FUNDAMENTALS OF
BIOCHEMISTRY

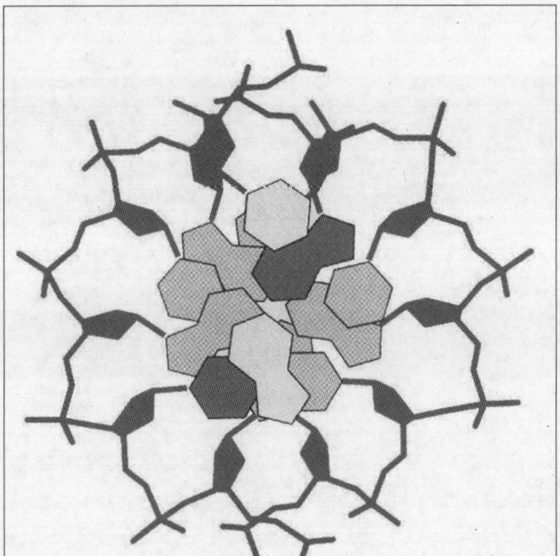

FUNDAMENTALS OF BIOCHEMISTRY

(With Life and Works of Over 100 Biochemists)

For University and College Students in India and Abroad

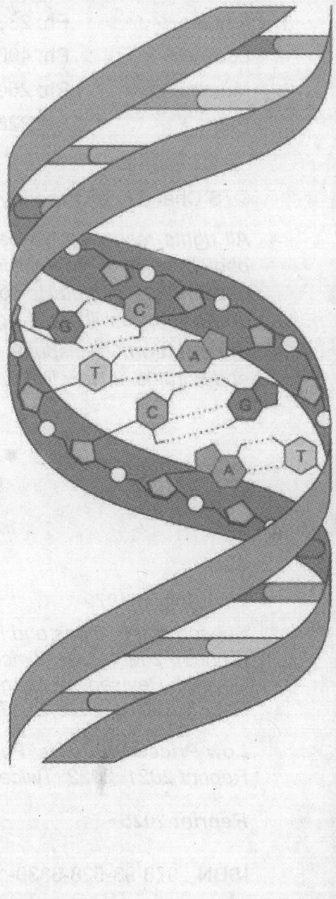

Dr. J.L. JAIN
MSc, PhD, FIES, FLS (London)
Former Reader in Botany Department
M S College, Saharanpur (UP), INDIA

Dr. SUNJAY JAIN
BSc, MBBS, MS (Ophth.), FCLI,
Affiliate, SEE Intl. (USA)

NITIN JAIN
BSc, PGDBA, OCP (USA)
PMP (USA)

S. CHAND
PUBLISHING

S Chand And Company Limited

(ISO 9001 Certified Company)

S Chand And Company Limited
(ISO 9001 Certified Company)
Head Office: D-92, Sector–2, Noida – 201301, U.P. (India), Ph. 91-120-4682700
Registered Office: A-27, 2nd Floor, Mohan Co-operative Industrial Estate,
New Delhi – 110 044, Phone: 011-49731800
www.**schandpublishing.com;** e-mail: **info@schandpublishing.com**

Marketing Offices:

Chennai	:	Ph: 23632120; chennai@schandpublishing.com
Guwahati	:	Ph: 2738811, 2735640; guwahati@schandpublishing.com
Hyderabad	:	Ph: 40186018; hyderabad@schandpublishing.com
Jalandhar	:	Ph: 4645630; jalandhar@schandpublishing.com
Kolkata	:	Ph: 23357458, 23353914; kolkata@schandpublishing.com
Lucknow	:	Ph: 4003633; lucknow@schandpublishing.com
Mumbai	:	Ph: 25000297; mumbai@schandpublishing.com
Patna	:	Ph: 2260011; patna@schandpublishing.com

First Edition 1979
Subsequent Editions and Reprints 1983, 88, 90, 92, 94, 95, 97, 98, 2000, 2001 (Twice), 2002, 2003, 2004, 2005 (Thrice), 2006, 2007 (Twice), 2008 (Twice), 2009, 2010, 2012, 2013
Seventh Revised & Enlarged Edition 2014
Reprints 2015, 2016, 2017, 2018

Low Priced Students' Paperback Edition (LPSPE), 2020
Reprint 2021, 2022 (Twice), 2024

Reprint 2025

ISBN : 978-93-528-3830-1 **Product Code :** H6BCH68BCHM10ENAG20L

PRINTED IN INDIA

By Vikas Publishing House Private Limited, Plot 20/4, Site-IV, Industrial Area Sahibabad, Ghaziabad – 201 010 and Published by S Chand And Company Limited, A-27, 2nd Floor, Mohan Co-operative Industrial Estate, New Delhi – 110 044.

Dedicated to
my reverend father

LATE SHRI HAR NARAYAN JAIN

(24-12-1899 — 11-1-1982)

Humanitarian, Philanthropist, Businessman and a True Father

—J. L. Jain

"His life was gentle, and the elements
So mix'd in him that Nature might stand up
And say to all the world, 'This was a man!' "

—William Shakespeare
Julius Caesar, Act V, Scene 5.

"We must trust in nothing but facts which are presented to us by nature and cannot deceive."

—Antoine Laurent Lavoisier (LT, 1743–1794)

ABOUT THE AUTHORS

Dr. J.L. Jain

Dr. J. L. Jain (born, 20.12.1938) has about 35 years of teaching experience of undergraduate and postgraduate classes. He did M.Sc. (Botany), with specialization in Cytogenetics and Plant Breeding, from Agra University, Agra and was conferred Ph.D. (Virology) from Meerut University, Meerut. He has published about a dozen research papers in various journals of international repute. He has also contributed articles of varied interest in many Magazines and Periodicals.

Dr. Jain has authored many text books pertaining to Botany, Chemistry and Biochemistry, catering to the needs of High School to postgraduate and research students and are being used in India and abroad. He has been an active member of the Leadership Project in Botany, sponsored by the University Grants Commission, New Delhi. He is presently associated with many universities and prestigious institutions such as Agra University, Agra; D.E.I. (Deemed University), Dayalbagh, Agra; Magadh University, Bodh-Gaya; Banasthali Vidyapeeth, Jaipur; and Madhya Pradesh Bhoj (Open) University, Bhopal, as an expert, counsellor and writer.

Dr. Jain is the recipient of **'Sahitya Bhooshan'** degree from Lucknow, in 1953. He has the rare distinction of being selected fellow of the coveted **'The Linnean Society of London'**, the oldest extant society in the world (founded in 1788) devoted to Life Sciences and patronized by Her Most Excellent Majesty Queen Elizabeth II and was awarded **F.L.S.** in 1994, in appreciation of his everlasting contributions in Life Sciences.

Dr. SUNJAY JAIN did his MBBS and MS (Ophth.), both from Sarojini Naidu Medical College, Agra, earning Gold Medal in Biochemistry and Ophthalmology. He has a vociferous appetite for exploring various human ailments. He has a vast collection of general and specialized books on diverse fields ranging from General Knowledge, Biology, Medical, Paramedical, Ayurveda, Naturopathy to Agriculture which keep him abreast with the recent developments in these fields.

NITIN JAIN has graduated in Information Technology from the Institute of Chartered Financial Analysts of India (rechristened as ICFAI University) and is now implementing ERP solutions for corporates in USA. He is also pursuing his interest on evolving Computer Biometrics which may eventually be used in human identification based on genetic prints used for authentication purposes. Presently, he is working on a concept of Biochemistry from information technology perspective.

PREFACE TO THE SEVENTH EDITION

More than 40 years have elapsed since one of us (JLJ) embarked on the intellectual journey of writing a textbook on Biochemistry at a time when the subject was in its infancy and the books on it were hardly available. Since then, Biochemistry has taken a long stride and has established itself in the forefront of scientific endeavour. As a matter of fact, it embraces all disciplines of Science and explains various ongoing phenomena in living beings in its own right to satisfy the human lust for knowledge and curiosity. Biochemistry has indeed become the very language of life itself.

In this latest Seventh Edition , five New Chapters (No. 28, 29, 33, 36 and 37) have been added to enhance the scope and utility of the book: three chapters pertain to **Bioenergetics and Metabolism (Biosynthesis of Nucleotides, Degradation of Nucleotides, Mineral Metabolism)** and two to **Nutrition Biochemistry (Principles of Nutrition, Elements of Nutrition**). In fact, all the previously-existing 35 chapters have been thoroughly revised, enlarged and updated in the light of recent advancements and the ongoing researches being conducted the world over.

It is hoped that this revised and enlarged seventh edition would serve in a better way, which the authors are keenly desirous of.

Gratitude is expressed to the students and teachers across the globe who have sent in their valuable suggestions which have been given due consideration. We are sincerely thankful to Mrs. Nirmala Gupta, the Chairperson and Managing Director of the Company for her wholehearted cooperation and goodwill gesture accorded to. The entire staff of the firm deserves appreciation for their unflinching support during the gestation period of this book. We are deeply indebted to our wives– Mrs. Mithlesh K. Jain, M.A.; Mrs. Shailja Jain, Ph.D. and Mrs. Neeti Jain, B.Com., respectively– for their patience, buoyancy and encouragement of this venture which was more arduous than anticipated. Finally, the authors express their appreciation for the help rendered by our own children, Miss Nikita Jain and Miss Shuchita Jain, both B. Tech.

The authors, however, still crave the indulgence of the readers of this book for pointing out any inadequacies and inaccuracies which they come across. Their healthy criticism shall be thankfully acknowledged. Suggestions for further improvement of the book are solicited.

114, Old Vijay Nagar Colony
AGRA - 282 004
Tel: 0562-4043940
Mob: 91-9997045122
e-mail: drjljain@yahoo.com

Dr. J.L. JAIN
Dr. SUNJAY JAIN
NITIN JAIN

'All the world knows me in my book, and my book in me.'
– Michael Montaigne

PREFACE

Biochemistry is a new branch of science which aims at answering, in chemical language, questions such as 'What is the life made of ?' and 'How does it work ?' Whereas the eye works at the gross level of visible objects, the microscope reaches down to the cellular level, exposing details of the various cell organelles, including nuclei and other particles. Biochemistry, however, works at a still finer level that cannot be accessed by the ultra-modern optical or phase-contrast microscopes. In other words, it operates at a molecular level and thus brings to light the hidden secrets of life. The rapid development and enormous expansion of every phase of biochemistry has not only markedly enriched our knowledge about the nature of life but has also made biochemistry the very language of life itself.

Realising the paramount importance of this discipline of science, degree courses in biochemistry are now offered in a good number of colleges and universities. But the students offering this subject at undergraduate level come as raw hand, as biochemistry is not taught to them at the school level. It is, in fact, to meet the requirements of such students that this book has been written. Besides, the book may also serve an useful purpose for higher studies.

The book contains the basic concept of biochemistry written in a manner suited to the broad spectrum of the college students. The matter has been lucidly presented, orderly arranged and profusely illustrated. References have been included at the end of each chapter in order to guide the reader to the classical and current literature. A few appendices are given at the end. These may, however, prove useful to the reader. Some illustrations have been adapted from standard texts, journals and research papers. The sources of all such figures have been duly acknowledged in their legends. The author feels greatly beholden to their authors and publishers.

I am also thankful to my publishers, specially to Shri Shyam Lal Gupta and Shri T.N. Goel for taking keen interest and in bringing out the book in its present form. Finally, I am grateful to my wife Mrs. Mithlesh K. Jain M.A., for inspiring me in proceeding from thoughts to print.

Helpful suggestions for improvement are welcome.

Saharanpur **J.L. JAIN**

LIST OF CHAPTERS

PART I: PREFATORY

PART II: BIOMOLECULES

PART III: BIOENERGETICS AND METABOLISM

PART IV : HORMONE BIOCHEMISTRY

PART V : NUTRITION BIOCHEMISTRY

PART VI : ANALYTICAL BIOCHEMISTRY

BIOCHEMISTRY

CONTENTS

PART I PREFATORY

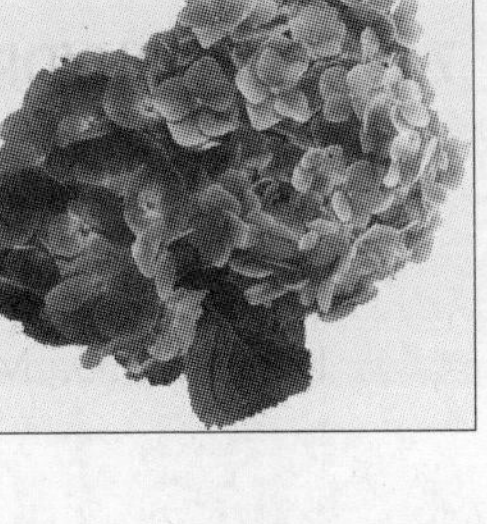

PART II BIOMOLECULES

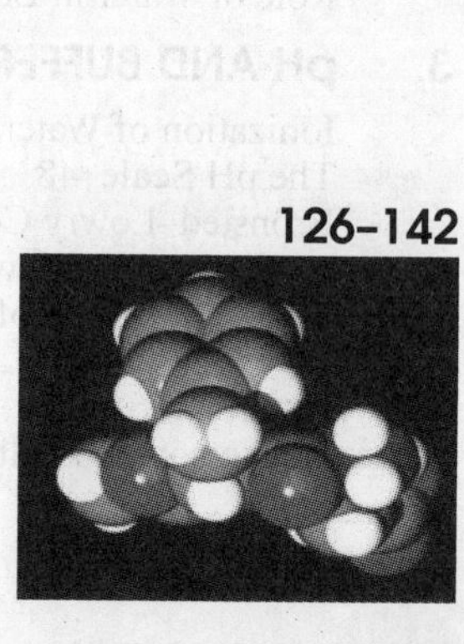

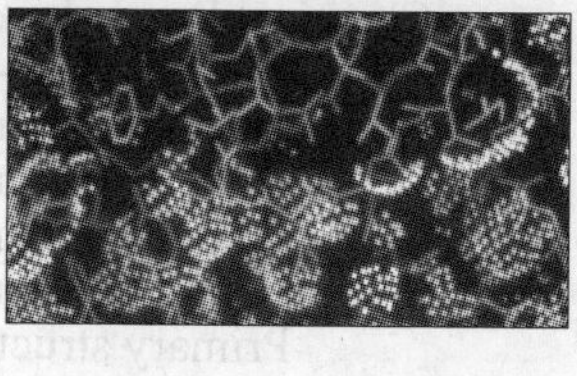

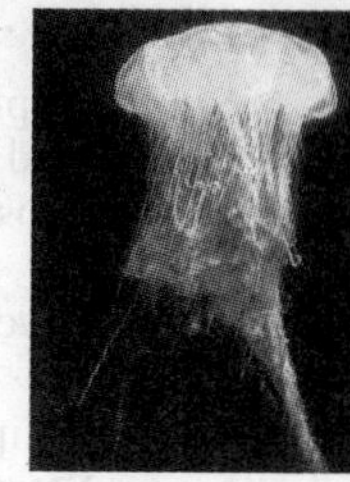

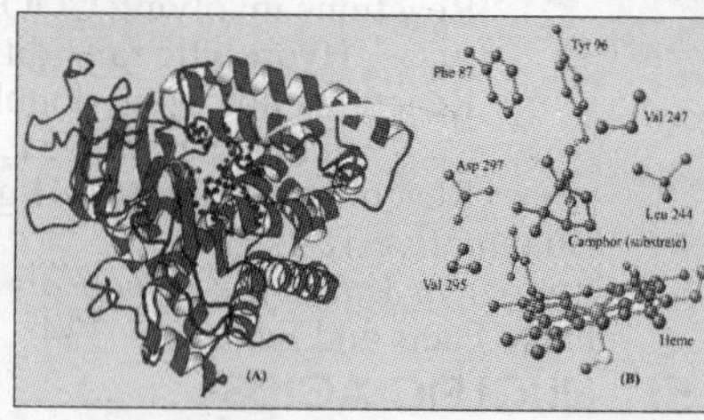

PART III BIOENERGETICS AND METABOLISM

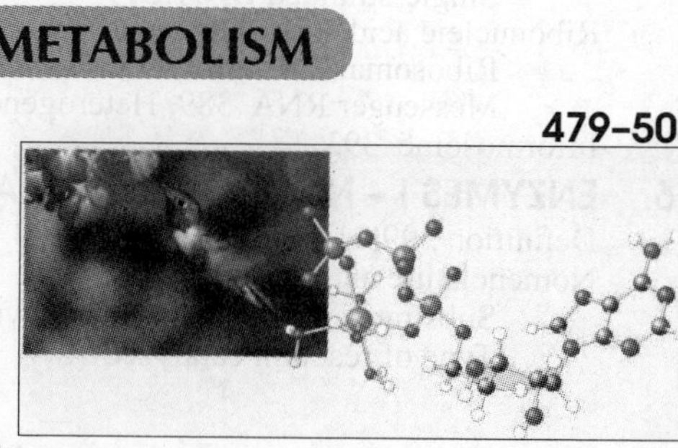

98
638
Mobil Mobil Mobil Mobil Mobil

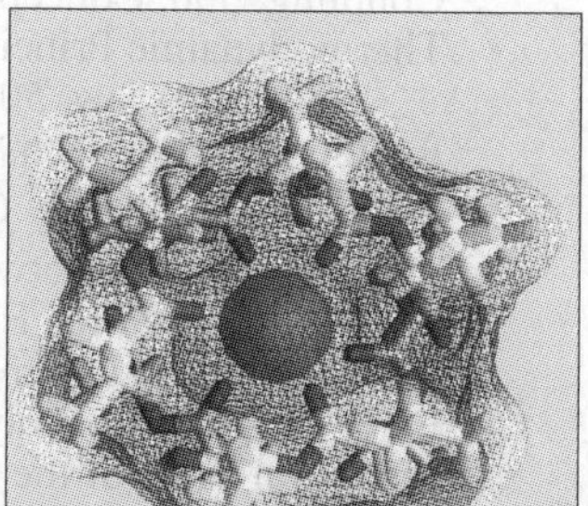

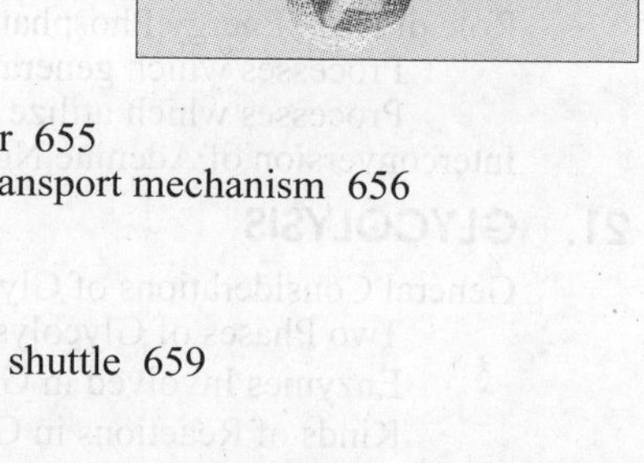

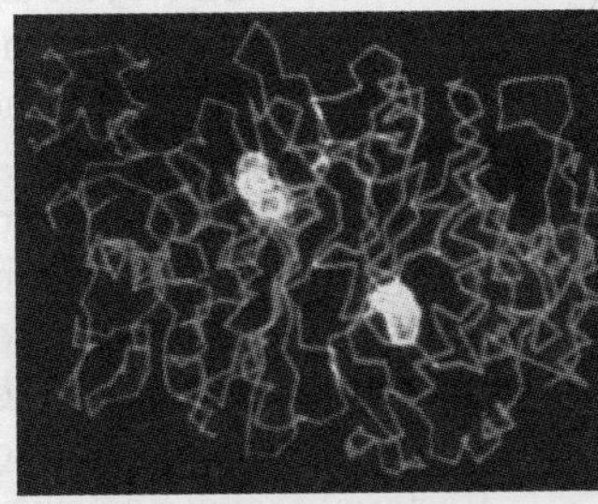

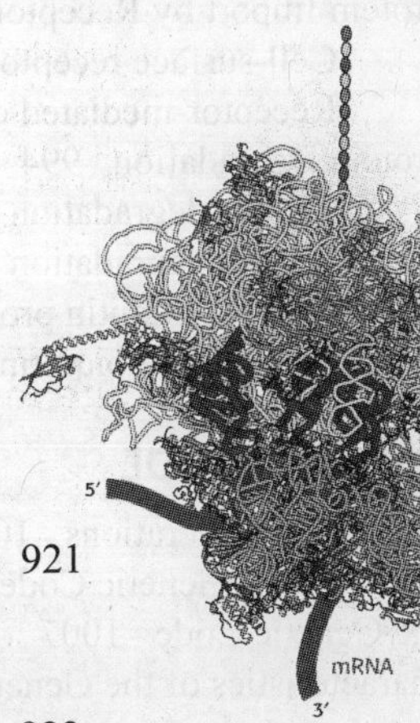
5′
mRNA
3′

31. PROTEIN TARGETING AND DEGRADATION 965–1003

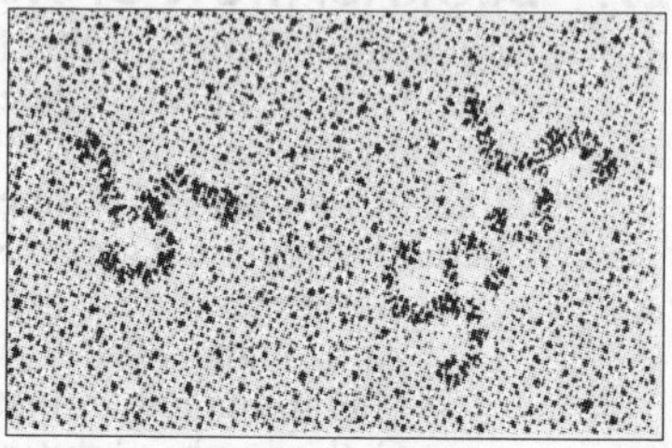

32. GENETIC CODE 1004–1032

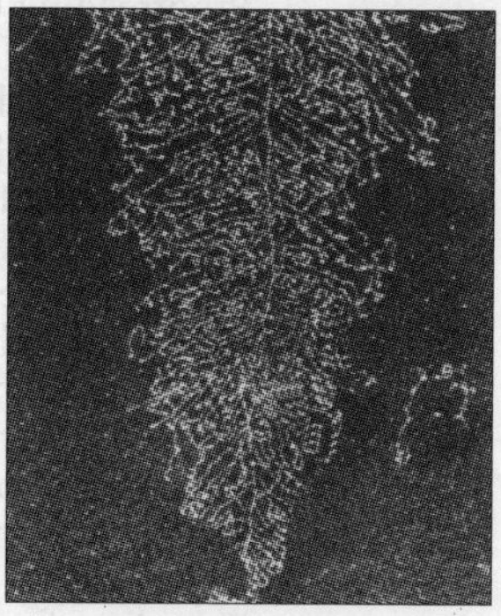

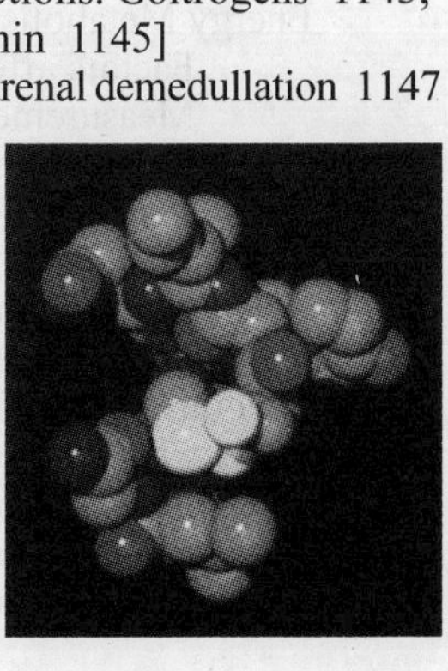

PART V NUTRITION BIOCHEMISTRY

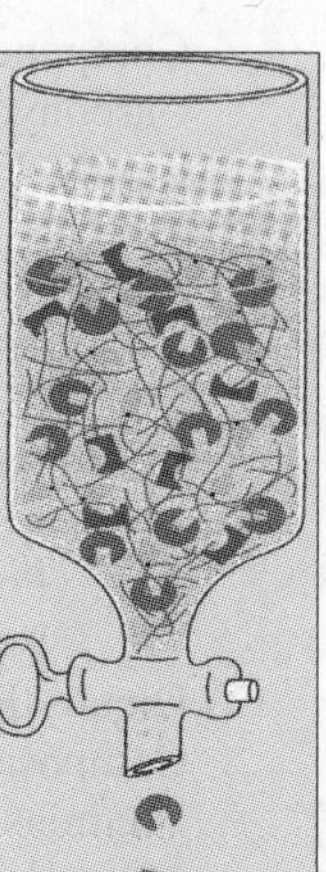

LIFE AND WORKS OF BIOCHEMISTS

CLINICAL IMPLICATIONS

PART

I

Prefatory

A large meal such as this breakfast contains representatives of most types of biological molecules, whose study forms the subject matter of Biochemistry.

> "There is no greater object of wonder, no greater thing of beauty, than the dynamic order, the organized complexity of life."
>
> —*Ariel G. Loewy and Philip Siekevitz: Cell structure and Function, 1969.*

CONTENTS

THE DIVERSITY OF LIVING SYSTEMS

The distinct morphology of the 3 organisms – a plant (*Amborella trichopoda*), an animal (frog, *Rana* sp.) and a microorganism (bacteria)– might suggest that they have little in common. Yet, biochemically they display a remarkable commonality that attests to a common ancestry.

CHAPTER 1

Introduction

"Anything found to be true of *E. coli* must also be true of elephants."

–Jacques Lucien Monod, LT, 1910–1976

DEFINITION

The term **Biochemistry** (*bios*[G] = life) was first introduced by a German chemist **Carl Neuberg** in 1903. Biochemistry may be defined as a science concerned with the chemical nature and chemical behaviour of the living matter. It takes into account the studies related to the nature of the chemical constituents of living matter, their transformations in biological systems and the energy changes associated with these transformations. Such studies have been conducted in the plant and animal tissues both. Broadly speaking, biochemistry may thus be treated as a discipline in which biological phenomena are analysed in terms of chemistry. The branch of biochemistry, for the same reason, has been variously named as **Biological Chemistry** or **Chemical Biology**.

In fact, biochemistry originated as an offshoot from human physiology when it was realized that the chemical analysis of urine, blood and other natural fluids can assist in the diagnosis of a particular disease. Hence in its infancy, biochemistry was accordingly known as **Chemical Physiology**. But physiology now covers the study of normal functions and phenomena of living beings. And biochemistry is concerned particularly with the chemical aspects of these functions and phenomena. In other words, biochemistry is but one of the many ways of studying physiology. The two may be compared by watching the monkeys in a zoo which means studying the physiology of behaviour. But if the behaviour of animal molecules is studied, rather than the whole animals, it would form the study of biochemistry.

Modern biochemistry has two branches, descriptive biochemistry and dynamic biochemistry. **Descriptive biochemistry** is concerned with the qualitative and quantitative characterization of the various cell components and the **dynamic biochemistry** deals with the elucidation of the nature and the mechanism of the reactions involving these cell components. While the former branch is more a concern of the organic chemist, the latter branch has now become the language of modern biochemistry.

However, as the knowledge of biochemistry is growing speedily, newer disciplines are emerging from the parent biochemistry. Some of the disciplines are **enzymology** (science of the study of enzymes), **endocrinology** (science dealing with the endocrine secretions or the hormones), **clinical biochemistry, molecular biochemistry** etc. Along with these branches certain link specialities have also come up such as agricultural biochemistry, pharmacological biochemistry etc.

HISTORICAL RESUME

In terms of history, biochemistry is a young science. It started largely as an offshoot of organic chemistry and later incorporated ideas obtained from physical chemistry. In fact, the science of biochemistry may be regarded to have begun with the writings of **Theophrastus Bombastus von Hohenheim,** better known as **Philippus Aureolus Paracelsus** (Lifetime or LT, 1493–1541), a Swedish physician and alchemist who laid the foundation of chemotherapy as a method of treating diseases and also the promoter of '*doctrine of signatures.*' He first acquired the knowledge of chemistry of his time and then entered the field of medicine to apply his knowledge of chemistry. He proclaimed, "Life processes are essentially of chemical nature and diseases can be cured by medicines". Later, his followers notably **Jan Baptist van Helmont** (LT, 1577–1644) amalgamated the science of chemistry with medicine which emerged under the name of 'medical chemistry' (or Iatrochemistry).

KARL (OR CARL) WILHELM SCHEELE
(LT, 1742–1786)

A Swedish chemist who discovered oxygen (ca 1773) perhaps a year before Joseph Priestley's similar discovery. He also discovered chlorine (1974). Schelle can probably be called the *unluckiest chemist in History*, for although he discovered several elements including the above two (such as barium, manganese, etc), he does not receive undisputed credit for having discovered even a single one. In some cases, chemists independently made the same discovery a little sooner. In others, Scheele did not quite carry matters far enough and other chemists took the last step and got the credit.

The basis of biochemistry was, in fact, laid down by chemists like Scheele and Lavoisier. **Karl Wilhelm Scheele** (LT, 1742–1786), a swedish pharmacist, discovered the chemical composition of various drugs and the plant and animal materials. He also isolated a number of substances such as citric acid from lime juice, lactic acid from sour milk, malic acid from apple and uric acid from urine. Scheele, thus, laid the foundation of descriptive biochemistry. Similarly, **Antoine Lavoisier** (LT, 1743–1794), a French chemist, who studied the composition of air and propounded the theory of conservation of matter, put dynamic biochemistry on firm standfootings. He developed the concept of oxidation and also clarified the nature of animal respiration. He had concluded that respiration could be equated with combustion and that it was slower but not essentially different from the combustion of charcoal. Lavoisier is often spoken of as '*father of modern biochemistry*'.

The earliest book relating to Biochemistry was '*Lectures in Animal Chemistry*' published by the famous Swedish chemist **Jöns Jacob Berzelius** (LT, 1779–1848), in 1806.

In 1828, **Friedrich Wöhler**, a German chemist, synthesized urea, a substance of biological origin, in the laboratory from the inorganic compound ammonium cyanate. This achievement was the unexpected result of attempts to prepare ammonium cyanates through the treatment of metal cyanates with ammonium salts. As Wöhler phrased it in a letter to a colleague, "I must tell you that

I can prepare urea without requiring a kidney or an animal, either man or dog." This was a shocking statement in its time, for it breached the presumed barrier between the living and the nonliving. Consequently, this rendered the vitalistic theory of organic materials untenable. *Vitalistic theory* (= *the doctrine of vitalism*) maintained that organic compounds could be synthesized only through the agency of a vital force, supposed to be present only in living tissues.

FRIEDRICH WÖHLER
(LT 1800–1882)

A German chemist who was born on July 1, 1800. He is well known for the historical synthesis of an organic compound, urea, from an inorganic salt, ammonium cyanate (in 1828):

$$\underset{\text{Ammonium cyanate}}{NH_4-O-C\equiv N} \xrightarrow{\text{Heat}} \underset{\text{Urea}}{O=C(NH_2)_2}$$

With **B.J. von Liebig**, he discovered the benzoyl radical. Wöhler earned worldwide acclaim for his teaching during his 46 years at the University of Göttingen. Between 1845 and 1866, he lectured to some 8,250 students. In his lifetime, 13 and 15 editions of his Organic and Inorganic texts, respectively, were published. He is also credited with the discovery of calcium carbide, preparation of metallic aluminium and isolation of beryllium and yttrium. He died on September 23, 1882.

Wöhler's work was followed by the synthesis of acetic acid from inorganic materials by another German chemist **Adolf Wilhelm Hermann Kolbe** (LT, 1818–1884), in 1845. However, the final blow to the theory of vital force was given by a French Chemist **Pierre Eugene Marcellin Berthellot** (LT, 1827–1907) who synthesized a host of organic compounds (such as methyl alcohol, ethyl alcohol, methane, benzene, acetylene) from inorganic compounds in 1850s. Vitalism was, thus, quietly laid to rest. Organic synthesis remains very much alive.

Justus von Liebig, a German chemist and discoverer of chloroform and who is often termed as '*father of agricultural chemistry*', arrived at the conclusion that "the nutritive materials of all green

BARON JUSTUS VON LIEBIG
(LT, 1803–1873)

A German organic chemist who was born on May 12, 1803 in Darmstadt. He worked at Giessen, near Frankfurt. He was the first to recognize that two different chemical compounds can have the same formula. He also developed a method of analyzing organic compounds by burning them and weighing the CO_2 and water produced. With **Friedrich Wöhler**, he proposed the *radical theory of organic structure.* This suggested that groups of atoms such as benzoyl radical ($C_6H_5CO^-$), now known as benzoyl group, remained unchanged in many chemical reactions. He was also one of the first to propose the use of mineral fertilizers for feeding plants. He is also credited with the discovery of chloroform. He also devised a condenser which goes by his name.

By age 36, Liebig headed the world's largest laboratory and school for training chemists. He was one of the early investigators of large-scale research. Today, his laboratory with its furnishings is preserved as a museum in Giessen, Germany. He wrote many books which provided an impetus in the early development of biochemistry. Of special interest is his book '*Organic Chemistry in Its Applications to Physiology and Pathology*', published in 1842. He is often termed as **'father of agricultural chemistry.'** He arrived at the conclusion that "the nutritive materials of all green plants are inorganic substances." Liebig's oft-repeated quotation, for which he is famous, reads as:

"The secret of all those who make discoveries is to look upon nothing as impossible."

plants are inorganic substances." He wrote many books which provided an impetus in the early development of biochemistry. Of special interest is his book '*Organic Chemistry in Its Applications to Physiology and Pathology*', published in 1842. **Michel Chevreul** (LT, 1786–1889) demonstrated through studies on saponification that fats were composed of glycerol and fatty acids, of which he isolated several. The excellent researches conducted by the great German biochemist **Hermann Emil Fisher** (LT, 1852–1919) may be regarded as landmark in the development of structural biochemistry. In the course of his studies, this remarkable man completely revolutionized research concerning the structures of carbohydrates, amino acids and fats. Although nucleic acids are the newest of the 4 great groups of biochemical materials, their discovery goes back to observations by **Friedrich Miescher** (LT, 1844–1895) in 1869. His discovery of nucleic acids in the nuclei of pus cells, obtained from discarded surgical bandages, led him to investigate the distribution and the properties of these compounds.

During the first half of the nineteenth century, the studies on heat conducted mainly by **Julius Robert von Mayer** (LT, 1814–1878) and **Ludwig von Helmholtz** (LT, 1821–1894) led to the formulation of the 'laws of thermodynamics' which are essential to the understanding of energy relations in biological systems.

Besides respiration, the other physiological process to attract the attention of biochemists was that of digestion. Main contributions towards this were made by **van Helmont** (LT, 1577–1644), **Abbé Lazaro Spallanzani** (LT, 1729–1799), **René Antoine Ferchault de Réaumur** (LT, 1683–1757), **William Beaumont** and Claude Bernard. **Claude Bernard** (LT, 1813–1878) of Paris was perhaps the greatest of these. His contributions included the discovery of liver glycogen and its relation to blood sugar in health and disease. He noted the digestive properties of pancreatic juice and began research in muscle and nerve physiology.

The process of fermentation is probably the single most important process around which the interest of biochemists persisted for a considerable period. By about 1780, fermentation had been recognized by **Theodor Schwann** (LT, 1810–1882) as a biological process. He established that yeast was a plant capable of converting sugar to ethanol and carbon dioxide. However, many of the leading chemists of the day, including Berzelius, Wöhler and Leibig, considered yeast to be nonliving and fermentation to be caused solely by oxygen. Ridicule by Liebig and Wöhler delayed acceptance of Schwann's views until the illustrious French microbiologist **Louis Pasteur** (LT, 1822–1895) presented evidence overwhelming all objections. He founded the useful branch of Microbiology in 1857 and identified several organisms that carried out various fermentations, including that leading to butyric acid, a type performed by organisms that function without oxygen. He defined fermentation as "*la vie sans l'air*" (life without air). Pasteur, thus, introduced the concept of aerobic and anaerobic organisms and their associated fermentations. These conclusions again aroused the ire of Liebig who again took sharp exception but this time with less effect. Such studies on fermentations climaxed by the demonstration in 1887 of **Eduard Buchner** (LT, 1860–1917) that sugars could be fermented by cell-free extracts of yeast. This discovery represents the cornerstone of much of the enzymological and metabolic study of the twentieth century since it led to techniques of isolation and identification that ultimately permitted the study of enzymes and the individual reactions concerned. It also expunged, yet once and for all, any traces of vitalism (or vitalistic theory) still lingering.

Researches conducted by pioneers such as Arrhenius, van't Hoff and Ostwald on electrolytic dissociation and osmotic pressure led physical chemists to turn their attention to biological phenomena. **Soren Sörensen** (LT, 1868–1939), a Danish chemist, developed our concept on pH, **Jacques Loeb** (LT, 1859–1924) studied the colloidal behaviour of proteins and their effect on the cell, **Leonor Michaelis** placed the concept of chemical compound formation between enzyme and substrate on an experimental basis and **Wendell Meredith Stanley** (LT, 1904–1971), a biochemist working at the Rockefeller Institute, New York, showed that viruses are nucleoproteins. For his fundamental research on TMV, Stanley was awarded a share of the 1946 Nobel Prize in

Chemisty. Later, many instruments were invented such as Van Slyke blood gas apparatus, the ultracentrifuge of **Theodore Svedberg** (LT, 1884–1971) and the electrophoresis apparatus of **Arne W.K. Tiselius** (LT, 1902–1971). The use of isotopes in biochemical research by **Harold Clayton Urey** (LT, 1883–1981) and **Schoenheimer** and the application of chromatography, first developed by **Martin** and **Richard Lawrence Synge** (LT, 1914–1994), opened a new chapter in modern biochemistry.

With the advent of twentieth century, biochemistry burst into full bloom. Important developments took place rapidly on several fronts including nutrition. The significance of unknown food factors was clearly recognized by **Frederick Gowland Hopkins** at Cambridge University and his associates, who developed the *concept of deficiency diseases*. Extensive series of feeding experiments utilizing synthetic diets were conducted mainly by **Babcook, E.V. McCollum, T.B. Osborne, L.B.Mendel** and **H.C. Sherman**. As a result, many deficiency diseases such as scurvy, rickets, beriberi and pellagra were recognized and their curative agents, which were called vitamins by a Polish biochemist **Casimer Funk** (LT, 1884–1967), were isolated and subsequently characterized.

Buchner's work on cell-free fermentation of sugars was actively extended in many laboratories, including those of **Harden** and **Young, Embden** and **Meyerhof,** with the result the complete biochemical pathway known as *Embden-Meyerhof-Parnas pathway* (or *glycolysis*) was elucidated. The researches conducted by **Otto Heinrich Warburg** (LT, 1883–1970), **Heinrich Otto Wieland** (LT, 1877–1957), **Keilin** and **Axel Hugo Teodor Theorell** (LT, 1903–1982) led to the discovery of enzymes and cofactors involved in cellular oxidation. Later, **Fritz Albert Lipmann** (LT, 1899–1986) and **Kurt Henseleit** made notable observations on the significance of the terminal pyrophosphate linkages of ATP as an energy storage reservoir. **Albert Szent-Györgyi** (LT, 1893–1986) and **Hans Adolf Krebs** (LT, 1900–1981) of England studied the fate of lactate (or pyruvate) during aerobic oxidation. This led to the development of a sequence of reactions known as *Krebs cycle* (or *citric acid cycle*). Later studies revealed that the fatty acids and amino acids, upon oxidation, also yield intermediates that are identical with those in the Krebs cycle, thus providing a common mechanism for the liberation of energy from all foodstuffs. **Frederick Sanger** (born, 1918) established the complete amino acid sequence of the protein hormone insulin and **Vincent du Vigneaud** (LT, 1901–1978) proved the structure of the nonapeptide hormones of posterior pituitary by direct synthesis.

The brilliant studies by **Linus Carl Pauling** (LT, 1901–1994) and **Robert Corey** (LT, 1897–1971) led to the concept of a secondary structure of protein molecules in the form of an α-helix. A similar kind of structure for the nucleic acids was also elucidated. **James Dewey Watson** and **Francis Harry Compton Crick,** in 1953, proposed that a double-stranded DNA molecule could be made by binding bases on adjacent strands to each other by hydrogen bonding. This *base-pairing hypothesis* was confirmed by the quantitative data of an Austrian refugee biochemist, **Erwin Chargaff** (LT, 1905–2002) and was soon followed by the enzymatic synthesis of DNA by **Arthur Kornberg.** These synthetic macromolecules have properties that suit the Watson-Crick hypothesis. Soon after, the base sequence of transfer RNA molecules specific for different amino acids was determined by many workers such as **Robert William Holley** (LT, 1922–1993), **Medison, Zachan** and others. A final and accurate list of the base sequences in messenger RNA that code for each of the amino acids was made available as a result of the brilliant researches by **Marshall Warren Nirenberg** (LT, 1927–2010), who used synthetic nucleotides as messenger molecules to identify the coded base sequences for each of the amino acids.

By 1835, **Jönes Jacob Berzelius,** a Swedish chemist, had clearly recognized the importance of catalysis in controlling the rates of chemical processes. He suggested that the formation of biological materials was controlled by catalytic actions and cited an enzyme from potato (*i.e.*, *potato diastase*) as an example of a biological catalyst effective in the hydrolysis of starch. He reasoned that all materials of living tissues are formed under the influence of catalytic action, a conclusion thoroughly established by subsequent work. Later, many catalysts were isolated, purified to some extent and the associated reactions investigated kinetically. However, the chemical nature of these so-called biocatalysts or enzymes remained unknown until 1926, when **James Batcheller Sumner** (LT, 1887–1955) at Cornell University, for the first time, crystallized the enzyme *urease* from the extracts of Jack bean and demonstrated its protein nature. For many years, Sumner's discovery was greeted with skepticism and derision, most particularly by the renowned German biochemist **Richard Willstätter**, an authoritative figure, who insisted that enzymes are low-molecular-weight compounds and that the protein found in urease crystals was merely a contaminant. Sumner was a tenacious man, armed with a body of convincing evidence and did not surrender to authority and even successfully repeated his experiments in Willstätter's laboratory. Subsequently, **Johan Howard Northrop** (LT, 1891–1987) and **Kunitz** crystallized a series of pancreatic and gastric enzymes. This work and subsequent isolation studies clearly confirmed that enzymes are proteins and established Sumner as the '*father of modern enzymology*'. Invaluable work in this field has been accomplished by **William Howard Stein** (LT, 1911–1980), **Stanford Moore** (LT, 1913–1982), **Max Ferdinand Perutz** (LT, 1914–2002), **John Cowdery Kendrew** (LT, 1917–1997) and **David C. Phillips.**

Perhaps the greatest impetus in biochemical researches is nowadays being given to the mechanisms for the regulation of the synthesis of cellular compounds. The phenomenon of feedback inhibition of enzyme activity by the end product of a reaction sequence illustrates a self-regulating mechanism. Enzyme induction and repression—the acceleration or inhibition of synthesis of an enzyme—have been described. These discoveries have led to a hypothesis proposed in 1961 by two Frenchmen, **François Jacob** (born, 1920) and **Jacques Lucien Monod** (LT, 1910–1976), suggesting that the DNA molecules consist of areas in which genes are maintained in an inactive state (by repressors) until they need to be activated for the production of messenger RNA molecules. In 1963, Jacob and Monod, with **Jean-Pierre Changeux**, also proposed a theory to explain the molecular aspects of regulation of the catalytic activity of enzymes. These two findings introduced scientific concepts about the control of genetic and metabolic functions of organisms which soon established a new field of scientific enquiry– the study of biological regulation.

At about this point, the strands of scientific development (shown in Fig. 1–1)– biochemistry, cell biology, and genetics– became inextricably woven, and the new science of **molecular biology** emerged. The distinction between molecular biology and biochemistry is not always clear, because both disciplines take as their ultimate aim the complete definition of life in molecular terms. The term *molecular biology* is often used in a narrower sense, to denote the study of nucleic acid structure and function and the genetic aspects of biochemistry, an area which one might more accurately call *molecular genetics*. Molecular biology and biochemistry are perhaps differentiated more easily by the orientations of their users than by the research problems being addressed. It may be said that while a biochemist thinks like chemist, a molecular biologist thinks like biologist. Even this distinction is somewhat artificial, since scientists in either field must use the approaches of all relevant disciplines, including chemistry, biology, and physics. In fact, 3 of the most powerful research techniques used by biochemists were developed by physicists: **electron microscopy (EM)**, which has revealed the minutest details of cellular structure, and **x-ray diffraction** and **nuclear magnetic resonance (NMR),** which have revealed the precise 3-dimensional structure of huge biological molecules (= biomolecules).

In the 1970s, molecular biology reached a new zenith with the introduction of recombinant DNA technology, which has brought about a revolution in biological research and has been instrumental in the emergence of a new branch of science called biotechnology. Use of this technique in science has allowed researchers to isolate eucaryotic genes to study their structure, regulation and expression.

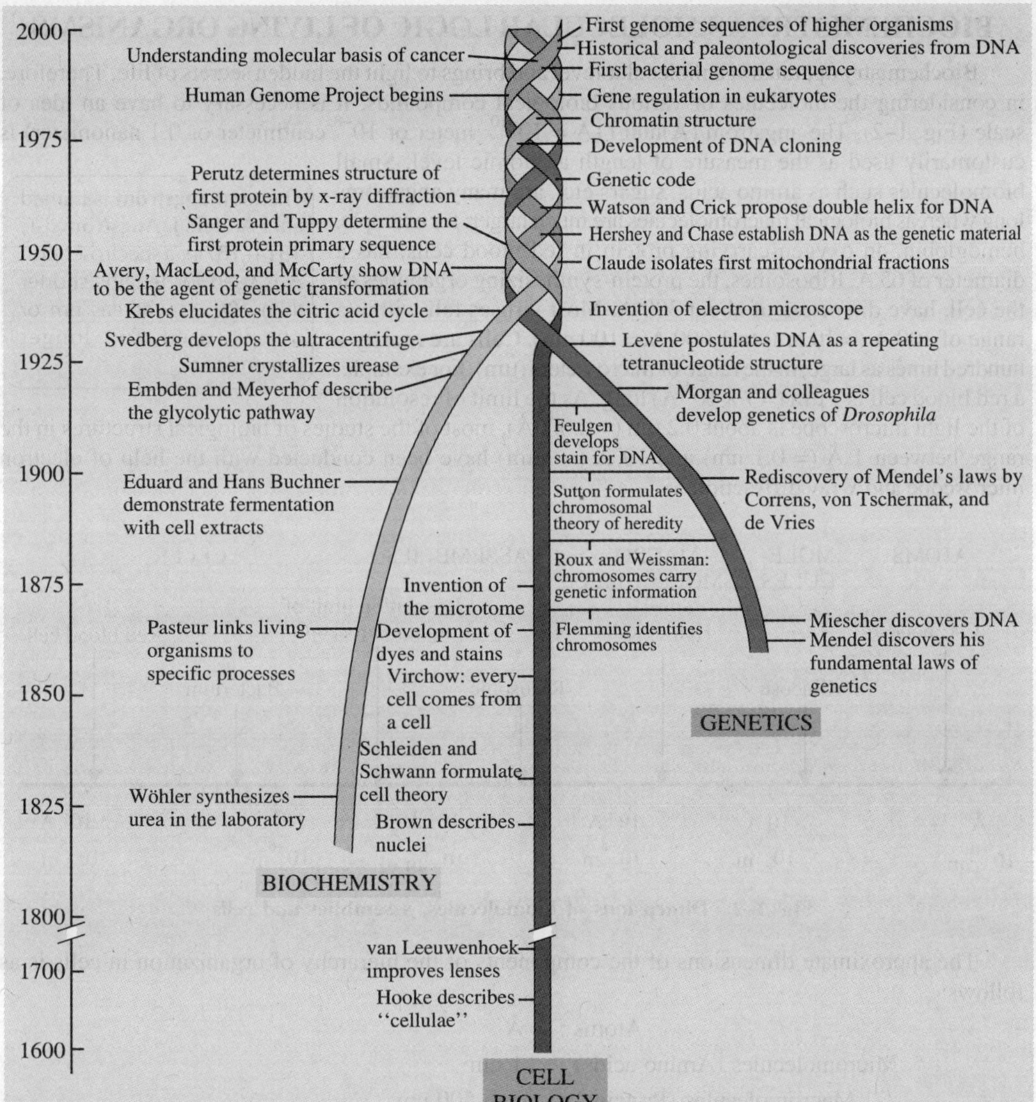

Fig. 1–1. Intertwining of historical traditions of biochemistry, cell biology and genetics

These 3 disciplines of science, originally considered quite distinct among each other, have become interweaved to yield a true molecular biology, which is the subject matter of present-day biochemistry.

(Adaped from Becker WM, Kleinsmith LJ and Hardin J, 2000)

In this era of biotechnology, biochemistry will continue to occupy a position of central importance. As a basic science, it is in the foreranks of many scientific endeavours that stand to make a reality of the statement that "the twentieth century belongs to the biologists." It was about 135 years ago that a German **Ernst Hoppe-Seyler,** one of the foremost chemists (and possibly the first biochemist) of the period, established a journal for those studies in physiological chemistry he believed should be recognized as the new discipline of biochemistry. If alive today, he would be justifiably proud of the science whose niche he helped define.

BIOCHEMISTRY AS MOLECULAR LOGIC OF LIVING ORGANISMS

Biochemistry operates at a molecular level and brings to light the hidden secrets of life. Therefore, in considering the molecules of various biological compounds, it is necessary to have an idea of scale (Fig. 1–2). The angström (Å)unit (1Å = 10^{-10} meter or 10^{-8} centimeter or 0.1 nanometer) is customarily used as the measure of length at atomic level. Small biomolecules such as amino acids, sugars etc., are many angströms long whereas biological macromolecules are much larger: for example, hemoglobin, an oxygen-carrying protein in red blood cells, has a diameter of 65 Å. Ribosomes, the protein-synthesizing organelles of the cell, have diameters of about 300 Å. Most viruses fall within a range of 100 Å (= 10 nm) to 1,000 Å(= 100 nm). Cells are mostly a hundred times as large, in the range of micrometers (µm). For example, a red blood cell is 7 µm (= 7 × 10^4 Å) long. As the limit of resolution of the light microscope is about 0.2 µm (= 2,000Å), most of the studies of biological structures in the range between 1 Å (= 0.1 nm) and 10^4 Å (= 1µm) have been conducted with the help of electron microscope and *x*-ray diffraction.

The term **angström** is named after **Anders J. Ångström** (LT, 1814–1874), a spectroscopist. Å is used for lengths shorter than 100 Å, whereas nm or *µ*m is used for longer dimensions.

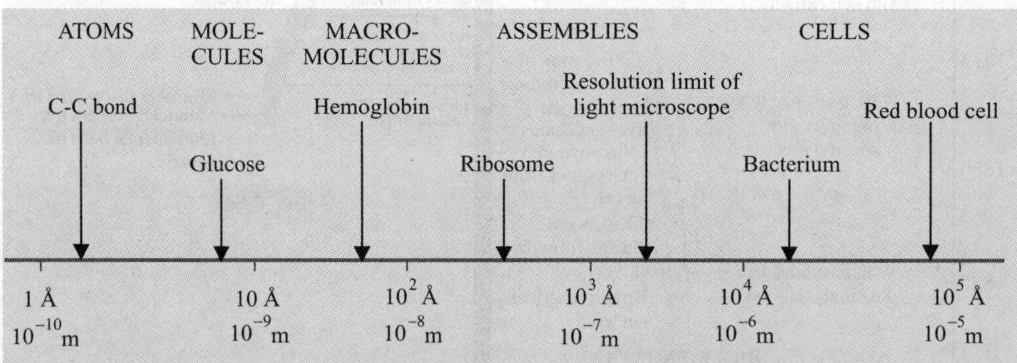

Fig. 1–2. Dimensions of biomolecules, assemblies and cells

The approximate dimensions of the components of the hierarchy of organization in cells is as follows:

Atoms :	Å
Micromolecules (Amino acids) :	1 nm
Macromolecules (Proteins) :	5 to 500 nm
Organelles :	nm to µm
Cells :	µm to cm

The biomolecules are in a state of flux (Fig. 1–3). The enzymes change their substrate into product in milliseconds (1 m sec = 10^{-3} second). Some enzymes are even more efficient and catalyze their substrate in even few microseconds (1 µ sec = 10^{-6} sec). The unwinding of the DNA double helix, which is essential for its replication, is completed in a microsecond. The rotation of one sphere of a protein with respect to another takes place in nanoseconds (1 n sec = 10^{-9} sec). It is remarkable to note that the primary event in vision— a change in structure of the light-absorbing group— occurs within a few picoseconds (1 p sec = 10^{-12} sec) after the absorption of a photon.

The molecular events are associated with energy changes (Fig. 1–4). The ultimate source of energy for the living objects is the sun. The energy of visible light, say green, is 57 kilocalories per mole (kcal/mol). ATP, which is the universal currency of energy, has a usable energy content of

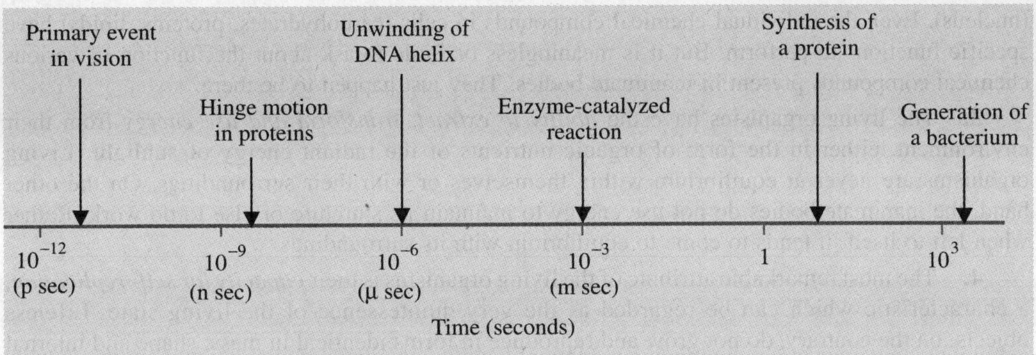

Fig. 1–3. Rates of some biological processes

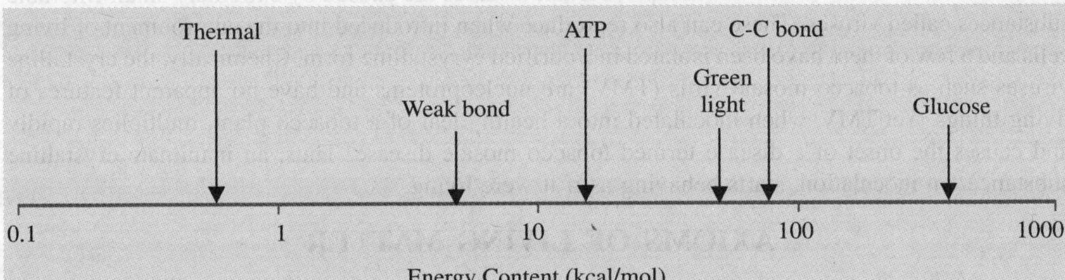

Fig. 1–4. Some biologically important energies

about 12 kcal/mol. The amount of energy needed for the cleavage of a covalent C—C bond is 83 kcal/mol. Hence, the covalent skeleton of biomolecules is stable in the absence of enzymes and inputs of energy. On the contrary, thermal energy is enough to make and break noncovalent bonds in living systems, which typically have energy contents of only a few kilocalories per mole.

NATURE

Biochemistry has been defined as **"the chemistry of living things"**. The living things are composed of lifeless molecules which, when examined individually, exhibit all the physical and chemical laws that are characteristic of inanimate bodies. Yet living organisms possess extraordinary attributes, not shown by inanimate molecules. At this stage, henceforth, it shall be useful to distinguish 'living' from 'nonliving'. Although the word 'life' is difficult to define but it can be associated with certain characteristics which are common to the living objects and are usually not found in the nonliving or lifeless objects. All **living objects** are characterized by their capacity of movement, power of growth, respiratory activity, sense of irritability and above all the power of reproduction.

Certain identifying characteristics of the living matter or the '*signs of life*', as they are also called, are enumerated below:

1. Living organisms are *highly complicated and organized structures* and contain a very large number of different organic molecules. For example, a single bacterial cell of *Escherichia coli* contains 5,000 different kinds of organic compounds, including as many as 3,000 different kinds of proteins and 1,000 kinds of nucleic acids. In contrast, the inanimate objects such as clay, sand, rock, sea water etc., consist of random mixtures of simpler chemical molecules.

2. Each component unit of a living object appears to have a *specific purpose or function*, whether it be a macroscopic structure (heart, lungs, brain) or a microscopic intracellular structure (nucleus). Even the individual chemical compounds in cells (carbohydrates, proteins, lipids) have specific functions to perform. But it is meaningless or futile to ask about the function of various chemical compounds present in inanimate bodies. They just happen to be there.

3. The living organisms have the *ability to extract, transform and use energy* from their environment, either in the form of organic nutrients or the radiant energy of sunlight. Living organisms are never at equilibrium within themselves or with their surroundings. On the other hand, the inanimate bodies do not use energy to maintain its sfurcture or else to do work. Rather when left to itself, it tends to come to equilibrium with its surroundings.

4. The most remarkable attribute of the living organisms is their *capacity for self- replication*, a characteristic which can be regarded as the very quintessence of the living state. Lifeless objects, on the contrary, do not grow and reproduce in forms identical in mass, shape and internal structure.

It should, however, be realized that the boundary between the living and nonliving objects is not always well demarcated. For example, certain diseases are caused by extremely small filterable substances called viruses. These can also reproduce when introduced into the environment of living cells and a few of them have been isolated in a purified cyrystalline form. Chemically, the crystalline viruses such as tobacco mosaic virus (TMV) are nucleoproteins and have no apparent features of living things. Yet TMV, when inoculated into a healthy leaf of a tobacco plant, multiplies rapidly and causes the onset of a disease termed tobacco mosaic disease. Thus, an inanimate crystalline substance, on inoculation, starts behaving as if it were living.

AXIOMS OF LIVING MATTER

The living objects are endowed with certain remarkable attributes. These attributes are due to the typical nature, function and interactions of the **biomolecules**, the kinds of molecules found in living organisms. **Albert L. Lehninger** (1984) lists some of the *axioms* or 'principles' which are characteristic of the living state. These are:

1. There is a basic simplicity in the structure of biological molecules.
2. All living organisms use the same kinds of building block molecules and thus appear to have a common ancestry.
3. The identities of each species or organism is preserved by its possession of distinctive sets of nucleic acids and of proteins.
4. All biomolecules have specific functions in cells.
5. Living organisms create and maintain their comnplex, orderly, purposeful structures at the expense of free energy from their environment to which they return energy in less useful forms.
6. Living cells are chemical engines that function at constant temperature.
7. The energy needs of all organisms are provided, directly or indirectly, by solar energy.
8. The plant and animal worlds– indeed,all living organisms– are dependent on each other through exchanges of energy and matter *via* the environment.
9. Living cells are self-regulating chemical engines, tuned to operate on the principle of maximum economy.
10. Genetic information is encoded in units that are submolecular in dimensions; these units are the four kinds of nucleotides, of which DNA is composed.
11. A living cell is self-assembling, self-adjusting, self-perpetuating isothermal system of organic molecules which extracts free energy and raw materials from its environment.
12. It carries out many consecutive organic reactions promoted by organic catalysts, which it produces itself.

13. It maintains itself in a dynamic steady state, far from equilibrium with its surroundings. It functions on the principle of maximum economy of parts and processes.
14. Its nearly precise self-replication through many generations is ensured by a self-repairing linear coding system.

MAJOR ORGANIC COMPOUNDS OF ANIMATE OBJECTS

Living beings contain a wide variety of organic compounds, besides the ubiquitous water and other inorganic compounds. Major organic molecules present in the living beings are: carbohydrates, proteins (of course, including enzymes), lipids and nucleic acids. Table 1–1 lists some details of these compounds.

Table 1–1. Major compounds of living beings

Organic compound	*Building block*	*Some major functions*	*Examples*
Carbohydrate			
Monosaccharide	—	Energy storage; physical structure	Glucose, fructose, galactose
Disaccharide	Monosaccharides	Energy storage; physical structure	Lactose, maltose, sucrose
Polysaccharide	Monosaccharides	Energy storage; physical structure	Starch, cellulose, chitin, inulin, pectin
Protein	Amino acids	Enzymes; toxins; physical structures	Antibodies; viral surface; flagella; pili
Lipid			
Triglycerides	Fatty acids and glycerol	Energy storage; thermal insulation; shock absorption	Fat, oil
Phospholipids	Fatty acids, glycerol, phosphate, and an R group*	Foundation for cell membranes	Plasma (cell membranes)
Steroids	Four-ringed structure†	Membrane stability	Cholesterol
Nucleic acid	Ribonucleotides; Deoxyri\bonucleotides	Inheritance; instructions for protein synthesis	DNA, RNA

* R group = a variable portion of a molecule.

† Technically, steroids are neither polymers nor macromolecules.

THE SCIENTIFIC METHOD OF APPROACH

Although biologists employ a variety of approaches in conducting research, the experimentally -oriented biologists such as biochemists and microbiologists often use the general approach known as the **scientific method.** They first gather **observations** of the process to be studied and then develop a tentative hypothesis—an educated guess—to explain the observations (Fig. 1–5). Thus, the **hypothesis** is simply a tentative explanation to account for observed phenomena. This step often is inductive and creative because there is no detailed, automatic technique for generating hypotheses. Next, they decide what information is required to test the hypothesis and collect this information through observation or carefully designed experiments. Then, they decide whether the hypothesis has been supported or falsified. If it has failed to pass the test, the hypothesis is rejected, and a new explanation or hypothesis is constructed. If the hypothesis passes the test, it is subjected to more severe testing. The procedure often is made more efficient by constructing and testing alternative hypotheses and then refining the hypothesis that survives testing. This general approach is often called the **hypothetico-deductive method.** One deduces predictions from the currently-accepted hypothesis and tests them. In *deduction,* the conclusion about specific cases follows logically from a general premise ("if . . ., then . . . ," reasoning). *Induction* is the opposite. A general conclusion is reached after considering many specific examples. Both types of reasoning are used by scientists.

When carrying out an experiment, it is essential to use a control group as well as an experimental group. The *control group* is treated precisely the same as the *experimental group* except that the experimental manipulation is not performed on it. In this way, one can be sure that any changes in the experimental group are due to the experimental manipulation rather than to some other factor not taken into account.

If a hypothesis continues to survive testing, it may be accepted as a valid theory. The term **theory** is applied to a hypothesis that has been extensively tested and that ties together and arranges the results of a number of observations and experiments. It provides a reliable, systematic, and rigorous account of an aspect of nature. It is important to note that *hypotheses and theories are never absolutely proven.* Scientists simply gain more and more confidence in their accuracy as they continue to survive testing, fit with new observations and experiments, and satisfactorily explain the observed phenomena.

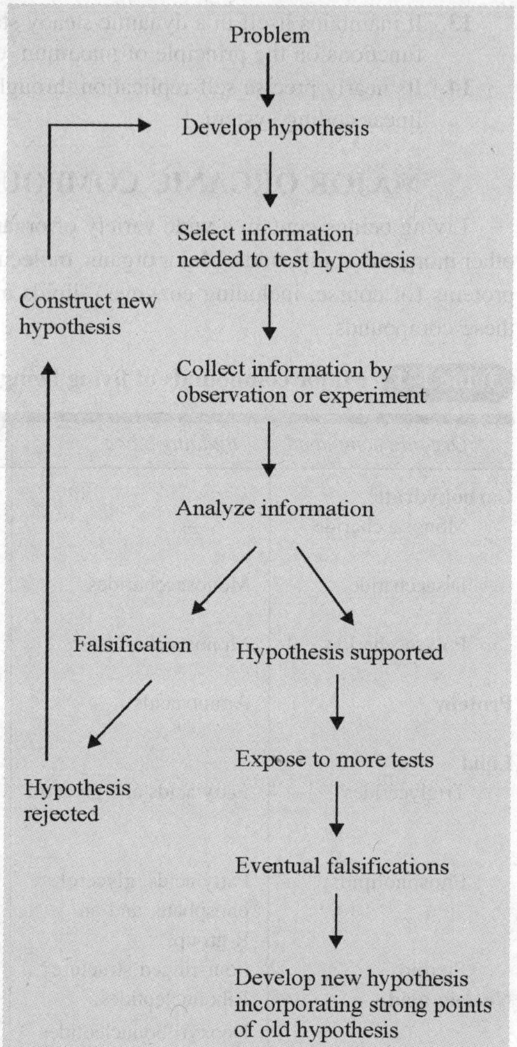

Fig. 1–5. The hypothetico-deductive method
This approach is most often used in scientific research.

EXPERIMENTAL TOOLS IN BIOCHEMICAL REVOLUTION

From its very inception, biochemistry has been much more an experimental science than a theoretical one. More knowledge has accumulated from attempts to explain laboratory explanations than from experiments designed to test theories. The **Buchner brothers (Eduard** and **Hans)**, for example, did not set out to prove that fermentation of sugar to ethanol could occur outside the cell. Rather, their accidental observation that this occurred spurred the developments that led to our understanding of carbohydrate metabolism.

The experimental nature of biochemistry has 2 important *corollaries*:

(*a*) first, in order to understand biochemistry, one must understand the critical experiments and experimental techniques on which our common understanding rests.

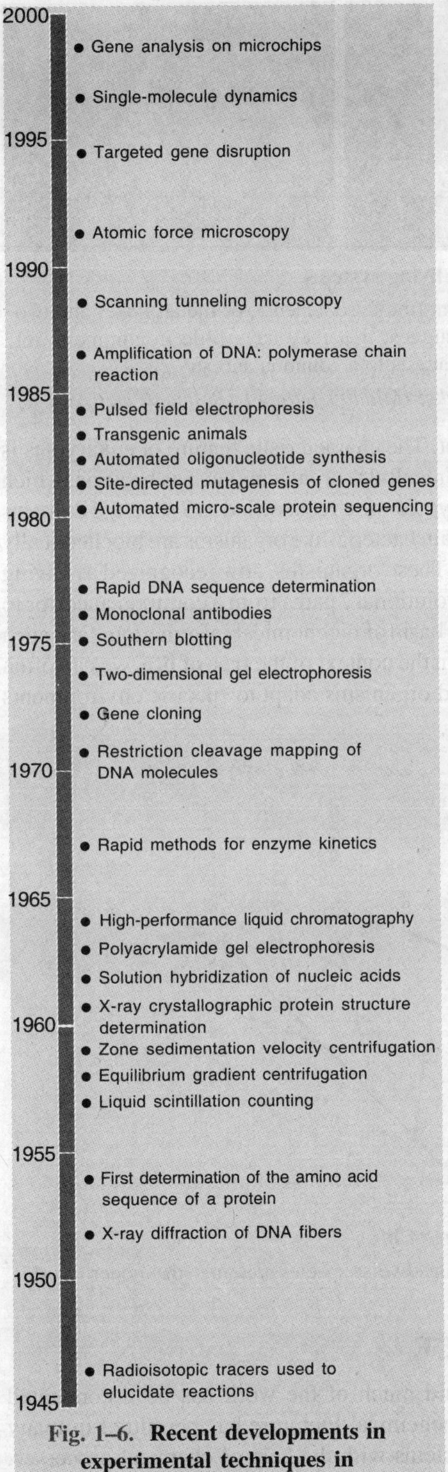

Fig. 1–6. Recent developments in experimental techniques in biochemistry

(*b*) second, the spectacular growth of biochemical knowledge over the past 50 years is due directly to the development of many powerful research methodologies – the armament of biological revolution. Some of this development is chronicled in Fig. 1–6.

A familiar example is **gene cloning**, which dates from about 1973. Our ability to isolate any desired gene, to read its information and to manipulate it so that any gene product can be formed in essentially unlimited amounts has transformed all of biology. However, gene cloning is but one of a long series of experimental advances that have transformed biochemistry and biology in equally fundamental ways. For example, the introduction of radioisotopes was essential to the illucidation of most metabolic pathways; x-ray crystallography permitted the structural localization of every atom within a protein structure; and 2-'D' gel electrophoresis permitted the visualization and quantitation of virtually all of the thousands of protein molecules present in a cell.

BIOCHEMICAL UNITY WITHIN BIOLOGICAL DIVERSITY

The remarkable variety of living systems (Fig. 1–7) belies a striking similarity. The common use of DNA and the genetic code by all organisms underlies one of the most important discoveries of the 20th century– namely, that **all organisms are remarkably uniform at the molecular level.** They are built from similar molecular components, distinguishable by relatively minor variations. This uniformity reveals that all organisms on Earth have arisen from a common ancestor. A core of the essential biochemical processes, common to all organisms, appeared early in the evolution of life. The diversity of life in the modern world has been generated by the evolutionary processes acting on these core processes through millions or even billions of years. As we will see repeatedly, the generation of diversity has very often resulted from the adaptation of existing biochemical components to new roles rather than the development of fundamentally new biochemical technology.

On the basis of their biochemical characteristics, the diverse organisms of the modern world can be divided into 3 fundamental groups called domains: **Eukarya** (eukaryotes), **Bacteria** (formerly Eubacteria), and **Archaea** (formerly Archaebacteria). Eukarya comprise all macroscopic organisms, including human beings as well

Fig. 1–7. The diversity of living systems

The distinct morphologies of the 3 organisms shown– a plant (the false hellebra, or Indian Poke) and two animals (sea urchins and a common house cat)– might suggest that they have little in common. Yet biochemically they display a remarkable commonality that attests to a common ancestry.

[Courtesy: (Left and Right) John Dudak/Phototake; (Middle) Jeffrey L. Rotman/Peter Arnold]

as many microscopic, unicellular organisms such as yeast. The characteristic feature of eukaryotes is the presence of a well-defined nucleus within each cell. Unicellular organisms such as bacteria, which lack a nucleus, are referred to as **prokaryotes**. The prokaryotes were reclassified as 2 separate domains in response to a discovery by **Carl Woese** in 1977 that certain bacteria-like organisms are biochemically quite distinct from better-characterized bacterial species. These organisms, now recognized as having diverged from bacteria early in evolution, are archaea. Evolutionary paths from a common ancestor to modern organisms can be developed and analyzed on the basis of biochemical information. One such path is depicted in Fig. 1–8. By examining biochemistry in the context of the **tree of life,** we can often understand how particular molecules or processes helped organisms adapt to specific environments or lifestyles.

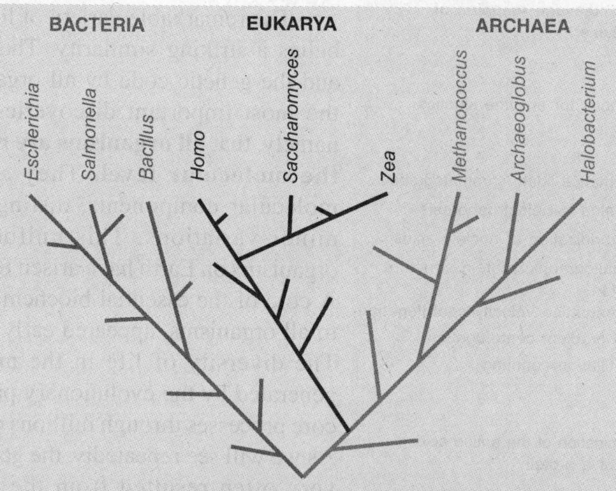

Fig. 1–8. The tree of life

A possible evolutionary path from a common ancestral cell to the diverse species present in the modern world can be deduced from DNA sequence analysis.

IMPORTANCE

Modern biochemistry is relatively a new branch and much of the work has been conducted during the present century. The enormous growth of biochemical literature has permitted in many cases the correlation of physical events in biological systems with the help of chemical processes. For a biochemist, therefore, a physiological process does not merely mean to elucidate the nature of the chemical substances involved in it but also to find out the physical relations among these substances and of these substances to the environment. Thus, the principal objective of biochemistry is to fill the wide gap between the highly integrated functions of the living cell and the various properties of its individual chemical constituents. A biochemist, therefore, has to perform an important arduous task of carrying the research work with utmost sincerity, patience and honesty. **Prof. Hopkins** (1931) has rightly remarked :

> "He (biochemist) should be bold in experiment but cautious in his claims. His may not be the last word in the description of life, but without his help the last word will never be said."

The *American Society of Biological Chemists* (1965) has, indeed, worked out a definition of a *biochemist* as a guideline for eligibility for membership in that society. The definition reads as follows :

> "A biochemist is an investigator who utilizes chemical, physical or biological technics to study the chemical nature and behaviour of living matter."

Ernest Baldwin (1937) entrusts the biochemist with the task of, "the study of physicochemical processes associated with the manifestations of what we call life– not the life of some particular animal or group of animals, but life in its most general sense."

APPLICATIONS OF BIOCHEMISTRY

Biochemistry is a research discipline, but the results of biochemical research are used extensively in the world outside the laboratory– in agriculture, medical sciences, nutrition etc. In clinical chemistry, biochemical measurements on people help diagnose illnesses and monitor responses to treatment. For example, liver disease is now routinely diagnosed and monitored by measurements of blood levels of enzymes called **transaminases** and of a hemoglobin breakdown product called **bilirubin.**

Pharmacology and toxicology are concerned with the effects of external chemical substances on metabolism. Drugs and poisons usually act by interfering with specific metabolic pathways. A good example is the antibiotic **penicillin,** which kills bacteria by inhibiting an enzyme that synthesizes an essential polysaccharide of the bacterial cell wall. Because animal cells do not synthesize these polysaccharides, they are not harmed by this inhibitor, which can henceforth be used therapeutically. A particularly exciting prospect in contemporary biochemistry is that of creating so-called designer drugs. If the target site for action of a drug is a protein enzyme or receptor, determining the detailed molecular structure of that target allows us to design inhibitors that bind to it with great selectivity.

Herbicides and pesticides, in many instances, act in similar ways– by blocking enzymes or receptors in the target organism. The first generations of these toxic agents (such as DDT) were so nonspecific in their effects that organisms other than the target populations were often affected, resulting in unforeseen and often severe damage to the environment. Furthermore, the indiscriminate use of these agents gave rise to resistant target populations, so that an ever-increasing number of toxic substances needed to be used. Biochemistry is involved in understanding the actions of herbicides and pesticides, in increasing their selectivity, and in understanding and dealing with mechanisms by which target organisms become resistant to them. Thus, biochemistry has become an important component of environmental science.

Finally, through gene engineering and cloning, remarkable progress in agriculture is being made. Already, genetically-engineered plants are in production, and whole-animal cloning has been successful with creatures as large as sheep.

SUMMARIZING BIOCHEMISTRY

Editors say that any text can be shortened, no matter how important the details. Below we attempt to condense this book to about 2 pages. This gives a very general overview but also creates a useful mental framework for further study. The highlighted terms take you through the contents of the following chapters.

1. **The major structural components of the body are proteins, carbohydrates and lipids.** Proteins are building blocks and catalysts; as structural units, they form the architectural

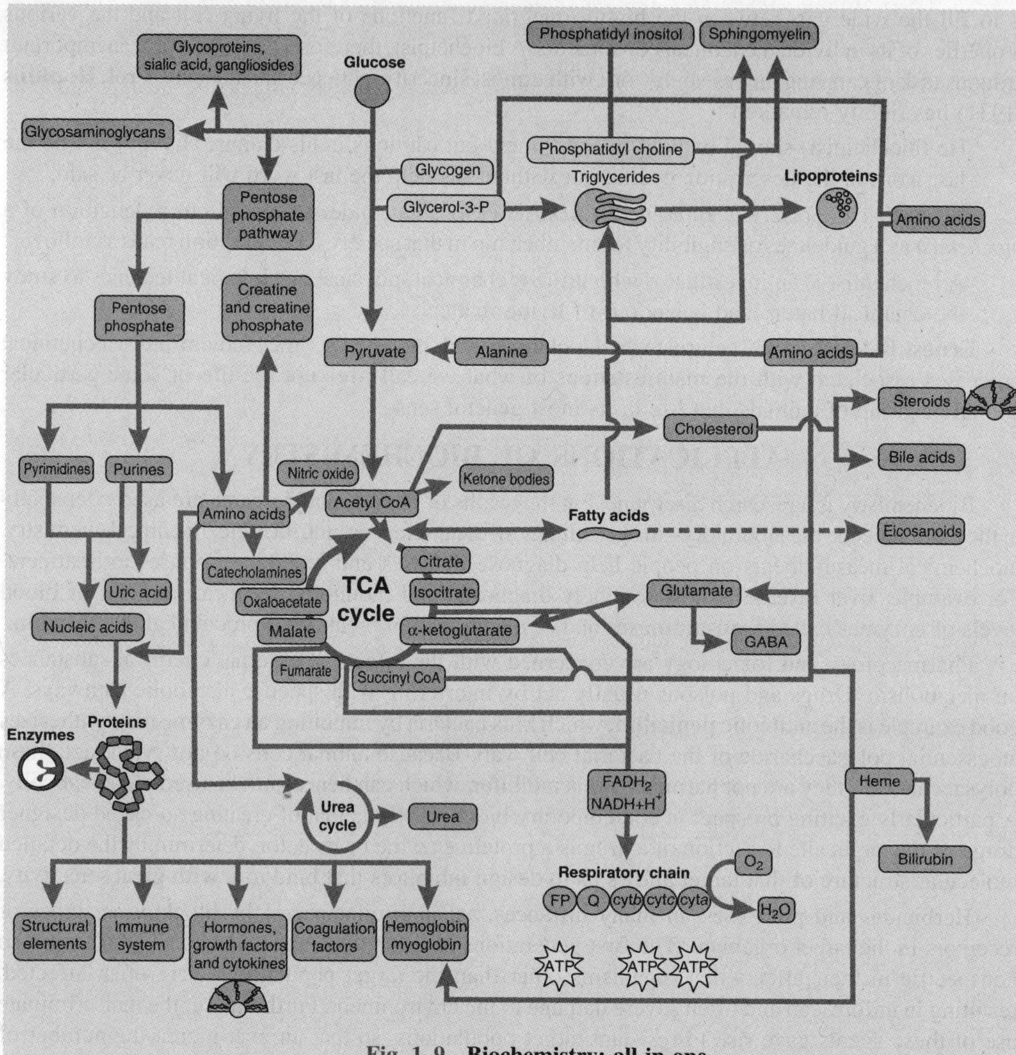

Fig. 1–9. Biochemistry: all in one

This figure gives the interrelationships between various biochemical pathways. This figure has been designed to give you a bird's-eye view of the field. It may help to structure your study or revision.

[GABA = γ-aminobutyrate; glycerol-3-P = glycerol-3-phosphate;
CoA = coenzyme A; TCA = tricarboxylic acid cycle; cyt = cytochrome;
FP = flavoprotein; Q = coenzyme Q_{10}; ATP = adenosine 5′-triphosphate]

(*After Baynes JW and Dominiczak MH, 2009*)

framework of our tissues; as enzymes together with helper molecules known as coenzymes and cofactors, they catalyze and control biochemical reactions. Carbohydrates and lipids are used primarily as energy sources. Their storage form in the body are glycogen and triglycerides. Carbohydrates are also present as glycoconjugates with proteins and lipids, and lipids form the backbone of biological membranes.

Chemical variables such as pH, oxygen tension, inorganic iron and buffer concentrations define the homeostatic environment in which metabolism takes place. Small changes in this environment, for example, less than one-tenth of a pH unit, can be life-threatening. All therapies, including emergency interventions, aim to maintain its stability. Blood is the medium for exchange of gases, fuels, metabolites and information between tissues. Blood plasma is an accessible 'window' on metabolism and serves as a source of clinical information for the diagnosis and management of the disease. The blood coagulation system and the immune system defend against disturbances in this environment.

Biological membranes compartmentalize metabolic pathways. They play a fundamental role in ion and metabolite transport, and also in transducing signals from one cell to another. Most of the body's energy is consumed to maintain ion and metabolite gradients across biological membranes; nerve and muscle function, and the red blood cell are critically dependent on membrane potentials, which are used for nerve transmission, muscle contraction and maintenance of cell shape.

2. **Energy released from nutrients is distributed throughout the cell in the form of ATP.** The recovery and utilization of energy in biological systems occur through oxidative phosphorylation which takes place in the mitochondria. This process involves oxygen consumption or respiration, by which we capture the energy of fuels, produce a hydrogen ion gradient and convert this energy to adenosine triphosphate (ATP). Biochemists call ATP the 'common currency of metabolism' for exchange of metabolic energy. It transduces energy from fuel metabolism for use in work, transport and biosyntheses.
3. **Central pathways of carbohydrate and lipid metabolism are access routes to other processes.** Sugars and fats are our primary sources of energy but our nutritional requirements also include amino acids or proteins and micronutrients such as vitamins and trace elements.

   Glucose is metabolized through glycolysis, a universal anaerobic pathway for energy production. Glycolysis transforms glucose into pyruvate, setting the stage for oxidative metabolism in the mitochondria. It also produces metabolites that are the starting points for the synthesis of amino acids, proteins, lipids and nucleic acids. Glycolytic enzymes are regulated by several mechanisms: small-molecule effectors, reversible and irreversible chemical modification of key enzymes, and by control of gene expression.

   The maintenance of a normal concentration of blood glucose is essential for our survival and is linked to the metabolism of glycogen, the short-term storage form of glucose. Glucose homeostasis is regulated by hormones (primarily insulin and glucagon, and also epinephrine) that coordinate metabolic activities among cells and organs.
4. **Oxygen is essential for survival but can also be toxic.** During aerobic metabolic, pyruvate is transformed into acetyl-coenzyme A (acetyl-CoA), which is the common intermediate in the metabolism of carbohydrates, lipids and amino acids. Acetyl-CoA enters the central metabolic engine of the cell, the tricarboxylic acid cycle (TCA cycle) in the mitochondrion. Acetyl-CoA is oxidized to carbon dioxide and reduces the important coenzymes nicotinamide adenine dinucleotide (NAD^+) and flavin adenine dinucleotide (FAD). Reduction of these nucleotides captures the energy from fuel oxidation, and they are, in turn, the substrates for the final pathway, the oxidative phosphorylation. They become oxidized by molecular oxygen through a chain of electron transport reactions, providing the energy for the synthesis of ATP.

   While oxygen is essential for aerobic metabolism, it can also cause oxidative stress and widespread tissue damage during inflammation. To protect us from the more damaging effects of oxygen, nature has provided us with powerful antioxidant defenses.
5. **Eating and fasting shift body metabolism between the anabolic and catabolic states.** Life can be divided into continuous cycle of eating and non-eating. The main pathways of carbohydrate and lipid metabolism are partially reversible and their direction changes in response to food intake. The direction of metabolism is constantly shifting during the feed-

fast cycle. In the fed state, the active pathways are glycolysis, glycogen synthesis, lipogenesis and protein synthesis, rejuvenating tissues and storing the excess of metabolic fuel. In the fasting state, the direction of metabolism is reversed: glycogen and lipid stores are degraded through glycogenolysis and lipolysis. Protein is converted into glucose by the pathway of gluconeogenesis, and other biosynthetic processes are slowed down. Derangements of fuel metabolism, which often develop as the person ages, lead to diabetes and dyslipidemias and are major contributors to atherosclerosis.

6. **Tissues perform specialized functions.** These functions include muscle contraction, nerve conduction, immune surveillance, hormonal signaling, maintenance of pH and electrolyte balance, detoxification of foreign substances and bone formation. Specialized microstructures, such as glycoconjugates (glyco-proteins, glycolipids and proteoglycans), play a crucial role in tissue organization, cell-cell interactions and in the structure and function of the extracellular matrix.
7. **Genome and cellular signaling underlie it all.** The genome provides the mechanism of conservation and transfer of genetic information, the regulation of the expression of constituent genes and the control of protein synthesis. Protein synthesis is controlled by information encoded in DNA and transcribed into RNA, which is then translated into peptides that fold into functional protein molecules. The spectrum of expressed proteins and the control of their temporal expression during development, adaptation and ageing are responsible for our protein make-up.

Applications of recombinant DNA and polymerase chain reaction (PCR) technology have revolutionized the work of clinical laboratories during the last decade. The recent ability to scan the entire genome and the potential of proteomics create an opportunity to gain new insights into the dynamics of gene-driven protein synthesis. Finally, cell growth, cellular signaling and repair mechanisms are important for survival, and the time-dependent decline in these systems leads to ageing and development of diseases such as cancer.

LITERATURE

A number of review publications and research journals appear periodically on various topics of biochemistry from different regions of the globe. A few of them are mentioned below:

REVIEWS

1. Advances in Carbohydrate Chemistry
2. Advances in Enzymology
3. Advances in Protein Chemistry
4. Annual Review of Biochemistry
5. Annual Review of Microbiology
6. Annual Review of Physiology
7. Annual Review of Plant Physiology
8. Biochemical Journal
9. Biological Reviews
10. Biochemical Society Symposia
11. British Medical Bulletin
12. Essays in Biochemistry
13. Harvey Lectures
14. Physiological Reviews
15. Vitamins and Hormones

JOURNALS

1. American Journal of Physiology
2. Analytical Biochemistry
3. Archives of Biochemistry and Biophysics
4. Biochemical Journal
5. Biochemische Zeitschriff
6. Biopolymers
7. Canadian Journal of Biochemistry
8. Comparative Biochemistry and Physiology
9. Endocrinology
10. Enzymologia
11. Indian Journal of Biochemistry
12. Indian Journal of Medical Research
13. Journal of Bacteriology
14. Journal of Biological Chemistry
15. Journal of Cell Biology
16. Journal of Cellular Physiology

17. Journal of Chromatography
18. Journal of Endocrinology
19. Journal of Lipid Research
20. Journal of Molecular Biology
21. Journal of Nutrition
22. Journal of Physiology
23. Nature
24. Plant Physiology
25. Proceedings of the Natural Academy of Sciences
26. Trends in Biochemical Sciences

REFERENCES

1. **Allen GE :** Life Sciences in The Twentieth Century. *John Wiley. 1975.*
2. **Bayliss WM :** Principles of General Physiology. *Longmans Green and Co., London. 1924.*
3. **Browne CA :** A Source Book of Agricultural Chemistry. *Chronica Botanica Co., Waltham. 1944.*
4. **Chittenden RH :** The Development of Physiological Chemistry in the United States. *Amer. Chem. Soc. Monograph No. 54, New York. 1930.*
5. **Chittenden RH :** The First Twenty-five Years of the American Society of Biological Chemists. 1945.
6. **Fruton JS :** Molecules and Life. Historical Essays on the Interplay of Chemistry and Biology. *Wiley Interscience, New York. 1972.*
7. **Fulton JF, Wilson LG :** Selected Readings in the History of Physiology. *Charles C. Thomas, Springfield, Illinois. 1966.*
8. **Henderson LJ :** The Fitness of the Environment. *Beacon Press, Boston, USA. 1958.*
9. **Ihde AJ :** The Development of Modern Chemistry. *Harper and Row, Publishers, Inc., New York. 1964.*
10. **Kornberg A :** The Two Cultures : Chemistry and Biology. *Biochem.* **26 :** *6888-6891, 1987.*
11. **Lipmann F :** Wanderings of a Biochemist. *Wiley, New York. 1971.*
12. **McCollum EV :** A History of Nutrition. *Houghton Mifflin Company, Boston, USA. 1957.*
13. **Needham J (editor) :** Chemistry of Life. Eight Lectures on the Histroy of Biochemistry. *Cambridge Univ., Press, New York. 1970.*
14. **Needham J, Baldwin E (editors) :** Hopkins and Biochemistry. *Heffer and Son Ltd., Cambridge. 1949.*
15. **Orgel LE :** The origin of life: A review of facts and speculations. *Trends Biochem. Sci.* **23 :** *491–495, 1998.*
16. **Orten JM, Neuhaus OW :** Biochemistry. *8th ed. The C.V. Mosby Co., Saint Louis. 1970.*
17. **Pace NR :** The universal nature of biochemistry. *Proc. Natl. Acad. Sci., USA.* **98 :** *805–808, 2000.*
18. **Partington JR :** A History of Chemistry. *Macmillan and Company Ltd., London. 1961, 1964.*
19. **Pirie NW, in Needham J, Green DE :** Perspectives in Biochemistry. *Cambridge Univ., Press, Cambridge. 1937.*
20. **Schrödinger E :** What is Life ? *Cambridge University Press, New York. 1944.* [Reprinted (1956) in What is Life ? and Other Scientific Essays. *Double-day Anchor Books, Garden City, New York.*]
21. **Solomons TWG :** Organic Chemistry. *2nd ed., John Wiley and Sons, New York. 1980.*
22. **Sumner JB :** The isolation and crystallization of the enzyme urease. *J. Biol. Chem.* **69 :** *435–441, 1926.*

23. **van Niel CB :** Bacteriological Reviews. **13 :** *161, 1949.*

24. **Vogel HJ, Vogel RH :** Some chemical glimpses of evolution. *Chem. Eng. News.* **45 :** *88, 1967.*

25. **Waksman SA :** Sergei N. Winogradsky. *Rutgers Univ., Press, New Brunswick. 1953.*

26. **Williams RJ, Lansford EM (editors) :** Encyclopedia of Biochemistry. *Reinhold Publishing Corp., New York. 1967.*

27. **Willstätter R :** From my life. *Benjamin. 1965.*

PROBLEMS

It is very difficult to learn something well without somehow participating in it. The chaper-end problems are therefore an asset of any textbook. This book contains few problems in each chapter which are of regurgitatory type. Rather they are designed to make one think and to offer insights not discussed in the text. The difficulties in solving these problems range from those that require only a few moments reflection to those that might take an hour or more of concentrated effort to work out.

1. Define biochemistry.
2. What are the goals of biochemistry?
3. Prove that biochemistry is an amalgamation mainly of chemistry, biology, physics and bioenergetics.
4. What type of scientific approach is normally used in conducting biochemical researches?
5. Which theory gave way to the theory of organic synthesis?
6. Who coined the word 'biochemistry'?
7. Who is considered the '*father of modern biochemistry*'?
8. Which is the oldest available literature in biochemistry?
9. Who is a biochemist?

ANSWERS TO PROBLEMS

1. Biochemistry is a science concerned with the chemical nature and chemical behaviour of the living matter. It takes into account the studies related to the nature of the chemical constituents of living matter, their transformations in biological systems and the energy changes associated with these transformations. Such studies have been conducted in the animal and plant tissues both. Broadly speaking, biochemistry may thus be treated as a discipline in which biological phenomena are analyzed in terms of chemistry. The branch of biochemistry, for the same reason, has been variously named as **Biological Chemistry** or **Chemical Biology.**
2. Biochemistry seeks to describe the structure, organization, and functions of the living matter in molecular terms. What are the chemical structures of the components of living matter? How do the interactions of these components give rise to organized supramolecular structures, cells, multicellular tissues, and then organisms? How does living matter extract energy from its surroundings in order to remain alive? How does an organism store and transmit the information it needs to growth and to reproduce itself accurately. What chemical changes accompany the reproduction, ageing, death of cells and organisms? How are chemical reactions controlled inside living cells? These are the kinds of questions being asked by biochemists; the search for the answers is the study of the chemistry of life.
3. The body of living organisms contains a number of molecules (biomolecules), most of which are of organic type. To study their behaviour within the organisms, it is essential first to know about their structure and properties which is usually studied under **chemistry**, *esp.*, organic chemistry. The transformations of these molecules to other molecules involves

the energy which is either taken in or given out; such studies are a part of another fascinating branch called **thermodynamics** or **bioenergetics**. The various transformations and the physiological processes (respiration, digestion, excretion, synthesis and degradation of biomolecules, photosynthesis etc) that go on in the living body are all governed by definite physical principles, which are integral part of the study of **physics**. For all this to study, a thorough knowledge of the external and internal structure of the organism, its behaviour and response to various stimuli, its genetic make-up and its niche are all required to be studied, which comes under the purview of **biology**. Thus, truly biochemistry is an amalgamation of chemistry, biology, physics and bioenergetics.

4. **Hypothetico-deductive type** of scientific approach.
5. **Vitalistic theory** (= the doctrine of vitalism).
6. A German chemist **Carl Neuberg,** in 1903.
7. A French chemist **Antoine Laurent Lavoisier** (LT, 1743–1794).
8. **'*Lectures in Animal Chemistry*'**, written by a Swedish chemist **Jöns Jacob Berzelius** (LT, 1779–1848) and published in 1806.
9. A biochemist is an investigator who utilizes chemical, physical or biological techniques to study the chemical nature and behaviour of living matter. He should be bold in experiment but cautious in his claims. His may not be the last word in the description of life, but without his help the last word will never be said.

CONTENTS

CHAPTER 2

Water : The Solvent of Life

"If there is magic on this planet, it is contained in water."

–Loren Eiseley: The Immense Journey, 1957

A striking example of water in both the solid and liquid states
Life as we know it depends on the properties of this simple molecule.

GENERAL CONSIDERATIONS

Water is the mother liquor of all forms of life. It is the vital essence, miracle of Nature, and the great sustainer of life. The essentiality of water for living systems is quite evident as without water, there is no life. The essentiality of water is a continuous reminder of the aqueous origin of life. It was in the solvent water that the chemical reactions of biological processes evolved. All aspects of cell structure and function are adapted to the physical and chemical properties of water. The strong attractive forces between water molecules result in water's solvent properties. A meagre tendency of water to ionize is of utmost importance to the structure and function of biomolecules. The water molecule and its ionization products (H^+ and OH^-) greatly influence the structure, assemblage and properties of all the cellular components, including enzymes and other proteins, nucleic acids and lipids. Water is needed not only for biochemical reactions, but also for transporting substances, across membranes, maintaining body temperature, dissolving waste products for excretion and producing digestive fluids.

Water is the medium in which the first cells arose, and the solvent in which most biochemical transformations take place. The properties of water have shaped the course of evolution and exert a decisive influence on the structure of biomolecules in aqueous solution. Many of the weak interactions within and between biomolecules are strongly influenced by the solvent properties of water. Even water-insoluble components, such as membrane lipids, interact with each other in ways dictated by the polar properties of water.

No other substance on earth is as abundant as water. It is almost everywhere in air, clouds, oceans, lakes, rivers, springs or glaciers. In the 5 km layer below the sea level on Earth, water is nearly 6 times as abundant as all other substances put together. And, none other occurs in 3 states– solid, liquid and gaseous– at the same time. It is water that had conditioned our climate. The water in oceans, seas and the atmosphere (vapour) acts as an accumulator of heat. In hot weather, it absorbs heat and in cold, it gives up heat; thus, it keeps the planet warm. Without water, our planet would be cooled to subzero temperatures long ago and all forms of life would have perished. It takes 10 litres of water to produce 1 litre of petroleum, 100 litres for 1 kg of paper, 4,500 litres for 1 kg of rice, 4,500 litres for l ton of cement and 20,000 litres to produce 1 ton of steel. This roughly gives us an idea of the extent of water usage. Water has deluded us into believing that it is abundant and in exhaustible. But it is merely an illusion rather than a reality. The volume of water on the earth is about 1,36,00,00,000 cubic kilometres which covers nearly 70% of the earth's surface. The global scenario is that 97% of water is in the sea, 2% locked up in the Arctic and Antarctic oceans, and 1% is fresh water. Of this 1% fresh water, 0.22% is in the underground acquifer and the remaining 0.78% of the world's water fills the rivers, streams, lakes and ponds. No doubt, water is a renewable source but it is finite too. So it has to be used judiciously. The daily water requirement of an average person is as follows:

15 litres – for cooking and drinking needs
35–40 litres – for toilet needs
65 litres – for bathing, washing, and other sundry needs.

Water accounts for ca 70% or more of the weight of most organisms. In human adults also, total body water accounts for about 70% of the lean body mass. Variations observed are mainly due to differences in fat contents. In obese males, water constitutes a lower percentage of body weight (45–60%) than in lean individuals (55–70%). Adult lean females have a low water content (45–60%) and the value in infants can be in the range 65–75%. Our normal body temperature of 37°C is maintained mainly because water is expelled by lungs and skin. However, a loss of 10% of water in our body is serious; a loss of 20% fatal. In humans, water is not evenly distributed between the two major compartments (intracellular and extracellular) of the body (Table 2–1). In an adult male of 70 kg body weight, 70% of water (*i.e.*, about 30 litres) is found in intracellular fluids. Of this, about 4 litres are found in bones which does not readily participate in fluid exchange. Nearly 30% of water (*i.e.*, about 12 litres) is found in extracellular fluid. Plasma (3 litres), interstitial fluid (8.5 litres) and transcellular fluid (l litre) come under this category.

Table 2–1. Water distribution in an adult male of 70 kg body weight

Compartment	*Total body water (in litres)*	
Intracellular fluids	30	
Extracellular fluids		
Plasma	3	12.5
Interstitial and lymph	8.5	
Transcellular fluid	1	

The maintenance of **water balance**, the equilibrium between water intake and output, is a critical aspect of metabolism. An adult in water balance generally takes in and loses about 2.8 litres of water per day in temperate climate and between 3.3 and 7.3 litres per day in tropical climate (Table 2–2). Besides the water obtained from foods and liquids, **metabolic water** is also made available through the oxidation of food in the body. Oxidation of 100 g each of fat, carbohydrate and protein yields 107, 55 and 41g of water, respectively. Water losses occur by evaporation (water vapour in exhaled air and perspiration) and by excretion of urine and feces.

Table 2–2. Daily water intake and output (or loss) from the human body

Intake/Output	*Temperate climate*		*Tropical climate*	
Intake				
As liquids	1.5		2 – 5	
In foods	1.0	2.8	1 – 2	3.3 – 7.3
Oxidation of foods	0.3		0.3	
Output (or Loss)				
Urine	1.5		1.0 – 1.5	
Feces	0.1	2.8	0.1 – 0.2	3.3 – 7.3
Evaporation through skin	0.8		1.8 – 5.2	
Evaporation through lung	0.4		0.4	

If the output of water significantly exceeds its intake, **dehydration** occurs due to diarrhea, vomiting, fever and high environmental temperatures. Dehydration is especially serious in young children because their body pool of water is small and hence can be readily depleted. If water accumulates in tissues, then **edema** occurs. Edema is often observed in children suffering from kwashiorkor, a nutritional disease caused by an inadequate intake of protein. In the Carribeans, children with kwashiorkor are called *sugar babies* because of their plump appearance (a deceiving indication of good health).

Water is the most important possession of humanity and keeping in view the importance of the role of water on our planet, NASA has launched a satellite named *Aqua* to study the effect of water in its various forms on climate. The aim is to find whether there are any changes in the pattern of water cycle-climate relationship as a result of human-generated factors. *Aqua* is second in a series of 3 satellites planned by NASA with a view to gathering comprehensive data on Earth and its surroundings. Information gathered by *Aqua* will complement that being gathered by an earlier launched satellite, *Terra,* which concentrates on land mass studies. A third satellite, *Aura,* is to be launched in 2004, which will focus on atmospheric studies. Mission *Aqua* is indeed a commendable initiative of NASA because it concerns water, the magic substance on earth.

CLINICAL IMPLICATIONS 2–1.

Kwashiorkor

Kwashiorkor is the Bantu word meaning displaced child. This refers to the fact that kwashiorkor appears in infants when they are no longer nursed by their mothers, *i.e.*, are displaced from the breast by the birth of a new baby. Kwashiorkor is the most widespread protein malnourishment syndrome that results from a diet adequate in calories but deficient in protein. It is characterized by degeneration of the liver, severe anemia, and inflammation of the skin. Kwashiorkor is especially common among children in industrially-underdeveloped countries where the diet consists primary of a single plant material, as in Indonesia, where rice is the main food, and in parts of Africa and Latin America, where corn (maize) is the principal staple. Worldwide, corn is the second or third largest crop and it is a staple for nearly half the world's malnourished people (For this reason, intense efforts have been directed at breeding stations for producing corn with a higher quality protein. The results have been marvellous; the quality protein maize (QPM) has a protein quality twice that of normal corn and almost equal to milk. If found fit on field trials, it will be good news for the impoverished populations in Mexico and Central America where maize constitutes about 85% of the grain consumed).

Kwashiorkor typically affects children in the early stages after weaning. Early clinical evidence of protein malnutrition is vague but does include lethargy, apathy or irritability. When well advanced, it results in retarded growth, lack of stamina, loss of muscular tissue, a secondary immunodeficiency, increased susceptibility to infections, and edema of dependent parts. The child may develop impaired appetite, anorexia, flabbiness of subcutaneous tissue and loss of muscle tone and may show little interest in play or in toys. The liver may enlarge early or late; fatty infiltration is common. Edema usually develops early; failure to gain weight may be masked by edema, which is often present in internal organs before it can be recognized externally in the face and limbs. The affected children are said to have a characteristic *"sugar baby" appearance* due to their bloated body. Renal plasma flow, glomerular filtration rate, and renal tubular function are decreased. The heart may be small in early stages, but is usually enlarged later.

Dermatitis is common. Darkening of the skin appears in irritated areas but not in those exposed to sunlight, a contrast to the situation in pellagra. Dyspigmentation (either hyper- or hypo-) may occur in these areas after desquamation (shedding of epithelial tissues, chiefly of the skin) or may be generalized. Thus, characteristic scaly erythematous rashes develop on the skin. The hair is often sparse, dry, thin, brittle, and hypopigmented (reddish-brown) with *"flag sign"* (alternating bands of black and brown hair). A decrease in the concentration of serum albumin is the most characteristic change. Bone growth is usually delayed. Growth hormone (GH) secretion may be enhanced. Stupor (partial unconsciousness), comma and death may ultimately follow.

PHYSICAL PROPERTIES OF WATER

The physical properties of water differ markedly from those of other solvents. For example, water as a hydride of oxygen (H_2O) has a higher melting point, boiling point, heat of vapourization and surface tension than do the comparable hydrides of sulfur (H_2S) and nitrogen (NH_3) and most other common liquids (Table 2–3). These unusual properties are a consequence of strong attractions between adjacent water molecules, which give liquid water great internal cohesion.

Table 2–3. Some physical properties of water and other common liquids

Substance	*Melting point, °C*	*Boiling point, °C*	*Heat of vapourization, cal/g*	*Heat capacity, cal/g*	*Heat of fusion, cal/g*	*Surface tension*	*Dielectric constant*
Water	0	100	540	1,000	79.7	72.8	80
Methanol	– 98	65	263	0.600	22	–	33
Ethanol	– 117	78	204	0.581	24.9		
Propanol	– 127	97					
Butanol	– 90	117					
Acetone	– 95	56	125	0.528	23	23.7	21.4
Butane	– 135	0.5					
Hexane	– 98	69					
Chloroform	– 63	61	59	0.226		27.1	5.1
Benzene	6	80	94	0.500	30	28.9	2.3
Hydrogen sulfide	– 85	– 60	132	–	16.7		
Ammonia	– 78	– 33	327	1.120	84		
Hydrofluoric acid	– 92	19	360	–	54.7		

The vapourization of ice is called *sublimation* and the vapourization of a liquid is called *evaporation*. Both processes occur more rapidly as the temperature increases. However, evaporation occurs most frequencly when water is heated above its boiling point, 100°C at sea level. Evaporation involves considerable amount of heat; 536 cal are needed to overcome the attraction between water molecules and to convert 1 g of water at 100°C into vapour. Heat lost at the point of evaporation returns at the point of condensation (the conversion from vapour to liquid). Such phenomena play a major role in meteorological cycles and the evaporative cooling of organisms.

Seawater behaves somewhat diferently. **Seawater** is defined as water with a minimum salinity of 24.7 0/00 (0/00 = parts per thousand). It's density, or rather its specific gravity ralative to that of an equal volume of pure water (sp. gra. = 1) at atmospheric pressure, is correlated with salinity. At 0°C, the density of seawater increases at a lower temperature and decreases at higher temperatures. No definite freezing point exists for seawater. Ice crystals begin to form at a temperature that varies with salinity. As pure water freezes out, the remaining unfrozen water becomes more salty, which further lowers its freezing point. If the temperature decreases further, a solid block of ice crystals and salt ultimately forms. Seawater has more salt concentration than our urine.

In humans, about 60% of red blood corpuscles (RBCs) and 92% of blood plasma are water. About 75% of most other tissues comprises of water; the only exceptions are such relatively inert tissues as hair, nails and the solid portion of bones. Water is also a principal constituent of the environment in which organisms thrive. Water has some unusual properties of physiological importance. These are described below:

1. Expansion on freezing. Most substances decrease in volume (and hence increase in density) as their temperature decreases. But in case of water, there is a temperature at which its density exceeds that at higher or lower temperatures. This temperature is 4 °C. In fact, water just above the freezing point is heavier than water at the freezing point. Therefore, it moves towards the bottom, freezing begins at the surface and the bottom is last to freeze. Organisms living at the bottoms of fresh-water lakes are, hence, protected from freezing. It may, however, be concluded that:

(*a*) While almost all substances contract on cooling, water expands. If water could contract on cooling, it would have become heavier and would have sunk. Thus, all water on earth would gradually have become ice.

(*b*) When temperature of water is raised above 0 °C, its volume decreases upto 4 °C and thereafter increases. Water thus has the minimum volume and hence the maximum density of 1.00 at 4 °C. The volume of water at 4 °C increases either on heating or cooling it.

(*c*) While all substances increase in volume when they are melted, the volume of ice decreases when melted : volume of 1g of ice is 1.09 cc but when it melts into water, it occupies only 1cc.

(*d*) Frozen water is less dense than liquid water.

2. Uniquely high surface tension. Like a stretched membrane, the surface of a liquid tends to contract as much as possible. This phenomenon is called *surface tension*. Water has the highest surface tension (of 72.8) of any known liquid. And it is the reason why water rises to unusually high levels in narrow capillary tubes. This has great significance in physiology.

3. Uniquely high heat capacity. There occurs a smaller temperature rise in water as compared to most other substances, when a given amount of heat is applied. Thus, water acts as a temperature buffer. It maintains its temperature more successfully than most other substances. We may, thus, say that has a high heat capacity (1,000 cal/g).

4. High solvent power. Water is a solvent for a great number of molecules which form ionized solutions in water. It may, thus, be called a universal solvent which facilitates chemical reactions both outside of and within biological systems.

Table 2–4 lists some of the unique features of water.

Table 2–4. Water and its features

	Property	*Chemistry*	*Result*
1.	Universal solvent	Polarity	Facilitates chemical reactions
2.	Adheres and is cohesive	Polarity; hydrogen bonding	Serves as transport medium
3.	Resists changes in temperature	Hydrogen bonding	Helps keep body temperature constant
4.	Resists change of state (from liquid to ice and from liquid to steam)	Hydrogen bonding	Moderates earth's temperature
5.	Less dense as ice than as liquid water	Hydrogen bonding changes	Ice floats on water

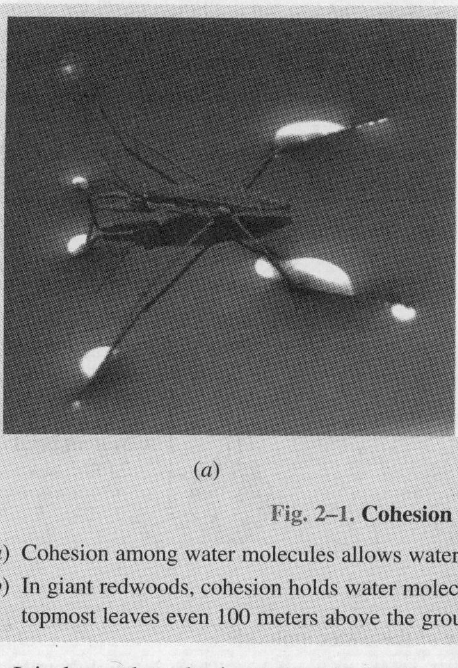

(*a*)

(*b*)

Fig. 2–1. Cohesion among water molecules

(*a*) Cohesion among water molecules allows water striders to skate across the surface of still waters.

(*b*) In giant redwoods, cohesion holds water molecules together in continuous strands from the roots to the topmost leaves even 100 meters above the ground.

It is due to the cohesive property of water molecules that allows water striders to skate across the surface of still waters and empowers giant redwoods to raise water about 100 metres above the ground (Fig. 2–1).

STRUCTURE OF WATER MOLECULE

In a water molecule (Figs. 2–2 and 2–3*a*, 3*b*), each hydrogen atom shares an electron pair with the oxygen atom. The geometry of the water molecule is dictated by the shapes of the outer electron orbitals of the oxygen atom, which are similar to the bonding orbitals of carbon. These orbitals describe a rough tetrahedron, with a hydrogen atom at each of the two corners and unshared electrons at the other two. The H—O—H bond angle is 104.5°, 5° less than the bond angle of a perfect tetrahedron which is 109.5°; the nonbonding orbitals of the oxygen atom slightly compress the orbitals shared by hydrogen.

The oxygen nucleus attracts electrons more strongly than does the hydrogen nucleus (*i.e.*, the proton); oxygen is more electronegative. The sharing of electrons between H and O is therefore unequal; the electrons are more often in the vicinity of the oxygen atom than of the hydrogen. This unequal electron sharing creates two electric dipoles in the water molecule, one along each of the H—O bonds. The oxygen atom bears a partial negative charge (δ^-), and each hydrogen a partial positive charge (δ^+). The resulting electrostatic attraction between the oxygen atom of one water molecule and the hydrogen of another water molecule (Fig. 2–3*c*) constitutes a **hydrogen bond.**

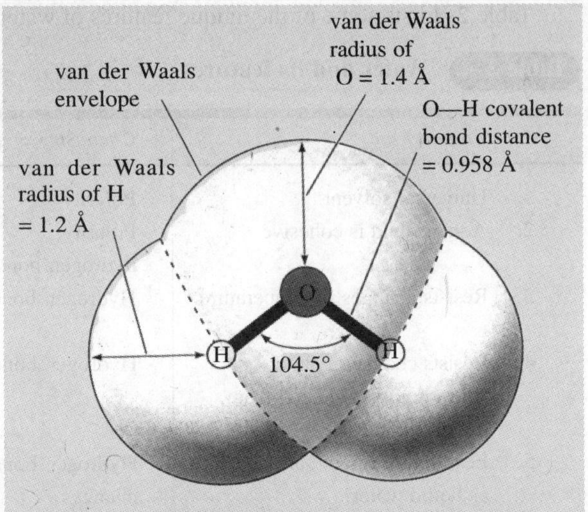

Fig. 2–2. The structure of a water molecule

The outline represents the van der Waals envelope of the molecule (where the attractive components of the van der Waals interactions balance the repulsive components). The skeletal model of the molecule indicates its covalent bonds. The H – O – H bond angle is 104.5°. Both hydrogen atoms carry a partial positive charge and the oxygen a partial negative charge, creating a dipole.

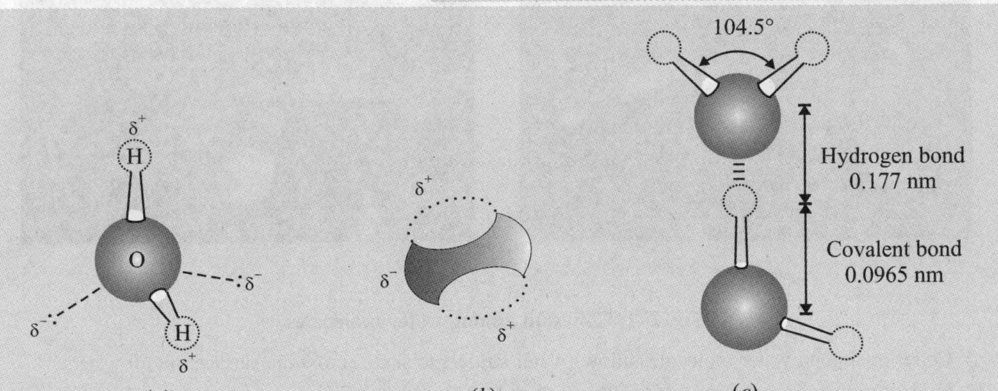

Fig. 2–3. The dipolar nature of the water molecule

(*a*) **Ball-and-stick model** (*dashed lines* represent the nonbonding orbitals)**.** There is nearly tetrahedral arrangement of the outer shell electron pairs around the oxygen atom; the two hydrogen atoms have localized partial positive charges and the oxygen atom has two localized partial negative charges.

(*b*) **Space-filling model.**

(*c*) **Two water molecules joined by a hydrogen bond** (designated by 3 thick *horizontal lines*) **between the oxygen atom of the upper molecule and a hydrogen atom of the lower one**. Hydrogen bonds are longer and weaker than covalent O — H bonds.

Polarity of water. Knowing the electronegativity of 2 atoms allows one to predict whether a bond between them will be covalent or ionic. The larger the difference in electronegativities of 2 atoms, the more likely they are to form an ionic rather than a covalent bond. Sodium and chlorine, for example, have a large difference in electronegativities and hence form ionic bonds. Carbon and nitrogen, on the contrary, have similar, moderate electronegativities and they usually form covalent bonds.

Even in a covalent bond, however, atoms may not share electrons equally. When atoms differ in electronegativity, they do not share electrons equally. Instead, the diffuse clouds of shared electrons tilt toward the more electronegative atoms. In a water molecule, for example, an oxygen atom shares electrons with 2 hydrogen atoms. But the shared electrons are more concentrated around the oxygen nucleus than around the 2 hydrogen nuclei (Fig. 2–4). Consequently, the oxygen atom has a slight negative charge and the 2 hydrogen atoms have a slight positive charge.

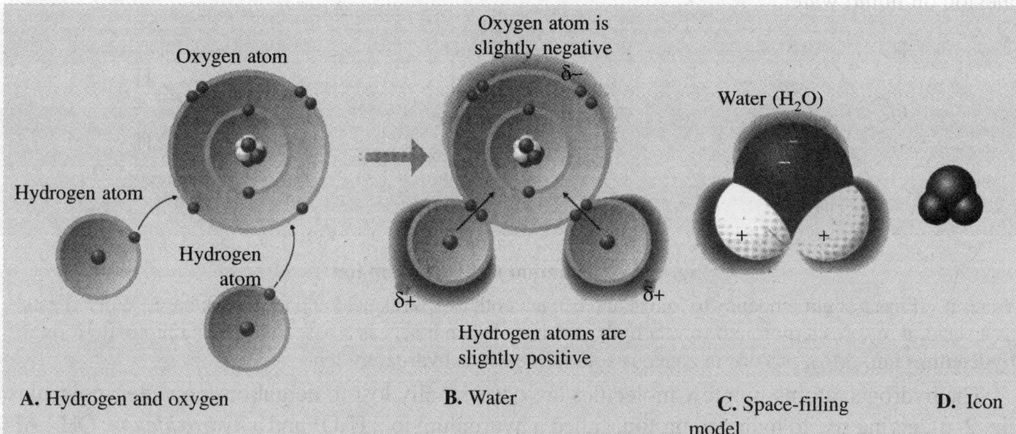

Fig. 2–4. The polarity of water

A. Two atoms of hydrogen and one atom of oxygen share electrons in two covalent bonds to form a molecule of water.

B. Because the oxygen atom is more electronegative than the hydrogen atoms, the electrons spend more time hovering around the oxygen end of the water molecule. As a result, the oxygen end has a slightly negative charge, while the hydrogens have a slight positive charge. Such a partial charge, less than one full electron, is symbolized by δ.

C. Space-filling model of water molecule.

D. Icon for water.

Molecules that have uneven distributions of electrical charge are said to be **polar**, since they have positive and negative poles in the same way that a magnet has 2 poles (Fig. 2–4). When a polar molecule, such as water, comes close to an ion or to another polar molecule, its negative pole points toward the other molecule's positive pole, and its positive pole toward a neighbouring negative pole. Molecules with approximately uniform charge distributions are said to be **nonpolar**.

Hydrogen bonds between water molecules. When a hydrogen atom attaches to a highly electronegative atom such as oxygen or nitrogen, the resulting covalent bond is polar. In this case, the hydrogen atom acquires a slight positive charge. Such a hydrogen atom can then participate in a hydrogen bond–a weak interaction to a negatively-charged atom in another molecule (Fig. 2–5). The most common hydrogen bonds are those between water molecules, but other hydrogen bonds also play a critical role in the structure of proteins and DNA.

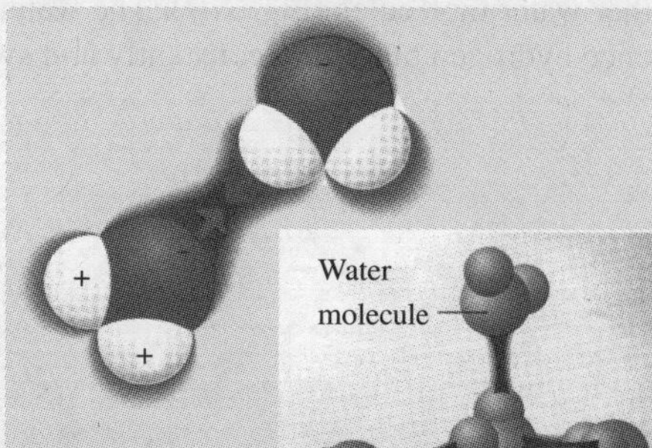

Fig. 2–5. Hydrogen bonds betweeen water molecules

The hydrogen atoms of one water molecule are attracted to the oxygen atoms of another water molecule.

Hydrogen bonds are weaker than covalent bonds. The hydrogen bonds in liquid water have a **bond energy** (the energy required to break a bond) of only about 20 kJ/mol, as compared to 460 kJ/mol for the covalent O—H bond. At room temperature, the thermal energy of an aqueous solution (*i.e.*, the kinetic energy resulting from the motion of individual atoms and molecules) is of the same order as that required to break hydrogen bonds. When water is heated, the resulting temperature increase causes the faster motion of individual water molecules. Although at any given time, most of the molecules in liquid water are hydrogen-bonded, the lifetime of each hydrogen bond is less than 1×10^{-9}s. Nevertheless, the very large number of hydrogen bonds between molecules confers great internal cohesion on liquid water.

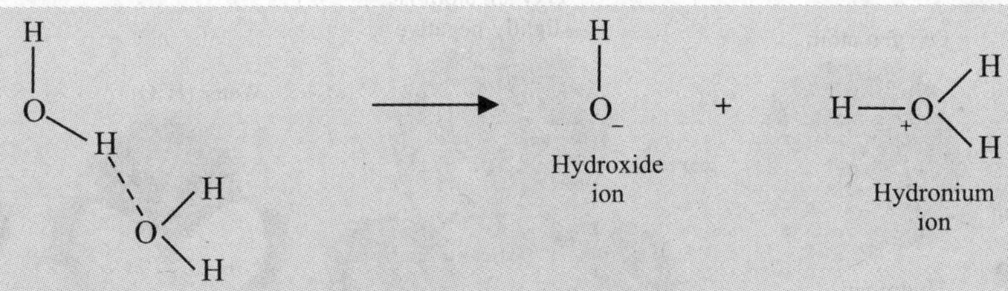

Fig. 2–6. Formation of hydronium ion

Water itself has a slight tendency to ionize and can act both as a weak acid and as a weak base. When it acts as an acid, it releases a proton to form a **hydroxyl ion.** When it acts as a base, it accepts a proton to form a **hydronium ion.** Most protons in aqueous solutions exist as hydronium ions.

The hydrogen atoms in a few molecules are occasionally lost to neighbouring water molecules (Fig. 2–6), giving rise to hydrated proton, called a hydronium ion, H_3O^+ and a *hydroxide ion*, OH^-. All protons in water are hydrated to some extent, but is unnecessary to write them as such. The symbol H^+ refers to any proton in water irrespective of its degree of hydration.

WEAK INTERACTIONS IN AQUEOUS SOLUTIONS

A. Hydrogen Bonding Between Water Molecules

The nearly tetrahedral arrangement of the oxygen electrons (bond angle 104.5°) allows each water molecule to form hydrogen bonds with 4 neighbouring water molecules. At any moment in liquid water at room temperature, each water molecule forms hydrogen bonds with an average of 3.4 other water molecules (Fig. 2–7). The water molecules are in continuous motion in the liquid state, hence hydrogen bonds are constantly and swiftly being broken and formed.

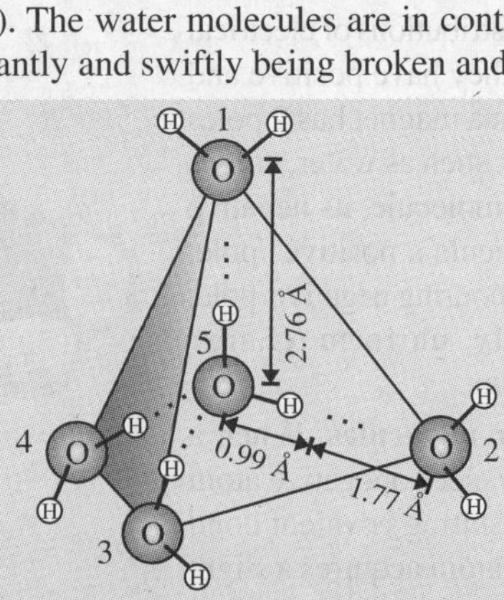

Fig. 2–7. Tetrahedral hydrogen bonding of a water molecule, in ice

Molecules 1, 2 and 5 are in the plane of the paper with molecule 3 above and molecule 4 below the plane. Molecules 1, 2, 3 and 4 are positioned at the corners of a regular tetrahedron.

In ice, however, each water molecule is fixed in space and forms hydrogen bonds with 4 other water molecules to produce a regular lattice structure (Fig. 2–8). Much thermal energy is needed to break the large number of hydrogen bonds in such a lattice and this is the reason for a relatively high melting point of water. When ice melts or water evaporates, heat is taken up by the system:

$$H_2O\ (s) \longrightarrow H_2O\ (l) \qquad \Delta H = +5.9\ kJ/mol$$
$$H_2O\ (l) \longrightarrow H_2O\ (g) \qquad \Delta H = +44.0\ kJ/mol$$

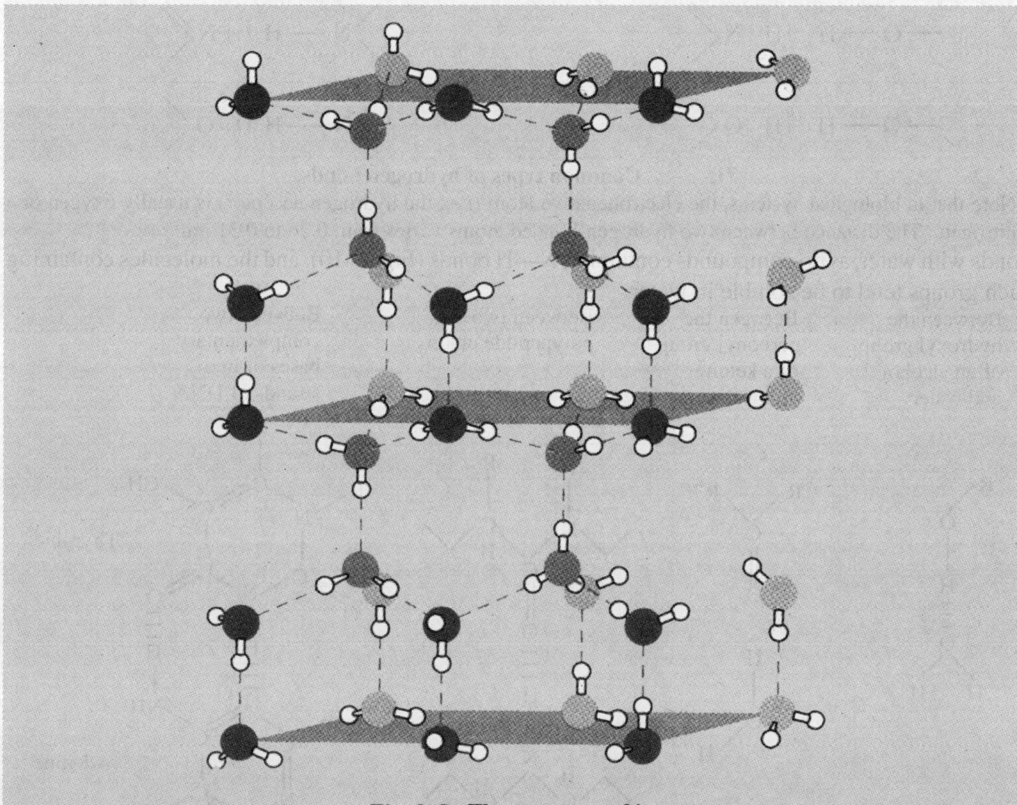

Fig. 2–8. The structure of ice

The tetrahedral arrangement of the water molecules is a consequence of the roughly tetrahedral disposition of each oxygen atom's sp^3- hybridized bonding and lone pair orbitals. Oxygen and hydrogen atoms are represented, respectively, by *red* and *white* spheres, and hydrogen bonds are indicated by *dashed lines*. Note the open structure that gives ice its low density relative to liquid water.

(*After Pauling LC, 1960*)

B. Hydrogen Bonding Between Water and Solute Molecules

Hydrogen bonding is not unique to water. They readily form between an electronegative atom (usually oxygen or nitrogen) and a hydrogen atom covalently bonded to another electronegative atom in the same or another molecule (Fig. 2–9). However, hydrogen atoms covalently bonded to carbon atoms (which are not electro-negative), do not participate in hydrogen bonding. The distinction explains why butanol ($CH_3\ CH_2\ CH_2\ CH_2.OH$) has a relatively high boiling point of 117°C in contrast to butane ($CH_3\ CH_2\ CH_2\ CH_3$) which has a boiling point of only –0.5°C. Butanol has a polar hydroxyl group and, hence, can form hydrogen bonds with other butanol molecules.

Uncharged but polar biomolecules such as sugars dissolve readily in water because of the stabilizing effect of the many hydrogen bonds that form between the hydroxyl groups or the carbonyl oxygen of the sugar and the polar water molecules. Alcohols, aldehydes and ketones all form hydrogen

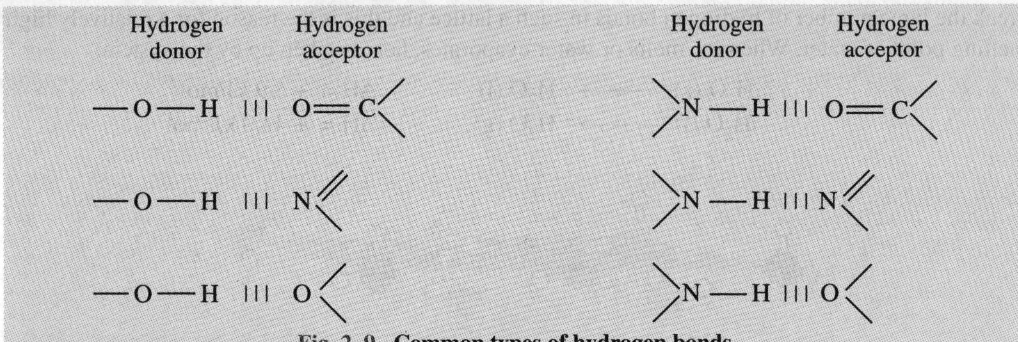

Fig. 2–9. Common types of hydrogen bonds

Note that in biological systems, the electronegative atom (*i.e.*, the hydrogen acceptor) is usually oxygen or nitrogen. The distance between two hydrogen-bonded atoms varies from 0.26 to 0.31 nm.

bonds with water, as do compounds containing N—H bonds (Fig. 2–10), and the molecules containing such groups tend to be soluble in water.

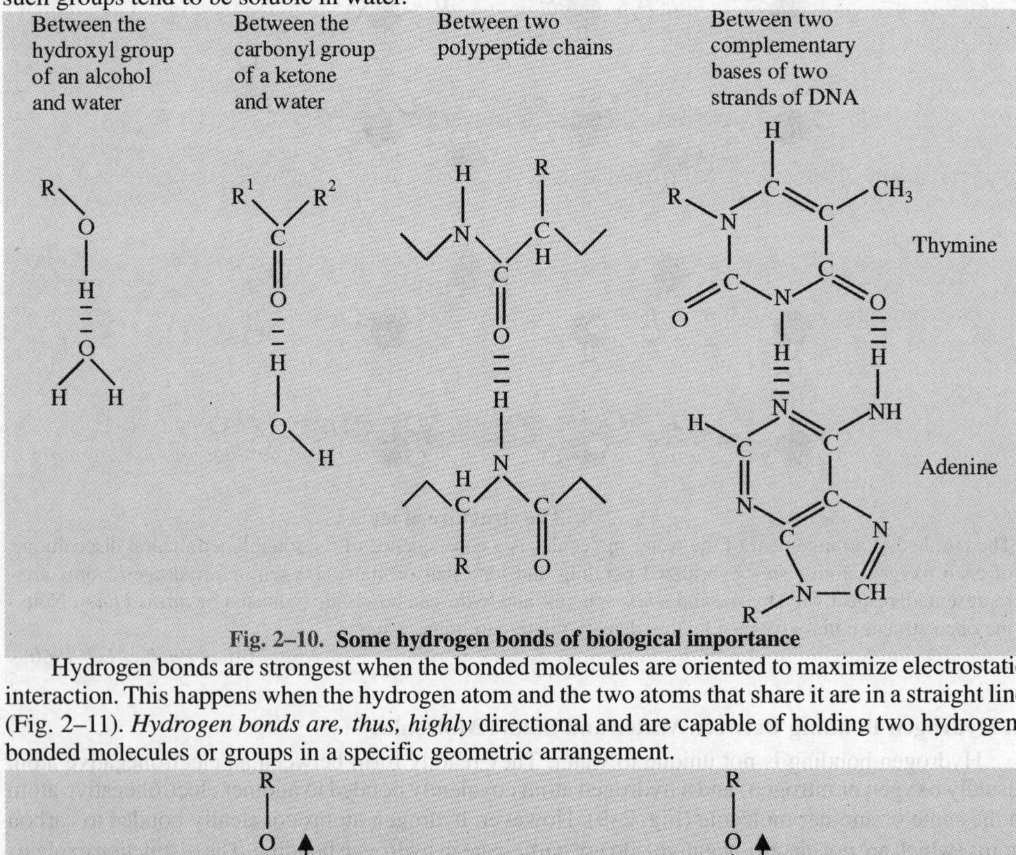

Fig. 2–10. Some hydrogen bonds of biological importance

Hydrogen bonds are strongest when the bonded molecules are oriented to maximize electrostatic interaction. This happens when the hydrogen atom and the two atoms that share it are in a straight line (Fig. 2–11). *Hydrogen bonds are, thus, highly* directional and are capable of holding two hydrogen-bonded molecules or groups in a specific geometric arrangement.

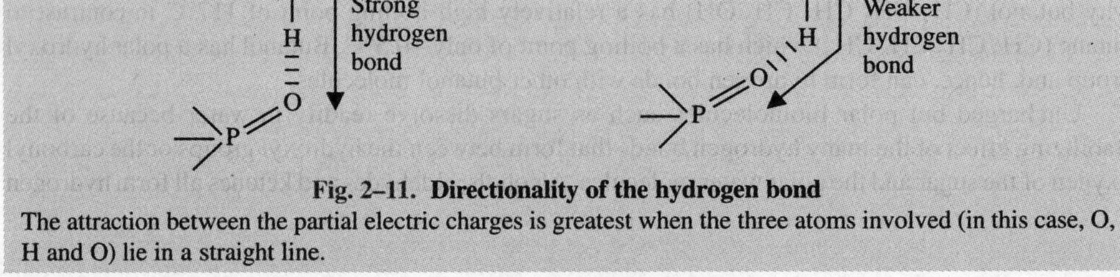

Fig. 2–11. Directionality of the hydrogen bond

The attraction between the partial electric charges is greatest when the three atoms involved (in this case, O, H and O) lie in a straight line.

C. Interaction Between Water and Charged Solutes

Water is a polar solvent. It readily dissolves most biomolecules, which are generally charged or polar compounds (Table 2–5). Compounds that dissolve readily in water are **hydrophilic** (*hudor*[G] = water; *philic*[G] = loving). In contrast, nonpolar solvents (such as chloroform and benzene) are poor solvents for polar biomolecules, but readily dissolve nonpolar biomolecules such as lipids and waxes.

Table 2–5. Some examples of polar, nonpolar and amphipathic biomolecules

Biomolecule	*Ionic form at pH 7*
Polar	
Glucose	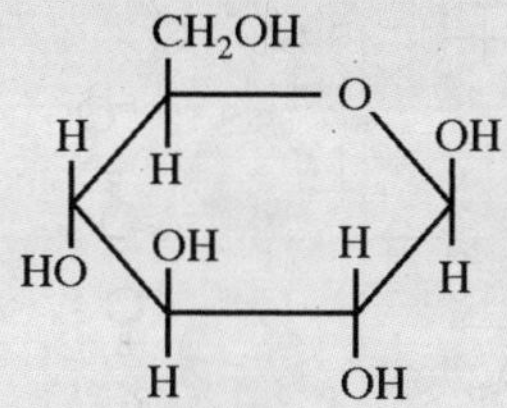
Glycine	$^{+}NH_3—CH_2—COO^-$
Aspartic acid	$^{-}COO—CH_2—CH(^{+}NH_3)—COO^-$
Lactic acid	$CH_3—CH(OH)—COO^-$
Glycerol	$HOCH_2—CH(OH)—CH_2OH$
Nonpolar	
Typical wax	$CH_3(CH_2)_7—CH{=}CH—(CH_2)_6—CH_2—C(=O)—O—CH_2—(CH_2)_7—CH{=}CH—(CH_2)_7—CH_3$
Amphipathic	
Phenylalanine	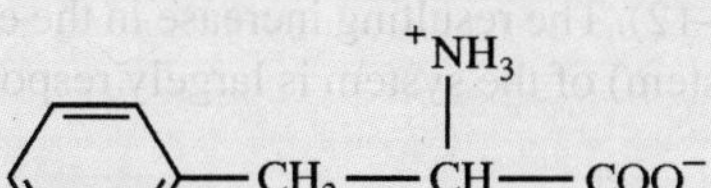
Phosphatidylcholine	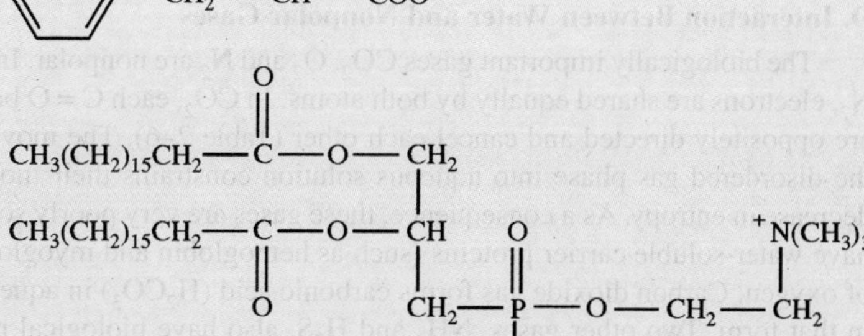

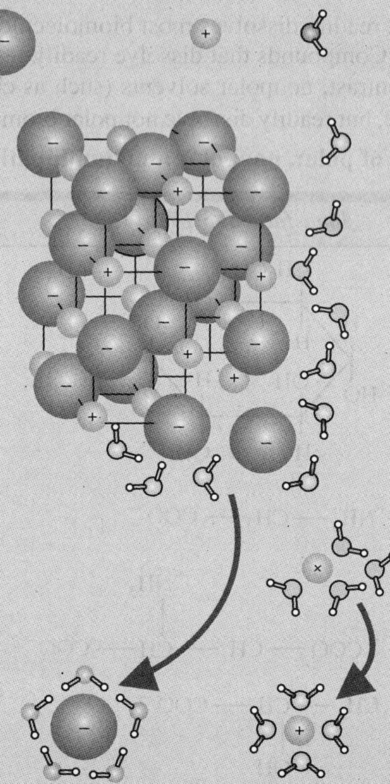

Fig. 2–12. Interaction between water and crystalline salts such as NaCl

Note that water dissolves many crystalline salts by hydrating their component ions (in this case, Na^+ and Cl^- ions). (Adapted fro*m AL Lehninger, DL Nelson and MM Cox, 1993*)

Water dissolves salts such as NaCl by hydrating and stabilizing the Na^+ and Cl^- ions, weakening their electrostatic interactions and thus counteracting their tendency to associate in a crystalline lattice. In fact, the NaCl crystal lattice is disrupted as water molecules cluster about the Cl^- and Na^+ ions. The ionic charges are, thus, partially neutralized and the electrostatic attractions necessary for lattice formation are weakened. As a result, the Na^+ and Cl^- ions leaving the crystal lattice acquire far greater freedom of motion (Fig. 2–12). The resulting increase in the entropy (a measure of the degree of randomness or disorder of a system) of the system is largely responsible for the ease of dissolving the salt in water.

D. Interaction Between Water and Nonpolar Gases

The biologically important gases CO_2, O_2 and N_2 are nonpolar. In the diatomic molecules O_2 and N_2, electrons are shared equally by both atoms. In CO_2, each C = O bond is polar, but the two dipoles are oppositely directed and cancel each other (Table 2–6). The movement of these molecules from the disordered gas phase into aqueous solution constrains their motion and therefore represents a decrease in entropy. As a consequence, these gases are very poorly soluble in water. Some organisms have water-soluble carrier proteins (such as hemoglobin and myoglobin) that facilitate the transport of oxygen. Carbon dioxide gas forms carbonic acid (H_2CO_3) in aqueous solution, and is transported in that form. Two other gases, NH_3 and H_2S, also have biological roles in some organisms. These gases are polar and dissolve easily in water.

Table 2–6. Solubilities of some gases in water

Gas	*Structure**	*Polarity*	*Solubility in water (g/L)*	*Temperature (°C)*
Nitrogen	N≡N	Nonpolar	0.018	40
Oxygen	O=O	Nonpolar	0.035	50
Carbon dioxide	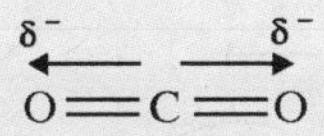	Nonpolar	0.970	45
Ammonia	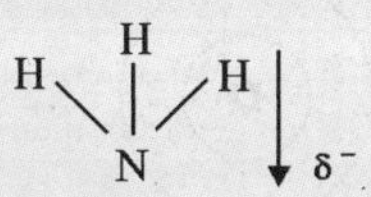	Polar	900	10
Hydrogen sulfide	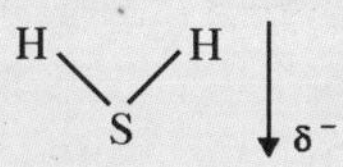	Polar	1,860	40

* The arrows represent electric dipoles; there is a partial negative charge (δ^-) at the head of the arrow, a partial positive (δ^+; not shown here) at the tail.

E. Interaction Between Water and Nonpolar Compounds

When water is mixed with a hydrocarbon such as benzene or hexane, two phases form: neither liquid is soluble in the other. Shorter hydrocarbons such as ethane have small but measurable solubility in water. Nonpolar compounds such as benzene, hexane and ethane are **hydrophobic** (*hudor*[G] = water; *phobic*[G] = hatred) as they are unable to undergo energetically favourable interactions with water molecules. In fact, they interfere with the hydrogen bonding among water molecules. All solute molecules or ions dissolved in water interfere with the hydrogen bonding of some water molecules in their immediate vicinity, but polar or charged solutes (such as NaCl) partially compensate for the lost hydrogen bonds by forming new solute-water interactions. The net change in enthalpy (the heat content of the reacting system) for dissolving these solutes is usually small. Hydrophobic solutes offer no such compensation. Addition of water to such solutes may hence result in a small gain of enthalpy; the breaking of hydrogen bonds requires the addition of energy to the system. Furthermore, dissolving hydrophobic solutes in water results in a measurable decrease in entropy.

Amphipathic compounds contain regions that are polar (or charged) and regions that are nonpolar (Table 2–5). When amphipathic compounds are mixed with water, the two regions of the solute molecule experience conflicting tendencies: the polar or charged, hydrophilic region interacts favourably with the solvent and tends to dissolve whereas the nonpolar hydrophobic region has the opposite tendency, to avoid contact with the water. The nonpolar regions of the molecules cluster together to present the smallest area to the solvent, and the polar regions are arranged to maximize their interaction with the aqueous solvent. These stable structures of amphipathic compounds in water, called **micelles**, may contain hundreds or thousands of molecules. The forces that hold the nonpolar regions of the molecules together are called **hydrophobic interactions.**

Many biomolecules are amphipathic (Table 2–5): proteins, pigments, the sterols and phospholipids of membranes and certain vitamins all have polar and nonpolar surface regions. Structures composed of these molecules are stabilized by hydrophobic interactions among the nonpolar regions. Hydrophobic interactions among lipids and between lipids and proteins, are the most important determinants of structure in biological membranes; also hydrophobic interactions between nonpolar amino acids stabilize the 3–'D' folding patterns of proteins.

Hydrogen bonding between water and polar solutes also causes some ordering of water molecules but the effect is less significant than with nonpolar solutes.

F. van der Waals Interactions

van der Waals interactions (named after a Dutch physicist **Johannes Diederik van der Waals,** LT, 1837–1923) are weak, nonspecific, interatomic attractions and come into play when any two uncharged atoms are 3 to 4 Å apart. Though weaker and less specific than electrostatic and hydrogen bonds, van der Waals interactions are no less important in biological systems. The basis of a van der Waals bond is that *the distribution of electronic charge around an atom changes with time.*

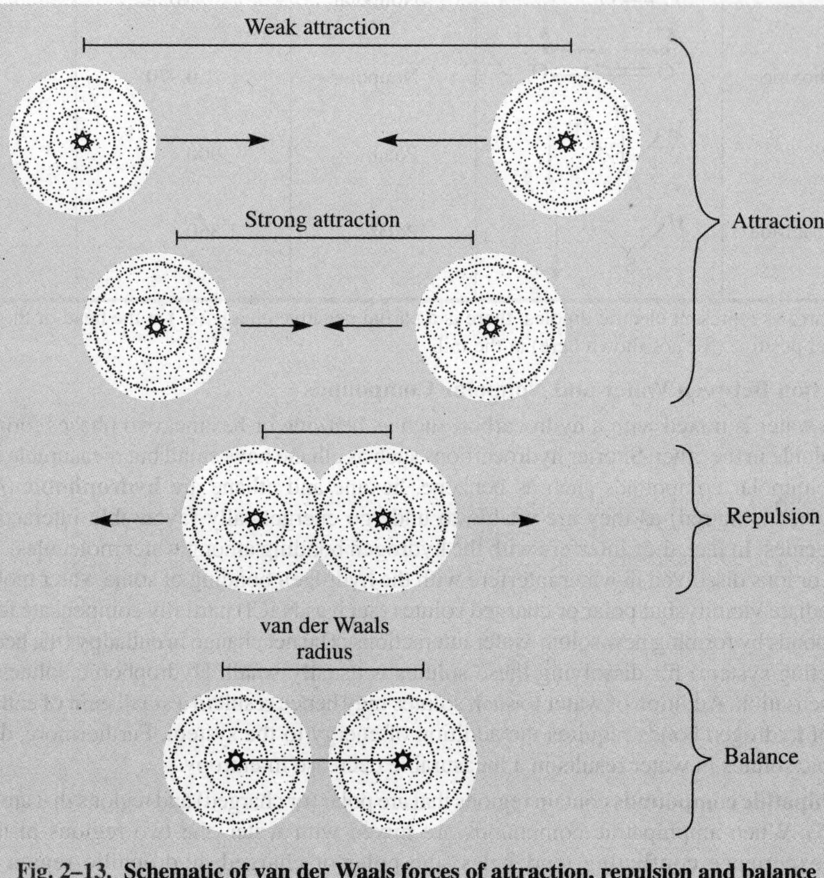

Fig. 2–13. Schematic of van der Waals forces of attraction, repulsion and balance between two atoms

All types of molecules exhibit van der Waals forces which arise from the attraction of the bound electrons of one atom for the nucleus of another. When two atoms are far apart, there is a very weak attraction which becomes stronger as the atoms move closer together (Fig. 2–13). However, if the atoms move close enough for their outer electron shells to overlap, then a force of repulsion ocurs. At a certain distance, defined as the *van der Waals contact radius*, there is a balance between the forces of attraction and those of repulsion. Each type of atom has a specific van der Waals contact radius (Table 2–7). At this point of balance, the two atoms are separated by the *van der Waals contact distance* (Fig. 2–14). The contact distance between an oxygen and carbon atom, for example, is 3.4 Å which is obtained by adding 1.4 and 2.0 Å, the contact radii of the O and C atoms, respectively.

Table 2–7. **van der Waals contact radii of atoms and group**

Atom/Group	*Radius (in Å)*
H	1.2
O	1.4
N	1.5
S	1.85
P	1.9
C	2.0
CH_3	2.0

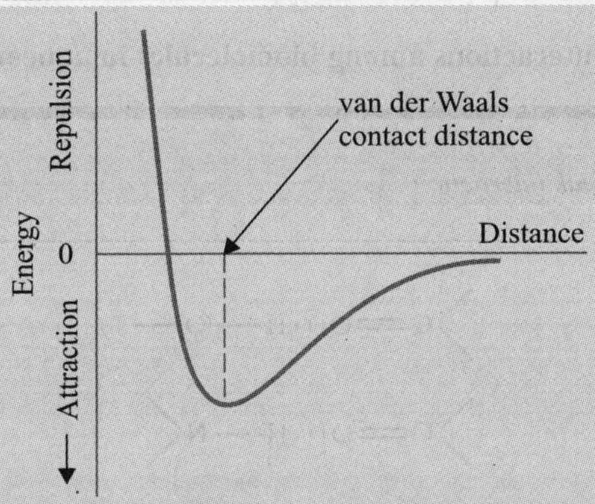

Fig. 2–14. Energy of a van der Waals interaction as two atoms approach each other

The energy is most favourable at the van der Waals contact distance. The energy rises rapidly owing to electron–electron repulsion as the atoms move closer together than this distance.

The van der Waals bonding energy between two average atoms is very small, *i.e.*, about 1 kcal/mol, which is only slightly greater than the average thermal energy (0.6 kcal/mol) of molecules at room temperature. It is considerably weaker than a hydrogen or electrostatic bond, which is in the range of 3 to 7 kcal/mol.

Table 2–8. **Strengths of bonds common in biomolecules**

Type of bond	*Bond dissociation energy (kJ/mol)*	*Type of bond*	*Bond dissociation energy (kJ/mol).*
Single bonds		**Double bonds**	
O—H	461	C = O	712
H—H	435	C = N	615
P—O	419	C = C	611
C—H	414	P = O	502
N—H	389	**Triple bonds**	
C—O	352	N ≡ N	930
C—C	348	C ≡ C	816
S—H	339	**Noncovalent bonds or interactions**	
C—N	293	Hydrogen bonds	4–20
C—S	260	van der Waals forces	4–20
N—O	222	Hydrophobic bonds	4–20
S—S	214	Ionic interactions	4–20

Since van der Waals interactions are very week, it is customary to call them as a force rather than a bond. The crystalline compounds possessing these forces are very soft and have low melting points. In fact, the more asymmetrical the molecule, the greater are the *van der Waals forces*. These forces are greater for compounds than for atoms and molecules of elements.

ROLE OF NONCOVALENT INTERACTIONS

The noncovalent interactions described above (hydrogen bonds and ionic interactions among charged groups and hydrophobic interactions and van der Waals interactions among nonpolar groups) are much weaker than covalent bonds (Table 2–8). The input of about 350 kJ of energy is required to break a mole (6×10^{23}) of C—C single bonds, and of about 410 kJ to break a mole of C—H bonds, but only 4-8 kJ is sufficient to disrupt a mole of typical van der Waals interactions (Table 2–9).

Table 2–9. Some weak interactions among biomolecules in aqueous solvent

Weak interaction		*Stabilization energy (kJ/mol)*
Hydrogen bonds		
Between neutral groups	>C=O···H—O—	8 – 21
Between peptide bonds	>C=O···H—N<	8 – 21
Ionic interactions		
Attraction	—$^{+}NH_3$ → ← $^{-}$O—C(=O)—	42
Repulsion	—$^{+}NH_3$ ← → H_3N^{+}—	≈ – 21
Hydrophobic interactions	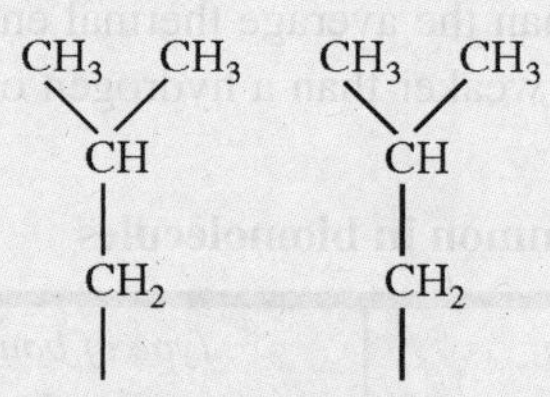	4 – 8
van der Waals interactions	Any two atoms in close proximity	4

Hydrophobic interactions are similarly weak, and ionic interactions and hydrogen bonds are only a little stronger; a typical hydrogen bond can be broken by the input of about 20 kJ/mol. In aqueous solvent at 25°C, the available thermal energy is of the same order as the strength of these weak interactions. Consequently, hydrogen bonds and ionic, hydrophobic and van der Waals interactions are continuously formed and broken.

Although these 4 types of interactions are individually weak relative to covalent bonds, the cumulative effect of many such interactions in a protein or nucleic acid can be very significant. For example, the noncovalent binding of an enzyme to its substrate may involve several hydrogen bonds and one or more ionic interactions, as well as hydrophobic and van der Waals interactions. The formation of each of these weak bonds contributes to a net decrease is free energy; this binding free energy is released as bond formation stabilizes the system.

The most stable (native) structure of most macromolecules is that in which weak-bonding possibilities are maximized. The folding of a single polypeptide or polynucleotide chain into its 3-dimensional shape is determined by this principle. The binding of an antigen to a specific antibody depends on the cumulative effects of many weak interactions. The energy released when an enzyme binds noncovalently to its substrate is the main source of catalytic power for the enzyme. The binding of a hormone or a neurotransmitter to its cellular receptor protein is the result of weak interactions.

ROLE OF WATER IN LIFE

A saying goes, 'there is no life without water'. Life and water are inextricably connected (Fig. 2–15). Life probably originated in water. Wherever life is found, there is water; and wherever liquid water is found, there is life. Water makes up more than 70% of the material of living organisms themselves and covers more than 75% of Earth's surface. It is the medium in which most cells are constantly bathed and the major component of cells themselves. Not only do most biochemical reactions occur in water, but water itself participates in many biochemical reactions. As the Hungarian 1937 Nobel Laureate **Albert von Szent-Györgyi** has rightly pointed out:

"Water is life's mater and matrix, mother and medium."

Fig. 2-15. Essentiality of water to organisms

Killer whale (*Orcinus orca*) spend their entire lives in the ocean.

(*Courtesy: Francois Gohier*)

REFERENCES

1. **Dick DAT :** Cell Water. *Butterworth Publishers, Inc., Stoneham, M.A. 1966.*
2. **Edsall JT, Wyman J :** Water and Its Biological Significance. *In Biophysical Chemistry. Vol.1. Academic Press, Inc., New York. 1958.*
3. **Eisenberg D, Kauzmann W :** The Structure and Properties of Water. *Oxford University Press, New York, 1969.*
4. **Franks R (editor) :** Water: A Comprehensive Treatise. *Vol. 4. Plenum Press, New York. 1975.*
5. **Franks R, Mathias SF (editors) :** Biophysics of Water. *John Wiley and Sons, Inc., New York. 1982.*

6. **Pullman A, Vesilescu V, Packer L (editors) :** Water and Ions in Biological Systems. *Plenum Press, New York. 1985.*
7. **Rupley JA, Gratton E, Careri G :** Water and glubular proteins. *Trends Biochem. Sci.* **8 :** *18, 1983.*
8. **Schrier RW:** Body water homeostasis: clinical disorders of urinary dilution and concentration. *J. Am. Soc. Nephrol.* **17 :** *1820–1832, 2006.*
9. **Snoeyink VL, Jenkins D :** Water Chemistry. *John Wiley and Sons, New York. 1980.*
10. **Wiggins PM :** Role of water in some biological processes. *Microbiol. Rev.* **54 :** *432–449, 1990.*

PROBLEMS

1. In a hospital laboratory, a 10.0 mL sample of gastric juice, obtained several hours after a meal, was titrated with 0.1 M NaOH to neutrality ; 7.2 mL of NaOH was required. The patient's stomach contained no ingested food or drink, thus assume that no buffers were present. What was the pH of the gastric juice?
2. One common description of the pK_a of an acid is that it represents the pH at which the acid is half ionized, that is, the pH at which it exists as a 50 : 50 mixture of the acid and the conjugate base. Demonstrate this relationship for an acid HA, starting from the equilibrium-constant expression.
3. The amino acid glycine is often used as the main ingredient of a buffer in biochemical experiments. The amino group of glycine, which has a pK_a of 9.6, can exist either in the protonated form ($—NH_3^+$) or as the free base ($—NH_2$) because of the reversible equilibrium

$$R—NH_3^+ \rightleftharpoons R—NH_2 + H^+$$

   (*a*) In what pH range can glycine be used as an effective buffer due to its amino group?

   (*b*) In a 0.1 M solution of glycine at pH 9.0, what fraction of glycine has its amino group in the $—NH_3^+$ form?

   (*c*) How much 5 M KOH must be added to 1.0 L of 0.1 M glycine at pH 9.0 to bring its pH to exactly 10.0?

   (*d*) In order to have 99% of the glycine in its $—NH_3^+$ form, what must the numerical relation be between the pH of the solution and the pK_a of the amino group of glycine?

4. Catechols substituted with long-chain alkyl groups are the components of poison ivy and poison oak that produce the characteristic itchy rash.

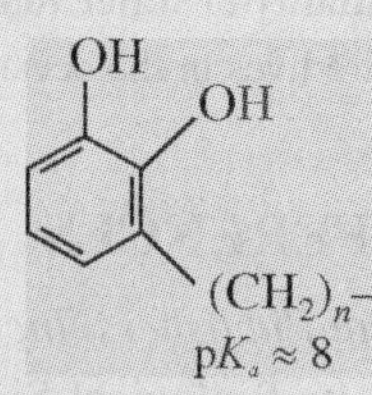

p$K_a \approx 8$

   If you were exposed to poison ivy, which of the treatments below would you apply to the affected area? Justify your choice.

   (*a*) Wash the area with cold water.

   (*b*) Wash the area with dilute vinegar or lemon juice.

   (*c*) Wash the area with soap and water.

   (*d*) Wash the area with soap, water, and baking soda (sodium bicarbonate).

5. Aspirin is a weak acid with a p*K*a of 3.5.

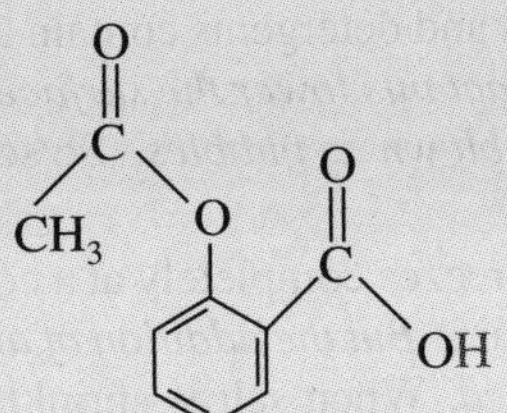

It is absorbed into the blood through the cells lining the stomach and the small intestine. Absorption requires passage through the cell membrane, which is determined by the polarity of the molecule: charged and highly polar molecules pass slowly, whereas neutral hydrophobic ones pass rapidly. The pH of the gastric juice in the stomach is about 1.5 and the pH of the contents of the small intestine is about 6. Is more aspirin absorbed into the bloodstream from the stomach or from the small intestine? Clearly justify your choice.

6. Why can we blow a bubble with soap water but not with plain water?
7. Why does ice melt when salt is sprinkled on it?
8. Why is salt not soluble in oil but soluble in water?
9. When a piece of paper is wet with water, it tears off easily. But the same paper if soaked in oil doesn't lose strength. Why?
10. Why does the water film move away when we touch a wet surface?
11. Why does water wet the glass while mercury does not?
12. Why does ice melt but wax harden when subjected to pressure?
13. Why are ice cubes preferred to water at 0°C to cool a soft drink?
14. When glucose is added and stirred, the water feels cooler. Why?
15. How can hydrogen, a combustible, and oxygen, a supporter of combustion when combined in the form of water extinguish fire?
16. Why is snow white although water of which it is made up of is transparent?
17. Why is water liquid but hydrogen sulfide a gas?
18. Why is sea water salty?
19. How do fishes survive in frozen lakes?
20. Why do we feel thirsty?
21. How is water purified?
22. What is the difference between ice and snow?
23. What are the water requirements of various animals?
24. How much of my body is comprised of water?
25. How can I tell if I am drinking enough water?

ANSWERS TO PROBLEMS

1. pH 1.1.
2. For the equilibrium HA $\rightleftharpoons H^+ + A^-$, the corresponding Henderson–Hasselbalch equation is: pH = pK_a + log [A^-]/[HA]. When the acid is half-ionized, [HA] = [A^-]. Hence [A^-]/[HA] = 1; log 1 = 0; and pK_a = pH.
3. (*a*) In a zone centred about pH 9.6 (*b*) 4/5 (*c*) 10^{-2} L (*d*) pH - pK_a = –2
4. (*d*); the weak base bicarbonate will titrate —OH to –O^-, making the compound more polar and more water-soluble.
5. Stomach; the neutral form of aspirin present at low pH is less polar, and therefore more membrane-permeant.

6. Surface tension of water is so strong that it does not allow water to form stable bubbles when blown through a pipe. Soaps and detergents contain a basic cleaning agent called surfactant or surface active agent. *Surfactants lower the surface tension of water sufficiently to enable a film of soap solution to be blown as bubbles.* So soapy water is used by children to blow bubbles.
7. A pure liquid such as water freezes completely at a constant temperature which is known as the freezing point for that liquid. *But the addition of any soluble impurity lowers the freezing point of the mixture so formed.* When salt is sprinkled on ice, a part of it dissolves in the water film that always covers the ice surface. As the freezing point of the solution so formed is much lower than the freezing point of water, it remains in a liquid state and continues dissolving more salt and melting more ice. *This phenomenon is used to clear snow from roads and side-walks in cold countries during winter.*
8. Water is a polar solvent, that is, its molecules behave like dipoles with two oppositely- charged ends. Common salt is a crystalline substance made up of sodium (Na^+) and chloride (Cl^-) ions, held together by strong electrostatic forces. When salt is put in water, the water molecules get attached to the sodium and chloride ions and neutralize the force of attraction between them. The negatively-charged chloride ions get surrounded by the positive ends of the water molecules, while the positively-charged sodium ions get surrounded by the negative ends of the water molecules. As a result, water molecules tend to pull the ions away from the solid crystal lattice, making salt dissolve in water. Oil, on the other hand, is not a polar solvent and its molecules are not able to neutralize the electrostatic force of attraction between ions. Salt, therefore, does not dissolve in oil.
9. Paper is made up of thin sheets of cellulosic fibres and other ingredients mixed together. It is sized to make it suitable for writing. The cellulosic fibres are held together by weak forces called the van der Waal's forces. When paper is dipped in water, the sizing material used for refining the paper being water-soluble becomes soft. Water being a polar solvent, weakens the van der Waal's forces that hold the fibres together. As a result, the fibres come off easily when pulled and so a wet paper tears off easily. On the other hand, oil being a nonpolar solvent does not affect the water-soluble sizing material. Nor does it weaken the forces holding the cellulosic fibres together. So paper soaked in oil does not tear off as easily as paper soaked in water does.
10. The surface of water behaves like any stretched membrane. In other words, when it is ruptured at any point it shrinks away. This property is known as surface tension. Some substances such as oil, grease or detergents have the property of reducing the surface tension when added to water. *When we touch a wet surface with our fingertips, the oil on our fingertips which is always present on the unwashed finger reduces the surface tension of water at that point and the water film moves away.* The area of the surface from which the water film moves away depends on how oily the finger is.
11. Wetting is a phenomenon which depends to a large extent on the difference between the adhesive force between a liquid and the surface in contact and the cohesive force between the molecules of the liquid which is also a measure of its surface tension. Mercury does not wet the surface of glass because of its higher surface tension which is six times that of water. In case of mercury, the cohesive force is much stronger than the adhesive force, that is the force of attraction between molecules of mercury and of glass. As a result, mercury neither spreads out nor wets the surface of glass. On the other hand, in case of water the adhesive force between water and glass is stronger than the cohesive force between water molecules which makes water to spread out and cling to the surface of glass wetting it.
12. Ice is nothing but solidified water in which H_2O molecules are bound together by the weak hydrogen bonds and form an open cage-like structure. When ice is subjected to pressure, the hydrogen bonds being the weaker ones break easily. Consequently, the H_2O molecules get

separated and ice melts into water. Whereas in the case of paraffin wax, which is hydrocarbon petroleum product, carbon atoms are held together by covalent bond. Also, since it is plastic solid, it usually deforms under pressure without the application of heat. When pressure is applied on soft wax, the molecular spaces decrease and molecules tend to come closer, thereby causing hardening of wax.

13. Ice cubes are preferred to water at 0°C to cool a soft drink because when ice melts, it absorbs a certain amount of heat known as latent heat until it is melted. *The heat absorbed does not raise the temperature of the drink but goes entirely into changing the ice into water. Thus ice cubes added to a drink chill it by absorbing heat from it.* On the other hand, chilled water at 0°C simply mixes up with the drink and the temperature attained by the mix will only be an average of the temperatures of the two. Hence, ice cubes are preferred over chilled water for cooling a soft drink.

14. Whenever a solute is dissolved in water, some energy change takes place. In most cases, the amount of this change is very small and is hardly noticeable. But in some cases, it is quite large and leads to large changes in temperature – a rise or fall depending upon whether the heat of solution is positive or negative. *Glucose has a negative heat of solution, i.e., heat is absorbed when glucose is dissolved in water. As a result, the temperature drops and the solution feels cool.*

15. Although hydrogen is a combustible gas and oxygen a supporter of combustion, when combined to form water they lose their individual properties. Water is not a mixture but a compound formed by the chemical reaction between hydrogen and oxygen in a definite proportion. Since the properties of a compound are entirely different from the properties of its constituent elements, water does not burn or support combustion. Rather because of its high heat capacity, it is an excellent coolant and fire-extinguisher.

16. Both snow and foam look white due to the same reason. While foam is a collection of large number of bubbles, snow is made up of large number of ice-crystals. In both snow and foam, the gaps between the bubbles or the crystals is filled by air. Light rays falling on such a collection of crystals gets bent and reflected by the crystals. However, all the crystals do not reflect or refract the light in the same direction. Thus, light gets scattered haphazardly. The presence of air adds to this scattering. The light reaching an observer looking at the snow, therefore, is just the scattered light and so the snow looks white. In case of water, the situation is different. Here since the water surface is smooth, very little light gets scattered although a part of it is reflected. Almost all of the light falling on a clear water surface passes through it to the other side making water look transparent.

17. Oxygen and sulphur belong to the same group in the periodic table, so theoretically the hydrides of these elements should have very similar properties. But in reality, there is a large difference in the physical properties of these hydrides. For example, water is a liquid and hydrogen sulphide is a gas. This unusual behaviour of water arises from its unique property of forming weak bonds called hydrogen bonds. In water, large number of H_2O molecules are held together by these bonds which form between O and H, thus effectively forming a giant molecule. As is the rule, with the increase in the molecular weight, the boiling point of water rises, compared to what it should be, had there been no hydrogen bonding, and it remains a liquid at room temperature. This does not happen with hydrogen sulphide and so it remains a gas.

18. Seawater is salty because it contains many dissolved salts, mostly sodium chloride with small proportions of potassium, magnesium, calcium and carbonates. The salt comes mainly with river water that flows through rocks eroded by frost and rain. The gradual wearing away of mountains releases chemicals which are carried down by rivers to the ocean as dissolved salts. Some salt also enters seawater from the rocks beneath the seabed. The river water carrying the salts to the sea does not taste salty because the salts are present in extremely small concentration. Whereas in the case of the oceans, water is continuously evaporating even as more and more salt continues to be added with river water. This, over millions of years has led to the high concentration of dissolved salts in seawater.

19. Lakes that appear frozen from the top never freeze completely. Only the top layer freezes and a thick layer of ice floats on water which remains liquid below and in which fishes and other aquatic animals survive. With the onset of winter, water starts cooling. As it cools, it shrinks in volume and becomes denser, but only until it reaches 4°C. Below that temperature, water expands and its density drops. This comparatively lighter water moves up and finally at 0°C it turns into ice. Since the density of ice is less than that of water, the ice floats. Also, ice being a very poor conductor of heat, further cooling of the water below it stops. Thus, the temperature of water below the ice layer remains at 4°C and fishes survive in it.

20. About 60 per cent of our body weight is water. Water is constantly lost from our body through exhalation, perspiration and urine. At the same time, we obtain a good amount of water from our food, but it is not enough to keep the balance of water our body needs. That is why we need to drink water from time to time. When the water balance is disturbed, either due to excessive loss by perspiration or due to insufficient intake, the salt content of the blood changes. This change activates the thirst centre in our brain which makes us feel thirsty

21. Water supplied to cities and towns is usually taken from rivers or lakes and contains various kinds of pollutants, and so it has to be purified. First, the raw water is passed through screens to separate coarse solids and it is then fed into settling tanks where the suspended solids settle out. Sometimes, substances such as alum are added to aid the settling process. From the settling tanks the water is sent to sand filter beds, which are large concrete tanks with thick layers of coarse and fine sand at the bottom. The sand filters remove the remaining suspended matter from the water. Finally, the clear water is treated with chlorine gas in a closed tank to kill bacteria. The purified water is then pumped into tanks for distribution. For industrial use, water is sometimes passed through ion exchangers to remove ions like magnesium and calcium present in hard water.

22. Though ice and snow are both made up of water, there is a difference between the two. Snow is nothing but the frozen atmospheric vapour which falls in winters on earth as light flakes whereas ice is simply frozen water. Another difference is that snowfall can be seen only in winters at areas of high altitudes and places near the polar regions whereas ice can be seen in any season of the year in our refrigerator.

23. All animals including man can produce water inside their bodies. Carbohydrates and fats from food, when oxidized inside the human body, release energy, CO_2 and water. An adult **human** produces about 300 ml of water every day by metabolism of carbohydrates and fats. Water generated thus is little but for some animals it is the only method to obtain water. A loss of 10% of water in our body is serious; a loss of 20% fatal.

Gebrils and other rodents of desert rarely drink water and depend on water that is released by the oxidation of fats. Unlike polar animals, which store fat under skin to keep their body warm, desert animals store fat in different places. For instance, camels store fat in their humps; gebrils and some sheep in their tails. **Sharks** do not drink water. Only **sea fish** drink water to compensate the water lost by exosmosis. **Whales** do not drink sea water. **Turtles**, like **crocodiles**, shed tears to excrete the excess salt. About 0.016% of ordinary water is heavy water (D_2O).

24. Your body is 2/3 water, but it is actually made up of 99% water molecules.

25. The colour of your urine will tell you if you are properly hydrated. Your urine should be a light yellow colour. If it is a deep, dark yellow, you are likely not drinking enough water. If you have not urinated in many hours, that is another indication that you need to drink more water.

CONTENTS

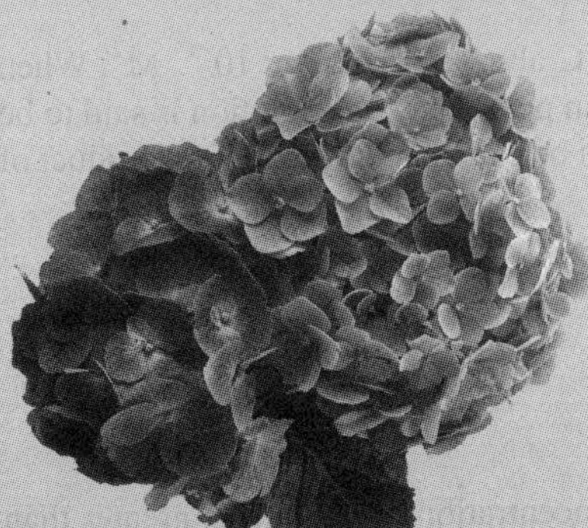

The compounds responsible for colour in plants are often sensitive to acids and alkalis.
Blue hydrangeas grow only in acidic soils; in neutral or alkaline soils, they will revert to pink.

CHAPTER 3

pH and Buffers

IONIZATION OF WATER, WEAK ACIDS AND WEAK BASES

Although many of the solvent properties of water can be explained in terms of the uncharged H_2O molecules,the small degree of ionization of water to hydrogen ions (H^+) and hydroxide ions (OH^-) must also be taken into account. *Like all reversible reactions, the ionization of water can be described by an equilibrium constant.* When weak acids or weak bases are dissolved in water, they can contribute H^+ by ionizing (if acids) or consume H^+ by being protonated (if bases). These processes are also governed by equilibrium constants. The total hydrogen ion concentration from all sources, which is experimentally measurable, is expressed as the pH of the solution.

Water molecules have a slight tendency to undergo reversible ionization to yield a hydrogen ion and a hydroxide ion:

$$H_2O \rightleftharpoons H^+ + OH^- \qquad ...(3.1)$$

This reversible ionization is crucial to the role of water in cellular functions. Henceforth, we must have a means of expressing the extent of ionization of water in quantitative terms. This is discussed below.

The position of equilibrium of any chemical reaction is given by its **equilibrium constant.** For the general reaction,

$$A + B \rightleftharpoons C + D \qquad ...(3.2)$$

an equilibrium constant can be defined in terms of concentrations of reactants (A and B) and products (C and D) present at equilibrium:

$$K_{eq} = \frac{[C]\,[D]}{[A]\,[B]} \quad ...(3.3)$$

Strictly speaking, the **bracketed terms** represent molar activity, rather than molar concentration.

The equilibrium constant is fixed and characteristic for any given chemical reaction at a specific temperature. It defines the composition of the final equilibrium mixture of that reaction, irrespective of the starting amounts of reactants and products. Conversely, one can calculate the equilibrium constant for a given reaction at a given temperature, if the equilibrium concentrations of all its reactants and products are known.

As mentioned, the degree of ionization of water at equilibrium (Eqn. 3.1) is small; at 25°C, only about one of every 10^7 molecules in pure water is ionized at any instant. The equilibrium constant for the reversible ionization of water (Eqn. 3.1) is:

$$K_{eq} = \frac{[H^+]\,[OH^-]}{[H_2O]} \quad ...(3.4)$$

In pure water at 25°C, the concentration of water is 55.5 M (*i.e.*, grams of H_2O in 1 litre divided by gram molecular weight or 1000/18 M = 55.5 M). This value is essentially constant in relation to the very low concentrations of H^+ and OH^-, namely 1×10^7 M. Accordingly, on substituting 55.5 M in the equilibrium constant expression (*i.e.*, Eqn. 3.4), we get:

$$K_{eq} = \frac{[H^+]\,[OH^-]}{55.5\ M} \quad ...(3.5)$$

which, on rearranging, becomes:

$$(55.5\ M)\,(K_{eq}) = [\,H^+\,]\,[\,OH^-\,] = K_w \quad ...(3.6)$$

where K_w designates the product (55.5 M) (K_{eq}), the **ion product of water** at 25°C.

The value for K_{eq} is 1.8×10^{-16} M at 25°C as calculated from electrical conductivity measurements. Substituting this value for K_{eq} in Eqn. 3.6 gives:

$$(55.5\ M)\,(1.8 \times 10^{-16}\ M) = [H^+]\,[OH^-]$$

$$99.9 \times 10^{-16}\ M^2 = [H^+]\,[OH^-]$$

$$1.0 \times 10^{-14}\ M^2 = [H^+]\,[OH^-] = K_w$$

Thus, the product $[H^+]\,[OH^-]$ in aqueous solutions at 25°C always equals $1 \times 10^{-14}\ M^2$. When there are exactly equal concentrations of both H^+ and OH^-, as in pure water, the solution is said to be at **neutral pH**. At this pH, the concentration of H^+ and OH^- can be calculated from the ion product of water as follows:

$$K_w = [H^+]\,[OH^-] = [H^+]^2$$

Solving for $[H^+]$ gives:

$$[H^+] = \sqrt{K_w} = \sqrt{1 \times 10^{-14} M^2}$$

$$[H^+] = [OH^-] = 10^{-7}\ M.$$

As the ion product of water is constant, whenever the concentration of H^+ ions is greater than 1×10^{-7} M, the concentration of OH^- must become less than 1×10^{-7} M, and vice versa. When the concentration of H^+ is very high, as in a solution of hydrochloric acid, the OH^- concentration must be very low.

THE pH SCALE

The ion product of water, K_w, is the basis for the **pH scale** (Table 3–1). It is a convenient means of designating the actual concentration of H^+ (and thus of OH^-) in any aqueous solution in the range between 1.0 M H^+ and 1.0 M OH^-. Biochemical reactions are often defined in terms of hydrogen ion (H^+) concentrations. In 1909, **Soren Sörensen**, a Danish biochemist, used a logarithmic scale for expressing the H^+ concentration. This scale was called pH, where *p* stands for power and H for hydrogen ion concentration. He defined pH of a solution as the negative logarithm of the concentration (in moles/litre) of hydrogen ions. Thus,

$$pH = \log \frac{1}{[H^+]} = -\log [H^+]$$

The symbol *p* denotes "negative logarithm of ".

For a precisely neutral solution at 25°C, in which the concentration of hydrogen ions is 1.0×10^{-7} M, the pH can be calculated as follows:

$$pH = \log \frac{1}{1 \times 10^{-7}} = \log (1 \times 10^7)$$
$$= \log 1.0 + \log 10^7$$
$$= 0 + 7.0$$
$$= 7.0$$

In fact, the term **pH** was introduced in 1909 by **Sörensen** who defined it as:

"Für die Zahl *p* schlage ich den Namen Wasserstoffionenexponent und die Schreibweise pH vor. Unter dem Wasserstoffionexponenten (pH) einer Lösung wird dann der Briggsche Logarithmus des reziproken Wertes des auf Wasserstoffionen bezagenen Normalitäts faktors de Lösung verstanden."

The translation into English is:

"For the sign *p* I propose the name 'hydrogen ion exponent' and the symbol pH. Then, for the hydrogen ion exponent (pH) of a solution, the negative value of the Briggsian logarithm of the related hydrogen ion normality factor is to be understood."

pH (short for "potential of hydrogen") is a symbol and denotes the relative concentration of hydrogen ions in a solution. pH values extend from 0 to 14; the lower the value, the higher the acidity or the more hydrogen ions the solution contains. Water at 25 °C has a concentration of H ion of 10^{-7}; the pH, therefore, is 7.

The value of 7.0 for the pH of a precisely neutral solution is not an arbitrarily chosen figure. *It is derived from the absolute value of the ion product of water at 25 °C, which by convenient coincidence is a round number.*

To calculate the pH of a solution:

1. Calculate the hydrogen ion concentration, $[H^+]$.
2. Calculate the base 10 logarithm of $[H^+]$.
3. pH is the negative of the value found in step 2.

For example, for pure water at 25°C:

$$pH = -\log [H^+] = -\log 10^{-7} = -(-7) = 7.0$$

The number of hydrogen ions present in a solution is a measure of the **acidity of the solution.** All acids do not ionize completely when dissolved in water, *i.e.*, all the molecules of acid do not ionize and exist in the solution as electrically-charged particles. The hydrogen ion concentration is a measure, therefore, of the amount of *dissociated acid* rather than of the anmount of acid present. Strong acids dissociate more freely than weak acids; hydrochloric acid, for example, dissociates freely into H^+ and Cl^- whereas carbonic acid, a weak acid, dissociates much less freely into H^+ and CO_3^-. The number of free hydrogen ions is a measure of its acidity rather than an indication of the type of molecule from which the hydrogen ions originated. The **alkalinity of a solution** is dependent upon the number of hydroxyl ions present. Water is a neutral solution because each molecule contains one H^+ and one OH^-. For each molecule of water dissociated, there is one H^+ and one OH^-, each one neutralizing the other.

The pH scale was developed taking water as the standard. It is an experimental fact that only 1 mole in 5,50,000,000 moles of water ionizes into a H^+ and OH^-. This is the same proportion as 1 gram hydrogen ion in 10,000,000 litres of water. Hence, 1 litre of water contains 1/10,000,000 (or $1/10^7 = 10^{-7}$) of a gram of H^+. Later, for every day use, only the 'power' figure was used and the symbol pH placed before it.

A neutral solution, such as water, where the number of hydrogen ions is balanced by the same number of hydroxyl ions, has a pH of 7.0. The range of the pH scale is from 0 to 14. If the pH is 0, it

Fig. 3–1. pH of washing soda indicates it is basic in nature.

would mean that 1 litre of water contained 1/1 = 1 gram hydrogen ion; or, at the other end of the scale, if there were no hydrogen ion present, it would be written $1/10^{14}$ or 10^{-14} or pH 14. The pH scale, thus, runs from 1 to 14; neutrality being at pH 7.0. Solutions having a pH lesser than 7 are acidic, *i.e.*, the concentration of H^+ is greater than that of OH^-. Conversely, solutions having a pH more than 7 are basic or alkaline, *i.e.*, denote an excess of OH^- over H^+ (Table 3–1). Washing soda for example, has a pH of ca 11 (Fig. 3–1).

Table 3–1. The pH scale

$[H^+]$ (M)	pH	$[OH^-]$ (M)	pOH*
$10^0(1)$	0	10^{-14}	14
10^{-1}	1	10^{-13}	13
10^{-2}	2	10^{-12}	12
10^{-3}	3	10^{-11}	11
10^{-4}	4	10^{-10}	10
10^{-5}	5	10^{-9}	9
10^{-6}	6	10^{-8}	8
10^{-7}	7	10^{-7}	7
10^{-8}	8	10^{-6}	6
10^{-9}	9	10^{-5}	5
10^{-10}	10	10^{-4}	4
10^{-11}	11	10^{-3}	3
10^{-12}	12	10^{-2}	2
10^{-13}	13	10^{-1}	1
10^{-14}	14	$10^0(1)$	0

* The expression of pOH is sometimes used to describe the basicity, or OH^- concentration, of a solution. pOH is defined by the expression pOH = – log $[OH^-]$, which is analogous to the expression for pH. *Note that for all cases, pH + pOH = 14.*

Note that *the pH scale is logarithmic, not arithmetic.* Thus, when the pH of a solution decreases one unit from 5 to 4, the H^+ concentration has increased tenfold from 10^{-5} to 10^{-4} M, since decimal logarithms are used for the pH scale. Similarly, when the pH has increased three-tenth of a unit from 6.0 to 6.3, the H^+ concentration has decreased from 10^{-6} M to 5×10^{-7} M. To say that two solutions differ in pH by 1 pH unit means that one solution has ten times the H^+ concentration of the other, but it does not tell us the absolute magnitude of the difference. Figure 3–2 gives the pH of some common aqueous fluids. A coca cola drink (pH 3.0) or red wine (pH 3.7) has an H^+ concentration approximately 10,000 times greater than that of blood (pH 7.4). If we now apply the term of pH to the ion product expression for pure water, we obtain another useful expression:

$$[H^+] \times [OH^-] = 1.0 \times 10^{-14}$$

On taking logarithmic of this equation:

$$\log [H^+] + \log [OH^-] = \log (1.0 \times 10^{-14}) = -14$$

and then multiplying by -1, we get:

$$-\log [H^+] - \log [OH^-] = 14$$

If we now define $-\log [OH^-]$as pOH, a definition similar to that of pH, we have an expression relating the pH and pOH in any aqueous solution:

$$pH + pOH = 14$$

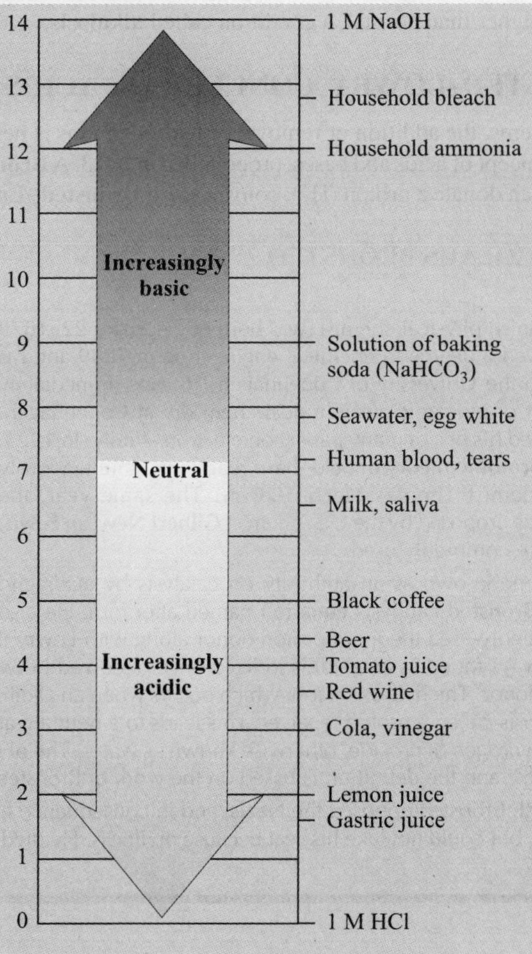

Fig. 3–2. The pH of some aqueous fluids

Measurement of pH is one of the most important and frequently used procedures in Biochemistry (Fig. 3–3). The pH affects the structure and activity of biological macromolecules; for example, the catalytic activity of enzymes. Measurements of the pH of the blood and urine are commonly used in

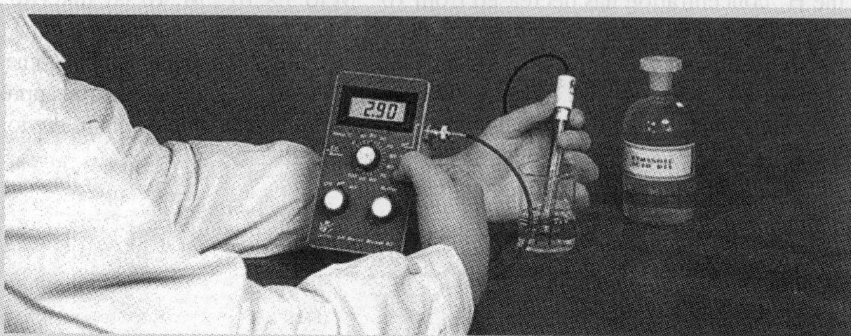

Fig. 3-3. Testing pH by pH meter

diagnosing disease. The pH of the blood plasma of severely diabetic people, for example, is often lower than the normal value of 7.4; this condition is called **acidosis.** In certain other disease states, the pH of the blood is higher than normal, a condition called **alkalosis.**

BRÖNSTED–LOWRY CONCEPT OF ACIDS AND BASES

In aqueous systems, the addition or removal of hydrogen ions is best understood in terms of the Brönsted–Lowry concept of acids and bases, propounded in 1923. A **Brönsted–Lowry acid** is defined as a substance that can donate a proton (H^+); conversely, a **Brönsted–Lowry base** is a substance that

JOHANNES NICOLAUS BRÖNSTED
(LT, 1879–1947)

Brönsted, a Danish physical chemist, was born on February 22, 1879 in Varde. He received a degree in chemical engineering in 1899 and his Ph.D. in 1908 from the University of Copenhagen. He was immediately appointed professor of inorganic and physical chemistry at Copenhagen. In 1906, he published his first of many papers on electron affinity. In 1923, he introduced the **protonic theory of acid–base reactions,** simultaneously with the British chemist **Thomas Martin Lowry.** The same year, the electronic theory was proposed by the U.S. chemist **Gilbert Newton Lewis,** but both theories are commonly used.

Brönsted became known as an authority on catalysis by acids and bases. He has the **Brönsted catalysis equation** named after him. He also came up with the highly-used theory of proton donor along with Lowry. Brönsted theorized that as a hydrogen atom (always found in an acid) is ionized once dissolved in water, it loses its electron and becomes a proton donor. The hydroxide ion, which occurs when an alkali is formed when a substance is dissolved in water is called a proton receiver. This leads to a neutralization reaction where the ions combine creating *hydrogen hydroxide,* otherwise known as *water.* The pH scale may be interpreted as "power of hydrogen", and the definition is based on the work of Brönsted and Lowry.

In World War II, Brönsted opposed the Nazis, and in consequence he was elected to the Danish parliament in 1947, but could not take his seat because of illness. He died shortly after the election on December 17, 1947.

THOMAS MARTIN LOWRY

(LT, 1874–1936)

Lowry, a physicochemical British scientist, was born on October 26, 1874 in Bradford (West Yorkshire), England. He studied chemistry under the direction of **Henry Armstrong,** and became his assistant in 1896. Although Armstrong's scientific interests were primarily organic chemistry, then the student was also initiated to study the nature of the ions in solution aqueous. He lectured at *Westminster Training College* from 1906 to 1912, before moving to *Guy's Hospital Medical School*. He became professor of Chemistry and Chairman, Department of Chemistry in 1913 and was the first professor of Chemistry in Medical School, London. He joined as a *Member of the Royal Society* the following year and was the first professor in physical chemistry from the University of Cambridge.

Lowry studied the changes in optical rotation caused by reactions catalyzed by acids and/or bases involving derivatives of camphor. It came to enable him to formulate the model for the characterization of protonic acids and bases in 1923– a work that was independent, but simultaneously also developed by fellow physical chemist Danish **Johannes Nicolaus Brönsted**. The new concept, because of the new model and the new acid–base theory on the basis of giving / receiving proton came to be named **conjugate acid–base theory of Brönsted–Lowry**. Lowry continued working in Cambridge until the end of his days. He died on November 2, 1936.

can accept a proton. A proton donor (*i.e.*, an acid) and its corresponding proton acceptor (*i.e.*, a base) make up a **conjugate** (*coniungere*[L] = to join together) **acid–base pair** (Table 3–2). This broad definition of acids and bases includes many substances that are not usually considered acidic or basic. In the general expressions:

$$HA \rightleftharpoons H^+ + A^-$$
$$HB^+ \rightleftharpoons H^+ + B:$$
$$HB^- \rightleftharpoons H^+ + B^{2-}$$

HA, HB^+ and HB^- represent Brönsted–Lowry acids, and the anion (A^-), neutral species (B:) and the negatively–charged species (B^{2-}), produced by the loss of a proton, are their respective conjugate bases. Note that each conjugate base can accept a proton to restore the corresponding acid. Because many biomolecules are capable of donating or accepting protons, they influence the H^+ concentration in biological systems. **G.N. Lewis**, also in 1923, proposed yet another definition of acids and bases. According to him, an acid is a compound which can accept a pair of electrons from a base. Such compounds are also called **electrophile** (electron-loving). A base is defined as a compound which can donate an electron pair to an acid. Such compounds are also called **nucleophile** (nucleus-loving).

A. Strong and Weak Acids

There are two general classes of acids– strong and weak. A **strong acid** is defined as a substance that has a greater tendency to lose its proton and therefore completely dissociates (or ionizes) in water, such as HCl and H_2SO_4. A **weak acid**, on the other hand, is a molecule that has a lesser tendency to lose its proton (or, in other words, displays a high affinity for its proton) and, therefore, does not readily dissociate in water, such as CH_3COOH.

Table 3–2. Some Brönsted–Lowry acids and their conjugate bases*

Acid	*Undissociated acid (HA)*	*Conjugate base (A^-)*
Acetic Acid	CH_3COOH	CH_3COO^-
Propionic acid	CH_3CH_2COOH	$CH_3CH_2COO^-$
Pyruvic acid	$CH_3COCOOH$	CH_3COCOO^-
Phosphoric acid	H_3PO_4	$H_2PO_4^-$
Dihydrogen phosphate	$H_2PO_4^-$	HPO_4^{2-}
Monohydrogen phosphate	HPO_4^{2-}	PO_4^{3-}
Ammonium ion	NH_4^+	NH_3
Carbonic acid	H_2CO_3	HCO_3^-
Bicarbonate ion	HCO_3^-	CO_3^{2-}
Water	HOH	OH^-
Glycine	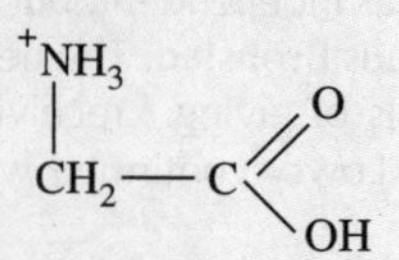	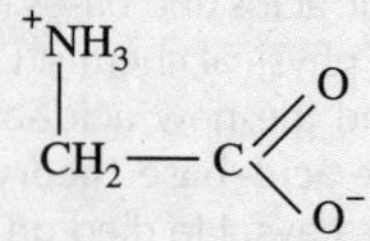
	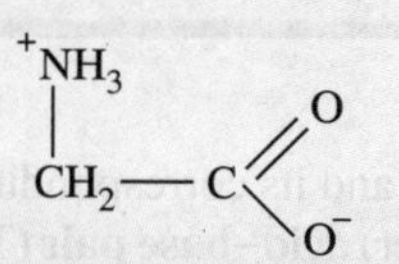	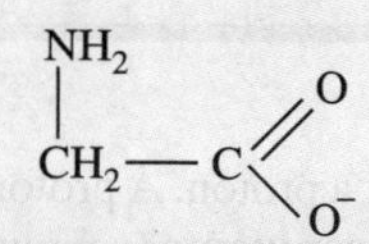

* Some compounds, such as acetic acid, are **monoprotic**, as they can give up only one proton. Others are **diprotic** (carbonic acid and glycine) or **triprotic** (phosphoric acid).

B. Ionization of Weak Acids

The selective dissociation of weak acids in water is a characteristic. It is of great importance to Biochemistry because of its role in influencing the H^+ concentration of an aqueous environment. The dissociation of the weak organic compound, acetic acid, is written as:

$$CH_3COOH \rightleftharpoons H^+ + CH_3COO^-$$

At a given temperature, the extent of ionization at equilibrium can be calculated by the following equation:

$$K_a' = [H^+]\,[CH_3\;COO^-]\,/\,[CH_3\;COOH]$$

The expression is comparable to that used to describe the dissociation of water, except that the symbol K_a' (apparent ionization constant) is substituted for K'_{eq} (equilibrium constant). The change recognizes the reaction as an ionization. The prime (′) symbol on the K_a' is used to identify the ionization constant as *apparent* (value based on the concentrations of reactants and products) rather than as a *true* value (K_a' corrected for deviation of the system from ideal behaviour).

The slight amount of ionization that a 1 M solution of acetic acid undergoes ($K_a' = 1.8 \times 10^{-5}$ M at 25°C) can be readily calculated by leting $x = [H^+]$ and $[CH_3COO^-]$ and not by correcting [CH_3 COOH], which is a relatively large concentration, for the small amount of x produced.

$$x^2 = 1.8 \times 10^{-5}\ \text{M}$$
$$x = 4.2 \times 10^{-3}\ \text{M or } 0.0042\ \text{M}$$

Thus, only slightly over 0.4 per cent of a 1 M solution of acetic acid is ionized at 25°C. The pH of the solution is 2.38 (*i.e.*, – log 0.0042).

C. Titration of a Weak Acid by a Strong Base

Titration is used to determine the amount of an acid in a given solution. In this procedure (Fig. 3-4), a measured volume of the acid is titrated with a solution of a strong base (usually NaOH) of known concentration. The NaOH is added in small increments until the acid is consumed (or neutralized), as determined with an indicator dye or with a pH meter. The concentration of the acid in the original solution can be calculated from the volume and concentration of NaOH added. A plot of the pH against the amount of NaOH added (a **titration curve**) reveals the pK_a of the weak acid.

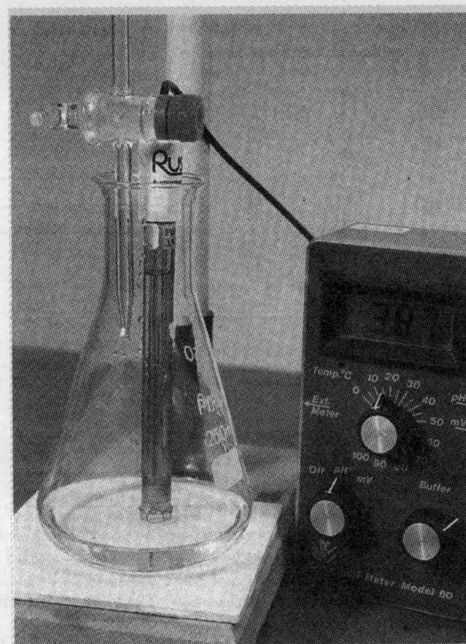

Fig. 3-4. A pH titration in action

Equilibrium constants for ionization reactions are more usually called *ionization or dissociation constants* and are often designated K_a. Table 3–3 lists dissociation constants of some weak acids. Stronger acids, such as formic and lactic acids, have higher dissociation constants; weaker acids, such as dihydrogen phosphate ($H_2PO_4^-$), have lower dissociation constants. Also included in the Table 3–3 are values of pK_a, which is analogous to pH and is defined by the equation:

$$pK_a = \log \frac{1}{K_a} = -\log K_a$$

The more strongly dissociated the acid, the lower is its pK_a.

A plot of the course of the titration reveals the property of Brönsted–Lowry acids and their conjugate bases that makes them useful in Biology. When acetic acid (CH_3COOH) is titrated with NaOH (Fig. 3–5), the greatest changes in pH take place at the beginning and end of the titration. The region of least change occurs at the midpoint of titration, when exactly 0.5 equivalents of base have been added. At this point, the concentration of undissociated acid (CH_3 COOH or HA) is equal to that of its anion (CH_3COO^- or A^-). At these particular concentrations of HA and A^-, the pH (4.76) is equal to the pK_a' value.

Table 3–3. Dissociation constant and pKa of some common weak acids (proton donors) at 25°C

Acid	*Formula*	K_a *(M)*	pK_a
Formic acid	$HCOOH$	1.78×10^{-4}	3.75
Acetic acid	CH_3COOH	1.74×10^{-5}	4.76
Propionic acid	CH_3CH_2COOH	1.35×10^{-5}	4.87
Lactic acid	$CH_3CH(OH)COOH$	1.38×10^{-4}	3.86
Phosphoric acid	H_3PO_4	7.25×10^{-3}	2.14
Dihydrogen phosphate	$H_2PO_4^-$	1.38×10^{-7}	6.86
Monohydrogen phosphate	HPO_4^{2-}	3.98×10^{-13}	12.40
Carbonic acid	H_2CO_3	1.70×10^{-4}	3.77
Biocarbonate	HCO_3^-	6.31×10^{-11}	10.20
Ammonium	NH_4^+	5.62×10^{-10}	9.25

The most important point about the titration curve of a weak acid is that it shows graphically that a weak acid and its anion— a conjugate acid-base pair— can act as a buffer.

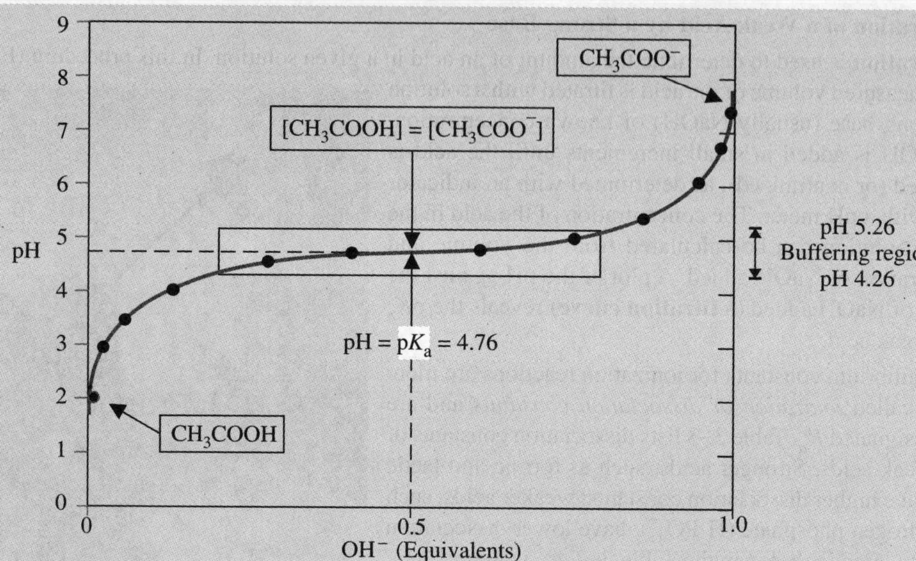

Fig. 3–5. The titration curve of acetic acid

After the addition of each increment of NaOH to the acetic acid solution, the pH of the mixture is measured. This value is plotted against the fraction of the total amount of NaOH required to neutralize the acetic acid (*i.e.*, to bring it to pH ≈ 7). The points so obtained yield the titration curve. Shown in the boxes are the predominant ionic forms at the points designated. At the midpoint of the titration, the concentration of the proton donor and proton acceptor are equal. The pH at this point is numerically equal to the pK_a of acetic acid. The screened zone is the useful region of buffering power.

Fig. 3–6 compares the titration curves of 3 weak acids with very different dissociation constants: acetic acid ($pK_a = 4.76$), dihydrogen phosphate ($pK_a = 6.86$) and ammonium ion ($pK_a = 9.25$). Although the titration curves of these acids have the same shape, they are displaced along the pH axis because these acids have different strengths. Acetic acid is the strongest and loses its proton most readily, since its K_a is highest (pK_a lowest) of the three. Acetic acid is already half dissociated at pH 4.76. $H_2PO_4^-$ loses a proton less readily, being half dissociated at pH 6.86. NH_4^+ is the weakest acid of the three and becomes half dissociated only at pH 9.25.

BUFFERS

A **buffer solution** is one that resists a change in pH on the addition of acid (H^+) or base (OH^-), more effectively than an equal volume of water. Most commonly, the buffer solution consists of a mixture of a weak Brönsted acid and its conjugate base; for example, mixtures of acetic acid and sodium acetate or of ammonium hydroxide and ammonium chloride are buffer solutions. A buffer system consists of a weak acid (the proton donor) and its conjugate base (the proton acceptor). As an example, a mixture of equal concentrations of acetic acid and acetate ion, found at the midpoint of the titration curve in Fig. 3–5, is a buffer system. The titration curve of acetic acid has a relatively flat zone extending about 0.5 pH units on either side of its midpoint pH of 4.76. In this zone, there is only a small change in pH when increments of either H^+ or OH^- are added to the system. This relatively flat zone is the buffering region of the acetic acid-acetate buffer pair. At the midpoint of the buffering region, where the concentration of the proton donor (acetic acid) exacts equals that of the proton acceptor (acetate), the buffering power of the system is maximal, *i.e.*, its pH changes least on addition of an increment of H^+ or OH^-. The pH at this point in the titration curve of acetic acid is equal to its pK_a. The pH of the acetate buffer system does change slightly when a small

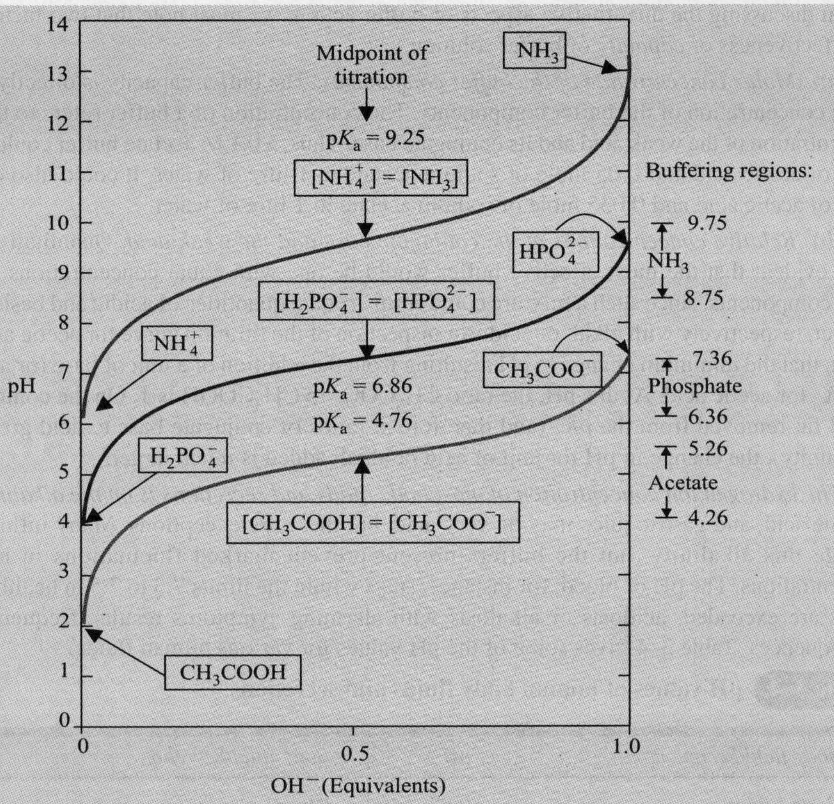

Fig. 3–6. Comparison of the titration curves of 3 weak acids, CH_3COOH, $H_2PO_4^-$ and NH_4^+

The predominant ionic forms at designated points in the titration are given in boxes. The regions of buffering capacity are indicated at the right. Conjugate acid-base pairs are effective buffers between approximately 25 and 75% neutralization of the proton-donor species.

amount of H^+ or OH^- is added, but this change is very small compared with the pH change that would result if the same amount of H^+ (or OH^-) were added to pure water or to a solution of the salt of a strong acid and strong base, such as NaCl, which have no buffering power.

Each conjugate-acid-base pair has a characteristic pH zone in which it is an effective buffer (Fig. 3–3). The $H_2PO_4^- / HPO_4^{2-}$ pair has a pK_a of 6.86 and thus can serve as a buffer system near pH 6.86; the NH_4^+/NH_3 pair, with a pK_a of 9.25, can act as a buffer near pH 9.25.

Let us consider the mechanism by which a buffer solution exerts control over large pH changes, by taking example of a buffering system consisting of acetic acid and sodium acetate.

$$CH_3COOH + CH_3COONa \longrightarrow Na^+ + H^+ + 2CH_3COO^-$$

If alkali (NaOH) is added to this system, it will form salt and no free H^+ or OH^- will be available:

$$CH_3COOH + CH_3COONa + NaOH \longrightarrow 2CH_3COONa + H_2O$$

If, however, acid (HCl) is added to this system, it will also form salt and no free H^+ or OH^- will be available:

$$CH_3COOH + CH_3COONa + HCl \longrightarrow NaCl + 2CH_3COOH$$

In either cases, there is no change in H^+ concentration. The buffer acts almost as if it were 'absorbing' the added free hydrogen or hydroxyl ions.

In discussing the quantitative aspects of buffer action, we must note that two factors determine the effectiveness or *capacity* of buffer solution :

(*a*) *Molar concentration of the buffer components.* The buffer capacity is directly proportional to the concentration of the buffer components. The concentration of a buffer refers to the sum of the concentration of the weak acid and its conjugate base. Thus, a 0.1 M acetate buffer could contain 0.05 mole of acetic acid and 0.05 mole of sodium acetate in 1 litre of water. It could also contain 0.065 mole of acetic acid and 0.035 mole of sodium acetate in 1 litre of water.

(*b*) *Relative concentrations of the conjugate base and the weak acid.* Quantitatively, it should seem evident that the most effective buffer would be one with equal concentrations of acidic and basic components, since such a mixture could furnish equal quantities of acidic and basic components to react, respectively with alkali or acid. An inspection of the titration curve for acetic acid (Fig. 3–3) shows that the minimum change in pH resulting from the addition of a unit of base (or acid) occurs at the pK_a for acetic acid. At this pH, the ratio CH_3COO^- to CH_3COOH is 1. On the contrary, at values of pH far removed from the pK_a (and therefore at ratios of conjugate base to acid greatly differing from unity), the change in pH for unit of acid or alkali added is much larger.

The hydrogen ion concentration of most body fluids and secretions is on the alkaline side. Urine may be acid, and gastric juice may be very acid, but these are exceptions. Many influences tend to change this alkalinity, but the buffers present prevent marked fluctuations in hydrogen ion concentrations. The pH of blood, for instance, stays within the limits 7.3 to 7.5 in health. When these limits are exceeded, acidosis or alkalosis with alarming symptoms results, frequently with dire consequences. Table 3–4 gives some of the pH values for various human fluids.

Table 3–4. pH values of human body fluids and secretions

Body fluid/Secretion	*pH*	*Body fluid/Secretion*	*pH*
Gastric juice	0.87	Blood	7.4
(parietal secretion)		Cerebrospinal fluid	7.4
Urine	6.0	Intestinal juice	7.7
Milk	6.6 – 6.9	Bile	7.8
Saliva	7.2	Pancreatic Juice	8.0
Aqueous humor of eye	7.2		

It is quite difficult to keep a solution at constant pH if no buffer is present because of the influence of the CO_2 of the air or the alkali of the glass container or because of other influences. Consequently, buffers are frequently required. Various mixtures, consisting of definite amounts of the acid or base and its respective salt have been prepared. Since such buffer sets maintain their pH indefinitely, they are used in the indicator method of determining pH. It should be noted that buffers can be prepared not only by titration of a weak acid with OH^- (or titration of a solution of A^- with H^+), but also by addition of A^-, the salt of HA, to a solution of HA.

HENDERSON–HASSELBALCH EQUATION

The quantitative relationship among pH, buffering action of a mixture of weak acid with its conjugate base, and the pK_a of the weak acid is given by a simple expression called **Henderson–Hasselbalch Equation.** The titration curves of acetic acid, $H_2PO_4^-$ and NH_4^+ (Fig. 3–6) have nearly identical shapes, suggesting that they all point towards a fundamental law or relationship. This is actually the case. The shape of the titration curve of any weak acid is expressed by Henderson–Hasselbalch equation. This equation is simply a useful way of restating the expression for the dissociation constant of an acid. For the dissociation of a weak acid HA into H^+ and A^-, the Henderson–Hasselbalch equation can be derived as follows:

$$K_a = \frac{[H^+][A^-]}{[HA]}$$

1. Rearrange the K_a equation to solve for (H^+):

$$[H^+] = K_a \frac{[HA]}{[A^-]}$$

2. Convert to logarithmic functions:

$$\log [H^+] = \log K_a + \log \frac{[HA]}{[A^-]}$$

3. Make the expression negative (or multiply by −1):

$$-\log [H^+] = -\log K_a - \log \frac{[HA]}{[A^-]}$$

4. Substitute pH for $-\log [H^+]$ and pK_a for $-\log K_a$

$$pH = pK_a - \log \frac{[HA]}{[A^-]}$$

5. Now, to remove the minus sign, invert the last term, *i.e.*, $-\log \frac{[HA]}{[A^-]}$ to obtain Henderson-Hasselbalch equation:

$$pH = pK_a + \log \frac{[A^-]}{[HA]}$$

The equation is expressed more generally as:

$$pH = pK_a + \log \frac{[\text{proton acceptor}]}{[\text{proton donor}]}$$

This equation fits the titration curve of all weak acids and enables one to deduce a number of important quantitative relationships. Henderson-Hasselbalch equation is of great predictive value in protonic equilibria as illustrated below:

A. When [A−] = [HA] or when an acid is exactly half neutralized : Under these conditions,

$$pH = pK_a + \log \frac{[A^-]}{[HA]} = pK_a + \log \frac{1}{1} = pK_a + 0 = pK_a.$$

Therefore, at half neutralization, pH = pK_a. The equation, thus, shows why the pK_a of a weak acid is equal to the pH of the solution at the midpoint of its titration.

B. When the ratio [A−]/[HA] = 100 to 1 :

$$pH = pK_a + \log \frac{[A^-]}{[HA]} = pK_a + \log \frac{100}{1} = pK_a + 2$$

C. When the ratio [A−]/[HA] = 1 to 10 :

$$pH = pK_a + \log \frac{[A^-]}{[HA]} = pK_a + \log \frac{1}{10} = pK_a + (-1)$$

If the equation is evaluated at several ratios of $[A^-] / [HA]$ between the limits 10^3 and 10^{-3}, and the calculated pH values plotted, the result obtained describes the titration curve for a weak acid as shown in Fig. 3–7.

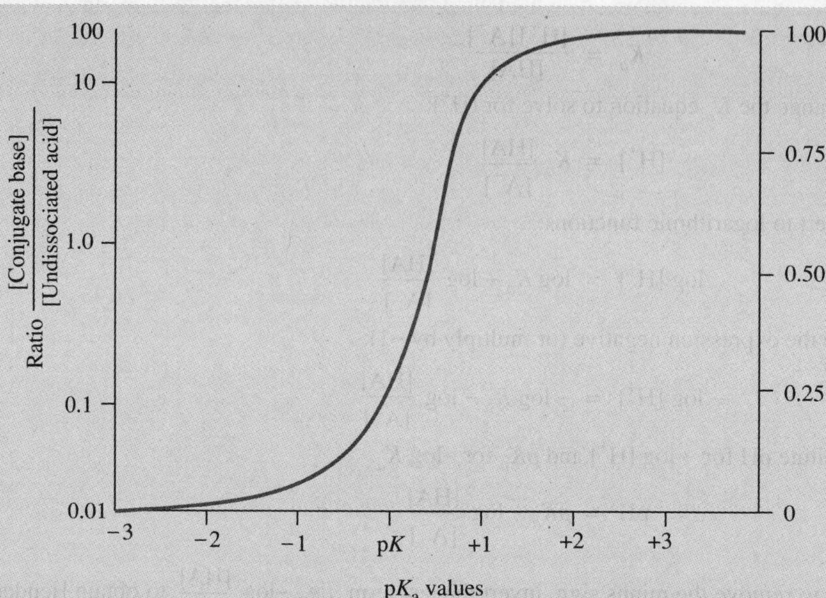

Fig. 3–7. General form of a titration curve calculated from the Henderson-Hasselbalch equation

BIOLOGICAL BUFFER SYSTEMS

Almost every biological process is pH-dependent; a small change in pH produces a large change in the rate of the process. This is true not only for the many reactions in which the H^+ ion is a direct participant, but also for those in which there is no apparent role for H^+ ions. The enzymes and many of the molecules on which they act, contain ionizable groups with characteristic pK_a values. The protonated amino ($-NH_3^+$) and carboxylic groups of amino acids and the phosphate groups of nucleotides, for example, function as weak acids; their ionic state depends upon the pH of the solution in which they are dissolved.

Cells and organisms maintain a specific and constant cytosolic pH, keeping biomolecules in their optimal ionic state, usually near pH 7. In multicelled organisms, the pH of the extracellular fluids (blood, for example) is also tightly regulated. Constancy of pH is achieved primarily by **biological buffers**: mixtures of weak acids and their conjugate bases. Table 3–5 lists some important buffering systems of body fluids which help maintaining pH. A certain amount of many of these is usually present in the body and cellular fluids, and so the maintenance of a constant pH depends on a complex system.

Table 3–5. Body fluids and their principal buffers

Body fluids	*Principal buffers*
Extracellular fluids	Biocarbonate buffer Protein buffer
Intracellular fluids	Phosphate buffer Protein
Erythrocytes	Hemoglobin buffer

A description of various buffering systems follows:

1. The Phosphate Buffer System

This system, which acts in the cytoplasm of all cells, consists of $H_2PO_4^-$ as proton donor and HPO_4^{2-} as proton acceptor:

$$H_2PO_4^- \rightleftharpoons H^+ + HPO_4^{2-}$$

The phosphate buffer system works exactly like the acetate buffer system, except for the pH range in which it functions. The phosphate buffer system is maximally effective at a pH close to its pK_a of 6.86 (see Table 3–3 and Fig. 3–6), and thus tends to resist pH changes in the range between 6.4 and 7.4. It is, therefore, effective in providing buffering power in intracellular fluids.

Since the concentration of phosphate buffer in the blood plasma is about 8% of that of the bicarbonate buffer, its buffering capacity is much lower than bicarbonate in the plasma. The concentration of phosphate buffer is much higher in intracellular fluid than in extracellular fluids. The pH of intracellular fluids (6.0 – 6.9) is nearer to the pK_a of the phosphate buffer. Therefore, the buffering capacity of the phosphate buffer is highly elevated inside the cells and the phosphate is also effective in the urine inside the renal distal tubules and collecting ducts.

In case the ratio of $[HPO_4^{2-}] / [H_2PO_4^-]$ tends to be changed by the formation of more $H_2PO_4^-$, there occurs the renal elimination of $H_2PO_4^-$ for which the ratio ultimately remains unaltered.

2. The Bicarbonate Buffer System

This is the main extracellular buffer system which (also) provides a means for the necessary removal of the CO_2 produced by tissue metabolism. The bicarbonate buffer system is the main buffer in blood plasma and consists of carbonic acid as proton donor and bicarbonate as proton acceptor:

$$H_2CO_3 \rightleftharpoons H^+ + HCO_3^-$$

This system has an equilibrium constant

$$K_1 = \frac{[H^+][HCO_3^-]}{[H_2CO_3]}$$

and functions as a buffer in the same way as other conjugate acid-base pairs. It is unique, however, in that one of its components, carbonic acid, is formed from dissolved (*d*) carbon dioxide and water, according to the reversible reaction:

$$CO_2(d) + H_2O \rightleftharpoons H_2CO_3$$

which has an equilibrium constant given by the expression:

$$K_2 = \frac{[H_2CO_3]}{[CO_2(d)][H_2O]}$$

Carbon dioxide is a gas under natural conditions, and the concentration of dissolved CO_2 is the result of equilibration with CO_2 of the gas phase (*g*):

$$CO_2(g) \rightleftharpoons CO_2(d)$$

This process has an equilibrium constant given by:

$$K_3 = \frac{[CO_2(d)]}{[CO_2(g)]}$$

The pH of a bicarbonate buffer system depends on the concentration of H_2CO_3 and HCO_3^-, the proton donor and acceptor components. The concentration of H_2CO_3, in turn, depends on the concentration of dissolved CO_2, which, in turn, depends on the concentration or partial pressure of CO_2 in the gas phase.

With respect to the bicarbonate system, a $[HCO_3^-] / [H_2CO_3]$ ratio of 20 to 1 is required for the pH of blood plasma to remain 7.40. The concentration of dissolved CO_2 is included in the $[H_2CO_3]$ value, *i.e.*,

$$[H_2CO_3] = [H_2CO_3] + [CO_2 \text{ (dissolved)}]$$

If there is a change in the ratio in favour of H_2CO_3, **acidosis** results. This change can result from a decrease in $[HCO_3^-]$ or from an increase in $[H_2CO_3]$. Most common forms of acidosis are metabolic or respiratory. *Metabolic acidosis* is caused by a decrease in $[HCO_3^-]$ and occurs, for example, in uncontrolled diabetes with ketosis or as a result of starvation. *Respiratory acidosis* is brought about when there is an obstruction to respiration (euphysema, asthma or pneumonia) or depression of respiration (toxic doses of morphine or other respiratory depressants). If acidosis is not treated promptly, the patient may go into a comma.

Alkalosis results when $[HCO_3^-]$ becomes favoured in the bicarbonate/carbonic acid ratio. *Metabolic alkalosis* occurs when the HCO_3^- fraction increases with little or no concomitant change in H_2CO_3. Severe vomiting (loss of H^+ as HCl) or ingestion of excessive amounts of sodium bicarbonate (bicarbonate of soda) can produce this condition. *Respiratory alkalosis* is induced by hyperventilation because an excessive removal of CO_2 from the blood results in a decrease in $[H_2CO_3]$. Hyperventilation can result in anxiety, hysteria, prolonged hot baths or lack of O_2 at high altitudes. Alkalosis can produce convulsive seizures in children and tetany in adults (characterized by sharp flexion of the wrist and ankle joints, muscle twitchings, and cramps).

The pH of blood is maintained at 7.4 when the buffer ratio $[HCO_3^-]/[H_2CO_3]$ becomes 20. If the bicarbonate neutralizes any acid or base, there may be the change of buffer ratio and the blood pH value. But the buffer ratio remains by the respiratory elimination of H_2CO_3 as CO_2 or the urinary elimination of HCO_3^-.

Since cells contain much lower amounts of HCO_3^-, the importance of bicarbonate buffer inside the cell is negligible.

3. The Protein Buffer Systems

The protein buffers are very important in the plasma and the intracellular fluids but their concentration is very low in cerebrospinal fluid, lymph and interstitial fluids. The proteins exist as anions serving as conjugate bases (Pr^-) at the blood pH 7.4 and form conjugate acids (HPr) accepting H^+. They have the capacity to buffer some H_2CO_3 in the blood.

$$H_2CO_3 + Pr^- \rightleftharpoons HCO_3^- + HPr$$

4. The Amino Acids Buffer System

This system also operates in humans. Amino acids contain in their molecule both an acidic (– COOH) and a basic (– NH_2) group. They can be visualized as existing in the form of a neutral zwitterion in which a hydrogen atom can pass between the carboxyl and amino groups. The glycine may, thus, be represented as:

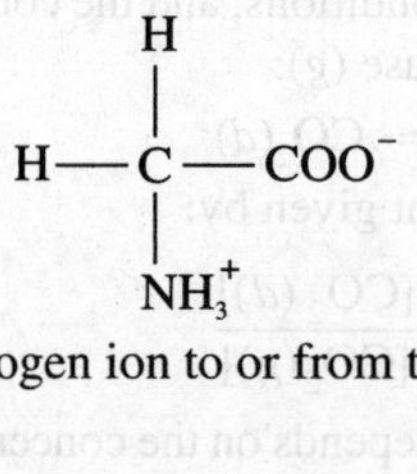

By the addition or subtraction of a hydrogen ion to or from the zwitterion, either the cation or anion form will be produced :

$$\underset{\text{Cation form}}{^+H_3N—CH_2—COOH} \xleftarrow[+H^+]{} \underset{\text{Zwitterion}}{^+H_3N—CH_2—COO^-} \rightleftharpoons \underset{\text{Anion form}}{H_2N—CH_2—COO^-} + H^+$$

Thus, when OH^- ions are added to the solution of amino acid, they take up H^+ from it to form water, and the anion is produced. If H^+ ions are added, they are taken up by the zwitterion to produce the cation form. In practice, if NaOH is added, the salt $H_2N—CH_2—COONa$ would be formed and the addition of HCl would result in the formation of amino acid hydrochloride, ClH —H_3N—CH_2—COOH, but these substances would ionize in solution to some extent to form their corresponding ions. Hemoglobin and plasma proteins act as buffers in a similar way.

Amino acids differ in the degree to which they will produce the cation or anion form. In other words, a solution of an amino acid is not neutral but is either predominantly acidic or basic, depending on which form is present in greater quantity. For this reason, different amino acids may be used as buffers for different pH values, and a mixture of them possesses a wide buffer range.

5. The Hemoglobin Buffer Systems

These buffer systems are involved in buffering CO_2 inside erythrocytes. The buffering capacity of hemoglobin depends on its oxygenation and deoxygenation. Inside the erythrocytes, CO_2 combines with H_2O to form carbonic acid (H_2CO_3) under the action of carbonic anhydrase. At the blood pH 7.4, H_2CO_3 dissociates into H^+ and HCO_3^- and needs immediate buffering. Oxyhemoglobin (HbO_2^-), on the other side, loses O_2 to form deoxyhemoglobin (Hb^-) which remains undissociated (HHb) by accepting H^+ from the ionization of H_2CO_3. Thus, Hb^- buffers H_2CO_3 in erythrocytes:

$$HbO_2^- \rightleftharpoons Hb^- + O_2$$
$$Hb^- + H_2CO_3 \rightleftharpoons HHb + HCO_3^-$$

Some of the HCO_3^- diffuse out into the plasma to maintain the balance between intracellular and plasma bicarbonates. This causes influx of some Cl^- into erythrocytes along the electrical gradient produced by the HCO_3^- outflow (*chloride shift*).

$HHbO_2$, produced in lungs by oxygenation of HHb, immediately ionizes into H^+ and HbO_2^-. The released hydrogen ions (H^+) are buffered by HCO_3^- inside erythrocyte to form H_2CO_3 which is dissociated into H_2O and CO_2 by carbonic anhydrase. CO_2 diffuses out of erythrocytes and escapes in the alveolar air. Some HCO_3^- return from the plasma to erythrocytes in exchange of Cl^- and are changed to CO_2.

$$HHb + O_2 \rightleftharpoons HHbO_2 \rightleftharpoons HbO_2^- + H^+$$
$$HCO_3^- + H^+ \rightleftharpoons H_2CO_3 \rightleftharpoons H_2O + CO_2$$

REFERENCES

1. **Bittar EE :** Cell pH. *Butterworth, Washington D.C. 1964.*
2. **Corey HE:** Bench-to-bedside review: fundamental principles of acid–base balance. *Crit. Care.* **9:** *184–192, 2005.*
3. **Davenport HW :** The ABC of Acid-Base Chemistry. *6th ed. University of Chicago Press, Chicago. 1974.*
4. **Dawson RMC, Elliott DC, Elliott WH, Jones KM (editors) :** Data for Biochemical Research. *3rd ed. Oxford University Press, New York. 1986.*
5. **Kellum:** Disorders of acid–base balance. *Crit. Care Med.* **35 :** *2630–2636, 2007.*
6. **Masoro EJ, Siegel PD :** Acid-Base Regulation *2nd ed. W.B. Saunders, Philadelphia. 1977.*
7. **Montgomery R, Swenson CA :** Quantitative Problems in Biochemical Sciences. *2nd ed. W.H. Freeman and Company, San Francisco, 1976.*
8. **Pullman A, Vasilescu V, Packer L (editors) BF :** Water and Ions in Biological Systems. *Plenum Press, New York. 1985.*
9. **Robinson JR :** Fundamentals of Acid-Base Balance. *Blackwell, Oxford. 1961.*
10. **Segel IH :** Biochemical Calculations. *2nd ed. John Wiley and Sons, Inc., New York. 1976.*

PROBLEMS

1. What is the pH of each of the following solutions?
   (*a*) 0.35 M hydrochloric acid
   (*b*) 0.35 M acetic acid
   (*c*) 0.035 M acetic acid
2. What is the pH of the following buffer mixtures?
   (*a*) 1 M acetic acid plus 0.5 M sodium acetate
   (*b*) 0.3 M phosphoric acid plus 0.8 M KH_2PO_4
3. You need to make a buffer whose pH is 7.0, and you can choose from the weak acids shown in Table 3–3. Briefly explain your choice.
4. It is possible to make a buffer that functions well near pH 7, using citric acid, which contains only carboxylate groups. Explain.

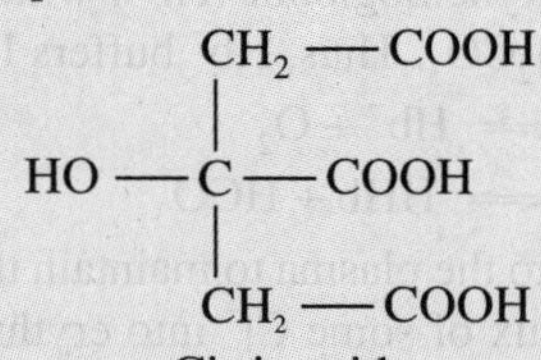

Citric acid

5. If a weak acid is 91% neutralized at pH 5.7, what is the p*K′* of the acid?
6. Metabolic alkalosis is a condition in which the blood pH is higher than its normal value of 7.40 and can be caused, among other reasons, by an increase in $[HCO_3^-]$. If a patient has the following blood values, pH = 7.45, $[CO_2]$ = 1.25 mM, what is the $[HCO_3^-]$? The p*K′* for the bicarbonate/carbonic acid system is 6.1. How does the calculated value of bicarbonate compare with the normal value?
7. Why does pure water have no taste or colour?
8. Red ant sting creates swelling due to which acid
   (*a*) Propionic acid
   (*b*) Butyric acid
   (*c*) Acetic acid
   (*c*) Formic acid

ANSWERS TO PROBLEMS

1. (*a*) pH = 0.456.
   (*b*) pH = 2.608 by approximation, 2.609 exact.
   (*c*) pH = 3.108 by approximation, 3.113 exact.
2. (*a*) pH = 4.46.
   (*b*) pH = 2.57.
3. The best choice would be a mixture of $H_2PO_4^-$ and HPO_4^{2-}, which has $pK_a = 6.86$.
4. Because of the proton-attracting power of the dianion produced when two of the citric acid protons have dissociated, the third is held anomalously strongly. In fact, it has a pK_a of 6.86.
5. If a weak acid is 91% neutralized, 91 parts are present as conjugate base and 9 parts remain as the undissociated acid. Thus, the conjugate base/acid ratio is approximately 10:1. Substituting this ratio into the Henderson–Hasselbalch equation gives 5.7 = p*K′* + log (10/ 1). Solving the equation for p*K′* gives an answer of 4.7. The acid could be β-hydroxybutyric acid, an important physiological acid, which has this p*K′*.

6. One need to simply substitute the given values into the Henderson–Hasselbalch equation:

$$7.45 = 6.1 + \log(x/1.25)$$
$$1.35 = \log(x/1.25)$$

The antilog of 1.35 is 22.39. Therefore

$$[HCO_3^-] = 22.39 \times 1.25 = 27.98 \text{ mM}$$
$$\text{Normal } [HCO_3^-] = 24.0 \text{ mM so there is an increase of 3.98 mmol (per liter) of } HCO_3^-.$$

7. Physical properties of a compound, such as colour, taste, touch etc, depend on its pH factor. The pH further depends upon the concentration of H^+ and OH^- ions in that compound. For example, compounds having higher concentration of H^+ ions are sour like acids and compounds having higher OH^- ion concentration are bitter like bases. Water has an equal concentration of H^+ and OH^- ions. Thus, it has neither sour nor a bitter taste and hence it is tasteless. Similarly, it is colourless and does not have an odour.
8. (*c*) Acetic acid.

CONTENTS

Element	Earth's crust	Human	Pumpkin
C	–	18	3.3
H	–	10	10.7
O	46.6	65	85
N	–	3	0.16
P	–	1.1	0.05
S	–	0.25	–
Na	2.8	0.15	0.001
K	2.6	0.35	0.34
Ca	3.6	2	0.02
Mg	2.1	0.05	0.01
Fe	5.0	0.004	0.008
Al	8.1	–	–
Si	27.7	–	–
Cl	–	0.15	–
I	–	0.0004	–
Zn	–	–	0.0002
Cu	–	–	0.0001
other elements	1.5	–	0.00005

Proportions of different elements in the earth's crust, the human body and a pumpkin

CHAPTER

4

The Atom and Chemical Bonds

Living organisms are composed of lifeless molecules. When these molecules are isolated and examined individually, they conform to all the physical and chemical laws that describe the behaviour of inanimate matter. Yet living organisms possess extraordinary attributes not shown by any random collection of molecules. The composition of living matter is strikingly different from that of the inanimate world. **Antoine Laurent Lavoisier** noted the relative chemical simplicity of the "mineral world" and contrasted it with the complexity of "plant and animal worlds"; the later he knew were composed of compounds rich in the elements carbon, oxygen, nitrogen, and phosphorus.

ELEMENTAL COMPOSITION OF LIVING MATTER

With the exception of certain metal ions, Biochemistry is, for the most part, related to the chemistry of the elements of the second and third periods of the Periodic Table. On earth, all cells, regardless of their origin (animal, plant, or microbial), contain the same elements in approximately the same proportions (Table 4–1). Thus, of the 107 known **elements**, only about 20 are essential for terrestrial life. Perhaps there is some logical chemical explanation for their selection.

Table 4–1. The elemental composition of living cells

Element	*Composition by weight (%)*	*Element*	*Composition by weight (%)*
O	65		
C	18	Cu, Zn, Se, Mo,	
H	10	F, Cl, I, Mn,	0.70
N	3	Co, Fe	
Ca	1.5		
P	1.0		
K	0.35		
S	0.25	Li, Sr, Al, Si,	
Na	0.15	Pb, V, As, Br	Traces*
Mg	0.05		

* Variable occurrences in cells. No known function in most cases.

Six nonmetals (O, C, H, N, P, S), which contribute almost 98% of the total mass of cells, provide the structural elements of protoplasm. From them, the functional components of cells (such as walls, membranes, genes, enzymes etc) are formed. These 6 elements all occur in the first three periods of the Periodic Table (refer Table 4–2).

Table 4–2. The structural elements of protoplasm*

Period	*Group* I	*II*	*III*	*IV*	*V*	*VI*	*VII*	*VIII*
1	(H)							He
2	Li	Be	B	(C)	(N)	(O)	F	Ne
3	Na	Mg	Al	Si	(P)	(S)	Cl	Ar

*The six common nonmetals have been shown *encircled*.

The relative abundance of these 6 lighter elements in the seas, crust and atmosphere of earth does not by itself explain their utilization for life. Aluminium, for example, is more abundant than carbon, but it performs no known function essential to life. By contrast, the intrinsic chemical properties of these elements suggest their unique suitability as building blocks for life. Desirable features of structural elements apparently are as follows (**Victor W. Rodwell**, 1979):

(*a*) Small atomic radius

(*b*) The versatility conferred by the ability to form 1−, 2−, 3− and 4− electron bonds

(*c*) The ability to form multiple bonds.

Carbohydrates and lipids contain C, H and O while proteins and nucleic acids contain in addition N and S, and N and P respectively. Na, K, Mg, Ca, and Cl are usually found in ionic forms whose concentrations must remain within narrow limits. The presence of Bo, Si, V, Mn, Fe, Co, Cu, Zn and Mb, though in ultratrace amounts, is essential for the functioning of enzymes. These elements are used as cofactors in the catalytic function of enzymes.

Small atoms form the tightest and most stable bonds — a distinct advantage for structural elements. **H, O, N and C are the smallest atoms capable of forming 1– , 2–, 3– and 4– electron bonds respectively.** Utilization of all possible types of electron bonds permits maximum versatility in molecular design. So also does the ability to form multiple bonds, a property confined almost entirely to P, S, and the elements of period 2. Advantages of carbon-based versus silicon-based life include the following :

1. Greater chemical stability of C—C bonds than the Si—Si bonds.
2. The ability of carbon, but not of silicon, to form multiple bonds; for example, the oxides of carbon are diffusible monomeric gases, whereas the oxides of silicon is a viscous polymer.
3. The stability of C—C bonds, but not of Si—Si bonds, to rupture by nucleophilic reagents (electron-rich elements or compounds) such as O_2, H_2O or NH_3.

Similar factors uniquely qualify P and S for utilization in energy transfer reactions. **Energy transfer is facilitated by bonds susceptible to nucleophilic attack.** A most common example is the nucleophilic attack of the 6-OH of glucose on the terminal P—O—P bond of ATP, forming ADP plus glucose-6-phosphate. P and S resemble Si in that P—O—P or S—O—S bonds, like Si—Si bonds, are susceptible to nucleophilic rupture by virtue of their unoccupied third orbitals. However, unlike Si, P and S form multiple bonds (more versatile), a consequence of their smaller atomic diameters. Most energy transfer reactions in Biochemistry may be visualized as resulting from attack of a nucleophil (N) on the unoccupied third orbital of a phosphorus atom :

Nucleophilic attack is the attack of an electron-rich centre upon an electron-deficient centre.

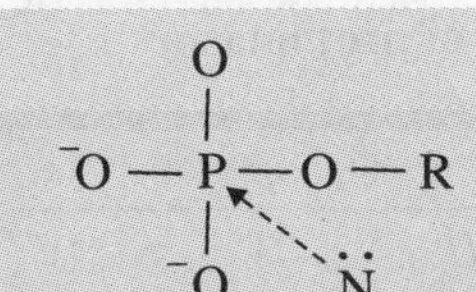

The characteristic chemical and physical properties of the chemical elements of life are the same throughout the known universe. It, thus, seems probable that if life exists elsewhere, the same elements are employed for the same or similar reasons.

STRUCTURE OF AN ATOM

Each matter is composed of very small particles called **atoms**, which cannot be created, destroyed or subdivided. Atoms of the same element are similar to one another and equal in weight. Atoms of different elements have different properties and weight. Although, at one time, the atoms were conceived to be the smallest particles, subatomic particles were later recognized in due course of time. The 3 fundamental subatomic particles are: proton, neutron and electron. Besides these fundamental particles, about 35 other atomic particles are also known to exist. Many of them are, however, extremely unstable and they merely represent a bundle of energy. Some of the stable particles other than fundamental particles, are positron, photon, neutrino, graviton and antiproton. These particles are, however, of little importance in the study of Biochemistry because their existence is rarely encountered in biological systems.

Regarding the arrangement of fundamental particles inside an atom, a British chemist **Ernest Rutherford** (1911) proposed the most satisfactory model which is accepted even today with some modifications (Fig. 4–1). Accordingly, an atom is made up of a central *nucleus* containing positively-charged protons and neutral neutrons, surrounded by negatively-charged electrons which move around it (the nucleus) in discrete, successive, concentric volumes in space known as *orbits* or *shells*. The

SIR ERNEST RUTHERFORD
(LT, 1871–1937)

Rutherford, a New Zealand-British chemist and physicist, was born on 30 August 1871 at Spring Grove (now Brightwater), near Nelson, New Zealand. His first name was mistakenly spelled Earnest when his birth was registered. He was the son of **James Rutherford**, a farmer and his wife **Martha Thompson**, originally from Essex, England. James had emigrated to New Zealand from Perth, Scotland, "to raise a little flax and a lot of children". He studied at Havelock School and then Nelson College and won a scholarship to study at Canterbury College, University of New Zealand. After gaining his BA, MA and BSc, and doing two years of research at the forefront of electrical technology, in 1895 Rutherford traveled to England for postgraduate study at the Cavendish Laboratory, University of Cambridge (1895–1898) and he briefly held the world record for the distance over which electromagnetic waves could be detected. In 1898, Rutherford was appointed to succeed **Hugh Longbourne Callendar** in the chair of Macdonald Professor of Physics at McGill University in Montreal, Canada. In 1900, he gained a DSc from the University of New Zealand. Also in 1900, he married **Mary Georgina Newton** (LT, 1876–1945); they had one daughter, **Eileen Mary**. Rutherford moved, in 1907, to Britain to take the chair of physics at the University of Manchester.

His early work was with **J.J. Thomson** on magnetism and thus on radio waves. Following Roentgen's discovery of x-rays (1895), they studied the conductivity of air bombarded by these rays and the rate at which the ions produced recombined. This led him to similar studies of the "rays" emitted by uranium. Rutherford worked in the UK at the Cavendish Laboratory. He identified that the radioactive elements including radium (Ra), uranium (U), and thorium (Th) emit 2 types of radiations:

(*a*) α *rays*, with a positive charge of 2 units, which was later identified as doubly-ionized helium atom, and

(*b*) β *rays*, with a negative charge of 1 unit, which was later identified as electrons.

In 1903, Rutherford realized that a type of radiation from radium discovered (but not named) by French chemist **Paul Villard** in 1900, must represent something different from a rays and b rays, due to its very much greater penetrating power. Rutherford gave this third type of radiation its name also: the γ *ray*. Gamma rays are electromagnetic radiation similar to x-rays.

Rutherford's studies showed that radioactive elements actually break down or disintegrate as they give off the different kinds of rays. He, along with one of his coinvestigators **Frederick Soddy**, brought forward, in 1902, the so-called "disintegration theory" to explain radioactivity. According to this theory, the radioactive elements disintegrate in such a way that in the space of every unit of time, a definite percentage of their atoms are broken up into one or more particles of radiation on the one hand, and a new atom on the other. He coined the term **half-life**. Though Rutherford was a physicist and his research was done mostly with the aid of physical methods, the importance of his discoveries for chemical investigation was far-reaching. That is why, for his researches concerning the disintegration of elements and the chemistry of radioactive substances, Rutherford was awarded the **Nobel Prize for Chemistry in 1908**.

Rutherford performed his most famous work after he had moved to the Victoria University of Manchester in the UK (in 1907) and he was already a Nobel Laureate. In 1911, he theorized that atoms have their positive charge concentrated in a very small nucleus, and thereby pioneered the **Rutherford model of the atom** through his discovery and interpretation of Rutherford scattering in his *gold foil experiment*. Based on his nuclear theory of the atom, **Niels H.D. Bohr** based his celebrated theory two years later. He is widely credited with first, "splitting the atom" in 1917 in a nuclear reaction between nitrogen and a particles, in which he also discovered (and named) the proton. This led to the first experiment to split the nucleus in a fully controlled manner, performed by two students working under his direction, **John Cockcroft** and **Ernest Walton**, in 1932.

Rutherford was knighted in 1914. In 1916, he was awarded the Hector Memorial Medal. In 1919, he returned to the Cavendish Laboratory as Director. Under him, Nobel Prizes were awarded to **James Chadwick** for discovering the neutron (in 1932), **John Cockcroft** and **Ernest Walton** for an experiment which was to be known as *splitting the atom* using a particle accelerator, and **Edward Appleton** for demonstrating the existence of the ionosphere. He was admitted to the **Order of Merit**, in 1925, and raised to the peerage as **Baron Rutherford of Nelson**, of Cambridge in the County of Cambridge, in 1931, a title that became extinct upon his unexpected death in 1937. In fact, for some time beforehand, Rutherford had a small hernia, which he had neglected to have fixed, and it became strangulated, causing him to be violently ill. Despite an emergency operation in London, he died 4 days afterwards (on October 19, 1937) of what physicians termed "intestinal paralysis". He was given the high honour of burial in Westminster Abbey near **Sir Issac Newton** and other illustrious British scientists. He remained **President of the Royal Society** from 1925 to 1930. He is commemorated by the naming of the element 104 as **rutherfordium**, in 1997.

Often pathbreaking discoveries in science have evolved from a *chance encounter* a scientist had. Rutherford's luckiest break in life was a Cambridge scholarship. He had applied for it and was ranked number two, but unluckily there was just one scholarship. Just when he had given up all hope of making it to Cambridge and had reconciled to spending the rest of his life in New Zealand away from the exciting researches going on in Europe, came the unexpected news. **Maclaurin**, the student who had won the scholarship, had declined it due to some personal problems. Ernest grabbed the chance, packaged his bags, moved over to Cambridge, worked with the likes of **J.J. Thompson**, and the rest is history.

Rutherford was famous for his ready wit. Once at a luncheon, he was asked what the population of South Island (New Zealand) was. It was his birthplace and Rutherford readily replied. "About 250,000." "Is that all?" the surprised questioner added, "Doesn't it mean, it is about the population of Stoke-on-Trent?" Rutherford rose admirably to the occasion:

> "May be the population is only that of Stoke-on-Trent. But let me tell you, Sir, that every single man in the South Island of New Zealand could eat up the whole population of Stoke-on-Trent, every day before breakfast, and still be hungry."

Some of his famous *statements* are as follows:

1. "The energy produced by the breaking down of the atom is a very poor kind of thing. Anyone who expects a source of power from the transformation of these atoms is talking moonshine." –1993
2. "All science is either physics or stamp collecting" (though he was in 1908 awarded the Nobel Prize in Chemistry).
3. "We haven't the money, so we've got to think."
4. "It your experiment needs statistics, you ought to have done a better experiment."
5. "You should never bet against anything in science at odds of more than about 1012 to 1."

model is analogous to the sun's planetary system but differs from it in having the subatomic particles, protons and electrons as charged.

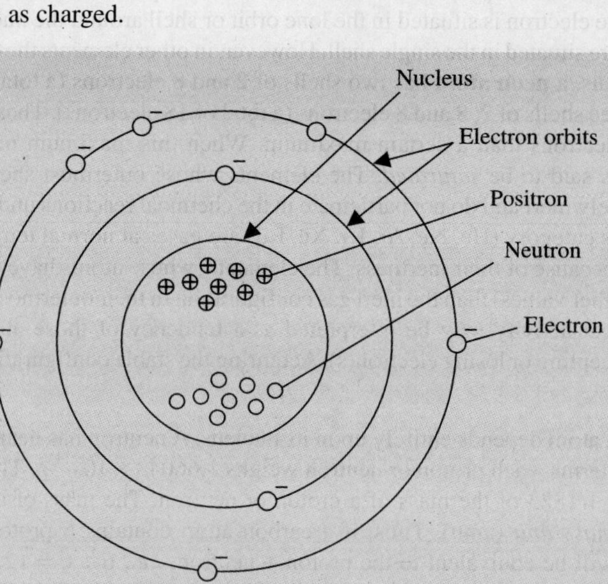

Fig. 4–1. Structure of an atom

The electron shells are numbered (from within) as 1, 2, 3, 4, 5, 6 and 7 and are designated by the letters *K, L, M, N, O, P, Q* respectively (Table 4–3). Each shell has a certain number of electrons and the maximum number of electrons for each shell is fixed. The maximum number is given by $2n^2$, where n is the serial number of the shell. Thus, the maximum number of electrons in *K, L, M, N, O, P, Q* shells will be 2, 8, 18, 32, 50, 72, 98 respectively. The maximum number of electrons in the outermost shell is 8 and in the penultimate shell is 18.

Table 4–3. Distribution of electrons in shells and subshells

Shell number (n)	1	2		3			4 etc			
Shell name	K	L		M			N			
Subshell number	0	0	1	0	1	2	0	1	2	3
Subshell name	*s*	*s*	*p*	*s*	*p*	*d*	*s*	*p*	*d*	*f*
Number of electrons in subshells	2	2	6	2	6	10	2	6	10	14
Total number of electrons in a subshell		2		18			32			

The shells are subdivided into subshells. The number of subshells in a shell is equal to the number of the shell from within. K shell has one subshell called *s*; second L shell has two subshells *s* and *p*; the third M shell has three subshells *s*, *p* and *d* and fourth N subshell has four subshells *s, p, d* and *f*. The subshells *s, p, d* and *f* can have a maximum of 2, 6, 10 and 14 electrons, respectively.

The position of electrons in the various shells and subshells are represented as follows. Major shells in which the electrons exist are indicated by the numbers 1, 2, 3 etc and the subshells designated by *s, p, d, f* etc. The superscript on *s, p, d* and *f* gives the number of electrons in the subshell. Thus, $1s^2$ indicates the presence of two electrons in the s subshell of the first major shell (*K*). Similarly, $4f^8$ indicates the presence of 8 electrons in the *f* subshell of fourth major shell (*N*).

Hydrogen atom is the simplest atom which consists of one proton, one neutron and one electron. In this atom, the lone electron is situated in the lone orbit or shell around the nucleus. In helium atom also, two electrons are situated in the single shell. However, in other elements the electrons are arranged in several shells. Thus, a **neon atom** has two shells of 2 and 8 electrons (a total of 10 electrons) and **argon atom** has three shells of 2, 8 and 8 electrons (a total of 18 electrons). These orbits or shells may never have more electrons than a certain maximum. When this maximum number of electrons is reached, the shell is said to be *saturated.* The elements whose outermost shells are saturated with electrons are relatively inert and do not participate in the chemical reactions under normal conditions. The elements in this category (He, Ne, Ar, Kr, Xe, Rd) are gases at normal temperature and they are called *noble gases* because of their inertness. The elements, whose atoms have electrons one more or one less (or even higher values) than the inert gas configuration in their outermost shell, are chemically active. This chemical activity may be interpreted as a tendency of those atoms to acquire noble configuration by accepting or losing electron(s). Acquiring the stable configuration results in lowering of energy.

The mass of an atom depends entirely upon its nucleus. A neutron has nearly the same mass as a proton. In absolute terms, each proton or neutron weighs 1.66043×10^{-24} g. The mass of an electron is negligible, about 1/1823 of the mass of a proton or neutron. The mass of a proton or neutron is called the *atomic mass unit* (amu). Thus, if a carbon atom contains 6 protons, 6 neutrons and 6 electrons, its amu will be equivalent to the proton + neutron, *i.e.*, 6 + 6 = 12. This amu is also the atomic weight of the element. Thus, *atomic weight* of an element may be defined as the combined weight of protons and neutrons, each taken as a unit weight. The number of protons on an atom is called the *atomic number* of the atom. Thus, the atomic number of the carbon is 6. Although the number of neutrons for a particular atom is fixed, it may vary, sometime giving rise to different species of the same atom. The different atomic species, having the same atomic number (as they have same proton number) but different atomic weight (as they have different neutron numbers) are called *isotopes*. Thus, a hydrogen atom with two neutrons (named as deuterium) is an isotope of normal hydrogen atom which contains only one neutron. The number of protons and electrons in both species of the hydrogen is same.

IONIZATION POTENTIAL

The minimum amount of energy required to remove the most loosely-held electron from a normal (uncharged) gaseous atom is called **ionization potential**. It may be expresed as:

$$A(g) \xrightarrow{EP} A^+(g) + e^-$$

where, g denotes gaseous form of the atom A, and EP denotes the amount of energy required to remove one electron (*i.e.*, ionization potential). It is now, however, defined as the energy to effect an ionization.

The removal of electron generates a positive charge on the atom. The magnitude of ionization potential depends upon several factors. If the **radius of the atom** is large, the orbiting electrons are farther from the positively-charged nucleus and it is easier to remove them. Therefore, the ionization potential of the atoms with larger radii is smaller. The **magnitude of the positive charge on the nucleus** also affects the ionization potential. If the charge (positive) on the nucleus is increased, it becomes more difficult to remove electrons and therefore the ionization potential increases. The potential also depends upon the **number of electron orbits.** The inner shells or orbits of the electrons act as a shield or screen between the nucleus and the outer shells. Therefore, each inner shell existing between the nucleus and the outermost shell produces a shielding effect, which decreases the force of attraction between the nucleus and the electrons of outer shells. Hence, *higher the number of inner shells, lesser the value of ionization potential.*

An atom can lose more than one electron during ionization. The amount of energy required to remove one electron from an uncharged atom is commonly called *first ionization potential*. The energy required for the removal of one electron from a charged ion, *i.e.*, a second electron from the original atom is called *second ionization potential.*

$$A^{+}(g) \xrightarrow{\text{II EP}} A^{2+}(g) + e^{-}$$

Similarly, *third ionization potential* represents the energy required to remove one electron from a 2^{+} ion, *i.e.*, third electron from the original atom.

$$A^{2+}(g) \xrightarrow{\text{III EP}} A^{3+}(g) + e^{-}$$

Ionization potentials can be determined by spectroscopy. It is conveniently measured in **electron volts,** eV (although this is not an SI unit) or **joules per mole.** It is synonymous with **ionization energy**.

NATURE OF CHEMICAL BONDING

A perusal of the periodic table shows that besides the inert gases (He, Ne, Ar, Kr, Xe, Rd–all of them have 8 electrons in their outermost shells except He where outermost shell is also the first shell (*K*) and cannot have more than 2 electrons), there are 3 types of elements:

(*a*) those having 1 to 3 electrons in the outermost shell,

(*b*) those having 5 to 7 electrons in the outermost shell, and

(*c*) those having 4 electrons in the outermost shell.

Since the driving force for chemical combination is the tendency to acquire inert gas stable configuration, **first types of elements** will try to stabilize themselves by losing 1, 2 or 3 electrons because it is difficult to gain 7, 6 or 5 electrons than to lose 1, 2 or 3 electrons for obtaining stability. Similarly, **elements of the second type** will try to stabilize themselves by gaining 3, 2 or 1 electrons as losing 5, 6, or 7 electrons will be more difficult. Thus, the elements of the first and second type usually combine by either losing or gaining electrons.

The **elements of the third type,** of which *carbon* is representative, have to lose 4 electrons or gain 4 electrons in order to obtain inert gas configuration and if the element has balanced elecropositive and electronegative characters then either process (of losing or gaining electrons) becomes still more difficult. In such cases, the element achieves the stable configuration by sharing of these electrons with other atoms. Besides carbon, one more element is in a similar position and that is *hydrogen* which may attain stable configuration by either losing or gaining one electron. It can also do so by sharing of electrons with other elements.

TYPES OF CHEMICAL BONDING

Several atoms are assembled and held together to form thousands of molecules which participate in the building and function of physical and biological systems. A **bond** is any force which holds two atoms together. The formation of bond between two atoms is due to some redistribution or regrouping of electrons to form a more stable configuration. The regrouping of electrons in the combining atoms may take place in either of the 3 ways:

(*a*) by a *transfer* of one or more electrons from one atom to another— **electrovalent bonding**

(*b*) by a *sharing* of one or more pairs of electrons between the combining atoms— **covalent bonding**

(*c*) by a combination of the two processes of *transfer and sharing*— **coordinate bonding.**

A. Electrovalent or Polar or Ionic Bond

Ionic bond formation takes place between atoms of strongly electropositive and strongly electronegative elements. An element preceding an inert gas in the periodic table is strongly electronegative and the element immediately following the inert gas is strongly electropositive. For example, chlorine is electronegative while sodium is electropositive. According to **W. Kossel** (1916), a transfer of electron(s) takes place from the outermost shell of the electropositive atom to the outermost shell of the electronegative atom, resulting in the formation of stable positive and negative ions respectively which are held together by electrostatic forces of attraction to form a *molecule* or more precisely an *ion pair*. Thus,

$$A^{e} + B \longrightarrow A^{+} + B^{-e} \quad \text{or} \quad A^{+} \equiv B^{-e}$$

The atoms involved are electrically neutral before combining. The element A which has lost its electrons is known as electropositive whereas element B which has gained the electrons is termed electronegative element. The compound formed by electron transfer is termed as *electrovalent* by **Langmuir** (1919) because the resulting compound is electrolyte. It is also called *polar* since the molecule develops a positive and a negative pole. The electrovalent compounds always exist in ionic form, are hard and nonvolatile, have high melting and boiling points because of stronger nature of the bond and are soluble in polar solvents (such as water and alcohol) and because of the presence of ions conduct electricity in solution or in the fused state. The electrovalent compounds having identical electronic configuration exhibit the phenomenon of isomorphism. For example, NaF (2, 8), magnesium oxide (2, 8), calcium chloride (2, 8, 8) and potassium sulfide (2, 8, 8) are isomorphous.

Electrovalent linkage is common in inorganic compounds like NaCl (Fig. 4–2), Na_2O, CaO etc. In the case of NaCl,

Fig. 4–2. Cubic crystals of sodium chloride
The bond involved is ionic bond.

$$\text{Na} + \cdot\ddot{\underset{\cdot\cdot}{\text{Cl}}}: \longrightarrow \text{Na}^+ + :\ddot{\underset{\cdot\cdot}{\text{Cl}}}:^-$$

The electronic configuraion of sodium atom is $1s^2\ 2s^2\ 2p^6\ 3s^1$ and that of chlorine is $1s^2\ 2s^2\ 2p^6\ 3s^2\ 3p^5$. After transfer of one electron from sodium to chlorine, the sodium atom contains a configuration ($1s^2\ 2s^2\ 2p^6$) similar to neon, and chlorine attains a configuration ($1s^2\ 2s^2\ 2p^6\ 3s^2\ 3p^6$) which is similar to that of argon. The Na^+ can be said to be isoelectronic with neon while Cl^- is isoelectronic with argon. The bond between the ions results from the attraction between oppositely-charged ions. This linking of the ionized atoms is the formation of ionic bonds. The strength of the bond is roughly proportional to the product of the charges on ions and inversely proportional to the square of the distance between the effective centres of ions.

Ionic bonds may be formed by transfer of more than one electrons also. For example, in the case of Na_2O, one oxygen atom can form an ionic bond by accepting one electron from each of the two sodium atoms:

$$\begin{matrix}\text{Na}\cdot \\ \text{Na}\cdot\end{matrix} + \ddot{\underset{\cdot\cdot}{\text{O}}}: \longrightarrow 2\text{Na}^+ + :\ddot{\underset{\cdot\cdot}{\text{O}}}:^{2-} \longrightarrow \text{Na}_2\text{O}$$

Oxygen can also form ionic bonds by accepting two electrons from a single atom of the element which has a low second ionization potential. For example, in the case of CaO, one calcium atom can link to one oxygen atom by transferring its two electrons:

$$\text{Ca}: + \ddot{\underset{\cdot\cdot}{\text{O}}}: \longrightarrow \text{Ca}^{2+} + :\ddot{\underset{\cdot\cdot}{\text{O}}}:^{2-} \longrightarrow \text{CaO}$$

The number of the charge (+ or –) or the number of electrons transferred during an ionic bond formation is called the *electrovalence number* or the *electrovalency* of the participating atoms. Thus, sodium has an electrovalency of 1 while calcium or oxygen has 2. The sign (+ or –) of this number depends upon whether the electron is donated or received. The element donating electron has a positive (+) electrovalency while that receiving electron has a negative (–) sign. Since the number of electrons transferred from an electropositive element to an elecronegative element during an ionic bond formation is equal to the number of electrons present in its outermost orbit, the electrovalency for such elements is the number of electrons present in outermost orbit. Similarly, the electrovalency of an electronegative element is the number of electrons which it can accept to achieve an inert gas configuration.

B. Covalent or Nonpolar Bond

Covalent bond formation, first suggested by **G.N. Lewis** (1916), consists in sharing or holding a pair of electrons in partnership between two combining atoms, so that the pair counts towards the electronic grouping of both atoms. By this mechanism also, the stability akin to the inert gas is attained by each atom. For each pair of electrons to be shared between two atoms, each of the constituent atom contributes one electron:

$$\text{A}\cdot + \text{B}\cdot \longrightarrow \text{A}:\text{B}$$

This type of linkage which is the result of equal contribution and equal sharing of electrons is known as covalent bond. The compound formed by electron sharing is termed as *covalent* or *nonpolar* by **Langmuir** (1919). The covalent compounds always exist in molecular form, are nonelectrolytes or nonionizable, soluble in organic solvents (such as benzene, ether, pyridine etc) and have low melting and boiling points because of weaker nature of the bond. They are usually liquids or gases and are generally soft, easily-fusible and volatile. They are nonconducting in the fused state or in solution.

GILBERT NEWTON LEWIS
(LT, 1875–1946)

Lewis, an American physical chemist, was born on October 23, 1875 in Weymouth, Massachusetts, where there exists a street named for him, *G.N. Lewis Way*, off of Summer Street. He was also raised here. Additionally, the wing of the new Weymouth High School Chemistry Department has been named in his honour. After earning his Ph.D. at Harvard under the direction of **Theodore Richards**, Lewis stayed as an instructor for a year before taking a traveling fellowship, studying under the physical chemist **Wilhelm Ostwald** at Leipzig and physicist **Walther Nernst** at Gottingen. While working in Nernst's lab, Nernst and Lewis apparently developed a lifelong enmity. A friend of Nernst, **Wilhelm Palmaer**, was a member of the Nobel Chemistry Committee. There is evidence that he used the Nobel nominating and reporting procedures to block a Nobel Prize for Lewis in thermodynamics by nominating Lewis for the prize 3 times, and then using his position as a committee member to write negative reports.

After his stay in Nernst's lab, Lewis returned to Harvard as an instructor for three more years, and in 1904 left to become Superintendent of Weights and Measures for the Beau Bureau of Science of the Philippine Islands in Manila. The next year, he returned to Cambridge, Massachusetts when the Massachusetts Institute of Technology (MIT) appointed him to a faculty position, in which he had a chance to join a group of outstanding physical chemists under the direction of **Arthur Amos Noyes**. He became an assistant professor in 1907, associate professor in 1908, and full professor in 1911. He left MIT in 1912 to become a professor of physical chemistry and dean of the College of Chemistry at the University of California, Berkeley. *Lewis Hall* at Berkeley, built in 1948, is named in his honour.

He is famous for his contributions to our knowledge of the electronic structure of organic molecules. In 1916, he introduced the idea of a stable octet of electrons and of a covalent bond being a shared pair of electrons in a paper published in the *Journal of the American Chemical Society*. He supposed that the outermost layer of electrons in the atoms of inert gases is 8, and tends to become 8 in atoms of other elements by 'sharing' electrons in pairs between atoms, forming valency bonds. The theory revolutionized chemistry, and it is in honour of this contribution that we refer to "electron dot" structures as *Lewis structures*. Of particular interest in this text is the extension of his theory of bonding to a generalized theory of acids and bases. His *theory of acids and bases* involved seeing *acids* (Lewis acids) as substances that are able to accept electron pairs from *bases* that are electron-pair donating species (Lewis bases). In 1933, Lewis became the first to prepare heavy water (D_2O). In addition to his pioneer work in describing chemical bonding through the symbolism named after him, Lewis was a driving force in the introduction of thermodynamics into the mainstream of chemistry.

Over the course of his career, Lewis published on many other subjects besides those mentioned in this entry, ranging from the nature of light quanta to economics of price stabilization. In the last years of his life, Lewis and his graduate student **Michael Kasha**, his last research associate, established that **phosphorescence** of organic molecules involves emission of light from one electron in an excited *triplet state* (a state in which two electrons have their spin vectors oriented in the same direction, but in different orbitals) and measured the magnetic nature of this triplet state, *paramagnetism*.

In 1946, a graduate student found Lewis's lifeless body under a laboratory workbench at Berkeley. Lewis had been working on an experiment with hydrogen cyanide (HCN), and deady fumes from a broken line had leaked into the laboratory. The coroner ruled that the cause of death was *coronary artery disease*, but some believe that it may have been a suicide. Berkeley **Emeritus Professor William Jolly**, who reported the various views on Lewis's death in his 1987 history of UC Berkeley's College of Chemistry, *From Retorts to Lasers*, wrote that a higher-up in the department believed that Lewis had committed suicide.

If Lewis's death was indeed a suicide, a possible explanation was depression brought on by a lunch with **Irving Langmuir**. Langmuir and Lewis had a long rivalry, dating back to Langmuir's extensions of Lewis's theory of the chemical bond. Langmuir had been awarded the 1932 Nobel Prize in Chemistry for his work on surface chemistry, while Lewis had not received the Prize, having been nominated 35 times! On the day of Lewis's death, Langmuir and Lewis had met for lunch at Berkeley, a meeting that Michael Kasha recalled only years later. Associates reported that Lewis came back from lunch in a dark mood, played a morose game of bridge with some colleagues, then went back to work in his lab. An hour later, he was found dead. Langmuir's papers at the *Library of Congress* confirm that he had been on the Berkeley campus that day to receive an honorary degree.

The covalent bond is rigid and directional and as such there is a possibility of position isomerism and stereoisomerism amongst these compounds.

Covalent linkage is common in organic compounds, although inorganic compounds also have it. In covalent compounds, one pair of shared electrons corresponds to a single bond, two pairs to double bond, three pairs of electrons or six shared electrons to triple bond. Some common examples from 3 categories are:

(*a*) One pair of electrons: single bond

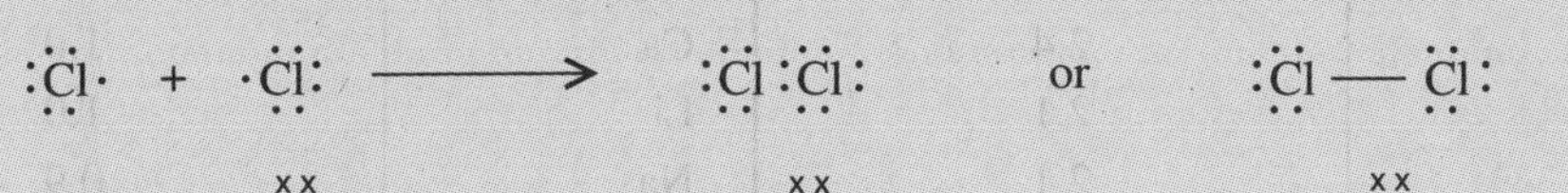

(*b*) Two pairs of electrons: double bond

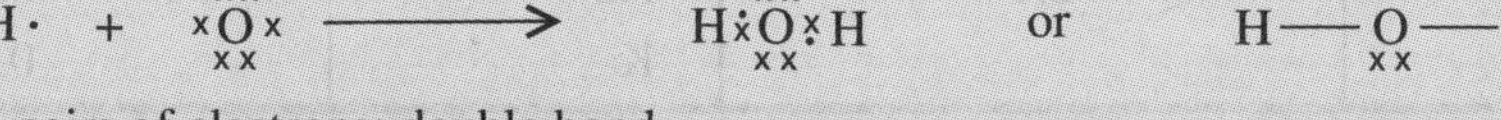

(*c*) Three pairs of electrons: triple bond

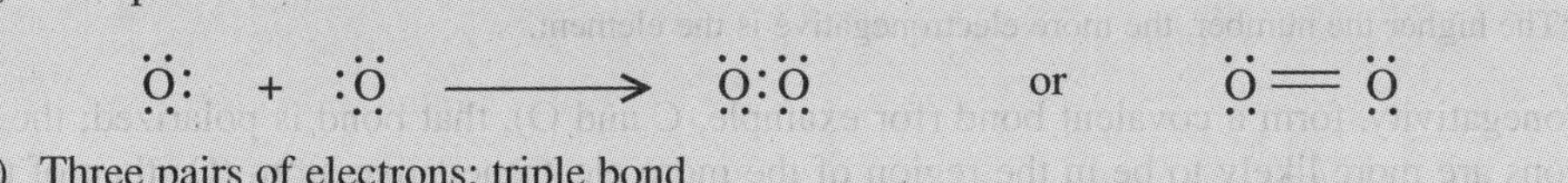

With the exception of hydrogen atom which by sharing one electron easily attains the helium configuration, it may be observed that only negative atoms can form covalent bonds by sharing electrons between them. Positive atoms contain an excess of electrons, which they have to lose to attain a stable inert gas structure; and therefore they form electrovalent bonds.

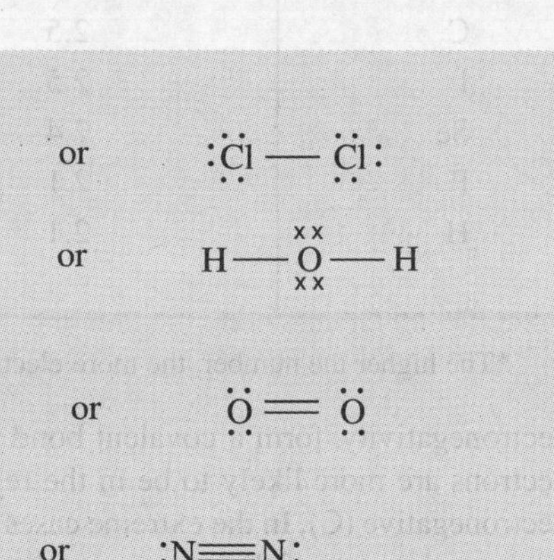

Fig. 4–3. Covalent sharing and bond formation between two atoms

The force of bond formation in covalent bonds is the same as that in ionic bonds, *i.e.*, electrostatic attraction between the two atoms, although this force of attraction develops in a different manner. When two atoms, destined to link through a covalent bond, come within a definite range, the wave function of the electrons of two atoms overlap each other (Fig. 4–3). This overlapping causes the accumulation of the negative charge between the two atomic nuclei. The accumulated negative charge, in turn, attracts the nuclei (with positive charges) of the two atoms and, thus, they are held together.

(*a*) Electron affinity or electronegativity

Electronegativity of an atom is a measure of its power to attract electrons that it is sharing in a covalent bond. Atoms of different elements have different electronegativities. Atoms with higher *atomic number* (magnitude of the positive charge on the nucleus) and less *atomic radius* (the distance of the valence electrons from the nucleus) have higher electronegativities. Electronegativity values of some elements commonly encountered are given in Table 4–4. When two elements, that differ in

Table 4–4. The electronegativities of some elements

Element	*Electronegativity**	*Element*	*Electronegativity**
F	4.0	Cu	1.9
O	3.5	Fe	1.8
Cl	3.0	Co	1.8
N	3.0	Ni	1.8
Br	2.8	Mo	1.8
S	2.5	Zn	1.6
C	2.5	Mn	1.5
I	2.5	Mg	1.2
Se	2.4	Ca	1.0
P	2.1	Li	1.0
H	2.1	Na	0.9
		K	0.8

*The higher the number, the more electronegative is the element.

electronegativity, form a covalent bond (for example, C and O), that bond is polarized; the shared electrons are more likely to be in the region of the more electronegative atom (O) than of the less electronegative (C). In the extreme cases of two atoms of very different electronegativity (Na and Cl, for example), one of the atoms actually gives up the electron(s) to the other atom, resulting in the formation of ions and ionic interactions such as those in solid NaCl.

(*b*) Polar and nonpolar covalent bonds

Polar bond. Polar covalent bonds may be characterized as transition state between ionic and covalent bonds. In this case, there is neither the complete transfer of electrons from one atom to the other nor equal sharing. When a covalent bond is formed between two atoms with *different electronegativities*, the electrons involved in the bond are not shared equally. The atom with higher electronegativity pulls the bonding electrons closer to it. In other words, the electron density in the molecular orbital would be greater around the atom with higher electronegativity. The result of this displacement of the molecular orbital toward the more electronegative atom will acquire a small negative charge and the less electronegative atom will acquire a small positive charge. For example, in the case of **C—Cl bond**, chlorine is more electronegative than carbon. As such, the electron density in the molecular orbital would be higher around chlorine atom than around the carbon atom. Thus, the chlorine atom acquires a small negative charge and the carbon atom acquires a small positive charge.

This may be indicated as in Fig. 4–4. Such a bond which appears to have a positive end and a negative end is said to be a **polar bond**.

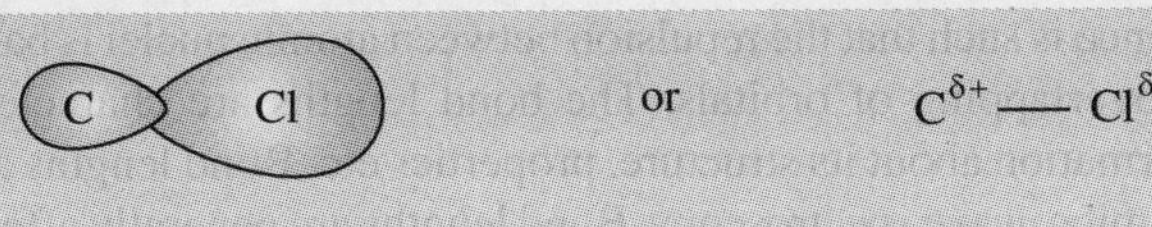

Fig. 4–4. C—Cl bond is a polar bond.
(The Greek letter delta, δ stands for *small*. The symbol δ^+ means a small positive charge and δ^- means a small negative charge.)

Water molecule is another good example of a polar bond. In this molecule, sharing of electrons between one oxygen and two hydrogen atoms is not equal. The electrons of hydrogen atoms are attracted more towards the nucleus of the oxygen atom. Consequently, oxygen develops a δ^- while hydrogen develops a δ^+ charge. Yet another common example of a polar bond is H—Cl, which may be depicted as $H^{\delta+}$—$Cl^{\delta-}$ (Fig. 4—5). A polar molecule like that of HCl having two centres of charge is called a **dipole** and the molecule becomes **dipolar**. Although the ends of molecule may have partial charges, the molecule itself is neutral because of the equal number of protons and electrons.

Fig. 4–5. A dipole for HCl molecule
d is the distance between the positive and negative centres and is called bond distance or bond length.

The degree of a polarity of a polar molecule (*i.e.*, a molecule having a polar covalent bond) is expressed in terms of its **dipole moment** (μ) which is equal to the product of electric charge (*e*) in **e.s.u.** and the distance (*d*) in Å between the positive and negative centres. Thus,

$$\mu = e \times d$$

As *e* is of the order of 10^{-10} e.s.u. and *d* of the order of 10^{-8} cm, the dipole moment (μ) is, therefore, of the order of 10^{-18} and this unit is known as **Debye** (D). Thus, $1D = 10^{-18}$ e.s.u. cm = 10^{-10} e.s.u. Å. The μ is a vector quantity and is often indicated by an arrow parallel to the line joining the points of charge and pointing towards the negative end. Thus, $H^{\delta+}$—$Cl^{\delta-}$ can also be represented as H—Cl. Greater the value of the dipole moment of a molecule, greater is the polarity of the bond between the atoms.

As expected, polar compounds have properties intermediate between ionic and covalent compounds. Because of the dipole attraction between the atoms, polar compounds have higher melting and boiling points than nonpolar covalent compounds. These compounds are hard but not as hard as ionic compounds.

Nonpolar bond. When a covalent bond is formed between two atoms with *same electronegativities* (C—C, H—H and F—F), the electrons involved in the bond are shared equally. Electron density in the molecular orbitals bonding the two atoms together is same around each atom. There is no positive and negative end. Such a bond is said to be **nonpolar bond** (Fig. 4–6).

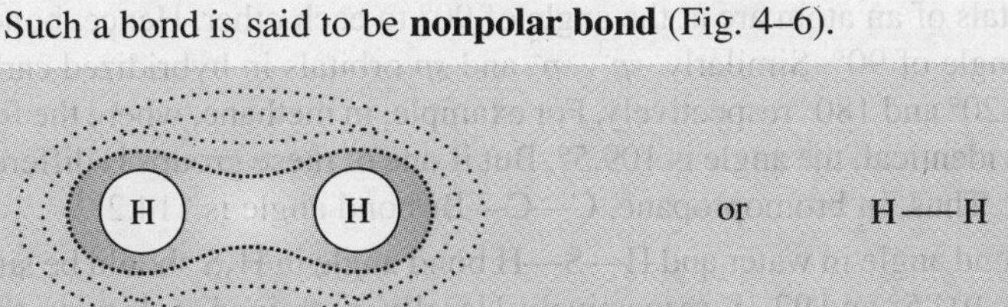

Fig. 4–6. H—H bond is a nonpolar bond.

(*c*) Bond length or bond distance

The distance between two atomic nuclei in a covalent molecule is called the **bond distance** or **bond length.** The distance is such that the repulsion between atomic nuclei is balanced by the stability acquired through the overlapping of orbitals. The bond lengths are characteristic properties of a molecule and give information about its structure, properties etc. Bond lengths are measured by x-ray crystallography and by microwave spectroscopy. Bond lengths are generally affected by hybridization, electronegativity, steric condition etc. Some typical bond lengths are given in Table 4–5.

Table 4–5. Bond lengths* of some common covalent bonds

Bond	*Bond length (in Å)*	*Typical compound*
C—C	1.54	Ethane, Propane
	1.48	Butadiene
	1.38	Butadiyne
C = C	1.34	Ethylene
	1.28	Butatriene
C ≡ C	1.20	Acetylene
C—H	1.11	Methane
	1.10	Benzene
	1.08	Acetylene
C—O	1.41	Ethanol
	1.34	Formic acid
C = O	1.20	Formaldehyde
	1.16	Carbon dioxide
C—N	1.47	Methylamine
	1.36	Formamide
C—Cl	1.78	Methyl chloride
C—Br	1.94	Methyl bromide
C—I	2.14	Methyl iodide
O—H	0.96	Methanol
N—H	1.01	Methyl amine

*Note that for most bonds, the value of bond length lies between 1 and 2 Å.

(*d*) Bond angle

The 3 *p* orbitals of an atom are at the angle of 90° to each other. Hence, bonds formed by these should have an angle of 90°. Similarly, sp^3, sp^2 and *sp* orbitals in hybridized carbon atoms have an angle of 109.5°, 120° and 180° respectively. For example, in methane, where the four groups attached to sp^3 orbitals are identical, the angle is 109.5°. But if one of these groups is different then usually the angle is distorted. Thus, in bromopropane, C—C—Br bond angle is 114.2°.

H—O—H bond angle in water and H—S—H bond angle in H_2S should be similar or 90° but the measured angle is 104.5° and 92.1°, respectively. Usually in strained molecules, the bond angles may be greatly distorted. An understanding of bond angle is essential for the study of the shape and geometry of the molecule. Bond angles in some molecules have been shown in Fig. 4–7.

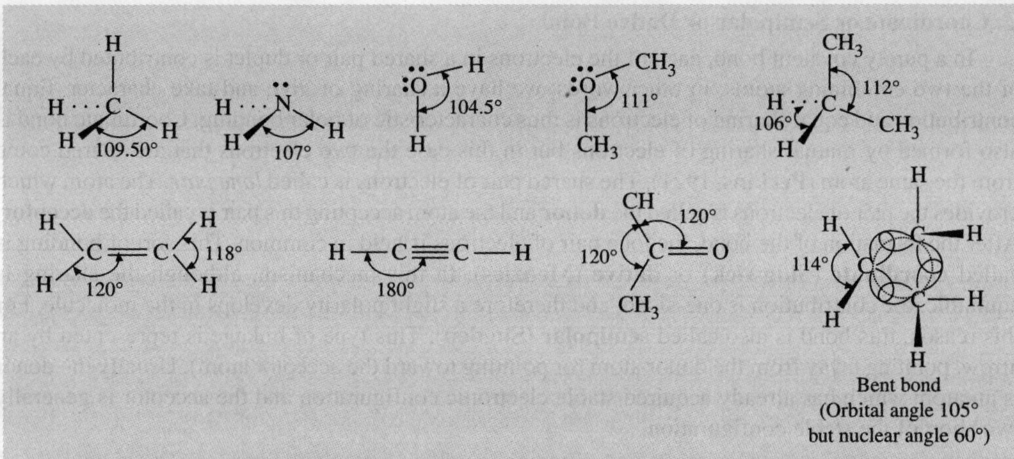

Fig. 4–7. A diagrammatic representation of bond angles in some molecules

(e) Bond strength and bond energy

The energy required to break a bond into its constituents is referred to as bond energy and is a measure of **bond strength**. The greater the energy required to break a bond, the greater will be its strength. Thus, bond energies provide a useful picture of the strengths of various bonds. Bond energy depends upon the relative electronegativities of the elements involved. The number of electrons shared also influences bond strength: double bonds are stronger than single bonds, and triple bonds are stronger yet. Bond strengths are usually determined by quantitative measurements of heats of chemical reactions (calorimetry) and by spectroscopic methods. The strength of a bond is expressed as bond energy, in joules. Bond energy can be thought of as either the amount of energy required to break a bond or the amount of energy gained by the surroundings when two atoms form the bond. Some of the important bond energies are given in Table 4–6.

> In biochemistry, **calories** have often been used as units of energy– bond energy and free energy. The **joule** is the unit of energy in the International System of Units. For conversions, 1 cal is equal to 4.18 J.

Table 4–6. Strength of bonds common in biomolecules

Type of bond	*Bond dissociation energy (kJ/mol)*	*Type of bond*	*Bond dissociation energy (kJ/mol)*
Single bonds		**Double bonds**	
O—H	461	C = O	712
H—H	435	C = N	615
P—O	419	C = C	611
C—H	414	P = O	502
N—H	389	**Triple bonds**	
C—O	352	N ≡ N	930
C—C	348	C ≡ N	816
S—H	339		
C—N	293	**Noncovalent bonds or interactions**	
C—S	260	Hydrogen bonds van der Waals interactions Hydrophobic interactions Ionic interactions	4–0
N—O	222		
S—S	214		

C. Coordinate or Semipolar or Dative Bond

In a purely covalent bond, each of the electrons in a shared pair or duplet is contributed by each of the two combining atoms; in other words, we have a sharing of give-and-take character. Equal contribution and equal sharing of electrons is thus characteristic of polar bonding. Coordinate bond is also formed by mutual sharing of electrons but in this case the two electrons that are shared come from the same atom (**Perkins**, 1921). The shared pair of electrons is called *lone pair*. The atom which provides the pair of electrons is called the **donor** and the atom accepting this pair is called the **acceptor**. After the formation of the bond, the lone pair of electrons is held in common. This sort of bonding is called **coordinate** (**Sidgwick**) or **dative** (**Menzies**). In this mechanism, although the sharing is equitable, the contribution is one-sided, and therefore a slight polarity develops in the molecule. For this reason, this bond is also called **semipolar** (**Sugden**). This type of linkage is represented by an arrow, pointing away from the donor atom (or pointing toward the acceptor atom). Usually the donor is an atom which has already acquired stable electronic configuration and the acceptor is generally two short of the stable configuration.

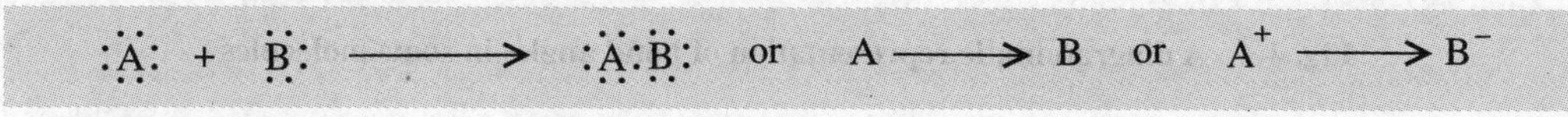

Because of slight increase in electron density near about the acceptor atom, the acceptor acquires a small negative charge (δ^-) and simultaneously the donor atom acquires a small positive charge (δ^+) because of slight decrease in electron density around it. Formation of O_3, SO_2, SO_3, boron hydride–ammonium complex and hydronium ion involves the coordinate bond.

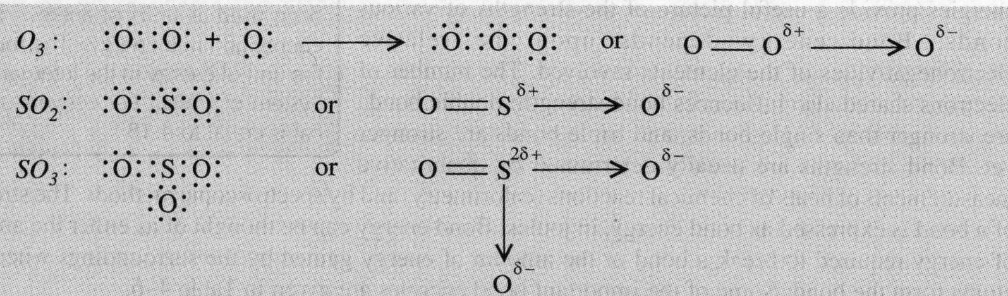

Boron hydride-ammonium complex :

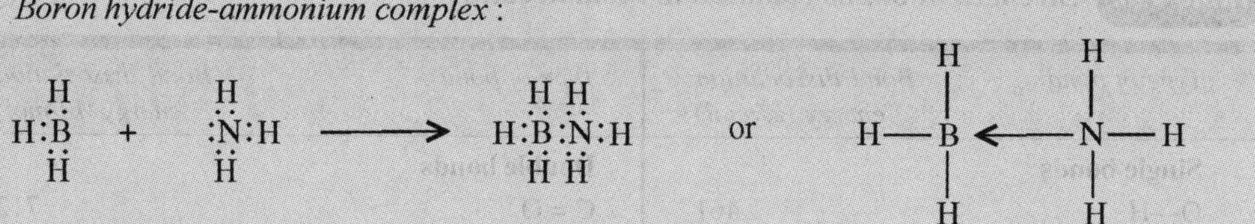

Hydronium ion :

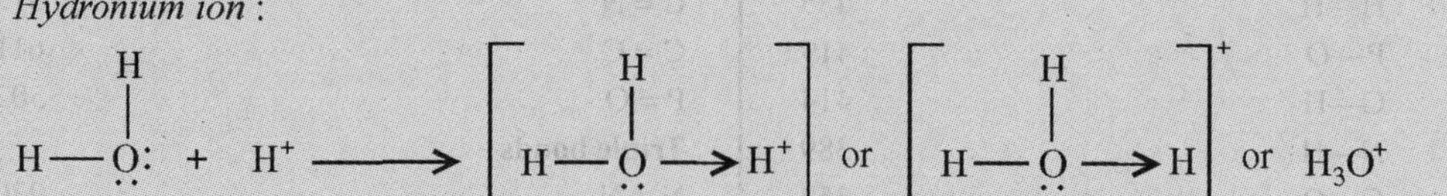

Coordinate compounds exhibit characteristics similar to covalent compounds:

1. They do not ionize in water and are poor conductors of electricity.
2. They are very sparingly soluble in water but dissolve readily in organic slovents.
3. Since a coordinate linkage is semipolar, the coordinate compounds possess melting and boiling points which are higher than those of purely covalent compounds but lower than those of ionic compounds.

4. The coordinate linkage is also rigid and directional. Therefore, such compounds exhibit space isomerism.
5. The coordinate linkage is easily broken when donor and acceptor are molecules which are capable of independent existence.

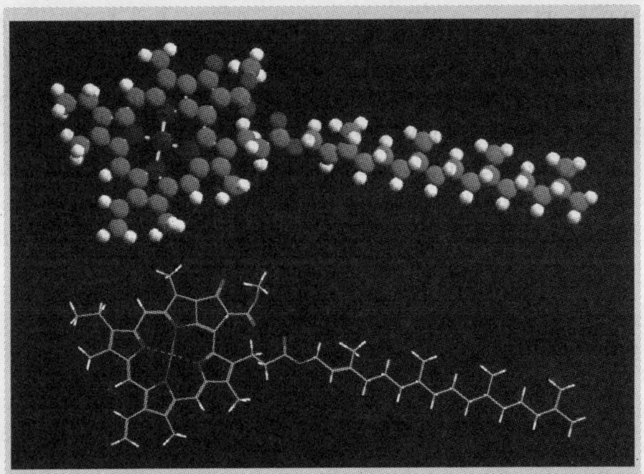

Fig. 4-8. Model of chlorophyll which contains coordinate covalent bond

Coordinate bonds are less common in organic compounds. However, *they are commonly found in complex compounds*. The two important biological molecules containing coordinate covalent bonds are **chlorophyll** (Fig. 4–8) and **heme** in which either Mg or Fe atom is linked coordinately to nitrogen atoms, respectively. In **vitamin** $\mathbf{B_{12}}$ also, a Co atom is linked to nitrogen atoms through coordinate bonds. The compound containing coordinate covalent bonds are called **coordinate complexes.**

D. Noncovalent Bonds or Interactions

In addition to covalent bonding, there are weaker forces of interaction that profoundly influence conformation of biomolecules and their function. These **noncovalent forces**, as they are called, play key roles in the faithful replication of DNA, the folding of proteins into intricate 3-dimensional forms, the specific recognition of substrates by enzymes, and the detection of signal molecules. *Indeed, all biological structures and processes depend on the interplay of noncovalent interactions as well as covalent ones*.

With respect to bonding, weak and strong are used to indicate the amount of energy in a bond. **Strong bonds** such as covalent bonds found in biomolecules require an average of 100 kilocalories/ mole or kcal mol^{-1}, to be cleaved and hence are stable and seldom break under physiological conditions. In contrast, **weak bonds,** such as hydrogen bonds, have energies of 2 to 7 kcal mol^{-1} and are easily broken. Weak bonds are transient; individually they form and break in small fractions of a second. The transient nature of noncovalent interactions confers flexibility on macromolecules, such as proteins and nucleic acids, that is critical to their function. Furthermore, the large number of noncovalent interactions in a single macromolecule makes it unlikely that at any given moment, all the interactions will be broken; thus macromolecular structures are stable over time. The four fundamental noncovalent bonds are : electrostatic (or ionic) bonds, hydrogen bonds, hydrophobic bonds and van der Waals forces or bonds. They differ in geometry, strength and specificity and are profoundly affected by the presence of water.

(*a*) Electrostatic or ionic bonds

Ionic bonds are formed due to the attraction between atoms or groups, of opposite charges (+ and −). A charged group on a substrate can attract an oppositely-charged group on an enzyme. The force (F) of such an electrostatic attraction is given by **Coulomb's law:**

$$F = \frac{q_1 q_2}{r^2 D}$$

where, q_1 and q_2 = charges of the two atoms or groups,

r = distance between the two atoms or groups, and

D = dielectric constant of the medium.

The attraction is strongest in a vaccum where D is 1 and is weakest in a medium such as water where D is 80. The distance between oppositely-charged atoms in an optimal electrostatic attraction is about

2.8 Å. The average bond energy of ionic bonds in aqueous solution is about 5 kcal mol^{-1}. This kind of attraction is also called *saline bond, salt linkage, salt bridge* or *ion pair*.

Ionic bonding occurs in crystals and salts that are ionized when dissolved in water. For example, NaCl is a salt composed of Na^+ (cation) and Cl^- (anion). Common examples of the salts of biomolecules are sodium acetate, potassium pyruvate and ethanolamine chloride (Fig. 4–9). This type of interaction even permits the bonding between two different molecules in heteroproteins (for example, in nucleoproteins, between the negatively-charged nucleic acid and the positively-charged basic proteins, esp., the histones).

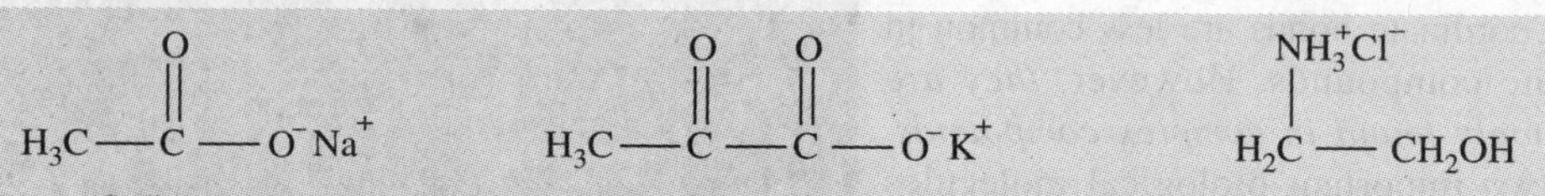

Sodium acetate   Potassium pyruvate   Ethanolamine chloride

Fig. 4–9. Some salts of biomolecules

(*b*) Hydrogen bonds

It is an electrostatic attraction forming relatively weak noncovalent bond between an electronegative atom (*e.g.,* O, N, F – the 3 highly electronegative elements) and a hydrogen atom attached to some other electronegative atom. Hydrogen bonds can be formed between uncharged molecules as well as charged ones. *In a hydrogen bond, a hydrogen atom is shared by two other atoms* (Fig. 4–10). The atom to which the hydrogen is more tightly linked is called the *hydrogen donor*, whereas the other atom is called as the *hydrogen acceptor*. The acceptor has a partial negative charge that attracts the hydrogen atom. In fact, a hydrogen bond can be considered as an intermediate in the transfer of a proton from an acid to a base.

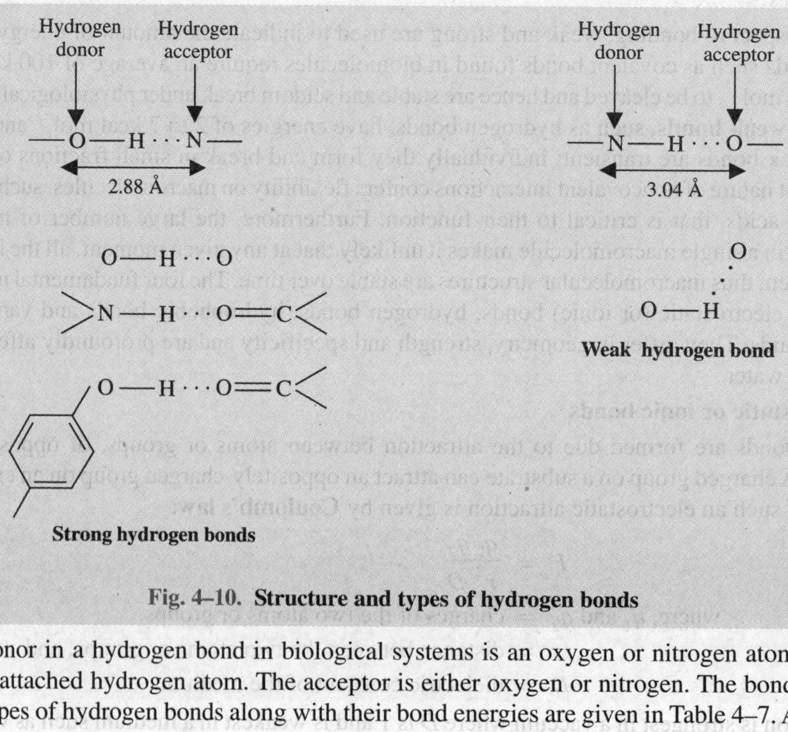

Fig. 4–10. Structure and types of hydrogen bonds

The donor in a hydrogen bond in biological systems is an oxygen or nitrogen atom that has a covalently-attached hydrogen atom. The acceptor is either oxygen or nitrogen. The bond lengths of different types of hydrogen bonds along with their bond energies are given in Table 4–7. A perusal of

Table 4–7. Bond lengths and bond energies of some hydrogen bonds

Bond	*Length (Å)*	*Energy (kcal mol^{-1})*
—O—H···N<	2.88	7
—O—H···O=	2.70	6
—O—H···O<		6
>N—H···N<	3.10	2 – 4
>N—H···O=	3.04	2 – 3

the table reveals that the length of a hydrogen bond is intermediate between that of a covalent bond and a van der Waals bond. The bond energies of hydrogen bonds range between 2 and 7 kcal/mol. Hydrogen bonds are stronger than the van der Waals but much weaker (about 1/10) than normal covalent bonds. The strongest hydrogen bonds are those in which the donor, hydrogen and acceptor atoms are colinear.

Hydrogen bonds in biomolecules (water, for example; Fig. 4–11) are also more specific than other weak bonds because they require particular complementary groups that donate or accept hydrogen. *An important feature of hydrogen bonds is that they are highly directional*. Hydrogen bonding is of 2 types: intramolecular (within a molecule) and intermolecular (between two molecules). Both types are common to many macromolecules such as proteins and DNA molecule. **Intramolecular bonding** gives rise to chelation, *i.e.*, ring formation and this normally occurs only with the formation of 5-, 6- or 7-membered rings. **Intermolecular bonding,** however, gives rise to association, thereby raising the boiling point; it also raises the surface tension and the viscosity, but lowers the dielectric constant. It may exist in compounds in the liquid or solid state and its formation is very much affected by the shape of the molecules, *i.e.*, by the steric factor.

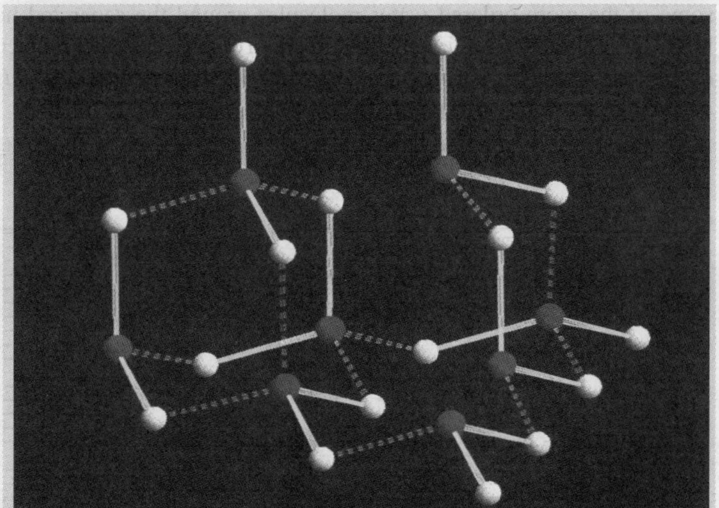

Fig. 4–11. Hydrogen bonds seen between water molecules

(c) Hydrophobic or nonpolar interactions

The essentiality of hydrophilic (water-loving) properties of biomolecules is obvious. That hydrophobic (water-fearing) characteristics can be valuable may not be as readily apparent . As in the formation of micelles, hydrophobic groups of macromolecules, if in proper spatial relation, will interact (*not bond*) to the exclusion of solvent molecules (water) and thereby reside in a hydrophobic environment. On the contrary, hydrophilic groups usually remain exposed to the aqueous environment where they interact with water molecules. Hydrophobic interactions are a major driving force in the folding of macromolecules, the binding of substrates to enzymes and the formation of membranes that define the boundaries of cells and their internal compartments.

In macromolecules such as proteins, the acceptance or rejection by the aqueous environment of the hydrophilic and hydrophobic moieties respectively exerts a dominant influence on their final conformation. Here the nonpolar side chains of neutral amino acids tend to be closely associated with one another. *The relationship is nonstoichiometric*; hence no true bond may be said to exist. *This clustering together of nonpolar molecules or groups in water is called hydrophobic interaction.* The familiar sight of dispersed oil droplets coming together in water to form a single large oil drop is an analogous process.

To understand the basis of hydrophilic attractions, let us take an example wherein a single nonpolar molecule, such as hexane, is introduced into some water (Fig. 4–12). A cavity in the water is created, which temporarily disrupts some hydrogen bonds between water molecules. The dispersed water molecules then reorient themselves to form a maximum number of hydrogen bonds. The water molecules around the hexane molecule are much more ordered than elsewhere in the solution. Now consider the arrangement of two hexane molecules in water. The two possibilities are : either they sit in two small cavities (Fig. 4–12A) or in a single larger one (Fig. 4–12B). The experimental fact is that the two hexane molecules come together and occupy a single large cavity. This association releases some of the more ordered water molecules around the separated hexanes. In fact, the basis of a hydrophilic attraction is this enhanced freedom of released water molecules. *Nonpolar solute molecules are driven together in water not primarily because they have a high affinity for each other but because water bonds strongly to itself.*

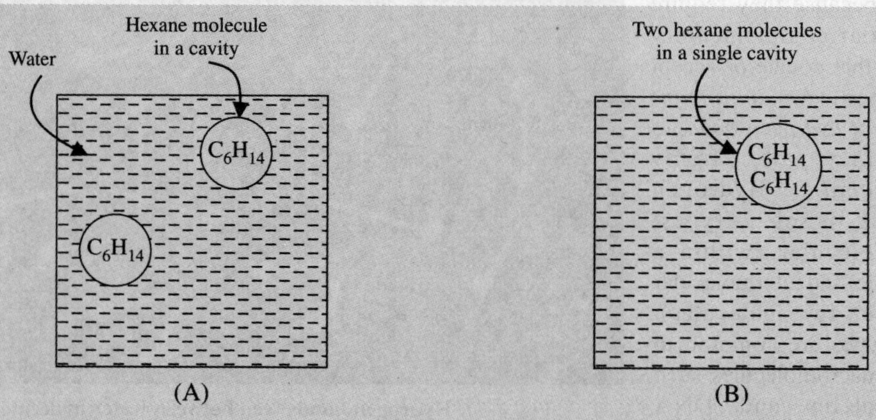

Fig. 4–12. Schematic of two molecules of hexane in a small volume of water

The hexane molecules occupy different cavities in the water structure (A), or they occupy the same cavity, which is energetically more favoured (B).

(d) van der Waals interactions

These have already been dealt with in detail in Chapter 2 (Water : The Solvent of Life) on page 38.

REFERENCES

1. **Chang R :** Physical Chemistry with Application to Biological Systems. *2nd ed. Clarendon Press, Oxford. 1979.*
2. **Fersht AR :** The hydrogen bond in molecular recognition. *Trends Biochem. Sci.* **12 :** *301–304, 1987.*
3. **Frieden E :** Non-covalent interactions: key to biological flexibility and specificity. *J. Chem. Educ.* **52 :** *754–761, 1975.*
4. **Fruton JS :** Molecules and Life. *Wiley, New York. 1972.*
5. **Gardner MLG :** Medical Acid-Base Balance: The Basic Principles. *Cassell, London. 1978.*
6. **Price NC, Dwek RA :** Principles and Problems in Physical Chemistry for Biochemists. *Clarendon Press, New York. 1978.*
7. **Tanford C :** The hydrophobic effect and the organization of living matter. *Science* **200 :** *1012–1018, 1975.*
8. **van Holde KE :** Physical Biochemistry. *2nd ed. Prentice-Hall, Englewood Cliffs. 1985.*

PROBLEMS

1. Rank the following in terms of expected dipole moment, and explain your choice:

   H_2S, CCl_4, $H_3\overset{+}{N}—CH_2—COO^-$, $H_3\overset{+}{N}—CH_2—CH_2—CH_2—COO^-$
2. Why has the element carbon maximum number of organic compounds?

ANSWERS TO PROBLEMS

1. CCl_4, H_2S, $H_3N^+—CH_2—COO^-$, $H_3N^+—CH_2—CH_2—CH_2—COO^-$. CCl_4 is symmetrical, $\mu = 0$; H_2S will be comparable to H_2O; the latter two involve separation of whole charges, and in the last, separation is greatest.
2. Carbon is an element which is most abundant in nature. It is found in 5 allotropic forms: graphite, diamond, fullerene, nanotubes and the recently-found nanofoam. Carbon atom is tetravalent, *i.e.*, it can combine with 4 atoms. The four bonds of carbon atom are directed towards the apices of a regular tetrahedron. Consequently, the compounds formed by carbon are very stable.

   The carbon–carbon skeleton forms a long chain or a ring, or both. Such compounds are found in nature in a large number and are the basis of life. They constitute a separate branch of chemistry– *Organic Chemistry*.

   Sometimes, the carbon–carbon structure also takes up atoms of another elements such as hydrogen, oxygen, nitrogen, sulfur etc. Compounds of such type are also in large number and are dealt with under *Inorganic Chemistry*.

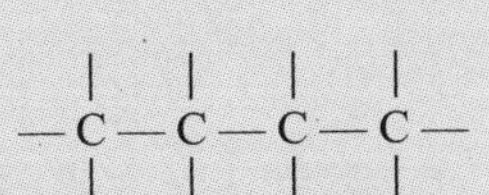
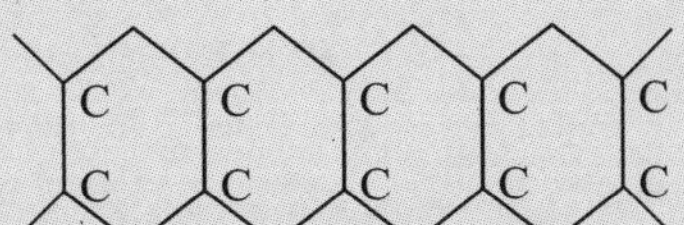

PART

II

Biomolecules

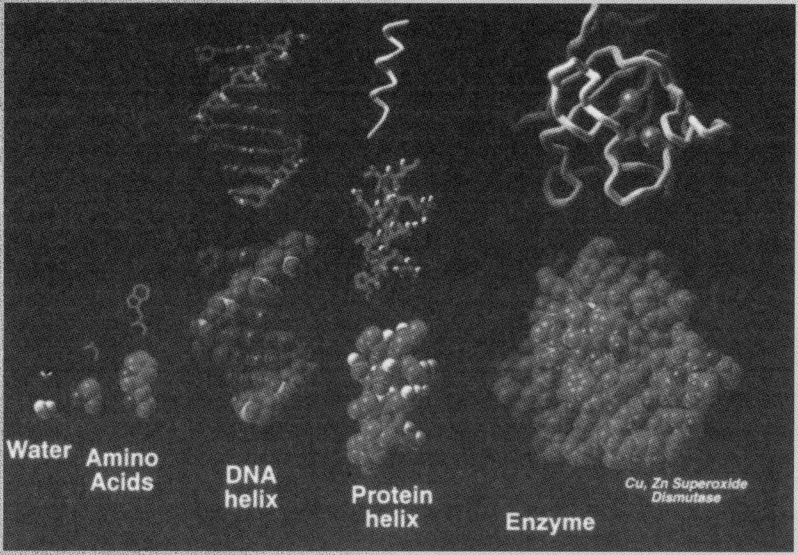

Water, Amino Acids, DNA and Protein Structure

Starting at the far left, we see a water molecule, two common amino acids, alanine and tryptophan, a segment of a DNA double helix, a segment of a protein single helix, and the folded polypeptide chain of the enzyme copper, zinc superoxide dismutase or SOD. With respect to the relative sizes of some of these molecules and structures, the water molecule is roughly half a nanometer (nm) across, the DNA and protein helices are about 2 nm and 1 nm in diameter, respectively, and the SOD, a small, globular protein of about 150 amino acids, is about 6 nm in width. SOD catalyzes the breakdown of harmful, negatively-charged oxygen radicals, thereby protecting people against neurodegenerative diseases such as Lou Gehrig's disease.

> "Life is a relationship among molecules and not a property of any one molecule."
>
> —*Linus Carl Pauling: The Chemical Bond, 1960*

CONTENTS

CHAPTER

5

Carbohydrates–I General Structure of Monosaccharides

Crystals of glucose, a key molecule in carbohydrate metabolism, viewed under polarized light.

IMPORTANCE

The carbohydrates (or saccharides), often termed as sugars, are the *'staff of life'* for most organisms. On the basis of mass, they are the most abundant class of biomolecules in nature. Carbohydrates are also known as *saccharides* (*sakcharon*[G] = sugar or sweetness) since many of those of relatively small molecular weight have a sweet taste, although this is not true of those with large molecules. They are widely distributed molecules (*moles*[L] = mass) in both plant and animal tissues. They are indispensable for living organisms, serving as skeletal structures in plants and also in insects and crustaceans. They also occur as food reserves in the storage organs of plants and in the liver and muscles of animals. In addition, they are an important source of energy required for the various metabolic activities of the living organisms; the energy being derived as a result of their oxidation. They also serve to lubricate skeletal joints, to provide adhesion between cells and to confer biological specificity on the surface of animal cells.

Plants are considerably richer (30 times) in carbohydrates in comparison to the animals. In fact, animal and plant tissues differ widely in the relative abundance of the various major classes of constituent chemicals (Table 5–1).

Table 5–1. Percentage composition of various major classes of constituent chemicals in plants and animals

Organism	*Water*	*Carbohydrates*	*Proteins*	*Lipids*	*Ash*
Animal	60	1	20	15	4
Plant	60	30	5	1	4

NOMENCLATURE AND DEFINITION

The term 'carbohydrate' was originally coined for this class of compounds as most of them were '**hydrates of carbon**' or could be represented by the general formula, $C_x(H_2O)_y$. Later, it was found that some of them, such as deoxyribose ($C_5H_{10}O_4$) and rhamnose ($C_6H_{12}O_5$) do not have the required ratio of hydrogen to oxygen. In addition, certain other carbohydrates are now known to possess nitrogen (*e.g.*, glucosamine, $C_6H_{13}O_5N$), phosphorus or sulfur also and obviously do not coincide with the above general formula. Furthermore, formaldehyde (H.CHO or CH_2O), acetic acid (CH_3.COOH or $C_2H_4O_2$) and lactic acid (CH_3.CHOH.COOH or $C_3H_6O_3$) which have C, H and O and the ratio of H : O is also the same as in water, but are not a carbohydrates. Hence, *the continued usage of the term 'carbohydrate' is for convenience rather than exactness.*

To accommodate a wide variety of compounds, the carbohydrates are nowadays broadly defined as **polyhydroxy aldehydes or ketones and their derivatives** or as substances that yield one of these compounds on hydrolysis.

CLASSIFICATION

Carbohydrates are usually classified in 3 groups :

A. **Monosaccharides** or **Monosaccharoses** (*mono*^{G} = one; *sakcharon*^{G} = sugar). The monosaccharides, often called *simple sugars*, are compounds which possess a free aldehyde (—CHO) or ketone (= CO) group and 2 or more hydroxyl (—OH) groups. They are, in fact, the simplest sugars and cannot be hydrolyzed into smaller units. Their **general formula** is $C_n(H_2O)_n$ or $C_nH_{2n}O_n$.

The monosaccharides may be subdivided into trioses, tetroses, pentoses, hexoses, heptoses etc., depending upon the number of carbon atoms they possess; and as aldoses or ketoses, depending upon whether they contain aldehyde or ketone group. Some important examples are:

Name	*Formula*	*Aldoses (Aldo sugars)*	*Ketoses (Keto sugars)*
Trioses	$C_3H_6O_3$	Glycerose	Dihydroxyacetone
Tetroses	$C_4H_8O_4$	Erythrose	Erythrulose
Pentoses	$C_5H_{10}O_5$	Ribose	Ribulose
Hexoses	$C_6H_{12}O_6$	Glucose	Fructose
Heptoses	$C_7H_{14}O_7$	Glucoheptose	Sodoheptulose

Both these characters (*i.e.*, the number of carbon atoms and the nature of functional group present) may also be combined into one. Thus, for example, glycerose (= glyceraldehyde) is an aldotriose; ribulose, a ketopentose and glucose, an aldohexose. It is noteworthy that, *except fructose, ketoses are not as common as aldoses*. The most abundant monosaccharide in nature is the 6-carbon sugar, D-glucose.

Sometimes, a distinction in naming between aldoses and ketoses is also maintained. The suffix -*oses* is kept reserved for the aldoses and the suffix -*uloses* is used for ketoses. Thus, glucose is a hexose and fructose, a hexulose. However, a few ketoses are named otherwise, such as fructose (*fructus*[L] = fruit) as fruits are a good source of this sugar.

> **Anselme Payen**, codiscoverer of diastase, also isolated a compound common to cell walls of higher plants, which he named `cellulose'. His naming of this polysaccharide, in fact, introduced the **-ose** suffix into the nomenclature of carbohydrates.

Aldoses and Ketoses: There are 2 trioses: glyceraldehyde and dihydroxyacetone. These molecules, as simple as they are, exhibit certain features which shall reoccur in our discussion on sugars. In fact, they represent two major classes of monosaccharides, as mentioned above: glyceraldehyde is an aldose and dihydroxyacetone is a ketose. Note that glyceraldehyde and dihydroxyacetone both have the same atomic composition. They are **tautomers** (*i.e.*, structural isomers differing in the location of their hydrogen atoms and double bonds) and can interconvert *via* an unstable *enediol* intermediate (Fig. 5–1). Such tautomeric interconversions occur to a certain extent between all such pairs of aldose and ketose monosaccharides, but the reactions are usually very slow unless catalyzed. Thus, glyceraldehyde and dihydroxyacetone can each exist as a quite stable compound.

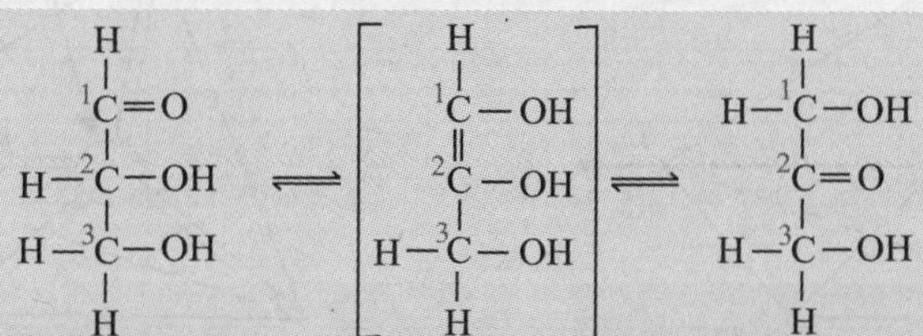

D-Glyceraldehyde (aldotriose) — Enediol intermediate — Dihydroxyacetone (ketotriose)

Fig. 5–1. Aldose–ketose interconversion *via* an enediol intermediate

The intermediate is unstable and cannot be isolated.

B. **Oligosaccharides** or **Oligosaccharoses** (*oligo*[G] = few). These are *compound sugars* that yield 2 to 10 molecules of the same or different monosaccharides on hydrolysis. Accordingly, an oligosaccharide yielding 2 molecules of monosaccharide on hydrolysis is designated as a dissaccharide, and the one yielding 3 molecules of monosaccharide as a trisaccharide and so on. The **general formula** of disaccharides is $C_n(H_2O)_{n-1}$ and that of trisaccharides is $C_n(H_2O)_{n-2}$ and so on. A few examples are:

Disaccharides	–	Sucrose, Lactose, Maltose, Cellobiose, Trehalose, Gentiobiose, Melibiose
Trisaccharides	–	Rhamninose, Gentianose, Raffinose (= Melitose), Rabinose, Melezitose
Tetrasaccharides	–	Stachyose, Scorodose
Pentasaccharide	–	Verbascose

The molecular composition of the 3 *legume oligosaccharides* (*viz.*, raffinose, stachyose and verbascose) is shown below:

Raffinose: α-Galactose (1–6) α-Glucose (1–2) β-Fructose

Stachyose: α-Galactose (1–6) α-Galactose (1–6) α-Glucose (1–2) β-Fructose

Verbascose: α-Galactose (1–6) α-Galactose (1–6) α-Galactose (1–6) α-Glucose (1–2) β-Fructose

C. **Polysaccharides** or **Polysaccharoses** (*poly*[G] = many). These are also compound sugars and yield more than 10 molecules of monosaccharides on hydrolysis. These may be further classified depending on whether the monosaccharide molecules produced as a result of the hydrolysis of polysaccharides are of the same type (homopolysaccharides) or of different types (heteropolysaccharides). Their *general formula* is $(C_6H_{10}O_5)_x$. Some common examples are:

Homopolysaccharides – Starch, Glycogen, Inulin, Cellulose, Pectin, Chitin

Heteropolysaccharides – "Specific soluble sugar" of pneumococcus type III, Hyaluronic acid, Chondrotin

ASYMMETRY

Jacobus H. van't Hoff, the first Nobel Laureate in Chemistry (1901) and **Joseph A. Le Bel**, in 1894, introduced the **concept of tetrahedral carbon atom**. It is now recognized that the carbon atom has the shape of a tetrahedron in which the carbon nucleus resides in the centre of this tetrahedron and the 4 covalent bonds extend out to the corners of it. The angle between any two covalent bonds is 109°28′ (Fig. 5–2).

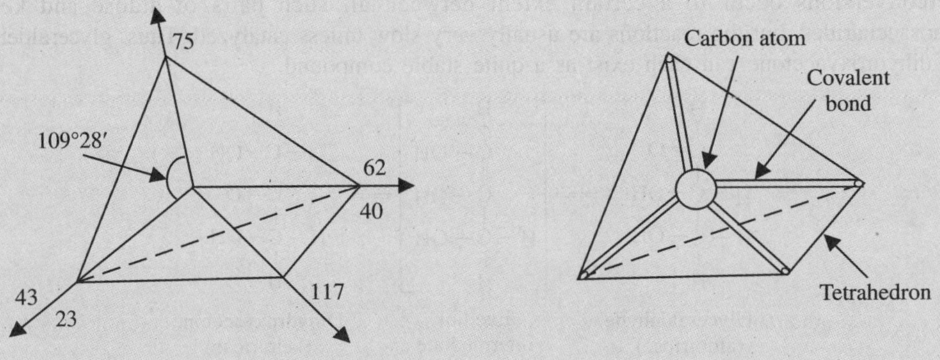

Fig. 5–2. Spatial arrangement of the valences of carbon

JACOBUS HENRICUS VAN'T HOFF
(LT, 1852–1911)

van't Hoff Jr., a Dutch physical and organic chemist, was born on August 30, 1852 in Rotterdam, the Netherlands. His father was **Jacobus Henricus van't Hoff, Sr.**, a physician and mother was **Alida Kolff van't Hoff**. From a young age, he was interested in science and nature, and frequently took part in botanical excursions. He considered **Lord Byron** to be his idol. Against the wishes of his father, van't Hoff chose to study chemistry. First, he enrolled at Delft University of Technology in September 1869, and studied until 1871, when he passed his final examination and obtained a degree of chemical technologist. Then he enrolled at the University of Leiden to study chemistry. Later, he studied in Bonn, Germany with **Friedrich Kekulé** and in Paris with **C.A. Wurtz**. He received his doctorate under **Eduard Mulder** at the University of Utrecht, in 1874.

Before he received his doctorate, van't Hoff had already published the first of his important contributions to the field of organic chemistry. In 1874, he accounted for the **phenomenon of optical activity** by assuming that the chemical bonds between carbon atoms and their neighbours were directed towards the corners of a regular tetrahedron. This 3-'D' structure accounted for the isomers found in nature. He shares credit for this with the French chemist **Joseph A. Le Bel**, who independently came up with the same idea.

van't Hoff published his work on **stereochemistry** in his book *La chimie dans l'espace*, in 1874. At the time, his theory was considered revolutionary and was strongly criticized by the scientific community. One such critic was the renowned editor of the German *Journal für praktische Chemie*, **Adolph Kolbe**, who stated:

> "A Dr. H. van't Hoff of the Veterinary School at Utrecht has no liking, apparently, for exact chemical investigation. He has considered it more comfortable to mount Pegasus (apparently borrowed from the Veterinary School) and to proclaim in his '*La chimie dans l'espace*' how the atoms appear to him to be arranged in space, when he is on the chemical Mt. Parnassus which he has reached by bold flight".

In 1884, van't Hoff published his research on **chemical kinetics** titled *Études de Dynamique chimique* ("Studies in Chemical Dynamics"), in which he described a new method for determining the order of a reaction using graphics, and applied the laws of thermodynamics to chemical equilibria. He also introduced the modern concept of chemical affinity. In 1886, he showed a similarity between the behaviour of dilute solutions and gases. In 1887, he and German chemist **Wilhelm Ostwald** founded an influential scientific magazine named *Zeitschrift für physikalische Chemie* ("Journal of Physical Chemistry"). He worked on Svante Arrhenius's theory of the dissociation of electrolytes and, in 1889, provided physical justification for the Arrhenius equation.

van't Hoff became a lecturer in chemistry and physics at the Veterinary College in Utrecht. He then worked as professor of chemistry, mineralogy, and geology at the University of Amsterdam for almost 18 years before eventually becoming the Chairman of the Chemistry Department. In 1896, van't Hoff moved to Germany where he finished his career at the University of Berlin in 1911.

In 1901, van't Hoff received the first-ever **Nobel Prize in Chemistry** for his work on rates of reactions, chemical equilibrium and osmotic pressure. His work on solutions showed that very dilute solutions follow mathematical laws that closely resemble the laws describing the behaviour of gases. van't Hoff received several honours and awards; a few noteworthy are as follows:

1. Member of the Royal Netherlands Academy of Sciences (1885)
2. Honorary Member of the British Chemical Society, London (1892)
3. Honorary Member of the Royal Dutch Academy of Sciences, Gottingen (1892)
4. Honorary Member of the American Chemical Society (1898)
5. Honorary Member of the Académie des Sciences, Paris (1905)

Of his numerous distinctions, van't Hoff regarded winning the first Nobel Prize in Chemistry as the culmination of his career.

JOSEPH A. LE BEL
(LT, 1847–1930)

Le Bel, a French chemist, was born on January 21, 1847 in Pechelbronn. He was educated at the École Polytechnique in Paris. In 1874, he announced his theory outlining the relationship between molecular structure and optical activity. This laid the foundation of the science of **stereochemistry**, which deals with the spatial arrangement of atoms in molecules. This hypothesis was put forward in the same year by the Dutch chemist **Jacobus Henricus van't Hoff Jr**. Also, he synthesized the first optically-active compound with asymmetric nitrogen atom. In 1929, Le Bel wrote *Cosmologie Rationelle* (Rational Cosmology).

The 5th Incarnation of Carbon

Carbon is one of the most important elements in the universe since *life itself*, as we know it, *is carbon-based.* The versatility of water to form a huge variety of compounds and their stability even under extreme conditions are the secret of carbon acting as the vehicle of life. Versatility of carbon compounds apart, the element itself has become a centre of great attention in recent years. For long, we knew carbon could manifest itself in vividly contrasting styles. It was known to assume a soft and black form as **graphite**; it was known equally well in its hard and sparkling - white incarnation of **diamond** which is the hardest material on earth. Then it was discovered (in 1985) to have had the shape of a soccer ball (called **fullerenes**), and of a cylindrical tube (called **nanotubes**). And now an announcement has been made that carbon can also assume a foamy shape, called **nanofoam**.

This new spongy form of carbon is also a modification of graphitic form. It is extremely light-weighted form and was created when some researchers, at the Australian National University in Canberra, bombarded a piece of graphite with a high-power laser beam. The graphite piece vapourized as temperature rose to more than 10,000°C. On cooling, the vapourized carbon atoms formed a web of tubes. But these tubes instead of being aligned in any particular direction (as happens in the case of nanotubes), were found to be highly dispersed, criss-crossing with one another to form a highly-porous mass. This web of nanotubes is seen as a nanofoam. Surprisingly, the nanofoam, unlike all the other forms of carbon, is *magnetic*. Although the magnetism is ephemeral at room temperature, it is an entirely new and unexpected property in the case of carbon. Because of light weight, foamy nature and magnetic property, nanofoam is being seen as useful for cancer and brain scans through magnetic resonance imaging (MRI). Besides, it might also be useful in the treatment of tumours. Being foam, it would be a bad conductor of heat and if injected into tumours and heated, it would retain the heat locally and kill tumours almost without affecting other body parts. But it needs confirmation through a series of studies.

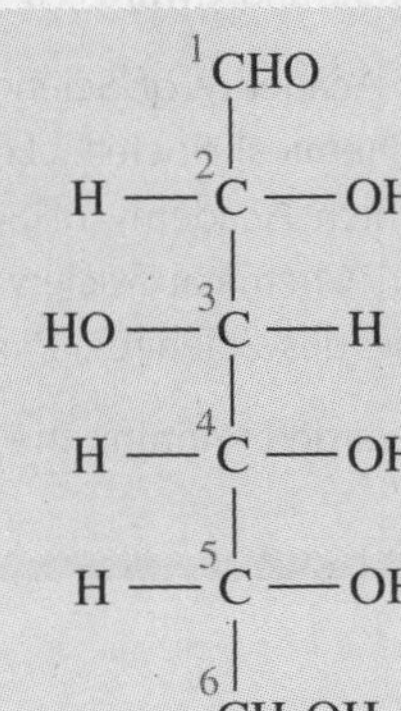

Glucose

[The carbon atoms of a sugar are numbered from the end closest to the aldehyde or ketone group.]

In the formula given above for glucose, it may be observed that a different group is attached to each of the 4 valences of carbon atoms 2 to 5. For example, the 4 groups attached at carbon atom 2 are:

— CHO, — H, — OH and 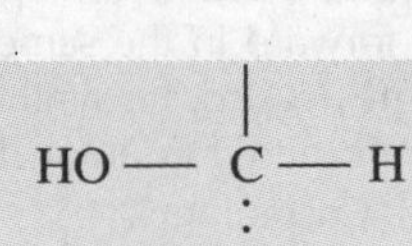

Thus, a carbon atom to which 4 different atoms or groups of atoms are attached is said to be **asymmetric** or **chiral** (*cheir*[G] = hand).

ISOMERISM

The term **isomer** (*isos*[G] = equal ; *meros*[G] = part) was originally applied by **Jönes Jacob Berzelius**, in 1827, to different compounds with the same molecular formula, and the phenomenon was called **isomerism**. The presence of asymmetric carbon atoms in the carbohydrates makes possible the formation of isomers in them. It shall, hence, be beneficial at this stage to discuss in brief about isomers.

The isomers are of 2 types: structural isomers and stereoisomers. **Structural isomers** have the same molecular formula but possess different structures. The difference in structure may be exhibited either in the length of the carbon chain (**chain isomers**) or in the position of a substituent group (**positional isomers**) or in possessing different functional groups (**functional-group isomers**).

Stereoisomers, on the other hand, have the same molecular formula and the same structure but differ only in spatial configuration. Stereoisomers are of 2 types: geometrical and optical. **Geometrical** or **cis-trans** (*cis*[L] = same side; *trans*[L] = across) **isomers** arise from peculiar geometry of compounds having a double bond within the carbon chain. It may be illustrated by *cis-trans* pair, maleic and fumaric acids.

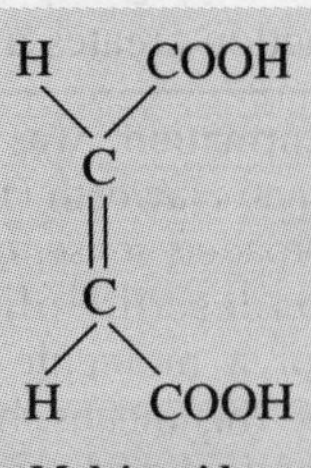

Maleic acid (*cis* form)

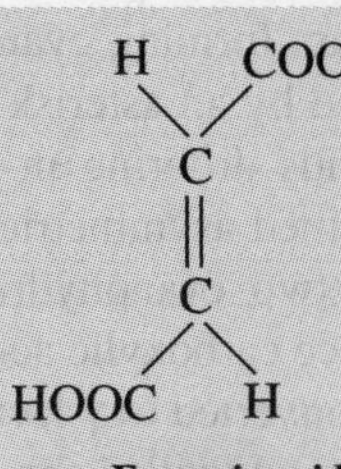

Fumaric acid (*trans* form)

Optical isomers (= enantiomers) differ from each other in the disposition of the various atoms or groups of atoms in space around the asymmetric carbon atom. These are, in fact, the mirror image of each other. These may also be likened to left- and right-handed gloves. For example, the glyceraldehyde has only one asymmetric carbon atom (numbered as 2) and it can, therefore, exist in 2 isomeric forms:

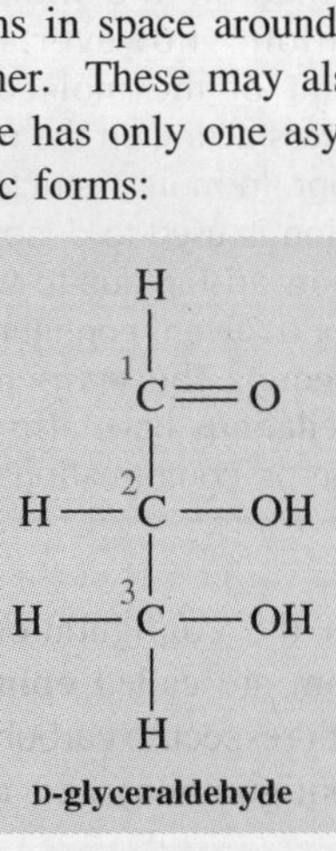

D-glyceraldehyde

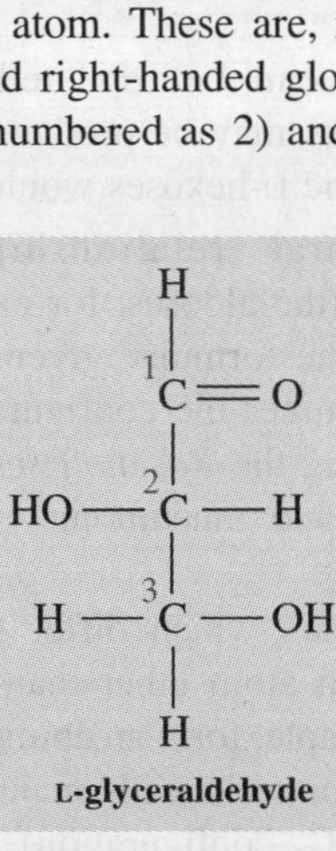

L-glyceraldehyde

One form in which H atom at carbon 2 is projected to the left side and OH group to the right is designated as **D-form** and the other form where H atom is projected to the right side and OH group to the left is called as **L-form** (note the use of small capital letters D and L) Similarly, a compound having 2 asymmetric carbon atoms (*e.g.*, a tetrose) might exist in 4 optical isomeric froms.

The number of possible isomers of any given compound, thus, depends upon the number of asymmetric carbon atoms present in its molecule. According to the **rule of n** (also called as

Le Bel–van't Hoff rule), the total number of optical isomers of a compound will be equal to 2^n, where n represents the number of asymmetric carbon atoms present in the molecule.

In fact, in sugars with 2 or more asymmetric carbon atoms, the designation of D- and L- forms depends upon the orientation of the H and OH groups around the lowermost asymmetric carbon atom (*i.e.*, the asymmetric carbon furthest from the aldehyde or keto group). The distribution of the H and OH groups on the other carbon atoms in the molecule is of no importance in this connection. It is interesting to note that *majority of the monosaccharides found in the human body are of D type*. However, some sugars do occur naturally in their L form such as L-arabinose, a pentaaldose.

KILIANI CYANOHYDRIN SYNTHESIS

A method for the synthesis of monosaccharides was first proposed by Heinrich Kiliani in 1886. It is, in fact, a method by which the chain length of a carbohydrate may be increased. The application of Kiliani synthesis to D-glyceraldeyde resulting in the production of 2 tetroses, D-erythrose and D-threose is shown in Fig. 5–3.

The process is based upon the addition of HCN to the carbonyl group of aldehydes (or ketones) of the sugars forming cyanohydrin (Reaction-1). This reaction creates a new asymmetric carbon atom, marked by an asterisk in the figure. Thus, two compounds differing in conformation about the newly-formed asymmetric carbon atom are formed. These two cyanohydrins are, then, hydrolyzed to produce carboxylic acids (Reaction-2) which are later converted to γ-lactones or inner esters (Reaction-3). Finally, the lactones are reduced to the corresponding aldoses, containing one carbon atom more than their parent sugar (Reaction-4).

The process can be repeated and the 4 isomeric D-pentoses may be produced; and from these the 8 isomeric D-hexoses would also result.

The structural relationships of the monosaccharides (the aldoses, for example) may be visualized by the formulae given in Fig. 5–4. It is easy to remember the configurations of the aldoses by recalling the *Rosanoff scheme*, drawn out on the chart and mneumonic– ET; RAXL; AAGMGIGT.

> The term **'conformation'** was first introduced by **Walter Norman Haworth** (1929), an English chemist. In its broadest sense, conformation has been used to describe different spatial arrangements of a molecule which are not superimposable. This means that, in effect, the terms *conformation* and *configuration* are equivalent. In the classical sense, the definition of conformation does not include the internal forces acting on the molecule. The term conformation, however, is the spatial arrangement of the molecule when all the internal forces acting on the molecule are taken into account. In more restricted sense, the term conformation is used to denote different spatial arrangements arising due to twisting or rotation of bonds of a 'given' configuration (used in the classical sense). The terms **rotational isomers** and **constellations** have also been used in the same sense as conformations.

Any two sugars which differ from each other only in the configuration around a single asymmetric carbon atom other than the carbonyl carbon atom are called **epimers.** Glucose and galactose, for example, form an epimeric pair as they differ with respect to carbon 4 only. Similarly, glucose is also epimeric with mannose (differing in C_2 configuration) and allose with altrose (differing also in C_2 configuration).

The two D- and L-forms of a compound constitute a pair of **enantiomers** (*enantios*[G] = opposite) or **enantiomorphs**. Thus, D-erythrose and L-erythrose are enantiomers; also D-glucose and L-glucose. The enantiomers (also called optical isomers or double image isomers) are nonsuperimposable mirror images of each other but are chemically identical in their reactions. They agree in their melting points, solubility etc., but differ in their ability to rotate the plane of polarized light in a polarimeter; a solution of one of the two enantiomers rotates the plane to the right, and a solution of the other to the left.

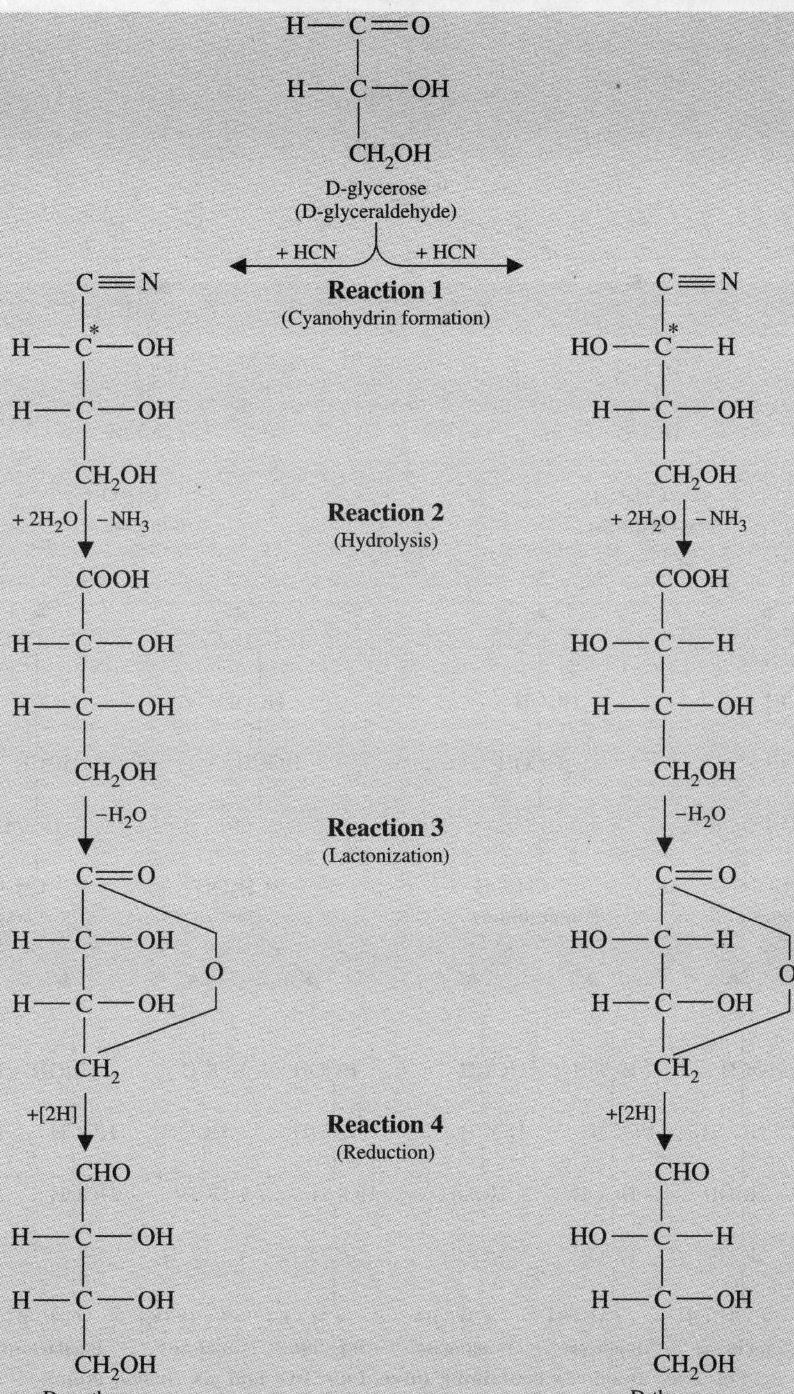

Fig. 5–3. Application of the cyanohydrin synthesis to D-glyceraldehyde

The conversion of D-glyceraldehyde into an aldotetrose yields D-erythrose and D-threose. These are called **diastereoisomers** *i.e.*, isomers but not mirror images of each other. In fact, diastereoisomers are different forms of a compound with two asymmetric centres. It may, hence,

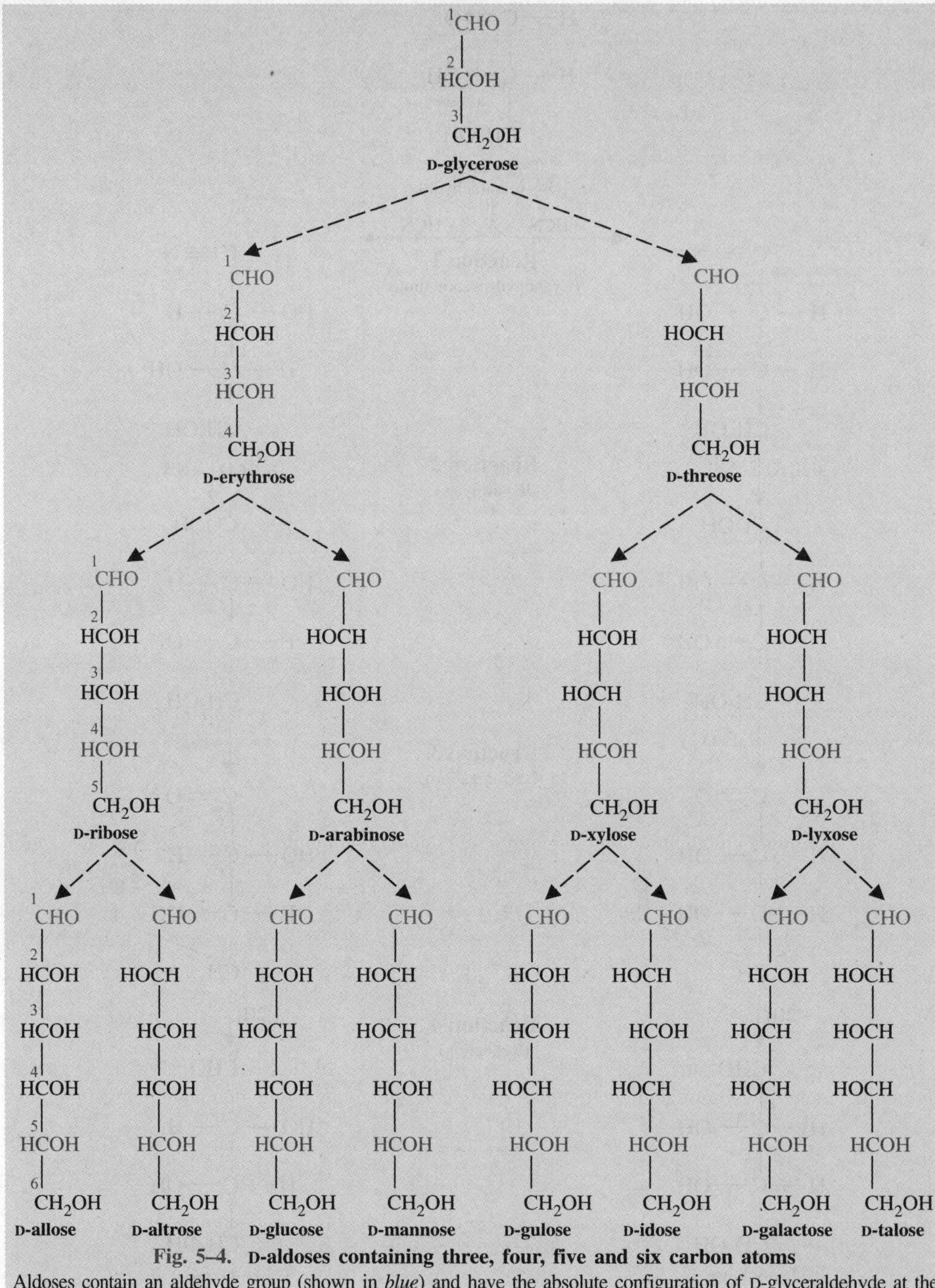

Fig. 5–4. D-aldoses containing three, four, five and six carbon atoms

Aldoses contain an aldehyde group (shown in *blue*) and have the absolute configuration of D-glyceraldehyde at the asymmetric centre (shown in *red*) farthest from the aldehyde group. The numbers indicate the standard designations for each carbon atom. The formulae of L-aldoses are in each case the mirror images of these structures.

be inferred that if two optical isomers are not enantiomers, they are related as diastereoisomers. They differ from each other in their melting points, solubilities and the chemical properties, in general.

OPTICAL ISOMERS OR ENANTIOMERS

As already pointed out, glyceraldehyde has 2 stereoisomers (because it has one carbon, C-2, which is asymmetric or chiral), which are nonsuperimposable mirror images of one another and are called **enantiomers**. The three-dimensional drawings of the 2 forms, designated as D- and L-glyceraldehyde, are shown in Fig 5–5.

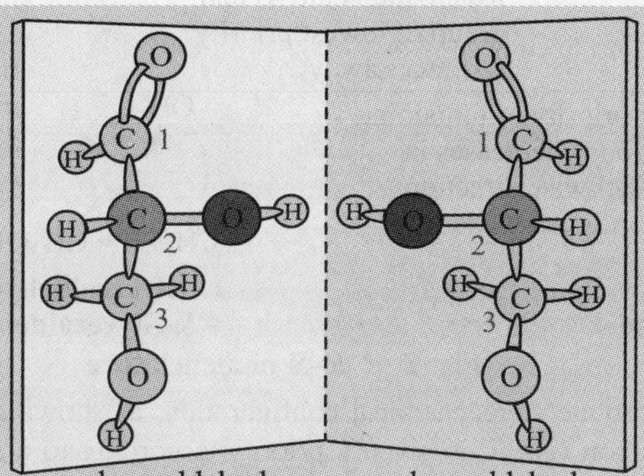

D-glyceraldehyde L-glyceraldehyde

Fig. 5–5. The enantiomers of glyceraldehyde

The configuration of groups around the chiral carbon 2 (shown in *darker grey*) distinguishes D-glyceraldehyde from L-glyceraldehyde. The 2 molecules are mirror images and cannot be superimposed on one another.

Such 3-'D' orientation about the asymmetric carbon can also be represented by the **bond convention:**

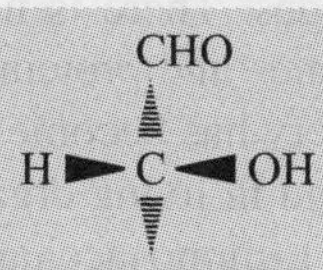

D-glyceraldehyde

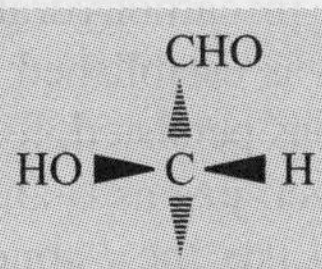

L-glyceraldehyde

Note that we do not need to draw the spatial orientation of atoms about carbons 1 or 3 because these carbons are not chiral centers.

The most compact way to represent enantiomers is to use a **Fischer projection.** In a Fischer projection, the bonds that are drawn *horizontally* are imagined as coming towards you; those drawn *vertically* are receding. Thus, we have:

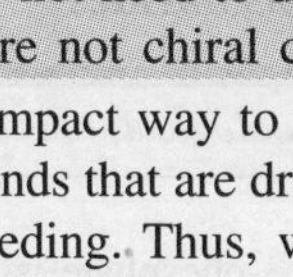

D-glyceraldehyde

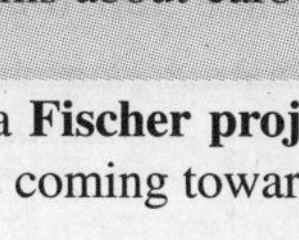

L-glyceraldehyde

Alternative Designations for Enantiomers: D–L and *R–S*. Originally, the terms D and L were meant to indicate the direction of rotation of the plane of polarization of polarized light: D for right (*dextro*) and L for left (*laevo*). It is true that a solution of D-glyceraldehyde does rotate the plane of polarization to the right, as do many other D-monosaccharides, but this correspondence does not always hold, because the magnitude and even the direction of optical rotation are a complicated function of the electronic structure surrounding the chiral center. Another disadvantage of the D–L nomenclature is that it is not absolute; the designation is always with respect to some reference compound. Accordingly, an *absolute* convention has been developed that allows us to assign a stereochemical designation to any compound from examination of its 3-'D' structure. An example of how this so-called *R–S* convention works is shown in Fig. 5–6. Although the *R–S* convention is more general, it is little used by biochemists. It becomes difficult to apply in the common situation in which a molecule contains more than one asymmetric carbon atom.

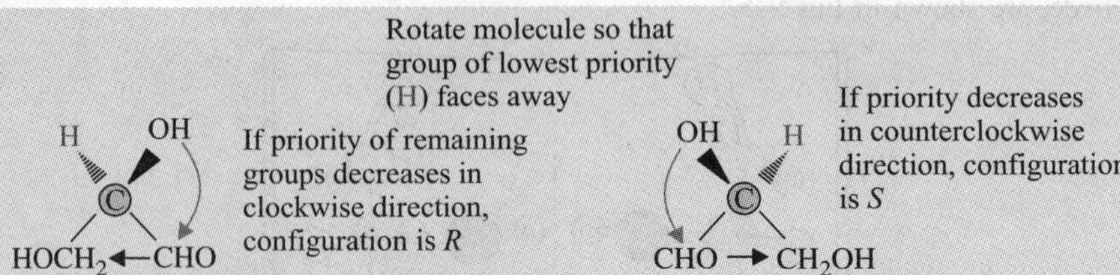

D-Glyceraldehyde = *R*-Glyceraldehyde **L-Glyceraldehyde = *S*-Glyceraldehyde**

Fig. 5–6. *R–S* nomenclature

The *R–S* system describes absolute stereochemical configuration, as shown in this example. Each type of group attached to a chiral carbon (*grey*) is given a priority, according to a set of defined rules. Priorities for groups common in carbohydrate chemistry are: OR > OH > NH_2 > CO_2H > CHO > CH_2OH > CH_3 > H. We view the molecule with the group of lowest priority away from us (H in our example). If the priority of the remaining 3 groups *decreases clockwise*, the absolute configuration is called *R* (from *rectus*[L] = right). If the priority *decreases counterclockwise*, the configuration is *S* (from *sinister*[L] = left). In this notation, D-glyceraldehyde is *R*-glyceraldehyde, and L-glyceraldehyde is *S*-glyceraldehyde.

Monosaccharide Enantiomers in Nature: One enantiomeric form of monosaccharides dominates in living organisms, just as in the case of amino acids (refer page 167). In proteins, it is the L-amino acids; in carbohydrates, it is the D-monosaccharides. There is no obvious reason why this preference was established in nature for both carbohydrates and proteins. But once fixed in early evolution, it has persisted, for most of the cellular machinery has become geared to operate with D-sugars. However, just as D-amino acids are sometimes found in living organisms, so are the L-monosaccharides. Like the "abnormal" D-amino acids, the L-monosaccharides play rather specialized roles. Table 5–2 includes, along with D-monosaccharides, the most commonly occurring L-monosaccharides, with examples of their occurrence and functions.

Table 5–2. Occurrence and biochemical roles of some monosaccharides

Monosaccharide	*Natural occurrence*	*Physiological role**
Trioses		
Glyceraldehyde	Widespread (as phosphate)	The 3-phosphate is an intermediate in glycolysis
Dihydroxyacetone	Widespread (as phosphate)	The 1-phosphate is an intermediate in glycolysis
Tetroses		
D-Erythrose	Widespread	The 4-phosphate is an intermediate in carbohydrate metabolism
Pentoses		
D-Arabinose	Some plants, tuberculosis bacilli	Plant glycosides, cell walls
L-Arabinose	Widely distributed in plants, bacterial cell walls	Constituent of cell walls, plant glycoproteins
D-Ribose	Widespread, in all organisms	Constituent of ribonucleic acid
2-D-Deoxyribose	Widespread, in all organisms	Constituent of deoxyribonucleic acid
D-Xylose	Woody materials	Constituent of plant polysaccharides
Hexoses		
D-Galactose	Widespread	Milk (as part of lactose); structural polysaccharides

L-Galactose	Agar, other polysaccharides	Polysaccharide structures
D-Glucose	Widespread	A major energy source for animal metabolism; role in cellulose
D-Mannose	Plant polysaccharides, animal glycoproteins	Polysaccharide structures
D-Fructose sucrose	A major plant sugar; part of esters)	Intermediate in glycolysis (phosphate
Heptoses		
D-Sedoheptulose	Many plants	Intermediate in Calvin cycle in photosynthesis and pentose phosphate pathway

* Some of these monosaccharides have additional roles that are not listed here.

OPTICAL ROTATION

The presence of asymmetric carbon atoms in the compound also confers optical activity to it. When a beam of polarized light is passed through an optically active solution, it may be rotated either to the right or to the left, depending upon the type of optical isomer present. A compound rotating the plane of polarized light to the right is called as *dextrorotatory* (*dexter*[L] = right, on the right side; *rotatus*[L] = pp of *rotare*[L], to turn) or clockwise and is designated as *d* or (+) type. If the compound causes rotation of polarized light to the left, it is said to be *levorotatory* (*laevus*[L] = left, on the left side) or anticlockwise and is denoted as *l* or (–) type. The direction in which the light is rotated (or in other words, the optical rotation) is a specific property of the molecule. It should, however, be emphasized that *the optical rotation is not at all related with the two* D *and* L *forms of a compound*. Thus, D-glucose is dextrorotatory and D-fructose is levorotatory. If it is desired to indicate the direction of rotation, the two names may be written as D(+)-glucose and D(–)-fructose. Similarly, D-erythrose is levorotatory and D-threose, dextrorotatory.

When equal amounts of dextrorotatory and levorotatory isomers are present, the resulting mixture becomes optically inactive because the optical activities of each isomer cancel each other. Such a mixture is called a **racemic** or **dl-mixture** or (±)-**conglomerate** and this process of converting an optically active compound into the racemic modification is known of **racemization** (*racemus*[L] = grape). Compounds produced synthetically are invariably racemic in nature because equal chances exist for the formation of two types of optical isomers or the *d* and *l* antipodes.

It may be concluded that all monosaccharides are optically active (except aldobiose and ketotriose) since they contain one or more asymmetric carbon atoms in their molecules.

Mutarotation

When a monosaccharide is dissolved in water, the optical rotatory power of the solution gradually changes until it reaches a constant value (**Dubrunfaut**, 1846). A freshly-prepared aqueous solution of α-D-glucose, for instance, has a specific rotation, $[\alpha]_D^{20}$ of +112.2°. And when this solution is allowed to stand, the rotation falls to +52.7° and remains constant at this value. The final stage can be attained more quickly either by heating the solution or by adding some catalyst which may be an acid or an alkali. This gradual change in specific rotation is known as **mutarotation** or **changing rotation**. In fact, this term reflects the discovery of the phenomenon by way of changes in the optical rotation of certain carbohydrates. The terms **multirotation** and **birotation** have sometimes also been used for mutarotation. The value of mutarotation for α-D-glucose is (+112.2°) – (+52.7°) or +59.5°. A fresh solution of β-D-glucose, on the other hand, has a rotation value of +18.7°; on standing, it also changes to the same value, +52.7°. *All reducing sugars (except* *a few ketoses) undergo mutarotation.* The specific rotations for the anomers of glucose and for the aqueous equilibrium mixture are:

$$[\alpha]_D^{20} = +112.2^\circ \longrightarrow [\alpha]_D^{20} = +52.7^\circ \longleftarrow [\alpha]_D^{20} = +18.7^\circ$$

α-D-glucose — Mutarotational equilibrium mixture — β-D-glucose

These substances (open and ring forms) interconvert due to a dynamic equilibrium. Although the ring form predominates, each molecule spends some time in the open form, when asymmetry is lost. When the open-chain form passes back to the ring form, it is possible for it to give rise to either anomer. Thus, a pure sample of an anomer only remains so when in the solid state. In solution, particularly at alkaline pH, a pure sample of one anomer will change to an equilibrium mixture of the two forms. There need not be equal quantities of the two forms at equilibrium because there are other centres for asymmetry and so anomers (like *allo* and *threo* pairs of amino acids) do not have identical physical properties, including ΔG° of formation. Equilibrium between the anomers is called mutarotation and it can be followed by observing optical rotation (Fig. 5–7).

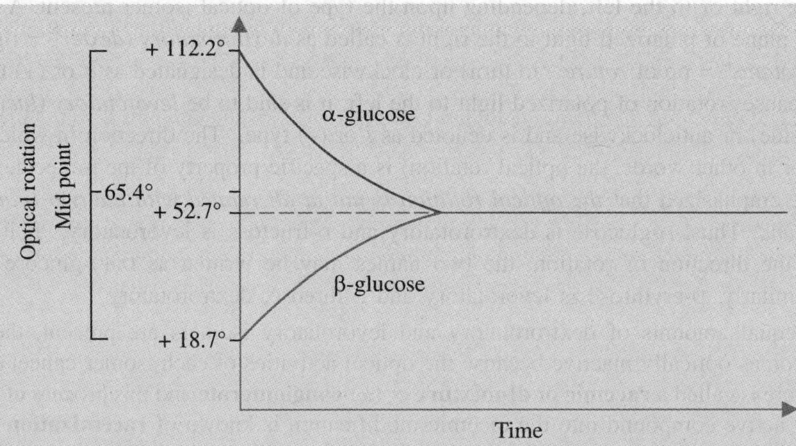

Fig. 5–7. The experimental detection of mutarotation

Mechanism of mutarotation. The phenomenon of mutarotation appears to be due to the changes of α- to β-forms and vice versa *via* the straight chain aldo or keto form (refer to the subsequent section). When an equilibrium is attained, the characteristic rotation is reached. According to **Lowry** (1925), mutarotation can take place only in the presence of an amphiprotic solvent (*i.e.*, a solvent which can act both as an acid and a base). Water is one such solvent. Thus, **Lowry** and **Faulkner** (1925) showed that mutarotation is checked in cresol solution (an acidic solvent) and in pyridine solution (a basic solvent). It is assumed that when mutarotation takes place, the ring opens and then recloses either in the inverted position or in the original position. The existence of the open chain form in mutarotation is proved by certain evidences. For example, the absorption spectra of fructose and sorbose in aqueous solution indicate the presence of open chain forms; the aldoses, however, give negative results (**Bednarczyk** *et al*, 1938).

FORMULATION OF MONOSACCHARIDES

Formula of Glucose— Linear Form

An organic chemist intends to incorporate as many characteristics as possible in writing the formulae (Fig. 5–8). When the molecular formula of glucose, $C_6H_{12}O_6$ is written in a form as propounded by **Fittig** and **Baeyer**, it tells us the presence of an aldehyde and 5 hydroxyl groups. Replacement of this formula by Fischer projection formula informs the reader about some more details. For example, the presence of 4 asymmetric carbon atoms in the molecule indicates towards the existence of 2^4 or 16 stereoisomers.

Ring Form

However, even Fischer formula fails to describe certain other properties of glucose. Hence, another method of representing the structure was searched for. The aldehyde or ketone group of a sugar can react with hydroxyl groups of alcohols forming **hemiacetals** or **hemiketals**, respectively. For the larger sugars (n = >4), this happens *within the same molecule* to form a 5- or 6-membered ring.

Fig. 5–8. Formulation of glucose

In fact, the angles of the tetrahedral carbon atom tend to bend the glucose molecule forming a ring. The aldehyde group and the alcohol group of carbon 5 or carbon 4 readily approach each

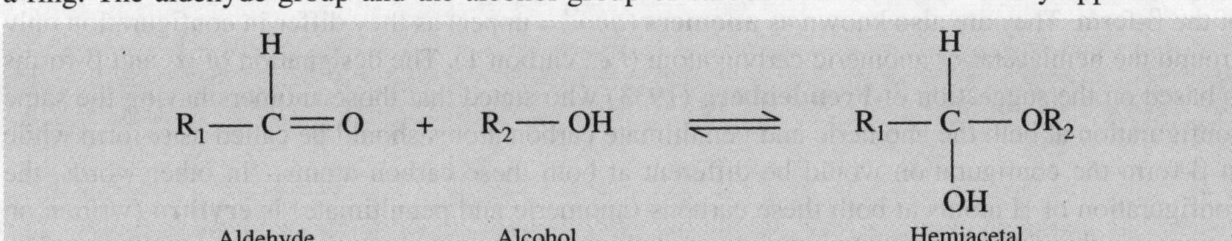

Formation of a hemiacetal

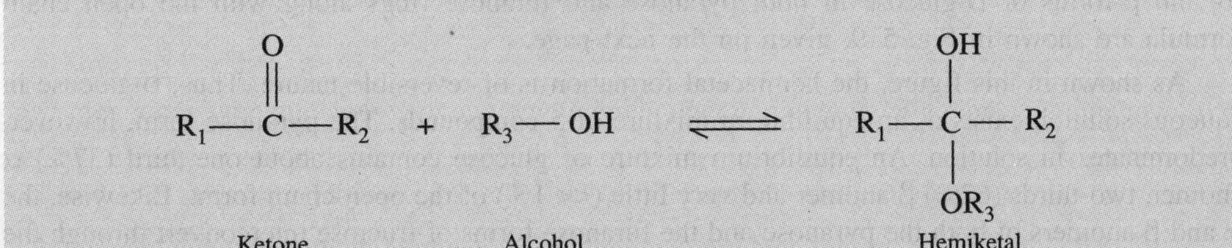

Formation of a hemiketal

other forming intramolecular *hemiacetals*. Similarly, the keto group of a ketose can also approach the alcohol group of either carbon 5 or carbon 4 forming intramolecular *hemiketals*. This results in the formation of either a 6-membered ring (**pyranose form**) or a 5-membered ring (**furanose form**). A 7-membered ring, however, becomes too strained to allow participation of the OH group of carbon 6 of aldohexoses with the CHO group in ring formation. The pyranose form has a lower $\Delta G°$ of formation than the furanose form.

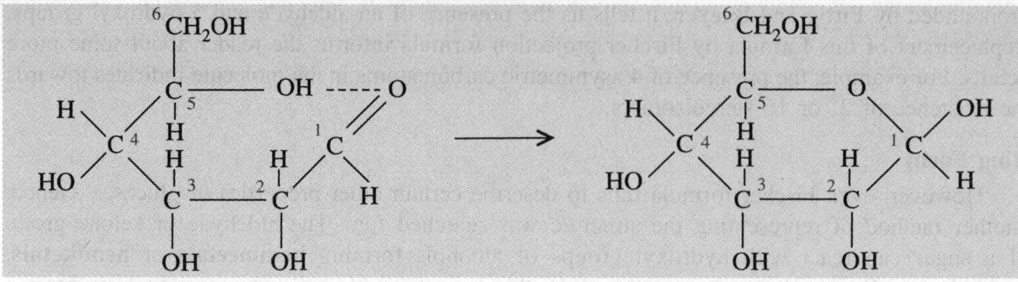

The terminology of such structures is, in fact, based on the 2 simple organic compounds, exhibiting a similar ring structure. These are **pyran** and **furan**. *The pyranose forms of the sugars are more stable than the furanose forms in solution.*

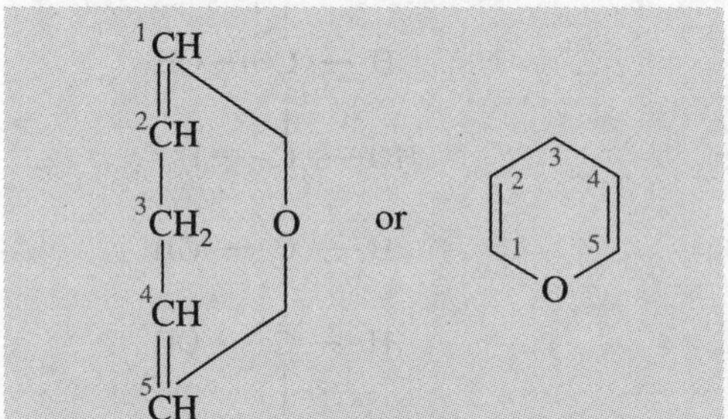

Pyran
(the parent compound of pyranoses)

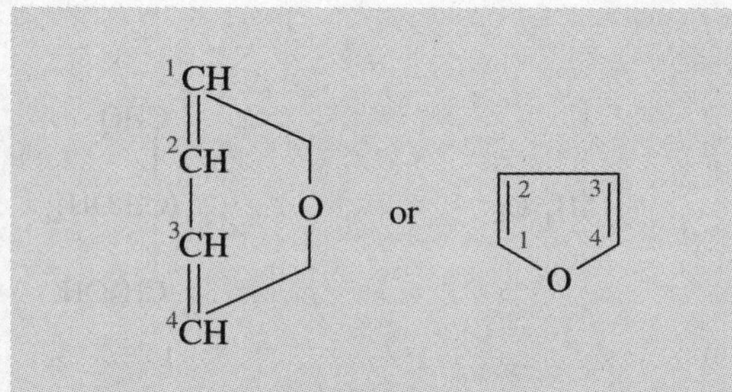

Furan
(the parent compound of furanoses)

The formation of a ring structure in the glucose molecule creates a new centre of asymmetry, *i.e.*, the carbon 1 (carbon 2 in ketoses). One may, therefore, find that the ring form of glucose can exist as either of the 2 isomeric forms, termed as α and β isomers. These two isomers are, in fact, diastereoisomers rather than enantiomers because the α-form of D-glucose is not the mirror image of the β-form. They are also known as **anomers** (*ano*[G] = upper) as they differ in configuration only around the hemiacetal or anomeric carbon atom (*i.e.*, carbon 1). The designation of α- and β-forms is based on the suggestion of **Freudenberg** (1933) who stated that those anomers having the same configuration at both the anomeric and penultimate carbon atoms should be called as α-form while in β-form the configuration would be different at both these carbon atoms. In other words, the configuration of H atoms at both these carbons (anomeric and penultimate) is **erythro** (written on the same side of the structure) in α-form and **threo** (written on opposite sides) in β-form. The α- and β-forms of D-glucose in both pyranose and furanose rings along with the open chain formula are shown in Fig. 5–9, given on the next page.

As shown in this figure, the hemiacetal formation is of reversible nature. Thus, D-glucose in aqueous solution exists as an equilibrium mixture of 5 compounds. The pyranose form, however, predominates in solution. An equilibrium mixture of glucose contains about one-third (37%) α anomer, two-thirds (62%) β anomer and very little (<< 1%) of the open-chain form. Likewise, the α and β anomers of both the pyranose and the furanose forms of fructose interconvert through the open-chain form. In fact, under physiologic conditions in solution, monosaccharides with 5 or more carbons are typically more than 99% in the ring forms. The distribution between pyranose and furanose forms depends on: (*a*) the particular sugar structure, (*b*) the pH, (*c*) the solvent composition, and (*d*) the temperature. Respresentative data obtained from nuclear magnetic resonance (NMR) studies are shown in Table 5–2. When the monomers are incorporated into polysaccharides, the structure of the polymer may also influence the ring form chosen. As an instance, as Table 5–2 depicts, D-ribose exists in solution as a mixture of the two ring forms. But in biological polysaccharides, specific forms are stabilized. Ribonucleic acid, for example, contains

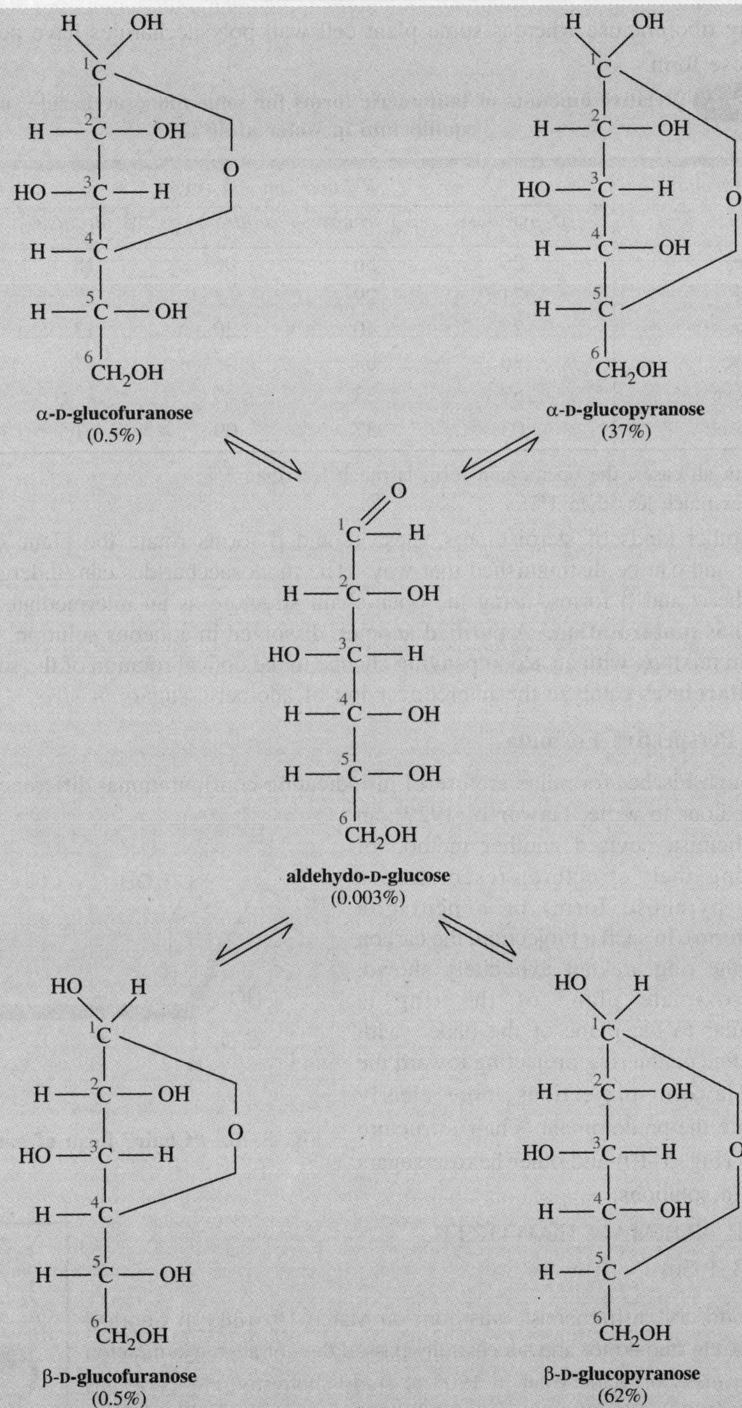

Fig. 5–9. Various forms of D-glucose present in an aqueous solution

[Note that in α-form, the OH group at anomeric carbon 1 is also on the right side like that at the penultimate carbon atom 5; whereas in β-form, the OH group at carbon 1 is on the left and that on carbon 5 is on the right side. Figures in parentheses indicate the molecular percentage present in an equilibrium mixture at pH 7.]

exclusively ribofuranose whereas some plant cell wall polysaccharides have pentoses entirely in the pyranose form.

Table 5–2. Relative amounts of tautomeric forms for some monosaccharide sugars at equilibrium in water at 40°C

Monosaccharide	*Relative amount* (in%)				
	α *pyranose*	β *pyranose*	α *furanose*	β *furanose*	*Total furanose*
Ribose	20	56	06	18	24
Lyxose	71	29	–[a]	–[a]	<1
Altrose	27	40	20	13	33
Glucose	36	64	–[a]	–[a]	<1
Mannose	67	33	–[a]	–[a]	<1
Fructose	03	57	09	31	40

Note that in all cases, the open-chain form is much less than 1%.
a designates much less than 1%.

Like other kinds of steroisomers, these α and β forms rotate the plane of polarized light differently and can be distinguished that way. The monosaccharides can undergo interconversion between the α and β forms, using the open-chain structure as an intermediate. This process is referred to as **mutarotation.** A purified anomer, dissolved in aqueous solution, will approach the equilibrium mixture, with an accompanying change in the optical rotation of the solution. Enzymes called **mutarotases** catalyze the interconversion of anomeric sugars *in vivo*.

Haworth Perspective Formula

Although Fischer formulae are useful in indicating configurational differences among sugars, they are tedious to write. **Haworth** (1929), an English chemist, devised another method of representing their structure, resembling a hexagon (pyranose form) or a pentagon (furanose form). In such a projection, the carbon atoms in the ring are not explicitely shown. The approximate plane of the ring is perpendicular to the plane of the paper, with the heavy line on the ring projecting toward the reader. Haworth projections more clearly approximate the predominant "chair" structure of glucose (Fig. 5–10) and other hexose sugars that exist in solutions.

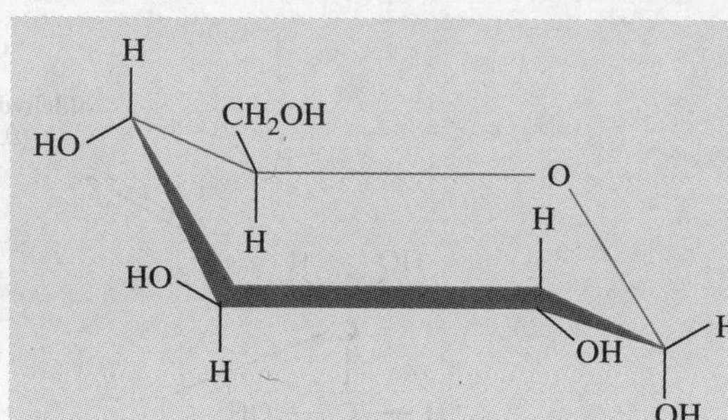

Fig. 5–10. "Chair" form of α-D-glucopyranose

WALTER NORMAN HAWORTH
(LT, 1883–1950)

Haworth, a British chemist, was born on March 19, 1883 in Chorley, Lancashire. He studied for and successfully passed the entrance examination to the University of Manchester in 1903 to study chemistry. He made this pursuit in spite of active discouragement by his parents who wanted him to join *Ryland's linoleum factory* which was being managed by his father. He gained his first-class honours degree in 1906. After gaining his master's degree under **William Henry Perkin Jr.**, he subsequently studied at the

University of Göttingen, earning his Ph.D. degree in Otto Wallach's lab after only one year of study. Then he obtained D.Sc. degree from the University of Manchester in 1911, after which he served a short time at the Imperial College of Science and Technology as senior demonstrator in Chemistry.

In 1912, Haworth became a lecturer at United College of the University of St. Andrews in Scotland and became interested in carbohydrate chemistry, which was being investigated at St. Andrews by **Thomas Purdie** (LT, 1843–1916) and **James Irvine** (LT, 1877–1952). Haworth began his work on simple sugars in 1915 and developed a new method for the preparation of the methyl ethers of sugars using methyl sulfate and alkali (now called *Haworth methylation*). He then began studies on the structural features of the disaccharides. He was appointed Professor of Organic Chemistry at the Armstrong College of Durham University, in 1920. The next year, Haworth was appointed Head of the Chemistry Department at the college. It was during this time in the North-east of England that he married **Violet Chilton Dobbie**. In 1925, he was appointed Mason Professor at the University of Birmingham, a position he held until 1948. Among his lasting contributions to science was the confirmation of a number of **structures of optically-active sugars:** by 1928, he had deduced and confirmed, among others, the structures of lactose, maltose, cellobiose, gentiobiose, melibiose (all disaccharides), gentianose, raffinose (both trisaccharides), as well as the glucoside ring tautomeric structure of aldose sugars.

In 1933, working with the then Assistant Director of Research (later **Sir**) **Edmund Hirst** and a team led by postdoctoral student **Maurice Stacey** (who in 1956 rose to the same Mason Chair), having properly deduced the correct structure and optical-isomeric nature of vitamin C, Haworth reported the **synthesis of vitamin C**. Haworth had been given his initial reference sample of "water-soluble vitamin C" or "hexuronic acid" (the previous name for the compound as extracted from natural products) by Hungarian physiologist **Albert Szent-Györgyi** who had codiscovered its vitamin properties along with **Charles G. King**, and had more recently discovered that it could be extracted in bulk from Hungarian paprika. In honour of the compound's antiscorbutic properties, Haworth and Szent-Györgyi now proposed the new name of **"a-scorbic acid"** for the molecule, with L-ascorbic acid as its formal chemical name.

Haworth also developed a simple method of representing on paper the '3'-dimensional structure of sugars. The representation, using perspective, now known as a **Haworth projection,** is still widely used in Biochemistry.

Haworth shared the **1937 Nobel Prize in Chemistry** with the Swiss chemist **Paul Karrer**, "for his investigations on carbohydrates and vitamin C". He was knighted in 1947. He died suddenly on 19 March 1950, his 67th birthday at Worcestershire, Birmingham.

In 1977, the Royal Mail issued a **postage stamp** (one of a series of four) featuring Haworth's achievement in synthesizing vitamin C and his Nobel Prize. Haworth is commemorated at the University of Birminghan in the *Haworth Building* which houses most of the University of Birmingham School of Chemistry. The School has a **Haworth Chair of Chemistry**, from 2007 held by **Prof. Nigel Simpkins**.

In the pyranose form, carbon atoms 1 to 5 of the aldohexoses and the O atom of the ring are represented in the form of a hexagon and the plane of this ring projects from the paper. Bonds that are nearer to the reader are thickened in this type of formula and the carbon atoms are numbered clockwise. Fischer formula may be converted to Haworth formula by twisting it around imaginatively or by constructing models. Certain '**assumptions**' have, however, been followed for conversion. These are:

1. Any group to the right of the carbon chain in Fischer formula is written *down* the plane of the ring in the Haworth formula ; those to the left are written *up*. A notable exception to this assumption is carbon atom 5 where H atom, present on the left side, is shown below the plane of the ring instead of showing it above the plane of the ring. This is as a result of torsion needed to bring about ring closure. The carbon atom 5 rotates so that the O atom of the OH group on the anomeric carbon comes in the plane of the first five carbon atoms. Consequently, the H atom on carbon 5 is shifted to the other side of the carbon chain by rotating through more than 90°.

2. Another important assumption is that the carbon atom (or atoms) not involved in ring formation (*e.g.*, carbon 6 in aldopyranses) will be shown above the plane of the ring, if the oxide ring in Fischer formula is to the right. Conversely, if the oxide ring is to the left, the extra carbon atom (or atoms) would be shown below the plane of the ring in Haworth projection. Two examples (α-D-glucopyranose and β-L-glucopyranose) will illustrate the mode of conversion (Fig. 5–11).

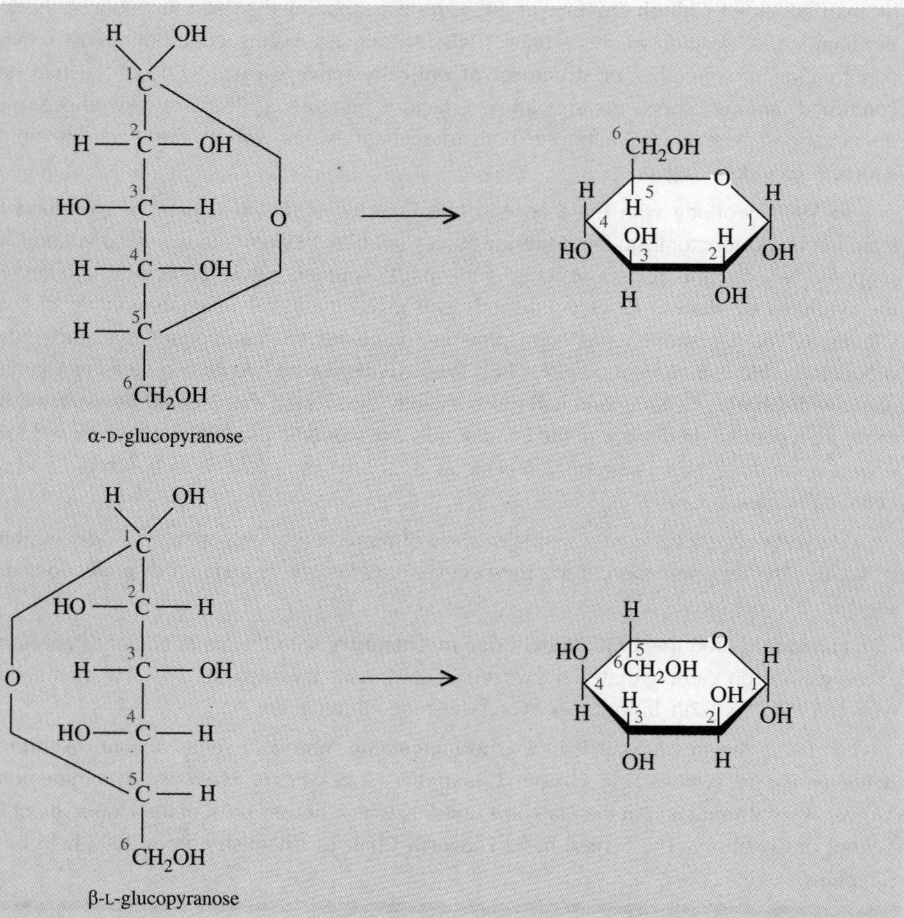

Fig. 5–11. Mode of conversion of two hexose sugars from Fischer projection formula to Haworth perspective formula

For the sake of simplicity, while writing Haworth formulae, the H and OH groups are not written and the bond indicating the single hydrogen atom is not shown. Thickened line in lower half of the hexagon is also omitted. The extra carbon atoms are, however, written as such. Thus, the above two compounds may be simply represented as in Fig. 5–12.

Fig. 5–12. Simplified Haworth formulae for two hexose sugars

The **6-membered pyranose ring**, like cyclohexane, cannot be planar because of the tetrahedral geometry of its saturated carbon atoms. Instead, pyranose rings adopt *chair* and *boat* conformations (Fig. 5–10). The substituents on the ring carbon atoms have two orientations : *axial* and *equatorial*. Axial bonds are nearly perpendicular to the average plane of the ring, whereas equatorial bonds are almost parallel to this plane. Axial substituents emerge above and below the average plane of the ring and sterically hinder each other if they emerge on the same side of the ring. By contrast, equatorial substituents emerge at the periphery and are less crowded. *The chair form of β-D-glucopyranose predominates because all axial positions are occupied by hydrogen atoms.* The bulkier –OH and $-CH_2OH$ groups emerge at the less hindered periphery. On the contrary, the boat form of glucose is highly disfavoured because it is much hindered sterically.

More generally, variations in the 3-'D' structure of biomolecules are described in terms of configuration and conformation. However, great care should be exercised in the use of these terms as they are not synonyms. **Conformation** refers to the spatial arrangement of substituent groups that are free to assume different positions in space, without breaking any bonds, because of the freedom of bond rotation. Usually no single conformation constitutes a discrete and isolable substance under usual conditions, in contrast to configurational isomers. In ethane ($H_3C—CH_3$), for example, there is nearly complete freedom of rotation around the single C—C bond. Many different interconvertible conformations of the ethane molecule are, therefore, possible, depending upon the degree of rotation. Two extreme conformations are of special interest: the *staggered* conformation, which is more stable than others and thus predominates, and the *eclipsed* form, which is least stable. It is not possible to isolate either of these conformational forms because *they are freely interconvertible without the breakage of any bonds and are in equilibrium with each other*. These 2 conformations are interconvertible by rotation about the C—C bond.

On the contrary, **configuration** denotes the spatial arrangement of an organic molecule that is conferred by the presence of (*a*) double bonds, around which there is no freedom of rotation, or (*b*) chiral centres, around which substituent groups are arranged in a specific sequence. The diagnostic feature of configurational isomers is that they *can be interconverted only by breaking one or more covalent bonds*. For example, maleic acid is a conformational isomer of fumaric acid.

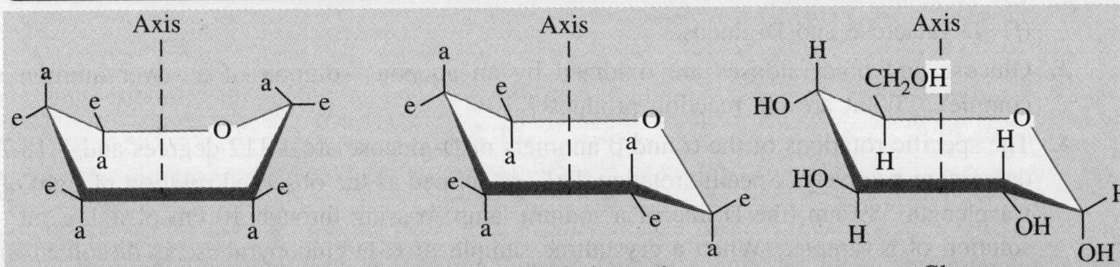

Fig. 5–13. Conformational formulae of the boat and chair forms of the pyranose ring

Substituents on the ring carbons may be either axial (a), projecting almost parallel with the vertical axis through the ring or equatorial (e) projecting almost perpendicular to this axis. For sugars with large equatorial substituents, the chair form is energetically more favourable because it is less hindered. The boat conformation is uncommon except in derivatives with very bulky substituents.

The **5-membered furanose rings,** like pyranose rings, are not planar. They can be puckered so that four atoms are nearly coplanar and the fifth is about 0.5Å away from this plane. This conformation is called *envelope form* because the structure resembles an opened envelope with the back flap raised (Fig. 5–11). In the ribose moiety of most biomolecules, either C-2 or C-3 is out of plane on the same side as C-5. These conformations are called C_2-*endo* and C_3-*endo*. [Note that the sugars in DNA double helix are in the C_2-*endo* form whereas the sugars in RNA are in the C_3-*endo* form.] Furanose rings can interconvert rapidly between different conformational states. *Furanose rings are more flexible than pyranose rings* and possibly this may be the reason for their selection as components of DNA and RNA.

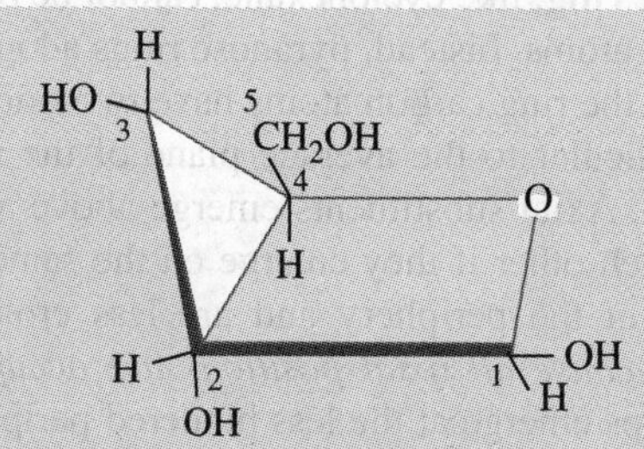

Fig. 5–14. An envelope form of β-d-ribose

The C_3-*endo* conformation is shown. C-3 is out of plane on the same side as C-5.

REFERENCES

See list following Chapter 8.

PROBLEMS

1. Account for the origin of the term *carbohydrate*.
2. Indicate whether each of the following pairs of sugars consists of anomers, epimers, or an aldose-ketose pair:
   (*a*) D-glyceraldehyde and dihydroxyacetone
   (*b*) D-glucose and D-mannose
   (*c*) D-glucose and D-fructose
   (*d*) α-D-glucose and β-D-glucose
   (*e*) D-ribose and D-ribulose
   (*f*) D-galactose and D-glucose
3. Glucose and other aldoses are oxidized by an aqueous solution of a silver–ammonia complex. What are the reaction products?
4. The specific rotations of the α and β anomers of D-glucose are + 112 degrees and + 18.7 degrees, respectively. Specific rotation, $[\alpha]_D$, is defined as the observed rotation of light of wavelength 589 nm (the D line of a sodium lamp) passing through 10 cm of a 1 g ml^{-1} solution of a sample. When a crystalline sample of α-D-glucopyranose is dissolved in water, the specific rotation decreases from 112 degrees to an equilibrium value of 52.7 degrees. On the basis of this result, what are the proportions of the α and β anomers at equilibrium? Assume that the concentration of the open-chain form is negligible.
5. Glucose reacts slowly with hemoglobin and other proteins to form covalent compounds. Why is glucose reactive? What is the nature of the product formed?
6. Compounds containing hydroxyl groups on adjacent carbon atoms undergo carbon–

carbon bond cleavage when treated with periodate ion (IO_4^-). How can this reaction be used to distinguish between pyranosides and furanosides?

7. Does the oxygen atom attached to C-1 in methyl α-D-glucopyranoside come from glucose or methanol?

8. Fructose in its β-D-pyranose form accounts for the powerful sweetness of honey. The β-D-furanose form, although sweet, is not as sweet as the pyranose form. The furanose form is the more stable form. Draw the two forms and explain why it may not always be wise to cook with honey.

9. (*a*) Compare the number of reducing ends to nonreducing ends in a molecule of glcogen.

(*b*) Glycogen is an important fuel storage form that is rapidly mobilized. At which end– the reducing or nonreducing– would you expect most metabolism to take place?

10. Draw Haworth projections for the following:

(*a*) in α-furanose form. Name the sugar.

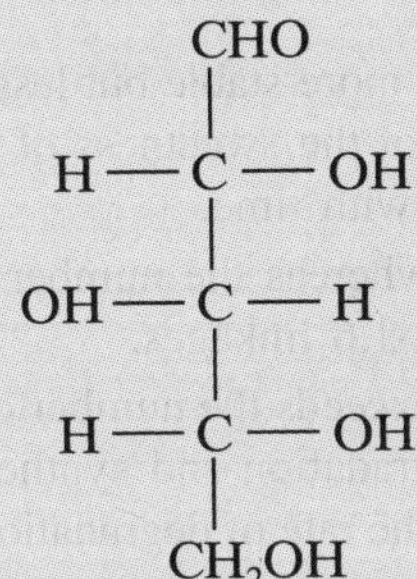

(*b*) The L isomer of (*a*)

(*c*) α-D-GlcNAc

(*d*) α-D-fructofuranose

11. α-D-galactopyranose rotates the plane of polarized light, but the product of its reduction with sodium borohydride (galactitol) does not. Explain the difference.

12. Provide an explanation for the fact that α-D-mannose is more stable than β-D-mannose, whereas the opposite is true for glucose.

13. What is the natural polysaccharide whose repeating structure can be symbolized by GlcUAβ(1→3)GlcNAc, with these units connected by β(1→4) links ?

14. A freshly prepared solution of the α form of D-galactose (1 g/mL in a 10 cm cell) shows an optical rotation of +150.7°. When the solution is allowed to stand for a prolonged period of time, the observed rotation gradually decreases and reaches an equilibrium value of + 80.2°. In contrast, a freshly prepared solution (1 g/mL) of the β form shows an optical rotation of only + 52.8°. Moreover, when the solution is allowed to stand for several hours, the rotation increases to an equilibrium value of +80.2°, identical to the equilibrium value reached by α-D-galactose.

(*a*) Draw the Haworth perspective formulas of the α and β forms of galactose. What features distinguishes the two forms?

(*b*) Why does the optical rotation of a freshly prepared solution of the α form gradually decrease with time ? Why do solutions of the α and β forms (at equal concentrations) reach the same optical rotation at equilibrium?

(*c*) Calculate the percentage composition of the two forms of galactose at equilibrium.

15. An unknown substance containing only C, H, and O was isolated from goose liver. A 0.423 g sample produced 0.620 g of CO_2 and 0.254 g of H_2O after complete combustion in excess oxygen. Is the empirical formula of this substance consistent with its being a carbohydrate? Explain.

ANSWERS TO PROBLEMS

1. Carbohydrates were originally regarded as *hydrates* of *carbon* because the empirical formula of many of them is $(CH_2O)_n$.

2. (*a*) aldose-ketose; (*b*) epimers; (*c*) aldose-ketose; (*d*) anomers; (*e*) aldose-ketose; (*f*) epimers.

3. Aldoses are converted into aldonic acids; the aldehyde group of the sugar is oxidized to a carboxylate.

4. The proporion of the α-anomer is 0.36, and that of the β-anomer is 0.64.

5. Glucose is reactive because of the presence of an aldehyde group in its open-chain form. The aldehyde group slowly condenses with amino groups to form Schiff-base adducts.

6. A pyranoside reacts with two molecules of periodate; formate is one of the products. A furanoside reacts with only one molecule of periodate; formate is not formed.

7. From methanol.

8. Heating converts the very sweet pyranose form to the more stable but less sweet furanose form. Consequently, it is difficult to accurately control the sweetness of the preparation, which also accounts for why honey loses sweetness with time.

9. (***a***) Each glycogen molecule has one reducing end, whereas the number of nonreducing ends is determined by the number of branches, or α-1,6 linkages.

(***b***) Because the number of nonreducing ends greatly exceeds the number of reducing ends in a collection of glycogen molecules, all of the degradation and synthesis of glycogen takes place at the nonreducing ends, thus maximizing the rate of degradation and synthesis.

10. (***a***)

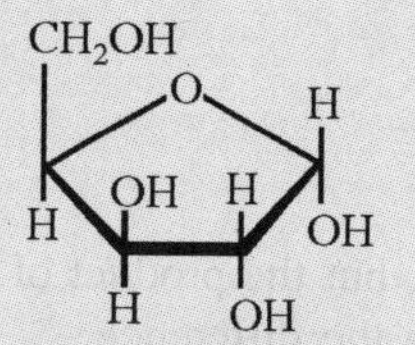

α-D-xylofuranose

(***b***)

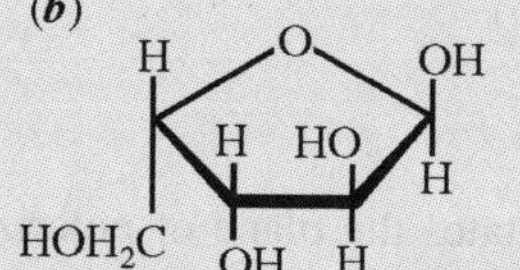

(***c***)

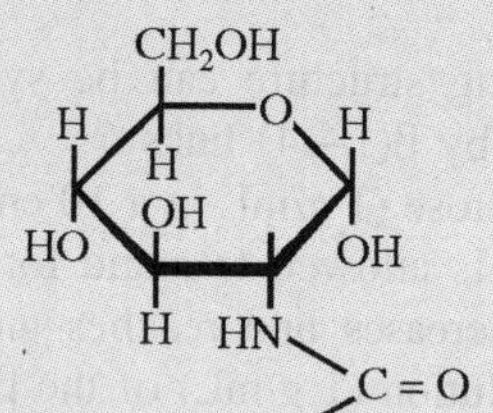

(***d***)

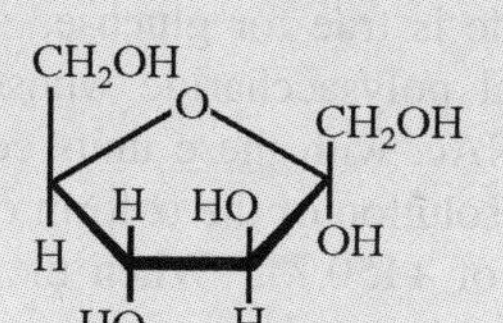

11. Galactitol has the following structure:

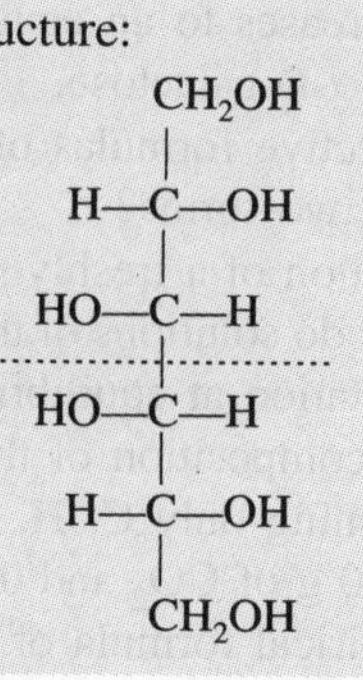

Since it has a plane of symmetry between C-3 and C-4, it is optically inactive. Such compounds, which contain asymmetric carbons but have no net optical activity, are called *meso* forms.

12. In the chair form, there is more steric clash between the 2-OH and the 1-OH in the α form of glucose and in the β form of mannose (compare glucose and mannose). Furthermore, dipole–dipole interactions will be more favourable in β-D-glucose and α-D-mannose.

13. Hyaluronic acid.

14. (*a*)

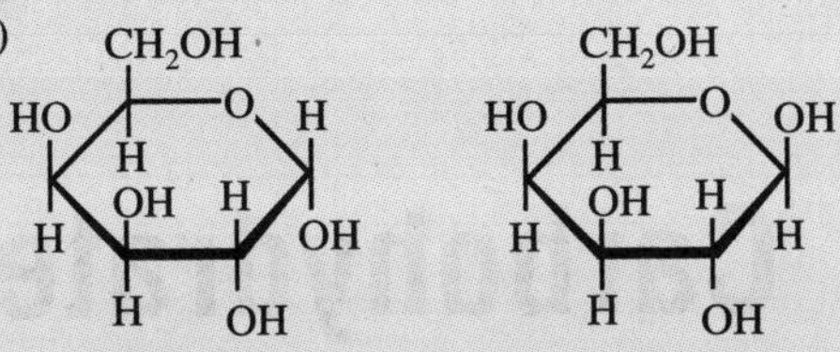

α-D-Galactose β-D-Galactose

(*b*) A freshly prepared solution of α-D-galactose undergoes mutaroation to yield an equilibrium mixure of α- and β-D-galactose. Mutarotation of either the pure α or the pure β form will yield the same equilibrium mixture.

(*c*) 72% β form; 28% α form.

15. Yes; the empirical formula is CH_2O, typical of a carbohydrate.

CONTENTS

Carbohydrates in food as important sources of energy
Starch, found in plant-derived food such as pasta, consists of chains of linked glucose molecules. These chains are broken down into individual glucose molecules for eventual use in generation of ATP and building blocks for other molecules.

CHAPTER

6

Carbohydrates–II

Properties of Monosaccharides

Carbohydrates possess active groups which are responsible for their chemical behaviour. These groups are glycosidic OH, alcoholic OH and CHO (or CO). Most of the reactions of mutarotating sugars take place in aqueous solution so that all the 3 forms become available for the reaction.

$$\alpha\text{-form} \rightleftharpoons \text{aldehydo/keto form} \rightleftharpoons \beta\text{-form}$$

The above equilibrium is shifted in favour of the group which reacts. Some reagents, however, react with more than one group.

REACTION OF GLYCOSIDIC OH GROUP

Reaction with alcohol. The glycosidic OH group of mutarotating sugars react with alcohols to form α- and β-**glycosides** or **acetals**. Thus, glucose forms glucosides and fructose forms fructosides.

The glucosides, or glycosides in general, do not exhibit mutarotation as the aldehyde group in them is converted to the acetal group. A number of glycosides occur in nature, *e.g.*, *phlorhizin* (glucose + phloretin) in rose bark; *digitonin* (4 galactose + xylose + digitogenin) in foxglove leaves; *amygdalin* (2 glucose + 2 mandelonitril) from bitter almonds and *saponin* (sugar + sapogenin) from soapwort. They are useful as medicaments.

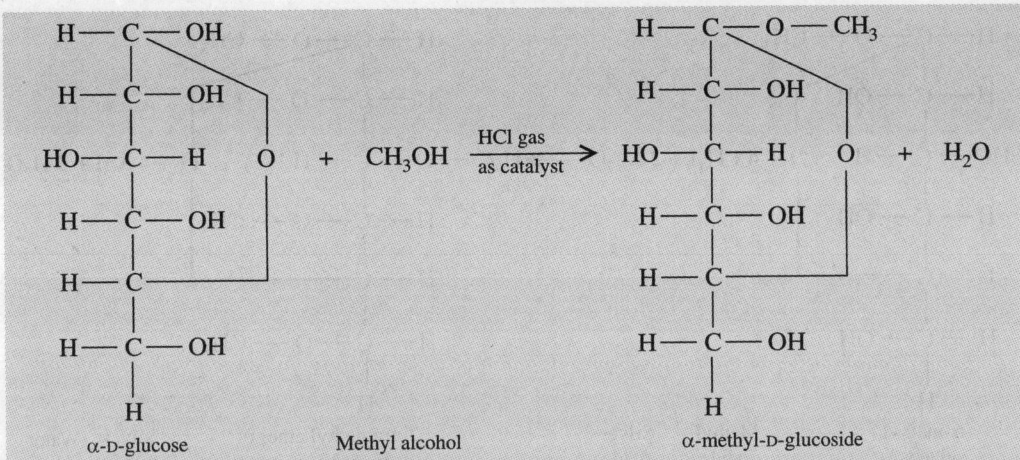

REACTION INVOLVING BOTH GLYCOSIDIC AND ALCOHOLIC OH GROUPS

Reaction with acetic anhydride (Esterification). The glycosidic and alcoholic OH groups of monosaccharides and disaccharides react with acetylating agents (like acetic anhydride in pyridine) to form acetate derivatives called **esters.**

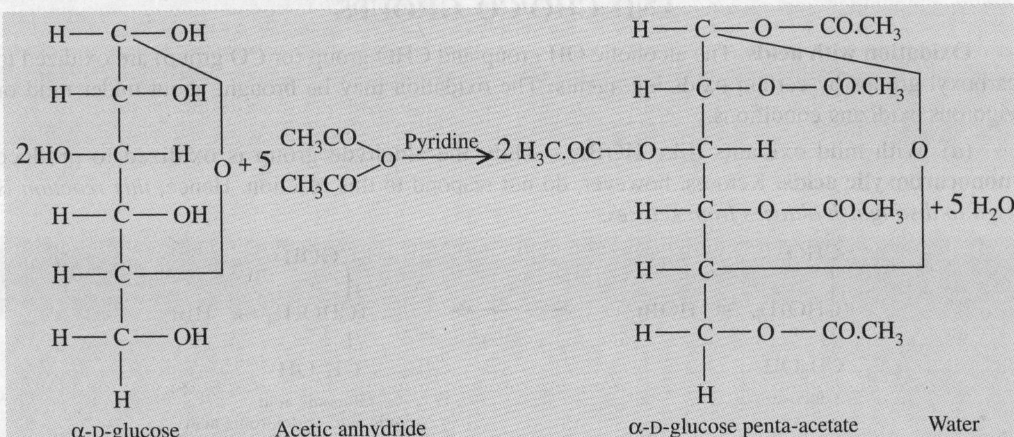

The ability of sugars to form esters indicates the presence of alcohol groups in their molecule. As glucose upon acetylation yields a penta-acetate derivative, it obviously contains 5 OH groups.

REACTION OF ALCOHOLIC OH GROUP

Reaction with methyl iodide (Etherification). The alcoholic OH groups of monosaccharides and disaccharides are converted to ether groups upon treatment with methylating agents.

[The glycosidic OH group of the sugars is, however, first protected with an R group by converting these sugars to glycosides which are, then, allowed to react with the methylating agent.]

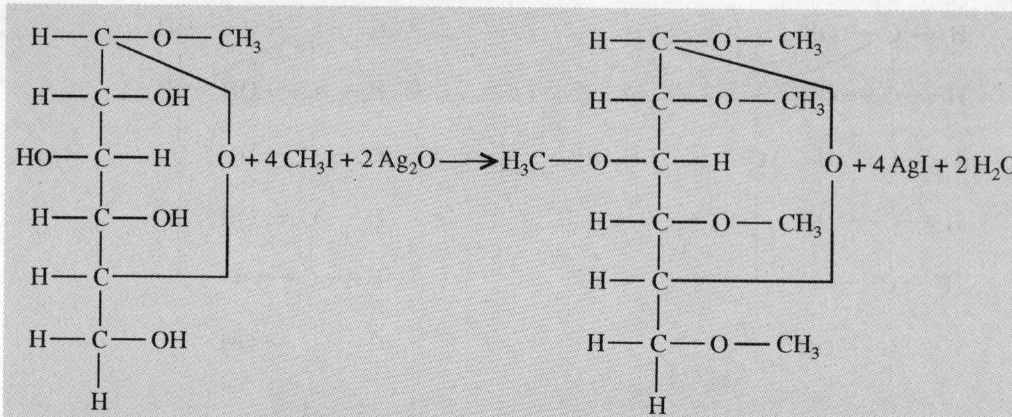

α-methyl glucose | Methyl iodide | Silver oxide | Tetramethyl ether of α-methyl glucoside | Silver iodide | Water

This reaction has been extensively used to ascertain which OH groups in a sugar are free and available for reaction. Thus, α-methyl glucoside is methylated at carbon 2, 3, 4 and 6 but not at carbon 5 which is involved in hemiacetal link.

REACTION INVOLVING BOTH ALCOHOLIC OH AND CHO/CO GROUPS

Oxidation with acids. The alcoholic OH group and CHO group (or CO group) are oxidized to carboxyl groups by certain oxidizing agents. The oxidation may be brought about under mild or vigorous oxidizing conditions.

(*a*) With mild oxidants (like HOBr) – Only the aldehyde group is oxidized to produce **monocarboxylic acids.** Ketoses, however, do not respond to this reaction. Hence, *this reaction is used to distinguish aldoses from ketoses.*

$$\underset{\text{Glucose}}{CHO-(CHOH)_4-CH_2OH} + HOBr \longrightarrow \underset{\text{Gluconic acid (= Aldonic acid or dextronic acid)}}{COOH-(CHOH)_4-CH_2OH} + HBr$$

(*b*) With strong oxidants (like conc. HNO_3) – Both the aldehyde group (or ketone group) and the primary alcohol group are oxidized to yield **dicarboxylic acids.** With aldoses, acids with same number of carbon atoms are obtained whereas ketoses react to produce acids with fewer number of carbon atoms (see reaction on the next page).

REACTIONS OF CHO/CO GROUP

1. **Oxidation with metal hydroxides.** Metal hydroxides like $Cu(OH)_2$, AgOH and Bi $(OH)_3$ oxidize the free aldehyde (or ketone) group of mutarotating mono- and di-saccharides and are, at the same time, themselves reduced to the lower oxides or to the free metals. Thus, $Cu(OH)_2$ is reduced to Cu_2O and AgOH and $Bi(OH)_3$ are reduced to the free metal, Ag and Bi. The reaction with cupric hydroxide is given below:

$$\text{Reducing sugar} + \underset{\text{Blue}}{2\,Cu^{++}} \longrightarrow \text{Oxidized sugar} + 2\,Cu^{+}$$

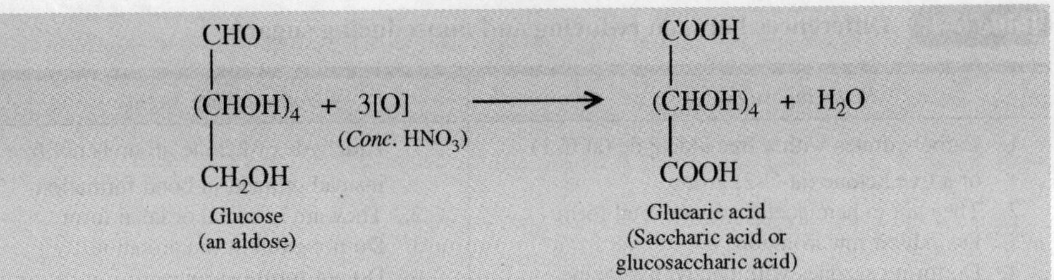

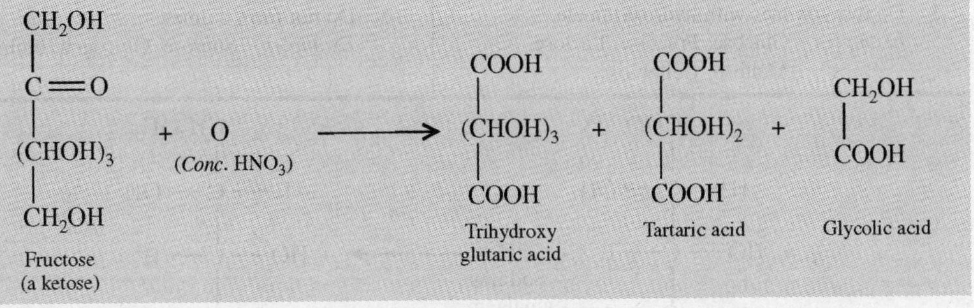

$$2\,Cu^{+} + 2\,OH^{-} \longrightarrow \underset{\text{Yellow}}{2\,Cu.OH} \longrightarrow \underset{\text{Red}}{Cu_2O} + H_2O$$

Cupric ion (Cu^{++}) is, in fact, the most common oxidizing agent of these. It is the active ingredient in Fehling's, Benedict's and Barfoed's reagents.

Fehling's reagent consists of 2 solutions: *solution A* contains 7% $Cu.SO_4$ and *solution B* contains 25% KOH and 35% sodium potassium tartarate. When the two solutions are mixed in equal amounts, a clear blue solution results for the tartarate forms a soluble complex with the copper hydroxide produced. This solution is widely used as an oxidizing agent preferably in quantitative sugar determinations.

With Fehling's reagent, the overall reaction occurs as follows:

$$\underset{\text{Glucose}}{C_6H_{12}O_6} + \underset{\text{Fehling's solution}}{2\,Cu(OH)_2} \longrightarrow \underset{\text{Gluconic acid}}{C_6H_{12}O_7} + \underset{\text{Cuprous oxide (Red ppt.)}}{Cu_2O} + 2H_2O$$

In clinical laboratories, however, a modification of Fehling's reagent is used for the qualitative determination of glucose in urine. This is known as **Benedict's reagent** and consists of a single solution containing $CuSO_4$, $Na_2.CO_3$ (in place of KOH) and sodium citrate (in place of sodium potassium tartarate). Sodium carbonate reduces alkalinity to such an extent that, unlike the Fehling's reagent, the two solutions ($Cu.SO_4$ and sodium citrate) here are kept mixed indefinitely as one solution.

Barfoed's reagent utilizes a weakly acid solution for oxidation. It contains 7% cupric acetate and 1% acetic acid. This reagent can be used to distinguish monosaccharides from disaccharides as the former give a red cuprous oxide precipitate in dilute acid more quickly than the latter. *Monosaccharides are, thus, more active reducing agents than the disaccharides.*

Carbohydrates with free carbonyl groups or in hemiacetal form give positive tests to these reagents without having been hydrolyzed first and are referred to as **'reducing'** sugars ; others (*i.e.*, the acetal types) are then **'nonreducing'** sugars (Table 6–1).

Table 6–1. Differences between reducing and nonreducing sugars

Reducing sugar	*Nonreducing sugar*
1. Carbohydrates with a free aldehyde (at C-1) or a free ketone (at C-2) group.	1. Aldehyde or ketone group is not free but instead utilized in bond formation.
2. They are in hemiacetal or hemiketal form.	2. They are in acetal or ketal form.
3. Do exhibit mutarotation.	3. Do not exhiblit mutarotation.
4. Do form osazones with phenyl hydrazine.	4. Do not form osazones.
5. Do form oximes with hydroxylamine.	5. Do not form oximes.
Examples – Glucose, Fructose, Lactose, Maltose, Cellobiose	*Examples* – Sucrose, Glycogen, Inulin

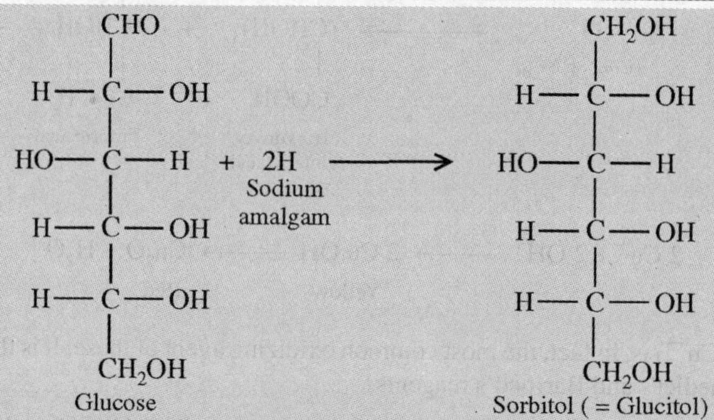

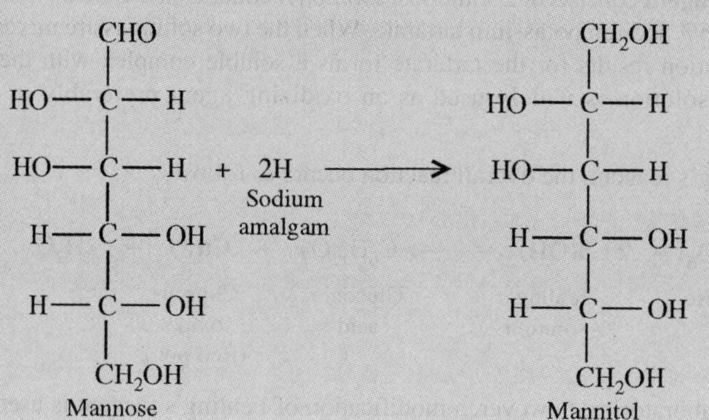

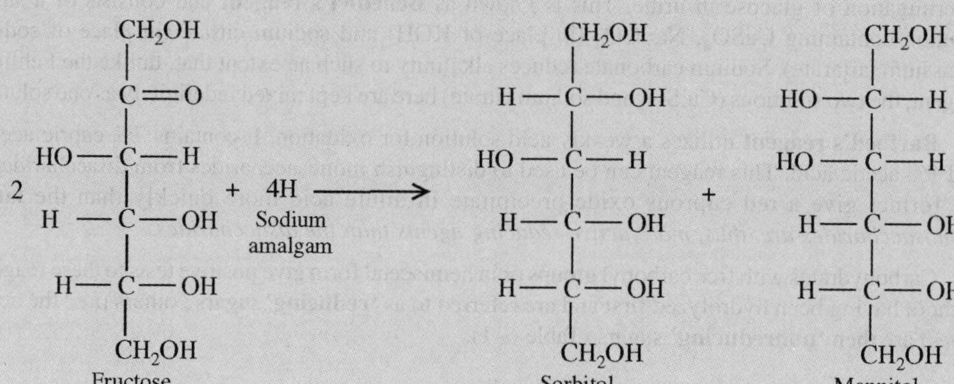

The **Molisch test** employs sulfuric acid for dehydration and α-naphthol as the phenol. The acid also hydrolyzes acetal groups present and thus makes this test probably the most general of all the tests for carbohydrates.

Since the alkali present in some of the above tests causes fragmentation of the sugar molecule, a mixture of hydroxy acids of different chain length is obtained upon oxidation of carbohydrates. Hence, these tests, although widely employed in analytical methods for sugars, are of little value in preparative chemistry.

2. **Reduction.** The sugars may be reduced in various ways depending upon the type of reducing agent used.

(*a*) With sodium amalgam – The monosaccharides are reduced to their corresponding **alcohols** by treating them with reducing agents like Na amalgam (refer reactions given on the previous page). Thus, glucose yields sorbitol (= glucitol), mannose yields mannitol, galactose yields dulcitol, fructose yields a mixture of sorbitol and mannitol, and glyceraldehyde yields glycerol.

Sorbitol and mannitol both are important and are used in the manufacture of surface-active agents and explosives, respectively.

(*b*) With strong mineral acids – With *hexoses,* the reduction leads to the formation of 5-hydroxymethylfurfural, which on further heating is transformed to levulinic acid.

This reaction forms the basis of some colour tests (*e.g.*, Molisch test and Bial's orcinol test) for sugars as the aldehyde products of these reactions condense with certain organic phenols to give characteristic coloured products.

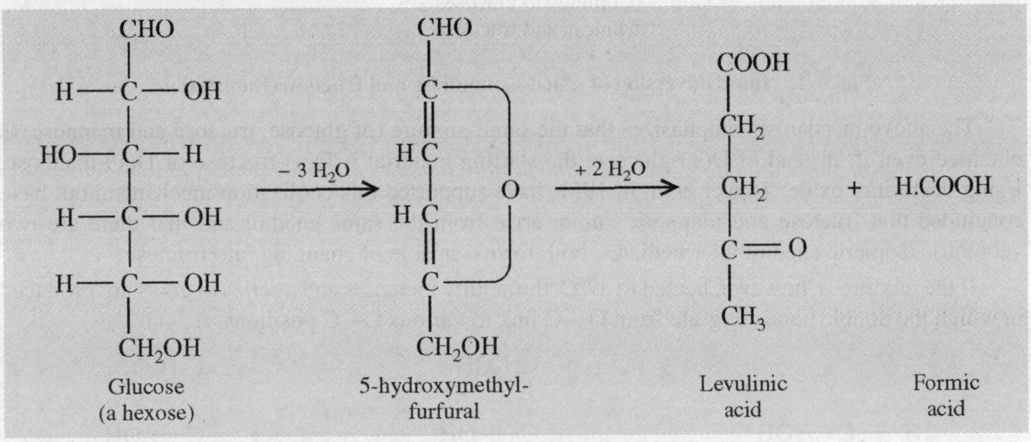

With *pentoses*, a product **furfural** is obtained on heating.

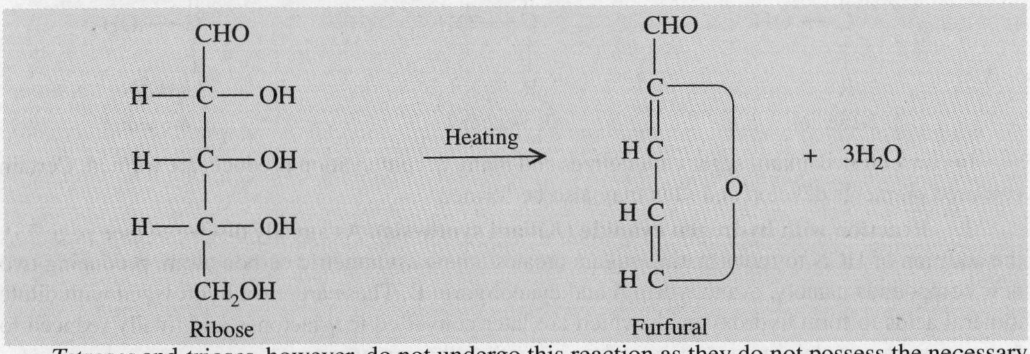

Tetroses and *trioses*, however, do not undergo this reaction as they do not possess the necessary minimum 5 carbon atoms for furfural formation.

(*c*) With dilute alkalies – Glucose, fructose and mannose are interconvertible in weak alkaline solutions such as $Ca(OH)_2$ and $Ba(OH)_2$ at low temperatures. This reaction is known as **Lobry de**

Bruyn–Alberda van Ekenstein transformation. This reaction, proposed in 1890, is of special significance as a similar reaction is supposed to take place in the human body. The mechanism of this reaction involves **enolization**, the migration of a proton from a carbon atom onto the oxygen of an adjacent carbonyl group, resulting in the formation of an unsaturated alcohol called **enol** (*ene* = unsaturation ; *ol* = alcohol). The interconversion between glucose, fructose and mannose may, in fact, be visualized as occurring through a common enediol form (Fig. 6–1).

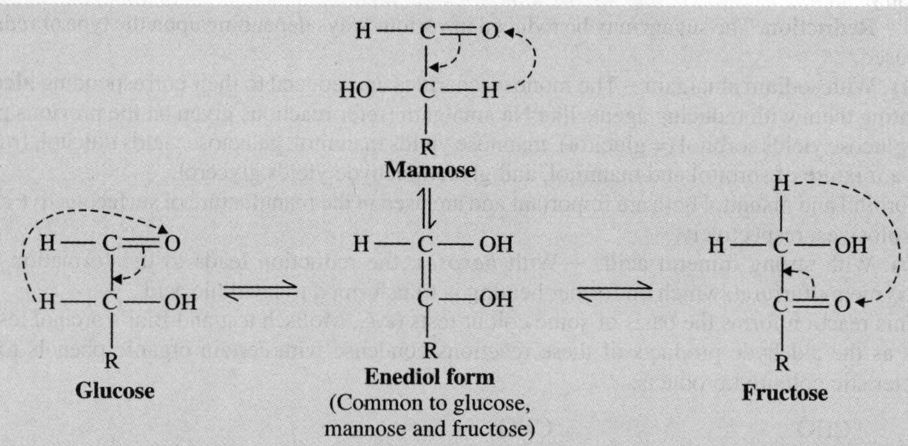

Fig. 6–1. Interconversion of glucose, mannose and fructose (Enolization)

The above mechanism emphasizes that the same mixture (of glucose, fructose and mannose) is obtained even if, instead of D(+)-glucose, the starting material is D(–)-fructose or D(+)-mannose. Using deuterium oxide, Topper et al, in 1951, have supported this enolization mechanism but have concluded that fructose and mannose cannot arise from the same enediol and that there are two geometric isomeric enediol intermediates, both forms capable of changing into fructose.

If the mixture is, however, heated to 37°C, the acidity increases and a series of enols are produced in which the double bonds migrate from O—C link to various C—C positions.

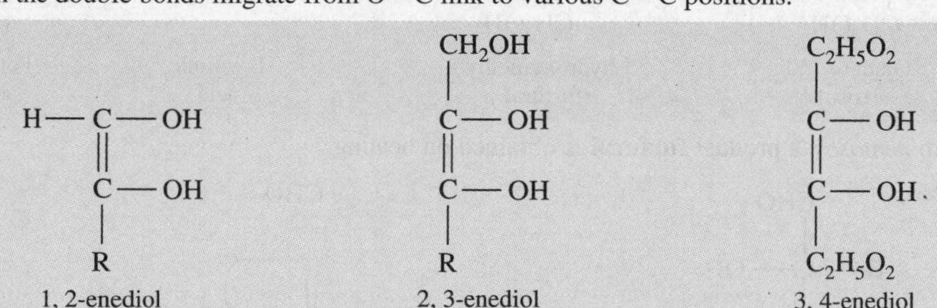

In concentrated alkali, sugar caramelizes and many decomposition products are formed. Certain coloured pigments develop and salts may also be formed.

3. **Reaction with hydrogen cyanide (Kiliani synthesis).** As already discussed (see page 73), the addition of HCN to mutarotating sugars creates a new asymmetric carbon atom, producing two new compounds namely, cyanohydrin A and cyanohydrin B. These are, then hydrolyzed with dilute mineral acids to form hydroxy acids which are later converted to γ-lactones and finally reduced to two new sugars containing one more carbon atom than the parent sugar.

4. **Reaction with alanine.** The aldehyde group of carbohydrates condenses with the amino group of alanine to form Schiff's base.

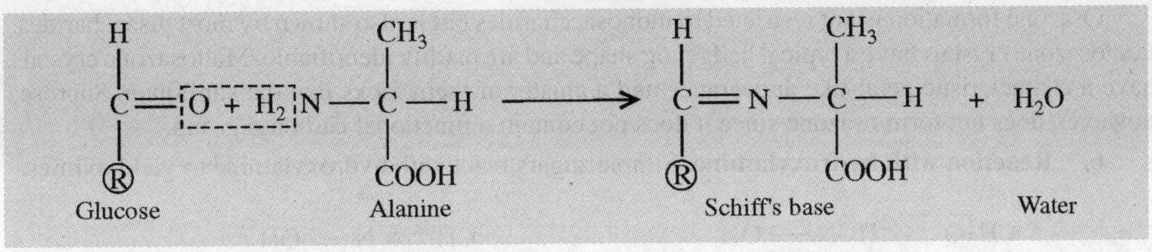

The browning reaction (Maillard's reaction) occurring during the baking of bread and other mixtures of carbohydrates and proteins is believed to be due to the formation of Schiff's base between the amino groups of proteins and the aldehyde groups of carbohydrates (Mertz, 1960).

5. Reaction with phenylhydrazine. Reaction with phenylhydrazine involves only 2 carbon atoms viz., the carbonyl (i.e., the aldehyde or ketone group) and the adjacent one. One mole of phenyl hydrazine reacts with one mole of aldose (or ketose) to form a mole of hydrazone. With a second mole of phenylhydrazine, the hydrazone is oxidized to **aldohydrazone** and the phenylhydrazine itself is reduced to aniline and ammonia. Finally, a third mole of phenylhydrazine reacts with the aldohydrazone to produce **osazone.** The hydrazone may, in fact, be regarded as a special type of Schiff's base and the osazone as a double Schiff's base (Fig. 6–2).

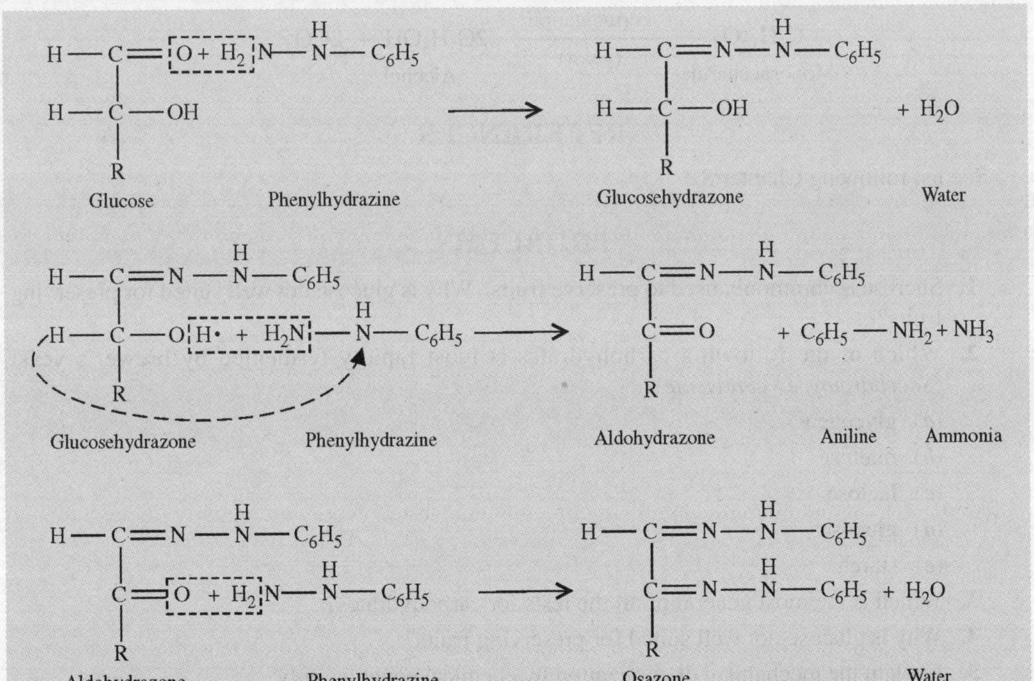

Fig. 6–2. Osazone formation

The reaction with ketose will also take place in a likewise manner.

The osazones of reducing (mutarotating) sugars are crystalline, yellow and usually insoluble compounds and hence may also be recovered. These have characteristic crystal structure, formation times and melting points and are, therefore, frequently used in the identification of sugars. It may, however, be noted that glucose, fructose, and mannose would yield the same osazone owing to their similar structure in the unaffected part of the molecule (from C_3 to C_6). Moreover, the asymmetry at C_2 is destroyed during the reaction. Obviously, galactose would, then, form a different osazone as it has a different configuration below C_2.

Osazone formation is not restricted to monosaccharides but is also shown by most disaccharides. Lactosazone crystals have a typical hedgehog shape and are readily identifiable. Maltosazone crystals have a characteristic petal-like appearance and a cluster of them looks like the sunflower. Sucrose, however, does not form osazone since it does not contain a functional carbonyl group.

6. Reaction with hydroxylamine. Simple sugars react with hydroxylamine to yield **oximes.**

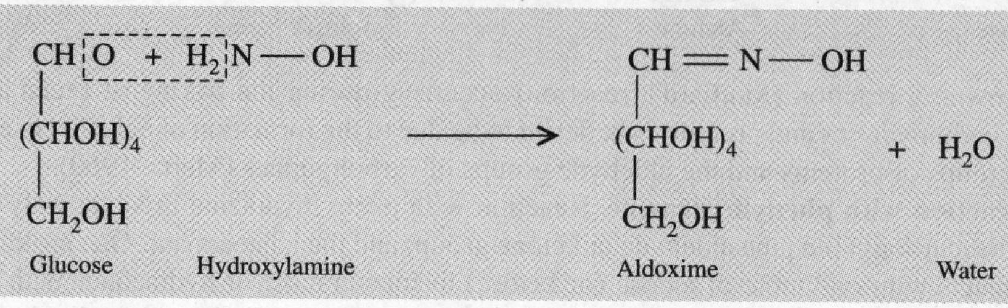

7. Fermentation. Monosaccharides such as glucose, fructose and mannose are readily fermented by yeast. The process of yeast fermentation is very complex. During this process, sugar phosphate and sugar acid phosphate are formed. Ordinarily, this process results in the formation of alcohol and carbon dioxide.

$$\underset{\text{Monosaccharide}}{C_6H_{12}O_6} \xrightarrow[\text{(yeast)}]{\text{Fermentation}} \underset{\text{Alcohol}}{2C_2H_5OH} + 2CO_2$$

REFERENCES

See list following Chapter 8.

PROBLEMS

1. Sucrose is commonly used to preserve fruits. Why is glucose not well suited for preserving fruits?
2. Which of the following carbohydrates is most rapidly fermented by brewer's yeast (*Saccharomyces cerevisiae*):
   (*a*) glycogen
   (*b*) maltose
   (*c*) lactose
   (*d*) glucose
   (*e*) starch
3. Which is the most general of all the tests for carbohydrates?
4. Why is glucose not well suited for preserving fruits?
5. Explain the mechanism of enolization by chemical equation only.
6. Why the reduction of cupric to cuprous is profuse with Benedict's reagent but not with Barfoed's reagent?
7. How can glucose and fructose be differentiated by their reactions?
8. What are glycosides?

ANSWERS TO PROBLEMS

1. Glucose is reactive because its open-chain form contains an aldehyde group.
2. (*d*) glucose

3. Molisch test
4. Glucose in its open-chain form contains an aldehyde group which is reactive and is, hence, unsuitable for preservation.
5.

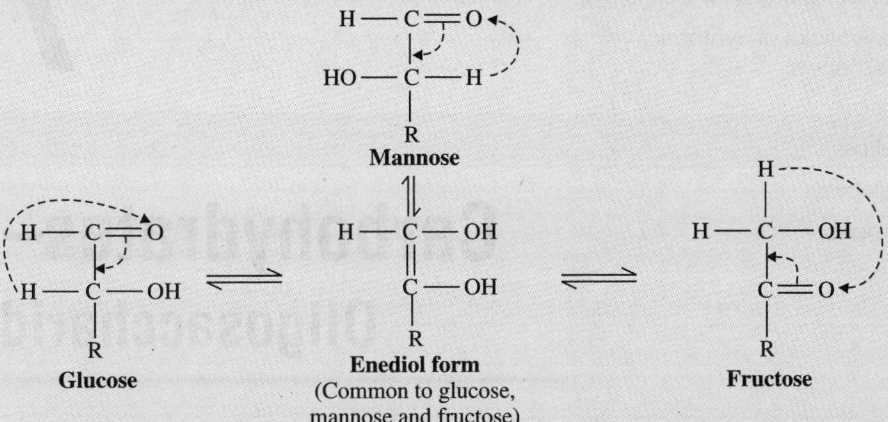

6. Benedict's reagent contains sodium carbonate (Na_2CO_3) which with its alkalinity produces enediol form. This is not possible with Barfoed's reagent which is a weakly acid solution of glacial acetic acid (CH_3COOH). As enediol is formed with Benedict's reagent, reduction is profuse.
7. Glucose and fructose can be differentiated by the following 2 tests:
   (*a*) *Seliwanoff's test.* When fructose is boiled with Seliwanoff's reagent (resorcinol in HCl), a cherry-red colour develops. However, glucose does not respond to this test.
   (*b*) *Foulger's test.* On boiling with Foulger's reagent (urea in H_2SO_4 with stannous chloride), fructose produces a blue colour while glucose gives a green colour.
8. Glycosides are plant products though they can be synthesized in the laboratory. In a glycoside, a sugar molecule is linked to a nonsugar with an oxygen bridge. The nonsugar is called aglycone. Some of the glycosides have profound biochemical effects. For example, cardiac glycosides like the derivative of *Digitalis* have effect on the heart. **Streptomycin**, the well-known antibiotic, **digitonin** which is used to precipitate cholesterol, and **quabain**, an inhibitor of sodium potassium ATPase, are all glycosides.

CONTENTS

Space-filling model of aspartame, a noncaloric sweetener, widely used in the drink and food industry

CHAPTER

7

Carbohydrates–III
Oligosaccharides

The oligosaccharides yield 2 to 10 monosaccharide molecules on hydrolysis. *Disaccharides are the most common oligosaccharides found in nature*. Most of the naturally-occurring representatives occur in plant rather than in animal sources. The three most important disaccharides found free in appreciable quantities are sucrose, lactose and maltose ; a few others are cellobiose and trehalose.

OLIGOSACCHARIDE STRUCTURES

The simplest and biologically most important oligosaccharides are the **disaccharides,** which are made up of 2 residues. The disaccharides play many roles in living organisms (Table 7–1). Some like **sucrose, lactose** and **trehalose** are soluble energy stores in plants and animals. Others like **maltose** and **cellobiose** can be regarded primarily as intermediate products in the degradation of much longer polysaccharides. Still others like **gentiobiose** are found principally as constituents of more complex, naturally occurring substances.

Distinguishing Features of Different Disaccharides: Four major features distinguish disaccharides from one another. These are:

1. *Two specific sugar monomers involved, and their stereo configuration.* The monomers may be of the same kind, as the 2 D-glucopyranose residues in maltose, or they may be different, as the D-glucopyranose and D-fructofuranose residues in sucrose.

Table 7-1 Occurrence and biochemical roles of some representative disaccharides

Disaccharide	*Structure*	*Natural occurrence*	*Physiological role*
Sucrose	Glcα(1→2)Fruβ	Many fruits, seeds, roots, honey	A final product of photosynthesis; used as primary energy source in many organisms
Lactose	Galβ(1→4)Glc	Milk, some plant sources	A major animal energy source
α,α-Trehalose	Glcα(1→1)Glcα	Yeast, other fungi, insect blood	A major circulatory sugar in insects; used for energy
Maltose	Glcα(1→4)Glc	Plants (starch) and animals (glycogen)	The dimer derived from the starch and glycogen polymers
Cellobiose	Glcβ(1→4)Glc	Plants (cellulose)	The dimer of the cellulose polymer
Gentiobiose	Glcβ(1→6)Glc	Some plants (*e.g.,* gentians)	Constituent of plant glycosides and some polysaccharides

2. *Carbons involved in the linkage.* Although many possibilities exist, the most common linkages are 1→1 (as in trehalose), 1→2 (as in sucrose), 1→4 (as in lactose, maltose, cellobiose), and 1→6 (as in gentiobiose). Note that all of these disaccharides involve the anomeric hydroxyl of at least one sugar as a participant in the bond.
3. *Order of the two monomer units, if they are different kinds.* The glycosidic linkage involves the anomeric carbon on one sugar, but in most cases the other is free. Thus, the two ends of the molecule can be distinguished by their chemical reactivity. For example, the glucose residue in **lactose**, having a free anomeric carbon and thus a potential free aldehyde group, could be oxidized by Fehling's solution; the galactose residue could not be. Lactose is, henceforth, a reducing sugar, and the glucose residue is at its *reducing end.* The other end is called the *nonreducing end.* In **sucrose,** neither residue has a potential free aldehyde group; both anomeric carbons are involved in the glycosidic bond. Hence, sucrose is a nonreducing sugar.
4. *Anomeric configuration of the hydroxyl group on carbon 1 of each residue.* This feature is especially important for the anomeric carbon(s) involved in the glycosidic bond. The configuration may be either α (as in trehalose, sucrose, maltose) or β (as in lactose, cellobiose, gentiobiose). This difference may seem small, but it has a major effect on the shape of the molecule, and the difference in shape is readily recognized by enzymes. For example, different enzymes are needed to catalyze the hydrolysis of **maltose** and **cellobiose**, even though both are dimers of D-glucopyranose.

NAMING OF DISACCHARIDES

To name disaccharides, certain rules are followed. First, the compound is written with its nonreducing end to the left. An *O* precedes the name of the first (left) monosaccharide unit, as a reminder that the sugar-sugar linkage is through an oxygen atom. The configuration at the anomeric carbon joining the first (left) monosaccharide unit to the second is then given (α or β). To distinguish 5- and 6-membered ring structures, "furanosyl" or "pyranosyl" is inserted in the name of each monosaccharide unit. The two carbon atoms joined by the glycosidic bond are then shown in parentheses, with an arrow connecting the two numbers; for example, (1 → 4) shows that C-1 of first sugar residue is joined to C-4 of the second. If there is a third residue, the second glycosidic bond is described next, by the same conventions. To shorten the description of a complex **polysaccharide**, 3-letter abbreviations for each monosaccharide are often used.

Following this convention for naming oligosaccharides, sucrose is therefore named *O*–α–D–glucopyranosyl–(1→2)–β–D–fructofuranoside. Because most sugars are the D-enantiomers, and the pyranose form predominates, a shorthand version of the formal name of such compounds is used, giving the configuration of the anomeric carbon and naming the carbons joined by the glycosidic bond. In this abbreviated nomenclature, sucrose is Glc (α1→2) Fru.

Based on the type of linkage present in their molecule, the disaccharides may by classified as follows:

Disaccharides

- Nonreducing (= Glycosylglycosides)
  - *C_1—C_1 glycosidic linkage*: Trehalose
  - *C_1—C_2 glycosidic linkage*: Sucrose
- Reducing (= Glycosylaldoses or glycosylketoses)
  - *C_1—C_4 glycosidic linkage*: Lactose, Maltose, Cellobiose
  - *C_1—C_6 glycosidic linkage*: Gentiobiose, Melibiose, Isomaltose

Synthesis of Disaccharides. Because different disaccharides (and oligosaccharides and polysaccharides) are distinguished by both the kinds of monomers involved and the precise glycosidic linkages between them, the enzymes needed for their synthesis and degradation must be very specific. For example, hydrolysis of the common nutritional disaccharides maltose, lactose and sucrose, which takes place in cells lining the wall of the small intestine, requires 3 different and specific enzymes, namely maltase, lactase and sucrase, respectively. None will substitute for another.

$$\text{Maltose} + H_2O \xrightarrow{\textit{Maltase}} 2\ \text{D-glucose}$$

$$\text{Lactose} + H_2O \xrightarrow{\textit{Lactase}} 2\ \text{D-galactose} + \text{D-glucose}$$

$$\text{Sucrose} + H_2O \xrightarrow{\textit{Sucrase}} 2\ \text{D-fructose} + \text{D-glucose}$$

Similarly, the synthesis of disaccharides, as exemplified by lactose (Fig. 7–1), requires enzymes that are specific not only for the pair of sugars to be joined, but also for the specific glycosidic linkage to be formed between them. Note that each galactose molecule has 5 different OH groups, each of which is potentially capable of forming a glycosidic linkage with each of the 5 OH groups on a glucose moiety. Even allowing for steric restriction, this still means that many different galactose – glucose disaccharides could conceivably be formed. Yet the enzymes involved in the reactions in Fig. 7–1 are so specific that only a single product is ever made.

The oligosaccharide (and polysaccharide) synthesis differs from the synthesis of nucleic acids and proteins in one important way. These sugar polymers are never copied from template molecules. Instead, in the formation of oligosaccharides or polysaccharides, a different enzyme is employed to catalyze the addition of each kind of monomer unit. Evidently, a vast array of plant and animal enzymes must be devoted to the synthesis and degradation of saccharide polymers.

SUCROSE OR SACCHAROSE
(Table sugar, Cane sugar, Beet sugar)

Occurrence. Sucrose is the common sugar of commerce and kitchen (hence, also called 'household' sugar) and is widely distributed in all photosynthetic plants. It is the chief constituent of sugarcane (*Saccharum officinarum*), beet (*Beta vulgaris*) and maple (*Acer saccharina*) and is also present in pineapple (*Ananas sativus*) and carrot (*Daucas carota*). Sucrose occurs in varying amounts in different plant organs such as fruits, seeds, flowers and roots. Nectar of flowers is particularly rich in sucrose. It is the raw material for honey. *Sucrose is probably the only foodstuff used in the crystalline form.*

Sucrose is something of a riddle in plant biochemistry. Although D-glucose is the major building block of both starch and cellulose, sucrose is a major intermediate product of photosynthesis. It is the predominant form in which sugar is transported in most plants from the leaves to other organs of plants *via* their vascular system.

Chemistry. It is derived commercially from either sugarcane or beet. Refinement removes the yellow-brown pigments of unrefined sugar to produce the white crystal form of table sugar, while leaving a syrup which does not crystallize. This syrup, known as *molasses,* is also a commercial product and is the least refined form of sucrose. In plants, sucrose is formed by elimination of a molecule of water from glycosidic hydroxyl groups (marked by an asterisk) of α-D-glucose and β-D-fructose (Fig. 7–2.)

It is noteworthy that fructose possesses the furanose ring structure in the sucrose molecule, although the pyranose ring is the dominant and more stable form in the free ketohexose. The linkage between the two disaccharide moieties (= units) of sucrose is a glycosidic one between C_1 of glucose and C_2 of fructose.

Properties. Sucrose is a white crystalline solid and is soluble in water. It exists in 2 crystalline forms:

(*a*) a stable form (*sucrose A*) – m.p. 184–185°C, and

(*b*) the unstable form (*sucrose B*) – m.p. 169–170°C.

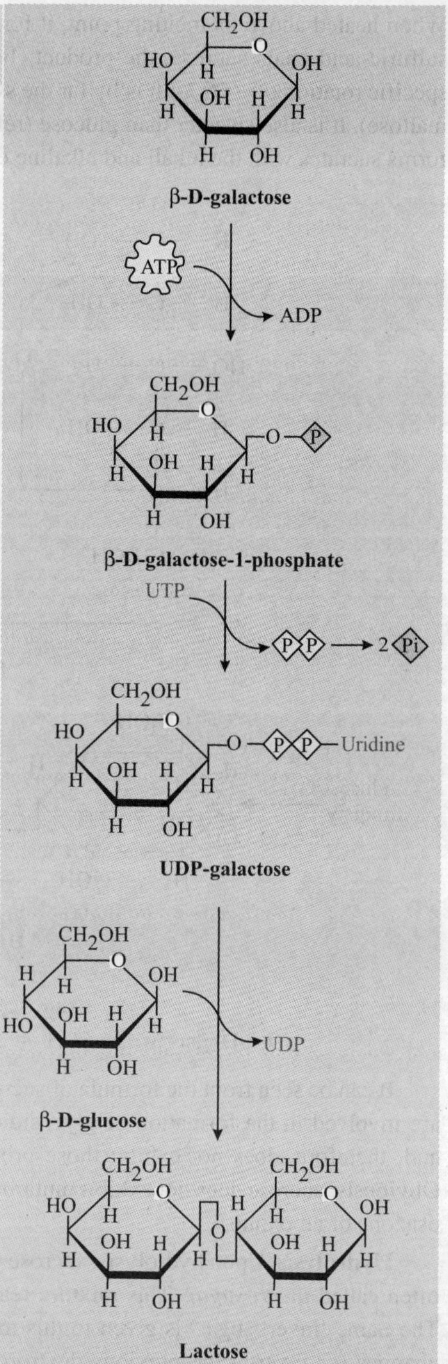

Fig. 7–1 Formation of lactose *in vivo*

The reaction shown occurs in the formation of milk in mammary tissue. Galactose is phosphorylated by ATP, then transferred to uridine diphosphate (UDP). UDP-galactose transfers galactose to glucose, with the simultaneous cleavage of a phosphate bond. The reaction is catalyzed by the enzyme *lactose synthase*.

When heated above its melting point, it forms a brown substance known as *caramel*. Concentrated sulfuric acid chars sucrose, the product being almost pure carbon. It is dextrorotatory and has a specific rotation of + 66.7°. It is by far the sweetest of the 3 common disaccharides (sucrose, lactose, maltose). It is also sweeter than glucose (refer Table 7–1). It is readily fermented to, *e.g.*, ethanol. It forms sucrates with the alkali and alkaline earth metals.

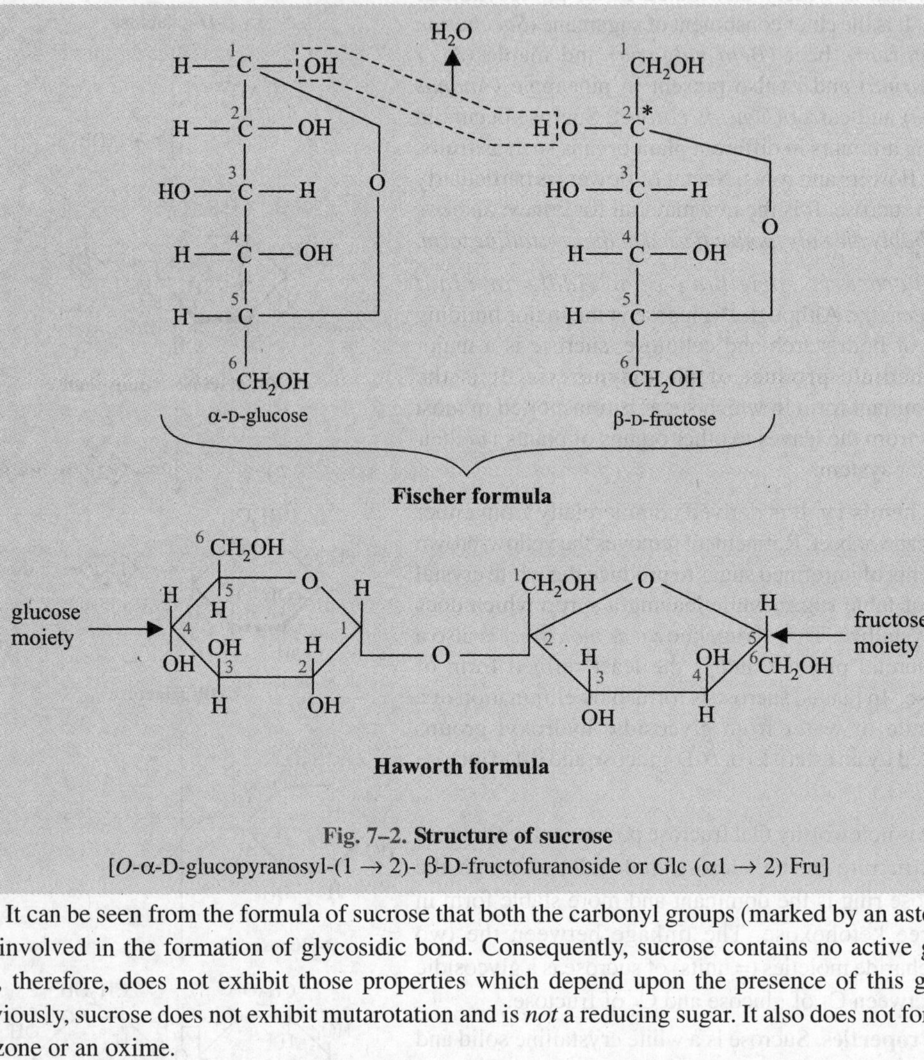

Fig. 7–2. Structure of sucrose

[*O*-α-D-glucopyranosyl-(1 → 2)- β-D-fructofuranoside or Glc (α1 → 2) Fru]

It can be seen from the formula of sucrose that both the carbonyl groups (marked by an asterisk) are involved in the formation of glycosidic bond. Consequently, sucrose contains no active group and, therefore, does not exhibit those properties which depend upon the presence of this group. Obviously, sucrose does not exhibit mutarotation and is *not* a reducing sugar. It also does not form an osazone or an oxime.

Hydrolysis. Upon hydrolysis, sucrose yields equimolar mixture of glucose and fructose which is often called *invert sugar*. This mixture readily reduces Fehling's solution and other alike reagents. The name 'invert sugar' is given to this mixture because the levorotatory fructose, thus produced, changes (or inverts) the previous dextrorotatory action of the sucrose (hence, the nomenclature–'invertase'). In fact, sucrose (which is dextrorotatory with a specific rotation, $[\alpha]_D^{20}$ of + 66.7°), upon hydrolysis, gives a mixture of equimolar quantities of D(+) glucose (dextrorotatory ; $[\alpha]_D^{20}$ of + 52.7°) and D(–) fructose (levorotatory; $[\alpha]_D^{20}$ of – 92.0°). And as the levorotation of fructose is greater than the dextrorotation of glucose, the mixture so obtained is levorotatory, contrary to the initial dextrorotatory sucrose.

Table 7–1. Sweetness index* of some sugars and noncaloric sweeteners

Sugar / Sweetener	*Relative sweetness*
Sugars	
Lactose	0.16
Raffinose	0.20
D-galactose	0.32
Maltose	0.32
D-xylose	0.40
Glycol	0.50
D-glucose	0.74
Sucrose	*1.00*
Invert sugar	1.23
D-fructose	1.73
Sweeteners	
Sucaryl sodium	30
Aspartame	180
Acesulfame-K	200
Stevia extract	300
Saccharin	400
Neohespiridine dihydrochalcone	1,000
Monellin	2,000

* Sweetness index is ascertained with respect to sucrose whose sweetness level is arbitrarily taken as one.

$$\underset{\text{Sucrose}}{C_{12}H_{22}O_{11}} + \underset{\text{Water}}{H_2O} \xrightarrow{\text{Invertase}} \underset{\text{Glucose}}{C_6H_{12}O_6} + \underset{\text{Fructose}}{C_6H_{12}O_6}$$

Sucrose: $[\alpha]_D^{20} = +66.7°$ dextrorotatory; Glucose: $[\alpha]_D^{20} = +52.7°$ dextrorotatory; Fructose: $[\alpha]_D^{20} = -92.0°$ levorotatory

Mixture of glucose and fructose is levorotatory with $[\alpha]_D^{20}$ value of (92.0 – 52.7)= – 39.3°.

This reaction which is called "inversion of sucrose" is not only catalyzed by the enzyme invertase but also by H^+ ions. Animals cannot absorb sucrose as such, but it is made available for absorption by the enzyme sucrase present in the intestinal mucosal cells. As already pointed out, this enzyme catalyzes the hydrolysis of sucrose to D-glucose and D-fructose, which are readily absorbed in the bloodstream.

It is interesting to note that the invert sugar is more sweeter than the sucrose itself owing to the presence of fructose in invert sugar which is sweetest of all the sugars. This also explains why honey which contains a large proportion of invert sugar is sweeter than sucrose. The invert sugar is used in candy both because of its sweetness and also because the monosaccharides do not crystallize. It is also used in making syrups for the confectionery trade.

Invertase is also called *sucrase, saccharase* or *6-D-fructosidase* the first name being the most preferred one. This enzyme has been found to occur in green leaves, fruits, grains, stems, potato tubers, some roots, pollen and such lower plants as fungi and bacteria. It is especially abundant in the yeast *Saccharomyces cerevisiae* where it is anchored in the periplasmic space (between a plasma membrane and cell wall). Salts of heavy metals (Ag, Cu, Hg) inhibit its action. Maximum activity is obtained with low concentrations of sucrose (5 –10%). Sucrase also hydrolyzes the sugars gentianose, raffinose, stachyose and, to some extent, inulin.

THE ARTIFICIAL OR SYNTHETIC SWEETENERS

Sweeteners are customarily grouped into two categories, based on the difference in the amount of energy provided by them. These two groupings are :

A. Nutritive or **Caloric Sweeteners.** These contain calories, hence have nutritive value. These include sugar sweeteners (*e.g.*, refined sugars, high fructose corn syrup, crystalline fructose, glucose, dextrose, corn sweeteners, honey, lactose, maltose, invert sugars, concentrated fruit juice) and sugar alcohols.

B. Nonnutritive or **Noncaloric Sweeteners.** These contain no calories, hence have no nutritive value. These have been developed artificially, hence also called as **artificial sweeteners.** As they sweeten with little volume, they may also be referred to as **high-intensity sweeteners.** They have been developed especially for obese or diabetics, for whom sugar consumption is harmful. Artificial sweeteners stimulate the taste buds of the tongue that are stimulated by sugars but, unlike sugars, they are of no food value and may have harmful effects, if used to excess. Some sweeteners are considered 'Generally Recognized As safe' (GRAS) ingredients and others are considered food additives. The safety limit of food additives or conditions of use is expressed as the acceptable daily intake (ADI), that is, the estimated amount per kilogram body weight that a person can safely consume every day over a lifetime without risk. ADI is a conservative level – it usually reflects an amount 100 times less than the maximum level at which no observed adverse effects occur in animals.

A description of some important artificial sweeteners (Fig. 7–3) follows:

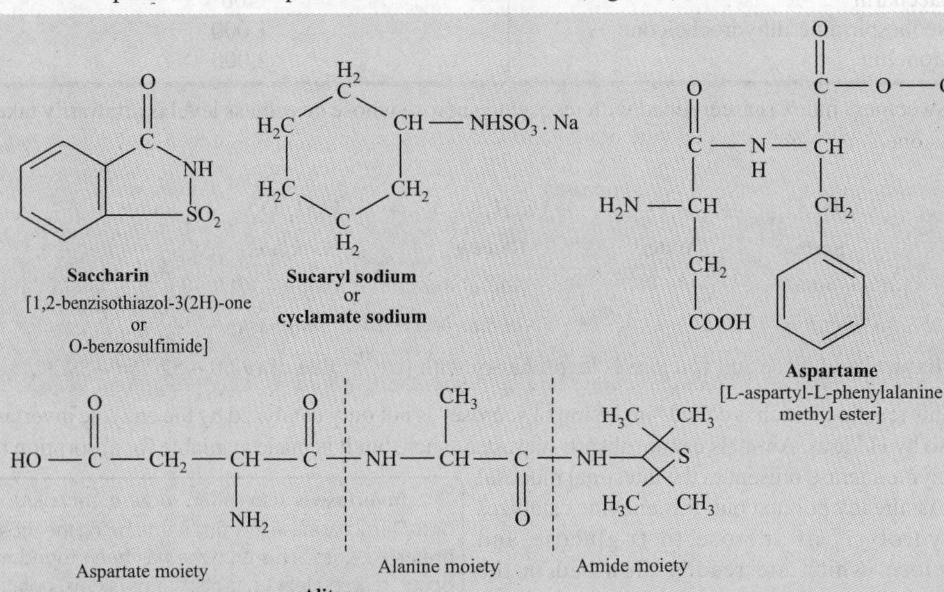

Fig. 7–3. Some artificial sweeteners

1. Saccharin. Saccharin is the first artificial sweetener which was discovered accidentally by **Constantine Fahlberg** in 1879. *It is the most-widely used artificial sweeteners*. In fact, saccharin was not used extensively as a sweetener until the advent of World War I when the use of sugar was strictly rationed. When World War II hit, sugar rationing started again, leading to another significant rise in saccharin use. Saccharin is sold under the tradename Sweetex™. It has a molecular formula $C_6H_4SO_2$ CONH, and its chemical structure is different from that of a carbohydrate. It possesses no OH group but rather a sulfonamide. It is manufactured from toluene, $C_6H_5.CH_3$. Saccharin is a white, crystalline, sparingly soluble solid with a m.p. 227 °C. It is heat-stable and is 400 times sweeter than sucrose by weight in aqueous solutions (Table 7–1). It is normally used as its soluble sodium salt. Not absorbed by the body, it is used by diabetics and in low-calorie dietetic foods. There is some concern that it may be carcinogenic. It is advised that pregnant women avoid saccharin.

> While working on new food preservatives, **Fahlberg** accidentally spilled a compound he had synthesized. When he ate his dinner that night, he noticed that the intense residual sweetness of the chemical still lingered on his hands. He named the compound saccharin after the Latin ***saccharum*** which means sugar, referring sweetness in this context.

2. Sucaryl sodium (or **cyclamate sodium).** It is the sodium salt of *N*-cyclohexylsulfanic acid, itself obtained by sulfonation of cyclohexylamine. It is 30 times sweeter than sucrose. It was formerly used as a sweetening agent in soft drink etc, until its use was banned when it was suspected of causing cancer.

3. Monellin. This protein sweetener is present in the sap of serendipity berries, fruit of a western African plant, *Dioscoreophyllum cumminsii*. This sweet protein is composed of two noncovalently-associated polypeptide chains (one having 42 amino acid residues and the other 50 residues) and is 2,000 times as sweet as sucrose. Monellin's sweetness requires the undissociated protein since neither protomer has a sweet taste. Presently, monellin is being studied as a nonfattening, nontoxic food sweetener for human use. In case of the African plant *Katemfe*, its intense sweetness is attributed to two proteins, which together are 1,600 times sweeter than sucrose.

4. Aspartame. This is an essentially noncaloric, sugar-free sweetener being commercially used (Technically, aspartame is nutritive as it contains 4 calories per gram, but since it is added in such tiny doses that it really makes little, if any, contribution to the total calorific intake). Aspartame is a synthesized dipeptide, containing aspartic acid (Asp) and the methyl ester of phenylalanine (Phe). Aspartame is marketed under the tradenames Nutrasweet™ (when included in food) and Equal™ (when sold as a powder). It is about 180 times sweeter than sucrose by weight (*i.e.*, about half as sweet than saccharin) and *does not have the bitter aftertaste, often associated with saccharin.* It, thus, has a better taste profile than saccharin. Interestingly, (S–S)-aspartame is a sweetener whereas its enantiomer, (S–R)-aspartame is bitter in taste. However, aspartame has a shelf life of about 6 months, after which it breaks down into its components and loses its sweetening power. Also, it splits at high temperatures (which means that it cannot be used in cooking or in hot or baked items). Intestinal enzymes hydrolyze aspartame to phenylalanine, aspartic acid and methanol in the ratio 5 : 4 : 1.

It is now widely used in the drink and food industry, *e.g.*, soft drinks, cold cereals, beverages, gelatin desserts, puddings, toppings and fillings for baked goods, and cookies. Aspartame has an additional advantage as it does not promote tooth decay, hence also used in gum and candies. Some users have reported suffered from adverse reactions to aspartame such as headache, dizziness, seizure and nausea. Such persons must abstain from using it. Because aspartame is made from phenylalanine, people with phenylketonuria (who cannot metabolize Phe) should avoid aspartame. Also, the pregnant women are advised to refrain from aspartame. Aspartame is approved for use in more than 100 nations.

5. Alitame. Alitame, also a peptide, is composed of *l*-aspartic acid, *d*-alanine and a novel C-terminal amide moiety. It is 2,000 times sweeter than sucrose without the bitter and metallic qualities of noncaloric sweeteners. Curiously, although alitame tastes sweet to most people, it tastes bitter to a miniscule part of population. This may be due to a genetic difference in taste perception. Alitame blends with other artificial sweeteners to maximize the quality of sweetness.

6. Neohespiridine dihydrochalcone. It is a carboxylic derivative product from bitter components of the grapefruit's skin. It is about 1,500–1,800 times sweeter than sucrose at threshold limit concentration but the intensity depends on many factors such as pH and the product to which it is added. It takes some time to reach maximum sweetness perception and a very light sweet methanol-like aftertaste. It has a synergistic sweetening effect when combined with sugar alcohols, and with aspartame, saccharin, acesulfame-K and cyclamate. This characteristic has many advantages:

(*a*) reduction in cost.

(*b*) reduction in the daily intake of any particular sweetener

(*c*) a more satisfying sucrose-like taste.

It may be used in beverages, juices, dairy products and alcoholic drinks such as beer. It is also used to sweeten diet and low-calorie food.

7. Sugar alcohols. Polyalcohols (or polyols) are part of a group of compounds called sugar alcohols. These include sorbitol, manitol and xylitol. Polyols can also be categorized as *'sugar replacers'* because they can replace sugar sweeteners, usually on a one-to-one basis. They offer less energy and offer potential health benefits (*e.g.*, reduced glycemic response, reduced dental caries risk, etc). Commercially, these sweeteners are synthesized and not extracted from natural sources. Sugar alcohols contain 2-3 calories per gram but are only slowly metabolized to glucose. These are found in sugar-free candy, gum, cake mix and syrup, as well as some medications. Like aspartame, they also do not promote tooth decay. However, sugar alcohols can cause gastrointestinal symptoms ranging from mild stomach discomfort to severe diarrhea. Hence, those products likely to cause ingestion of 50 g or more of sorbitol are required to carry a level reading,

"Excess consumption may have a laxative effect".

8. Acesulfame potassium (=Acesulfame-K). It is the newest alternative sweetener approved in the US and is marketed under the tradenames Sunette™ and SweetOne™. It was formulated in the late 1960s and was evaluated for safety in 1983. Food and Drugs Administration (FDA) approved acesulfame-K in 1988. Acesulfame-K is about 200 times sweeter than sucrose. Blends of acesulfame-K with other nutritive and nonnutritive sweeteners can synergize the sweetness potential and the bitter taste. It has a shelf life of about 3-4 years as it is heat-stable (hence can be used for baked items). It does not provide any energy as it is not metabolized by the body and is excreted in the urine unchanged. Acesulfame-K finds applications in gum, powdered drink mixes, gelatins, puddings, candy, throat lozenges, tabletop sweeteners, yogurt and nondairy creamers. It is also used as a sweetener in soft drinks, chewing gum, coffee, tea, flavouring, salted food and dairy items.

9. *Stevia* extract. The plant sweet honey leaf *(Stevia rebaudiana)* is a herbaceous (60–70 cm in height), highly branched perennial and grows best between 11 and 41°C. It is native to Brazil, Venezuela, Columbia and Paraguay where the native Guarani have used it for over 1,500 years to sweeten otherwise unpalatable medicinal drinks. The plant is now being cultivated commercially in Japan, Korea, Taiwan and the United States; in India, *Stevia* is being grown in the state of Maharashtra. *Stevia* is planted in the months of February and March or else in September, although the seedlings may be developed by tissue culture. Once planted, the produce (which consists of leaves) can be harvested continuously for 5 years, cutting the leaves 5 times in a year. It is a natural herbal sweetener. The sweetness of *Stevia* is attributed to 2 compounds, *stevioside* and *rebaudioside*, which can be up to 300 times sweeter than sucrose. *Stevia* has an advantage over artificial sweeteners in that it is stable at high temperatures and a pH range of 3 to 9. *Stevia* extract is used as sweetener or flavour enhancer in many countries such as China, Japan, Korea, Israel, Brazil and Paraguay. It is also used in soft drinks, ice creams, cookies, pickles, chewing gum, tea and skin care products. *Stevia* plant and its extract both are used in weight-loss programmes because of their ability to reduce the cravings for sweet and fatty foods. Sweet honey leaf is, however, banned in food products in the US because a derivative of stevioside may be harmful to humans.

Thus, mankind's search for the ultimate sweetener goes on. In fact, sweetness is rather subjective measure of taste. The sensation of sweetness is believed to be due to the interaction between molecules of the sweetening agent and receptor sites of the tongue. Furthermore, perception of sweet taste can be influenced by genetic behaviour, health status and aging. **Robert Margolskee** of the Mount Sinai of Medicine in New York hopes to have definite proof that gene TIR3 is truly the "sweet tooth" receptor gene. He feels that the present artificial sweeteners were developed at random and by chance. However, by having the receptor protein that normally binds sugars and sweeteners, one should be able to target more specifically better-designed molecules that will bind and activate the receptor.

LACTOSE
(Milk sugar)

Occurrence. Lactose is solely of animal origin and is found in the milk of mammals. Human milk contains about 6.8 % (and 0.3% of higher oligosaccharides); cows' milk, about 4.8% (refer Table 7-2); hence, mother's milk is about 1.5 times sweeter than cow's milk. It is an unique product of mammary glands and is not found in other parts of the animal body. A minute quantity of lactose is also present in the milk as *neuramin lactose*. However, during pregnancy, it may be found in urine. Interestingly and curiously enough, there is no lactose in the milk of seals and other close relatives.

> **Kuhn** (1949) has, however, found lactose to be present in the pollen of *Forsythia* plant.

Table 7–2. Percentage composition of milk

Contents	*Human milk*	*Cow's milk*
Water	88.5	87.0
Fat	3.3	3.5
Lactose	6.8	4.8
Casein	0.9	2.7
Lactalbumin + other proteins	0.4	0.7
Ash (containing minerals)	0.2	0.7

Chemistry. Lactose is prepared commercially from whey by evaporation to crystallization. Whey is obtained as a by-product in the manufacture of cheese. Lactose may be formed by elimination of a molecule of water from the glycosidic OH group of β-D-galactose and the alcoholic OH group on carbon atom 4 of D-glucose (Fig. 7-3). As D-glucose exists in 3 forms (α, *aldehydo* and β), it is obvious to have 3 forms of lactose *viz.*, α form, *aldehydo* form and β-form. The aldehydo form is present in traces only. The formula shown in Fig. 7–3 is that of α-lactose.

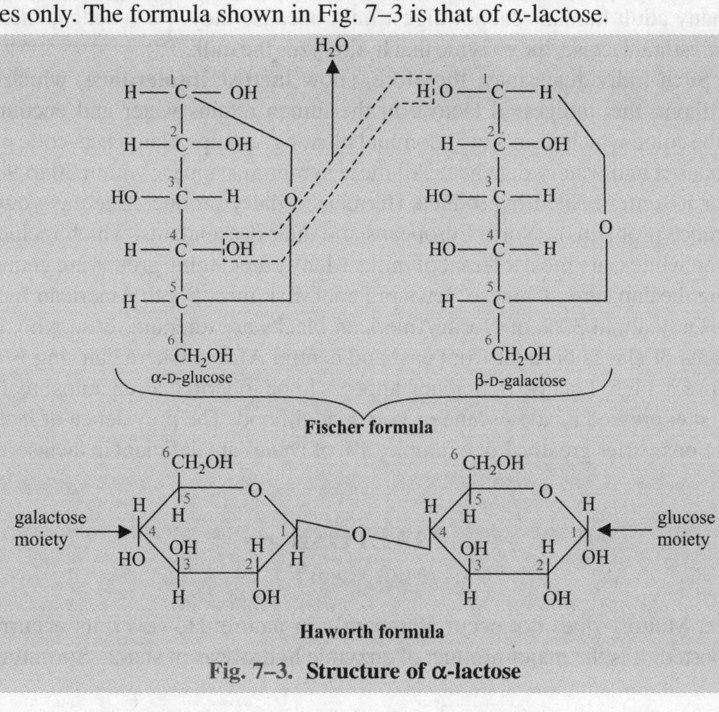

Fig. 7–3. Structure of α-lactose

Lactose, in solution, consists of an equilibrium mixture of all the 3 forms. Using the simplified scheme of representing the Haworth formula, the 3 forms may be shown as below:

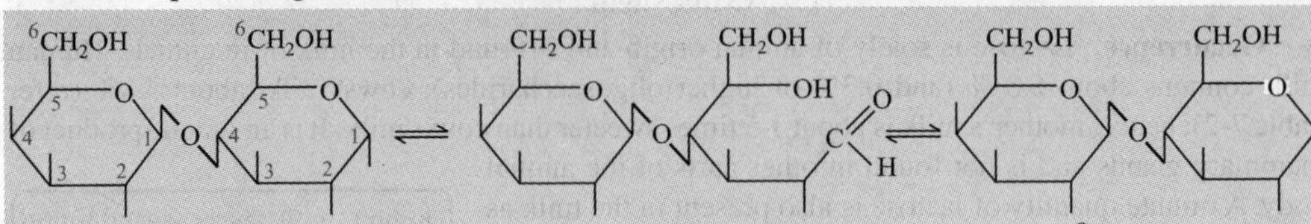

It may be emphasized that the α- and β-forms of lactose differ from each other only in hemiacetal configuration; the linkage in both the forms being β-1,4-galactoside.

Properties. Lactose is a white, crystalline solid with a melting point 203°C (with decomposition) and is also dextrorotatory. The α- and β-forms have a specific rotation of + 90° and +35° respectively. The equilibrium mixture has a specific rotation +52.5°. It is less soluble in water and much less sweet than sucrose. The α-diastereoisomer is less soluble than the β-form. Sometimes, the α-isomer crystallizes in ice-cream, making the product seem sandy in texture. The more soluble β isomer is used in diets for infants. Since lactose has a free carbonyl group on carbon atom 1 of glucose unit, it is a reducing sugar. Hence, it exhibits mutarotation, reduces Cu^{++} ions to Cu^{+} ions and forms osazone and oxime. It crystallizes in rhombic prisms with a mole of water.

Hydrolysis. Upon hydrolysis with the enzyme, *lactase*, it yields an equimolar mixture of glucose and galactose. It should be noted that lactase is a β-glycosidase, *i.e.*, splits β-glycosides (hence, it is identical with emulsin). Lactose itself cannot be absorbed from the intestine into the bloodstream unless it is first hydrolyzed to the component monosaccharide units. The enzyme *lactase* is actively secreted by the intestinal mucosal cells in suckling infants and therefore the lactose, in the form of milk suckled by them, is easily hydrolyzed to yield glucose and galactose, thus effecting absorption by the intestinal tract. Although milk is the universal food of newborn mammals and one of the most complete human foods, many adult humans cannot digest milk because they are deficient in intestinal lactase, the enzyme that hydrolyzes the milk sugar, lactose. Such individuals may, therefore, show **lactose intolerance,** which is genetically determined. In them, the undigested lactose in the lumen retains water and encourages bacterial fermentation in the colon which results in abdominal bloating, cramps, flatulence, colic pains, abnormal intestinal flow, nausea and watery diarrhea. All these symptoms appear within 30 to 90 minutes after ingesting milk or its unfermented byproducts (fermented dairy products such as yogurt and cheese create no intolerance problems). North Europeans and their descendants, which include the majority of North American whites, are most tolerant of milk. Many other ethnic groups are generally intolerant to milk, including the Japanese, Chinese, Jews in Israel, Eskimos, South American Indians and most African blacks. Only about 30% of North American blacks are tolerant; those who are tolerant are mostly descendants of slaves brought from east and central Africa, where dairying is traditional and tolerance to lactose is high. Lactase deficiency appears to be inherited as an autosomal recessive trait and is usually first expressed in adolescene or young adulthood. The prevalence of lactase deficiency in human populations varies greatly. For example, 3% of Danes are deficient in lactase, compared with 97% of Thais.

> **Lactose intolerance** should not be confused with the genetic *galactosemia*.

MALTOSE
(Malt sugar)

Occurrence. Maltose does not occur abundantly in nature. However, its occurrence has been occasionally reported. It is the major product of enzymic hydrolysis of starch. Sprouting cereal grains are rich in *amylases* which split the starch present to dextrins and maltose. Malt, prepared from sprouting barely, is an excellent source of maltose.

Chemistry. Maltose is the simplest of the disaccharides and is produced by the action of malt (which contains the enzyme *diastase*) on starch:

$$(C_6H_{10}O_5)_n + \frac{n}{2}H_2O \xrightarrow{\text{Diastase}} \frac{n}{2}C_{12}H_{22}O_{11}$$

Maltose may be considered as originating by splitting out a molecule of water from the glycosidic OH group of α-D-glucose and the alcoholic OH group on carbon atom 4 of D-glucose (Fig. 7–4).

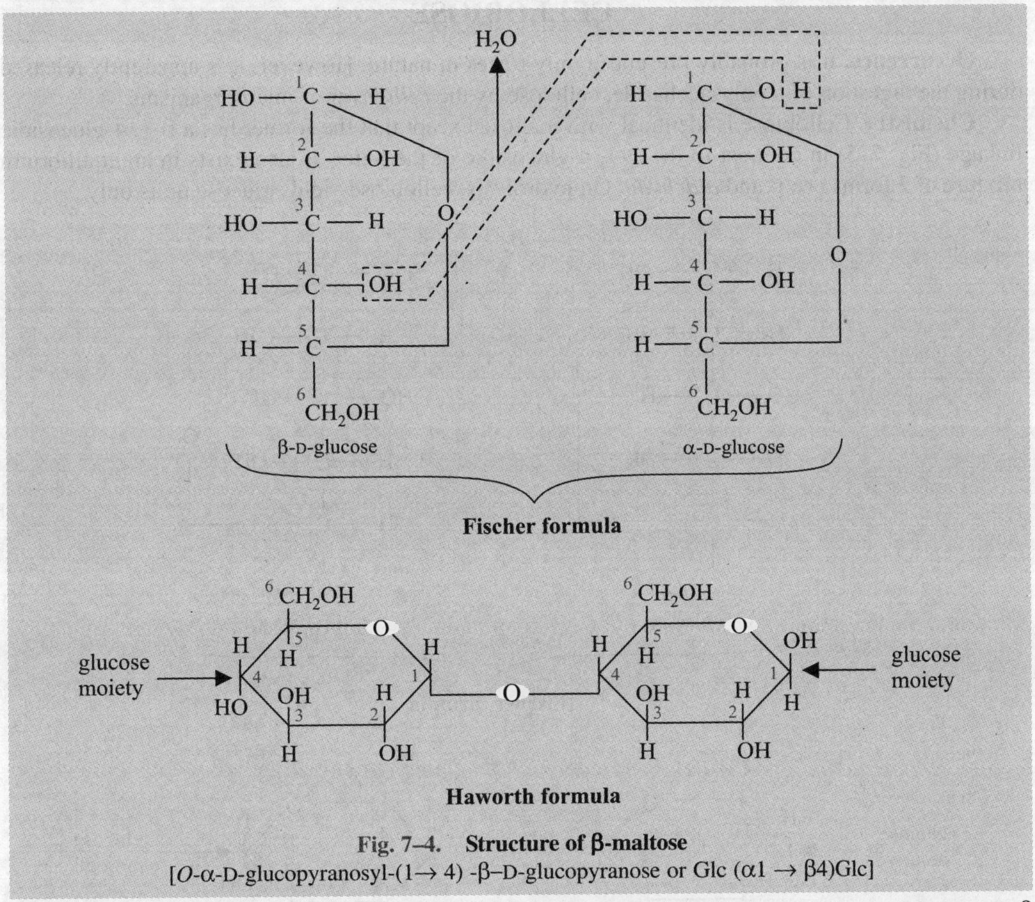

Fig. 7–4. Structure of β-maltose
[*O*-α-D-glucopyranosyl-(1→ 4) -β–D-glucopyranose or Glc (α1 → β4)Glc]

Maltose, like lactose, has one free hemiacetal group. Consequently, it too exists in 3 forms ; α, β and *aldehydo*. Maltose is usually found in β-form. Maltose, in solution, also exists as an equilibrium mixture of all the 3 forms. It may be noted that the linkage in both α- and β-forms of maltose is always an α-1,4-glucoside.

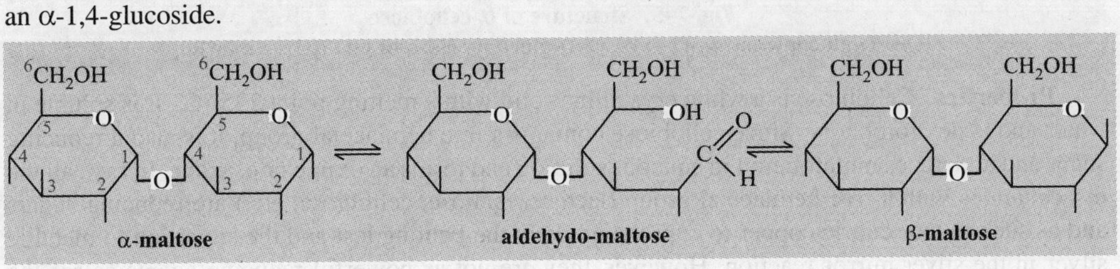

Properties. Maltose is a white crystalline solid, with a melting point 160–165 °C. It is soluble in water and is dextrorotatory. Because of the free aldehyde group, maltose is also a reducing sugar and forms osazone with phenyl hydrazine. It is also capable of exhibiting mutarotation.

Hydrolysis. Maltose is easily hydrolyzed into 2 identical units of glucose by dilute acids or by the enzyme, *maltase*, found in the intestine. It should also be noted that the enzyme maltase hydrolyzes or splits only α-glycoside linkages and has no action upon β-glycosides. It has, hence, been used to ascertain the presence of this bond in disaccharides and polysaccharides. Hydrolysis may also be brought about by the enzyme, *diastase* (= *amylase*), found in sprouting barley.

CELLOBIOSE

Occurrence. It is probably present in only traces in nature. However, it is apparently released during the digestion of a polysaccharide, cellulose by the *cellulases* of microorganisms.

Chemistry. Cellobiose is identical with maltose except that the former has a β-1, 4-glucosidic linkage (Fig. 7–5) in contrast to the α-1, 4-glucosidic of the latter. It also exists in an equilibrium mixture of 3 forms : α, β and *aldehydo*. On hydrolysis, cellobiose yields glucose units only.

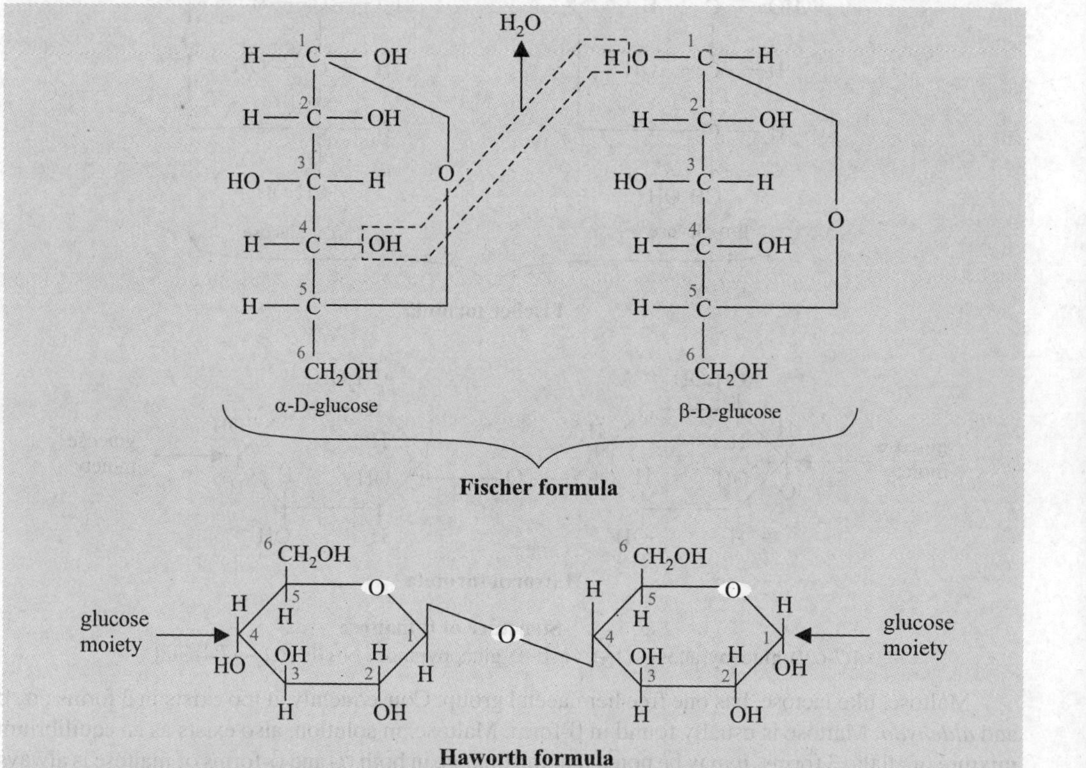

Fig. 7–5. Structure of α-cellobiose
[*O*-β-D-glucopyranosyl-(1→ 4) -α-D-glucopyranose or Glc (β1 → α4)Glc]

Properties. Cellobiose is a white crystalline solid with a melting point 225 °C. It is soluble in water and is dextrorotatory. Since cellobiose contains a free hemiacetal group, it is also a reducing sugar and undergoes mutarotation in aqueous solution and forms an oxime or osazone. In fact, all the disaccharides with a free hemiacetal group (lactose, maltose, cellobiose, etc.) are reducing sugars and as such reduce cupric copper to cuprous oxide in the Fehling test and the silver ion to metallic silver in the silver mirror reaction. However, they are not as powerful reducing agents as are the monosaccharides glucose, galactose and fructose, and do not readily reduce Barfoed's reagent to cuprous copper. The reason for this difference is that while C-1 of one of the two glucose residues is free to revert back to the carbonyl (= reducing) form, that on the second residue is stabilized in the cyclic form by the existence of the glycosidic linkage.

Hydrolysis. When hydrolyzed with dilute acids or by the enzyme *emulsin,* cellobiose yields identical units of D (+)-glucose. It may be noted that the enzyme emulsin splits only β-glycosidic linkages (*cf* maltase).

ISOMALTOSE

It is similar to maltose except that it has an α-1, 6-glucoside linkage (Fig. 7–6).

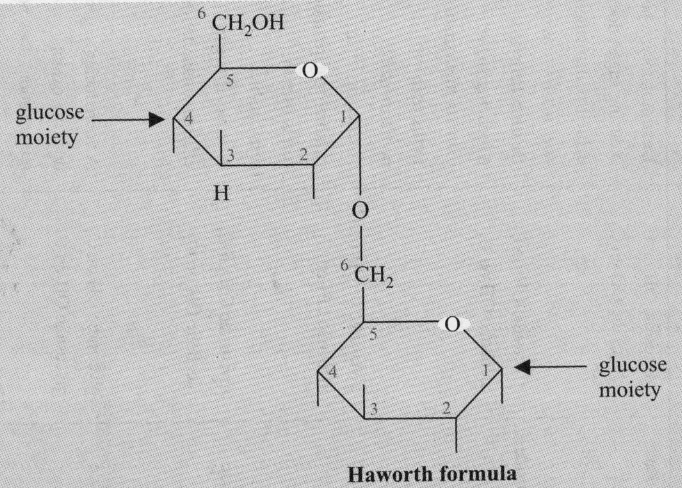

Fig. 7–6. Structure of isomaltose

[*O*-α-D-glucopyranosyl-(1→ 6) -α–D-glucopyranose or Glc (α1 → α6)Glc]

TREHALOSE

Trehalose is a major constituent of the circulating fluid (hemolymph) of insects, in which it serves as an energy storage compound. It is also found in yeasts and young mushrooms (*e.g., Amanita muscaria*). Here, the two anomeric carbon atoms of the two α-D-glucose moieties connect each other, the linkage thus being α-Glc(1→1)α-Glc (Fig. 7–7). Consequently, it resembles sucrose in being a nonreducing sugar as it has no free aldehyde group. Trehalose does not form an osazone. On hydrolysis with a special disaccharidase *trehelase*, it yields glucose. It is hydrolyzed by HCl to 2 moles of D-glucose. It generally replaces sucrose in plants lacking chlorophyll and starch.

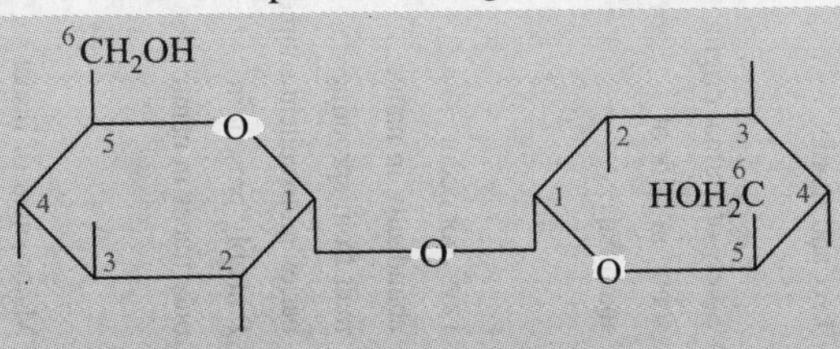

Fig. 7–7. Structure of trehalose

[*O*-β-D-glucopyranosyl-(1→4)-α-D-glucopyranose or Glc(β1→α4)Glc]

Two isomeric forms of trehalose are recognized:

(*a*) *Isotrehalose* (β-D-glucopyranosyl-β-D-glucospyranoside)

(*b*) *Neotrehalose* (α-D-glucopyranosyl-β-D-glucopyranoside)

A summary chart of the important disaccharides (Table 7–3) is given on the next page.

Table 7-3. Summary of occurrence, structure and properties of some important disaccharides

Name	*Occurrence*	*Constituent monosaccharides**	*Oxide linkage involved*	*Groups involved in linkage*	*Properties*	*Hydrolysis*
Sucrose (Cane sugar)	Solely of plant origin; found in sugaracane, beet and maple	α-D-glucose and β-D-fructose	glycosidic linkage between C_1 of glucose and C_2 of fructose	glycosidic OH on C_1 glycosidic OH on C_2	Dextrorotatory; a non-reducing sugar; does not exhibit mutarotation; does not form osazone;	Yields an equimolar mixture of glucose and fructose on hydrolysis with invertase
Lactose (Milk sugar)	Solely of animal origin; found in milk of mammals	β-D-galactose and D-glucose	β-1, 4-galactoside	glycosidic OH on C_1 alcoholic OH on C_4	Dextrorotatory; a reducing sugar; exhibits mutarotation; forms osazone; nonfermentable.	Yields an equimolar mixture of galactose and glucose on hydrolysis with lactase
Maltose (Malt sugar)	Does not occur abundantly in nature; malt from sprouting barley an excellent source	α-D-glucose and D-glucose	α-1, 4-glucoside	glycosidic of OH on C_1 alcoholic OH on C_4	A reducing sugar; exhibits mutarotation; forms osazone; fermentable	Yields identical units of glucose on hydrolysis with maltase or diastase
Cellobiose	Probably present in only traces in nature	β-D-glucose and D-glucose	β-1, 4-glucoside	glycosidic OH on C_1 alcoholic OH on C_4	A reducing sugar; exhibits mutarotation	Yields identical units of glucose on hydrolysis with β-glycosides or with dilute acids
Trehalose	Major sugar of insect hemolymph; also occurs in fungi and yeasts	α-D-glucose and α-D-glucose	α-1, 1-glucoside	glycosidic OH on C_1 glycosidic OH on C_1	A nonreducing sugar; does not exhibit mutarotation; does not form osazone	Yields glucose units on hydrolysis

* All the constituent monosaccharides are present in pyranose ring form except β-D-fructose which has a furanose ring structure.

REFERENCES

See list following Chapter 8.

PROBLEMS

1. The hydrolysis of sucrose (specific rotation + 66.5°) yields an equimolar mixture of D-glucose (specific rotation + 52.5°) and D-fructose (specific rotation –92°).
   (*a*) Suggest a convenient way to determine the rate of hydrolysis of sucrose by an enzyme preparation extracted from the lining of the small intestine.
   (*b*) Explain why an equimolar mixture of D-glucose and D-fructose formed by hydrolysis of sucrose is called invert sugar in the food industry.
   (*c*) The enzyme invertase (its preferred name is now sucrase) is allowed to act on a solution of sucrose until the optical rotation of the solution becomes zero. What fraction of the sucrose has been hydrolyzed?
2. The manufacture of chocolates containing a liquid centre is an interesting application of enzyme engineering. The flavored liquid centre consists largely of an aqueous solution of sugars rich in fructose to provide sweetness. The technical dilemma is the following : the chocolate coating must be prepared by pouring hot melted chocolate over a solid (or almost solid) core, yet the final product must have a liquid, fructose-rich centre. Suggest a way to solve this problem. (**Hint:** The solubility of sucrose is much lower than the solubility of a mixture of glucose and fructose.)
3. Although lactose exists in two anomeric forms, no anomeric forms of sucrose have been reported. Why?
4. A sample of disaccharide is either lactose or sucrose. No reddish precipitate forms in Fehling's reaction, unless the compound is first warmed in dilute acid. Is it lactose or sucrose? Explain.
5. The carbohydrate portion of some glycoproteins may serve as a cellular recognition site. In order to perform this function, the oligosaccharide moiety of glycoproteins must have the potential to occur in a large variety of forms. Which can produce a larger variety of structures: oligopeptides composed of five different amino acid residues or oligosaccharides composed of five different monosaccharide residues? Explain.
6. Sucrose is commonly used to preserve fruits. Why?
7. Although the disaccharide lactose exists in two anomeric forms, no anomeric forms of the disaccharide sucrose have been reported. Why?
8. In oriental and black adults, symptoms such as abdominal distention, cramps, pain and diarrhea develop on consumption of milk. Explain.
9. Why sucrose, which contains in its molecule glucose and fructose which are reducing sugars, is itself a nonreducing sugar?
10. Does sugar contain any nutrients?
11. Can sugar help heal a wound?
12. What are the benefits of using Stevia as a sweetener?

ANSWERS TO PROBLEMS

1. (*a*) Measure the change in optical rotation with time.
   (*b*) The optical rotation of the mixture is negative (inverted) relative to that of the sucrose solution.
   (*c*) 0.63.

2. Prepare a slurry of sucrose and water for the core; add a small amount of invertase; immediately coat with chocolate.
3. Sucrose is not a reducing sugar.
4. Lactose; in sucrose, the anomeric carbons of both glucose and fructose are involved in the glycosidic bond and are not available to reduce Fehling's reagent. Lactose is a reducing sugar that converts Fe^{3+} to Fe^{2+}, which precipitates as the red oxide.
5. Oligosaccharides; their subunits can be combined in more ways than the amino acid subunits of oligopeptides. Each of the several hydroxyl groups can participate in glycosidic bonds, and the configuration of each glycosidic bond may be either α or β. Furthermore, the polymer may be linear or branched.
6. In sucrose, both the carbonyl groups of its 2 moieties (*i.e.*, α-D-glucose and β-D-fructose) are involved in the formation of glycosidic bond. Consequently, sucrose contains no active group and is, hence, well suited for preservation.
7. Both the active carbonyl groups (present on C_1 of α-D-glucose and C_2 of β-D-glucose) are involved in the formation of glucosidic bond. Hence, no anomeric forms of sucrose are available. Thus, sucrose is not a reducing sugar because of the absence of the active group.
8. This is due to a substance, lactose. The ability to secrete the enzyme *lactase*, which hydrolyzes lactose, decreases with age. The unabsorbed lactose passes through the small intestine to the large intenstine, where it is fermented by intestinal bacteria causing distention, pain and diarrhea.
9. In the formation of a sucrose molecule, the reducing aldehyde group of glucose and the ketonic group with adjacent hydroxyl group of fructose are both involved and result in the formation of an oxide linkage. Hence, sucrose is a nonreducing sugar. But the hydrolysis of sucrose breaks the glycosidic oxygen bond, releasing free glucose and fructose.
10. Refined white sugar is devoid of any nutrients. *Molasses* is a natural sweetener that is very nutritious. It contains B vitamins, calcium, iron, magnesium, potassium and chromium as well as other micronutrients. It may be tried as a substitute for sugary pancake syrups.
11. Eating sugar will not help heal a wound, of course. But sprinkling it on the wound will. Sugar is hygroscopic, which means that it soaks up moisture in the wound. Bacterial growth is suppressed without moisture, thereby assisting in healing the wound.
12. Stevia comes from a natural source– the leaf of the *Stevia* plant, native to South America. Extensive testing has shown Stevia to be safe. It does not contain calories and does not raise insulin levels, so it is also safe for diabetics. Stevia may lower blood pressure and blood sugar. And it is safe for teeth.

CONTENTS

The glistening outer skeleton of this grasshopper is primarily made up of chitin, one of the most abundant biopolymer on the Earth.
(Courtesy: Robert and Linda Mitchell)

CHAPTER

8

Carbohydrates–IV

Polysaccharides

Polysaccharides (variously called as *glycanes* or *polyholosides* or *polyosides*) are high molecular weight carbohydrates which, on hydrolysis, yield mainly monosaccharides or products related to monosaccharides. They may also be regarded as **polymeric anhydrides of simple sugars**. D-glucose is the commonest component of polysaccharides. However, D-and L-galactose, D-mannose, D-xylose, L-arabinose as well as D-glucuronic, D-galacturonic, D-mannuronic acids, D-glucosamine, D-galactosamine and amino uronic acids also occur as constituents of polysaccharides. The various polysaccharides differ from one another not only in the composition of the constituent monosaccharide but also in the molecular weight, in the nature of the chain (whether linear or branched), in the type of glycosidic bond (whether α or β) and in the type of linkage (whether 1→2, 1→3, 1→4, or 10→6) involved in the respective monosaccharide units. *A great majority of carbohydrates of nature occur as polysaccharides.* Chemically, the polysaccharides may be distinguished into **homopolysaccharides** (or **homoglycanes**), which yield, on hydrolysis, a single monosaccharide and **heteropolysaccharides** (or

> The **small arrow** shows the direction of the glycosidic bond which is always from the hemiacetal hydroxyl to some alcohol or hemiacetal group of the following sugar.

heteroglycanes), which produce a mixture of monosaccharides on hydrolysis. They form ong, often branched molecules which are insoluble, osmotically inactive and unsweat.

Based on their functional aspect, the polysaccharides may be grouped under two heads :

(*a*) **Nutrient** (or **digestible**) **polysaccharides.** These act as metabolic reserve of monosaccharides in plants and animals, *e.g.*, starch, glycogen and inulin.

(*b*) **Structural** (or **indigestible**) **polysaccharides.** These serve as rigid mechanical structures in plants and animals, *e.g.*,cellulose, pectin and chitin and also hyaluronic acid and chondroitin.

The sequence of monomer residues in a polysaccharide, like that in polypeptides and polynucleotides, defines its **primary structure.** Whereas proteins usually have complicated sequences, polysaccharides often have rather simple primary structures. In some cases (***e.g.,*** cellulose), the polymer is made from only one kind of monomer residue (β-D-glucose, in case of cellulose); these kinds of polymers are, as pointed out earlier, referred to as homopolysaccharides. If 2 or more kinds of residues are involved, the polymer is called a heteropolysaccharide. Even those nutrient and structural polysaccharides that are heteropolymers are rarely complex; usually no more than two kinds of residues are involved. In further contrast to protein and nucleic acid molecules, which are almost always of defined length, polysaccharide chains grow to random lengths.

The functional reasons for these differences are not hard to find. A storage material, such as starch, needs neither to convey information nor to adopt a complicated 3-'D' form. It is simply a bin in which to put away glucose molecules for future use. Many structural polysaccharides (like structural proteins) form extended, regular **secondary structures,** well suited to the formation of fibres or sheets. Often a regular repetition of some simple monosaccharide or disaccharide motif will serve this function (See for comparison, the simple and repetitive amino acid sequences of collagen and silk fibroin, to be described in Chapter 9). The only saccharide polymers in which well-defined and complex sequences are found are some of the oligosaccharides attached to cell surfaces or those attached to specific glycoproteins. Because these oligomers serve to identify cells or molecules, they must convey information. This function requires precisely defined "words" in the polysaccharide language, just as nucleic acid sequences spell out information in their own language.

HOMOPOLYSACCHARIDES

These yield, on hydrolysis, a single monosaccharide. They serve as both storage (starch, glycogen, inulin) and structural (cellulose, pectin, chitin) polysaccharides.

STARCH

Occurrence. Starch occurs widely in plants, especially in roots, tubers, seeds and fruits, as a carbohydrate storage product and energy source. It is present in cereals, legumes, potatoes and other vegetables. It is found in colourless plastids (leucoplasts), in storage tissue and in the stroma of chloroplasts in plants. *More than half the carbohydrate ingested by humans is starch*. Sago starch is obtained from sago palm, *Meteroxylon rumphii* ; the arrowroot from *Maranta arundinacea* and the tapioca, a starchy food, from *Manihot utillissima*. It is usually present inside the plant cells as compact insoluble granules which may be spherical, lens-shaped or ovoid, and which are laid down in a series of concentric layers.

Chemistry. Natural starches consist of two components : *amylose* (15–20%), a long unbranched straight-chain component and *amylopectin* (80–85%), a branched chain polysaccharide. Potato and cereal starches are 20–30% amylose and 70–80% amylopectin. Starch from waxy corn is notable as it consists practically of amylopectin component, there being no amylose. Starch is made from corn, wheat, potatoes, rice and other cereals by various physical processes such as steeping, milling and sedimentation. Starch can be split into amylose and amylopectin by a commercial process based on selective solubilities.

α-amylose or simply **amylose** (Fig. 8–1) has a molecular weight range of 10,000 to 50,000. It may be formed in plant cells by elimination of a molecule of water from glycosidic OH group of one α-D-glucose molecule and alcoholic OH group on carbon 4 of the adjacent α-D-glucose molecule. The linkage in amylose is, thus, an α-1,4-glucoside, like that in maltose. Enzymic hydrolysis of amylose with *amylase*, henceforth, yields maltose units mainly. Amylose may be considered as an anhydride of α-D-glucose units.

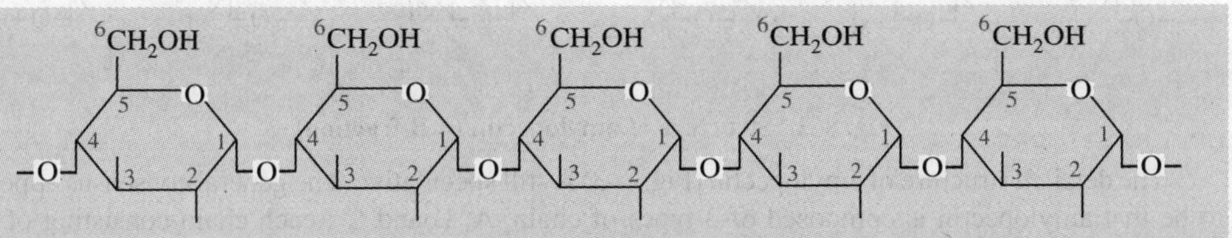

Fig. 8–1. Structure of amylose (= A-fraction)

Various types of starches differ in the amount of the amylose component present in them (refer Table 8–1).

Table 8–1. Amylose contents of some starches

Starch	*Amylose component (in %)*
Sugary mutant corn	70
Steadfast pea	67
Alderman pea	65
Buckwheat	28
Barley	27
Sorghum	27
Commercial corn	26
Wheat	25
White potato	23
Arrowroot	21
Sweet potato	20
Tapioca	18
Waxy barley	3
Waxy corn	0–6

(Adapted from Whistler RD and Smart CL, 1953)

β-amylose or **isoamylose** or **amylopectin** (Fig. 8–2) has a high molecular weight range of 50,000 to 1,000,000, thus indicating the presence of 300–5,500 linked glucose units per molecule. This possesses the same basic chain of α-1,4-glucoside linkage like that of amylose but has, in addition, many side chains attached to the basic chain by α-1,6-glucoside linkages, similar to those in isomaltose. It may, thus, be seen that the glucose unit, present at each point of branching, has substituents not only on carbon atoms 1 and 4 but also on carbon atom 6. In other words, these glucose units have 3 points of attachment to serve as branching points. The average chain length is about 24 glucose units. Amylopectin, upon incomplete hydrolysis, yields the disaccharide isomaltose.

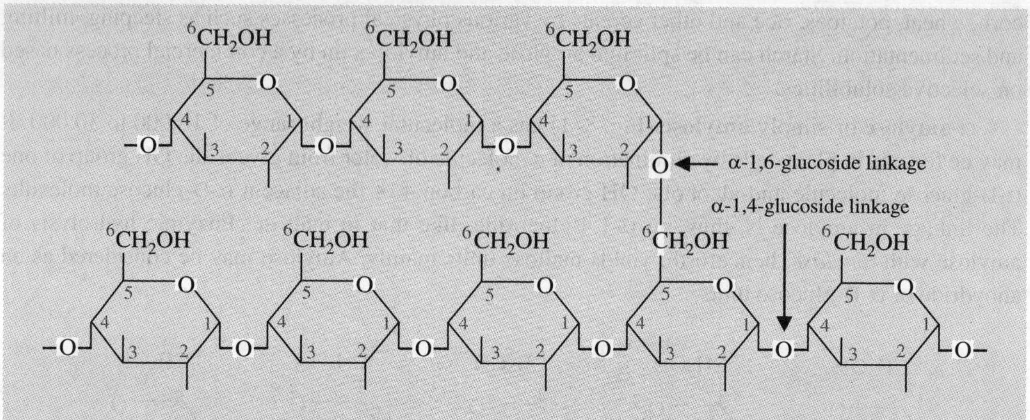

Fig. 8–2. **Structure of amylopectin (= B-fraction)**

The detailed structure of amylopectin (Fig. 8–3) is still speculative. The general consensus appears to be that amylopectin is composed of 3 types of chain, A, B and C ; each chain consisting of 24 glucose units. *A-chains* are linked to *B-chains* in 1, 6, or 1, 3 manner. These B-chains are further linked to other B-chains, the terminal one of which is linked to a single *C-chain*, which characteristically possesses a free reducing group. There are 2 different types of combinations of these chains which explain reasonably well the properties of amylopectin : the laminated structure as proposed by **Haworth** in 1937 and the randomly highly-branched structure as given by **Meyer** in 1940.

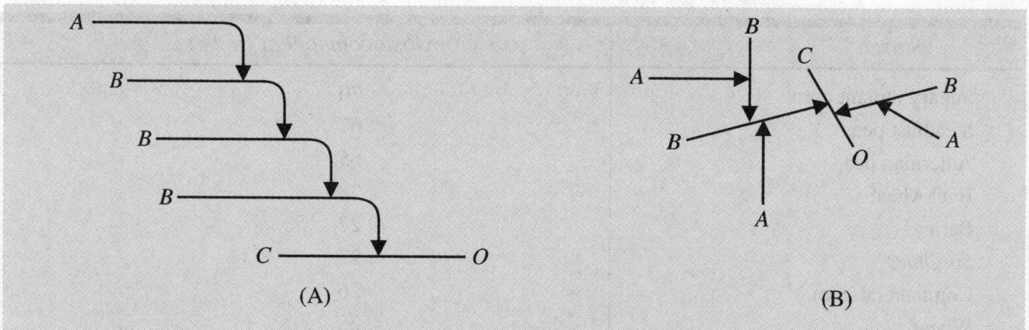

Fig. 8–3. **Detailed structure of amylopectin**

(A) Laminated structure (B) Randomly-branched structure

Properties. Starch is a white soft amorphous powder and lacks sweetness. It is insoluble in water, alcohol and ether at ordinary temperature. The specific rotation of starch, $[\alpha]_D^{20}$ is + 196°. The microscopic form of the starch grains is characteristic of the source of starch. Starch, on partial hydrolysis by boiling with water under pressure at about 250°C, breaks down into large fragments called **dextrins.** The resulting dextrins then confer stiffness to clothes that have been starched and ironed. Starch molecule is highly hydrated since it contains many exposed hydroxyl groups. With the result, starch, when extracted from granules with hot water, forms turbid colloidal solutions. In common with other polysaccharides, starch is a nonreducing carbohydrate since carbonyl groups of all units (except one of the two terminal ones) participate in the glycosidic linkages.

The two components of starch differ in following points:

I. Amylose has a simpler structure and hence more soluble in water than amylopectin. Because of this difference in solubility, the two components may be separated partially by keeping starch in water for a prolonged period. When potatoes are boiled, amylose is extracted by the hot water turning it milky and opalescent. The amylopectin left behind accounts for most of the starch in boiled potatoes.

II. Amylose is soluble without swelling in hot water whereas amylopectin swells in water.

III. Amylose is readily dispersed in water but does not form the characteristic gel or starch paste. However, when dispersed in sufficient concentrations, the amylopectin forms the typical starch gel.

IV. Amylose produces a typical blue colour with iodine whereas amylopectin gives a purplish colour with iodine. In fact, amylose is an open, helical molecule. The inside diameter of the helix is just large enough to accommodate an iodine molecule and it is the consequent change in the light-absorbing properties of the halogen that is responsible for the blue colour given in the starch-iodine test. By contrast, most other polysaccharides, including amylopectin, give only dull, reddish-brown colours.

Hydrolysis. Starch forms hydrated micelles in water, giving the impression of solubility. An enzyme, α-*amylase*, found in the digestive tract of animals (in saliva and the pancreatic juice) hydrolyzes the linear **amylose** chain by attacking α (1→ 4) linkages at random throughout the chain to produce a mixture of maltose and glucose. β-*amylase*, an enzyme found in plants, attacks the nonreducing end of amylose to yield successive units of maltose (The prefixes α and β used with the amylases do not refer to glycosidic linkage, but simply designate these two enzymes). **Amylopectin** can also be attacked by α- and β-amylases, but the α (1 → 4) glycosidic bonds near the branching point in amylopectin and the α (1 → 6) bond itself are not hydrolyzed by these enzymes. A separate "debranching" enzyme, an *α(1→6) glucosidase* can hydrolyze the bond at the branch point. Therefore, the combined action of α-amylase and the α (1 →6) glucosidase will hydrolyze amylopectin ultimately to a mixture of glucose and maltose. Starch is readily hydrolyzed by mineral acid with the final production of glucose. The course of hydrolysis may be followed by the gradual change in colour when treated with iodine (blue ⟶ purple ⟶ red ⟶ none) and also by the increase in concentration of reducing sugar. A compound which yields only glucose molecules on hydrolysis is called a **glucosan.** Hence, starch is a glucosan.

> The suffix **-an** is used for designating polymers; for example, starch is a *glucosan* as it yields only glucose molecules on hydrolysis; similarly, inulin is a *fructosan* as it yields only fructose molecules on hydrolysis.

$$\underset{\substack{\text{Starch or Glycogen}\\ \text{(insoluble)}}}{(C_6H_{10}O_5)_n} + \underset{\substack{(n-1)\text{ water}\\ \text{molecules}}}{(n-1)\,H_2O} \xrightarrow{\text{Hydrolysis}} \underset{\substack{n\text{ glucose molecules}\\ \text{(soluble)}}}{n(C_6H_{12}O_6)}$$

If hydrolysis of starch is interrupted early, dextrins are formed. The dextrins are white powders, and are used for making adhesives and confectionery, for sizing paper, etc. These materials possess adhesive properties and are used as adhesives on paper products.

GLYCOGEN

Occurrence. It is the major reserve food in animals (hence, often called **'animal starch'** and also in many fungi and bacteria due to its being osmotically inactive. Glycogen is stored in the liver and muscles of animals. It is especially abundant in the liver where it may attain up to 7% of the wet weight. It occurs in animal cells as particles much smaller than the starch grains. It can be isolated from tissues by digesting them with hot KOH solutions.

> Although of animal origin, a **glycogen-like polysaccharide** has also been described in *maize*.

Chemistry. Glycogen is a branched-chain polysaccharide and resembles amylopectin very much in structure, rather than amylose (for which there is no 'animal' equivalent), but has somewhat more glucose residues per molecule and about one-and-a-half times as many branching points. Also the chains are shorter (10–20 glucose units), and hence the molecule is even more highly branched and more compact. These differences, however, do not alter the functional behaviour of the molecule to any significant extent. Its molecular weight is $1 - 2 \times 10^7$.

Properties. It is a white, amorphous, tasteless powder and is more soluble in water than amylopectin. Hence, it readily forms suspension even in cold water. Glycogen is precipitated from aqueous solutions by addition of ethyl alcohol and is fairly stable in hot alkali. It is a nonreducing sugar and gives red colour with iodine. The red colour, however, disappears on boiling and reappears on cooling.

Hydrolysis. On incomplete hydrolysis with α-1,4-*glucanmaltohydrolase*, glycogen yields maltose. However, the acids completely hydrolyze it into glucose.

INULIN

Occurrence. It is a complex polysaccharide which is the storage form of carbohydrate in the members of the family– Compositae, such as dahlias (*Dahlia sp.*), artichokes (*Helianthus tuberosus*) and dandelions (*Taraxacum officinale*). Inulin is stored in the tubers of the dahlia and artichoke and in the roots of dandelion. It is also found in onion and garlic. Animals lack it.

Chemistry. Inulin (Fig. 8–4) has a molecular weight of about 5,000 and consists of about 30–35 fructose units per molecule. It is formed in the plants by eliminating a molecule of water from the glycosidic OH group on carbon atom 2 of one β-D-fructose unit and the alcoholic OH group on carbon atom 1 of the adjacent β-D-fructose unit. In inulin, the fructose residues (in furanose form) are, thus, joined together in a straight chain by β-2, 1-fructosidic linkages. Inulin is, therefore, 2→1-β-polyfructose.

On hydrolysis, however, inulin also yields a small amount of glucose besides fructose molecules. It is now thought that there are 2 glucose units in the inulin molecule, one somewhere in the centre and the other at the reducing end of the chain.

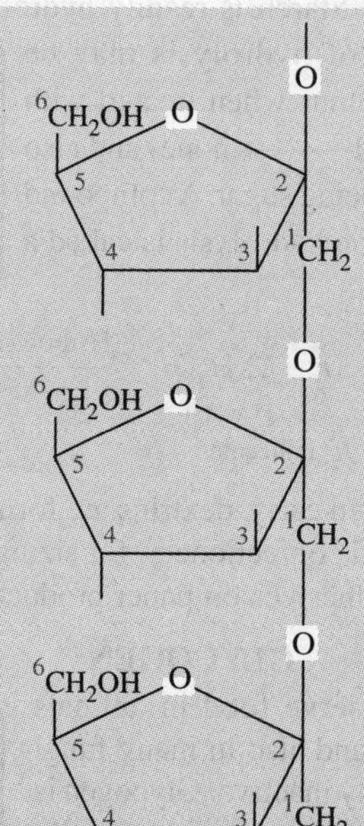

Fig. 8–4. Structure of inulin

[The two glucose units have not been shown for the sake of simplification.]

Properties. It is a white tasteless powder and is insoluble in cold water. In warm water, it easily dissolves forming a colloidal solution which does not form a gel on cooling. Inulin solution does not give any colour with iodine. It is a nonreducing sugar.

Hydrolysis. Inulin, on complete hydrolysis by dilute acids, yields D-fructose and hence it is a **fructosan** or **levan**.

CELLULOSE

Occurrence. Cellulose is not only the most abundant extracellular structural polysaccharide of the plant world but is also *undoubtedly the most abundant of all biomolecules in the biosphere*. It is present in all land plants, but is completely lacking in meat, egg, fish and milk. It is, however, not metabolized by the human system. Some 10^{15} kg of cellulose is synthesized and degraded on earth each year! It shares as much as 50% of all the carbon in vegetation. It is the most widely distributed carbohydrate of the plants, although certain tunicates also possess it. Cellulose occurs in the cell walls of plants where it contributes in a major way to the structure of the organism. Lacking a skeleton of bone onto which organs and tissues may be organized, the higher plant relies on its cell walls to bear its own weight, whether it is the smallest *Wolffia* flower or the tallest *Sequoia* tree. Cellulose is often found associated with other structural substances such as lignin. Plant residues in soil consist of 40-70% cellulose. This high cellulose content emphasizes the importance of cellulolytic organisms in the mineralization process and in the carbon cycle. It occurs in nearly pure form in cotton (94%), flax (90%) and in the wood of many plants (60%). It is not usually found in bacterial cell walls, but it is the substance which keeps the cells of *Sarcina ventriculi* together in large packets or bunches.

Chemistry. The molecular weight of cellulose (Fig. 8–5) ranges between 200,000 and 2,000,000, thus corresponding to 1,250–12,500 glucose residues per molecule. It may be formed by taking out a molecule of water from the glycosidic OH group on carbon atom 1 of one β-D-glucose molecule and the alcoholic OH group on carbon atom 4 of the adjacent β-D-glucose molecule. It, thus, resembles in structure with amylose except that the glucose units are linked together by β-1, 4-glucoside linkages. Cellulose may, henceforth, be regarded as an anhydride of β-D-glucose units.

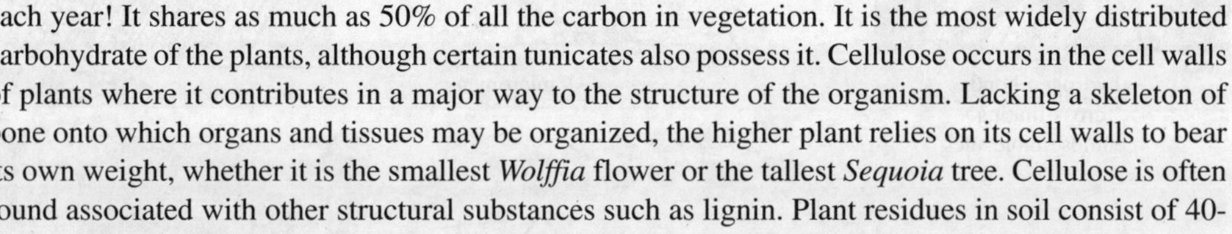

Fig. 8–5. Structure of cellulose

The correlation between the structure of cellulose molecule and the function it perform is illustrated in Fig. 8–6.

Properties. It is a fibrous, tough, white solid, insoluble in water but soluble in ammoniacal cupric hydroxide solution (Schweitzer's reagent). It gives no colour with iodine and lacks sweetness. Although insoluble in water, cellulose absorbs water and adds to the bulk of the fecal matter and facilitates its removal. As it passes through the digestive tract unchanged, cellulose is helpful only in stimulating movement of the intestines.

Because of the lack of chemical reactivity, the cellulose is of no nutritive value unlike starch. However, the same characteristics have made it so useful as fibres for paper and cloth. In man, the cellulose is not digested since it does not possess an enzyme capable of catalyzing the cleavage of β-glucoside bonds. Hence, it serves as an important source of "bulk" in the diet. However, the ruminants (cattles, sheep, goats, camels, giraffes) and certain wood-eating insects are able to digest it because the microorganisms, present in their digestive tract, do possess such digesting enzymes. Also certain wood-eating insects such as termites readily digest cellulose because their intestinal tract harbours a parasitic protozoan, *Trichonympha*. The latter secretes *cellulase*, a cellulose-hydrolyzing enzyme which enables termites to digest wood. Wood-rot fungi and bacteria also produce cellulase.

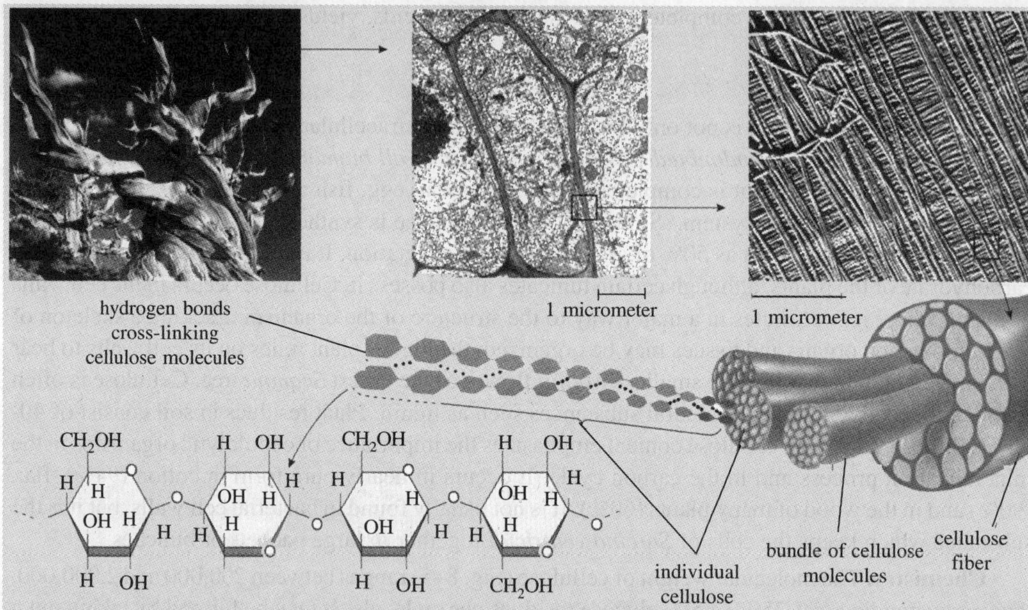

Fig. 8–6. Interrelation between cellulose structure and function

Cellulose, like starch, is composed of glucose monomers but the orientation of the bond between monomers in cellulose is different such that every other glucose molecule is "upside down." Unlike starch, cellulose has great structural strength, due partly to the difference in bonding and partly to the arrangement of parallel molecules of cellulose into long, cross-linked fibres. Plant cells often deposit cellulose fibres in layers that run at angles to each other, resulting in resistance to tearing in both directions. The final product can be incredibly tough, as this 3000-year-old bristlecone pine testifies.

Hydrolysis. Cellulose is a relatively inert material and is completely degraded only under the most drastic conditions. For example, it may be hydrolyzed to glucose when treated with conc. H_2SO_4 or HCl or with conc. NaOH.

Biological importance. Huge quantities of cellulose are made annually by the plants of the globe. It is estimated that every day some 50 kg of cellulose is synthesized by the plant world for each human being on the earth. Cellulose is also a very useful substance. Wood, cotton, paper and cardboard are all largely cellulose. Dried wood contains about 65% cellulose, 30% lignin and 2% pentosans. In the manufacutre of **paper**, wood is cooked with water, lime and SO_2 to remove lignin leaving behind the cellulose. These fibres are then spread into mats and rolled into sheets which may be coated to provide a smooth writing surface. When heavy paper is treated with H_2SO_4, the translucent parchment paper is obtained.

Cotton treated with alkali becomes somewhat translucent, acquires a silky lustre, and is called **mercerized cotton**. Rayon is made from cellulose by dissolving the material in NaOH and CS_2. The viscous material is forced through small jets into an acid solution of sodium bisulfite, regenerating the cellulose in small threads.

Cellulose nitrates are used in explosives, lacquers, celluloid and collodion. **Cellulose acetate** is widely used in photographic film and packing materials.

It may, thus, be easily visualized that the way the constituent monomeric units of various polysaccharides (glycogen, starch and cellulose, for example) are linked together has a profound bearing on the physical and chemical properties of these molecules, although the polysaccharides are made up of identical glucose units (Fig. 8–7).

The amount of **lignin** in wood varies in different species and with age in the same species. It makes up about 23–33% of softwoods, but only about 16–25% of hardwoods.

When wood passes from the green pliable stage, it is hardened by the interpenetration of a highly complex non-polysaccharide polymeric substance called *lignin*. This substance is formed by the cross-linking of a number of *non-sugar aromatic compounds,* such as coniferyl alcohol, vanillin, syringaldehyde etc. The process of lignification has some chemical resemblance to that which strengthens the proteins of the insect cuticle.

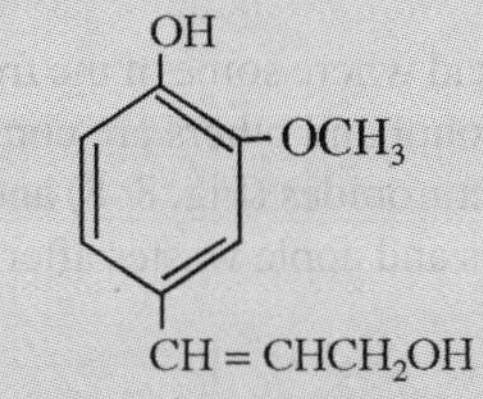

Coniferyl alcohol

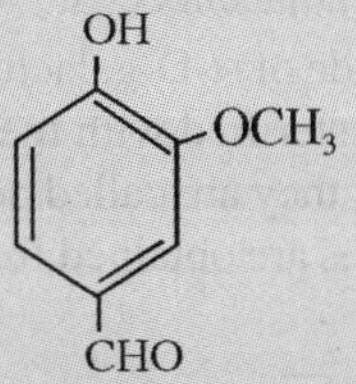

Vanillin

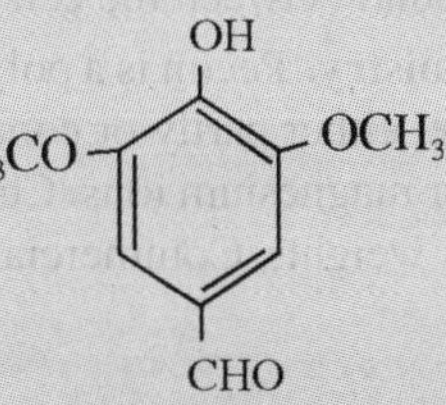

Syringaldehyde

Fruits with seeds and lignified cells of pear are important sources of lignin. Woody nondigestible parts of vegetables too are rich in lignin. Lignin is water-insoluble and non-digestible. It adsorbs organic substances like bile acid from the intestinal lumen and thus facilitates its fecal excretion. Lignin is an intractable material that is difficult to remove from the wood.

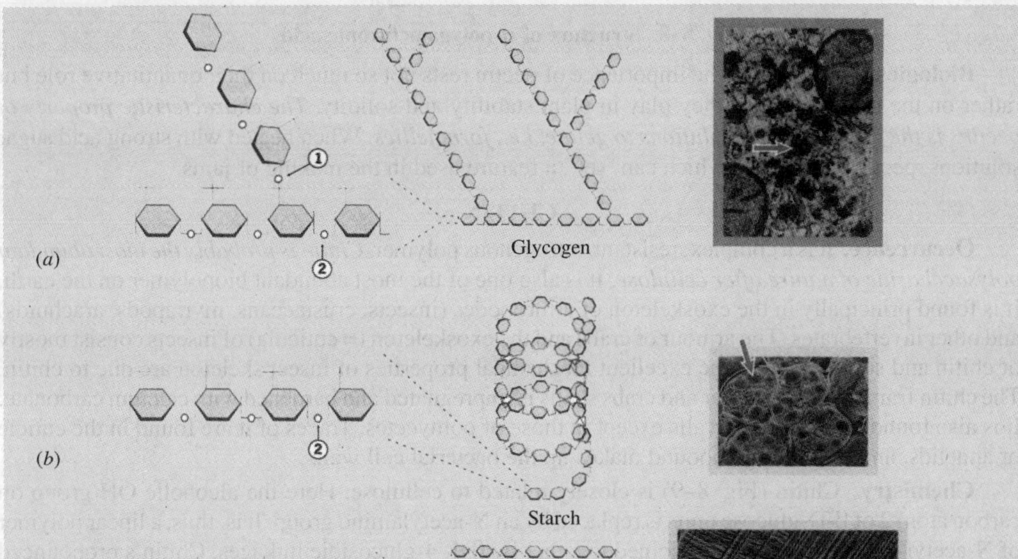

Fig. 8–7. Three polysaccharides with identical sugar monomers but dramatically different properties

Gycogen (*a*), starch (*b*), and cellulose (*c*) are each composed entirely of glucose subunits, yet their physical and chemical properties are very different due to the distinct ways that the monomers are linked together (three different types of linkages are indicated by the circled numbers). *Glycogen* molecules are the most highly branched, *starch* molecules assume a helical arrangement, and *cellulose* molecules are unbranched and highly extended. Whereas glycogen and starch are energy stores, cellulose molecules are bundled together into tough fibres that are suited for their structural role. Colourized electron micrographs show glycogen granules in a liver cell, starch grains (amyloplasts) in a plant seed, and cellulose fibres in a plant cell wall; each is indicated by an *arrow*.

[*Photo insets:* (*Top*) *Don Fawcett;* (*Centre*) *Jeremy Burgess;* (*Bottom*) *Cabisco*]

PECTIN

Occurrence. It is a complex carbohydrate consisting of a mixture of polysaccharides that holds cellulose fibrils together. Pectins are found as intercellular substances in the tissues of young plants and are especially abundant in ripe fruits such as guava, apples and pears. Normally, they are present in an insoluble form, but in ripening fruits and in tissues affected by certain diseases, they change into a soluble form, which is evidenced by softening of the tissues. They are components of the middle lamella found between the cell wall and adjacent cells.

Chemistry. Pectin is a polysaccharide of α-D-galacturonic acid where some of the free carboxyl groups are, either partly or completely, esterified with methyl alcohol and others are combined with calcium or magnesium ions. Chemically, they are called polygalacturonides (Fig. 8–8) and have high molecular weights. Commercial quantities are obtained from citrus and apple wastes after removal of the juice.

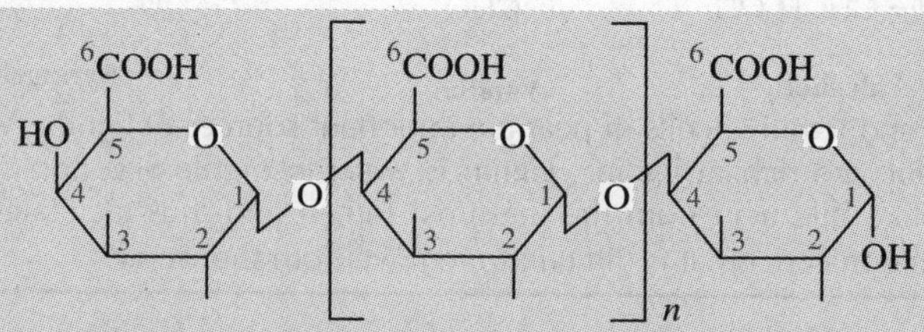

Fig. 8–8. Structure of α-polygalacturonic acid

Biological properties. The importance of pectin rests not so much on their quantitative role but rather on the functional part they play in plant stability and solidity. *The characteristic property of pectins is the ability of their solutions to gelate, i.e., form jellies.* When heated with strong acid sugar solutions, pectin forms a gel which can 'set', a feature used in the making of jams.

CHITIN

Occurrence. It is a complex, resistant, nitrogenous polymer. *Chitin is probably the most abundant polysaccharide of nature after cellulose.* It is also one of the most abundant biopolymer on the earth. It is found principally in the exoskeleton of arthropodes (insects, crustaceans, myriapods, arachnids) and other invertebrates. The armour of crabs and the exoskeleton (= cuticula) of insects consist mostly of chitin and some protein. The excellent mechanical properties of insect skeleton are due to chitin. The chitin framework of lobster and crabs shells is impregnated and hardened with calcium carbonate. It is also found in fungal cell walls except in those of oomycetes. Traces of it are found in the cuticle of annelids, and a similar compound makes up the bacterial cell walls.

Chemistry. Chitin (Fig. 8–9) is closely related to cellulose. Here the alcoholic OH group on carbon atom 2 of β-D-glucose units is replaced by an N-acetylamino group. It is, thus, a linear polymer of N-acetyl-D-glucosamine units joined together by β-1, 4-glucosidic linkages. Chitin's pronounced stability is based on the hydrogen bonding of the *N*-acetyl side chains. Like cellulose, chitin is indigestible by vertebrate animals.

Hydrolysis. Chitin is a tough, horny, insoluble polysaccharide of high molecular weight. On hydrolysis with mineral acids, chitin yields 2 final end products, namely, glucosamine and acetic acid. Glucosamine is an important component of some glycoproteins (=mucoproteins) such as mucin of saliva. *Chitinases* (from the gastric juice of snails or from bacteria), however, decompose the chitin to *N*-acetylglucosamine.

HEMICELLULOSES

Xylan and other related substances are collectively referred to as hemicelluloses. They are not structurally related to cellulose, nor do they contain the same building blocks but they are, at least partially, soluble in water or alkali. The hemicelluloses consist of either pentoses (xylose, arabinose) or hexoses (glucose, mannose, galactose) as well as uronic acids. Unrefined cereals, some fruits and

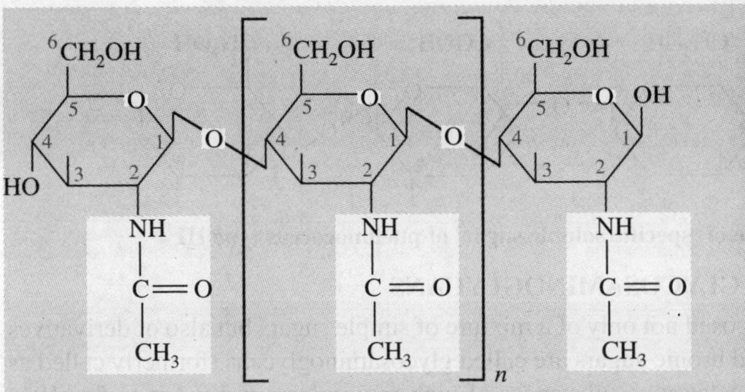

Fig. 8–9. Structure of chitin, a unique polysaccharide

Chitin has the same "alternating upside down" bonding of glucose molecules as cellulose has, but in chitin the glucose subunits are modified by the replacement of one of the hydroxyl groups with a nitrogen-containing functional group (yellow). Tough, slightly flexible chitin supports the otherwise soft bodies of arthropods (insects, spiders, and their relatives) and fungi.

vegetables and whole wheat are rich sources of hemicelluloses. They adsorb water and are partially digestible. They function as storage and supporting substances in plants. The designation 'hemicelluloses' has been discontinued since a large number of similar polysaccharides were discovered in fungi and bacteria.

Xylan is representative of the group and is, therefore, being discussed below.

XYLAN

Occurrence. It is a complex polysaccharide belonging to the group hemicelluloses, that occurs in close association with cellulose in lignified cell walls of plants. *Xylan is the next most abundant and widely distributed carbohydrate after cellulose.* Straw and bark consist of upto 30% xylan; residues of sugarcane also contain about 30% xylan. It also occurs in conifer wood (7–12%) and deciduous wood (20–25%).

Chemistry. Xylan is a linear homopolymer of D-xylose (an aldopentose) in β-1→4 linkage, bearing side chains of 4–*O*–methylglucuronic acid and/or arabinose. Xylan can be derived from a cellulose chain by substituting hydrogen atoms for the —CH_2OH groups, but its polymer size (number of units per polymer) is considerably lower (30–100).

Properties. Xylan is more rapidly degraded by a large number of microorganisms than cellulose. Many cellulose-degrading organisms (such as *Sporocytophaga myxococcoides*) also produce *xylanase*. The ability to utilize xylans is very common among fungi. Xylan is even an excellent substrate for the cultivation of mushrooms. In bacteria, xylanase is formed constitutively in some organisms (*e.g.*, Clostridia) and in others it is inducible by xylan. Xylan is used therapeutically.

HETEROPOLYSACCHARIDES

These yield, on hydrolysis, a mixture of monosaccharides. They are numerous in both plants and animals such as hyaluronic acid, chondroitins, heparin etc.

The 'specific soluble sugar' of pneumococcus type III (Fig. 8–10) is one of the simplest heteropolysaccharides. It contains repeating units of a mixed disaccharide consisting of glucose and glucuronic acid. Apparently, on hydrolysis, it yields equimolar amounts of glucose and glucuronic acid.

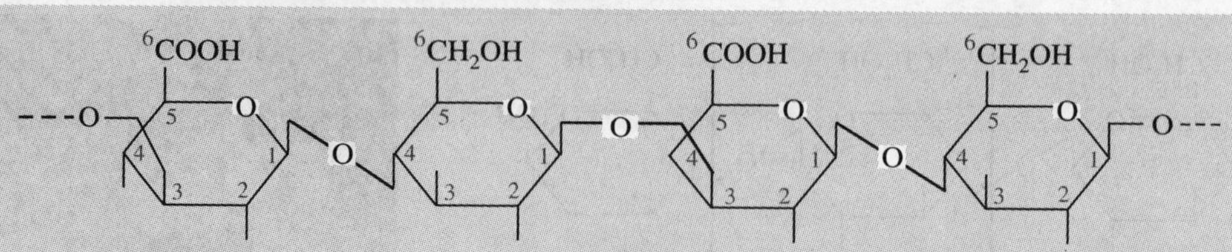

Fig. 8–10. Structure of 'specific soluble sugar' of pneumococcus type III

GLYCOSAMINOGLYCANS

Polysaccharides that are composed not only of a mixture of simple sugars but also of derivatives of sugars such as amino sugars and uronic sugars are called glycosaminoglycans (formerly called as ***mucopolysaccharides***). They are gelatinous substances of high molecular weights (up to 5×10^6). Most of these act as structural support material for connective tissue or mucous substances of the body. They serve both as a lubricant and a cementing substance.

Structurally, they have a common feature. They consist of disaccharide units in which a uronic acid is bound by a glycosidic bond to the C_3 of an acetylated amino acid ($1 \rightarrow 3$ linkage). These disaccharide residues are polymerized by $1 \rightarrow 4$ linkages to give a linear macromolecule. The uronic and sulfuric acid residues impart a strong acidic character to these substances. Table 8–2 lists the structural features of some common mucopolysaccharides.

Table 8–2. Some common glycosaminoglycans

Glycosaminoglycan	*Two components of the disaccharide units*	*Linkages**	
Hyaluronic acid	D-glucuronic acid + N-acetyl-D-glucosamine	β-1 → 3	β-1 → 4
Chondroitin sulfate A	D-glucuronic acid + N-acetyl-D-galactosamine-4-sulfate	β-1 → 3	β-1 → 4
Chondroitin sulfate C	D-glucuronic acid + N-acetyl-D-galactosamine-6-sulfate	β-1 → 3	β-1 → 4
Dermatan sulfate (Chondroitin sulfate B)	L-iduronic acid + N-acetyl-D-galactosamine-4-sulfate	α-1 → 3	β-1 → 4
Keratosulfate	D-galactose + N-acetyl-D-glucosamine-6-sulfate	β-1 → 4	β-1 → 3

* Linkage of the first column represents the linkage involved between the two monosaccharides of the disaccharide unit whereas the linkage of the second column is the one involved between the repeating disaccharide units.

HYALURONIC ACID

Occurrence. It is the most abundant member of glycosaminoglycans and is found in higher animals as a component of various tissues such as the vitreous body of the eye, the umbilical cord and the synovial fluid of joints. The high viscosity of the synovial fluid and its role as biological lubricant is largely due to the presence of its hyaluronic acid content (about 0.03%). Frequently, it is prepared from umbilical cord.

Chemistry. Hyaluronic acid (Fig. 8–11) has the least complicated structure among mucopolysaccharides. It is a straight-chain polymer of D-glucuronic acid and N-acetyl- D-glucosamine (NAG) alternating in the chain. Its molecular weight approaches approximately, 5,000,000. Here, apparently two linkages are invloved, β-1 → 3 and β-1 → 4. Hyaluronic acid is an acidic substance, because the carboxyl groups are largely ionized at cellular pH.

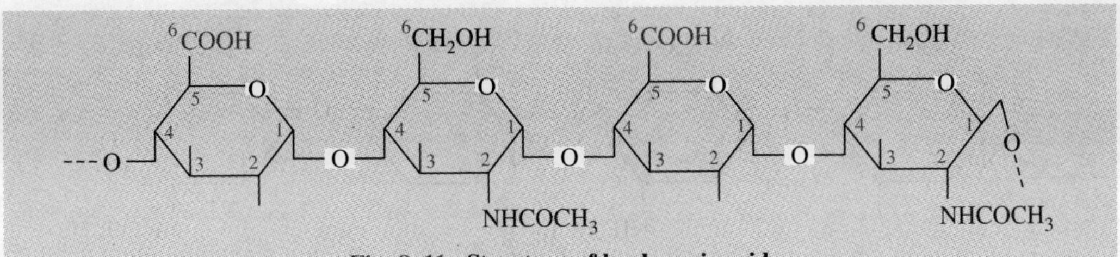

Fig. 8–11. Structure of hyaluronic acid

The Ca^{2+} polyanionic form of hyaluronic acid is depicted in Fig. 8-12. It is an extended left-handed helix, which is stabilized by hydrogen bonds.

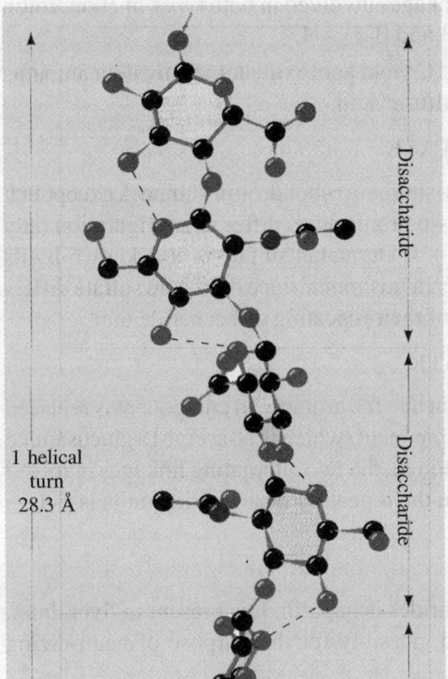

Fig. 8–12. The x-ray fibre structure of Ca^{2+} hyaluronate

The hyaluronate polyanion forms an extended left-handed single-stranded helix with three disaccharide units per turn that is stabilized by intramolecular hydrogen bonds (dashed lines). H atoms and Ca^{2+} ions are omitted for clarity.

(*After Winter WT and Arnott S. 1977*)

Hydrolysis. Hyaluronic acid, upon hydrolysis, yields an equimolar mixture of D-glucuronic acid, D-glucosamine and acetic acid.

It is noteworthy that hyaluronic acid is split swiftly by the enzyme *hyaluronidase*. The enzyme brings about depolymerization of hyaluronic acid (leading to a drop in its viscosity) and cleavage to smaller fragments. Hyaluronidase is a 'spreading factor' of skin and connective tissue. The depolymerization effect allows any foreign bodies (such as pigments, pathogenic bacteria) to penetrate the tissue, since the cementing substance is being dissolved. The enzyme also has a physiologic role in fertilization. The sperm is rich in hyaluronidase and hence can advance better in the cervical canal and finally fertilize the ovum.

CHONDROITIN

Occurrence. Chondroitin is of limited distributioin. It is found in cartilage and is also a component of cell coats. It is a parent substance for two more widely distributed mucopolysaccharides, chondroitin sulfate A and chondroitin sulfate B.

Chemistry. Chondroitin is similar in structure to hyaluronic acid except that it contains galactosamine rather than glucosamine. It is, thus, a polymer of β-D-glucuronido-1, 3-N-acetyl-D-galactosamine joined by β-1 → 4 linkages.

Hydrolysis. On hydrolysis, chondroitin produces equimolar mixture of D-glucuronic acid, D-galactosamine and acetic acid.

CHONDROITIN SULFATES

Occurrence. The two chondroitin sulfate A and C are widely distributed and form major structural components of cartilage, tendons and bones. They are very often associated with collagen and probably with other proteins too.

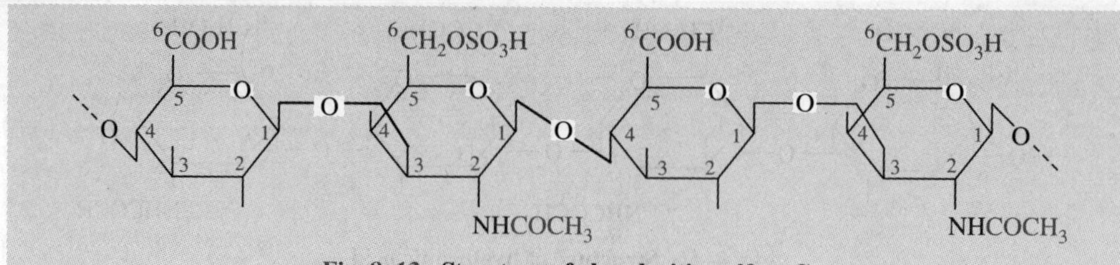

Fig. 8–13. Structure of chondroitin sulfate C

Chemistry. Chondroitin sulfates may be regarded as derivatives of chondroitin where, in the galactosamine moiety, a sulfate group is esterified either at carbon 4 as in chondroitin sulfate A or at carbon 6 as in chondroitin sulfate C (Fig. 8–13). The two linkages involved in both types of chondroitin sulfate would, obviously, be the same. These are β-1 → 3 and β-1 → 4.

Hydrolysis. On hydrolysis, chondroitin sulfates A and C yield approximately equivalent amounts of D-glucuronic acid, D-galactosamine, acetic acid and sulfuric acid.

DERMATAN SULFATE

Dermatan sulfate is a mucopolysaccharide structurally similar to chondroitin sulfate A except that the D-glucuronic acid is replaced by L-iduronic acid (the two uronic acids differ in configuration only at C_5). The two linkages involved are α-1 → 3 and β-1 → 4. Dermatan sulfate is also known by its conventional name, **chondroitin sulfate B**. This is, however, a misnomer since dermatan sulfate differs from both chondroitin sulfate A and C in the composition of their repeating disaccharide unit.

KERATOSULFATE

Keratosulfate differs from other mucopolysaccharides in that the uronic acid component is replaced by D-galactose. Here, the second acetylated amino sugar component (which is N-acetyl-D-glucosamine in this case) is esterified by a sulfate group at carbon 6. Although, the two alternating linkages involved are β-1 → 4 and β-1 → 3, in this case the linkage between the repeating disaccharide units is β-1 → 3 rather than β-1 → 4.

HEPARIN

Occurrence. Related to the sulfated mucopolysaccharides is heparin. It is present in liver, lung, arterial walls and, indeed, wherever mast cells are found, possibly for the purpose of neutralizing biogenic amines (*e.g.*, histamine).

Chemistry. Heparin (Fig. 8–14) is a heteropolysaccharide composed of D-glucuronic acid units, most of which (about 7 out of every 8) are esterified at C_2 and D-glucosamine-N-sulfate (= sulfonylaminoglucose) units with an additional O-sulfate group at C_6. Both the linkages of the polymer are alternating α-1 → 4. Thus, the sulfate content is very high and corresponds to about 5–6 molecules per tetrasaccharide repeating unit. The relative positions of the sulfate residues may also vary. Its molecular weight ranges between 17,000 and 20,000.

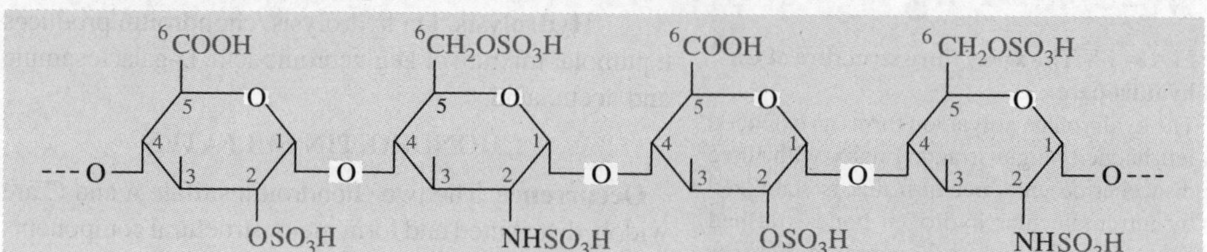

Fig. 8–14. Structure of heparin

Heparin acts as an anticoagulant. It prevents coagulation of blood by inhibiting the transformation of prothrombin to thrombin. This eliminates the effect of thrombin on fibrinogen.

Anticoagulant Effect of Heparin

Heparin is a naturally-occurring, heterogeneous (3,000–30,000 kDa), polyonionic sulfated polysaccharide that is used to reduce the clotting tendency of patients. Both *in vivo* and *in vitro,* heparin prevents the activation of clotting factors but does not act directly on the clotting factors. Rather, the anticoagulant activity of heparin is brought about by the binding interaction of heparin with an inhibitor of the coagulation process, called *antithrombin III* (AT). Heparin activates antithrombin III which is a plasma protein inhibitor of serine proteases. AT is a slow but quantitatively important inhibitor of thrombin (factor X) and other factors (IX, XI, XII) in the blood-clotting cascade. When heparin binds to AT, it converts from a slow inhibitor to a rapid inhibitor of coagulating enzymes. Heparin interacts with a lysine (Lys) residue in AT and induces a conformational change that promotes covalent binding of AT to the active serine (Ser) centres of coagulating enzymes inhibiting their procoagulant activity. Heparin then detaches from the ternary complex and can be recycled for anticoagulation. The smallest, most active component of heparin is a pentasaccharide: GlcN-(*N*-sulfate-6-*O*-sulfate)-α1, 4-GlcUA-β1, 4-GlcN-(*N*-sulfate-3,6-di-*O*-sulfate)-α1, 4-IdUA-(2-*O*-sulfate)-α-1, 4-GlcN-(*N*-sulfate-6-*O*-sulfate). Heparin has an average half-life of 30 min in the circulation, hence it is commonly administered by infusion. Heparin does not have fibrinolytic activity; therefore, it will not lyse existing clots.

The heteropolysaccharides described above contain two different sugars in their repeating units or monomers. However, there are many heteropolysaccharides which contain more than 2 carbohydrates in their repeating units. Vegetable 'gums' and agar-agar are two notable examples.

VEGETABLE 'GUMS'. These contain as many as 4 different monosaccharide units. Most common of these are D-glucuronic acid, D-mannose, D-xylose (*the second most abundant sugar in the biosphere*) and L-rhamnose.

AGAR-AGAR. It is a gelatinous polysaccharide produced by certain marine red algae (= rhodophycean members) such as species of *Gelidium, Gracilaria, Gigartina, Eucheuma, Campylaephora, Ahnfeltia, Hypnea,* etc. Japan is the major producer of agar-agar. It consists of D- and L-galactose, predominantly with 1 → 3 bonds and always contains some amount of sulfuric acid. On hydrolysis, it yields D- and L-galactose in a ratio of 9 : 1. It forms highly viscous gels and has melting point between 90 and 100°F. At lower temperature, it solidifies. It is insoluble in cold but soluble in hot water.

Agar-agar is of great value in the preparation of foodstuffs, particularly as articles of diet for invalids. It is used extensively in biological laboratories as the base material for the preparation of culture media, especially for bacteria and fungi. It is also used in the preparation of some medicines, and in cosmetics and leather industry. It is largely used as a solidifying agent in desserts, as a laxative and as a sizing material for textiles. It also finds application as an emulsifier in dairy products. Agar-agar is not utilized by man and hence adds to the bulk of the feces and helps in its propulsion.

The use of **agar** as a solidifying agent was first suggested, in 1882, by **Fannie Eilshemius Hesse** (LT, 1850–1934), the wife of **Walther Hesse** (LT, 1846–1911), one of Robert Koch's assistants. She had been using it successfully to make jellies for sometime.

A wide variety of heteropolysaccharides are found in animals and microbes. These, on hydrolysis, yield various types of monosaccharides. These more complex polysaccharides are found bound to proteins (**glycoproteins**) and lipids (**glycolipids**). Carbohydrates occur in them as an oligosaccharide moiety.

ANALYSIS OF CARBOHYDRATES

Mixtures of carbohydrates can be resolved into their individual components (Fig. 8–15) by many techniques (such as differential centrifugation, ion exchange chromatography and gel filtration) which are also used in protein and amino acid separation. Each carbohydrate separated in the first stage of analysis is subjected to the following 3 analytic routes for complete characterization. The 3 routes are as follows:

1. Hydrolysis in strong acid yields a mixture of monosaccharides, which after conversion to suitable volatile derivatives may be separated, identified and quantified by gas-liquid chromatography to yield the overall composition of the polymer.
2. For simple linear polymers such as amylose, the position of the glycosidic bond between monosaccharides is determined by treating the intact polysaccharide with methyl iodide to convert all free hydroxyls to acid-stable methyl esters. When the methylated polysaccharide is hydrolyzed in, the only free hydroxyls present in the monosaccharides produced are those that were involved in glycosidic bonds.
3. To determine the stereochemistry at the anomeric carbon, the intact polymer is tested for sensitivity to purified glycosidases known to hydrolyze only α- or only β-glycosides. Total structure determination for complex heteropolysaccharides is much more difficult. Stepwise degradation with highly specific glycosidases, followed by isolation and identification of the products, is often helpful. Mass spectral analysis and high-resolution nuclear magnetic resonance (NMR) spectroscopy are highly powerful analytic tools for carbohydrates.

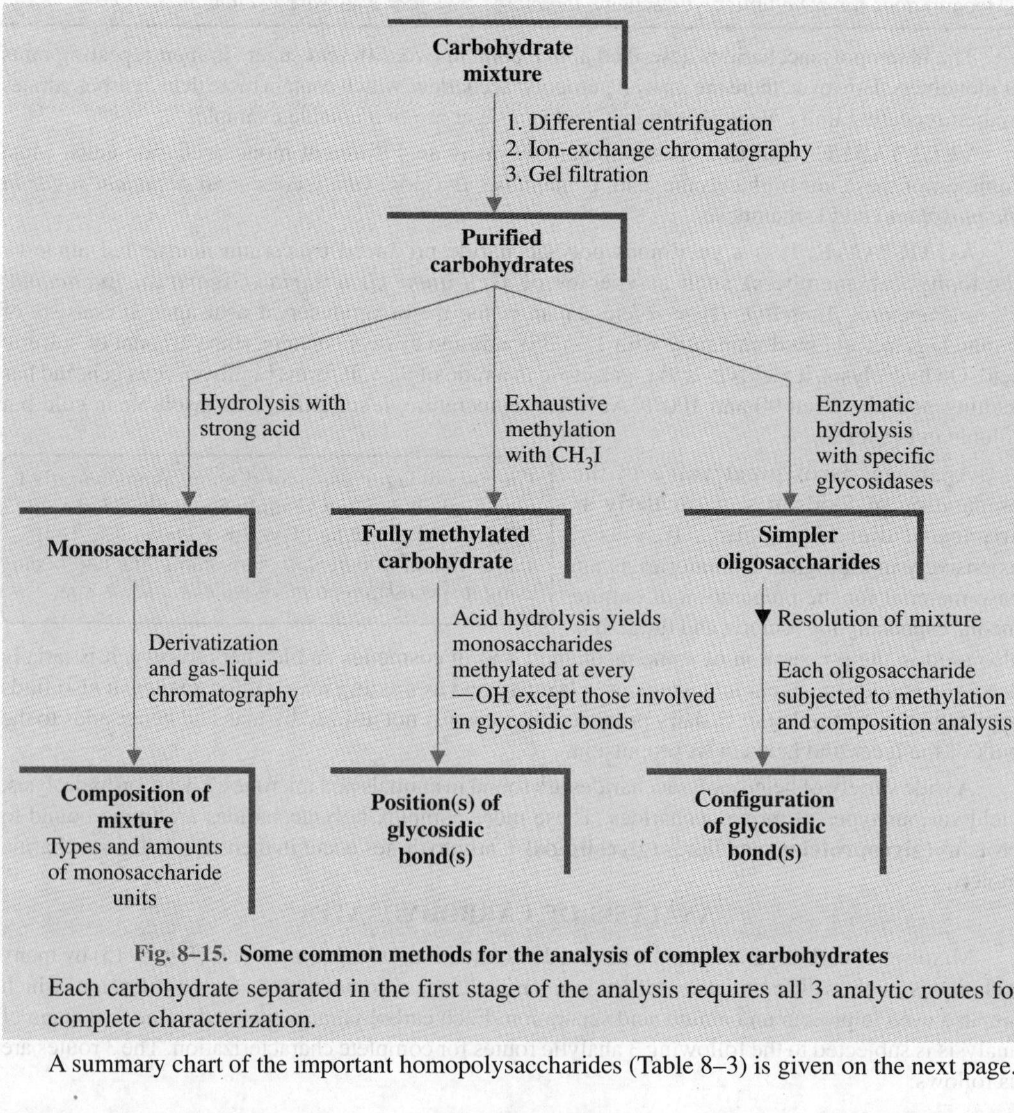

Fig. 8–15. Some common methods for the analysis of complex carbohydrates

Each carbohydrate separated in the first stage of the analysis requires all 3 analytic routes for complete characterization.

A summary chart of the important homopolysaccharides (Table 8–3) is given on the next page.

Table 8–3. Summary of occurrence, structure and properties of some important homopolysaccharides

Name	Occurrence	Constituent monosaccharides*	Oxide linkage(s) involved	Groups involved in linkage	Properties	Hydrolysis
Starch	Major reserve food of higher plants; found in cereals, legumes, potatoes and other vegetables	α-D-glucose	(In amylose) α-1, 4-glucoside (In amylopectin) α-1, 4-glucoside and α-1, 6-glucoside The averange chain length is about 24 glucose units.	glycosidic OH on C_1 alcoholic OH on C_4 glycosidic OH on C_1 alcoholic OH on C_4 alcoholic OH on C_6	amylose gives blue colour with iodine; amylopectin gives purple colour with iodine; dextrorotatory	Yields glucose units only on hydrolysis, hence it is a glucosan
Glycogen	Major reserve food of animals; stored in liver and muscles of animals	α-D-glucose	α-1, 4-glucoside and α-1, 6-glucoside Chains are shorter (10–20 glucose units) and much more branched than amylopectin.	glycosidic OH on C_1 alcoholic OH on C_4 alcoholic OH on C_6	a nonreducing sugar; gives red colour with iodine	Yields glucose units only on complete hydrolysis, hence a glucosan
Inulin	Found in tubers and roots of certain composite plants such as dahlias, artichokes and dandelions	β-D-fructose	β-2, 1-fructoside	glycosidic OH on C_2 alcoholic OH on C_1	a nonreducing sugar; gives no colour with iodine	Yields fructose units only on complete hydrolysis, hence a fructosan
Cellulose	Most abundant organic compound found in world; widely distributed in plants such as cotton and flax	β-D-glucose	β-1, 4-glucoside	glycosidic OH on C_1 alcoholic OH on C_4	gives no colour with iodine	Yields glucose units only on complete hydrolysis, hence a glucosan
Pectin	Fruits of many plants	α-D-galacturonic acid	α-1, 4-galactoside	glycosidic OH on C_1 alcoholic OH on C_4	—	—
Chitin	Most abundant in nature after cellulose; found in fungi and arthropods	N-acetyl-D-glucosamine	β-1, 4-glucoside	glycosidic OH on C_1 alcoholic OH on C_4	a nonreducing sugar	Yields N-acetyl-glucosamine on hydrolysis with chitinases

* All the constituent monosaccharides are present in pyranose ring form except β-D-fructose which has a furanose ring structure.

REFERENCES

1. **Annual Review :** Advances in Carbohydrate Chemistry. *Academic Press, Inc., New York. 1945-current.*
2. **Ashford DA :** Carbohydrate analysis. *Curr. Opin. Biotechnol.* **3(1)** *: 45, 1992.*
3. **Aspinall GO (editor) :** The Polysaccharides. *Vols.1-3, Academic Press, Inc., New York. 1982, 1983, 1985.*
4. **Bailey RW :** Oligosaccharides. *Pergamon Press, Oxford, 1964.*
5. **Biermann CJ, McGinnis GD (editors) :** Analysis of Carbohydrates by GLC and MS. *CRC Press, Inc., Boca Raton, FL. 1989.*
6. **Binkley RW :** Modern Carbohydrate Chemistry. *Marcel Dekker, Inc., San Diego, CA. 1988.*
7. **Davidson EA :** Carbohydrate Chemistry. *Holt, Rinehart and Winston, New York. 1967.*
8. **El Khadem HS :** Carbohydrate Chemistry. Monosaccharides and Their Oligomers. *Academic Press, Inc., New York. 1988.*
9. **Florkin M, Stotz E :** Comprehensive Biochemistry ; Carbohydrates *Sec. II, vol, 5. Elsevier Pub. Co., New York, 1963.*
10. **Goodwin TW, Mercer EI:** Introduction to Plant Biochemistry. *Pergamon, Oxford. 1983.*
11. **Guthrie RD :** Introduction to Carbohydrate Chemistry. *4th ed., Clarendon Press, Oxford. 1974.*
12. **Heath EC :** Complex polysaccharides *Ann. Rev. Biochem.,* **40 :** *29, 1971.*
13. **Hughes RC:** Glycoproteins. *Chapman and Hall, London. 1983.*
14. **Hudson CS :** Emil Fischer's discovery of the configuration of glucose. *J. Chem. Ed.* **18 :** *353, 1941.*
15. **Jaques LB : Heparin :** An old drug with a new paradigm. *Science.* **206 :** *528–533, 1979.*
16. **Kretchner N :** Lactose and lactase. *Scientific American.* **227(4) :** *270, 1971.*
17. **Lennarz WJ (editor) :** The Biochemistry of Glycoproteins and Proteoglycans. *Plenum, New York. 1980.*
18. **Lundblad RL *et al* :** Chemistry and Biology of Heparin. *Elsevier. 1980.*
19. **Mcllroy RJ :** Introduction to Carbohydrate Chemistry. *Butterworths, Reading, Mass. 1967.*
20. **Morrison RT, Boyd RN :** Organic Chemistry. *4th ed., Allyn and Bacon, Boston, 1983.*
21. **Northcote DH :** Polysaccharides. *Ann. Rev. Biochem.,* **33 :** *51, 1964.*
22. **Oosta GM *et al* :** Multiple functional domains of the heparin molecule. *Proc. Nat. Acad. Sci., U.S.* **78 :** *829, 1981.*
23. **Pearl IA :** Lignin Chemistry. *Chem. Eng. News. p. 81, July 6, 1964.*
24. **Percival EGV, Pericival E :** Structural Carbohydrate Chemistry. *Prentice-Hall. 1962.*
25. **Pigman WW, Horton D (editors) :** The Carbohydrates : Chemistry and Biochemistry. *2nd ed., Academic Press Inc., New York. 1970, 1972, 1980 (4 volumes).*
26. **Preston RD :** Cellulose. *Sci. Amer.* **197 :** *157, 1957.*
27. **Rees DA :** Polysaccharide Shapes. *Halsted Press. 1977.*
28. **Sharon N :** Polysaccharides. *Ann. Rev. Biochem.,* **35 :** *485—520, 1966.*
29. **Sharon N :** Complex Carbohydrates : Their Chemistry, Biosynthesis, and Functions. *Addison-Wesley Publishing Co., Inc., Reading, MA. 1975.*
30. **Sharon N :** Carbohydrates. *Scientific American.* **243 (5)** *: 90–116, 1980.*
31. **Speck Jr. CJ :** The Lobry de Bruyn–Alberda van Eckenstein transformation. *Adv. Carbohydrate Chemistry.* **13,** *63, 1958.*
32. **Spiro RG :** Glycoproteins. *Adv. Protein Chem.* **27 :** *349, 1973.*

33. **Stanek J *et al* :** The Monosaccharides. *Academic Press, Inc., New York. 1963.*
34. **Stumpf PK, Conn EE (editors) :** Carbohydrates : Structure and Function. Vol. 3 : **Preiss J (editor) :** The Biochemistry of Plants : A Comprehensive Treatise. *Academic Press, Inc., New York. 1980.*
35. **Whelan WJ :** Biochemistry of Carbohydrates. *Butterworth, London. 1975.*
36. **Whistler RL, Wolfrom ML (editors) :** Methods in Carbohydrate Chemistry. *Academic Press, Inc., New York . 1962–1980 (8 volumes).*
37. **Whistler RL, Smart CL :** Polysaccharide Chemistry. *Academic Press, Inc., New York. 1953.*
38. **Wolfrom ML, Tipson RS :** Advances in Carbohydrate Chemistry and Biochemistry. *Academic Press, Inc., New York. 1945–1984 (42 volumes).*
39. **Wood WA :** Carbohydrate metabolism. *Ann. Rev. Biochem.,* **35** *: 521-558, 1966.*

PROBLEMS

1. α-amylose is an unbranched glucose polymer. Why would this polymer not be as effective a storage form for glucose as glycogen?
2. The stems of bamboo, a tropical grass, can grow at the phenomenal rate of 0.3 m/d under optimal conditions. Given that the stems are composed almost entirely of cellulose fibres oriented in the direction of growth, calculate the number of sugar residues per second that must be added enzymatically to growing cellulose chains to account for the growth rate. Each D-glucose unit in the cellulose molecule is about 0.45 nm long.
3. Both cellulose and α-amylose consist of (1→4)-linked D-glucose units and can be extensively hydrated. Despite this similarity, a person on a diet consisting predominantly of α-amylose (starch) will gain weight, whereas a person on a diet of cellulose (wood) will starve. Why?
4. The practically pure cellulose obtained from the seed threads of the plant genus *Gossypium* (cotton) is tough, fibrous, and completely insoluble in water. In contrast, glycogen obtained from muscle or liver disperses readily in hot water to make a turbid solution. Although they have markedly different physical properties, both substances are composed of (1→4)-linked D-glucose polymers of comparable molecular weight. What features of their structures cause these two polysaccharides to differ in their physical properties ? Explain the biological advantages of their respective properties.
5. A polysaccharide of unknown structure was isolated, subjected to exhaustive methylation, and hydrolyzed. Analysis of the products revealed three methylated sugars: 2, 3, 4-tri-*O*-methyl-D-glucose, 2, 4-di-*O*-methyl-D-glucose, and 2, 3, 4, 6-tetra-*O*-methyl-D-glucose, in the ratio 20 : 1 : 1. What is the structure of the polysaccharide?
6. One of the critical functions of chondroitin sulfate is to act as a lubricant in skeletal joints by creating a gel-like medium that is resilient to friction and shock. This function appears to be related to a distinctive property of chondroitin sulfate: the volume occupied by the molecule is much greater in solution than in the dehydrated solid. Why is the volume occupied by the molecule so much larger in solution?
7. Why is it not possible to write on an oily paper?
8. Why does an apple turn orangish when kept in the air?
9. What is gun cotton?
10. In what forms carbohydrates are stored in plants and animals?
11. If one has diabetes, should he avoid potatoes because they raise one's blood pressure?
12. Why starch forms turbid colloidal solution, when extracted from granules with hot water?
13. Why do we need to fold a paper to tear it along a straight line?
14. Why do the smooth pages of paper, after being dipped in water and dried, get wrinkled?

15. Cellulose is soluble in:
(*a*) Ammoniacal $Cu(OH)_2$ solution (*c*) Water
(*b*) Organic solvents (*d*) None of these

ANSWERS TO PROBLEMS

1. As an unbranched polymer, α-amylose has only one nonreducing end. Therefore, only one glycogen phosphorylase molecule could degrade each α-amylose molecule. Because glycogen is highly branched, there are many nonreducing ends per molecule. Consequently, many phosphorylase molecules can release many glucose molecules per glycogen molecule.
2. 7,840 residues/s.
3. The human intestinal enzyme that splits (1 →4) linkages between glucose residues is absolutely specific for the (α1→4) linkage; cellulose, with its (β1→4) linkages, cannot be digested.
4. Native cellulose consists of glucose units linked by (β1→4) glycosidic bonds. The β linkage forces the polymer chain into an extended conformation. A parallel series of these extended chains can form intermolecular hydrogen bonds, aggregating into long, tough, insoluble fibres. Glycogen consists of glucose units linked together by (α1→4) glycosidic bonds. The α linkage causes a bend in the chain and prevents the formation of long fibres. In addition, glycogen is highly branched. Because many of its hydroxyl groups are exposed to water, glycogen is highly hydrated, and can be extracted as a dispersion in hot water.

   The physical properties of these two polymers are well suited for their biological roles. Cellulose serves as a structural material in plants, consistent wih its side-by-side aggregation into insoluble fibres. Glycogen serves as a storage fuel in animals. The highly hydrated glycogen granules with their abundance of nonreducing ends can be rapidly hydrolyzed by glycogen phosphorylase to release glucose-1-phosphate.
5. Chains of (1→6)-linked D-glucose residues with occasional (1→3)-linked branches, with about one branch for every 20 residues in the polymer.
6. The negative charges on chondroitin sulfate repel each other and force the molecule into an extended conformation. Furthermore, the polar structure attracts many water molecules (water of hydration), which increases the molecular volume of the chondroitin sulfate.
7. Writing with ink on a surface depends on both the ink and the surface. Ink contains a colouring substance (dye) dissolved in water or a volatile solvent. As paper is made of cellulose, when we write with a fountain pen, the colouring substance in the ink enters the fine capillaries through capillary action. When the solvent evaporates, dry ink remains in the capillaries, thus making the visible mark on the paper. If the surface of the paper is oily, the oil repels the ink which as a result does not stick to the paper to produce a mark. So, it is not possible to write on oily paper with a fountain pen.
   On the other hand, a ball-point pen does not write on oily paper for a different reason. When drawn over the oily surface, the ball on the tip of a ball-point pen slips and does not rotate as there is no friction. As a result, the thick ink does not flow out and does not write.
8. Apples contain tannins which act as *astringent agents i.e., is they demoisturise the fruit.* Their concentration in the apple depends upon the season and the stage of maturity of the fruit. When the apple is green, the concentration of tannins is high and it decreases as the fruit matures. When the apple is sliced or cut, tannins get exposed to the air. In presence of some enzymes that apples contain, tannins combine with oxygen in the air to form oxides which impart an orangish hue to the sliced part.

9. Gun cotton is a highly inflammable, explosive chemical called cellulose nitrate. If is prepared by treating cotton, a natural almost pure (94%) cellulose, with concentrated nitric acid and hence the name gun cotton. Due to its inflammable nature and with the arrival of better plastic material, its use has been discontinued. However, it is still used for making certain kinds of explosives.
10. In the form of starch in plants and in the form of glycogen in animals.
11. No, any starch will increase one's blood sugar some. One should always eat a well-balanced diet that includes various starches.
12. Because starch molecule is highly hydrated since it contains many exposed hydroxyl groups.
13. Paper is made of cellulose fibres found in plant cell walls. The fibres are beaten into pulp and joined together by the action of water to form sheets of paper. It is these fibres which break when we try to tear a sheet of paper. As these fibres are randomly placed with respect to each other, they can break in any direction when one pulls a paper apart. However, when we fold a paper, the fibres get uniformly weakened along a straight line. So, when this folded paper is pulled apart to tear it, those weakened fibres give way easily and the paper gets torn along the straight line.
14. Paper is made of cellulose fibres and other ingredients mixed together. The smooth surface of finished paper results from the process of paper making itself. During manufacture, the wet film of pulp is drained, dried and finally calendered by pressing it between steam-heated rollers which set the fibres in place in a somewhat stretched condition. When paper is dipped in water, the sizing material gets loosened, the fibres soak water and get deformed. If this wet paper is allowed to dry, it gets wrinkled because the fibres dry in a deformed state.
15. (*a*) Ammoniacal $Cu(OH)_2$ solution

CONTENTS

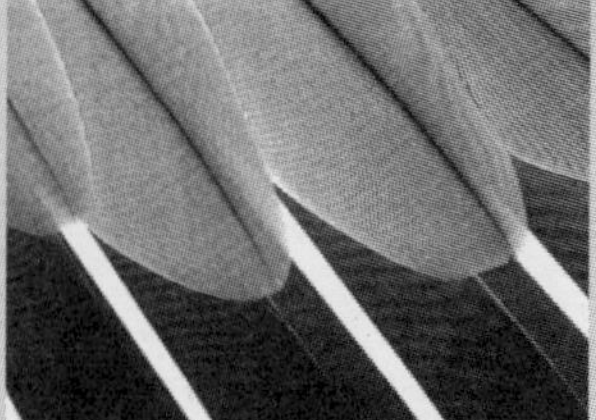

Two examples of the thousands of biological structures composed predominantly of protein

Lenses of eyes, as in this net-casting spider, which focus light rays

Feathers, which are adaptations in birds for thermal insulation, flight, and sex recognition

[*Courtesy: (Upper) Mantis wildlife Films and (Lower) Frans Lanting*]

CHAPTER

9

Proteins–I
General Structure

"I believe that as the methods of structural chemistry are further applied to physiological problems, it will be found that the significance of the hydrogen bond for physiology is greater than that of any other single structural feature."

–*Pauling LC (in the Nature of the Chemical Bond, 1939)*

IMPORTANCE

The name protein (*proteios*^{G} = pre-eminent or first) was first suggested, in 1838, by a Swedish chemist **Berzelius** to a Dutch chemist **Mulder**, who referred it to the complex organic nitrogenous substances found in the cells of the living beings. *They are most abundant intracellular macromolecules and constitute over half the dry weight of most organisms.* "Proteins occupy a central position in the architecture and functioning of living matter. They are intimately connected with all phases of chemical and physical activity, that constitute the life of the cell. Some proteins serve as important structural elements of the body, for example, as hair, wool and collagen, an important constituent of connective tissue ; other proteins may be enzymes, hormones or oxygen- carriers. Still other proteins participate in muscular contraction, and some are associated with the genes, the hereditary factors."(**White A, Handler P** and **Smith EL**, 1964).

They are, therefore, essential to cell structure and cell function. The proteins with catalytic activity (enzymes) are largely responsible for determining the phenotype or

JÖNS JACOB BERZELIUS

(LT, 1779–1848)

Berzelius, a Swedish chemist, was born in Sorgard, on August 20, 1779. He was Professor of Chemistry at Stockholm, Sweden. He is famous for developing the modern system of chemical symbols. He isolated several elements and is also credited with the discovery of *selenium* (Se), a nonmetallic element, closely related to sulfur. The story of the discovery of selenium goes like this. Once, the housekeeper of the house, where Berzelius lived, complained that he had eaten too much garlic. But it was due to his having examined a selenium deposit in a sulfuric acid chamber. He had inhaled fumes which were probably those of *dimethyl selenide* (Garlicky odour is actually due to the secretion of dimethyl selenide in the breath. In the body, various selenium compounds can actually get converted to this compound to give rise to garlicky odour). This set Berzelius thinking and he went on to discover selenium.

It is also noteworthy that the suggestion to name the element lithium (Li), which was discovered by **Johan August Arfwedson** in 1817, came from his fellow scientist Berzelius, who called this as **lithia** (from *lithos*^{G} = stone or rock) because it was obtained from the mineral petatite which was stony in nature (However, later the British chemist **Humphrey Davy** changed the name to *lithium,* which stuck). Berzelius also coined the name '**polymer**' in 1827. He developed the **concept of atomic weight**. With Davy, he laid the foundations of **electrochemistry**. He is also credited with publishing the earliest book on Biochemistry entitled '*Lectures in Animal Chemistry*' in 1806. Berzelius also suggested the term "protein", in 1838, to a Dutch chemist **G. J. Mulder** by writing as follows:

> "The word protein that I propose to you... I would wish to derive from *proteios,* because it appears to be the primitive or principal substance of animal nutrition that plants prepare for the herbivores, and which the latter then furnish to the carnivores."
>
> *(J.J. Berzelius, letter to G.J. Mulder, 1838)*

Berzelius died on August 7, 1848.

GERARDUS JOHANNES MULDER

(LT, 1802–1880)

Mulder, a Dutch chemist, studied some complex organic nitrogenous compounds found in the cells of living beings. They were most abundant intracellular macromolecules and constituted over half the dry weight of most organisms. Mulder, at the suggestion of a Swedish chemist **J.J. Berzelius** (in 1838), named them **proteins**. His comment about proteins reads as follows:

> "There is present in plants and animals a substance which...is without doubt the most important of the known substances in living matter, and without it, life would be impossible on our planet. This material has been named Protein."

properties of a cell in a particular environment. The total hereditary material of the cell or genotype dictates which type of protein the cell can produce. In fact, the proteins have built into their structure the information that instructs them in "what to do" (catalytic activity), "where to go" in the cell (intracellular organization) and "when and how to perform" (control of function through bacterial interactions with other proteins, small activators or inhibitors).

The bacterium *Escherichia coli* is estimated to contain about 5,000 different types of compounds which include some 3,000 different kinds of proteins and 1,000 nucleic acids. In humans, there may be 1,00,000 different kinds of proteins, each with a unique structure. None of the proteins of *E. coli* is identical with any of the human proteins. Thus, in about 1.5 million species of living organisms, there are probably 10^{10} to 10^{12} different kinds of protein molecules and about 10^{10} different kinds of nucleic acids.

Animals are 4 times richer in proteins than plants whose composition of proteins is merely 5%. The constituent elements of proteins are carbon, hydrogen, oxygen, nitrogen and very rarely sulfur also. In certain complex proteins, phosphorus occurs as well. The elemental composition of proteins in plants and animals presents a great deal of variation (Table 9–1).

Table 9–1. Elemental composition of plant and animal proteins

Protein	*Carbon*	*Hydrogen*	*Oxygen*	*Nitrogen*
Green plants	54	7	38	0.003
Mammals	21	10	62	3.0

Most animal proteins contain from 0.5 to 2.0 per cent sulfur. Insulin is, however, a notable exception to this in possessing about 3.4% sulfur.

AMINO ACIDS

Structure

All proteins are macromolecules because of their very high molecular weights. These are the polymers, *i.e.*, chain-like molecules produced by joining a number of small units of amino acids called monomers. The amino acids are, therefore, regarded as **'building blocks of proteins'**. The general formula of an amino acid along with its space-filling model is presented in Fig, 9–1.

> The term **macromolecule** was coined by **Schaudinger** in 1920s

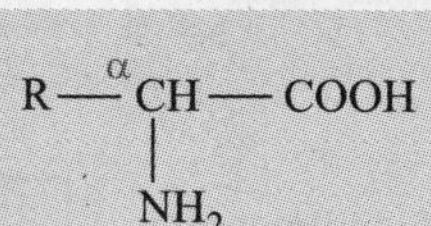

General formula

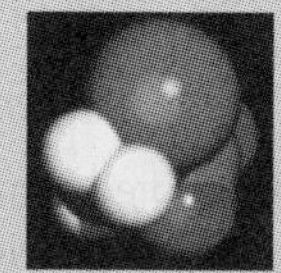

Space-filling model
The large green sphere represents the R group.

Fig. 9–1. General structure of an amino acid

Each amino acid is a nitrogenous compound having both an acidic carboxyl (—COOH) and a basic amino (—NH_2) group. R stands for the side chains that are different for each amino acid. R can be as simple as a hydrogen atom (H) as in glycine or a methyl group (—CH_3) as in alanine or a more complex structure. The first carbon is the part of the carboxyl group. The second carbon, to which is attached the amino group, is called the *α-carbon*. The α-carbon of most amino acids is joined by covalent bonds to 4 different groups. Thus, the α-carbon in all the amino acids is asymmetric except in glycine where the α -carbon is symmetric. Because of this asymmetry, the amino acids (of course, except glycine) exist in two optically active forms : those having —NH_2 group to the right are designated as D-forms and those having —NH_2 group to the left as L-forms (Fig. 9–2). However, the two amino acids, threonine and isoleucine have two asymmetric carbon atoms each and thus have $2^n = 2^2 = 4$ optical isomers. At pH 7.0, both the carboxyl and amino groups are ionized.

D-amino acid L-amino acid

Fig. 9–2. Two optical isomers of amino acids

Specific Rotation

It is interesting to note that *the amino acids found in the proteins belong to the L-series*. Many of the naturally occurring L-amino acids rotate the plane of polarized light to the right (*i.e.*, they are dextrorotatory) while others rotate the plane of polarized light to the left (*i.e.*, they are levorotatory). Table 9–2. shows the specific rotation of some L-amino acids.

Table 9–2. Specific rotation of some amino acids isolated from proteins

Amino acid	*Specific rotation,* $[\alpha]_D^{25°}$
L-alanine	– 86.2
L-histidine	– 38.5
L-phenylalanine	– 34.5
L-threonine	– 28.5
L-serine	– 7.5
L-alanine	+ 1.8
L-glutamic acid	+ 12.0
L-isoleucine	+ 12.4
L-arginine	+ 12.5
L-lysine	+ 13.5

Thus, it is evident that the symbols D and L do not identify the property of light rotation, *i.e.*, D-isomers can be either dextrorotatory (*d*) or levorotatory (*l*); similarly, L-isomers can be either (*d*) or (*l*). However, to minimize confusion, the symbols *d* and *l* are usually not used nowadays. Moreover, the **DL nomenclature** has limitations because it describes the asymmetry of only one carbon atom in a compound and many biomolecules contain two or more asymmetric carbon atoms. The R and S classification or **RS notation** of isomers, introduced in 1956 by **Robert Cahen**, **Christopher Ingold** and **Vladimir Prelog** and being currently used in chemistry, is more useful for defining the asymmetry of biomolecules because its accounts for all asymmetric carbons in an isomer. If any atom (other than H) or group on the asymmetric carbon is on the right side, that asymmetric carbon is designated as R (from *rectus*[L] = right) ; conversely, if any atom (other than H) or group is on the left side, the asymmetric carbon is then designated as S (from *sinister*[L] = left). As an illustration, isoleucine (with two asymmetric carbon atoms, C–2 and C–3) will have four stereoisomers. And the configuration of the biological isomer, L-isoleucine is more completely described as (2 S), (3 S)–isoleucine (Fig. 9–3).

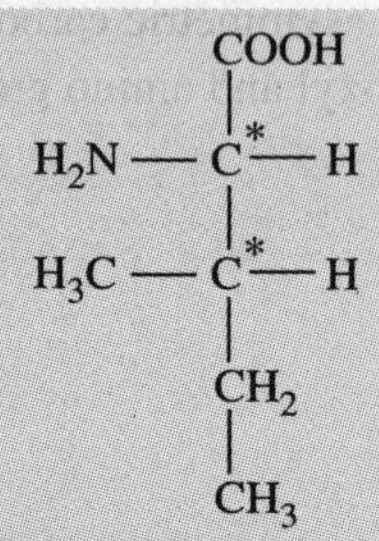

Fig. 9–3. Fisher projection formula of L-isoleucine of (2S), (3S)-isoleucine

The two asymmetric carbon atoms are indicated by an *asterisk* each.

Distribution in Proteins

The distribution of the 20 amino acids is not uniform in all proteins. Nearly 40% by weight of fibroin and 25% by weight of collagen are accounted for by glycine. Fibroin is all rich in alanine (30% by weight). Serine and threonine predominate in casein and phosvitin. Collagen (in connective tissue), gliadin (in wheat) and zein (in corn) are rich in proline. Human serum albumin with 585 amino acid residues has only one tryptophan moiety.

The **pulses** are notable as they lack S-containing amino acid, methionine (Met) but contain good amount of the basic amino acid, lysine (Lys); whereas **cereals** lack lysine but have sufficient quantity of methionine. When combined, these make good the deficiency of each other through mutual supplementation and are therefore better utilized in human body.

Location in Proteins

Amino acids with uncharged polar side chains are relatively hydrophilic and are usually on the outside of the proteins, while the side chains on nonpolar amino acids tend to cluster together on the inside. Amino acids with acidic or basic side chains are very polar, and they are nearly always found on the outside of the protein molecules.

Physical Properties

Regarding their physical characteristics, the amino acids are colourless crystalline substances. The crystal form may vary from slender needles (tyrosine) to thick hexagonal plates (cystine). They may be either tasteless (tyrosine), sweet (glycine and alanine) or bitter (arginine). It is interesting to note that (S)-phenylalanine is bitter in taste whereas its enantiomer (R)-phenylalanine is sweet. They have high melting points (above 200°C) and often result in decomposition. Amino acids are soluble in polar solvents such as water and ethanol but they are insoluble in nonpolar solvents such as benzene and ether. Some amino acids like tryptophan, tyrosine, histidine and phenylalanine absorb ultraviolet rays at 260–290 m μ . This property enables the identification of not only these amino acids but also the proteins which contain them.

The order of hydrophilicity and hydrophobicity of various amino acids is as follows:

Hydrophilic (= polar) amino acids:

Tyr > Ser > Asp > Glu > Asn > Gln > Arg

Hydrophobic (= apolar) amino acids:

Phe > Ala > Val > Gly > Leu > Cys

Although over 100 amino acids have been shown to be present in various plants and aminals, only 20 of them (L-isomers) are found as constituent of most proteins. These 20 amino acids of proteins are often referred to as **standard, primary** or **normal amino acids**, to distinguish them from others. The almost universal use of these 20 amino acids for the synthesis of various protein molecules is "one of nature's enigmatic rules." These have, therefore, rightly been called as the "**magic 20**", a phrase coined by **Francis Crick**.

Electrochemical Properties

All molecules possessing both acidic and basic groups might exist as uncharged molecules or in ionic form or as a mixture of both. Calculations have revealed that an aqueous solution of most amino acids, glycine for example, can have only one part of uncharged molecules in one lakh parts of the ionic form. Hence, it is more accurate to write the structure of glycine as

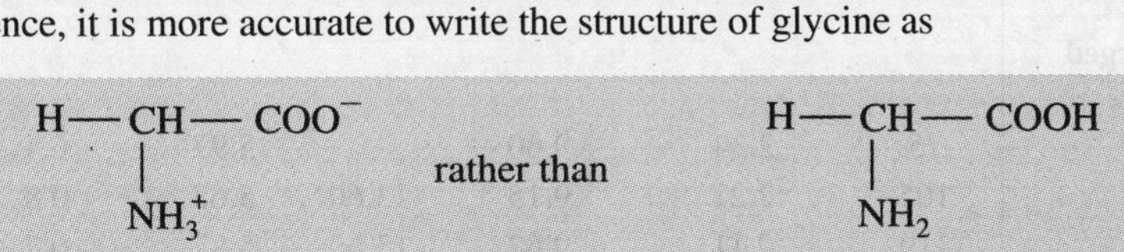

In the ionic form, the proton migrates from the carboxyl to amino group, thus producing carboxylate (COO^-) and ammonium (NH_3^+) ions.

Amino acids react with both acids and bases. Hence, they are **amphoteric** in nature.

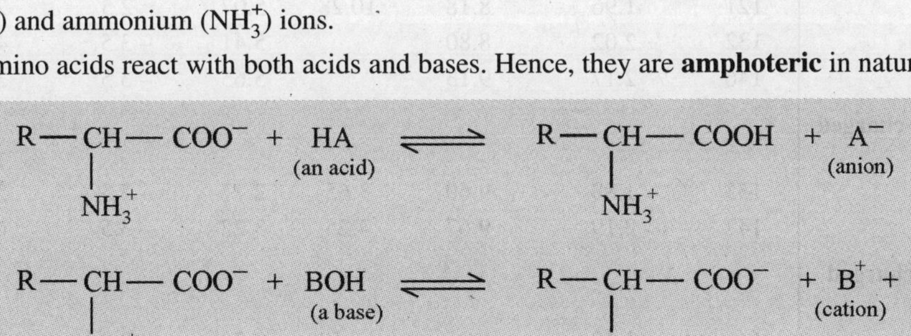

Thus, in acid solution, the COO^- ion acquires a proton and the amino acid becomes an ammonium salt of the acid. Conversely, in alkaline solution, the NH_3^+ ion loses a proton and the amino acid becomes the anion of a salt. These reactions are of reversible nature and depend on the pH of the medium. Henceforth, amino acids serve as buffers and tend to prevent pH change when an acid or a base is added.

The α–COOH and α–NH_2 groups of amino acids are ionized in solutions at physiological pH, with the deprotonated carboxyl group (—COO^-) bearing a negative charge and the protonated amino group (—NH_3^+) a positive charge. An amino acid in its dipolar state is called a *zwitterion*. The dissociable α-COOH and α-NH_3^+ groups are responsible for the two characteristic pK values (pK_1 for —COOH and pK_2 for —NH_3^+) of α-amino acids. An amino acid with a third dissociable group in its side chain (*i.e.*, R group) exhibits an additional pK_R value (refer Table 9–3). However, pK values depend on temperature, ionic strength and the microenvironment of the ionizable group.

Table 9–3. Some properties associated with the standard amino acids

Amino acid	*Mr*	pK_1 (*—COOH*)	pK_2 (*—NH_3^+*)	p*KR* (*R group*)	*pI*	*Hydropathy index**	*Occurrence in proteins (%)†*
Nonpolar R groups							
Alanine	89	2.34	9.69		6.01	1.8	9.0
Valine	117	2.32	9.62		5.97	4.2	6.9
Leucine	131	2.36	9.60		5.98	3.8	7.5
Isoleucine	131	2.36	9.68		6.02	4.5	4.6
Proline	115	1.99	10.96		6.48	– 1.6	4.6
Phenylalanine	165	1.83	9.13		5.48	2.8	3.5
Tryptophan	204	2.38	9.39		5.89	– 0.9	1.1
Methionine	149	2.28	9.21		5.74	1.9	1.7
Polar, uncharged R groups							
Glycine	75	2.34	9.60		5.97	– 0.4	7.5
Serine	105	2.21	9.15	13.60	5.68	–0.8	7.1
Threonine	119	2.11	9.62	13.60	5.87	– 0.7	6.0
Tyrosine	181	2.20	9.11	10.07	5.66	– 1.3	3.5
Cysteine	121	1.96	8.18	10.28	5.07	2.5	2.8
Asparagine	132	2.02	8.80		5.41	– 3.5	4.4
Glutamine	146	2.17	9.13		5.65	– 3.5	3.9
Negatively-charged R groups							
Aspartate	133	1.88	9.60	3.65	2.77	– 3.5	5.5
Glutamate	147	2.19	9.67	4.25	3.22	– 3.5	6.2
Positively-charged R groups							
Lysine	146	2.18	8.95	10.53	9.74	– 3.9	7.0
Arginine	174	2.17	9.04	12.48	10.76	– 4.5	4.7
Histidine	154	1.82	9.17	6.00	7.59	– 3.2	2.1

* A scale combining hydrophobicity and hydrophilicity can be used to predict which amino acid will be found in an aqueous environment (-values) and which will be found in a hydrophobic environment (+ values).

† Average occurrence in over 200 proteins

Note that in nature, the commonest amino acid is alanine and the rarest amino acid is tryptophan.

(Adapted from Klapper MH, 1977)

As an instance, ionization of the amino acid, **alanine** (Fig. 9–4) is discussed.

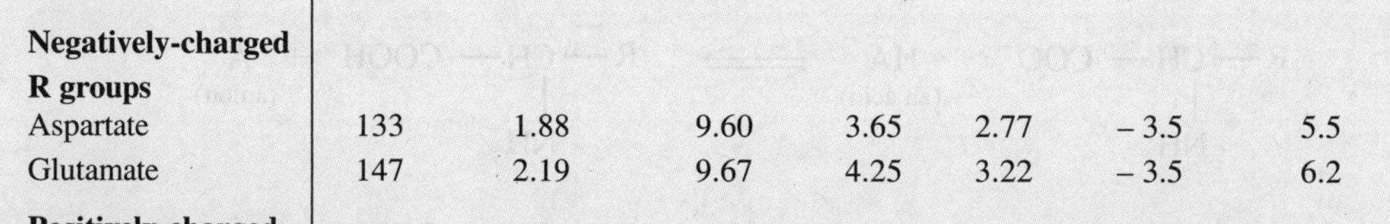

Fig. 9–4. The two dissociations of L-alanine

Titration involves the gradual addition or removal of protons. Fig. 9–5 shows the titration curve of the diprotic form of alanine. Each molecule of added base results in the net removal of one proton from one molecule of amino acid. *The plot has two distinct stages*, each corresponding to the removal of one proton from alanine. At low pH values, molecules of alanine bear a net charge of + 1 because both functional groups are protonated ; for example, at pH 0.35, 99% molecules are positively-charged and are present as ^+H_3N—$CH(CH_3)$—COOH. At the midpoint in the first stage of titration, in which the —COOH group of alanine loses it proton, equimolar concentrations of proton-donor, ^+H_3N—$CH(CH_3)$—COOH and proton-acceptor, ^+H_3N —CH(CH3)—COO^- species are present. At the midpoint of a titration, the pH is equal to the pK_a of the protonated group being titrated. For alanine, the pH at the midpoint is 2.34, thus its —COOH group has a pK_a of 2.34.

> The **pH and pK_a** are simply convenient notations for proton concentration and equilibrium constant for ionization, resepectively. The pK_a is a measure of the tendency of a group to give up a proton, with that tendency decreasing tenfold as the pK_a increases by one unit.

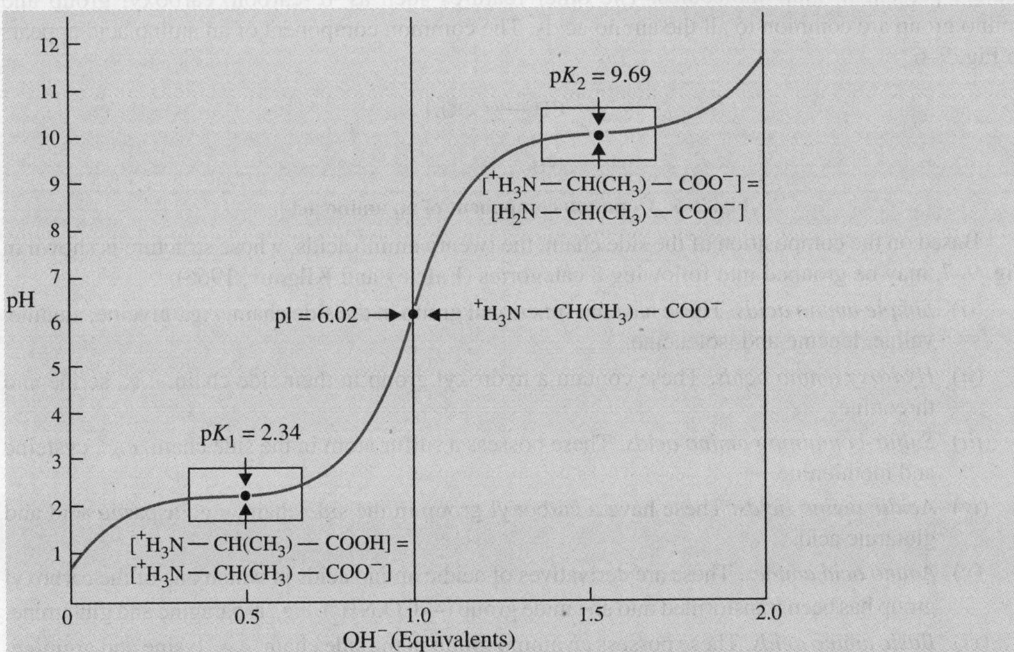

Fig. 9–5. The titration curve of alanine

The ionic species predominating at key points in the titration are shown above the graph. The shaped boxes, centred about pK_1 = 2.34 and pK_2 = 9.69, indicate the regions of maximum buffering power.

As the titration proceeds, another important point is reached at pH **6.02.** Here there is a point of inflexion, at which removal of the first proton is essentially complete, and removal of the second has just begun. At this pH, the alanine is present as its dipolar form, fully ionized (^+H_3N—$CH(CH_3)$—COO^-) but with no *net* electric charge.

The second stage of titration corresponds to the removal of a proton from the $-NH_3^+$ group of alanine. The pH at the midpoint of this stage is **9.69** which is equal to the pK_a for the $-NH_3^+$ group. The titration is complete at a pH of **11.69**, at which point the molecules of alanine bear a net charge of −1 because both functional groups are deprotonated. And 99% of the molecules are negatively charged and are present as H_2N—$CH(CH_3)$— COO^-. It may, however, be noted that the characteristic pH value of 6.02, at which the average net charge is zero (called the isoelectric point or isoelectric pH), is midway between the two p*Ka* values [(2.35 + 9.69)/2].

The study of titration curve provides three useful informations:

1. It gives a quantitative measure of the pK_a of each of the two ionizing groups, 2.34 for the —COOH group and 9.69 for the —NH_3^+ group in case of alanine.
2. It gives information that alanine, say for example, has *two* regions of buffering power : one centred around the first pK_a of 2.34 and the second centred around pH 9.69.
3. It also provides information regarding the relationship between the net electric charge of the amino acid and the pH of the solution.

Classification

Three systems of classifying amino acids are in vogue.

A. On the basis of the composition of the side chain or R group. There are 20 different amino acids which regularly appear in proteins. These possess a side chain which is the only variable feature present in their molecules. The other features such as α -carbon, carboxyl group and amino group are common to all the amino acids. The common component of an amino acid appears in Fig. 9–6.

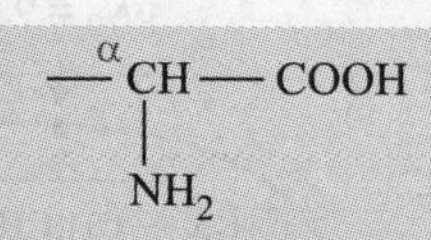

Fig. 9–6. Common component of an amino acid

Based on the composition of the side chain, the twenty amino acids, whose structure is shown in Fig. 9–7, may be grouped into following 8 categories (**Fairley** and **Kilgour**, 1966):

(*i*) *Simple amino acids.* These have no functional group in the side chain, *e.g.*, glycine, alanine, valine, leucine and isoleucine.

(*ii*) *Hydroxy amino acids.* These contain a hydroxyl group in their side chain, *e.g.*, serine and threonine.

(*iii*) *Sulfur-containing amino acids.* These possess a sulfur atom in the side chain, *e.g.*, cysteine and methionine.

(*iv*) *Acidic amino acids.* These have a carboxyl group in the side chain, *e.g.*, aspartic acid and glutamic acid.

(*v*) *Amino acid amides.* These are derivatives of acidic amino acids in which one of the carboxyl group has been transformed into an amide group (—CO.NH_2), *e.g.*, asparagine and glutamine.

(*vi*) *Basic amino acids.* These possess an amino group in the side chain, *e.g.*, lysine and arginine.

(*vii*) *Heterocyclic amino acids.* These amino acids have in their side chain a ring which possesses at least one atom other than the carbon, *e.g.*, tryptophan, histidine and proline.

(*viii*) *Aromatic amino acids.* These have a benzene ring in the side chain, *e.g.*, phenylalanine and tyrosine.

The classification given above is only a practical one and can conveniently be followed. It does not, however, strictly delimit the various categories. For example, tryptophan may also be included under aromatic amino acids and similarly, histidine under basic amino acids.

B. On the basis of the number of amino and carboxylic groups. McGilvery and **Goldstein** (1979) have classified various amino acids as follows:

I. *Monoamino-monocarboxylic amino acids:*

1. Unsubstituted — Glycine, Alanine, Valine, Leucine, Isoleucine
2. Heterocyclic — Proline
3. Aromatic — Phenylalanine, Tyrosine, Tryptophan
4. Thioether — Methionine

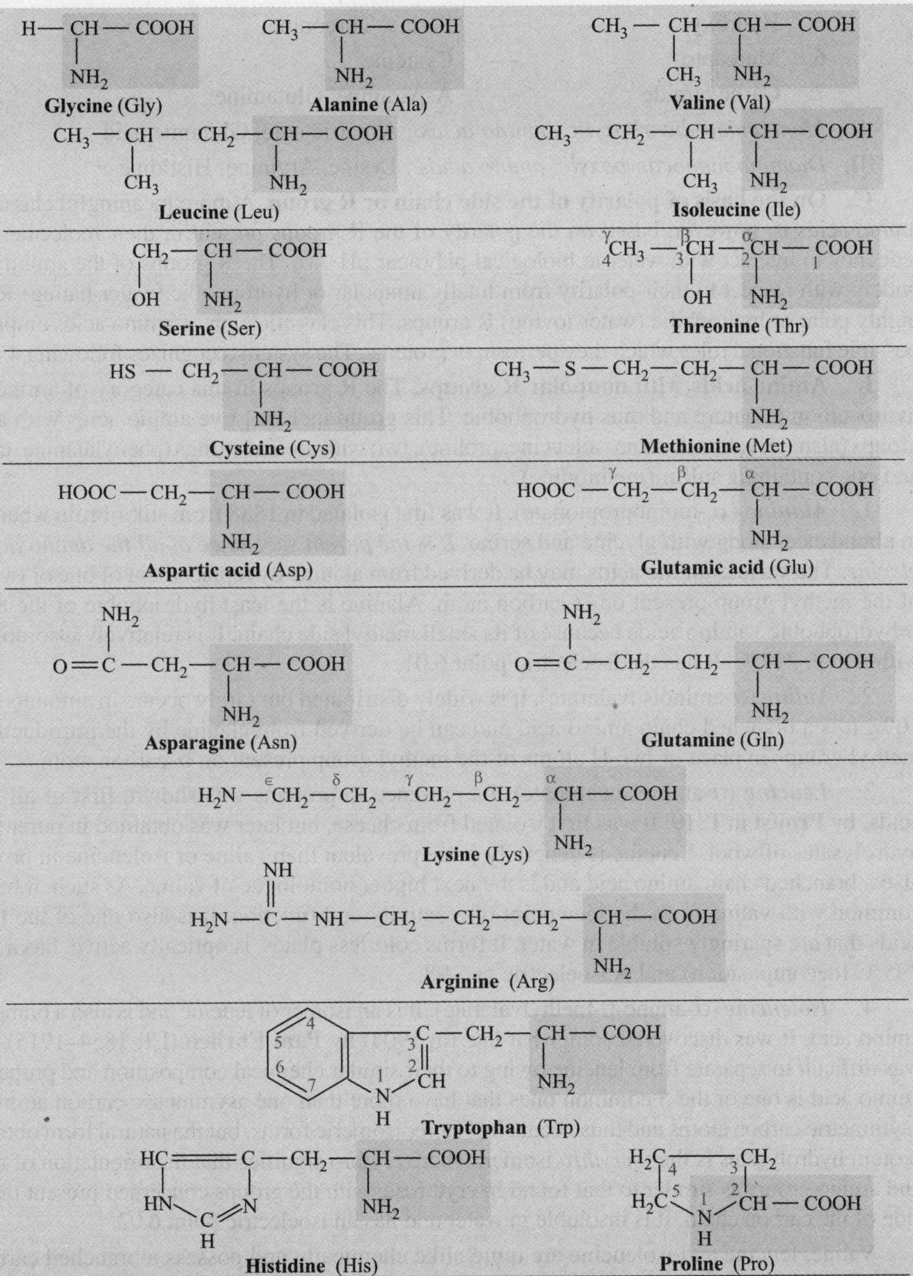

Fig. 9–7. The twenty amino acids ("Magic 20") found in proteins

The standard three-letter abbreviation for each amino acid is written in brackets.
The common component is shown in a *blue enclosure*.

5. Hydroxy — Serine, Threonine
6. Mercapto — Cysteine
7. Carboxamide — Asparagine, Glutamine

II. *Monoamino-dicarboxylic amnino acids* : Aspartic acid, Glutamic acid

III. *Diamino-monocaryboxylic amino acids* : Lysine, Arginine, Histidine

C. On the basis of polarity of the side chain or R group. A more meaningful classification of amino acids is, however, based on the polarity of the R groups present in their molecules, *i.e.*, their tendency to interact with water at biological pH (near pH 7.0). The R groups of the amino acids vary widely with respect to their polarity from totally nonpolar or hydrophobic (water-hating) R groups to highly polar or hydrophilic (water-loving) R groups. This classification of amino acids emphasizes the possible functional roles which they perform in proteins. The system recognizes following 4 categories:

I. Amino acids with nonpolar R groups. The R groups in this category of amino acids are hydrocarbon in nature and thus hydrophobic. This group includes five amino acids with aliphatic R groups (alanine, valine, leucine isoleucine, proline), two with aromatic rings (phenylalanine, tryptophan) and one containing sulfur (methionine).

1. ***Alanine*** (α -aminopropionate). It was first isolated in 1888 from silk fibroin where it occurs in abundance, along with glycine and serine. *It is the parent substance of all the amino acids except glycine.* The various amino acids may be derived from alanine by replacement of one or two H atoms of the methyl group present on α-carbon atom. Alanine is the least hydrophobic of the 8 nonpolar (=hydrophobic) amino acids because of its small methyl side chain. It is relatively insoluble in water with a m.p. 297°C. It has the isoelectric point 6.01.

2. ***Valine*** (α-aminoisovalerate). It is widely distributed but rarely occurs in amounts exceeding 10%. It is a branched chain amino acid and can be derived from alanine by the introduction of two methyl groups in place of two H atoms of the methyl group present on α-carbon atom.

3. ***Leucine*** (α-aminoisocaproate). Its presence in proteins was shown, first of all the amino acids, by **Proust** in 1819. It was first isolated from cheese, but later was obtained in purer form from hydrolysates of wool. Leucine is generally more prevalent than valine or isoleucine in proteins. It is also a branched chain amino acid and is the next higher homologue of valine. As such, it has much in common with valine from the viewpoint of reactivity and function. It is also one of the few amino acids that are sparingly soluble in water. It forms colorless plates, is optically active, has a m.p. 293–295°C (decomposition) and is isoelectric at 5.98.

4. ***Isoleucine*** (α-amino-β-methylvalerate). It is an isomer of leucine and is also a branched chain amino acid. It was discovered somewhat late (in 1904) by **Paul Ehrlich** (LT, 1854–1915) because it was difficult to separate from leucine owing to their similar chemical composition and properties. This amino acid is one of the 5 common ones that have more than one asymmetric carbon atoms. It has 2 asymmetric carbon atoms and thus occurs in 4 stereoisomeric forms, but the natural form obtained from protein hydrolysates is the L-*erythro* isomer, where erythro signifies that the orientation of the methyl and amino groups is similar to that found in erythrose, with the groups concerned present on the same side of the carbon chain. It is insoluble in water and has an isoelectric point 6.02.

Valine, leucine and isoleucine are quite alike chemically and possess a branched carbon chain. As most animals cannot synthesize the branched carbon chain, these amino acids are, therefore, indispensable or essential in the diet.

5. ***Proline*** (2-pyrrolidinecarboxylate). *It is present in almost all proteins that have been studied*, but zein from corn and gelatin are relatively high in proline content. It is also found in wheat gluten (12%). It is a cyclized derivative of glutamic acid. Its α -amino group is not free but is substituted by a portion of its nonpolar R group to yield a cyclic structure. The secondary amino group in proline is held in a rigid conformation. This reduces the structural flexibility of the protein because of the maximal steric hindrance of its side chain. It forms colourless crystals and has a m.p. 220–222°C. It is relatively insoluble in water and has an isoelectric point 6.48. Unlike other amino acids, proline is very soluble in

ethanol. Moreover, it does not give many of the characteristic amino acid tests which are generally based on the presence of the unsubstituted α-amino group. Hence, *strictly, it is not an amino acid.*

6. ***Phenylalanine*** (α-amino-β-phenylpropionate). It is one of the two common, clearly benzenoid amino acids. It is of widespread distribution and closely resembles tyrosine in structure. It also cannot be formed in animals because of its aromatic ring. Interestingly, (S)-phenylalanine is bitter whereas its stereoisomer, (R)-phenylalanine is sweet.

7. ***Tryptophan*** (α-amino-β-3-indolepropionate or β-indolylalanine). It was discovered in 1901 in the laboratory of **F.G. Hopkins**, one of the pioneers in biochemistry. *Tryptophan is the most complex amino acid found in proteins.* It is a heterocyclic amino acid and is a derivative of indole (with one N atom). Although widespread, it is usually limited in quantity. Tryptophan is the only amino acid of proteins which is nearly completely destroyed upon acid hydrolysis. *Tryptophan is so weakly polar, however, that it is often classified as a nonpolar amino acid.*

8. ***Methionine*** (α-amino-β-methylmercapto-butyrate). It is the only common amino acid possessing an ether linkage. Cereals have sufficient quantity of methionine whereas pulses lack in it. It is a methylation product of homocysteine. Apart from its role as a protein constituent and as an essential amino acid, *methionine is also important as a donor of active methyl groups.*

It may, however, be seen that **proline** (and also its 4-hydroxylated derivate, **hydroxyproline**) do not contain a primary amino group (—NH_2). The amino group in them is, in fact, utilized in ring formation. Both proline and hydroxyproline are actually the *substituted α-amino acids* as they have a secondary amino group $\left(\begin{smallmatrix}C\\C\end{smallmatrix}\!>\!NH\right)$ in their molecule (By contrast, all other 19 protein amino acids are designated as *α-amino acids*). Since they are an integral part of the protein molecule and exhibit similar reactions, they are usually inadvertently referred to as amino acids.

> Recently (1996), **homocysteine** has been regarded in the same category as cholesterol, in inducing heart attacks. In 1980s, a study on 15,000 doctors in the U.S. called "physicians health study", showed that those with homocysteine higher than 12% had a risk of heart attack that was three-and-a-half times higher than those with lower levels. In fact, homocysteine, though not a well-known name, badly clogged the arteries. However, researchers have pointed out that high homocysteine levels are easier to treat than high cholesterol levels and can be effected by consuming abundant amounts of a B-vitamin, folic acid.

> Some authors use the term **"imino acids"** for proline and hydroxyproline which is not correct, because an imine is characterized by a $>CH = NH$ and not a $\begin{smallmatrix}C\\C\end{smallmatrix}\!>\!NH$.

II. Amino acids with polar but uncharged R groups. The R groups of these amino acids are more soluble in water *i.e.*, more hydrophilic than those of the nonpolar amino acids because they contain functional groups that form hydrogen bonds with water. This category includes 7 amino acids, *viz.*, glycine, serine, threonine, tyrosine, cysteine, asparagine and glutamine. The polarity of these amino acids may be due to either a hydroxyl group (serine, threonine, tyrosine) or a sulfhydryl group (cysteine) or an amide group (asparagine, glutamine). The R group of glycine, a single hydrogen atom, is too small to influence the high degree of polarity of the α-amino and α-carboxyl groups.

9. ***Glycine** or **glycocoll*** (α-aminoacetate). *It is the simplest amino acid and the only one lacking an asymmetric α-carbon atom.* Where it is present in a protein, the minimal steric hindrance of the glycine side chain allows much more structural flexibility than the other amino acids. Glycine represents the opposite structural extreme in comparison to proline. It is, therefore, unique in being optically inactive. It is one of the first to be isolated from proteins and has a characteristic sweet taste, hence so named (*glykos*[G] = sweet). *Glycine is present in abundance in scleroproteins.* This occurs in especially large amounts in the protein gelatin (25% of the total amino acids) and in silk fibroin (40%). In addition, many nonprotein compounds also contain glycine as a component part, for instance glutathione (a common tripeptide), sarcosine (= N-methylglycine) and hippuric acid (= benzoylglycine). Methylation of glycine yields sarcosine which is found in many peptides. Further

methylation produces betaine, a substance which exists exclusively in the zwitterionic or salt form (for the same reason, zwitterionic formulae are sometimes called as *betaine structures*).

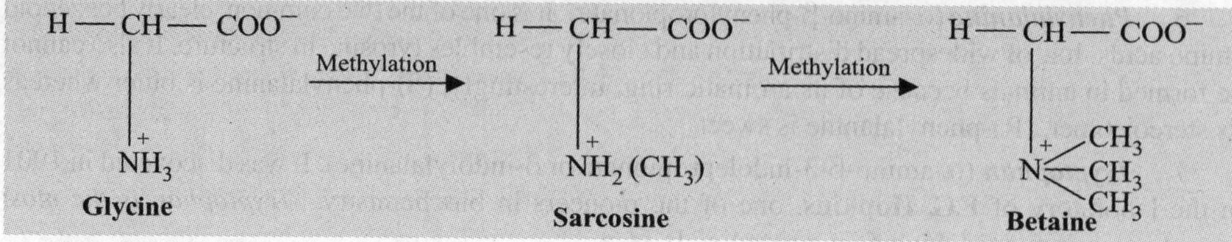

It crystallizes in colourless prisms; has a m.p. 260°C (decomp.), turning brown at 228°C. Glycine, the nonsymmetric amino acid, is sometimes considered nonpolar. However, glycine's small R group (a hydrogen atom) exerts essentially no effect on the hydrophilicity of the molecule.

10. ***Serine*** (α -amino-β-hydroxypropionate; derived from the word, *serum*). It was first obtained from silk protein, sericin. It is found in all proteins but occurs in unusually high proportions in silk proteins, fibroin and sericin. It forms colourless crystals with a m.p. 228°C (decomp.) and is soluble in water. It has an isoelectric point 5.68. Serine contains an alcoholic hydroxyl group which participates in ester formation. Esters with phosphoric acid have physiological significance as components of nucleotides and proteins. The unesterified serine residue appears to conduct special functions in many enzymes. The hydroxyl group on serine (and also on threonine) makes it much more hydrophilic and reactive than alanine and valine.

11. ***Threonine*** (α-amino-β-hydroxybutyrate). It was the last of the common amino acids to be discovered in proteins (**Meyer** and **Rose**, 1936). Its very name points towards its relationship with the sugar threose. Threonine is the next higher homologue to serine. It has two asymmetric carbon atoms and thus occurs in 4 stereoisomeric forms, namely D and L threonine and D and L allothreonine. The natural form is the L-*threo* isomer, hence its name. The *erythro* (unnatural) form of threonine is commonly referred to as allothreonine. Threonine is less abundant than serine in most proteins.

12. ***Tyrosine*** (α-amino β-(*p*-hydroxypyhenyl) propionate). It is the other aromatic amino acid normally found as a component of proteins. It was first isolated in 1857 from cheese and hence so named (*tyros*[G] = cheese). The phenolic group of tyrosine is weakly acidic and loses its proton above pH 9. It is sparingly soluble in water. It is destroyed during acid hydrolysis of proteins but may be isolated after enzymic hydrolysis. Some proteins (*e.g.*, protamines) contain almost no tyrosine.

13. ***Cysteine*** (α-amino-β-mercaptopropionate). It is the sulfur analogue of serine and is *one of the most reactive amino acids found in proteins*. It contains sulfhydryl (SH) group which is quite reactive and esp., easily dehydrogenated. When it is dehydrogenated (*i.e.*, oxidized), two molecules of it join to form a molecule of the amino acid cystine, which on being reduced brings back 2 cysteine molecules (see reaction on page 187). Fibrous proteins such as keratin from hair are especially rich in cystine (12%). *Cysteine and its oxidation product cystine are together counted as one of the twenty protein amino acids.* **Alberts** *et al* (1994) have included cysteine (and glycine, as well) among amino acids with nonpolar side chains.

Cystine, from which the name 'cysteine' was later derived, was isolated in 1843 from urinary stones and was hence so named (*cystos*[G] = bladder).

14. ***Asparagine*** (β-amide of α-aminosuccinate). It was the first amino acid to be discovered in 1806 and was first isolated in 1813 from a plant, *Asparagus*, for which it was so named. This is the β-amide of aspartic acid and has been isolated from proteins after enzymic hydrolysis. Asparagine has long been known as a constituent of plant tissues, being found widely in all the Leguminosae and Gramineae, and in many other seeds, roots and buds. L-asparagine can be prepared from lupin seedlings. It is water-soluble and has the m.p. 234°C. It has an isoelectric point 5.41.

15. ***Glutamine*** (γ-amide of α-aminoglutarate). It is the homologue of asparagine. This is the γ-amide of glutamic acid and has been isolated from proteins after enzymic hydrolysis. Free glutamine is found in many animal and plant tissues. It is even less stable toward hydrolysis than is asparagine. *Glutamine occurs widely in proteins.*

Amino acids of categories I and II are jointly referred to as *neutral amino acids* because each one of them contains one acidic and one basic group.

III. Amino acids with negatively-charged (= acidic) R groups. These are monoamino-dicarboxylic acids. In other words, their side chain contains an extra carboxyl group with a dissociable proton. The resulting additional negative charge accounts for the electrochemical behaviour of proteins. The two amino acids which belong to this category are aspartic and glutamic.

16. ***Aspartic acid*** (α -aminosuccinate). Its presence in protein was discovered by **Ritthausen** in 1868. It is the parent compound of asparagine. It has an extra carboxyl group and is, hence, acidic in solution. It has a m.p. 271°C and an isoelectric point 2.77. It is used as dispersing agent and in forming aspartame.

17. ***Glutamic acid*** (α-aminoglutarate). It is an acidic amino acid which is the homologue of aspartic acid. Its presence in protein was discovered also by **Ritthausen** in 1866. It was found in wheat gluten (33%), hence so named. It is the parent compound of glutamine and *occurs widely in proteins.*

Both aspartic and glutamic acids in almost all proteins are linked to other amino acids only through their α -carboxyl and α-amino groups, not through their side chain carboxyl groups. Aspartic and glutamic acids are usually called aspartate and glutamate respectively to emphasize that their side chains are nearly always negatively charged at physiological pH.

IV. Amino acids with positively-charged (=basic) R groups. These are diamino-monocarboxylic acids. In other words, their side chain contains an extra amino group which imparts basic properties to them. Lysine, arginine and histidine belong to this category.

18. ***Lysine*** (α , ε-diaminocaproate). It has a second amino group at ε-position. Lysine is generally abundant in animal proteins but present in limited amounts in plant proteins, such as those of corn and wheat. Ionic charges are provided by protonation of the amino group of the ε-carbon. It is nutritionally important since it is not synthesized by higher animals. Pulses contain good amount of lysine whereas cereals lack in it.

19. ***Arginine*** (α-amino-δ-guanidinovalerate). It is abundant in highly basic proteins of the cell nucleus (histones) and in sperm proteins, such as the protamines from salmon and herring sperm. In the latter, arginine may be as much as 80% of the total amino acids. It is found generally in all proteins, although in lesser amounts. Arginine is unique in possessing the guanidinium group, and due to which *it is more strongly basic than lysine*. Protonation of the guanidinium group provides ionic charges. It has a m.p. 207°C and has the isoelectric point 10.76. It plays an important role in the production of urea as an excretory product and in nitric oxide cell signaling.

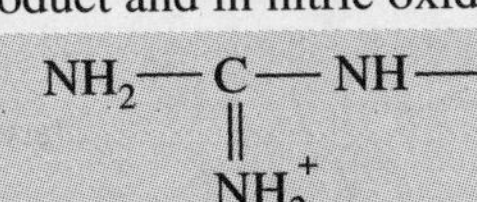

Guanidinium group (charged form) of arginine

This group is very basic because its positive charge is stabilized by resonance.

20. ***Histidine*** (α-amino-β-imidazolopropionate). *Histidine is the last entry in the list of "magic 20"*. It contains a weakly basic imidazolium R group with pK'_a value of 6.0, and, therefore, is less than 10% protonated at pH 7. In many enzyme proteins, it functions as a proton donor or acceptor. *Histidine is the only amino acid which has a proton that dissociates in the neutral pH range*. It is this property which allows certain histidine residues to play an important role in the catalytic activities of

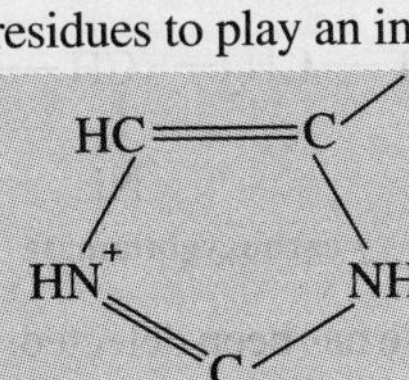

Imidazolium group (charged form) of histidine

some enzymes. Histidine occurs in limited quantities in most proteins, but hemoglobin, protamines and histones contain relatively large amounts. It is also a component in the simple peptides carnosine and anserine. Histidine carries an extra basic group, making it alkaline but its basic properties are clearly marginal. The two ring nitrogen atoms have a relatively weak affinity for an H^+ and are only partly positive at neutral pH. It is soluble in water. It has a m.p. 277°C and an isoelectric point 7.59. Histidine is an essential constituent of the food animals.

> **Carnosine** ($C_9H_{14}O_3N_4$) is an optically active crystalline dipeptide; m.p. 260°C; found in muscle tissue.

> In addition to the 20 common amino acids found in proteins, a 21st amino acid, named **selenocysteine,** has been discovered and shown to be an active site amino acid in several enzymes, including the antioxidant enzyme *glutathione peroxidase* and *5′-deiodinases*. Selenocysteine is derived from serine and has unique chemical properties; substitution of selenocysteine with cysteine may have a pronounced inhibitory effect on enzyme activity. It is because of the need for selenocysteine that trace amounts of selenium are required in the diet.

NONSTANDARD PROTEIN AMINO ACIDS

In addition to the above-mentioned twenty standard amino acids which are building blocks of proteins and have a wide range of distribution in proteins, several other amino acids exists. These have a limited distribution but may be present in high amounts in a few proteins and hence deserve mention. As an example, **hydroxyproline** has a limited distribution in nature but constitutes as much as 12% of the composition of collagen, an important structural protein of animals. Similarly, **hydroxylysine** is also a component of collagen, where it accounts for about 1% of the total amino acids. **N-methyllysine** is found in myosin, a contractile protein of muscle. Another important nonstandard or less common amino acid is **γ-carboxyglutamate**, which is found in the blood-clotting protein, prothrombin as well as in certain other proteins that bind Ca^{2+} in their biological function.

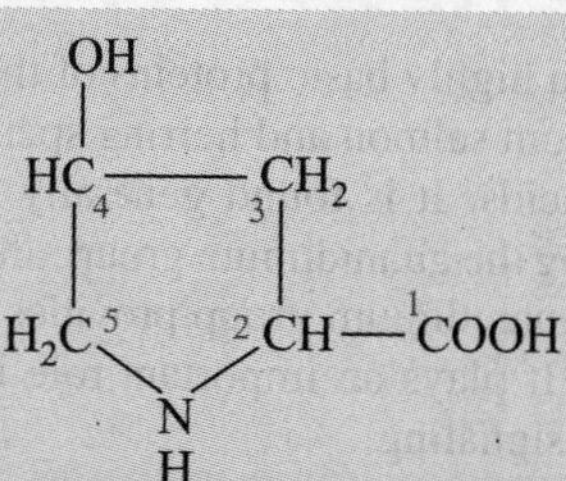

L-hydroxyproline
(erythro-4-hydroxy-L-proline)

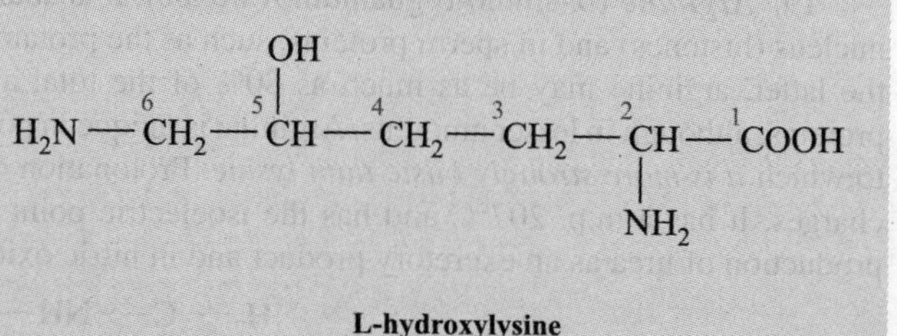

L-hydroxylysine
(erythro-5-hydroxy-L-lysine)

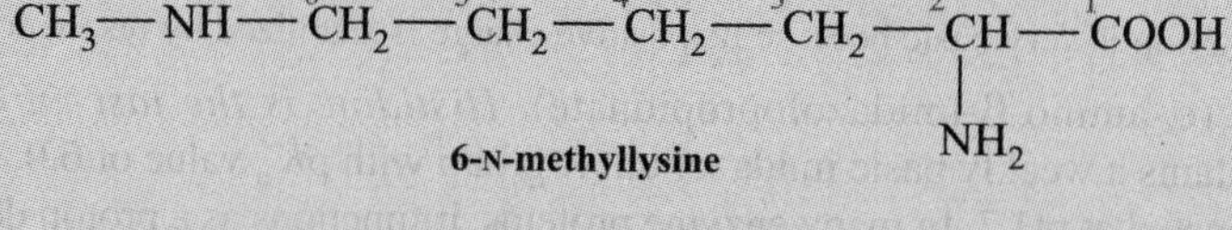

6-N-methyllysine

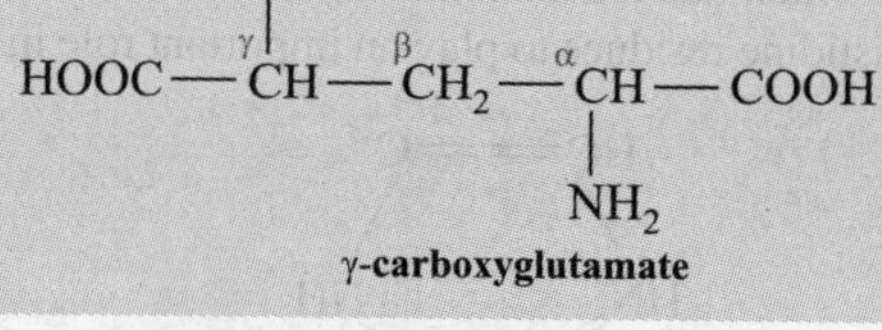

γ-carboxyglutamate

In protein from corn, **α-aminoadipate** has been detected. It has one CH_2 group more than glutamic acid. Another amino acid, **α-ε-diaminopimelate** has been found in bacterial protein.

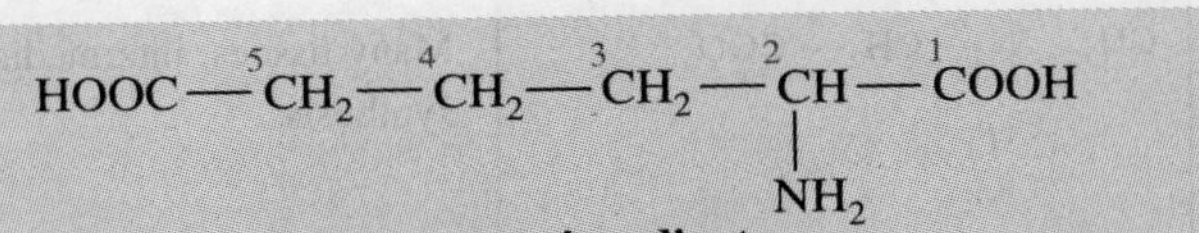

L-α-aminoadipate

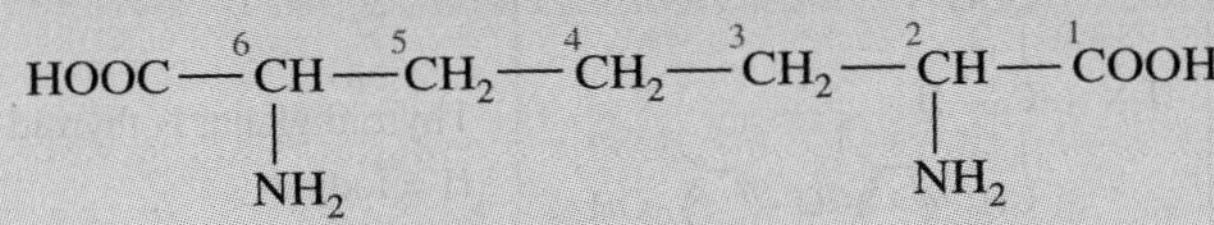

L, L-α, ε-diaminopimelate

NONPROTEIN AMINO ACIDS

There are some 300 additional amino acids which are never found as constituents of proteins but which either play metabolic roles or occur as natural products in various animal, plant and microbial systems (Table 9–4).

Table 9-4 Some biologically important amino acids not found in proteins

Name	*Formula*	*Biochemical source, Function*
β-alanine	$H_3\overset{+}{N}-CH_2-CH_2-COO^-$	Found in the vitamin pantothenic acid and in some important natural peptides
D-alanine	$H-C(COO^-)(CH_3)-\overset{+}{N}H_3$	In polypepetides in some bacterial cell walls
γ-aminobutyric acid	$H_3\overset{+}{N}-CH_2-CH_2-CH_2-COO^-$	Brain, other animal tissues; functions as neurotransmitter
D-glutamic acid	$H-C(COO^-)(CH_2-CH_2-COO^-)-\overset{+}{N}H_3$	In polypeptides in some bacterial cell walls
L-homoserine	$H_3\overset{+}{N}-C(COO^-)(CH_2-CH_2OH)-H$	Many tissues; an intermediate in amino acid metabolism
L-ornithine	$H_3\overset{+}{N}-C(COO^-)(CH_2-CH_2-CH_2\overset{+}{N}H_3)-H$	Many tissues; an intermediate in arginine synthesis

Sarcosine	$CH_3-N(H)-CH_2-COO^-$	Many tissues; intermediate in amino acid synthesis
L-thyroxine	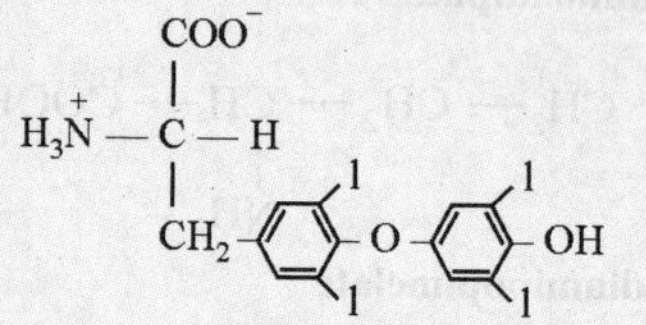	Thyroid gland; is thyroid hormone (I = iodine)

Among the important nonprotein amino acids, which play metabolic roles, are L-ornithine, L-citrulline, β -alanine, creatine and γ-aminobutyrate. **L-ornithine** and **L-citrulline** occur in free state in the animal tissues and are metabolic intermediates in the urea cycle. L-ornithine possesses one CH_2 group less than its homologue, lysine. **β-alanine**, an isomer of alanine, occurs free in nature and also as a constituent of an important vitamin pantothenic acid and of coneyzme A. It is also found in the naturally occurring peptides, carnosine and anserine. The quaternary amine **creatine**, a derivative of glycine, plays an important role in the energy storage process in vertebrates where it is phosphorylated and converted to creatine phosphate. Lastly, γ-aminobutyrate is found in free form in the brain.

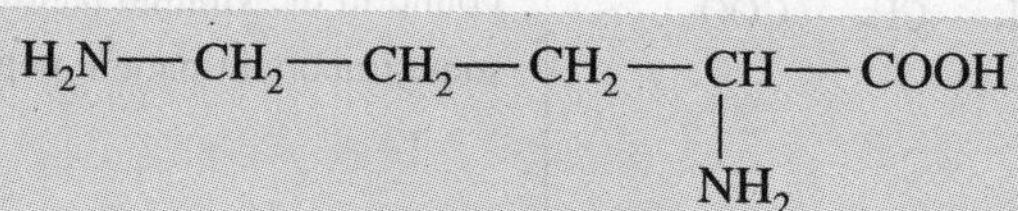

L-ornithine

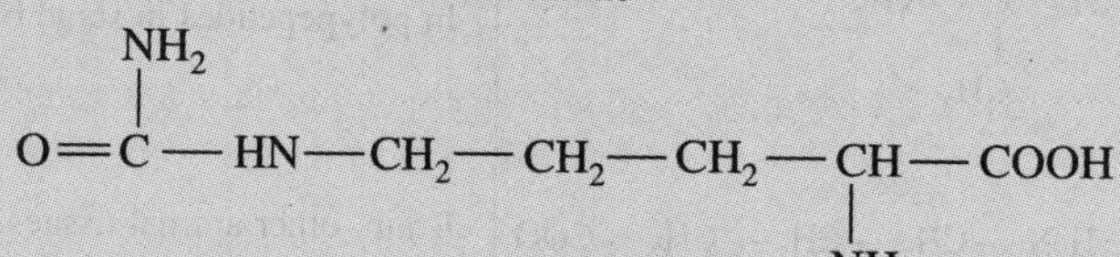

L-citrulline

β-alanine

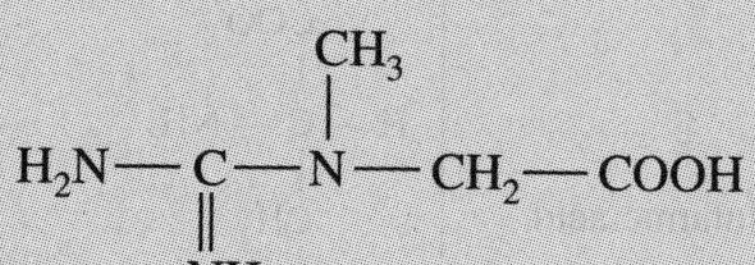

Creatine

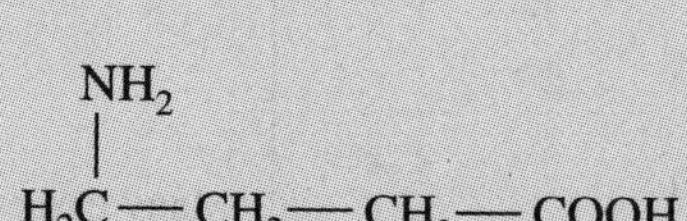

γ-aminobutyrate

Higher plants are especially rich in nonprotein amino acids. These nonprotein amino acids are usually related to the protein amino acids as homologues or substituted derivatives. They have a limited distribution, sometimes to a single species even. Thus, **L-azetidine-2-carboxylic acid**, a homologue of proline, accounts for 50% of the nitrogen present in the rhizome of Solomon's seal, *Polygonatum multiflorum*. **Orcylalanine** is found in the seed of cornocockle, *Agrostemma githago*. It may be considered as a substituted phenylalanine. Furthermore, in the toxic polypeptides of *Amanita phalloides*, in addition to **hydroxyleucine**, ***allo*-threonine** is also found.

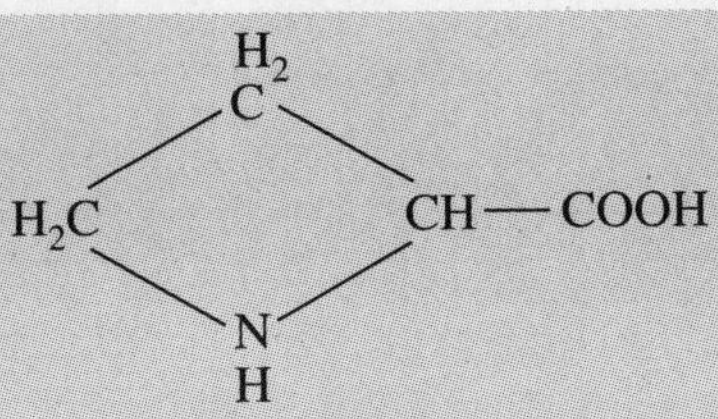

L-azetidine-2-carboxylic acid

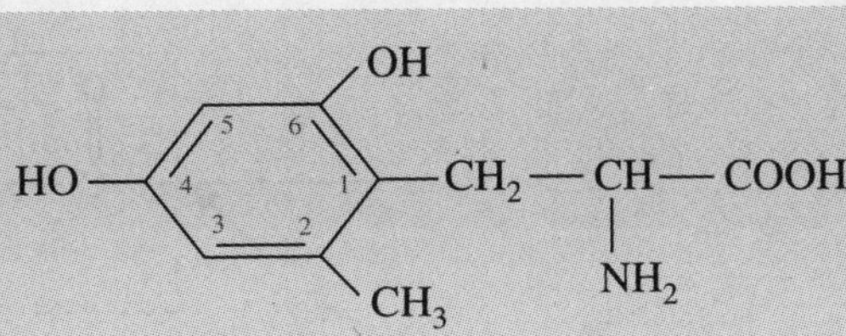

Orcyl-L-alanine
(4, 6-dihydroxy-2-methyl phenyl–L-alanine)

It is interesting to note that *all of the amino acids incorporated by organisms into proteins are of the L form.* Although it is evident that random mixtures of D and L amino acids could not form protein structures as precise as that shown for myoglobin in Fig. 9–35, the absolute preference for the L isomer is puzzling. Indeed, we see that each of the 3 major classes of biological macromolecules exhibits a strong preference for one stereoisomer class or the other. Most naturally-occurring polysaccharides involve D-sugars, as do DNA and RNA. It may be that mutual interaction between these substances has established these relationships very early in the evolution of life. But why was a particular set of enantiomers chosen at all? It is hard to see how L-amino acids have any inherent superiority over their D-isomers for biological function. Indeed D-amino acids exist in nature, and some play important biochemical roles (some examples are given in Table 9–4), but they are never found in proteins.

Many scientists have attempted to provide explanations for this "handedness preference" in biology. Most point to an intrinsic asymmetry in the behaviour of subnuclear particles, a kind of asymmetry that gives the electrons emitted in β decay a preferential left-hand spin. Such influences are very weak but might, in a competition between primitive organisms using L or D protein, give a slight advantage to one or the other. After billions of generations, even a small advantage can become overwhelming.

PEPTIDES

Peptide Bond

The amino acid units are linked together through the carboxyl and amino groups to produce the primary structure of the protein chain. The bond between two adjacent amino acids is a special type of amide bond, in which the hydrogen atom of amino (-NH_2) group is replaced by an R radical.Such a substituted amide bond is known as the **peptide bond**.

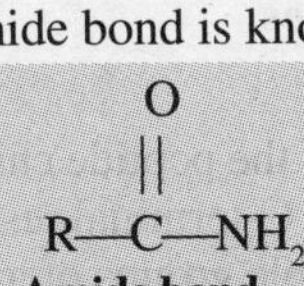

Amide bond
(an unsubstituted amide)

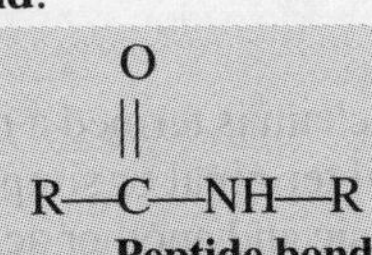

Peptide bond
(an *N*-substituted amide)

And the chain, thus formed by linking together of many amino acid units, is called a **peptide chain** (Fig. 9–8).

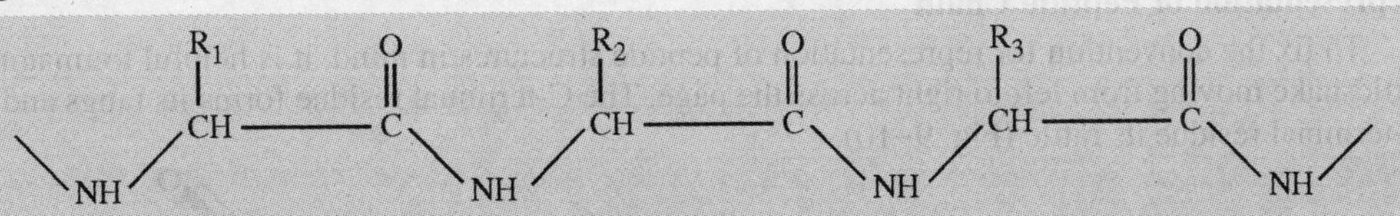

Fig. 9–8. Peptide chain

The characteristic structure of the peptide bond is shown is Fig. 9–9.

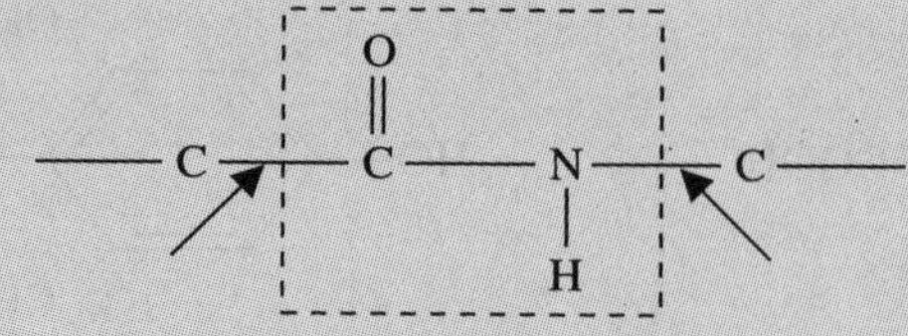

Fig. 9–9. Peptide bond

The peptide bond is shown enclosed in the *dashed box*. The four atoms (C, O, N, H) of the peptide bond form a rigid planar unit. There is no freedom of rotation about the C—N bond. On the contrary, the 2 single bonds (shown with *arrows*) on either side of the rigid peptide unit, exhibit a high degree of rotational freedom.

Each peptide chain is of considerable length and may possess from 50 to millions of amino acid units. Depending on the number of amino acid molecules composing a chain, the peptides may be termed as a **dipeptide** (containing 2 amino acid units), a **tripeptide** (containing 3 amino acid units) and so on. If a peptide is made up of not more than 10 amino acids, it is called an **oligopeptide**; beyond that it is a **polypeptide**. Polypeptides when they are made up of over 100 amino acids are, sometimes, called as **macropeptides**. *Strictly speaking, the proteins are polypeptides with more than 100 amino acids.* All naturally-occurring important peptides, however, possess a shorter individual name, such as glutathione etc. Proteins differ widely in amino acid content. Various types of proteins in an organism may have varied amounts of a particular amino acid. Some amino acids are in abundance in one protein, may be in meagre amounts in others and may even be lacking in the rest. Tryptophan, for instance, lacks in certain proteins. However, *most of the proteins contain all the 20 amino acids.* As the number and manner in which the amino acids are grouped is highly variable, the number of proteins approaches almost to infinity.

> Sometimes, the word **'proteinoids'** is used for short polypeptides containing up to 18 amino acids (**Fried GH**, 1990).

It is analogous to the indefinite number of words that can be formed with the 26 letters of English alphabet. While the words have to be restricted in length, there is no such restriction regarding the number of amino acids that may form a protein. According to an estimate given by **Erlene B. Cunningham** (1978), if each protein molecule were to consist of only 250 amino acid residues (The term "residue" reflects the loss of the elements of water when one amino acid is joined to the other), the utilization of all the 20 different monomers would permit the formation of 10325 different protein molecules! In reality, protein molecules often contain more than 250 aminoacyl units, and hence there is possibility for an even greater number of different protein molecules.

N- and C-terminals

Each amino acid in the chain is termed a *residue*. The two ends of the peptide chain are named as amino terminal and carboxyl terminal or simply as an N-terminal and C-terminal respectively. These two terminal groups, one basic and another acidic, are the only ionizable groups of any peptide chain except those present in the side chain. The terminal amino acid with the free amino group is called as the *N-terminal amino acid* and the one with the free carboxyl group at the other end as *C-terminal amino acid*.

Representation of Peptide Chain

To fix the convention for representation of peptide structures in mind, it is helpful to imagine a rattlesnake moving from left to right across the page. The C-terminal residue forms its fangs and the N-terminal residue its rattle (Fig. 9–10).

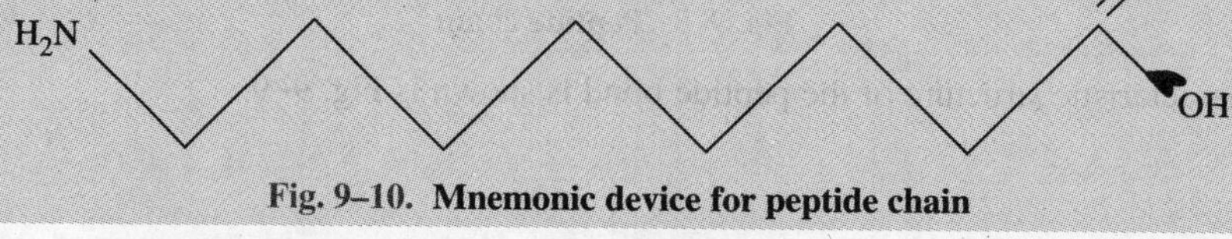

Fig. 9–10. Mnemonic device for peptide chain

Naming of Peptide Chain

In naming a polypeptide, the convention is that the N-terminal residue (which is shown at the left hand part of the structure) is written first and the C-terminal residue in the formation of each peptide

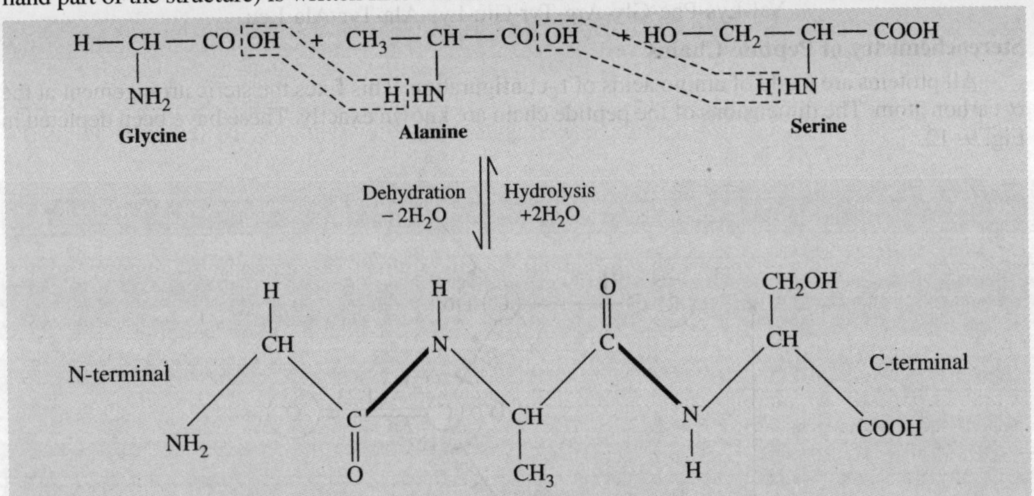

Glycyl-L-alanyl-L-serine

Fig. 9–11. Construction of a tripeptide chain from three different amino acids

The formation of each peptide bond (indicated by bold line) involved splitting of a molecule of water. Thus, the process of synthesis of proteins may be deemed as essentially a dehydration synthesis.

at the end. The names of various intermediary amino acid residues are written in the same sequence as they are placed. Further, the names of all the amino acid residues, except the last one, are written by adding the suffix *-yl* because all these are the acyl groups. The name of the last amino acid, however, is written as such. For example, a tripeptide containing glycine, alanine and serine (structure shown in Fig. 9–11) is named as glycyl-L-alanyl-L-serine and abbreviated as Gly-Ala-Ser. If the sequence of amino acid in such a tripeptide is not known, the abbreviation would be (Gly, Ala, Ser), the parenthesis and commas indicating that only the compostiton of the tripeptide is known.

Similarly, glutathione or GSH (a tripeptide containing glutamic acid, cysteine and glycine) is named as γ-glutamyl-cysteyl-glycine and abbreviated as Glu-Cys-Gly. Glutathione is a naturally occurring and widely distributed polypeptide.

Determination of the Amino Acid Sequence of a Polypeptide

This can be explained by taking the example of a **dodecapeptide** whose composition was found to be Ala_2, Arg, Glu, Gly, Leu, Lys_2, Phe, Tyr_2, Val. It was determined that the N-terminal amino acid of the dodecapeptide was valine and the C-terminal amino acid, leucine. Hydrolysis of the dodecapeptide by trypsin yielded four peptides whose structures were determined and found to be those given in A to D.

Tyr-Glu-Lys	Phe-Gly-Arg	Val-Lys	Ala-Tyr-Ala-Leu
A	**B**	**C**	**D**

Since valine was the N-terminus and leucine the C-terminus of the dodecapeptide, it is apparent that peptide C must represent the amino acid sequence at the N-terminal end, and peptide D the amino acid sequence at the C-terminal end of the dodecapeptide. To establish the order of the A and B peptides in the interior of the dodecapeptide, another sample of the dodecapeptide was hydrolyzed by *chymotrypsin*, the four peptides formed were sequenced, and their structures were found to be those given in **E** to **H**.

Ala-Leu	Glu-Lys-Ala-Tyr	Val-Lys-Phe	Gly-Arg-Tyr
E	**F**	**G**	**H**

The sequences Gly-Arg-Tyr in peptide **H** and Glu-Lys-Ala-Tyr in peptide **F** clearly establish that peptide **B** must precede peptide **A** in the dodecapeptide. Hence, the structure of the dodecapeptide is unambiguously determined to be that as shown belew:

Val-Lys-Phe-Gly-Arg-Tyr-Glu-Lys-Ala-Tyr-Ala-Leu.

Stereochemistry of Peptide Chains

All proteins are made of amino acids of L-configuration. This fixes the steric arrangement at the α-carbon atom. The dimensions of the peptide chain are known exactly. These have been depicted in Fig. 9–12.

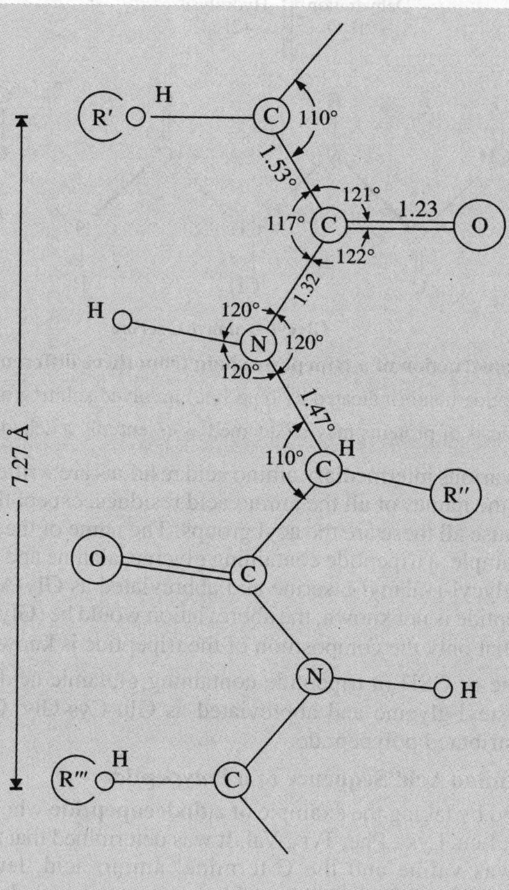

Fig. 9–12. Dimensions within the peptide chain

The peptide bond, which is an imide (substituted amide) bond, has a planar structure. The 6 atoms within the plane are related to each other by bond lengths and angles that vary little from amino acids residue to amino acid residue. Only 3 of these bonds are part of the peptide chain *per se* : the α-carbon to carbonyl carbon bond, the C–N bond, and the imide nitrogen to α-carbon bond. Since the double bond character of the C–N bond limits rotation about it, only the first and the last allow rotation. The rotation angles ϕ and ψ establish the relative positions of any 2 successive amide planes along the polypeptide chain. The α-carbon atoms can be thought of as swivel (a mechanism of two parts which can revolve the one on the other) centres for the adjacent amide planes.

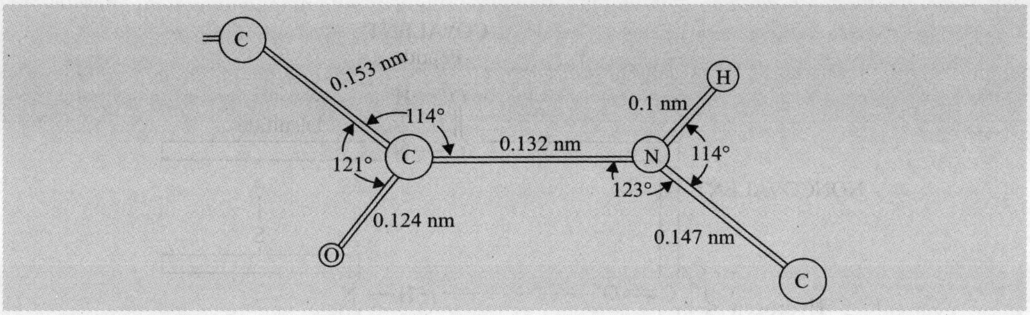

Biological Roles

Peptides participate in a number of biological activities.

1. They serve as intermediates in the formation of **proteins.**
2. They appear as constituents in a group of compounds called **alkaloids.** Majority of these have been isolated from fungi, although they are also found in higher plants. Ergotamine is a peptide alkaloid from rye ergot and has pronounced pharmacological properties. The four components of this alkaloid are lysergic acid, alanine, proline and phenylalanine.
3. Many of them possess **antibacterial activities** and are usually present in fungi and bacteria. Penicillin G with 3 components (valine, cysteine and phenylacetic acid) is a common antibiotics.
4. Certain other peptides serve as **growth factors.** Folic acid, a water-soluble vitamin, is a noteworthy example of it (see Chapter 39). Another group of peptides serving as growth factor for a variety of microorganisms is streptogenins.
5. Higher animals do synthesize certain peptides serving as **hormones** (see Chapter 34).
6. Certain peptides like glutathione participate in controlling the **oxidation-reduction potential** of the cell. This may also serve as a key intermediate in electron-transfer systems.
7. A direct correlation has been found to exist between the amount of peptides in the urine of the patients and their **mental state of disturbance.** A group of Norwegian doctors have found an excess of peptides in urine specimens from patients with psychiatric disturbances. The peptides have been shown to induce in animals some of the conditions for the development of psychiatric disorders which lead to mania, depression or schizophrenia. The urine tests can, thus, indicate if a person is suffering from mental illness. The cause of this hypersecretion of peptides is not yet well established. It may exist from birth in organic genetic derangements or may be induced in healthy people by environmental factors.

CHEMICAL BONDS INVOLVED IN PROTEIN STRUCTURE

Given a full assortment of amino acids, a cell can synthesize all of its protein components. *Protein synthesis is a multiple dehydration process* (refer Fig. 9–11). The net structure of a protein becomes possible as a result of linking together of various amino acid units. The union of these amino acids to each other forming a chain and also among various amino acid residues of different chains involves various types of chemical bonds (Fig. 9–13). These are described below.

A. Primary Bond

The principal linkage found in all proteins is the covalent **peptide bond,** —CO—NH— (**Fischer**, 1906). It is a specialized amide linkage where C atom of —COOH group of one amino acid is linked with the N atom of —NH_2 group of the adjacent amino acid. *Peptide bond is, in fact, the backbone of the protein chain.*

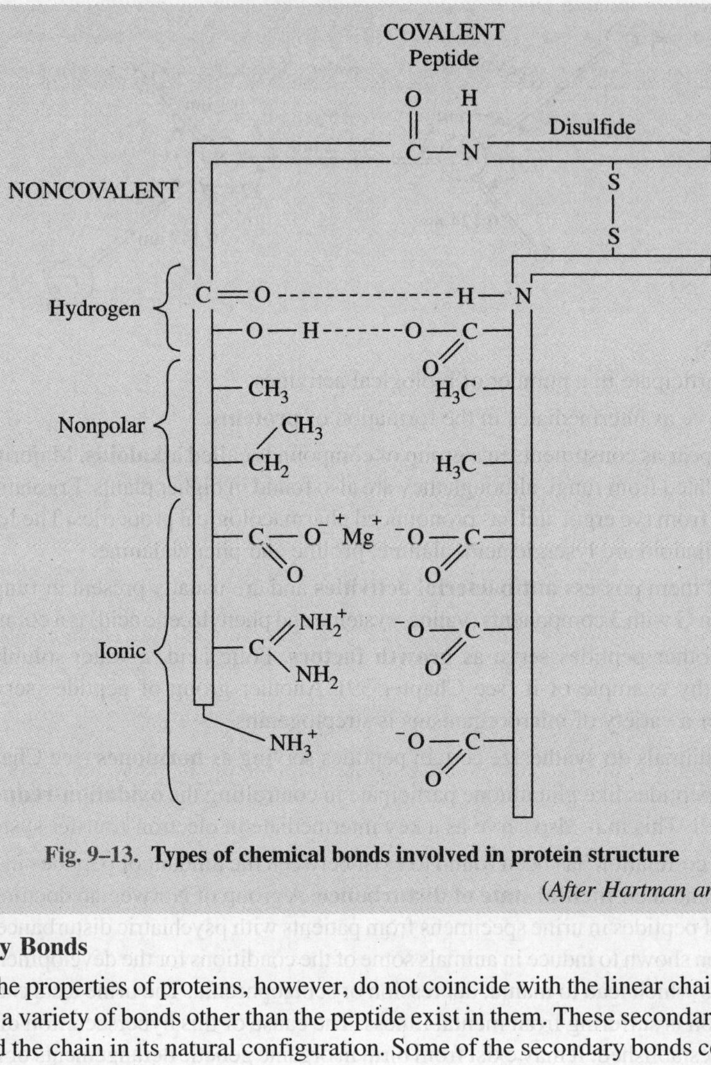

Fig. 9–13. Types of chemical bonds involved in protein structure

(After Hartman and Suskind, 1969)

B. Secondary Bonds

Many of the properties of proteins, however, do not coincide with the linear chain structure, thus indicating that a variety of bonds other than the peptide exist in them. These secondary bonds, as they are called, hold the chain in its natural configuration. Some of the secondary bonds commonly found in proteins are listed below:

1. Disulfide Bond (—S—S—). In addition to the peptide bond, a second type of covalent bond found between amino acid residues in proteins and polypeptides is the disulfide bond, which is formed by the oxidation of the thiol or sulfhydryl (—SH) groups of two cysteine residues to yield a mole of **cystine**, an amino acid with a disulfide bridge (Fig. 9–15). In generalized form, the above reaction may be written as:

$$2R{-}SH + 1/2\ O_2 \rightleftharpoons R{-}S{-}S{-}R + H_2O$$

A disulfide bond is characterized by a bond strength of approximately 50 kilocalories per mole and a bond length of about 2Å between the two sulfur atoms. Hence, disulfide bond formation between 2 cysteine residues located some distance apart in the polypeptide chain requires that the polypeptide chain be folded back on itself to bring the sulfur groups close together. Although disulfide bridges are very strong, when compared to the strength of noncovalent bonds they are very short-range since, as with all covalent bonds, even a slight extension breaks them completely. They, therefore, only stabilize the tertiary structure when it has reached something approximating to its final form.

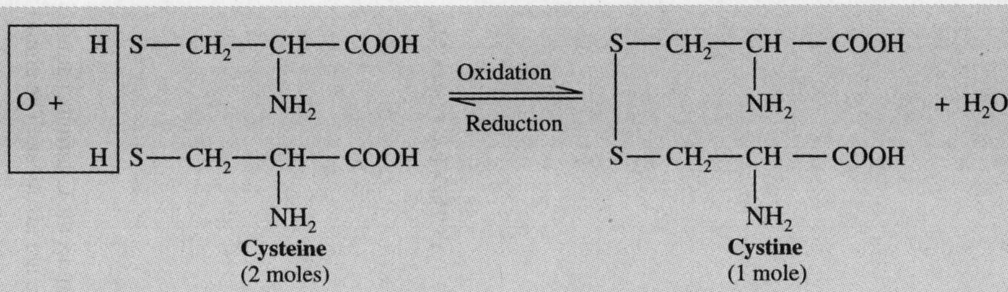

Fig. 9–14. **Formation of cystine (= dicysteine)**

Ocytocin (Fig. 9–15), a hormone stimulating the contraction of smooth muscles especially during childbirth, is an example where an internal disulfide bond is present between two cysteine units separated from each other in the peptide chain by 4 other amino acid units.

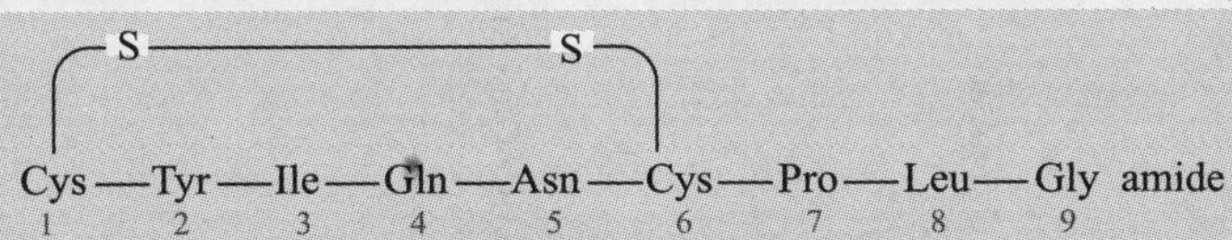

Fig. 9–15. **Structure of ocytocin**

Insulin (Fig. 9–16) is another excellent example where two peptide chains are linked together by 2 disulfide bonds. The presence of an internal disulfide bond in the glycyl (or A) chain between residues 6 and 11 is noteworthy.

These two chemical bonds, namely, peptide and disulfide, are relatively stable. Both these bonds, collectively or individually, maintain the linear form (or the *primary structure*) of the protein molecule.

2. Hydrogen Bond (>CO......HN<). When a group containing a hydrogen atom, that is covalently-bonded to an electronegative atom, such as oxygen or nitrogen, is in the vicinity of a second group containing an electronegative atom, an energetically favourable interaction occurs which is referred to as a hydrogen bond. The formation of a hydrogen bond is due to the tendency of hydrogen atom to share electrons with two neighbouring atoms, *esp.*, O and N. For example, the carbonyl oxygen of one peptide bond shares its electrons with the hydrogen atom of another peptide bond. Thus,

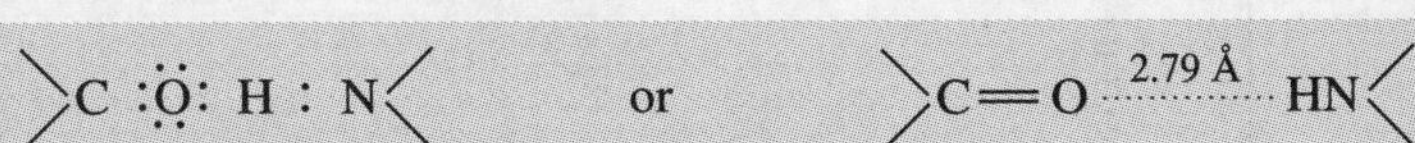

An interaction sets in between a C=O group and the proton of an NH or OH group if these groups come within a distance of about 2.8 Å. This secondary valence bond is symbolized by a dotted line, ... (The solid line, however, represents the normal covalent bond). The strength of the hydrogen bond is only 5 to 8 kilocalories per mole and is maximal when the bond is linear (Fig. 9–17). Hydrogen bonding between amides or peptides of the type depicted in Fig. 9–17 (*c*) plays an important role in stabilizing some conformations of the polypeptide chain.

Silk fibroin, composed mainly of glycine, alanine and serine units, is an example of the presence of hydrogen bonds involving the imide (>NH) and carbonyl (>C=O) groups of the peptide bonds. Here the hydrogen bonds link the vicinal peptide chains (refer Fig. 9–28).

In other proteins like **keratin of wool**, however, the hydrogen bonds link the side chains so that a single peptide chain is held in a coiled or helical form.

Since the binding energy of a hydrogen bond amounts to only 1/10th of that of a primary valence, the hydrogen bonds are relatively weak linkages but many such bonds collectively exert considerable

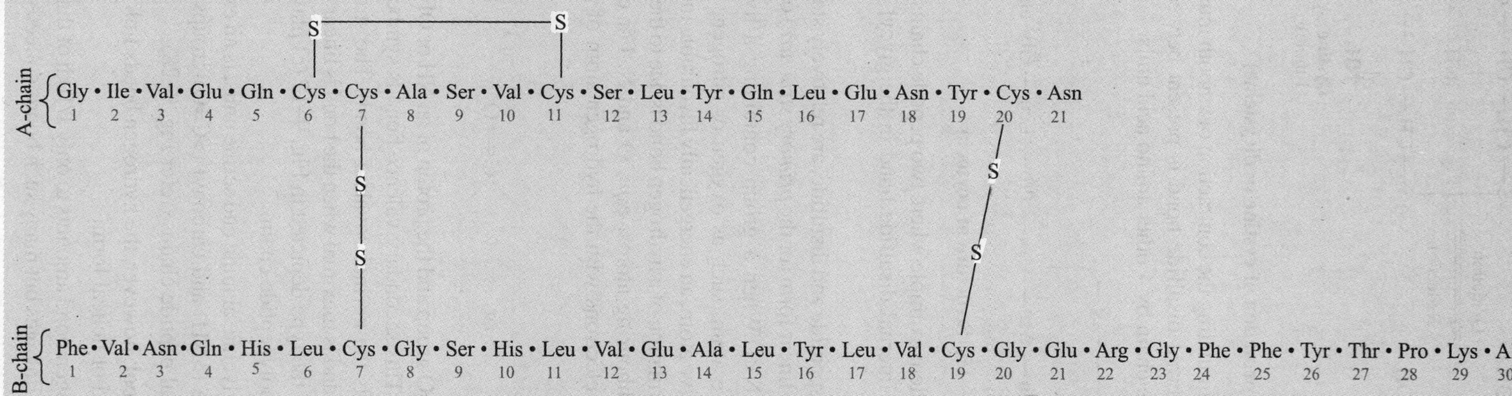

Fig. 9–16. Structure of bovine insulin

Numbers (in *red*) refer to specific amino acid residues. Glycine and phenylalanine are the N-terminal amino acids and the asparagine and alanine, the C-terminal amino acids. The A-chain of the insulin of man, pig, dog, rabbit and sperm whale are identical. The B-chains of the cow, pig, dog, goat and horse are identical. The amino acid replacements in the A-chain usually occur in positions 8, 9 and 10 (For details, refer Chapter: Animal Hormones).

force and help in maintaining the helical structure (or the *secondary structure*), characteristic of many proteins (**Mirsky** and **Pauling,** 1936).

3. Nonpolar or **Hydrophobic Bond.** Many amino acids (like alanine, valine, leucine, isoleucine, methionine, tryptophan, phenylalanine and tyrosine) have the side chains or R groups which are essentially hydrophobic, *i.e.*, they have little attraction for water molecules in comparison to the strong hydrogen bonding between water molecules. Such R groups can unite among themselves with elimination of water to form linkages between various segments of a chain or between different chains. This is very much like the coalescence of oil droplets suspended in water.

The association of various R groups in this manner leads to a relatively strong bonding. It also serves to bring together groups that can form hydrogen bonds or ionic bonds in the absence of water. Each type linkage, thus, helps in the formation of the other; the hydrophobic bonds being most efficient in this aspect. The hydrophobic bonds also play important role in other protein interactions, for example, the formation of enzyme-substrate complexes and antibody-antigen interactions.

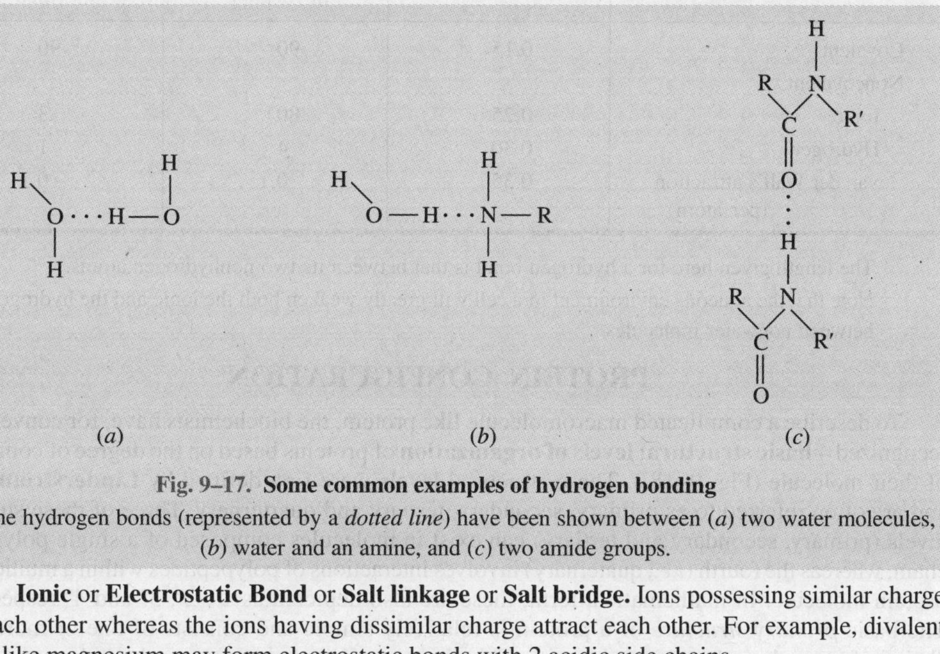

Fig. 9–17. Some common examples of hydrogen bonding

The hydrogen bonds (represented by a *dotted line*) have been shown between (*a*) two water molecules, (*b*) water and an amine, and (*c*) two amide groups.

4. Ionic or **Electrostatic Bond** or **Salt linkage** or **Salt bridge.** Ions possessing similar charge repel each other whereas the ions having dissimilar charge attract each other. For example, divalent cations like magnesium may form electrostatic bonds with 2 acidic side chains.

Another instance of ionic bonding may be the interaction between the acidic and basic groups of the constituent amino acids shown at the bottom of Fig. 9–13. The R groups of glutamic acid and aspartic acid contain negatively charged carboxylate groups, and the basic amino acids (arginine, histidine, lysine) contain positively charged amino groups in the physiological pH range. Thus, these amino acids contribute negatively charged and positively charged side chains to the polypeptide backbone. When two oppositely charged groups are brought close together, electrostatic interactions lead to a strong attraction, resulting in the formation of an electrostatic bond. In a long polypeptide chain containing a large number of charged side chains, there are many opportunities for electrostatic interaction. Intramolecular ionic bonds are rather infrequently used in the stabilization of protein structure but when they are so used, it is often with great effect. In fact, ionized groups are more frequently found stabilizing interactions between protein and other molecules. Thus, ionic bonds between positively charged groups (side chains of lysine, arginine and histidine) and negatively charged groups (COO^- group of side chain of aspartic and glutamic acids) do occur.

These ionic bonds, although weaker than the hydrogen bonds, are regarded as responsible for maintaining the folded structure (or the *tertiary structure*) of the globular proteins.

CHARACTERISTICS OF CHEMICAL BONDS

Table 9–4 lists some characteristics of the 2 types of chemical bonds : covalent and non cavalent. The strength of a bond or the **bond strength** can be measured by the energy required to break it, here in the table given in kilocalories per mole (kcal/mole). One kilocalorie is the quantity of energy needed to raise the temperature of 1,000 g of water by 1°C. An alternative unit in wide use is the kilojoule, kJ which is equal to 0.24 kcal. Individual bonds vary a great deal in strength, depending on the atoms involved and their precise environment, so that *the values are only a rough guide*. The **bond length** is the centre-to-centre distance between the two interacting atoms.

Table 9–4. Characteristics of chemical bonds

Bond Type	*Length (nm)**	*Strength (kcal/mole)†*	
		In vacuum	*In water*
Covalent	0.15	90	90
Noncovalent			
Ionic	0.25	80	3
Hydrogen	0.30	4	1
van der Wall's attraction (per atom)	0.35	0.1	0.1

* The length given here for a hydrogen bond is that between its two nonhydrogen atoms.

† Note that the aqueous environment in a cell will greatly weaken both the ionic and the hydrogen bonds between nonwater molecules.

PROTEIN CONFIGURATION

To describe a complicated macromolecule like protein, the biochemists have, for convenience, recognized 4 **basic structural levels of organization** of proteins based on the degree of complexity of their molecule (Fig. 9–18). These structural levels were first defined by **Linderström–Lang** and are often referred to as primary, secondary, tertiary and quarternary. Three of these structural levels (primary, secondary and tertiary) can exist in molecules composed of a single polypeptide chain, whereas the fourth (*i.e.*, quaternary) involves interactions of polypeptides within a multichained protein molecule. In mathematical term, these are also depicted as 1°, 2°, 3° and 4° respectively. The basic *primary structure* of a protein is relatively simple and consists of one or more linear chains of a number of amino acid units. This linear, unfolded structure or the polypeptide chain often assumes a helical shape to produce the *secondary structure*. This, in turn, may fold in certain specific patterns to produce the twisted three-dimensional or the *tertiary structure* of the protein molecule. Finally, certain other proteins are made up of subunits of similar or dissimilar types

IRVING GEIS

(LT, 1908–1997)

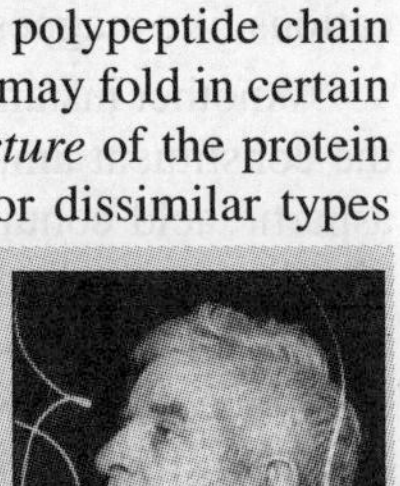

Irving Geis is well known for his lucid visualizations of molecular structures, particularly proteins and nucleic acids. These have appeared in *Scientific American* for the past thirty five years and in major chemistry, biology, and biochemistry textbooks. He is a co-author with **R.E. Dickerson**, Director of the Molecular Biology Institute of UCLA, of 3 books:

1. Chemistry, Matter and the Universe
2. The Structure and Action of Proteins
3. Hemoglobin: Structure, Function, Evolution and Pathology

In addition to drawing, painting, and writing, Irving Geis was a frequent lecturer at universities and medical schools on protein structure and function. A recent Guggenheim fellowship made possible the assembly and cataloging of his drawings and paintings into *The Geis Archives of molecular structure.*

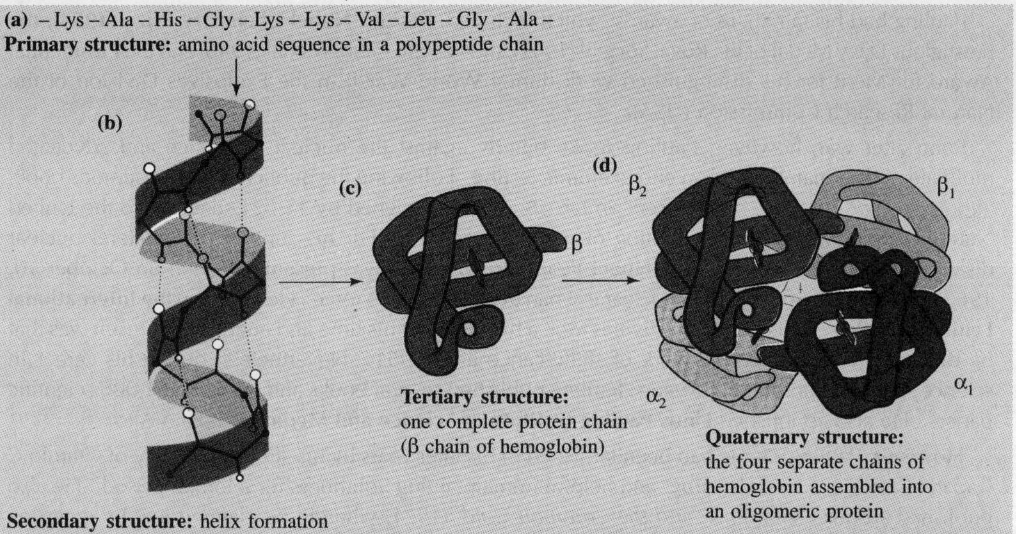

An α-helix is a secondary structure found in proteins. This structure emphasizes the polypeptide backbone (shown as a *blue ribbon*), which is composed of amino acids linked together in a linear fashion. Hydorgen bonding between hydrogen and oxygen atoms stabilizes the helical conformation. (*Courtesy: Irving Geis*)

Fig. 9–18. The four levels of structural hierarchy in proteins

of the polypeptide chains. These subunits interact with each other in a specific manner to give rise to the so-called *quaternary structure* of the protein. This, in fact, defines the degree of polymerization of a protein unit.

1. Primary Structure : Amino Acid Sequence

The *primary structure* of a protein refers to the number and sequence of amino acids, the constituent units of the polypeptide chain. The main mode of linkage of the amino acids in proteins is the peptide bond which links the α-carboxyl group of one amino acid residue to the α-amino group of the other. The proteins may consist either of one or of more peptide chains.

Rigid and Planar Peptide Bond

Linus Pauling and **Robert Corey**, in the late 1930s, demonstrated that the α-carbons of adjacent

LINUS CARL PAULING

(LT, 1901–1994)

Linus Pauling (with his α-helix ball-and-stick model in the photograph) , the son of a German father and English mother, was an American physical chemist. He was born on February 28, 1901 at Portland, Oregon, USA. Although he had initially an undistinguished school education, he graduated from the Oregon Agriculture College in 1922 (bachelor of science in chemical engineering). Pauling obtained his doctorate degree in 1925 from California Institute of Technology (Caltech). He was one of the exceptional men of his times and carried the message of Quantum Mechanics to the New World. His pioneering application of principles of Quantum Physics to explain chemical properties, his investigations of molecular structures of proteins and, above all, his work on the nature of chemical bond rank among the outstanding pieces of chemical research of the century. To the graduate students of Chemistry in the 1940s and 50s, his book, *'The Nature of Chemical Bond'* (1939) became a Bible. At age 30, Pauling had become a full professor in 1931 and Chairman of Chemistry and Chemical Engineering some 6 years later. With **Robert Corey** (LT, 1867–1971), he investigated amino acids and polypeptides. Around 1940s, he and his colleagues conducted researches on antibody-antigen reactions. Pauling was the first one to devise a scale for comparing the electronegativity of different elements.

Pauling had his fair share of awards, which included, besides **Nobel Chemistry Prize (1954),** the prestigious Davy Medal of the Royal Society (1947), the Willard Gibbs Medal (1946) and the Presidential Award for Merit for his distinguished work during World War II in the Explosives Division of the Natural Research Commission (USA).

Soon after war, however, Pauling spoke bitterly against the nuclear arms race and advocated multilateral disarmament and an end to atomic testing. Following the publication of his famous book, *'No More War !'* (1958), he then sent, in Jan'58, a petition signed by 11,021 scientists to the United Nations, urging an end to the testing of nuclear weapons. For his support of unilateral nuclear disarmament, he earned the **1962-Nobel Peace Prize,** which was presented to him on October 10, 1963, the day a US-Soviet partial nuclear test ban treaty came into force. He also won the **International Lenin Peace Prize**. His political activities took a heavy toll of his time and energy. The result was that he did not produce scientific work of significance after 1951. Nevertheless, during his career in science spanning more than 60 years, Pauling published several books and more than 1,000 scientific papers. He also established **Linus Pauling Institute of Science and Medicine** at Palo Alto.

However, Pauling's fame had been tarnished in his later years by his strong advocacy of vitamin C (ascorbic acid) as a 'wonder drug' and helpful in maintaining youthness for a longer period. He also published a book, *'Vitamin C and the Common Cold'* (1971) wherein he claimed that by ingesting 1,000 mg daily of vitamin C (instead of the 60 mg daily minimum as recommended by the U.S. National Research Council), a person would catch 45% fewer colds and suffer 60% fewer days of illness. His findings were bitterly criticized experimentally by **Terene Anderson** of the University of Toronto in Canada. But this should not minimize the magnitude of his earlier contributions. Till today, he holds the distinction of being **the only person to have won two Nobel Prizes on his own.** Pauling died on August 19, 1994 at Big Sur, California. His life fell into 3 distinct phases: the first evoked reverences, the second love and the third ridicule. Truly, Pauling ranks among the most versatile scientists of more than one generation of 20th century.

amino acids are separated by three covalent bonds, arranged C_α—C—N—C_α. They also demonstrated that the amide C—N bond in a peptide is somewhat shorter (1.32 — or 0.132 nm) than the C—N bond in a simple anine (1.49 Å or 0.149 nm) and that the atoms associated with the bond are coplanar. This indicated a **resonance** or partial sharing of two pairs of electrons between the carbonyl oxygen and the amide nitrogen [Fig. 9–19 (*a*)]. The oxygen has a partial negative charge and the nitrogen a partial positive charge, setting up a small electric dipole. The 4 atoms of the peptide group (C, H, O, N) lie in a single plane, in such a way that the oxygen atom of the carbonyl group and the hydrogen atom of the amide nitrogen are *trans* to each other. *Virtually, all peptide bonds in proteins occur in trans configuration.* From these studies, Pauling and Corey concluded that the amide C—N bonds are unable to rotate freely because of their **partial double-bond character.** The backbone of a polypeptide chain can thus be separated by substituted methylene groups —CH(R)— [Fig. 9–19 (*c*)]. The rigid peptide bonds limit the number of conformations that can be assumed by a polypeptide chain.

However, rotation is permitted about the bond between the nitrogen and α-carbon atoms of the main chain (N—C_α) and between the α carbon and carbonyl carbon atoms (C_α—C). By convention, the degree of rotation at the N—C_α bond is called *phi* (ϕ) and that between C_α—C bond is called psi (ψ). Again, by convention, both ϕ and ψ are defined as 0° in the conformation in which the two peptide bonds connected to a single a carbon are in the same plane [Fig. 9–19 (*d*)]. In principle, ϕ and ψ can have any value between −180° and + 180°, but many values of ϕ and ψ are prohibited by steric interference between atoms in the polypeptide backbone and the amino acid side chains. The conformation in which ϕ and ψ are both 0° is prohibited for this reason.

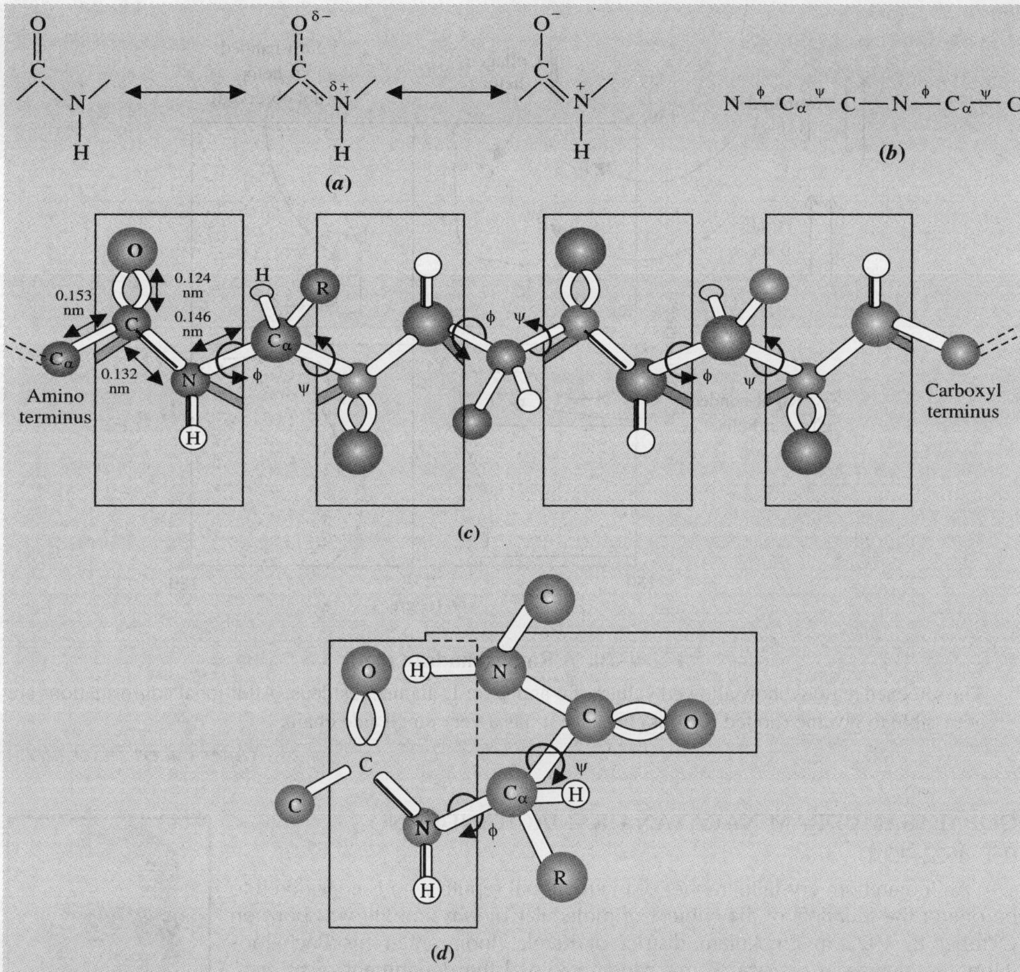

Fig. 9–19. The details of the planar peptide bond

(*a*) **The planar peptide group.** Note that the oxygen and hydrogen atoms are on opposite sides of the C—N bond. This is *trans* configuration.

(*b*) **The three bonds between the sequential C_α carbons in a polypeptide chain.** The N—C_α and C_α—C bonds can rotated, with bond angles designated φ and ψ respectively.

(*c*) **Limited rotation around two (N—C_α and C_α—C) of the three types of bond in a polypeptide chain.** The third type *i.e.*, C—N bonds in the planar polypeptide groups, which make one-third of all the backbone bonds, are not free to rotate.

(*d*) **The two coplanar peptide bonds flanking an α carbon.** By convention, φ and ψ are both defined as 0° when the two peptide bonds flanking an α carbon are in the same plane. In a fully stretched out polypeptide chain, φ = ψ = 180°.

Ramachandran Plot

The conformation of the main polypeptide chain can be completely determined if the values φ and ψ for each amino acid residue in the chain are known. In a fully stretched polypeptide chain, φ = ψ = 180°. **G.N. Ramachandran** (1963) recognized that an amino acid residue in a polypeptide chain cannot have just *any* pair of values of φ and ψ. By assuming that atoms behave as hard spheres, allowed ranges of φ and ψ can be predicted and visualized in steric contour diagram called **Ramachandran plots.** Such a plot for poly-L-alanine (or *any amino acid except glycine and proline*) shows three separate allowed ranges (the screened regions in Fig. 9–20). One of them contains φ–ψ

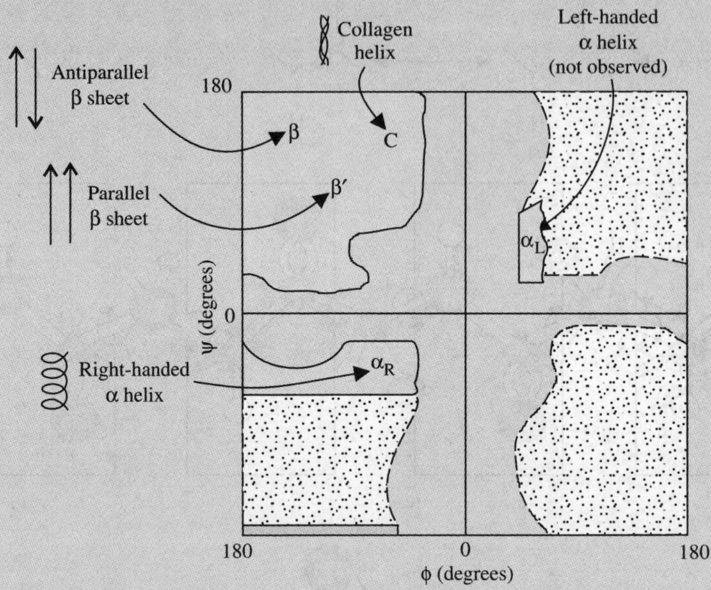

Fig. 9–20. A Ramachandran plot

The screened regions show allowed values of ϕ and ψ for L-alanine resideus. Additional conformations are accessible to glycine (dotted regions) because it has a very small side chain.

(*After Lubert Stryer, 1995*)

GOPALASAMUDRAM NARAYANA RAMACHANDRAN

(LT, 1922–2001)

An Indian-born crystallographer of international repute who is considered to be one of the founders of the subject of molecular biophysics. He was born on October 8, 1922 in Ernakulam district of Kerala, India. After his Bachelor's (Honours) degree in physics, Ramachandran joined the department of physics, Indian Institute of Science, Bangalore, to work with legendary **Sir C.V. Raman**. This proved to be the beginning of his passionate involvement in the pursuit of science culminating in a large number of poineering contributions. He did Master's degree in Electrical Engineering in 1940. He got his D.Sc. from the University of Madras in 1948 and completed his Ph.D. from the University of Cambridge, England, in 1949. In 1951, Ramachandran went back to Cambridge, to work with **Lawrence Bragg** (who along with his father **Sir William Henry Bragg** shared 1915 Nobel Prize in Physics and was the youngest scientist, at age 25, so honoured) and **W.A. Wooster**, and returned home with another Ph.D.

Ramachandran established the Department of Physics at the University of Madras, in 1952. He began work on modelling collagen from fibre diffraction during 1952–53 when **J.D. Bernal** (who did pioneering work in x-ray crystallography) visited him and suggested that the collagen structure is a major unsolved problem. In 1954, Ramachandran, along with **Gopinath Kartha**, arrived at the triple helix model for collagen, that had 2 hyrogen bonds per unit. **Crick**, the codiscoverer of DNA structure, had independently deduced one hydrogen bond per collagen unit. More recent single crystal work has shown the result to be 1.5 hydrogen bonds per collagen unit, a compromise between these 2 proposals.

Ramachandran became distinguished leader and brought a galaxy of luminaries from around the globe, including **Lawrence Bragg, Linus Pauling, Dorothy Hodgkin, Stanford Moore, Severo Ochoa, Maurice Wilkins, Paul Florey, David Phillips, John Edsall** and **Harold Scheraga** who visited the young Madras School of Biophysics during the International Meets he organized in 1963 and 1967. In 1968, Ramachandran was made the Jawaharlal Nehru Fellow of the University of Madras.

In 1971, Ramachandran was invited by the Indian Institute of Science, Bangalore to establish the department of Molecular Biophysics. Very soon, he raised the status of the department to international fame. At this department, research was carried out on several giant biomolecules such as nucleic acids and polysaccharides.

In 1970s, Ramachandran turned to problems of mathematical logic and proposed a new method for structural analysis, in 1990. His work with **A.V. Lakshminarayanan** has been acclaimed as the starting point of the CAT scan technique in radiography.

Ramachnadran single-handedly set up an active school, working simultaneously in many disciplines that embraced construction of analogue computers, theoretical physics, theoretical crystallography and biomolecular structure determination. His background in physics and mathematics enabled him to venture into diverse fields.

Prof. Ramachandran was honoured with the prestigious **Ewald Prize** in 1999 for his outstanding contribution in the field of crystallography. He died in Chennai (erstwhile Madras) on April 7, 2001 after a protracted illness.

values that generate the antiparallel β sheet, the parallel Beet and the collagen helix. A second region has ϕ–ψ values that produce the right-handed α helix: a third, the left-handed α helix. Though sterically allowed, left-handed α helices are not found in proteins because they are energetically much less favoured.

For **glycine**, these three allowed regions are larger, and a fourth appears (shown as dotted in Fig. 9–20) because a hydrogen atom causes less steric hindrance than a methyl group. *Glycine enables the polypeptide backbone to make turns that would not be possible with another residue.*

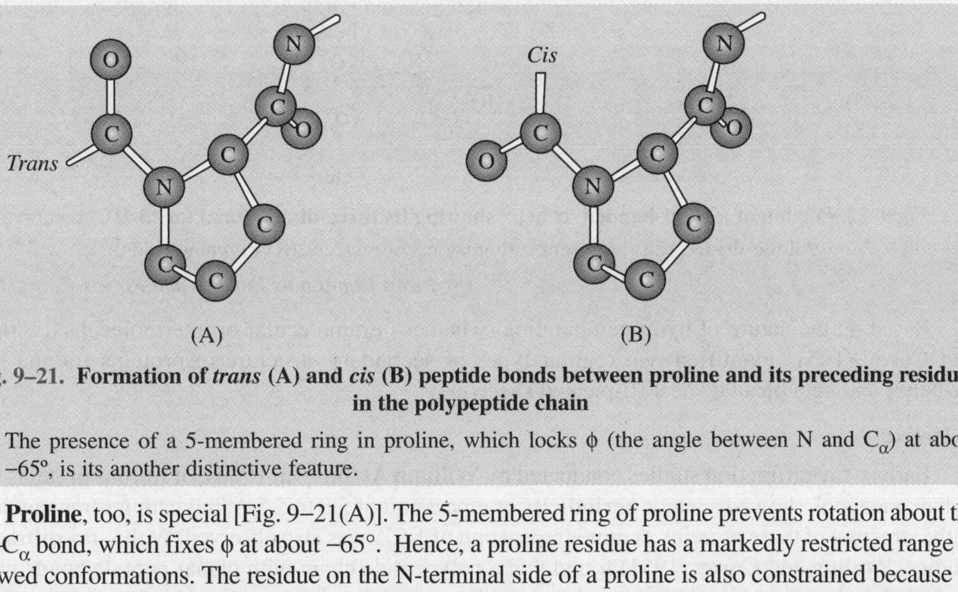

Fig. 9–21. Formation of *trans* (A) and *cis* (B) peptide bonds between proline and its preceding residue in the polypeptide chain

The presence of a 5-membered ring in proline, which locks ϕ (the angle between N and C_α) at about –65°, is its another distinctive feature.

Proline, too, is special [Fig. 9–21(A)]. The 5-membered ring of proline prevents rotation about the N—C_α bond, which fixes ϕ at about –65°. Hence, a proline residue has a markedly restricted range of allowed conformations. The residue on the N-terminal side of a proline is also constrained because of steric hindrance imposed by the 5-membered ring. Proline also disfavours α-helix formation because it lacks an amide H atom for hydrogen bonding.

2. Secondary Structure : Helix Formation or **Local Folding**

If the peptide bonds were the only type of linkage present in proteins, these molecules would have behaved as irregularly coiled peptide chains of considerable length. But the globular proteins, however, do show some regular characteristic properties, indicating the presence of a regular coiled structure in these molecules. This involves the folding of the chain which is mainly due to the presence of hydrogen bonds. Thus, folding and hydrogen bonding between neighbouring amino acids results in the formation of a rigid and tubular structure called a **helix**. This constitutes the *secondary structure* of proteins, which refers to the steric or spatial relationship of amino acids that are near to each other in the amino acid sequence.

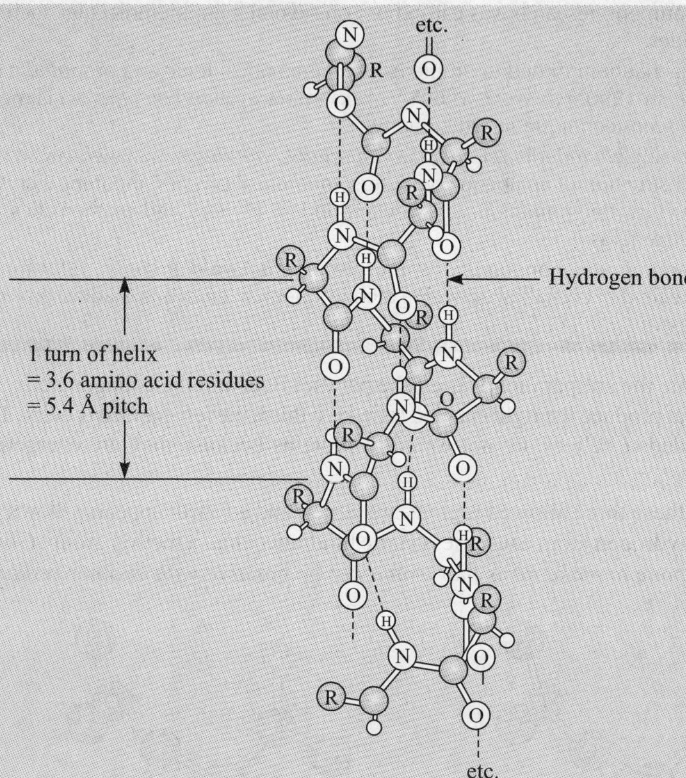

Fig.9–22. Portion of a right-handed α helix showing its three-dimensional (or '3-D') structure

All the shaded balls represent carbon atoms or residues (R) of amino acids.

(Redrawn from Linus Pauling and Robert Corey, 1955)

Based on the nature of hydrogen bonding (whether intramolecular or intermolecular), **Pauling** and **Corey** (1951) identified two regular types of secondary structure in proteins : alpha helix (α-helix) and beta pleated sheet (β-pleated sheet).

α-Helix:

Early x-ray diffraction studies conducted by **William Astbury** in 1930s, of fibrous proteins such as hair and wool, showed a major **periodicity** or repeat unit of 5.0 to 5.5 Å, indicating some regularity in the structure of these proteins. A minor repeat unit of 1.5 Å was also observed. With x-ray diffraction studies, **Pauling** and **Corey** (1951) found that a polypeptide chain with planar peptide bonds would form a right-handed helical structure by simple twists about the α-carbon-to-nitrogen and the α-carbon-to-carboxyl carbon bonds. They called this helical structure as **α-helix.** The helix is so named because of the mobility of α-carbon atoms.

The α-helix is a rodlike structure. The tightly coiled polypeptide main chain forms the inner part of the rod, and the side chains extend outward in a helical array (Figs. 9–22 and 9–23). The α-helix is stabilized by hydrogen bonds between the NH and CO groups of the main chain. The CO group of each amino acid is hydrogen-bonded to the NH group of the amino acid that is situated four residues ahead in the linear sequence (Fig. 9–24). Thus, all the main chain CO and NH groups are hydrogen-bonded.

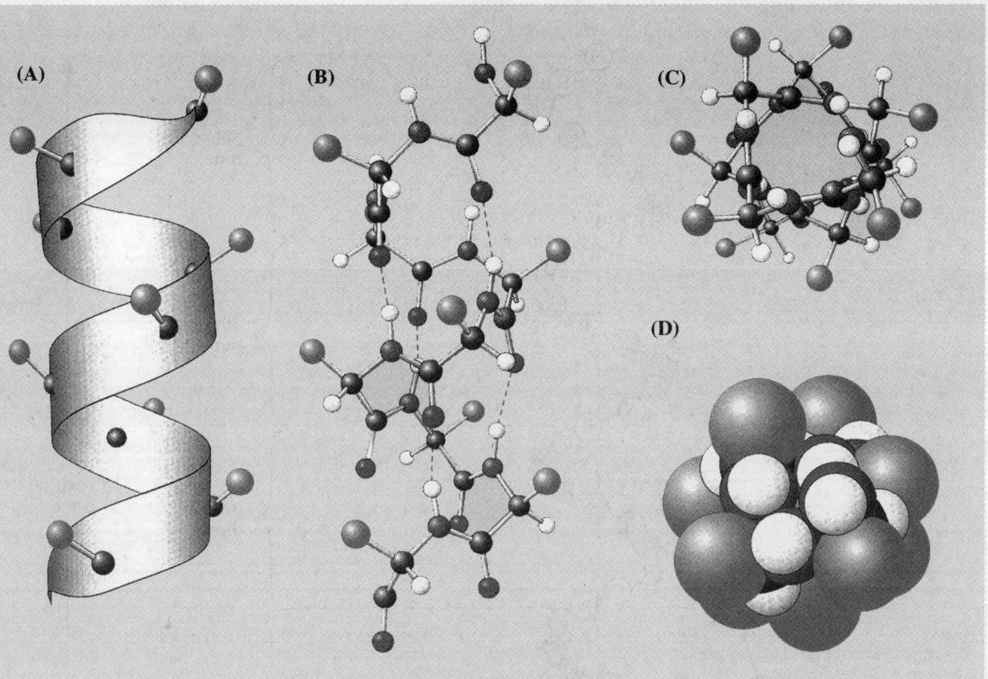

Fig. 9–23. Structure of the α helix

(A) A ribbon depiction with the α-carbon atoms and side chains (*green*) shown.

(B) A side view of a ball-and-stick version depicts the hydrogen bonds (*dashed lines*) between NH and CO groups.

(C) A cross-sectional view of the α-helix. The end view shows the coiled backbone as the inside of the helix and the side chains projecting outward. The van der Waals radii of the atoms are larger than shown here for the sake of clarity. Hence, there is actually no free space inside the helix.

(D) A space-filling view of part (C) shows the tightly-packed interior core of the helix.

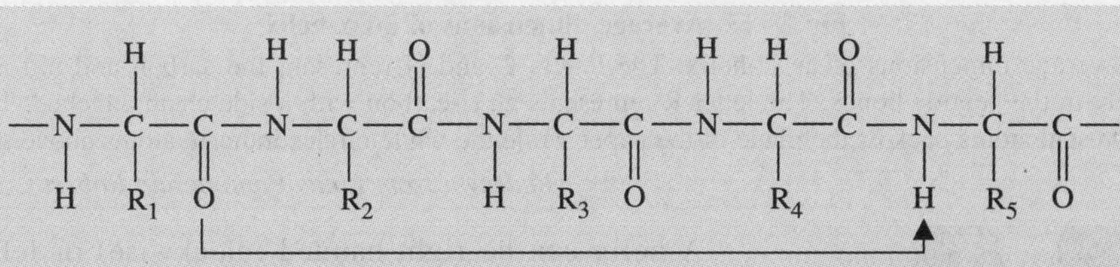

Fig. 9–24. Diagram showing that in the α-helix, the CO group of residue *n* is hydrogen-bonded to the NH group of residue (*n* + 4).

It is, thus, apparent that the α-helical structure depends on the intramolecular (= intrachain) hyrogen bonding between the NH and CO groups of peptide bonds. The hydrogen bonding occurs spontaneously and, as a result, a polypeptide can assume a rod-like structure with well-defined dimensions (Fig. 9–25). The α-helix (or α-conformation, as it is also called) has a pitch of 5.4 Å (= 0.54 nm) and contains 3.6 amino acids per turn of the helix, thereby giving a rise per residue of 5.4/3.6 = 1.5 Å (= 0.15 nm), which is the *identity* period of α-helix. The amino acid residues in an α-helix have conformations with $\phi = -60°$ and $\psi = -45°$ to $-50°$.

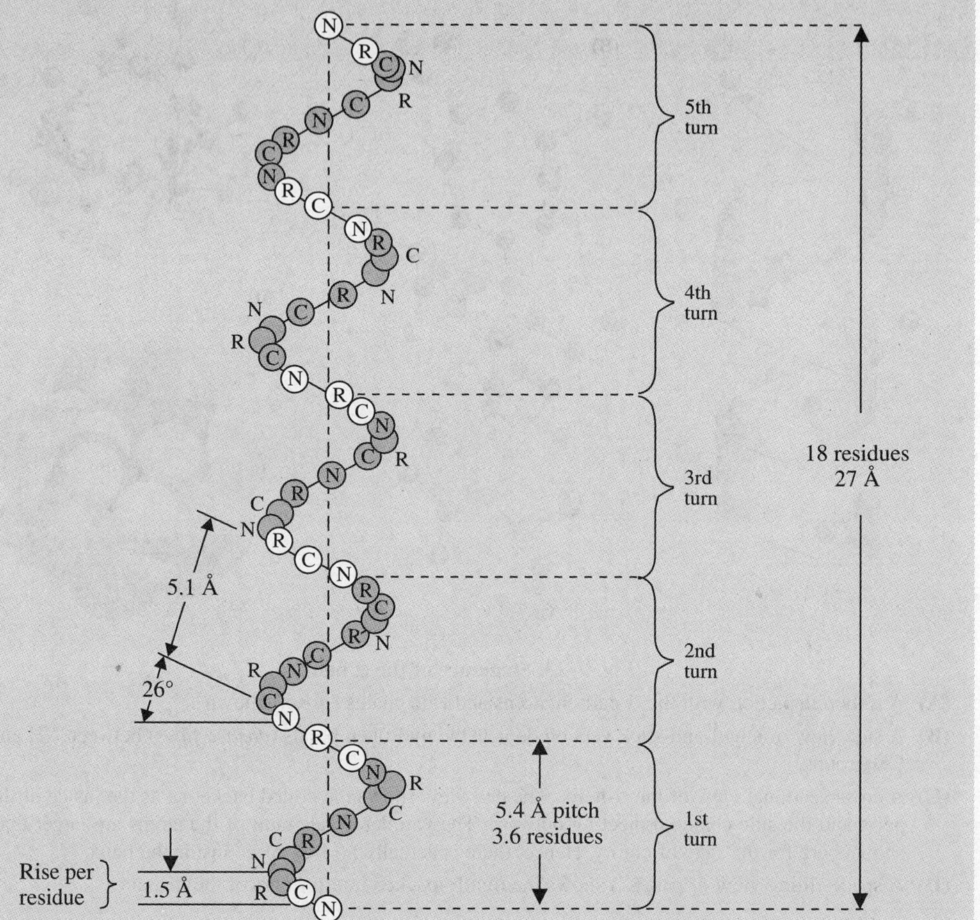

Fig. 9–25. Average dimensions of an α-helix

Note the average dimensions of an α-helix. The letters C and N represent the carbon and nitrogen atoms respectively in the peptide bonds. The letter R represents an α-carbon with a side chain attached. The shaded circles represent atoms back of the plane of the paper while the white circles indicate atoms above the plane.

(Redrawn from Linus Pauling and Robert Corey, 1955)

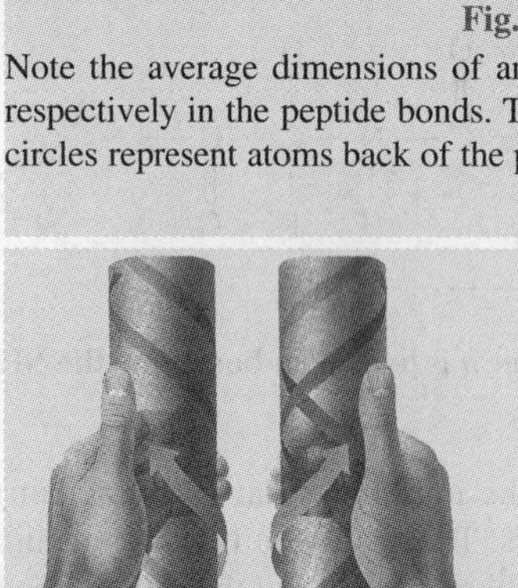

Fig. 9–26. Two types of helices

Right- and left-handed helices are related to each other in the same way as right and left hands.

A helix can be right-handed (clockwise) or left-handed (anticlockwise) (refer Fig 9–26); *α-helices of known polypeptides (i.e., L-amino acids) are right-handed*. Biologically functional proteins do not usually exhibit cent per cent α-helical structure. Some have a high percentage of their residues in α-helical structures, *e.g.*, myoglobin and hemoglobin ; others have a low percentage, *e.g.*, chymotrypsin and cytochrome *C*. A very interesting feature of the α-helix, besides the periodicity, is the fact that the carbonyl group (>CO═O) of every peptide bond is in a position to form a hydrogen bond with the >N—H group of the peptide bond in the next turn of the helix, thereby contributing to the ***stability*** of the α-helix. Long polypeptides chains in the α-helical conformation, forming structures are reminiscent of a multistranded rope. The protofibril of hair, for example, is 20 Å in diameter and appears to be made up of 3 right-handed helices wound about each other to form a left-handed supercoil.

The model-building experiments have shown that an α-helix can form with either D- or L-amino acids. However, the residues must all be of one stereoisomer ; a d-amino acid will disrupt any regular structure consisting of L-amino acids, and vice versa. The preferential formation of such a helix over many other possible conformations in nature is due, in part, to the fact that it (α-helix) makes optimal use of internal hydrogen bonds. The helical twist of the α-helix found in all proteins is right-handed. The α-helix proved to be the predominant structure in α-keratins. More often, about 1/4th of all amino acid residues in polypeptides are found in α helices, the exact fraction varying greatly from one protein to the next.

Knowing the Right Hand from the Left

There is a simple method for determining whether a helical structure is right-handed or left-handed. Make fists of your two hands with thumbs outstretched and pointing straight up. Looking at your right hand, think of a helix spiraling up your right thumb in the direction in which the other four fingers are curled as shown (counterclockwise) in the following figure. The resulting helix is *right-handed*. Your left hand will demonstrate a *left-handed* helix, which rotates in the clockwise as it spirals up your thumb.

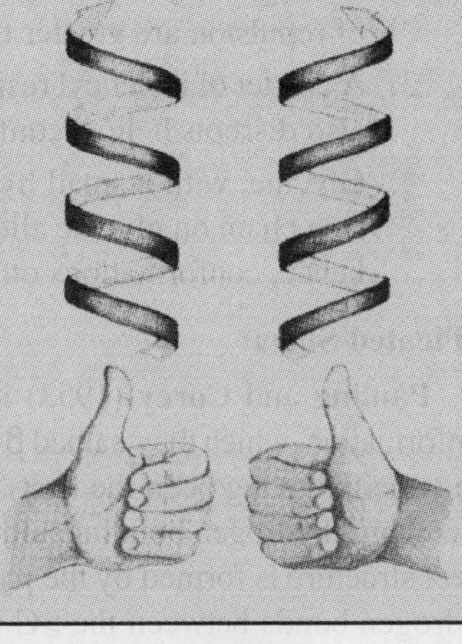

Many globular proteins contain short regions of such α-helices, and those portions of a transmembrane protein that cross the lipid bilayer are usually α-helices because of the constraints imposed by the hydrophobic lipid environment. In aqueous environment, an isolated α-helix is usually not stable on its own. Two identical α-helices that have a repeating arrangement of nonpolar side chains, however, will twist around each other gradually to form a particularly stable structure known as a coiled coil. Long rod-like *coiled coils* are found in many fibrous proteins, such as intracellular α-keratin fibres that reinforce skin and its appendages.

α-helix occurs in the protein **α-keratin**, found in skin and its appendages such as hair, nails and feathers and constitutes almost the entire dry weight of hair, wool, feathers, nails, claws, quills, scales, horns, hooves, tortoise shell, and much of the outer layer of skin. The basic structural unit of α-keratin usually consists of 3 right-handed helical polypeptides in a left-handed coil that is stabilized by crosslinking disulfide bonds (Fig. 9–27).

Destabilization of α-helical conformation by certain amino acid residues (Table 9–5) can occur in various ways:

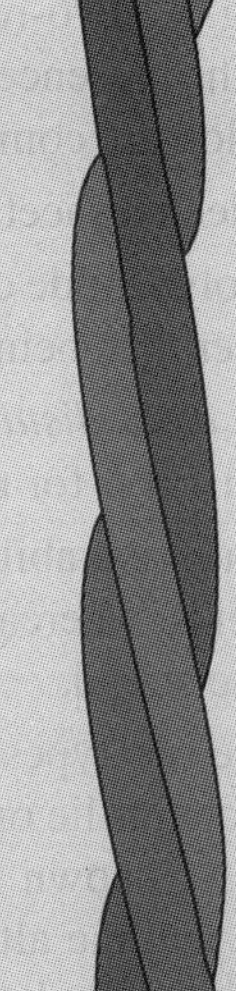

Fig. 9–27. Schematic of the 3 α-helical coils in keratin

Table 9–5. Amino acids affecting α-helical structure

Destabilize α-helix		*Create bends in α-helix*
Aspartic acid	Glycine	Proline
Glutamic acid	Serine	Hydroxyproline
Lysine	Isoleucine	
Arginine	Threonine	

1. A prolyl reisdue has its α-N atom in a rigid ring system and cannot participate in α-helical structure; instead, it creates a sharp bend in the helix.
2. A sequence of aspartyl and/or glutamyl residues can destabilize α-helical structure because the negatively-charged side chains repel one another (electrostatic repulsion), and the forces of repulsion are greater than those of hydrogen bonding.
3. A cluster of isoleucyl residues, because of steric hindrance imposed by their bulky R groups, also disrupts helical conformation.
4. Glycine, with a small hydrogen atom as an R group, is another destabilizer. The lack of a side chain on glycine allows for a great degree of rotation about the amino acid's α carbon; hence, conformations other than a helical bond angles are possible.

β-Pleated Sheet:

Pauling and **Corey** (1953) identified a second type of repetitive, minimum-energy or stable conformation, which they named **β-pleated sheet** (β because it was the second structure they elucidated, the α-heilx having been the first). The formation of β-pleated sheets depends on ***intermolecular*** (= interchain) hydrogen bonding, although intramolecular hydrogen bonds are also present. The pleated sheet structure is formed by the parallel alignment of a number of polypeptide chains in a plane, with hydrogen bonds between the >C = O and —N—H groups of adjacent chains. The R groups of the constituent amino acids in one polypeptide chain alternately project above and below the plane of the sheet, leading to a two-residue repeat unit. The β sheet structures are quite common in nature and are favoured by the presence of amino acids, glycine and alanine. Silk and certain synthetic fibres such as nylon and orlon are composed of β-structures.

The β-pleated sheet differs markedly from the rodlike α-helix:

1. A polypeptide chain in a β-pleated sheet, called a β-strand, has *fully extended conformation*, rather than being tightly coiled as in the α-helix.
2. The *axial distance* between adjacent amino acids in β-pleated sheets is 3.5 Å, in contrast with 1.5 Å for the α-helix.
3. β-sheet is stabilized by hydrogen bonds between NH and CO groups in *different* polypeptide strands, whereas in the α-helix, the hydrogen bonds are between NH and CO groups in the same strand.

There are two types of β-pleated sheet structures. If the N-terminal ends of all the participating polypeptide chains lie on the same edge of the sheet, with all C-terminal ends on the opposite edge, the structure is known as a **parallel β-pleated sheet.** In contrast, if the direction of the chains alternates so that the alternating chains have their N-terminal ends on the same side of the sheet, while their C-terminal ends lie on the opposite edge, the structure is known as the **antiparallel β-pleated sheet**. In other words, the hydrogen-bonded neighbouring polypeptides are aligned in the same N-to-C terminus direction in the parallel pleated sheets and in the opposite N-to-C direction in the antiparallel pleated sheets. Both parallel and antiparallel β-pleated sheets have similar structures, although the repeat period is shorter (6.5 Å 0.65 nm) for the parallel conformation in comparison to antiparallel conformation (7 Å or 0.7nm). Pleated-sheet structures can also be formed from a single polypeptide chain if the chain repeatedly folds back on itself. Although a β-pleated sheet structure is usually associated with structural proteins, it is also known to occur in the 3-dimensional structures of certain globular proteins, *e.g.*, the enzymes lysozyme and carboxypeptidase A.

Silk fibroin is one example of a protein that has the antiparallel pleated sheet structure that is very stable. It is a member of a class of fibrillar proteins called α-keratins. Silk fibroin (Fig. 9–28) is composed mainly of glycine, alanine and serine units that are linked together by the peptide bonds.

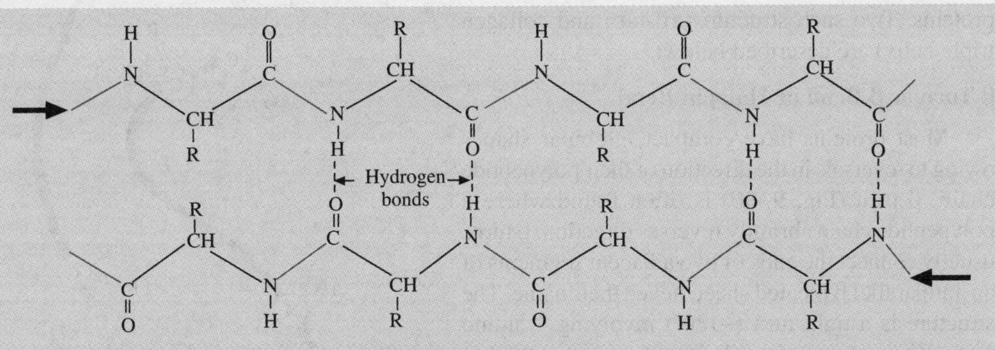

Fig. 9–28. A portion of two chains of silk fibroin

Note that the two chains run in opposite direction. Each *red line* represents a single peptide bond.

Glycine constitutes approximately 45% of the total amino acid residues, and alanine plus serine compose another 42%. Hence, the R groups extending above and below the plane of the pleated sheet are small and allow the pleated sheets to stack. Here the two chains which run in opposite directions are linked by hydrogen bonds (shown by dotted lines). It may be noted that each of the two chains can also form hydrogen bonds with still other protein and so on. Very large protein aggregates are, thus, formed provided the R groups are relatively small. If the R groups are too large, the hydrogen bonding may not occur as the chains are held far apart for union. An interesting phenomenon occurs when hair or wool (α-keratins) is treated with moist heat and stretched because β-keratin structures with parallel β pleated sheets are produced.

α-keratin pattern $\xrightarrow{\text{stretch}}$ β-keratin pattern

(slow relaxation: β-keratin pattern → α-keratin pattern)

Random Coil:

Besides the α-helix and β-pleated sheet structures of proteins, which were recognized by Pauling and Corey in 1950s, there also exist a third type of secondary structure in proteins called **random coil**. When a polypeptide contains adjacent bulky residues such as isoleucine or charged residues such as glutamic acid and aspartic acid, repulsion between these groups causes the polypeptide to assume a random coil configuration (Fig. 9–29).

Thus, we see that the R groups distributed along the polypeptide backbone determine the secondary structure adopted by different portions of the polypeptide (α-helix or β-pleated sheet), or the lack of a well-defined structure (random coil).

Other Secondary Structures

The α-helix and β-conformation are the major repetitive secondary structures easily recognized in a wide variety of proteins. The **γ-helix**, however, is an example of a highly hydrogen-bonded structure that is insufficiently stable to be used in proteins because of the lack of interatomic contacts. In addition to the α-helix and β-pleated sheet structures, other repetitive structures also exist in one or a few specialized proteins. Two such structures (β-turn and collagen triple helix) are described below:

β Turn or β Bend or Hairpin Bend

Most proteins have compact, globular shapes owing to reversals in the direction of their polypeptide chain. β turn (Fig. 9–30) is often found where a polypeptide chain abruptly reverses direction. β turns usually connect the ends of two adjacent segments of an antiparallel β pleated sheet, hence their name. The structure is a tight turn (~180°) involving 4 amino acids. The essence of this hairpin turn is that the CO group of residue n of a polypeptide is hydrogen-bonded to the NH group of residue $n + 3$. In other words, the hydrogen bond between the peptide groups of the first and fourth amino acid residues is involved in bend formation. Glycine and proline residues often occur in β turns; the former because it is small and flexible and the latter because peptide bonds involving the imino nitrogen of proline readily assume the *cis* configuration, a form that is particularly amenable to a tight turn. β turns are often found near the surface of a protein. They are also called as **reverse turns.**

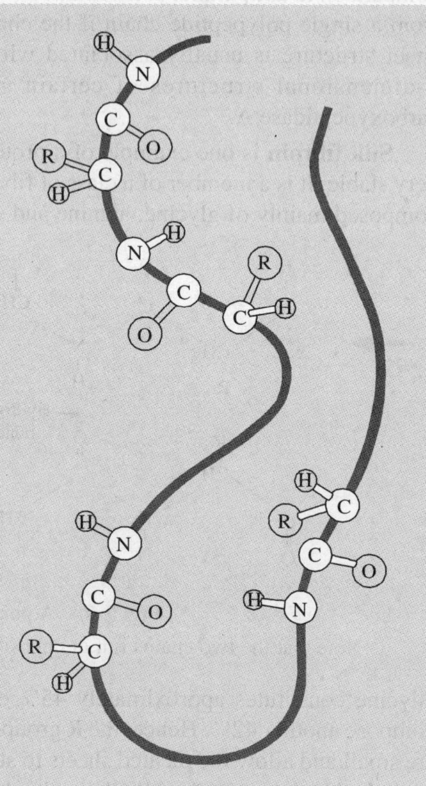

Fig. 9–29. A random coil

The backbone of the polypeptide is drawn, but only a few amino acid residues are shown.

(Adapted from Francisco J. Ayala and John A. Kiger Jr., 1980)

Collagen Triple Helix

Collagen is the most abundant protein of mammals. It is the principal structural element of the human body and makes up 25–33% of all the body protein. It is found in connective tissues such as tendons, cartilage, the organic matrix of bones and the cornea of the eye. This extracellular protein contains three helical polypeptide chains, each nearly 1,000 residues long. The amino acid sequence of collagen is remarkably regular: *nearly every third residue is glycine.* Proline is also present to a much greater extent than in most other proteins. Furthermore, collagen contains 4-hydroxyproline (Hyp), which is rarely found elsewhere. The percentage composition of predominant amino acids found in collagen is: Gly (35%), Ala (11%) and Pro + Hyp (25%). The amino acid sequence in collagen is generally a repeating tripeptide unit, Gly–X–Pro or Gly–X–Hyp, where X can be any amino acid. This repeating tripeptide sequence adopts a left-handed helical structure with 3 residues per turn.

Collagen is a rod-shaped molecule, about 3,000 Å (= 300 nm) long and only 15 Å (=1.5 nm) in diameter. The helical motif of its 3 chains is entirely different from that of the α helix. Hydrogen bonds within a strand (*i.e.*, intrachain hydrogen bonds) are absent. Instead, each of these 3 helices is stabilized by steric repulsion of the pyrrolidone rings of the proline and hydroxyproline residues.

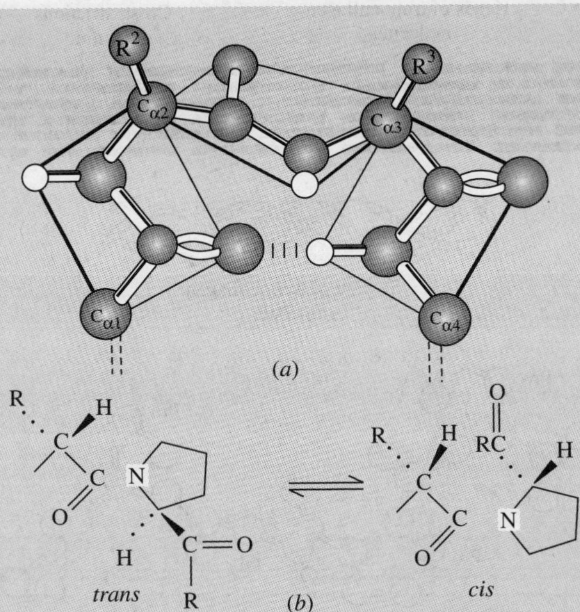

Fig. 9–30. Structure of a β turn or hairpin bend

(*a*) Hydrogen bonding in a tetrapeptide involving first and fourth amino acid residues. Here the NH and CO groups of residue 1 of the tetrapeptide are hydrogen-bonded, respectively, to the CO and NH groups of residue 4.

(*b*) The *trans* and *cis* isomers of a peptide bond involving the imino nitrogen of proline.

The 3 strands wind around each other in a cable fashion (Fig. 9–31), to form a **superhelix**. The superhelical twisting is right-handed (cf α-keratin). The rise (axial distance) per residue in this superhelix is 2.9 Å, and there are nearly 3 residues per turn. The three strands are stabilized by the interchain formation of hydrogen bonds between the > C = O group of one chain and the >N—H group of another chain. Proline does not have a hydrogen atom attached to its nitrogen when participating in a peptide bond and therefore cannot participate in this interchain hydrogen-bonding. This 3-stranded superhelix is known as the **collagen triple helix**. The amino acid residue on either side of glycine is located on the outside of the cable, where there is room for the bulky rings of Pro and Hyp residues. The tight wrapping of the collagen triple helix provides great tensile strength with no capacity to stretch. *Collagen fibres can support up to 10,000 times their own weight and are said to have greater tensile strength than a steel wire of equal cross section.*

The collagen fibrils consist of recurring 3-stranded polypeptide units called **tropocollagen** (MW 3,00,000), arranged head-to-tail in parallel bundles (Fig. 9–31). In some collagens, all three chains are identical in amino acid sequence and thus are homotrimers; while in others, two chains are identical and the third differs and thus are heterotrimers. The heads of adjacent molecules are staggered, and the alignment of the head groups of every fourth molecule produces characteristic cross-striations 640 Å (= 64 nm) apart. A series of complex covalent cross-links are formed within and between the tropocollagen molecules in the fibril, leading to the formation of strong mature collagen. The rigid, brittle character of the connective tissue in older people is the result of an accumulation of covalent cross-links in collagen as they age. The electron micrograph of human collagen fibres in given in Fig 9–32.

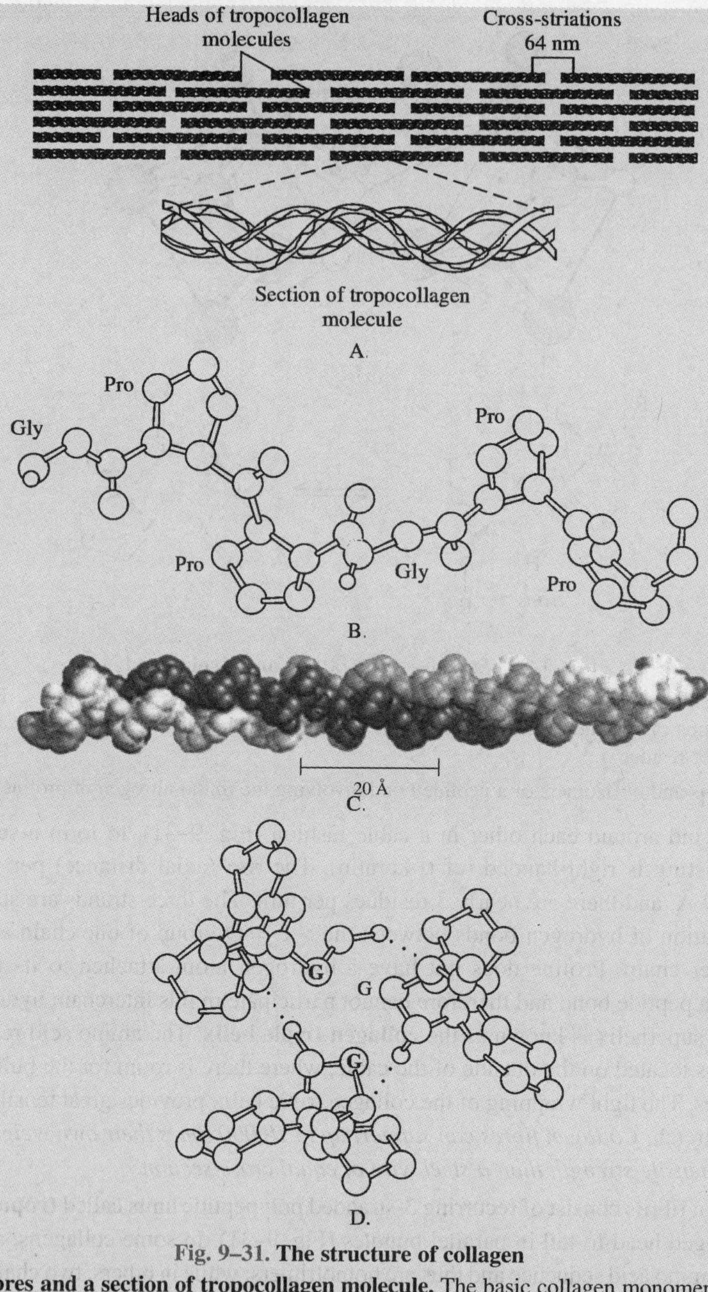

Fig. 9–31. The structure of collagen

A. **Collagen fibres and a section of tropocollagen molecule.** The basic collagen monomer is a triple helix, composed of 3 helical α chains. The collagen monomers become aligned in rows in which the molecules in one row are staggered relative to those in the neighbouring rows.

B. **Conformation of a single strand of the collagen triple helix.** The sequence shown here is Gly-Pro-Pro-Gly-Pro-Pro.

C. **Space-filling model of the collagen triple helix.** The three strands are shown in differnt shades.

D. **Cross section of a model of the triple-stranded helix of collagen.** Each strand is hydrogen-bonded to other two (... denotes a hydrogen bond). The α-carbon atom of a glycine residue in each strand is labelled G. Every third residue must be glycine because there is no space near the helix axis (centre) for a larger amino acid residue. Note that the pyrrolidone rings are on the outer side.

The food product **gelatin** is derived from collagen. Although it is a protein, it has little nutritional value because collagen lacks significant amounts of many amino acids that are essential in the human diet.

Human genetic defects involving collagen illustrate the close relationship between amino acid sequence and three-dimensional structure in this protein. **Osteogenesis imperfecta** results in abnormal bone formation in human babies. **Ehlers–Danlos syndrome** is characterized by loose joints. Both can be lethal and both result if a different glycine residue in each case is replaced by a cysteine (in case of osteogenesis imperfecta) or a serine (in case of Ehlers-Danlos syndrome) residue. These seemingly small replacements have a catastrophic effect on collagen function because they disrupt the Gly–X–Pro (or Hyp) repeat that gives collagen its unique helical structure.

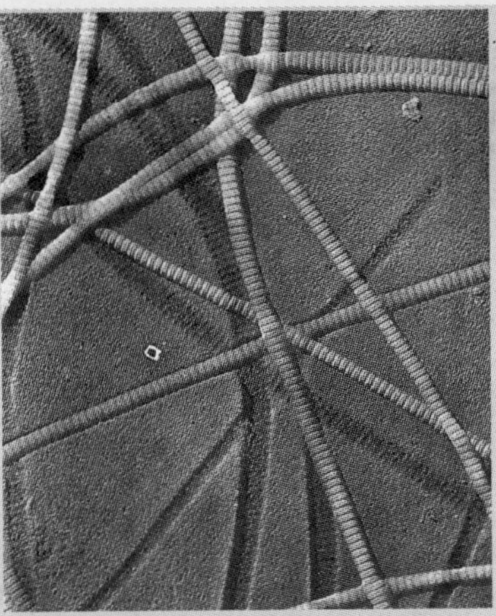

Fig. 9–32. An electron micrograph of human collagen fibres showing their characteristic banding pattern

The bands repeat along the fibre with a periodicity of 64 to 70 nm.

(Courtesy : Jesome Gross and Francis O. Schmitt)

CLINICAL IMPLICATIONS 9–1.

Osteogenesis Imperfecta

Osteogenesis imperfecta (OI), also called **brittle bone diease,** is a congenital disease caused by multiple genetic defects in the synthesis of type I collagen, one of the fibril-forming collagens. It is characterized by fragile bones, thin skin, abnormal teeth and weak tendons. It has a frequency of 1 in 30,000–50,000 live births. The majority of the individuals with OI have mutations in genes encoding a1(l) and a2(l) collagen chains. Many of these mutations are single-base substitutions that convert glycine in the Gly-X-Y repeat to bulky amino acids, preventing the correct folding of the collagen chains into a triple helix and their assembly to form collagen fibrils. The dominance of type I collagen in bone explains why bones are predominantly affected. However, there is a remarkable clinical variability characterized by bone fragility, osteopenia, short stature, and progressive skeletal deformities. The most common form of OI, with a presentation that is sometimes mistaken for child abuse, has a good prognosis, with fractures decreasing after puberty, though the general reduction in bone mass ensures lifetime risk remains high. Patients frequently develop deafness due to osteosclerosis, partly from recurrent fractures of the stapes. *Biphosphonate drugs,* which inhibit osteoclast activity and thereby inhibit normal bone turnover, have reduced the incidence of fractures.

CLINICAL IMPLICATIONS 9–2.

Ehlers–Danlos Syndrome

Ehlers–Danlos syndrome is a group of at least 10 disorders that are clinically, genetically and biochemically, distinguishable but that share manifestations of structural weaknesses in connective tissue. The usual problems are with fragility and hyperextensibility of skin and hypermobility of the joints. The weaknesses result from defects in collagen structure. For example, *type IV Ehlers Danlos syndrome* is caused by defects in type III collagen, which is particularly important in skin, arteries, and hollow organs. Characteristics include thin, translucent skin through which veins are easily seen, marked bruising, and sometimes an appearance of ageing in the hands and skin. Clinical problems arise from arterial rupture, intestinal perforation, and rupture of the uterus during pregnancy or labour. Surgical repair is difficult because of tissue fragility. The basic defects in type IV Ehlers–Danlos syndrome are due to changes in the primary structure of type III chains. These arise from point mutations that result in replacement of glycine residues and thus disruption of the collagen triple helix, and from exon-skipping, which shortens the polypeptide and can result in inefficient secretion and decreased thermal stability of the collagen and in abnormal formation of type III collagen fibrils.

Table 9–6 summarizes the important differences between the structures of α-keratin of hair and collagen of bones.

Table 9–6. Structural differences between α-keratin and collagen

α-keratin	*Collagen*
1. Found in skin and its appendages such as hair, nails and feathers.	1. Found in connective tissues such as tendons, cartilage, bones and the cornea of the eye.
2. The polypeptide chain is a right-handed helix.	2. The unique polypeptide chain is a left-handed helix.
3. Hydrophobic amino acids (Phe, Ile, Val, Met and Ala) predominate in the helix.	3. Gly, Pro, Hyp and Ala predominate in the helix.
4. The 3 helical strands wrap together into a superhelix, called protofibril.	4. The 3 helical strands wrap together into a repeating superhelical structure, called tropocollagen.
5. The superhelical twisting is left-handed.	5. The superhelical twisting is right-handed.
6. The covalent cross-links, between polypeptide chains within the triple-helical ropes and between adjacent ones, are contributed by disulfide bonds.	6. The covalent links are contributed by an unusual type of covalent link between two Lys residues that creates a nonstandard amino acid residue called lysinonorleucine.

Elastin

Elastin is found in elastic connective tissues such as ligaments and blood vessels. It resembles collagen in some of its properties, but differs in other. The polypeptide subunit of elastin fibrils is

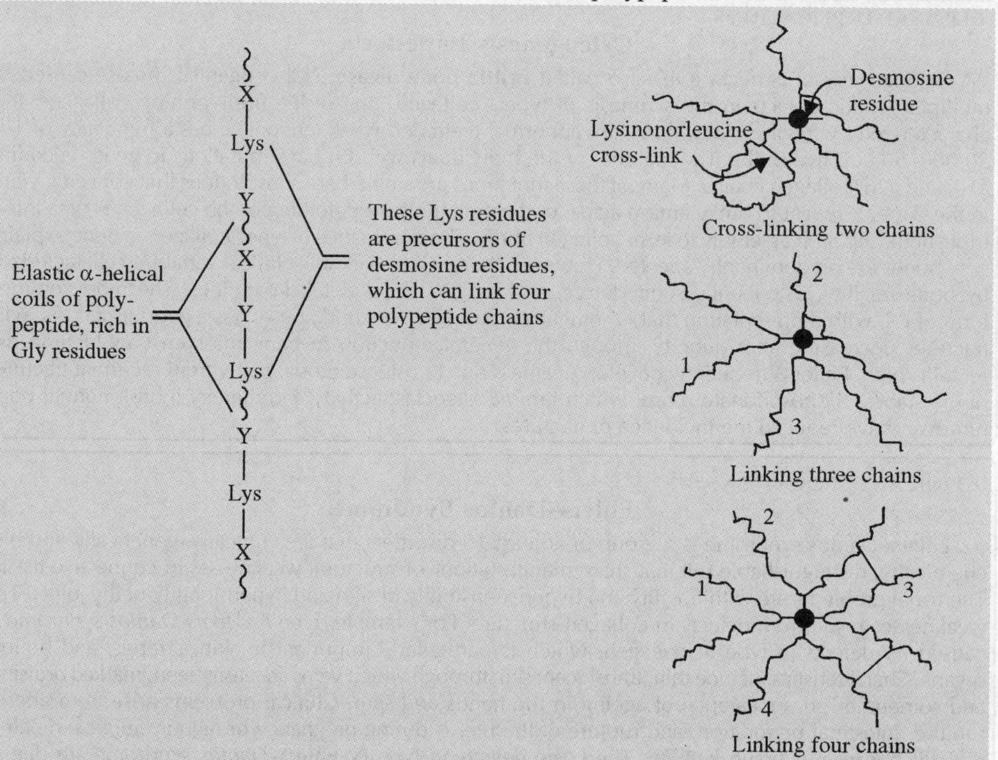

Fig. 9–33. Tropoelastin molecules and their linkage to form a network of polypeptide chains in elastin
As shown, desmosine residues can link two, three or four tropoelastin molecules. In addition, other types of linkages such as lycinonorleucine are also present.

(Adapted from Lehninger AL, Nelson DL and Cox MM, 1993)

tropoelastin (MW 72,000), containing about 800 amino acid residues. Tropoelastin differs from tropocollagen in having many lysine but few proline residues. Also, the helix which it forms is quite different from a helix and the collagen helix.

Tropoelastin consists of α helical portions of polypeptides rich in Gly residues, separated by short regions containing Lys and Ala residues. The helical regions stretch on applying tension but regain their original length when tension is released. The regions containing Lys residues form covalent cross-links. Four lysine side chains come together and are enzymatically converted into **desmosine** and a related compound, **isodesmosine** ; these amino acids are found exclusively in elastin. Like collagen, elastin also contains lysinonorleucine. These nonstandard amino acids are capable of joining tropoelastin chains into arrays that can be stretched reversibly in all directions (Fig. 9–33).

Hydrogen-bonding Potentiality of Proteins

All reversible molecular interactions in biological systems are mediated by 3 kinds of forces: electrostatic bonds, hydrogen bonds and van der Waals bonds. We have already seen that the hydrogen bonds between main-chain NH and CO groups work in forming α helices, β sheets and collagen fibrils. In fact, side chains of 11 of the 20 fundamental amino acids also can participate in hydrogen-bonding. These have been separated into 3 groups, according to their hydrogen- bonding potentialities:

1. **Tryptophan and Arginine.** Their side chains can serve *as hydrogen bond donors only.*

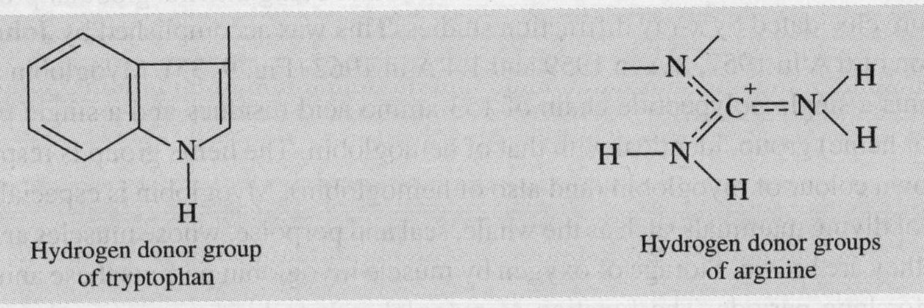

2. **Asparagine, Glutamine, Serine and Threonine.** Like the peptide unit itself, their side chains can serve *as hydrogen bond donors and acceptors.*
3. **Lysine, Aspartic acid, Glutamic acid, Tyrosine and Histidine.** The hydrogen-bonding capabilities of lysine (and the terminal amino group), aspartic and glutamic acids (and the terminal carboxyl group), tyrosine and histidine vary with pH. These groups can *serve as both acceptors and donors* within a certain pH range and *as acceptors* or *donors* (but not both) at other pH values (Fig. 9–34).

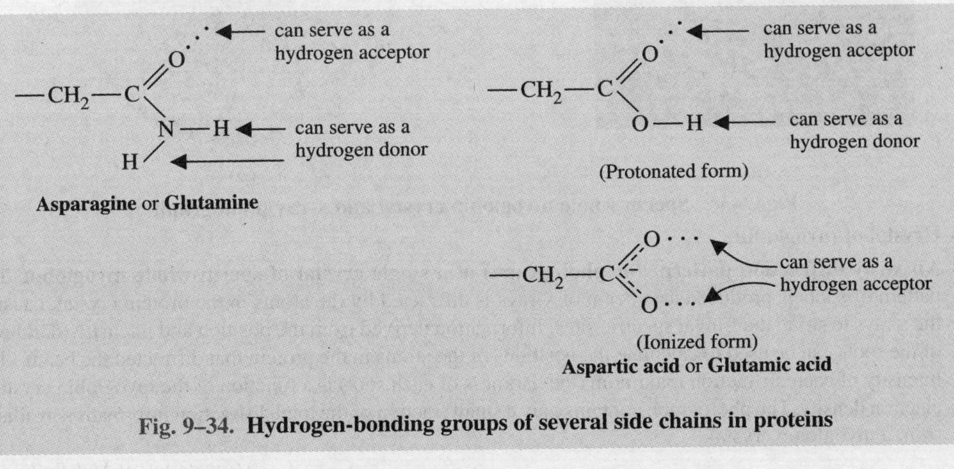

Fig. 9–34. Hydrogen-bonding groups of several side chains in proteins

3. Tertiary Structure : Folding of the Chain or Overall Folding

If the globular proteins consisted only of a small helix, these molecules would have been elongated structures with considerable length and a small cross-sectional area (*i.e.*, a large axial ratio). But as we now know about the existence of globular proteins, the helix must, therefore, possess many other types of bonds placed at regular intervals. These additional bonds include disulfide, hydrogen, hydrophobic and ionic. In such globular proteins (including enzymes, transport proteins, some peptide hormones and immunoglobulins), polar groups because of their hydrophobicity are most often located on the molecule's exterior and nonpolar R groups in the interior, where their interactions create a hydrophobic environment. The *tertiary structure*, thus, involves the folding of the helices of globular proteins. It refers to the spatial arrangement of amino acids that are far apart in linear sequence and to the pattern of disulfide bonds. The dividing line between secondary and tertiary structure is, hence, a matter of taste. X-ray crystallographic studies have revealed the detailed 3-'D' structures of more than 300 proteins.

Myoglobin (Mb): An Oxygen Storage Protein

Myoglobin (*myo*^{G} = muscle; *globin*^{G} = a type of protein) is a relatively small, oxygen-binding heme protein, found in muscle cells. It has the distinction of being the first globular protein to have its 3-'D' structure elucidated by x-ray diffraction studies. This was accomplished by **John C. Kendrew** at a resolution of 6 Å in 1957, 2 Å in 1959 and 1.4 Å in 1962 (Fig. 9–35). Myoglobin molecule (Fig. 9–36) contains a single polypeptide chain of 153 amino acid residues and a single prosthetic iron-porphyrin (or heme) group, identical with that of hemoglobin. The heme group is responsible for the deep red-brown colour of myoglobin (and also of hemoglobin). Myoglobin is especially abundant in the muscles of diving mammals such as the whale, seal and porpoise, whose muscles are so rich in this protein that they are brown. Storage of oxygen by muscle myoglobin permits these animals to remain submerged for long periods. The function of myoglobin is to bind oxygen in the muscles and to enhance its transport to the mitochondria, which consume oxygen during respiration.

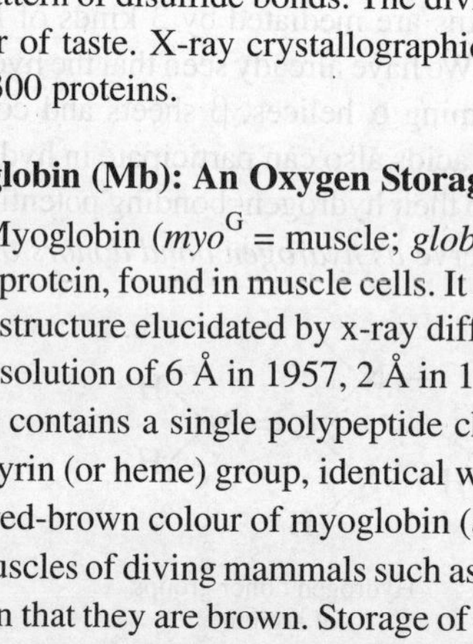

(*a*)

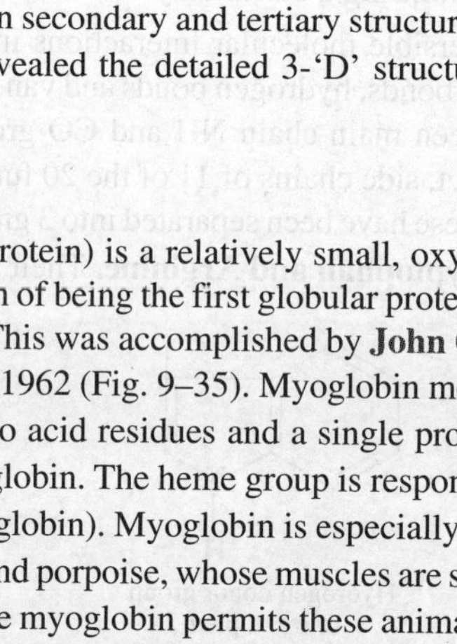

(*b*)

Fig. 9–35. Sperm whale myoglobin crystal and x-ray photograph

(*a*) **Crystal of myoglobin.**

(*b*) **An x-ray diffraction pattern (or photograph) of a single crystal of sperm whale myoglobin.** The pattern of spots is produced as a beam of x-rays is diffracted by the atoms in the protein crystal, causing the x rays to strike the film at specific sites. Information derived from the position and intensity (darkness) of the spots can be used to calculate the positions of the atoms in the protein that diffracted the beam. The intensity of each diffraction maximum (the darkness of each spot) is a function of the myoglobin crystal's electron density. The photograph contains only a small fraction of the total diffraction information available from a myoglobin crystal.

(*Courtesy: (a) Mel Pollinger*)

SIR JOHN COWDERY KENDREW

(LT, 1917–1997)

Kendrew, a British biochemist and crystallographer, was born on March 24, 1917 in Oxford. His father **Wilford George Kendrew** was a reader in climatology in the University of Oxford and mother **Evelyn May Graham Sandburg**, an art historian. After prep school at the Dragon School in Oxford, he was educated at Clifton College in Bristol, from 1930 to 1936. He attended Trinity College, Cambridge in 1936, as a Major Scholar, graduating in chemistry in 1939. He spent the early months of World War II doing research on reaction kinetics, and then became a member of the Air Ministry Research Establishment, working on radar. In 1940, he became engaged in operational research at the Royal Air Force (R.A.F.) headquarters, holding the honorary rank of *Wing Commander R.A.F.*

In 1945, he approached **Dr. Max Perutz** in the Cavendish Laboratory in Cambridge to work on the structure of proteins. **Joseph Barcroff**, a respiratory physiologist, suggested he might make a comparative protein crystallographic study of adult and fetal sheep hemoglobin and he started that work. In 1947, Kendrew became a Fellow of *Peterhouse*, and the Medical Research Council (MRC) agreed to create a research unit for the Study of the Molecular Structure of Biological Systems, under the direction of **Sir Lawrence Bragg**. The original studies were on the structure of sheep hemoglobin, but when this work had progressed as far as possible using the resources then available, Kendrew embarked on the study of **myoglobin**, a molecule only one-fourth the size of the hemoglobin molecule. His initial source of raw material was horse heart, but the crystals thus obtained were too small for x-ray analysis. Kendrew realized that the oxygen-conserving tissue of *diving mammals* could offer a better prospect, and a chance encounter led to his acquiring a large chunk of whale meat from Peru. Whale myoglobin did give large crystals with clean x-ray diffraction patterns. However, the problem still remained insurmountable, until 1953 **Max Perutz** discovered that phase problem in analysis of the diffraction patterns could be solved by **multiple isomorphous replacement**– comparison of patterns from several crystals; one from the native protein, and others that had been soaked in solutions of heavy metals and had metal ions introduced in different well-defined positions. An *electron density map* at 6 Ångström (0.6 nanometer) resolution was obtained by 1957, and by 1959 an atomic model could be built at 2 Ångström (0.2 nm) resolution. In 1954, Kendrew had already become a Reader at the Davy–Faraday Laboratory of the Royal Institution in London.

Kendrew was awarded with **M.F. Perutz** the **1962 Nobel Prize for Chemistry** for determining for the first time the structure of a globular protein, myoglobin– the protein that transports O_2 from the lungs to the tissues *via* blood cells. The structure was accomplished by Kendrew at a resolution of 6Å in 1957, 2Å in 1959 and 1.4 Å in 1962. Regarding myoglobin, Kendrew wrote:

> "Perhaps the more remarkable features of [myoglobin] are its complexity and its lack of symmetry. The arrangement seems to be almost totally lacking in the kind of regularities which one instinctively anticipates, and it is more complicated than has been predicted by any theory of protein structure."
>
> *(John C. Kendrew, article in Nature, 1958)*

Their work was done at what is now the MRC Laboratory of Molecular Biology in Cambridge. On Saturday, the 20th October, 1962, the award of Nobel Prizes to **John Kendrew** and **Max Perutz**, and to **Crick**, **Watson** and **Wilkins** was satirized in a short sketch in the BBC TV programme *That was The Week That Was* with the Nobel Prizes being referred to as 'The Alfred Nobel Peace Pools'.

In 1963, Kendrew became one of the founders of the *European Molecular Biology Organization;* as well, he founded and was for many years editor-in-chief of the *Journal of Molecular Biology*. He became Fellow of the American Society of Biological Chemists in 1967 and honorary member of the International Academy of Science. From 1974 to 1979, he was a Trustee of the *British Museum,* and from 1974 to 1988 he was successively Secretary-General, Vice-President, and President of the *International Council of Scientific Unions.* After his retirement from the European Molecular Biology Laboratory, Kendrew became President of St. John's College at Oxford University, a post he held from 1981 to 1987. He died on August 23, 1997.

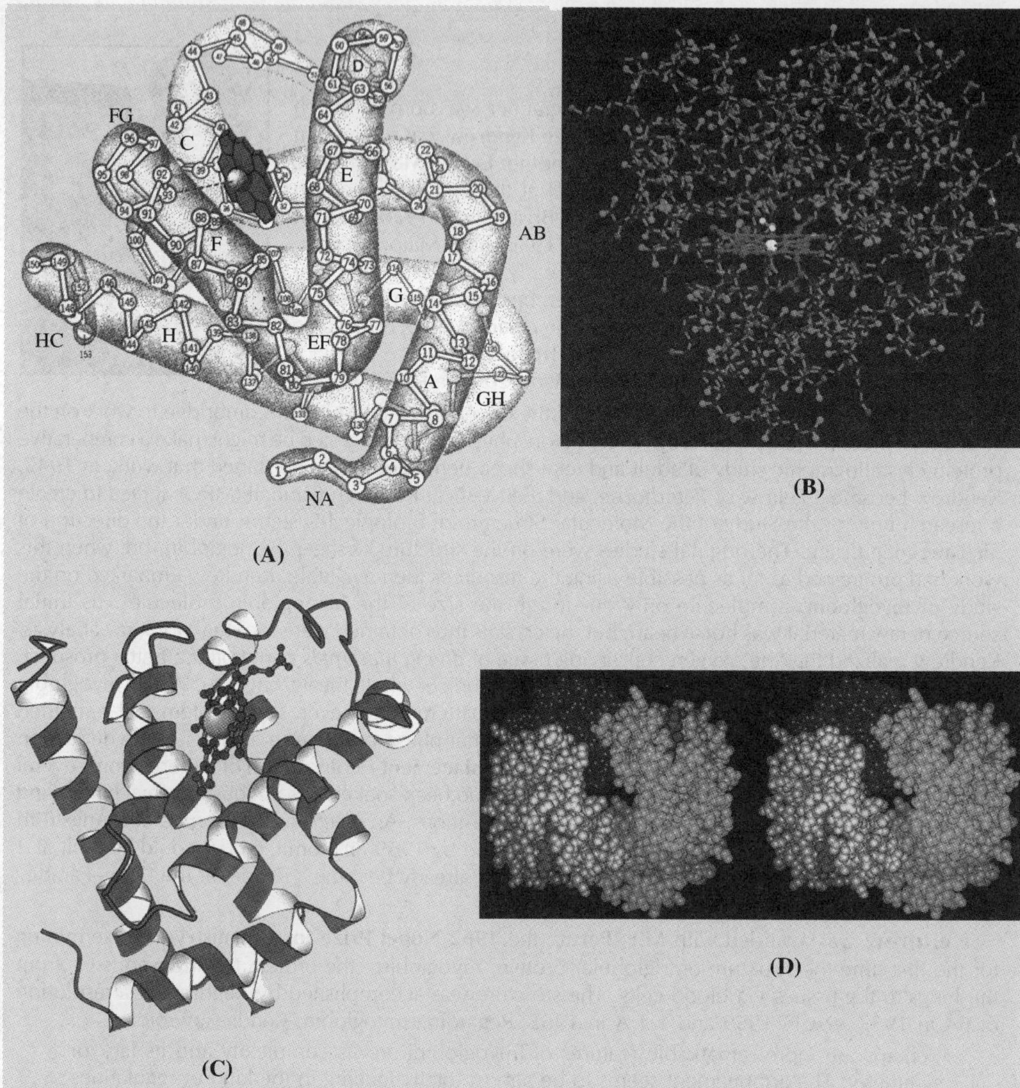

Fig. 9–36. The structure of sperm whale myoglobin

(A) **The tertiary structure**. Its 153 C_α positions are numbered from the N-terminus and its eight helices are sequentially labeled A through H. The last half of the EF corner is now regarded as a turn of helix and is therefore designated the F′ helix. The heme group is shown in *red*. Most of the amino acids are part of α helices. The nonhelical regions occur primarily as turns, where the polypeptide chain changes direction.

(B) **The ball-and-stick model.** The three-dimensional structure of myoglobin shows the position of all of the molecule's atoms other than hydrogen and reveals many interactions between the amino acids. The heme group is indicated in red.

(C) **Ribbon model.** A schematic view shows that the protein consists largely of α helices. The heme group is shown in *black* and the iron atom is shown as a *purple* sphere.

(D) **A stereo, space-filling model showing the quaternary structure of hemoglobin.** The model shows the α_1, α_2, β_1, and β_2 subunits as colored *yellow*, *green*, *light blue*, and *dark blue*, respectively, Heme groups are *red*. The protein is viewed along its molecular twofold rotation axis which relates the $\alpha_1\beta_1$ protomer to the $\alpha_2\beta_2$ protomer.

[*Courtesy : (A) Irving Geis (B) Ken Eward*]

Myoglobin (MW = 16,700) is an extremely compact macromolecule with oblate, spheroid shape and leaves little empty space in its interior. Its overall molecular dimensions are 45 × 35 × 25 Å. The sausagelike outline of the tertiary structure of myoglobin chain is folded into an odd, irregular form. The backbone of the molecule is made up of 8 almost-straight α-helical segments, designated from the N-terminus as A through H (Fig 9–37). The first residue in helix A is designated A_1, the second A_2 and so forth. Interspersed among the helical segments are 5 nonhelical regions, each identified by the two helical segments it joins, *e.g.*, AB is located between helical sections A and B. There are also two other nonhelical regions : two residues at the N-terminus (named NA_1 and NA_2) and five residues at the C-terminus (named HC_1 through HC_5). The longest helical segment has 23 amino acid residues and the shortest only 7. All the helical segments are of α-type and right-handed; there being no β-structure (refer Table 9–7).

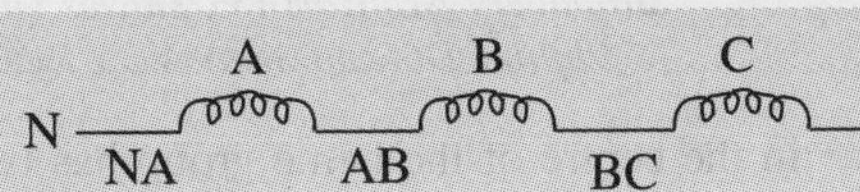

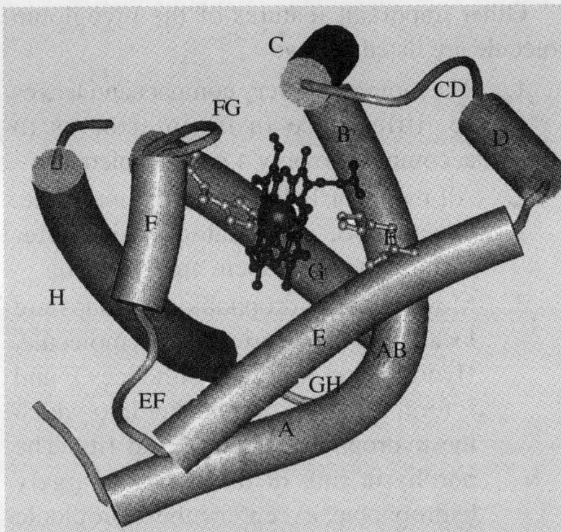

Fig. 9–37. The eight α helical segments of myoglobin molecule

The eight α helical segments (shown here as cylinders) are labeled A through H. Nonhelical residues in the bends that connect them are labeled AB, CD, EF, and so forth, indicating the segments they interconnect. A few bends, including BC and DE, are abrupt and do not contain any residues; these are not normally labeled. (The short segment visible between D and E is an artifact of the computer representation.) The heme is bound in a pocket made up largely of the E and F helices, although amino acid residues from other segments of the protein also participate.

Table 9–7. Approximate amounts of α-helix and β-structure in some single-chain proteins*

Protein	*Total residues*	*Residues, %*	
		α-helix	*β-structure*
Cytochrome C	104	39	0
Ribonuclease	124	26	35
Lysozyme	129	40	12
Myoglobin	153	78	0
Chymotrypsin	247	14	45
Carboxypeptidase	307	38	17

* Portions of polypeptide chains, that are not accounted for by α-helix or β-structure, consist of bends, reverse turns and irregularly-coiled stretches.

(*Adapted from Cantor and Schimmel, 1980*)

Of the 153 amino acid residues, 121 (*i.e.*, 79%) are present on the helical regions and the remaining 32 amino acid residues are distributed over the nonhelical areas. The nonhelical areas possess various types of bonding such as hydrogen and nonpolar linkages. The flat heme group is tightly but noncovalently bound to the polypeptide chain. A notable feature of whale myoglobin is the absence of a disulfide bridge, since both cysteine and cystine residues are lacking. The electron density map of sperm whale myoglobin is presented in Fig. 9–38.

Other important features of the myoglobin molecule are listed below:

1. The molecule is very compact and leaves so little space in its interior as to accommodate only 4 water molecules.
2. All the polar R groups of the molecule, except two, are located on its outer surface and all of them are hydrated.
3. Most of the hydrophobic R groups are located in the interior of the molecule. Hydrophobic R groups of helices E and F form the sides of a pocket into which the hydrophobic heme group fits. The porphyrin ring of the heme is largely hydrophobic, except for the 2 propionic acid side chains which stick out of the pocket and into the aqueous environment.
4. Each of thc 4 proline residues occurs at a turn. Other turns or bends contain serine, threonine and asparagine.
5. All the peptide bonds of the polypeptide chain are planar, with the carbonyl and amide groups being *trans* to each other.
6. The heme group is flat and rests in a crevice in the molecule. The iron atom in the centre of the heme group has two coordination bonds that are perpendicular to the plane of the heme group. One bond is attached to the R group of histidine (93), whereas the other bond is the site to which an O_2 molecule is bound. Upon oxygenation, the iron atom descends into the heme plane and the oxygen bound to myoglobin is stabilized by hydrogen-bonding to the imidazole ring of His E_7 (Fig. 9–39). Reversible oxygenation requires that the iron atom be in the ferrous state (Fe^{2+}) and myoglobin with or without oxygen bound to the Fe^{2+} of heme is called oxyhemoglobin and deoxyhemoglobin respectively.
7. The inside and outside are well defined. There is little empty space inside. *The interior consists almost entirely of nonpolar residues* such as Leu, Val, Met, and Phe. On the contrary, Glu, Asp, Gln, Asn, Lys and Arg are absent from the interior of the

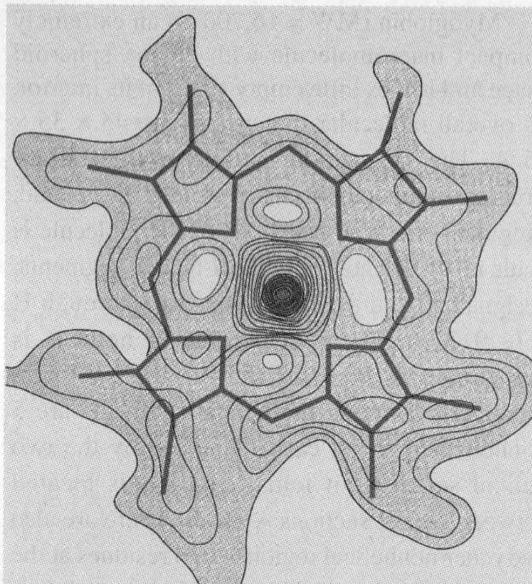

Fig. 9–38. A section through the 2.0-Å-resolution electron density map of sperm whale myoglobin

The heme group is represented by *red*. The large peak the centre of the map represents the electron-dense Fe atom.

(After Kendrew JC, Dickerson RE, Strandberg BE, Hart RG, Davies DR, Phillips DC and Shore VC, 1960)

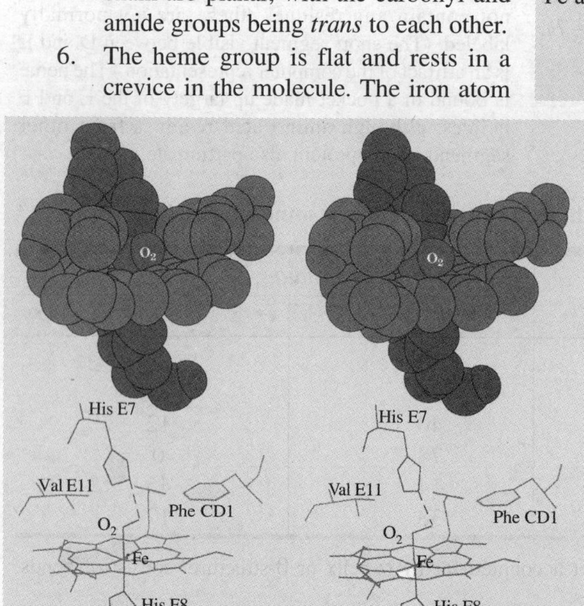

Fig. 9–39. Stereo drawings of the heme complex in oxymyoglobin

In the upper drawing, atoms are represented as spheres of van der Waals radii. The lower drawing shows the corresponding skeletal model with a dashed line representing the hydrogen bond between the distal His and the bound O_2.

(After Phillips SEV, 1980)

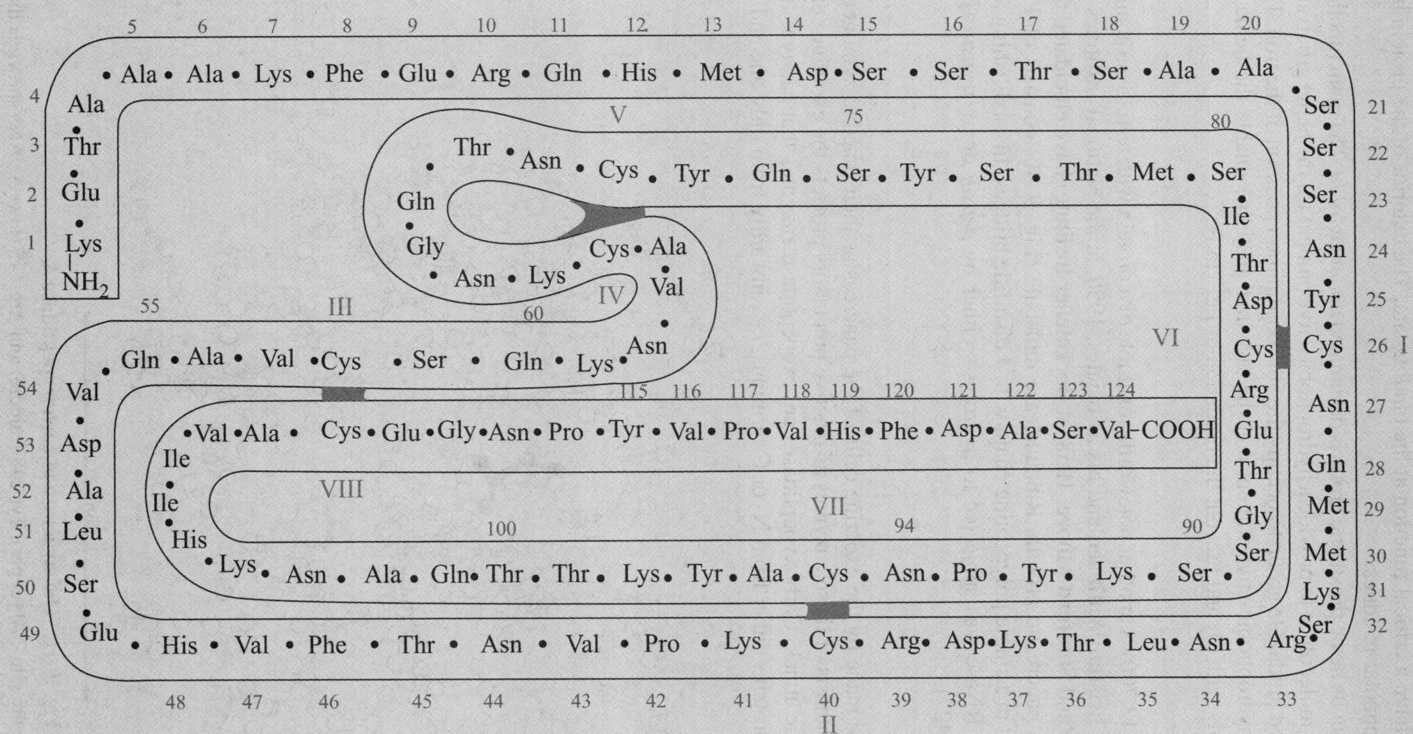

Fig. 9–40. The amino acid sequence of the enzyme ribonuclease from bovine

This enzyme molecule consists of 124 amino acid residues most of which are *numbered*. The 4 dark blocks between cysteines represent 4 disulfide bonds or bridges. Note that this illustration is diagrammatic; the polypeptide chain is actually folded to give a complex 3-dimensional configuration.

(*After Smyth DG, Stein WH and Moore S, 1963*)

protein. Residues with polar and nonpolar parts within them (Thr, Tyr, Trp) are so oriented that their nonpolar portions point inward. The only polar residues inside myoglobin are two histidines which have a critical function at the binding site. The exterior of the protein has both polar and nonpolar residues.

8. Myoglobin without its heme prosthetic group is designated *apomyoglobin*. The main function of apomyoglobin molecule is to provide a hydrophobic environment for the heme group and a properly oriented imidazole group to occupy the 5th coordination position of the iron. The presence of the hydrophobic environment and the proximal histidine enable the heme to combine reversibly with O_2 and prevent the oxidation of Fe^{2+} to Fe^{3+}.

Ribonuclease (RNase)

Ribonuclease (MW = 13,700) is another noteworthy example of a protein with tertiary structure. Two groups of workers (**Christian Anfinsen** and his associates, 1950, at the National Institutes of Health ; **William H. Stein** and **Stanford Moore**, 1958, at Rockefeller Institute) have elucidated the complete structure of this pancreatic protein. Ribonuclease molecule (Fig. 9–40) consists of 124 amino acid units contained in a single polypeptide chain with 4 disulfide bridges. In establishing the amino acid sequence, the RNase was subjected to hydrolysis either by *pepsin* or by *trypsin* (for further details, refer page 427).

Carboxypeptidases

Carboxypeptidases are made by the exocrine cells of the pancreas as their inactive zymogens, *procarboxypeptidases*. Synthesis of these enzymes as inactive precursors protects the exocrine cells from destructive proteolytic attack. Carboxypeptidases are exopeptidase enzymes that catalyze the hydrolysis of (or cleave) proteins either from N- or C-terminus, *i.e.*, inwardly from only one end of

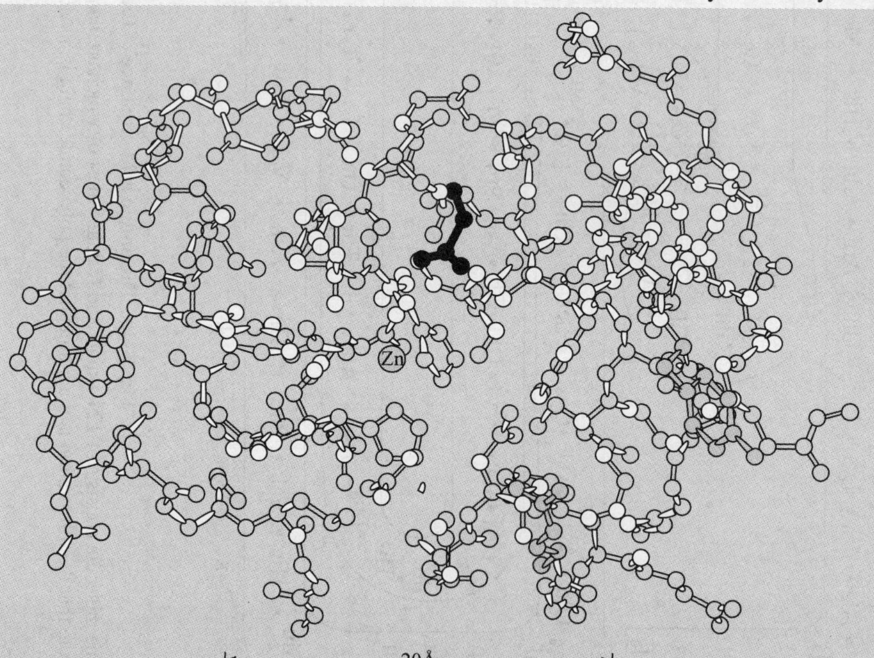

Fig. 9–41. The structure of carboxypeptidase A

The three shaded amino acids are Glu 270 (upper), Tyr 248 (middle) and Arg 145 (lower). Also shown is the zinc ion (Zn^{2+}) located in a groove near the surface of the molecule.

(*After William N. Lipscomb, 1971*)

the chain (By contrast, endopeptidases are enzymes that cleave internal peptide bonds.). Different forms of carboxypeptidases are usually *substrate-specific*. *Carboxypeptidase A*, the enzyme used initially in C-terminal residue determinations, readily hydrolyzes (or cleaves) a C-terminal residue with an aromatic or a bulky aliphatic side chain. *Carboxypeptidase B* cleaves C-terminal lysine or arginine residues while *carboxypeptidase C* cleaves a C-terminal proline residue. However, *carboxypeptidase Y* has the capacity to cleave all C-terminal amino acids.

Carboxypeptidase A (MW = 34,600) is a zinc-containing proteolytic enzyme. Its 3-'D' structure (Fig. 9–41) was accomplished at a resolution of 2 Å by **William Nunn Lipscomb** in 1967. This enzyme is a globular protein which contains 307 amino acid residues in a single polypeptide chain. Carboxypeptidase A molecule is a compact ellipsoid with dimensions of 50 × 42 × 38Å. This molecule is considerably larger than myoglobin and shows more variety in its secondary structures. Of the 307 amino acid residues, 38% are involved in the formation of 8 helical segments (designated A through H) and 17% are involved in β pleated sheet structure that extends through the molecule, forming a core. Helix D is connected to helix E by a segment containing 52 amino acids. The folding of this segment is very complex and contains no recognizable regular secondary structure. A tightly bound zinc ion (which is essential for enzymatic activity) is located in a groove near the surface of the molecule, where it coordinates with 2 histidine side chains, a glutamate side chain and a water molecule. A large pocket near the zinc ion accommodates the side chain of the terminal residue of the peptide substrate.

Fig. 9–42 gives the ribbon model of the structure of bovine carboxypeptidase A, along with its central β sheet.

Fig. 9–42. Bovine carboxypeptidase A, showing its central β sheet

(Photo courtesy: Irving Geis, 1986)

Two aspects of catalytic mechanism of carboxypeptidase A, are noteworthy:

(A) **Binding of substrate:** For instance, the binding of glycyltyrosine (a slowly-hydrolyzed substrate) is accompanied by a large structural rearrangement of the active site. The phenolic hydroxyl group of tyrosine 248 moves a distance of 12 Å from the surface of the molecule to the vicinity of the terminal carboxylate of the substrate. Consequently, the active site cavity closes so that the water molecule extrudes from it.

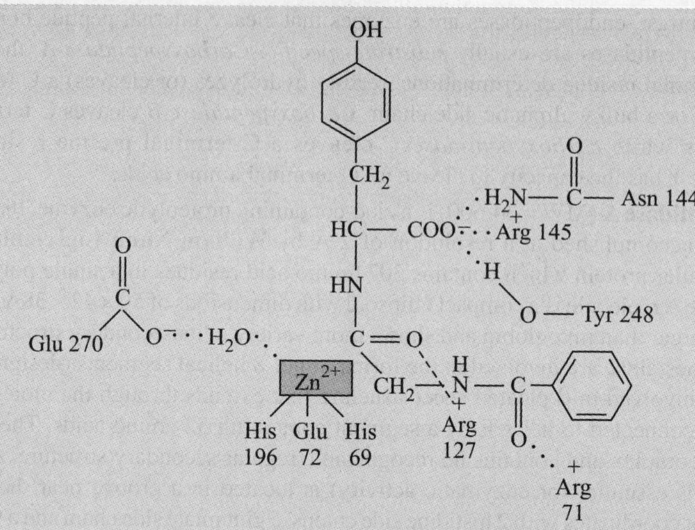

Fig. 9–43. Mode of binding of N-benzoylglycyltyrosine to carboxypcptidase A

(*After Christianson DW and Lipscomb WN, 1989*)

How does carboxypeptidase A recognize the terminal carboxylate group of a peptide substrate? Three interactions explain this (Fig. 9–43). The induced-fit movement of Tyr 248 enables its phenolic OH group to become hydrogen-bonded to the terminal carboxylate group. Further, the guanidium of Arg 145 moves 2 Å in this transition so as to enable it to form a salt bridge with the terminal COO^-, which is hydrogen-bonded to a side chain Asn 144. Besides, the terminal side chain of the substrate sits in a hydrophobic pocket of the enzyme (this is the reason why carboxypeptidase A requires an aromatic or bulky nonpolar residues at this position). Two interactions involving the penultimate amino acid residue are also noteworthy. The carbonyl oxygen of the scissile peptide bond interacts with the guanidium group of Arg 127. Finally, Tyr 248 is also hydrogen-bonded to the peptide NH of penultimate residue.

(B) **Activation of bound water:** Proteins and peptides are stable at neutral pH in the absence of a protease because water does not readily attack peptide bonds. In fact, carboxypeptidase A also activates a water molecule. This is accomplished by the bound zinc ion with the help of adjacent carboxylate group of Glu 270. Thus, *the zinc-bound water behaves muck like an* OH^- *ion*. The activation of water by zinc ion involves two steps:

1. First step in catalysis is the attack of activated water molecule on the carbonyl group of the scissile peptide bond (Fig. 9–44). Specifically, the nucleophilic oxygen atom of activated water attacks the carbonyl carbon atom. Concurrently, Glu 270 accepts a proton from the water. A negatively-charged tetrahedral intermediate is, thus, formed. This intermediate is stabilized by electrostatic interactions with Zn^{2+} and the positively-charged side chain of Arg 127.

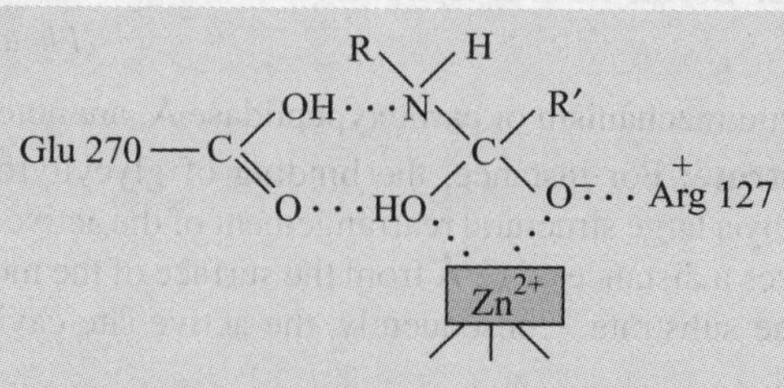

Fig. 9–44. Proposed tetrahedral transition state in the peptide bond hydrolysis by carboxypeptidase A

2. The final step is the transfer of a proton from the COOH group of Glu 270 to the peptide NH group. The peptide bond is simultaneously cleaved and the reaction products diffuse away from the active site. In fact, the bound substrate is surrounded on all sides by catalytic groups of the enzyme. And the substrate-induced structural changes in the active site promote catalysis in 3 ways:
   (*a*) activation of H_2O by Zn^{2+},
   (*b*) proton abstraction and donation by Glu 270, and
   (*c*) electrostatic stabilization by Arg 127.

It is, thus, apparent that a substrate could not enter such an array of catalytic groups (nor could a product leave) unless the enzyme were flexible.

4. Quaternary Structure: Protein-Protein Interactions or **Multichain Association**

A fourth degree of complexity in protein structure has recently been recognized to be of great value in many proteins. Some globular proteins consist of 2 or more interacting peptide chains. Each peptide chain in such a protein is called a *subunit*. These chains may be identical or different in their primary structure. This specific association of a number of subunits into complex large-sized molecules is referred to as the *quaternary structure*. In other words, quaternary structure refers to the spatial arrangement of subunits and the nature of their contact. The same forces (disulfide, hydrogen, hydrophobic and ionic bonds) involved in the formation of tertiary structure of proteins are also involved here to link the various polypeptide chains.

Tobacco Mosaic Virus (TMV)

TMV, with 158 amino acid residues, is an instance of the protein-protein interactions. TMV particles are rod like, tubular structures measuring 3,000 Å length and 180 Å diameter. The tube is made up of a single RNA molecule (30,000 Å long) which is coiled round the central axis. The RNA molecule is a right-handed helical filament and is embedded in a protein matrix or the coat which is made up of 2,130 identical molecules of a single protein that interact to form a cylinder enclosing the RNA genome. The amino acid sequence of this coat protein (Fig. 9–45) was elucidated in 1962 independently by **H. Fraenkel-Conrat** and **H. Wittmann**. The coat protects the RNA molecule from disintegration by various environmental changes.

AcSer-Tyr-Ser-Ile-Thr-Thr-Pro-Ser-Gln-Phe-Val-Phe-Leu-Ser-Ser-Ala-Trp-Ala-Asp-Pro-
1 5 10 15 20

Ile-Glu-Leu-Ile-Asn-Leu-Cys-Thr-Asn-Ala-Leu-Gly-Ans-Gln-Phe-Gln-Thr-Gln-Gln-Ala-
25 30 35 40

Arg-Thr-Val-Val-Gln-Arg-Gln-Phe-Ser-Glu-Val-Trp-Lys-Pro-Ser-Pro-Gln-Val-Thr-Val-
45 50 55 60

Arg-Phe-Pro-Asp-Ser-Asp-Phe-Lys-Val-Tyr-Arg-Tyr-Asn-Ala-Val-Leu-Asp-Pro-Leu-Val-
65 70 75 80

Thr-Ala-Leu-Leu-Gly-Ala-Phe-Asp-Thr-Arg-Asn-Arg-Ile-Ile-Glu-Val-Glu-Asn-Gln-Ala-
85 90 95 100

Asn-Pro-Thr-Thr-Ala-Glu-Thr-Leu-Asp-Ala-Thr-Arg-Arg-Val-Asp-Asp-Ala-Thr-Val-Ala-
105 110 115 120

Ile-Arg-Ser-Ala-Ile-Asn-Asn-Leu-Ile-Val-Glu-Leu-Ile-Arg-Gly-Thr-Gly-Ser-Tyr-Asn-
125 130 135 140

Arg-Ser-Ser-Phe-Glu-Ser-Ser-Ser-Gly-Leu-Val-Trp-Thr-Ser-Gly-Pro-Ala-Thr
145 150 155 158

Fig. 9–45. Amino acid sequence of the coat protein of tobacco mosaic virus (TMV)

Note that the serine in position 1 is acetylated (AcSer). Also note that there are no extended repetitions of an amino acid sequence, that there are no clusters of either polar or nonpolar amino acids, and that the frequencies of occurrence of different amino acids vary widely.

Hemoglobin (Hb): An Oxygen Transport Protein

Hemoglobin (*hemo*[G] = blood ; *globin*[G] = a protein, belonging to the myoglobin-hemoglobin family), the oxygen transporter in erythrocytes, constitutes about 90% of the protein of red blood cells. It is a tetrameric protein, *i.e.*, it contains 4 polypeptide chains. The four chains (of which two are of one kind and two of another) are held together by noncovalent interactions. Each chain contains a heme group and a single oxygen-binding site. *Hemoglobin A*, the principal hemoglobin in adults, consists of two alpha (α) chains and two beta (β) chains. Adults also have a minor hemoglobin (~ 2% of the total hemoglobin) called hemoglobin A_2, which contains two delta (δ) chains in place of two β chains of hemoglobin A. Thus, the subunit composition of hemoglobin A is $\alpha_2\beta_2$, and that of hemoglobin A_2 is $\alpha_2\delta_2$. HbA and HbA_2 are the postnatal forms of hemoglobin.

Prior to birth, embryonic and fetal hemoglobins are used as oxygen carrier. The first hemoglobin to appear in embryonic development has the subunit composition $\zeta_2\ \varepsilon_2$: the two zeta (ζ) chains are analogous to the α chains and the two epsilon (ε) chains are analogous to the β chains. When, after about 6 weeks, ζ chain production ceases, the tetramer $\alpha_2\ \varepsilon_2$ appears, *i.e.*, ζ is replaced by α. A third embryonic hemoglobin, $\zeta_2\ \gamma_2$ has also been identified where ε is replaced by γ. These last two hemoglobins represent transition phases leading to the appearance of *fetal hemoglobin* (*HbF*), whose tetrameric composition is $\alpha_2\ \gamma_2$. The α and ζ chains contain 141 residues each. The β, γ and δ chains contain 146 residues each and have homologous (similar but not identical) amino acid sequences. The γ and δ chains differ from the β chain at 39 and 10 amino acid residues respectively.

Hemoglobin (MW = 64,500) provides an example of the interaction of unidentical protein subunits. Since hemoglobin is 4 times as large as myoglobin, much more time and effort were required to solve its 3-'D' structure, finally achieved by **Max Perutz**, **John C. Kendrew** and their

SIR MAX FERDINAND PERUTZ

(LT, 1914–2002)

Perutz, an Austrian-born British molecular biologist, was born on May 19, 1914 in Vienna. Both his parents, **Hugo Perutz** and **Dely Goldschmidt**, came from families of textile manufacturers who had made their fortune in the 19th century by the introduction of mechanical spinning and weaving into the Austrian monarchy. Max was sent to school at the *Theresianum*, a grammar school derived from an officers academy of the days of the **Empress Maria Theresa**, and his parents suggested that he should study law in preparation for entering the family business. However, a good schoolmaster awakened his interest in chemistry, and he had no difficulty in persuading his parents to let him study the subject of his choice. In 1932, he entered Vienna University, where he, in his own words, "wasted five semesters in an exacting course of inorganic analysis". His curiosity was aroused, however, by organic chemistry, and especially by an organic chemistry course, given by **F. von Wessely**, in which Sir Frederick Gowland Hopkin's work at Cambridge was mentioned. He therefore decided to work at Cambridge for his Ph.D. Thesis.

So after completing his first university degree at the University of Vienna, Perutz in 1936 became a research student at the University of Cambridge's Cavendish Laboratory, in a crystallography research group under the direction of **J.D. Bernal**. He completed his Ph.D. under **William Lawrence Bragg**. Here, he started to work on hemoglobin, which was to occupy him for most of his professional career. As a research student, Perutz became a member of *Peterhouse*, where he was elected an Honorary Fellow in 1962. He took a keen interest in the Junior Members, and was a regular and popular speaker at the Kelvin Club, the College's scientific society.

After Hitler's invasion in Austria and Czechoslovakia, the family business was expropriated, his parents became refugees, and his own funds were soon exhausted. He was saved by being appointed research assistant to **Sir Lawrence Bragg**, under a grant from the Rockefeller Foundation on January 1, 1939. The grant continued, with various interruptions due to WW II, until 1945, when Perutz was given an Imperial Chemical Industries Research Fellowship. In October 1947, he was made Head of the newly-constituted Medical Research Council Unit for Molecular Biology, with **Sir J.C. Kendrew** representing its entire staff. He continued holding this post until he was made Chairman of the Medical Research Council Laboratory of Molecular Biology, in March 1962. His collaboration with **Sir L. Brag** continued through all these years.

In 1953, Perutz showed that diffracted x-rays from protein crystals could be phased by comparing the patterns from crystals of the protein with and without heavy atoms attached. In 1959, he employed this method to determine the molecular structure of the protein hemoglobin, which transports oxygen in the blood. He shared with his compatriot **J.C. Kendrew** the **1962 Nobel Prize for Chemistry** for his x-ray diffraction analysis of the structure of hemoglobin, a globular protein with quaternary structure and 4 times as large as myoglobin.

After 1959, Perutz and his colleagues went on to determine the structure of oxyhemoglobin and deoxyhemoglobin at high resolution. As a result, in 1970, he was able to suggest how it works as a molecular machine: how it switches between its deoxygenated and its oxygenated states, in turn triggering the uptake of oxygen and then its release to the muscles and other organs. In addition, Perutz studies the structural changes in a number of hemoglobin diseases and how these might affect oxygen binding. In his final years, Perutz turned to the study of changes in protein structures implicated in Huntington and other neurodegenerative diseases. He demonstrated that the onset of **Huntington disease** is related to the number of glutamine repeats as they bind to form what he called a *polar zipper*.

In his later years, Perutz was a regular reviewer / essayist for *The New York Review of Books* on biomedical subjects. Many of these essays are reprinted in his 1998-book *I wish I had made you angry earlier*. Perutz's flair for writing was a late development. His relative **Leo Perutz**, a distinguished writer, told Max when he was a boy that he would never be a writer, an unwarranted judgement, as demonstrated by Perutz's remarkable letters written as an undergraduate. They are published in 'What A Time I Am Having: Selected Letter of Max Perutz'. Perutz was delighted to win the **Lewis Thomas Prize** for *Writing about science*, in 1997.

Perutz attacked the theories of philosophers **Sir Karl Popper** and **Thomas Kuhn** and biologist **Richard Dawkins** in a lecture given at Cambridge on 'Living Molecules', in 1994. He criticized Popper's notion that science progresses through a process of hypothesis formation and refutation, saying that hypotheses are not necessarily the basis of scientific research and, in molecular biology at least, they are not necessarily subject to revision either. For Perutz, Kuhn's notion that science advances in paradigm shifts that are subject to social and cultural pressures is an unfair representation of modern science. These criticisms extended to scientists who attack religion, in particular to Dawkins. Statements which offend religious faith were for Perutz tactless and simply damage the reputation of science. They are of a quite different order to criticism of the demonstrably false *theory of creationism*. He concluded that,

> "even if we do not believe in God, we should try to live as though we did".

Within days of the September 11 attacks in 2001, Perutz wrote to **British P.M. Tony Blair**, appealing to him not to respond with military force:

> "I am alarmed by the American cries for vengeance and concerned that President Bush's retaliation will lead to the death of thousands more innocent people, driving us into a world of escalating terror and counter-terror. I do hope that you can use your restraining influence to prevent this happening."

Besides the Nobel Prize, Perutz also received a number of other honours:

1. Commander of the Order of the British Empire (1963)
2. The Royal Medal (1971)
3. The Companions of Honour (1975)
4. The Copley Medal (1979)

colleagues in 1959. They determined the 3-'D' structure of adult hemoglobin (HbA) of horse, which is very similar to that of human hemoglobin (HbA). The x-ray analysis has revealed that the hemoglobin molecule is nearly spherical with a diameter of 55 Å (5 .5 nm).

Human hemoglobin protein (Fig. 9–46) consists of 4 polypeptide chains of two types, two α-chains and two β-chains. The polypeptide portion is collectively called as *globin*. The α-chain has valine at the N-terminal and arginine at the C-terminal whereas in the β-chain, valine is situated at the N-terminal and histidine at the C-terminal. Each α chain is in contact with both β chains. In contrast, there are few interactions between the two α-chains or between the two β–chains. Each chain has a heme prosthetic group in a crevice near the exterior of the molecule. The heme groups are involved in the binding of oxygen. The α-chain has 141 residues and the β-chain, which is more acidic, has 146 residues. The protein, thus, has a total (141 × 2 + 146 × 2) 574 amino acid residues. Each of the 4 chains has a characteristic *tertiary structure*, in which the chain is folded. Like myoglobin, the α- and β-chains of hemoglobin contain several segments of α-helix, separated by bends. The α- and β-chains are held together as a pair by ionic and hydrogen bonds. The 2 pairs are then joined to each other by additional ionic bonds, hydrogen bonds and the hydrophobic forces. Thus, the 4 polypeptide chains fit together almost tetrahedrally to produce the characteristic *quaternary structure*. The hemes are 2.5 nm apart from each other and tilted at different angles. Each heme is partly burried in a pocket lined with hydrophobic R groups. It is bound to its

The **designations α and β,** which are common labels in Biochemistry, are used simply to identify two different polypeptide chains and are not to be confused with α and β secondary structures or α and β amino acids.

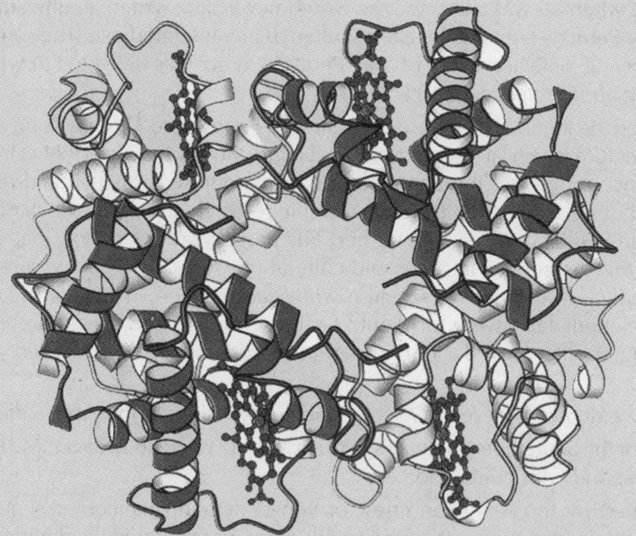

Fig. 9–46. Quaternary structure of hemoglobin molecule, showing interaction of 4 polypeptide chains
Hemoglobin, which is composed of 2 α chains and 2 β chains, functions as a pair of αβ dimers. The structure of the two identical α subunits (*red*) is similar to but not identical with that of the two identical β subunits (*yellow*). The molecule contains 4 heme groups (*black*) with the iron atom shown in *purple*.

polypeptide chain through a coordination bond of the iron atom to the R group of a histidine residue. The sixth coordination bond of the iron atom of each heme is available to bind a molecule of oxygen. Myoglobin and the α and β chains of hemoglobin have nearly the same tertiary structure. Both have well over 70% α-helical nature, both have similar lengths of α-helical segments and the bends have about the same angles.

Similarity in 3- 'D' Structure of Hemoglobin and Myoglobin

The 3-'D' structures of myoglobin and the α and β chains of human hemoglobin are strikingly similar (Fig. 9–47). The 8 helices in each chain of hemoglobin are virtually superposable on those of myoglobin. This close resemblance in the folding of their main chains was unexpected because their amino acid sequences are rather different. In fact, these 3 chains are identical at only 24 or 141 positions.

The amino acid sequence of hemoglobins from more than 60 species, ranging from lamprey eels to humans, are known. A comparison of these sequences shows considerable variations at most positions. However, 9 positions have the same residues in most species studied so far (Table 9–8). These highly conserved 9 residues are especially important for the functioning of the hemoglobin molecule. For example, the invariant ***F8 histidine*** is directly bonded to the heme iron; proline ***C2*** may be essential because it defines one end of the C helix; and tyrosine ***HC2*** stabilizes the structure by forming a hydrogen bond between the H and F helices.

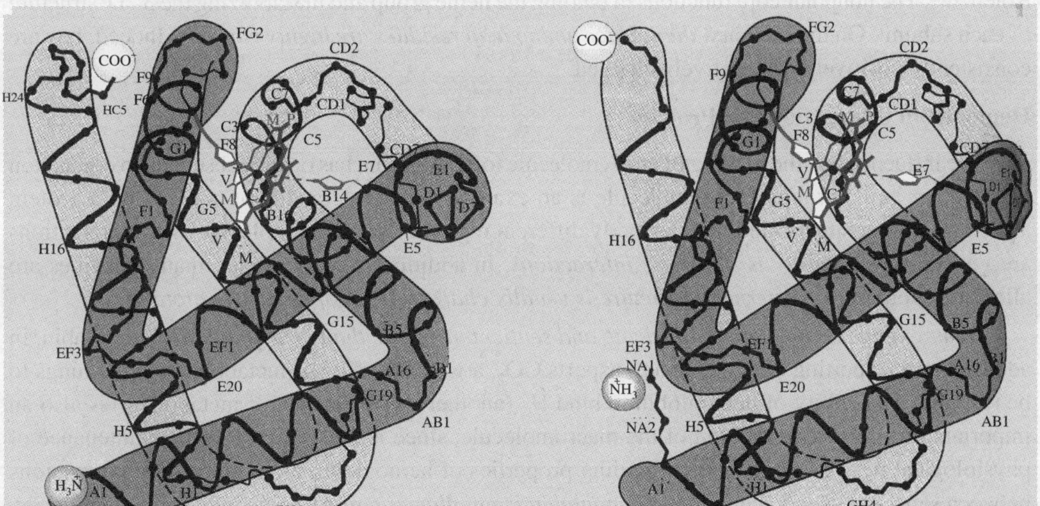

Fig. 9–47. Comparison of the conformation of the main chain of myoglobin (*a*) and the β chain of hemoglobin A_1 (*b*)

Proximal His F8, distal His E7 and Val E11 side chains are shown. Other amino acids of polypeptide chain are represented by α-carbon positions only; the letters M, V and P. refer to the methyl, vinyl and propionate side chains of the heme. The overall structures are very similar, except at NH_2-terminal and COOH-terminal ends.

(*Courtesy: Fersht A, 1977*)

Table 9–8. The nine highly conserved (or invariant) amino acid residues in hemoglobins

Position	*Amino acid*	*Role*
F8	Histidine	Proximal heme-linked histidine
E7	Histidine	Distal histidine near the heme
CD1	Phenylalanine	Heme contact
F4	Leucine	Heme contact
B6	Glycine	Allows close approach between B and E helices
C2	Proline	Helix formation
HC2	Tyrosine	Cross-links the F and H helices
C4	Threonine	Uncertain
H10	Lysine	Uncertain

(*Adapted from Lubert Stryer, 1993*)

The amino acid residues in the interior of hemoglobin are strikingly nonpolar but vary considerably. However, the change is always of one nonpolar residue for another as from alanine to isoleucine. The nonpolar core functions in binding the heme group and in stabilizing the 3-'D' structure of each subunit. On the contrary, *the surface amino acid residues are highly variable.* Indeed, few are consistently positively or negatively charged.

Hemoglobin as an Allosteric Protein

The influence that the binding of an O_2 molecule to hemoglobin has on the subsequent oxygenation of the other heme groups of the molecule is an example of an interaction at one site on a protein affecting another site located in a distinctly different region of the same molecule. Such interactions are generally referred to as *allosteric interactions*. In addition to hemoglobin, many enzymes are allosteric proteins. *A quaternary structure is usually characteristic of allosteric proteins.*

Hemoglobin is a much more intricate and sentient molecule than is myoglobin. Hemoglobin, in addition to transporting oxygen, also transports CO_2, a waste product of metabolism, to the lungs to be respired. The ability of hemoglobin to bind H^+ (another waste product of metabolism) is also an important physiological function of the macromolecule, since it is essential for the maintenance of physiological pH. Also, the oxygen-binding properties of hemoglobin are regulated by interactions between separate, nonadjacent sites. *Hemoglobin is an allosteric protein, whereas myoglobin is not.* This difference is expressed in 3 ways.

1. The binding of O_2 to hemoglobin enhances the binding of extra O_2 to the same hemoglobin molecule. In other words, *oxygen binds cooperatively to hemoglobin.* In contrast, the binding of O_2 to myoglobin is not cooperative.
2. *The affinity of hemoglobin for O_2 depends on pH* (*i.e., pH-dependent*), whereas that of myoglobin is independent of pH.
3. The oxygen affinity of hemoglobin is further regulated by organic phosphates such as 2,3-bisphosphoglycerate (BPG), whereas that of myoglobin is not. Thus, *hemoglobin has a lower affinity for oxygen than does myoglobin.*

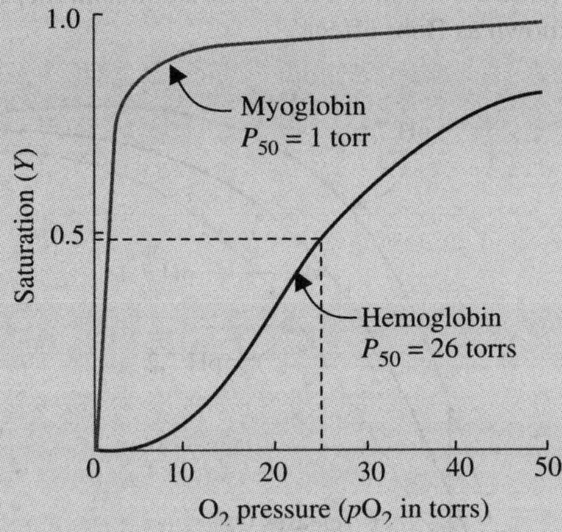

Fig. 9–48. Oxygen dissociation curves of myoglobin and hemoglobin in whole blood

Cooperativity in Oxygen-binding of Hemoglobin

> **Torr** is a unit of pressure equal to that exerted by a column of mercury 1 mm high at 0°C and standard gravity; named after **Evangelista Torricelli** (LT, 1608–1647), the inventor of the mercury barometer.

The differences between hemoglobin and myoglobin as oxygen carriers are readily apparent when their oxygen dissociation curves are compared. An **oxygen dissociation curve** is a plot of the binding sites *i.e.*, heme groups with oxygen (expressed as *Y*) versus the partial pressure of oxygen (expressed as pO_2 in torr) to which the protein solution is exposed, *i.e.*, $Y = [HbO_2] / [HbO_2] + [Hb]$. The oxygen dissociation curves of myoglobin and hemoglobin (Fig. 9–48) differ in two ways:

A. For any given pO_2, *Y* is higher for myoglobin than for hemoglobin. This means that *myoglobin has a higher affinity for O_2 than does hemoglobin*. Oxygen affinity can be characterized by a quantity P_{50}, which is the partial pressure of oxygen at which 50% of sites are filled (*i.e.*, at which $Y = 50$). For myoglobin, P_{50} is typically 1 torr, whereas for hemoglobin, P_{50} is 26 torrs.

B. Secondly, *the oxygen dissociation curve of myoglobin is hyperbolic, whereas that of hemoglobin is sigmoidal (S-shaped).*

Because myoglobin has only a single heme group, the phenomenon of cooperative binding is not possible, as evidenced by its hyperbolic curve. It has been calculated that the cooperative binding of oxygen by hemoglobin enables it to deliver 1.83 times as much O_2 under typical physiological conditions as it would if the sites were independent.

The Bohr Effect : H^+ and CO_2 promote the release of O_2.

Hemoglobin not only furnishes O_2 to tissues but also transports the waste products of metabolism ($H^+ + CO_2$). Since the same biomolecule is responsible for both transport systems, it is not surprising that there is a regulatory interplay between the two functions. It has been observed that increased concentrations of CO_2 and H^+ (*i.e.*, lowering of pH) decrease the O_2 affinity of hemoglobin; conversely, increased concentrations of O_2 lower the affinity for CO_2 and H^+ (Fig. 9–49). By contrast, increased concentrations of CO_2 and H^+ have almost no effect on myoglobin's O2 affinity. Increasing the concentrations of CO_2 (at constant pH) also lowers the O_2 affinity. In actively metabolizing tissue, such as contracting muscle, much CO_2 and acid are produced. The presence of higher levels of H^+ and CO_2 in the capillaries of such metabolically active tissue promotes the release of O_2 from oxyhemoglobin. This important mechanism for meeting the higher O_2 needs of metabolically active tissues was discovered by **Christian Bohr** in 1904. This metabolic interdependence of the two transport functions of hemoglobin is known as Bohr effect.

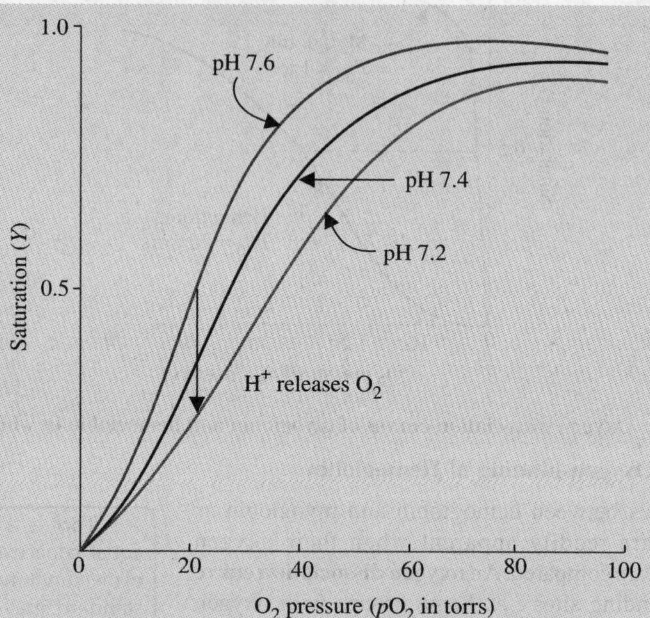

Fig. 9–49. Effect of pH on the oxygen affinity of hemoglobin

The three O_2 dissociation curves of hemoglobin at different pH have been shown. Lowering the pH from 7.5 to 7.2 results in the release of O_2 from oxyhemoglobin.

J.B.S. Haldane, in 1914, discovered the reciprocal effect which occurs in the alveolar capillaries of the lungs. The high concentration of O_2 there unloads H^+ and CO_2 from hemoglobin, just as the high concentrations of H^+ and CO_2 in active tissues drives off O_2 (Fig. 9–50).

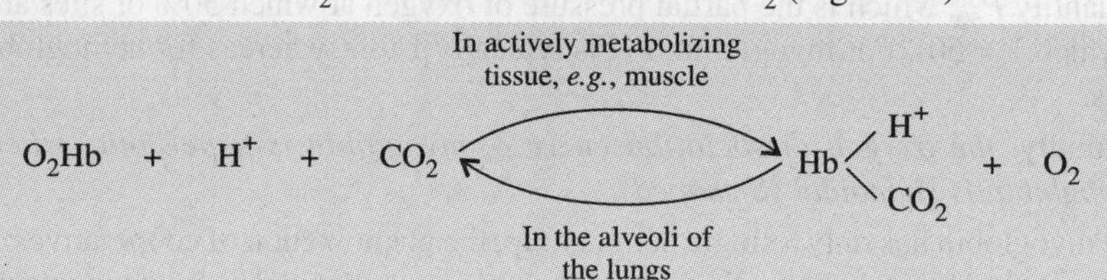

Fig. 9–50. Summary of the Bohr effect

BPG as a Hemoglobin Regulator

Joseph Bancroft, as early as 1921, observed that the O_2 affinity of hemoglobin within erythrocytes is lower than that of hemoglobin in free solution. Much later in 1967, **Reinhold Benesch** and **Ruth Benesch** showed that an anionic organic phosphate present in human red cells, that is 2, 3-*bisphosphogylcerate, BPG* (also known as *2,3-diphosphoglycerate, DPG*) binds to hemoglobin and thereby lowering the O_2 affinity of hemoglobin. In the absence of BPG, the P_{50} of hemoglobin is 1 torr, like that of myoglobin. In its presence, P_{50} becomes 26 torrs (Fig. 9–51). Thus, *BPG lowers the* O_2 *affinity of hemoglobin by a factor of 26, which is essential in enabling hemoglobin to unload oxygen in tissue capillaries*. BPG reduces the oxygen affinity of hemoglobin by binding in the central cavity of deoxyhemoglobin and not to the oxygenated form. As oxygenation occurs, the accompanying conformational changes in hemoglobin make the central cavity too small to accommodate BPG, which is then expelled.

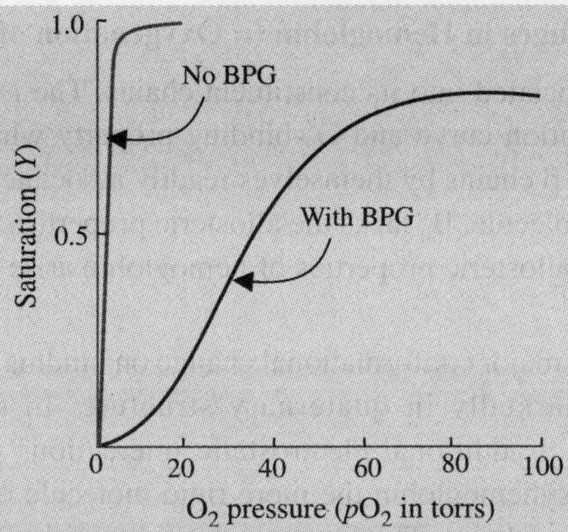

Fig. 9–51. BPG regulation of O_2 affinity of hemoglobin

Note that 2,3-bisphosphoglycerate (BPG) decreases the oxygen affinity of hemoglobin molecule.

Oxygen Affinity of Fetal Hemoglobin

Hemoglobin F has a higher O_2 affinity under physiological conditions than does hemoglobin A (Fig. 9–52). The higher O_2 affinity of hemoglobin F optimizes the transfer of oxygen from the maternal to the fetal circulation. Hemoglobin F is oxygenated at the expense of hemoglobin A on the other side of the placental circulation. In fact, hemoglobin F does not bind BPG as strongly as does hemoglobin A because the γ chain (analogous to the β chain of hemoglobin A) has a seryl residue in the H21 position instead of a positively charged histidine. Because of this particular amino acid difference, hemoglobin F has a higher affinity for oxygen than does hemoglobin A. In the absence of BPG, the O_2 affinity of hemoglobin F is actually lower than that of hemoglobin A. Thus, we see that different forms of a protein, called **isoforms** or **isotypes**, in different tissues have a clear-cut biological advantage, as beautifully illustrated by diverse forms of hemoglobin.

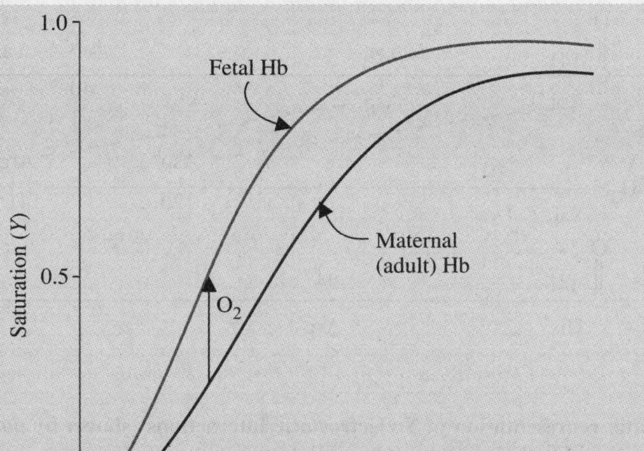

Fig. 9–52. Oxygen dissociation curves for fetal hemoglobin (HbF) and adult hemoglobin (HbA)

In the presence of biphosphoglycerate (BPG), the O_2 affinity of fetal hemoglobin is higher than that of maternal hemoglobin. The arrow represents transfer of oxygen from maternal oxyhemoglobin to fetal deoxyhemoglobin.

Quaternary Structural Changes in Hemoglobin (= Oxygenation of Hemoglobin)

Hemoglobin can be dissociated into its constituent chains. The isolated α chain has a high O_2 affinity, a hyperbolic dissociation curve and O_2-binding property which is indifferent to pH, CO_2 concentration and BPG level. β chains by themselves readily associate to form a tetramer (B_4). Like the α chain and myoglobin molecule, B_4 lacks the allosteric properties of hemoglobin and has a high oxygen affinity. In short, the allosteric properties of hemoglobin arise from interactions between its subunits.

Hemoglobin undergoes a major conformational change on binding oxygen, and as such oxy- and deoxyhemoglobin differ markedly in quaternary structure. In the quaternary structure of deoxyhemoglobin, there are 8 additional electrostatic interactions (salt linkages), not found in oxyhemoglobin, making deoxyhemoglobin the more rigid molecule of the two (Fig. 9–53). Six of these 8 interactions are between chains. The C-terminal residue of the 4 chains of deoxy-hemoglobin are also involved in salt linkages. As a consequence of oxygenation, a hemoglobin molecule undergoes conformational changes which disrupt 8 salt linkages and the cooperative binding observed with oxygen is the result of these structural changes.

In the oxyhemoglobin molecule, the distance between the iron atoms of the β chains decreases from 40 to 33 Å, thus making the molecule more compact. During the transition phase from oxy- to deoxyhemoglobin, large structural changes take place at two of the four contact regions (the $\alpha_1 \beta_2$ contact and the identical $\alpha_2 \beta_1$ contact) but not at the others (the $\alpha_1 \beta_1$ contact and the identical $\alpha_2 \beta_2$ contact). In fact, the $\alpha_1 \beta_2$ contact region is designated to act as a switch between two alternative

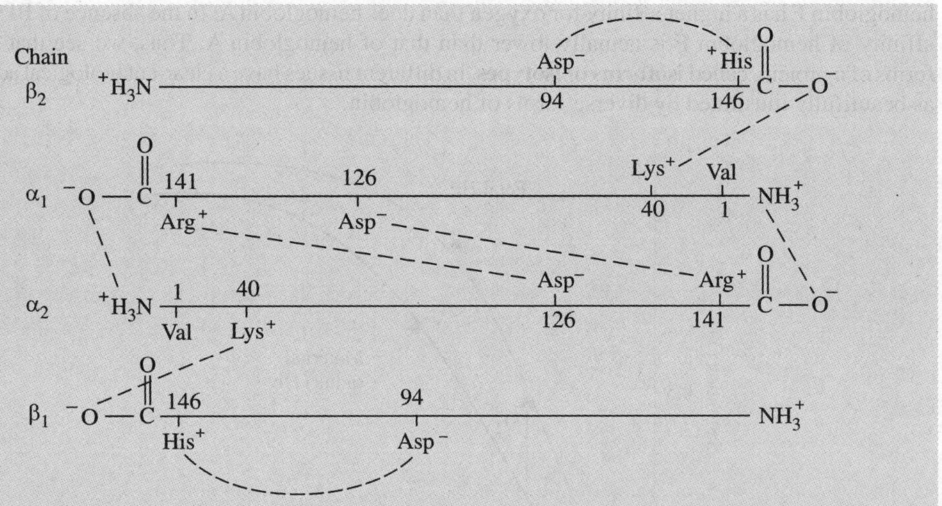

Fig. 9–53. Schematic representation of 8 electrostatic interactions (shown by *dotted lines*) that occur between α and β chains of deoxyhemoglobin

(Adapted from Frank B. Armstrong, 1989)

structures. In oxyhemoglobin, the C-terminal residues of all 4 chains have almost complete freedom of rotation. By contrast, in deoxyhemoglobin, these terminal groups are anchored. Deoxyhemoglobin is a tauter, more constrained molecule than oxyhemoglobin because of the presence of additional salt links. The quaternary structure of deoxyhemoglobin is termed the T (tense or tout) form; that of oxyhemoglobin, the R (relaxed) form.

> The **designations R and T** are generally used to describe alternative quaternary sturctures of an allosteric protein, the T form having a lower affinity for the substrate.

The above-mentioned conformational changes take place at some distance from the heme. But changes also take place at the heme group itself on oxygenation. In deoxyhemoglobin, the iron atom is about 0.4 Å out of the porphyrin plane toward the proximal histidine, so that the heme group domed (convex) in the same direction (Fig. 9–54). On oxygenation, the iron atom moves into the plane of the porphyrin to form a strong bond with O_2, and the heme becomes more planar. Besides, His F8, because of its bonding to the iron atom, also moves toward the heme plane. This slight displacement of the histidine residue initiates a sequential series of conformational changes that ultimately results in the disruption of some of the subunit interactions in the quaternary structure of deoxyhemoglobin.

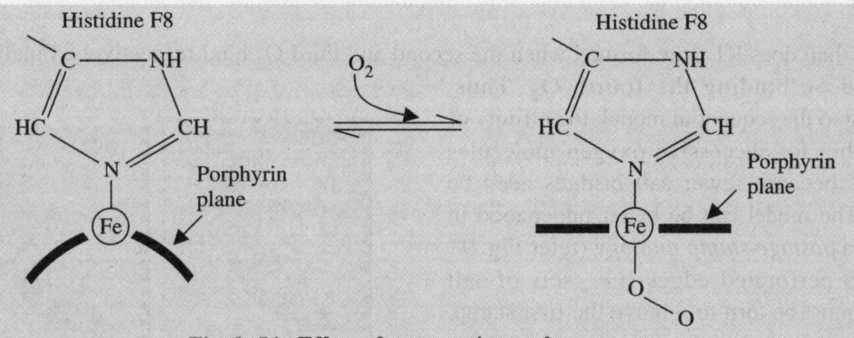

Fig. 9–54. Effect of oxygenation on heme group

The iron atom moves into the plane of the heme group on oxygenation. The proximal histidine (F8) is pulled along with the iron atom and becomes less tilted.

Mechanism of Allosteric Interactions of Hemoglobin

The structures of fully deoxygenated and fully oxygenated hemoglobin provide insight into how the binding of O_2, CO_2, H^+ and BPG influence each other. How does allosteric mechanism take place? How does hemoglobin switch from the deoxy to the oxy structure when it binds successive O_2 molecules? Two models– the sequential model and the concerted model– have been proposed to explain the allosteric changes in multi-subunit protein. Both these models are based upon the ideas that the proteins are flexible in conformation and that a protein subunit can exist in only two states.

A. **Sequential Model.** Developed by **Daniel E. Koshland Jr.** in 1966, the sequential model, in its simplest form, makes 3 *assumptions*:

1. Only two conformational states, T and R, are accessible to any subunit.
2. The binding of the ligand switches the conformation of the subunit to which it is bound but not that of its neighbours.
3. The conformational change elicited by the binding of substrate in one subunit can increase or decrease the binding affinity of the other subunit in the same molecule. A T subunit with an R neighbour has higher affinity for ligand than does a T subunit with a T neighbour because the subunit interfaces are different in TR and TT.

The sequential model (Fig. 9–55), in its essence, proposes that *symmetry in not conserved in allosteric transitions and that the subunits change conformation one at a time*. The T state is symbolized by a square, and the R state by a circle. Deoxyhemoglobin is in the T_4 state. The binding of O_2 to one of the subunits changes its conformation from T to R, but leaves the other subunits in the T form. The oxygen-binding affinity of unoccupied sites in RT_3 is higher than in T_4 because some salt links have been broken on binding the first O_2. R_2T_2 and R_3T, which have higher oxygen

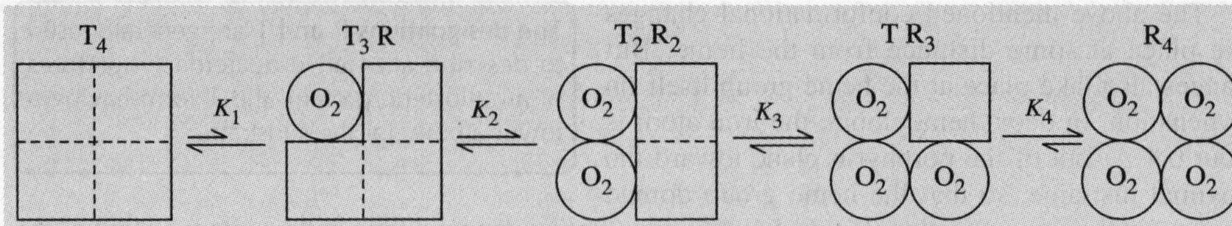

Fig. 9–55. Simple sequential model for a tetrameric allosteric protein

The binding of a ligand to a subunit changes the conformation of that particular subunit from the T (square) to the R (circle) form. This transition affects the affinity of the other subunits for the ligand.

affinities than does RT_3, are formed when the second and third O_2 bind respectively. Finally, R_4 is produced on binding the fourth O_2. Thus, according to the sequential model, the affinity of hemoglobin for successive oxygen molecules increases because fewer salt bridges need be broken. The model can be better understood in terms of a ***postage-stamp analogy*** (refer Fig. 9–56). Two perforated edges (*i.e.*, sets of salt bridges) must be torn to remove the first stamp. Only one perforated edge must be torn to remove the second stamp, and one edge again to remove the third stamp. The fourth stamp is then free.

Fig. 9–56. Postage-stamp analogy of the simple sequential model

B. Concerted Model (or **Symmetry Model**). Developed by **Jacques Monod, Jeffries Wyman** and **Jean-Pierre Changeux** (hence, also called **MWC model**) in 1965, the concerted model takes a different view of allosteric interactions. The essence of this elegant and incisive model (Fig. 9–57) is that *symmetry is conserved in allosteric transitions* and that *all the subunits change conformation together.* This model also is based on 3 *assumptions*:

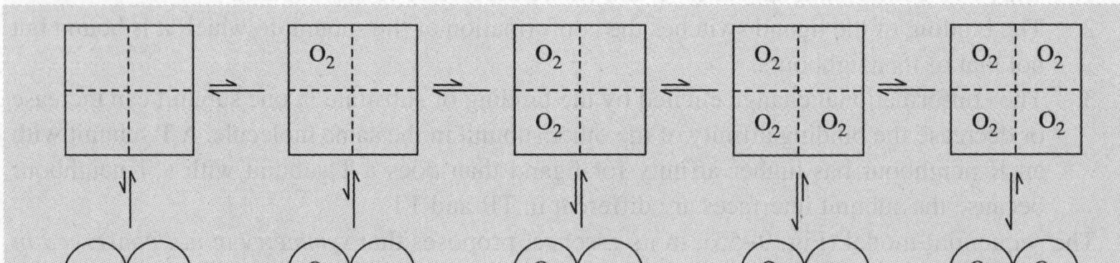

Fig.9–57. Concerted model (or MWC model) for a tetrameric allosteric protein

The *squares* denote the T form, and the *circles* denote the R form.

1. The protein interconverts between two conformations, T and R. All the subunits of a particular molecule must be in the T form, or all must be in R form. Hybrids such as TR are forbidden.
2. Ligands bind with low affinity to the T form, and with high affinity to the R form.
3. The binding of each ligand increases the probability that all subunits in that molecule are in the R form. The allosteric transition is said to be *concerted* because subunits change in unison from T to R or vice versa. Stated in terms of the postage-stamp analogy, either all four perforated edges or none are broken.

It is, thus, apparent that the two models offer contrasting views regarding the mode of cooperative interactions in multi-subunit protein. The models differ in many points through:

1. The symmetry model assumes that the active and inactive states are in equilibrium, whereas sequential model assumes that the transition from inactive state to active state in *induced* by substrate binding.
2. The symmetry model states that if one subunit changes its form, the other subunit necessarily changes to that form also because, unless symmetry is preserved, the two subunits cannot interact. By contrast, the sequential model allows interaction between subunits in *different* states.
3. The symmetry model assumes that binding of the first molecule necessarily enhances binding of a second molecule of the same type, whereas the sequential model allows *either an increase or decrease* of affinity for the second molecule.
4. The symmetry model has few intermediate states, whereas the sequential model has *more* intermediate states.

In fact, the actual allosteric mechanism of hemoglobin is more complex than visualized by either the sequential or the concerted model. These models should be regarded as limiting cases. Actual allosteric processes combine, of course in varying degrees, elements of both.

The *protein-protein interactions*, thus, serve many important functions:

(*a*) protect nucleic acids of virus particles from destruction,

(*b*) conserve genetic information, and

(*c*) construct the various enzyme-complexes found in the cells.

The foregoing discussion and also the evidences presently available confirm the view that "the primary structure of the protein, which dictates the limitations of secondary and tertiary structure, also delineates the correct quaternary structure necessary for biological activity of the molecule." (**Hartman** and **Suskind**, 1969).

DYNAMICS OF GLOBULAR PROTEIN STRUCTURE

The precise pictures produced by x-ray diffraction give the impression of a rigid and inflexible structure for the globular proteins. Recent researches, however, indicate that globular proteins fold *via* complex kinetic pathways and that even the folded structure, once acquired, is a dynamic structure. Some of the dynamic aspects of globular proteins are described below :

1. Kinetics of Protein Folding

The folding of globular proteins from their denatured states (or conformations) is a remarkably rapid process and often completes in less than a second. This viewpoint was first expressed in 'Levinthal's paradox', first propounded by **Cyrus Levinthal** in 1968. A rough estimate reveals that about 10^{50} conformations are possible for a polypeptide chain such as ribonuclease (124 residues). And if the molecule tries a new conformation every 10^{-13} second, it would still take about 10^{30} years to try a significant fraction of them. In actuality however, ribonuclease folds experimentally *in vitro* in about 1 minute. To explain this anomaly, it has been suggested that folding takes place through a series of intermediate states. These observations led to the **pathway model of folding,** as depicted below:

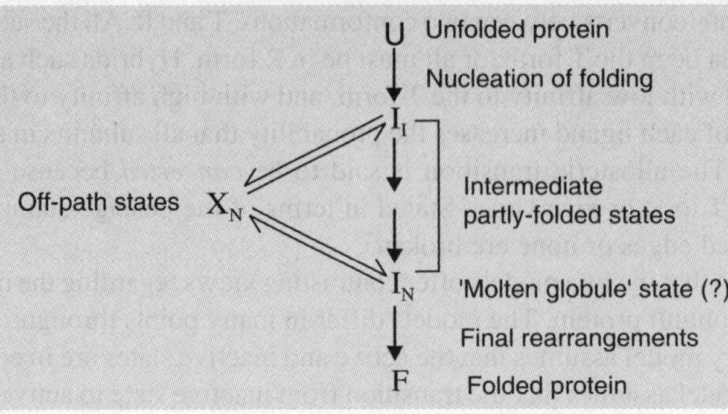

The nucleation step is critical because it is much more difficult to begin an α helix than to extend it (note that at least 4 residues must fold properly to make the first stabilizing H bond). On the contrary, it is now recognized that nucleation may begin at many places and that all of these partially folded structures will be 'funneled' by energy minimizations toward the final state. The funneling model proposes that there is not just one but *many* possible paths from the denatured state to the folded state, and each path leads downhill in energy. During the descent toward the free energy minimum, there may be pauses corresponding to metastable intermediates, as incorporated in the pathway model. An important folding intermediary for many proteins appears to be what is called as the "molten globule" state. It is a compact structure in which much of the secondary and tertiary folding has occurred, but the internal hydrophobic residues have not yet settled into their final packing.

Evidence suggests for the presence of "off-path" states– those in which some key element is incorrectly folded. However, the cell has ways to assist incorrectly-folded proteins to find their proper conformations. Larger complex proteins may never achieve 100% folding success *in vivo* but the cell can identify those which are incorrect and dispose of them. One such common folding error occurs *via* the incorrect *cis-trans* isomerization of the amide bond adjacent to a proline residue:

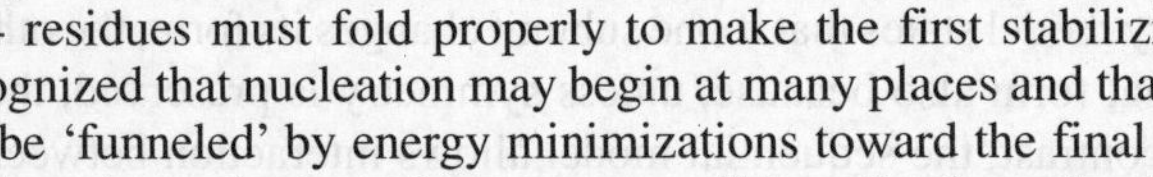

Unlike other peptide bonds in proteins, for which the *trans* isomer is highly favoured (by a factor of about 1,000), proline residues favour the *trans* form in the preceding bond by a factor of only 4. Hence, there are more chances that the 'wrong' isomer (in terms of the functional protein conformation) will form first; this will later be converted to the correct configuration involving chain rearrangement enzymatically to speed up *in vivo* folding.

2. Kinetics of Disulfide Bond Formation

If a protein is folded from a state in which disfulfide bonds have been disrupted and must hence be re-formed, the process is more complicated and slower– often taking many minutes. Some disulfide bonds that are missing in the native structure are formed in intermediate stages of the folding. Obviously, the protein can utilize a number of alternative pathways to fold but ultimately finds both its proper tertiary structure and the correct set of disulfide bonds. This process is aided *in vivo* by enzymatic catalysis of —S—S— bond rearrangement.

3. Chaperonins

It has been discovered that protein folding and assembly *in vivo* sometimes requires the aid of special proteins called *chaperonins* or *molecular chaperones*. As the name signifies, the function of these chaperones is to keep the newly-formed protein away from improper folding or aggregation. Improper folding may correspond to being trapped in a deep local minimum on the energy scenario.

Aggregation is often a danger because the protein, released from the ribosome in an unfolded state, will have groups exposed. These will be tucked inside in normal folding, but when exposed they stand the chance of making hydrophobic interactions with *other* polypeptide strands and thereby aggregating.

Many chaperonin systems have been discovered but the GroEL-ES complex from *E. coli* is the best-studied one. The structure of this enormous complex has recently been elucidated by x-ray diffraction [Fig. 9–58(*a*)]. It consists of 2 basic portions– GroEL and GroES. GroEL is made up of 2 rings, each consisting of 7 protein molecules; the centre of each ring is an open cavity, accessible to the solvent at the ends. Either cavity can be 'capped' with GroES, which is again a 7-membered ring of smaller subunits.

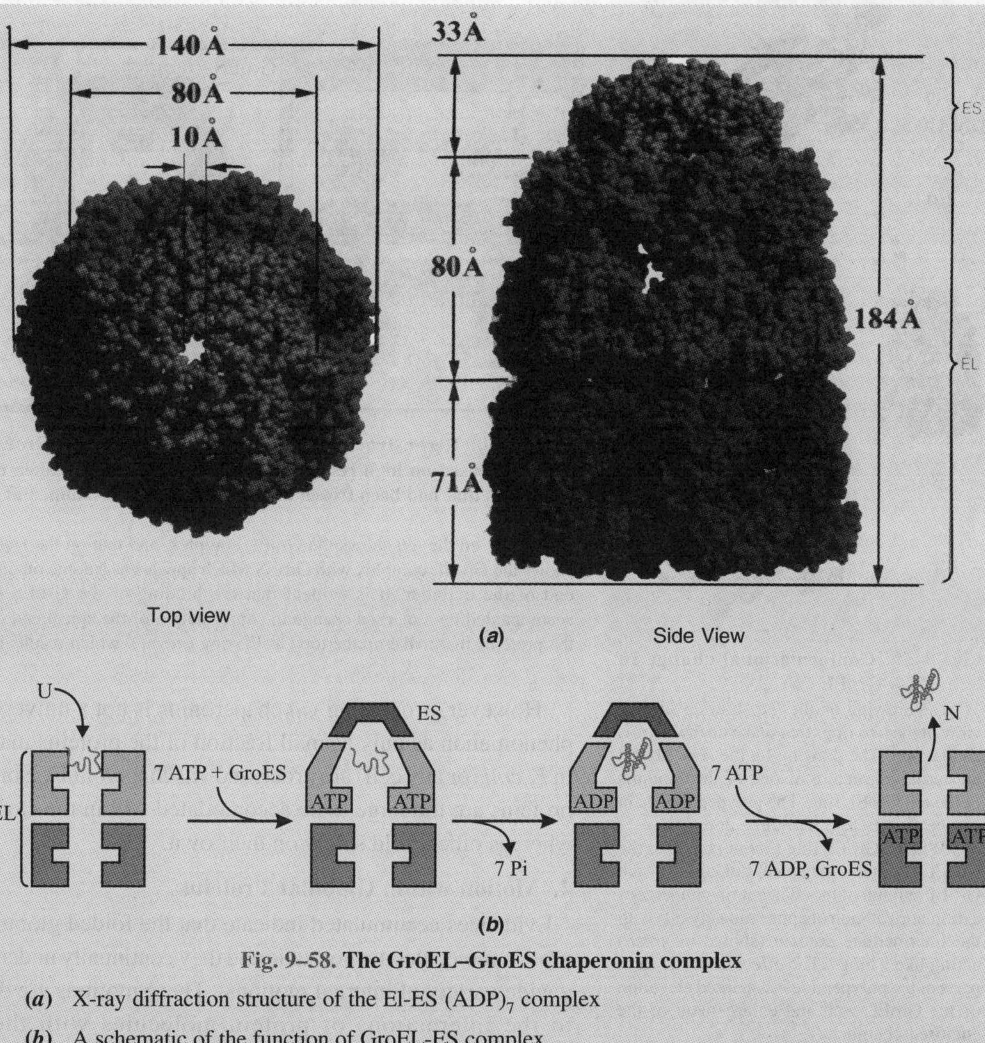

Fig. 9–58. The GroEL–GroES chaperonin complex

(*a*) X-ray diffraction structure of the El-ES $(ADP)_7$ complex

(*b*) A schematic of the function of GroEL-ES complex

The unfolded protein enters into a sort of cavity lined with hydrophobic residues. Then the cavity changes, presenting a hydrophilic lining. This releases the protein from the walls and it folds and is then released. Note that ATP is required, probably to drive the process in one direction.

(Adapted from (a) Xu Z, Horwich AL and Sigler. PB, 1997 and (b) Netzer and Hartl, 1998)

It is argued that basically, the cavities provide 'shelters' in which nascent protein chains can be 'incubated' until they have folded properly. The GroEL-ES complex does not stipulate the folding pattern *i.e.*, upto the protein itself to do. But insulation from the environment prevents chances of aggregation or misfolding. The cycle experienced by a protein molecule is schematically shown in [Fig. 9–58(*b*).]

The conformational changes in GroEL are diagrammatically represented in Fig 9–59.

The reconstructions of the GroEL and GroEL–GroES complexes, based on high resolution electron micrographs, have been presented in Fig. 9–60.

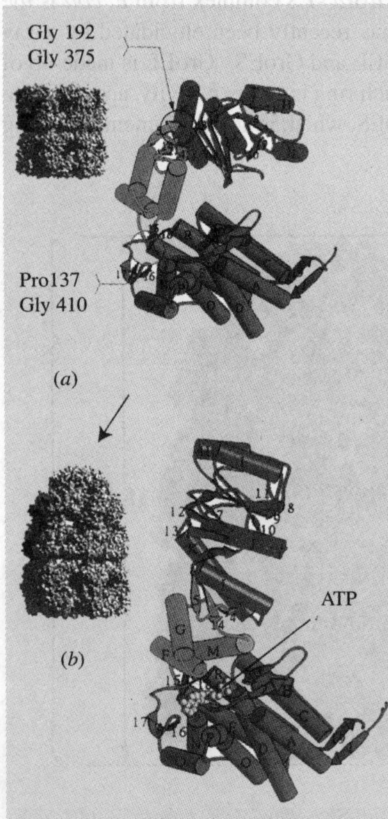

Fig. 9–59. Conformational change in GroEL

(*a*) The drawing on the *left* shows a surface view of the two rings that make up the GroEL chaperonin. The drawing on the *right* shows the tertiary structure of one of the subunits of the top GroEL ring. The polypeptide chain can be seen to fold into three domains

(*b*) When a GroES ring (*arrow*) binds to the GroEL cylinder, the apical domain of each GroEL subunit of the adjacent ring undergoes a dramatic rotation of approximately 60° with the intermediate domain (shown in *green*) acting like a hinge. The effect of this shift in parts of the polypeptide is a marked elevation of the GroEL wall and enlargement of the enclosed chamber.

(Adapted from Xu Z, Horwich AL and Sigler PB, 1997)

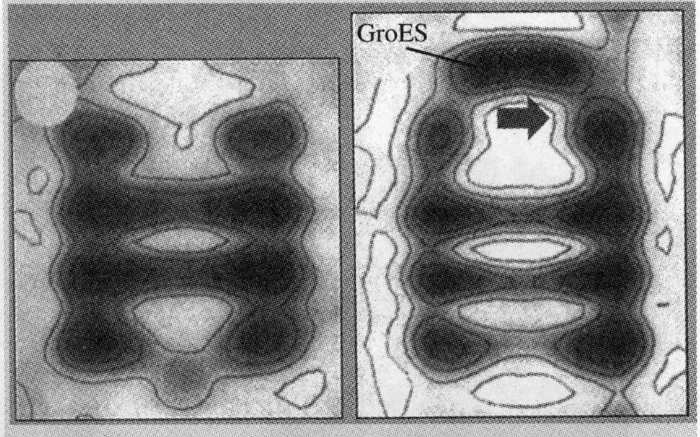

Fig. 9–60. Reconstructions of the GroEL and GroEL-GroES complex, based on high resolution electron micrographs, taken of specimens that had been frozen in liquid ethane and examined at –170°C

The image on the *left* shows the GroEL complex, and that on the *right* shows the GroEL complex with GroES which appears as a dome on one end of the cylinder. It is evident that the binding of the GroES is accompanied by a marked change in conformation of the apical end of the proteins that make up the top GroEL ring (*arrow*), which results in

However, processing *via* chaperonins is not a universal phenomenon as only a small fraction of the proteins made in *E. coli* (or any cell) are processed *via* chaperonins. Some proteins are too large to be accomodated within the cavity whereas others fold safely on their own.

4. Motion within Globular Proteins

Evidences accumulated indicate that the folded globular protein molecules are not static and they continually undergo a wide variety of internal motions. These motions are due to the interactions of protein molecules with their environment. The resulting motions can be roughly grouped into 3 classes (Table 9–9).

Table 9–9. Motions within globular protein molecules

Class	*Type of motion*	*Approximate amplitude (nm)*	*Range time(s)*
1	Vibrations and oscillations of individual atoms and groups	0.2	10^{-15}–10^{-12}
2	Constructed motions of structural elements, like α helices and groups of residues	0.2–1	10^{-12}–10^{-8}
3	Motions of whole domains; opening and closing of clefts	1–10	$\geq 10^{-8}$

Class 1 motions occur even within protein molecules in crystals and account, in part, for the limits of resolution obtainable in x-ray diffraction studies. In **class 2** and **class 3**, the motions are larger in magnitude and slower in rate. These are more likely to occur in solution. Some of the motions, like the opening and closing of clefts in molecules, are probably involved in the enzymatic functions. Binding or release of a small molecule from a protein depends on the time required to open or close a shaft. Likewise, the protein 'gates' that pass molecules and ions through membranes rapidly change from open to closed states.

STANLEY BEN PRUSINER

(born, 1942)

Prusiner, an American neurologist and biochemist, was born on May 28, 1942 in Des Moines, Iowa. He spent his childhood in Des Moines and Cincinnatio, Ohio, where he attended Walnut Hills High School. He graduated in chemistry from the University of Pennsylvania and later received his M.D. from the University of Pennsylvania School of Medicine. Prusiner then completed an internship in medicine at the University of California, San Francisco (UCSF). Later, he moved to the National Institutes of Health (NIH), where he studied *glutaminases* in *E. coli* in the laboratory of **Earl Stadtman**.

After 3 years at NIH, Prusiner returned to UCSF to complete a residency in neurology. Upon completion of the residency in 1974, Prusiner joined the faculty of UCSF Neurology Department. Currently, he is the Director of the Institute for Neurodegenerative Diseases at UCSF.

Prusiner won the **1997 Nobel Prize in Physiology or Medicine** for his work in proposing an explanation for the cause of bovine spongiform encephalopathy, BSE (more commonly called *mad cow disease*) and its human equivalent, Creutzfeldt–Jakob disease, CJD. The explanation given by him goes by the name **prion theory**. *Prions* are a class of infectious self-reproducing pathogens, primarily or solely composed of proteins.

Prusiner was elected to the *National Academy of Science* in 1992 and to its governing council in 2007. He is also an elected member of the *American Academy of Arts and Sciences* (1993), the *Royal Society* (1996), the American Philosophical Society (1998), and the Serbian Academy of Sciences and Arts (2003). He has won several awards, some of which are as follows:

1. The Gairdner Foundation International Award (1993)
2. The Albert Lasker Award for Basic Medical Research (1994)
3. The Paul Ehrlich and Ludwig Darmstaedter Prize from the Federal Republic of Germany (1995)
4. The Franklin Medal (1998)
5. The National Medal of Science (2010)

5. Prions

Until very recently, scientists believed that the diseases could be transmitted from one organism to another *via* viruses or microorganisms. After all, DNA and RNA were the carriers of genetic information. But now evidences have gathered to point out that some diseases are transmitted by a protein and nothing more. Table 9–10 lists some such diseases (called **prion diseases**) along with their host and geographic distribution.

Table 9–10. Prion disease in different species

Host	*Disease*	*Geographic distribution*	*First clinical observation*
Sheep	Scrapie	Cosmopolitan except Australia, New Zealand and some European countries	1730
Goat	Scrap	—	—
Man	Kuru ('Laughing death')	Papua New Guinea	1900
	Creutzfeldt-Jacob disease	Cosmopolitan	1920
	Gertsmann-Straussler–Scheinker syndrome	Cosmopolitan	1926
	Fatal familial insomnia	—	—
Mink	Transmissible mink encephalopathy	North America; Europe	1947
Mule deer	Chronic wasting disease	North America	1967
Cattle	Bovine spongiform encephalopahy, BSE (= Mad cow disease)	U.K., Ireland and some other European countries	1985

In 1900, a fatal neurodegenerative disease called **Kuru** was reported only in a tribe of Papua New Guinea. **Carleton Gajdusek**, a virologist at the US National Institute of Health, discovered and understood the basis of the disease and the mode of its transmission. He pointed out that a peculiar ritual of the tribal people of eating the brain of dead relatives was the main mode of transmission of this disease. The frequency of Kuru gradually declined since the practice was abandoned. **Gajdusek** was awarded 1976 Nobel prize in medicine or physiology for this basic work, along with **Baruch S. Blumberg**, a virologist of the Institute of Cancer Research, Philadelphia.

But **Stanley B. Prusiner**, an Americal biochemist, was not convinced about the nature of the infectious agent which was described by Gajdusek and others as 'unconventional slow virus'. However, **Prusiner** was against the so-called virus theory because of the following 3 reasons:

(*a*) The infective agent is extremely resistant to UV and ionizing radiations whereas all viruses are sensitive.

(*b*) The infectivity of the scrapie agent is not changed by treatment with nucleases (DNAase and RNAase enzymes). That means the infective agent does not carry DNA or RNA as genetic material which is the most essential component of a virus particle.

(*c*) Absence of any agent-specific antibody titre, strong tendency to aggregate and hydrophobicity go against the nature of a true virus.

Backed by these evidences, **Prusiner** and his colleagues in 1982, extracted the infectious material from hamster brains, and they gave the first blow to the virus theory. **Prusiner** suggested that the infectious agent, causing certain degenerative disorders of central nervous system (CNS) in animals and more rarely in humans, is a small proteinaceous infectious particle, which he called *prion* (pronounced as 'preeon') and the protein believed to be responsible for infection was called *prion-related protein* or *PrP.* Soon he found that there were 2 isoforms of PrP. The first one is the normal cellular **prion-related protein, PrP^{c}** (the superscript '*c*' denotes cellular), which is the nonpathological form; the PrP^{c} may act as an acetyl-choline receptor inducer and plays an important role in the transmission of nerve signal. The other isoform of PrP is infective and called **prion-related protein scrapie, PrP^{sc}** (the superscript 'sc' denotes scrapie, which is now used to refer to all infectious form of prion causing scrapie- like diseases in animals and humans). It is this form, in which the disordered N-terminal portion appears to fold into a β-sheet, that wrecks havoc with the nervous system. It is postulated that when PrP^{sc} is ingested into the body, it induces the conversion of PrP^{c} in the recipient to PrP^{sc}; thus the disease is transmitted. It is something like a bad guy, who converts a good guy into one of its 'bad' form. How this conversion is catalyzed is unknown, but it strongly suggests that PrP^{c} represents an especially stable "off-path" folding of the type hypothesized in the preceding section. Fig. 9–61 presents the computer-generated images of the two forms of human prion protein.

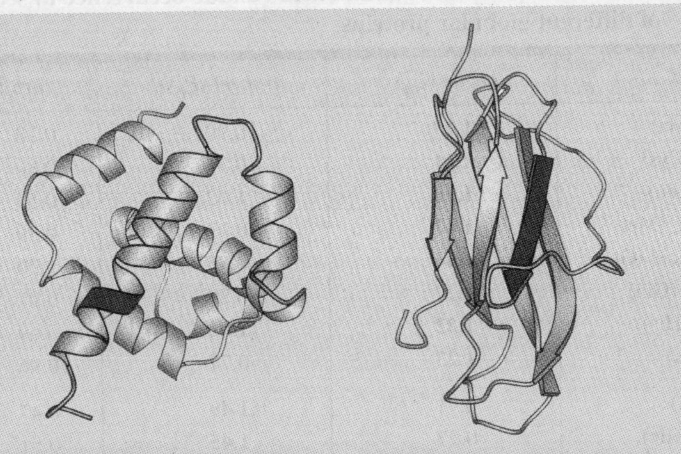

(a)
Normal prion protein, PrP^{c}

(b)
Abnormal prion protein, Pr P^{sc}

Fig. 9–61. Computer-generated images of the two forms of human prion protein

In 1988, Prusiner and his team reported that human prion diseases can certainly be inherited, *i.e.*, they could be heritable and communicable. Later studies on transmission of human prion to transgenic mice threw more light on prion diseases in 1994. When researchers tried to infect transgenic mice (carrying the human PrP gene) with human prion, there was no development of CNS dysfunction as expected. The mice could become susceptible to human prions only after removal of mouse PrP gene which is called **'gene knockout'** in genetics terminology.

Unfortunately, prion diseases have already played havoc among cattle population in some European countries. **Bovine spongiform encephalopathy (BSE),** or more commonly known as **mad cow disease,** took its toll in cattle population in Britain and elsewhere. Epidemiological studies revealed that the most probable cause of the epidemic was the inclusion of ruminant-derived meat and bone meal (which utilizes sheep brain, spinal cord etc.) in the cattle feed. Things became worse as reports of 12 persons afflicted with Creutzfeldt-Jakob disease came in. It was reported that these new cases occurred due to consumption of BSE-infected beef. These reports sent a shock wave throughout the world, apprehending the possible impact of prion diseases.

The recognition of the relationship of PrP to diseases won Prusiner the Nobel prize in physiology or medicine in 1997. His discovery of a new biological principle of infection goes by the name **'prion theory'.**

PREDICTION OF SECONDARY AND TERTIARY PROTEIN STRUCTURES

A. Prediction of Secondary Structure

The protein amino acids arrange in innumerable ways, with the help of a variety of chemical bonds, to produce a definite pattern of secondary structure in proteins, which may be either an α helix, or a β sheet or a turn. Acquisition of any one, of these 3 forms (α helix, β sheet, turn) by a protein depends upon the frequency of occurrence of particular amino acid residues in these secondary structures. Table 9–11. lists the relative frequencies (P_α, P_β, P_t) of amino acids in producing secondary structures, *i.e.*, α helix, β sheet and turn, respectively.

Table 9–11. Relative frequencies of amino acid residue occurrence in secondary structures of different globular proteins

Amino acid	*α helix* (P_α)	*β sheet* (P_β)	*Turn* (P_t)	
Alanine (Ala)	**1.29**	0.90	0.78	Favour α-helices
Cysteine (Cys)	**1.11**	0.74	0.80	
Leucine (Leu)	**1.30**	1.02	0.59	
Methionine (Met)	**1.47**	0.97	0.39	
Glutamic acid (Glu)	**1.44**	0.75	1.00	
Glutamine (Gln)	**1.27**	0.80	0.97	
Histidine (His)	**1.22**	1.08	0.69	
Lysine (Lys)	**1.23**	0.77	0.96	
Valine (Val)	0.91	**1.49**	0.47	Favour β sheets
Isoleucine (Ile)	0.97	**1.45**	0.51	
Phenylalanine (Phe)	1.07	**1.32**	0.58	
Tyrosine (Tyr)	0.72	**1.25**	1.05	
Tryptophan (Trp)	0.99	**1.14**	0.75	
Threonine (Thr)	0.82	**1.21**	1.03	
Glycine (Gly)	0.56	0.92	**1.64**	Favour turns
Serine (Ser)	0.82	0.95	**1.33**	
Aspartic acid (Asp)	1.04	0.72	**1.41**	
Asparagine (Asn)	0.90	0.76	**1.28**	
Proline (Pro)	0.52	0.64	**1.91**	
Arginine (Arg)	0.96	0.99	0.88	

* Amino acids are grouped according to their preference for α helices (first or top group, highlighted by showing the relative frequencies in *boldface*), β sheets (second group) or turns (third group). Note that arginine shows no significant preference for any of the structures.

(*Adapted from Creighton TE, 1992*)

A perusal of the above table reveals that Met, Glu, Ala, and Leu residues tend to be present in α helices, whereas Val, Ile, Phe and Tyr tend to be present in β strands. Pro, Gly, Asp, and Ser have a propensity (or inclination) to lie in turns.

There are some obvious reasons for these preferences:

1. The α helix can be regarded as the default conformation. Branching at the β carbon atom (as in **Val**, **Thr** and **Ile**) tends to destabilize α helices because of steric clashes. These residues are easily accomodated in β strands, where their side chains project out of the plain containing the main chain.
2. **Ser, Asp** and **Asn** tend to disintegrate α helices because their side chains contain H bond donors or acceptors in close proximity to the main chain, where they compete for the main-chain NH and CO groups.
3. **Pro** tends to disrupt both α helices and β strands because it lacks an NH group and because its ring structure restricts its ϕ value to near -60 degrees.
4. **Gly** readily fits into all structures and hence does not favour helix formation, in particular.

It is worthmentioning that the conformational preferences of amino acid residues are not toppled over all the way to one structure, as seen in Table 9–11. As an instance, glutamic acid, one of the strongest helix formers, prefers α helix to β strand by only a factor of (1.44/0.75) about 2. The preference ratios of most other amino acid residues are smaller; for example, for methionine it is (1.47/0.97) about 1.5 and for alanine it is (1.29/0.90) about 1.25. Indeed, some penta- and hexapeptide sequences have been found to adopt one structure in one protein and an entirely different structure in another protein (refer Fig. 9–52). Hence, some amino acid sequences do not uniquely determine secondary structure of proteins. Tertiary interactions between residues that are far apart in the sequence may be decisive in specifying the secondary structure of some segments.

It is now possible to predict the protein secondary structure with moderate accuracy. Based on the relative frequency values of the constituent amino acids, **P.Y. Chou** and **G.D. Fasman** (1974) have framed certain rules for the prediction of globular protein secondary structures. Following are the 3 **Chou–Fasman rules** for prediction:

1. Any segment of 6 residues or more, with $(P_\alpha) \geq 1.03$, as well as $(P_\alpha) > (P_\beta)$, and not including Pro, is predicted to be α helix.
2. Any segment of 5 residues or more, with $(P_\beta) \geq 1.05$, and $(P_\beta) > P_\alpha$, is predicted to be β sheet.
3. Examine the sequence for tetrapeptides with $(P_\alpha) < 0.9$, $(P_t) > (P_\beta)$. They have a good chance of being turns. The actual rules for predicting β turns are more complex, but this method will work in most cases.

B. Prediction of Tertiary Structure

Prediction of tertiary structure of proteins is much more difficult because the higher-order folding depends so critically on specific side chain interactions, often between residues far removed from one another in the sequence. However, recent recognition of overall patterns in tertiary folding has given some success. For example, predictions of secondary structure have been used to predict an α/β barrel structure for the enzyme **tryptophan synthase**, that is in excellent harmony with x-ray result.

The method for predicting the tertiary structure of proteins depends on the fact that, in their spontaneous folding, proteins are seeking a free energy minimum. A random-coil chain is allowed, in computer simulation, to undergo a large number of small permutations in its configuration, through rotation about individual bonds. The computer programme keeps track of the total energy, in terms of possible interactions, and seeks an energy minimum. This approach is still in its infancy.

CLEAVAGE OF A PROTEIN

Certain proteolytic enzymes and chemical reagents cleave protein between specific amino acid residues (Table 9–12). The enzyme *trypsin*, for example, cuts on the carboxyl side of lysine or arginine, whereas the chemical *cyanogen bromide* splits peptide bonds next to methionine residues. Since these enzymes and chemicals cleave at relatively few sites in a protein molecule, they tend to create relatively large and relatively few peptides. If such a mixture of peptides is separated by biochemical procedures (chromatography, electrophoresis etc), a characteristic pattern or **peptide map** would be obtained which will be diagnostic of the protein from which the peptides were generated. This peptide map is, sometimes, referred to as the **protein's fingerprint.**

Table 9–12. Some reagents frequently used to cleave peptide bonds in proteins*

Reagent	*Amino acid 1*	*Amino acid 2*
Enzymes		
Trypsin	Lys or Arg	Any
Chymotrypsin	Phe, Trp or Tyr	Any
V8 protease	Glu	Any
Chemicals		
Cyanogen bromide	Met	Any
2-nitro-5-thiocyano-benzoate	Any	Cys

* The specificity for the amino acids on either side of the cleaved bond is indicated. The carboxyl group of amino acid 1 is released by the cleavage; this amino acid is to the left of the peptide bond as normally written.

ANALYSIS OF PROTEIN STRUCTURE

The purification of a protein is first done before ascertaining its structure. The typical steps in the purification of a protein are summarized in Fig. 9–62. The primary structure of the protein is determined by automated Edman degradation of peptides. The 3-'D' structure of the protein may be determined by x-ray crystallography.

Once purified, for the determination of its amino acid composition (*i.e.*, the primary structure), a protein is subjected to hydrolysis, commonly in 6 mol/L HCl at 110°C in a sealed and evacuated tube for 24–48 h. Under these conditions, tryptophan, cysteine and most of the cystine are destroyed, and glutamine and asparagine are quantitatively deaminated to yield glutamate and aspartate, respectively. Recovery of serine and threonine is incomplete and decreases with increasing time of hydrolysis. Alternative hydrolysis procedures may be used for measurement of tryptophan, while cysteine and cystine may be converted to an acid-stable cysteic acid prior to hydrolysis. Following hydrolysis, the free amino acids are separated on an automated amino acid analyzer using an ion exchange column or by reversed-phase high-performance liquid chromatography (HPLC). The free amino acids fractionated by ion exchange chromatography are detected by reaction with a chromogenic or fluorogenic reagent such as ninhydrin or dansyl chloride, Edman reagent (see below) or α-phthalaldehyde. These techniques allow the measurement of as little as 1p mol of each amino acid. A typical elution pattern of amino acids in a purified protein is shown in Fig. 9–63.

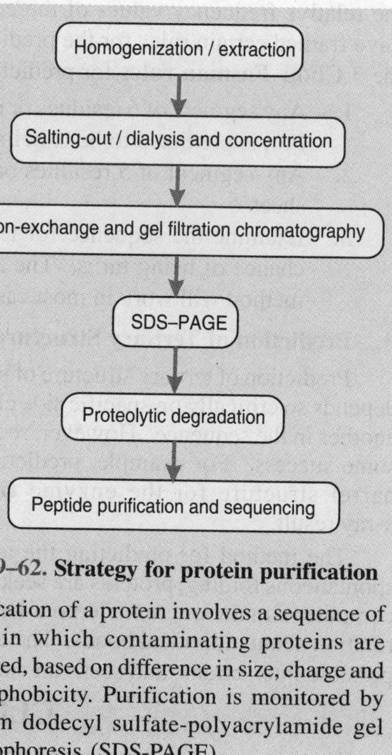

Fig. 9–62. Strategy for protein purification

Purification of a protein involves a sequence of steps in which contaminating proteins are removed, based on difference in size, charge and hydrophobicity. Purification is monitored by sodium dodecyl sulfate-polyacrylamide gel electrophoresis (SDS-PAGE).

A. Determination of the Primary Structure of Proteins

For determining the amino acid sequence of a protein, it is cleaved first by digestion by specific endoproteases, such as trypsin, V8 protease or lysyl endopeptidase, to obtain peptide fragments. Trypsin cleaves peptide bonds on the C-terminal side of arginine and lysine residues, provided the next residue is not proline. Lysyl endopeptidase is also frequently used to cleave at the C-terminal side of lysine. Cleavage by chemical reagents such as cyanogen bromide is also useful. *Cyanogen bromide* cleaves on the C-terminal side of methionine residues. Before cleavage, proteins with cysteine and cystine residues are reduced by 2-mercaptoethanol and then treated with iodoacetate to form carboxymethylcysteine residues. This avoids spontaneous formation of inter- or intramolecular disulfides during analyses.

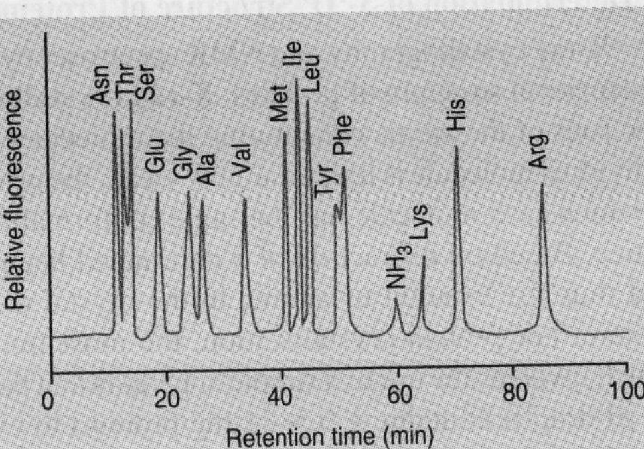

Fig. 9–63. Typical chromatogram from an amino acid analysis by cation exchange chromatography

A protein hydrolysate is applied to the cation exchange column in a dilute buffer at acidic pH (~3.0), at which all amino acids are positively charged. The amino acids are then eluted by a gradient of increasing pH and salt concentrations. The most anionic (acidic) amino acids elute first, followed by the neutral and basic amino acids. Amino acids are derived by post-column reaction with a fluorogenic compound, such as *o-pthalaldehyde*.

The cleaved peptides are then subjected to reverse-phase HPLC to purify the peptide segments, and then sequenced on an automated protein sequencer, using the **Edman degradation technique** (Fig. 9–64). The sequence of overlapping peptides is then used to obtain the primary structure of the protein. Mass spectrometry (MS) is more commonly used these days to obtain both the molecular mass and sequence of polypeptides simultaneously. Both techniques can be applied directly to proteins or peptides recovered from SDS-PAGE or 2-'D' electrophoresis (IEF + SDS-PAGE). Once the partial amino acid sequence is obtained, one can determine the nucleotide sequence of the DNA that encodes this polypeptide segment. After chemically synthesizing this DNA, it can be used to identify and isolate the gene containing its nucleotide sequence.

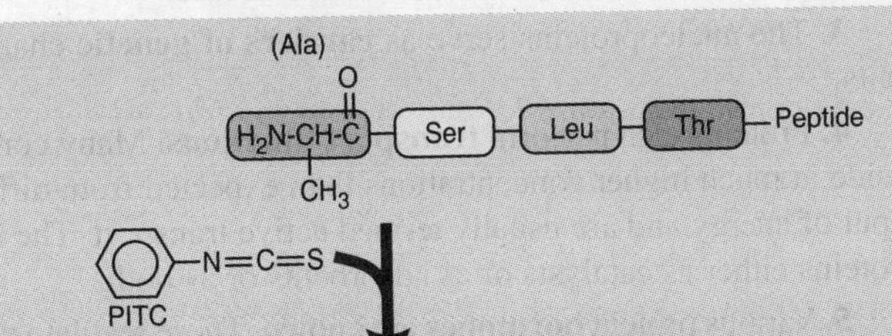

Fig. 9–64. Steps in Edman degradation

The Edman degradation method sequentially removes one residue at a time from the amino end of a peptide. Phenyl isothiocyanate (PITC) converts the N-terminal amino group of the immobilized peptide to a phenylthiocarbamyl derivative (PTC amino acid) in alkaline solution. Acid treatment removes the first amino acid as the phenylthiohydantoin (PTH) derivative, which is identified by HPLC.

B. Determination of 3-'D' Structure of Proteins

X-ray cystallography and NMR spectroscopy are usually used for the determination of the three-dimensional structure of proteins. **X-ray crystallography** depends on the diffraction of x-rays by the electrons of the atoms constituting the molecule. However, since the x-ray diffraction caused by an individual molecule is immeasurably weak, the protein must exist in the form of a well-ordered crystal, in which each molecule has the same conformation in a specific position and orientation on a 3-'D' lattice. Based on diffraction of a collimated beam of electrons, the distribution of electron density, and thus the location of atoms, in the crystal can be calculated to determine the structure of the protein. For protein crystallization, the most frequently-used method is the *hanging drop method* which involves the use of a simple apparatus that permits a small portion of a protein solution (typically 10 μl droplet containing 0.5—1 mg/protein) to evaporate gradually to reach the saturating point at which the protein begins to crystallize. **NMR spectroscopy** is usually used for structural analysis of small organic compounds, but high-field NMR is also useful for determination of the structure of a protein in solution and complements information obtained by x-ray crystallography.

BIOLOGICAL ROLES OF PROTEINS

Proteins are of utmost significance to biological systems. These are most critical to life and perform various functions. Some of their roles are given below.

1. Many proteins act as **catalysts**, thus usually enhancing the rate of chemical reactions to such extents as needed by the living cells.

2. The fibrous proteins serve as **components of the tissues** holding the skeletal elements together. Collagen is a structural unit of connective tissues.

3. The nucleoproteins serve as **carriers of genetic** characters and hence govern inheritance of traits.

4. Proteins also perform **transport functions.** Many compounds enter the cells and accumulate inside at much higher concentrations than expected from diffusion alone. These changes require the input of energy and are usually termed active transport. The mechanism of active transport involves proteins either as catalysts or as adsorbents or as both.

5. Various protein **hormones** are known. These regulate the growth of plants and animals, besides controlling many other physiological functions.

6. Under conditions of non-digestion and no chances for denaturation, the proteins accumulate inside the cells and produce **toxicity**. Venoms of snakes and insects are injected by biting into the blood. Certain of the foreign proteins present in venom are actually enzymes. These enzymes attack body tissue causing destruction of blood cells leading ultimately to death.

In other cases, the toxicity from foreign proteins results from responses by the affected animals. Allergic reactions exemplify this category. These reactions occur when an animal is exposed to a foreign protein to which it has been sensitized by prior exposure. The toxic disorders include skin blisters, swelling of limbs and respiratory congestion, leading sometimes to death.

7. Blood plasma, which is obtained after removal of the blood cells by centrifugal action, is essentially a solution of proteins in water. It is used for the **treatment of shock** produced by serious injuries and operations.

8. Interferon (IF or **IFN)** is a generic term which applies to a number of (about over 20) related low molecular weight, regulatory glycoproteins produced by many eukaryotic cells in response to numerous inducers : a virus infection, double-stranded RNA, endotoxins, antigenic stimuli, mitogenic agents, and many parasitic organisms capable of intracellular growth (*Listeria monocytogenes*, chlamydiae, rickettsias, protozoa). They are effective in treating viral diseases and cancer, and in eliminating its side effects. The most widely-studied property of interferons is their ability to 'interfere' (hence, their nomenclature) with the replication of viruses. *They are usually species-specific but virus-nonspecific.* Interferon was discovered in 1957 in London by the British virologist **Alick Issacs** and a visiting scientist from Switzerland **Jean Lindenmann**, both of the National Institute for Medical Research, London. They found an agent responsible for such viral interference ; a protein released by cells exposed to a virus, that enables other cells to resist viral infection. They called it interferon (IF). Since then, several different classes of interferons have been identified. Human interferons have been classified into 3 classes, depending on which type of they are produced by:

(*a*) Leukocytes (L) or alpha interferons (IFN-α)

(*b*) Fibroblasts (F) or beta interferons(IFN-β)

(*c*) Immune type (I) or gamma interferons (IFN-γ)

In 1978, a Tokyo metropolitan medical team has cultivated this "wonder substance" using cells from the placenta taken at the time of birth (or parturition). Large-scale production of this protein is possible because the human cells reproduce in tissue culture. Because of their antiviral and antiproliferative activities, interferons have been considered important therapeutically in treating viral diseases such as hepatitis, encephalitis, cancer and even common cold. The IFs may not be a magic bullet like penicillin against all viruses, yet it may be a very useful addition to the armamentarium already available against viruses in general, and cancer in particular.

9. Peptides from humans called *defensins* have been found to be **antibiotic** in nature. Produced by the immune system, these cells smother and kill the invading pathogens. They are secreted by the epithelial cells lining the moist body surface of mammals and serve as the body's own disinfectants.

10. Another group of peptides called *endorphins* are found in the brain and are involved in the **suppression of pain**, creation of euphoric highs and feelings of joy. **Candace Pert**, a codiscoverer of the body's opiates, says "They reveal the harmony as well as the economy inherent in nature. Nothing is wasted". She adds "Once a molecule is found effective, it tends to be used over and over again in the ladder of life, in newer and ever more exciting configurations."

11. The frog's secretion *adenoregulin*, a 33-amino-acid-long peptide, works on the receptors in the brain which handle adenosine. Adenosine is a fundamental component in all human cell fuel. And the frog peptide seems to **enhance the binding of adenosine** by subtly altering the receptor in the brain.

12. The American scientists at the Chicago Medical School and the Harvard Medical School, in 1982, have isolated about 30 g of a glycopeptide, known as *S factor*, from 4.5 tones of urine from healthy males. The S factor is composed of alanine, glutamic acid, diaminopimelic acid, and muramic acid. The S factor acts **as a soporific,** *i.e.*, as a sleep-promoting agent and has little side effects. The researchers infused a very light concentration of the S factor into the brain of the rabbit and it was found that it induced a 50% increase in what is known as *slow wave sleep*– a deep dream-free sleep that occurs in animals and humans after sleep derivation and is normal as judged by various criteria.

REFERENCES

See list following Chapter 11.

PROBLEMS

1. (*a*) Tropomyosin, a 70-kd muscle protein, is a two-stranded α-helical coiled coil. Estimate the length of the molecule? (*b*) Suppose that a 40-residue segment of a protein folds into a two-stranded antiparallel β structure with a 4-residue hairpin turn. What is the longest dimension of this motif?

2. Glycine is a highly conserved amino acid residue in the evolution of proteins. Why?
3. Identify the groups in protein that can form hydrogen bonds or electrostatic bonds with an arginine side chain at pH 7.
4. The shape of hair is determined in part by the pattern of disulfide bonds in keratin, its major protein. How can curls be induced?
5. Proteins are quite stable. The lifetime of a peptide bond in aqueous solution is nearly 1000 years. However, the $\Delta G^{\circ\prime}$ of hydrolysis of proteins is negative and quite large. How can you account for the stability of the peptide bond in light of the fact that hydrolysis releases much energy?
6. For an amino acid such as alanine, the major species in solution at pH 7 is the zwitterionic form. Assume a pK_a value of 8 for the amino group and a pK_a value of 3 for the carboxylic acid and estimate the ratio of the concentration of neutral amino acid species (with the carboxylic acid protonated and the amino group neutral) to that of the zwitterionic species at pH 7.
7. All L amino acids have an *S* absolute configuration except L-cysteine, which has the *R* configuration. Explain why L-cysteine is designated as the *R* absolute configuration.
8. Translate the following amino acid sequence into one-letter code: Leu-Glu-Ala-Arg-Asn-Ile-Asn-Gly-Ser-Cys-Ile-Glu-Cys-Glu-Ile-Ser-Gly-Arg-Glu-Ala-Thr.
9. Would you expect Pro–X peptide bonds to tend to have *cis* conformations like those of X–Pro bonds? Why or why not?
10. A protein was purified to homogeneity. Determination of the molecular weight by molecular exclusion chromatography yields 60 kd. Chromatography in the presence of 6 M urea yields a 30-kd species. When the chromatography is repeated in the presence of 6 M urea and 10 mM β-mercaptoethanol, a single molecular species of 15 kd results. Describe the structure of the molecule.
11. The three-dimensional structure of biomolecules is more conserved evolutionarily than is sequence. Why is this the case?
12. The sequences of three proteins (A, B, and C) are compared with one another, yielding the following levels of identity:

	A	B	C
A	100%	65%	15%
B	65%	100%	55%
C	15%	55%	100%

Assume that the sequence matches are distributed relatively uniformly along each aligned sequence pair. Would you expect protein A and protein C to have similar three-dimensional structures? Explain.
13. Would a homopolymer of alanine be more likely to form an α helix in water or in a hydrophobic medium? Explain.
14. Using the data in Table 9–3, calculate the *average* amino acid residue weight in a protein of typical composition. This is a useful number to know for approximate calculations.
15. The melanocyte-stimulating peptide hormone *α-melanotropin* has the following sequence:

Ser-Tyr-Ser-Met-Glu-His-Phe-Arg-Trp-Gly-Lys-Pro-Val

(*a*) Write the sequence using the one-letter abbreviations.

(*b*) Calculate the molecular weight of α-melanotropin, using data in Table 9–3. Why is this result not *exactly* correct at neutral pH?

16. A protein has been sequenced after destruction of —S—S— bonds. It is known to contain 3 Cys residues, located as shown below. However, only one of these is a free —SH; two are involved in an S—S bond.

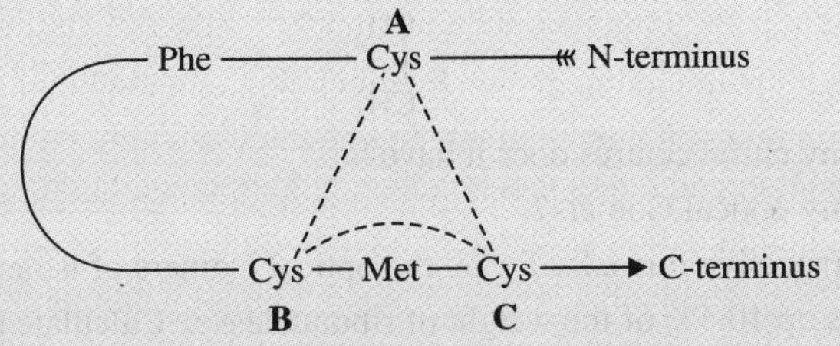

The only methionine and the only aromatic amino acid (Phe) in this protein are in the positions indicated. Cleavage of the *intact* protein (with —S—S— bridge intact) by either cyanogen bromide or chymotrypsin does *not* break the protein into two peptides. Where is the —S—S— bridge (AB, BC, or AC)?

17. *Apamine* is a small protein toxin present in the venom of the honeybee. It has the sequence

CNCKAPETALCARRCQQH

(*a*) It is known that apamine does not react with iodoacetate. How many disulfide bonds are present?

(*b*) Suppose trypsin cleavage gave two peptides. Where is (are) the S—S bond(s)?

18. In the protein *adenylate kinsase,* the C-terminal region is α-helical, with the sequence

Val-Asp-Asp-**Val-Phe**-Ser-Gln-**Val**-Cys-Thr-His-**Leu**-Asp-Thr-**Leu**-Lys-

The hydrophobic residues in this sequence are presented in boldface type. Suggest a possible reason for the periodicity in their spacing.

19. Consider a small protein containing 101 amino acid residues. The protein will have 200 bonds about which rotation can occur. Assume that three orientations are possible about each of these bonds.

(*a*) Based on these assumptions, about how many *random-coil* conformations will be possible for this protein?

(*b*) The estimate obtained in (*a*) is surely too large. Give one reason why.

20. It has been postulated that the normal (noninfectious) form of prion differs from the infectious form only in secondary/tertiary structure.

(*a*) How might you show that changes in secondary structure occur?

(*b*) How might you check for changes in quaternary structure?

(*c*) If this model is correct, what are the implications for structural prediction schemes like that of **Chou** and **Fasman**?

21. It is observed that chloride ion acts as a negative allosteric effector for hemoglobin. Suggest a possible explanation for why this should be so.

22. Is citrulline isolated from watermelons (shown below) a D- or L-amino acid? Explain.

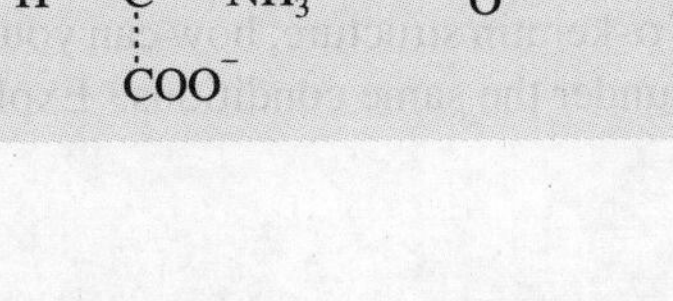

23. The structure of the amino acid isoleucine is:

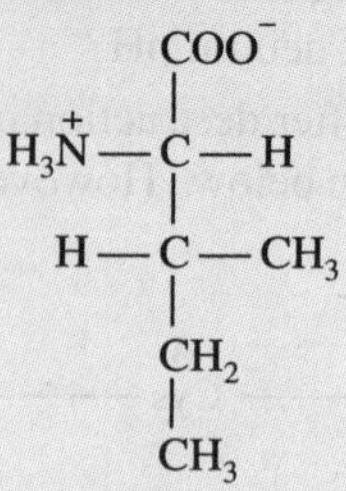

(*a*) How many chiral centres does it have?

(*b*) How many optical isomers?

(*c*) Draw perspective formulas for all the optical isomers of isoleucine.

24. Lysine makes up 10.5% of the weight of ribonuclease. Calculate the minimum molecular weight of ribonuclease. The ribonuclease molecule contains ten lysine residues. Calculate the molecular weight of ribonuclease.

25. What is the approximate molecular weight of a protein containing 682 amino acids in a single polypeptide chain?

26. Pepsin of gastric juice (pH ≈ 1.5) has a pI of about 1, much lower than that of other proteins. What functional groups must be present in relatively large numbers to give pepsin such a low pI? What amino acids can contribute such groups?

27. One method for separating polypeptides makes use of their differential solubilities. The solubility of large polypeptides in water depends upon the relative polarity of their R groups, particularly on the number of ionized groups; the more ionized groups there are, the more soluble the polypeptide. Which of each pair of polypeptides below is more soluble at the indicated pH?

(*a*) $(Gly)_{20}$ or $(Glu)_{20}$ at pH 7.0

(*b*) $(Lys–Ala)_3$ or $(Phe–Met)_3$ at pH 7.0

(*c*) $(Ala–Ser–Gly)_5$ or $(Asn–Ser–His)_5$ at pH 6.0

(*d*) $(Ala–Asp–Gly)_5$ or $(Asn–Ser–His)_5$ at pH 3.0

28. **William Astbury** discovered that the x-ray pattern of wool shows a repeating structural unit spaced about 0.54 nm along the direction of the wool fiber. When he steamed and stretched the wool, the x-ray pattern showed a new repeating structural unit at a spacing of 0.70 nm. Steaming and stretching the wool and then letting it shrink gave an x-ray pattern consistent with the original spacing of about 0.54 nm. Although these observations provided important clues to the molecular structure of wool, Astbury was unable to interpret them at the time. Given our current understanding of the structure of wool, interpret Astbury's observations.

29. A number of natural proteins are very rich in disulfide bonds, and their mechanical properties (tensile strength, viscosity, hardness, etc.) are correlated with the degree of disulfide bonding. For example, glutenin, a wheat protein rich in disulfide bonds, is responsible for the cohesive and elastic character of dough made from wheat flour. Similarly, the hard, tough nature of tortoise shell is due to the extensive disulfide bonding in its α-keratin. What is the molecular basis for the correlation between disulfide-bond content and mechanical properties of the protein?

30. When wool sweaters or socks are washed in hot water and/or dried in an electric dryer, they shrink. From what you know of α-keratin structure, how can you account for this? Silk, on the other hand, does not shrink under the same conditions. Explain.

31. In the following polypeptide, where might bends or turns occur? Where might intrachain disulfide cross-linkages be formed?

1 2 3 4 5 6 7 8 9 10 11 12 13 14 15 16 17 18
Ile–Ala–His–Thr–Tyr–Gly–Pro–Phe–Glu–Ala–Ala–Met–Cys–Lys–Trp–Glu–Ala–Gln–

19 20 21 22 23 24 25 26 27 28
Pro–Asp–Gly–Met–Glu–Cys–Ala–Phe–His–Arg–

32. Both myoglobin and hemoglobin consist of globin (protein) bound to a heme prosthetic group. One heme group binds one O_2. Why is the oxygen saturation curve (saturation versus pO_2) of myoglobin a rectangular hyperbola while that of hemoglobin is sigmoidal?

33. Using known endo- and exopeptidases, suggest a pathway for the complete degradation of the following peptide:

His-Ser-Lys-Ala-Trp-Ile-Asp-Cys-Pro-Arg-His-His-Ala

34. How do depilatory creams remove hair?

35. Which of the following characteristics are associated with myoglobin, hemoglobin, both of them or neither of them?

(*a*) majority of structure in α-helical conformation
(*b*) oxygen carrier
(*c*) carbon dioxide carrier
(*d*) not an allosteric protein
(*e*) heme group(s) in a polar crevice(s)
(*f*) protoporphyrin IX
(*g*) sigmoidal oxygen dissociation curve
(*h*) tertiary structure
(*i*) quaternary structure
(*j*) blocked N-terminal residue

36. What is homocysteine?

37. Why do some people have curly hair while in others hair grows straight?

38. Why does our skin become wrinkled in old age?

39. What does the blood test for homocysteine signify?

40. Why do some wounds leave scars while some do not?

41. Why don't we feel pain when cutting hair / nails?

42. What is the fraction of glycine (Gly) present as the completely uncharged form at the isoelectric point (pI)?

43. Suppose you want to synthesize tripeptides using glycine, alanine and serine as building blocks.

(*a*) How many different tripeptides can you prepare using any of the three amino acids in any of the 3 positions, if you can use an amino acid more than once?

(*b*) How many different tripeptides can you prepare using each amino acid only once?

44. Proteins in a pea plant differ from those of a dog, yet they are built from similar amino acids. How can the difference be explained?

45. Why amino acids are called ampholytes?

46. What are the evidences for the existence of zwitterion structure of amino acids.

47. How many amino acids are involved in five complete turns of α-helix?

48. Why L-phenylalanine is an essential amino acid whereas L-alanine is not?

ANSWERS TO PROBLEMS

1. (*a*) Each strand is 35 kd and hence has about 318 residues (the mean residue mass is 110 daltons). Because the rise per residue in an α helix is 1.5 Å, the length is 477 Å. More precisely, for an α-helical coiled coil the rise per residue is 1.46 Å so that the length will be 464 Å.

   (*b*) Eighteen residue in each strand (40 minus 4 divided by 2) are in a β-sheet conformation. Because the rise per residue is 3.5 Å, the length is 63 Å.
2. Glycine has the smallest side chain of any amino acid. Its size often is critical in allowing polypeptide chains to make tight turns or to approach one another closely.
3. Glutamate, aspartate, and the terminal carboxylate can form salt bridges with the guanidinium group of arginine. In addition, this group can be a hydrogen-bond donor to the side chains of glutamine, asparagine, serine, threonine, aspartate, and glutamate, and to the main-chain carbonyl group.
4. Disulfide bonds in hair are broken by adding a thiol and applying gentle heat. The hair is curled, and an oxidizing agent is added to re-form disulfide bonds to stabilize the desired shape.
5. The energy barrier that must be crossed to go from the polymerized state to the hydrolyzed state is large even though the reaction is thermodynamically favourable.
6. Using the Henderson-Hasselbach equation, we find the ratio of alanine-COOH to alanine-COO^- at pH 7 to be 10^{-4}. The ratio of alanine-NH_2 to alanine-NH_3^+, determined in the same fashion, is 10^{-1}. Thus, the ratio of neutral alanine to zwitterionic species is $10^{-4} \times 10^{-1} = 10^{-5}$.
7. The assignment of absolute configuration requires the assignment of priorities to the four groups connected to a tetrahedral carbon. For all amino acids except cysteine, the priorities are: (1) amino group; (2) carbonyl group; (3) side chain; (4) hydrogen. For cysteine, because of the sulfur atom in its side chain, the side chain has a greater priority than does the carbonyl group, leading to the assignment of an *R* rather than *S* configuration.
8. LEARNINGSCIE(N)CEISGREAT.
9. No, Pro–X would have the characteristics of any other peptide bond. The steric hindrance in X–Pro arises because the R group of Pro is bonded to the amino group. Hence, in X–Pro, the proline R group is near the R group of X. This would not be the case in Pro–X.
10. Treatment with urea will disrupt noncovalent bonds. Thus, the original 60-kd protein must be made of two 30-kd subunits. When these subunits are treated with urea and mercaptoethanol, a single 15-kd species results, suggesting that disulfide bonds link the 30-kd subunits.
11. Because three-dimensional structure is much more closely associated with function than is sequence, tertiary structure is more evolutionarily conserved than is primary structure. In other words, protein function is the most important characteristic, and protein function is determined by structure. Thus, the structure must be conserved, but not necessarily a specific amino acid sequence.
12. Protein A is clearly homologous to protein B, given 65% sequence identity, and so A and B are expected to have quite similar three-dimensional structures. Likewise, proteins B and C are clearly homologous, given 55% sequence identity, and so B and C are expected to have quite similar three-dimensional structures. Thus, one can conclude that proteins A and C are likely to have similar three-dimensional structures, even though they are only 15% identical in sequence.

13. In a hydrophobic environment, the formation of intrachain hydrogen bonds would stabilize the amide hydrogen atom and carbonyl oxygen atoms of the polypeptide chain; so an α helix would form. In an aqueous environment, these groups would be stabilized by interaction with water, so there would be no energetic reason to form an α helix. Thus, the α helix would be most likely to form in an hydrophobic environment

14. 109.5.

15. (*a*) SYSMEHFRWGKPV.

(*b*) 1624. This is not exactly correct because it does not take into account dissociation of some protons.

16. AC.

17. (*a*) 2.

(*b*) Between the first and second, and third and fourth cysteines.

18. They are spaced about three to four residues apart. Therefore, they will all lie on the same side of the α helix. This suggests that this side of the helix may face the interior of the protein.

19. (*a*) $3^{200} = 2.7 \times 10^{95}$.

(*b*) Not all of these conformations will be sterically possible. But even if only 0.1% of these are allowed, there are still 2.7×10^{92}, a very large number.

20. Assuming that x-ray diffraction is not practical, we have:

(*a*) Analysis by CD or by NMR.

(*b*) Sedimentation equilibrium.

(*c*) More than one secondary/tertiary folding can be observed for the same sequence. Therefore, sequence alone cannot dicate folding in all cases, and sequence-based predictions must sometimes fail badly.

21. Chloride ions must interact with positively charged groups so as to stabilize the deoxy state. Thus, the higher the Cl^- concentration, the lower the O_2 affinity of hemoglobin will be.

22. L; determine the absolute configuration of the α carbon and compare with D- and L-glyceraldehyde.

23. (*a*) 2

(*b*) 4

(*c*)

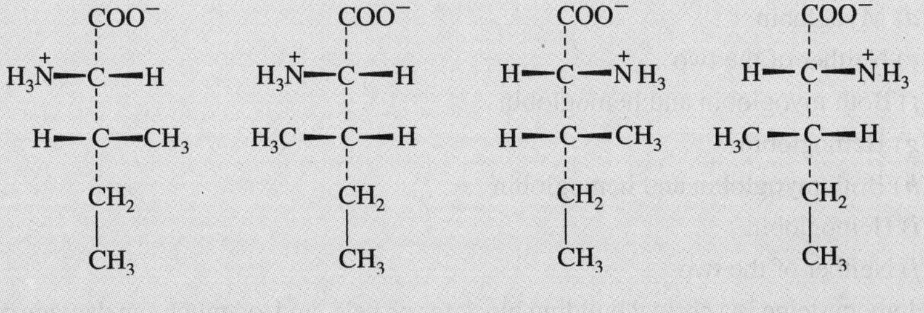

24. 1,220; 12,200

25. 75,020

26. Carboxyl groups; Asp and Glu

27. (*a*) $(Glu)_{20}$ (*b*) $(Lys–Ala)_3$ (*c*) $(Asn–Ser–His)_5$ (*d*) $(Asn–Ser–His)_5$

28. The principal structural units in the wool fibre polypeptides are successive turns of the α helix, which are spaced at 0.54 nm intervals. Steaming and stretching the fibre yields an extended polypeptide chain with the β conformation, in which the distances between the R groups are about 0.70 nm. On shrinking, the fibre reassumes an α-helical structure.

29. The disulfide bonds are covalent bonds, which are much stronger than the noncovalent bonds that stabilize most proteins. They serve to cross-link protein chains, increasing their stiffness, mechanical strength, and hardness.

30. Wool shrinks when polypeptide chains are converted from the extended conformation (β-pleated sheet) to the α-helical conformation in the presence of heat. The β-pleated sheets of silk, with their small, closely packed amino acid side chains, are more stable than those of wool.

31. At residues 7 and 19; prolines are often but not always found at bends. Between residues 13 and 24.

32. Myoglobin is a monomer with one globin and one heme binding one O_2. O_2 binding is a simple equilibrium with a Hill coefficient of 1. Hemoglobin is a tetramer with four O_2 bound to the four hemes. Hemoglobin without oxygen is in the T conformation in which binding of O_2 is difficult. Thus, the curve starts with a slow increase as pO_2 increases. Binding of the first O_2 shifts the conformation of the whole molecule to favour the R form, which binds O_2 more readily. Thus, the slope rises steeply giving it a sigmoidal shape. This is an example of cooperativity. The Hill coefficient for hemoglobin is 2.8.

33. Trypsin cleavage gives (a) His-Ser-Lys + (b) Ala-Trp-Ile-Met-Cys-Gly-Pro-Arg + (c) His-His-Ala. Further degradation of (a) is accomplished by elastase and dipeptidase. Further degradation of (b) would start with chymotrypsin and also use dipeptidases, tripeptidase, and carboxypeptidase B. To degrade (c), carboxypeptidase A and dipeptidase would be enough. The point is that several peptidases with varying specificities are required.

His-Ser-Lys-Ala-Trp-Ile-Asp-Cys-Pro-Arg-His-His-Ala

34. Methods of removal of hair are classified as *depilatory* (removal of hair from the surface of the skin) or *epilatory* (removal of intact hair with the root). Depilatory creams are substituted mercaptans, such as thioglycollic acid, which reduce the disulfide bond in the peptide chains of keratin, a hair protein. As a result, the hair fibre swells and softens and can be washed from the skin.

35. (***a***) Both myoglobin and hemoglobin

(***b***) Both myoglobin and hemoglobin

(***c***) Hemoglobin

(***d***) Myoglobin

(***e***) Neither of the two

(***f***) Both myoglobin and hemoglobin

(***g***) Hemoglobin

(***h***) Both myoglobin and hemoglobin

(***i***) Hemoglobin

(***j***) Neither of the two

36. Homocysteine is a normal building block for protein, and too much can damage one's arteries. Vitamins B_6, B_9 and B_{12} can lower it.

37. Scalp hair is not the same among all the people in the world. The Negro has tightly curled hair while the American Indians have straight, lank and coarse hair. The Caucasians have hair that may be straight or wavy or curly. Australian natives have "frizzly" hair. So the type of hair one has depends on one's racial inheritance and is genetically determined.

The reason why someone's hair is curly while another's is not depends on the structure of the strand of hair. If you cut a shaft of hair transversely and observe it under microscope, the difference would become clear. *The cut section of curly hair is flat.* The flatter the hair, the more easily it bends and hence the curlier it becomes. *Straight hair has a round cross-section.* Thus what kind of hair you have would be genetically determined and stable for a life-time. Cosmetic treatment can only temporarily curl or straighten hair as per our wishes.

38. Wrinkling of skin is a sign of old age. Our skin becomes wrinkled because of changes in the underlying structure. There are some connective tissues just under the outer layer of the skin which are made up of two types of protein fibres—collagen and elastin. The collagen provides the material for the tissue and the elastin fibres, which are smaller in number, give elasticity and suppleness to the skin. *With time, as one ages, the amount of elastin in the skin diminishes and the collagen fibres become disorganized. The cross-linking between collagen fibres also increases. Consequently, the tissue gradually loses its elasticity.* Thus, wrinkles appear on the face and other parts of the body. Another factor responsible for the development of wrinkles is sunlight. *The ultraviolet part of the sunlight accelerates the changes in the skin which cause wrinkles.*

39. Homocysteine is a substance made in the body. Too much of it means one has an increased risk of heart attack and stroke.

40. A deep wound or cut usually leaves a scar whereas a superficial one seldom does so. A wound in which collage fibres – situated deep in the dermis layer of the skin – are damaged leaves a scar. This is because the collagen fibres are arranged in the skin in a specific sequencial manner. To restore the same sequence when wound is healing is not possible. As a result of a change in the deposition pattern of collagen fibres, a scar is formed on the skin.

41. Pain is felt only when a nerve end is stimulated by a cut or injury. Since hair and nails do not have nerve endings, cutting them does not produce pain. However, cells at the base of a nail and hair are living cells and have nerve endings. So, if a nail is cut from the base or if a hair is plucked off, pain is felt.

42. 1 part in 10^7 (one ten-millionth)

43. (*a*) 27 (*b*) 6

44. Proteins contain about 20 different amino acids. The number and manner in which the amino acids may be grouped is enormously high. Hence, the number of possible proteins is correspondingly large. From 20 different amino acids, it is theoretically possible to make 400 dipeptides and as many as 8,000 tripeptides.

45. Since amino acids contain at least one amino group and at least one carboxyl group, they are amphoteric. *In dry state*, amino acids form lower salts which are called dipolar ions or zwitterions. In the dipolar ion, the amino group remains as $-NH_3^+$ group and the carboxyl group remains as $-COO^-$ group. However, *in aqueous medium*, an equilibrium exists involving the dipolar ion, and the cationic and anionic forms of the amino acids :

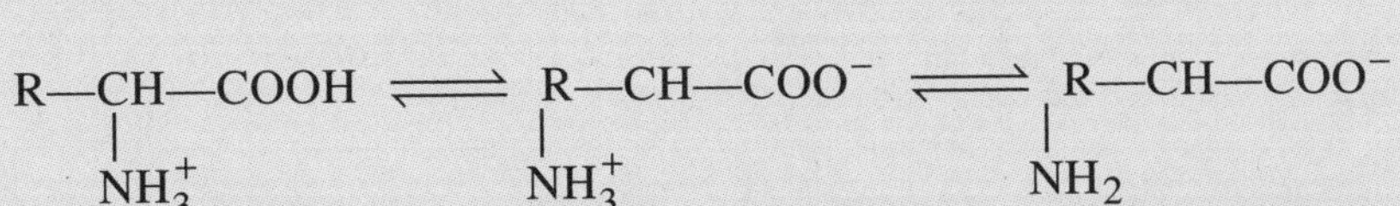

Cationic form (predominant at pH <2) | Zwitterionic form (Predominant at isoelectric point) | Anionic form (predominant at pH >11)

Hence, amino acids act as an ampholyte by donating a proton and also by accepting a proton.

46. Amino acids in solid crystalline form exist having inner salt structure, in which the $-NH_2$ group exists as $-NH_3^+$ and $-COOH$ group exists as $-COO^-$:

$$\underset{\displaystyle \underset{NH_3^+}{|}}{R—CH—COO^-}$$

Following are the evidences for their existence:

(*a*) Amino acids are soluble in polar solvent like water.

(*b*) Amino acids, dissolved in water, have pH near the neutral region, which is in contrary to the carboxylic acids.

(*c*) Amino acids have high melting point (about 260°C) which is possible for ionic salts.

47. Per turn of helix = 3.6 amino acids

Five turns of helix = 3.6×5

= 18.0 amino acids

48. L-alanine can be formed within the body by metabolic activity of pyruvate. On the contrary, L-phenylalanine cannot by synthesized within the body and has to be supplied from outside in the diet.

$$\text{Glutamate + pyruvate} \rightleftharpoons \alpha\text{-ketoglutarate + alanine}$$

CONTENTS

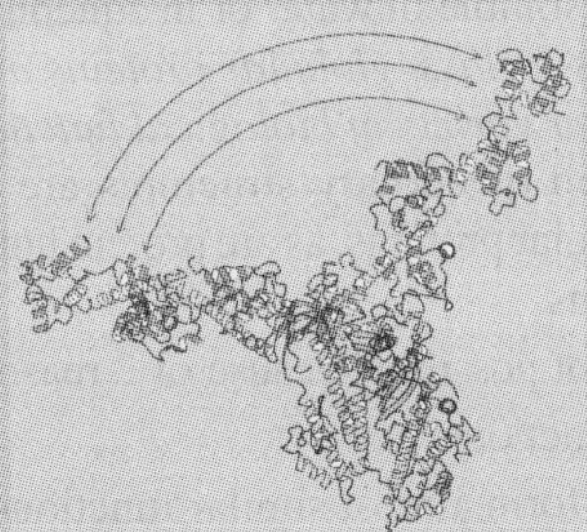

The horse, like all animals, Is powered by the molecular motor protein, myosin.

A portion of myosin moves dramatically (as shown *above*) in response to ATP binding, hydrolysis, and release, propelling myosin along an actin filament. This molecular movement is translated into movement of the entire animal, excitingly depicted in da Vinci's rearing horse (as shown *below*).

[(Lower) Leonardo da vinci's "Study of a rearing horse" for the battle of Anghiari (C. 1504) from the Royal Collection ® Her Royal Majesty Queen Elizabeth II)]

CHAPTER

10

Protein–II Classification

The proteins are all remarkably similar in structure insofar as they contain amino acids. But as little is known so far regarding their structure, a classification based totally on this criterion is not possible. However, a few systems based on one or the other cirterion are given below.

CLASSIFICATION BASED ON THE SOURCE OF PROTEIN MOLECULE

Since long, the proteins have been traditionally divided into two well-defined groups: animal proteins and plant proteins. **Animal proteins** are the proteins derived from animal sources such as eggs, milk, meat and fish. They are usually called *higher-quality proteins* because they contain (and hence supply) adequate amounts of all the essential amino acids. On the other hand, **plant proteins** are called *lower-quality proteins* since they have a low content (limiting amount) of one or more of the essential amino acids. The *four most common limiting amino acids are methionine, lysine, threonine and tryptophan* (Table 10–1). Although plant proteins have limiting amounts of some (but not all) amino acids, it should not be construed that they are poor protein sources.

Table 10–1. Limiting amino acids in some plant proteins

Food	*Amino acid(s)*
Cereal grains and millets	Lysine, Threonine
Rice and soybeans	Methionine
Legumes (peas and beans)	Methionine, Tryptophan
Groundnuts	Methionine, Lysine, Threonine
Sunflower seeds	Lysine
Green leafy vegetables	Methionine

CLASSIFICATION BASED ON THE SHAPE OF PROTEIN MOLECULE

On the basis of the shape of protein molecule, the proteins have been grouped under two categories: globular and fibrous.

1. Globular or Corpuscular Proteins

These have an *axial ratio* (length : width) of less than 10 (usually not over 3 or 4) and, henceforth, possess a relatively spherical or ovoid shape. These are usually soluble in water or in aqueous media containing acids, bases, salts or alcohol, and diffuse readily. *As a class, globular proteins are more complex in conformation than fibrous proteins, have a far greater variety of biological functions and are dynamic rather than static in their activities.* Tertiary and quaternary structures are usually associated with this class of proteins. Nearly all enzymes are globular proteins, as are protein hormones, blood transport proteins, antibodies and nutrient storage proteins.

A simple functional classification of globular proteins is not possible because of 2 reasons:

(*a*) Firstly, these proteins perform a variety of different functions.

(*b*) Secondly, many widely-differing globular proteins perform almost similar functions.

However, **Conn** and **Stumpf** (1976) have classified globular proteins as follows:

Globular Proteins
- Cytochrome C
- Blood proteins
  - Serum albumin
  - Glycoproteins
  - Antibodies (= Immunoglobulins)
  - Hemoglobin
  - Hormones
- Enzymes
- Nutrient proteins

2. Fibrous or Fibrillar Proteins

These have axial ratios (length : width) greater than 10 and, henceforth, resemble long ribbons or fibres in shape. *These are mainly of animal origin and are insoluble in all common solvents* such as water, dilute acids, alkalies and salts and also in organic solvents. Most fibrous proteins serve in a structural or protective role. The fibrous proteins are extremely strong and possess two important properties which are characteristic of the elastomers. These are:

(*a*) They can *stretch* and later recoil to their original length.

(*b*) They have a tendency to *creep*, *i.e.*, if stretched for a long time, their basic length increases and equals the stretched length but, if the tension on the two ends of the fibril is relaxed, they creep to their shorter and shorter length. A large scar, for example, creeps to a smaller size if there is no tension on the scar. On the contrary, if the scar is in a region of high tension, the scar becomes larger and larger as happens in the skin of a person gradually becoming obese.

It is a heterogeneous group and includes the proteins of connective tissues, bones, blood vessels, skin, hair, nails, horns, hoofs, wool and silk. The important examples are:

I. **Collagens** (pronounced as KAHL-uh-jehns or KAHL-uh-gens). These are of mesenchymal origin and form the major proteins of white connective tissues (tendons*, cartilage) and of bone. *More than half the total protein in mammalian body is collagen*; acted upon by boiling in water, dilute acids or alkalies to produce the soluble gelatins; *unique in containing high contents (12%) of hydroxyproline*; poor in sulfur since cysteine and cystine are lacking.

Tendons connect muscle to bone, whereas **ligaments** attach one bone to another. Tendons are nonelastic but ligaments can be stretched, because they contain, in addition to collagen fibres, the protein elastin.

II. **Elastins.** Also of mesenchymal origin; form the major constituents of yellow elastic tissues (ligaments, blood vessels); differ from collagens in not being converted to soluble gelatins.

III. **Keratins.** These are of ectodermal origin; form the major constituents of epithelial tissues (skin, hair, feathers, horns, hoofs, nails); *usually contain large amounts of sulfur in the form of cystine*–human hair has about 14% cystine.

IV. **Fibroin.** It is the principal constituent of the fibres of silk; composed mainly of glycine, alanine and serine units.

CLASSIFICATION BASED ON COMPOSITION AND SOLUBILITY

This is nowadays the most accepted system of classification and is based on the proposals made by the committees of *British Physiological Society* (1907) and the *American Physiological Society* (1908). The system divides the proteins into 3 major groups, based on their composition, *viz*., simple, conjugated and derived.

A. Simple Proteins or Holoproteins.

These are of globular type except for scleroproteins which are fibrous in nature. This group includes proteins containing only amino acids, as structural components. On decomposition with acids, these liberate the constituent amino acids.

These are further classified mainly on their solubility basis as follows:

1. Protamines and histones. These are basic proteins and occur almost entirely in animals, mainly in sperm cells; possess simplest structure and lowest molecular weight (approximately 5,000); soluble in water; unlike most other proteins, not coagulated by heat; strongly basic in character owing to high content of basic amino acids (lysine, arginine); form salts with mineral acids and nucleic proteins. Protamines are virtually devoid of sulfur and aromatic amino acids. Histones are somewhat weaker bases and are, therefore, insoluble in NH_4OH solution, whereas the protamines are soluble.

e.g., protamines—*clupeine* from herring sperm, *salmine* from salmon sperm, *sturine* from sturgeon and *cyprinine* from carp.

histones—*nucleohistones* of nuclei; *globin* of hemoglobin.

2. Albumins. These are widely distributed in nature but more abundant in seeds; soluble in water and dilute solutions of acids, bases and salts; precipitated with a saturated solution of an acid salt like $(NH_4)_2.SO_4$ or a neutral salt like $Na_2.SO_4$; coagulated by heat.

e.g., leucosine in cereals, *legumeline* in legumes, *ovalbumin* from white of egg, *serum albumin* from blood plasma, *myosin* of muscles and *lactalbumin* of milk whey.

Pseudoglobulins and euglobulins are differentiated on the basis of solubility behaviour. A pseudoglobulin is soluble at very low ionic strength, whereas euglobulin remains sparingly soluble until the ionic strength is raised. Euglobulins remain soluble at the isoelectric point (pI), whereas pseudoglobulins do not.

3. Globulins. These are of two types— pseudoglobulins and euglobulins. Euglobulins are more widely distributed in nature than the pseudoglobulins; either soluble (pseudoglobulins) or insoluble (euglobulins) in water; precipitated with half saturated solution of $(NH_4)_2.SO_4$; coagulated by heat.

e.g., pseudoglobulins— *pseudoglobulin* of milk whey.

euglobulins—*serum globulin* from blood plasma, *ovoglobulin* from eggwhite; *myosinogen* from muscle; globulins of various plant seeds like hemp (*edestin*), soybeans (*glycinine*), peas (*legumine*), peach (*amandine*), oranges (*pomeline*); also potato (*tuberin*).

4. **Glutelins.** These have been isolated only from plant seeds; insoluble in water, dilute salt solutions and alcohol solutions but soluble in dilute acids and alkalies; coagulated by heat.

e.g., *glutenin* from wheat, *glutelin* from corn, *oryzenin* from rice, etc.

5. **Prolamines.** These have also been isolated only from plant seeds; insoluble in water and dilute salt solutions but soluble in dilute acids and alkalies and also in 60–80% alcohol solutions; not coagulated by heat

e.g., *gliadin* from wheat, *zein* from corn, *hordein* from oat, etc.

Some biochemists like **Karlson** (1968) are of the viewpoint that glutelins and prolamines should not be granted the status of exclusive classes since they are small groups of vegetable proteins occurring in grain kernels.

6. **Scleroproteins or Albuminoids.** These occur almost entirely in animals and are, therefore, commonly known as the '*animal skeleton proteins*'; insoluble in water, dilute solution of acids, bases and salts and also in 60–80% alcohol solutions; not attacked by enzymes.

e.g., collagen of bones, *elastin* in ligaments, *keratin* in hair and horry tissues and *fibroin* of silk.

B. Conjugated or Complex Proteins or Heteroproteins.

These are also of globular type except for the pigment in chicken feathers which is probably of fibrous nature. These are the proteins linked with a separable nonprotein portion called *prosthetic group*. The prosthetic group may be either a metal or a compound. On decomposition with acids, these liberate the constituent amino acids as well as the prosthetic group.

Their further classification is based on the nature of the prosthetic group present. The various divisions are metalloproteins, chromoproteins, glycoproteins, phosphoproteins, lipoproteins and nucleoproteins. (Instead of metalloproteins, chromoproteins etc, the terms metalloproteids, chromoproteids etc, are sometimes used.)

1. **Metalloproteins.** These are the proteins linked with various metals. These may be of stable nature or may be more or less labile. Based on their reactivity with metal ions, the metalloproteins may be classified into 3 groups:

(*a*) *Metals strongly bound by proteins*. Some heavy metals (Hg, Ag, Cu, Zn) become strongly binded

> **Siderophilin**, also called as **transferrin**, is an important metalloprotein and constitutes about 30% of the total plasma protein. It has a molecular weight of about 90,000 and is capable of binding 2 atoms of iron per mole. It facilitates iron transport.

> **Ceruloplasmin** is an important blue copper-binding protein in the blood of humans and other vertebrates. This protein contains about 90% of copper in serum. It has a molecular weight of about 150,000 and contains 8 atoms of copper per mole. Ceruloplasmin is only one of the many sialoglycoproteins whose removal from the bloodstream is triggered by the loss of sialic acid units. It probably functions by reversibly releasing and binding copper at various sites in the body, whereby regulating copper absorption. In its deficiency, the Wilson's disease develops in man which is characterized by hepatolenticular degeneration.

with proteins like *collagen, albumin, casein* etc., through the —SH radicals of the side chains. Some other proteins have strong binding affinities for Fe (*siderophilin**) and Cu (*ceruloplasmin***). In these cases, the following pattern of binding may be present:

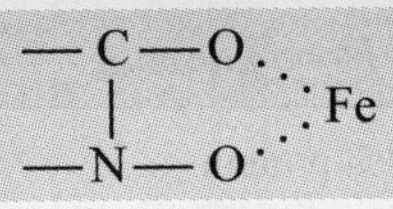

(*b*) *Metals bound weakly by proteins.* Ca belongs to this category. Here the binding takes place with the help of radicals possessing the electron charge.

(*c*) *Metals which do not couple with proteins.* Na and K belong to this group. These form compounds with nucleic acids where apparently electrostatic bonds are present.

2. **Chromoproteins.** These are proteins coupled with a coloured pigment. Such pigments have also been found among the enzymes like catalase, peroxidase and flavoenzymes. Similarly, chlorophyll is present in leaf cells in the form of a protein, the chloroplastin. The chloroplastin dissolves in water as a colloid and is readily denatured.

e.g., myoglobin, hemoglobin, hemocyanin, hemoerythrin, cytochromes, flavoproteins, catalase, etc.

3. **Glycoproteins and Mucoproteins.** These are the proteins containing carbohydrate as prosthetic group. Glycoproteins contain small amounts of carbohydrates (less than 4%), whereas mucoproteins contain comparatively higher amounts (more than 4%).

e.g., glycoproteins— *egg albumin*, *elastase* (Fig. 10–1.), certain *serum globulins* and also certain *serum albumins.*

mucoproteins— *ovomucoid* from egg-white, *mucin* from saliva and *Dioscorea* tubers, *osseomucoid* from bone and *tendomucoid* from tendon.

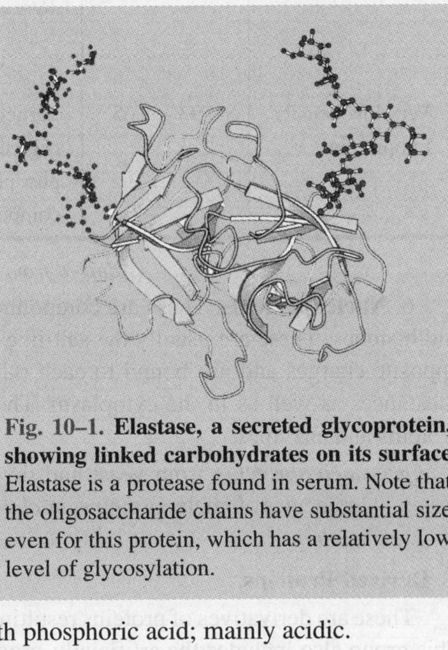

Fig. 10–1. Elastase, a secreted glycoprotein, showing linked carbohydrates on its surface
Elastase is a protease found in serum. Note that the oligosaccharide chains have substantial size even for this protein, which has a relatively low level of glycosylation.

4. **Phosphoproteins.** These are proteins linked with phosphoric acid; mainly acidic.

e.g., casein from milk and *ovovitellin* from egg yolk.

5. **Lipoproteins.** Proteins forming complexes with lipids (cephalin, lecithin, cholesterol) are called lipoproteins; soluble in water but insoluble in organic solvents.

e.g., lipovitellin and *lipovitellenin* from egg yolk; lipoproteins of blood.

The lipoproteins are in reality the temporary intermediates in the process of transfer of lipids from the site of absorption to the site of utilization. The classification of lipoproteins is frequently based on an operational definition, *i.e.*, the migration of the fraction in a density gradient separation. On this basis, the lipoproteins have been classified into following 4 categories (Table 10–2):

(*a*) Very high density lipoproteins (VHDLs). These have densities greater than 1.21.

(*b*) High density lipoproteins (HDLs). These possess density range of 1.063 to 1.21.

(*c*) Low density lipoproteins (LDLs). Their densities range between 1.05 and 1.063.

(*d*) Very low density lipoproteins (VLDLs). Their density range is from 0.93 to 1.05.

Table 10–2 also lists the compositions of the fractions floated at the respective densities, along with proposed functions.

Table 10–2. Classification of lipoproteins based on density

Name	*Density range*	*Composition by weight (in blood plasma)*	*Half-life*	*Function*
Very high density lipoprotein	> 1.21	62% protein 28% lipid	> 5 days	Phospholipid transport (contains 8–15% of serum phospholipids)
High density lipoprotein	1.063 – 1.21	50% protein 50% lipid	3 – 5 days	Cholesterol transport Lipid transport
Low density lipoprotein	1.05 – 1.063	22% protein 12% triglyceride 20% phospholipids 46% cholesterol	3 – 5 days	Lipid transport (congenital absence accompanied by intestinal malabsorption and distorted shape of erythrocyte)
Very low density lipoprotein	0.93 – 1.05	Largely protein, triglyceride and phospholipid; composition variable	3 – 4 hours	Triglyceride transport (related to chylomicron units)

(Adapted from Mallette MF, Clagett CO, Phillips AT and McCarl RL, 1979)

6. **Nucleoproteins.** These are compounds containing nucleic acid and protein, esp., protamines and histones. These are usually the salt-like compounds of proteins since the two components have opposite charges and are bound to each other by electrostatic forces. They are present in nuclear substances as well as in the cytoplasm. These may be considered as the sites for the synthesis of proteins and enzymes.

e.g., *nucleoproteins* from yeast and thymus and also viruses which may be regarded as large molecules of nucleoproteins; *nucleohistones* from nuclei-rich material like glandular tissues; *nuclein*.

C. Derived Proteins.

These are derivatives of proteins resulting from the action of heat, enzymes or chemical reagents. This group also includes the artificially-produced polypeptides.

I. *Primary derived proteins.* These are derivatives of proteins in which the size of protein molecule is not altered materially.

1. **Proteans.** Insoluble in water; appear as first product produced by the action of acids, enzymes or water on proteins.

e.g., edestan derived from edestin and *myosan* derived from myosin.

2. **Metaproteins or Infraproteins.** Insoluble in water but soluble in dilute acids or alkalies; produced by further action of acid or alkali on proteins at about 30–60°C.

e.g., *acid* and *alkali metaproteins.*

3. **Coagulated Proteins.** Insoluble in water; produced by the action of heat or alcohol on proteins.

e.g., *coagulated eggwhite.*

II. *Secondary derived proteins.* These are derivatives of proteins in which the hydrolysis has certainly occurred. The molecules are, as a rule, smaller than the original proteins.

1. **Proteoses.** Soluble in water; coagulable by heat; produced when hydrolysis proceeds beyond the level of metaproteins; *primary proteoses* are salted out by half saturation with $(NH_4)_2.SO_4$ and precipitated by HNO_3 and picric acid; *secondary proteoses* are salted out only by complete saturation with $(NH_4)_2.SO_4$ but are not precipitated by HNO_3 or picric acid.

e.g., albuminose from albumin; *globulose* from globulin.

2. **Peptones.** Soluble in water; noncoagulable by heat; produced by the action of dilute acids or enzymes when hydrolysis proceeds beyond proteoses; neither salted out by $(NH_4)_2SO_4$ nor precipitated by HNO_3 or picric acid.

3. **Polypeptides.** These are combinations of two or more amino acid units. In fact, *the proteins are essentially long chain polypeptides*.

Drawbacks. Although widely accepted, the system outlined above has certain discrepancies:

1. The classification is arbitrary.
2. The criterion of solubility is not well demarcated as some globulins (pseudoglobulins) are also soluble in water.
3. Protamines and histones should have been kept under derived proteins.
4. The group metaproteins is an artificial assemblage.

The two (second and third) systems of classification, described above, may be merged into one, as shown on page 258.

CLASSIFICATION BASED ON BIOLOGICAL FUNCTION

Depending upon their physical and chemical structure and location inside the cell, different proteins perform various functions. As such diverse proteins may be grouped under following categories, based on the metabolic functions they perform (Table 10–3).

Table 10–3. Classification of proteins on the basis of their biological functions

Class of protein	*Function*	*Examples*
Enzymic proteins	Biological catalysts	Urease, Amylase, Catalase, Cytochrome C, Alcohol dehydrogenase.
Structural proteins	Strengthening or protecting biological structures	Collagen, Elastin, Keratin, Fibroin
Transport or carrier proteins	Transport of ions or molecules in the body	Myoglobin, Hemoglobin, Ceruloplasmin, Lipoproteins
Nutrient and storage proteins	Provide nutrition to growing embryos and store ions	Ovalbumin, Casein, Ferritin
Contractile or motile proteins	Function in the contractile system	Actin, Myosin, Tubulin
Defense proteins	Defend against other organisms	Antibodies, Fibrinogen, Thrombin
Regulatory proteins	Regulate cellular or metabolic activities	Insulin, G proteins, Growth hormone
Toxic proteins	Hydrolyze (or degrade) enzymes	Snake venom, Ricin.

1. **Enzymic proteins.** The most varied and most highly specialized proteins are those with catalytic activity– the enzymes. Virtually, all the chemical reactions of organic biomolecules are catalyzed by the enzymes. *Nearly all enzymes are globular proteins.* Chemically, some enzymes are

PROTEINS

- Fibrous
  - **SIMPLE** *(Insoluble in water)*
    - **Scleroproteins**
      Collagen
      Elastin
      Keratin
      Fibroin
  - **CONJUGATED** *(Insoluble in water)*
    - Pigment in chicken feather?
- Globular
  - **SIMPLE**
    - *(Soluble in water)*
      - **Protamines** clupeine, salmine, sturine, cyprinine
      - **Histones** nucleohistones of nuclei, globin
      - **Albumins** leucosine, legumeline, ovalbumin, lactalbumin
      - **Pseudoglobulins** pseudoglobulin from milk whey
    - *(Insoluble in water)*
      - **Euglobulins** ovoglobulin, myosinogen, glycinine, legumine
      - **Glutelins** glutenin, oryzenin
      - **Prolamines** gliadin, zein, hordein
  - **CONJUGATED**
    - **Metalloproteins** siderophilin ceruloplasmin
    - **Chromoproteins** hemoglobin hemocyanin cytochromes flavoproteins
    - **Glycoproteins** egg albumin
    - **Mucoproteins** ovomucoid, mucin
    - **Phosphoproteins** casein, ovovitellin
    - **Lipoproteins** lipovitellin
    - **Nucleoproteins** nucleohistones from nuclei, nuclein

simple proteins, containing only amino acid residues; others are complex proteins, containing a major protein part (apoenzyme) and a small nonprotein part (prosthetic group) associated with the protein unit. Enzymes catalyze a variety of reactions. *Urease, amylase, catalase, cytochrome C, alcohol dehydrogenase* are some of the examples of enzymic proteins.

2. **Structural proteins.** The structural proteins are usually inert to biochemical reactions. They maintain the native form and position of the organs. The cell wall and primary fibrous constituents of the cell have structural proteins. *Collagen*, which has very high tensile strength, is the most abundant protein of animals. It is found in connective tissue such as tendons, cartilage, matrix of bones and cornea of the eye. *Leather is almost pure collagen.* Ligaments contain *elastin*, a structural protein capable of stretching in two dimensions. α *keratin* constitutes almost the entire dry weight of hair, wool, feathers, nails, claws, quills, scales, horns, hooves, tortoise shell and much of the outer layer of skin. The major component of silk fibres and spider webs is *fibroin*. The wing hinges of some insects are made of *resilin*, which has nearly perfect elastic properties.

3. **Transport or carrier proteins.** Certain proteins, specially in the animals, are involved in the transport of many essential biological factors to various parts of the organisms. *Hemoglobin* of erythrocytes binds O_2 as the blood passes through the lungs, carries it to the peripheral tissues, and there releases it to participate in the oxidation of nutrients. The blood plasma contains *lipoproteins*, which carry lipids from the liver to other organs. Other kinds of transport proteins are present in the plasma membranes and intracellular membranes of all organisms. *Ceruloplasmin* transports copper in blood.

4. **Nutrient and storage proteins.** *Ovalbumin* is the major protein of eggwhite. The milk protein, *casein* stores amino acids. The seeds of many plants store nutrient proteins, required for the growth of the germinating seedlings. *Ferritin*, found in some bacteria and in plant and animal tissues, stores iron.

5. **Contractile or motile proteins.** Some proteins endow cells and organisms with the ability to contract, to change shape, or to move about. *Actin* and *myosin* function in the contractile system of skeletal muscle and also in many nonmuscle cells. *Tubulin* is the protein from which microtubules are built.

6. **Defense proteins.** Many proteins defend organism against invasion by other species or protect them from injury. The *antibodies* (or immunoglobulins), the specialized proteins made by the lymphocytes of vertebrates, can precipitate or neutralize invading bacteria, viruses or foreign proteins from another species. *Fibrinogen* and *thrombin*, although enzymic, are blood-clotting proteins that prevent loss of blood when the vascular system is injured.

7. **Regulatory proteins.** Some proteins help regulate cellular or physiological activity. Among them are many hormones, such as *insulin* which regulates sugar metabolism, and *growth hormone* which is required for bone growth. The cellular response to many hormonal signal is often mediated by a class of GTP-binding proteins called *G proteins*. Other regulatory proteins bind to DNA and regulate the biosynthesis of enzymes and RNA molecules involved in cell division.

8. **Toxic proteins.** Some proteins act as toxic substances, such as *snake venom, bacterial toxins* and toxic plant proteins like *ricin*. These toxic proteins also have defensive functions.

EGG PROTEINS

The eggs and milk are consumed as food by an appreciable portion of the global population. Hence, the egg and milk proteins deserve a special mention. Eggs contain 2 types of proteins: the egg white protein and egg yolk protein. The various components present in the egg white protein are listed in Table 10–4.

Table 10–4. Protein composition of egg white

Protein	*Percentage*
Total protein	10-11% (on wet basis) 82.8% (on dry basis)
Ovalbumin	70% of total proteins
Conalbumin	9%
Ovomucoid	13%
Globulins	
Lysozyme (G_1)	2.6%
(G_2) (G_3)	7%
Mucin	2%
Avidin	0.06%

(*Adapted from Fevold HL, 1951*)

When egg yolk is diluted with water, proteins precipitate. When the yolk is heated, the proteins undergo heat denaturation and precipitate. Egg yolk contains 2 phosphoproteins, lipovitellin and lipovitellenin, which differ from each other in the following characteristics:

	Lipovitellin	***Lipovitellenin***
Lipid concentration:	17-18%	36-41% (mainly lecithins)
Phosphorus concentration:	1%	0.29%
Protein present:	Vitellin	Vitellenin

Egg yolk also contains water-soluble protein that does not precipitate on dilution of the yolk. This fraction is called ***livetin***. The enzymatic activity of yolk lies in this fraction.

Whole egg is an excellent food because it is a very rich source not only of protein and lipid but also of most of the vitamins (except vitamin C) and most of the required minerals (except calcium).

MILK PROTEINS

Milk (Fig. 10–2) contains about 0.6–0.7% protein which is not precipitated on acidification to pH 4.7. This represents about 20% of the protein contained in skim milk. These ***whey proteins*** are separated into 2 fractions: ***lactalbumin*** and ***lactoglobulin***.

The name ***casein*** is assigned to the fraction precipitated by acidifying milk to a pH of 4.7. It is present in cow's milk (3–3.5%) and human milk (0.3–0.6%). Casein may be further purified by redissolving and precipitating again. It is of 3 types: α, β and γ. These differ one another in their molecular weights, their rate of migration in an electric field and their phosphorus content.

Fig. 10–2. Milk, a source of nourishment for all mammals, is composed, in part, of a variety of proteins.

REFERENCES

See list following Chapter 11.

PROBLEMS

1. Which amino acids are usually present in low amount in plant proteins?
2. Which group of proteins is a heterogeneous group and has properties characteristic of the elastomers?
3. How tendons are distinguished from ligaments?
4. Why is a classification of globular proteins totally based on their structure not possible?
5. Why are scleroproteins best suited for animal skeleton?
6. What is casein?
7. Following is the average composition of milk:

Water	87.30	Ash	0.73
Fat	3.67	Solids	12.69
Total protein	3.42	Solids not fat	8.77
Lactose	4.78		

(*After Jacobs MB, 1951*)

Some people maintain that milk is the perfect food and must be included in the diet of all people for proper nutrition. Is this statement correct? Provide a biochemical basis for your answer.

8. In Oriental and Black adults, symptoms such as abdominal distention, cramps, pain and diarrhea develop on consumption of milk. Explain.

ANSWERS TO PROBLEMS

1. Threonine (Thr), Methionine (Met), Lysine (Lys), Tryptophan (Trp).
2. Fibrous (or fibrillar) proteins
3. Tendons connect muscle to bone, whereas ligaments attach one bone to another. Tendons are nonelastic but ligaments can be stretched, because they contain, in addition to collagen fibres, the protein elastin.
4. A simple functional classification of globular proteins is not possible because of 2 reasons:
   (a) these proteins perform a variety of different functions.
   (b) many widely-differring globular proteins perform almost similar functions.
5. Scleroproteins (albuminoids) are simple fibrous proteins; they have the axial ratios greater than 10 and henceforth resemble long ribbons or fibres in shape. Moreover, these are insoluble in all common solvents such as water, dilute acids, alkalis and salts and also in organic solvents. Also, these are not attacked by enzymes. Hence, scleroproteins are best suited for animal skeleton and serve in a structural or protective role.
6. The name casein is assigned to the fraction precipitated by acidifying milk to a pH of 4.7. It is present in cow's milk (3–3.5%) and human milk (0.3–0.6%). Casein may further be purified by redissolving and precipitating again. It is of 3 types : α, β, and γ. These differ one another in their molecular weights, their rate of migration in an electric field and their phosphorus content.
7. From a nutritional standpoint, people need a balanced diet to meet their biochemical requirements for energy, and to provide building blocks and cofactors for biosynthesis. These requirements can be met by dietary intake of a large variety of sources; there is no evidence that a specific food (*e.g.*, milk) is required for normal nutrition.
8. This is due to a substance, lactose. The ability to secrete the enzyme *lactase*, which hydrolyzes lactose, decreases with age. The unabsorbed lactose passes through the small intestine to the large intestine, where it is fermented by intestinal bacteria causing distention, pain, and diarrhea.

CONTENTS

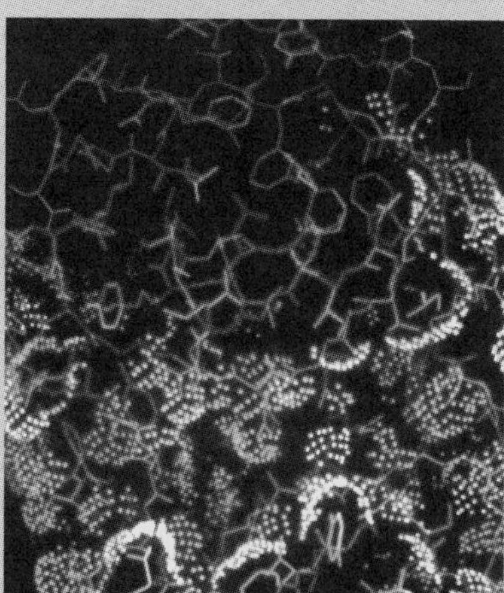

Computer-generated model of the protein pepsin

This enzyme protein helps in the digestion of food ingested by living beings.

CHAPTER

11

Proteins–III

General Properties

The general properties of proteins are reminiscent of those of the amino acids.

PHYSICAL PROPERTIES

1. Colour and Taste. Proteins are colourless and usually tasteless. These are homogeneous and crystalline.

2. Shape and Size. As already discussed, the proteins range in *shape* from simple crystalloid spherical structures to long fibrillar structures. Two distinct patterns of shape have been recognized :

A. Globular proteins—These are spherical in shape and occur mainly in plants, esp., in seeds and in leaf cells. These are bundles formed by folding and crumpling of protein chains.

e.g., pepsin, edestin, insulin, ribonuclease etc.

B. Fibrillar proteins—These are thread-like or ellipsoidal in shape and occur generally in animal muscles. Most of the studies regarding protein structure have been conducted using these proteins.

e.g., fibrinogen, myosin etc.

Each protein molecule is characterized for its specific size (Fig. 11–1). For example:

(*a*) Hemoglobin has a diameter of 55 Å.

(*b*) Edestin has a diameter of 80 Å.

(*c*) Catalase has dimensions of 80 × 64 × 54 Å (of the axes).

(*d*) Human fibrinogen has a diameter of 38 Å and a length of 700 Å.

(*e*) Collagen is one of the longest proteins, with a length of 3,000 Å.

In general, the protein molecules are always very large, as can be seen in the following examples:

(a) *Gliadin* (from wheat)— $C_{685}H_{1068}N_{196}O_{211}S_5$

(b) *Zein* (from corn)— $C_{736}H_{1161}N_{184}O_{208}S_3$

(c) *Casein* (from milk)— $C_{708}H_{1130}N_{180}O_{224}S_4P_4$

(d) *Beta-lactoglobulin* (from milk)— $C_{1642}H_{2652}O_{492}N_{420}S_{18}$; mol.wt. 36,684

3. Molecular Weight. The extraordinary size, poor stability, specific solubility conditions and high reactivity have rendered the determination of molecular weight of proteins as a difficult task (**Edsall**, 1953). However, the proteins generally have large molecular weights ranging between 5×10^3 and 1×10^6 (see Table 11–1). It might be noted that the values of molecular weights of many proteins lie close to or multiples of 35,000 and 70,000. Previously, this was interpreted as a regularity under the name *Svedberg's rule*. Also, it was then assumed that proteins are composed of units of molecular weight 17,500. This corresponds to about 145–150 amino acid residues, since the average molecular weight of an amino acid residue amounts to about 115–120. The discovery in recent times of too many exceptions to this rule, however, finally forced its abandonment.

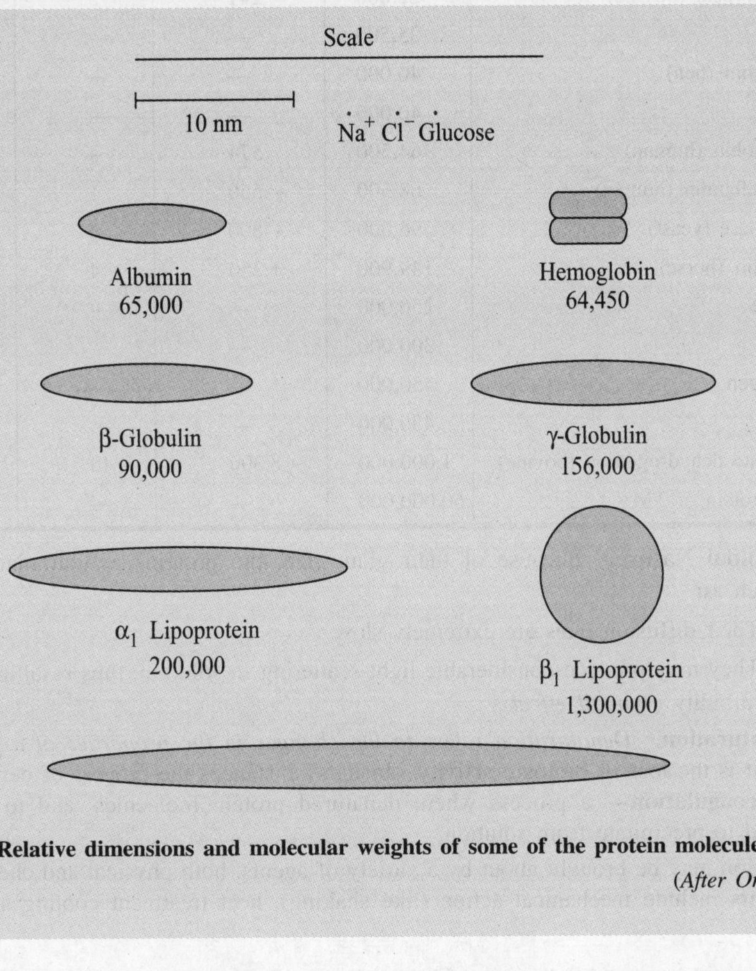

Fig. 11–1. Relative dimensions and molecular weights of some of the protein molecules in the blood
(*After Oncley JL, 1949*)

The approximate number of amino acid residues in a simple protein having no prosthetic group can be calculated by dividing its molecular weight by 110. The average molecular weight of the 20 amino acids is about 138. But as the smaller amino acids predominate in most proteins, hence the average molecular weight of an amino acid is nearer 128. Since a molecule of water (MW = 18) is eliminated to produce each peptide bond, the average molecular weight of the amino acid residue is about 128–18 = 110. Table 11–1 also gives the number of amino acid residues present in different proteins.

Table 11–1. Exact / Approximate molecular weights and the isoelectric points of some important proteins

S.N.	*Protein*	*Molecular weight (MW)*	*Number of residues*	*Number of chains*	*Isoelectric point (pI)*
1.	Insulin (bovine)	5,733	51	2	5.4
2.	Cytochrome C	12,500	104	—	9.8
3.	Ribonuclease (bovine)	14,000	124	1	7.8/9.5
4.	Lysozyme (eggwhite)	14,600	129	1	11.0
5.	Myoglobin (horse)	16,700	153	1	7.0
6.	Chymotrypsin (bovine)	22,600	241	3	—
7.	Pepsin	35,500	—	—	2.7
8.	Ovalbumin (hen)	40,000	—	—	4.6
9.	Zein	40,000	—	—	—
10.	Hemoglobin (human)	64,500	574	4	—
11.	Serum albumin (human)	68,500	~ 550	1	4.9
12.	Hexokinase (yeast)	96,000	~ 800	4	—
13.	γ-globulin (horse)	149,900	~ 1,250	4	6.6
14.	Catalase	250,000	—	—	5.6
15.	Edestin	300,000	—	—	6.9
16.	Fibrinogen	450,000	—	—	5.5
17.	Urease	480,000	—	—	5.0
18.	Glutamate dehydrogenase (bovine)	1,000,000	~ 8,300	~ 40	—
19.	Virus protein of TMV	60,000,000	—	—	—

4. Colloidal Nature. Because of their giant size, the proteins exhibit many colloidal properties, such as:

I. Their diffusion rates are extremely slow.

II. They may produce considerable light-scattering in solution, thus resulting in visible turbidity (*Tyndall effect*).

5. Denaturation. *Denaturation refers to the changes in the properties of a protein.* In other words, it is the loss of biologic activity. In many instances the process of denaturation is followed by **coagulation**— a process where denatured protein molecules tend to form large aggregates and to precipitate from solution.

Denaturation may be brought about by a variety of agents, both physical and chemical. The **physical agents** include mechanical action (like shaking), heat treatment cooling and freezing operations, rubbing, high hydrostatic pressures, (5,000 to 10,000 atm.), ultraviolet rays, etc. The **chemical agents**, that cause denaturation, are many ionizing radiations (like X-rays, radioactive and ultrasonic radiations), organic solvents (acetone, alcohol), aromatic anions (salicylates), some anionic detergents (like sodium dodecyl sulfate), etc. A common example of protein easily denatured by shaking or heat is the albumin of eggwhite.

$$H_3C—(CH_2)_{10}—CH_2OSO^-Na^+$$

Sodium dodecyl sulfate, SDS

It was suggested by **Wu** (1931) that denaturation leads mainly to the unfolding of the peptide chain, thus causing disorganization of the internal structure of protein (Fig. 11–2). This is evidenced by the fact that the denatured proteins are more easily hydrolyzed (Mirsky, 1935).

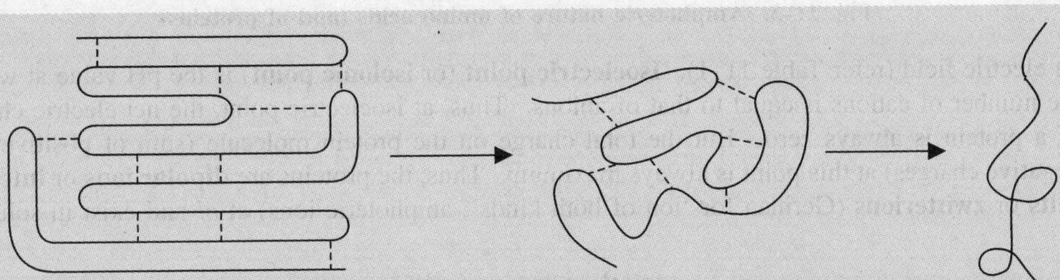

Fig. 11–2. Representation of denaturation of a protein

The process of denaturation breaks bonds.

When the peptide chains or the protein molecules are unrolled, certain bonds split and new sites of bundles are exposed to the action of certain proteolytic enzymes causing hydrolysis. Thus, the H-bonds linking the 2 peptide chains are partly freed and the disulfide (—S—S—) bonds also linking the two peptide chains split open to yield the free sulfhydryl (—SH) groups.

According to **Putnam** (1953), the proteins, on denaturation, undergo following changes :

1. Decrease in their solubility.
2. Cessation of their biochemical activity as enzymes or hormones.
3. Decrease in size and shape of the molecule.
4. Increased activity of some radicals present in the molecule such as —SH group of cysteine, —S—S— bond of cystine and phenolic group of tyrosine.

Further, on denaturation, new ionizable groups become available for acid-base titration (**Steinhardt** and **Zaiser**, 1955) and also there occurs a change in optical rotation in the direction of increased levorotation (**Simpson** and **Kauzmann**, 1953). Denaturation also leads to alteration in surface tension and loss of antigenecity.

Some proteins, when denatured, cannot be brought back to their original state. In that case denaturation is described as of '*irreversible*' type. On the other hand, denaturation in other proteins is of '*reversible*' type. For example, if trypsin is exposed to a temperature of 80–90°C, it denatures and when this solution is cooled at 37°C, the solubility and the activity of this protein-enzyme is regained. The process of regaining normal protein properties by a denatured protein is called **renaturation** or **refolding**. During renaturation, certain antibodies may cause a re-rolling of the protein bundles so that most of the original bonds are recovered (**Pauling**, 1940). The recovery of the renatured protein is, however, never complete.

6. Amphoteric Nature. Like amino acids, the proteins are *amphoteric, i.e.*, they act as acids and alkalies both (Fig. 11–3). These migrate in an electric field and the direction of migration depends upon the net charge possessed by the molecule. The net charge is influenced by the pH value. Each protein has a fixed value of isoelectric point (*pI*) at which it will move in

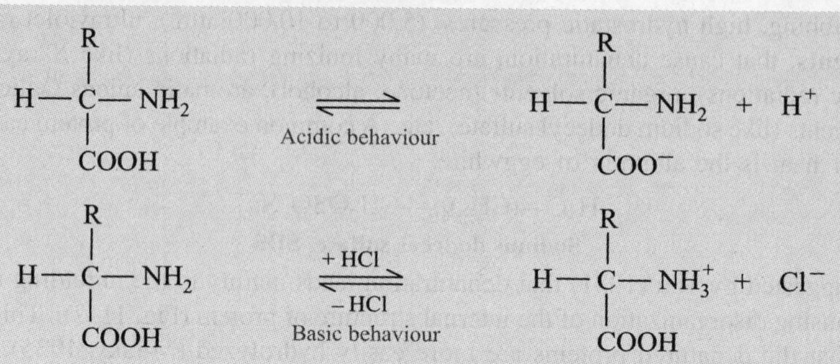

Fig. 11–3. Amphoteric nature of amino acids (and of proteins)

an electric field (refer Table 11–1). **Isoelectric point** (or **isoionic point**) is the pH value at which the number of cations is equal to that of anions. Thus, at isoelectric point, the net electric charge of a protein is always zero. But the total charge on the protein molecule (sum of positive and negative charges) at this point is always maximum. Thus, the proteins are **dipolar ions** or **internal salts** or **zwitterions** (German for 'ion of both kinds'; amphoteric ions) at *pl* and exist in solution as:

$$(H_3N^+)_m—R—(COO^-)_n$$

This general structure of amino acids, represented as an inner salt, was originally proposed by **N. Bjerrum** in 1923. The formula depicts the carboxyl group as being dissociated while the amino group is protonated.

At pH values lower than *pl*, the protein will have a net positive charge and, as a cation, will migrate towards negative pole (cathode). Similarly, at pH values higher than *pl*, the protein will have a net negative charge and, as an anion, will move towards positive pole (anode).

It may be stated in general that those proteins having an excess of carbonyl groups will tend to have a lower *pl* while those having an excess of amino groups will tend to have a higher *pl*.

The osmotic pressure and viscosity of the protein solution are a minimum at the isoelectric point. Also at the isoelectric point, proteins are found to be least soluble and can be precipitated most easily.

7. Ion Binding Capacity. Being amphoteric in nature, the proteins can form salts with both cations and anions based on their net charge. In fact, a mixture of different proteins at a given pH (except at *pl*) will include cations and anions both and the salts of protein-protein combinations will be formed. This occurs in tissues since both acidic and basic proteins are present.

Many ions form insoluble salts with proteins and serve as excellent precipitating agents for proteins. For example, anions of some acids like phosphotungstic, trichloroacetic, picric etc., form insoluble salts with proteins when the latter behave **as cations** (acid side of their *pl*).

Heavy metals are used for precipitating proteins on the alkaline side of their *pl*, the proteins behaving **as anions**. Ions of Hg, Cu, Ag, Zn etc., are frequently used for this purpose. Many acid dyes find practical use for colouring the insoluble proteins like silk and wool.

8. Solubility. The solubility of proteins is markedly influenced by pH. *Solubility is lowest at isoelectric point and increases with increasing acidity or alkalinity.* This is because when the protein molecules exist as either cations or anions, repulsive forces between ions are high, since all the molecules possess excess charges of the same sign. Thus, they will be more soluble than in the isoelectric state.

A. *'Salting-in' effect.* Globulins are sparingly soluble in water but their solubility is greatly increased by the addition of neutral salts like NaCl. This phenomenon is commonly described as 'salting-in' effect.

B. '*Salting-out*' *effect*. Proteins are precipitated from aqueous solution by high concentrations of neutral salts. This is the 'salting-out' process. Divalent and trivalent ions are more effective than univalent ions. The salts commonly used for this purpose are $Na_2.SO_4$, $(NH_4)_2.SO_4$, magnesium salts and phosphates.

C. *Isoelectric precipitation*. Some proteins like casein of milk, however, are readily precipitated at or near their isoelectric point. This process is, therefore, described as isoelectric precipitation.

9. Optical Activity. All protein solutions rotate the plane of polarized light to the left, *i.e.*, these are *levoratotory*. For example, the specific rotation $[\alpha]_D$ for ovalbumin is near —30° over the *p*H range between 3.5 and 11. However, at lower or higher pH values the rotation becomes more negative, *e.g.*, at pH 13, the $[\alpha]_D$ is about —60°. The rotation is further increased by subjecting proteins to high temperatures.

CHEMICAL PROPERTIES

A. HYDROLYSIS

Proteins are hydrolyzed by a variety of hydrolytic agents.

1. By acidic agents. Proteins, upon hydrolysis with conc. HCl (6–12N) at 100–110°C for 6 to 20 hrs, yield amino acids in the form of their hydrochlorides. Undesirable side-effects of acid hydrolysis include the following:

(*a*) Tryptophan, serine and threonine are destroyed during acid hydrolysis and as such this reaction is not used for their isolation.

(*b*) Asparagine and glutamine are deamidated to aspartate and glutamate respectively.

(*c*) Glutamic acid undergoes *intramolecular dehydration* to pyrollidone 5-carboxylic acid.

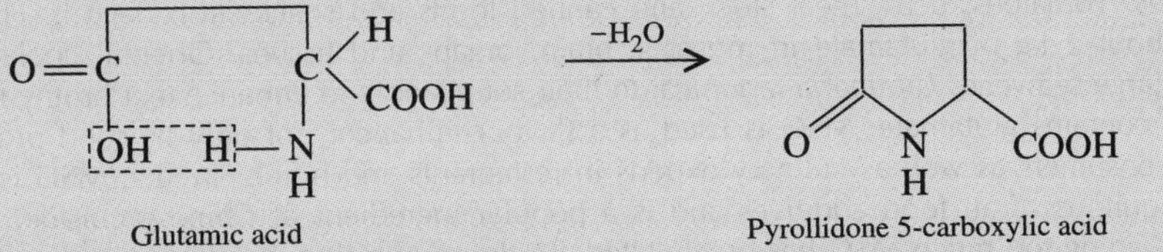

(*d*) Other amino acids may undergo *intermolecular dehydration* forming cyclic anhydrides or diketopiperazines.

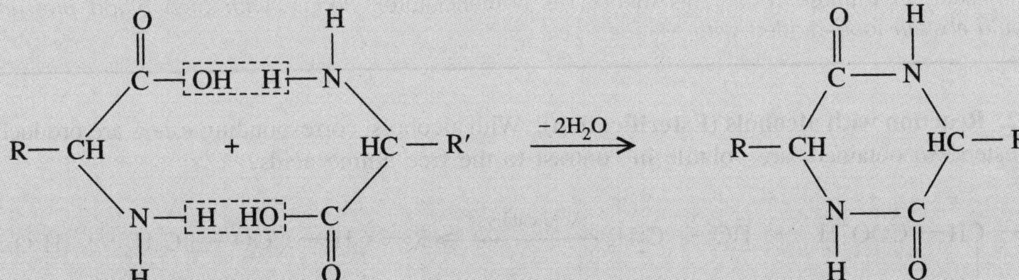

2. By alkaline agents. Proteins may also be hydrolyzed with 2N NaOH. Alkaline hydrolysis is, however, less used as it is highly disadvantageous:

(*a*) It leads to the destruction of certain amino acids like arginine, cysteine, cystine, serine, threonine etc.

(*b*) It also causes loss of optical activity (or racemization) of the amino acids.

3. By proteolytic enzymes. Under relatively mild conditions of temperature and acidity, certain proteolytic enzymes like pepsin and trypsin hydrolyze the proteins. Enzyme hydrolysis is used for the isolation of certain amino acids like tryptophan. Two important drawbacks with this type of hydrolysis are:

(*a*) It requires prolonged incubation.
(*b*) Hydrolysis is incomplete.

B. REACTIONS INVOLVING COOH GROUP

1. Reaction with alkalies (Salt formation). The carboxylic group of amino acids can release a H^+ ion with the formation of carboxylate (COO^-) ions. These may be neutralised by cations like Na^+ and Ca^{2+} to form salts. Thus, amino acids react with alkalies to form *salts*.

$$\underset{\displaystyle NH_2}{\underset{|}{R-CH}}-COO\boxed{H\ +\ HO}Na \longrightarrow \underset{\displaystyle NH_2}{\underset{|}{R-CH}}-COO\cdot Na + H_2O$$

Sodium salt of glutamic acid (monosodium glutamate) is used commercially as a flavouring agent. It imparts a meat-like flavour to soups, for example.

> **Monosodium glutamate (MSG)** or 'ajinomoto' in common parlance, or the 'taste powder' as some people call it, has a long history of use in China. Like sugar or salt, it is regarded as a fairly standard part of cuisine. It makes a great difference to the way the food tastes. MSG actually represents a new taste - the meaty, broth-like taste, common to foods as diverse as steak, lobster and tomato. *Virtually, all proteins contain glutamates, including mother's milk.* Most glutamates are `bound' and have no effect on the flavour but 'free' glutamates increase food's palatability. Every 100 g of Parmesan cheese, for instance, contains 9,847 mg of bound and 1,200 mg of free glutamates. Tomatoes, peas, meat stock and canned foods and snacks also have it. Human body carries about 2 kg of glutamate in muscles, brain, organs and tissues. Oriental cooks have long added a large seaweed, *Laminaria japonica*, to their soup stocks to enhance that brothy flavour. The seaweed contains glutamate. MSG is used as a flavour enhancer not only by food processors but also by consumers as well as culinary experts in restaurants and hotels. In fact, MSG is one of the most ubiquitous of all food additives and is a popular ingredient of Chinese cuisine. Consuming food containing too much MSG has precipitated attacks of sweating, headaches and gastrointestinal disorders in individuals, sensitive to the chemical. The disease characterized by the above symptoms is called **Kwok's disease** or the '**Chinese Restaurant Syndrome (CRS)**. Kwok, a patron of eateries, described people experiencing numbness of the back of the neck, general weakness and palpitations after eating in Chinese restaurants (hence, its nomenclature). *People with high blood pressure should abstain foods added with MSG.*

2. Reaction with alcohols (Esterification). With alcohols, corresponding *esters* are produced. The esters, so obtained, are volatile in contrast to the free amino acids.

$$\underset{\text{Amino Acid}}{\underset{\displaystyle NH_2}{\underset{|}{R-CH}}-COO}\boxed{H\ +\ HO}-\underset{\text{Ethyl alcohol}}{C_2H_5} \xrightarrow{\text{Acid catalyst}} \underset{\text{Ethyl ester of amino acid}}{\underset{\displaystyle NH_2}{\underset{|}{R-CH}}-COO-C_2H_5} + \underset{\text{Water}}{H_2O}$$

The reaction was, for the first time, used by **Emil Fischer** for the isolation of amino acids in pure form from protein hydrolysates by the fractional distillation in vacuum of their ethyl esters.

3. Reaction with amines. Amino acids react with amines to form *amides*.

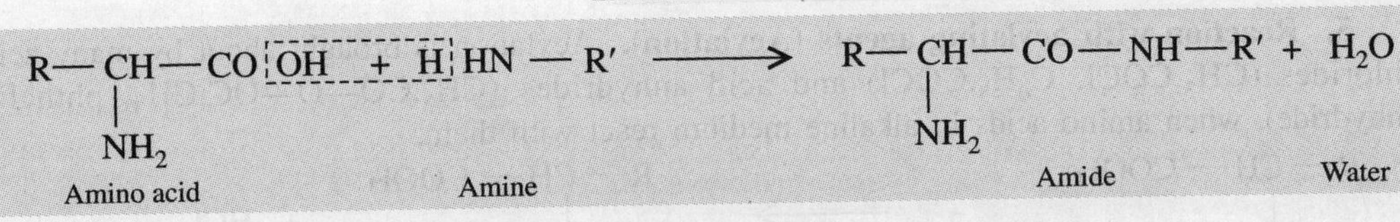

C. REACTIONS INVOLVING NH_2 GROUP

1. Reaction with mineral acids (Salt formation). When either free amino acids or proteins are treated with mineral acids like HCl, the *acid salts* are formed.

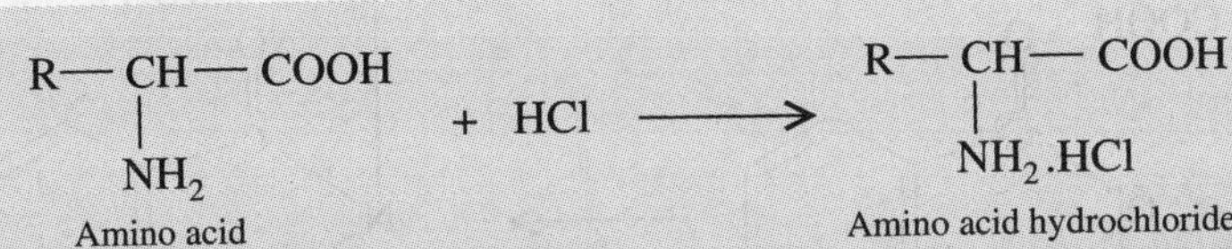

The basic amino acids, arginine and lysine react with CO_2 in the presence of air to form carbonate salts. Because of this property, these are usually stored and also sold in the form of their monochlorides.

2. Reaction with formaldehyde. With formaldehyde, the *hydroxy-methyl* derivatives are formed. These derivatives are insoluble in water and resistant to attack by microorganisms. Because of this action, formaldehyde is the principal reagent in embalming fluids and is used to harden and preserve certain fibres (Aralac, Vicara) obtained from globular proteins.

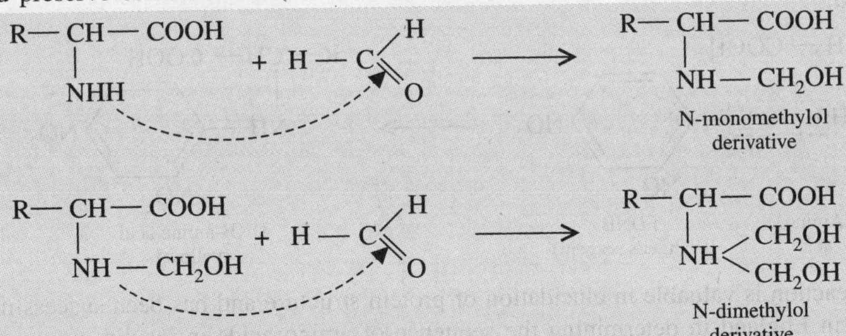

This reaction is the basis of the Sorensen titration method for determining the purity of the individual amino acids.

3. Reaction with benzaldehyde. *Schiff's* bases are formed.

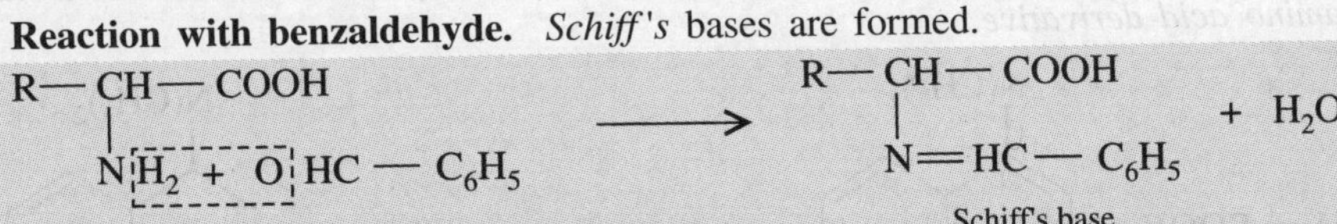

4. Reaction with nitrous acid (Van Slyke reaction). The amino acids react with HNO_2 to liberate N_2 gas and to produce the corresponding *α-hydroxy acids*.

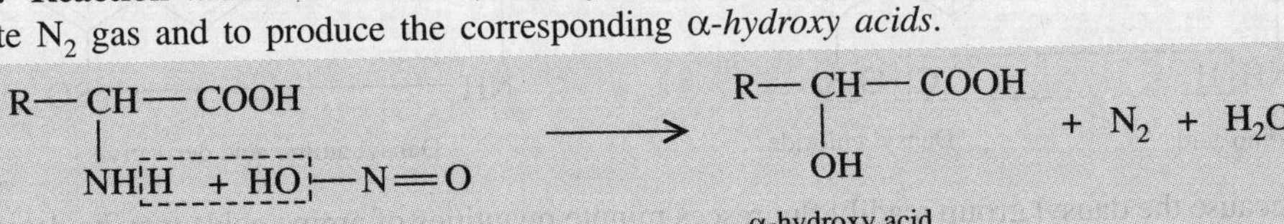

This reaction is characteristic of aliphatic primary amines and has been utilized by **Van Slyke** (1912) as the basis for his *`nitrous acid' method* for the estimation of amino acids by measuring the volume of N_2 gas liberated. The imino acids proline and hydroxyproline, however, do not respond to this reaction.

5. Reaction with acylating agents (Acylation). Acylation is brought about by many acid chlorides ($CH_3.COCl$, $C_6H_5.COCl$) and acid anhydrides ($CH_3.CO—O—OC.CH_3$, phthalic anhydride), when amino acids in alkaline medium react with them.

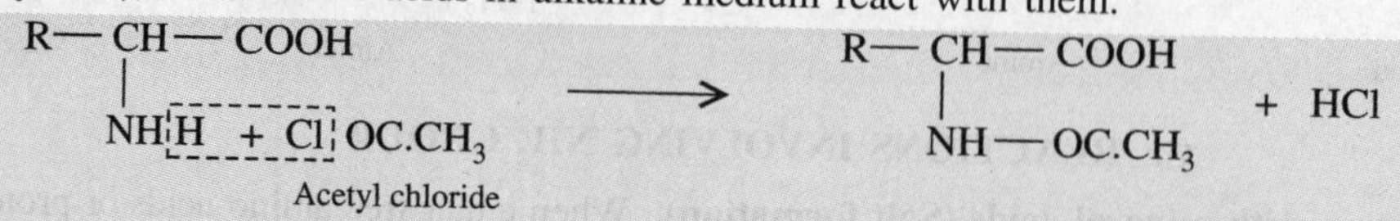

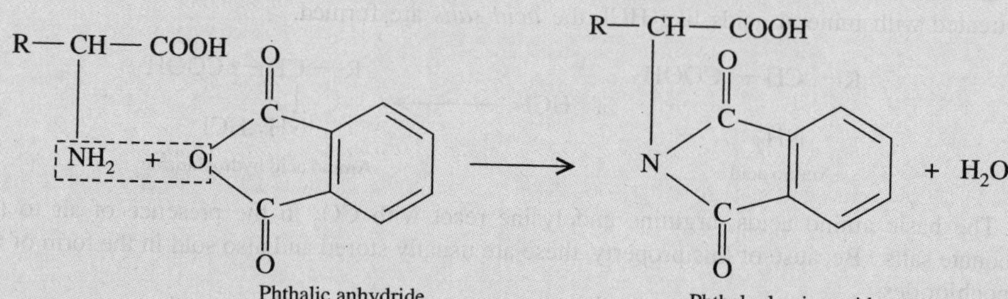

6. Reaction with FDNB or **Sanger's reagent.** In mildly alkaline solution, FDNB (1-fluoro- 2, 4-dinitrobenzene) reacts with α-amino acids to produce yellow coloured derivative, *DNB-amino acid.*

> FDNB is also called as **DNFB** (2, 4-dinitrofluorobenzene).

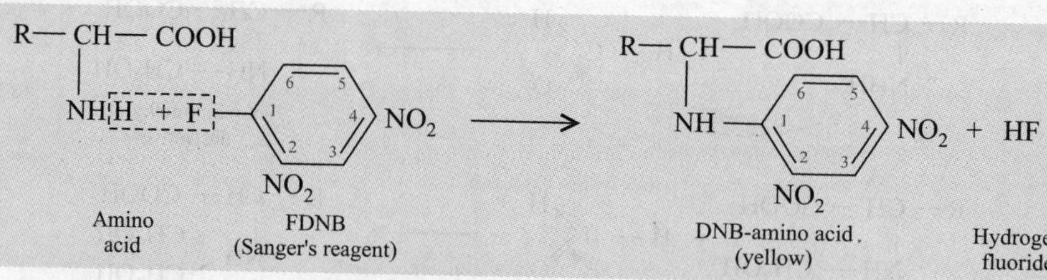

This reaction is valuable in elucidation of protein structure and has been successfully uitlized by Sanger in England in determining the sequence of amino acids in insulin.

7. Reaction with dansyl chloride. The N-terminal amino acid of the protein also combines with 1-dimethylaminonaphthalene-5-sulfonyl chloride (or dansyl chloride) to form a fluorescent *dansyl amino acid derivative.*

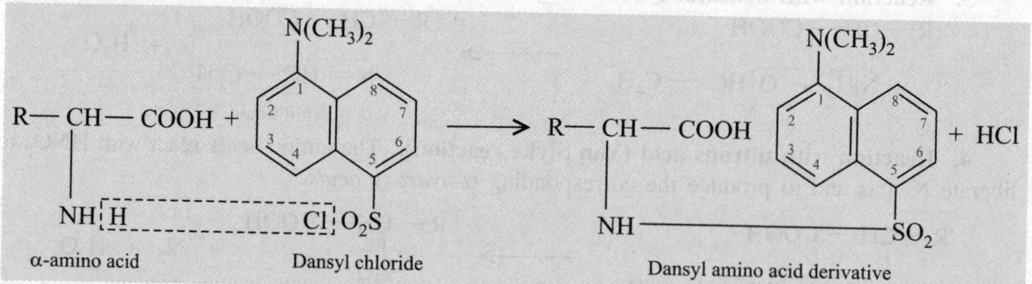

Because the dansyl group readily fluoresces minute quantities of amino acids may be determined by this procedure.

D. REACTIONS INVOLVING BOTH COOH AND NH_2 GROUPS

1. Reaction with triketohydrindene hydrate (Ninhydrin reaction). Ninhydrin (= indane-1,2,3,-trione hydrate) is a powerful oxidizing agent and causes oxidative decarboxylation of α-amino acids producing CO_2, NH_3 and an aldehyde with one less carbon atom than the parent amino acid. The reduced ninhydrin (or hydrindantin) then reacts with the liberated NH_3 and a mole of ninhydrin, forming blue-coloured Ruheman's complex.

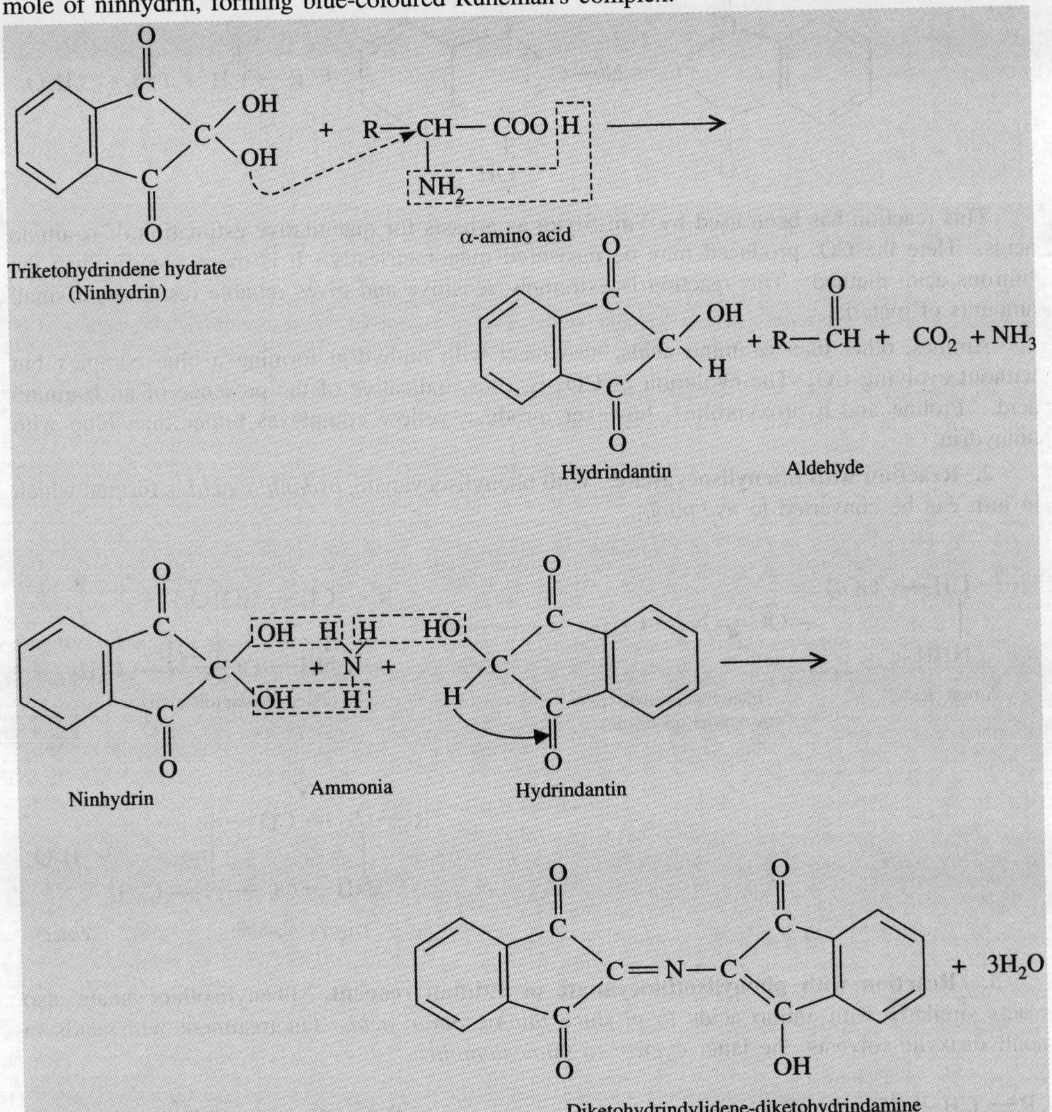

The net equation may, thus, be written as follows:

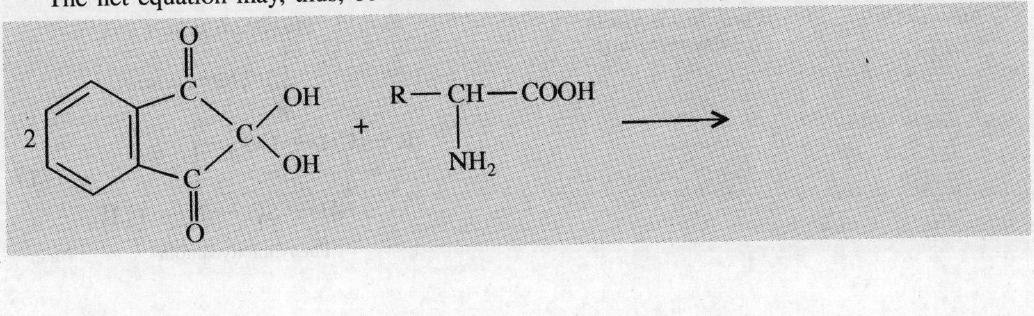

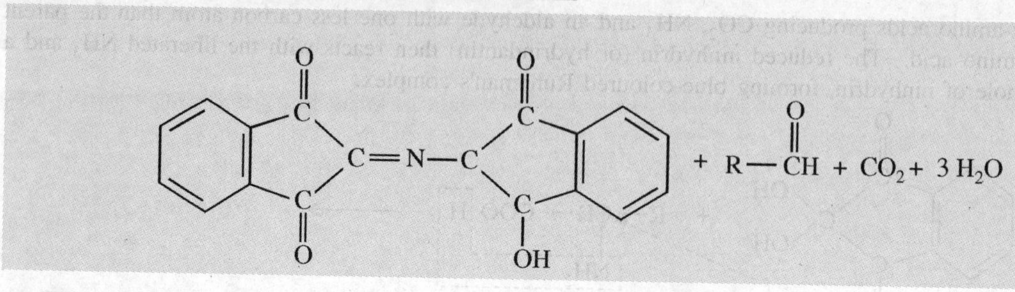

This reaction has been used by **Van Slyke** as a basis for quantitative estimation of α-amino acids. Here the CO_2 produced may be measured manometrically. It is more specific than the 'nitrous acid' method. This reaction is extremely sensitive and gives reliable results with small amounts of material.

Amines, other than α-amino acids, also react with ninhydrin forming a blue complex but without evolving CO_2. The evolution of CO_2 is, thus, indicative of the presence of an α-amino acid. Proline and hydroxyproline, however, produce yellow complexes rather than blue with ninhydrin.

2. Reaction with phenylisocyanate. With phenylisocyanate, *hydantoic acid* is formed which in turn can be converted to *hydantoin*.

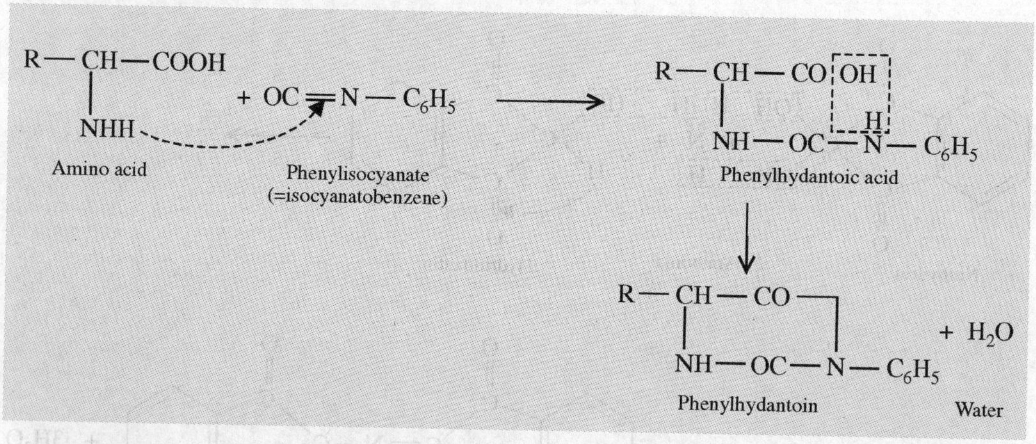

3. Reaction with phenylisothiocyanate or Edman reagent. Phenylisothiocyanate also reacts similarly with amino acids to produce *thiohydantoic acid*. On treatment with acids in nonhydroxylic solvents, the latter cyclize to *thiohydantoin*.

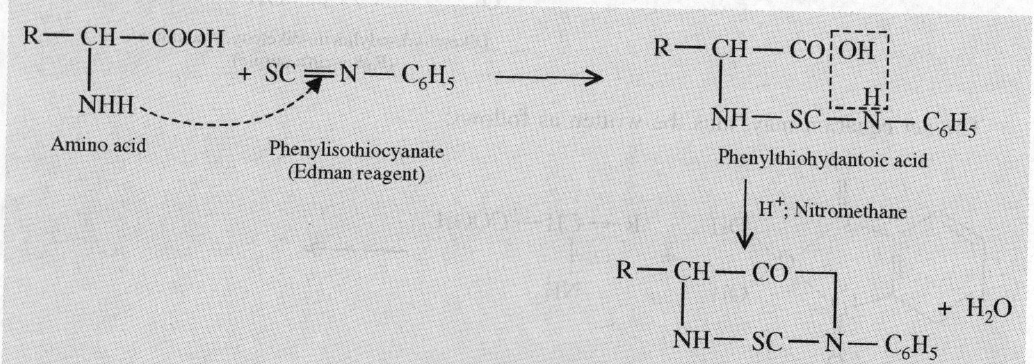

This reaction has proved useful in the studies of protein structure.

4. Reaction with phosgene. With phosgene, *N-carboxyanhydride is formed.*

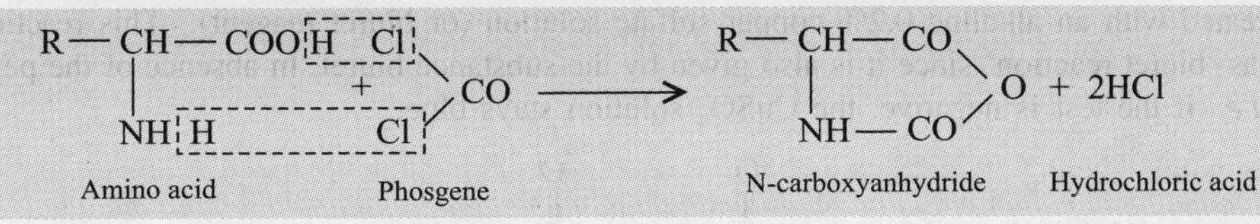

Phosgene, $COCl_2$ (first synthesized by **Sir Humphrey Davy**, in 1812, by passing CO and chlorine through charcoal) is an acidic chloride which does not occur in nature. Phosgene can form by burning of several compounds, most important of which is polyurethane. It is a colourless gas with a musty penetrating odour resembling fresh brown hay or green corn. It liquefies at 8°C and has m.p. –127.9°C. It is a highly poisonous gas which is lethal at 50 ppm if inhaled for a few minutes and is second only to arsine (AH_3) gas which kills at 10 ppm concentration (the equivalent figures for 2 other deadlier gases, chlorine and carbon monoxide, are 400 and 1,000, respectively). It gives urea (NH_2CONH_2) and ammonium chloride (NH_4Cl) with ammonia. It is stored as triphosgene, *bis*(trichloromethyl)carbonate which decomposes to phosgene on heating. Phosgene is an important industrial chemical being used as an intermediate in the synthesis or manufacture of isocyanates, polyurethane, polycarbonate resins, aniline dyes, pharmaceuticals, plastics and insecticides. It is also used in the 'uranium enrichment' process. Because of its widespread use in industry, there is always a danger of workers getting dangerously exposed to it. Currently, it is generally agreed that a maximum allowable concentration (MAC) of phosgene should be 1.0 ppm of gas, which means that during a 10-hour continuous working period, there should be no more than 0.1 ppm of the gas in the environment (0.1 ppm x 10 hours = 1 ppm). This gas represents one of the many hazards to fire-fighting personnel and fire victims for similar reasons. Welding in a poorly-ventillated, confined space in the presence of chlorinated solvents or other halides can cause phosgene to form rapidly in lethal concentrations. Since phosgene is heavier (3.43 times) than air, it was used extensively in gas warfare in World War I, and caused 80% of the deaths by gas in that conflict. Luckily, gas warfare was sparingly used in World War II, thus reducing the number of casualties.

Phosgene was also implicated in the Bhopal tragedy that occurred in India on the night between 2nd and 3rd December 1984. In fact, methyl isocyanate (MIC) was one of the main gases involved in the tragedy. MIC, in turn, is made from phosgene and methylamine and is used as an intermediate product in the manufacture of carbamate insecticide Carbaryl or Sevin, the chemical which Union Carbide Industry Limited (UCIL), Bhopal used to make. On that fateful night, nearly 50,000 pounds of stored MIC was released in vapour and liquid forms over a period of 2 hours. Because of the hypothesized reactions that took place within the storage tank and in the surrounding atmosphere, it is thought that MIC, phosgene, and hydrogen cyanide all played a significant role in this disaster.

Phosgene damages the body in several ways. It slowly hydrolyzes in mucus-membrane water to produce CO_2 and HCl. Because this hydrolysis tends to occur slowly, phosgene is less irritating to the mucus membrane of the upper airway and eyes and therefore penetrates more deeply into the lungs. Formation of HCl within the lungs causes damage to lung cells. This causes water to ooze out of the blood into the lungs. This, in turn, causes lungs to become waterlogged, a condition medically known as *pulmonary edema.* Different concentrations of phosgene affect humans in diverse ways. In concentrations of 3 ppm, phosgene causes dryness and burning of the throat. At 4 ppm, it causes irritation in, and redness and watering of, eyes. At 5–10 ppm, it causes coughing. Exposure to 25 ppm for 30 minutes is very dangerous and even brief exposures to 50 ppm or higher is rapidly fatal. About 80% of mortalities occur within 48 hours of initial exposure, which may be considered as its fatal period.

It is variously called *carbon oxychloride, carbonic dichloride, carbonyl chloride.*

5. Reaction with carbon disulfide. With carbon disulfide, *2-thio-5-thiozolidone* is produced.

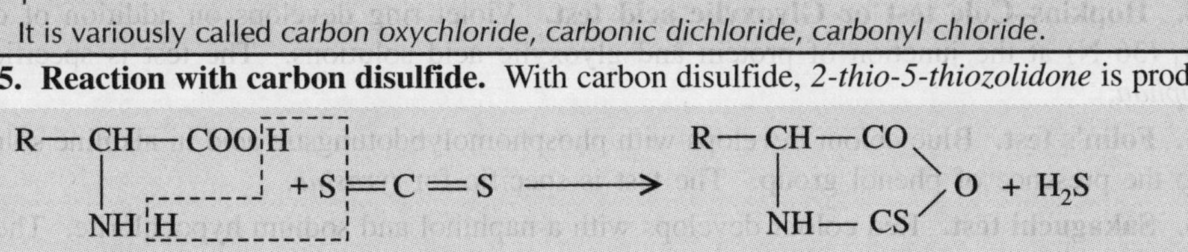

E. REACTIONS INVOLVING R GROUP OR SIDE CHAIN

1. Biuret test. Compounds containing peptide bonds produce a characteristic purple colour when treated with an alkaline 0.2% copper sulfate solution (or biuret reagent). This reaction is termed as 'biuret reaction' since it is also given by the substance biuret. In absence of the peptide bonds, *i.e.*, if the test is negative, the $CuSO_4$ solution stays blue.

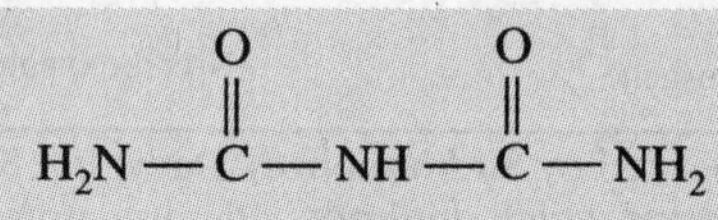

Biuret

The colour deepens as the number of peptide bonds is increased and the proteins produce a deep blue-violet colour due to the probable formation of a coordination complex whose structure is given below:

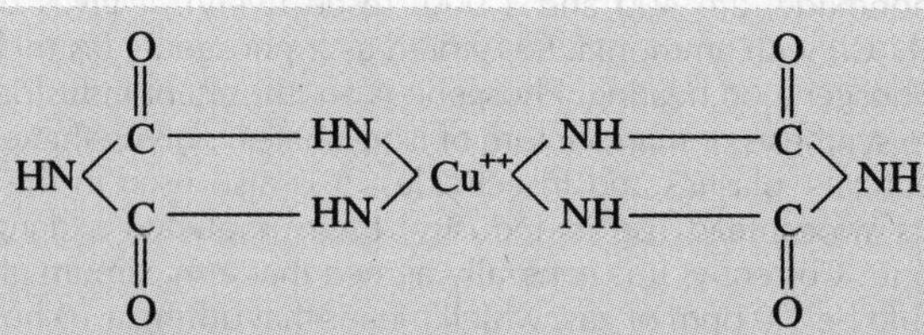

The test is, in fact, given by biuret as well as any similar structure having 2 amide or peptide bonds linked directly or through an intermediate carbon atom. The required unit is shown below between the two broken lines:

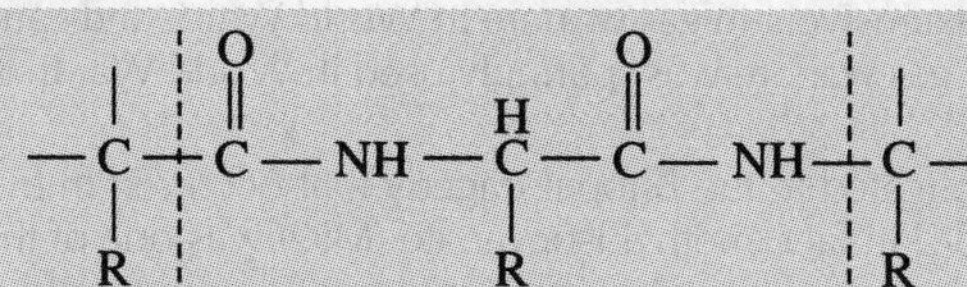

All proteins except dipeptides, therefore, respond to this reaction. This reaction is widely used both as a qualitative test for the detection of proteins and as a quantitative measure of protein concentration.

2. Xanthoproteic test. Yellow colour develops on boiling proteins with conc. HNO_3 due to the presence of benzene ring. This reaction is due to the nitration of the phenyl rings (of *tyrosine, tryptophan* and *phenylalanine*) to yield yellow substitution products, which turn orange upon addition of alkali due to formation of salts of nitro derivatives.

$$\text{Protein} + HNO_3 \xrightarrow{\text{Xanthoproteic test}} \underset{\text{(= Picric acid)}}{\text{Trinitrophenol}} + NO\uparrow$$

3. Millon's test. Red colour develops when proteins are heated with $Hg.NO_3$ in HNO_2. The reaction is specific for *tyrosine* and takes place between mercuric and mercurous nitrates and tyrosine residues of the protein. *Tryptophan* also responds to this reaction.

4. Hopkins–Cole test or Glyoxylic acid test. Violet ring develops on addition of conc. H_2SO_4 (36 N) at the junction of protein and glyoxylic acid solutions. The test is specific for *tryptophan.*

5. Folin's test. Blue colour develops with phosphomolybdotungstic acid in alkaline solution due to the presence of phenol group. The test is specific for *tyrosine.*

6. Sakaguchi test. Red colour develops with a-naphthol and sodium hypochlorite. The test is applied for the detection of *arginine.*

7. Pauly test. Red colour develops with diazotized sulfanilic acid in alkaline solution. The reaction is specific for *tyrosine* and *histidine*.

8. Ehrlich test. With *p*-dimethylaminobenzaldehyde in 12 N HCl, *tryptophan* develops a blue colour.

F. REACTIONS INVOLVING SH GROUP

1. Nitroprusside test. Red colour develops with sodium nitroprusside in dilute NH_4OH. The test is specific for *cysteine*.

2. Sullivan test. *Cysteine* develops red colour in the presence of sodium 1, 2-naphthoquinone-4-sulfonate and sodium hydrosulfite.

The various colour reactions for different amino acids are summarized in Table 11–2.

Table 11–2. Colour reactions for specific amino acids

S.N.	*Test*	*Reagent*	*Colour*	*Amino acid(s)*
1.	Biuret test	Alkaline $Cu.SO_4$ soln.	Violet	Peptide bonds
2.	Xanthoproteic test	Conc. HNO_3	Yellow	Tyrosine Tryptophan Phenylalanine
3.	Millon's test	$Hg.NO_3$ in HNO_2; heat	Red	Tyrosine Tryptophan
4.	Hopkins–Cole test	Glyoxylic acid + conc. H_2SO_4	Violet	Tryptophan
5.	Folin's test	Alkaline phosphomolybdotungstic acid	Blue	Tyrosine
6.	Sakaguchi test	α-naphthol + sodium hypochlorite	Red	Arginine
7.	Pauly test	Diazotized sulfanilic acid in alkaline soln.	Red	Tyrosine Histidine
8.	Ehrlich test	*p*-dimethylaminobenzaldehyde in 12 N HCl	Blue	Tryptophan
9.	Nitroprusside test	Sodium nitroprusside + dil. NH_4OH	Red	Cysteine
10.	Sulivan test	Sodium 1,2-naphthoquinone-4-sulfonate + sodium hydrosulfite	Red	Cysteine

REFERENCES

1. **Agard DA :** To fold or not to fold. *Science*. **260 :** *1903-1904, 1993.*
2. **Anfinsen CB (editor) :** Aspects of Protein Biosynthesis. *Academic Press, Inc., New York. 1970.*
3. **Anfinsen CB :** Principles that govern the folding of polypeptide chains. *Science.* **181:** *223–230, 1973.*
4. **Annual Review :** Advances in Protein Chemistry. *Academic Pres, Inc., New York. 1944-current.*
5. **Barrett GC (editor) :** Chemistry and Biochemistry of the Amino Acids. *Chapman and Hall, New York. 1985.*

6. **Bell JE, Bell ET :** Proteins and Enzymes. *Prentice-Hall Inc., New Jersey. 1988.*
7. **Blake CCF, Koenig DF, Mair GA, North ACT, Phillips DC, Sarma VR :** Structure of the Hen Egg-white Lysozyme. A Three Dimensional Fourier Synthesis at 2 Å Resolution. *Nature* **206 :** *760, 1965.*
8. **Bosch L :** Mechanism of Protein Synthesis and its Regulation. *Elsevier Publishing Co., New York. 1972.*
9. **Boulter D :** Protein synthesis in plants. *Ann. Rev. Plant Physiol.* **21** *: 1, 1970.*
10. **Branden C, Tooze J :** Introduction to Protein Structure. *Garland, New York. 1991.*
11. **Bunn HF, Forget BG :** Hemoglobin: Molecular, Genetic and Clinical Aspects. *W.B. Saunders, Philadelphia, Pa. 1986.*
12. **Burke DC :** The status of interferon. *Sci. Amer.* **236** *(April) : 42, 1977.*
13. **Cantor CR, Schimmel PR :** Biophysical Chemistry. Pt 1. The Conformation of Biological Molecules. *Freeman, San Francisco. 1980.*
14. **Carlier M-F :** Actin: protein structure and filament dynamics. *J. Biol. Chem.* **266 :** *1–4, 1991.*
15. **Clark BFC, Marcker KA :** How Proteins Start. *Sci. Amer.,* **218 (1) :** *36–42, 1968.*
16. **Creighton TE :** Disulfide bonds and protein stability. *Bioessays.* **8 :** *57-63, 1988.*
17. **Creighton TE :** Proteins: Structure and Molecular Properties . *2nd ed., W.H. Freeman, New York. 1993.*
18. **Croft LR :** Handbook of Protein Sequences. *2nd ed., John Wiley, New York. 1980.*
19. **Dayhoff MO (editor) :** Atlas of Protein Structure and Sequence. *Vol. 5. Supple. 1Ä3. National Biomedical Research Foundation, Washington D.C. 1972-79.*
20. **Dickerson RE, Geis I :** Hemoglobin : Structure, Function, Evolution, and Pathology. *Benjamin/Cummings, Menlo Park, California. 1982.*
21. **Dickerson RE, Geis I :** Proteins : Structure, Function, and Evolution. *2nd ed., Benjamin/ Cummings, California. 1983.*
22. **Doolittle RF :** Proteins. *Scientific American.* **253 (4) :** *88-99, 1985.*
23. **Fermi G, Perutz MF :** Atlas of Molecular Structures in Biology. Vol. 2 Hemoglobin and Myoglobin. *Oxford Univ. Press, New York. 1981.*
24. **Fox JL, Deyl Z, Blazej A (editors) :** Protein Structure and Evolution. *Marcel Dekker, New York. 1976.*
25. **Fraser RDB :** Keratins. *Sci. Amer.* **221(2):** *86-96, 1969.*
26. **Goldberg ME :** Protein folding and assembly. *Trends in Biochemical Sciences.* **10 :** *388, 1985.*
27. **Goldenberg D :** Genetic studies of protein stability and mechanisms of folding. *Ann. Rev. Biophys. Chem.,* **17 :** *481, 1988.*
28. **Greenstein JP, Winitz M :** Chemistry of the Amino Acids. *Vol. 3. Wiley. 1961.*
29. **Gross J :** Collagen. *Scientific American* **204 (5) :** *120, 1961.*
30. **Hartwig JH, Kwiatkowski DJ :** Actin-binding proteins. *Curr. Opin. Cell Biol.* **3 :** *87–97, 1991.*
31. **Haschemeyer RH, Haschemeyer AEV :** Proteins: A Guide to Study by Physical and Chemical Methods. *John Wiley and Sons Inc., New York. 1973.*
32. **Haurowitz F :** The Chemistry and Function of Proteins. *2nd ed., Academic Press, Inc., New York. 1963.*

33. **Hughes RC :** Glycoproteins. *Chapman and Hall, London. 1983.*
34. **Hulmes DJ :** Building collagen molecules, fibrils and supra-fibrillar structures. *J. Struct. Biol.* **137 :** *2–10, 2002.*
35. **Kadler KE, Baldock, Bella J, Boot-Handford RP :** Collagens at a glance. *J. Cell Science.* **120 :** *1955–1958, 2007.*
36. **Kendrew JC :** The Three-Dimensional Structure of a Protein Molecule. *Scientific American* **205 (6) :** *96–110, 1961.*
37. **Kendrew JC :** Myoglobin and the structure of Proteins. *Science.* **139 :** *1259, 1963.*
38. **Kilmartin JV :** The Bohr effect of human hemoglobin. *Trends in Biochemical Sciences.* **2 :** *247, 1977.*
39. **Koshland DE Jr :** Protein Shape and Biological Control. *Sci. Amer.* **229 :** *52–64, 1973.*
40. **Kovacs GG, Budka H :** Prion diseases: from protein to cell pathology. *Am. J. Pathol.* **172 :** *555–565, 2008.*
41. **Kyte J :** Structure in Protein Chemistry. *Garland. 1994.*
42. **Linderström-Lang K :** Proteins and Enzymes. *Lane Lectures No. 6, p.93. Univ., Press, Stanford, California. 1952.*
43. **Meister A :** Biochemistry of the Amino Acids. *2nd ed., Vols. 1 and 2. Academic Press, Inc., New York. 1965.*
44. **Monod J, Wyman J, Changeux J :** On the nature of allosteric transitions: A plausible model. *J. Mol. Biol.* **12 :** *88–118, 1965.*
45. **Neurath H, Hill RL (editors) :** The Proteins. *3rd ed., Academic Press, Inc., New York. 1975–1977.*
46. **Pauling LC, Corey RB, Hayward R :** The structure of protein molecules. *Scientific American.* **191(1) :** *51, 1954.*
47. **Pauling LC, Itano HA, Singer SJ, Wells IC :** Sickle cell anemia: A molecular disease. *Science.* **110 :** *543–548, 1949.*
48. **Perutz MF :** Proteins and Nucleic Acids: Structure and Function. *Elsevier Publishing Co., New York. 1962.*
49. **Perutz MF :** The hemoglobin molecule. *Scientific American.* **211(5) :** *64, 1964.*
50. **Perutz MF :** Hemoglobin Structure and Respiratory Transport. *Sci. Amer.* **239(6) :** *92-125, 1978.*
51. **Perutz MF :** Regulation of oxygen affinity of hemoglobin: Influence of the globin on the heme iron *Ann. Rev. Biochem.* **48 :** *327, 1979.*
52. **Perutz MF :** Molecular anatomy, physiology and pathology of hemoglobin. In Stamatayonnopoulos C (editor) : Molecular Basis of Blood Diseases. Saunders. 1987.
53. **Perutz MF :** Mechanisms of cooperativity and allosteric regulation in proteins. *Quart. Rev. Biophys.* **22 :** *139–237, 1989.*
54. **Perutz MF, Fermi G, Luisi B, Shaanan B, Liddington RC :** Stereochemistry of cooperative mechanisms in hemoglobin. *Acc. Chem. Res.* **20 :** *309–321, 1987.*
55. **Ramachandran GN (editor) :** Aspects of Protein Structure. *Academic Press, Inc., New York. 1963.*
56. **Richards FM :** The protein of folding problem. *Sci. Amer.* **264 (1) :** *54–63, 1991.*
57. **Richardson JS :** The Anatomy and Taxonomy of Protein Structure. *Adv. Protein Chem.* **34 :** *168, 1981.*

58. **Robson B :** Protein folding. Trends in Biochemical *Sciences.* **1 :** *49, 1976.*
59. **Rossman MG, Argos P :** Protein Folding. *Ann. Rev. Biochem* **50 :** *497, 1981.*
60. **Sanger F :** Sequences, sequences and sequences. *Ann. Rev. Biochem.* **57 :** *1–28, 1988.*
61. **Scheraga HA :** Protein Structure. *Academic Press, Inc., New York. 1963.*
62. **Schroder E, Lubke K :** The Peptides. *Vols.1 and 2. Academic Press, Inc., New York. 1965–66.*
63. **Schrosder WA :** The Primary Structure of Proteins. *Harper and Row, New York. 1968.*
64. **Schultz GE, Schirmer RH :** Principles of Protein Structure. *Springer-Verlag, New York. 1990.*
65. **Scott JE :** Molecules for strength and shape. Trends Biochem. *Sci.* **12 :** *318–321, 1987.*
66. **Shkundina IS, Ter-Avanesyan MD :** Prions. *Biochemistry (Moscow).* **72 :** *1519–1536, 2007.*
67. **Srinivasan PR, Fruton JS, Edsall JT (editors) :** The origins of modern biochemistry: A retrospective on proteins. *Ann. N.Y. Acad. Sci. 325, 1979.*
68. **Stein WH, Moore S :** The Chemical Structure of Proteins. *Scientific American.* **204(2) :** *81, 1961.*
69. **Tschesche H (editor) :** Modern Methods in Protein Chemistry. *de Gruyter, New York. 1983.*
70. **Wilson AC :** The molecular basis of evolution. *Sci. Amer.* **253 :** *164–173, 1985.*
71. **Wyman J, Gill SJ :** Binding and Linkage. Functional Chemistry of Biological Molecules. *University Books. 1990.*
72. **Zuckerlandl E :** The evolution of hemoglobin. *Sci. Amer.* **212(5) :** *110, 1965.*

PROBLEMS

1. After proper purification, the Edman reaction was used to sequence a dodecapeptide. The following data were obtained. The C-terminal amino acid is isoleucine; N-terminal amino acid is methionine; peptide fragments are Ala–Ala–Ile, Leu–Arg–Lys–Lys–Glu–Lys–Glu–Ala, Met–Gly–Leu, and Met–Phe–Pro–Met. What is the sequence of this peptide?
2. Why is pepsin useful as a digestive aid in precooked foods?
3. Why eggs become harder on boiling?
4. What is Edman reagent? Describe its use in the determination of amino acid sequence in protein.
5. Why do eggs solidify after boiling while 'ghee' melts after heating?

ANSWERS TO PROBLEMS

1. You must have the correct N- and C-terminal amino acids and the correct total number (12) of amino acids in the peptide. Keeping these criteria in mind and overlapping the fragments, you should get the sequence: Met–Phe–Pro–Met–Gly–Leu–Arg–Lys–Glu–Ala–Ala–Ile. Note that the Lys–Glu fragment doesn't give any additional information for the sequencing.
2. Pepsin partially digests (hydrolyzes) the protein in foods.
3. The eggs become harder on boiling due to coagulation of natural proteins, present in egg.

4. Phenylisothiocyanate (C_6H_5–N=C=S) is Edman reagent. It is used in the determination of N-terminal amino acid in a protein. Phenylisothiocyanate reacts with the $-NH_2$ group of amino acid to produce phyenylthiohydantoic acid which, on treatment with mild acids, results in the formation of phenylthiohydantoin (see reaction on page 272). Phenylthiohydantion can be separated and compared with the corresponding known derivative of a known amino acid. This method is quite useful because it results in the removal and identification of the NH_2^- terminoid amino acid together with the production of a polypeptide containing one amino acid less.
5. *Eggs* contain rich proteins composed of many amino acids. On heating eggs, their proteins coagulate into a solid mass. *Ghee* is a mixture of more than one fatty acids, which may be saturated or unsaturated. These fatty acids soften at high temperatures and subsequently melt. In the case of eggs, the reaction is irreversible, but it is reversible in the case of ghee.

CONTENTS

CHAPTER

12

Lipids–I
General Structure

Triacylglycerols fuel the long migration flights of the golden plover (*Pluvialis dominica*).

(*Courtesy : Gerard Fuehrer*)

IMPORTANCE

The lipids are important constituents of the diet because of their high energy value and also because of the fat-soluble vitamins and the essential fatty acids found with the fat of the natural foodstuffs. In the body, the fats serve as efficient source of energy which is stored in the adipose tissues. They also serve as an insulating material in the subcutaneous tissues and around certain organs. Fats combined with proteins (lipoproteins) are important constituents of the cell membranes and mitochondria of the cell. Animals are considerably richer (15 times) in lipids than the plants.

Proteins, polysaccharides, DNA and RNA are macromolecules. *Lipids are not generally classed as macromolecules, even though they share some of their features*: for example, most are synthesized as linear polymers of a smaller molecule (the acetyl group on acetyl-CoA), and they self-assemble into larger structures (membranes). It is noteworthy that water and protein comprise most of the mass of both mammalian and bacterial cells (Table 12–1).

Table 12–1. Approximate chemical composition of a typical bacterium and a typical mammalian cell

Component	Percent of total cell weight	
	E. coli bacterium	*Mammalian cell*
H_2O	70	70
Inorganic ions (Na^+, K^+, Mg^{2+}, Ca^{2+}, Cl^- etc)	1	1
Miscellaneous small metabolites	3	3
Proteins	15	18
RNA	6	1.1
DNA	1	0.25
Phospholipids	2	3
Other lipids	—	2
Polysaccharides	2	2
Total cell volume:	2×10^{-12} cm^3	4×10^{-9} cm^3
Relative cell volume:	1	2,000

DEFINITION

The lipids (*lipos*[G] = fat) are a heterogeneous group of compounds related to fatty acids and include fats, oils, waxes and other related substances. These are oily or greasy organic substances, relatively insoluble in water and considerably soluble in organic solvents like ether, chloroform and benzene. They are, thus, **hydrophobic** in nature. These are variously called as *lipins* or *lipoids*. The latter term is, however, sometimes used to refer "fat-like" substances which may not actually be related to the fatty acids. The term 'lipid' was first used by the German biochemist **Bloor** in 1943 for a major class of tissue components and foodstuffs. The structure of some common fats and fatty acids is presented in Fig.12–1.

Chemical Abstracts and some biochemical journals use the spelling **'lipide'**.

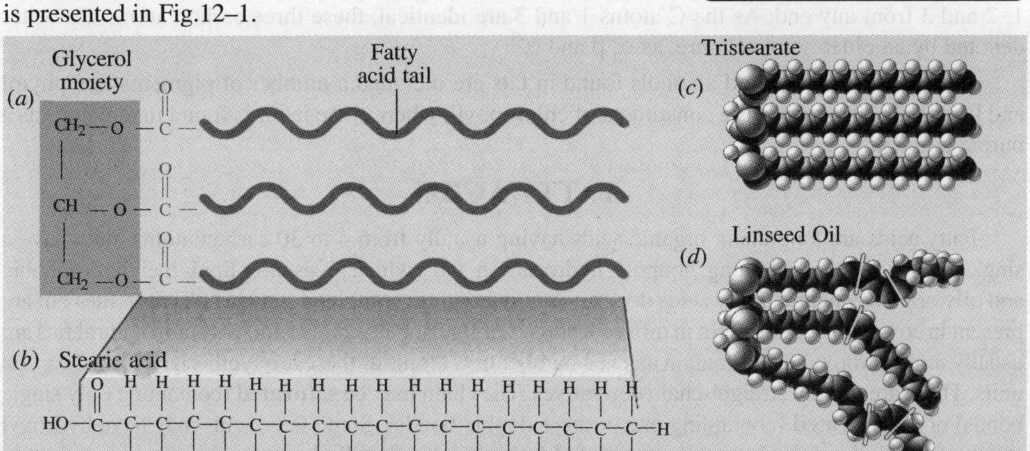

Fig. 12–1. Fats and fatty acids

(*a*) **Basic structure of a triacylglycerol (also called a triglyceride or a neutral fat)**. The glycerol moiety, indicated in *orange*, is linked by three ester bonds to the carboxyl groups of three fatty acids whose tails are indicated in *green*.

(*b*) **Stearic acid.** It is an 18-carbon saturated fatty acid that is common in animals fats.

(*c*) **Space-filling model of tristearate.** It is a triacylglycerol containing three identical stearic acid chains.

(*d*) **Space-filling model of linseed oil.** It is a triacylglycerol containing two different unsaturated fatty acids, derived from flax seeds. The sites of unsaturation, which produce kinks in the molecule, are indicated by the *yellow-orange* bars.

Chemically, the fats are defined as the esters of glycerol and fatty acids or as the triglycerides of fatty acids.

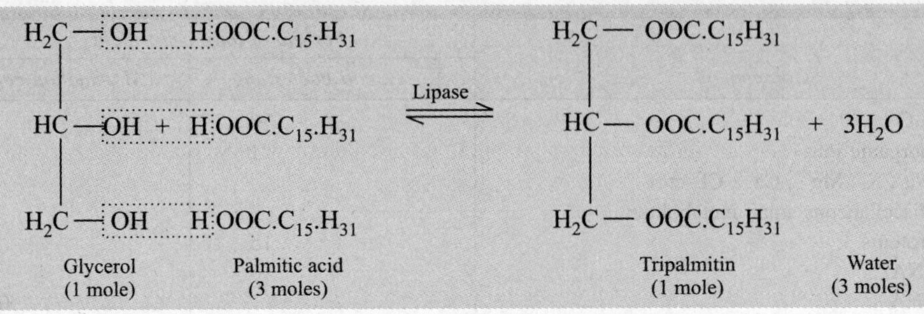

Both these components are described below for a better understanding of the lipids.

ALCOHOLS

Alcohols found in lipid molecules may be **saturated.** These commonly include glycerol, cholesterol and higher alcohols such as cetyl alcohol and myricyl alcohol.

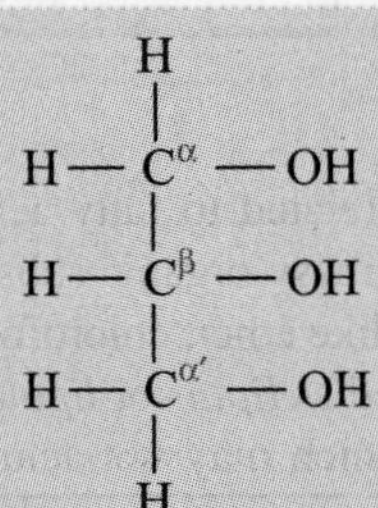

Glycerol

$CH_3(CH_2)_{14}CH_2OH$

Cetyl alcohol
(=Hexadecanol)

$CH_3(CH_2)_{28}CH_2OH$

Myricyl alcohol
(=Triacontanol)

In the structural formula of glycerol, the C atoms are designated by *stereospecific numbers* (*sn*), 1, 2 and 3 from any end. As the C atoms 1 and 3 are identical, these three carbon atoms are better denoted by an older nomenclature, as α, β and α'.

Among the **unsaturated** alcohols found in fats are included a number of pigments like phytol and lycophyll. The former is a constituent of chlorophyll, whereas the latter is found in tomatoes as a purple pigment.

FATTY ACIDS

Fatty acids are long-chain organic acids having usually from 4 to 30 carbon atoms; they have a single carboxyl group and a long, nonpolar hydrocarbon 'tail', which gives most lipids their hydrophobic and oily or greasy nature. Fatty acids do not occur in free or uncombined state in cells or tissues but are present in covalently bound form in different classes of lipids. Fatty acids which occur in natural fats are usually monocarboxylic and contain an even number of C atoms as these are synthesized from 2 carbon units. These are usually straight-chain derivatives. The chain may be **saturated** (containing only single bonds) or **unsaturated** (containing one or more double bonds). Some fatty acids may have hydroxyl group(s) in the chain (**hydroxy** or **oxygenated** fatty acids) and still others may possess ring structure (**cyclic** fatty acids). Fatty acids are stored as an energy reserve (fat) through an ester linkage to glycerol to form *triglycerides*. If free, the *carboxyl group* of a fatty acid will be ionized.

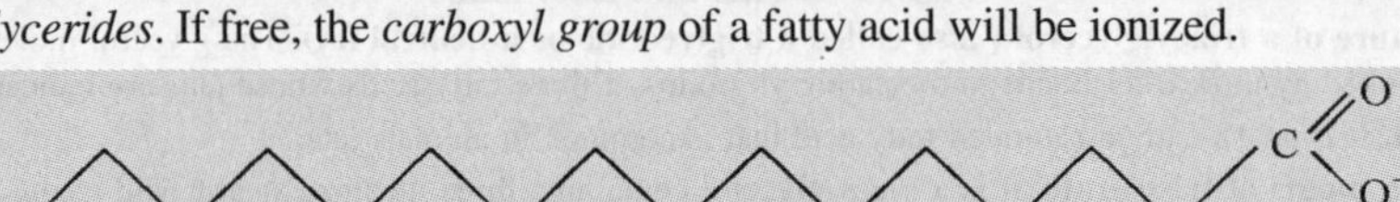

But more usually, it is linked to other groups to form either *esters*

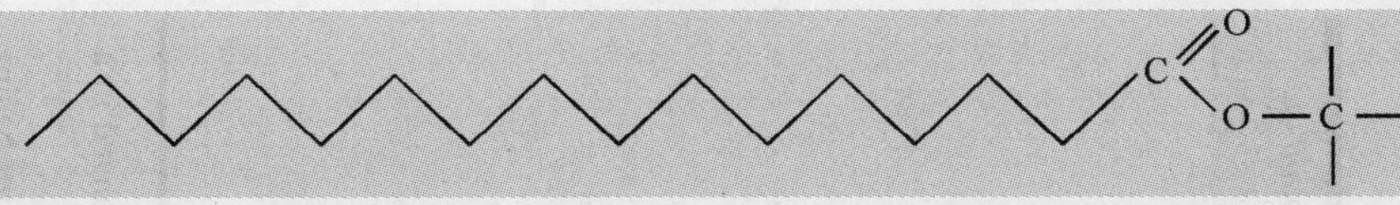

or *amides.*

Nomenclature. The systematic nomenclature of the fatty acids is based on the **Genevan system.** According to this system, the fatty acid is named after the hydrocarbon with the same number of carbon atoms, the suffix *-oic* is written in place of the final letter *e* in the name of the hydrocarbon. The names of saturated fatty acids end with the suffix *-anoic* and those of unsaturated acids with the suffix *-enoic*.

Generally, the numbers are used with the International Union of Pure and Applied Chemistry (IUPAC) **system of nomenclature** and the Greek letters when the common names are employed. For example :

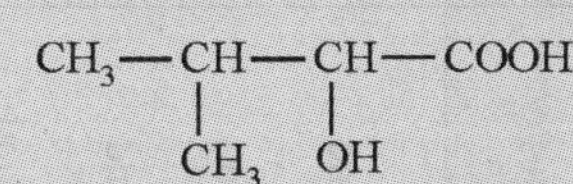

2-hydroxy-3-methylbutanoic acid
or
α-hydroxy-*iso*-valeric acid

The position of carbon atoms in the fatty acid chain is indicated either by **numbering** (in which case the carboxyl carbon is numbered as C1, the carbon adjacent to C1 as C2 and so on) or by the **use of Greek letters** (in which case C2 is denoted as α-carbon, C3 as β-carbon and so on). Thus:

$$R-\underset{\varepsilon}{\overset{6}{CH_2}}-\underset{\delta}{\overset{5}{CH_2}}-\underset{\gamma}{\overset{4}{CH_2}}-\underset{\beta}{\overset{3}{CH_2}}-\underset{\alpha}{\overset{2}{CH_2}}-\overset{1}{COOH}$$

A widely-used convention to indicate the number and position of the double bond(s) in the case of unsaturated fatty acids is to write the number of carbon atoms, the number of double bond(s) and the position of the double bonds(s) below the name of the acid. For example, oleic acid having 18 carbon atoms and a double bond between carbon atoms 9 and 10 is written as 18:1; 9. Similarly, linoleic acid (18 carbon atoms and 2 double bonds at C 9 and C 12) is written as 18:2; 9, 12.

The **position of a double bond** is indicated by the lower number of the two carbon atoms involved in double bonding.

An alternative method to write the name of an unsaturated fatty acid is to write first the position of double bond(s) in numerals and then the total number of carbon atoms in Roman followed by the suffix *-enoic* acid. Thus, oleic acid may be written as 9-octadecenoic acid and linoleic acid as 9, 12-octadecadienoic acid.

A description of the various categories of fatty acids involved in lipid formation follows.

1. Saturated fatty acids

The general formula for these acids is $C_nH_{2n+1}COOH$. Table 12–2. lists some even-numbered straight chain saturated fatty acids, found distributed in both plant and animal worlds. In addition, lipids from all sources contain small amounts of saturated fatty acids with an odd number of carbon atoms (C 5 through C 17). Generally, these odd-carbon acids account for less than 1% of the total fatty acids.

In animal fats, palmitic and stearic acids (C_{16} and C_{18}) are the most abundantly found saturated fatty acids, next in order are shorter chain fatty acids (C_{14} and C_{12}) and longer chain fatty acids (C_{20}, C_{22} and C_{24}). Fatty acids of 10 carbon atoms or less are present in limited amounts in animal lipids except the milk fat which contains appreciable amounts of lower molecular weight fatty acids. The preponderance of these acids may, thus, be shown in the descending order as below:

$$C_{16}, C_{18} > C_{14}, C_{12}, C_{20}, C_{22}, C_{24} > C_{10} \text{ and less.}$$

Table 12–2. Straight-chain saturated fatty acids, commonly found in natural fats

Trivial name	*Systematic name**	*Carbon skeleton*	*Structure†*	*Common source*	*Melting point (in °C)*	*Boiling point (in °C)*
Butyric	*n*-butanoic	4 : 0	$CH_3(CH_2)_2COOH$	Butter	–7.9	162
Caproic	*n*-hexanoic	6 : 0	$CH_3(CH_2)_4COOH$	Coconut and palm oils	–3.4	206
Caprylic	*n*-octanoic	8 : 0	$CH_3(CH_2)_6COOH$	Coconut and palm oils	16.7	237
Capric	*n*-decanoic	10 : 0	$CH_3(CH_2)_8COOH$	Coconut and palm oils	31.6	270
Lauric (*laurus*[L] = laurel plant)	n-dodecanoic	12 : 0	$CH_3(CH_2)_{10}COOH$	Laurel oil and spermaceti	44.2	299
Myristic (*Myristica*[L] = nutmeg genus)	*n*-tetradecanoic	14 : 0	$CH_3(CH_2)_{12}COOH$	Butter and wool fats	53.9	149[a]
Palmitic (*palma*[G] = palm tree)	*n*-hexadecanoic	16 : 0	$CH_3(CH_2)_{14}COOH$	Animal and plant fats	63.1	167[a]
Stearic (*stear* = hard fat)	*n*-octadecanoic	18 : 0	$CH_3(CH_2)_{16}COOH$	Animal and plant fats	69.6	184[a]
Arachidic (*Arachis*[L] = legume genus)	*n*-eicosanoic	20 : 0	$CH_3(CH_2)_{18}COOH$	Groundnut oil	76.5	204[a]
Behenic	*n*-docosanoic	22 : 0	$CH_3(CH_2)_{20}COOH$	Groundnut oil	79.9	306[b]
Lignoceric (*lignum*[L] = wood + *cera*[L] = wax)	*n*-tetracosanoic	24 : 0	$CH_3(CH_2)_{22}COOH$	Groundnut and rapeseed oils	86.0	—
Cerotic	*n*-hexacosanoic	26 : 0	$CH_3(CH_2)_{24}COOH$	Wool fat	87.7	—
Montanic	*n*-octacosanoic	28 : 0	$CH_3(CH_2)_{26}COOH$	—	—	—

* The **prefix *n***- indicates the "normal" unbranched structure. For instance, "octadecanoic" simply indicates 18 carbon atoms, which could be arranged in a variety of branched forms. Thus, *n*-octadecanoic specifies the linear, unbranched form.

† All **fatty acids** are shown in their unionized form. At pH 7, all free fatty acids have an ionized carboxylate. Note that the numbering of carbon atoms begins at the carboxyl-group carbon.

[a] at 1 mm Hg

[b] at 60 mm Hg

In addition to the straight chain fatty acids, a number of branched chain fatty acids having either an even or an odd number of carbon atoms have been identified as minor components of natural fats and oils. Table 12-3 lists a representative sample of these fatty acids.

Table 12–3. Some branched chain saturated fatty acids, identified in natural fats

Trivial	*Systematic name*	*No. of C atoms*	*Structure*	*Common source*
Isopalmitic	Isohexa-decanoic	16	CH_3 ╲ / CH_3 ╱ $CH(CH_2)_{12}COOH$	Wool fat
Anteiso-palmitic	14-methyl-hexadecanoic	17	CH_3CH_2 ╲ / CH_3 ╱ $CH(CH_2)_{12}COOH$	Wool fat
Tuberculo-stearic	D-(–) 10-methylocta-decanoic	19	$CH_3(CH_2)_7CH(CH_3)(CH_2)_8COOH$ (with CH_3 on the CH)	Bacteria

2. **Unsaturated fatty acids**

These may be classified, based on the *degree of unsaturation.*

A. Monoethenoid acids — These contain one double bond and conform to the general formula, $C_nH_{2n-1}COOH$. The common example is oleic acid.

B. Diethenoid acids — Two double bonds;
$C_nH_{2n-3}COOH$; Linoleic acid.

C. Triethenoid acids — Three double bonds;
$C_nH_{2n-5}COOH$; Linolenic acid.

D. Tetraethenoid acids — Four double bonds;
$C_nH_{2n-7}COOH$; Arachidonic acid

Monoethenoid acids are commonly called as *monounsaturated fatty acids (MUFAs)* and the remaining ones as *polyunsaturated fatty acids (PUFAs).* A few important unsaturated acids are listed in Table 12–4. A perusal of the table indicates that in most of the unsaturated fatty acids, there is a double bond (designated Δ^9) between carbon atoms 9 and 10. This is particularly true of the unsaturated fatty acids, commonly found in the plant world. If there are additional bonds, they usually occur between Δ^9 and the methyl-terminal end of the chain. It may, however, be generalized that in mammals, polyunsaturated fatty acids can have up to 22 carbon atoms and 6 double bonds, but in plants these acids do not exceed 18 carbon atoms and 4 double bonds.

Human body can convert stearic acid to oleic acid by inserting a double bond but is incapable of inserting further double bonds so that the oleic acid cannot be converted to either linoleic, linolenic or arachidonic acid. For normal cell functioning esp., of skin tissues any one of these acids is needed. Since they cannot be synthesized by the cells, they must be obtained from diet. On account of the important physiological role, these 3 acids are collectively called as *essential fatty acids (EFAs),* a term introduced by **Burr** and **Burr** in 1930.

> Linoleic acid was once known as **vitamin F**. Its function in this capacity is now discredited. It is a yellow oily liquid with a b.p. 229°C.

Vegetable oils contain two types of polyunsaturated fatty acids (PUFAs)– linoleic acid (*lin* with 2 double bonds) and α-linolenic acid (*len* with 3 double bonds). In chemical jargon, *lin* and longer chain fatty acids derived from it are referred to as **n-6 fatty acids**, since the first double bond in

Table 12–4. **Unsaturated fatty acids*, commonly found in natural fats**

Trivial name	*Systematic name*	*Carbon skeleton*	*Structure*	*Common source*	*Melting point (in °C)*	*Boiling point (in °C)*
Crotonic	2-butenoic	4 : 1 ; 2	$CH_3CH{=}CHCOOH$	Croton oil	—	—
Myristoleic	9-tetradecenoic	14 : 1 ; 9	$CH_3(CH_2)_3CH{=}CH(CH_2)_7COOH$	*Pycnanthyus*	—	—
Palmitoleic	9-hexadecenoic	16 : 1 ; 9	$CH_3(CH_2)_5CH{=}CH(CH_2)_7COOH$	Animal and plant fats	0.5–1	—
Oleic (*oleum*[L] = oil)	9-octadecenoic	18 : 1 ; 9	$CH_3(CH_2)_7CH{=}CH(CH_2)_7COOH$	Animal and plant fats	13.4	234
Vaccenic	11-octadecenoic	18 : 1 ; 11	$CH_3(CH_2)_5CH{=}CH(CH_2)_9COOH$	Bacterial fat	44	—
Linoleic (*linon*[G] = flax)	9,12-octadecadienoic	18 : 2 ; 9, 12	$CH_3(CH_2)_4CH{=}CHCH_2CH{=}CH(CH_2)_7COOH$	Linseed and cottonseed oils	–5	202[a]
Eleostearic	9,11,13-octadecatrienoic	18 : 3 ; 9, 11, 13	$CH_3(CH_2)_3CH{=}CH{-}CH{=}CH{-}CH{=}CH(CH_2)_7COOH$	Tung oil	48	—
Linolenic	9,12,15-octadecatrienoic	18 : 3 ; 9, 12, 15	$CH_3CH_2CH{=}CHCH_2CH{=}CHCH_2CH{=}CH(CH_2)_7COOH$	Linseed oil	–11.0	157[b]
Arachidonic	5,8,11,14-eicosatetraenoic	20 : 4 ; 5, 8, 11, 14	$CH_3(CH_2)_4CH{=}CHCH_2CH{=}CHCH_2CH{=}CHCH_2CH{=}CH(CH_2)_3COOH$	Animal fat	–49.5	—
Nervonic	15-tetracosenoic	24 : 1 ; 15	$CH_3(CH_2)_7CH{=}CH(CH_2)_{13}COOH$	–	—	—

* All of the unsaturated fatty acids in this table are of the *cis* form with the exception of **eleostearic acid,** which is of the *cis*-9-*trans*-11-*trans*-13 form.

[a] at 1.4 mm Hg

[b] at 0.001 mm Hg

their molecule occurs on carbon no. 6 from the methyl end of the molecule.. *Len* and fatty acids derived from it are referred to as **n-3 fatty acids** since the first double bond in their molecule occurs on carbon no. 3 from the methyl end of the molecule. ***Lin*** *is more abundant in nature and is present in all vegetable oils, whereas* ***len*** *is present only in some vegetable oils such as mustard oil, rapeseed and soybean.* Fish oils are, however, good sources of both *lin* and *len* fatty acids and hence are considered nutritionally quite rich. *Hydrogenated fat* or *'vanaspati'* is manufactured from vegetable oils by the process of hydrogenation. It is a saturated fat and, though derived from vegetable oils, behaves like other saturated fats of animal origin.

All foods contain small quantities of bound fat or invisible fat. Such bound fat is found in green leafy vegetables and pulses in plenty and is a rich source of *n*-3 fatty acids. About half of the daily human requirement of fat (40 g) can be derived through such bound fat. Thus, the minimum daily requirement of visible fat (oil etc) is only 20 g. For long, the 2 types of PUFAs, *n*-6 and *n*-3, were thought to have similar effects. But recent study shows that *n*-6 and *n*-3 PUFAs have different types of effects on blood lipids and blood clotting and that they should be present in the diet in a certain proportion. Too much of *n*-6 with very little *n*-3 is not desirable. A healthy cooking practice, therefore, would be to blend vegetable oils containing both *lin* and *len*, instead of patronizing any one type of oil, including even the much-publicized unsaturated oils like safflower or rice bran which are very effective in reducing blood cholesterol.

A most unusual unsaturated fatty acid, *nemotinic acid*, is excreted in the growth medium by a citrivorium mould. This fatty acid is unique in that it contains the single, double and triple C—C linkages. *Nemotinic acid is one of the few naturally-occurring compounds containing the allene group.*

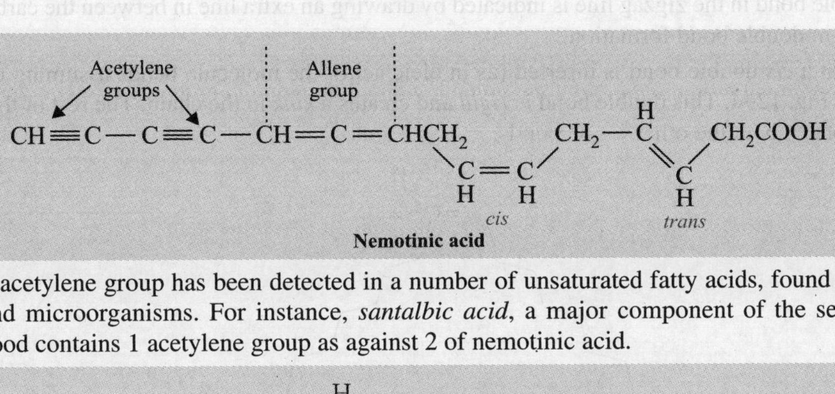

Nemotinic acid

The acetylene group has been detected in a number of unsaturated fatty acids, found in higher plants and microorganisms. For instance, *santalbic acid*, a major component of the seed oil of sandlewood contains 1 acetylene group as against 2 of nemotinic acid.

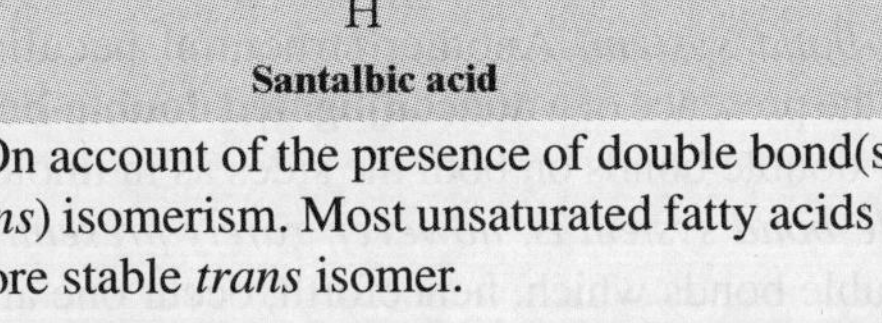

Santalbic acid

Geometric isomerism. On account of the presence of double bond(s), the unsaturated fatty acids exhibit geometric (or *cis-trans*) isomerism. Most unsaturated fatty acids are found as the **unstable *cis* isomer** rather than as the more stable *trans* isomer.

Oleic acid **Elaidic acid**

The hydrocarbon chain of saturated fatty acids (stearic acid, for example) has a zigzag configuration with the C—C bond forming a bond angle of 109° (Fig. 12–2).

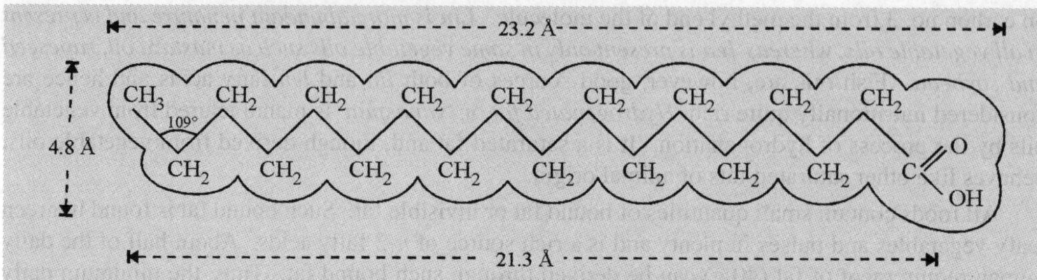

Fig. 12–2. Stearic acid

For the sake of simplicity, the long chains of CH_2 groups are represented by zigzag lines where each corner represents a C atom and the hydrogen atoms are left out. *The zigzag line represents the most stable configuration of such carbon chains.* The free end of this line, if not indicated otherwise, obviously represents the presence of a methyl group. The simplified formula for stearic acid would, thus, be as represented in Fig. 12–3.

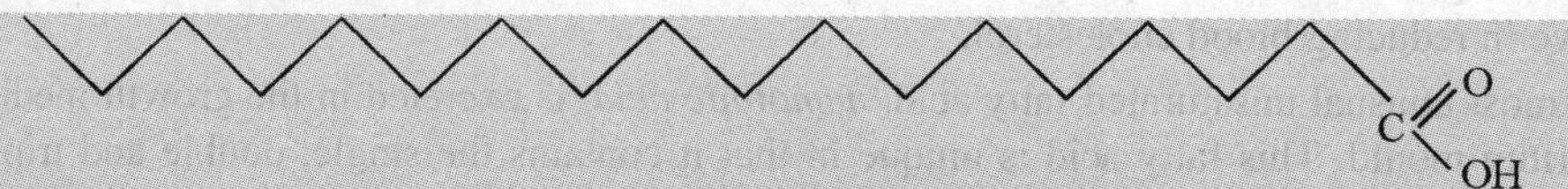

Fig. 12–3. Stearic acid (simplified formula)

The double bond in the zigzag line is indicated by drawing an extra line in between the carbon atoms involved in double bond formation.

When a *cis* double bond is inserted (as in oleic acid), the molecule bends assuming the shape, shown in Fig. 12–4. This double bond is *rigid* and creates a *kink* in the chain. The rest of the chain is free to rotate about the other C—C bonds.

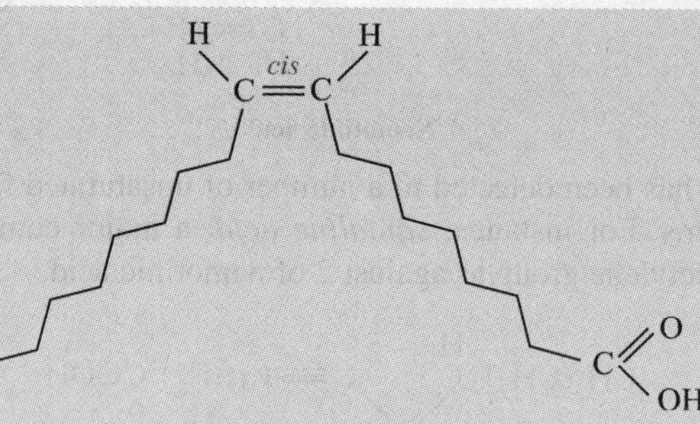

Fig. 12–4. Oleic acid

Nonconjugated double-bond system. Another structural peculiarity of naturally occurring polyunsaturated fatty acids is the presence of a **nonconjugated double-bond system.** It has a methylene group (—CH_2—) flanked by double bonds on both the sides as in linoleic, linolenic and arachidonic acids. *The conjugated double-bond system is, however, rarely present.* In it the methylene group is not found in between the double bonds which, henceforth, occur one after the other as in eleostearic acid. This acid has valuable properties as a drying oil since it polymerizes readily. These two double-bond systems (Fig. 12–5) have different chemical reactivities.

$$-CH{=}CH-CH_2-CH{=}CH-CH_2-CH{=}CH- \quad \text{Nonconjugated system}$$

$$-CH{=}CH-CH{=}CH-CH{=}CH-CH{=}CH- \quad \text{Conjugated system}$$

Fig. 12–5. Two double-bond systems

3. Hydroxy or oxygenated fatty acids

One of these found in castor oil is *ricinoleic acid* (87%). It is a C-18 acid with a double bond at C_9 and an OH group on C_{12}.

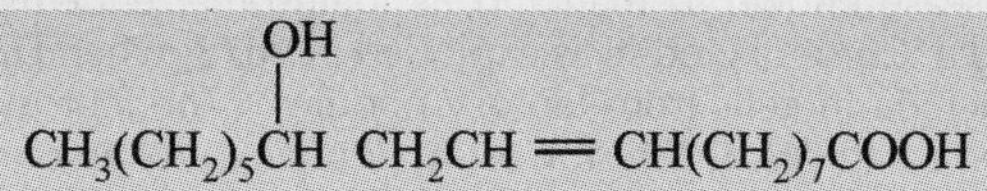

Ricinoleic acid
(12-hydroxyoctadec-9-enoic acid)

Cerebronic acid, a C-24 acid obtained from animal lipid, is another important hydroxy acid with an OH group on C_2.

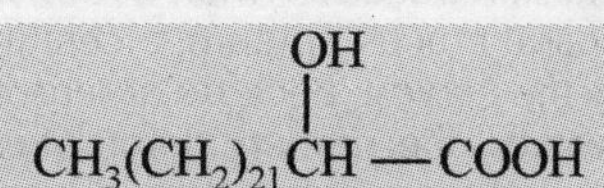

Cerebronic acid
(2-hydroxytetracosanoic acid)

A common oxygenated fatty acid, isolated from plants and bacterial lipids, is *9, 10-dihydroxystearic acid.*

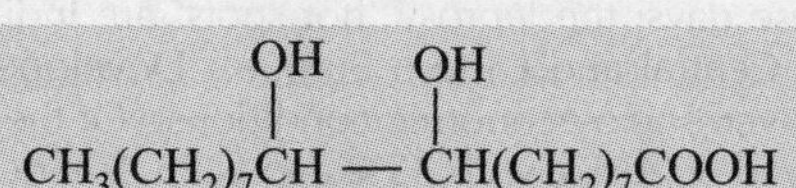

9,10-dihydroxystearic acid
(9,10-dihydroxyoctadecanoic acid)

Similarly, 9, 10-*epoxystearic acid* is isolated from rust spore lipids (20%).

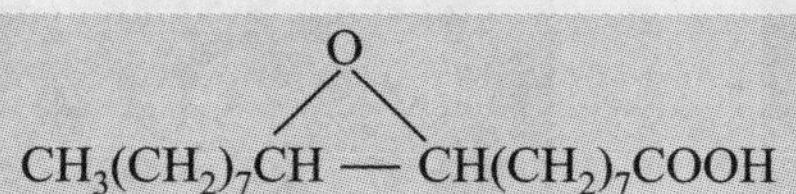

9,10-epoxystearic acid
(9,10-epoxyoctadecanoic acid)

4. Cyclic fatty acids

These are of rare occurrence. Chaulmoogra oil, obtained from the plant *Hydnocarpus kurzil* and used in the treatment of leprosy, contains 2 such acids *hydnocarpic* and *chaulmoogric*. Chaulmoogric acid has a cyclopentenyl ring in its 18-carbon structure.

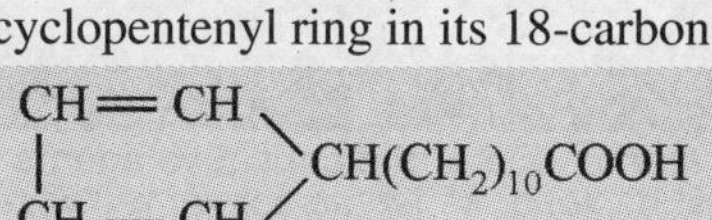

Hydnocarpic acid

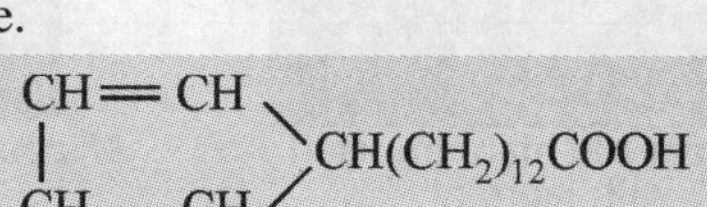

Chaulmoogric acid

Lipids from the lactobacilli contain a fatty acid, *lactobacillic acid*, with a cyclopropyl group. This fatty acid may result from the addition of a methylene group across the double bond of vaccenic acid (18:1; 11).

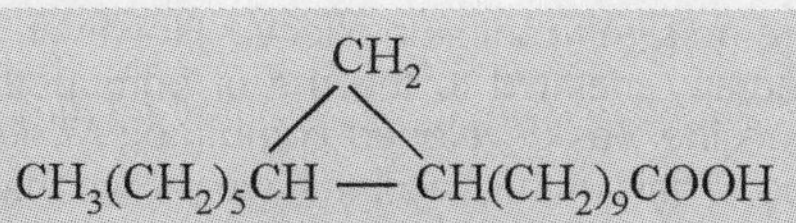

Lactobacillic acid

Similarly, *sterculic acid* from plant sources has a comparable structure, with a suggested relationship to oleic acid. It may be derived from oleic acid by the addition of a methylene group across the double bond in a manner that the unsaturated nature is not altered, unlike the lactobacillic acid.

CLINICAL IMPLICATIONS 12–1.

Leprosy

Leprosy, the oldest disease known to mankind, is loathsome, endemic disease. The myth that leprosy was hereditary and a 'Curse of God' was exploded in 1873, when **Dr. G. A. Hansen** in Norway discovered rod-shaped bacteria in cells from leprosy nodules. *The leprosy bacillus* (*Mycobacterium leprae*), *also called Hansen's bacillus, is the slowest multiplying prokaryote known and takes about 2 weeks to double*. Once infected, it takes a long time for the disease to develop : from less than a year to as long as 30 years, though normally 3–7 years. About 20% of the infected people develop the disease, on an average. In patients with high immune resistance, the disease is relatively mild (*tuberculoid form*); those with low resistance develop highly infectious and more severe conditions (*lepromatous leprosy*), affecting not only nerves and skin but also lymph nodes, eyes, nose, mouth, larynx, spleen etc and the bacilli may appear in tears, nasal mucus, sputum, ulcerating nodules etc. *M. leprae is the only bacterium that can enter and destroy the nerves*. Because of sensory impairment due to nerve damage, the person does not feel pain. However, the primary deformities are pronounced on the face. The infected, untreated individual is the only source of infection. There is no known natural reservoir other than man. Leprosy is only weakly contagious: usually prolonged and close contact is considered necessary. The nose is the main portal of exit of the bacilli. Although leprosy affects individuals of both sexes, *males are more susceptible than females,* often in the ratio of 2 : 1. Although leprosy had assumed epidemic proportions in Europe in 13th and 14th century A.D., at present, Western Europe is virtually free from it. And these days, the leprosy 'hot spots' are India, Indonesia, Myanmar, Brazil, Columbia and few pockets of Central and Eastern Europe. Developed countries are, however, free from leprosy due to their improved socio-economic conditions. Of the estimated disease burden, Asia accounts for about 60%, followed by Africa with 35% and Latin America with 4%; the rest of the world accounting for about 1% of the disease load. India alone accounts for about 1/5th of the total lepers in the world.

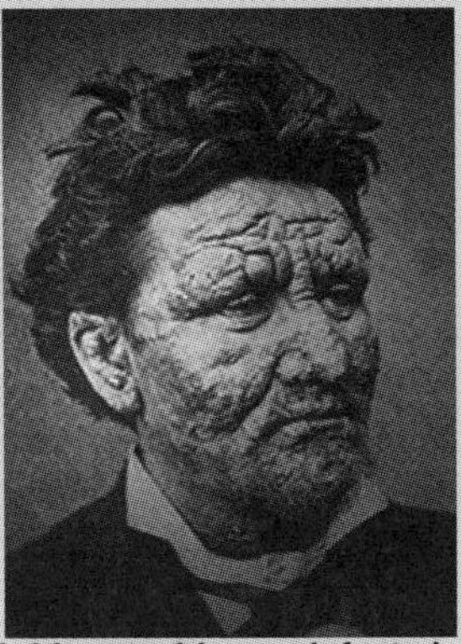

A 24-year-old man infected with leprosy

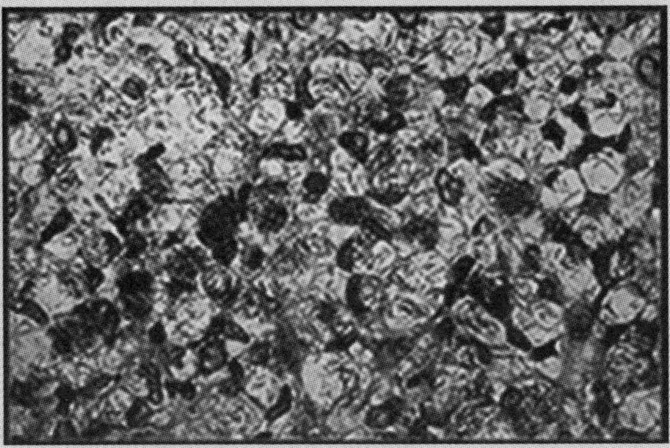

***Mycobacterium leprae*, one of the causative agents of leprosy**
Acid-fast bacteria, *M. leprae* appear red when a Ziehl-Neelsen stain is used.

Under chemotherapy, the drugs most commonly used against leprosy were *dapsone, rifampicin* and *clofazimine,* the last one being the best alternative as it had the much-desired anti-inflammatory effect and produced least resistance. However, in 1982, the WHO recommended the multi-drug therapy (MDT), combining all the 3 above drugs, as it has a very high potential for control of leprosy. The duration of treatment in most cases is about 3 years or more. A single dose of MDT kills about 95% of bacilli in a patient and there are no significant side effects. However, the overall long duration of the treatment regimen in chemotherapy, has led to the use of immunotherapy, using leprosy vaccines. Under the aegis of WHO, a vaccine was developed combining *M. leprae* (inactivated) derived from armadillo, and the BCG vaccine. Trials of this vaccine have shown that its application can restore cell-mediated immunity in patients and thus it could remarkably suppress the disease in about 18 months only.

Leprosy eradication programme has made spectacular progress worldover. From a long list of 122 endemic countries in 1982, the disease is now endemic only in 24 of them. Ninety per cent of leprosy is now confined to only 12 countries, including India, Brazil, Indonesia, Myanmar, Nepal, Mosambique, Guinea, and Angola.

Table 12–5. Fatty acid content of some animal and plant fats and oils

Fat/Oil	*Butyric*	*Caproic*	*Caprylic*	*Capric*	*Lauric*	*Myristic*	*Palmitic*	*Stearic*	*Arachidic*	*Palmi-toleic*	*Oleic*	*Linoleic*	*Linolenic*	*Others*
Buffer fat	3.6	2.3	0.5	1.0	2.5	11.1	29.0	9.2	2.4	4.6	26.7	3.6	–*	–
Beef tallow	–	–	–	–	–	6.3	27.4	14.1	–	–	49.6	2.5	–	–
Human depot fat	–	–	–	–	–	2.7	24.0	8.4			46.9	10.2		
Pork fat (lard)	–	–	–	–	–	1.3	28.3	11.9	–	2.7	47.5	6.0		
Castor oil	–	–	–	–	Trace	Trace	Trace	–		–	7.4	3.1		88†
Corn oil	–	–	–	–	–	1.4	10.2	3.0	–	1.5	49.6	34.3	–	–
Soybean oil	–	–	–	–	0.2	0.1	9.8	2.4	–	0.4	28.9	50.7	6.5	
Linseed oil	–	–	–	–	–	–	6.8	2.5	0.5	–	19.0	24.1	47.4	
Herring oil						7.3	13.0	Trace	–	4.9	–	–	20.7	30.1[a] 23.2[b]
Whale oil					0.2	9.3	15.6	2.8	–	14.4	35.2	–	–	13.6[a] 5.9[b]

* Not detected
† Ricinoleic acid
[a] C-20 polyunsaturated
[b] C-22 polyunsaturated

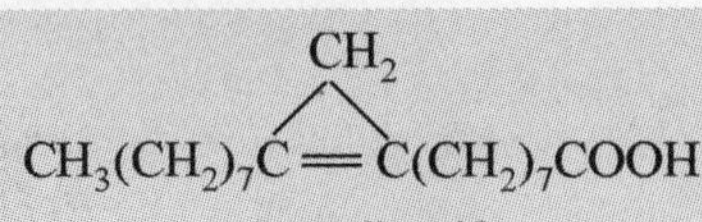

Sterculic acid

The percentage composition of various types of fatty acids in some typical fats and oils is presented in Table 12–5.

BIOLOGICAL ROLES OF LIPIDS

The lipids perform a wide variety of functions (Fig. 12–6). These are briefly listed below:

1. **Food material.** Lipids provide food, highly rich in calorific value. One gram lipid produces 9.3 kilocalories of heat.
2. **Food reserve.** Lipids provide are insoluble in aqueous solutions and hence can be stored readily in the body as a food reserve.
3. **Structural component.** Lipids are an important constituent of the cell membrance.

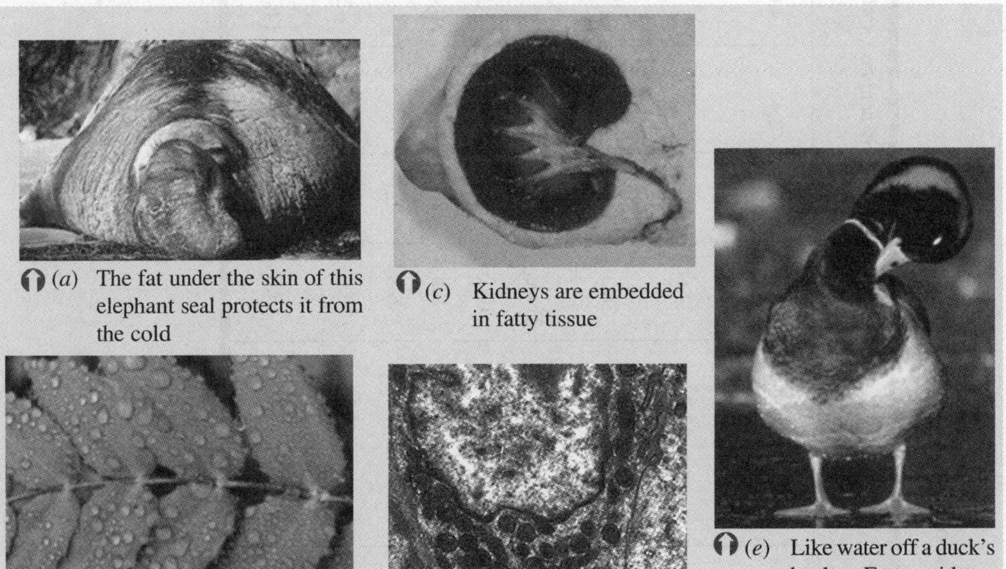

(*a*) The fat under the skin of this elephant seal protects it from the cold

(*b*) The wax surface of leaves repels water, but also reduces water loss from the plant

(*c*) Kidneys are embedded in fatty tissue

(*d*) Every cell membrane contains lipids

(*e*) Like water off a duck's back.... Fatty acids on a duck's feathers repel water

Fig. 12–6. **Various functions performed by lipids**

4. **Heat insulation.** The fats are characterized for their high insulating capacity. Great quantities of fat are deposited in the subcutaneous layers in aquatic mammals such as whale and in animals living in cold climates.
5. **Fatty acid absorption.** Phospholipids play an important role in the absorption and transportation of fatty acids.
6. **Hormone synthesis.** The sex hormones, adrenocorticoids, cholic acids and also vitamin D are all synthesized from cholesterol, a steroidal lipid.
7. **Vitamin carriers.** Lipids act as carriers of natural fat-soluble vitamins such as vitamin A, D and E.

8. **Blood cholesterol lowering.** Chocolates and beef, especially the latter one, were believed to cause many heart diseases as they are rich in saturated fatty acids, which boost cholesterol levels in blood and clog the arterial passage (Fig. 12–7). But researches conducted at the University of Texas by **Scott Grundy** and **Andrea Bonanome** (1988) suggest that at least one saturated fatty acid stearic acid, a major component of cocoa butter and beef fat, does not raise blood cholesterol level at all. The researchers placed 11 men on three cholesterol-poor liquid diets for three weeks each in random order. One formula was rich in palmitic acid, a known cholesterol booster; the second in oleic acid; and the third in stearic acid. When compared with the diet rich in palmitic acid, blood cholesterol levels were 14% lower in subjects put on the stearic acid diet and 10% lower in those on the oleic acid diet.

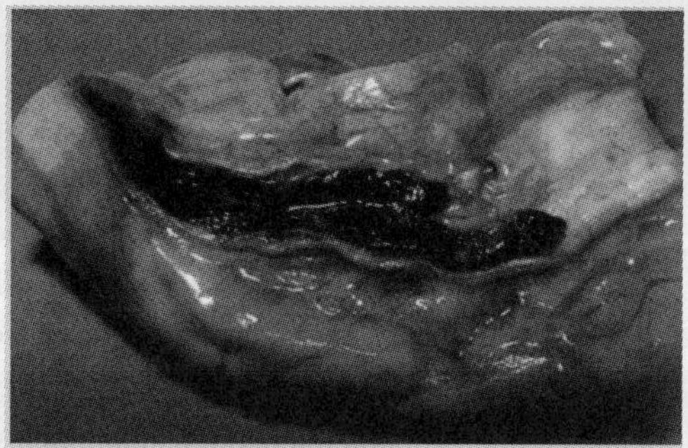

Fig. 12–7. A cholesterol deposit, (known as *atheroma*) filling almost all the space inside the artery.

9. **Antibiotic agent.** *Squalamine*, a steroid from the blood of sharks, has been shown to be an antibiotic and antifungal agent of intense activity. This seems to explain why sharks rarely contract infections and almost never get cancer.

REFERENCES

See list following Chapter 14.

PROBLEMS

1. Why are the most unsaturated fatty acids found in phospholipids in the *cis* rather than the *trans* conformation? Draw the structure of a 16-carbon fatty acid as saturated, *trans* monounsaturated, and *cis* monounsaturated.
2. Given these molecular components– glycerol, fatty acid, phosphate, long-chain alcohol, and carbohydrate– answer the following :
   (*a*) Which two are present in both waxes and sphingomyelin?
   (*b*) Which two are present in both fats and phosphatidylcholine?
   (*c*) Which are present in a ganglioside but not in a fat?
3. The melting points of a series of 18-carbon fatty acids are stearic acid, 69.6°C, oleic acid, 13.4°C; linoleic acid –5°C; and linolenic acid, –11°C. What structural aspect of these 18-carbon fatty acids can be correlated with the melting point? Provide a molecular explanation for the trend in melting points.
4. How is the definition of "lipid" different from the definitions of other types of biomolecules that we have considered, such as amino acids, nucleic acids, and proteins?
5. **Johann Thudichum**, who practiced medicine in London about 100 years ago, also dabbled in lipid chemistry in his spare time. He isolated a variety of lipids from neutral tissue, and characterized and named many of them. His carefully sealed and labeled vials of isolated lipids were rediscovered many years later. How would you confirm, using techniques available to you but not to him, that the vials he labeled "sphingomyelin" and "cerebroside" actually contain these compounds?

6. How would you distinguish sphingomyelin from phosphatidylcholine by chemical, physical, or enzymatic tests?
7. Do all kinds of fibre help lower one's cholesterol?
8. How much fat does one need in his diet?
9. Is a high-protein diet good for one's health?
10. Why is tea oil better for frying than olive oil?
11. Why soaps cannot be manufactured from mineral oil or kerosene oil?
12. How would you differentiate between hydrophilic and hydrophobic groups?
13. Why fatty acids are called amphipathic molecules?
14. Differentiate between prokaryotic and eukaryotic cells in terms of their lipids.

ANSWERS TO PROBLEMS

1. The presence of a *cis* double bond introduces a kink that prevents packing of the fatty acid chains. *Cis* double bonds maintain fluidity. *Trans* fatty acids have no structural effect, relative to saturated fatty acids, and so they are rare.

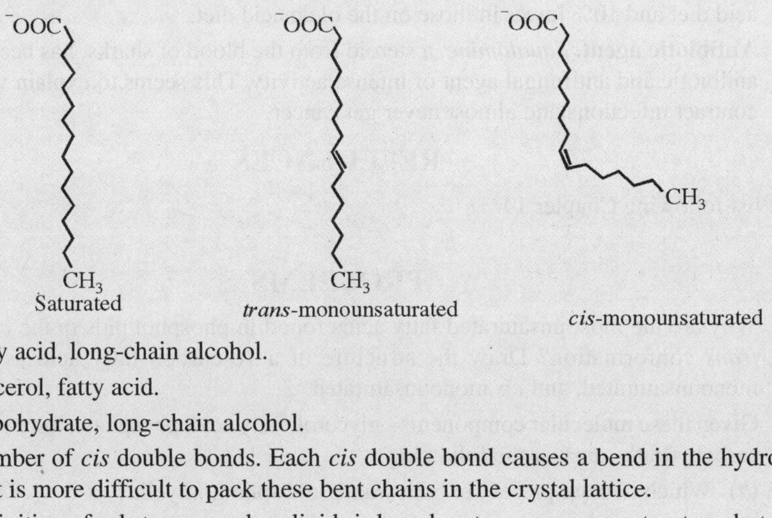

Saturated *trans*-monounsaturated *cis*-monounsaturated

2. (*a*) Fatty acid, long-chain alcohol.
   (*b*) Glycerol, fatty acid.
   (*c*) Carbohydrate, long-chain alcohol.
3. The number of *cis* double bonds. Each *cis* double bond causes a bend in the hydrocarbon chain; it is more difficult to pack these bent chains in the crystal lattice.
4. The definition of substances such as lipids is based, not on a common structure, but on their solubilities in water and in nonpolar solvents.
5. Both compounds should yield sphingosine and a fatty acid upon strong alkaline hydrolysis. Phosphocholine is produced only from sphingomyelin; cerebroside produces one or more sugars, but no phosphate.
6. Strong alkaline hydrolysis of phosphatidylcholine released 2 fatty acids per phosphate in PC; the same treatment of sphingomyelin produces 1 fatty acid per phosphate, as well as sphingosine.
7. No, only soluble fibre (in oat bran, fruits and vegetables) does. Insoluble fibre (in wheat bran and beans) does not.
8. Not much. About one tablespoon of oil a day will supply all the fat one needs.
9. No. This diet is high in fat and low in complex carbohydrates. Just the opposite of good nutrition.
10. Tea oil, which is cold-pressed from tea seeds, is unrefined and low in saturated fat like olive oil. But tea oil is a better choice for frying because it resists breaking down into *trans* fats when heated. It also has a naturally neutral flavour.

11. The mineral and kerosene oils do not have glycerides and, hence, cannot be converted into soaps.

12. Colloids are of 2 types, when taking water as the dispersion medium:

(*a*) Hydrophilic (water-loving) or lyophilic–These are capable of binding water and as such readily yield colloidal solutions when brought in contact with water. They are, therefore, also called *reversible colloids*. Agar-agar, gelatin, starch, cellulose, gums, and proteins are examples of hydrophilic colloids.

(*b*) Hydrophobic (water-hating) or lyophobic–They are incapable of binding water and as such do not readily yield colloidal solutions when brought in contact with water. They are, henceforth, also called *irreversible colloids*. Hydrophobic colloids include metals, lipids, etc.

13. A fatty acid molecule has a polar end ($-COO^-$), and a nonpolar end of long carbon chain. The polar end is water-soluble, and is hydrophilic. The nonpolar end is water-insoluble and is, thus, hydrophobic. Molecules like these are called amphipathic, *i.e*., they have both polar and nonpolar ends.

14. Prokaryotic cells (*e.g*., bacterial cells) have over 95% of its total components associated with its cell membrane and remaining 5% distributed between cytoplasm and cell wall. They are unable to synthesize steroid ring structure and are devoid of sterols. Barring mycobacteria, triacylglycerols are also missing in bacteria. Polyunsaturated fatty acids (PUFAs) are not synthesized by bacteria (with the exception of bacilli) and mostly saturated and branched-chain fatty acids are synthesized.

Eukaryotic cells, such as plant cells, always contain triacylglycerol, PUFAs and other fatty acids like digalactosyl diacylglycerol, phosphatidyl glycerol etc. Animal cells, on the other hand, cannot synthesize PUFAs. However, phospholipids, sphingolipids, glyceryl ethers and plasmalogens play a vital role in maintaining the nervous system.

CONTENTS

(a)

(b)

Lipids

(*a*) **Fat.** A European brown bear ready to hibernate. Fat is an efficient way to store energy. If this bear stored the same amount of energy in carbohydrates instead of fat, she probably would be unable to walk! (*b*) **Wax.** Wax is a highly saturated lipid that remains very firm at normal outdoor temperatures. Its rigidity allows it to be used to form the strong but thin-walled hexagons of this honeycomb.

CHAPTER

13

Lipids–II Classification

Bloor (1943) has proposed the following classification of lipids based on their chemical composition.

A. Simple lipids or **Homolipids.** These are esters of fatty acid with various alcohols.

1. *Fats and oils (triglycerides, triacylglycerols).* These are esters of fatty acids with a trihydroxy alcohol, glycerol. A fat is solid at ordinary room temperature wheras an oil is liquid.
2. *Waxes.* These are esters of fatty acids with high molecular weight monohydroxy alcohols.

B. Compound lipids or **Heterolipids.** These are esters of fatty acids with alcohol and possess additional group(s) also.

1. *Phospholipids (phosphatids)*, These are compounds containing, in addition to fatty acids and glycerol, a phosphoric acid, nitrogen bases and other substituents.
2. *Glycolipids (cerebrosides).* These are the compounds of fatty acids with carbohydrates and contain nitrogen but no phosphoric acid. The glycolipids also include certain structurally-related compounds comprising the groups, gangliosides, sulfolipids and sulfatids.

C. Derived lipids. These are the substances derived from simple and compound lipids by hydrolysis. These include fatty acids, alcohols, mono- and diglycerides, steroids, terpenes and carotenoids.

Glycerides and cholesterol esters, because of their uncharged nature, are also called *neutral lipids.*

However, **Conn** and **Stumpf** (1976) have traditionally classified lipids into following 6 classes:

1. Acyl glycerols
2. Waxes
3. Phospholipids
4. Sphingolipids
5. Glycolipids
6. Terpenoid lipids including carotenoids and steroids

HARRY JAMES DEUEL JR.

(LT, 1897–1956)

Deuel was born on October 15, 1897 in St. Paul, Minnesota. He did his graduation from Carleton College. He was appointed junior chemist under **Dr. Langworthy**. He also published some papers on the digestibility of fats in collaboration with **Dr. Langworthy** and **Dr. Holmes**. In 1920 , he joined Yale University as a graduate student and worked with **Oscar Bandish, Arthur H. Smith** and **George R. Cowgill**, leading to the publication of many research papers. He received his Ph.D. degree, in 1993, under **Prof. Lafayette B. Medel** (LT, 1872–1935). Then, he was appointed as instructor in the Department of Physiology of Cornell University Medical College and later as Assistant Professor, the post he held till 1928. He valued very much his association with **Dr. Graham Lusk,** with whom he published 7 research papers. In 1925, *Deuel carried out a study on the minimum endogenous nitrogen loss on himself.* This has become a classic piece of work. In 1928, Deuel was appointed Professor of Physiology in the University of Maryland Medical School. After a year, he took up the post of Professor of Biochemistry in the School of Medicine of the University of Southern California. During the course of next 20 years, he developed the department into an active school of research.

For the first 8 to 9 years, Deuel made important contributions in the field of **carbohydrate and fat metabolism in relation to ketosis**. In 1939, Deuel initiated a multigeneration experiment to study the **comparative nutritive value of margarine and butter fat**. His studies also demonstrated the high nutritive value of vegetable fats and their partial hydrogenation products. He further demonstrated that the intestines, rather than liver, are the principal site of the conversion of carotene to vitamin A. In collaboration with **Dr. L. Zechmeister**, he published a series of important papers on the vitamin A activity of different isomers of carotenoids. In 1939, Deuel was appointed Dean of the Graduate School, but he retained his professorship and continued his research work. During the 1950s, he published his monumental book 'The Lipids' in 3 volumes which appeared in 1951, 1955 and 1956. His health deteriorated swiftly and he died on April 17, 1956 at an early age of 59. Besides his great contributions in the field of **lipids,** Deuel was also an inspiring teacher and a great writer.

SIMPLE LIPIDS

FATS AND OILS

(= Triglycerides or Triacylglycerols)

The triglycerides are the most abundant of all lipids. They constitute about 98% of total dietary lipids; the remaining 2% consists of phospholipids and cholesterol and its esters. They are the major components of storage or depot fats in plant and animal cells but are not normally found in membranes. They are nonpolar, hydrophobic molecules since they contain no electrically charged or highly polar functional groups. In animals, the *fat cells* or *adipocytes* contain very large quantities of triglycerides in the form of fat droplets, which fill almost the entire cell volume

(Fig. 13-1). Adipocytes are abundantly found under the skin, in the abdominal cavity and in the mammary glands. Triglycerides can be stored in quantities, sufficient to supply the energy needs of the body for many months, as in the case of obese persons. On the contrary, the body can store the carbohydrate glycogen in meagre amounts, sufficient to supply energy need of a day only. *Triglycerides are much better adapted than glycogen to serve as storage form of energy.* They are not only stored in large amounts but also yield over twice as much energy as carbohydrates. This is because triglycerides have a greater proportion of hydrogen atoms in their molecules. Since fats tend to remain in the stomach longer than carbohydrates and are digested more slowly, they also have greater satiety value than carbohydrates. The arctic and antarctic animals (Fig. 13-2) such as whales, seals, walruses and penguins are amply padded with triglycerides to serve both as energy storage depots and as an insulation against very low temperatures. Most fats and oils, upon hydrolysis, yield several fatty acids as well as glycerol. However, the milk of spiny anteater is an exception in that it comprises almost pure triolein. Human body contains enough fat to make 7 bars of soap!

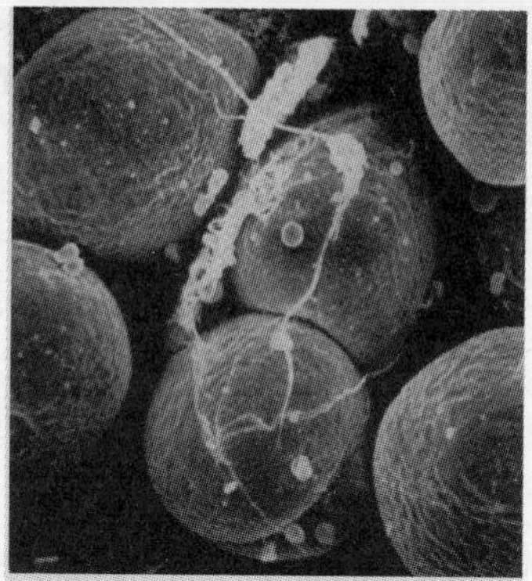

Fig 13–1. A scanning electron micrograph of adipocytes

Each adipocyte contains a fat globule that occupies nearly the entire cell.

(*Courtesy* : *Fred E. Hossler*)

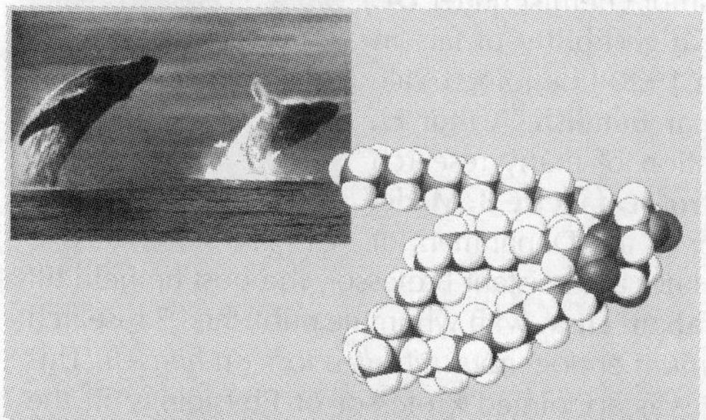

Fig 13–2. Insulating blubber of whales and the fat molecules

Fats such as the triacylglycerol molecule (*lower*) are widely used to store excess energy for later use and to fulfill other purposes, illustrated by the insulating blubber of whales. The natural tendency of fats to exist in nearly water-free forms makes these molecules well-suited for these roles.

[*Courtesy* : (*Upper*) *Francois Cohier*)

In a normal man, weighing 70 kg, at least 10-20% of the body weight is lipid, the bulk of which is triacylglycerol (TAG). Triacylglycerol is found in all organs of the human body, particularly in adipose tissue, in which droplets of triacylglycerols may represent more than 90% of the cytoplasm of the cells. Body lipid is a reservoir of potential chemical

Sperm Whales: Fatheads of the Deep Waters

The usefulness of the triacylglycerols is explicitly revealed by the study of sperm whales. The sperm whale's head is very large, accounting for over 1/3 rd of its body weight. About 90% of the weight of the head is made up of the *spermaceti organ*, a blubbery mass that contains up to 3,600 kg (about 4 tons) of spermaceti oil, which is a mixture of triacylglycerols and waxes rich in unsaturated fatty acids. This mixture is liquid at the normal resting body temperature of the whale (about 37°C) and becomes solid when the temperature drops appreciably. The probable biological function of spermaceti oil has been deduced from research on the anatomy and feeding behaviour of the sperm whale. These mammals feed almost exclusively on squid in very deep waters. In their feeding dives, they descend 1,000 m or more; the deepest recorded dive is 3,000 m (about 2 miles). At these depths, there are no competitors for the very plentiful squids; the sperm whale rests quietly, waiting for schools of squid to pass.

For a marine animal to remain at a given depth, without a constant swimming effort, it must have the same density as the surrounding water. The sperm whale undergoes changes in buoyancy to match the density of its surroundings– from the tropical ocean surface to great depths where the water is much colder and thus denser. The key is the freezing point of spermaceti oil. When the temperature of the oil is lowered several degrees during a deep dive, it congeals or crystallizes and becomes denser. Thus, the buoyancy of the whale changes to match the density of seawater. Various physiological mechanisms promote rapid cooling of the oil during a dive. During the return to the surface, the congealed spermaceti oil warms and melts, decreasing its density to match that of the surface water. Thus, *we see in the sperm whale a remarkable anatomical and biochemical adaptation.* The triacylglycerols and waxes synthesized by the sperm whale contain fatty acids of the necessary chain length and degree of unsaturation to give the spermaceti oil the proper melting point for the animal's diving habits.

Unfortunately for the sperm whale population, spermaceti oil was at one time considered the finest lamp oil and continues to be commercially valuable as a lubricant. Several centuries of intensive hunting of these mammals have driven sperm whales onto the endangered species list.

energy. About 100 times more energy is stored as mobilizable lipid than as mobilizable carbohydrate in the normal human being. TAG is stored in a relatively water-free state in the tissue, in comparison to carbohydrate, which is heavily hydrated.

Chemically, triglycerides are esters of glycerol with 3 fatty acid molecules. Their generic formula is shown in Fig. 13–3.

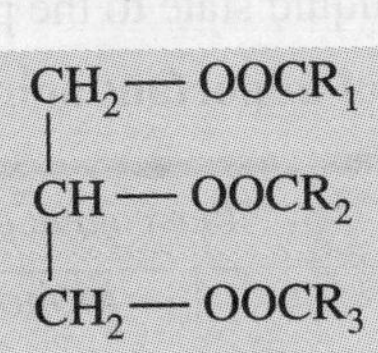

Fig. 13–3. Generic formula of triglycerides

Obviously, when the groups attached to carbon 1 and 3 differ, a centre of asymmetry is created at C 2. The 2 optical isomers may, thus, be represented as shown in Fig. 13–4. *The naturally-occurring fats are of L-type.*

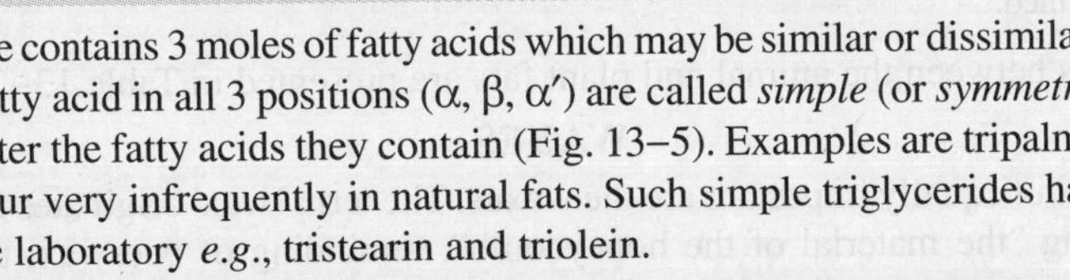

D-isomer L-isomer

Fig. 13–4. Two optical isomers of triglycerides

A fat molecule contains 3 moles of fatty acids which may be similar or dissimilar. Those containing a single kind of fatty acid in all 3 positions (α, β, α′) are called *simple* (or *symmetrical*) *triglycerides*; they are named after the fatty acids they contain (Fig. 13–5). Examples are tripalmitin, tristearin, and triolein. They occur very infrequently in natural fats. Such simple triglycerides have, however, been synthesized in the laboratory *e.g.*, tristearin and triolein.

Tristearin (tristearoylglycerol) **Triolein** (trioleylglycerol)

Fig. 13–5. Two simple triglycerides

Most of the triglycerides of nature are mixed (or asymmetrical) triglycerides, i.e., they contain 2 or 3 different fatty acid units in the molecule. Representatives of such mixed triglycerides are oleodipalmitin and oleopalmitostearin (Fig. 13–6).

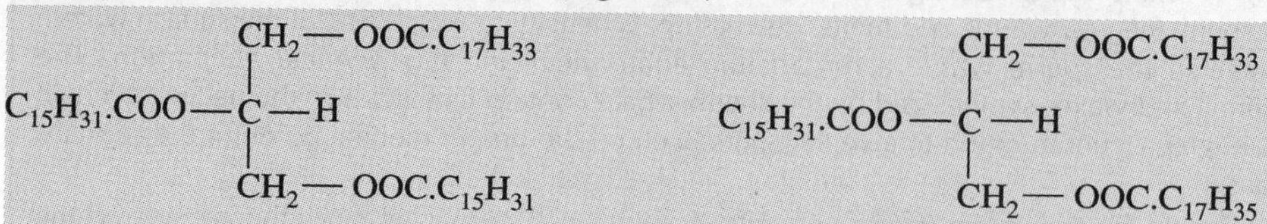

Oleodipalmitin (α-oleo-β, α′-dipalmitin)

Oleopalmitostearin (α-oleo-β, palmito- α′-stearin)

Fig. 13–6. Two mixed triglycerides

Most **animal fats** such as those from meat, milk and eggs are relatively rich in saturated fatty acids but contain a rather low content of polyunsaturated fatty acids (Table 13–1) ; two exceptions are chicken fat and fish fat. The large proportion of saturated fatty acids (*esp.*, palmitic and stearic) with high melting point confers solid state to the animal fats. The **plant fats,** on the other hand, contain a large proportion of unsaturated fatty acids (*esp.*, polyunsaturated). The unsaturated fatty acids have low melting point and confer liquid state to the plant fats.

Table 13–1. Fatty acid composition of important animal and plant fats

Fat/Oil	*Percentage of total fatty acids*		
	Saturated	*Monounsaturated*	*Polyunsaturated*
Animal fats			
Butter fat	60	36	4
Pork fat	59	39	2
Beef fat	53	44	2
Chicken fat	39	44	21
Plant fats			
Olive oil*	20	26	54
Corn oil	15	31	53
Soybean oil	14	24	53
Soft margarine	23	22	52

* data not confirmed.

The differences between the animal and plant fats are presented in Table 13–2.

WAXES

Far less spread but equally important are the waxes. The term `wax' originates in the Old English word *weax,* meaning "the material of the honeycomb", reminding of beeswax, the honeycomb is made of. Beeswax is used for candles, for artificial fruits and flowers, and for modeling wax. It is also an ingredient in the manufacture of furniture and floor waxes, leather dressings, shoe polishes, waxed paper, lithographic inks, cosmetics and ointments. Waxes are esters of long-chain saturated and unsaturated fatty acids with long-chain monohydroxy alcohols. The fatty acids range in between C_{14} and C_{36} and the alcohols range from C_{16} to C_{36}. In vertebrates, waxes are secreted by cutaneous glands as a protective coating to keep the skin pliable, lubricated and water-proof. Hair, wool and fur are also coated with wax. Birds, particularly waterfowl, secrete waxes in their preen glands to make their feathers water-repellent. The leaves of many plants such as *Rhododendron, Calotropis* etc., are shiny because of the deposition of protective waxy coating. Waxes also serve as the chief storage form of fuel in planktons. Since marine organisms (whale, herring, salmon) consume planktons in large quantities, waxes act as major food and storage lipids in them.

Table 13–2. Differences between animal and plant fats

Animal fat	*Plant fat*
1. Relatively rich in saturated fatty acids, *esp*., C_{16} and C_{18} acids.	1. Relatively rich in unsaturated fatty acids, *esp*., polyunsaturated acids.
2. Solid at ordinary room temperature.	2. Liquid at ordinary room temperaure.
3. These have usually low iodine number.	3. These have usually high iodine number.
4. These have usually high Reichert-Meissl . number	4. These have usually low Reichert-Miessl number.
5. These are stored mainly in liver and bone marrow	5. These are stored mainly in seeds and fruits.
6. Oxidative rancidity is observed more frequently.	6. Oxidative rancidity is observed less frequently.
Examples– Butterfat, Beef fat, Tallow	***Examples–*** Olive oil, Castor oil, Soybean oil, Corn oil

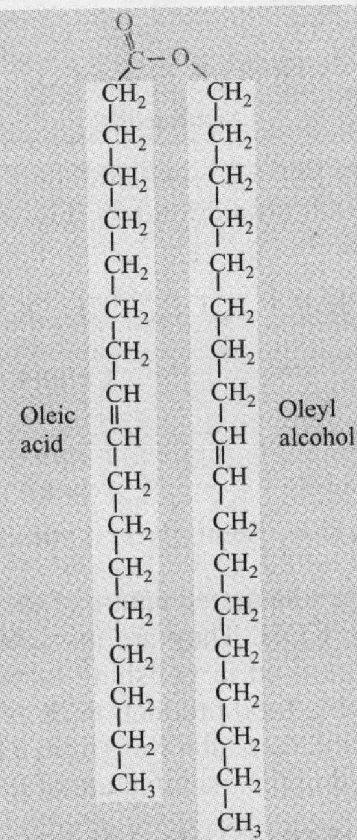

Fig. 13– Structure of a typical wax

Waxes are formed by esterification of fatty acids and long-chain alcohols. This yields a small head group that is only weakly hydrophilic, attached to the 2 long hydrocarbon chains which are, in contrast, significantly hydrophobic. As a consequence, the waxes are completely water-insoluble. In fact, they are so hydrophobic that they often serve as water repellents as in the feathers of some birds and the leaves of some plants. As with the triacylglycerols, the hardness of waxes increases with chain length and degree of hydrocarbon saturation.

Most of the waxes are mixtures of esters. Thus, *sperm whale wax* (spermaceti) and *beeswax* are composed mainly of palmitic acid esterified with either hexacosanol, $CH_3(CH_2)_{24}CH_2OH$ or triacontanol (=myricyl alcohol or melissyl alcohol), $CH_3(CH_2)_{28}.CH_2OH$.

$$\underset{\text{Palmitic acid}}{CH_3(CH_2)_{14}COO}\,[H + HO]\,\underset{\text{Hexacosanol}}{CH_2(CH_2)_{24}CH_3} \xrightarrow{-H_2O} \underset{\text{Hexacosanyl palmitate}}{CH_3(CH_2)_{14}COO{-}CH_2(CH_2)_{24}CH_3}$$

Moreover, sperm whale wax is also rich in cetyl palmitate and the beeswax in myricyl palmitate.

$CH_3(CH_2)_{14}COO—CH_2(CH_2)_{14}CH_3$
Cetyl palmitate

$CH_3(CH_2)_{14}COO—CH_2(CH_2)_{28}CH_3$
Myricyl palmitate

Carnauba wax, the hardest known wax, consists mainly of fatty acids esterified with tetracosanol, $CH_3(CH_2)_{22}.CH_2OH$ and tetratriacontanol, $CH_3(CH_2)_{32}.CH_2OH$.

The waxes from the conifers contain polymers formed by the ester-linking of many ω-hydroxy acids, such as juniperic acid, with each other.

$HOOC.(CH_2)_{14}.CH_2OH$
Juniperic acid

There are certain waxes which have a characteristic odour. This is due to the presence of hydroxy acids in the form of lactones in them. For example, a wax named *ambretolide*, which is extracted from the seeds of lady's finger, *Abelmoschus esculentus*, has a characteristic musky smell.

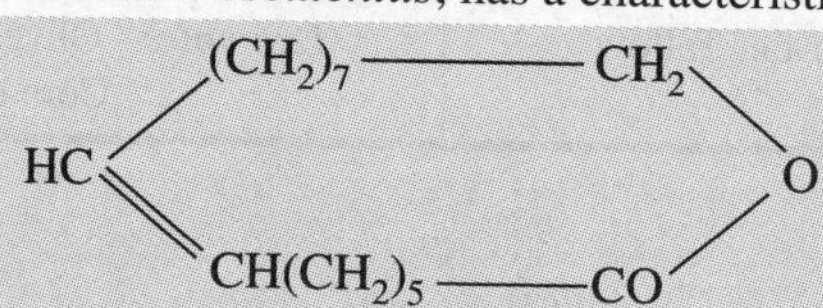

Ambretolide

Lipids of marine organisms such as starfish, squid and shark contain fatty alcohols that are long-chain alkyl ethers of glycerol. Three such *glyceryl ethers* (Fig. 13–7), as they are called, have been isolated from shark oil.

Chimyl alcohol	Batyl alcohol	Selachyl alcohol
$CH_2OCH_2(CH_2)_{14}CH_3$	$CH_2OCH_2(CH_2)_{16}CH_3$	$CH_2OCH_2(CH_2)_7CH{=}CH(CH_2)_7CH_3$
CHOH	CHOH	CHOH
CH_2OH	CH_2OH	CH_2OH

Fig. 13–7. Some glyceryl ethers

Waxes are unusually inert due to their saturated nature of the hydrocarbon chain. However, they can be split slowly with hot alcoholic KOH. They are insoluble in water and highly resistant to atmospheric oxidation. Hence, these are used in polishing furnitures and automobiles and also in wax-coated paper used to wrap perishable food products such as biscuits, cakes, etc. Lanolin (from lamb's wool), beeswax (from honeycomb), carnauba wax (from a Brazilian palm tree) and spermaceti oil (from sperm whales) are widely used in the manufacture of lotions, ointments and polishes.

COMPOUND LIPIDS

PHOSPHOLIPIDS
(= Phosphatids)

Phospholipids are the most abundant membrane lipids. They serve primarily as structural components of membranes and are never stored in large quantities. As their name implies, phospholipids contain phosphorus in the form of phosphoric acid groups. They differ from triglycerides in possessing usually one hydrophilic polar "head" group and usually two hydrophobic nonpolar "tails". For this reason, they are often called *polar lipids*. Thus, *phospholipids are amphipathic, whereas the storage lipids (triglycerides and waxes) are not.* In phospholipids, two of the OH groups in glycerol are linked to fatty acids while the third OH group is linked to phosphoric acid. The phosphate is further linked to one of a variety of small polar head groups (alcohols). **Folch** and **Sperry** (1955) have classified phospholipids into phosphoglycerides, phosphoinositides and phosphosphingosides.

A. Phosphoglycerides

These are the major phospholipids found in membranes and contain two fatty acid molecules or "tails" esterified to the first and second hydroxyl groups of glycerol. The third hydroxyl group of glycerol forms an ester linkage with phosphoric acid. In addition, phosphoglycerides contain a second alcohol, which is also esterified to the phosporic acid. This is referred to as 'head alcohol group' as it is present at one end ('head') of the long phosphoglyceride molecule.

The various phosphoglycerides differ in their head alcohol groups. However, all of them contain two nonpolar tails, each consisting of a long chain (usually C_{16} or C_{18}) fatty acid. Usually one of the fatty acids is saturated and the other unsaturated ; the latter is always esterified to the middle or β-hydroxy group of glycerol. A noteworthy feature of the phosphoglycerides is that they contain an asymmetric carbon atom at position 2 in the glycerol part of their molecule. It has the L-configuration since it is related to L-glyceraldehyde. All phosphoglycerides have a negative charge on phosphoric group at pH 7. In addition, the head alcohol group may also have one or more electric charges at pH 7.

1. *Lecithins (= phosphatidylcholines)* – Lecithins (*likithos*[G] = yolk) are widely distributed in nature. They are found in every animal and plant cell and are an essential constituent of cells, especially membranes. Various oil seeds like soybean and the yeasts are important sources from plant world. In animals, the glandular and nervous tissues are rich in these lipids. The lecithins are required for the normal transport and utilization of as other lipids *esp*., in the liver of animals. Lecithins swell up with water to give slimy emulsions or colloidal solutions. They melt at about 60°C. In their absence, accumulation of lipids occurs in the liver to as much as 30% against a normal value of 3-4%, giving rise to a condition called "*fatty liver*". This fatty infiltration may lead to fibrotic changes, characteristic of the liver disease *cirrhosis*.

In addition to glycerol and 2 moles of fatty acids, the lecithins (Fig. 13–8) also contain phosphoric acid and a nitrogen base choline at either the end or middle carbon atom of glycerol unit. Accordingly, two forms of lecithins, α and β are recognized.

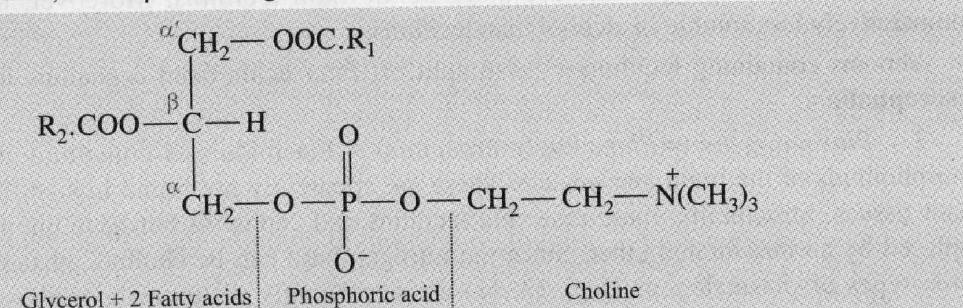

Fig. 13–8. Structure of α-lecithin

On complete hydrolysis, lecithin yields choline, phosphoric acid, glycerol and 2 moles of fatty acids. But partial hydrolysis of lecithins by *lecithinases* (active principles found in snake venoms) causes removal of only one fatty acid to yield substances called **lysolecithins** (Fig. 13–9). These, therefore, contain only one acyl radical. When subjected into the blood stream by sting as a result of snake bite or by needle, the lysolecithins cause rapid rupture (hemolysis) of the red blood corpuscles.

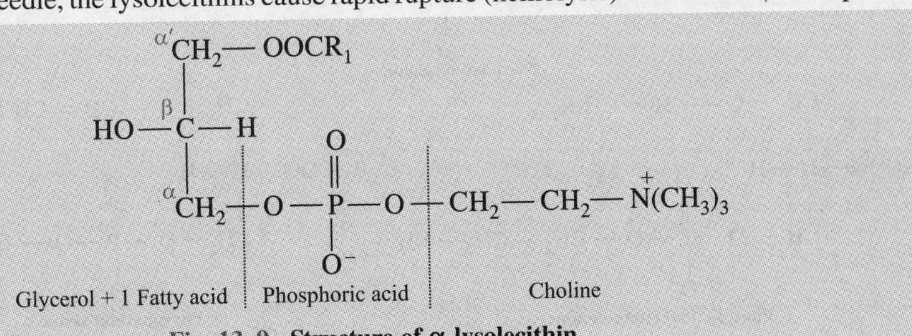

Fig. 13–9. Structure of α-lysolecithin

2. *Cephalins* – The cephalins (*kephalus*[G] = head) are closely associated with lecithins in animal tissues. These have also been identified from soybean oil. These are similar in structure to the lecithins except that the choline is replaced by either ethanolamine or serine. Serine is the biochemical precursor of ethanolamine.

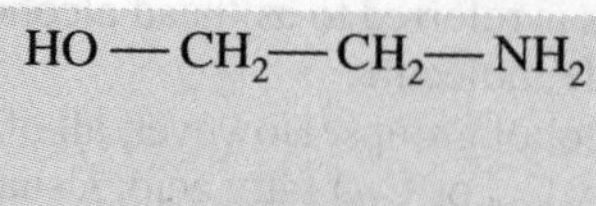

Ethanolamine

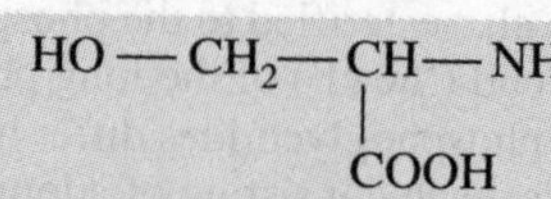

Serine

Accordingly, two types of cephalins are recognized, phosphatidyl ethanolamine and phosphatidyl serine. Like lecithins, the cephalins (Fig. 13–10) also exist in 2 forms, α and β, depending upon the relative positions of the two substituent fatty acids.

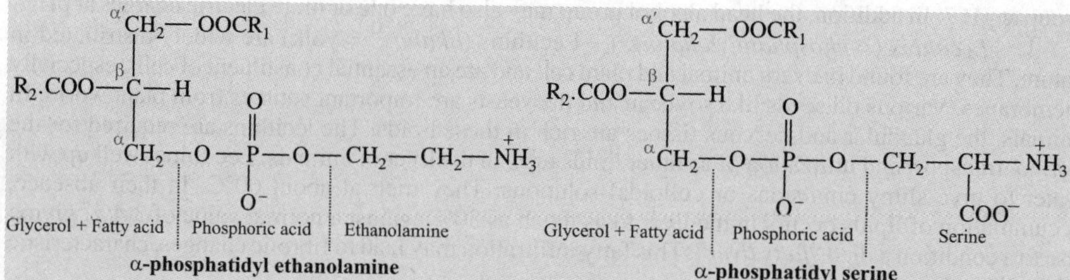

Fig. 13–10. The two cephalins

Since the primary amino group of ethanolamine is a weaker base than the quaternary ammonium group of choline, *the cephalins are more acidic than lecithins.* Moreover, the cephalins are comparatively less soluble in alcohol than lecithins.

Venoms containing lecithinases also split off fatty acids from cephalins, leaving hemolytic **lysocephalins.**

3. *Plasmalogens (=Phosphoglyceracetals)* – Plasmalogens constitute about 10% of the phospholipids of the brain and muscle. These are apparently not found in significant quantities in plant tissues. Structurally, these resemble lecithins and cephalins but have one of the fatty acids replaced by an unsaturated ether. Since the nitrogen base can be choline, ethanolamine or serine, three types of plasmalogens (Fig. 13–11) are accordingly distinguished: phosphatidal choline, phosphatidal ethanolamine and phosphatidal serine.

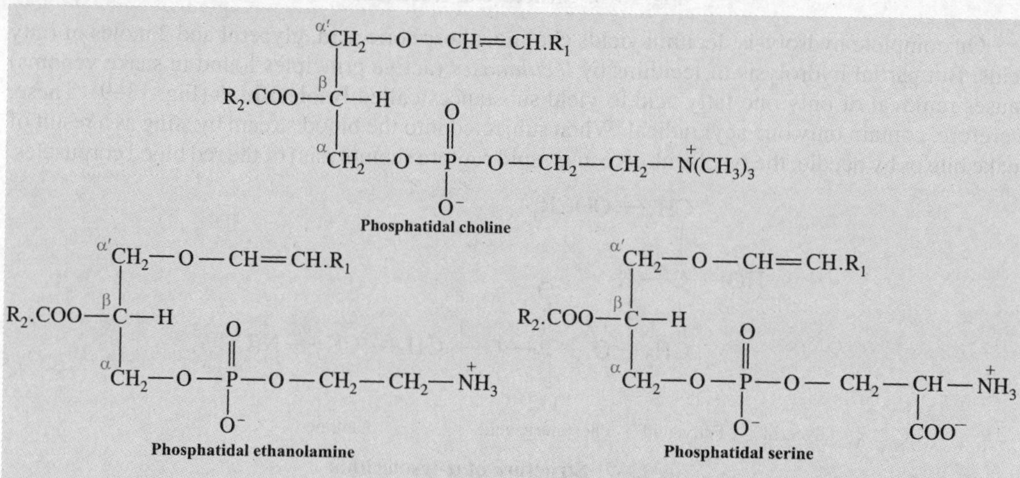

Fig. 13–11. The three plasmalogens

B. Phosphoinositides (=Phosphatidyl inositols)

Phosphoinositides (Fig. 13–12) have been found to occur in phospholipids of brain tissue and of soybeans and are of considerable importance because of their role in transport processes in cells. These are phospholipids where a cyclic hexahydroxy alcohol called inositol replaces base. The inositol is present as the stereoisomer, *myo*-inositol. On hydrolysis, the phosphoinositides yield 1 mole of glycerol, two moles of fatty acid, 1 mole of inositol and 1, 2, or 3 moles of phosphoric acid. Accordingly, mono-, di- or triphosphoinositides are found.

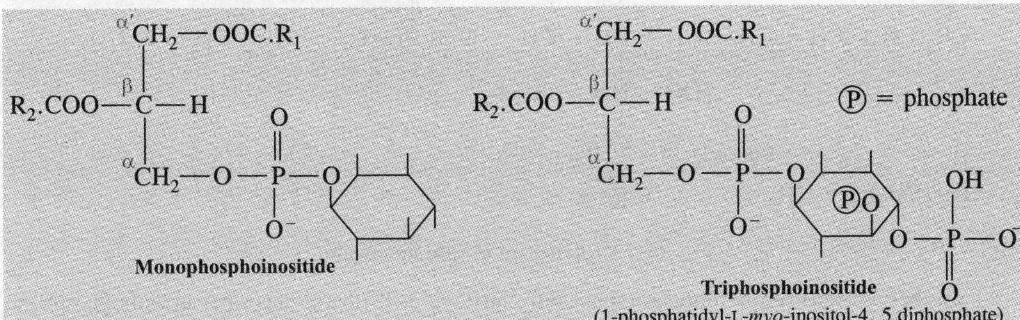

Monophosphoinositide

Triphosphoinositide
(1-phosphatidyl-L-*myo*-inositol-4, 5 diphosphate)

Fig. 13–12. The two phosphoinositides

Phosphoinositides are also classified as glycolipids, in as much as they contain carbohydrate residue.

C. Phosphosphingosides (=Sphingomyelins)

> A **syndrome** (*syn*[G] = together; *dramein*[G] = to run) is a group of symptoms associated with a particular disease or abnormality, although not every symptom is necessarily shown by every patient with the disease.

These compounds are commonly found in nerve tissue *esp.*, in the myelin sheath of the nerve (hence their name, sphingomyelins) and apparently lack in plants and the microorganisms. In a syndrome called *Niemann–Pick disease*, the sphingomyelins are stored in the brain in large quantities. These differ from other phospholipids in their lack of glycerol and the presence of another nitrogenous base sphingosine or a closely related dihydrosphingosine, besides choline, in place of glycerol. Sphingomyelins are electrically charged molecules and contain phosphocholine as their polar head groups.

CLINICAL IMPLICATIONS 13–1.

Niemann–Pick Disease

Niemann-Pick disease (=sphingomyelin lipidosis) is a rare genetic disorder, inherited as an autosomal recessive condition. The disease is caused by a deficiency of the enzyme *sphingomyelinase*, which cleaves ceramide-phosphocholine bond of sphingomyelin. As a result, sphingomyelin accumulates in large amounts in the reticuloendothelial system since their synthesis is normal in rate but their degradation is interrupted. Niemann–Pick cells (prototype of 'foam cells') are found in the bone marrow, spleen, lymphoid tissues, liver, lung and tissues, of virtually any organ. Clinically, the disease manifests itself in two forms:

(a) *Early or infantile form*. This is characterized by hepatos-plenomegaly, macular degeneration, cherry-red retinal spot (=macula) in nearly half of the cases, mental retardation and blindness. Death ensues by two years of age.

(b) *Late or adult form*. This perhaps is the most common of the two forms and reveals intellectual impairment during late infancy, slow evolution of the disease, no changes in the ocular fundi and survival beyond the age of five. However, death of the patient occurs by the second decade.

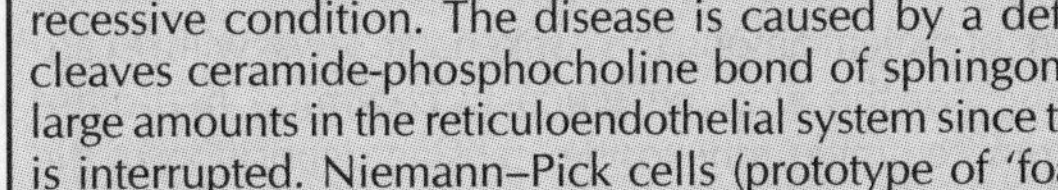

Sphingosine or **4-sphingenine**
(1, 3-dihydroxy-2-amino-*trans*-octadec-4-ene)

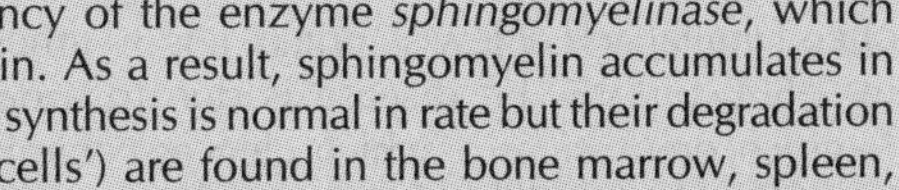

Dihydrosphingosine or **sphinganine**

On hydrolysis, the phosphosphingosides yield equimolar amounts of fatty acid, phosphoric acid, choline and sphingosine or dihydrosphingosine but no glycerol. Thus, in these compounds the atomic ratio N/P is 2, in contrast to phosphoglycerides where this ratio equals unity. Phosphoinositides, however, do not contain a nitrogen base.

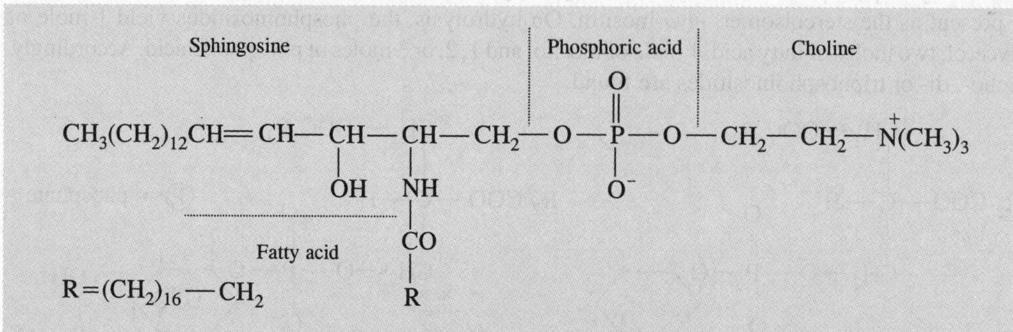

Fig. 13–13. Structure of sphingomyelin

It may be observed from the formula of sphingomyelin (Fig. 13–13) that sphingosine carries the phosphoric acid on its primary alcohol group and the fatty acid by amide linkage on its primary amino group.

GLYCOLIPIDS
(= Cerebrosides or Glycosphingosides)

Glycolipids are substances whose molecules consist of a lipid covalently attached with a simple or complex sugar residue. They are important constituent of brain where they amount to about 8% of the solid matter (hence, their alternative nomenclature *cerebrosides*). These may also occur in tissues other than brain. Since the head group characteristically consists of one or more sugar units, cerebrosides are often called glycosphingosides. Like phospholipids, glycolipids are composed of a hydrophobic region, containing 2 long hydrocarbon "tails", and a hydrophilic polar "head" region, which now contains one or more sugar residues and no phosphate. Both phospholipids and glycolipids form self-sealing lipid bilayers that are the basis for all cellular membranes. In *Gaucher disease*, the cerebrosides appear in relatively large amount in the liver and the spleen. They are also present in large amount in the brain in Niemann–Pick disease. These are present in much higher concentration in medullated than in nonmedullated nerve fibres. There is evidence that they also occur in some plant organs.

CLINICAL IMPLICATIONS 13–2.

Gaucher Disease

Gaucher Disease(= glucocerebrosidosis) is a not-so-rare, autosomal recessive, hereditary syndrome, often familial and of Jewish origin. It is 30 times more prevalent among the Ashkenazic Jews, with 1 case in 2,500 births. The disease is due to a deficiency of the enzyme *glucosyl ceramide hydrolase* (=*glucocerebrosidase*), which cleaves ceramide from ceramide trihexoside, causing deposition of the later. Gaucher's cells (pale lipid-containing cells) are found in abundance in all the organs specifically in the lungs and bone marrow, resulting in marked hepatosplenomegaly. Thus, the disease affects the lungs and causes destruction of bones. Clinically, two patterns of the disease have been reported:

(*a*) *Infantile* or *acute form.* In this, the symptoms appear early with hepatosplenomegaly, mental retardation, hypertonicity and signs of respiratory involvement. Most patients die during infancy.

(*b*) *Adult* or *chronic form.* This is more common a form of this disease. There is slow evolution of the disease with hepatosplenomegaly, hypersplenism, pathological fractures and often no neurological deficit.

The structure of cerebrosides (Fig. 13–14) is somewhat similar to that of sphingomyelins (phosphosphingosides). They contain a high molecular weight fatty acid, sphingosine and either galactose or glucose instead of choline but no phosphoric acid. They have no electric charge since their polar head groups are neutral. In general properties, they resemble sphingomyelins.

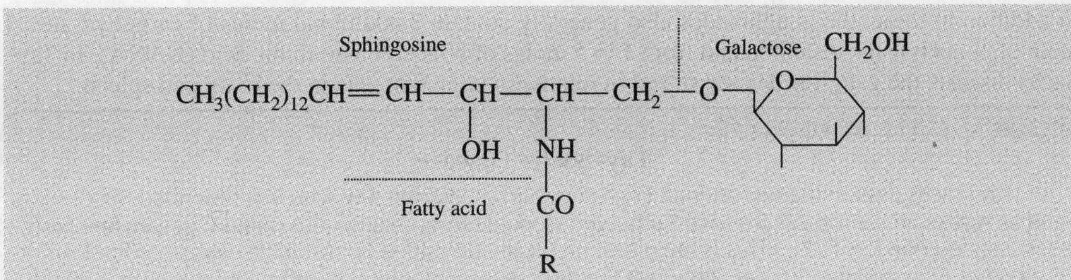

Fig. 13–14. Structure of a cerebroside

The name **ceramide** is commonly used to designate the sphingosine–fatty acid (or N-acylsphingosine) portion of the cerebrosides. Here again the sphingosine carries the galactose by glycosidic linkage on its primary alcohol group and the fatty acid by an amide linkage on its primary amino group.

Individual cerebrosides are differentiated on the basis of their fatty acid component (Fig. 13–15). The various classes, so differentiated, are as follows:

$CH_3(CH_2)_{22}COOH$ — **Lignoceric acid**

$CH_3(CH_2)_{21}CH(OH)COOH$ — **Cerebronic acid**

$CH_3(CH_2)_7CH=CH(CH_2)_{13}COOH$ — **Nervonic acid**

$CH_3(CH_2)_7CH=CH(CH_2)_{12}CH(OH)COOH$ — **Oxynervonic acid**

Fig. 13–15. Characteristic fatty acids of cerebrosides

(*a*) Kerasin– contains saturated C 24 lignoceric acid.

(*b*) Phrenosin (cerebron)– contains a 2-hydroxy derivative of lignoceric acid called cerebronic acid.

(*c*) Nervon– contains an unsaturated homologue of lignoceric acid called nervonic acid,

(*d*) Oxynervon– contains a 2-hydroxy derivative of nervonic acid called oxynervonic acid.

A. Gangliosides

In 1955, **Klenk** isolated a new type of glycolipid from brain tissue and named it as **ganglioside.** These are found in significant concentrations in ganglion cells of nervous tissue (hence so named) and also in most parenchymatous tissues like spleen and erythrocytes. They make up about 6% of the membrane lipids in the gray matter of the brain. They are also found in lesser amounts in the membranes of most nonneural tissues.

The structure of gangliosides (Fig. 13–16) is complex and related to that of cerebrosides in that they contain a ceramide (N-acylsphingosine) linked to a carbohydrate (galactose or glucose).

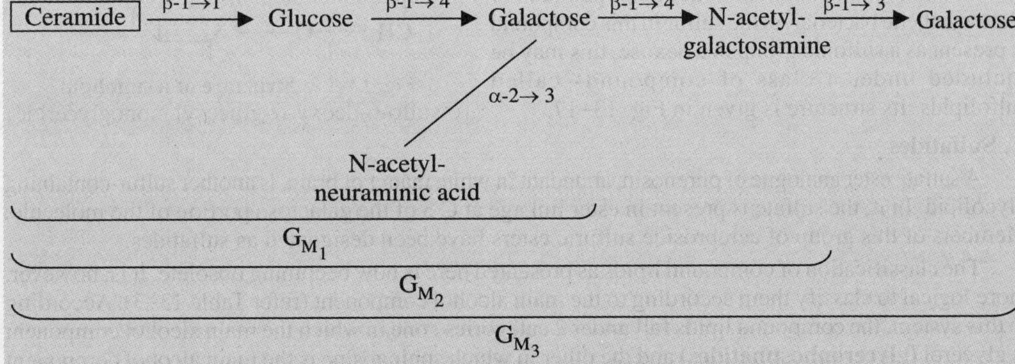

Fig. 13–16. The three gangliosides

In addition to these, the gangliosides also generally contain 2 additional moles of carbohydrates, 1 mole of N-acetylgalactosamine and from 1 to 5 moles of N-acetylneuraminic acid (NANA). In Tay–Sachs disease, the gangliosides are stored in relatively large amounts in the brain and spleen.

CLINICAL IMPLICATIONS 13–3.

Tay–Sachs Disease

Tay–Sachs disease (named after an English physician **Warren Tay** who first described the disease and an American neurologist **Bernard Sachs** who worked out its details), also called **G_{M2} gangliosidosis,** was first described in 1881. This is the oldest medically described lipid storage disease or lipidosis. It is a common hereditary disorder. Although the disease is rare in the population at large (1 in 3,00,000 births), it has a very high incidence (1 in 3,600 births) in Ashkenazic Jews, who make up about 90% of the Jewish population in America. One in 30 Ashkenazic Jews carries the defective gene in recessive form (*i.e.*, the carrier rate is 1/30), whereas the carrier rate is 1/300 in non-Jewish Americans. Consequently, the incidence of the disease is about 100 times higher in Jewish Americans. When both parents are carrier, their children are likely to develop Tay-Sachs disease even though the parents show no symptoms of the disease. Because the disease is irreversible and needs intensive care, genetic counseling of parents has become important in averting its consequences. The disease is caused due to a deficiency of the enzyme N-*acetylgalactosaminidase* (= *hexosaminidase A*), which cleaves a specific bond (β-8→4) between an N-acetyl-D-galactosamine and a D-galactose residue in the polar head of the ganglioside, G_{M2}. In effect, G_{M2} is not degraded to G_{M1}. With the result, G_{M2}, which is also called as Tay–Sachs ganglioside, is accumulated in large amounts in the lysosomes, particularly in the brain cells to the tune of 100-300 times the normal G_{M2} content, causing degeneration of the nervous system. In Tay–Sachs disease, the symptoms are evident before an affected infant is a year old. Weakness, retardation in development and difficulty in eating are typical early symptoms. Mental retardation and blindness usually follow a few months later; death, between two and five years, is inevitable. Over 90% of the patients have a characteristic cherry-red spot in the retina. Tay–Sachs disease can be diagnosed during fetal development. Amniotic fluid is obtained by amniocentesis and assayed for b-N-acetylgalactosaminidase activity. As mentioned, Tay–sachs disease shares the tragic element of Niemann–Pick and Gaucher diseases in that the affected child seldom lives beyond age three.

More than 15 different gangliosides have been characterized and their structures determined. These are commonly abbreviated by the letter G, with a subscript $_{M}$, $_{D}$ or $_{T}$ to designate that they contain one, two or three sialic acid and (N-acetylneuraminate or N-glycolyl neuraminate) residues respectively, and a number or letter to distinguish different members of a group from one another. In essence, they consist of an oligosaccharide chain attached to ceramide (N-acylsphingosine) by a sugar residue which is usually glucose. They lack phosphoric acid. Gangliosides are thought to act as receptors for toxic agents like the pathogens, *Vibrio cholerae* influenza virus and tetanus toxin. They are also implicated to play a role in cell-to-cell interaction.

B. Sulfolipids

A glycolipid that contains sulfur is widely distributed in plants. It is localized in the chloroplasts but is also found in the chromatophores of photosynthetic bacteria. As the sulfur in this compound is present as a sulfonic group in a hexose, this may be included under a class of compounds called sulfolipids. Its structure is given in Fig. 13–17.

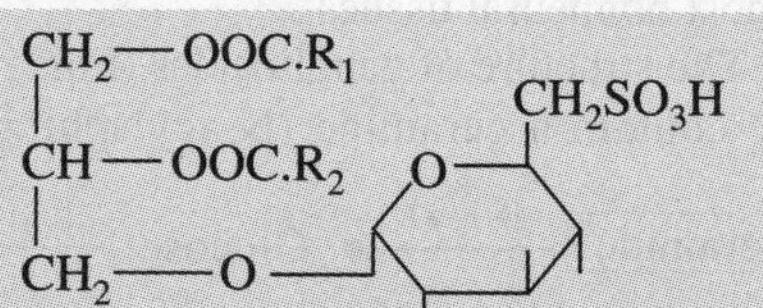

Fig. 13–17. Structure of a sulfolipid (6-sulfo-6-deoxy-α-glucosyl monoglyceride)

C. Sulfatides

A sulfate ester analogue of phrenosin, abundant in white matter of brain, is another sulfur-containing glycolipid. In it, the sulfate is present in ester linkage at C 3 of the galactose portion of the molecule. Members of this group of cerebroside sulfuric esters have been designated as sulfatides.

The classification of compound lipids as presented here is now becoming obsolete. It is, however, more logical to classify them according to the main alcohol component (refer Table 13–3). According to this system, the compound lipids fall under 2 categories : one in which the main alcohol component is glycerol **(glycerophosphatides)** and the other in which sphingosine is the main alcohol component **(sphingolipids).** The further classification of these two categories depends on the nature of the other alcohol components.

Table 13–3. Classification of compound lipids

Name	*Main alcohol component(s)*	*Other alcohol component(s)**	*N:P ratio*
Glycerophos-phatides			
Phosphatidic acids	Diglyceride (= glycerol diester)	—	0 : 1
Lecithins	Diglyceride (= glycerol diester)	Choline	1 : 1
Cephalins	Diglyceride (= glycerol diester)	Ethanolamine, serine	1 : 1
Inositides	Diglyceride (= glycerol diester)	Inositol	0 : 1
Plasmalogens	Monoglyceride (= glycerol ester) + Enol ether	Choline, ethanolamine, serine	1 : 1
Sphingolipids			
Sphingomyelins	N-acylsphingosine	Choline	2 : 1
Cerebrosides	N-acylsphingosine	Galactose, glucose	1 : 0
Sulfatides	N-acylsphingosine	Galactose	($1H_2SO_4$)
Gangliosides	N-acylsphingosine	Galactose, glucose + N-acetylgalactosamine + N-acetylneuraminic acid	No P

* The other alcohol components in all the sphingolipids except sphingomyelins are not present as phosphoric esters, but rather in glycosidic linkages. For this reason, these sphingolipids are grouped under glycolipids.

DERIVED LIPIDS

The group derived lipids is a **"catch all" group** in Bloor's classification. It includes the hydrolysis products of simple and compound lipids and also various other compounds such as steroids, terpenes, fatty acids, alcohols, fatty aldehydes, ketones etc.

STEROIDS

The steroids (*stereos*^{G} = solid) are one of the most studied classes of biological compounds and are often found in association with fat. Since they contain no fatty acids, they are *nonsaponifiable*, *i.e.*, cannot be hydrolyzed by heating with alkali to yield soaps of their fatty acid components. Fats, on the other hand, are saponifiable and form soaps when hydrolyzed with alkali. The steroids may be separated from the fat after the latter is saponified since they occur in 'nonsaponifiable residue'. All steroids may be considered as derivatives of a fused and fully saturated ring system called **cyclopentanoperhydrophenanthrene** or **sterane.** This system consists of 3 cyclohexane rings (A, B and C) fused in nonlinear or phenanthrene manner and a terminal cyclopentane ring (D). The sterane nucleus along with the conventional numbering of the carbon atoms is shown in Fig. 13–18.

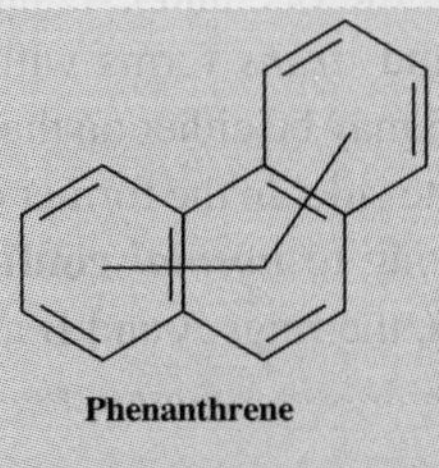

Phenanthrene

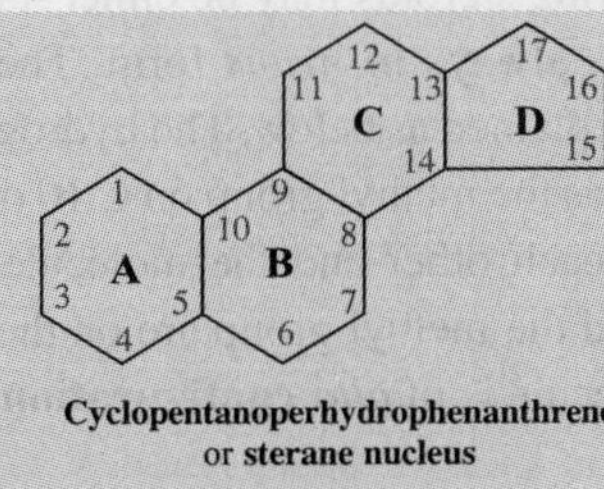

Cyclopentanoperhydrophenanthrene or sterane nucleus

Fig. 13–18. Phenanthrene and sterane nucleus

It may, however, be emphasized that in steroids the hexagonal rings are not the benzene rings as in them the valences of C atoms are fully satisfied by hydrogen bonds, unless shown otherwise. The double bonds, if present, are shown as such. The 'angular' methyl groups occur typically at positions 10 and 13 and constitute carbon atoms 19 and 18 respectively. The steroids may have one or more hydroxyl groups, one OH group being present usually on C 3. A side chain at C 17 is usual. This side chain serves as a convenient basis for classification of steroids. For example, the side chain contains 8, 9 or 10 C atoms in sterols, 5 C atoms in bile acids, 2 C atoms in adrenal cortical steroids and in progesterone and none in naturally-occurring estrogens and androgens.

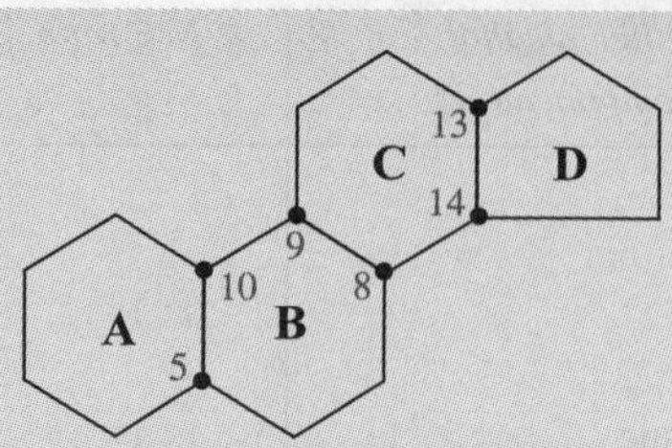

Fig. 13–19. Asymmetric centres in sterane (= gonane) nucleus
(shown by *solid black circles*)

Stereochemical considerations of steroids. Without even considering the substituent groups, there are 6 possible asymmetric centres present in the sterane nucleus (Fig. 13–19). Consequently, the steroids have many potential stereoisomers. Each of 3 cyclohexane rings, on the basis of the tetrahedral theory, is capable of existing in 3-dimensional conformation either of a **'chair'** or of a **'boat'** (Fig. 13–20). These 2 forms are aplanar and strainless and were first proposed by Sachse in 1890. The 'chair' form is *rigid* (in the sense that it resists distortion) and when it is changed to the 'boat' form, some angular deformity invariably takes place. The 'chair' conformation of cyclohexane itself is more *stable* than the 'boat' conformation. The 'chair' and 'boat' forms are both free from angle strain but because of differences in steric strain and bond opposition strain, the two forms differ in energy content (**Finar IL**, 1975).

> Steroids containing hydroxyl groups (hydroxy steroids) are often referred to as **sterols**. Broadly speaking, the various sterols fall under 2 groups based on their source : the animal sterols or *zoosterols* (cholesterol, cholestanol, coprostanol) and the plant sterols or *phytosterols* (stigmasterol, spinasterol, sitosterol). A third group of sterols, which are obtained from fungi including yeasts, is sometimes separately referred to as *mycosterols* (*e.g.*, ergosterol). This classification, however, is not rigid as some sterols are obtained from more than one of these sources.

Fig. 13–20. Two types of conformation of a cyclohexane ring

In naturally occurring steroids, all the rings are in 'chair' form. With respect to each other, the rings can be either *-cis* or *-trans* (Fig. 13–21). The junction between the rings A and B (A/B junction) in naturally-occurring steroids may be either *-cis* or *-trans*. But the B/C and C/D junctions are, with certain exceptions, always in *-trans* form. The idea of *-cis* and *-trans* forms with respect to A/B junction may be explained as below. The hydrogen atom on C 5 may be either on the same side of the plane of molecule as the methyl group on C 10 or on the opposite side. In the former case, rings A and B will be *-cis* to each other and the molecule is said to belong to ***normal*** **configuration.** If the hydrogen atom and the methyl group are on the opposite sides, the rings A and B are *-trans* to each other and the molecule is of ***allo*** **configuration.**

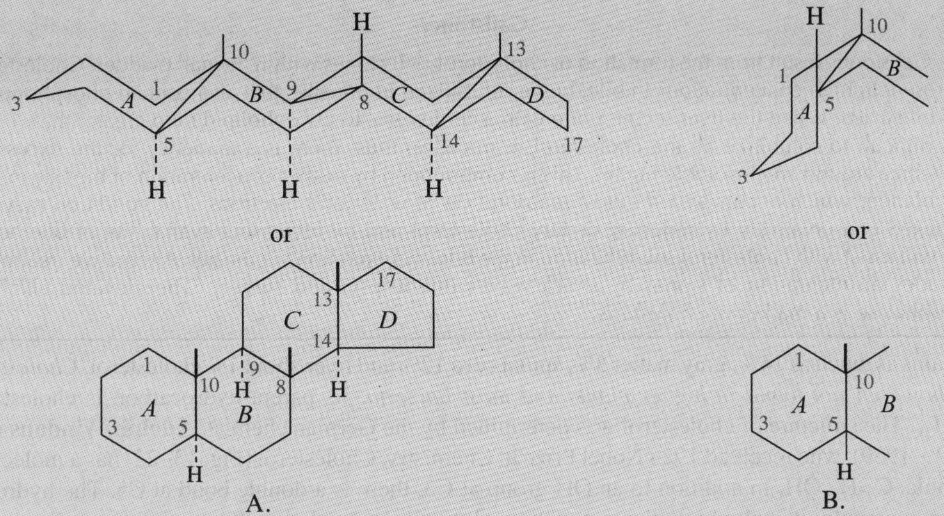

Fig. 13–21. Stereochemical aspects of steroid nucleus, A. all - *trans* configuration between adjacent rings, B. a -*cis* configuration between rings A and B

In the structural representation of the steroids, substituents that are above the plane of the ring (or β-**oriented**) are shown with solid lines, while those which are below the plane of the ring (or α-**oriented**) are represented with broken lines. The methyl groups at C 10 and C 13, the hydroxyl group on C 3 and the side chain at C 17 are all β-oriented.

Certain common steroids are described below.

C 29, C 28 and C 27 Steroids

1. Cholesterol (*chole*^{G} = bile). Cholesterol is undoubtedly the most publicized lipid in nature, because of the strong correlation between high levels of cholesterol in the blood and the incidence of diseases of the cardiovascular system in humans. It is not only an important component of some cell membranes and of plasma lipoproteins but also the precursor of many other biologically important steroids, such as bile acids and various steroid hormones. It is the principal sterol of higher animals and is especially abundant in nerve tissues and in gallstones. It occurs either free or as fatty esters in all animal cells. It was first isolated in 1784, from human **gallstones** which consist almost entirely of cholesterol and hence so named (cholesterol literally means 'solid alcohol from bile'). Human body contains 1/4 lb of cholesterol! Its main sources are fish liver oils and the brain and spinal cord of cattle. White matter

> **Cholesterol** is generally believed to be notorious as a major cause of heart disease. There are 2 types of cholesterol, the low-density lipoprotein cholesterol. (LDL-C) and the high-density lipoprotein cholesterol (HDL-C). The LDL-C grows with intake of eggs and dairy products and triggers heart trouble. However, in a study conducted by **Dr. Ian Craig** of the Goteborg University, Sweden in the early 70s showed that patients with heart trouble such as myocardial infarction had low levels of HDL-C, as compared to other people. He pointed out that most of the preventive heart attack studies have so far been aimed at lowering the LDL-C levels through drugs and diet, but it seemed that normal variations in diet made little difference in HDL-C levels. Also, HDL-C levels were generally found higher in women than men. Perhaps this is the reason why fewer women than men have heart attacks. There are, thus, hopeful signs that incidence of heart disease may be lowered by increasing the HDL-C level in heart patients.

CLINICAL IMPLICATIONS 13–4.

Gallstones

Gallstones result from the formation of cholesterol-rich stones within the gall bladder. Cholesterol is present in high concentrations in bile, being solubilized in micelles that also contain phospholipids and bile acids. When the liver secretes bile with a cholesterol to phospholipid ratio greater than 1 : 1, it is difficult to solubilize all the cholesterol in micelles; thus, there is a tendency for the excess to crystallize around any insoluble nuclei. This is compounded by further concentration of the bile in the gall bladder which occurs as a result of reabsorption of water and electrons. The condition may be managed conservatively by reducing dietary cholesterol and by increasing availability of bile acids that will assist with cholesterol solubilization in the bile and excretion *via* the gut. Alternative treatment includes disintegration of stones by shock waves (lithotripsy) and surgery. The elevated alkaline phosphatase is a marker of *cholestasis*.

contains as much as 14%, gray matter 5%, spinal cord 12% and liver about 1% cholesterol. *Cholesterol is , however, not found in higher plants and most bacteria.* Its parent hydrocarbon is cholestane, $C_{27}H_{48}$. The structure of cholesterol was determined by the German chemist, **Adolph Windaus** (LT, 1879 – 1959), who received 1928 Nobel Prize in Chemistry. Cholesterol (Fig. 13–22) has a molecular formula, $C_{27}H_{45}OH$. In addition to an OH group at C3, there is a double bond at C5. The hydroxyl group constitutes its polar head, the rest of the molecule is hydrophobic. It is a white crystalline solid and is optically active, $[\alpha]_D$ 39°. The crystals are rhombic plates with one of the angles broken. It has a melting point of 149°C. **Konrad E. Bloch**, a Germany-born American biochemist, elucidated the biosynthetic pathway of cholesterol synthesis, one of the most complex known, for which he received the coveted Nobel Prize in Chemistry along with **Feodor Lynen**, a German and **John Cornforth**, a Briton, in 1964. **Bloch** combined work and pleasure by traveling to Bimini for collection of shark livers.

ADOLF OTTO REINHOLD WINDAUS

(LT, 1876–1959)

Windaus, a German chemist, was born on December 25, 1876, in Berlin. His interest in chemistry was raised by the lectures of the 1902 Nobel Laureate in Chemistry, **Hermann E. Fischer** of Germany. He started studying medicine and chemistry in Berlin and later in Freiburg. He got his Ph.D. in early 1900 and focused on cholesterol and other sterols at the University of Freiburg. In 1913, he became Professor of Chemistry at the University of Innsbruck and, in 1915, he changed to the University of Göttingen where he stayed until his retirement in 1944. He was involved in the discovery of the **transformation of cholesterol** through several steps to vitamin D_3 (cholecalciferol), for which work he earned the **1929 Nobel Prize in Chemistry.** He gave his patents to Merck and Bayer and they brought the medical *Vigantol* in 1927. Windaus was the doctoral advisor of his compatriot **Adolf Butenandt** who also won a Nobel Prize in Chemistry in 1939 along with **Leopold S. Ružička**, a Swiss.

KONRAD EMIL BLOCH

(LT, 1912–2000)

Bloch, a German-born U.S. biochemist, was born an January 21, 1912, in Niesse (Nysa) in the German Empire's Prussian Province of Silesia. From 1930 to 1934, he studied chemistry at the Technical University of Munich. In 1934, due to the Nazi persecutions of Jews, he fled to the *Schweizerische Forschungsinstitut* in Davos, Switzerland, before moving to the United States in 1936. Later, he was appointed to the Department of Biological Chemistry at Yale Medical School. In the US, Bloch enrolled at Columbia University, and received a Ph.D. in Biochemistry in 1938. He taught at Columbia from 1939 to 1946. From there, he went to the University of Chicago and then to Harvard University as Higgins Professor of Biochemistry in 1954, a post he held until 1982. After retirement at Harvard, he served as the Mack and Effie Campbell Tyner Eminent Scholar Chair in the College of Human Sciences at Florida State University.

Bloch shared the **1964 Nobel Prize in Physiology or Medicine** with **Feodor Lynen**, a German, for their discoveries concerning the mechanism and regulation of the cholesterol and fatty acid metabolism. He combined work and pleasure by traveling to Bimini for collection of shark livers. He was elected **Fellow of the Royal Society** (F.R.S.) in 1985. In 1988 he was awarded the **National Medal of Science.** Bloch died on October 15, 2000 in Lexington, Massachusetts of congestive heart failure (C.H.F.), at age 88.

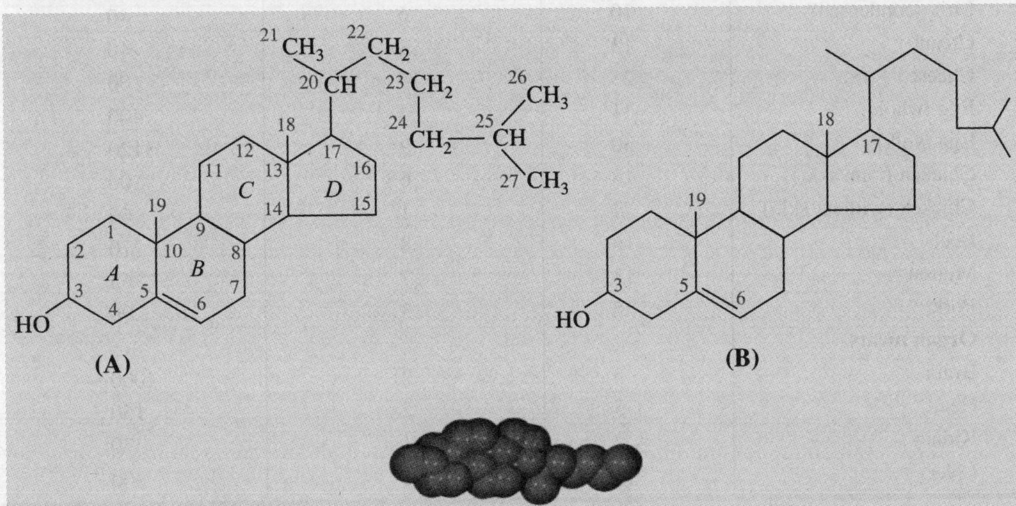

(A) (B) (C)

Fig. 13–22. Structure of cholesterol

(3-hydroxy-5,6-cholestane)

(A) Molecular representation. The individual rings in the fused-ring system are designated A through D.

(B) Simplified version. It is customary to write steroid structure in the manner shown here, without, however, the methyl substituents.

(C) Space-filling model. The C is shown with *grey*, vinyl C with *yellow*, O with *red*, N with *blue* and P with *green*. H is not shown.

[*Courtesy : (C) Richard Pastor and Richard Venable*]

On receiving the Nobel Prize in 1985 for their work on cholesterol, **Michael Stuart Brown** and **Joseph Goldstein** recounted in their lecture the extraordinary history of cholesterol:

"Cholesterol is the most highly decorated small molecule in biology. **Thirteen Nobel Prizes** have been awarded to scientists who devoted major parts of their careers to cholesterol. Ever since it was isolated from gallstones in 1784, cholesterol has exerted an almost hypnotic fascination for scientists from the most diverse areas of science and medicine.... Cholesterol is a Janus-faced molecule. The very property that makes it useful in cell membranes, namely its absolute insolubility in water, also makes it lethal."

(Courtesy: The Nobel Foundation, 1985)

Table 13-4 provides data on fat and cholesterol contents of some notable animal foods.

Table 13-4. Fat and cholesterol content of some animal foods*

Food Item	*Fat g/100 g*	*Saturated fatty acids g/100 g*	*Cholesterol mg/100 g*
Butter	80	50	250
Ghee	100	65	300
Milk (cow)	4	2	14
Milk (buffalo)	8	4	16
Milk (skimmed)	0.1	—	2
Milk (condensed)	10	6	40
Cream	13	8	40
Cheese	25	15	100
Egg (whole)	11	4	400
Egg (yolk)	30	9	1120
Chicken (with skin)	18	6	100
Chicken (without skin)	4	1	60
Beef	16	8	70
Mutton	13	7	65
Pork	35	13	90
Organ meats			
Brain	6	2	2000
Heart	5	2	150
Kidney	2	1	370
Liver	9	3	300

* The data are based on per 100 g edible portion of the food item.

No vegetable oil contains any cholesterol; hence, it is misleading for oil manufacturers to advertise any oil as 'low cholesterol' oil. Saturated fats of animals origin (such as butter, ghee and lard) do contain cholesterol but *vanaspati,* which is derived from vegetable oils, does not contain cholesterol. Only a little portion of the body cholesterol is derived from diet. The bulk of it is synthesized in the body. Unsaturated fatty acids lower blood cholesterol levels by controlling cholesterol's synthesis as well as its elimination from the body. Thus, vegetable oils when consumed within the recommended amounts help to maintain proper levels of lipids in the blood than do the saturated fats like butter, ghee, lard and vanaspati. These saturated fats elevate blood lipids (cholesterol and triglycerides) and reduce the ratio of good to bad cholesterol even when consumed within the recommended limits. High intake of dietary cholesterol can probably raise blood cholesterol levels and hence should be avoided, esp., by the elderly people and those having high blood cholesterol.

The level of cholesterol in the blood is measured in milligrams per decilitre (mg/dl, which is equivalent to parts per 1,00,000). The levels range from less than 50 in infants to an average of 215 in adults to 1,200 or more in individuals suffering from a rare, inherited disease called **familial cholesterolemia**. For those persons in the normal range, about two-thirds of their cholesterol is transported as LDLs. Most of the rest is carried by HDLs.

An often-posed question relates to the consumption of eggs and meat, and sometimes even milk, because of the relatively high content of cholesterol in these foods. Milk is no doubt a very good and safe food. Persons aspiring to control their cholesterol intake (and calories) must use defatted or skimmed milk. Egg and meat proteins are of high grade and desirable especially for the growing children. Those desirous of reducing their cholesterol intake should eat only the white of egg and lean on white meat like chicken and fish. Nutritionally, fish is a very good food since it has high-quality protein (fish fat is rich in *n*-3 PUFA) and its flesh has very little cholesterol. Organ meats like brain and liver are rich in cholesterol and should hence be avoided by those wanting to restrict cholesterol.

The notion that cholesterol is a poisonous substance in the body is misleading. It is the parent hydrocarbon from which many important hormones (corticosteroids and sex steroids, for example) are synthesized. The problem creates when blood cholesterol level rise and the level of LDL-cholesterol, which carries cholesterol to the tissues, goes below the desired level. HDL-cholesterol or good cholesterol removes cholesterol from the tissues and helps eliminating it from the body. High blood cholesterol results in the deposition of cholesterol in the arteries and narrows the passage, a condition called *atherosclerosis.* Also, the arterial surface roughens due to cholesterol deposition, which leads to frequent blood clotting or *thrombosis.* Both are major health hazards. Studies, however, point out that even very low levels of blood cholesterol are too not desirable. *The thumb rule, therefore, is moderation.* Both too much and too little of fat may be harmful.

In blood about two-thirds of the cholesterol is esterified mainly to unsaturated fatty acids, the remaining portion occurring as the free alcohol. Reduction of the double bond gives rise to 2 products, coprostanol and cholestanol.

2. Coprostanol (= coprosterol). It occurs in feces and is produced in the intestine as a result of bacterial action on the double bond of cholesterol. The A/B junction is *-cis* in coprostanol in contrast to *-trans* in cholesterol.

3. Cholestanol. It occurs as a minor constituent of the sterols of blood and other tissues. Here the A/B junction is in *-trans* form.

4. Ergosterol. It is present in ergot (hence its nomenclature), yeast and the mould *Neurospora.* Its parent hydrocarbon is ergostane, $C_{28}H_{50}$. Ergosterol (Fig. 13–23) has a molecular formula, $C_{28}H_{43}OH$ with one OH group at C_3 and 3 double bonds at C_5, C_7 and C_{22}. It is also optically active, $[\alpha]_D$ -135°.

Rupture of the ring B by UV radiation produces vitamin D_2 which is chemically known as ergocalciferol. A similar compound cholecalciferol (or vitamin D_3) is, however, obtained from 7-dehydrocholesterol on irradiation with UV light (refer Chapter 32).

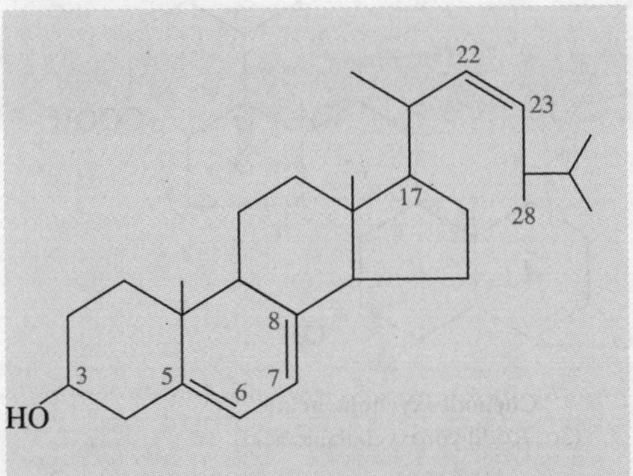

Fig. 13–23. Structure of ergosterol

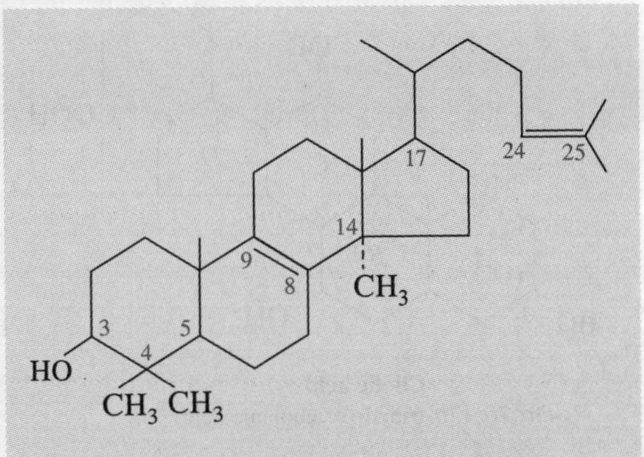

Fig. 13–24. Structure of lanosterol

5. Lanosterol (= cryptosterol). It is a major constituent of wool fat and is also present in minor quantities in liver and yeast. Lanosterol (Fig. 13–24) is a C 30 compound with twin methyl groups at C 4 and a third angular methyl group on C 14. There are 2 double bonds at C_8 and C_{24}. It is an intermediate in the biosynthesis of cholesterol.

A number of sterols have also been obtained from various plant and animal sources, for example **stigmasterol** (from soybean and wheat germ oils), **spinasterol** (spinach and cabbage), **sitosterol** (many higher plants), **ostreasterol** (oysters) and **chondrillasterol** (marine sponges).

C 24 Steroids or Bile Acids

The bile acids are the important end products of cholesterol metabolism in higher plants. About 20 natural bile acids have been characterized (Table 13–4). All these are derived from a C 24 parent steroid, cholanic acid and resemble coprostanol in having rings A and B in cis form. The most abundant bile acids in human bile are cholic acid (25-60% of the total bile acids), chenodeoxycholic acid (30-50%) and deoxycholic acid (5-25%). Various bile acids differ from each other in the number and position of OH groups which are all in a configuration. In the bile acids, the number of OH group(s) may be 1, 2, or 3 and the position of OH group(s) may be any of the following : 3, 6, 7, 11, 12 and 23. The side chain is usually made up of 5 carbon atoms and bears the carboxyl group.

Table 13–5. Some important natural bile acids

Name	*Formula*	*Melting point, 0ºC*	*Position of OH group(s)*	*Source*
Monohydroxycholanic acid	$C_{23}H_{38}(OH).COOH$			
Lithocholic acid		186	3	Man, Ox
Dihydroxycholanic acids	$C_{23}H_{37}(OH)_2.COOH$			
Deoxycholic acid		172	3, 12	Man, Ox
Chenodeoxycholic acid		140	3, 7	Man, Ox, Hen
α-hydrodeoxycholic acid		197	3, 6	Pig
Trihydroxycholanic acid	$C_{23}H_{36}(OH)_3.COOH$			
Cholanic acid		195	3, 7, 12	Man, Ox

The structure of two principal bile acids is given in Fig. 13–25.

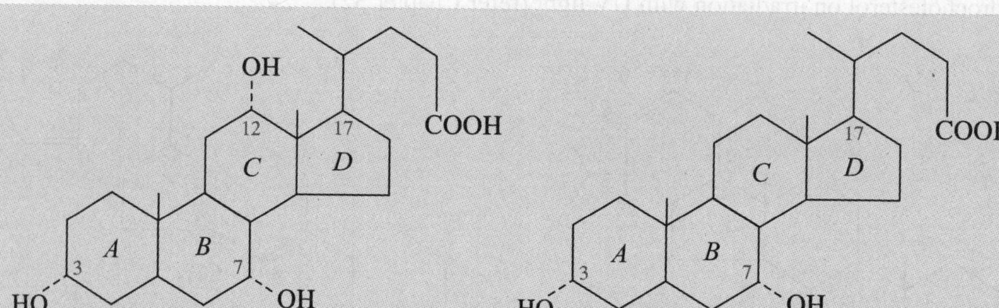

Cholic acid
(3α,7α,12α-trihydroxycholanic acid)

Chenodeoxycholic acid
(3α,7α-dihydroxycholanic acid)

Fig. 13–25. Two principal bile acids

These acids are coupled in amide linkage to the amino acids glycine (NH_2—CH_2—COOH) and taurine (NH_2—CH_2—CH_2—SO_3H) to form glycocholic and taurocholic acids respectively (Fig. 13–26). The reaction is characteristic of COOH group and involves the side chain only. The salts of these conjugated acids have high surface activity and hence promote the intestinal absorption of lipids like cholesterol. They are water-soluble and are powerful detergents.

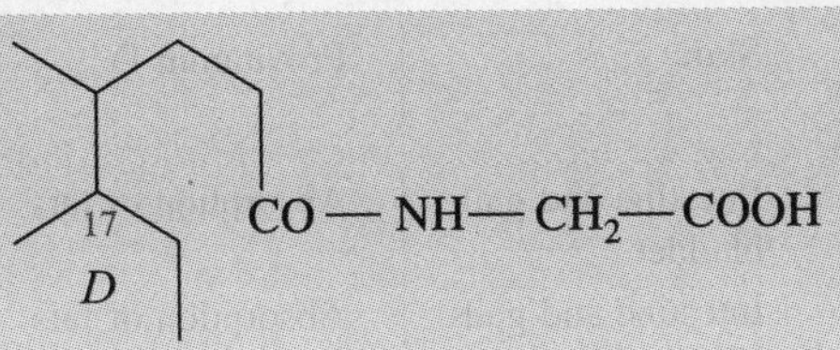

Glycocholic acid
(=cholylglycine)

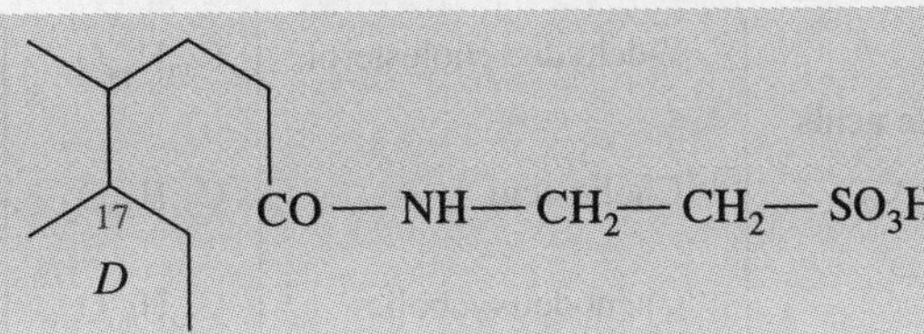

Taurocholic acid
(=cholylglycine)

Fig. 13–26. Side chain of two bile-acid amides

C 21, C 19 and C 18 Steroids

On account of their hormonal role, these have been described in Chapter 34.

Several important poisons are based on the steroid structure. **Digitoxin** or **digitalin** (Fig. 13–27 *a*, *b*) is present in the foxglove plant (*Digitalis purpurea*) and has a powerful heart stimulating action. Another cardiac glycoside, **Quabain**, (Fig. 13–27 *c*) is isolated from the East African Quabio tree and has the sugar, rhamnose, attached to a modified sterol nucleus. It is of interest to note that it is a powerful inhibitor of the 'sodium pump', a device which normally ensures that the cell content of potassium is higher and that of sodium is lower than in the circumambient fluid.

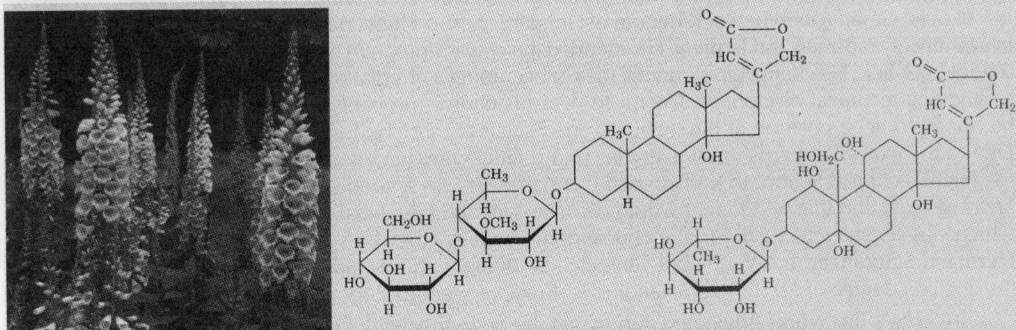

(*a*) (*b*) (*c*)

Fig. 13–27. The foxglove plant and two most-commonly prescribed cardiac drugs

(*a*) **The foxglove plant, *Digitalis purpurea*.** The leaves of purple foxglove plant are the source of the popular heart muscle stimulant, digitalis.

(*b*) **Digitoxin** or **digitalin.** It is extracted from the dried leaves of foxglove plant and is the major component of digitalis, one of the most-widely used cardiac muscle stimulant.

(*c*) **Quabain.** It is a cardiac glycoside, isolated from the East African Quabio tree.

(Courtesy : (a) Derek Fell)

Table 13–6 lists the occurrence and biological role of some important steroids.

Table 13–6. Some steroids and their biological roles

Group	*Representative*	*Formula*	*Occurrence*	*Biological role*
Sterols				
C_{30}	Lanosterol	$C_{30}H_{51}O$	Wool fat	Biosynthesis of cholesterol
C_{28}	Ergosterol	$C_{28}H_{44}O$	Yeast	Provitamin D
C_{27}	Cholesterol	$C_{27}H_{46}O$	Animal fats	Structural component and precursor

	7-dehydro-cholesterol	$C_{27}H_{44}O$	Skin	Provitamin D
Bile acids				
C_{24}	Cholic acid	$C_{24}H_{40}O_5$	Intestine and gall-bladder	Absorption of fats
	Chenodeoxycholic acid	$C_{24}H_{40}O_4$	Intestine and gall-bladder	Absorption of fats
Hormones				
C_{21}	Progesterone	$C_{21}H_{30}O_2$	Corpus luteum	Hormone
	Deoxycorticosterone	$C_{21}H_{30}O_3$	Adrenal cortex	Hormone
	Cortisol	$C_{21}H_{30}O_5$	Adrenal cortex	Hormone
C_{19}	Testosterone	$C_{19}H_{28}O_2$	Testes	Hormone
C_{18}	Estradiol	$C_{18}H_{24}O_2$	Ovaries	Hormone

LEOPOLD STEPHEN RUŽIČKA

(LT, 1887–1976)

Ružička, a Croatian-born Swiss chemist, was born on September 13, 1887 in Vukovar, Croatia, then part of Kingdom of Hungary, Austro-Hungarian Empire in a family of craftsmen and farmers. He attended the classics-program secondary school in Osijek. He changed his original idea of becoming a priest and switched to studying technical disciplines. Chemistry was his choice, probably because he hoped to get a position at the newly-opened sugar refinery built in Osijek. Due to the excessive hardship of everyday and political life, he left and chose the High Technical School in Karlsruhe in Germany. He was a good student in areas he liked and that he thought would be necessary and beneficial in future, which was *organic chemistry*. In the course of his studies, Ružička set up excellent cooperation with **Hermann Staudinger** (a 1953-Nobel Laureate). Studying within Staudinger's department, he obtained his doctor's degree in 1910. With Staudinger, Ružička went to Zürich and was his assistant.

Ružička's first works originated during that period in the field of chemistry of natural compounds. He remained in this field of research all his life. He investigated the ingredients of the Dalmatian insect powder (*Pyrethrum cinerariifolium*), a highly esteemed insecticide. In this way, he came into contact with the chemistry of turpene, a fragrant oil of vegetable origin, interesting to the perfume industry. He intended to start individual research and even started successful and productive cooperation with the *Chuit & Naef Company* (later known as *Firmenich*) in Geneva. During 1916–1917, he received the support of the oldest perfume manufacturer in the world, *Haarman & Reimer*, of Holzminden in Germany. With expertise in the terpene field, he became senior lecturer in 1918, and, in 1923, honorary professor at the ETH (*Eidgenössische Technische Hochschule*) as well the University of Zürich. Here, with a group of his doctoral students, he proved the structure and existence of the compounds of *muskone* and *civetone*, the scents derived from the musk deer and the civet cat. The **Ružička large ring synthesis** is a method in organic chemistry for the organic synthesis of these type of compounds.

In 1921, the Geneva perfume manufacturers Chuit and Naef asked him to collaborate. Working here, Ružička achieved financial independence, but not as big as he did plan, so he left Zürich to start working for the Ciba, Basel-based company. In 1927, he took over the organic chemistry chair at Utrecht University in Netherlands. In Netherlands, he remained for 3 years and then returned to Switzerland, which was superior in its chemical industry. Back to Zürich, at ETH he became professor of organic chemistry and started the most brilliant period of his professional career. Here, he researched on higher terpenes and steroids. After the successful synthesis of sex hormones (testosterone and androsterone), his laboratory became the global centre of organic chemistry.

Ružička shared the **1939 Nobel Prize in Chemistry** with **Adolf F.J. Butenandt** for his work on terpenes and for his demonstration that the ring compounds muskone and civetone had respectively 16 and 17 carbon atoms in the ring: previously it had been thought that rings with more than 8 carbon atoms would be unstable. In 1940, following the award, he was invited by the *Croatian Chemical*

Association, where he delivered a lecture to an overpacked hall of dignitaries on the topic *From the Dalmatian Insect Powder to Sex Hormones.* During WW II, some of his excellent collaborators were lost, but Ružička restructured his laboratory with new, younger and promising people; among them was young scientist **Vladimir Prelog.**

Following 1950, Ružička returned to chemistry, which had entered a new era of research. Now he turned to the field of biochemistry, the problems of evolution and genesis of life, particularly to the biogenesis of terpenes. He published his hypothesis, **Biogenetic Isoprene Rule**, which was the peak of his scientific career. The rule states that many organic compounds contain multiples of 5 carbons. Ružička retired in 1957, handing over the charge of running the laboratory to his assistant and future Nobel Laureate **Vladimir Prelog.**

Ružička dedicated significant efforts to the problems of education. He insisted on a better organization of academic education and scientific work in the new Yugoslavia, and established the *Swiss-Yugoslav Society.* He became an honorary academician at the then *Yugoslav Academy of Sciences and Arts* in Zagreb. In Switzerland, the **Ružička Award** was established, for young chemists working in Switzerland. In his native Vukovar, a museum was opened in his honour, in 1977.

Ružička married twice: to **Anna Hausmann** in 1912, and to **Gertrud Acklin** in 1951. He died on September 26, 1976 in Mammern, Switzerland, a village on Lake Constance.

TERPENES

Among the nonsaponifiable lipids found in plants are many hydrocarbons known as terpenes (from *turpentine*). In general, these hydrocarbons and their oxygenated derivatives have lesser than 40 carbon atoms. The simplest terpenes are called *monoterpenes* and conform to the formula $C_{10}H_{16}$ (equivalent to 2 isoprene units, Fig. 13–28), those with the formula $C_{15}H_{24}$ are called as *sesquiterpenes,* with $C_{20}H_{32}$ as *diterpenes* and with $C_{30}H_{48}$ as *triterpenes.* Terpenes with 40 carbon atoms (or *tetraterpenes*) include compounds called carotenoids (refer Chapter 32).

$$\overset{1}{C}H_2{=}\overset{2}{C}(CH_3){-}\overset{3}{C}H{=}\overset{4}{C}H_2$$

or simply as

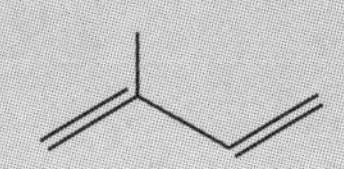

Fig. 13–28. Structure of isoprene
(2-methyl-1,3-butadiene)

In fact, **O. Wallach** (1910 Nobel Laureate in Chemistry) was the first to point out, in 1887, that nearly all the terpenoids are made of varying number of repetitive units (C_5H_8), called **isoprene units**. His finding later came to be known **isoprene rule.** Structurally, isoprene is a 5-carbon diene. The carbon skeletons of open-chain monoterpenoids and sesquiterpenoids are:

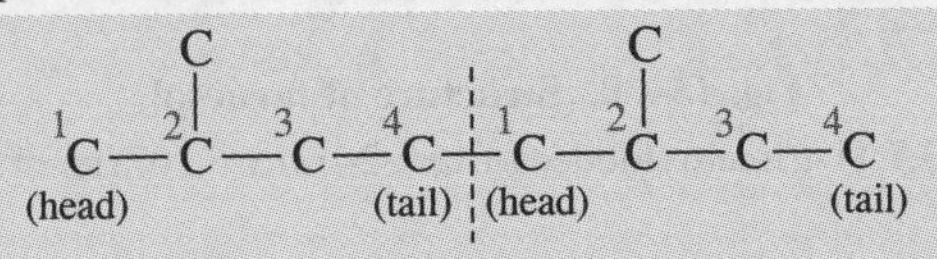

Monoterpenoid

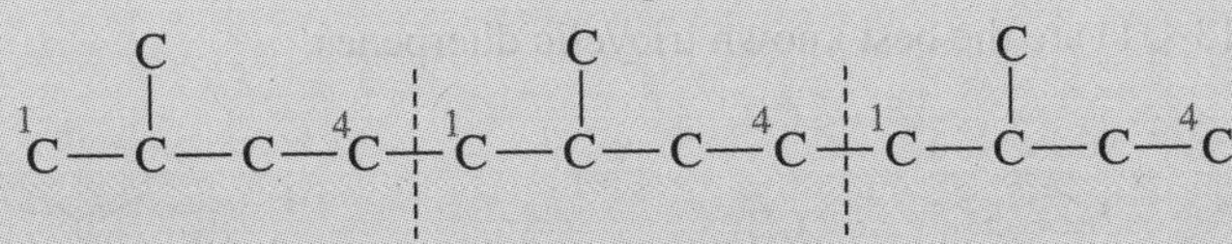

Sesquiterpenoid

Even not only the presence of isoprene units but their special type of arrangement is found to be present in nearly all the terpenoids. **Ingold**, in 1925, formulated this observation under another rule, the **special isoprene rule**, according to which the isoprene units in terpenoids are usually joined in *head-to-tail linkages* or 1,4 linkages (the branched end of the isoprene unit was considered as the head).

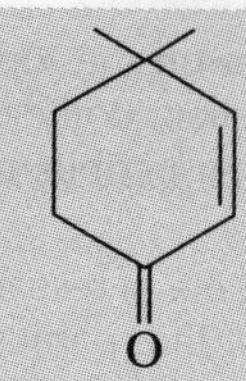

Cryptone

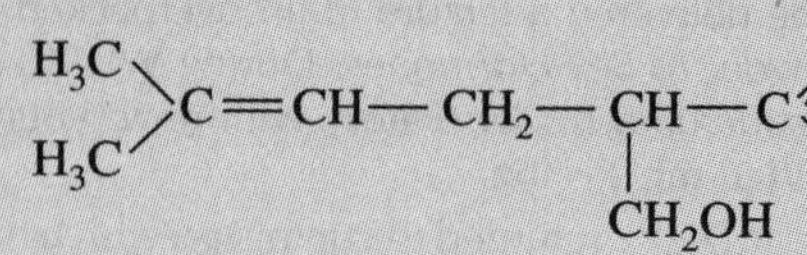

Lavandulol

Exceptions, however, do occur for these two isoprene rules. For example, *cryptone*, a natural terpenoid, contains only 9 carbon atoms instead of 10. Also, in *lavandulol*, the 2 isoprene units are not joined in head-to-tail manner, and in *carotenoids*, the two halves (each with 4 isoprene units) are linked with each other by *tail-to-tail* at the centre.

Monoterpenes and Sesquiterpenes

The fragrances of many plants arise from volatile C_{10} and C_{15} terpenes. These and their oxygenated derivatives occur as components of the essential oils, some of which are used as perfumes. Important monoterpenes (Fig. 13–29) are *myrcene* (from oil of bay), *geraniol* (from rose oil), *limonene* (from lemon oil) and *menthol* (from peppermint oil).

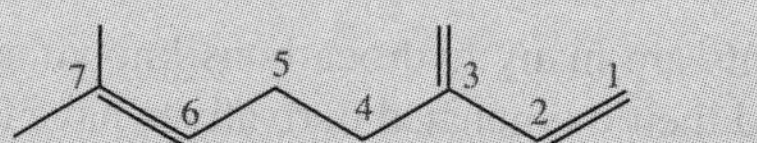

Myrcene, $C_{10}H_{16}$

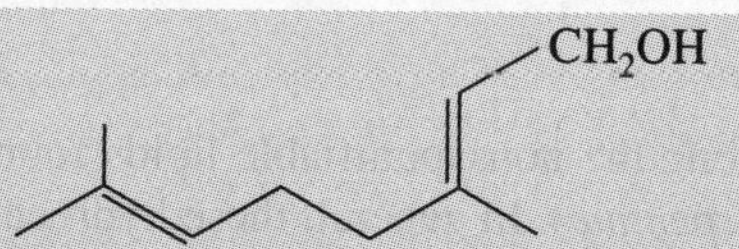

Geraniol, $C_{10}H_{18}O$

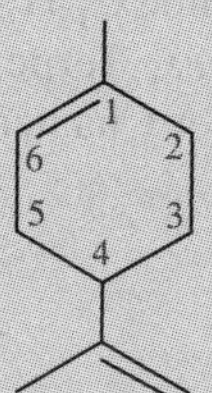

Limonene, $C_{10}H_{16}$

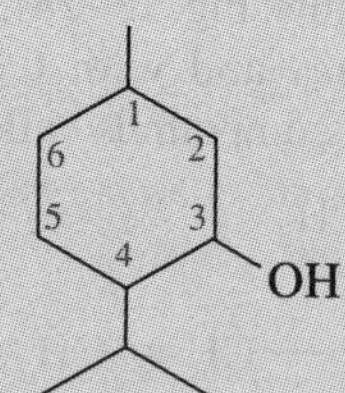

Menthol, $C_{10}H_{20}O$

Fig. 13–29. Four important monoterpenes

Among the sesquiterpenes, farnesol (Fig. 13–30) is widely distributed but is present in essential oils in small amounts.

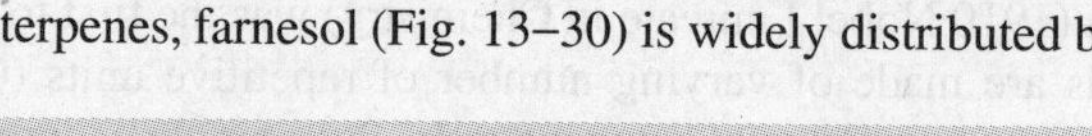

Fig. 13–30. Structure of farnesol

(3,7,11-trimethyldodeca-*trans*-2-*trans*-6,10-trienol)

Diterpenes

These are usually found as substituents of the resins and balsams. The two resin acids, **abietic** and **sapietic** (Fig. 13–31) are the best known tricyclic diterpenes.

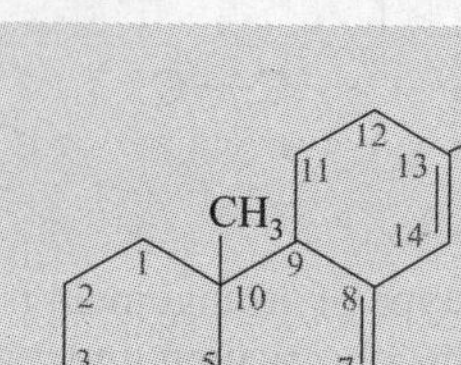

Abietic acid, $C_{20}H_{30}O_2$

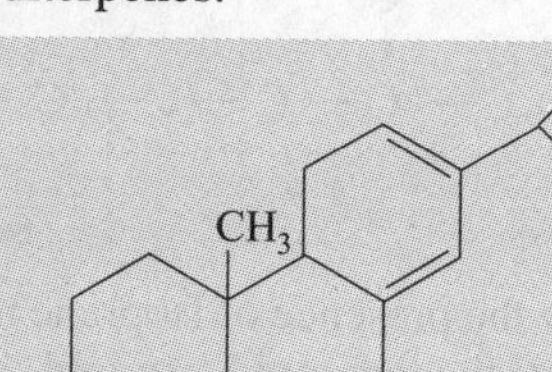

Sapietic acid

Fig. 13–31. Two best-known diterpenes

Vitamins A_1 and A_2 and their aldehydes called retinenes are important monocyclic derivatives of diterpenes.

Phytol ($C_{20}H_{40}O$; b.p. 145°C) is an acyclic diterpene and is obtained from the hydrolysis of chlorophyll. It was isolated from nettles by **Paul Karrer** *et al* in 1943. The phytol molecule (Fig. 13–32) has 2 chiral (= asymmetric) centres, 7 and 11. Natural phytol is very weakly dextrorotatory.

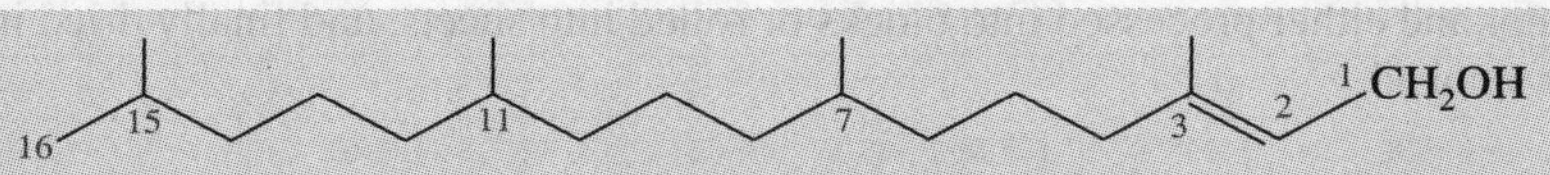

Fig. 13–32. Structure of phytol

PAUL KARRER

(LT, 1889–1971)

Karrer, a Russian-born Swiss chemist, was born on April 21, 1889 in Moscow to **Paul Karrer Sr.** and **Julie Lerch**, both Swiss nationals. In 1892, Karrer's family returned to Switzerland where he was educated at Wildegg and at the Grammar School in Lenzburg, Aarau, where he matriculated in 1908. He studied chemistry at the University of Zürich under **Alfred Werner** and after gaining his Ph.D. in 1911, he spent a further year as assistant in the Chemical Institute. He then took a post as chemist with **Paul Ehrlich** at the Georg Speyer Haus, Frankfurt-am-Main. In 1914, Karrer married **Helena Froelich**. In 1919, he became Professor of Chemistry and Director of the Chemical Institute.

Karrer's early research concerned complex metal compounds but his important work was on plant pigments, particularly the yellow **carotenoids**. He elucidated their chemical structure and showed that some of these substances are transformed in the body into vitamin A. His work led to the establishment of the correct constitutional formula for β-carotene, the precursor of vitamin A; the first time that the structure of a vitamin or provitamin had been established. **George Wald** worked briefly in Karrer's lab while studying the role of vitamin A in the retina. Later, Karrer **confirmed the structure of ascorbic acid** (vitamin C) and extended his researches into the vitamins B_2 (riboflavin) and E (tocopherol). His important contributions to the chemistry of the flavins led to **identification of lactoflavin** as part of the complex originally thought to be vitamin B_2.

Karrer was awarded the **1937 Nobel Prize for Chemistry** for his researches into the constitution of carotenoids, flavins and vitamins A and B, along with **Walter N. Haworth** of England. His textbook *Lehrbuch der Organischem Chemie* (Textbook of Organic Chemistry) was published in 1927, went through 13 editions, and was published in 7 languages. He died on June 18, 1971, at the age of 82, in Zürich.

Triterpenes

These are not widespread in nature but are significant in that some of them are intermediates in the biosynthesis of cholesterol. Two such compounds are : a tetracyclic alcohol, **lanosterol** (see page 260) and an acyclic hydrocarbon, **squalene**. Squlaene (C_{30} H_{50} ; b.p. 240–242°C) was first isolated from the liver of sharks (genus *Squalus*), hence so named. The olive oil and several other vegetable oils are other sources. It has also been detected in the leaves. Each squalene molecule (Fig. 13–33) has 6 double bonds. The double-bond system present is of nonconjugated type. The conjugated double bonds are, however, absent from the molecule.

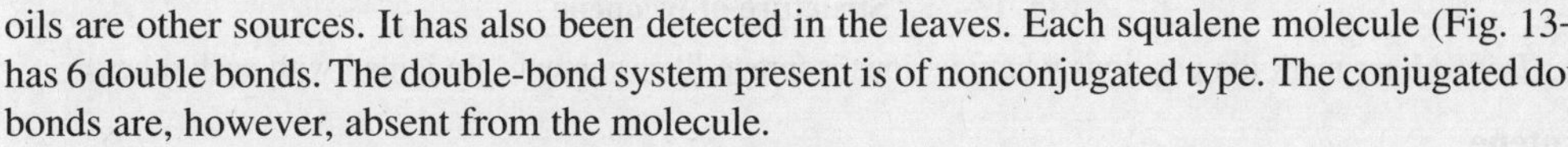

Fig. 13–33. Structure of squalene

[E-(trans-2,6,10,15,19,23-hexamethyl-2,6,10,14,18,22-tetracosane-hexaene]

Polyterpene

Mention, however, should also be made of **rubber** (MW = ca 3,00,000), a polyterpene present in the latex of many tropical plants. A molecule of rubber (Fig. 13–34) is composed of about 500 to 5,000 isoprene units joined in a long straight chain. When the bark of the rubber tree is cut, latex slowly exudes from the cut. Addition of acetic acid coagulates the rubber, which is then separated from the liquor and either processed into blocks or rolled into sheets, and finally dried in a current of warm air, or smoked.

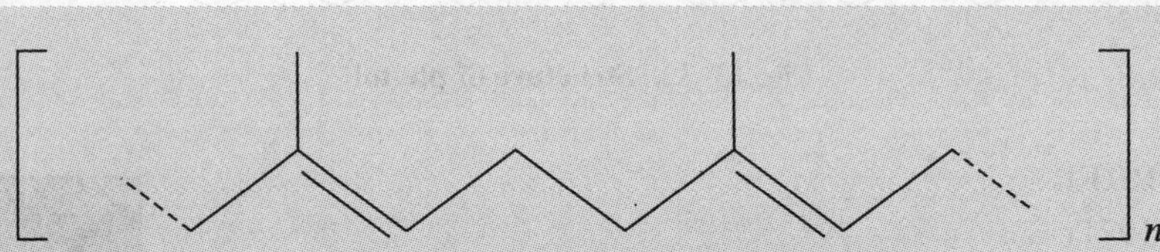

Fig. 13–34. Structure of rubber

(*cis* configuration about double bonds)

CAROTENOIDS

The carotenoids are tetraterpenes. These are widely distributed in both the plant and animal kingdoms but are exclusively of plant origin. These occur in unsaponifiable residue of plant and animal lipids. They are isoprene derivatives with a high degree of unsaturation. Because of the presence of many conjugated double bonds, they are coloured red or yellow. As an example, the pigment of tomato (*lycopene*) and that of carrot (α- and β-*carotene*) are red while many oxygen-containing carotenoids are yellow (*xanthophylls*). Since the olefinic (= double) bonds permit *cis-trans* isomerism, numerous forms are possible. *Most carotenoids, however, exist in all-trans form.* The presence of long hydrocarbon chain in carotenoids makes them lipid-soluble; they are hence also called *lipochromes* or *chromolipids*.

Lycopene

Lycopene ($C_{40}H_{56}$) is the main pigment of tomato, paprika and many other fruits. It is a highly unsaturated, unbranched, long chain hydrocarbon (a polyene) and is composed of two identical units ($C_{20}H_{28}$), joined by a double bond between C_{15} and $C_{15'}$. Each of these units may, in turn, be considered to have been derived from 4 isoprene units (C_5H_8). A molecule of lycopene (Fig. 13–35) in all contains as many as 13 double bonds, of which 11 are conjugated.

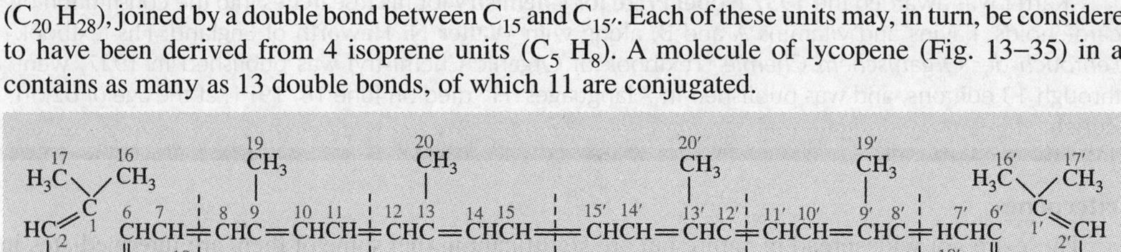

Fig. 13–35. Structure of lycopene

(*Dotted lines* mark the positions where various isoprene-like segments are joined with each other.)

Carotene

Another group of naturally-occurring carotenoids, with the same molecular formula as that of lycopene, is **carotene**. Carotene was first isolated by **Wackenroder** (1831) from carrots (this was the origin of the name *carotin*, which was later on modified to *carotene*). The 3 types of carotenes are:

1. α-carotene– violet crystals; m.p. 187°C; optically active (dextrorotatory).
2. β-carotone– red crystals; m.p. 183°C; optically inactive.

3. γ-carotene– dark red crystals ; m.p. 152–154°C; optically inactive.

(The prefixes α, β and γ in the carotenes were assigned not on the basis of structural properties but rather as a result of observations of their elution patterns on a $CaCO_3$ chromatographic column, eluted with petroleum ether).

The carotenes are obtained commercially by chromatography. The two best sources are carrots and alfalfa. Their relative proportion, however, varies with the source; for example carrots contain 15% α-, 85% β- and 0.1% γ-form. These are unstable to air, heat, acids and alkalies.

It is to be noted that the ends of the lycopene chain can easily close up to form rings, with the disappearance of the double bonds. Ring closure on only one end results in the production of γ-carotene while closure of the ring at both ends produces β- and α-carotenes which differ from each other in the position of the double bonds in the rings. The ring is structurally related to the ionones. (Fig. 13–36). In β-carotene, both annular double bonds are in conjugation with the system of double bonds of the long chain (β-ionone structure). In α-carotene, one of the annular double

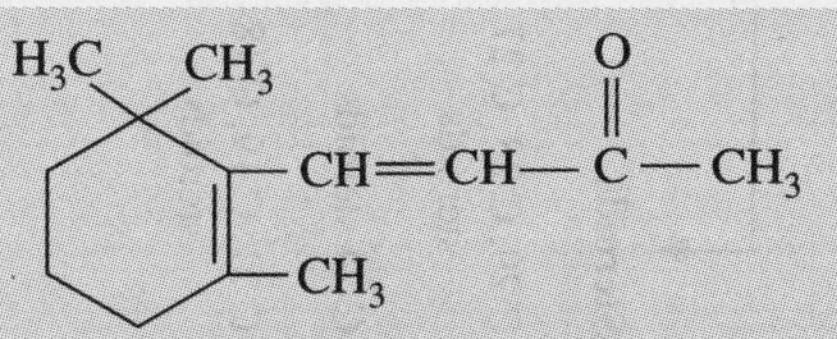

Fig. 13–36. Structure of β-ionone

bonds is removed from the system of conjugation by one position (α-ionone structure). Of particular interest is β-carotene (Fig. 13–37), which is precursor of vitamin A.

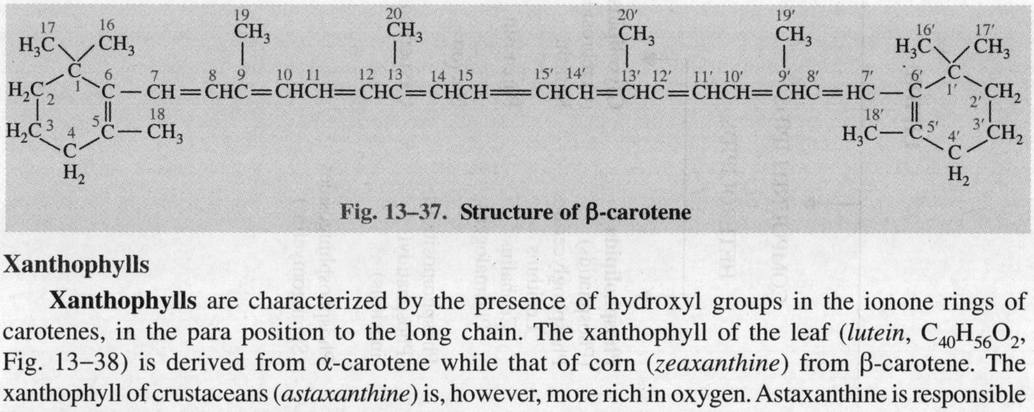

Fig. 13–37. Structure of β-carotene

Xanthophylls

Xanthophylls are characterized by the presence of hydroxyl groups in the ionone rings of carotenes, in the para position to the long chain. The xanthophyll of the leaf (*lutein*, $C_{40}H_{56}O_2$, Fig. 13–38) is derived from α-carotene while that of corn (*zeaxanthine*) from β-carotene. The xanthophyll of crustaceans (*astaxanthine*) is, however, more rich in oxygen. Astaxanthine is responsible for the appetizing redness of boiled lobsters.

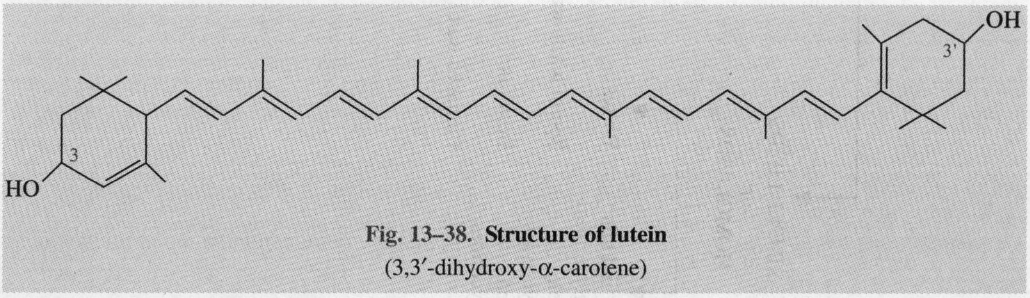

Fig. 13–38. Structure of lutein

(3,3′-dihydroxy-α-carotene)

The classification of lipids described above may be summarized as depicted on the next page.

LIPIDS

- SIMPLE LIPIDS or HOMOLIPIDS
  - **Fats and oils** (Triglycerides)
    - Simple triglycerides
    - Mixed triglycerides
  - **Waxes**
    - Sperm whale wax
    - Beeswax
    - Carnauba wax
- COMPOUND LIPIDS or HETEROLIPIDS
  - **Phospholipids** (Phosphatids)
    - Phosphoglycerides
      - Lecithins
      - Cephalins
      - Plasmalogens
    - Phosphoinositides (Phosphatidyl inositols)
    - Phosphosphingosides (Sphingomyelins)
  - **Glycolipids** (Cerebrosides)
    - Kerasin
    - Phrenosin
    - Nervon
    - Oxynervon
- DERIVED LIPIDS
  - **Steroids**
    - C 29, C 28, C 27 steroids
    - C 24 steroids
    - C 21, C 19, C 18 steroids
  - **Terpenes**
    - Monoerpenes
    - Sequiterpenes
    - Diterpenes
    - Triterpenes
    - Tetraterpenes
    - Polyterpenes
  - **Carotenoids**
    - Lycopene
    - Carotenes
    - Xanthophylls

REFERENCES

See list following Chapter 14.

PROBLEMS

1. A common structural feature of membrane lipid molecules is their amphipathic nature. For example, in phosphatidylcholine, the two fatty acid chains are hydrophobic and the phosphocholine head group is hydrophilic. For each of the following membrane lipids, name the components that serve as the hydrophobic and hydrophilic units:
   (*a*) phosphatidylethanolamine
   (*b*) sphingomyelin
   (*c*) galactosylcerebroside
   (*d*) ganglioside
   (*e*) cholesterol
2. Which of the following could not be classified as a lipid?
   (*a*) steroids
   (*b*) fats and oils
   (*c*) waxes
   (*d*) photosynthetic pigments
   (*e*) anthocyanin pigments
3. What food has the most lycopene?
4. At pH 7.0, what is the electrical charge of:
   (*a*) α-lecithin
   (*b*) α-lysolecithin
   (*c*) phosphatidyl ethanolamine
   (*d*) phosphatidyl serine
   (*e*) phosphatidal choline
   (*f*) phosphatidal ethanolamine
   (*g*) phosphatidal serine
   (*h*) sphingomyelin
5. Do tomato products lose their lycopene while being stored in the pantry?
6. How sphingomyelin can be differentiated from a diacylglyceride?
7. Is a glass of canned tomato juice equal to a serving of vegetables?
8. Is cholesterol beneficial?
9. What is an effective nutrient for battling cancer?

ANSWERS TO PROBLEMS

1. *Hydrophobic units*: (*a*) 2 fatty acids; (*b*), (*c*) and (*d*) 1 fatty acid and the hydrocarbon chain of sphingosine; (*e*) steroid nucleus and acyl side chain.
   Hydrophilic units: (*a*) phosphoethanolamine; (*b*) phosphocholine; (*c*) D-galactose; (*d*) several sugar molecules; (*e*) alcohol group (–OH)
2. (*e*) anthocyanin pigments

3. Contrary to popular belief, watermelon has 40% more lycopene than raw tomatoes. One-and-a-half cup of watermelon contains 9–13 mg of lycopene. If the tomatoes are cooked, their lycopene content is similar to that of watermelon.

4. (*a*) 0 (*b*) 0 (*c*) 0 (*d*) –1 (*e*) 0 (*f*) 0 (*g*) –1 (*h*) 0

5. Even after a year, tomato juice, spaghetti sauce and other tomato products maintain the same level of lycopene content.

6. The differences between a sphingomyelin and a diacylglyceride are given in tabulated form, as shown below:

Sphingomyelin	*Diacylglyceride*
1. This is a compound lipid and belongs to the category of phospholipids.	1. This is a simple lipid and is neutral in nature
2. This is also phosphodiester-like diacylglyceride but the first substituent is not a diglyceride but a sphingosine.	2. Here the glycerol forms ester with fatty acids.
3. Sphingomyelin can exist as a zwitterion over a wide range of pH	3. Zwitterionic form is not possible.

7. Yes. The lycopene in tomato juice helps fight prostate and breast concers. One glass of tomato juice raises blood lycopene by about 45%. The more veggies you eat, the lower your risk of strokes, heart attack and cancer.

8. Cholesterol is a critical part of the body's fundamental building materials. Your liver and your brain produce cholesterol. It is the raw material for all of your steroid hormones and vitamin D. Cholesterol is essential for optimum health.

9. Lycopene is a key nutrient in the battle against cancer. This amazing supplement has been shown to be effective in reducing the risk of prostate, breast, ovarian, cervical and colon cancers.

CONTENTS

The surface of a soap bubble is a bilayer formed by detergent molecules.

The polar heads (*red*) pack together leaving the hydrophobic groups (*green*) in contact with air on the inside and outside of the bubble. Other bilayer structures define the boundary of a cell.

[*Courtesy : (Upper) Photonica*]

CHAPTER

14

Lipids–III Properties of Fats and Oils

The fats and oils owe the manifestation of their properties to the fatty acids and alcohols, the two component units which they possess.

PHYSICAL PROPERTIES

1. State. Fats containing saturated fatty acids are solid at ordinary room temperature. The *animal fats* belong to this category. Most *plant fats*, on the contrary, possess unsaturated fatty acids and are, henceforth, liquid at room temperature.

2. Colour, Odour and Taste. When pure, the fats are *colourless,* virtually *odourless* and *possess* an extremely *bland taste*. They are capable of absorbing a variety of odours and hence flavour during storage. For the same reason, a housewife knows that the flavour of an onion quickly permeates butter that is stored with it in a refrigerator. In some cases, however, this absorbing property of fats is of advantage. For example, the perfumes of some flowers can be isolated by placing their petals in contact with the fat for a certain period, then extracting the fat with alcohol and concentrating the essence.

3. Solubility. The fats are, however, only sparingly soluble in water. These are, therefore, described as *hydrophobic* in contrast to the water-soluble or hydrophilic substances like many carbohydrates and proteins.

However, these are freely soluble in organic solvents like chloroform, ether, acetone and benzene. These solvents, as they dissolve fats in them, are also known as *'fat solvents'*. The solubility of the fatty acids in organic solvents, in fact, decreases with the increase of chain length. *The introduction of hydroxyl groups, however, increases solubility.*

4. Melting point. The melting point of fats depends on the chain length of the constituent fatty acid and the degree of unsaturation. Fats containing saturated fatty acids from C 4 to C 8 are liquid at room temperature but those containing C 10 or higher saturated fatty acids are solid and their melting points increase with increasing chain length. With the introduction of double bond in the fat molecule, the melting point lowers considerably. It may be stated, in general, that *greater the degree of unsaturation* (*or higher the number of double bonds*) *of the constituent fatty acid, the lower is the melting point of the fat.* This may be easily visualized in terms of constituent fatty acids from the Table 14–1. In fact, short chain length and unsaturation enhance the fluidity of fatty acids and of their derivatives.

Table 14–1. Melting points of the common fatty acids*

Fatty acid	*No. of carbon atoms*	*No. of double bonds*	*M.P. (in °C)*
Saturated			
Caprylic	8	0	16.7
Capric	10	0	31.6
Lauric	12	0	44.2
Myristic	14	0	53.9
Palmitic	16	0	63.1
Stearic	18	0	69.6
Arachidic	20	0	76.5
Behenic	22	0	79.9
Lignoceric	24	0	86.0
Unsaturated			
Palmitoleic	16	1	11.0
Oleic	18	1	13.4
Linoleic†	18	2	– 5.0
Linolenic	18	3	–11.0
Arachidonic	20	4	– 49.5

* Here mention has been made only of even numbered fatty acids as they alone are one of the two constituents of the saturated fats. If, however, the entire homologous series of the saturated fatty acids is taken into consideration, the fact emerges that this series shows an alternation or oscillation of melting points, *i.e.*, the melting point of an 'even' acid is higher than that of the 'odd' acid immediately below and above it. Thus, the fatty acid series follows what is known as the **saw-tooth rule.** However, in many other homologous series the melting points of the n-members rise continuously, tending towards a maximum value.

† Once known as vitamin E, now the name is discredited.

5. Specific gravity. The specific gravity of the fats is less than 1 (about 0.86) and, therefore, they float on water surface. Solid fats are lighter than the liquid fats. Oils spread on water to form thin monomolecular layers. In general, either unsaturation of the fatty acid chains increase or increase in chain length of the fatty acid residues tend to increase the specific gravity.

6. Geometric isomerism. As stated earlier, the presence of double bond(s) in the unsaturated fatty acid part of the fat molecule produces geometric (or *cis-trans*) isomerism.

7. Insulation. The fats possess *high insulating power*, *i.e.*, they are bad conductor of heat. A layer of fat below the skin provides a sort of blanket for warm-blooded animals (or *homoiotherms*). This is especially important for whales and seals which have to maintain a high temperature in cold waters. The fishes are cold-blooded animals (*or poikilotherms*) and, therfore, do not require maintenance of high temperature and so have very little subcutaneous fat.

8. Emulsification. It is the process by which a lipid mass is converted into a number of small lipid droplets. The fats may be emulsified by shaking either with water or with emulsifying agents like soaps, gums, proteins etc. An emulsifying agent helps in the production of a finely divided suspension of a fat in an aqueous medium. The hydrocarbon portions of the two (the emulsifier and the fat) tend to aggregate. This leaves the water-soluble group of the emulsifier projecting into the aqueous phase. A fat droplet will associate with a number of molecules of the emulsifier, thus producing a new water-soluble surface. Water molecules, henceforth, tend to be held in a layer or '*cloud*' around each droplet, thus disallowing the aggregation of the fat droplets (Fig. 14–1).

The process of emulsification is of great metabolic significance. In fact, *the fats have to be emulsified before they can be absorbed by the intestinal wall.* The process is accomplished by the bile juice secreted from liver.

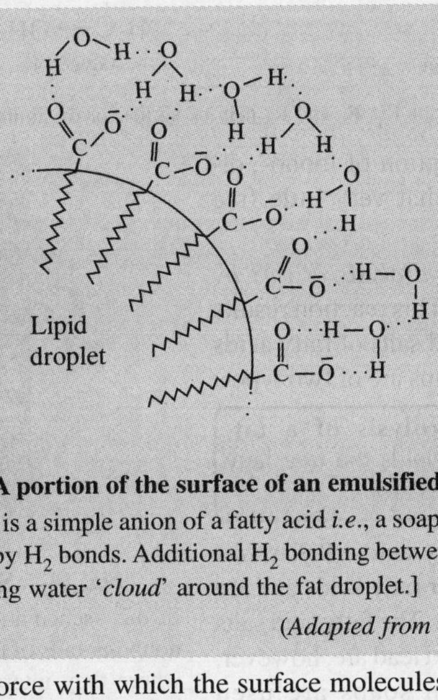

Fig. 14–1. A portion of the surface of an emulsified fat globule

[Here, the emulsifier selected is a simple anion of a fatty acid *i.e.*, a soap. The projecting COOH groups are linked with water molecules by H_2 bonds. Additional H_2 bonding between the water molecules leads to the further stability of the resulting water '*cloud*' around the fat droplet.]

(*Adapted from Fairley JL and Kilgour GL, 1966*)

9. Surface tension. The force with which the surface molecules are held together is called the surface tension. When liquid fat is poured on water, it spreads uniformly over the surface of water in the form of a unimolecular layer and thus reduces the surface tension of water.

CHEMICAL PROPERTIES

The chemical reactions of the fats reflect the reactivities of the ester linkage and the degree of unsaturation of the hydrocarbon chain.

REACTIONS INVOLVING COOH GROUP

1. Hydrolysis. The fats are hydrolyzed by the enzymes *lipases* to yield fatty acids and glycerol. The lipases catalyze this reaction at a slightly alkaline *p*H (7.5 – 8.5) in a stepwise manner. The fats first split to produce *diglycerides*, part of these are then split to *monoglycerides*. Finally, part of the monoglycerides split to yield *fatty acid* and *glycerol*.

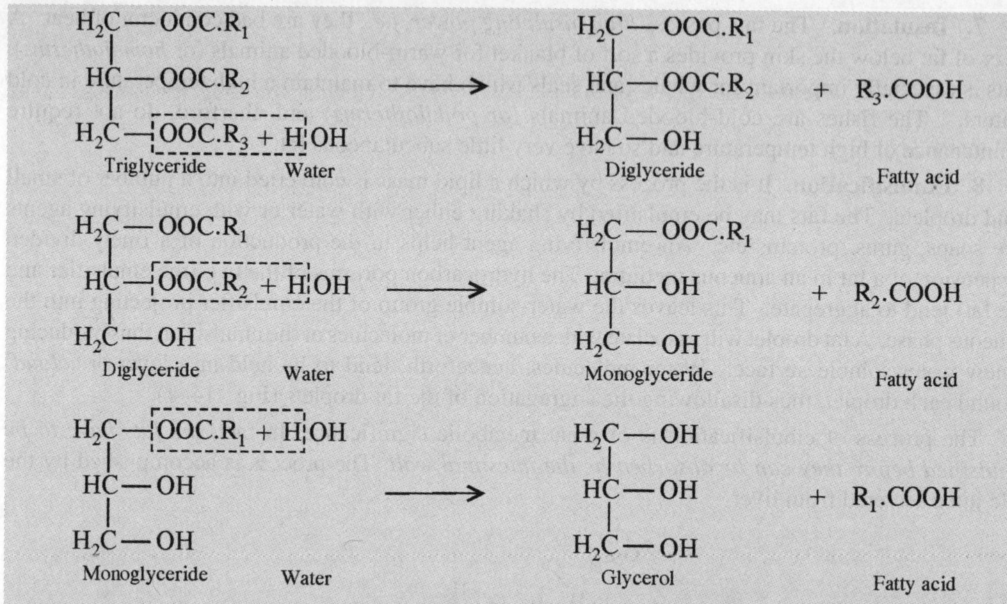

[Radicals R_1, R_2 and R_3 may be similar or dissimilar.]

In the intestine, the absorption of mono-, di- and triglycerides is so rapid that very little free glycerol is formed.

2. Saponification. *The hydrolysis of fats by alkali is called saponification.* This reaction results in the formation of glycerol and salts of fatty acids which are called *soaps*. The soaps are of two types: hard and soft. *Hard soaps* such as the common bar soaps are the sodium salts of the higher fatty acids. *Soft soaps* are the potassium salts of higher fatty acids and are marketed as semisolids or pastes. The fatty acid salts of calcium, magnesium, zinc and lead are, however, insoluble in water. Calcium soaps are used industrially as lubricating greases. Zinc soaps are employed in the manufacture of talcum powder and other cosmetics. Lead and magnesium soaps are used in paints industry to hasten the process of drying.

> **Acid hydrolysis of a fat**, however, yields the free fatty acid and glycerol.

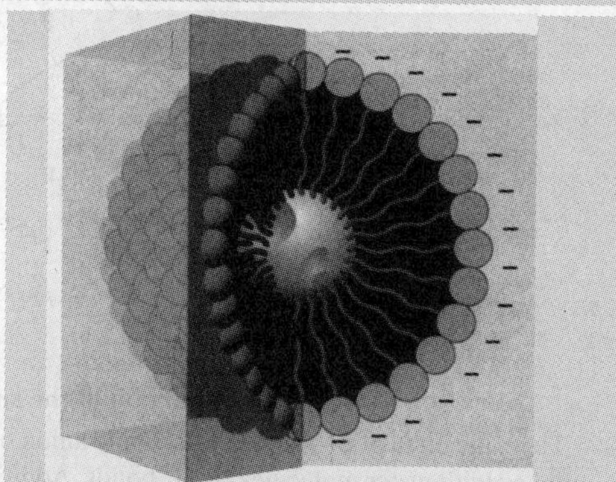

Fig. 14-2. Soaps consist of fatty acids.
In this schematic drawing of a soap micelle, the nonpolar tails of the fatty acids are directed inward, where they interact with the greasy matter to be dissolved. The negatively-charged heads are located at the surface of the micelle, where they interact with the surrounding water. Membrane proteins, which also tend to be insoluble in water, can also be solubilized in this way by extraction of membranes with detergents.

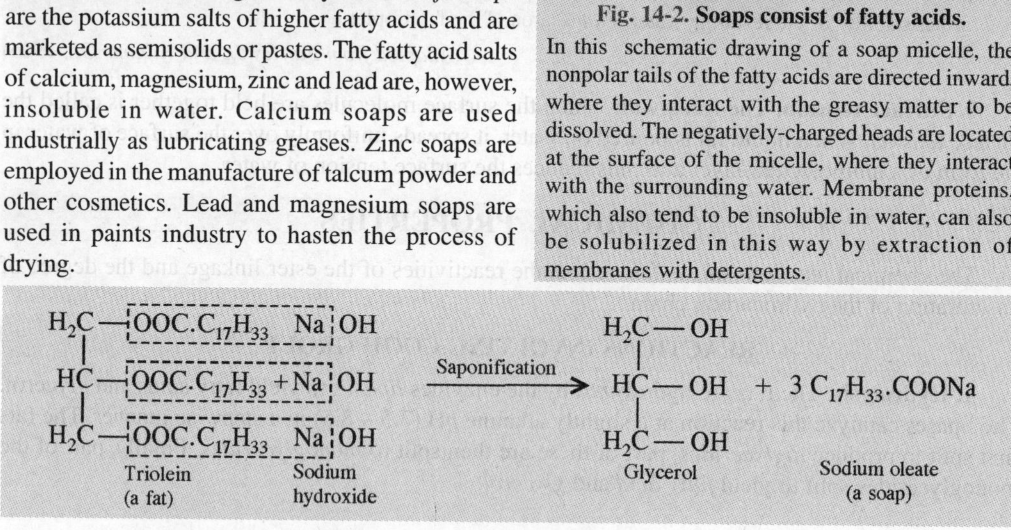

Soaps are important cleansing agents. Their cleansing property is due to their *emulsifying action* (*i.e.*, capacity to render more prolonged the mixing of oil and water). This is accomplished by means of negative charge the soap anion confers on oil droplets (Fig. 14–2). The electrostatic repulsion then prevents the coalescence of soap and oil droplets into an oil phase.

3. Hydrolytic rancidity. When butter or other fats are stored, they often become rancid and hence unpalatable. Rancidity is caused by the growth of microorganisms which secrete enzymes like *lipases*. These split the fats into glycerol and free fatty acids. The fatty acids impart unpleasant odour and flavour to the fat. However, butter may be prevented from becoming rancid by refrigeration or by exclusion of water.

REACTIONS INVOLVING DOUBLE BOND

1. Hydrogenation. Unsaturated fatty acids, either free or combined in lipids, react with gaseous hydrogen to yield the *saturated fatty acids*. The reaction is catalyzed by platinum, palladium or nickel. The addition of hydrogen takes place at the C—C double bond (*s*). Thus, 1 mole of oleic, linoleic or linolenic acid reacts with 1, 2 or 3 moles of hydrogen respectively to form stearic acid.

$$\underset{\text{Oleic acid}}{CH_3(CH_2)_7\overset{10}{C}H{=}\overset{9}{C}H(CH_2)_7COOH} + H_2 \xrightarrow[\text{as catalyst}]{\text{Pt, Pd or Ni}} \underset{\text{Stearic acid}}{CH_3(CH_2)_7\overset{10}{C}H_2{-}\overset{9}{C}H_2(CH_2)_7COOH}$$

This reaction is of great commercial importance since it permits transformation of inexpensive and unsaturated liquid vegetable fats into solid fats. The latter are used in the manufacture of candles, vegetable shortenings and of oleomargarine.

2. Halogenation. Unsaturated fatty acids and their esters can take up halogens like Br_2 and I_2 at their double bond (*s*) at room temperature in acetic acid or methanol solution.

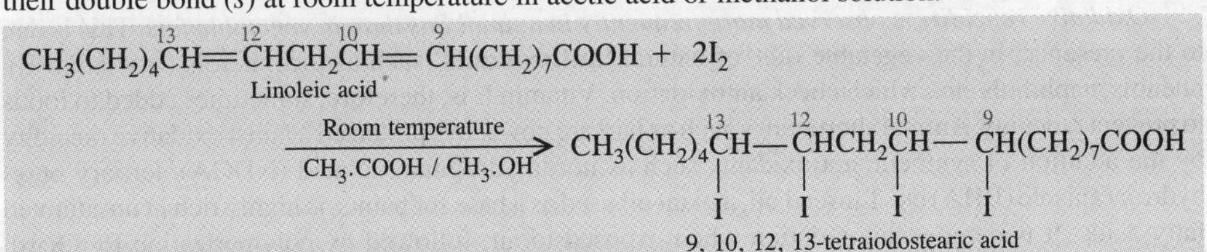

This reaction is the basis of the 'iodine number determination'.

3. Oxidation. Unsaturated fatty acids are susceptible to oxidation at their double bonds. Oxidation may be carried with ozone or $KMnO_4$.

(*a*) With ozone – An unstable *ozonide* is formed which later cleaves by water to give rise to 2 *aldehydic groups*.

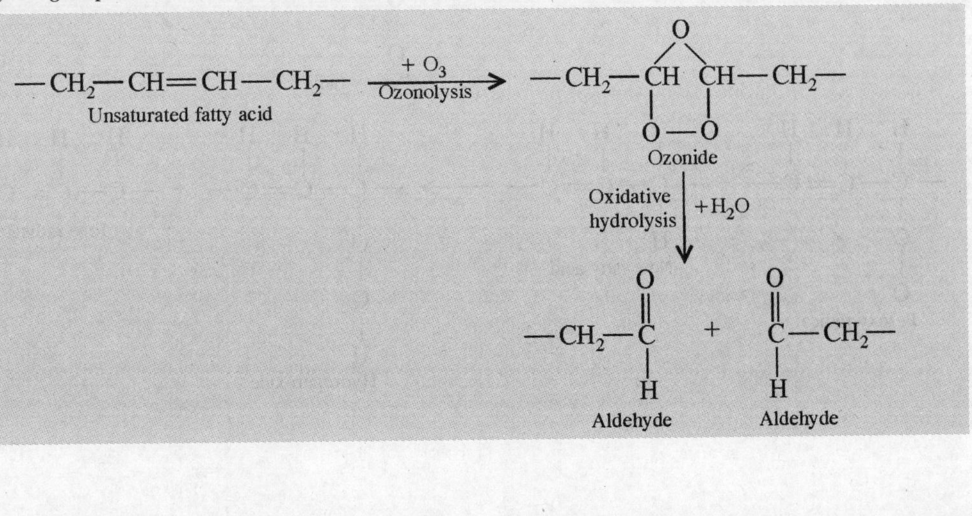

(*b*) With $KMnO_4$ – Under mild conditions, the *glycols* are formed at the sites of double bonds.

$$\underset{\text{Oleic acid}}{CH_3(CH_2)_7\overset{10}{C}H{=}\overset{9}{C}H(CH_2)_7COOH} \xrightarrow{KMnO_4} \underset{\text{9, 10-dihydroxystearic acid}}{CH_3(CH_2)_7\overset{10}{\underset{OH}{C}}H{-}\overset{9}{\underset{OH}{C}}H(CH_2)_7COOH}$$

Under vigorous conditions, the same reagent cleaves the molecule at the double bond and oxidizes the terminal portions to the *carboxyl group*.

$$\underset{\text{Oleic acid}}{CH_3(CH_2)_7\overset{10}{C}H{=}\overset{9}{C}H(CH_2)_7COOH} \xrightarrow{KMnO_4} \underset{\text{Pelargonic acid}}{CH_3(CH_2)_7COOH} + \underset{\text{Azelaic acid}}{HOOC(CH_2)_7COOH}$$

The oxidation reactions have been extensively used in establishing the position of double bond(s) in the fatty acid chain. This gives important clues regarding lipid structure.

4. Oxidative rancidity. Oils containing highly unsaturated fatty acids are spontaneously oxidized by atmospheric oxygen at ordinary temperatures. The oxidation takes place slowly and results in the formation of *short chain fatty acids* (*C_4 to C_{10}*) and *aldehydes* which give a rancid taste and odour to the fats. This type of rancidity or rancidification is called 'oxidative rancidity' and is due to a reaction called `*autoxidation*'. Autoxidation proceeds by a free radical mechanism in which the a- methylene group is primarily attacked. A hydrogen atom is removed from an α-methylene group. This initiates a chain of reactions leading to oxidation (**Holman**, 1954).

Oxidative rancidity is observed more frequently in animal fats than in vegetable fats. This is due to the presence, in the vegetable oils, of natural **'antioxidants'** such as tocopherols (= vitamin E), phenols, naphthols etc., which check autoxidation. Vitamin E is, therefore, sometimes added to foods to prevent rancidity. Animal shortenings such as lard are nowadays protected against oxidative rancidity by the addition of synthetic antioxidants such as nordihydroguiaretic acid (NDGA), tertiary butyl hydroxy anisole (BHA) etc. Linseed oil, a plant oil used as a base for paints, is highly rich in unsaturated fatty acids. It undergoes autoxidation when exposed to air, followed by polymerization to a hard, resinous coating as it 'dries' or oxidizes.

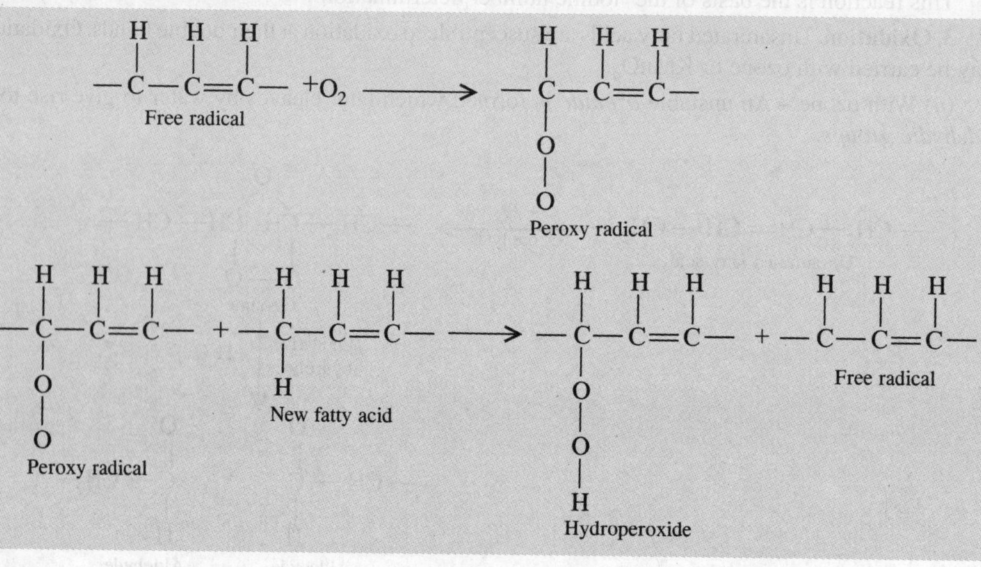

The action of antioxidants is opposed by a group of compounds present in the fats and oils. These accelerate the oxidation of the parent compound and are called *pro-oxidants.* Majority of these substances are formed during the processing and refining of fats. Among the noteworthy pro-oxidants are the copper, iron and nickel salts of organic acids like lactic, etc.

REACTION INVOLVING OH GROUPS

1. Dehydration (Acrolein test). Fats, when heated in the presence of a dehydrating agent, $NaHSO_4$ or $KHSO_4$ produce an unsaturated aldehyde called acrolein from the glycerol moiety. Acrolein is easily recognized by its pungent odour and thus forms the basis of the test for the presence of glycerol in fat molecule.

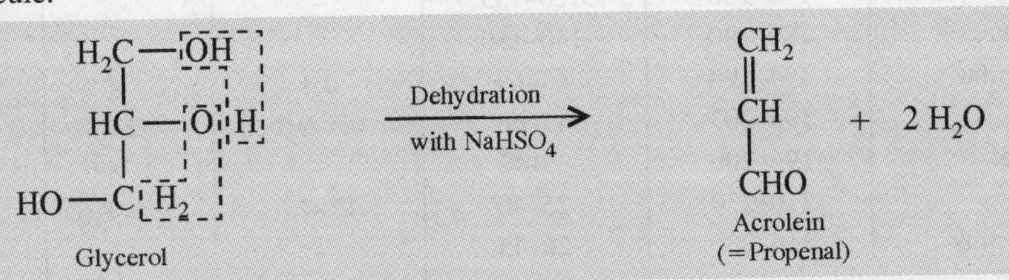

QUANTITATIVE TESTS

The reactions described above give valuable information about the chemical nature of fatty acids and the number of hydroxyl groups present in the fat molecule. Such chemical determinations involve various analytical tests. These are called *chemical constants* and include the following:

1. Acid value. It is the number of milligrams of KOH required to neutralize the free fatty acids present in 1 gm of fat. The acid number, thus, tells us of the quantity of free fatty acid present in a fat. Obviously, a fat which has been both processed and stored properly has a very low acid number.

2. Saponification number. It is the number of milligrams of KOH required to saponify 1 gm of fat. The saponification number, thus, provides information of the average chain length of the fatty acids in the fat. It varies inversely with the chain length of the fatty acids. *The shorter the average chain length of the fatty acids, the higher is the saponification number.*

3. Iodine value (or Koettstorfer number). It is the number of grams of iodine absorbed by 100 g of fat. The iodine number is, thus, a measure of the degree of unsaturation of the fatty acids in the fat. Oils like soybean, corn and cottonseed have higher iodine numbers (133, 127 and 109, respectively) than the solid fats such as beef fat or tallow (42) because the former possess more unsaturated fatty acids in the fat molecule. However, the iodine number gives no indication as to the number of double bonds present in the fatty acid molecule.

4. Polenske number. It is the number of millilitres of 0.1N KOH required to neutralize the insoluble fatty acids (*i.e.*, those which are not volatile with steam distillation) obtained from 5 gm of fat.

5. Reichert-Meissl number. It is the number of millilitres of 0.1N KOH required to neutralize the soluble, volatile fatty acids derived from 5 g of fat. The Reichert-Meissl number, thus, measures the quantity of short chain fatty acids (up to C 10 inclusive) in the fat molecule. The Reichert-Meissl numbers of coconut and palm oils range between 5 and 8. Butterfat is exceptional in having a high Reichert-Meissl number, ranging from 17 to 35. This high value makes possible the detection of any foreign fats which are, sometimes, adulterated in the manufacture of butter.

6. Acetyl number. It is the number of milligrams of KOH required to neutralize the acetic acid obtained by saponification of 1 gm of fat after it has been acetylated (The treatment of fat or fatty acid mixture with acetic anhydride results in acetylation of all alcoholic OH groups). The acetyl number is, thus, a measure of the number of OH groups in the fat. For example, the castor oil has a high acetyl number (146) because of high content of a hydroxy acid, ricinoleic acid, in it.

Some examples of the value of these numbers in defining fats are given in Table 14–2.

Table 14–2. Analytical values for some fats and oils

Fat/Oil	*Saponification number (mg KOH/g)*	*Iodine value (g I_2/100 g)*	*Rechert-Meissl number*	*Acetyl number*	*Acid value (mg KOH/g)*
Animal fats					
Halibut liver oil	170–180	120–135	–	–	1.0
Cod liver oil	180–190	140–180	–	–	0.5–5
Human fat	194–198	65–69	0.4	–	–
Lard	195–203	47–66	0.5–0.8	2.6	–
Beef fat	196–200	35–42	–	2.78.6	–
Butter	210–230	25–40	17–35	1.9–8.6	0.5–30
Cow's milk	216–235	26–45	–	–	–
Plant fats					
Castor oil	175–187	80–90	1.4	146–150	0.2–4
Corn oil	187–195	104–128	–	–	1.0–2
Linseed oil	188–195	170–195	1.0	4.0	1.0–4
Olive oil	190–195	80–90	0.6–1.5	10–11	0.2–3
Coconut oil	254–262	7–10	0.6–7.5	2.0	2.5–6

FATS– FACTS AND FANTASIES

Irrespective of the source, fat is the most concentrated form of dietary energy. While sugars (carbohydrates) and proteins provide 4 Calories (kilocalories) of energy per gram, fat of any type provides 9 Calories. *There is no such thing as low-calorie fat*. Thus, by replacing fats of animal origin like `ghee' and butter with vegetable oils like groundnut, sunflower, safflower, mustard, palm oil or any other, weight cannot be reduced. For weight reduction, the only rule is *to eat less and exercise more*. Despite the same Calorie content, the nutritional quality of different fats does differ. As already stated, the fats are composed of an alcohol component (usually glycerol) and a fatty acid component which may be saturated or unsaturated. Nutritionally, polyunsaturated fatty acids (PUFAs) are considered superior. While the human body can synthesize saturated fatty acids (SFAs) and monounsaturated fatty acids (MUFAs), it cannot synthesize the polyunsaturated fatty acids (PUFAs) and hence PUFAs are referred to as essential

Butter or Margarine

There is a continuing debate among nutritionists about the health benefits of butter versus margarine in foods. **Butter** is rich in both cholesterol and triglycerols containing saturated fatty acids, which are dietary risk factors for atherosclerosis. **Margarine** contains no cholesterol and is richer in unsaturated fatty acids. However, the unsaturated fatty acids in margarine are mostly the unnatural *trans*-fatty acids formed during the partial hydrogenation of vegetable oils. *Trans*-fatty acids affect plasma lipids in the same fashion as saturated fatty acids, suggesting that there are comparable risks associated with the consumption of butter or margarine. The resolution of this issue becomes complicated by the fact that various forms of margarine, for example soft-spread and hard-block types, vary significantly in their content of *trans*-fatty acids. Partially hydrogenated oils are more stable than the natural oils during heating; when used for deep-frying, they need to be changed less frequently. Despite the additional expense, the food industries have gradually shifted to the use of natural oils, rich in unsaturated fatty acids and without *trans*-fatty acids, for cooking and baking.

fatty acids (EFAs). Fats of animal origin (lard, cream, butter and 'ghee') are saturated whereas vegetable oils, except coconut, have varying degree of unsaturation. Fish oils too are unsaturated and have long-chain-length fatty acids. Table 14–3. lists fatty acid composition of some common fats and oils.

Table 14–3. **Fatty acid composition of some common fats and oils (in g/100g)**

Fat / Oil	*Saturated*	*Mono-unsaturated*	*Polyunsaturated*		*Predominant fatty acids**
			Linoleic	*α-linolenic*	
Coconut[a]	90	7	2	< 0.5	SFAs
Palm kernel	82	15	2	< 0.5	SFAs
'Ghee'[a,b]	65	32	2	< 1.0	SFAs
'Vanaspati'[b]	24	19	3	< 0.5	SFAs
Red palm oil	50	40	9	< 0.5	SFAs + MUFAs
Palm oil	45	44	10	< 0.5	SFAs + MUFAs
Olive	13	76	10	< 0.5	MUFAs
Groundnut	24	50	25	< 0.5	MUFAs
Rape / Mustard[c]	8	70	12	10	MUFAs
Sesame	15	42	42	1.0	MUFAs + PUFAs
Rice bran	22	41	35	1.5	MUFAs + PUFAs
Cottonseed	22	25	52	1.0	PUFAs
Corn	12	32	55	1.0	PUFAs
Sunflower	13	27	60	< 0.5	PUFAs
Safflower	13	17	70	< 0.5	PUFAs
Soybean	15	27	53	5.0	PUFAs

* SFAs = Saturated fatty acids
MUFAs = Monounsaturated fatty acids
PUFAs = Polyunsaturated fatty acids

a : Mainly short medium-chain fatty acids (Coconut 77%; 'Ghee' 25%)
b : *Trans* fatty acids ('Ghee' 2% ; 'Vanaspati' 53%)
c : Long-chain monounsaturated fatty acids (50% Erucic acid; 5% Eicosenoic acid)

Note that Rape / Mustard and Soybean are good sources of α-linolenic acid.

It is long known that saturated fats raise blood cholesterol level whereas unsaturated fats lower blood cholesterol and also ensure a more favourable ratio of 'good' to 'bad' cholesterol. It is for this reason that the vegetable oils such as groundnut, mustard, rice bran, sunflower, safflower etc (except coconut oil, which is saturated) are regarded to be healthy, rather health-friendly fats. Therefore, these are prescribed for those suffering with high levels of blood cholesterol or those suffering from coronary artery disease.

REFERENCES

1. **Ansell GB, Dawson RMC, Hawthorne JN (editors) :** Form and Function of Phospholipids, *2nd ed., Elsevier, Amsterdam. 1973.*
2. **Asselineau J, Lederer E :** Chemistry of Lipids. *Ann. Rev. Biochem.* **30** *: 71–92, 1961.*
3. **Bittar EE (editor) :** Membrane Structure and Function. *5 vols. Wiley-Interscience, New York. 1980–1984.*
4. **Bloch K (editor) :** Lipide Metabolism. *John Wiley and Sons, Inc., New York. 1960.*

5. **Bloor WR :** Biochemistry of the Fatty Acids and Their Compounds, the Lipids. *Reinhold Publishing Corp., New York. 1943.*
6. **Burton RM, Guerra FC (editors) :** Fundamentals of Lipid Chemistry. *Webster Groves, Missouri : Bi-Science Publication Division. 1972.*
7. **Chapman D (editor) :** Biological Membranes. *5 vols. Academic Press, Inc., New York. 1968–1984.*
8. **Chapman D :** Introduction to Lipids. *McGraw-Hill, New York. 1969.*
9. **Chapman DC :** The Structure of Lipids. *John Wiley and Sons, Inc., New York. 1965.*
10. **Christie WW :** Lipid Analysis. *Pergamon Press, Oxford. 1973.*
11. **Colowick SP, Kaplan NO :** Methods of Enzymology. *Academic Press, Inc., New York. Numerous annual volumes.*
12. **Cook RP (editor) :** Cholesterol : Chemistry, Biochemistry, and Pathology. *Academic Press, Inc., New York. 1958.*
13. **Dawson RMC, Rhoades RN :** Metabolism and Physiological Significance of Lipids. *John Wiley and Sons, Inc., London. 1964.*
14. **Deuel HJ Jr. :** The Lipids : Their Chemistry and Biochemistry. *3 vols. John Wiley and Sons, Inc., New York. 1951, 1955, 1957.*
15. **Fasman GD (editor) :** Handbook of Biochemisty and Molecular Biology. *3rd ed. Sec. C. Lipids, Carbohydrates and Steroids. Vol. 1. CRC Press, Boca Raton Fla. 1976.*
16. **Fieser LF, Fieser M :** Steroids. *Reinhold Pub. Corp., New York. 1959.*
17. **Finean JB, Michell RH (editors) :** Membrane Structure. *Elsevier North Holland, New York. 1981.*
18. **Fishman PH, Brady RO :** Biosynthesis and Function of Gangliosides. *Science.* **194 :** *906–915, 1976.*
19. **Grabowski G :** Phenotype, diagnosis and treatment of Gaucher's disease. *Lancet.* **372 :** *1263–1271, 2008.*
20. **Gunstone FD :** An Introduction to the Chemistry and Biochemistry of Fatty Acids and Their Glycerides. *Halstead Press, New York. 1975.*
21. **Gurr MI, Hardwood JL :** Lipid Biochemistry. An Introduction. *4th ed. Chapman and Hall, London. 1990.*
22. **Haber E :** The Cell Membrane. *Plenum Press, New York. 1984.*
23. **Hakomori S :** Glycosphingolipids. *Sci. Amer.* **254 (5) :** *44–53, 1986.*
24. **Hanahan DJ** : Lipid Chemistry. *John Wiley and Sons, Inc., New York. 1960.*
25. **Hansen HS :** The essential nature of linoleic acid in mammals. Trends Biochem. *Sci.* **11 :** *263, 1985.*
26. **Hanson JR :** Introduction to Steroid Chemistry. *Pergamon, New York. 1968.*
27. **Harrison R, Lunt GG :** Biological Membranes : Their Structure and Function. *2nd ed., Halstead, New York. 1980.*
28. **Harwood JL, Russell NJ :** Lipids in Plants and Microbes. *George Allen and Unwin, Ltd., London. 1984.*
29. **Hawthorne JN, Ansell GB :** Phospholipids. *Elsevier, New York. 1982.*
30. **Heftman E :** Steroid Biochemistry. *Academic Press Inc., New York. 1970.*
31. **Jain MH :** Introduction to Biological Membranes. *2nd ed., Wiley, New York. 1988.*
32. **Johnson AR, Davenport JB :** Biochemistry and Methodology of Lipids. *John Wiley and Sons, Inc., New York. 1971.*
33. **Kates M :** Techniques of Lipidology : Isolation, Analysis and Identification of Lipids. 2nd ed., Laboratory Techniques in Biochemistry and Molecular Biology. Vol.3, Part 2 (Burdon RH, van Knippenberg PH, editors), *Elsevier Science Publishing Co., Inc., New York. 1986.*

34. **Klyne W :** The Chemistry of the Steroids. *Methuen and Co. Ltd., London. 1965.*
35. **Martonosi AN :** Membranes and Transport. *Plenum. 1982.*
36. **Marx JL :** Liposomes : Research Application Grow. *Science.* **199 :** *1056–1128, 1978.*
37. **Mead JF, Alfin-Slater RB, Howton DR, Popjak G :** Lipids : Chemistry, Biochemistry and Nutrition. *Plenum Press, New York. 1986.*
38. **Morton RA (editor) :** Biochemistry of Quinones. *Academic Press, Inc., New York. 1965.*
39. **Olson JA :** Lipid Metabolism. *Ann. Rev. Biochem.* **35 :** *559–598, 1966.*
40. **Porter JW, Anderson DG :** Biosynthesis of Carotenes. *Ann. Rev. Plant Physiol.* **18 :** *197–228, 1967.*
41. **Quinn PJ :** The Molecular Biology of Cell Membranes. *Macmillan, London, 1976.*
42. **Richards JH, Hendrickson JB :** The Biosynthesis of Steroids, Triterpenes and Acetogenins. *W.A. Benzamin, New York. 1964.*
43. **Ryman BE, Tyrell DA :** Liposomes– Bags of Potential. *Essays Biochem.* **16 :** *49, 1980.*
44. **Shoppee CW :** Chemistry of the Steroids. *2nd ed., Academic Press, Inc,. New York. 1964.*
45. **Stumpf PK :** Metabolism of Fatty Acids. *Ann. Rev. Biochem.* **38 :** *159, 1969.*
46. **Vance DE, Vance JE (editors) :** Biochemistry of Lipids, Lipoproteins and Membranes. New Comprehensive Biochemistry. *Vol. 20, Elsevier Science Publishing Co. Inc., New York. 1991.*
47. **Weigandt H :** The Gangliosides. *Adv. Neurochem.* **4 :** *149–223, 1982.*

PROBLEMS

1. Some fats used in cooking, such as olive oil, spoil rapidly upon exposure to air at room temperature, whereas others, such as solid shortening, remain unchanged. Why?
2. Name the products of mild hydrolysis of the following lipids with dilute NaOH:
   (*a*) 1-stearoyl-2,3-dipalmitoylglycerol
   (*b*) 1-palmitoyl-2-oleoylphosphatidylcholine
3. A common procedure for cleaning the grease trap in a sink is to add a product that contains sodium hydroxide. Explain why this works.
4. Draw all the possible triacylglycerols that you could construct from glycerol, palmitic acid, and oleic acid. Rank them in order of increasing melting point.
5. Rank, in order of increasing solubility in water, a triacylglycerol, a diacylglycerol, and a monoacylglycerol, all containing only palmitic acid.
6. Why do eggs solidify after boiling while 'ghee' melts after heating?
7. How can exercise make one thinner even though one does not lose any weight?
8. Which type of diet is the more likely to lead to overconsumption and obesity, one rich in sugar or one rich in fat?
9. Is it true that you will get rid of more fat if you exercise less strenuously?
10. Which oils have greatest amount of the desirable monounsaturated fatty acid?
11. Is margarine better for you than butter?
12. Why can we blow a bubble with soap water but not with plain water?
13. Why is it not possible to write on an oily paper?
14. Why does butter turn rancid if left unrefrigerated for some time?
15. Why do soap bubbles last longer than water bubbles?
16. How is cooking oil refined?

17. Why do phenyls and dettol turn water milky-white when mixed with it?
18. Name two lipids which exist in oil as well as fat form.
19. Why fats are found in the form of globules in cytoplasm of cells?
20. Why fats float on water surface?
21. Why fats are synthesized in a cell in which they occur?
22. Why coconut oil and not mustard oil solidifies during winter season?
23. Which oil is best for cooking?
24. Why do different soaps irrespective of their colour produce only white foam?
25. By which process the vegetable oils can be converted into vegetable 'ghee' (vanaspati 'ghee')?
26. Milk contains water, sugars, fats and proteins. However, the colour of the milk is white. Which of these components is responsible for the colour?
27. When a piece of paper is wet with water, it tears off easily. But the same paper if soaked in oil does not lose strength. Why?
28. Why does the surface of boiling milk rise?
29. Is bread fattening?
30. How the purity of 'desi ghee' is tested?
31. Is white part of the orange peel edible?

ANSWERS TO PROBLEMS

1. Unsaturated fats (*e.g.*, olive oil) are susceptible to oxidation by molecular oxygen.
2. (*a*) The sodium salts of palmitic and stearic acids, plus glycerol
(*b*) The sodium salts of palmitic and oleic acids, plus glycerol-3-phosphorylcholine.
3. The triacylglycerols of animal fats (grease) are hydrolyzed by NaOH, in a process known as saponification, to form soaps, which are much more soluble in water than are the triacylglycerols.
4. Eight different triacylglycerols can be constructed. All saturated (palmitic) fatty acids, all unsaturated (oleic) fatty acids, or any combination of oleic and palmitic acids can be used. Furthermore, positional isomers are possible, as the three carbons of glycerol are not equivalent. In order of increasing melting point: OOO, OOP and OPO; PPO and POP; PPP, where O = oleic, P = palmitic.
5. In order of increasing solubility: triacylglycerol; diacylglycerol; monoacylglycerol.
6. Eggs contain rich proteins composed of many amino acids. On heating eggs, their proteins coagulate into a solid mass. 'Ghee' is a mixture of more than one fatty acids, which may be saturated or unsaturated. These fatty acids soften at high temperatures and subsequently melt. In the case of eggs, the reaction is irreversible, but it is reversible in the case of 'ghee'.
7. Exercise gets rid of fat but increases muscle which weighs more. Muscles are denser and take less space than fat.
8. Obesity is the result of caloric intake in excess of body needs. On a per-gram-of-diet basis, a diet rich in fat is more likely to lead to obesity than a diet rich in sugar because the former contains a higher caloric content per grain.
9. No. The amount of fat that you lose depends on the number of calories you burn, not the intensity of the workout.
10. Olive, canola, peanut and sesame-seed oils. Olive oil has the most.
11. A little better, provided you use the tub or liquid type and use it sparingly.

12. Surface tension of water is so strong that it does not allow water to form stable bubbles when blown through a pipe. Soaps and detergents contain a basic cleaning agent called *surfactant* or surface-active agent. Surfactants lower the surface tension of water sufficiently to enable a film of soap solution to be blown as bubbles. So soapy water is used by children to blow bubbles.

13. As paper is made of cellulose, when we write with a fountain pen, the colouring substance in the ink enters the fine capillaries through capillary action. When the solvent evaporates, dry ink remains in the capillaries, thus making the visible mark on the paper. If the surface of the paper is oily, the oil repels the ink which as a result does not stick to the paper to produce a mark. So it is not possible to write on oily paper with a fountain pen.

In the case of a ballpoint pen, when drawn over the oily surface, the ball on the tip of a ballpoint pen slips and does not rotate as there is no friction. As a result, the thick ink in it does not flow out and does not write.

14. Butter is nothing but solidified milk fat obtained from cream or whole milk by churning. Butter turns rancid if left unrefrigerated for some time because of the formation of butyric acid by the action of oxygen of air or due to bacterial action. Butyric acid is present in butter in the form of glycerol ester. This ester breaks down into butyric acid and butyl alcohol by the action of oxygen of air or due to bacterial action when kept at room temperature for a long time. It is the butyric acid that imparts the strong unpleasant smell to rancid butter. Refrigeration reduces the bacterial action and good packaging, the action of oxygen.

15. Bubbles are spherical in shape because of the surface tension of liquids. Soap bubbles are formed by blowing air into a film of soap water. *Soap is a surfactant, a chemical which when added to water reduces its surface tension. In other words, it makes the film to stretch more without breaking.* Therefore, the bubble formed from soap solution lasts longer. On the other hand, a water bubble bursts as soon as it is formed because of the higher surface tension of water.

16. Vegetable oils are mixtures of chemicals called glycerides or esters of glycerides and long chain fatty acids. Oil is obtained by crushing oil-bearing seeds such as groundnut, sunflower seeds, or rapeseed, and then pressing the oil through expellers. These mechanically-extracted oils contain impurities like gums and free fatty acids (FFAs), which have to be removed to make the oils suitable cooking media. Refining of oil is done by first treating it with alkali, which forms soap with the FFAs. The soaps settle out carrying with them some colouring matter. The colour and odour are removed by treating the oil with some absorbing material like Fuller's earth. Refining can also be done by extracting the pure oil with a suitable organic solvent like hexane and then removing the solvent by distillation.

17. Any substance which is soluble in water will completely dissolve in it and form a clear and transparent solution. *But oil-based substances are immiscible with water and so they form a fine emulsion. Since dettol and phenyls contain substances that are immiscible with water, they do not mix when added to the water. However, they also contain some chemicals which act as surfactants or detergents that help in breaking up of the oily layer into small droplets.* This makes dettol or phenyl to form a fine emulsion when added to water and make it appear cloudy. Since the emulsion is formed of fine droplets of the immiscible liquid and the amount of dettol or phenyl added is also quite small, the emulsion stays and does not separate out into a different oily layer on standing.

18. Coconut oil and vegetable 'ghee'.

19. It is because the fats are insoluble in water.

20. It is because the specific gravity of fats is less than one (*i.e.*, around 0.86).

21. Being insoluble in water, the fats cannot diffuse from cell to cell. Hence, they remain in the cell in which they are synthesized.

22. The melting point of lipids depend on the degree of unsaturation. The more is the degree of unsaturation, less is its melting point. Coconut oil has greater degree of unsaturation compared to mustard oil and hence the melting point of the former falls rapidly with fall in temperature. Consequently, it solidifies in winter season.

23. Olive oil and canola oil are best for cooking. Although flaxseed oil is the best source of essential fatty acids (EFAs), it should not be used for cooking. Heat changes flaxseed oil into a toxic derivative called *lipid peroxide*.

24. Foam or lather is nothing but a large collection of small soap bubbles. A soap bubble is, in turn, a round ball of many thin film of soap solution enclosing some air. Because of the low surface tension of the soap solution, the film can stretch and spread. Thus, the total surface area in the lather is larger than that could be formed by water of equal volume. Because of this, whatever slight tint present in the thin film of the coloured soap solution gets subdued. Although the soap film is more or less transparent, the foam or lather looks white because the light striking this large collection of bubbles bounces in several different directions or becomes scattered. This is the reason why all kinds of foam look white.

25. Hydrogenation

26. Fats

27. Paper is made of thin sheets of cellulosic fibres and other ingredients mixed together. It is sized to make it suitable for writing. The cellulosic fibres are held together by weak forces of attraction called the van der Waal's forces. When paper is dipped in water, the sizing material used for refining the paper being water-soluble becomes soft. Water being a polar solvent weakens the van der Waal's forces that hold the fibres together. As result, the fibres come off easily when pulled and so a wet paper tears off easily.

On the contrary, oil being a nonpolar solvent does not affect the water-soluble sizing material. Nor does it weaken the forces holding the cellulosic fibres together. So paper soaked in oil does not tear off as easily as paper soaked in water does.

28. Milk is a fine suspension of fat and protein globules in a watery liquid containing milk sugar (lactose) and other dissolved substances. Normally, the fat is dispersed uniformly in the form of fine droplets and hence remain as an emulsion. But when milk is heated, these fine droplets tend to rise up and at a temperature of about 50°C they form a thick layer of cream on the surface. When milk comes to boiling, bubbles of steam formed at the bottom of the vessel rise up and force the thick creamy layer upwards, causing the milk to spill.

29. No, a slice of bread has only 75 calories, but butter, margarine or mayonnaise on bread is fattening.

30. Take 5 ml 'ghee' (to be tested) in a test tube and heat it. When the ghee melts, add 5 ml HCl and shake the tube well. Then add 5 to 6 drops of furfural solution (2% solution of furfural in alcohol). The test tube is again shaken for 5 minutes. If there is no change in colour, ghee is pure.

31. Yes. The white membrane in the orange peel is rich in antioxidant compounds called *limonoids*. They have been shown to offer cancer protection.

CONTENTS

CHAPTER

15

Nucleic Acids

"A structure this pretty just had to exist."
–James Dewey Watson
(on the discovery of DNA structure,
in *The Double Helix, 1968*)

Model of DNA built by James D. Watson and Francis H.C. Crick at Cambridge University, 1953
[*Courtesy: Science and Society Picture Library, Science Museum, London.*]

INTRODUCTION

The final class of biomolecules to be considered, the nucleotides plus molecules derived from them (*i.e.*, nucleic acids), represent a clear case in which last is not least. "Nucleotides themselves participate in a plethora of crucial supporting roles in cell metabolism, and the nucleic acids provide the script for everything that occurs in a cell" (**Lehninger**, **Nelson** and **Cox**, 1993).

Nucleotides are energy-rich compounds that drive metabolic processes (*esp.*, biosynthetic) in all cells. They also serve as chemical signals, key links in cellular systems that respond to hormones and other extracellular stimuli, and are structural components of a number of enzyme cofactors and metabolic intermediates.

The **nucleic acids** (DNA and RNA) are the molecular repositories for genetic information and are jointly referred to as the 'molecules of heredity'. The structure of every protein, and ultimately of every cell constituent, is a product of information programmed into the nucleotide sequence of a cell's nucleic acids.

HISTORICAL RESUME

The nucleic acids have been the subject of biochemical investigations since 1869 when **Friedrich Miescher**, a 25-year old Swiss chemist, isolated nuclei from pus cells (white blood corpuscles) and found that they contained a hitherto unknown phosphate-rich substance, which he named *nuclein*. This substance was quite different from the carbohydrates, proteins and fats. He continued his studies on salmon sperm, a prime source of nuclein and isolated it as a nucleoprotein complex in 1871, when he prophetically wrote:

> "It seems to me that a whole family of such phosphorus-containing substances, differing somewhat from each other, will emerge, as a group of nuclein substances, which perhaps will deserve equal consideration with the proteins."

FRIEDRICH MIESCHER

(LT, 1844–1895)

Miescher, a Swiss and a student of the eminent German chemist, **Felix Hoppe-Seyler**, became interested in the chemistry of the nucleus in 1860s. He chose to work with white blood corpuscles (WBCs) *because they contained a large and easily observable nucleus.* A hospital at Tubingen supplied him with surgical bandages that had been peeled off purulent wounds. From the pus on the bandages, he obtained the WBCs, which he then treated with gastric juice (we now know that gastric juice contains an enzyme, *pepsin*, that digests protein). He observed microscopically that only the shrunken nuclei were left after the pepsin treatment; the remainder of the cell had dissolved. These nuclei were then analyzed and found to have different composition from any cellular component then known. It was called **nuclein.**

Miescher continued his studies when he returned to his native city, Basel, in Switzerland. He soon found that a more convenient source of this material was salmon sperm. With this favourable material, he further purified nuclein. When all the protein was completely removed, it became clear that the new material was an acid. It was then referred to as **nucleic acid.**

It was not until 1874, that he (Miescher) isolated pure nucleic acid from the DNA-protamine complex in salmon sperm nuclei. Later, it was found that the nuclein had acid properties and hence **Altmann**, in 1899, introduced the term *nucleic acid* to replace nuclein.

However, in 1880s, **Fischer** discovered purine and pyrimidine bases in nucleic acids. In 1881, **Zacharis** identified nuclein with chromatin. In 1882, **Sachs** stated that nucleins of sperm and egg are different and in 1884, **Hertwig** claimed that nuclein is responsible for the transmission of hereditary characteristics. The same year (*i.e.*, in 1894), **Geheimrat Albrecht Kossel**, of the University of Heidelberg, Germany, recognized that histones and protamines are associated with nucleic acids and also found that the histones were basic proteins. He was honoured with Nobel Prize for Physiology or Medicine in 1910 for demonstrating the presence of two purine and two pyrimidine bases in nucleic acids. In 1894, cytosine was identified and in 1909, uracil was isolated. **Phoebus Levene**, a Russian-born biochemist, recognized the 5-carbon ribose sugar in 1910 and later also discovered deoxyribose in nucleic acids. In 1914, **Robert Feulgen**, another German chemist, demonstrated a colour test known as *Feulgen test* for the deoxyribonucleic acid.

P.A. Levine (1931) stressed that there are 2 types of nucleic acids, *viz.*, deoxyribonucleic acid and ribonucleic acid. In 1941, **Caspersson** and **Brachet**, independently, related that nucleic acids were connected to protein synthesis. **Oswald T. Avery, Colin M. MacLeod** and **Maclyn McCarty**, in 1944, first of all demonstrated that DNA is directly involved in inheritance. **Alfred D. Hershey** and **Martha J. Chase** of Cold Spring Harbor Lab., New York, in 1952, demonstrated that only the DNA of T4 bacteriophage enters the host, the bacterium *Escherichia coli*, whereas the protein (*i.e.*, capsid) remains behind. They, thus, confirmed that DNA is the genetic material of most living organisms. **Matthew S. Meselson** and **Franklin H. Stahl** (1957), at California Institute of Technology, presented evidence that nucleic acid forms the genetic material.

In 1953, **James D. Watson** and **Francis H.C. Crick** constructed the double helical model for the DNA molecule which could successfully explain DNA replication. In 1957, **Arthur Kornberg** proved the Watson-Crick model in the cell-free system. In 1967, he also synthesized a molecule of DNA from the 6,000 nucleotides.

DEFINITION

Although the name nucleic acid suggests their location in the nuclei of cells, certain of them are, however, also present in the cytoplasm. The nucleic acids are the hereditary determinants of living orgainsms. They are the macromolecules present in most living cells either in the free state or bound to proteins as nucleoproteins. Like the proteins, *the nucleic acids are biopolymers of high molecular weight with mononucleotide as their repeating units*, just as amino acids are the repeating units of proteins. As regards their elemental composition, the nucleic acids contain carbon, hydrogen, oxygen, nitrogen and, strangely enough, phosphorus ; the percentage of the last two elements being about 15 and 10, respectively.

TYPES

There are two kinds of nucleic acids, deoxyribonucleic acid (DNA) and ribonucleic acid (RNA). Both types of nucleic acids are present in all plants and animals. Viruses also contain nucleic acids; however, unlike a plant or animal, a virus has either RNA or DNA, but not both. The previously held view, that the DNA occurred only in animals and the RNA only in plants, is now known to be incorrect. DNA is found mainly in the chromatin of the cell nucleus whereas most of the RNA (90%) is present in the cell cytoplasm and a little (10%) in the nucleous. It may be added that extranuclear DNA also exists; it occurs, for example, in mitochondria and chloroplasts. Upon hydrolysis, under different set of conditions (Fig. 15–1), the two nucleic acids yield 3 components : phosphoric acid, a pentose sugar and nitrogenous bases. Table 15–1. summarizes the structural components of the two types of nucleic acids, DNA and RNA.

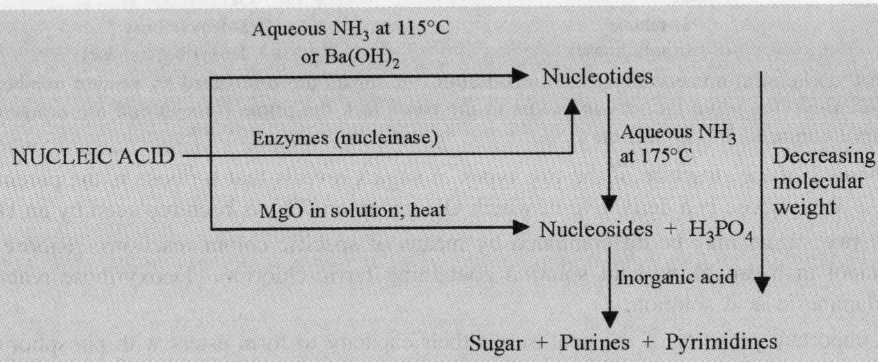

Fig. 15–1. Hydrolytic products of nucleic acid

Table 15–1. **Structural components of RNA and DNA**

Components	*Ribonucleic acid*	*Deoxyribonucleic acid*
Acid	Phosphoric acid	Phosphoric acid
Pentose sugar	D-ribose	D-2-deoxyribose
Nitrogenous bases		
Purines	Adenine	Adenine
	Guanine	Guanine
Pyrimidines	Cytosine	Cytosine
	Uracil	Thymine

Each of the 3 components of nucleic acids is discussed sepatately.

PHOSPHORIC ACID

The molecular formula of phosphoric acid is H_3PO_4. It contains 3 monovalent hydroxyl groups and a divalent oxygen atom, all linked to the pentavalent phosphorus atom

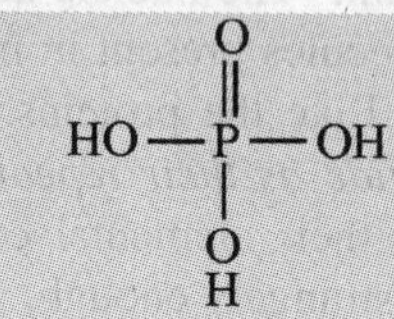

Phosphoric acid

PENTOSE SUGAR

The two types of nucleic acids are distinguished primarily on the basis of the 5-carbon (pentose) aldo sugar which they possess. One possesses D-2-deoxyribose, hence the name deoxyribose nucleic acid or deoxyribonucleic acid, while the other contains D-ribose, hence the name ribose nucleic acid or ribouncleic acid. *Both these sugars in nucleic acids are present in the furanose form and are of β configuration.*

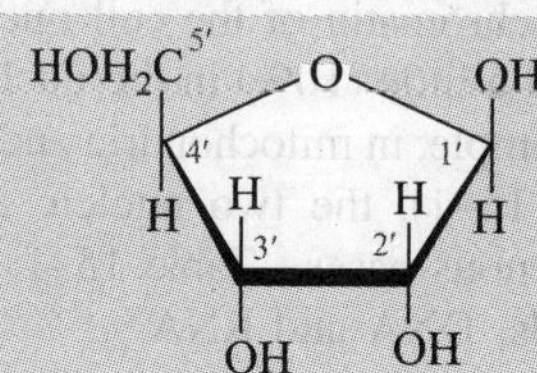

D-ribose
(β-D-ribofuranose)

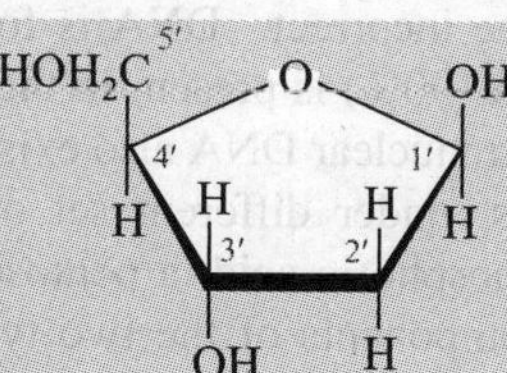

D-2-deoxyribose
(β-D-2-deoxyribofuranose)

[Note that in chemical nomenclature, the carbon atoms of sugars are designated by primed numbers, *i.e.*, C-1′, C-2′, C-3′ etc., while the various atoms in the bases lack the prime (′) sign and are designated by the cardinal numbers, *i.e.*, 1, 2, 3 etc.]

A perusal of the structure of the two types of sugars reveals that D-ribose is the parent sugar while D-2-deoxyribose is a derivative in which OH group on C2 has been replaced by an H atom.

The two sugars may be differentiated by means of specific colour reactions. Ribose reacts with orcinol in hydrochloric acid solution containing ferric chloride. Deoxyribose reacts with diphenylamine in acid solution.

An important property of the pentoses is their capacity to form esters with phosphoric acid. In this reaction the OH groups of the pentose, especially those at C3 and C5, are involved forming a 3′, 5′- phosphodiester bond between adjacent pentose residues. This bond, in fact, is an integral part of the structure of nucleic acids.

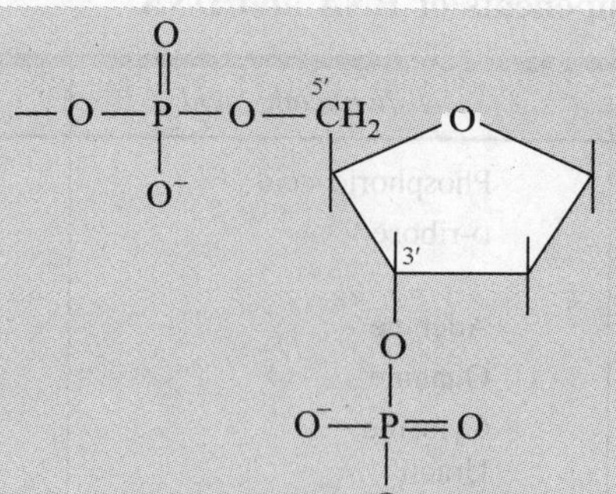

3′, 5′-phosphodiester bond

NITROGENOUS BASES

Two types of nitrogenous bases are found in all nucleic acids. The base is linked to the sugar moiety by the same carbon (C1) used in sugar-sugar bonds. The nitrogenous bases are derivatives of pyrimidine and purine. Owing to their π electron clouds, both the pyrimidine and purine bases are planar molecules.

Pyrimidine Derivatives

These are all derived from their parent heterocyclic compound *pyrimidine*, which contains a six-membered ring with two nitrogen atoms and three double bonds. It has a melting point of 22°C and a boiling point of 123.5°C.

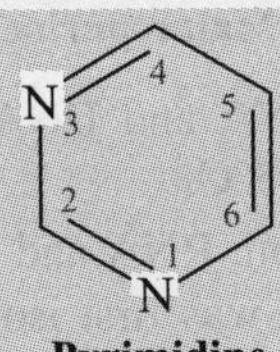

Pyrimidine

[For pyrimidine, the Chemical Abstracts use the numbering system, as depicted in the formula. This system is proposed by IUPAC. However, an older convention used a system, analogous to that indicated for the pyrimidine portion of purine, whose structural formula is given under the next subheading 'Purine Derivatives'.]

The common pyrimidine derivatives found in nucleic acids are *uracil*, *thymine* and *cytosine* (Fig. 15–2).

Uracil ($C_4H_4O_2N_2$), found in RNA molecules only, is a white, crystalline pyrimidine base with MW = 112.10 daltons and a m.p. 338°C. Only rarely does uracil occur in DNA.

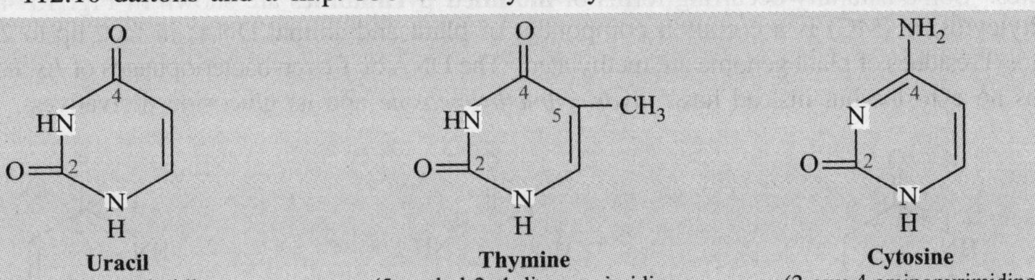

Uracil (2,4-dioxypyrimidine or pyrimidinedione) | **Thymine** (5-methyl-2, 4-dioxypyrimidine or 5-methyluracil) | **Cytosine** (2-oxy-4-aminopyrimidine or aminopyrimidone)

Fig. 15–2. Major pyrimidine derivatives

Thymine ($C_5H_6O_2N_2$), found in DNA molecules only, has MW = 126.13 daltons. It was first isolated from thymus, hence so named. Only rarely does thymine occcur in RNA.

Cytosine ($C_4H_5ON_3$), found in both RNA and DNA, is a white crystalline substance, with MW = 111.12 daltons and a m.p. 320-325°C.

Purine Derivatives

These are all derived from their parent compound *purine*, which contains a six-membered pyrimidine ring fused to the five-membered imidazole ring and is related to uric acid. It has a melting point of 216°C.

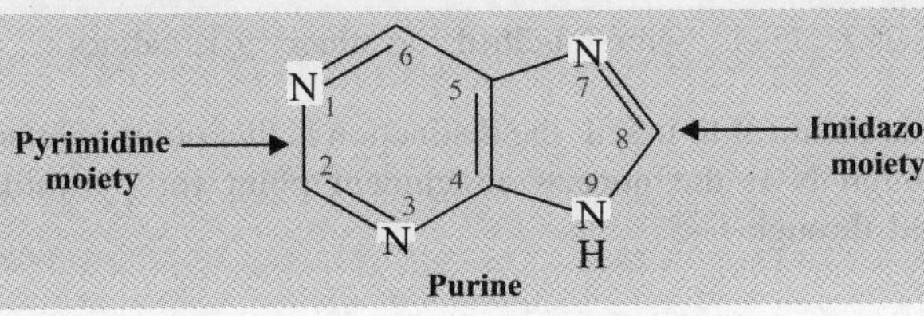

Purine

The prevalent purine derivatives found in nucleic acids are *adenine* and *guanine* (Fig. 15–3).

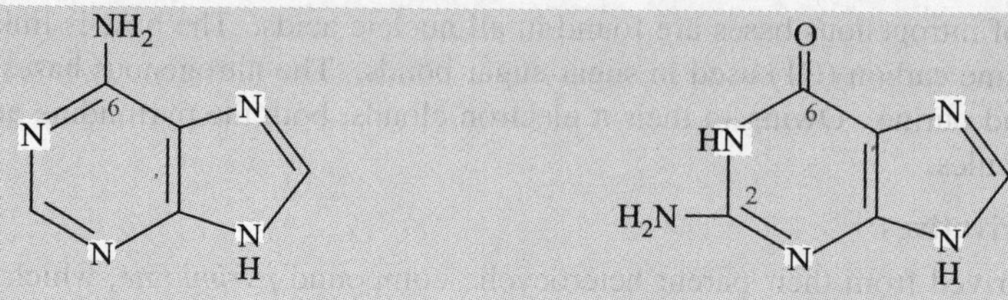

Adenine
(6-aminopurine)

Guanine
(2-amino-6-oxypurine)

Fig. 15–3. Major purine derivatives

Adenine ($C_5H_5N_5$), found in both RNA and DNA, is a white crystalline purine base, with MW = 135.15 daltons and a m.p. 360-365°C.

Guanine ($C_5H_5ON_5$), also found in both RNA and DNA, is a colourless, insoluble crystalline substance, with MW = 151.15 daltons. It was first isolated from guano (bird manure), hence so named.

Other purine derivatives are also found in plants. For example, *caffeine* (1,3,7-trimethylxanthine) is present in coffee and tea and *theobromine* (3, 7-dimethylxanthine) is found in tea and cocoa. Both caffeine and theobromine have important pharmacological properties.

Modified Nitrogenous Bases

The above 5 nitrogenous bases were once believed to account for the total base composition of animal and plant nucleic acids and were hence designated as **major nitrogenous bases**. It is now known that other minor bases called **modified nitrogenous bases** also occur in polynucleotide structures. Some naturally-occurring forms of **modified pyrimidines** are shown in Fig. 15–4. The 5-methylcytosine (MC) is a common component of plant and animal DNA; in fact, up to 25% of the cytosyl residues of plant genome are methylated. The DNA of T-even bacteriophages of *Escherichia coli* has no cytosine but instead has 5-*hydroxylmethylcytosine* and its glucoside derivatives.

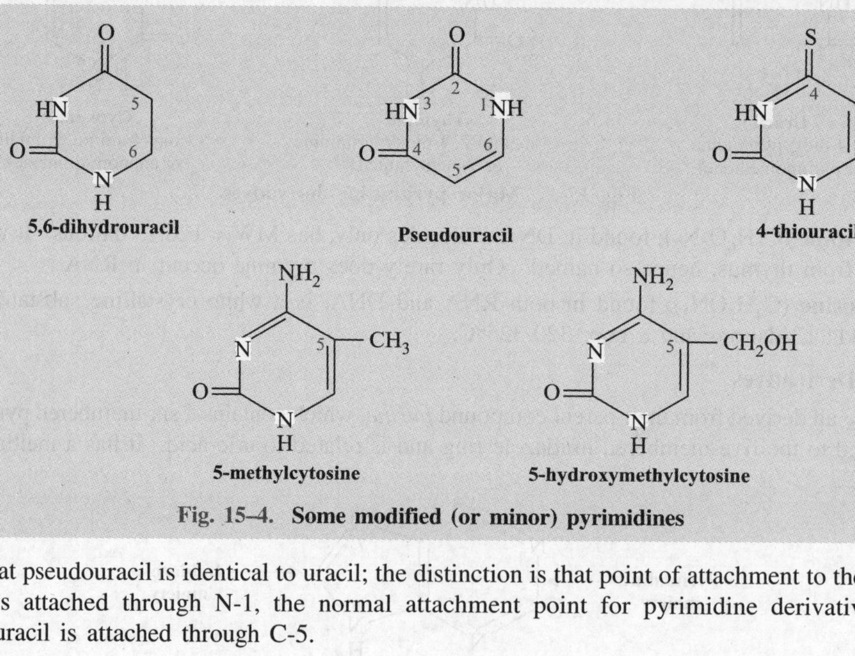

5,6-dihydrouracil **Pseudouracil** **4-thiouracil**

5-methylcytosine **5-hydroxymethylcytosine**

Fig. 15–4. Some modified (or minor) pyrimidines

Note that pseudouracil is identical to uracil; the distinction is that point of attachment to the ribose-uracil is attached through N-1, the normal attachment point for pyrimidine derivatives, and pseudouracil is attached through C-5.

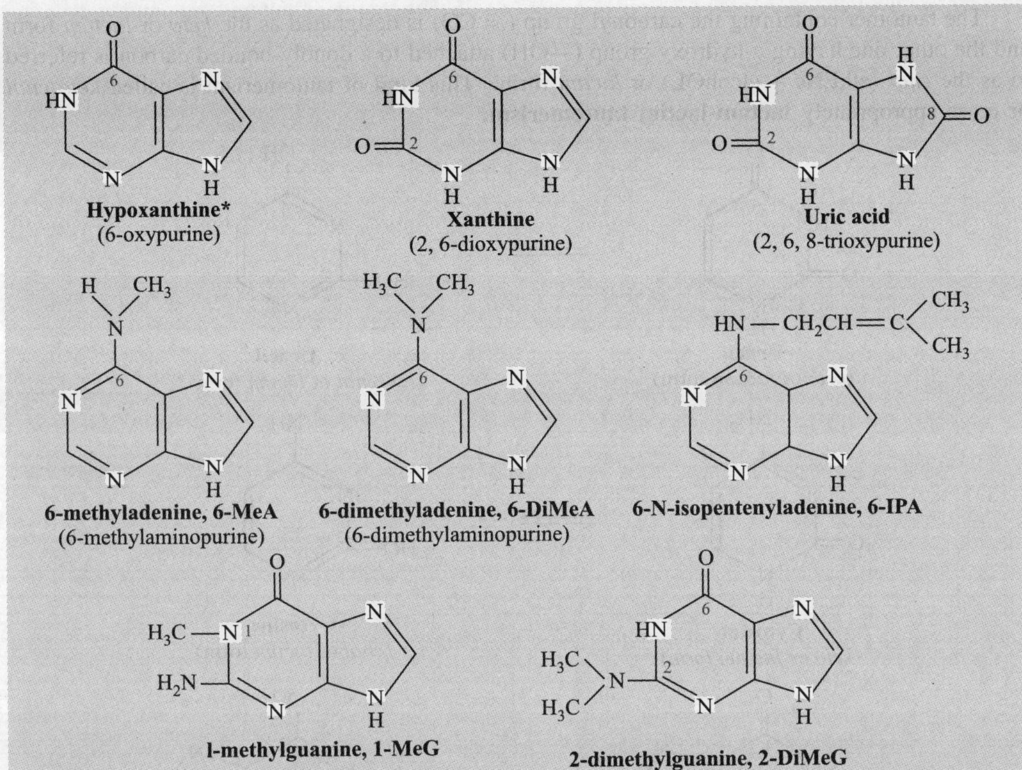

Fig. 15–5. Some modified (or minor) purines

* While names of nucleosides and nucleotides are generally derived from the corresponding bases, we have here an exception to this rule: the base corresponding to the nucleoside called inosine (and the derived nucleotides) is called **hypoxanthine**.

Among the **modified purines** (Fig. 15–5), some of them are found in transfer RNAs (tRNAs), which are a class of nucleic acids. *Methylation is the most common form of purine modification.* Methylation of purines (particularly of adenine) in DNA is now known to occur in the genetic material of microorganisms, and it is believed that plant genomes will also be shown to have methylated purines. The 6-methyladenine is found in bacterial DNA.

Note that pseudouracil is identical to uracil; the distinction is the point of attachment to the ribose: uracil is attached through N–1, the normal attachment point for pyrimidine derivatives, and pseudouracil is attached through C–5

Tautomerism in Nitrogenous Bases

Compounds that exist in 2 structural isomeric forms which are mutually interconvertible and exist in dynamic equilibrium are called **tautomers** and the phenomenon is termed **tautomerism**. The term tautomer (*tauto*^{G} = same ; *meros*^{G} = part) was first coined by **Laar** in 1885. *Aldehydes, ketones and other carbonyl compounds such as esters exhibit tautomerism.* Tautomerism, which expresses wandering tendency of an H atom, involves migration of a proton (H^+) from α-carbon to carbonyl oxygen by the following mechanism:

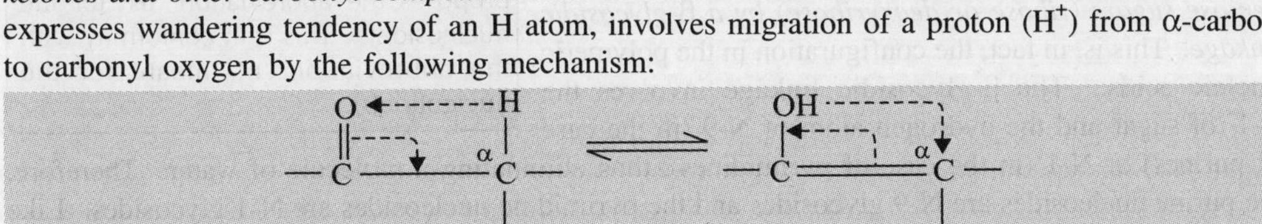

The tautomer containing the carbonyl group (= CO) is designated as the *keto* or *lactam* form and the other one having a hydroxy group (—OH) attached to a doubly-bonded carbon is referred to as the *enol* (alkENe + alcohOL) or *lactim* form. This kind of tautomerism is called *keto-enol* or more appropriately **lactam-lactim tautomerism.**

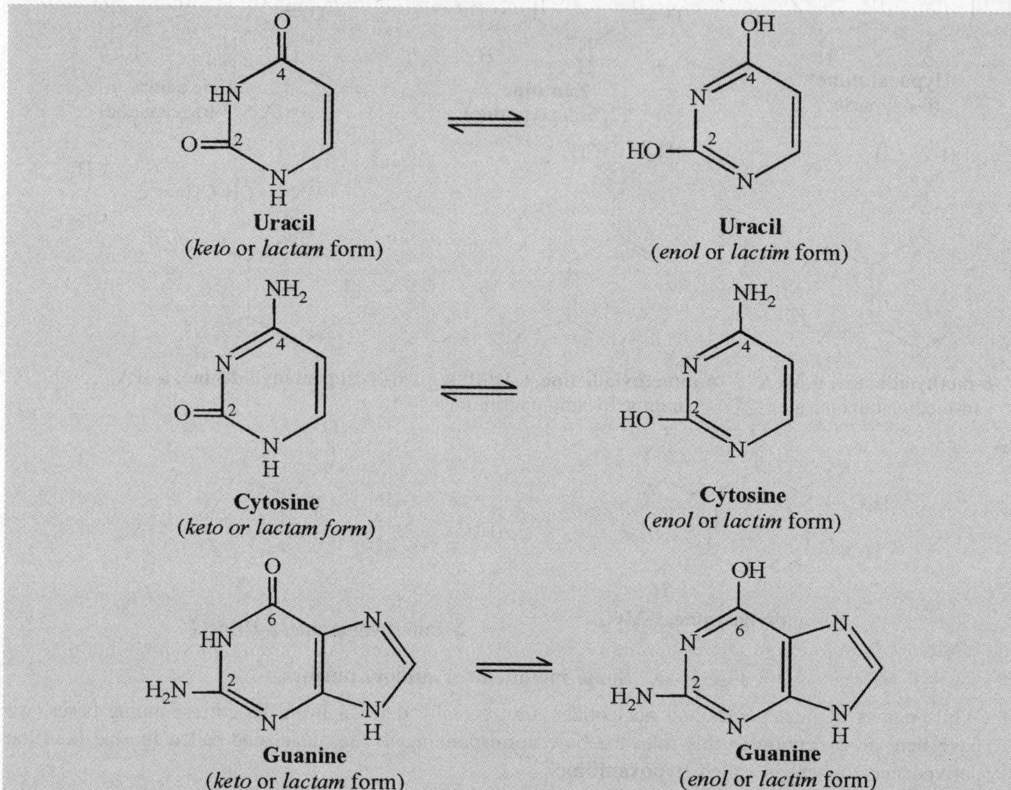

Fig. 15–6. Tautomeric forms of uracil, cytosine and guanine

The oxygen-containing nitrogenous bases, for example, uracil, thymine and cytosine among the pyrimidine derivatives and guanine, hypoxanthine, xanthine, uric acid, 1-methylgunanine, 2-dimethylguanine etc., among the purine derivatives exist in keto-enol (= lactam-lactim) forms. Three such examples are given in Fig. 15–6 and may be written in either form. However, *it is the keto (= lactam) form which predominates at neutral and acid pH values which are of physiological importance* ; the enol (= lactim) form becomes more prominent as pH decreases (*i.e.*, at acidic values).

NUCLEOSIDES

The nucleosides are compounds in which nitrogenous bases (purines and pyrimidines) are conjugated to the pentose sugars (ribose or deoxyribose) by a β-glycosidic linkage. This is, in fact, the configuration in the polymeric nucleic acids. The β-glycosidic linkage involves the C-1′ of sugar and the hydrogen atom of N-9 (in the case of purines) or N-1 (in the case of pyrimidines), thus eliminating a molecule of water. Therefore, the purine nucleosides are N-9 glycosides and the pyrimidine nucleosides are N-1 glycosides. Like the O-glycosides, the nucleosides are stable in alkali. Purine nucleosides are readily hydrolyzed

> One of the present authors (**JL Jain**) proposes the abbreviation '**ns**' for the nucleoside(s). This is in conformity with the abbreviation 'nt' meant for the nucleotide(s).

by acid whereas pyrimidine nucleosides are hydrolyzed only after prolonged treatment with concentrated acid.

The nucleosides are generally named for the particular purine or pyrimidine present. Nucleosides containing ribose are called *ribonucleosides*, while those possessing deoxyribose as *deoxyribonucleosides*. Table 15–2 lists the trival names of the purine and pyrimidine nucleosides which are related to the bases that occur in RNA and DNA.

Table 15–2. The nucleosides

Base	*Sugar*	*Nucleoside*	*Trivial name**	*Abbreviation*
Ribonucleosides				
Adenine	Ribose	Adenine ribonucleoside	Adenosine	AR
Guanine	Ribose	Guanine ribonucleoside	Guanosine	GR
Cytosine	Ribose	Cytosine ribonucleoside	Cytidine	CR
Thymine	Ribose	Thymine ribonucleoside	Thymidine	TR
Uracil	Ribose	Uracil ribonucleoside	Uridine	UR
Deoxyribo-nucleosides				
Adenine	Deoxyribose	Adenine deoxyribonucleoside	Deoxyadenosine	AdR
Guanine	Deoxyribose	Guanine deoxyribonuceloside	Deoxyguanosine	GdR
Cytosine	Deoxyribose	Cytosine deoxyribonucleoside	Deoxycytidine	CdR
Thymine	Deoxyribose	Thymine deoxyribonucleoside	Deoxythymidine**	TdR
Uracil	Deoxyribose	Uracil deoxyribonucleoside	Deoxyuridine	UdR

* Note that the trivial names for the pyrimidine nucleosides end with the suffix *-dine*, whereas those of purine nucleosides end with the suffix *-sine*.

** The deoxyribonucleoside of thymine is usually termed as thymidine instead of deoxythymidine since this pyrimidine is primarily found in DNA. However, as thymine also occurs in one type of RNA termed transfer RNA (tRNA), a growing tendency amongst the biochemists is to call this deoxyribonucleoside as deoxythymidine in conformity to the nomenclature and to restrict the use of the term thymidine to the ribonucleoside of thymine only.

The structure of two nucleosides, one ribonucleoside (also called riboside) and another deoxyribonucleoside (or deoxyriboside) is given in Fig. 15–7.

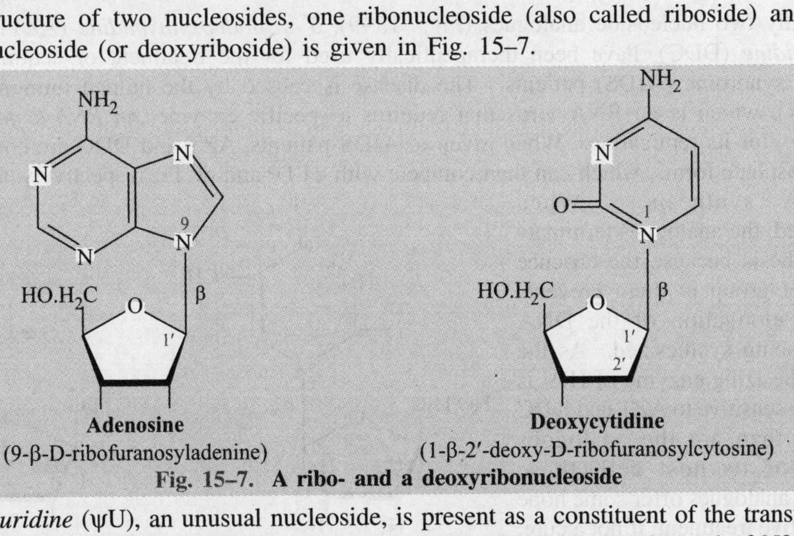

Adenosine
(9-β-D-ribofuranosyladenine)

Deoxycytidine
(1-β-2′-deoxy-D-ribofuranosylcytosine)

Fig. 15–7. A ribo- and a deoxyribonucleoside

Pseudouridine (ψU), an unusual nucleoside, is present as a constituent of the transfer RNAs. In it, the β-glycoside linkage occurs between the C1 of ribose and C5 (instead of N3) of uracil (Fig. 15–8). The uracil moiety at physiological pH has one oxy group in the *keto* form and the other in the *enol* form.

Table 15–3. The common nucleotides

*Nucleotide**	*Trivial name*	*Abbreviations†*	
Ribouncleotides			
Adenosine-5′-monophosphate	Adenylic acid	AMP	Ado-5′-P
Guanosine-5′-monophosphate	Guanylic acid	GMP	Guo-5′-P
Cytidine-5′-monophosphate	Cytidylic acid	CMP	Cyd-5′-P
Uridine-5′-monophosphate	Uridylic acid	UMP	Urd-5′-P
2′-deoxyribonucleotides			
Deoxyadenosine-5′-monophosphate	Deoxyadenylic acid	dAMP	dAdo-5′-P
Deoxyguanosine-5′-monophosphate	Deoxyguanylic acid	dGMP	dGuo-5′-P
Deoxycytidine-5′-monophosphate	Deoxycytidylic acid	dCMP	dCyd-5′-P
Deoxythymidine-5′-monophosphate	Deoxythymidylic acid	dTMP	dThd-5′-P

* The table lists only the monophosphate derivatives at C5′ position of both ribonucleotides and 2′-deoxyribonucleotides. Among the other monophosphates of **ribonucleotides,** those at C2′ position are called as adenosine-2′-monophosphate (abbreviated as 2′-AMP or Ado-2′-P), cytidine-2′-monophosphate (2′-CMP or Cyd-2′-P) etc., and those at C3′ position as adenosine-3′-monophosphate (3′-AMP or Ado-3′-P), cytidine-3′-monophosphate (3′-CMP or Cyd-3′-P) etc. Similarly, in the case of **2′-deoxyribonulceotides,** the 3′-monophosphate derivatives are called as adenosine- 3′-monophosphate (3′-dAMP or dAdo-3′P), cytosine-3′-monophosphate (3′-dCMP or dCyd-3′-P) etc.

† In each case, the first abbreviation is more generally used. Note that in the case of 5′-monophosphate derivatives, the primed number 5′ is not written in the first system of abbreviation as these derivatives (as well as the 5′-diphosphates and 5′-triphosphates) are more important physiologically and are very frequently used in the literature. While adopting the second sytem of abbreviation, the diphosphates are written as Ado-5′-PP, Cyd-5′-PP etc., and the triphosphates as Ado-5′-PPP, Cyd-5′-PPP etc.

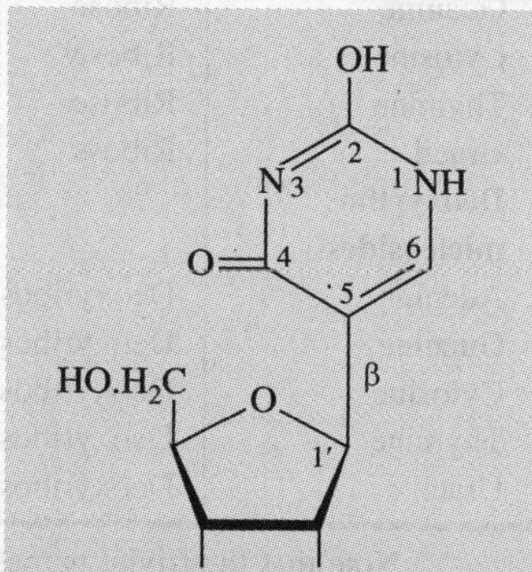

Fig. 15–8. Pseudouridine, ψU (= 5-ribosyluracil)

Note the absence of a true nucleosidic bond.

Nucleoside Analogues as Drugs

Recently two nucleoside analogues (Fig. 15–9), 3′-*azidodeoxythymidine* (AZT) and 2′, 3′-*dideoxycytidine* (DDC), have been therapeutically used for the treatment of acquired immune deficiency syndrome (AIDS) patients. The disease is caused by the human immunodeficiency viurs (HIV), which is an RNA virus that requires a specific enzyme, *an RNA-dependent DNA polymerase*, for its replication. When given to AIDS patients, AZT and DDC are converted into their triphosphate forms, which can then compete with dTTP and dCTP, respectively, as substrates for DNA synthesis. When incorporated, the analogues terminate DNA synthesis because the absence of a 3′-OH group in them prevents continued elongation of the DNA molecule being synthesized. As the DNA-synthesizing enzyme of HIV is much more sensitive to AZT and DDC inhibition than are the analogous enzymes of its host cell, these nucleoside analogues offer some hope as an effective treatment, if not a cure, for AIDS patients.

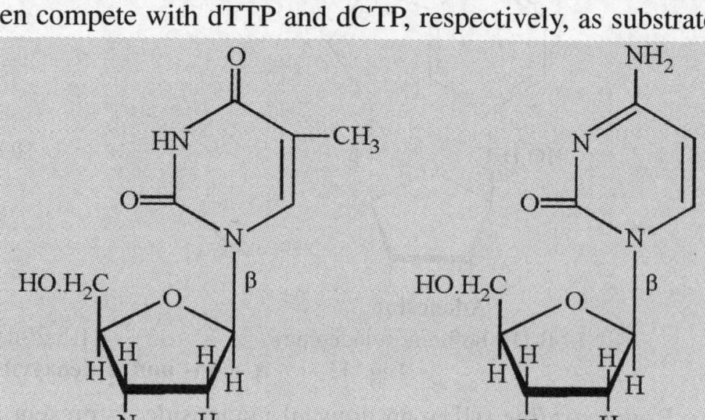

3′-azidodeoxythymidine, AZT **2′, 3′-dideoxycytidine, DDC**

Fig. 15–9. Structure of two nucleoside analogues

NUCLEOTIDES

Nucleotides are the phosphoric acid esters of nucleosides. These occur either in the free form or as subunits in nucleic acids. As mentioned earlier, the phosphate is always esterified to the sugar moiety. The component units of nucleotides are shown in Fig 15-10.

In the ribose moiety of a **ribonucleoside**, phosphorylation is possible only at *three* positions (C2′, C3′, C5′) since C1′ and C4′ are involved in the furanose ring formation. In other words, the

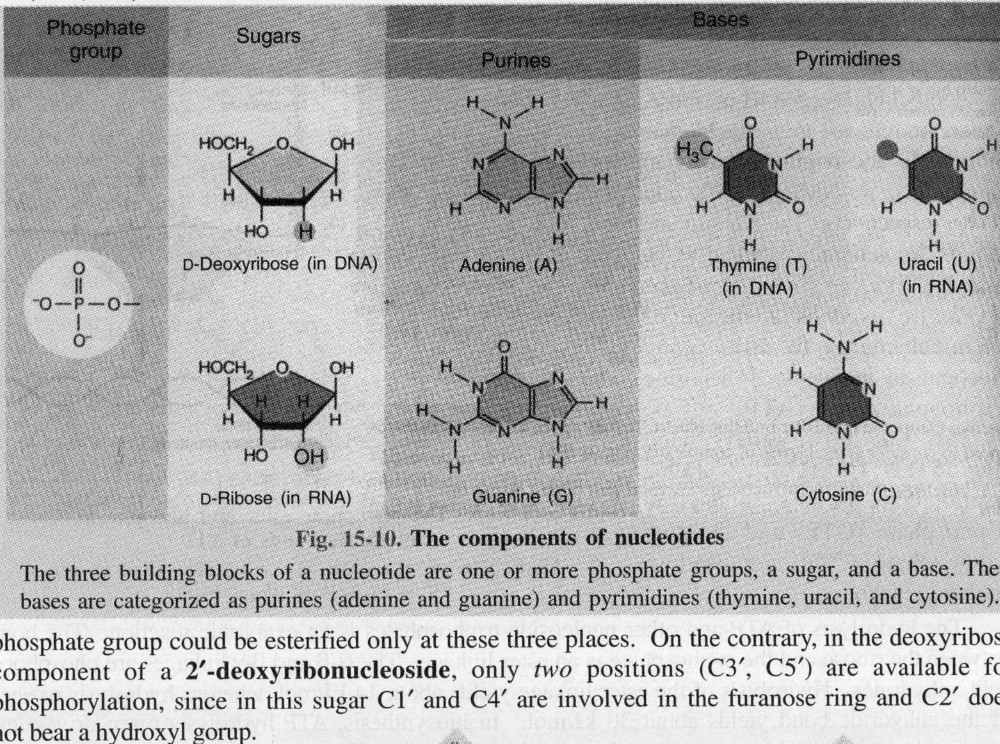

Fig. 15-10. The components of nucleotides

The three building blocks of a nucleotide are one or more phosphate groups, a sugar, and a base. The bases are categorized as purines (adenine and guanine) and pyrimidines (thymine, uracil, and cytosine).

phosphate group could be esterified only at these three places. On the contrary, in the deoxyribose component of a **2′-deoxyribonucleoside**, only *two* positions (C3′, C5′) are available for phosphorylation, since in this sugar C1′ and C4′ are involved in the furanose ring and C2′ does not bear a hydroxyl gorup.

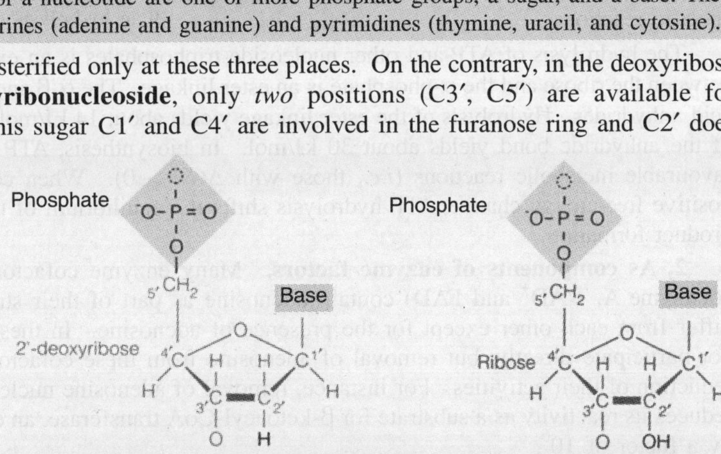

(*a*) Repeating unit of deoxyribonucleic and (DNA)

(*b*) Repeating unit of ribonucleic acid (RNA)

Fig. 15-11. The structure of nucleotides found in (*a*) DNA and (*b*) RNA

DNA contains deoxyribose as its sugar, and the bases A, T, G, and C. RNA contains ribose as its sugar, and the bases A, U, G, and C. In a DNA or RNA strand, the oxygen on the 3′ carbon is linked to the phosphorus atom of phosphate in the adjacent nucleotide. This bond formation removes the hydrogen attached to the 3′-oxygen atom, and it removes an oxygen (shown with a *dashed line*) attached to phosphorus.

Accordingly, hydrolysis of the two nucleic acids, RNA and DNA, by various methods and under different set of conditions gives rise to isomeric nucleotides of 3 types and 2 types respectively. Table 15–3 lists some important ribonucleotides (also called *ribotides*) and deoxyribonucleotides (or *deoxyribotides*). While names of nucleosides and nucleotides are generally derived from the corresponding bases, there is one exception to this rule : the base corresponding to the nucleoside called inosine (and the derived nucleotides) is called hypoxanthine. Fig. 15–11 depicts the structure of nucleotides found in DNA and RNA both.

Functions of Nucleotides

In addition to their roles as the subunits of nucleic acids, nucleotides perform some other functions. These are enumerated below:

1. As carriers of chemical energy. Nucleotides may have one, two or three phosphate groups covalently linked at 5′-OH of ribose. These are referred to as **nucleoside mono-, di-** and **triphos-phates** and abbreviated as NMPs, NDPs and NTPs, respectively. The 3 phosphate groups are generally labelled as α, β and γ, *starting from the ribose.* NTPs are used as a source of chemical energy to drive many biochemical reactions. Adenosine triphosphate (ATP) is, by far, the most widely used (Fig. 15–12). Others such as uridine triphosphate (UTP), guanosine triphosphate (GTP) and cytidine triphosphate (CTP) are used in specific reactions.

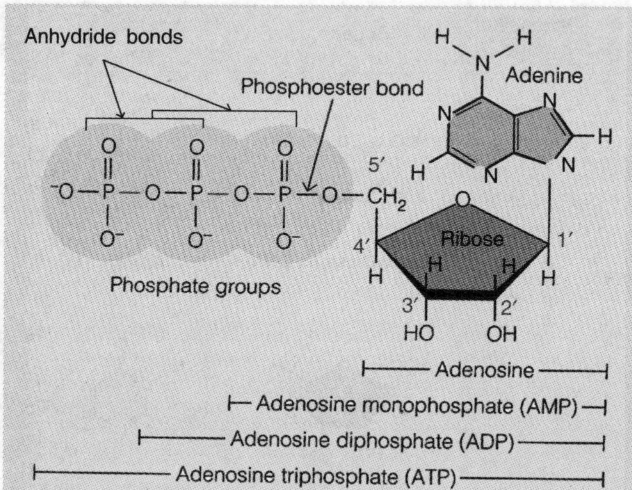

Fig. 15–12. The phosphate ester and phosphoric acid anhydride bonds of ATP
Hydrolysis of an anhydride bond yields more energy than hydrolysis of the ester.

The hydrolysis of ATP and other nucleoside triphosphates is an exergonic reaction. The bond between the ribose and the α-phosphate is an ester linkage. The α-β and β-γ linkages are phosphoric acid anhydrides. Hydrolysis of the ester linkage yields about 14 kJ/mol, whereas hydrolysis of each of the anhydride bond yields about 30 kJ/mol. In biosynthesis, ATP hydrolysis often drives less favourable metabolic reactions (*i.e.*, those with $\Delta G^{o\prime} > 0$). When coupled to a reaction with a positive free-energy change, ATP hydrolysis shifts the equilibrium of the overall process to favour product formation.

2. As components of enzyme factors. Many enzyme cofactors and coenzymes (such as coenzyme A, NAD^+ and FAD) contain adenosine as part of their structure (Fig. 15–13). They differ from each other except for the presence of adenosine. In these cofactors, adenosine does not participate directly, but removal of adenosine from these cofactors usually results in drastic reduction of their activities. For instance, removal of adenosine nucleotide from acetoacetyl-CoA reduces its reactivity as a substrate for β-ketoacyl-CoA transferase, an enzyme of lipid metabolism, by a factor of 10^6.

3. As chemical messengers. The cells respond to their environment by taking cues from hormones or other chemical signals in the surrounding medium. The interaction of these chemical signals (first messengers) with receptors on the cell surface often leads to the formation of **second messengers** (Fig. 15–13) inside the cell, which in turn lead to adaptive changes inside the cell. *Often, the second messenger is a nucleotide.*

One of the most common second messengers is the nucleotide **adenosine 3′, 5′- cyclic monophosphate (cyclic AMP or cAMP)**, formed from ATP in a reaction catalyzed by *adenylate cyclase*, associated with the inner face of the plasma membrane. Cyclic AMP serves regulatory functions in virtually every cell outside the plant world. **Guanosine 3′, 5′- cyclic monophosphate (cGMP)** occurs in many cells and also has regulatory functions.

DEOXYRIBONUCLEIC ACID

A nucleic acid may be visualized as a polymer of a nucleotide monomer. In other words, it may be considered as a polynucleotide.

DNA as .... (base-deoxyribose-phosphate)$_n$
RNA as .... (base-ribose-phosphate)$_n$

> The name **'polymer'** was coined by **Jöns Jacob Berzelius**, in 1827.

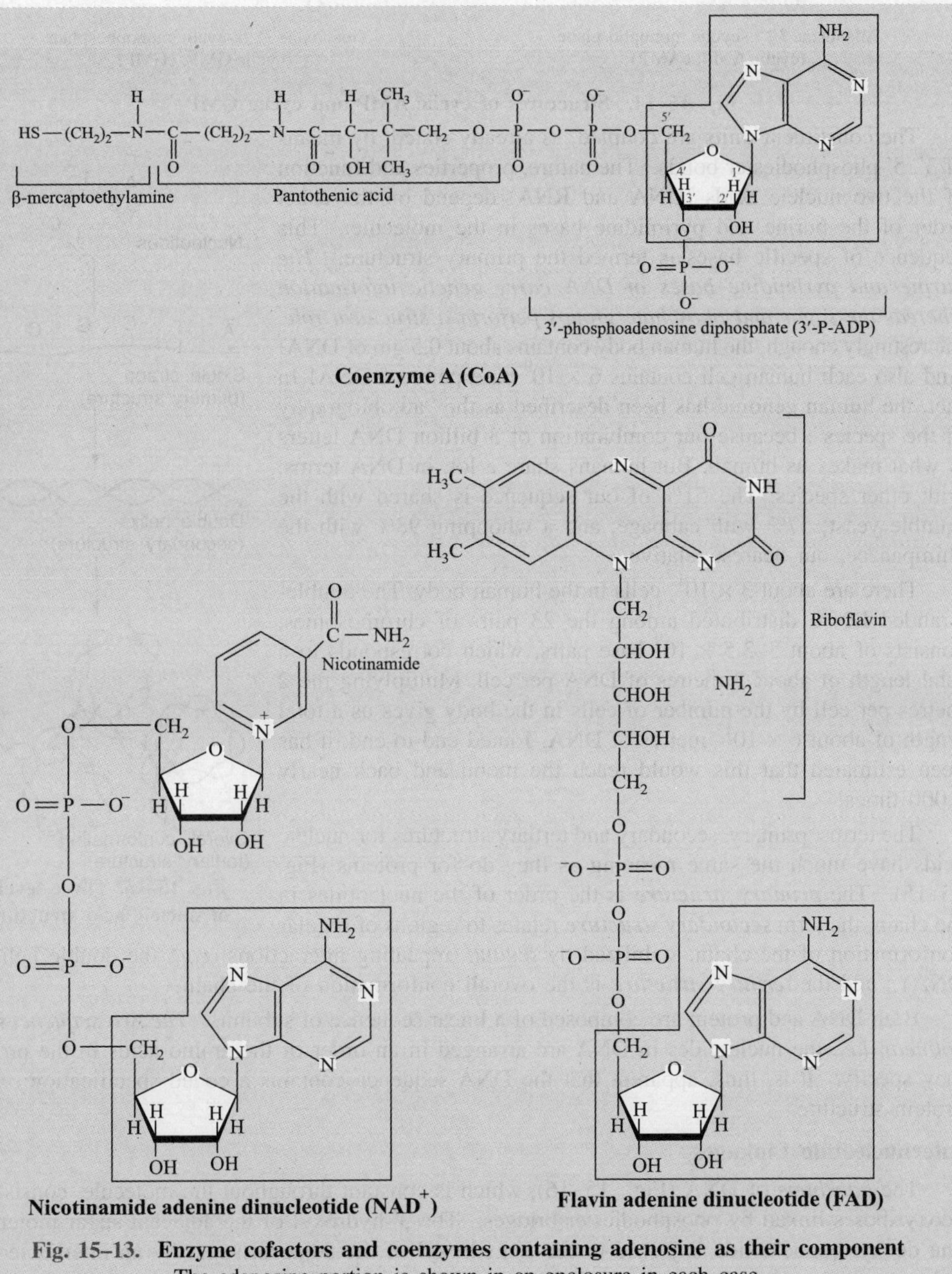

Fig. 15–13. Enzyme cofactors and coenzymes containing adenosine as their component
The adenosine portion is shown in an enclosure in each case.

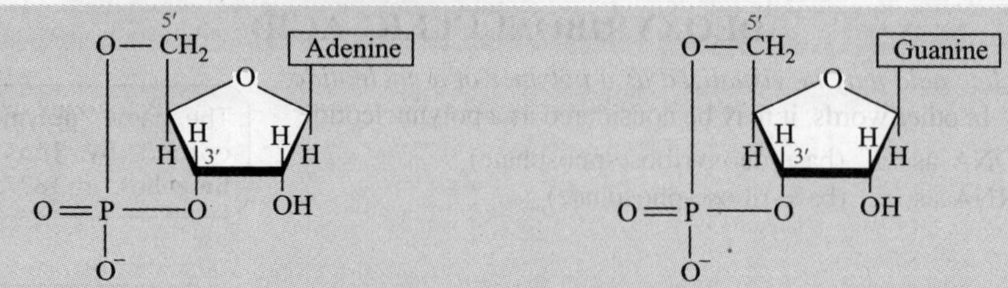

Adenosine 3′, 5′-cyclic monophosphate (cyclic AMP; cAMP)

Guanosine 3′, 5′-cyclic monophosphate (cyclic GMP; cGMP)

Fig. 15–14. Structures of cyclic AMP and cyclic GMP

The constituent units are coupled, as already stated, by means of 3′, 5′-phosphodiester bonds. The nature, properties and function of the two nucleic acids (DNA and RNA) depend on the exact order of the purine and pyrimidine bases in the molecule. This sequence of specific bases is termed the primary structure. *The purine and pyrimidine bases of DNA carry genetic information whereas the sugar and phosphate groups perform a structural role.* Interestingly enough, the human body contains about 0.5 gm of DNA! And also each human cell contains 6 × 10^9 base pairs of DNA! In fact, the human genome has been described as the "autobiography of the species", because our combination of 3 billion DNA letters is what makes us human. But humans share a lot, in DNA terms, with other species. The 51% of our sequence is shared with the humble yeast; 57% with cabbage; and a whopping 98% with the chimpanzee, our nearest relative.

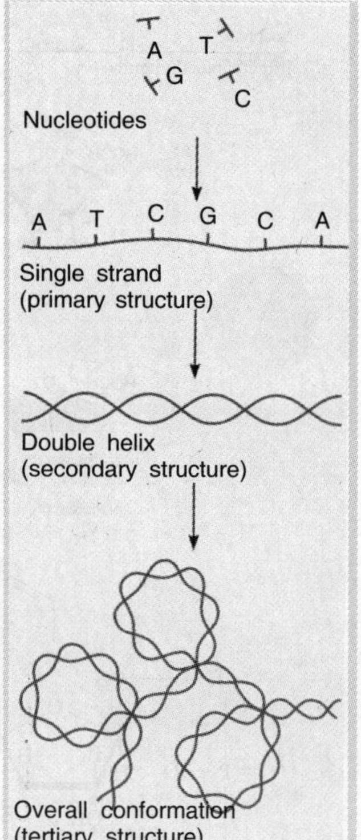

Fig. 15–15. Three levels of nucleic acid structure

There are about 3 × 10^{12} cells in the human body. The double-stranded DNA, distributed among the 23 pairs of chromosomes, consists of about 3–3.5 × 10^9 base pairs, which corresponds to a total length of about 2 metres of DNA per cell. Multiplying the 2 metres per cell by the number of cells in the body gives us a total length of about 6 × 10^{12} metres of DNA. Joined end-to-end, it has been estimated that this would reach the moon and back nearly 8,000 times!

The terms, primary, secondary and tertiary structures for nucleic acids have much the same meaning as they do for proteins (Fig. 15–15). The ***primary structure*** is the order of the nucleotides in the chain; the term ***secondary structure*** relates to regions of regular conformation of the chain, stabilized by regular, repeating interactions (*e.g.*, the double helix of DNA) ; and the ***tertiary structure*** is the overall conformation of the chain.

Both DNA and protein are composed of a linear sequence of subunits. *The two sequences are colinear, i.e.*, the nucleotides in DNA are arranged in an order of the amino acids in the protein they specify. It is, thus, apparent that the DNA sequence contains a coded specification of the protein structure.

Internucleotide Linkages

The *backbone* of DNA (Fig. 15–16), which is constant throughout the molecule, consists of deoxyriboses linked by phosphodiester bridges. The 3′-hydroxyl of the adjacent sugar moiety of one deoxyribonucleotide is joined to the 5′-hydroxyl of the adjacent sugar by an internucleotide linkage called a phosphodiester bond. *The variable part of DNA is its sequence of bases.*

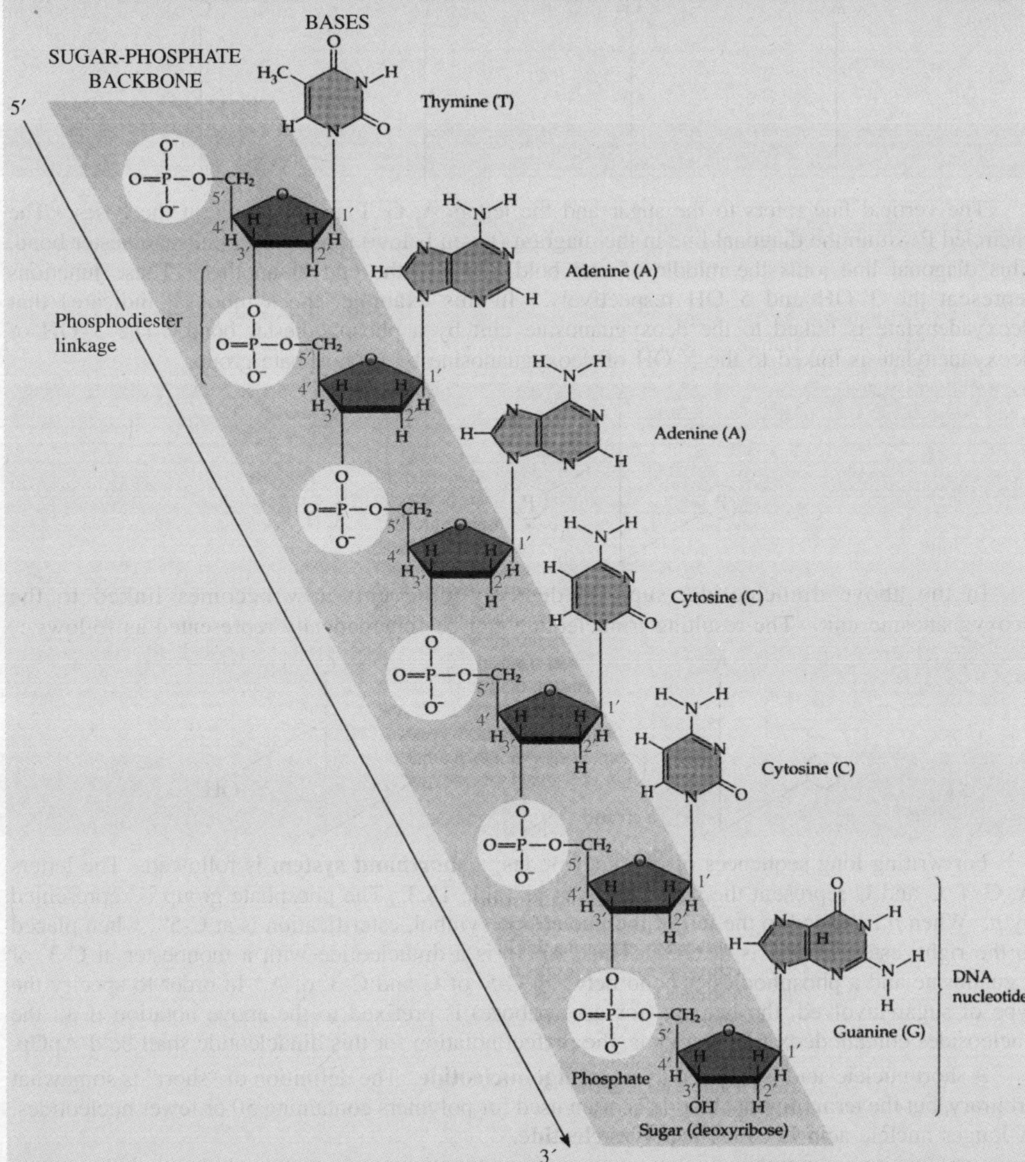

Fig. 15–16. A short strand of DNA containing six nucleotides

Nucleotides are covalently linked together to form a strand of DNA. Note the position of the internucleotide linkage between C3′ and C5′.

Representation of Nucleic Acid Backbones. The writing of the molecular structure of DNA chain, as depicted in Fig. 15–16, is tedious and time-taking. Henceforth, a system of schematic representation has been evolved. According to this system, the symbols for the 4 principal deoxyribonucleosides are:

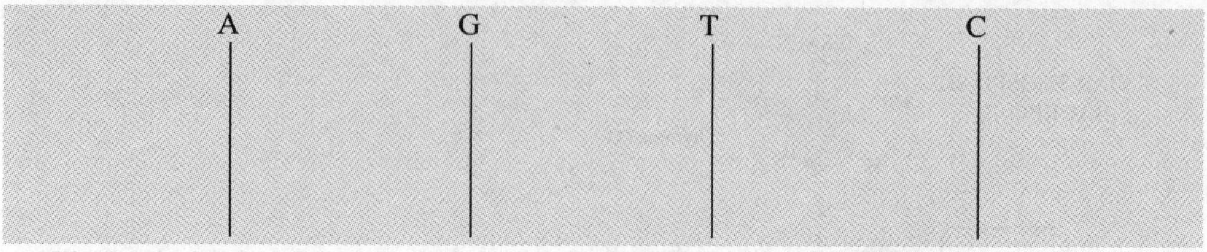

The vertical line refers to the sugar and the letters A, G, T and C represent the bases. The encircled P within the diagonal line in the diagram (given below) represents a phosphodiester bond. This diagonal line joins the middle of one bold line and the end of another. These junctions represent the 3′-OH and 5′-OH respectively. In this example, the symbol Ⓟ indicates that deoxyadenylate is linked to the deoxyguanosine unit by a phosphodiester bond. The 3′-OH of deoxyadenylate is linked to the 5′-OH of deoxyguanosine by a phosphate group.

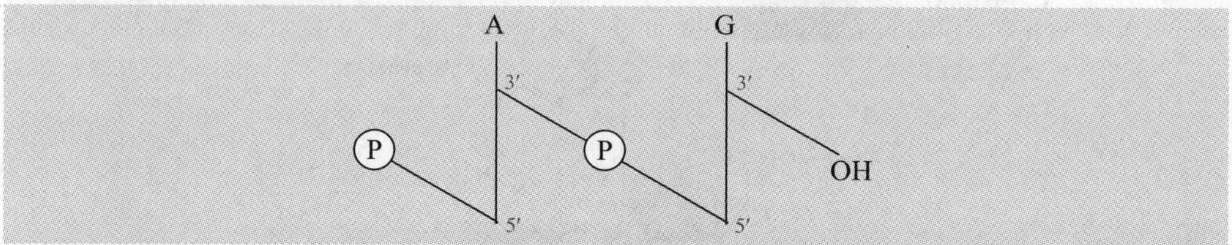

In the above dinucleotide, suppose deoxycytidine unit now becomes linked to the deoxyguanosine unit. The resulting trinucleotide can be schematically represented as follows :

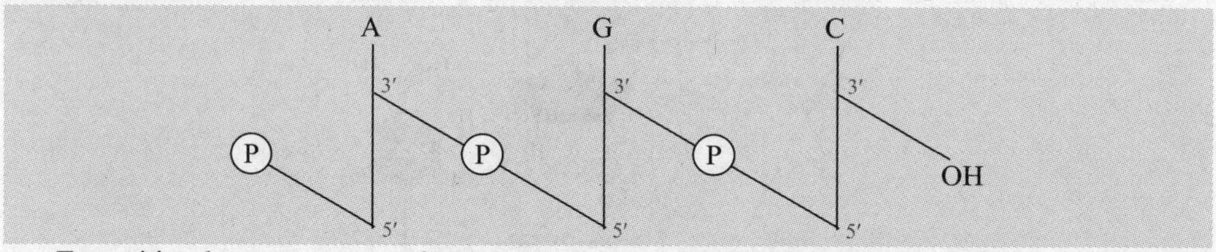

For writing long sequences of polynucleotides, a **shorthand system** is followed. The letters A, G, T, C and U represent the nucleosides as in Table 15.3. The phosphate group is represented by *p*. When *p* is placed to the left of the nucleoside symbol, esterification is at C-5′ ; when placed to the right, esterification is at C-3′. Thus, ApGp is a dinucleotide with a monoester at C-3′ of a guanosine and a phosphodiester bond between C-5′ of G and C-3′ of A. In order to specify the type of sugar involved, the letter *d* (for deoxyribose) is prefixed to the above notation if all the nucleosides contain deoxyribose. Thus, the perfect notation for this dinucleotide shall be d-ApGp.

A short nucleic acid is referred to as an **oligonucleotide**. The definition of "short" is somewhat arbitrary, but the term oligonucleotide is often used for polymers containing 50 or fewer nucleotides. A longer nucleic acid is called a **polynucleotide**.

Base Composition of DNA

The base composition of DNA from a number of sources has been worked out by a number of investigators. Table 15–4 lists the DNA composition of various species. These analyses show a characteristic regularity of pattern as first noted by **Erwin Chargaff** and his coworkers in 1950. The important conclusions drawn by him are:

1. The sum of purines (Pu) is equal to the sum of pyrimidines (Py), *i.e.*, Pu/Py = 1. In other words, A + G = T + C (+ MC) where MC stands for methylcytosine, where it occurs.

Table 15–4. DNA composition of various species

Species	Base proportions, moles%*					Pu/Py ratio			Dissymmetry ratio
	Adenine A	Guanine G	Thymine T	Cytosine C	Methyl-cytosine MC	$\frac{A+G}{T+C(+MC)}$	$\frac{A}{T}$	$\frac{G}{C(+MC)}$	$\frac{A+T}{G+C(+MC)}$
1. *Sarcina lutea*	13.4	37.1	12.4	37.1	—	1.02	1.08	1.00	0.35
2. *Acaligenes faecalis*	16.5	33.9	16.8	32.8	—	1.02	0.98	1.03	0.50
3. *Brucella abortus*	21.0	20.0	21.1	28.9	—	1.00	1.00	1.00	0.73
4. *Escherichia coli K 12*	26.0	24.9	23.9	25.2	—	1.08	1.09	0.99	1.00
5. *Salmonella paratyphi A*	24.8	24.9	25.3	25.0	—	0.99	0.98	1.00	1.00
6. Wheat germ	28.1	21.8	27.4	16.8	5.9	0.99	1.02	0.96	1.25
7. Rat bone marrow	28.6	21.4	28.4	20.4	1.1	1.00	1.00	0.99	1.33
8. Ox	29.0	21.2	28.7	19.9	1.3	1.00	1.01	1.00	1.36
9. *Staphyllococcus aureus*	30.8	21.0	29.2	19.0	—	1.07	1.05	1.11	1.50
10. Human thymus	30.9	19.9	29.4	19.8	—	1.03	1.05	1.01	1.52
11. Human liver	30.3	19.5	30.8	19.9	—	0.99	1.00	0.98	1.54
12. *Saccharomyces cerevisiae*	31.7	18.3	32.6	17.4	—	1.00	0.97	1.05	1.80
13. *Paracentrus lividus*	32.8	17.1	32.1	16.2	1.1	1.01	1.02	0.98	1.85
14. *Pasteurella tularensis*	32.4	17.6	32.9	17.1	—	1.00	0.98	1.03	1.88
15. *Clostridium perfringens*	36.9	14.0	36.3	12.8	—	1.04	1.02	1.09	2.70

* The mole per cent is the number of molecules of one base divided by the total number of molecules of all 4 bases, multiplied by 100. The data are accurate to ± 1 per cent.

ERWIN CHARGAFF

(LT, 1905–2002)

An Austrian refugee biochemist who was born on August 11, 1905 in Czernowitz, Austria (now Chernovtsy, the Ukraine). He received his PhD from the University of Vienna in 1928, and then after periods at Yale, Berlin and Paris, joined Columbia University, New York, in 1935. He worked in many fields of biochemistry and his research papers include studies on subjects as diverse as lipid metabolism and blood coagulation. However, his most influential contribution was the demonstration during 1945–50 that *the base ratios in DNA are constant*. Using the simple but sensitive technique of paper chromatography, he along with his collaborators at Columbia University analyzed the base composition of DNA from various sources. Although this was one of the seminal discoveries that led to the structure of DNA, Chargaff never claimed any allegiance to molecular biology, and indeed had become one of the sternest critics. He was the source of many pungent comments about contemporary science and scientists. He described **Watson** and **Crick** as two pitchmen in the search of a helix and explained his objection to molecular biology as 'by its claim to be able to explain everything, it actually impedes the flow of scientific explanation'. His autobiography is *Heraclitean Fire: Sketches From a Life Before Nature,* Rockefeller University Press, New York, 1978. He believed in the dictum, 'we can only seek what we have already found'. Chargaff is also famous for his following quote:

"What counts... in science is to be not so much the first as the last."

MAURICE HUGH FREDERICK WILKINS

(born, 1916)

A British biophysicist who was born on December 15,1916 at Pongaroa, New Zealand. He moved to England 6 years later with his parents. Wilkins did BA from Cambridge (1938) and PhD from Birmingham (1940). He began as a physicist but his experiences on the atomic bomb project during the WWII turned him away from physical science and towards biology. He joined King's College, London in 1947 and began studying the structure of chromosomes and genes using physical methods. In May 1950, he was given by **Rudolf Signer** of Bern a sample of what was probably the purest DNA available at the time, and from it obtained extremely detailed x-ray diffraction patterns. Wilkins continued with DNA analysis for a number of years and shared with **J.D. Watson** and **F.H.C. Crick** the **1962 Nobel Prize for Physiology or Medicine** for his *x-ray diffraction studies of DNA*, which enabled Watson and Crick to deduce DNA's molecular structure. In 1969, he became the founding President of the British Society for Social Responsibility in Science.

ROSALIND FRANKLIN

(LT, 1920–1958)

A British biophysicist who was born on July 25, 1920 in a prominent London banking family. She graduated from Cambridge University and in 1942 quit her studies to accept a position as a research officer in the British Coal Utilisation Research Association. After the war, she studied x-ray diffraction techniques in Paris. In January 1951, she returned to England accepting a position to develop an x-ray diffraction unit in the biophysics department at King's College. Here, she took over analysis of the Signer DNA (actually, **Rudolf Signer** of Bern handed over a sample of what was probably the purest DNA available at the time to Franklin's senior colleague at King's,

M.H.F. Wilkins), producing pictures that showed clearly that DNA is a helix with phosphate group on the outside of the molecule. Unfortunately, there was a personality clash between Franklin and Wilkins and the intellectual discussion needed to solve the structure of DNA (which Watson and Crick enjoyed) never took place at King's. Nevertheless, Franklin was very close to a double helix structure when Watson and Crick announced their results.

Franklin moved to Birbeck College to work on virus structure in 1953 and remained there until she died of ovarian cancer on April 16, 1958, at the premature age of 37, without knowing the significance her work had played in determining the structure of DNA, as is evident by the statement made by **J.D. Watson** on the elucidation of the structure of DNA, "actually it was a matter of 5 persons: Maurice Wilkins, Rosalind Franklin, Linus Pauling, F.H.C. Crick and me." Franklin's untimely death resulted in her not sharing the 1962 Nobel Prize in Chemistry (along with Watson, Crick and Wilkins), as the Nobel Prizes are not awarded posthumously. Most scientists agree that she had played a crucial but underappreciated role in elucidation of the structure of DNA. Her portrayal in *The Double Helix* as 'Rosie' is now considered a fiction that it is clear from accounts of her contemporaries that, had she lived, she would probably have become one of the most eminent British scientists of her day [see *Rosalind Franklin and DNA* by A. Sayre (1978), W.W. Norton & Co., Inc., New York].

2. The ratio of adenine to thymine is also one, *i.e.*, A/T = 1.

3. The ratio of guanine to cytosine (plus methylcytosine, where it occurs) is also one, *i.e.*, G/C(+ MC) = 1.

4. Bases with 6-amino groups are equal to bases with 6-keto (hydroxyl) groups, *i.e.*, A + C (+ MC) = G + T.

5. The ratio of A + T/G + C(+ MC), known as *dissymmetry ratio*, varies greatly from one species of DNA to the other and is characteristic of that species. When the dissymmetry ratio exceeds one, such a DNA is called *AT type* ; when the value is less than one, such a DNA is designated as *GC type*. In bacteria, both AT and GC types of DNA are found. However, in the higher organisms, the range of dissymmetry ratio is more limited : in most animals the value ranges from 1.3 to 2.2 and in higher plants from 1.1 to 1.7. The value of dissymmetry ratio in human beings is 1.4 and that in *Mycobacterium tuberculosis* is 0.60.

Chargaff's data suggest that A is always paired with T and G is always paired with C. A-paired bases occur in any order:

①	②	③	④
A...T	T...A	C...G	T...A
T...A	C...G	A...T	A...T
C...G	A...T	T...A	C..G

The variability that can be obtained is overwhelming. For example, it has been calculated that human chromosome contains on the average about 140 million base pairs. Since any of the 4 possible nucleotides can be present at each nucleotide position, the total number of possible nucleotide sequences is $4^{140} \times 10^{6}$ or $4^{14,00,00,000}$. No wonder each species has its own base percentages!

Only the 2 purines (adenine and guanine) and the 2 pyrimidines (thymine and cytosine) have been found in DNA molecule of most microorganisms such as bacteria, actinomycetes, algae, fungi and protozoa. However, the presence of methylcytosine, a pyrimidine, is a characteristic feature of DNA of higher plants and animals ; *the DNA of plants being richer in methylcytosine than the DNA of animals*. The richest source of methylcytosine, found as yet, is wheat germ which contains 6 moles of methylcytosine per 100 moles of bases (refer Table 15–4).

Evolution of Watson–Crick Model

W. T. Astbury was the first person to give any thought to the 3-'D′ structure of DNA. By his x-ray crystallographic studies on DNA molecule, conducted in 1940s, he concluded that because DNA has a high density, so its polynucleotide was a stack of flat nucleotides, each of which was oriented perpendicularly to the long axis of the molecule and was placed 3.4 Å apart from each other, a concept given the name a **"3.4 Å repeat".** The x-ray crystallographic studies of **Astbury** were continued in early 1950s by **Maurice Wilkins** and his student **Rosalind Franklin,** both of King's College, London, who obtained a superior x-ray diffraction photograph of a DNA fibre (Fig. 15-17) which confirmed Astbury's earlier inference of 3.4 Å internucleotide distance and suggested a helical configuration for DNA molecule. They also proposed a **"34 Å repeat"**, *i.e.*, the helix is folded into many turns and each turn causes a vertical rise of 34 Å. Furthermore their findings indicated a helical structure containing two or more strands of polydeoxyribonucleotides. **Erwin Chargaff**, in late 1940s, independently had provided the crucial observation that, in DNA obtained from a wide variety of organisms, the molar ratio of adenine to thymine and that of guanine to cytosine were very close to unity. These results indicated that a specific relationship must exist between the two bases within each of the ratios. These and some other generalizations (discussed earlier) were collectively termed as **"Chargaff's equivalence rules".** Later, the results of titration studies (= analytical studies) also suggested that the long polynucleotide chains were held together by hydrogen bonding between the base residues.

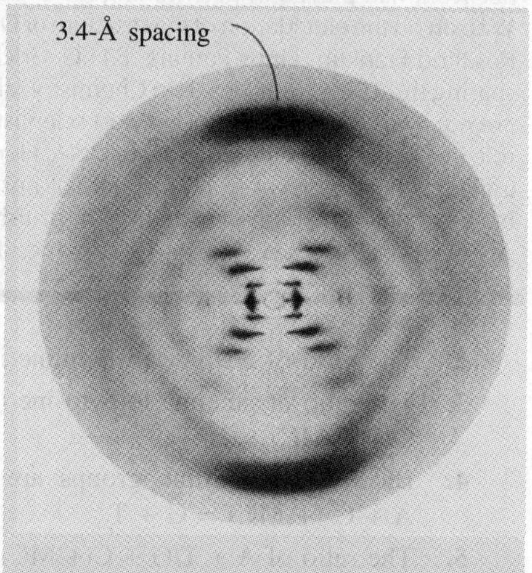

Fig. 15–17. X-ray diffraction photograph of a hydrated DNA fibre

The spots forming a cross in the centre denote a helical structure. The strong arcs at the top and bottom correspond to the stack of base pairs, which are 3.4 Å apart.

(*Courtesy: Dr. Maurice Wilkins*)

Until the 1950s, tetranucleotides with the four different bases in varying arrangements were believed to be the basic repeating unit of RNA structure, and the analogous deoxytetranucleotides the basic repeating unit of DNA. This so-called "**tetranucleotide hypothesis**" conferred nucleic acids a monotonous structure of repeating tetranucleotides which, however, could not account for the idea of DNA as genetic material as was first proposed by **Avery** *et al* in 1944. Later findings that DNA bases are not arranged in repetitive tetranucleotide sequences, dispelled doubts about the hereditary role of the nucleic acids. **Linus Pauling** suggested that the nitrogenous bases projected outwards from a central core formed by the backbone phosphate group. But Franklin felt she had evidence that phosphate pointed towards outside and that the bases were in the centre.

This was the situation around 1951 when **James Dewey Watson**, a 22-year-old American postdoctoral research fellow arrived in Cambridge and met **Francis Harry Compton Crick,** a physicist working for his Ph.D. degree in Biophysics. Although they were to work on different problems, they decided to collaborate on the study of DNA. Using intelligently all the available information, Watson and Crick, in April, 1953, published an article proposing the double helical structure for DNA molecule in the same issue of journal 'Nature', in which **Wilkins** *et al* presented the x-ray evidence for

In fact, **Watson** and **Crick** shared the Nobel Prize with **Wilkins**, although elucidation of the double helical structure of DNA was a concerted effort on the part of 5 scientists as **Watson** had, truly, once said, "actually it was a matter of 5 persons: **Maurice Wilkins, Rosalind Franklin, Linus Pauling, F.H.C. Crick** and me."

that structure. Thus, we see that Watson-Crick structural model of DNA is the outcome of 3 types of studies (x-ray diffraction, base-pairing and analytical), admirably reasoned, interwoven and adjusted into a hypothetical double helical structure. In fact, their discovery of the double helix represents an elegant case of scientific sleuthing in the literature, thinking and master model building. This brilliant accomplishment ranks as one of the most significant in the history of biology because it led the way to an understanding of gene function in molecular terms. For this epoch-making discovery, **Watson** and **Crick** were awarded the prestigious Nobel Prize in medicine or physiology in 1962. In fact, their model of DNA structure was radically different from that suggested by **Linus Pauling** and **Robert Corey** in February, 1953 which was a **triple helix**. The latter model, however, failed to explain the process by which a DNA molecule copies itself (*i.e.*, DNA replication) whereas Watson and Crick's model admirably explained this phenomenon.

Double Helical Structure of DNA (Watson-Crick Model)

The *salient features* of the **Watson-Crick model** (Fig. 15–18) for the commonly-found DNA (known as the B-DNA) are:

1. DNA molecule consists of two helical polynucleotide chains which are coiled around (or wrapped about) a common axis in the form of a right-handed double helix (refer Fig. 15–19). The two helices are wound in such a way so as to produce 2 interchain spacings or grooves, a major or wide groove (width 12 Å, depth 8.5 Å) and a minor or narrow groove (width 6 Å, depth 7.5 Å). The two grooves arise because the glycosidic bonds of a base pair are not diametrically opposite each other. The minor groove contains the pyrimidine O-2 and the purine N-3 of the base pair, and the major groove is on the opposite side of the pair. Each groove is lined by potential hydrogen bond donor and acceptor atoms. It is noteworthy that the major groove displays more distinctive features than does the minor groove. Also, the larger groove is more accessible to proteins interacting with specific sequences of DNA.

Sometimes, one chain or strand of DNA is called the **Watson strand** (W strand) and its complement the **Crick strand** (C strand) as a gesture to the propounders of this model.

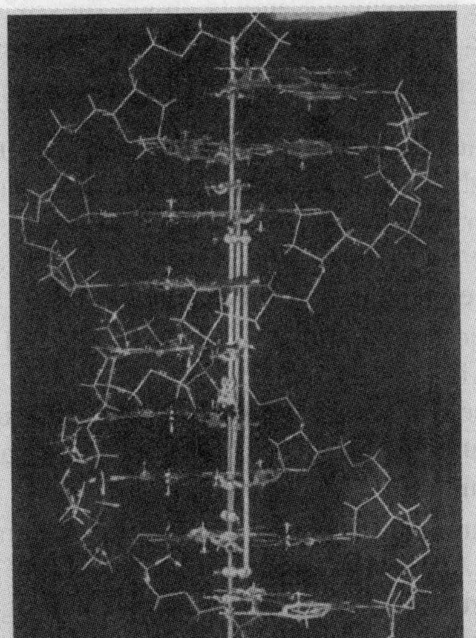

Fig. 15–18. The original molecular model of DNA double helix as proposed by Watson and Crick

The model shows A hydrogen-bonded to T and G hydrogen-bonded with C.

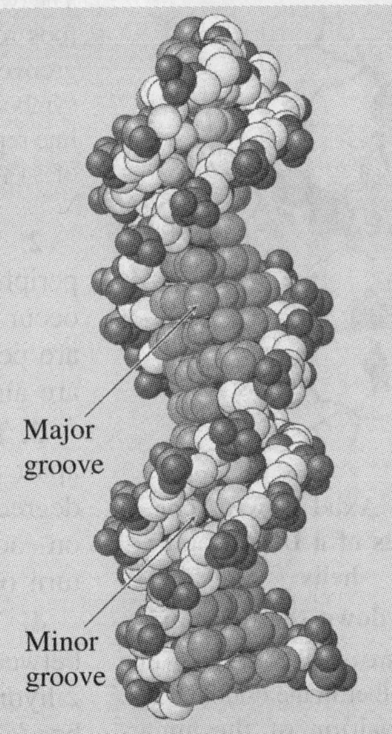

Fig. 15–19. Space-filing model of B-form DNA showing major and minor grooves

The major groove is depicted in *orange*, and the minor groove is depicted in *yellow*. The carbon atoms of the backbone are shown in *white*. Note that the bases fill the space between the two backbones. Atoms are shown at a size that includes their van der Waals radii.

The two helices wind along the molecule parallel to the phosphodiester backbones. These may be visualized as to have formed when two wires are wound about a pencil or a common axis (Fig. 15–20). In these grooves, specific proteins interact with DNA molecules. Such double helices cannot be pulled apart and can be separated only by an unwinding process. They are called as *plectonemic coils*, *i.e.*, coils which are interlocked about the same axis. The two chains run in opposite direction, *i.e.*, they are *antiparallel* which means that in one strand, the ring oxygen of sugar moieties face upwards while in the other strand, the ring oxygen faces downwards. Also they are not identical but, because of base pairing, are *complementary* to each other.

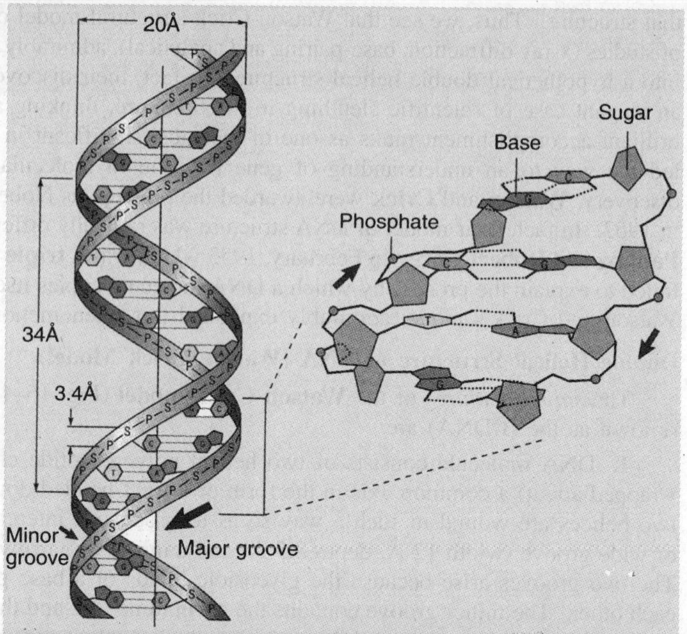

Fig. 15–20. Schematic representation of the Watson-Crick ribbon model of the B-DNA double helix

[P = phosphate; S = deoxyribose sugar; A = adenine; G = guanine; T = thymine; C = cytosine]

The two ribbons represent the phosphate-sugar chains and the horizontal rods represent the bonding between the pair of bases. Two kinds of grooves, a major and a minor, are evident. The horizontal parallel lines symbolize hydrogen bonding between complementary bases. The vertical line represents the fibre axis. Note that the structure repeats at intervals of 34 Å, which corresponds to 10 residues on each chain.

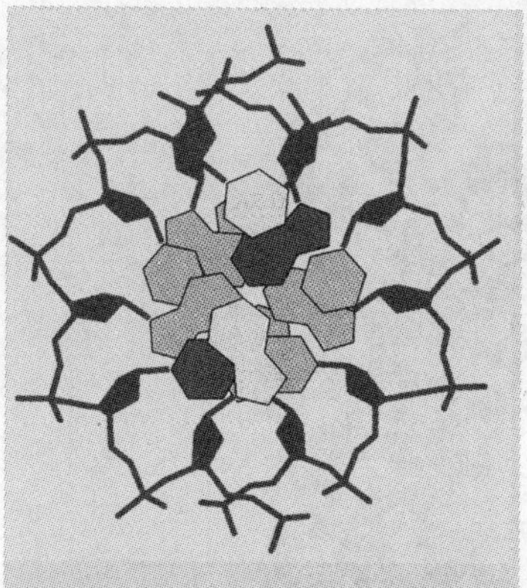

Fig. 15–21. Axial view of one of the strands of a DNA double helix

(as viewed down the helix axis)

Base pairs are stacked nearly one on top of another in the double helix. Note the position of the sugar-phosphate backbone and bases which is outward and inward respectively. The 10-fold symmetry is evident.

(Adapted from Lubert Stryer, 2000)

2. The phosphate and deoxyribose units are found on the periphery of the helix, whereas the purine and pyrimidine bases occur in the centre (refer Fig. 15–21). The planes of the bases are perpendicular to the helix axis. The planes of the sugars are almost at right angles to those of the bases.

3. The diameter of the helix is 20 Å. The bases are 3.4 Å apart along the helix axis and are related by a rotation of 36 degrees. Therefore, the helical structure repeats after 10 residues on each chain, *i.e.*, at intervals of 34 Å. In other words, each turn of the helix contains 10 nucleotide residues.

4. The two chains are held together by hydrogen bonds between pairs of bases. Adenine always pairs with thymine by 2 hydrogen bonds and guanine with cytosine with 3 hydrogen bonds. This specific positioning of the bases is called *base complementarity*. The individual hydrogen bonds are weak in nature but, as in the case of proteins, a large number of them involved in the DNA molecule confer stability to it. It is now thought that the stability of the DNA molecule is primarily a consequence of van der Waals forces between the planes of stacked bases.

JAMES DEWEY WATSON

(born, 1928)

An American biochemist who was born in 1928 at Chicago, USA. He was a child prodigy who entered the University of Chicago at the age of 15. He graduated in 1947 and obtained his PhD from the University of Indiana in 1950. He was then awarded a fellowship to study in Copenhagen but after a year there, he moved to the Cavendish Laboratory, Cambridge, with the specific objective of studying the gene. There, he met **Francis H.C. Crick**, who was working on protein structure for his PhD after starting as a physicist but becoming diverted to biology after the War. The famous fusion of minds that led Watson and Crick to deduce that DNA is a double helix has been described many times, notably by Watson in *The Double Helix* (1968) which contains his personalized account of the research and became a bestseller. The two shared the **1962 Nobel Prize for Physiology or Medicine** with **Maurice H.F. Wilkins**. Athough Watson and Crick shared the Nobel Prize with Wilkins, elucidation of the double helical structure of DNA was, in fact, a concerted effort on the part of 5 scientists as Watson had, truly, once said, " actually it was a matter of 5 persons : Maurice Wilkins, Rosalind Franklin, Linus Pauling, F.H.C. Crick and me." The molecular details of DNA that continue to unfold are reminiscent of a statement once made by the great **Albert Einstein** (Nobel Laureate in Physics, 1921) that, in science, for every milestone reached there is a signpost pointing to yet another.

After the double helix, Watson and Crick went separate ways. Watson returned to the USA in 1953 and eventually took up a professorship at Harvard before moving in 1976 to the Cold Spring Harbor Laboratory on Long Island, where he became Director (Crick stayed in Cambridge until moving to the Salk Institute in Southern California in 1977). Watson has worked on RNA synthesis, protein synthesis and the role of viruses in cancer. Recently, he has been one of the prime motivators behind the *Human Genome Project* (HGP), the ambitious research programme that has as its goal the complete nucleotide sequence on the human genome.

Regarding the structure of DNA molecule, Watson had famously written:

"A structure this pretty just had to exist."

(*in The Double Helix, 1968*)

"Francis and I were running against God, in the sense that we wanted to know what made us human. Both of us had been subjected to religious truths which came by revelation, and we didn't have much acceptance of truths by revelation. We wanted to know really what we were. We were both very curious about what life was. What is life? And that thinking led you to "What is a gene?"

FRANCIS HARRY COMPTON CRICK

(LT, 1916–2004)

An English biochemist-cum-biophysicist who is largely known (along with **James Dewey Watson**) for the *discovery of the structure of DNA*. But, in addition to this, Crick made several other contributions during a productive career in science. He was born in a middle class family to a shoe-factory owner **Harry Crick** and his wife **Anne Elizabeth Wilkins** in Northampton, England on June 8, 1916. He attended University College, London and obtained a Bachelor of Science degree, in 1937, with second class honours in physics and then joined Medical Research Council (MRC) unit at Cambridge, for PhD. But the outbreak of the World War II disrupted his studies, and he was drafted into the British Admiralty Research Establishment where he worked designing circuits for magnetic acoustic mines. He left Admiralty in 1947 and joined Strangeways Research Laboratory in Cambridge to work on the effects of magnetism in cells, and later, the Cavendish Laboratory's MRC unit in 1949, completing his PhD in 'X-ray diffraction: polypeptides and proteins' in 1954. The lab was headed by **Sir Max Ferdinand Perutz** (who later also received the Nobel Prize) under the general direction of **Sir Lawrence Bragg**, the formulator of Bragg's law for x-ray diffraction (Unfortunately, his loud laugh would often annoy Prof. Bragg, his supervisor).

With the brash 23-year-old American, James Dewey Watson in 1951, who had just arrived at the Cavendish Laboratory, led to an instantaneous realization of the common interests (in fact, both shared a fascination with DNA and both were intrigued by Linus Carl Pauling's theories) and a lifelong friendship, and most importantly, the discovery of the DNA structure. Both Watson and Crick worked on the structure of DNA in their spare time. Neither Watson nor Crick performed experiments on DNA, concentrating instead on building structural models that theoretically should fit the known data about DNA, *thus relying mainly on abstract thinking*. Using laboratory clamps and pieces of metal, they set about building a giant model of a section of DNA, which is now preserved and displayed in the London Science Museum. As Crick stated "chance prepares the favoured mind", and when they hit upon the correct structure, they immediately recognized it for what it was (The famous fusion of minds that led Watson and Crick to deduce that DNA is a double helix has been described many times, notably by Watson in *The Double Helix*, published by Atheneum, in 1968). For this epoch-making discovery, the two shared the 1962 Nobel Prize for Medicine or Physiology with M.H.F. Wilkins. The molecular details of DNA that continue to unfold are reminiscent of a statement once made by the great **Albert Einstein** (1921 Physics Nobel Laureate) that, in science, for every milestone reached there is a signpost pointing to yet another.

In fact, it was Crick, who on February 28, 1953 walked into the Eagle Pub and stunned his colleagues by announcing that he and his American colleague Watson "had found the secret of life" (He did remember his wife telling him that he was always making such apparently extravagant claims). They had discovered that morning the "double helix" structure of DNA, one which could replicate itself, confirming theories that it carried life's hereditary information. And in his famous paper of April 25, 1953, Crick noted, "the specific pairing of purine and pyrimidine bases that we have postulated immediately suggests a possible copying mechanism for the genetic material" Later, in one of his books, he wrote:

> "One of the striking generalizations of biochemistry is that the 20 amino acids and four bases are, with minor exceptions, the same throughout nature.".

Following the discovery of DNA structure, Crick went on to postulate the characteristics of the genetic code in detail. He also coined the term "codon" for the individual base-pair triplets that were soon accepted as the specific codes for particular amino acids. In 1956, Crick wrote a paper suggesting the existence of an "adaptor molecule" that could physically interact with DNA to read the code, as amino acids are incapable of directly reading DNA base sequences (This paper was never published but was widely circulated). Again with Watson, who had spent a couple of years at the California Institute of Technology (CALTECH), Crick investigated the RNA of viruses, and together they proposed models of viral structure, which carried molecular biology into a new realm. Most of this work was carried out in California, at MRC. In 1977, Crick moved to take up a specially-created research post at the Salk Institute for Biological Studies and, in parallel, as professor of biology, chemistry and psychology at the University of California in San Diego, California, and carried out brain research. Collaborating with Caltech cognitive scientist **Christof Koch**, he set out to find what he called "the neural correlates of consciousness" arguing that brain chemistry— as opposed to anything mystical— was responsible for human thought, memory, identity and free will. Also at Caltech, Crick collaborated with **Leslie E. Orgell** to write a stimulating paper on selfish DNA in 1980. In 1988, Crick wrote about his experiences in his famous autobiography "What Mad Pursuit" (*Weidenfield and Nicolson, London, 1989*), a personal view of scientific discovery.

Besides winning the Nobel Prize, Crick received many other awards and honours. He was elected **a fellow of the Royal Society** in 1959, a Foreign Honorary Member of the American Academy of Arts and Sciences, and a fellow of the University College, London, both in 1962. He was appointed to the **Order of Merit** in 1991.

Crick appropriately named his house at Cambridge "The Golden Helix". He had hung on the wall behind his desk in the Cavendish Lab the following motto: 'Reading Rots the Mind'. He often used to say:

"It is very difficult to explain something scientifically when you cannot explain what it is."

Crick died of colon cancer in San Diego on July 29, 2004 at the ripe old age of 88, after having witnessed some breathtaking advancements that could be credited to his discovery of the structure of DNA. **Steve Jones**, Professor of Genetics at University College, London on his demise said,

"Francis Crick was the Charles Darwin of the 20th Century."

He will, in fact, be remembered for eons to come. He had once stated:

"The view of ourselves as 'persons' is just as erroneous as the view that the Sun goes round the Earth... this sort of language will disappear in a few hundred years... In the fullness of time, educated people will believe there is no soul independent of body, and hence no life after death."

5. The sequence of bases along a polynucleotide chain is not restricted in any way. *The precise sequence of bases carries the genetic information.*

As a corollary, the entire structure of a DNA molecule resembles a *winding staircase*, with the sugar and phosphate molecules of the nucleotides forming the railings and the linked nitrogen base pairs (A-T and G-C) forming the steps.

Antiparallelity of the polynucleotide chain. The procedure of *nearest neighbour analysis* has shown that the two complementary chains do not run in the same direction with respect to their internucleotide linkages, but rather are *antiparallel, i.e.*, one strand runs in the 5′ → 3′ direction and the other in the 3′ → 5′ direction (refer Fig. 15–22). Thus, of the two chains, one ascends and the other descends. They have reverse polarity, that is to say the sequence of the atoms in the two chains is opposite to each other. This is analogous to 2 parellel streets, each running one way but carrying traffic in opposite directions.

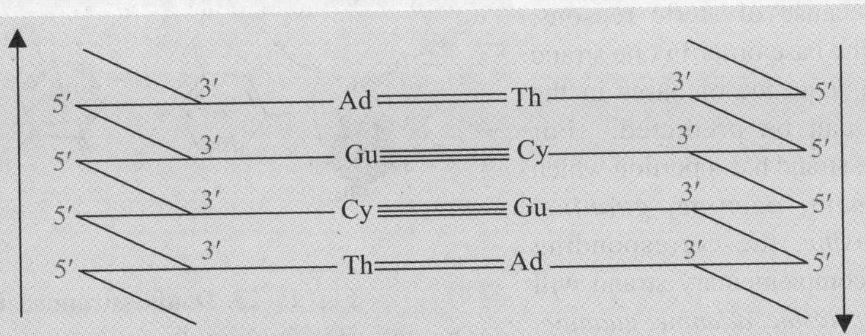

Fig. 15–22. Antiparallelity of the two polynucleotide chains of DNA molecule

Base complementarity of the polynucleotide chain. An important feature of the double helix is the specificity of the pairing of bases. *Pairing always occurs between adenine and thymine and between guanine and cytosine* (Fig. 15–23). This means that a 6-aminopurine will always bond with a 6-ketopyrimidine, and a 6-ketopurine with a 6-aminopyrimidine. Consequently, the two helices of a DNA molecule are complementary to each other and not identical (Fig 15–23). Base-pairing is due to steric and hydrogen-bonding factors.

Complementarity is a term introduced into quantum theory by a physicist **Niels Bohr** (LT, 1885–1962), implying that evidence relating to atomic systems, that has been obtained under different experimental conditions, cannot necessarily be comprehended by one single model. Thus, for example, the wave model of the electron is *complementary* to the particle model.

(A) ***Steric factor*****.** The steric restriction is imposed by the regular helical nature of the sugar-phosphate backbone of each polynucleotide chain. The geometry of the base pair has some special consequences in that the distances between the glycosidic bonds are the same for both the base pairs and also because the bond angle between the glycosidic bond direction and the line joining the C 1 atoms is the same in each pair. With the result, the glycosidic bonds of all nucleotides are arranged in an identical manner in relation to the axis of the helix, despite the differences in the bases.

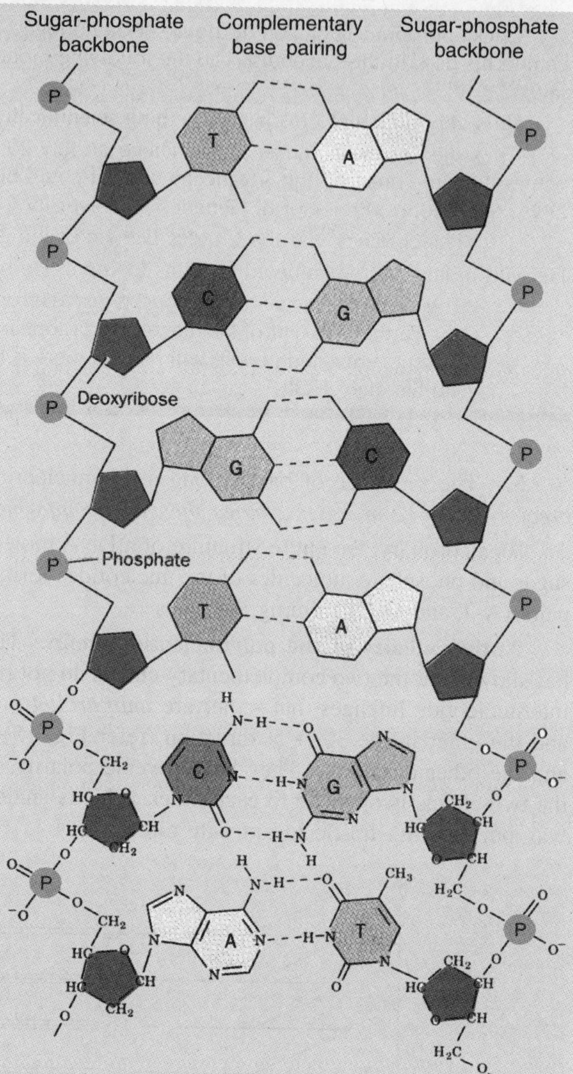

Fig. 15–23. Double-stranded DNA

The two polynucleotide chains associate through complementary hydrogen bonding.

The glycosidic bonds that are attached to a base pair are always 10.85 Å apart. A purine-pyrimidine base pair fits perfectly in this space. In contrary to this, there is insufficient space for two purines, whereas there is more than enough space for two pyrimidines so that they would be far apart to form hydrogen bonds. Hence, one member of a base pair in a DNA helix must invariably be a purine and the other a complementary pyrimidine because of steric reasons. Obviously, if the base order in one strand is known, the sequence of bases in the other strand can be predicted. For instance, if one strand has a portion which has arrangement, *adenine*, *guanine*, *thymine*, *cytosine*, the corresponding region in the complementary strand will have *thymine, cytosine, adenine, guanine*.

(B) ***Hydrogen-bonding factor*****.** The base pairing is further restricted by hydrogen-bonding requirements. The hydrogen atoms in purine and pyrimidine bases have well-defined positions. Adenine cannot pair with cytosine because there would be two hydrogen atoms near one of the bonding positions and none at the other. Similarly, guanine cannot pair with thymine. In contrast, *adenine forms 2 hydrogen bonds with thymine whereas guanine forms 3 with cytosine* (refer Fig. 15–24). Thus, *the G–C bond is stronger by 50% than the A–T bond*. The higher the G–C content of a DNA moelcule, the greater is its buoyant density.

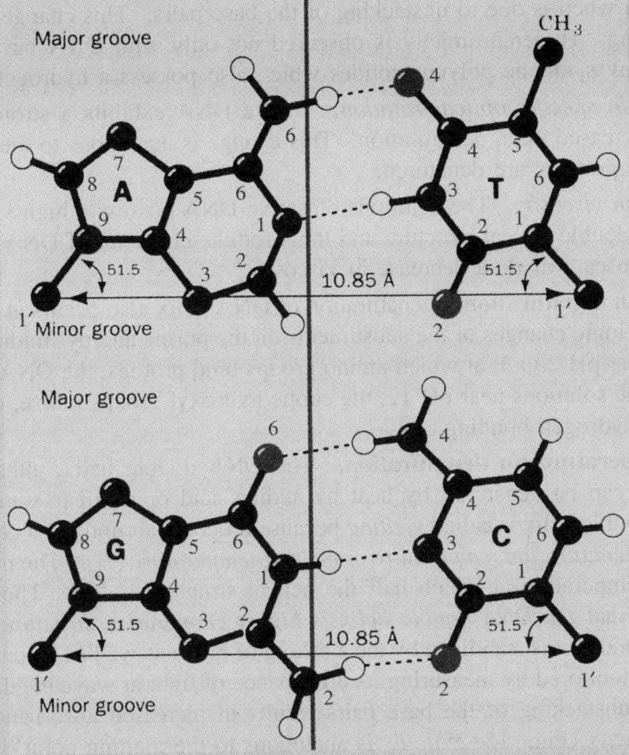

Fig. 15–24. Structure of the two Watson–Crick base pairs

The line joining the C1′ atoms is the same length in both base pairs and makes equal angles with the glycosidic bonds to the bases. This gives DNA a series of pseudo-twofold symmetry axes (often referred to as **dyad axes)** that pass through the centre of each base pair (*red line*) and are perpendicular to the helix axis. Note that A. T base pairs associate *via* two hydrogen bonds, whereas C. G base pairs are joined by three hydrogen bonds.

(*After Arnott S, Dover SD and Wonacott AJ, 1969*)

Denaturation and Renaturation of DNA Helix

Denaturation of DNA is a loss of biologic activity and is accompanied by cleavage of hydrogen bonds holding the complementary sequences of nucleotides together. This results in a separation of the double helix into the two constituent polynucleotide chains. In it, the firm, helical, two-stranded native structure of DNA is converted to a flexible, single-stranded `denatured' state. The splitting of DNA molecule into its two strands or chains, during denaturation, is obvious because of the fact that the hydrogen bonds holding the bases are weaker than the bonds holding these bases to the sugar-phosphate groups. The transition from native to a denatured form is usually very abrupt and is accelerated by reagents such as urea and formamide, which enhance the aqueous solubility of the purine and pyrimidine groups. Denaturation involves the following changes:

1. *Increase in absorption of ultraviolet light* (= *Hyperchromic effect*). As a result of resonance, all of the bases in nucleic acids absorb ultraviolet light. And all nucleic acids are characterized by a maximum absorption of UV light at wavelengths near 260 nm. When the native DNA (which has base pairs stacked similar to a stack of coins) is denatured, there occurs a marked increase in

optical absorbancy of UV light by pyrimidine and purine bases, an effect called **hyperchromicity** or **hyperchromism** whch is due to unstacking of the base pairs. This change reflects a decrease in hydrogen-bonding. Hyperchromicity is observed not only with DNA but with other nucleic acids and with many synthetic polynucleotides which also possess a hydrogen-bonded structure.

2. *Decrease in specific optical rotation.* Native DNA exhibits a strong positive rotation which is highly decreased upon denaturation. This change is analogous to the change in rotation observed when the proteins are denatured.

3. *Decrease in viscosity.* The solutions of native DNA possess a high viscosity because of the relatively rigid double helical structure and long, rodlike character of DNA. Disruption of the hydrogen bonds causes a marked decrease in viscosity.

Effect of pH on denaturation. Denaturation of DNA helix also occurs at acidic and alkaline pH values at which ionic changes of the substituents on the purine and pyrimidine bases can occur. In acid solutions near pH 2 to 3, at which amino groups bind protons, the DNA helix is disrupted. Similarly, in alkaline solutions near pH 12, the enolic hydroxyl groups ionize, thus preventing the keto-amino group hydrogen bonding.

Effect of temperature on denaturation. The DNA double helix, although stabilized by hydrogen bonding, can be denatured by heat by adding acid or alkali to ionize its bases. The unwinding of the double helix is called *melting* because it occurs abruptly at a certain characteristic temperature called *denaturation temperature* or *melting temperature* (T_m). The melting temperature is defined as the temperature at which half the helical structure is lost. The abruptness of the transition indicates that *the DNA double helix is highly cooperative structure*, held together by many reinforcing bonds ; it is stabilized by the stacking of bases as well as by pairing. The melting of DNA is readily monitored by measuring its absorbance of light at wavelength near 260 nm. As already stated, the unstacking of the base pairs results in increased absorbance, a phenomenon termed hyperchromicity (Fig. 15–25). T_m is analogous to the melting point of a crystal. It can be lowered by the addition of urea which disrupts hydrogen bonds. In 8M urea, T_m is decreased by nearly 20°C. DNA can be completely denatured (*i.e.*, separated into a single-stranded structure) by 95% formamide at room temperature only.

Since the G-C base pair has 3 hydrogen bonds as compared to 2 for A-T, it follows that DNAs

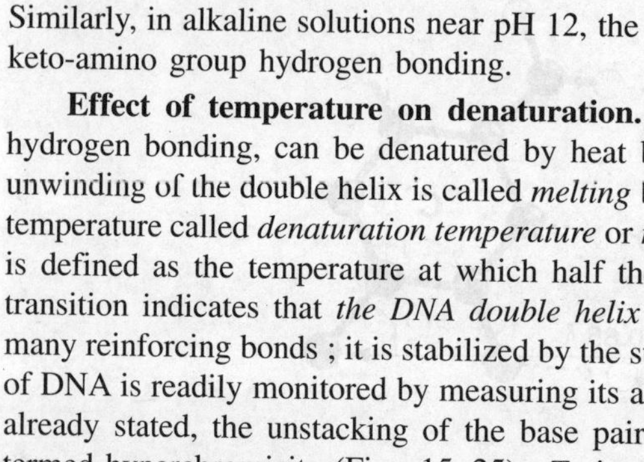

Fig. 15–25. The absorbance spectra of a DNA solution at 260 nm and pH 7.0

When mixtures of mucleotides are present, the wavelength at 260 nm (*dashed vertical line*) is used for measurement.

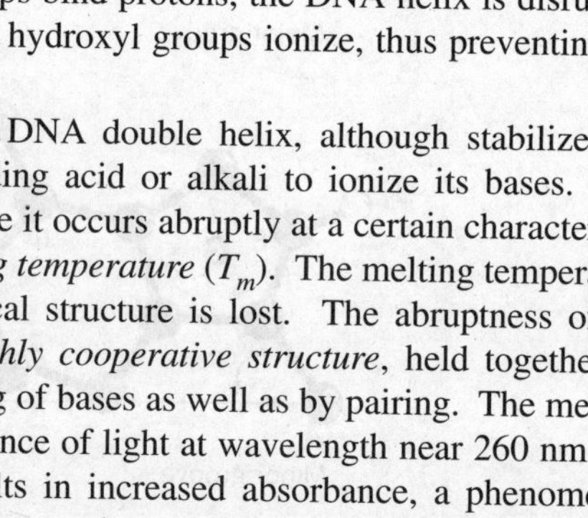

Fig. 15–26. DNA melting curves

The absorbance relative to that at 25°C is plotted against temperature. The wavelength of the incident light was 260 nm. The Tm is about 72°C for *Escherichia coli* DNA (50% G-C pairs) and 79°C for the bacterium, *Pseudomonas aeruginosa* DNA (66% G-C pairs).

with high concentrations of G and C might be more stable and have a higher T_m than those with high concentrations of A and T (Fig. 15–26). In other words, the DNA molecules containing less G-C bond denature first as G-C bond has higher thermal stability. In fact, the Tm of DNA from many species varies linearly with G-C content, rising from 77 to 100°C as the fraction of G-C pairs increases from 20% to 78%. This measurement may, hence forth, be used as an index of heterogeneity of nucleic acid molecules. For instance, viral DNA has a much sharper thermal transition than that prepared from animal sources.

Complete rupture of the two-stranded helix by heating is not a readily reversible process. However, if a solution of denatured DNA, prepared by heating, is cooled slowly to room temperature, some amount of DNA is renatured. Maximum reversibility (50-60%) is usually attained by *annealing* (= slow cooling) the denatured DNA, *i.e.*, holding the solution at a temperature about 25°C below T_m and above a concentration of 0.4 M Na^+ for several hours. Fast cooling will not reverse denaturation, but if the cooled solution is again heated and then cooled slowly, renaturation takes place. The restoration occurs because the complementary bases reunite by hydrogen bonds and the double helix again forms.

Renaturation of complementary single strands to produce fully double-stranded moelcule requires 2 separate reactions (Fig. 15–27):

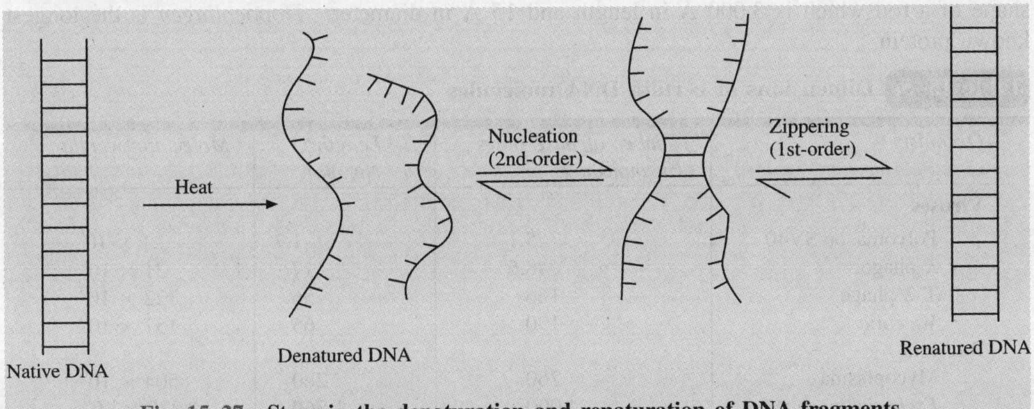

Fig. 15–27. Steps in the denaturation and renaturation of DNA fragments

I. **Nucleation reaction.** In this hydrogen bonds form between two complementary single strands ; this is a bimolecular, second-order reaction.

II. **Zippering reaction.** In this hydrogen bonds form between all the bases in the complementary strands; this is a unimolecular, first-order reaction.

Molecular Weight of DNA

DNA molecules are among the largest known. It is difficult to isolate DNA without fragmentation. The entire DNA of bacteria such as *Escherichia coli* has a molecular weight of 2.6 $\times 10^9$. Viral DNAs range in size from about 1 to 350 $\times 10^6$. The residue weight of a single nucleotide is 300 to 350. Thus, there are about 3,000 nucleotides per million molecular weight of DNA.

Length of DNA Molecule

The amount of DNA is usually measured by the microunit of the weight known as the **picogram, pg** (1 pg = 10^{-12} g). The amount of DNA has been found to be constant from cell to cell and species. *A piece of double helical DNA, 31 cm in length, weighs one picogram.* So by taking the weight of DNA amount in the nucleus, the length of the DNA molecule can be calculated easily. By employing this method, **Dupraw** and **Bahr** (1968) have calculated that the human diploid cells contain 5.6 pg DNA amount which is equivalent to (5.6 × 31) or 173.6 cm-long DNA molecule. Similarly, they also calculated that a diploid cell of *Trillium* has dsDNA of 37.2 m (or 3,720 cm) length, which is equivalent to (3,720/31) or 120 pg DNA by weight. Also, the polytenic chromosomes of *Drosophila*, with 293 pg DNA contents, have (293 × 31) or 90.83 m-long DNA molecule.

Shape and Size of DNA Molecules

The DNA molecules are characteristically highly **elongate** structures. The *Escherichia coli* chromosome is a single molecule of double helical DNA, containing about 4 million base pairs. This DNA is highly asymmetric. Its contour length is 14 × 10^6 Å and its diameter is 20 Å. Whereas its contour length (1.4 mm) corresponds to a macroscopic dimension, its width (20 Å) is on the atomic scale. However, the largest chromosome of *Drosophila melanogaster* contains a single DNA molecule with 6.2 × 10^7 base pairs. These have a contour length of 2.1 cm. Such highly asymmetric DNA molecules are very susceptible to fragmentation by shearing forces.

Table 15–5 lists the dimensions of the various viral, bacterial and eukaryotic DNA molecules. A perusal of the table indicates that *even the smallest DNA molecules are highly elongate.* For instance, the DNA from polyoma virus contains 5,100 base pairs and has a contour length of 1.7 μm (or 17,000 Å). Furthermore, tropocollagen, the basic structural unit of collagen, has the shape of a rod which is 3,000 Å in length and 15 Å in diameter. *Tropocollagen* is the longest known protein.

Table 15–5. Dimensions of certain DNA molecules

Organism	*Number of base pairs (in thousands or kb)*	*Length* (in μm)*	*Molecular weight*
Viruses			
Polyoma or SV40	5.1	1.7	3.1 × 10^6
λ phage	48.6	17	31 × 10^6
T 2 phage	166	56	122 × 10^6
Vaccinia	190	65	157 × 10^6
Bacteria			
Mycoplasma	760	260	504 × 10^6
Escherichia coli	4,000	1,360	2,320 × 10^6
Eukaryotes			
Yeast	13,500	4,600	—
Drosophila	1,65,000	56,000	—
Human	2,900,000	9,90,000	—

* 1 μm of double helix = 2.94 × 10^3 base pairs = 1.94 × 10^6 daltons.

(*Modified from A. Kornberg and T.A. Baker, 1992*)

The existence of a **circular** form of DNA was first demonstrated by John Cairns (1963) in *Escherichia coli* which was grown in the presence of tritium-labelled thymidine. The chromosomal material was then visualized by radioautography. This revealed a tangled circle that, when fully extended, had a circumference of 1,100 to 1,400 μ (refer Fig. 15–28). In fact, the finding that *E. coli* has a circular chromosome was anticipated by genetic studies that revealed that the gene-linkage map of this bacterium is circular. *The term 'circular' refers to the continuity of the DNA chain, and not to its geometric form.*

DNA molecule from the T 7 bacteriophage is, however, **linear** in shape. Interestingly, the DNA molecules of some viruses such as λ bacteriophage interconvert between linear and circular forms. The linear form is present inside the virus particle whereas the circular form is present in the host cell.

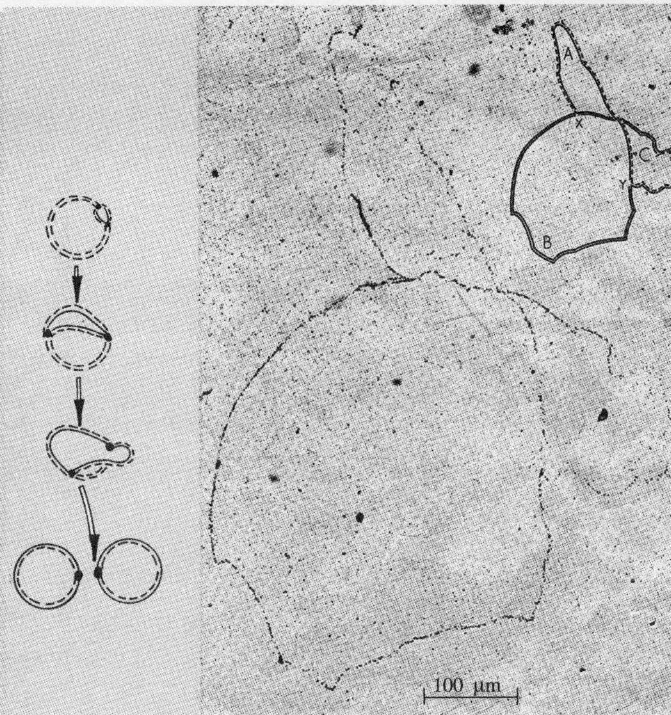

Fig. 15–28. **Autoradiograph of the chromosome of *Escherihia coli***

Right: *E. coli DNA in the process of replication*. The inset is a schematic representation of the autoradiogram. The dashed line in regions A and C represents one parent strand, one-fourth tritiated (solid portion) from the previous replication. The other parent strand and the new daughter strands (solid lines) are fully tritiated. X and Y are the "replication forks."

Left: The replication scheme suggested by autoradiography. Here the addition of tritium is synchronized with the beginning of replication.

[*Courtesy of John Cairns, from Cold Spring Harbor Symp. Quant. Biol.,* **28** : *43, 1963.*]

Variants of Double Helical DNA

Under physiologic conditions, most of the DNA in a bacterial or eukaryotic genome is in the classic Watson-Crick form which is called the **B-form DNA** or simply **B-DNA**. The B-form is the most stable structure for a random-sequence DNA molecule, and is therefore the standard point of reference in any study of the properties of DNA. X-ray analyses of DNA crystals at atomic resolution have revealed that DNA exhibits much more structural diversity than formerly envisaged. Two such variations are:

(*a*) A DNA chain can be rotated about 6 bonds in each monomer—the glycosidic bond between the base and sugar, the $C_{4'}$ — $C_{5'}$, bond of the sugar and four bonds in the phosphodiester bridge joining $C_{3'}$ of one sugar and $C_{5'}$ of the next one — compared with just two for polypeptide chains.

(*b*) The DNA helix can be smoothly bent into an arc or it can be supercoiled with rather little change in local structure. This ease of deformation enables circular DNA to be formed and allows DNA to be wrapped around proteins. Deformability of DNA also allows compacting of DNA into a much smaller volume. *DNA can also be kinked, i.e.*, bent at discrete sites. Kinking can be induced by specific base sequences, such as a segment of at least 4 adenine residues or by the binding of a protein.

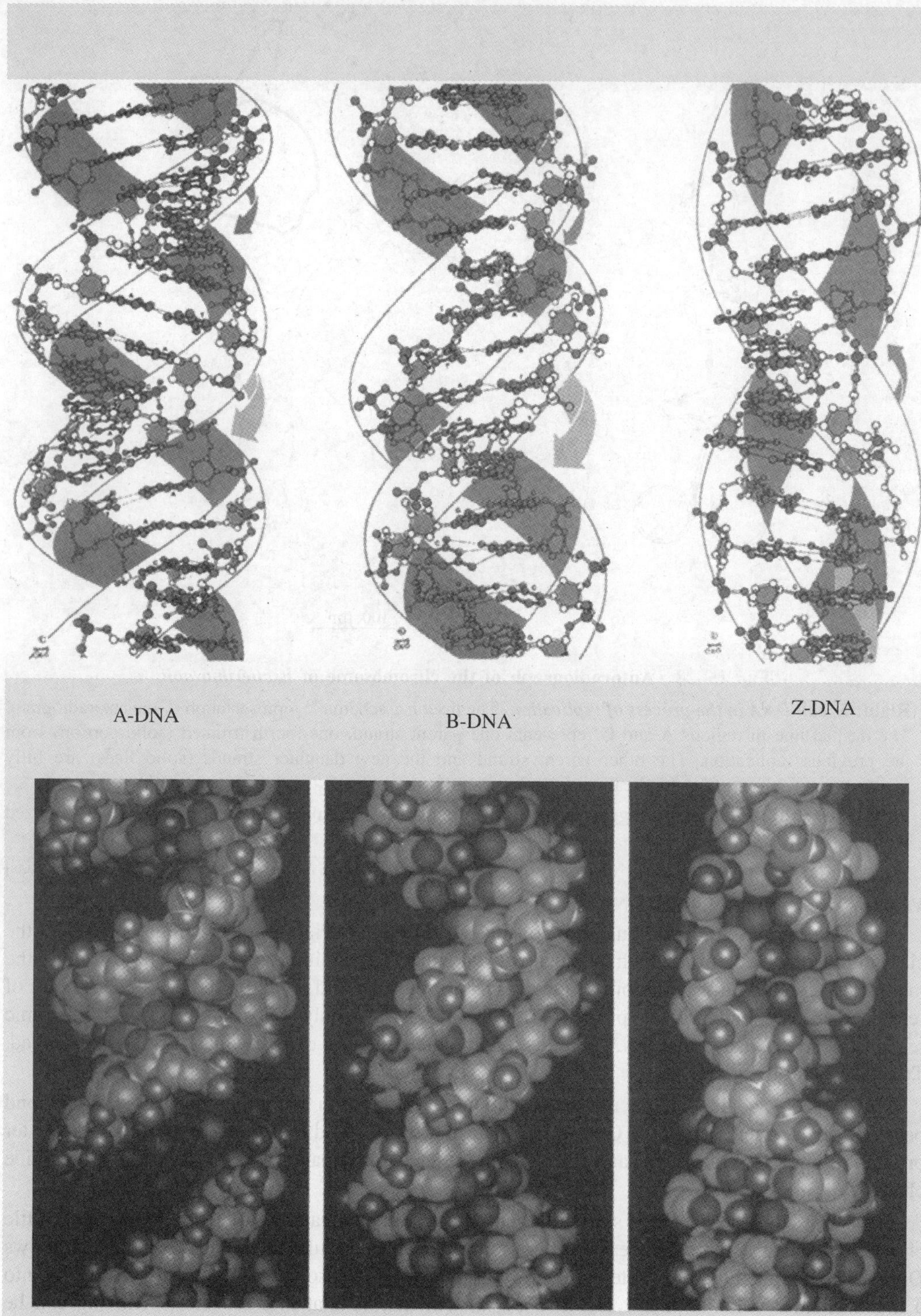

Part (a)

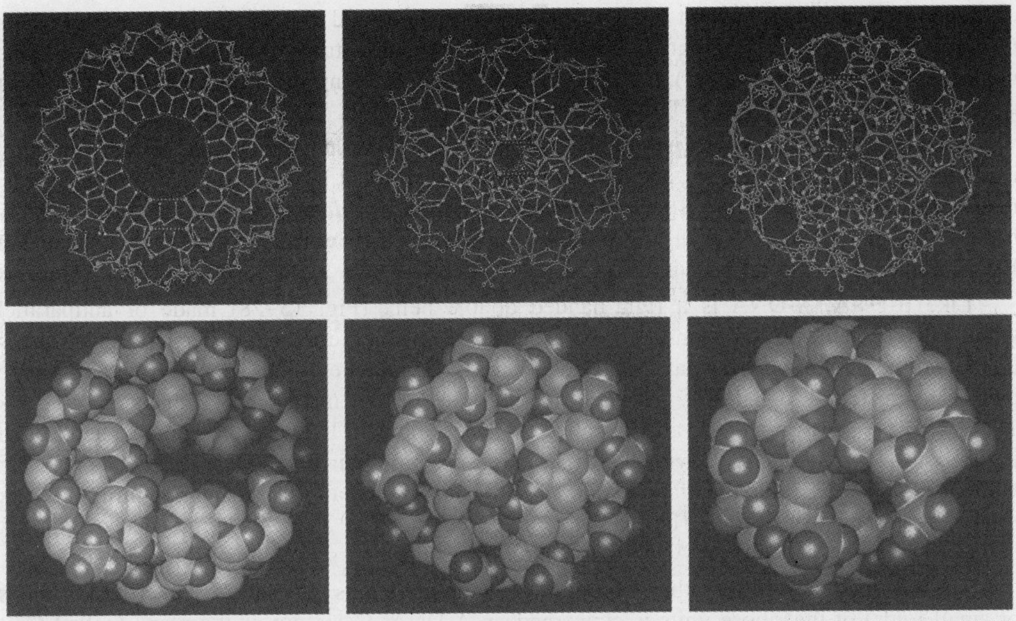

A-DNA B-DNA Z-DNA

Part (b)

Fig. 15–29. A comparison of the structures of 3 important forms of DNA– A-DNA, B-DNA and Z-DNA

Part (a). Views perpendicular to the helix axis (*i.e.*, axial or side views) of all the 3 forms

Part (b). Views down the helix axis (*i.e.*, radial or top views) of all the 3 forms

The top half of both parts shows ball-and-stick drawings; the bottom half of both parts include computer-generated space-filling models. The H atoms have been omitted for the sake of clarity in both drawings. The highly detailed structures shown here were deduced by x-ray crystallography performed on short segments of DNA, rather than the less detailed structures obtained from DNA wet fibres. The diffraction pattern obtained from the crystallization of short segments of DNA provides much greater detail concerning the exact placement of atoms within a double helical structure. **Alexander Rich** at MIT, **Richard Dickerson** then at the California Institute of Technology, and their colleagues were the first reseachers to crystallize a short piece of DNA.

In the **A-DNA**, the based pairs have a marked propeller twist with respect to the helix axis. Note that the base pairs are inclined to the helix axis and that the helix has a hollow core. In the **B-DNA**, the base pairs lie in a plane that is close to perpendicular to the helix axis. Note that the helix axis passes through the base pairs so that the helix has a solid core. In the **Z-DNA** , the helix is left-handed and in this respect differs from A-DNA and B-DNA, both of which are right-handed. Note that in Z-DNA, the sugar, phosphate chains follow a zigzag course [alternate ribose residues lie at different radii in Part (b)] indicating that Z-DNA's repeating motif is a dinucleotide. *Under physiologic conditions, all DNA molecules are in B form.*

It may, thus, be inferred that "DNA is a remarkably flexible molecule. Considerable rotation is possible around a number of bonds in the sugar-phosphate backbone and thermal fluctuations can produce bending, stretching and unpairing (melting) in the structure." (**Lehninger**, **Nelson** and **Cox**, 1993).

However, these structural variations do not affect the key properties of the DNA molecule as defined by **Watson** and **Crick**: strand complementarity, antiparallel nature of strands and base pairing specificity of A combining with T (A = T) and G combining with C (G ≡ C).

The DNA can, in fact, adopt several 3-'D' conformations and *can change reversibly from one form to the other.* These alternative forms of DNA differ in features like nature of the helix, the number of residues per turn ('n'), the spacing of the residues along the helix axis ('h') etc. Some significant alternative forms of DNA are described below (Fig. 15–29):

A-DNA : A-DNA appears when the DNA fibre (B-DNA) is dehydrated, *i.e.*, relative humidity is reduced from 92 to 75% and Na^+, K^+ and Cs^+ ions are present in the medium. In other words, *in solution, DNA assumes the B form and under conditions of dehydration, the A form.* This is because the phosphate groups in the A-DNA bind fewer water molecules than do phosphates in B-DNA.

Like B-DNA, A-DNA is a right-handed double helix (Fig. 15–28) made of antiparallel strands held together by Watson-Crick base pairing. But the vertical rise per base pair (*bp*) is 2.3 Å and the number of base pairs per helical turn is 11, relative to the 3.4 Å rise and 10.4 base pairs per turn found in B-DNA. As such, the helix in A-DNA moves a total distance of (2.3 × 11) or 25.30 Å per turn which value is called the *helix pitch* (In B-DNA, the value of helix pitch is (3.4 × 10.4) or 35.36 Å. Obviously, the rotation per base pair in A-DNA is 360/11 or 32.72° while in B-DNA the value of rotation is 360/10.4 or 34.61° (Recall that during one turn, the helix rotates by an angle of 360°). The diameter of A-DNA helix is 25.5 Å whereas that of B-DNA helix is 23.7 Å. The helix of A-DNA is, therefore, wider and shorter than that of B-DNA.

The base pairs are tilted rather than normal to the helix axis. As far as puckering (wrinkles or small folds) of the ribose units is concerned, in A-DNA, $C_{3'}$ is out of the plane (called **C3′-endo**) formed by the other 4 atoms of the furanose ring of sugar. By contrast, in B-DNA, $C_{2'}$ is out of plane (called **C2′-endo**). In A-DNA, the base pairs are inclined at 19° with the axis of the helix but in B-DNA, the base pairs lie almost perpendicular (1° tilt) to the helix axis. Moreover, the minor groove is practically nonexistent in A-DNA, whereas it is quite deep in B-DNA.

C-DNA : C-DNA is formed at 66% relative humidity in the presence of Li^+ ions. This form of DNA is also right-handed, with an axial rise of 3.32 Å per base pair. There are 9.33 base pair per turn of the helix ; the value of helix pitch is, therefore, 3.32 × 9.33 Å or 30.97 Å. The rotation per base pair in C-DNA is 360/9.33 or 38.58°. The C-helix has a diameter of 19 Å, smaller than that of both B- and A-helix. The tilt of the base pairs is 7.8°.

D-DNA : D-DNA is an extremely rare variant with only 8 base pairs per helical turn. This form of DNA is found in some DNA molecules devoid of guanine. By contrast, *A-, B- and C-forms of DNA are found in all DNA molecules, irrespective of their base sequence.* There is an axial rise of 3.03 Å per base pair, with a tilting of 16.7° from the axis of the helix.

Z-DNA : Z-DNA is the more radical departure from B-DNA and is characterized by a *left-handed helical rotation.* It was discovered by **Rich**, **Nordheim** and **Wang** in 1984. They found that a hexanucleotide, CGCGCG, forms a duplex of antiparallel strands held together by Watson-Crick base pairing, as expected. Surprisingly, they found that this double helix was left-handed and the phosphates in the DNA backbone were in a *zigzag* manner ; hence, they termed this new form as Z-DNA. By contrast, other forms of DNA are right-handed, with a regular phosphate backbone. Another remarkable characteristic of Z-DNA is that in it the adjacent sugar residues have *alternating* orientation and it is because of this reason that in Z-DNA, the repeating unit is a *dinucleotide* as against the B-DNA, where the adjacent sugar residues have same orientation so that the repeating unit in B-DNA is a mononucleotide (Fig. 15–30). Unlike A- and B-DNAs, Z-DNA contains only one deep helical groove. There are 12 base pairs (or six repeating dinucleotide units) per helical turn, with an axial rise of 3.8 Å per base pair; the bases are inclined at 9° with the axis of the helix. Because 12 base pairs are accomodated in one helix in Z-DNA, as against 10.4 in B-DNA, the angle of twist per repeating unit (dinucleotide) is (360/12 × 2) or 60° as against 34.61° per nucleotide in B-DNA. One complete helix is 45.60 Å in length in contrast to 35.36 Å in B-DNA. Since the bases get more length to spread out in Z-DNA and since the angle of tilt is 60°, they are more closer to the axis ; hence, the diameter of Z-DNA molecule is 18.4 Å hereas it is 23.7 Å in B-DNA.

The Z-DNA form occurs in short oligonucleotides that have sequences of alternating pyrimidine and purine bases. Methylation of C-5 of cytosyl residues in alternating CG sequences (as for example in the sequence CGCGCG) facilitates the transition of B-DNA to Z-DNA, because the added hydrophobic methyl groups stabilize the Z-DNA structure. The alternance of purines and pyrimidines permits at the glycosidic bonds an alternance of conformations, *syn* (base and sugar are near one another on the same side) and *anti* (base and sugar are distant, because they are on opposite sides). While in A- and B-DNA, all glycosidic bonds are *anti*, in Z-DNA, the purines are *syn* and the pyrimidines are *anti*.

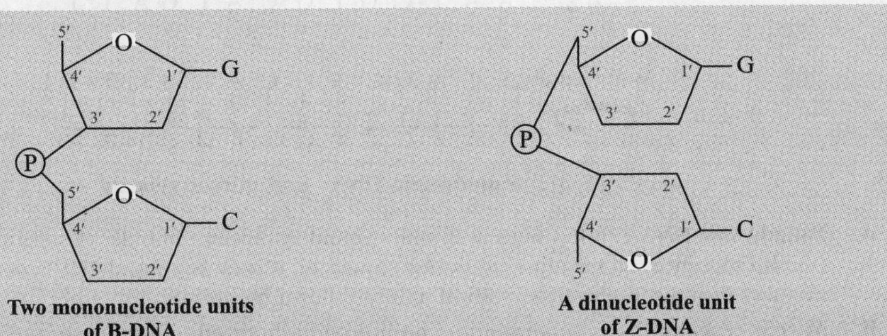

Fig. 15–30. Different orientation patterns of adjacent sugar residues in B-DNA and Z-DNA

Note that in B-DNA, the orientaion of adjacent sugar residues is same, while it is opposite in Z-DNA. The result is that in B-DNA, the repeating unit is a mononucleotide, while in Z-DNA, the repeating unit is a dinucleotide.

General characteristics of various forms of DNA are summarized in Table 15–6.

Table 15–6. Comparison of different forms of DNA

Characteristics	*A-DNA*	*B-DNA*	*C-DNA*	*Z-DNA*
Conditions	75% relative humidity; Na^+, K^+, Cs^+ ions	92% relative humidity; Low ion strength	60% relative humidity; Li^+ ions	Very high salt concentration
Shape	Broadest	Intermediate	Narrow	Narrowest
Helix sense	Right-handed	Right-handed	Right-handed	Left-handed
Helix diameter	25.5 Å	23.7 Å	19.0 Å	18.4 Å
Rise per base pair ('h')	2.3 Å	3.4 Å	3.32 Å	3.8 Å
Base pairs per turn of helix ('n')	11	10.4	9.33	12 (= 6 dimers)
Helix pitch (h × n)	25.30 Å	35.36 Å	30.97 Å	45.60 Å
Rotation per base pair	+ 32.72°	+ 34.61°	+ 38.58°	–60° (per dimer)
Base pair tilt	19°	1°	7.8°	9°

Glycosidic bond	*anti*	*anti*	—	*anti* for C, T *syn* for A, G
Major groove	Narrow and very deep	Wide and quite deep	—	Flat
Minor groove	Very broad and shallow	Narrow and quite deep		Very narrow and deep

DNAs with Unusual Structures

A number of other sequence-dependent structural variations in DNA have been detected that may serve locally important functions in DNA metabolism. These are:

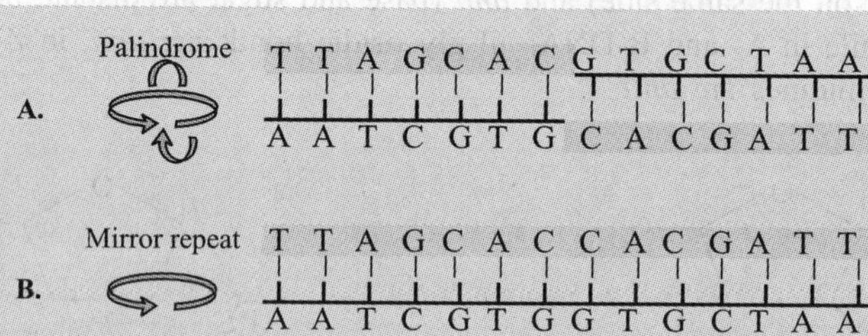

Fig. 15–31. Palindromic DNAs and mirror repeats

A. Palindromic DNA : It has sequences with twofold symmetry. In order to superimpose one repeat (*shaded* sequence) on the other (*unshaded* sequence), it must be rotated 180° around the horizontal axis and then again about the vertical axis, as shown by arrows.

B. Mirror repeat : It has a symmetric sequence on each strand. Superimposing one repeat on the other requires only a single 180° rotation about the vertical axis.

1. **Bent DNA.** Some sequences cause bends in the DNA helix. Bends are produced whenever 4 or more adenine residues appear sequentially in one of the two strands. Six adenines in a row produce a bend of about 18°. Bending may be important in the binding of some proteins to DNA.

2. **Palindromic DNA.** The term palindromic DNA is applied to regions of DNA in which there are *inverted repetitions* of base sequence with twofold symmetry occuring over two strands (Fig. 15–31A). Such sequences are self-complementary within each of the strands and therefore have the potential to form hairpin (when only a single strand is involved) or cruciform (when both strands of a duplex DNA are involved) structures, involving intrastrand base pairing (Fig. 15–32).

A **palindrome** (*palindromos*[G] = running back again) is a name, word, phrase, sentence, or verse that reads the same in either direction; for examples names such as NITIN, words such as LEVEL and ROTATOR, and sentences such as SLAP NO GAG ON PALS and ABLE WAS I ERE I SAW ELBA. A spectacular, but less pure, example that ignores spaces and punctuation has been devised by **Alastair Reid** : "T. Eliot, top bard, notes putrid tang emanating, is sad. I'd assign it a name: gnat dirt upset on drab pot toilet." [Quoted by **Brendan Gill** in *Here at the New Yorker*, Random House (1975).]

When the inverted sequence occurs within each individual strand of the DNA, the sequence is called a **mirror repeat** (Fig. 15–31B). Mirror repeats do not have complementary sequences within the same strand and cannot form hairpin or cruciform structures. Sequences of these types are found in virtually every large DNA molecule and can involve a few or up to thousands of base pairs.

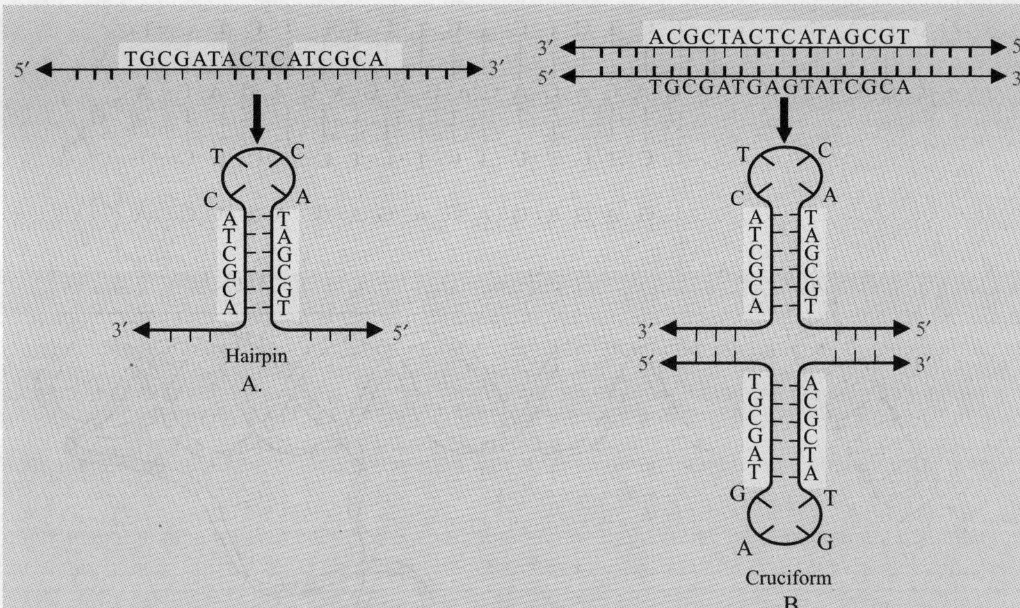

Fig. 15–32. Hairpins and cruciforms

Bases (enclosed in *rectangle*) are the asymmetric sequences that can pair alternatively with a complementary sequence in the same or opposite strand.

A. Hairpin: When only a single strand of palindromic DNA (or RNA) is involved, a hairpin is formed.

B. Cruciform: When both the strands of a double helical DNA are involved, a cruciform is formed.

3. H-DNA. H-DNA is usually found in polypyrimidine or polypurine segments that contain within themselves a mirror repeat. One simple example is a long stretch of alternating T and C residues (Fig. 15–33). A striking feature of H-DNA is the pairing and interwinding of 3 strands of DNA to form a triple helix. Triple-helical DNA is produced spontaneously only within long sequences containing only pyrimidines (or purines) in one strand. Two of the three strands in the H-DNA triple helix contain pyrimidines and the third contains purines.

Single-stranded DNA

The DNA molecules are not always a double-helical structure. **Robert Sinsheimer**, in 1959, discovered that the DNA in ϕ X174, a small virus that infects *Escherichia coli*, is single-stranded. His finding was based on the following evidences:

1. The base ratios (A/T and G/C) of ϕ X174 DNA do not approach unity.
2. The ϕ X174 DNA behaves as a randomly coiled polymer whereas the DNA double helix behaves as a rigid rod.
3. The amino groups of the bases of ϕ X174 DNA react readily with formaldehyde. In contrast, the bases in the double-helical DNA do not react with this reagent.

The discovery of this single-stranded DNA raised doubts as to whether the semiconservative scheme of replication as proposed by Watson and Crick, is universally true. However, it was soon found that ϕ X174 DNA is single-stranded for only a part of its life span and the infected *E. coli* cells contain a double helical form of ϕ X174 DNA. This double helical DNA is called *replicative form*, since it acts as a template for the synthesis of DNA of the progeny virus. Viruses that contain single-stranded RNA also replicate *via* a double-stranded replicative form.

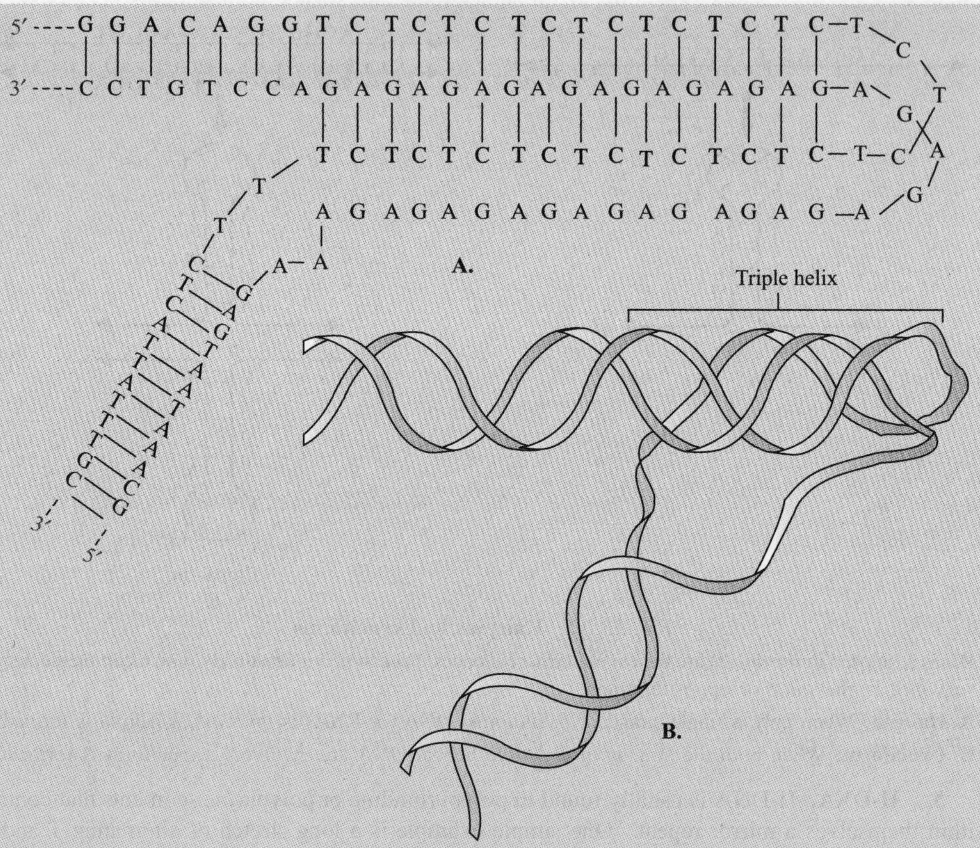

Fig. 15–33. H-DNA

A. A sequence of alternating T and C residues can be considered a mirror repeat centred about one of the central T or C residues.

B. The sequences of alternating T and C residues form an unusual structure in which the strands in one half of the mirror repeat are separated and the pyrimidine-containing strand folds back on the other half of the repeat to form a triple helix.

Table 15–7. Differences between double-stranded and single-stranded DNAs

dsDNA	*ssDNA*
1. It is linear or filamentous types.	1. It is somewhat star-shaped (stellate).
2. DNA double helix behaves as a rigid rod.	2. φ X 174 DNA behaves as a randomly coiled polymer.
3. A/T ratio approaches unity.	3. A/T ratio is 0.77.
4. G/C ratio also approaches unity.	4. G/C ratio is 1.3.
5. It is resistance to the action of formaldehyde.	5. The amino groups of the bases of φ X 174 DNA react readily with formaldehyde.
6. Ultraviolet absorption increases from 0 to 80°C	6. Ultraviolet absorption increases from 20 to 90°C.

RIBONUCLEIC ACID

Ribonucleic acid (RNA), like DNA, is a long, unbranched macromolecule consisting of nucleotides joined by 3′ → 5′ phosphodiester bonds (refer Fig. 15–34). The number of ribonucleotides in RNA ranges from as few as 75 to many thousands.

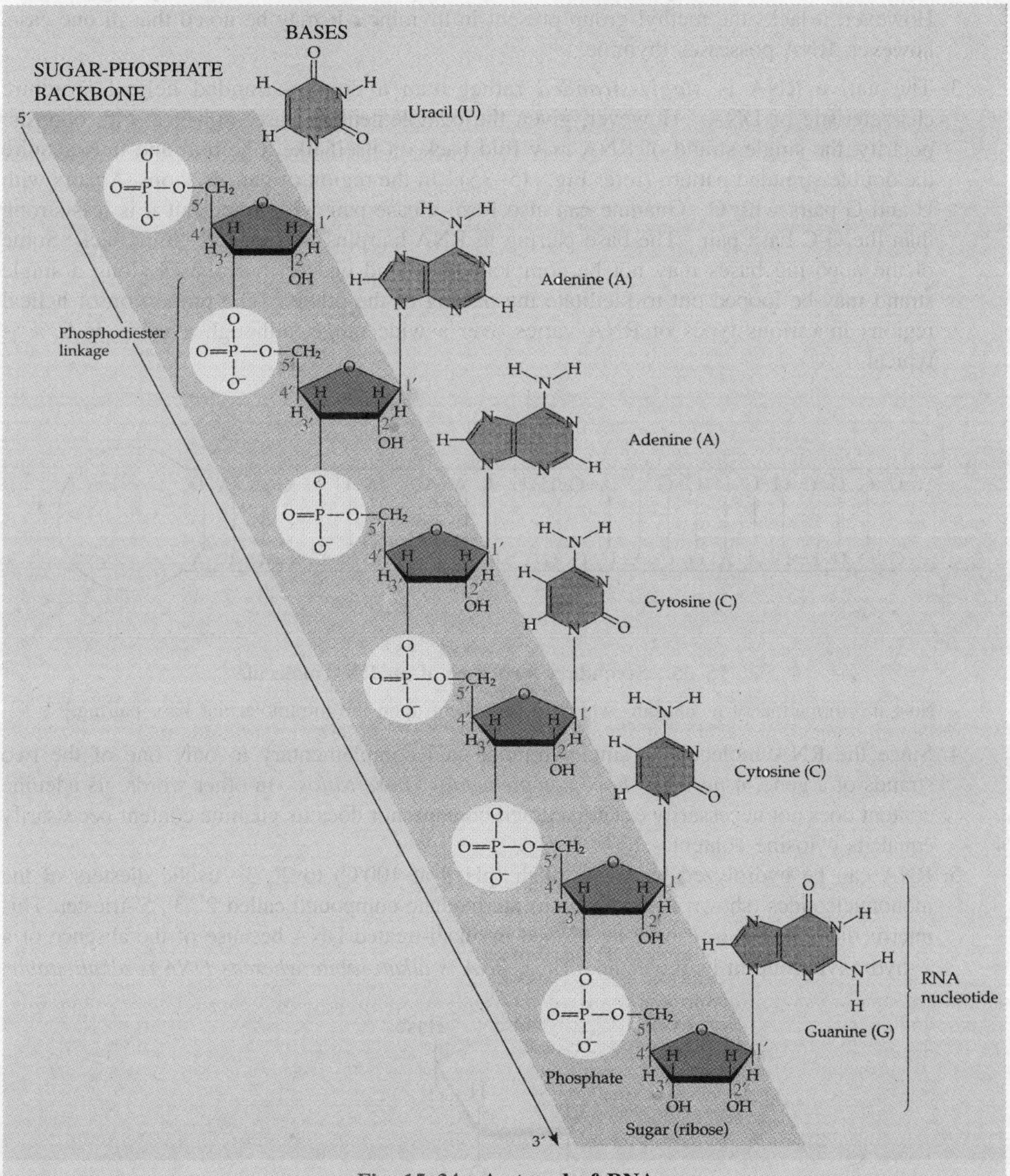

Fig. 15–34. A strand of RNA

This structure is very similar to a DNA strand (see Figure 15–16), except that the sugar is ribose instead of deoxyribose and uracil is substituted for thymine.

Differences with DNA

Although sharing many features with DNA, the RNA molecules possess several sepcific differences:

1. As apparent from its name, the sugar moiety in RNA, to which the phosphate and the nitrogen bases are attached, is *ribose* rather than 2′-deoxyribose of DNA. Ribose contains a 2′-hydroxyl group not present in deoxyribose.
2. RNA contains the pyrimidine *uracil* (U) in place of thymine which is characteristic of DNA molecule. Uracil, like thymine, can form a base pair with adenine by 2 hydrogen bonds. However, it lacks the methyl group present in thymine. It may be noted that in one case, however, RNA possesses thymine.
3. The native RNA is *single-stranded* rather than a double-stranded helical structure characteristic of DNA. However, given the complementary base sequence with opposite polarity, the single strand of RNA may fold back on itself like a hairpin and thus acquire the double-stranded pattern (refer Fig. 15–35). In the region of hairpin loops, A pairs with U and G pairs with C. Guanine can also form a base pair with uracil but it is less strong than the G-C base pair. The base pairing in RNA hairpins is frequently imperfect. Some of the apposing bases may not be complementary and one or more bases along a single strand may be looped out to facilitate the pairing of the others. The proportion of helical regions in various types of RNA varies over a wide range, although a value of 50% is typical.

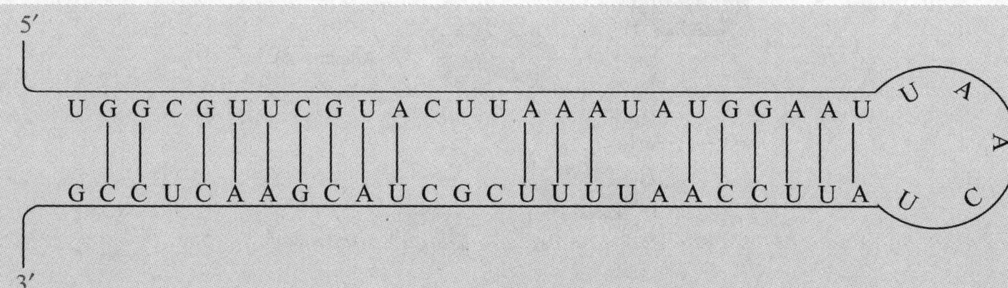

Fig. 15–35. Secondary structure of an RNA molecule

[Note the formation of a 'hairpin' which is dependent upon the intramolecular base pairing.]

4. Since the RNA molecule is single-stranded and complementary to only one of the two strands of a gene, *it need not have complementary base ratios*. In other words, its adenine content does not necessarily equal its uracil content, nor does its guanine content necessarily equal its cytosine content.
5. RNA can be hydrolyzed by weak alkali (pH 9 at 100′C) to 2′, 3′- cyclic diesters of the mononucleotides (shown below) *via* an intermediate compound called 2′, 3′, 5′-triester. This intermediate, however, cannot be formed in alkali-treated DNA because of the absence of a 2′-hydroxyl group in its molecule. Thus, *RNA is alkali-labile whereas DNA is alkali-stable*.

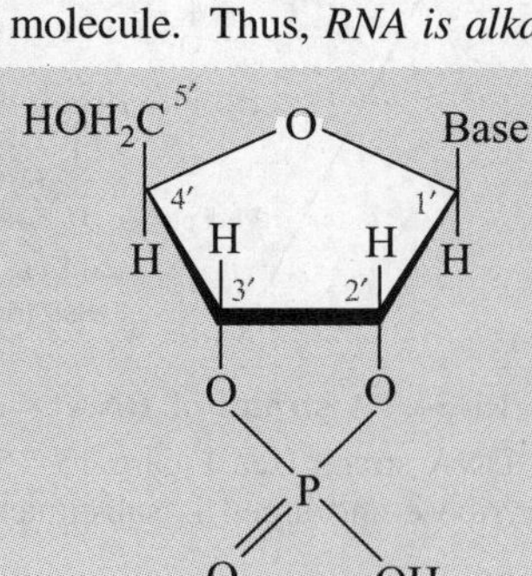

A 2′, 3′-cyclic monophosphate nucleotide

Table 15–8. Comparison between DNA and RNA

DNA	*RNA*
1. Found mainly in the chromatin of the cell nucleus	1. Most of RNA (90%) is present in the cell cytoplasm and a little (10%) in the nucleolus.
2. Never present in free state in cytoplasm	2. May be present in free state.
3. Normally double-stranded and rarely single-stranded.	3. Normally single-stranded and rarely double-stranded.
4. DNA has both 'sense' and 'antisense' strands.	4. The sequence of an RNA molecule is the same as that of the 'antisense' strand.
5. Sugar moiety in DNA is 2′-deoxyribose (hence the nomenclature) which contains an H atom at C-2.	5. Sugar moiety in RNA is ribose (hence the nomenclature) which contains a 2′-hydroxyl group.
6. Sugars in DNA are in the $C_{2'}$ - endo form.	6. Sugars in RNA are in the $C_{3'}$ - endo form.
7. The common nitrogenous bases are adenine, guanine, cytosine and thymine (but not uracil).	7. The common nitrogenous bases are adenine, guanine, cytosine and uracil (but not thymine)
8. Base pairing is inevitable during which adenine pairs with thymine and guanine with cytosine.	8. In case pairing takes place, adenine pairs with uracil and guanine with cytosine.
9. The base ratios (A/T and G/C) are necessarily around one.	9. It need not have complementary base ratios.
10. Base pairing involves the entire length of DNA molecule.	10. Base pairing takes place in only the helical regions of RNA molecule, which amount to roughly half (50%) of the entire RNA molecule.
11. DNA contains only few unusual bases.	11. RNA contains comparatively more unusual bases.
12. DNA is of 3 types: filamentous (double helical or duplex), circular or single-stranded.	12. RNA is of 5 types: viral RNA, rRNA, tRNA, mRNA and double-stranded RNA.
13. It consists of a large number of nucleotides (up to 3–4 million) and has, therefore, high molecular weight.	13. It consists of fewer nucleotides (up to 12,000) and has, therefore, low molecular weight.
14. DNA is alkali-stable.	14. RNA is alkali-labile.
15. DNA stains blue with azurephthalate.	15. RNA stains red with azurephthalate.
16. DNA acts as a template for its synthesis.	16. RNA does not act as a template for its synthesis.
	17. Usually RNA does not replicate or transcribe.
17. DNA on replication forms DNA and on transcription forms RNA.	18. During biosynthesis, exonuclease is not needed.
18. During replicatin, exonuclease is needed.	19. RNA exhibits complete and practically instantaneous reversibility of the process of melting.
19. DNA is partially reversible only under certain conditions of slow cooling (= annealing).	20. RNA does not undergo mutation.
20. DNA undergoes mutation.	21. RNA is the genetic material of some viruses only.
21. DNA is the usual genetic material.	22. RNA is stained red with pyronin.
22. DNA is stained green with a dye, pyronin.	

Types of RNA

In all procaryotic and eucaryotic organisms, 3 general types of RNAs are found: *ribosomal*, *transfer* and *messenger RNAs*. Each of these polymeric forms serves as extremely important informational links between DNA, the master carrier of information and proteins. The 3 types of RNA molecules differ from each other by size, function and general stability. Their properties are summarized in Table 15–9.

Table 15–9. RNA molecules in Escherichia coli

Type	*Relative amount*	*Sedimentation coefficient*	*Molecular weight*	*Number of nucleotides*
Ribosomal RNA (rRNA)	80%	23 s	1.2×10^6	3,700
		16 s	0.55×10^6	1,700
		5 s	3.6×10^4	120
Transfer RNA (tRNA)	15%	4 s	2.5×10^4	75
Messenger RNA (mRNA)	5%	—	Heterogeneous	

Ribosomal RNA (rRNA) or Insoluble RNA

It is the most stable form of RNA and is found in ribosomes. It has the highest molecular weight and is sedimented when a cell homogenate containing 10^{-2} M of Mg^{2+} is centrifuged at high speed (100,000 gravity for 120 minutes). In the bacterium, *Escherichia coli*, there are 3 kinds of RNA called 23 s, 16 s, and 5 s RNA because of sedimentation behaviour. These have molecular weights of 1,200,000, 550,000 and 36,000 respectively. One molecule of each of these 3 types of rRNA is present in each ribosome. *Ribosomal RNA is most abundant of all types of RNAs* and makes up about 80% of the total RNA of a cell. Ribosomal RNA represents about 40-60% of the total weight of ribosomes.

Table 15–10. Ribosomes and their RNAs

Ribosomes	*rRNA*
Procaryotic ribosomes	
30 s	16 s
50 s	5 s, 23 s
Eucaryotic ribosomes	
40 s	18 s
60 s	5 s, 28 s

The ribosomes of procaryotic and eucaryotic cells possess different species of RNA which have been summarized in Table 15–10. The ribosomes of procaryotes and those found in the plastids and mitochondria of eucaryotes contain 3 different types of RNA: 23 s RNA and 5 s RNA in the larger (50 s) subunit and 16 s RNA in the smaller (30 s) subunit. The ribosomes of the eucaryotic cells also contain 3 kinds of RNA: 28 s RNA and 5 s RNA in the larger (60 s) subunit and 18 s RNA in the smaller (40 s) subunit.

The mammalian ribosomes deserve special mention (refer Table 15–11). They also contain 2 major nucleoprotein subunits : a larger one of 2.7 megadaltons (60 s) and a smaller one of 1.3 megadaltons (40 s). The 60 s subunit contains a 5 s rRNA, a 5.8 s rRNA (formerly called 7 s rRNA) and a 28 s rRNA. There are also probably more than 50 specific polypeptides. The 40 s subunit contains a single 18 s rRNA and about 30 polypeptides chains. All types of rRNAs, except the 5 s rRNA, are processed from a single 45 s precursor RNA molecule in the nucleolus. The 5 s rRNA apparently has its own precursor which is independently transcribed.

Table 15–11. RNA components of mammalian ribosomes

Subunit		*RNA*	
Size	Mol. wt.	Size	Mol. wt.
60 s (> 50 polypeptides)	2.7×10^6	5 s 5.8 s 28 s	3.5×10^4 4.5×10^4 1.5×10^6
40 s (> 30 polypeptides)	1.3×10^6	18 s	7.5×10^5

It is remarkable to note that *rRNA from all sources has G-C contents more than 50%.* The rRNA molecule appears as a single unbranched polynucleotide strand (= *primary structure*). At low ionic strength, the molecule shows a compact rod with random coiling. But at high ionic strength, the molecule reveals the presence of compact helical regions with complementary base pairing and looped outer region (= *secondary structure*). The helical structure (Fig. 15–36) results from a folding back of a single-stranded polymer at areas where hydrogen bonding is possible because of short lengths of complementary structures. The double helical secondary structures in RNA can form within a single RNA molecule or between 2 separate RNA molecules (Fig. 15–37). RNAs can often assume even more complex shapes as in bacteria (Fig. 15–38). The function of rRNA molecules in the ribosomes is not yet fully understood, but they are necessary for

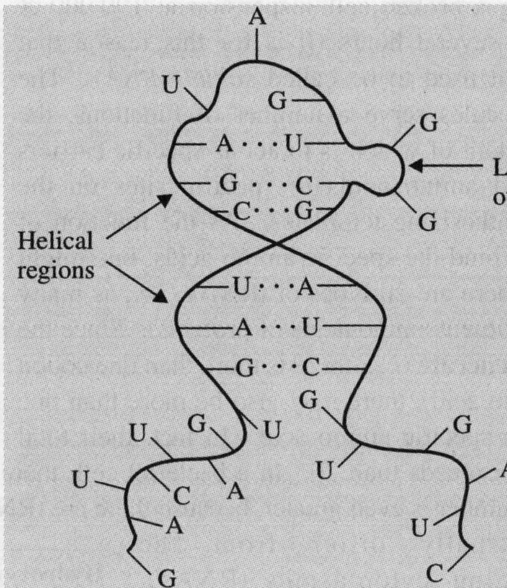

Fig. 15–36. Molecular strucutre of rRNA

Note the helical regions with complementary base pairs and a looped outer region.

(*After Davidson JN, 1972*)

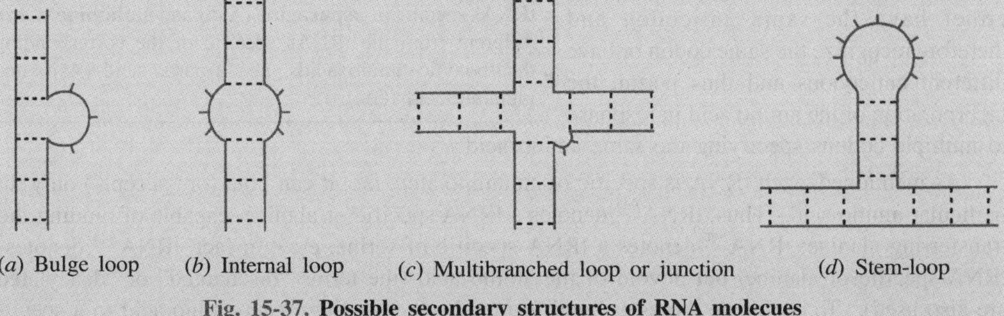

(*a*) Bulge loop (*b*) Internal loop (*c*) Multibranched loop or junction (*d*) Stem-loop

Fig. 15-37. Possible secondary structures of RNA molecues

The double-stranded regions are depicted by connecting hydrogen bonds. Loops are noncomplementary regions that are not hydrogen bonded with complementary bases. Double-stranded RNA structures can form within a single RNA molecule or between two separate RNA molecules.

(*Modified from Jaeger et al, 1993*)

ribosomal assembly and seem to provide a specific sequence to which the messenger RNA moelcule can bind in order to be translated.

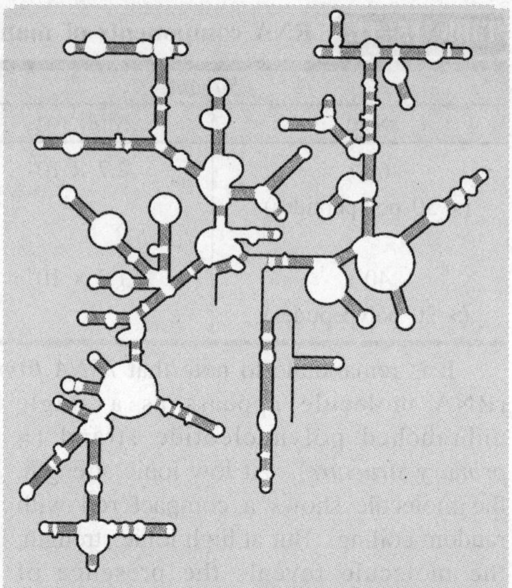

Fig. 15–38. An unusual complex shape of RNA
This ribosomal RNA is an integral component of the small ribosomal subunit of a bacterium. The RNA strand is folded back on itself in a highly ordered pattern so that most of the molecule is double-stranded.

Transfer RNA (tRNA) or Soluble RNA (sRNA)

Transfer RNA is the smallest polymeric form of RNA. These molecules seem to be generated by the nuclear processing of a precursor molecule. In abundance, the tRNA comes next to rRNA and amounts to about 15% of the total RNA of the cell. The tRNA remains dissolved in solution after centrifuging a broken cell suspension at 100,000 X gravity for several hours (It is for this reason that previously it used to be called *soluble RNA*). The tRNA molecules serve a number of functions, the most important of which is to act as specific carriers of activated amino acids to specific sites on the protein- synthesizing templates. As the function of tRNA is to bind the specific amino acids, one might think that there are 20 types of tRNAs (*i.e.*, as many as the constituent amino acids of proteins). Since the code is **degenerate** (*i.e.*, there is more than one codon for an amino acid), there may also be more than one tRNA for a specific amino acid. In fact, their total number far exceeds than 20. In a bacterial cell, there are more than 70 tRNAs and in eukaryotic cells, this number is even greater, because there are tRNAs specific of mitochondria and chloroplasts (which usually differ from the corresponding cytoplasmic tRNAs). Therefore, there are generally several tRNAs specific of the same amino acid (sometimes up to 4 or 5); they are called **isoacceptor tRNAs.** These various tRNAs, capable of binding the same amino acid, differ in their nucleotide sequence; they can either have the same anticodon and therefore recognize the same codon or have different anticodons and thus permit the incorporation of the amino acid in response to multiple codons specifying the same amino acid.

> **Hydroxyproline (Hyp)** is a constituent amino acid of some proteins, but it is formed by a modification of proline after its incorporation into the polypeptide chain, so that there is no tRNA specific of hydroxyproline. There is no tRNA specific of **cystine** either, because the latter is formed by the union of 2 cysteine residues present in the same chain at places sometimes rather distant from one another or even in different polypeptide chains. But there are tRNAs specific of **asparagine (Asn)** and **glutamine (Gln)**, different from the tRNAs specific of the corresponding dicarboxylic amino acids, *i.e.*, aspartic acid (Asp) and glutamic acid (Glu).

As mentioned, each tRNA is specific of an amino aicd, *i.e.*, it can bind (or "accept") only that particular amino acid. Thus, $tRNA^{Ala}$ denotes a tRNA specific of alanine, capable of binding then transferring alanine; $tRNA^{Ser}$ denotes a tRNA specific of serine, etc. In fact, $tRNA^{Ala}$ denotes a tRNA specific of alanine, but devoid of the amino acid (the terms "discharged" or "deacylated" are also used). To designate the ester bond formed by the binding of an amino acid to a specific tRNA, one must write alanyl-$tRNA^{Ala}$ or simply alanyl-tRNA.

Because of its small size, tRNA has lent itself more readily to isolation procedures and to studies of nucleotide sequence. *Although tRNAs are quite stable in procaryotes, they are somewhat less stable in eucaryotic organisms.*

Primary structure (or base sequence) of a tRNA molecule. The base sequence of a

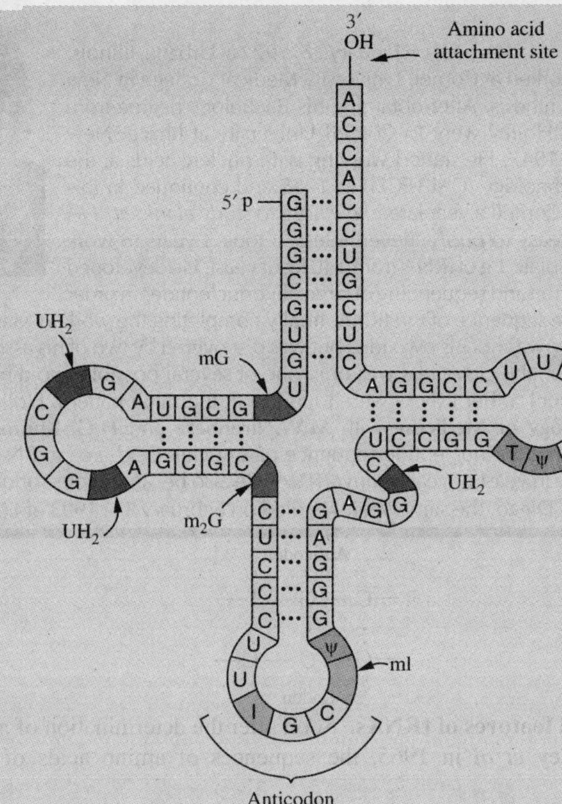

Fig. 15–39. Secondary structure, showing the base sequence of yeast alanine-specific tRNA, drawn in the cloverleaf form

The abbreviations used for modified nucleosides stand for I = inosine, mI = methylinosine, T = ribothymidine, UH$_2$ = dihydrouridine, ψ = pseudouridine, mG = methylguanosine and m$_2$G = dimethylguanosine. Note that a molecule of tRNA^{Ala} contains a total of 77 amino acid residues, with the break-up: 8A + 11U + 25G + 23C + 10 modified bases.

(*Adapted from Lubert Stryer, 1995*)

tRNA molecule specific for the amino acid alanine (refer Fig. 15–39) was first determined by **Robert W. Holley** and coworkers in 1965, the culmination of seven years of effort. In fact, this was the first determination of the complete sequence of a nucleic acid. **Holley** first suggested the '*cloverleaf*' model (*i.e.*, consisting of 3 folds) which is based on a secondary structure which would contain a maximum of intramolecular hydrogen bonding.

The alanyl-tRNA of yeast consists of an unbranched chain of 77 ribonucleotides including 8A, 11U, 25G, 23C and 10 unusual nucleosides which are inosine, ribothymidine, dihydrouridine, pseudouridine and methylated derivatives of inosine and guanosine. The 77 residues represent a molecualr weight of 26,600 as the sodium salt. The 5′ terminus is phosphorylated whereas the 3′ terminus has a free hydroxyl group. While writing down the nucleotide sequence, the usual conventions are followed in that the *5′-phosphate is shown at the left hand and the 3′-hydroxyl on the right side.* The attachment site for the amino acid alanine is 3′-hydroxyl group of the adenosine residue at 3′ terminus of the molecule. The sequence *inosine-guanine-cytosine* in the middle of the molecule is the anticodon. It is complementary to *guanine-cytosine-cytosine*, one of the codons for alanine.

ROBERT WILLIAM HOLLEY

(LT, 1922–1993)

Holley, a US biochemist, was born on January 28, 1922 at Urbana, Illinois. From 1944 to 1946, he worked at Cornell University Medical College in New York City on penicillin synthesis. After obtaining his Bachelor's degree from the University of Illinois, Holley went to Cornell University at Ithaca, New York, gaining his PhD in 1947. He started working with nucleic acids at the California Institute of Technology (CALTECH) in 1955 and continued in this area when he returned to Cornell a year later. *He chose to study alanine tRNA because it was relatively easy to purify*; nevertheless, it took 3 years to work out new procedures and isolate 1 g of tRNA from 140 kg of yeast. He developed new methods for the isolation and sequencing of large oligonucleotides in order to work out the nucleotide sequence of the tRNA, finally completing the work in March 1965. He stated that the cloverleaf structure was not his own idea but was discovered by two of his associates, **Betty Keller** and **John Penswick**. In fact, the cloverleaf was only one of several possible two-dimensional structures and was not confirmed until other tRNAs were found to fit the same pattern. Holley shared the **1968 Nobel Prize for Physiology or Medicine** with **M.W.Nirenberg** and **H.G. Khorana** for his work in establishing for the first time the nucleotide sequence of a nucleic acid, *i.e.*, a tRNA molecule specific for the amino acid alanine in yeast or *yeast alanyl-tRNA*. He also became Professor of Molecular Biology at the Salk Institute in San Diego, the same year. He died on February 11, 1993 at Los Gatos, California.

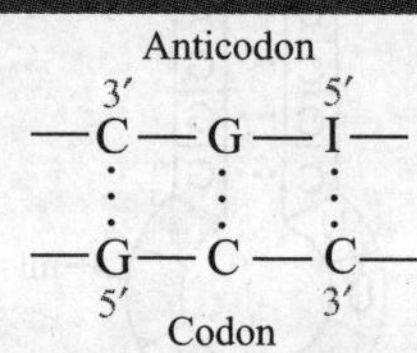

Common structural features of tRNAs. Soon after the determination of amino acid sequence of alanyl-tRNA by Holley *et al* in 1965, the sequences of amino acids of many other tRNA

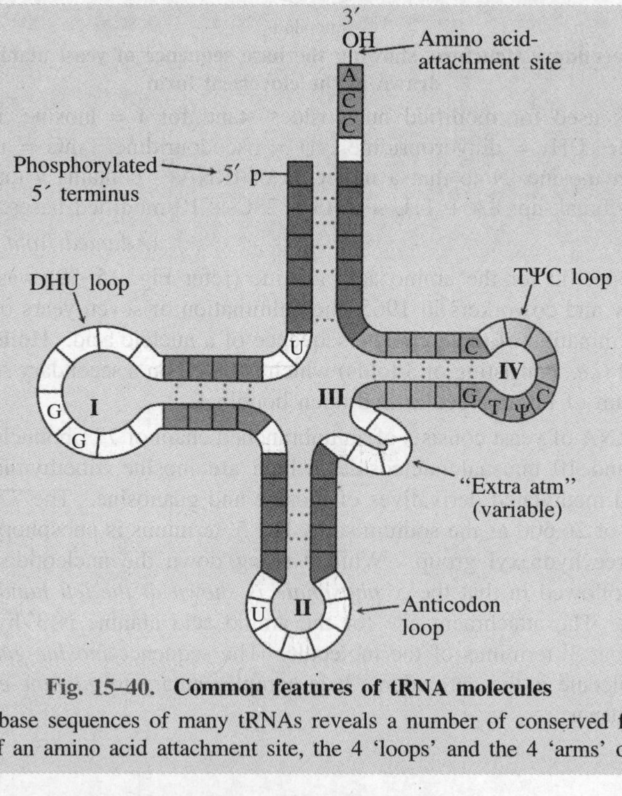

Fig. 15–40. Common features of tRNA molecules

Comparison of the base sequences of many tRNAs reveals a number of conserved features. Note the presence of an amino acid attachment site, the 4 'loops' and the 4 'arms' or stems.

molecules were determined by many workers. As a result, about 50 sequences are now known. These have been found to possess certain *common structural features* (refer Fig. 15–40). These are enumerated below:

1. All tRNA molecules have a common design and consist of 3 folds giving it a shape of the cloverleaf with four arms (refer Fig. 15–41); the longer tRNAs have a short fifth or extra arm. The actual 3-dimensional structure of a tRNA looks more like a twisted L than a cloverleaf (Fig. 15–41).

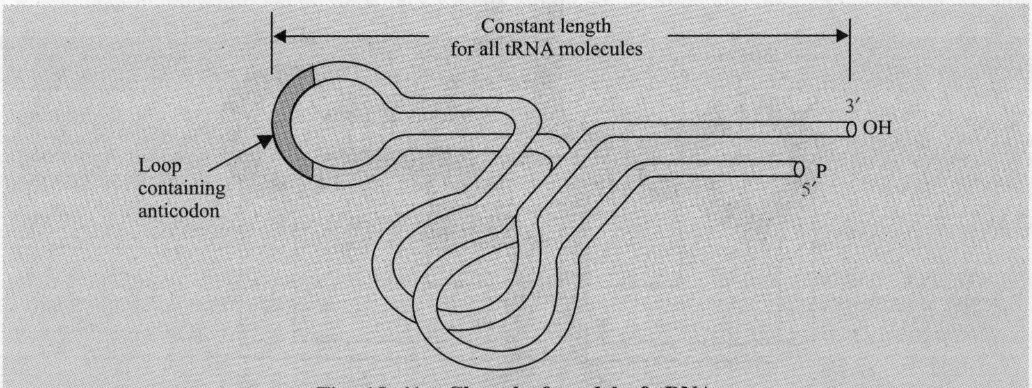

Fig. 15–41. Cloverleaf model of tRNA

2. All tRNA molecules are unbranched chains containing from 73 to 93 ribonucleotide residues, corresponding to molecular weights between 24,000 and 31,000 (Mitochondria contain distinct RNAs that are somewhat smaller). For example, $tRNA^{Ala}$ contains 77 nucleotides (**Holley** *et al*, 1965), $tRNA^{Tyr}$ contains 78 nucleotides (**Medison** *et al*, 1967) $tRNA^{Ser}$ contains 85 nucleotides (**Zachan** *et, al,* 1967).
3. They contain from 7 to 15 unusual modified bases. Many of these unusual bases are methylated or dimethylated derivatives of A, U, G and C. These include nucleotides of pseudouridine, various methylated adenines and guanines, methylated pyrimidines such as thymine and 5-methylcytosine and others. Not all these are present in any one source of tRNA but *pseudouridine* (*ϕU*) *is the most abundant and universally distributed.* Although, the role of these bases is uncertain, yet two roles seem certain:
   a. Methylation prevents the formation of certain base pairs so that some of the bases become accessible for other interactions.
   b. Methylation imparts hydrophobic character to some portions of tRNA molecules which may be important for their interaction with the *synthetases* and with ribosomal proteins.

   In fact, the modified bases of tRNAs prompted **Crick** to remark:

   > "It almost appears as if tRNAs were nature's attempt to make an RNA molecule play the role of a protein."

   What he meant by this statement is that the unusual bases are probably needed to stabilize a nucleic acid molecule that has the intricate folding of the 3-'D' structure of a globular protein.
4. The 5′ end of tRNAs is phosphorylated. The 5′ terminal residue is usually guanylate (pG).
5. The base sequence at the 3′ end of all tRNAs is CCA. All amino acids bind to this terminal adenosine *via* the 3′-OH group of its ribose.

   (5′ end) G........................tRNA.........................CCA (3′ end)
6. About 50% of the nucleotides in tRNAs are base-paired to form double helices.

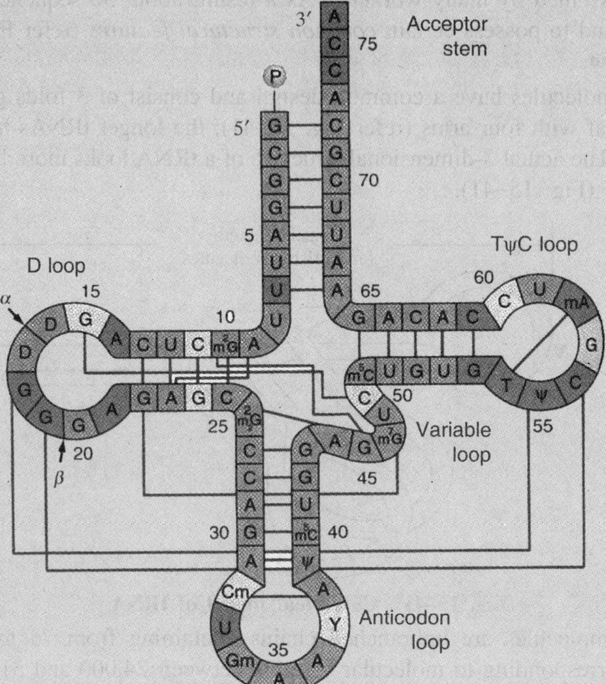

Fig. 15-42. Secondary or two-dimensional structure of yeast phenylalanine tRNA in the form of a cloverleaf

The conventional numbering of nucleotides begins at the 5′ end and proceeds toward the 3′ end. In all tRNAs, the nucleotides at the 3′ end contains the sequence CCA. Certain locations can have additional nucleotides that are not found in all tRNA molecules. The entire nucleotide sequence of yeast $tRNA^{Phe}$ is shown. *Red* lines connecting nucleotides indicate hydrogen-bonded bases. *Rose* squares indicate constant nucleotides; *tan* squares indicate a constant purine or pyrimidine. Insertion of nucleotides in the D loop occurs at positions α and β for different tRNAs.

(*Courtesy: Quigley GJ and Rich A, 1976*)

7. There are, however, 5 groups of bases which are not base-paired. These 5 groups, of which 4 form 'loops', are:
   a. the 3′ CCA terminal region,
   b. the ribothymine-pseudouracil-cytosine (= T ϕ C) loop,
   c. the 'extra arm' or little loop, which contains a variable number of residues,
   d. the dihydrouracil (= DHU) loop, which contains several dihydrouracil residues, and
   e. the anticodon loop, which consists of 7 bases with the sequence,
   5′ —pyrimidine—pyrimidine—X—Y—Z—modified purine—variable base — 3′

   This loop contains a triplet of bases which allows the tRNA to hydrogen-bond to a complementary sequence on mRNA attached to a ribosome.
8. The 4 loops are *recognition sites.* Each tRNA must have at least two such recognition sites : one for the activated amino acid-enzyme complex with which it must react to form the aminoacyl-tRNA and another for the site on a messenger RNA molecule which contains the code (codon) for that particular amino acid. It is interesting to note that the former involves recognition by bases of amino acid residues (either of the activated amino acid or of a site on the enzyme molecule) whereas the latter involves recognition by bases of bases (hydrogen bonding).
9. A unique similarity among all tRNA molecules is that *the overall distance from CCA at one*

end to the anticodon at the other end is constant (refer Fig. 15–33). The difference in nucleotide numbers in various tRNA molecules is, in fact, compensated for by the size of the "extra arm", which is located between the anticodon loop and T ϕ C loop.

Tertiary structure (or three-dimensional structure) of a tRNA molecule. Alexander Rich and Aaron Klug (1960s), on the basis of their x-ray crystallographic studies, have elucidated the 3-'D' structure of the phenylalanine-accepting tRNA from yeast (refer Figs. 15–42 and 15-43). This

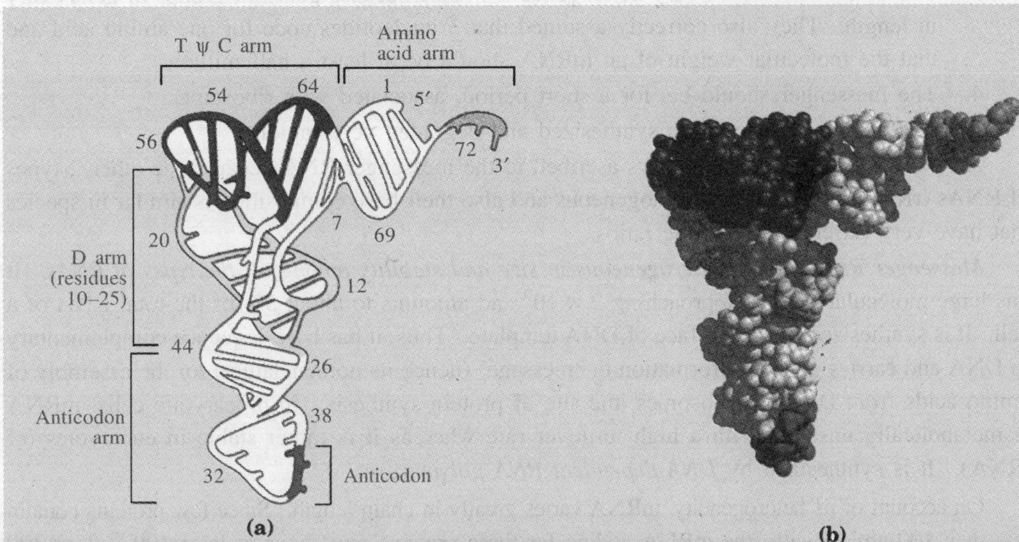

Fig. 15–43. The 3-dimensional tertiary structure of yeast phenylalanine tRNA, as deducted from diffraction analysis

The specified amino acid is attached at the CCA 3′ terminus. The amino acid (or amino acceptor) arm and TΨC) arm form a continuous double helix, and the anticodon (AC) arm and the DHU (D) arm form the other partially continuous double helix. These 2 helical columns meet to form a twisted L-shaped molecule.

is probably closer to reality than the postulated cloverleaf structure. The important features observed by them are:

1. The molecule is L-shaped.
2. There are 2 segments of the double helix. Each of these helices contains about 10 base pairs which correspond to one turn of the helix. These helical segments are perpendicular to each other, thus forming an L.
3. The CCA terminus containing the attachment site for the amino acid is at one end of the L. The other end of the L is occupied by the anticodon loop. The DHU and TϕC loops form the corner of the L.
4. The CCA terminus and the adjacent helical region do not interact strongly with the rest of the molecule. This part of the molecule may change conformation during amino acid activation and also during protein synthesis on the ribosome.

Messenger RNA (mRNA) or Template RNA

The abundance of RNA in the cytoplasm and its role in protein synthesis suggested that the genetic information of nuclear DNA is transmitted to an RNA which functions at the sites of protein synthesis. In 1961, the two Nobel laureates, **François Jacob** and **Jacques Monod** postulated that control of protein formation, at least in certain microorganisms, is determined by the rate of synthesis of templates. This requires that the templates do not accumulate in contrast of the constant presence of DNA, rRNA and tRNA. They, therefore, suggested the transient existence of an RNA, which they called "the messenger" RNA (mRNA). Later, they proposed that "the messenger" should have the following properties:

1. The messenger should be a polynucleotide.
2. The base composition of the messenger should reflect the base composition of the DNA that specifies it.
3. The mRNA should be very heterogeneous in size because genes or groups of genes vary in length. They also correctly assumed that 3 nucleotides code for one amino acid and that the molecular weight of an mRNA should be at least a half million.
4. The messenger should be, for a short period, associated with ribosomes.
5. The messenger should be synthesized and degraded very rapidly.

All these properties are nowadays ascribed to the messenger RNA because the other 2 types of RNAs (rRNA and tRNA)are homogeneous and also their base composition is similar in species that have very different DNA base ratios.

Messenger RNA is most heterogeneous in size and stability among all the types of RNAs. It has large molecular weight approaching 2×10^6 and amounts to about 5% of the total RNA of a cell. It is synthesized on the surface of DNA template. Thus, it has base sequence complementary to DNA and carries genetic information or 'message' (hence its nomenclature) for the assembly of amino acids from DNA to ribosomes, the site of protein synthesis. In procaryotic cells, mRNA is metabolically unstable with a high turnover rate whereas it is rather stable in eucaryotes (cf tRNA). It is synthesized by *DNA-dependent RNA polymerase*.

On account of its heterogeneity, mRNA varies greatly in chain length. Since few proteins contain less than 100 amino acids, the mRNA coding for these proteins must have at least 100×3 or 300 nucleotide residues. In *E. coli*, the average size of mRNA is 900 to 1,500 nucleotide units.

If mRNA carries the codes for the synthesis of simple protein molecule, it is called *monocistronic type* and if it codes for more than one kind of protein, it is known as *polycistronic type* as in *Escherichia coli*.

The mRNAs are unstable in the bacterial systems with a half-life from a few seconds to about 2 minutes. In mammalian systems, however, mRNA molecules are more stable with a half-life ranging from a few hours to one day.

Half-life is the period of time after which half is degraded.

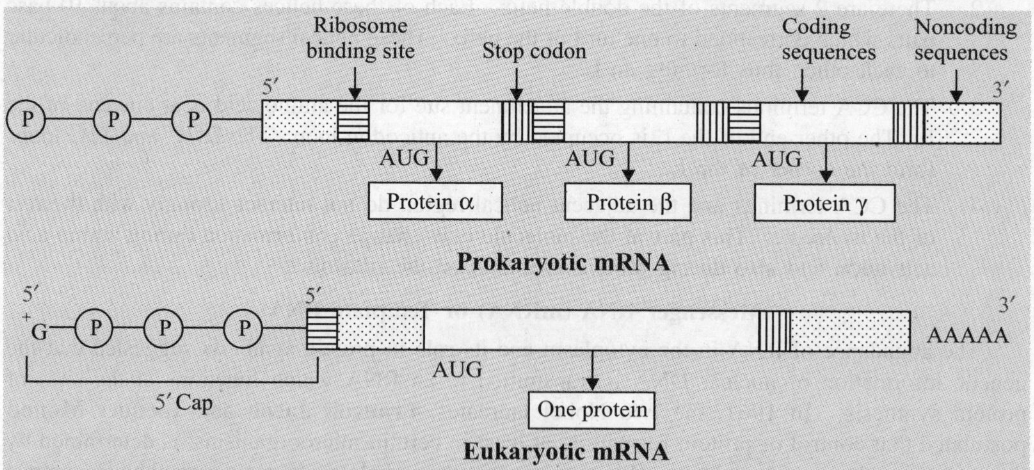

Fig. 15–44. A comparison of the structures of prokaryotic and eukaryotic messenger RNA molecules

The mRNAs are single-stranded and complementary to the sense strand of their respective structural genes. Although both types of mRNA molecules (prokaryotic and eukaryotic) are synthesized with a triphosphate group at the 5′ end, there is a basic difference between the two, as depicted in Fig. 15–44. In eukaryotes, the mRNA molecule immediately acquires a 5′ cap, which is part of the structure recognized by the small ribosomal subunit. Protein synthesis, therefore, begins at the start codon near the 5′ end of the mRNA. On the contrary, in prokaryotes the 5′end has no special significance, and there can be many ribosome-binding sites (called Shine–Dalgarno sequences, see page 739) in the interior of an mRNA chain, each resulting in the synthesis of a different protein.

As stated above, the eukaryotic mRNA molecules, especially those of mammals, have some peculiar characteristics. The 5′ end of mRNA is 'capped' by a 7-methylguanosine triphosphate which is linked to an adjacent 2′-O-methylribonucleoside at its 5′-hydroxyl through the 3 phosphates (Fig. 15–45). Although the function of this capping of mRNA is not well understood, the cap is probably involved in the recognition of mRNA by the translating machinery. The translation of mRNA into proteins begins at the capped 5′ end. The other end of most mRNA molecules, the 3′ hydroxyl end, has attached a polymer of adenylate residues, 20–250 nucleotides in length. The specificity of the poly A "tail" at the 3′ hydroxyl end of mRNA is not understood. It probably serves to maintain the intracellular stability of the specific mRNA.

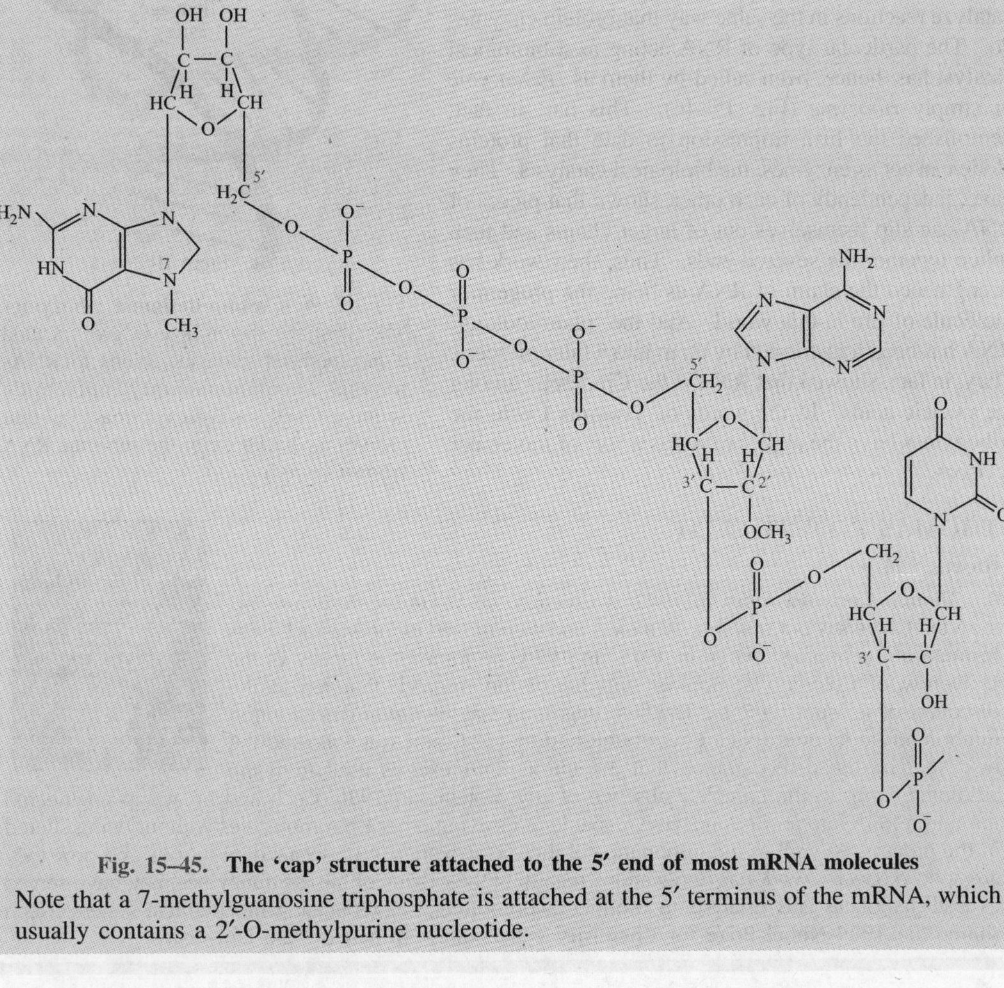

Fig. 15–45. The 'cap' structure attached to the 5′ end of most mRNA molecules
Note that a 7-methylguanosine triphosphate is attached at the 5′ terminus of the mRNA, which usually contains a 2′-O-methylpurine nucleotide.

Heterogeneous Nuclear RNA(hnRNA)

In mammalian cells including those of human beings, a precursor RNA is first synthesized in the nucleoplasm by *DNA-dependent RNA polymerase*. This precursor is then degraded by a *nuclear nuclease* to mRNA that is then translocated to the cytoplasm where it becomes associated to the ribosomal system. This precursor RNA constitutes the fourth class of RNA molecules and is designated as heterogeneous nuclear RNA (*hn*RNA). The *hn*RNA molecules may have molecular weights exceeding 10^7 daltons whereas the mRNA molecules are generally smaller than 2×10^6 daltons. Most mammalian mRNA molecules are 400–4,000 nucleotides in length whereas a *hn*RNA molecule possesses 5,000–50,000 nucleotides. Some uncertainty still exists concerning the precursor-product relationship between *hn*RNA and mRNA, the former being 10–100 times longer than the latter. Thus, the hnRNA molecules appear to be processed to generate the mRNA molecules which then enter the cytoplasm to serve as templates for protein synthesis.

The painstaking researches conducted by the two 1989 Nobel Laureates in Chemistry, **Thomas Cech** from the University of Colorado and **Sidney Altman** from the Yale University, have recently shown that RNA not only carries the blueprint message from the imperious DNA to the factories that churn out proteins but it could catalyze reactions in the same way that protein enzymes do. The particular type of RNA acting as a biological catalyst has, hence, been called by them as *riboenzyme* or simply *ribozyme* (Fig. 15–46). This has, in fact, demolished the firm impression to date that proteins alone can act as enzymes, the biological catalysts. They have, independently of each other, shown that pieces of RNA can slip themselves out of larger chains and then splice together the severed ends. Thus, their work has strengthened the claim of RNA as being the progenitor molecule of life in this world. And the 'plain-looking' RNA has been transformed by them into a fairy princess. They, in fact, showed that RNA is the Cinderella among the nucleic acids. In the words of **Thomas Cech**, the 'ribozymes have the ability to act as a sort of molecular scissors.'

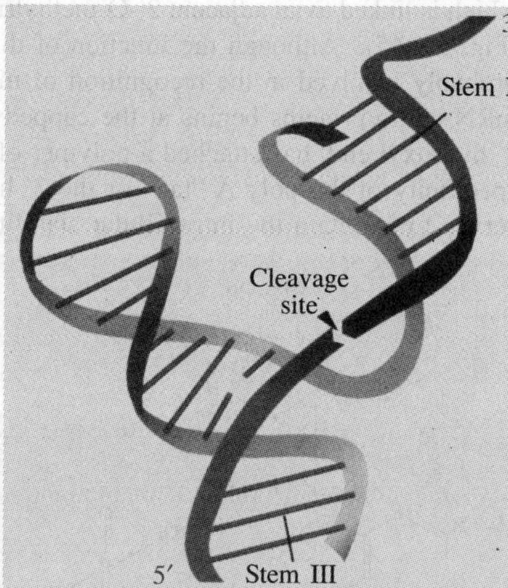

Fig. 15-46. Custom-designed ribozymes
The ribozyme shown here in *green*, called a hammerhead ribozyme, binds to RNAs having a complementary nucleotide sequence and catalyzes a reaction that cleaves the backbone of the substrate RNA (shown in *red*).

THOMAS ROBERT CECH
(born, 1947)

Thomas Cech was born in 1947 at Chicago, USA. He received his PhD from the University of California, Berkeley, and then moved to the Massachusetts Institute of Technology (MIT), in 1975. In 1978, he joined the faculty of the University of Colorado at Boulder and began the research that led to the discovery of self-splicing RNA. The first suggestion that the *Tetrahymena* intron might mediate its own splicing was published in 1981, and was followed the next year by the demonstration that the intron RNA excises itself from the adjoining exons in the complete absence of any proteins. In 1986, Cech and his group engineered the intron RNA into an RNA enzyme, capable of cleaving other RNA molecules without being altered in the process. As well as the importance of these discoveries in understanding introns and how they are spliced, Cech's work has implications regarding the origins of life, as ribozymes may have served as both templates and catalysts to produce a primordial, self-replicating biochemical system. Cech shared the **1989 Nobel Prize for Chemistry** with **Sidney Altman** of Yale University.

In fact, the discovery of ribozymes has led to the concept of an **"RNA world",** in which the primordial biomolecule was RNA. Later researches have revealed that RNA cannot only catalyze excision of introns and splicing of exons, it can even break bonds between amino acids and nucleic acids. Cech has suggested that if RNA can break the bonds, it may also make them. And evidences indicate that RNA-driven protein synthesis may occur even now in certain bacterial systems. Nevertheless, an intricate problem with the notion of an RNA world is the origin of RNA in the first place. RNA is a complex molecule, notoriously difficult to synthesize under ideal conditions in the laboratory. How could it have arisen spontaneously under conditions prevailing in the prebiotic soup ?

INFORMOSOME

Spirin, Beltisina and **Lerman** (1965) have investigated that in certain eucaryotic cells, the mRNA does not enter the cytoplasm as a naked RNA strand but often remains encased by certain proteins. Spirin has proposed the term informosome for this mRNA-protein complex. The proteins bound to informosomes provide stability to mRNA molecule and also protect it from degradation by the enzyme *ribonuclease*.

REFERENCES

1. **Adams RLP, Burdon RH, Cambell AM, Leader DP, Smellie RMS :** The Biochemistry of the Nucleic Acids. *9th ed., Chapman and Hall, London. 1981.*
2. **Bloomfield VA, Crothers DM, Tinoco I :** Physical Chemistry of Nucelic Acids. *Harper and Row, New York. 1974.*
3. **Borst P :** Structure and function of mitochondrial DNA. *Trends Biochem. Sci.* **2 :** *31, 1977.*
4. **Brawerman G :** Eukaryotic messenger RNA. *Ann. Rev. Biochem.*, **43 :** *621, 1974.*
5. **Burhans WC, Weinberger M:** DNA replication stress, genome instability and aging. *Nucleic Acids Res.* **25:** *2045–2056, 2007.*
6. **Cantor CR, Schimmel PR :** Biophysical Chemistry. 3 vols. *W.H.Freeman. 1980.*
7. **Chargaff E, Davidson JN (editors) :** The Nucleic Acids : Chemistry and Biology. *3 vols. Academic Press, Inc., New York. 1955, 1955, 1960.*
8. **Crick FHC :** The structure of the hereditary material. *Sci. Amer.* **191 (4) :** *54-61, 1954.*
9. **Crick FHC :** Nucleic acids. *Sci. Amer.* **197(3) :** *188, 1957.*
10. **Crick FHC, Klug A :** Kinky Helix. *Nature* **255 :** *530, 1975.*
11. **Darnel JE Jr :** RNA. *Sci. Amer.* **253 (4) :** *68-78, 1985.*
12. **Davidson JN :** The Biochemistry of the Nucleic Acids. *8th ed., Chapman and Hall, London. 1976.*
13. **Dickerson RE :** The DNA helix and how it is read. *Sci. Amer.* **249 (6) :** *94-111, 1983.*
14. **Dickerson RE :** DNA structure from A to Z. *Methods Enzymol.* **211 :** *67-111, 1992.*
15. **Doctor et al :** Nucleotide sequence in *E. coli* tyrosine-tRNA. *Science.* **163 :** *693, 1969.*
16. **Falsenfeld G :** DNA. *Sci. Amer.* **253 (4) :** *58-67, 1985.*
17. **Felsenfeld G, Miles HT :** The physical and chemical properties of nucleic acids. *Ann. Rev. Biochem.*, **36 :** *407-448, 1967.*
18. **Fraenkel-Conrat H :** The Chemistry and Biology of Viruses. *Academic Press, Inc., New York. 1969.*
19. **Friedberg EC :** DNA Repair. *W.H. Freeman and Company, New York. 1985.*
20. **Gesteland RF, Atkins JF (editors) :** The RNA World. *Cold Spring Harbour, New York. 1993.*

21. **Grivell LA :** Mitochondrial DNA. *Sci. Amer.* **248 (3) :** *60-73, 1983.*
22. **Hartman SC :** Purines and Pyrimidines. *In Greenberg DM (editor), Metabolic Pathways. (3rd editon) vol. IV, 1-68, Academic Press, Inc., New York. 1970.*
23. **Henderson JF, Paterson ARP :** Nucleotide Metabolism : An Introduction. *Academic Press, Inc., New York. 1973.*
24. **Holley RW :** The nucleotide sequence of a nucleic acid. *Sci. Amer.* **214 (2) :** *30-39, 1966.*
25. **Holley RW** ***et al*** **:** Structure of a ribonucleic acid. *Science.* **147 :** *1462-1465, 1965; also Science.* **153 :** *531-534, 1966.*
26. **Htun H, Dalhberg JE :** Topology and formation of triple-stranded H-DNA. *Science.* **243 :** *1571-1576, 1989.*
27. **Hurst DT :** An Introduction to the Chemistry and Biochemistry of Pyrimidines, Purines, and Pteridines. *Wiley, New York. 1980.*
28. **Hutchinson DW:** Nucleotides and Coenzymes. *Wiley, New York. 1973.*
29. **Jordan :** The Chemistry of Nucleic Acids. *Butterworth, 1960.*
30. **Khan AU, Lal SK:** Ribozymes: a modern tool in medicine. *J. Biomed. Sci.* **10:** *457– 467, 2003.*
31. **Kornberg A :** Biologic synthesis of deoxyribonucleic acid. *Science.* **131 :** *1503-1508, 1960.*
32. **Kornberg A :** DNA Synthesis. *W.H. Freeman and Company, New York. 1974.*
33. **Kornberg A, Baker TA :** DNA Replication. 2nd ed., *W.H. Freeman and Company, New York. 1991.*
34. **Lewin B:** Gene Expression, vol. 2. – Eukaryotic Chromosomes. *2nd ed., 428–440. Wiley, New York. 1980.*
35. **Lewin B:** Genes. *4th ed., Wiley, New York. 1990.*
36. **Mainwaring WIP, Parish JH, Pickering JD, Manna NH :** Nucleic Acid Biochemistry and Molecular Biology. *Blackwell Scientific Publications, Boston. 1982.*
37. **McCarty M :** The Transforming Principle : Discovering that Genes Are Made of DNA. *Norton. 1985.*
38. **McClain WH :** Transfer RNA identity. *FASEB J.* **7 (1) :** *72-78, 1993.*
39. **Michelson AM :** The Chemistry of Nucleosides and Nucleotides. *Academic Press, Inc., New York. 1963.*
40. **Missky AE :** The discovery of DNA. *Sci. Amer.* **218 (6) :** *78, 1968.*
41. **Olby R :** The Path to the Double Helix. *University of Washington Press, Seattle. 1974.*
42. **Orgell LE, Crick FHC :** Selfish DNA : the ultimate parasite. *Nature.* **284 :** *604-607, 1980.*
43. **Perutz MF :** Proteins and Nucleic Acids. *Elsevier Publishing Co., New York. 1962.*
44. **Portugal FH, Cohen JS :** A Century of DNA : A History of the Discovery of the Structure and Function of the Genetic Substance. *MIT Press. 1977.*
45. **Ptashne MA:** Genetic Switch, *2nd ed., Cell Press and Blackwell Press, Cambridge, M.A. 1992.*
46. **Rich A, Kim H :** The three-dimensional structure of transfer RNA. *Sci. Amer. January, 1978.*
47. **Rich A, Nordheim A, Wang AH-J :** The chemistry and biology of left-handed Z-DNA. *Ann. Rev. Biochem.* **53 :** *791-846, 1984.*
48. **Rich A, Raj Bhandari UL :** Transfer RNA : Molecular structure, sequence and properties. *Ann. Rev. Biochem.,* **45 :** *805, 1976.*
49. **Robertus JD** ***et al*** **:** Structure of yeast phenylalanine tRNA at 3 Å resolution. *Nature.* **250 :** *546-551, 1974.*

50. **Saenger W :** Principles of Nucleic Acid Structure. *Springer-Verlag, New York. 1984.*
51. **Sayre A :** Rosalind Franklin and DNA. *W.W. Norton & Co., Inc., New York. 1978.*
52. **Scheit KH :** Nucleotide Analogs : Synthesis and Biological Function. *John Wiley and Sons, New York. 1980.*
50. **Schimmel PRD, Soll D, Abelson JN (editors) :** Transfer RNA : Structure, Properties and Recognition. *Cold Spring Harbour, New York. 1979.*
51. **Singer MF, Leder P :** Messenger RNA : An evaluation. *Ann. Rev. Biochem.,* **35** *: 195-230, 1966.*
52. **Smith MM :** Histone structure and function. *Curr. Opin. Cell Biol.* **3** *: 429-437, 1991.*
53. **Spencer JH :** The Physics and Chemistry of DNA and RNA. *W.B. Saunders, Philadelphia. 1972.*
54. **Stewart PR, Letham DS (editors) :** The Ribonucleic Acids. *2nd ed., Springer-Verlag, New York 1977.*
55. **Venter JC, Adams MD, Myers EW :** The sequence of the human genome. *Science.* **291:** *1304–1351, 2001.*
56. **Watson JD:** A passing for DNA. Genes, genomes and society. *Oxford university Press, Oxford. 2000.*
57. **Watson JD:** The double helix: a personal account of the discovery of the structure of DNA. *WW Norton, New York. 1980.*
58. **Watson JD :** The Double Helix. *Atheneum. 1968.*
59. **Watson JD, Crick FHC :** Molecular structure of nucleic acid. A structure for deoxyribose nucleic acid. *Nature.* **171** *: 737-738, 1953 a.*
60. **Watson JD, Crick FHC :** Genetic implications of the structure of deoxyribonucleic acid. *Nature,* **171** *: 964-967, 1953 b.*
61. **Watson JD, Hopkins NH, Roberts JW, Steitz JA, Weiner AM :** Molecular Biology of the Gene. *4th ed., The Benjamin/Cummings Publishing Company, Menlo Park, CA. 1987.*
62. **Wells RD, Harvey SC (editors) :** Unusual DNA Structures. *Springer-Verlag, New York. 1988.*
63. **Wilkins M :** Molecular Configuration of Nucleic Acids. *Science.* **140** *: 941, 1963.*
64. **Younghusband HB, Inman RB :** The electron microscopy of DNA. *Ann. Rev. Biochem.* **43** *: 605, 1974.*

PROBLEMS

1. Write the complementary sequence (in the standard 5′ → 3′ notation) for (*a*) GATCAA, (*b*) TCGAAC, (*c*) ACGCGT, and (*d*) TACCAT.
2. The composition (in mole-fraction units) of one of the strands of a double-helical DNA molecule is [A] = 0.30 and [G] = 0.24. (*a*) What can you say about [T] and [C] for the same strand? (*b*) What can you say about [A], [G], [T], and [C] of the complementary strand?
3. (*a*) Write the sequence of the mRNA molecule synthesized from a DNA template strand having the sequence:

   5′-ATCGTACCGTTA-3′

   (*b*) What amino acid sequence is encoded by the following base sequence of an mRNA molecule? Assume that the reading frame starts at 5′ end.

   5′-UUGCCUAGUGAUUGGAUG-3′

   (*c*) What is the sequence of the polypeptide formed on addition of poly(UUAC) to a cell-free protein-synthesizing system?

4. RNA is readily hydrolyzed by alkali, whereas DNA is not. Why?
5. The transition from B-DNA to Z-DNA occurs over a small change in the superhelix density, which shows that the transition is highly cooperative.
   (*a*) Consider a DNA molecule at the midpoint of this transition. Are B- and Z-DNA regions frequently intermingled or are there long stretches of each?
   (*b*) What does the finding reveal about the energetics of forming a junction between the two kinds of helices?
   (*c*) Would you expect the transition from B- to A-DNA to be more or less cooperative than the one from B- to Z-DNA? Why?
6. The sequence of part of an mRNA is:
   5′-AUGGGGAACAGCAAGAGUGGGGCCCUGUCCAAGGAG-3′
   What is the sequence of the DNA coding strand? Of the DNA template strand?
7. Why is RNA synthesis not as carefully monitored for errors as is DNA synthesis?
8. Why is it advantageous for DNA synthesis to be more rapid than RNA synthesis?
9. A viral DNA is analyzed and found to have the following base composition, in mole present: A = 32, G = 16, T = 40, C = 12.
   (*a*) What can you immediately conclude about this DNA?
   (*b*) What kind of secondary structure do you think it would have?
10. Given the following sequence for one strand of a double-strand oligonucleotide:
    $^{5'}$ACCGTAAGGCTTTAG$^{3'}$
    (*a*) Write the sequence for the complementary DNA strand.
    (*b*) Suppose you knew that the strand shown above had phosphate on both ends. Using an accepted nomenclature, write the sequence so as to show this.
    (*c*) Write the sequence of the RNA complementary to the strand shown above.
11. What positions in a purine ring have the potential to form hydrogen bonds, but are not involved in the hydrogen bonds of Watson–Crick base pairs?
12. Write the base sequence of the complementary strand of double-helical DNA in which one strand has the sequence (5′)ATGCCCGTATGCATTC(3′).
13. Calculate the weight in grams of a double-helical DNA molecule stretching from the earth to the moon (~320,000 km). The DNA double helix weighs about 1×10^{-18} g per 1,000 nucleotide pairs; each base pair extends 0.34 nm. For an interesting comparison, your body contains about 0.5 g of DNA!
14. Hairpins may form at palindromic sequences in single strands of either RNA or DNA. How is the helical structure of a hairpin in RNA different from that of a hairpin in DNA?
15. In the cells of many eukaryotic organisms, there are highly specialized systems that specifically repair G–T mismatches in DNA. The mismatch is repaired to form a G≡C base pair (not A=T). This G–T mismatch repair system occurs in addition to a more general system that repairs virtually all mismatches. Can you think of a reason why cells require a specialized system to repair G–T mismatches?
16. Explain why there is an increase in the absorption of UV light (hyperchromic effect) when double-stranded DNA is denatured.
17. In samples of DNA isolated from two unidentified species of bacteria, adenine makes up 32 and 17%, respectively, of the total bases. What relative proportions of adenine, guanine, thymine, and cytosine would you expect to find in the two DNA samples? What

assumptions have you made ? One of these bacteria was isolated from a hot spring (64 °C). Which DNA came from this thermophilic bacterium ? What is the basis for your answer?

18. In the coding strand of DNA for the alpha gene of normal hemoglobin (HbA), the three bases that correspond to position 142 of the mRNA synthesized are TAA and the alpha chain has 141 amino acids. In the coding strand of the gene for the alpha chain of hemoglobin Constant Spring, the three bases in the same position as above are CAA and the alpha chain produced contains 172 amino acids. Explain the mutation that has occurred. Is the mutation a frameshift or a point mutation?

19. How would you measure the turnover of DNA?

20. Write the complementary sequence (in the standard 5′ → 3′ notation) for:

(*a*) GATCAA

(*b*) TCGAAC

(*c*) ACGCGT

(*d*) TACCAT

21. Write the sequence of the mRNA molecule synthesized from a DNA template strand having the sequence:

5′-ATCGTACCGTTA-3′

22. Which of the following DNA shows left-handed double helix?

(*a*) Z-DNA (*c*) A-DNA

(*b*) B-DNA (*d*) A-DNA and B-DNA

23. Stability of a DNA is determined by its

(*a*) Number of adenine-guanine bonds

(*b*) Number of guanine-cytosine bonds

(*c*) Number of phosphodiester bonds

(*d*) None of the above

24. Who produced the first synthetic DNA in a test tube?

25. Who synthesized a complete gene with regulatory sequences and without any template?

26. The structure presented by **Watson** and **Crick** is of which type of DNA.

ANSWERS TO PROBLEMS

1. (*a*) TTGATC; (*b*) GTTCGA; (*c*) ACGCGT; (*d*) ATGGTA.

2. (*a*) [T] + [C] = 0.46, (*b*) [T] = 0.30, [C] = 0.24, and [A] + [G] = 0.46.

3. (*a*) 5′ -UAACGGUACGAU-3′

(*b*) Leu-Pro-Ser-Asp-Trp-Met

(*c*) Poly(Leu-Leu-Thr-Tyr)

4. The 2′ -OH group in RNA acts as an intramolecular nucleophil. In the alkaline hydrolysis of RNA, it forms a 2′-3′ cyclic intermediate.

5. (*a*) There are long stretches of each because the transition is highly cooperative.

(*b*) B–Z junctions are energetically highly unfavourable.

(*c*) A–B transitions are less cooperative than B–Z transitions because the helix stays right-handed at an A–B junction but not at a B–Z junction.

6. The sequence of the coding (+, sense) strand is:

5′–ATGGGGAACAGCAAGAGTGGGGCCCTGTCCAAGGAG–3′

and the sequence of template (–, antisense) strand is:

3′–TACCCCTTGTCGTTCTCACCCCGGGACAGGTTCCTC–5′

7. An error will only affect one molecule of mRNA of many synthesized from a gene. In addition, the errors do not become a permanent part of the genomic information.

8. At any given instant, only a fraction of the genome (total DNA) is being transcribed. Consequently, speed is not necessary.

9. (*a*) It must be a single-strand DNA, since Chargaff's rules are not obeyed.

(*b*) It should be mostly random coil, with perhaps some self-bonding into hairpins, etc.

10. (*a*) $^{3'}$TGGCATTCCGAAATC$^{5'}$.

(*b*) $^{5'}$pApCpCpGpTpApApGpGpCpTpTpTpApGp$^{3'}$.

(*c*) $^{3'}$UGGCAUUCCGAAAUC$^{5'}$.

11. N-3, N-7, and N-9.

12. (3′) TACGGGCATACGTAAG (5′).

13. 0.94×10^{-3} g

14. The RNA helix will be in the A conformation; the DNA helix will generally be in the B conformation.

15. In eukaryotic DNA, about 5% of cytosine residues are methylated. 5-methylcytosine can spontaneously deaminate to form thymine, and the resulting G–T pair is one of the most common mismatches in eukaryotic cells.

16. Base stacking in nucleic acids tends to reduce the absorption of UV light. Denaturation of DNA involves the loss of base stacking, and UV absorption increases.

17. One DNA contains 32% A, 32% T, 18% G, and 18% C; the other 17% A, 17% T, 33% G, and 33% C. This assumes that both are double-stranded. The DNA containing 33% G and 33% C almost certainly came from the thermophilic bacterium; its higher G ≡ C content makes it much more stable to heat.

18. The coding strand of the gene has the same sequence as the mRNA (except U replaces T in the RNA). In HbA, the codon at position 142 of mRNA is a stop codon so the last amino acid added is 141. In Hb Constant Spring, a point mutation has mutated the DNA so that the mRNA codon at 142 now codes for an amino acid instead of stop. Translation continues until a stop codon appears at position 173 (so 172 amino acids). This could be a transition mutation– pyrimidine for pyrimidine.

19. Thymine is derived almost exclusively from DNA. The degradation of thymine leads to the unique product β-aminoisobutyrate, which is excreted. This can be measured in urine.

20. (*a*) TTGATC

(*b*) GTTCGA

(*c*) ACGCGT

(*d*) ATGGTA

21. 5′–UAACGGUACGAU–3′

22. (*a*) Z-DNA

23. (*b*) Number of guanine-cytosine bonds

24. **Arthur Kornberg**

25. **Har Gobind Khorana**

26. Right-handed B-DNA

CONTENTS

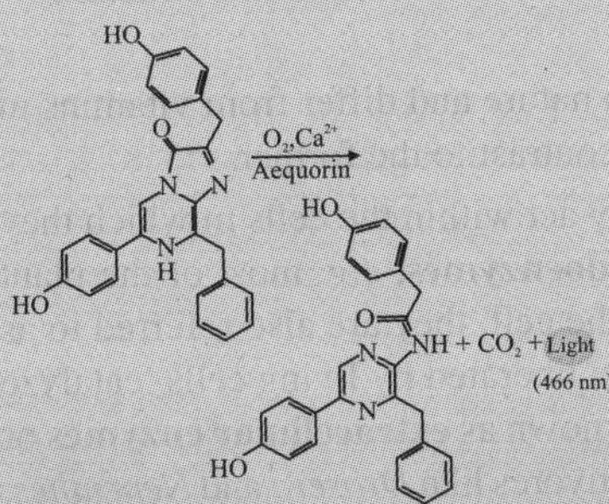

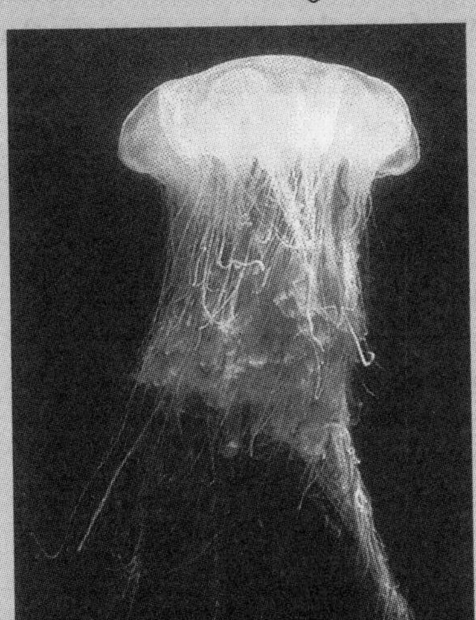

The activity of an enzyme (shown *above*) is responsible for the glow of the luminescent jellyfish (shown *below*).
The enzyme *aequorin* catalyzes the oxidation of a compound by oxygen in the presence of calcium to release CO_2 and light.
[*Courtesy: (Lower) Fred Bavendam/ Peter Arnold*]

CHAPTER

16

Enzymes–I Nomenclature and Classification

DEFINITION

Life is an intricate meshwork involving a perfect coordination of a vast majority of chemical reactions. Some of these reactions result in synthesizing large molecules, others in cleaving large molecules and all of them either utilize energy or liberate energy. All these reactions occur very slowly at the low temperatures and the atmospheric pressures– the conditions under which living cells carry on their life processes. Yet in the living cells these reactions proceed at extremely high rates. This is due to the presence of some catalysts produced and synthesized inside the body of the organisms. The term 'enzyme' was coined in 1878 by **Friedrich Wilhelm Kuhne** to designate these **'biological catalysts'** that had previously been called 'ferments'. As they quicken most of the chemical reactions occurring in the body, the enzymes have been designated as "*the manifestations of nature's impatience*".

> In fact, **Kuhne** intended for the name **enzyme** to apply to both yeast ferment and the extracellular catalysts of more complex animals, *e.g., pepsin and trypsin*. Prophetically, his definition specifically implied that lower and higher forms of life are not so fundamentally different.

The name **'enzyme'** (en^G = in ; $zyme^G$ = yeast) literally means 'in yeast'. This was referred to denote one of the most noteworthy reactions wherein the production of ethyl alcohol and carbon dioxide through the agency of an enzyme, the *zymase*, present in yeast takes place. This reaction is most popularly known as *alcoholic fermentation*. **Sumner** and **Myrbäck** (1950) have beautifully defined the enzymes as 'simple or combined proteins acting as specific catalysts'. They are soluble, colloidal molecules which are produced by living cells. All enzymes are globular proteins with a complex 3-'D' structure, capable of binding substrate molecules to a part of their surface. They affect the life of an organism to such an extent that life has aptly been called as "an orderly function of enzymes."

As an analogy, if one considers a living cell as a factory, the individual enzymes might be compared to the machines that work together to cause the transformation of a raw material (like steel) into parts of a finished product (like an automobile).

Difference from catalysts. Like catalysts, the enzymes do not alter the chemical equilibrium point of a reversible reaction but only the speed of the reaction is changed. They, however, differ from catalysts in being the biological products, *i.e.*, produced from the living cells. Moreover, the enzymes are all protein and, unlike catalysts, cannot last indefinitely in a reaction system since they, being colloidal in nature, often become damaged or inactivated by the reactions they catalyze. Henceforth, they must be replaced constantly by further synthesis in the body. Furthermore, unlike catalysts, most individual enzymes are very specific in that they act either on a single or at the most on some structurally-related substrates.

Difference from vitamins. All enzymes are proteinaceous in nature and differ from vitamins in the fact that the latter are not synthesized by the animal cells, in contrast to the former.

Endoenyzmes and exoenyzmes. Most of the enzymes usually act within the cells in which they are produced and hence are called **intracellular enzymes** or **endoenzymes**, *e.g.* most of the plant enzymes. As these enzymes catalyze the metabolic reactions of the cell, they are also referred to as *metabolic enzymes*. On the other hand, certain enzymes which are liberated by living cells, catalyze useful reactions outside the cell in its environment and hence are known as **extracellular enzymes** or **exoenzymes**, *e.g.*, enzymes found in bacteria, fungi and some insectivores like *Drosera* and *Nepenthes*. They act chiefly as digestive enzymes, catalyzing the breakdown of complex substances to simpler ones which can readily be absorbed by the cell.

LOUIS PASTEUR
(LT, 1822–1895)

Louis Pasteur, a French microbiologist and chemist, was born on December 27,1822 in Dole France into a family of poor tanners. His father Joseph Pasteur was a sergent-major in Nepoleon's army and his valour had earned him the Legion of Honour. Louis did not seem to be particularly outstanding at school but was an excellent artist and made numerous portraits of his family. He studied at the Paris school, Ecole Normale Superieure, a school founded by Napoleon to train professors, and graduated as a Ph.D. in Sciences. In 1848, he achieved distinction in organic chemistry for his discovery that tartaric acid, a C-4 compound, forms two different types of crystals and successfully separated the two forms while looking through a microscope. He formulated a fundamental law: asymmetry differentiates the organic world from the mineral world. His work became the basis of a new science– **stereochemistry**. In January 1849, he became an Assistant Professor of Chemistry at the University of Strasborough and soon after in May, 1849 married Marie Laurent, daughter of the University principal. She often discussed his work, spurred his thinking and was one of the best scientific collaborators. The couple had 5 children, but only 2 survived to adulthood. And in September 1854, at the age of 32, he was appointed full-fledged Professor of Chemistry and Dean at the University of Lille in Northern France.

Three years later (*i.e.*, in 1857), he was called back and appointed as Research Director at the Ecole Normal on the Rue d'Ulm in Paris. As a researcher, he showed remarkable observational skill which allowed him to discover what other people looked at but did not see.

The same year, 1857, he unfolded the mystery of why local wines were turning sour. The prevailing theory held that wine fermentation results from the chemical breakdown of grape juice to alcohol. Pasteur, however, saw yeast cells under microscope and believed that yeasts played a major role in fermentation. In a classic series of experiments, he classified the role of **yeasts** in fermentation and showed that the sticks and rod-like structures (now known as bacteria) were responsible for making the wine sour. Pasteur's work also indicated that the bacteria could be a cause of disease, for if they could sour the wine, perhaps they could also make the body ill or diseased. He, henceforth, held the view that microorganisms are responsible for infectious diseases. He, thus, set down the foundation for the **germ theory of disease** and also founded the science of **Bacteriology.**

Pasteur also recommended a practical solution to the sour wine problem. He suggested that grape juice be heated to 55°C for several minutes to destroy all the evidence of life, after which yeasts could be added to begin the fermentation. Acceptance of this technique, known as **pasteurization** (which is also applied to milk), gradually ended the problem. Pasteur's elation was tempered with sadness as his daughter Jeanne died of typhoid fever in 1859.

Finally, in an elegant series of experiments using swan-neck flasks, Pasteur successfully performed his meticulous experiments in public for the Academy of Science and silenced all but the ardent supporters of spontaneous generation hypothesis. By now, Pasteur became a national celebrity. And the theory of spontaneous generation was given a burial after holding sway for 20 centuries! In 1862, he was elected to the Academy. Once again, tragedy struck him as his two-year-old daughter **Camille** developed a tumour and died of blood poisoning in September 1865. Pasteur, thus, realized that he was no closure to solving the riddle of disease.

The same year (*i.e.*, 1865) **cholera** engulfed Paris, killing about 200 people daily! Pasteur tried to capture the responsible bacterium by filtering the hospital air and trapping the bacterium in cotton. Unfortunately, he was unable to cultivate one bacterium apart from the others as he was using broth (Later, **Robert Koch** used solid culture media, instead of broth media).

In order to help the ailing French industry, Pasteur turned to pebrine, the disease of silkworms. Late in 1865, he identified a protozoan infesting the sick silkworms and the mulberry leaves fed to them. Then, he separated the healthy silkworms from the diseased ones and their food, and he managed to check the spread of disease. His success further endorsed the germ theory of disease. It was again a terrible blow to Pasteur when twelve-year-old **Cecille**, another of his daughters died of typhoid fever in May 1866. This diverted his attention again to the study of human diseases, and he worked still harder. He himself suffered a brain hemorrhage which left him paralyzed on the left side of his body. But all this could not keep him down for long. In 1873, he was elected to the Academy of Medicine. Later, he developed **vaccines** against chicken cholera in poultry (1880), anthrax in sheep (1881) and most notably rabies in dogs (1885). He also discovered 3 bacteria responsible for human illnesses: *Staphyllococcus, Streptococcus* and *Pneumococcus.*

In fact, Pasteur's genius reached a peak with his development of a vaccine for rabies ('rage' or 'madness' in Latin). To prepare the vaccine, he inoculated rabbits with the brain tissue of rabid animals. He then removed the spinal cords and dried them. Next, he inoculated experimental animals with 15-day-old cord tissue and followed this up the next day with 14-day-old cord tissue and so on for two weeks. Animals so treated did not develop rabies.

Pasteur was a sturdy and unitiring worker who drove himself and his subordinates mercilessly. At heart, however, he was quite sentimental as the following event displayed. On July 6, 1885, a 9-year-old boy named Joseph Meister was brought to Pasteur. Only two days ago, the boy had suffered as many as 18 bites from a rabid dog, and physicians assured Pasteur of the imminent horrible death, to the boy. The boy's only hope was the vaccine. But Pasteur's vaccine had never been tested on humans, and the scientist, now 63, could not bring himself to make an immediate decision. In his writing, he recalls the incident:

> "The child's death appeared inevitable. I decided not without acute and harrowing anxiety, as may be imagined, to apply to Joseph Meister the method I had found consistently successful with dogs."

The next day, Pasteur began the treatment. He turned the first sample of vaccine over to two physicians from the Academy of Medicine and watched as they gave 12 successive injections in 10 days of the vaccine to the terrified, crying child. Each day his concern lessened as the injections proceeded smoothly. The vaccine appeared to be working. Meister survived his ordeal and the scientist heaved a sigh of relief. When Meister left Paris for his village, Pasteur gave him stamped envelopes so that he could write often. Pasteur's viewpoint in dealing with a disease was well known. In his own words:

> "When meditating over a disease, I never think of finding a remedy for it, but, instead, a means of preventing it."

The defeat of rabies was Pasteur's crowing glory. After Pasteur saved the boy Joseph Meister from the dreaded rabies in 1885, French government on November 4,1888, established the **Pasteur Institute** in Paris to treat cases of rabies, one of the world's foremost scientific establishments. The institute was inaugurated by the then President of France Sadi Carnot. Many monetary rewards were floated towards the institute, including a generous gift from the Russian government after Pasteur immunized 20 Russians against rabies. Louis continued working till 1887, when he suffered another paralytic stroke which prevented him from personally doing experimental work but his dialogue with his pupils and his collaborators never ceased. Pasteur presided over the Institute until his death on September 28, 1895 in Villeneuvel Etang. He was given a State Funeral at the Cathedral of Notre Dame and his body placed in a permanent crypt at the Pasteur Institute. He would teach his disciples:

> "Do not put forward anything that you cannot prove by experimentation."

For nearly half a century, he had dominated the scientific world; for a quarter of a century, he had surged ahead despite a half-paralyzed body. Such was his genius! In later years, a grateful Meister returned to Paris as caretaker of the Pasteur Institute and in 1940 preferred commiting suicide rather than obey the demands of some occupying Hitler's Nazi soldiers to open Pasteur's crypt.

Pasteur was perhaps best known to the French nation as the **'saviour of the wine industry'** because his pasteurization salvaged an ailing industry beset with problems of microbial contamination. He is truly called as the **'father of microbiology'**.

The life of Pasteur stands as a supreme testimony to the fact that the best and most far-reaching applications of science come from passionate studies of seemingly esoteric subjects. He evolved into a prophet whose vision accelerated the progress of science. He was a true **benefactor of mankind**.

Pasteur's famous quote reads as follows:

"Dans les champs de l'observation, le hasard ne favorise que les esprits prepares."

The English transliteration is:

"In the field of observation, chance favours only prepared minds."

HISTORICAL RESUME

The history of enzymes may be regarded as commencing with the work of **Dubrunfaut** (1830) who prepared malt extract from germinating barley seeds. This extract possessed the power of converting starch into sugar. Later in 1833, **Payen** and **Persoz** prepared an enzyme, the *diastase* (now known as *amylase*), from malt extract by precipitation from alcohol. The same year, **Horace de Saussure** prepared a substance from germinating wheat which acted like diastase, *i.e.*, converted starch into sugar. Within the next few years, **Theodor Schwann** succeeded in extracting *pepsin*, which digests meat (protein), from gastric juice and he later identified *trypsin*, a peptidase in digestive fluids. Hence, the notion of diastases (the early name for enzymes) was soon extended to animals. By 1837, the famous chemist **Jönes Jacob Berzelius** recognized with remarkable foresight the catalytic nature of these biological diastases.

> **Anselme Payen**, a chemist-industrialist was the owner of the sugar factory in Paris. When he and **Jean-Francois Persoz** called the first known biological catalyst "diastase", they introduced the now historical use of the suffix **-ase** in the naming of most enzymes.

Later, **Pasteur** (1857) demonstrated that alcoholic fermentation was brought about by the action of living yeast cells. It, thus, became apparent that such catalytic actions could be induced by the action of living cells (as in alcoholic fermentation) or by nonliving substances (*as diastase* and *emulsin*). Obviously, such catalysts or ferments were regarded as forming two classess : **organized ferments** (like yeasts and certain bacteria) which were living cells and **unorganized ferments** (like diastase and emulsin) which acted independently of living cells.

In fact, the term 'enzyme' was later proposed for the unorganized ferments by **Kuhne**. On the contrary, **Eduard Buchner** and his brother **Hans Buchner** (1897) showed that, besides the living yeast cells, even their extract (which Eduard named as *zymase*) could bring about alcoholic fermentation. In fact, the Buchners discovered cell-free fermentation when they attempted to preserve their yeast extracts

EDUARD BUCHNER
(LT, 1860–1917)

A German biochemist, studied under Prof. Adolf von Baeyer and later became his assistant. He received the **Nobel Prize in Chemistry in 1907** "for his biochemical researches and his discovery of cell-less fermentation." In fact, Buchner, in 1897, reported that he had prepared from brewer's yeast, a cell-free press juice that caused CO_2 and ethyl alcohol to form in solutions of various sugars (sucrose, glucose, fructose, maltose). And he concluded:

> "the initiation of the fermentation process does not require so complicated an apparatus as is represented by the yeast cell. The agent responsible for the fermenting action of the press juice is rather to be regarded as a dissolved substance, doubtless a protein; this will be denoted **zymase.**"

In fact, his discovery of zymase was the first proof that fermentation was caused by enzymes and did not require the presence of living cells. Eduard, a major in the German army, died in action on the Romanian front (1917) during World War I.

JAMES BATCHELLER SUMNER
(LT, 1887–1955)

After Sumner lost his left arm as a result of hunting at age 17, he was discouraged by his teachers from pursuing a career in Chemistry. They felt that Sumner was too handicapped for the profession. However, his illustrious career belied all this. He is truly called as the **'father of modern enzymology'.** Sumner received the **1946 Nobel Prize in Chemistry** for crystallizing the first enzymes, along with his compatriots **John H. Northrop** and **W.M. Stanley** for his work on preparation of enzymes and virus proteins in pure form.

with sugar, the preservative of jellies and jams. The words– ferment and enzyme– thus became synonym and the latter is now frequently used in the literature. The use of the term ferment (*fermentare*[L] = to agitate) is, however, misleading as it also denotes the fermenting microorganisms.

James B. Sumner (1926), at Cornell University, for the first time isolated and purified an enzyme, the *urease*, in crystalline form from jack bean (*Canavalia ensiformis*), thus confirming the proteinaceous nature of the enzymes. Since then, some 250 enzymes have been obtained in pure crystalline form such as :

(*a*) Trypsin from beef pancreas by **John Howard Northrop** and **Kunitz** (1936)
(*b*) Catalase from beef liver by **Sumner** and **Dounce** (1937)
(*c*) RNase from beef pancreas by **Kunitz** (1940)
(*d*) Pepsin from swine stomach by **John Howard Northrop** (1946)
(*e*) DNase from beef pancreas by **Kunitz** (1950)

NOMENCLATURE AND CLASSIFICATION

With the continuous increase in our knowledge of enzymology, various systems have evolved to name and classify the enzymes, using one or the other criterion as the basis. However, many of the enzymes were known before these systems of naming enzymes were adopted. The names of such enzymes were not changed under the new systems. In this category belong: bromelin, chymotrypsin, diastase, emulsin, papain, pepsin, ptyalin, rennin, trypsin etc.

A. Substrate acted upon by the enzyme. The substance upon which an enzyme acts is called the substrate. **Duclaux** (1883) named the enzymes by adding the suffix *-ase* in the name of the substrate catalyzed. For example, enzymes acting upon carbohydrates were named as *carbohydrases*, upon proteins as *proteinases*, upon lipids as *lipases*, upon nucleic acids as *nucleases* and so on. A few of the names were even more specific like *maltase* (acting upon maltose), *sucrase* (upon sucrose), *urease* (upon urea), *lecithinase* (upon lecithin), *tyrosinase* (upon tyrosine) etc.

B. Type of reaction catalyzed. The enzymes are highly specific as to the reaction they catalyze. Hence, this has necessitated their naming by adding the suffix-*ase* in the name of the reaction; for example *hydrolases* (catalyzing hydrolysis), *isomerases* (isomerization), *oxidases* (oxidation), *dehydrogenases* (dehydrogenation), *transaminases* (transamination), *transaldolases* (transaldolation), *transketolases* (transketolation), *phosphorylases* (phosphorylation) etc.

In fact, the consensus at that time in Europe was that **enzymes** were not proteins, because of the findings of the influential German chemist **Richard Willstätter** (Nobel Laureate, 1915), who, in 1920s, reported that he could not detect protein in purified enzyme preparations from yeast. With hindsight, it is realized that the protein assays used in that era were not sensitive enough to detect the small amounts present in Willstätter's purified preparations. The nonprotein nature of enzyme was so entrenched in scientific thinking that decades passed before their polypeptidyl composition was unequivocally accepted.

Although these two systems are quite simple and easy to follow, there are certain discrepancies present in them. The former system does not take into account the type of the reaction catalyzed, whereas in the latter system no idea can be derived regarding the nature of the substrate acted upon by the enzyme.

C. Substrate acted upon and type of reaction catalyzed. The names of some enzymes give clue of both the substrate utilized and the type of reaction catalyzed. For example, the enzyme *succinic dehydrogenase* catalyzes the dehydrogenation of the substrate succinic acid. Similarly, *L-glutamic dehydrogenase* indicates an enzyme catalyzing a dehydrogenation reaction involving *L-glutamic* acid.

D. Substance that is synthesized. A few enzymes have been named by adding the suffix *-ase* to the name of the substance synthesized, *viz.*, *rhodonase* that forms rhodonate irreversibly from hydrocyanic acid and sodium thiosulphate, and also *fumarase* that forms fumarate irreversibly from L-malate.

E. Chemical composition of the enzyme. Based on their chemical composition, the enzymes have been classified into following three categories:

1. Enzyme molecule consisting of protein only— *e.g.*, pepsin, trypsin, urease, papain, amylase etc.
2. Enzyme molecule containing a protein and a cation— *e.g.*, carbonic anhydrase (containing Zn^{2+} as cation), arginase (Mn^{2+}), tyrosinase (Cu^{2+}) etc.
3. Enzyme molecule containing a protein and a nonprotein organic compound known as prosthetic group—**Tauber** (1950) has further subdivided them, on the basis of the nature of prosthetic group involved:
   (*a*) Iron prophyrin enzymes— catalase, cytochrome *c* peroxidase I and II.
   (*b*) Flavoprotein enzymes— glycine oxidase, pyruvate oxidase, histamine.
   (*c*) Diphosphothiamin enzymes — β-carboxylase, pyruvate mutase.
   (*d*) Enzymes requiring other coenzymes— phosphorylase, amino acid decarboxylase.

F. Substance hydrolyzed and the group involved.

1. Carbohydrate-hydrolyzing enzymes
   (*a*) Glycosidases—cellulase, amylase, sucrase, lactase, maltase
   (*b*) β-glucorinidase
2. Protein-hydrolyzing enzymes
   (*a*) Peptide bonds
      I. Endopeptidases
      Animals— pepsin, trypsin, rennin
      Plants—papain ficin, bromelin
      II. Exopeptidases—dipeptidase, tripeptidase
   (*b*) Nonpeptide C—N linkages (amidases)
   urease, arginase, glutaminase
3. Lipid-hydrolyzing enzymes
   lipases, esterases, lecithinases
4. Other ester-hydrolyzing enzymes
   (*a*) Phosphatases
   (*b*) Cholinesterases
   (*c*) Chlorophyllases
   (*d*) Sulfatases
   (*e*) Pectinesterases
   (*f*) Methylases
5. Oxidation-reduction enzymes
   hydrases, mutases, oxidases, dehydrogenases, peroxidases
6. Miscellaneous enzymes
   catalase, carboxylase, carbonic anhydrase, thiaminase, transpeptidase

> The I.U.B. system is based on the report of the Commission on Enzymes, International Union of Biochemistry Symposium, Vol. 20, Pergamon Press, New York, 1961. In fact, the **international symbolic language of chemistry** was originally developed by **Jönes Jacob Berzelius**, who proposed that an element be identified by the initial letter or first two letters of its Latin name.

G. Over-all chemical reaction taken into consideration. The chemical reaction catalyzed is the specific property which distinguishes one enzyme from another. In 1961, *International Union of Biochemistry* (*I.U.B.*) used this criterion as a basis for the classification and naming of enzymes. Although complicated, the I.U.B. system is precise, descriptive and informative.

The *major features* of this system of classification of enzymes are as follows:

(*a*) The reactions and the enzymes catalyzing them are divided into 6 major classes, each with 4 to 13 subclasses.

(*b*) Each enzyme name has 2 parts—the first part is the name of the substrate(s) and the second part which ends in the suffix *-ase*, indicates the type of reaction catalyzed.

(*c*) Additional information regarding the nature of the reaction, if needed, is given in parenthesis. For example, the enzyme *malate dehydrogenase* catalyzes the following reaction:

$$\text{L-malate} + \text{NAD}^+ \longrightarrow \text{Pyruvate} + \text{CO}_2 + \text{NADH} + \text{H}^+$$

This enzyme has now been designated as L-malate : NAD oxidoreductase (decarboxylating).

(*d*) Each enzyme has been alloted a systemic code number called Enzyme Commission (E.C.) number. The E.C. number for each enzyme consists of a series of numbers at 4 places : the first place numbers representing the major class to which the enzyme belongs, the two median numbers denoting the subclass and the sub-subclass of the enzyme within the major class. The last place number or the fourth digit represents the serial number of the enzyme within the sub-subclass. Thus E.C. 2.7.1.1 represents class 2 (a transferase), subclass 7 (transfer of phosphate), sub-subclass 1 (an alcohol group as phosphate acceptor). The final digit denotes the enzyme, *hexokinase* or *ATP*: *D-hexose-6-phosphotransferase*. This enzyme catalyzes the transfer of phosphate from ATP to the hydroxyl group on carbon 6 of glucose.

$$\text{ATP} + \text{D-hexose} \xrightarrow{\text{Hexokinase}} \text{ADP} + \text{Hexose-6-phosphate}$$

(*e*) Where no specific category has been created for an enzyme, it is listed with a final figure of 99 in order to leave space for new subdivisions. For example, 4.2.99 refers to "other carbon-oxygen lyases."

The 6 major classes of enzymes with some important examples from some subclasses are described below:

1. Oxidoreductases. This class comprises the enzymes which were earlier called dehydrogenases, oxidases, peroxidases, hydroxylases, oxygenases etc. The group, in fact, includes those enzymes which bring about oxidation-reduction reactions between two substrates, A and B.

$$\text{A}_{\text{reduced}} + \text{B}_{\text{oxidized}} \longrightarrow \text{A}_{\text{oxidized}} + \text{B}_{\text{reduced}}$$

More precisely, they catalyze electron transfer reactions. In this class are included the enzymes catalyzing oxidoreductions of CH—OH, C=O, CH—CH, CH—NH_2 and CH=NH groups. Some important subclasses are:

1.1 Enzymes acting on CH—OH group of electron donor. For example:

1.1.1.1 Alcohol : NAD oxidoreductase

[common or trivial name, Alcohol dehydrogenase]

This enzyme catalyzes the following reaction:

$$\text{Alcohol} + \text{NAD} \longrightarrow \text{Aldehyde or Ketone} + \text{NADH} + \text{H}^+$$

1.3 Enzymes acting on CH—CH group of electron donor. For example:

1.3.2.2 Acyl-CoA : cytochrome c oxidoreductase

[Acyl-CoA dehydrogenase]

Acyl-CoA + oxidized cytochrome c ⟶ 2,3-dehydroacyl-CoA + reduced cytochrome c

1.9 Enzymes acting on the heme groups of electron donors. For example:

1.9.3.1 Cytochrome c : O_2 oxidoreductase

[Cytochrome oxidase]

4. reduced cytochrome + O_2 + $4H^+$ ⟶ 4 oxidized cytochrome c + $2H_2O$

1.11 Enzymes acting on H_2O_2 as electron acceptor. For example:

1.11.1.6 H_2O_2: H_2O_2 oxidoreductase

[Catalase]

$$H_2O_2 + H_2O_2 \rightleftharpoons 2H_2O + O_2$$

> Although not an oxidizing enzyme, **catalase** is usually classified with oxidases because its action is closely connected with physiological oxidation.

2. Transferases. Enzymes which catalyze the transfer of a group, G (other than hydrogen) between a pair of substrates, A and B are called transferases.

$$A—G + B \longrightarrow A + B—G$$

In these are included the enzymes catalyzing the transfer of one-carbon groups, aldehydic or ketonic residues and acyl, glycosyl, alkyl, phosphorus or sulfur-containing groups. Some important subclasses are:

2.3 Acyltransferases. For example:

2.3.1.6 Acetyl-CoA : choline O-acetyltransferase

[Choline acetyltransferase]

Acetyl-CoA + choline ⟶ CoA + O-acetylcholine

2.4 Glycosyltransferases. For example:

2.4.1.1 α-1, 4-Glucan : orthophosphate glycosyl transferase

[Phosphorylase]

$(\alpha\text{-1, 4-Glucosyl})_n$ + orthophosphate ⟶ $(\alpha\text{-1, 4-Glycosyl})_{n-1}$ + α-D-glucose-1-phosphate

2.7 Enzymes catalyzing the transfer of phosphorus-containing groups. For example:

2.7.1.1 ATP : D-hexose-6-phosphotransferase

[Hexokinase]

ATP + D-hexose ⟶ ADP + D-hexose-6-phosphate

3. Hydrolases. These catalyze the hydrolysis of their substrates by adding constituents of water across the bond they split. The substrates include ester, glycosyl, ether, peptide, acid-anhydride, C—C, halide and P—N bonds.

$$A—B + H_2O \longrightarrow A—H + B—OH$$

Representative subclasses are:

3.1 Enzymes acting on ester bonds. For example:

3.1.1.3 Glycerol ester hydrolase

[Lipase]

A triglyceride + H_2O ⟶ A diglyceride + a fatty acid

3.2 Enzymes acting on glycosyl compounds. For example:

3.2.1.23 β-D-galactoside galactohydrolase

[β-galactosidase]

A β-D-galactoside + H_2O ⟶ An alcohol + D-galactose

3.4. Enzymes acting on peptide bonds

Here the classical trivial names (pepsin, trypsin, thrombin, plasmin etc.) have been largely retained due to their consistent long usage and also due to dubious specificities which make systematic nomenclature almost impractical at this time.

3.5 Enzymes acting on C—N bonds, other than peptide bonds. For example:

3.5.3.1 L-arginine ureohydrolase

[Ariginase]

L-arginine + H_2O ⟶ L-ornithine + urea

4. Lyases (= Desmolases). These are those enzymes which catalyze the removal of groups from substrates by mechanisms other than hydrolysis, leaving double bonds.

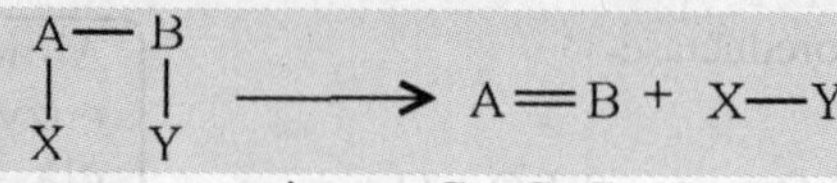

In these are included the enzymes acting on C—C, C—O, C—N, C—S and C—halide bonds. Important subclasses include:

4.1 Carbon-carbon lyases. For example:

4.1.2.7 Ketose-1-phosphate aldehyde-lyase

[Aldolase]

A ketose-1-phosphate ⟶ Dihydroxyacetone phosphate + an aldehyde

4.2 Carbon-oxygen lyases. For example:

4.2.1.2 L-malate hydro-lyase

[Fumarase]

L-malate ⟶ Fumarate + H_2O

4.3 Carbon-nitrogen lyases. For example:

4.3.1.3 L-histidine ammonia-lyase

[Histidase]

L-histidine ⟶ Urocanate + NH_3

5. Isomerases. These catalyze interconversions of optical, geometric or positional isomers by intramolecular rearrangement of atoms or groups.

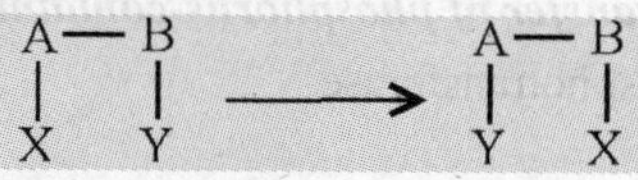

Important subclasses are:

5.1 Racemases and epimerases. For example:

5.1.1.1 Alanine racemase

L-alanine ⟶ D-alanine

5.2 Cis-trans isomerases. For example:

5.2.1.3 *All trans*-retinene 11-*cis-trans* isomerase

[Retinene isomerase]

All *trans*-retinene ⟶ 11-*cis*-retinene

5.3 Intramolecular oxidoreductases. For example:

5.3.1.9 D-glucose-6-phosphate keto-isomerase

[Glucosephosphate isomerase]

D-glucose-6-phosphate ⟶ D-fructose-6-phosphate

6. Ligases (*ligare*[L] = to bind) **or Synthetases.** These are the enzymes catalyzing the linking together of two compounds utilizing the energy made available due to simultaneous breaking of a pyrophosphate bond in ATP or a similar compound. This category includes enzymes catalyzing reactions forming C—O, C—S, C—N and C—C bonds.

A + B + ATP ⟶ A—B + ADP + P_i

Important subsclasses are:

6.2 Enzymes catalyzing formation of C—S bonds. For example :

6.2.1.1 Acetate : CoA ligase (AMP)

[Acetyl-CoA synthetase]

ATP + acetate + CoA ⟶ AMP + pyrophosphate + acetyl-CoA

6.3 Enzymes catalyzing formation of C—N bonds. For example:

6.3.1.2 L-glutamate : ammonia ligase (ADP)

[Glutamine synthetase]

$$\text{ATP + L-glutamate + NH}_3 \longrightarrow \text{ADP + orthophosphate + L-glutamine}$$

6.4 Enzymes catalyzing formation of C—C bonds. For example:

6.4.1.2 Acetyl-CoA : CO_2 ligase (ADP)

[Acetyl-CoA carboxylase]

$$\text{ATP + acetyl-CoA + CO}_2 + \text{H}_2\text{O} \longrightarrow \text{ADP + orthophosphate + malonyl-CoA}$$

To date, over 2,000 different enzymes are known, of which the oxidoreductases, transferases and hydrolases predominate. Because official names are often lengthy, the trivial names of enzymes are generally used after initial identification.

Proportion of Enzyme Genes in Whole Human Genome

About a quarter of genes encode enzymes. Names of enzyme groups with number and proportion (percentage in parenthesis) in a total of 26,383 human genes were as follows:

oxidoreductases – 652 (2.1); transferases – 610 (2.0)
hydrolases – 1,227 (4.0); lyases – 117 (0.4)
isomerases – 163 (0.5); ligases – 56 (0.2)
kinases – 868 (2.8); synthases and synthetases – 313 (1.0)
nucleic acid enzymes – 2,308 (7.5)

[Original data (from **Venter** *et al* Science **291:** 1335, 2001.) are quoted here, and so classification does not exactly match the nomenclature in the Enzyme Classification, described here.]

ISOENZYMES OR ISOZYMES

Many enzymes occur in more than one molecular form in the same species, in the same tissue or even in the same cell. In such cases, the different forms of the enzyme catalyze the same reaction but since they possess different kinetic properties and different amino acid composition, they can be separated by appropriate techniques such as electrophoresis. Such multiple forms of the enzymes are called **isoenzymes** or **isozymes**. Isozymes are of widespread nature. Over a hundred enzymes are now known to be of isozymic nature and consequently occur in two or more molecular forms. *Lactic dehydrogenase*, LDH (E.C. No 1.1.1.27), for example, is an enzyme which exists in 5 possible forms in various organs of most vertebrates. LDH catalyzes the reversible oxidation-reduction reaction:

Isoenzymes is, in fact, the preferable spelling according to the International Union of Biochemistry.

$$\text{Lactate + NAD}^+ \rightleftharpoons \text{Pyruvate + NADH + H}^+$$

Two basically-different types of LDH are found:

(*a*) *heart LDH*. This predominates in the heart and is active at low levels of pyruvate. This has 4 identical subunits called *H* subunits (H for heart).

(*b*) *muscle LDH*. This is characteristic of many skeletal muscles and maintains its activity in much higher concentrations of pyruvate. This also has 4 identical subunits called *M subunits* (M for muscle) which are enzymatically inactive.

The two types of subunits, H and M, have the same molecular weight (35,000) but differ in amino acid composition and in immunological properties. The two subunits are produced by two separate genes. LDH can be formed from two types of polypeptide chains designated as H and M subunits to yield a pure H tetramer and a pure M tetramer. Combinations of H and M subunits will, however, produce 3 additional types of hybrid enzymes, thus making the total number of possible forms as five (refer Fig. 16–1). This is confirmed by the fact that when the 2 subunits are mixed in equal proportions, a sequence of 5 bands is obtained by electrophoresis. The 5 different LDH enzyme forms are designated as H_4, H_3M, H_2M_2, HM_3 and M_4. The various isozyme forms of LDH differ significantly in the maximum activities V_{max}, in the Michaelis constant Km for their substrates, especially for pyruvate and in the degree of their allosteric inhibition by pyruvate.

The relative concentrations of the LDH enzymes differ from one type of tissue to another, and the relative composition is characteristic of the tissue. For example, LDH-1 plus LDH-2 makes up about 60% of the LDH of cardiac muscle, while LDH-4 plus LDH-5 constitutes about 80% of the LDH of liver. Thus, a diagnosis based on elevated serum LDH can be made more specific with respect to the organ involved by an electrophoretic analysis of the LDH isozymes. For example, the elevated serum level of LDH in a myocardial infarct is primarily due to an increase in LDH-1 and LDH-2; while in liver disease the elevated serum LDH is primarily due to increases in LDH-4 and LDH-5.

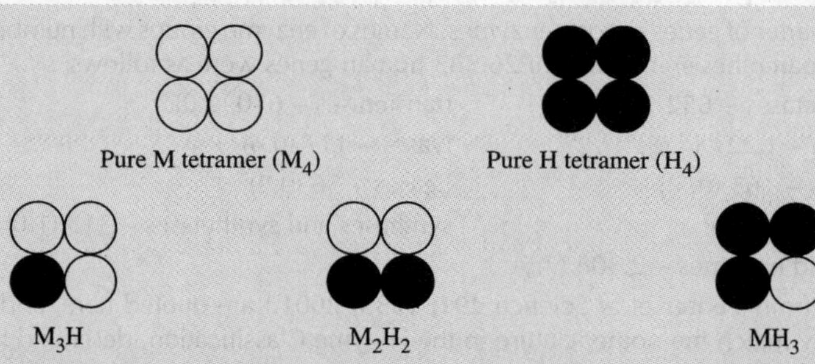

Fig. 16–1. Possible combinations of H and M subunits of lactic dehydrogenase

Damages to tissues and cell death both result in a release of intracellular enzymes from damaged cells into the blood. Therefore, the concentrations of these enzymes increase in the blood serum and can serve as a valuable diagnostic aid in a number of diseases, including myocardial infarction, pancreatitis, liver disease and prostatic cancer.

The serum concentrations of 3 enzymes are often used in diagnosis of myocardial infarction (MI). These are creatinine phosphokinase (CPK), glutamate oxaloacetate transaminase (GOT), and lactate dehydrogenase (LDH). In more than 95% of MI patients, serum levels of GOT rise rapidly and then return to normal in 4—5 days. The serum levels of CPK characteristically rise and fall even more rapidly than GOT levels. However, the serum levels of LDH typically rise and fall more slowly, returning to normal levels in about 10 days.

There seems to be an interesting relation between LDH and the physiological role of flight in birds. In those birds grouped as *short flyers*, over 90% muscle LDH is present in breast muscle. In *long flyers*, breast muscle contains over 95% heart LDH which suggests a possible relation to sustained heart contraction. Those birds classed as *intermediate flyers* contain mixtures of both heart LDH and muscle LDH.

Other noteworthy examples of isozymes are *malic dehydrogenase* (*MDH*), *hexokinase, esterase and glycol dehydrogenase*.

MULTIENZYME SYSTEMS

A few examples of complex enzyme systems are known to exist. These are not independent molecules but occur as aggregates in a mosaic pattern involving several different enzymes. *Pyruvic acid dehydrogenase* of *E. coli* is one such example. This complex molecule has a molecular weight 4,800,000 and consists of 3 enzymes: 24 moles of pyruvate decarboxylase (90,000), 24 moles of pyruvate decarboxylase (90,000), 24 moles of dihydrolipoic dehydrogenase (55,000) and 8 subunits of lipoyl reductase transacetylase (120,000). Each component of this complex enzyme is so arranged as to provide an efficient coupling of the individual reactions catalyzed by these enzymes. In other words, the product of the first enzyme becomes the substrate of the second and so on.

BIOLOGICAL ROLES OF ENZYMES

The enzymes find many applications in our daily life. Enzymatic processes such as baking, brewing and tanning have been known from antiquity. The manifold applications of the enzymes are described below:

1. **Wine manufacturing.** Much of the early interest in enzymology was developed by scientists like **Pasteur**, **Payen** and **Persoz**, who were associated with food, wine, and beer industries. Pasteur was perhaps best known to the French nation as the "*saviour of the wine industry*" because his pasteurization process salvaged an ailing industry beset with problems of microbial contamination. *Papain* is used in brewing industry as a stabilizer for chill-proof beer, because it removes small amounts of protein that cause turbidity in chilled beer.
2. **Cheese making.** Since long, the animal *rennin* (the impure form being referred to as *rennet*) is employed in making cheese. The enzyme rennet is obtained on a commercial scale from the fourth or true stomach of the unweaned calves which are specifically slaughtered for this purpose. One calf produces only 5 to 10 gm of rennet. The enzyme helps in coagulating the casein of milk. Certain preservatives (boric acid, benzoic acid or sodium chloride) are, sometimes, added to prevent decomposition of the enzyme preparations by bacteria. An enzyme *lipase* is added to cheese for imparting flavour to it. Many vegetarians are unaware that the cheese made in india contains animal rennet. However, an international charitable trust concerned with the welfare of animals, the *Beauty Without Cruelty* (BWC) has, with the help of Aurey Dairy, Mumbai, undertaken successful experimental trials in cheese making using nonanimal rennet.

> **Rennin** (= chymase or chymosin) is a milk-curdling enzyme found in the stomach of ruminant mammals and is probably widely distributed among other mammals. By clotting and precipitating milk proteins, it apparently slows the movement of milk through the stomach. Human infants, who lack rennin, digest milk proteins with acidic pepsin, just as adults do. Rennin should not be confused with another phonetically similar enzyme called **renin** which is produced by the kidney. Renin catalyzes the synthesis of angiotensins which cause vasoconstriction in the kidneys, thereby causing electrolyte and water retention in the body.

3. **Candy making.** An enzyme, *invertase* helps preventing granulation of sugars in soft-centred candies. Another enzyme, *lactase* prevents formation of lactose crystals in ice cream which would otherwise not allow the product seem sandy in texture.
4. **Bread whitening.** *Lipoxygenase* is used for whitening the bread.
5. **Clarifying fruit-juices.** The enzymes are being used in processing of fruit juices such as apple juice and grape juice. The juices are clarified by adding a mixture of *pectic enzymes* which hydrolyze the pectic substances causing turbidity.
6. **Tenderizing meat.** Because hydroxyprolyl residues create bends in collagen helices, which contribute to the tough and rubbery texture often associated with cooked meat, treating the meat with a protease (*bromelin* or *papain*) prior to its cooking hydrolyzes peptide bonds, and thus tenderizes it.
7. **Desizing fabrics.** The woven fabrics are sized by applying starch to the warp (lengthwise) threads to strengthen the yarn before weaving. But when these fabrics are printed or dyed, the sizing should be removed. Desizing may be done by acids, alkalies or enzymes. Enzymatic desizing is, however, preferred as it does not weaken the fabrics. Enzymes for this purpose are obtained from a variety of sources including bacteria, fungi and malt.

8. **Destaining fabrics.** In drycleaning, the stains due to glue, gelatin or starch are removed by employing certain enzymes, such as *alcalase*.
9. **Dehairing hide.** In the manufacture of leather, the hide is made free from hair. This is done by *employing pancreatic enzymes* which hydrolyze the proteins of the hair follicles, thus freeing the hair so that it may be easily scraped off from the hide.
10. **Recovering silver.** *Pepsin* is used to digest gelatin in the process of recovering silver from photographic films.
11. **Correcting digestion.** When the enzymes are present insufficiently in the body, certain digestive disorders come up. These may be cured by supplying the lacking enzymes. *Pepsin, papain* and *amylases* aid digestion in the stomach while *pancreatic enzymes* act in the duodenum.
12. **Wound healing.** *Proteolytic enzymes* from pig pancreas are used to alleviate skin diseases, bed sores and sloughing wounds. These enzymes act by destroying proteolytic enzymes of man, that prevent the healing of such wounds. The enzymes commonly used for wound debridement are the proteases such as *streptodornase*, *ficin* and *trypsin*.
13. **Analyzing biochemicals.** Certain enzymes are used in clinical analysis. For example, *uricase* and *urease* are employed in the determination of uric acid and urea respectively in blood. Besides, sucrose and raffinose contents in sugar mixtures are determined by polarimetry before and after treating the solutions with the enzymes, *sucrase* and *melibiase*.
14. **Dissolving blood clot.** The enzyme *urokinase*, which is manufactured from urine, is being used effectively in Japan in the treatment of blood clot in brain, artery and other circulatory diseases. A team of Soviet scientists led by **Yevgeni Chazov**, Director of the National Cardiological Research Centre, Moscow, have, in 1982, developed an effective enzyme *streptodekase*, which can dissolve blood clots in vessels. The new enzyme is particularly useful in preventing heart attacks as clots are responsible for 9 out of 10 fatal cases of cardiac arrests.
15. **Changing the blood type.** In 1981, **Prof. Ken Furukawa** and his associates of the Gunma University's Medical School, Japan have successfully employed several types of specific enzymes in an epoch-making experiment to freely change human blood types. They found that the composition of polysaccharide on the surface of blood corpuscles determines each person's type of blood. Different kinds of sugar characteristics of each blood type form on the surface of RBCs due to the function of a synthetic enzyme. If the sugar is separated from the surface of RBCs by using a specific decomposition enzyme, type A blood and type B blood can be reverted to type O, the prototype of the two blood types. If this breakthrough can be put to practical use, it will fulfill a long-cherished dream of doctors to administer blood transfusions irrespective of the type of blood a patient has by merely changing the patient's blood type to match the blood available.
16. **Diagnosing hypertension.** A new method called radioimmunoassay (RIA) procedure for diagnosing cases of hypertension has been developed by *Bhabha Atomic Research Centre* (BARC). In it, the activity of *renin*, a proteolytic enzyme secreted by the kidneys, is calculated indirectly by measuring angiotensin-I which is formed by the action of renin. Renin acts as part of a complex feedback mechanism for regulating blood volume and pressure.
17. **Augmenting surgery.** A technique using the enzyme *trypsin* as an adjunct to cataract surgery has been developed in 1980 by **Dr. Joseph Spina Jr.** of Philadelphia. With older techniques it required an incision about 2.5 cm long in the white of the eye to remove the clouded lens. Modern microsurgery has, however, reduced this cut to only 0.3 cm. But Dr. Spina's method involves a still smaller cut wide enough for a needle 0.025 cm wide. The hollow needle is used to inject a microscopic amount of trypsin, a digestive enzyme secreted by the pancreas.

Trypsin digests and liquefies the semisolid interior of the lens without harming other parts of the eye. Once the enzyme liquefies the lens— which takes from a few hours to overnight— the lens is removed by suction through the same hollow needle. This eliminates the necessity of intervention in the eye, the constant passing in and out of the instruments and suturing. The lesser the tissue is wounded, the quicker it recovers. This enzyme surgery for cataracts could be done as an outpatient operation. The patients would come in one day to have the enzyme injected and return the next day to have the cataract removed.

> A **cataract** is a clouding of eye lens which can cause blindness. It is the most common eye disease and comes with age. The lens of the eye is made of protein and as time goes by, there occurs denaturation process and the protein gets cloudy and one cannot see. India, at present, has 18 million blind people, out of which 12 million are cataract patients.

18. **Breaking down chemicals.** Recently, in 1993, a group of scientists from the Netherlands led by Han G. Brunner have found a tiny genetic defect that appears to predispose some men toward aggression, impulsiveness and violence. The afflicted persons often react to the most mildly stressful occasions with aggressive outbursts, cursing or assaulting the persons they deem a threat. The researchers have linked the abnormal behaviours to mutations in the gene responsible for the body's production of *monamine oxidase*-a, an enzyme critical for breaking down chemicals that allow brain cells to communicate. It is proposed that lacking the enzyme, the brains of afflicted men end up with excess deposits of potential signalling molecules like serotonin, dopamine and noradrenaline. Those surplus neurotransmitters, in turn, stimulate often hostile conduct. The erratic behaviour is due to point mutation. The gene is on the X chromosome, which explains why only males, with their single copy of the X chromosome, can suffer from the enzyme deficiency. Women can serve as carriers of the genetic defect, but are themselves protected from its symptoms by their possession of a second, good copy of the gene, sitting on their second X chromosome. Although the number of persons afflicted with this disease is not known but based on other types of hereditary disorders, the researchers estimate that the illness is likely to be quite rare in the general population, *i.e.*, no more than one in 1,00,000 people.
19. **Destroying acids.** Sprouts are the natural health boosters. They are basically the young new plants and are, in fact, the organic answer to simple natural health. Almost any edible seed can be sprouted. Far from being the invention of food faddists, sprouting dates back to 2939 B.C. in China. Sprouts are found in all shapes and colours and it is best to choose these from legume plants. Sprouting greatly improves the safety and nutritional quality of all pulses, seeds and grains. The enzymes which go into action during sprouting not only neutralize trypsin-inhibiting factors but also destroy harmful acids like phytic acid. Phytic acid, an integral constituent of grains, tends to bind minerals, making them unavailable to the body.
20. **Syrup manufacturing**. An *immobilized enzyme* is one that is physically entrapped or covalently-bonded by chemical means to an insoluble matrix, *e.g.*, glass beads, polyacrylamide or cellulose. Immobilization of an enzyme often greatly enhances its stability, which makes its prolonged catalytic life a valuable industrial trait. These days, *immobilized glucose isomerase* is being successfully used in the production of high-fructose corn syrup, *esp.*, in the United States.

Table 16–1 outlines some of industrial applications of enzymes. This is a rapidly changing field, and new applications of enzyme technology appear all the time.

Table 16–1. Some applications of enzymes

Enzyme	*Reaction*	*Source of enzyme*	*Application*
Industrial applications			
α-amylase	breaks down starch	bacteria	converts starch to glucose in the food industry
Glucose isomerase	converts glucose to fructose	fungi	production of high fructose syrups
Proteases	digest protein	bacteria	washing powder
Rennin	clots milk protein	animal stomach linings; bacteria	cheese making
Catalase	splits hydrogen peroxide into $H_2O + O_2$	bacteria; animal livers	turns latex into foam rubber by producing gas
β-galactosidase	hydrolyzes lactose	fungi	in dairy industry, hydrolyzes lactose in milk or whey
Medical applications			
L-asparginase	removes L-asparagine from tissues - this nutrient is needed for tumour growth	bacteria (*E. coli*)	cancer chemotherapy- particularly leukaemia
Urokinase	breaks down blood clots	human urine	removes blood clots, *e.g.*, in heart disease patients
Analytical applications			
Glucose oxidase	oxidizes glucose	fungi	used to test for blood glucose, *e.g.*, in Clinistix™ diabetics
Luciferase	produces light	marine bacteria; fireflies	binds to particular chemicals indicating their presence, *e.g.*, used to detect bacterial contamination of food
Manipulative applications			
Lysozyme	breaks 1-4 glycosidic bonds	hen egg white	disrupts bacterial cell walls
Endonucleases	break DNA into fragments	bacteria	used in genetic manipulation techniques, *e.g.*, gene transfer, DNA fingerprinting

(*Source: Biochemical Society Guidance Notes 3, Enzymes and their role in biotechnology.*)

IDENTIFICATION AND TREATMENT OF AN ENZYME DEFICIENCY

The enzyme deficiencies usually lead to an increased accumulation of specific intermediary metabolites in plasma, and hence in urine. Recognition of the intermediates that accumulate in biological fluids is useful in pinpointing possible enzyme defects. After the enzyme deficiency is established, metabolites that normally occur in the pathway but are distal to the block may be supplied exogenously in order to overcome the metabolic effects of the enzyme deficiency.

In *hereditary orotic aciduria,* there is a double enzyme deficiency in the pyrimidine biosynthetic pathway, leading to accumulation of orotic acid. Both orotate phosphoribosyltransferase and orotidine 5′-phosphate decarboxylase are deficient, causing decreased *in vivo* levels of CTP and TTP. The two activities are deficient because they reside in separate domains of a bifunctional polypeptide of 480 amino acids. dCTP and dTTP, which arise from CTP and TTP, are required for cell division. In these enzyme deficiency diseases, the patients are pale, weak, and fail to thrive. Administration of the missing pyrimidines as uridine or cytidine promotes growth and general well-being, and also decreases orotic acid excretion. The latter occurs because the TTP and CTP formed from the supplied uridine and cytidine repress carbamoyl-phosphate synthetase, the committed step, by feedback inhibition, resulting in a decrease in orotate production.

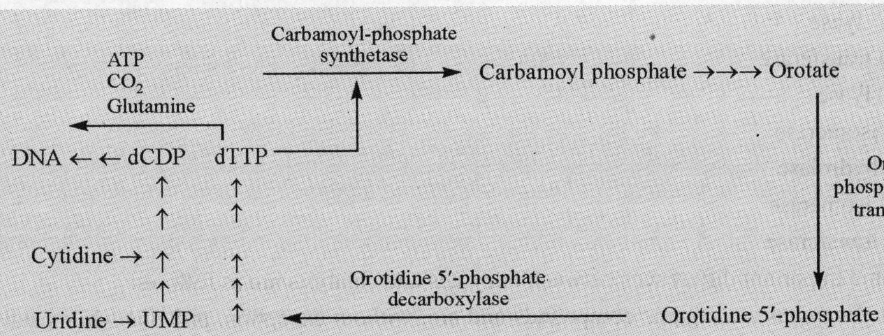

REFERENCES

See list following Chapter 18 .

PROBLEM

1. Which of the reactions listed below are catalyzed by an isomerase, lyase, hydrolase or transferase?
   (*a*) Protein ⟶ Amino acids
   (*b*) Histidine ⟶ Histamine + CO_2
   (*c*) Glucose + ATP ⟶ Glucose-6-phosphate + ADP
   (*d*) Glucose-6-phosphate ⟶ Glucose + H_3PO_4
   (*e*) $CH_3COCOOH + 2H^+ \longrightarrow CH_3CHOHCOOH$
   (*f*) $CH_3COCOOH \longrightarrow CH_3CHO + CO_2$
   (*g*) 3-phosphoglycerate ⟶ 2-phosphoglycerate
   (*h*) Tryptophan ⟶ Tryptamine + CO_2
   (*i*) L-lysine ⟶ D-lysine
   (*j*) Acetycholine ⟶ Acetic acid + choline
   (*k*) $H_3C{-}\underset{\displaystyle H}{\overset{\displaystyle OH}{C}}{-}COOH \longrightarrow H_3C{-}\underset{\displaystyle OH}{\overset{\displaystyle H}{C}}{-}COOH$
   (*l*) Glutaric acid + oxaloacetate ⟶ α-ketoglutarate + aspartic acid
2. How are enzymes different from catalysts?
3. Why does our mouth water when we see or smell something spicy?
4. Why is pepsin useful as a digestive aid and in precooked foods?

ANSWERS TO PROBLEMS

1. (*a*) hydrolase
   (*b*) lyase
   (*c*) transferase
   (*d*) hydrolase
   (*e*) reductoisomerase
   (*f*) lyase
   (*g*) transferase
   (*h*) lyase
   (*i*) isomerase
   (*j*) hydrolase
   (*k*) isomerase
   (*l*) transferase
2. Some important differences between enzymes and catalysts are as follows:
   (*a*) Enzymes are organic compounds and are, without exception, proteins while catalysts are inorganic in nature.
   (*b*) Some of the enzyme molecules are inactivated or destroyed during the reaction, while catalysts are not.
3. Our mouth starts watering on seeing or smelling a spicy food because of what is known as *reflex action*. The salivary glands in our mouth become active when we eat. The enzymes present in the saliva help in digestion of food. But sometimes the glands start secreting copious amounts of saliva even when there is no food in the mouth. This happens in anticipation of eating a tasty or spicy food when we see or smell it. On seeing or smelling a spicy food, our nerve cells in the brain are stimulated and signals are sent which activate the salivary glands. The latter start secreting *enzymes* and our mouth starts watering.
4. Pepsin partially digests (hydrolyzes) the protein in foods.

CHAPTER

17

Enzymes–II Characteristics and 3 'D' Structure

CONTENTS

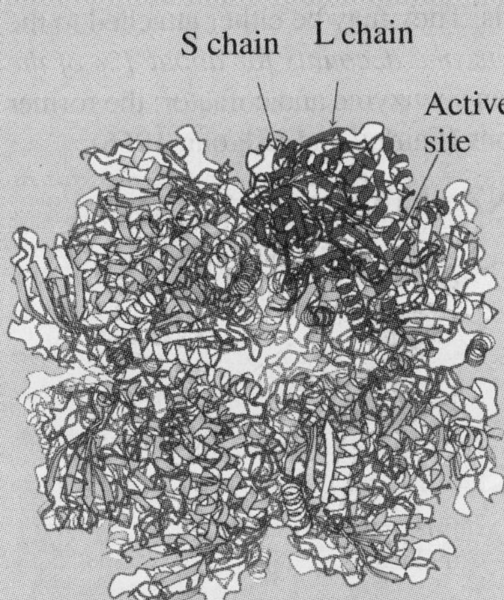

Ribbon model of the tertiary structure of ribulose 1,5-bisphosphate carboxylase/oxygenase (rubisco)

The enzyme comprises eight large subunits (one shown in the *red* and the others in *yellow*) and eight small subunits (one shown in *blue* and the others in *white*). The active sites lie in the large subunits.

CHEMICAL NATURE OF ENZYMES

All the enzymes are essentially proteins and possess properties characteristic to these. **Dixon** and **Webb** (1964) have stressed the protein nature of an enzyme by defining it as "*a protein with catalytic properties due to its power of specific activation*".

Evidences Proving the Protein Nature of the Enzymes

(*a*) **Elementary composition.** In their elementary composition, the enzymes show the usual proportion of C, H, N and S, as found in the proteins. Some crystalline enzymes, however,also contain minute quantities of P or metal ions such as Cu^{2+}, Mg^{2+}, Zn^{2+} etc. On hydrolysis, the crystalline enzymes yield the amino acids.

(*b*) **Identical action of some enzymes over other enzymes and the proteins.** Enzymes are subjected to the action of those enzymes which are specifically meant for the breakdown of peptide bonds of proteins.

(*c*) **Amphoteric nature.** Like other proteins, the enzymes behave as ampholytes in an electric field. The isoelectric point (*pI*) for various enzymes has also been determined.

(*d*) **Denaturation.** Enzymes, like other proteins, also undergo denaturation. If the crystalline *proteinase chymotrypsin* is subjected to an unfavourable *p*H, some part of protein becomes denatured. This percentage of denatured protein is usually found to be equal to the per cent loss in enzymic activity, thus proving a sort of correlation between the enzymes and the proteins..

(*e*) **Formation of antibodies.** Many purified enzymes, on injection into animal body, produce the specific antibodies. Since many nonprotein materials have been shown to serve as antigens, this cannot be treated as an evidence in support of the protein nature of enzymes but simply a further support to it.

Chemically, the enzymes may be divided into 2 categories:

1. **Simple-protein enzymes.** These contain simple proteins only *e.g., urease, amylase, papain* etc.
2. **Complex-protein enzymes.** These contain conjugated proteins *i.e.,* they have a protein part called *apoenzyme* (*apo*[G] = away from) and a nonprotein part called *prosthetic group* associated with the protein unit. The two parts constitute what is called a *holoenzyme, e.g., catalase, cytochrome c* etc.

The activity of an enzyme depends on the fact that the non-proteinaceous prosthetic group is intimately associated with the proteinaceous apoenzyme. But sometimes the prosthetic group is loosely bound to the protein unit and can be separated by dialysis and yet indispensable for the enzyme activity. In that case, this dialyzable prosthetic group is called as a coenzyme or cofactor. Thus:

$$\text{Conjugated-protein enzyme} \rightleftharpoons \text{Protein part + Prosthetic group}$$

or

$$\text{Holoenzyme} \rightleftharpoons \text{Apoenzyme + Coenzyme}$$

Coenzymes are thermostable, dialyzable organic compounds. They may be either attached to the protein molecules or may be present in the cytoplasm. *The coenzyme accounts for about 1% of the entire enzyme molecule.* Sometimes, a distinction is made between coenzyme and cofactor: the former includes the organic prosthetic groups and the latter the metal ions (Fairley and Kilgour, 1966).

CHARACTERISTICS OF ENZYMES

The enzymes possess many outstanding characteristics. These are enumerated below :

1. Colloidal Nature. Enzyme molecules are of giant size. Their molecular weights range from 12,000 to over 1 million. They are, therefore, very large compared with the substrates or functional group they act upon (Fig. 17–1).

It has been observed that the molecular weights of many enzymes prove to be approximately an *n*-fold multiple (where *n* is an integer) of 17,500 which is found to be an unit in most proteins (Table. 17–1).

On account of their large size, the enzyme molecules possess extremely low rates of diffusion and form colloidal systems in water. Being colloidal in nature, the enzymes are nondialyzable although some contain dialyzable or dissociable component in the form of coenzyme.

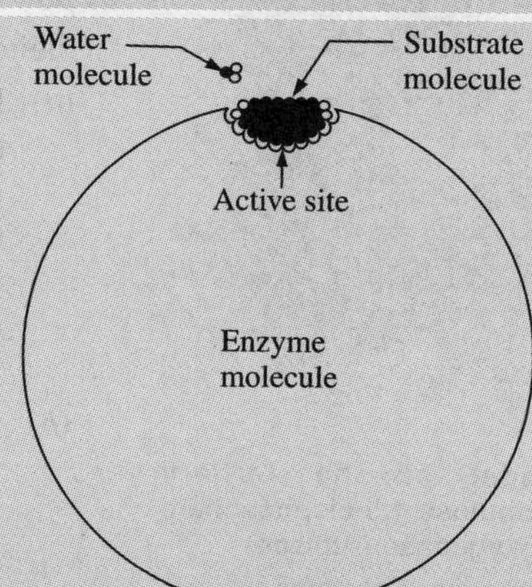

Fig. 17–1. Relative dimensions of a medium-sized enzyme molecule (MW 1,00,000 ; diameter 7 nm) and a typical substrate molecule (MW 250; diameter 0.8 nm)

[Note that the active site occupies only a small fraction of the surface area of the enzyme molecule. Also shown for comparison is a water molecule.]

2. Catalytic Nature or Effectiveness. *An universal feature of all enzymatic reactions is the virtual absence of any side products.* Therefore, just as hemoglobin is precisely tailored to transport oxygen, an enzyme is precisely adapted to catalyze a particular reaction. They act catalytically and accelerate the rate of chemical reactions occurring in plant and animal tissues. They do not normally participate in these reactions or if they do so, at the end of the reaction, they are recovered as such without undergoing any qualitative or quantitative change. This is the reason why they, in very small amounts, are capable of catalyzing the transformation of a large quantity of substrate. Thus, *the catalytic potency of enzymes is exceedingly great.*

Table 17–1. Molecular weight of some enzymes

Enzyme	*Molecular weight*	*n**
Pepsin	35,500	2
Catalase	250,000	14
Urease	480,000	27

**n* = an integer, which is a multiple of 17,500

The catalytic power of an enzyme is measured by the "**turnover number**" (a term devised by Wechselzahl) or **molecular activity** (a term devised by **Norman Arthur Edwards** and **Kenneth Arnold Hassall**, 1980) which is defined as– *the number of substrate molecules converted into product per unit time, when the enzyme is fully saturated with substrate.* For example, a single molecule of *catalase* can convert 50,00,000 H_2O_2 molecules into H_2O and O_2 in a minute (Sumner and Somers, 1947). The value of turnover number varies with different enzymes and depends upon the conditions in which the reaction is taking place. However, *for most enzymes, the turnover numbers fall between 1 to 10^4 per second* (refer Table 17–2). The turnover number of 600,000 sec^{-1} for *carbonic anhydrase* is one of the largest known. Carbonic anhydrase (Fig. 17-2) catalyzes the hydration of carbon dioxide to produce 3,60,00,000 molecules of carbonic acid per minute. This catalyzed reaction is 6×10^7 times faster than the uncatalyzed one.

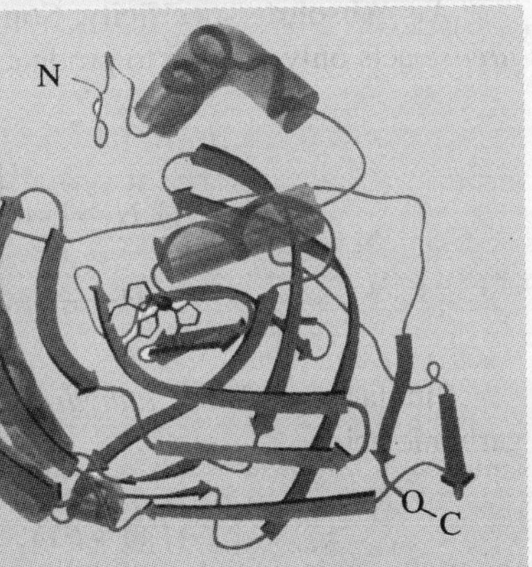

Fig. 17-2. Ribbon model of the tertiary structure of human carbonic anhydrase

The α helices are represented as cylinders and each strand of β sheet is drawn as an *arrow* pointing towards the polypeptide's C-terminus. The *grey* ball in the middle represents a Zn^{2+} ion that is coordinated by three His side chains (*blue*). Note that the C-terminus is tucked through the plane of a surrounding loop of polypeptide chain so that carbonic anhydrase is one of the rare native proteins in which a polypeptide chain forms a knot.

(Courtesy: Kannan KK et al, 1971)

$$CO_2 + H_2O \xrightarrow{\text{Carbonic anhydrase}} H_2CO_3$$

Table 17–2. Maximum turnover numbers of some enzymes

Enzyme	*Turnover number (per second)*
1. Lysozyme	0.5
2. Tryptophan synthetase	2
3. DNA polymerase I	15
4. Phosphoglucomutase	20.5

5. Chymotrypsin	100
6. β–galactosidase	208
7. Lactate dehydrogenase	1,000
8. Penicillinase	2,000
9. β–amylase	18,333
10. Acetylcholinesterase	25,000
11. Carbonic anhydrase	600,000

3. Specificity of Enzyme Action. With few exceptions, the enzymes are specific in their action. Their specificity lies in the fact that they may act (*a*) on one specific type of substrate molecule or (*b*) on a group of structurally-related compounds or (*c*) on only one of the two optical isomers of a compound or (*d*) on only one of the two geometrical isomers. Accordingly, four patterns of enzyme specificity have been recognized:

A. Absolute specificity. Some enzyme are capable of acting on only one substrate. For example, *urease* acts only on urea to produce ammonia and carbon dioxide.

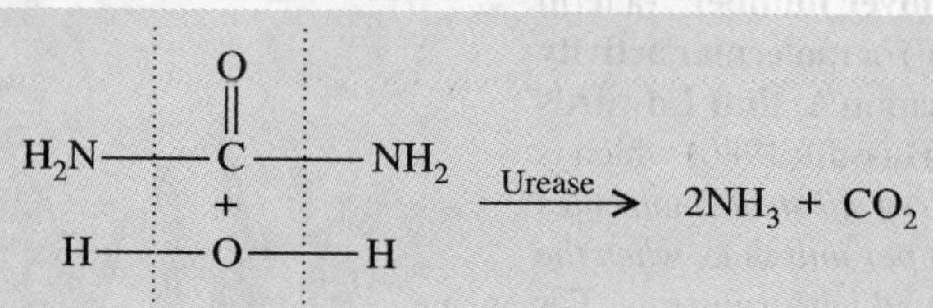

Similarly, *carbonic anhydrase* brings about the union of carbon dioxide with water to form carbonic acid.

$$H_2O + CO_2 \xrightarrow{\text{Carbonic anhydrase}} H_2CO_3$$

B. Group specificity. Some other enzymes are capable of catalyzing the reaction of a structurally-related group of compounds. For example, *lactic dehydrogenase* (LDH) catalyzes the interconversion of pyruvic and lactic acids and also of a number of other structurally-related compounds.

$$\underset{\text{Pyruvic acid}}{CH_3.CO.COOH} + NADH + H^+ \underset{}{\overset{\text{Lactic dehydrogenase}}{\rightleftharpoons}} \underset{\text{Lactic acid}}{CH_3.CHOH.COOH} + NAD^+$$

C. Optical specificity. The most striking aspect of specificity of enzymes is that a particular enzyme will react with only one of the two optical isomers. For example, *arginase* acts only on L-arginine and not on its D-isomer. Similarly, D-*amino acid oxidase* oxidizes the D-amino acids only to the corresponding keto acids.

Although, the enzymes exhibit optical specificity, some enzymes, however, interconvert the two optical isomers of a compound. For example, *alanine racemase* catalyzes the interconversion between L- and D-alanine.

$$\text{L-alanine} \overset{\text{Alanine racemase}}{\rightleftharpoons} \text{D-alanine}$$

D. Geometrical specificity. Some enzymes exhibit specificity towards the *cis* and *trans* forms. As an example, *fumarase* catalyzes the interconversion of fumaric and malic acids:

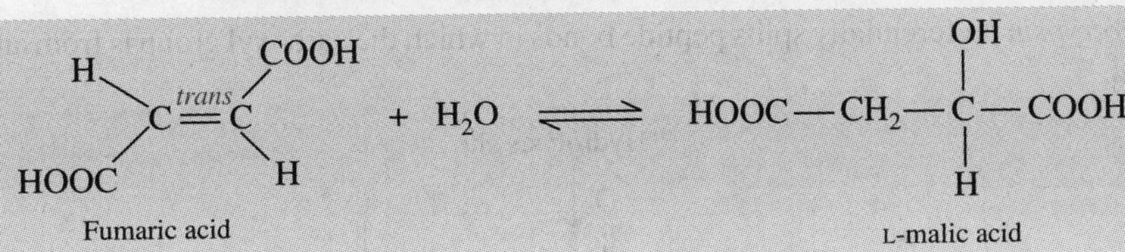

It does not react with maleic acid which is the *cis* isomer of fumaric acid or with D-malic acid.

The degree of specificity of the enzymes for substrate is usually high and sometimes virtually absolute. Proteolytic enzymes, for instance, catalyze the hydrolysis of a peptide bond:

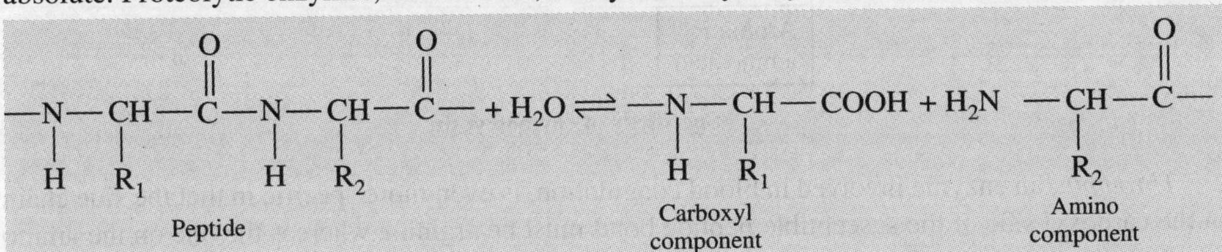

Many proteolytic enzymes (pepsin, trypsin, chymotrypsin) catalyze a different but related reaction, namely the hydrolysis of an ester bond.

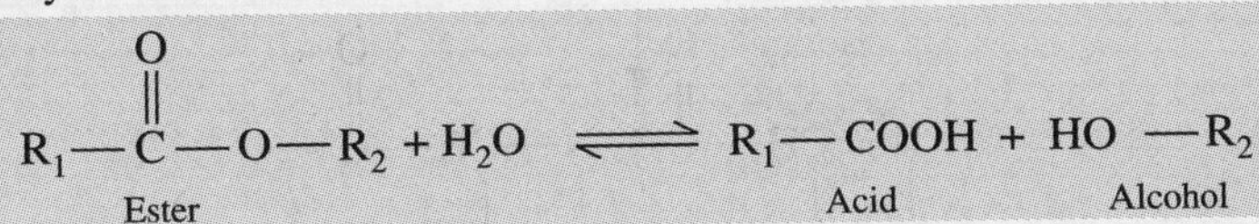

These enzymes vary markedly in their degree of specificity. For example, *subtilisin*, a bacterial enzyme, does not discriminate the nature of the side chains adjacent to the peptide bond to be cleaved. Another enzyme *pepsin* prefers bonds involving the carboxyl and amino groups of dicarboxylic and aromatic amino acids respectively. Since the bonds attached are usually located in the interior of the protein substrate, pepsin is called an *endopeptidase*.

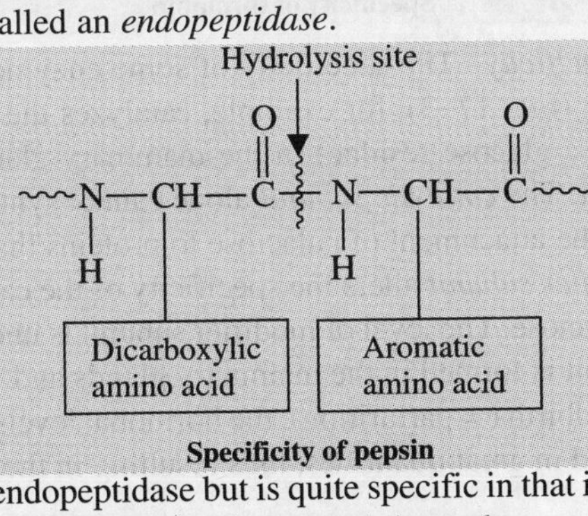

Specificity of pepsin

Trypsin, likewise, is an endopeptidase but is quite specific in that it splits peptide bonds in which carboxylic group is contributed by either lysine or arginine only.

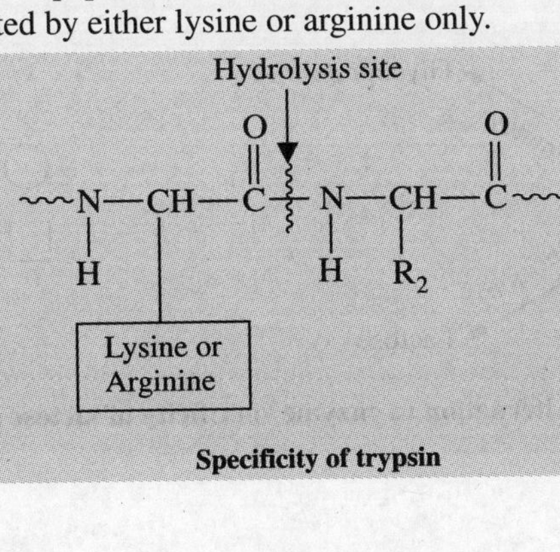

Specificity of trypsin

Chymotrypsin preferentially splits peptide bonds in which the carboxyl group is from an aromatic amino acid.

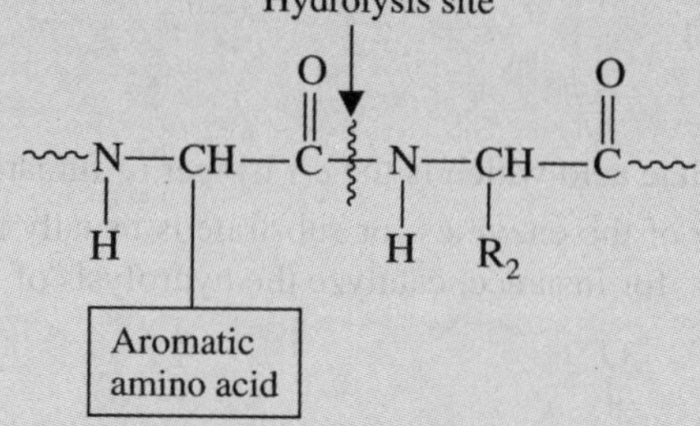

Specificity of chymotrypsin

Thrombin, an enzyme involved in blood coagulation, is even more specific in that the side chain on the carboxyl side of the susceptible peptide bond must be arginine whereas the one on the amino side must be glycine.

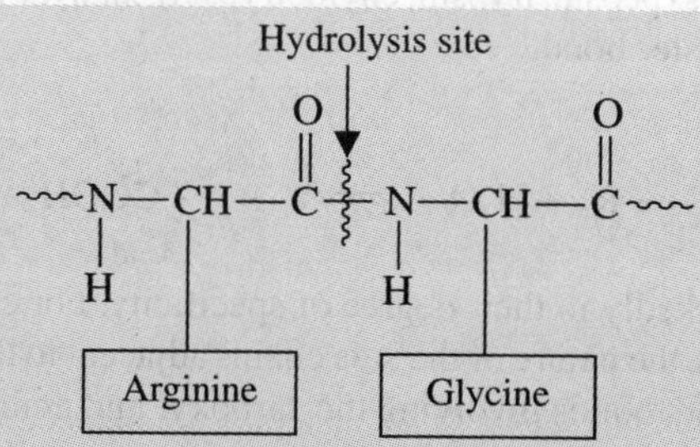

Specificity of thrombin

Alteration of enzyme specificity– The specificity of some enzymes is altered by physiological behaviour. *Lactose synthetase* (Fig. 17–3), for example, catalyzes the synthesis of lactose (a sugar consisting of a galactose and a glucose residue) in the mammary glands. It consists of a catalytic subunit and a modifier subunit. The *catalytic subunit* alone cannot synthesize lactose. Instead, it has a different role of catalyzing the attachment of galactose to proteins that contain a covalently linked carbohydrate chain. The *modifier subunit* alters the specificity of the catalytic subunit so that it links galactose to glucose to form lactose. The level of modifier subunit is under hormonal control. During pregnancy, the catalytic subunit is formed in the mammary glands and very little modifier subunit is formed. But at the time of childbirth (= parturition), the hormonal levels change significantly and the modifier subunit is synthesized in great quantities, thus resulting in the production of large amounts of lactose.

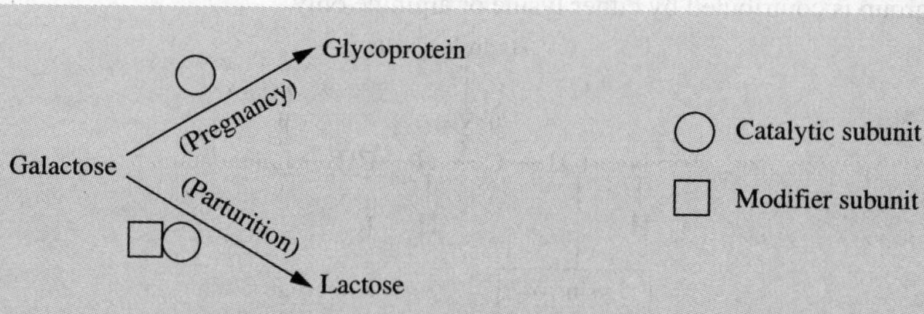

Fig. 17–3. Alternation in enzyme specificity of lactose synthetase

There are, however, instances where one enzyme acts on more than one substrate and conversely a substrate may also be catalyzed by more than one enzyme (Fig. 17–4). For example,

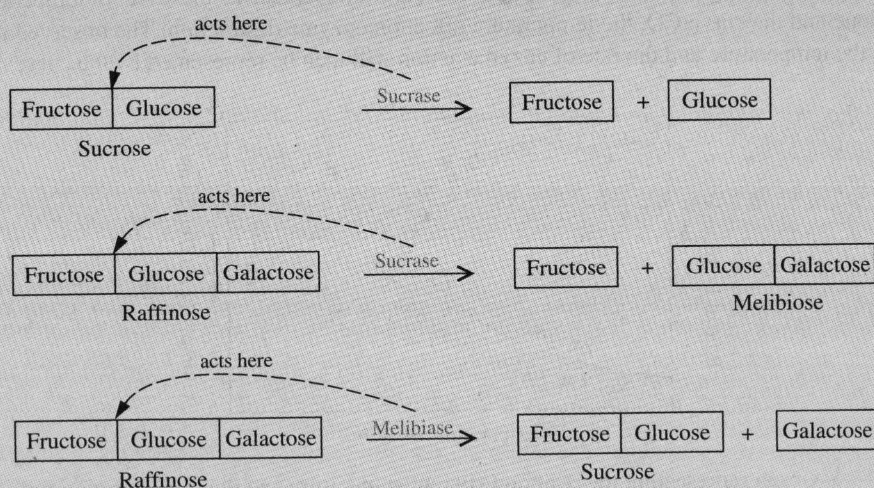

Fig. 17–4. Enzyme specificity of sucrase and melibiase

sucrase acts on both sucrose (a disaccharide sugar, containing one mole of glucose and fructose each) and raffinose (a trisaccharide sugar, containing one mole each of glucose, fructose and galactose). But in both these cases, the enzyme sucrase attacks only the glucose-fructose linkage resulting in the production of fructose and glucose (in the case of sucrose) or fructose and melibiose (in the case of raffinose). However, raffinose is also acted upon by another enzyme, the *melibiase*. But this enzyme, unlike sucrase, breaks up glucose-galactose linkage so that at the end of the reaction sucrose and galactose are produced.

The coenzymes possess much less specificity. For example, among the *hydrolases*, NAD^+ and $NADP^+$ act as common coenzymes. The relative nonspecificity of the coenzymes, in contrast to the absolute specificity of the enzymes, can be visualized by comparing coenzyme to a common hammer, used equally by various apoenzyme workers (ironworker, watchmaker, shoemaker or electrician). Although some differences may occur in the nature of hammers (*e.g.*, between NAD^+ and $NADP^+$), the apoenzyme worker is strictly specific, corresponding to the nature of the substrate concerned.

4. Thermolability (= Heat sensitivity). Being proteinaceous in nature, the enzymes are very sensitive to heat. The rate of an enzyme action increases with rise in temperature, the rate being frequently increased 2 to 3 times for a rise in temperature of 10°C, *i.e.*, the value of temperature quotient or Q_{10} is 2 to 3. But at higher temperatures, the value of coefficient does not remain constant and decreases rapidly. Above 60°C, the enzymes coagulate and thus become inactivated, because there occurs an irreversible change in their chemical structure. The enzymes of dry tissues like seeds and spores, however, can endure still higher temperatures of about 100° to 120°C.

A temperature difference of 10 °C has become a standard that is used to measure the temperature sensitivity of a biological function. This value, called the **temperature quotient (Q_{10})**, is determined (for temperature intervals of exactly 10 °C) simply by dividing the value of a rate function (such as metabolic rate or the rate of an enzymatic reaction) at the higher temperature by the value of the rate function at the lower temperature. In general, metabolic reactions have Q_{10} values about 2 to 3. Purely physical processes, such as diffusion, have much lower Q_{10} values, usually close to 1.

The observed effect of temperature on enzyme action is the net result of the effect of temperature on the rate of enzyme action and their destruction as well. There will, thus, be obtained an optimum temperature for the enzyme action (Fig. 17–5). The curve AB represents the effect of temperature on action alone and the curve CD, the temperature effect on enzyme destruction. The observed relation between the temperature and the rate of enzyme action will then be represented by the curve AE.

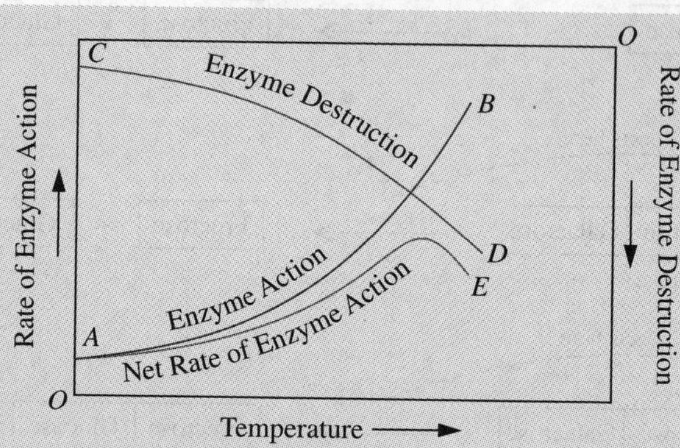

Fig. 17–5. Graph representing the relation between temperature and the rate of enzyme activity
(After Duclaux, 1883)

If, however, the effect of increasing temperature (in terms of three ill-demarcated categories of low, medium and high) on enzyme activity is studied (Fig. 17–6), it may be observed that the initial velocity of the reaction (given by the shape of the curves at $t = 0$) steadily increases with temperaure. However, after a certain temperature is passed, the cessation of activity comes earlier and earlier with the result that less product is formed. There is, thus, a somewhat ill-defined optimum region of temperature which is that at which the two factors of increased initial rate and decreased active life of the enzyme are balanced to produce the most product in a reasonable time. It is not easy to determine the exact value for the optimum temperature because it is somewhat vague concept, and will depend on the length of time over which the measurements are made. However, the approximate values obtained often show a distinct correlation with the body temperatures of the organisms from which the enzyme came. Thus, mammalian enzymes often have optimum temperatures in the range 35–45°C, while the enzymes from the bacteria, that live in volcanic hot springs, may have optima of 80°C.

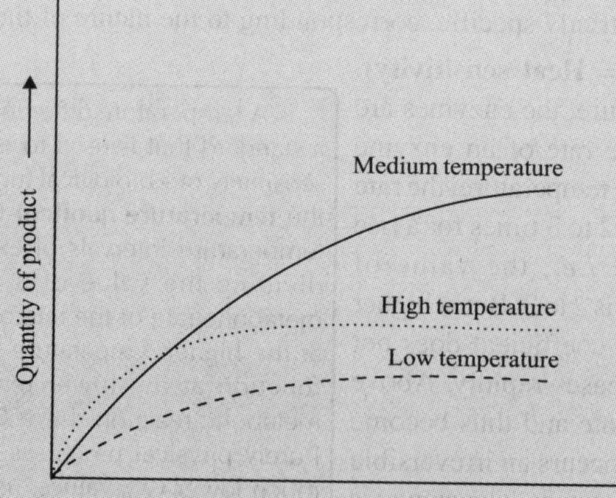

Fig. 17–6. Effect of temperature (coupled with time) on enzyme activity

In general, the optimum temperature range for most enzymes varies between 30 and 45ºC. For instance, it is 30ºC for *catalase*. Because enzymes are globular proteins, most are thermolabile and begin to denature (indicated by loss of enzyme activity) at temperatures between 45° and 50ºC (Fig. 17–7).

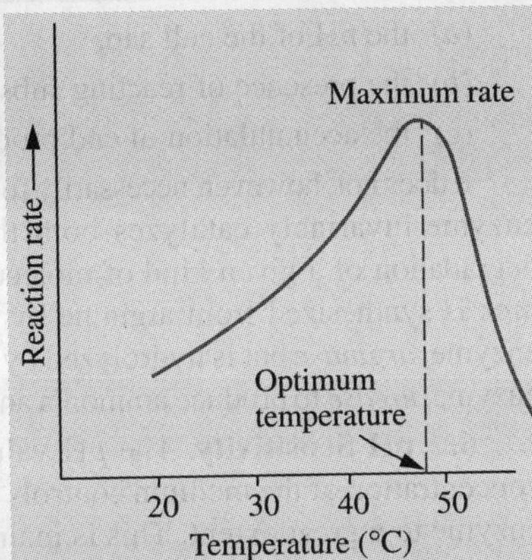

Fig. 17–7. Hypothetical temperature activity profile of an enzyme

At low temperatures, the catalytic activity of the enzyme predominates, although some thermal denaturation takes place during this period. Decreasing temperatures to near or below 0ºC although inactivate the enzyme but this is a reversible type of change and the enzyme regains its catalyzing power upon increasing the temperature to optimum. At higher temperaure, although the catalytic activity of the enzyme increases, yet its denaturation predominates. Henceforth, all the enzyme is denatured in a very short time. The enhanced enzyme activity with the rise in temperature is due to the fact that the energy of molecule becomes greater which, in turn, enhances the inherent reactivity of the molecules and the frequency of their collisions. It is because of the high rate of enzyme destruction at increasing temperatures that an enzyme is stable for weeks at 0ºC, for days at 10ºC, for hours at 30ºC but for fraction of seconds at 70ºC.

The effect of heat also mainfests itself in the preservation of enzyme activity during storage. *The best preservation of enzyme preparations is by refrigeration or quick freezing*. This has been shown by **Nord** (1932) in the case of *zymase*.

5. **Reversibility of a Reaction.** The enzymes are capable of bringing about reversion in a chemical reaction. The digestive enzymes catalyze the hydrolytic reactions which are reversible. For instance, *lipase*, which catalyzes the synthesis of fat from glycerol and fatty acid, can also hydrolyze them into their component units.

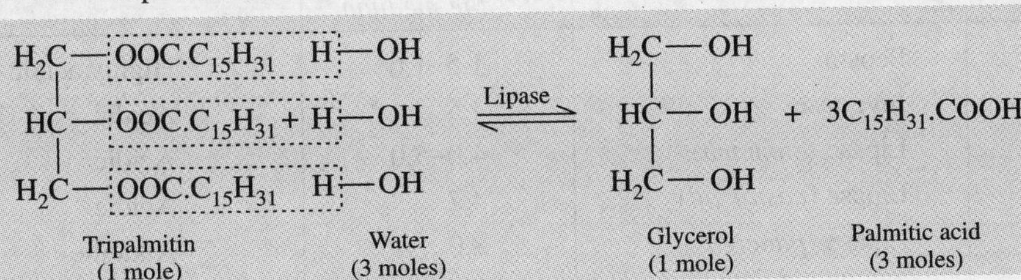

Fig. 17–8 shows the results of experiments on the action of *lipase* from castor on a fat, triolein. The final equilibrium mixture is the same whether one starts with the ester or with its individual components.

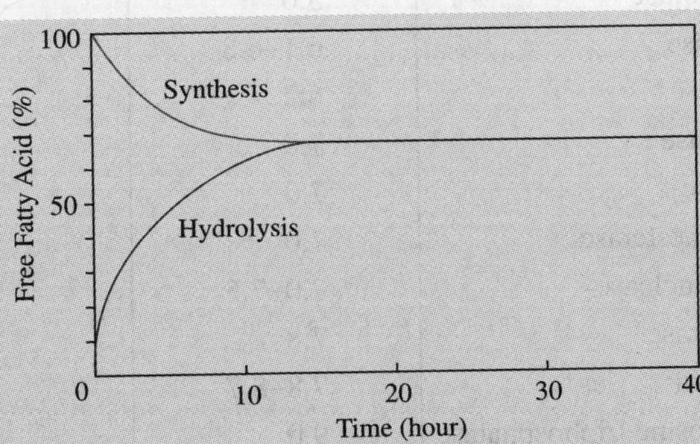

Fig. 17–8. Reversibility of action of lipase (from castor) on triolein

The direction in which the reaction proceeds depends upon many factors, such as:

(*a*) the *p*H of the cell sap,
(*b*) the presence of reacting substances, and
(*c*) the accumulation of end products.

It does not, however, necessarily follow that the same enzyme invariably catalyzes both the synthesis and degradation of a given kind of molecule. For instance, urea is synthesized from arginine by the action of the enzyme, *arginase* but is hydrolyzed by action of another enzyme, *urease* to produce ammonia and carbon dioxide.

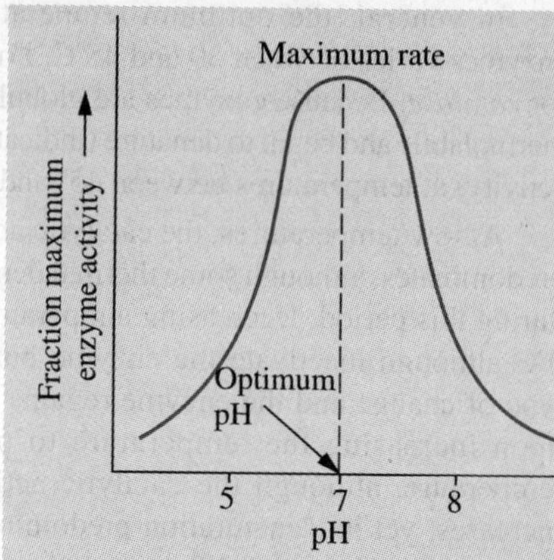

Fig. 17–9. Hypothetical pH activity profile of an enzyme

6. pH Sensitivity. The *p*H value or the H^+ ion concentration of the medium controls the activity of an enzyme to a great extent. This is mainly related to the degree of dissociation, to the electric charge of the enzyme and, through this, to the formation of the enzyme-substrate complex (a discussion of which will follow in the succeeding chapter). Each enzyme, thus, acts best in a certain pH value which is specific to it and its activity slows down with any appreciable change (increase or decrease) in the H^+ ion concentration. In fact, the *p*H will affect the efficiency of an enzyme and usually there will be a *p*H at which the activity is at a maximum. The activity will fall off on either side of this value. Fig. 17–9 depicts the effect of *p*H on an enzyme-catalyzed reaction.

Table 17–3. pH optima for various enzymes

S.N.	*Enzyme*	*Optimum pH of the medium*	*Nature of the medium*
1.	Pepsin	1.5–1.6	Highly acidic
2.	Invertase	4.5	Acidic
3.	Lipase (*stomach*)	4.0–5.0	Acidic
4.	Lipase (*castor oil*)	4.7	Acidic
5.	Lipase (*pancreas*)	8.0	Alkaline
6.	Amylase (*malt*)	4.6–5.2	Acidic
7.	Amylase (*pancreas*)	6.7–7.0	Acidic–neutral
8.	Cellobiase	5.0	Acidic
9.	Maltase	6.1–6.8	Acidic
10.	Sucrase	6.2	Acidic
11.	Catalase	7.0	Neutral
12.	Urease	7.0	Neutral
13.	Cholinesterase	7.0	Neutral
14.	Ribonuclease	7.0–7.5	Neutral
15.	Fumarase	7.8	Alkaline
16.	Trypsin	7.8–8.7	Alkaline
17.	Adenosine triphosphatase	9.0	Alkaline
18.	Arginase	10.0	Highly alkaline

Some optimum *p*H values for various enzymes are given in Table. 17–3. A perusal of the table indicates that *the approximate optimum pH value for most enzymes lies near neutrality.* This value depends on many factors such as:

(*a*) the nature of buffer system,
(*b*) the presence of other colloids, activators or inhibitors,
(*c*) the age of the cell tissue, and
(*d*) the nature of the substrate.

Usually maximum enzyme activity is obtained at or near the isoelectric point (*p*I) of the enzymes. Thus *trypsin*, whose *p*I value is 10.1, shows maximum activity at *p*H range between 7 and 9.

The correction between the enzymic activity and the *p*H value for 3 enzymes has been graphically represented in Fig. 17–10.

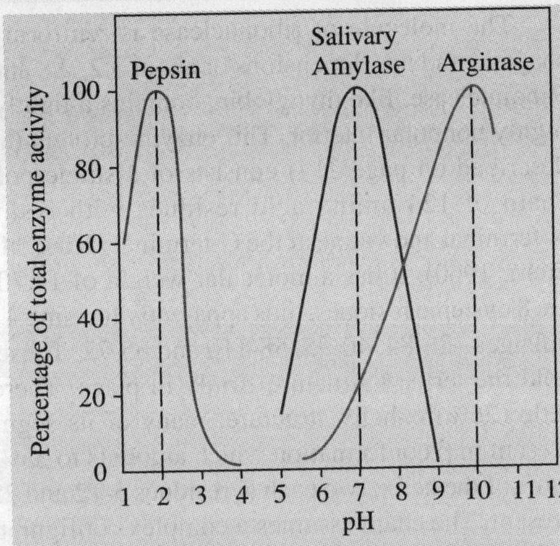

Fig. 17–10. Effect of pH on enzyme action
(Modified from Fruton and Simmonds, 1958)

THREE DIMENSIONAL STRUCTURE OF THE ENZYMES

A single crystal of protein or the protein fibres will deflect x-rays and the resultant image formed on a photographic plate can give certain important clues regarding the structure of the crystal or the fibres. This techinque is called **x-ray crystallography** and has been widely used for the elucidation of protein structure at micro level. X-ray crystallography has so far revealed the structure of many enzymes, namely, *ribonuclease, lysozyme, chymotrypsin, trypsin* etc. The structure of four of them is described below:

1. Ribonuclease (RNase)

Ribonuclease (Fig. 17–11), a small globular protein, is an enzyme secreted by the pancreas into the small intestine, where it catalyzes the hydrolysis of certain bonds in ribonucleic acids present in ingested food.

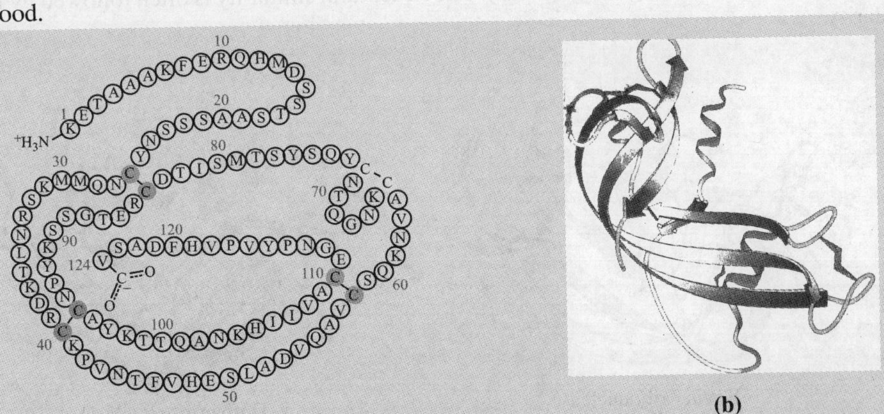

Fig. 17–11. Structure of bovine ribonuclease as determined from x-ray diffraction studies

(a) The amino acid sequence: The 4 disulfide bonds are shown in *colour.* The polypeptide chain is actually folded to give a complex 3'D' configuration.

(b) A ribbon model: The regions of α helix are depicted as *spirals* and β strands as *flattened ribbons* with the arrows indicating the N-terminal → C-terminal direction of the polypeptide. Those segments of the chain that do not adopt a regular secondary structure (*i.e.*, an α helix or β strand) consist largely of loops and turns and are shown in *lime green.* The disulfide bonds are shown in *blue.*

[*Courtesy: (a) Hirs CHW, Moore S and Stein WH, 1960 and (b) Richardson JS*]

The molecule of ribonuclease is reniform (kidney-shaped) and has dimensions of about 3.2, 2.8 and 2.2 nm. Ribonuclease, like myoglobin, contains a tightly packed, highly nonpolar interior. This enzyme-protein (as already described on page 214) consists of a single polypeptide chain of 124 amino acid residues with lysine at the N-terminal and valine at the C-terminal (**Hirs, Moore** and **Stein**, 1960). It has a molecular weight of 13,700. There are 8 cysteine residues, thus apparently forming 4 disulfide linkages– 26-84, 40-95, 58-110 and 65-72. These serve to hold the tertiary structure firmly in place. There is very little (26%) α-helix structure; many of its segments are present in β conformation which amounts to 35%. Only 4 turns of the helix, two each at residues 5-12 and 28-35, are present. The chain assumes a complex configuration with a deep depression in the middle of one side. The active site is believed to be on the edge of this depression and the residues forming the active site are 6-8, 11, 12, 41, 42, 46-48 and 117-119. A phosphate ion is associated directly with the active site of the enzyme. The amino acid residues 12 and 119 (both histidine) are nearest to the phosphate ion.

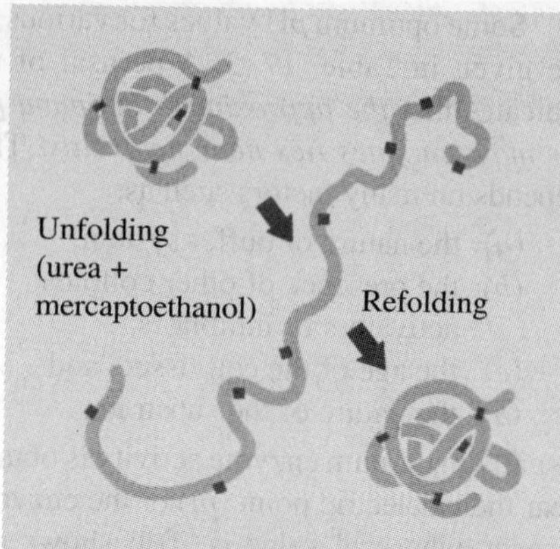

Fig. 17–12. Denaturation and refolding of ribonuclease

A native ribonuclease molecule (with intramolecular disulfide bonds indicated) is reduced and unfolded with β-mercaptoethanol and 8 M urea. After removal of these reagents, the protein undergoes spontaneous refolding.

(From: CJ Epstein, RF Goldberger and CB Anfinsen, 1963)

The bacterial enzyme, *subtilisin*, cleaves the chain into 2 inactive fragments : the shorter one (S-peptide) consisting of first 21 amino acids from N-terminal and the longer one (S-protein) with remaining residues. On reunion of the two segments, the enzyme molecule regains its full activity upon treatment with 8 M urea and mercaptoethanol, the native ribonuclease molecule become reduced and unfolded. After removal of these reagents, the enzyme undergoes spontaneous refolding (Fig. 17–12).

It is interesting to note that there is remarkable similarity in structure between the ribonucleases from cows and humans beings (Fig. 17–13). The strucutural similarity is often followed by functional similarity.

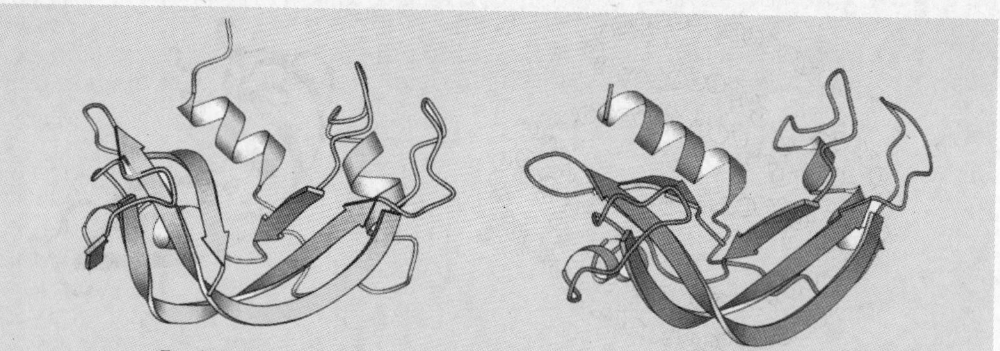

Fig. 17–13. Ribbon diagrams of the structure of ribonucleases from cows and human beings

Structural similarity often follows functional similarity.

2. Lysozyme (= Muramidase)

Lysozyme was discovered by a British bacteriologist **Sir Alexander Fleming**, in 1922. Lysozyme (Figs. 17–14 and 17–15), another small globular protein, is an enzyme present in body fluids and secretions including tears, nasal mucus, saliva, gastric secretions, milk and egg white. Lysozyme is so

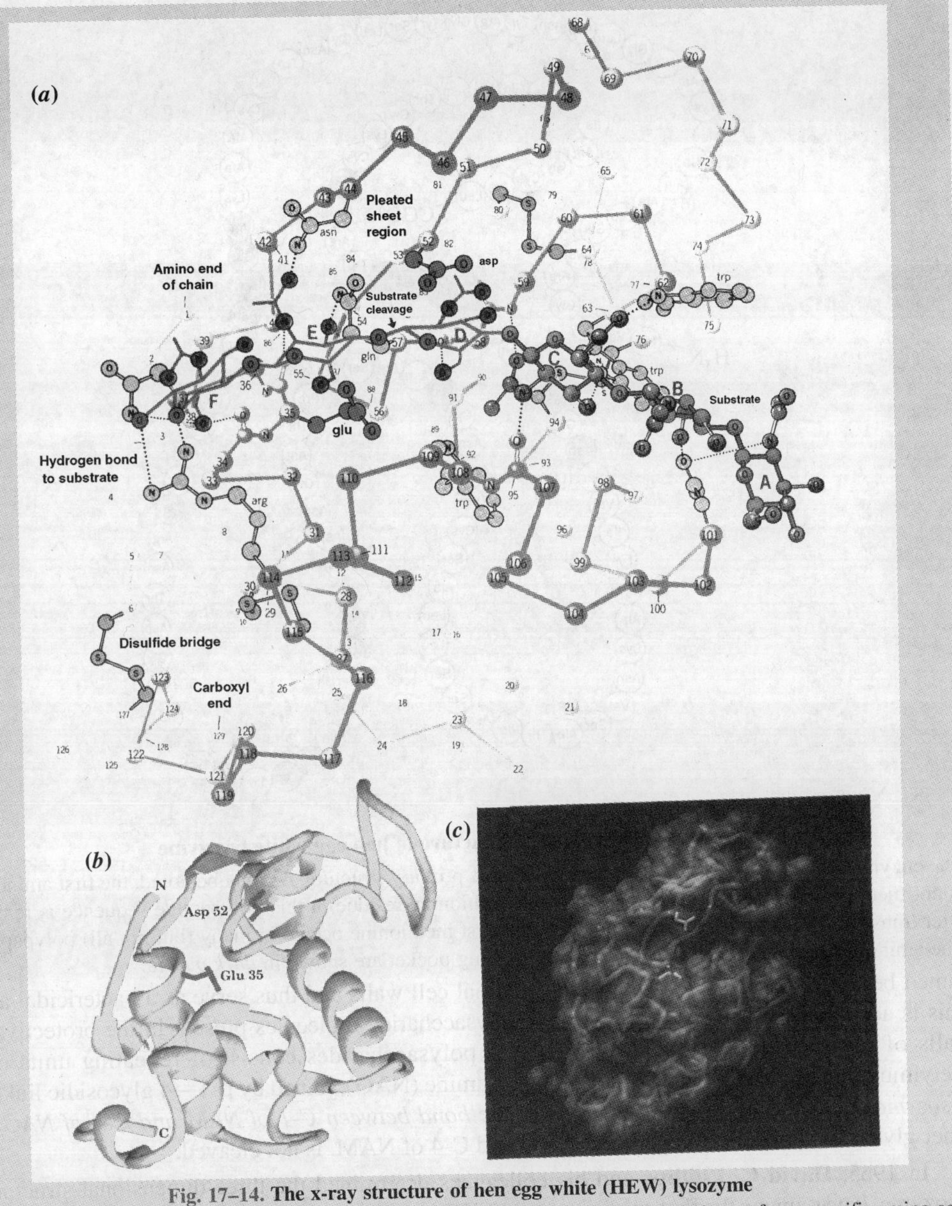

Fig. 17–14. The x-ray structure of hen egg white (HEW) lysozyme

(*a*) **The ball-and-stick model.** Each circle represents a single amino acid residue. Numbers refer to specific amino acid residues. The segment between residues 43 and 54 has a pleated sheet structure. The polypeptide chain is shown with a bound $(NAG)_6$ substrate (*green*). The positions of the backbone C_α atoms are indicated together with those of the side chains that line the substrate-binding site and form disulfide bonds. The substarate's sugar rings are designated A, at its nonreducing end (*right*), through F, at its reducing end (*left*). Lysozyme catalyzes the hydrolysis of the glycosidic bond between residues D and E. Rings A, B, and C are observed in the x-ray structure of the complex of $(NAG)_3$ with lysozyme; the positions of rings D, E, and F were inferred from model studies.

(*b*) **The ribbon model.** It highlights the protein's secondary structure and indicates the positions of its catalytically important side chains. The letters N and C represent the amino and carboxyl terminals of the protein molecule, respectively.

(*c*) **A computer-generated model.** It shows the protein's molecular envelope (*purple*) and C_α backbone (*blue*). The side chains of the catalytic residues, Asp 52 (*above*) and Glu 35 (*below*) are coloured *yellow*. Note the enzyme's prominent substrate-binding cleft. Parts (*a*), (*b*) and (*c*) have approximately the same orientation .

(Courtesy of (a) Irving Geis, (b) & (c) Arthur Olson)

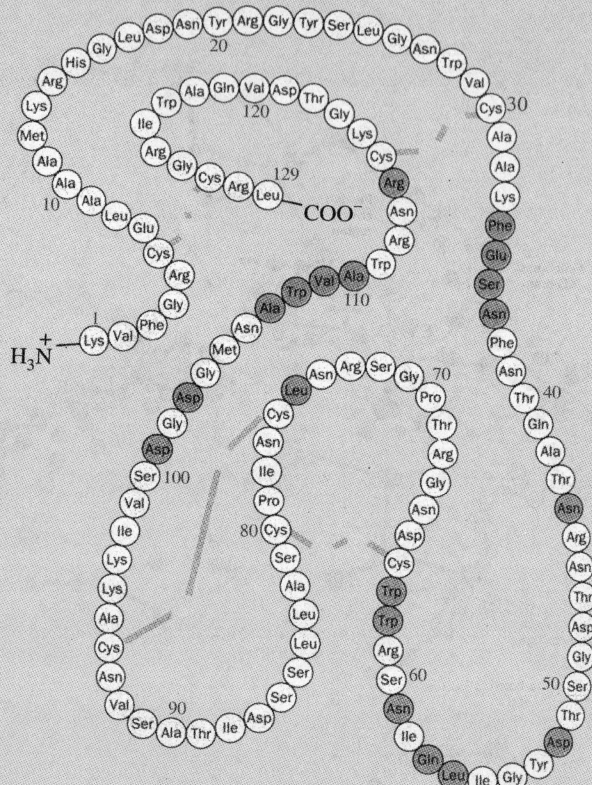

Fig. 17–15. The primary structure of hen egg white lysozyme

The enzyme lysozyme contains 129 amino acids in its primary structure. As may be noted, the first amino acid is not methionine; instead, it is lysine. The first methionine residue in this polypeptide sequence is removed after translation is completed. The removal of the first methionine occurs in many (but not all) polypeptides. The amino acid residues that line the substrate-binding pocket are shown in *dark purple*.

named because it can 'lyse', or dissolve, bacterial cell walls and thus serve as a bactericidal agent. This is accomplished by disrupting certain polysaccharide molecules present in the protective cell walls of many gram-positive bacteria. These polysaccharides consist of repeating units of N-acetylmuramic acid (NAM) and N-acetylglucosamine (NAG), joined by β-1→4 glycosidic linkages. *Lysozyme catalyzes the hydrolysis of glycosidic bond between C-1 of NAM and C-4 of NAG.* The other glycosidic bond, between C-1 of NAG and C-4 of NAM, is not cleaved.

In 1965, **David C. Phillips** and his colleagues determined the three-dimensional structure of lysozyme. It was among the first proteins whose 3-'D' structure was elucidated. Lysozyme is a relatively small, compact molecule, roughly ellipsoidal in shape and with dimensions 45 ×30 × 30 Å. It has a molecular weight of 14,600. Its molecule (Fig. 17–15) consists of a single polypeptide chain of 129 amino acid residues with 4 intrachain disulfide linkages– 6–127, 30–115, 64–80 and 76–94. It has lysine at the N-terminal and leucine at the C-terminal. *Lysozyme is devoid of coenzyme or metal ion cofactors* and thus lacks a built-in marker at its active site, in contrast with such proteins as myoglobin and hemoglobin. Like myoglobin and cytochrome c, lysozyme has a compactly-folded conformation and has most of its hydrophobic R groups inside the globular structure, shielded from water, and its hydrophilic R groups outside, facing the aqueous medium. The enzyme has only 12% β conformation and 40% α-helical segments which line a long deep cleft in the side of the molecule. This central cleft is the active site of the enzyme molecule. The interior of lysozyme, like that of myoglobin and hemoglobin, is almost entirely nonpolar. Hydrophobic interactions evidently play an important role in the folding of lysozyme, as they do for most proteins.

The active site has 6 *subsites* (A to F ; Fig. 17–16) which bind various substrates or inhibitors. The amino acid residues located at the active sites are 35, 52, 59, 62, 63 and 107. It is, thus, apparent that the active site may include amino acid residues which are distantly placed, as shown in Fig. 17–17. The residues which bring about bond cleavage lie between the subsites D and E, close to the COOH groups of glutamic acid (35) and aspartic acid (52). It is thought that glutamic acid protonates the acetal bond of the substrate while the aspartic acid stabilizes the resulting carbonium ion from the back side (**Harper** and **Rodwell**, 1973).

Lysozyme binds 6 of the monomeric units of its polysaccharide substrate, and the strain induced by the binding facilitates the formation of the unstable carbonium ion, $>C^{+}-$, intermediate. The proposed mechanism of lysozyme catalysis employs (**Charles J. Flickinger** *et al*, 1979):

1. orientation and approximation through formation of ES complex,
2. strain,
3. general acid-base catalysis, and
4. electrostatic stabilization of a carbonium ion intermediate.

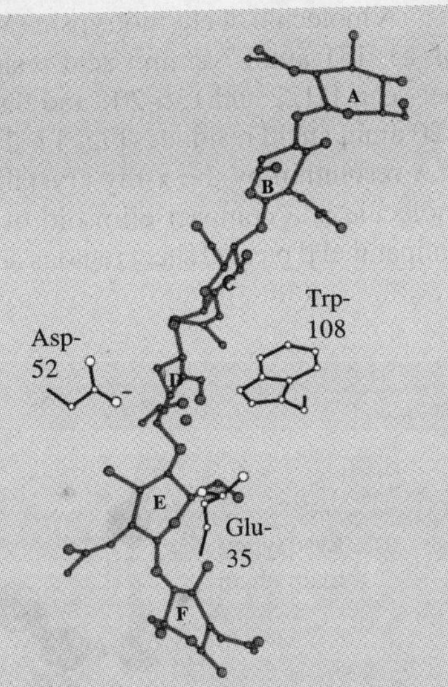

Fig. 17–16. Simplified model of a lysozyme molecule showing a hexasaccharide bound in the cleft of the enzyme

A to F represent the subsites of the enzyme. The locations of key amino acid residues of the enzyme are indicated.

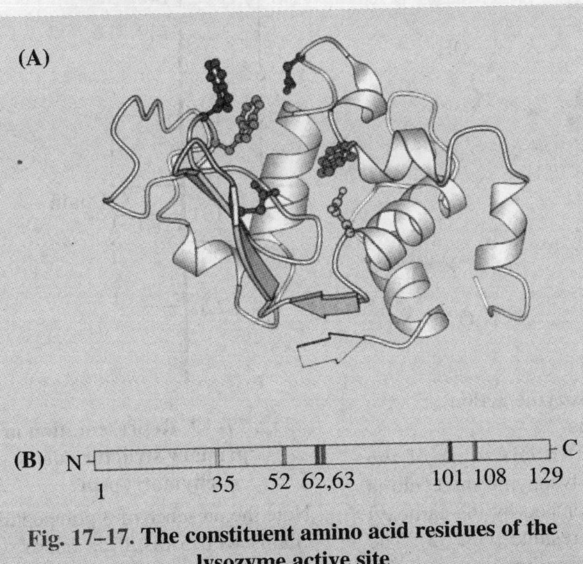

Fig. 17–17. The constituent amino acid residues of the lysozyme active site

(A) Ribbon diagram of the enzyme lysozyme with several components of the active site shown in color.
(B) A schematic of the primary structure of lysozyme showing that the active site is composed of residues that come from different parts of the polypeptide chain.

The importance of the ability of proteins to structure precisely a volume of space is evident.

The mechanism for lysozyme action, as proposed by **David C. Phillips** (1965), is presented in Fig. 17–18.

3. Chymotrypsin

Chymotrypsin, like carboxypeptidase, is a mammalian digestive enzyme which catalyzes the hydrolysis of proteins in the small intestine. Chymotrypsin is highly selective in its action as it catalyzes the hydrolysis of only those peptide bonds which are on the carboxyl side of amino acids with aromatic (phenylalanine, tyrosine, tryptophan) or bulky hydrophobic (methionine) R groups, irrespective of the length or amino acid sequence of the polypeptide chain. Chymotrypsin is synthesized by the exocrine cells of the pancreas as its inactive precursor or zymogen form called **chymotrypsinogen**.

A molecule of chymotrypsin (MW = 25,000) consists of 3 short polypeptide chains (A, B, and C) of 13, 131 and 97 amino acid residues respectively, connected by two *interchain* disulfide bonds between 1-122 and 136-201 and three *intrachain* disulfide bonds between 42-58, 168-182 and 191-220 amino acid residues (Fig. 17–19). The 3-dimensional structure of the enzyme was elucidated at 2 Å resolution by the x-ray crystallographic studies of **David Blow** (Figs. 17–20 and 17–21). The molecule is a compact ellipsoid of dimensions 51× 40 × 40 Å. Chymotrypsin consists of several antiparallel β pleated sheet regions and, unlike myoglobin and hemoglobin, has little α helical structure.

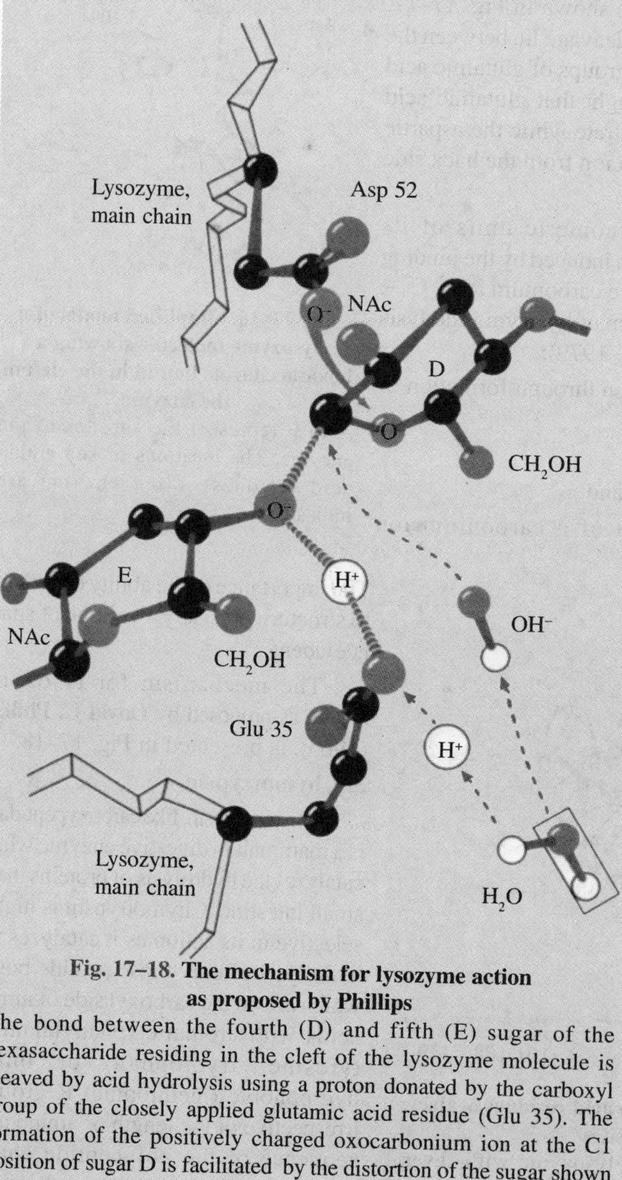

Fig. 17–18. The mechanism for lysozyme action as proposed by Phillips

The bond between the fourth (D) and fifth (E) sugar of the hexasaccharide residing in the cleft of the lysozyme molecule is cleaved by acid hydrolysis using a proton donated by the carboxyl group of the closely applied glutamic acid residue (Glu 35). The formation of the positively charged oxocarbonium ion at the C1 position of sugar D is facilitated by the distortion of the sugar shown in the figure and stabilized by the nearby aspartic acid residue (Asp52) of the enzyme. In the final step, the oxocarbonium ion reacts with an OH^- group from the solvent.

(From D Voet and JG Voet, 1995)

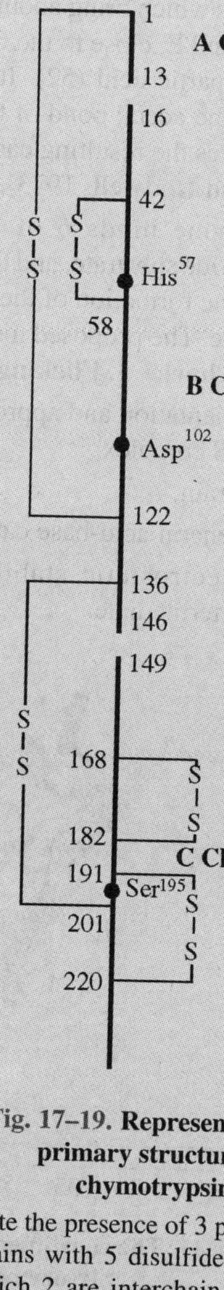

Fig. 17–19. Representation of primary structure of chymotrypsin

Note the presence of 3 polypeptide chains with 5 disulfide bonds, of which 2 are interchain and 3 are intrachain. Location of 3 amino acid residues forming catalytic triad is shown. The active-site amino acids are found grouped together in the 3-'D' structure.

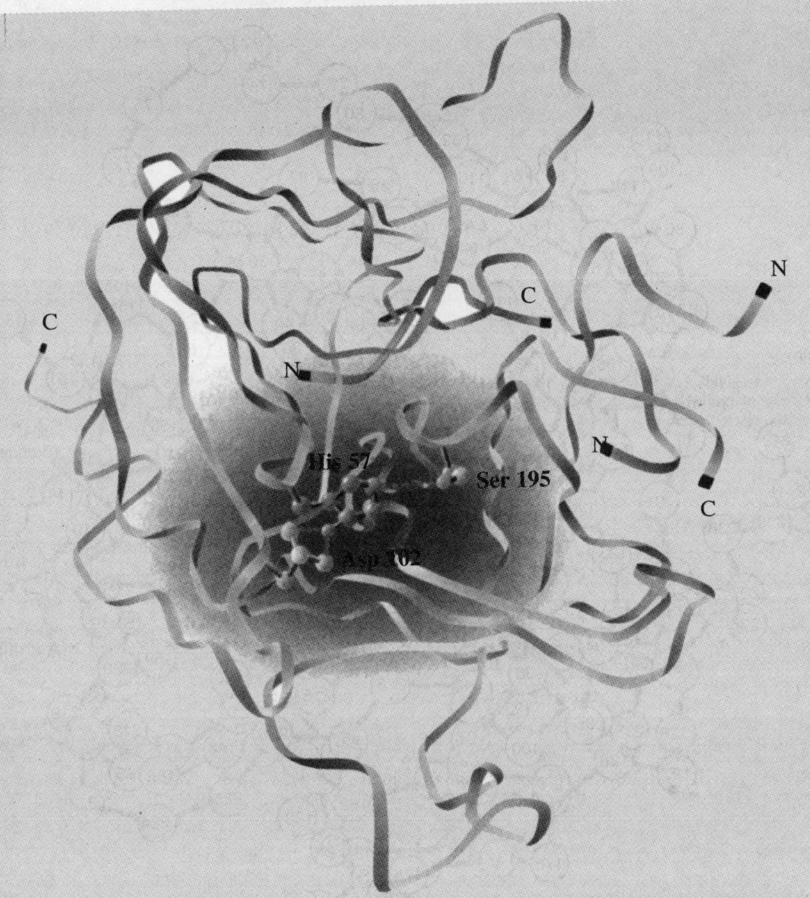

17–20. Ribbon model of the tertiary structure of chymotrypsin

The amino- and carboxyl-terminals of the 3 constituent chains are labelled as N and C respectively. The tertiary structure of chymotrypsin places the essential amino acid residues close to one another. They are shown as ball-and-stick representations in *red* and *blue*, respectively.

(From D Voet and JG Voet, 1995)

All charged groups are on the surface of the molecule except for three (His^{57}, Asp^{102} and Ser^{195}) that play a critical role in catalysis.

Double Displacement Mechanism. Chymotrypsin, like many proteases, hydrolyzes *ester bonds*, in addition to peptide bonds. The hydrolysis (of peptide or *ester bonds*) takes place by a two-step displacement with an amine being produced first, followed by production of an acid. The two steps of this double displacement mechamism are:

First step. *Acylation: Formation of the acetyl-enzyme complex.*

p-nitrophenylacetate (*p*-NPA) combines with chymotrypsin to form an enzyme-substrate (ES) complex. The ester bond of the substrate then cleaves. One of the products, *p*-nitrophenol is released from the enzyme, whereas the acetyl group of the substrate becomes covalently attached to the enzyme.

Second step. *Deacylation: Hydrolysis of the acetyl-enzyme complex.*

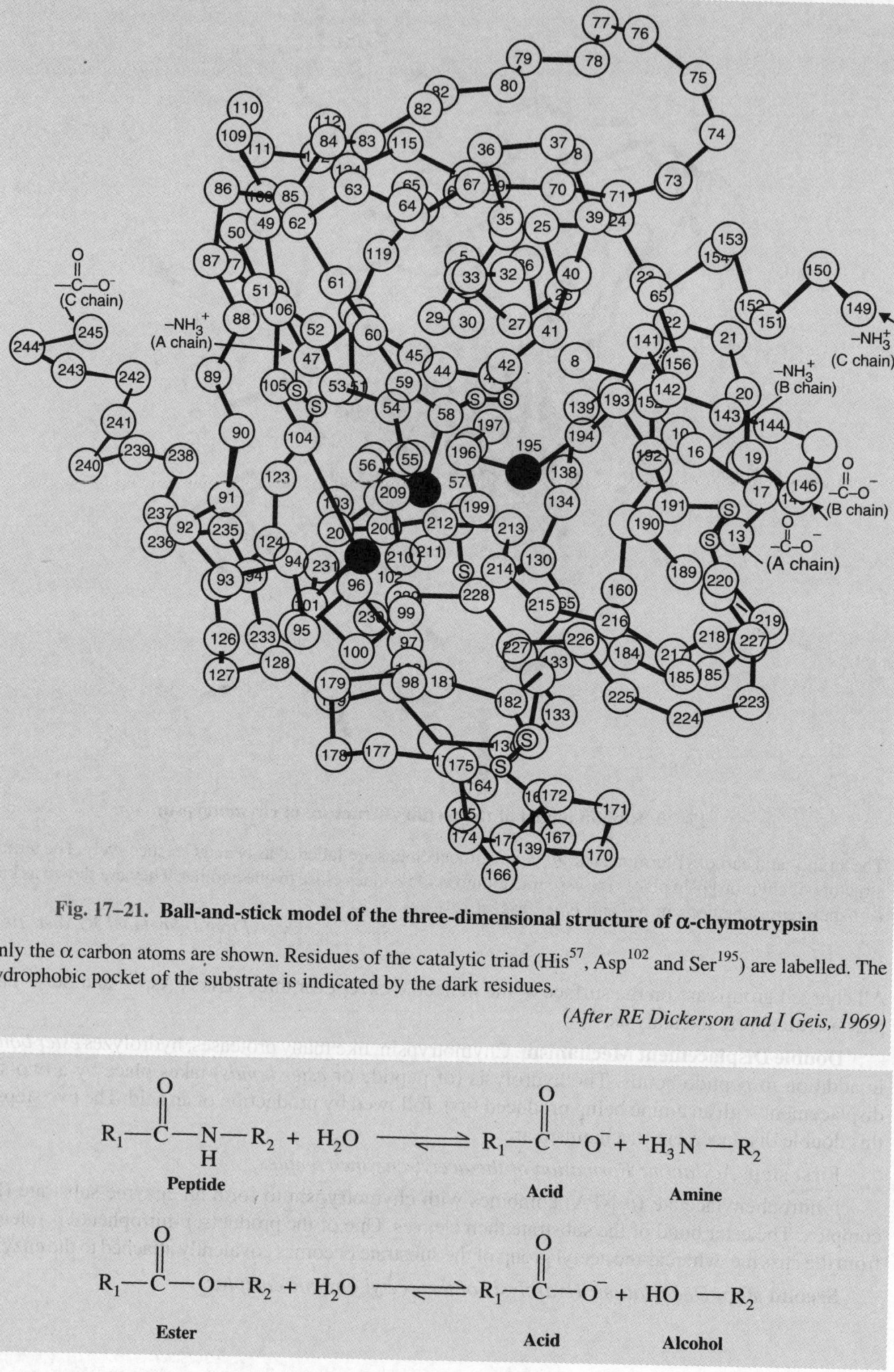

Fig. 17–21. Ball-and-stick model of the three-dimensional structure of α-chymotrypsin

Only the α carbon atoms are shown. Residues of the catalytic triad (His^{57}, Asp^{102} and Ser^{195}) are labelled. The hydrophobic pocket of the substrate is indicated by the dark residues.

(After RE Dickerson and I Geis, 1969)

$$\underset{\textbf{Peptide}}{R_1-\overset{O}{\overset{\|}{C}}-\underset{H}{N}-R_2} + H_2O \rightleftharpoons \underset{\textbf{Acid}}{R_1-\overset{O}{\overset{\|}{C}}-O^-} + \underset{\textbf{Amine}}{{}^+H_3N-R_2}$$

$$\underset{\textbf{Ester}}{R_1-\overset{O}{\overset{\|}{C}}-O-R_2} + H_2O \rightleftharpoons \underset{\textbf{Acid}}{R_1-\overset{O}{\overset{\|}{C}}-O^-} + \underset{\textbf{Alcohol}}{HO-R_2}$$

Water then attacks the acetyl-enzyme complex to yield acetate ion and regenerate the enzyme.

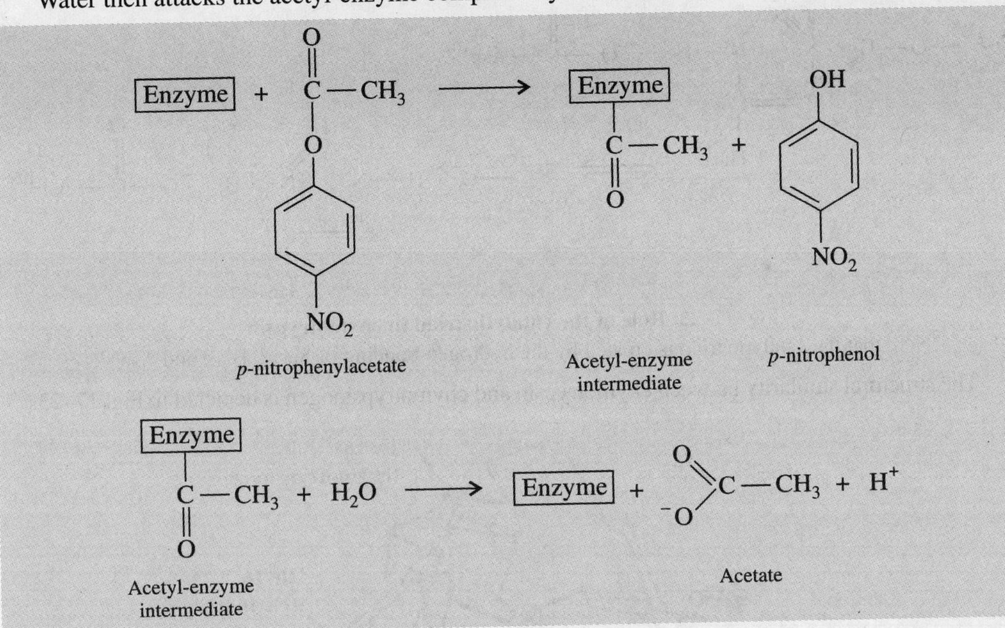

The second step (deacylation) is much slower than the first step (acylation), so that it determines the overall rate of hydrolysis of esters by chymotrypsin. The acetyl-enzyme complex is sufficiently stable to be isolated under proper conditions. The catalytic mechanism of chymotrypsin can, thus, be represented by,

$$E + S \rightleftharpoons ES \xrightarrow[P_1 \text{ (an amine)}]{} E{-}P_2 \xrightarrow[P_2 \text{ (an acid)}]{} E$$

where P_1 is the amine (or alcohol) component of the substrate, E-P_2 is the covalent intermediate, and P_2 is the acid component of the substrate. A distinct feature of this mechanism is the appearance of a covalent intermediate. In the first-step-reaction, an acetyl group is covalently bonded to the enzyme and the group attached to chymotrypsin at E-P_2 stage is an acyl group. Thus, E-P_2 is an acyl-enzyme intermediate.

Catalytic Triad. Proteolytic enzymes containing a highly reactive serine residue are known as **serine proteases**. These enzymes are readily identifiable by their rapid inactivation by DIPF. Chymotrypsin, trypsin and thrombin are noteworthy examples of this clan.

Chymotrypsin contains 28 seryl residues but only one of them (Ser^{195}) is a strong nucleophile. This is due to a specific spatial relationship between three amino acid residues (His^{57}, Asp^{102} and Ser^{195}) which constitute a ***catalytic triad*** and is based on hydrogen bonding (Fig. 17–22). The hydrogen bonding, that occurs between the buried Asp^{102} and His^{57} and between His^{57} and Ser^{195}, establishes an equilibrium that allows for the loss of the proton of the OH group of Ser^{195} (at the catalytic site) to His^{57}. This loss makes the oxygen atom of Ser^{195} residue a strong nucleophile, *i.e.*, makes serine^{195} an active serine. The proton gained by His^{57} converts the side chain of that residue into a positive imidazolium ion that forms an **ion pair** with the negative carboxylate ion of Asp^{102}. Thus, loss of a proton by Ser^{195} and formation of an ion pair by His^{57} and Asp^{102} explain the mechanism behind the functioning of catalytic triad.

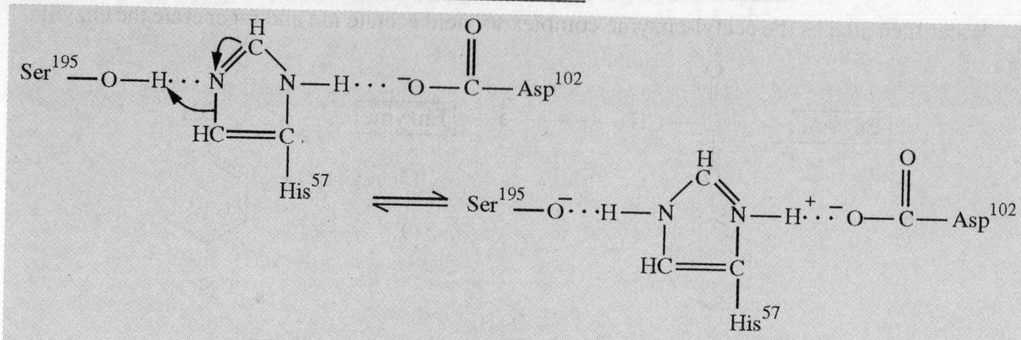

Fig. 17–22. Role of the catalytic triad in chymotrypsin

Note that the catalytic triad is created by the hydrogen-bonding of Ser^{195}, His^{57} and Asp^{102}.

The structural similarity between chymotrypsin and chymotrypsinogen is depicted in Fig. 17–23.

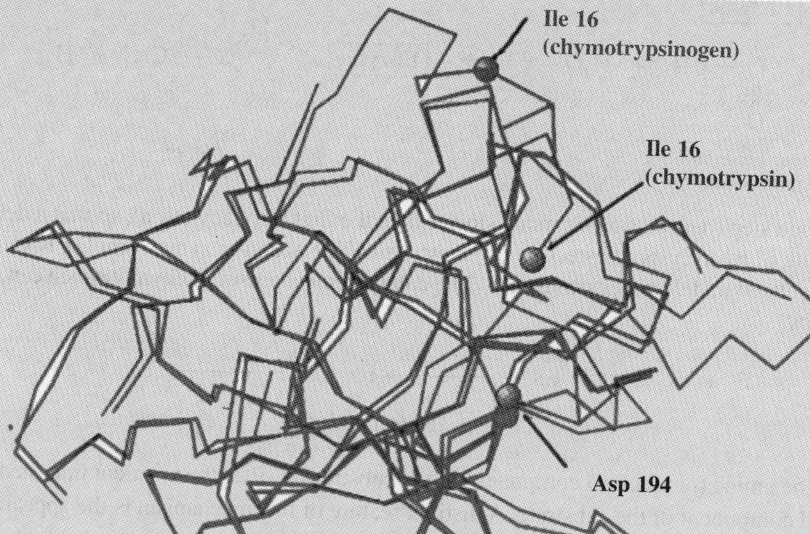

Fig. 17–23. Conformations of chymotrypsinogen (*red*) and chymotrypsin (*blue*)

The electrostatic interaction between the carboxylate of aspartate 194 and the α-amino group of isoleucine 16, essential for the structure of active chymotrypsin, is possible only in chymotrypsin.

4. Trypsin

Trypsin (and another enzyme *elastase*) are homologues of chymotrypsin. These two have catalytic triads similar to that discovered in chymotrypsin. The catalytic triad in trypsin consists of His 57, Asp 102 and Ser 195. (Fig. 17–24). Their sequences are approximately 40% identical with that of

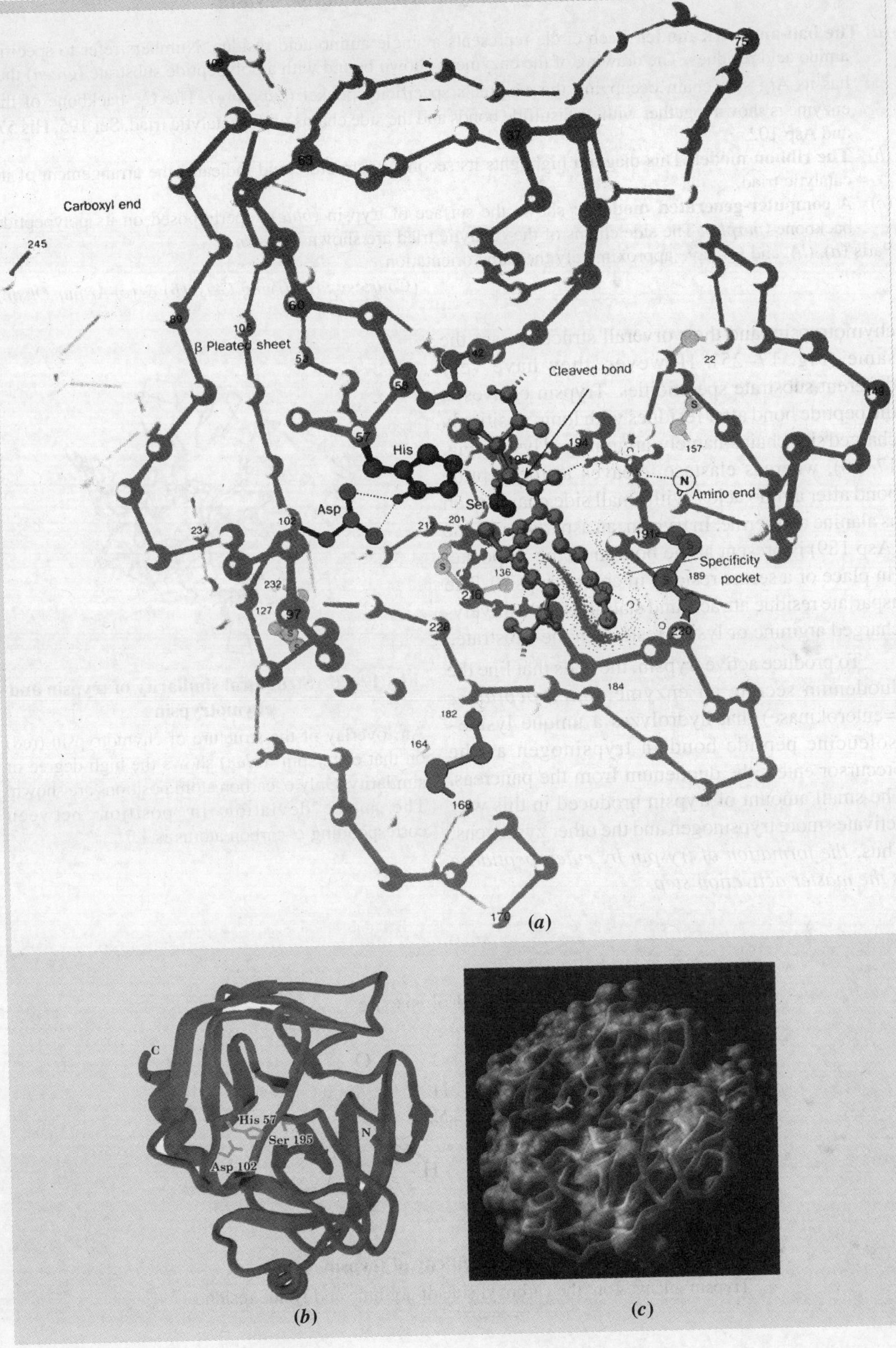
Carboxyl end
245
β Pleated sheet
Cleaved bond
His
Asp
Ser
Amino end
Specificity
pocket
His 57
Ser 195
Asp 102
C
N

(a)

(b)

(c)

Fig. 17–24. The x-ray structure of bovine trypsin

(*a*) **The ball-and-stick model.** Each circle represents a single amino acid residue. Numbers refer to specific amino acid residues. The drawing of the enzyme is shown bound with a polypeptide substrate (*green*) that has its Arg side chain occupying the enzyme's specificity pocket (*stippling*). The C_α backbone of the enzyme is shown together with its disulfide bonds and the side chains of the catalytic triad, Ser 195, His 57, and Asp 102.

(*b*) **The ribbon model.** This diagram highlights its secondary structure and indicates the arrangement of its catalytic triad.

(*c*) **A computer-generated model.** It shows the surface of trypsin (*blue*) superimposed on its polypeptide backbone (*purple*). The side chains of the catalytic triad are shown in *green*.

Parts (*a*), (*b*), and (*c*) have approximately the same orientation.

(Courtesy : (a) Irving Geis, (b) & (c) Arthur Olson)

chymotrypsin, and their orverall structures are the same (Fig. 17–25). However, they have very different substrate specificities. Trypsin cleaves at the peptide bond after residues with long, positively charged side chains-namely, arginine and lysine (Fig. 17–26), whereas elastase, cleaves at the peptide bond after amino acids with small side chains-such as alanine and serine. In trypsin, an aspartate residue (Asp 189) is present at the bottom of the S_1 pocket (in place of a serine residue in chymotrypsin). The aspartate residue attracts and stabilizes a positively-charged arginine or lysine residue in the substrate.

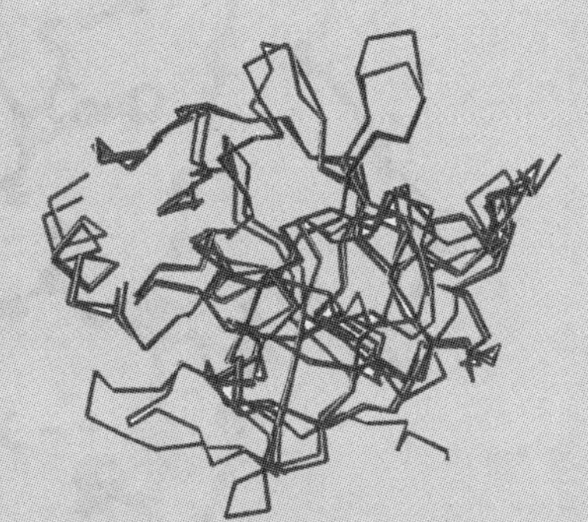

Fig. 17–25. Structural similarity of trypsin and chymotrypsin

An overlay of the structure of chymotrypsin (red) on that of tryspin (*blue*) shows the high degree of similarity. Only α-carbon atom positions are shown. The mean deviation in position between corresponding α-carbon atoms is 1.7.

To produce active trypsin, the cells that line the duodenum secrete an enzyme, *enteropeptidase* (=enterokinase) that hydrolyzes a unique lysine–isoleucine peptide bond in trypsinogen as the precursor enters the duodenum from the pancreas. The small amount of trypsin produced in this way activates more trypsinogen and the other zymogens. Thus, *the formation of tryspin by enteropeptidase is the master activation step.*

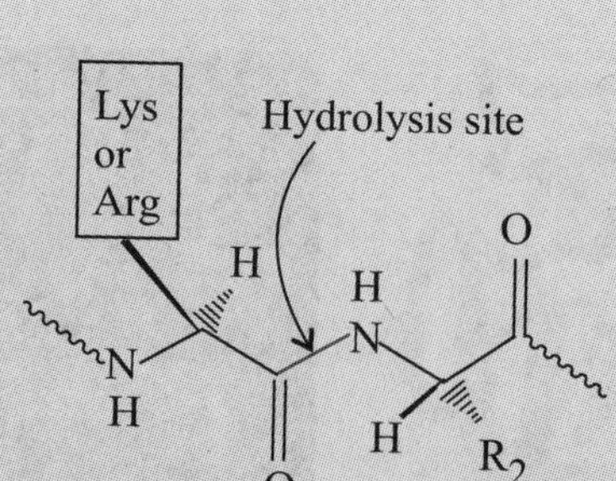

Fig. 17–26. Specificity of trypsin

Trypsin cleaves on the carboxyl side of arginine and lysine residues.

It is to be noted that in all the four enzymes described above, the active site is present in the groove or depression. The groove is, in fact, the ideal place for the active site as it provides a nonpolar microenvironment in which alone the van der Waal's forces and the hydrogen bond formation can operate between the polar groups of the active site and the substrate. In the case of an `exposed' active site, on the contrary, the water molecules will interfere with such activity.

From the foregoing discussion, it may be concluded that the chemical nature of an enzyme depends much on the active site of the enzyme molecule. Apart from the various kinds of constituent amino acid residues comprising an enzyme, its properties further depend on the pattern of '3-D' structure. The typical folding of the polypeptide chain, for example, ensures that the constituent amino acids of the active site are brought closer in the grooves.

REFERENCES

See list following Chapter 18.

PROBLEMS

1. The sweet taste of freshly picked corn is due to the high level of sugar in the kernels. Store-bought corn (several days after picking) is not as sweet because about 50% of the free sugar of corn is converted into starch within one day of picking. To preserve the sweetness of fresh corn, the husked ears are immersed in boiling water for a few minutes ("blanched") and then cooled in cold water. Corn processed in this way and stored in a freezer maintains its sweetness. What is the biochemical basis for this procedure?
2. The enzymatic activity of lysozyme is optimal at pH 5.2.

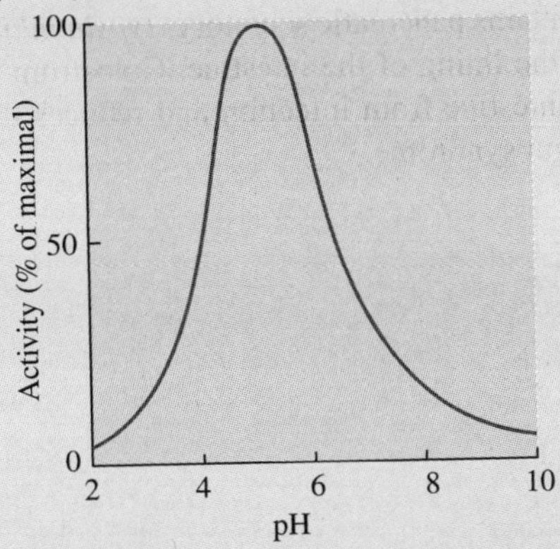

The active site of lysozyme contains two amino acid residues essential for catalysis : Glu^{35} and Asp^{52}. The pK_a values of the carboxyl side chains of these two residues are 5.9 and 4.5, respectively. What is the ionization state (protonated or deprotonated) of each residue at the pH optimum of lysozyme ? How can the ionization states of these two amino acid residues explain the pH-activity profile of lysozyme shown above?
3. Which of the following statements is not universally applicable to enzymes:
   (*a*) they generally work very rapidly.
   (*b*) they can catalyze a reaction in both directions
   (*c*) they are not used up during a reaction.
   (*d*) they will bind one substrate only.
   (*e*) they are proteins.
4. Why does a wound on tongue heal faster than those on other parts of the body ?
5. Cite the factors that contribute to the catalytic power of enzymes.

6. How can one get rid of pain naturally?
7. What can one take for digestive problems?

ANSWERS TO PROBLEMS

1. The activity of the enzyme that converts sugar to starch is destroyed by heat denaturation.
2. Glu^{35}: protonated; Asp^{52}: deprotonated. The pH-activity profile suggests that maximum catalytic activity occurs when Glu^{35} is protonated and Asp^{52} is deprotonated.
3. (*d*) They will bind one substrate only.
4. Cuts on the tongue heal fast because *the tongue is highly vascularized, that is, has plenty of blood vessels.* The rich supply of blood means that white blood corpuscles can easily reach the site of the wound and carry out defense functions preventing infection by bacteria. The secondary infection can prolong healing. *The saliva too helps in this as it bathes the tongue in a chemical called lysozyme that has mildly antiseptic properties.* Besides, *the epithelial cells of the tongue divide very rapidly* and so the wounds in the mouth and on the tongue heal fast.
5. Precise positioning of catalytic residues and substrates, geometrical strain (distortion of the substrate), electronic strain and desolvation of the substrate.
6. Try *proteolytic enzymes,* which are also known as systemic enzymes. After age 25, one's body's natural production of these enzymes declines, lowering his body's defenses against inflammation, which causes pain. Replacing proteolytic enzymes reduces inflammation and pain. Look for them in Health Food Stores.
7. Bovine colostrum contains pancreatic secretory *trypsin inhibitor,* a powerful compound that protects and repairs the lining of the intestine. Colostrum (or 'the first milk', as it is also called) protects the intestine from infection and reduces the toxicity of infections. It also helps prevent leaky gut syndrome.

CONTENTS

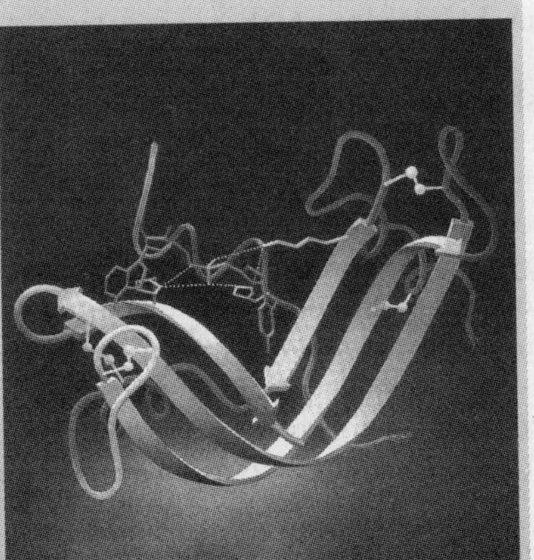

Bovine pancreatic ribonuclease S in complex with a non-hydrolyzable substrate analogue, the dinucleotide phosphonate UpcA.

CHAPTER

18

Enzymes–III Mechanism of Enzyme Action

ENERGY MECHANICS OF ENZYMATIC REACTIONS

A simple enzymatic reaction might be written:

$$E + S \rightleftharpoons ES \rightleftharpoons EP \rightleftharpoons E + P$$

where, E, S and P represent enzyme,substrate and product, respectively. ES and EP are complexes of the enzyme with substrate and with the product, respectively.

For a better understanding of the kinetics of a chemical reaction, the two terms, reaction equilibria and reaction rates must be differentiated. The function of a catalyst is to increase the *rate* of a reaction. Catalysts do not affect reaction *equilibria*. Any reaction, such as $S \rightleftharpoons P$, can be described by a **reaction coordinate diagram** (Fig. 18–1). Energy in biological systems is described in terms of free energy, G. In the coordinate diagram, the free energy of the system is The free energy of the system is plotted against the progress of the reaction. Energy diagrams describe the energetic course of the reaction (vertical axis) and the progressive chemical changes (horizontal axis) as the substrate is converted to product.

S = free energy of the substrate

P = free energy of the product

≠ = transition state

$\Delta G^{\neq}$ = activation energies for the two reactions, $S \rightarrow P$ and $P \rightarrow S$

$\Delta G^{o\prime}$ = overall standard free-energy change in moving from S to P

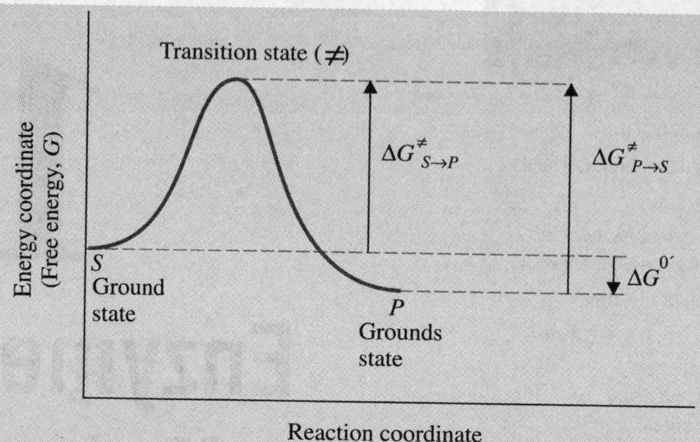

Fig. 18–1. Free energy diagram for a simple chemical reaction, S ⇌ P

described in terms of free enery, *G*. In the coordinate diagram, the free energy of the system is plotted against the progress of the reaction (or reaction coordinate). In its normal stable form or **ground state**, any molecule contains a specific amount of free energy. Chemists express the free energy change for this reacting system under standard set of conditions (temperature, 298 K; partial pressure of gases, each 1 atm or 101.3 kPa ; pH = 0; concentration of solutes, each 1M) and call this as **standard free-energy change, $\Delta G°$**. Because biochemical systems commonly involve H+ concentrations far from 1M, biochemists define a constant $\Delta G°'$, the standard free-energy change at pH 7.0.

The equilibrium between S and P reflects the difference in the free energy of their ground states. In the example shown in Fig., the free energy of the ground state of P is lower than that of S, hence $\Delta G°'$ for the reaction is negative and the equilibrium favours P. This equilibrium is *not* affected by any catalyst. A favourable equilibrium, however, *does not* mean that S → P conversion is fast. The rate of a reaction, in fact, depends on an entirely different parameter. There exists an energetic barrier between S and P that represents the energy required for alignment of reacting groups, bond rearrangements and other changes needed for the reaction to occur in either direction. To undergo reaction, the molecules must overcome this barrier or "energetic hill" and therefore must be raised to a higher energy level. At the top of energy hill is a point at which decay to the S or P state is equally probable, which is downhill either way. This is called the **transition state** and should not be confused with a reaction intermediate. It is simply a moment of fleeting molecules in which certain events (bond breakage, bond formation, charge development etc) have proceeded to decide the future course of the reaction, *i.e.*, a collapse to either substrate or product. The difference between the energy levels of the ground state and the transition state is called the **Gibbs free energy of activation** or simply **activation energy** and is symbolized by $\Delta G^{\neq}$. The double dagger (≠) denotes a thermodynamic quantity of a transition state. The rate of reaction reflects this activation energy; *a higher activation energy corresponds to a slower reaction.* Reaction rates can be increased by raising the temperature, thereby increasing the number of molecules with higher energy to overcome this energy barrier. As an alternate, the activation energy can be lowered by adding a catalyst (Fig. 18–2). Thus, *catalysts enhance reaction rates by lowering activation energies.*

The ES and EP intermediates occupy minima in the energic process curve of the enzyme-catalyzed reaction. The terms $\Delta G_{NE}^{\neq}$ and $\Delta_{GE}^{\neq}$ correspond to the activation energies for the nonenzymatic and enzymatic reactions. The activation energy for the overall process is lower when the enzyme catalyzes the reaction. Note that for both reactions, the free energy for S and P, called the $\Delta G°'$ of the reaction, is the same.

The catalysts affect the reaction rates, not the reaction equilibria. And so are the enzymes, the bidirectional arrows put in the equation on page 441 make the point clear: any enzyme that catalyzes the reaction, S → P also catalyzes the reverse reaction, P → S. Its only role is to accelerate the

Enzyme Inhibition: Transition State Inhibition and Suicide Substrate

Enzymes catalyze reactions by inducing the transition state of the reaction. It should, hence, be possible to construct molecules that bind very tightly to the enzyme by mimicking the transition state of the substrate. *Transition states* themselves cannot be isolated, because they are not a stable arrangement of atoms, and some bonds are only partially formed or broken. But for some enzymes, analogues can be synthesized that are stable, but still have some of the structural features of the transition state.

Penicillin is a good example of a *transition state analogue*. It inhibits the transpeptidase that crosslinks bacterial cell wall peptidoglycan strands, the last step in cell-wall synthesis in bacteria. It has a strained 4-membered lactam ring that mimicks the transition state of the normal substrate. When penicillin binds to the active site of the enzyme, its lactam ring opens, forming a covalent bond with a serine residue at the active site. Penicillin is a potent irreversible inhibitor of bacterial cell-wall synthesis, making the bacterium osmotically fragile and unable to survive in the body.

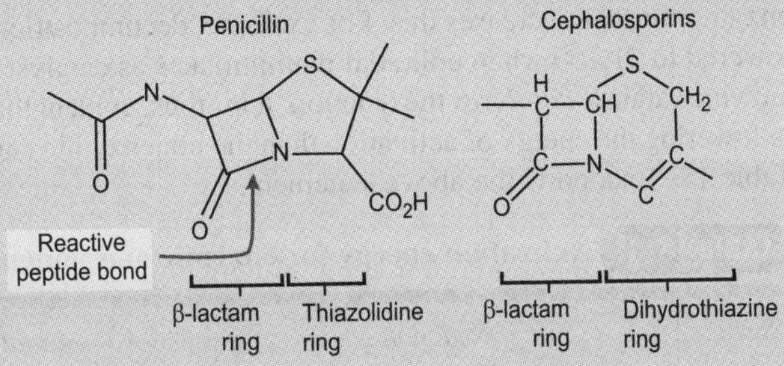

Structure of penicillin showing the reactive peptide bond in the β-lactam ring and core structure of cephalosporins

Penicillins contain β-lactam ring in conjunction with thiazolidine ring. Cephalosporins are another class of compounds containing the β-lactam ring fused to a 6-membered dihydrothiazine ring. Because of their effectiveness and lack of toxicity, β-lactam compounds are widely-used antibiotics. Bacteria with β-*lactamase*, which breaks the β-lactam ring, are resistant to these antibiotics.

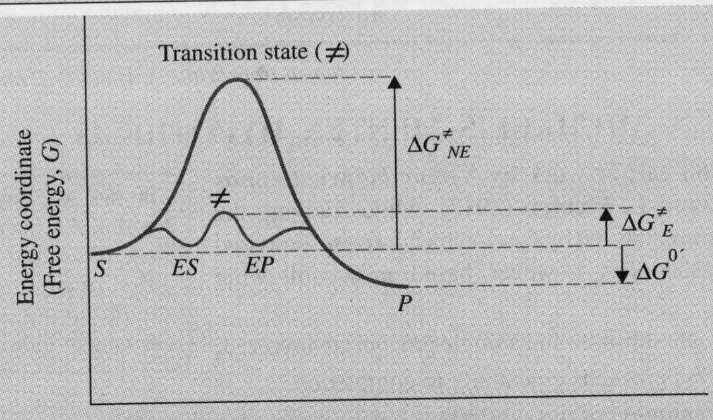

Fig. 18–2. Energy diagram, comparing the nonenzymatic and enzymatic reactions, S → P

interconversion of S and P. The enzyme is not consumed in the process, and the equilibrium point remains unaffected. However, the reaction reaches equilibrium much faster when the appropriate enzyme is present because the rate of the reaction is increased.

Practically, any reaction may have several steps involving the formation and decay of transient (unstable) chemical compounds called **reaction intermediates.** When the S → P reaction is catalyzed by an enzyme, the ES and EP complexes are intermediates and they occupy valleys in the reaction coordinate diagram (Fig. 18–2). When several steps occur in a reaction, the overall rate is determined by the step(s) with the maximum activation energy. This is called the **rate limiting step.** In practice, the rate limiting step can vary with reaction conditions. It may, however, be inferred from the above discussion that reaction *equilibria* are intimately linked with $\Delta G^{o\prime}$ and reaction *rates* are linked with $\Delta G^{\neq}$.

The energy of activation is a measure of the energy needed for the conversion of molecules to the reactive state. An inorganic catalyst lowers the activation energy and the organic catalyst (i.e., an enzyme) further decreases this. For example, decomposition of H_2O_2 requires 18,000 cal/mol; this is lowered to 11,700 when colloidal platinum acts as catalyst and is further lowered to 2,000 when an enzyme catalase *catalyzes* the reaction. It is , thus, evident that enzymic catalysts are far more efficient in lowering the energy of activation than the nonenzymic catalysts for the same reaction. A perusal of Table 18–1 supports the above statement.

Table 18–1. Activation energy for enzymic and nonenzymic catalyses

S.No.	*Reaction*	*Catalyst*	*Activation Energy (expressed as cal/mol)*
1.	Decomposition of hydrogen peroxide	None	18,000
		Colloidal platinum	11,700
		Catalase	2,000
2.	Hydrolysis of ethyl butyrate	Hydrogen ion	16,800
		Hydroxyl ion	10,200
		Lipase	4,500
3.	Hydrolysis of sucrose	Hydrogen ion	25,600
		Invertase	10,000
4.	Hydrolysis of casein	Hydrogen ion	20,600
		Trypsin	12,000

(Adapted from White A, Handler P and Smith EL, 1964)

MICHAELIS–MENTEN HYPOTHESIS

Building upon earlier work by **Victor Henri, Leonor Michaelis** and **Maud L. Menten** (1913), while studying the hydrolysis of sucrose catalyzed by the enzyme *invertase*, proposed this theory. Their theory is, however, based on the following assumptions:

> The discovery by **Michaelis** that **keratin**, the major ingredient of hair, is soluble in thioglycolic acid made possible the development of the home permanent industry.

1. Only a single substrate and a single product are involved.
2. The process proceeds essentially to completion.
3. The concentration of the substrate is much greater than that of the enzyme in the system.
4. An intermediate enzyme-substrate complex is formed.
5. The rate of decomposition of the substrate is proportional to the concentration of the enzyme-substrate complex.

The theory postulates that the enzyme (E) forms a weakly-bonded complex (ES) with the substrate (S). This enyzme-substrate complex, on hydrolysis, decomposes to yield the reaction product (P) and the free enzyme (E). These reactions may be symbolically represented as follows :

$$E + S \rightleftharpoons ES \longrightarrow E + P$$

Although one may not feel any difficulty (at least theoretical) in describing the kinetics of these reactions, yet the difficulty is encountered when one starts to determine the concentration of ES or even S practically. The same difficulty was experienced by **Michaelis** and **Menten,** who devised an equation where these immeasurable quantities were replaced by those which could be easily measured experimentally.

Following symbols may be used for deriving Michaelis-Menten equation:

$$(E_t) = \text{total concentration of enzyme}$$
$$(S) = \text{total concentration of substrate}$$
$$(ES) = \text{concentration of enzyme-substrate complex}$$
$$(E_t) - (ES) = \text{concentration of free enzyme}$$

The rate of appearance of products (*i.e.*, velocity, V) is proportional to the concentration of the enzyme-substrate complex.

$$V = k \neq (ES) \qquad ...(1)$$

The maximum reaction rate, V_m will occur at a point where the total enzyme E_t is bound to the substrate. Then the maximum concentration of ES will be equal to the total enzyme concentration, E*t*. Thus:

$$V_m = k \times (Et) \qquad ...(2)$$

Dividing equation (1) by (2,) we get:

$$\frac{V}{V_m} = \frac{(ES)}{(E_t)} \qquad ...(3)$$

With the help of this equation, one can easily measure the immeasurable quantities, (ES) and (E_t), in terms of the reaction rates experimentally.

Now coming back to the reversible reaction, $E + S \rightleftharpoons ES$, one can write the equilibrium constant for dissociation of ES as K_m which is equal to:

$$K_m = \frac{(E_t) - (ES) \times (S)}{(ES)} \qquad ...(4)$$

or $$(ES) \times K_m = (E_t) \times (S) - (ES) \times (S)$$

or $$(ES) \times K_m + (ES) \times (S) = (E_t) \times (S)$$

or $$(ES) \times [K_m + (S)] = (E_t) \times (S)$$

or $$\frac{(ES)}{(E_t)} = \frac{(S)}{K_m + (S)} \qquad ...(5)$$

Substituting the value of $\frac{(ES)}{(E_t)}$ from equation (3) to equation (5), we get:

$$\frac{V}{V_m} = \frac{S}{K_m + (S)}$$

or $$V = \frac{V_m \times (S)}{K_m + (S)} \qquad ...(6)$$

or $$K_m = (S)\left[\frac{V_m}{V} - 1\right] \qquad ...(7)$$

Equation (7) is called as **Michaelis–Menten equation.** This can be used to calculate K_m after experimentally determining the reaction rates at various substrate concentrations. This equilibrium constant, K_m, is usually called **Michaelis constant.** *It is a measure of the affinity of an enzyme for its substrate.* Referring to the equation (4), the greater the concentration of ES complex, the lower is the concentration of free enzyme and consequently the lower is the value of K_m.

For experimental determination of K_m, the velocity of the reaction (relative activity of the enzyme) is measured as a function of substrate concentration. When $V = ½V_m$, it will be seen from equation (6) that K_m is numerically equal to the substrate concentration or in other words K_m is equal to the concentration of the substrate which gives half the numerical maximal velocity, V_m. Thus, it is possible to determine K_m. The K_m, shown in Fig. 18–3, is indicated to be 0.017 M. It is noteworthy that *for any enzyme-substrate system, K_m has a characteristic value which is independent of the enzyme concentration.*

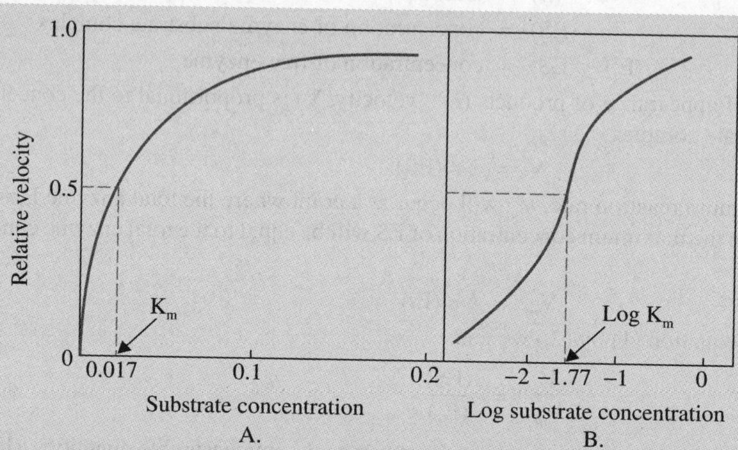

Fig. 18–3. Relative initial velocity as a function of substrate concentration (A) and as a function of the logarithm of the substrate concentration (B) for the action of yeast invertase on sucrose

(Adapted from White A, Handler P and Smith EL, 1964)

Lineweaver–Burk equation. The method described above for the detertmination of K_m is somewhat complex and, therefore, simpler methods have been devised. Two such methods (Fig. 18–4) are given below:

First method. A convenient means of evaluating K_m and V_m is to plot kinetic data as the reciprocals of V and [S]. Such a double reciprocal plot was proposed by **Hans Lineweaver** and **Dean Burk** in 1934. If one takes the reciprocal of Michaelis–Menten equation, the following equation is obtained:

$$\frac{1}{V} = \frac{K_m + (S)}{V_m \times (S)}$$

or

$$\frac{1}{V} = \frac{K_m}{V_m} \times \frac{1}{(S)} + \frac{1}{V_m} \qquad ...(8)$$

This is known as **Lineweaver–Burk equation.** This equation is of the form, $y = mx + b$, if one considers the variables to be $\frac{1}{V}$ and $\frac{1}{(S)}$. When one plots the graph against these two variables, a straight line is obtained. The slope of this line corresponds to $\frac{K_m}{V_m}$ and the $\frac{1}{V}$ intercept corresponds to $\frac{1}{V_m}$. Since V_m can be determined from the intercept, the K_m may also be calculated.

Second Method. Another graphical method for the measurement of K_m from experimental data on V as a measure of (S) makes use of the above Lineweaver-Burk equation. Multiplication of both sides of this equation by (S) gives:

$$\frac{(S)}{V} = \frac{K_m}{V_m} + \frac{(S)}{V_m} \qquad ...(9)$$

A plot of $\frac{(S)}{V}$ versus (S) gives a straight line. The intercept of the line on $\frac{(S)}{V}$ axis is $\frac{K_m}{V_m}$ and the slope is $\frac{1}{V_m}$.

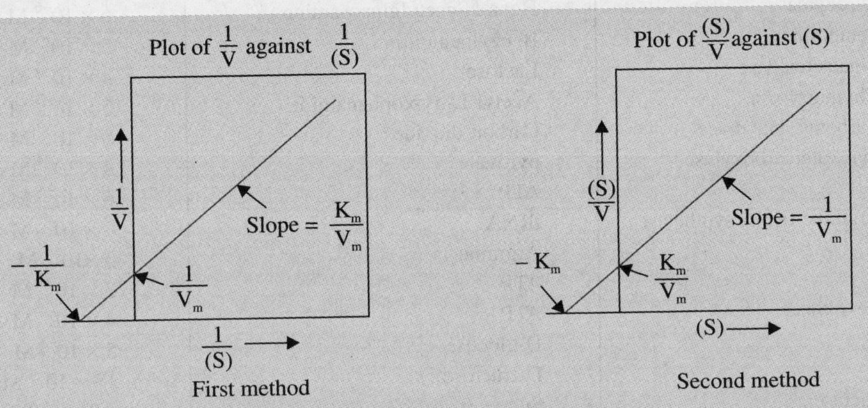

Fig. 18–4. Two methods of obtaining linear plots involving measurement of initial velocity, V and substrate concentration, (S)

V_m and K_m can be obtained from the intercept on the slope and the abscissa.

(Adapted from Fruton JS and Simmond S, 1960)

A Lineweaver-Burk plot provides a quick test for adherence to Michaelis-Menten kinetics and allows easy evaluation of the critical constants. As we shall see, it also allows discrimination between different kinds of enzyme inhibition and regulation. A disadvantage of Lineweaver-Burk plot is that a long extrapolation is often required to determine K_m, with corresponding uncertainty in the result. Consequently, other ways of plotting the data are sometimes used. One such alternative is to rearrange equation (6) into the form,

$$V = V_m - K_m \frac{V}{[S]} \quad \text{...(10)}$$

and graph V versus V/[S]. This yields what is called an **Eadie-Hofstee plot** (Fig. 18–5).

Significance of K_m and V_m Values

The **Michaelis constant (K_m)** values of the enzymes differ greatly. However, for most of the enzymes, the general range is between 10^{-1} and 10^{-6} M (refer Table 18–2). The K_m value depends on the particular substrate and on the environmental conditions such as temperature and ionic concentration.

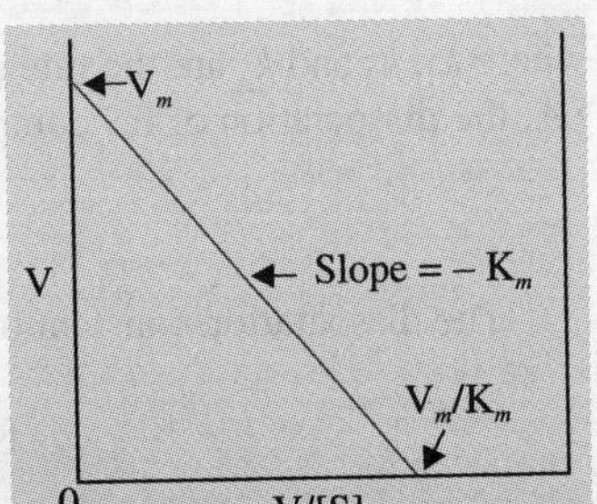

Fig. 18–5. An Eadie-Hofstee plot

Graphing V versus V/[S], one obtains V_m at (V/[S]) = 0 and K_m.

The K_m value signifies two meanings:

(*a*) K_m is the concentration of substrate at which half the active sites are occupied. If K_m is known, the fraction of the sites filled, f_{ES}, at any substrate concentration, can be calculated from:

$$f_{ES} = \frac{V}{V_m} + \frac{(S)}{K_m + (S)} \quad \text{...(11)}$$

(*b*) K_m is related to the rate constants of the individual steps in the catalytic scheme,

Table 18–2. Values of Michaelis constant (Km) of some enzymes

Enzyme	*Substrate*	K_m*
1. Lysozyme	Hexa-N-acetylglucosamine	6×10^{-6} M
2. Penicillinase	Benzylpenicillin	5×10^{-5} M
3. β-galactosidase	Lactose	4×10^{-3} M
4. Chymotrypsin	Acetyl-L-tryptophanamide	5×10^{-3} M
5. Carbonic anhydrase	Carbon dioxide	8×10^{-3} M
6. Pyruvate carboxylase	pyruvate	4×10^{-4} M
	ATP	6×10^{-5} M
7. Arginine-tRNA synthetase	tRNA	4×10^{-7} M
	Arginine	3×10^{-6} M
	ATP	3×10^{-4} M
8. Hexokinase	ATP	4×10^{-4} M
	D-glucose	5×10^{-5} M
	D-fructose	15×10^{-4} M
9. Sucrase	Sucrose	0.016 – 0.04 M
10. α-glucosidase	Methyl-α -D-glucoside	0.037 – 0.075 M
11. β-glucosidase	Methyl-β-D-glucoside	0.060 – 1.12 M
12. Pepsin	Ovalbumin	4.5%
13. Trypsin	Casein	2.0%
14. Urease	Urea	0.025 M
15. Catalase	Hydrogen peroxide	0.025 M
16. Lipase (Esterase)	Ethyl butyrate	> 0.03 M
17. Phosphatase	Glycerophosphate	< 0.003 M
18. Xanthine oxidase	Xanthine	4×10^{-7} M
19. Succinic dehydrogenase	Succinate	5×10^{-7} M
20. Amylase	Starch	0.8 – 0.25%
21. Dipeptidase	Glycylleucine	0.02 – 0.07 M

* K_m values are not absolute constants but depend on the temperature, substrate and source of enzyme.

$$E + S \underset{k_2}{\overset{k_1}{\rightleftharpoons}} ES \xrightarrow{k_3} E + P$$

where k_1, k_2 and k_3 are the rate constants for the 3 reactions. If k_2 is much greater than k_3 ($K_2 >> k_3$), *i.e.*, the dissociation of ES complex is more rapid than formation of P and regeneration of E, then

$$K_m = \frac{k_2}{k_1} \qquad ...(12)$$

The dissociation constant of ES complex is given by,

$$K_{ES} = \frac{(E)\,(S)}{(ES)} = \frac{k_2}{k_1} \qquad ...(13)$$

In other words, K_m is equal to the dissociation constant of the ES complex if k_3 is much smaller than k_2 ($k_3 << k_2$). When this condition is reached, K_m is a measure of the strength of ES complex. The high K_m value indicates weak binding whereas the low K_m value signifies strong binding. It may, however, be emphasized that K_m indicates the affinity of the ES complex only when $k_2 >> k_3$.

The **maximal rate (V_m)** represents the turnover number of an enzyme, if the concentration of the active sites (E*t*) is known, since,

$$V_m = k_3\,(E_t) \qquad ...(14)$$

A 10^{-6} M solution of *carbonic anhydrase*, for instance, catalyzes the formation of 0.6 M carbonic acid per second when it is fully saturated with substrate. Hence, k_3 is 6×10^5 sec^{-1}. *The kinetic constant, k_3 is known as the turnover number.*

HERMANN EMIL FISCHER

(LT, 1852–1919)

Emil Fischer, a German, was the son of a wealthy merchant. He graduated from the Gymnasium of Bonn in 1869. After an abortive foray into the business world, he entered the University of Bonn in 1871 to study chemistry under **Frederick August Kekule** (LT, 1829-1867), master of structural chemistry. After receiving his doctoral degree from the University of Strasbourg in 1874, he taught at the Universities of Erlangen and Wurzburg, eventually becoming Professor of Chemistry at the University of Berlin in 1892 while succeeding Hofmann. He specialized in synthetic organic chemistry and published his discovery of phenylhydrazine in 1875. Later in 1899, he turned to a detailed study of proteins. It was in the study of structural chemistry to the study of proteins that Fischer saw the possible future collaboration of biology and chemistry. He developed the **theory of the peptide bond**, the chemical linkage by which all amino acids are joined together to form their respective proteins. He also chemically synthesized an *octadecapeptide*, composed of 3 leucine + 15 glycine residues. Frequently referred to as the **father of biochemistry**, Fischer received the second-ever **Nobel Prize in Chemistry (1902)** for his work on the synthesis of purines, peptides and sugars. Fischer's collection of 9,000 reference compounds is housed in the Department of Biochemistry, the University of California at Berkeley. The prized collection was a gift of **H.O.L. Fischer**, a biochemist and the only one of Fischer's three sons to survive World War I. Following the deaths of his two sons in World War I, Emil Fischer committed suicide.

Following incident reflects the importance of Fischer's works. **Claude Hudson**, an American chemist, did pioneering research in carbohydrate chemistry. He was a meticulous writer, and as a reviewer of books his criticism was often tinged with humour. Once, an editor was too eager to get a mediocre book on carbohydrates reviewed from Hudson. To this Hudson's reply was:

"Dear Sir, I have the book you refer to and I have sandwiched it between the volumes of Emil Fischer's collected works, in the hope that by the process of 'osmosis' there may result something worth reviewing!"

ACTIVE SITE

As the substrate molecules are comparatively much smaller than the enzyme molecules, there should be some specific regions or sites on the enzyme for binding with the substrate. Such sites of attachment are variously called as **'active sites'** or **'catalytic sites'** or **'substrate sites'**.

Although the enzymes differ widely in their properties, the active site present in their molecule possesses some common features (Fig. 18–6). These are litsed below:

1. The active site occupies a relatively small portion of the enzyme molecule.
2. The active site is neither a point nor a line or even a plane but is a 3-dimensional entity. It is made up of groups that come from different parts of the linear amino acid sequence. For example, as already stated, *lysozyme* (refer Fig. 17–16) has 6 subsites in the active site. The amino acid residues located at the active site are 35, 52, 59, 62, 63 and 107.
3. Usually the arrangement of atoms in the active site is well defined, resulting in a marked specificity of the enzymes. Although cases are known where the active site changes its configuration in order to bind a substance which is only slightly different in structure from its own substrate.
4. The active site binds the substrate molecule by relatively weak forces.

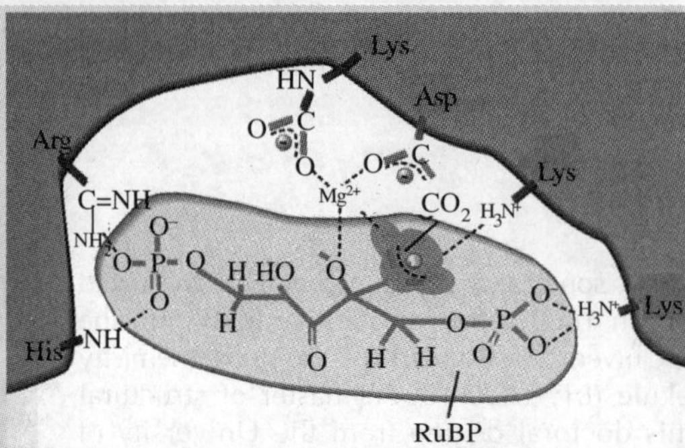

Fig. 18–6. Diagrammatic representation of the active site of the enzyme ribulose bisphosphate carboxylase (Rubisco) showing the various sites of interaction between the bound substrates (RuBP and CO_2) and certain amino acid side chains of the enzyme

In addition to determining the substrate-binding properties of the active site, these noncovalent interactions alter the properties of the substrate in ways that accelerate its conversion to products.

(After Harris DA, 1995)

5. The active sites in the enzyme molecules are grooves or crevices from which water is largely excluded. It contains amino acids such as aspartic acid, glutamic acid, lysine serine etc. The side chain groups like —COOH, —NH_2, —CH_2OH etc., serve as catalytic groups in the active site. Besides, the crevice creates a micro-environment in which certain polar residues acquire special properties which are essential for catalysis.

Remember that the active site is only a small part of the total enzyme structure. Yet, for catalytically perfect enzymes, every encounter between enzyme and substrate is productive. In these cases, there may be attractive electrostatic forces on the enzyme that entice the substrate to the active site. These forces are sometimes referred to poetically as *Circe effects*.

Circe effect is the utilization of attractive forces to lure a substrate into a site in which it undergoes a transformation of structure, as defined by **William P. Jencks**, an enzymologist, who coined the term. A goddess of Greek mythology, Circe lured Odysseus's men to her house and then transformed them into pigs.

Fischer's Lock and Key Model

Previously, the interaction of substrate and enzyme was visualized in terms of a lock and key model (also known as *template model*), proposed by **Emil Fischer** in 1898. According to this model, the union between the substrate and the enzyme takes place at the active site more or less in a manner in which a key fits a lock and results in the formation of an enzyme–substrate complex (Fig. 18–7).

To quote **Fischer** (1898) himself, "There is a relation between the unknown structure of an active enzyme and that of substrate, they are complementary and the one may be said to fit the other as a key fits a lock."

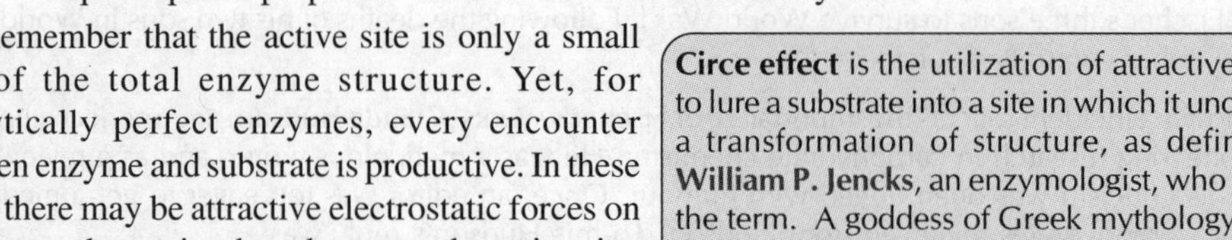

Fig. 18–7. Formation of an enzyme-substrate complex according to Fischer's lock and key model

In fact, the enzyme-substrate union depends on a *reciprocal fit* between the molecular structure of the enzyme and the substrate. And as the two molecules (that of the substrate and the enzyme) are involved, this hypothesis is also known as the **concept of intermolecular fit.** The enzyme-substrate complex (Fig. 18–8) is highly unstable and almost immediately this complex decomposes to produce the end products of the reaction and to regenerate the free enzyme. The enzyme-substrate union results in the release of energy. It is this energy which, in fact, raises the energy level of the substrate molecule, thus inducing the *activated state*. In this activated state, certain bonds of the substrate molecule become more susceptible to cleavage.

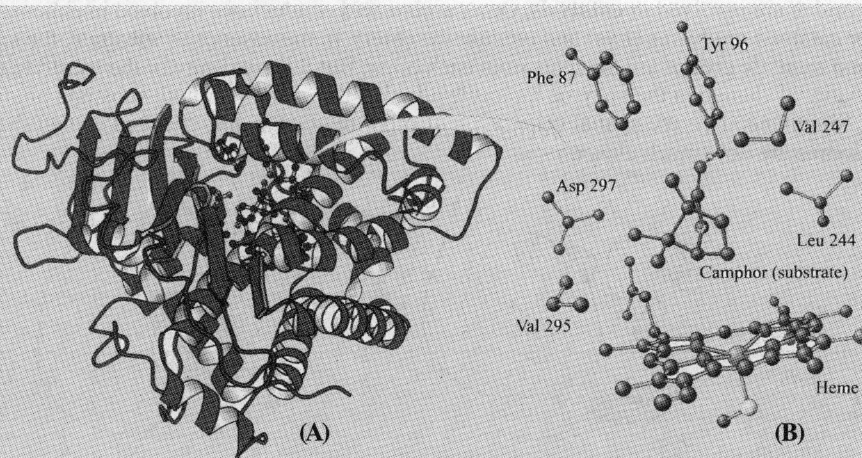

Fig. 18–8. Structure of an enzyme–substrate complex

(A) The enzyme cytochrome P-450 is illustrated bound to its substrate, camphor.

(B) In the active site, the substrate is surrounded by residues from the enzyme. Note also the presence of a heme cofactor.

Evidences Proving the Existence of an ES Complex:

The existence of an ES complex during enzymatically-catalyzed reaction has been shown in many ways:

1. The ES complexes have been directly observed by electron microscopy and x-ray crystallography.
2. The physical properties of enzymes (*esp.*, solubility, heat sensitivity) change frequently upon formation of an ES complex.
3. The spectroscopic characteristics of many enzymes and substrates change upon formation of an ES complex. It is a case parallel to the one in which the absorption spectrum of deoxyhemoglobin changes markedly, when it binds oxygen or when it is oxidized to ferric state.
4. Stereospecificity of highest order is exhibited in the formation of ES complexes. For example, D-serine is not a substrate of tryptophan synthetase. As a matter of fact, the D-isomer does not even bind to the enzyme.
5. The ES complexes can be isolated in pure form. This may happen if in the reaction, $A + B \rightleftharpoons C$, the enzyme has a high affinity for the substrate A and also if the other reactant B is absent from the mixture.
6. A most general evidence for the existence of ES complexes is the fact that at a constant concentration of enzyme, the reaction rate increases with increase in the substrate concentration until a maximal velocity is reached.

Koshland's Induced Fit Model

An important but unfortunate feature of Fischer's model is the rigidity of the active site. The active site is presumed to be pre-shaped to fit the substrate. In order to explain the enzyme properties more efficiently, **Koshland**, in 1958, modified the Fischer's model. **Koshland** presumed that the enzyme molecule does not retain its original shape and structure. But the contact of the substrate *induces* some configurational or geometrical changes in the active site of the enzyme molecule. Consequently, the enzyme molecule is made to *fit* completely the configuration and active centres of the substrate. At the same time, other amino acid residues may become buried in the interior of the molecule. Koshland's hypothesis has been confirmed by **Lipscomb**.

To explain the theory, a hypothetical illustration may be given (Fig. 18–9). The hydrophobic and charged groups both are involved in **substrate binding.** A phosphoserine (-P) and the -SH group of cysteine residue are involved in **catalysis.** Other amino acid residues not involved in either substrate binding or catalysis are lysine (Lys) and methionine (Met). In the absence of substrate, the substrate binding and catalytic groups are far apart from each other. But the proximity of the substrate induces a conformational change in the enzyme molecule aligning the groups for both substrate binding and catalysis. Simultaneously, the spatial orientation of other regions is also changed so that the lysine and methionine are now much closer.

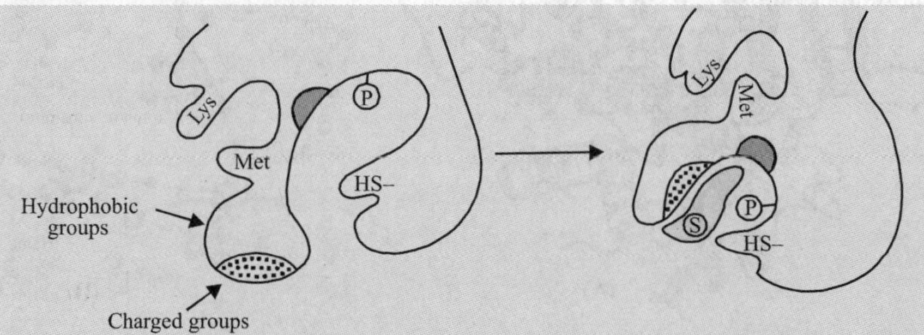

Fig. 18–9. Conformational changes brought about by induced fit in an enzyme molecule

(Modified from Daniel E. Koshland Jr., 1964)

An illustration of the competitive inhibitor or substrate analogue may also be given (Fig. 18–10). On contact with the true substrate, all groups are brought into correct spatial orientation. But attachment of a competitive inhibitor, which is either too "slim" or too "bulky", induces incorrect alignment.

As to the sequence of events during the conformational changes, 3 possibilities exist (refer Fig. 18–11). The enzyme may first undergo a conformational change (A), then bind substrate (B). An alternative pathway is that the substrate may first be bound (C) and then a conformational change (D) may occur. Thirdly, both the processes may occur simultaneously (E) with further isomerization (F) to the final conformation.

Originally little more than an attractive hypothesis, Koshland's model has now gained much experimental support. Conformational changes during substrate binding and catalysis have been demonstrated for various enzymes such as *phosphoglucomutase, creatine kinase, carboxypeptidase* etc.

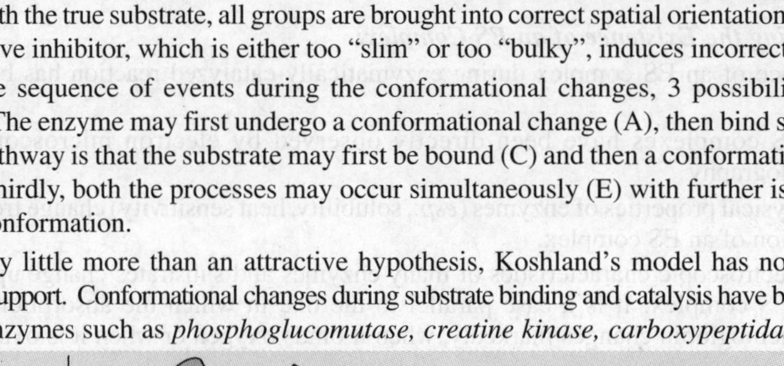

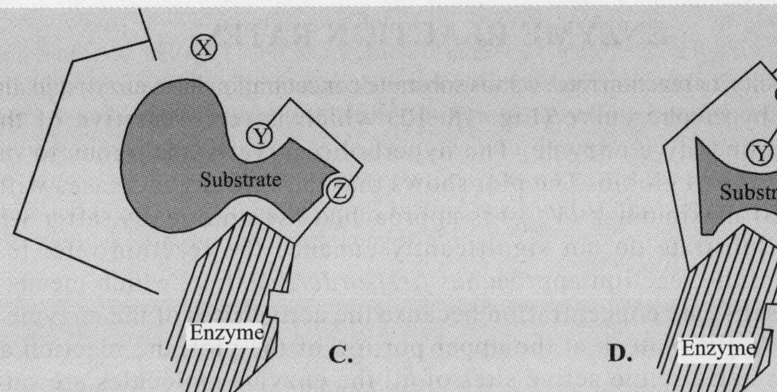

Fig. 18–10. Schematic model of flexible active site mechanism

Black lines indicate protein chains containing catalytic groups X and Y and binding groups Z.

A. Substrate and enzyme dissociated.

B. Substrate with induced change of protein chains to bring X and Y into proper alignment for reaction.

C. Bulky groups added to substrate prevent proper alignment of X and Y.

D. Deleted group eliminates buttressing action on chain containing X so that the complex has incorrect alignment of X and Y.

(From Daniel E. Koshland Jr., 1964)

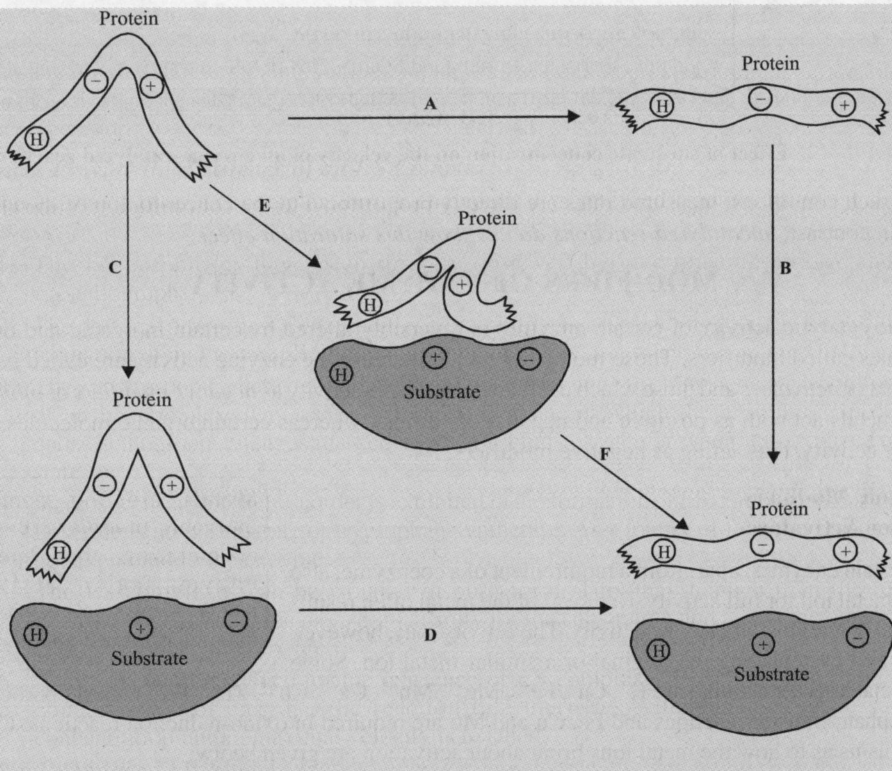

Fig. 18–11. Alternative pathways for a substrate-induced conformational change

(Adapted from Koshland and Neet, 1968)

ENZYME REACTION RATES

A plot of *V* (velocity, or reaction rate) versus substrate concentration in an enzymatically- catalyzed reaction produces a hyperbolic curve (Fig. 18–12) which is representative of the reaction kinetics of a noncooperative enzyme. The hyperbolic curve is analogous to the oxygen-dissociation curve of myoglobin. The plot shows that the velocity increases with substrate concentration until maximum *V* (V_{max}) is approached asymptotically, after which larger concentrations of substrate do not significantly enhance the reaction rate. In the lower region of the curve, the reaction approaches *first-order kinetics*, which means that *v* is a direct function of substrate concentration because the active sites of the enzyme molecules are not saturated. At the plateau at the upper portion of the plot, the reaction approaches *zero-order kinetics* because the active sites of all the enzyme molecules are saturated and the reaction rate is, therefore, independent of further increases in substrate concentration. For the intermediate portion of the curve, as the enzyme approaches substrate saturation, kinetics are mixed zero and first order in substrate concentration. Routine enzyme assays are designed to follow zero-order kinetics to avoid the influence of substrate concentration on reaction velocity.

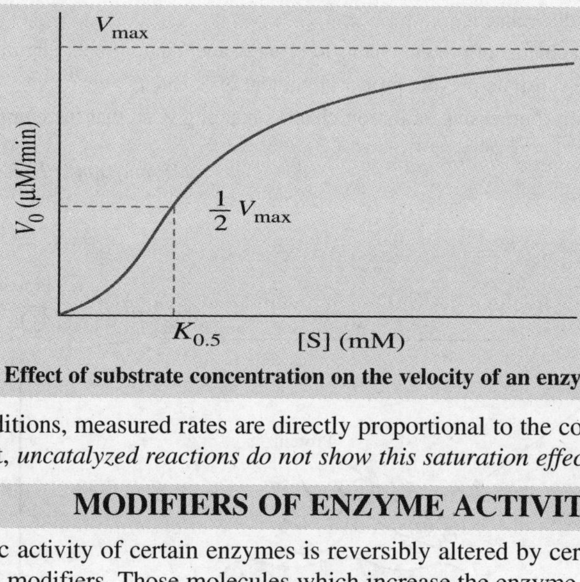

Fig. 18–12. Effect of substrate concentration on the velocity of an enzyme-catalyzed reaction

Under such conditions, measured rates are directly proportional to the concentration of the enzyme itself. In contrast, *uncatalyzed reactions do not show this saturation effect.*

MODIFIERS OF ENZYME ACTIVITY

The catalytic activity of certain enzymes is reversibly altered by certain inorganic and organic molecules called modifiers. Those molecules which increase the enzyme activity are called *positive modifiers* or *activators* and those which decrease the enzyme activity as *negative modifiers* or *inhibitors*. Many metals act both as postitive and negative modifiers, whereas certain organic molecules retard enzyme activity, thus acting as negative modifiers.

Inorganic Modifiers (Enzyme Activators)

> **Metals** are the common inorganic modifiers. Besides accelerating the rate of enzymatically-catalyzed reactions, metals also inhibit the rate of such reactions.

Certain enzymes, apart from a requirement of a coenzyme, also need a metal ion for full activity. Removal of the metal often results in partial or total loss of enzyme activity. The activity may, however, be restored by replacing the original or a similar metal ion. Some such metal ions (or cations) are K^+, Cu^+, Fe^{2+}, Mg^{2+}, Mn^{2+}, Ca^{2+}, Cu^{2+}, Zn^{2+}, Fe^{3+} etc. Mg participates in phosphate-transfer reactions and Fe, Cu and Mo are required in oxido-reduction reactions. Certain mechanisms as to how the metal ions bring about activation are given beolw:

1. **Direct participation in catalysis.** Certain metals may directly participate in the oxidoreduction reactions by undergoing a valence change and thus function in

electron transport system. Fe, for exmaple, functions similarly in *cytochromes* or in *catalase*.

2. **Formation of a metallosubstrate.** Sometimes a metal combines with the substrate to form a metallosubstrate (MS) which, in fact, is the true substrate for the enzyme and forms an enzyme-metal-substrate (EMS) complex. This complex, later on, decomposes to produce the reaction product (P) and regenerate the enzyme and the metal.

$$S + M \rightleftharpoons MS \xrightarrow{+E} EMS \longrightarrow E + M + P$$

3. **Formation of a metalloenzyme.** A metal ion may first combine with an enzyme to form a metalloenzyme (ME) which then combines with the substrate forming an enzyme-metal-substrate (EMS) complex.

$$E + M \rightleftharpoons EM \xrightarrow{+S} EMS \longrightarrow E + M + P$$

4. **Alteration of equilibrium constant.** Metals may also change the nature of the reactants so that the apparent equilibrium constant of the reaction is also altered.
5. **Conformational change in the enzyme.** Metal ions may also bring about conformational change in the enzyme molecule, converting it into an active form. In such a case, the metal may be linked at a point far remote from the substrate and may serve to maintain an active tertiary or quaternary structure.

Regulation of Enzyme Activity

The activity of the enzymes is controlled by certain mechanisms which are enumerated below :

1. ***Zymogen activation.*** Some enzymes (esp., digestive and coagulating ones) are synthesized in inactive forms. These inactive forms of enzymes are called **proenzymes** or **zymogens**. Thus, zymogens are precursors of enzymes. They are not active on their own but can be active after conversion (Table 18–3). Generally, prefix *pro-* or suffix *-ogen* is added to enzyme's name to denote

Table 18–3. Conversion of zymogens to active enzymes

Zymogen	*Activating agent*	*Active enzyme*		*Inactive peptide*
Chymotrypsinogen	Trypsin + Chymotrypsin →	Chymotrypsin	+	Amino acid residues
Trypsinogen	Enterokinase or Trypsin →	Trypsin	+	Hexapeptide
Pepsinogen	H^+ or Pepsin →	Pepsin	+	Fragments
Procarboxypeptidase A	Trypsin →	Carboxypeptidase A	+	Fragments
Proelastase	Trypsin →	Elastase	+	Fragments

its zymogen such as prothrombin, proelastase, trypsinogen, pepsinogen, fibrinogen etc. It is difficult to argue as to why cells should make precursors for some enzymes and not others, rather than merely forming the active enzymes in all cases. Perhaps, the precursors are evolutionary anachronisms or else they are required at times or places when the active enzymes would injure cells.

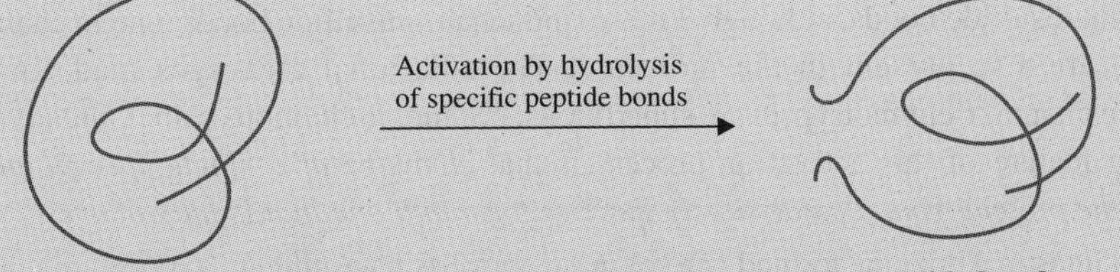

Fig. 18–13. Zymogen activation by hydrolysis of specific peptide bonds

The *precursors* should be converted to the active forms before they can catalyze a chemical reaction. *This type of activation usually involves cleavage of the peptide bond* (Fig. 18–13). Often, a part of the zymogen is also removed. Examples of some common precursor systems are given below:

(A) Chymotrypsin is synthesized by the exocrine cells of the pancreas in its precursor form, *chymotrypsinogen*. Hydrolysis by trypsin converts chymotrypsinogen into the active form, chymotrypsin.

$$\text{Chymotrypsinogen} \xrightarrow{\text{Trypsin}} \text{Chymotrypsin}$$

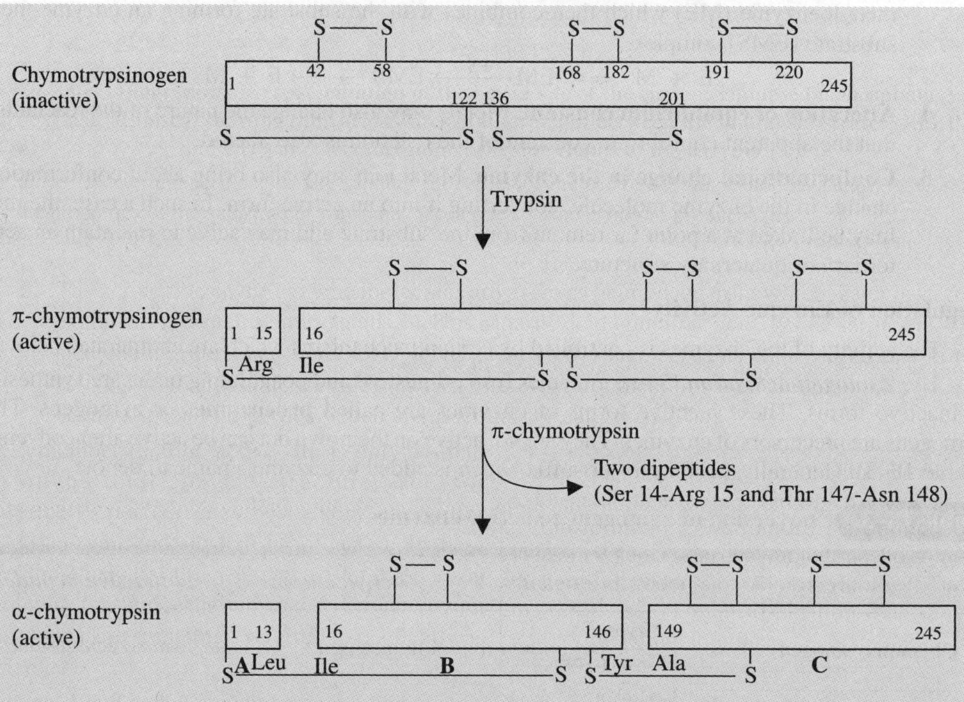

Fig. 18–14. Proteolytic activation of chymotrypsinogen

The 3 chains of α-chymotrypsin are linked by two interchain disulfide bonds (A to B and B to C).

Chymotrypsinogen is composed of a single polypeptide chain with 245 amino acid residues (representing all the 20 standard amino acids) and 5 intrachain disulfide bridges (Fig. 18–14). It is virtually devoid of enzymatic activity and is converted into a fully active enzyme when the peptide bond linking Arg^{15} and Ile^{16} is cleaved by trypsin. The resulting enzyme, called **π-chymotrypsin**, then acts on other π-chymotrypsin molecules. Two dipeptides (between residues 14 and 15 and residues 147 and 148) are removed to yield **α-chymotrypsin**, the stable form of the active enzyme. The 3 resulting chains (designated A, B and C) in α-chymotrypsin remain covalently linked to each other by two interchain disulfide bonds, although 3 more (intrachain) disulfide bonds, one in chain B and two in chain C, are also present in the molecule. The additional cleavages made in converting π-chymotrypsin into α-chymotrypsin are superfluous, because π-chymotrypsin is already fully active. The striking feature of this activation process is that *cleavage of a single specific peptide bond transforms the protein from a catalytically inactive form into one that is fully active.*

(B) An intestinal enzyme termed enterokinase converts *trypsinogen*, a proenzyme also secreted by pancreas, to trypsin. Trypsin acts upon peptide linkages involving the carboxyl group of arginine and lysine.

$$\text{Trypsinogen} \xrightarrow{H^+ \text{ or Pepsin}} \text{Trypsin}$$

(C) Another instance is the proteinase of the gastric juice called as pepsin. This is derived from its proenzyme *pepsinogen*, secreted by gastric mucosa. Pepsinogen is converted to the active form, pepsin, both by the acidity of the gastric juice and by pepsin itself. This process is, therefore, called as **autocatalysis.** During this process, a polypeptide is liberated from the proenzyme.

$$\text{Pepsinogen} \xrightarrow{\text{Enterokinase}} \text{Pepsin} + \text{a polypeptide}$$

Autocatalytic activity of an ezyme precursor may be graphically represented. If a graph of time versus enzyme activity is plotted, a sigmoid (S-shaped) curve is obtained (Fig. 18–15).

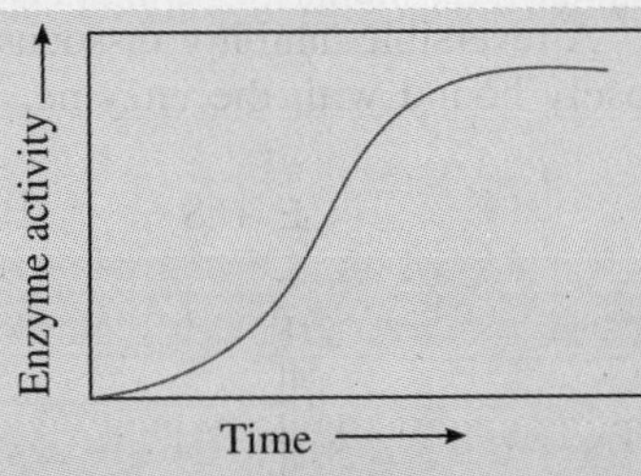

Fig. 18–15. Autocatalytic activation of an enzyme precursor

2. ***Covalent modification.*** The activity of serine-containing enzymes is regulated by the covalent insertion of a small group on an enzyme molecule. For example, the activity of the enzymes, that synthesize and cleave glycogen, is regulated by the attachment of a phosphoryl group to a specific serine residue on these enzymes. This modification can be reversed by hydrolysis.

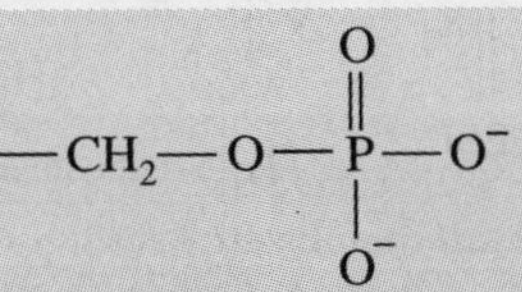

Phosphorylated derivative of a serine residue

3. ***Feedback inhibition.*** Some enzymes catalyze the synthesis of small molecules (such as amino acids) in a number of steps. The enzyme catalyzing the first step in this biosynthesis is inhibited by the end product of the reaction. Such type of regulatory mechanism, which is called feedback inhibition (Fig. 18–16), is beautifully illustrated by the biosynthesis of isoleucine from threonine.

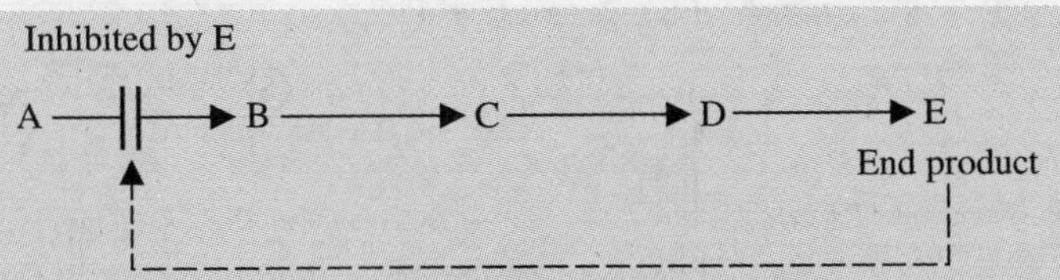

Fig. 18–16. Feedback inhibition of the first enzyme in a multistep reaction by reversible binding of the end product

The reaction completes in 5 steps. The first step reaction is catalyzed by the enzyme *threonine aminase.* The activity of this enzyme is inhibited upon accumulation of high quantities of isoleucine. Isoleucine binds to a different site from threonine. This is called *allosteric interaction.* However, when the level of isoleucine drops sufficiently, the enzyme reactivates and isoleucine is resynthesized.

Organic Modifiers (Enzyme Inhibitors)

Compounds which convert the enzymes into inactive substances and thus adversely affect the rate of enzymatically-catalyzed reaction are called as *enzyme inhibitors.* Such a process is known as *enzyme inhibition*. Two broad classes of enzyme inhibitors are generally recognized: reversible and nonreversible, depending on whether the enzyme-inhibitor (EI) complex dissociates rapidly or very slowly.

Reversible Enzyme Inhibition

A reversible inhibitor dissociates *very rapidly* from its target enzyme because it becomes very loosely bound with the enzyme. Three general types of reversible inhibition (Fig. 18–17) are

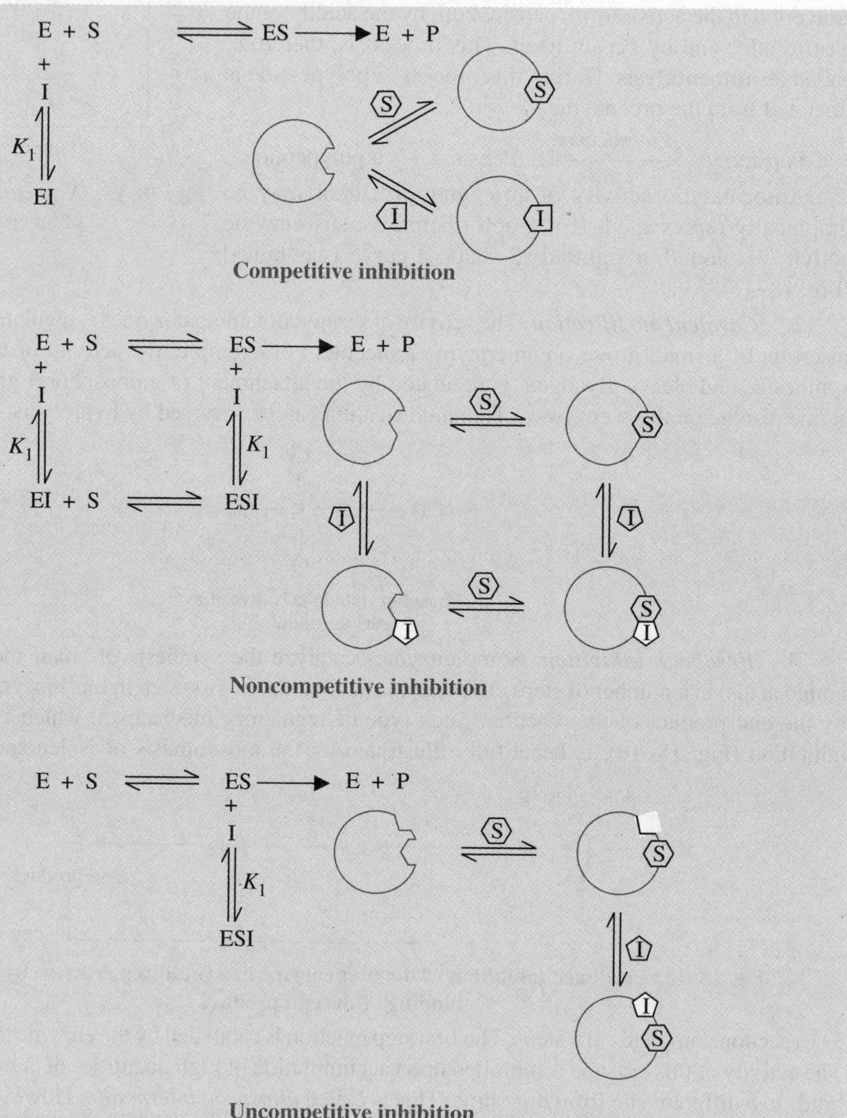

Fig. 18–17. Three types of reversible enzyme inhibition

Competitive inhibitors bind to the enzyme's active site. Noncompetitive inhibitors generally bind at a separate site termed allosteric site. Uncompetitive inhibitors also bind at an allosteric site but they bind only to the ES complex.

distinguished: competitive, noncompetitive and uncompetitive, depending on three factors:

a. whether the inhibition is or is not overcome by increasing the concentration of the substrate,

b. whether the inhibitor binds at the active site or at allosteric site, and

c. whether the inhibitor binds either with the free enzyme only, or with the enzyme substrate complex only or with either of the two.

A. Competitive or Substrate analogue inhibition. This type of competition occurs at the active site (Fig. 18-18). Here the structure of the inhibitor (I) closely resembles with that of the substrate (S). It may, thus, combine with the enzyme (E), forming an enzyme-inhibitor (EI) complex rather than an ES complex. The inhibitor, thus, competes with the substrate to combine with the enzyme. *The degree of inhibition depends upon the relative concentrations of the substrate and the inhibitor.* Thus, by increasing the substrate concentration and keeping the inhibitor concentration constant, the amount of inhibition decreases and conversely a decrease in substrate concentration results in an increased inhibition. It may, however, be noted that in competitive inhibition, the enzyme can bind substrate (forming an ES complex) or inhibitor (EI), but not both (ESI). Thus, we see that a competitive inhibitor diminishes the rate of the reaction by reducing the proportion of the enzyme molecules bound to a substrate.

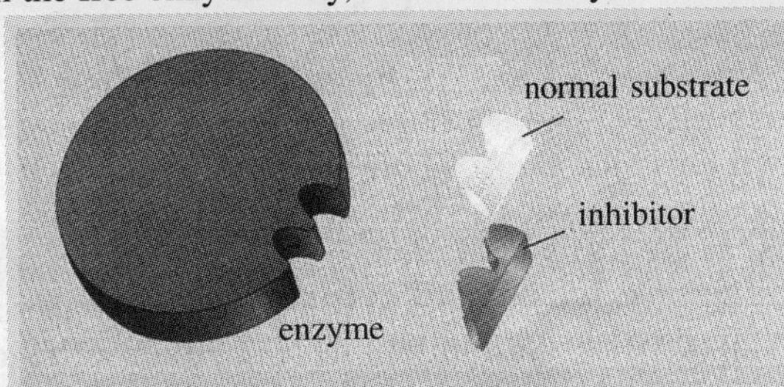

Fig. 18–18. Competitive or substrate analogue inhibition

A competitive inhibitor fits into the active site of the enzyme, preventing the real substrate from gaining access. The inhibitor cannot be converted to the products of the reaction and so the overall rate of reaction is slowed down. If an inhibitor is present in equal concentrations to the substrate, and if both types of molecule bind to the active site equally well, the enzyme can only work at half its normal rate.

Competitive inhibition can be analyzed quantitatively by steady-state kinetics. Because the inhibitor binds reversibly to the enzyme, the competition can be biased to favour the substrate only simply by adding more substrate. When more substrate is present, the probability that an inhibitor molecule will bind is minimized, and the reaction exhibits a normal V_{max}. However, the [s] at which $V_0 = 1/2\ V_{max}$, the K_m will increase in the presence of inhibitor.

Some well-known examples of competitive inhibition are given below:

1. An enzyme, *succinic acid dehydrogenase* (= *succinodehydrogenase*) catalyzes the conversion of succinic acid to fumaric acid.

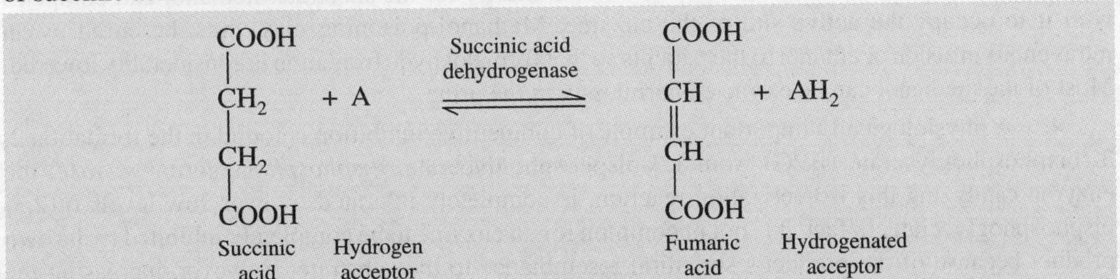

Many organic compounds, which are structurally related to succinic acid, combine with the enzyme, thus inhibiting the reaction. A few inhibitors of this reaction are:

Malonic acid is most efficient of all these inhibitors. When the inhibitor : enzyme ratio is 1 :50, the enzyme is inhibited 50%. Malonic acid differs from succinic acid in having one rather than two methylene groups.

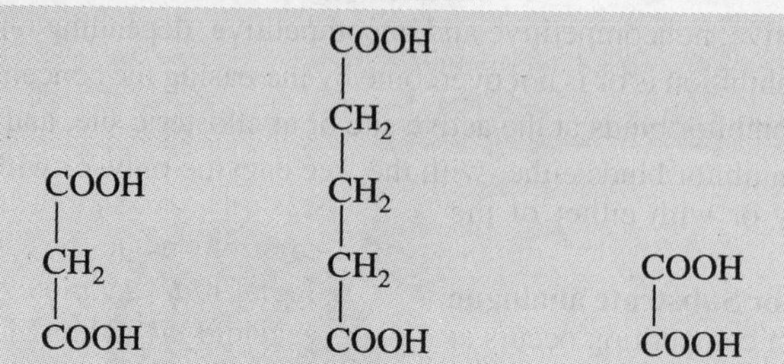

Malonic acid — Glutaric acid — Oxalic acid

2. Many microorganisms, like bacteria, synthesize the vitamin folic acid from para-aminobenzoic acid (PABA). Sulfanilamide and other sulfa drugs are structural analogues of PABA.

***p*-aminobenzoic acid, PABA** (Substrate) — ***p*-aminobenzenesulfonamide** or **Sulfanilamide** (Inhibitor)

Hence, sulfa drugs act as enzyme inhibitor and occupy the active site of some *bacterial enzymes* catalyzing this reaction. Failure of PABA to combine with the bacterial enzymes at the active site results in blocking off the folic acid synthesis. The resulting deficiency of this vitamin is fatal to these microorganisms. Since man lacks the enzymes necessary for folic acid synthesis from PABA, folic acid is needed as a vitamin in the diet. Thus, the sulfa drugs inhibit growth of the bacteria in man by competing with PABA for the active centres of the bacterial enzymes.

3. Competitive inhibition is used therapeutically to treat patients who have ingested methanol ($CH_3.OH$), a solvent found in gas-line antifreeze. Methanol is converted to formaldehyde by the action of the enzyme alcohol dehydrogenase (E.C. No. 1.1.1.1). Formaldehyde damages many tissues esp., the optic ones, causing blindness. Ethanol competes effectively with methanol as a substrate for alcohol dehydrogenase. Ethanol, thus, acts as an inhibitor for the substrate methanol and competes with it to occupy the active site of the enzyme. Methanol-poisoning may, thus, be cured by an intravenous infusion of ethanol to the patients so that formaldehyde formation is considerably lowered. Most of the methanol can be extracted harmlessly in the urine.

4. A physiologically important example of competitive inhibition is found in the formation 2, 3- bisphosphoglycerate (BPG) from 1,3-bisphosphoglycerate. *Bisphosphoglycerate mutase*, the enzyme catalyzing this isomerization reaction, is completely inhibited by even low levels of 2,3-bisphosphoglycerate. In fact, it is not uncommon for an enzyme to be completely inhibited by its own product because of the product's structural resemblance to the substrate. However, increasing the concentration of substrate checks the inhibitory effect.

Michaelis-Menten equation may also be applied to the competitive inhibition of enzymes. Here, besides the normal equation:

$$E + S \rightleftharpoons ES \longrightarrow E + P$$

one must also consider the equilibrium state between the enzyme and the inhibitor, I as follows:

$$E + I \rightleftharpoons EI$$

In the presence of the competitive inhibitor, henceforth, the concentration of the free enzyme would be expressed as:

$$[(E) — (ES) — (EI)]$$

The dissociation of the enzyme-inhibitor compound, K_i would then be defined as:

$$K_i = \frac{[(E) - (ES) - (EI)]\,(1)]}{(EI)}$$

Deriving the Michaelis-Menten equation for this case of inhibition, one obtains as follows:

$$V = \frac{V_m \times (S)\, K_i}{K_m K_i + K_m\,(I) + K_i (S)} \qquad ...(15)$$

Reversal of this equation gives the modified form of Lineweaver-Burk equation:

$$\frac{1}{V} = \frac{K_m K_i + K_m (I) + K_i\,(S)}{V_m \times (S)\, K_i}$$

or

$$\frac{1}{V} = \frac{K_m}{V_m}\left(1 + \frac{(I)}{K_i}\right) \times \frac{1}{(S)} + \frac{1}{V_m} \qquad ...(16)$$

When $\frac{1}{V}$ is plotted against $\frac{1}{(S)}$ (Fig. 18–19), the intercept $\frac{1}{V_m}$ remains the same as in the case of non-inhibited reaction but the slope, which is now $\frac{K_m}{V_m}\left(1 + \frac{(I)}{K_i}\right)$, is increased by the

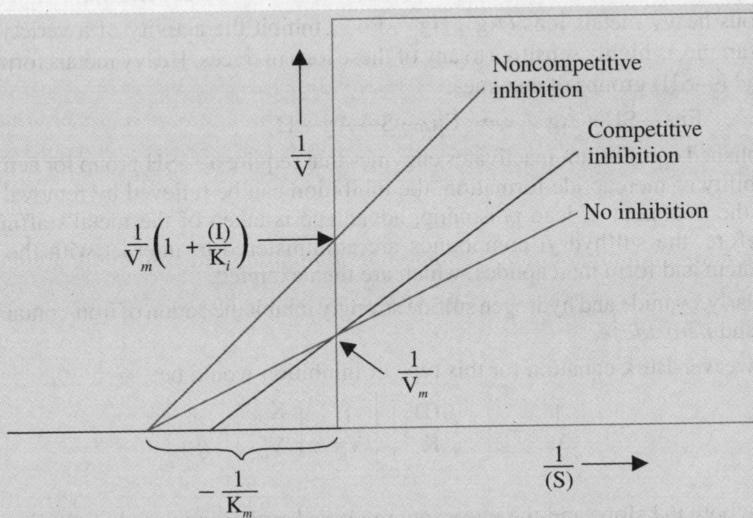

Fig. 18–19. Double reciprocal graph (Lineweaver-Burk plots)

A plot of $\frac{1}{V_m}$ versus $\frac{1}{(S)}$ for an enzyme reaction with (*i*) no inhibitor (*ii*) a competitive inhibitor and (*iii*) a noncompetitive inhibitor.

(Modified from Fairley JL and Kilgour GL, 1966)

factor, $1 + \frac{(I)}{K_i}$. Thus, if the substrate concentration is large enough, the effect of the competitive inhibitor can be overcome and V_m may be reached.

The phenomenon of competitive inhibition is of practical value. It may be employed to prevent the growth of one organism in the presence of other, for example: (*a*) bacterial growth inhibited in the presence of an animal, (*b*) growth of the insect inhibited on a fruit tree.

B. Noncompetitive inhibition. Here no competition occurs between the substrate, S and the inhibitor, I (Fig. 18-20). The inhibitor has little or no structural resemblance with the substrate and it binds with the enzyme at a place other than the active site (*i.e.*, at the allosteric site). Since I and S may combine at different sites, formation of both EI and ESI complexes takes place. Both ES and ESI may break down to produce the reaction product (P). It may, however, be noted that in noncompetitive inhibition, the inhibitor and substrate can bind simultaneously to an enzyme molecule since their binding sites are different and hence do not overlap. The enzyme is inactivated when inhibitor is bound, whether or not substrate is also present. Thus, it is apparent that a noncompetitive inhibitor acts by lowering the turnover number rather than by decreasing the proportion of enzyme molecules that are bound to the substrate. Noncompetitive inhibition, in contrast to competitive inhibition, cannot be overcome by increasing substrate concentration. The inhibitor effectively lowers the concentration of active enzyme and hence lowers V_{max}. There is almost no effect on K_m, however.

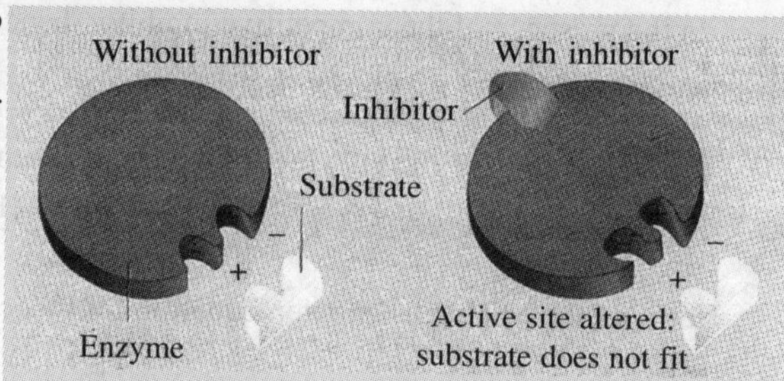

Fig. 18–20. Noncompetitive inhibition

Noncompetitive inhibitors attach to enzyme molecules and alter the overall shape, so that the active site cannot function. Although the substrate may still bind, forming an *enzyme–inhibitor–substrate* complex, the substrate cannot be turned into product. When the inhibitor molecule is removed, normal function is restored.

Certain noteworthy examples of noncompetitive inhibition are as follows:

1. Various heavy metals ions (Ag^+, Hg^{2+}, Pb^{2+}) inhibit the activity of a variety of enzymes. *Urease,* for example, is highly sensitive to any of these ions in traces. Heavy metals form mercaptides with sulfhydryl (—SH) groups of enzymes:

$$\text{Enz—SH} + Ag^+ \rightleftharpoons \text{Enz—S—Ag} + H^+$$

The established equilibrium inactivates enzymes that require a—SH group for activity. Because of the reversibility of mercaptide formation, the inhibition can be relieved by removal of the heavy metal ion. In the treatment of lead poisoning, advantage is taken of the metal's affinity for —SH groups. Therefore, the sulfhydryl compounds are administered to interact with the metal in the circulatory system and form mercaptides, which are then excreted.

2. Similarly, cyanide and hydrogen sulfide strongly inhibit the action of iron-containing enzymes like *catalase* and *peroxidase*.

The Lineweaver-Burk equation for this type of inhibition would be:

$$\frac{1}{V} = \left[1 + \frac{(I)}{K_i}\right]\left[\frac{1}{V_m} + \left(\frac{K_m}{V_m}\right) \times \frac{1}{(S)}\right] \quad ...(17)$$

Thus, here both the slope and the intercept are altered, rather increased by the factor $1 + \frac{(I)}{K_i}$ in contrast to the competitive inhibition where only the slope is changed. Moreover, the maximal velocity attained is less than that found in noninhibited case.

C. Uncompetitive inhibition. An uncompetitive inhibitor also binds at an allosteric site (like the noncompetitive inhibitors) but the binding takes place only with the enzyme-substrate (ES) complex, and not the free enzyme molecule.

Irreversible Enzyme Inhibition

Although irreversible inhibition was once categorized and tested as noncompetitive inhibition, it is now recognized as a distinct type of inhibition. Irreversible inhibitors are those that combine with or destroy a functional group on the enzyme that is essential for its activity. In fact, an irreversible inhibitor dissociates *very slowly* from its target enzyme because it becomes very tightly bound to its active site, thus inactivating the enzyme molecule. The bonding between the inhibitor and enzyme may be covalent or noncovalent.

Two common examples of irreversible inhibition are discussed below:

1. Alkylating reagents, such as **iodoacetamide,** irreversibly inhibit the catalytic activity of some enzymes by modifying cysteine and other side chains. Iodoacetamide is a widely-used agent for the detection of sulfhydryl group.

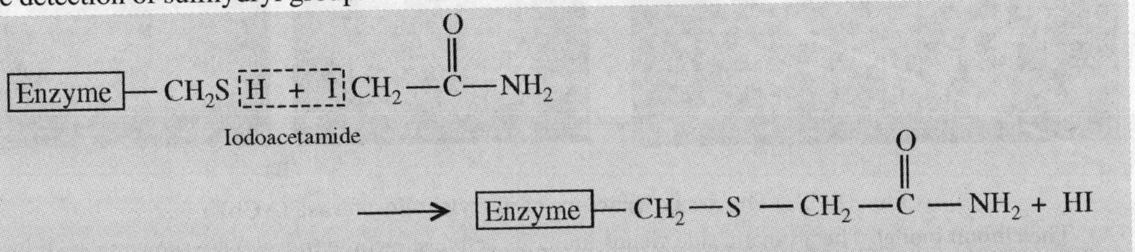

Inactivation of an enzyme (with a critical cysteine residue) by iodoacetamide

2. Organophosphorus compounds, such as diisopropylphosphofluoridate, DIPF, are potent irreversible inhibitors of enzymes that have active seryl residues at their catalytic sites. DIPF is closely

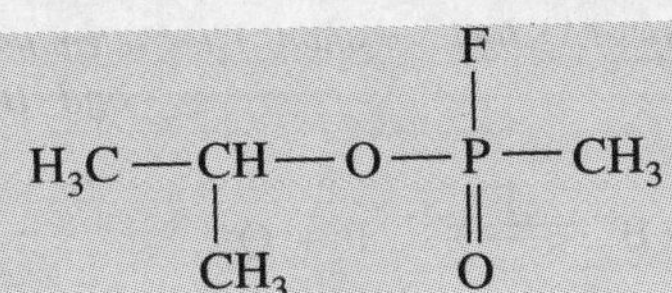

Sarin, a nerve gas

related chemically to nerve gas, whose lethality is due to the inactivation of *acetylcholinesterase*, an enzyme critical for the transmission of nerve impulses. Acetylcholinesterase (Fig. 18–21) cleaves the neurotransmitter acetylcholine, an essential step in normal functioning of nervous system.

DIPF is also called **diisopropylfluorophosphate, DIFP** (AL Lehninger, DL Nelson and MM Cox, 1993)

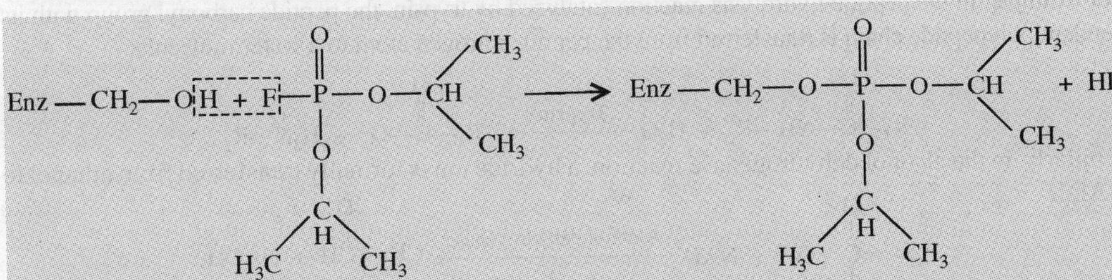

Diisopropylphosphofluoridate (DIPF)

Diisopropylphosphoryl-enzyme (DIP-enzyme)

Inactivation of acetylcholinesterase by diisopropylphosphofluoridate, DIPF

A special class of irreversible inhibitors are the **suicide inhibitors.** These compounds are relatively unreactive until they bind to the active site of a specific enzyme. A suicide inhibitor is designed to carry out the first few chemical steps of the normal enzyme reaction. Instead of being transformed into the normal product, however, the inhibitor is converted to a very reactive compound that combines irreversibly with the enzyme. These are also called **mechanism-based inactivators**, because they utilize the normal enzyme reaction mechanism to inactivate the enzyme.

BISUBSTRATE REACTIONS

We have heretofore been concerned with reactions involving enzymes that require only one substrate. Yet, enzymatic reactions involving two substrates and yielding two products account for ~60% of known biochemical reactions.

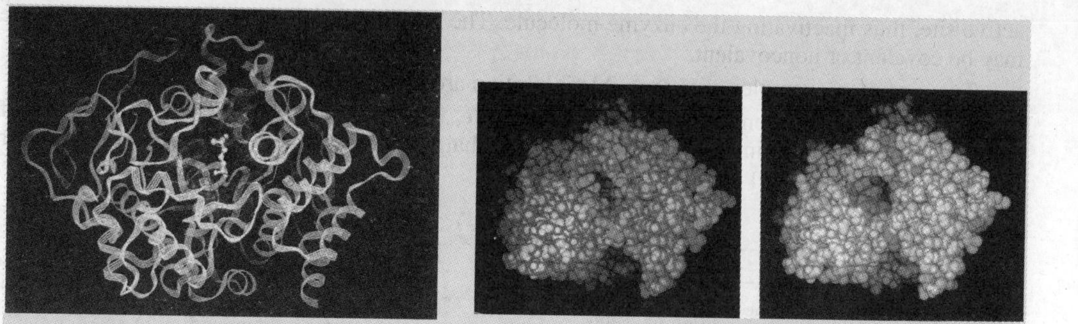

(A) (B)

Fig. 18–21. The x-ray structure of acetylcholinesterase (AChE)

(A) The ribbon model. The aromatic side chains lining its active site gorge (*purple*) are shown in stick form surrounded by their van der Waals dot surface. The ACh substrate, which was modeled into the active site (the enzyme was crystallized in its absence), is shown in ball-and-stick form with its atoms *gold* and its bonds *light blue*. The entrance to the gorge is at the top of the figure.

(B) Space-filling stereo view looking down into the active site. Aromatic residues are in *green*, Ser ^{200} is *red*. Glu ^{199} is *cyan*, and other residues are *grey*.

[Courtesy : (A) Joel Sussman, The Weizmann Institute of Science, Israel and (B) Joel Sussman et al, 1991].

$$A + B \xrightleftharpoons{E} P + Q$$

Almost all of these so-called ***bisubstrate reactions*** are either transferase reactions in which the enzyme catalyzes the transfer of a specific functional group, X, from one of the substrates to the other:

$$P{-}X + B \xrightleftharpoons{E} P + B{-}X$$

or oxidation-reduction reactions in which reducing equivalents are transferred between the two substrates. For example, in the peptide hydrolysis reaction catalyzed by trypsin, the peptide carbonyl group with its pendent polypeptide chain is transferred from the peptide nitrogen atom to a water molecule:

$$R_1{-}\overset{\overset{\displaystyle O}{\|}}{C}{-}NH{-}R_2 + H_2O \xrightarrow{\text{Trypsin}} R_1{-}\overset{\overset{\displaystyle O}{\|}}{C}{-}O^- + H_3\overset{+}{N}{-}R_2$$

Similarly, in the alcohol dehydrogenase reaction, a hydride ion is formally transferred from ethanol to NAD^+:

$$CH_3{-}\underset{\underset{\displaystyle H}{|}}{\overset{\overset{\displaystyle H}{|}}{C}}{-}OH + NAD^+ \xrightarrow[\searrow H^+]{\text{Alcohol dehydrogenase}} CH_3{-}\overset{\overset{\displaystyle O}{\|}}{C}H + NADH$$

A. Terminology

In such reactions, the sequence of binding of the substrate to the enzyme molecule may be of different types. Depending on the sequence of binding of the substrate to the enzyme, **W. Wallace Cleland** (1967) propounded 4 types of mechanisms and explained them with the help of what are commonly called **'Cleland diagrams'** (Fig. 18–22). Cleland's nomenclatural concept is based on the following 5 *conventions*:

1. The substrates are designated by the letters A, B, C and D in the order that they add to the enzyme.
2. The products are denoted by the letters P, Q, R and S in the order that they leave the enzyme.
3. The enzyme is indicated by the letter E. The intermediary stable forms of enzymes, that are produced and then disappear during the course of reaction, are represented by the letters F, G, H etc.
4. The enzyme-substrate complexes are shown in parenthesis.

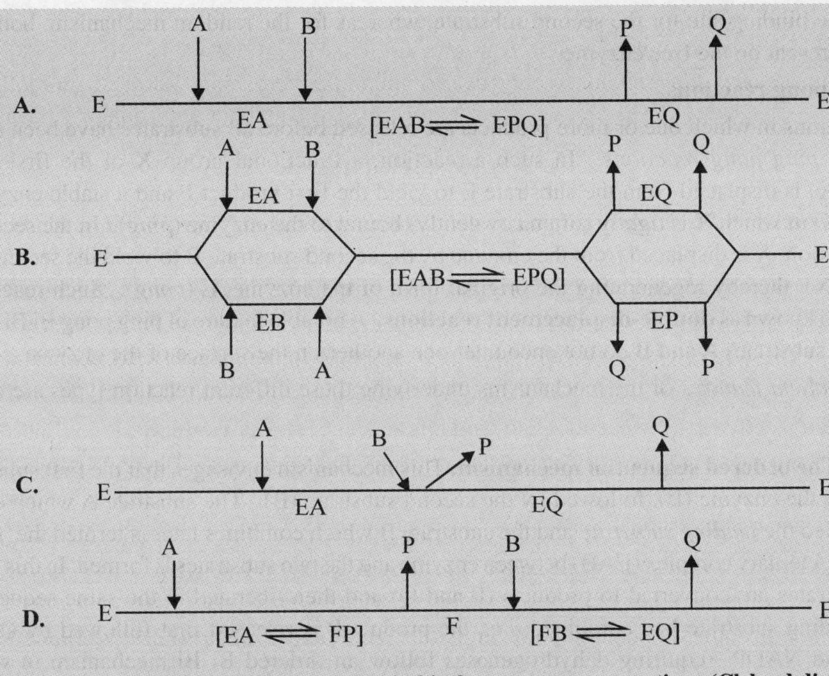

Fig. 18–22. Four substrate binding patterns in bisubstrate enzyme reactions (Cleland diagrams)

A. The ordered sequential mechanism **C.** Theorell–Chance mechanism

B. The random sequential mechanism **D.** The ping pong mechanism

For describing the enzymatic reactions using Cleland's shorthand notation, certain conventions are followed. The enzyme is represented by a horizontal line and successive additions of substrates and release of products are denoted by vertical arrows. Enzyme forms are placed under the line and rate constants, if given, are to the left of the arrow or on top of the line for forward reactions.

5. The number of reactants and products in a given reaction are specified, in order by the terms *Uni* (one), *Bi* (two), *Ter* (Three), and *Quad* (four). For example, a reaction requiring one substance and yielding three products is designated a *Uni Ter reaction*.

B. Types of Bi Bi reactions

In this section, we shall deal with reactions that require two substrates and yield two products, that is *Bi Bi reactions.* It may, however , be remembered that there are numerous examples of even more complex reactions. Enzyme-catalyzed group-transfer Bi Bi reactions fall under two major mechanistic classifications.

1. Sequential reactions

Reactions in which all substrates combine with the enzyme before a reaction can occur and products be released are known as *sequential reactions.* In such reactions, the group being transferred, X, is directly passed from A (= P-X) to B, yielding P and Q (= B -X). Hence, such reactions are also called **single-displacement reactions.**

Sequential reactions can be further classified into 2 types:

(*a*) those with a compulsory order of substrate addition to the enzyme, which are said to have an **ordered mechanism,** and

(*b*) those with no preference for the order of substrate addition, which are described as having a **random mechanism.**

In the ordered mechanism, the binding of the first substrate is apparently required for the enzyme to form the binding site for the second substrate, whereas for the random mechanism, both binding sites are present on the free enzyme.

2. Ping pong reactions

Reactions in which one or more products are released before all substrates have been added are known as *ping pong reactions.* In such a reaction, a functional group X of the first substrate A (= P—X) is displaced from the substrate E to yield the first product P and a stable enzyme form F (= E—X) in which X is tightly (often covalently) bound to the enzyme (*ping*). In the second stage of the reaction, X is displaced from the enzyme by the second substrate B to yield the second product Q (= B—X), thereby regenerating the original form of the enzyme, E (*pong*). Such reactions are, hence, also known as **double-displacement reactions.** A notable feature of ping pong Bi Bi reactions is that the substrates A and B do not encounter one another on the surface of the enzyme.

The *salient features* of the mechanisms underlying these different reaction types are described hereunder:

1. The ordered sequential mechanism. This mechanism envisages that the first substrate (A) binds with the enzyme (E), followed by the second substrate (B). The substrate A which combines first is called the *leading substrate* and the substrate B which combines later is termed the *following substrate.* A ternary complex (EAB) between enzyme and the two substrates is formed. In this complex, both substrates are converted to products (P and Q) and then liberated in the same sequence their corresponding substrate had combined, *i.e.*, the product P is released first followed by Q. Many NAD^+- and $NADP^+$-requiring dehydrogenases follow an ordered Bi Bi mechanism in which the coenzyme is the leading substrate.

2. The random sequential mechanism. In this mechanism, there is no definite sequence of substrate association to the enzyme. They can bind in any order and the products from the ternary complex can also be released in any order. Some dehydrogenases and kinases operate through this mechanism.

3. Theorell–Chance mechanism. It is a variant of the ordered sequential mechanism in which the ternary complex (EAB) is not at all formed. This mechanism operates in the reaction catalyzed by *alcohol dehydrogenase* (E.C. No. 1.1.1.1).

4. The ping pong mechanism. This mechanism envisages that one substrate binds and one product is released before the second substrate can bind to the enzyme and may release the second substrate. Thus, here the enzyme acts as a board and the substrates act as ping pong balls. Many enzymes, including glutamate dehydrogenase (E.C. No. 1.4.1.2), chymotrypsin, transaminases and some flavoenzymes obey ping pong mechanism.

C. Kinetics of Bi Bi Reactions

Steady state kinetic measurements can be utilized to distinguish among the foregoing bisubstrate mechanisms. For doing so, one must first derive their *rate equations.* This can be done in much the same way as for single-substrate-enzymes, *i.e.*, solving a set of simultaneous linear equations consisting of an equation expressing the steady state condition for each kinetically distinct enzyme complex plus one equation representing the conservation condition for the enzyme. This, indeed, is a more complex situation for bisubstrate enzymes than it is for single-substrate or monosubstrate enzymes.

The rate equations for the above-described bisubstrate mechanisms in the absence of products are given below in double reciprocal form.

For ordered Bi Bi reactions:

$$\frac{1}{v_0} = \frac{1}{V_m} + \frac{K_M^A}{V_m[A]} + \frac{K_M^B}{V_m[B]} + \frac{K_S^A K_M^B}{V_m[A][B]} \quad ...(18)$$

For random Bi Bi reactions:

The rate equation for the *general* random Bi Bi reaction is quite complicated. However, in the special case that both substrates are in *rapid and independent equilibrium* with the enzyme; that is, the EAB–EPQ interconversion is rate determining, the initial rate equation reduces to the following relatively simple form. This mechanism is known as the *rapid equilibrium random Bi Bi mechanism*:

$$\frac{1}{v_0} = \frac{1}{V_m} + \frac{K_S^A K_M^B}{V_m K_S^B [A]} + \frac{K_M^B}{V_m [B]} + \frac{K_S^A K_M^B}{V_m [A][B]} \quad ...(19)$$

For ping pong Bi Bi reactions:

$$\frac{1}{v_0} = \frac{K_M^A}{V_m [A]} + \frac{K_M^B}{V_M [B]} + \frac{1}{V_m} \quad ...(20)$$

Physical Significance of the Bisubstrate Kinetic Parameters

The parameters in the equations describing bisubstrate reactions have meanings similar to those for single-substrate reactions. V_m is the maximal velocity of the enzyme obtained when both A and B are present at saturating concentrations; K_M^A and K_M^B are the respective concentrations of A and B, necessary to achieve $\frac{1}{2}V_m$ in the presence of a saturating concentration of the other; and K_S^A and K_S^B are the respective dissociation constants of A and B from the enzyme, *E*.

ALLOSTERIC ENZYMES

A **modulator** is a metabolite which, when bound to the allosteric site of an enzyme, alters its kinetic characteristics. The modulators for allosteric enzymes may be either stimulatory or inhibitory. A stimulator is often the substrate itself. The regulatory enzymes for which substrate and modulator are identical are called **homotropic.** When the modulator has a structure different than the substrate, the enzyme is called **heterotropic.** Some enzymes have more than one modulators.

The term **allosteric** (*allos*[G] = other; *stereos*[G] = space or site) **site** has been introduced by the two Nobel Laureates, **Monod** and **Jacob** to denote an enzyme site, different from the active site, which noncompetitively binds molecules other than the substrate and may influence the enzyme activity.

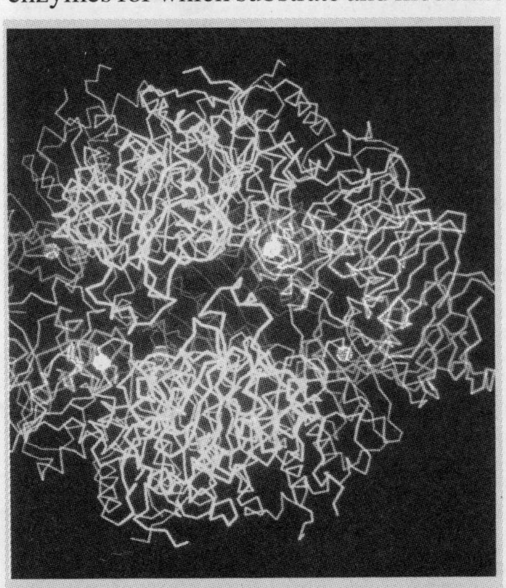

Fig. 18–23. The structure of the enzyme aspartate transcarbamoylase (ATCase) in the T form

Catalytic subunits are shown in *yellow* and regulatory subunits in *green.*

The allosteric enzymes are structurally different from the simple nonregulatory enzymes. Besides the presence of an active or catalytic site, the allosteric enzymes also have one or more regulatory or allosteric sites for binding the modulator. Just as an enzyme's active site is specific for its substrate, the allosteric site is specific for its modulator. Enzymes with several modulators generally have different specific binding sites for each. By contrast, in homotropic enzymes, the active site and allosteric site are the same. Most of them have two or more polypeptide chains or subunits, for example, *aspartate transcarbamoylase,* ATCase (Fig. 18–23) has 12 subunits.

FRANÇOIS JACOB

(born, 1920)

Jacob, a French cellular geneticist, was born on June 17, 1920 at Nancy. He served heroically during World War II (1939–1945) as a member of the Free French Forces, and got injured seriously in Normandy during August 1944. Consequently, he had to wait until 1947 to gain his MD from the University of Paris, but his physical disabilities prevented him from fulfilling his original desire to practice surgery. He spent the major part of his career at the Pasteur Institute, the famous Paris research centre set up around **Louis Pasteur** in the late 19th century and still one of the most influential European laboratories. Jacob contributed to some of the early work on bacteriophages, with **André M. Lwoff** and **Elie Wollman**. Later, he worked in collaboration with **J.C. Monod** that led to the *operon theory* and the *concept of messenger RNA*, in 1961. Both Jacob and Monod shared the **1965 Nobel Prize for Physiology or Medicine** for their work on regulatory gene action in bacteria, along with Lwoff. Jacob has published 2 important books:

1. *The Logic of Life*– which is an illuminating account of changes in scientific thinking over the past three centuries.
2. *The Statue Within*– which is Jacob's autobiography, published in 1987.

JACQUES LUCEIN MONOD

(LT, 1910–1976)

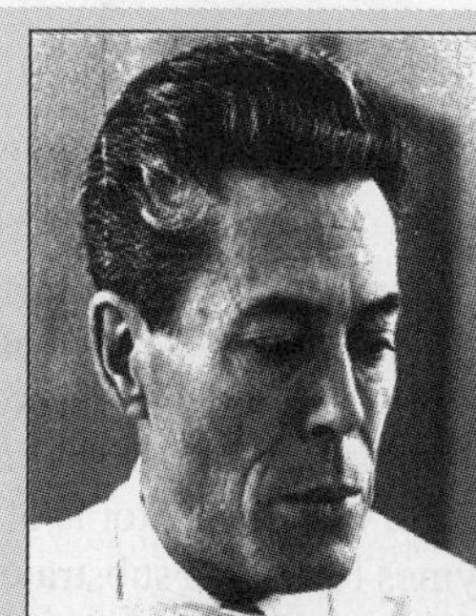

Monod, a French biochemist, was born in Paris on February 9, 1910 to **Sharlie Monod** (formerly, Sharlie MacGregor from Milwaukee) and **Lucein Monod**, a painter, engraver and art historian. Monod lived in Paris most of his time, receiving his advanced degrees from the Sorbonne and the Pasteur Institute, both very distinguished institutions for teaching and research. It was at the Pasteur Institute that Monod developed an intense interest in microbiology. However, after concluding his formal education, Monod was unable to carry out research because Paris was occupied by the invading Nazis. Monod joined the underground Paris Resistance Movement and was arrested by the Gestapo, but he managed to escape.

After the war, Monod began his long career at the Pasteur Institute in Paris where later he became director. Here, he had collaborations with many great scientists, including **F.H.C. Crick** and **François Jacob**. During the late 1940s and early 1950s, Monod's work was directed toward understanding the induction of β-galactosidase in *Escherichia coli*. This led to the development of *operon hypothesis*.

The second great contribution of Monod to science was the *concept of allostery* which he proposed in 1961. This concept, which was not accepted by the enzymologists of the day, asserted that the activity of an enzyme is dependent upon the conformation of the protein. The conformation was controlled at the attachment-detachment site by the effector molecule. He envisioned that the inducer (lactose) would be bound to the repressor which, in turn, would change the binding properties of the repressor so that it could no longer bind to the operator site. Monod was awarded the **1965 Nobel Prize for Physiology or Medicine,** along with **F. Jacob** and **A. M. Lwoff**, for their work on regulatory gene action in bacteria. Monod has published an important book: *Chance and Necessity*, which is a powerful vindication of natural selection. He died on May 31, 1976 from infectious hepatitis at the southern resort town of Cannes in France.

Other differences between nonregulatory enzymes and allosteric enzymes involve kinetic properties. The allosteric enzymes often display sigmoidal plots (Fig. 18–24) of the reaction velocity, V versus substrate concentration, [S], rather than the hyperbolic plots predicted by Michaelis-Menten equation. Recall that the oxygen-binding curve of myoglobin is hyperbolic, whereas that of

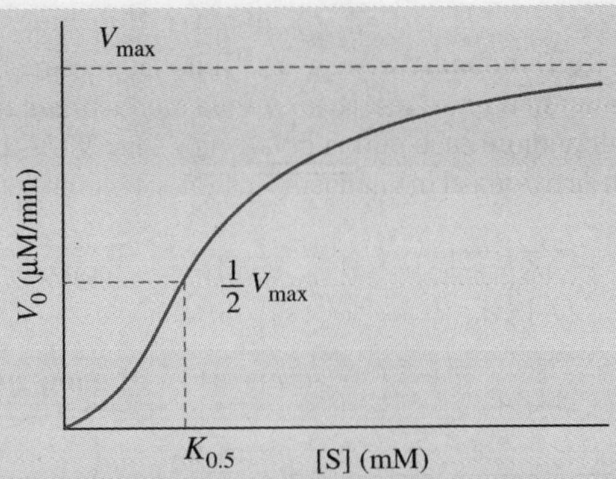

Fig. 18–24. Schematic of an allosteric enzyme's activity by its substrate

Note the sigmoid curve given by a homotropic enzyme, in which the substrate also serves as a positive (stimulatory) modulator. The analogous curve for the nonregulatory enzymes is hyperbolic, as also predicted by the Michaelis–Menten equation. The allosteric enzymes, however, do not obey the Michaelis–Menten kinetics. Also note that a relatively small increase in [S] in the steep part of the curve can cause a very large increase in V_o.

hemoglobin is sigmoidal. The binding of enzymes to substrates is analogous. In allosteric enzymes, the binding of substrate to one active site can affect the properties of other active sites in the same enzyme molecule. A possible outcome of this interaction between subunits is that the binding of substrate becomes cooperative. In addition, the activity of an allosteric enzyme may be altered by regulatory molecules that are bound to the allosteric sites, just as oxygen binding to hemoglobin is affected by H^+, CO_2 and BPG.

With heterotropic enzymes, however, it is difficult to generalize about the shape of the substrate-saturation curve (Fig. 18–25). An activator may cause the substrate-saturation curve to become more nearly hyperbolic with a decrease in $K_{0.5}$ but no change in V_{max}, thus resulting in an increased reaction velocity at a fixed substrate concentration. On the contrary, a negative modulator (= an inhibitor) may produce a more sigmoid substrate-saturation curve, with an increase in $K_{0.5}$. In other words, allosteric activators shift the substrate-saturation curve to the left (to higher saturation), whereas allosteric inhibitors shift it to the right (to lower saturation).

[Although one can find a value of [S] on the sigmoid-saturation curve at which V_0 is half-maximal, one cannot put the designation K_m to it because the enzyme does not follow the hyperbolic Michaelis–Menten relationship. Instead, the symbol $[S]_{0.5}$ or $K_{0.5}$ is often used to represent the substrate concentration giving half-maximal velocity of the reaction catalyzed by an allosteric enzyme.]

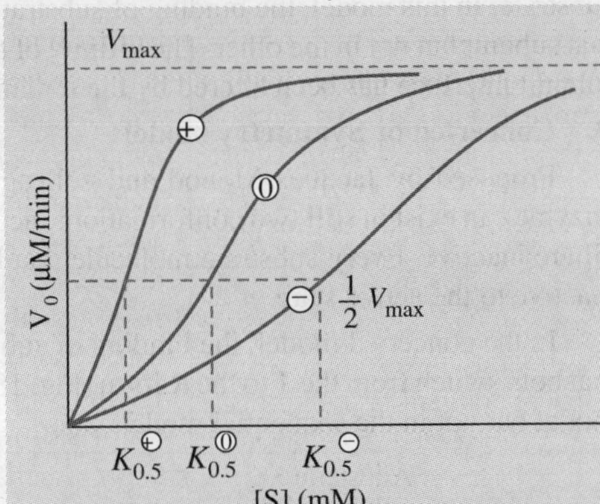

Fig. 18–25. Effects of an activator ⊕, and inhibitor ⊖ and no modulator ◎ on an allosteric enzyme in which $K_{0.5}$ is modulated without a change in V_{max}

Mechanism of Kinetic Behaviour of Allosteric Enzymes

Two general models for the interconversion of inactive and active forms of allosteric enzymes have been proposed.

A. Simple sequential model

Proposed by **Daniel E. Koshland Jr.** in 1966, the model envisages that the allosteric enzyme can exist in only two conformational change individually. Binding of substrate increases the probability of the conformational change. A conformational change in one subunit makes a similar change in an adjacent subunit, as well as the binding of a second substrate molecule, more likely. There are more potential intermediate states in this model than in the other (symmetry) model.

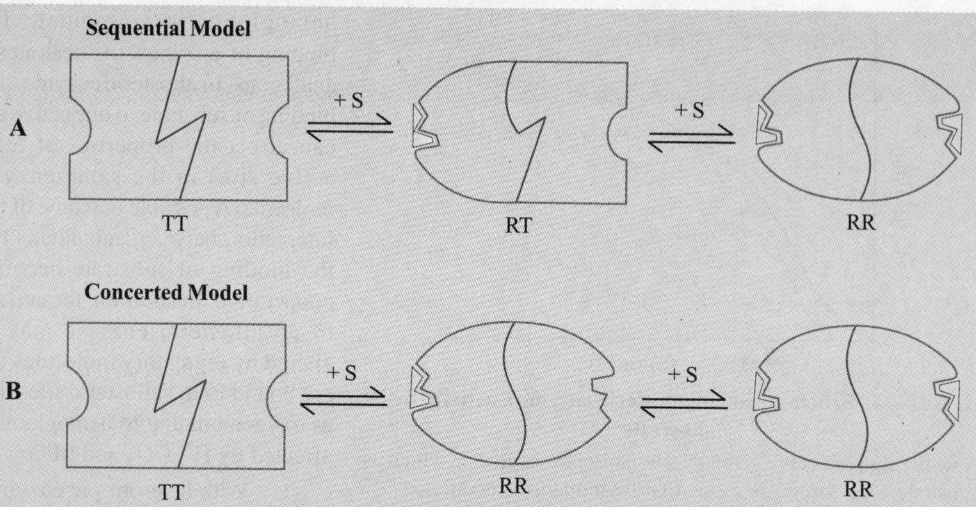

Fig. 18–26. Comparison of the two allosteric models

A. Simple sequential model

B. Concerted or symmetry model

(Adapted from Lubert Stryer, 1995)

Consider an allosteric enzyme consisting of two identical subunits, each containing an active site (Fig. 18–26A). The T (tense) form has low affinity and the R (relaxed) form has high affinity for substrate. In this model, the binding of substrate to one of the subunits induces a T $\rightarrow$ R transition in that subunit but not in the other. The affinity of the other subunit for substrate is increased because the subunit interface has been altered by the binding of the first substrate molecule.

B. Concerted or Symmetry model

Proposed by Jacques Monod and colleagues in 1965, the model envisages that an allosteric enzyme can exist in still two conformations, active and inactive. All subunits are in the active form or all are inactive. Every substrate molecule that binds increases the probability of transition from the inactive to the active state.

In the concerted model, the binding of substrate to one of the subunits increases the probability that both switch from the T to the R form (Fig. 18–26B). Thus, symmetry is conserved in the concerted model but not in the sequential model.

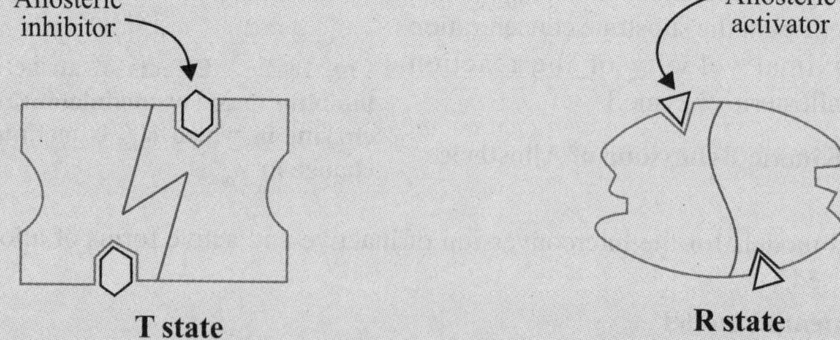

Fig. 18–27. Effect of allosteric activator and allosteric inhibitor on substrate binding

Note that in the concerted model, an allosteric inhibitor stabilizes the T state, whereas an allosteric activator stabilizes the R state.

The effects of allosteric activators and inhibitors can be explained quite easily by the concerted model. An allosteric inhibitor binds preferably to the T form, whereas an allosteric activator binds preferentially to the R form (Fig. 18–27). Consequently, *an allosteric inhibitor shifts the R → T conformational equilibrium toward T, whereas an allosteric activator shifts it toward R*. The result is that an allosteric activator increases the binding of substrate to the enzyme, whereas an allosteric inhibitor decreases substrate binding.

REFERENCES

1. **Abelson JN, Simon MI (editors) :** Methods in Enzymology. *Academic Press Inc., New York. 1992.*
2. **Annual Review :** Advances in Enzymology. *Interscience Publishers, Inc., New York. 1941–Current.*
3. **Atkinson DE :** Regulation of Enzyme Action. *Ann Rev. Biochem.* **35 :** *85–124, 1966.*
4. **Bell JE, Bell ET :** Proteins and Enzymes. *Prentice-Hall Inc., New Jersey. 1988.*
5. **Bender ML, Brubacher LJ:** Catalysis and Enzyme Action. *McGraw-Hill, New York. 1973.*
6. **Bernhard S :** The Structure and Functions of Enzymes. *W.A. Benzamin Inc., New York. 1968.*
7. **Blow DM:** Structure and Mechanism of Chymotrypsin. *Acc. Chem. Res.* **9:** *145–152, 1976.*
8. **Boyer PD, Lardy H, Myrbäck K (editors) :** The Enzymes. *3rd ed. 16 vols. Academic Press, Inc., New York. 1970–1983.*
9. **Brookhaven Symposia in Biology :** Enzyme Models and Enzyme Structure. *No. 15, Brookhaven National Laboratory, Upton, New York.1962.*
10. **Cech TR :** RNA as an enzyme. *Sci. Amer.* **255 (5)** *: 64–75, 1986.*
11. **Chipman DM, Sharon N :** Mechanism of Lysozyme Action. *Science* **165 :** *454–465, 1969.*
12. **Cleland WW :** Enzyme Kinetics. *Ann. Rev. Biochem.* **36 :** *77, 1967.*
13. **Colowick SP, Kaplan NO (editors) :** Methods in Enzymology. *52 vols. Academic Press, Inc., New York. 1955–1978.*
14. **Dische Z :** The discovery of feedback inhibition. *Trends Biochem. Sci.* **1 :** *269–270, 1976.*
15. **Dixon M, Webb EC :** Enzymes. *3rd ed. Academic Press, Inc., New York. 1979.*
16. **Dugas H, Penney C :** Bioorganic Chemistry : A Chemical Approach to Enzyme Action. *Springer-Verlag. 1981.*
17. **Editorial :** Evergeen Enzyme. *Nature.* **218 :** *1202, 1968.*
18. **Fersht A :** Enzyme Structure and Mechanism. *2nd ed. W.H. Freeman and Co., New York. 1985.*
19. **Fersht AR, Leatherbarrow RJ, Wells TNC :** Binding energy and catalysis. A lesson from protein engineering of the tyrosyl-tRNA synthetase. *Trends Biochem. Sci.* **11 :** *321–325, 1986.*
20. **Friedman H (editor) :** Benchmark Papers in Biochemistry. Vol. 1 : Enzymes. *Hutchinson Ross Publishing Company, Stroudsburg, PA. 1981.*
21. **Goodwin TW, Harris JI, Hartley BS (editors) :** Structure and Activity of Enzymes. *Academic Press, Inc., New York. 1964.*
22. **Gutfruend H :** Enzyme : Physical Principles. *Wiley-Intescience, New York. 1972.*
23. **Haldane JBS :** The Enzymes. *MIT Press, Cambridge, Mass. 1965.*
24. **Hammes GG :** Enzyme Catalysis and Regulation. *Academic Press, Inc., New York. 1982.*
25. **Hansen DE, Raines RT :** Binding energy and enzyme catalysis. *J. Chem. Educ.* **67:** *483–489, 1990.*

26. **Henning U :** Multienzyme Complexes. *Angew. Chem. Intern. Ed. Engl.* **5 :** *785–790, 1966.*
27. **Himori K:** Kinetics of Fast Enzyme Reactions. *Halstead, New York. 1979.*
28. **Ingraham LL :** Biochemical Mechanism. *John Wiley and Sons, Inc., New York. 1962.*
29. **International Union of Biochemistry :** Enzyme Nomenclature. *Academic Press, Inc., New York. 1984.*
30. **Jencks WP :** Mechanism of enzyme action. *Ann. Rev. Biochem.* **32 :** *639–676, 1963.*
31. **Jencks WP :** Binding energy, specificity and enzymic catalysis. The Circe effect. *Adv. Enzymol. 43 : 219–410, 1975.*
32. **Jencks WP :** Catalysis in Chemistry and Enzymology. *Dover Publications, Inc., New York. 1987.*
33. **Kirsch JF :** Mechanism of enzyme action. *Ann. Rev. Biochem.* **42 :** *205–234, 1973.*
34. **Knowles JR :** Enzyme catalysis : not different, just better. *Nature.* **350 :** *121–124, 1991.*
35. **Knowles JR, Albery WJ :** Evolution of enzyme function and the development of catalytic efficiency. *Biochemistry.* **15 :** *5631–5640, 1976.*
36. **Koshland DE Jr :** Correlation of structure and function in enzyme action. *Science.* **142 :** *1533–1541, 1963.*
37. **Koshland DE Jr. :** Control of enzyme activity and metabolic pathways. *Trends Biochem. Sci.* **9 :** *155–159, 1984.*
38. **Koshland DE Jr. :** Evolution of catalytic function. *Cold Spring Harbor Symp. Quant. Biol.* **52 :** *1–7, 1987.*
39. **Koshland DE Jr., Neet KE :** The catalytic and regulatory properties of enzymes. *Ann. Rev. Biochem.* **37 :** *359–410, 1968.*
40. **Kraut J :** How do enzymes work ? *Science. 242 : 533–540, 1988.*
41. **Kurganov BI :** Allosteric Enzymes. *John Wiley and Sons, New York. 1982.*
42. **Leinhard GE :** Enzyme catalysis and transition-state theory. *Science.* **180 :***149–154, 1973.*
43. **Mildvan AS :** Mechanism of Enzyme Action. *Ann. Rev. Biochem.* **43 :** *357, 1974.*
44. **Monod J, Changeux J-P, Jacob F :** Allosteric proteins and cellular control systems. *J. Mol. Biol.* **6 :** *306–329, 1963.*
45. **Moss DW :** Isoenzymes. *Chapman and Hall, London. 1982.*
46. **Moss DW, Butterworth PJ:** Enzymology and Medicine - *Pitman Medical, London. 1974.*
47. **Osserman EF, Canfield RE, Beychok S (editors) :** Lysozyme. *Academic Press, Inc., New York. 1974.*
48. **Page MI, Williams A (editors) :** Enzyme Mechanisms. *Royal Society of Chemistry. 1987.*
49. **Palmer T :** Understanding Enzymes. *John Wiley and Sons, New York. 1985.*
50. **Phillips DC :** The three-dimensional structure of an enzyme molecule. *Scientific American.* **215 (5) :** *78–90, 1966.*
51. **Philips DC, Richards FM:** Atlas of Molecular Structures in Biology. I. Ribonuclease. *S. Oxford. 1973.*
52. **Price NC, Stevens L :** Fundamentals of Enzymology. *Oxford University Press, Oxford. 1982.*
53. **Rose IA :** Mechanism of enzyme action. *Ann. Rev. Biochem.* **35 :** *23–56, 1966.*
54. **Schultz PG :** The interplay of chemistry and biology in the design of enzymatic catalysts. *Science.* **240 :** *426–433, 1988.*
55. **Segel IH :** Enzyme Kinetics: Behaviour and Analysis of Rapid Equilibrium and Steady State Enzyme Systems. *John Wiley and Sons, Inc., New York. 1975.*

56. **Segel IH :** Biochemical Calculations. *2nd ed., John Wiley and Sons, Inc., New York. 1976.*
57. **Sigman DS, Mooser G :** Chemical studies of enzyme active sites. *Ann. Rev. Biochem.* **44 :** *889, 1975.*
58. **Sumner JB, Somers GF :** Chemistry and Methods of Enzymes. *Academic Press, Inc., New York. 1953.*
59. **Walsh C :** Enzymic Reaction Mechanisms. *W.H. Freeman and Company, San Francisco, California. 1979.*
60. **Warshel A:** Energetics of enzyme catalysis. *Proc. Nat. Acad. Sci.* **75:** *5250 – 5254, 1978.*
61. **Weeb JL :** Enzyme and Metabolic Inhibitors. *Vols. 1 to 3. Academic Press, Inc., New York. 1963–1966.*
62. **Welch GR (editor) :** Organized Multienzyme Systems : Catalytic Properties. *Academic Press, Inc., New York. 1985.*
63. **Wharton CW, Eisenthalo R :** Molecular Enzymology. *Blackie, London. 1981.*
64. **Wilkinson JH :** Isoenzymes. *2nd ed., Chapman and Hall, London. 1970.*
65. **Zeffren E, Hall PL :** The Study of Enzyme Mechanisms. *Wiley, New York. 1973.*

PROBLEMS

1. For enzyme that follows simple Michaelis-Menten kinetics, what is the value of V_{max} if V_0 is equal to 1 μmol/minute at 1/10 K_M?
2. A simple Michaelis-Menten enzyme, in the absence of any inhibitor, displayed the following kinetic behaviour. The expected value of V_{max} is shown on the *y*-axis.

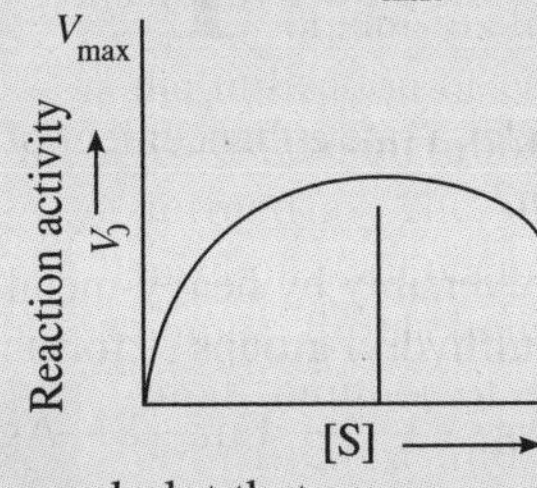

   (*a*) Draw a double-reciprocal plot that corresponds to the velocity-versus-substrate curve.
   (*b*) Provide an explanation for the kinetics results.
3. The active site of an enzyme usually consists of a pocket on the enzyme surface lined with the amino acid side chains necessary to bind the substrate and catalyze its chemical transformation. Carboxypeptidase, which sequentially removes the carboxyl-terminal amino acid residues from its peptide substrates, consists of a single chain of 307 amino acids. The two essential catalytic groups in the active site are furnished by Arg^{145} and Glu^{270}.
   (*a*) If the carboxypeptidase chain were a perfect α helix, how far apart (in nanometers) would Arg^{145} and Glu^{270} be?
   (*b*) Explain how it is that these two amino acids, so distantly separated in the sequence, can catalyze a reaction occurring in the space of a few tenths of a nanometer.
   (*c*) If only these two catalytic groups are involved in the mechanism of hydrolysis, why is it necessary for the enzyme to contain such a large number of amino acid residues?
4. (*a*) At what substrate concentration will an enzyme having a k_{cat} of 30 s^{-1} and a K_m of 0.005 M show one-quarter of its maximum rate?
   (*b*) Determine the fraction of V_{max} that would be found in each case when $[S] = ½K_m$, $2\ K_m$, and $10\ K_m$.

5. The following experimental data were collected during a study of the catalytic activity of an intestinal peptidase capable of hydrolyzing the dipeptide glycylglycine:

$$\text{Glycylglycine} + H_2O \longrightarrow 2\ \text{glycine}$$

[S] (mM)	Product formed (μmol/min)
1.5	0.21
2.0	0.24
3.0	0.28
4.0	0.33
8.0	0.40
16.0	0.45

From these data, determine by graphical analysis the values of K_m and V_{max} for this enzyme preparation and substrate.

6. Carbonic anhydrase of erythrocytes (M_r 30,000) is among the most active of known enzymes. It catalyzes the reversible hydration of CO_2:

$$H_2O + CO_2 \rightleftharpoons H_2CO_3$$

which is important in the transport of CO_2 from the tissues to the lungs.

(***a***) If 10 μg of pure carbonic anhydrase catalyzes the hydration of 0.30 g of CO_2 in 1 min at 37°C under optimal conditions, what is the turnover number (k_{cat}) of carbonic anhydrase (in units of min^{-1})?

(***b***) From the answer in (*a*), calculate the activation energy of the enzyme-catalyzed reaction (in kJ/mol).

(***c***) If carbonic anhydrase provides a rate enhancement of 10^7, what is the activation energy for the uncatalyzed reaction?

7. Many enzymes are inhibited irreversibly by heavy-metal ions such as Hg^{2+}, Cu^{2+}, or Ag^+, which can react with essential sulfhydryl groups to form mercaptides:

$$\text{Enz—SH} + Ag^+ \longrightarrow \text{Enz—S—Ag} + H^+$$

The affinity of Ag^+ for sulfhydryl groups is so great that Ag^+ can be used to titrate —SH groups quantitatively. To 10 mL of a solution containing 1.0 mg/mL of a pure enzyme was added just enough $AgNO_3$ to completely inactivate the enzyme. A total of 0.342 μmol of $AgNO_3$ was required. Calculate the *minimum* molecular weight of the enzyme. Why does the value obtained in this way give only the minimum molecular weight?

8. When enzyme solutions are heated, there is a progressive loss of catalytic activity with time. This loss is the result of the unfolding of the native enzyme molecule to a randomly coiled conformation, because of its increased thermal energy. A solution of the enzyme hexokinase incubated at 45°C lost 50% of its activity in 12 min, but when hexokinase was incubated at 45°C in the presence of a very large concentration of one of its substrates, it lost only 3% of its activity. Explain why thermal denaturation of hexokinase was retarded in the presence of one of its substrates.

9. An experiment measuring velocity versus substrate concentration was run, first in the absence of substance A, and then in the presence of substance A. The following data were obtained:

$[S]$ (μM)	*Velocity in Absence of A* (μ*mol min*$^{-1}$)	*Velocity in Presence of A* (μ*mol min*$^{-1}$)
2.5	0.32	0.20
3.3	0.40	0.26
5.0	0.52	0.36
10.0	0.69	0.56

Is substance A an activator or an inhibitor? If it is an inhibitor, what kind of inhibitor is it?

10. For the experiment in Question 9, calculate the K_m and V_{max} both in the absence and in the presence of substance A. Are these results consistent with your answer for Question 9?

11. Bacterial growth is inhibited by sulfanilamide and related sulpha drugs, and there is a concomitant accumulation of 5-aminoimidazole-4-carboxamide ribonucleotide. This inhibition is reversed by the addition of *p*-aminobenzoate. Propose a mechanism for the inhibitory effect of sulfanilamide.

ANSWERS TO PROBLEMS

1. 11μmol minute^{-1}.

2. (*a*)

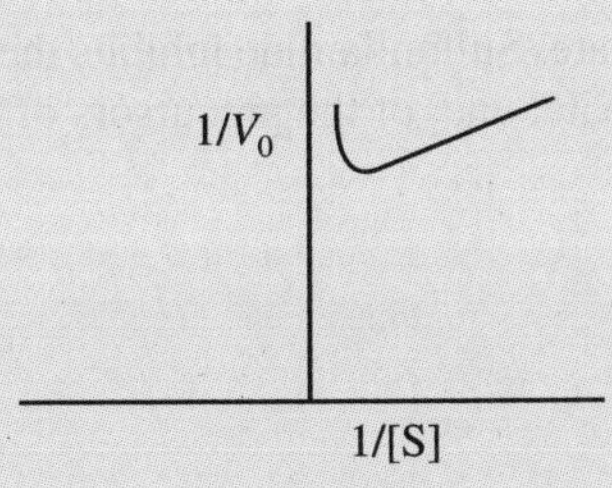

(*b*) This behaviour is substrate inhibition— at high concentrations, the substrate forms unproductive complexes at the active site. The drawing below shows what might occur, Substrate normally binds in a defined orientation, shown in the drawing as red to red and blue to blue. At high concentrations, the substrate may bind at the active site such that the proper orientation is met for each end of the molecule, but two different substrate molecules are binding.

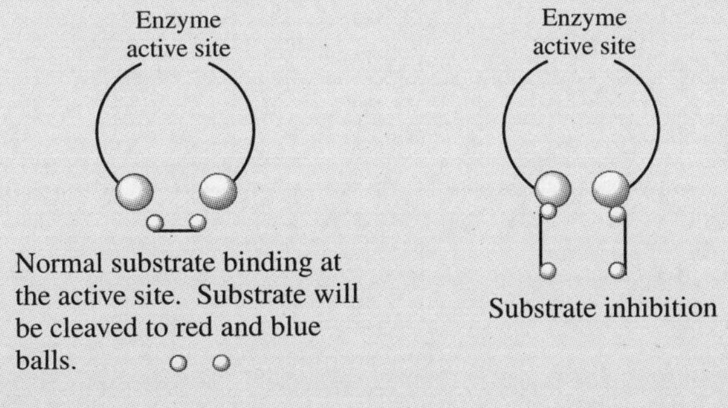

3. (*a*) 18.8 nm

(*b*) Three-dimensional folding of the enzyme brings these amino acid residues into close proximity.

(*c*) The protein serves as "scaffolding" to keep the catalytic groups in a precise orientation. Also, many other interactions occur between the enzyme and its substrate, and some of the binding energy derived from these interactions contributes to catalysis.

4. (*a*) 1.7×10^{-3} M (*b*) 0.33; 0.67; 0.91.
5. $K_m = 2.2$ mM; $V_{max} = 0.51$ μmol/min.
6. (*a*) $k_{cat} = 2.0 \times 10^7\ min^{-1}$
   (*b*) $\Delta G^{\ddagger} = 43.3$ kJ/mol (at 37 °C or 310 K) (*c*) $\Delta G^{\ddagger} = 83.2$ kJ/mol
7. 29,000; it is assumed that each enzyme molecule contains only one titratable sulfhydryl group.
8. The enzyme–substrate complex is more stable than the enzyme alone.
9. The best way to handle the data is to take the reciprocals of both [S] and *v* and construct a Lineweaver–Burk plot. You should find that the two curves cross the *y*-axis at the same point but the curve in the presence of A crosses the *x*-axis closer to the origin. This pattern indicates that A is a competitive inhibitor.
10.

	$-1/K_m$	K_m	$1/V_{max}$	V_{max}
Absence of A	−0.14	7.1	0.8	1.25
Presence of A	−0.08	12.5	0.8	1.25

With a competitive inhibitor V_{max} remains constant (be sure you understand why) but the apparent K_m is larger. It takes more substrate to reach a given velocity because the substrate has to compete with the inhibitor

11. There is a deficiency of N^{10}-formyltetrahydrofolate. Sulfanilamide inhibits the synthesis of folate by acting as an analogue of *p*-aminobenzoate, one of the precursors of folate.

PART

III

Bioenergetics and Metabolism

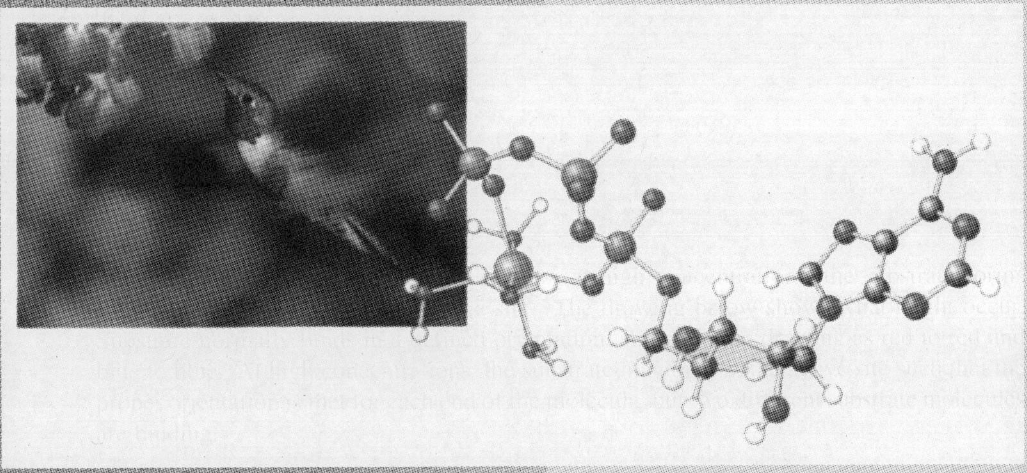

Prodigious feats of endurance exhibited by hummingbirds

The tiny ruby-throated hummingbird can store enough fuel to fly across the Gulf of Mexico, a distance of some 800 kilometres, without resting. This achievement is possible because of the ability to convert fuels into the cellular energy currency, ATP, represented by the model at the right.

(Courtesy: (Left) K.D. McGraw)

> "All the vital mechanisms, varied as they are, have only one object, that of preserving constant the conditions of life in the internal environment."
>
> **– *Claude Bernard: Memoirs, 1855.***

CONTENTS

Like motor traffic, metabolic pathways flow more efficiently when regulated by signals.

CHAPTER

19 Metabolic Concepts

DEFINITION OF METABOLISM

The term metabolism applies to the assembly of biochemical reactions which are employed by the organisms for the synthesis of cell materials and for the utilization of energy from their environments. In other words, the **metabolism** of an organism (or of a cell) may be defined as 'the sum total of all the enzyme-catalyzed reactions that occur in an organism (or in a cell)'. The large number of reactions in a cell are organized into a relatively small number of sequences or pathways. It is a highly coordinated and purposeful cell activity, in which multienzyme systems cooperate. This obviously points out to the fact that the metabolism of even a simple unicellular organism is time-variant, *i.e.*, all its aspects are not actually expressed at any given point.

The magnitude of metabolism may be appreciated by taking the examples of microorganisms and the human beings :

(*a*) Microorganisms like bacteria (*e.g.*, *Escherichia coli*) can double in number every 40 minutes in a culture medium containing only glucose and inorganic salts, or in 20 minutes in a rich broth. The components of the medium are depleted and very little is added to the medium by the cells. Each cell contains hundreds to thousands of molecules of each of about 2,500 different proteins, about 1,000 types of organic compounds and a variety of nucleic acids. It is, thus, apparent that the bacterial cells participate in a variety of metabolic activities in a remarkable way.

(*b*) Human adults maintain a constant weight for about 40 years, during which period a total of about 60 quintals of solid food and 45,000 litres of water are metabolized. And yet both body weight and body composition remain almost constant.

TERMINOLOGY OF METABOLISM

The various processes constituting metabolism may be divided, somewhat arbitrarily, into catabolism and anabolism. Those processes, whose major function is the generation of chemical energy in forms suitable for the mechanical and chemical processes of the cells, are termed as **catabolism** (*cata*[G] = down ; *ballein*[G] = to throw) ; whereas those processes, which utilize the energy generated by catabolism for the biosynthesis of cell components, are termed as **anabolism** (ana[G] = up). The various activities powered by catabolism include mechanical movement, growth, reproduction, accumulation of foods, elimination of wastes, generation of electricity, maintenance of temperature etc. The various anabolic activities may be exemplified by food manufacture, protein synthesis, fat synthesis etc.

The terms **catabolism** and **anabolism** were coined by the physiologist **Gasket**, in 1886.

Some processes can be either catabolic or anabolic, depending on the energy conditions in the cell. These are referred to as **amphibolism.**

$$\text{Fuels (carbohydrates, fats)} \xrightarrow{\text{Catabolism}} CO_2 + H_2O + \text{Useful energy}$$

$$\text{Useful energy + Small molecules} \xrightarrow{\text{Anabolism}} \text{Complex molecules}$$

The relationship among catabolism, anabolism and other terms of metabolic importance is summarized in Table 19–1.

Table 19–1. Classification of general metabolic terms

Processes are	degradative:	biosynthetic:
	Catabolism	Anabolism
	Dissimilation	Assimilation
Energy	yielding:	consuming:
	Exergonic	Endothermic
	Exothermic	Endergonic
Terminal electron acceptor is	oxygen:	not oxygen:
	Aerobic	Anaerobic
	Respiration	Fermentation

(Adapted from MF Mallette, CO Clagett, AT Phillips and RL McCarl, 1979)

In all cells, metabolism enables the cell to perform its vital functions. Metabolism performs following 4 specific functions:

(*a*) to obtain chemical energy from the degradation of energy-rich nutrients or from captured solar energy.

(*b*) to convert nutrient molecules into precursors of cell macromolecules.

(*c*) to assemble these precursors into proteins, lipids, polysaccharides, nucleic acids and other cell components.

(*d*) to form and degrade biomolecules required in specialized functions of cells.

These features of metabolism are closely interrelated since the synthesis of the molecules, that are a component of cell, requires an input of energy, while at the same time it is obvious that the cell components (such as those that make up the cell membrane and its constituent transport proteins) are needed to provide the energy supply and to control intracellular solute concentrations. The specialized functions (such as movement, the secretion of a particular type of molecule or the stimulation of an adjacent cell) also require biosynthetic processes as well as a supply of energy.

Terms, *catabolism* and *dissimilation* are synonyms and refer to the pathways or routes breaking down food materials into simpler compounds and resulting in the release of energy contained in them. The processes of *anabolism* or *assimilation* (also synonyms) utilize food materials (or intermediates from catabolism) and energy to synthesize cell components.

In dealing with the energy relations of the biological processes, the term *exergonic* is used to denote a chemical reaction which liberates chemical-free energy. The term *exothermic* refers to the total energy liberated including heat. As the magnitude of heat energy is small and also that it cannot drive biological reactions, the biochemists are more interested in free energy changes and often use the term exergonic. The corresponding energy-consuming term *endergonic* refers to the processes which require an input of free energy while the term *endothermic* denotes a total energy requirement including heat.

The organisms which reduce oxygen are said to be *aerobic*. The route or pathway by which this reaction is accomplished (*esp*., its terminal steps) is called *respiration*. The organisms which reduce not oxygen but other compounds are said to be *anaerobic*. The term *fermentation* is complementary to respiration. It is to be noted that the term respiration is included twice in the above table. According to the modern trend among the biochemists, respiration is defined as the terminal processes involved in the reduction of molecular oxygen. The previous trend, as depicted in the older literature, used the term respiration for any process transferring electrons and releasing energy, whether terminating at molecular oxygen or not.

FUNCTIONS OF METABOLISM

An arrays of enzyme-catalyzed chemical reactions, that bring about transformations of certain organic compounds vital to the organism, constitute *metabolic pathways* or *metabolic routes*. The molecules or the compounds which participate in these reactions are called as *metabolic intermediates* or *metabolites*. The synthesis of cell constituents begins with the metabolic pathways which supply the building blocks. The major cell constituents are complex carbohydrates, the proteins, the lipids, and the nucleic acids. They are formed from relatively simple starting materials (Fig. 19–1). In mammalian cells, these key metabolites come from one of the following 3 sources.

(*a*) Absorption of the metabolites *from outside the cell* and by implication, if we consider the organism as a whole, by absorption of the metabolites as dietary constituents from the alimentary tract.

(*b*) Release of the metabolite *from a source stored with the cell*. The metabolite may be released either from a molecule used solely for storage (such as glycogen or triacylglycerol), or from a molecule that has another function within the cell (such as an enzyme or a membrane lipid).

(*c*) The metabolite may be formed *by the metabolism of a simple precursor*. However, the precursor must be absrobed by the cell or derived from a source stored within the cell.

Following are the key functions of metabolism:

A. Metabolism enables the cell to convert some of the energy found in nutrients into a form which will support biosynthesis, the maintenance of homeostasis and the cell's other energy-requiring processes.

Biosynthetic reaction sequences or pathways in the cell require an input of energy and this will normally be in the form of ATP or a reduced coenzyme (NADH, NADPH). It is noteworthy that the usefulness of ATP as an energy source is not a consequence of any special or 'high energy' form of the phosphate bond, but is merely a function of how far the hydrolysis of ATP is displaced from equilibrium.

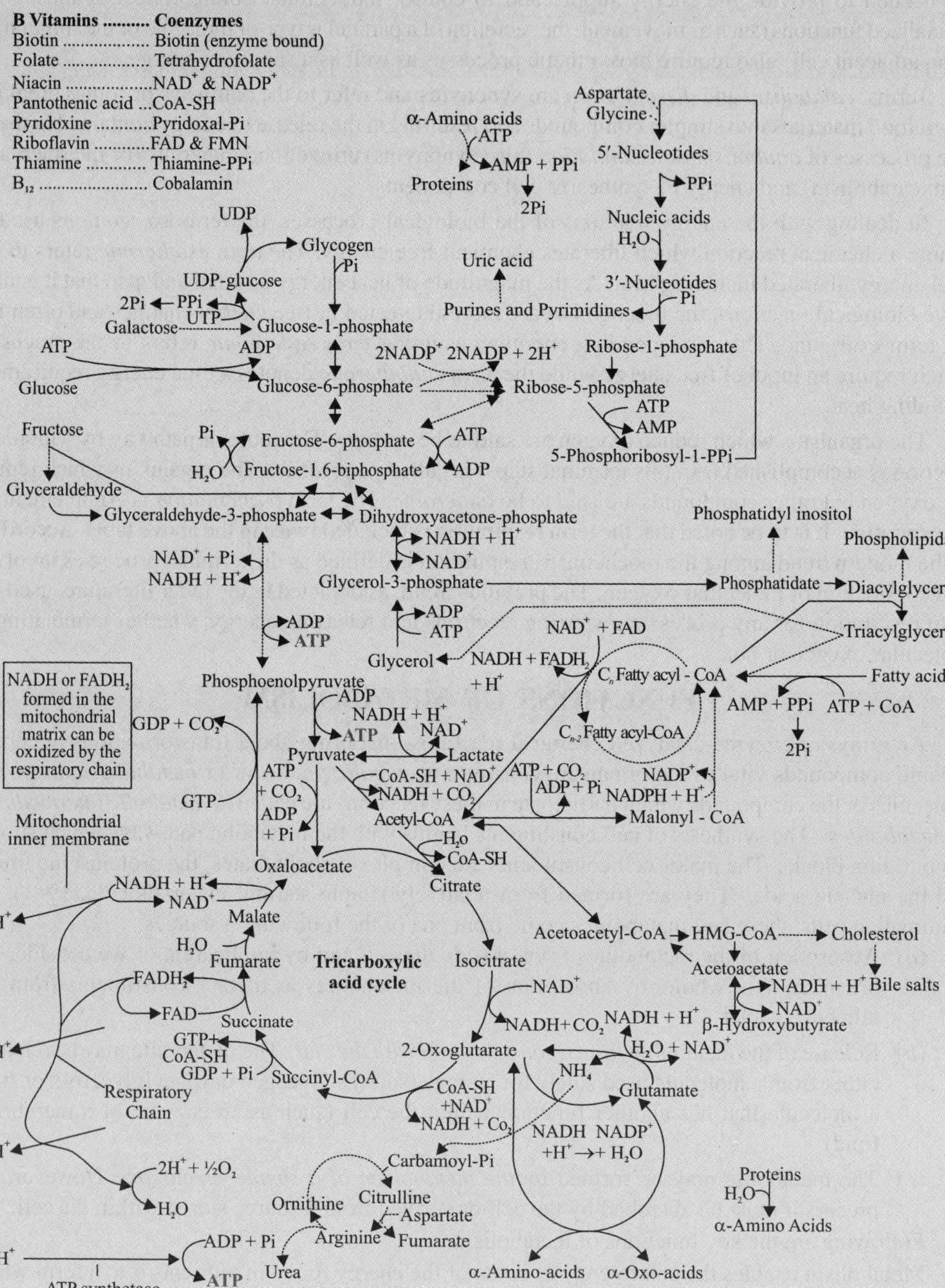

Fig. 19–1. Overview of metabolism

The points of ATP formation are in *red*.

Mammalian cells have 5 reaction schemes which are capable of supplying cells with adequate amounts of energy in one of these forms. These energy conserving pathways (or energy-supplying processes) are listed below:

1. The ***glycolytic pathway*** **:** This converts glucose into 3-c compounds, pyruvate or lactate.
2. The ***citric acid cycle (CAC)*** **:** This converts 2-c acetate units into carbon dioxide.
3. The ***pentose phosphate pathway (PPP)*** **:** This converts glucose-6-phosphate into pentose phosphate, and reduces the $NADP^+$ coenzyme.
4. The ***β oxidation of fatty acids*** **:** This converts fatty acids into 2-c units, and reduces the coenzymes FAD and NAD^+.
5. ***Oxidative phosphorylation*** **:** This is the electron transport phosphorylation process in which molecular oxygen is used to oxidize the coenzymes, which are reduced in the other 4 pathways, with the production of water and the conversion of ADP plus phosphate into ATP.

One of the 5 energy-supplying processes, the glycolytic pathway, can be distinguished from the others because it is capable of operating independently of a supply of oxygen. Unluckily, this has led people to believe that it is an anaerobic pathway, which is misleading since the pathway can (and does!) operate in an oxygen-independent mode even oxygen is plentifully available. The other 4 energy-supplying pathways lack this flexibility since they can only function in the presence of oxygen.

B. Metabolism provides key metabolites which are required for the synthesis of many essential cell components.

The central metabolic pathways (Fig. 19–1) include all 5 of the energy-conserving pathways listed in the preceding section. In addition to supplying the cell with energy, these central pathways provide precursors required for the formation of the carbohydrates, proteins, lipids and nucleic acids which the cell is composed of. Most of the pathways, shown in Fig. 19–1, are found in almost all mammalian cells, although some reactions which occur only in specialized cells have been incorporated *for the sake of completeness* (for example, the metabolism of galactose and fructose, the formation of urea and glycogen and the hydrolysis of glucose-6-phosphate).

Glycolysis, which is responsible for the conversion of glucose into 3-C compounds, illustrates the dual role of these central metabolic pathways. The glycolytic pathway supplies the cell with ATP and also provides the acetyl-coenzyme A (= acetyl-CoA) and glycerol-3-phosphate required to form the phospholipids needed for the cell membranes.

The pentose phsophate pathway provides a different sort of example since it is responsible for supplying energy (as NADPH) for fatty acid synthesis, but it also supplies the phosphoribosyl-1-pyrophosphate needed for the formation of nucleotides, the precursors of the nucleic acids.

CLASSICAL SUBDIVISIONS OF METABOLISM

The concept of metabolism must be viewed as an integrated set of pathways within the cell, rather than emphasizing the artificial subdivisions between the different areas of metabolism. However, the metabolism of the cell is too complex to consider all at once and there are some advantages in using the 3 classical subdivisions of metabolism: the carbohydrate, lipid and nitrogen metabolisms.

Fig. 19–2 shows how the metabolic pathways, depicted in Fig. 19–1, are divided into the 3 main areas of metabolism.

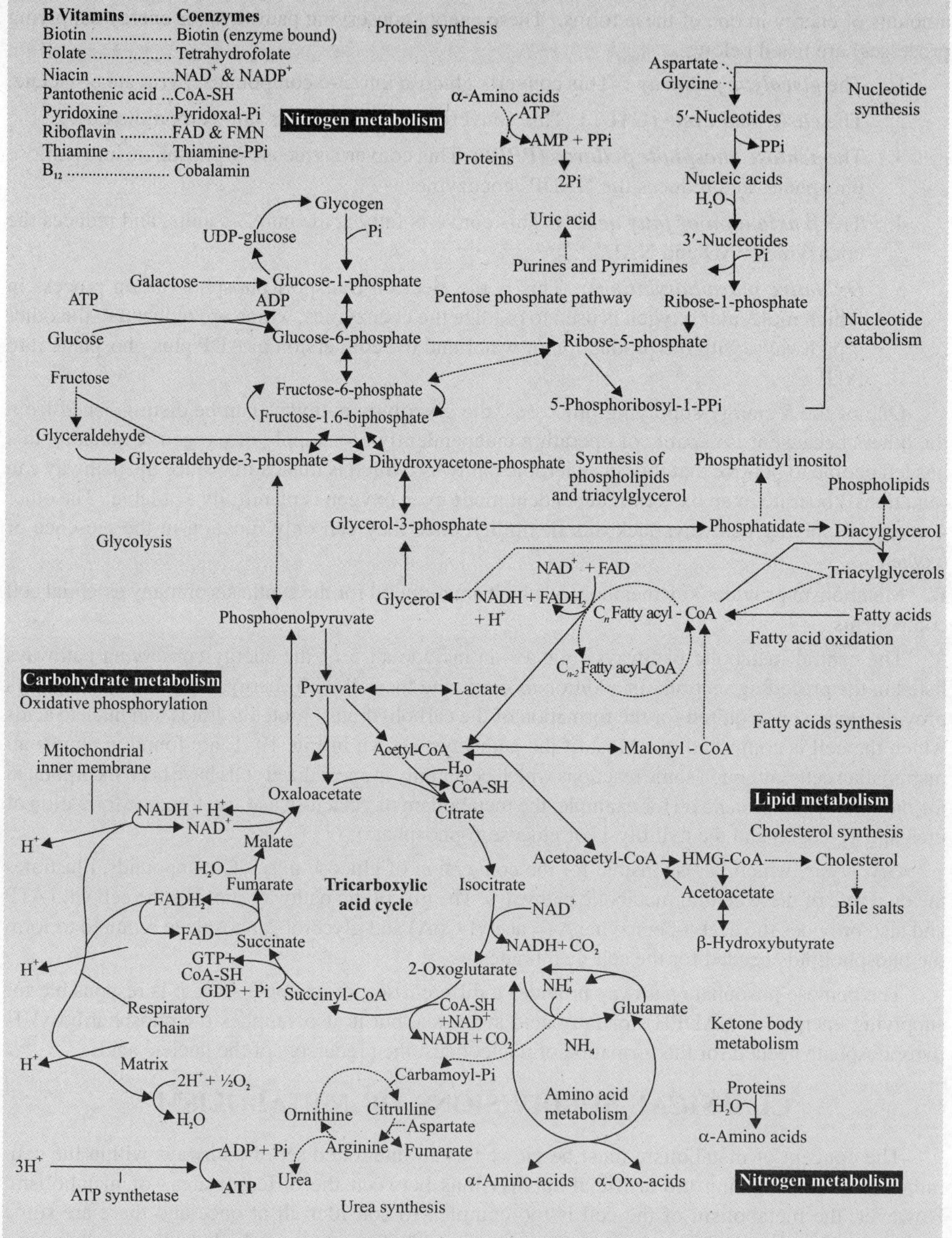

Fig. 19–2. Classical subdivisions of metabolism

Nitrogen metabolism and lipid metabolism are shown. The remaining areas show carbohydrate metabolism plus the final common pathways of citric acid cycle and oxidative phosphorylation.

Within each of these major areas, Fig. 19–2 also identifies the figure that gives details of particular pathways with well-defined functions, such as the energy-supplying pathways mentioned earlier. This is done for carbohydrate metabolism in Fig. 19–3, but a comparison of Figs. 19.2 and 19.3. illustrates the artificiality of the 3 classical subdivisions. Most of the energy conserving pathways fall within carbohydrate metabolism, except the β oxidation scheme from lipid metabolism which is not included in Fig. 19–3. Furthermore, the carbon skeletons of amino acids provide an important source of substrate for CAC, although they are part of nitrogen metabolism.

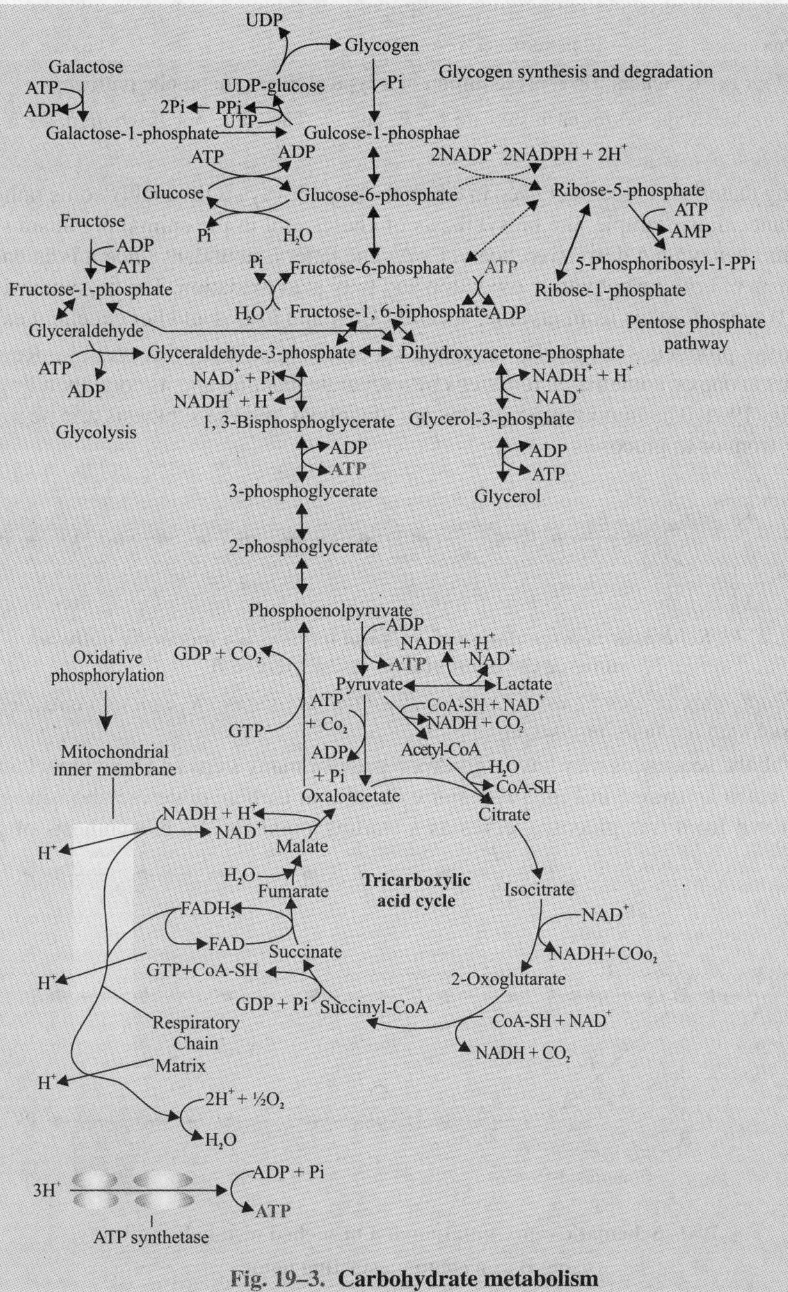

Fig. 19–3. Carbohydrate metabolism

The points of ATP formation are in *red*.

METABOLIC PATHWAYS

Since metabolic processes take place in a series of progressive, stepwise individual reactions, they can be described conveniently in the form of metabolic pathways. A linear metabolic pathway is shown in Fig. 19–4.

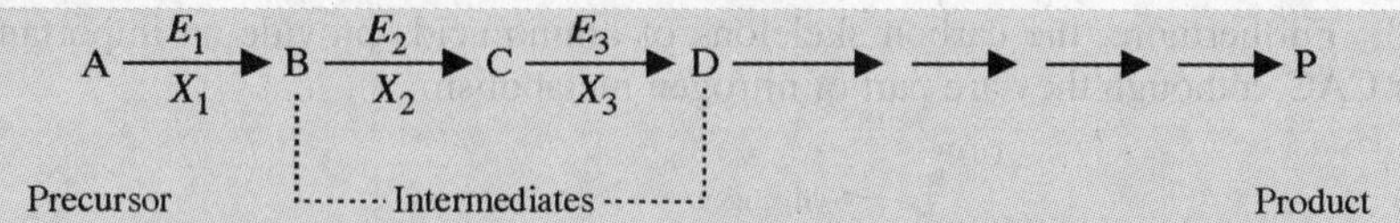

Fig. 19–4. Schematic representation of a typical linear metabolic pathway

[The specific enzymes required for each step are E_1, E_2 and E_3. The cofactors, if required, are X_1, X_2 and X_3.]

The starting material or precursor used in the anabolic pathways is invariably some rather simple, plentiful substance. For example, the biosynthesis of cholesterol in the animal organism starts with acetic acid as its coenzyme A derivative, acetyl CoA. The latter is abundant since it is an end product in the metabolism of both carbohydrate oxidation and fatty acid oxidation. The biosynthesis of heme in some 9 or 10 separate steps from glycine, succinyl CoA and iron is another excellent example.

To obtain the product, a metabolic sequence should be essentially irreversible. Reversibility, however, occurs at one or, sometimes, few steps by a separate reaction and its corresponding enzyme, as shown in Fig. 19–5. The important examples are glycolysis and the synthesis and degradation of liver glycogen from or to glucose.

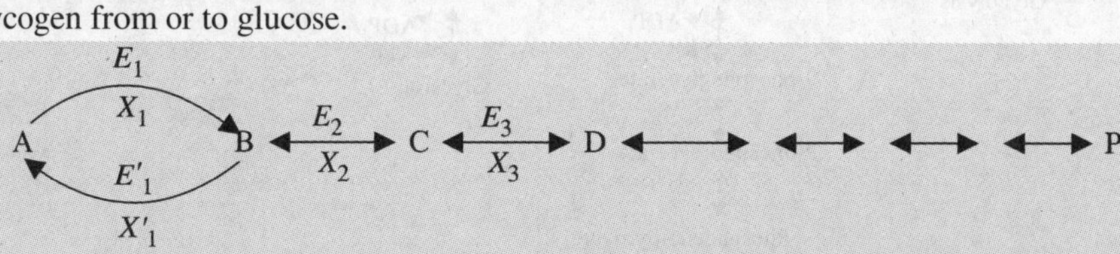

Fig. 19–5. Schematic representation of a typical irreversible metabolic pathway, showing the point of irreversibility, A to B

[For reversion, different enzymes E_1 and E_1', and usually different cofactors X_1 and X_1' are required for the forward and backward reactions, respectively.]

Some metabolic sequences may have a common path for many steps and then branch into two or more separate paths as shown in Fig. 19–6. For example, in carbohydrate metabolism, glucose-6-phosphate, derived from free glucose, serves as a starting point for the biosynthesis of glycogen,

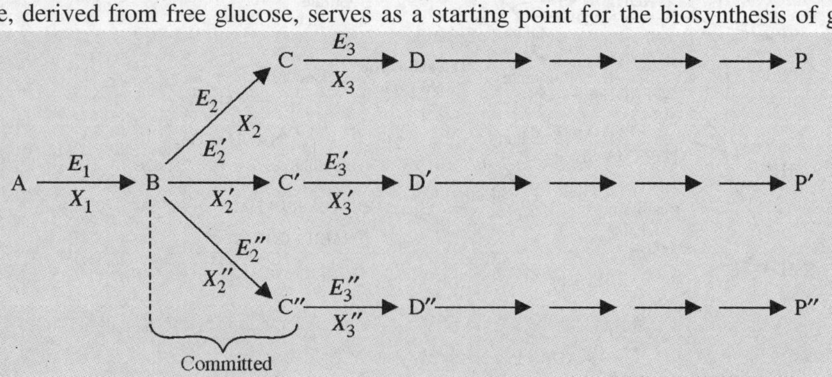

Fig. 19–6. Schematic representation of a branched metabolic pathway, using B as a common starting point

[Deficiency of an enzyme E_2, E_3 etc., or a cofactor X_2, X_3 etc., required for any of the path shown may result in a diversion of B into one or both of the other paths.]

the Embden-Meyerhof-Parnas (E.M.P.) pathway and the pentose phosphate pathway. The first step after such a branch point is often termed a **committed step** since its metabolite and succeeding ones have no role in metabolism other than the formation of the specific product of that path. This step often involves a great loss of free energy and hence is essentially irreversible. In case of blockage of one branch due to hereditary deficiency of an enzyme, the metabolism of the initial metabolite will be diverted into one or more of the `alternate' pathways at the branching point. Phenylketonuria (PKU) affords a good example. The diversion of phenylalanine (Phe) into other pathways occurs and there is formation of excessive amounts of phenylpyruvic acid and other related metabolites. Their increased amounts are toxic and apparently cause abnormalities in the affected children.

The metabolic pathways have been classically divided into two types, *catabolic* and *anabolic*. Both these pathways take place simultaneously in cells and their rates are regulated independently.

(A) Catabolic (= Degradative) Pathways

These comprise pathways in which large organic nutrient molecules (*e.g.*, carbohydrates, proteins and lipids) are broken down to smaller simpler compounds (*e.g.*, CO_2, NH_3 and lactic acid), frequently involving the participation of oxidation reactions, and result in the release of chemical free energy contained in the large organic molecules. This energy is then utilized by the organism for growth, movement, replication and also for transduction into other forms of energy, such as mechanical, thermal or electrical. At certain steps in the catabolic pathway, much of the free energy is conserved in the form of energy-carrying molecule adenosine triphosphate (ATP). Some may be conserved as energy-rich hydrogen atoms carried by the coezyme, NADP in its reduced form, NADPH (refer Fig. 19–7).

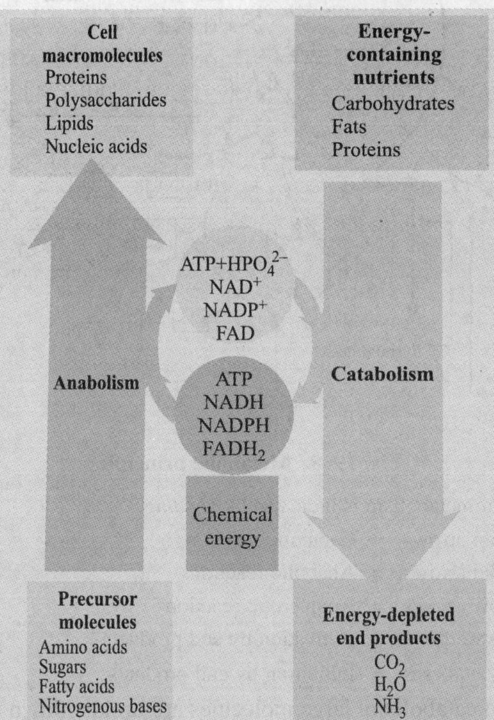

Fig. 19–7. Energy relations between catabolic and anabolic pathways

Catabolic pathways deliver chemical energy in the form of ATP, NADH, NADPH and $FADH_2$. These energy carriers are used in anabolic pathways to convert small precursor molecules into cell macromolecules.

(B) Anabolic (= Biosynthetic) Pathways

These include pathways in which complex organic compounds are produced from simpler precursors, usually involving the participation of reduction reactions, and require an input of chemical free energy which is furnished by the breakdown of ATP to ADP and phosphate. Biosynthesis of some cell components also requires high-energy hydrogen atoms, which are donated by NADPH (Fig. 19–7).

CENTRAL PATHWAYS

In spite of much diversity in the structure and function of the metabolites, there is a remarkable order and simplicity in various metabolic routes. This can be well illustrated by describing a central area of metabolism which is common to the two pathways (catabolic and anabolic) and also provides a direct link between them (refer Fig. 19–8).

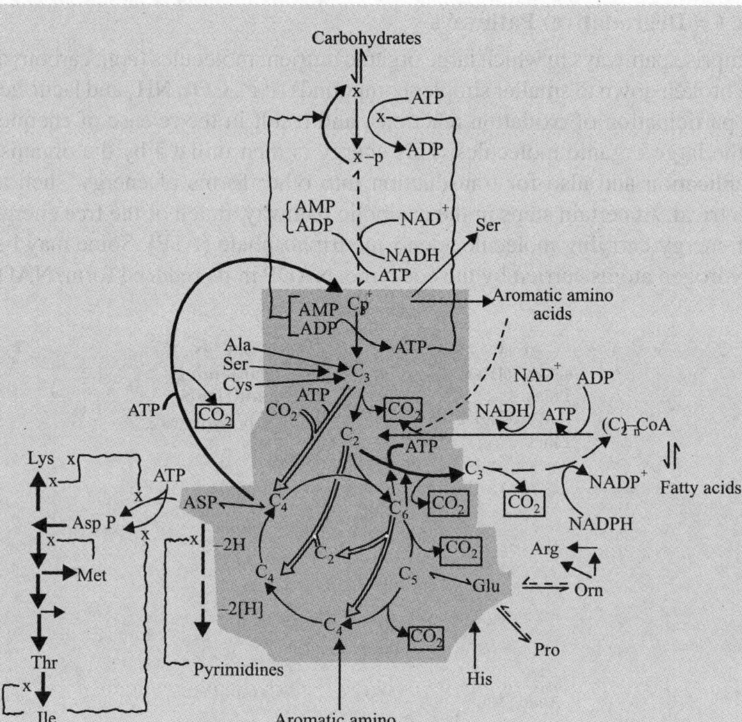

Fig. 19–8. Metabolic principles

Central area of metabolism is indicated by *shading*.

Thin arrows	=	Catabolic reactions
Thick solid arrows	=	Anabolic reactions
Thick hollow arrows	=	Anaplerotic reactions
Wavy arrows	=	Activation by end products
Wavy crosses	=	Inhibition by end products

During early stages of catabolism, large molecules are broken down to yield CO_2 and H_2O and also a 'restricted group' of small organic molecules, liberating about 1/3rd of the available free energy. These organic molecules are as follows:

(*a*) for carbohydrates– triose phosphates and/or pyruvate

(*b*) for proteins– acetyl-CoA, oxaloacetate, α-ketoglutarate, fumarate and succinate (*c*) for fats– acetyl-CoA, propionyl-CoA and glycerol

The same compounds are also produced even when organisms like bacteria utilize exotic carbon sources such as aliphatic (*e.g.*, γ-aminobutyric or itaconic acids), aromatic (*e.g.*, benzoic or mandelic acids) or heterocyclic (*e.g.*, purines or uric acid) compounds.

It is interesting to note that one and the same set of reactions is involved in 3 crucial phases of metabolism:

(*a*) the *interconversion* of the various products of catabolism, mentioned above.

(*b*) their *complete combustion* to CO_2 and H_2O, thus providing the organism with the remaining 2/3rd of its energy supply.

(*c*) the supply of *crucial intermediates* for the anabolic processes.

The central pathways are composed of relatively few reactions. In nutshell, they consist of following steps:

1. triose phosphate $\rightleftharpoons$ pyruvate
2. pyruvate $\longrightarrow$ acetyl-CoA
3. oxaloacetate $\rightleftharpoons$ aspartate
4. α-keto- (or o × o) glutarate $\rightleftharpoons$ glutamate
5. cyclic set of reactions catalyzing the complete combustion of acetyl-CoA (a C_2 compound) to CO_2 plus H_2O.

CATABOLISM VERSUS ANABOLISM

There are some fundamental differences between the catabolic and anabolic processes which have been shown in Fig. 19–9. These are:

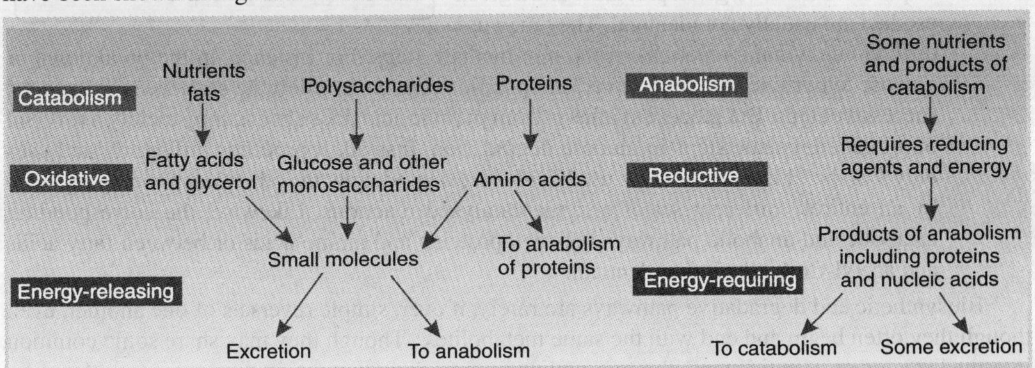

Fig. 19–9. A comparison of catabolism and anabolism

A. The catabolic pathways have clearly defined beginnings but no clearly distinguishable end products. On the contrary, the anabolic pathways lead to clearly identifiable end products from diffuse beginnings. The catabolic routes affect the conversion of nutrient carbon sources into the intermediates of central pathways; the anabolic routes comprise of various enzymatic steps which result in synthesizing the macromolecules from these intermediates.

B. Catabolism is converging process since it begins with many diverse cell components but ends in a final common pathway with only a few end products. Anabolism, on the other hand, is a diverging process since it begins with a few simple precursor molecules from which a large number of different macromolecules are made.

C. The catabolic and anabolic sequences or routes mostly follow altogether different pathways in detail. This is evident from the fact that the product of catabolism is quite different from the carbon source in anabolism. Following examples support this statement:

1. The amino acids, serine and cysteine are catabolized to pyruvate but their biosynthesis begins with glycerophosphate and aspartate respectively.
2. The aromatic amino acids are catabolically degraded to acetyl-CoA plus fumarate (or succinate) and their synthesis begins with phosphoenolpyruvate and an aldotetrose phosphate.
3. The catabolism of fatty acids terminates in the production of acetyl-CoA and its biosynthesis although commences with acetyl-CoA but follows a different path. It is first converted to the more reactive malonyl-CoA, which is not an intermediate in the catabolic chain, and then to other long-chain acetyl-CoAs.
4. In glucose metabolism, which is characterized by a reversal of enzymatic reactions, biosynthetic and degradative steps differ at the two most critical points, i.e., at either end. For instance, glucose is converted to glucose-6-phosphate during catabolism using ATP ; yet it is formed in anabolism from phosphate ester by hydrolysis. Pyruvate is produced catabolically from its enol phosphate by transphosphorylation to ADP ; it is, however, used anabolically by virtue of two linked reactions (carboxylation to oxaloacetate and then its transformation to enolphosphate).

The above-mentioned examples point out to a generalization that the initial and final reactions in most catabolic and anabolic pathways are virtually irreversible thermodynamically, *i.e.*, they have $\Delta G^{\circ\prime}$ values equal to ≤ -4 kcal/mole.

> The free energy change of a reaction occurring when reactants and products are at unit activity is called the standard free energy change and is denoted by the symbol ΔG°. However, biochemists usually use $\Delta G^{\circ\prime}$, which is the **standard free energy change at pH 7.**

D. The catabolic and the corresponding anabolic sequences between a given precursor and a given product are usually not identical. They may use different enzymatic reactions in the intermediate steps. For instance, in the breakdown of glucose to pyruvic acid in the liver, 11 specific enzymes in a definite sequence catalyze the successive steps. But glucose synthesis from pyruvic acid does not occur by merely a reversal of all the enzymatic steps in glucose degradation. Instead, it proceeds differently and uses only 9 of the 11 enzymatic steps used in glucose degradation, the other 2 steps being replaced by an entirely different set of enzyme-catalyzed reactions. Likewise, the corresponding catabolic and anabolic pathways between proteins and amino acids or between fatty acids and acetyl-CoA are also unidentical.

Biosynthetic and degradative pathways are rarely, if ever, simple reversals of one another, even though they often begin and end with the same metabolites. Though they may share some common intermediates or enzymatic steps, the two pathways are separate reaction sequences, regulated by distinct mechanisms and with different enzymes catalyzing their regulated reactions. Although seemingly, it appears wasteful to have different catabolic and anabolic pathways between a substrate and the product, there are at least two weighty reasons for the different corresponding catabolic and anabolic pathways:

I. ***A pathway can be exergonic in only one direction.*** Firstly, to proceed in a particular direction, a pathway must be exergonic in that direction. If a pathway was strongly exergonic then reversal of that pathway is just as strongly endergonic under the same conditions. Thus, the pathway adopted for breakdown of a biomolecule may not be possible for its synthesis because of high energy requirements. This can be easily explained by the famous *hill-and-boulder example* (Fig. 19–10). A boulder dislodged at the top of a hill will roll downhill, losing energy as it descends. At many points, the boulder may lose a large amount of energy as it falls over highly steep places. A tractor, however, may not be able to haul the boulder back up the hill by the path of its descent, but may succeed if it bypasses the steep slopes by taking a gradual

> A **boulder** is a large stone, worn smooth and round by weather and water.

zigzag path. Thus, degradation of a complex organic molecule is essentially a 'downhill' process, resulting in a loss of free energy; whereas biosynthesis of organic molecule is an 'uphill' process, requiring an input of energy.

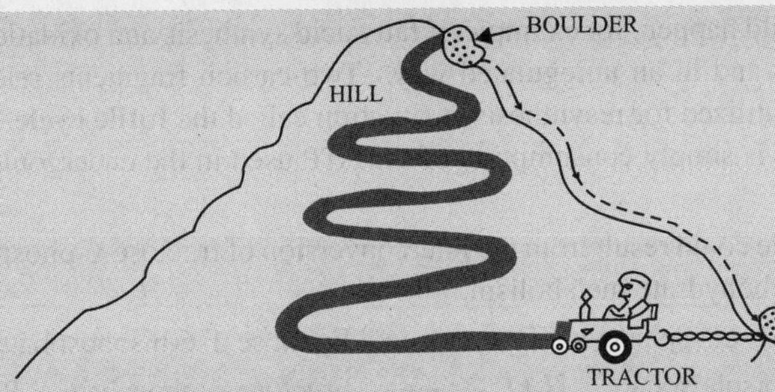

Fig. 19–10. The hill-and-boulder analogy

[*Dashed lines* indicate gradual fall while the *solid arrows* represent abrupt fall of the boulder.]

II. ***Pathways must be separately regulated to avoid futile cycles.*** The catabolic and anabolic pathways must be independently regulated. If only one pathway were used reversibly for both degradation and biosynthesis, slowing down the degradative pathway by inhibiting one of its enzymes would also slow down the corresponding biosynthetic pathway. Independent regulation is possible only when the two pathways are either entirely different or at least the rate-controlling step(s) should have different enzymes (refer Fig. 19–11). An important aspect of the metabolic pathways is the need to control the flow of metabolites in relation to the bioenergetic status of a cell. When ATP is in plenty, there is less need for carbon to be oxidized in the citric acid cycle. At such times, the cell can store carbon as carbohydrates and fats, so that gluconeogenesis, fatty acid synthesis and related pathways come into play. When ATP levels lower down, the cell must mobilize stored carbon to generate substrates for the citric acid cycle, so carbohydrate and fat breakdown must occur. Using distinct pathways for the biosynthetic and degradative processes is crucial for control, so conditions that activate one pathway tend to inhibit the opposed pathway and *vice versa*.

Sometimes, the degradative and biosynthetic pathways occur simultaneously in different parts of the cell. For instance, the oxidation of fatty acids to acetyl-CoA in the liver takes place in mitochondria where different enzymes catalyzing the various steps are present; whereas the synthesis of fatty acids

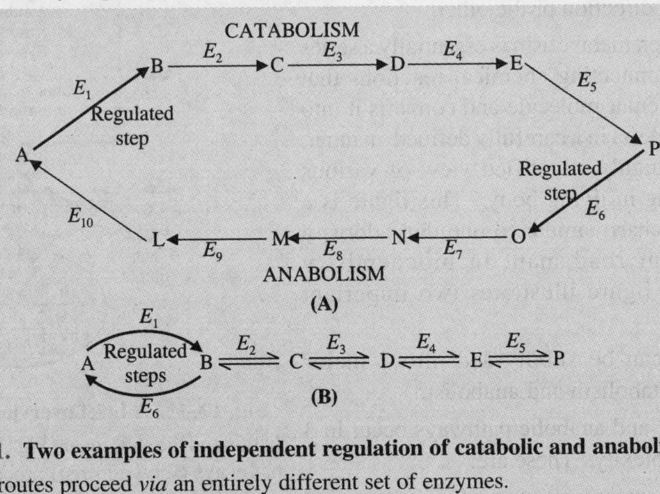

Fig. 19–11. Two examples of independent regulation of catabolic and anabolic pathways

(A) Parallel routes proceed *via* an entirely different set of enzymes.

(B) The two pathways differ in only one enzyme.

[Regulated steps are designated by *thick arrows*.]

from acetyl-CoA takes place in cytosol where a different set of enzymes catalyzing these steps are localized. Thus, *oxidative events (i.e., fatty acid oxidation) are favoured in mitochondria and reductive events (i.e., fatty acid synthesis) in cytosol.*

Consider what would happen, for example, if fatty acid synthesis and oxidation took place in the same cell compartment and in an unregulated way. Two-carbon fragments released by oxidation would be immediately utilized for resynthesis, a situation called the **futile cycle**. No useful work is done, and the net result is simply consumption of the ATP used in the endergonic reactions of fatty acid synthesis.

A similar futile cycle could result from the interconversion of fructose-6-phosphate with fructose-1,6- bisphosphate in carbohydrate metabolism.

$$\text{Fructose-6-phosphate} + \text{ATP} \longrightarrow \text{Fructose-1-6-bisphosphate} + \text{ADP}$$
$$\text{Fructose-1,6-bisphosphate} + H_2O \longrightarrow \text{Fructose-6-phosphate} + P_i$$
$$\text{Net:} \quad \text{ATP} + H_2O \longrightarrow \text{ADP} + P_i$$

The first reaction occurs in glycolysis, and the second participates in a biosynthetic pathway, glyconeogenesis. Both processes occur in the cytosol. The net effect of carrying out both reactions simultaneously would be the wasteful hydrolysis of ATP to ADP and P*i*. However, enzymes catalyzing both of the above reactions respond to allosteric effectors, such that one enzyme is inhibited by conditions that activate the other. Thus, effective control prevents the futile cycle from operating, even though the two enzymes occupy the same cell compartment. Henceforth, it is more appropriate to call this situation– two seemingly opposed cellular reactions that are independently controlled– a **substrate cycle**. Evidences suggest that a substrate cycle represents an efficient regulatory mechanism, because a small change in the activity of either or both enzymes can have immense effect on the flux of metabolites in one direction or the other.

As stated earlier, metabolism is essentially a series of coupled, interconnecting chemical reactions that begins with a particular molecule and converts it into some other molecule(s) in a carefully defined manner. Fig. 19–12 is a highly simplified view of various processes occurring in living body. This figure is a sort of metabolic chart, similar to one that adorns a wall, like a giant road map, in biochemistry laboratories. The figure illustrates two important principles:

(*a*) Metabolism can be subdivided into 2 major categories– catabolism and anabolism.

(*b*) Both catabolic and anabolic pathways occur in 3 stages of complexity. These are:

Stage 1 : the interconversion of polymers and complex lipids with monomeric intermediates

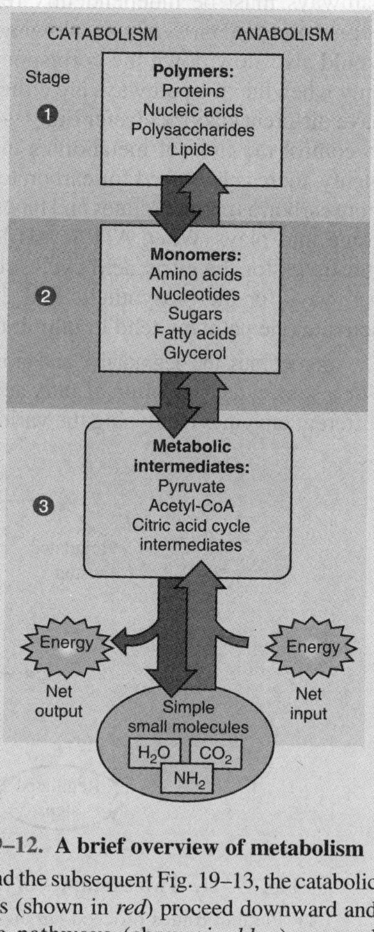

Fig. 19–12. A brief overview of metabolism

In this and the subsequent Fig. 19–13, the catabolic pathways (shown in *red*) proceed downward and anabolic pathways (shown in *blue*) proceed upward. Electron flow is shown by narrow arrows. Note the 3 stages of metabolism.

Stage 2 : the interconversion of monomeric sugars, amino acids, and lipids with still simpler organic compounds

Stage 3 : the ultimate degradation to, or synthesis from, inorganic compounds, including CO_2, H_2O and NH_3

Fig. 19–13 presents metabolism in more detail. Shown in this figure are the central metabolic pathways and some key intermediates. It may, however, appear from Figs. 19–12 and 19–13 that

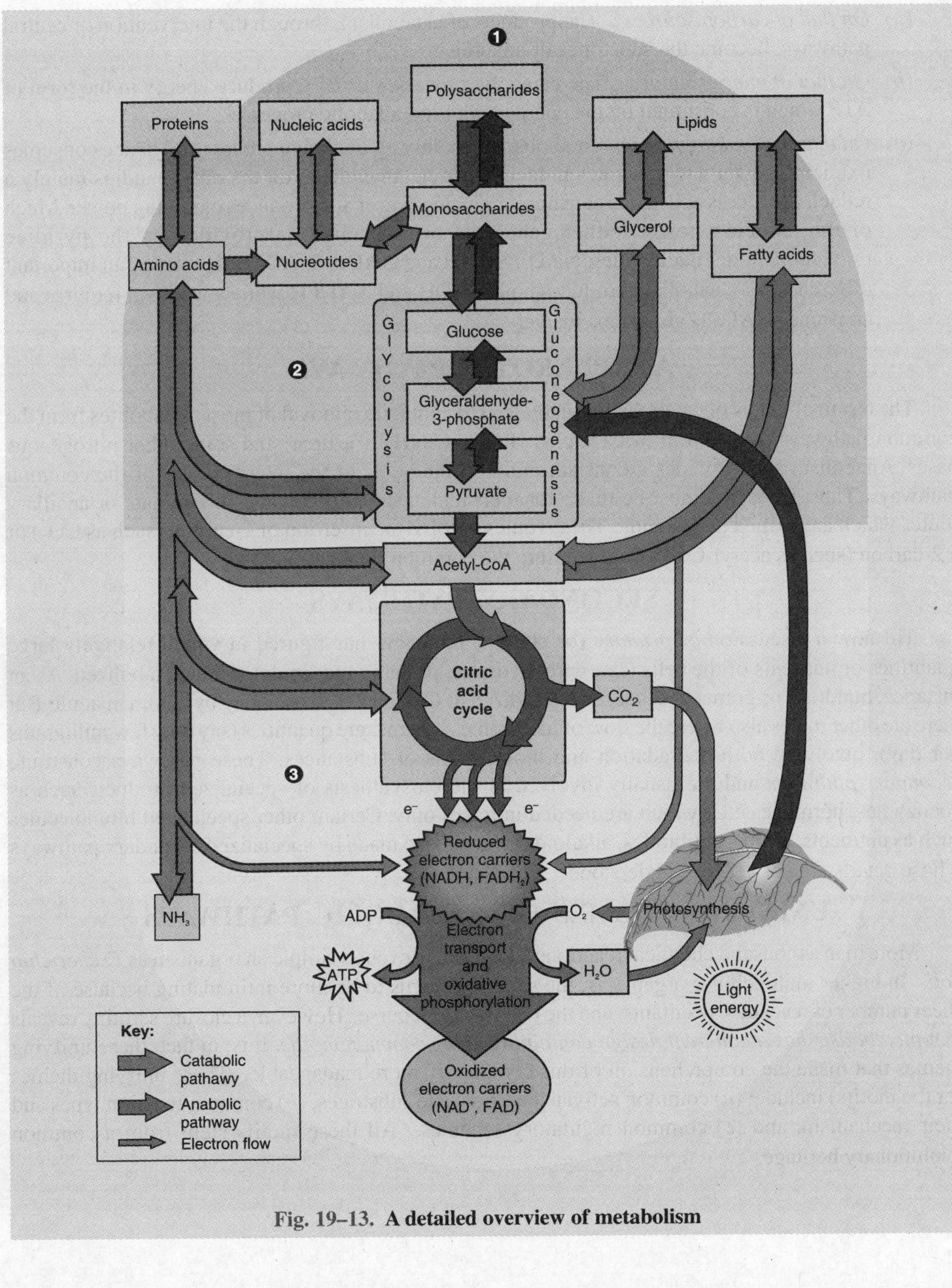

Fig. 19–13. A detailed overview of metabolism

some pathways operate simply as the reversal of other pathways. For example, fatty acids are synthesized from acetyl-CoA, but they are also converted to acetyl-CoA by β oxidation. Similarly, glucose-6-phosphate is synthesized from pyruvate in gluconeogenesis, which looks at first glance like a simple reversal of glycolysis. It is important to realize that in these cases, the opposed pathways are quite distinct from one another.

In spite of these well-marked differences, there is a close connection between catabolism and anabolism at 3 levels:

(*a*) *On that of carbon sources* : The products of catabolism, through the intervention of central pathways, become the substrates of anabolism.

(*b*) *On that of energy supply* : The catabolic processes usually produce energy in the form of ATP which is consumed by the energy-requiring anabolic processes.

(*c*) *On that of reducing power*: Catabolism is mainly an oxidative process and hence consumes oxidizing power and generates reducing power. Anabolism, on the other hand, is mainly a reductive process and hence consumes reducing power and produces oxidizing power. Much of the reducing power, either generated or consumed, is provided by the pyridine (= nicotinamide) nucleotides, NAD^+/NADH and $NADP^+$/NADPH. However, an important difference is : catabolism produces both NADH and NADPH while anabolism requires and consumes NADPH almost exclusively.

ANAPLEROTIC PATHWAYS

The terminal stages of catabolism usually lead to complete removal of most metabolites from the common pathways in the form of CO_2 plus H_2O plus NH_4^+ (or urea) and some other nitrogenous bases. Anabolism also provides a constant drain on the pools of the intermediates of the common pathways. Thus, provision must be made for their replenishment by some subordinate or ancillary routes which are termed *anaplerotic*. These routes involve the insertion of 1-carbon (such as CO_2) or a 2-carbon (such as acetyl-CoA) fragment into the common pool.

SECONDARY PATHWAYS

Till now a discussion of *primary* (or *central*) *pathways* has figured in which relatively large quantities of nutrients of the cells (*i.e.*, carbohydrates, proteins and lipids) are metabolized. As an instance, hundreds of grams of glucose are oxidized to CO_2 and H_2O each day by a human adult. But there are other routes also where the flow of metabolites is in meagre quantities (say only few milligrams per day), involving both degradation and biosynthesis of substances. These pathways constitute *secondary pathways* and are usually involved in the biosynthesis of specialized products such as coenzymes, hormones etc., which are needed in traces only. Certain other specialized biomolecules such as pigments, toxins, antibiotics, alkaloids etc., are also made by specialized secondary pathways whose details are not yet well understood.

UNIFYING THEMES OF METABOLIC PATHWAYS

More than a thousand chemical reactions take place in even as simple an organism as *Escherichia coli*. In higher multicellular organisms, the situation seems to be more intimidating because of the sheer number of reactants, resultants, and the reactions, of course. However, a closure scrutiny reveals that *metabolism has a coherent design containing many common motifs*. It is, in fact, these unifying themes that make the comprehension of this complexity more manageable. These unifying themes (or the motifs) include: (*a*) common activated carriers and substrates, (*b*) common reaction types and their mechanism, and (*c*) common regulatory schemes. All these motifs stem from a common evolutionary heritage.

1. Activated Carriers

There is repeated appearance of a limited number of activated intermediates in all metabolic process. Table 19–2 lists the names of some activated carriers used in metabolism. The phosphoryl transfer can be used to drive otherwise endergonic reactions. The phosphoryl-group donor in all such reactions is ATP. Many such activated carriers are described below:

Table 19–2. Some activated carriers in metabolism

Carrier molecule in activated form	*Group carried*	*Vitamin precursor*
ATP	Phosphoryl	
NADH and NADPH	Electrons	Niacin (vitamin B_5)
$FADH_2$	Electrons	Riboflavin (vitamin B_2)
$FMNH_2$	Electrons	Riboflavin (vitamin B_2)
Coenzyme A	Acyl	Pantothenate (vitamin B_3)
Lipoamide	Acyl	
Thiamine pyrophosphate	Aldehyde	Thiamine (vitamin B_1)
Biotin	CO_2	Biotin
Tetrahydrofolate	One-carbon units	Folate
S-adenosylmethionine	Methyl	
Uridine diphosphate glucose	Glucose	
Cytidine diphosphate diacylglycerol	Phosphatidate	
Nucleoside triphosphates	Nucleotides	

Note that many of the activated carriers are coenzymes that are derived from water-soluble vitamins.

A. *Activated carriers of electrons for fuel oxidation.* In aerobic organisms, the ultimate electron acceptor in the oxidation of fuel molecules is O_2. As electrons are not transferred directly to O_2, the fuel molecules transfer them to special carriers, which are either pyridine nucleotides or flavins. The reduced forms of thse carriers then transfer their high potential electrons to O_2.

Nicotinamide adenine dinucleotide, NAD⁺ (Fig. 19–14) is a major electron carrier in the oxidation of fuel molecules. The reactive part of NAD^+ is its nicotinamide ring, a pyridine derivative

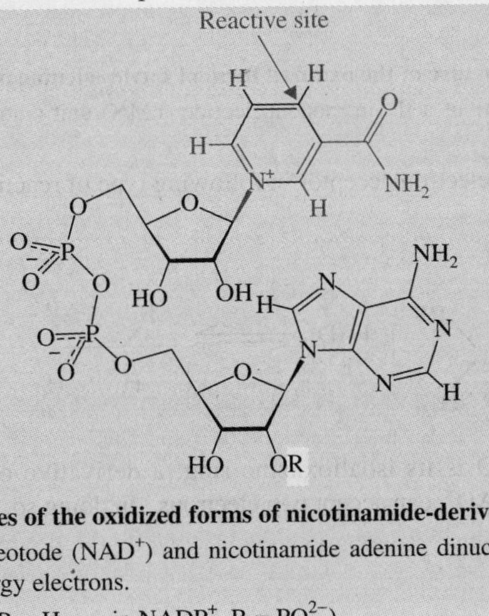

Fig. 19–14. Structures of the oxidized forms of nicotinamide-derived electron carriers

Nicotinamide adenine dinucleotode (NAD^+) and nicotinamide adenine dinucleotide phosphate ($NADP^+$) are main carriers of high-energy electrons.

(In NAD^+, R = H; in $NADP^+$, R = PO_3^{2-})

synthesized from the vitamin B_5 (niacin). During the oxidation of a substrate, the nicotinamide ring of NAD^+ accepts a hydrogen ion and two electrons, which are equivalent to a hydride ion. The reduced form of NAD^+ is called NADH. In the oxidized form, the nitrogen atom carries a positive charge, as indicated by NAD^+. NAD^+ is the electron acceptor in following type of reactions:

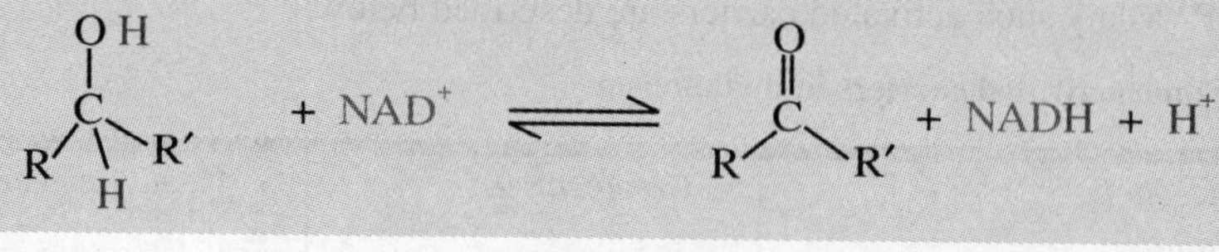

In this dehydrogenation, one H atom of the substrate is directly transferred to NAD^+, while the other appears in the solvent as a proton.

The other major electron carrier in the oxidation of fuel molecules is the coenzyme **flavin adenine dinucleotide, FAD** (Fig. 19–15). FAD is the oxidized form of this carrier and $FADH_2$, the

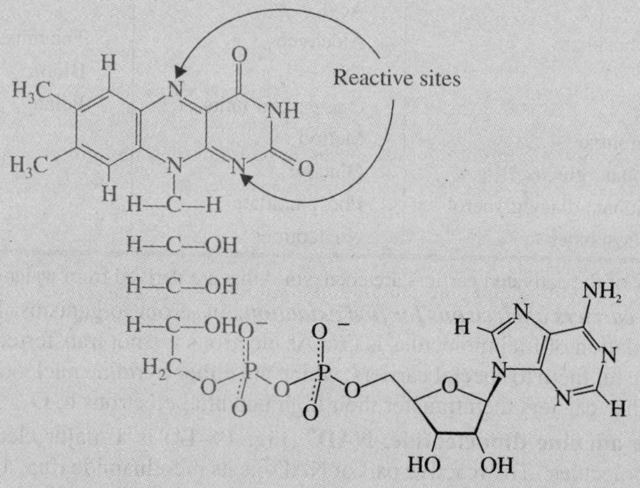

Fig. 19–15. Structure of the oxidized form of flavin adenine dinucleotide (FAD)
This electron carrier consists of a flavin mononucleotide (FMN) unit (shown in *blue*) and an AMP unit (shown in *black*).

reduced form. FAD is the electron acceptor in following type of reactions:

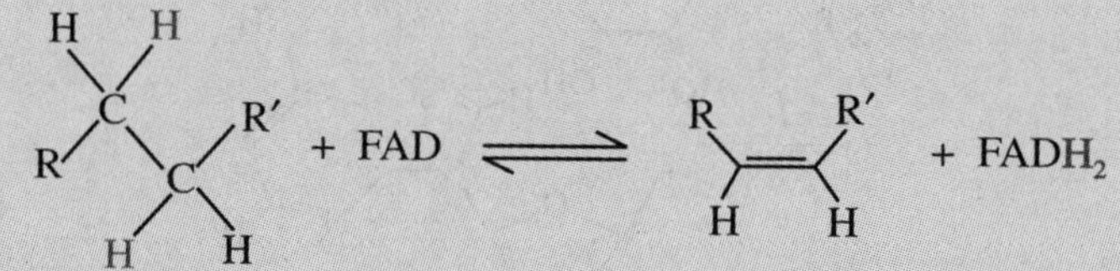

The reactive part of FAD is its isoalloxazine ring, a derivative of the vitamin B_2 riboflavin (Fig. 19–16). FAD, like NAD^+, can accept two electrons. In doing so, FAD, unlike NAD^+, takes up two protons.

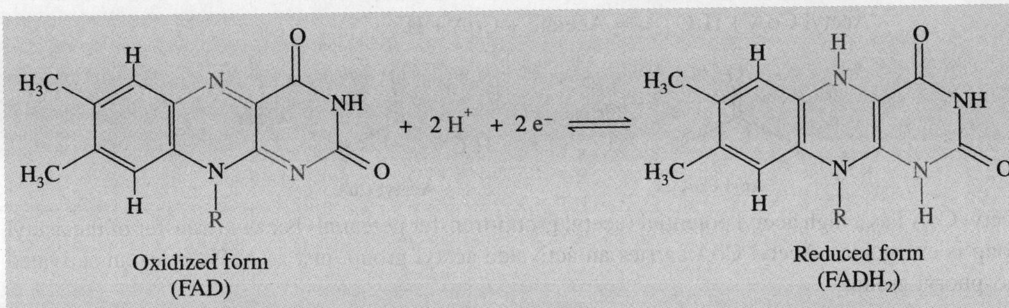

Fig. 19–16. Structures of reactive parts of FAD and $FADH_2$

The electrons and protons are carried by the isoalloxazine ring component of FAD and $FADH_2$.

B. Activated carriers of electrons for reductive biosynthesis. High potential electrons are required in most biosyntheses because the precursors are more oxidized than the products. Hence, reducing power is needed besides ATP. As an illustration, in the biosynthesis of fatty acids, the keto group of an added 2-C unit is reduced to a methylene (—CH_2) group in many steps. This requires an input of 4 electrons.

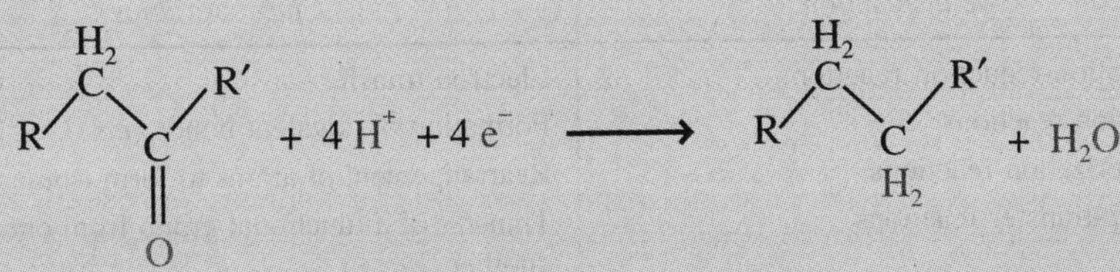

The electron donor in most reductive biosyntheses is NADPH, the reduced form of nicotinamide adenine dinucleotide phosphate ($NADP^+$; see Fig. 19–17). NADPH differs from NADH in that the 2′-hydroxyl group of its adenosine moiety is esterified with phosphate. NADPH carries electrons in the same way as NADH. *Whereas NADPH is used almost exclusively for reductive biosyntheses, NADH is used primarily for ATP production.*

C. An activated carrier of 2-carbon fragments. Coenzyme A is a carrier of acyl groups (Fig. 19– 16). Acyl groups are important constituents both in catabolism (as in the oxidation of fatty acids)

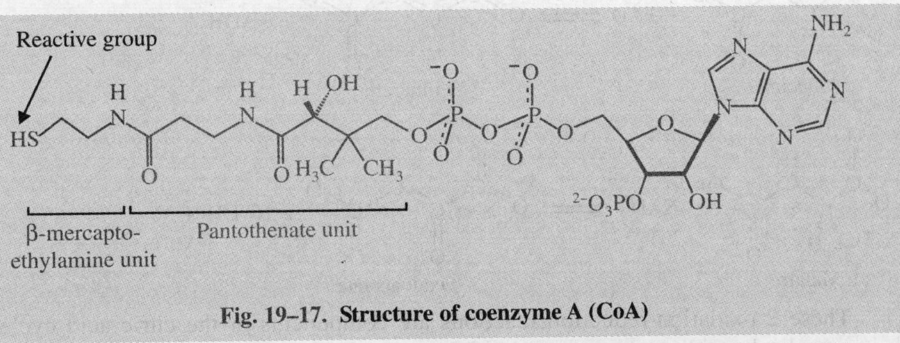

Fig. 19–17. Structure of coenzyme A (CoA)

and in anabolism (as in the synthesis of membrane lipids). The terminal sulfhydryl (—SH) group in CoA is the reactive site. Acyl groups are linked to CoA by thioester bonds. The resultant derivative is called an acyl CoA. An acyl group often linked to CoA is the acetyl unit and the resultant derivative is called *acetyl CoA*. The hydrolysis of acetyl CoA proceeds as follows:

$$\text{Acetyl CoA} + H_2O \rightleftharpoons \text{Acetate} + \text{CoA} + H^+$$

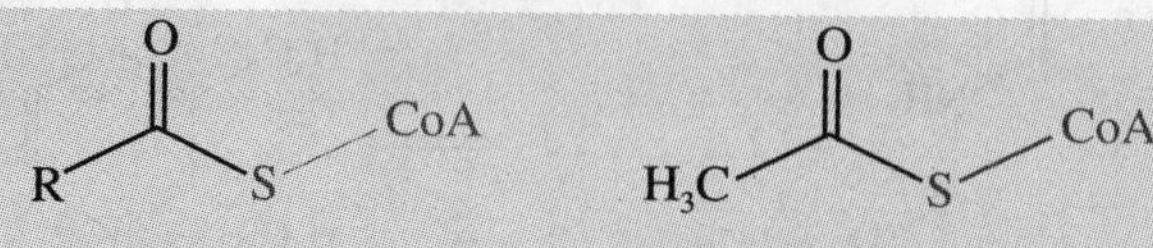

Acyl CoA **Acetyl CoA**

Acetyl CoA has a high acetyl potential (acetyl group-transfer potential) because transfer of the acetyl group is exergonic. Acetyl CoA carries an activated acetyl group, just as ATP carries an activated phosphoryl group.

2. Key Reactions

Just as there is an economy of design in the use of activated carriers, so is there an economy of design in biochemical reactions. Although the number of reactions in metabolism is bewilderingly large, the number of *kinds* of reactions is amazingly small. There are only 6 types of key reactions which are reiterated throughout the metabolism (Table 19–3). These reaction types are discussed below:

Table 19–3. Types of chemical reactions in metabolism

Type of reaction	*Characteristics*
1. Oxidation-reduction reactions	Electron transfer
2. Ligation reactions	Formation of covalent bonds (*i.e.*, C-C bonds)
3. Isomerization reactions	Rearrangement of atoms to form isomers
4. Group-transfer reactions	Transfer of a functional group from one molecule to another
5. Hydrolytic reactions	Cleavage of bonds by the addition of water
6. Addition or removal of functional groups	Addition of functional groups to double bonds or their removal to form double bonds

1. **Oxidation-reduction reactions.** These are essential components of many pathways and the energy is often derived from the oxidation of carbon compounds. Following 2 reactions may be considered:

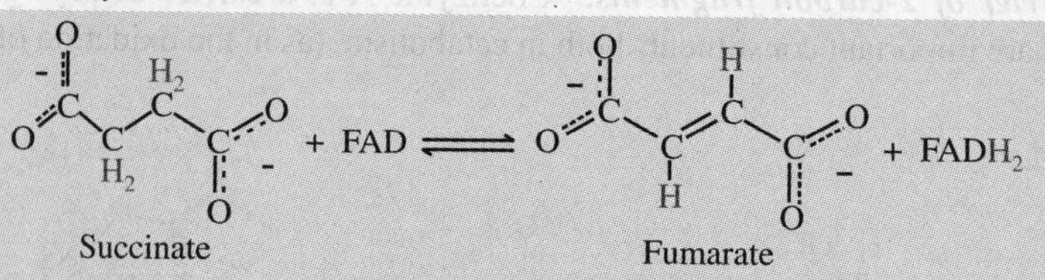

... (1)

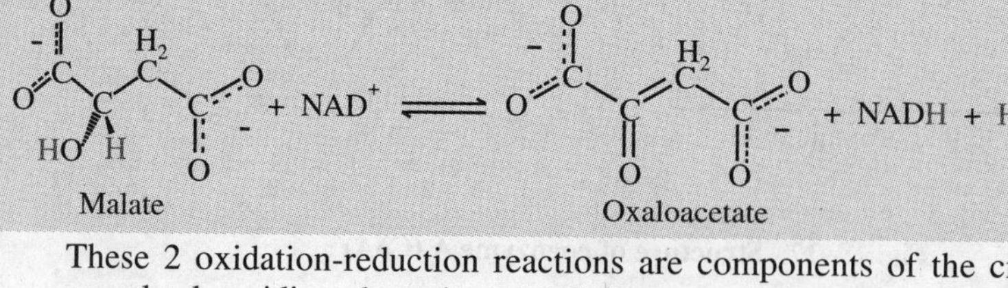

... (2)

These 2 oxidation-reduction reactions are components of the citric acid cycle, which completely oxidizes the activated 2-C fragment of acetyl-CoA to two molecules of CO_2. In reaction 1, $FADH_2$ carries the electrons whereas in Reaction 2, NADH acts as electron carrier. In biosynthetic oxidation-reduction reactions, NADPH is the reductant.

2. **Ligation reactions.** These reactions form bonds by using free energy from cleavage of ATP molecule. Reaction 3 illustrates the ATP-dependent formation of a C—C bond, required to

combine smaller molecules to form larger molecules. Oxaloacetate (OAA) is formed from pyruvate and CO_2. OAA can be used in the citric acid cycle or converted into amino acids, such as aspartic acid (Asp).

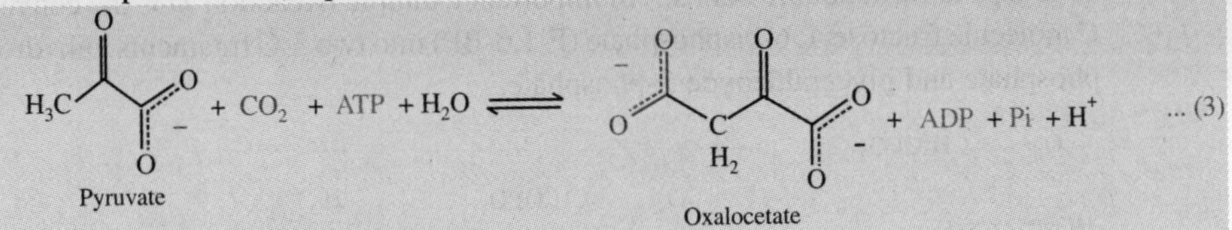

3. **Isomerization reactions.** These rearrange particular atoms within the molecule. Their role is often to prepare a molecule for subsequent reactions.

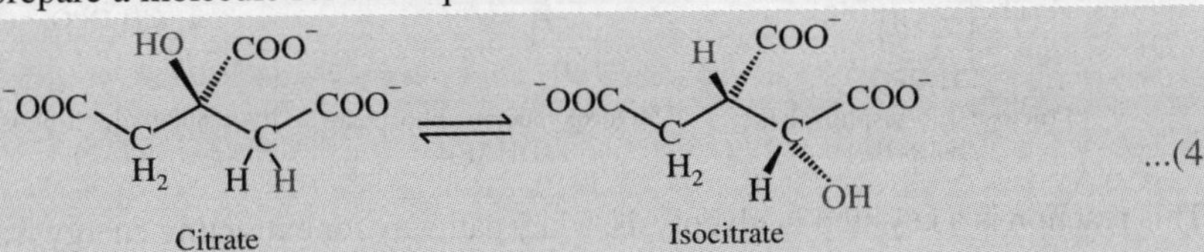

Reaction 4 is again a component of the citric acid cycle. This isomerization prepares the molecule for subsequent oxidation and decarboxylation by removing the OH group of citrate from a tertiary to a secondary position.

4. **Group-transfer reactions.** These reactions play a variety of roles. In one such reaction (Reaction 5), a phosphoryl group is transferred from the activated phosphoryl-group carrier, ATP, to glucose. This reaction traps glucose in the cell so that further catabolism can take place.

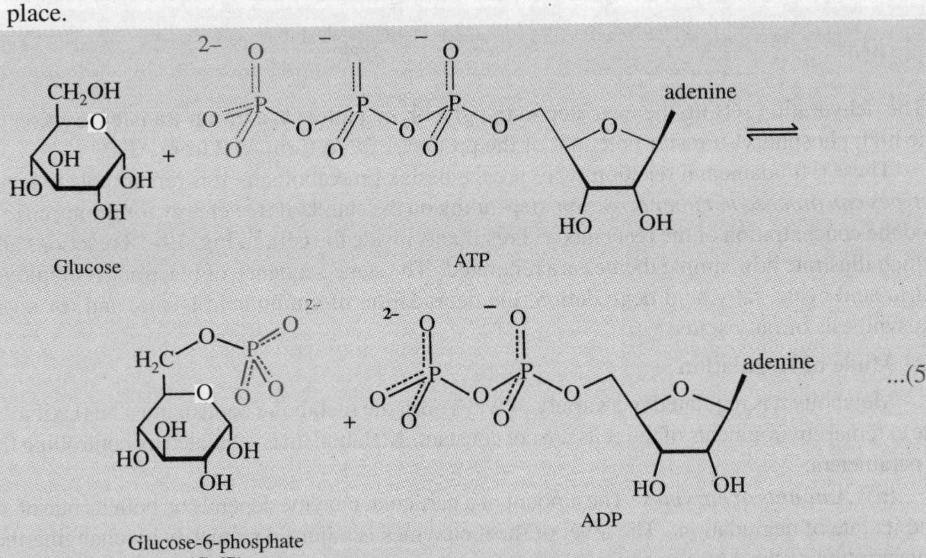

5. **Hydrolytic reactions.** These cleave bonds by the addition of water. Hydrolysis is a common means employed to break down large molecules. Proteins are digested by hydrolytic cleavage. Reaction 6 illustrates the hydrolysis of a peptide to yield two smaller peptides.

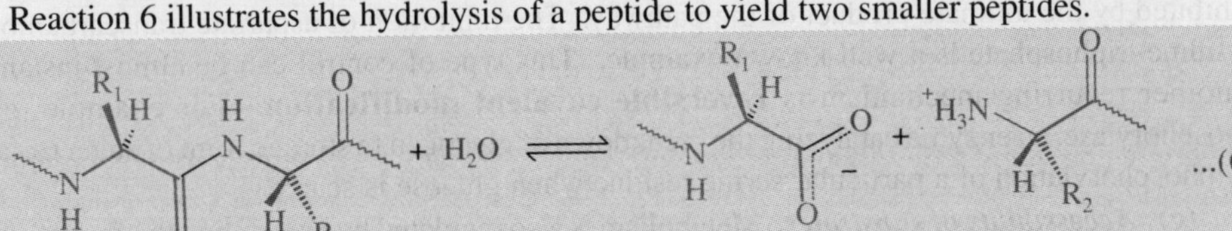

6. **Addition or removal of functional groups.** These types of reactions are catalyzed by *lyases* and consist either in the addition of functional groups to double bonds or the removal of groups to form double bonds. An important example (Reaction 7) is the conversion of 6-C molecule fructose-1, 6-bisphosphate (F-1,6-BP) into two 3-C frgaments, dihydroxyacetone phosphate and glyceraldehyde-3-phosphate.

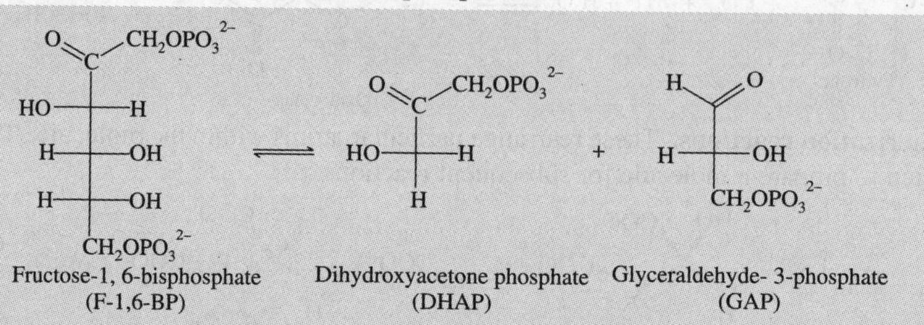

...(7)

This reaction is a key step in glycolysis, a key pathway for extracting energy from glucose. Another important reaction of this type is the formation of phosphoenolpyruvate (PEP) from 2-phosphoglycerate (2-PG).

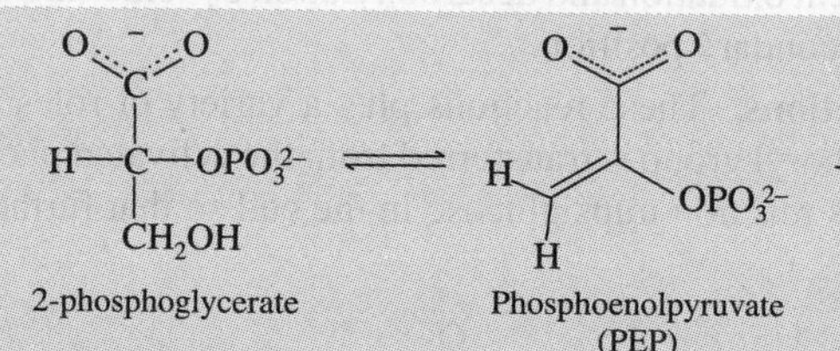

...(8)

The dehydration sets up the next step in the glycolytic pathway, a group-transfer reaction that use the high phosphoryl-transfer potential of the product PEP to form ATP from ADP.

These 6 fundamental reaction types are the basis of metabolism. It is remarkable to note that *all 6 types can proceed in either direction,* depending on the standard free energy for the specific reaction and the concentration of the reactants and resultants inside the cell. In Fig. 19–18 reactions are shown which illustrate how simple themes are reiterated. The same sequence of reactions is employed in the citric acid cycle, fatty acid degradation, the degradation of amino acid lysine, and (*in reverse*) the biosynthesis of fatty acids.

3. Mode of Regulation

Metabolism is regulated in a variety of ways. But the metabolic control must be flexible, because the external environments of the cells are not constant. Metabolism is regulated by controlling following 3 parameters:

(a) ***Amounts of enzymes.*** The amount of a particular enzyme depends on both its rate of synthesis and its rate of degradation. The level of most enzymes is adjusted primarily by changing the rate of transcription of the genes encoding them.

(b) ***Catalytic activities of enzymes.*** The catalytic activities of many enzymes are regulated in many ways. In **feedback inhibition,** the first reaction in many biosynthetic pathways is allosterically inhibited by the ultimate product of the pathway. The inhibition of aspartate transcarbamoylase by cytidine triphosphate is a well-known example. This type of control can be almost instantaneous. Another recurring mechanism is **reversible covalent modification.** For example, glycogen phosphorylase, an enzyme catalyzing the breakdown of glycogen (a storage form of sugar), is activated by phosphorylation of a particular serine residue when glucose is scarce.

(c) ***Accessibility of substrates.*** Metabolism is also regulated by controlling the flux of substrates.

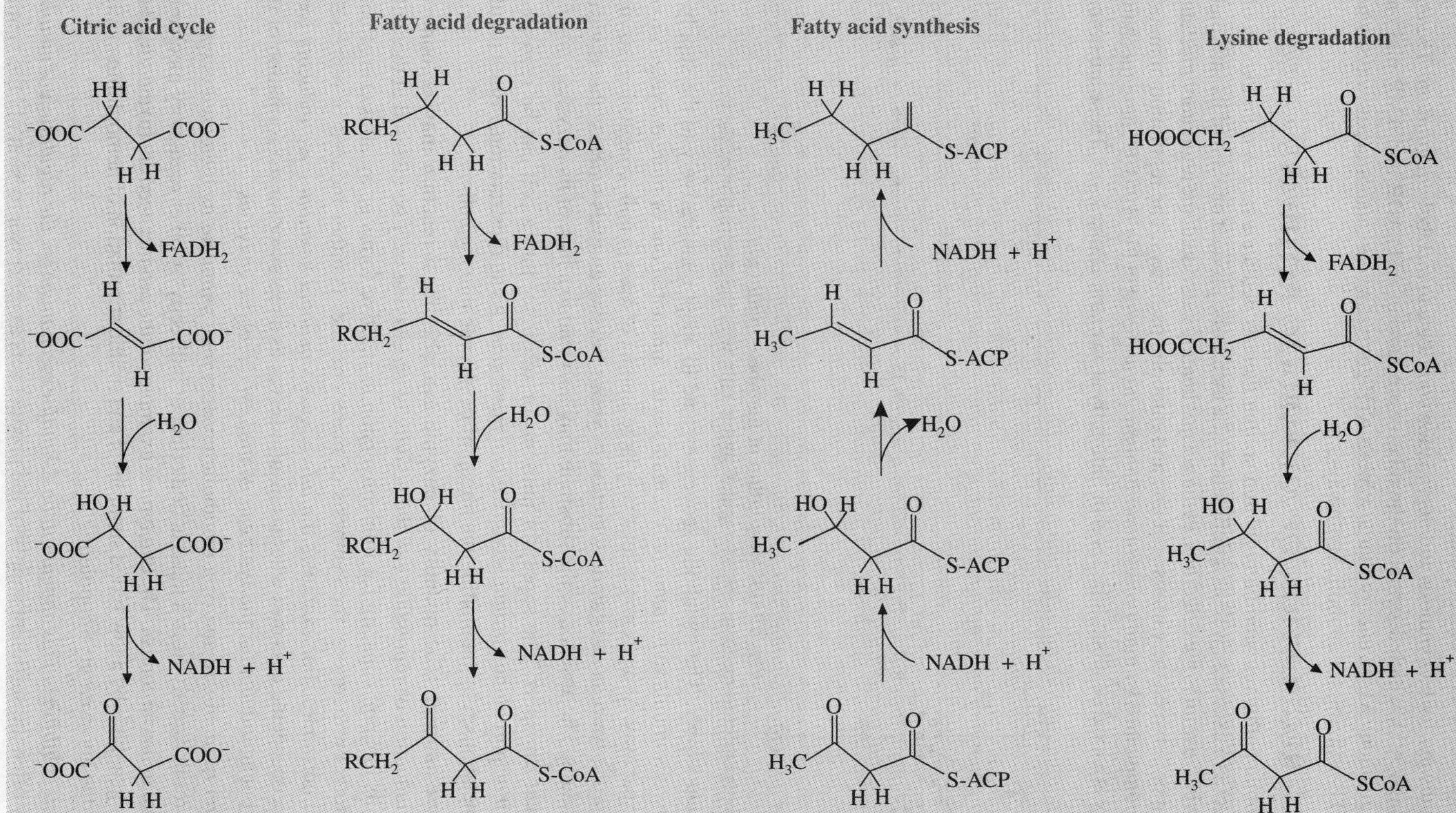

Fig. 19–18 . Metabolic motifs

Some metabolic pathways have similar sequences of reactions in common– in this case, an oxidation, the addition of a functional group (from a water molecule) to a double bond, and another oxidation.
(ACP = Acyl carrier protein)

The transfer of substrates from one compartment of a cell to another (for example, from the cytosol to mitochondria) can serve as a control point.

Distinct pathways for biosynthesis and degradation contribute to metabolic regulation. The energy charge (refer page 591), which depends on the relative amounts of ATP, ADP, and AMP, plays a role in metabolic regulation. A high energy charge inhibits ATP-generating (or catabolic) pathways, whereas it stimulates ATP-utilizing (or anabolic) pathways.

REGULATION OF METABOLIC PATHWAYS

The metabolic pathways need be regulated so that there is neither a lack nor an excess of any essential product. An excess as well as a deficiency of a metobolic product or even of its intermediate metabolites can be harmful to the cell. Thus, in the normal healthy individual, the regulatory mechanisms maintain a balance between the various anabolic and catabolic pathways. The regulation of metabolic pathways is accomplished by many control mechanisms, as depicted in Fig. 19–19. These mechanisms may act directly at a local or subcellular level or indirectly at an extracellular level. These are described below:

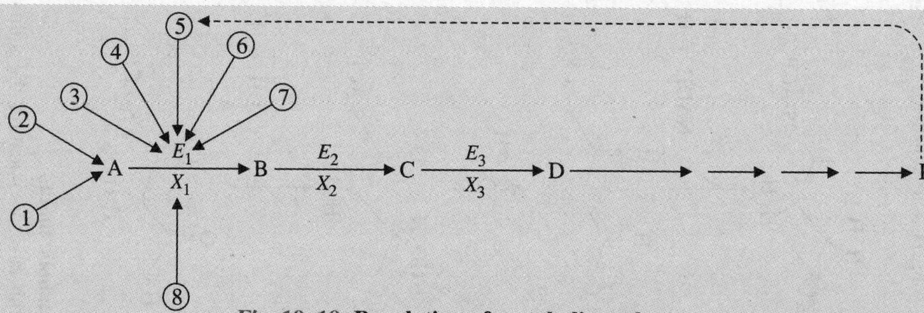

Fig. 19–19. **Regulation of metabolic pathways**
[Numbers refer to mechanisms of regulation that tally with the description of the text.]

1. *Nutrient supply.* The metabolic sequences tend to adapt quantitatively to the supply of a nutrient. This usually entails an increase (or decrease) in the amount of one or more enzymes involved in the metabolic pathway. As an example, relatively large amount of renin (a milk-coagulating enzyme) is present in the gastric mucosa and gastric secretion of young nursing animals whereas this is virtually lacking in the adults. The absence of the substrate may also lead to a lack of its enzyme.

2. *Nutrient transport.* The supply of nutrient (or substrate) into a cell can be regulated by controlling the transport of the nutrient across the cell membrane. As an illustratation, *insulin* regulates carbohydrate metabolism by facilitating the transport of glucose into the cells.

3. *Enzyme amount.* The quantity of enzyme available for a reaction may be controlled genetically by induction or repression ; or the activity of an enzyme may be affected at the cellular level either by its inherent capacity for interconversion to inactive forms or by allosteric effects or even by inhibitors. Furthermore, the synthesis of many enzymes is either induced or repressed by certain specific hormones. For example, the *adrenoglucocorticoid hormones* are inducers for the synthesis of gluconeogenic enzymes whereas insulin serves as a repressor for the formation of these same enzymes but an inducer for the synthesis of the key glycolytic enzymes.

4. *Product need.* A demand of a metabolic product may stimulate its increased output. This may, however, result directly from a mass action effect or indirectly by other regulatory mechanisms such as hormonal or neural control. The two typical examples of the product need are (a) the stimulation of hepatic glucogenolysis by a low blood sugar level and (b) the stimulation of hemoglobin synthesis in anaemias by the hormone erythropoietin.

5. *Product inhibition.* *This appears to be the major mechanism for the regulation of metabolic pathways.* Very often the sufficient supply of the product acts as repressor to 'shut off' the synthesis of usually the first of the series of enzymes of a particular pathway. This is particularly advantageous as accumulation of the later metabolites in the pathway, which may be deleterious to the cell, is prevented. This could occur if a later enzyme in the pathway were the one suppressed. Such type of inhibition is also called *feedback* (or *end product*) *inhibition*. A classic example is the inhibition of porphyrin heme biosynthesis by the end product heme. Here the biosynthesis of the first enzyme in the metabolic sequence, δ-*aminolevulinic acid synthetase is repressed.*

6. *Endocrine control.* The metabolic pathways are controlled by hormonal secretions in many ways. Conversion of an inactive form of enzyme to the active form (*e.g.*, conversion of phosphorylase b to active phosphorylase a by epinephrine and glucagon *via* cyclic-AMP) is one such mechanism. Other mechanisms are the regulation of transport through the cell membrane (*e.g.*, insulin), the enzyme induction and repression etc.

7. *Neural control.* The effects of nerve stimulation on metabolic pahtways are probably indirect by hormonal or other mechanisms. For example, the effect of psychologic factors such as fright or anger on carbohydrate metabolism, specifically in increasing the blood sugar level, apparently results from an increased secretion of epinephrine which, in turn, accelerates glucogenolysis by the mechanism described under `Product Need'. Neural effects by way of an increased secretion of neurohormones are still other examples of indirect means of affecting metabolic pathways.

8. *Cofactor availability.* The enzyme-catalyzed reactions, requiring a cofactor, are affected by the amount of the cofactor available. As an example, relatively large amount of nicotinamide adenine dinucleotide (NAD^+) as cofactor is required for the metabolism of ethanol. Evidently, during ethanol metabolism, lesser NAD^+ may then be available for other metabolic pathways requiring NAD^+ as cofactor. These pathways may, therefore, become partially or even completely blocked. However, metabolic sequences requiring $NADH_2$ (which is formed from NAD^+ during alcohol oxidation) as a cofactor may then be stimulated. Conversion of pyruvic acid to lactic acid is one such example.

EVOLUTION OF METABOLIC PATHWAYS

Until the 1980s, all biological catalysts, termed *enzymes*, were believed to be proteins. Then,

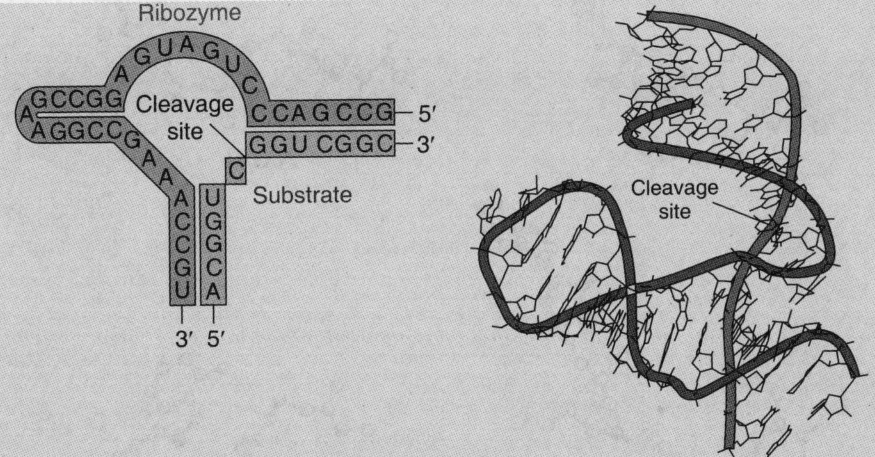

Fig. 19–20. The catalytic RNA

A. The base-pairing pattern of a 'hammerhead' ribozyme and its substrate

B. The folded conformation of the complex

The ribozyme cleaves the bond at the cleavage site. The paths of the nucleic acid backbones are highlighted in *red* and *blue*.

(*Adapted from Berg JM, Tymoczko JL and Stryer L, 2002*)

Sidney Altman and **Thomas R. Cech** (Nobel Laureates, 1997) independently discovered that certain RNA molecules can be effective catalysts. These RNA catalysts have come to be known as **ribozymes**. The discovery of ribozymes suggested the possibility that catalytic RNA molecules could have played fundamental roles early in the evolution of life.

The catalytic ability of RNA molecules is related to their ability to adopt specific yet complex structures. This principle is illustrated by a *'hammerhead' ribozyme,* an RNA structure first identified in plant viruses (Fig. 19–20). This RNA molecule promotes the cleavage of specific RNA molecules at specific sites (This cleavage is necessary for certain aspects of the viral life cycle). The ribozyme requires Mg^{2+} ion or other ion for the cleavage step and forms a complex with its substrate RNA molecule, that can adopt a reactive conformation. The existence of RNA molecules that possess specific binding and catalytic properties makes plausible the idea of an early **'RNA World'**, inhabited by life forms dependent on RNA molecules to play all major roles, including those important in biosynthesis and energy metabolism. Thus, RNA served as catalysts and information-storage molecules.

The activated carriers such as ATP, NADH, FAD and coenzyme A all contain adenosine diphosphate units (Fig. 19–21). What can be the reason behind it ? A possible explanation is that these molecules evolved from the early RNA catalysts. Non-RNA units such as the isoalloxazine ring may have been recruited to serve as efficient carriers of activated electrons and chemical units, a function not readily performed by RNA itself. The adenine ring of $FADH_2$ binds to a uracil unit in a niche of a ribozyme by base-pairing, whereas the isoalloxazine ring bulges out and functions as an electron carrier. When the more versatile proteins replaced RNA as the major catalysts, the

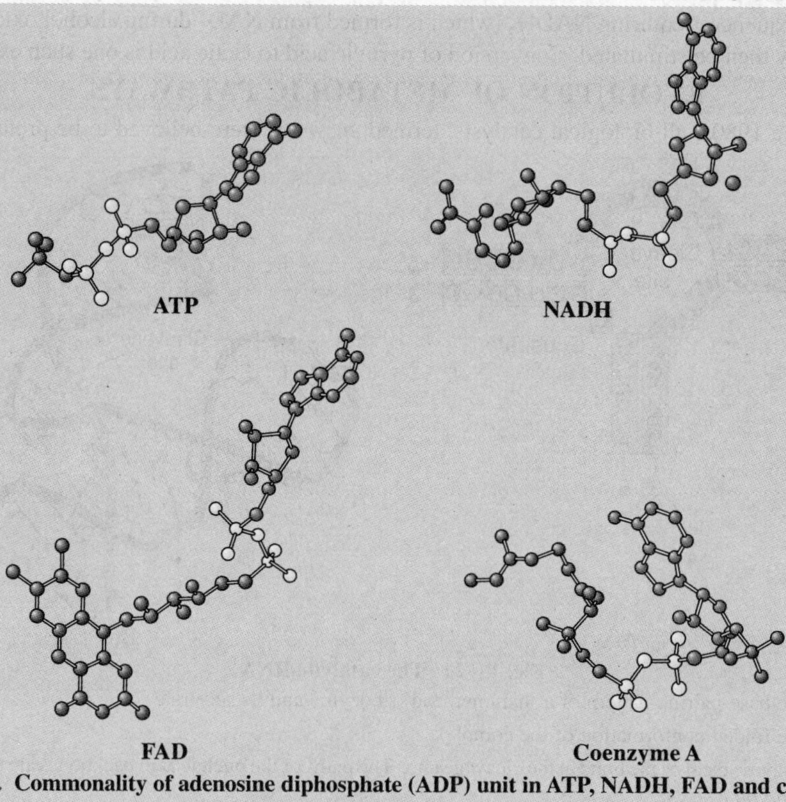

Fig. 19–21. Commonality of adenosine diphosphate (ADP) unit in ATP, NADH, FAD and coenzyme A

The adenine unit is shown in *blue*, the ribose unit in *red*, and the diphosphate unit in *yellow*.

ribonucleotide coenzymes stayed essentially unchanged because they were already well adapted to their metabolic roles. For example, the nicotinamide unit of NADH can readily transfer electrons, regardless of whether the adenine unit interacts with a base in an RNA enzyme or with amino acid residues in a protein enzyme. With the advent of protein enzymes, these important cofactors evolved as free molecules, without losing the adenosine disphosphate vestige of their RNA-world ancestry. That modules and motifs of metabolism are common to all forms of life testifies to their common origin and to the retention of functioning modules through billions of years of evolution.

REFERENCES

1. **Atkinson DE :** Cellular Energy Metabolism and Its Regulation. *Academic Press Inc., New York. 1977.*
2. **Dagley S, Nicholson DE :** Introduction to Metabolic Pathways. *Wiley, NewYork. 1970.*
3. **Godsell DS :** Inside a living cell. *Trends Biochem. Sci.* **16** *: 203–206, 1991.*
4. **Greenberg DM (editor) :** Metabolic Pathways. 3rd ed. vol. 1. *Academic Press Inc., New York 1967.*
5. **Krebs HA, Kornberg HL :** Energy Transformations in Living Matter. *Springer-Verlag. 1957.*
6. **Linder MC (editor) :** Nutritional Biochemistry and Metabolism *2nd ed. Elsevier, New York. 1991.*
7. **Martin BR :** Metabolic Regulation. A Molecular Approach. *Blackwell Scientific Publications. 1987.*
8. **Newholme EA, Start C :** Regulation in Metabolism, *Wiley, New York. 1973.*
9. **Nicholls DG, Ferguson SJ :** Bioenergetics 2. *Academic Press, Inc., New York. 1992.*
10. **Ochs RS, Hanson RW, Hall J (editor) :** Metabolic Regulation. *Elsevier Science Publishing Co. Inc., New York. 1985.*
11. **Salway JG :** Metabolism at a Glance. *Blackwell Scientific. 1994.*
12. **Wood WB :** The Molecular Basis of Metabolism *Unit 3 in Biocore, McGraw-Hill, 1974.*

PROBLEM

1. Write a balanced equation for the complete oxidation of each of the following, and calculate the respiratory quotient for each substance.
   (*a*) Ethanol
   (*b*) Acetic acid
   (*c*) Stearic acid
   (*d*) Oleic acid
   (*e*) Linoleic acid
2. How does metabolism slow as we grow older?

ANSWERS TO PROBLEMS

1. *Equation* — *RQ*
   (*a*) $C_2H_5OH + 3\ O_2 \longrightarrow 2\ CO_2 + 3\ H_2O$ — 0.67*
   (*b*) $CH_3COOH + 2\ O_2 \longrightarrow 2\ CO_2 + 2\ H_2O$ — 1.0
   (*c*) $CH_3(CH_2)_{16}COOH + 26\ O_2 \longrightarrow 18\ CO_2 + 18\ H_2O$ — 0.69
   (*d*) $CH_3(CH_2)_{14}\ (CH)_2COOH + 25\frac{1}{2}\ O_2 \longrightarrow 18\ CO_2 + 17\ H_2O$ — 0.71
   (*e*) $CH_3(CH_2)_{12}\ (CH)_4COOH + 25\ O_2 \longrightarrow 18\ CO_2 + 16\ H_2O$ — 0.71
   * Example: RQ = $2\ CO_2/3O_2$.
2. It is true that metabolism slows down as we age, contributing to weight gain. But another contributing factor is the lack of exercise as we grow older. This causes a reduction in lean muscle mass.

CONTENTS

The potential energy of the water at the top of a waterfall is transformed into kinetic energy in spectacular fashion.

CHAPTER

20

Bioenergetics

You can't win.

—First Law of Thermodynamics

You can't even break even.

—Second Law of Thermodynamics

You can't stay out of the game.

—Third Law of Thermodynamics

COUPLING

Thermodynamics (*therme*[G] = heat; *dynamis*[G] = power) is the branch of physical chemistry that deals with the energy changes. And **biochemical thermodynamics** (or **biochemical energetics** or **bioenergetics**, as it is also called) is the field of biochemistry concerned with the transformation and use of energy by living cells. The chemical reactions occurring in living beings (or biochemical reactions) are associated with the liberation of energy as the reacting system moves from a higher to a lower energy level. *Most often, the energy is liberated in the form of heat*. In nonbiologic systems, heat energy may be transformed into mechanical or electrical energy. Since the biologic systems are isothermic, the heat energy cannot

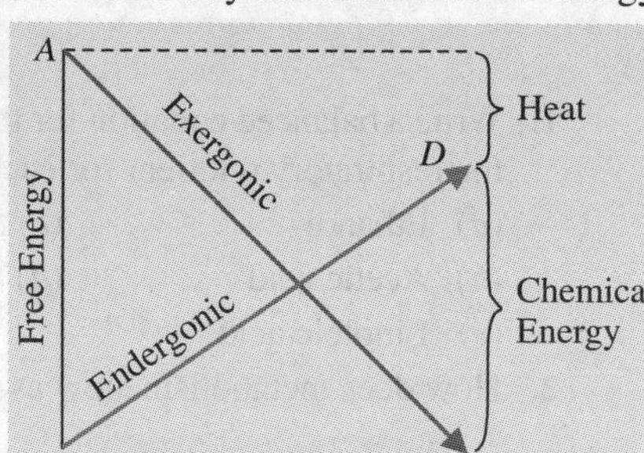

$$A + C \longrightarrow B + D + \text{Heat}$$

Fig. 20–1. Coupling of an exergonic to an endergonic reaction

be used to drive the vital processes (such as synthesis, active transport, nerve conduction, muscular contraction etc.) obtain energy by chemical linkage (or coupling) to oxidation reactions. The simplest type of coupling (Fig. 20–1) may be represented by the equation.

$$A + C \longrightarrow B + D + \text{Heat}$$

The conversion of metabolite A to metabolite B occurs with the release of energy. It is coupled to another reaction, wherein energy is required to convert metabolite C to metabolite D.

CONCEPT OF ENERGY

Energy is defined as the capacity to do work, which is the product of a given force acting through a given distance:

$$\text{Work} = \text{Force} \times \text{Distance}$$

It is one of the fundamental components of any system. Energy exists in a variety of forms, such as electrical, mechanical, chemical, heat and light energy. These different forms of energy are interconvertible, for example, an electric motor converts electric into mechanical energy, a battery changes chemical into electrical energy and a steam engine transforms heat into mechanical energy. Some of the common examples of transformations of energy in biological systems are given in Fig. 20–2. Besides, these various forms of energy are also interrelated quantitatively, for example,

$$1.0 \text{ calorie of heat energy} \equiv 4.185 \times 10^7 \text{ ergs of mechanical energy}$$

Furthermore, during conversion of one form of energy to the other, there is always some loss.

(*a*) (*b*) (*c*)

Fig. 20–2. Transformation of energy in biological systems

(*a*) The running horse represents conversion of chemical energy to mechanical energy.

(*b*) The electric fish (*Torpedinidae*) converts chemical energy to electrical energy.

(*c*) The phosphorescent bacteria convert chemical energy into light energy.

As an example, when an electric motor transforms electric into mechanical energy, the output of useful energy is always less than the input. This is due to friction in the motor which generates heat. The heat, in turn, is dissipated in the environment and is no longer useful. Thus, when a work is done or when one form of energy is changed to the other, there is a loss of useful energy.

THERMODYNAMIC PRINCIPLES

Many quantitative observations made on the interconversion of various forms of energy have led scientists to the formulation of two fundamental laws of thermodynamics, the first and second. These laws help us understanding:

(*a*) the direction of a reaction, whether from left to right or *vice versa*,

(*b*) the accomplishment of work, whether useful or not, and
(*c*) whether the energy for driving a reaction must be delivered from an external source.

The First Law : Principle of Conservation of Energy

In thermodynamics, a *system* is a matter within a defined region. The matter in the rest of the universe is called the *surroundings*. Thus, the system plus the surroundings constitute the universe (Fig. 20–3), which literally includes the entire earth, rather even the outer space. Some physical or chemical processes can be made to take place in isolated systems which are unable to exchange energy with their surroundings. But in the biological world, the reacting system do exchange energy, and often matter also with their surroundings.

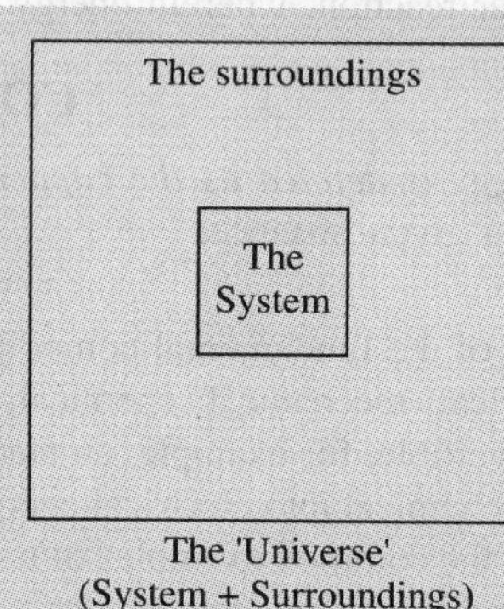

Fig. 20–3. Schematic representation of a reacting system and its surroundings
Energy can be exchanged between the system and the surroundings but only according to the laws of thermodynamics.

The principle of conservation of energy was first formulated by **V. Mayer** in 1841 as a result of a study of energy transformation in the inanimate world, but it is equally applicable to living systems. The first law of thermodynamics states that *the total amount of energy in the universe (i.e., the system + surroundings) remains constant*. Paraphrased, it says that energy cannot be created or destroyed. To date, there is no known exception to this law. Thus, whenever energy is used to do work or is converted from one to the other form, the total amount of energy is unchanged. The mathematical expression of the first law is:

$$\Delta E = E_B - E_A = Q - W \quad \text{...(1)}$$

where,
ΔE = change in internal energy
ΔE_A = energy of a system at the start of a process
ΔE_B = energy of a system at the end of a process
Q = heat absorbed by the system
W = work done by the system.

A noteworthy point about equation 1 is that the change in energy of a system depends only on the initial and the final stages and not on the path of transformation.

The Second Law

The first law of thermodynamics cannot be used to predict whether a reaction can occur spontaneously, although ΔE is positive. In such cases, the system absorbs heat from its surroundings so that the sum of the energies of the system and its surroundings remains the constant. It is evident that a function other than ΔE is required. One such function is entropy (*en*[G] = in; *trope*[G] = to warm in), which is denoted by the symbol *S*. *Entropy is energy in a state of randomness or disorder*. It is unavailable, useless energy. In other words, entropy is a measure of the degree of randomness or disorder of a system. The entropy of a system increases (*i.e.*, ΔS is positive) when it becomes more disordered. Entropy becomes maximum in a system as it approaches equilibrium. When equilibrium is attained, no further change can occur spontaneously unless additional energy is supplied from outside the system.

The term **entropy** was first used in 1851 by **Rudolf Julius Emanuel Clasius** (LT, 1822–1888) of Germany, one of the propounders of the second law of thermodynamics. The importance of entropy was pronounced by **John A. Wheeler** in following words: "No one is considered scientifically literate today who does not know what a Gaussian distribution is, or the meaning and scope of the concept of entropy."

Thus, according to the second law, at each transfer or transformation of energy, part of that energy assumes a form that cannot be passed on any further because it is randomly dispersed, often as heat. And it is this randomly dispersed energy which is called *entropy.* Ultimately closed systems run down. Open systems, such as ecosystems, however, with their constant input and outflow, maintain a steady state in spite of the second law of thermodynamics. Biological systems, thus, do not seem to conform to the second law, for the tendency of life is to produce order out of disorder, to decrease rather than increase entropy.

Although it is true that energy is conserved in the universe, the first law of the thermodynamics does not account for many physical and chemical processes. A simple example may help make this clear. Suppose a full gas cylinder is connected to an empty cylinder by a tube with a valve (Fig 20–4). If the valve is opened, gas flows from the full to the empty cylinder until the gas pressure is equal on both sides. Energy has not only been redistributed but also conserved. The expansion of gas is explained by the second law of thermodynamics and a condition of matter called entropy which, as already stated, is a measure of the randomness or disorder of a system. The greater the disorder of a system, the greater is its entropy. The second law states that physical and chemical processes proceed in such a way that the randomness or disorder of the universe (the system and its surroundings) increases to the maximum possible. Gas will always expand into an empty cylinder.

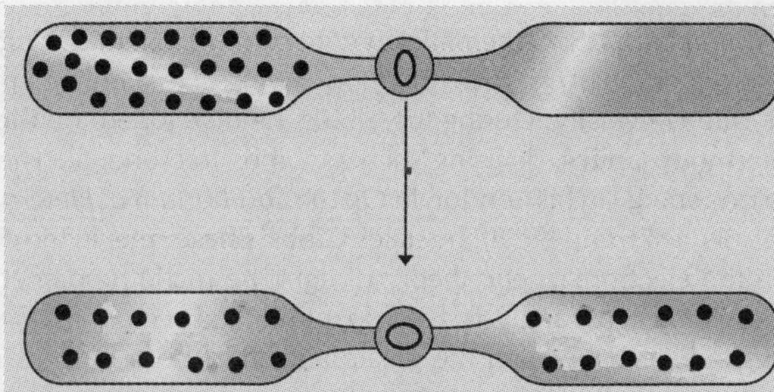

Fig. 20–4. A second law process

The expansion of gas into an empty cylinder simply redistributes the gas molecules until equilibrium is reached. The total number of molecules remains unchanged.

The second law of thermodynamics states that a process can occur spontaneously only if the sum of the entropies of the system and its surroundings increases. This can be represented as:

$$(\Delta S_{system} + \Delta S_{surroundings}) > 0 \quad \text{for a spontaneous process} \quad ...(2)$$

Thus, the total entropy of a system must increase if a process is to occur spontaneously. However, the entropy of a system can decrease even during a spontaneous process, provided the entropy of the surroundings increases to such extent that their sum is positive. When water freezes, its entropy decreases because the water molecules of ice exist in a more ordered state with less freedom of movement than in a liquid state. The decrease of entropy is particularly apparent in the formation of a snowflake (Fig. 20–5).

The difference between the first and second laws of thermodynamics is that the *first law* is concerned with the accounting of the various kinds of energy involved in a given process, whereas the *second law* is concerned with the availability of the energy of a given system for doing useful work.

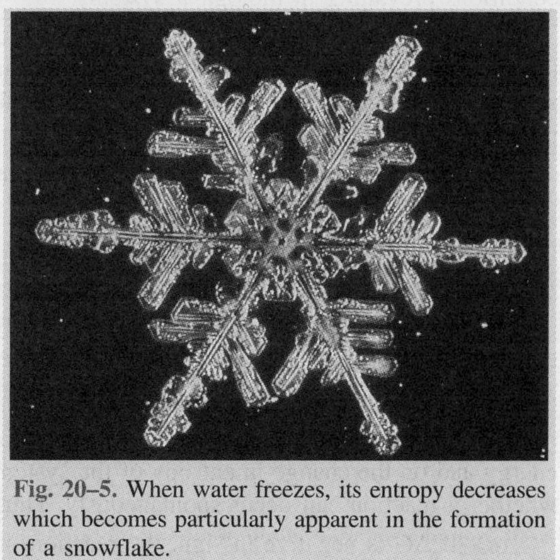

Fig. 20–5. When water freezes, its entropy decreases which becomes particularly apparent in the formation of a snowflake.

(Courtesy: Nuridsany and Perennou)

Combining the Two Laws

Since the entropy changes of chemical reactions are not readily measurable, the entropy is not used as a criterion whether a biochemical process can occur spontaneously or not. Furthermore, for spontaneity, both the entropy changes (that of the surroundings and of the system) should be known. These difficulties are alleviated by using a different thermodynamic function, called the *free energy*, which is denoted by the symbol G (In older literature, however, the symbol F is used in its place).

JOSIAH WILLARD GIBBS
(LT, 1839–1903)

Gibbs, a US mathematician-cum-physical chemist, single-handedly created a large portion of chemical thermodynamics. He spent his entire academic career at Yale University. During the 1870s, he developed the **theory of biochemical thermodynamics**, devising such as Gibbs free energy. The symbol G is given to free energy in his honour. In *On the Equilibrium of Heterogeneous Substances* (2 vols, 1876 and 1878), he states **Gibbs' phase rule.** In the course of his research on the electromagnetic theory of light, he made fundamental contributions to the **art of vector analysis.** He also introduced *vector notations*. He is considered by some to have been the greatest scientist born in the United States.

In 1878, **Gibbs** created the free energy function by combining the first and second laws of thermodynamics in the form of following equation:

$$\Delta G = \Delta H - T\Delta S \qquad ...(3)$$

where, ΔG = the change in free energy of a reacting system,

ΔH = the change in heat content or enthalpy of this system,

T = the absolute temperature at which the process is taking place, and

ΔS = the change in entropy of this system.

In fact, this equation represents quantitative interrelationship between the changes in free energy, heat and entropy in chemical reactions at constant temperature (T) and pressure (P), the conditions prevailing in biological systems.

The term $T\Delta S$ is that fraction of ΔH which cannot be put to useful work. The ΔG indicates the free energy change or the theoretically available useful work. Naturally, in most cases the system is 'inefficient' and not all of the theoretically available work can be utilized. However, the properties of the surroundings do not enter into this equation.

The enthalpy (*enthalpein*[G] = to warm in) change, ΔH is given by the following equation:

$$\Delta H = \Delta E + P\Delta V \qquad ...(4)$$

where, ΔE = the change in internal energy of a reaction, and

ΔV = the change in volume of this reaction.

As the volume change, ΔV is small for nearly all biochemical reactions, hence ΔH is nearly equal to the change in internal energy, ΔE. Therefore, equation 3 modifies as follows :

$$\Delta G \cong \Delta E - T\Delta S \qquad ...(5)$$

Thus, the change in free energy of a reaction, ΔG depends both on the change in internal energy and on the change in entropy of the system. The ΔG is a valuable criterion in determining whether a reaction can occur spontaneously. Thus,

(*a*) If ΔG is negative in sign, the reaction proceeds spontaneously with loss of free energy, *i.e.*, it is **exergonic**. If, in addition, ΔG is of great magnitude, the reaction goes virtually to completion and is essentially irreversible.

(*b*) If, however, ΔG is positive, the reaction proceeds only if free energy can be gained, *i.e.*, it is **endergonic**. If, in addition, ΔG is of high magnitude, the system is stable with little or no tendency for a reaction to occur.

(*c*) If ΔG is zero, the reaction system is at equilibrium and no net change takes place.

With regard to the free energy change, ΔG of a reacting system two more points need to be emphasized:

I. Firstly, the ΔG of a reaction depends only on the free energy of the products minus that of the reactants. *The ΔG of a reaction is independent of the path of transformation.* Obviously, the mechanism of a reaction has no effect on ΔG. As an instance, the value of ΔG is the same for the oxidation of glucose to CO_2 and H_2O whether it takes place by combustion or by a series of enzyme-catalyzed reactions.

II. Secondly, *the value of ΔG provides no information about the rate of a reaction.* A negative ΔG indicates that a reaction can occur spontaneously, but it does not signify that it will occur at a perceptible rate. As already pointed out, the rate of a reaction rather depends on the free energy of activation ($\Delta G^{\neq}$), which is unrelated to ΔG.

To get a feeling for the magnitude of changes in various forms of energy, an actual example of aerobic oxidation of glucose may be cited. The living cells carry out oxidation of glucose in the presence of oxygen to CO_2 and H_2O at constant temperature and pressure.

$$C_6H_{12}O_6 + 6O_2 \longrightarrow 6CO_2 + 6H_2O$$

Assuming that the temperature is 25°C (or 298 K) and the pressure is 1.0 atm. (or 760 mm Hg), which are standard conditions in thermodynamic calculations, the following energy changes take place per molecule of glucose oxidized:

$$\Delta G = -\ 686{,}000 \text{ cal/mol}$$ (*i.e.*, the free energy of the reacting molecules has decreased)

$$\Delta H = -\ 673{,}000 \text{ cal/mol}$$ (*i.e.*, the reacting molecules have released heat)

The equation 3 may also be written as:

$$\Delta S = \frac{\Delta H - \Delta G}{T} \qquad \text{...(6)}$$

Substituting the above values in equation 6, we get:

$$\Delta S = \frac{-673{,}000 - (-686{,}000)}{298}$$

$$= 44 \text{ cal/deg}$$ (*i.e.*, the entropy of the universe has increased)

RELATIONSHIP BETWEEN STANDARD FREE ENERGY CHANGE AND EQUILIBRIUM CONSTANT

In a model reaction,

$$A + B \rightleftharpoons C + D \qquad \text{...(7)}$$

The free energy change, ΔG of this reaction is given by

$$\Delta G = \Delta G^\circ + RT \log_e \frac{[C][D]}{[A][B]} \qquad \text{...(8)}$$

where,
ΔG° = Standard free energy change,
R = Gas constant,
T = Absolute temperature, and
$[A]$, $[B]$, $[C]$ and $[D]$ = Molar concentrations (*i.e.*, activities) of the reactants

To elaborate, ΔG° is the free energy change for this reaction under standard conditions, *i.e.*,when each of the reactants *A, B, C* and *D* is present at a concentration of 1.0 M. Thus, the ΔG of a reaction depends on the nature of the reactants (expressed in ΔG° term) and on their concentrations (expressed in logarithmic terms), as shown in equation (8).

At equilibrium, $\Delta G = 0$. Equation (8) then becomes,

$$O = \Delta G^\circ + RT \log_e \frac{[C][D]}{[A][B]} \quad ...(9)$$

or

$$\Delta G^\circ = -RT \log_e \frac{[C][D]}{[A][B]} \quad ...(10)$$

The equilibrium constant under standard conditions, K'_{eq} for the reaction $A + B \rightleftharpoons C + D$, is given by

$$K'_{eq} = \frac{[C][D]}{[A][B]} \quad ...(11)$$

However, in reactions in which more than one molecule of any reactant or product participates, the general reaction"

$$aA + bB \rightleftharpoons cC + dD$$

where *a, b, c* and *d* are the number of molecules of A, B, C and D participating, the equilibrium constant is given by

$$K'_{eq} = \frac{[C]^c\,[D]^d}{[A]^a\,[B]^b} \quad ...(12)$$

Substituting equation 11 into equation 10 gives

$$\Delta G^\circ = -RT \log_e K'_{eq} \quad ...(13)$$

or

$$\Delta G^\circ = -2.303\ RT \log_{10} K'_{eq} \quad ...(14)$$

or

$$K'_{eq} = 10^{-\Delta G^\circ/(2.303RT)} \quad ...(15)$$

Substituting $R = 1.98 \times 10^{-3}$ kcal mol^{-1} degree^{-1} and $T = 298°K$ (corresponding to 25°C) gives

$$K'_{eq} = 10^{-\Delta G^\circ/1.36} \quad ...(16)$$

when ΔG° is expressed in kcal/mol. Thus, the standard free energy, $\Delta G^\circ$and the equilibrium constant, K'_{eq} are related by a simple expression. For example, a change in equilibrium constant by a factor of 10 results in a change in standard free energy of –1.36 kcal/mol at 25°C (refer Table 20–1). At 37°C, however, the change in standard free energy would be of 1.42 kcal/mol. Values of ΔG° may be expressed in joules or calories per mole.

Table 20–1. Numerical relationship between equilibrium constants of chemical reactions and their standard free energy changes at 25°C

K'_{eq}	ΔG° *(kcal/mol)*
10^{-5} or 0.00001	+ 6.82
10^{-4} or 0.0001	+ 5.46
10^{-3} or 0.001	+ 4.09
10^{-2} or 0.01	+ 2.73
10^{-1} or 0.1	+ 1.36
1	0
10^{1} or 10	– 1.36
10^{2} or 100	– 2.73
10^{3} or 1,000	– 4.09
10^{4} or 10,000	– 5.46
10^{5} or 1,00,000	– 6.82

$\Delta G°$ may also be defined as the difference between the free energy content of the reactants and that of the products under standard conditions (*i.e.*, 298 K pressure and 1.0 atmospheric pressure) when the reactants and products are present in their standard concentrations, namely 1.0 *M*. When $\Delta G°$ is negative (*i.e.*, the products contain less free energy than the reactants), the reaction will proceed to form the products under standard conditions, since all chemical reactions will tend to go in that direction resulting in decrease in the free energy of the system. When $\Delta G°$ is positive (*i.e.*, the products of the reaction contain more free energy than the reactants), the reaction will tend to go in the reverse direction if we start with 1 *M* concentration of all components. To be more explicit, reactions with a negative $\Delta G°$ proceed forward in the direction written when they start with all reactants and products at 1.0 *M*, until they reach equilibrium. Reactions with a positive $\Delta G°$ will proceed in the reverse of the direction written when they start with all components at 1.0 *M*. Table 20–2 summarizes these relationships.

Table 20–2. Relationship among K′eq, ΔG° and the direction of chemical reactions under standard conditions

When K'_{eq} *is:*	$\Delta G°$ *is:*	*Starting with 1 M components, the reaction:*
> 1.0	Negative	proceeds forward
1.0	Zero	remains at equilibrium
< 1.0	Positive	proceeds backward

Two more points need be emphasized:

I. Biochemical reactions take place near *p*H 7.0. Hence, *p*H 7.0 is conventionally designated the standard pH in biochemical energetics. The standard free energy change at *p*H 7.0 in biochemical energetics is designated by the symbol, $\Delta G°'$, which shall be used in further discussions.

II. The basic SI (Systéme International) unit of energy is joule (J). However, in biology and medicine, heat and energy measurements are equally expressed in calories (cal). The interrelationship between calories and joules is:

$$1.000 \text{ cal} = 4.184 \text{ J}$$

A **calorie (cal)** is equivalent to the amount of heat needed to raise the temperature of 1 gram of water from 14.5 to 15.5°C. A calorie is most oftenly used by biologists. A **kilocalorie (kcal)** is equal to 1,000 cal. Also, 1 kcal = 4.184 kJ.

JAMES PRESCOTT JOULE
(LT, 1818–1889)

Joule, a British physicist-cum-engineer, was born at Salford near Manchester on December 24, 1818. His father was the proprietor of a large brewery in Manchester, England where he got opportunities of carrying on with his experiments. He had a very delicate health and had to be educated at home. Fortunately, **John Dalton** was employed as his private tutor from whom the young **Joule** got the inspiration for Science.

He spent first few years of his student life in electrical researches due to the influence of **Michael Faraday** (LT, 1791–1867), a British scientist who discovered the principle of electromagnetic induction and built the first transformer. Joule designed an electromagnetic engine. Later, his attention was concentrated upon the manner in which work was done by an engine. He found that in all engines, work is done at the disappearance of certain quantity of heat. Then he wanted to show the relation between work done and heat necessary to do the required amount of work. He determined the

mechanical equivalent of heat in 1843. To complete his investigations of the mechanical equivalent of heat, he devoted 35 years from 1843 to 1878. He experimentally proved the first law or the *principle of interconvertibility of various forms of energy* and also proposed the **kinetic theory of gases**. The energy unit Joule is named after him. A *Joule* (J) is the amount of energy required to apply a 1 newton force over a distance of 1 metre. The joule is normally used by chemists and physicists. A *kilojoule* (kJ) is equal to 1,000 J. Joule also invented various apparatuses including electric meters and speed-counter of hydrogen molecules. Faraday's belief in the unity of nature was vindicated in the work by **Joule, Thomson, Helmholtz, Clasius** and **Maxwell.** In 1872, Joule's health gave away and he was forced to retire. He died on October 11, 1889 at Sale, Cheshire, UK.

STANDARD FREE ENERGY CHANGES AT pH 7.0 OR $\Delta G^{\circ\prime}$

To calculate $\Delta G^{\circ\prime}$, an example of isomerization of dihydroxyacetone phosphate (DHAP) to glyceraldehyde 3-phosphate (G-3-P) may be taken. This is one of the reactions of glycolysis.

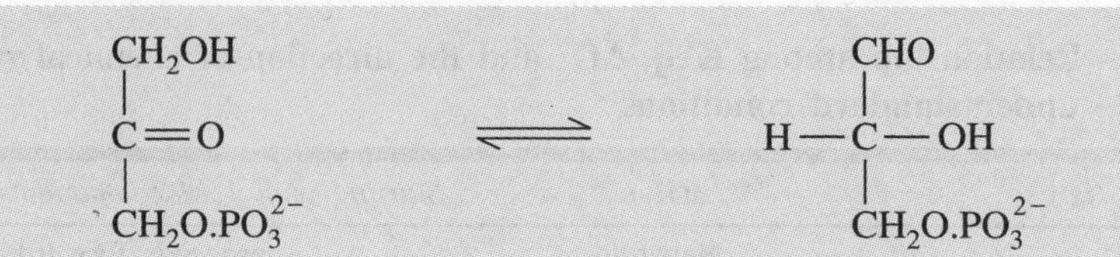

Dihydroxyacetone phosphate   Glyceraldehyde 3-phosphate

At equilibrium, the ratio of glyceraldehyde 3-phosphate to dilydroxyacetone phosphate is 0.0475 at 25°C (298°K) and pH 7.0. Hence,

$$K'_{eq} = 0.0475$$

The standard free energy change for this reaction is, then, calculated from equation 14, as follows:

$$\begin{aligned}\Delta G^{\circ} &= -2.303\ RT \log_{10} K'_{eq} \\ &= -2.303 \times 1.98 \times 10^{-3} \times 298 \times \log_{10}(0.0475) \\ &= +1.8 \text{ k cal/mol}\end{aligned}$$

When the initial concentration of DHAP is $2 \times 10^{-4}\ M$ and the initial concentration of G-3-P is $3 \times 10^{-6}\ M$, the ΔG can be calculated from equation 8, as follows:

$$\begin{aligned}\Delta G &= \Delta G^{\circ} + RT \log_e \frac{[C][D]}{[A][B]} \\ &= 1.8 \text{ kcal/mol} + 2.303\ RT \log_{10} \frac{3\times10^{-6}\ M}{2\times10^{-4} M} \\ &= 1.8 \text{ kcal/mol} - 2.5 \text{ kcal/mol} \\ &= -0.7 \text{ kcal/mol}\end{aligned}$$

The negative value for ΔG points out that isomerization of dihydroxyacetone phosphate to glyceraldehyde 3-phosphate can occur spontaneously, when these compounds are present at concentrations mentioned above. Although ΔG is negative for this reaction, the value of $\Delta G^{\circ\prime}$ is positive. It is important to note that the magnitude of ΔG for a reaction (whether smaller, larger or the same as $\Delta G^{\circ\prime}$) depends on the concentrations of the reactants .The criterion of spontaneity for a reaction is ΔG and not $\Delta G^{\circ\prime}$. Table 20–3 lists the standard free energy changes for some chemical reactions.

A perusal of the table (given on the next page) indicates that hydrolysis of esters, amides, peptides and glycosides, as well as rearrangements, and elimination proceed with relatively small standard free energy changes. Hydrolysis of acid anhydrides proceeds with relatively large decreases in standard free energy, whereas oxidation of organic compounds to CO_2 and H_2O takes place with rather huge decreases in standard free energy.

Difference Between ΔG and ΔG°′

The difference between the free energy change, ΔG and the standard free energy change, $\Delta G^{\circ\prime}$ of a chemical reaction should be clearly understood. The actual free energy change, ΔG of a chemical reaction is a function of the conditions of concentration, *p*H and temperature under which the reaction is taking place. Moreover, the ΔG of an ongoing chemical process is always negative, becomes smaller (*i.e.*, less negative) as the reaction proceeds and is zero at the point of equilibrium, indicating that no more work can be done by the reaction. The value of ΔG declines with time as the reaction proceeds because the actual concentrations of the reactants will be getting smaller and those of the resultants getting larger.

Table 20–3. Standard free energy changes for representative chemical reactions under standard conditions (*i.e.*, at 25°C and pH 7.0)

Reaction Types and Reactions	*ΔG°′ (in kcal/mol)*
Elimination of Water	
Malate ⟶ Fumarate + H_2O	+0.75
Rearrangements	
Glucose 1-phosphate ⟶ Glucose 6-phosphate	–1.74
Fructose 6-phosphate ⟶ Glucose 6-phosphate	–0.40
Hydrolysis	
Esters:	
Ethyl acetate + H_2O ⟶ Ethanol + Acetate	–4.7
Glucose 6-phosphate + H_2O ⟶ Glucose + Phosphate	–3.3
Amides and peptides:	
Glutamine + H_2O ⟶ Glutamate + NH_4^+	–3.4
Glycylglycine + H_2O ⟶ 2 Glycine	–2.2
Glycosides:	
Maltose + H_2O ⟶ 2 Glucose	–3.7
Lactose + H_2O ⟶ Glucose + Galactose	–3.8
Acid anhydrides:	
Acetic anhydride + H_2O ⟶ 2 Acetate	–21.8
ATP + H_2O ⟶ ADP + Phosphate	–7.3
Oxidations with Molecular Oxygen	
Glucose + $6O_2$ ⟶ $6CO_2$ + $6H_2O$	–686
Palmitic acid + $23O_2$ ⟶ $16CO_2$ + $16H_2O$	–2,338

Thus, in every spontaneous chemical or physical process, the free energy of the reacting system always decreases, *i.e.*, ΔG is negative. On the contrary, the value of standard free energy change, $\Delta G^{\circ\prime}$ for a chemical reaction is characteristic and unchanging, and may be positive, negative or zero, depending on the equilibrium constant of the reaction. $\Delta G^{\circ\prime}$ is, thus, an immutable constant and tells us in which direction and how far a given reaction will go in order to reach equilibrium when it occurs under standard conditions, *i.e.*, when the initial concentration of all components is 1.0 *M*, *p*H is 7.0 and the temperature is 25°C.

Standard Free Energy Value of Chemical Reactions are Additive.

A noteworthy thermodynamic fact is that the overall free energy change for a series of reactions is equal to the sum of the free energy changes of the individual steps. Consider the two consecutive reactions where each reaction has its own equilibrium constant and a characteristic standard free energy change, $\Delta G_1{}^{\circ\prime}$ and $\Delta G_2{}^{\circ\prime}$.

$$A \longrightarrow B \qquad \Delta G_1{}^{\circ\prime} = -\ 8 \text{ kcal/mol}$$
$$B \longrightarrow C \qquad \Delta G_2{}^{\circ\prime} = +\ 5 \text{ kcal/mol}$$

Since the two reactions are sequential, the intermediate product *B* cancels out and the overall reaction with its standard free energy change, $\Delta Gs^{\circ\prime}$ may be written as:

$$A \longrightarrow C \qquad \Delta Gs^{\circ\prime} = \Delta G_1{}^{\circ\prime} + \Delta G_2{}^{\circ\prime} = -8 + (+5) = -3 \text{ kcal/mol}$$

Under standard conditions, *A* can be spontaneously converted into *B* because ΔG is negative. However, the conversion of *B* into *C*, under standard conditions is thermodynamically not feasible. But as the free energy changes are additive, the conversion of *A* into *C* has a $\Delta G^{\circ\prime}$ value of –3 kcal/mol, which obviously means that A can be converted into C spontaneously under standard conditions. Thus, the above two sequential reactions are coupled by the intermediate product, *B*. In other words, *a thermodynamically unfavourable reaction can be driven by a thermodynamically favourable reaction.*

Two sequential steps from glycogen breakdown in muscles will illustrate this fact more clearly.

$$\text{Glucose 1-phosphate} \xrightarrow[\text{mutase}]{\text{Phosphogluco-}} \text{Glucose 6-phosphate} \qquad \Delta G_1{}^{\circ\prime} = -1.74 \text{ kcal/mol}$$

$$\text{Glucose 6-phosphate} \xrightarrow[\text{isomerase}]{\text{Glucose phosphate}} \text{Fructose 6-phosphate} \qquad \Delta G_2{}^{\circ\prime} = +0.40 \text{ kcal/mol}$$

On adding the two reactions, we get:

$$\text{Glucose 1-phosphate} \longrightarrow \text{Fructose 6-phosphate}$$

This has a standard free energy change value,

$$\Delta Gs^{\circ\prime} = \Delta G_1{}^{\circ\prime} + \Delta G_2{}^{\circ\prime} = -1.74 + (+\ 0.40) = -1.36 \text{ kcal/mol}$$

Because $\Delta Gs^{\circ\prime}$ is negative, glucose 1-phosphate is converted into fructose 6-phosphate in the muscles.

ATP AS UNIVERSAL CURRENCY OF FREE ENERGY IN BIOLOGICAL SYSTEMS

The living objects require a continuous supply of free energy mainly for the following 4 purposes:

(*a*) to synthesize macromolecules from simpler and smaller precursors,

(*b*) to transport molecules and ions across membranes against gradients,

(*c*) to perform mechanical work, as in the muscle contraction, and

(*d*) to ensure fidelity of information transfer.

The free energy in these processes is derived from the environment. The phototrophs obtain this energy by trapping light energy from the sun. On the other hand, the chemotrophs obtain it by the oxidation of foodstuffs. This free energy (derived from light or from the oxidation of foodstuffs) is partly transformed into a special form before it is used for biosynthesis, transport, motion and fidelity. This special carrier of free energy is adenosine triphosphate (ATP). ATP plays a central role in the transference of free energy from the exergonic (= *energy-yielding*) to the endergonic (= *energy-requiring*) processes in the cells. During breakdown of energy-rich foodstuffs or fuel molecules, some of the free energy is harnessed to make ATP from adenosine diphosphate (ADP) and inorganic phosphate (P_i), a process that requires input of free energy. ATP then donates much of its chemical energy to energy-requiring processes (biosynthesis, transport etc.) by undergoing a breakdown to ADP and P_i (Fig. 20–6).

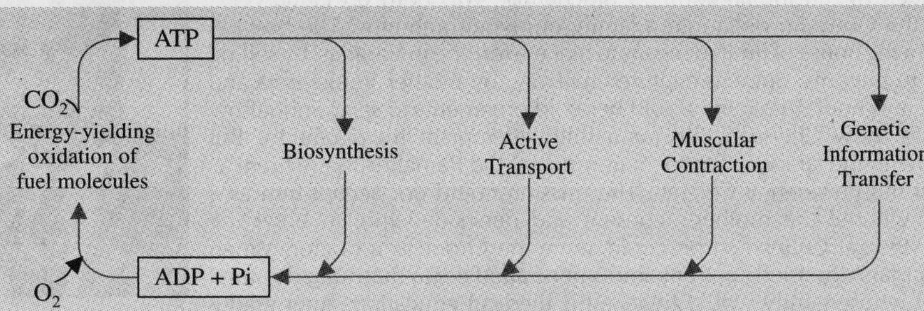

Fig. 20–6. The ATP cycle in cells

An interesting use of the cleavage of ATP to AMP and PPi occurs in the firefly, which uses ATP as an energy source to produce light flashes.

Firefly Flashes Reveal Glowing of ATP

The phenomenon of **bioluminescence** requires considerable amounts of energy. In the firefly, a beetle of the Lampyridae family, ATP is used in a set of reactions that converts chemical energy into light energy. In the 1950s, from many thousands of fireflies collected by children in and around Baltimore, **William McElroy** and his colleagues at The John Hopkins University isolated the principal biochemical components: **luciferin**, a complex carboxylic acid and **luciferase**, an enzyme. The generation of a light flash requires activation of luciferin by an enzymatic reaction involving pyrophosphate cleavage of ATP to form luciferyl adenylate. In the presence of molecular oxygen and luciferase, the luciferin undergoes a multistep oxidative decarboxylation to oxyluciferin. This process is accompanied by emission of light. The colour of the light flash differs with the firefly species and appears to be determined by differences in the structure of the luciferase. Luciferin is regenerated from oxyluciferin in a subsequent series of reactions.

In the laboratory, pure firefly luciferin and luciferase are used to measure minute quantities of ATP by the intensity of the light flash produced. As little as a few picomoles (10^{-12} mol) of ATP can be measured in this way. An enlightening extension of the studies in luciferase was the cloning of the luciferase gene in tobacco plants. When watered with a solution containing luciferin, the plants glowed in the dark.

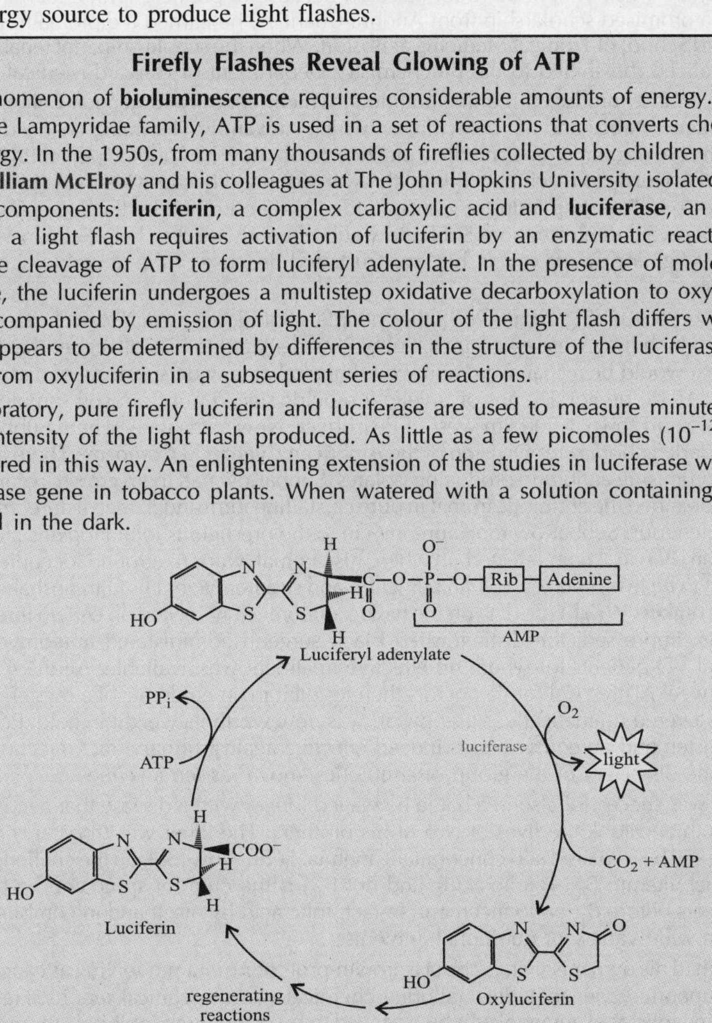

Important components in the firefly bioluminescence cycle

YELLAPRAGADA SUBBAROW

(LT, 1895–1948)

SubbaRow, an Indian-born US biochemist, was born on January 12, 1895 in the Godavari delta into a family of niyogi brahmins. The boy ran away from the house of limited means to make a fortune in Varanasi by selling bananas to pilgrims, but was captured halfway by mother **Venkamma** and put back in school. Venkamma sold her gold ornaments to send SubbaRow to Madras (now Chennai) city for a third attempt at matriculation. But SubbaRow in the faraway city spent as much at the Ramakrishna Ashram as he did at the Presidency College. The mission could not accept him as a 'sanyasi' without the mother's consent and persuaded him to enter the Madras Medical College so he could serve the Order as a doctor. When private sholarships dried up, Venkamma persuaded her to marry a girl named **Seshagiri** whose family would finance his medical education. After graduation, he took admission in Madras Ayurvedic College, whose Vice Principal **Achanta Laxmi Pathi** had saved his life when the allopaths could not treat his bout with sprue. Soon he realized that the college atmosphere was not conducive for research because of politics. With a loan from his father-in-law against a promised scholarship from Andhra Charities, he earned a diploma from Dr. Richard Strong's Harvard School of Tropical Medicine at Boston. When the scholarship, not tenable for medical studies, materialized, he shifted to the Biochemistry Department of Harvard Medical School.

SubbaRow, in fact, produced the first antibiotic which, contrary to general belief, is *gramicidin* discovered by **René Dubos** and not penicillin on which **Alexander Flemming** chanced upon. Its side effects barred the use of gramicidin in medical practice. SubbaRow had purifed penicillin but was beaten in race with **H.W. Florey** to reach pencillin to the war-wounded because his management would not let him join the "pool" of drug industry giants, funded and controlled by the "creeping socialists" of the Franklin Roosevelt administration. SubbaRow subsequently tried to help **Selman Waksman** with his turberculosis-fighting *Streptomycin* but was frustrated because of the microbiologist's tieup with a rival drug firm.

Having fed up of running about with other people's antibiotics which were limited in the range of infections in which they were efficacious, SubbaRow declared to his colleagues that he would find an antibiotic that would be a "panacea" and would neutralize a wide spectrum of disease organisms. And in August, 1945, he got a series of 6 yellow moulds sieved out from soil samples of a field in Columbia, a Missouri town 1,500 km west of Pearl river, New York state, where SubbaRow headed research for Lederle Laboratories Division of the American Cyanamid Company. There in the test tube were the most promising antibiotic-producing isolates. SubbaRow had them grown in corn steep liquor and ultimately obtained the antibiotic from it in pure crystalline form and named it *aureomycin*, because of its golded splendour. SubbaRow took aureomycin to the prestigious John Hopkins Hospital. There, famed physician **Perrin Long** rejected it when his animal warren (ground occupied by rabbits) demonstrated it to be less spectacular than penicillin and streptomycin. His faith unshaken, SubbaRow showed **Louis Tomkins Wright** the data on the havoc aureomycin wrecked on certain infectious viruses in animal hosts. Impressed, the distinguished black surgeon administered aureomycin at Harlem Hospital to viral VD patients for whom no effective treatment was available. All the 4 patients were cured within a week. Three California doctors then tested it in an epidemic of a mysterious viral fever and patients recovered dramatically. Aureomycin was now worth its weight in gold. Perrin Long was now ready to listen and aureomycin was indeed effective against urinary tract infections. He is also credited with the discovery of the group of antibiotics known as *tetracyclines*.

SubbaRow was successful also in a battle he waged longer with a disease that nearly cut short his own life in youth and took the lives of two of his brothers. The killer was *tropical sprue*, a vitamin deficiency, marked by impaired and concomitant anemia. SubbaRow took up the challenge and proved conclusively that vitamin B_9 (= folic acid), and not B_{12}, is the cure for sprue and he had wrested it from liver, 5 years before B_{12} announcement, In fact, folic acid by itself and in combination with B_{12} is effective in a wide variety of nutritional ailments.

SubbaRow had his own misgivings about a growth-promoting vitamin serving as cancer cell-growth inhibitor but supported cancer studies, tailoring chemical work to clinical reports. He soon had his "team" producing folic acid antagonists which proved to be real growth inhibitors of cancer cells. One of them *aminopterin*, brought about miraculous recovery of children suffering from acute lukaemia or blood cancer. A derivative, methotrexate, is one of the most potent drugs against cancer and helped bring about a revolution in chemotherapy. Conquest of cancer became a magnificent obsession with SubbaRow after reports of aminopterin success came in. He planned a cancer research institute but before his plans could get off the ground, SubbaRow died in his sleep the night of August 8–9, 1948. He was merely 53.

Thus, in less than 10 years, SubbaRow had produced an antibiotic, a vitamin and an anticancer drug, besides an antifilarial drug diethycarbamazine, still the best guarantee against deforming elephantiasis. He had also broken the old vitamin monopolies by devising alternative processes for several members of B complex. How then has such a wizard of drugs, hailed as *"miracle man of miracle drugs"* remained an unsung hero? Especially when intimates believe he was a fame hunter? In fact, SubbaRow never recognized any of his accomplishments as worthy of the fame he sought as one who changed fortunes in man's fight against disease. He was always on a back seat in the auditorium when the accolades were handed out on the stage.

Not only this, with **Cyrus Fiske** as his guide, he devised a colour measurement of phosphorus in blood and urine. In diagnosis of rickets, renal diseases, parathyroid hormone deficiency and overactivity, and bone calcification, the **Fiske–SubbaRow method** is comparable in value to the sugar test for diagnosing and monitoring diabetes. SubbaRow, in joint research with Fiske, **discovered phosphocreatine** and **adenosine triphosphate** (ATP) which have unravelled a mystery of life: the source of muscular energy which keeps the world moving.

> "You have probably never heard of Yellapragada SubbaRow. Yet because he lived you may be alive and well today. Because he lived you may live longer."

Such was the tribute offered by an American magazine to SubbaRow. He is truly regarded by many scientists as 'one of the most eminent medical minds of the (twentieth) century.'

FRITZ ALBERT LIPMANN
(LT, 1899–1986)

Lipmann was a German-born American biochemist. After obtaining M.D. degree, he started research career in 1927 as an unpaid graduate student under the famous biochemist **Otto Meyerhof.** He was awarded the **1953 Nobel Prize in Medicine or Physiology** for his discovery, in 1947, of *coenzyme A* and revealing its importance in intermediary metabolism, along with **Sir Hans Adolf Krebs** of Great Britain. Lipmann is popularly called as the **'father' of ATP cycle.** He introduced the "*squiggle*" notation (~) to designate the energy-rich bonds of biomolecules such as ATP and ADP. He was professionally active at Rockefeller University until his death at age 87. His famous quote (1949) reads as follows:

> "It seems that in the field of biosynthesis we have a rare example of progress leading to simplification."

Adenosine triphosphate (ATP) was discovered in extracts of skeletal muscles by **Karl Lohmann** in Germany and by **Cyrus Fiske** and **Yellapragada SubbaRow** in the United States, simultaneously in 1929. Later, it was found to be present in all types of cells– animal, plant and microbial. Although first thought to be concerned with muscular contraction only, ATP has now been assigned many cell activities. In 1941, **Fritz A. Lipmann** postulated that ATP is the primary and universal carrier of chemical energy in cells. He adequately stated that the small chemical units from which the material of organisms is built "are glued together by an enormously versatile condensing reagent, ATP." He also first proposed the ATP cycle, shown in a present-day form in Fig. 20–6.

Adeosine triphosphate (ATP) and its successive hydrolysis products, adenosine diphosphate (ADP) and adenosine monophosphate (AMP) are nucleotide, consisting of an adenine, a ribose and a 3, 2 or 1 phosphate group(s) respectively (Fig. 20–7). ATP, ADP and AMP occur not only in cell cytosol but also in mitochondria and the nucleus. In normal respiring cells, ATP makes up about 75% or more of the sum of all 3 adenine ribonucleotides (refer Table 20–4).

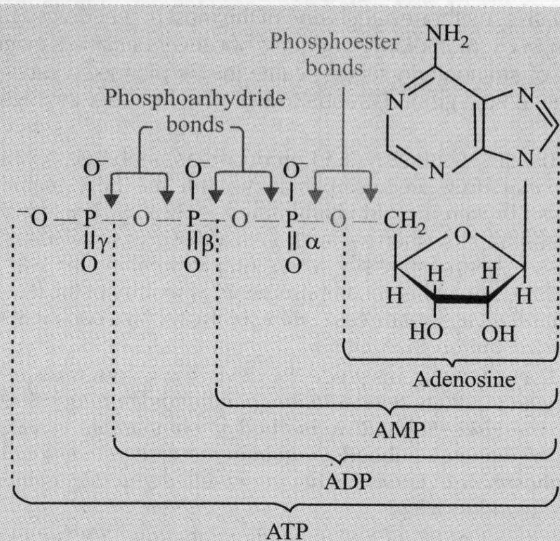

Fig. 20–7. The structure of ATP, indicating its relationship to ADP, AMP and adenosine

The phosphoryl groups of ATP, starting with that on AMP, are designated as α, β and γ phosphates. At pH 7.0, the phosphate groups are almost completely ionized, hence so shown here. The terminal phosphate group (*i.e.*, γ) can be enzymatically transferred to various phosphate acceptors. Note the differences between phosphoester and phosphoanhydride bonds.

Table 20–4. Concentrations (in mM) of 3 adenine ribonucleotides, phosphate (Pi) and phosphocreatine (PCr) in some cells*

Cells type	*ATP*	*ADP*	*AMP*	*Pi*	*PCr*
Rat muscle	8.05	0.93	0.04	8.05	28
Rat liver	3.38	1.32	0.29	4.80	0
Rat brain	2.59	0.73	0.06	2.72	4.7
Escherichia coli	7.90	1.04	0.82	7.90	0
Human erythrocytes	2.25	0.25	0.02	1.65	0

* These data are for the entire cell contents except in the case of erythrocytes where the concentrations are those of cytosol, since erythrocytes are absent from nucleus and mitochondria both.

(*Adapted from Lehninger AL, 1984*)

At *p*H 7.0, the phosphate group(s) of ATP, ADP and AMP are almost fully ionized, so that they occur as multiple charged anions ATP^{4-} ADP^{3-} and AMP^{2-}. But as the cell fluid contains high concentrations of Mg^{2+}, the ATP and ADP both exist largely as Mg ATP^{2-} and $MgADP^{-}$ complexes (Fig. 20–8). In fact, in phosphate transfer reactions, ATP participates as its complex form. However, ATP can also form complex with Mn^{2+}.

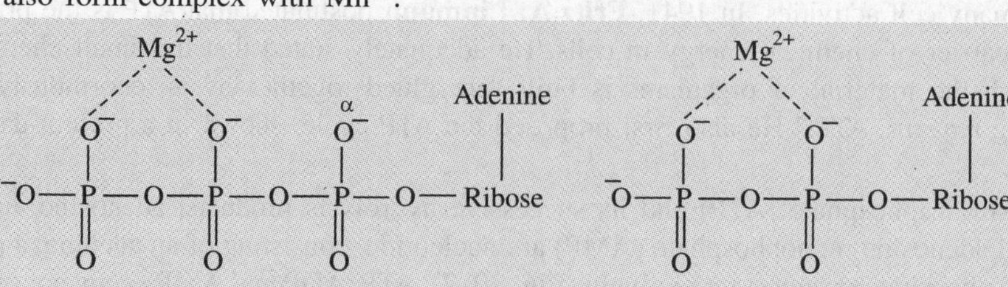

Fig. 20–8. Magnesium ion complexes of ATP and ADP

ATP serves as the principal immediate donor of free energy in biological systems rather than as a storage form of energy. In a typical cell, an ATP molecule is consumed within a minute of its formation. *The turnover of ATP is very high.* For instance, a resting human consumes about 40 kg ATP in a day. During strenuous labour, the ATP is consumed at the rate of even 0.5 kg per minute. The endergonic processes such as biosynthesis, active transport etc., can occur only if ATP is continuously regenerated from ADP (refer Fig. 20–9). Phototrophs harvest the free energy in light to regenerate ATP whereas chemotrophs form ATP by the oxidation of foodstuffs.

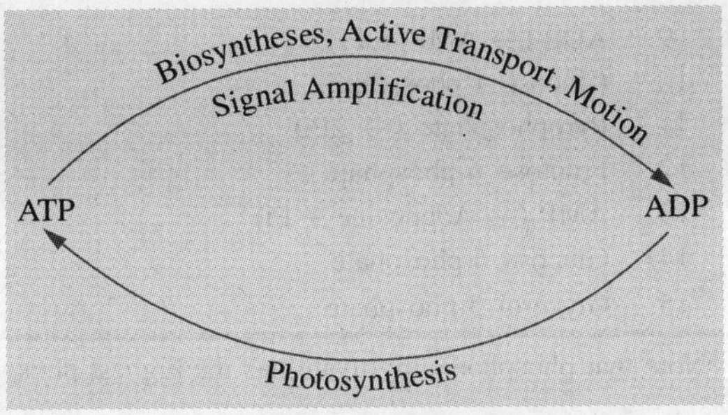

Fig. 20–9. The ATP-ADP cycle

The cycle is the fundamental mode of energy exchange in biological systems.

FREE ENERGY OF HYDROLYSIS OF ATP AND OTHER ORGANOPHOSPHATES

When ATP is hydrolyzed, it loses its terminal γ phosphate group to form ADP and orthophosphate or inorganic phosphate (Pi).

$$ATP + H_2O \longrightarrow ADP + Pi$$

The standard free energy change, $\Delta G^{\circ\prime}$ for this reaction is –7.3 kcal/mol. Standard free energy changes have also been determined for the hydrolysis of other phosphorylated compounds or organophosphates (Table 20–5). Some phosphates yield more and some yield less free energy than ATP upon hydrolysis, under standard conditions. This intermediate position enables ATP to function efficiently as a carrier of phosphoryl groups. Thus, with respect to the $\Delta G^{\circ\prime}$ value of hydrolysis of ATP, two classes of *organophosphates* are recognized: high energy phosphates exemplified by enolphosphates (*e.g.*, phosphoenolpyruvate), phosphoguanidines (*e.g.*, creatine phosphate and arginine phosphate) etc., which have $\Delta G^{\circ\prime}$ values larger than that of ATP and low energy phosphates, exemplified by ester phosphates found in the intermediates of glycolysis, which have $\Delta G^{\circ\prime}$ values smaller than that of ATP. However, the designations 'high' and 'low' do not clearly indicate that there are 3 classes of phosphates. Such phosphates as phosphoenolpyruvate, creatine phosphate etc., whose $\Delta G^{\circ\prime}$ values are higher than that of ATP, should better be designated as 'super' high energy phosphates, the ATP then bedesignatedas'high'energyphosphate and theester phosphatesas 'low' energy phosphates.

Table 20–5. Standard free energy of hydrolysis of some phosphorylated compounds

Compound	$\Delta G^{\circ\prime}$ *(kcal/mol)*
1. Phosphoenolyruvate*	–14.8
2. Carbamoyl phosphate	–12.3
3. 3-phosphoglyceroyl phosphate or 1, 3-diphosphoglycerate (→ 3-phosphoglycerate + P*i*)	–11.8
4. Creatine phosphate or phosphocreatine	–10.3
5. Acetyl phosphate	–10.1
6. Arginine phosphate	–8.0
7. ATP (→ AMP + PPi)	–7.7
8. ATP (→ ADP + Pi)	–7.3

9.	ADP (→ AMP + Pi)	–7.3
10.	Glucose 1-phosphate	–5.0
11.	Pyrophosphate (→ 2Pi)	–4.6
12.	Fructose 6-phosphate	–3.8
13.	AMP (→ Adenosine + Pi)	–3.4
14.	Glucose 6-phosphate	–3.3
15.	Glycerol 3-phosphate	–2.2

*Note that phosphoenolpyruvate has the highest phosphate group transfer potential of the compounds listed.

A persual of Table 20–5 indicates that when ADP is hydrolyzed to AMP and inorganic phosphate, the $\Delta G^{o\prime}$ value of this reaction is the same as that of the reaction, ATP ⟶ ADP + P*i*, that is – 7.3 kcal/mol. Thus, the two terminal phosphate groups of ATP (β and γ) are both high energy groups. On the contrary, the $\Delta G^{o\prime}$ value of hydrolysis of AMP to yield adenosine and phosphate is much lower, that is only –3.4 kcal/mol. Thus, the phosphate group of AMP (*i.e.*, the a phosphate group of ATP) is in the low energy class. The hydrolysis of ATP to yield AMP plus PPi proceeds with $\Delta G^{o\prime}$ –7.7 kcal/mol, slightly greater than the $\Delta G^{o\prime}$ for the hydrolysis of the terminal or γ phosphate bond.

$$\text{ATP} + H_2O \longrightarrow \text{AMP} + \text{PP}i$$

The inorganic pyrophosphate is subsequently hydrolyzed by the enzyme pyrophosphatase to yield 2 moles of inorganic orthophosphate. The $\Delta G^{o\prime}$ value of this reaction is –6.9 kcal/mol.

$$\text{PPi} + H_2O \longrightarrow 2\text{Pi}$$

The overall reaction has a $\Delta G^{o\prime}$ value –14.6 kcal/mol which is the sum of the $\Delta G^{o\prime}$ values of the two sequential component reactions.

$$\text{ATP} + 2\ H_2O \longrightarrow \text{AMP} + 2\text{P}i$$

It is noteworthy that the $\Delta G^{o\prime}$ of the overall reaction is exactly twice the $\Delta G^{o\prime}$ of the terminal phosphate groups of ATP and ADP.

Some biosynthetic reactions are, however, driven by nucleotides that are analogous to ATP, namely guanosine triphosphate (GTP), cytidine triphosphate (CTP) and uridine triphosphate (UTP). They are present in all cells but in much lower concentrations than ATP. Also found in the cells in low concentrations are the corresponding deoxyribonucleoside 5′-triphosphates denoted as *d*ATP, *d*GTP, *d*CTP and *d*TTP. Although ATP is the mainstream carrier of phosphate groups in the cell, the other types of nucleotides serve certain specific biosynthetic pathways. They acquire their terminal phosphate groups mainly from ATP in reactions catalyzed by Mg^{2+}-

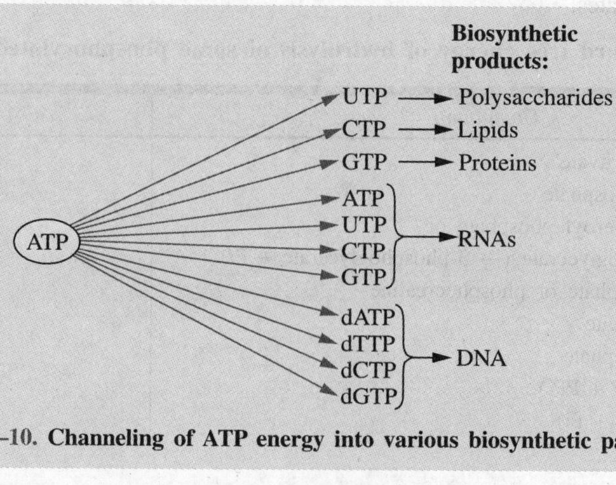

Fig. 20–10. Channeling of ATP energy into various biosynthetic pathways

dependent enzymes called *nucleoside diphosphokinases*. These enzymes promote the following types of reversible reactions:

$$\text{ATP} + \text{GDP} \rightleftharpoons \text{ADP} + \text{GTP}$$
$$\text{ATP} + \text{CDP} \rightleftharpoons \text{ADP} + \text{CTP}$$
$$\text{ATP} + \text{UDP} \rightleftharpoons \text{ADP} + \text{UTP}$$
$$\text{GTP} + \text{UDP} \rightleftharpoons \text{GDP} + \text{UTP}$$
$$\text{ATP} + d\text{CDP} \rightleftharpoons \text{ADP} + d\text{CTP}$$

The energy contained in ATP may be channelized into different biosynthetic pathways *via* different nucleoside and deoxynucleoside triphosphates (refer Fig. 20–10).

STRUCTURAL BASIS OF THE HIGH GROUP TRANSFER POTENTIAL OF ATP

The ATP molecules deliver considerably more free energy on hydrolysis of the terminal phosphate group than the hydrolysis of ester phosphates, say glucose 6-phosphate. The $\Delta G°'$ values of both these reactions are –7.3 kcal/mol and –3.3 kcal/mol respectively (refer Table 20–5). This means that ATP has a stronger tendency to transfer its terminal phosphate group to water than does glucose 6-phosphate. In other words, *ATP has a higher phosphate group transfer potential than glucose 6-phosphate*.

The standard free energy change is a measure of the difference in the free energy content of the reactants and the products. Therefore, the reason for the high $\Delta G°'$ value for ATP should be sought in the properties of both the substrate and products. There are 3 main reasons of high $\Delta G°'$ value (or high group transfer potential) for ATP:

1. Degree of ionization. At pH 7.0, ATP is almost completely ionized as the ATP^{4-} ion. This, on hydrolysis, yields 3 products, namely ADP^{3-}, HPO^{2-} and H^+.

$$ATP^{4-} + H_2O \longrightarrow ADP^{3-} + HPO_4^{2-} + H^+$$

At pH 7.0 (the standard *p*H for $\Delta G°'$ hydrolysis), the hydrogen ion concentration is only 10–7 M. This means that by the law of mass action, the equilibrium for ATP hydrolysis tends to be pulled far to the right, since $[H^+]$ at pH 7.0 is very small as compared with the standard concentration of 1.0 M. In contrast, when glucose 6–phosphate is hydrolyzed at pH 7.0, no significant extra H^+ ion is formed and, therefore, no such pulling force takes place.

$$\text{Glucose 6-phosphate}^{2-} + H_2O \longrightarrow \text{Glucose} + HPO_4^{2-}$$

2. Electrostatic repulsion. At *p*H 7.0, the ATP molecule carries 4 negative charges. These charges repel each other strongly because they are in close proximity. However, when ATP is hydrolyzed, the electrostatic repulsion between these negatively-charged groups is reduced because of the separation of the negatively-charged products, ADP^{3-} and HPO_4^{2-}. These products have relatively little tendency to approach each other to react in the backward direction to form ATP again. On the contrary, when glucose 6-phosphate is hydrolyzed, two products glucose and HPO_4^{2-} are formed. Because glucose has no charge, the two products do not repel each other and tend to recombine more readily.

3. Resonance stabilization. The two products of ATP hydrolysis, ADP^{3-} and HPO_4^{2-} are resonance hybrids, *i.e.*, special stable forms in which certain electrons are in a configuration having much less energy than in their original positions in the ATP molecule. When ATP is hydrolyzed, the electrons in the two products (ADP^{3-} and HPO_4^{2-}) can sink to lower energy levels than in the nonhydrolyzed ATP. This causes ADP^{3-} and HPO_4^{2-} ions to contain less free energy than when they were still combined as ATP^{4-}.

ADP and Pi enjoy greater resonance stabilization than does ATP. For example, orthophosphate (Pi) has a number of resonance forms of similar energy (refer Fig. 20–11), whereas the terminal phosphate portion of ATP has fewer resonance forms per phosphate group.

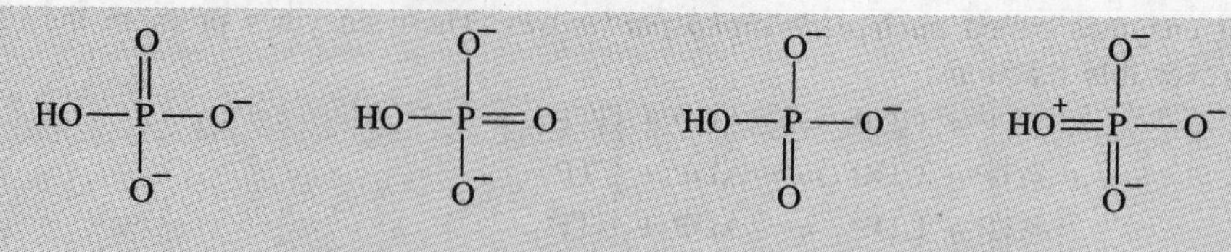

Fig. 20–11. Significant resonance forms of orthophosphate

To indicate the presence of high energy phosphate group, **Lipmann** introduced the symbol ~P indicating high energy phosphate bond. Although in long usage among biochemists, the term '*high energy phosphate bond*' is misleading since it wrongly suggests that the bond itself contains high amount of energy. What the symbol ~P indicates is that the group attached to the bond, on transfer to an appropriate acceptor, results in transfer of large amount of free energy. Hence, the term '*group transfer potential*' is preferred to '*high energy bond*'. In fact, the free energy released by hydrolysis of phosphate esters does not come from the specific bond that is broken but results from the fact that the products of the reaction have a smaller free energy content than the reactants. Therefore, the use of the term '*high energy phosphate compound*' be preferred when referring to ATP and other phosphate compounds having a large negative $\Delta G^{o\prime}$ value of hydrolysis. Thus, whereas ATP contains two high energy phosphate groups and one low energy phosphate group, ADP contains one high energy and one low energy phosphate groups. AMP, however, contains only one phosphate group of low energy type (Fig. 20–12).

Adenosine –P~P~P	Adenosine –P~P	Adenosine –P
ATP	**ADP**	**AMP**

Fig. 20–12. Structure of adenine nucleotides, showing the position and number of high energy bonds (~)

ATP HYDROLYSIS AND EQUILIBRIA OF COUPLED REACTIONS

For a better understanding of the role of ATP in energy coupling, the follownig reaction with a positive $\Delta G^{o\prime}$ value (say + 4 kcal/mol) is taken as an example:

$$A \rightleftharpoons B \qquad \Delta G^{0\prime} = +4 \text{ kcal/mol}$$

This reaction is thermodynamically unfavourable without an input of free energy. The equilibrium constant, K'_{eq} of this reaction at 25°C as per equation 16 is:

$$\frac{[B]_{eq}}{[A]_{eq}} = K'_{eq} = 10^{-\Delta G^{o\prime}/1.36} = 1.15 \times 10^{-3}$$

Thus, A cannot be spontaneously converted into B when the molar ratio of B/A is equal to or greater than 1.15×10^{-3}. However, if the reaction is coupled to the hydrolysis of ATP (whose $\Delta G^{o\prime}$ value is –7. 3 kcal/mol), A can be converted into B when $[B]/[A]$ ratio is higher than 1.15×10^{-3}. The new overall reaction shall be:

$$A + ATP + H_2O \rightleftharpoons B + ADP + Pi + H+$$

Because the $\Delta G^{o\prime}$ values of the sequential reactions are additive, the value of standard free energy change of the above reaction shall be:

$$\begin{aligned} \Delta G^{o\prime} &= \Delta G_1{}^{o\prime} + \Delta G_2{}^{o\prime} \\ &= +4 + (-7.3) \\ &= -3.3 \text{ kcal/mol} \end{aligned}$$

The equilibrium constant of this coupled reaction will be:

$$K'_{eq} = \frac{[B]_{eq}}{[A]_{eq}} \times \frac{[ADP]_{eq}\,[P_i]_{eq}}{[ATP]_{eq}}$$
$$= 10^{3.3/1.36}$$
$$= 2.67 \times 10^2$$

At equilibrium, the ratio $[B]/[A]$ will be:

$$\frac{[B]_{eq}}{[A]_{eq}} = K'_{eq} \times \frac{[ATP]_{eq}}{[ADP]_{eq}\,[P_i]_{eq}}$$

The $[ATP]/[ADP]\,[Pi]$ ratio in the cells generating ATP is maintained at a high level, typically of the order of 500. Thus,

$$\frac{[B]_{eq}}{[A]_{eq}} = 2.67 \times 10^2 \times 500$$
$$= 1.34 \times 10^5$$
$$= 10^8 \text{ approx.}$$

This means that the hydrolysis of ATP enables A to be converted to B until the $[B]/[A]$ ratio reaches a value of 1.34×10^5. This equilibrium is quite different from the value of 1.15×10^{-3} that does not include ATP hydrolysis. In other words, the coupled hydrolysis of ATP has changed the equilibrium ratio of B to A by a factor of about 10^8. Thus, the cells maintain a high level of ATP by using light or substrates as sources of free energy. The hydrolysis of one ATP molecule then changes the equilibrium ratio of products to reactants of a coupled reaction by a very large factor of about 10^8. More generally, the hydrolysis of n ATP molecules changes the equilibrium ratio of a coupled reaction (or sequence of reactions) by factor of 10^{8n}. For example, the hydrolysis of 3 ATP molecules in a coupled reaction changes the equilibrium ratio by a factor of $10^{8\times3}$ or 10^{24}. Thus, a thermodynamically unfavourable reaction sequence can be converted into a favourable one by coupling it to the hydrolysis of a sufficient number of ATP molecules.

ROLE OF HIGH ENERGY PHOSPHATES AS THE 'ENERGY CURRENCY' OF THE CELL

ATP has a position midway down the list of standard free energies of hydrolysis of organophosphates (refer Table 20–5). Therefore, ATP acts as a donor of high energy phosphates to the compounds below it in the table.

Likewise, ADP can accept high energy phosphate, in the presence of enzyme, to form ATP from the compounds above ATP in the table.

Thus, an ATP/ADP cycle connects these two types of processes, *i.e.*, those which generate ~P and those which utilize ~P.

A. Processes which generate ~P. The processes that feed ~P into ATP/ADP cycle fall mainly within 4 groups:

I. ***Oxidative phosphorylations.*** The greatest quantitative source of ~P in aerobes is from reactions catalyzed by the enzyme *ATP synthetase* (Fig. 20–13.), which in fact reverses the hydrolysis of ATP. The free energy to drive this process is obtained from respiratory chain oxidation within mitochondria, often called oxidative phosphorylation.

II. ***Catabolism of glucose to lactic acid.*** Energy is also captured during catabolism of glucose to lactic acid in a sequence of reactions commonly called as Embden–Meyerhof–Parnas (E.M.P.) pathway of glycolysis. In this pathway, there is a net formation of 2 high energy phosphate groups which produce 2 ATP molecules from 2 ADP molecules.

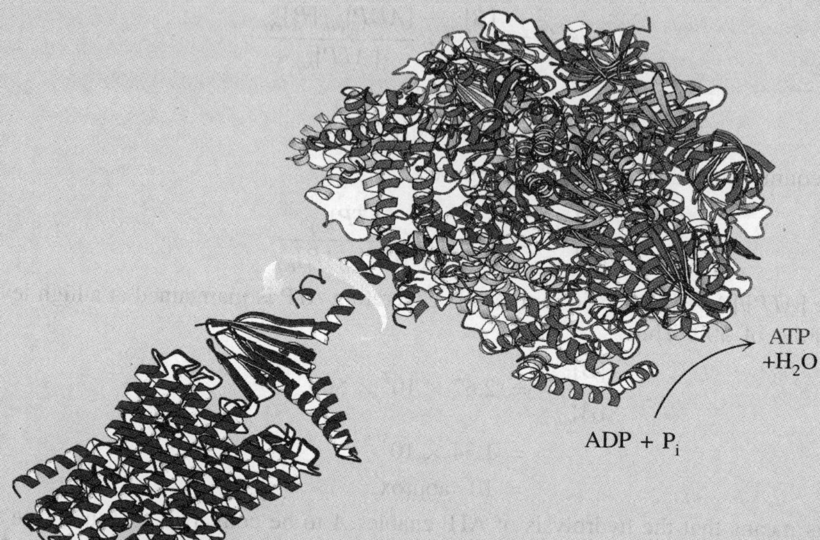

Fig. 20–13. Ribbon model of the x-ray structure of ATP synthetase

This enzyme is a molecular assembly that transduces the free energy associated with a proton gradient to the chemical energy associated with ATP. The proton gradient drives the rotation of one component of the assembly within the other. This rotational motion, in turn, drives the synthesis and release of ATP.

One ATP molecule is formed when 1, 3 diphosphoglycerate, one of the intermediates of the pathway, is enzymatically catalyzed to 3-phosphoglycerate.

$$\text{1,3-diphosphoglyceric acid} + \text{ADP} \xrightarrow{\text{Phospholglycerate kinase}} \text{3-phosphoglyceric acid} + \text{ATP}$$

Another ATP molecule results when another intermediate, 2-phosphoenolpyruvate changes enzymatically to pyruvate.

$$\text{2-phosphoenolpyruvic acid} + \text{ADP} \xrightarrow{\text{Pyruvate kinase}} \text{Pyruvic acid} + \text{ATP}$$

III. ***Oxidation of pyruvic acid.*** Further energy capture occurs at the *succinyl thiokinase* step of the Krebs' Cycle (= citric acid cycle) wherein succinyl CoA, one of the intermediates of the cycle, transforms to succinate. Here again, one ATP molecule is formed from one ADP molecule.

IV. ***The muscle compounds.*** Another group of compounds acting as storage forms of high energy phosphate within muscles include creatine phosphate, occurring in vertebrate muscle and arginine phosphate, occurring within invertebrate muscle. A substantial free energy is lost in the reaction,

$$\text{Creatine} \sim \text{P} \underset{\text{ADP} \quad \text{ATP}}{\rightleftharpoons} \text{Creatine} \quad -3.0 \text{ kcal/mol}$$

Under physiologic conditions, the above reaction allows ATP concentrations to be maintained in muscle whereas ATP is being used as a source of energy for muscular contraction. On the other hand, when ATP is in sufficient quantity, the reaction occurs in reverse direction and allows the concentration of creatine phosphate to increase substantially and thus acts as a store of high energy phosphate.

B. Processes which utilize ~P. When adenosine triphosphate (ATP) acts as a phosphate donor, the phosphate group is invariably converted to one with low energy, producing phosphoric esters of alcohols. Such reactions are catalyzed by kinases. Among them is *hexokinase*, which catalyzes transfer of phosphate group from ATP to D-glucose. This reaction proceeds by an *orthophosphate cleavage* of ATP, in which ATP loses a single orthophosphate group.

$$\text{D–glucose} + \text{Adenosine–P\~P\~P} \xrightarrow{\text{Hexokinase}} \text{D–glucose 6-phosphate} + \text{Adenosine–P\~P}$$

Another enzyme is glycerol kinase, which catalyzes transfer of phosphate group from ATP to glycerol. This reaction also proceeds by an orthophosphate cleavage of ATP.

$$\text{Glycerol} + \text{Adenosine–P\~P\~P} \xrightarrow{\text{Glycerol kinase}} \text{Glycerol 3-phosphate} + \text{Adenosine–P\~P}$$

In both the cases, a hydroxyl group of the acceptor molecule is phosphorylated to yield a phosphoric ester. Since the $\Delta G°'$ values for the hydrolysis of glucose 6-phosphate and glycerol 3-phosphate are smaller than for ATP, the above reactions proceed to the right as written, if we begin with 1.0 *M* concentrations of the substrates and reactants. Glucose 6-phosphate and glycerol 3-phosphate may be considered as energized forms of glucose and glycerol respectively. These compounds can now undergo further enzymatic reactions in which they serve as activated building blocks for the synthesis of larger molecules, *i.e.*, glycogen and lipid respectively. Fig. 20–14 shows the flow sheet of enzymatic phosphate transfer reactions in the cell.

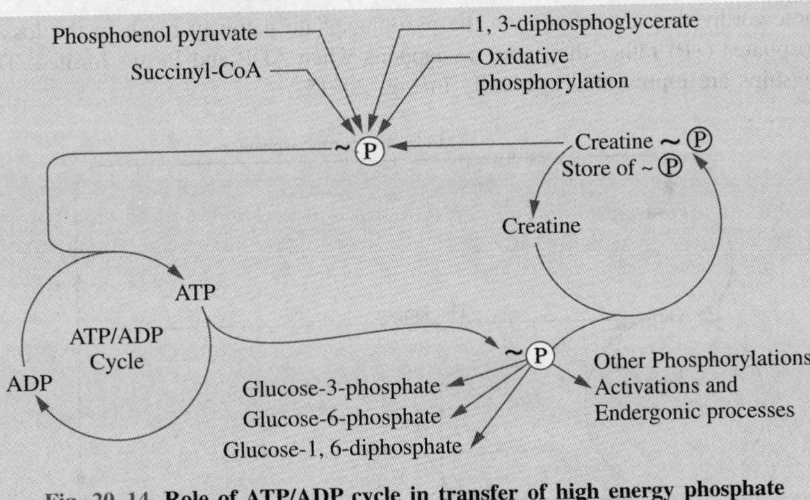

Fig. 20–14. Role of ATP/ADP cycle in transfer of high energy phosphate

An important feature of this cycle is that all super high-energy phosphate compounds must pass their phosphate groups *via* ATP to various acceptor molecules to form their low energy phosphate derivatives.

(*Adapted from Harper HA, Rodwell VW and Mayes PA, 1979*)

INTERCONVERSION OF ADENINE NUCLEOTIDES

The enzyme adenylate kinase (= myokinase) is present in most cells. It catalyzes the interconversion of ATP and AMP on the one hand and ADP on the other.

$$\underset{\text{(1 ATP)}}{\text{Adenosine–P\~P\~P}} + \underset{\text{(1 AMP)}}{\text{Adenosine–P}} \xrightarrow{\text{Adenylate kinase}} \underset{\text{(2 ADP)}}{\text{2 Adenosine–P\~P}}$$

This reaction performs 3 main functions:

(*a*) It permits the high energy phosphate of ADP to be used in the formation of ATP.

(*b*) It is a means by which AMP can be rephosphorylated to form ADP.

(*c*) Under stress of ATP depletion, AMP concentration increases which acts as a metabolic signal to increase the rate of catabolic reactions so that more ATP molecules are regenerated.

When ATP reacts enzymatically to form AMP, inorganic pyrophosphate (symbolized PPi) is formed. An example is the enzymatic activation of a fatty acid to form its coenzyme A ester, a reaction that 'energizes' the fatty acid and converts it into an activated precursor, fatty acyl-CoA, for the biosynthesis of lipids. In this reaction, the two terminal groups of ATP (*i.e.*, β and γ groups) are removed in *one piece* as inorganic pyrophosphate, leaving adenosine monophosphate.

$$\underset{\text{}}{\text{ATP}} + \underset{\text{(Fatty acid)}}{\text{R.COOH}} + \text{CoA–SH} \xrightarrow{\text{Thiokinase}} \text{AMP} + \text{PPi} + \underset{\text{(Fatty acyl-CoA)}}{\text{R.CO\sim S–CoA}}$$

Thus, this activation reaction proceeds by a pyrophosphate cleavage of ATP, in contrast to the usual orthophosphate cleavage (refer page 362). This reaction, whose $\Delta G^{\circ\prime}$ value is + 0.2 kcal/mol, results in a loss of free energy as heat which ensures that the reaction will proceed to the right as written. Further, the hydrolytic splitting of PPi has also a large $\Delta G^{\circ\prime}$ valaue of –6.9 kcal/mol. This also aids the above reaction to go to the right.

$$\underset{\text{(Pyrophosphate)}}{\text{PPi}} + H_2O \xrightarrow{\text{Inorganic pyrophosphatase}} \underset{\text{(Phosphate)}}{\text{2Pi}}$$

It is noteworthy that activations via the pyrophosphate pathway result in the loss of 2 high energy phosphates (~P) rather than one, as happens when ADP and P*i* are formed. The various interrelationships are represented cyclically in Fig. 20–15.

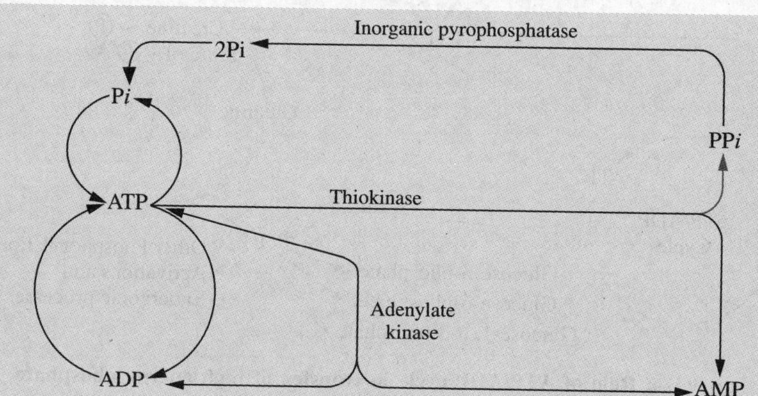

Fig. 20–15. Phosphate cycles and interconversion of adenine nucleotides

REFERENCES

1. **Atkins PW :** The Second Law. *Scientific American Books, Inc., New York. 1984.*
2. **Becker WM :** Energy and the Living Cell. *Harper and Row, New York.* 1977.
3. **Bray HG, White K :** Kinetics and Thermodynamics in Biochemistry. *2nd ed., Academic Press Inc., New York.* 1966.
4. **Bridger WA, Henderson JF :** Cell ATP. *John Wiley and Sons, New York. 1983.*
5. **Dickerson RE :** Molecular Thermodynamics. *W.A. Benjamin, Inc., Menlo Park, CA. 1969.*
6. **Edsall JT, Gutfreund H :** Biothermodynamics : The Study of Biochemical Processes at Equilibrium. *John Wiley and Sons, Inc., New York. 1983.*
7. **Ernster L (editor) :** Bioenergetics. *Elsevier, New York, 1984.*

8. **Ernster L (editor) :** Molecular Mechanisms in Bioenergetics. *Elsevier, New York, 1992.*
9. **Gates DM :** The flow of energy, in the bioshpere. *Sci. Amer.,* **224 :** *88–100, 1971.*
10. **Hanson RW :** The role of ATP in Metabolism. *Biochem. Educ.* **17 :** *86–92, 1989.*
11. **Harold FM :** The Vital Force : A Study of Bioenergetics. *W.H. Freeman and Company, New York. 1986.*
12. **Hill TL :** Free Energy Transduction in Biology. *Academic Press Inc., New York. 1977.*
13. **Ingraham LL, Pardee AB :** Free Energy and Entropy in Metabolism. In Greenberg, DM (editor) : *Metabolic Pathways. 3rd ed.,* **vol. 1 :** *1–46. Academic Press Inc., New York. 1967.*
14. **Interunion Commission on Biothermodynamics :** Recommendations for measurement and presentation of biochemical equilibrium data. *Quart. Rev. Biophys.* **9 :** *439, 1976.*
15. **Jencks WP :** How does ATP make work ? *Chemtracts Biochem. Mol. Biol. 1 : 1–3, 1993.*
16. **Kalckar HM :** Biological Phosphorylations : Development of Concepts. *Prentice-Hall, Inc., Englewood Cliffs, NJ. 1969.*
17. **Kaplan NO, Kennedy EP (editors) :** Current Aspects of Biochemical Energetics. *Academic Press Inc., New York. 1966.*
18. **Krebs HA, Kornberg HL :** Energy Transformations in Living Matter. *Springer, New York. 1967.*
19. **Lehninger AL :** Energy Transformation in the Cell. *Sci. Amer.* **202 (5) :** *102, 1960.*
20. **Lehninger AL :** Bioenergetics : The Molecular Basis of Biological Energy Transformations. *2nd ed., Benjamin, Menlo Park, California. 1971.*
21. **Lipman F :** Wanderings of a Biochemist. *Wiley-Interscience, New York. 1971.*
22. **Racker E :** A New Look at Mechanisms in Bioenergetics. *Academic Press, Inc., New York. 1976.*
23. **Segel IH :** Biochemical Calculations. 2nd ed., *Wiley-Interscience, New York. 1976.*
24. **Stumpf PK :** ATP. *Sci. Amer. 188 (4) : 85, 1953.*
25. **Szent-Györgyi A :** Bioenergetics. *Academic Press Inc., New York. 1957.*
26. **Westheimer FH :** Why nature chose phosphates? *Science.* **235 :** *1173–1178, 1987.*

PROBLEMS

1. What is the direction of each of the following reactions when the reactants are initially present in equimolar amounts ? Use the data given in Table 20-5.
   (*a*) ATP + creatine $\rightleftharpoons$ creatine phosphate + ADP
   (*b*) ATP + glycerol $\rightleftharpoons$ glycerol 3-phosphate + ADP
   (*c*) ATP + pyruate $\rightleftharpoons$ phosphoenolpyruate + ADP
   (*d*) ATP + glucose $\rightleftharpoons$ glucose 6-phosphate + ADP
2. What information do the $\Delta G^{o\prime}$ data given in Table 20-5 provide about the relative rates of hydrolysis of pyrophosphate and acetyl phosphate?
3. Consider the following reactions:
   ATP + pyruate $\rightleftharpoons$ phosphoenolpyruate + ADP
   (*a*) Calculate $\Delta G^{o\prime}$ and K'_{eq} at 25ºC for this reaction, by using the data given in Table 20–5.
   (*b*) What is the equilibrium ratio of pyruate to phosphoenolpyruate if the ratio of ATP to ADP is 10?
4. The formation of acetyl CoA from acetate is an ATP-driven reaction:
   Acetate + ATP + CoA $\rightleftharpoons$ acetyl CoA + AMP + PP_i
   (*a*) Calculate $\Delta G^{o\prime}$ for this reaction by using data given in this chapter.

(*b*) The PP_i formed in the preceding reaction is rapidly hydrolyzed *in vivo* because of the ubiquity of inorganic pyrophosphatase. The $\Delta G^{o\prime}$ for the hydrolysis of PPi is –4.6 kcal mol^{-1}. Calculate the $\Delta G^{o\prime}$ for the overall reaction. What effect does the hydrolysis of PP_i have on the formation of acetyl-CoA?

5. The p*K* of an acid is a measure of its proton group-transfer potential.

(*a*) Derive a relation between $\Delta G^{o\prime}$ and p*K*.

(*b*) What is the $\Delta G^{o\prime}$ for the ionization of acetic acid, which has a p*K* of 4.8?

6. What is the structural feature common to ATP, FAD, NAD^+, and CoA?

7. Glycolysis is a series of 10 linked reactions that convert one molecule of glucose into two molecules of pyruvate with the concomitant synthesis of two molecules of ATP. The $\Delta G^{o\prime}$ for this set of reactions is –8.5 kcal mol^{-1} (– 35.6 kJ mol^{-1}), where as ΔG is –18.3 kcal mol^{-1} (– 76.6 kJ mol^{-1}). Explain why the free-energy release is so much greater under intracellular conditions than under standard conditions.

8. The combustion of glucose to CO_2 and water is a major source of energy in aerobic organisms. It is a reaction favoured mainly by a large negative enthalpy change.

$$C_6H_{12}O_6(s) + 6\ O_2(g) \longrightarrow 6\ CO_2(g) + 6\ H_2O(l)$$

$$\Delta H^o = -2816 \text{ kJ/mol} \qquad \Delta S^o = +\ 181 \text{ J/K·mol}$$

(*a*) At 37ºC, what is the value for $\Delta G^{o\prime}$?

(*b*) In the overall reaction of aerobic metabolism of glucose, 38 moles of ATP are produced from ADP for every mole of glucose oxidized. Calculate the standard state free energy change for the *overall* reaction when glucose oxidation is coupled to the formation of ATP.

(*c*) What is the *efficiency* of the process in terms of the percentage of the available free energy change captured in ATP?

9. In another key reaction in glycolysis, dihydroxyacetone phosphate (DHAP) is isomerized into glyceraldehyde-3-phosphate (G3P):

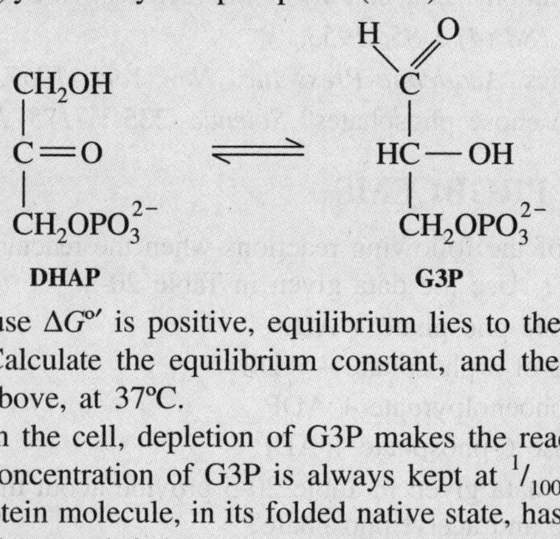

$\Delta G^{o\prime} = +7.5$ kJ/mol

DHAP **G3P**

Because $\Delta G^{o\prime}$ is positive, equilibrium lies to the left.

(*a*) Calculate the equilibrium constant, and the equilibrium fraction of G3P from the above, at 37ºC.

(*b*) In the cell, depletion of G3P makes the reaction proceed. What will ΔG be if the concentration of G3P is always kept at $^1/_{100}$ of the concentration of DHAP?

10. A protein molecule, in its folded native state, has *one* favoured conformation. But when it is denatured, it becomes a random coil, with *many* possible conformations.

(*a*) What must be the sign of ΔS for the change native $\longrightarrow$ denatured?

(*b*) Will the contribution of ΔS to the free energy change be + or –?

What requirement does this impose on ΔH if proteins are to be stable structures?

11. We give a value of the standard state free energy change for ATP hydrolysis to be –31 kJ/mol. Would you expect the same value for the reaction $ATP^{4-} + H_2O \longrightarrow ADP^{2-} + P_i^{2-}$? Explain.

12. For the reaction A $\rightleftharpoons$ B, $\Delta G^{o\prime}$ = –7.1 kcal mol^{-1}. At 37ºC, –2.303 *RT* = –1.42 kcal mol^{-1}. What is the equilibrium ratio of B/A?

13. Organophosphate compounds are irreversible inhibitors of acetylcholinesterase. What effect does an organophosphate inhibitor have on the transmission of nerve impulses ?

14. Calculate the standard free-energy changes of the following metabolically important enzyme-catalyzed reactions at 25ºC and pH 7.0 from the equilibrium constants given.

(*a*) Glutamate + oxaloacetate $\xrightleftharpoons[]{\text{aspartate aminotransferase}}$ aspartate + α-ketoglutarate $K'_{eq} = 6.8$

(*b*) Dihydroxyacetone phosphate $\xrightleftharpoons[]{\text{triose phosphate isomerase}}$ glyceraldehyde-3-phosphate $K'_{eq} = 0.0475$

(*c*) Fructose-6-phosphate + ATP $\xrightleftharpoons[]{\text{phosphofructokinase}}$ fructose-1,6-bisphosphate + ADP $K'_{eq} = 254$

15. Calculate the equilibrium constants K'_{eq} for each of the following reactions at pH 7.0 and 25ºC, using the $\Delta G^{o\prime}$ values of Table 20–3:

(*a*) Glucose-6-phosphate + H_2O $\xrightleftharpoons[]{\text{glucose-6-phosphatase}}$ glucose + P*i*

(*b*) Lactose + H_2O $\xrightleftharpoons[]{\beta\text{-galactosidase}}$ glucose + galactose

(*c*) Malate $\xrightleftharpoons[]{\text{fumarase}}$ fumarate + H_2O

16. Consider the following interconversion, which occurs in glycolysis :

Fructose-6-phosphate $\rightleftharpoons$ glucose-6-phosphate $K'_{eq} = 1.97$

(*a*) What is $\Delta G^{o\prime}$ for the reaction (assuming that the temperature is 25ºC) ?

(*b*) If the concentration of fructose-6-phosphate is adjusted to 1.5 M and that of glucose-6-phosphate is adjusted to 0.5 M, what is ΔG?

(*c*) Why are $\Delta G^{o\prime}$ and ΔG different?

17. The free energy released by the hydrolysis of ATP under standard conditions at pH 7.0 is –30.5 kJ/mol. If ATP is hydrolyzed under standard conditions but at pH 5.0, is more or less free energy released? Why?

18. Glucose-1-phosphate is converted into fructose-6-phosphate in two successive reactions:

Glucose-1-phosphate ⟶ glucose-6-phosphate

Glucose-6-phosphate ⟶ fructose-6-phosphate

Using the $\Delta G^{o\prime}$ values in Table 20–3, calculate the equilibrium constant, K'_{eq}, for the sum of the two reactions at 25ºC:

Glucose-1-phosphate ⟶ fructose-6-phosphate

19. Calculate the physiological ΔG (not $\Delta G^{o\prime}$) for the reaction,

Phosphocreatine + ADP ⟶ creatine + ATP

at 25ºC as it occurs in the cytosol of neurons, in which phosphocreatine is present at 4.7 mM, creatine at 1.0 mM, ADP at 0.20 mM, and ATP at 2.6 mM.

20. The ATP concentration in muscle tissue (approximately 70% water) is about 8.0 mM. During strenuous activity, each gram of muscle tissue uses ATP at the rate of 300 μmol/min for contraction.

(*a*) How long would the reserve of ATP last during a 100 meter dash?

(*b*) The phosphocreatine level in muscle is about 40.0 mM. How does this help extend the reserve of muscle ATP?

(*c*) Given the size of the reserve ATP pool, how can a person run a marathon?

21. The synthesis of the activated form of acetate (acetyl-CoA) is carried out in an ATP-dependent process:

Acetate + CoA + ATP ⟶ acetyl-CoA + AMP + PP_i

(*a*) The $\Delta G^{o\prime}$ for the hydrolysis of acetyl-CoA to acetate and CoA is –32.2 kJ/mol and that for hydrolysis of ATP to AMP and PP_i is –30.5 kJ/mol. Calculate $\Delta G^{o\prime}$ for the ATP-dependent synthesis of acetyl-CoA.

(*b*) Almost all cells contain the enzyme inorganic pyrophosphatase, which catalyzes the hydrolysis of PP_i to P_i. What effect does the presence of this enzyme have on the synthesis of acetyl-CoA? Explain.

22. Electron transfer in the mitochondrial respiratory chain may be represented by the net reaction equation:

$$NADH + H^+ + ½O_2 \rightleftharpoons H_2O + NAD^+$$

(*a*) Calculate the value of $\Delta E'_0$ for the net reaction of mitochondrial electron transfer.

(*b*) Calculate $\Delta G^{o\prime}$ for this reaction.

(*c*) How many ATP molecules can *theoretically* be generated by this reaction if the standard free energy of ATP synthesis is 30.5 kJ/mol ?

23. List the following substances in order of increasing tendency to accept electrons:

(*a*) α-ketoglutarate + CO_2 (yielding isocitrate)

(*b*) oxaloacetate

(*c*) O_2

(*d*) $NADP^+$

24. Distinguish between anabolic and catabolic reactions.

25. How is the energy generated during metabolic processes usually stored for later use?

26. DNA, RNA and ATP are all composed of:

(*a*) nucleotides (*c*) nucleic acids

(*b*) purines (*d*) pentose sugars

27. The role played by ATP in biochemical reactions is that of :

(*a*) a reducing agent (*c*) a coenzyme

(*b*) an energy-donor substance (*d*) an energy-donor substance or a coenzyme

28. ATP synthesis occurs in:

(*a*) chloroplasts

(*b*) mitochondria

(*c*) chloroplasts and mitochondria both

(*d*) all cell organelles

ANSWERS TO PROBLEMS

1. Reactions in parts *a* and *c*, to the left; reactions in parts *b* and *d*, to the right.

2. None whatsoever.

3. (*a*) $\Delta G^{0\prime}$ = +7.5 kcal mol^{-1} (+31.4 kJ mol^{-1}) and $K'_{eq} = 3.2 \times 10^{-6}$ (*b*) 3.28×10^4

4. (*a*) Acetate + CoA + H^+ goes to acetyl-CoA + H_2O,
$\Delta G^{0\prime}$ = +7.5 kcal mol^{-1} (+31.4 kJ mol^{-1}). ATP hydrolysis,
$\Delta G^{0\prime}$ = –10.9 kcal mol^{-1} (–45.6 kJ mol^{-1}). Overall reaction,
$\Delta G^{0\prime}$ = –3.4 kcal mol^{-1} (–14.2 kJ mol^{-1}).

(*b*) With pyrophosphate hydrolysis, $\Delta G^{0\prime}$ = –8.0 kcal mol^{-1} (–33.4 kJ mol^{-1}).

5. (*a*) For an acid AH, $AH \rightleftharpoons A^- + H^+$, $K = \dfrac{[A^-][H^+]}{[AH]}$.

The pK is defined as pK = $-\log_{10} K$. $\Delta G^{0'}$ is the standard free energy change at pH 7. Thus, $\Delta G^{0'} = -RT \ln K = -2.303 \log_{10} K = -2.303$ (pK −7) kcal mol^{-1} since $[H^+] = 10^{-7}$ M.

(*b*) $\Delta G^{0'} = -2.303\ (4.8 - 7) = -5.1$ kcal mol^{-1} (−21.3 kJ mol^{-1}).

6. An ADP unit.

7. Recall that $\Delta G = \Delta G^{0'} + RT \ln$ [products/reactants]. Altering the ratio of products to reactants will cause ΔG to vary. In glycolysis, the concentrations of the components of the pathway result in a value of ΔG greater than that of $\Delta G^{0'}$.

8. (*a*) $\Delta G^0 = -2872$ kJ/mol

(*b*) $\Delta G^0 = -1694$ kJ/mol

(*c*) 41%

9. (*a*) $K = 5.44 \times 10^{-2}$; $(f_{G3P})_{eq} = 0.052$

(*b*) $\Delta G = -4.37$ kJ/mol

10. (*a*) ΔS must be positive, because the increase in available states corresponds to an increase in entropy.

(*b*) Since $\Delta G = \Delta H - T\Delta S$, a positive ΔS yields a negative contribution to ΔG (T is always a positive number). Thus, for proteins to be stable, which requires ΔG for the above to be positive, denaturation must involve a large positive ΔH and/or an additional negative contribution to ΔS. In fact, both occur.

11. No, because the repulsion between ADP^{2-} and P_i^{2-} is less than that between ADP^{3-} and ADP^{2-}, and the release of an H^+ will change the value as well.

12. $\Delta G^{0'} = -RT \ln K'_{eq} = -2.303\ RT \log K'_{eq}$. Substitution gives $\log K'_{eq} = 5$. K'_{eq} then is 1,00,000 so the B/A = 1,00,000/1.

13. Acetylcholine is an excitatory neurotransmitter, causing channels to open, Na^+ to enter, and a depolarization of the membrane. Hydrolysis of acetylcholine allows repolarization. The presence of the inhibitor keeps the channel open and prevents repolarization, and thus continued transmission of nerve impulses.

14. (*a*) 4.75 J/mol (*b*) −7.6 J/mol (*c*) −13.7 kJ/mol

15. (*a*) 261 M (*b*) 609 M (*c*) 0.29

16. (*a*) −1.7 kJ/mol (*b*) −4.4 kJ/mol

(*c*) At a given temperature, the value of $\Delta G^{0'}$ for any reaction is fixed and is defined for standare conditions (both fructose-6-phosphate and glucose-6-phsophate at 1M). In contrast, ΔG is a variable that can be calculated for any set of reactant and product concentrations.

17. Less. The overall equation for ATP hydrolysis can be approximated as:

$$ATP^{4-} + H_2O \longrightarrow ADP^{3-} + HPO_4^{2-} + H^+$$

(This is only an approximation because the ionized species shown here are the major, but not the only, forms present). Under standard conditions (*i.e.*, [ATP] = [ADP] = $[P_i]$ = 1 M), the concentration of water is 55 m and does not change during the reaction. Because H^+ ions are produced in the reaction, at a higher $[H^+]$ (pH 5.0) the equilibrium would be shifted to the left and less free energy would be released.

18. 9.6

19. 10.0 kJ/mol

20. (*a*) 1.1 s

(*b*) Phosphocreatine + ADP → creatine + ATP.

(*c*) ATP synthesis coupled to the catabolism of glucose, amino acids, and fatty acids.

21. (*a*) 0.8 kJ/mol

(*b*) Inorganic pyrophosphatase catalyzes the hydrolysis of pyrophosphate and drives the net reaction toward the synthesis of acetyl-CoA.

22. (*a*) 1.14 V (*b*) –220 kJ/mol (*c*) About 7

23. In order of increasing tendency: (*a*); (*d*); (*b*); (*c*)

24. An anabolic reaction consumes ATP; a catabolic reaction generates ATP.

25. In ATP and other compounds containing high-energy phosphates, and in glycogen.

26. (*a*) nucleotides

27. (*d*) an energy-donor substance or a coenzyme

28. (*c*) chloroplasts and mitochondria both

CONTENTS

Fermentation

(*Above*) **Sprinters at the end of a race.** A runner's respiratory and circulatory systems cannot supply oxygen to her leg muscles fast enough to keep up with the demand for energy, so glycolysis and lactate fermentation must provide the ATP. Panting after the race brings in the oxygen needed to remove the lactate through cellular respiration.

(*Below*) **Bread**

Bread rises as CO_2 is liberated by fermenting yeast, which converts glucose to ethanol *via* the alcoholic fermentation pathway.

CHAPTER

21

Glycolysis And Pentose Phosphate Pathway

Carbohydrates are the first cellular constituents formed by photosynthetic organisms and result from the fixation of CO_2 on absorption of light. The carbohydrates are metabolized to yield a vast array of other organic compounds, many of which are subsequently utilized as dietary constituents by animals. The animals ingest great quantities of carbohydrates that can be either stored, or oxidized to obtain energy as ATP, or converted to lipids for more efficient energy storage or used for the synthesis of many cellular constituents.

The major function of carbohydrates in metabolism is as a fuel to be oxidized and provide energy for other metabolic processes. *The carbohydrate is utilized by cells mainly as glucose*. The 3 principal monosaccharides resulting from digestive processes are glucose, fructose and galactose. Much of the glucose is derived from starch which accounts for over half of the fuel in the diets of most humans. Glucose is also produced from other dietary components by the liver and, to a lesser extent, by the kidneys. *Fructose* results on large intake of sucrose while *galactose* is produced when lactose is the principal carbohydrate of the diet. Both fructose and galactose are easily converted to glucose by the liver. It is thus apparent that glucose is the major fuel of most organisms and that it can be quickly metabolized from glycogen stores when there arises a sudden need for energy.

Pentose sugars such as arabinose, ribose and xylose may be present in the diet. But their fate after absorption is, however, obscure.

GENERAL CONSIDERATIONS OF GLYCOLYSIS

Glycolysis (*glycos*^{G} = sugar (sweet); *lysis*^{G} = dissolution or splitting) is a ubiquitous, catabolic pathway. These 2 words ('sweet' and 'splitting') are literally correct terms, for glycolysis is the pathway by which six-carbon sugars (which are sweet) are split, yielding a three-carbon compound, pyruvate. It consists of a sequence of 10 enzyme-catalyzed reactions that converts glucose into pyruvate with the simultaneous production of usable energy in the form of ATP. Moreover, glycolysis also includes the formation of lactate from pyruvate. The glycolytic sequence of reactions differs from one species to the other only in the mechanism of its regulation and in the subsequent metabolic fate of the pyruvate formed. In aerobic organisms, glycolysis is the prelude to the citric acid cycle and the electron transport chain which together harvest most of the energy contained in glucose. In fact, glycolysis is the central pathway of glucose catabolism.

Glycolysis takes place in the extramitochondrial part of the cell (or the soluble cytoplasm) and is common to both aerobic respiration and anaerobic respiration. It is frequently referred to as **Embden-Meyerhof-Parnas** or **EMP pathway**, in honour of these poineer workers in the field, and still represents one of the greatest achievements in the field of biochemistry. Other illustrious investigators, who contributed significantly to the final elucidation of glycolytic pathway, include **Fritz A. Lipmann**, **Harden** and **Young**, **A.V. Hill**, **Carl Neuberg**, **Otto Warburg**, and **Carl F. Cori** and his wife **Gerty T. Cori**.

GUSTAV GEORG EMBDEN

(LT, 1874–1933)

Embden, a German chemist, was born on November 10, 1874. He conducted studies on carbohydrate metabolism and muscle contraction. He had also studied the liver's metabolic processes, laying a foundation for understanding **diabetes.** He was the first to discover and link together all the steps involved in the conversion of glycogen to lactic acid. In 1918, **Otto F. Meyerhof** explained, together with **Jakub K. Parnas,** cellular metabolism by showing that it involved the breakdown of glucose to lactic acid. Embden worked out the precise steps in the **breakdown of glucose**. He died on July 25, 1933.

OTTO FRITZ MEYERHOF

(LT, 1884–1951)

Meyerhof, a German-born physician and biochemist, was born in Hildesheim, on April 12, 1884 (Because Hildensheim is a smaller city, about half-an-hour south of Hannover, a number of biographies mistakenly list his birthplace as Hannover). He was the son of wealthy Jewish parents. He spent most of his childhood in Berlin, where he started his study of medicine. He continued these studies in Strasbourg and Heidelberg, from which he graduated in 1909, with a work titled "Contributions to the Psychological Theory of Mental Illness". In Heidelberg, he met **Hedwig Schallenberg**, who later became his wife. They had a daughter **Bettina Meyerhof** and two sons, **Gottfried** and **Walter**.

In 1912, Meyerhot moved to the University of Kiel, where he became professor in 1918. In 1918 also, he along with **Jakub K. Parnas** explained cellular metabolism by showing that it involved the breakdown of glucose to lactic acid. He was awarded the **1922 Nobel Prize in Physiology or Medicine**, with **Archibald Vivian Hill**, for his work on muscle metabolism, including **glycolysis**. In 1929, he became one of the directors of the *Kaiser Wilhelm Institute for Medical Research*, a position he held until 1938. Fleeing the Nazi regime he moved to Paris in 1928. He than sought refuge in the United States in 1940, where he became a guest professor at the University of Pennsylvania in Philadelphia. He died on October 6, 1951 in Philadelphia.

JAKUB KAROL PARNAS

(LT, 1884–1949)

Parnas, a prominent Jewish-Polish-Soviet biochemist, was born on January 16, 1884 in Mokriany near Drohobych, at that time part of the Austro-Hungarian Empire, in the province of Galicia (now split between Poland and Ukraine). He graduated from the Royal Technical College of Charlottenburg in 1904. From 1920 to 1941, he was Head of the Institute of the Medical Chemistry at Lviv University. He traveled across Europe collaborating with universities in Cambridge, Naples, Strasbourg, Ghent and Zürich.

Parnas's major work was the study of the mechanisms of **carbohydrate metabolism in muscle tissue**. Together with Wladyslaw Baranowski, he discovered the process of **phosphorolysis**. He also made a major contribution to the theoretical analysis of **glycolysis**.

After the Soviet invasion of Poland and annexation of Western Ukraine in 1939 by the Soviet Union, Parnas remained in Lviv to continue his work in the Institute. He also started collaborating with the Soviet authorities by taking on a political role in the communist *District Soviet Worker's Delegation*. In 1941, after the German invasion of the Soviet Union, Parnas was evacuated deeper into the USSR and remained there for the rest of his life. Only a few days after his departure, Lviv was occupied by the Nazi Germany army, who massacred about 45 Lwów professors. In the Soviet Union, Parnas met **Joseph Stalin**, convinced him of his usefulness and shortly thereafter received his own laboratory. He became an Academician of the *Academy of Sciences of the USSR* and a founding member of the *Academy of Medical Sciences of the USSR*. Despite his achievements and popularity, Parnas was falsely accused of being a spy of the West and arrested by the KGB on January 28, 1949, and was murdered during the Stalinist Doctor's Plot purge in 1949 (According to KGB's archives, he died during his first interrogation at Lubyanka prison "from a heart attack" on January 29, 1949).

Parnas is honoured by the *Polish–Ukrainian Parnas Conference* organized by the Polish and Ukrainian Biochemical Societies, which has been held biannually since 1999.

There are 3 important routes taken by pyruvate after glycolysis, depending on the organism and the metabolic conditions (refer Fig. 21–1):

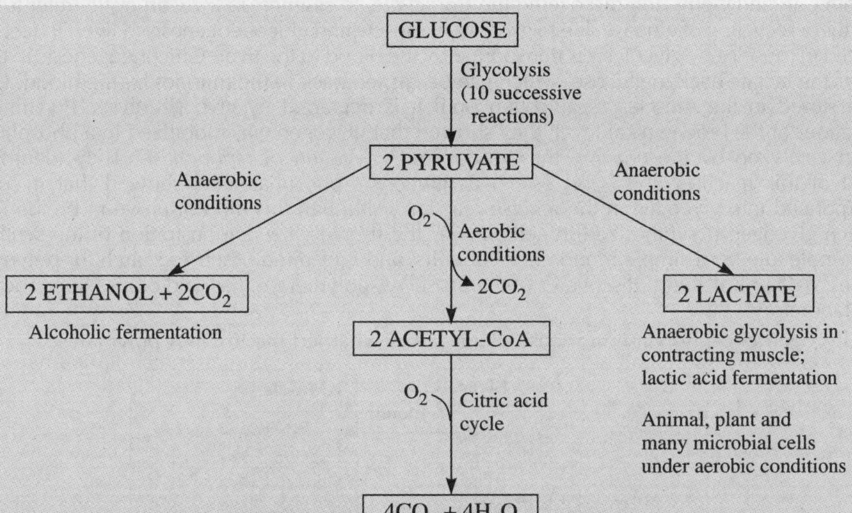

Fig. 21–1. Some important fates of glucose

(*a*) In aerobic organisms, the pyruvate so formed then enters mitochondria where it is oxidized, with the loss of its carboxyl group as CO_2, to form the acetyl group of acetyl-coenzyme A. Later, the acetyl group is completely oxidized to CO_2 and H_2O by the citric acid cycle with the intervention of molecular oxygen. This pathway is followed by aerobic animal and plant cells.

CARL FERDINAND CORI
(LT, 1896–1984)

A Czech-born American biochemist at Washington State University who shared the **1947 Nobel Prize for Medicine or Physiology** with his wife **Gerty Theresa Radnitz Cori** (LT, 1896–1957) and **B.A. Houssay** from Argentina for their joint elucidation of the processes by which glycogen is broken down and re-formed in the body. Both Carl and Gerty Cori were trained in medicine in Europe at the end of World War I (Gerty completed premedical studies and medical school in one year!). They left Europe together in 1922 to establish research laboratories in the United States, first for 9 years in Buffalo, New York, at what is now the Roswell Park Memorial Institute, then from 1931 until the end of their lives at Washington University in St. Louis. They showed that the cells used an enzyme called *phosphorylase* to convert the stored glycogen to glucose, the sugar form which normally meets energy requirements of all cells in the body, including muscle cells. They also found that the enzyme exists in either of its 2 forms: active or inactive. *The breakdown of glycogen to glucose is, in fact, a classic biochemical reaction that has bred three separate Nobel Prizes: first* by **Carl** and **Gerty Coris** along with **B.A. Houssay** in 1947, *second* by **Earl W. Sutherland Jr** in 1971 and *third* by **Edwin G. Krebs** and **Edmund H. Fischer**, both Emeritus professors at the University of Washington, Seattle in 1992, although 35 years after the work was accomplished by them in 1957.

Both Carl and Gerty Cori discovered the *Cori cycle*. In essence, in the Cori cycle (refer following figure) there is cycling of glucose due to glycosis in muscle and gluconeogenesis in liver. In fact, lactate produced in muscles by glycolysis is transported by the blood to the liver. Gluconeogenesis in the liver converts the lactate back to glucose, which can be carried back to the muscles by the blood. Glucose can be stored in the muscles as glycogen until it is degraded by glycogenolysis. Pursuing these observations at the biochemical level, they showed that glycogen was mobilized in a phosphorolysis reaction catalyzed by the enzyme they discovered, *glycogen phosphorylase*. They identified the product of this reaction (the "Cori ester") as glucose 1-phosphate and showed that it could be reincorporated into glycogen in the reverse reaction. Although this did not prove to be the reaction by which glycogen is synthesized in cells, it was the first *in vitro* demonstration of the synthesis of a macromolecule from simple monomeric subunits, and it inspired others to search for polymerizing enzymes. **Arthur Kornberg**, discoverer of the first DNA polymerase, has said of his experience in the Coris' laboratory:

"Glycogen phosphorylase, not base pairing, was what led me to DNA polymerase."

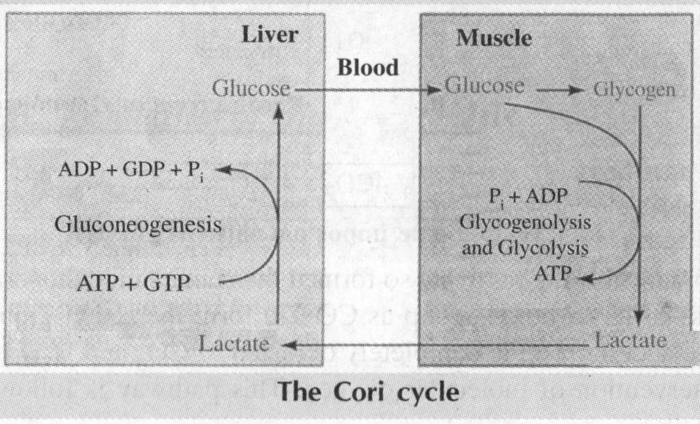

The Cori cycle

The Cori laboratories in St. Louis became an international centre of biochemical research in the 1940s and 1950s, and at least 6 scientists who trained with the Coris became Nobel Laureates! These are:

1. **Arthur Kornberg** (for DNA synthesis, 1959)
2. **Severo Ochoa** (for RNA synthesis, 1959)
3. **Luis Federico Leloir** (for the role of sugar nucleotides in polysaccharide synthesis, 1970)
4. **Earl Wilbur Sutherland Jr** (for the discovery of cAMP in the regulation of carbohydrate metabolism, 1971)
5. **Christian de Duve** (for subcellular fractionation, 1974)
6. **Edwin G. Krebs** (for the discovery of phosphorylase kinase, 1992)

In fact, much of what is written in present-day biochemistry textbooks about the metabolism of glycogen was discovered between about 1925 and 1950 by this remarkable husband-and-wife team of **Carl F. Cori** and **Gerty T. Cori.**

(*b*) If the supply of oxygen is insufficient, as in vigorously contracting skeletal muscles, the pyruvate cannot be oxidized further for lack of oxygen. Under such conditions, it is then reduced to lactate, a process called *anaerobic glycolysis*. Lactate is also produced from glucose in anaerobic microorganisms that carry out *lactic acid fermentation.*

(*c*) In some microorganisms (*e.g.*, brewer's yeast), the pyruvate formed from glucose is transformed anaerobically into ethanol and CO_2, a process called *alcoholic fermentation.* Since living organisms first arose in an atmosphere devoid of oxygen, anaerobic breakdown of glucose is the most ancient type of biological mechanism for obtaining energy from organic fuel molecules (**Lehninger AL**, 1984).

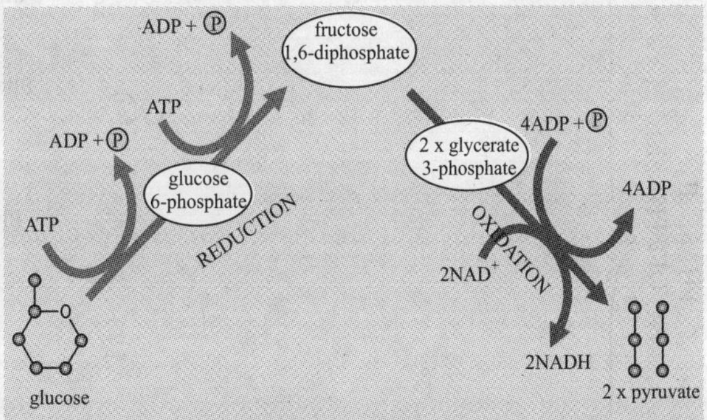

Fig 21–2. Glycolysis is a two-stage process.

The 'uphill' part involves raising glucose to a higher energy level by using ATP. In the 'downhill' part, the products are oxidized, yielding 2 molecules of pyruvate, 2 molecules of reduced coenzyme and a net gain in ATP.

Two Phases of Glycolysis

During glycolysis, the 6-carbon glucose is broken down into two moles of 3-carbon pyruvate *via* 10 enzyme-catalyzed sequential reactions. These reactions are grouped under 2 phases, phase I and II (refer Figs. 21–2 and 21–3).

A. **Phase I** or **Preparatory Phase.** It consists of the first 5 steps. In these reactions, glucose is enzymatically

OTTO HEINRICH WARBURG

(LT, 1883–1970)

A German scientist who is considered by many the pre-eminent biochemist of the first half of the 20th century. He made seminal contributions to many areas of biochemistry including respiration, photosynthesis, and the enzymology of intermediary metabolism. He was the first to show, as early as 1928, that tumours have a higher rate of glucose metabolism than other tissues. With his associates, Warburg purified and crystallized 7 of the enzymes of glycolysis. In these studies, he developed and used an experimental tool that revolutionized biochemical studies of oxidative metabolism: the *Warburg manometer,* which measured directly the consumption of oxygen by monitoring changes in gas volume, and therefore allowed quantitative measurement of any enzyme with oxidase activity.

Trained in carbohydrate chemistry in the laboratory of the great **Emil Fischer** (1902 Nobel Laureate in Chemistry), Warburg himself won the **1931 Nobel prize in Physiology or Medicine** for his work elucidating the *chemistry of cell respiration*. A number of Warburg's students and colleagues also were awarded Nobel prizes: **Otto Meyerhof** (in 1922), **Hans A. K rebs** and **Fritz A. Lipmann** (in 1953), and **Axel Hugo T. Theorell** (in 1955), incidentally all in physiology or Medicine. His first publication (with his teacher **Emil Fischer**) appeared in 1904, and his last in 1970, the year of his death at age 87.

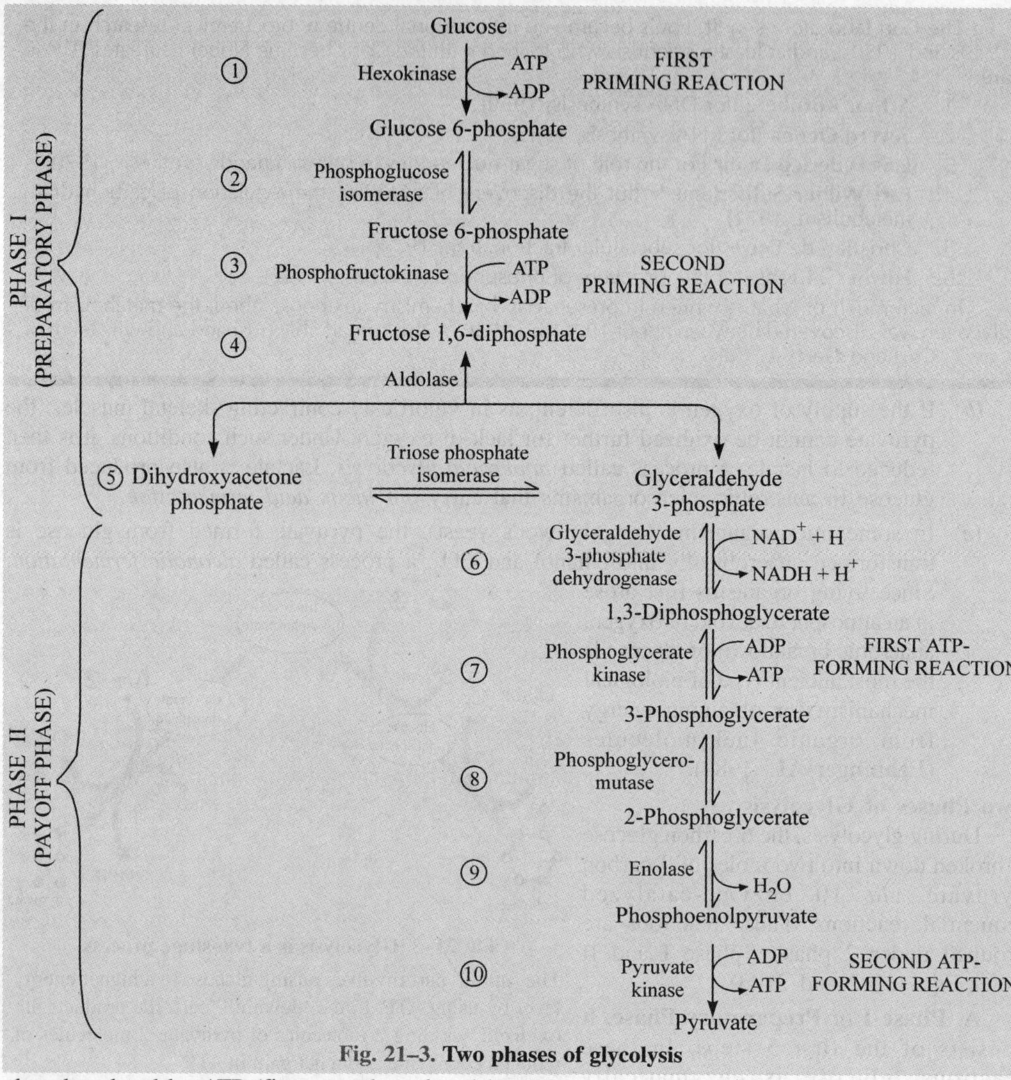

Fig. 21–3. Two phases of glycolysis

phosphorylated by ATP (first at carbon 6 and later at carbon 1) to yield fructose 1,6-diphosphate which is then split in half to yield 2 moles of the 3-carbon compound, glyceraldehyde 3-phosphate. The first phase of glycolysis, thus, results in cleavage of the hexose chain. This phase requires an investment of 2ATP moles to activate (or *prime*) the glucose mole and prepare it for its cleavage into two 3-carbon pieces. Besides glucose, other hexoses such as D-fructose, D-galactose and D-mannose may also convert into glyceraldehyde 3-phosphate

B. Phase II or **Payoff Phase.** The last 5 reactions of glycolysis constitute this phase. This phase represents the payoff of glycolysis, in which the energy liberated during conversion of 2 moles of glyceraldehyde 3-phosphate to 2 moles of pyruvate is converted by the coupled phosphorylation of 4 moles of ADP to ATP. Although 4 moles of ATP are formed in phase II, the net overall yield is only 2 moles of ATP per mole of glucose oxidized, since 2 moles of ATP are invested in phase I. The phase II is, thus, energy conserving.

A noticeable feature of glycolysis is that each of the 9 metabolic intermediates between glucose and pyruvate is a phosphorylated compound. The phosphoryl groups in these compounds are in either *ester* or *anhydride* likage. The phosphoryl or phosphate groups perform the following 3 functions:

1. The phosphate groups are completely ionized at *p*H 7, so that each of the 9 intermediates of glycolysis gains a net negative charge. Since cell membranes are, in general, impermeable to charged molecules, the glycolytic intermediates cannot escape from the cell. Only glucose can enter cells and pyruvate or lactate can leave cells because cell membranes have specific transport systems that allow these molecules to pass.
2. The phosphate groups are essential components in the conservation of energy since they are ultimately transferred to ADP to produce ATP.
3. The phosphate groups act as recognition or binding groups required for the proper fit of the glycolytic intermediates to the active site of their corresponding enzymes.

Enzymes Involved in Glycolysis

In most kinds of cells, the enzymes that catalyze glycolytic reactions (refer Table 21–1) are present in the extramitochondrial soluble fraction of the cell, the *cytosol*. On the other hand, the enzymes involved in citric acid cycle are located in the mitochondrial membrance in eukaryotes and in the plasma membrane in prokaryotes. *A remarkable feature of the glycolytic enzymes is that nearly all of them require* Mg^{2+} *for activity.*

Kinds of Reactions in Glycolysis

For a better understanding of the various reactions of glycolysis, some of the kinds of reactions that occur in glycolysis are listed below:

1. **Phosphoryl transfer.** A phosphate group is transferred from ATP to a glycolytic intermediate or *vice versa*.

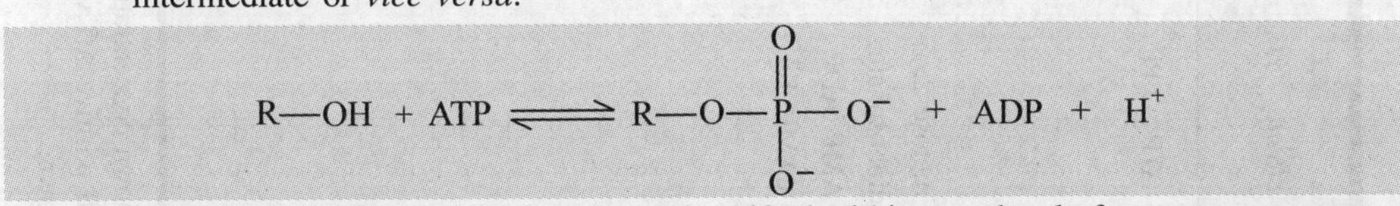

2. **Phosphoryl shift.** A phosphate group is shifted within a molecule from one oxygen atom to another.

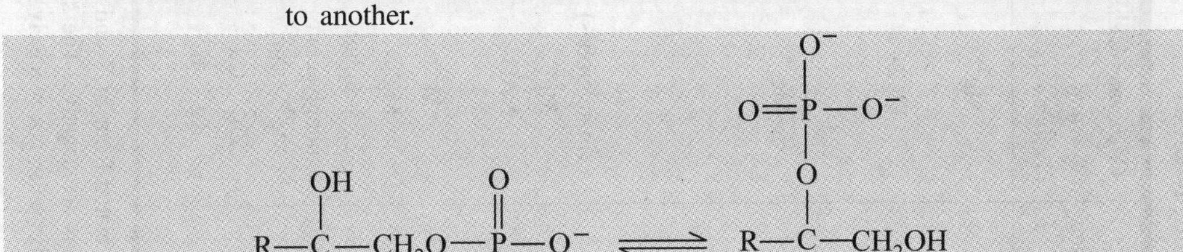

3. **Isomerization.** An aldose is converted into a ketose or *vice versa*.

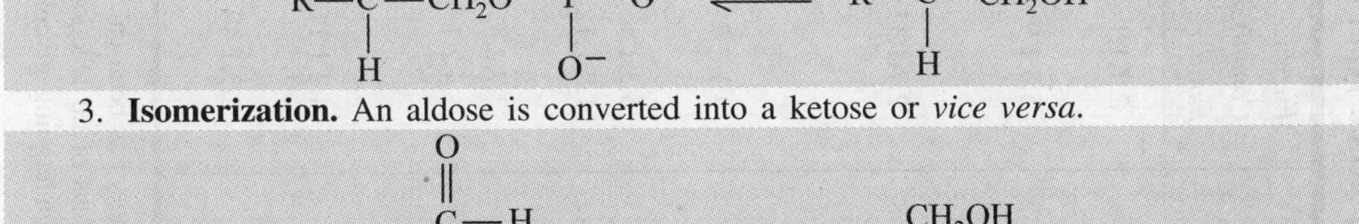

4. **Dehydration.** A molecule of water is eliminated.

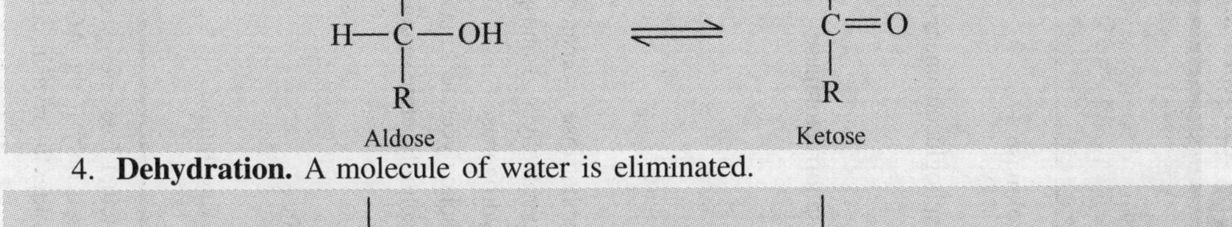

Table 21-1. Enzymes and reaction types of glycolysis

Step No.	*Enzyme*	*Enzyme Commission number*	*Coenzyme (s) and cofactor (s)*	*Activator (s)*	*Inhibitor (s)*	*Kind of reaction catalyzed*	*$\Delta G^{o\prime}$ kcal/mol*	*ΔG^{o}* kcal/mol*
1	Hexokinase	2.7.1.1	Mg^{2+}	ATP4–, Pi	Glucose 6-phosphate	Phosphoryl transfer	– 4.00	–8.0
2	Phosphoglucoisomerase	5.3.1.9	Mg2+	—	2-deoxyglucose 6-phosphate	Isomerization	+ 0.40	–0.6
3	Phosphofructokinase	2.7.1.11	Mg^{2+}	Fructose 2, 6-diphosphate, AMP, ADP, cAMP, K^{+}	ATP4–, citrate	Phosphoryl transfer	– 3.40	–5.3
4	Aldolase	4.1.2.7	Zn^{2+} (in microbes)	—	Chelating agents	Aldol cleavage	+ 5.73	–0.3
5	Phosphotriose isomerase	5.3.1.1	Mg^{2+}	—	—	Isomerization	+ 1.83	+ 0.6
6	Glyceraldehyde 3-phosphate dehydrogenase	1.2.1.12	NAD	—	Iodoacetate	Phosphorylation coupled to oxidation	+ 1.50	+ 0.6
7	Phosphoglycerate kinase	2.7.2.3	M^{2+}	—	—	Phosphoryl transfer	– 4.50	+ 0.3
8	Phosphoglycerate mutase	5.4.2.1	Mg^{2+} 2, 3-diphosphoglycerate	—	—	Phosphoryl shift	+ 1.06	+ 0.2
9	Enolase	4.2.1.11	Mg^{2+}, Mn^{2+} Zn^{2+}, Cd^{2+}	—	Fluoride + phosphate	Dehydration	+ 0.44	– 0.8
10	Pyruvate kinase	2.7.1.40	Kg^{2+}, K^{+}	—	Acetyl CoA, alanine, Ca^{2+}	Phosphoryl transfer	– 7.50	– 4.0

* ΔG, the actual free energy change, has been calculated from $\Delta G^{o\prime}$ and known concentrations of reactants under typical physiologic conditions. Glycolysis can proceed only if the ΔG values of all reactions are negative. The small positive ΔG values of 3 of the above reactions indicate that the concentrations of metabolites *in vivo* in cells undergoing glycolysis are not precisely known.

5. **Aldol cleavage.** A C–C bond is split in a reversal of an aldol condensation.

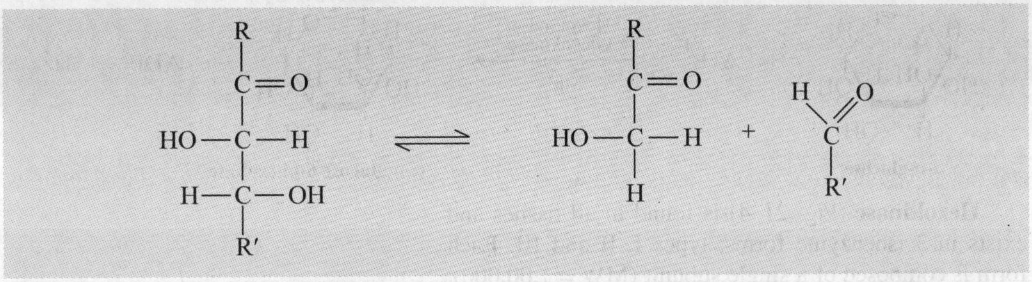

REACTION STEPS OF GLYCOLYSIS

The various reaction steps of glycolysis are schematically represented in Fig. 21–3. The details of these reactions and those of the enzymes, which catalyze them, are given below:

Step 1 : *Phosphorylation of Glucose*

In the first step, glucose is activated (or *primed*) for subsequent reactions by its phosphorylation at C_6 to yield glucose 6-phosphate, using ATP as phosphate donor. It is phosphoryl transfer type of reaction and is catalyzed by the inducible enzyme *hexokinase*, found in most animal, plant and microbial cells, and by an additional enzyme in the liver, *glucokinase*. The reaction is accompanied by considerable loss of free energy as heat. It is physiologically irreversible reaction because of the relatively low energy character of glucose 6-phosphate and the lower stability of Mg^{2+}.ADP compared to Mg^{2+}.ATP. Glucose 6-phosphate is an important compound, being at the junction of many metabolic pathways such as glycolysis, glycogenolysis, gluconeogenesis and the hexose monophosphate shunt.

> The terms used for glyceric acid, pyruvic acid and lactic acid are respectively, glycerate, pyruvate and lactate. These terms are used to emphasize that at pH of the cell, the acid involved in the reaction is largely in **dissociated form.**

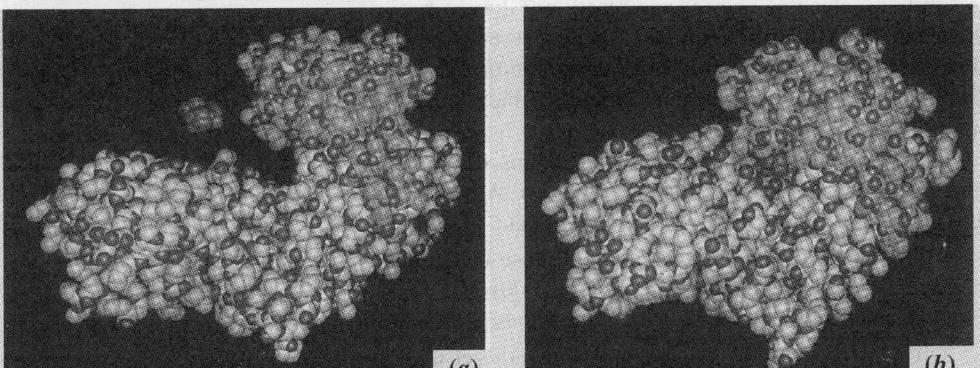

Fig. 21–4. A space-filling model of a subunit of yeast hexokinase in the "open" (*a*) and "closed" (*b*) conformations

The glucose, with which the enzyme complexes, is shown in *purple*. Note the prominent bilobal appearance of the free enzyme (the C atoms in the small lobe are shaded *green*, whereas those in the large lobe are light *grey*; the N and O atoms are *blue* and *red*). In the enzyme-substrate complex, these 2 lobes have swung together so as to engulf the substrate. Hexokinase also has been crystallized with a bound analogue of ATP. In the absence of glucose, the enzyme with the bound ATP analogue remains in the open conformation. The structural change caused by glucose results in the formation of additional contacts between the enzyme and ATP. This can explain why the binding of glucose enhances the binding of ATP.

(*Courtesy: Dr. Thomas A. Steitz, Yale University*)

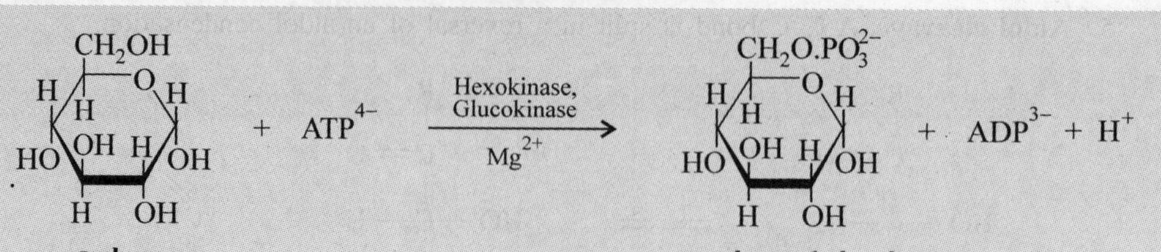

Hexokinase (Fig. 21–4) is found in all tissues and exists in 3 isoenzyme forms, types I, II and III. Each form is composed of a single subunit (MW = 1,00,000). Brain contains chiefly type I and skeletal muscle, the type II. *All the 3 types are present in human liver and fat.* Type I is found in the cytosol or bound to the mitochondria, whereas type II exists primarily in the cytosol. Hexokinase, like all other kinases, requires Mg^{2+} (or other divalent metal ions such as Mn^{2+}, Ca^{2+} etc) for activity. The 2 lobes of hexokinase remain separate in the absence of its substrate molecule, *i.e.*, glucose. However, the conformation changes markedly on binding with glucose and the 2 lobes of the enzyme come together and surround the substrate. This induced fit is shown in Fig, 21–5. Hexokinase not only acts on glucose but also on some other common hexoses such as fructose and mannose. The activity of hexokinase is inhibited by the product of the reaction (*i.e.*, glucose 6-phosphate) which binds the enzyme at an allosteric site. Hexokinase has a high affinity (*i.e.*, low Km value of about 1.0 mM) for its substrate, glucose. The reverse reaction (Glucose 6-phosphate → Glucose) requires a different enzyme, glucose 6-phosphatase with Mg^{2+} as cofactor. This reaction occurs in liver but not in muscle which lacks glucose 6-phosphatase. Yeast hexokinase, however, differs somewhat from the mammalian forms. It is a dimer of identical subunits (MW = 50,000). Also, its activity is not affected by glucose 6-phosphate.

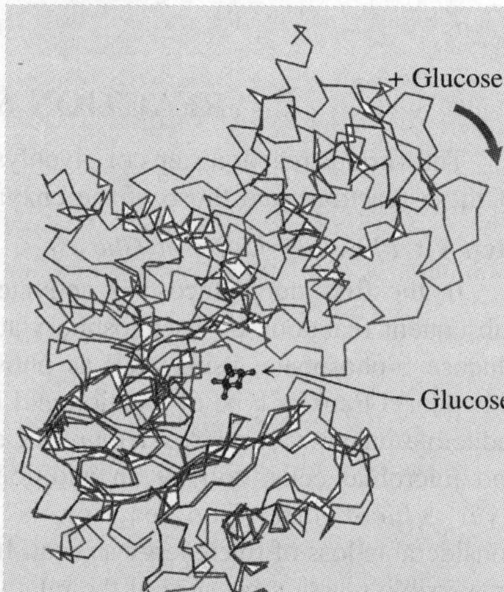

Fig. 21–5. Computer graphics of the induced fit in hexokinase

As shown in *blue*, the two lobes of hexokinase are separated in the absence of glucose. The conformation of hexokinase changes markedly on binding glucose, as shown in *red*. The two lobes of the enzyme come together and surround the substrate. Except around the binding site, the remaining portion of the enzyme in both the conditions (free and bound) is almost superimposed as seen in the figure.

(*Courtesy: Dr. Thomas A. Steitz*)

Glucokinase (often designated **hexokinase IV**) is a monomeric inducible enzyme (MW = 48,000) and is found almost exclusively in the liver. Glucokinase differs from mammalian hexokinase in 3 respects:

(*a*) It is specific for glucose and does not act on other hexoses.

(*b*) It is not inhibited by glucose 6-phosphate.

(*c*) It has a low affinity (*i.e.*, a much higher Km value of about 10 *m*M) for glucose than hexokinase.

The function of glucokinase is to remove glucose from the blood following a meal and to trap it in the liver cells, thereby allowing storage of glucose as glycogen or, after further metabolism, as fatty acids.

Isoenzymes: Hexokinase and Glucokinase

Hexokinase catalyzes the first step in glucose metabolism in all cells, namely the phosphorylation reaction of glucose by ATP to form glucose 6-phosphate (Glc-6-P):

Glucose + ATP → Glucose 6-phosphate + ADP

This enzyme has a low K_m for glucose (0.2 mmol/L) and is inhibited allosterically by its product, Glc-6-P. Since normal glucose levels in blood are about 5 mmol/L and intracellular levels are 0.2–2 mmol/L, hexokinase efficiently catalyzes this reaction (55–90% of V_{max}) under normal conditions, e.g., in muscle.

Hepatocytes, which store glucose as glycogen, and pancreatic β cells, which regulate glucose consumption in tissues and its storage in liver by secreting insulin, contain an enzyme, glucokinase. **Glucokinase** catalyzes the same reaction as hexokinase, but has a higher K_m for glucose (10 mmol/L) and is *not* inhibited by the product, Glc-6-P. Since glucokinase has a much higher K_m than hexokinase, glucokinase phosphorylates glucose with increasing efficiency as blood glucose levels increase following a meal. One of the physiological roles of glucokinase in the liver is to provide Glc-6-P for the synthesis of glycogen, a storage form of glucose. In the pancreatic b cell, glucokinase functions as the glucose sensor, determining the threshold for insulin secretion.

Step 2 : *Isomerization of Glucose 6-phosphate*

Glucose 6-phosphate is reversibly isomerized to frucose 6-phosphate by *phosphoglucoisomerase*. Thus, the 6-membered pyranose ring of glucose 6-phosphate is converted into the 5-membered furanose ring of fructose 6-phosphate. This reaction involves a shift in the carbonyl oxygen from C_1 to C_2, thus converting an aldose into a ketose. At equilibrium, the ratio of aldose to ketose is 7 : 3, *i.e.*, glucose 6-phosphate predominates, having concentration over twice that of fructose 6-phosphate. The reaction proceeds readily in either direction because of relatively small standard free energy change. Fructose 6-phosphate has metabolic fates other than glycolysis.

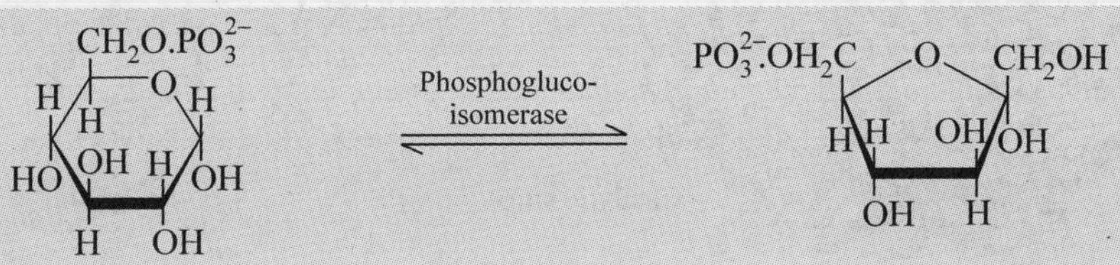

α-D-glucose 6-phosphate (Robinson's ester) — **α-D-fructose 6-phosphate** (Newberg's ester)

Human phosphoglucoisomerase (MW = 134,000) is a dimer of identical subunits and requires Mg^{2+} for activity. It is specific for glucose 6-phosphate and fructose 6-phosphate. An interesting sidelight of this enzyme is that it binds the α- pyranose form of glucose 6-phosphate, but the open chain form of fructose 6-phosphate. The α- and β-pyranose forms are interconvertible, so all of the glucose 6-phosphate is available to the enzyme.

Step 3 : *Phosphorylation of Fructose 6-phosphate*

This is the second of the two priming or activating reactions of glycolysis (the first one being Step 1). Fructose 6-phosphate is phosphorylated by ATP to produce fructose 1, 6-diphosphate in the presence of another inducible allosteric enzyme, *phosphofructokinase* (abbreviated as PFK). The enzyme catalyzes the transfer of a phosphate group from ATP to fructose 6-phosphate at C1 to yield fructose 1, 6-diphosphate. Since the reaction proceeds with ΔG° = – 3.4 kcal/mol, it is essentially irreversible. *It is considered to be the committed step in glycolysis* since the PFK action 'commits' the cell to metabolizing glucose rather than storing or converting it to some other hexose. In addition to being a key step, it is an important control point of glycolysis.

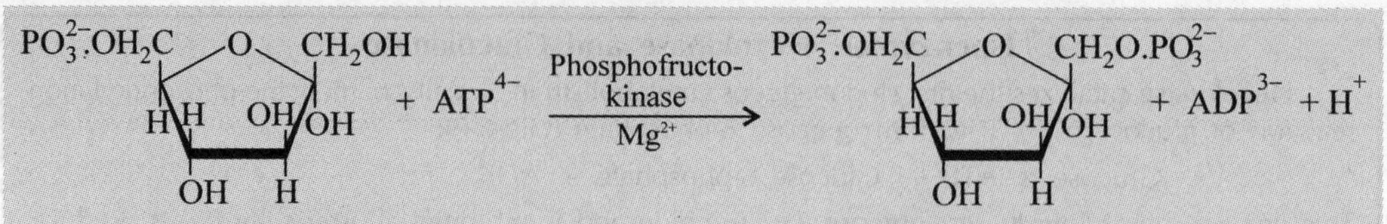

D-fructose 6-phosphate (Newberg's ester) → **D-fructose 1, 6-diphosphate** (Harden-Young ester)

The muscle **phosphofructokinase** (MW = 3,20,000) *is one of the most complex known enzymes*. It is a tetramer of 4 identical subunits (Fig. 21–6)but dissociates into inactive dimers in the presence of citrate. Fructose 6-phosphate, however, promotes their reunion to form the tetramer. It is a major regulatory enzyme in muscle glycolysis because of its allosteric nature. The activity of phosphofructokinase is accelerated whenever the cell is deficient in ATP or there is an excess of ATP breakdown products (*i.e.*, ADP and AMP). The activity is, however, inhibited whenever the cell has plentiful ATP (and other fuels such as citrate or fatty acids) which lowers the affinity of the enzyme for fructose 6-phosphate. For the same reason, PFK is rightly regarded as the '*pace maker*' of glycolysis.

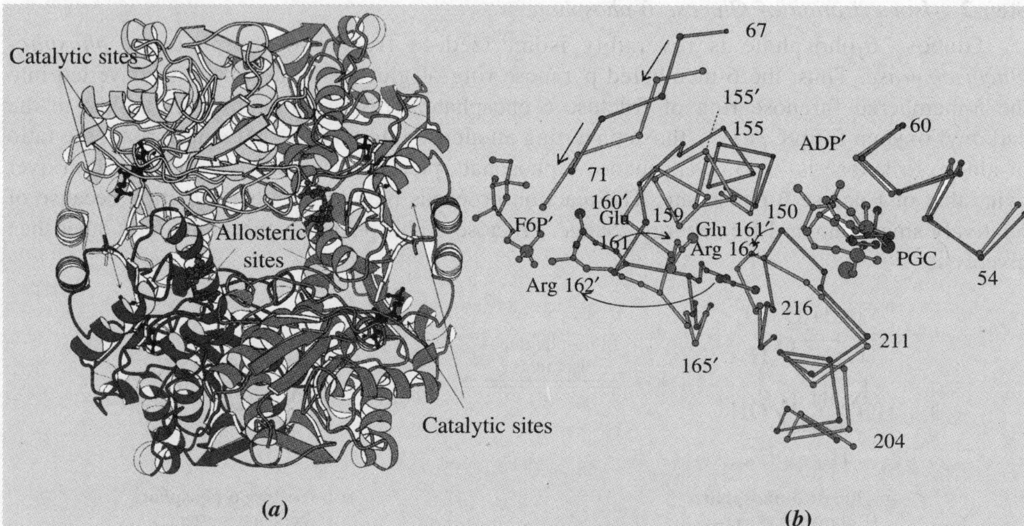

Fig. 21–6. The x-ray structure of phosphofructokinase (PFK)

(*a*) **Ribbon model.** Phosphofructokinase in the liver is a tetramer of four identical subunits. The positions of the catalytic and allosteric sites are indicated.

(*b*) **A superposition of those segments of the T state (*blue*) and R state (*red*) enzymes that undergo a large conformational rearrangement upon the T→ R allosteric transition** (*indicated by the arrows*). Residues of the R state structure are marked by a prime. Also shown are bound ligands: the nonphysiological inhibitor 2-phosphoglycolate (PGC; a PEP analogue) for the T state, and the cooperative substrate F6P and the activator ADP for the R state.

(*Courtesy : (b) Schirmer T and Evans PR, 1990*)

Step 4 : *Cleavage of Fructose 1,6-diphosphate*

This is a unique C–C bond scission reaction. Since fructose 1, 6-diphosphate is a molecule with phosphate group on both ends, it splits in the middle into two different triose phosphates, glyceraldehyde 3-phoshpate (an aldose) and dihydroxyacetone phosphate (a ketose). This reaction is catalyzed by the enzyme *fructose diphosphate aldolase* (often simply called *aldolase*) which cleaves the fructose 1, 6-diphosphate molecule between C_3 and C_4. Carbon atoms 4, 5 and 6 appear in glyceraldehyde 3-phosphate and 1, 2 and 3 in dihydroxyacetone phosphate. Although the aldolase reaction has a highly positive standard free energy change, it can readily proceed in either direction under the *p*H and concentration conditions existing in cells. Thus, this is a reversible aldol condensation type of reaction. The remaining steps in glycolysis involve 3 carbon units, rather than 6 carbon units.

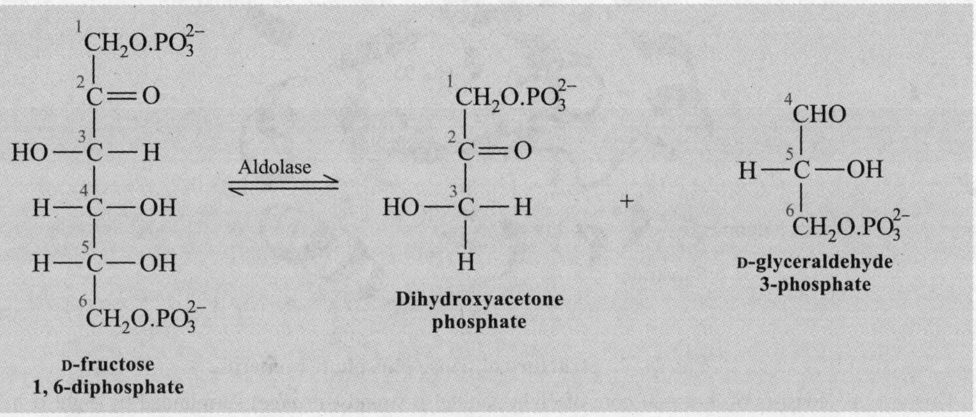

The **aldolase** (name derived from the nature of the reaction it catalyzes) from all animal tissues is a tetramer (MW = 160,000) but various tissues contain primarily one of the 3 different forms, characteristic of muscle, liver and brain respectively. All forms catalyze the above reaction but at different rates. The aldolase of animal tissues does not require Mg^{2+} but in many microbes (yeast), aldolase needs Zn^{2+} for its activity.

Step 5: *Isomerization of Dihydroxyacetone phosphate*

Glyceraldehyde 3-phosphate (an aldose) can be directly degraded in the subsequent reaction steps of glycolysis but dihydroxyacetone phosphate (a ketose) cannot be. However, dihydroxyacetone phosphate can be readily and reversibly converted into glyceraldehyde 3-phosphate by the enzyme *triose phosphate isomerase* (also called *phosphotriose isomerase*) in the same way that glucose and fructose phosphates are interconverted by phosphoglucoisomerase. This is an isomerization reaction and occurs very rapidly. At equilibrium, about 95% of the triose phosphate is dihydroxyacetone phosphate. However, the reaction proceeds towards glyceraldehyde 3-phosphate formation because of the efficient removal of this product. It may be noted that by this reaction, carbon atoms 1, 2 and 3 of the starting glucose now become indistinguishable from carbon atoms 6, 5 and 4 respectively. This reaction completes the first phase of glycolysis.

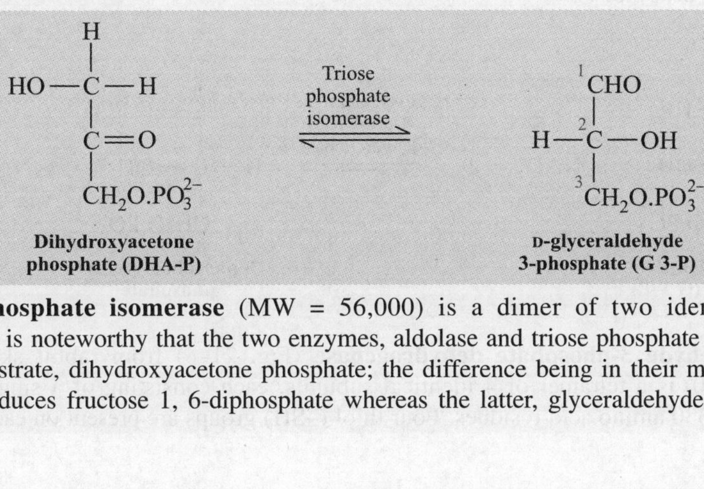

Triose phosphate isomerase (MW = 56,000) is a dimer of two identical subunits (Fig. 21–7). It is noteworthy that the two enzymes, aldolase and triose phosphate isomerase have a common substrate, dihydroxyacetone phosphate; the difference being in their mode of action – the former produces fructose 1, 6-diphosphate whereas the latter, glyceraldehyde 3-phosphate.

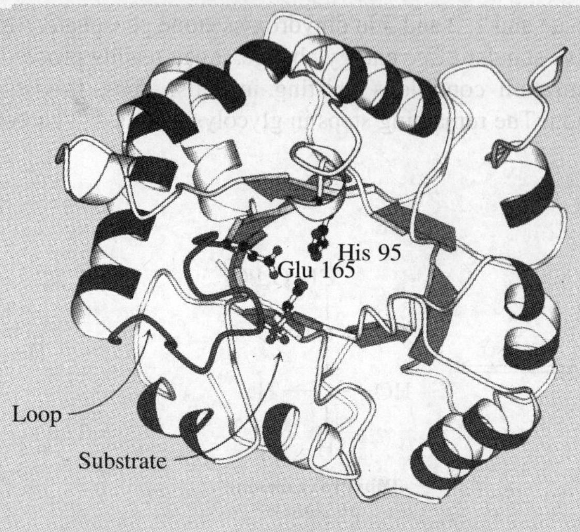

Fig. 21–7. Structure of triose phosphate isomerase

This enzyme consists of a central core of eight parallel β strands (*orange*) surrounded by eight α helices (*blue*). This structural motif, called an αβ barrel, is also found in the glycolytic enzymes, aldolase, enolase, and pyruvate kinase. Histidine 95 and glutamate 165, essential components of the active site of triose phosphate isomerase, are located in the barrel. A loop (*red*) closes off the active site on substrate binding.

Step 6 : *Oxidative Phosphorylation of Glyceraldehyde 3-phosphate*

This is first of the two energy-conserving reactions of glycolysis (the second one being Step 9) in which a high energy phosphate compound, 3-phosphoglyceroyl phosphate is formed. Here, glyceraldehyde 3-phosphate is converted into 3-phosphoglyceroyl phosphate (= 1,3-diphosphoglycerate, 1,3-DPG) by the enzyme *glyceraldehyde* 3-phosphate *dehydrogenase* (= phosphoglyceraldehyde dehydrogenase), which is NAD^+-dependent. In this complex and reversible reaction, the aldehyde group of glyceraldehyde 3-phosphate is dehydrogenated, not to a free carboxyl group, but to a carboxylic anhydride with phosphoric acid, the 3-phosphoglyceroyl phosphate. This type of anhydride called an acyl phosphate has a very high standard free energy of hydrolysis ($\Delta G^{o\prime}$ = –11.8 kcal/mol) and is thus a super high energy phosphate (refer Table 20–5). The acyl phosphate or 1,3–diphosphoglyceraldehyde conserves much of the free energy liberated during oxidation of the aldehyde group of glyceraldehyde 3-phosphate. The value of $DG^{o\prime}$ for this reaction is +1.5 kcal/mol.

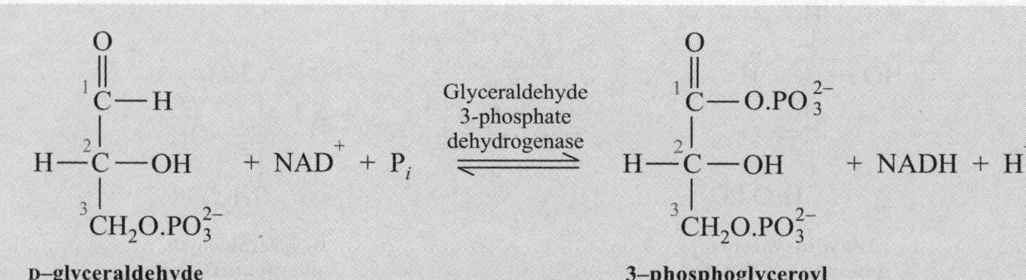

Glyceraldehyde 3-phosphate dehydrogenase (Fig. 21–8) from rabbit skeletal muscles (MW = 146,000) is a tetramer of 4 identical subunits, each consisting of a single polypeptide chain of about 330 amino acid residues. Four thiol (–SH) groups are present on each polypeptide, probably derived from cysteine residues within the polypeptide chain. One of the –SH groups is found at the active site on which an acyl-enzyme complex is formed. Each mole of enzyme also contains 4 moles of NAD^+. The enzyme is, however, inactivated by the –SH poison, iodoacetate which combines with the essential –SH group of the enzyme, thus preventing its participation in catalysis.

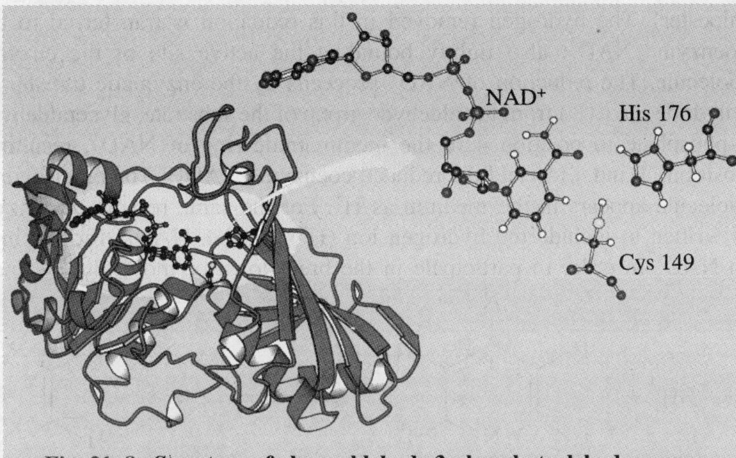

Fig. 21–8. Structure of glyceraldehyde 3-phosphate dehydrogenase

The active site includes a cysteine residue and a histidine residue adjacent to a bound NAD^+.

$$\underset{\text{Active enzyme}}{\text{Enzyme—SH}} + \underset{\text{Iodoacetate}}{ICH_2COO^-} \longrightarrow \underset{\text{Inactive enzyme}}{\text{Enzyme—S—}CH_2COO^-} + HI$$

The mechanism of action of glyceraldehyde 3-phosphate dehydrogenase is rather complex and resembles that of a-ketoglutarate dehydrogenase. It involves 3 steps (refer Fig 21–9):

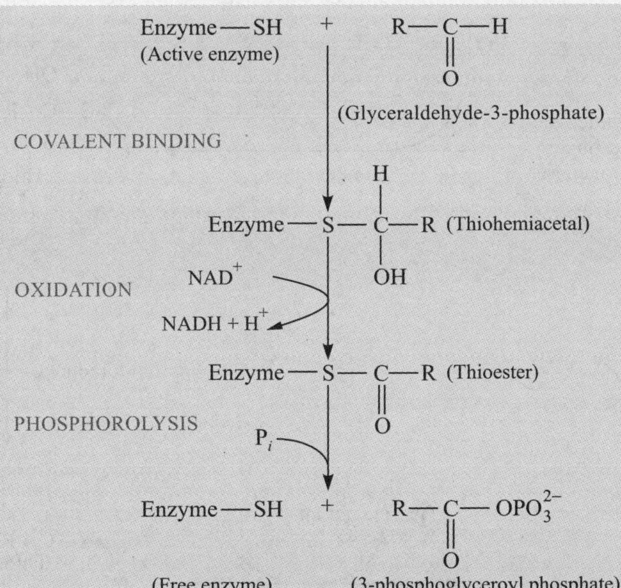

Fig. 21–9. Mechanism of action of glyceraldehyde 3-phosphate dehydrogenase, represented here as Enzyme—SH

(*a*) ***Covalent binding of substrate to —SH group.*** The substrate first combines with an —SH group of an essential cysteine residue present at the active site of the enzyme, forming a thiohemiacetal.

(*b*) ***Oxidation of thiohemiacetal and reduction of NAD^+.*** Thiohemiacetal (= hemithioacetal) is then oxidized to produce a high energy covalent acyl-enzyme complex, called thiol ester (=

thioester). The hydrogen removed in this oxidation is transferred to the coenzyme NAD^+, also tightly bound at the active site of the enzyme molecule. The reduction of NAD^+ proceeds by the enzymatic transfer of a hydride ion (:H^-) from the aldehyde group of the substrate, glyceraldehyde 3-phosphate to position 4 of the nicotinamide ring of NAD^+, resulting in its reduction at ring positions 1 and 4 to yield the reduced coenzyme NADH. The other hydrogen atom of the substrate molecule appears in the medium as H^+. For the same reason, the enzymatic reduction of NAD^+ is written to include the hydrogen ion (H^+) formed. NADH formed in the reaction is reoxidized to NAD^+ in order to participate in the breakdown of more glucose molecules to pyruvate.

> **Hydride ion** is a hydrogen nucleus and two electrons.

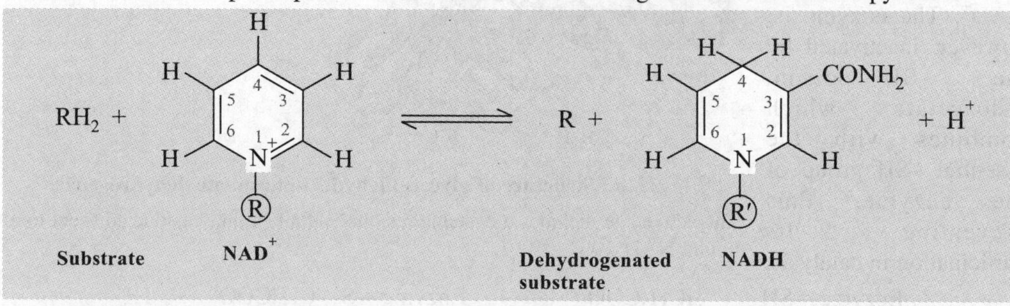

(*c*) ***Phosphorolysis of thioester.*** Finally, the acyl-enzyme (or thioester) reacts with an inorganic phosphate (Pi) forming an acyl phosphate called 3-phosphoglyceroyl phosphate and the free enzyme with a reconstituted —SH group is liberated. A diagrammatic representation of the mechanism of action of glyceraldehyde 3-phosphate dehydrogenase is given in Fig. 21–10.

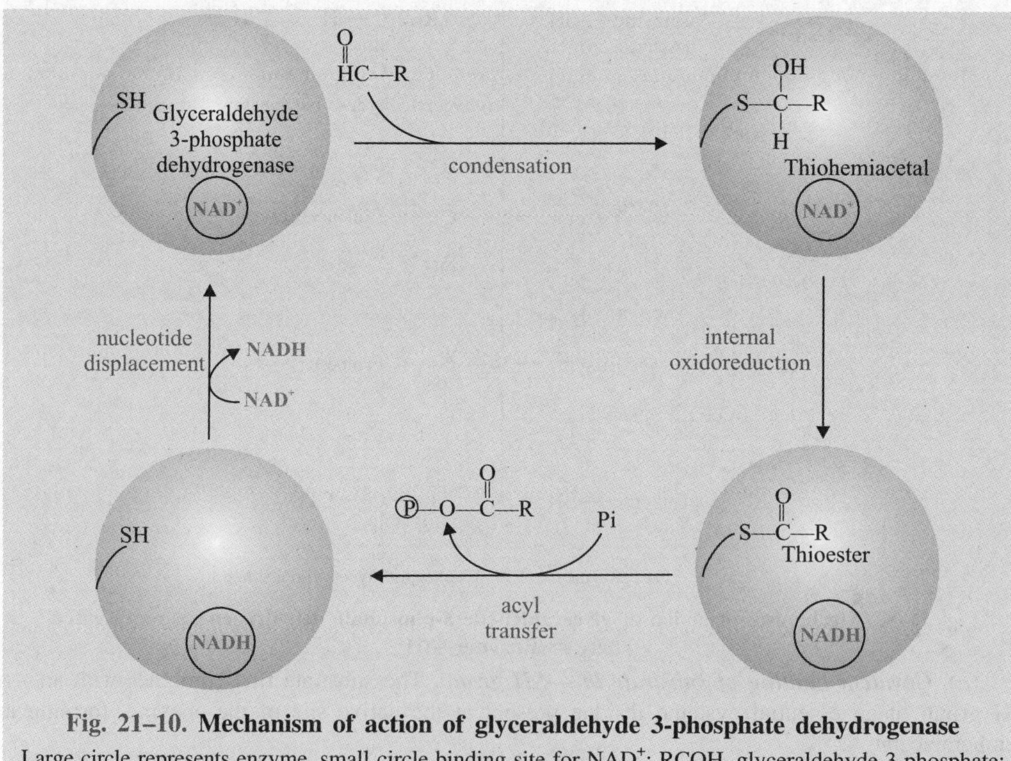

Fig. 21–10. Mechanism of action of glyceraldehyde 3-phosphate dehydrogenase

Large circle represents enzyme, small circle binding site for NAD^+; RCOH, glyceraldehyde 3-phosphate; —SH, the sulfhydryl group of the cysteine residue located at the active site; and ~P, the high-energy phosphate bond of 1,3-bisphosphoglycerate.

Step 7 : *Transfer of Phosphate from 1,3-DPG to ADP*

This is the first ATP-generating reaction in glycolysis (the second one being Step 10). It involves the transfer of high-energy phosphate group from the carboxylic group of 3-phosphoglyceroyl phosphate (= 1,3-diphosphoglycerate or 1,3-DPG) to ADP by the enzyme phosphoglycerate kinase, thus producing ATP and leaving 3-phosphoglycerate. Since 2 moles of triose phosphate are produced per mole of glucose, 2 moles of ATP are generated at this stage per mole of glucose oxidized. The value of $\Delta G°'$ for this essentially reversible reaction is –4.5 kcal/mol.

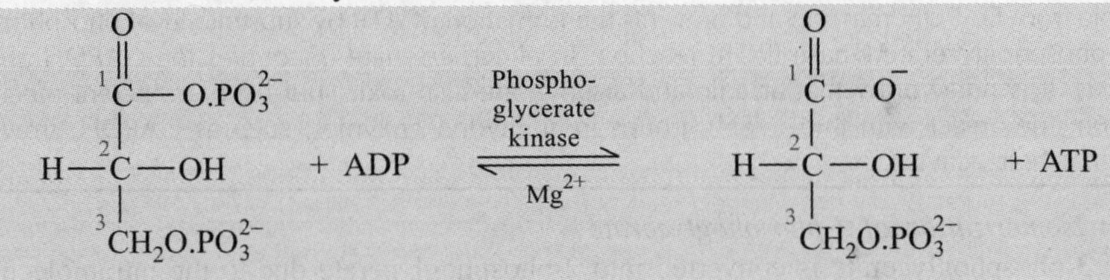

1, 3-diphosphoglycerate **3-phosphoglycerate**

Phosphoglycerate kinase (Fig. 21–11) has a molecular weight of about 45,000. As with other enzymes of this type, there is an absolute requirement for a divalent metal cofactor (Mg^{2+}, Mn^{2+} or Ca^{2+}). The metal interacts with the ADP or ATP to form the reactive complex. This and the preceding reaction (*i.e.*, Steps 6 and 7) together constitute an energy-coupling process. The sum of these two sequential reactions would give:

$$\text{Glyceraldehyde 3-phosphate} + NAD^+ + Pi + ADP \rightleftharpoons \text{3-phosphoglycerate} + ATP + NADH + H^+$$

$$\Delta G°' = (+1.5) + (-4.5) = -3.0 \text{ kcal/mol}$$

Thus, the net outcome of Steps 6 and 7 is:

(*a*) Glyceraldehyde 3-phosphate, an aldehyde, is oxidized to 3-phosphoglycerate, a carboxylic acid.

(*b*) NAD^+ is reduced to NADH.

(*c*) ATP is produced from ADP and Pi.

In other words, the net result of Steps 6 and 7 is that the energy released on oxidation (or dehydrogenation) of an aldehyde to a carboxylate group is conserved by the coupled formation of ATP from ADP and Pi. Such reactions are called *substrate-level phosphorylations* because the energy required to bring them about arises from the dehydrogenation of an organic substrate molecule (for example, glyceraldehyde 3-phosphate in the present case). This term distinguishes these phosphorylations from *oxidative phosphorylation*, which is the formation of ATP coupled to the oxidation of NADH and $FADH_2$ by oxygen. As the oxidative phosphorylation is coupled to electron transport, it is also called as *respiratory-chain phosphorylation*.

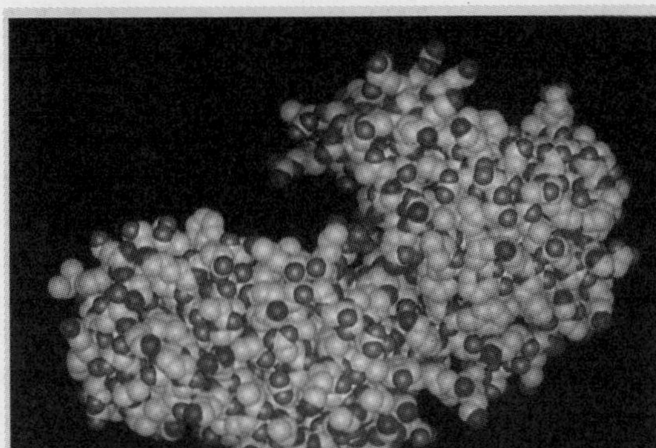

Fig. 21–11. A space-filling model of yeast phosphoglycerate kinase showing its deeply clefted bilobal structure

The substrate-binding site is at the bottom of the cleft as marked by the P atom (*purple*) of 3PG. Compare this structure with that of hexokinase (Fig. 21–4).

(*Courtesy: Herman Watson, University of Bristol, U.K.*)

Inhibition of Substrate-level Phosphorylation by Arsenate

Arsenic is just below phosphorus in the Periodic Table of the elements, and it might be expected to share some of the properties and reactivity of phosphate. In fact, arsenate has pK_a values similar to those of phosphate and can actually be used by glyceraldehyde-3-phosphate dehydrogenase (GAPDH), producing 1-arsenato-3-phosphoglycerate. However, the acyl–arsenate bond is unstable and hydrolyzes rapidly, and ATP is not generated y substrate-level phosphorylation. While arsenate does not inhibit any of the enzymes of glycolysis, it dissipates the redox energy available from GAPDH reaction and prevents the formation of ATP by substrate-level phosphorylation at the phosphoglycerate kinase (PGK) reaction. In effect, *arsenate uncouples the GAPDH and PGK reactions.* It is noteworthy that arsenic and arsenite are also toxic, but have a different mechanism of action: they react with thiol (–SH) groups in sulfhydryl enzymes, such as GAPDH, irreversibly inhibiting their activity.

Step 8 : *Isomerization of 3-phosphoglycerate*

The 3-phosphoglycerate is converted into 2-phosphoglycerate due to the intramolecular shift of phosphoryl group from C_3 to C_2, by the enzyme *phosphoglycerate mutase* (= *phosphoglyceromutase*). This is a reversible reaction with a $\Delta G°'$ value = +1.06 kcal/mol.

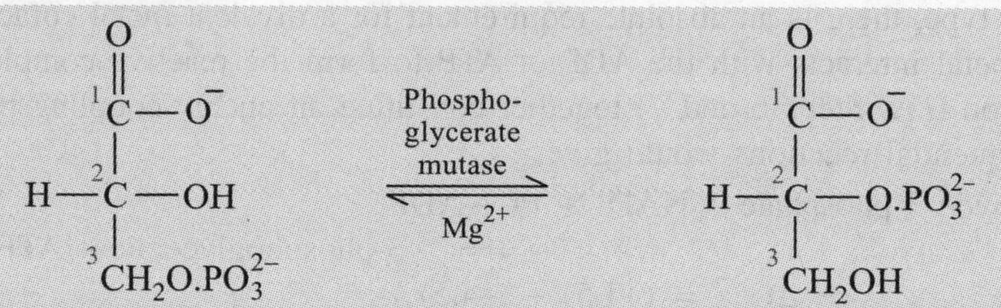

3-phosphoglycerate **2-phosphoglycerate**

Phosphoglycerate mutase (MW = 65,000) is a dimer of identical subunits. Mg^{2+} is essential for this reaction. The enzyme requires 2, 3-diphosphoglycerate as cofactor for its action and combines with it to give a phosphoenzyme and either the 2-phosphoglycerate or the 3-phosphoglycerate:

$$\text{Enzyme—OH} + \text{2,3-diphosphoglycerate} \rightleftharpoons \text{Enzyme—O.PO}_3^{2-} + \text{2- or 3-phosphoglycerate}$$

At high concentration of 3-phosphoglycerate, the 3-phosphoglycerate and phosphoenzyme react to produce free enzyme and 2, 3-diphosphoglycerate, which in turn yields the 2-phosphoglycerate. The reaction occurs in reverse direction at high concentration of 2-phosphoglycerate. At equilibrium, the ratio of 2- to the 3-phosphate is 50 : 1. The 2,3-diphosphoglycerate is produced from 3-phosphoglycerate by the enzyme *2,3-diphosphoglycerate kinase* as follows:

$$\text{3-phosphoglycerate} + \text{ATP} \longrightarrow \text{2,3 diphosphoglycerate} + \text{ADP}$$

Step 9 : *Dehydration of 2-phosphoglycerate*

This is the second reaction of glycolysis in which is a high-energy phosphate compound (*i.e.*, phosphoenolpyruvate) is formed. The 2-phosphoglycerate is dehydrated by the action of enolase (= *phosphopyruvate hydratase*) to phosphoenolypyruvate (abbreviated as PEP), which is the phosphate ester of enol tautomer of pyruvate. This is a reversible reaction and has a relatively small free energy change value of + 0.44 kcal/mol. At equilibrium, the ratio of 2-phosphoglycerate to phosphoenolpyruvate is 2 : 1. However, phosphoenolpyruvate, and not 2-phosphoglycerate, contains a super high-energy phosphate bond. The loss of water from 2-phosphoglycerate causes a redistribution of energy within the molecule, raising the phosphate on position 2 to the high energy state, thus forming PEP. The reaction is freely reversible since there is little free energy change. The $\Delta G°'$ value for the hydrolysis of PEP is –14.8 kcal/mol. There is more than sufficient energy to allow synthesis of ATP from PEP in the next step of glycolysis.

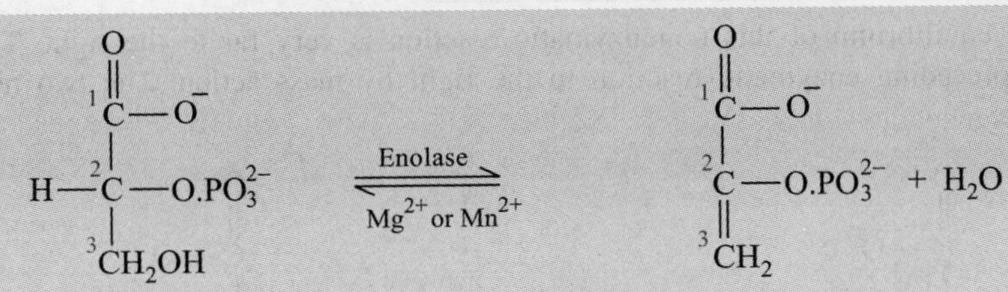

2-phosphoglycerate **Phosphoenolpyruvate**

Enolase (MW = 88,000) is a dimer with identical subunits. It requires Mg^{2+}, Mn^{2+}, Zn^{2+} or Cd^{2+} as cofactor in its active site which forms a complex with the enzyme before the substrate is bound. Enolase is inhibited by the simultaneous presence of fluoride and phosphate. In fact, the fluorophosphate ion, which binds with Mg^{2+} forming magnesium fluorophosphate, is the true inhibitor.

Glycolysis and Dental Caries

Streptococcus mutans and *Lactobacillus* are anerobic bacteria that colonize the oral cavity and contribute to the development of dental caries. These bacteria grow optimally on refined, fermentable carbohydrates in the diet and excrete organic acids, such as lactate. They thrive in acidic, anerobic microenvironments in fissures in the teeth and in gingival pockets. The organic acids gradually erode tooth enamel and dentin, and the chronic dissolution of the calcium phosphate (hydroxyappetite) matrix of the teeth sets the state for cavity formation. Fluoride, provided either topically or in toothpaste, at levels too low to inhibit **enolase**, integrates into the tooth surface, forming fluoroappatite, which is more resistant to demineralization.

Step 10 : ***Transfer of Phosphate from PEP to ADP***

This is the second ATP-generating reaction in glycolysis. Here, phosphoenolpyruvate (PEP) is converted into pyruvate in enol form (*i.e.*, enolpyruvate) by the inducible allosteric enzyme *pyruvate kinase* (abbreviated as *PK*). The enzyme catalyzes the transfer of a phosphoryl group from PEP to ADP, thus forming ATP. *This phosphorylation reaction is nonoxidative* in contrast with the one catalyzed by glyceraldehyde 3-phosphate dehydrogenase (*i.e.*, Step 6). This is another physiologically irreversible step in glycolysis (the first one being Step 1) and proceeds with $\Delta G^{\circ\prime}$ = −7.5 kcal/mol.

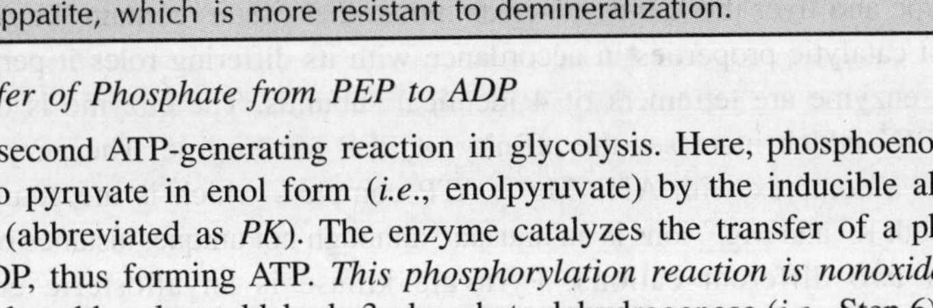

Phosphoenolpyruvate **Enolpyruvate**

The enolpyruvate, however, rearranges rapidly and *nonenzymatically* to yield the keto form of pyurvate (*i.e.*, ketopyruvate). The keto form predominates at *p*H 7.0.

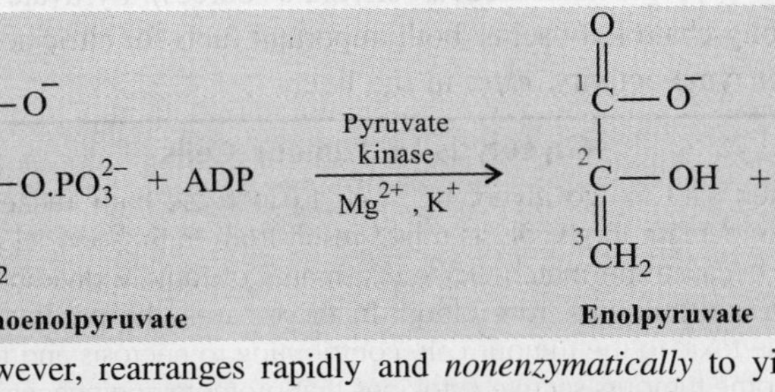

Enolpyruvate **Ketopyruvate**

The point of equilibrium of this nonenzymatic reaction is very far to the right. Therefore, it 'drives' the preceding enzymatic reaction to the right by mass action. The two reactions, on addition, give:

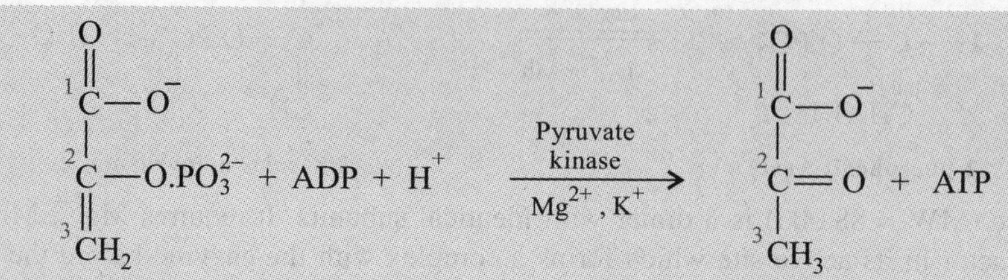

Phosphoenol-pyruvate **Ketopyruvate**

The overall reaction has a very large negative $\Delta G^{\circ\prime}$ value due to the spontaneous conversion of enol form of pyruvate to the keto form. The $\Delta G^{\circ\prime}$ value for hydrolysis of PEP is – 14.8 kcal/mol. About half of this energy is recovered as ATP ($\Delta G^{\circ\prime}$ = – 7.3 kcal/mol) and the rest (–7.5 kcal/mol) constitutes a large driving force, pushing the reaction far to the right. Since 2 moles of PEP are formed per mole of glucose oxidized, 2 moles of ATP are also produced per mole of glucose. The conversion of phosphoenolpyruvate into pyruvate is the second example of *substrate-level phosphorylation* in glycolysis.

Pyruvate kinase (MW = 1,90,000–2,50,000) is found in 3 major forms: muscle and brain contain the M type and liver the L type whereas most other tissues contain the A type. Each type controls different catalytic properties in accordance with its differing roles it performs. However, all forms of the enzyme are tetramers of 4 identical subunits. The enzyme is dependent on the concentration of K^+ which increases the affinity of PEP for enzyme. The enzyme also requires Mg^{2+} as it forms a complex with ATP, Mg^{2+}. ATP complex, which is the actual substrate. The participation of both K^+ and Mg^{2+} ions is an unusual, although not unique, instance of complementary requirement for two different cations. Pyruvate kinase is an allosteric enzyme and like phosphofructokinase, its activity is regulated by several means. In general, its activity is high when a net flux of glucose to pyruvate or lactate is required and low during gluconeogenesis (*i.e.*, during formation of glucose or glycogen from noncarbohydrate sources). Pyruvate kinase is also inhibited by acyl-CoA and by long-chain fatty acids, both important fuels for citric acid cycle. Certain amino acids also modulate enzyme activity, *esp.*, in the liver.

Glycolysis in Tumour Cells

Tumours are often said to 'go glycolytic', *i.e.*, to increase their reliance on glycolysis as a source of energy. The increase in glycolysis might result from mitochondrial dysfunction as a result of hypoxia, possibly because the metabolic requirements of rapidly dividing tumour cells exceed the supply of oxygen and nutrients from blood. In these cases, the production and accumulation of lactate may become toxic to the tumour cell, contributing to necrosis and formation of a necrotic core in the tumour. Some tumours secrete cytokines that promote *angiogenesis* (neovascularization), thereby increasing their fuel and oxygen supply and enhancing tumour growth. Angiogenesis inhibitors, designed to inhibit the vascularization of the tumour, are being evaluated as a nonsurgical approach to tumour therapy. The ability to survive by relying on glycolysis in hypoxic environments may be an important factor in tumour survival and growth.

In metabolic pathways, the enzymes catalyzing essentially irreversible reactions are the key sites of control. In glycolysis, the steps 1, 3 and 10 which are catalyzed by hexokinase (or glucokinase), phosphofructokinase and pyruvate kinase respectively are virtually irreversible. Hence, they perform regulatory as well as catalytic functions.

STOICHIOMETRY OF GLYCOLYSIS

Overall Balance Sheet

> The word stoichiometry is pronounced as **'stoy-ke-om'–etry'**. It combines two Greek words: *stoicheion* = element and *metron* = measure.

Keeping in mind that each molecule of glucose yields 2 molecules of glyceraldehyde 3-phosphate, the total inputs and the outputs of all the 10 glycolytic reactions may be written as follows:

$$\text{Glucose} + 2\ \text{ATP} + 2\ \text{Pi} + 2\ \text{NAD}^+ + 2\ \text{H}^+ + 4\ \text{ADP} \longrightarrow$$
$$2\ \text{pyruvate} + 2\ \text{H}^+ + 4\ \text{ATP} + 2\ \text{H}_2\text{O} + 2\ \text{NADH} + 2\ \text{H}^+ + 2\text{ADP}$$

On cancelling the common terms, we get the net equation for the transformation of glucose into pyruvate:

$$\text{Glucose} + 2\ \text{Pi} + 2\ \text{ADP} + 2\ \text{NAD}^+ \longrightarrow 2\ \text{pyruvate} + 2\ \text{ATP} + 2\ \text{NADH} + 2\ \text{H}^+ + 2\ \text{H}_2\text{O}$$

Thus, three things happen simultaneously in glycolysis:

(*a*) Glucose is oxidized to pyruvate.

(*b*) NAD+ is reduced to NADH.

(*c*) ADP is phosphorylated to form ATP.

There can be no EMP pathway without all 3 events which means that NAD^+, ADP and Pi, as well as glucose, must be present.

Energy Yield

Further, 2 moles of ATP are generated in glycolysis. A summary of the steps in which ATP is consumed or formed is given in Table 21–2.

Table 21–2. Energy yield of glycolysis

Step	*Reaction*	*Consumption of ATP*	*Gain of ATP*
1	Glucose ⟶ Glucose 6-phosphate	1	
3	Fructose 6-phosphate ⟶ Fructose 1, 6-diphosphate	1	
7	1, 3-diphosphoglycerate ⟶ 3-phosphoglycerate		1 × 2 = 2
10	Phosphoenolpyruvate ⟶ Pyruvate		1 × 2 = 2
		2	4
		Net gain of ATP =	4 – 2 = 2

MUSCLE (OR ANAEROBIC) GLYCOLYSIS AND HOMOLACTIC FERMENTATION

In plant and animal tissues, under aerobic conditions, the pyruvate is the product of glycolysis and NADH, formed by the dehydrogenation of glyceraldehyde 3-phosphate, is then reoxidized to NAD^+ by oxygen. However, under anaerobic conditions, in actively contracting skeletal muscles, the NADH generated in glycolysis cannot be reoxidized by oxygen but must be reoxidized to NAD^+ by the pyruvate itself, converting pyruvate into lactate. Such type of glycolytic sequence occurring under anaerobic conditions in the muscle tissues is commonly spoken of as *muscle glycolysis or anaerobic glycolysis*. Besides the skeletal muscles, a large number of microorganisms, the lactic acid bacteria (esp., species of *Lactobacilli, Bacilli, Streptococci and Clostridia*) also follow the same path for the reduction of pyruvate to lactate. Such type of fermentation that yields lactate as the sole product is termed *homolactic fermentation*.

The reduction of pyruvate by NADH to form lactate is catalyzed by lactate dehydrogenase (abbreviated as LDH) which forms the L-isomer of lactate. As the reaction has a large negative

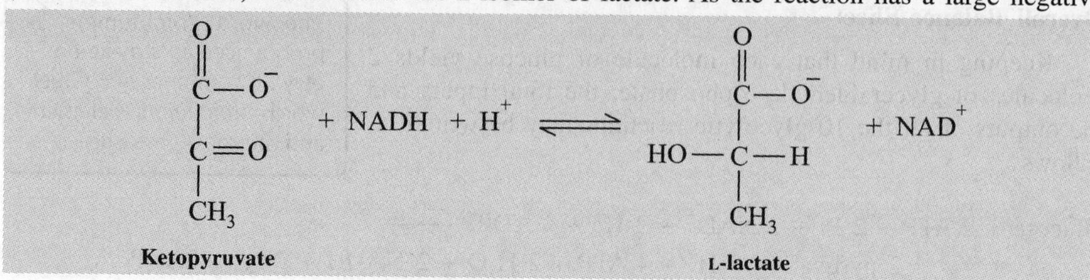

value of $\Delta G^{\circ\prime}$ (–6.0 kcal/mol), it proceeds far to the right. The reoxidation of NADH via lactate formation allows glycolysis to proceed in the absence of O_2 by regenerating sufficient NAD^+ for the reaction Step 6 of glycolysis, catalyzed by glyceraldehyde 3-phosphate dehydrogenase. Thus, the reduction of pyruvate to lactate is coupled to the oxidation of glyceraldehyde 3-phosphate to 3-phosphoglyceroyl phosphate as shown below:

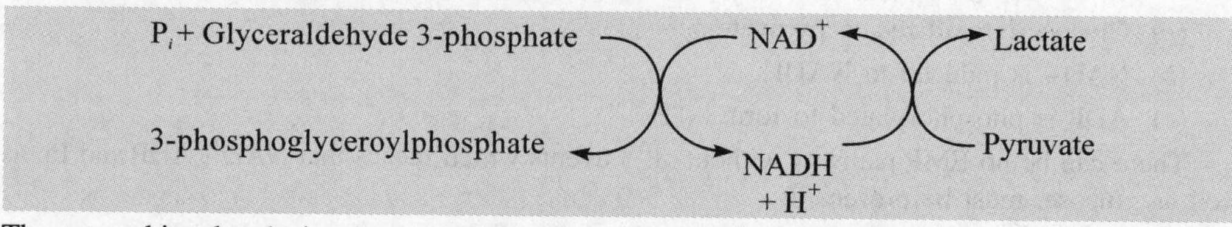

The anaerobic glycolysis, thus, results in the accumulation of 2 moles of lactate per mole of glucose utilized. *Lactate is one of the blind alleys or "dead-end streets" in metabolism and, once formed, it can only be reconverted to pyruvate.* The reconversion is, however, accomplished in the liver cells into which lactate is transported by circulation from muscle cells.

Lactate dehydrogenase (MW = 140,000) is a tetramer consisting of 4 subunits which are of 2 types. These are designated as M (for skeletal muscle) and H (for heart muscle) and differ in sequence. The skeletal muscle contains LDH of M_4 type mainly whereas heart muscle LDH is largely H_4. However, various tissues contain all possible hybrids, M_1H_3, M_2H_2 and M_3H_1. The exact amount of each of these 5 isozyme forms, as they are called, differs with the tissue. The various isozyme forms differ with respect to their K_m value for pyruvate, their turnover numbers or V_{max} and the degree of their allosteric inhibition by pyruvate. The H_4 has a low K_m for pyruvate and is strongly inhibited by pyruvate. It is, thus, better adapted for a more highly aerobic organ which removes lactate and oxidizes it to pyruvate mainly in the mitochondria. In contrast, the M_4 has a higher K_m for pyruvate and is not inhibited by pyruvate. Hence, it is useful in a tissue in which a more anaerobic environment may predominate as in skeletal muscle. *M_4 is catalytically more active than H_4.* Lactate dehydrogenase is characteristically inhibited by oxamate.

The *net equation* for anaerobic glycolysis in muscles and lactate fermentation in some microbes would then be:

$$\text{Glucose} + 2\ \text{Pi} + 2\ \text{ADP} \longrightarrow 2\ \text{lactate} + 2\ \text{ATP} + 2\ \text{H+} + 2\ H_2O$$

No oxygen is consumed in anaerobic glycolysis. Two steps involve oxidation-reduction, the oxidation of glyceraldehyde 3-phosphate and the reduction of pyruvate to lactate. NAD^+ participates in both reactions. Hence, the two cancel out and there is no net oxidation or reduction.

In anaerobic glycolysis, however, there occurs no net change in the oxidation state of carbon. This becomes evident by comparing the empirical formula of glucose ($C_6H_{12}O_6$) with that of lactic acid ($C_3H_6O_3$). The ratio of C to H to O is the same (*i.e.*, 1 : 2 : 1) for both, showing that no net oxidation of carbon has occurred. Nevertheless, some of the energy of glucose is extracted by anaerobic glycolysis, sufficient to give a net gain of 2 ATP per glucose utilized. This is equivalent to an estimated 15 kcal. Since some 56 kcal per mole are produced when glucose is degraded to lactate under standard conditions, the overall efficiency of glycolysis is 15/56 × 100 or approximately 25%– a rather high figure!

ALCOHOLIC FERMENTATION

In yeast and other microorganisms, the reactions of glycolysis up to pyruvate formation are identical to those described for anaerobic glycolysis and the difference occurs only in its terminal steps. In contrast to animals, which utilize lactate dehydrogenase reaction for the reoxidation of NADH to generate NAD^+, the yeast cells utilize two enzymatic reactions for the purpose, as lactate dehydrogenase is not found in them:

(*a*) In the first step, the pyruvate resulting from glucose breakdown is decarboxylated by the action of *pyruvate decarboxylase* (= *2-oxo-acid carboxylase*) to produce acetaldehyde and carbon dioxide. This is an irreversible reaction and does not involve the net oxidation of pyruvate.

$$\underset{\textbf{Pyruvate}}{CH_3-\overset{\displaystyle O}{\overset{\|}{C}}-COO^-} + H^+ \xrightarrow[Mg^{2+}]{\text{Pyruvate decarboxylase}} \underset{\substack{\textbf{Acetaldehyde}\\ \textbf{(= Ethanal)}}}{CH_3-\overset{\displaystyle O}{\overset{\|}{C}}-H} + CO_2$$

Pyruvate decarboxylase (E.C. 4.1.1.1) requires the usual Mg^{2+} for activity although certain trivalent ions such as Al^{3+}, Fe^{3+} may also satisfy the need. It has a tightly bound coenzyme, thiamine pyrophosphate, TPP (= cocarboxylase). TPP functions as a transient carrier of acetaldehyde group. In fact, the carboxyl group of pyruvate is lost as CO_2 and the rest of the molecule (sometimes referred to as *active acetaldehyde*) is simultaneouly transferred to the position 2 of the thiazole ring of TPP, to yield its hydroxyethyl derivative. This is unstable since the hydroxyethyl group quickly dissociates from the coenzyme to yield free acetaldehyde.

$$\text{Pyruvate} + H_2O + \text{TPP-E} \longrightarrow \alpha\text{-hydroxyethyl-TPP-E} + HCO_3^-$$

$$\alpha\text{-hydroxyethyl-TPP-E} \longrightarrow \text{Acetaldehyde} + \text{TPP-E}$$

(*b*) In the second and final step, acetaldehyde is reduced to ethanol by NADH, derived from glyceraldehyde 3-phosphate dehydrogenase reaction, *i.e.*, Step 6 of Glycolysis, through the catalytic action of *alcohol dehydrogenase* (abbreviated as ADH). This is a reversible oxidation-reduction reaction.

$$\underset{\textbf{Acetaldehyde}}{CH_3-\overset{\displaystyle O}{\overset{\|}{C}}-H} + NADH + H^+ \xrightleftharpoons{\text{Alcohol dehydrogenase}} \underset{\substack{\textbf{Ethyl alcohol}\\ \textbf{(= Ethanol)}}}{CH_3-CH_2OH} + NAD^+$$

The yeast **alcohol dehydrogenase** (E.C.1.1.1.1; MW = 151,000) is an NAD-dependent enzyme and contains a zinc ion at its active site.

The conversion of glucose into ethanol is called alcoholic fermentation. Thus, ethanol and CO_2, instead of lactate, are the end products of this process. Alcoholic fermentation is prevented by $NaHSO_3$ which combines with acetaldehyde to give a bisulfite addition compound that is not a substrate for alcohol dehydrogenase.

The *net equation* for alcoholic fermentation would then be:

$$\text{Glucose} + 2\ \text{Pi} + 2\ \text{ADP} \longrightarrow 2\ \text{Ethanol} + 2\ CO_2 + 2\ \text{ATP} + 2\ H_2O$$

It may be pointed out that NAD^+ and NADH do not appear in this equation because NAD^+ generated during reduction of acetaldehyde is used in the reaction catalyzed by glyceraldehyde

3-phosphate dehydrogenase, *i.e.*, Step 6 of Glycolysis. Thus, there is no net oxidation-reduction, as in anaerobic glycolysis. Another noteworthy point of alcoholic fermentation is that there is no net change in the ratio of hydrogen to carbon atoms when D-glucose (H/C ratio = 12/6 = 2) is fermented to 2 molecule each of ethanol and CO_2 (H/C ratio = 12/6 = 2). In fact, *in all anaerobic fermentations, the H/C ratio of the reactants and products remains the same.*

GLUCONEOGENESIS

(= Neoglucogenesis or **Neoglycogenesis)**

The central role of **glucose** in metabolism arose early in evolution, and this sugar remains the nearly universal fuel and building block in modern organisms, from microbes to humans. In mammals, some tissues depend almost completely on glucose for their metabolic energy. The glucose requirements of the human brain are relatively enormous– about 120 g per day out of the total body's need of about 160 g. And the amount of glucose that can be generated from the body's glycogen reserves at any time is about 190 g, and the total amount of glucose present in body fluids is about 20 g. Thus, the readily available glucose reserves amount to about one day's supply. During periods of fasting for more than one day, glucose must be formed from other precursors. The same is true during intense exertion, for example during a marathon run, when the glucose reserves are rapidly depleted. For these times, organisms need a method of synthesizing glucose from noncarbohydrate precursors. This is accomplished by a pathway called **gluconeogenesis**– literally meaning, "formation of new sugar glucose or glucose". *Gluconeogenesis is defined as the biosynthesis of carbohydrate from 3-C or 4-C precursors, generally noncarbohydrate in nature.*

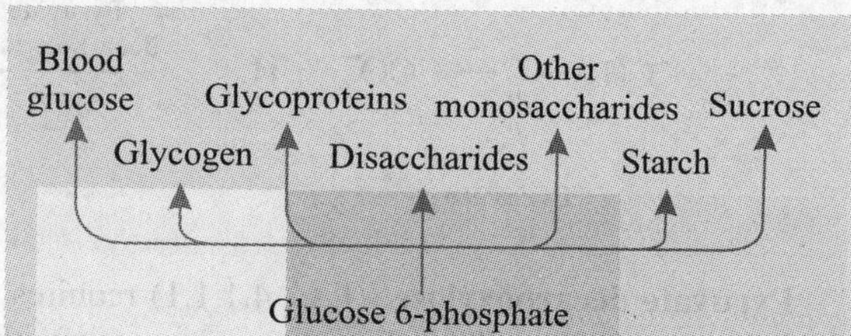

Fig. 21–12. Carbohydrate synthesis from simple precursors

The pathway from phosphoenolpyruvate to glucose 6-phosphate is common to the biosynthetic conversion of many different precursors of carbohydrates in animals and plants. Plants and photosynthetic bacteria are uniquely able to convert CO_2 to carbohydrates. The fatty acids released during lipid breakdown are mostly converted to acetyl-CoA and cannot be used for carbohydrate synthesis, except in organisms that have a functioning glyoxylate cycle.

Gluconeogenesis occurs in all animals, plants, fungi and microbes. The reactions are essentially the same in all tissues and all species. The principal precursors (or substrates) of gluconeogenesis (Fig. 21–12) are: *lactate*, produced primarily from glycolysis in skeletal muscle and erythrocytes; *amino acids*, generated from dietary protein or from the breakdown of muscle protein during starvation; the specific amino acid *alanine*, produced in muscle through the glucose–alanine cycle; *propionate*, derived from the breakdown of some fatty acids and amino acids; and *glycerol*, derived from the catabolism of fats

Gluconeogenesis occurs primarily in the cytosol, although some precursors are generated in mitochondria and must be transported to the cytosol to be utilized. In animals, especially mammals, gluconeogenesis takes place mainly in the liver and to a lesser extent in kidney (renal) cortex (Fig. 21–13). In fact, gluconeogenesis in the liver and kidney helps to maintain the glucose level in the blood so that brain and muscle can extract sufficient glucose from it to meet their metabolic demands. The major fates of glucose formed by gluconeogenesis are catabolism by nervous tissue and utilization by skeletal muscle. In addition, glucose is the primary precursor for all other carbohydrates, including amino sugars, complex polysaccharides, and the carbohydrate components of glycoproteins and glycolipids. In plant seedlings, stored fats and proteins are converted, *via* paths that include gluconeogenesis, to the disaccharide sucrose for transport throughout the developing plant. Glucose and its derivatives are precursors for the synthesis of plant cell walls, nucleotides and coenzymes, and a variety of other essential metabolites. In many microorganisms, gluconeogenesis starts from simple organic compounds to 2 or 3 carbons, such as acetate, lactate and propionate, in their growth medium.

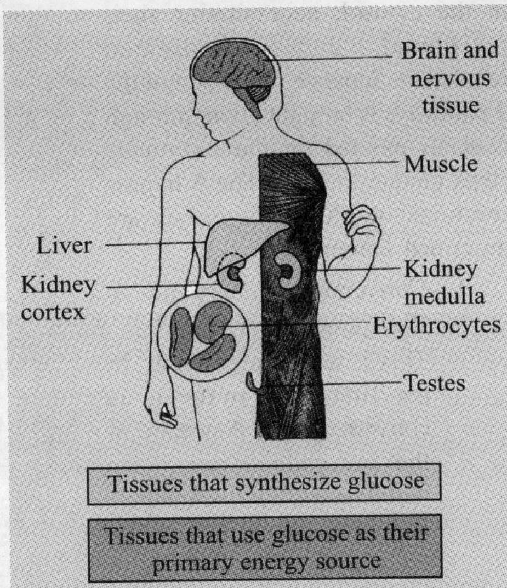

Fig. 21–13. Synthesis and use of glucose in the human body

Liver and kidney cortex are the primary gluconeogenic tissues. Brain, skeletal muscle, kidney medulla, erythrocytes and testes use glucose as their sole or primary energy source, but they lack the enzymes needed to synthesize it.

Glycogenolysis, that is, formation of glucose or glucose 6-phosphate from glycogen, should be differentiated from gluconeogenesis; glycogenolysis refers to:

$$\text{Glycogen or (glucose)}_n \longrightarrow n \text{ molecules of glucose}$$

and thus does not correspond to *de novo* synthesis of glucose, the hallmark of the process of gluconeogenesis.

In glycolysis, glucose is converted into pyruvate; in gluconeogenesis, pyruvate is converted into glucose. Yet, the two are not identical pathways running in opposite direction, although they do share several steps (Fig. 21–14); 7 of the 10 enzymatic reactions of gluconeogenesis are the reverse of glycolytic reactions. However, 3 reactions of glycolysis are essentially irreversible *in vivo* and cannot be used in gluconeogenesis; these are:

1. the conversion of glucose to glucose 6-phosphate by hexokinase (*Step 1*; refer page 543),
2. the phosphorylation of fructose 6-phosphate to fructose 1, 6-bisphosphate by phosphofructokinase-1 (*Step 3*; refer page 545), and
3. the conversion of phosphoenolpyruvate to pyruvate by pyruvate kinase (*Step 10*; refer page 553).

In cells, these 3 reactions are characterized by a large negative free-energy change, ΔG, whereas other glycolytic reactions have a ΔG near 0. In gluconeogenesis, the 3 irreversible steps are bypassed by a separate set of enzymes, catalyzing reactions that are sufficiently exergonic to be effectively irreversible in the direction of glucose synthesis. Thus, both glycolysis and gluconeogenesis are irreversible processes in the cells. In animals, both pathways occur largely in the cytosol, necessitating their reciprocal and coordinated regulation. Separate regulation of the 2 pathways is brought about through controls exerted on the enzymatic steps unique to each. The 3 bypass reactions of gluconeo-genesis are described hereunder:

1. **Conversion of Pyruvate to Phosphoenolpyruvate**

   This is a 2-step reaction. In the first step, pyruvate is converted to oxaloacetate in the presence of *pyruvate carboxylase*, a mitochondrial enzyme that requires the coenzyme biotin; the reaction occurs at the expense of a molecule of ATP, as follows:

$$\text{Pyruvate} + CO_2 + \text{ATP} + H_2O \longrightarrow \text{Oxaloacetate} + \text{ADP} + P_i + 2H^+ \quad ...\Delta G^{0'} = -2.1 \text{ kJ/mol}$$

   This is one of the anaplerotic reactions used to maintain levels of citric acid cycle intermediates. Pyruvate carboxylase generates oxaloacetate (OAA) in the mitochondrial matrix, where it can be oxidized in the citric acid cycle. To be used for gluconeogenesis, OAA must move out of the mitochondrion to the cytosol, where the remaining reactions of the pathway occur.

Pyruvate carboxylase (Fig. 21–15) is of special interest because of its structural, catalytic and allosteric properties. The N-terminal 300 to 350 amino acids form an ATP-grasp domain, which is widely used *ATP-activating domain* (see nucleotide biosynthesis). The C-terminal 80

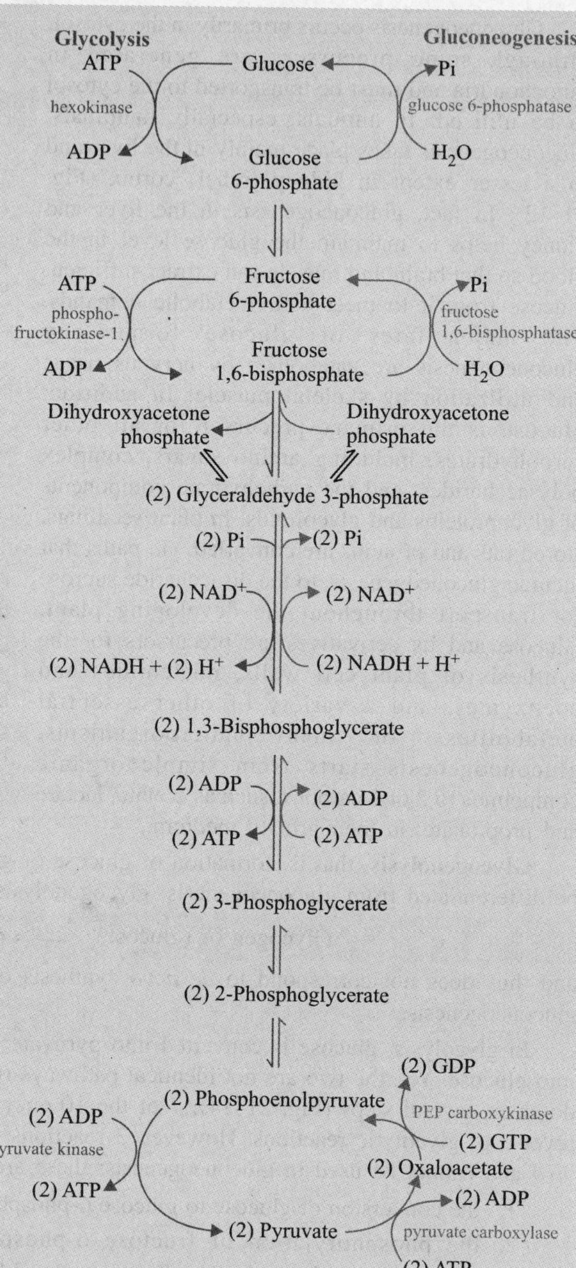

Fig. 21–14. The opposing pathways of glycolysis and gluconeogenesis

The reactions of glycolysis are shown (on the left) in *red*; the opposing pathway of gluconeogenesis is shown (on the right) in *blue*. The major sites of regulation of gluconeogenesis shown here are discussed later in this chapter.

Fig. 21–15. Domain structure of pyruvate carboxylase

The ATP-grasp domain activates HCO_3^- and transfers CO_2 to the biotin-binding domain. From there, the CO_2 is transferred to pyruvate, generated in the central domain.

amino acids constitute a *biotin-binding domain* (see fatty acid synthesis). Biotin is a covalently-attached prosthetic group, which serves as a carrier of activated CO_2. The caboxylate group of biotin is linked to the E-amino group of a specific lysine residue by an amide bond (Fig. 21–16). Note that biotin is attached to pyruvate carboxylase by a long, flexible chain.

Once in the cytosol, OAA is simultaneously decarboxylated and phosphorylated to yield phosphoenolpyruvate in the presence of *phosphoenolpyruvate carboxykinase* (PEPCK); the reaction occurs at the expense of a molecule of GTP, as follows:

$$\text{Oxaloacetate} + \text{GTP} \rightleftharpoons \text{Phosphoenolpyruvate} + \text{GDP} + CO_2$$

$$...\Delta G^{0'} = +\ 2.9 \text{ kJ/mol}$$

Note the use of GTP, rather than ATP, as an energy donor. Note also that the CO_2 that was fixed by pyruvate carboxylase is released in this reaction, so that no net fixation of CO_2 occurs. The PEPCK reaction requires Mg^{2+} or Mn^{2+} and is readily reversible. In the reaction, the carboxyl group formed from the transferred CO_2 provides electrons to facilitate O–P bond formation:

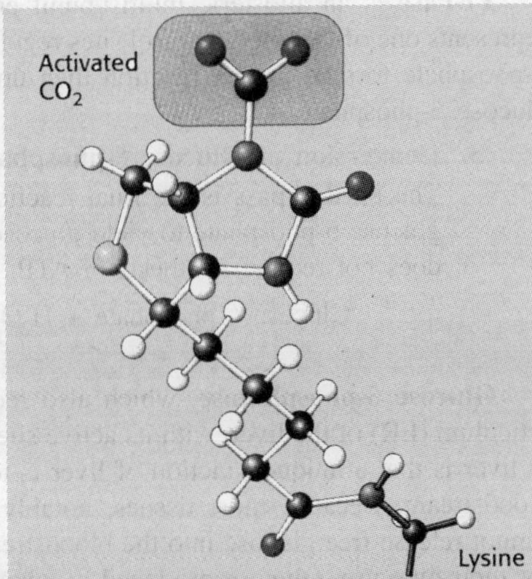

Fig. 21–16. Structure of carboxybiotin

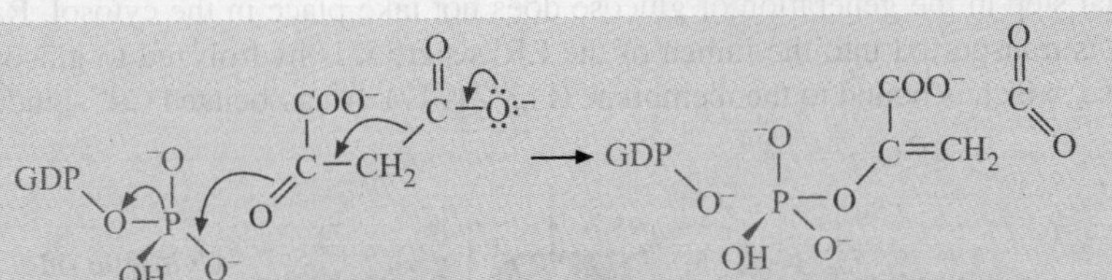

PEPCK is localized to the cytosol of most animal cells, but in some species it is found in both cytosol and mitochondria. The cytosolic form is more important in gluconeogenesis, as shown by the fact that its intracellular concentration is regulated by hormonal conditions known to control gluconeogenesis.

The overall equation for the bypass of pyruvate kinase reaction is:

$$\text{Pyruvate} + \text{ATP} + \text{GTP} + H_2O \rightleftharpoons \text{Phosphoenolpyruvate} + \text{ADP} + \text{GDP} + P_i + 2\ H^+$$

$$...\Delta G^{0'} = +\ 20.8 \text{ kJ/mol}$$

The $\Delta G^{0'}$ for the 2 reactions combined is slightly positive. However, under intracellular conditions the sequence is quite exergonic, with $\Delta G'$ of about –25 kJ/mol. As shown in the *summary equation*,

2 high-energy phosphates (ATP and GTP) must be invested for the synthesis of one super-high-energy phosphoenolpyruvate. After this bypass, PEP is converted into fructose 1, 6-bisphosphate by glycolytic enzymes acting in reverse.

2. Conversion of Fructose 1, 6-bisphosphate to Fructose 6-phosphate

The second glycolytic reaction that cannot participate in gluconeogenesis is the phosphorylation of fructose 6-phosphate by phosphofructokinase-1; this is because this reaction is highly exergonic and therefore irreversible in intact cells. A bypass reaction in gluconeogenesis is the generation of fructose 6-phosphate from fructose 1,6-bisphosphate. This simple hydrolytic reaction is catalyzed by *fructose 1,6-bisphosphatase* (FBPase), which promotes the essentially-irreversible hydrolysis of the C-1 phosphate (*not* phosphoryl group transfer to ADP):

$$\text{Fructose 1,6-bisphosphate} + H_2O \longrightarrow \text{Fructose 6-phosphate} + P_i$$

$$...\Delta G^{0'} = -16.3 \text{ kJ/mol}$$

FBPase is an allosteric multisubunit enzyme and requires Mg^{2+} for activity. This enzyme represents one of the major control sites regulating the overall gluconeogenic pathway. The fructose 6-phosphate formed in this reaction then undergoes isomerization by phosphoglucoisomerase to glucose 6-phosphate.

3. Conversion of Glucose 6-phosphate to Glucose

The third bypass is the final reaction of gluconeogenesis, *i.e.,* the dephosphorylation of glucose 6-phosphate to yield glucose. This reaction, catalyzed by *glucose 6-phosphatase*, does not require synthesis of ATP; it is a simple hydrolysis of a phosphate ester:

$$\text{Glucose 6-phosphate} + H_2O \longrightarrow \text{Glucose} + P_i$$

$$...\Delta G^{0'} = -13.8 \text{ kJ/mol}$$

Glucose 6-phosphatase, which also requires Mg^{2+}, is found primarily in the endoplasmic reticulum (ER) of the liver, with its active site on the cytosolic side. The significance of its location in liver is that a unique function of liver is to synthesize glucose for export to the tissues *via* the bloodstream. Because most tissues, notably brain and skeletal muscle, lack this enzyme, they cannot release free glucose into the bloodstream, although they do synthesize glucose 6-phosphate for their own use. Glucose produced by gluconeogenesis in the liver or kidney or ingested in the diet is delivered to brain and muscle through the bloodstream.

This final step in the generation of glucose does not take place in the cytosol. Rather, glucose 6-phosphate is transported into the lumen of the ER, where it is hydrolyzed to glucose by glucose 6-phosphatase, which is bound to the membrane (Fig. 21–17). An associated Ca^{2+}-binding stabilizing

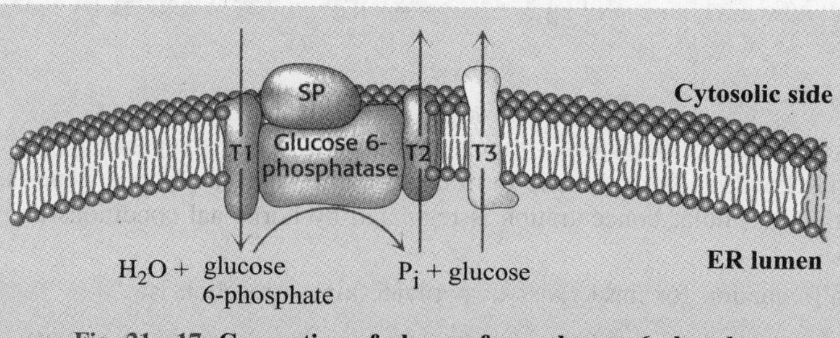

Fig. 21– 17. Generation of glucose from glucose 6-phosphate

Several ER proteins play a role in the generation of glucose from glucose 6-phosphate. T1 transports glucose 6-phosphate into the lumen of the ER, whereas T2 and T3 transport Pi and glucose, respectively, back into the cytosol. Glucose 6-phosphatase is stabilized by a Ca^{2+}-binding protein (SP).

(Courtesy: Buchell A and Waddel ID, 1991)

protein is essential for phosphatase activity. Glucose and Pi are then shuttled back to the cytosol by a pair of transporters. The glucose transporter in the ER membrane is like those found in the plasma membrane. It is striking that 5 proteins are needed to transform cytosolic glucose 6-phosphate into glucose.

STOICHIOMETRY OF GLUCONEOGENESIS

The stoichiometry of gluconeogenesis is:

$$2 \text{ Pyruvate} + 4 \text{ ATP} + 2 \text{ GTP} + 2 \text{ NADH} + 6 \text{ H}_2\text{O} \longrightarrow \text{Glucose} + 4 \text{ ADP} + 2 \text{ GDP} + 6 \text{ P}_i + 2 \text{ NAD}^+ + 2 \text{ H}^+$$

$$...\Delta G^{0'} = -38 \text{ kJ/mol}$$

It is a well known fact that catabolic pathways generate energy, whereas anabolic pathways carry an energy cost, *i.e.,* consume energy. The overall conversion of 2 moles of pyruvate to 1 mole of glucose during gluconeogenesis is quite exergonic, as shown in Table 21–3. The $\Delta G^{0'}$ for the overall process is about –47.6 kJ/mol. But the synthesis of glucose is costly to the cell in an energetic sense. For each molecule of glucose formed from pyruvate, six high-energy phosphate groups are consumed (4 ATPs and 2 GTPs). In addition, two moles of NADH are also required, which are energetically equivalent to 6 more ATPs (because mitochondrial oxidation of 1 mole of NADH generates 3 moles of ATP). Evidently, the equation given above for gluconeogenesis is not simply the reverse of the equation for glycolysis, which requires only 2 moles of ATP.

Table 21–3. Sequential reactions in gluconeogenesis from pyruvate to glucose

Reaction		$\Delta G^{0'}$ *(kJ/Mol)*
Pyruvate + CO_2 + ATP + H_2O ⟶ Oxaloacetate + ADP + P_i + 2 H^+	× 2	– 4.2
Oxaloacetate + GTP ⇌ Phosphoenolpyruvate + CO_2 + GDP	× 2	+5.8
Phosphoenolpyruvate + H_2O ⇌ 2-phosphoglycerate	× 2	–3.4
2-Phosphoglycerate ⇌ 3-phosphoglycerate	× 2	–9.2
3-Phosphoglycerate + ATP ⇌ 1,3-bisphosphoglycerate + ADP	× 2	+37.6
1,3-bisphosphoglycerate + NADH + H^+ ⇌ Glyceraldehyde-3-phosphate + NAD^+ + P_i	× 2	–12.6
Glyceraldehyde-3-phosphate ⇌ Dihydroxyacetone phosphate		–7.6
Glyceraldehyde-3-phosphate + dihydroxyacetone phosphate ⇌ Fructose-1,6-bisphosphate		–23.9
Fructose-1,6-bisphosphate + H_2O ⟶ Fructose-6-phosphate + P_i		–16.3
Fructose-6-phosphate ⇌ Glucose-6-phosphate		–1.7
Glucose-6-phosphate + H_2O ⟶ Glucose + P_i		–12.1
Pyruvate + 4 ATP + 2 GTP + 2 NADH + 6 H_2O ⟶ Glucose + 4 ADP + 2 GDP + 6 P_i + 2 NAD^+ + 2 H^+		–47.6

Note: The reactions in boldface type are those that bypass irreversible glycolytic reactions; the remaining reactions are reversible reactions of glycolysis. The first six reactions are multiplied by 2, because 2 three-carbon precursors are required to make one molecule of glucose.

By contrast, if glycolysis could operate *in reverse*, the net equation would show far less energy input, *i.e.,* 2 moles of NADH and 2 moles of high-energy phosphate:

$$2 \text{ Pyruvate} + 2 \text{ ATP} + 2 \text{ NADH} + 2 H_2O \longrightarrow \text{Glucose} + 2 \text{ ADP} + 2 P_i + 2 \text{ NAD}^+ \quad ...\Delta G^{0'} = +84 \text{ kJ/mol}$$

This process would be highly endergonic, however. Therefore, it is evident that the investment of 4 extra high-energy phosphate bond is essential if the net synthesis of glucose is to occur as an irreversible process. In other words, much of this high-energy cost is necessary to ensure the irreversibility of gluconeogenesis. This validates the fact that *gluconeogenesis is energetically expensive but essential process.*

COORDINATED REGULATION OF GLYCOLYSIS AND GLUCONEOGENESIS

Regulation of gluconeogenesis is crucial for many physiological functions, but especially so for the proper functioning of nervous tissue. Although other organs can use a variety of energy sources, the well-being of the central nervous system (CNS) demands maintenance of blood glucose levels within narrow limits. Gluconeogenic control is important also as an animal adjusts to muscular exertion or to cycles of feeding and fasting. The flux through the pathway must increase or decrease, based on the availability of lactate produced by the muscles, of glucose from the diet, or of other gluconeogenic precursors.

Gluconeogenesis is also controlled in large part by the diet. Animals fed a high-carbohydrate diet show low rates of gluconeogenesis, whereas fasted animals or those fed carbohydrate-poor diets show enhanced rates of gluconeogenesis. These hormonal effects, mediated primarily through insulin and glucagon, involve both control of the synthesis of phosphoenolpyruvate carboxykinase and regulation effected through control of cyclic AMP levels.

In mammals, gluconeogenesis occurs primarily in the liver, where its role is to provide glucose for export to other tissues when glycogen stores are exhausted. Gluconeogenesis employs most of the enzymes that act in glucolysis, but it is not simply the reversal of glycolysis. Seven of the 10 glycolytic reactions are freely reversible, and the enzymes that catalyse these reactions also function in gluconeogenesis (Fig. 21–18). Three reactions of glycolysis are so exergonic as to be essentially irreversible: those catalyzed by hexokinase, phosphofructokinase-1 (PFK-1), and pyruvate kinase. All these 3 reactions have a large, negative $\Delta G'$. Gluconeogenesis uses detours (longer routes) around each of these irreversible steps; for example, the conversion of fructose 1,6-bisphosphate to fructose 6-phosphate is catalyzed by fructose 1,6-bisphosphatase (FBPase-1). Note that each of these bypass reactions also has a large negative $\Delta G'$.

At each of these 3 points where glycolytic reactions are bypassed by alternative gluconeogenic reactions, simultaneous operation of both pathways would consume ATP without performing any chemical or biological work. As an example, PFK-1 and FBPase-1 catalyze opposing reactions:

$$\text{ATP} + \text{fructose 6-phosphate} \xrightarrow{\text{PFK-1}} \text{ADP} + \text{fructose1,6-bisphosphate}$$

$$\text{Fructose 1,6-bisphosphate} + H_2O \xrightarrow{\text{FBPase-1}} \text{Fructose 6-phosphate} + P_i$$

The *sum* of these 2 reactions is:

$$\text{ATP} + H_2O \longrightarrow \text{ADP} + P_i + \text{Heat}$$

which means hydrolysis of ATP without any useful metabolic work being done. Evidently, if these 2 reactions were allowed to proceed simultaneously at a high rate in the same cell, a large amount of chemical energy would be dissipated as heat. This cyclic process has been called a **futile cycle** as this cycling was regarded uneconomical and as an imperfection in metabolic control. However, such cycles may provide advantages for controlling pathways, and the term

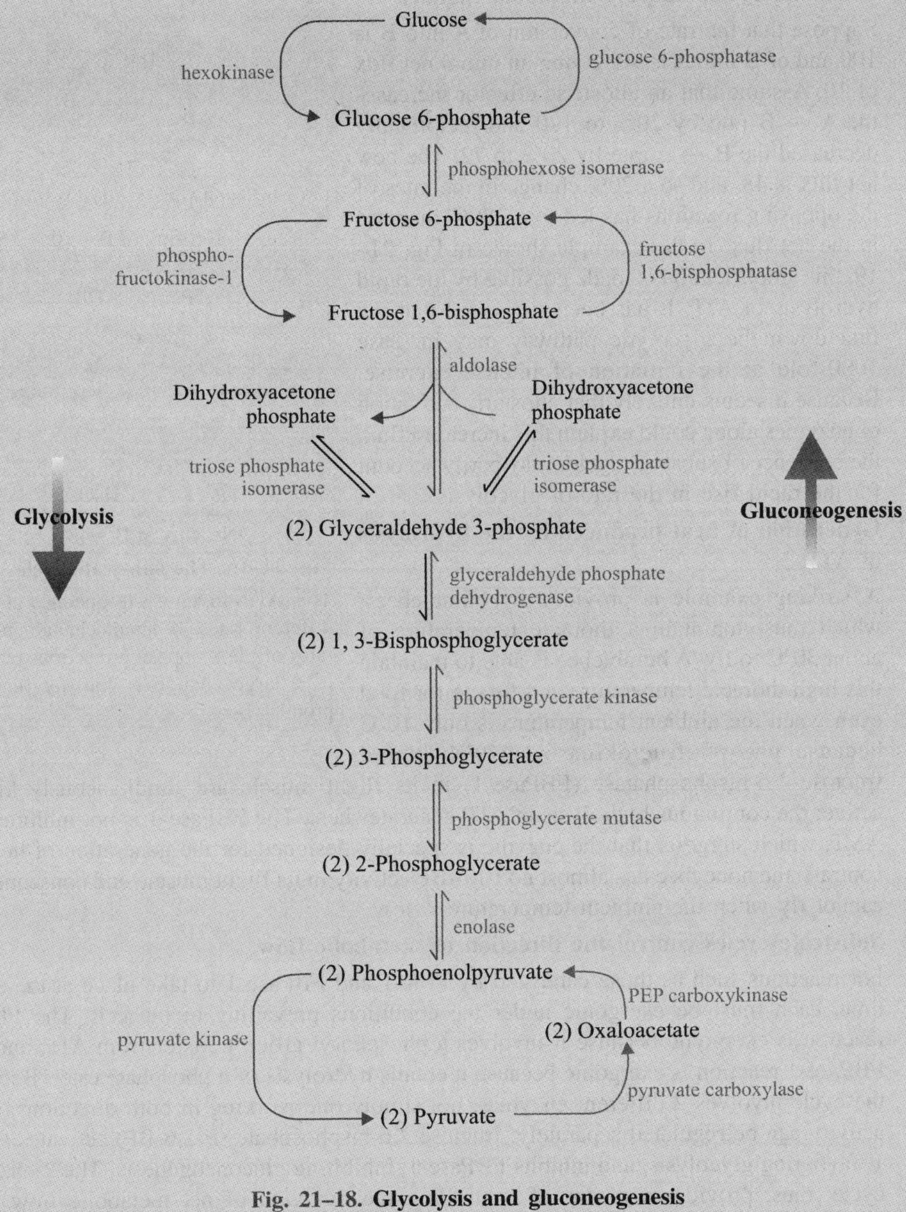

Fig. 21–18. Glycolysis and gluconeogenesis

The opposing pathways of glycolysis (*pink*) and gluconeogenesis (*blue*) are shown in rat liver. Three steps are catalyzed by different enzymes in gluconeogenesis (the "bypass reactions") and glycolysis; seven steps are catalyzed by the same enzymes in the 2 pathways. *Cofactors have been omitted for simplicity*.

substrate cycle is a better alternative name. Similar substrate cycles also occur with the other 2 sets of bypass reactions of gluconeogenesis (see Fig. 21–18).

Despite extraordinary circumstances, it now seems likely that substrate cycles are biologically important for the following reasons:

(a) Substrate cycles amplify metabolic signals.

Suppose that the rate of conversion of A into B is 100 and of B into A is 90, giving an initial net flux of 10. Assume that an allosteric effector increases the A → B rate by 20% to 120 and reciprocally decreased the B → A rate by 20% to 72. The new net flux is 48, and so a 20% change in the rates of the opposing reactions has led to a 380% increase in the net flux. In the example shown in Fig. 21–19, this amplification is made possible by the rapid hydrolysis of ATP. It has been suggested that the flux down the glycolytic pathway may increase 1000-fold at the initiation of intense exercise. Because it seems unlikely that allosteric activation of enzymes alone could explain this increased flux, the existence of substrate cycles may partly account for the rapid rise in the rate of glycolysis.

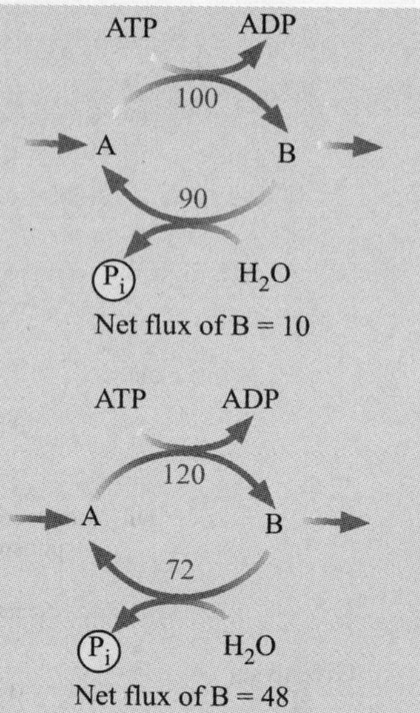

Fig. 21–19. The substrate cycle

This ATP-driven cycle operates at two different rates. A small change in the rates of the 2 opposing reactions results in a large change in the *net* flux of product B.

(b) Generation of heat produced by the hydrolysis of ATP.

A striking example is provided by bumblebees, which must maintain a thoracic temperature of about 30°C to fly. A bumblebee is able to maintain this high thoracic temperature and forage for food even when the ambient temperature is only 10°C because phosphofructokinase-1 (PFK-1) and fructose 1,6-bisphosphatase (FBPase-1) in its flight muscle are simultaneously highly active; the continuous hydrolysis of ATP generates heat. The FBPase-1 is not inhibited by AMP, which suggests that the enzyme is specially designed for the generation of heat. It contrast, the honeybee has almost no F1,6-BP activity in its flight muscle and consequently cannot fly when the ambient temperature is low.

(c) Substrate cycles control the direction of metabolic flow.

For reactions such as those catalyzed by PFK-1 and FBPase-1 to take place at the same time, each must be exergonic under the conditions prevailing in the cell. The PFK-1 reaction is exergonic because it involves a phosphoryl group transfer from ATP, and the FBPase-1 reaction is exergonic because it entails hydrolysis of a phosphate ester. Because the cycle involves 2 different enzymes, not simply one working in both directions, each activity can be regulated separately: fructose 2,6-bisphosphate (F-2,6-BP) activates PFK-1, favouring glycolysis, and inhibits FBPase-1, inhibiting gluconeogenesis. The 2-enzyme cycle, thus, provides a means of controlling the *direction* of net metabolic flow. The apparent energetic disadvantage of the "futile" cycle is evidently outweighed by the advantage of allowing this type of control of pathway direction.

Reciprocal Regulation of Gluconeogenesis and Glycolysis

Gluconeogenesis and glycolysis both proceed largely in the cytosol. Because gluconeogenesis synthesizes glucose and glycolysis degrades glucose, it is evident that *gluconeogenesis and glycolysis must be controlled in a reciprocal way*. In other words, intracellular conditions that activate one pathway tend to inhibit the other pathway. If it were not for reciprocal control, glycolysis and gluconeogenesis would operate together as a giant futile cycle. Reciprocal regulation is related in large part to the adenylate energy charge. Conditions of low energy charge tend to activate the rate-controlling steps in glycolysis while inhibiting carbon flux through gluconeogenesis. Conversely, gluconeogenesis is stimulated at high energy charge, under conditions where catabolic flux rates are adequate to maintain sufficient ATP levels.

Recall that glycolysis is controlled primarily by regulation of the 3 strongly exergonic reactions – those catalyzed by hexokinase (or glucokinase in liver), phosphofructokinase, and pyruvate kinase. The opposing reactions in gluconeogenesis– those catalyzed by glucose-6-phosphatase, fructose 1, 6-bisphosphatase, and the combination of pyruvate carboxylase and phosphoenolpyruvate carboxykinase – are also strongly exergonic and represent the chief targets for control of this pathway. In other words, the 3 substrate cycles that differentiate glycolysis from gluconeogenesis (see Fig. 21–14) represent the primary sites for reciprocal regulation of these pathways. Fig. 21–20 identifies the major allosteric activators and inhibitors of the key exergonic reactions in glycolysis and gluconeogenesis.

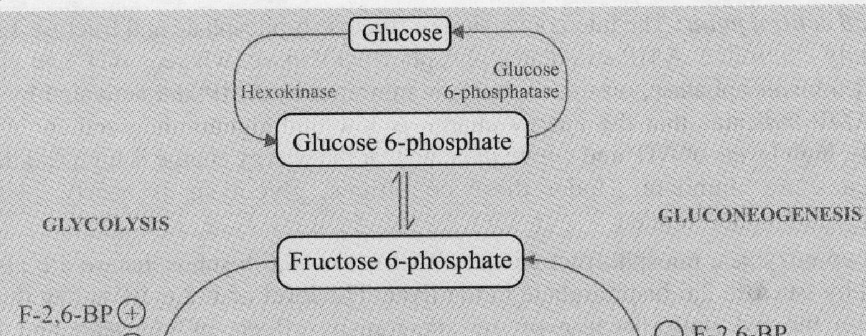

Fig. 21– 20. Reciprocal regulation of gluconeogenesis and glycolysis in the liver

The first control point is the inhibition of pyruvate kinase by phosphorylation during starvation. The second control point, the level of 2,6-bisphosphate is high in the fed state and low in starvation.

***First control point*:** The interconversion of phosphoenolpyruvate and pyruvate is precisely regulated. The enzyme pyruvate kinase is controlled by allosteric effectors and by phosphorylation. High levels of ATP and the amino acid alanine, which signal that the energy charge is high and that the building blocks are abundant, inhibit the enzyme in the liver. Conversely, pyruvate carboxylase, which catalyzes the first step in gluconeogenesis from pyruvate, is activated by acetyl-CoA and inhibited by ADP. Likewise, ADP inhibits phosphoenolpyruvate carboxykinase. Hence, gluconeogenesis is favoured when the cell is rich in biosynthetic precursors and ATP.

The energy charge, however, does control one of the 2 isoenzyme forms of pyruvate kinase. In liver and in other gluconeogenic tissues, the so-called L form of pyruvate kinase predominates, whereas the M form is found largely in muscle. The L form is inhibited both by ATP and alanine. This relationship allows the inhibition of glycolysis, with consequent activation of gluconeogenesis, specifically in gluconeogenic tissues, when ample energy and substrates are available. Control at the pyruvate kinase step allows conservation of high-energy phosphate in the phosphoenolpyruvate molecule.

Acetyl-CoA can also be seen as a reciprocal regulator of glycolysis and gluconeogenesis, acting on the enzymes that interconvert pyruvate and phosphoenolpyruvate. Acetyl-CoA is a required activator of pyruvate carboxylase and an inhibitor of pyruvate kinase and pyruvate dehydrogenase complex. It can thus signal, when its levels rise, that adequate substrates are available to provide energy through the citric acid cycle and that more carbon can be shuttled into gluconeogenesis and ultimately stored as glycogen.

***Second control point*:** The interconversion of fructose 6-phosphate and fructose 1,6-bisphosphate is stringently controlled. AMP stimulates phosphofructokinase, whereas ATP and citrate inhibit it. Fructose 1,6-bisphosphatase, on the contrary, is inhibited by AMP and activated by citrate. A high level of AMP indicates that the energy charge is low and signals the need for ATP generation. Conversely, high levels of ATP and citrate indicate that the energy charge is high and that biosynthetic intermediates are abundant. Under these conditions, glycolysis is nearly switched off and gluconeogenesis is accelerated.

The two enzymes, phosphofructokinase and fructose 1,6-bisphosphatase are also reciprocally controlled by fructose 2,6-bisphosphate in the liver. The level of F-2,6-BP is low during starvation and high in the fed state, because of the antagonistic effects of glucagon and insulin on the production and degradation of this signal molecule. *Fructose 2,6-bisphosphate strongly stimulates phosphofructokinase and inhibits fructose 1,6-bisphosphatase*. Hence, glycolysis is accelerated and gluconeogenesis is diminished in the fed state. During starvation, gluconeogenesis predominates because the level of F-2,6-BP is very low. Glucose formed by the liver under these conditions is essential for the well-being and viability of the brain and muscle.

***Third control point*:** Glucose-6-phosphatase is not known to be allosterically controlled, but its K_m for glucose-6-phosphate is far higher than intracellular concentrations of this metabolite. Thus, intracellular activity is largely controlled in first-order fashion by the concentration of this substrate.

The interplay between glycolysis and gluconeogenesis is summarized in Fig. 21–21, which shows how these pathways help meet the energy needs of different cell types.

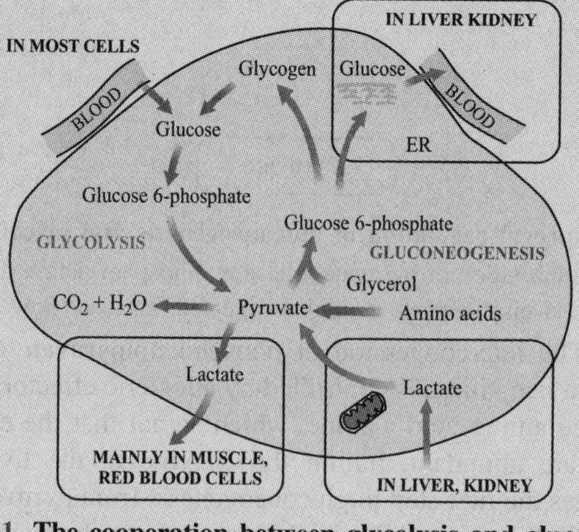

Fig. 21– 21. The cooperation between glycolysis and gluconeogenesis

Glycolysis and gluconeogenesis are coordinated, in a tissue-specific fashion, to ensure that the glucose-dependent energy needs of all cells are met

Ethanol Consumption and Gluconeogenesis

Although one can visualize of a pathway by which ethanol could be converted rather oxidized to glucose, ethanol is actually a poor gluconeogenic precursor. *In fact, ethanol strongly inhibits gluconeogenesis by liver.* It is oxidized primarily in liver, by alcohol dehydrogenase, with production of a large load of reducing equivalents (NADH) that must be transported into the mitochondria by the malate–aspartate shuttle. This excess NADH in the cytosol elevates the [NADH]/[NAD^+] ratio in the liver, which in turn shifts the equilibrium of lactate dehydrogenase, so that pyruvate tends to be reduced to lactate; it also shifts the equilibrium of cytosolic malate dehydrogenase, so that oxaloacetate tends to be reduced to malate.

$$\text{Ethanol} + NAD^+ \rightleftharpoons \text{Acetaldehyde} + NADH + H^+$$

$$\text{Pyruvate} + NADH + H^+ \longrightarrow \text{Lactate} + NAD^+$$

$$[\textit{Sum:}\ \text{Ethanol} + \text{pyruvate} \longrightarrow \text{Acetaldehyde} + \text{lactate}]$$

$$\text{Oxaloacetate} + NADH + H^+ \longrightarrow \text{Malate} + NAD^+$$

$$[\textit{Sum:}\ \text{Ethanol} + \text{Oxaloacetate} \longrightarrow \text{Acetaldehyde} + \text{malate}]$$

Thus, both pyruvate and oxaloacetate become unavailable for glucose synthesis through gluconeogenesis.

The resultant hypoglycemia can affect the parts of the brain concerned with temperature regulation. This response, in turn, can lower the body temperature by as much as 2°C. Therefore, the time-tested practice of feeding brandy or whiskey to those rescued from cold or wet conditions is counterproductive. To be sure, alcohol creates a sense of warming through vasodilation, but this peripheral vasodilation causes further heat loss. From metabolic viewpoint, glucose would be far more effective in raising body temperature.

Evolutionary Relationship between Glycolysis and Gluconeogenesis

The metabolism of glucose originated in ancient times. Organisms living in early biosphere depended on the anaerobic generation of energy till appreciable amounts of oxygen began to accumulate about 2 billion year ago. The fact that glycolytic enzymes with similar properties do not have similar amino acid sequences also provides a clue as to how the pathway originated. Although there are 4 kinases and 2 isomerases in the pathway, both sequence and structural comparisons do not suggest that these sets of enzymes are related to one another by divergent evolution. The absence of such similarities implies that glycolytic enzymes were derived independently, rather than by gene duplication. The common dinucleotide-binding domain found in the dehydrogenases and the αβ barrels are the only major recurring elements.

Speculations regarding the relationship between glycolysis and gluconeogenesis can be made if we think of glycolysis as consisting of 2 segments: the metabolism of hexoses (the upper segment) and the metabolism of trioses (the lower segment). The enzymes from the upper segment are different in some species and are missing entirely in some archaea, whereas enzymes from the lower segment are quite conserved. In fact, 4 enzymes of the lower segment are present in all species. *This lower part of the pathway is common to glycolysis and gluconeogenesis.* This common part of the 2 pathways may be the oldest part, constituting the core to which the other steps were added. The upper part would vary according to the sugars that were available to evolving organisms in particular niches. Interestingly, this core part of carbohydrate metabolism can generate triose precursors for ribose sugars, a component of RNA and a critical requirement for the RNA world.

PENTOSE PHOSPHATE PATHWAY

(= Hexose Monophosphate Pathway)

The predominant pathway for glucose breakdown is glycolysis to yield pyruvate, followed by oxidation to CO_2 in the citric acid cycle. An alternative process, the pentose phosphate pathway, is a remarkable, multipurpose pathway that operates to varying extents in different cells and tissues. *The role of this pathway is primarily anabolic rather than catabolic.* But it does involve oxidation of glucose and in certain modes it can operate to oxidize glucose completely to CO_2 and water.

The **pentose phosphate pathway** is a cytosolic pathway present in all cells, so named because it is the primary pathway for formation of pentose phosphates for the synthesis of nucleotides for incorporation into DNA and RNA. This pathway branches from glycolysis at the level of glucose 6-phosphate (Glc-6-P), thus its alternative designation, the **hexose monophosphate shunt.** The pentose phosphate pathway is sometimes described as a **shunt,** rather than a pathway, because when pentoses are not needed for biosynthetic reactions, the pentose phosphate intermediates are recycled to the mainstream of glycolysis by conversion into fructose-6-phosphate (Fru-6-P) and glyceraldehyde-3-phosphate (Gly-3-P). This rerouting is especially important in the red blood cell (RBC) and in nondividing or quiescent cells, where there is limited need for synthesis of DNA and RNA. The pathway is also called the **phosphogluconate pathway** (after the name of one of the intermediates of the cycle). The pathway also goes by the name **Dickens–Horecker pathway,** in view of the contribution of these two authors to the study of this cycle.

It may be said that the advantage of this pathway is fourfold:

(*a*) It provides pentose phosphates which are indispensable for the synthesis of nucleotides found in various coenzymes and in nucleic acids.

(*b*) It permits the formation of NADPH required for various reactions, particularly for the synthesis of fatty acids and steroids.

(*c*) It can perform the oxidation of glucose into CO_2 with the formation of ATP in appreciable quantity, if transhydrogenation (from NADPH to NAD^+) can take place and provided the NADPH is not used to carry out reactions requiring this reduced coenzyme.

(*d*) In photosynthetic organisms, it permits the synthesis of carbohydrates (and therefore of other organic compounds) from CO_2.

In this oxidative pathway, NAD^+ is the electron acceptor, yielding NADPH. Rapidly dividing cells, such as those of bone marrow, skin and intestinal mucosa, use the pentoses to make RNA, DNA and such enzymes as ATP, NADH, $FADH_2$ and coenzyme A. In other tissues, the essential product of this pathway is not the pentoses but the electron donor NADPH, needed for reductive biosynthesis or to counter the damaging effects of oxygen radicals. Tissues that carry out extensive fatty acid synthesis (liver, adipose, lactating mammary gland) or very active synthesis of cholesterol and steroid hormones (liver, adrenal gland, gonads) require the NADPH provided by the pathway. Erythrocytes, and the cells of the lens and cornea are directly exposed to oxygen and thus to the damaging free radicals generated by oxygen. By maintaining a reducing atmosphere (a high ratio of NADPH to $NADP^+$ and a high ratio of reduced to oxidized glutathione), they can prevent or undo oxidative damage to proteins, lipids, and other sensitive molecules. In erythrocytes, the NADPH produced by the pentose phosphate pathway is so important in preventing oxidative damage that a genetic defect in glucose-6-phosphate dehydrogenase, the first enzyme of the pathway, can have serious medical consequences.

The oxidative pentose phosphate pathway provides a means for cutting the carbon chain of a sugar molecule, one carbon at a time. However, in contrast to the citric acid cycle, the operation of this pathway does not occur as a consecutive set of reactions leading directly from glucose-

6-phosphate (G-6-P) to 6 molecules of CO_2. The pathway operates in 2 phases: the oxidative phase 1 and the nonoxidative phase 2.

Phase 1: Oxidative Phase. Oxidation and Decarboxylation of Glucose 6-phosphate to a Pentose Phosphate.

This phase consists of 3 reactions, of which the two are oxidations and the third is hydrolysis (Fig. 21–22). The first reaction of this pathway is the oxidation of glucose 6-phosphate to an intramolecular ester **6-phosphoglucono-δ-lactone**; the reaction being catalyzed by *glucose 6-phosphate dehydrogenase* (G6PD). This enzyme is highly specific for $NADP^+$; the K_m for NAD^+ is about a thousand times as great as that for $NADP^+$. $NADP^+$ is the electron acceptor, and the overall equilibrium lies far in the direction of NADPH formation. In fact, this reaction is rate limiting under physiological conditions and serves as the control site. The most important regulatory factor is the level of $NADP^+$. The inhibitory effect of low levels of $NADP^+$ is exacerbated by the fact that NADPH competes with $NADP^+$ in binding to the enzyme. The ratio of $NADP^+$ to NADPH in the cytosol of a liver cell from a well-fed rat is about 0.014, several orders of magnitude lower than the ratio of NAD^+ to NADH, which is 700 under the same conditions. The marked effect of the $NADP^+$ level on the rate of the oxidative phase ensures that NADPH generation is tightly coupled to its utilization in reductive biosyntheses. The nonoxidative phase of the pentose phosphate pathway is controlled primarily by the availability of substrates. Special interest in this enzyme stems from the severe hemolytic anemia that may result from the absence of glucose 6-phosphate dehydrogenase in erythrocytes or from the presence of one of over 300 genetic variants of this enzyme that contains 516 amino acids.

In the second reaction, the 6-phosphoglucono-δ-lactone is hydrolyzed by a specific *lactonase* to give a free acid **6-phosphogluconate.** In the third reaction, 6-phosphogluconate undergoes oxidative decarboxylation by *6-phosphogluconate dehydrogenase* to form the ketopentose **ribulose 5-phosphate.** $NADP^+$ is again the electron acceptor.

In the final step, ribulose 5-phosphate is converted to its aldose isomer, **ribose 5-phosphate** by *phosphopentose isomerase*. The net result of the oxidative phase is generation of 2 moles of NADPH, oxidation of one carbon to CO_2, and synthesis of one mole of ribulose 5-phosphate.

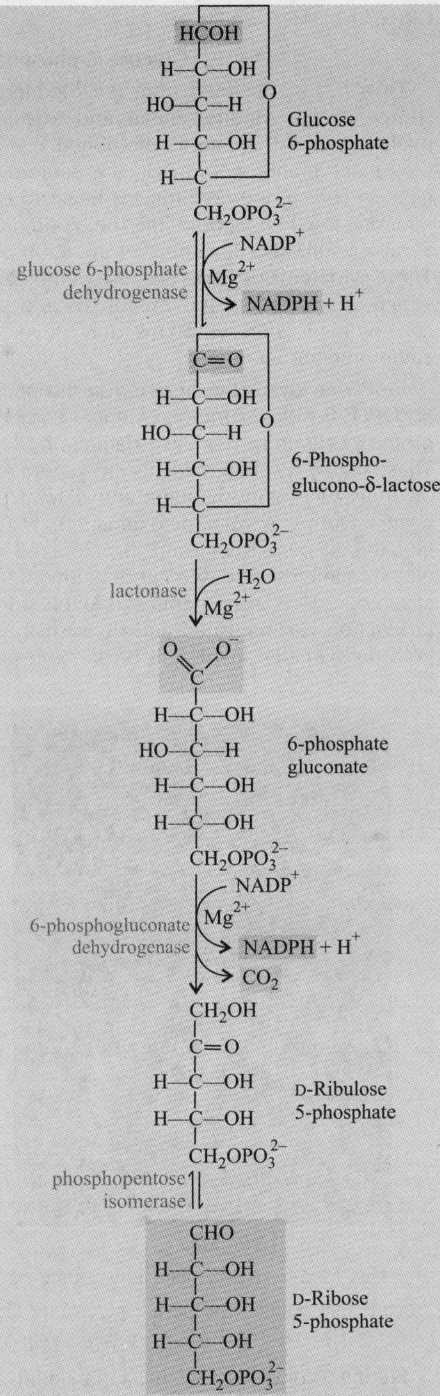

Fig. 21–22. Reactions of the oxidative phase of the pentose phosphate pathway

CLINICAL IMPLICATIONS 21-1.

Glucose 6-phosphate Dehydrogenase Deficiency

Fava beans, obtained from the Mediterranean plant *Vicia faba,* have been an important food source in the Mediterranean and Middle East since antiquity. The Greek philosopher and mathematician **Pythagoras** prohibited his followers from dining on fava beans, perhaps because they make many people sick, the pathologic condition being called **favism,** which may even be fatal. In favism, the erythrocytes begin to disintegrate 48 to 96 hours after ingestion of the beans, releasing free hemoglobin into the blood. Jaundice, and sometimes kidney damage, may also result. Similar results also develop after ingestion of the antimalarial drug *primaquine* or of sulfa antibiotics. These symptoms develop because of the deficiency of *glucose 6-phosphate dehydrogenase* (G6PD), which affects about 400 million people across the globe. Most G6PD-deficient individuals are asymptomatic but symptoms may develop if the G6PD-deficiency is associated with certain environmental factors.

G6PD catalyzes the first step in the pentose phosphate pathway which produces the reductant NADPH. Besides being an essential component in many biosynthetic pathways, it (NADPH) also protects cells from oxidative damage by hydrogen peroxide (H_2O_2) and superoxide free radicals. These highly reactive oxidants are generated as metabolic byproducts and also through the actions of drugs such as primaquine and natural products such as *divicine*– the toxic ingredient of fava beans. During normal detoxification, H_2O_2 is converted to H_2O by reduced glutathione and glutathione peroxidase, and the oxidized glutathione is converted back to the reduced form by glutathione reductase (see figure below). H_2O_2 is also broken down to H_2O and O_2 by the enzyme catalase, which also requires NADPH for activity. In G6PD-deficient individuals, the NADPH production is diminished and detoxification of H_2O_2 is inhibited. Cellular damage, thus, results: lipid peroxidation leading to breakdown of erythrocyte membranes and oxidation of proteins and DNA.

Vicia faba

This Mediterranean plant is a source of fava beans that contain the purine glycoside, *vicine*.

(*Courtesy: Visuals Unlimited*)

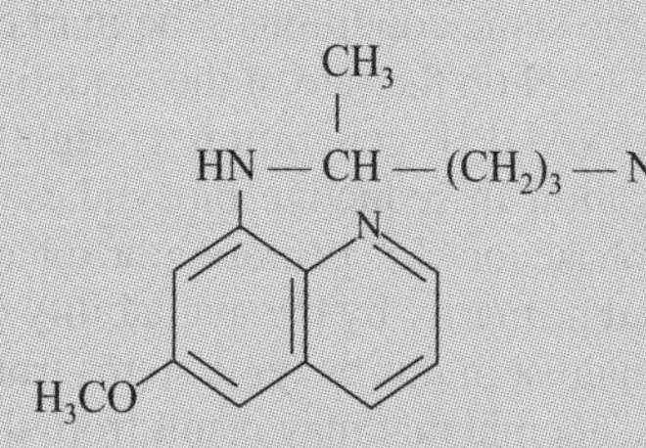

***Primaquine*, the antimalarial drug**

During World War II, the antimalarial drug **primaquine** was prophylactically administered to members of the armed forces. As a result, significant proportion of servicemen suffered hemolytic anemia (massive destruction of red blood cells). They were also sensitive to many other compounds that, like primaquine, generate oxidative stress, as manifested by the appearance of H_2O_2 and organic peroxides in RBCs. These individuals were later found to be deficient in *glucose 6-phosphate dehydrogenase*.

With regard to G6PD-deficiency, frequencies as high as 25% occur in tropical Africa, parts of Middle East and Southeast Asia, areas where malaria is prevalent. In addition to such epidemiological observations, *in vitro* studies show that growth of one malaria parasite, *Plasmodium falciparum*, is inhibited in G6PD-deficient erythocytes. The parasite is very sensitive to oxidative stress and is killed by a level of oxidative stress that is tolerable to a G6PD-deficient human host. This explains why the G6PD-deficiency is seen most frequently among individuals of malaria-infested regions of the world, *i.e.*, African and Mediterranean areas. This enzymatic deficiency, which is usually undetected until administration of certain drugs, illustrates the interplay of heredity and environment on the production of disease. Because the advantage of resistance to malaria balances the disadvantage of lowered resistance to oxidative damage, natural selection sustains the G6PD-deficient genotype in human populations where malaria is prevalent. Only under overwhelming oxidative stress, caused by drugs or divicine, does G6PD-deficiency cause serious medical problems.

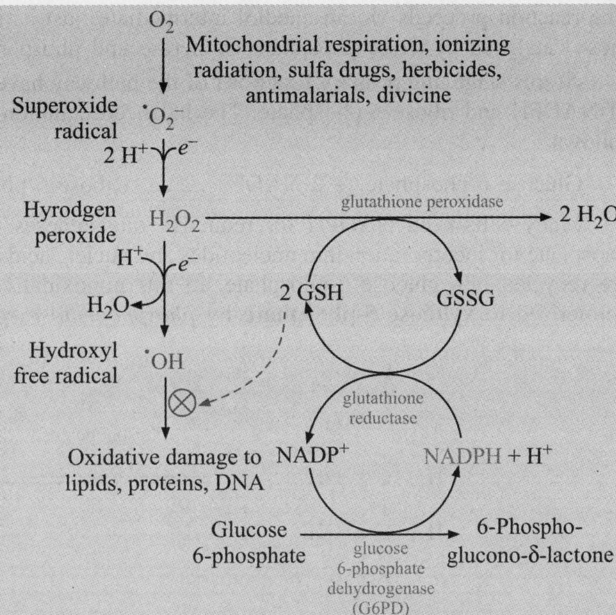

Role of NADPH and glutathione in protecting cells against highly reactive oxygen derivatives

Reduced glutathione (GSH) protects the cell by destroying hydrogen peroxide and hydroxyl free radicals. Regeneration of GSH from its oxidized form (GSSG) requires the NADPH produced in the glucose 6-phosphate dehydrogenase reaction.

It is ironic to see that the antimalarial drugs can cause illnesses through the same biochemical mechanism that provides resistance to malaria. Divicine also acts as an antimalarial drug, and ingestion of fava beans may protect against malaria. By refusing to eat falafel (whose fava bean is an ingredient), many Pythagoreans with normal G6PD activity may have unwittingly increased their risk of malaria!

Phase 2: Nonoxidative Phase. Interconversions of Pentose Phosphates to Glycolytic Intermediates.

In the nonoxidative phase, some of the ribulose 5-phosphate produced in the oxidative phase, depending on the prevailing conditions in the cell, is converted to **ribose 5-phosphate** by *phosphopentose isomerase*.

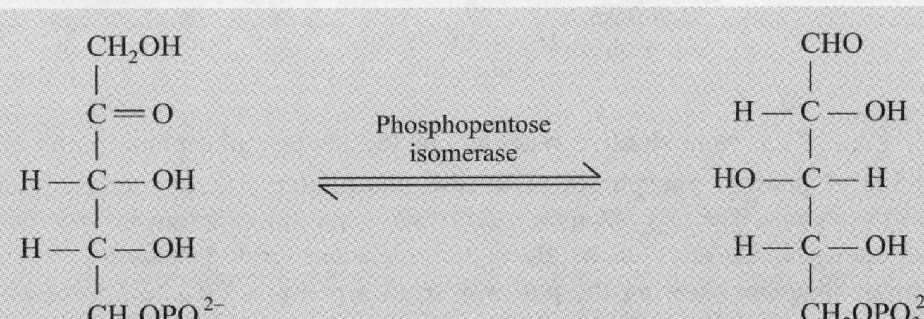

Ribulose 5-phosphate | Ribose 5-phosphate

The reaction proceeds *via* an enediol intermediate, just as in 2 different reactions of glycolysis– those catalyzed by triose phosphate isomerase and phosphoglucoisomerase.

At this stage, the primary functions of the pathway have been fulfilled, namely the generation of NADPH and ribose 5-phosphate. The balanced equation up to this stage of the pathway is as follows:

$$\text{Glucose 6-phosphate.} + 2\ NADP^+ \longrightarrow \text{Ribose 5-phosphate.} + CO_2 + 2\ NADPH + 2\ H^+$$

Many cells need NADPH for reductive biosyntheses much more than they need ribose 5-phosphate for incorporation into nucleotides and nucleic acids. In such cases, the pentose phosphates are recycled into glucose 6-phosphate. In this nonoxidative phase, ribulose 5-phosphate is first epimerized to **xylulose 5-phosphate** by *phosphopentose epimerase*:

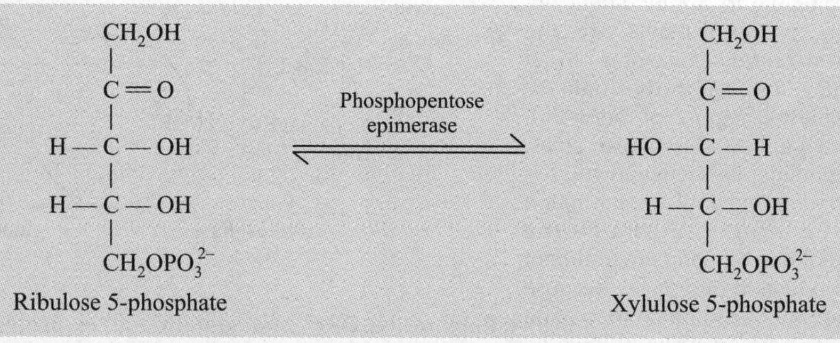

Then in a series of rearrangements of the carbon skeletons (Fig. 21–23), 6 five-carbon sugar phosphates are converted to 5 six-carbon sugar phosphates, completing the cycle and allowing continued oxidation of glucose 6-phosphate with production of NADPH. Continued recycling leads ultimately to the conversion of glucose 6-phosphate to 6 CO_2.

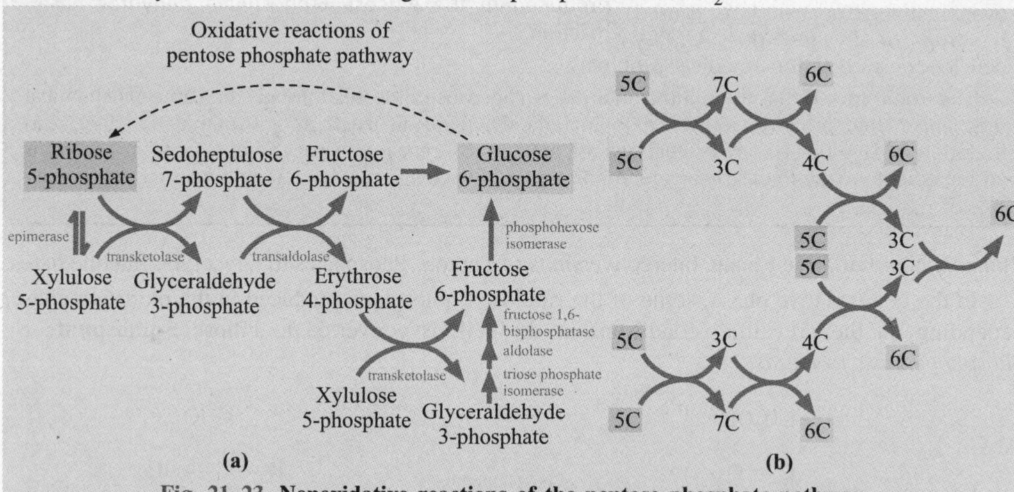

Fig. 21–23. Nonoxidative reactions of the pentose phosphate pathway

(*a*) **Conversion of pentose phosphates to hexose phosphates:** These reactions allow the oxidative reactions to continue. The two enzymes, *transketolase* and *transaldolase* are specific to this pathway; the other enzymes also serve in the glycolytic or gluconeogenic pathways.

(*b*) **A schematic diagram showing the pathway from 6 pentoses (5C) to 5 hexoses (6C):** Note that this involves 2 sets of interconversions shown in (*a*). Every reaction shown here is reversible; the unidirectional arrows are used only to make clear the direction of the reactions during continuous oxidation of glucose 6-phosphate. In the light-independent reactions of photosynthesis, the direction of these reactions is reversed.

(Adapted from: Lehninger AL, Nelson DL, Cox MM, 2005)

This so-called *sugar rearrangement* system forms triose, tetrose, hexose, and heptose sugars from the pentoses, thus creating a **disposal mechanism** for ribose 5-phosphate, and providing a reversible link between the pentose phosphate pathway and glycolysis *via* intermediates common to both pathways. For the interconversions, another pentose phosphate, xylulose 5-phosphate, is formed, as already mentioned, through isomerization of ribulose 5-phosphate by the action of phosphopentose epimerase. Consequently, ribulose 5-phosphate, ribose 5-phosphate, and xylulose 5-phosphate exist as an equilibrium mixture and can then undergo 3 more transformations by the two unique enzymes, transketolase and transaldolase (Fig. 21–24).

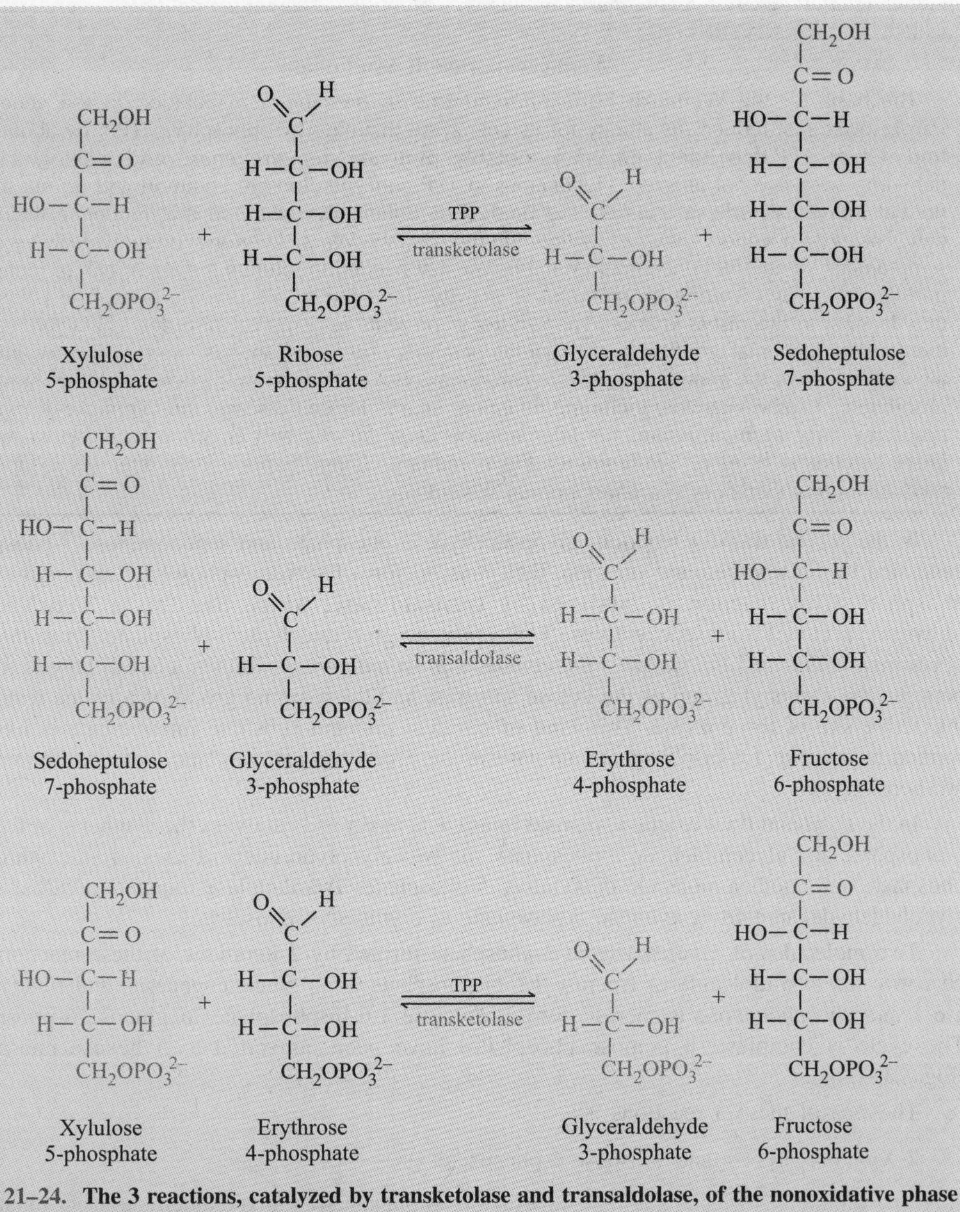

Fig. 21–24. The 3 reactions, catalyzed by transketolase and transaldolase, of the nonoxidative phase of the pentose phosphate pathway

Note that the first and third reactions are catalyzed by the same enzyme *transketolase* whereas the second reaction is catalyzed by another enzyme *transaldolase*.

In the *first* of the 3 reactions, one mole of xylulose 5-phosphate reacts with one mole of ribose 5-phosphate. The reaction is catalyzed by **transketolase,** which transfers a *2 carbon unit* (carbanion) from xylulose 5-phosphate to ribose 5-phosphate to produce a triose phosphate, glyceraldehyde 3-phosphate and a heptose sugar, sedoheptulose 7-phosphate. The 2-carbon unit is bound transiently to the carbon 2 of the thiazole ring of the required coenzyme thiamine pyrophosphate (TPP). Transketolase is homologous to the E_1 subunit of the pyruvate dehydrogenase complex (PDC) and the reaction mechanism is similar. Alterations in transketolase can lead to **Wernicke-Korsakoff syndrome.**

CLINICAL IMPLICATION 21–2.

Wernicke–Korsakoff Syndrome

The basis for the Wernicke–Korsakoff syndrome in humans is a mutation in the gene for transketolase that reduces its affinity for its coenzyme thiamine pyrophosphate (TPP) by about 10-fold. Other TPP-dependent enzymes, notably pyruvate dehydrogenase and a-ketoglutarate dehydrogenase, are not affected. Fluctuations in TPP concentration are common and do not affect normal individuals whose transketolase binds TPP sufficiently tightly so that moderate thiamine deficiencies do not affect enzyme function. On the contrary, Wernicke–Korsakoff individuals become symptomatic when TPP concentrations fall below that needed to saturate the abnormal (or mutated) transketolase. The lowering of transketolase activity slows the whole pentose phosphate pathway, thus leading to the diseased state. The syndrome presents as a mental disorder, characterized by memory loss, mental confusion and partial paralysis. The syndrome is more common among alcoholics than in the general population; chronic alcohol consumption interferes with the intestinal absorbtion of some vitamins, including thiamine. Like sickle-cell disease, the Wernicke–Korsakoff syndrome once again illustrates the interdependence of genetic and environmental factors in the onset of clinical disease. Symptoms of the hereditary change express only after some kind of moderate stress that does not affect normal individuals.

In the *second* transfer reaction, glyceraldehyde 3-phosphate and sedoheptulose 7-phosphate, generated in the transketolase reaction, then react to form fructose 6-phosphate and erythrose 4-phosphate. This reaction is catalyzed by **transaldolase,** which transfers a *3-carbon unit* (dihydroxyacetone) from sedoheptulose 7-phosphate to glyceraldehyde 3-phosphate. *Transaldolase, in contrast with transketolase, does not contain a prosthetic group*. Rather, a Schiff base is formed between the carbonyl group of the ketose substrate and the ε-amino group of a lysine residue at the active site of the enzyme. This kind of covalent enzyme–substrate intermediate is like that formed in fructose 1,6-bisphosphate aldolase in the glycolytic pathway and, indeed, the enzymes are homologous.

In the *third* and final reaction, **transketolase** acts again and catalyzes the synthesis of fructose 6-phosphate and glyceraldehyde 3-phosphate, the two glycolytic intermediates, from erythrose 4-phosphate and another molecule of xylulose 5-phosphate. Transketolase transfers a carbanion (= glycolaldehyde) unit from xylulose 5-phosphate to erythrose 4-phosphate.

Two molecules of glyceraldehyde 3-phosphate formed by 2 iterations of these reactions can be converted to a molecule of fructose 1,6-bisphosphate as in gluconeogenesis, and finally FBP ase-1 and phosphohexose isomerase convert fructose 1,6-bisphosphate to glucose 6-phosphate. The cycle is complete: 6 pentose phosphates have been converted to 5 hexose phosphates (Fig. 21–23b).

The *sum* of these 3 reactions is:

$$\text{2 Xylulose 5-phosphate + ribose 5-phosphate} \rightleftharpoons \text{2 Fructose 6-phosphate + glyceraldehyde 3-phosphate}$$

Since xylulose 5-phosphate is derived from ribose 5-phosphate by the sequential action of phosphopentose isomerase and phosphopentose epimerase, the *net reaction* starting from ribose 5-phosphate is:

$$\text{3 Ribose 5-phosphate} \rightleftharpoons \text{2 Fructose 6-phosphate + glyceraldehyde 3-phosphate}$$

Therefore, excess ribose 5-phosphate, whether it arises from the first stage of the pentose phosphate pathway or from the degradative metabolism of nucleic acids, is effectively scavenged by conversion to intermediates that can enter the carbon flow of glycolysis. Moreover, any ribose ingested in the diet can also be processed into glycolytic intermediates by this pathway. It is evident that the carbon skeletons of sugars can be extensively rearranged to meet physiologic needs (Table 21–4).

Table 21–4. The pentose phosphate pathway

Reaction	*Enzyme*
Oxidative phase	
Glucose 6-phosphate + $NADP^+$ $\longrightarrow$ 6-phosphoglucono-δ-lactone + NADPH + H^+	Glucose 6-phosphate dehydogenase
6-phosphoglucono-δ-lactone + H_2O $\longrightarrow$ 6-phosphogluconate + H^+	Lactonase
6-phosphogluconate + $NADP^+$ $\longrightarrow$ Ribulose 5-phophate + CO_2 + NADPH	6-phosphogluconate dehydrogenase
Nonoxidative phase	
Ribulose 5-phosphate $\rightleftharpoons$ Ribose 5-phosphate	Phosphopentose isomerase
Ribulose 5-phosphate $\rightleftharpoons$ Xylulose 5-phosphate	Phosphopentose epimerase
Xylulose 5-phosphate + ribose 5-phosphate $\rightleftharpoons$ Sedoheptulose 7-phosphate + glyceraldehyde 3-phosphate	Transketolase
Sedoheptulose 7-phosphate + glyceraldehyde 3-phosphate $\rightleftharpoons$ Fructose 6-phosphate + erythrose 4-phosphate	Transaldolase
Xylulose 5-phosphate + erythrose 4-phosphate $\rightleftharpoons$ Fructose 6-phosphate + glyceraldehyde 3-phosphate	Transketolase

ALTERNATIVE PENTOSE PHOSPHATE PATHWAYS

We can write a balanced equation for the entire pentose phosphate pathway as follows:

$$\text{3 Glucose 6-phosphate} + 6\,NADP^+ + 3\,H_2O \longrightarrow \text{2 Fructose 6-phosphate + glyceraldehyde 3-phosphate} + 6\,NADPH + 6\,H^+ + 3\,CO_2$$

In the equation for the overall reaction, 3 hexose phosphates yield 2 hexose phosphates, 1 triose phosphate, and 3 molecules of CO_2, for a net oxidation of 3 carbons to CO_2. In the formal sense, therefore, the pathway can be seen as a means to oxidize the 6 carbons glucose 6-phosphate to CO_2, just as occurs in glycolysis plus the citric acid cycle.

However, as noted above, the pentose phosphate pathway is not primarily an energy-generating pathway. The actual fate of the sugar phosphates depends on the metabolic needs of the cell in which the pathway is occurring. The following 4 *metabolic situations* or *modes* are recognized (Fig. 21–25):

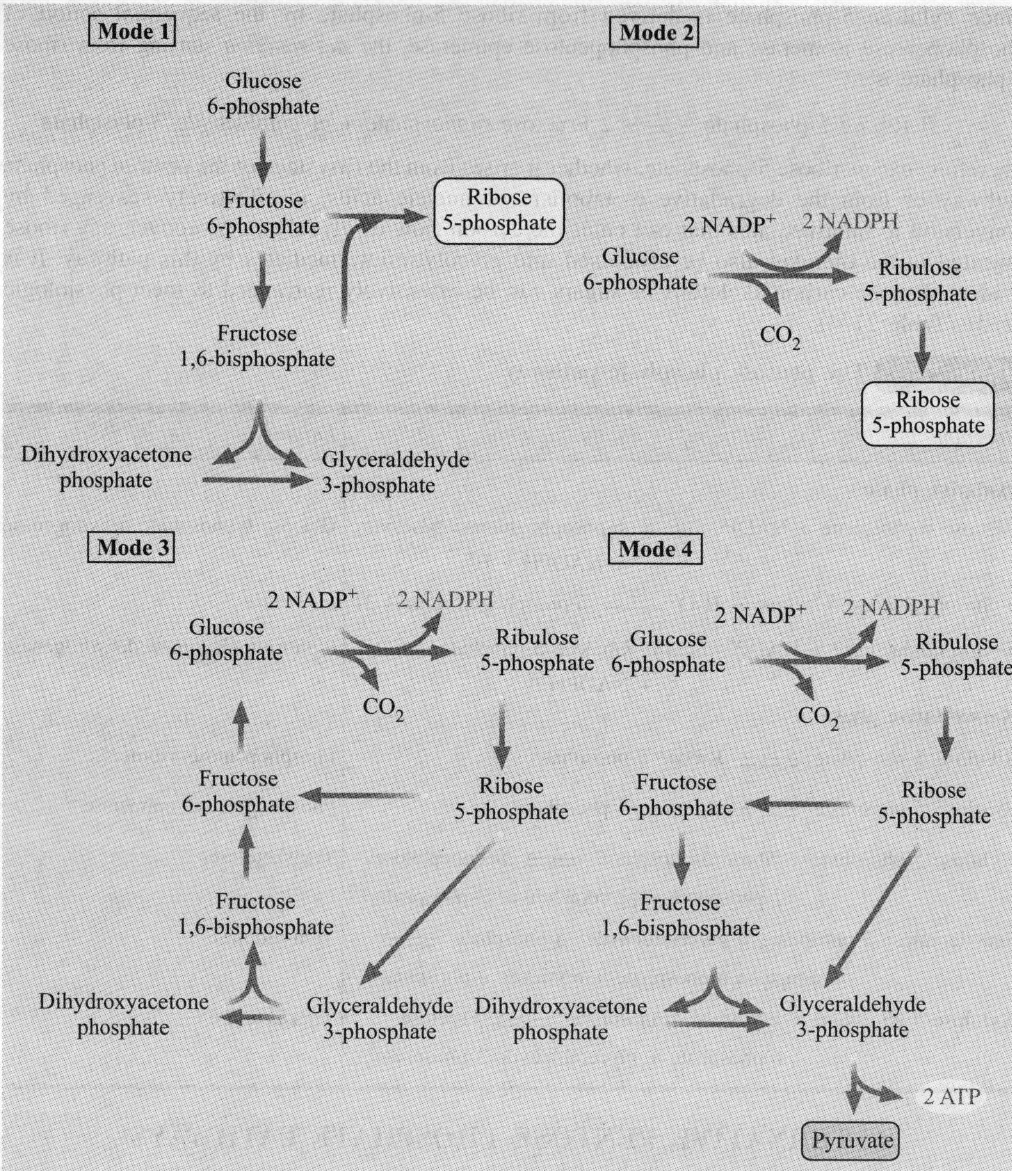

Fig. 21–25. Four modes of the pentose phosphate pathway

The pentose phosphate pathway has different modes of operation to meet varying metabolic needs.

Mode 1: When the primary need is for ribose 5-phosphate to synthesize nucleotide precursors, this mode utilizes NADPH to reduce ribonucleotides to deoxyribonucleotides.

Mode 2: When there is a balanced need for both NADPH and ribose 5-phosphate, which are derived from glucose 6-phosphate.

Mode 3: When the primary need is for the reducing power (NADPH), fructose phosphates are reconverted to glucose 6-phosphate for reoxidation in the oxidative phase.

Mode 4: When both NADPH and ATP are required, this pathway is used to supply energy, with the reaction products being oxidized through glycolysis and the citric acid cycle.

Major products are shown in *yellow*, *green, red and blue* colour.

***Mode 1*: When much more ribose 5-phosphate than NADPH is required.** In such a situation, for example in rapidly dividing cells, there is a need for ribose 5-phosphate (R5P) for the synthesis of nucleotide precursors of DNA. Most of the glucose 6-phosphate is converted to fructose 6-phosphate (F6P) and glyceraldehyde 3-phosphate (G3P) by the glycolytic pathway. The two enzymes, **transketolase** and **transaldolase,** then convert 2 moles of F6P and one mole of G3P into 3 moles of R5P by a reversal of the reactions described earlier. The stoichiometry of Mode 1 is:

$$\text{5 Glucose 6-phosphate + ATP} \longrightarrow \text{6 Ribose 5-phosphate + ADP + H}^+$$

***Mode 2*: When the needs for NADPH and ribose 5-phosphate are balanced.** In such a situation, the predominant reaction is the formation of 2 moles of NADPH and 1 mole of R5P from 1 mole of G6P in the oxidative phase of the pentose phosphate pathway. The stoichiometry of Mode 2 is:

$$\text{Glucose 6-phosphate + 2 NADP}^+ + \text{H}_2\text{O} \longrightarrow \text{Ribose 5-phosphate + 2 NADPH + 2 H}^+ + \text{CO}_2$$

***Mode 3*: When more NADPH than ribose 5-phosphate is required.** In such a situation, for example in adipose tissue, there is a need for high level of NADPH for the synthesis of fatty acids. In this case, G6P is completely oxidized to O_2.

Three groups of reactions are active in this situation. *First*, the oxidative phase of the pentose phosphate pathway forms 2 moles of NADPH and 1 mole of R5P. *Second*, the R5P is converted into F6P and G3P by transketolase and transaldolase. *Third*, the G6P is resynthesized from F6P and G3P by the gluconeogenic pathway. The stoichiometries of these 3 sets of reactions are:

$$\text{Glucose 6-phosphate + 12 NADP}^+ + \text{6 H}_2\text{O} \longrightarrow \text{6 Ribose 5-phosphate + 12 NADPH + 12 H}^+ + \text{6 CO}_2$$

$$\text{6 Ribose 5-phosphate} \longrightarrow \text{4 Fructose 6-phosphate + 2 glyceraldehyde 3-phosphate}$$

$$\text{4 Fructose 6-phosphate + 2 glyceraldehyde 3-phosphate + H}_2\text{O} \longrightarrow \text{5 Glucose 6-phosphate + P}_\text{i}$$

The *sum* of the Mode 3 reactions is:

$$\text{Glucose 6-phosphate + 12 NADP}^+ + \text{7 H}_2\text{O} \longrightarrow \text{6 CO}_2 + \text{12 NADPH + 12 H}^+ + \text{P}_\text{i}$$

Thus, the equivalent of glucose 6-phosphate can be completely oxidized to CO_2 with the concomitant generation of NADPH. In essence, ribose 5-phosphate produced by the pentose phosphate pathway is recycled into glucose 6-phosphate by transketolase and transaldolase, and some of the enzymes of gluconeogenic pathway.

***Mode 4*: When both NADPH and ATP are required.** As an alternative, ribose 5-phosphate formed by the oxidative phase of the pentose phosphate pathway can be converted into pyruvate. Fructose 6-phosphate and glyceraldehyde 3-phosphate, derived from ribose 5-phosphate, enter the glycolytic pathway, rather than reverting to glucose 6-phosphate. In this mode, ATP and NADPH are concomitantly generated, and 5 of the 6 carbons of glucose 6-phosphate emerge in pyruvate.

$$\text{3 Glucose 6-phosphate + 6 NADP}^+ + \text{5 NAD}^+ + \text{5 P}_\text{i} + \text{8 ADP} \longrightarrow \text{5 Pyruvate + 3 CO}_2 + \text{6 NADPH + 5 NADH + 8 ATP + 2 H}_2\text{O + 8 H}^+$$

Pyruvate formed by these reactions can be oxidized to generate more ATP or it can be used as a building block in a variety of biosyntheses.

Because of the multiple metabolic needs of a cell for biosynthesis, it is unlikely that any one of these 4 modes operates exclusively in any one cell.

Distribution of Pentose Phosphate Pathway in Tissues

The distribution of the pentose phosphate pathway in tissues is consistent with its functions (Table 21–5). It is present in *erythrocytes* for the production of NADPH, required to generate reduced glutathione, which is essential for maintenance of normal red cell structure. It is also active in the *liver*, *mammary gland*, *testis* and *adrenal cortex*, sites of fatty acid or steroid synthesis that also require the reducing power of NADPH. In contrast, in mammalian *striated muscle*, which exhibits little fatty acid or steroid synthesis, all catabolism proceeds *via* glycolysis and the citric acid cycle and no direct oxidation of glucose 6-phosphate occurs through the pentose phosphate pathway. In some other tissues like *liver*, 20–30% of the CO_2 produced may arise from the pentose phosphate pathway, and the balance between glycolysis and the pentose phosphate pathway depends on the metabolic requirements of the cell.

Table 21–5. Tissues with active pentose phosphate pathways

Tissue	*Function*
Adipose tissue	Fatty acid synthesis
Mammary gland	Fatty acid synthesis
Liver	Fatty acid and cholesterol syntheses
Adrenal gland	Steroid synthesis
Testes	Steroid synthesis
Ovary	Steroid synthesis
Erythrocytes	Maintenance of reduced glutathione

Partitioning of Glucose 6-phosphate Between Glycolysis And Pentose Phosphate Pathway

Whether glucose 6-phosphate enters glycolysis or the pentose phosphate pathway depends on the current needs of the cell and on the concentration of $NADP^+$ in the cytosol. Without $NADP^+$ (which is an electronic acceptor), the first reaction of the pentose phosphate pathway (*i.e.*, the glucose 6-phosphate dehydrogenase reaction) cannot proceed. When a cell is rapidly converting NADPH to $NADP^+$ in biosynthetic reductions, the level of $NADP^+$ rises, glucose 6-phosphate dehydrogenase is allosterically stimulated so that there occurs an increase in the flux of glucose 6-phosphate through the pentose phosphate pathway. (Fig. 21–26). When the demand for NADPH decreases, the level of $NADP^+$ drops, the pentose phosphate pathway shows, and glucose 6-phosphate is instead used to start glycolysis.

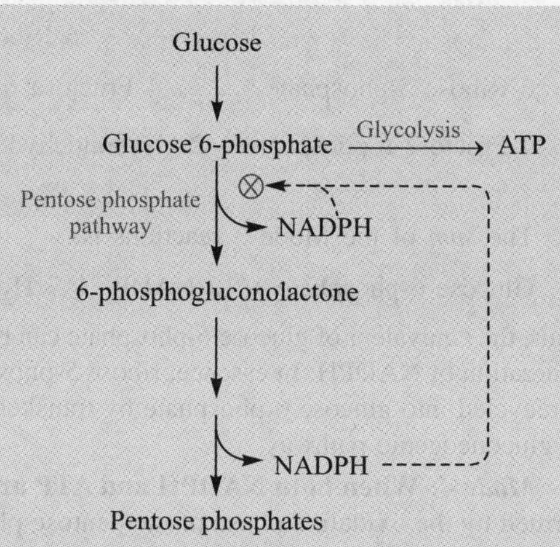

Fig. 21–26. Role of NADPH in regulating the partitioning of glucose 6-phosphate between glycolysis and the pentose phosphate pathway

When NADPH is forming faster than it is being used for biosynthesis and glutathione reduction, the concentration of NADPH rises and inhibits the first enzyme in the pentose phosphate pathway (*i.e.*, glucose 6-phosphate dehydrogenase, G6PD). As a result, more glucose 6-phosphate is available for glycolysis.

REFERENCES

1. **Ashwell G :** Carbohydrate Metabolism. *Ann, Rev. Biochem.,* **33** *: 101, 1964.*
2. **Atkinson DE :** Cellular Energy Metabolism and Its Regulation. *Academic Press Inc., New York. 1977.*
3. **Axelord B :** Glycolysis. *In Greenberg DM (editor) : Metabolic Pathways. 3rd ed., vol 1, Academic Press Inc., New York. 1967.*
4. **Barker HA :** Bacterial Fermentations. *John Wiley and Sons, Inc., New York. 1956.*
5. **Bennett WS Jr, Steitz TA :** Glucose-induced conformational change in yeast hexokinase. *Proc. Nat. Acad. Sci.* **75** *: 4848–4852, 1978.*
6. **Boiteux A, Hess B :** Design of glycolysis. *Phil. Trans. Roy. Soc., London* **B 293** *: 5–22, 1981.*
7. **Caputto R, Barra HS, Cumar FA :** Carbohydrate metabolism. *Ann. Rev. Biochem.* **36** *: 211–246, 1967.*
8. **Clark MG, Lardy HA :** Regulation of intermediary carbohydrate metabolism. *Intern. Rev. Sci. (Biochem.)* **5** *: 223–266, 1975.*
9. **Colowick SP :** The Hexokinases. *In Boyer PD (editor): The Enzymes 3rd ed. vol. IX part B, 1–48. Academic Press, Inc., New York. 1976.*
10. **Davison EA :** Carbohydrate Chemistry. *Holt, Rinehart and Winston. 1967.*
11. **Dickens F, Randle PJ, Whelan WJ :** Carbohydrate Metabolism and Its Disorders. vols. I and II. *Academic Press, Inc., New York. 1968.*
12. **Everse J, Kaplan NO :** Lactate dehydrogenases : Structure and function. *Adv. Enzymol.* **37** *: 61–134, 1973.*
13. **Fothergill-Gilmore LA :** The evolution of the glycolytic pathway. *Trends Biochem. Sci.* **11** *: 47, 1986.*
14. **Fruton JS :** Molecules and Life : Historical Essays on the Interplay of Chemistry and Biology. *Wiley-Interscience. 1972.*
15. **Gatenby RA, Gillies RJ :** Glycolysis in cancer: a potential target for therapy. *Cell Biol.* **39** *: 1358–1366, 2007.*
16. **Goldhammer AR, Paradies HH :** Phosphofructokinase : Structure and function. *Curr. Top. Cell. Regul.* **15** *: 109–141, 1979.*
17. **Harris JI, Waters M :** Glyceraldehyde 3-phosphate Dehydrogenase. *In Boyer PD (editor); The Enzymes. 3rd ed. vol. XIII,. 1–49, Academic Press, Inc., New York. 1976.*
18. **Hochachka P :** Living Without Oxygen. *Harward Univ., Press, Cambridge, Mass., 1980.*
19. **Hofmann E :** Phosphofructokinase– A favourite of enzymologists and of students of metabolic regulation. *Trends Biochem. Sci.* **3** *: 145, 1978.*
20. **Horecker BL :** Fermentation mechanisms. *Ciba Lectures Microbial Biochem.* **6** *: 36–64, 1962.*
21. **Horecker BL :** Transketolase and Transaldolase. *In Florkin M, Stotz EH (editor) : Comprehensive Biochemistry.* **vol. 15.** *48–70. Elsevier Publishing Co., Amsterdam. 1964.*
22. **Kalckar HM (editor) :** Biological Phosphorylations : Development of Concepts. *Prentice-Hall, Englewood Cliffs, N.J., 1969.*
23. **Knowles JR, Albery WJ :** Perfection in Enzyme Catalysis : The Energetics of Triosephosphate Isomerase. *Acc. Chem. Res.,* **10** *: 105–111, 1977.*

24. **Middleton RJ :** Hexokinases and glucokinases. *Biochem. Soc. Trans.* **18 :** *180–183, 1990.*
25. **Newsholme EA, Start C :** Regulation in Metabolism. *Wiley, New York. 1973.*
26. **Ottaway JH, Mowbray J :** The role of compartmentation in the control of glycolysis. *Curr. Top. Cell. Regul.* **12 :** *107–195, 1977*
27. **Phillips D, Blake CCF, Watson HC (editors) :** The enzymes of glycolysis : Structure, activity and evolution. *Phil. Trans. R. Soc. London (Biol.)* **293 :** *1–214, 1981.*
28. **Pigman WW, Horton D (editors) :** The Carbohydrates. Chemistry and Biochemistry. *Academic Press, Inc., New York. 1972.*
29. **Pilkis SJ (editor) :** Fructose-2,6-bisphosphate. *CRC Press, Inc., Boca Raton, FL. 1990.*
30. **Rose ZB :** The glucose bisphosphate family of enzymes. *Trends Biochem. Sci.* **11 :** *253, 1986.*
31. **Uyeda K :** Phosphofructokinase, *Adv. Enzymol.* **48 :** *193, 1979.*
32. **Weinhouse E :** Regulation of glucokinase in liver. *Curr. Top. Cell. Regul.* **11 :** *1–50, 1976.*
33. **Wilson JE :** Brain hexokinase, the prototype. Ambiquitous enzyme. *Curr. Top. Cell. Regul.* **16 :** *1–54, 1980.*
34. **Wood WA :** Carbohydrate metabolism. *Ann. Rev. Biochem.* **35 :** *521–558, 1966.*
35. **Zanella A, Fermo E, Bianchi P, Chiarelli LR, Valentini G :** Pyruvate kinase deficiency: the genotype-phenotype association. *Blood Rev.* **21 :** *217–231, 2007.*

PROBLEMS

1. Xylose has the same structure as that of glucose except that it has a hydrogen atom at C-6 in place of a hydroxymethyl group. The rate of ATP hydrolysis by hexokinase is markedly enhanced by the addition of xylose. Why?
2. The intravenous infusion of fructose into healthy volunteers leads to a two- to fivefold increase in the level of lactate in the blood, a far greater increase than that observed after the infusion of the same amount of glucose.
   (*a*) Why is glycolysis more rapid after the infusion of fructose?
   (*b*) Fructose has been used in place of glucose for intravenous feeding. Why is this use of fructose unwise?
3. What are the likely consequences of a genetic disorder rendering fructose 1,6-bisphosphatase in liver less sensitive to regulation by fructose 2,6-bisphosphate?
4. If cells synthesizing glucose from lactate are exposed to CO_2 labeled with ^{14}C, what will be the distribution of label in the newly-synthesized glucose?
5. In the conversion of glucose into two molecules of lactate, the NADH generated earlier in the pathway is oxidized to NAD^+. Why is it not to the cell's advantage to simply make more NAD^+ so that the regeneration would not be necessary ? After all, the cell would save much energy because it would no longer need to synthesize lactic acid dehydrogenase.
6. People with galactosemia display central nervous system abnormalities even if galactose is eliminated from the diet. The precise reason for it is not known. Suggest a plausible explanation.
7. Write a pathway leading from glucose to lactose in mammary gland, and write a balanced equation for the overall pathway.
8. How many high-energy phosphates are generated in (*a*) converting 1 mole of glucose to lactate? (*b*) converting 2 moles of lactate to glucose?

9. Write balanced equations for all of the reactions in the catabolism of D-glucose to two molecules of D-glyceraldehyde-3-phosphate (the preparatory phase of glycolysis). For each equation, write the standard free-energy change. Then write the overall or net equation for the preparatory phase of glycolysis, including the net standard free-energy change.

10. In working skeletal muscle under anaerobic conditions, glyceraldehyde-3-phosphate is converted into pyruvate (the payoff phase of glycolysis), and the pyruvate is reduced to lactate. Write balanced equations for all of the reactions in this process, with the standard free-energy change for each. Then write the overall or net equation for the payoff phase of glycolysis (with lactate as the end product), including the net standard free-energy change.

11. The concentration of glucose in human blood plasma is maintained at about 5 mM. The concentration of free glucose inside muscle cells is much lower. Why is the concentration so low in the cell? What happens to the glucose upon entry into the cell ?

12. Glycerol (see below) obtained from the breakdown of fat is metabolized by being converted into dihydroxyacetone phosphate, an intermediate in glycolysis, in two enzyme-catalyzed reactions. Propose a reaction sequence for the metabolism of glycerol. On which known enzyme-catalyzed reactions is your proposal based? Write the net equation for the conversion of glycerol to pyruvate based on your scheme.

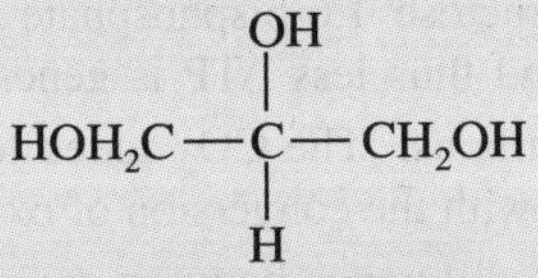

Glycerol

13. In muscle tissue, the rate of conversion of glycogen to glucose-6-phosphate is determined by the ratio of phosphorylase *a* (active) to phosphorylase *b* (less active). Determine what happens to the rate of glycogen breakdown if a muscle preparation containing glycogen phosphorylase is treated with **(*a*)** phosphorylase *b* kinase and ATP; **(*b*)** phosphorylase *a* phosphatase; **(*c*)** epinephrine.

14. The intracellular use of glucose and glycogen is tightly regulated at four points. In order to compare the regulation of glycolysis when oxygen is plentiful and when it is depleted, consider the utilization of glucose and glycogen by rabbit leg muscle in two physiological settings: a resting rabbit, whose leg-muscle ATP demands are low, and a rabbit who has just sighted its mortal enemy, the coyote, and dashes into its burrow at full speed. For each setting, determine the relative levels (high, intermediate, or low) of AMP, ATP, citrate, and acetyl-CoA and how these levels affect the flow of metabolites through glycolysis by regulating specific enzymes. In periods of stress, rabbit leg muscle produces much of its ATP by anaerobic glycolysis (lactate fermentation) and very little by oxidation of acetyl-CoA derived from fat breakdown.

15. Unlike the rabbit with its short dash, migratory birds require energy for extended periods of time. For example, ducks generally fly several thousand miles during their annual migration. The flight muscles of migratory birds have a high oxidative capacity and obtain the necessary ATP through the oxidation of acetyl-CoA (obtained from fats) *via* the citric acid cycle. Compare the regulation of muscle glycolysis during short-term intense activity, as in the fleeing rabbit, and during extended activity, as in the migrating duck. Why must the regulation in these two settings be different?

16. If a cell is forced to metabolize glucose anaerobically, how much faster would glycolysis have to proceed to generate the same amount of ATP as it would get if it metabolized glucose aerobically?

17. Which metabolic pathway is common to both fermentation and cellular respiration?
(*a*) Krebs cycle
(*b*) Glycolysis
(*c*) Electron transport chain
(*d*) Synthesis of acetyl-CoA from pyruvate

ANSWERS TO PROBLEMS

1. Hexokinase has a low ATPase activity in the absence of a sugar because it is in a catalytically inactive conformation. The addition of xylose closes the cleft between the two lobes of the enzyme. However, xylose lacks a hydroxymethyl group, and so it cannot be phosphorylated. Instead, a water molecule at the site normally occupied by the C-6 hydroxymethyl group acts as the acceptor of the phosphoryl group from ATP.

2. (*a*) The fructose-1-phosphate pathway forms glyceraldehyde 3-phosphate.
(*b*) Phosphofructokinase, a key control enzyme, is bypassed. Furthermore, fructose 1-phosphate stimulates pyruvate kinase.

3. Fructose 2,6-bisphosphate, present at high concentration when glucose is abundant, normally inhibits gluconeogenesis by blocking fructose 1,6-bisphosphatase. In this genetic disorder, the phosphatase is active irrespective of the glucose level. Hence, substrate cycling is increased. The level of fructose 1,6-bisphosphate is consequently lower than normal. Less pyruvate is formed and thus less ATP is generated.

4. There will be no labeled carbons. The CO_2 added to pyruvate (formed from the lactate) to form oxaloacetate is lost with the conversion of oxaloacetate into phosphoenolpyruvate.

5. This example illustrates the difference between *stoichiometric* and *catalytic* utilization of a molecule. If cells used NAD^+ stoichiometrically, a new molecule of NAD^+ would be required each time a lactate is produced. As we will see, the synthesis of NAD^+ requires ATP. On the other hand, if the NAD^+ that is converted into NADH could be recycled and reused, a small amount of the molecule could regenerate a vast amount of lactate. This is the case in the cell. NAD^+ is regenerated by the oxidation of NADH and reused. NAD^+ is thus used catalytically.

6. Galactose is a component of glycoproteins. Possibly, the absence of galactose leads to the improper formation or function of glycoproteins required in the central nervous system. More generally, the fact that the symptoms arise in the absence of galactose suggests that galactose is required in some fashion.

7. Glc ⟶ G6P ⟶ G1P ⟶ UDP-Glc ⟶ UDP-Gal $\xrightarrow{\text{Glc}}$ lactose

2 Glucose + ATP + UTP ⟶ lactose + ADP + UDP + PPi

8. (*a*) 2 (glycolysis).
(*b*) 6 (gluconeogenesis requires ATP).

9. Net equation: Glucose + 2ATP ⟶ 2 glyceraldehyde-3-phosphate + 2 ADP + 2 H^+

$\Delta G^{0\prime} = 2.34$ kJ/mol

10. Net equation:

Glyceraldehyde-3-phosphate + 2 ADP + Pi + H^+ ⟶ lactate + 2 ATP + H_2O

$\Delta G^{0\prime} = -63$ kJ/mol

11. The glucose in cells is phosphorylated to glucose-6-phosphate. Because the equilibrium of this reaction strongly favours the product, glucose that enters the cell is rapidly and irreversibly converted to glucose-6-phosphate.

12. Net equation: Glycerol + 2 NAD^+ + ADP + P_i → pyruvate + 2 NADH + ATP + 2 H^+
13. (*a*) Increases (*b*) Decreases (*c*) Increases.
14. *Resting*: [ATP] high; [AMP] low; [acetyl-CoA] and [citrate] intermediate. *Running:* [ATP] intermediate; [AMP] high; [acetyl-CoA] and [citrate] low. Glucose flux through glycolysis increases during the anaerobic sprint, because: (1) the ATP inhibition of glycogen phosphorylase and PFK-1 is partially relieved, (2) AMP stimulates both enzymes, and (3) lower [citrate] and [acetyl-CoA] relieves their inhibitory effects on PFK-1 and pyruvate kinase, respectively.
15. The migrating bird relies on the highly efficient aerobic oxidation of fats, rather than the anaerobic metabolism of glucose used by a sprinting rabbit. The bird reserves its muscle glycogen for short bursts of energy during emergencies.
16. Anaerobically, there is a net of 2 mol ATP/mol glucose. Aerobically, the same net of 2 ATP is obtained plus 2 NADH because pyruvate is the product. Let us assume the cell uses the malate-aspartate shuttle where each NADH yeilds 2½ ATP. Thus, there is a net of 7 mol ATP/mol glucose. Each pyruvate is converted to AcCoA and the AcCoA is oxidized by the tricarboxylic acid cycle. Each mole of pyruvate then yields 12.5 mol ATP or 25 mol ATP for the 2 pyruvates. This gives a total of 32 mol ATP/mol glucose aerobically. Therefore, glycolysis must proceed 16 times as rapidly under anaerobic conditions to generate the same amount of ATP as occurs aerobically.
17. (*b*) Glycolysis.

CONTENTS

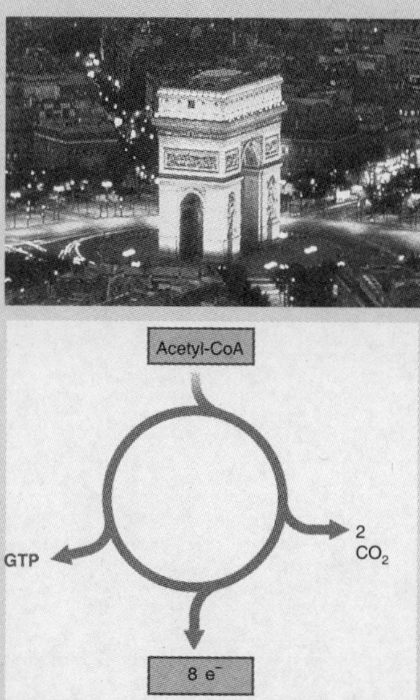

Roundabouts, or traffic circles, function as hubs to facilitate traffic flow.
The citric acid cycle is the biochemical hub of the cell, oxidizing carbon fuels, usually in the form of acetyl-CoA, as well as serving as a source of precursors for biosynthesis.
(Courtesy : (Upper) Chris Warren)

CHAPTER 22

Pyruvate Oxidation and Citric Acid Cycle

Under aerobic conditions, the cells obtain energy from ATP, produced as a result of breakdown of glucose. However, most plant and animal cells are aerobic and hence oxidize their organic fuels (carbohydrates etc.) completely to CO_2 and H_2O. Under these conditions, the pyruvate formed during glycolysis is not reduced to lactate or ethanol and CO_2 as occurs in anaerobic conditions but instead oxidized to CO_2 and H_2O aerobically. This process is called respiration by biochemists. The biochemists use the term in a *microscopic sense* and define **respiration** as *a sequence of molecular processes involved on O_2 consumption and CO_2 formation by cells.*

THREE STAGES OF CELL RESPIRATION

Respiration in cells occurs in 3 stages (refer Fig. 22–1):

1. First stage: *Oxidative decarboxylation of pyruvate to acetyl CoA and CO_2*

In this stage, the organic fuels such as carbohydrates, fatty acids and also some amino acids are oxidized to yield the 2-carbon fragments, the acetyl groups of the acetyl-coenzyme A.

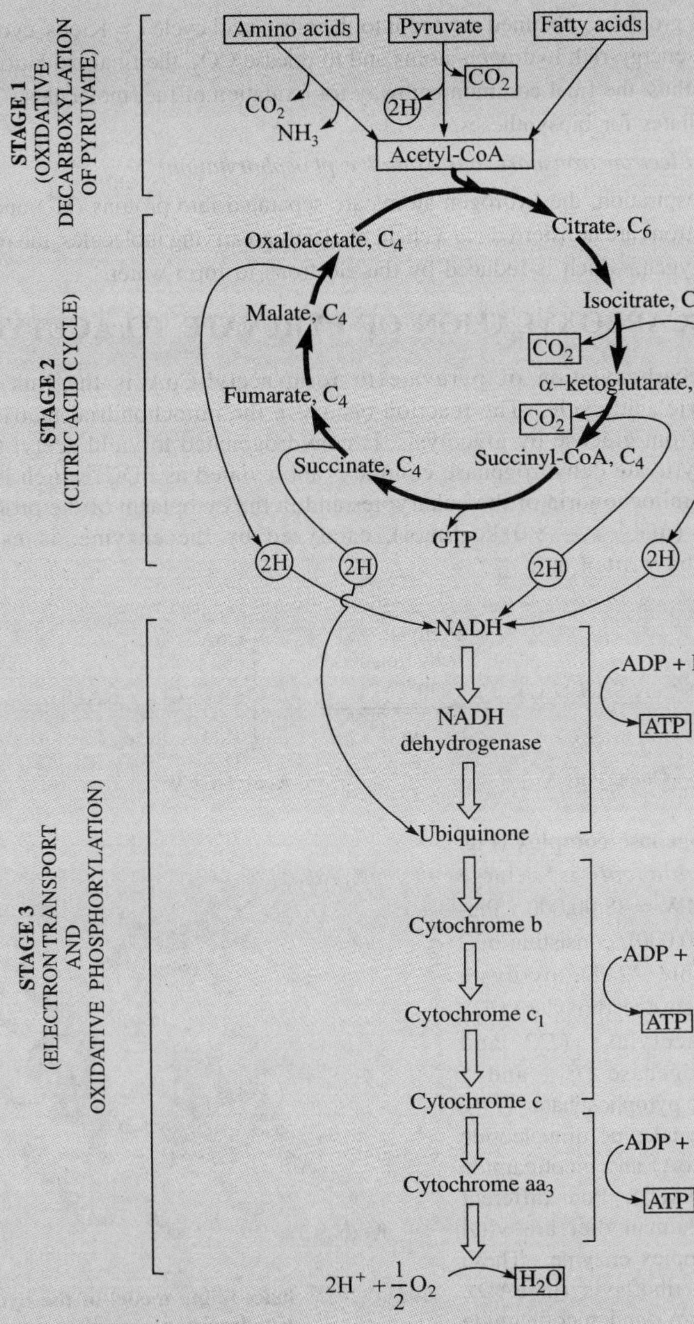

Fig. 22–1. Three stages in cell respiration

(*Adapted from Lehninger AL, 1984*)

2. Second stage: *Citric acid cycle or Acetyl CoA catabolism*

In this stage, the acetyl groups so obtained are fed into the citric acid cycle (= Krebs' cycle) which degrades them to yield energy-rich hydrogen atoms and to release CO_2, the final oxidation pruduct of organic fuels. It is, thus, the final common pathway for oxidation of fuel molecules. The cycle also provides intermediates for biosyntheses.

3. Third stage: *Electron transport and oxidative phosphorylation*

In this final stage of respiration, the hydrogen atoms are separated into protons (H^+) and energy-rich electrons. The electrons are transferred *via* a chain of electron-carrying molecules, the respiratory chain, to molecular oxygen, which is reduced by the electrons to form water.

OXIDATIVE DECARBOXYLATION OF PYRUVATE TO ACETYL-CoA

The oxidative decarboxylation of pyruvate to form acetyl-CoA is the link between glycolysis and the citric acid cycle. The reaction occurs in the mitochondrial matrix. Here, the pyruvate derived from glucose by glycolysis, is dehydrogenated to yield acetyl CoA and CO_2 by the enzyme pyruvate dehydrogenase complex (abbreviated as PDC) which is located in the matrix space of mitochondria of the eukaryotes and in the cytoplasm of the prokaryotes. The overall reaction ($\Delta G°' = -8.0$ kcal/mol), catalyzed by the enzyme, is essentially irreversible and may be written as :

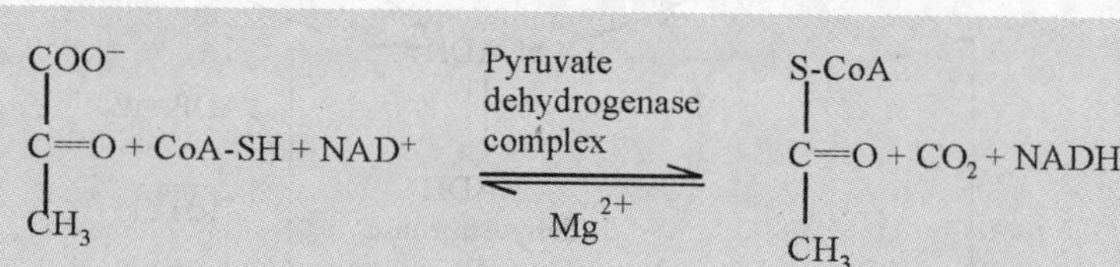

Pyruvate **Coenzyme A** **Acetyl-CoA**

Pyruvate dehydrogenase complex (Fig. 22–2) from *Escherichia coli* is a large multienzyme cluster (MW = 48,00,000 ; Pig-heart has MW = 1,00,00,000), consisting of 3 component units (Table 22–1): pyruvate dehydrogenase or pyruvate decarboxylase (E_1), dihydrolipoyl transacetylase (E_2) and dihydrolipoyl dehydrogenase (E_3), and 5 coenzymes *viz*., thiamine pyrophosphate (TPP), lipoic acid (LA), flavin adenine dinucleotide (FAD), coenzyme A (CoA) and nicotinamide adenine dinucleotide (NAD^+). Four different vitamins required in human diet are vital components of this complex enzyme. These are thiamine (in TPP), riboflavin (in FAD), pantothenic acid (in CoA) and nicotinamide (in NAD^+). Lipoic acid, however, is an essential vitamin or growth factor for many microorganisms but not so for higher animals, where it can be made from readily available precursors.

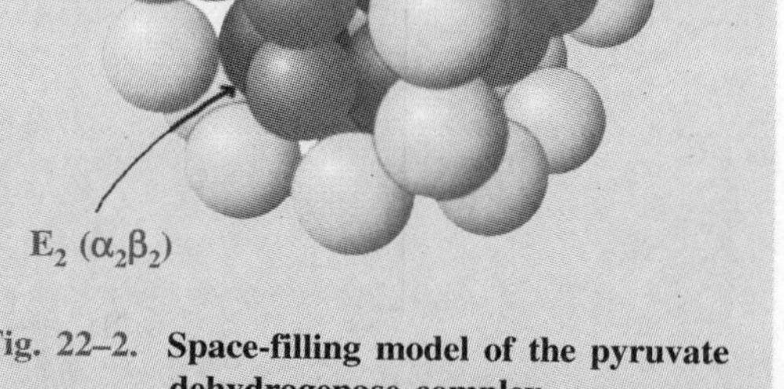

Fig. 22–2. Space-filling model of the pyruvate dehydrogenase complex

The pyruvate dehydrogenase component (E_1) is shown in *yellow*, the transacetylase core (E_2) in *red*, and the dihydrolipoyl dehydrogenase (E_3) in *green*.

Table 22-1. Pyruvate dehydrogenase complex of *Escherichia coli*

Enzyme Component	*Abbreviations*	*No. of chains*	*Prosthetic group*	*Reaction catalyzed*
Pyruvate dehydrogenase	A or E_1	24	TPP	Oxidative decarboxylation of pyruvate
Dihydrolipoyl transacetylase	B or E_2	24	Lipoamide	Transfer of the acetyl group to CoA
Dihydrolipoyl dehydrogenase	C or E_3	12	FAD	Regeneration of the oxidized form of lipoamide

This complex enzyme from *Escherichia coli* was first isloated and studied in detail by **Lester J. Reed** and **R. M. Oliver** of the University of Texas in 1974. Its molecule exhibits a distinct polyhedral appearance with a diameter of 350 ± 50 Å and a height of 225 ± 25 Å. The 'core' of the multienzyme cluster is occupied by dihydrolipoyl transacetylase (E_2) which consists of 24 polypeptide chain subunits, each containing two lipoic acid groups in amide linkage with the ε-amino groups of specific lysine residues in the active sites of the subunits. The 3-'D' structure of dihydrolipoyl transacetylase (E_2) catalytic domain is presented in Fig. 22–3. The other two enzyme components (pyruvate dehydrogenase and dihydrolipoyl dehydrogenase) are attachced to the outside of dihydrolipoyl transacetylase core. Pyurvate dehydrogenase contains bound TPP

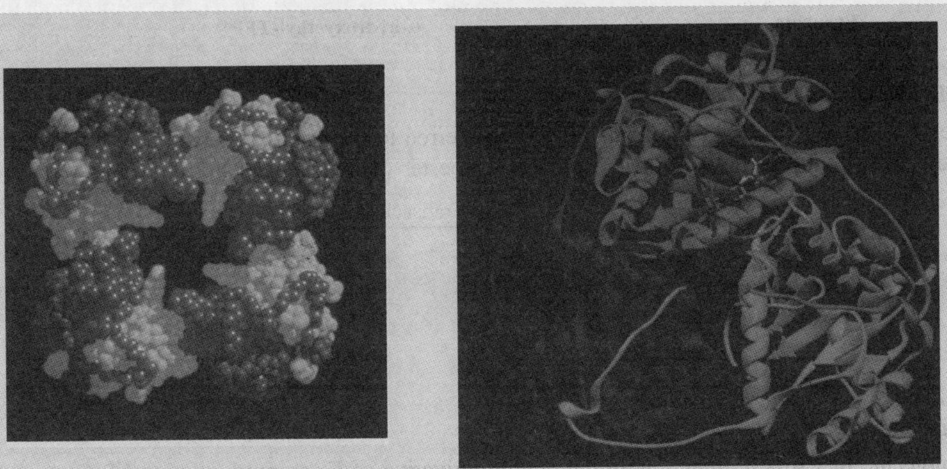

(*a*) (*b*)

Fig. 22–3. The x-ray structure of the *A. vinelandii* dihydrolipoyl transacetylase (E_2) catalytic domain

(*a*) **A space-filling drawing.** Each residue is represented by a sphere centred on its Cα atom. The 8 trimers (24 identical subunits) are arranged at the corners of a cube as viewed along one of the cube's 4-fold axes (only the forward half of the complex is visible). The edge length of the cube is ~ 125 Å. Note that the subunits in a trimer are extensively associated, but that the interactions between contacting trimers are relatively tenuous.

(*b*) **Ribbon diagram of a trimer as viewed along its 3-fold axis (along the cube's body diagonal) from outside the complex.** Coenzyme A (*purple*) and lipoamide (*light blue*), in skeletal form, are shown bound in the active site of the red subunit. Note how the N-terminal "elbow" of each subunit extends over a neighbouring subunit; its deletion greatly destabilizes the complex.

(From Wim Hol, University of Washington)

and dihydrolipoyl dehydrogenase contains bound FAD. The lipoyllysyl groups of the 'core' enzyme serve as `swinging or flexible arms' that can transfer hydrogen atoms and acetyl groups from one enzyme molecule to another in the multienzyme complex. All these enzymes and coenzymes are organized into a cluster to keep the prosthetic groups close together, thus allowing the reaction intermediates to react quickly with each other and also minimizing the side reactions.

The constituent polypeptide chains of the complex are held together by noncovalent forces. At alkaline pH, the complex dissociates into the pyurvate dehydrogenase component and a subcomplex of the other two enzymes. The transacetylase can then be separated from the dehydrogenase at neutral pH in the presence of urea. These 3 enzyme components associate to form the pyruvate dehydrogenase complex when they are mixed at neutral pH in the absence of urea.

There are 5 successive stages in the conversion of pyruvate into acetyl CoA. These are as follows:

Stage 1:

Pyruvate loses its carboxyl group as it reacts with the bound TPP of pyruvate dehydrogenase (E_1) to form the hydroxyethyl derivative of thiazole ring of TPP.

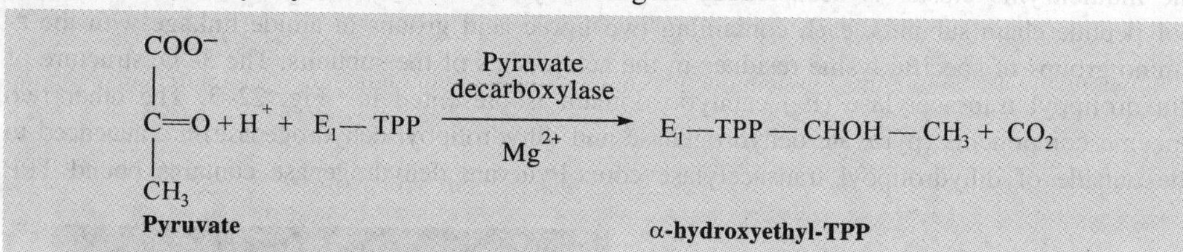

Stage 2:

The H atoms and acetyl group from TPP is transferred to the oxidized form of lipoyllysyl groups of the 'core' enzyme E_2 to form the 6-acetyl thioester of the reduced lipoyl groups.

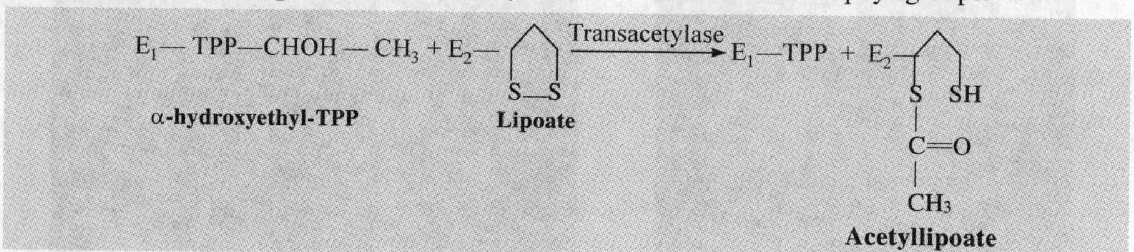

Stage 3:

A molecule of CoA-SH reacts with the acetyl derivative of E_2 to produce acetyl-S-CoA and the fully reduced (or dithiol) form of lipoyl groups.

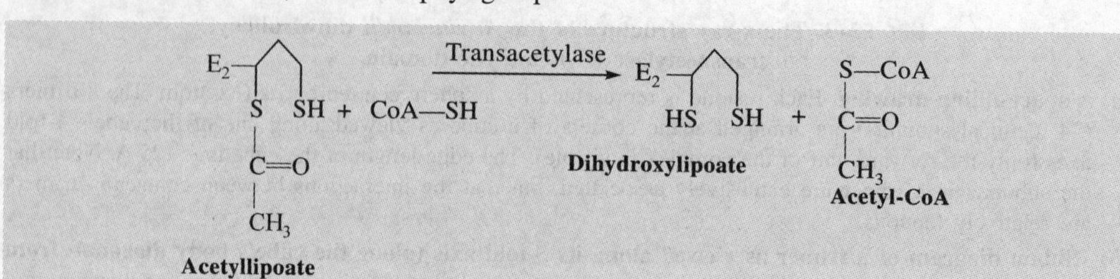

Stage 4:

The fully reduced form of E_2 is acted upon by E_3 which promotes transfer of H atoms from the reduced lipoyl groups to the FAD prosthetic group of E_3.

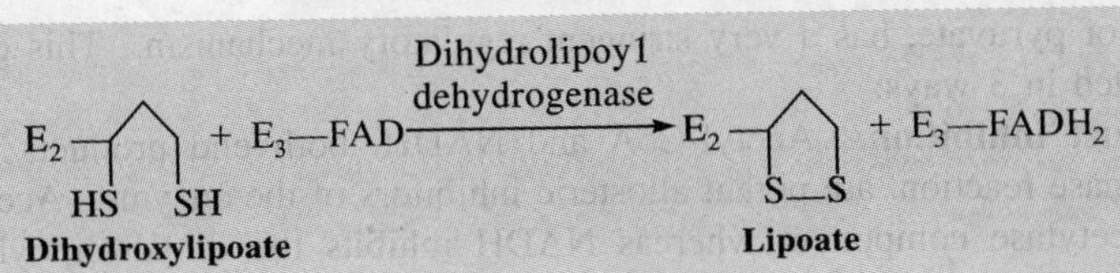

Stage 5:

In this last stage, the reduced FAD group of E_3 transfers hydrogen to NAD^+, forming NADH.

$$E_3\text{—}FADH_2 + NAD^+ \xrightarrow[\text{dehydrogenase}]{\text{Dihydrolipoyl}} E_3\text{—}FAD + NADH + H^+$$

Regulation of Oxidative Decarboxylation of Pyruvate

The conversion of pyruvate into acetyl-CoA is a key irreversible step in the metabolism of animals becaause the animals cannot convert acetyl-CoA into glucose. The carbon atoms of glucose has two fates : (*a*) oxidation of CO_2 *via* the citric acid cycle and (*b*) incorporation into lipid. Therefore, it seems that pyruvate dehydrogenase complex (PDC) which catalyzes oxidative

Energy Charge

Energy charge (reference on the next page), a term coined by **Daniel Atkinson** in 1970, is the energy status of the cell. In mathematical terms, **Atkinson** defined energy charge as that fraction of the adenylic system (ATP + ADP + AMP) which is composed of ATP. In other words, it is proportional to the mole fraction of ATP plus half the mole fraction of ADP, given that ATP contains two and ADP contains one anhydride bond. Hence, the energy charge is given by the equation:

$$\text{Energy charge} = \frac{[\text{ATP}] + \frac{1}{2}[\text{ADP}]}{[\text{ATP}] + [\text{ADP}] + [\text{AMP}]}$$

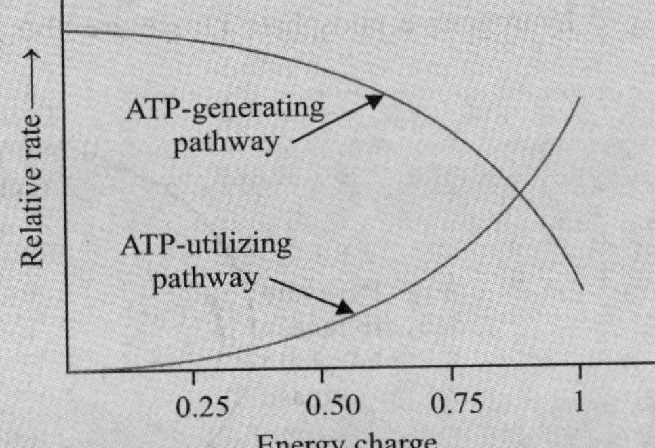

Energy charge and the metabolic rate

High concentrations of ATP inhibit the relative rates of a typical ATP-generating (= catabolic) pathway and stimulate that of the typical ATP-utilizing (= anabolic) pathway.

The energy charge is one if the total adenine nucleotide pool is fully phosphorylated to ATP (*i.e.*, it is all ATP) and zero if the adenine nucleotides are fully `empty' and present only as AMP (*i.e.*, it is all AMP). The energy charge may have values ranging between these two extremes. Normally, the energy charge of cells ranges between 0.80 and 0.95 which means that adenylate system is almost completely charged. A high energy charge inhibits all ATP-generating (*i.e.*, catabolic) pathways but stimulates the ATP-utilizing (*i.e.*, anabolic) pathways.

In plots of the reaction rates of such pathways versus the energy charge, the curves are steep near an energy charge of 0.9, where they usually intersect (adjoining figure). It is evident that the control of these pathways has evolved to maintain the energy charge within rather narrow limits. In other words, *the energy charge, like the pH of a cell, is buffered.*

An alternative index of the energy status is the ***phosphorylation potential,*** which is given by the equation:

$$\text{Phosphorylation potential} = \frac{[\text{ATP}]}{[\text{ADP}][\text{Pi}]}$$

The phosphorylation potential, contrary to the energy charge, depends on the concentration of Pi and is directly related to the free energy-storage available from ATP.

decarboxylation of pyruvate, has a very stringent regulatory mechanism. This complex enzyme system is regulated in 3 ways:

A. End-product inhibition. Acetyl-CoA and NADH, both end products of the pyruvate dehydrogenase reaction, are potent allosteric inhibitors of the enzyme. Acetyl-CoA inhibits the transacetylase component whereas NADH inhibits the dihydrolipoyl dehydrogenase component. The inhibitory effects are reversed on addition of coenzyme A and NAD^+ respectively.

B. Feedback regulation. The activity of PDC is controlled by the *energy charge*. The pyruvate dehydrogenase component is specifically inhibited by GTP and activated by AMP.

C. Covalent modification. Under conditions of high concentrations of ATP, acetyl-CoA and those of the cycle intermediates, further formation of acetyl-CoA is slowed down. This is accomplished by what is called *covalent modification*. ATP inactivates pyruvate dehydrogenase complex (PDC) by phosphorylating 3 different serine residues of its pyruvate dehydrogenase component in the presence of an auxiliary enzyme, *pyruvate dehydrogenase kinase*. NADH, ATP and acetyl-CoA stimulate the rate of phosphorylation whereas pyruvate, Ca^{2+} and K^+ inhibit phosphorylation. However, if the demand for ATP increases causing ATP level to decline, the inactive enzyme complex becomes active again by the hydrolytic removal of phosphate group from its pyruvate dehydrogenase component in the presence of another enzyme, *pyruvate dehydrogenase phosphate kinase*. Dephosphorylation is enhanced by high levels of pyruvate, Ca^{2+} and Mg^{2+}. Both pyruvate dehydrogenase kinase and pyruvate dehydrogenase phosphate kinase are also present in PDC.

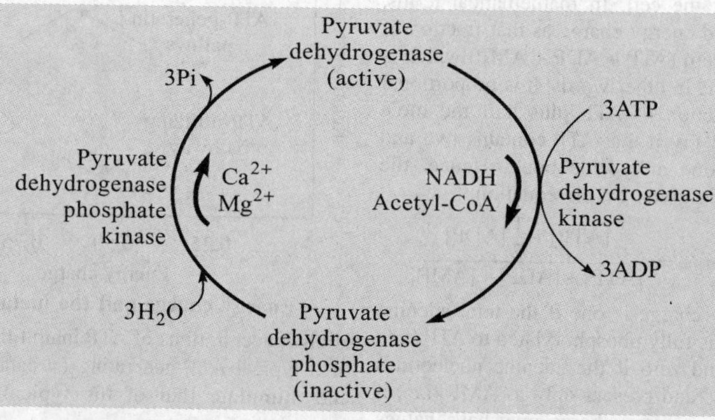

CITRIC ACID CYCLE OR KREBS CYCLE OR ACETYL-CoA CATABOLISM

The most nearly universal pathway for aerobic metabolism is the cyclic series of reactions, termed *citric acid cycle* (CAC) or *Krebs cycle*. The first name has been applied because citric acid (Fig. 22–4) is the first intermediate formed in this cycle. The second name has been given in honour of its most illustrious proponent, **Sir Hans A. Krebs**, who first postulated it in 1937 and the cycle has since then been slightly modified in a form we know today. However, a third name *tricarboxylic acid cycle* (TCA) was given to it some years after Krebs postulated the cycle because it was then not certain whether citric acid or some other tricarboxylic acid (*e.g.*, isocitric acid) was first product of the cycle. Since we now know with certainty that citric acid is indeed the first tricarboxylic acid formed, the use of the term `tricarboxylic acid cycle' is not appropriate and hence be discouraged. *This cycle forms the hub of metabolism of almost all cells* and has truly been regarded as the most important single discovery in the history of metabolic biochemistry.

In fact, more of the details have been worked out and more of the ramifications explored for the citric acid cycle than for any other pathway, except perhaps glycolysis.

The citric acid cycle is a series of reactions in mitochondria that bring about the complete oxidation of acetyl-CoA to CO_2 and liberate hydrogen equivalents which ultimately form water. This cyclic sequence of reactions provides electrons to the transport system, which reduces oxygen while generating ATP. *The citric acid cycle is the final common pathway for the oxidation of fuel molecules– amino acids, fatty acids and carbohydrates.* Most fuel molecules enter the cycle as acetyl-CoA. Two fundamental *differences* exist between glycolysis and citric acid cycle:

1. Glycolysis takes place by a *linear* sequence of 10 enzyme-catalyzed reactions. In contrast, citric acid cycle proceeds in a *cyclic* way by 8 enzyme-catalyzed reactions.
2. The reactions of glycolysis occur in the *cytosol* in contrast with those of citric acid cycle which occur inside *mitochondria*.

Fig. 22–4. Crystals of citric acid, as viewed under polarized light

SIR HANS ADOLF KREBS

(LT, 1900–1981)

A German born and trained British biochemist, **Krebs** was one of the outstanding scientists of the twentieth century. From 1926 to 1930, he worked in Berlin with **Otto Warburg**, himself one of the great pioneers of modern biochemistry. In 1932, Krebs worked out the outlines of the *urea cycle* with a medical student **Kurt Henseleit** at the University of Freiburg. In 1933, he emigrated to England to the University of Cambridge. Later, he moved to the University of Shefield, where he worked out a major part of the citric acid cycle (or CAC) using mainly the pigeon breast muscles, a *very actively respiring tissue*. In 1954, he became head of the Biochemistry Department at Oxford. On his retirement from that position in 1967, he engaged himself in examining the regulation of metabolism in the Department of Medicine, at Oxford until his death. Krebs shared the **1953 Nobel Prize for Physiology or Medicine** for his discovery of the citric acid cycle or "Krebs cycle", with **Fritz Albert Lipmann**, the 'father' of ATP cycle. Krebs died on November 22, 1981 at Oxford, UK. It is of interest to note that when Krebs' original manuscript on tricarboxylic acid cycle (or CAC cycle) was submitted for publication to *Nature*, it was rejected because of a lack of publishing space. It was subsequently published in *Enzymologia*. Dr. Krebs proudly displayed the rejection letter throughout his career as encouragement for young scientists.

"June 1937

The editor of NATURE presents his compliments to Dr. H. A. Krebs and regrets that as he has already sufficient letters to fill the correspondence columns of NATURE for seven or eight weeks, it is undesirable to accept further letters at the present time on account of the time delay which must occur in their publication.

If Dr. Krebs does not mind much delay the editor is prepared to keep the letter until the congestion is relieved in the hope of making use of it.

He returns it now, in case Dr. Krebs prefers to submit it for early publication to another periodical."

ENZYMES INVOLVED IN THE CITRIC ACID CYCLE

A landmark discovery was made in 1948 by **Eugene P. Kennedy** and **Albert L. Lehninger** when they found that rat liver mitochondria could catalyze the oxidation of pyruvate and all the intermediates of the citric acid cycle by molecular oxygen. Since only Mg^{2+} and an adenylic acid (ATP, ADP or AMP) had to be added, this finding meant that mitochondria contain not only all the enzymes and coenzymes required for the citric acid cycle but also those needed to transport the electrons from the substrate to molecular oxygen. Later work has shown that some enzymes (such as aconitase, fumarase and malate dehydrogenase), required in the cycle, are also found in the cytoplasm, but the reactions they catalyze are independent of the mitochondrial oxidation process. The different enzymes involved in the citric acid cycle are located either in the inner membrane or in the matrix space of the mitochondria (Fig. 22–5).

ALBERT L. LEHNINGER
(LT, 1917–1986)

An American biochemist who along with **Eugene P. Kennedy** discovered, in 1948, that the enzymes involved in the citric acid and respiratory cycles of energy transformation in the cell are located in mitochondria. With regard to the regulation of enzyme activity, he held the view:

"Living cells are self-regulating chemical engines, tuned to operate on the principle of maximum economy."

He also wrote a very famous, self-explanatory and highly informative text *Principles of Biochemistry*, first published in 1982 which transformed the teaching of Biochemistry.

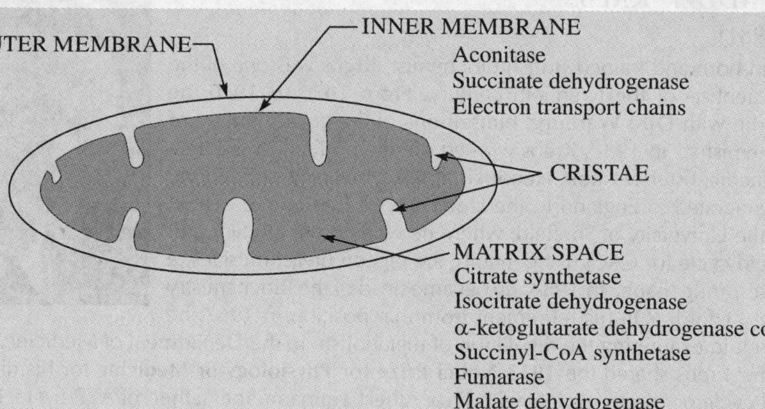

Fig. 22–5. Location of citric acid cycle enzymes in mitochondria

The various enzymes involved in different reactions of the cycle are presented in Table 22–2 along with their characteristics.

OVERVIEW OF THE CITRIC ACID CYCLE

An overall pattern of the citric acid cycle is represented in Fig. 22–6. To begin with, acetyl-CoA donates its 2-carbon acetyl group to the 4-carbon tricarboxylic acid, **oxaloacetate** to form a 6-carbon tricarboxylic acid, **citrate.** Citrate is then transformed into another 6-carbon tricarboxylic acid, **isocitrate.** Isocitrate is then dehydrogenated with the loss of CO_2 (*i.e.*, oxidatively decarboxylated) to yield the 5-carbon dicarboxylic acid, **α-ketoglutarate.** The latter compound

Table 22–2. Enzymes of the citric acid cycle

*Step No.**	*Enzyme*	*Location (in mitochondria)*	*Coenzymes(s) and Cofactor(s)*	*Inhibitor(s) (kcal/mol)*	*Type of Reaction Catalysed*	*$\Delta G^{o\prime}$ (kcal/mol)*
1	Citrate synthase	Matrix space	CoA	Monofluoroacetyl-CoA	Condensation	–7.7
2	Aconitase	Inner membrane	Fe^{2+}	Fluoroacetate	Isomerization	+1.59
3	Isocitrate dehydrogenase	Matrix space	NAD^+, $NADP^+$ Mg^{2+}. Mn^{2+}	ATP	Oxidative decarboxylation	–2.0
4	α-ketoglutarate dehydrogenase complex	Matrix space	TPP, LA, FAD, CoA, NAD^+	Aresenite, Succinyl-CoA NADH	Oxidative decarboxylation	–8.0
5	Succinyl-CoA synthetase	Matrix space	CoA	—	Substrate-level phosphorylation	–7.0
6	Succinate dehydrogenase	Inner membrane	FAD	Malonate, Oxaloacetate	Oxidation	~0
7	Fumarase	Matrix space	None	—	Hydration	≈0
8	Malate dehydrogenase	Matrix space	NAD^+	NADH	Oxidation	+7.1

* The numbers in this column correspond to the reactions as described in the text and catalyzed by the enzyme specified.

undergoes further oxidative decarboxylation to yield the 4-carbon dicarboxylic acid, succinate and to release a second molecule of CO_2. Succinate then undergoes three successive enzyme-catalyzed reactions to regenerate oxaloacetate, with which the cycle began. Thus, oxalocetate is regenerated after one turn of the cycle and is now ready to react with another molecule of acetyl-CoA to start a second turn.

> As per Circular No 200 of the Committee of Editors of Biochemical Journals Recommendations (1975), there is a standard biochemical convention, according to which the names of the various carboxylic acids are written by adding the suffix **-ate** in the name of the acid mentioned, as it is taken to mean any mixture of free acid and the ionized form(s) (according to pH) in which the cations are not specified.

In each turn of the cycle, one acetyl group (*i.e.*, 2 carbons) enters as acetyl-CoA and 2 moles of CO_2 are released. In each turn, one mole of oxaloacetate is used to form citrate but after a series of reactions, the oxaloacetate is regenerated again. Therefore, no net removal of oxaloacetate occurs when the citric acid cycle operates and one mole of oxaloacetate can, theoretically, be sufficient to bring about oxidation of an indefinite number of acetyl groups.

Another diagram (Fig 22–7, akin to Fig. 22–6) shows the positions of various intermediate products, whether drawn off or replenished in the citric acid cycle.

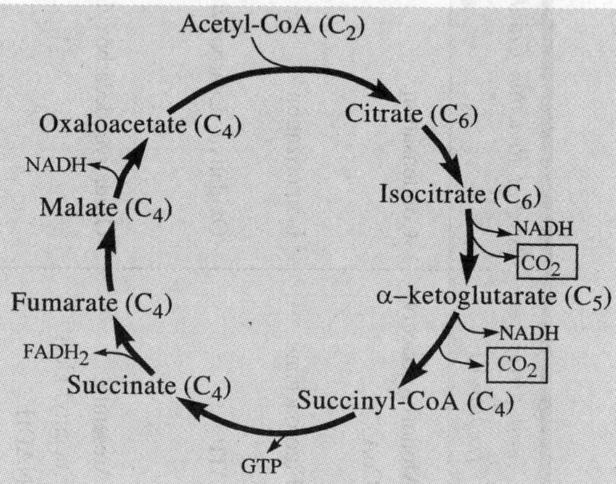

Fig. 22–6. An outline of the citric acid cycle

In bracket, the number of carbon atoms in each intermediate is shown. However, succinyl-CoA has 4 carbon atoms in its succinyl groups, the portion of the molecule which is converted into free succinate.

REACTION STEPS OF THE CITRIC ACID CYCLE

The citric acid cycle proper consists of a total of 8 successive reaction steps,. each of which is catalyzed by an enzyme. The details of these reactions and those of the enzymes which catalyze them are given below:

Step 1 : *Condensation of acetyl-CoA with oxaloacetate*

> For oxaloacetate, some biochemical journals use the spelling, **'oxalacetate'.**

The cycle starts with the joining of a 4-carbon unit, oxaloacetate (OAA) and a 2-carbon unit,the acetyl group of acetyl-CoA. Oxaloacetate reacts with acetyl-CoA plus water to yield a C_6 compound, citrate plus coenzyme A in the presence of a regulatory enzyme, **citrate synthase** (variously called as *condensing enzyme* or *citrate oxaloacetate lyase* (*CoA acetylating or citrogenase*). The enzyme condenses the methyl carbon of acetyl group of acetyl-CoA and the carbonyl carbon of oxaloacetate forming a transient intermediate compound, citryl-CoA (which is the coenzyme A thiol-ester of citric acid) on the active site of the enzyme. Citryl-CoA then undergoes hydrolysis so that its thiol ester bond is cleaved to set free coenzyme A and to form citrate. Both coenzyme A and citrate are then released from the active site of the enzyme. Thus, this reaction is an aldol condensation followed by hydrolysis. The overall reaction has a large negative standard free energy of hydrolysis (−7.7 kcal/mol). Therefore, the reaction proceeds far to the right, *i.e.*, in the direction of citrate synthesis. The coenzyme A formed in this reaction is now free to participate in the oxidative decarboxylation of another molecule of acetyl-CoA for entry into the citric acid cycle.

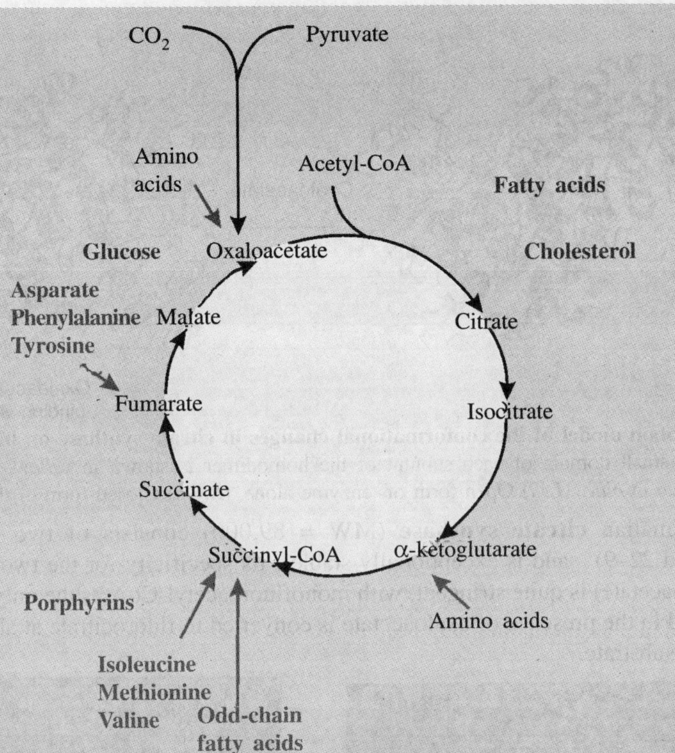

Fig. 22–7. A diagram of the citric acid cycle, indicating the positions at which intermediates are drawn off for use in anabolic pathways *(red arrows)* and the points where anaplerotic reactions replenish depleted cycle intermediates *(green arrows)*.

Reactions involving amino acid transamination and deamination are reversible, so their direction varies with metabolic demand.

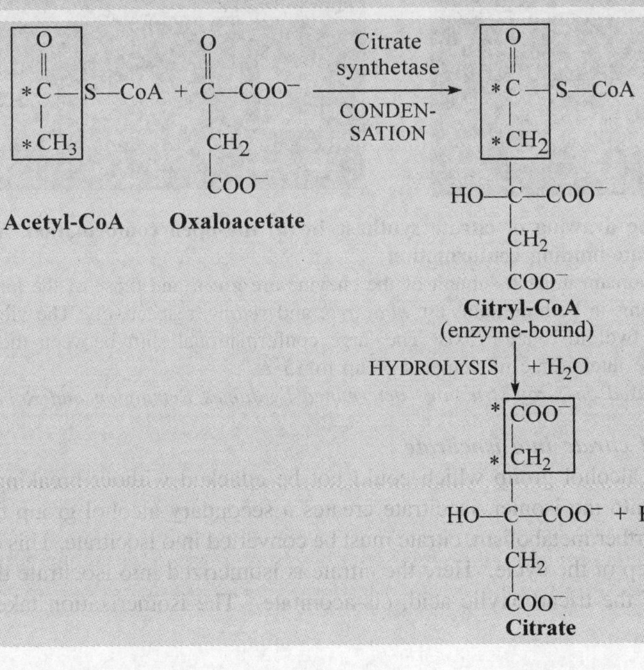

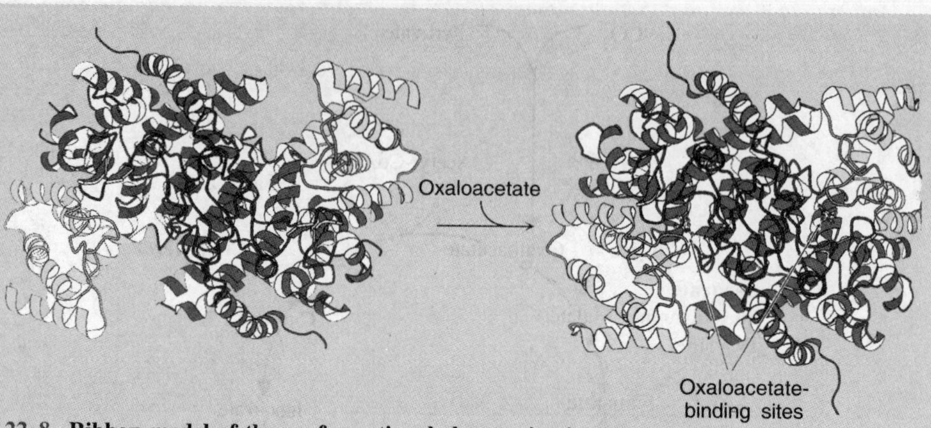

Fig. 22–8. Ribbon model of the conformational changes in citrate synthase on binding oxaloacetate
The small domain of each subunit of the homodimer is shown in *yellow*, the large domain is shown in *blue*. (*Left*) Open form of enzyme alone. (*Right*) Closed form of the liganded enzyme.

The mammalian **citrate synthase** (MW = 89,000) consists of two identical subunits (Figs. 22–8 and 22–9) and is exceptionally stable. Its specificity for the two substrates (acetyl-CoA and oxaloacetate) is quite stringent, with monofluoroacetyl-CoA as the only alternate reactant. This compound in the presence of oxaloacetate is converted to fluorocitrate at about 1/10th the rate of the natural substrate.

(*a*) 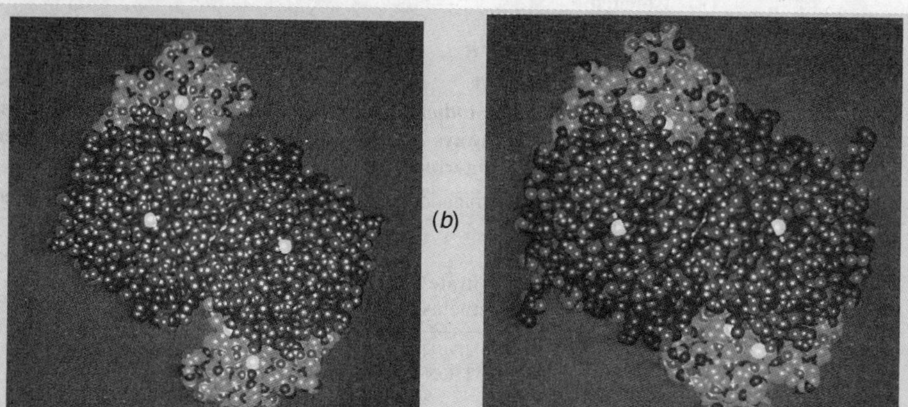 (*b*)

Fig. 22–9. A space-filling drawing of citrate synthase in (*a*) the open conformation and (*b*) the closed, substrate-binding conformation
The C atoms of the small domain in each subunit of the enzyme are *green* and those of the large domain are *purple*, N, O, and S atoms in both domains are *blue, red,* and *yellow*, respectively. The view is along the homodimeric protein's twofold rotation axis. The large conformational shift between the open and closed forms entails relative interatomic movements of up to 15 Å.
(Based on x-ray structures determined by James Remington and Robert Huber)

Step 2 : *Isomerization of citrate into isocitrate*

Citrate has a tertiary alcohol group which could not be attacked without breaking a carbon bond, but rearrangement into the isomer, isocitrate creates a secondary alcohol group that can be oxidized. Therefore, for further metabolism, citrate must be converted into isocitrate. This conversion is accomplished by this step of the cycle. Here the citrate is isomerized into isocitrate through the intermediary formation of the tricarboxylic acid, cis-aconitate. The isomerization takes place in 2 stages:

(*a*) *dehydration* of citrate to cis-aconitate, which remains bound to the enzyme, and

(*b*) *rehydration* of cis-aconitate to isocitrate.

The result is an interchange of an H and OH. Both the reactions are reversible and are catalyzed by the same enzyme, *aconitase* (= *aconitate hydratase*). When aconitase catalyzes the addition of water to the double bond of cis-aconitate, the OH group may be added to either carbon; in one case citrate is formed and in the other, isocitrate. The enzyme also catalyzes the removal of water from either citrate or isocitrate to form the intermediate, *cis*-aconitate.

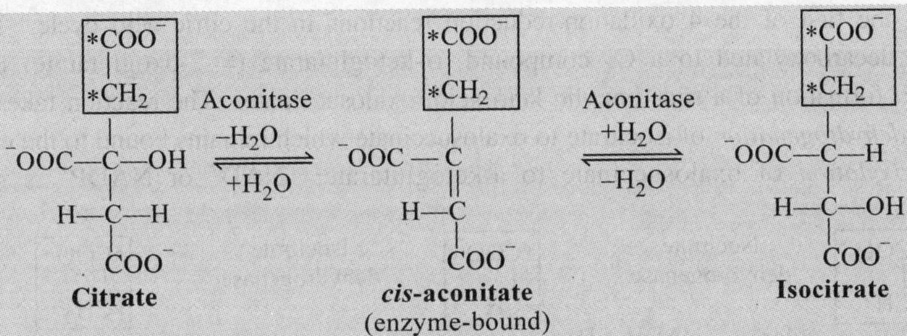

It is possible that *cis*-aconitate may not be an obligatory intermediate between citrate and isocitrate but may, in fact, be a side branch from the main pathway.

Experiments using ^{14}C-labelled intermediates indicate that citrate, although being a symmetrical molecule, reacts in an asymmetrical manner and that the *aconitase always acts on that part of the citrate molecule which is derived from oxaloacetate.* The standard free energy change of the overall reaction (+1.59 kcal/mol) is small enough so that the reaction can go in either direction. Although the equilibrium mixture (at pH 7.4 and 25°C) contains about 90% citrate, 4% *cis*-aconitate and 6% isocitrate, the reaction in the cell is driven to the right (*i.e.*, towards isocitrate formation) because the product isocitrate is rapidly transformed in the subsequent step of the cycle.

Aconitase is a rather complex enzyme. Pig heart aconitase (MW = 89,000) is a dimer of identical units, each containing iron and sulfur atoms arranged in a cluster called iron-sulfur centre. However, aconitase differs from other enzymes of its own category, *i.e.*, hydratases (= enzymes catalyzing the reversible hydration of double bonds) in that they lack such an iron-sulfur centre.

Toxicity of Fluoroacetate: A Suicide Substrate

Fluoroacetate, originally obtained from plants, is a potent toxin. It is activated as fluoroacetyl-CoA and then condenses with oxaloacetate (OAA) to form fluorocitrate:

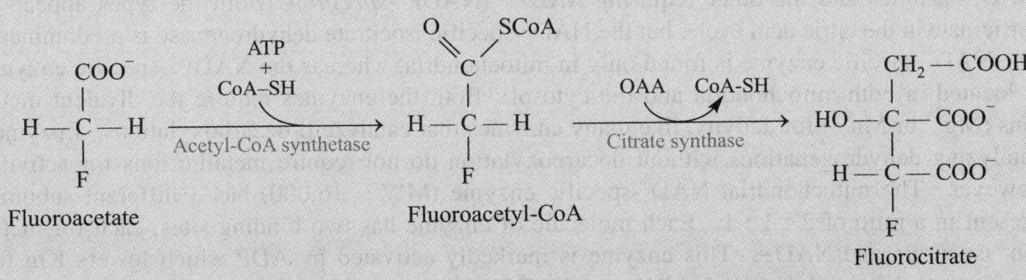

Death results from inhibition of the citric acid cycle by 2-fluorocitrate, a strong inhibitor of *aconitase.* Fluoroacetate is an example of a 'suicide substrate', a compound that is not toxic *per se* but is metabolically activated to a toxic product. Thus, the cell is said to commit suicide by converting an apparently harmless substrate to a lethal toxin. Similar processes are involved in the activation of many environmental procarcinogens to carcinogens that induce mutations in DNA.

The iron-sulfur centre presumably acts as a prosthetic group, although its precise function is still not known. The presence of such a centre, however, is baffling because there occurs no transfer of electrons during the reaction. The reaction is inhibited by fluoroacetate which, in the form of fluoroacetyl-CoA, condenses with oxaloacetate to form fluorocitrate. The latter inhibits aconitase and thus prevents utilization of citrate. Fluoroacetate occurs naturally in the leaves of a South African plant, *Dichopetalum cymosum* which is toxic to animals that feed on it.

Step 3 : *Oxidative decarboxylation of isocitrate*

This is the first of the 4 oxidation-reduction reactions in the citric acid cycle. Isocitrate is oxidatively decarboxylated to a C_5 compound, α-ketoglutarate (= 2-oxoglutarate) through the intermediary formation of a tricarboxylic keto acid, oxalosuccinate. The reaction takes place in 2 stages : (*a*) *dehydrogenation* of isocitrate to oxalosuccinate which remains bound to the enzyme, and (*b*) *decarboxylation* of oxalosuccinate to a-ketoglutarate. NAD^+ or $NADP^+$ is required as

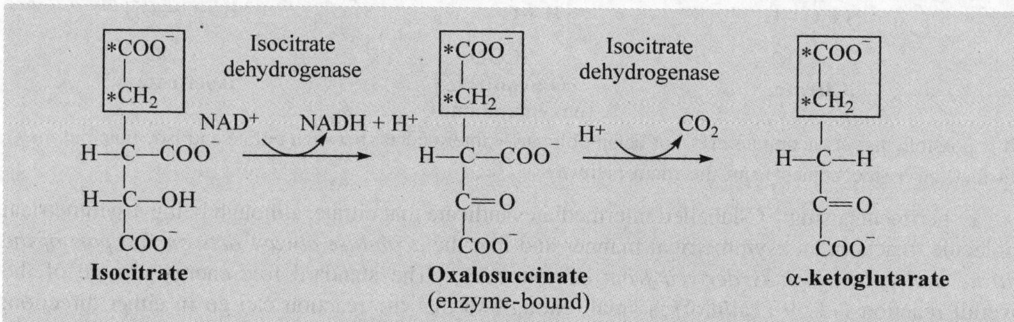

electron acceptor in the first stage. Both the reactions are irreversible and are catalyzed by the same enzyme, *isocitrate dehydrogenase*. Equilibrium favours α-ketoglutarate formation as, under physiologic conditions, the $\Delta G°'$ is equal to –5 kcal/mol (*i.e.*, a large negative value). This is the first '*committed step*' in the Krebs cycle as it has no other function than to participate in the cycle.

It may be noticed that the carbonyl group of the intermediate oxalosuccinate is in the β position relative to the middle carboxylate group (The carbonyl group is also in the α-position with respect to the top carboxylate group, but that is not relevant to our point). And we know from the knowledge of organic chemistry that β-keto acids are easily decarboxylated and oxalosuccinate does it.

There are 2 types of **isocitrate dehydrogenase**, one requiring NAD^+ as electron acceptor (*NAD^+-specific*) and the other requiring *$NADP^+$* ($NADP^+$-*specific*). Both the types appear to participate in the citric acid cycle, but the NAD^+-specific isocitrate dehydrogenase is predominant. The NAD^+-specific enzyme is found only in mitochondria, whereas the $NADP^+$-specific enzyme is located in both mitochondria and the cytosol. Both the enzymes require the divalent metal ions (Mg^{2+} or Mn^{2+}) for activity, like many enzymes that catalyze β-decarboxylations. Enzymes catalyzing dehydrogenations without decarboxylation do not require metallic ions for activity, however. The mitochondrial NAD^+-specific enzyme (MW = 16,000) has 3 different subunits present in a ratio of 2 : 1 : 1. Each molecule of enzyme has two binding sites, each for metal ion, isocitrate and NAD+. This enzyme is markedly activated by ADP which lowers Km for isocitrate and is inhibited by NADH and NADPH.

The $NADP^+$-specific (or $NADP^+$-dependent) enzymes show none of these peculiar properties. They also differ from the NAD^+-dependent enzymes in their size and in the nature of the reaction catalyzed. Furthermore, the $NADP^+$-specific enzyme also oxidizes oxalosuccinate when added to the system. On the contrary, the $NADP^+$-specific enzyme, which functions as an integral component of the citric acid cycle, does not decarboxylate added oxalosuccinate. The metabolic significance of $NADP^+$-specific enzyme lies not in the operation of the citric acid cycle but as a source of reducing equivalents.

Step 4 : *Oxidative decarboxylation of a-ketoglutarate*

One of the peculiarities of the citric acid cycle is that *it contains two successive oxidative decarboxylation steps* (*Steps 3 and 4*) *of quite different reaction types*. In this step, α-ketoglutarate is oxidatively decarboxylated, in a manner analogous to the oxidative decarboxylation of pyruvate, to form a C_4 thiol ester, succinyl-CoA and CO_2 by the enzyme *α-ketoglutarate dehydrogenase complex* (α-KDC) which is located in the mitochondrial space. The reaction has a high negative value of $\Delta G^{\circ\prime}$ (– 8.0 kcal/mol), and is therefore physiologically irreversible and proceeds far to the right. The high negative $\Delta G^{\circ\prime}$ value is sufficient for the creation of a high energy bond in succinyl-CoA.

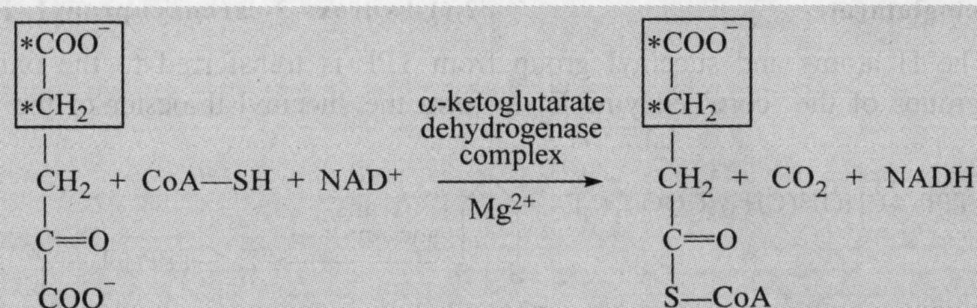

α-ketoglutarate **Succinyl-CoA**

The reaction is virtually identical to the pyruvate dehydrogenase complex (PDC) reaction in that both promote the oxidation of an a-keto acid with loss of the carboxyl group as CO_2. However, there exists an important difference between the two : *the α-KDC system does not have so elaborate a regulatory mechanism as PDC system.*

The immediately preceding isocitrate dehydrogenase reaction (*i.e.*, Step 3) is also an oxidative decarboxylation, but of a 3-hydroxycarboxylate. Moreover, the standard free energy change value of this reaction (–5 kcal/mol) is too less to support the formation of an extra high-energy bond (which is formed in Step 4) even at low CO_2 concentration in the tissues.

The **α-ketoglutarate dehydrogenase complex** from pig heart (MW = 33,00,000) is much smaller than the PDC from the same source (MW = 1,00,00,000). The pig heart α-KDC is a large multienzyme cluster consisting of 3 enzyme components, *viz.*, α-ketoglutarate dehydrogenase or α-ketoglutarate decarboxylase (12 moles per mole of the complex), transsuccinylase (24 moles) and dihydrolipoyl dehydrogenase (12 moles). The complex also requires the same 5 coenzymes, as required by pyruvate dehydrogenase complex, for activity, *viz.*, thiamine pyrophosphate (6 moles), lipoic and (6 moles), flavine adenine dinucleotide (8 moles), coenzyme A and nicotinamide adenine dinucleotide. The compelx enzyme molecule is comparable in size to ribosomes.

The transsuccinylase (B′) component, like the transacetylase of PDC, forms the 'core' of the multienzyme complex while the α-ketoglutarate dehydrogenase (A′) and dihydrolipoyl dehydrogenase (C′) components are arranged on the periphery,. Again, A′ binds to B′ and B′ binds to C′ but A′ does not bind directly to C′ . The α-ketoglutarate dehydrogenase (A′) and transsuccinylase (B′) components are different from the corresponding components (A and B) in the pyruvate dehydrogenase complex. However, the dihydrolipoyl dehydrogenase components (C and C′) of the two enzyme complexes are similar. Lipoic acid is attached to the core transsuccinylase by forming an amide bond with lysine side chains. This places the reactive disulfide groups (—S—S—) at the end of a long flexible chain. The ability of the chain to swing the disulfide group in contact with the different proteins is an important feature of the enzyme complex. As in the case of pyruvate oxidation, *arsenite* inhibits the reaction, causing the substrate α-ketoglutarate to accumulate. The α-KDC is inhibited by both succinyl-CoA and NADH, the former being more effective.

There are 5 successive stages in the conversion of a-ketoglutarate to succinyl-CoA. These are as follows:

Stage 1. α-ketoglutarate loses its carboxyl group as it reacts with the bound TPP of a-ketoglutarate dehydrogenase (E_1) to form the hydroxy-carboxypropyl derivative of thiazole ring of TPP.

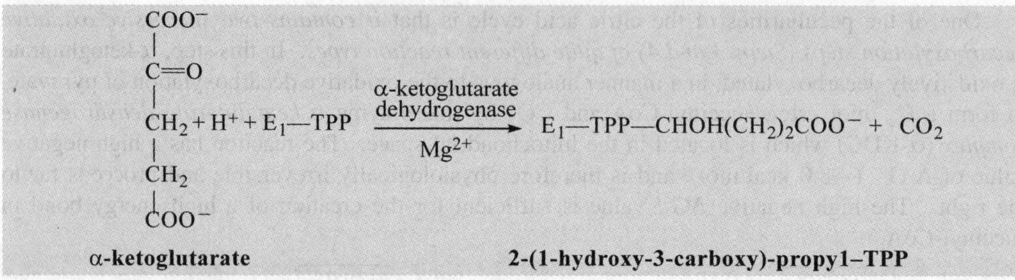

Stage 2. The H atoms and succinyl group from TPP is transferred to the oxidized form of lipoyllysyl groups of the 'core' enzyme E_2 to form the succinyl thioester of the reduced lipoyl groups.

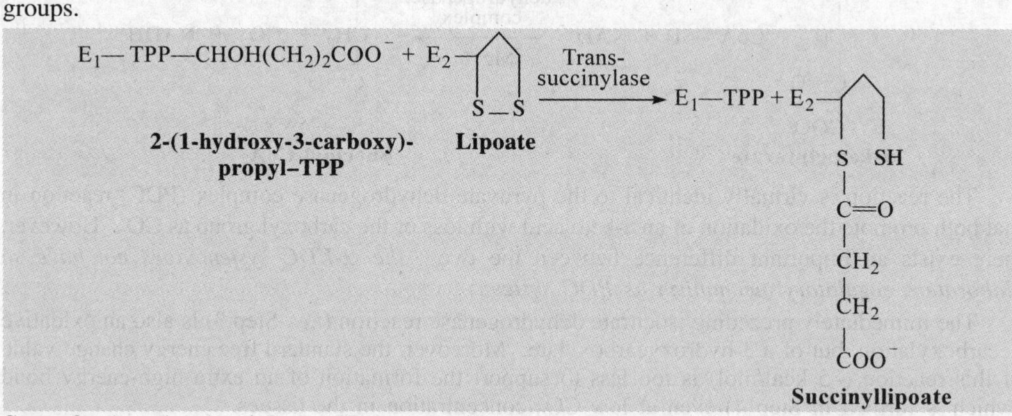

Stage 3. A molecule of CoA—SH reacts with the succinyl derivative of E_2 to produce succinyl-S-CoA and the fully reduced (or dithiol) form of lipoyl groups.

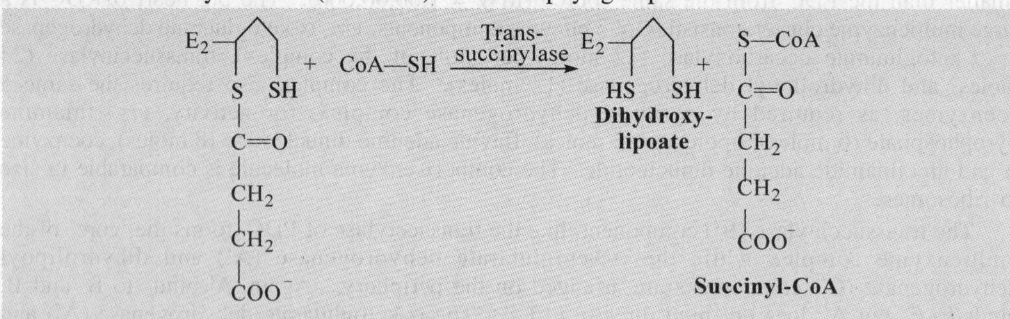

Stage 4. The fully reduced form of E_2 is acted upon by E_3 which promotes transfer to H atoms from the reduced lipoyl groups to the FAD prosthetic group of E_3.

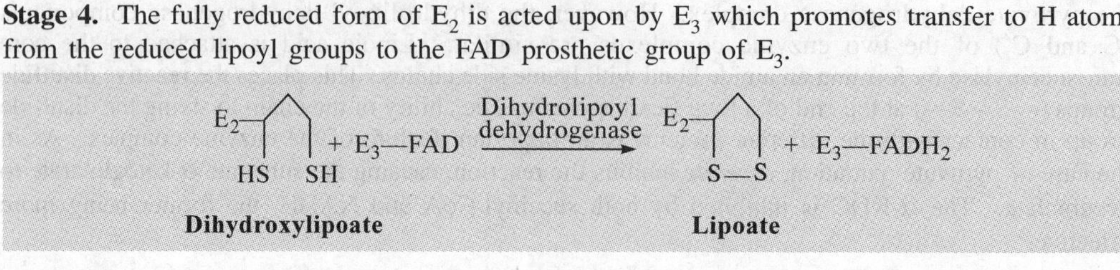

Stage 5. In this last stage, the reduced FAD group of E_3 transfers hydrogen to NAD^+, forming NADH.

$$E_3\text{—}FADH_2 + NAD^+ \xrightarrow[\text{dehydrogenase}]{\text{Dihydrolipoyl}} E_3\text{—}FAD + NADH + H^+$$

Step 5 : *Conversion of succinyl-CoA into succinate*

While acetyl-CoA can undergo a wide variety of metabolic reactions, the fate of succinyl- CoA is much more limited. Its main route is to continue in the cycle. Succinyl-CoA is a high- energy compound. Like acetyl-CoA, it has a strong negative $\Delta G^{o\prime}$ value (–8.0 kcal/mol) for hydrolysis of the thioester bond:

$$\text{Succinyl-S-CoA} + H_2O \longrightarrow \text{Succinate} + \text{CoA—SH} + H^+$$

But such a simple hydrolysis of succinyl-CoA in cells does not occur, as it would mean wastage of free energy. Instead, succinyl-CoA undergoes an energy-conserving reaction in which the cleavage of its thioester bond is accompanied by the phosphorylation of guanosine diphosphate (GDP) to guanosine triphosphate (GTP). The reaction is catalyzed by **succinyl-CoA synthase (= succinic thiokinase)**. Its structural details are presented in Fig. 22–10. The enzyme involves the formation of an intermediate, succinyl phosphate. The phosphate is transferred, first onto the imidazole side chain of a histidine residue in the enzyme and then onto GDP, producing GTP. Inosine diphosphate (IDP), another energy-phosphate acceptor, also functions as an alternate cosubstrate, in place of guanosine diphosphate. In that case, inosine triphosphate (ITP) is produced. This is a readily reversible reaction with $\Delta G^{o\prime}$ value –0.7 kcal/mol.

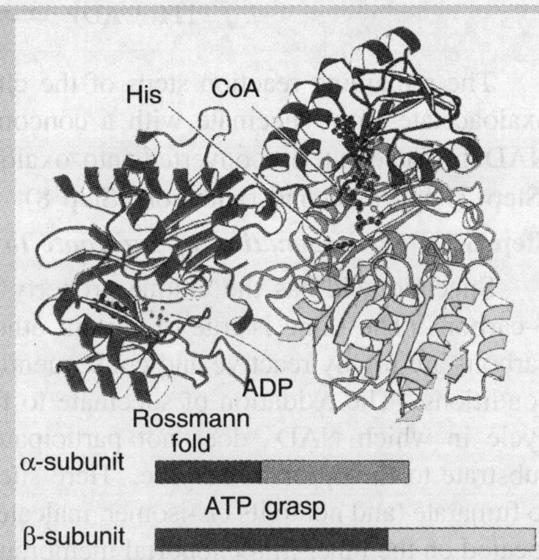

Fig. 22–10. Ribbon model of the structure of succinyl CoA synthetase

The enzyme is composed of two subunits. The α subunit contains a Rossmann fold that binds the ADP component of CoA, and the β subunit contains a nucleotide-activating region called the ATP-grasp domain. The ATP-grasp domain is shown here binding a molecule of ADP. The histidine residue picks up the phosphoryl group from near the CoA and swings over to transfer it to the nucleotide bound in the ATP-grasp domain.

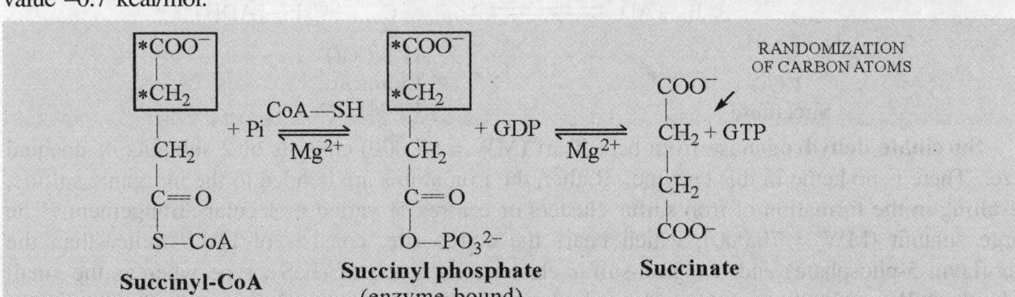

The generation of a high-energy phosphate from succinyl-CoA is an example of a *substrate-level phosphorylation*. In fact, this is the only reaction in the citric acid cycle that directly yields a high-energy phosphate.

The GTP (or ITP) formed by succinyl-CoA synthase then readily donates its terminal phosphate group to ADP to form ATP by the action of Mg^{2+}-dependent enzyme, *nucleoside diphosphokinase* present in the interspace membrane of mitochondria. This is a reversible reaction ($\Delta G^{o\prime}$ = 0.0 kcal/mol).

$$GTP + ADP \underset{Mg^{2+}}{\overset{\text{Nucleoside diphosphokinase}}{\rightleftharpoons}} GDP + ATP$$

$$ITP + ADP \underset{Mg^{2+}}{\overset{\text{Nucleoside diphosphokinase}}{\rightleftharpoons}} IDP + ATP$$

The remaining reaction steps of the citric acid cycle are concerned with the regeneration of oxaloacetate from succinate with a concomitant trapping of energy in the form of $FADH_2$ and NADH. Succinate is converted into oxaloacetate in 3 steps: an oxidation (Step 6), a hydration (Step 7) and a second oxidation (Step 8).

Step 6 : *Dehydrogenation of succinate to fumarate*

Succinic acid has the unique property of possessing two carbon atoms that are both α- and β-carbons (The same is true of certain other C_4 dicarboxylic acids). It is well known that such carbons are highly reactive and consequently would be expected to react rapidly under favourable conditions. The oxidation of succinate to fumarate is the only dehydrogenation in the citric acid cycle in which NAD^+ does not participate. Rather hydrogen is directly transferred from the substrate to flavoprotein enzyme. Here succinate, formed from succinyl-CoA, is dehydrogenated to fumarate (and not to its *cis*-isomer, maleate) by the flavoprotein enzyme, *succinate dehydrogenase*, located on the inner mitochondrial membrane. The enzyme contains the reducible prosthetic group flavin adenine dinucleotide (FAD) as the coenzyme. FAD functions as the hydrogen acceptor in this reaction, rather than NAD^+ (which is used in the other 3 oxidation- reductions in the cycle), because the free energy change is insufficient to reduce NAD^+. In succinate dehydrogenase, the isoalloxazine ring of FAD is covalently linked to a histidine side chain of the enzyme. The enzyme may, thus, be represented by E–FAD. Isotopic experiments have shown that *succinate dehydrogenase is specific for the trans hydrogen atoms of the 2 methylene carbons of succinate*, thus producing fumarate, which is in *trans* form. This points out to high geometrical specificity revealed by the enzyme. The reaction is readily reversible with a free energy change in the neighbourhood of 0.

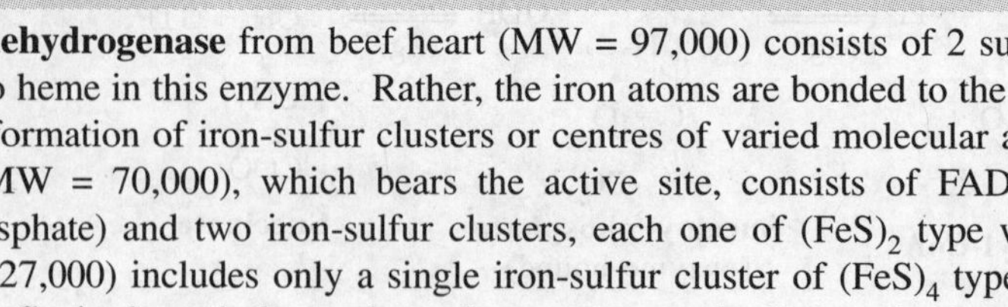

Succinate **Fumarate** *(trans* form*)*

Succinate dehydrogenase from beef heart (MW = 97,000) consists of 2 subunits of unequal size. There is no heme in this enzyme. Rather, the iron atoms are bonded to the inorganic sulfide, resulting in the formation of iron-sulfur clusters or centres of varied molecular arrangement. The large subunit (MW = 70,000), which bears the active site, consists of FAD (rather than the riboflavin 5-phosphate) and two iron-sulfur clusters, each one of $(FeS)_2$ type whereas the small subunit (MW = 27,000) includes only a single iron-sulfur cluster of $(FeS)_4$ type. Electrons flow from substrate to flavin through the nonheme iron groups on the large subunit to that on the small subunit. Succinate dehydrogenase differs from other enzymes in the citric acid cycle in that *it is an integral part of the inner mitochondrial membrane* and is tightly bound to it. In fact, the enzyme is directly linked to the electron-transport chain.

The $FADH_2$, produced by the oxidation of succinate, does not dissociate from the enzyme, in contrast with NADH. Rather two electrons from $NADH_2$ are transferred directly to the Fe^{3+} atoms of the enzyme. The enzyme transfers electrons directly to ubiquinone (*formerly* known as coenzyme Q) bypassing the first phosphorylation site in the electron transfer scheme. Thus, the overall transfer of electrons from succinate to oxygen results in the generation of only 2 high-energy phosphates. Addition of malonate or oxaloacetate inhibits succinate dehydrogenase competitively, resulting in accumulation of succinate.

Step 7 : *Hydration of fumarate to malate*

Fumarate is hydrated to form L-malate in the presence of *fumarate hydratase* (formerly known as *fumarase*). This is a freely reversible reaction ($\Delta G^{\circ\prime} \approx 0.0$ kcal/mol) and involves hydration in malate formation and dehydration in fumarate formation. *Fumarate hydratase is highly specific and catalyzes trans addition and removal of H and OH*, as shown by deuterium-labelling studies. The enzyme hydrates *trans* double bond of fumarate but does not act on maleate, the *cis*-isomer of fumarate. In the reverse direction, fumarase dehydrates L-malate and does not act on D-malate. Thus, there is absolute specificity for the *trans* decarboxylic unsaturated acid and the L-hydroxy dicarboxylic acid.

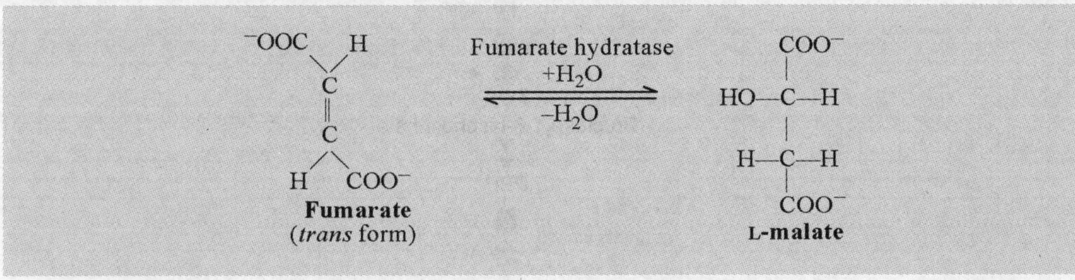

Fumarate hydratase from pig heart (MW = 2,20,000) is a tetramer of identical polypeptide subunits.The enzyme requires no cofactors but the participation of an acidic (*i.e.*, protonated) and a basic (*i.e.*, deprotonated) residue has been clearly implicated.

Step 8 : *Dehydrogenation of malate to oxaloacetate*

This is the fourth oxidation-reduction reaction in the citric acid cycle (the other 3 reactions being Steps 3, 4 and 6). Here L-malate is dehydrogenated to oxaloacetate in the presence of *l L-malate dehydrogenase*, which is present in the mitochondrial matrix. NAD^+, which remains linked to the enzyme molecule, acts as a hydrogen acceptor. This is a reversible reaction ($\Delta G^{o\prime} = +7.1$ kcal/mol). Although the equilibrium of this reaction is far to the left (*i.e.*, it favours malate formation), the reaction proceeds to the right since oxaloacetate and NADH, the two reaction products, are removed rapidly and continuously in further reactions. Oxaloacetate, so regenerated, allows repetition of the cycle and NADH participates in oxidative phosphorylation. This reaction, thus, completes the cycle.

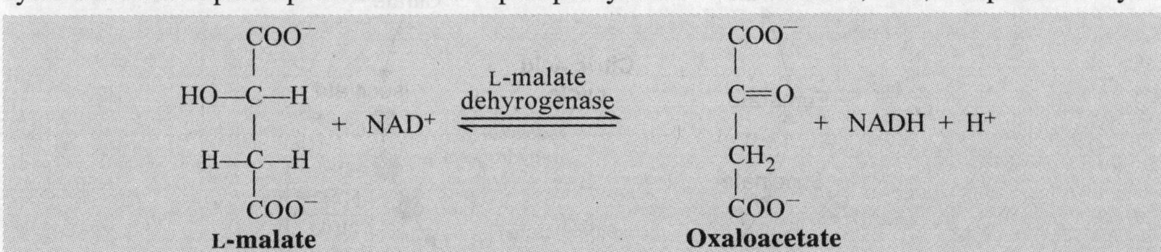

L-malate dehydrogenase (MW = 66,000) consists of 2 subunits. It is a highly stereospecific NAD-requiring enzyme. Mammalian cells appear to contain 2 isozymes of malate dehydrogenase, one species probably being mitochondrial in localization. A high NAD^+/NADH ratio is stimulatory and a low ratio is inhibitory.

Fig 22–11 represents the coordinated control of glycolysis and the citric acid cycle by ATP, ADP, AMP and P.

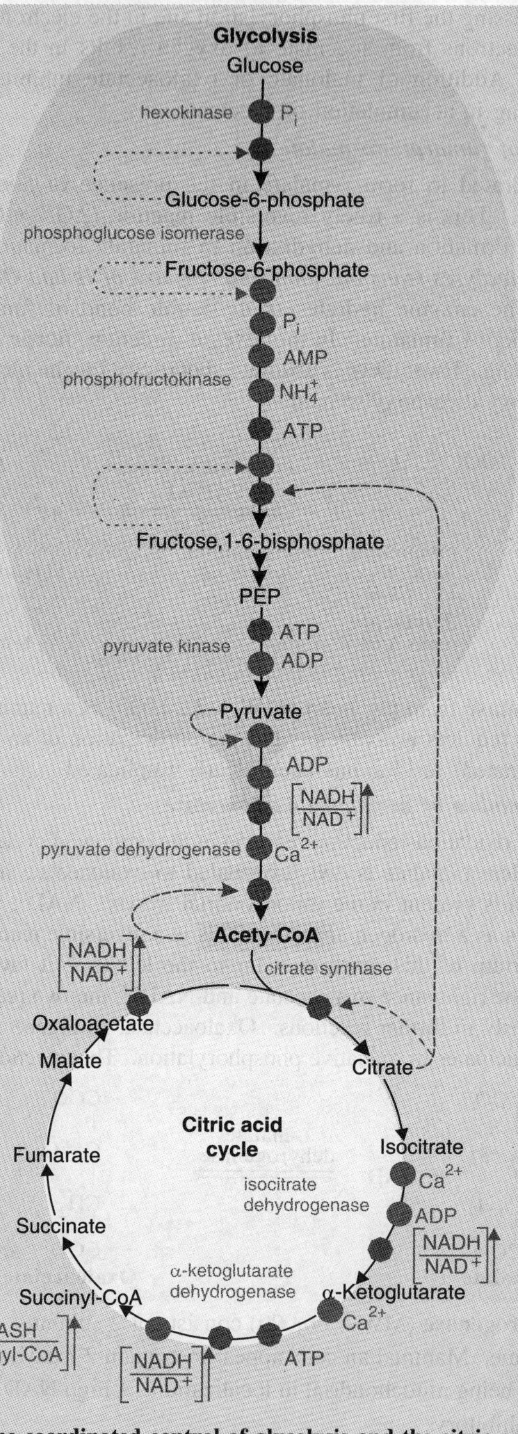

Fig. 22–11. A schematic of the coordinated control of glycolysis and the citric acid cycle by ATP, ADP, AMP, P_i Ca^{2+}, and the [NADH]/[NAD$^+$] ratio

The *vertical arrows* indicate increases in this ratio. Here a *green dot* signifies activation and a *red octagon* represents inhibition.

The Malonate Block

The malate dehydrogenase reaction played an important role in the elucidation of the *cyclic nature of the citric acid cycle.* Addition of tricarboxylic acids (citrate, aconitate) and α-ketoglutarate was known to catalyze pyruvate metabolism– we now know that this is the result of formation of catalytic amounts of oxaloacetate from these intermediates. In 1937, **Sir Hans Adolf Krebs** found that malonate, the C-3 dicarboxylic acid homologue of succinate and competitive inhibitor of succinate dehydrogenase, blocked metabolism of pyruvate by minced muscle preparations. He also showed that malonate inhibition of pyruvate metabolism led to accumulation not only of succinate but also of citrate and α-ketoglutarate, suggesting that succinate was a product of pyruvate metabolism and that the tricarboxylic acids might be intermediate in this process. Interestingly, fumarate and oxaloacetate also stimulated pyruvate oxidation and led to accumulation of citrate and succinate during malonate block, suggesting that the C-3 and C-4 acids might combine to form the tricarboxylic acids. The experiments with fumarate indicated that there were 2 paths between fumarate and succinate– *one* involving reversal of the succinate dehydrogenase reaction, which was inhibited during malonate block, and the *other* involving conversion of fumarate to succinate through a series of organic acids. These observations, combined with Krebs' experience a few years earlier in characterization of the urea cycle, led to his description of the citric acid cycle.

STOICHIOMETRY OF THE CITRIC ACID CYCLE

A. Overall Balance Sheet

We have just seen that one turn of the citric acid cycle involves 8 enzyme-catalyzed reactions and leads to the conversion of one mole of acetyl-CoA to CO_2 plus H_2O. The net reaction of the cycle may be written as:

$$\underset{\text{(Acetyl–CoA)}}{CH_3CO\text{—}SCoA} + 3\ NAD^+ + FAD + GDP + Pi + 2\ H_2O \xrightarrow{\text{Krebs cycle}} 2CO_2 + \underset{\text{(Coenzyme A)}}{CoA\text{—}SH} + 3\ NADH + FADH_2 + GTP + 2\ H^+$$

The $\Delta G^{o\prime}$ for this overall reaction is –14.3 kcal/mol.

The net result of the Krebs cycle may be recapitulated as follows:

1. Two carbon atoms in the form of acetyl unit of acetyl-CoA enter the cycle and condense with oxaloacetate. Two carbon atoms emerge from the cycle as CO_2 in the two successive decarboxylation reactions (Steps 3 and 4) catalyzed by isocitrate dehydrogenase and α-ketoglutarate dehydrogenase respectively. It is noteworthy that the two C atoms that leave the cycle are different from the ones that entered in that round. However, additional turns around the cycle are required before the C atoms that entered as an acetyl group finally leave the cycle as CO_2.
2. Four pairs of hydrogen atoms are removed from the four cycle intermediates by enzymatic dehydrogenation (Steps 3, 4, 6 and 8). Three pairs of hydrogen are used to reduce 3 moles of NAD^+ to NADH and one pair to reduce the FAD of succinate dehydrogenase to $FADH_2$.
3. As would be clear in the folowing chapter, the four pairs of electrons, derived from the four pairs of hydrogen atoms removed in Steps 3, 4, 6 and 8, pass down the electron-transport chain and ultimately reduce 2 molecules of oxygen to form 4 molecules of water.

$$3\ NADH + FADH_2 + 2\ O_2 \longrightarrow 3\ NAD^+ + FAD + 4\ H_2O$$

4. One high-energy phosphate bond (in the form of GTP) is generated from the energy-rich thioester linkage in succinyl-CoA (Step 5). GTP then donates its terminal phosphate group to ADP to produce a mole of ATP, which is thus a by-product of the cycle.

5. Two molecules of water are consumed : one in the synthesis of citrate (Step 1) and the other in the hydration of fumarate (Step 7).
6. *The citric acid cycle does not involve the net production or consumption of oxaloacetate (OAA) or of any other constituent of the cycle itself.* The cycle does not provide a route for making additional OAA from acetyl groups. Failure to appreciate this point has led intelligent men into serious error in the past. However, at the end of each round, a mole of OAA is regenerated and the same condenses with the acetyl group of acetyl-CoA to continue the next round.
7. It is interesting to note that molecular oxygen deos not participate directly in the citric acid cycle. However, the cycle operates only under aerobic conditions because NAD^+ and FAD can be regenerated from their reduced forms (NADH and $FADH_2$) in the mitochondrion only by electron transfer to molecular oxygen. Thus, *whereas glycolysis has an aerobic and an anaerobic nature, the citric acid cycle is strictly aerobic in nature.*

B. Energy Yield

The main purpose of the citric acid cycle is not just the disposition of the carbon and hydrogens of all those compounds that can generate acetyl-CoA or any of the cycle members, but is *the conversion of potential chemical energy into metabolic energy in the form of ATP*. A grand total of 12 ATP molecules is formed during complete oxidation of one mole of acetyl-CoA, *i.e.*, during one turn of the citric acid cycle (refer Table 22–3). (However, a total of 15 ATP moles is formed when we start with the oxidation of 1 mole of pyruvate. This corresponds to a total of 38 ATP moles for the oxidation of 1 mole of glucose). Out of these 12 ATP molecules, only one ATP is produced directly at the level of the cycle itself (*i.e.*, at the substrate level) in the reaction catalyzed by succinyl-CoA synthase (Step 5). And the rest 11 ATP molecules are generated as a consequence of oxidation-reduction reactions of the cycle (Steps 3, 4, 6 and 8). Three molecules of NADH (one each in Steps 3, 4 and 8) and one mole of $FADH_2$ (Step 6) are produced in one turn of the cycle. The oxidation of one mole of NADH by O_2 results in the formation of 3 ATP molecules from ADP plus P_i, thus producing a total of 9 ATP molecules. However, the oxidation of $FADH_2$ by O_2 produces 2 molecules of ATP, rather than 3. Thus, a total of 12 ATP moles are generated utilizing 12 ADP and 12 P*i*.

Since 2 molecules of oxygen are consumed in the citric acid cycle, the P : O ratio (the ratio of high-energy phosphates to atoms of oxygen consumed) is 12/4 or 3.0. This value has a bearing on the relative energy production obtained from various fuels.

Table 22–3. Energy yield of the citric acid cycle

Step No.	*Reaction*	*Method of ATP formation*	*ATP yield per mole*
3	Isocitrate ⟶ α-ketoglutarate + CO_2	Respiratory chain oxidation of NADH	3
4	α-ketoglutarate ⟶ Succinyl-CoA + CO_2	Respiratory chain oxidation of NADH	3
5	Succinyl-CoA + ADP + P*i* ⟶ Succinate + ATP	Oxidation at substrate level	1
6	Succinate ⟶ Fumarate	Respiratory chain oxidation of $FADH_2$	2
8	Malate ⟶ Oxaloacetate	Respiratory chain oxidation of NADH	3
		Total gain of ATP = 12	

ROLE OF WATER IN THE CITRIC ACID CYCLE

Water is an extremely important metabolite for the various biologic processes. It functions as a solvent for the enzymes and most intermediates of the cycle. In addition, it also participates either as a reactant or as a product in many reactions of this path. *This ubiquitous character is nowhere more apparent than in the citric acid cycle.*

Water serves as a reactant in Steps 1 and 7. Moreover, it is both first eliminated and then utilized in Step 2 of the cycle. Furthermore, water is formed when the electron transport system (ETS) reoxidizes the NADH and FADH2, which are generated by the oxidoreduction reactions, to NAD^+ and FAD. Thus, water also appears as a product at 4 additional sites which are not directly the parts of the citric acid cycle. Finally, the elements of water are consumed when GTP is restored to GDP asnd P_i. Thus, *the citric acid cycle and the connected reactions consume 4 moles of water and release 5 moles.*

STEREOSPECIFICITY OF THE CITRIC ACID CYCLE

It has long been known that *aconitate hydratase* (catalyzing isomerization of citrate– Step 2), *succinate dehydrogenase* (catalyzing dehydrogenation of succinate– Step 6) and *fumarate hydratase* (catalyzing hydration of fumarate– Step 7) produce stereospecific products.

A. Biological Asymmetry of Citrate

Citrate (a C_6 compound) is a symmetric molecule but it behaves asymmetrically. This may be understood by tracing the fate of a particular carbon atom in the cycle. Suppose, in oxaloacetate (C_4), the carboxyl carbon furthest from the keto group is labelled with radioactive isotope of carbon *i.e.*, with ^{14}C. Analysis of ketoglutarate (C_5), formed later in the cycle, reveals that it contains the radioactive carbon and a mole of normal or nonradioactive CO_2, derived from the middle carboxylate group of citrate molecule, is given off along with it (Step 3). But succinate (C_4), formed at a latter stage in the cycle (Step 5), shows no radioactivity and all the radioactive carbon (^{14}C) has appeared in CO_2 released in the preceding reaction (*i.e.*, Step 4). This was quite a surprising result because citrate is a symmetric molecule and therefore it was assumed that the 2 —CH_2COO^- groups in the citrate mole would react identically. Thus, it was thought that if one citrate molecule reacts in a way as written above (refer Fig. 22.12, Path 1), another molecule of citrate would react in a way shown in Path 2, so that only half of the released CO_2 molecules should have labelled CO_2. But such an expectation did not come true.

This peculiar behaviour of citrate can be explained on the basis of a **3-point attachment hypothesis** developed by **Alexander G. Ogston** (1948) of Oxford University. According to him, aconitase always acts on that part of citrate molecule which is derived from oxaloacetate (shown by boldface in Path 1). He further pointed out that the two carboxymethyl (—CH_2COO^-) groups of citrate are not truly geometrically equivalent and this nonequivalence becomes apparent upon its attachment on the enzyme surface by 3 points.

An illustration will make the point clear (Fig. 22–13). Consider a molecule which contains a centrally-located tetrahedral carbon atom. To this carbon atom are bonded 2H atoms (which are labelled as H_A and H_B), a group X and a different group Y. Suppose an enzyme binds 3 groups of this citrate, X, Y and H_A. It should, however, be noted that X, Y and H_B cannot be bound to this active site of the enzyme; two of these 3 groups can be bound but not all three. Thus, H_A and H_B are geometrically not equivalent and have different fates. Similarly, the two —CH_2COO^- groups in citrate molecule are geometrically not equivalent even though the citrate is optically inactive. This is due to the fact that the enzyme holds the substrate in a specific orientation.

An organic molecule, which has handedness and hence optically active, is called **chiral** molecule. Contrary

> A compound is said to be **chiral** and to possess **chirality** if it cannot be superimposed on its mirror image, either as a result of simple reflection or after rotation and reflection. If superposition can be achieved, then the molecule is said to be **achiral**. The term chirality is equivalent to Pasteur's dissymmetry.

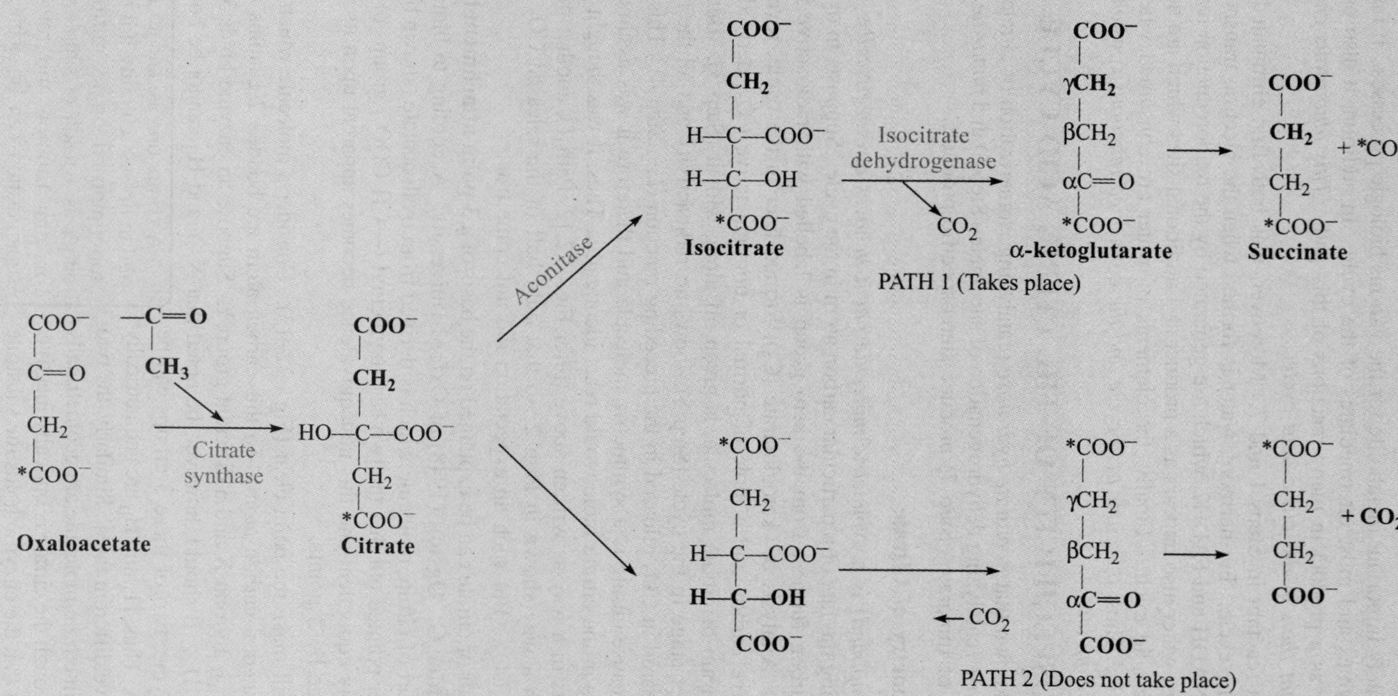

Fig. 22–12. Incorporation of isotopic carboxylate carbon (furthest from the keto group) of oxaloacetate into α-ketoglutarate

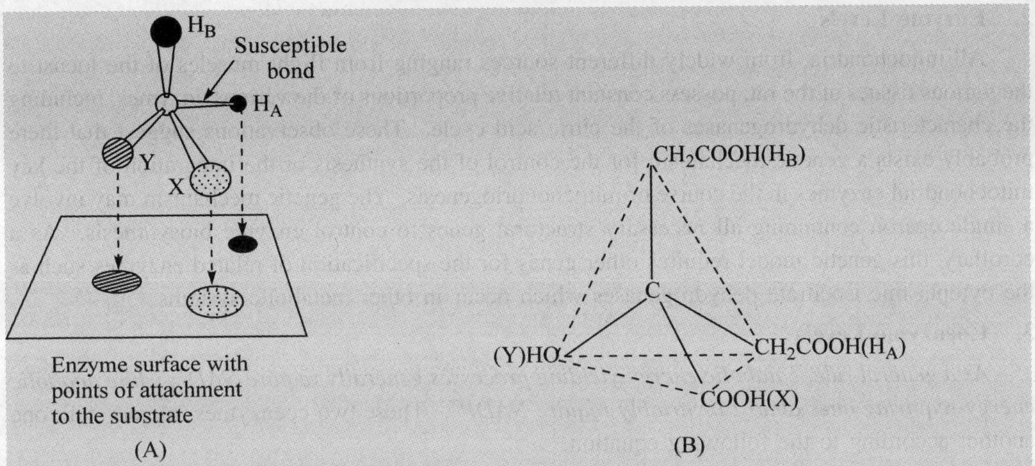

Fig. 22–13. Asymmetry of citrate and the action of aconitase

(A) Ogston's 3-point attachment hypothesis

(B) Structure of citrate

to this, a **prochiral** molecule, (such as citrate or CXY H_2) lacks handedness and is hence optically inactive. *The prochiral molecule can become chiral in one step*. For instance, CXY H_AH_B, which is a prochiral molecule, is transformed into a chiral one, CXYZ H_B, when one of its identical atoms or groups (H_A in this case) is replaced by another (Z in this case).

In citric acid cycle, one mole of acetyl-CoA is consumed and 2 moles of CO_2 are released. But neither of the carbon atoms lost as CO_2 is derived from that acetyl-CoA on the first turn of the cycle. However, at the level of free succinate, randomization does occur and the carboxyl carbon of acetate group of acetyl-CoA will be symmetrically distributed about the plane of symmetry. Hence, on the subsequent rounds of the cycle, with reappearance and reutilization of oxaloacetate, the acetyl carbon will be liberated as CO_2.

B. Geometrical Specificity of Succinate Dehydrogenase

It has already been discussed in Step 6 of the citric acid cycle.

C. Geometrical Specificity of Fumarate Hydratase

It has also been discussed in Step 7 of the same cycle.

REGULATION OF THE CITRIC ACID CYCLE

Since the citric acid cycle is one of the major routes of fuel consumption in most cells, there must be some control on the rate at which it proceeds. It would not do to have the cycle going full swing like a runaway boiler at times of rest nor would it do to have it sluggishly going when there is an immediate demand for ATP.

Several factors serve to control the rate of this sequence of reactions. These are described below:

1. Substrate Levels

One of the controlling features for any reaction sequence is the availability of the various substrates involved in it. It is known that the steady-state concentrations of the half-lives of most substrates of the cycle are of the order of a few seconds. *The outstanding exception is oxaloacetate (OAA), whose half-life is of the order of a tenth of a second.* This relatively restricted concentration of OAA puts it in high demand and also emphasizes its role in controlling the input of acetyl-CoA into the cycle. Regulation of the rate of this reaction would control activity in the enzyme cycle.

2. Enzyme Levels

All mitochondria, from widely different sources ranging from flight muscles of the locust to the various tissues of the rat, possess constant relative proportions of the various enzymes, including the characteristic dehydrogenases of the citric acid cycle. These observations suggest that there probably exists a genetic mechanism for the control of the synthesis or the integration of the key mitochondrial enzymes in the course of mitochondriogenesis. The genetic mechanism may involve a single operon containing all necessary structural genes to control enzyme biosynthesis. As a corollary, this genetic model requires other genes for the specification of related enzymes such as the cytoplasmic isocitrate dehydrogenases which occur in other metabolic systems.

3. Coenzyme Levels

As a general rule, catabolic energy-yielding processes generally require NAD$^+$, while anabolic energy-requiring ones almost invariably require NADP$^+$. These two coenzymes interact with one another according to the following equation.

$$\text{NADH} + \text{NADP}^+ \xrightleftharpoons[]{\text{NAD(P) transhydrogenase}} \text{NAD}^+ + \text{NADPH}$$

It is apparent that such a system would represent a sensitive control point for the regulation of the levels of these coenzymes. The relative concentrations of NAD$^+$ and NADH are important in regulating metabolic pathways. When NAD$^+$/NADH ratio is high, the rate of citric acid cycle becomes rapid. However, the activity of this cycle is retarded when NAD$^+$/NADH ratio is low because of (*a*) insufficient NAD$^+$ concentration for otherwise normal enzymatic function, and (*b*) reoxidation of NADH coupled to ATP formation.

4. Respiratory Control

Respiration rate depends, not only on the nature and concentration of the substrates to be oxidized, but also on the coupling of respiration to phosphorylation. Intact mitochondria are usually 'tightly' coupled, so that their rate of respiration is actually controlled by the ratio [ADP]/[ATP]. When this ratio is high, respiration is promoted ("State 3"). In contrast, low ratios (*i.e.*, high ATP concentrations) decline respiration ("State 4"). Added ATP can even inhibit respiration because they bring about **reversed electron flow.** These phonomena are now known as respiratory control.

The term **'reversed electron flow'** is used to describe certain oxidation-reduction reactions not ordinarily observed because of their unfavourable equilibrium which, however, can be 'reversed' by the addition of ATP.

5. Accessibility of Cycle Intermediates

The activity of the citric acid cycle is also controlled by its accessibility to acetyl-CoA of intermediates of the cycle. This problem consists of a combination of permeability barriers and geometry. The mitochondrial membrane itself provides a means for the admission of some substrates and the exclusion of others. A few examples are given below:

(*a*) Intact mitochondria do not allow NADH, which is produced on the 'outside' by oxidation of 3-phosphoglyceraldehyde during glycolysis, to enter 'inside' and with the result NADH is not connected directly to the electron transport system. Instead, NADH is reoxidized by a substrate. The reduced substrate then penetrates the mitochondrion and is reoxidized. The reduced substrate, thus, serves as a shuttle for electrons between NADH in cytoplasm and electron carriers in the motochondria.

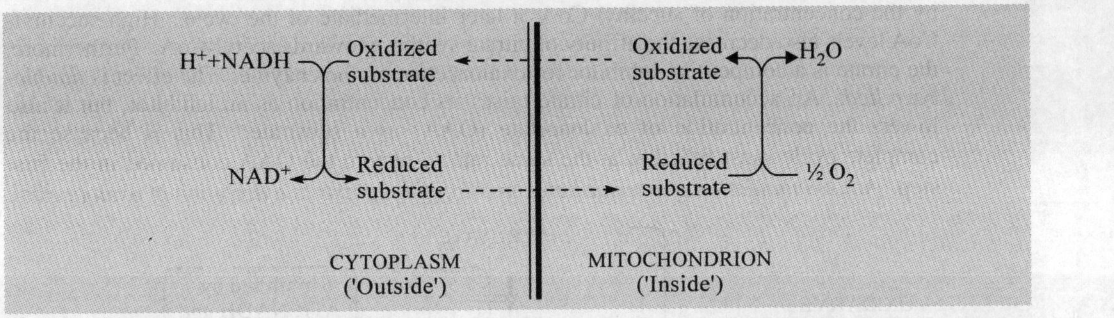

(*b*) Similarly, acetyl-CoA generated in mitochondria does not readily diffuse out. Rather, it is first converted to citrate which is then cleaved in the cytoplasm to generate acetyl-CoA for various reactions taking place there.

(*c*) Another example of mitochondrial compartmentation is that of succinate dehydrogenase. Mitochondrial succinate dehydrogenase is freely available to succinate from outside the mitochondria but not to fumarate.

(*d*) Furthermore, added fumarate is also not freely accessible to the mitochondrial fumarate hydratase (= fumarase).

6. Ketosis

The accumulation of ketone bodies, acetoacetate and acetone formed by the liver in diabetics results from the production of more acetyl-CoA than can be cyclized via the Krebs cycle or other synthetic reactions. Under these conditions, the rate of Krebs cycle slows down probably due to hormonal action since ketone body formation (*i.e.*, ketosis) is affected by hormones of the hypophysis and adrenal cortex.

The CoA group of succinyl-CoA may be transferred to acetoacetate in mitochondria of muscle and kidney:

$$\text{Succinyl-CoA} + \text{Acetoacetate} \longrightarrow \text{Succinate} + \text{Acetoacetyl-CoA}$$

This is the point at which interaction between Krebs cycle and ketosis occurs, since acetoacetyl-CoA can be further degraded to acetyl-CoA.

7. Control of Enzyme Activity (= Regulation by Effectors)

Some of the reactions in the citric acid cycle require individual regulation (refer Fig. 22–14) because they are essentially irreversible under physiological conditions. There is always a danger from such reactions because they may continue till they have consumed the available supply of substrate or coenzyme. In general, at many points, stimulation or inhibition of the cycle is determined by the relative levels of NADH/NAD$^+$, ATP/ADP, acetyl-CoA/CoA or succinyl-CoA/CoA. When these ratios are high, the cell has ample supply of energy and flow through the cycle is slowed. When these ratios are low, the cell is in need of energy, and flow through the cycle quickens.

(*a*) **Citrate synthase reaction.** As in most metabolic cycles, the initial steps of the citric acid cycle are believed to be rate-setting for the cycle as a whole. The rate of citrate synthesis from oxaloacetate and acetyl-CoA is controlled by the concentration of acetyl-CoA itself which is, in turn, governed by the activity of pyruvate dehydrogenase complex. The concentration of oxaloacetate is also most important factor since its concentration in mitochondria is very low and depends upon metabolic conditions. ATP is an allosteric inhibitor of citrate synthase. The effect of ATP is to increase the K_m for acetyl-CoA. Thus, as the level of ATP increases, the affinity of the enzyme towards its substrate (acetyl-CoA) decreases and so less citrate is formed. The activity of citrate synthase is also regulated by the concentration of succinyl-CoA, a later intermediate of the cycle. High succinyl-CoA levels also decrease the affinity of citrate synthase towards acetyl-CoA. Furthermore, the citrate is a competitive inhibitor for oxaloacetate on the enzyme. The effect is *double-barrelled.* An accumulation of citrate raises its concentration as an inhibitor, but it also lowers the concentration of oxaloacetate (OAA) as a substrate. This is because the complete cycle must function at the same rate to restore the OAA consumed in the first step. *Any accumulation of intermediates in the cycle represents a depletion of oxaloacetate.*

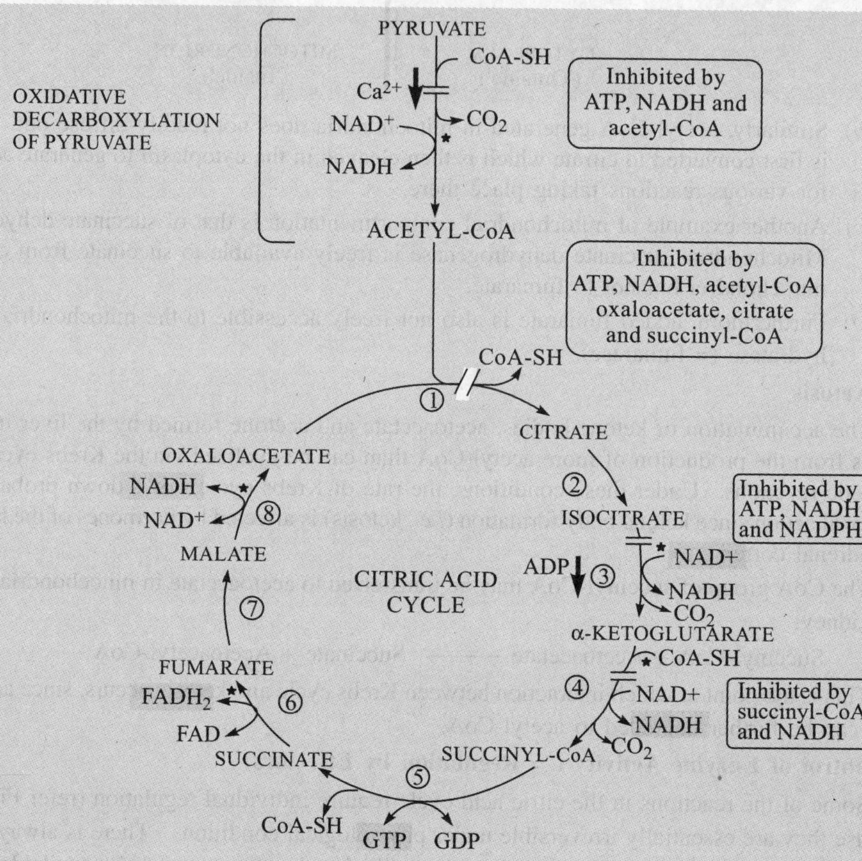

Fig. 22–14. Control of the oxidative decarboxylation of pyruvate and the citric acid cycle
Circled numbers correspond to the reaction steps of the citric acid cycle. Thick arrows indicate the stimulatory effect of the molecule or ion mentioned. Starred reactions indicate steps that require an electron acceptor (NAD^+ or FAD) that is regenerated by the respiratory chain.

(*b*) **Isocitrate dehydrogenase reaction.** *This reaction appears to be the rate-limiting step of the citric acid cycle.* Mammalian isocitrate dehydrogenase is allosterically stimulated by ADP which enhances its activity for the substrate whereas NADPH is an allosteric inhibitor. NADH and ATP are the competitive inhibitors for the NAD^+ site. NADH and NADPH bind to different sites on the enzymes. However, in the the case of yeast and *Neurospora*, AMP rather than ADP is the main stimulator (or positive effector) of the enzyme.

(*c*) **α-ketoglutarate dehydrogenase reaction.** The activity of a-ketoglutarate dehydrogenase complex is inhibited by its two products, succinyl-CoA and NADH, the former being more effective. In fact, succinyl-CoA is a competitive inhibitor for one of its substrate, coenzyme A. Here again is a *double-barrelled effect*. A rise in succinyl-CoA concentration in itself inhibits but it also represents a depletion of coenzyme A, leading to still more

Table 22–4. Some controls on the citric acid cycle and connected systems

Enzyme system	*Location*	*Reactant(s)*	*Products(s)*	*Activator(s)*	*Inhibitor(s)*	*Remarks*
1. ETS-oxidative phosphorylation	Microconidia	ATP, P_i	ATP, CO_2 (by virture of respiration)	—	ATP uncouplers	Coupled ATP production
2. Pyruate carboxylase	Mitochondria	ATP, O_2	ADP	Acetyl-CoA	—	Controls carbohydrate synthesis
3. Acetyl-CoA carboxylase	Cytoplasm	ATP, CO_2	ADP	Citrate etc	Long-chain acyl-CoAs	Controls fatty acid synthesis
4. Citrate synthase (= condensing enzyme)	Mitochondria	Acetyl-CoA, Oxaloacetate	Citrate, CoA—SH	—	Long-chain acyl-CoAs ATP or NADH*	Controls diversion of acetyl-CoA to other pathways
5. Citrate lyase (ATP-requiring)	Cytoplasm	Citrate (from acetyl-CoA)	Acetyl-CoA, Oxaloacetate	—	—	Makes available extramitochondiral acetyl-CoA for lipid synthesis
6. Isocitrate lyase	Cytoplasm	Citrate (from acetyl-CoA)	Succinate	—	Phosphoenol-pyruate	Controls combination of C_2 units (in bacterial and plants only)
7A. Isocitrate dehydrogenase (NAD-specific)	Mitochondria	NAD^+	NADH, CO_2	ADP	ATP, NADH	Oxidizes isocitrate
7B. Isocitrate dehydrogenase (NADP-specific)	Cytoplasm Mitochondria	$NADP^+$	NADPH, CO_2	Oxaloacetate ?	—	Generates NADPH
8. Glutamate dehydrogenase	Mitochondria	NADH or NADPH, NH_3	NAD^+ or $NADP^+$	ADP	GDP + NADH	Precise nature of control not known

* Mammalian and yeast enzymes are inhibited by ATP whereas *Escherichia coli* enzyme is inhibited by NADH. Both inhibition controls are of the feedback type.

effective inhibition. *The α-ketoglutarate dehydrogenase reaction represents a threat to the supply of coenzyme A which is needed for other reactions.* Indeed 70% of the coenzyme A supply in some tissues is present as succinyl-CoA under some conditions, even though the enzyme is regulated.

The various controls of the citric acid cycle have been listed in Table 22–4.

AMPHIBOLIC ROLES OF THE CITRIC ACID CYCLE

The citric acid cycle has a dual or amphibolic (*amphi*[G] = both) nature. The cycle functions not in the oxidative catabolism of carbohydrates, fatty acids and amino acids but also as the first stage in many biosynthetic (= anabolic) pathways, for which it provides precursors. Under certain conditions, one or more intermediates of the cycle may be drawn off for synthesis of other metabolites. In order to avoid cutting the cycle, the intermediates of the cycle are, however, replenished at the same or different locus by other reactions called as anaplerotic reactions. Under normal conditions, the reactions by which the cycle intermediates are drained away and those by which they are replenished are in dynamic balance so that at any time the concentration of the citric acid cycle intermediates in mitochondria usually remains relatively constant. Thus, for example, if a cell over a period of time needs to make 4.25 m moles of glutamic acid from α-ketoglutarate, it must provide 4.25 μ moles of acetyl-CoA and 4.25 μ moles of oxaloacetate (the two compounds from which synthesis will occur) to "balance the books", so to speak. The biosynthetic and anaplerotic functions of the cycle (refer Fig. 22–15) are discussed below.

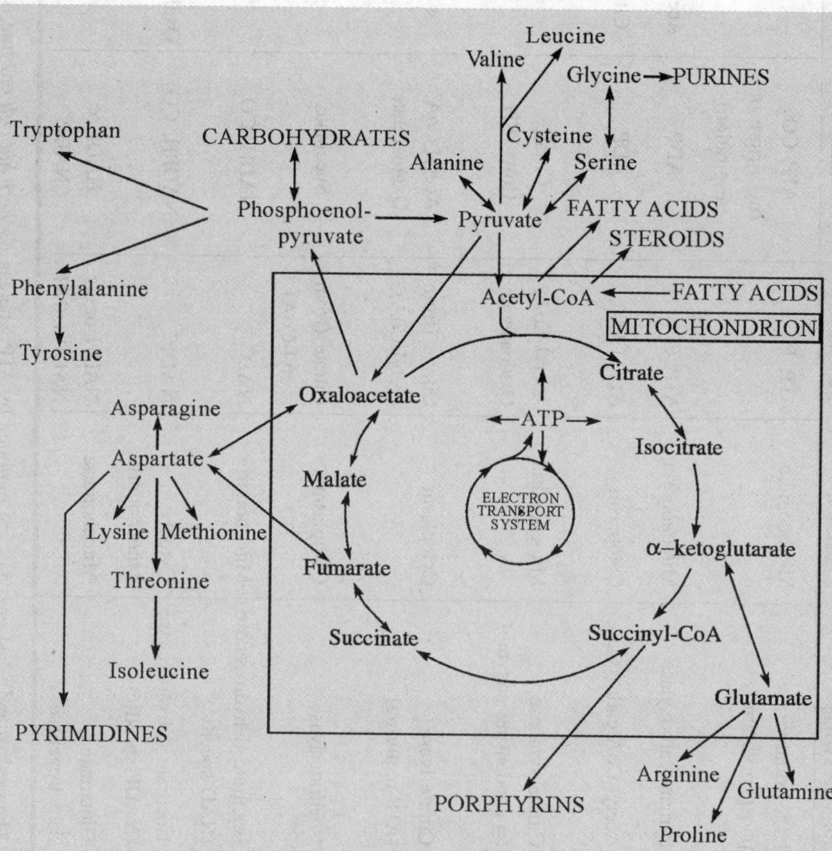

Fig. 22–15. **Biosynthetic and anaplerotic functions of the citric acid cycle**

A. Biosynthetic Roles

Citric acid cycle is the primary source of some key metabolites of the cell as it provides intermediates for their biosyntheses. In fact, intermediates from the cycle serve as biosynthetic precursors of all 4 major classes of compounds *viz.*, carbohydrates, lipids, proteins and nucleic acids.

1. Carbohydrates. The intermediates of the citric acid cycle are, as such, probably not important in the biosynthesis of carbohydrates by most animal cells. However, when such cells have ample supply of C_4 acids or C_3 compounds, certain mitochondrial reactions involving oxaloacetate become important routes to carbohydrate synthesis. *The general process appears to be the conversion of such C_3 or C_4 compounds to oxaloacetate.* For example, if sufficient succinate is available, it passes through a portion of cycle to oxaloacetate. However, other compounds can be first converted to pyruvate. The pyruvate is, then, carboxylated to oxaloacetate in the presence of pyruvate carboxylase to which biotin is covalently attached (The reaction is discussed at length in the following subheading). Although, *Escherichia coli* might convert pyruvate to phosphoenolpyruvate (PEP) with a single ATP per unit, it appears that the animals require the equivalent of two ATP's. The carbohydrate is, then, obtained from PEP by reversal of the EMP pathway.

2. Lipids. Acetyl-CoA is the key intermediate in the biosynthesis of lipids. It is generated by just 2 main processes:

(*a*) the thiolytic cleavage of acetoacetyl-CoA, which is generated by the oxidation of fatty acids and certain amino acids.

(*b*) the oxidative decarboxylation of pyruvate.

Yet the biosynthesis of fatty acids requires an obligatory first step, the carboxylation of acetyl-CoA to malonyl-CoA which is probably catalyzed by an extramitochondrial enzyme complex. Acetyl-CoA is first converted to citrate within the mitochondrion by condensation with oxaloacetate. The citrate, so formed, then passes out to the cytoplasm where it is cleaved back to oxaloacetate plus acetyl-CoA by the ATP-requiring enzyme, citrate lyase which is also known as citarate cleavage enzyme.

$$\text{Citrate} + \text{ATP} + \text{CoA—SH} + \text{Enz} \underset{\text{Condensation}}{\overset{\text{Citrate lyase}}{\rightleftharpoons}} [\text{Enz-citryl}\sim\text{SCoA}] + \text{ADP} + \text{P}i$$

$$[\text{Enz—citryl}\sim\text{SCoA}] \xrightarrow[\text{Hydrolysis}]{+\,H_2O} \text{Acetyl-CoA} + \text{OAA} + \text{Enz}$$

This reaction, unlike that responsible for the synthesis of carbohydrates, requires only catalytic amounts of OAA and synthesis of fatty acids, therefore, does not require the obligatory participation of C_4 dicarboxylic acids.

Fatty acids obtained by the hydrolysis of lipids are catabolized in the mitochondrion and contribute acetyl-CoA for its breakdown by the citric acid cycle. Acetyl-CoA, from this source or synthesized by the PDC system in the outer membrane of the mitochondrion, is transferred outside for the biosynthesis of fatty acids and steroids.

3. Proteins. Protein synthesis, like all the other synthetic processes, requires a supply of monomeric units or precursors. In the case of proteins, the monomeric units are some 20 L-amino acids. Most higher animals are unable to synthesize about half of these amino acids (Arg, His, Ile, Leu, Lys, Met, Phe, Thr, Trp and Val) in sufficient amounts to meet their demands. Plants and most microorganisms, on the contrary, are able to synthesize almost all the amino acids. Again, the citric acid cycle provides a means for the generation of 'nonessential' amino acids in animals and the bulk of all amino acids in other cells and organisms.

Two of these amino acids (glutamate and aspartate) are obtained directly from intermediates of the citric acid cycle and one of these (alanine) is made directly from pyruvate. Glutamate is made directly from a-ketoglutarate by all cells. Aspartate is made by most cells from oxaloacetate. Bacterial cells have got the additional capacity of producing aspartate from fumarate, although animals cannot. All other amino acids are made indirectly from glutamate, aspartate and alanine by series of reactions. Thus, there exist 3 'families' of amino acids that originate from the 3 -keto acids, namely a-ketoglutarate, oxaloacetate and pyruvate. These are:

(*a*) **Glutamic Family**— This group consists of glutamate, glutamine, arginine and proline. These amino acids are formed throughout the living organisms.

(*b*) **Aspartic Family**— This group includes aspartate, asparagine, lysine, methionine, threonine and isoleucine. These are simply not produced at all by mammals and perhaps not by other animals.

(*c*) **Pyruvic Family**— This group comprises of alanine, valine, leucine, cysteine, serine, glycine, histidine, tryptophan, phenylalanine and tyrosine. Only some of these are synthesized in higher organisms.

4. Purines and pyrimidines. Purines and pyrimidines are important constituents of the coenzymes and nucleic acids. Aspartate provides the carbon skeleton of pyrimidines whereas glycine from pyruvate contributes the carbon skeleton for purines.

5. Porphyrins. Porphyrins are essential components of the respiratory pigments and enzymes. Animals utilize succinyl-CoA as one of the components for the synthesis of porphyrins.

B. Anaplerotic Roles

Many intermediates of the citric acid cycle are used for the synthesis of other substances (as seen in the preceding subsection). The drain of these intermediates would ultimately prevent operation of the cycle were they not replenished. **H. L. Kornberg** (1966) has proposed the term **anaplerotic** for these replenishing or "filling-up" or "building up" reactions. *Anaplerosis is defined as any reaction that can restore the concentration of a crucial but depleted intermediate.* Some important anaplerotic reactions of the citric acid cycle are listed below :

1. Pyruvate carboxylase reaction. The citric acid cycle will cease to operate unless oxaloacetate is formed *de novo* because acetyl-CoA cannot enter the cycle unless it condenses with oxaloacetate. Mammals lack the enzyme system needed to convert acetyl-CoA into oxaloacetate or another citric acid cycle intermediate. However, in animal tissues *esp.*, liver and kidney, oxaloacetate is formed by the carboxylation of pyruvate through the action of *pyruvate carboxylase*, a mitochondrial complex regulatory enzyme. Since the standard free-energy change of the reaction is very small ($\Delta G^{o\prime}$ = 0.5 kcal/mole), it is a readily reversible reaction.

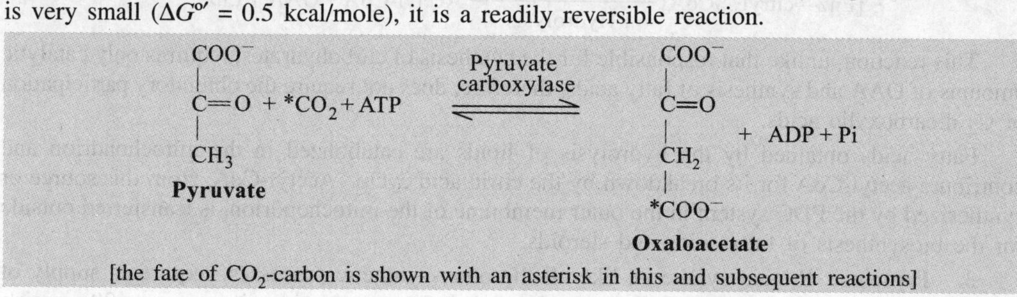

[the fate of CO_2-carbon is shown with an asterisk in this and subsequent reactions]

Pyruvate carboxylase (MW, of that from chicken liver = 6,60,000) is a tetramer of 4 identical subunits. Each subunit contains an active site, including tightly-bound Mn^{2+} (or Mg^{2+}) and covalently-bound biotin, and also an allosteric site, which binds acetyl-CoA. Biotin is covalently attached through an amide linkage with the ϵ-NH_2 group of a specific lysine residue in the active site. Pyruvate kinase has an unusual metal requirement : a monovalent cation (K^+) and a divalent cation (Mg^{2+} or Mn^{2+}). The enzyme has nearly absolute requirement for acetyl-coenzyme A as an activator.

The carboxylation of pyruvate occurs in 2 steps, each of which is catalyzed by a distinct subsite of the active site:

(*a*) **First step :** Free CO_2, the precursor of new carboxyl group of oxaloacetate, is first, 'energized' through its covalent union with a ring N atom of biotin, forming 1′-N-carboxybiotinyl enzyme. The energy needed is provided by ATP. The rate of this reaction is accelerated by acetyl-CoA which acts as an allosteric modulator.

$$ATP + CO_2 + \text{Enz-biotin} + H_2O \rightleftharpoons ADP + P_i + \text{Enz-biotin-COO}^- + 2H^+$$

(*b*) **Second step :** The new carboxyl group covalently bound to the biotin prosthetic group is transferred to pyruvate to form oxaloacetate.

$$\text{Enz-biotin-COO}^- + \text{Pyruvate} \rightleftharpoons \text{Enz-biotin} + \text{Oxaloacetate}$$

2. **Phosphoenolpyruvate carboxykinase reaction.** This is another anaplerotic reaction that feeds the Krebs cycle and occurs in animal tissues, *esp.*, heart and muscles. This reversible reaction brings about carboxylation of phosphoenolpyruvate (PEP) to produce oxaloacetate (OAA) with the concomitant use of GDP (or IDP) as phosphate acceptor, by the enzyme *phosphoenolpyruvate carboxykinase* (= PEP *carboxykinase*). The breakdown of phosphoenolpyruvate, a super-energy compound, furnishes the energy for the carboxylation to yield oxaloacetate and also for the phosphorylation of GDP to yield GTP. However, the affinity of the enzyme for oxaloacetate is very high ($K_m = 2 \times 10^{-6}$) while that for CO_2 is low. Hence, the enzyme favours PEP formation.

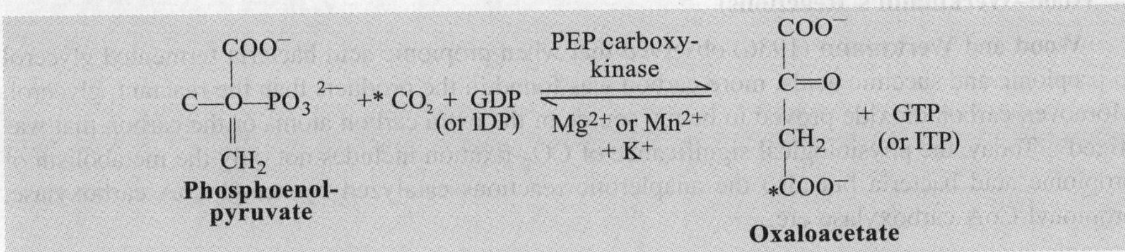

PEP carboxykinase (MW = 75,000) is found primarily in cytosol and to less extent in mitochondria. It differs in mechanism from that of pyruvate carboxylase as it does not involve biotin and also CO_2 is not 'activated'.

3. **Phosphoenolpyruvate carboxylase reaction.** This is another replenishing reaction of the citric acid cycle and occurs chiefly in higher plants, yeast and bacteria (except pseudomonads) but not in animals. In this reaction, phosphoenolpyruvate (PEP) is irreversibly carboxylated to oxaloacetate *by phosphoenolpyruvate carboxylase (= PEP carboxylase).*

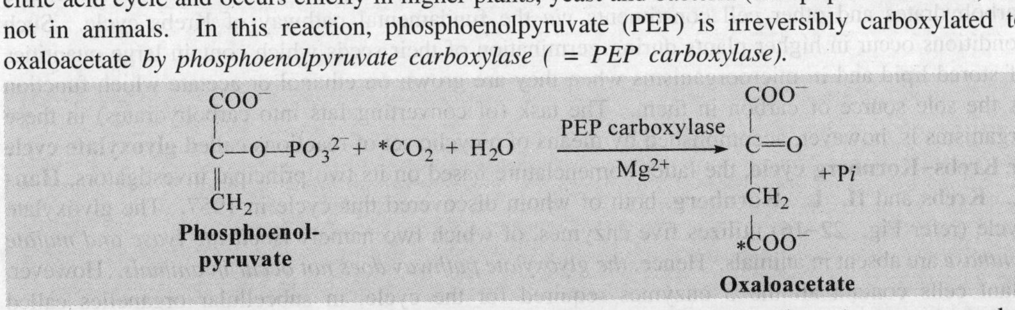

PEP carboxylase has the same function as that of pyruvate carboxylase, *i.e.*, to ensure that the citric acid cycle has an adequate supply of oxaloacetate. The enzyme requires Mg^{2+} for activity. PEP carboxylase is activated by fructose 1, 6-diphosphate and inhibited by aspartate. The inhibitory effect is understandable because oxaloacetate is the direct precursor of aspartate by transamination. The above biosynthetic sequence is a simple means for synthesizing aspartate from PEP, and aspartate can control its own formation by inhibiting the Ist step of this sequence.

$$\text{Phosphoenolpyruvate} \xrightarrow[\text{STEP 1}]{\text{Aspartate} \dashv} \text{Oxaloacetate} \xrightarrow[\text{STEP 2}]{} \text{Aspartate}$$

4. Malic enzyme reaction. The reaction occurs in plants, in several animal tissues and in some bacteria grown on malic acid. Malic enzyme catalyzes the reversible formation of L-malate from pyruvate and CO_2

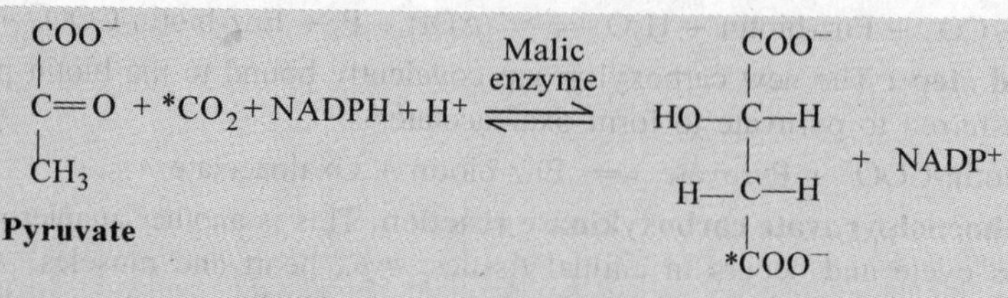

Pyruvate **L-malate**

Malic enzyme is found in two different forms, one in cytosol and another in mitochondria. Its major function is probably the formation of NADPH required for biosynthetic processes, or the reverse of the reaction as written. Mitochondrial malic enzyme, but not the cytosolic form, is an allosteric enzyme. Succinate is a positive effector (*i.e.*, activator) and decreases the K_m for malate.

Carbon Dioxide-fixation Reactions
(= Wood–Werkmann's Reactions)

Wood and **Werkmann** (1936) observed that when propionic acid bacteria fermented glycerol to propionic and succinic acids, more carbon was found in the products than the reactant, glycerol. Moreover, carbon dioxide proved to be the source of the extra carbon atoms or the carbon that was 'fixed'. Today, the physiological significance of CO_2-fixation includes not only the metabolism of propionic acid bacteria but also the anaplerotic reactions catalyzed by acetyl-CoA carboxylase, propionyl-CoA carboxylase etc.

MODIFICATION OF THE CITRIC ACID CYCLE : GLYOXYLATE CYCLE

Higher plants and some microorganisms (bacteria, yeast and molds), under certain specific conditions, face the problem of converting fats into 2-carbon compounds (such as acetyl-CoA) into carbohydrates and other cell constituents *via* the fundamental pathway of Krebs cycle. Such conditions occur in higher plants during germination of their seeds which contain large quantities of stored lipid and in microorganisms when they are grown on ethanol or acetate which function as the sole source of carbon in them. The task (of converting fats into carbohydrates) in these organisms is, however, accomplished by means of a cyclic set of reactions called **glyoxylate cycle** or **Krebs–Kornberg cycle**, the latter nomenclature based on its two principal investigators, **Hans A. Krebs** and **H. L. Kornberg**, both of whom discovered this cycle in 1957. The glyoxylate cycle (refer Fig. 22–16) utilizes five enzymes, of which two namely *isocitrate lyase and malate synthase* are absent in animals. Hence, *the glyoxylate pathway does not occur in animals.* However, plant cells contain all the 5 enzymes required for the cycle, in subcellular organelles called *glyoxysomes*, hence this cycle is operative in them. It is interesting to note that the glyoxysomes appear in the cotyledons of lipid-rich seeds (*e.g.*, groundnut, castor, bean) shortly after germination begins and at a time when lipids are being utilized as the major source of carbon for carbohydrate synthesis.

In effect, the glyoxylate cycle bypasses steps 3 to 7 of the citric acid cycle, thereby omitting the two oxidative decarboxylation reactions (Steps 3 and 4) in which CO_2 is produced. The bypass consists of 2 reactions, namely:

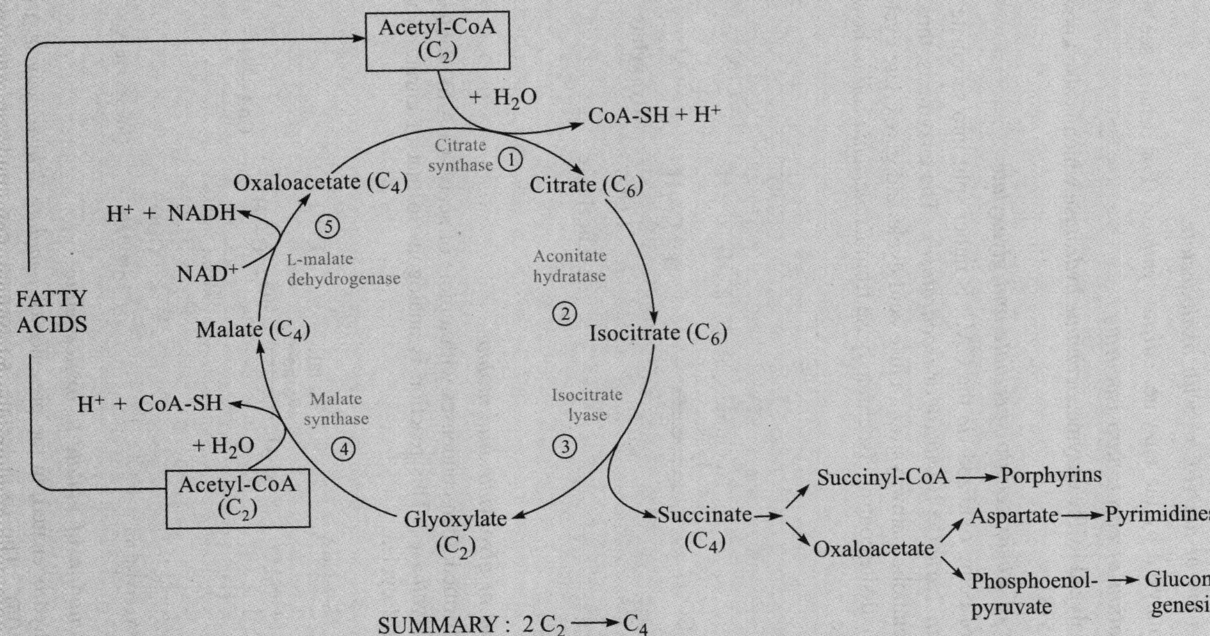

Fig. 22–16. The glyoxylate cycle: an anabolic variant of the citric acid cycle

Oxidation of fatty acids can provide acetyl-CoA, which enters the cycle at two points. Each turn of the cycle results in generation of one mole of succinate from 2 moles of acetyl-CoA. The net result of the bypass is the formation of C–C bonds between C_2 subunits.

(*a*) splitting of isocitrate into succinate and glyoxylate, and

(*b*) conversion of glyoxylate into malate.

Thus, the glyoxylate cycle consists of 5 enzyme-catalyzed steps (Fig. 22–16), of which 3 steps are the same as in Krebs cycle, *viz*., Steps 1, 2 and 8. The various steps are sequentially described as follows:

Step 1 : ***Condensation of acetyl-CoA with oxaloacetate***

This reaction is catalyzed by the enzyme, *citrate synthase* (For details, see page 596).

Step 2 : ***Isomerization of citrate into isocitrate***

This reaction is catalyzed by the enzyme, *aconitate hydratase* which is also known as *aconitase* (For details, see page 598).

Step 3 : ***Cleavage of isocitrate into succinate and glyoxylate***

Isocitrate, instead of being oxidized *via* the CAC, is rather split into succinate and *glyoxylate* by the enzyme, *isocitrate lyase* (= *isocitrase* or *isocitratase*). The enzyme is, thus, present at the branch point of two metabolic pathways (*i.e.*, citric acid cycle and glyoxylate cycle) and appears to be susceptible to second-site control. Isocitrate lyase from *Escherichia coli* is, however, inhibited by phosphoenolpyruvate.

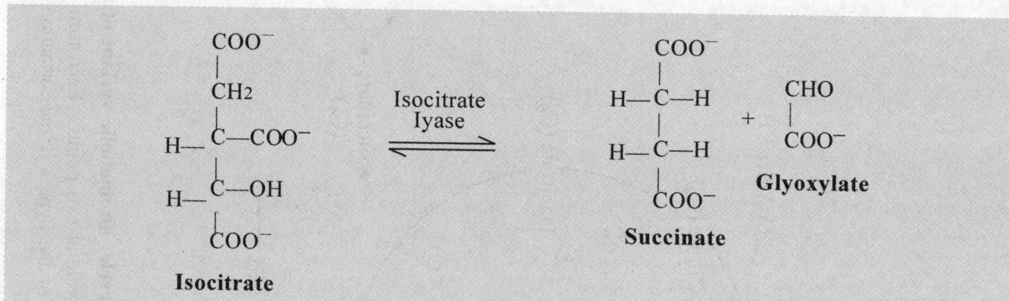

Step 4 : ***Conversion of glyoxylate into malate***

The glyoxylate, so formed, then condenses with a mole of acetyl-CoA to produce L-malic acid by the enzyme, *malate synthase*. This reaction is analogous to that of citrate synthase reaction (Step 1) of the citric acid cycle.

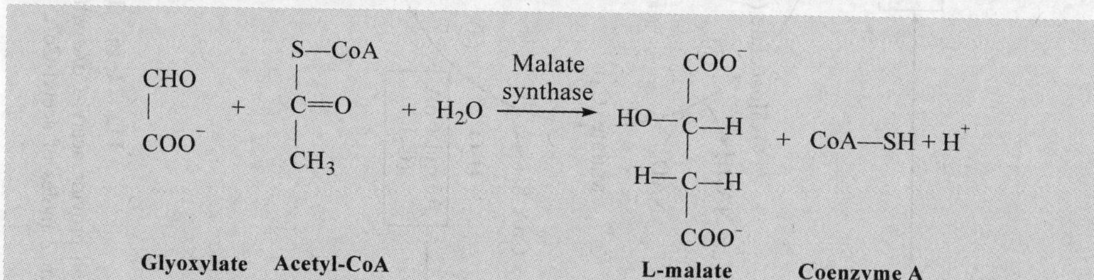

Step 5 : ***Dehydrogenation of malate to oxaloacetate***

L-malate is then dehydrogenated, in the presence of *L-malate dehydrogenase*, to oxaloacetate (For details, see page 605). The oxaloacetate, so formed, can condense with another mole of acetyl-CoA to start another turn of the cycle. It may, however, be pointed out that *whereas in the citric acid cycle the conversion of isocitrate to malate is an aerobic process, in glyxylate cycle the conversion takes place anaerobically.*

To summarize, operation of glyoxylate cycle, in addition to the 2 novel reactions, also requires the participation of 3 of the enzymes of the citric acid cycle, *viz*., citrate synthase, aconitate hydratase and malate dehydrogenase. Also required is the electron transport chain for the oxidation of NADH (produced in Step 5) by molecular oxygen. This plus the malate synthase reaction (Step 4) provide the essential driving force for the cycle. One turn of the cycle leads to the oxidation of 2 moles of acetyl-CoA to yield 1 mole of succinate with the simultaneous removal of two reducing equivalents. The overall reaction for the glyoxylate cycle may be written as:

$$\underset{\text{Acetyl-CoA}}{2\,CH_3CO\text{—}S\text{—}CoA} + NAD^+ + 2\,H_2O \xrightarrow{\textbf{Glyoxylate cycle}} \underset{\text{Succinate}}{\begin{array}{l} CH_2\text{—}COO^- \\ | \\ CH_2\text{—}COO^- \end{array}} + \underset{\text{Coenzyme A}}{2\,CoA\text{—}SH} + NADH + 3\,H^+$$

The succinate, so formed, may be used for biosynthetic purposes. For example,

(*a*) Succinate can be converted to succinyl-CoA (Step 5 of the citric acid cycle) and serve as a precursor of porphyrins.

(*b*) Succinate can be oxidized to oxaloacetate (*via* Steps 6, 7 and 8 of the citric acid cycle) which can be utilized for the synthesis of aspartate and other compounds (such as pyrimidines), derived from aspartate.

(*c*) Oxaloacetate can also be converted into phosphoenolpyruvate (For details, see page 413) which is then used for the reactions of gluconeogenesis (*i.e.*, formation of glucose or glycogen from noncarbohydrate sources such as phosphoenolpyruvate).

(*d*) Finally, the oxaloacetate can also condense with acetyl-CoA (Step 1 of the citric acid cycle) and thus initiate the reactions of the citric acid cycle.

Thus, succinate may be *anaplerotic*, giving rise to oxaloacetate, or *gluconeogenic*, giving rise to phosphoenolpyruvate. In this way, net carbohydrate synthesis from fatty acids *via* acetylCoA, which is not accomplished in animals, is achieved in the higher plants and microorganisms. ".......the glyoxylate cycle provides an example par excellence of an anaplerotic sequence in operation" (**Mahler** and **Cordes**, 1968).

As may be visualized in Fig. 22–17, the glyoxylate pathway makes use of both mitochondrial and glyoxysomal enzymes.

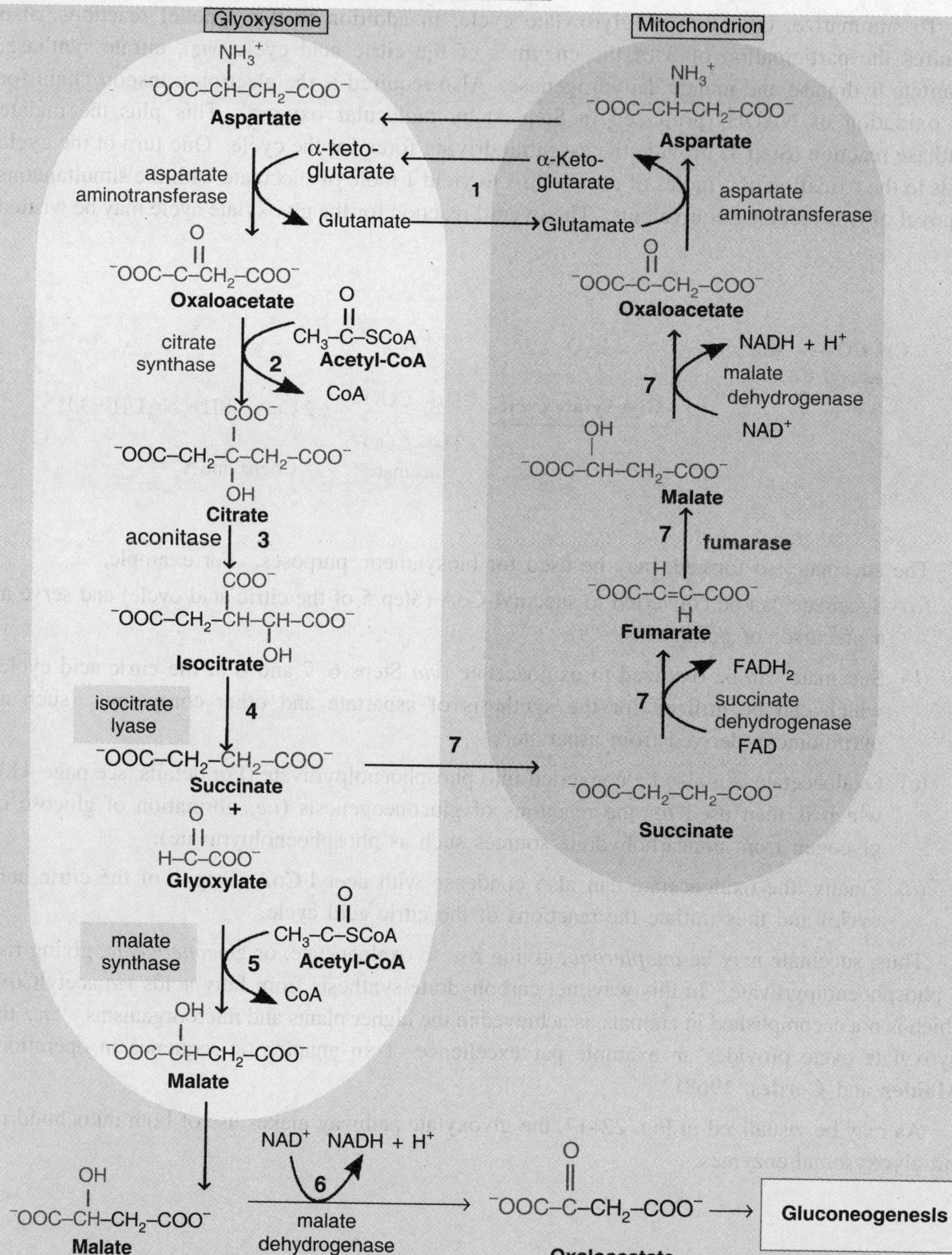

Fig. 22–17. Participation of both mitochondrial and glyoxysomal enzymes in the glyoxylate pathway
Isocitrate lyase and malate synthase, enzymes unique to plant glyoxysomes, are boxed. The pathway results in the net conversion of two acetyl-CoA to oxaloacetate. **(1)** Mitochondrial oxaloacetate is converted to aspartate, transported to the glyoxysome, and reconverted to oxaloacetate **(2)** Oxaloacetate is condensed with acetyl-CoA to form citrate. **(3)** Aconitase catalyzes the conversion of citrate to isocitrate. **(4)** Isocitrate lyase catalyzes the cleavage of isocitrate to succinate and glyoxylate **(5)** Malate synthase catalyzes the condensation of glyoxylate with acetyl-CoA to form malate. **(6)** After transport to the cytosol, malate dehydrogenase catalyzes the oxidation of malate to oxaloacetate, which can then be used in gluconeogenesis **(7)** Succinate is transported to the mitochondrion, where it is reconverted to oxaloacetate *via* the citric acid cycle.

REFERENCES

1. **Ashwell G :** Carbohydrate metabolism. *Ann. Rev. Biochem.* **33 :** *101, 1964.*
2. **Atkinson DE :** Regulation of enzyme activity. *Ann. Rev. Biochem.* **35 :** *85, 1966.*
3. **Baldwin JE, Krebs H :** The evolution of metabolic cycles. *Nature.* **291 :** *381–382, 1981.*
4. **Bentley R :** Molecular Asymmetry in Biology. *vol. 1 and 2. Academic Press Inc., New York. 1969.*
5. **Boyer PD (editor) :** The Enzyme : *3rd ed. Academic Press Inc., New York. 1971.*
6. **De Meirleir L:** Defects of pyruvate metabolism and the Krebs cycle. *J. Child Neurol.* **17 (Suppl. 3):** *3S26 – S33; discussion 3S33–34, 2002.*
7. **Denton RM, Pogson CI :** "Metabolic Regulation", Outline Studies in Biology. *John Wiley and Sons,. New York. 1970.*
8. **Gluster JP :** Aconitase. In Boyer PD (editor)'s The Enzymes. *3rd ed. vol. 5, 413–439, 1971.*
9. **Goodwin TW (editor) :** The Metabolic Roles of Citrate. *Academic Press Inc., New York. 1968.*
10. **Gottschalk G :** Bacterial Metabolism. *2nd ed. Springer-Verlag, New York. 1986.*
11. **Greenberg DM (editor) :** Metabolic Pathways. *vol. 1. Academic Press Inc., New York. 1960.*
12. **Gribble GW :** Fluoroacetate toxicity. *J. Chem. Ed.* **50 :** *460–462, 1973.*
13. **Kay J, Weitzman PDJ (editors) :** Krebs' Citric Acid Cycle : Half a Century and Still Turning. *Biochemical Society Symposium 54, The Biochemical Society, London. 1987.*
14. **Krebs HA :** The History of the Tricarboxylic Acid Cycle. *Perspect. Biol. Med.* **14 :** *154–170, 1970.*
15. **Krebs HA, Lowenstein JM :** The Tricarboxylic Acid Cycle. In Greenberg DM (editor)'s Metabolic Pathways. *vol. 1 : 129-203. Academic Press Inc., New York. 1960.*
16. **Krebs HA, Martin A :** Reminiscences and Reflections. *Clerendon Press. 1981.*
17. **Kornberg HL :** Anaplerotic Sequences and Their Role in Metabolism. *In Campbell PN and Greville GD (editor)'s Essays in Biochemistry. vol.* **2 :** *1–31, Academic Press Inc., New York. 1966.*
18. **LaPorte DC :** The isocitrate dehydrogenase phosphorylation cycle : Regulation and enzymology. *J. Cell Biochem.* **51 :** *14–18, 1993.*
19. **Lehninger AL :** The Mitrochondrion– Structure and Function. *W.A. Benjamin Inc., New York. 1964.*
20. **Lowenstein JM ;** The Tricarboxylic Acid Cycle : *In Greenberg DM (editor)'s Metabolic Pathway 3rd ed.* **vol. 1 :** *146–270, Academic Press Inc., New York. 1967.*
21. **Lowenstein JM (editor) :** Citric Acid Cycle : Control and Compartmentation. *Marcel Dekker, New York. 1969.*
22. **Lowenstein JM (editor) :** In Colowick SP and Kaplan NO's Methods in Enzymology. *vol. 13, Academic Press Inc., New York. 1969.*
23. **Lowenstein JM (editor) :** The Pyruvate Dehydrogenase Complex and the Citric Acid Cycle. *Compre. Biochem.* **185 :** *1–55, 1971.*
24. **Munn EA :** The Structure of Mitochondria. *Academic Press Inc., New York. 1974.*
25. **Newsholme EA, Start C :** Regulation in Metabolism. *Wiley, London. 1973.*
26. **Owen OE, Kalhan SC, Hanson RW:** The key role of anaplerosis and cataplerosis for citric acid cycle function. *J. Biol. Chem.* **277:** *30409–30412, 2002.*
27. **Parsons DS (editor) :** Biological Membranes. *Oxford, London. 1975.*

28. **Popjak G :** Stereospecificity of Enzyme Reactions. *In Boyer PD (editor)'s The Enzymes, 3rd ed. vol.* **2** *: 115–215, Academic Press Inc., New York. 1970.*
29. **Randall DD, Miernyk JA, Fang TK, Budde RJ, Schuller KA :** Regulation of the pyruvate dehydrogenase complexes in plants. *Ann. N. Y. Acad. Sci.* **573** *: 192–205, 1989.*
30. **Randle PJ :** Pyruvate dehydrogenase complex : Meticulous regulator of glucose disposal in animals. *Trends Biochem. Sci.,* **3** *: 217–219, 1978.*
31. **Reed LJ :** Multienzyme Complexes. *Acc. Chem. Res.* **7** *: 40–46, 1974.*
32. **Remington SJ :** Structure and mechanism of citrate synthetase. *Curr. Top. Cell. Regul.* **33** *: 209–229, 1992.*
33. **Srere PA :** The enzymology of the formation and breakdown of citrate. *Adv. Enzymol.* **43** *: 57–101, 1975.*
34. **Srere PA :** The molecular physiology of citrate. *Curr. Top. Cell. Regul.* **33** *: 261–275, 1992.*
35. **Tchen TT, van Milligan H :** The stereospecificity of the succinic dehydrogenase reaction. *J. Amer. Chem. Soc.,* **82** *: 4115–4116, 1960.*
36. **Tzagoloff A :** Mitochondria. *Plenum, New York. 1982.*
37. **Williamson DH :** Sir Hans Krebs, the First 80 Years. *Trends Biochem. Sci.* **5** *: vi–viii 1980.*
38. **Williamson JR, Cooper RH :** Regulation of the citric acid cycle in mammalian systems. *FEBS Lett.* **117** *(Suppl) : K73–K85, 1980.*

PROBLEMS

1. What is the $\Delta G^{\circ\prime}$ for the complete oxidation of the acetyl unit of acetyl-CoA by the citric acid cycle?
2. The citric acid cycle itself, which is composed of enzymatically catalyzed steps, can be thought of essentially as the product of a supramolecular enzyme. Explain.
3. Patients in shock will often suffer from lactic acidosis due to a deficiency of O_2. Why does a lack of O_2 lead to lactic acid accumulation ? One treatment for shock is to administer dichloroacetate, which inhibits the kinase associated with the pyruvate dehydrogenase complex. What is the biochemical rationale for this treatment?
4. The oxidation of malate by NAD^+ to form oxaloacetate is a highly endergonic reaction under standard conditions ($\Delta G^{\circ\prime}$ = + 7 kcal mol^{-1} (+ 29 kJ mol^{-1})]. The reaction proceeds readily under physiological conditions.
   (*a*) Why?
   (*b*) Assuming an [NAD^+]/[NADH] ratio of 8 and a pH of 7, what is the lowest [malate]/[oxaloacetate] ratio at which oxaloacetate can be formed from malate?
5. Propose a reaction mechanism for the condensation of acetyl-CoA and glyoxylate in the glyoxylate cycle of plants and bacteria.
6. The citrate (or any other symmetric compound) cannot be an intermediate in the formation of α-ketoglutarate, because of the asymmetric fate of the label. This view seemed compelling until Alexander Ogston incisively pointed out in 1948 that "it is possible that *an asymmetric enzyme which attacks a symmetrical compound can distinguish between its identical groups.*" For simplicity, consider a molecule in which two hydrogen atoms, a group X, and a different group Y are bonded to a tetrahedral carbon atom as a model for citrate. Explain how a symmetric molecule can react with an enzyme in an asymmetric way.
7. Consider the fate of pyruvate labeled with ^{14}C in each of the following positions : carbon 1 (carboxyl), carbon 2 (carbonyl), and carbon 3 (methyl). Predict the fate of each labeled carbon during one turn of the citric acid cycle.
8. Which carbon or carbons of glucose, if metabolized *via* glycolysis and the citric acid cycle, would be most rapidly lost as CO_2?

9. Would you expect NAD^+ or CoA-SH to affect the activity of pyruvate dehydrogenase kinase? Briefly explain your answer.

10. Given what you know about the function of the glyoxylate cycle and the regulation of the citric acid cycle, propose control mechansims that might regulate the glyoxylate cycle.

11. Write a balanced equation for the conversion in the glyoxylate cycle of two acetyl units, as acetyl-CoA, to oxaloacetate.

12. FAD is a stronger oxidant than NAD^+; FAD has a higher standard reduction potential than NAD^+. Yet in the last reaction of the pyruvate dehydrogenase complex, $FADH_2$ bound to E_3 is oxidized by NAD^+. Explain this apparent paradox.

13. Given the roles of NAD^+/NADH in dehydrogenation reactions and NADPH/$NADP^+$ in reductions, would you expect the intracellular ratio of NAD^+ to NADH to be high or low? What about the ratio of $NADP^+$ to NADPH? Explain your answers.

14. Oxaloacetate is formed in the last step of the citric acid cycle by the NAD^+-dependent oxidation of L-malate. Can a net synthesis of oxaloacetate take place from acetyl-CoA using only the enzymes and cofactors of the citric acid cycle, without depleting the intermediates of the cycle? Explain. How is the oxaloacetate lost from the cycle (to biosynthetic reactions) replenished?

15. α-ketoglutarate plays a central role in the biosynthesis of several amino acids. Write a series of known enzymatic reactions that result in the net synthesis of α-ketoglutarate from pyruvate. Your proposed sequence must not involve the net consumption of other citric acid cycle intermediates. Write the overall reaction for your proposed sequence and identify the source of each reactant.

16. Although oxygen does not participate directly in the citric acid cycle, the cycle operates only when O_2 is present. Why?

17. Two of the steps in the oxidative decarboxylation of pyruvate (Steps 4 and 5, Fig. 22-) do not involve any of the three carbons of pyruvate yet are essential to the operation of the pyruvate dehydrogenase complex. Explain.

18. Using pyruvate, labeled with ^{14}C in its keto group, *via* the pyruvate dehydrogenase reaction and the TCA cycle, where would the carbon label be at the end of one turn of the TCA cycle? Where would the carbon label be at the end of the second turn of the cycle?

19. Explain why, when glucose is the sole carbon source, bacteria grow much more slowly in the absence of O_2 than in the presence of O_2.

20. In which region of a mitochondria are most enzymes of the citric acid cycle located?

(*a*) outer membrane

(*b*) inner membrane

(*c*) intermembrane space

(*d*) matrix

21. In which of the following organisms would you never find mitochondria?

(*a*) a bacterium

(*b*) *Mucor*

(*c*) an yeast

(*d*) *Amoeba*

22. Irrespective of what we eat– carbohydrates, proteins and fats, the body breaks it all down through various biochemical reactions during the process of metabolism. There is one biochemical pathway which is the final common pathway for the oxidation of carbohydrates, proteins and lipids. Which is it?

ANSWERS TO PROBLEMS

1. -9.8 kcal mol^{-1} (-41.0 kJ mol^{-1}).
2. Enzymes or enzyme complexes are biological catalysts. Recall that a catalyst facilitates a chemical reaction without the catalyst itself being permanently altered. Oxaloacetate can be thought of as a catalyst because it binds to an acetyl group, leads to the oxidative decarboxylation of the two carbon atoms, and is regenerated at the completion of a cycle. In essence, oxaloacetate (and any cycle intermediate) acts as a catalyst.
3. A decrease in the amount of O_2 will necessitate an increase in anaerobic glycolysis for energy production, leading to the generation of a large amount of lactic acid. Under conditions of shock, the kinase inhibitor is administered to ensure that pyruvate dehydrogenase is operating maximally.
4. (*a*) The steady-state concentrations of the products are low compared with those of the substrates.

   (*b*) The ratio of malate to oxaloacetate must be greater than 1.75×10^4 for oxaloacetate to be formed.
5. The enol intermediate of acetyl-CoA attacks the carbonyl carbon atom of glyoxylate to form a C–C bond. This reaction is like the condensation of oxaloacetate with the enol intermediate of acetyl-CoA in the reaction catalyzed by citrate synthase. Glyoxylate contains a hydrogen atom in place of the – CH_2COO^- group of oxaloacetate; the reactions are otherwise nearly identical.
6. Call one hydrogen atom A and the other B. Now suppose that an enzyme binds three groups of this substrate—X, Y, and H—at three complementary sites. The adjoining diagram shows X, Y, and H_A bound to three points on the enzyme. In contrast X, Y, and H_B cannot be bound to this active site; two of these three groups can be bound, but not all three. Thus, H_A and H_B will have different fates.

   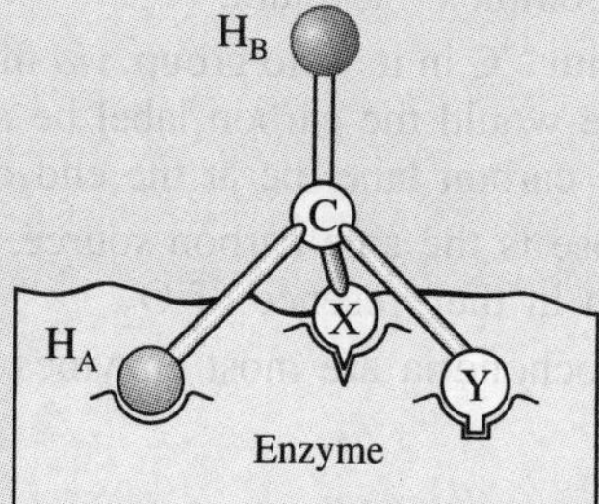

   Sterically nonequivalent groups such as H_A and H_B will almost always be distinguished in enzymatic reactions. The essence of the differentiation of these groups is that the enzyme holds the substrate in a specific orientation. Attachment at three points, as depicted in the diagram, is a readily visualized way of achieving a particular orientation of the substrate, but it is not the only means of doing so.
7. C-1: all released as CO_2. C-2 and C-3: all retained in oxaloacetate.
8. C-3 and C-4, since these become the carboxyl group of pyruvate, which is lost in the pyruvate dehydrogenase reaction.
9. Because NADH and acetyl-CoA activate the enzyme, it makes metabolic sense to expect that NAD^+ and CoA-SH would be inhibitory, and these inhibitions are observed.
10. Some possible mechanisms: substrate-level control of citrate synthase, activation of citrate lyase by acetyl-CoA or fatty acids, inhibition of isocitrate lyase by succinate (to ensure adequate flux through the citric acid cycle).

11. 2 Acetyl-CoA + 2 NAD^+ + FAD + 3 H_2O $\longrightarrow$
oxaloacetate + 2 NADH + $FADH_2$ + 2 CoA-SH + 4 H^+

12. NAD^+ + $FADH_2$ $\rightleftharpoons$ NADH + FAD + H^+
NAD^+ can oxidize $FADH_2$ if its local concentration is much higher than that of $FADH_2$.

13. [NAD^+]/ [NADH] should be high, so that it can promote the oxidation of substrates, *e.g.*, malate + NAD^+ $\rightleftharpoons$ oxaloacetate + NADH + H^+. Conversely, since NADPH and $NADP^+$ usually promote reduction of substrates, we expect [$NADP^+$]/[NADPH] to be low.

14. No. For every two carbons that enter as acetate, two leave the cycle as CO_2; thus there is no net synthesis of oxaloacetate. Net synthesis of oxaloacetate occurs by the carboxylation of pyruvate, an anaplerotic reaction.

15. Net reaction:
2 Pyruvate + ATP + 2 NAD^+ + H_2O $\longrightarrow$
α-ketoglutarate + CO_2 + ADP + P_i + 2 NADH + 3 H^+

16. Oxygen is needed to recycle NAD^+ from the NADH produced by the oxidative reactions of the citric acid cycle. Reoxidation of NADH occurs during mitochondrial oxidative phosphorylation.

17. Steps 4 and 5 are essential in the reoxidation of the enzyme's reduced lipoamide cofactor.

18. The labeled *keto* carbon of pyruvate becomes the labeled *carboxyl* carbon of acetyl-CoA. After condensation with oxaloacetate, the first carboxyl group of citrate is labeled. This label is retained through subsequent reactions to succinate. However, succinate is a symmetrical compound to the enzyme so, in effect, both carboxyl groups of succinate are labeled. This means that the oxaloacetate regenerated is labeled in both carboxyl groups at the end of one turn (actually half the molecules are labeled in one carboxyl and half in the other but this can't be distinguished experimentally). Note that CO_2 is not labeled. In the second turn, the same carboxyl labeled acetyl-CoA is added but this time to labeled oxaloacetate. Both carboxyl groups of the oxaloacetate are released as CO_2 so it will be labeled, as will the regenerated oxaloacetate.

19. In the absence of oxygen, the citric acid cycle is inactive, and less ATP is generated per molecule of glucose.

20. (*d*) matrix

21. (*a*) a bacterium

22. Citric acid cycle. Essentially, it comprises the pathway for metabolizing glucose, many amino acids and fatty acids to acetyl-CoA.

CONTENTS

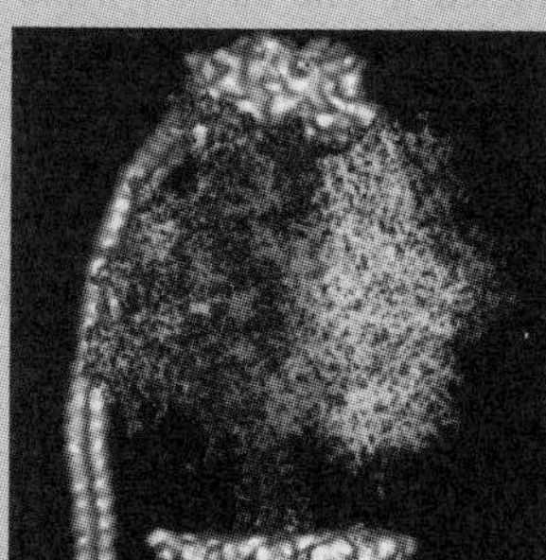

A model for mitochondrial F_0F_1-ATP synthetase, a rotating molecular motor
ATP synthesis occurs on F_1 domain, while F_0 domain contains a proton channel. The a, b, α, β, and δ subunits constitute the stator while the c, γ, and ε subunits provide the rotor. Protons flow through the structure causing the rotor to turn, resulting in conformational changes in the β subunits where ATP is synthesized.

(Courtesy : Drs. Peter L. Pedersen, Young Hee Ko, and Sangjin Hong)

CHAPTER

23

Electron Transport and Oxidative Phosphorylation

All the enzyme-catalyzed steps in the oxidative degradation of carbohydrates, fats and amino acids in aerobic cells converge into electron transport and oxidative phosphorylation, the final stage of cell respiration. This stage consists of flow of electrons from organic substrates to oxygen with the simultaneous release of energy for the generation of ATP molecules. The importance of this final stage of respiration in the human body can be realized by the fact that a normal adult businessman with a 70 kg weight requires about 2,800 kcal of energy per day. This amount of energy can be produced by the hydrolysis of about 2,800/7.3 = 380 mole or 190 kilograms of ATP. However, the total amount of ATP present in his body is about 50 grams. In order to provide chemical energy for the body need, the 50 g of ATP must be broken down into ADP and phosphate and resynthesized thousands of times in a day. *i.e.*, 24 hours.

ELECTRON FLOW AS SOURCE OF ATP ENERGY

Fig. 23–1 shows the electron transport chain in abbreviated form. In each turn of citric acid cycle, 4 pairs

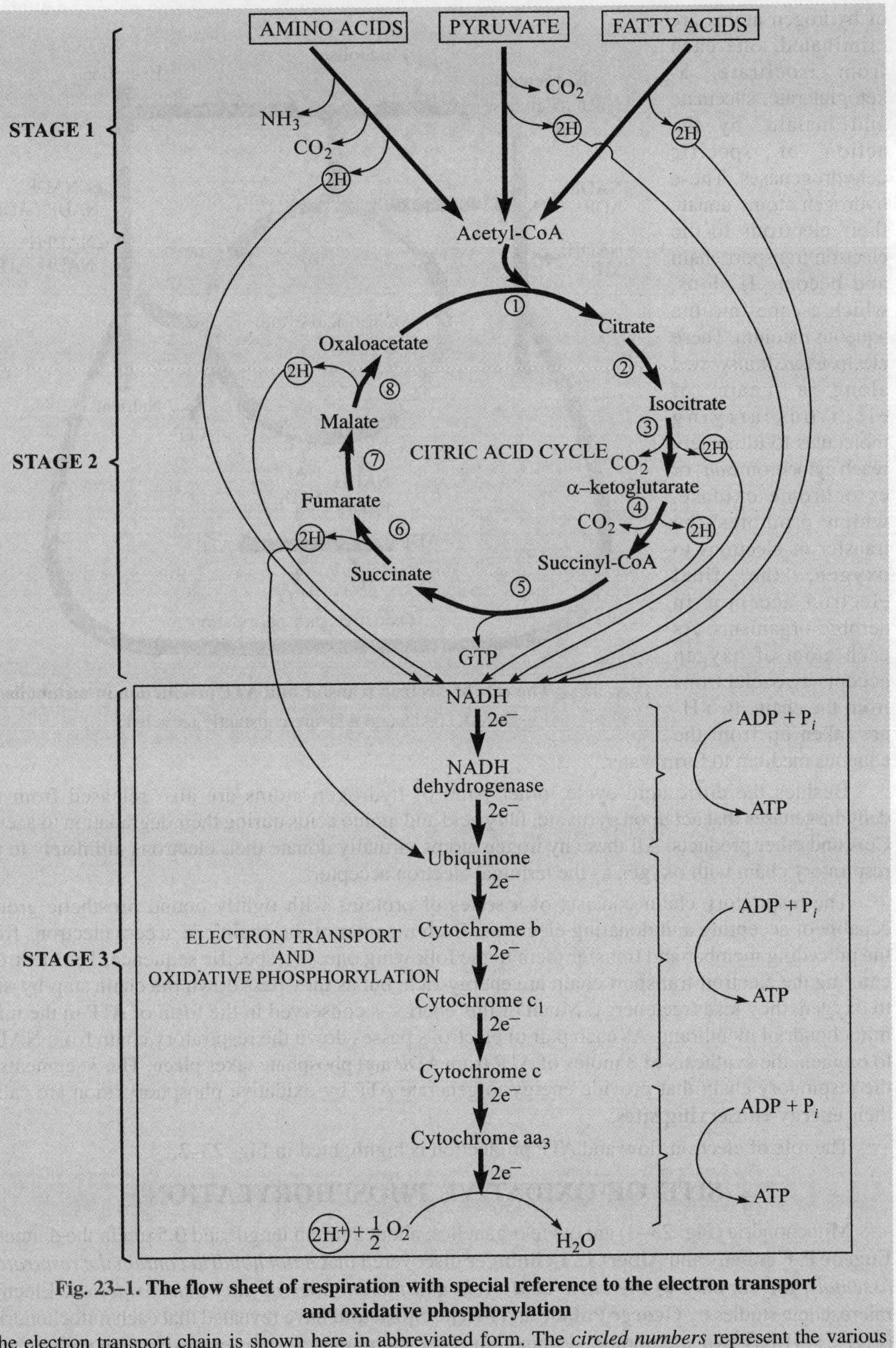

Fig. 23–1. The flow sheet of respiration with special reference to the electron transport and oxidative phosphorylation

The electron transport chain is shown here in abbreviated form. The *circled numbers* represent the various steps of the citric acid cycle.

of hydrogen atoms are eliminated, one each from isocitrate, a-ketoglutarate, succinate and malate, by the action of specific dehydrogenases. These hydrogen atoms donate their electrons to the electron transport chain and become H^+ ions, which escape into the aqueous medium. These electrons are transported along a chain of electron-carrying molecules to ultimately reach cytochrome aa_3 or cytochrome oxidase, which promotes the transfer of electrons to oxygen, the final electron acceptor in aerobic organisms. As each atom of oxygen accepts two electrons from the chain, two H^+ are taken up from the aqueous medium to form water.

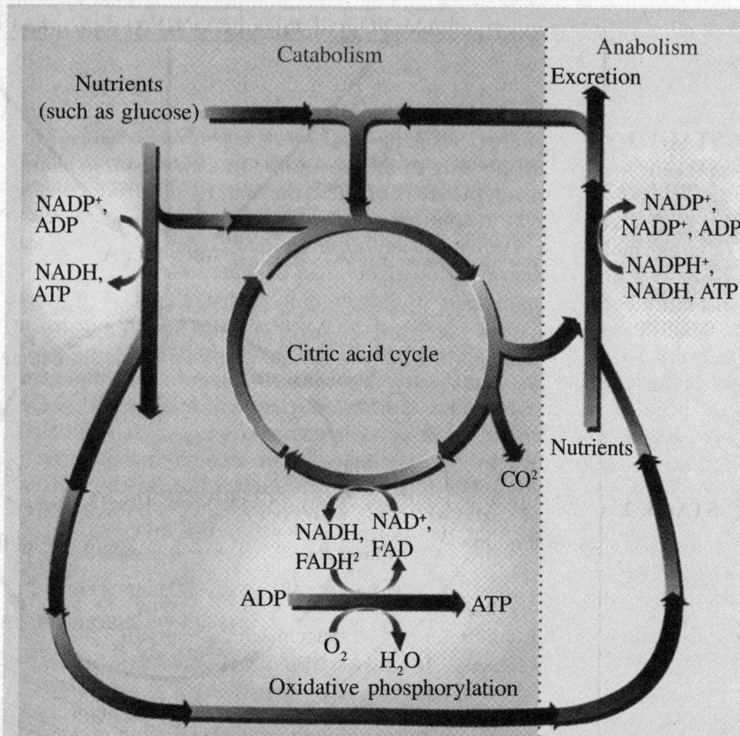

Fig. 23–2. The role of electron transfer and ATP production in metabolism
NAD^+, FAD, and ATP are constantly recycled.

Besides the citric acid cycle, other pairs of hydrogen atoms are also released from the dehydrogenases that act upon pyruvate, fatty acid and amino acids during their degradation to acetyl-CoA and other products. All these hydrogen atoms virtually donate their electrons ultimately to the respiratory chain with oxygen as the terminal electron acceptor.

The respiratory chain consists of a series of proteins with tightly bound prosthetic groups capable of accepting and donating electrons. Each member of the chain can accept electrons from the preceding member and transfer them to the following one, in a specific sequence. The electrons entering the electron-transport chain are energy-rich, but as they pass down the chain step-by-step to oxygen, they lose free energy. Much of this energy is conserved in the form of ATP in the inner mitochondrial membrane. As each pair of electrons passes down the respiratory chain from NADH to oxygen, the synthesis of 3 moles of ATP from ADP and phosphate takes place. The 3 segments of the respiratory chain that provide energy to generate ATP by oxidative phosphorylation are called their **energy-conserving sites**.

The role of electron flow and ATP production is highlighted in Fig- 23–2.

SITE OF OXIDATIVE PHOSPHORYLATION

Mitochondria (Fig. 23–3) are ovate organelles, about 2 μm in length and 0.5 μm in the diameter. **Eugene P. Kennedy** and **Albert L. Lehninger** discovered that *mitochondria contain the respiratory assembly, the enzymes of the citric acid cycle and the enzymes of fatty acid oxidation.* Electron microscopic studies by **George Palade** and **Fritjof Sjöstrand** have revealed that each mitochondrion (Fig. 23–4) has two membrane systems: an outer membrane and an extensive *inner membrane*, which is highly-folded into a series of internal ridges called *cristae*. Obviously, there are two compartments in motochondria : the intermembrane space between the outer and inner membranes and the *matrix*, which is bounded by the inner membrane.

The **outer membrane** is freely permeable to most small molecules and ions and contains some enzymes. In contrast, the **inner membrane** is impermeable to nearly all ions and most uncharged molecules and contains the electron-transport chains, succinate dehydrogenase and ATP-synthesizing enzymes. The inner membrane of a single liver mitochondrion may have over 10,000 sets of electron-transport chains and ATP synthetase molecules. The heart mitochondria have profuse cristae and therefore contain about 3 times more sets of electron-transport chains than that of liver mitochondria. The **intermembrane space** contains adenylate kinase and some other enzymes whereas the **matrix** compartment contains most of the citric acid cycle enzymes, the pyruvate dehydrogenase system and the fatty acid oxidation system. It also contains ATP, ADP, AMP, phosphate, NAD, NADP, coenzyme A and various ions such as K^+, Mg^{2+} and Ca^{2+}.

Fig. 23–3. Mitochondria, shown here, are the sites of the citric acid cycle, electron transport, and oxidative phosphorylation.

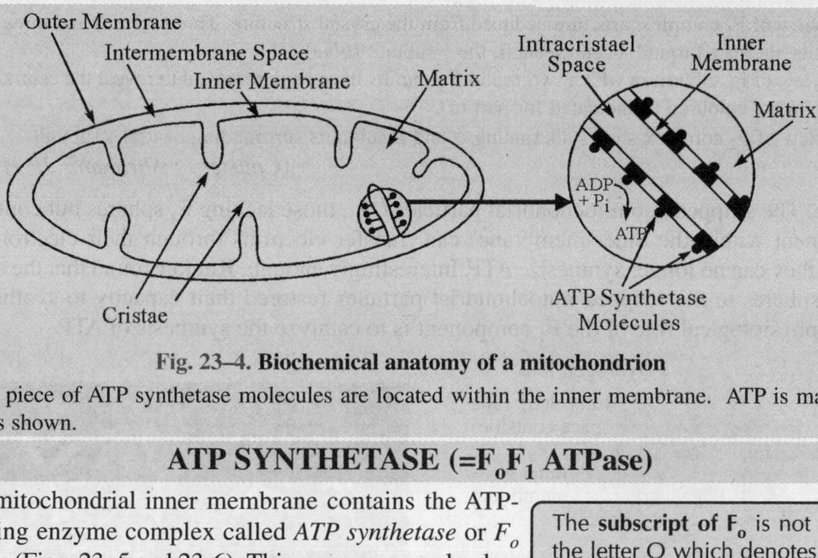

Fig. 23–4. Biochemical anatomy of a mitochondrion

The base piece of ATP synthetase molecules are located within the inner membrane. ATP is made in the matrix, as shown.

ATP SYNTHETASE (=F_oF_1 ATPase)

The mitochondrial inner membrane contains the ATP-synthesizing enzyme complex called *ATP synthetase* or *F_o F_1 ATPase* (Figs. 23–5 and 23-6). This enzyme complex has 2 major components F_o and F_1 (F for factor). The F_1 component is like a doorknob protruding into the matrix from the inner membrane. It is attached by a stalk to Fo component, which is embedded in the inner membrane and extends across it.

> The **subscript of F_o** is not zero but the letter O which denotes that it is the portion of the ATP synthetase which binds the toxic antibiotic, oligomycin. Oligomycin is a potent inhibitor of this enzyme and thus of oxidative phosphorylation.

The F_1 components was first extracted from the mitochondrial inner membrane and purified by **Efraim Racker** and his collaborators in 1960. During the sonic treatment of the inner mitochondrial membrane (Fig. 23–7), the cristae membranes are fragmented which, later on, reseal to form vesicles called *submitochondrial particles*, in which the F_1 spheres are on the outside, rather than the inside. In other words, the vesicles are "inside out". When these inverted vesicles are treated with urea or trypsin, the F_1 spheres become detached. The isolated F_1 alone cannot make ATP from ADP and inorganic phosphate but it can hydrolyze ATP to ADP and phosphate. It is, therefore, also called as

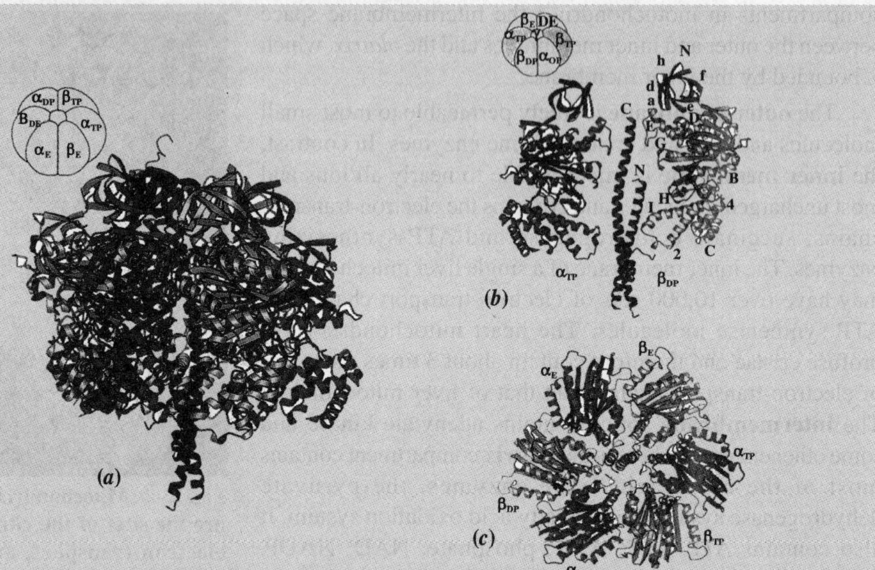

Fig. 23–5. Ribbon model of the 3-dimensional structure of mitochondrial ATP synthase complex (=F_0F_1 ATPase)

(*a*) Side view of F_1 complex structure deduced from the crystal structure. Three α (*red*) and three β (*yellow*) subunits alternate around a central shaft, the γ subunit (*blue*).

(*b*) Side view of F_1 subunit in which two α and β subunits have been removed to reveal the central γ subunit. Subunits are coloured as indicated for part (*a*).

(*c*) Top view of F_1 complex shows alternating α and β subunits surrounding central γ subunit.

(*Courtesy : Abrahams JP et al, 1994*)

F_1 ATPase. The stripped submitochondrial particles (*i.e.*, those lacking F_1 spheres but containing the F_o component within the inner membrane) can transfer electrons through their electron-transport chain, but they can no longer synthesize ATP. Interestingly enough, **Racker** found that the addition of these F_1 spheres to the stripped mitochondrial particles restored their capacity to synthesize ATP. Thus, the physiological role of the F_1 component is to catalyze the synthesis of ATP.

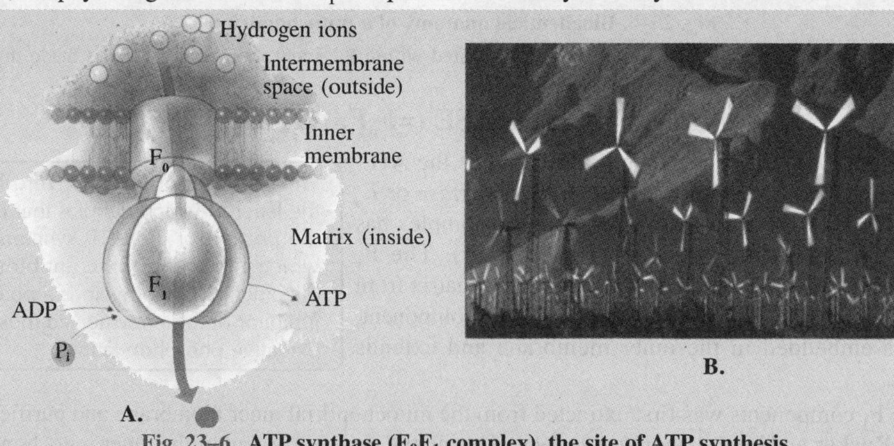

Fig. 23–6. ATP synthase (F_0F_1 complex), the site of ATP synthesis

A. The F_oF_1 complex harnesses energy stored in the proton gradient across the inner mitochondrial membrane. F_o serves as a channel for protons flowing back into the matrix, and F_1 is an enzyme that makes ATP as the protons pass.

B. A windmill's blades spin around as it generates electricity. In the same way, the F_1 blades of the F_oF_1 complex spin around as the complex generates ATP.

(*Courtesy : (B) Alan Reininger*)

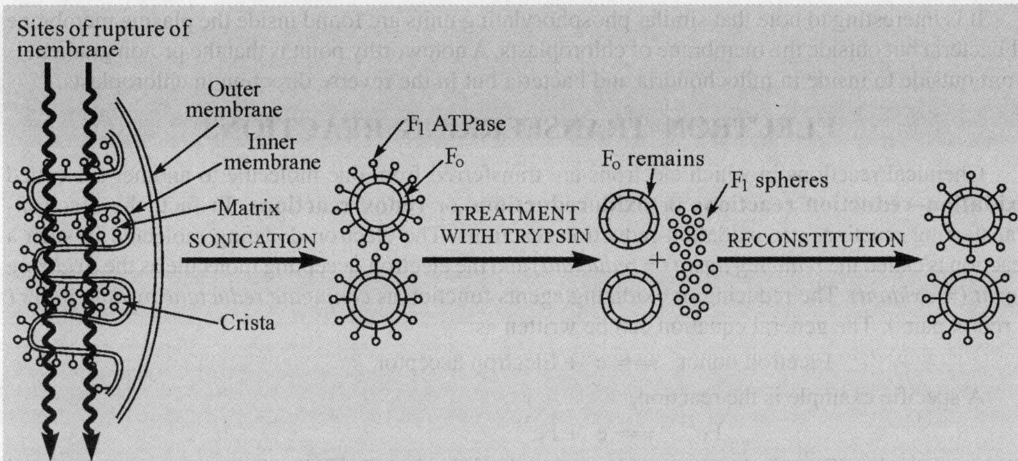

Fig. 23–7. Sonication of the mitochondrial inner membrane

(After Lehninger AL, 1984)

Table 23–1 lists some characteristics of the various components of the mitochondrial ATP-synthesizing complex.The spheric **F_1 component** (MW = 360 kdal) contains 9 polypeptide chain subunits of five kinds (designated as α, β, γ, δ and ∈) arranged into a cluster. It has many binding sites for ATP and ADP. The cuboidal **F_o component** is a hydrophobic segment of 4 polypeptide chains. It acts as a base piece and normally extends across the inner membrane. *F_o is the proton channel of the enzyme complex.* The cylindric stalk between F_o and F_1 includes many other proteins. One of them renders the enzyme complex sensitive to oligomycin, an antibiotic that blocks ATP synthesis by interferring with the utilization of the proton gradient. *The stalk is the communicating portion of the enzyme complex.* F_oF_1 ATPase is called an ATPase because, in isolated form, it hydrolyzes ATP to ADP plus Pi. However, since its major biological role in intact mitochondria is to produce ATP from ADP and Pi, it is better called ATP synthetase.

Table 23–1. Components of the mitochondrial ATP synthetase

Subunits	*Mass (in kcal)*	*Role*	*Location*
F_1	360	Contains catalytic site for ATP synthesis	Spherical headpiece on matrix side
α	53		
β	50		
γ	33		
δ	17		
∈	7		
F_o	29	Contains proton channel	Transmembrane
	22		
	12		
	8		
F_1 inhibitor	10	Regulates proton flow and ATP synthesis	Stalk between F_0 and F_1
Oligomycin-sensitivity conferring protein (OSCP)	18		
Fc_2 (F_6)	8		

(Adapted from De Pierre JW and Ernster L, 1977)

It is interesting to note that similar phosphorylating units are found inside the plasma membrane of bacteria but outside the membrane of chloroplasts. A noteworthy point is that the proton gradient is from outside to inside in mitochondria and bacteria but in the reverse direction in chloroplasts.

ELECTRON–TRANSFERRING REACTIONS

Chemical reactions in which electrons are transferred from one molecule to another are called **oxidation–reduction reactions** or **oxidoreductions** or **redox reactions**. In fact, the electron-transferring reactions are oxidation-reduction reactions. The electron-donating molecule in such a reaction is called the *reducing agent* (= *reductant*) and the electron-accepting molecule as the *oxidizing agent* (= *oxidant*). The reducing or oxidizing agents function as *conjugate reductant-oxidant pairs* (= redox pairs). The general equation can be written as:

$$\text{Electron donor} \rightleftharpoons e^- + \text{Electron acceptor}$$

A specifie example is the reaction,

$$Fe^{2+} \rightleftharpoons e^- + Fe^{3+}$$

where ferrous ion (Fe^{2+}) is the electron donor and the ferric ion (Fe^{3+}) the electron acceptor. Fe^{2+} and Fe^{3+} together constitute a conjugate redox pair.

Electrons are transferred from one molecule to another in one of the following ways (**Lehninger AL**, 1984):

1. *Directly in the form of electrons*— For example, the $Fe^{2+} - Fe^{3+}$ redox pair can transfer an electron to the $Cu^+ - Cu^{2+}$ pair.

$$Fe^{2+} + Cu^{2+} \longrightarrow Fe^{3+} + Cu^+$$

2. *In the form of hydrogen atoms*— A hydrogen atoms consists of a proton (H^+) and a single electron (e^-). The general equation can be written as:

$$AH_2 \rightleftharpoons A + 2e^- + 2H^+$$

in which AH_2 is the hydrogen (or electron) donor, A is the hydrogen acceptor and AH_2 and A together constitute a conjugate redox pair. This redox pair can reduce the electron acceptor B by the transfer of H atoms as follows:

$$AH_2 + B \longrightarrow A + BH_2$$

3. *In the form of a hydride ion*— The hydride ion (: H^-) bears two electrons, as in the case of NAD-linked dehydrogenases."

4. *During direct combination of an organic reductant with oxygen*— In such reactions a product is formed in which the oxygen is covalently incorporated, for example in the oxidation of a hydrocarbon to an alcohol,

$$R—CH_3 + \frac{1}{2}O_2 \longrightarrow R—CH_2—OH$$

where the hydrocarbon is the electron donor and the oxygen atom, the electron acceptor.

All four types of electron transfer occur in cells. The neutral term reducing equivalent is commonly used to designate a single electron equivalent participating in redox reactions, whether it is in the form of an electron *per se*, a hydrogen atom, a hydride ion, or as a reaction with oxygen to yield an oxygenated product. Because biological fuel molecules usually undergo dehydrogenation to lose two reducing equivalents at a time and also because each oxygen atom can accept two reducing equivalents, it is customary to treat the unit of biological oxidations as a pair of reducing equivalents passing from substrate to oxygn.

STANDARD OXIDATION–REDUCTION POTENTIAL

The tendency of a given conjugate redox pair to lose an electron can be specified by a constant variously called as the *standard oxidation-reduction potential, standard redox potential* or simply *standard potential* and is denoted by the symbol, E_0'. This is defined as the electromotive force (e.m.f.) in volts given by a responsive electrode placed in solution containing both the electron donor and its conjugate electron acceptor at 0.1 M concentration, 25°C and 7.0 pH. Although standard potentials are given in unts of *volts*, they are often expressed in *millivolts* for convenience. Each conjugate redox pair has a characteristic standard oxidation-reduction potential. By convention, the standard potentials of conjugate redox pairs are expressed as reduction potentials, which assign increasingly negative values to systems having an increasing tendency to lose electrons, and increasingly positive values to systems having an increasing tendency to accept electrons. In other words, the more negative the E_0', the lower is the affinity of the system for electrons and conversely, the more positive the E_0' of a system, the greater is its *electron affinity*. Thus, electrons tend to flow from one redox couple to another in the direction of the more positive system.

Table 23–2 gives the standard redox potentials of some systems useful in biological electron transport. They are listed in order of increasing potential, *i.e.*, in the order of decreasing tendency to lose electrons. Thus, conjugate redox pairs having relatively negative standard potential tend to lose electrons to those lower in the table. For example, when the isocitrate/α-ketoglutarate + CO_2 couple is present in 1.0 concentration, it has a standard potential E_0' of –0.38 V. This redox couple tends to pass electrons to the redox couple NADH/NAD^+, which has a relatively more positive potential in the

Table 23–2. Standard redox potentials, E_0' of some redox pairs participating in oxidative metabolism*

Redox couple	E_o' (in volts)
Some substrate couples	
Acetyl-CoA + CO_2 + $2H^+$ + $2e^-$ ⟶ Pyruvate + CoA	– 0.48
α-ketoglutarate + CO_2 + $2H^+$ + $2e^-$ ⟶ Isocitrate	– 0.38
3-phosphoglyceroyl phosphate + $2H^+$ + $2e^-$ ⟶ Glyceraldehyde 3-phosphate + Pi	– 0.29
Pyruvate + $2H^+$ + $2e^-$ ⟶ Lactate	– 0.19
Oxaloacetate + $2H^+$ + $2e^-$ ⟶ Malate	– 0.18
Fumarate + $2H^+$ + $2e^-$ ⟶ Succinate	+ 0.03
Components of the electron-transport chain	
$2H^+$ + $2e^-$ ⟶ H_2	**– 0.41**
NAD^+ + H^+ + $2e^-$ ⟶ NADH	–0.32
$NADP^+$ + H^+ + $2e^-$ ⟶ NADPH	– 0.32
NADH dehydrogenase (FMN form) + $2H^+$ + $2e^-$ ⟶ NADH dehydrogenase ($FMNH_2$ form)	– 0.30
Ubiquinone + $2H^+$ + $2e^-$ ⟶ Ubiquinol	+ 0.04
Cytochrome *b* (oxi.) + e^- ⟶ Cytochrome *b* (red.)	+ 0.07
Cytochrome c_1 (oxi.) + e^- ⟶ Cytochrome c_1 (red.)	+ 0.23
Cytochrome *c* (oxi.) + e^- ⟶ Cytochrome *c* (red.)	+ 0.25
Cytochrome *a* (oxi.) + e^- ⟶ Cytochrome *a* (red.)	+ 0.29
Cytochrome a_3 (oxi.) + e^- ⟶ Cytochrome a_3 (red.)	+ 0.55
$1/2O_2$ + $2H^+$ + $2e^-$ ⟶ H_2O	**+ 0.82**

* Assuming 1 M concentrations of all compontents, pH = 7.0 and temperature = 25°C.

E_0' refers to the partial reaction written as:

Oxidant + e^- ⟶ Reductant

Note the two landmark potentials (in boldface) which are for the $H_2/2H^+$ and the $H_2O/\frac{1}{2}O_2$ couple.

presence of isocitrate dehydrogenase. Conversely, the strongly positive standard potential of H_2O/O_2 couple, 0.82 V, indicates that water molecule has very little tendency to lose electrons to form molecular oxygen. In other words, molecular oxygen has a very high affinity for electrons or hydrogen atoms.

In oxidation systems, the electrons will tend to flow from a relatively electronegative conjugate redox pair, such as $NADH/NAD^+$ ($E_0' = -0.32$ V), to the more electropositive pair, such as reduced cytochrome *c*/oxidized cytochrome *c* ($E_0' = +0.25$ V). Likewise, they will also tend to flow from the cytochrome *c* redox pair to the water/oxygen pair ($E_0' = +0.82$ V). The greater the difference in the standard potentials between two redox pairs, the greater is the free-energy loss as electrons pass from the electronegative to the electropositive pair. Therefore, when electrons flow down the complete electron-transport chain from NADH to oxygen *via* several electron-carrying molecules, they lose a large amount of free energy.

The amount of free-energy which becomes available as a pair of electrons passes from NADH to O_2 can also be calculated. The standard-free-energy change of an electron-transferring reaction is given by the equation,

$$\Delta G^{\circ\prime} = -nF\ \Delta E_0'$$

where, $\Delta G^{\circ\prime}$ is the standard-free-energy change in calories, n is the number of electrons transferred, F is the caloric equivalent of the constant, the faraday (23.062 kcal/V. mol) and $\Delta E_0'$ is the difference between the standard potential of the electron-donor system and that of the electron-acceptor system. The standard-free-energy change as a pair of electron equivalents passes from the $NADH/NAD^+$ pair ($E_0' = -0.32$ V) to the H_2O/O_2 pair ($E_0' = +0.82$ V) is

$$\begin{aligned}\Delta G^{\circ\prime} &= -2\ (23.062)\ [0.82 - (-0.32)] \\ &= -2 \times 23.062 \times 1.14 \\ &= -52.58\ \text{kcal/mol}\end{aligned}$$

This amount of free energy (*i.e.*, –52.58 kcal) is more than sufficient to bring about the synthesis of 3 moles of ATP, which requires an input of 3 (7.3) = 21.9 kcal under standard conditions.

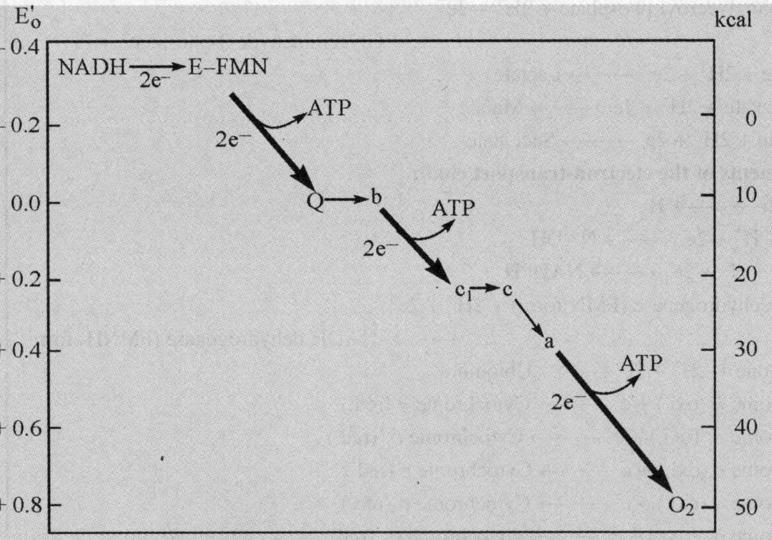

Fig. 23–8. The energy relationships in the respiratory chain of mitochondria

[E-FMN = NADH dehydrogenase; *Q* = ubiquinone; *b*, c_1, *c* and *a* = cytochromes]

Note that there are 3 steps (shown with boldface arrows) in the electron-transport chain in which relatively large decreases in free-energy occur as electrons pass. These are, in fact, the steps that provide free energy for ATP synthesis.

Likewise using the expression $\Delta G^{\circ\prime} = -nF\,\Delta E_0'$, the free-energy changes for individual segments of the electron-transport chains can be calculated from the differences in the standard potentials of the electron-donating redox pair and the electron-accepting pair. Fig. 23–8 is an energy diagram showing (*a*) the standard potentials of the electron carriers of the respiratory chain, (*b*) the direction of electron flow, which is always "downhill" toward oxygen, and (*c*) the relative free-energy change at each step.

ELECTRON CARRIERS

Electrons are transferred from substrates to oxygen through a series of electron carriers— flavins, iron-sulfur complexes, quinones and hemes (Fig. 23–9). The 3 *noteworthy features* of the electron carriers are enumerated below:

1. Although the figure shows the respiratory chain to have 17 electron carriers, there are some 15 or more chemical groups in the electron-transport chain that can accept or transfer reducing equivalents (hydrogen and electrons) in sequence.
2. All electron carriers, except for the quinones, are prosthetic groups of proteins. They include nicotinamide adenine dinucleotide (NAD), active with various dehydrogenases; flavin mononucleotide (FMN), in NADH dehydrogenases; ubiquinone or coenzyme Q, which functions in association with one or more proteins; two different kinds of iron-containing proteins, the iron-sulfur centres (Fe–S) and the cytochromes; and copper of cytochrome aa_3.
3. Almost all the electron-carrying proteins are water insoluble and are embeded in the inner mitochondrial membrane.

1. Pyridine Nucleotides

Most of the electron pairs entering the respiratory chains arise from the action of dehydrogenases that utilize the coenzymes NAD^+ or $NADP^+$ (Table 23–3) as electron acceptors. As a group, they are designated as NAD(P)-linked dehydrogenases. They catalyze the reversible reactions of the following general types:

$$\text{Reduced substrate} + NAD^+ \rightleftharpoons \text{Oxidized substrate} + NADH + H^+$$

$$\text{Reduced substrate} + NADP^+ \rightleftharpoons \text{Oxidized substrate} + NADPH + H^+$$

Most of the pyridine-linked dehydrogenases are specific for NAD^+ (refer Table 23–3). However, certain others require $NADP^+$ as electron acceptor, such as glucose 6-phosphate dehydrogenase. Very few, such as glutamate dehydrogenase can react with either NAD^+ or $NADP^+$. Some pyridine-linked dehydrogenases are located in the mitochondria, some in the cytosol and still others in both.

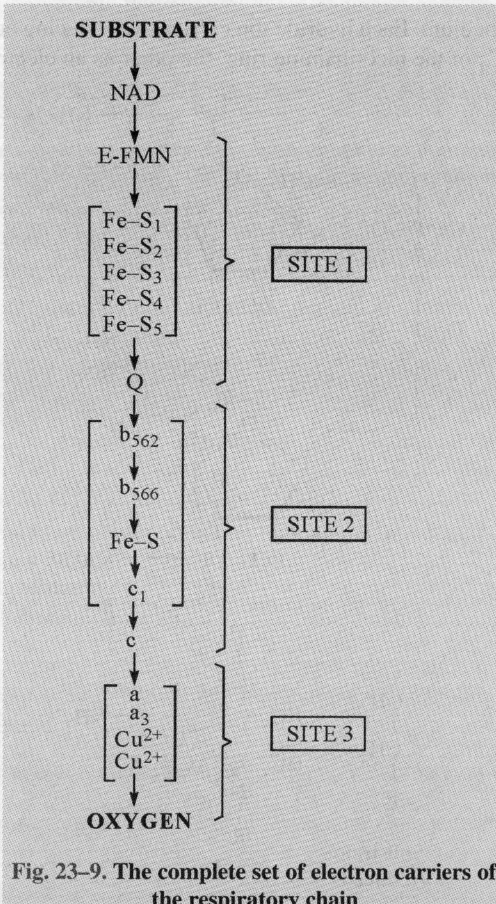

Fig. 23–9. The complete set of electron carriers of the respiratory chain

The precise sequence and function of all the oxidation-reduction centres is not exactly known.

The pyridine-linked dehydrogenases remove 2 hydrogen atoms from their substrates. One of these is transferred as a hydride ion (: H^-) to the NAD^+ or $NADP^+$; the other appears as H^+ in the

Table 23–3. Some important reactions catalyzed by NAD (or NADP)-linked dehydrogenases

Reaction	*Location*
NAD-linked:	
Isocitrate + NAD^+ ⇌ α-ketoglutarate + CO_2 + NADH + H^+	Mitochondria
α-ketoglutarate + CoA + NAD^+ ⇌ Succinyl-CoA + CO_2 + NADH + H^+	Mitochondria
L-malate + NAD^+ ⇌ Oxaloacetate + NADH + H^+	Mitochondria and cytosol
Pyruvate + CoA + NAD^+ ⇌ Acetyl-CoA + CO_2 + NADH + H^+	Mitochondria
Glyceraldehyde 3-phosphate + Pi + NAD ⇌ 1, 3-diphosphoglycerate + NADH + H^+	Cytosol
Lactate + NAD^+ ⇌ Pyruvate + NADH + H^+	Cytosol
NADP-linked:	
Isocitrate + $NADP^+$ ⇌ α-ketoglutarate + CO_2 + NADPH + H^+	Mitochondria and cytosol
Glucose 6-phosphate + $NADP^+$ ⇌ 6-phosphogluconate + NADPH + H^+	Cytosol
NAD or NADP:	
L-glutamate + H_2O + NAD^+ ($NADP^+$) ⇌ α-ketoglutarate + NH_3 + NADH(NADPH) + H^+	Mitochondria

medium. Each hydride ion carries two reducing equivalents: one is transferred as a hydrogen atom to C_4 of the nicotinamide ring, the other as an electron to the ring nitrogen (Fig. 23–10).

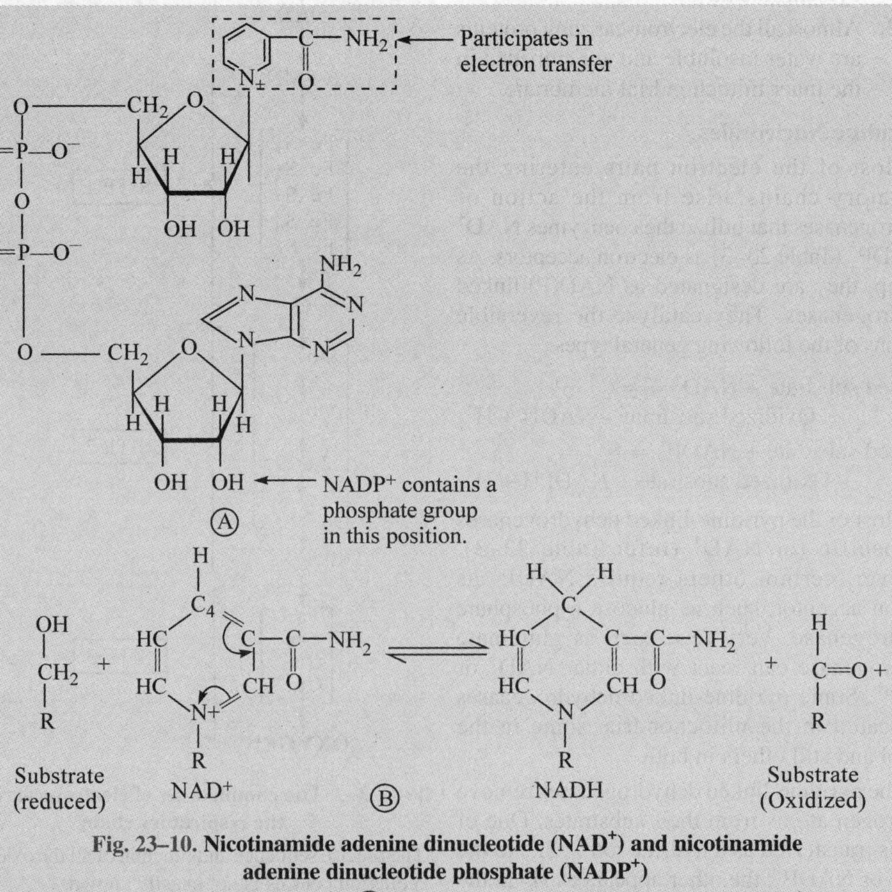

Fig. 23–10. Nicotinamide adenine dinucleotide (NAD^+) and nicotinamide adenine dinucleotide phosphate ($NADP^+$)

(A) Oxidized forms of NAD^+ and $NADP^+$; (B) Reduction of the nicotinamide ring of NAD^+ by substrate.

Since most dehydrogenases in cells transfer H atoms from their substrates to NAD^+, this coenzyme *collects* pairs of reducing equivalents from many different substrates, in one molecular form, NADH (Fig. 23–11). Ultimatley, NAD^+ can also collect reducing equivalent from substrates acted upon by NADP-linked dehydrogenases in the presence of *pyridine nucleotide transhydrogenase.*

$$NADPH + NAD^+ \rightleftharpoons NADP^+ + NADH$$

2. NADH Dehydrogenase (=NADH-Q Reductase)

The transfer of electrons from NADH in the mitochondrial matrix to ubiquinone in the membrane core, and the accompanying pumping of protons, is catalyzed by a highly organized enzyme complex, *NADH dehydrogenase*. This complex includes a flavoprotein and iron-sulfide proteins as electron carriers. In the next step of electron transfer, a pair of reducing equivalents is transferred from NADH to NADH dehydrogenase. In this reaction, the tightly bound prosthetic group of NADH dehydrogenase becomes reduced. This prosthetic group is flavin mononucleotide (FMN), which contains a molecule of vitamin B_2 or riboflavin.

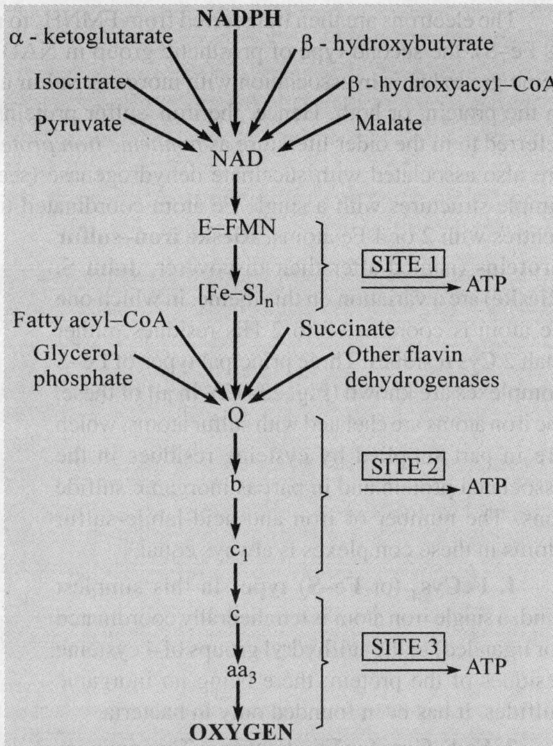

Fig. 23–11. The collecting function of nicotinamide adenine dinucleotide (NAD) and ubiquinone (Q)

Pairs of reducing equivalents collected from flavin dehydrogenases do not pass through the first phosphorylation site and thus give rise to only two ATP. molecules.

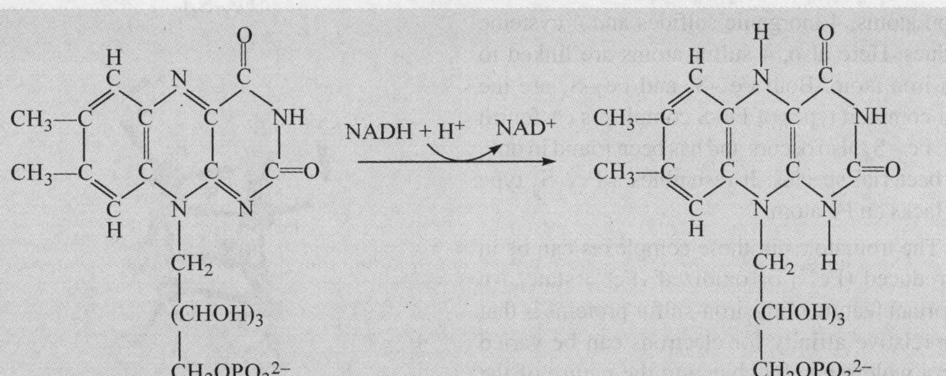

Oxidized form (FMN) **Reduced form ($FMNH_2$)**

NADH dehydrogenase is a complex and highly organized flavoprotein enzyme and consists of at least 16 polypeptide chains. It is located in the inner mitochondrial membrane. Transfer of two reducing equivalents from NADH to NADH dehydrogenase (here designated as E-FMN) reduces the FMN to $FMNH_2$ as follows:

$$NAD + H^+ + E\text{—}FMN \rightleftharpoons NAD^+ + E\text{—}FMNH_2$$

The electrons are then transferred from $FMNH_2$ to a series of iron-sulfur complexes (abbreviated as Fe–S), the second type of prosthetic group in NADH dehydrogenase. The iron is not a part of a heme group but is in association with inorganic sulfur atoms or with the sulfur atoms of Cys residues in the protein, or both. Hence, the iron–sulfur proteins, first discovered by **Helmut Beinert,** were referred to in the older literature as *nonheme iron proteins* (NHI proteins). Recall that Fe—S centres are also associated with succinate dehydrogenase (see page 604). These Fe–S centres range from simple structures with a single Fe atom coordinated to 4 Cys–SH groups to more complex Fe—S centres with 2 or 4 Fe atoms. **Rieske iron–sulfur proteins** (named after their discoverer, **John S. Rieske**) are a variation on this theme, in which one Fe atom is coordinated to 2 His residues, rather than 2 Cys residues. Three principal types of Fe–S complexes are known (Fig. 23–12). In all of these, the iron atoms are chelated with sulfur atoms, which are in part supplied by cysteine residues in the associated protein and in part as inorganic sulfide ions. The number of iron and acid-labile-sulfur atoms in these complexes is always equal.

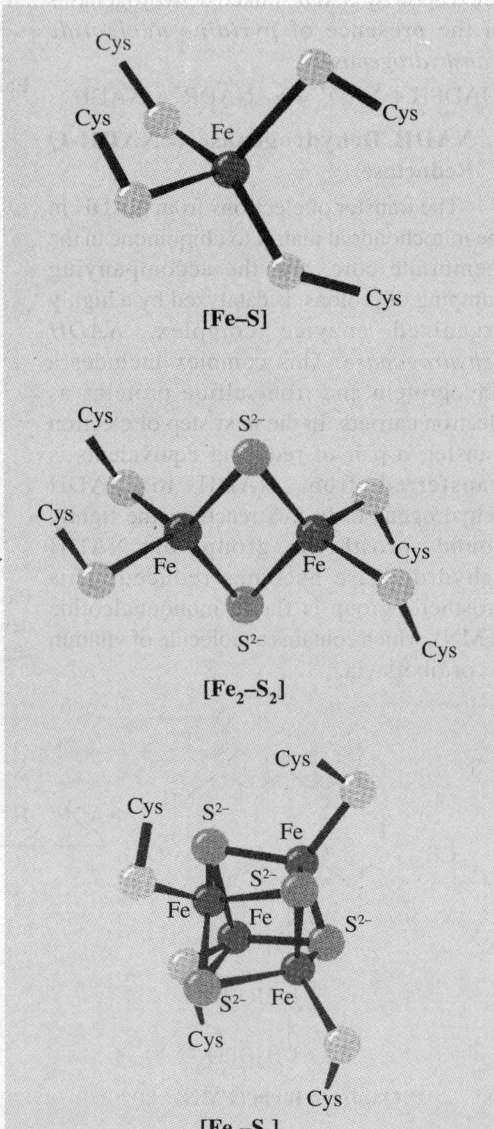

Fig. 23–12. Three principal forms of iron–sulfur (Fe–S) proteins

1. $FeCys_4$ (or Fe–S) type: In this simplest kind, a single iron atom is tetrahedrally coordinated (or liganded) to the sulfhydryl groups of 4 cysteine residues of the protein; there being no inorganic sulfides. It has been founded only in bacteria.

2. $Fe_2S_2Cys_4$ (or Fe_2–S_2) type: This contains 2 iron atoms and 2 inorganic sulfides, in addition to 4 cysteine residues. The bonding occurs in such a manner that a total of 4 sulfur atoms is linked to each iron atom.

3. $Fe_4S_4Cys_4$ (or Fe_4–S_4) type: This contains 4 iron atoms, 4 inorganic sulfides and 4 cysteine residues. Here also, 4 sulfur atoms are linked to each iron atom. Both Fe_2–S_2 and Fe_4–S_4 are the most common types of Fe–S complexes (A fourth type, Fe_3–S_4 also occurs and has been found in only one bacterial species. It resembles a Fe_4–S_4 type that lacks an Fe atom.

The iron atoms in these complexes can be in the reduced (Fe^{2+}) or oxidized (Fe^{3+}) state. An important feature of the iron-sulfur proteins is that their relative affinity for electrons can be varied over a wide range by changing the nature of the polypeptide chain. Some are relatively strong oxidizing agents; others are powerful reducing agents– even stronger than NADH. NADH dehydrogenase contains both the $Fe_2S_2Cys_4$ and $Fe_4S_4Cys_4$ types of complexes. One means of identifying these clusters utilizes the fact that their sulfide ions are acid-labile: they are released as H_2S near pH 1. *The oxidized and reduced states of all iron–sulfur clusters differ by one formal charge regardless of their number of Fe atoms.* This is because the Fe atoms in each cluster form a conjugatged system and thus can have oxidation states between the +2 and +3 values possible for individual Fe atoms. For example, each of the two [Fe_4–S_4] clusters in the protein **ferredoxin** contains one Fe(II) and three Fe(III)'s in its oxidized form and two Fe(II)'s and two Fe(III)'s in its reduced form. Iron–sulfur proteins also occur in photosynthetic electron–transport chains of plants and bacteria; indeed, photosynthetic electron–transport chains are thought to be the evolutionary precursors of oxidative electron–transport chains.

3. Ubiquinone (= Coenzyme Q)

The next carrier of reducing equivalents in the respiratory chain is ubiquinone (UQ), a name reflecting its ubiquitous nature, as it occurs virtually in all cells. It was formerly called as **coenzyme Q** (Q for quinone) and abbreviated as CoQ or simply Q.

Ubiquinone is actually a group of compounds, all containing the same quinone structure but substituted with a long side chain composed of varying numbers (from 6 to 10) of isoprene units (=prenyl groups), linked head to tail. For example, certain microorganisms contain 6 isoprene units, and in which case the compound is referred to as Q_6 or CoQ_6 or also as UQ_{30}, where the subscript number 30 represents the total number of carbon atoms in the side chain. However, the most common form in mammals contains 10 isoprene units, and when its designation is Q_{10} or CoQ_{10} or UQ_{50}. The isoprenoid tail makes ubiquinone highly nonpolar which enables it to diffuse readily in the hydrocarbon phase of the inner mitochondrial membrane. *Ubiquinone is the only electron carrier in the respiratory chain that is not tightly bound or covalently attached to a protein.* In fact, it serves as a highly mobile carrier of electrons between the flavoproteins and the cytochromes of the electron-transport chain.

The closely related plastoquinones, which function as an analogous carrier of electrons in photosynthesis, differ from ubiquinones in the alkyl substituents of the benzene ring : two $-CH_3$ groups instead of two $-OCH_3$ and H instead of $-CH_3$. Plastoquinones B and C carry one hydroxyl group in the side chain.

Like other quinones, the ubiquinones may be reduced one electron at a time through the semiquinone free radical or may be reduced directly to the quinol by two electrons (Fig. 23–13).

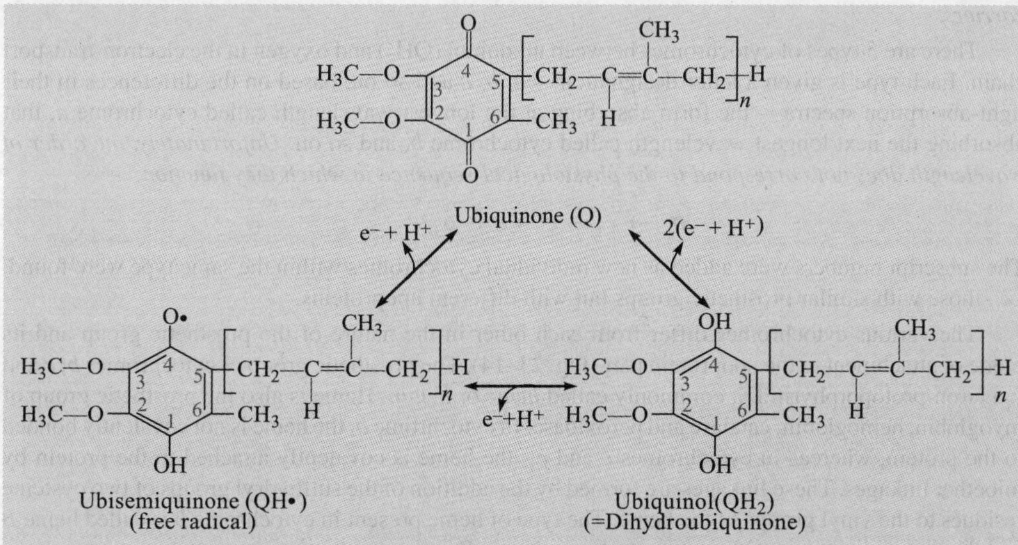

Fig. 23–13. Reduction of ubiquinone

When reduced NADH dehydrogenase (E–$FMNH_2$) donates its reducing equivalents via the Fe—S centres to ubiquinone, the latter becomes reduced to ubiquinol (QH_2) and the oxidized form of NADH dehydrogenase is regenerated.

$$E\text{—}FMNH_2 + Q \rightleftharpoons E\text{–}FMN + QH_2$$

The function of ubiqunione is to collect reducing equivalents not only from NADH dehydrogenase but also from other flavin-linked dehydrogenases of mitochondria (see Fig. 23–10).

4. Cytochromes

Cytochromes, found in mitochondria and endoplasmic reticulum, are proteins that contain heme groups, but which are not involved in oxygen transport. The core structure of these heme groups is a tetrapyrrole ring similar to that of hemoglobin, sometimes differing only in the composition of the side chains. The heme group of cytochromes *b* and c_1 is known as iron protoporphyrin IX and is the same heme that is found in hemoglobin, myoglobin, and catalase. Cytochrome *c* contains heme C that is covalently bound to the protein through cysteine residues. Cytochromes *a* and a_3 contain heme A which, in common with ubiquinone, contains a hydrophobic isoprene side chain. In hemoglobin and myoglobin, heme must remain in the ferrous (Fe^{2+}) state; in cytochromes, the heme iron is reversibly reduced and oxidized between the Fe^{2+} and Fe^{3+} states as electrons are shuttled from one protein to another.

The enzyme complex catalyzing the oxidation of ubiquinol (QH_2) contains an iron-sulfide protein of Fe_2S_2 Cys_4 type and 2 types of cytochromes. The cytochromes are electron-transferring, red or brown proteins that contain a heme prosthetic group and act in sequence to carry electrons from ubiquinone to molecular oxygen. They are, thus, *heme proteins* (or *hemoproteins*) like hemoglobin but, unlike hemoglobin, their iron atoms are oxidized and reduced to transfer electrons between other compounds. The cytochromes were discovered in 1886 by **MacMunn**, a Scottish physician who also named them as *histohematin* and *myohematin*, since they appeared to him to be related to heme and hemin. But their role in biologic oxidation was first shown by **David Keilin** in 1925. He also renamed these pigments as 'cytochromes'. The iron atom in cytochromes alternates between a reduced ferrous (Fe^{2+}) state and an oxidized ferric (Fe^{3+}) state during electron transport. *A heme group, like an Fe–S centre, is one-electron carrier, in contrast with NADH, flavins and ubiquinone, which are two-electron carriers.*

There are 5 types of cytochromes between ubiquinol (QH_2) and oxygen in the electron-transport chain. Each type is given a letter designation— *a, b, c* and so on, based on the differences in their light-absorption spectra— the form absorbing at the longest wavelength called cytochrome *a*, that absorbing the next longest wavelength called cytochrome *b*, and so on. *Unfortunately, the order of wavelength does not correspond to the physiological sequence in which they function:*

$$b \rightarrow c_1 \rightarrow c \rightarrow a \rightarrow a_3$$

The subscript numbers were added as new individual cytochromes within the same type were found, *i.e.*, those with similar prosthetic groups but with different apoproteins.

The various cytochromes differ from each other in the nature of the prosthetic group and its mode of attachment to the apoprotein part (Fig. 23–14). The prosthetic group of cytochromes *b*, *c* and c_1 is iron-protoporphyrin IX, commonly called *heme* or *hemin*. Heme is also the prosthetic group of myoglobin, hemoglobin, catalase and peroxidase. In cytochrome *b*, the heme is not covalently bonded to the protein, whereas in cytochromes *c* and c_1, the heme is covalently attached to the protein by thioether linkages. These linkages are formed by the addition of the sulfhydryl groups of two cysteine residues to the vinyl groups of the heme. The type of heme present in cytochrome *b* is called heme *B* and the one present in cytochromes *c* and c_1 as heme *C*.

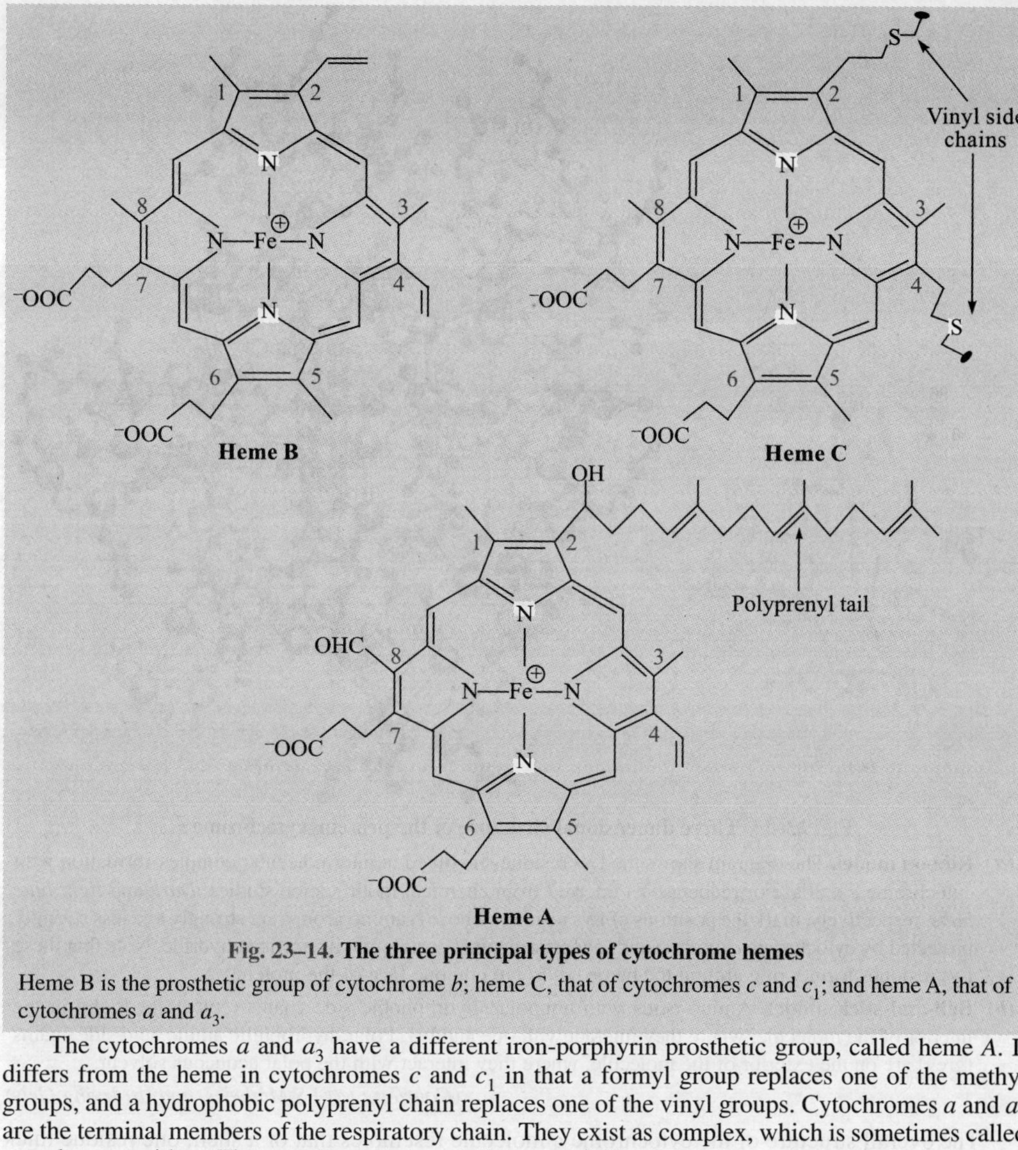

Fig. 23–14. The three principal types of cytochrome hemes

Heme B is the prosthetic group of cytochrome *b*; heme C, that of cytochromes *c* and c_1; and heme A, that of cytochromes *a* and a_3.

The cytochromes *a* and a_3 have a different iron-porphyrin prosthetic group, called heme *A*. It differs from the heme in cytochromes *c* and c_1 in that a formyl group replaces one of the methyl groups, and a hydrophobic polyprenyl chain replaces one of the vinyl groups. Cytochromes *a* and a_3 are the terminal members of the respiratory chain. They exist as complex, which is sometimes called cytochrome oxidase. The cytochrome aa_3 complex, thus, differs from other cytochromes as it contains 2 moles of highly-bound heme *A*. Moreover, cytochrome aa_3 also contains 2 essential copper atoms. It is the terminal member of the electron-transport chain and was first identified by **Warburg** as *Atmungsferment*.

Cytochrome *c* is the best known of cytochromes. It is the only electron-transport protein that can be separated from the inner mitochondrial membrane by gentle treatment. The solubility of this peripheral membrane protein in water has facilitated its purification and crystallization. It is a small protein (MW = 12,500) with an iron-porphyrin group (or *heme C*) covalently attached to its single polypeptide chain, containing about 100 amino acid residues (104 in a fish called tuna) in most species. Cytochrome *c* from tuna (Fig. 23–15) is roughly spherical, with a diameter of 34 Å. The heme group is surrounded by many closely-packed hydrophobic side chains. The iron atom is bonded to the sulfur atom of a methionine residue and to the nitrogen atom of a histidine residue. The hydrophobic nature of the heme environment makes the redox potential of cytochrome *c* more positive, corresponding to a higher electron affinity.

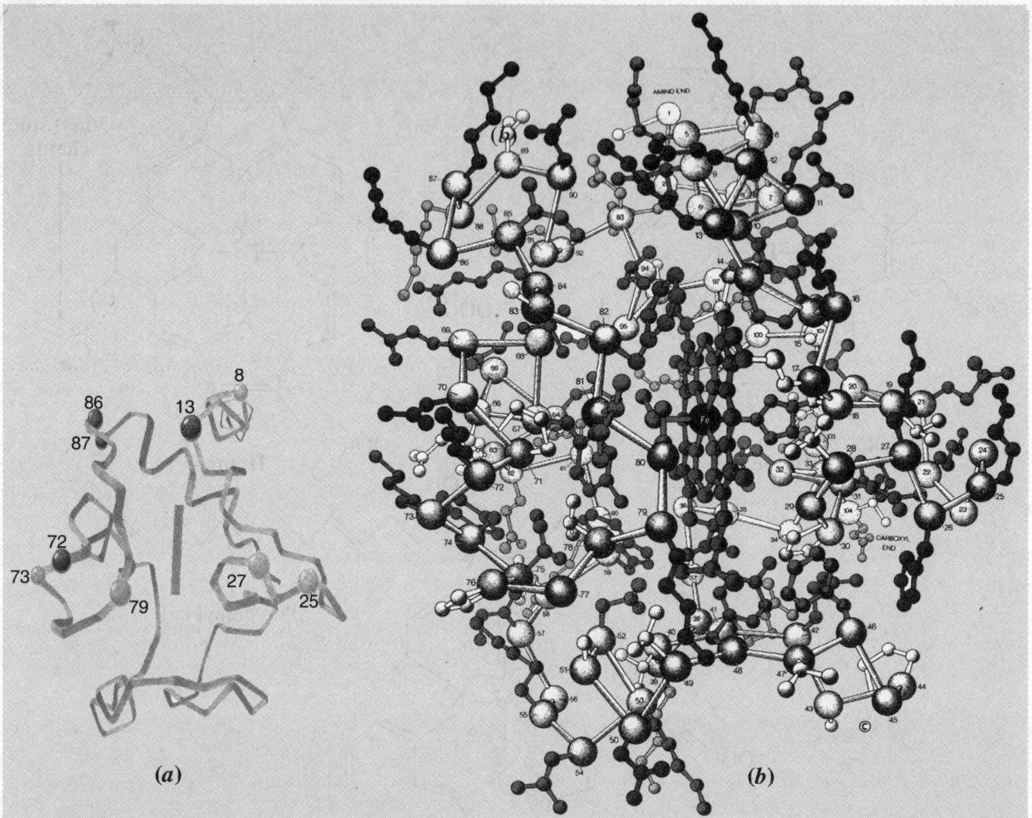

(*a*) (*b*)

Fig. 23–15. Three dimensional structure of the protein, cytochrome *c*

(*a*) **Ribbon model.** The diagram shows the Lys residues involved in intermolecular complex formation with cytochrome *c* oxidase or reductase as inferred from chemical modification studies. *Dark* and *light blue balls*, respectively, mark the positions of Lys residues whose ε-amino groups are strongly and less strongly protected by cytochrome *c* oxidase and reductase against acetylation by acetic anhydride. Note that these Lys residues form a ring around the heme (*solid bar*) on one face of the protein.

(*b*) **Ball-and-stick model.** Amino acids with nonpolar, hydrophobic side chains (color) are found in the interior of the molecule, where they interact with one another. Polar, hydrophilic amino acid side chains (grey) are on the exterior of the molecule, where they interact with the polar acqueous solvent.

(*Courtesy: (a) FS Mathews, and (b) Irving Geis*)

The overall structure of the cytochrome *c* molecule resembles that of a shell, one residue thick surrounding the heme. The side chains make up the interior of the shell. The main polypeptide chain comes next, followed by charged side chains on the surface. *There is a very little a-helix and no* β-*pleated sheet*. In essence, the polypeptide chain is coiled around the heme. Residues 1 to 47 are on the histidine-18 side of the heme (called the right side), and residues 48 to 91 are on the methionine-80 side (called the left side). The remaining 92 to 104 residues come back across the heme to the right side. Cytochorme *c* is an ancient protein, since its amino acid sequence has many points of similarity in all eukaryotes– microbes, plants and animals.

ELECTRON–TRANSPORT COMPLEXES
(= Complexes of the Respiratory Chain)

It is now a well-established fact that the electron carriers in the respiratory chain function in a specific sequence. The following evidences support the statement:

1. Firstly, as expected their standard redox potentials (refer Table 23–2 and Fig. 23–6) are more positive going toward oxygen, since electrons tend to flow from electronegative to electropositive systems, causing a decrease in free energy.
2. Secondly, each member of the chain is specific for a given electron donor and acceptor. For instance, NADH can transfer electrons to NADH dehydrogenase but cannot transfer them directly to cytochrome b or to cytochrome *c*.
3. Lastly, four structured complexes of functionally related electron carriers have been isolated from mitochondrial membrane (refer Fig. 23–16). These are:

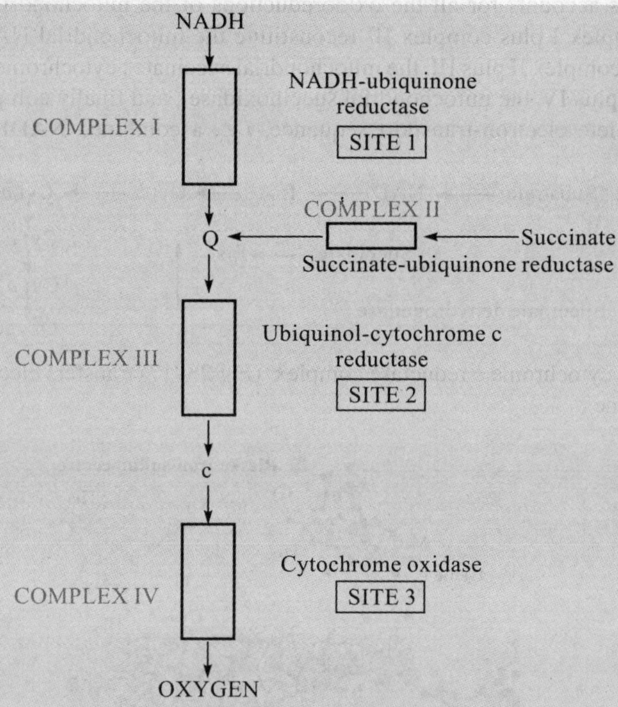

Fig. 23–16. The electron transport complex

The various complexes can be isolated as functional assemblies.

Complex I: NADH to Ubiquinone. It consists of NADH dehydrogenase and its iron-sulfur centres, which are closely linked in their function. Complex I carries out the following characteristic reaction:

$$\text{NADH} + \text{H}^+ + \text{CoQ} \xrightarrow[\text{Complex I}]{\text{NADH : CoQ (oxido) reductase}} \text{NAD}^+ + \text{CoQH}_2$$

Complex II: Succinate to Ubiquinone. It consists of succinate dehydrogenase and its iron-sulfur centres. This complex carries out the following characteristic reaction:

$$\text{Succinate} + \text{CoQ} \xrightarrow[\text{Complex II}]{\text{Succinate : CoQ (oxido) reductase}} \text{Fumarate} + \text{CoQH}_2$$

Complex III: Ubiquinone to Cytochrome c. It consists of cytochromes *b* and *c*, and a specific iron-sulfur centre. This brings about the following characteristic reaction:

$$\text{CoQ H}_2 + 2\text{ cyt } c\ (\text{Fe}^{3+}) \xrightarrow[\text{Complex III}]{\text{Hydro-CoQ : Cytochrome (oxido) reductase}} \text{CoQ} + 2\text{ cyt } c\ (\text{Fe}^{2+})$$

Complex IV: Cytochrome c to Oxygen. It consists only of cytochromes *a* and a_3, which is sometimes called as cytochrome oxidase. This brings about the following characteristic reaction:

$$4 \text{ cyt } c\ (Fe^{2+}) + O_2 \xrightarrow[\text{Complex IV}]{\text{Cytochrome oxidase (Cytochrome c : } O_2 \text{oxido reductase)}} 4 \text{ cyt } c\ (Fe^{3+}) + H_2O$$

Ubiquinone is the connecting link between complexes I , II and III, and cytochrome *c* connects the complexes. III and IV. It is evident that in the presence of 2 mobile carriers (CoQ and cytochrome *c*), this accounts for all the oxidoreductions of the mitochondrial electron transport system. Thus, complex I plus complex III reconstitute the mitochondrial NADH : cytochrome *c* (oxido) reductase; complex II plus III, the mitochondrial succinate : cytochrome *c* (oxido) reductase; complex I plus III plus IV, the mitochondrial succinoxidase ; and finally complex I plus II plus III plus IV, the complete electron-transport sequence, *i.e.*, a combined NADH and succinoxidase equation:

Substrate ⟶ NAD ⟶ fp_D ⟶ Cyt *b* ⟶ Cyt c_1
↓
Succinate ⟶ fps ↑ Cyt *c*
↓
Cyt *a* plus a_3
↓
fps =Succinate dehydrogenase Oxygen

The ubiquinol-cytochrome *c* reductase complex (Fig 23–17) transfers electrons from ubiquinol (QH_2) to cytochrome *c*.

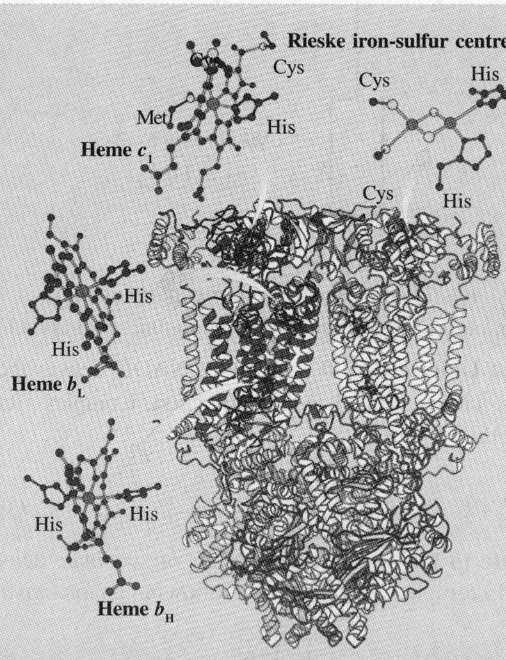

Fig 23–17. Structure of Q-cytochrome *c* oxidoreductase or cytochrome bc_1 complex or complex III
This enzyme is a dimer of identical monomers (*i.e.*, homodimer), each with 11 distinct polypeptide chains. The major prosthetic groups, three hemes and a 2Fe-2S cluster, mediate the electron-transfer reactions between quinones in the membrane and cytochrome *c* in the intermembrane space. The enzyme protrudes 75 Å into the matrix and 38 Å into the intermembrane space.

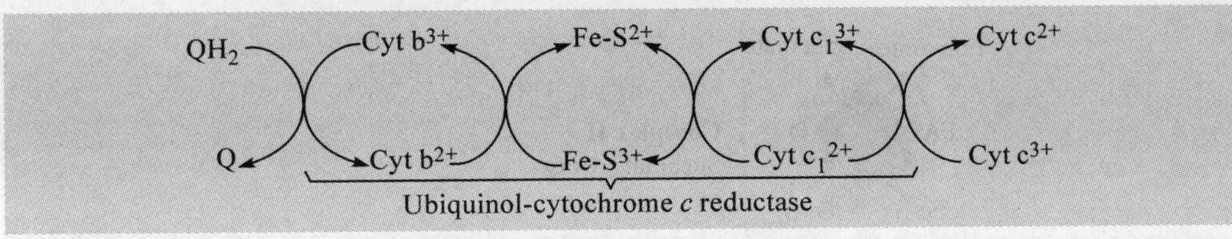

Reduced cytochrome *c* then transfers its electrons to the cytochrome *c* oxidase complex (Fig. 23-18). The role of cytochrome *c* is analogous to that of coenzyme Q, *i.e.*, it is a mobile carrier of electrons between different complexes in the respiratory chain.

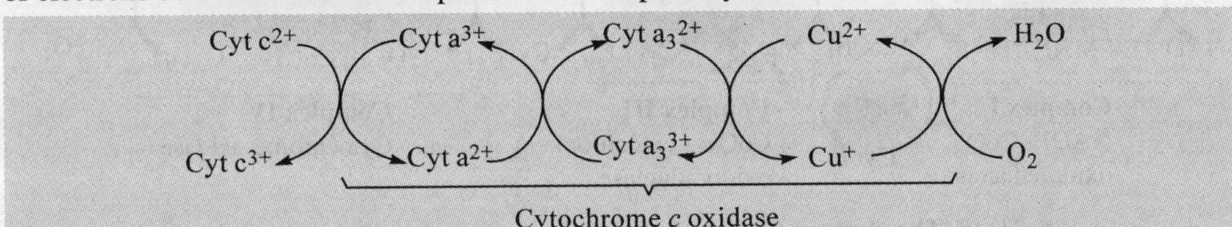

Electrons are transferred to the cytochrome *a* moiety of the complex, and then to cytochrome a_3, which contains copper. This copper atom alternates between an oxidized (2+) form and a reduced (1+) form as it transfers electrons from cytochrome a_3 to molecular oxygen. *The formation of water is a four-electron process,* whereas heme groups are one-electron carriers. It is not yet certain as to how four electrons converge to reduce a molecule of oxygen.

$$O_2 + 4H^+ + 4e^- \longrightarrow 2\,H_2O$$

After the cytochrome *a* component receives electrons from cytochrome *c* and thus becomes reduced to the ferrous (Fe^{2+}) form, it passes its electrons to cytochrome a_3. Reduced cytochrome a_3 then passes electrons to molecular oxygen. Participating with the two heme groups in this process are 2 bound copper atoms, which undergo cuprous—cupric redox changes ($Cu^+ - Cu^{2+}$) in their function. This is an important and complex step in electron transport, since 4 electrons must be passed almost simultaneously to O_2 to yield two H_2O molecules, with uptake of 4 H^+ from the aqueous medium. *Of all the members of the electron-transport chain, only cytochrome aa_3 can react directly with oxygen.* The complete electron transport chain along with the respiratory complexes is presented in Fig. 23–19.

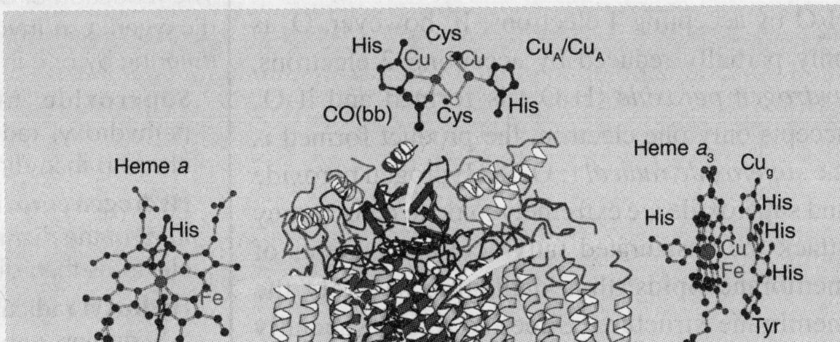

Fig. 23–18. Three dimensional structure of cytochrome *c* oxidase (complex IV)

This enzyme consists of a dimer in which each monomer is composed of 13 different polypeptide chains. The major prosthetic groups include Cu_A/Cu_A, heme *a*, and heme a_3-Cu_B. Heme a_3-Cu_B is the site of the reduction of oxygen to water. CO(bb) is a carbonyl group of the peptide backbone.

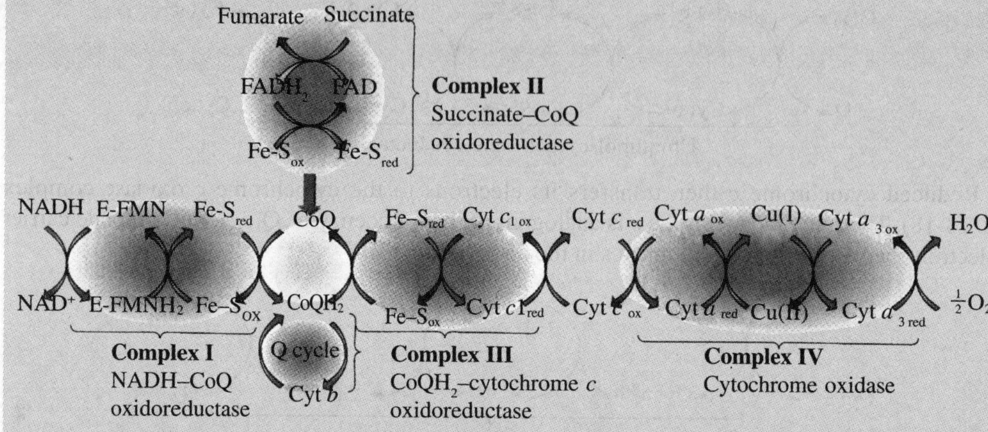

Fig. 23–19. The electron transport chain, showing the respiratory complexes

In the reduced cytochromes, the iron is in the Fe(II) oxidation state, while in the oxidized cytochromes, the oxygen is in the Fe(III) oxidation state.

INCOMPLETE REDUCTION OF OXYGEN (=Toxic Metabolites of Oxygen)

It is quite important for the cell that the O_2 molecule be completely reduced to 2 molecules of H_2O by accepting 4 electrons. If, however, O_2 is only partially reduced by accepting 2 electrons, *hydrogen peroxide* (H_2O_2) is formed and if O_2 accepts only one electron, the product formed is the *superoxide radical* ($:O_2^-$). Hydrogen peroxide and superoxide are extremely toxic to cells as they attack the unsaturated fatty acid components of membrane lipids, thus damaging severely the membrane structure. Superoxide is especially dangerous. It does not itself react readily with most cellular constituents, but it will spontaneously combine with peroxides to form *hydroxyl radicals* (OH) and *singlet oxygen* ($^1O_2$), which are disruptively reactive:

> The reduction of oxygen, which is also called as dioxygen, can involve a variety of intermediates differing by one in electrons contents:
>
> **Superoxide** ($:O_2^-$) is the anion of the perhydroxyl radical, HO_2 and has one more electron than dioxygen.
>
> **Hydrogen peroxide** (H_2O_2) is the undissociated form of the dianion, O_2^{2-}, which has two more electrons than dioxygen.
>
> **Hydroxyl radical** (·OH) is equivalent to half of a hydrogen peroxide molecule, but is much more reactive. Addition of one electron to each hydroxyl radical results in the formation of hydroxide ions, OH^-, the completely reduced form of oxygen.

$$:O_2^- + H_2O_2 \longrightarrow HO + {}^1O_2 + OH^-$$

Almost without exception, the aerobic cells protect themselves against : O_2^- and H_2O_2 by the action of *superoxide dismutase* and *catalase*, which these cells do contain, respectively. Superoxide dismutase converts superoxide radical into H_2O_2 while catalase transforms H_2O_2 into water and molecular oxygen.

> **Singlet oxygen** ($^1O_2$) is a form of dioxygen in which the electrons are in less stable orbitals with antiparallel spins.

$$2\ O_2^- + 2\ H^+ \xrightarrow{\text{Superoxide dismutase}} H_2O_2 + O_2$$

$$2\ H_2O_2 \xrightarrow{\text{Catalase}} 2\ H_2O + O_2$$

Although toxic, H_2O_2 may be useful to some organisms such as **bombardier beetle**. This insect generates a concentrated solution of H_2O_2 in one sac of its spray gland and a solution of hydroquinone in the other sac. When threatened, the insect frightens (and also poisons) its enemy by firing a hot (100°F) spray of toxic quinone, which is produced from the oxidation of hydroquinone by H_2O_2.

MECHANISMS OF OXIDATIVE PHOSPHORYLATION

One of the most challenging and difficult problems in biochemical research is that how does the electron-transport chain cooperate with the ATP synthetase to bring about oxidative phosphorylation of ADP to ATP? One of the reasons is that the enzymes concerned in electron-transport and oxidative phosphorylation are very complex and they are embedded in the inner mitochondrial membrane, rendering the detailed study of their interactions difficult. However, 3 principal hypotheses have been advanced to account for the coupling of oxidation and phosphorylation. In other words, these hypotheses explain how the energy transfer between electron transport and ATP synthesis takes place.

1. Chemical Coupling Hypothesis

This is the oldest of the 3 hypotheses and proposes that electron transport is coupled to ATP synthesis by a sequence of consecutive reactions in which a high-energy covalent intermediate is formed by electron transport and subsequently is cleaved and donates its energy to make ATP. The hypothesis, thus, postulates direct chemical coupling at all stages of the process. It is similar to the concept in glycolysis which states that the ATP produced in oxidative phosphorylation results from an energy-rich intermediate encountered in electron transport. Specifically, when an oxidoreduction reaction occurs between A_{red} and B_{oxi}, the factor I is incorporated into the formation of an energy-rich structure $A_{oxi} \sim I$, where the ~ indicates a linkage having an energy-rich nature:

$$A_{red} + I + B_{oxi} \rightleftharpoons A_{oxi} \sim I + B_{red}$$

In subsquent reactions, an enzyme (E) replaces A_{oxi} in the compound $A_{oxi} \sim I$ to form an energy-rich E ~ I. Later, inorganic phosphate reacts with E ~ I to form phosphoenzyme complex E ~ P containing the energy-rich enzyme-phosphate bond:

$$A_{oxi} \sim I + E \rightleftharpoons A_{oxi} + E \sim I$$

$$E \sim I + Pi \rightleftharpoons E \sim P + I$$

The enzyme-phosphate component finally reacts with ADP to form ATP.

$$E \sim P + ADP \rightleftharpoons E + ATP$$

Although suggestions have been made regarding the nature of E ~ I and E ~ P, such compounds have not been identified in mitochondria.

Oxidative phosphorylation occurs in certain reactions of glycolysis, in the citric acid cycle and in the respiratory chain. However, it is only in those phosphorylations occurring at the substrate level in glycolysis and the citric acid cycle that the chemical mechanisms involved are known. Three such equations are given below:

3-phosphoglyceraldehyde + NAD^+ + Pi ⟶ 1 ~ 3-biphosphoglycerate + NADH + H^+

1 ~ 3-biphosphoglycerate + ADP ⟶ 3-phosphoglycerate + ATP

2-phosphoglycerate ⟶ 2 ~ phosphoenolpyurvate

2-phosphoenolpyruvate + ADP ⟶ Pyruvate + ATP

α-ketoglutarate + NAD+ + CoA ⟶ Succinyl ~ CoA + NADH + H^+

Succinyl ~ CoA + GDP + Pi ⟶ Succinate + GTP

Some key differences are evident in these equations. In equation I, phosphate is incorporated into the product of the reaction after the oxidoreduction. In equation II, phosphate is incorporated into the substrate before the internal arrangement or redox change. In equation III, the redox reaction leads to the generation of a high-energy compound other than a phosphate, which in a subsequent reaction leads to the formation of high-energy phosphate.

It is presumed that oxidative phosphorylations in the respiratory chain follow the pattern shown in equations I and 3, the latter being an extension of reaction I, to which an extra nonphosphorylated high-energy intermediate stage is added. Of the possible mechanisms shown in Fig. 23–20, mechanism Ⓒ is favoured since in the presence of uncouplers (see page 452) such as 2, 4-dinitrophenol, oxidoreduction in the respiratory chain is independent of Pi. However, at present, the identities of the hypothetical high-energy carrier (Car ~ I), and the postulated intermediates I and X are not known. In recent years, several so-called "coupling factors" have been isolated that restore phosphorylation when added to disrupted mitochondria.

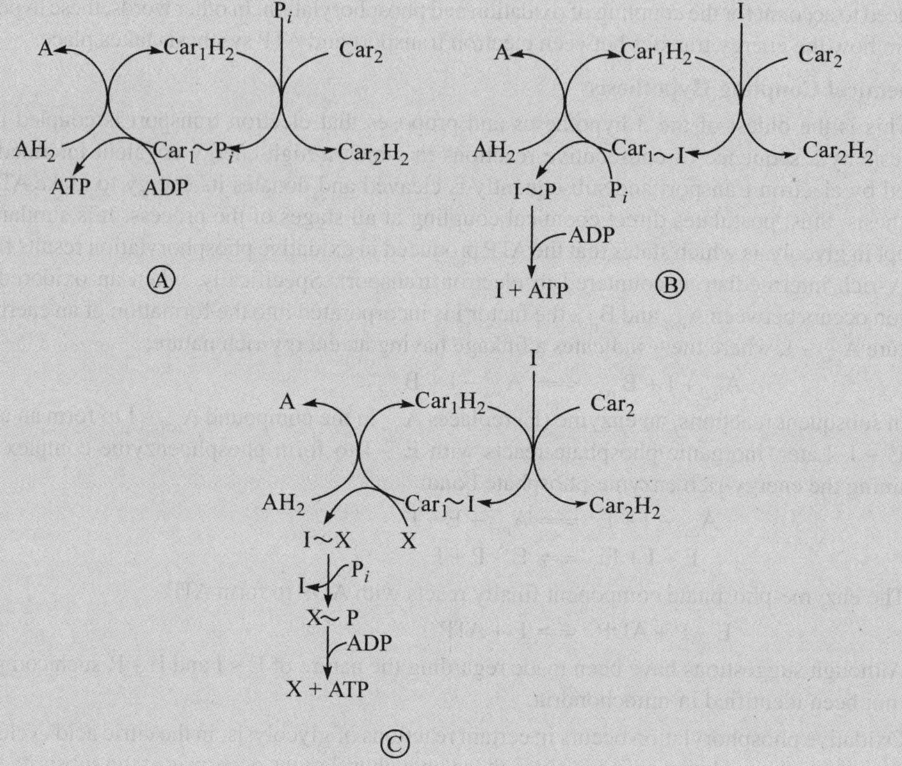

Fig. 23–20. Possible mechanisms for the chemical coupling of oxidation and phosphorylation in the respiratory chain

2. Conformational Coupling Hypothesis

In mitochondria, that are actively phosphorylating in the presence of an excess of ADP, the inner membrane pulls away from the outer membrane and assumes a "*condensed state*". In the absence of ADP, the mitochondria have the normal structure or the "*swollen state*", in which the cristae project into the large matrix. The propounders of this hypothesis believe that the energy released in the transport of electrons along the respiratory chain causes the conformational changes, just described, in the inner mitochondrial membrane and that this energy-rich condensed structure, in turn, is utilized for ATP synthesis as it changes to the energy-poor swollen conformation. However, the mode of the conformational changes that take place in the inner mitochondrial membrane is not yet clearly understood.

3. Chemiosmotic Coupling Hypothesis

Salient features. This is a simpler radically different and novel mechanism and was postulated by **Peter Mitchell**, a British biochemist, in 1961. He proposed that *electron trasnsport and ATP synthesis are coupled by a proton gradient, rather than by a covalent high-energy intermediate or an activated protein.* According to this model (Fig. 23–21), the transfer of electrons through the respiratory chain results in the pumping of protons (H^+) from the matrix side (M-side) to the cytosol or cytoplasmic side (C-side) of inner mitochondrial membrane. The concentration of H^+ becomes higher on the cytoplasmic side, thus creating an electrochemical potential difference. This consists of a chemical potential (difference in pH) and a membrane potential, which becomes positive on the cytoplasmic side. The hypothesis further proposes that the H^+ ions, ejected by electron transport, flow back into the matrix through a specific H^+ channel or 'pore' in the F_oF_1 ATPase molecule, driven by the concentration gradient of H^+. The free energy released, as proton (H^+) flows back through the ATPase, causes the coupled synthesis of ATP from ADP and phosphate by ATP synthetase (Fig. 23–22).

PETER DENNIS MITCHELL

(LT, 1920–1992)

Mitchell, a British biochemist, was born in Mitcham, Surrey on September 29, 1920. His parents were **Christopher Gibbs Mitchell,** a civil servant and **Kate Beatrice Dorothy (nee) Taplin.** He was educated at Queen's College, Taunton, and at Jesus College, Cambridge where he studied the Natural Sciences Tripos specializing in Biochemistry. He accepted a research post in the Department of Biochemistry, Cambridge in 1942 and received his Ph.D. in 1951 for his work on the *mode of action of penicillin*. In 1955, he was invited by **Prof. Michael Swann** to set up a biochemical research unit called the Chemical Biology Unit, in the Department of Zoology, Edinburgh University, where he was appointed a Senior Lecturer in 1961 and then a Reader in 1962, although ill health led to his resignation in 1963.

From then to 1965, he supervised the restoration of a Regency-fronted Mansion, known as Glynn House, near Bodmin, Cornwall– adapting a major part of it for use as a research laboratory. He and his former research colleague, **Jennifer Moyle**, founded a charitable company, known as **Glynn Research Ltd.,** to promote fundamental biological research at Glynn House and they embarked on a programme of research on chemiosmotic reactions and reaction systems. Thus, Mitchell was a rare example of the truly independent scientist since, in his native England, he was not affiliated with a university, industry, or government.

In the 1960s, ATP was known to be the energy currency of life, but the mechanism by which ATP was created in the mitochondria was assumed to be by substrate-level phosphorylation. Mitchell's **chemiosmotic coupling hypothesis** was the basis for understanding the actual process of *oxidative phosphorylation.* At the time, the biochemical mechanism of ATP synthesis by oxidative phosphorylation was unknown. Mitchell realized that the movement of ions across an electrochemical membrane potential could provide the energy needed to produce ATP. His hypothesis was derived from information that was well known in the 1960s. He knew that living cells had a *membrane potential*; interior negative to the environment. The movement of charged ions across a membrane is, thus, affected by the electrical forces (the attraction of + to – charges). Their movement is also affected by *thermodynamic forces,* the tendency of substances to diffuse from regions of higher to lower concentrations. He went on to show that ATP synthesis was coupled to this electrochemical gradient. He, thus, proposed that electron transport and ATP synthesis are coupled by a proton gradient, rather than by a covalent high-energy intermediate or an activated protein. His hypothesis was confirmed by the discovery of *ATP synthase,* a membrane-bound protein that uses the potential energy of the electrochemical gradient to make ATP.

Mitchell received the **1978 Nobel Prize in Chemistry** "for his contribution to the understanding of biological energy transfer through the formulation of chemiosmotic theory. His work was on the coupling of oxidation and phosphorylation. He died on April 10, 1992.

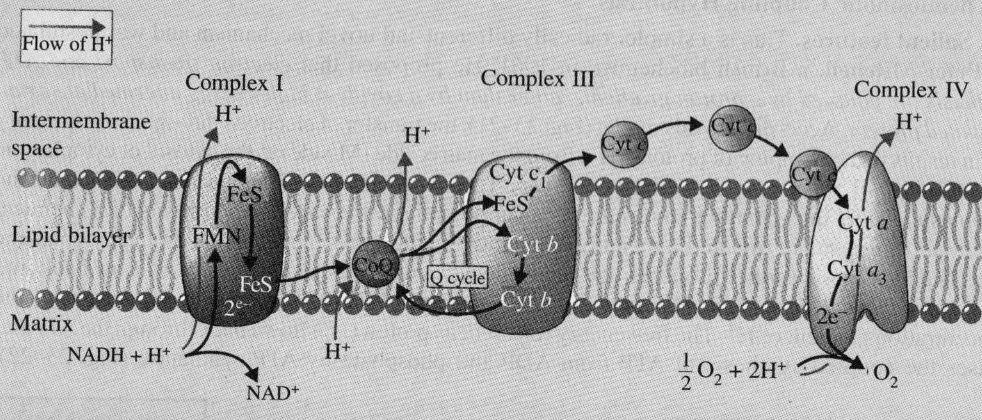

Fig. 23–21. The compositions and locations of respiratory complexes in the inner mitochondrial membrane, showing the flow of electrons from NADH to O_2

Complex II is not involved and not shown. NADH has accepted electrons from substrates such as pyruvate, isocitrate, α-ketoglutarate, and malate. Note that the binding site for NADH is on the matrix side of the membrane. Coenzyme Q is soluble in the lipid bilayer. *Complex III* contains two *b*-type cytochromes, which are involved in the Q cycle. Cytochrome *c* is loosely bound to the membrane, facing the intermembrane space. In *Complex IV*, the binding site for oxygen lies on the side toward the matrix.

The overall effect of the electron transport reaction series is to move protons (H^+) out of the matrix into the intermembrane space, creating a difference in pH across the membrane.

The model requires that the electron carriers in the respiratory chain and the ATP synthetase be anisotropically (= vectorially) organized, i.e., they must be oriented with respect to the two faces of the coupling membrane (= inner mitochondrial membrane). Further, the inner mitochondrial membrane must be intact in the form of a completely closed vesicle (either in intact mitochondria or in submitochondrial vesicles produced during sonication of the inner membrane), since, an H^+ gradient across the inner membrane could not otherwise exist. If, however, a 'leak' of proton across the membrane is induced by uncouplers, the proton gradient would be discharged and consequently energy-coupling would fail.

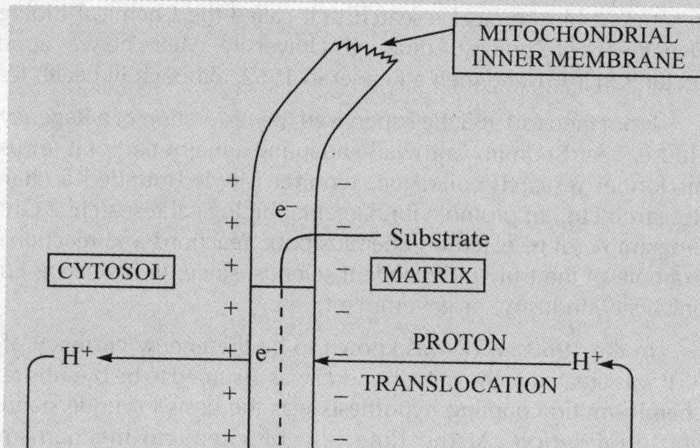

Fig. 23–22. Diagram illustrating principles of the chemiosmotic coupling hypothesis

ATP synthestase (=F_oF_1 ATPase) is responsible for oxidative phosphorylation.

Summarily, according to the chemiosmotic hypothesis, the high-energy chemical intermediates are replaced by a link between chemical processes ("chemi") and transport process ("osmotic" – from the *osmos*[G] = push) – hence the phrase, **chemiosmotic coupling**. As the high-energy electrons from the hydrogens of NADH and $FADH_2$ are transported down the respiratory chain in the mitochondrial inner membrane, the energy released as they pass from one carrier molecule to the next is used to pump protons (H^+) across the inner membrane from the mitochondrial matrix into the innermembrane space. This creates an *electrochemical proton gradient* across the mitochondrial inner membrane, and the backflow of H^+ down this gradient is, in turn, used to drive the membrane-bound enzyme *ATP synthase*, which catalayzes the conversion of oxidative phosphorylation.

Evidences in Favour. Mitchell's hypothesis that oxidation and phosphorylation are coupled by a proton gradient is supported by a wealth of evidences:

1. No hypothetical 'high-energy' intermediates, linking electron transport to ATP synthesis, have been found to date.
2. Oxidative phosphorylation requires a closed compartment, *i.e.*, the inner mitochondrial membrane should be intact. Breaks or holes in the inner membrane do not allow oxidative phosphorylation, although electron transport from substrates to oxygen may still continue. ATP synthesis coupled to electron transfer does not also occur in soluble cell preparations.
3. The inner mitochondrial membrane is impermeable to H^+, K^+, OH^- and Cl^- ions. If the membrane is damaged in order to pass through such ions readily, oxidative phosphorylation will not take place. However, evidences available indicate the existence of specific transport systems which enable ions to penetrate the inner mitochondrial membrane.
4. Both the respiratory chain and the ATPase are vectorially organized in the coupling membrane.
5. A proton gradient across the mitochondrial inner membrane is generated during electron transport. The pH inside is 1.4 units higher than outside, and the membrane potential is 0.14 V, the outside being positive. The total electrochemical potential Δp (in volts) consists of a membrane potential contribution ($\Delta\psi$) and a H^+ concentration-gradient contribution (ΔpH). Taking R as the gas constant, T as the absolute temperature and F as the caloric equivalent of Faraday, the value of total electrochemical potential (Δp) can be written as:

$$\begin{aligned}\Delta p &= \Delta\psi - \frac{RT}{F}\,\Delta p\text{H}\\ &= \Delta\psi - 0.06\,\Delta p\text{H}\\ &= 0.14 - 0.06\,(-1.4)\\ &= 0.224 \text{ V}\end{aligned}$$

This total proton-motive force of 0.224 V corresponds to a free energy of 5.2 kcal per mole of protons.

6. ATP is synthesized when a *p*H gradient is imposed on mitochondria or chloroplasts in the absence of electron transport.
7. Oxidative phosphorylation can be checked by uncouplers and certain ionophores (see page 453). Uncouplers such as 2,4-dinitrophenol increase the permeability of mitochondria to protons, thus reducing the electrochemical potential and short-circuiting the vectorial ATP synthetase system for the production of ATP.
8. Addition of acid to the external medium, establishing a proton gradient, leads to the synthesis of ATP.

Oxidation–reduction loop. Perhaps the most debated issue of the chemiosmotic hypothesis is the manner in which the process of electron transport in the inner mitochondrial membrane pumps protons (H^+) from the matrix to the exterior. **Mitchell** has proposed a fantastic scheme which is based on the fact that reducing equivalents are transferred as H atoms by some of the electron carriers (such as ubiquinone), and as electrons by others (such as Fe—S centre and cytochromes). He opined that hydrogen-carrying and electron-carrying proteins alternate in the respiratory chain to form 3 'functional loops' called the *oxidation-reduction loops* (= *o/r loops*). Each loop corresponds functionally to the coupling sites I, II and III of the chemical hypothesis respectively. An ideal single o/r loop consists of a hydrogen carrier and an electron carrier, as depicted in Fig. 23–23. In each loop, two H^+ are carried outward through the inner membrane and deliver two H^+ to the cytosol; the corresponding pair of electrons is then carried back from the outer to the inner surface of the membrane. Thus, each pair of reducing equivalents passing through such a loop carries two H^+ from the matrix to the exterior. Each loop is thought to provide the osmotic energy to make one mole of ATP.

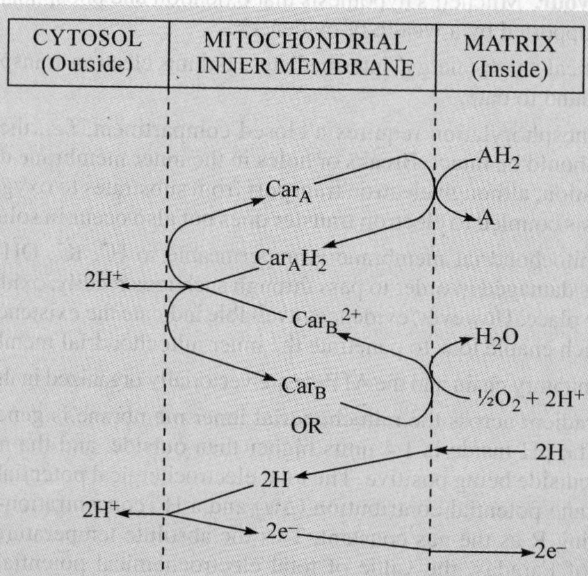

Fig. 23–23. An ideal oxidation-reduction loop as envisaged in chemiosmotic hypothesis

The oxidation-reduction (or o/r) loop translocates protons.

(After Peter A. Mayes, 1979)

Proton Transport Mechanism. In the overall scheme of electron transport, as envisaged by Peter Mitchell, the respiratory chain is folded into 3 oxidation-reduction (o/r) loops (Fig. 23–24). It is assumed that the members of the respiratory chain are organized in the membrane to provide the necessary sidedness. Each pair of electrons transferred from NADH to oxygen causes 6 protons to be translocated from inside to the outside of the coupling membrane. NADH first donates one H^+ and 2 electrons which, together another H^+ from the internal medium, reduce FMN to $FMNH_2$. FMN extends the full width of the membrane, so that it can release $2H^+$ to the exterior of the membrane and then return 2 electrons to the inside *via* Fe—S proteins which become reduced. Each reduced Fe—S complex donates one electron to ubiquinone (Q) which, after accepting a proton (H^+) from inside the membrane, is reduced to QH_2. QH_2, being a small lipid-soluble molecule, moves to the exterior of the membrane to discharge a pair of protons into the cytosol and donates 2 electrons to 2 moles of cytochrome *b*, the next carrier of electrons in the respiratory chain. Cytochrome *b* is thought to extend the mitochondrial membrane enabling the electrons to join another molecule of ubiquinone along with 2 more protons from the internal medium. The QH_2, so produced, moves to the outer surface to liberate 2 protons and the 2 electrons are passed onto the 2 moles of cytochrome *c*. These electrons, passing through cytochrome *a*, then traverse the membrane to reach cytochrome a_3, which is located on the inner face of the membrane. Here, 2 electrons combine with 2 protons from the internal medium and an oxygen atom to form a molecule of water.

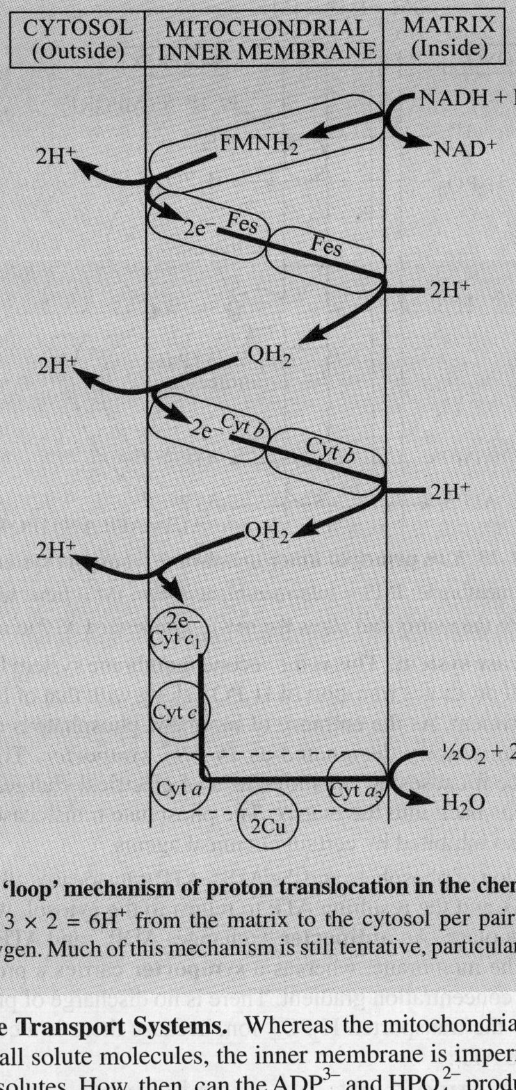

Fig. 23–24. The 'loop' mechanism of proton translocation in the chemiosmotic hypothesis

The 3 o/r loops carry 3 × 2 = $6H^+$ from the matrix to the cytosol per pair electrons passing from the substrate NADH to oxygen. Much of this mechanism is still tentative, particularly around the Q/cytochrome *b* region.

Inner Membrane Transport Systems. Whereas the mitochondrial outer membrane is freely permeable to most small solute molecules, the inner membrane is impermeable to H^+, OA^-, K^+ and also many other ionic solutes. How, then, can the ADP^{3-} and HPO_4^{2-} produced in the cytosol enter the matrix and how can the newly-formed ATP^{4-} leave again, since oxidative phosphorylation takes place within the inner matrix space ? Two of the many specific transport systems present in the inner mitochondrial membrane make these events possible (Fig. 23–25):

1. **Adenine nucleotide translocase system.** This system consists of a specific protein that extends across the inner membrane. It translocates one molecule of ADP^{3-} inward in exchange for one molecule of ATP^{4-} coming out. As the entrance of ADP is coupled to the exit of ATP, this system is better called as *ADP—ATP antiporter*. Obviously, this transport system is moving more negative charges out than it is bringing in, so it is effectively discharging the outside-positive electrical potential across the inner membrane. The adenine nucleotide transport system is specific since its carries only ATP and ADP and not AMP or any other nucleotides, such as GDP or GTP. Adenine nucleotide translocase is specifically inhibited by a toxic glycoside, *atractyloside*.

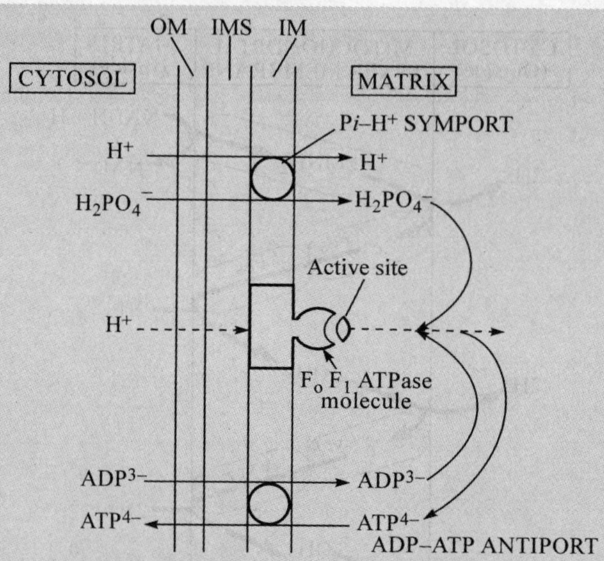

Fig. 23–25. Two principal inner membrane transport systems

[OM = Outer membrane; IMS = Intermembrane space; IM = Inner membrane]

These bring P*i* and ADP into the matrix and allow the newly-synthesized ATP to move out of the matrix.

2. **Phosphate translocase system.** This is the second membrane system functioning in oxidative phosphorylation. It promotes transport of $H_2PO_4^-$ along with that of H^+ from the cytosol into the matrix compartment. As the entrance of inorganic phosphate is coupled to the entrance of H^+, this system is aptly designated as *Pi—H*$^+$ *symporter*. The Pi–H^+ symporter is electroneutral since it causes no net movement of electrical charge, rather it is effectively transporting protons back into the matrix. The phosphate translocase system is specific for phosphate. It is also inhibited by certain chemical agents.

Thus, the combined action of phosphate and the ADP–ATP translocases allows external phosphate and ADP to enter the matrix and the resulting ATP to return to the cytosol, where most of the ATP-requiring cell activities take place. An **antiporter** exchanges ADP^{3-} and ATP^{4-} and is driven by the electrical potential across the membrane, whereas a **symporter** carries a proton and monobasic P*i* and is driven by the proton concentration gradient. There is no discharge of proton gradient from the release of a proton (H^+) by the transported $H_2PO_4^-$ ion, because there is a counterbalancing uptake when the P*i* is used to make ATP:

$$ADP^{3-} + HPO_4^{2-} + H^+ \longrightarrow ATP^{4-} + H_2O$$

OXIDATION OF EXTRAMITOCHONDRIAL NADH (=NADH Shuttle Systems)

Although NADH cannot penetrate the mitochondrial inner membrane, it is produced continuously in the cytosol by *3-phosphoglyceraldehyde dehydrogenase*, a glycolytic enzyme. However, under aerobic conditions, extramitochondrial NADH does not accumulate and is presumably oxidized by the mitochondrial respiratory chain. How, then, does this occur ? Special shuttle systems have been proposed which are based on the fact that the electrons from cytosolic NADH, rather than cytosolic NADH itself, are carried across the mitochondrial inner membrane by an indirect route. Two shuttle systems, which explain these events, are described below:

1. Malate-oxaloacetate-aspartate shuttle

This shuttle is of comparatively more universal occurrence and operates in heart, liver and kidney mitochondria. This shuttle (Fig. 23–26), is mediated by two membrane carriers and four enzymes. In this shuttle, the reducing electrons are first transferred from cytosolic NADH to cytosolic oxaloacetate to yield malate by the enzymatic action of *cytosolic malate dehydrogenase*. The malate, carrying the electrons, then passes through the inner membrane into the matrix by a dicarboxylate-transport system (A). Here the malate donates its electrons to the matrix NAD^+, reducing it to NADH in the presence of *matrix-malate dehydrogenase*. The NADH then passes its electrons directly to the respiratory chain in the inner membrane. Three moles of ATP are generated as this pair of electrons passes to oxygen. The oxaloacetate, so formed, cannot pass through the mitochondrial inner membrane from the matrix back into the cytosol but is converted by *transaminase* into aspartate which can pass *via* the amino acid-transport system (C). The function of transport system B is to regenerate oxaloacetate into the cytosol. Transport system B makes possible the exchange of glutamate for aspartate. The dicarboxylate-transport system A carries α-ketoglutarate out in exchange for malate passing inward. The net reaction of *malate-aspartate shuttle*, as it is also called, is:

$$\underset{\text{Cytosolic}}{NADH} + \underset{\text{Mitochondrial}}{NAD^+} \rightleftharpoons \underset{\text{Cytosolic}}{NAD^+} + \underset{\text{Mitochondrial}}{NADH}$$

As evident, this is a readily reversible shuttle, *i.e.*, can operate in both directions either into or out of the mitochondria. The complexity of this system is due to the impermeability of the mitochondrial membrane to oxaloacetate. However, the other anions are not freely permeable and require specific transport systems for passage across the membrane.

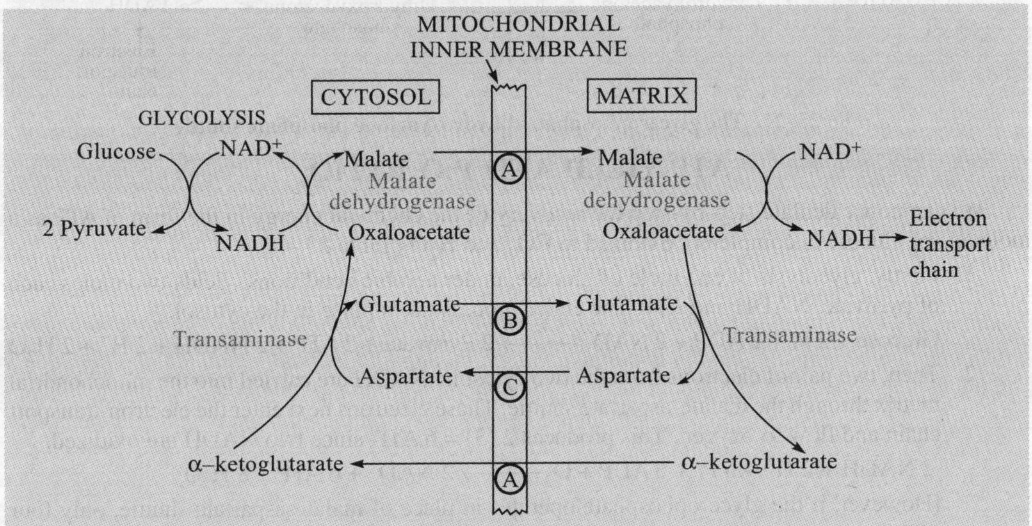

Fig. 23–26. The malate-oxaloacetate-aspartate shuttle

2. Glycerophosphate-dihydroxyacetone phosphate shuttle

This shuttle is not so common and operates prominently in the insect flight muscle and in the brain. This NADH shuttle is medicated by membrane carriers and two enzyme systems, the cytosolic and mitochondrial glycerol 3-phosphate dehydrogenase. *Glycerol 3-phosphate dehydrogenase* is NAD-linked in the cytosol whereas the enzyme found in the matrix is a flavoprotein enzyme. In this shuttle (Fig. 23–27), first of all, the electrons from cytosolic NADH are transferred to cytosolic dihydroxyacetone phosphate (DHAP) to form glycerol 3-phosphate (G-3-P). This reaction is catalyzed by cytosolic glycerol 3-phosphate dehydrogenase. Glycerol 3-phosphate then enters matrix, where it is reoxidized to dihydroxyacetone phosphate by the matrix glycerol 3-phosphate dehydrogenase which is FAD-bound. The DHAP, so formed, then diffuses out of the mitochondria into the cytosol to complete one turn of the shuttle. It is to be noted that as the mitochondrial enzyme is linked to the respiratory chain via a flavoprotein rather than NAD, only 2 rather than 3 mole of ATP are formed per atom of oxygen consumed, as also happens with succinate. The use of FAD enables electrons from cytoplasmic NADH to be transported into mitochondria against an NADH concentration gradient. The price of this transport is one ATP per electron pair. Evidently, if more reducing equivalents are passed through G-3-P—DHAP shuttle, as it is abbreviated, oxygen consumption must increase to maintain ATP production. This mechanism may, therefore, account for at least part of the extra oxygen consumption of hyperthyroid individuals (the administration of thyroxine increases the activity of the flavin-bound glycerol 3-phosphate dehydrogenase). The net reaction of G-3-P—DHAP shuttle is:

$$\underset{\text{Cytosolic}}{\text{NADH} + \text{H}^+} + \underset{\text{Mitochondrial}}{\text{E-FAD}} \rightleftharpoons \underset{\text{Cytosolic}}{\text{NAD}^+} + \underset{\text{Mitochondrial}}{\text{E-FADH}_2}$$

A remarkable feature of this shuttle is that it is irreversible or unidirectional. *i.e.*, can transfer reducing equivalents only from cytosol to matrix and not *vice versa*.

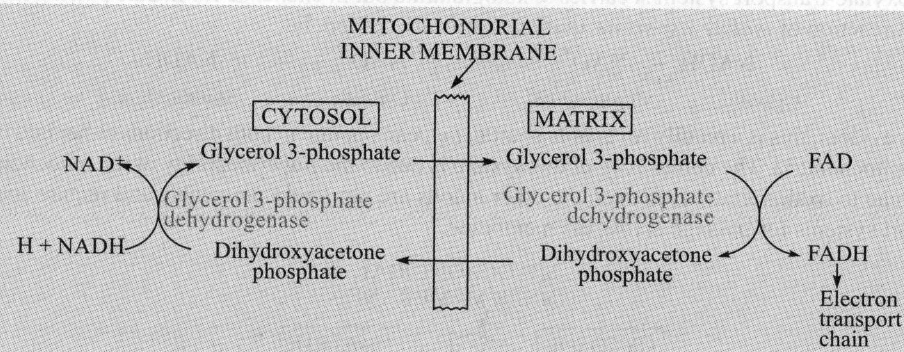

Fig. 23–27. The glycerophosphate-dihydroxyaetone phosphate shuttle

ATP YIELD AND P:O RATIO

We can now calculate step-by-step the recovery of the chemical energy in the form of ATP as a molecule of glucose is completely oxidized to CO_2 and H_2O (Table 23–4):

1. Firstly, glycolysis of one mole of glucose, under aerobic conditions, yields two moles each of pyruvate, NADH and ATP. The entire process takes place in the cytosol.

   $$\text{Glucose} + 2\,\text{P}i + 2\,\text{ADP} + 2\,\text{NAD}^+ \longrightarrow 2\,\text{Pyruvate} + 2\,\text{ATP} + 2\,\text{NADH} + 2\,\text{H}^+ + 2\,\text{H}_2\text{O}$$

2. Then, two pair of electrons from the two cytosolic NADH are carried into the mitochondrial matrix through the malate-aspartate shuttle. These electrons next enter the electron-transport chain and flow to oxygen. This produces 2 (3) = 6 ATP, since two NADH are oxidized.

   $$2\,\text{NADH} + 2\,\text{H}^+ + 6\,\text{P}i + 6\,\text{ADP} + \text{O}_2 \longrightarrow 2\,\text{NAD}^+ + 6\,\text{ATP} + 8\,\text{H}_2\text{O}$$

   [However, if the glycerophosphate operates in place of malate-aspartate shuttle, only four ATP are generated, instead of six.

   $$2\,\text{NADH} + 2\,\text{H}^+ + 4\,\text{P}i + 4\,\text{ADP} + \text{O}_2 \longrightarrow 2\,\text{NAD}^+ + 4\,\text{ATP} + 6\,\text{H}_2\text{O}]$$

3. Then, two moles of pyruvate are dehydrogenated to yield two moles each of acetyl-CoA and CO_2 in the mitochondria. This results in the formation of two NADH. The two electron pairs from two NADH are carried to O_2 via the electron-transport chain, each mole providing three moles of ATP.

   $$2\,\text{Pyruvate} + 2\,\text{CoA} + 6\,\text{P}i + 6\,\text{ADP} + \text{O}_2 \longrightarrow 2\,\text{Acetyl-CoA} + 2\,\text{CO}_2 + 6\,\text{ATP} + 8\,\text{H}_2\text{O}$$

4. Ultimately, two moles of acetyl-CoA are oxidized to CO_2 and H_2O *via* the citric acid cycle, along with the oxidative phosphorylation coupled to electron transport from isocitrate, α-ketoglutarate and malate to O_2, each of which yields 3 moles of ATP. The oxidation of succinate, however, yields 2 ATP and another two ATPs are generated from succinyl-CoA *via* GTP.

   $$2\,\text{Acetyl-CoA} + 24\,\text{P}i + 24\,\text{ADP} + 4\,\text{O}_2 \longrightarrow 2\,\text{CoA—SH} + 4\,\text{CO}_2 + 24\,\text{ATP} + 26\,\text{H}_2\text{O}$$

Table 23–4. ATP yield from the complete oxidation of glucose

Reaction Sequence	*ATP Yield per Glucose*
Glycolysis (in the cytosol)	
Phosphorylation of glucose	– 1
Phosphorylation of fructose 6-phosphate	– 1
Dephosphorylation of 2 moles of 1, 3-DPG	+ 2
Dephosphorylation of 2 moles of PEP	+ 2
2 NADH are formed in the oxidation of 2 moles of G-3-P	
Conversion of pyruvate into acetyl-CoA (inside mitochondria)	
2 NADH are formed	
Citric acid cycle (inside mitochondria)	
2 moles of GTP are formed from 2 moles of succinyl-CoA	+ 2
6 NADH are formed in the oxidation of 2 moles each of isocitrate, α-ketoglutarate and malate	
2 $FADH_2$ are formed in the oxidation of 2 moles of succinate	
Oxidative phosphorylation (inside mitochondria)	
2 NADH formed in glycolysis ; each yields 2 ATP (assuming transport of NADH by malate-oxaloacetate-aspartate shuttle)	+ 6*
2 NADH formed in oxidative decarboxylation of pyruvate ; each yields 3 ATP	+ 6
2 FADH formed in the citric acid cycle ; each yields 2 ATP	+ 4
6 NADH formed in the citric acid cycle ; each yields 3 ATP	+ 18
Net yield per glucose	+ 38

* If, however, 2 NADH are transported by the glycerophosphate-dihydroxyacetone phosphate shuttle, there would be an yield of 4 ATP rather than 6. In such a case, the net yield of ATP per glucose oxidized would be 36, instead of 38.

Summing up the 4 equations. On adding the above 4 equations, we get the overall equation for the process of respiration as a whole.

$$\text{Glucose} + 38\ Pi + 38\ \text{ADP} + 6\ O_2 \longrightarrow 6\ CO_2 + 38\ \text{ATP} + 44\ H_2O$$

Thus, one mole of glucose on complete oxidation to CO_2 and H_2O in the heart, liver and kidney, *where the malate-aspartate shuttle operates*, leads to the production of 38 moles of ATP. The P:O ratio in such cases in 3.16, since 38 ATP are produced and 12 atoms of O_2 are consumed.

> The **P:O ratio** is defined as the number of moles of inorganic phosphate incorporated into organic form per atom of oxygen consumed. In other words, the P : O ratio may be taken to mean the number of moles of high-energy phosphate generated per atom of oxygen consumed.

[However, in the skeletal muscles, *where glycerophosphate shuttle operates*, a total of 36 ATP moles is generated per mole of glucose oxidized. In that case, the overall equation is modified to:

$$\text{Glucose} + 36\ Pi + 36\ \text{ADP} + 6\ O_2 \longrightarrow 6\ CO_2 + 36\ \text{ATP} + 42\ H_2O$$

Here the P:O ratio is 36/12= 3.0.]

Under standard conditions (1.0 M), the theoretical recovery of free energy in the complete oxidation of glucose is:

$$38\ (7.3/686)\ (100) = 40\%$$

[or 36 (7.3/686) (100) = 38%, when glycerophosphate shuttle operates]. But in the intact cell, the efficiency of this process is high (over 70%) because the cellular concentration of glucose, oxygen, P*i*, ADP and ATP are unequal and much lower than the standard concentration of 1.0 M. The trapping of this amount of energy is a noteworthy achievement for the living cell.

ROLES OF ELECTRON TRANSPORT ENERGY

The main function of electron transport in mitochondria is to provide energy for the synthesis of ATP during oxidative phosphorylation. But the energy generated during electron transport is also used for other biological purposes (Fig. 23–28), which are listed below:

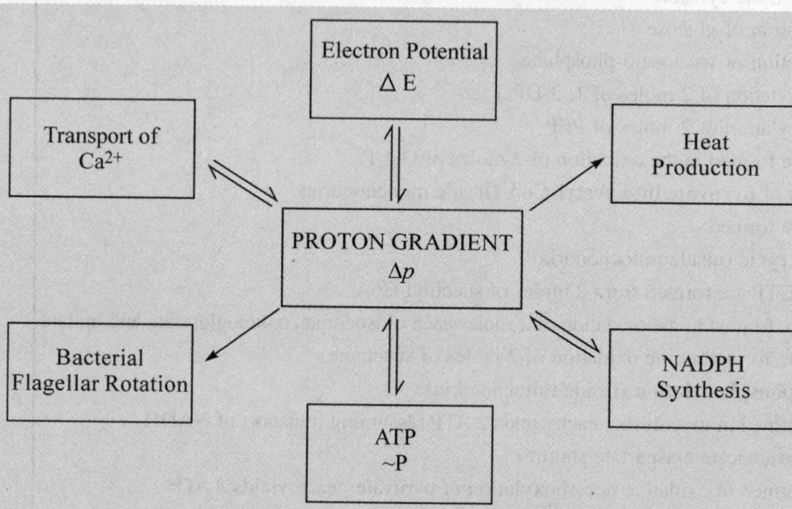

Fig. 23–28. Some key roles of the transmembrane proton gradient

1. The proton gradient generated by electron transport can be used to generate heat. For example, human infants, other mammals born hairless and some hibernating animals have a special type of **brown fat** in the neck and upper back. The brown fat is so named because it contains profuse mitochondria which, in turn, are rich in the red-brown cytochromes. These specialized brown-fat mitochondria do not usually produce ATP, rather they dissipate the free energy of electron transport as heat in order to maintain the body temperature of the young ones. This is because the brown-fat mitochondria have special *proton pores* in their inner membrane that allow the protons, pumped on by electron transport, to flow back into the matrix, rather than through the F_oF_1 ATPase or ATP synthetase. Consequently, the free energy of electron transport is diverted from ATP synthesis into heat production.
2. The electron transport energy is also used to transport Ca^{2+} from the cytosol into the matrix of animal mitochondria. In fact, the inner membrane contains two transport systems for Ca^{2+} : one transports Ca^{2+} inward and the other transports Ca^{2+} outward. The concentration of external Ca^{2+} is maintained at a very low level (about 10^{-7} M). This is due to a balance between the rates of Ca^{2+} influx and efflux. Thus, the inward transport of Ca^{2+} is counterbalanced by Ca^{2+} efflux whose rate is regulated. High Ca^{2+} concentrations initiate or promote many cell functions such as muscle concentration, glycogen breakdown and the oxidation of pyruvate ; low Ca^{2+} concentrations have inhibitory effects on these functions.
3. Rotation of bacterial flagella is also controlled by the proton gradient generated across the membrane.
4. Transfer fo electrons from NADH to NADPH is also powered by the proton gradient.
5. The entry of some amino acids and sugars is also governed by the energy generated during electron transport.

RESPIRATORY INHIBITORS

The electron transport chain contains 4 complexes (I, II, III and IV) which participate in the transfer of electrons. In general, it may be stated that the structural integrity of these complexes appears essential for its interaction with most inhibitors, since the soluble, phospholipid-free enzymes do not exhibit the characteristic inhibitory pattern. Specific inhibitors are summarized in Table 23–5.

Table 23–5. Different classes of respiratory inhibitors

Class with examples	*Affected complex*or reaction*	*Concentrations employed*
Inhibitors of electron transport		
Rotenone, Piericidin A	I FMN ⟶ Q and /or cyt b (Fe^{3+})	Stoichimetric with f_{PN}
Amytal, Seconal	I same as above	$\geq 10^{-3}$ M
Thenoyltrifluoroacetone	II Peptide-FAD ⟶ Q and/or cyt *b*	$\geq 10^{-4}$ M
Antimycin A, Dimercaprol	III cyt *b* (Fe^{2+}) ⟶ cyt c1 (Fe^{3+}) (perhaps at NH1)	Stoichiometric with cytochromes
HCN, H_2S	IV cyt *c* (Fe^{2+}) ⟶ O_2	$\leq 10^{-4}$ M (CN^-)
HN_3, CO	IV same as above	Higher than above
Inhibitors of oxidative phosphorylation		
Oligomycin		
Rutamycin	Inhibit respiration (NADH ⟶ O_2	
Atractylate	or Succinate O_2) when coupled to phosphorylation	
Bongkrekate		
Uncouplers of oxidative phosphorylation		
2,4 dinitrophenol	Stimulate respiration when rate is	
Dicoumarol	limited by phosphorylation or blocked by oligomycin.	
Valinomycin †		
Gramicidin A†		

* The complexes are indicated by Roman numerals.
† Now classed under 'Ionophores of oxidative phosphorylation'.

A. Inhibitors of Electron Transport

(= Inhibitors of Respiratory Chain)

These are the inhibitors that arrest respiration by combining with members of the respiratory chain, rather than with the enzymes that may be involved in coupling respiration with ATP synthesis. They appear to act at 3 loci that may be identical to the energy transfer sites I, II and III (Fig. 23–29). The various inhibitors of this category are described below:

1. Rotenone. Rotenone (Fig. 23–30) is a compound extracted from the roots of tropical plants such as *Derris elliptica* and *Lonchoncarpus nicou*. It complexes avidly with NADH dehydrogenase and acts between the Fe—S proteins and ubiquinone. Only 30 nanomoles per gram of mitochondrial protein are effective for blocking site I. Rotenone is relatively nontoxic to mammals because it is absorbed poorly, although exposure of the lungs to dust is a little more dangerous. However, the compound is intensely toxic to the fishes and insects as it readily passes into their gills and breathing tubes respectively. The Amazonian Indians hence, use rotenone to poison fishes and kill insects.

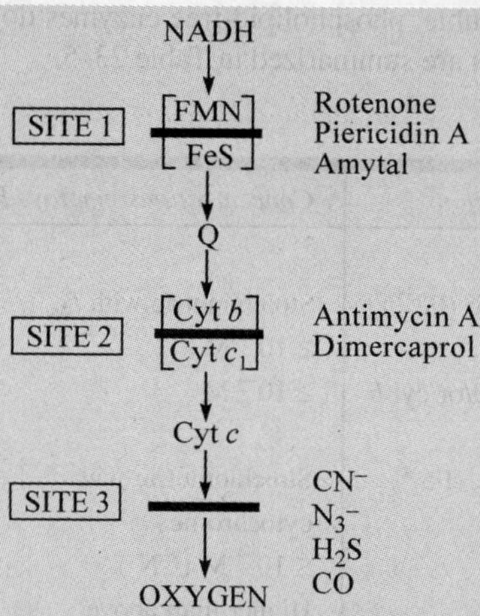

Fig. 23–29. Sites of action of some inhibitors of electron transport

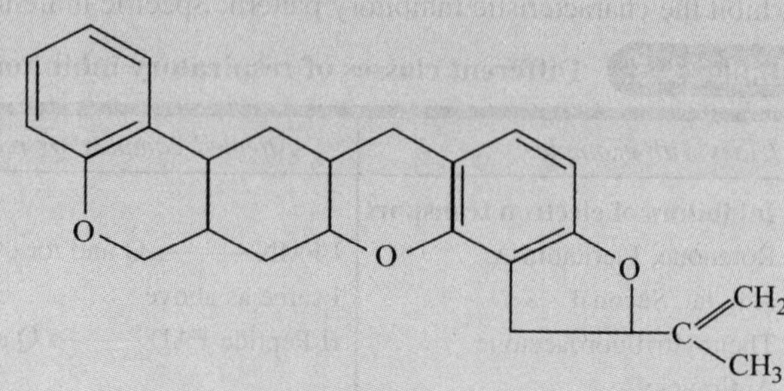

Fig. 23–30. Rotenone

2. **Piericidin A.** It is an antibiotic, produced by species of *Streptomyces*. It has an action similar to that of rotenone.

3. **Barbiturates (Amytal, seconal).** These also block NADH dehydrogenase, but are required in much higher concentrations for the purpose. The sedative actions of these compounds appear to depend on their actions on neural membranes, but inhibition of respiration may assist the effect. In contrast, amytal (and also rotenone and piericidin A) do not interfere with the oxidation of succinate, because the electrons of these substrates enter the electron-transport chain beyond the block of coenzyme Q.

4. **Antimycins.** These are also antibiotics, also produced by *Streptomyces*. They inhibit the respiratory chain at or around site II and block electron flow between cytochromes *b* and c_1, which prevents ATP synthesis coupled to the generation of a proton gradient at site II. This block can be bypassed by the addition of ascorbate, which directly reduces cytochrome *c*. Electrons then flow from cytochrome *c* to O_2, with the concomitant synthesis of ATP coupled to a proton gradient at site III. About 0.07 micromole of antimycin A per gram of mitochondrial protein is effective.

5. **Dimercaprol.** It is identical in action to the antimycins.

6. **Cyanides.** These are among the poisons better known by the general public. Although they are not extraordinarily potent (the minimum lethal dose for human beings is between 1 and 3 millimoles), they enter the tissues very rapidly so that a sufficient quantity becomes lethal within a

> **Cyanide Poisoning**
>
> Inhalation of hydrogen cyanide (HCN) gas or ingestion of potassium cyanide (KCN) causes a rapid and extensive inhibition of the mitochondrial electron transport chain at the cytochrome oxidase step. **Cyanide is one of the most potent and rapidly acting poisons known.** It binds to the Fe^{3+} in the heme a_3 of cytochrome c oxidase that catalyzes the terminal step in the electron transport chain. Cyanide thus prevents the binding of oxygen to the binuclear center and the role of oxygen as the final electron acceptor. Mitochondrial respiration and energy production cease, and cell death occurs rapidly. Death due to cyanide poisoning occurs from tissue asphyxia, most notably of the central nervous system (CNS). An antidote to cyanide poisoning, if the poisoning is diagnosed rapidly, is the administration of various nitrites that convert oxyhemoglobin to methemoglobin by oxidizing Fe^{2+} of hemoglobin to Fe^{3+}. Methemoglobin (Fe^{3+}) competes with cytochrome a_3 (Fe^{3+}) for cyanide, forming a methemoglobin–cyanide complex. Administration of thiosulfate causes the cyanide to react with the enzyme rhodonase, forming the nontoxic thiocyanate.

few minutes. It is this quick effect that has gained the cyanides so much respect. The cyanide ion (CN^-) combines tightly with cytochrome oxidase, leading to the cessation of transfer of electrons to oxygen. The previous electron carriers in the chain accumulate in their reduced state, and the generation of high-energy phosphate stops. In fact, the effect of cyanide is as fundamental as deprivation of oxygen, and like the latter causes rapid damage to the brain.

7. Azide. It also blocks the electron flow between the cytochrome oxidase complex and oxygen. Azide (N_3^-), as also cyanides, react with the ferric form (Fe^{3+}) of this carrier.

8. Hydrogen sulfide. Few realize that H_2S is as toxic as HCN. However, its disagreeing odour gives more warning. It is a lethal menace in all drilling operations, especially so on oil-drilling platforms at sea. *In vitro* tests reveal that 0.1 mM sulfide inhibits cytochrome oxidase more than does 0.3 mM cyanide– 96% against 90%.

9. Carbon monoxide. It also attacks between cytochrome oxidase and O_2 but, unlike cyanide and azide, CO inhibits the ferrous form (Fe^{2+}) of the electron carrier.

Phosphorylation coupled to the generation of a proton gradient at site III does not occur in the presence of these inhibitors (*i.e.*, cyanides, azide, H_2S and CO) because electron flow is blocked.

B. Inhibitors of Oxidative Phosphorylation

These compounds inhibit electron transport in cells in which such transport is coupled with ATP synthesis, but they do not affect electron transport in cells in which such transport is not coupled with phosphorylation. Thus, these compounds inhibit both electron transport and oxidative phosphorylation.

1. Oligomycins. These polypeptide antibiotics are obtained from various species of *Streptomyces*. They inhibit the transfer of high-energy phosphate to ADP and, therefore, also inhibit electron transfers coupled to phosphorylation. However, they do not affect those redox reactions that are not coupled. Henceforth, they are widely employed as experimental tools for differentiating between the two kinds of reactions. The F_o component of ATP synthetase binds oligomycin, which is a potent inhibitor of this enzyme and, thus, of oxidative phosphorylation. Apparently, oligomycin appears to block one of the primary phosphorylation steps.

2. Rutamycin. This antibiotic also inhibits both electron transport and oxidative phosphorylation.

3. Atractylate (= Atractyloside). This toxic glycoside is extracted from the rhizomes of *Atractylis gummifera*, a plant native to Italy. It blocks oxidative phosphorylation by competing with ATP and ADP for a site on the ADP—ATP antiport of the inner mitochondrial membrane. Hence, it checks renewal of ATP supply in the cytosol. Evidently, atractylate inhibits at a step beyond that blocked by oligomycin, *i.e.*, one specifically concerned with the entry and exit of ADP/ATP into a mitochondrial compartment.

4. Bongkrekate. It is a toxin formed by a bacteria (*Pseudomonas*) in a coconut preparation (called 'bongkrek') from Java. It also blocks the ADP—ATP antiport. Only 2 micromoles per gram of mitochondrial protein are effective.

C. Uncouplers of Oxidative Phosphorylation

Uncoupling agents are compounds which dissociate (or 'uncouple') the synthesis of ATP from the transport of electrons through the cytochrome system. This means that the electron transport continues to function, leading to oxygen consumption but phosphorylation of ADP is inhibited. In the intact mitochondria, these two processes are intimately associated. When they are uncoupled, the transport of electrons speeds up, thereby pointing out that the phosphorylation of ADP has been a rate limiting process. In the presence of uncouplers, the free energy released by electron transport appears as heat, rather than as newly-made ATP. Uncoupling agents greatly enhance the permeability of the inner membrane to H^+. They are lipophilic and bind H^+ from one side of the membrane and carry it through the membrane toward the side with the lower H^+ concentration. In Mitchell's hypothesis, uncouplers are agents that are capable of destroying the vectorial, anisotropic structure of the membrane, leading to elimination of the pH gradient. As the uncouplers bind and carry protons, they are also called **protonophores**. Uncoupling can be distinguished from inhibition. Uncoupling causes an increased oxygen consumption in the absence of increased utilization of ATP, whereas inhibition of phosphorylation (or inhibition of ADP—ATP antiport) diminishes oxygen consumption in normal coupled mitochondria. Some uncouplers commonly employed are:

1. 2,4-dinitrophenol (DNP). Introduced by Loomis and Lipmann, dinitrophenol (Fig. 23–31) is one of the most effective agents for uncoupling respiratory-chain phosphorylation. It does not have any effect on the substrate-level phosphorylations that take place in glycolysis. It acts at a concentration of 10 micromolar. The uncoupling action of DNP is due to the fact that both the phenol and the corresponding phenolate ion are significantly soluble in the lipid core of the inner mitochondrial membrane (Fig. 23–32). The phenol diffuses through the core toward the matrix, where it loses a proton ; the phenolate ion then diffuses back toward the cytosol side, where it picks up a proton to repeat the process.

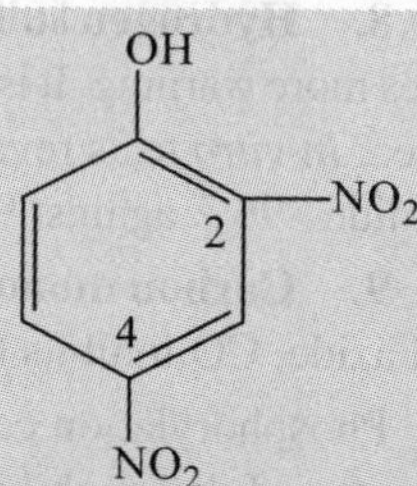

Fig. 23–31. 2,4-dinitrophenol, DNP

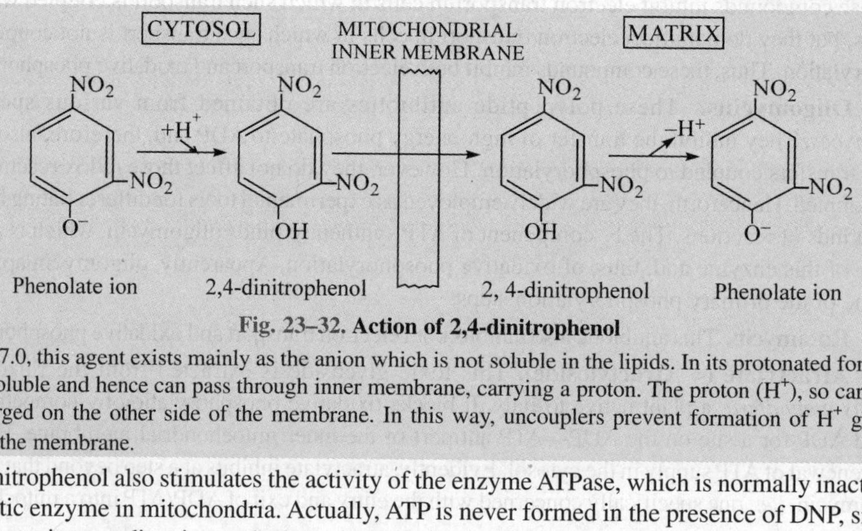

Fig. 23–32. Action of 2,4-dinitrophenol

At pH 7.0, this agent exists mainly as the anion which is not soluble in the lipids. In its protonated form, it is lipid-soluble and hence can pass through inner membrane, carrying a proton. The proton (H^+), so carried, is discharged on the other side of the membrane. In this way, uncouplers prevent formation of H^+ gradient across the membrane.

Dinitrophenol also stimulates the activity of the enzyme ATPase, which is normally inactive as a hydrolytic enzyme in mitochondria. Actually, ATP is never formed in the presence of DNP, since the high-energy intermediate is attacked *i.e.*, it acts prior to the step of ATP synthesis.

2. Dicoumarol. Dicoumarol (Fig. 23–33) arises from the action of microorganisms on coumarin, a natural constituent of sweet clover, *Melilotus indica*. It has an action identical to that of 2,4-dinitrophenol. Dicoumarol is also an antagonist of vitamin K function.

Dicoumarol is also spelt as **dicumarol.**

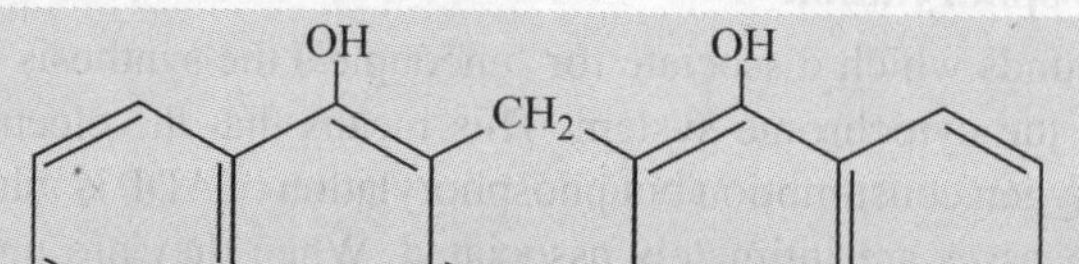

Fig. 23–33. Dicoumarol
(3, 3′-methylene-bishydroxycoumarin)

3. m-chlocarbonyl cyanide phenylhydrazone (CCCP). Its action is also similar to that of 2, 4-dinitrophenol but it is about 100 times more active than the latter.

2,4-dinitrophenol Poisoning

Dinitrophenol (DNP) is an uncoupler of oxidative phosphorylation. It was first discovered to induce weight-loss during World War I when it was noticed that French munitions workers who were exposed to DNP during the synthesis of dynamite (trinitrotoluene, TNT) rapidly lost weight. In the 1930s, it was prescribed by physicians for weight-loss and was also available OTC (over the counter), but because people suffered significant side effects, such as cataract, blindness, kidney and liver damage and death, it was banned for medical use in the U.S. after a congressional investigation. DNP is currently used industrially in the manufacture of dyes, explosives, herbicides, insecticides and lumber preservatives. DNP kills bacteria and fungi by uncoupling phosphorylation. Unluckily, DNP has resurfaced as an illegal weight-loss product. It radically increases consumption of oxygen and metabolic fuels, and nearly all metabolic energy is wasted as heat. Cells die because of both– excess temperature and lack of ATP.

D. Ionophores of Oxidative Phosphorylation

Ionophores ("ion carriers") are lipophilic substances, capable of binding and carrying specific cations through the biologic membranes. *They differ from the uncouplers in that they promote the transport of cations other than H^+ through the membrane.*

1. Valinomycin. This toxic antibiotic (Fig. 23–34)is synthesized by *Streptomyces*. Valinomycin is a repeating macrocyclic molecule made up of four kinds of residues (L-lactate, L-valine, D-hydroxyisovalerate and D-valine) taken 3 times. The four residues are alternately joined by ester and peptide bonds. It contains 6 peptide and 6 ester bonds with side chains consisting of hydrophobic alkyl radicals. The antibiotic forms a lipid-soluble complex with K^+ which readily passes through the inner mitochondrial membrane, whereas K^+ alone in the absence of valinomycin penetrates only very

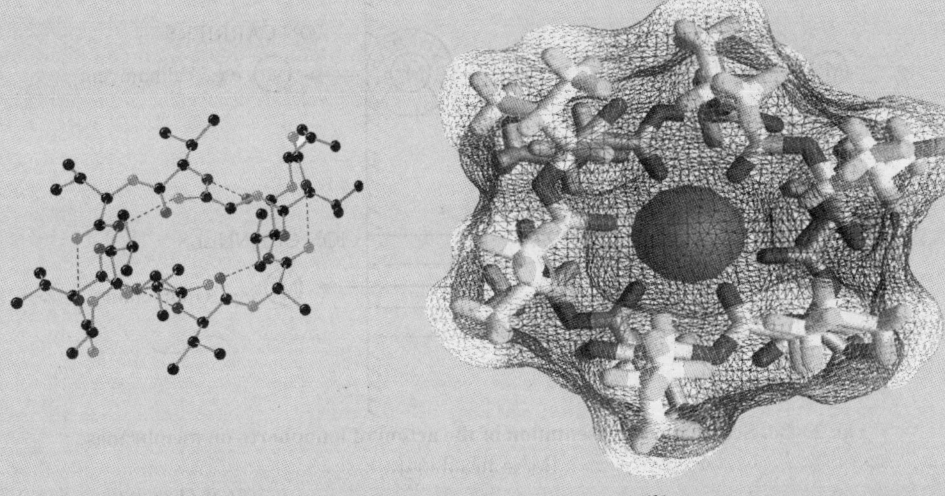

(*a*) (*b*)

Fig. 23–34. Valinomycin, a peptide hormone that binds K^+

(*a*) **Ball-and-stick model.** For the sake of clarity, hydrogen atoms are not shown.

(*b*) **Computer graphics.** In this image, the surface contours are shown as a transparent mesh through which a stick structure of the peptide and a K^+ atom (*green*) are visible. The oxygen atoms (*red*) that bind K^+ are part of a central hydrophilic cavity. Hydrophobic amino acid side chains (*yellow*) coat the outside of the molecule. Because the exterior of the K^+-valinomycin complex is hydrophobic, the complex readily diffuses through membranes, carrying K^+ down its concentration gradient. The resulting dissipation of the transmembrane ion gradient kills microbial cells, making valinomycin a potent antibiotic.

(*Courtesy : (a) Smith GD et al, 1975*)

slowly. It has a high degree of selectivity for K^+ as compared to Na^+. In fact, valinomycin binds K^+ about a thousand times as strongly as Na^+ because *water has less attraction for K^+ than for Na^+* and it is energetically more costly to pull Na^+ away from water. Thus, valinomycin interferes with oxidative phosphorylation in mitochondria by making them permeable to K^+. The result is that mitochondria use the energy generated by electron transport to accumulate K^+ rather than to make ATP.

2. Gramicidin A. Gramicidin A (Fig. 23–35) is a linear polypeptide consisting of 15 amino acid residues. Two noteworthy features of the molecule are that (a) the L-and D-amino acids alternate and that (*b*) both the N– and C–terminals of this polypeptide are modified. Gramicidin promotes penetration not only of K^+ but also of Na^+ and several other monovalent cations through the inner membrane. *Unlike valinomycin, gramicidins do not complex cations.* Rather they induce ion permeability of membranes at concentrations as low as 10^{-10} M by forming dimers which in effect provide `tubes' or 'channels' which span the membrane and through which the cation passes (Fig. 23–36). They are, therefore, better called as 'ion channels' in contrast to valinomycin and allied substances which are classed as 'ion carriers'.

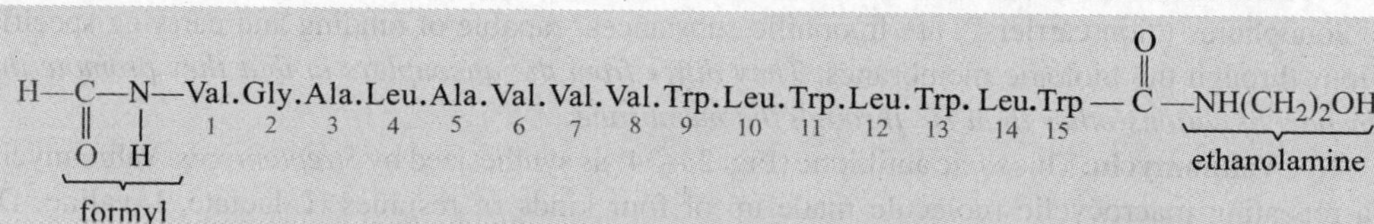

Fig. 23–35. Gramicidin A

Gramicidin A consists of 15 alternating D- and L-amino acid residues and is blocked at both its N- and C-termini.

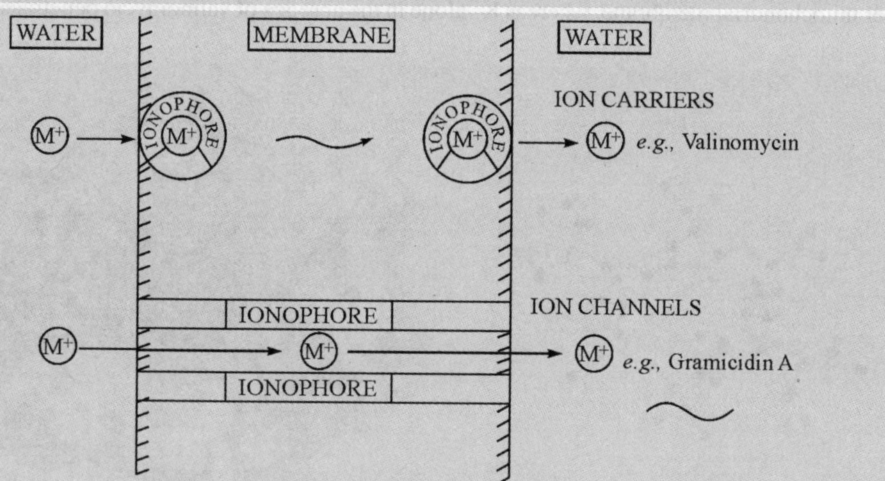

Fig. 23–36. Schematic representation of the action of ionophores on membranes

[M^+ = metal ion]

(*After Ovchinnikov YA, 1979*)

3. Nigericin. Like valinomycin, it also acts as an ionophore for K^+ but in exchange for H^+. It, therefore, abolishes the pH gradient across the membrane. In the presence of both valinomycin and nigericin, both the membrane potential and the pH gradient are eliminated, and phosphorylation is, therefore, completely inhibited.

REGULATORY CONTROLS AMONG GLYCOLYSIS, THE CITRIC ACID CYCLE AND OXIDATIVE PHOSPHORYLATION

The 3 energy-yielding stages in carbohydrate metabolism are glycolysis, the citric acid cycle and oxidative phosphorylation. Each stage is so regulated as to satisfy the time-to-time need of the cell for its products. These 3 stages are coordinated with each other in such a way that they function

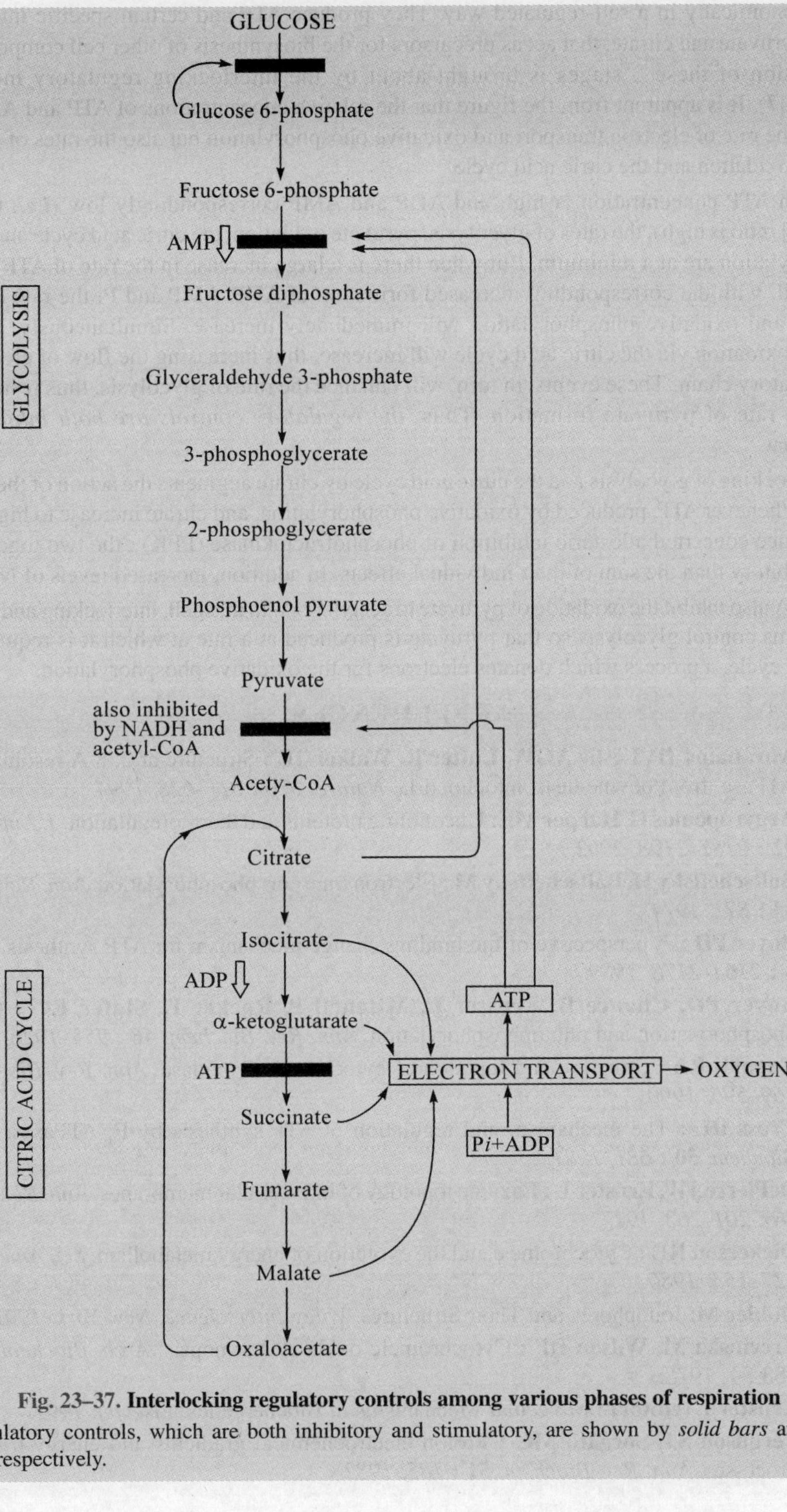

Fig. 23–37. Interlocking regulatory controls among various phases of respiration

The regulatory controls, which are both inhibitory and stimulatory, are shown by *solid bars* and *hollow arrows*, respectively.

most economically in a self-regulated way. They produce ATP and certian specific intermediates such as pyruvate and citrate, that act as precursors for the biosynthesis of other cell components. The coordination of these 3 stages is brought about by the interlocking regulatory mechanisms (Fig. 23–37). It is apparent from the figure that the relative concentrations of ATP and ADP control not only the rate of electron transport and oxidative phosphorylation but also the rates of glycolysis, pyruvate oxidation and the citric acid cycle.

When ATP concentration is high and ADP and AMP correspondingly low (*i.e.*, the [ATP]/[ADP][P*i*] ratio is high), the rates of glycolysis, pyruvate oxidation, the citric acid cycle and oxidative phosphorylation are at a minimum. But when there is a large increase in the rate of ATP utilization by the cell, with the corresponding increased formation of ADP, AMP and Pi the rate of electron transport and oxidative phosphorylation will immediately increase. Simultaneously, the rate of pyruvate oxidation via the citric acid cycle will increase, thus increasing the flow of electrons into the respiratory chain. These events, in turn, will enhance the rate of glycolysis, thus resulting in an increased rate of pyruvate formation. Thus, *the regulatory controls are both inhibitory and stimulatory.*

Interlocking of glycolysis and the citric acid cycle by citrate augments the action of the adenylate system. Whenever ATP, produced by oxidative phosphorylation, and citrate increase to higher levels, they produce concerted allosteric inhibition of phosphofructokinase (PFK) ; the two together being more inhibitory than the sum of their individual effects. In addition, increased levels of NADH and acetyl-CoA also inhibit the oxidation of pyruvate to acetyl-CoA. In nutshell, interlocking and regulatory mechanisms control glycolysis so that pyruvate is produced at a rate at which it is required by the citric acid cycle, a process which donates electrons for the oxidative phosphorylation.

REFERENCES

1. **Abrahams JP, Leslie AGW, Lutter R, Walker JE :** Structure at 2. 8 Å resolution of F_1-ATPase, from bovine heart mtochondria. *Nature.* **370 :** *621–628, 1994.*
2. **Argyropoulos G, Harper ME:** Uncoupling proteins and thermoregulation. *J. Appl. Physiol.* **92 :** *2187–2198, 2002.*
3. **Baltscheffsky H, Baltscheffsky M :** Electron transport phosphorylation. *Ann. Rev. Biochem.* **43 :** *871, 1974.*
4. **Boyer PD :** A perspective of the binding change mechanism for ATP synthesis. *FASEB J.* **3 :** *2164–2178, 1989.*
5. **Boyer PD, Chance B, Ernster L, Mitchell P, Racker E, Slater EC :** Oxidative phosphorylation and photophosphorylation. *Ann. Rev. Biochem.* **46 :** *955–1026, 1977.*
6. **Capaldi RA** ; Structure and function of cytochrome *C* oxidase. *Ann. Rev. Biochem.* **59 :** *569–596, 1990.*
7. **Cross RL :** The mechanism and regulation of ATP synthesis by F_1-ATPases. *Ann. Rev. Biochem.* **50 :** *681, 1981.*
8. **DePierre JW, Ernster L :** Enzyme topology of intracellular membranes. *Ann. Rev. Biochem.* **46 :** *201–262, 1977.*
9. **Dickerson RE :** Cytochrome *c* and the evolution of energy metabolism. *Sci. Amer.* **242(3) :** *137–153, 1980.*
10. **Dobler M:** Ionophores and Their Structures. *Wiley-Interscience, New York. 1991.*
11. **Erecinska M, Wilson DF :** Cytochromeic oxidase. A synopsis. *Arch. Biochem. Biophys.* **188 :** *1, 1978.*
12. **Ernster L (editor) :** Molecular Mechanisms in Bioenergetics. *Elsevier. 1992.*
13. **Ferguson SJ, Sorgato MC :** Proton electrochemical gradients and energy-transduction processes. *Ann. Rev. Biochem.* **51 :** *185, 1982.*

14. **Fillingame RH :** The proton-translocating pumps of oxidation phosphorylation. *Ann. Rev. Biochem.* **49 :** *1079–1114, 1980.*
15. **Fridovich I :** Superoxide dismutase. *Ann. Rev. Biochem.* **44 :** *147–159, 1975.*
16. **Green DE, MacLennan DH :** Structure and function of the mitochondrial cristael membranes. *Bioscience.* **19 :** *213–222, 1969.*
17. **Hall DO, Rao KK, Cammack R :** The iron-sulfur proteins : Structure, function and evaluation of a ubiquitous group of proteins. *Sci. Prog.* **62 :** *285–316, 1975.*
18. **Harold FM :** The Vital Force : A study of Bioenergetics. *Academic Press, Inc., NewYork. 1986.*
19. **Hinckle PC, McCarty RE :** How cells make ATP. *Sci. Amer.* **238(3) :** *104–123, 1978.*
20. **Kadenbach B:** Intrinsic and extrinsic uncoupling of phosphorylation. *Biochim Biophys Acta–Bioenergetics.* **1604 :** *77–94, 2003.*
21. **Kalckar HM (editor) :** Biological Phosphorylations : Development of Concepts. *Prentice-Hall, 1969.*
22. **Kalckar HM :** Fitfy years of biological research– From oxidative phosphorylation to energy requiring transport and regulation. *Ann. Rev. Biochem.* **60 :** *1–37. 1991.*
23. **Keilin D :** The History of Cell Respiration and Cytochromes. *Cambridge Univ. Press, 1966.*
24. **Klingenberg M :** Structure-function of the ADP/ATP carrier. *Biochem. Soc. Trans.* **20 :** *547–550, 1992.*
25. **Krebs HA, Kornberg HL :** Energy Transformation in Living Matter, A Survey. *Springer, Berlin. 1957.*
26. **Lehninger AL :** How cells transform energy. *Sci. Amer.* **205 (3) :** *62, 1961.*
27. **Lehninger AL :** The Mitochondrion : Molecular Basis of Structure and Function. *Benjamin, New York. 1965.*
28. **Lehninger AL, Wadkins CL :** Oxidative phosphorylation. *Ann. Rev. Biochem.* **31 :** *47–78, 1962.*
29. **Lemberg R, Barrett J :** Cytochromes. *Academic Press Inc., New York. 1973.*
30. **Lovenberg W (editor) :** Iron-sulfur Proteins. vol 3. *Academic Press Inc., New York. 1977.*
31. **Malmström BG :** Enzymology of oxygen. *Ann. Rev. Biochem.* **51 :** *21, 1982.*
32. **Martonösi AN :** Membranes and Transport. *Plenum, New York. 1982.*
33. **Massey V, Veeger C :** Biological oxidations. *Ann. Rev. Biochem.* **32 :** *579, 1963.*
34. **Mitchell P :** Chemiosmotic Coupling and Energy Transduction. *Glynn Research, Bodmin, England. 1968.*
35. **Mitchell P :** Keilin's respiratory chain concept and its chemiosmotic consequences. *Science.* **206 :** *1148–1159, 1979.*
36. **Munn EA :** The Structure of Mitochondria. *Academic Press Inc., New York. 1974.*
37. **Nicholls DG:** An Introduction to the Chemiosmotic Hypothesis. *2nd ed., Academic Press, San Diego, C.A. 1981.*
38. **Nicholls DG, Ferguson SJ :** Bioenergetics 2. *Academic Press, Inc., New York. 1992.*
39. **Parsons DS :** Biological Membranes. *Oxford, London. 1975.*
40. **Pederson PL, Amzel LM :** ATP synthases. Structure, reaction centre, mechanism, and regulation of one of nature's most unique machines. *J. Biol. Chem.* **268 :** *9937–9940, 1993.*
41. **Racker E :** Mechanisms in Bioenergetics. *Academic Press Inc., New York. 1965.*
42. **Racker E :** The two faces of the inner mitochondrial membrane. *Essays in Biochemistry.* **6 :** *1–22, 1970.*

43. **Racker E :** A New Look at Mechanisms in Bioenergetics. *Academic Press Inc., New York. 1976.*
44. **Racker E :** From Pascher to Mitchell. A hundred years of Bioenergetics. *Fed. Pro. 39 210–215, 1980.*
45. **Salemme SR :** Structure and function of cytochromes *c*. *Ann. Rev. Biochem.* **46 :** *299–329, 1977.*
46. **Senior AE :** ATP synthesis by oxidative phosphorylation. *Physiol. Rev.* **68 :** *177–231, 1988.*
47. **Siekevitz P :** Powerhouse of the cell. *Sci. Amer.* **197(1) :** *131, 1957.*
48. **Singer TP (editor) :** Flavins and Flavoproteins. *Elsevier, Amsterdam. 1976.*
49. **Singer TP (editor) :** Biological Oxidations. *Interscience, 1968.*
50. **Skulachev VP :** Energy transformations in the respiratory chain. *Curr. Top. Bioenerg.* **4 :** *127–190, 1971.*
51. **Skulachev VP :** The laws of cell energetics. *Eur. J. Biochem.* **208 :** *203–209, 1992.*
52. **Spiro TG :** Iron-Sulfur Proteins. *Wiley-Interscience. 1982.*
53. **Stayman CL (editor) :** Electronic ion pumps. *Curr. Top. Membr. Transp. vol. 16, 1982.*
54. **Sund H (editor) :** Pyridine Nucleotide Dependent Dehydrogenases. *Springer-Verlag, New York. 1970.*
55. **Sweeney WV, Rabinowitz JC :** Proteins containing 4Fe-4S clusters. *An overview. Ann. Rev. Biochem.* **49 :** *131–162, 1980.*
56. **Tedeschi H :** Mitochondria : Structure, Biogenesis and Transducing Functions. *Springer-Verlag, New York. 1976*
57. **Thomas PJ, Bianchet M, Garboczi DN, Hullihen J, Amzel LM, Pederson PL :** ATP synthase : Structure-function relationships. *Biochem. Biophys. Acta.* **1101 :** *228–231, 1992.*
58. **Tzagoloff A :** Mitochondria. Plenum, *New York. 1982.*
59. **Von Jagow G, Sebald W :** *b*-type cytochromes. *Ann. Rev. Biochem.* **49 :** *281–314, 1980.*
60. **Wainio WW :** The Mammalian Mitochondrial Respiratory Chain. *Academic Press Inc., New York. 1970.*
61. **Whittaker DA, Danks SM :** Mitochondria ; Structure, Function and Assembly. *Longman, London. 1978.*
62. **Wikström M, Krebs K, Suraste M :** Proton-translocating cytochrome complexes. *Rev. Biochem.* **50 :** *623–655, 1981.*

PROBLEMS

1. What is the yield of ATP when each of the following substrates is completely oxidized to CO_2 by a mammalian cell homogenate? Assume that glycolysis, the citric acid cycle, and oxidative phosphorylation are fully active.
   (*a*) Pyruvate
   (*b*) Lactate
   (*c*) Fructose 1,6-bisphosphate
   (*d*) Phosphoenolpyruvate
   (*e*) Galactose
   (*f*) Dihydroxyacetone phosphate
2. The standard oxidation–reduction potential for the reduction of O_2 to H_2O is given as 0.82 V in Table 23–2. However, the value given in textbooks of chemistry is 1.23 V. Account for this difference.
3. What is the effect of each of the following inhibitors on electron transport and ATP formation by the respiratory chain?

(*a*) Azide (*d*) DNP
(*b*) Atractyloside (*e*) Carbon monoxide
(*c*) Rotenone (*f*) Antimycin A

4. The number of molecules of inorganic phosphate incorporated into organic form per atom of oxygen consumed, termed the P : O ratio, was frequently used as an index of oxidative phosphorylation.

(*a*) What is the relation of the P : O ratio to the ratio of the number of protons translocated per electron pair ($H^+/2e^-$) and the ratio of the number of protons needed to synthesize ATP and transport it to the cytosol (P/H^+)?

(*b*) What are the P : O ratios for electrons donated by matrix NADH and by succinate?

5. The immediate administration of nitrite is a highly effective treatment for cyanide poisoning. What is the basis for the action of this antidote? (**Hint :** Nitrite oxidizes ferrohemoglobin to ferrihemoglobin.)

6. Suppose that the mitochondria of a patient oxidizes NADH irrespective of whether ADP is present. The P : O ratio for oxidative phosphorylation by these mitochondria is less than normal. Predict the likely symptoms of this disorder.

7. Years ago, it was suggested that uncouplers would make wonderful diet drugs. Explain why this idea was proposed and why it was rejected. Why might the producers of antiperspirants be supportive of the idea?

8. You are asked to determine whether a chemical is an electron-transport-chain inhibitor or an inhibitor of ATP synthase. Design an experiment to determine this.

9. Years ago there was interest in using uncouplers such as dinitrophenol as weight control agents. Presumably, fat could be oxidized without concomitant ATP synthesis for re-formation of fat or carbohydrate. Why was this a bad idea?

10. The NADH dehydrogenase complex of the mitochondrial respiratory chain promotes the following series of oxidation–reduction reactions, in which Fe^{3+} and Fe^{2+} represent the iron in iron–sulfur centres, UQ is ubiquinone, UQH_2 is ubiquinol, and E is the enzyme:

(1) $NADH + H^+ + E\text{–}FMN \longrightarrow NAD^+ + E\text{–}FMNH_2$

(2) $E\text{–}FMNH_2 + 2Fe^{3+} \longrightarrow E\text{–}FMN + 2Fe^{2+} + 2H^+$

(3) $2Fe^{2+} + 2H^+ + UQ \longrightarrow 2Fe^{3+} + UQH_2$

Sum: $NADH + H^+ + UQ \longrightarrow NAD^+ + UQH_2$

For each of the three reactions catalyzed by the NADH dehydrogenase complex, identify (*a*) the electron donor, (*b*) the electron acceptor, (*c*) the conjugate redox pair, (*d*) the reducing agent, and (*e*) the oxidizing agent.

11. The standard reduction potential of any redox couple is defined for the half-cell reaction (or half-reaction):

$$\text{Oxidizing agent} + n \text{ electrons} \longrightarrow \text{reducing agent}$$

The standard reduction potentials of the NAD^+/NADH and pyruvate/lactate redox pairs are – 0.320 and – 0.185 V, respectively.

(*a*) Which redox pair has the greater tendency to lose electrons? Explain.

(*b*) Which is the stronger oxidizing agent? Explain.

(*c*) Beginning with 1 M concentrations of each reactant and product at pH 7, in which direction will the following reaction proceed?

$$\text{Pyruvate} + NADH + H^+ \rightleftharpoons \text{lactate} + NAD^+$$

(*d*) What is the standard free-energy change, $\Delta G^{\circ\prime}$, at 25 °C for this reaction?

(*e*) What is the equilibrium constant for this reaction at 25 °C?

12. Electron transfer functions to translocate protons from the mitochondrial matrix to the external medium to establish a pH gradient across the inner membrane, the outside more acidic than the inside. The tendency of protons to diffuse from the outside into the matrix, where $[H^+]$ is lower, is the driving force for ATP synthesis *via* the ATP synthase. During oxidative phosphorylation by a suspension of mitochondria in a medium of pH 7.4, the internal pH of the matrix has been measured as 7.7.

(*a*) Calculate $[H^+]$ in the external medium and in the matrix under these conditions.

(*b*) What is the outside : inside ratio of $[H^+]$? Comment on the energy inherent in this concentration.

(*c*) Calculate the number of protons in a respiring liver mitochondrion, assuming its inner matrix compartment is a sphere of diameter 15 μm.

(*d*) From these data would you think the pH gradient alone is sufficiently great to generate ATP?

(*e*) If not, can you suggest how the necessary energy for synthesis of ATP arises?

13. ATP production in the flight muscles of the fly *Lucilia sericata* results almost exclusively from oxidative phosphorylation. During flight, 187 ml of O_2/h • g of fly body weight is needed to maintain an ATP concentration of 7 μmol/g of flight muscle. Assuming that the flight muscles represent 20% of the weight of the fly, calculate the rate at which the flight-muscle ATP pool turns over. How long would the reservoir of ATP last in the absence of oxidative phosphorylation ? Assume that reducing equivalents are transferred by the glycerol-3-phosphate shuttle and that O_2 is at 25 °C and 101.3 kPa (1 atm). (**Note:** Concentrations are expressed in micromoles per gram of flight muscle.)

14. Iron-containing compounds that act as hydrogen acceptors in the respiratory chain are:

(*a*) flavoproteins

(*b*) dehydrogenases

(*c*) cytochromes

(*d*) oxidases

ANSWERS TO PROBLEMS

1. (*a*) 12.5; (*b*) 14; (*c*) 32; (*d*) 13.5; (*e*) 30; (*f*) 16.

2. Biochemists use E_0', the value at pH 7, whereas chemists use E_0, the value in 1 M H^+. The prime denotes that pH 7 is the standard state.

3. (*a*) Blocks electron transport and proton pumping at Complex III.

(*b*) Blocks electron transport and ATP synthesis by inhibiting the exchange of ATP and ADP across the inner mitochondrial membrane.

(*c*) Blocks electron transport and proton pumping at Complex I.

(*d*) Blocks ATP synthesis without inhibiting electron transport by dissipating the proton gradient.

(*e*) Blocks electron transport and proton pumping at Complex III.

(*f*) Blocks electron transport and proton pumping at Complex II.

4. (*a*) The P:O ratio is equal to the product of $(H^+/2e^-)$ and $(\sim P/H^+)$. Note that the P:O ratio is identical with the $(P: 2e^-)$ ratio,

(*b*) 2.5 and 1.5, respectively.

5. Cyanide can be lethal because it binds to the ferric form of cytochrome oxidase and thereby inhibits oxidative phosphorylation. Nitrite converts ferrohemoglobin into ferrihemoglobin, which also binds cyanide. Thus, ferrihemoglobin competes with cytochrome oxidase for cyanide. This competition is therapeutically effective because the amount of ferrihemoglobin that can be formed without impairing oxygen transport is much greater than the amount of cytochrome oxidase.

6. Such a defect (called **Luft syndrome**) was found in a 38-year-old woman who was incapable of performaing prolonged physical work. Her basal metabolic rate was more than twice normal, but her thyroid function was normal. A muscle biopsy showed that her mitochondria were highly variable and atypical in structure. Biochemical studies then revealed that oxidation and phosphorylation were not tightly coupled in these mitochondria. In this patient, much of the energy of fuel molecules was converted into heat rather than ATP.

7. If oxidative phosphorylation were uncoupled, no ATP could be produced. In a futile attempt to generate ATP, much fuel would be consumed. The danger lies in the dose. Too much uncoupling would lead to tissue damage in highly aerobic organs such as the brain and heart, which would have severe consequences for the organism as a whole. The energy that is normally transformed into ATP would be released as heat. To maintain body temperature, sweating might increase, although the very process of sweating itself depends on ATP.

8. Add the inhibitor with and without an uncoupler, and monitor the rate of O_2 consumption. If the O_2 consumption increases again in the presence of inhibitor and uncoupler, the inhibitor must be inhibiting ATP synthase. If the uncoupler has no effect on the inhibition, the inhibitor is inhibiting the electron-transport chain.

9. Because the energy not used for ATP was dissipated as heat, and the subjects developed uncontrollable fevers.

10. *Reaction (1)* : **(*a*), (*d*)** NADH; **(*b*)** E-FMN; **(*c*)** NAD^+/ NADH and E-FMN/$FMNH_2$
Reaction (2) : **(*a*), (*d*)** E–$FMNH_2$; **(*b*), (*e*)** Fe^{3+}; **(*c*)** E-FMN/$FMNH_2$ and Fe^{3+}/Fe^{2+}
Reaction (3) : **(*a*), (*d*)** Fe^{2+}; **(*b*), (*e*)** UQ; **(*c*)** Fe^{3+}/Fe^{2+} and UQ/UQH_2.

11. **(*a*)** NAD^+/NADH **(*b*)** Pyruvate/lactate **(*c*)** Lactate formation **(*d*)** −25 kJ/mol **(*e*)** 2.58×10^4

12. **(*a*)** External medium: 4.0×10^{-8} M; Matrix 2.0×10^{-8} M
(*b*) 2 : 1 **(*c*)** 21 **(*d*)** No **(*e*)** From the transmembrane potential.

13. About 70 μ mol/s • g. With a steady state [ATP] of 7 μmol/g, this is equivalent to 10 turnovers of the ATP pool per second; the reservoir would last about 0.1 s.

14. (*c*) cytochromes

CONTENTS

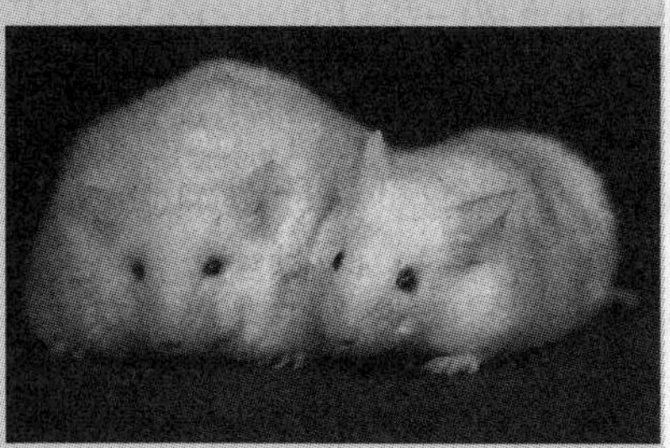

Fats provide an efficient means for storing energy for later use.
The processes of fatty acid synthesis (preparation for energy storage) and fatty acid degradation (preparation for energy use) are, in many ways, the reverse of each other. Studies of mice are revealing the interplay between these pathways and the biochemical bases of appetite and weight control.
[*Courtesy* : *Jackson/Visuals Unlimited*]

CHAPTER 24

Oxidation of Fatty Acids

INTRODUCTION

The lipids of metabolic significance in the mammalian organisms include triacylglycerols (= triglycerides, neutral fats), phospholipids and steroids, together with products of their metabolism such as long-chain fatty acids, glycerol and ketone bodies. An overview of their metabolic interrelationships and their relationship to carbohydrate metabolism is depicted in Fig. 24–1.

At least 10 to 20% of the body weight of a normal animal is due to the presence of lipids, a major part of which is in the form of triglycerides which are uncharged esters of glycerol. Body lipids are distributed in varying amounts in all organs and stored in highly specialized connective tissues called *depot*. In these depots, a large part of the cytoplasm of the cell is replaced by droplets of lipids. Body lipids serve as an important source of chemical potential energy.

Fats (or triacylglycerols) are highly concentrated stores of metabolic energy. They are the best heat producers of the three chief classes of foodstuffs. Carbohydrates and proteins each yield 4.1 kilocalories (4.1 kcal) of heat for every gram oxidized in the body; whereas fats yield 9.3 kcal more than twice as much. The basis of this large difference in caloric yield is that fats contain relatively

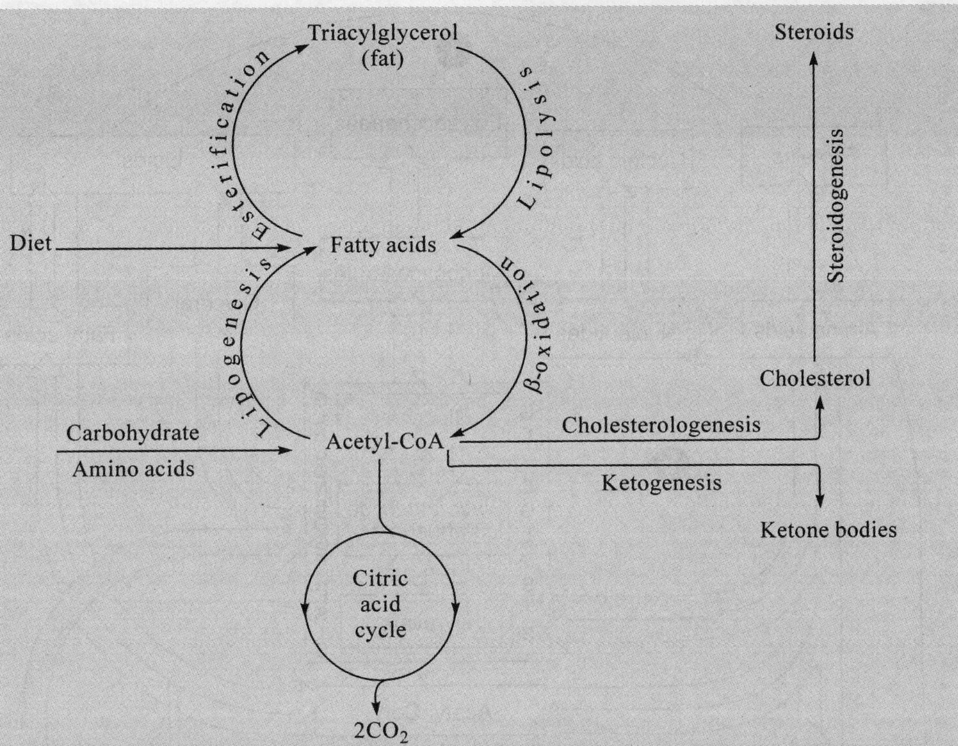

Fig. 24–1. An overview of the principal pathways of lipid metabolism

(Adapted from Harper HA, Rodwell VW and Mayes PA, 1977)

more carbon and hydrogen in relation to oxygen as compared to proteins or carbohydrates. In other words, fats are compounds that are less completely oxidized to begin with and therefore can be oxidized further and yield more energy. Furthermore, triacylglycerols are very nonpolar and so they are stored in a nearly anhydrous form, whereas carbohydrates and proteins are much more polar and more highly hydrated. In fact, a gram of dry glycogen (a carbohydrate) binds about 2 grams of water. Consequently, a gram of nearly anhydrous fat stores more than 6 times as much energy as a gram of hydrated glycogen, which is the reason that triacylglycerols, rather than glycogen, were selected in evolution as the major energy reservoir. A normal man, weighing 70 kg, possesses fuel reserves of 10,000 kcal in triacylglycerols, 25,000 kcal in proteins (mostly in muscles), 600 kcal in glycogen and 40 kcal in glucose. As mentioned, triacylglycerols constitute about 11 kg of his total body weight. If this amount of energy were stored in glycogen, his total body weight would be 55 kg greater.

An overview of the intermediary metabolism with special emphasis on fatty acids and triglycerides is given in Fig. 24–2.

The normal animal contains a greater quantity of easily mobilized lipids than of carbohydrates or proteins. About 100 times more energy is stored as mobilizable lipids than as mobilizable carbohydrate in the normal human being. In times of caloric insufficiencies, an animal can meet the endogenous requirements necessary for the maintenance of life by drawing on its lipid depots. In addition, neutral lipids serve as insulators of delicate internal organs of the body. This function is best exemplified in marine animals, whose water environment is both colder than body temperature and a far better thermal conductor than air. Lipids also serve as shock absorbers in protecting joints, nerves and other organs against mechanical trauma.

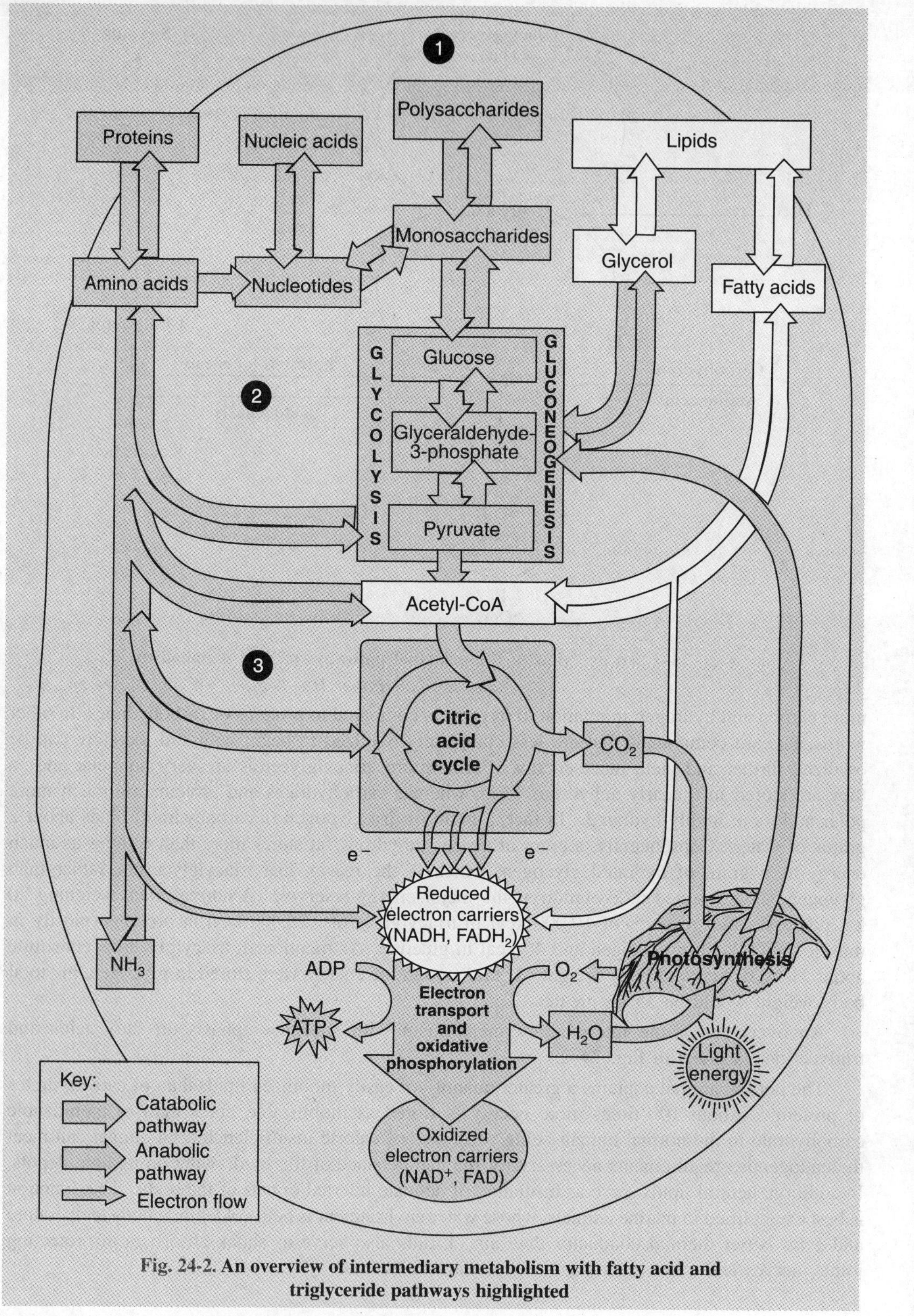

Fig. 24-2. An overview of intermediary metabolism with fatty acid and triglyceride pathways highlighted

When food intake exceeds caloric utilization, the excess energy is invariably stored as fat for the body cannot store any other form of food in such large amounts. The capacity of the animal to store carbohydrates (such as glycogen) is strictly limited, and there is no provision for the storage of excess proteins. Moreover, in an adult organism in which active growth has ceased, nitrogen output is more or less geared to nitrogen intake, and the organism shows no tendency to store surplus proteins from the diet. Plants differ from animals in that the energy reserves needed for reproduction are stored in the form of carbohydrates (as in corn or wheat) or as a combination of reserve proteins and oils (as in oil seeds, flax seed, safflower seed or sunflower seed).

In mammals, the major site of accumulation of triacylglycerols is the cytoplasm of adipose cells (= fat cells). Droplets of triacylglycerol coalesce to form a large globule, which may occupy most of the cell volume. Adipose cells are specialized for the synthesis and storage of triacylglycerols and for their mobilization into fuel molecules that are transported to other tissues by the blood. More than 99% of the lipid of human adipose tissue is triacylglycerol, regardless of anatomical location. In general, depot lipid is richer in saturated fatty acids than liver lipid. The more nearly saturated a sample of lipid, the higher the energy yield available from oxidation.

Fig. 24–3 presents, schematically, the flow of lipids in the body. Three important compartments are the liver, blood and adipose tissue. Both liver and adipose tissue are the principal sites of metabolic activity while the blood serves as a transport system. Other compartments, such as cardiac and skeletal muscle, are important utilizers of fatty acids and ketone bodies.

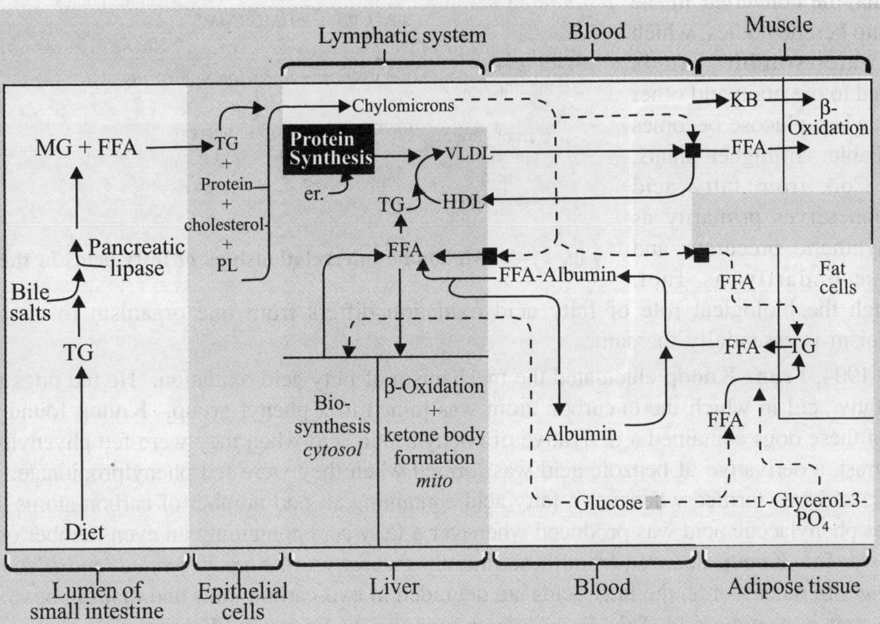

Fig. 24–3. Scheme depicting role of compartments in the utilization of lipids in the animals

TG = triacylglycerol, MG = monoacylglycerol, FFA = free fatty acids, PL = phospholipids, KB = ketone bodies, VLDL = very low-density lipoprotein, HDL = high density lipoprotein, ■ = lipoprotein lipase

(Adapted from Conn EE, Stumpf PK and Doi RH, 1997)

A diagrammatic representation of the metabolic interrelationships of fatty acids is presented in Fig. 24–4.

OXIDATION OF FATTY ACIDS

General Considerations

The importance of oxidation of fatty acids is not limited to the obese or to devotees of greasy foods ; it is a critical part of the metabolic economy in the lean as well as the lardy. The oxidation of long-chain fatty acids to acetyl-CoA is a central energy-yielding pathway in animals, many protists and some bacteria. Complete combustion or oxidaton of a typical fatty acid, palmitic acid, yields 2,380 kcal per mole.

CH_3—$(CH_2)_{14}$—COOH + 23 O_2 →→→→ 16 CO_2 + 16 H_2O + 2380 kcal/mole

In some organisms, acetyl-CoA produced by fatty acid oxidation has alternative fates. In vertebrate animals, acetyl-CoA may be converted in the liver into ketone bodies, which are water-soluble fuels exported to the brain and other tissues when glucose becomes unavailable. In higher plants, acetyl-CoA from fatty acid oxidation serves primarily as a biosynthetic precursor and only secondarily as fuel.

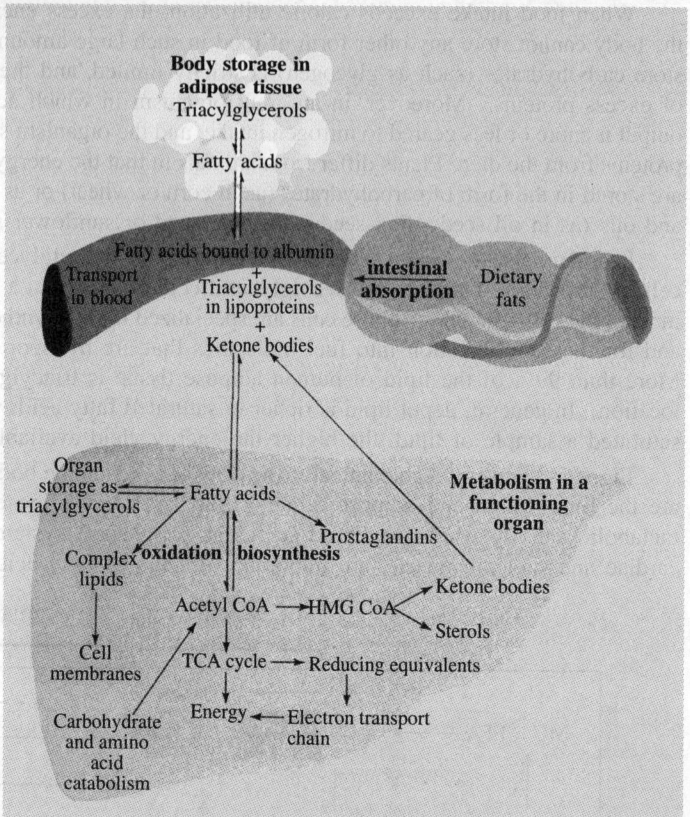

Fig. 24–4. Metabolic interrelationships of fatty acids in the human

Although the biological role of fatty acid oxidation differs from one organism to another, the mechanism is essentially the same.

In 1904, **Franz Knoop** elucidated the mechanism of fatty acid oxidation. He fed dogs straight-chain fatty acid in which the ω-carbon atom was joined to a phenyl group. **Knoop** found that the urine of these dogs contained a derivative of phenylacetic acid when they were fed phyenylbutyrate. In contrast, a derivative of benzoic acid was formed when they were fed phenylpropionate. In fact, benzoic acid was formed whenever a fatty acid containing an odd number of carbon atoms was fed, whereas phenylacetic acid was produced whenever a fatty acid containing an even number of carbon atoms was fed. **Knoop** deduced from these findings that *fatty acids are degraded by oxidation at the β-carbon.* In other words, the fatty acids are degraded in two-carbon units and that the obvious two-carbon unit is acetic acid. This finding later came to be known as **Knoop's**

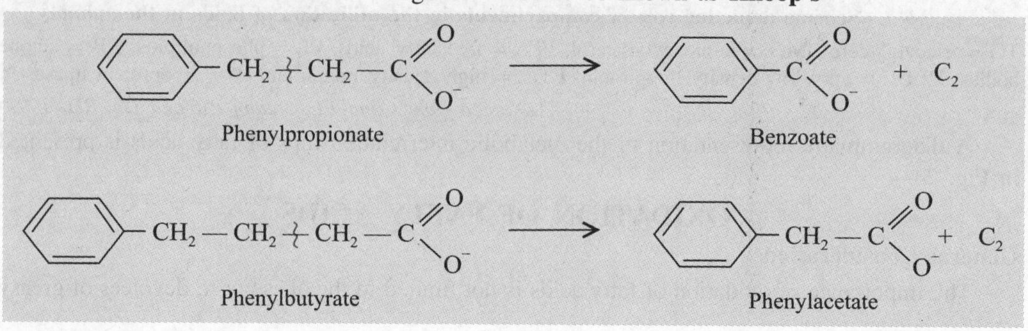

hypothesis. These experiments are a landmark in Biochemistry because they were the first to use a synthetic level to elucidate synthetic mechanisms. Deuterium and radioisotopes came into Biochemistry several decades later.

The complete combustion of fatty acids to CO_2 and H_2O occurs in the mitochondria, where the transfer of electrons from the fatty acids to oxygen can be used to generate ATP. The combustion occurs in 2 stages:

(*a*) the fatty acid is sequentially oxidized so as to convert all of its carbons to acetyl-coenzyme A, and

(*b*) the acetyl-coenzyme A is oxidized by the reactions of the citric acid cycle.

Both stages generate ATP by oxidative phosphorylation.

Activation of a Fatty Acid

Because of their hydrophobicity and extreme insolubility in water, triacylglycerols are segregated into lipid droplets, which do not raise osmolarity of the cytosol and, unlike polysaccharides, do not contain extra weight as water of solvation. The relative chemical inertness of triacylglycerols allows their extracellular storage in large quantities without the risk of undesired chemical reactions with other cellular constituents.

But the same properties that make triacylglycerols good storage compounds present problems in their role as fuels. Because of their insolubility in water, the ingested triacylglycerols must be emulsified before they can be digested by water-soluble enzymes in the intestine, and triacylglycerols absorbed in the intestine must be carried in the blood by proteins that counteract their insolubility. The relative stability of the C—C bonds in a fatty acid is overcome by activation of the carboxyl group at C-1 by attachment to coenzyme A, which allows stepwise oxidation of the fatty acyl group at the C-3 position. This later carbon atom is also called the *beta* (β) carbon in common nomenclature, from which the oxidation of fatty acids gets its common name: β **oxidation.**

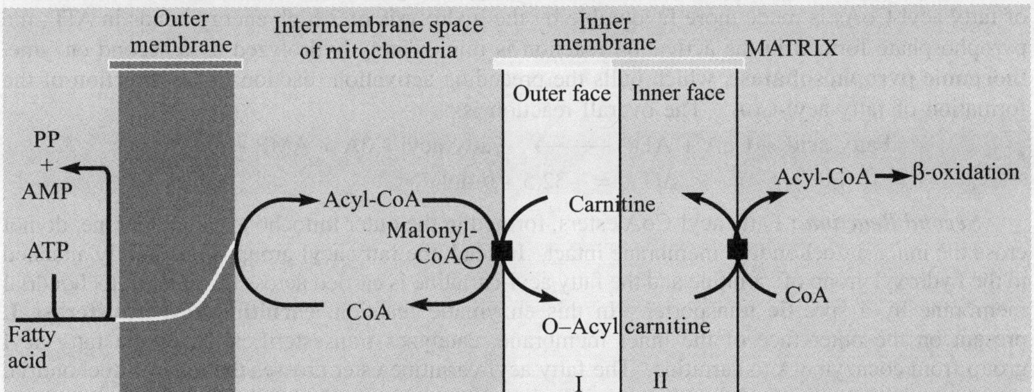

Fig. 24–5. Transport mechanism of fatty acids from cytosol to the β-oxidation site in the mitochondrion

■, Carnitine : acyl-CoA transferase I (outer face) and carnitine : acyl-CoA transferase II (inner face), the two distinct enzymes that catalyze the same reaction; malonyl-CoA ⊖ indicates inhibition of transferase I.

(*Redrawn from Conn EE, Stumpf PK and Doi RH, 1997*)

An unusual property of liver and other tissue mitochondria is their inability to oxidize fatty acids or fatty acyl-CoA's unless (–)- carnitine (3-hydroxy-4-trimethyl ammonium butyrate) is added in catalytic amounts. Evidently, free fatty acids or fatty acyl-CoA's cannot pentrate the inner membranes of liver and other tissue mitochondria, whereas acyl carnitine readily passes through the membrane and is then converted to acetyl-CoA in the matrix. Fig. 24–5 outlines the translocation of acetyl-CoA from outside the mitochondrion to the internal site of the β oxidation system. The key enzyme is *carnitine acetyl-CoA transferase.*

Reactions of Fatty Acid Oxidation

The free fatty acids that enter the cytosol from the blood cannot pass directly through the mitochondrial membranes, but must first undergo a series of 3 enzymatic reactions. These are described as under:

First Reaction : It is catalyzed by a series of family of isozymes present in the outer mitochondrial membrane, **acyl-CoA synthetases** (also called **fatty acid thiokinases**), which promote the general reaction.

$$\text{Fatty acid} + \text{CoA} + \text{ATP} \rightleftharpoons \text{Fatty acyl-CoA} + \text{AMP} + \text{PP}i$$
$$\Delta G = \sim \text{o kcal/mole}$$

Three different acyl-CoA synthetases occur in the cell and act on fatty acids of short, intermediate and long carbon chains, respectively. One type of synthetase activates acetate and propionate to corresponding thioesters, another activates medium chain fatty acids from C_4 to C_{11}, and the third activates fatty acids from C_{10} to C_{20}. Acetyl-CoA synthetase catalyzes the formation of thioester linkage between the fatty acid carboxyl group and the thiol (—SH) group of coenzyme A to yield a fatty acyl-CoA ; simultaneously, ATP undergoes cleavage to AMP and PP*i*. The reaction actually takes place in 2 steps:

$$\text{Fatty acid} + \text{ATP} + \text{Enzyme} \rightleftharpoons \text{Enzyme-acyladenylate} + \text{PP}i$$

$$\text{Enzyme-acyladenylate} + \text{CoA-SH} \rightleftharpoons \text{Enzyme} + \text{Fattyl-acyl-S-CoA} + \text{AMP}$$

Fatty acyl-CoAs, like acetyl-CoA, are high-energy compounds ; their hydrolysis to free fatty acid and CoA has a large negative standard free-energy change ($\Delta G°' \approx -31$ kJ/mol). The formation of fatty acyl-CoAs is made more favourable by the hydrolysis of 2 high energy bonds in ATP; the pyrophosphate formed in the activation reaction is immediately hydrolyzed by a second enzyme, **inorganic pyrophosphatase**, which pulls the preceding activation reaction in the direction of the formation of fatty acyl-CoA. The overall reaction is:

$$\text{Fatty acid} + \text{CoA} + \text{ATP} \longrightarrow \text{Fatty acyl-CoA} + \text{AMP} + 2\ P_i$$
$$\Delta G°' = -32.5 \text{ kJ/mol}$$

Second Reaction : Fatty acyl-CoA esters, formed in the outer mitochondrial membrane, do not cross the inner mitochondrial membrane intact. Instead, the fatty acyl group is transiently attached to the hydroxyl group of carnitine and the fatty acyl-carnitine is carried across the inner mitochondrial membrane by a specific transporter. In this enzymatic reaction, **carnitine acyl-transferase I**, present on the outer face of the inner membrane, catalyzes transesterification of the fatty acyl group from coenzyme A to carnitine. The fatty acyl-carnitine ester crosses the inner mitochondrial membrane into the matrix by facilitated diffusion through the acyl-carnitine/carnitine transporter.

Third Reaction : In this final step of the entry process, the fatty acyl group is enzymatically transferred from carnitine to intramitochondrial coenzyme A by **carnitine acyltransferase II.** This isozyme is located on the inner face of the inner mitochondrial membrane, where it regenerates fatty acyl-CoA and releases it, along with free carnitine, into the matrix. Carnitine reenters the space between the inner and outer mitochondrial membranes *via* the acyl-carnitine/carnitine transporter. Once inside the mitochondrion, the fatty acyl-CoA is ready for the oxidation of its fatty acid component by a set of enzymes in the mitochondrial matrix.

OXIDATION OF EVEN-CHAIN SATURATED FATTY ACIDS (= KNOOP'S β OXIDATION PATHWAY)

Mitochondrial oxidation of fatty acids takes place in 3 stages (Fig. 24–6):

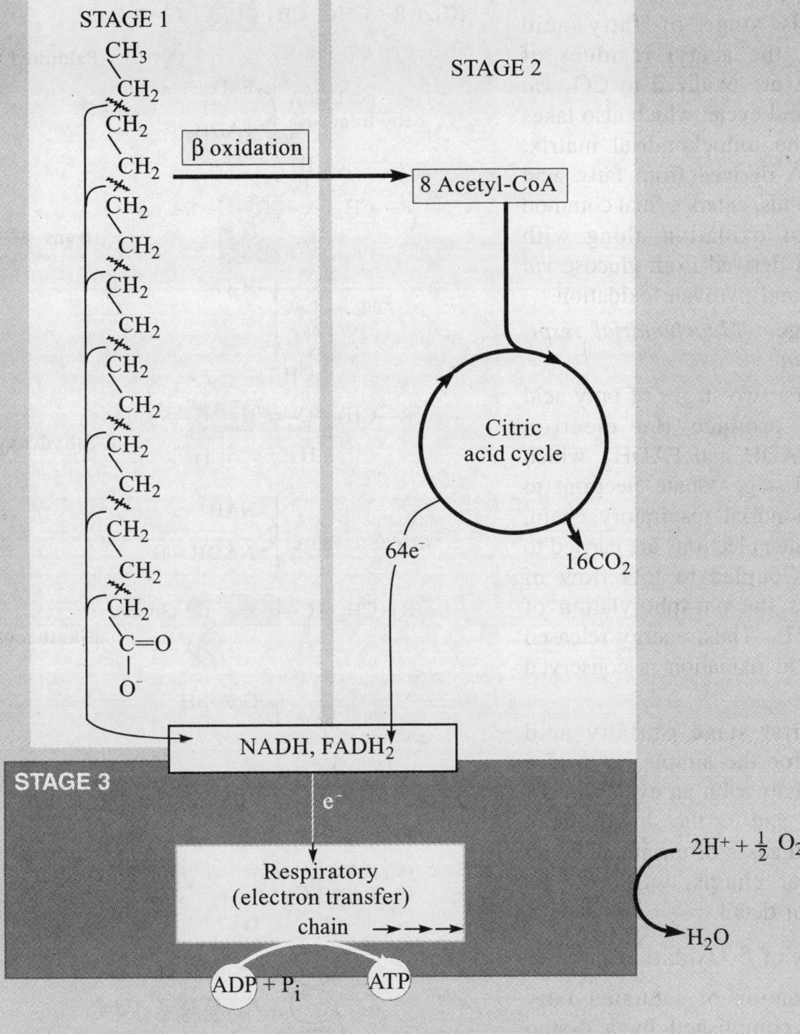

Fig. 24–6. Three stages of fatty acid oxidation

Stage 1 : A long-chain fatty acid is oxidized to yield acetyl residues in the form of acetyl-CoA.

Stage 2 : The acetyl residues are oxidized to CO_2 *via* the citric acid cycle.

Stage 3 : Electrons derived from the oxidations of Stages 1 and 2 are passed to O_2 *via* the mitochondrial respiratory chain, providing the energy for ATP synthesis by oxidative phosphorylation.

(Adapted from Lehninger AL, Nelson DL and Cox MM, 1993)

First Stage : β *oxidation pathway*

In this stage, the fatty acids undergo oxidative removal of successive two-carbon units in the form of acetyl-CoA, starting from the carboxyl end of the fatty acyl chain. For example, the C-16 fatty acid palmitic acid (palmitate at pH 7) undergoes 7 passes through this oxidative sequence, in each pass losing two carbons as acetyl-CoA. At the end of seven cycles, the last two carbons of palmitate (originally C-15 and C-16) are left as acetyl-CoA. The overall result is the conversion of 16-carbon chain of palmitate to 8 two-carbon acetyl-CoA molecules. Formation of each molecule of acetyl-CoA requires removal of 4 hydrogen atoms (two pairs of electrons and 4 H^+) from the fatty acyl moiety by the action of dehydrogenases.

Second Stage : *Citric acid cycle.*

In this stage of fatty acid oxidation, the acetyl residues of acetyl-CoA are oxidized to CO_2 *via* the citric acid cycle, which also takes place in the mitochondrial matrix. Acetyl-CoA derived from fatty acid oxidation, thus, enters a final common pathway of oxidation along with acetyl-CoA derived from glucose *via* glycolysis and pyruvate oxidation.

Third Stage : *Mitochondrial respiratory chain*

The first two stages of fatty acid oxidation produce the electron carriers, NADH and $FADH_2$, which in the third stage donate electrons to the mitochondrial respiratory chain, through which electrons are carried to oxygen. Coupled to this flow of electrons is the phosphorylation of ADP to ATP. Thus, energy released by fatty acid oxidation is conserved as ATP.

The first stage of fatty acid oxidation for the simple case of a saturated chain with an even number of carbons, and for the slightly more complicated cases of unsaturated and odd-number chains, will now be described in detail.

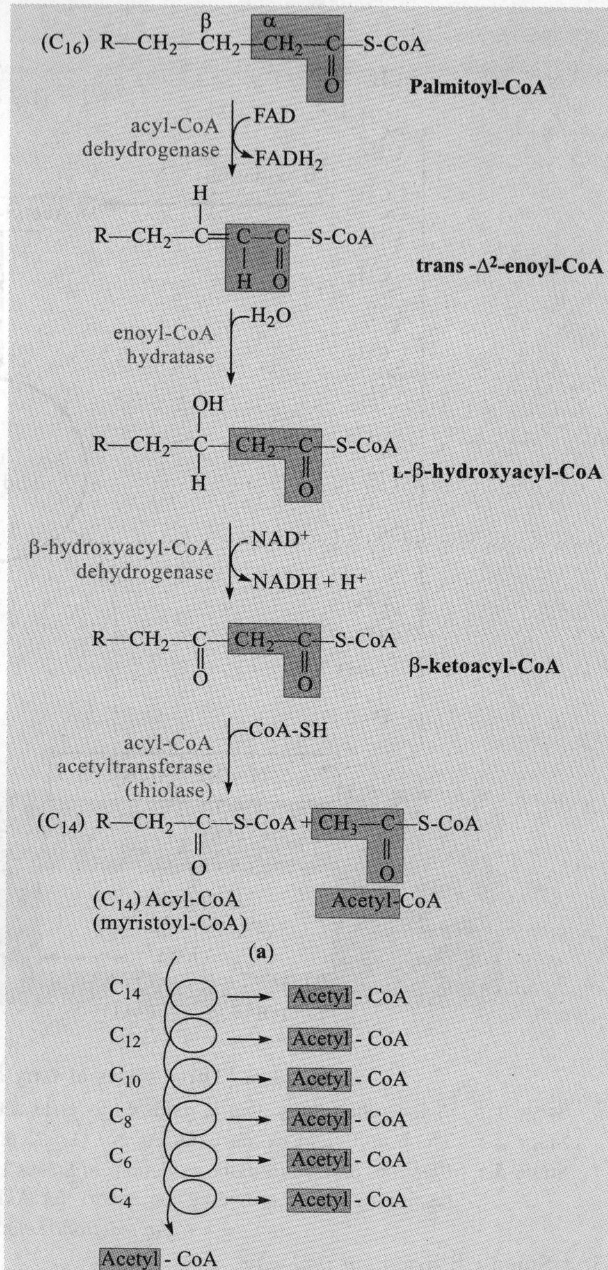

Fig. 24–7. The fatty acid oxidation pathway (= β oxidation cycle)

(a) In each pass through this 4-step sequence, one acetyl residue (shaded in pink) is removed in the form of acetyl-CoA from the carboxyl end of the fatty acyl chain– in this example, palmitate (C_{16}), which enters as palmitoyl-CoA.

(b) Six more passes through the pathway yield 7 more molecules of acetyl-CoA, the seventh arising from the last 2 carbon atoms of the 16-carbon chain. In all, 8 molecules of acetyl-CoA are formed.

Four Steps of β Oxidation

β Oxidation of saturated fatty acids is accomplished by a 4-step mechanism, illustrated in Fig. 24–7. The four steps of the **fatty acid spiral**, as it is also called, are described below:

First Step : α, β *dehydrogenation of acyl-CoA*

In this step, acyl-CoA is oxidized by an acyl-CoA dehydrogenase to produce an enoyl-CoA with *a trans* double bond between α and β carbon atoms (C-2 and C-3). It is thus, better written as *trans*-Δ^2-enoyl-CoA (Recall that naturally occurring unsaturated fatty acids normally have their double bonds in the *cis* configuration).

$$\text{R—C}^{\gamma}\text{H}_2\text{—C}^{\beta}\text{H}_2\text{—C}^{\alpha}\text{H}_2\text{—}\overset{\text{O}}{\overset{\|}{\text{C}}}\text{—S—CoA + FAD} \longrightarrow \text{R—CH}_2\text{—CH = CH—}\overset{\text{O}}{\overset{\|}{\text{C}}}\text{—S—CoA + FADH}_2$$

$$\Delta G' = -4.8 \text{ kcal/mole}$$

Three *acyl-CoA dehydrogenases* (E.C. No. 1.3.99.3) are found in the matrix of mitochondria. They all have FAD as a prosthetic group. The first has a specificity ranging from C_4 to C_6 acyl-CoAs, the second from C_6 to C_{14} and the third from C_6 to C_{18}. The $FADH_2$ is not directly oxidized by oxygen but traces the following path:

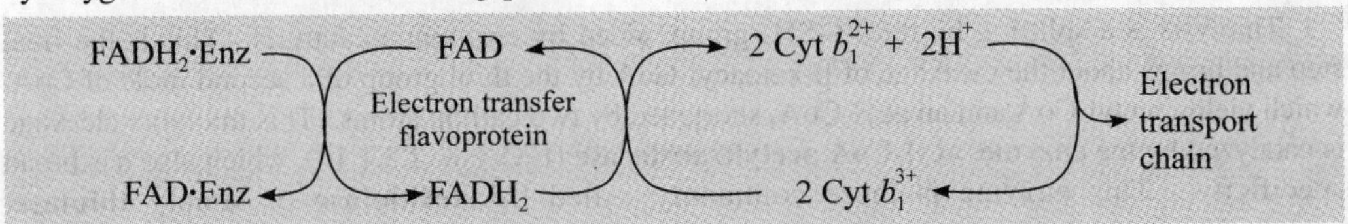

The oxidation catalyzed by acetyl-CoA dehydrogenase is analogous to succinate dehydrogenation in the citric acid cycle (see page 604), as in both the reactions:

(*a*) the enzyme is bound to the inner membrane,

(*b*) a double bond is introduced into a carboxylic acid between the α and β carbons,

(*c*) FAD is the electron acceptor, and

(*d*) electrons from the reaction ultimately enter the respiratory chain and are carried to O_2 with the concomitant synthesis of 2 ATP molecules per electron pair.

Second Step : *Hydration of* α, β*-unsaturated acyl-CoAs*

In this step, a mole of water is added to the double bond of the *trans*-Δ^2-enoyl-CoA to form the L-stereoisomer of β-hydroxyacyl-CoA (also called 3-hydroxyacyl-CoA). The reaction is catalyzed by **enoyl-CoA hydratase** or **crotonase** (E.C. No. 4.2.1.17), and has broad specificity with respect to the length of the acyl group. However, its activity decreases progressively with increasing chain length of the substrate (It may be noted that the enzyme will also hydrate α, β-*cis* unsaturated acyl-CoA, but in this case D (–)-β-hydroxyacyl-CoA is formed).

$$\underset{trans}{\text{R}-\text{CH}_2-\overset{\beta}{\text{C}}\text{H}=\overset{\alpha}{\text{C}}\text{H}-\overset{\text{O}}{\overset{\|}{\text{C}}}-\text{S}-\text{CoA}}+\text{H}_2\text{O} \rightleftharpoons \underset{\text{L (+) -}\beta\text{-hydroxyacyl-CoA}}{\text{R}-\text{CH}_2-\underset{\text{OH}}{\underset{|}{\text{CH}}}-\underset{\text{H}}{\underset{|}{\text{CH}}}-\overset{\text{O}}{\overset{\|}{\text{C}}}-\text{S}-\text{CoA}}$$

$$\Delta G' = -0.75 \text{ kcal/mole}$$

This reaction catalyzed by enoyl-CoA hydratase is formally analogous to the fumarase reaction in the citric acid cycle, in which water adds across an α-β double bond (see page 401). The hydration of enoyl-CoA is, in fact, the prelude to the second oxidation reaction, *i.e.*, Step 3.

Third Step : *Oxidation of* β*-hydroxyacyl-CoA*

In this step of fatty acid oxidation cycle, the L-β-hydroxyacyl-CoA is dehydrogenated (or oxidized) to form β-ketoacyl-CoA by the action of an enzyme, **β-hydroxyacyl-CoA dehydrogenase** (E.C. No. 1.1.1.35), which in absolutely specific for the L stereoisomer of the hydroxyacyl substrate. NAD^+ is the electron acceptor in this reaction and the NADH, thus formed, donates its electrons to NADH dehydrogenase (complex I), an electron carrier of the respiratory chain. Three ATP molecules are generated from ADP per pair of electrons passing from NADH to O_2 *via* the respiratory chain.

$$\text{R}-\text{CH}_2-\underset{\text{OH}}{\underset{|}{\text{CH}}}=\underset{\text{H}}{\underset{|}{\text{CH}}}-\overset{\text{O}}{\overset{\|}{\text{C}}}-\text{S}-\text{CoA}+\text{NAD}^+ \rightleftharpoons \underset{\beta\text{-ketoacyl-CoA}}{\text{R}-\text{CH}_2-\text{CO}-\text{CH}_2-\overset{\text{O}}{\overset{\|}{\text{C}}}-\text{S}-\text{CoA}}+\text{NADH}+\text{H}^+$$

$$\Delta G' = +3.75 \text{ kcal/mole}$$

This reaction, catalyzed by β-hydroxyacyl-CoA dehydrogenase, is closely analogous to the malate dehydrogenase reaction of the citric acid cycle (see page 401). Thus, we see that the first three reactions in each round of fatty acid oxidation closely resemble the last steps in the citric acid cycle:

Aceyl-CoA	⟶	Enoyl-CoA	⟶	Hydroxyacyl-CoA	⟶	Ketoacyl-CoA
Succinate	⟶	Fumarate	⟶	Malate	⟶	Oxaloacetate

The net result of the first three reactions is the oxidation of methylene group at β (or C-3) position to a keto group of the substrate, acyl-CoA.

Fourth Step : *Thiolysis or Thioclastic scission*

Thiolysis is a splitting by thiol (–SH) group, aided by enzymatic catalysis. This is the final step and brings about the cleavage of β-ketoacyl-CoA by the thiol group of a second mole of CoA, which yields acetyl-CoA and an acyl-CoA, shortened by two carbon atoms. This thiolytic cleavage is catalyzed by the enzyme, **acyl-CoA acetyltransferase** (E.C. No. 2.3.1.16), which also has broad specificity. This enzyme is more commonly called β-**ketothiolase** or simply **thiolase**.

$$\underset{\substack{\beta\text{-ketoacyl-CoA} \\ \text{(n carbons)}}}{R-CH_2-CO-CH_2-\overset{\overset{\displaystyle O}{\|}}{C}-S-CoA} + \underset{\text{Coenzyme A}}{CoA-SH} \xrightleftharpoons[\Delta G' = -6.65 \text{ kcal/mole}]{} \underset{\substack{\text{Acyl-CoA} \\ \text{(n-2 carbons)}}}{R-CH_2-CO-S-CoA} + \underset{\text{Acetyl-CoA}}{CH_3-\overset{\overset{\displaystyle O}{\|}}{C}-CoA}$$

Although the overall reaction is reversible, the equilibrium position is greatly in the direction of cleavage.

As to the mechanism of thiolase action, the enzyme protein has a reactive thiol (–SH) group on a cysteinyl residue that is involved in the following series of reactions:

$$\underset{\beta\text{-ketoacyl-CoA}}{R-CH_2-CO-CH_2-\overset{\overset{\displaystyle O}{\|}}{C}-S-CoA} + \underset{\text{Thiolase}}{Enz-SH} \rightleftharpoons \underset{\text{Acyl-S-Enz}}{R-CH_2-CO-S-Enz} + \underset{\text{Acetyl-CoA}}{CH_3-\overset{\overset{\displaystyle O}{\|}}{C}-S-CoA}$$

$$R-CH_2-CO-S-Enz + CoA-SH \rightleftharpoons \underset{\text{Acyl-CoA}}{R-CH_2-CO-S-CoA} + Enz-SH$$

In summary, the shortening of a fatty acyl-CoA derivative by two carbon atoms can be represented by the equation:

$$R—CH_2—CH_2—CH_2—CO—S—CoA + FAD + NAD + CoA—SH$$
$$\longrightarrow R—CH_2—CO—S—CoA + CH_3—CO—S—CoA + FADH_2 + NADH + H^+$$

The shortened acyl-CoA then undergoes another cycle of oxidation, starting with the reaction catalyzed by acyl-CoA dehydrogenase. Beta-ketothiolase, hydroxyacyl dehydrogenase and enoyl-CoA hydratase all have broad specificity with respect to the length of the acyl group. Thus, by repeated turns of the cycle, a fatty acid is degraded to acetyl-CoA molecules with one being produced every turn until the last cycle, wherein two are produced. The β-oxidation of fatty acids is presented in a cyclical manner in Fig 24–8.

The β oxidative system is found in all organisms. However, in bacteria grown in the absence of fatty acids, the β oxidative system is practically absent but is readily induced by the presence of fatty acids in the growth medium. *The bacterial* β *oxidation system is completely soluble and hence is not membrane-bound.* Curiously, in germinating seeds possessing a high lipid content, the β oxidation system is exclusively located in microbodies called glyoxysomes, but in seeds with low lipid content, the enzymes are associated with mitochondria.

Stoichiometry of β Oxidation

The energy yield derived from the oxidation of a fatty acid can be calculated. In each reaction

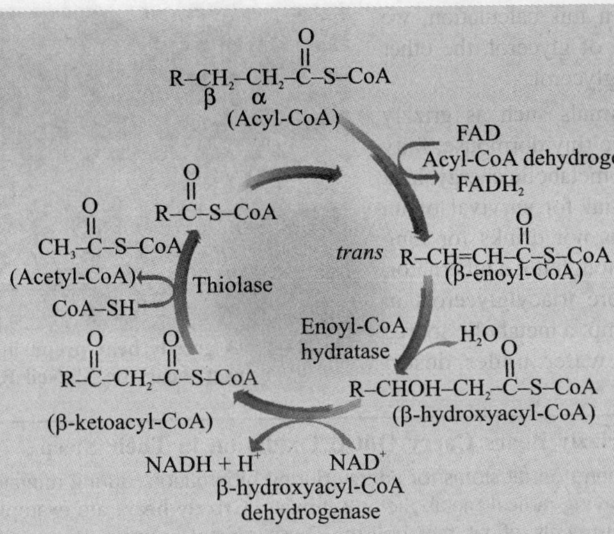

Fig. 24–8. The β-oxidation cycle for fatty acids

cycle, an acyl-CoA is shortened by two carbons and one mole each of $FADH_2$, NADH and acetyl-CoA are formed.

$$C_n\text{–acyl–CoA} + \text{FAD} + \text{NAD}^+ + \text{CoA} + H_2O \longrightarrow C_n\text{–2–acyl–CoA} + FADH_2 + \text{NADH} + \text{Acetyl-CoA} + H^+$$

The degradation of palmitoyl-CoA (C_{16}-acyl-CoA), for example, requires 7 reaction cycles. In the seventh cycle, the C_4-ketoacyl-CoA is thiolyzed to 2 moles of acetyl-CoA. Hence, the stoichiometry of oxidation of palmitoyl-CoA is:

$$\text{Palmitoyl-CoA} + 7\ \text{FAD} + 7\ \text{NAD}^+ + 7\ \text{CoA} + 7\ H_2O \longrightarrow 8\ \text{Acetyl-CoA} + 7\ FADH_2 + 7\ \text{NADH} + 7\ H^+$$

Three ATP are generated when each of these NADH is oxidized by the respiratory chain, whereas two ATP are formed for each $FADH_2$ because their electrons enter the chain at the level of coenzyme Q. Recall that the oxidation of one mole of acetyl-CoA by the citric acid cycle yields 12 ATP molecules. Hence, the number of ATP moles formed in the oxidation of palmitoyl-CoA is 14 from the 7 $FADH_2$, 21 from the 7 NADH, and 96 from the 8 moles of acetyl-CoA. This totals to 131. Two high-energy phosphate bonds are consumed in the activation of palmitate, in which ATP is split into AMP and 2 P*i*. Thus, *the net yield from the complete oxidation of a mole of palmitate is 131–2 = 129 ATP molecules.*

The efficiency of energy conservation in fatty acid oxidation can be estimated from the number of ATP formed and from the free energy of oxidation of palmitic acid to CO_2 and H_2O, as determined by calorimetry. The standard free energy of hydrolysis of 129 ATP is 129 × –7.3 kcal = –941.7 kcal or roughly – 942 kcal. The standard free energy of oxidation of palmitic acid is –2,340 kcal.

$$CH_3(CH_2)_7\text{—COOH} + 23O_2 \longrightarrow 16CO_2 + 16H_2O$$

$$\Delta G^{\circ\prime} = -\ 2{,}340\ \text{kcal/mole}$$

Hence, the efficiency of energy conservation of the *in vivo* oxidation of fatty acids, under standard conditions, is 942/2340 × 100 = 42.56%, a surprisingly high figure. This value is similar to those of glycolysis, the citric acid cycle and oxidative phosphorylation. In other words, the 129 ATP produced account for a conservation of 942 kcal of the 2,340 kcal released by the oxidation of one mole of palmitic acid, *i.e.*, roughly 42% efficiency of energy conservation. The remaining energy is lost probably as heat. It, hence, becomes clear why, as a food, fat is an effective source of available energy. In this calculation, we neglect the combustion of glycerol, the other component of a triacylglycerol.

In hibernating animals such as grizzly bear (Fig. 24–9) and the tiny dormouse, fatty acid oxidation provides metabolic energy, heat and water— all essential for survival of an animal that neither eats nor drinks for long periods. The camel, although not a hibernator, can synthesize and store triacylglycerols in large amounts in its hump, a metabolic source of both energy and water under desert conditions.

Fig. 24–9. A grizzly bear preparing its hibernation nest, near the McNeil River in Canada

Grizzly Bears Carry Out β Oxidation in Their Sleep

Many animals depend on fat stores for energy during hibernation, during migratory periods, and in other situations involving radical metabolic adjustments. **Grizzly bears** are examples of one of the most pronounced adjustments of fat metabolism. These animals remain in a continuous state of dormancy for periods as long as 6–7 months. Unlike most hibernating species, the bear maintains a body temperature of 32–35°C, close to the normal (nonhibernating) level. Although expending about 25,000 kJ/day (6,000 kcal/day), the bear does not eat, drink, urinate, or defecate for months at a time.

It has been experimentally shown that hibernating grizzly bears use body fat as their sole fuel. Fat oxidation yields sufficient energy for maintenance of body temperature, active synthesis of amino acids and proteins, and other energy-requiring activities, such as membrane transport. Fat oxidation also releases large amounts of water which replenishes water lost in breathing. The glycerol released by degradation of triacylglycerols is converted into blood glucose by gluconeogenesis. Urea formed during breakdown of amino acids is reabsorbed in the kidneys and recycled, the amino groups reused to make new amino acids for maintaining body proteins.

Bears store an enormous amount of body fat in preparation for their long sleep. An adult grizzly bear consumes about 38,000 kJ/day during the late spring and summer, but as winter approaches it feeds 20 hours a day, consuming up to 84,000 kJ daily. This change in feeding is a response to a seasonal change in hormone secretion. Large amounts of triacylglycerols are formed from the huge intake of carbohydrates during the fattening-up period.

OXIDATION OF UNSATURATED FATTY ACIDS

The fatty acid oxidation scheme described above operates only when the incoming fatty acid is a saturated one (having only single bonds) and possesses an even number of carbon atoms. However, most of the fatty acids in the triacylglycerols and phospholipids of animals and plants are unsaturated, having one or more double bonds in its carbon chain. These bonds are in *cis* configuration and cannot be acted upon by the enzyme, enoyl-CoA hydratase which catalyzes the addition of H_2O to the *trans* double bond of the Δ^2 - enoyl-CoA generated during β oxidation. However, by the action of two auxiliary enzymes, the fatty acid oxidation sequence described above can also break down the common unsaturated fatty acids. The action of these two enzymes, one an isomerase and the other a reductase, will be illustrated by the following two examples:

(*a*) Oxidation of Monounsaturated Fatty Acids

This requires only one additional enzyme, enoyl-CoA isomerase. Oleate, an abundant C–18 monounsaturated fatty acid with a *cis* double bond between C–9 and C–10 (denoted *cis*-Δ^9) is aken as an example (Fig. 24–10). Oleate is converted into oleoyl-CoA which is transported through the mitochondrial membrane as oleoyl carnitine and then converted back into oleoyl-CoA in the matrix. Oleoyl-CoA then undergoes 3 passes through the β oxidation cycle to yield 3 moles of acetyl-CoA and the coenzyme A ester of a Δ^3, 12-carbon unsaturated fatty acid, *cis*-Δ^3- dodecenoyl-CoA (Fig. 24–10). This product cannot be acted upon by the next enzyme of the β oxidation pathway, *i.e.*, enoyl-CoA hydratase, which acts only on trans double bonds. However, by the action of the auxiliary enzyme, **enoyl-CoA isomerase**, the *cis*-Δ^3-enoyl-CoA is isomerized to yield the *trans*-Δ^2-enoyl-CoA. The latter compound is now converted by enoyl-CoA hydratase into the corresponding L-β-hydroxyacyl-CoA (*trans*-Δ^2-dodecenoyl-CoA). This intermediate is now acted upon by the remaining enzymes of β oxidation to yield acetyl-CoA and a C–10 saturated fatty acid as its coenzyme A ester (decanoyl-CoA). The latter undergoes 4 more passes through the pathway to yield altogether 9 acetyl-CoAs from one mole of the C–18 oleate.

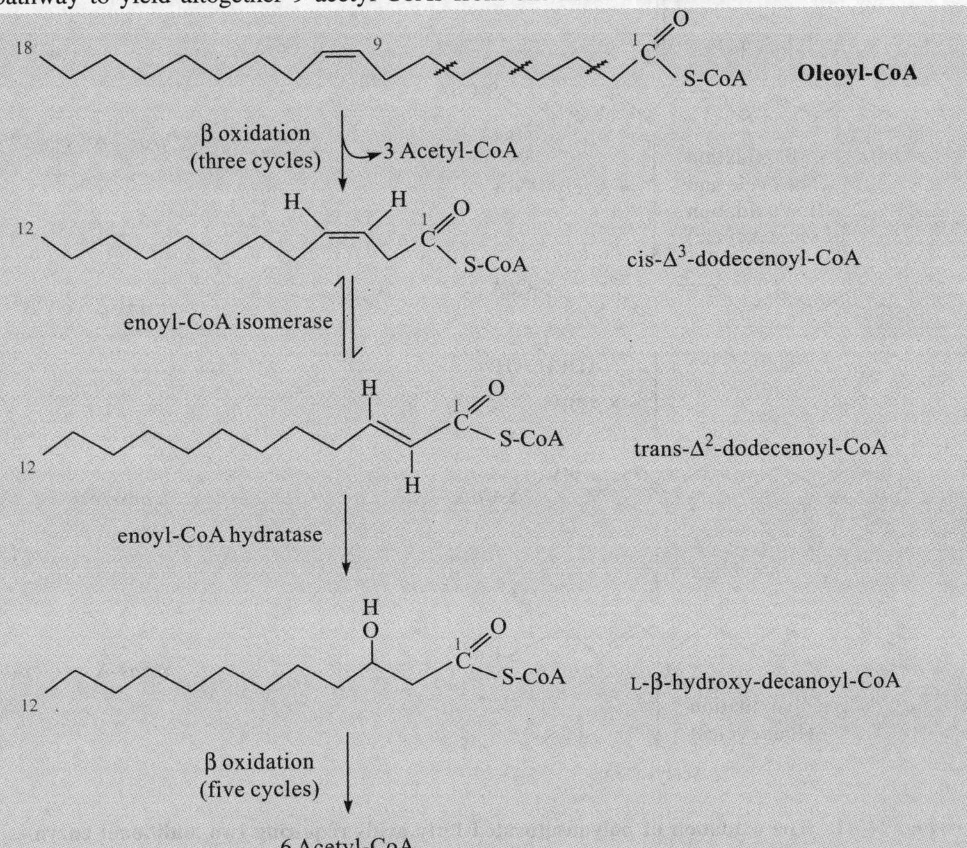

Fig. 24–10. The oxidation of a monounsaturated fatty acyl-CoA such as oleoyl-CoA (Δ9) requiring an additional enzyme, enoyl-CoA isomerase

Note that the enzyme repositions the double bond converting the *cis* isomer to a *trans* isomer, a normal intermediate in β oxidation. Thus, both position and configuration of the double bond are shifted by the action of the enzyme.

(*b*) Oxidation of Polyunsaturated Fatty Acids

This process requires two auxiliary enzymes, enoyl-CoA isomerase and 2, 4-dienoyl-CoA-reductase. The mechanism is illustrated by taking linoleate, a C-18 polyunsaturated fatty acid with 2 *cis* double bonds at C_9 and C_{12} (denoted *cis*-Δ^9, *cis*-Δ^{12}), as an example. Linoleoyl-CoA undergoes 3 passes through the typical β oxidation sequence to yield 3 moles of acetyl-CoA and the coenzyme A ester of a C-12 unsaturated fatty acid with a *cis*-Δ^3, *cis*-Δ^6 configuration. This intermediate cannot be used by the enzymes of the β oxidation pathway ; its double bonds are in the wrong position and have the wrong configuration (*cis*, not *trans*). However, the combined action of **enoyl-CoA isomerase** and **2,4-dienoyl-CoA reductase** (Fig. 24–11) allows re-entry of

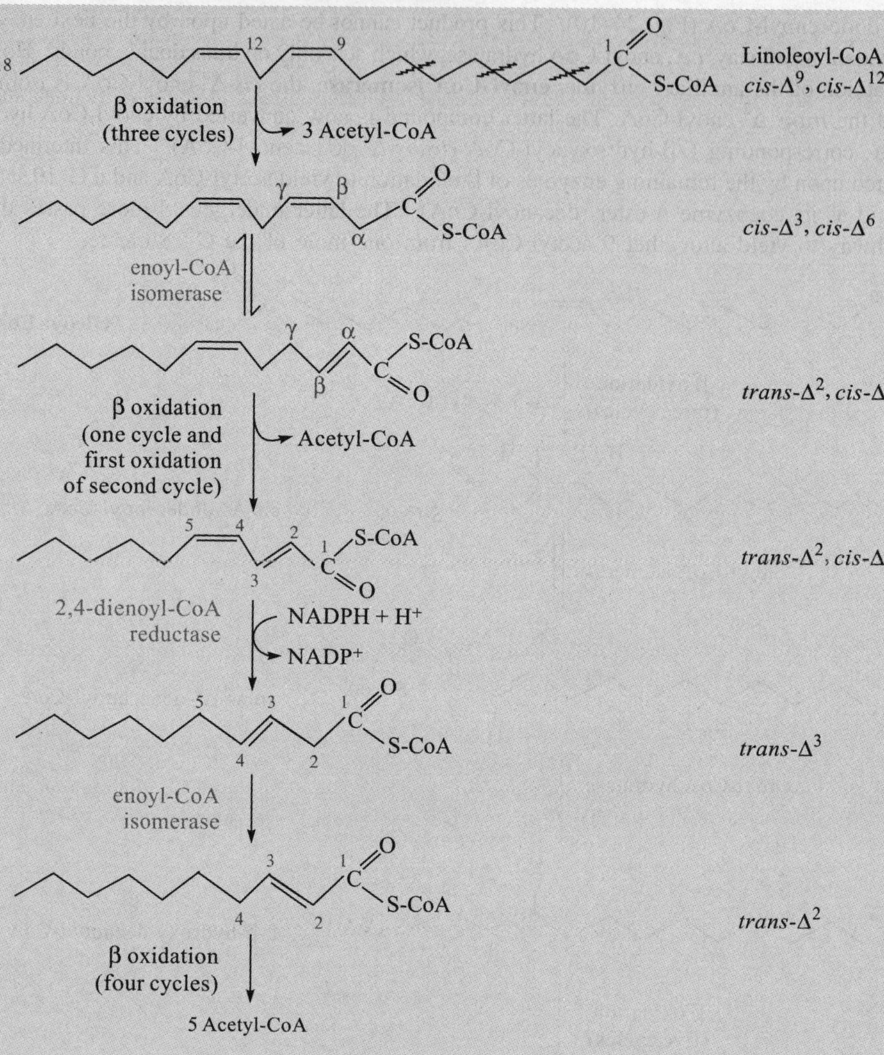

Fig. 24–11. The oxidation of polyunsaturated fatty acids requiring two additional enzymes, enoyl-CoA isomerase and 2, 4-dienoyl-CoA reductase

Note that the combined action of these two enzymes converts a *trans*-Δ^2, *cis* Δ^4-dienoyl-CoA intermediate into the *trans*-Δ^2-enoyl-CoA substrate, necessary for β oxidation.

this intermediate into the typical β oxidation pathway and its degradation to 6 acetyl-CoAs. The overall result is the conversion of linoleate to 9 moles of acetyl-CoA.

Here is an excellent example of the beautiful economy of organization of metabolism.The introduction of 2 additional types of enzymes (an enoyl-CoA isomerase and a 3-hydroxyacyl-CoA racemase) makes it possible to handle any combination of double bonds found in an unsaturated chain through the same route used for saturated fatty acids.

The roles of the 3 additional enzymes which are necessary for the oxidation of a dienoic (or polyenoic) acid may be shown in outline below, where A is enoyl-CoA isomerase; B, enoyl-CoA hydratase; and C, 3-hydroxyacyl-CoA epimerase. Monoenoic and dienoic acids are oxidized at comparable rates.

$$C_{18:2}\,(9\,cis,\,12\,cis) \xrightarrow{-3C_2} C_{12:2}\,(3\,cis,\,6\,cis) \xrightarrow{A} C_{12:2}\,(2\,trans,\,6\,cis)$$

$$\xrightarrow{-C_2} C_{10:1}\,(4\,cis) \xrightarrow{-C_2} C_{8:1}\,(2\,cis) \xrightarrow{B} \text{3-D-hydroxy } C_{8:0}$$

$$\xrightarrow{C} \text{3-L-hydroxy } C_{8:0} \xrightarrow{-C_2} C_{6:0} \longrightarrow 3C_2$$

OXIDATION OF ODD-CHAIN FATTY ACIDS

Most naturally-occurring lipids contain fatty acids with an even number of carbon atoms, yet fatty acids with an odd number of carbon atoms are found in significant amounts in the lipids of many plants and some marine animals. Small quantities of C-3 propionate are added as a mould inhibitor to some breads and cereals, and thus propionate enters the human diet. Besides, cattle and other ruminants form large amounts of propionate during fermentation of carbohydrates in the rumen. The propionate so formed is absorbed into the blood and oxidized by the liver and other tissues.

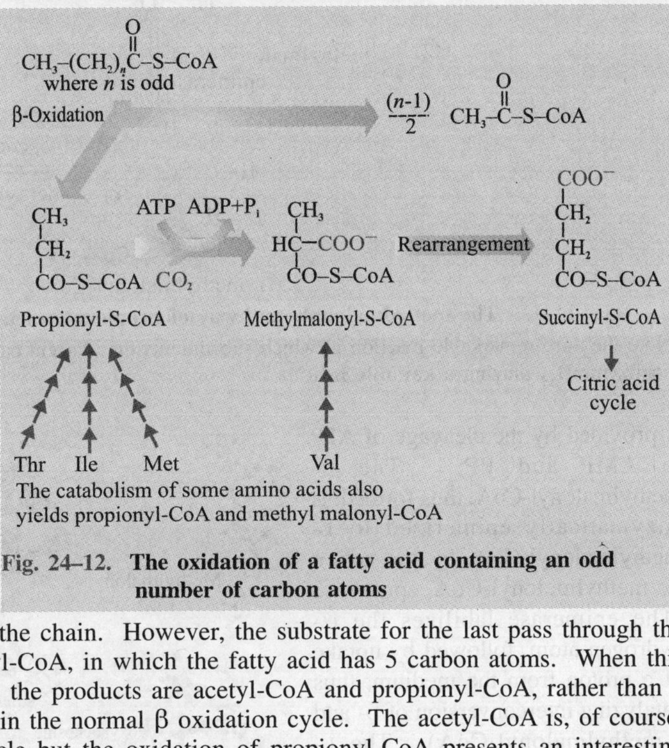

Fig. 24–12. The oxidation of a fatty acid containing an odd number of carbon atoms

A generalized scheme of the oxidation of an odd-chain fatty acid is presented in Fig. 24–12. The odd-carbon long-chain fatty acids are oxidized by the same pathway as the even-carbon fattty acids, starting at the carboxyl end of the chain. However, the substrate for the last pass through the β oxidation cycle is a fatty acyl-CoA, in which the fatty acid has 5 carbon atoms. When this is oxidized and finally cleaved, the products are acetyl-CoA and propionyl-CoA, rather than 2 moles of acetyl-CoA produced in the normal β oxidation cycle. The acetyl-CoA is, of course, oxidized *via* the citric acid cycle but the oxidation of propionyl-CoA presents an interesting problem, since at first glance the propionic acid (or propionyl-CoA) appears to be a substrate unsuitable for β oxidation. However, the substrate is held by two strikingly dissimilar pathways: methylmalonate pathway and β-hydroxy-propionate pathway.

(*a*) Methylmalonate Pathway

This pathway is found only in animals and occurs in the mitochondria of liver, cardiac and skeletal muscles, kidney and other tissues. Propionate (or propionyl-CoA) is also produced by the oxidation of isoleucine, valine, methionine and threonine. Propionate is catalyzed by acetyl-CoA synthetase to produce propionyl-CoA (Fig. 24–13). The propionyl-CoA is carboxylated to form the D stereoisomer of methylmalonyl-CoA by an enzyme propionyl-CoA carboxylase, which contains the cofactor biotin. In this reaction, as in pyruvate carboxylase reaction (see page 413), the CO_2 (or its hydrated ion, HCO_3^-) is activated by attachment to biotin before its transfer to the propionate moiety. The formation of the carboxybiotin intermediate requires energy, which

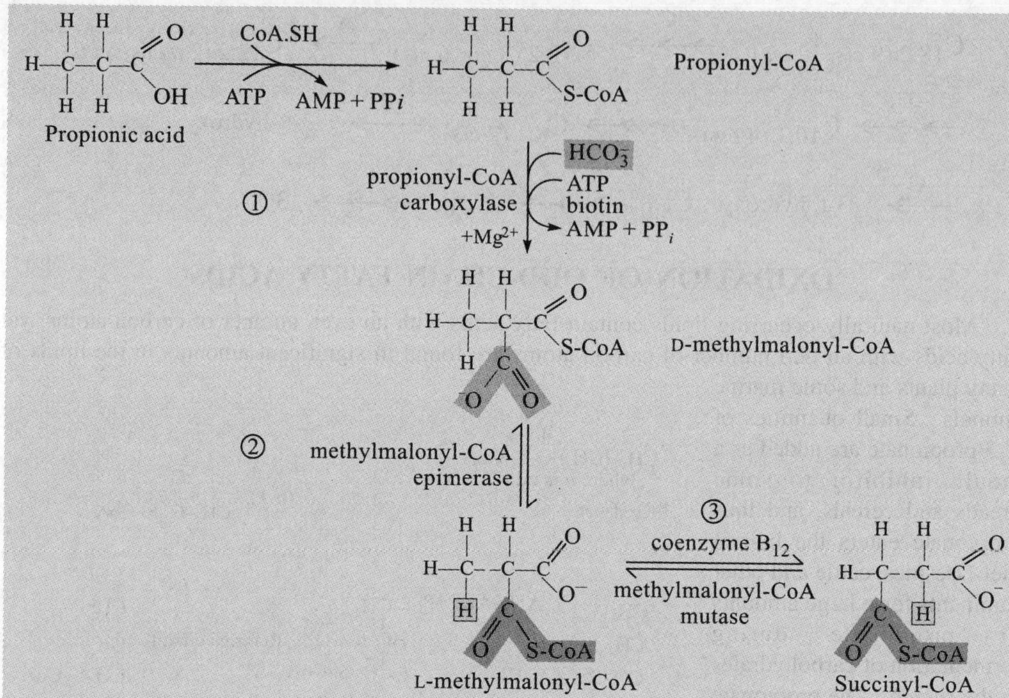

Fig. 24–13. The methylmalonate pathway of propionate metabolism, as found in animals

Note the third remarkable reaction in which substituents on adjacent carbon atoms exchange positions; the coenzyme B_{12} playing a key role in it.

is provided by the cleavage of ATP to AMP and PP*i*. The D-methylmalonyl-CoA, thus formed, is enzymatically epimerized to L-methylmalonyl-CoA, by the action of methylmalonyl-CoA epimerase (The epimerase labilizes the α-hydrogen atom, followed by uptake of a proton from the medium, thus catalyzing interconversion of D- and L-methylmalonyl-CoA). The L-methylmalonyl-CoA undergoes an intramolecular rearrangement to form succinyl-CoA by the enzyme methylmalonyl-CoA mutase (Fig 24–14), which requires as its coenzyme deoxyadenosyl-cobalamin or coenzyme B_{12}. When [2–^{14}C] methyl-malonyl-CoA was converted by the mutase enzyme, the label (marked by an asterisk, below) was found in the 3 position of succinyl-CoA, thus indicating an intramolecular transfer of the entire thioester group, –CO–S–CoA, rather than migration of the carboxyl carbon.

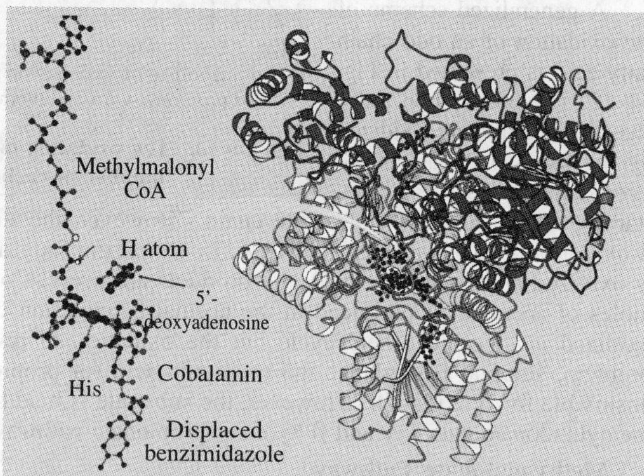

Fig. 24–14. Active site of methylmalonyl CoA mutase

The arrangement of substrate and coenzyme in the active site facilitates the cleavage of the cobalt-carbon bond and the subsequent abstraction of a hydrogen atom from the substrate.

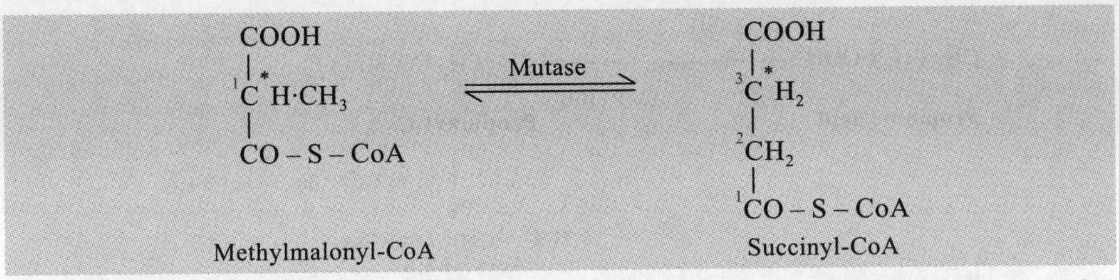

The role of the coenzyme B_{12} is to remove a hydrogen from one carbon atom by transferring it directly to an adjacent carbon atom, simultaneously effecting the exchange of a second (R) substituent. The H and R are not released into solution.

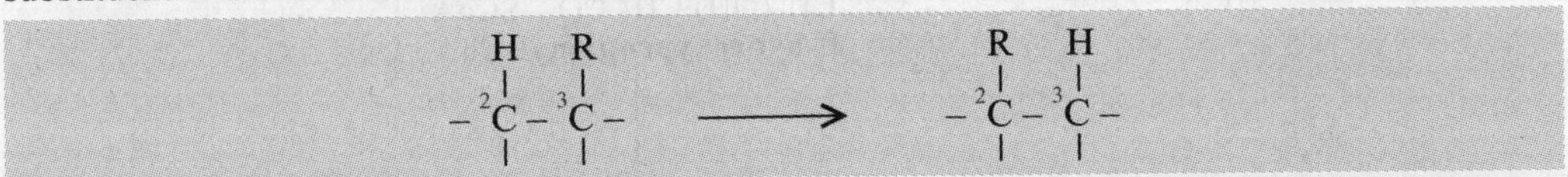

At equilibrium, formation of succinyl-CoA favoured by a ratio of 20 : 1 over methylmalonyl-CoA. The succinyl-CoA can then be oxidized via succinate and the citric acid cycle to CO_2 and H_2O. In patients with vitamin B_{12} deficiency, both propionate and methylmalonate are excreted in the urine in abnormally large amounts.

The odd-chain fatty acids are only a small fraction of the total, and only the terminal 3 carbons appear as propionyl-CoA. The metabolism of propionyl-CoA is, therefore, not of quantitative significance in fatty acid oxidation.

Two inheritable types of **methylmalonic acidemia** (and **aciduria**) are associated in young children with failure to grow and mental retardness. In *one type*, the mutase protein is absent or defective since addition of coenzyme B_{12} to liver extracts does not restore the activity of the mutase. In the *other type*, feeding large doses of vitamin B_{12} relieves the acidemia and aciduria, and addition of coenzyme B_{12} to liver extracts restores the activity of the mutase ; in these cases, there is limited ability to convert the vitamin to the coenzyme.

Another inheritable disorder of propionate metabolism is due to a defect in propionyl-CoA carboxylase, resulting in **propionic acidemia** (and **aciduria**). Such individuals, as well as those with methylmalonic acidemia, are capable of oxidizing some propionate to CO_2, even in the absence of propionyl-CoA carboxylase.

(b) β-hydroxypropionate Pathway

This pathway is ubiquitous in plants and is a modified form of β oxidation scheme. It nicely resolves the problem of how plants can cope with propionic acid by a system not involving vitamin B_{12} as cobamide coenzyme. Since plants have no B_{12} functional enzymes, the methylmalonate pathway does not operate in them. This pathway (Fig. 24–15), thus, bypasses the B_{12} barrier in an effective way.

α OXIDATION OF FATTY ACIDS

Although β oxidation is major pathway for the oxidation of fatty acids, two other types of oxidation also occur, α and ω oxidation. α oxidation is the removal of one carbon atom (*i.e.*, α carbon) at a time from the carboxyl end of the molecule. α oxidation was first observed in seeds and leaf tissues of plants. α oxidation of long-chain fatty acids to 2-hydroxy acids and then to fatty acids with one carbon atom less than the original substrate have been demonstrated in the microsomes of brain and other tissues also. Long-chain α hydroxy fatty acids are constituents of brain lipids, *e.g.*, the C_{24} cerebronic acid (= 2 hydroxylignoceric acid), $CH_3(CH_2)_{21}.CH(OH).COOH$. These hydroxy fatty acids can be converted to the 2-keto acids, followed by oxidative decarboxylation, resulting in the formation of long-chain fatty acids with an odd number of carbon atoms:

$$CH_3{\cdot}CH_2{\cdot}COOH \xrightarrow[\text{ATP} \quad \text{AMP+PP}i]{\text{CoA–SH}} CH_3{\cdot}CH_2{\cdot}CO\text{-}SCoA$$

Propionic acid → **Propionyl-CoA**

↓ Acyl-CoA dehydrogenase

$H_2C{=}CH{\cdot}CO–SCoA$
Acrylyl-CoA

↓ Enoyl-CoA hydratase

$CH_2(OH){\cdot}CH_2{\cdot}CO—SCoA$
β-hydroxypropionyl-CoA

↓ Acyl-CoA thioesterase

$CH_2(OH){\cdot}CH_2{\cdot}COOH$
β-hydroxypropionic acid

↓ β-hydroxypropionic dehydrogenase

$CHO{\cdot}CH_2{\cdot}CO–SCoA$
Malonyl semialdehyde-CoA

↓ Malonyl semialdehyde dehydrogenase

$HOOC{\cdot}CH_2{\cdot}CO–SCoA$
Malonyl –CoA

↓ Malonyl-CoA decarboxylase ($\rightarrow CO_2$)

CH3·CO–SCoA
Acetyl-CoA

Fig. 24–15. The β-hydroxypropionate pathway of propionate metabolism, as found in plants

$$R{\cdot}CH_2\,CH_2\,CH_2\,COOH \xrightarrow[\text{Monoxygenase}]{\text{Hydroxylation}} R{\cdot}CH_2\,CH_2\,CHOH—COOH \longrightarrow$$

Long-chain fatty acid (n carbon) → 2-hydroxy fatty acid

$$R{\cdot}CH_2\,CH_2\,CO–COOH \xrightarrow{\text{Oxidative decarboxylation}} R{\cdot}CH_2\,CH_2\,COOH + CO_2$$

2-keto fatty acid → Long-chain fatty acid (n–1 carbon)

The initial hydroxylation reaction is catalyzed by a mitochondrial enzyme, monoxygenase that requires O_2, Mg^{2+}, NADPH and a heat-stable cofactor. Conversion of the α hydroxy fatty acid to CO_2 and the next lower unsubstituted acid appears to occur in the endoplasmic reticulum and to require O_2, Fe^{2+} and ascorbate.

The *salient features* of α oxidation are as follows:

1. Only free long-chain fatty acids serve as substrates.
2. Molecular oxygen is indirectly involved.
3. It does not require CoA intermediates.
4. It does not lead to generation of high-energy phosphates.

This mechanism explains the occurrence of α hydroxy fatty acids and of odd-numbered fatty acids in the biomolecules. The latter may, in nature, also be synthesized *de novo* from propionate.

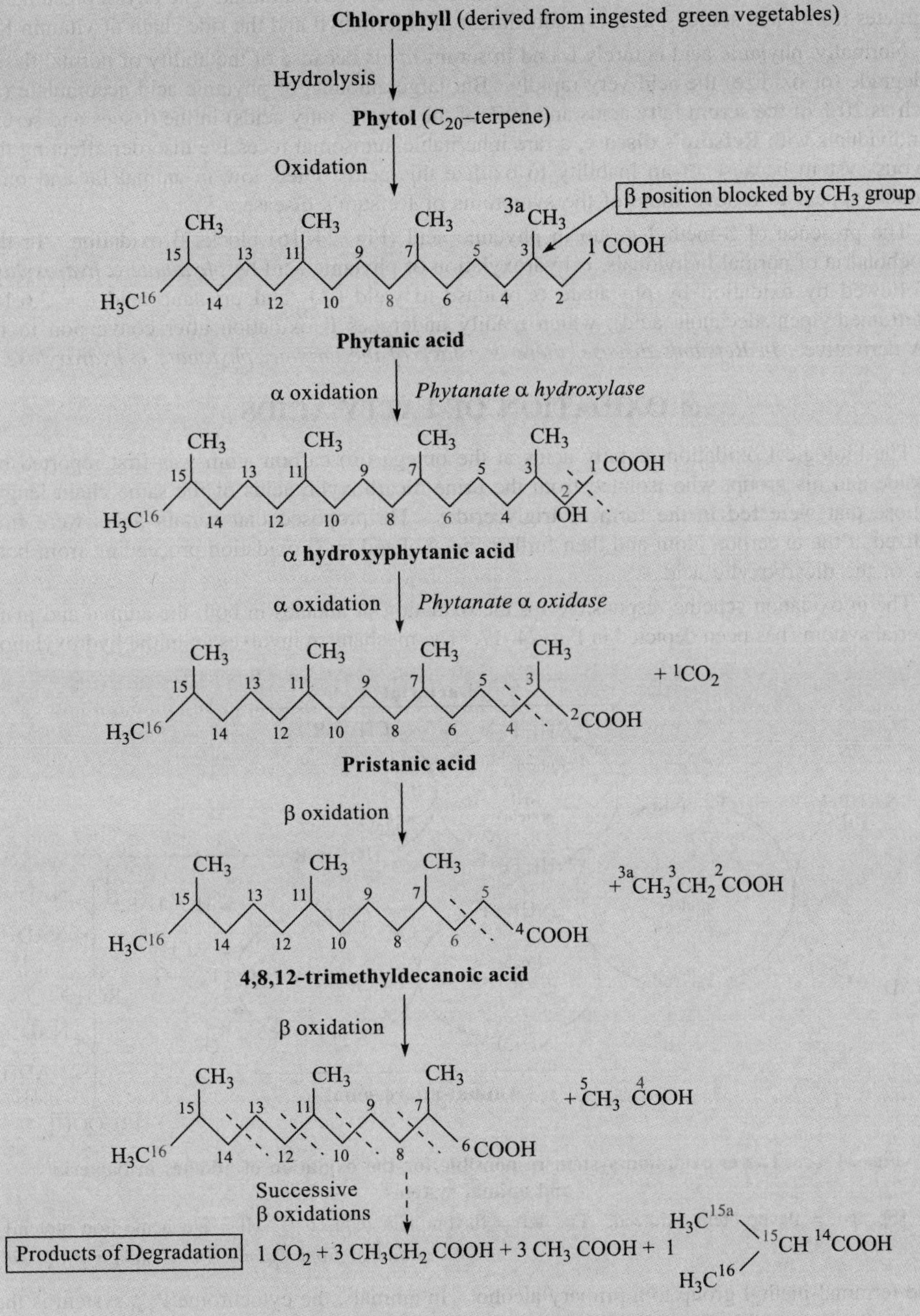

Fig. 24–16. The metabolism of phytanic acid by the normal animal cell

Dashed lines show the successive points of cleavage of the molecule.

The α oxidation system plays a key role in the capacity of mammalian tissues to oxidize **phytanic acid** (= 3,7,11,15-tetramethylhexadecanate). Phytanic acid is an oxidation product of phytol and is present in animal fat, cow's milk and foods derived from milk. The phytol presumably originates from plant sources, as it is a substituent of chlorophyll and the side chain of vitamin K_2.

Normally, phytanic acid is rarely found in serum lipids because of the ability of normal tissue to degrade (or oxidize) the acid very rapidly. But large amounts of phytanic acid accumulate (as much as 20% of the serum fatty acids and 50% of the hepatic fatty acids) in the tissues and serum of individuals with **Refsum's disease**, a rare inheritable autosomal recessive disorder affecting the nervous system because of an inability to oxidize this acid. Diets low in animal fat and milk products appear to relieve some of the symptoms of Refsum's disease.

The presence of 3-methyl group in phytanic acid (Fig. 24–16) blocks β oxidation. In the mitochondria of normal individuals, α hydroxylation of phytanic acid by *phytanate α hydroxylase* is followed by oxidation by phytanate α oxidase to yield CO_2 and pristanic acid (= 2,6,10, 14-tetramethylpentadecanoic acid), which readily undergoes β oxidation after conversion to its CoA derivative. *In Refsum's disease, there is a lack of the enzyme, phytanate α hydroxylase.*

ω OXIDATION OF FATTY ACIDS

The biological oxidation of fatty acids at the omega (ω) carbon atom was first reported by Verkade and his group, who isolated from the urine dicarboxylic acids of the same chain length as those that were fed in the form of triglycerides. He proposed that certain acids were first oxidized at the ω carbon atom and then further metabolized by β oxidation proceeding from both ends of the dicarboxylic acid.

The ω oxidation scheme responsible for the oxidation of alkanes in both the animal and plant bacterial systems has been depicted in Fig. 24–17. The mechanism involves an initial hydroxylation

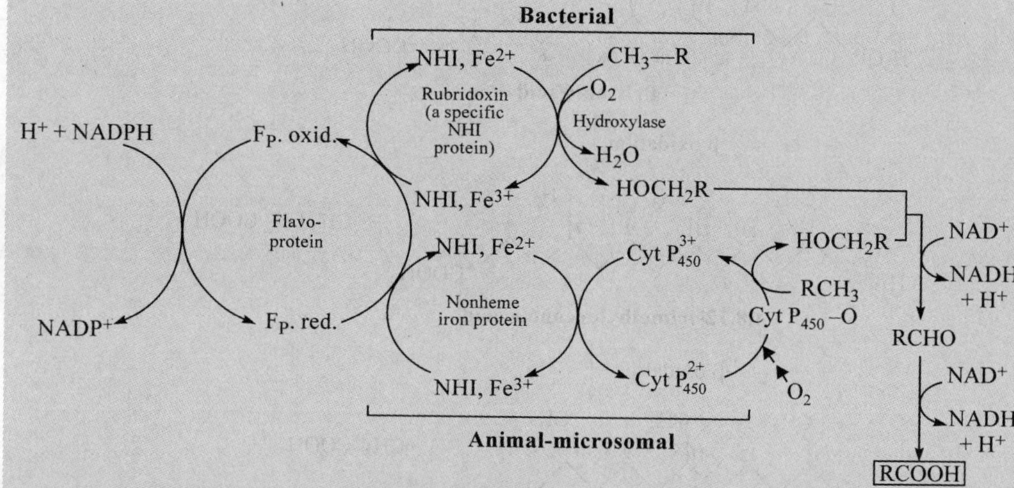

Fig. 24–17. The ω oxidation system responsible for the oxidation of alkanes in bacteria and animal systems

[Fp. oxi. = flavoprotein oxidized, Fp. red. = flavoprotein reduced, NHI = Nonheme iron protein]

(Adapted from Conn EE and Stumpf PK, 1976)

of the terminal methyl group to a primary alcohol. In animals, the cytochrome P_{450} system is the hydroxylase responsible for this alkane hydroxylation ; whereas in bacteria, rubridoxin is the intermediate electron carrier which feeds electrons to ω hydroxylase system. The immediate product, RCH_2OH is oxidized to an aldehyde by an *alcohol dehydrogenase*, which in turn is oxidized to a carboxylic acid by an *aldehyde dehydrogenase* in both systems. Summarily, the

—CH_3 group is converted to a —CH_2OH group which subsequently is oxidized to —COOH, thus forming a dicarboxylic acid. Once formed, the dicarboxylic acid may be shortened from either end of the molecule, by the β oxidation sequence, to form acetyl-CoA.

These series of reactions now have assumed an extremely important scavenging role in the bacterial biodegradation of both detergents derived from fatty acids and even more important the large amounts of oil spilled over the ocean surface. The rate of bacterial oxidation of floating oil under aerobic conditions is estimated as high as 0.5g/day per square metre of oil surface. The bacterial oxidation of oils is brought about primarily by ω oxidation mechanism.

KETOGENESIS

General Considerations

The term **ketogenesis** means formation of ketone bodies. The acetyl-CoA, formed in fatty acid oxidation, enters the citric acid cycle if fat and carbohydrate degradation are approximately balanced. The molecular basis of the adage that *fats burn in the flame of carbohydrates* is now evident. The entry of acetyl-CoA into citric acid cycle depends on the availability of oxaloacetate for the formation of citrate. However, if fat breakdown predominates, the acetyl-CoA undergoes a different fate. This is because the concentration of oxaloacetate is lowered if carbohydrate is not available or else poorly utilized. Also, in fasting or in diabetes, the oxaloacetate is utilized in the formation of glucose and is thus unavailable for condensation with acetyl-CoA. Under these conditions, acetyl-CoA is diverted to the formation of acetoacetate (**3-oxobutyrate**, in systematic nomenclature) and **D-3-hydroxybutyrate**. Acetoacetate continually undergoes spontaneous decarboxylation to yield acetone, which is exhaled. The reaction is slow, but if the concentration of acetoacetate becomes high, enough acetone may be formed to make its characteristic odour detectable in the breath of the individuals. This is the part of the reason that the 3 substances (acetoacetate, D-3-hydroxybutyrate and acetone) were collectively but inaccurately called the "**ketone bodies**" (or "**acetone bodies**") by early investigators even though acetone is the minor part of the total. The term now seems quaint, but it is still in use. And an increase in blood concentrations, of these compounds is called **ketonemia**.

Biosynthesis and Utilization of Ketone Bodies

Biosynthesis. Ketone bodies are formed by a series of unique reactions (Fig. 24–18), primarily in the liver and kidney mitochondria. The enzymes involved in the synthesis of ketone bodies are localized primarily in liver and kidney mitochondria. *Ketone bodies cannot be utilized in the liver since the key utilizing enzyme, β-ketoacyl : CoA transferase (= 3-oxoacid : CoA transferase) is absent in the tissue* but is present in all tissues metabolizing ketone bodies, namely red muscle, cardiac muscle, brain and kidney.

Acetoacetate is produced from acetyl-CoA in the liver and kidneys by a simple three-step process. The first step in the formation of acetoacetate, one of the 3 ketone bodies, is the reversible head-to-tail condensation of 2 moles of acetyl-CoA, to produce a mole of acetoacetyl-CoA enzymatically by *thiolase* (**Step 1**). This steps is simply the reversal of the last step of β oxidation, *i.e.*, thiolysis. The acetoacetyl-CoA then condenses with acetyl-CoA and water to form β-hydroxy-β-methylglutaryl-CoA, HMG-CoA (**Step 2**). The unfavourable equilibrium in the formation of acetoacetyl-CoA is compensated for by the favourable equilibrium of this reaction, which is due to the hydrolysis of a thioester linkage.

If we think of acetoacetyl-coenzyme A as being analogous to oxaloacetate, we find that the condensation reaction (Step 2) in exactly analogous to the formation of citrate in the first step of the citric acid cycle. However, it is catalyzed by a quite different enzyme, 3-hydroxy-3-methylglutaryl-CoA synthetase. HMG-CoA is ultimately cleaved at a different point to yield acetoacetate and acetyl-CoA in an irreversible reaction (**Step 3**). The sum of these reactions is:

$$2 \text{ Acetyl-CoA} + H_2O \longrightarrow \text{Acetoacetate} + 2 \text{ CoA} + H^+$$

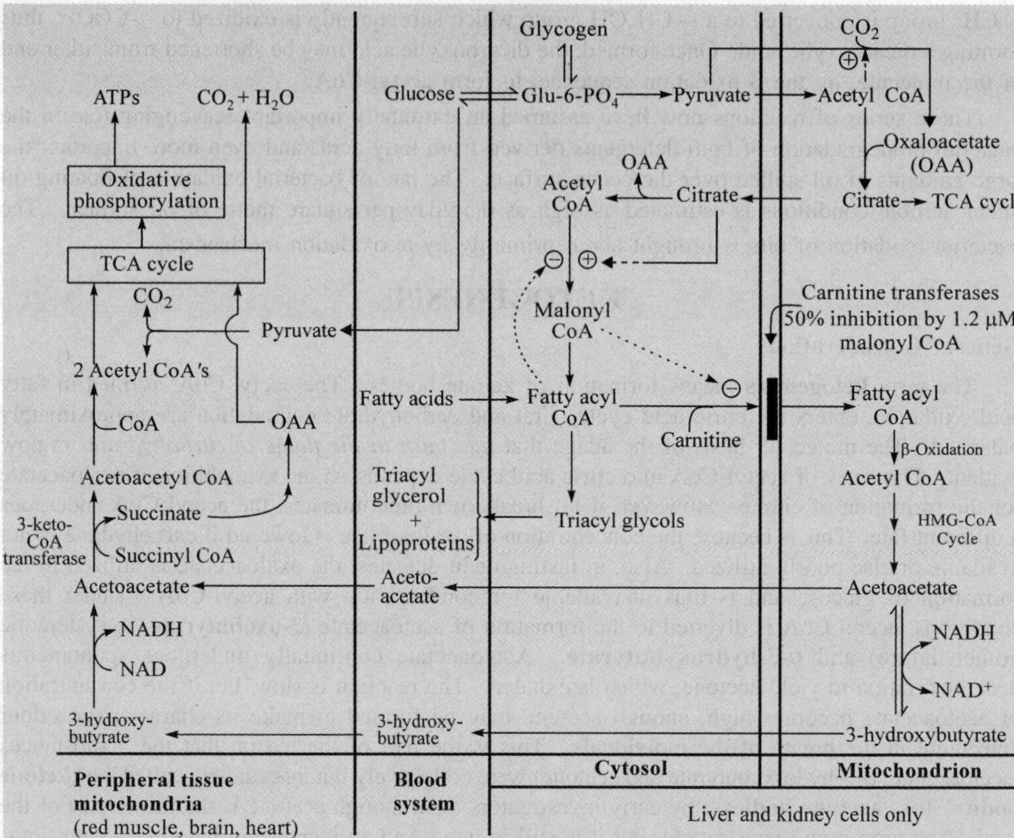

Fig. 24–18. Biosynthesis of ketone bodies and their utilization

Note that the C-6 compound, β-methylglutaryl-CoA (HMG-CoA) is also an intermediate of sterol biosynthesis, but the enzyme that forms HMG-CoA in that pathway is cytosolic. HMG-CoA lyase is present in the mitochondrial matrix but not in the cytosol.

The acetoacetate so produced is reversibly reduced by a mitochondrial enzyme, D-β-*hydroxybutyrate dehydrogenase* to produce D-β-hydroxybutyrate (Fig. 24–18). This enzyme is specific for the D-stereoisomer; it does not act on L-β-hydroxyacyl-CoAs and is not to be confused with L-β-hydroxyacyl-CoA dehydrogenase, which participates in the β-oxidation pathway. The ratio of hydroxybutyrate to acetoacetate depends on the $NADH/NAD^+$ ratio inside mitochondria. Accetoacetate is also easily decarboxylated to acetone, either spontaneously or by the action of *acetoacetate decarboxylase*.

Utilization. The acetoacetate and D-β-hydroxybutyrate diffuse from the liver mitochondria into the blood and are transported to peripheral tissues. George Cahill and others have shown that these two acetyl-CoA products are important molecules in energy production. *Acetoacetate and D-β-hydroxybutyrate are normal fuels of respiration and are quantitatively important as source of energy*. In fact, heart muscles and renal cortex use acetoacetate in preference to glucose. On the contrary, glucose is the major fuel for the brain in well-nourished persons on a balanced diet. However, brain adapts to utilization of acetoacetate during fasting, starvation and diabetes. In prolonged starvation, 75% of the fuel needs of the brain are met by acetoacetate.

Fatty acids are released by adipose tissue and converted into acetyl units by the liver, which then exports them as acetoacetate. *Acetoacetate can, thus, be regarded as a water-soluble transportable form of acetyl units.*

To sum up, ketone bodies are alternative substrates to glucose, for energy sources in muscle and brain. The precursors of ketone bodies, namely free fatty acids are toxic in high concentrations, have very limited solubility, and readily saturate the carrying capacity of the plasma membrane. On the other hand, the ketone bodies are low in toxicity, and tolerated at high concentrations, are very soluble, diffuse rapidly through membranes, and are rapidly metabolized to CO_2 and H_2O.

The importance of this pathway is indicated by the fact that the normal human liver is potentially capable of making the equivalent of half its weight as acetoacetate each day!

In the extrahepatic tissues, D-β-hydroxybutyrate is oxidized to acetoacetate by an enzyme, D-β-*hydroxybutyrate dehydrogenase*. Acetoacetate is activated to form its coenzyme A ester by transfer of CoA from succinyl-CoA, an intermediate of the citric acid cycle, in a reaction catalayzed by b-ketoacyl-CoA transferase. The acetoacetyl-CoA is then cleaved by thiolase to yield 2 moles of acetyl-CoA, which enter the citric acid cycle.

Ketogenic and Antiketogenic Substances

The *ketogenic substances* are, of course, all the fatty acids. In addition, at least 3 amino acids belong to this category : leucine, phenylalanine and tyrosine. *Antiketogenic substances*, in the sense of preventing the accumulation of ketone bodies, are the carbohydrates, the glycerol fraction of fat, and the following amino acids : glycine, alanine, valine, serine, threonine, cysteine, methionine, aspartic acid, glutamic acid, arginine, histidine, proline and ornithine. These are antiketogenic because their non-nitrogenous residues are convertible to glucose.

Regulation of Ketogenesis

Ketosis arises as a result of deficiency in available carbohydrate. This leads to the enhanced rate of ketogenesis. There are 3 crucial steps in the pathway of metabolism of free fatty acids (FFAs) that determine the magnitude of ketogenesis (Fig. 24–19). These are:

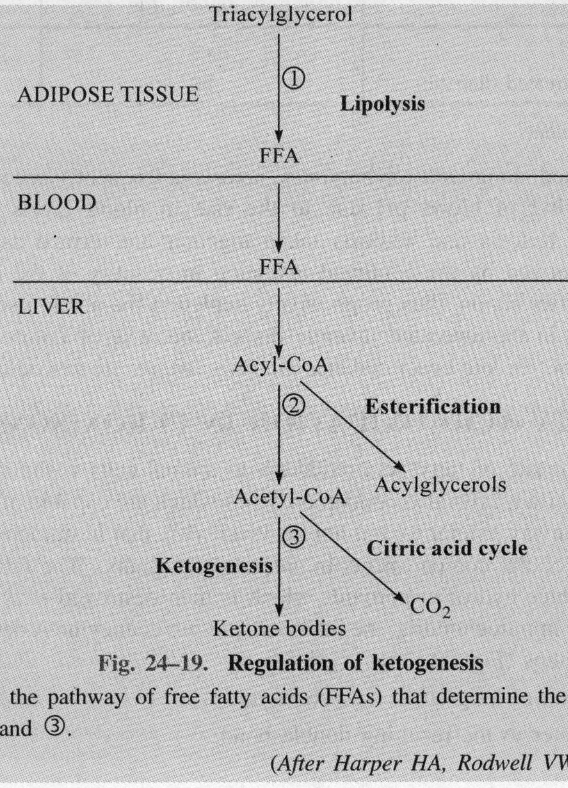

Fig. 24–19. Regulation of ketogenesis

The 3 crucial steps in the pathway of free fatty acids (FFAs) that determine the rate of ketogenesis are numbered as ①, ②, and ③.

(After Harper HA, Rodwell VW and Mayes PA, 1977)

1. It causes an imbalance between esterification and lipolysis in adipose tissue, with consequent of free fatty acids into the circulation. FFAs are the principal substrates for ketone body formation in the liver and therefore all factors, metabolic or endocrine, affecting the release of free fatty acids from adipose tissues, influence ketogenesis.

2. Upon entry of free fatty acids into the liver, the balance between esterification and oxidation of FFAs is influenced by the hormonal state of the liver and possibly by the availability of glycerol-3-phosphate or by the redox state of the tissue.

3. As the quantity of fatty acids presented for oxidation increases, more form ketone bodies and less form CO_2, regulated in such a manner that the total energy production remains constant.

Under circumstances of limited utilization of carbohydrates and/or substantial mobilization of fatty acids to the liver, there is a markedly diminished rate of operation of two of the three pathways for metabolizing acetyl-CoA, *viz.*, the citric acid cycle and fatty acid synthesis. The result is a channeling of acetyl-CoA into β-hydroxy-β-methylglutaryl-CoA (HMG-CoA). This results in increased formation of acetoacetate, D-β-hydroxybutyrate and acetone, whereby elevating the concentration of the total ketone bodies in the blood and urine above normal (Table 24–1). Higher than normal quantities of the ketone bodies present in the blood or urine constitute ketonemia (= hyperketonemia) and ketonuria, respectively. D-β-hydroxybutyrate is quantitatively the predominant ketone body present in the blood and urine in ketosis. Whenever a marked degree of ketonemia and ketonuria exists, the odour of acetone is likely to be detected in the exhaled air. The triad of ketonemia, ketonuria and acetone odour of the breath is collectively termed **ketosis.**

Table 24–1. Accumulation of ketone bodies in untreated diabetics with severe ketosis

State of individual	*Blood concentration* (mg/100 mL)*	*Urinary excretion (mg/24 hr)*
Normal	<3	≤ 125
Extreme ketosis (untreated diabetic)	90	5,000

* as acetone equivalents

Since H^+ is produced along with oxybutyrates, ketosis is frequently accompanied by acidosis. **Acidosis** is the lowering of blood pH due to the rise in blood levels of acetoacetate and D-β-hydroxybutyrate. Ketosis and acidosis taken together are termed as **ketoacidosis.** The ketoacidosis is characterized by the continual excretion in quantity of the ketone bodies which entails some loss of buffer cation, thus progressively depleting the alkali reserve. Ketoacidosis is of frequent occurrence in the untreated juvenile diabetic because of failure to secrete sufficient insulin and may be fatal. In late-onset diabetes after age 40, severe ketoacidosis is less frequent.

FATTY ACID OXIDATION IN PEROXISOMES

Although the major site of fatty acid oxidation in animal cells is the mitochondrial matrix, other compartment in certain cells also contain enzymes which are capable of oxidizing fatty acids to acetyl-CoA, by a pathway similar to, but not identical with, that in mitochondria. **Peroxisomes** are membrane-bound cellular compartments in animals and plants. The fatty acids are oxidized in peroxisomes to produce hydrogen peroxide which is then destroyed enzymatically. As in the oxidation of fatty acids in mitochondria, the intermediates are coenzyme A derivatives. The whole process consists of 4 steps (Fig. 24–20):

(*a*) dehydrogenation of long-chain acyl-coenzyme A,

(*b*) addition of water to the resulting double bond,

(*c*) oxidation of the β-hydroxyacyl-CoA to a ketone, and

(*d*) thiolytic cleavage by coenzyme A.

The fatty acid oxidation in **glyoxysomes** occurs by the peroxisomal pathway.

The peroxisomal pathway differs from the mitochondrial pathway in 3 respects:

1. In peroxisomal pathway, the flavoprotein dehydrogenase, that introduces the double bond, passes electrons directly to O_2, producing H_2O_2. This strong and potentially damaging oxidant is immediately cleaved to H_2O and O_2 by *catalase*. Whereas in mitochondrial pathway, the electrons removed in first oxidative step pass through the respiratory chain to O_2, forming H_2O. This process is accompanied by ATP synthesis. In peroxisomes, the energy released in the first oxidative step of fatty acid breakdown is dissipated as heat.
2. The NADH formed in β oxidation cannot be reoxidized, and the peroxisomes must export reducing equivalent to the cytosol (These equivalents are eventually passed on to mitochondria).
3. In mitochondria, the acetyl-CoA is further oxidized *via* the citric acid cycle. Acetyl-CoA produced by peroxisomes (and also glyoxysomes) is exported. The acetate from glyoxysomes serves as a biosynthetic precursor of polysaccharides, amino acids, nucleotides and some metabolic intermediates.

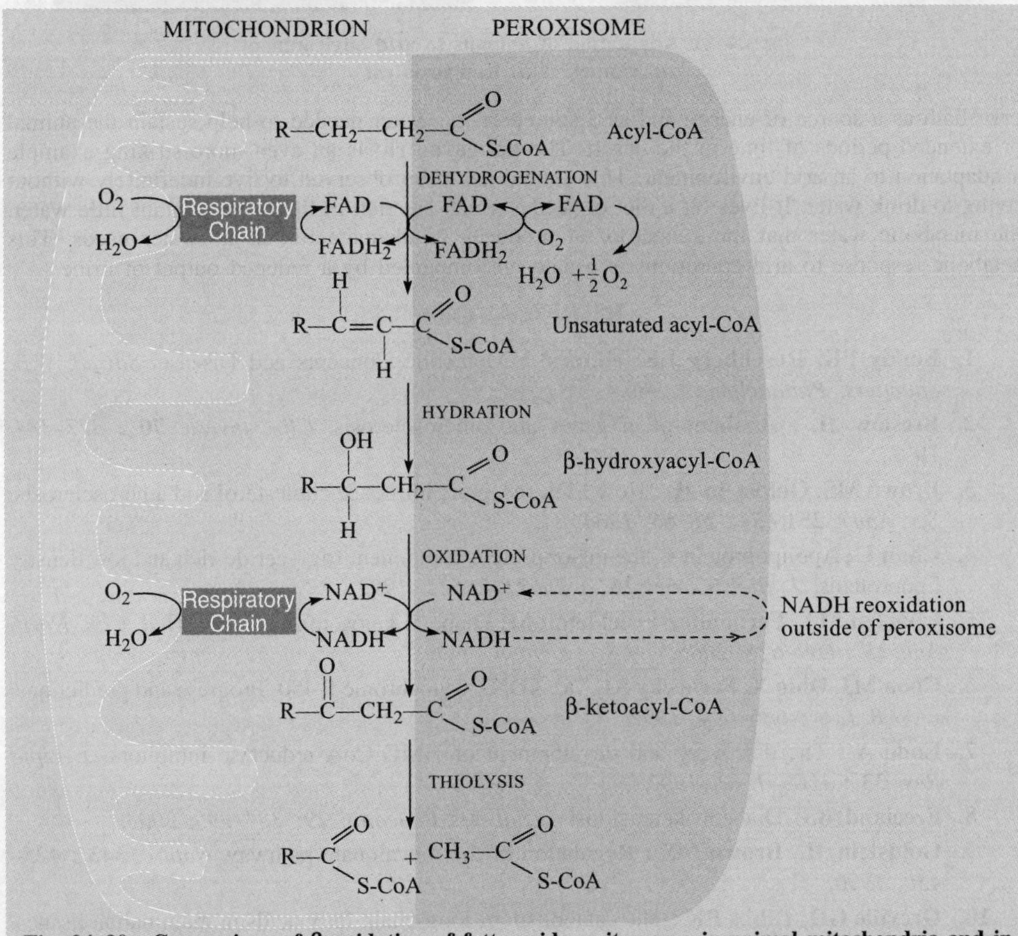

Fig. 24–20. Comparison of β oxidation of fatty acids as it occurs in animal mitochondria and in animal and plant peroxisomes

METABOLIC WATER

Another important biological feature of fatty acid oxidation (and of aerobic respiration, in general) is the production of metabolic water. For example, a mole of palmitic acid upon oxidation produces 16 moles of H_2O. This metabolic production of water is of significant importance to many organisms (Fig. 24–21). In the case of a **camel**, for example, the lipids stored in its humps

(a)

(b)

Fig. 24–21. Adaptation of animals to arid environment
(*a*) Camel (*b*) Kangaroo rat

serve both as a source of energy and as a source of the water needed to help sustain the animal for extended periods of time in the desert. The **kangaroo rat** is an even more striking example of adaptation to an arid environment. This animal has been observed to live indefinitely without having to drink water. It lives on a diet of seeds, which are rich in lipids but contain little water. The metabolic water that the kangaroo rat produces is adequate for all its water needs. This metabolic response to arid conditions is usually accompanied by a reduced output of urine.

REFERENCES

1. **Bondy PK, Rosenberg EE (editors) :** Metabolic Concepts and Disease. *8th ed. W.B. Saunders, Philadelphia. 1980*
2. **Breslow JL :** Apolipoprotein genes and atherosclerosis. *Clin. Invesh.* **70 :** *377–384, 1992.*
3. **Brown MS, Goldstein JL :** How LDL receptors influence cholesterol and atherosclerosis. *Sci. Amer.* **251 (5) :** *58–66, 1984.*
4. **Chan L :** Apolipoprotein B, the major protein component triglyceride-rich and low density lipoproteins. *J. Biol. Chem.* **267 :** *25621–25624, 1992.*
5. **Charfen MA, Fernández-Frackleton M:** Diabetic ketoacidosis. *Emer. Med. Clin. North Am.* **23 :** *609–628, 2005.*
6. **Coon MJ, Ding X, Pernecky SJ, Vaz ADN :** Cytochrome P 450: Progress and predictions. *FASEB J.* **6 :** *686–694, 1992.*
7. **Endo A :** The discovery and development of HMG-CoA reductase inhibitors. *J. Lipid Res.* **33 :** *1569–1582, 1992.*
8. **Freeland BS:** Diabetic ketoacidosis. *Diabetes Educator.* **29:** *384–395, 2003.*
9. **Goldstein JL, Brown MS :** Regulation of the mevalonate pathway. *Nature.* **343 :** *425–430, 1990.*
10. **Greville GD, Tubbs PK :** The catabolism of long-chain fatty acids in mammalian tissues. *Essays Biochem.* **4 :** *155–212, 1968.*

11. **Gurr MI, Harwood JL :** Lipid Biochemistry : An Introduction. *4th ed. Chapman & Hall, London. 1991.*
12. **Hakomori S :** Glycosphingolipids. *Sci. Amer.* **254 (5) :** *44–53, 1986.*
13. **Hardwood JL :** Fatty acid metabolism. *Ann. Rev. Plant Physiol. Plant Mol. Biol.* **39 :** *101–138, 1988.*
14. **Hawthorne JN, Ansell GB (editors) :** Phospholipids. *Elsevier. 1982.*
15. **Hayaishi O (editor) :** Molecular Mechanisms of Oxygen Activation. *Academic Press Inc., New York. 1974.*
16. **Jain MH :** Introduction to Biological Membranes. *2nd ed. Wiley, New York. 1988.*
17. **Majerus PW :** Inositol phosphate biochemistry. *Ann. Rev. Biochem.* **61 :** *225–250, 1992.*
18. **Masters C, Crane D :** The role of peroxisomes in lipid metabolism. *Trends Biochem. Sci.* **9 :** *314, 1984.*
19. **Numa S (editor) :** Fatty Acid Metabolism and its Regulation. *Elsevier, Armsterdam. 1984.*
20. **Osumi T, Hashimoto T :** The inducible fatty acid oxidation system in mammalian peroxisomes. *Trends Biochem. Sci.* **3 :** *317, 1984.*
21. **Owen JS, McIntyre N :** Plasma lipoprotein metabolism and lipid transport. *Trends Biochem Sci.* **7 :** *92, 1982.*
22. **Raetz CR :** Bacterial endotoxins : Extraordinary lipids that activate eucaryotic signal transduction. *J. Bacteriol.* **175 :** *5745–5753, 1993.*
23. **Ross R :** The pathogenesis of atherosclerosis : A perspective for the 1990s. *Nature.* **362 :** *801–809, 1993.*
24. **Russell DW :** Cholesterol biosynthesis and metabolism. *Cardiovasc Drugs Therapy.* **6 :** *103–110, 1992.*
25. **Schulz H :** Oxidation of fatty acids. In Biochemistry of Lipids and Membranes (Vance DE & Vance JE, editors). *116–142. The Benjamin/Cummings Publishing Company, Menlo Park, CA. 1985.*
26. **Schulz H, Kunau WH :** Beta-oxidation of unsaturated fatty acids : A revised pathway. *Trends Biochem. Sci.* **12 :** *403–406, 1987.*
27. **Scriver CR, Beaudet AL, Sly WS, Valle D (editors) :** The Metabolic Basis of Inherited Disease. *6th ed. McGraw-Hill, New York. 1989.*
28. **Snyder F (editor) :** Lipid Metabolism in Mammals. *Vols. 1 and 2. Plenum, New York. 1977.*
29. **Vance DE, Vance JE (editors) :** Biochemistry of Lipids, Lipoproteins and Membranes. New Comprehensive Biochemistry. *Vol. 20. Elsevier Science Publishing Co., Inc., New York. 1991.*
30. **Vanden BT, Cronan JJ :** Genetics and regulation of bacterial lipid metabolism. *Ann. Rev. Microbiol.* **43 :** *317–343, 1989.*
31. **Vaz AD, Coon MJ :** On the mechanism of action of cytochrome P 450 : Evaluation of hydrogen abstraction in oxygen-dependent alcohol oxidation. *Biochemistry.* **33 :** *6442–6449, 1994.*
32. **Wakil SJ (editor) :** Lipid Metabolism. *Academic Press Inc., New York. 1970.*
33. **Wanders RJ, Waterham HR:** Peroxisomal disorders: the single peroxisomal enzyme deficiencies. *Biochim. Biophys. Acta.* **1763 :** *1707–1720, 2006.*
34. **Wierzbicki AS, Lloyd MD, Schofield CJ, Feher MD, Gibberd FB:** Refsum's disease: a peroxisomal disorder affecting phytanic acid alpha-oxidation. *J. Neurochem.* **80 :** *727–735, 2002.*
35. **Wood PA:** Defects in mitochondrial beta-oxidation of fatty acids. *Curr. Opin. Lipidol.* **10 :** *107–112, 1999.*

PROBLEMS

1. A number of genetic deficiencies in acyl CoA dehydrogenases have been described. This deficiency presents early in life after a period of fasting. Symptoms include vomiting, lethargy, and sometimes coma. Not only are blood levels of glucose low (hypoglycemia), but starvation-induced ketosis is absent. Provide a biochemical explanation for these last two observations.
2. High blood levels of triacylglycerides are associated with heart attacks and strokes. Clofibrate, a drug that increases the activity of peroxisomes, is sometimes used to treat patients with such a condition. What is the biochemical basis for this treatment?
3. Suppose that, for some bizarre reason, you decided to exist on a diet of whale and seal blubber, exclusively.
   (*a*) How would lack of carbohydrates affect your ability to utilize fats?
   (*b*) What would your breath smell like?
   (*c*) One of your best friends, after trying unsuccessfully to convince you to abandon this diet, makes you promise to consume a healthy dose of odd-chain fatty acids. Does your friend have your best interests at heart? Explain.
4. An animal is fed stearic acid that is radioactively labeled with [^{14}C]carbon. A liver biopsy reveals the presence of ^{14}C-labeled glycogen. How is this possible in light of the fact that animals cannot convert fats into carbohydrates?
5. Adipose tissue cannot resynthesize triacylglycerols from glycerol released during lipolysis (fat breakdown). Why not? Describe the metabolic route that is used to generate a glycerol compound for triacylglycerol synthesis.
6. (*a*) Briefly describe the relationship between intracellular malonyl-CoA levels in the liver and the control of ketogenesis.
   (*b*) Describe how the action of glucokinase helps the liver to buffer the level of blood glucose.
7. The action of glucagon on liver cells leads to inhibition of pyruvate kinase. What is the most probable mechanism for this effect?
8. What is the approximate maximum net synthesis of ATP-associated high-energy phosphate bonds (~P) for complete oxidation of 1 mol of oleic acid in a muscle cell?.
9. How does oxidation of a 17-carbon fatty acid (from plants) lead to the production of propionyl-CoA? Be specific in your answer.
10. On a per-carbon basis, where does the largest amount of biologically-available energy in triacylglycerols reside: in the fatty acid portions or the glycerol portion ? Indicate how knowledge of the chemical structure of triacylglycerols provides the answer.
11. Triacylglycerols have the highest energy content of the major nutrients.
    (*a*) If 15% of the body mass of a 70 kg adult consists of triacylglycerols, calculate the total available fuel reserve, in both kilojoules and kilocalories, in the form of triacylglycerols. Recall that 1.00 kcal = 4.18 kJ, and that 1.0 kcal = 1.0 nutritional Calorie.
    (*b*) If the basal energy requirement is approximately 8,400 kJ/day (2,000 kcal/day), how long could this person survive if the oxidation of fatty acids stored as triacylglycerols were the only source of energy?
    (*c*) What would be the weight loss per day in pounds under such starvation conditions (1 lb = 0.454 kg)?

12. Cells often follow the same enzyme reaction pattern for bringing about analogous metabolic reactions. For example, the steps in the oxidation of pyruvate and α-ketoglutarate to acetyl-CoA and succinyl-CoA, although catalyzed by different enzymes, are very similar. The first stage in the oxidation of fatty acids follows a reaction sequence closely resembling one in the citric acid cycle. Show by equations the analogous reaction sequences in the two pathways.

13. Free palmitate is activated to its coenzyme A derivative (palmitoyl-CoA) in the cytosol before it can be oxidized in the mitochondrion. If palmitate and [^{14}C] coenzyme A are added to a liver homogenate, palmitoyl-CoA isolated from the cytosolic fraction is radioactive, but that isolated from the mitochondrial fraction is not. Explain.

14. Contrary to legend, camels do not store water in their humps, which actually consist of a large fat deposit. How can these fat deposits serve as a source of water ? Calculate the amount of water (in litres) that can be produced by the camel from 1 kg (0.45 lb) of fat. Assume for simplicity that the fat consists entirely of tripalmitoylglycerol.

15. When the acetyl-CoA produced during β oxidation in the liver exceeds the capacity of the citric acid cycle, the excess acetyl-CoA reacts to form the ketone bodies acetoacetate, D-β-hydroxybutyrate, and acetone. This condition exists in cases of severe diabetes because the patient's tissues cannot use glucose ; they oxidize large amounts of fatty acids instead. Although acetyl-CoA is not toxic, the mitochondrion must divert the acetyl-CoA to ketone bodies. Why? How does this diversion solve the problem?

16. Suppose you had to subsist on a diet of whale and seal blubber with little or no carbohydrate.

(*a*) What would be the effect of carbohydrate deprivation on the utilization of fats for energy?

(*b*) If your diet were totally devoid of carbohydrate, would it be better to consume odd- or even-numbered fatty acids? Explain.

17. Write the net equation for the complete oxidation of D-β-hydroxybutyrate in the kidney. Include any required activation steps and all oxidative phosphorylations.

18. The complete oxidation of palmitate to carbon dioxide and water is represented by the overall equation:

$$\text{Palmitate} + 23\ O_2 + 129\ P_i + 129\ \text{ADP} \longrightarrow 16\ CO_2 + 129\ \text{ATP} + 145\ H_2O$$

The 145 H_2O molecules come from two separate reactions. What are they, and how many H_2O molecules are produced in each?

ANSWERS TO PROBLEMS

1. The absence of ketone bodies is due to the fact that the liver, the source of blood-ketone bodies, cannot oxidize fatty acids to produce acetyl-CoA. Moreover, because of the impaired fatty acid oxidation, the liver becomes more dependent on glucose as an energy source. This dependency results in a decrease in gluconeogenesis and a drop in blood-glucose levels, which is exacerbated by the lack of fatty acid oxidation in muscle and a subsequent increase in glucose uptake from the blood.

2. Peroxisomes enchance the degradation of fatty acids. Consequently, increasing the activity of peroxisomes could help to lower levels of blood triacylglycerides. In fact, clofibrate is rarely used because of serious side effects.

3. (*a*) Fats burn in the flame of carbohydrates. Without carbohydrates, there would be no anaplerotic reactions to replenish the TCA-cycle components. With a diet of fats only, the acetyl-CoA from fatty acid degradation would build up.

(*b*) Acetone from ketone bodies.

(*c*) Yes. Odd-chain fatty acids would lead to the production of propionyl CoA, which can be converted into succinyl CoA, a TCA-cycle component. It would serve to replenish the TCA cycle and mitigate the halitosis.

4. A labeled fat can enter the citric acid cycle as acetyl-CoA and yield labeled oxaloacetate, but only after two carbon atoms have been lost as CO_2. Consequently, even though oxaloacetate may be labeled, there can be no net synthesis in the amount of oxaloacetate and hence no net synthesis of glucose or glycogen.

5. Adipose tissue lacks glycerol kinase. Glycolysis generates dihydroxyacetone phosphate, which is reduced to glycerol 3-phosphate.

6. (*a*) Malonyl-CoA at high levels inhibits carnitine acyltransferase I, and this inhibits ketogenesis by blocking the transport of fatty acids into mitochondria, both for β-oxidation and for ketogenesis.

(*b*) The high K_M of glucokinase, a liver-specific enzyme, allows the liver to control the rate of glucose phosphorylation over a wide range of glucose concentrations. Accumulation of glucose-6-phosphate activates the D form of glycogen synthase and promotes glycogen deposition. By several mechanisms, the liver also senses when blood glucose levels are low and mobilizes its glycogen reserves accordingly.

7. Phosphorylation of pyruvate kinase by cyclic AMP-dependent protein kinase. The phosphorylated form of the enzyme is far less active than the dephosphorylated form.

8. Oleic acid has 18 carbons and 1 double bond. Nine Acetyl CoA × 10 = 90 ATP. Seven reduced flavoproteins × 1.5 = 10.5 ATP. (**Note:** In one step the flavoprotein reduction is not necessary because the double bond is not present. The isomerase to convert the *cis*-3-enoyl bond to the *trans*-2-enoyl bond does not involve ATP.) Eight NADH × 2.5 = 20 ATP. Total = 120.5 ATP – 2 ~ P (for activation) = 118.5 ~ P/oleic acid.

9. β-oxidation proceeds normally but the final thiolase cleavage yields acetyl CoA and propionyl CoA. Propionyl CoA is *not* a substrate for SCAD (short-chain acyl-CoA dehydrogenase) so β-oxidation terminates.

10. The fatty acid portion; the carbons in fatty acids are more reduced than those in glycerol.

11. (*a*) 4.0×10^5 kJ (9.5×10^4 kcal)

(*b*) 48 days (*c*) 0.5 lb/day

12. The first step in fatty acid oxidation is analogous to the conversion of succinate to fumarate; the second step, to the conversion of fumarate to malate; the third step, to the conversion of malate to oxaloacetate.

13. Fatty acyl groups condensed with CoA in the cytosol are first transferred to carnitine, releasing CoA, then transported into the mitochondrion, where they are again condensed with CoA. The cytosolic and mitochondrial pools of CoA are thus kept separate, and no radioactive CoA from the cytosolic pool enters the mitochondrion.

14. Oxidation of fats releases metabolic water; 0.49 L of water per pound of tripalmitin.

15. Because the mitochondrial pool of CoA is small, CoA must be recycled from acetyl-CoA *via* the formation of ketone bodies. This allows the operation of the β-oxidation pathway, necessary for energy production.

16. (*a*) Glucose yields pyruvate *via* glycolysis, and pyruvate is the main source of oxaloacetate. Without glucose in the diet, [oxaloacetate] drops and the citric acid cycle slows.

(*b*) Odd-numbered; propionate conversion to succinyl-CoA provides intermediates for the citric acid cycle.

17. $$\underset{}{CH_3}\text{—}\overset{\overset{\displaystyle OH}{|}}{CH}\text{—}CH_2COO^- + 4\tfrac{1}{2}\ O_2 + 25\ ADP + 25\ P_i + 25\ H^+ \longrightarrow$$
$$4\ CO_2 + 29\ H_2O + 25\ AT$$

18. (1) The reduction of O_2 in the respiratory chain:

$$CH_3(CH_2)_{14}COO^- + 23\ O_2 \longrightarrow 16\ CO_2 + 16\ H_2O$$

(2) The formation of the anhydride bond in ATP:

$$129\ Pi + 129\ ADP + 129\ AH^+ \longrightarrow 129\ ATP + 129\ H_2O$$

CONTENTS

CHAPTER 25

Biosynthesis of Lipids

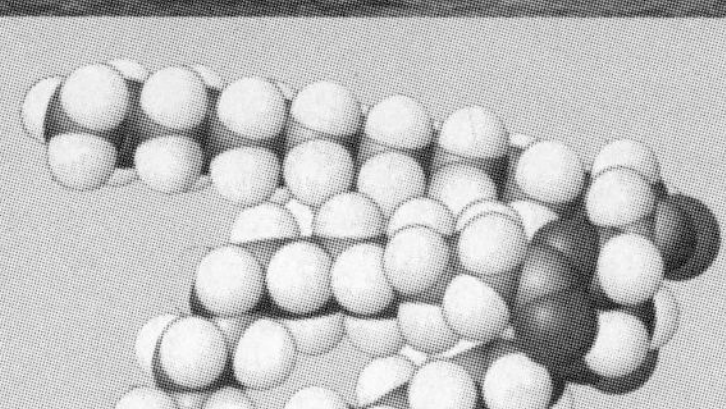

Fats such as the triacylglycerol molecule (*below*) are widely used to store excess energy for later use and to fulfill other purposes, illustrated by the insulating blubber of whales (*above*).
The natural tendency of fats to exist in nearly water-free forms makes these molecules well-suited for these roles.
[*Courtesy : Upper Francois Cohier*]

Lipids play a variety of roles. "They are the principal form of stored energy in most organisms, as well as major constituents of cell membranes. Specialized lipids serve as pigments (retinal), cofactors (vitamin K), detergents (bile salts), transporters (dolichols), hormones (vitamin D derivatives, sex hormones), extracellular and intracellular messengers (eicosanoids and derivatives of phosphatidylinositol) and anchors for membrane proteins (covalently attached fatty acids, phenyl groups and phosphatidylinositol)." (**Lehninger**, **Nelson** and **Cox**, 1993). All organisms are, thus, able to synthesize a variety of lipids which are essential to them. It is, therefore, imperative to deal with biosynthetic pathways for some of the principal lipids present in most cells which will illustrate the strategies that are employed in assembling these water-insoluble products from simple, water-soluble precursors such as acetate. *Like other biosynthetic pathways, these reaction sequences are endergonic and reductive*. They use ATP as a source of metabolic energy and a reduced electron carrier (usually NADPH) as a reductant.

NATURE AND DISTRIBUTION OF FAT STORES

Fat is a fuel for long-term storage. Glycogen and starch are fuel meant for *long-term storage* or for the maintenance of organisms in the presence of limited amounts of oxygen. The human epitomizes this dual storage; the ordinary adult has only enough glycogen to maintain activity for one day or less but he can live from his fat for nearly a month. A human being is built for a daily routine in which he oxidizes glucose residues for energy immediately after meals while rebuilding glycogen reserves and converting any excess glucose to fatty acids. As the time of the last meal recedes, and glycogen supply again becomes depleted, more and more of his energy is obtained by oxidizing fatty acids previously stored as triglycerides. Even the overnight fast is sufficient to cause the amount of oxygen used for the oxidation of fatty acids to be twice that for the oxidation of glucose from glycogen at rest.

The fat is an ideal stored fuel because it is light in weight and the initial appearance on earth of organisms with large fat deposits evidently coincided with the development of the ability to move over relatively long distances without an intake of food. Salmon and ducks, for instance, are alike in building up large stores of fat before they begin their long migrations, but vertebrates of more fixed domicile, along with many insects, also can store fat for less dramatic exertion.

The importance of fat deposition began to be realized with the evolution of vertebrates and *the liver was the initial site of deposition.* Modern sharks frequently have massive livers containing cells loaded with triglycerides. With the appearance of bony fish, fat began to be deposited mainly in and around the muscle fibres, creating the oily flesh we see in salmon and sardines. Insects followed a different route and created a multipurpose organ with many of the functions of vertebrate liver, but which contains so much fat that it is known as the **fat body**. The advanced vertebrates, starting with some fish, developed a discrete **adipose tissue** by modifying the same kind of cells that produced blood cells. These adipocytes contain globules of triglyceride which constitutes 90% or more of the mass of the cell. Adipose tissue is especially abundant in subcutaneous tissues and can become the largest in the body, comprising 50% or more of the total mass of some individuals. Human can become tubs of lard. Such people are objects of humour, disdain or concern in our society. But the advantage with them is that in societies subject to famine they may be happily living on their own fat while burying the last of their formerly lean companions. In fact, the different parts of an animal cell are the sites of fatty acid metabolism : cytosol for fatty anabolism and formation of acyl-CoA and mitochondria for catabolism of acetyl-CoA by β-oxidation process (Fig. 25–1).

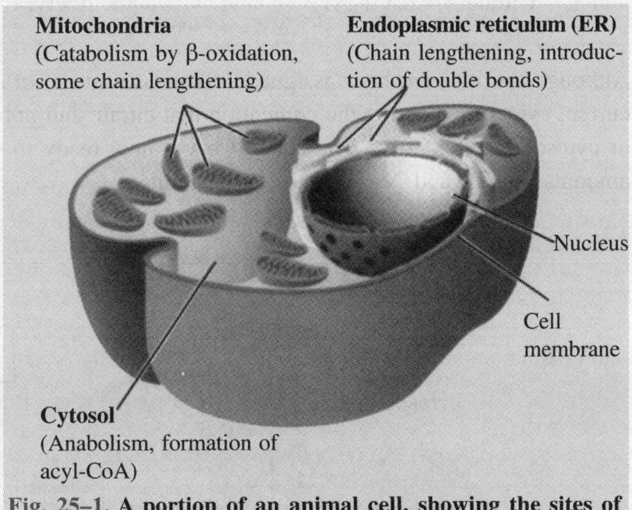

Fig. 25–1. A portion of an animal cell, showing the sites of various aspects of fatty acid metabolism

The cytosol is the site of fatty anabolism. It is also the site of formation of acyl-CoA, which is transported to the mitochondria for catabolism by the β-oxidation process. Some chain-lengthening reactions (beyond C_{16}) take place in the mitochondria. Other chain-lengthening reactions take place in the endoplasmic reticulum (ER), as do reactions that introduce double bonds.

Plants do store fat, especially in the tissues surrounding the embryos in seeds. The light weight of fuel no doubt aids in the dispersal of small seeds, but fat is also the predominant stored fuel in large seeds, where its hydrophobic nature may be of primary importance in protecting the embryo until time for development.

BIOSYNTHESIS OF FATTY ACIDS

When fatty acid oxidation was found to occur by oxidative removal of successive two-carbon (acetyl-CoA) units, biochemists thought that the biosynthesis of fatty acid might proceed by simple reversal of the same enzymatic steps used in their oxidation. However, *fatty acid synthesis and breakdown occur by different routes, are catalyzed by different sets of enzymes and take place in different parts of the cell.*

Acetyl-CoA Transport into the Cytosol

Acetyl-CoA serves as a key intermediate between lipid and carbohydrate metabolism (Fig. 25–2). For the production of fatty acids, acetyl-CoA (which is produced in mitochondria) must first be transported across the organelle's membrane into the cytosol. Since acetyl-CoA itself cannot traverse the membrane, this transfer relies on the transport of the acetyl moiety as citrate (produced from acetyl-CoA and oxaloacetate). After citrate is transferred *via* the tricarboxylate transport system from mitochondria into the cytosol, it is cleaved by **ATP-citrate lyase** to produce acetyl-CoA by the following reaction:

$$\text{Citrate + CoA + ATP} \xrightarrow[\Delta G' = -3{,}400 \text{ cal/mol}]{\text{ATP-citrate lyase}} \text{Acetyl-CoA + Oxaloacetate + ADP + } P_i$$

Although carnitine has been assigned the role as a carrier of acetyl groups, as well as of fatty acids, current evidence supports the contention that citrate and not acetylcarnitine is the principal source of cytosolic acetyl-CoA. The acetyl-CoA is now ready to serve as a substrate with the required amounts of ATP and NADPH to form palmitate.

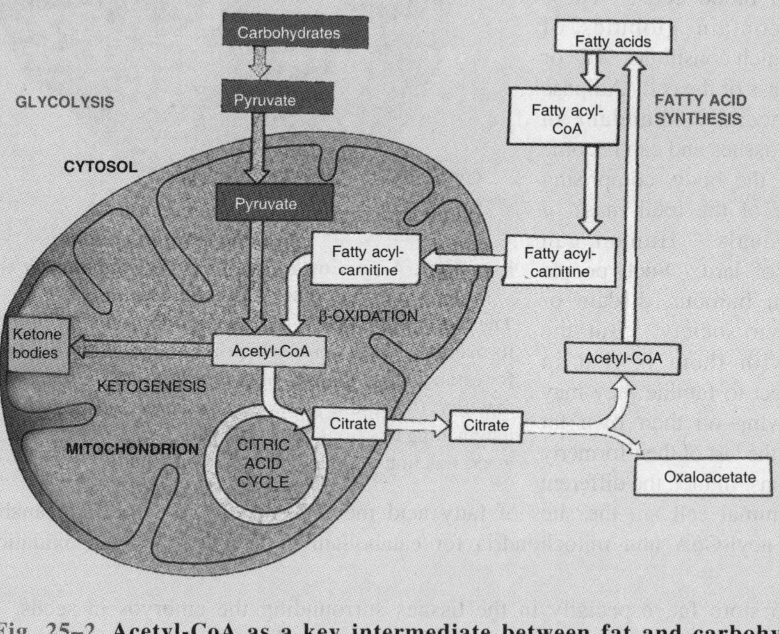

Fig. 25–2. Acetyl-CoA as a key intermediate between fat and carbohydrate metabolism

Arrows identify major routes of formation or utlization of acetyl-CoA. Citrate serves as a carrier to transport acetyl units from the mitochondrion to the cytosol for fatty acid synthesis.

Production of Malonyl-CoA : The Initiation Phase

The production of malonyl-CoA is the initial committed step in the fatty acid synthesis. In 1961, Salih Wakil's observation that CO_2 greatly stimulates the incorporation of acetyl-CoA into fatty acid structures was an important finding in the elucidation of this process. In fact, his studies revealed that acetyl-CoA must be converted, rather carboxylated, into **malonyl-CoA** prior to its utilization for fatty acid synthesis. This irreversible two-step reaction is the committed step in fatty acid synthesis and, as would be expected, it is also the primary rate-limiting reaction of the process. The reaction is catalyzed by the enzyme, acetyl-CoA carboxylase.

The **acetyl-CoA carboxylase** from bacteria is a multienzyme complex and consists of 3 separate, functional polypeptide subunits (Fig. 25–3):

(*a*) *biotin carboxyl carrier protein*, BCP (MW = 45,000), containing two identical subunits each of which has one mole of biotin as its prosthetic group, covalently bound in amide linkage to an ε-amino group of a lysine residue,

(*b*) *biotin carboxylase*, BC (MW = 98,0000), an enzyme with two identical subunits and which catalyzes carboxylation of the biotin unit in biotin carboxyl carrier protein in an ATP-dependent reaction, and

(*c*) *transcarboxylase*, TC (MW = 1,30,000), an enzyme with two pairs of subunits of molecular weight 35,000 and 30,000 respectively, and which catalyzes the transfer of activated CO_2 unit from carboxybiotin to acetyl-CoA, producing malonyl-CoA.

In yeast, higher plants and animals, the activities of all the three subunits are present in a single biotin-containing polypeptide chain (MW ≈ 2,20,000).

Malonyl-CoA is synthesized in two steps by the action of two enzymes, each of which employs the biotin carrier protein as one substrate. The two steps are:

First Step : The biotin carboxylase (BC) catalyzes carboxylation of biotin carboxyl carrier protein (BCP) to yield carboxybiotin carboxyl carrier protein ($BCP–COO^-$); the carboxyl group being derived from bicarbonate (HCO_3^-). This is an ATP-dependent reaction.

$$\underset{\text{(Biotinyl-enzyme)}}{BCP} + HCO_3^- + ATP \xrightarrow{\text{Biotin carboxylase}} \underset{\text{(Carboxybiotinyl enzyme)}}{BCP–COO^-} + ADP + Pi \quad ...(1)$$

Second Step : The transcarboxylase transfers the "bound" CO_2 from $BCP–COO^-$ to acetyl-CoA, forming malonyl-CoA and regenerating BCP.

$$BCP–COO^- + CH_3CO–SCoA \xrightarrow{\text{Transcarboxylase}} BCP + {}^-OOC–CH_2CO–SCoA \quad ...(2)$$

The free energy of cleavage of carboxybiotin protein, ΔG° = 4.7 kcal/mole at pH 7.0, is sufficient to allow the compound to act as a carboxylating agent in reaction (2) as well as in other reactions with suitable acceptors. The exergonic nature of the cleavage also explains the requirement for ATP for formation of the carboxybiotin protein.

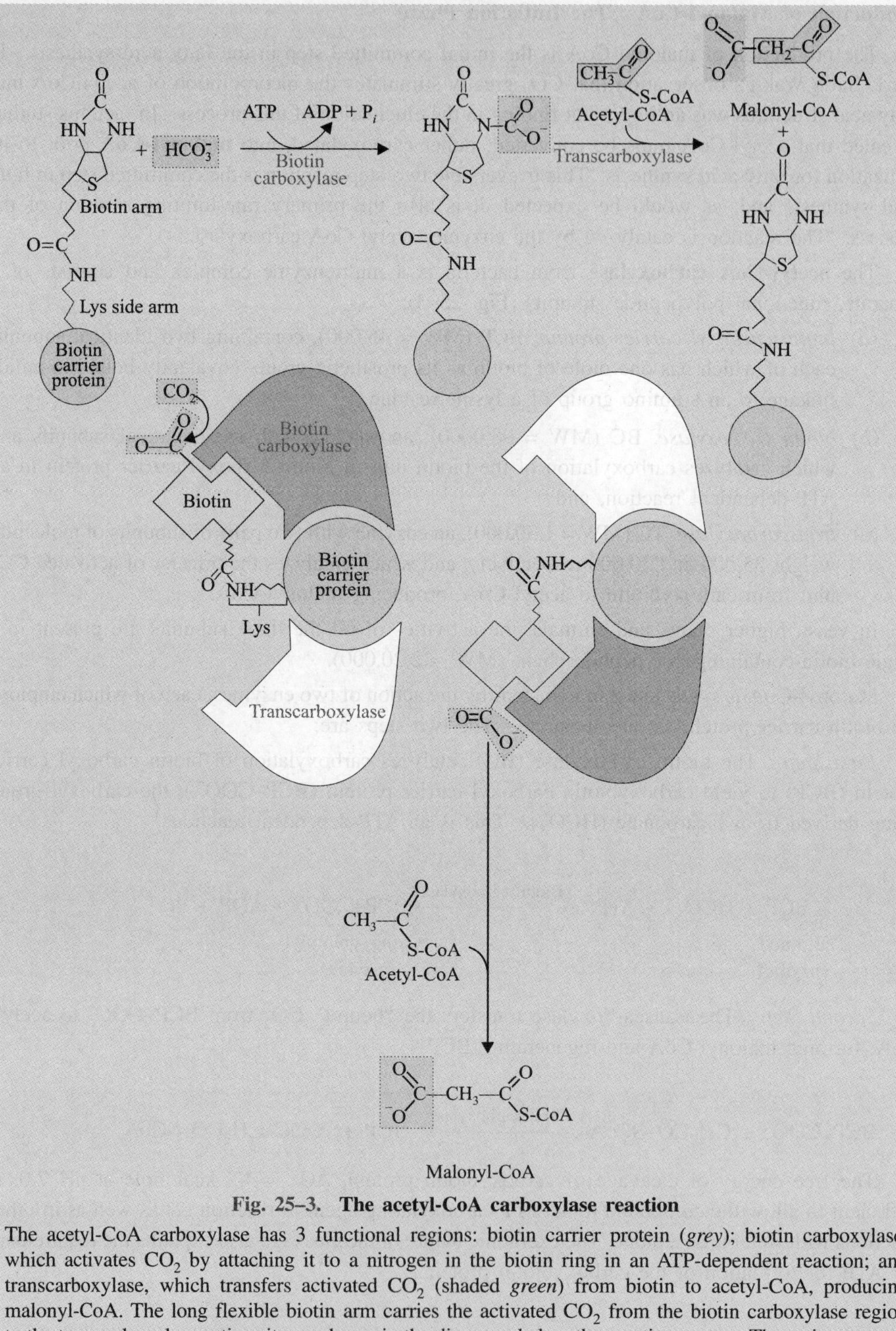

Malonyl-CoA

Fig. 25–3. The acetyl-CoA carboxylase reaction

The acetyl-CoA carboxylase has 3 functional regions: biotin carrier protein (*grey*); biotin carboxylase, which activates CO_2 by attaching it to a nitrogen in the biotin ring in an ATP-dependent reaction; and transcarboxylase, which transfers activated CO_2 (shaded *green*) from biotin to acetyl-CoA, producing malonyl-CoA. The long flexible biotin arm carries the activated CO_2 from the biotin carboxylase region to the transcarboxylase active site, as shown in the diagrams below the reaction arrows. The active enzyme in each step is shaded *blue*.

(*Redrawn from Lehninger AL, Nelson DL and Cox MM, 2005*)

Thus, the substrates are bound to acetyl-CoA carboxylase and products are released in a specific sequence (Fig. 25–4). *Acetyl-CoA carboxylase exemplifies a ping-pong reaction mechanism* in which one or more products are released before all the substrates are bound.

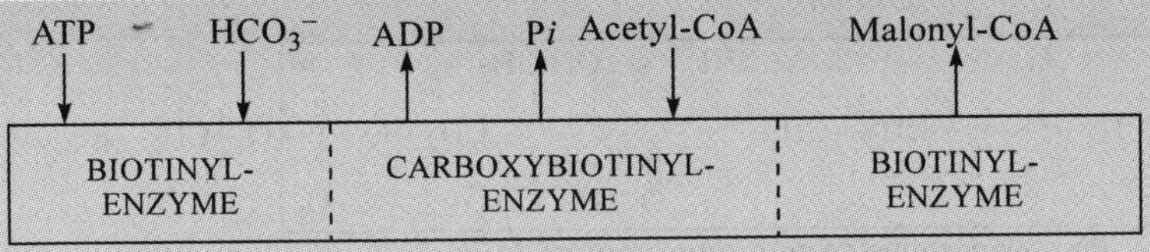

Fig. 25–4. **The reaction sequence of acetyl-CoA carboxylase**

Note that these reactions are "CO_2-fixation" processes in which inorganic CO_2 is used, even by animals, to form organic compounds. The overall result of these two reactions would be the production of a mole of malonyl-CoA by the addition of a mole of CO_2 (actually as HCO_3^-) to a mole of acetyl-CoA; the ATP mole providing energy for driving the reaction. The net equation then would be:

$$\underset{\text{(Acetyl-CoA)}}{CH_3CO - SCoA} + HCO_3^- + ATP \xrightarrow[\text{carboxylase}]{\text{Acetyl-CoA}} \underset{\text{(Malonyl-CoA)}}{^-OOC - CH_2CO - SCoA} + ADP + Pi$$

The malonyl-CoA provides 14 out of 16 carbon atoms of palmitate.

This reaction is very similar to other biotin-dependent carboxylation reactions, such as those catalyzed by pyruvate carboxylase and propionyl-CoA carboxylase.

Acetyl-CoA carboxylase is also important because it is a regulatory step; citrate acts as an allosteric activator for the animal enzyme, but not in plant or microbial systems. The high degree of structural organization of the animal carboxylases, which are absent in their counterparts in plants, yeast and *Escherichia coli*, suggests a possible structural role, in addition to their known catalytic and regulatory functions. It could serve as an organizing matrix for a supramolecular (multienzyme) complex with other enzymes which take part in lipid biosynthesis.

Intermediates in Fatty Acid Synthesis and the ACP

Vagelos (1964) discovered that *the intermediates in fatty acid synthesis are linked to an acyl carrier protein, ACP (MW = # 9,000)*. Specifically, the intermediates are attached to the sulfhydryl (–SH) terminus of a phosphopantetheine group (Fig. 25–5). In the degradataion of fatty acids,

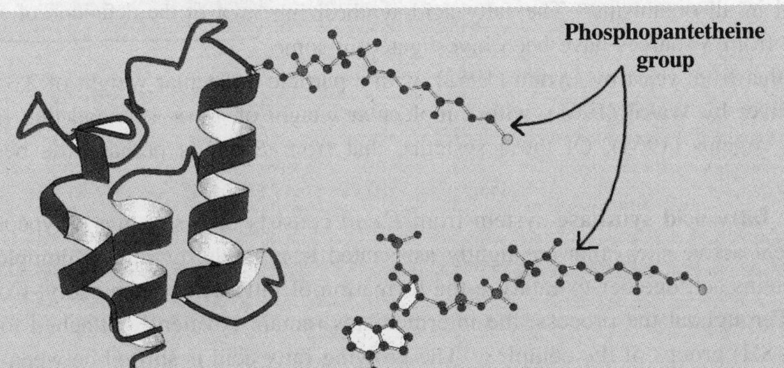

Fig. 25–5. **Phosphopantetheine**

Both acyl carrier protein and CoA include phosphopantetheine as their reactive units.

this unit is part of the CoA; whereas in synthesis, it is attached to a serine residue of the ACP (Fig.25–6). This single polypeptide chain of 77 residues can be regarded as a giant prosthetic group, a "*macro-CoA*". The molecule apparently contains no cysteine.

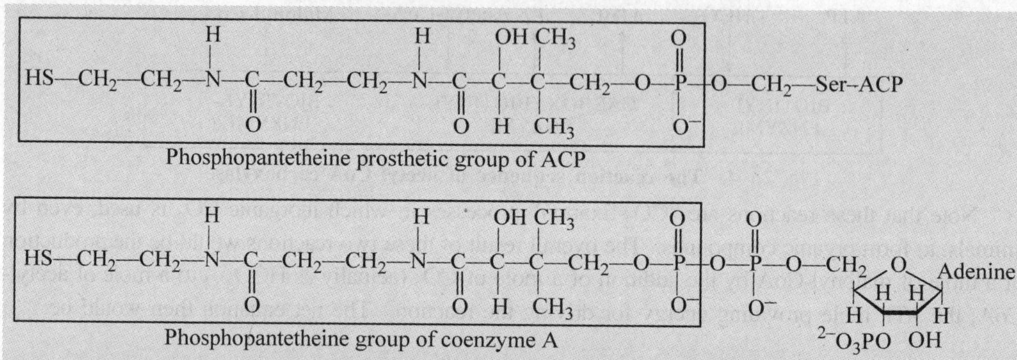

Fig. 25–6. Phosphopantetheine unit of ACP and CoA

Note that in the upper figure, the fatty acid binds to the prosthetic group by forming a thioester bond with the sulfhydryl group. In other words, the —SH group is the site of entry of malonyl groups during fatty acid synthesis.

Acyl carrier protein (ACP) of *Escherichia coli* is a small protein (Relative molecular mass, Mr = 8,860) containing the prosthetic group 4′–phosphopantetheine (Pn), an intermediate in the synthesis of coenzyme A. The thioester bond that links ACP to the fatty acyl group has a high free energy of hydrolysis. And when this bond is broken, energy is released which makes the first reaction in fatty acid synthesis (*i.e.*, condensation reaction) thermodynamically favourable. The 4′-phosphopantetheine prosthetic group of ACP serves as a flexible arm, tethering the growing fatty acyl chain to the surface of the fatty acid synthase complex and carrying the reaction intermediates from one enzyme active site to the other.

The Fatty Acid Synthase Complex

All of the reactions in the biosynthesis of fatty acids are catalyzed by a multienzyme complex, the fatty acid synthase. The detailed structure of this multienzyme complex and its location in the cell differ from one species to another, but the reaction sequence is identical in all organisms. The fatty acid-synthesizing systems from 3 sources have been investigated in some detail : that from yeast by **Lynen** (1952), with a particle molecular weight of 2.3×10^6; that from pigeon liver by **Wakil** (1961), with a molecular weight of 4.5×10^5; and that from *Escherichia coli* by **Vagelos** (1964). Of these systems, that from *E.coli* is perhaps the best understood at present.

Fatty acid synthase is frequently named **fatty acid synthetase**, but its action does not fit the definition of a synthetase.

The **fatty acid synthase** system from *E.coli* consists of 7 separate polypeptides (and hence 7 different active sites) that are tightly associated in a single, organized complex (Table 25–1). The proteins act together to catalyze the formation of fatty acids from acetyl-CoA and malonyl-CoA. Throughout the process, the intermediates remain covalently attached to one of the two thiol (—SH) groups of the complex. The growing fatty acid is shifted between these two —SH groups. *One* is relatively fixed in position because it is on a cysteine residue. It acts as a parking place for acyl groups that are to be lengthened. The *other* —SH group carries the extended chain while it undergoes the reactions necessary for reduction to a saturated acyl group, and it also accepts the acetyl and malonyl groups from which the fatty acid is built. This —SH group can swing across the 7 different catalytic sites because it is located in a residue of phosphopantetheine.

Table 25–1. Seven components* of the fatty acid synthase complex from *Escherichia coli*

Component	*Abbreviation*	*E.C No.*	*Role*
Acyl carrier protein	ACP	6.4.1.2	Carries acyl groups in thioester linkage
Acyl-CoA–ACP transacetylase	AT	2.3.1.9	Transfers acyl group from CoA to cysteine residue of KS
Malonyl-CoA–ACP transferase	MT	2.8.3.3	Transfers malonyl group from CoA to ACP
β-ketoacyl–ACP synthase	KS	2.3.1.16	Condenses acyl and malonyl groups
β-ketoacyl–ACP reductase	KR	1.1.1.36	Reduces β-keto group to β-hydroxy group
β-hydroxyacyl–ACP dehydratase	HD	4.2.1.17	Removes H_2O from β-hydroxyacyl-ACP, creating double bond
Enoyl–ACP reductase	ER	1.3.99.2	Reduces double bond, forming saturated acyl-ACP

* ACP has the specific task of binding the acyl intermediates during fatty acid synthesis. Of the 7 components, *ACP is not an enzyme* while the remaining are enzymatic in behaviour.

The two thiol groups are designated as 'central' and 'peripheral'. The 'central' one is the —SH group of acyl carrier protein (ACP), with the intermediates of fatty acid synthesis form a thioester and the 'peripheral' one is the —SH group of a cysteine residue in β-ketoacyl-ACP synthase, one of the 7 proteins of the multienzyme complex.

Thus, we see that the bacteria contain separate proteins to catalyze the individual reactions of fatty acid synthesis; even the formation of malonyl-CoA occurs in 2 stages (carboxylation of biotin and transfer of the —COO^- group to acetyl-CoA). This lucky circumstance made it much easier to discover the sequence of reactions, since each reaction could be studied separately.

The Fatty Acid Synthase From Some Organisms

It may, however, be noted that the 7 active sites for fatty acid synthesis (6 enzymes + ACP) reside in 7 separate polypeptides in the fatty acid synthase of *Escherichia coli* ; the same holds good for the enzyme complex from **higher plants** (Fig. 25–7). In these complexes, each enzyme is positioned with its active site near that of the preceding and succeeding ezymes of the sequence. The flexible pantetheine arm of ACP can reach all of the active sites, and it carries the growing fatty acyl chain from one site to the next ; the intermediates are not released from the enzyme complex until the finished product is obtained.

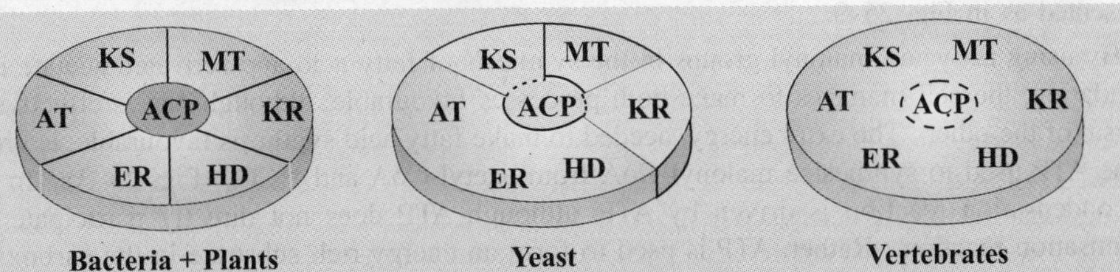

Fig. 25–7. A comparison among the fatty acid synthase complexes from different sources

Note that the fatty acid synthase from **bacteria** and **plants** is a complex where all seven activities reside in seven separate polypeptides. In **yeast**, all 7 activities reside in only 2 polypeptides. And in **vertebrates**, the 7 activities reside in a single large polypeptide.

The fatty acid synthases of yeast and of vertebrates are also multienzyme complexes, but their integration is even more complete than in *E.coli* and higher plants. In **yeast**, the seven distinct active sites reside in only two large, multifunctional polypeptides, and in **vertebrates**, a single large polypeptide (Relative molecular mass, M_r = 2,40,000) contains all seven enzymatic activities as well as a hydrolytic activity that cleaves the fatty acid from the ACP-like part of the enzyme complex. The active form of this multienzyme protein is a dimer (M_r = 4,80,000).

The organized structure of the fatty acid synthases of yeast and higher organisms enhances the efficiency of the overall process because of the following reasons:

1. The intermediates are directly transferred from one active site to the next.
2. The intermediates are not diluted in the cytosol.
3. The intermediates do not have to find each other by random diffusion.
4. The covalently-bound intermediates are secluded and protected from competing reactions.

Priming of the Fatty Acid Synthesis by Acetyl-CoA : The Priming Phase

The sequence of events that occurs during synthesis of a fatty acid is listed in Fig. 25–8. Before the condensation reactions, that build up the fatty acid chain, can begin, the two —SH groups on the enzyme complex must be charged with the correct acyl groups. The 'priming' of the system, as it is called, takes place in 2 steps:

In the *first step*, the acetyl group of acetyl-CoA is tranferred to the cysteine—SH group of the β-ketoacyl-ACP synthase. This reaction is catalyzed by **acetyl-CoA-ATP transacetylase.**

In the *second step*, the malonyl group from malonyl-CoA is transferred to the —SH group of ACP by the enzyme **malonyl-CoA-ACP transferase**, also part of the complex.

Growth of the Fatty Acyl Chain by Two Carbons : The Elongation Phase

First Round : In the charged synthase complex, the acetyl and malonyl groups are very close to each other and are activated for the chain-lengthening process, which consists of the following four steps (or reactions):

1. Condensation. The first step in the formation of a fatty acid chain is condensation of the activated acetyl and malonyl groups to form an acetoacetyl group bound to ACP through the phosphopantetheine —SH group, thus producing **acetoacetyl-ACP**; simultaneously, a mole of CO_2 is eliminated from the malonyl group. In this reaction, catalyzed by β-*ketoacyl-ACP synthase*, the acetyl group is transferred from the cysteine —SH group of this enzyme to the malonyl group on the —SH of ACP, becoming the methyl-terminal two-carbon unit of the new acetoacetyl group. The carbon atom in the CO_2 formed in this reaction is the same carbon atom that was originally introduced into malonyl-CoA from HCO_3^- by the acetyl-CoA carboxylase reaction. Thus, CO_2 is only transiently in covalent linkage during fatty acid biosynthesis; it is removed as such as each two-carbon unit is inserted. Thus, the net effect of condensation reaction is the extension of the acyl chain by 2 carbon atoms. Thus, the first condensation reaction in the biosynthesis of a fatty acid may be diagrammatically represented as in Fig. 25–9.

By using activated malonyl groups in the synthesis of fatty acid and activated acetate in their degradation, the cell manages to make both processes favourable, although one is effectively the reversal of the other. The extra energy, needed to make fatty acid synthesis favourable, is provided by the ATP used to synthesize malonyl-CoA from acetyl-CoA and HCO_3^- (Fig. 25–3). In effect, the condensation reaction is driven by ATP, although ATP does not directly participate in the condensation reaction. Rather, ATP is used to form an energy-rich substrate in the carboxylation of acetyl-CoA to malonyl-CoA. The free energy stored in malonyl-CoA in the carboxylation reaction is released in the decarboxylation accompanying the formation of acetoacetyl-ACP. Although HCO_3^- is required for fatty acid synthesis, its carbon does not appear in the product. Rather, all of the carbon atoms of even-chain fatty acids are derived from acetyl-CoA.

The next 3 steps in fatty acid synthesis reduce the keto (> CO) group at C-3 to a methylene (—CH_2—) group, the result being the conversion of acetoacetyl-ACP into butyryl-ACP.

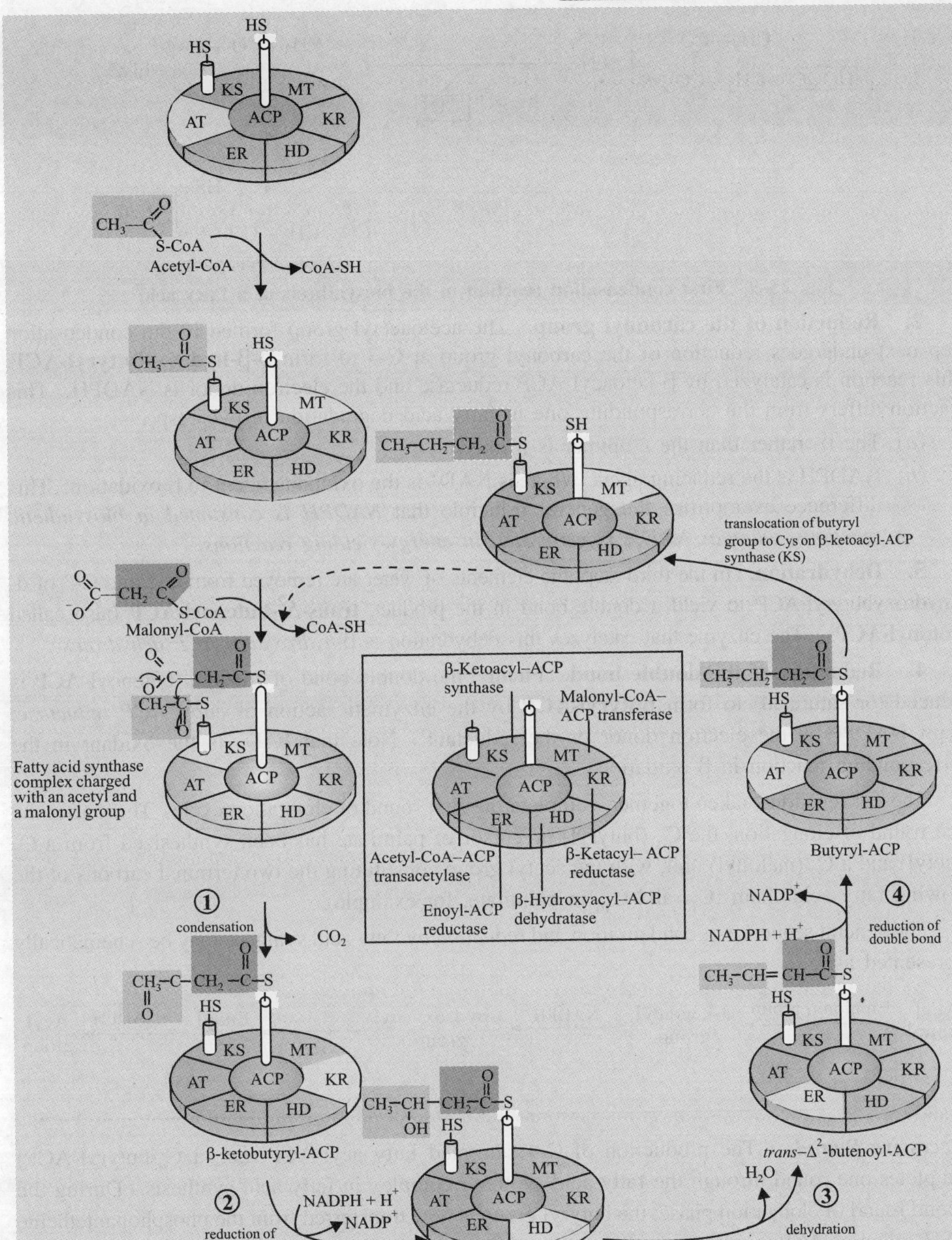

Fig. 25–8. The sequence of events occuring during fatty acid synthesis

The fatty acid synthase complex is shown schematically. Each segment of the disc represents one of the 6 enzymatic activities of the complex. At the centre is acyl carrier protein (ACP), with its phosphopantetheine arm ending in an —SH. The enzyme shown in *blue* is the one that will act in the active step. The initial acetyl group is shaded *yellow*, the C-1 and C-2 of malonate are shaded *pink*, and the carbon released as CO_2 is shaded *green*.

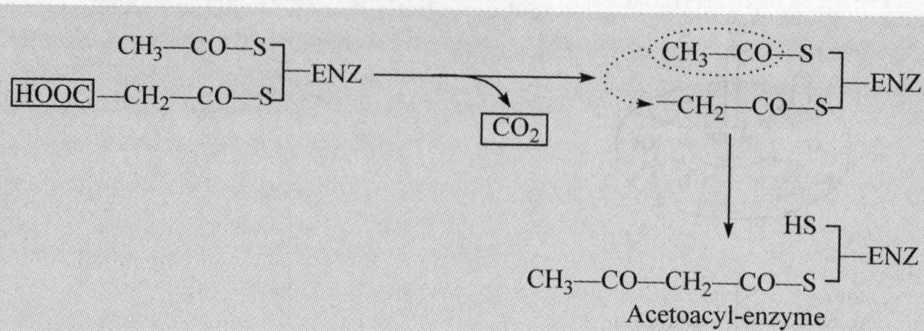

Fig. 25–9. First condensation reaction in the biosynthesis of a fatty acid

2. **Reduction of the carbonyl group.** The acetoacetyl group formed in the condensation step next undergoes reduction of the carbonyl group at C-3 to form **D-β-hydroxybutyryl-ACP.** This reaction is catalyzed by β-ketoacyl-ACP reductase and the electron donor is NADPH. This reaction differs from the corresponding one in fatty acid degradation in two respects:

(*a*) The D- rather than the L-epimer is formed.

(*b*) NADPH is the reducing agent, whereas NAD^+ is the oxidizing agent in β oxidation. This difference exemplifies the general principle that *NADPH is consumed in biosynthetic reactions, whereas NADH is generated in energy-yielding reactions.*

3. **Dehydration.** In the third step, the elements of water are removed from C-2 and C-3 of d-β-hydroxybutyryl-ACP to yield a double bond in the product, **trans-Δ^2-butenoyl-ACP** (also called **crotonyl-ACP**). The enzyme that catalyzes this dehydration is *β-hydroxyacyl-ACP dehydratase.*

4. **Reduction of the double bond.** Finally, the double bond of *trans*-Δ^2-butenoyl-ACP is reduced (or saturated) to form **butyryl-ACP** by the enzymatic action of *enoyl-ACP reductase*; again NADPH is the electron donor or the reductant. Note that FAD^+ is the oxidant in the corresponding reaction in β oxidation.

These 4 reactions, taken together, complete the first round of elongation cycle. Thus, after the first round of elongation, the C_4 (butyryl) precursor of palmitate has been synthesized from a C_2 (acetyl) and a C_3 (malonyl) unit, with the acetyl group constituting the two terminal carbons of the growing fatty acid chain (C_{15} and C_{16} in palmitate, for example).

The general sequence of condensation and reduction by fatty acid synthase may be schematically represented as:

Acyl groupl $\xrightarrow[-\,CO_2]{+\text{ malonyl group}}$ 3-ketoacyl group $\xrightarrow{+\text{ NADPH}}$ 3-hydroxyacyl group $\xrightarrow[-\,H_2O]{}$ Enoyl group $\xrightarrow{+\text{ NADPH}}$ Acyl group

n carbons ← —— *(n+2) carbons* —— →

Successive Rounds **:** The production of C-4 saturated fatty acyl-ACP (*i.e.*, a C_4-butyryl-ACP) completes one round through the fatty acid synthase complex in fatty acid synthesis. During the second round of elongation phase, the butyryl group is now transferred from the phosphopantetheine —SH group of ACP to cysteine —SH group of β-ketoacyl-ACP synthase (KS). To start the next cycle of 4 reactions, that lengthens the chain by 2 more carbons, another malonyl group is linked to the now vacant phosphopantetheine —SH group of ACP. Condensation occurs as the butyryl group, acting exactly as did the acetyl group in the first round, is linked to two carbons of the malonyl-ACP with simultaneous release of a mole of CO_2. The product of this condensation is a C-6 acyl group, covalently bound to the phosphopantetheine —SH group of ACP (*i.e.*, a **C_6-β-ketoacyl-ACP**). Its β-keto group is reduced in the next 3 reactions of the second round of synthesis cycle to yield the C-6 saturated fatty acyl-ACP (*i.e.*, a **C_6-fatty acyl-ATP**), exactly as in the first round of reactions.

The C_6-fatty acyl-ACP is now ready for a third round of elongation. Seven such cycles of condensation and reduction produce the **C-16 saturated palmitoyl group**, still bound to ACP. This intermediate is not a substrate for the condensing enzyme, β-ketoacyl-ACP synthase (KS) and the chain elongation generally stops at this point. Rather, it is hydrolyzed to yield **palmitate** and ACP. Small amounts of longer-chain fatty acids such as stearate (18 : 0) are also formed. In certain plants (coconut and palm, for example), chain termination occurs earlier; a majority of the fatty acids (up to 90%) in the oils of these plants contain between 8 and 14 carbon atoms. Thus, we see that the fatty acid synthase reactions are repeated to form palmitate. The origin of carbon atoms in palmitic acid is as shown in Fig. 25–10.

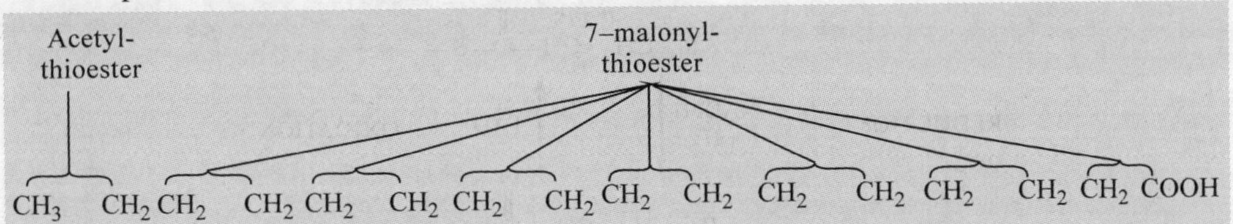

Fig. 25–10. The origin of carbon atoms in palmitic acid

Stoichiometry of Fatty Acid Synthesis

The overall reaction for the synthesis of palmitate from acetyl-CoA can be broken down into 2 parts:

(*a*) *the formation of 7 malonyl-CoA molecules*

$$7 \text{ Acetyl-CoA} + 7\, CO_2 + 7 \text{ ATP} \longrightarrow 7 \text{ Malonyl-CoA} + 7 \text{ ADP} + 7 \text{ P}i \quad ...(i)$$

(*b*) *the 7 cycles of condensation and reduction*

$$\text{Acetyl-CoA} + 7 \text{ Malonyl–CoA} + 14 \text{ NADPH} + 14\, H^+ \longrightarrow \text{Palmitate} + 7\, CO_2 + 8 \text{ CoA} + 14 \text{ NADP}^+ + 6\, H_2O \quad ...(ii)$$

Hence, the overall process (the sum of Equations *i* and *ii*) for the synthesis of palmitate is:

$$8 \text{ Acetyl-CoA} + 7 \text{ ATP} + 14 \text{ NADPH} + 14\, H^+ \longrightarrow \text{Palmitate} + 8 \text{ CoA} + 7 \text{ ADP} + 7 \text{ P}i + 14 \text{ NADP}^+ + 6\, H_2O \quad ...(iii)$$

[Note that the CO_2 utilized (formation of malonyl-CoA) and the CO_2 produced (condensation reaction) cancel each other when the overall stoichiometry is tabulated.]

The biosynthesis of fatty acids such as palmitate, thus, requires acetyl-CoA and the input of chemical energy in 2 forms: the group transfer potential of ATP and the reducing power of NADPH. The ATP is required to attach CO_2 to acetyl-CoA to produce malonyl-CoA ; the NADPH is required to reduce the double bonds to form the corresponding saturated fatty acyl group.

Comparison of Fatty Acid Synthesis and Degradation

Although fatty acid synthesis and degradation represent two independent cellular mechanisms, the two systems share many similarities. The one consistent feature of all synthetic and catabolic reactions is that they have a strict specificity for fatty acyl derivatives, either those of CoA or those of ACP. As illustrated in Fig. 25–11, the last 3 steps in synthesis are the biochemical reversal of the 3 key catabolic reactions of β oxidation.

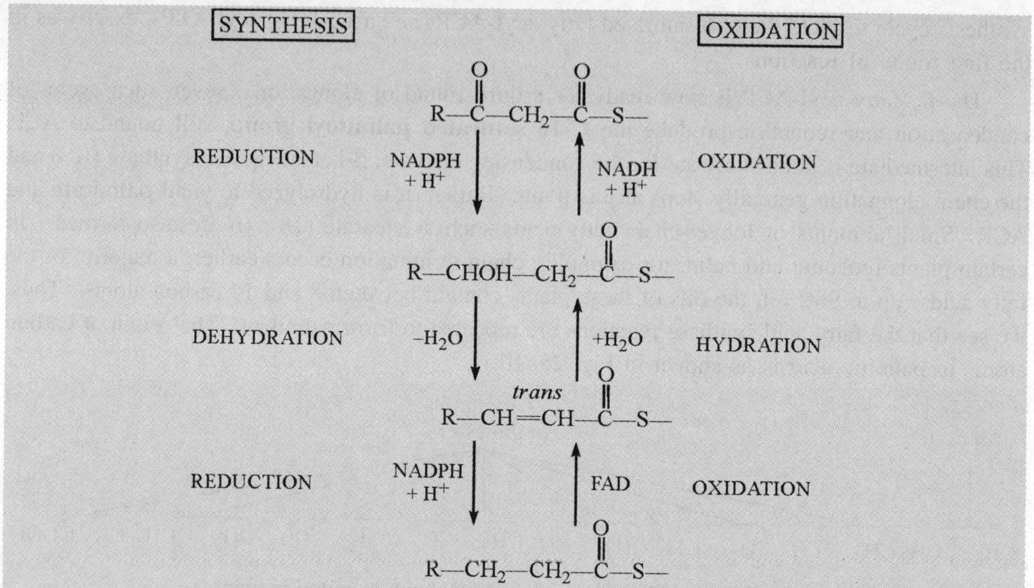

Fig. 25–11. Comparison of fatty acid synthesis and β oxidation

Since each step in the enzymatic pathway responsible for the oxidation of fatty acids is reversible, and also that acetyl-CoA is the starting material, the proposition was advanced that fatty acid biosynthesis might occur by simple reversal of fatty acid oxidation. Rather, fatty acid oxidation consists of a new set of reactions, which once again exemplifies the principle that *synthetic and degradative pathways in biological systems are usually distinct*. The two processes (fatty acid synthesis and oxidation) are distinct from each other in following respects:

1. Fatty acid synthesis (lipogenesis) takes place in the soluble *cytosol* fraction of many tissues, including liver, kidney, brain, lung and adipose tissue, in contrast with fatty acid degradation which occurs principally or entirely in the mitochondrial matrix.
2. The intermediates in fatty acid synthesis are covalently linked to the sulfhydryl groups of an *acyl carrier protein* (ACP), whereas the intermediates in the fatty acid breakdown are bonded to coenzyme A.
3. Many of the enzymes of fatty acid synthesis in higher organisms are organized into a multienzyme complex called the *fatty acid synthase*. In contrast, the degradative enzymes do not seem to be associated.
4. The growing fatty acid chain is elongated by the sequential addition of two-carbon units, derived from acetyl-CoA. The activated donor of two-carbon units in the elongation step is *malonyl*-ACP. The elongation reaction is driven by the release of CO_2.
5. Elongation of the fatty acid synthase complex stops upon formation of *palmitate* (C_{16}). In other words, free palmitate is the main end product. Further elongation and the insertion of double bonds are carried out by other enzyme systems.
6. The source of CO_2 in fatty acid synthesis is *bicarbonate* (HCO_3^-) whereas bicarbonate has no effect upon fatty acid oxidation.
7. Also, C-3 intermediate, *malonyl-CoA* participates in the biosynthesis of fatty acids but not in their breakdown.
8. The reductant in fatty acid synthesis is *NADPH*, whereas NAD^+ and FAD serve as electron acceptors in the oxidative pathway. This difference is an excellent example of the concept that *NAD^+ and NADPH coenzymes are used in catabolic and biosynthetic reactions, respectively.*

9. The activating groups in the synthetic sequence are two different enzyme-bound thiol (–SH) groups; whereas in the degradative process, the activating group is the –SH group of coenzyme A.
10. During fatty acid synthesis, the substrates are *ACP-S derivatives*, whereas fatty acid oxidation involves the action of the enzymes on CoA derivatives as substrates.
11. Also, during fatty acid synthesis, it is the D(–)β-hydroxyacyl–S–ACP derivative that is substrate, whereas during fatty acid oxidation, it is the CoA derivative of the L(+) β-hydroxyacyl compound, that is the substrate. Thus, the configurations of the β-hydroxyacyl intermediates differ in the two systems.

Table 25–2. lists a comparison of compounds involved in fatty acid metabolism.

Table 25–2. Comparison of compounds involved in fatty acid metabolism

Step / Component	*Degradation*	*Synthesis*
SH component	CoA	Acyl carrier protein
Intemediate SH derivative	Acetyl-CoA	Malonyl ACP + acetyl ACP
Keto ⟷ hydroxy	NAD, L-β-hydroxybutyryl CoA	NADPH, D-β-hydroxybutyryl ACP
Crotonyl ⟷ butyryl	FAD, electron transport system	NADPH, fatty acyl-ACP

BIOSYNTHESIS OF LONG-CHAIN FATTY ACIDS

(= Mitochondrial And Microsomal Fatty Acid Synthesis)

Palmitate (a C_{16} fatty acid) is the major product of the fatty acid synthase system. This system is also called as the *de novo system* in that the palmitate is constructed from acetyl-CoA (ACP) and malonyl-CoA (ACP). In plants and animals, the most important fatty acids are the C_{18} fatty acids, namely stearic, oleic, linoleic and linolenic. These C_{18} fatty acids are synthesized by *elongation systems* which differ markedly from the *de novo system.* Palmitate acts as a precursor of other long-chain fatty acids (Fig. 25–12). It may be lengthened to form stearate (18 : 0) or even larger saturated fatty acids by further addition of acetyl groups through the action of fatty acid elongation systems present in the endoplasmic reticulum (= microsomes) and mitochondria in the case of animals and in the soluble cytosol in the case of plants.

In animals:

Endoplasmic reticulum membrane

$$\text{Palmitoyl-CoA} \xrightarrow[\text{+ NADPH}]{\text{Malonyl-CoA}} \text{Stearyl-CoA}$$

Mitochondrial outer and inner membranes

$$\text{Palmitoyl-CoA} \xrightarrow[\text{+ NADPH}]{\text{Acetyl-CoA}} \text{Stearyl-CoA}$$

In plants:

Soluble cytosolic system

$$\text{Palmitoyl-ACP} \xrightarrow[\text{+ NADPH}]{\text{Malonyl-ACP}} \text{Stearyl-ACP}$$

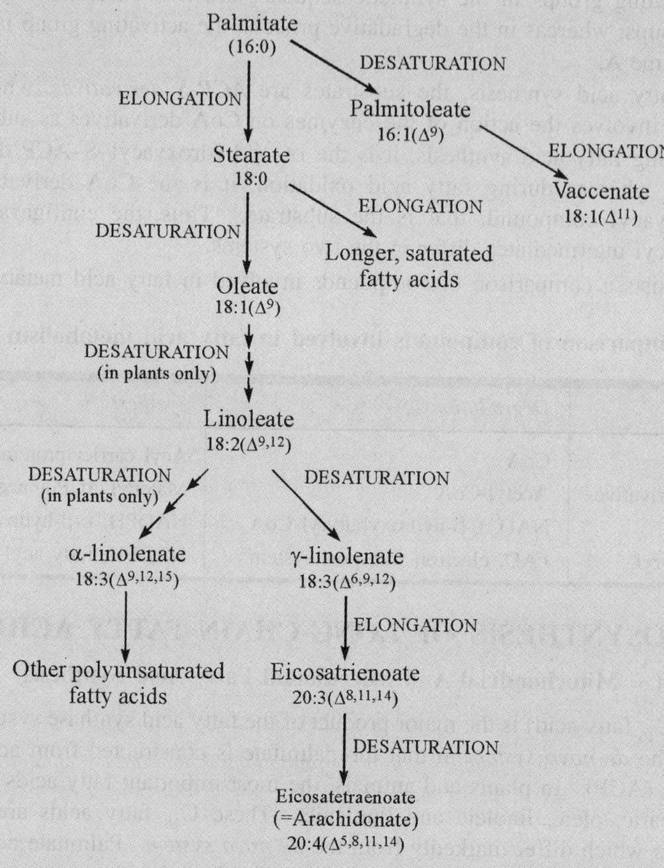

Fig. 25–12. Biosynthesis of some fatty acids in mammals

Palmitate is the precursor of stearate and longer-chain saturated fatty acids, as well as the monounsaturated fatty acids, palmitoleate and oleate. Mammals cannot convert oleate into linoleate or a-linolenate, hence required in the diet as essential fatty acids (EFAs). Conversion of linoleate into other polyunsaturated fatty acids and eicoandnoids is also outlined.

Although different enzyme systems are involved and coenzyme A, rather than ACP, is the acyl carrier directly involved in animal fatty acid synthesis, the mechanism of elongation is otherwise identical with that employed in palmitate synthesis: donation of two carbons by malonyl-ACP, followed by reduction (with NADH), dehydration and another reduction (with NADPH) to the saturated C_{18} product, stearyl-CoA. *Fasting largely abolishes chain elongation.* Elongation of stearyl-CoA in brain increases rapidly during myelination in order to provide C_{22} and C_{24} fatty acids that are present in sphingolipids.

BIOSYNTHESIS OF UNSATURATED FATTY ACIDS

Palmitate and stearate serve as precursors of the two most common monounsaturated fatty acids of animal tissues, palmitoleate and oleate. Both of them possess a single *cis* double bond between C–9 and C–10. The double bond is introduced into the fatty acid chain by an oxidative reaction catalyzed by **fatty acyl-CoA desaturase** (Fig. 25–13). The enzyme is an example

> The name **mixed-function oxidase** indicates that the enzyme oxidizes two different substrates simultaneously.

of a mixed-function oxidase. Two different substrates, a fatty acyl-CoA and NADPH, simultaneously undergo two-electron oxidations. The path of electron flow includes a cytochrome (cytochrome b_5) and a flavoprotein (cytochrome b_5 reductase), both of which, like fatty acyl-CoA desaturase itself, are present in the smooth endoplasmic reticulum.

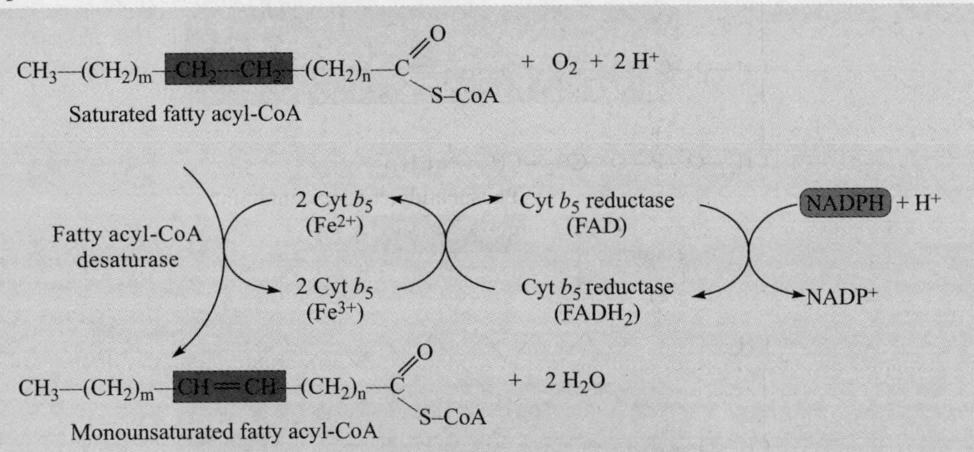

Fig. 25–13. The pathway of electron transfer in the desaturation of fatty acids by a mixed-function oxidase in vertebrates

Two different substrates– a fatty acyl-CoA and NADPH– undergo oxidation by molecular oxygen. These reactions occur on the lumenal face of the smooth ER. A similar pathway, but with different electron carriers, exists in plants.

Mammalian hepatocytes can readily introduce double bonds at Δ^9 position of fatty acids but cannot introduce additional double bonds in the fatty acid chain between C–10 and the methyl-terminal end. Thus, linoleate, 18 : 2 ($\Delta^{9, 12}$) and α-linolenate, 18 : 3 ($\Delta^{9, 12, 15}$) cannot be synthesized by mammals, but plants can synthesize both. The plant desaturases that introduce double bonds at Δ^{12} and Δ^{15} positions are located in the endoplasmic reticulum. These enzymes, in fact, act not on free fatty acids but on a phospholipid called phosphatidylcholine which contains at least one oleate linked to glycerol (Fig. 25–14). Because linoleate and linolenate are necessary precursors for the synthesis of other products, they are *essential fatty acids* (EFAs) for mammals and must be obtained from plant material in the diet. Once ingested, linoleate may be converted into certain other polyunsaturated fatty acids, particularly γ linolenate, eicosatrienoate and eicosatetraenoate (= arachidonate), which can be made only from linoleate. Arachidonate, 20 : 4 ($\Delta^{5, 8, 11, 14}$) is an essential precursor of regulatory lipids, the eicosanoids.

Families of Fatty Acids **:** Animal tissues contain a variety of polyunsaturated fatty acids. Of these, one series can be fabricated by the animal *de novo*. These are those fatty acids where all the double bonds lie between the 7th carbon from the terminal methyl group and the carboxyl group. Such fatty acids can be made by desaturation and chain elongation, starting with oleic acid (Fig. 25–12). Thus, oleate (which is produced from its corresponding saturated fatty acid, *i.e.*, stearate) can be elongated to a 20 : 1, *cis*- Δ^{11} fatty acid. Alternatively, a second double bond can be inserted to yield an 18 : 2 , *cis*-Δ^6, Δ^9 fatty acid. However, polyunsaturated fatty acids, in which one or more double bonds are situated within the terminal 7 carbon atoms, cannot be made *de novo*. Such polyunsaturated fatty acids are essential in the diet.

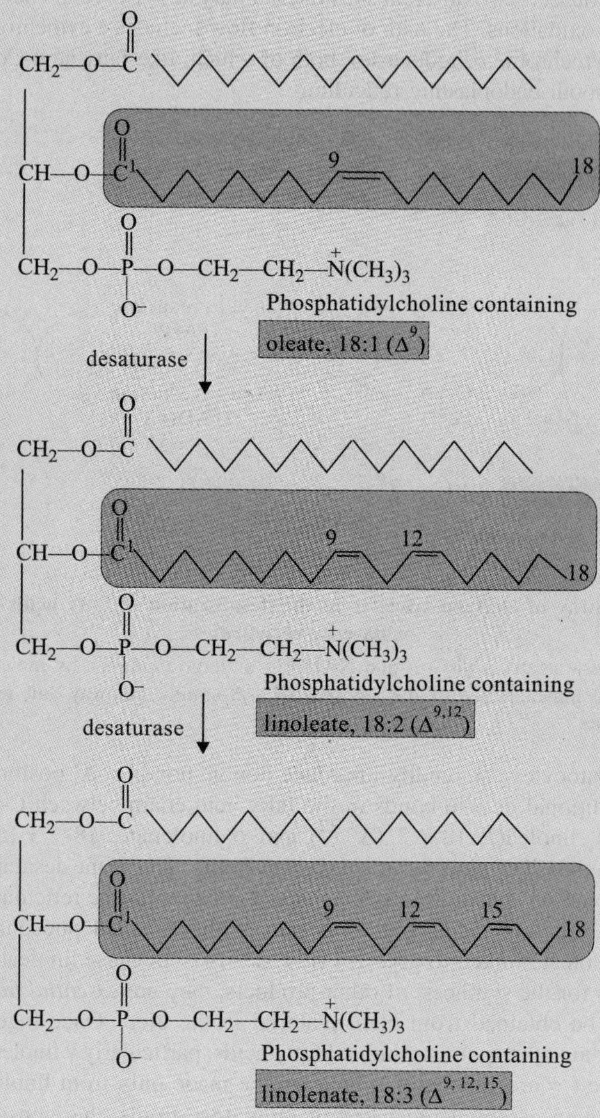

Fig. 25–14. Action of plant desaturases

Desaturases in plants oxidize phosphatidylcholine-bound oleate to polyunsaturated fatty acids. Some of the products are released from the phosphatidylcholine by hydrolysis.

There are 4 series of polyunsaturated fatty acids in the mammals (Table 25–3): two are derived from dietary linoleate and linolenate and two are sythesized from the monounsaturated fatty acids, oleate and palmitoleate, which in their turn are formed from the corresponding saturated fatty acids. *Conjugated double bonds are not formed in animal tissues*. The four series can be recognized by the distance between the terminal methyl group (ω) and the nearest double bond. Desaturation and elongation reactions occur more extensively in liver than in extrahepatic tissues. Fasting and diabetes are marked by an inhibition of desaturation pathways.

Table 25–3. **The four series of polyunsaturated fatty acids**

Precursor family	*Systemic code*	*Formula*	*Series**
Linolenate	18 : 3 ; 9, 12, 15	$CH_3–(CH_2)_1$— CH = CH — R	ω-3†
Linoleate	18 : 2 ; 9, 12	$CH_3–(CH_2)_4$ — CH = CH — R	ω-6
Palmitoleate	16 : 1 ; 9	$CH_3–(CH_2)_5$ — CH = CH — R	ω-7
Oleate	18 : 1 ; 9	$CH_3–(CH_2)_7$ — CH = CH — R	ω-9

* Note that the family of precursor and its products generated in animals through elongation and desaturation can be identified by subtracting the number designating the last double bond from the total number of carbon atoms. The result is the same within a family. For example, with linoleate (18 : 2; 9, 12) and arachidonate (20 : 4; 5, 8, 11, 14): 18–12 = 6 = 20–14.

† Nutritionists number fatty acids backward from biochemists, so the term ω-3 (in case of C-18 linolenic acid, for example) refers to a double bond on the 3rd carbon from the terminal methyl group what we would call the bond between C-15 and C-16.

BIOSYNTHESIS OF EICOSANOIDS

Eicosanoids is a family of very potent biological signalling molecules that act as short-range messengers affecting tissues *near* the cells that produce them. All 3 subclasses of eicosanoids (prostaglandins, thromboxanes and leukotrienes) are unstable and insoluble in water. These signalling molecules generally do not move far from the tissue that produced them and act primarily on cells very near their point of release. Eicosanoids are derivatives of the C-20 polyunsaturated fatty acid, arachidonate.

In the beginning, a specific *phospholipase*, present in most of the mammalian cells, attacks the membrane phospholipids and releases **arachidonate**. Arachidonate is then converted into **PGH_2**, the immediate precursor of many **prostaglandins** and thromboxanes (Fig. 25–15 A), by the enzymes of the smooth endoplasmic reticulum. These two reactions which lead to the formation of PGH_2 involve the addition of molecular oxygen and are both catalyzed by a bifunctional enzyme, *prostaglandin endoperoxide synthase*. Aspirin (= acetylsalicylate) irreversibly inactivates this enzyme by acetylating a serine residue essential to catalytic activity (Fig. 25–15 B). Thus, the synthesis of prostaglandins and thromboxanes is blocked. *Ibuprofen* (Fig. 25–15 C), a widely-used nonsteroidal antiinflammatory drug, also inhibits this step, probably by mimicking the structure of the substrate or an intermediate in the reaction. Another drug *naproxen* also has the inhibitory action identical to that of ibuprofen.

PGH_2 is enzymatically converted into **thromboxane A_2**, from which **other thromboxanes** are derived. This reaction is catalyzed by *thromboxane synthase*, an enzyme present in blood platelets (= thrombocytes). Thromboxanes induce blood vessel constriction and platelet aggregation, the early steps in blood clotting. Thromboxanes and prostaglandins both contain a ring of 5 or 6 atoms, and the pathway that leads from arachidonate to these two classes of compounds is sometimes called the **'cyclic'** pathway, to differentiate it from the **'linear'** pathway that results in the synthesis of the 'linear' molecules of leukotrienes from arachidonate (Fig. 25–17).

Leukotriene synthesis begins with the incorporation of molecular oxygen into arachidonate by the enzymatic action of the mixed-function oxidases called *lipoxygenases*. The lipoxygenases (Fig. 25–16) are found in leukocytes and in heart, brain, lung and spleen, and utilize cytochrome P-450 for their acitvity. The various leukotrienes differ in the position of the peroxide that is introduced by these lipoxygenases. Unlike the cyclic pathway, this linear pathway for leukotriene synthesis is not inhibited by aspirin or ibuprofen.

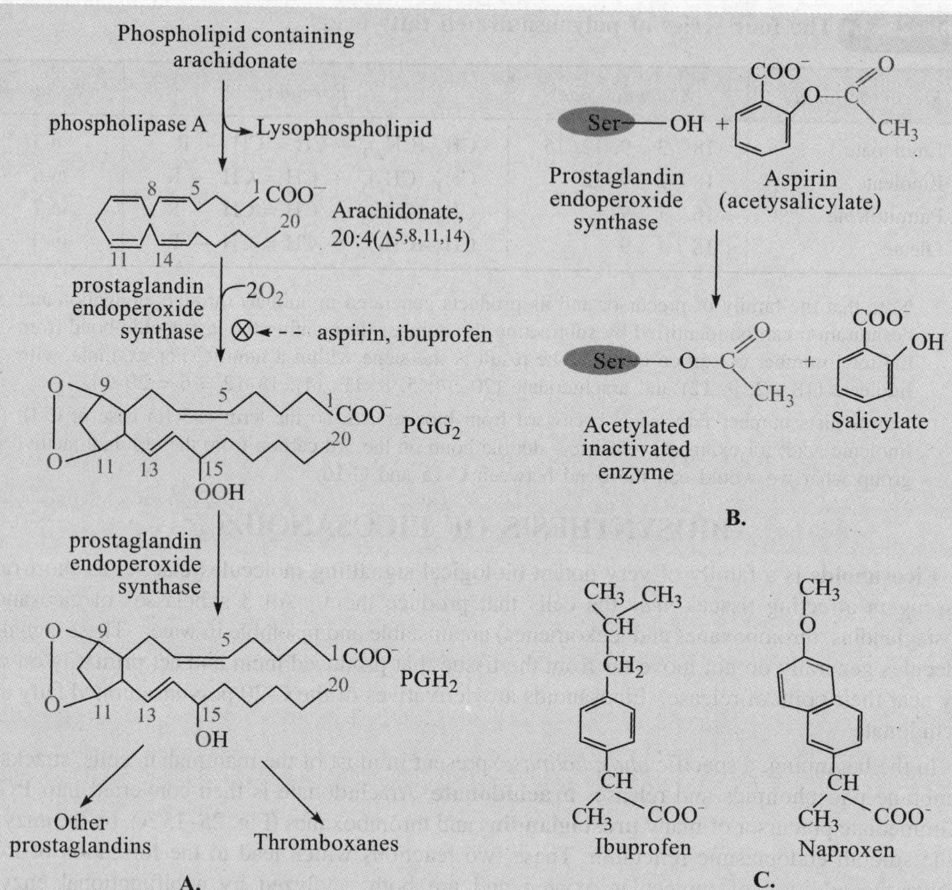

Fig. 25–15. The "cyclic" pathway from arachidonate to prostaglandins and thromboxanes

A. Production of PGH_2 from arachidonate by the action of prostaglandin endoperoxide synthase
B. Aspirin inhibits the first reaction by acetylating an essential Ser residue on the enzyme.
C. Ibuprofen and naproxen inhibit the same step, probably by mimicking the structure of the substrate or an intermediate in the reaction.

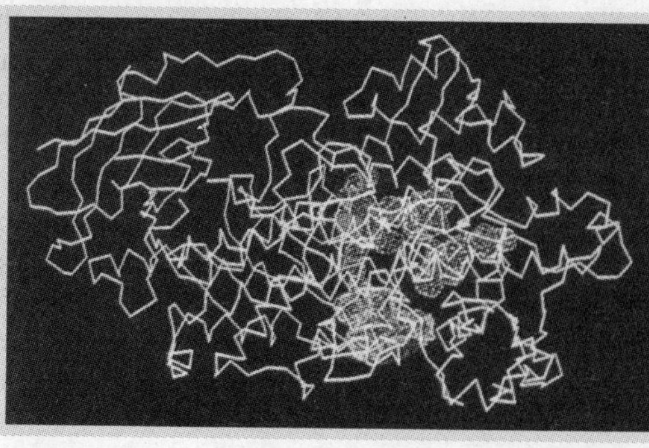

Fig. 25–16. The x-ray structure of soybean lipoxygenase-I

The protein is represented by its C_α diagram (*yellow*). Its internal cavities are outlined by dot surfaces with cavity I in *green* and cavity II in *pink*. The active site Fe atom is represented by an *orange* sphere.

(Courtesy: Mario Amzel, The Johns Hopkins University)

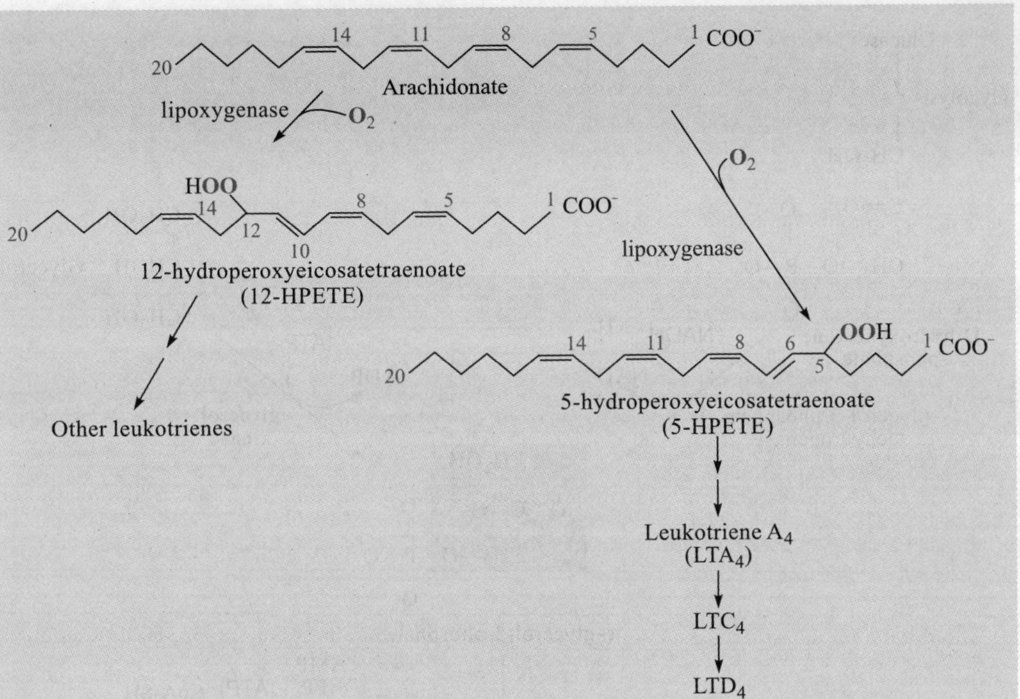

Fig. 25–17. The "linear" pathway from arachidonate to leukotrienes

BIOSYNTHESIS OF TRIACYLGLYCEROLS

Depending upon the requirements of organisms, most of the fatty acids synthesized or ingested by an organism are either incorporated into triacylglycerols for the storage of metabolic energy or incorporated into the phospholipid components of membranes. The organisms, that are not actively growing but have abundant food supply, do not convert most of their fatty acids into storage fats; but during rapid growth, the synthesis of new cell membranes requires membrane phospholipid synthesis. It is of interest to note that *both the pathways (triacylglycerol and phospholipid syntheses) begin at the same point: the formation of fatty acyl esters of glycerol*.

Humans store only about half a kilo of glycogen in their liver and muscles, which is hardly enough to supply the body's energy needs for 12 hours. In contrast, the total amount of stored triacylglycerol in a 70 kg man is about 15 kg, which is enough to supply his basic energy needs for as long as 12 weeks. When carbohydrate is ingested in excess of the capacity to store glycogen, it is converted into triacylglycerols and stored in adipose tissues. Plants also do manufacture triacylglycerols as an energy-rich fuel and store them in fruits and seeds.

Triacylglycerols and glycerophospholipids (phosphatidylethanolamine, for example) share two precursors (fatty acyl-CoAs and glycerol-3-phosphate) and many enzymatic steps in their biosynthesis in animal tissues. Glycerol-3-phosphate can be formed in two ways (Fig. 25–18):

(*a*) either from dihydroxyacetone phosphate generated during glycolysis by the action of cytosolic NAD-linked *glycerol-3-phosphate dehydrogenase*, or

(*b*) from glycerol by the action of glycerol kinase in liver and kidney.

Phosphatidate, which occurs in traces in cells, is a key intermediate in lipid biosynthesis. It can be converted either to a triacylglycerol or to a glycerophospholipid. In the pathway to triacylglycerols, phosphatidate is hydrolyzed by *phosphatidate phosphatase* to form

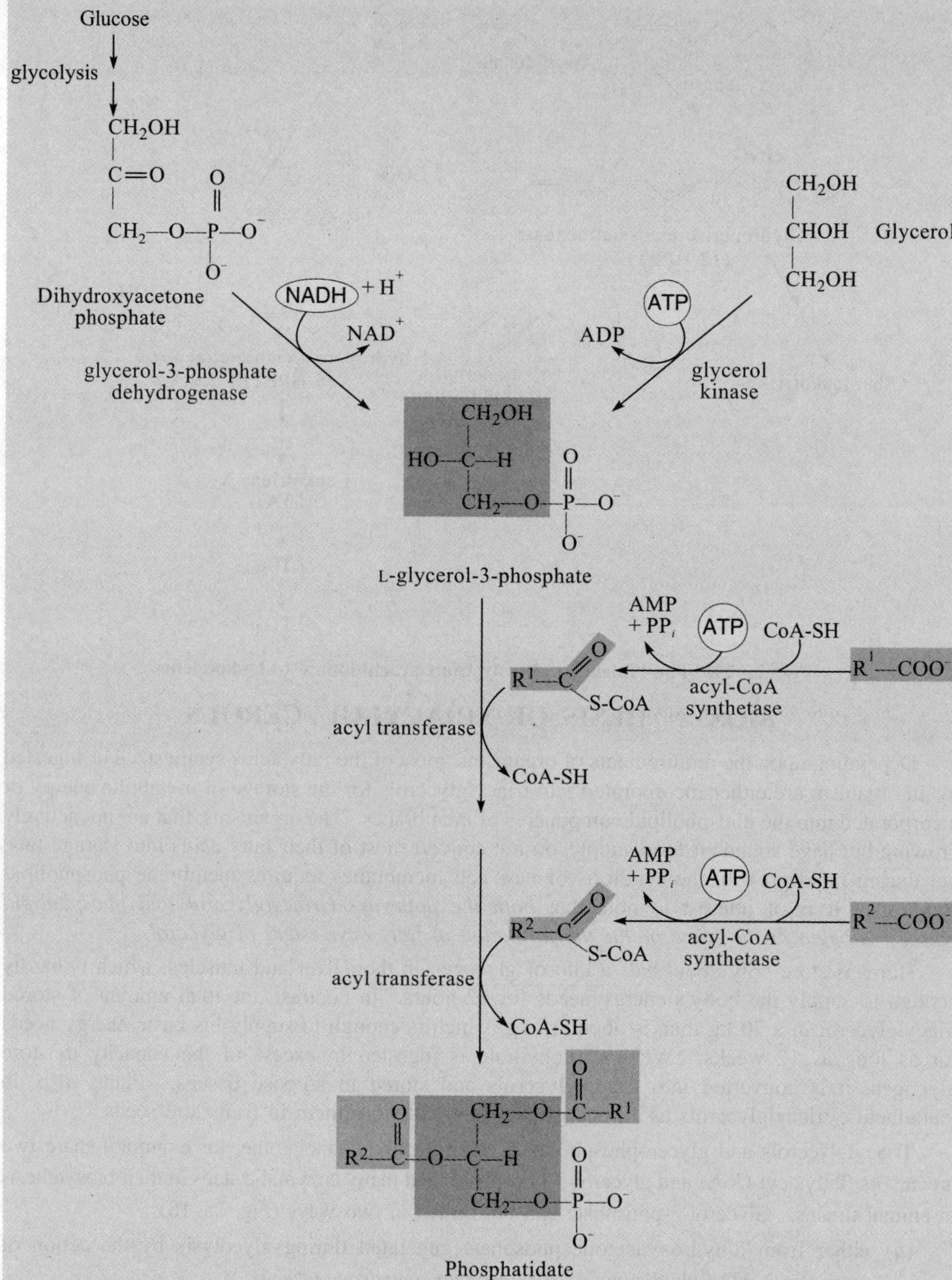

Fig. 25–18. The biosynthetic pathway to phosphatidate

Fatty acyl groups are first activated by formation of fatty acyl–CoA molecules, then they are transferred to ester linkage with L-glycerol-3-phosphate, formed in either of the 2 ways shown. Phosphatidate is shown here with the correct stereochemistry at C-2 of the glycerol molecule.

1, 2-diacylglycerol (Fig. 25–19). **Diacylglycerols** are then converted to **triacylglycerols** by transesterification with a third fatty acyl-CoA.

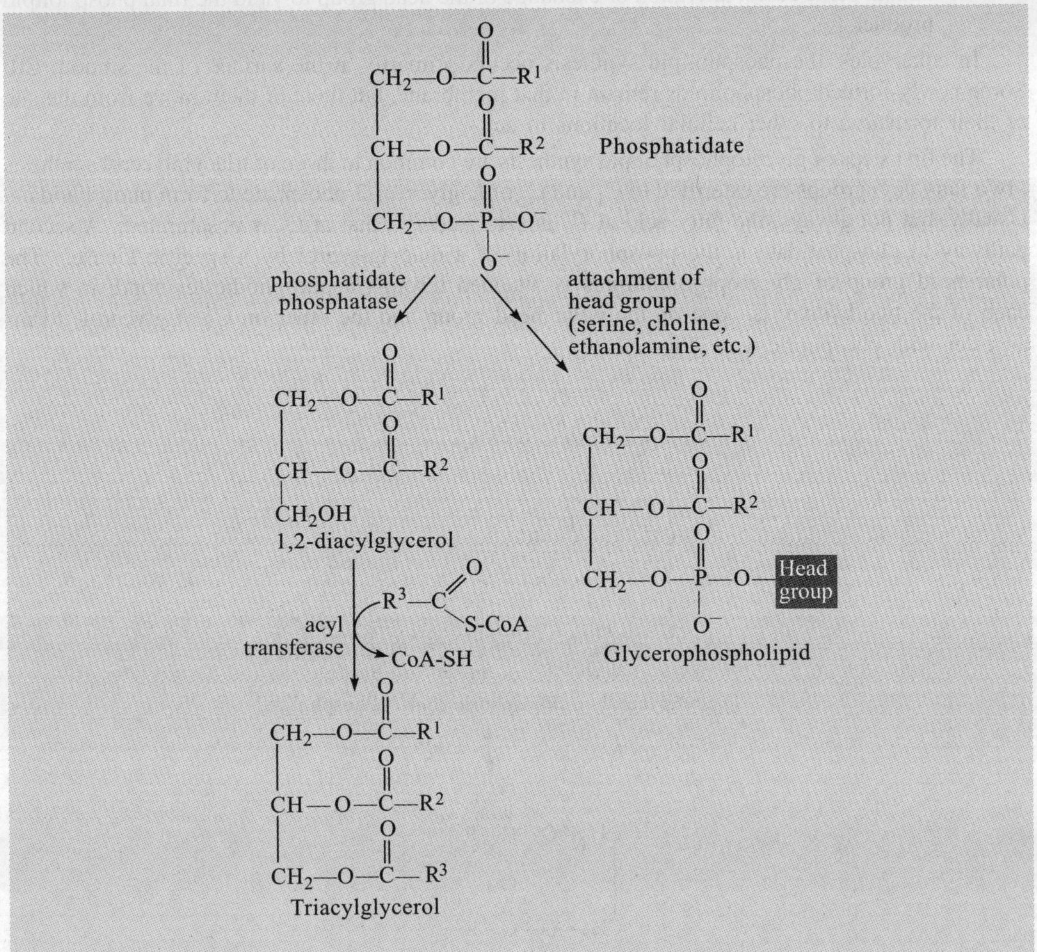

Fig. 25–19. Biosynthesis of triacylglycerols and glycerophospholipids from a common precursor phosphatidate

In men, the amount of body fat remains almost constant over long periods. However, if carbohydrate, protein or fat is consumed in excess of their energy needs, the surplus is stored in the form of triacylglycerols. This stored fat can be utilized for energy during normal body activity or during fasting. Triacylglycerol metabolism is influenced by hormones such as insulin and glucagon and also by the growth hormone and adrenalocorticoids.

BIOSYNTHESIS OF MEMBRANE PHOSPHOLIPIDS

Phospholipid biosynthesis results in the production of a large number of end products. But all of these diverse products are produced according to some basic patterns. In general, the assembly of phospholipids from simple precursors requires (**Lehninger**, **Nelson** and **Cox**, 1993):

1. synthesis of backbone molecule (glycerol or sphingosine),
2. attachment of fatty acid(s) to the backbone in ester or amide linkage,

3. addition of a hydrophilic head group, joined to the backbone through a phosphodiester linkage, and
4. and, in some cases, alteration or exchange of the head group to yield the final phospholipid product.

In eukaryotes, the phospholipid synthesis occurs primarily at the surface of the smooth ER. Some newly-formed phospholipids remain in that membrane, but most of them move from the site of their formation to other cellular locations to act.

The first steps of glycerophospholipid synthesis are common to those of triacylglycerol synthesis : two fatty acyl groups are esterified to C_1 and C_2 of L-glycerol-3-phosphate to form phosphatidate. Usually, but not always, the fatty acid at C_1 is saturated and that at C_2 is unsaturated. A second pathway to phosphatidate is the phosphorylation of a diacylglycerol by a specific kinase. The polar head group of glycerophospholipids is attached through a phosphodiester bond, in which each of the two hydroxyls, (one on the polar head group and the other on C_3 of glycerol) forms an ester with phosphoric acid (Fig. 25–20).

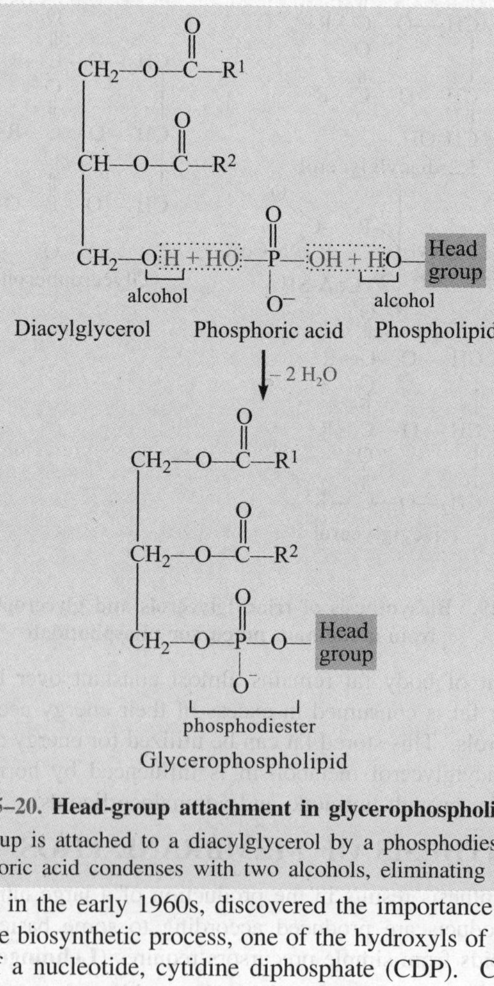

Fig. 25–20. Head-group attachment in glycerophospholipid

The phosphlipid head group is attached to a diacylglycerol by a phosphodiester bond. The bond is formed when phosphoric acid condenses with two alcohols, eliminating 2 moles of water.

Eugene P. Kennedy, in the early 1960s, discovered the importance of cytidine nucleotides in lipid biosynthesis. In the biosynthetic process, one of the hydroxyls of the phospholipid is first activated by attachment of a nucleotide, cytidine diphosphate (CDP). Cytidine monophosphate (CMP) is then displaced in a nucleophilic attack by the other hydroxyl (Fig. 25–21). The CDP is attached either to the diacylglycerol, forming an activated phosphatidate, CDP-diacylglycerol (Strategy 1), or to the hydroxyl of the head group (Strategy 2).

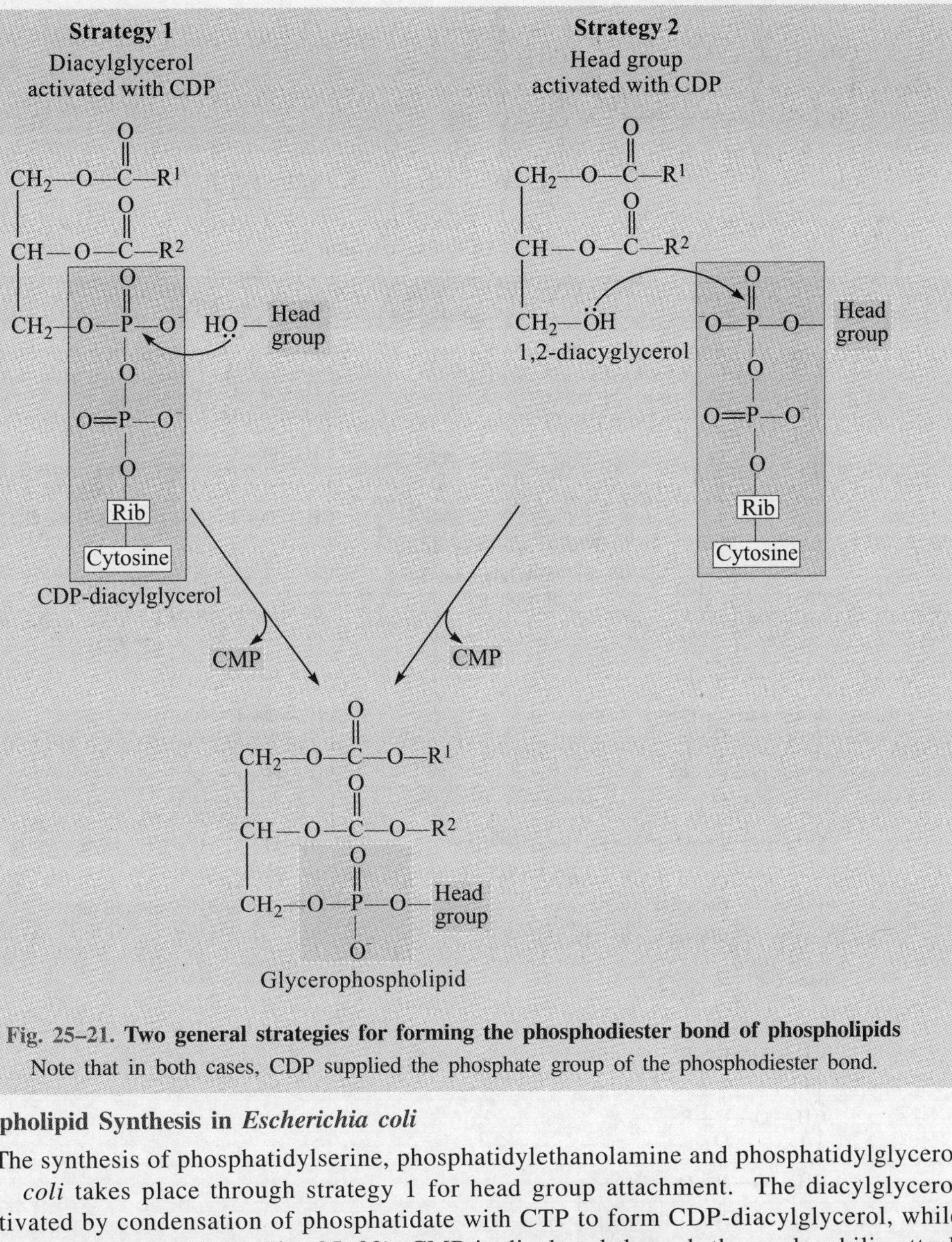

Fig. 25–21. Two general strategies for forming the phosphodiester bond of phospholipids
Note that in both cases, CDP supplied the phosphate group of the phosphodiester bond.

Phospholipid Synthesis in *Escherichia coli*

The synthesis of phosphatidylserine, phosphatidylethanolamine and phosphatidylglycerol in *E. coli* takes place through strategy 1 for head group attachment. The diacylglycerol is activated by condensation of phosphatidate with CTP to form CDP-diacylglycerol, while eliminating pyrophosphate (Fig. 25–22). CMP is displaced through the nucleophilic attack either by the hydroxyl group of serine to yield **phosphatidylserine** or by the C-1 hydroxyl of glycerol-3-phosphate to yield phosphatidylglycerol-3-phosphate. The latter is processed further by cleavage of the phosphate monoester to yield **phosphatidylglycerol**, while freeing inorganic phosphate, P*i*.

The two products (phosphatidylserine and phosphatidylglycerol) can serve as precursors of other membrane lipids in bacteria (Fig. 25–22). Decarboxylation of the serine moiety in phosphatidylserine by the enzyme phosphatidylserine decarboxylase yields **phosphatidylethanolamine.** Condensation of two moles of phosphatidylglycerol produces

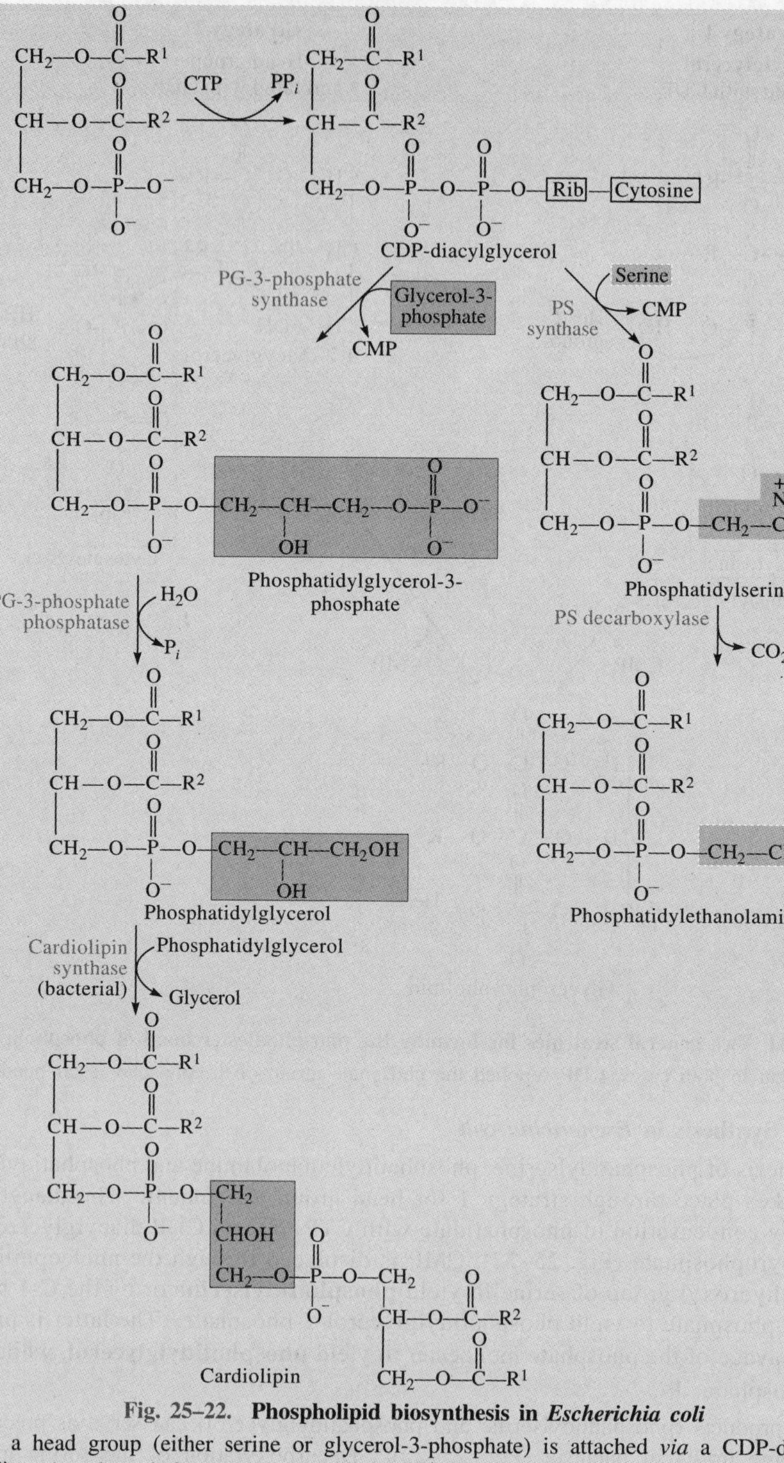

Fig. 25–22. Phospholipid biosynthesis in *Escherichia coli*

Initially, a head group (either serine or glycerol-3-phosphate) is attached *via* a CDP-diacylglycerol intermediate, *i.e.*, through Strategy 1. For phospholipids other than phosphatidylserine, the head group is further modified, as shown in the diagram.

[PG = Phosphatidylglycerol; PS = Phosphatidylserine]

cardiolipin, and a mole of glycerol is eliminated in the process. In cardiolipin, the two diacylglycerol moles are joined through a common head group.

Phospholipid Synthesis in Eukaryotes

In eukaryotes, the synthesis of acidic phospholipids (phosphatidylglycerol, cardiolipin and the phosphatidylinositols) takes place by the same strategy as employed for phospholipid synthesis in bacteria. **Phosphatidylglycerol** is made exactly like that in bacteria. However, **cardiolipin** synthesis differs slightly as phosphatidylglycerol condenses with CDP-diacylglycerol (Fig. 25–23) and not with another mole of phosphatidylglycerol as in *E. coli* (Fig. 25–22).

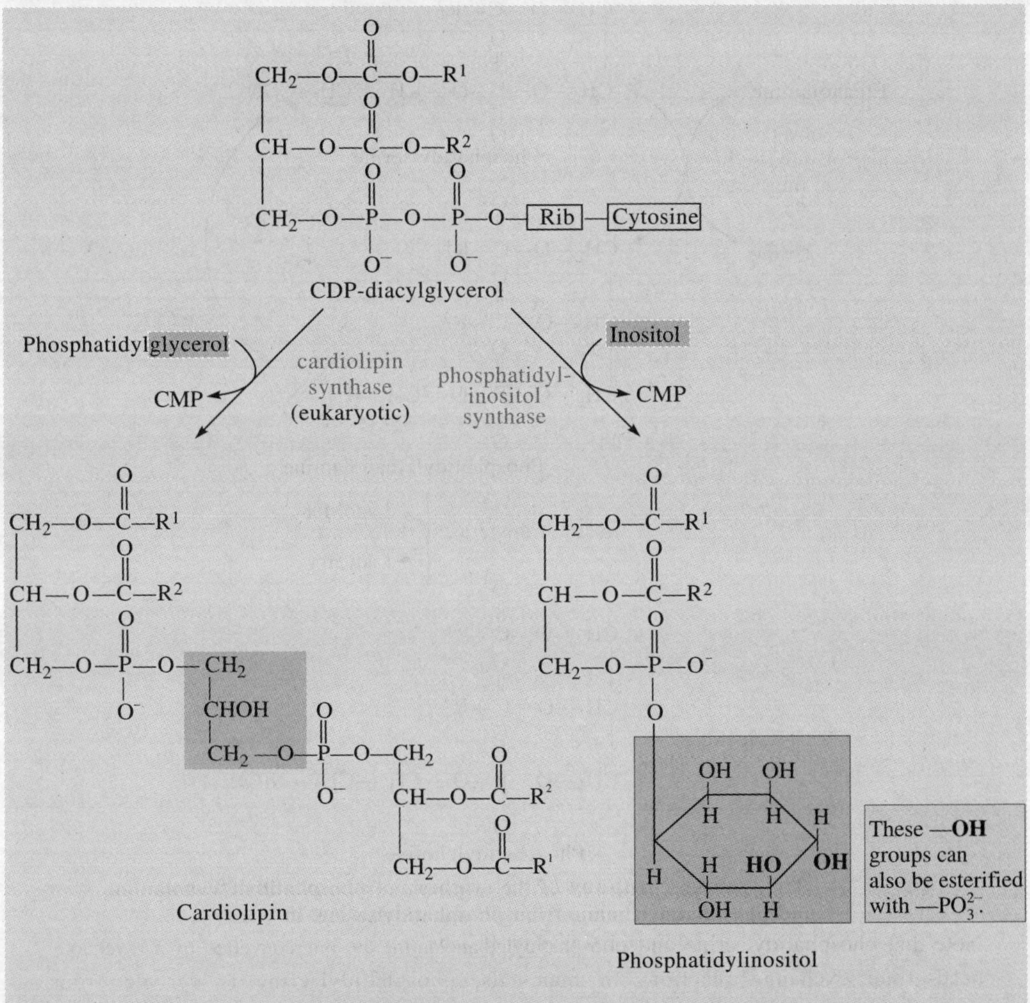

Fig. 25–23. Biosynthesis of cardiolipin and phoshatidylinositol in eukaryotes
These glycerophospholipids are synthesized as per Strategy I, shown in Fig. 25–21. Phosphatidylglycerol is synthesized as in bacteria (see Fig. 25–22).

The synthesis of **phosphatidylinositol** takes place by the condensation of CDP-diacylglycerol with inositol. Phosphatidylinositol and its phosphorylated derivatives play a central role in signal transduction in eukaryotic individuals.

Interrelationship among the Eukaryotic Pathways to Phosphatidylserine, Phosphatidylethanolamine and Phosphatidylcholine

In yeast, as also in bacteria, phosphatidylserine may be formed by condensation of CDP-diacylglycerol and serine. And phosphatidylethanolamine can be synthesized from phosphatidylserine in the reaction catalyzed by phosphatidylserine decarboxylase (Fig. 25–24).

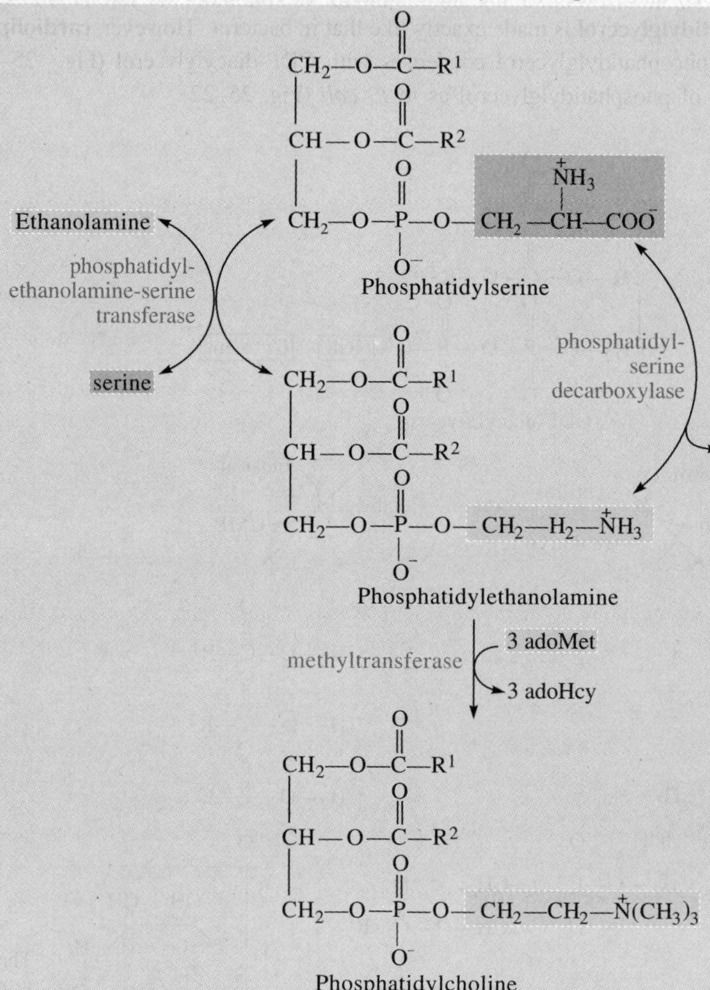

Fig. 25–24. The "salvage" pathway of the synthesis of phosphatidylethanolamine and phosphatidylcholine from phosphatidylserine, in yeast

Note that phosphatidylserine and phosphatidylethanolamine are interconverted by a reversible head-group exchange reaction. In mammals, phosphatidylserine is derived from phosphatidylethanolamine by a reversal of this reaction.

[adoMet = S-adenosylmethionine; adoHcy = *S*-adenosylhomocysteine]

An alternative pathway to phosphatidylserine is a head group exchange reaction, in which free serine displaces ethanolamine. Phosphatidylethanolamine may also be converted into **phosphatidylcholine** (lecithin) by the addition of 3 methyl groups to its amino group.

Mammals, however, do not synthesize phosphatidylserine from CDP-diacylglycerol. Instead, they derive it from phosphatidylethanolamine *via* the head group exchange reaction, as depicted in Fig. 25–24. *In mammals, the synthesis of all nitrogen-containing phospholipids occurs by*

Strategy 2: phosphorylation and activation of the head group followed by condensation with diacylglycerol. As an instance, choline is resused (or "salvaged") by first being phosphorylated and then converted into CDP-choline by condensation with CTP. A diacylglycerol displaces CMP from CDP-choline, forming phosphatidylcholine (Fig. 25–25).

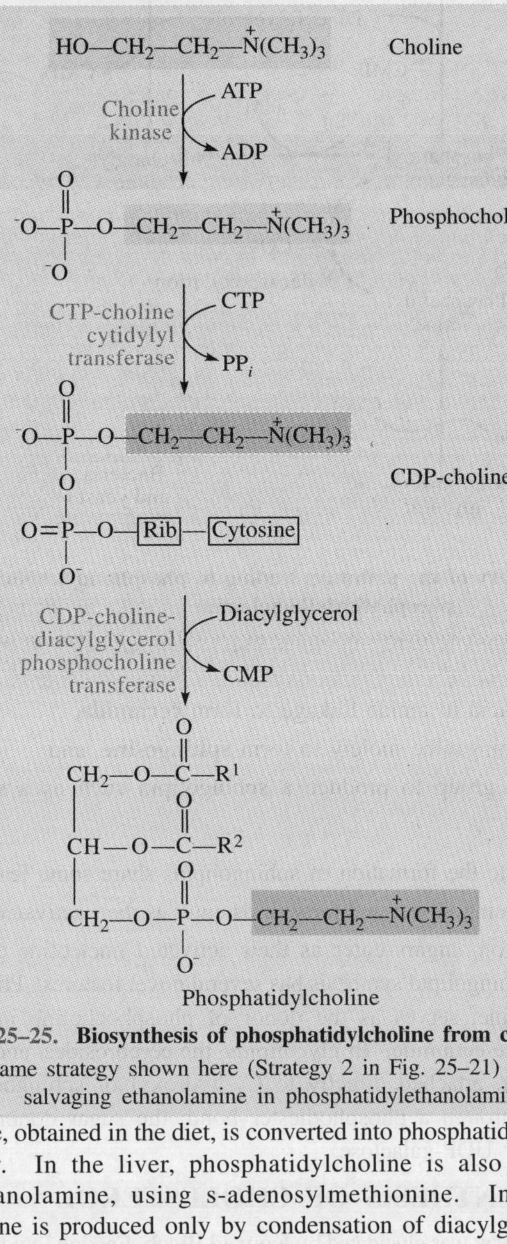

Fig. 25–25. Biosynthesis of phosphatidylcholine from choline in mammals
The same strategy shown here (Strategy 2 in Fig. 25–21) is also employed for salvaging ethanolamine in phosphatidylethanolamine synthesis.

Ethanolamine, obtained in the diet, is converted into phosphatidylethanolamine by an analogous salvage pathway. In the liver, phosphatidylcholine is also produced by methylation of phosphatidylethanolamine, using s-adenosylmethionine. In all other tissues, however, phosphatidylcholine is produced only by condensation of diacylglycerol and CDP-choline. Fig. 25–26 summarizes the pathways leading to the formation of phosphatidylcholine and phosphatidylethanolamine in various organisms.

Biosynthesis of Sphingolipids

The biosynthesis of sphingolipids occurs in 4 stages (Fig. 25–27):

1. synthesis of a C_{18} amine, **sphinganine** from palmitoyl-CoA and serine,

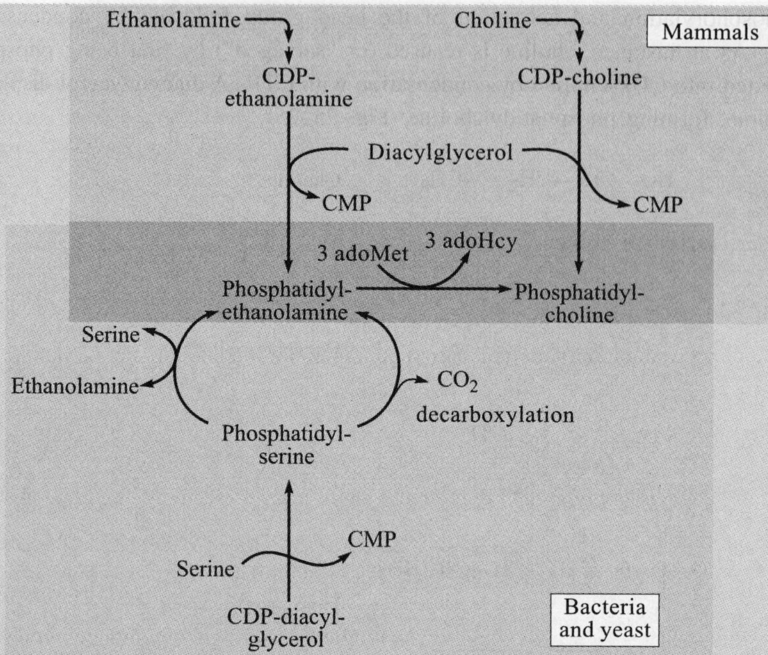

Fig. 25–26. Summary of the pathways leading to phosphatidylcholine and phosphatidylethanolamine

Note that the conversion of phosphatidylethanolamine to phosphatidylcholine in mammals occurs exclusively in the liver.

2. attachment of a fatty acid in amide linkage to form **ceramide**,
3. desaturation of the sphinganine moiety to form **sphingosine**, and
4. attachment of a head group to produce a sphingolipid such as a **sphingomyelin** or **cerebroside.**

The pathways leading to the formation of sphingolipids share some features in common:

1. NADPH provides reducing power and fatty acids enter as their activated CoA derivatives.
2. In cerebroside formation, sugars enter as their activated nucleotide derivatives. Head group attachment in sphingolipid synthesis has several novel features. Phosphatidylcholine, rather than CDP-choline, serves as the donor of phosphocholine in the synthesis of sphingomyelin from the ceramide. In glycolipids, the cerebrosides, and gangliosides, the head group is a sugar, attached directly to C_1 hydroxyl of sphingosine in glycosidic linkage, rather than through a phosphodiester bond; the sugar donor is a UDP-sugar, either UDP-glucose or UDP-galactose.

BIOSYNTHESIS OF CHOLESTEROL

This extremely difficult pathway was elucidated by **Konrad Bloch, Feodor Lynen, John Kornforth** and **George Popjáck** in the late 1950s. Cholesterol, an essential molecule in many animals including man, is not required in the mammalian diet because the liver can synthesize it from simple precursors. Like long-chain fatty acids, cholesterol is made from acetyl-CoA but the assembly plan is quite unlike each other.

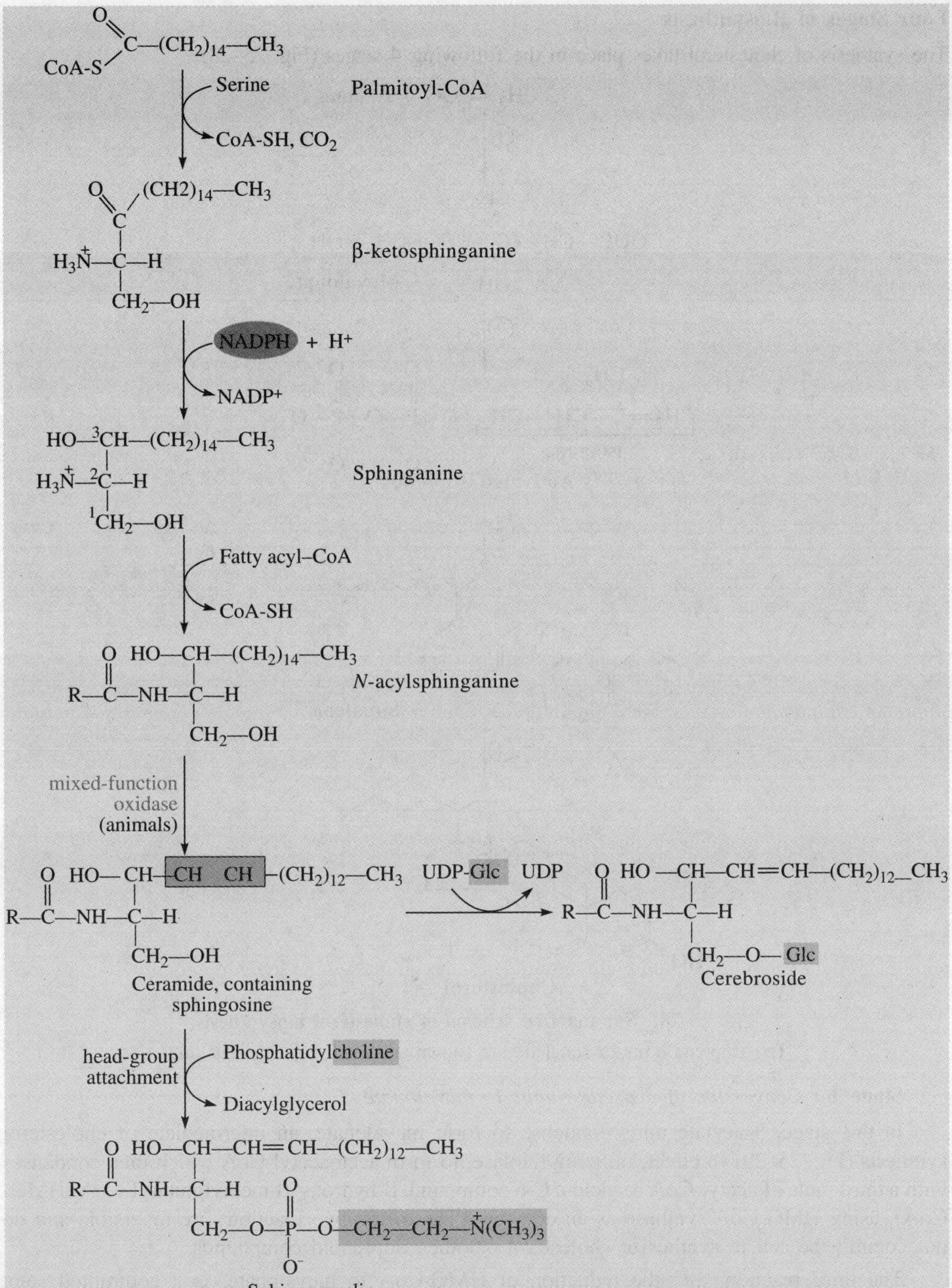

Fig. 25–27. Biosynthesis of sphingolipids

The synthesis involves first the condensation of palmitate and serine to produce sphinganine, followed by acylation of sphinganine to produce *N*-acylsphinganine (a ceramide). In animals, a double bond (shown in *pink rectangle*) is then created by a mixed-function oxidase. At last, a head group is added: phosphatidylcholine, to form sphingomyelin; or glucose, to form a cerebroside.

Four Stages of Biosynthesis

The synthesis of cholesterol takes place in the following 4 stages (Fig. 25–28):

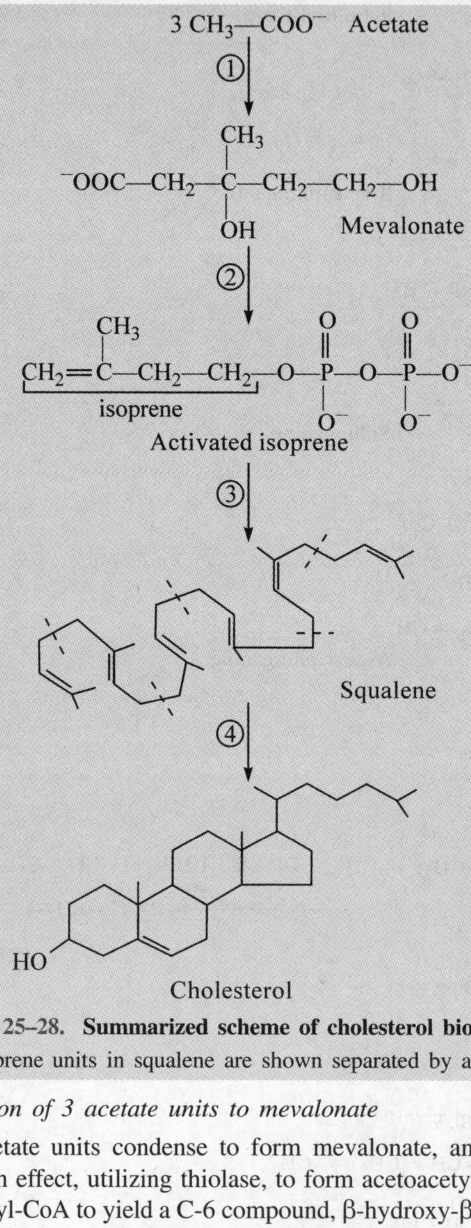

Fig. 25–28. Summarized scheme of cholesterol biosynthesis

The isoprene units in squalene are shown separated by a *dashed line.*

Stage 1 : *Conversion of 3 acetate units to mevalonate*

In this stage, 3 acetate units condense to form mevalonate, an intermediate of cholesterol synthesis (Fig. 25–29) In effect, utilizing thiolase, to form acetoacetyl-CoA which then condenses with a third mole of acetyl-CoA to yield a C-6 compound, β-hydroxy-β-methylglutaryl-CoA (HMG-CoA), using HMG-CoA synthase as an enzyme. These first two reactions are reversible and do not commit the cell to synthesize cholesterol or other isoprenoid compounds.

The third reaction, *i.e.*, the reduction of HMG-CoA to mevalonate, is a committed step, wherein 2 moles of NADPH each donate two electrons. This reaction is catalyzed by HMG-CoA reductase (Fig. 25–30), an integral membrane protein of the smooth endoplasmic reticulum. It is the major point of regulation for cholesterol synthesis. It may, however, be pointed out that HMG-CoA is converted into mevalonate in the cytosol, but it is converted into acetyl-CoA and acetoacetate in the mitochondria (Fig. 25–31).

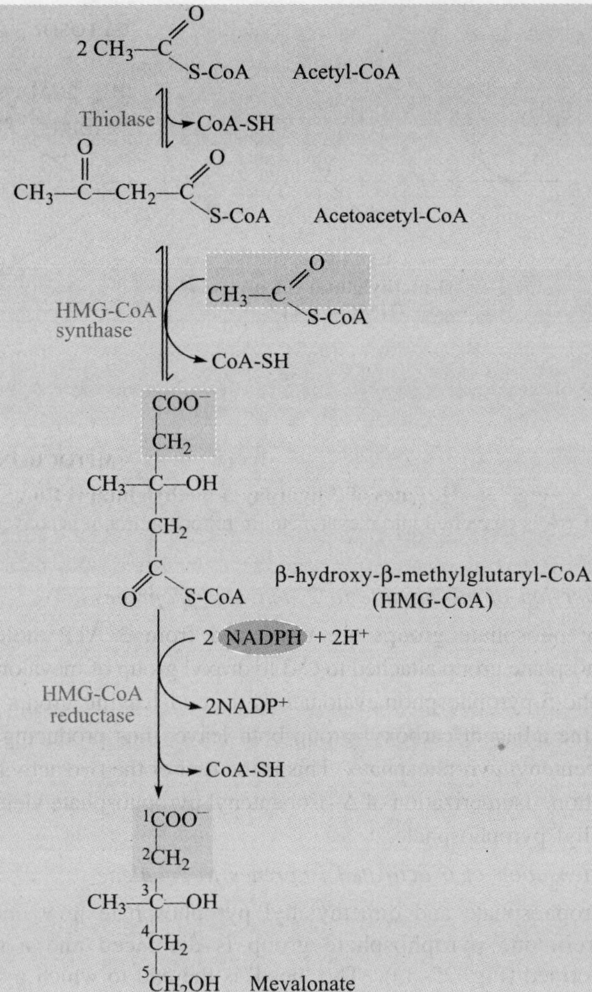

Fig. 25–29. Formation of mevalonate from acetyl-CoA

The origin of C-1 and C-2 in mevalonate from acetyl-CoA is shown enclosed in *shaded rectangles*.

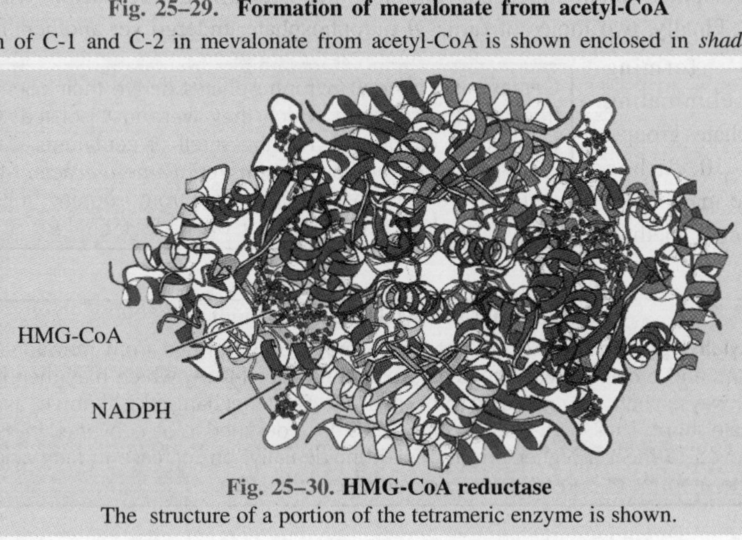

Fig. 25–30. HMG-CoA reductase

The structure of a portion of the tetrameric enzyme is shown.

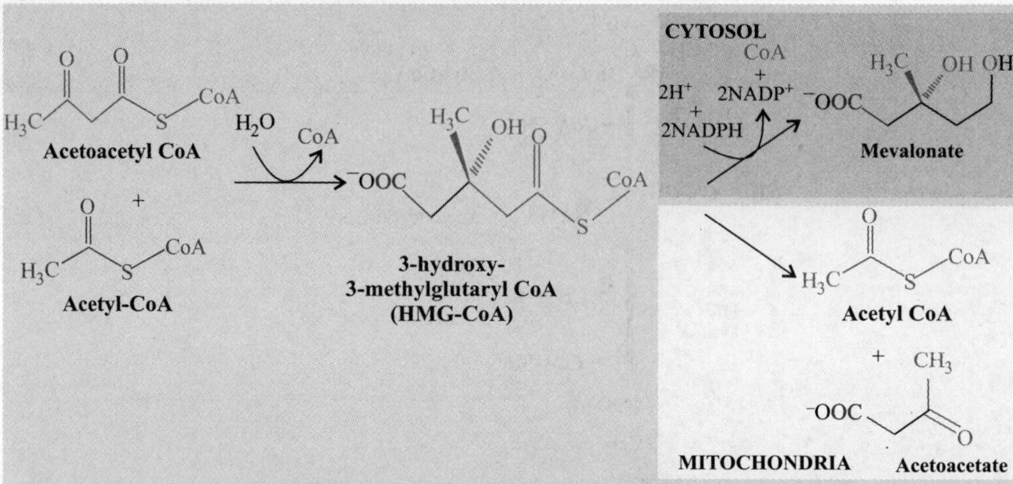

Fig. 25–31. Fates of 3-hydroxy-3-methylglutaryl CoA
In the cytosol, HMG-CoA is converted into mevalonate. In mitochondria, it is converted into acetyl-CoA and acetoacetate

Stage 2 : *Conversion of mevalonate to 2 activated isoprenes*

Here, the three phosphate groups are tranferred from 3 ATP molecules to mevalonate (Fig. 25–32). The phosphate group attached to C-3 hydroxyl group of mevalonate in the intermediate compound, 3-phospho-5-pyrophosphomevalonate is a good leaving group. In the next reaction, this phosphate and the adjacent carboxyl group both leave, thus producing a double bond in the C-5 product, Δ^3-isopentenyl pyrophosphate. This is the first of the two activated isoprenes essential to cholesterol formation. Isomerization of Δ^3-isopentenyl pyrophosphate yields the second activated isoprene, dimethylallyl pyrophosphate.

Stage 3 : *Condensation of 6 activated isoprenes to squalene*

Isopentenyl pyrophosphate and dimethylallyl pyrophosphate now undergo a 'head-to-tail' condensation wherein one pyrophosphate group is displaced and a C-10 chain, **geranyl pyrophosphate** is formed (Fig. 25–33). The 'head' is the end to which pyrophosphate is joined. Geranyl pyrophosphate then undergoes another 'head-to-tail' condensation with **farnesyl pyrophosphate**. Finally, two moles of farnesyl pyrophosphate undergo yet another 'head-to tail' condensation, forming **squalene** and eliminating both pyrophosphate groups. Squalene has 30 carbon atoms, 24 in the main chain and 6 in the form of methyl-group branches.

Geranyl and farnesyl pyrophosphates derive their common names from the sources from which they were first isolated. **Geraniol**, a component of rose oil, has the smell of geraniums. **Farnesol** is a scent present in the flowers of a tree, *Farnese acacia*. Many natural scents of plant origin are synthesized from isoprene units. **Squalene** was first isolated from the liver of sharks (*Squalus*).

The Trans-methylglutaconate Shunt

The **dimethylallyl pyrophosphate,** one of the isoprene units formed from mevalonate, can be dephosphorylated and broken down into acetoacetate and acetyl-CoA, which may then be diverted into other pathways, such as fatty acid biosynthesis. This mechanism is known as the trans-methylglutaconate shunt. Thus, high-energy compounds once destined to be converted into cholesterol may be redeployed to meet a higher priority need. Incidentally, an increase in fatty acid synthesis will increase the amount of substrate for cholesterol esterification.

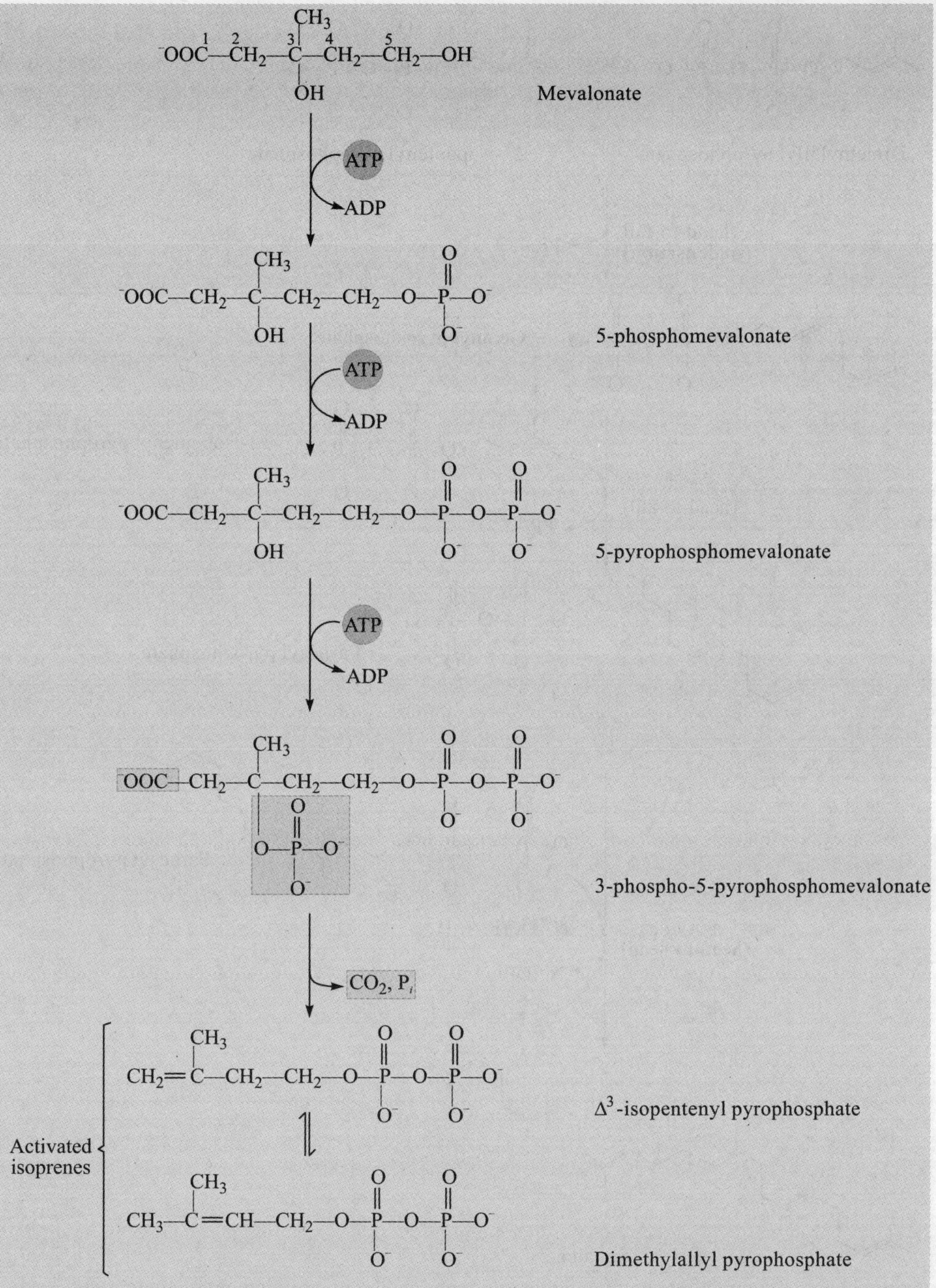

Fig. 25–32. Conversion of mevalonate into activated isoprene units

Note that 6 isoprene units combine to form squalene. The leaving groups of 3-phosho-5-pyrophosphomevalonate are shown in *light-screened areas*.

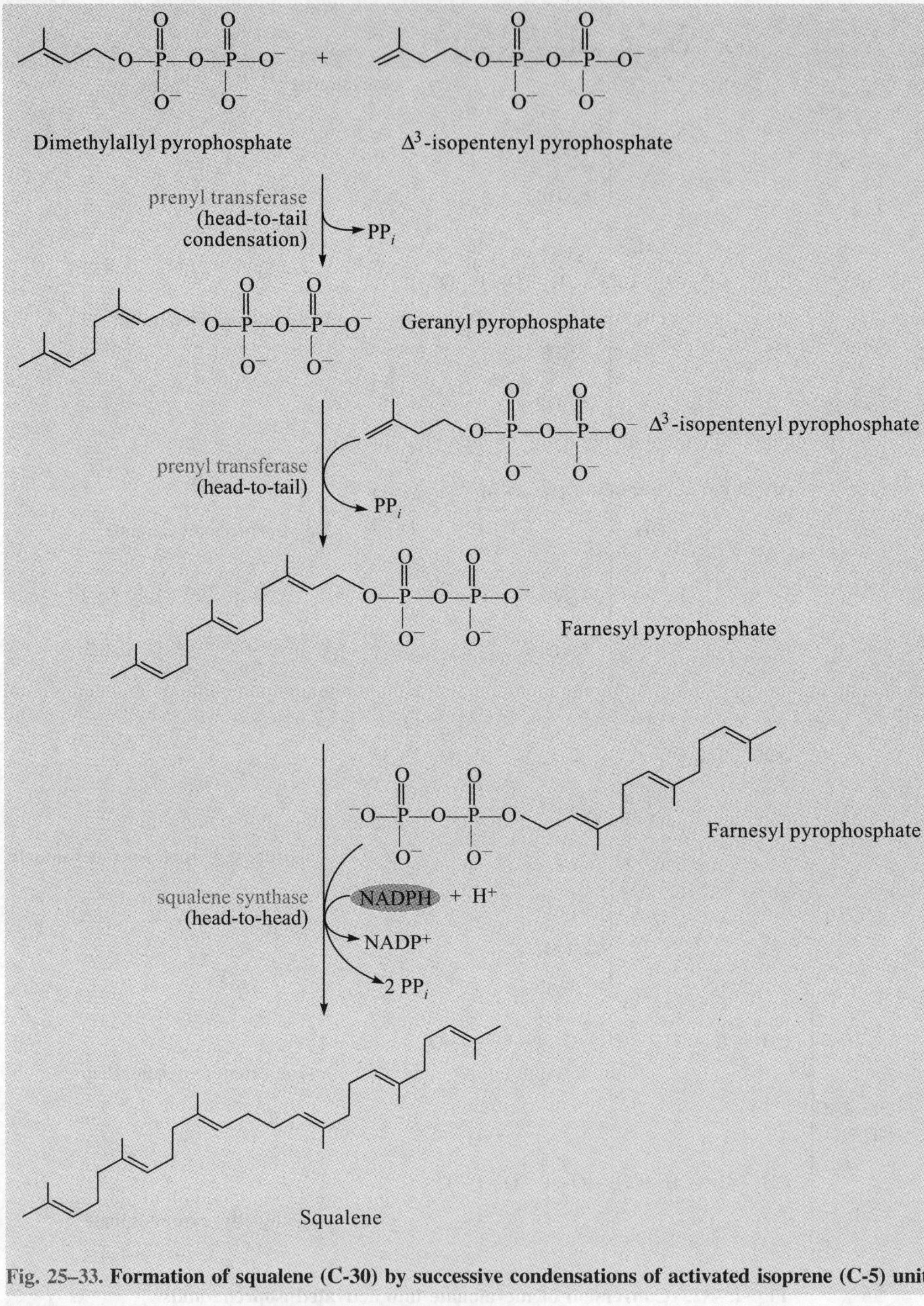

Fig. 25–33. Formation of squalene (C-30) by successive condensations of activated isoprene (C-5) units

Stage 4 : *Conversion of squalene to the steroid nucleus*

All of the sterols are alcohols with a OH group at C-3 and have a steroid nucleus containing 4 fused rings (hence, so named). One oxygen atom from oxygen is added to the end of squalene chain forming an epoxide, **squalene-2,3-epoxide** (Fig. 25–34). The reaction is catalyzed by squalene monooxygenase, another mixed-function oxidase enzyme. NADPH reduces the other oxygen atom of oxygen to water. The double bonds of the product, squalene-2,3-epoxide are positioned so that a concerted reaction can convert the linear squalene epoxide into a cyclic structure. In animals, this cyclization results in the formation of **lanosterol**, which contains the 4 rings characteristic of the steroid nucleus. Lanosterol is finally converted into **cholesterol** in a series of about 20 reactions, including the migration of some methyl groups and the removal of others.

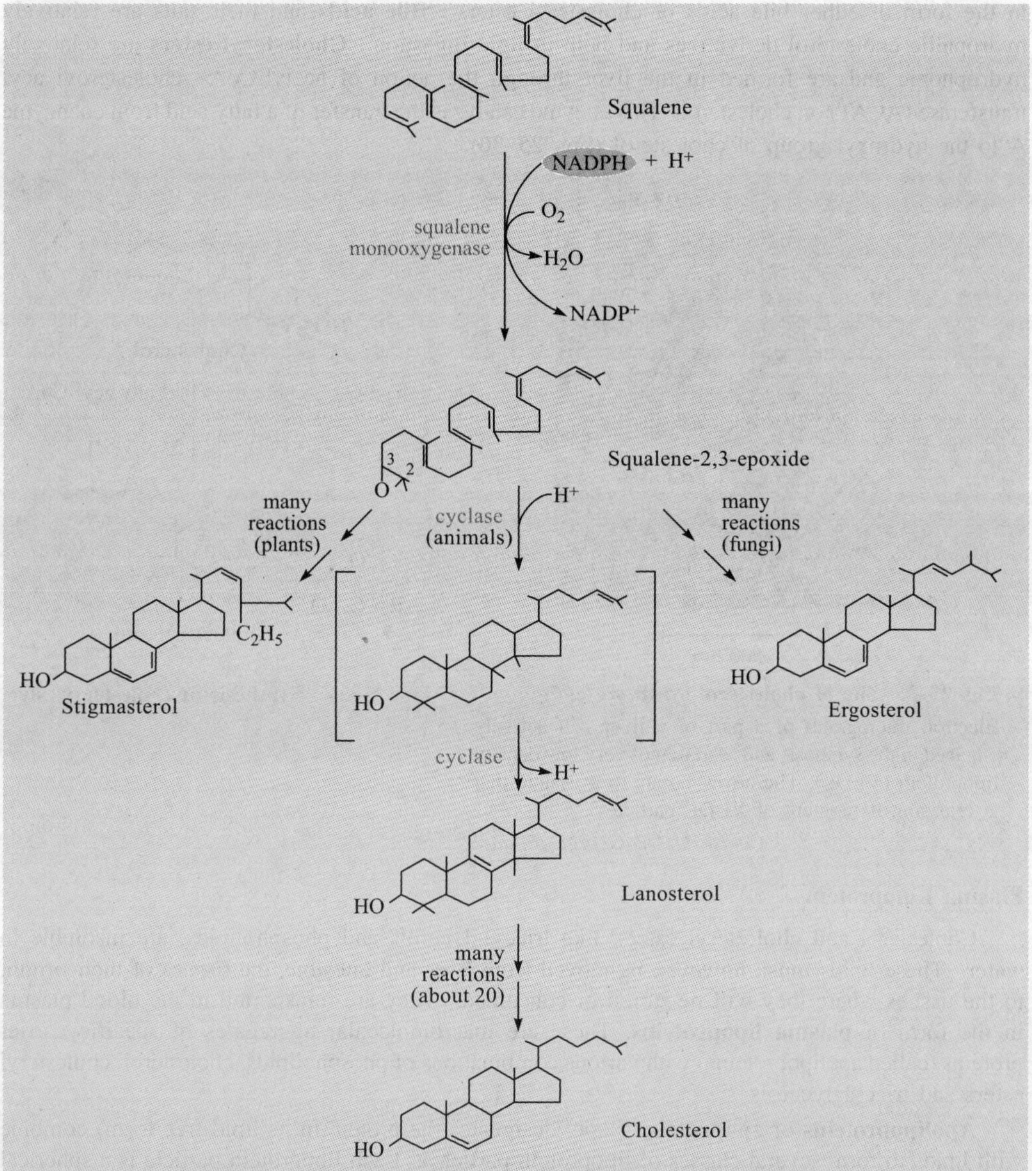

Fig. 25–34. Conversion of linear squalene into cyclic steroid nucleus
The first step in this sequence is catalyzed by a mixed-function oxidase (a monooxygenase), for which the cosubstrate is NADPH. The product is an epoxide, which in the next step is cyclized to the steroid nucleus. Note that the final product of these reactions in animal cells is cholesterol, but in other organisms slightly different sterols are produced.

In other organisms apart from animals, **closely-related sterols**, instead of cholesterol, are formed. In such cases, after the formation of squalene-2,3-epoxide, the synthetic pathways diverge slightly, yielding other sterols. As an instance, stigmasterol is formed in many plants and ergosterol in fungi (Fig. 25–34).

Most of the cholesterol synthesis takes place in the vertebrate liver cells (Fig. 25–35). While a small fraction of this is incorporated into the membranes of hepatocytes, most of it is exported in the form of either bile acids or cholesteryl esters. **Bile acids** and their salts are relatively hydrophilic cholesterol derivatives and help in lipid digestion. **Cholesteryl esters** are relatively hydrophobic and are formed in the liver through the action of acetyl-CoA—cholesteroyl acyl transferase (ACAT) on cholesterol. This enzyme catalyzes the transfer of a fatty acid from coenzyme A to the hydroxyl group of cholesterol (Fig. 25–36).

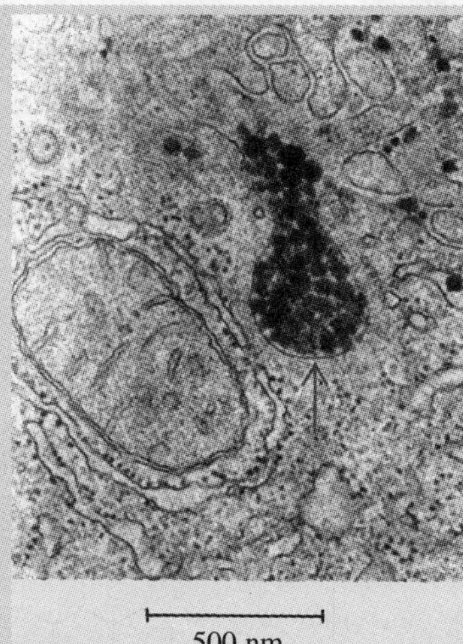

Fig. 25–35. Site of cholesterol synthesis

Electron micrograph of a part of a liver cell actively engaged in the synthesis and secretion of very low density lipoprotein (VLDL). The arrow points to a vesicle that is releasing its content of VLDL particles.

(Courtesy: Dr. George Palade)

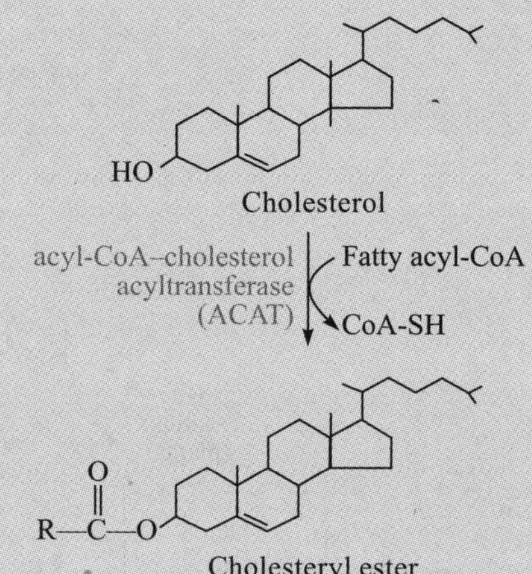

Fig. 25–36. Synthesis of cholesteryl ester

Plasma Lipoproteins

Cholesterol and cholesteryl esters, like triacylglycerols and phospholipids, are insoluble in water. These lipids must, however, be moved from liver and intestine, the tissues of their origin, to the tissues where they will be stored or consumed. They are transported in the blood plasma in the form of **plasma lipoproteins**. These are macromolecular aggregates of specific carrier proteins (called apolipoproteins) with various combinations of phospholipids, cholesterol, cholesteryl esters and triacylglycerols.

Apolipoproteins or **apoproteins** ('apo' designates the protein in its lipid-free form) combine with lipids to form several classes of lipoprotein particles. Each lipoprotein particle is a spherical aggregate, consisting of a core of hydrophobic lipids surrounded by a shell of polar lipids and apoproteins (Fig. 25–37). Lipoproteins are classified according to increasing density (Table 25–4) into chylomicrons, very low-density lipoproteins (VLDL), low-density lipoproteins (LDL) and high-density lipoproteins (HDL), which may be separated by ultracentrifugation and visualized by electron microscopy. (Fig. 25–38)

Table 25–4. Properties of the major classes of human plasma lipoproteins

Liproprotein	*Density (g/mL)*	*Composition (wt %)*				
		Protein	*Free cholesterol*	*Cholesteryl esters*	*Phospho-lipids*	*Triacyl-glycerols*
Chylomicrons	< 0.95	2	1	3	9	85
VLDL	0.95 — 1.006	10	7	12	18	50
LDL	1.007 — 1.063	23	8	37	20	10
HDL	1.064 — 1.210	55	2	15	24	4

(Modified from D. Kritchevsky, 1986)

Each class of lipoprotein performs a specific function which is determined by its point of synthesis, lipid composition and apoprotein content. Lipoproteins of human plasma contain at least 9 types of apolipoproteins (Table 25–5), based on their size, their reaction with specific antibodies and their distribution in the lipoprotein classes. These protein components act as signals, targeting lipoproteins to specific tissues or activating enzymes that act on the lipoproteins. They also solubilize hydrophobic lipids.

Table 25–5. Apolipoproteins of the human plasma lipoproteins

Apolipoprotein	*Mol. Wt.*	*Lipoprotein association*	*Functions (if known)*
ApoA–I	28,331	HDL	Activates LCAT*
ApoA–II	17,380	HDL	—
ApoB–48	2,40,000	Chylomicrons	
ApoB–100	5,13,000	VLDL, LDL	Binds to LDL receptor
ApoC–I	7,000	VLDL, HDL	—
ApoC–II	8,837	Chylomicrons,VLDL, HDL	Activates lipoprotein lipase
ApoC–III	8,751	Chylomicrons, VLDL, HDL	Inhibits lipoprotein lipase
ApoD	32,500	HDL	—
ApoE	34,145	Chylomicrons,VLDL, HDL	Triggers clearance of VLDL and chylomicron remnants

* LCAT = lecithin-cholesterol acyl transferase

(Modified from Vane DE and Vane JE, 1985)

Chylomicrons : The chylomicrons (Fig. 25–37), the first major lipoprotein type, are the largest of the lipoproteins and are between 180 and 500 nm in diameter. They have very low density (d < 0.95 g/cm^3) because of a high triacylglycerol contents (about 85%) in them (Table 25–4). Chylomicrons are synthesized in the smooth endoplasmic reticulum of epithelial cells lining the small intestine. They then move through the lymphatic system to enter the bloodstream. The apoproteins of chylomicrons consist of apoB-48, apoI, apoC-II. ApoC-II activates lipoprotein lipase in the capillaries of adipose, heart, skeletal muscle and lactating mammary tissues, allowing the release of free fatty acids to these tissues. Chylomicrons, thus, carry fatty acids obtained in the diet to the tissue in which they will be consumed or stored as fuel.

Very low-density lipoproteins (VLDLs) : Triacylglycerols and cholesterol in excess of the liver's own requirements are exported into the blood in the form of very low-density lipoproteins ($0.95 < d < 1.006$ g/cm^3). Excess carbohydrate in the diet can also be converted into triacylglycerols in the liver and exported as VLDLs. Besides triacylglycerols, VLDLs contain some cholesterol, cholesteryl esters, as well as apoB–100, apoC–I, apoC–II, apoC–III and apoE. These lipoproteins are transported in the blood from the liver to adipose tissue, where activation of lipoprotein lipase by apoC–II causes the release of free fatty acids from the triacylglycerols of the VLDL. Thus, triacylglycerols in VLDL, as in chylomicrons, are hydrolyzed by lipases on capillary surfaces.

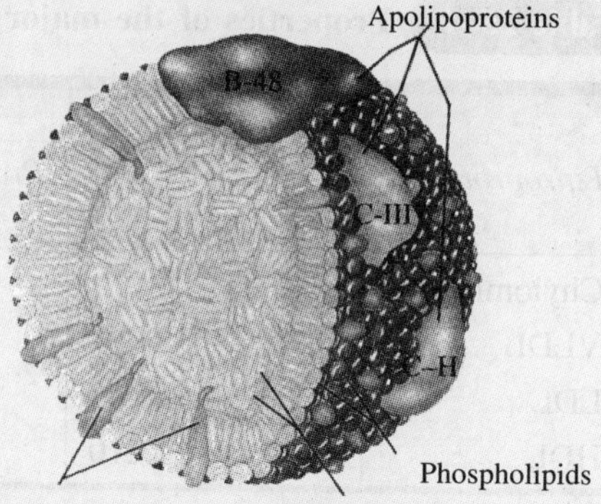

Fig. 25–37. Molecular structure of a chylomicron
The surface is a layer of phospholipids with head groups facing the aqueous phase. Triacylglycerols sequestered in the interior (*yellow*) make more than 80% of the mass. Several apolipoproteins that protrude from the surface (B-48, C-II, C-III) act as signals in the uptake and metabolism of chylomicron contents. The diameter of chylomicrons ranges between 100 and 500 nm.

Low-density lipoproteins (LDLs) : The loss of triacylglycerols converts VLDL to low-density liprotein. LDL (Fig. 25–39) is the major carrier of cholesterol in blood. It has a diameter of 22 nm and a mass of 3,000 kd. It contains a core of about 1,500 esterified cholesterol molecules; linoleate is the most common fatty acyl chain in these esters. This highly hydrophobic core is surrounded by a shell of phospholipids and unesterified cholesterols. The shell also contains a single copy of apoB–100, which is recognized by target cells. *The role of LDL is to transport cholesterol to peripheral tissues and regulate do novo cholesterol synthesis at these sites.*

High-density lipoproteins (HDLs) : The fourth major lipoprotein type, high-density lipoproteins, has a high density ($1.063 < d < 1.210$ g/cm^3) and is synthesized in the liver as small, protein-rich particles containing relatively little cholesterol and cholesteryl esters. HDLs

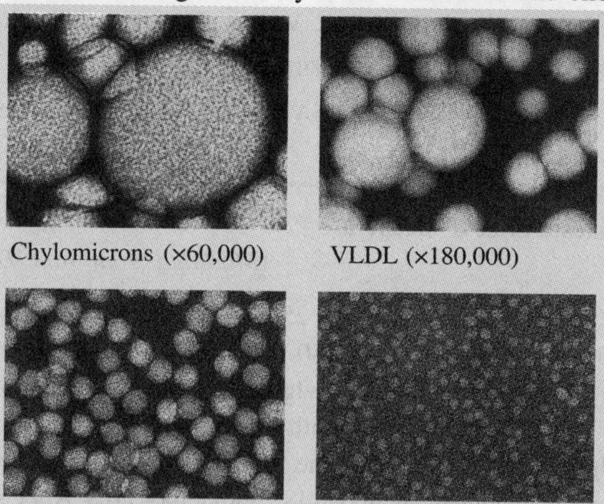

Fig. 25–38. Four classes of lipoproteins, as visualized in the electron microscope after negative staining
Clockwise from top left: Chylomicrons, 50–200 nm in diameter; VLDL, 28–70 nm; HDL, 8–11 nm; and LDL, 20–25 nm.

contain apoC–I and apoC–II, among other apolipoproteins, as well as the enzyme, lecithin-cholesterol acyl transferase (LCAT). The enzyme catalyzes the formation of cholesteryl esters from lecithin (phosphatidylcholine) and cholesterol (Fig. 25–40). After release into the bloodstream, the nascent (or newly synthesized) HDL collects cholesteryl esters from other circulating lipoprotiens. Chylomicrons and VLDLs, after the removal of their triacylglycerols by lipoprotein lipase, are rich in cholesterol and phosphatidylcholine. LACT on the surface of nascent HDL converts this phosphatidylcholine and cholesterol to cholesteryl esters, which enter the interior of the nascent HDL, converting it from a flat disc to a sphere– a mature HDL. This cholesterol-rich lipoprotein now returns to the liver, where the cholesterol is unloaded. Some of this cholesterol is converted into bile salts.

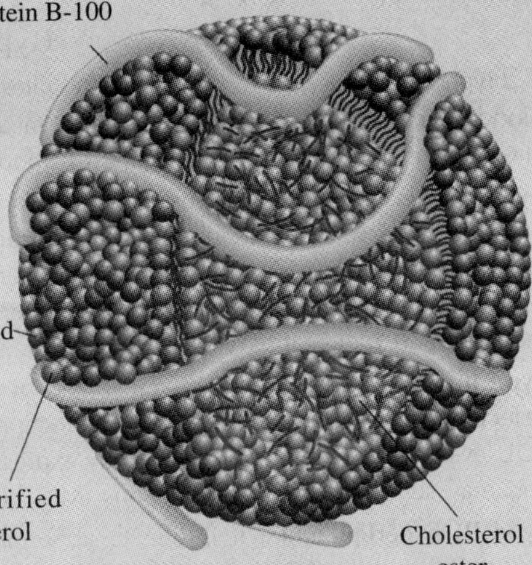

Fig. 25–39. LDL cholesterol

Each particle is approximately 22 nm (220 Å) in diameter and consists of approximately 1500 cholesterol ester molcules, surrounded by a mixed monomolecular layer of phospholipids and cholesterol, and a single molcule of the protein apolipoprotein B, which interacts specifically with the LDL receptor projecting from the plasma membrane. Apolipoprotein B-100 (apoB-100) is one of the largest single polypeptide chains known with 4,636 amino acid residues and has a molecular weight of 5,13, 000 daltons.

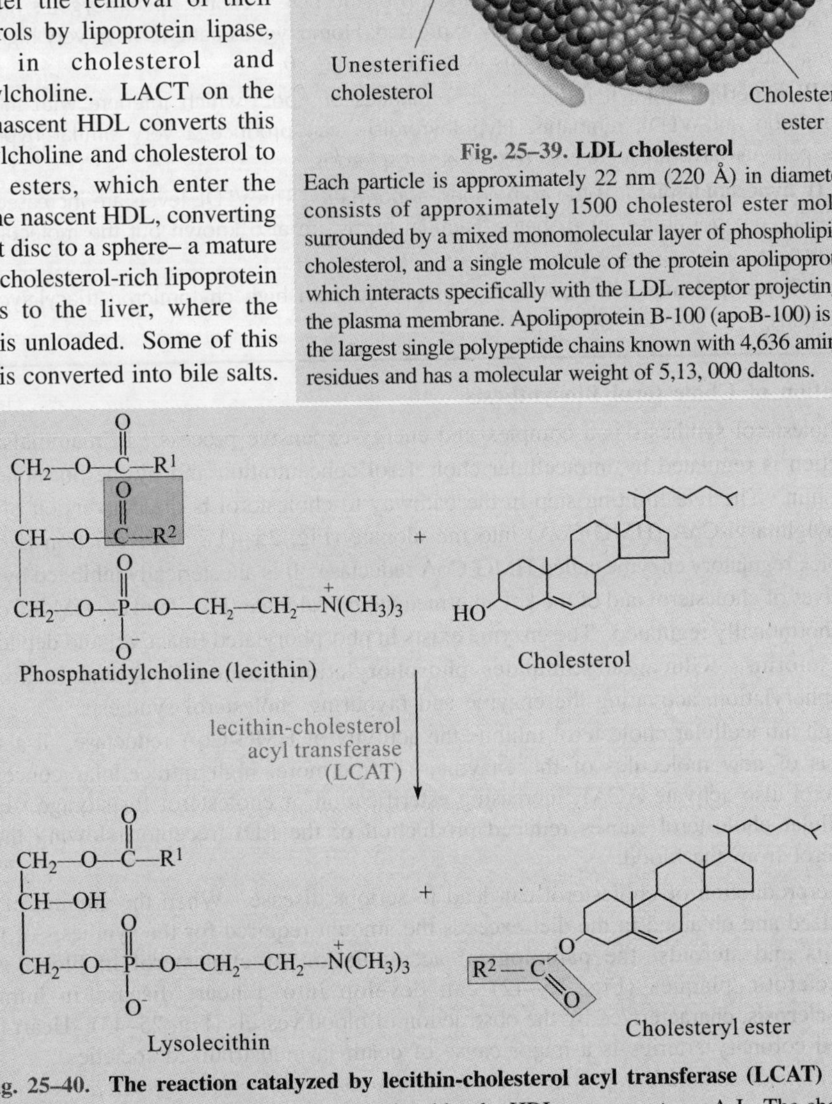

Fig. 25–40. The reaction catalyzed by lecithin-cholesterol acyl transferase (LCAT)

LCAT is present on the surface of HDL and is stimulated by the HDL component, apoA-I. The cholesteryl esters accumulate within nascent HDLs, converting them to mature HDLs.

CLINICAL IMPLICATIONS 25–1.

Hyperlipidemias

Hyperlipidemias are disorders of the rates of synthesis or clearance of lipoproteins from the bloodstream. Usually they are detected by measuring plasma triacylglycerol and cholesterol and are classified on the basis of which class of lipoproteins is elevated.

Type I hyperlipidemia: It is due to accumulation of chylomicrons. Two genetic forms are known: lipoprotein lipase deficiency and ApoC-II deficiency. ApoC-II is required by lipoprotein lipase for full activity. Patients with type I hyperlipidemia have exceedingly high plasma triacylglycerol concentrations (over, 1000 mg dL^{-1}) and suffer from eruptive *xanthomas* (triacylglycerol deposits in the skin) and *pancreatitis*.

Type II hyperlipidemia: It is characterized by elevated LDl levels. Most cases are due to genetic defects in the synthesis, processing or function of the LDL receptor. Heterozygotes have elevated LDL levels, hence the trait is dominantly expressed. Homozygous patients have very high LDL levels and may suffer myocardial infarctions (MIs) before age 20.

Type III hyperlipidemia: It is due to abnormalities of ApoE, which interfere with the uptake of chylomicron and VLDL remnants. Hypothyroidism can produce a very similar hyperlipidemia. These patients have an increased risk of atherosclerosis.

Type IV hyperlipidemia: *It is the commonest abnormality*. The VLDL levels are increased, often due to obesity, alcohol abuse, or diabetes. Familial forms are also known but the molecular defect is unknown.

Type V hyperlipidemia: Like type I, it is associated with high chylomicron triacylglycerol levels, pancreatitis, and eruptive xanthomas.

Regulation of Cholesterol Biosynthesis

Cholesterol synthesis is a complex and energy-expensive process. In mammals, cholesterol production is regulated by intracellular cholesterol concentration and by the hormones, glucagon and insulin. The rate-limiting step in the pathway to cholesterol is the conversion of β-hydroxy-β-methylglutaryl-CoA (HMG-CoA) into mevalonate (Fig. 25–41). This reaction is catalyzed by a complex regulatory enzyme called HMG-CoA reductase. It is allosterically inhibited by unidentified derivatives of cholesterol and of the key intermediate mevalonate (Fig. 25–41). HMG-CoA reductase is also hormonally regulated. The enzyme exists in phosphorylated (inactive) and dephosphorylated (active) forms. Glucagon stimulates phosphorylation (inactivation) and insulin promotes dephosphorylation, activating the enzyme and favouring cholesterol synthesis.

High intracellular cholesterol inhibits the acitivity of HMG-CoA reductase. It also slows the synthesis of new molecules of the enzyme. Furthermore, high intracellular concentrations of cholesterol also activate ACAT, increasing esterification of cholesterol for storage. Finally, high intracellular cholesterol causes reduced production of the LDL receptor, slowing the uptake of cholesterol from the blood.

Overproduction of cholesterol can lead to serious disease. When the amount of cholesterol synthesized and obtained in the diet exceeds the amount required for the synthesis of membranes, bile salts and steroids, the pathological accumulation of cholesterol in blood vessels *i.e.*, atherosclerotic plaques (Fig. 25–42) can develop into a heart disease in humans called **atherosclerosis**, characterized by the obstruction of blood vessels (Fig, 25–43). Heart failure from occluded coronary arteries is a major cause of death in industrialized societies.

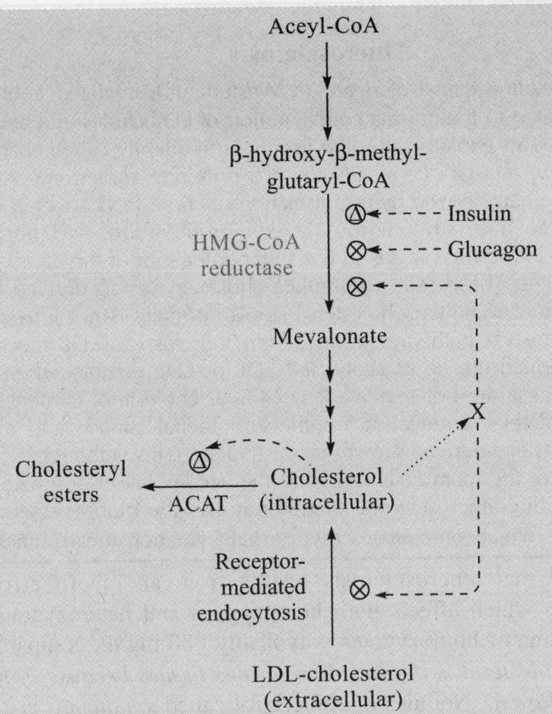

Fig. 25–41. Regulation of cholesterol biosynthesis

Glucagon promotes phosphorylation (inactivation) of HMG-CoA reductase, insulin promotes dephosphorylation (activation). X represents unidentified metabolites of cholesterol that stimulate proteolysis of HMG-CoA reductase.

(*Adapted from Lehninger AL, Nelson DL and Cox MM, 2005*)

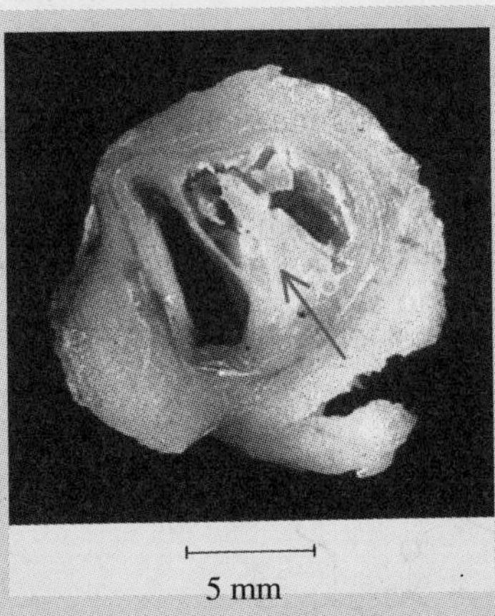

Fig. 25–42. An atherosclerotic plaque

A plaque (marked by an *arrow*) blocks most of the lumen of this blood vessel. The plaque is rich in cholesterol.

[Courtesy: Dr. Jeffrey Sklar]

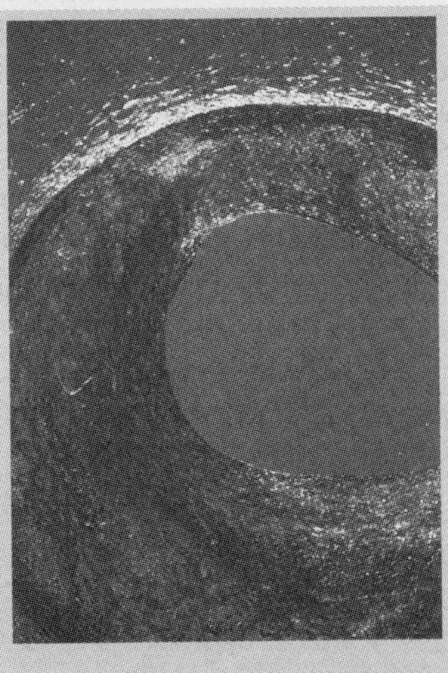

Fig. 25–43. Atherosclerosis

A human heart artery considerably blocked by cholesterol as veiwed under polarized light.

CLINICAL IMPLICATIONS 25–2.

Atherosclerosis

Atherosclerosis is the leading cause of death in Western industrialized countries. The risk of developing it is directly related to the plasma concentration of LDL cholesterol and inversely related to that of HDL cholesterol. This explains why the former is frequently called "bad" cholesterol and the latter "good" cholesterol, *though chemically there is only one cholesterol*. Atherosclerosis is a disorder of the arterial wall characterized by accumulation of cholesteryl esters in cells derived from the monocyte– macrophage line, smooth muscle cell proliferation, and fibrosis. The earliest abnormality is migration of blood monocytes to the subendothelium of the artery. Once there, they differentiate into macrophages. These cells accumulate cholesteryl esters derived from plasma LDL. Why these cells do not regulate cellular cholesterol stores normally is not fully understood. Some of the LDL may be taken up *via* pathways distinct from the classical LDL receptor pathway. For instance, receptors that mediate uptake of acetylated LDL or LDL complexed with dextran sulfate have been described and these are not regulated by cellular cholesterol content. Distortion of the subendothelium leads to platelet aggregation on the endothelial surface and release of platelet-derived mitogens such as platelet-derived growth factor (PDGF). This is thought to stimulate smooth muscle cell growth. Death of the foam cells results in the accumulation of a cellular lipid that can stimulate fibrosis. The resulting atherosclerotic plaque narrows the blood vessel and serves as the site of thrombus formation, which precipitates myocardial infarction (heart attack).

Studies have revealed that atherosclerosis results from an inborn error called **familial hypercholesterolemia** (FH) which affects both homozygous and heterozygous individuals. The cholesterol level in the plasma of homozygotes is typically 680 mg/dl, compared with 300 mg/dl in heterozygotes. *In FH, cholesterol is deposited in various tissues because of high concentration of LDL-cholesterol in the plasma.* Nodules of cholesterol called *xanthomas* are prominent in skin and tendons. More harmful is the deposition of cholesterol in atherosclerotic plaques, causing arterial narrowing and leading to heart attacks. In reality, most homozygotes die of coronary artery disease in childhood. The disease in heterozygotes has a milder and more variable clinical manifestation. *The molecular defect in most cases of familial hypercholesterolemia is an absence or deficiency of functional receptors for LDL.* Homozygotes have almost no receptors for LDL, whereas heterozygotes have about half the normal number. As a result, the entry of LDL into liver and other cells is impaired, leading to an increased plasma level of LDL. Thus, cholesterol obtained in the diet is not cleared from the blood; it accumulates and contributes to the formation of atherosclerotic plaques. This genetic disorder results from a mutation at a single autosomal locus.

Two natural products derived from fungi, **lovastatin** (also called **mevinolin**) and **compactin** (Fig. 25–44) are being widely used to lower the plasma cholesterol level. Both are potent competitive inhibitors of HMG-CoA reductase, the key control point in the biosynthetic pathway. The consequent increase in the number of LDL receptors on liver cells leads to a decrease in the LDL level in blood.

Clinical assay results are often expressed in mg/dl, which is equal to mg per 100 ml.

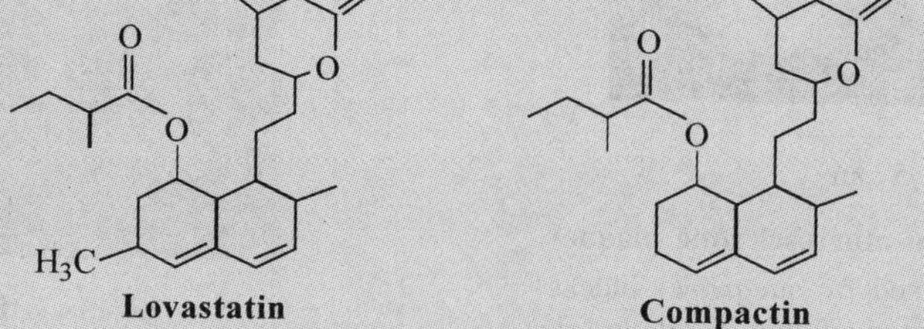

Lovastatin **Compactin**

Fig. 25–44. The two competitive inhibitors of HMG-CoA reductase

Note that their structure resembles 3-hydroxy-3-methylglutaryl-CoA, the substrate of the reaction catalyzed by HMG-CoA reductase.

BIOSYNTHESIS OF STEROID HORMONES

There are five major classes of steroid hormones: glucocorticoids, mineralocorticoids, androgens, estrogens and progestagens. All these steroid hormones in humans are derived from cholesterol. Glucocorticoids and mineralocorticoids are synthesized in the cortex of the adrenal gland; androgens in the testis; estrogens in the ovary; and progestagens in the corpus luteum. **Glucocorticoids** (such as corticosterone) promote gluconeogenesis and the formation of glycogen and enhance the degradation of fat and protein; they also enable animals to respond to stress– indeed, the absence of glucocorticoids can be fatal. **Mineralocorticoids** (primarily aldosterone) act on the distal tubules of the kidney to increase the reabsorption of Na^+ and the excretion of K^+ and H^+, which lead to an increase in blood volume and blood pressure. **Androgens** (such as testosterone) influence the development of secondary sex characteristics in males. **Estrogens** (such as estradiol) are responsible for the development of female secondary sex characteristics.

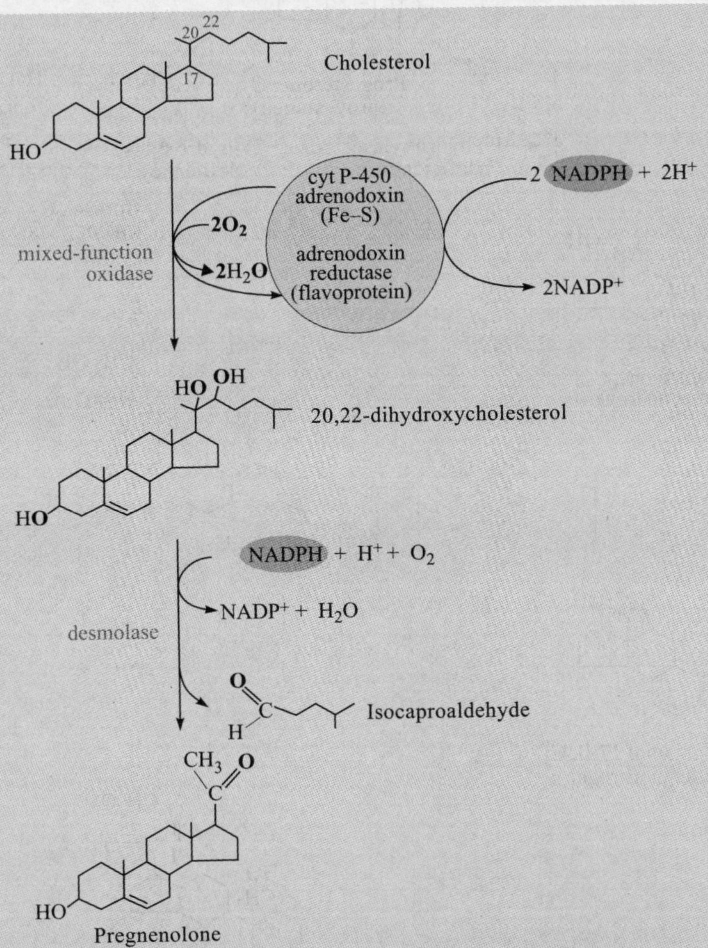

Fig. 25–45. Side chain cleavage in the synthesis of steroid hormones

This involves oxidation of adjacent carbons. Cytochrome P-450 acts as electron carrier in this mixed-function oxidase system, which also requires the electron-transferring proteins, adrenodoxin and adrenodoxin reductase. This side-chain cleaving system is found in mitochondria of the adrenal cortex, where active steroid production occurs. Note that pregnenolone is produced by the cleavage of 6 carbon atoms from the side chain of cholesterol. Pregnenolone is the precursor of all other steroid hormones.

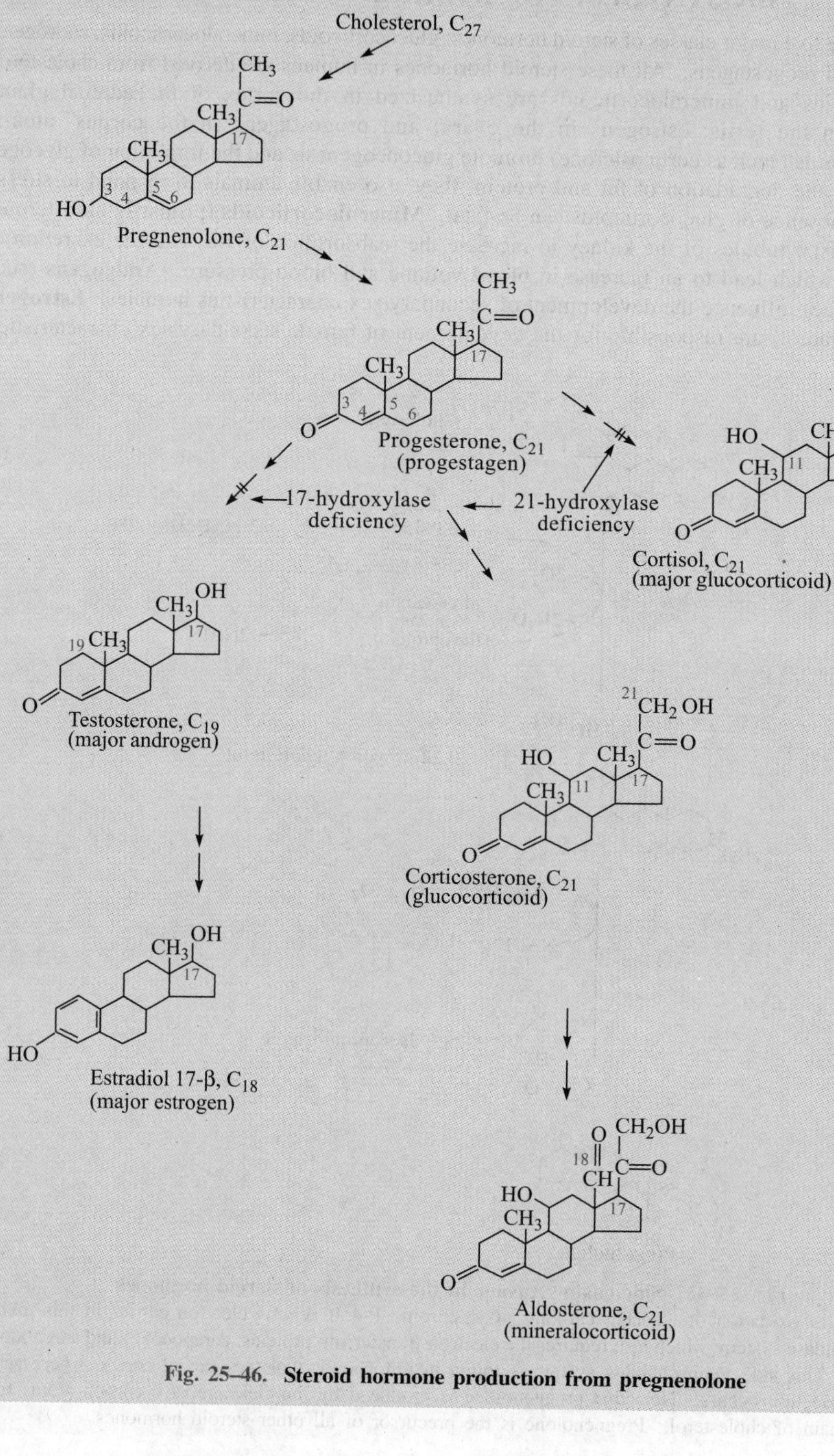

Fig. 25–46. Steroid hormone production from pregnenolone

Progesterone, a **progestagen**, prepares the lining of the uterus for implantation of an ovum ; it is also essential for the maintenance of pregnancy; it also participates, along with estrogens, in the ovarian cycle. The steroid hormones are effective at very low concentrations and are hence synthesized in relatively small quantities. Little cholesterol is consumed in their production, in contrast to bile salts.

The synthesis of steroid hormones requires removal of some or all of the carbons in the "side chain" that projects from C-17 of the D ring of cholesterol. Side chain removal takes place in the mitochondria of tissues that make steroid hormones. It involves first the hydroxylation of two adjacent carbons in the side chain (C-20 and C-22) and then cleavage of the bond between them (Fig. 25–45). Formation of individual hormones also involves the introduction of oxygen atoms. All of the hydroxylation and oxygenation reactions in steroid biosynthesis are catalyzed by monooxygenases (mixed-function oxidases of mitochondria) that require molecular oxygen and NADPH. Mitochondrial cytochrome P-450 acts as an electron carrier in this monooxygenase system, which also requires the electron-transferring proteins adrenodoxin and adrenodoxin reductase.

The removal of 6 carbons from C_{27} **cholesterol** yields the C_{21} structure, **pregnenolone**, which is common to many steroid hormones. Pregnenolone is then converted into C_{21} **progesterone** (a progestagen) in a two-step process that includes oxidation of the OH group of the C-3 to a keto group and isomerization of the double bond from C-5,6 to C-4,5 (Fig. 25–46).

The 3 principal steroid hormones are cortisol, corticosterone (both glucocorticoids) and aldosterone (a mineralocorticoid). Hydroxylation of C-11, C-17 and C-21 converts progesterone into cortisol (Fig. 25–46) and, in this reaction series, hydroxylation of C-17 must occur prior to that of C-21. The C-11 hydroxylation can take place at any stage of the conversion. For aldosterone synthesis, progesterone is first hydroxylated at C-21 and then at C-11 to produce **corticosterone**. The C-18 methyl group of corticosterone is then oxidized to an aldehyde in the last step, which yields **aldosterone**.

Progesterone is also the precursor of androgens, which in turn are precursors of estrogens (Fig. 25–46). Synthesis of testosterone (C_{19}) involves loss of C-20 and C-21 from progesterone (C_{21}); the steroid product of these reactions is another androgen, **androstenedione** (C_{19}) which has a keto group at C-17. This keto group is then reduced to a hydroxyl group to produce **testosterone.** **Estrogen** (C_{18}) is synthesized from testosterone by the loss of its C-19 methyl group and conversion of A ring into an aromatic structure. **Estrone** (C_{18}), another estrogen, is produced in an analogous manner from androstenedione.

REFERENCES

1. **Angulo P :** Nonalcoholic fatty liver disease. *N. Engl. J. Med.* **346 :** *1221–1231, 2002.*
2. **Benveniste P :** Sterol Biosynthesis. *Ann. Rev. Plant Physiol.* **37 :** *275–308, 1986.*
3. **Bloch K :** The biological synthesis of cholesterol. *Science.* **150 :** *19–28, 1965.*
4. **Bloch K :** Sterol structure and membrane function. *Crit. Rev. Biochem.* **14 :** *47–92, 1983.*
5. **Bloch K, Vance D :** Control mechanisms in the synthesis of saturated fatty acids. *Ann. Rev. Biochem.* **46 :** *263, 1977.*
6. **Bonomini F, Tengattini S, Fabiano A, Bianchi R, Rezzani R :** Atherosclerosis and oxidative stress. *Histol. Histopathol.* **23 :** *381–390, 2008.*
7. **Brown MS, Goldstein JL :** How LDL receptors influence cholesterol and atherosclerosis. *Sci. Amer.* **251 (November) :** *58–66, 1984.*

8. **Browse J, Somerville C :** Glycerolipid synthesis : Biochemistry and regulation. *Ann. Rev. Plant Physiol. Plant Mol. Biol.* **42** : *467–506, 1991.*
9. **Capdevila JH, Falck JR, Estabrook RW :** Cytochrome P 450 and the arachidonate cascade. *FASEB J.* **6** : *731–736, 1992.*
10. **Chan L :** Apolipoprotein B, the major protein component of triglyceride-rich and low density lipoproteins. *J. Biol. Chem.* **267** : *25621–25624, 1992.*
11. **Charlton-Menys V, Durrington PN :** Human cholesterol metabolism and therapeutic molecules. *Exp. Physiol.* **93** : *27–42, 2007.*
12. **DeLuca HF :** New concepts of vitamin D functions. *Ann. N.Y. Acad. Sci.* **669** : *59–68, 1992.*
13. **Dempsey ME :** Regulation of steroid biosynthesis. *Ann. Rev. Biochem.* **43** : *967, 1974.*
14. **Durrington P :** Dyslipidaemia. *Lancet.* **362** : *717–731, 2003.*
15. **Endo A :** The discovery and development of HMG-CoA reductase inhibitors. *J. Lipid Res.* **33** : *1569–1582, 1992.*
16. **Gibbons GF, Mitropoulos KA, Myant NB :** Biochemistry of Cholesterol. *Elsevier Biomedical, New York. 1982.*
17. **Goldstein JL, Brown MS :** Regulation of the mevalonate pathway. *Nature.* **343** : *425–430, 1990.*
18. **Hakomori S :** Glycosphingolipids. *Sci. Amer.* **254(5)** : *44–53, 1986.*
19. **Hardwood JL :** Fatty acid metabolism. *Ann. Rev. Plant Physiol. Plant Mol. Biol.* **39** : *101–138, 1988.*
20. **Harrison GK :** Inflammation, atherosclerosis and coronary artery disease. *N. Engl. J. Med.* **352** : *1685–1695, 2005.*
21. **Hawthorne JN, Ansell GB (editors) :** New comprehensive Biochemistry. Vol. 4 (Neuberger A & van Deenen LLM, Series editors). *Elsevier Biomedical Press, Amsterdam. 1982.*
22. **Heftman E :** Steroid Biochemistry. *Academic Press, Inc., New York. 1970.*
23. **Hobkisk R :** Steroid Biochemistry. *Vols. 1 and 2. CRC Press, Boca Raton, Fla. 1979.*
24. **Kennedy EP :** The metabolism and function of complex lipids. *Harvey Lectures.* **57** : *143–171, 1962.*
25. **Kent C, Carman GM, Spence MW, Dowhan W :** Regulation of eukaryotic phospholipid metabolism. *FASEB J.* **5** : *2258–2266, 1991.*
26. **Kleinig H :** The role of plastids in isoprenoid biosynthesis. *Ann. Rev. Plant Physiol. Plant Mol. Biol.* **40** : *39–59, 1989.*
27. **Lands WEM :** Biosynthesis of prostaglandins. *Ann. Rev. Nutr.* **11** : *41–60, 1991.*
28. **Lenz A, Diamond FB :** Obesity: the hormonal milieu. *Curr. Opin. Endocrinol. Diabetes Obesity.* **15** : *9–20, 2008.*
29. **LaRosa JC :** Low-density lipoprotein cholesterol reduction: the end is more important than the means. *Am. J. Cardiol.* **100** : *240–242, 2007.*
30. **McCarthy AD, Hardie DG :** Fatty acid synthase–an example of protein evolution by gene fusion. *Trends Biochem. Sci.* **9** : *60–63, 1984.*
31. **Mead JF, Alfin-Slater RB, Howton DR, Popják G :** Lipids : Chemistry, Biochemistry, and Nutrition. *Plenum Press, New York. 1986.*

32. **Myant NB :** Cholesterol Metabolism, LDL, and the LDL Receptor. *Academic Press, Inc., New York. 1990.*

33. **Raetz CRH, Dowhan W :** Biosynthesis and function of phospholipids in *Escherichia coli. J. Biol. Chem.* **265 :** *1235–1238, 1990.*

34. **Russell DW :** Cholesterol biosynthesis and metabolism. *Cardiovasc. Drugs Therapy.* **6 :** *103–110, 1992.*

35. **Schroepfer GJ Jr.** Sterol biosynthesis. *Ann. Rev. Biochem.* **51 :** *555–585, 1982.*

36. **Synder F (editor) :** Lipid Metabolism in Mammals. *Vols. 1 and 2. Plenum, New York. 1977.*

37. **Vance DE, Vance JE (editors) :** Biochemistry of Lipids, Lipoproteins and Membranes. *Elsevier. 1991.*

38. **Vegiopoulos A, Herzig S :** Glucocorticoids, metabolism and metabolic diseases. *Mol. Cell Endocrinol.* **275 :** *43–61, 2007.*

39. **Wakil SJ, Stoop JK, Joshi VC :** Fatty acid synthesis and its regulation. *Ann. Rev. Biochem.* **52 :** *537–579, 1983.*

40. **Walsh C :** Enzymatic Reaction Mechanisms. *W.H. Freeman. 1979.*

PROBLEMS

1. Compare and contrast fatty acid oxidation and synthesis with respect to
   (*a*) site of the process.
   (*b*) acyl carrier.
   (*c*) reductants and oxidants.
   (*d*) stereochemistry of the intermediates.
   (*e*) direction of synthesis or degradation.
   (*f*) organization of the enzyme system.

2. For each of the following unsaturated fatty acids, indicate whether the biosynthetic precursor in animals is palmitoleate, oleate, linoleate, or linolenate.
   (*a*) 18 : 1 cis-Δ^{11}
   (*b*) 18 : 3 cis-Δ^{6}, Δ^{9}, Δ^{12}
   (*c*) 20 : 2 cis-Δ^{11}, Δ^{14}
   (*d*) 20 : 3 cis-Δ^{5}, Δ^{8}, Δ^{11}
   (*e*) 22 : 1 cis-Δ^{13}
   (*f*) 22 : 6 cis-Δ^{4}, Δ^{7}, Δ^{10}, Δ^{13}, Δ^{16}, Δ^{19}

3. What is the role of decarboxylation in fatty acid synthesis ? Name another key reaction in a metabolic pathway that employs this mechanistic motif.

4. The serine residue in acetyl-CoA carboxylase that is the target of the AMP-dependent protein kinase is mutated to alanine. What is a likely consequence of this mutation ?

5. What is a potential disadvantage of having many catalytic sites together on one very long polypeptide chain?

6. Write a balanced equation for the synthesis of a triacylglycerol, starting from glycerol and fatty acids.

7. Write a balanced equation for the synthesis of phosphatidyl serine by the *de novo* pathway, starting from serine, glycerol, and fatty acids.

8. What is the activated reactant in each of the following biosyntheses ?

(*a*) Phosphatidyl serine from serine

(*b*) Phosphatidyl ethanolamine from ethanolamine

(c) Ceramide from sphingosine

(*d*) Sphingomyelin from ceramide

(*e*) Cerebroside from ceramide

(*f*) Ganglioside G_{M1} from ganglioside G_{M2}

(*g*) Farnesyl pyrophosphate from geranyl pyrophosphate

9. Would you expect the reaction catalyzed by cardiolipin synthase to be strongly exergonic or strongly endergonic? Explain your reasoning.

10. Identify a pathway for utilization of the four carbons of acetoacetate in cholesterol biosynthesis. Carry your pathway as far as the rate-determining reaction in cholesterol biosynthesis.

11. Explain why a deficiency of steroid 21-hydroxylase leads to excessive production of sex steroids (androgens and estrogens).

12. After a person has consumed large amounts of sucrose, the glucose and fructose that exceed caloric requirements are transformed to fatty acids for triacylglycerol synthesis. This fatty acid synthesis consumes acetyl-CoA, ATP, and NADPH. How are these substances produced from glucose?

13. Write the net equation for the biosynthesis of palmitate in rat liver, starting from mitochondrial acetyl-CoA and cytosolic NADPH, ATP, and CO_2.

14. In the condensation reaction catalyzed by β-ketoacyl-ACP synthase (Fig. 25–8), a four-carbon unit is synthesized by the combination of a two-carbon unit and a three-carbon unit, with the release of CO_2. What is the thermodynamic advantage of this process over one that simply combines two two-carbon units?

15. The biosynthesis of palmitoleate (Fig. 25–12), a common unsaturated fatty acid with a *cis* double bond in the Δ^9 position, uses palmitate as a precursor. Can this be carried out under strictly anaerobic conditions? Explain.

16. Use a net equation for the biosynthesis of tripalmitoylglycerol (tripalmitin) from glycerol and palmitate to show how many ATPs are required per molecule of tripalmitin formed.

17. Write the sequence of steps and the net reaction for the biosynthesis of phosphatidylcholine by the salvage pathway from oleate, palmitate, dihydroxyacetone phosphate, and choline. Starting from these precursors, what is the cost in number of ATPs of the synthesis of phosphatidylcholine by the salvage pathway?

18. The rate-limiting step in the early stages of cholesterol biosynthesis is the conversion of β-hydroxy-β-methylglutaryl-CoA to mevalonate, catalyzed by HMG-CoA reductase (Fig. 25–31). The liver of a fasting animal has decreased reductase activity. When the flow through this reaction is reduced, what is the effect on the formation of ketone bodies from acetyl-CoA? How does this explain increased ketosis during fasting?

19. Cells from a patient with familial hypercholesterolemia (FH) and cells from an individual without that disease were incubated with LDL particles containing radioactively-labeled cholesterol. After incubation, the incubation medium was removed and the radioactivity of the cells measured. The cells were treated to remove any bound material, lysed, and internal cholesterol content measured. Results are given below. What mutation of the gene for the LDL receptor protein could account for the results?

Cell type	*Radioactivity of cell*	*Cholesterol content*
Normal	3,000 cpm/mg cells	Low
FH	3,000 cpm/mg cells	High

20. The combination of bile salt-binding resin and an HMG CoA reductase inhibitor is very effective in reducing serum cholesterol for most patients with high cholesterol. Why is this treatment much less effective for patients with familial hypercholesterolemia?

21. What is the difference between 'good' and 'bad' cholesterol?

ANSWERS TO PROBLEMS

1. (*a*) Oxidation in mitochondria; synthesis in the cytosol.

(*b*) Acetyl-CoA in oxidation ; acyl carrier protein for synthesis.

(*c*) FAD and NAD^+ in oxidation; NADPH for synthesis.

(*d*) L isomer of 3-hydroxyacyl CoA in oxidation; D isomer in synthesis.

(*e*) From carboxyl to methyl in oxidation; from methyl to carboxyl in synthesis.

(*f*) The enzymes of fatty acid synthesis, but not those of oxidation, are organized in a multienzyme complex.

2. (*a*) palmitoleate

(*b*) linoleate

(*c*) linoleate

(*d*) oleate

(*e*) oleate

(*f*) linolenate

3. Decarboxylation drives the condensation of malonyl ACP and acetyl ACP. In contrast, the condensation of two molecules of acetyl ACP is energetically unfavourable. In gluconeogenesis, decarboxylation drives the formation of phosphoenolpyruvate from oxaloacetate.

4. The mutant enzyme would be persistently active because it could not be inhibited by phosphorylation. Fatty acid synthesis would be abnormally active. Such a mutation might lead to obesity.

5. The probability of synthesizing an error-free polypeptide chain decreases as the length of the chain increases. A single mistake can make the entire polypeptide ineffective. In contrast, a defective subunit can be spurned in forming a noncovalent multienzyme complex; the good subunits are not wasted.

6. Glycerol + 4 ATP + 3 fatty acids + 4 $H_2O \longrightarrow$

triacylglycerol + ADP + 3 AMP + 7 Pi + 4 H^+

7. Glycerol + 3 ATP + 2 fatty acids + 2 H_2O + CTP + serine $\longrightarrow$

phosphatidyl serine + CMP + ADP + 2 AMP + 6 Pi + 3 H^+

8. (*a*) CDP-diacylglycerol

(*b*) CDP-ethanolamine

(*c*) acyl CoA

(*d*) CDP-choline

(*e*) UDP-glucose or UDP-galactose

(*f*) UDP-galactose

(*g*) geranyl pyrophosphate

9. Probably K_{eq} is close to unity, because the bond broken is identical to the bond created.

10. Acetoacetate + succinyl-CoA $\longrightarrow$ acetoacetyl-CoA + succinate

Acetoacetyl-CoA + acetyl-CoA $\longrightarrow$ HMG-CoA + CoA-SH

HMG-CoA + 2 NADPH + 2 H^+ $\longrightarrow$ mevalonate + 2 $NADP^+$ + CoA-SH

11. By shutting down the pathway leading to aldosterone, this deficiency increases the supply of progesterone available for conversion to sex steroids.

12. Both glucose and fructose are degraded to pyruvate in glycolysis. The pyruvate is converted to acetyl-CoA by the pyruvate dehydrogenase complex. Some of this acetyl-CoA enters the citric acid cycle, which produces reducing equivalents (NADH and NADPH). Mitochondrial electron transfer to O_2 yields ATP.

13. 8 Acetyl-CoA + 15 ATP + 14 NADPH + 9 H_2O $\longrightarrow$

palmitate + 8 CoA + 15 ADP + 15 P_i + 14 $NADP^+$ + 2 H^+

14. By using the three-carbon unit malonyl-CoA, the activated form of acetyl-CoA (recall that malonyl-CoA synthesis requires ATP), metabolism is driven in the direction of fatty acid synthesis by the exergonic release of CO_2.

15. The double bond in palmitoleate is introduced by an oxidation catalyzed by fatty acyl–CoA desaturase, a mixed-function oxidase that requires O_2 as a cosubstrate.

16. 3 Palmitate + glycerol + 7 ATP + 4 H_2O $\longrightarrow$ tripalmitin + 7 ADP + 7 P_i + 7 H^+

17. Net reaction:

Dihydroxyacetone phosphate + NADH + palmitate +

oleate + 3 ATP + CTP + choline + 4 H_2O $\longrightarrow$

phosphatidylcholine + NAD^+ + 2 AMP + ADP + H^+ + CMP + 5 *Pi*;

7 ATP per molecule of PC.

18. The pathways to mevalonate and ketone bodies have steps in common. A decrease in HMG-CoA reductase results in the accumulation of HMG-CoA. This is a substrate for HMG-CoA lyase, which produces acetoacetate. Thus, a decreased HMG-CoA reductase activity leads to increased production of acetoacetate.

19. FH cells have LDL receptors with normal binding properties as indicated by the bound radioactivity being the same as that of normal cells. They are unable to internalize the receptor-LDL complex so cholesterol synthesis is not inhibited as it is with normal cells. The mutation is most likely on the carboxy terminus of the protein, which is involved in internalization. Make sure you understand why mutation in other regions would not lead to the observed results.

20. By binding bile salts, forcing increased excretion, liver has to convert more cholesterol to bile salts. If liver synthesis of cholesterol is inhibited, liver synthesizes more LDL receptors, removing increased LDL particles (and thus cholesterol) from blood. Patients with FH either have no receptors or have non-functioning LDL receptors so liver cannot increase uptake of LDL from blood. Synthesis is reduced but this does not have as dramatic an effect on blood cholesterol.

21. 'Bad' cholesterol (or LDL) carries cholesterol to arteries where it embeds and participates in plaque growth. 'Good' cholesterol (or HDL) carries cholesterol away from arteries to the liver where it is disposed of.

CONTENTS

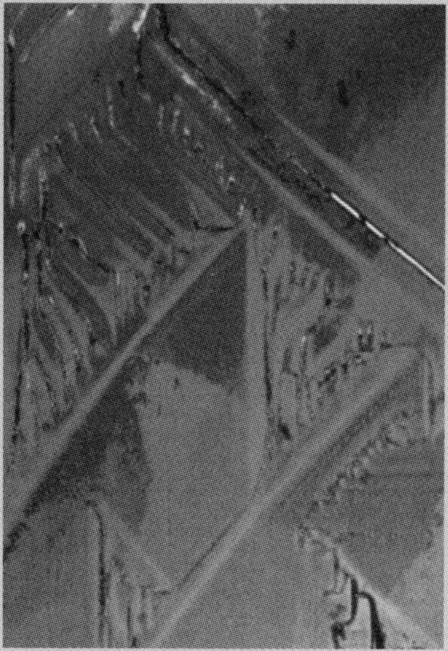

Crystals of urea, viewed under polarized light

Most of the ureotelic animals including man and shark secrete the excess ammonia as urea.

CHAPTER

26

Oxidation of Amino Acids

INTRODUCTION

Amino acids are the final class of biomolecules whose oxidation makes a significant contribution towards generation of metabolic energy. The fraction of metabolic energy derived from amino acids varies greatly with the type of organism and with the metabolic situation in which an organism finds itself. **Carnivores** may derive up to 90% of their energy requirements from amino acid oxidation. **Herbivores**, on the other hand, may obtain only a small fraction of their energy needs from this source. Most **microorganisms** can scavenge amino acids from their environment if they are available; these can be oxidized as fuel when the metabolic conditions so demand. **Photosynthetic plants,** on the contrary, rarely oxidize amino acids to provide energy. Instead they convert CO_2 and H_2O into the carbohydrate glucose that is used almost exclusively as an energy source. *Amino acid metabolism does occur in plants, but it is generally concerned with the production of metabolites for other biosynthetic pathways.*

In animals, the amino acids can be oxidatively degraded in 3 different metabolic conditions:

(*a*) *During normal protein synthesis:* Some of the amino acids released during protein breakdown will undergo oxidative degradation.

(*b*) *During protein-rich diet:* The surplus may be catabolized and amino acids cannot be stored.

(*c*) *During starvation* or *in diabetes mellitus:* Body proteins are used as fuel.

Under these different circumstances, amino acids lose their amino groups, and the resulting α-keto acids may undergo oxidation to produce CO_2 and H_2O.

Pathways leading to amino acid degradation are quite alike in most organisms. *As is the case for sugar and fatty acid catabolic pathways, the processes of amino acid degradation converge on the central catabolic pathways for carbon metabolism.* However, one major factor distinguishes amino acid degradation from the catabolic processes described till now, *i.e.*, every amino acid contains an amino group. As such every degradative pathway passes through a key step in which α-amino group is separated from the carbon skeleton and shunted into the specialized pathways for amino group metabolism (Fig. 26–1). This biochemical fork in the road is the point around which this chapter is centered.

Amino acids are needed for the synthesis of proteins and other biomolecules. The excess amount of amino acids, in contrast with glucose and fatty acids, cannot be stored; nor are they excreted. Rather surplus amino acids are used as metabolic fuel. The *α-amino group of the amino acids is removed and the resulting carbon skeleton is converted into a major metabolic intermediate.* Most of the amino groups of surplus amino acids are converted into urea whereas their carbon skeletons are transferred to acetyl-CoA, acetoacetyl-CoA, pyruvate, or one of the intermediates of the citric acid cycle. It follows that *amino acids can form glucose, fatty acids and ketone bodies.* The major site of amino acid degradation in mammals is the liver. The fate of the α-amino groups will be dealt with first, followed by that of the carbon skeleton.

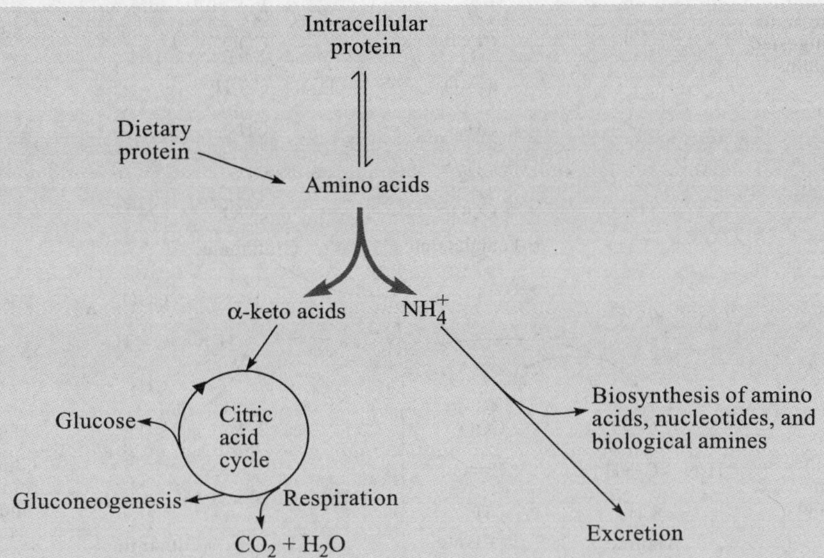

Fig. 26–1. An overview of the catabolism of amino acids

The *thick bifurcated red arrow* indicates the separate paths taken by the carbon skeleton and the amino groups.

AMINO GROUP METABOLISM (= Metabolic Fates of Amino Groups)

Nitrogen is the fourth most important contributor (after carbon, hydrogen and oxygen) to the mass of living cells. Atmospheric nitrogen is abundant but is too inert for use in most biochemical processes. Only a few microorganisms have the capacity to convert into biologically useful forms (such as NH_3) and as such amino groups are used with great economy in biological systems.

The catabolism of ammonia and amino groups in vertebrates is presented in Fig. 26–2. Amino acids derived from dietary proteins are the source of most amino groups. *Most of the amino groups are metabolized in the liver.* Some of the ammonia that is generated is recycled and used in a variety of biosynthetic processes; the excess is either excreted directly or converted to uric acid or urea for excretion. Excess ammonia generated in extrahepatic (*i.e.*, other than liver) tissues is transported to the liver in the form of amino groups, as described below, for conversion to the appropriate excreted form. The coenzyme pyridoxal phosphate (PLP or PALP) participates in these reactions.

Two amino acids, glutamate and its amide form glutamine, play crucial roles in these pathways. Amino groups from amino acids are generally first transferred to a α-ketoglutarate in the cytosol of liver cells (= hepatocytes) to form glutamate. Glutamate is then transported into the mitochondria. In muscle, excess amino groups are generally transferred to pyruvate to form alanine. Alanine is another important molecule in the transport of amino groups, transporting them from muscle to the liver.

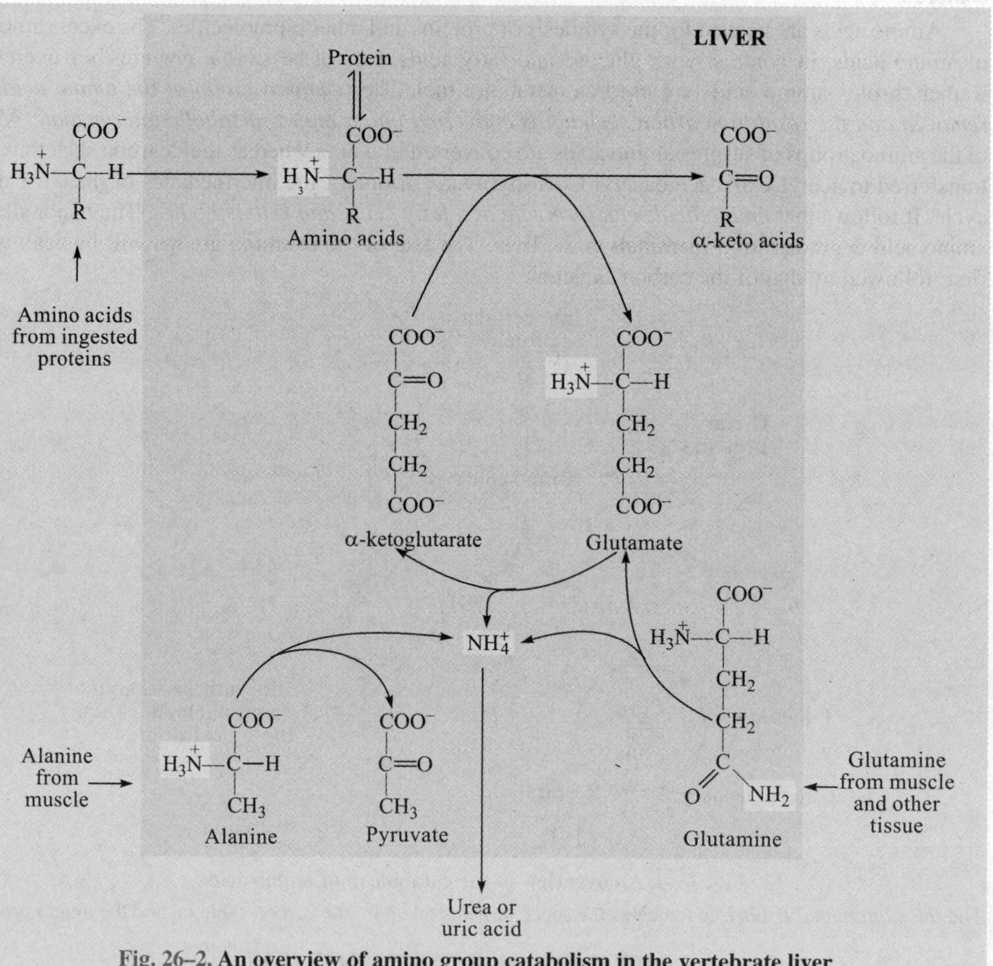

Fig. 26–2. An overview of amino group catabolism in the vertebrate liver

Note that the excess NH_4^+ is excreted as urea or uric acid

A. Transfer of Amino Groups to Glutamate

The α-amino groups of the 20 l-amino acids, commonly found in proteins, are removed during the oxidative degradation of the amino acids. If not reused for the synthesis of new amino acids, these amino groups are channelled into a single excretory product (Fig. 26–3). Many aquatic organisms simply release ammonia as NH_4^+ into the surrounding medium. Most terrestrial vertebrates first convert ammonia into either urea (*e.g.*, humans, other mammals and adult amphibians) or uric acid (*e.g.*, reptiles, birds).

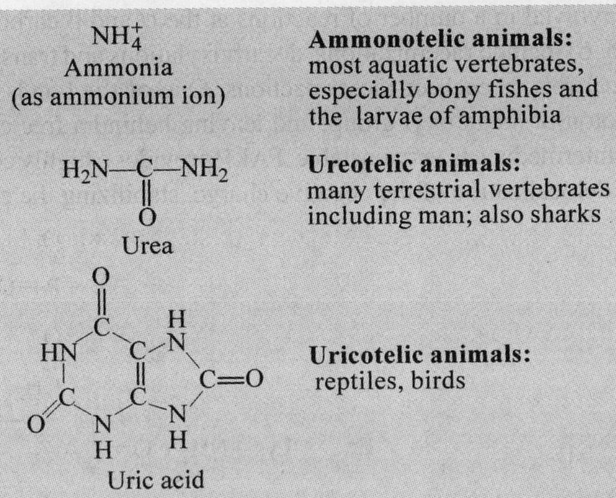

Fig. 26.–3. Excretory forms of amino-group nitrogen in different forms of life

Excess NH_4^+ is excreted as ammonia (microbes, bony fishes), urea (most terrestrial vertebrates), or uric acid (birds and terrestrial reptiles). Note that the C atoms of urea and uric acid are highly oxidized; the organism discards carbon only after extracting most of its available energy of oxidation.

The first step in the catabolism of most of the L-amino acids is the removal of the α-amino group (*i.e.*, transamination) by a group of enzymes called **aminotransferases** (= **transaminases**). In these reactions, the α-amino group is transferred to the α-carbon atom of α-ketoglutarate, leaving behind the corresponding α-keto acid analogue of the amino acid (Fig. 26–4). There is no net deamination in such reactions because the α-ketoglutarate becomes aminated as the α-amino acid is deaminated. *The effect of transamination reactions is to collect the amino groups from many different amino acids in the form of only one, namely L-glutamate.* Cells contain several different aminotransferases, many of which are specific for α ketoglutarate as the amino group acceptor. The amino-transferases differ in their specificity for the other substrate (*i.e.*, the L-amino acid that donates the amino group) and are named for the amino group donor. The reactions catalyzed by the aminotransferases are freely reversible, having an equilibrium constant of about 1.0 ($\Delta G^{o\prime} \simeq 0$ kJ/mol)).

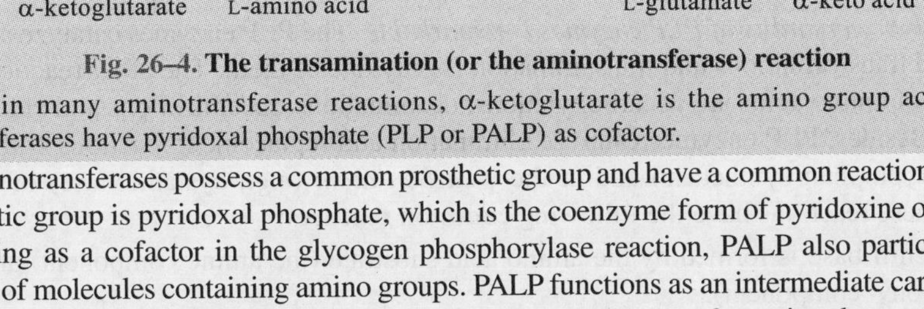

Fig. 26–4. The transamination (or the aminotransferase) reaction

Note that in many aminotransferase reactions, α-ketoglutarate is the amino group acceptor. All aminotransferases have pyridoxal phosphate (PLP or PALP) as cofactor.

All aminotransferases possess a common prosthetic group and have a common reaction mechanism. The prosthetic group is pyridoxal phosphate, which is the coenzyme form of pyridoxine or vitamin B_6. Besides acting as a cofactor in the glycogen phosphorylase reaction, PALP also participates in the metabolism of molecules containing amino groups. PALP functions as an intermediate carrier of amino groups at the active site of aminotransferases. It undergoes reversible transformations between its aldehyde form (pyridoxal phosphate, PALP) which can accept an amino group, and its aminated form (pyridoxamine phosphate, PAMP) which can donate its amino group to an α-keto acid (Fig. 26–5a). PALP is generally bound covalently to the enzyme's active site through an imine (Schiff-base) linkage to the ε-amino group of a lysine (Lys) residue (Fig. 26–5b).

Pyridoxal phosphate is involved in a number of reactions at the α and β carbons of amino acids. Reactions at α carbon (Fig. 26–6) include racemizations, decarboxylations and transaminations as well. PALP plays the same chemical role in each of these reactions. One of the bonds to the α carbon is broken, removing either a proton or a carboxyl group, and leaving behind a free electron pair on the carbon, *i.e.*, a carbanion. This intermediate is very unstable. PALP provides a highly conjugated structure (an *electron sink*) that allows delocalization of the negative charge, stabilizing the carbanion.

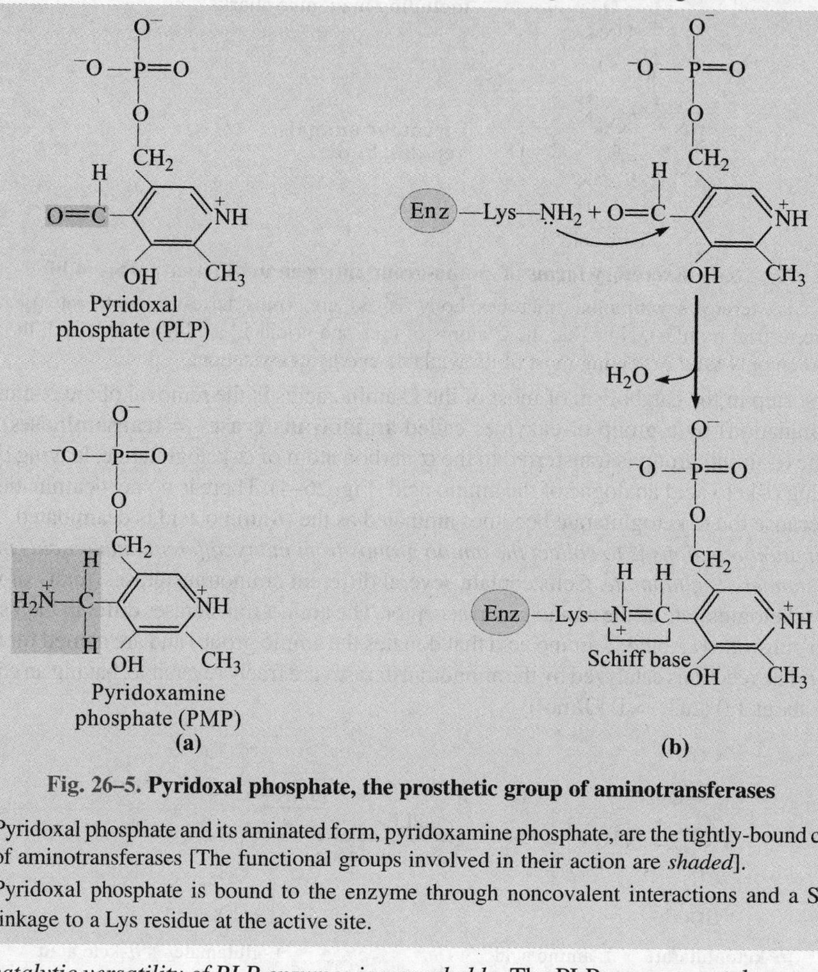

Fig. 26–5. Pyridoxal phosphate, the prosthetic group of aminotransferases

(a) Pyridoxal phosphate and its aminated form, pyridoxamine phosphate, are the tightly-bound coenzymes of aminotransferases [The functional groups involved in their action are *shaded*].

(b) Pyridoxal phosphate is bound to the enzyme through noncovalent interactions and a Schiff-base linkage to a Lys residue at the active site.

The catalytic versatility of PLP enzymes is remarkable. The PLP enzymes catalyze a wide range of amino acid transformations and transamination is only one of them. The other reactions at the *α-carbon atom* of amino acids are decarboxylations, deaminations, racemizations and aldol cleavages (Fig. 26–7). Besides, PLP enzymes catalyze elimination and replacement reactions at the *β-carbon atom* (*e.g.*, tryptophan synthetase) and at the *γ-carbon atom* (*e.g.*, cystathionase) of amino acid substrates. These reactions have the following **common features**:

1. A Schiff base is formed by the amino acid substrate (the amine component) and PLP (the carbonyl component).
2. Pyridoxal phosphate acts as an electron sink to stabilize catalytic intermediates that are negatively charged. The ring nitrogen of PLP attracts electrons from the amino acid substrate. In other words, *PLP is an electrophilic catalyst.*

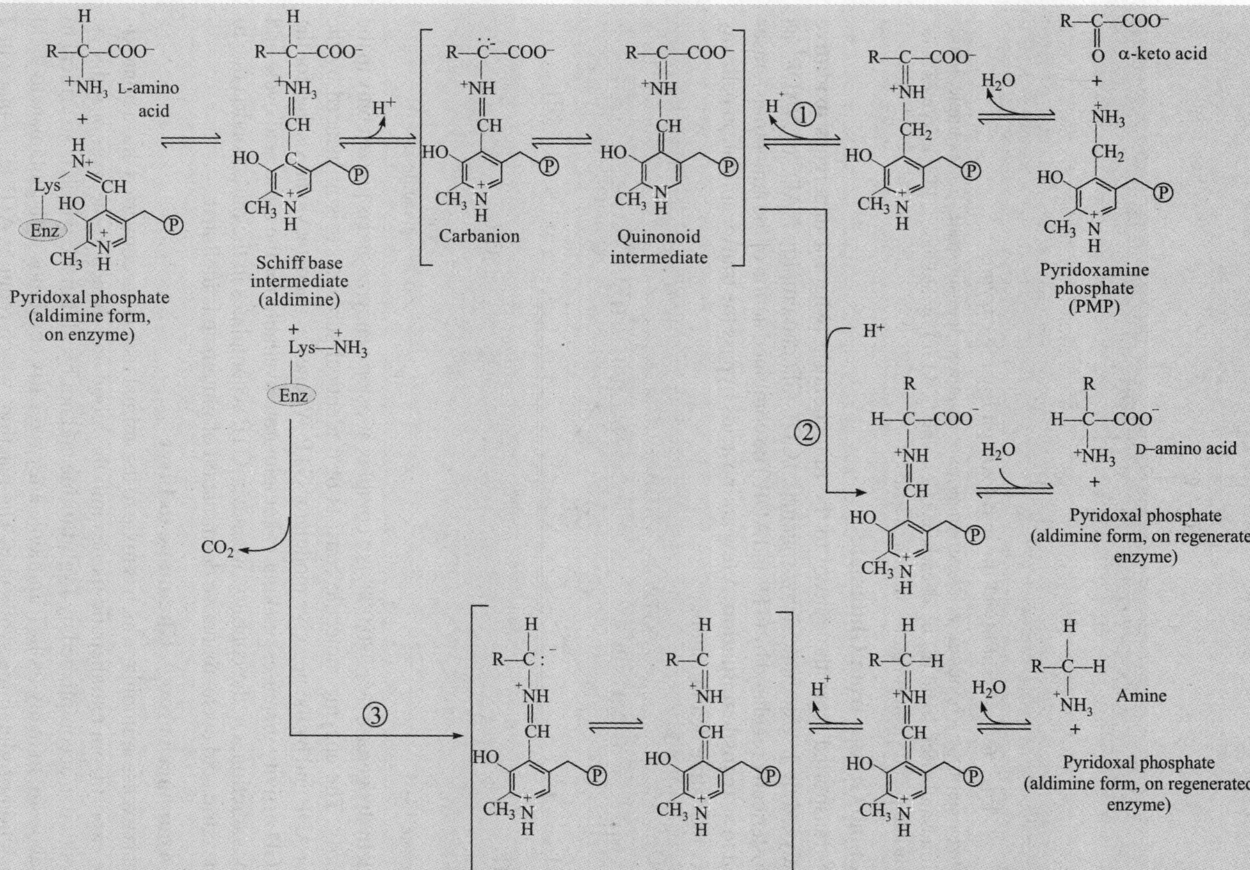

Fig. 26–6. Some amino acid transformations at the α carbon that are facilitated by pyridoxal phosphate (PLP or PALP)

PALP is usually bound to the enzyme through a Schiff base. Reactions begin (*top left*) with the formation of a new Schiff base (aldimine) between the α-amino group of the amino acid and PALP, which substitutes for the enzyme-PALP linkage. The amino acid then can undergo 3 alternative fates: (1) transamination, (2) racemization, and (3) decarboxylation. The Schiff base formed between PALP and the amino acid is in conjugation with the pyridine ring, which acts as an electron sink, allowing delocalization of the negative charge of the carbanion (as shown within the *large-sized brackets*). A quinonoid intermediate is involved in all 3 types of reactions.

(Adapted from Lehninger AL, Nelson DL and Cox MM, 1993)

3. The product Schiff base is then hydrolyzed.

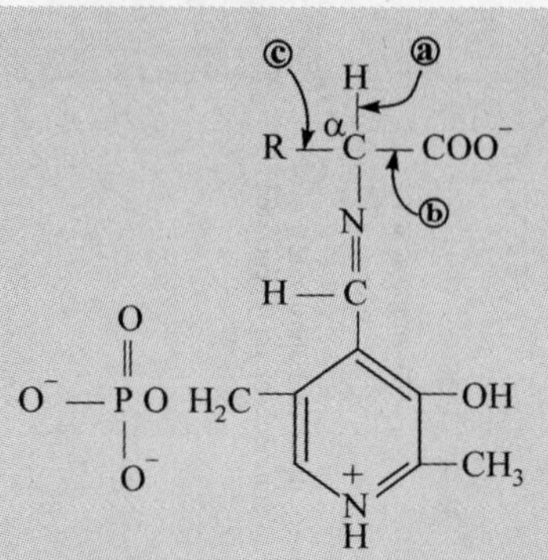

Fig. 26–7. Versatility of the pyridoxal phosphate enzymes

PLP enzymes labilize one of the 3 bonds at the α-carbon atom of an amino acid substrate. For example, bond **(a)** is labilized by transaminases, bond **(b)** by decarboxylases, and bond **(c)** by aldolases. PLP enzymes also catalyze reactions at the β and γ carbon atoms of amino acids.

B. Removal of Amino Groups from Glutamate

Glutamate is transported from the cytosol to the mitochondria, where it undergoes **oxidative deamination** catalyzed by L-glutamate dehydrogenase (GD). GD can employ NAD^+ or $NADP^+$ as cofactor and is allosterically regulated by GTP and ADP. The combined action of the aminotransferases and GD is referred to as **transdeamination**. A few amino acids bypass the transdeamination pathway and undergo direct oxidative deamination.

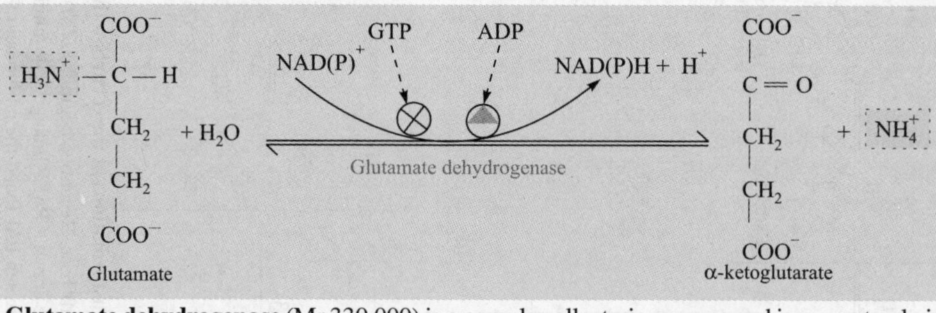

Glutamate dehydrogenase (Mr 330,000) is a complex allosteric enzyme and is present only in the mitochondrial matrix. The enzyme molecule consists of 6 identical subunits. It is influenced by the positive modulator ADP and by the negative modulator GTP. Whenever a hepatocyte needs fuel for the citric acid cycle, GD activity increases, making α-ketoglutarate available for the citric acid cycle and releasing NH_4^+ for excretion. On the contrary, whenever GTP accumulates in the mitochondria due to high activity of the citric acid cycle, oxidative deamination of glutamate is inhibited.

C. Transport of Ammonia Through Glutamine to Liver

In most animals excess ammonia, which is toxic to the animal tissues, is converted into a nontoxic compound before export from extrahepatic tissues into the blood and thence to the liver or kidneys. This transport function is accomplished by **L-glutamine** and not by glutamate which is so critical to amino group metabolism. In many tissues, ammonia enzymatically combines with glutamate to yield glutamine by the action of glutamine synthetase. The reaction, which requires ATP, takes place in 2 steps:

I. *First step:* glutamate and ATP react to form ADP and γ-glutamyl phosphate

II. *Second step:* γ-glutamyl phosphate reacts with ammonia to produce glutamine and inorganic phosphate.

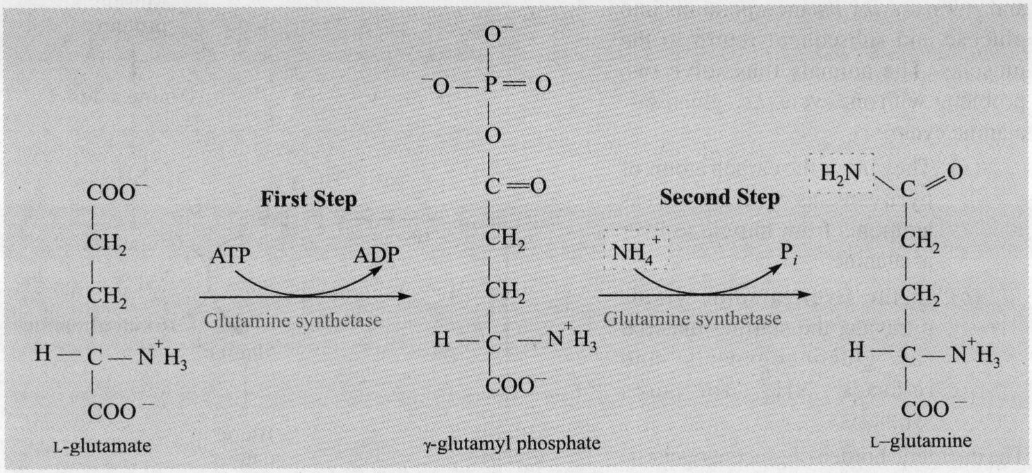

Glutamine is a nontoxic, neutral compound that can readily pass through cell membranes, whereas glutamate which bears a negative net charge, cannot. In most terrestrial animals, glutamine is carried through blood to the liver. The amide nitrogen of glutamine, like the amino group of glutamate, is

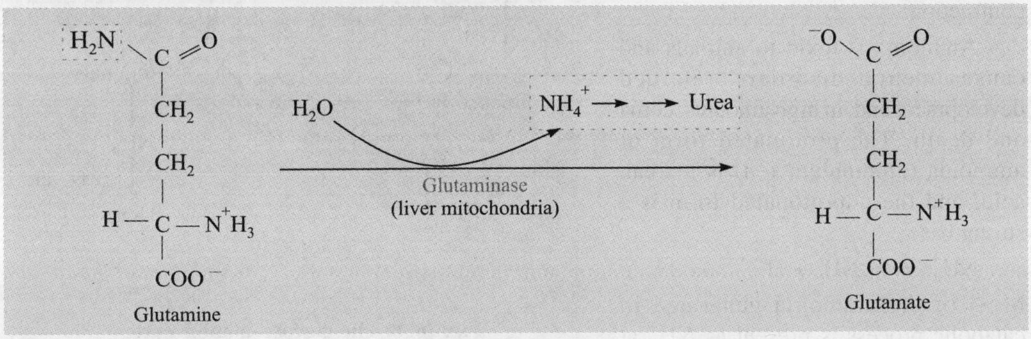

released as ammonia only within liver mitochondria, where the enzyme glutaminase converts glutamine to glutamate and NH_4^+. *Glutamine is, thus, a major transport form of ammonia.* It is normally present in blood in much higher concentrations than other amino acids.

D. Transport of Ammonia from Muscle to Liver Through Alanine (= Glucose–Alanine Cycle)

Alanine also plays a special role in transporting amino groups to the liver in a nontoxic form by **glucose–alanine cycle** (Fig. 26–8). In muscle and certain other tissues that degrade amino acids for fuel, amino groups are collected in glutamate by transamination (refer Fig. 26–2). Glutamate may then either be converted to glutamine for transport to the liver, or it may transfer its α-amino group to pyruvate, a readily-available product of muscle glycolysis, by the action of **alanine aminotransferase**. Alanine passes into the blood and is carried to the liver. As in the case of glutamine, excess nitrogen carried to the liver as alanine is ultimately delivered as ammonia in the mitochondria. During a reversal of this alanine aminotransferase reaction, alanine transfers its amino group to α-ketoglutarate, forming glutamate in the cytosol. Some of this glutamate is transported into the mitochondria and acted upon by glutamate dehydrogenase, releasing NH_4^+. Alternatively, transamination with oxaloacetate moves amino groups from glutamate to aspartate, another nitrogen donor in urea synthesis.

Vigorously contracting skeletal muscles operate anaerobically and produce not only ammonia from protein degradation but also large amounts of pyruvate from glycolysis. Both these products must find their way to the liver– *ammonia* for its conversion into urea for excretion and *pyruvate* for its incorporation into glucose and subsequent return to the muscles. The animals thus solve two problems with one cycle (*i.e.*, glucose—alanine cycle):

(*a*) They move the carbon atoms of pyruvate, as well as excess ammonia from muscle to liver as alanine.

(*b*) In the liver, alanine yields pyruvate, the starting material for gluconeogenesis and releases NH_4^+ for urea synthesis.

The energetic burden of gluconeogenesis is, thus, imposed on the liver rather than on the muscle, so that the available ATP in the muscle can be devoted to muscle contraction.

Ammonia is toxic to animals and causes mental disorders, retarded development and, in high amounts, coma and death. The protonated form of ammonia (ammonium ion) is a weak acid, and the unprotonated form is a strong base:

$$NH_4^+ \rightleftharpoons NH_3 + H^+$$

Most of the ammonia generated in catabolic process is present as NH_4^+ at neutral pH. Although most of the reactions that produce ammonia yield NH_4^+ a few reactions produce NH_3. Excessive amounts of ammonia cause alkalization of cellular fluids, which has multiple effects on cellular metabolism.

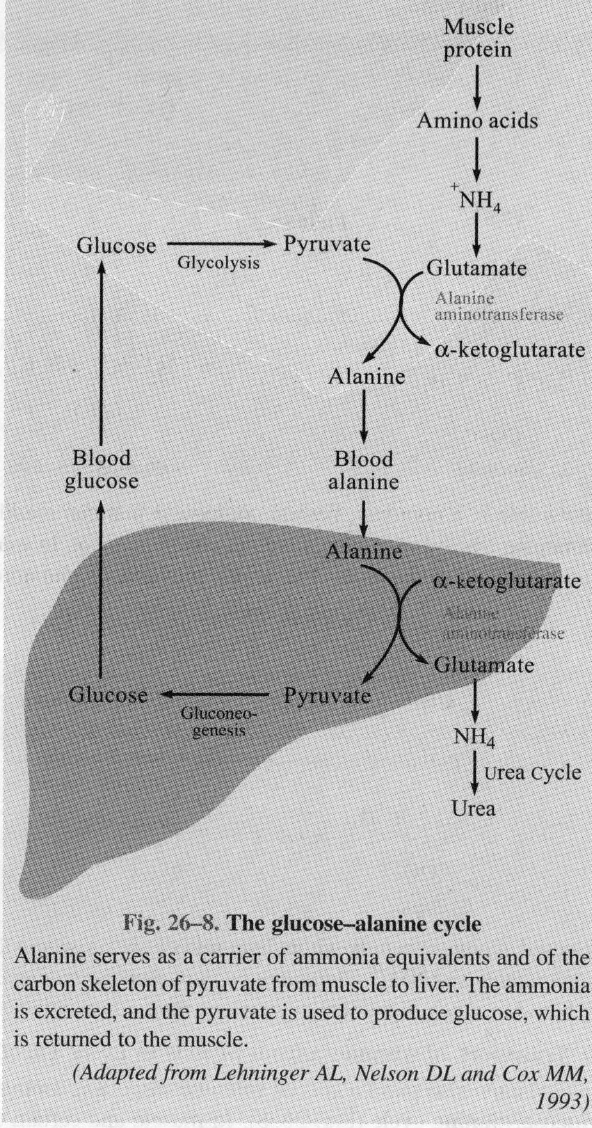

Fig. 26–8. The glucose–alanine cycle

Alanine serves as a carrier of ammonia equivalents and of the carbon skeleton of pyruvate from muscle to liver. The ammonia is excreted, and the pyruvate is used to produce glucose, which is returned to the muscle.

(Adapted from Lehninger AL, Nelson DL and Cox MM, 1993)

NITROGEN EXCRETION

The amino nitrogen is excreted in 3 different forms in various types of life-forms (Fig. 26–3).

(*a*) as *ammonia* in most aquatic vertebrates, bony fishes and amphibian larvae (**ammonotelic animals**)

(*b*) as *urea* in many terrestrial vertebrates including man, also sharks and adult amphibian (**ureotelic animals**)

(*c*) as *uric acid* in reptiles and birds (**uricotelic animals**)

Plants, however, recycle virtually all amino groups, and nitrogen excretion occurs only under highly unusual circumstances. *There is no general pathway for nitrogen excretion in plants.*

In ureotelic organisms, the ammonia in the mitochondria of liver cells (= hepatocytes) is converted to urea *via* the **urea cycle**. This pathway was discovered by **Hans Adolf Krebs** and a medical student associate, **Kurt Henseleit** in 1932, five years before the elucidation of the citric acid cycle. In fact, *the urea cycle was the first cyclic metabolic pathway to be discovered.* They found that the rate of urea formation from ammonia was greatly accelerated by adding any one of the 3 α-amino acids: ornithine, citrulline or arginine. Their structure suggests that they might be related in a sequence. This finding led them to deduce that a cyclic process occurs (Fig. 26–9), in which ornithine plays a role resembling that of oxaloacetate in the citric acid cycle. A mole of ornithine combines with one mole of ammonia and one of CO_2 to form citrulline. A second amino group is added to citrulline to form arginine. Arginine is then hydrolyzed to yield urea with concomitant regeneration of ornithine. Ureotelic animals have large amounts of the enzyme arginase in the liver. This enzyme catalyzes the irreversible hydrolysis of arginine to urea and ornithine. The ornithine is then ready for the next turn of the urea cycle. The urea is passed *via* the bloodstream to the kidneys and is excreted into the urine.

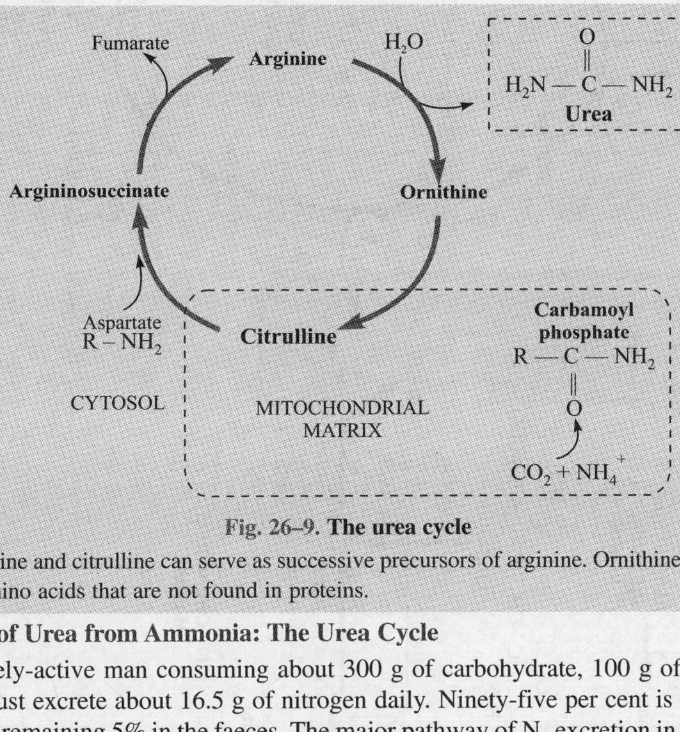

Fig. 26–9. The urea cycle

Note that ornithine and citrulline can serve as successive precursors of arginine. Ornithine and citrulline are nonstandard amino acids that are not found in proteins.

A. Production of Urea from Ammonia: The Urea Cycle

A moderately-active man consuming about 300 g of carbohydrate, 100 g of fat and 100 g of protein daily must excrete about 16.5 g of nitrogen daily. Ninety-five per cent is eliminated by the kidneys and the remaining 5% in the faeces. The major pathway of N_2 excretion in humans is as urea which is synthesized in the liver, released into the blood, and cleared by the kidney. In humans eating an occidental diet, urea constitutes 80-90% of the nitrogen excreted. The urea cycle spans two cellular compartments (Figs 26–9 and 26–10). It begins inside the mitochondria of liver cells (= hepatocytes), but 3 of the steps occur in the cytosol. The first amino group to enter the urea cycle is derived from ammonia inside the mitochondria, arising by multiple pathways described above. Whatever its source, the NH_4^+ generated in liver mitochondria is immediately used, together with HCO_3^- produced by mitochondrial respiration, to form **carbamoyl phosphate** in the matrix. This ATP-dependent reaction is catalysed by **carbamoyl phosphate synthetase I**, a regulatory enzyme present in liver mitochondria of all ureotelic organisms including man. In bacteria, glutamine rather than ammonia serves as a substrate for carbamoyl phosphate synthesis.

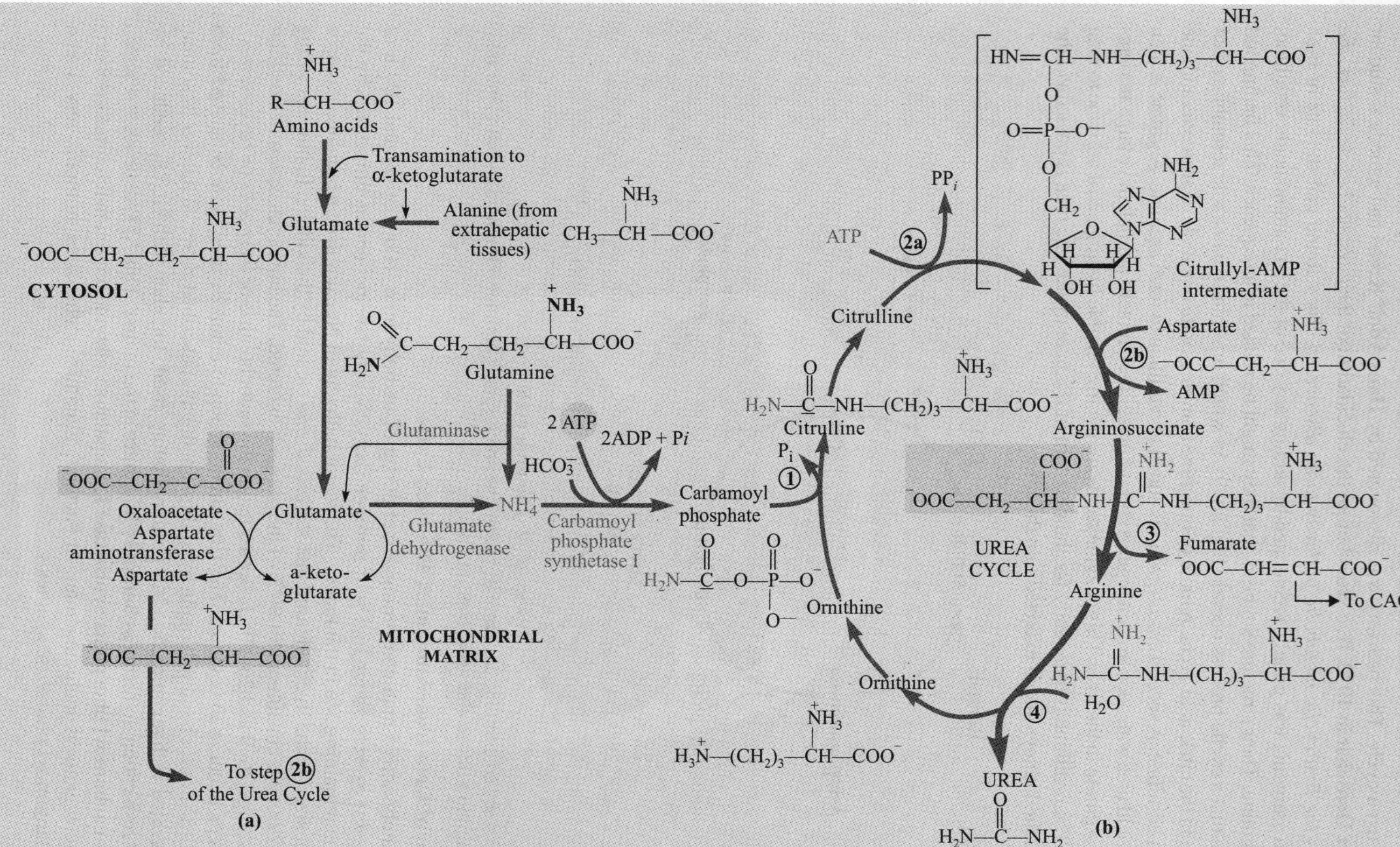

Fig. 26–10. The urea cycle and the reactions that feed amino groups into it

Note that one of the nitrogen atoms of the urea synthesized by this pathway is transferred from an amino acid, aspartate. The other nitrogen atom is derived from NH_4^+ and the carbon atom comes from CO_2. Ornithine, a nonprotein amino acid, is the carrier of these carbon and nitrogen atoms. Also note that the enzymes catalyzing these reactions are distributed between the mitochondrial matrix and the cytosol. One amino group enters the urea cycle from carbamoyl phosphate (Step①), formed in the matrix; the other (entering at step②) is derived from aspartate, also formed in the matrix *via* transamination of oxalocetate and glutamate in a reaction catalyzed by aspartate aminotransferase. Also note that the arginase reaction (Step④) also regenerates the starting compound in the cycle, *i.e.*, ornithine.

The carbamoyl phosphate now enters the urea cycle, which itself consists of 4 enzymatic steps. These are:

Step 1: *Synthesis of citrulline*

Carbamoyl phosphate has a high transfer potential because of its anhydride bond. It, therefore, donates its carbamoyl group to ornithine to form citrulline and releases inorganic phosphate. The reaction is catalyzed by **L-ornithine transcarbamoylase** of liver mitochondria. The citrulline is released from the mitochondrion into the cytosol.

Step 2: *Synthesis of argininosuccinate*

The second amino group is introduced from aspartate (produced in the mitochondria by transamination and transported to the cytosol) by a condensation reaction between the amino group of aspartate and the ureido (= carbonyl) group of citrulline to form argininosuccinate. The reaction is catalyzed by **argininosuccinate synthetase** of the cytosol. It requires ATP which cleaves into AMP and pyrophosphate and proceeds through a citrullyl-AMP intermediate.

Step 3: *Cleavage of argininosuccinate to arginine and fumarate*

Argininosuccinate is then reversibly cleaved by **argininosuccinate lyase (= argininosuccinase)**, a cold-labile enzyme of mammalian liver and kidney tissues, to form free arginine and fumarate, which enters the pool of citric acid cycle intermediates. These two reactions, which transfer the amino group of aspartate to form arginine, preserve the carbon skeleton of aspartate in the form of fumarate.

Step 4: *Cleavage of arginine to ornithine and urea*

The arginine so produced is cleaved by the cytosolic enzyme **arginase**, present in the livers of all ureotelic organisms, to yield urea and ornithine. Smaller quantities of arginase also occur in renal tissue, brain, mammary gland, testes and skin. Ornithine is, thus, regenerated and can be transported into the mitochondrion to initiate another round of the urea cycle.

In the urea cycle, mitochondrial and cytosolic enzymes appear to be clustered and not randomly distributed within cellular compartments. The citrulline transported out of the mitochondria is not diluted into the general pool of metabolites in the cytosol. Instead, each mole of citrulline is passed directly into the active site of a molecule of argininosuccinate synthetase. This channeling continues for argininosuccinate, arginine and ornithine. Only the urea is released into the general pool within the cytosol. Thus, the compartmentation of the urea cycle and its associated reactions is noteworthy. The formation of NH_4^+ by glutamate dehydrogenase, its incorporation into carbamoyl phosphate, and the subsequent synthesis of citrulline occur in the *mitochondrial matrix*. In contrast, the next 3 reactions of the urea cycle, which lead to the formation of urea, takes place in the cytosol.

A perusal of the urea cycle reveals that of the 6 amino acids involved in urea synthesis, one, N-acetylglutamate functions as an enzyme activator rather than as an intermediate. The remaining 5 amino acids— aspartate, arginine, ornithine, citrulline and argininosuccinate— all function as carriers of atoms which ultimately become urea. Two (aspartate and arginine) occur in proteins, while the remaining three (ornithine, citrulline and argininosuccinate) do not. The major metabolic roles of these latter 3 amino acids in mammals is urea synthesis. Note that urea formation is, in part, a cyclical process. The ornithine used in Step 1 is regenerated in Step 4. There is thus no net loss or gain of ornithine, citrulline, argininosuccinate or arginine during urea synthesis; however, ammonia, CO_2, ATP and aspartate are consumed.

B. The "Krebs Bicycle"

The citric acid cycle and urea cycle both are linked together (Fig. 26–11). The fumarate produced in the cytosol by argininosuccinate lyase reaction is also an intermediate of the citric acid cycle. Fumarate enters the citric acid cycle in the the mitochondrion where it is first hydrated to malate by fumarase which, in turn, is oxidized to oxaloacetate by malate dehydrogenase. The oxaloacetate

CLINICAL IMPLICATIONS 26–1.

Deficiencies of Urea Cycle Enzymes

Ornithine transcarbamoylase: The most common deficiency involving urea cycle enzymes is the lack of ornithine transcarbamoylase. Mental retardation and death often result. The gene for ornithine transcarbamoylase is on the X chromosome, and males are generally more seriously affected than heterozygotic females. In addition to ammonia and amino acids appearing in the blood in increased amounts, orotic acid also increases, presumably because carbamoyl phosphate that cannot be used to form citrulline diffuses into the cytosol, where it condenses with aspartate, ultimately forming orotate.

Arginosuccinate synthetase and **Arginosuccinate lyase:** The inability to condense citrulline with aspartate results in accumulation of citrulline in blood and its excretion in the urine (**citrullinemia**). Therapy for this normally benign disease requires specific supplementation with arginine for protein synthesis and for formation of creatine. Impaired ability to split arginosuccinate to form arginine resembles arginosuccinate synthetase deficiency in that the substrate, in this case arginosuccinate, is excreted in large amounts. Severity of symptoms in this disease varies greatly so that it is hard to evaluate the effect of therapy, which includes dietary supplementation with arginine.

Arginase: Arginase deficiency is rare but causes many abnormalities in development and function of the central nervous system (CNS). Arginine accumulates and is excreted. Precursors of arginine and products of arginine metabolism may also be excreted. Unexpectedly, some urea is also excreted. A diet including essential amino acids but excluding arginine has been used effectively.

accepts an amino group from glutamate by transamination, and the aspartate thus formed leaves the mitochondrion and donates its amino group to the urea cycle in the argininosuccinate synthetase reaction; the other product of this transamination is α-ketoglutarate, another intermediate of the citric acid cycle. Because the reactions of the urea and citric acid cycles are inextricably interwined, cumulatively they have been called "Krebs bicycle".

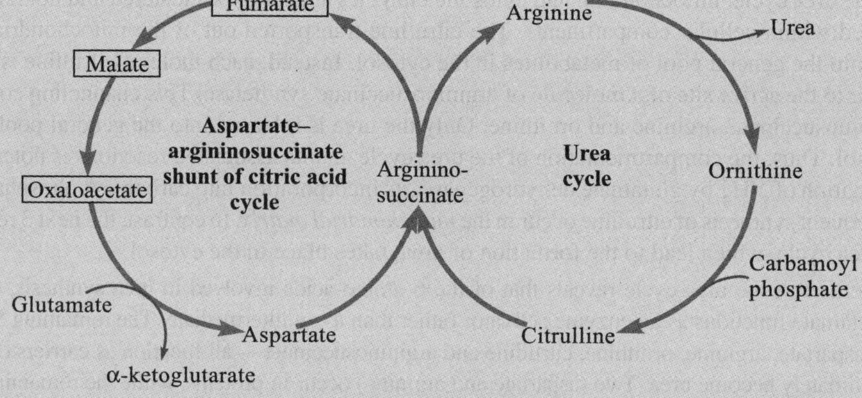

Fig. 26–11. The Krebs bicycle

It is composed of the urea cycle on the *right*, which meshes with the aspartate–argininosuccinate shunt of the citric acid cycle on the *left*. Note that the urea cycle, the citric acid cycle and the transamination of oxaloacetate are linked by fumarate and aspartate. Intermediates in the citric acid cycle are *boxed*

C. Energetics of the Urea Cycle

The urea cycle is energetically expensive. It brings together two amino groups and HCO_3^- to form a mole of urea which diffuses from the liver into the bloodstream. The overall reaction of the urea cycle is:

$$2\,NH_4^+ + HCO_3^- + 3\,ATP^{4-} + H_2O \longrightarrow \text{Urea} + 2\,ADP^{3-} + 4\,Pi^{2-} + AMP^{2-} + 5\,H^+$$

The synthesis of one mole of urea requires four high-energy phosphate groups. Two ATPs are required to make carbamoyl phosphate, and one ATP is required to make argininosuccinate. In the latter reaction, however, the ATP undergoes a pyrophosphate cleavage to AMP and pyrophosphate which may be hydrolyzed to yield two P_i.

> **Carbamoyl phosphate** is the official nomenclature for the –CO–NH_2 group, but carbamyl is sometimes used.

D. Genetic Defects in the Urea Cycle

The synthesis of urea in the liver is the major pathway of the removal of NH_4^+. A blockage of carbamoyl phosphate synthesis or any of the 4 steps of the urea cycle has serious consequences because there is no alternative pathway for the synthesis of urea. *They all lead to an elevated level of NH_4^+ in the blood* (**hyperammonemia**). Some of these genetic defects become evident a day or two after birth, when the afflicted infant becomes lethargic and vomits periodically. Coma and irreversible brain damage may ensue. The high levels of NH_4^+ are toxic *probably* because elevated levels of glutamine, formed from NH_4^+ and glutamate, lead directly to brain damage:

People cannot tolerate a protein-rich diet because amino acids ingested in excess of the minimum daily requirements for protein synthesis would be deaminated in the liver, producing free ammonia in the blood. As we have seen, ammonia is toxic to humans. Human beings are incapable of synthesizing half of the 20 amino acids, and these **essential amino acids** (Table 26–1) must be provided in the diet.

Table 26-1. Nonessential and essential amino acids for humans

Nonessential amino acids		*Essential amino acids*	
Name	*Abbreviation*	*Name*	*Abbreviation*
Alanine	Ala	Arginine	Arg
Asparagine	Asn	Histidine	His
Aspartate	Asp	Isoleucine	Ile
Cysteine	Cys	Leucine	Leu
Glutamate	Glu	Lysine	Lys
Glutamine	Gln	Methionine	Met
Glycine	Gly	Phenylalanine	Phe
Proline	Pro	Threonine	Thr
Serine	Ser	Tryptophan	Trp
Tyrosine	Tyr	Valine	Val

Patients with defects in the urea cycle are often treated by substituting in the diet the **α-keto acid analogues** of the essential amino acids, which are the indispensable parts of the amino acids. The α-keto acid analogues can then accept amino groups from excess nonessential amino acids by aminotransferase action (Fig. 26–12).

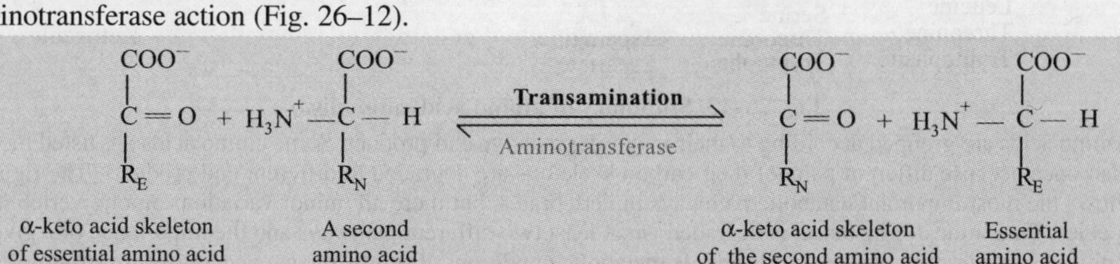

Fig. 26–12. Transamination reaction for the synthesis of essential amino acids from the corresponding α-keto acids

The dietary requirement for essential amino acids can, hence, be met by the α-keto acid skeletons. [R_E and R_N represent R groups of the essential and nonessential amino acids, respectively].

PATHWAYS OF AMINO ACID CATABOLISM

Twenty standard amino acids, with a variety of carbon skeletons, go into the composition of proteins. As such, there are 20 different pathways for amino acid degradation. All these pathways taken together, in human beings, account for only 10-15% of the body's energy production. Therefore, *the individual amino acid degradative pathways are not nearly as active as glycolysis and fatty acid oxidation.* Moreover, the activity of the catabolic pathways varies greatly from one amino acid to the other. For this reason, these will not be examined in detail. *The 20 catabolic pathways converge to form only 5 products, all of which enter the citric acid cycle.* From here, the carbons can be diverted to gluconeogenesis or ketogenesis, or they can be completely oxidized to CO_2 and H_2O (Fig. 26–13).

All or part of the carbon skeletons of *ten* of the amino acids are finally broken down to yield acetyl-CoA. *Five* amino acids are converted into α-ketoglutarate, *four* into succinyl-CoA, *two* into fumarate and *two* into oxaloacetate. The individual pathways for the 20 amino acids will be summarized by means of **flow diagrams**, each leading to a specific point of entry into the citric acid cycle. Note that some amino acids appear more than once which means that their carbon skeleton is broken down into different fragments and each of which enters the citric acid cycle at a different point. *The strategy of amino acid degradation is to form major metabolic intermediates that can be converted into glucose or be oxidized by the citric acid cycle.* In fact, the carbon skeletons of the diverse set of 20 amino acids are funneled into only 7 molecules (refer Fig. 26–13), viz., pyruvate, acetyl-CoA, acetoacetyl-CoA, α-ketoglutarate, succinyl-CoA, fumarate and oxaloacetate. Thus, *here we have an example of the remarkable economy of metabolic conversions.*

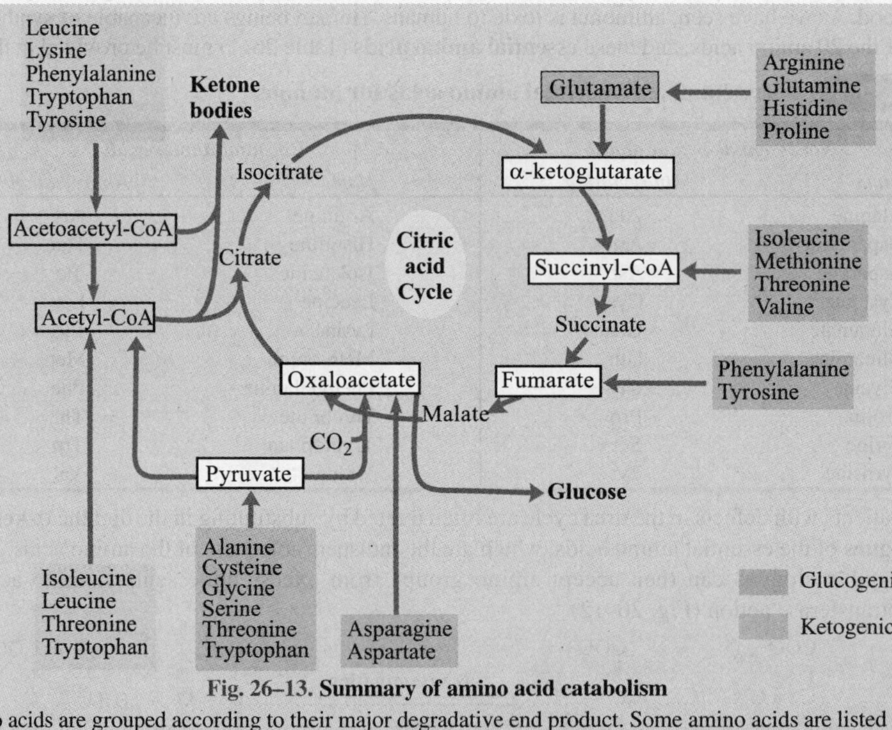

Fig. 26–13. Summary of amino acid catabolism

Amino acids are grouped according to their major degradative end product. Some amino acids are listed more than once because different parts of their carbon skeletons are degraded to different end products. The figure shows the most important catabolic pathways in vertebrates, but there are minor variations among vertebrate species. Threonine, for instance, is degraded *via* at least two different pathways and the importance of a given pathway can vary with the organism and its metabolic conditions. The glucogenic and ketogenic amino acids are also delineated in the figure, by colour shading. Notice that five of the amino acids are both glucogenic and ketogenic. The amino acids degraded to pyruvate are also potentially ketogenic. Only two amino acids, leucine and lysine, are exclusively ketogenic.

Based on their catabolic products, amino acids are classified as glucogenic or ketogenic (Table 26–2):

(*a*) those amino acids that generate precursors of glucose, *e.g.*, pyruvate or citric acid cycle intermediate (*i.e.*, α-ketoglutarate, succinyl-CoA, fumarate or oxaloacetate), are referred to as **glucogenic**. Ala, Arg, Asn, Asp, Cys, Gln, Glu, Gly, His, Met, Pro, Ser, Thr and Val belong to this category. Net synthesis of glucose from these amino acids is feasible because these glucose precursors can be converted into phosphoenolpyruvate (PEP) and then into glucose.

(*b*) those amino acids that are degraded to acetyl-CoA or acetoacetyl-CoA are termed as **ketogenic** because they give rise to ketone bodies. Their ability to form ketones is particularly evident in untreated diabetes mellitus, in which large amounts of ketones are produced by the liver, not only from fatty acids but from the ketogenic amino acids. Leucine (Leu) exclusively belongs to this category.

The remaining 5 amino acids (Ile, Lys, Phe, Trp and Tyr) are both glucogenic and ketogenic. Some of their carbon atoms emerge in acetyl-CoA or acetoacetyl-CoA, whereas others appear in potential precursors of glucose. Thus, the division between glucogenic and ketogenic amino acids is not sharp. Whether an amino acid is regarded as being glucogenic, ketogenic or both depends partly on the eye of the beholder.

Table 26–2. Classification of amino acids as glucogenic or ketogenic

Glucogenic		*Ketogenic*	*Glucogenic and ketogenic*
Alanine	Glycine	Leucine	Isoleucine
Arginine	Histidine		Lysine
Asparagine	Methionine		Phenylalanine
Aspartate	Proline		Tryptophan
Cysteine	Serine		Tyrosine
Glutamate	Threonine		
Glutamine	Valine		

Table 26–3 lists the catabolic end products of the 20 protein (or standard) amino acids. It may be seen that nearly all of the amino acids yield on breakdown either an intermediate of the citric acid cycle, pyruvate or acetyl-CoA. Five amino acids (Leu, Lys, Phe, Trp and Tyr) are exception to this since they give rise to acetoacetic acid. Since, however, this compound also forms acetyl-CoA, carbon skeletons of all of the amino acids are ultimately oxidized *via* the citric acid cycle.

Table 26–3. End products of amino acid metabolism

Amino acid(s)	*End product*
Alanine, Cysteine (Cystine), Glycine, Serine and Threonine (2)	Pyruvic acid
Leucine (2)	Acetyl-CoA
Leucine (4), Lysine (4), Phenylalanine (4), Tryptophan (4) and Tyrosine (4)	Acetoacetic acid (or its CoA-ester)
Arginine (5), Glutamic acid, Glutamine, Histidine (5) and Proline	α-ketoglutaric acid
Isoleucine (4), Methionine and Valine (4)	Succinyl-CoA
Phenylalanine (4) and Tyrosine (4)	Fumarate
Asparagine and Aspartic acid	Oxaloacetic acid

* The figures in parentheses specify the number of carbon atoms in the amino acid that are actually converted to the end product listed.

A. Ten Amino Acids are Degraded to Acetyl-CoA.

The carbon skeletons of 10 amino acids yield acetyl-CoA, which enters the citric acid cycle directly. Five of the ten are degraded to acetyl-CoA *via* pyruvate. Alanine, cysteine, glycine, serine and tryptophan belong to this category. In some organisms, threonine is also degraded to form acetyl-CoA as shown in Fig. 26–14; in men, it is degraded to succinyl-CoA, as described later. The other five amino acids are converted into acetyl-CoA and/or acetoacetyl-CoA which is then cleaved to form acetyl-CoA. Leucine, lysine, phenylalanine, tryptophan and tyrosine come under this category.

Degradation to acetyl-CoA via pyruvate

Alanine yields pyruvate directly on transamination with α-ketoglutarate. The side chain of **tryptophan** is cleaved to yield alanine and thus pyruvate. **Cysteine** is converted to pyruvate in 2 steps: one to remove the S atom and the other a transamination. Serine is converted (or deaminated) to pyruvate by serine dehydratase. Both the α-amino and the β-hydroxyl groups of serine are removed in this single PLP-dependent reaction (an analogous reaction with threonine is shown in Fig. 26–23).

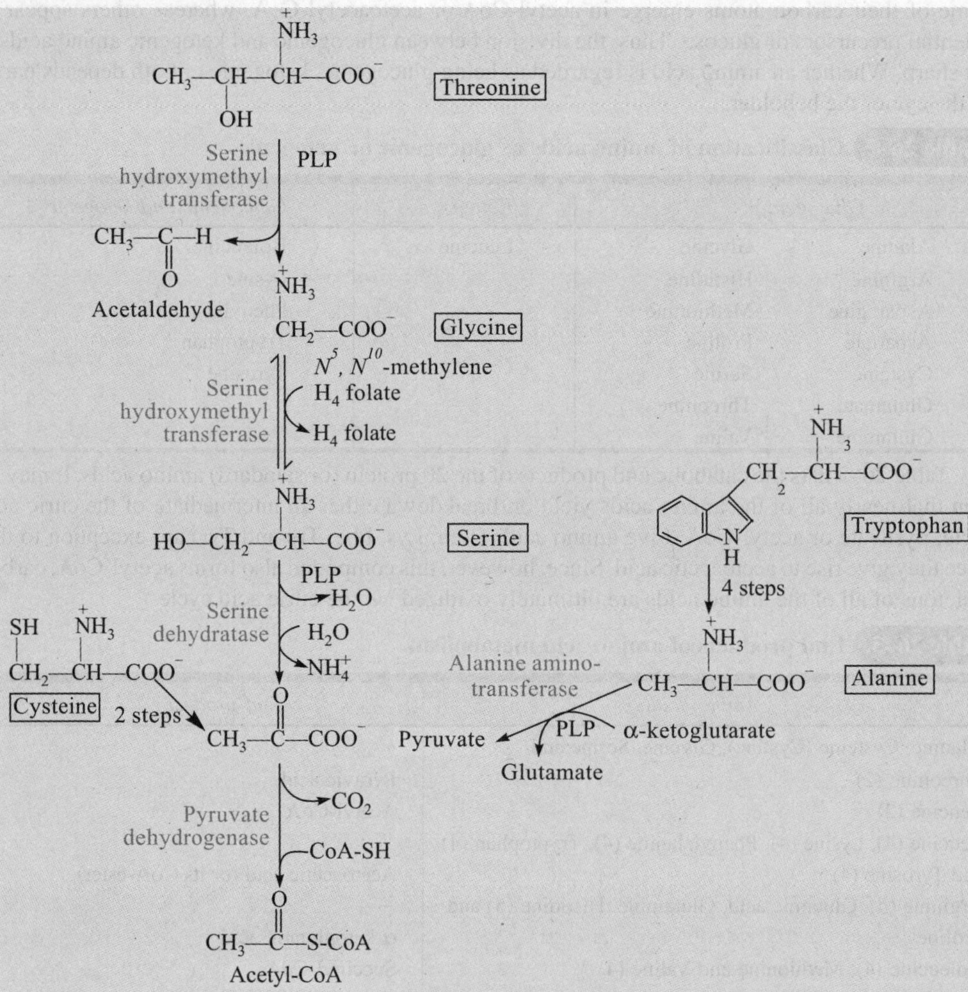

Fig. 26–14. Catabolic fates of glycine, serine, cysteine, tryptophan, alanine and also threonine
The fate of the indole group of tryptophan is shown in Fig. 26–16. Details of the two pathways for glycine are shown in Fig. 26–15.

Glycine has 2 routes. It can be converted to serine by enzymatic addition of a hydroxymethyl group (Fig. 26–15a). This reaction, catalyzed by serine hydroxymethyl transferase, requires two coenzymes, tetrahydrofolate and pyridoxal phosphate. The other route for glycine, which predominates in animals, involves its oxidative change into CO_2, NH_4^+ and a $-CH_2-$ group (Fig. 26–15b). This readily reversible reaction, catalyzed by glycine synthase, also requires tetrafolate which accepts the methylene ($-CH_2-$) group. In this oxidative cleavage pathway, the 2 carbon atoms of glycine do not enter the citric acid cycle. One is lost as CO_2, and the other becomes the methylene group of N^5,N^{10}-methylene-tetrahydrofolate.

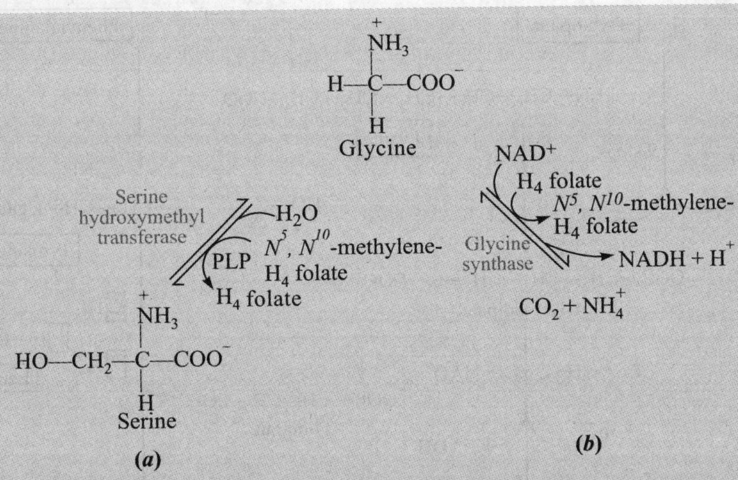

Fig. 26–15. Two metabolic fates of glycine

(*a*) Conversion to serine (*b*) Breakdown to CO_2 and ammonia

Note that the cofactor tetrahydrofalate carries one-carbon units in both of these reactions.

Degradation to acetyl-CoA via acetoacetyl-CoA

The portions of the carbon skeleton of six amino acids– tryptophan, lysine, phenylalanine, tyrosine, leucine and isoleucine– yield acetyl-CoA and/or acetoacetyl-CoA; the latter is then converted into acetyl-CoA (Fig. 26–16). It may be noted that some of the final steps in the degradative pathways for leucine, lysine and tryptophan resemble steps in the oxidation of fatty acids. The breakdown of 2 of these six amino acids– tryptophan and phenylalanine– deserves special mention.

The degradation of tryptophan is the most complex of all the pathways of amino acid catabolism in animal tissues. Portions of **tryptophan** yield acetyl-CoA by 2 different pathways: one *via* pyruvate and the other *via* acetoacetyl-CoA. Some of the intermediates in tryptophan catabolism are required as precursors for biosynthesis of other important biomolecules (Fig. 26–17), such as nicotinate (a precursor of NAD and NADP), indoleacetate (a plant growth factor) and serotonin (a neurotransmitter).

The pathway for the degradation of **phenylalanine** and its oxidation product **tyrosine** has some remarkable features. *This series of reactions shows how molecular oxygen is used to break an aromatic ring* (Fig. 26–19).

The first step in phenylalanine degradation is the hydroxylation of phenylalanine to tyrosine, a reaction catalyzed by **phenylalanine hydroxylase** (= phenylalanine 4-monooxygenase). This enzyme is called a mixed-function oxygenase or a monooxygenase because only one atom of O_2 appears in the product tyrosine as hydroxyl group and the other is reduced to water by NADH. The reaction requires an unusual coenzyme called tetrahydrobiopterin, which carries electrons from NADH to O_2 in the

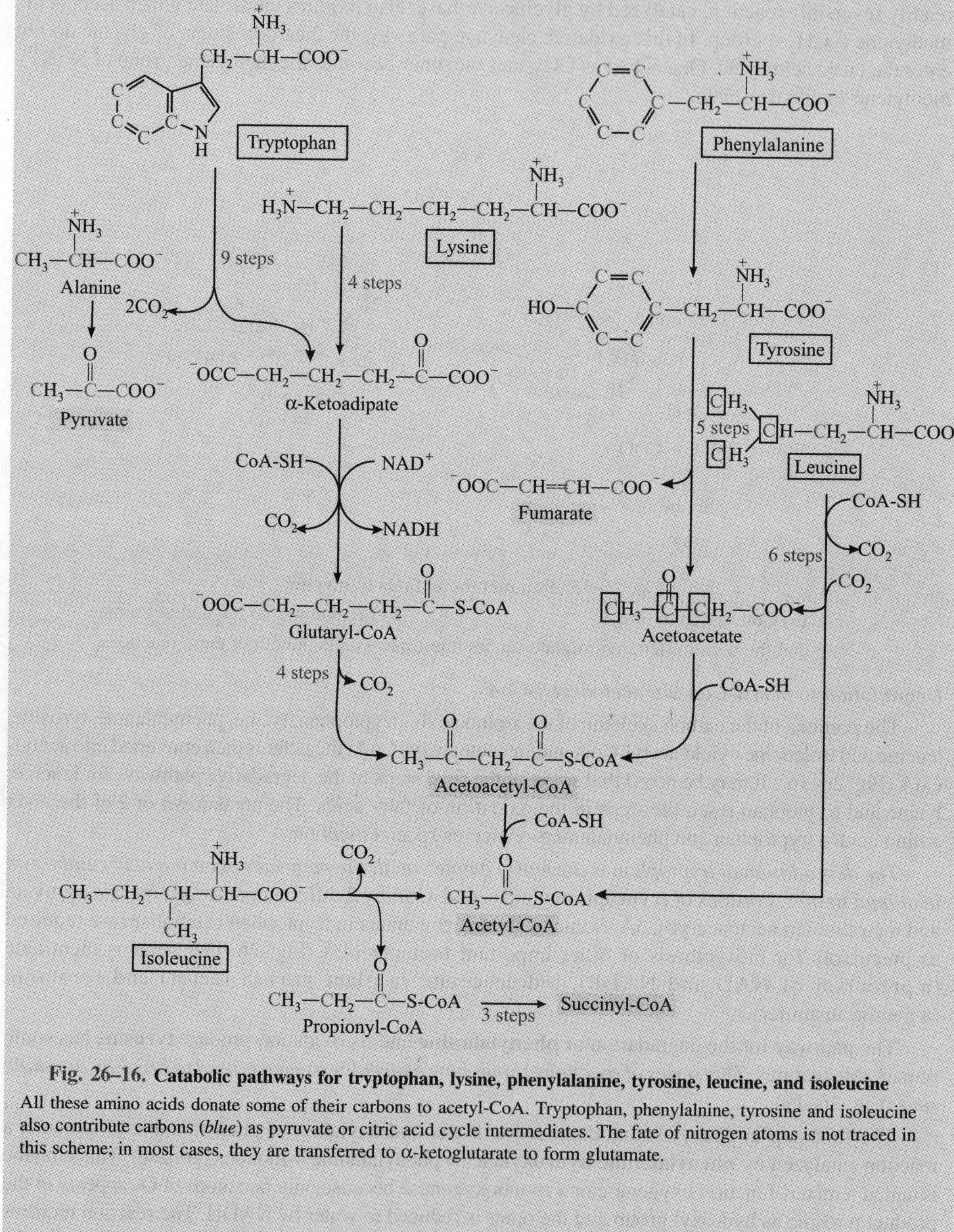

Fig. 26–16. Catabolic pathways for tryptophan, lysine, phenylalanine, tyrosine, leucine, and isoleucine

All these amino acids donate some of their carbons to acetyl-CoA. Tryptophan, phenylalnine, tyrosine and isoleucine also contribute carbons (*blue*) as pyruvate or citric acid cycle intermediates. The fate of nitrogen atoms is not traced in this scheme; in most cases, they are transferred to α-ketoglutarate to form glutamate.

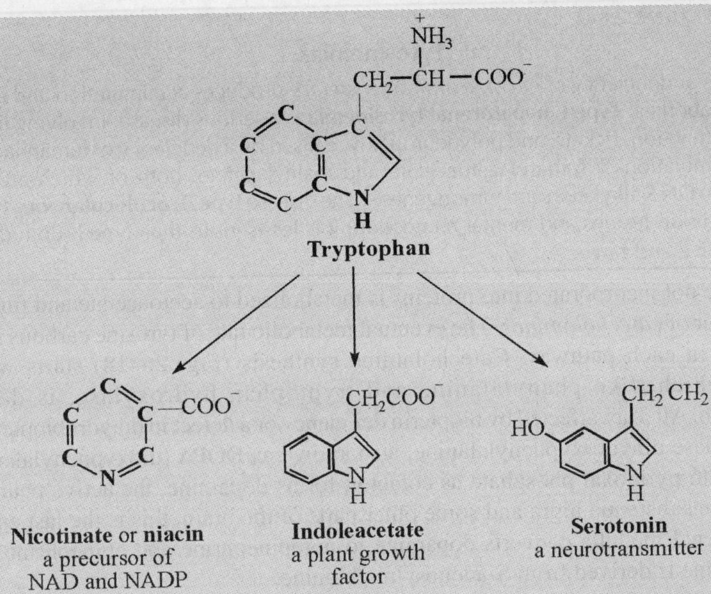

Fig. 26–17. Some of the intermediates of tryptophan catabolism

Atoms in *boldface* are used to trace the source of the ring atoms in nicotinate.

hydroxylation of phenylalanine. The oxidized form of this electron carrier is dihydrobiopterin. Tetrahydrobiopterin is initially formed by reduction of dihydrobiopterin by NADPH; the reaction being catalyzed by **dihydrofolate reductase** (Fig. 26–20). The quinonoid form of dihydrobiopterin, produced in the hydroxylation of phenylalanine, is reduced back to tetrahydrobiopterin by NADH in a reaction catalyzed by **dihydropteridine reductase.** The sum of the reactions catalyzed by phenylalanine hydroxylase and dihydropteridine reductase is:

$$\text{Phenylalanine} + O_2 + \text{NADH} + H^+ \longrightarrow \text{Tyrosine} + \text{NAD}^+ + H_2O$$

The next step (Fig. 26–19) is the transamination of tyrosine to produce an α-keto acid, *p*-hydroxyphenylpyruvate in a reaction catalyzed by **tyrosine aminotransferase**. *p*-hydroxyphenylpyruvate then reacts with O_2 to form homogentisate. In this reaction, a carbon atom is lost as CO_2 and a hydroxyl group is added to the phenyl ring to yield homogentisate. The enzyme catalyzing this complex reaction, ***p*-hydroxyphenylpyruvate hydroxylase** is called a **dioxygenase** because both atoms of O_2 become incorporated into the product. Next follows the oxidative cleavage of the phenyl ring to produce 4-maleylacetoacetate. This reaction is catalyzed by another dioxygenase called **homogentisate oxidase**. In fact, *nearly all cleavages of the aromatic rings in biological systems are catalyzed by dioxygenases*, a class of enzymes discovered by **Osamu Hayaishi**. The 4-maleylacetoacetate is then isomerized by **4-maleylacetoacetate isomerase** to produce its *trans* isomer, 4-fumarylacetoacetate. In the last step, 4-fumarylacetoacetate is hydrolyzed by **fumarylacetoacetate** to generate fumarate and acetoacetate. The latter product, acetoacetate can subsequently be activated by **β-ketoacyl-CoA transferase** (with succinyl-CoA serving as the second substrate) to produce acetoacetyl-CoA, which can be oxidized aerobically.

Thus, we see that phenylalanine and tyrosine are both degraded into 2 fragments, each of which can enter the citric acid cycle, but at different points. Four of the 9 carbon atoms of phenylalanine and tyrosine yield acetoacetate, which is converted into acetoacetyl-CoA. A second four-carbon fragment of phenylalanine and tyrosine is recovered as fumarate. Eight of the 9 carbon atoms of these two amino acids thus enter the citric acid cycle; the remaining carbon is lost as CO_2.

CLINICAL IMPLICATIONS 26–2.

Tyrosinemias

The absence or deficiency of *tyrosine aminotransferase* produces accumulation and excretion of tyrosine and metabolites. **Type I, hepatorenal tyrosinemia** is a serious disease, involving liver failure, renal tubular dysfunction, rickets, and polyneuropathy, caused by a deficiency of fumarylacetoacetate hydrolase. Accumulation of fumarylacetoacetate and maleylacetate, both of which are alkylating agents, can lead to DNA alkylation and tumorigenesis. The disease **type II, oculocutaneous tyrosinemia** results in eye and skin lesions and mental retardation; it is less serious than type I. Both diseases are autosomal recessive and rare.

Most tyrosine not incorporated into proteins is metabolized to acetoacetate and fumarate. Some is used as precursor of *catecholamines*. The eventual metabolic fate of tyrosine carbons is determined by the first step in each pathway. Catecholamine synthesis (Fig. 26–18) starts with **tyrosine hydroxylase,** which, like phenylalanine and tryptophan hydroxylase, is dependent on tetrahydrobiopterin. All 3 are affected by biopterin deficiency or a defect in dihydrobiopterin reductase. Tyrosine hydroxylase dihydroxyphenylalanine, also known as DOPA (dioxyphenylalanine). DOPA decarboxylase, with pyridoxal phosphate as cofactor, forms dopamine, the active neurotransmitter, from DOPA. In the substantia nigra and some other parts of the brain, this is the last enzyme in this pathway. The adrenal medulla converts dopamine to norepinephrine and epinephrine. The methyl group of epinephrine is derived from S-adenosylmethionine.

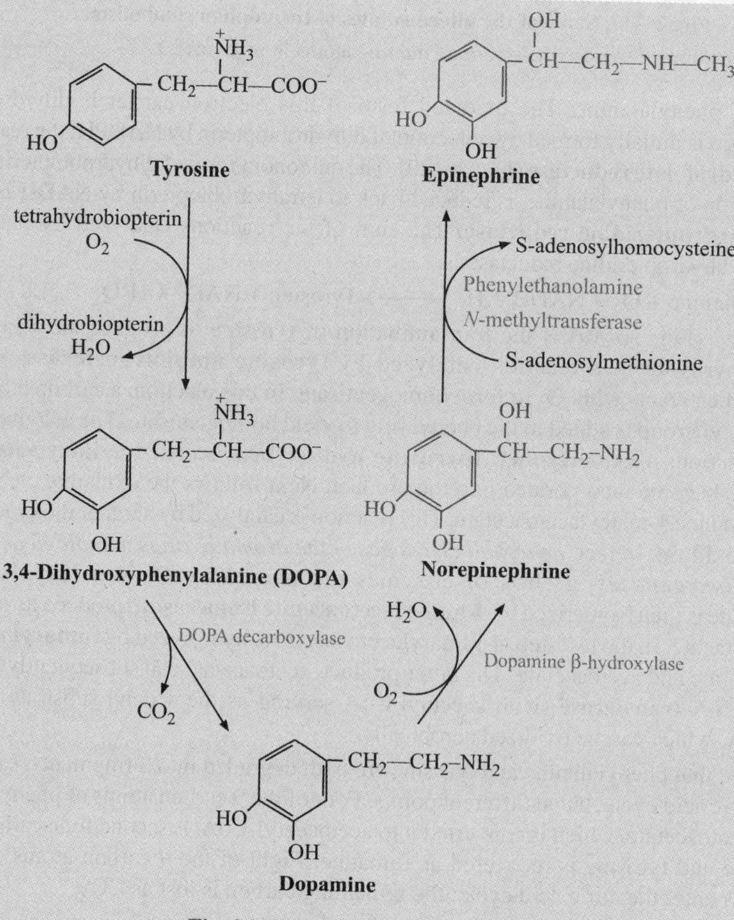

Fig. 26–18. Synthesis of catecholamines

Brain plasma tyrosine regulates norepinephrine formation. Estrogens decrease tyrosine concentration and increase tyrosine aminotransferase activity, diverting tyrosine into the catabolic pathway. Estrogen sulfate competes for the pyridoxal phosphate site on DOPA carboxylase. These 3 effects combined may help explain mood variations during the menstrual cycle. Tyrosine is therapeutic in some cases of depression and stress. Its transport appears to be reduced in skin fibroblasts from schizophrenic patients, indicating other roles for tyrosine derivatives in mental disorders.

CLINICAL IMPLICATIONS 26–3.

Parkinson's Disease

This condition is named for the physician who described "shaking palsy" in 1817. In it, usually elderly people over-60 develop tremors that gradually interfere with motor function of various muscle groups. The disease is caused by the degeneration of dopamine-producing cells in the substantia nigra and the locus ceruleus of the brain. These cells normally produce dopamine as a neurotransmitter, the amount released being proportional to the number of surviving cells. A dramatic outbreak of Parkinsonism, as the disease is also called, occurred in young adult drug addicts using a derivative of pyridine (methylphenyl-tetrahydropyridine, MPTP). It (or a contaminant produced during its manufacture) appears to be directly toxic to dopamine-producing cells of substantia nigra. Symptomatic relief, often dramatic, is obtained by administering DOPA, the precursor of dopamine. Clinical problems developed when DOPA (levodopa, L-DOPA) was used for treatment of many people who have Parkinson's disease. The side effects included nausea, vomiting, hypotension, cardiac arrhythmias, and various central nervous system (CNS) symptoms. These were explained as effects of dopamine produced outside the CNS. Administration of DOPA analogues that inhibit DOPA decarboxylase but are unable to cross the blood–brain barrier has been effective in decreasing side effects and increasing effectiveness of the DOPA. The interactions of the many brain neurotransmitters are very complex, cell degeneration continues even after treatment, and elucidation of the major biochemical abnormality has not yet led to complete control of the disease.

Although medication can markedly reduce the symptoms, the disease is progressive and may result in severe disability. Recently, a controversial treatment by transplantation of the dopaminergic fetal adrenal medullary tissue into the brain has been tried. The adrenal tissue synthesizes dopamine and improves the movement disorder.

B. Five Amino Acids are Converted into α-ketoglutarate.

The carbon skeletons of 5 amino acids— arginine, histidine, proline, glutamate, glutamine— enter the citric acid cycle at α-ketoglutarate (Fig. 26–21). Proline, glutamate and glutamine have 5-carbon skeletons. The cyclic structure of **proline** is opened by oxidation of the carbon furthest from the carboxyl group to create a Schiff base and hydrolysis of the Schiff base to a linear semialdehyde, glutamate γ-semialdehyde. This is further oxidized at the same carbon to produce **glutamate**. By the action of glutaminase, **glutamine** donates its amide nitrogen to some acceptor and gets converted to glutamate. Transamination or deamination of glutamate produces the citric acid cycle intermediate.

Arginine and histidine contains 5 adjacent carbons and a 6th carbon attached through a nitrogen atom. The catabolic conversion of these two amino acids to glutamate is, hence, slightly more complex than the path starting from proline or glutamine to glutamate. Arginine is converted to the 5-carbon skeleton of ornithine by arginase in the urea cycle (see Fig. 26–9), and the ornithine is transaminated to glutamate semialdehyde.

The conversion of **histidine** to the 5-carbon glutamate occurs in a 4-step pathway (Fig. 26–22). Histidine is first converted to urocanate by **histidine ammonia lyase**. Urocanate is then converted into 4-imidazolone 5-propionate enzymatically by the action of **urocanate hydratase**. The amide bond in the ring of this intermediate is hydrolyzed to the *N*-formimino derivative of glutamate enzymatically by **imidazolonepropionase**. The *N*-formiminoglutamate is then converted into glutamate by transfer of its formimino group to tetrahydrofolate, a carrier of activated one-carbon units. The enzyme **glutamate formimino transferase** catalyzes the reaction. The extra carbon is, thus, removed in this last step.

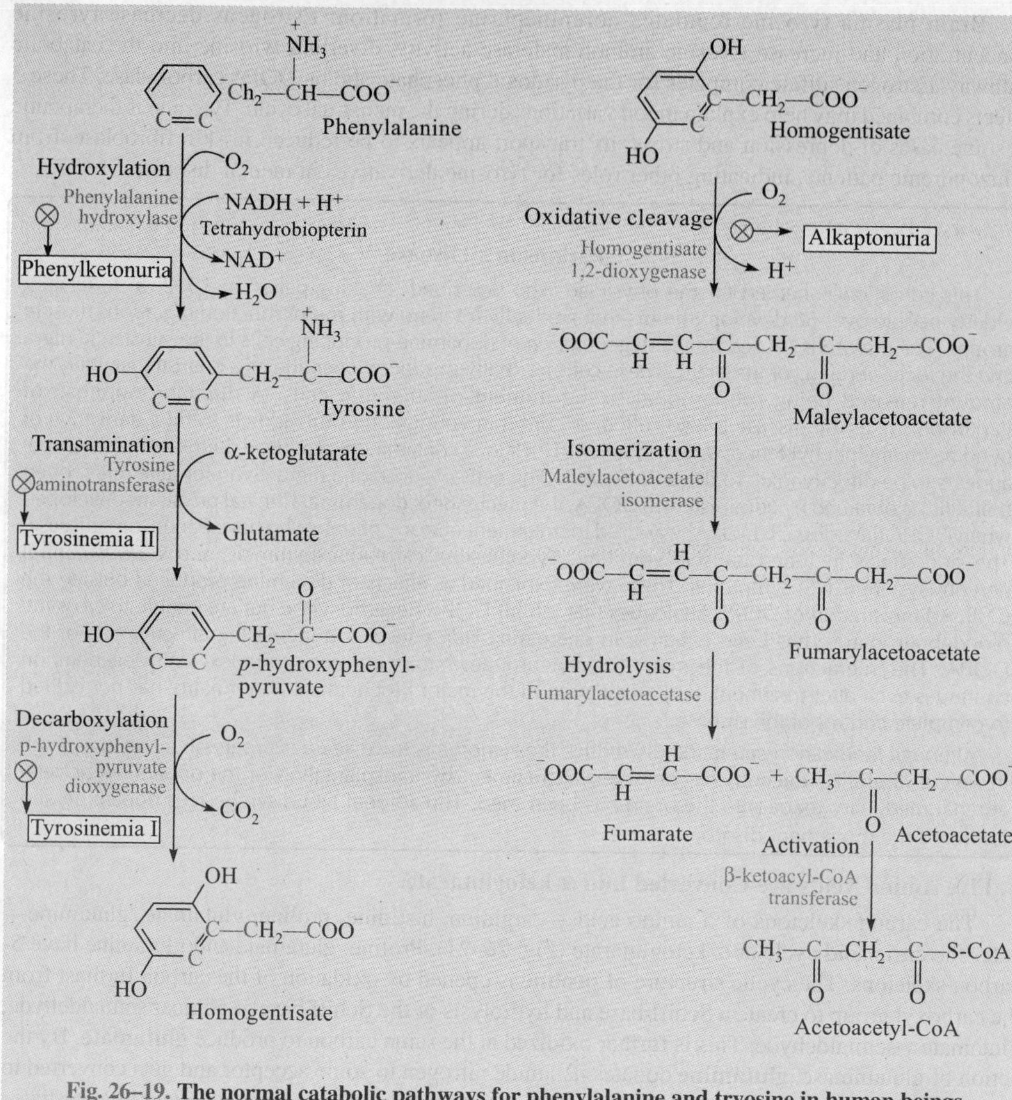

Fig. 26–19. The normal catabolic pathways for phenylalanine and tryosine in human beings

Note that the pathway ultimately results in the production of fumarate and acetoacetyl-CoA. The genetic defects in each of the first four enzymes in this pathway are known to cause inheritable human diseases, shown here shaded *yellow*. The path of the 4 carbon atoms of acetoacetyl-CoA from phenylalanine downwards is traced by depicting them in *red*. The enzymes are shown in *blue* print.

C. Four Amino Acids are Converted into Succinyl-CoA.

The carbon skeletons of methionine, isoleucine, threonine and valine are degraded by pathways that produce succinyl-CoA (Fig. 26–23). Methylmalonyl-CoA is an intermediate in the breakdown of these 4 amino acids.

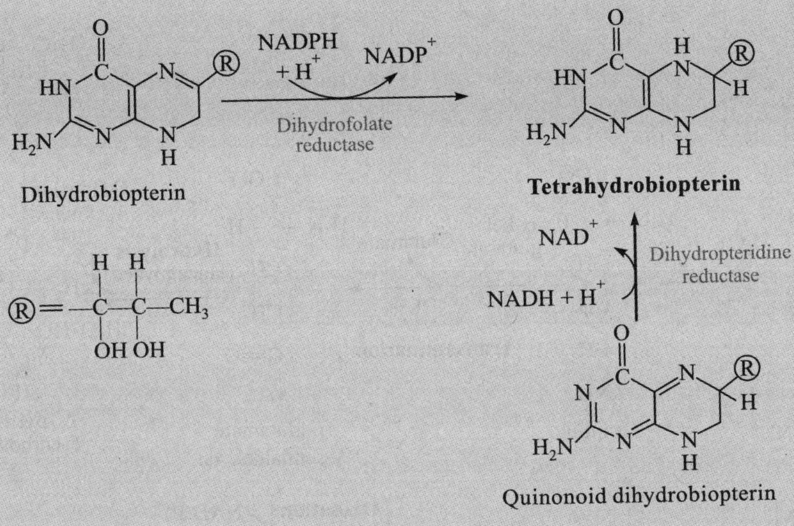

Fig. 26–20. Formation of tetrahydrobiopterin by reduction of either of the two forms of dihydrobiopterin

Methionine donates its methyl group to one of the many possible acceptors through *S*-adenosylmethionine and 3 of the four remaining atoms of its carbon skeleton are converted into those of propionate as propionyl-CoA. **Isoleucine** first undergoes transamination and then oxidative decarboxylation of the resulting α-keto acid. The remaining 5-carbon skeleton derived from isoleucine undergoes further oxidation, producing acetyl-CoA and propionyl-CoA. In human beings, **threonine** is also converted into propionyl-CoA. The degradation of **valine** follows a path similar to those for methionine, isoleucine and threonine. After transamination, decarboxylation and a series of oxidation reactions, valine is converted to propionyl-CoA.

The pathway from propionyl-CoA to succinyl-CoA is especially interesting. Propionyl-CoA is carboxylated at the expense of an ATP to yield the D-isomer of methylmalonyl-CoA. This carboxylation reaction is catalyzed by **propionyl-CoA carboxylase**, a biotin enzyme that has a catalytic action like that of acetyl-CoA carboxylase and pyruvate carboxylase. The D-isomer of methylmalonyl-CoA is unusually racemized to the L-isomer. Lastly, L-methylmalonyl-CoA is converted to succinyl-CoA by an *intramolecular rearrangement*, using the enzyme **methylmalonyl-CoA mutase**, which is one of the two mammalian enzymes known to possess a derivative of vitamin B_{12} as its coenzyme. The –CO–S–CoA group migrates from C–2 to C–3 in exchange for a hydrogen atom.

D. Three Branched-chain Amino Acids are Degraded in Extrahepatic Tissues.

Although most of the amino acids are mainly catabolized in liver, the 3 amino acids with branched side chains (*viz.*, leucine, isoleucine and valine) are oxidized as fuels primarily in the extrahepatic tissues such as muscle, adipose, kidney and brain tissue. These 3 amino acids share the first two enzymes in their catabolic pathways, which occur in extrahepatic tissues. The first of these 2 enzymes, **aminotransferase**, not present in liver, acts on all 3 branched-chain amino acids to produce the corresponding α-keto acid (Fig. 26–24). The second enzyme, the **branched-chain α-keto acid dehydrogenase** catalyzes oxidative decarboxylation of the corresponding α-keto acid, releasing the carboxyl group as CO_2 and producing the acyl-CoA derivative. This dehydrogenase complex is formally analogous to the pyruvate and α-ketoglutarate dehydrogenases. In fact, all 3 enzymes are closely homologous in structure and the reaction

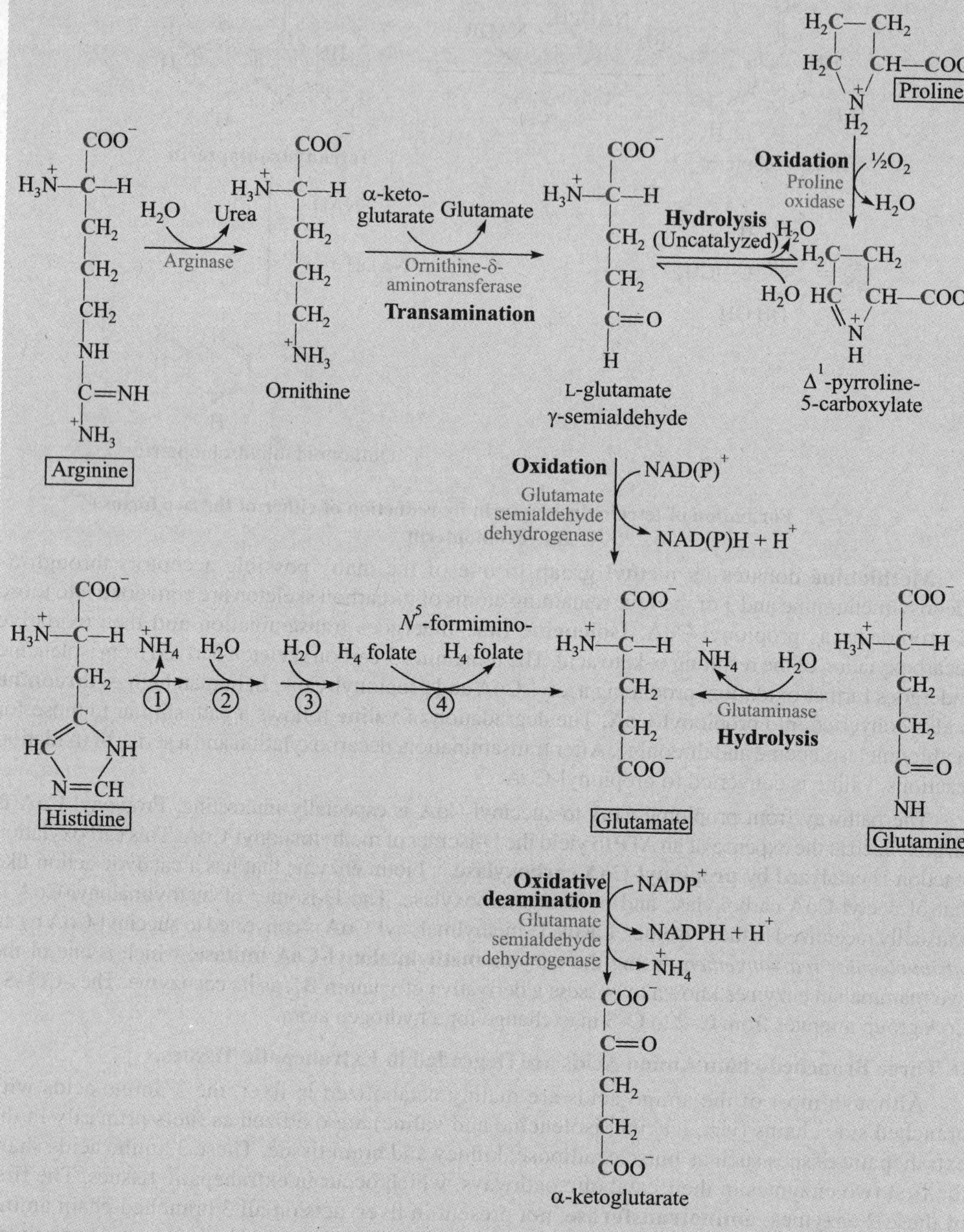

Fig. 26–21. Catabolic pathways for arginine, histidine, proline, glutamate and glutamine
Note that all these amino acids are converted to α-ketoglutarate. For the details of the breakdown of histidine to glutamate, which is a 4-step mechanism, refer Fig. 26–22. The numbered steps in the histidine pathway are catalyzed by (1) histidine ammonia lyase, (2) urocanate hydratase, (3) imadazolonepropionase, and (4) glutamate formiminotransferase.

mechanism is essentially the same for all. Five cofactors (TPP, FAD, NAD, lipoate and coenzyme A) participate and the 3 proteins in each of these complexes catalyze homologous reactions. The subsequent reactions are like those of fatty acid oxidation. Isoleucine yields acetyl-CoA, whereas valine yields methylmalonyl-CoA.

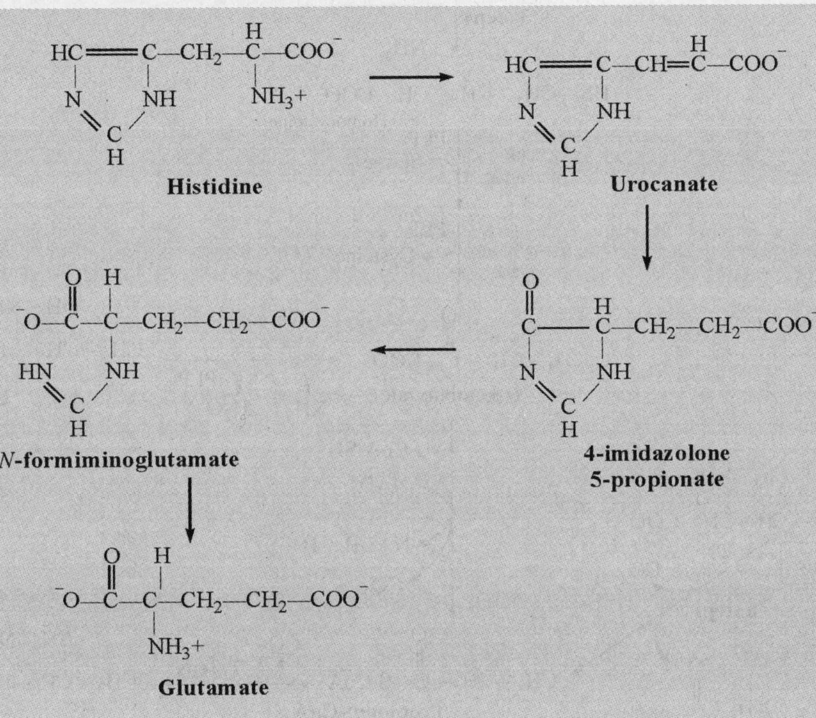

Fig. 26–22. Conversion of histidine into glutamate

E. Two Amino Acids are Degraded to Oxaloacetate.

The carbon skeletons of asparagine and aspartate ultimately enter the citric acid cycle via oxaloacetate. **Asparagine** is hydrolyzed by the enzyme **asparaginase** to NH_4^+ and aspartate (Fig. 26–25). **Aspartate** then undergoes a transamination reaction with α-ketoglutarate to yield glutamate and oxaloacetate. The latter enters the citric acid cycle. Recall that aspartate can also be converted into fumarate by the urea cycle (refer page 769). Fumarate is also a point of entry for half the carbon atoms of phenylalanine and tyrosine.

Thus, we have now seen that how the 20 different protein amino acids, after losing their nitrogen atoms, are degraded through dehydrogenation, decarboxylation and other reactions to yield portions of their carbon skeleton in the form of 5 central metabolites that can enter the citric acid cycle. Here they are completely oxidized to CO_2 and H_2O. During electron transfer, ATP is generated by oxidative phosphorylation, and in this way, the amino acids contribute to the total energy supply of the organism.

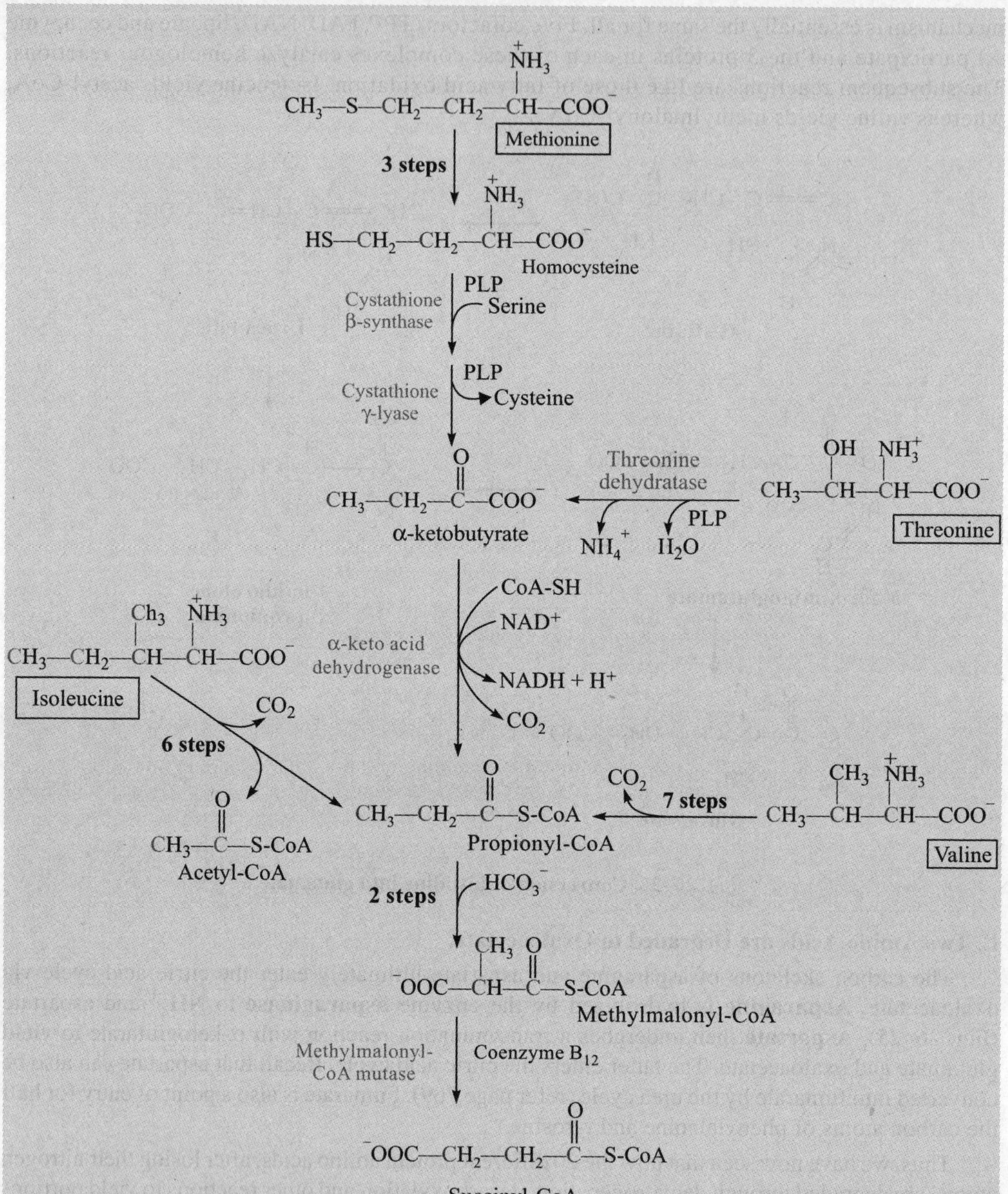

Fig. 26–23. The catabolic pathways for methionine, isoleucine, threonine and valine

These amino acids are converted to succinyl-C0A; isoleucine also contributes two of its carbon atoms to acetyl-CoA. The threonine degradation pathway shown here occurs in human beings; a pathway found in other organisms is shown in Fig. 26–14.

INBORN ERRORS OF AMINO ACID CATABOLISM

Many different genetic defects in amino acid metabolism have been identified in humans (see Table 26–4 on page 788). Most such defects cause specific intermediates to accumulate a condition that can cause defective neural development and mental retardation.

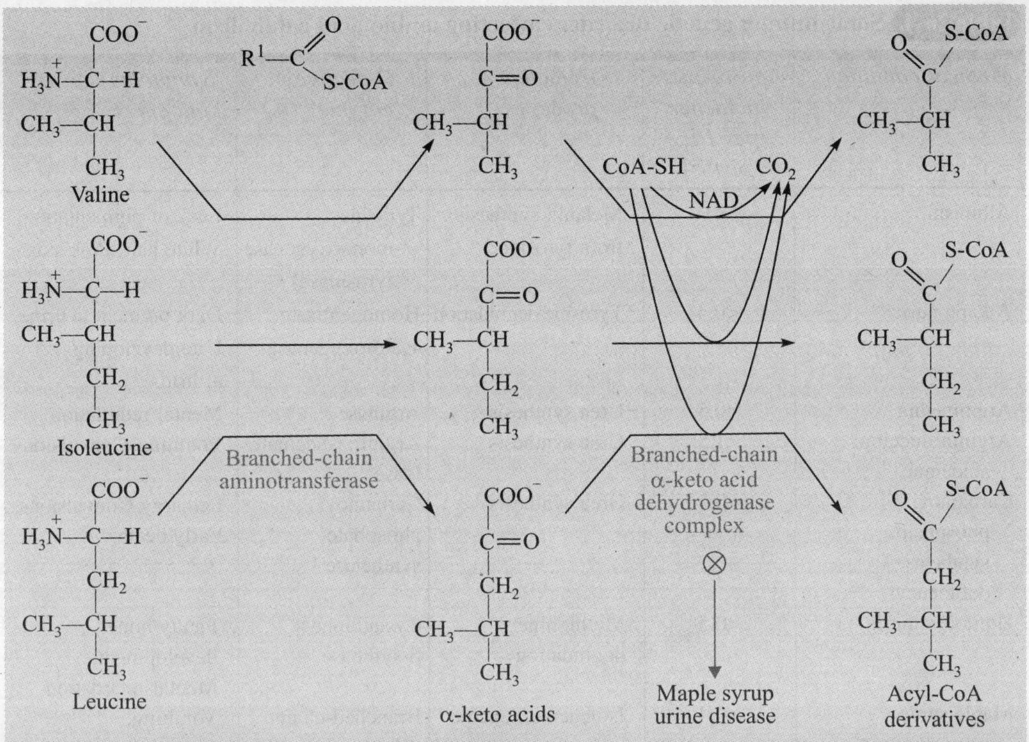

Fig. 26–24. The catabolic pathways for the 3 branched-chain amino acids: valine, isoleucine and leucine

The 3 pathways, which occur in extrahepatic tissues, share the first two enzymes, as shown here. The second enzyme, the branched-chain α-keto acid dehydrogenase complex is analogous to the pyruvate and α-ketoglutarate dehydrogenase complexes and requires the same 5 cofactors, some of which are not shown here. This enzyme is defective in people suffering from maple syrup urine disease.

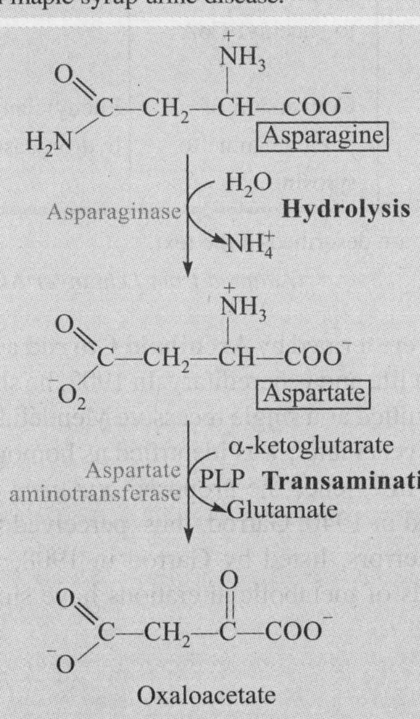

Fig. 26–25. The catabolic pathway for asparagine and aspartate

Both amino acids are converted to oxaloacetate.

Table 26-4. Some human genetic disorders affecting amino acid catabolism

Medical condition	*Approximate incidence (per 1 lac births)*	*Defective process*	*Defective enzyme*	*Symptoms and effects*
Albinism*	<3	Melanin synthesis from tyrosine	Tyrosine 3-monooxygenase (= tyrosinase)	Lack of pigmentation, White hair, Pink skin
Alkaptonuria*	<0.4	Tyrosine degradation	Homogentisate 1,2-dioxygenase	Dark pigment in urine, Late developing arthritis
Argininemia	<0.5	Urea synthesis	Arginase	Mental retardation
Argininosuccinic acidemia	<1.5	Urea synthesis	Argininosuccinate lyase	Vomiting, Convulsions
Carbamoyl phosphate synthetase I deficiency	<0.5	Urea synthesis	Carbamoyl phosphate synthetase I	Lethargy, Convulsions, Early death
Homocystinuria*	<0.5	Methionine degradation	Cystathione β-synthase	Faulty bone development, Mental retardation
Maple syrup urine disease* (= Branched-chain ketoaciduria)	<0.4	Isoleucine, leucine and valine degradation	Branched-chain α-keto acid dehydrogenase complex	Vomiting, Convulsions, Mental retardation, Early death
Methylmalonic acidemia	<0.5	Conversion of propionyl-CoA to succinyl-CoA	Methylmalonyl-CoA mutase	Vomiting, Convulsions, Mental retardation, Early death
Phenylketonuria* (= Hyperphenyl-alaninemia)	<8	Conversion of phenylalanine to tyrosine	Phenylalanine hydroxylase	Neonatal vomiting, Mental retardation

* Disorders marked with an *asterisk* are described in the text.

(Adapted from Lehninger AL, Nelson DL and Cox MM, 1993)

Several metabolic disorders were termed by **Archibald Garrod** as **inborn errors of metabolism** because each is present throughout life and is hereditary. In 1902, he showed that alkaptonuria, one of the metabolic syndromes, is transmitted as a single recessive Mendelian trait and is due to the absence of a metabolic enzyme which, 45 years later, was identified as homogentisate 1,2-dioxygenase with the help of liver biopsy analysis. In essence, he proposed *one gene-one enzyme hypothesis*, which **Beadle** and **Tatum** later elucidated in 1940. Garrod, thus, perceived the direct relationship between genes and enzymes. The inborn errors, listed by Garrod in 1908, were cystinuria, alkaptonuria, pentosuria and albinism. Hundreds of metabolic alterations have since been described that have a genetic basis.

SIR ARCHIBALD EDWARD GARROD

(LT, 1857–1936)

An English pediatrician and often called the 'father of biochemical genetics', is an example of a scientist ahead of his time. Garrod was born on November 25, 1857 in London as the 4th son of **Sir Alfred Baring Garrod,** a physician at King's college Hospital, who discovered the abnormal uric acid metabolism associated with gout. Archibalad was educated at Marlborough College and Christ Church, University of Oxford. He graduated with honours in Natural Science, in 1880. He received further medical training at St. Bartholomew's Hospital in London. In 1885, he obtained his Bachelor of Medicine and Master of Arts from Oxford, and became a member of the Royal College of Physicians, London. In 1886, married **Laura Elizabeth**. They had 3 sons and a daughter, **Dorothy Garrod**.

During the First World War (WWI), Garrod served as medical consultant to the army, primarily in Malta and in 1916 was appointed Knight Commander of the Order of St. Michael and St. George in recognition of his wartime services. Two of his three sons were killed on warfront, **Thomas Martin Garrod** aged 20 in 1915 and **Alfred Noel Garrod** aged 28 in 1916. In 1919, his 3rd son **Basil Rahere Garrod** died, aged 21, in Cologne during the great Spanish Flu pandemic.

Over the next 20 years, he served on the attending staff of several hospitals in London: Marylebone General Dispensary, West London Hospital, St. Bartholomew's Hospital, Great Ormond Street Hospital for Children, and Alexandria Hospital for Children with Hip Disease. Garrod was a proponent of scientific research as the foundation of medical practice, and published on a variety of diseases and topics throughout his career, including *An Introduction to the Use of the Laryngoscope* (1886) and *A Treatise on Rheumatism and Rheumatoid Arthritis* (1890). He helped found the *Quarterly Journal of Medicine* to provide a forum for more fundamental research into the processes of disease. He helped edit a pediatrics textbook, *Diseases of Children* (1913), with **FE Batten** and **Hugh Thursfield**.

Garrod was acquainted with the newly-rediscovered laws of Mendel and was also familial with certain human traits that occur as distinct attributes rather than as extremes of a continuous curve. These were some rare diseases– alkaptonuria, cystinuria, pentosuria and albinism– that were present apparently from birth and that often affected two or more members of a sibship in spite of the rarity among the population as a whole. Parents ordinarily were not affected. Garrod reasoned that these might be recessive traits, and, at least with alkaptonuria, he proposed that an essential enzyme was missing. His studies on alkaptonuria, published in 1902, state clearly the hypothesis that *alkaptonuria is inherited as a Mendelian recessive trait*. Garrod proposed that the enzyme that catalyzes the breakdown of homogentistic acid to acetoacetic acid is nonfunctional in alkaptonurics. This results in a build up of the acid in the urine, which turns brown then black on exposure to air, and is readily observed in infant's nappies. It was not, however, until 1958 that the absence of a functional *homogentisic acid oxidase* (= homogentisate 1,2-dioxygenase) enzyme was demonstrated in the liver of a patient with alkaptonuria. In later life, individuals with alkaptonuria develop *arthritis* characterized by deposition of brown pigment in joint cartilage and connective tissue. Garrod studied the recurrence patterns in several families, realized it followed an *autosomal recessive pattern of inheritance,* and postulated that it was caused by a mutation in a gene encoding an enzyme (*homogentisic acid oxidase*) involved in the metabolism of a class of compounds called *alkaptans.* He published 'The Incidence of Alkaptonuria: a Study in Chemical Individuality', in 1902. Garrod further provided a clear explanation of the effects of consanguinity on the appearance of recessive traits

Although Garrod had developed his hypothesis by 1902, it is of interest to note that the word 'gene' was not coined until 1911, and the first enzyme was not crystallized until 1926. His contributions were summarized in the Croonian lectures in 1908. His classic and prescient book, "Inborn Errors of Metabolism" (1909) was a most imaginative and important contribution to Biology and Medicine. Garrod expanded his metabolic studies to cover cystinuria, pentosuria, and albinism. These 3 inborn errors, along with alkaptonuria are collectively called **Garrod's tetrad**. In 1923, he summarized these studies in an expanded edition of his best-known work, *Inborn Errors of Metabolism.* As it became clearer that he had pioneered a new field of medicine, Garrod was increasingly honoured in England and abroad. He succeeded **William Osler** as Regius Professor of Medicine at Oxford. He was elected a **Fellow of the Royal Society** in 1910, was appointed to the Medical Research Council (MRC), and was made an honorary member of the American Association of Physicians and of the Ärtzlicher Verein is Munich. In 1935, he was awarded the **Gold Medal of the Royal Society of Medicine.** The Canadian Association of Centres for the Management of Hereditary Metabolic Diseases is commonly referred to as the *Garrod Association* to honour his contributions to the field of inborn errors of metabolism. He died at the Cambridge home of his daughter after a brief illness on March 28, 1936. As in the case of **Gregor Johann Mendel**, Garrod's perception was well ahead of his time and consequently had little impact. But some 4 decades after, Garrod's findings were well appreciated and acknowledged. Truly, Garrod was one of the pioneers of human genetics.

His famous quote reads as follows:

'... scientific method is not the same as the scientific spirit. The scientific spirit does not rest content with applying that which is already known, but is a restless spirit, ever pressing forward towards the regions of the unknown, ... it acts as a check, as well as a stimulus, sifting the value of the evidence, and rejecting that which is worthless, and restraining too eager flights of the imagination and too hasty conclusions.'

(Archibald Garrod: The Scientific Spirit in Medicine: *Inaugural Sessional Address to the Abernethian Society. St. Bartholomew's Hospital Journal.* **20:** *19, 1912.*)

1. Alkaptonuria

Alkaptonuria is an inherited metabolic disorder found in infants (approximately one in every 2,00,000 live births) and caused by the absence of **homogentisate 1,2-dioxygenase** which is required for the complete breakdown of the amino acids, tyrosine and phenylalanine. Homogentisate is a normal intermediate in the degradation of phenylalanine and tyrosine (refer Fig. 26–19) and it accumulates in alkaptonurics because its breakdown is blocked. Homogentisate is then excreted in the urine which turns dark on standing open to the atmosphere as homogentisate is oxidized and polymerized to a melanin-like substance. Alkaptonuria is a relatively benign condition and results in no serious ill effects. But it has historical importance for being the first such disease to be associated with an inborn error. The alkaptonurics live until well into reproductive age with no difficulty other than whatever esthetic offense the darkening urine may represent. Many in their fourth or fifth decade develop arthritis. The degeneration of the connective tissue in the joints is apparently associated with a deposition of pigment (a condition called **ochronosis**, because of the ochre colour of the deposits), presumably resulting from further oxidation of homogentisate in cartilage.

> Alkaptonuria is also spelt as **alcaptonuria.**

2. Albinism

Another inborn error associated with phenylalanine and tyrosine metabolism is albinism, an autosomal recessive trait. The biochemical defect involves melanin production, for which tyrosine is the precursor (Fig. 26–26). One type of albinism is believed to be due to a deficiency of **tyrosinase (= tyrosine 3-monooxygenase,)** a copper-containing enzyme needed for melanin synthesis. Albinos have normal eyesight, in spite of the lack of pigmentation, but are generally very sensitive to bright light.

> **Melanin** (*melan*[G] = black) is a black pigment present in the skin and hair. This polymeric pigment is formed in granules called *melanosomes* that are rich in *tyrosinase,* a monooxygenase enzyme.

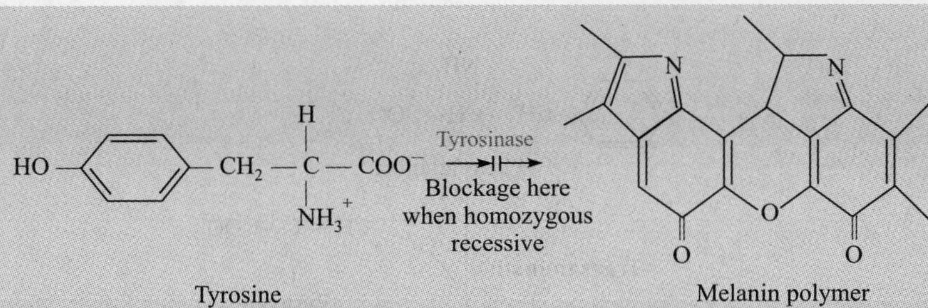

Fig. 26–26. Production of melanin

3. Phenylketonuria (PKU) or Hyperphenylalaninemia

Phenylketonuria (PKU), the commonest inborn error of metabolism, is so named because of high levels of phenylpyruvate, a type of phenylketone, in the urine of individuals afflicted with this disease. It is also known as **phenylpyruvic oligophrenia** because the condition results in early neurological damage preventing normal intellectual development. It was among the first human genetic defects of metabolism discovered. Phenylketonuria can have devastating effects in contrast with alkaptonuria. Untreated individuals with PKU are nearly always *severely mentally retarded* (a mean I.Q. of 20). In fact, about 1% of the institutionalized patients (those admitted in mental institutions) have phenylketonuria. The weight of the brain of these individuals is below normal, myelinization of their nerves is defective and their reflexes are hyperactive. The life expectancy of untreated phenylketonurics is drastically shortened. Nearly half are dead by age 20, and three-quarters by age 30. Very fair skin and light blonde hair are two characteristics of phenylketonuria.

Phenylketonuria is caused by an absence or deficiency of **phenylalanine hydroxylase (= phenylalanine 4-monooxygenase)** or, more rarely, of its tetrahydrobiopterin cofactor. Phenylalanine cannot be converted to tyrosine and so there is an accumulation of phenylalanine in all body fluids. Under such conditions, a secondary pathway of phenylalanine metabolim, normally little used, comes into play. In this minor pathway (Fig. 26–27), phenylalanine undergoes transamination with pyruvate to form **phenylpyruvate**. Much of the phenylpyruvate is either decarboxylated to produce **phenylacetate** or reduced to form **phenyllactate**. Phenylacetate imparts a characteristic "mousey" odour to the urine that has been used by nurses to detect PKU in infants.

Phenylalanine Hydroxylase

The hydroxylation of phenylalanine is a critical step in its catabolism and its conversion to tyrosine, thyroid hormone, and the catecholamine hormones. *Phenylalanine hydroxylase* is an example of a **mixed function oxidase,** an enzyme that uses a reduced cofactor and molecular oxygen to carry out a hydroxylation reaction. The cofactor is tetrahydrobiopterin, which is oxidized to dihydrobiopterin during the hydroxylation reaction. The residual hydrogens and oxygen are released as water. In order for this reaction to continue, the dihydrobiopterin must be reduced to tetrahydrobiopterin, and this requires a second enzyme, *dihydrobiopterin reductase*, which uses NADH to drive the reduction. The further hydroxylation of tyrosine, in the pathway that leads to catecholamines, requires a similar mixed-function oxidase, *tyrosine hydroxylase*. Phenylketonuria may develop as a result of mutations in either phenylalanine hydroxylase or enzymes involved in the synthesis or recycling of tetrahydrobiopterin.

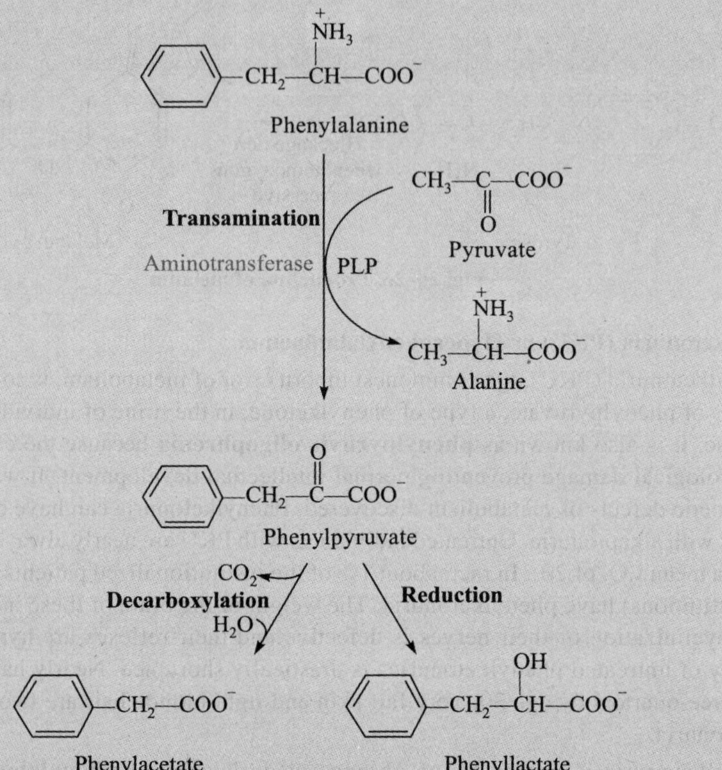

Fig. 26–27. Alternative pathways for catabolism of phenylalanine in phenylketonurics
Phenylpyruvate accumulates in the tissues, blood and urine. Both phenyacetate and phenyllactate can also be found in the urine.

At a first glance, it might be assumed that the inability to synthesize tyrosine, the precursor of melanin, is responsible for the observed lack of pigmentation. However, tyrosine is not lacking since food proteins (*e.g.*, casein of milk) furnish adequate amounts of the amino acid. Melanin production is impaired because high levels of phenylalanine effectively compete with tyrosine as substrate, *i.e.*, competitive inhibition.

Phenylketonurics appear normal at birth but are severely defective by age one, if untreated. PKU can be cured by giving low phenylalanine diet, just enough to meet the needs for growth and development. Foods that are rich in proteins must be avoided, rather curtailed. Proteins with low phenylalanine content, such as casein of milk, must first be hydrolyzed and much of the phenylalanine removed by adsorption to provide an appropriate diet for phenylketonurics, at least through childhood to prevent irreversible brain damage. A phenylketonuric child must be maintained on this restricted low-protein diet (with no more than 200-500 mg/day of phenylalanine), and continued as long as possible or until at least by age 3, by which time brain development is complete. Although severe mental retardation can be prevented by this treatment, the phenylketonurics exhibit other physical and/or emotional problems, many of which remain biochemically unexplained. It should be noted that a high blood level of phenylalanine in a pregnant mother can result in abnormal development of the fetus. *This is a striking example of maternal-fetal relationship at the molecular level.*

Phenylketonuria is inherited as an autosomal recessive and the approximate incidence of phenylketonuria in individuals is 80 out of 10,00,000 live births. Data on the low frequency of this and other genetic disorders can mislead the unwary on the prevalence of the altered genes that cause them. In the present instance of PKU, for example, both chromosomes carrying specifications for the protein must be defective because the disease is recessive and as such both parents must have contributed defective versions of the genes. In such cases, the probability that a child will have two defective genes is the product of the probabilities for the presence of one defective gene in each chromosome. Since the gene is **autosomal**, *i.e.*, it does not occur on a sex chromosome, the probabilities are the same for males as they are for females in large populations. Therefore, the estimated gene frequency for PKU would be $(80 \times 10^{-6})^{1/2}$, which is 8.9×10^{-3}. Approximately, one out of every 57 humans has a gene for phenylketonuria in one of his paired chromosomes, an impressive incidence. Over 170 mutations in the gene have been reported.

4. Maple syrup urine disease (MSUD) or Branched-chain ketoaciduria

Maple syrup urine disease is a rare autosomal, recessive, hereditary disorder in which the 3 branched-chain amino acids (leucine, isoleucine, valine) accumulate in the blood and "spill over" into the urine. MSUD presents acutely in the neonatal period. The disease is so named because of the characteristic odour of the urine which resembles that of maple syrup or burnt sugar and is imparted to the urine by the α-keto acids. In afflicted individuals, both plasma and urinary levels of 3 branched-chain amino acids and their corresponding α-keto acids are greatly elevated. For this reason, the disease has also been termed **branched-chain ketonuria**. Smaller quantities of branched-chain - hydroxy acids, formed by reduction of α-keto acids, also are present in the urine.

Although the afflicted newborn infant initially appears to be normal, the characteristic symptoms of the disease are evident by the end of the first week of extrauterine life. Besides the biochemical abnormalities mentioned above, the infant is difficult to feed, may vomit and may show alternate periods of hypertonicity and flaccidity. The patient may be highly lethargic. Extensive brain damage and demyelinization occur in surviving children. If left untreated, death ensues by the end of first year of life.

The biochemical defect is the absence or highly reduced activity of the **branched-chain α-keto acid dehydrogenase complex,** which catalyzes conversion of all 3 branched-chain keto acids to CO_2 plus acetyl-CoA derivatives. All 3 branched-chain α-keto acids are competitive inhibitors of L-glutamate dehydrogenase activity.

Maple syrup urine disease is usually fatal until the patient is placed on a diet in which protein is supplied by a mixture of all the purified amino acids except leucine, isoleucine and valine. When the plasma levels of these amino acids fall within the normal range, they are restored to the diet in the form of milk and other foods in amounts adequate to supply, but not to exceed, the requirements for these branched-chain amino acids.

5. Homocystinuria

Homocystinuria is a relatively rare, autosomal, recessive condition with an approximate incidence of 1 in every 2,00,000 live births. The disease results in a variety of symptoms including mental retardation, vision problems, and thrombotic strokes and coronary artery disease at a young age. The condition is caused by lack of an enzyme which catalyzes the transfer of sulfur from homocysteine to serine through the formation of a cystathionine intermediate. Some of these patients respond to vitamin supplementaion. Moderately elevated levels of homocysteine in plasma are implicated in the development of cardiovascular disease (CVD) and cerebrovascular ischemic episodes (stroke). Cross-sectional and retrospective studies suggest that even moderately elevated levels of homocysteine may be correlated with increased incidence of heart disease and stroke, but the jury is still out as to whether lowering homocysteine levels will reduce the development of these serious illnesses.

REFERENCES

1. **Bender DA :** Amino Acid Metabolism. *2nd ed., John Wiley and Sons, New York, 1985.*
2. **Bondy PK, Resenberg LE (editors) :** Metabolic Control and Disease. *8th ed., W.B. Saunders, Philadelphia. 1980.*
3. **Cederbaum S :** Phenylketonuria: an update. *Curr. Opin. Pediatr.* **14 :** *702–706, 2002.*
4. **Childs B :** Sir Archibald Garrod's conception of chemical individuality: A modern appreciation. *New Engl. J. Med.* **282 :** *71–78, 1970.*
5. **Christen P, Metzler DE :** Transaminases. *Wiley-Interscience, Inc., New York, 1985.*
6. **Cooper AJL :** Biochemistry of sulfur-containing amino acids. *Ann. Rev. Biochem.* **52 :** *187–222, 1983.*
7. **Dakshinamurti K (editor) :** Vitamin B_6; Annals of the New York Academy of Sciences. *Vol. 585, 1990.*
8. **Eisensmith RC, Woo SLC :** Phenylketonuria and the phenylalanine hydroxylase gene. *Mol. Biol. Med.* **8 :** *3–18, 1991.*
9. **Garrod AE :** Inborn Errors in Metabolism. *Oxford University Press (Reprinted in 1963 with a supplement by H. Harris).*
10. **Grisolia S, Báguena R, Mayor F (editors) :** The Urea Cycle. *Wiley, New York, 1976.*
11. **Gropman AL, Summar M, Leonard JV :** Neurological implications of urea cycle disorders. *J. Inherit. Metab. Dis.* **30 :** *865–879, 2007.*
12. **Holmes FL :** Hans Krebs and the discovery of the ornithine cycle. *Fed. Proc.* **39 :** *216–225, 1980.*
13. **Kaufman S (editor) :** Amino Acids in Health and Disease: New Perspectives. *ULCA Symposium on Molecular and Cellular Biology. New Series. Vol. 55. Fox CF (editor), Alan R. Liss, 1987.*
14. **King J (editor) :** Protein and Nucleic Acid Structure and Dynamics. *Benjamin/Cummings, Reading, Massachusetts, 1985.*
15. **Levy HL :** Nutritional therapy for selected inborn errors of metabolism. *J. Amer. Coll. Nutrit.* **8 :** *54S–60S, 1989.*
16. **Lippard SJ, Berg JM :** Principles of Bioenergetic Chemistry. *University Science Books, 1994.*
17. **Livesey G :** Methionine degradation: Anabolic and catabolic. *Trends Biochem. Sci.* **9 :** *27, 1984.*
18. **Mc Phalen CA, Vincent MG, Jansonius JN :** X-ray structure refinement and comparison of three forms of mitochondrial aspartate aminotransferase. *J. Mol. Biol.* **225 :** *495–517, 1992.*
19. **Mc Phalen CA, Vincent MG, Picot D, Jansonius JN, Lesk AM, Chothia C :** Domain closure in mitochondrial aspartate aminotransferases. *J. Mol. Biol.* **227 :** *197–213, 1992.*
20. **Mehler A :** Amino acid metabolism I: general pathways; Amino acid metabolism II: metabolism of the individual amino acids. [In Textbook of Biochemistry with Clinical Correlations. *3rd ed., TM Devlin, editor], 475–528, Wiley-Liss, New York, 1992.*
21. **Meister A :** Biochemistry of the Amino Acids. 2nd ed., *Vols. 1 and 2. Academic Press, Inc., New York, 1965.*
22. **Morris SM Jr :** Regulation of enzymes of the urea cycle and arginine metabolism. *Annu. Rev. Nutr.* **22 :** *87–105, 2002.*
23. **Nichol CA, Smith GK, Duch DS :** Biosynthesis and metabolism of tetrahydrobiopterin and molybdopterin. *Ann. Rev. Biochem.* **54 :** *729–764, 1985.*

24. **Nyhan WL (editor) :** Abnormalities in Amino Acid Metabolism in Clinical Medicine. *Appleton-Century-Crofts, Norwalk, CT., 1984.*
25. **Powers-Lee SG, Meister A :** Urea synthesis and ammonia metabolism. [In The Liver: Biology and Pathobiology, 2nd ed., IM Arias, WB Jackoby, H Popper, D Schachter, DA Shafritz, editors] *317–329, Raven Press, New York. 1988.*
26. **Schander P, Wahren J, Paoletti R, Bernardi R, Rinetti M (editors) :** Branched-Chain Amino Acids: Biochemistry, Physiopathology and Clinical Sciences. *Raven Press, New York. 1992.*
27. **Scriver CR, Beaudet AL, Sly WS, Valle D (editors) :** The Metabolic Basis of Inherited Diseases. *6th ed. Part 4: 493–771, McGraw-Hill Book Company, New York. 1989.*
28. **Scriver CR, Kaufman S, Woo SLC :** Mendelian hyperphenylalaninemia. *Ann. Rev. Genet.,* **22 :** *301–321, 1988.*
29. **Singh RH :** Nutritional management of patients with urea cycle disorders. *J. Inherit. Metab. Dis.* **30 :** *880–887, 2007.*
30. **Snell EE, DiMari SJ :** Schiff base intermediates in enzyme catalysis [In PD Boyer's (editor) The Enzymes. *3rd ed., Vol. 2] 335–370, Academic Press, Inc., New York. 1970.*
31. **Stanbury JB, Wyngaarden JB, Fredrickson DS, Goldstein JL, Brown MS (editors) :** The Metabolic Basis of Inherited Disease. 5th ed., Part 3: Disorders of Amino Acid Metabolism. *McGraw-Hill Book Company, New York. 1983.*
32. **Torchinsky YM :** Transamination: Its discovery, biological and chemical aspects. *Trends Biochem. Sci.* **12 :** *115–117, 1989.*
33. **Walsh C :** Enzymatic Reaction Mechanisms. *W.H. Freeman and Company, San Francisco. 1979.*
34. **Wellner D, Meister A :** A survey of inborn errors of amino acid metabolism and transport in man. *Ann. Rev. Biochem.* **50 :** *911, 1981.*

PROBLEMS

1. Name the α-ketoacid that is formed by transamination of each of the following amino acids:
   (*a*) Alanine
   (*b*) Aspartate
   (*c*) Glutamate
   (*d*) Leucine
   (*e*) Phenylalanine
   (*f*) Tyrosine
2. Compound A has been synthesized as a potential inhibitor of a urea-cycle enzyme. Which enzyme do you think compound A might inhibit ?

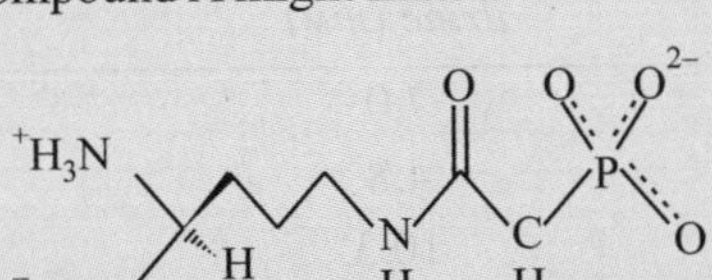

Compound A

3. How would you treat an infant who is deficient in argininosuccinate synthetase? Which molecules would carry nitrogen out of the body ?
4. Why should phenylketonurics avoid using aspartame, an artificial sweetener? (**Hint:** Aspartame is L-aspartyl-L-phenylalanine methyl ester.)
5. Within a few days after a fast begins, nitrogen excretion accelerates to a higher-than-normal level. After a few weeks, the rate of nitrogen excretion falls to a lower level and continues at

this low rate. However, after the fat stores have been depleted, nitrogen excretion rises to a high level.

(*a*) What events trigger the initial surge of nitrogen excretion?

(*b*) Why does nitrogen excretion fall after several weeks of fasting?

(*c*) Explain the increase in nitrogen excretion when the lipid stores have been depleted.

6. Isoleucine is degraded to acetyl-CoA and succinyl-CoA. Suggest a plausible reaction sequence, based on reactions discussed in the text, for this degradation pathway.

7. Use numbers 1 to 5 to identify each carbon atom in the product of this reaction. What is the coenzyme ?

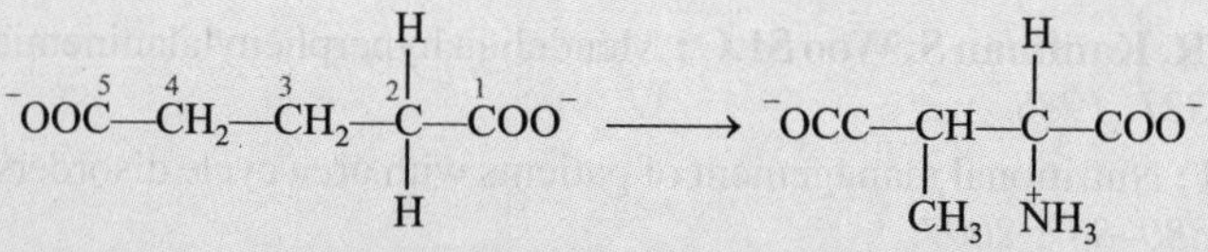

8. Explain the basis for the following statement: As a coenzyme, pyridoxal phosphate is covalently bound to enzymes with which it functions, yet during catalysis the coenzyme is not covalently bound.

9. Briefly discuss how a yeast cell might contain two glutamate dehydrogenases, one specialized for nitrogen assimilation and one for amino acid catabolism, and not dissipate energy in the futile cycle,

$$\text{glutamate} \rightleftharpoons \alpha\text{-ketoglutarate.}$$

10. Draw the structure and give the name of the α-keto acid resulting when the following amino acids undergo transamination with α-ketoglutarate:

(*a*) Aspartate
(*b*) Glutamate
(*c*) Alanine
(*d*) Phenylalanine

11. If your diet is rich in alanine but deficient in aspartate, will you show signs of aspartate deficiency ? Explain.

12. A two-year-old child was brought to the hospital. His mother indicated that he vomited frequently, especially after feedings. The child's weight and physical development were below normal. His hair, although dark, contained patches of white. A urine sample treated with ferric chloride ($FeCl_3$) gave a green color characteristic of the presence of phenylpyruvate. Quantitative analysis of urine samples gave the results shown in the table below.

Substance	*Concentration in patient's urine (mM)*	*Normal concentration in urine (mM)*
Phenylalanine	7.0	0.01
Phenylpyruvate	4.8	0
Phenyllactate	10.3	0

(*a*) Suggest which enzyme might be deficient. Propose a treatment for this condition?

(*b*) Why does phenylalanine appear in the urine in large amounts?

(*c*) What is the source of phenylpyruvate and phenyllactate ? Why does this pathway (normally not functional) come into play when the concentration of phenylalanine rises?

(*d*) Why does the patient's hair contain patches of white?

13. The three carbons in lactate and alanine have identical states of oxidation, and animals can use either carbon source as a metabolic fuel. Compare the net ATP yield (moles of ATP per mole of substrate) for the complete oxidation (to CO_2 and H_2O) of lactate versus alanine when the cost of nitrogen excretion as urea is included.

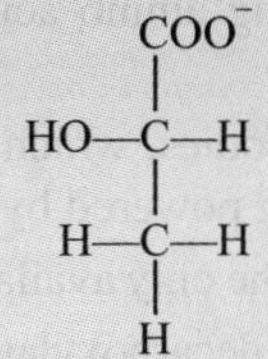

Lactate

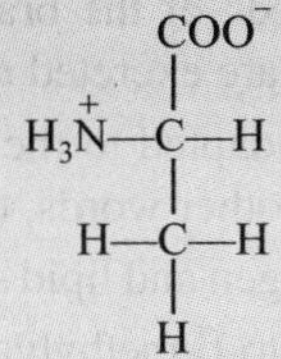

Alanine

14. Aspartate aminotransferase has the highest activity of all the mammalian liver aminotransferases. Why?

15. A weight-reducing diet, heavily promoted some years ago, required the daily intake of "liquid protein" (soup of hydrolyzed gelatin), water, and an assortment of vitamins. All other food and drink were to be avoided. People on this diet typically lost 10 to 14 lb in the first week.
   (*a*) Opponents argued that the weight loss was almost entirely water and would be regained almost immediately when a normal diet was resumed. What is the biochemical basis for the opponent's argument?
   (*b*) A number of people on this diet died. What are some of the dangers inherent in the diet and how can they lead to death?

16. Blood plasma contains all the amino acids required for the synthesis of body proteins, but they are not present in equal concentrations. Two amino acids, alanine and glutamine, are present in much higher concentrations in normal human blood plasma than any of the other amino acids. Suggest possible reasons for their abundance.

17. An inability to generate tetrahydrobiopterin would have what specific effects on the metabolism of phenylalanine, tyrosine, and tryptophan?

18. Untreated phenylketonuria patients, in addition to mental retardation, have diminished production of catecholamines and light skin and hair. If the defect is in phenylalanine hydroxylase itself, a diet lacking phenylalanine but including tyrosine alleviates these conditions. If the defect is in the ability to produce tetrahydrobiopterin, the same dietary treatment may alleviate the mental retardation and light hair but not the diminished catecholamine production. What is the rationale explaining these findings?

19. Albinism is caused due to the deficiency of the enzyme:
   (*a*) Hydroxylase
   (*c*) Zymase
   (*b*) Cellulase
   (*d*) Tyrosinase

ANSWERS TO PROBLEMS

1. (*a*) Pyruvate
   (*b*) oxaloacetate
   (*c*) α-ketoglutarate
   (*d*) α-ketoisocaproate
   (*e*) phenylpyruvate
   (*f*) hydroxyphenylpyruvate.
2. Ornithine transcarbamoylase (analogous to *N*-(phosphonacetyl)-L-aspartate, PALA)
3. Benzoate, phenylacetate, and arginine would be given to supply a protein-restricted diet. Nitrogen would emerge in hippurate, phenylacetylglutamine, and citrulline.
4. Aspartame, a dipeptide ester (L-aspartyl-L-phenylalanine methyl ester), is hydrolyzed to

L-aspartate and L-phenylalanine. High levels of phenylalanine are harmful in phenylketonurics.

5. (*a*) Depletion of glycogen stores. When they are gone, proteins must be degraded to meet the glucose needs of the brain. The resulting amino acids are deaminated, and the nitrogen atoms are excreted as urea.

   (*b*) The brain has adapted to the use of ketone bodies, which are derived from fatty acid catabolism. In other words, the brain is being powered by fatty acid breakdown.

   (*c*) When the glycogen and lipid stores are gone, the only available energy source is protein.

6. Deamination to α-keto-β-methylvalerate; oxidative decarboxylation to α-methylbutyryl CoA; oxidation to tiglyl CoA; hydration,oxidation and thiolysis yields acetyl-CoA and propionyl CoA; propionyl CoA to succinyl CoA.

7. $^{-}OO{}^{5}C-{}^{4}CH-{}^{2}CH-{}^{1}COO^{-}$, with ${}^{3}CH_3$ on C-4 and NH_3^+ on C-2

8. PLP forms a covalent Schiff base between the aldehyde carbon of the coenzyme and an ε-amino group of a lysine residue. Obviously, this bond must be broken for the coenzyme to form a Schiff base with an amino acid substrate.

9. The most straightforward answer is to place the two enzymes in different cell compartments– the catabolic enzyme in mitochondria and the assimilative enzyme in cytosol.

10. (*a*) $^{-}OOC-CH_2-C(=O)-COO^{-}$ Oxaloacetate

    (*b*) $^{-}OOC-CH_2-CH_2-C(=O)-COO^{-}$ α-ketoglutarate

    (*c*) $CH_3-C(=O)-COO^{-}$ Pyruvate

    (*d*) 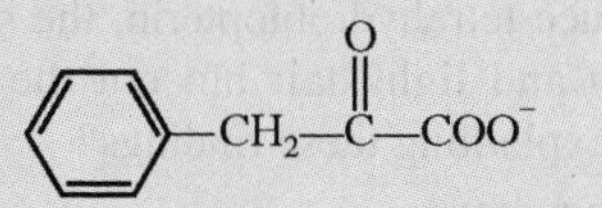 Phenylpyruvate

11. No; the nitrogen in Ala can be transferred to oxaloacetate *via* transamination, to form Asp.

12. (*a*) Phenylalanine-4-monooxygenase; a low-phenylalanine diet.

    (*b*) The normal route of Phe metabolism *via* hydroxylation to Tyr is blocked, and Phe accumulates.

    (*c*) Phe is transformed to phenylpyruvate by transamination, and then to phenyllactate by reduction. The transamination reaction has an equilibrium constant of 1.0, and phenylpyruvate is formed in significant amounts when phenylalanine accumulates.

    (*d*) Because of the deficiency in production of Tyr, a precursor of melanin, the pigment normally present in hair.

13. 17–18 moles of ATP per mole of lactate, depending on which shuttle moves the electrons from NADH produced by the dehydrogenases into the mitochondrion. 15–16 ATP per alanine when nitrogen removal *via* the urea cycle is included.

14. The second amino group introduced into urea is transferred from Asp, which is generated during the transamination of Glu to oxaloacetate, a reaction catalyzed by aspartate aminotransferase. Approximately one half of all the amino groups excreted as urea must

pass through the aspartate aminotransferase reaction, making this the most highly active aminotransferase.

15. (*a*) A person on a diet consisting only of protein must use amino acids as the principal source of metabolic fuel. Because the catabolism of amino acids requires the removal of nitrogen as urea, the process consumes abnormally large quantities of water to dilute and excrete the urea in the urine. Furthermore, electrolytes in the "liquid protein" must be diluted with water and excreted. If the daily water loss through the kidney is not balanced by a sufficient water intake, a net loss of body water results.

(*b*) When considering the nutritional benefits of protein, one must keep in mind the total amount of amino acids needed for protein synthesis and the distribution of amino acids in the dietary protein. *Gelatin contains a nutritionally unbalanced distribution of amino acids.* As large amounts of gelatin are ingested and the excess amino acids are catabolized, the capacity of the urea cycle may be exceeded, leading to ammonia toxicity. This is further complicated by the dehydration that may result from excretion of large quantities of urea. A combination of these two factors could produce coma and death.

16. Ala and Gln play special roles in the transport of amino groups from muscle and from other nonhepatic tissues, respectively, to the liver.

17. Tetrahydrobiopterin is a necessary component of phenylalanine, tyrosine, and tryptophan hydroxylases. Its deficiency would inhibit normal degradation of both phenylalanine and tyrosine because their degradative pathways begin with the respective hydroxylases. Catecholamine formation (norepinephrine and epinephrine) begin with the formation of DOPA from tyrosine *via* tyrosine hydroxylase so catecholamine synthesis would be inhibited. The initial step in the conversion of tryptophan to serotonin is catalyzed by tryptophan hydroxylase.

18. Mental retardation seems to be caused by the elevated levels of phenylalanine in the blood so removing phenylalanine from the diet is beneficial. Light skin and hair is a secondary effect of phenylalanine competing with tyrosine for tyrosinase, not an enzyme defect, which is necessary to form melanins. Low tyrosine because of the inability to convert phenylalanine to tyrosine could also be a factor. Since tyrosine is low, the production of catecholamines is affected. If the defect is in the production of tetrahydrobiopterin, diet is less effective because tyrosine hydroxylase leading to DOPA (the first step in catecholamine formation) is tetrahydrobiopterin-dependent.

19. (*d*) Tyrosinase

CONTENTS

CHAPTER 27

Biosynthesis of Amino Acids

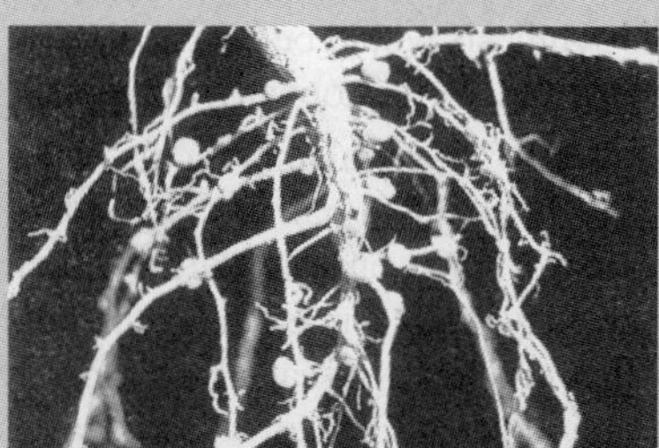

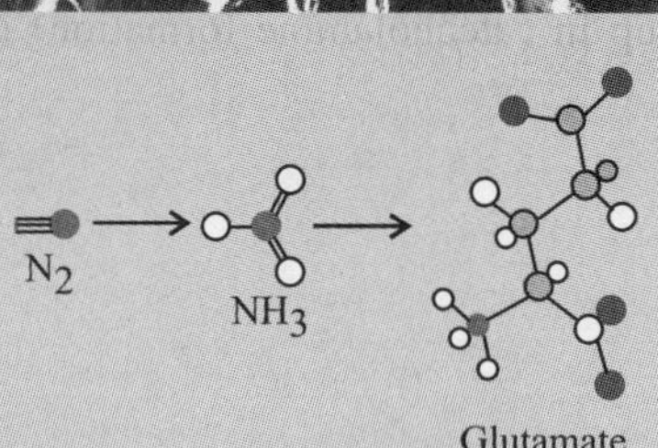

Nitrogen as a key component of amino acids

The atmosphere is rich in nitrogen gas (N_2), a very unreactive molecule. Certain organisms such as bacteria that live in the root nodules of bird's foot trefoil (photo *above*) can convert nitrogen gas into ammonia. Ammonia can then be used to synthesize first glutamate and then other amino acids.

[*Courtesy : (Upper) Vu/Cabisco*]

AN OVERVIEW OF AMINO ACID BIOSYNTHESIS

General Considerations

The broad subject area of amino acid metabolism includes several processes such as protein synthesis and degradation, conversion of carbon skeletons of the amino acids to amphibolic intermediates, urea synthesis and the formation of a wide variety of physiologically active compounds such as serotonin. This concept is diagrammed in Fig. 27–1.

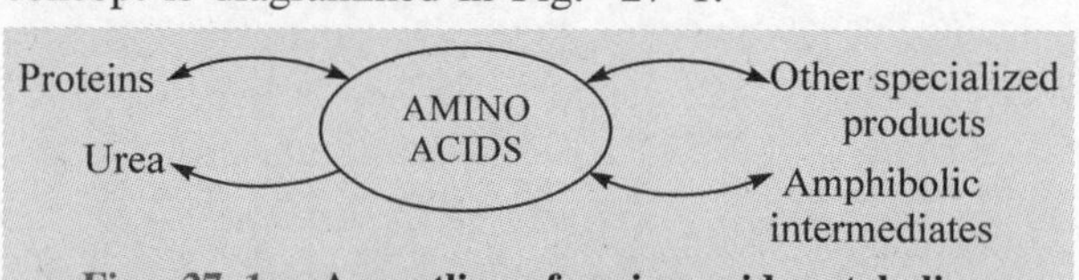

Fig. 27–1. An outline of amino acid metabolism

Note that the indicated process except urea formation proceed reversibly in intact cells. However, the enzymes and intermediates in biosynthetic and degradative process frequently differ.

Every living cell must have a supply of amino acids constantly available for diverse synthetic processes, especially protein synthesis. These amino acids may be derived by synthesis from simple substances, *i.e.*, from glucose as a carbon source and ammonium ion as a nitrogen source, or by their absorption from the surrounding environment. In the first case, where there is *de novo* synthesis of the compounds from simple substances, they are often described as being of **endogenous origin**. The second case, where the compounds are obtained preformed from the environment, would be termed an **exogenous** source. Respectively, these two terms literally mean 'formed within' and 'formed without'.

Amino acids rely upon the diet for the provision of some amino acids. Indeed, it might be questioned why animals synthesize any of the amino acids, since all 20 are obtained from the food. However, we know that some of the amino acids have important metabolic roles, which in themselves involve a constant synthesis and degradation of the compounds in quantities much higher than those needed for protein synthesis.

It is of interest to note that the catabolic breakdown of amino acids produces intermediates of citric acid cycle whereas the anabolic formation of amino acids utilizes citric acid cycle intermediates as precursors. Fig 27–2 (on the next page) highlights the relationship between amino acid metabolism and the citric acid cycle intermediates.

Essential and Nonessential Amino Acids

The organisms differ markedly in terms of their ability to carry out the *de novo* synthesis of the protein amino acids from inorganic substances. Most microorganisms and plants are competent in all such syntheses, but most animals lack about half of these synthetic capacities. For the latter organisms, amino acids may then be classified as 'essential' or 'nonessential'. On the basis of the most frequently employed experimental criteria, an amino acid is considered essential if it is included in the diet for:

(*a*) optimal growth, or

(*b*) the maintenance of nitrogen balance.

An animal is said to be in **nitrogen balance**, if its daily intake of nitrogen is just balanced by its daily excretion of nitrogen.

The 'essentiality' of an amino acid is not only a function of the criteria employed for its determination but is a function of many other variables, including the presence of metabolically related substances in the diet. Besides, the amino acid requirement may also vary with the physiological state of the animal (*i.e.*, in pregnancy, lactation and disease), with age, and probably with the nature of the intestinal flora. It is also noteworthy that the 'essential' feature of the essential amino acids is their **carbon skeleton**; most of these may be derived from the corresponding keto acids through transamination :

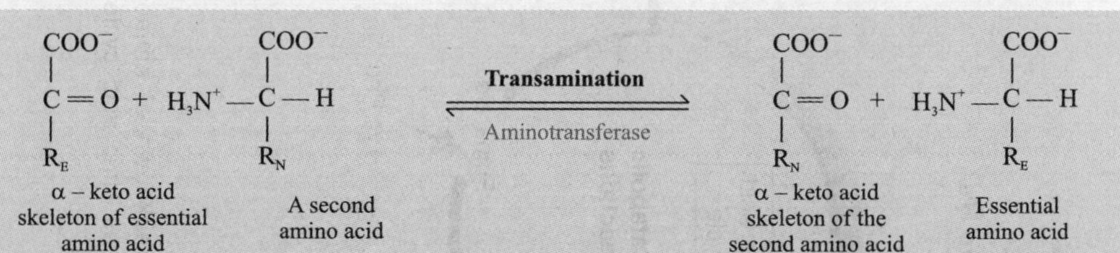

[R_E and R_N represent R groups of the essential and nonessential amino acids, respectively.]

In fact, all the 20 amino acids are essential to the organism in the sense that all must be present in order for protein synthesis and, therefore, life to occur. As already stated, some life forms (*e.g.*, bacteria and plants) can form all the 20 amino acids from amphibolic intermediates, whereas other forms including human and other animals, can biosynthesize only half of those required. These amino acids are termed **nutritionally-nonessential amino acids** (refer Table 27–1). The remaining

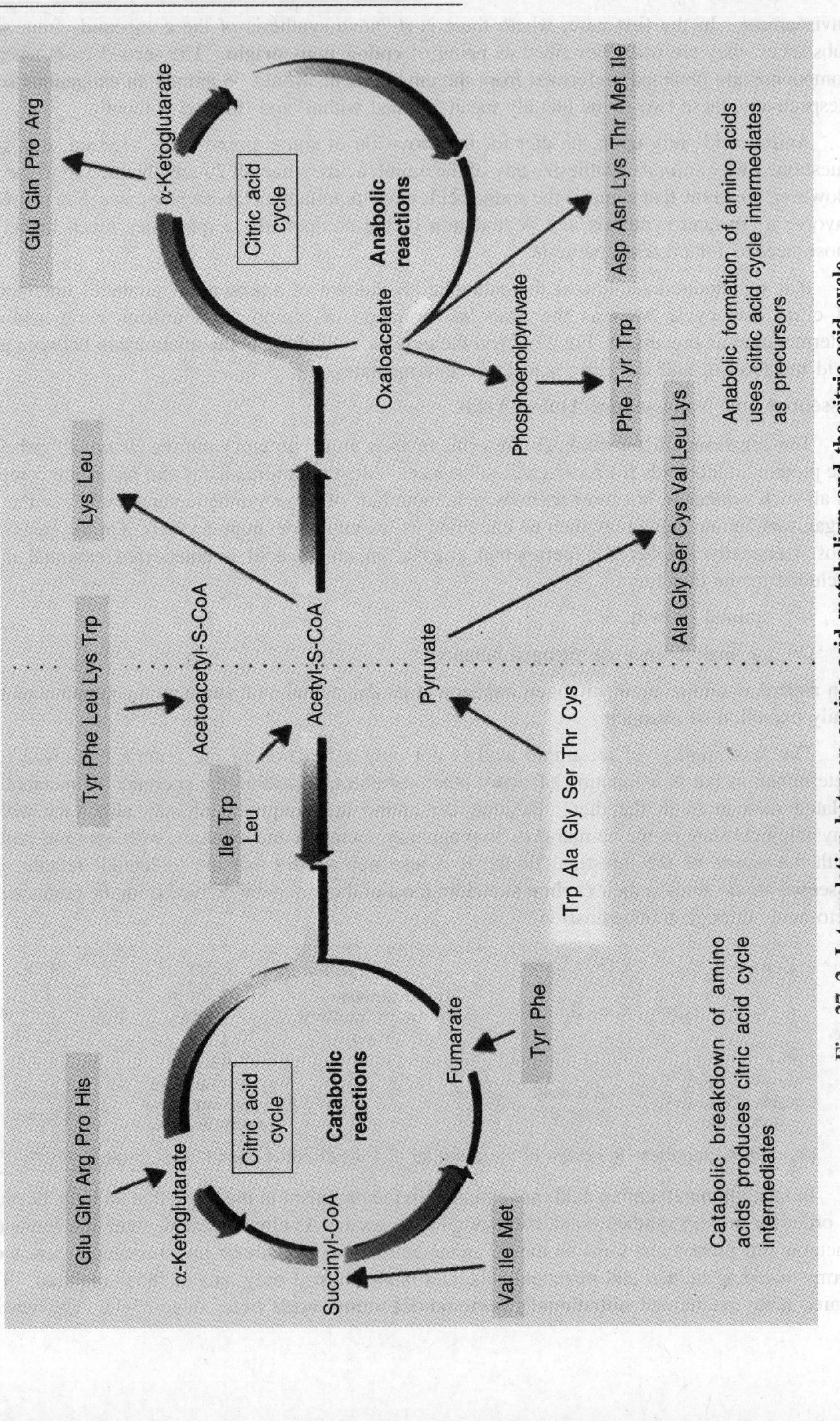

Fig. 27–2. Interrelation between amino acid metabolism and the citric acid cycle

Table 27–1. **Essential and nonessential amino acids and the number of enzymes required for their synthesis**

Nutritionally essential amino acids		*Nutritionally nonessential amino acids*	
Arginine*, [a]	7	Alanine	1
Histidine*	6	Asparagine[b]	1
Isoleucine	8 (6 shared)	Aspartic acid	1
Leucine	3 (7 shared)	Cysteine[d]	2
Lysine	8	Glutamic acid	1
Methionine	5 (4 shared)	Glutamine[a]	1
Phenylalanine	1 (9 shared)	Glycine[c]	1
Threonine	6	Proline[a]	3
Tryptophan	5 (8 shared)	Serine	3
Valine	+1 (7 shared)	Tyrosine	+10
	50		**24**

* Arginine and histidine, though belonging to the nutritionally essential category, are sometimes classified as **nutritionally-semiessential amino acids** because they may be synthesized in tissues at rates inadequate to support growth of children.

[a]From glutamic acid [b]From aspartic acid [c]From serine [d]From serine plus S^{2-}

amino acids, which must therefore be supplied by the diet, are termed **nutritionally-essential amino acids**. It may be emphasized that a given amino acid may be nutritionally essential for one form of life but nutritionally nonessential for another. These differences reflect their different genetic heritages. Confusion may arise because nutritionists frequently refer to the nutritionally essential amino acids as 'essential' or 'indispensable' amino acids and the nutritionally nonessential amino acids as 'nonessential' or 'dispensable' amino acids. Although in a nutritional context these terms are correct, it is unfortunate that they obscure the biologically essential nature of all the 20 amino acids.

A perusal of the Table 27–1 indicates that the number of enzymes required by cells to synthesize the nutritionally essential amino acids is large in relation to the number of enzymes required to synthesize the nutritionally nonessential amino acids. This suggests that there is a positive survival advantage in retaining the ability to manufacture 'easy' amino acids while losing the ability to make 'difficult' amino acids.

Metabolic Precursors of Amino Acids

The pathways for the biosynthesis of amino acids are diverse. However, they have an important common feature: *their carbon skeletons come from intermediates in glycolysis, tricarboxylic acid cycle or pentose phosphate pathway* (Fig. 27–3). The nonessential amino acids are synthesized by quite simple reactions, whereas the pathways for the formation of essential amino acids are quite complex. Ten of the amino acids are only one or a few enzymatic steps removed from their precursors. The pathways for others, such as the aromatic amino acids, are more complex. A useful way to organize the amino acid biosynthetic pathways is to group them into families corresponding to the metabolic precursor of each amino acid (Table 27–2).

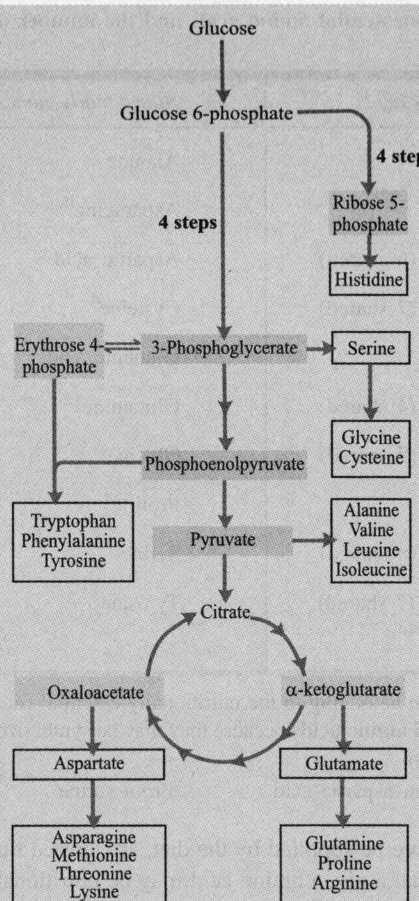

Fig. 27–3. An overview of amino acid biosynthesis

The carbon skeleton precursors derive from 3 sources: glycolysis (*pink*), the citric acid cycle (*blue*), and the pentose phosphate pathway.

Table 27–2. Six biosynthetic families of amino acids, based on different metabolic precursors (shown in *boldface*)

α-ketoglutarate	**Oxaloacetate**	**Phosphoenolpyruvate and erythrose-4-phosphate**
Glutamate	Aspartate	Phenylalanine*
Glutamine	Asparagine	Tyrosine
Proline	Methionine*	Tryptophan*
Arginine†	Threonine*	
	Isoleucine*	
	Lysine*	
3-phosphoglycerate	**Pyruvate**	**Ribose-5-phosphate**
Serine	Alanine	Histidine*
Cysteine	Valine*	
Glycine	Leucine*	

† Although essential in young, growing animals but not in adults, the status of arginine is rather ambiguous since the entire arginine molecule can be synthesized by human and other animals.

* Essential amino acids; rest are nonessential.

In addition to these, there is a notable intermediate that recurs in several pathways: **phosphoribosyl pyrophosphate (PRPP)**. PRPP is synthesized from ribose-5-phosphate derived from the pentose phosphate pathway, in a reaction catalyzed by **ribose phosphate pyrophosphokinase**:

$$\text{Ribose-5-phosphate} + \text{ATP} \rightleftharpoons \text{5-phosphoribosyl-1-pyrophosphate} + \text{AMP}$$

Ribose phosphate pyrophosphokinase is allosterically regulated by many of the biomolecules for which PRPP is a precursor. PRPP is an intermediate in tryptophan and histidine biosyntheses, with the ribose ring contributing many of its carbons to the final structure of these amino acids. It is also of fundamental importance in the biosynthesis of nucleotides.

AN OVERVIEW OF NITROGEN METABOLISM

The pathways leading to the biosynthesis of amino acids and nucleotides both share a requirement for nitrogen. But as soluble and biologically-useful nitrogen compounds are scarce in natural environment, therefore ammonia, amino acids and nucleotides are used economically by most organisms. Nitrogen from the environment is introduced into biological systems by many pathways which are described below.

Nitrogen Cycle

The atmospheric air contains 79% (*i.e.*, about four-fifths) molecular nitrogen (N_2). However, only a relatively few species can convert atmospheric nitrogen into forms useful to living organisms. Therefore, different organisms display an interdependent manner to salvage and reuse biologically available nitrogen in a vast **nitrogen cycle** (Fig. 27–4).

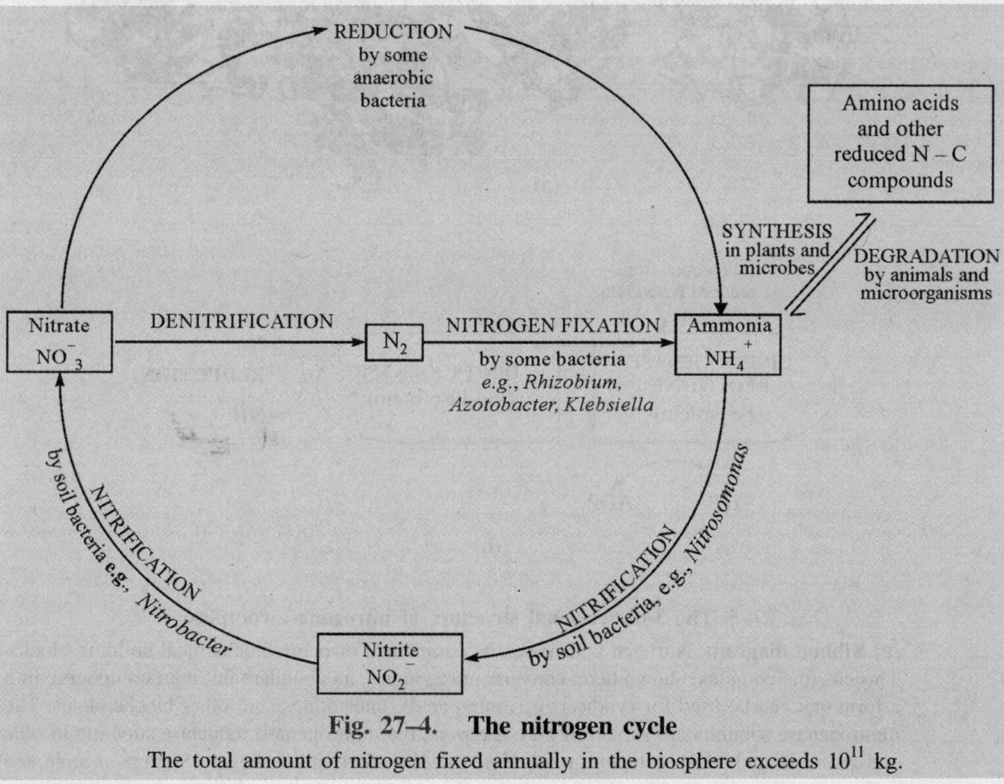

Fig. 27–4. The nitrogen cycle

The total amount of nitrogen fixed annually in the biosphere exceeds 10^{11} kg.

The first step in nitrogen cycle is **nitrogen fixation,** *i.e.*, reduction of N_2 to ammonia (NH_3 or NH_4^+) which is accomplished only by nitrogen-fixing bacteria. Plants cannot reduce but live symbiotically with the bacteria (*usually Rhizobium*) and thereby enrich the nitrogen content of the soil. Thus, the symbiotic and photosynthetic N_2-fixing systems fix a major amount of the total of approximately 10^8 tons of N_2 fixed annually. On the contrary, the free-living and nonphotosynthetic bacteria (*Azotobacter, Klebsiella, Clostridium*) contribute relatively little to the total.

Although ammonia can be used by most living organisms, the soil bacteria, which are so abundant and active, oxidize soil ammonia first to NO_2^- and then to NO_3^-, a process called **nitrification**. Plants and many bacteria can easily reduce NO_3^- to ammonia by the action of nitrate reductases. Ammonia so formed can be synthesized into amino acids by plants which are then used up by animals as source of amino acids, both essential and nonessential, to built animal proteins. When organisms die, the microbes degrade their body proteins to ammonia in the soil, where nitrifying bacteria convert it into nitrite and nitrate again.

Certain denitrifying bacteria, however, convert nitrate to N_2 under anaerobic conditions, so that a balance is maintained between fixed nitrogen and atmospheric nitrogen. In this process of **denitrification**, these soil bacteria utilize NO_3^-, rather than O_2, as the ultimate electron acceptor.

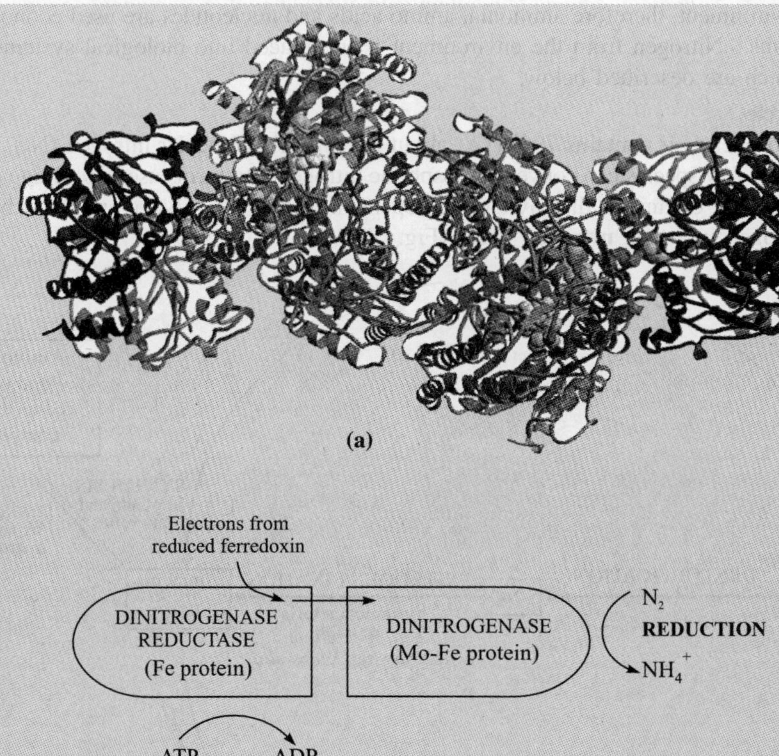

Fig. 27–5. The 3-dimensional structure of nitrogenase complex

(a) Ribbon diagram. Nitrogen is an essential component of many biochemical building blocks. This enzyme complex, shown here, converts nitrogen gas, an abundant but inert compound, into a form that can be used for synthesizing amino acids, nucleotides, and other biochemicals. The dinitrogenase subunits are shown in *grey* and *pink*, the dinitrogenase reductase subunits in *blue* and *green*. Bound ADP is shown in *red*. Note the 4Fe-4S complex (Fe and S atoms *orange* and *yellow*, respectively), and the Fe-Mo cofactor (Mo *black*, homocitrate *light grey*). The P clusters (bridged pairs of the 4Fe-4S complexes) are also shown.

(b) Schematic diagram. The dinitrogenase reductase component dissociates from the nitrogenase component before N_2 is converted into NH_4^+.

Nitrogenase Complex

The conversion of nitrogen to ammonia is a reduction reaction which is exergonic in nature:

$$N_2 + 3H_2 \longrightarrow 2NH_3 \quad \Delta G^{0\prime} = -33.5 \text{ kJ/mol}$$

The N ≡ N triple bond, which has a bond energy of 942 kJ/mol, is highly resistant to chemical attack. Indeed, Lavoisier named it "azote", meaning "without life", because it is quite unreactive. The industrial process for nitrogen fixation, devised by Fritz Haber in 1910 and currently used in fertilizer factories, is typically carried out over an iron catalyst at about 500°C and a pressure of 300 atm of N_2 and H_2 to provide the necessary activation energy. Biological nitrogen fixation must occur at 0.8 atm of N_2, and the high activation barrier is partly overcome by the binding and hydrolysis of ATP. The stoichiometry of the overall process of nitrogen fixation can be written as:

$$N_2 + 10\ H^+ + 8\ e^- + 16\ ATP \longrightarrow 2\ NH_4^+ + 16\ ADP + 16\ Pi + H_2$$

Biological fixation of nitrogen is carried out by a highly conserved complex of metal-containing proteins called **nitrogenase complex** (Fig. 27–5.). The nitrogenase complex comprises 2 protein components: *component I* or dinitrogenase and component *II* or dinitrogenase reductase; neither of the two components is active in the absence of the other and their sizes vary with the microbial source. **Dinitrogenase** or **Mo-Fe protein** (MW ≈ 2,10,000 to 2,40,000) is a tetramer with 4 identical peptide chains, each with an Mo-Fe coenzyme of unknown structure and an Fe_4–S_4 group. Its redox centres have a total of 2 Mo, 32 Fe and 30 S per tetramer. **Dinitrogenase reductase** or **Fe protein** (MW ≈ 55,000 to 60,000) is a dimer with 2 identical chains, each having an Fe_4 – S_4 group and can be oxidized and reduced by one electron. It also has two binding sites for ATP. The nitrogenase complex has one or two units of component II for each component I. The reductase component provides electrons with high reducing power and the nitrogenase component uses these electrons to reduce N_2 to NH_4^+.

Nitrogen fixation is carried out by a highly reduced form of dinitrogenase, and it needs 8 electrons : 6 for the reduction of N_2 and 2 to produce one mole of H_2 as an obligate part of the reaction mechanism (Fig. 27–6). Dinitrogenase is reduced by the transfer of electrons from dinitrogenase reductase. Dinitrogenase has 2 binding sites for the reductase and the required 8 electrons are transferred to dinitrogenase one at a time, with the reduced reductase binding and the oxidized reductase dissociating from dinitrogenase in a cycle. This cycle requires the hydrolysis of ATP by the reductase. The immediate source of electrons to reduce dinitrogenase reductase varies, although in at least one instance, the ultimate source of electrons is pyruvate, as shown in the figure.

Two important characteristics of the nitrogenase complex are:

(*a*) The *ATP seems to play a catalytic role, rather than thermodynamic*. It may be recalled that ATP, besides contributing chemical energy through the hydrolysis of one or more of its phosphodiester bonds, can also contribute binding energy through noncovalent interactions that can be used to lower the activation energy. In the reaction carried out by dinitrogenase reductase, both ATP binding and ATP hydrolysis bring about protein conformational changes that help overcome the high activation energy of nitrogen fixation.

(*b*) Another characteristic of nitrogenase complex is that it is *extremely labile when oxygen is present*. The reductase is inactivated in air, with a half-life of 30 s. The dinitrogenase has a half-life of 10 min in air. Free-living bacteria, that fix nitrogen, avoid or solve this problem by diverse means.

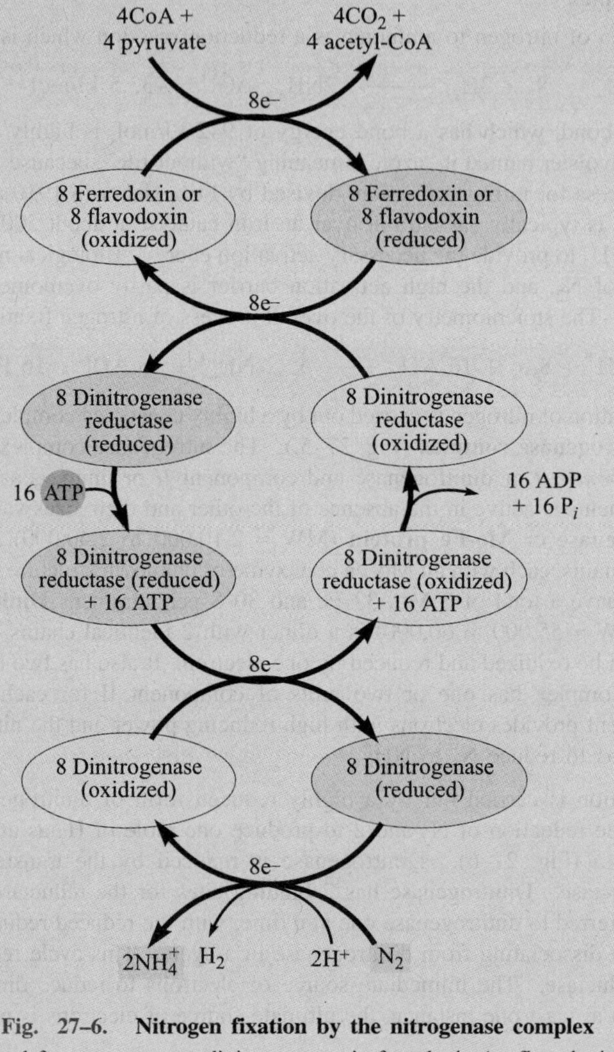

Fig. 27–6. Nitrogen fixation by the nitrogenase complex

Electrons are transferred from pyruvate to dinitrogenase via ferredoxin (or flavodoxin) and dinitrogenase reductase. Dinitrogenase is reduced one electron at a time by dinitrogenase reductase, and must be reduced by at least 6 electrons to fix one mole of N_2. An additional 2 electrons (thus a total of 8) are used to reduce $2H^+$ to H_2 in a process that accompanies nitrogen fixation in anaerobes.

Both components of the nitrogenase complex are irreversibly poisoned by oxygen. The apparent function of the leghemoglobins (= legume hemoglobins) supplied to the nodules of *Rhizobium* in the legumes, is to bind O_2 and maintain pO_2 below 0.001 mm Hg so that it cannot interfere with nitrogen fixation. In the cyanophycean members (blue-green algae), the walls of the heterocysts separate the nitrogenase from the oxygen produced by photosynthesis.

The activity of nitrogenase complex is controlled in 2 ways: one is a *coarse control*, in which enzyme synthesis is repressed by excess of ammonia; the other is a *fine control*, in which the activity of the nitrogenase is regulated by ADP. Should the amount of ADP increase to about twice that of ATP, further utilization of ATP by nitrogenase is completely inhibited.

It may, thus, be inferred from the foregoing discussion that the **basic features** of biological nitrogen fixation are (Fig. 27–7):

1. nitrogenase, the cardinal enzyme complex in the process,
2. a strong reductant, such as reduced ferredoxin or flavodoxin,
3. ATP,
4. a system for regulation of the rate of ammonia production and one for assimilation, since biosynthesis of the nitrogenase complex ceases when ammonia accumulates, and
5. protection of the nitrogen fixation system from molecular oxygen, which inactivates nitrogenase and competes for reductant (in aerobic bacteria).

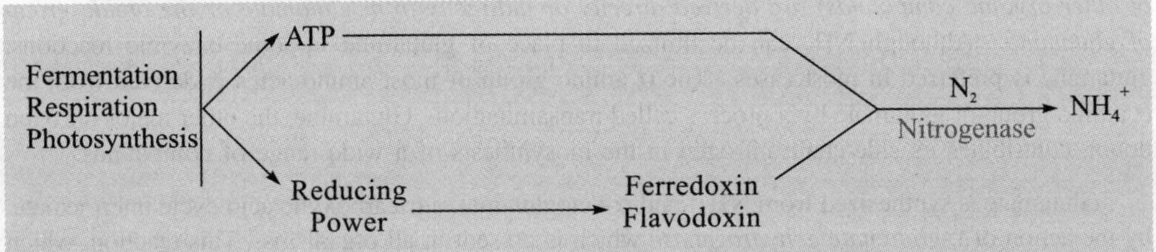

Fig. 27–7. Overall representation of nitrogen fixation

Reduction of Nitrate and Nitrite

Whereas nitrogen-fixing plants can obtain their nitrogen requirements from N_2, nonfixing plants acquire most of their nitrogen from nitrate (NO_3^-), the most abundant form of nitrogen. Most anaerobic and facultative aerobes (those that can also grow anaerobically) and the algae are capable of converting NO_3^- into biological nitrogen. This utilization of NO_3^- entails its reduction to ammonia, which is then incorporated into nitrogenous biomolecules. Reduction of nitrate to ammonia in diverse microorganisms and most higher plants is accomplished by the catalytic action of two enzymes, *nitrate reductase and nitrite reductase*.

Nitrate reductase. This enzyme catalyzes the two-electron reduction of nitrate to nitrite, with NADH (bacterial and plant cells) or NADPH (fungal cells) as electron donor. Nitrate reductase (MW = 8,00,000) is a tetrameric aggregation of 4 dimers, each possessing one protomer of MW ≈ 1,50,000 and another of MW ≈ 50,000. The mechanism of nitrate reduction is analogous to that of nitrogen fixation in that nitrate reductase also contains an electron pathway with one FAD, one Mo atom and many Fe_4–S_4 centres per basic dimer, serving as electron carriers. The flow of electrons derived from NADH in plants, that results in reduction of NO_3^- to NO_2^-, appears to proceed as follows:

$$\text{NADH} \xrightarrow{e} \text{FAD} \xrightarrow{e} Fe_4\text{–}S_4 \xrightarrow{e} \text{Mo} \xrightarrow{e} NO_3^- \longrightarrow NO_2^-$$

The overall reaction catalyzed is:

$$NO_3^- + \text{NADH} + H^+ \longrightarrow NO_2^- + \text{NAD}^+ + H_2O$$

Molybdenum undergoes cyclic changes from Mo^{5+} to Mo^{4+} during reduction of nitrate. The nitrate reductase of plants is induced by nitrate and is believed to be repressed by ammonia.

Nitrite reductase. This enzyme catalyzes the six-electron reduction of nitrite to ammonia, with NADH (bacterial and plant cells) or NADPH (fungal cells) as reductant. The nitrite reductase of plants has an Fe_2–S_2 centre and siroheme, the immediate reductant of nitrogen, as carriers of the electrons furnished to the enzyme by reduced ferredoxin. The siroheme iron is the binding site for NO_2^-; no intermediates dissociate NH_3 appears. Some fungal nitrite reductases also possess an FAD or FMN as an electron carrier. Presumably, the electron pathway is similar to that in sulfite reductase. The overall reaction for the reduction of NO_2^- to NH_4^+ is:

> **Siroheme** is an iron tetrahydroporphyrin of the isobacteriochlorin type, *viz.*, two adjoining pyrrole rings are fully reduced, and each pyrrole bears one propionic and one acidic side chain. Hence, it is a derivative of uroporphyrin.

$$NO_2^- + 6e^- + 8H^+ \longrightarrow NH_4^+ + 2H_2O$$

FIXATION OF AMMONIA INTO AMINO ACIDS

Ammonia, of whatever origin, can be combined into organic linkage by 3 major reactions that occur in all organisms. These reactions result in the formation of glutamate, glutamine and carbamoyl phosphate. Utilization of the nitrogen of carbamoyl phosphate is limited to two pathways, one contributing a single nitrogen atom in the synthesis of pyrimidines and the other donating one nitrogen atom to the synthesis of arginine. *Essentially all other nitrogen atoms of amino acids (or of other organic compounds) are derived directly or indirectly from glutamate or the amide group of glutamine.* Although NH_3 can be utilized in place of glutamine in some enzymic reactions, glutamine is preferred in most cases. The α amino group of most amino acids is derived from the α amino group of glutamate by a process called transamination. Glutamine, the other major nitrogen donor, contributes its side-chain nitrogen in the biosynthesis of a wide range of compounds.

Glutamate is synthesized from $NH_4{}^+$ and α-ketoglutarate, a tricarboxylic acid cycle intermediate, by the action of L-*glutamate dehydrogenase* which is present in all organisms. This reaction, which is termed reductive amination, has already been dealt with in the catabolism of amino acids. NADPH acts as a reductant in glutamate biosynthesis, whereas acts as an oxidant in its degradation.

$$\alpha\text{-ketoglutarate} + NH_4^+ + NADPH \rightleftharpoons \text{L-glutamate} + NADP^+ + H_2O$$

Glutamate dehydrogenase from *Escherichia coli* (MW = 50,000) is a hexamer of 6 identical subunits. In eukaryotic cells, l-glutamate dehydrogenase is located in the mitochondrial matrix. The equilibrium for the reaction favours reactants and the K_m for NH_4^+ (~ 1 mM) is so high that a modest NH_4^+ assimilation takes place. Soil bacteria and plants rarely have sufficiently high concentrations of NH_4^+ and as such not enough glutamate is formed and these organisms have to generally rely on the two-enzyme pathway, described below.

The pathway for the conversion (or assimilation) of NH_4^+ into glutamate requires 2 reactions. First, glutamate and NH_4^+ react to produce glutamine by the action of *glutamine synthetase*, which is found in all organisms.

$$\text{Glutamate} + NH_4^+ + ATP \longrightarrow \text{Glutamine} + ADP + Pi + H^+$$

In fact, this is a two-step reaction, with enzyme-bound γ-glutamyl phosphate as an intermediate.

$$\text{Glutamate} + ATP \rightleftharpoons \gamma\text{-glutamyl phosphate} + ADP$$

$$\gamma\text{-glutamyl phosphate} + NH_4{}^+ \rightleftharpoons \text{Glutamine} + Pi + H^+$$

Besides its importance for NH_4^+ assimilation in bacteria, this is a central reaction in amino acid metabolism. It is the main pathway for converting toxic free ammonia into the nontoxic glutamine for transport in the blood.

Glutamine synthetase (Fig 27–8), (MW from that of *E. coli* = 6,00,000), consists of 12 identical subunits and is inactivated by the transfer from ATP of a 5′-adenylyl group to form a phosphodiester linkage with the phenolic hydroxyl group of a specific tyrosine residue of each subunit. At least 6 end products of glutamine metabolism plus alanine and glycine are allosteric inhibitors of the enzyme (Fig. 27–9) and each subunit (MW = 50,000) has binding sites for all 8 inhibitors as well as an active site for catalysis. Each inhibitor alone gives only partial inhibition. The effects of the different inhibitors, however, are more than additive, and all 8 together virtually shut down the enzyme. Thus, *glutamine synthetase is a primary regulatory point in nitrogen metabolism.*

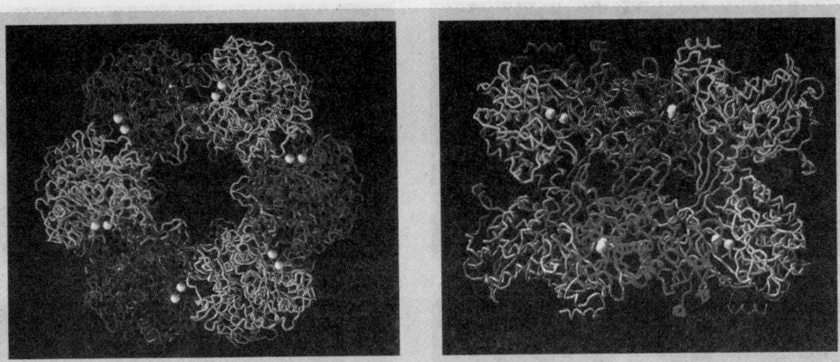

(*a*) Top view (*b*) Side view

Fig. 27–8. **The structure of glutamine synthetase from *Salmonella typhimurium***
The enzyme consists of twelve identical subunits arranged in two rings of six subunits, here represented by their C_α backbones, arranged with D_6 symmetry (the symmetry of a hexagonal prism).

(a) Top view of the enzyme down the sixfold axis of symmetry. The top ring of monomers are alternately colored light and dark *blue* and the bottom ring of monomers light and dark *red*. The subunits or monomers of the bottom ring are roughly directly below those of the top ring. The protein, including its side chains (not shown), has a diameter of 143 Å. The active sites of each monomer shown are marked by pairs of Mn^{2+} ions (*white* spheres).

(b) Side view of the enzyme along one of the twofold axes of symmetry showing only the six nearest subunits. The molecule extends 103 Å along the sixfold axis which is vertical in this view.

(*Courtesy: Michael Pique/Scripps Research Institute*)

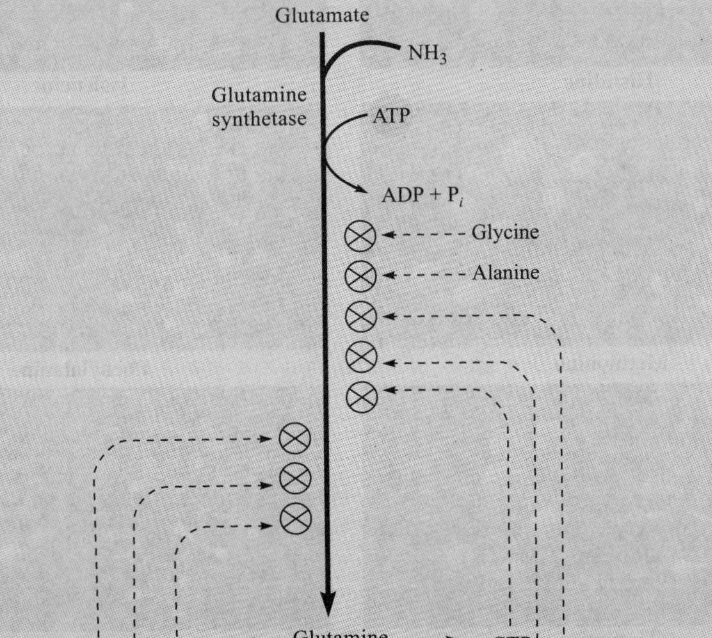

Fig. 27–9. **Cumulative allosteric inhibition of glutamine synthetase by 6 end products of glutamine metabolism**

Alanine and glycine probably serve as indicators of the general status of cellular amino acid metabolism.

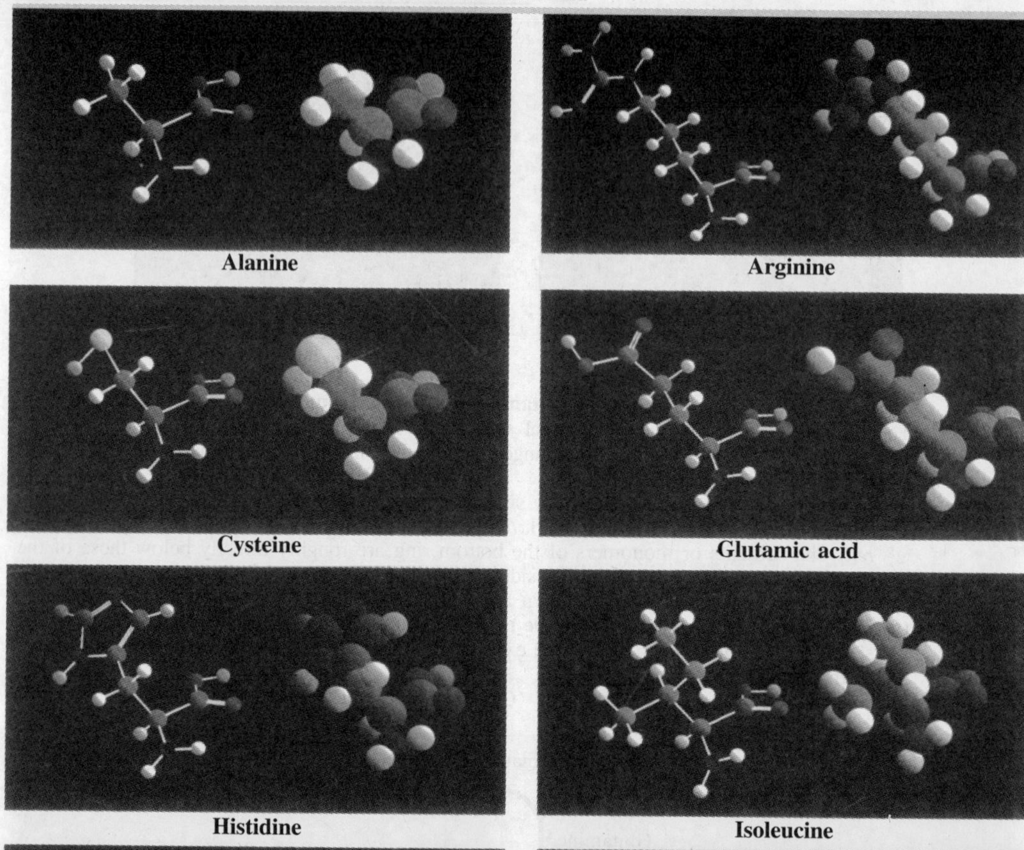

Fig. 27–10. (Cont'd.)

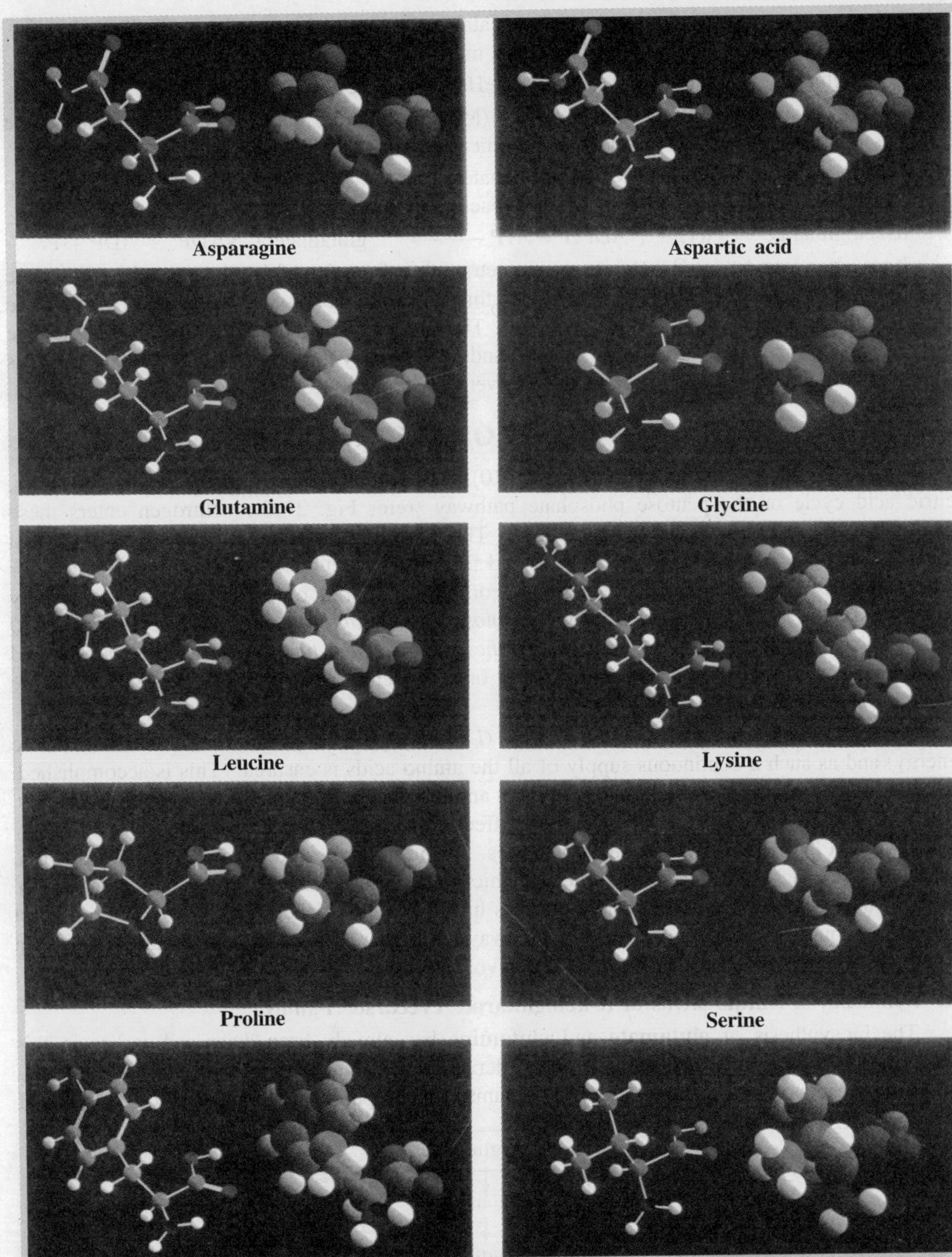

Fig. 27–10. The structure of the 20 protein amino acids

(For each amino acid, diagram on the *left* represents the ball-and-stick model and the diagram on the *right* represents the space-filling model.)

The various proteins found in the living beings are made of these 20 amino acids which are synthesized by discrete pathways, utilizing intermediates of glycolysis, citric acid cycle and pentose phosphate pathway.

In bacteria, an enzyme *glutamate synthase* catalyzes the reductive amination of α-ketoglutarate, using glutamine as nitrogen donor. Thus, two moles of glutamate are produced.

$$\alpha\text{-ketoglutarate} + \text{Glutamine} + \text{NADPH} + \text{H}^+ \longrightarrow 2\ \text{Glutamate} + \text{NADP}^+$$

Glutamate synthase obtained from *E. coli* (MW = 8,00,000) has two types of subunits: one contains nonheme iron, FAD and FMN; the other subunit binds NADPH.

When NH_4^+ is limiting, most of the glutamate is made by the sequential action of glutamine synthetase and glutamate synthase. The net reaction of these two enzymes in bacteria is:

$$\alpha\text{-ketoglutarate} + NH_4^+ + \text{NADPH} + \text{ATP} \longrightarrow \text{L-glutamate} + \text{NADP}^+ + \text{ADP} + \text{P}i$$

It may, however, be noted that this stoichiometry differs from that of the glutamate dehydrogenase reaction in that here an ATP is hydrolyzed. At this stage, a querry may arise as to why this more expensive pathway sometimes is used by *E. coli*? The answer lies in the fact that the K_m of glutamate dehydrogenase for NH_4^+ is high (~ 1 mM), and so this enzyme is not saturated when NH_4^+ is limiting. In contrast, glutamine synthetase has very high affinity for NH_4^+.

BIOSYNTHESIS OF AMINO ACIDS

All the 20 protein amino acids (Fig. 27–10) are derived from intermediates in glycolysis, citric acid cycle or the pentose phosphate pathway (refer Fig. 27–3). Nitrogen enters these pathways by way of glutamate and glutamine. The pathways for 10 amino acids are simple and are only one or a few enzymatic steps removed from their precursors, whereas the pathways for others (such as aromatic amino acids) are more complex. Different organisms have varied capacity to synthesize these 20 amino acids. *Whereas most bacteria and plants can synthesize all the 20 amino acids, mammals including man can synthesize only about half of them.* These are termed as nonessential amino acids and the remaining ones, which must be obtained from food, as the essential amino acids (refer Table 27–1).

Most pathways are essentially irreversible (*i.e.*, they proceed with a substantial loss of free energy) and as such a continuous supply of all the amino acids is ensured. This is accomplished, in general, by reactions in which ATP is utilized and in effect hydrolyzed to ADP + P*i*. The cases, where ATP splits into AMP + PP*i*, are more effective because the pyrophosphate is irreversibly hydrolyzed (PP*i* ⟶ 2 P*i*). In other instances, synthesis is ensured by a reductive amination, usually employing a pyridine nucleotide, in which equilibrium strongly favours such reduction. Most of the pathways for amino acid syntheses have been established mainly for bacteria, yeast and other moulds. It is believed that the pathways in higher plants and animals are similar, but comparatively less is known of the enzymes involved.

A. Syntheses of Amino Acids of α-ketoglutarate Precursor Family

The biosynthesis of **glutamate** and **glutamine** has already been discussed earlier in this chapter. The synthesis of **proline**, a cyclized derivative of glutamate, is depicted in Fig. 27–11. In the first reaction, the γ-carboxyl group of glutamate is phosphorylated using ATP to form an acyl

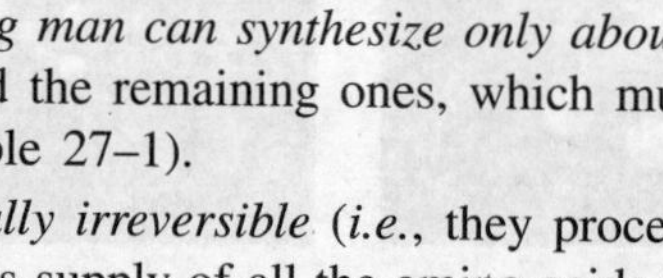

phosphate, which is then reduced by NADPH to form glutamate γ-semialdehyde. This intermediate ultimately undergoes cyclization and further reduction to form proline.

Arginine is synthesized from glutamate *via* ornithine and the urea cycle (refer page 770). Ornithine could also be synthesized from glutamate γ-semialdehyde by transamination but cyclization of the semialdehyde in the proline pathway is a rapid spontaneous reaction so that only a little

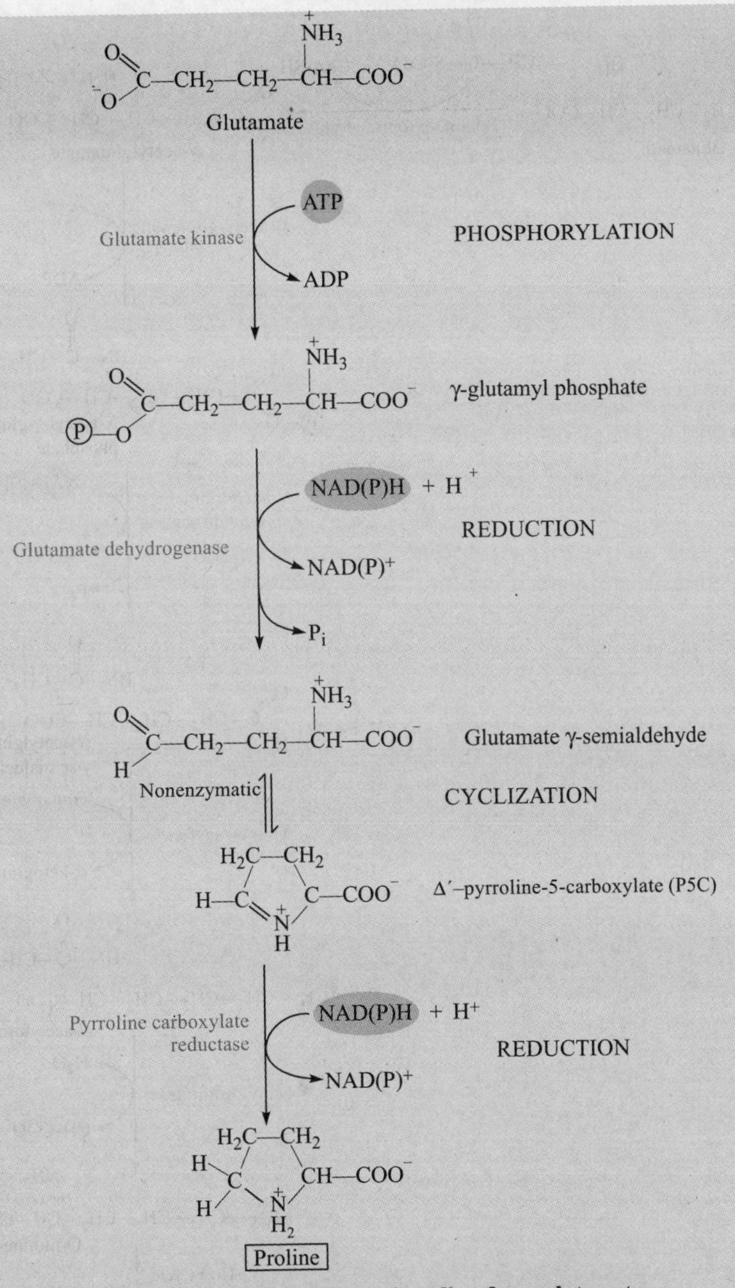

Fig. 27–11. Biosynthesis of proline from glutamate

Note that all 5 carbon atoms of proline arise from glutamate. In many organisms, glutamate dehydrogenase is unusual in that it uses either NADH or NADPH as a cofactor. The same may be true of other enzymes in these pathways. The nonenzymatic cyclization of glutamate γ-semialdehyde is so rapid that the γ-semialdehyde cannot give rise to ornithine *via* transamination.

amount of this intermediate is left for ornithine synthesis. The biosynthetic pathway for ornithine therefore parallels some steps of the proline pathway but includes 2 additional steps to chemically block the amino group of glutamate γ-semialdehyde and prevent cyclization (Fig. 27–12). To begin with, the α-amino group of glutamate is blocked by acetylation using

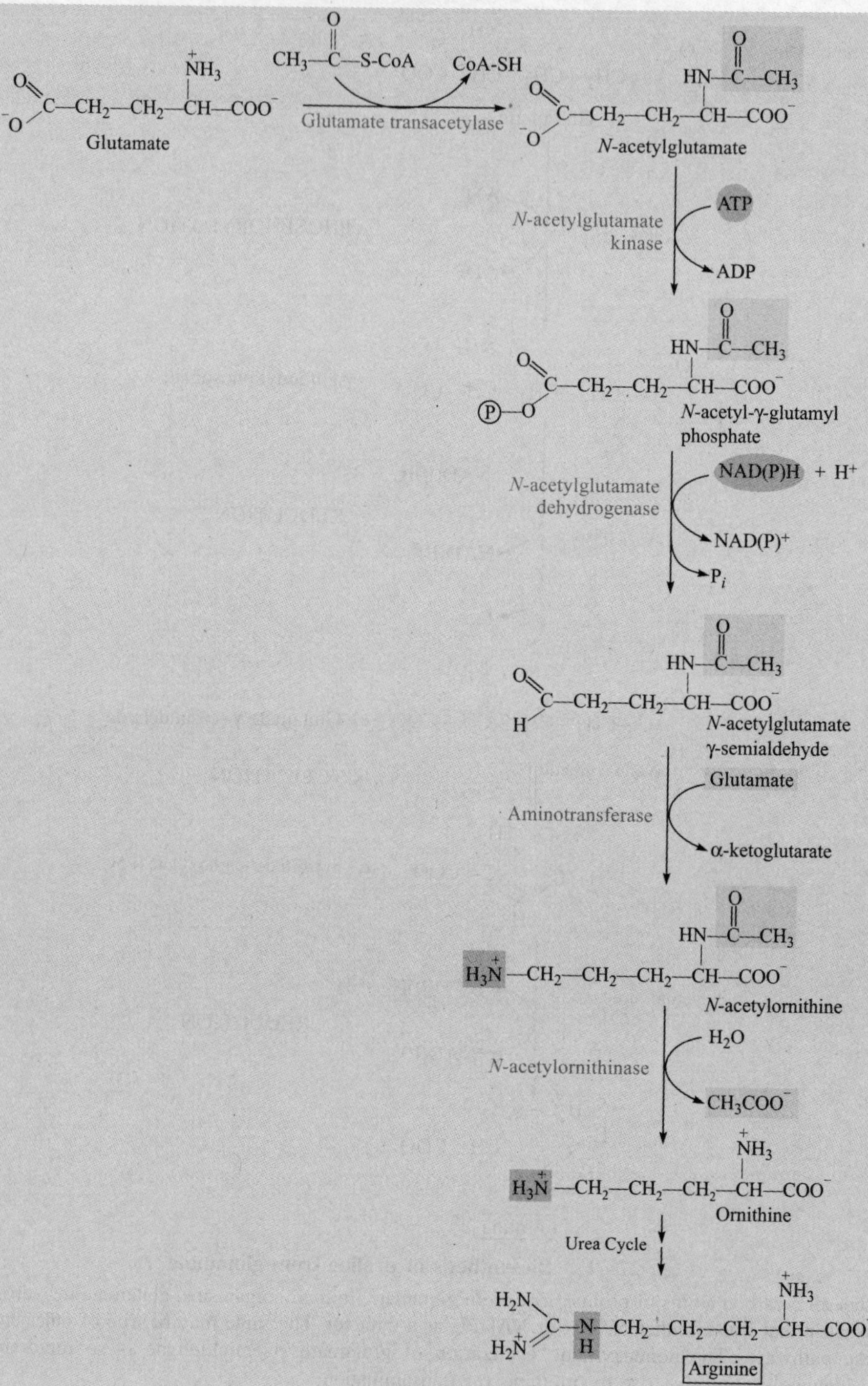

Fig. 27.12. Biosynthesis of arginine from glutamate

Note that, in contrast to proline pathway, cyclization is averted in ornithine/arginine pathway by acetylation of the α-amino group of glutamate in the first step and removal of the acetyl group after the transamination. Arginine is synthesized from ornithine *via* the urea cycle, as shown in Fig. 26–10.

acetyl-CoA, and after the transamination step the acetyl group is removed to yield ornithine. Most of the arginine formed in mammals is cleaved to form urea so that the available arginine is depleted. This makes arginine an essential amino acid in young animals that need higher amounts of amino acids for growth.

B. Syntheses of Amino Acids of 3-phosphoglycerate Precursor Family

Fig. 27–13 outlines the serine pathway. In the first step, the hydroxyl group of 3-phosphoglycerate is oxidized by NAD^+ to produce an 3-phosphohydroxypyruvate. This is the

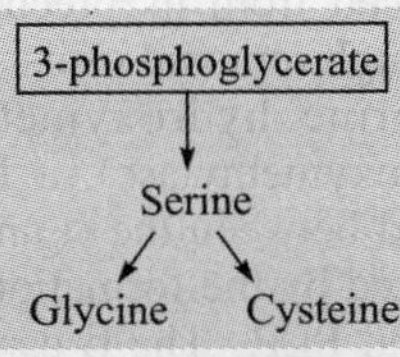

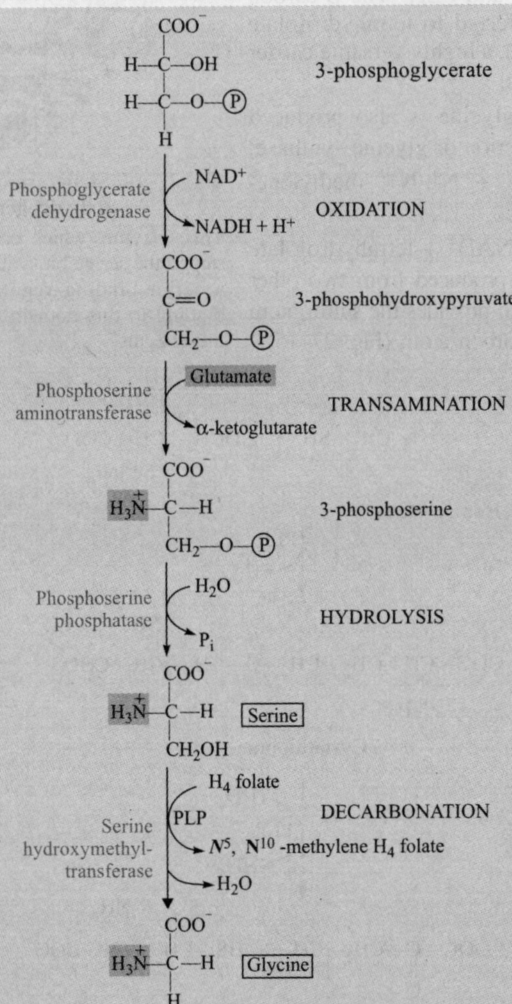

Fig. 27–13. Biosynthesis of serine from 3-phosphoglycerate and the subsequent conversion of serine into glycine

Glycine is also made from CO_2 and NH_4^+ by the action of glycine synthase, with N^5,N^{10}-methylene-tetrahydrofolate as methyl group donor.

committed step in the serine biosynthetic pathway and is catalyzed by the enzyme **3-phosphoglycerate dehydrogenase** (Fig. 27–14). Transamination of 3-phosphohydroxypyruvate (an α-keto acid) from glutamate produces 3-phosphoserine, which upon hydrolysis by phosphoserine phosphatase yields free serine.

Serine is the precursor of glycine and cysteine. The 2-carbon **glycine** is produced from its precursor 3-carbon amino acid serine through removal of its side chain β carbon atom by serine hydroxymethyl transferase (= serine transhydroxymethylase), a PLP enzyme (Fig. 27–15). The bond between the α and β carbon atoms of serine is labilized by the formation of a Schiff base between serine and PLP. The β carbon atom of serine is then transferred to tetrahydrofolate (= tetrahydropteroylglutamate), a highly versatile carrier of activated one-carbon units.

In the vertebrate liver, glycine is also produced from CO_2 and NH_4^+ by the action of glycine synthase:

$$CO_2 + NH_4^+ + NADH + H^+ + N^5, N^{10}\text{-methylene-tetrahydrofolate} \rightleftharpoons \text{Glycine} + NAD^+ + \text{tetrahydrofolate}$$

In mammals, **cysteine** is produced from two other amino acids, namely Met which provides the sulfur atom and Ser which furnishes the carbon atom (Fig. 27–15).

Fig. 27–14. Structure of 3-phosphoglycerate dehydrogenase

This enzyme, which catalyzes the committed step in the serine biosynthetic pathway, includes a serine-binding regulatory domain. Serine binding to this domain reduces the activity of the enzyme.

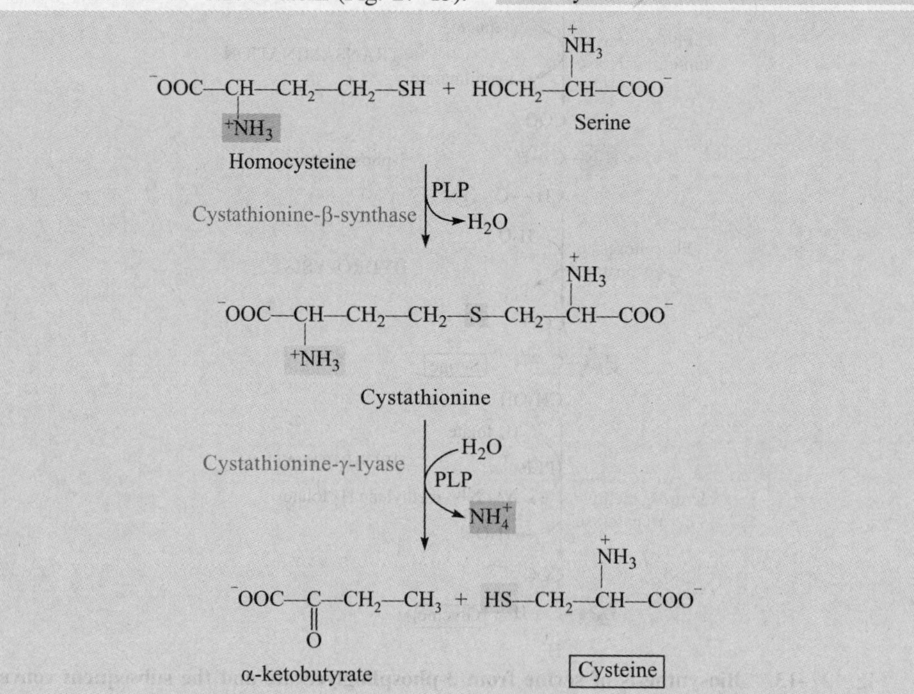

Fig. 27–15. Biosynthesis of cysteine from homocysteine and serine in mammals

The homocysteine is formed from methionine.

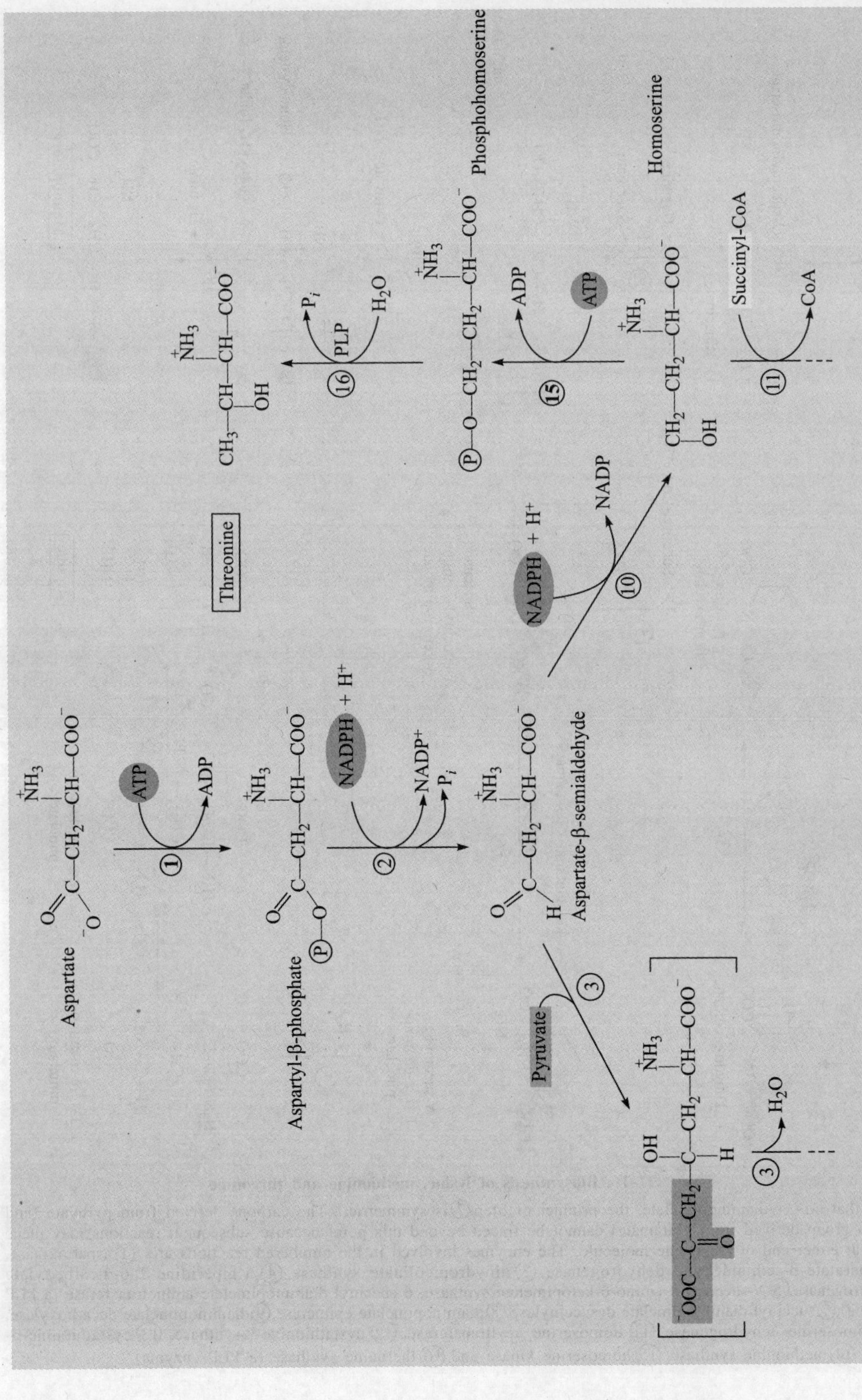

Fig. 27–16. (Cont'd)

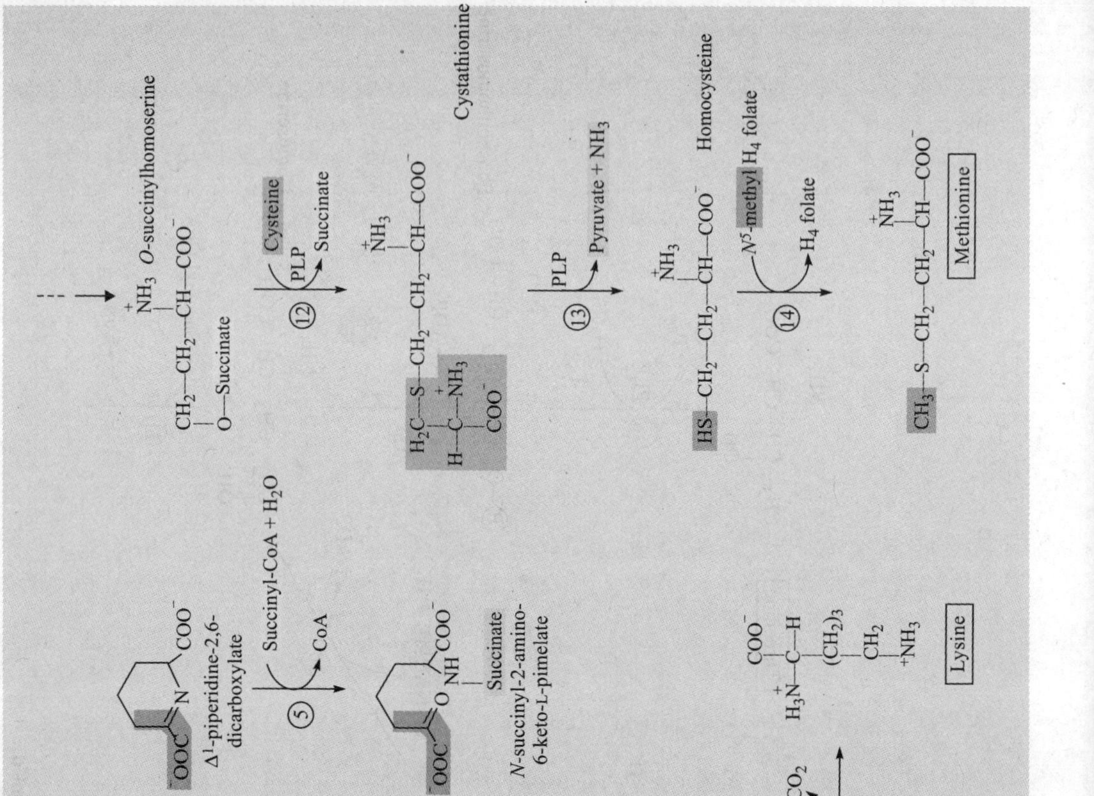

Fig. 27–16. **Biosynthesis of lysine, methionine and threonine**

Note that L-L-, ε-diaminopimelate, the product of Step ⑦ is symmetric. The carbons derived from pyruvate (and amino group derived from glutamate) cannot be traced beyond this point because subsequent reactions may place them at either end of the lysine molecule. The enzymes involved in the numbered reactions are: ① aspartokinase, ② asparatate-β-semialdehyde dehydrogenase, ③ dihydropicolinate synthase, ④ Δ^1piperidine-2, 6-dicarboxylate dehydrogenase, ⑤ *N*-succinyl-2-amino-6-ketopimelate synthase, 6 succinyl diaminopimelate aminotransferase (a PLP enzyme), ⑦ succinyl diaminopimelate desuccinylase, ⑧ diaminopimelate epimerase, ⑨ diaminopimelate decarboxylase, ⑩ homoserine dehydrogenase, ⑪ homoserine acyltransferase, ⑫ cystathionine-γ-synthase, ⑬ cystathionine-β-lyase, ⑭ methionine synthase, ⑮ homoserine kinase and ⑯ theronine synthase (a PLP enzyme).

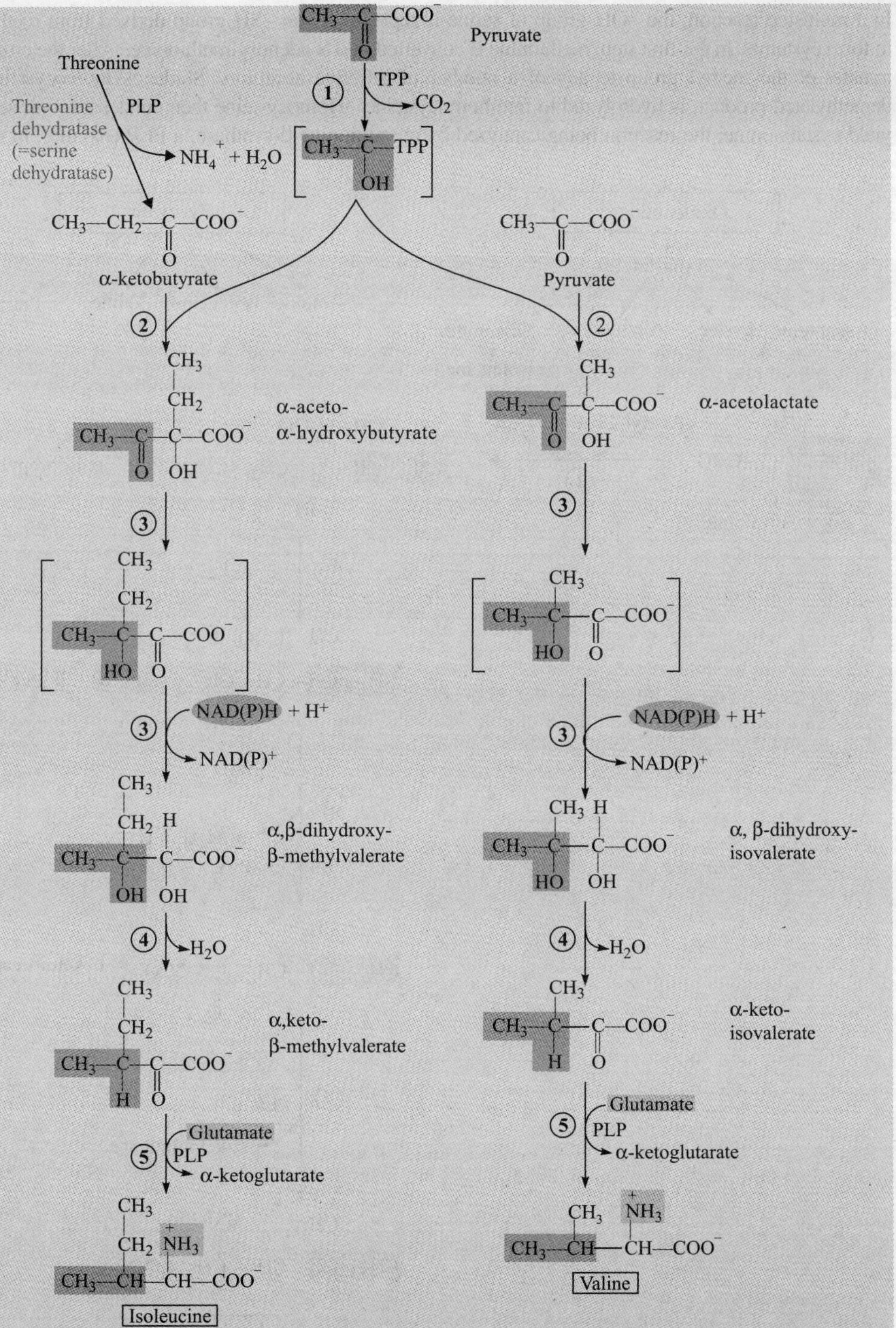

Fig. 27–17. Biosynthesis of isoleucine and valine

Note that pyruvate is the metabolic precursor for both isoleucine and valine pathways. The pathway enzymes for the numbered reactions are: ① and ② acetoacetate synthase (a TPP enzyme), ③ acetohydroxy acid isomeroreductase, ④dihydroxy acid dehydratase and ⑤valine aminotransferase (a PLP enzyme)

In a multistep reaction, the –OH group of serine is replaced by an –SH group derived from methionine to form cysteine. In the first step, methionine is converted into S-adenosylmethionine. After the enzymatic transfer of the methyl group to any of a number of different acceptors, S-adenosylhomocysteine, the demethylated product, is hydrolyzed to free homocysteine. Homocysteine then condenses with serine to yield cystathionine; the reaction being catalyzed by cystathionine-β-synthase, a PLP enzyme. In the last

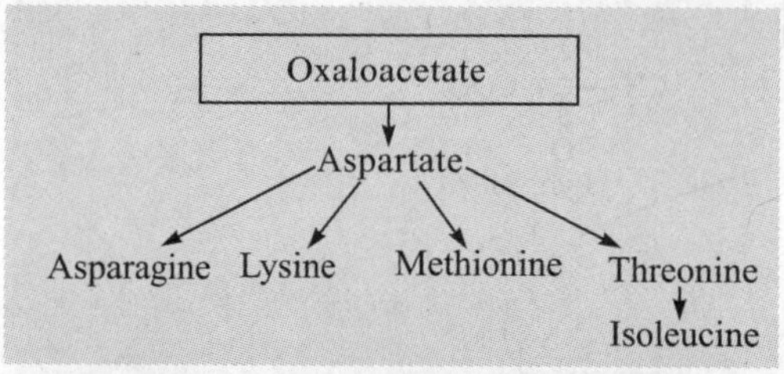

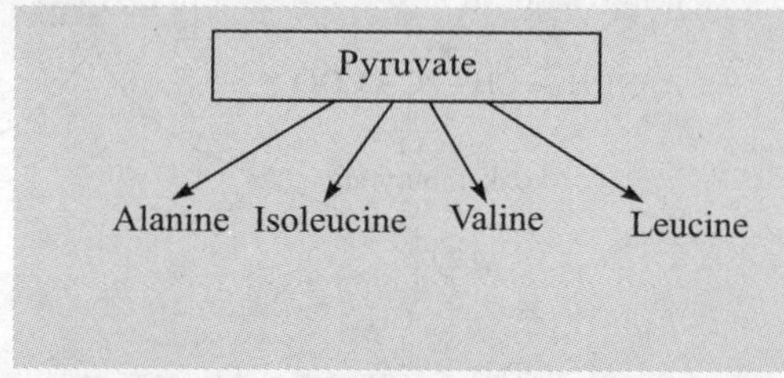

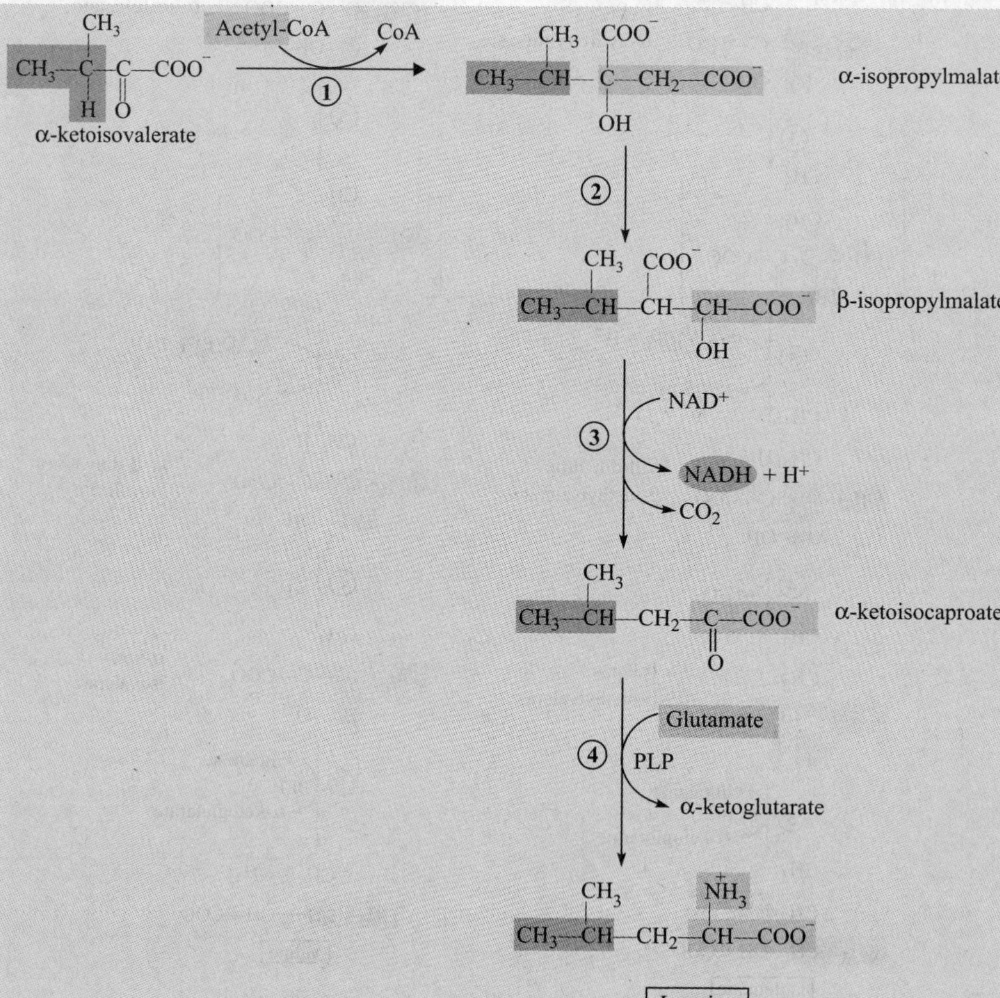

Fig. 27–18. Biosynthesis of leucine

Note that α-ketoisovalerate, an intermediate in valine pathway, is the starting point for leucine synthesis. The enzymes involved in the numbered reactions are: ①α-isopropylmalate synthase,② isopropylmalate β-isopropylmalate isomerase,③ dehydrogenase, and ④ leucine aminotransferase (a PLP enzyme).

step, cystathionine is then deaminated and cleaved to cysteine plus α-ketobutyrate, by cystathionine-γ-lyase (= cystathioninase), another PLP enzyme. *Note that the sulfur atom of cysteine is derived from homocysteine, whereas the carbon skeleton comes from serine.*

C. Syntheses of Amino Acids of Oxaloacetate and Pyruvate Precursor Families

Alanine and **aspartate** are synthesized from pyruvate and oxaloacetate respectively, by transamination from glutamate. **Asparagine** is then synthesized by amidation of aspartate; the NH_4^+ being donated by glutamine. These are all nonessential amino acids and their simple biosynthesis occurs in all organisms. The biosynthetic pathways for the 6 of the essential amino acids, *viz.*, methionine, threonine, lysine, isoleucine, valine and leucine, are complex and interlinked. In some cases, there are marked differences in the pathways found in bacteria, fungi and higher plants. The bacterial pathways have been outlined in Figs. 27–16, 27–17 and 27–18.

Aspartate gives rise to **lysine, methionine** and **threonine**. Branch points occur at aspartate β-semialdehyde, an intermediate in all three pathways and at homoserine, a precursor of methionine and threonine. Threonine, in turn, is one of the precursors of isoleucine.

The **isoleucine** and **valine** pathways (Fig. 27–17) have 4 common enzymes. These two pathways begin with the condensation of 2 carbons of pyruvate with either another mole of pyruvate (*valine path*) or with α-ketobutyrate (*isoleucine path*). The α-ketobutyrate is derived from threonine in a reaction that requires pyridoxal phosphate (PLP).

The α-ketoisovalerate, an intermediate in the valine pathway, is the starting point for a 4-step branch pathway which leads to the production of **leucine** (Fig. 27–18).

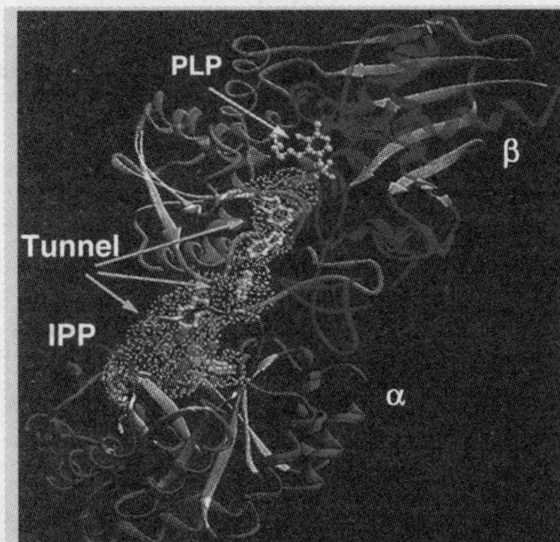

Fig. 27–19. A ribbon diagram of the bifunctional enzyme tryptophan synthase from *Salmonella typhimurium*

Only one αβ unit of this twofold symmetric αββα heterotetramer is shown. The 268-residue α subunit is *blue*, the 397-residue β subunit's N-terminal domain is *yellow*, its C-terminal domain is *red*, and all β sheets are *tan*. The active site of the α subunit is located by its bound competitive inhibitor, **indolpropanol phosphate** (IPP; *red* ball-and-stick model), whereas that of the β subunit is marked by its PLP coenzyme (*yellow* ball-and-stick model). The solvent-accessible surface of the ~25-Å-long "tunnel" connecting the active sites of the α and β subunits is outlined by *yellow* dots. Several indole molecules (*green* ball-and-stick models) have been modeled into the tunnel in head-to-tail fashion, thereby demonstrating that the tunnel has sufficient width to permit the indole product of the α subunit to pass through the tunnel to the β subunit's active site.

[*Courtesy : Craig Hyde, National Institutes of Health*]

D. Synthesis of Amino Acids of Phosphoenolpyruvate–erythrose-4-phosphate Precursor Family

The synthesis of tryptophan, tyrosine and phenylalanine takes place by pathways that share a number of early steps. The *first 4 steps* lead to the production of shikimate whose 7 carbon atoms are derived from phosphoenolpyruvate and erythrose-4-phosphate. Shikimate is then converted to chorismate through *3 more steps* that include the addition of 3 more carbon atoms from another molecule of phosphoenolpyruvate (Fig. 27–20). Chorismate is the first branch point, with one branch leading to tryptophan and the other to tyrosine and phenylalanine through prephenate, the other branch point.

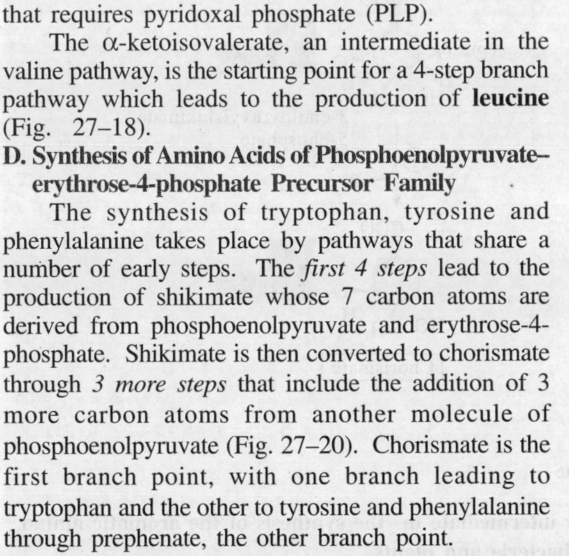

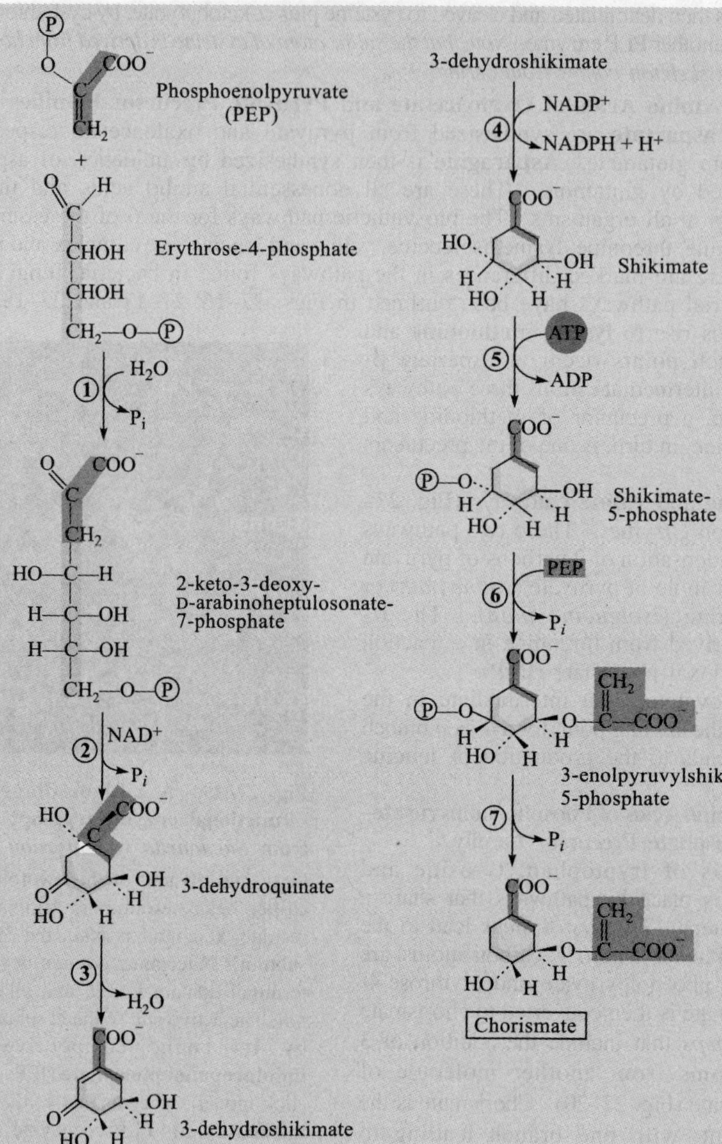

Fig. 27–20. Synthesis of chorismate, a key intermediate in the synthesis of the aromatic amino acids in bacteria and plants

Note that all carbons are derived from either erythrose-4-phosphate or phosphoenolpyruvate. The pathway enzymes for the numbered reactions are: ① 2-keto-3-deoxy-D-arabinoheptulosonate-7-phosphate synthase, ② dehydroquinate synthase, ③ 3-dehydroquinate dehydratase, ④ shikimate dehydrogenase, ⑤ shikimate kinase, ⑥ 5-enolpyruvylshikimate-3-phosphate synthase, and ⑦ chorismate synthase. Note that the NAD^+ required as a cofactor in Step ② is released unchanged; it may be transiently reduced to NADH during the reaction, with the formation of an oxidized reaction intermediate. Step ⑥ is competitively inhibited by glyphosate ($^-OOC-CH_2-NH-CH_2-PO_3^{2-}$), the active ingredient in the widely-used herbicide *Roundup*. The herbicide is relatively nontoxic to mammals, which lack this biosynthetic pathway. The chemical names quinate, shikimate, and chorismate are derived from the name of plants in which these intermediates have been found to accumulate.

Tryptophan is synthesized from chorismate in a 5-step process (Fig. 27–21). Chorismate acquires an amino group from the side chain of glutamine and releases pyruvate to form anthranilate.

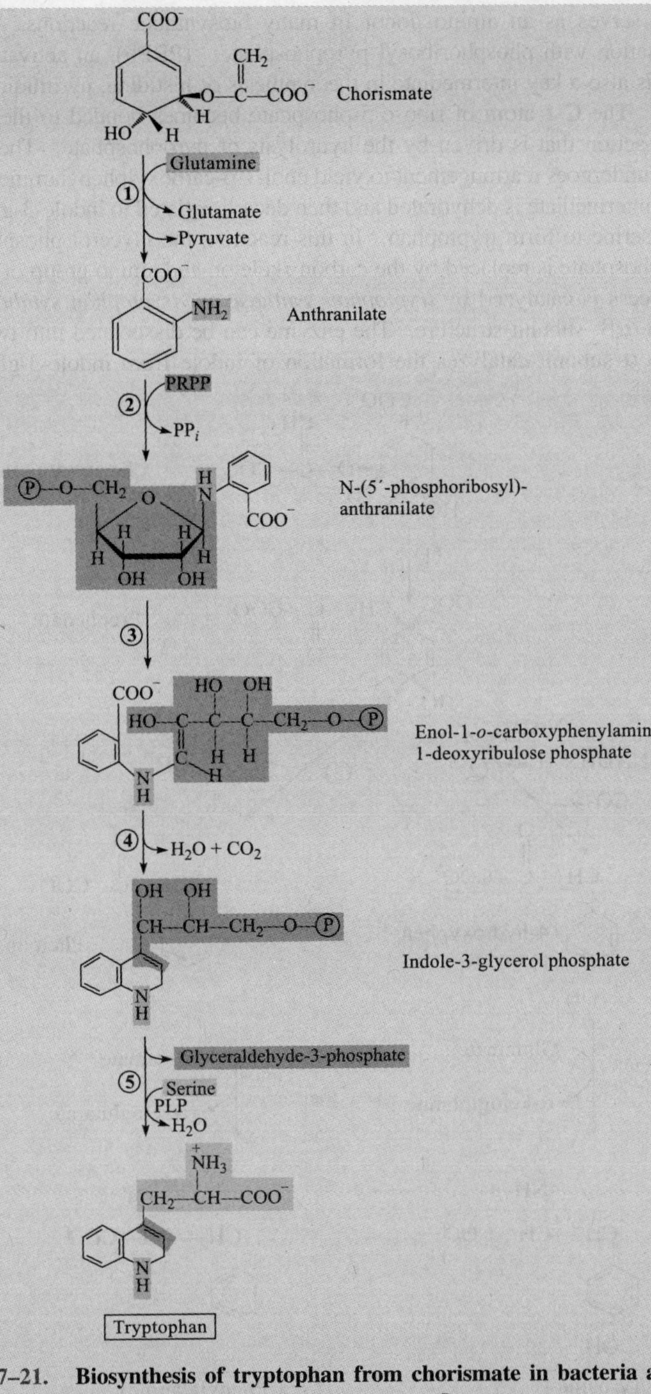

Fig. 27–21. Biosynthesis of tryptophan from chorismate in bacteria and plants

The pathway enzymes for the numbered reactions are: ① anthranilate synthase, ② anthranilate phosphoribosyltransferase, ③ *N*-(5'-phosphoribosyl)- anthranilate isomerase, ④ indole-3-glycerol phosphate synthase, and ⑤ tryptophan synthase. In *Escherichia coli*, the two enzymes, anthranilate synthase and anthranilate phosphoribosyltransferase, are subunits of a single complex.

In fact, glutamine serves as an amino donor in many biosynthetic reactions. Anthranilate then undergoes condensation with phosphoribosyl pyrophosphate (PRPP), an activated form of ribose phosphate. PRPP is also a key intermediate in the synthesis of histidine, pyrimidine nucleotides and purine nucleotides. The C-1 atom of ribose 5-phosphate becomes bonded to the nitrogen atom of anthranilate in a reaction that is driven by the hydrolysis of pyrophosphate. The ribose moiety of ribosylanthranilate undergoes rearrangement to yield enol-1-o-carboxylphenylamino-1-deoxyribulose-5-phosphate. This intermediate is dehydrated and then decarboxylated to indole-3-glycerol phosphate, which reacts with serine to form tryptophan. In this reaction, the glycerol phosphate side chain of indole-3-glycerol phosphate is replaced by the carbon skeleton and amino group of serine. This final reaction in the process is catalyzed by *tryptophan synthase* or *tryptophan synthetase* (Fig 27–19), an enzyme with an $\alpha_2\beta_2$ subunit structure. The enzyme can be dissociated into two α subunits and a β_2 subunit. The α subunit catalyzes the formation of indole from indole-3-glycerol phosphate,

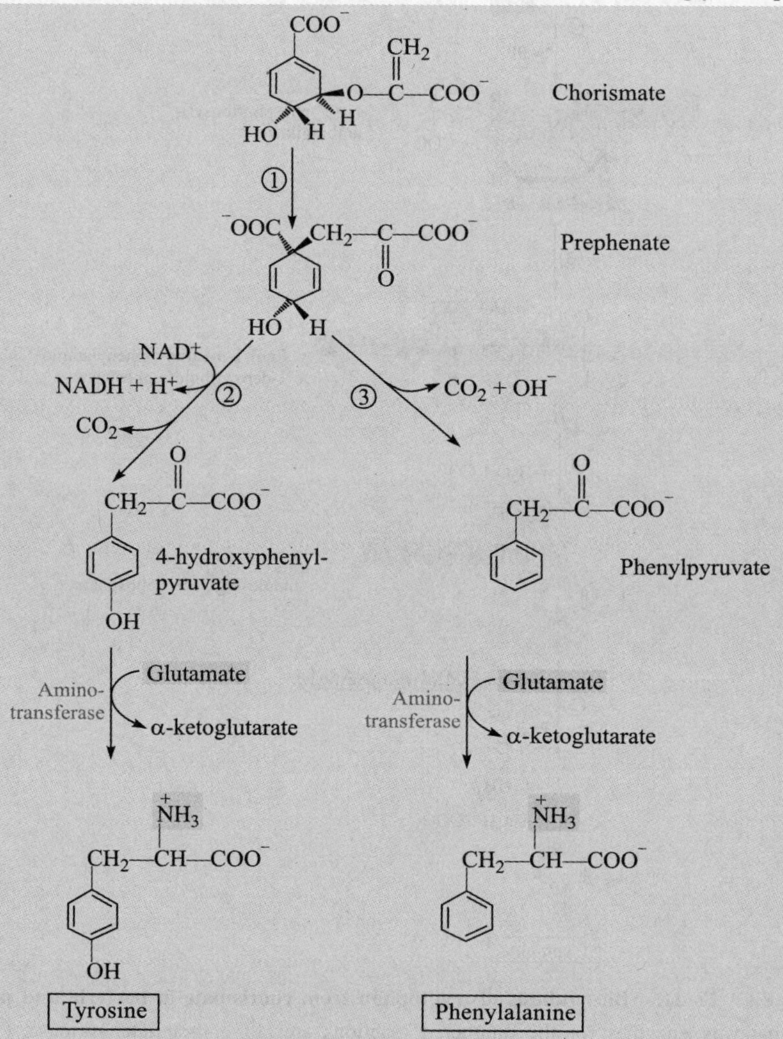

Fig. 27–22. Biosynthesis of tyrosine and phenylalanine from chorismate in bacteria and plants. The enzymes inovled in the numbered reactions are: ① chorismate mutase, ② prephenate dehydrogenase, and ③ prephenate dehydratase. Conversion of chorismate to prephenate is a rare biological example of a *Claisen rearrangement*.

whereas the β_2 subunit *catalyzes* the condensation of indole and serine to form tryptophan.

$$\text{Indole-3-glycerol phosphate} \xrightarrow{\alpha\text{-subunit}} \text{Indole + Glyceraldehyde-3-phosphate}$$

$$\text{Indole + Serine} \xrightarrow{\beta_2\text{-subunit}} \text{Tryptophan + Water}$$

The second reaction requires a pyridoxal phosphate as cofactor. Indole is rapidly channeled from the α-subunit active site to the β-subunit active site, where it undergoes condensation with a Schiff base intermediate, derived from serine and PALP.

Tyrosine and **phenylalanine** are synthesized from chorismate in plants and microorganisms *via* simpler pathways. A mutase converts chorismate into prephenate, the immediate precursor of the aromatic ring of tyrosine and phenylalanine (Fig. 27–22). Prephenate is oxidatively decarboxylated to *p*-hydroxyphenylpyruvate. Alternatively, dehydration followed by decarboxylation yields phenylpyruvate. These α-keto acids are then transaminated, with glutamate as amino group donor, to form tyrosine and phenylalanine, respectively.

Tyrosine can also be made by animals directly from phenylalanine *via* hydroxylation at C-4 of the phenyl group by **phenyl hydroxylase**, which also participates in the degradation of phenylalanine Tyrosine is considered a nonessential amino acid only because it can be synthesized from the essential amino acid phenylalanine.

E. Synthesis of Amino Acid of Ribose-5-phosphate Precursor Family

The biosynthetic pathway for histidine in all plants and bacteria is unique in many respects:

1. Histidine is derived from 3 precursors (Fig. 27–23):
   (*a*) PRPP contributes 5 carbon atoms
   (*b*) the purine ring of ATP contributes a nitrogen and a carbon
   (*c*) glutamine contributes the second ring nitrogen.
2. The key steps of histidine biosynthesis are:
   (*a*) the condensation of ATP and PRPP (step ①)
   (*b*) purine ring opening that ultimately leaves N-1 and C-2 linked to the ribose (Step ②)
   (*c*) formation of the imidazole ring in a reaction during which glutamine donates a nitrogen (Step ③).
3. The use of ATP as a metabolite rather than a high-energy cofactor is unusual, but not wasteful because it dovetails with the purine biosynthetic pathway.
4. The remnant of ATP that is released after the transfer of N-1 and C-2 is 5-amino- imidazole-4-carboxamide ribonucleotide, an intermediate in the biosynthesis of purines that can rapidly be recycled to ATP.

Ribose-5-phosphate

↓

Histidine

REGULATION OF AMINO ACID BIOSYNTHESIS

The amino acid biosynthesis is regulated by feedback inhibition. The rate of synthesis depends mainly on the *amount* and *activity* of the biosynthetic enzymes.

The first irreversible reaction in a biosynthetic pathway, called the committed step, is usually an important regulatory site. In a linear (*i.e.*, unbranched) biosynthetic pathway, the final product (Z) often inhibits the enzyme that catalyzes the committed step (A ⟶ B). This kind of control is essential for the conservation of building blocks and metabolic energy.

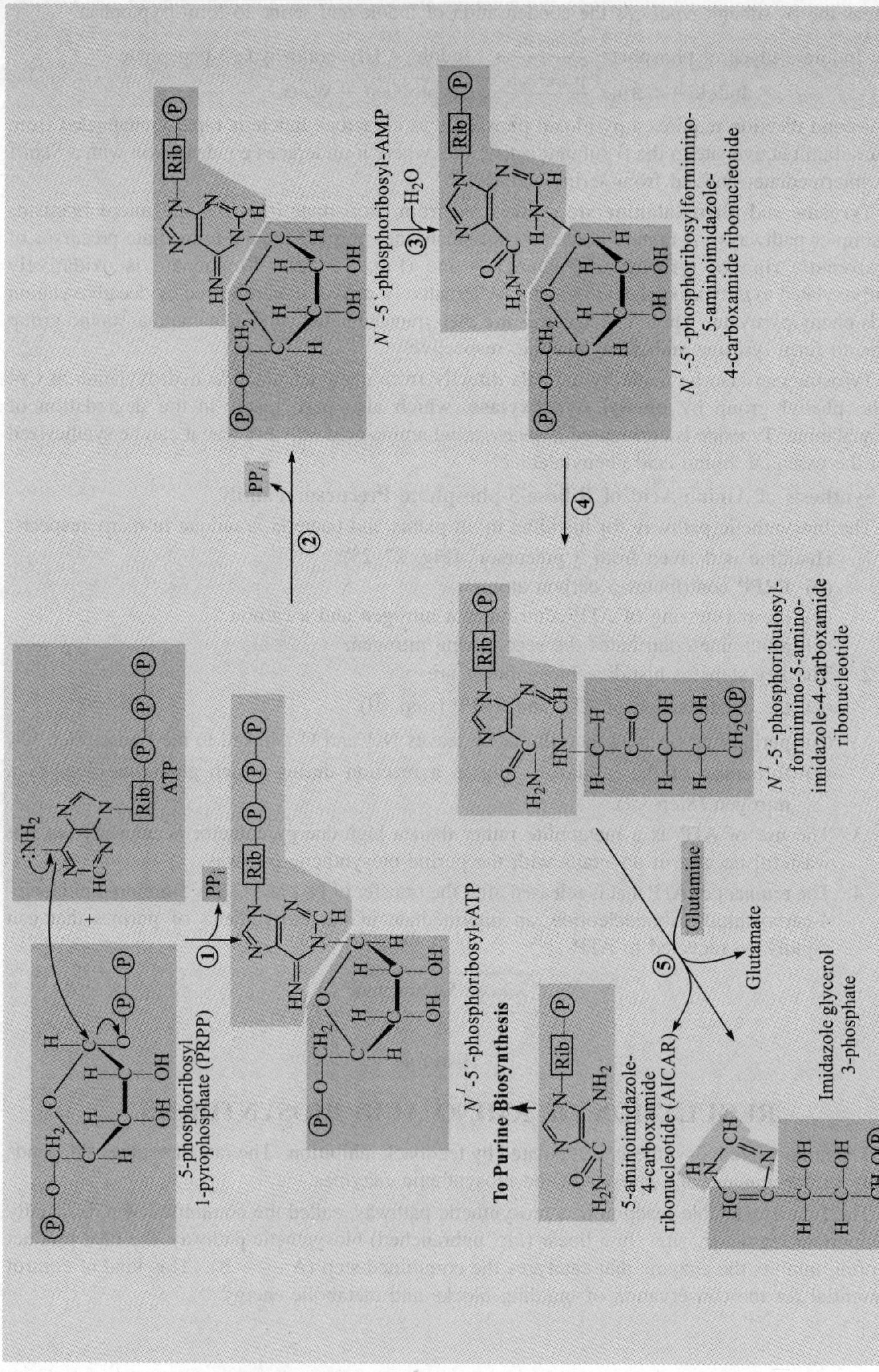

Fig. 27–23. (Cont'd)

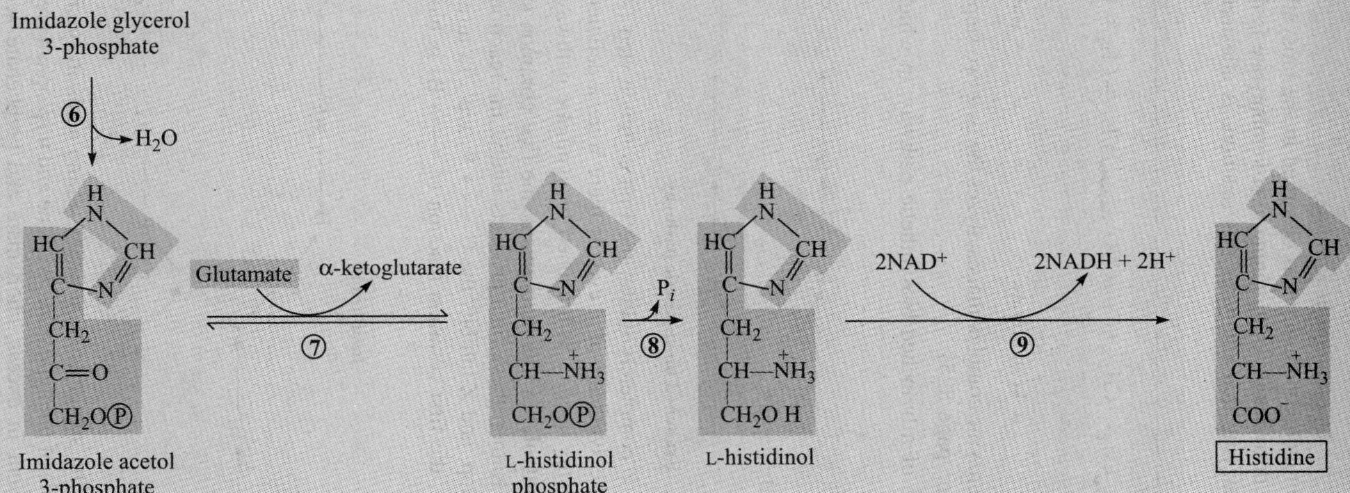

Fig. 27–23. Biosynthesis of histidine in bacteria and plants

Atoms derived from PRPP and ATP are shaded *red* and *blue*, respectively. Two of the histidine nitrogens are derived from glutamine and glutamate (*green*). The pathway enzymes for the numbered reactions are: ① ATP phosphoribosyl transferase, ② pyrophosphohydrolase, ③ phosphoribosyl-AMP cyclohydrolase, ④ phosphoribosylformimino-5-aminoimidazole-4-carboxamide ribonucleotide isomerase, ⑤ glutamine amidotransferase, ⑥ imidazole glycerol 3-phosphate dehydratase, ⑦ L-histidinol phosphate aminotransferase, ⑧ histidinol phosphate phosphatase, and ⑨ histidinol dehydrogenase. Note that the derivative of ATP remaining after Step ⑤ (AICAR) is an intermediate in purine biosynthesis, so ATP is rapidly regenerated.

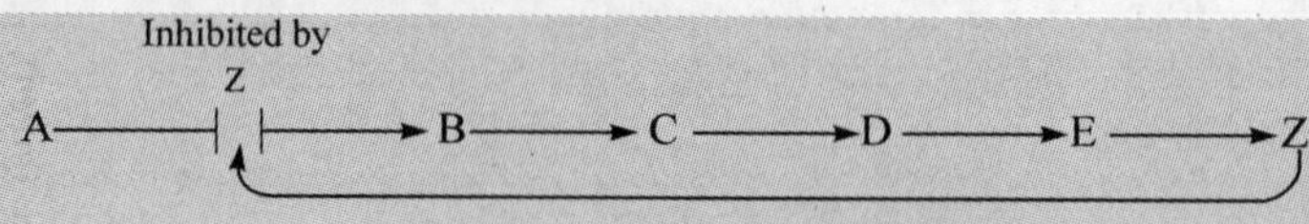

Linear biosynthetic pathway

The first example of this unbranched pathway was observed in the biosynthesis of isoleucine in *E. Coli.* The dehydration and deamination of threonine to α-ketobutyrate is the committed step. *Threonine deaminase*, the PLP enzyme that catalyzes this reaction, is allosterically inhibited by isoleucine:

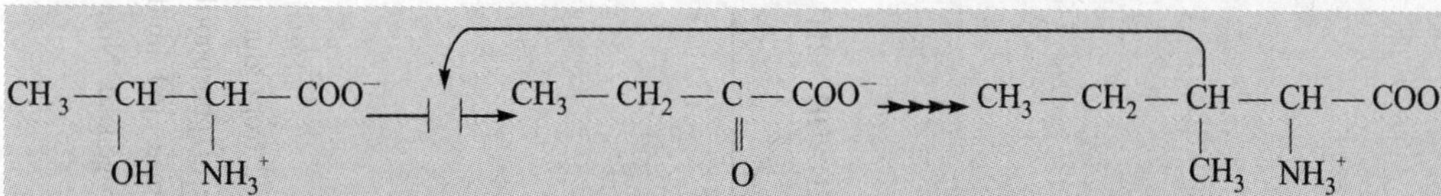

Threonine **α-ketobutyrate** **Isoleucine**

Likewise, tryptophan inhibits the enzyme complex that catalyzes the first two steps in the conversion of chorismate into tryptophan (see page 825).

Take a hypothetical example of a branched biosynthetic pathway, in which *Y* and *Z* are the final products:

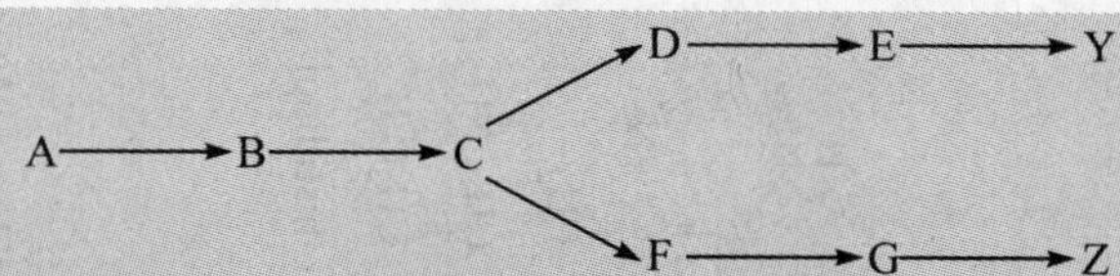

Branched biosynthetic pathway

Suppose that high levels of *Y* or *Z* completely inhibit the first common step (A ⟶ B). Then, high levels of *Y* would prevent the synthesis of *Z* even if there were a deficiency of *Z*. In fact, several complex control mechanisms operate in branched biosynthetic pathways:

1. Sequential feedback control. In this mechanism, the first common step (A ⟶ B) is not inhibited directly by *Y* or *Z*. Rather, these final products inhibit the reactions after the branch point: *Y* inhibits the C ⟶ D step, and *Z* inhibits the C ⟶ F step. In turn, high levels of C inhibit the A ⟶ B step. Thus, the first common reaction (A ⟶ B) is blocked only if both *Y* and *Z* are present in excess.

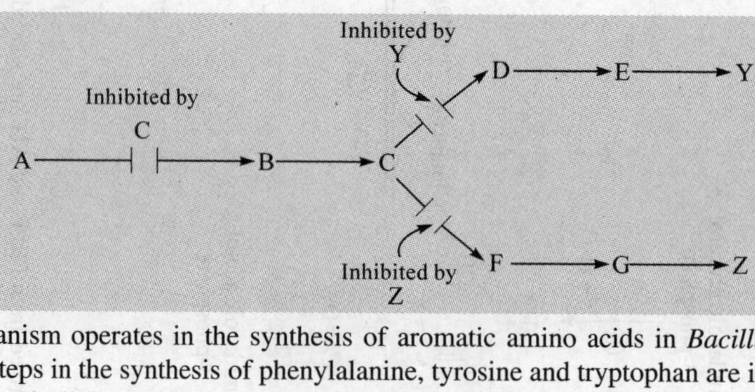

This mechanism operates in the synthesis of aromatic amino acids in *Bacillus subtilis*. The first divergent steps in the synthesis of phenylalanine, tyrosine and tryptophan are inhibited by their final products. If all 3 are present in excess, chorismate and prephenate accumulate. These branch-point intermediates, in turn, inhibit the first common step in the overall pathway, *i.e.*, the condensation of phosphoenolpyruvate and erythrose-4-phosphate.

2. Differential multiple enzyme control (= Enzyme multiplicity). Here, the first common step (A ⟶ B) is catalyzed by either of 2 different enzymes. One of them is directly inhibited by *Y*, and the other by Z. A high level of either *Y* or *Z* partially blocks the first step. In fact, both *Y* and *Z* must be present in sufficient quantities to prevent the conversion of *A* into *B* completely. In the rest of this control mechanism, like in sequential feedback control, Y inhibits the C ⟶ D step and Z inhibits the C ⟶ F step.

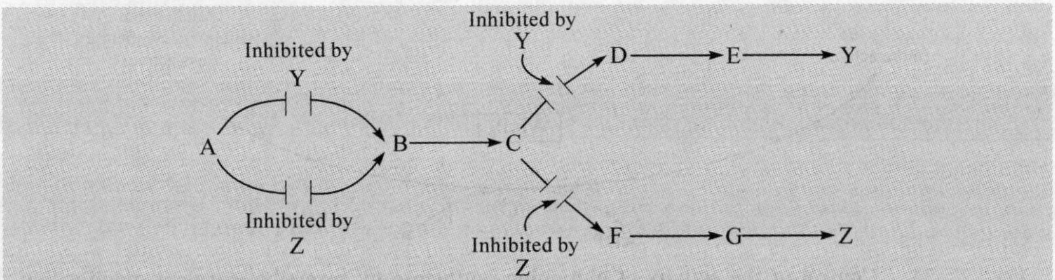

Enzyme multiplicity mechanism operates in a variety of biosynthetic pathways in microbes. In *E. coli*, the condensation of phosphoenolpyruvate and erythrose-4-phosphate is catalyzed by 3 different enzymes. One is inhibited by phenylalanine, another by tyrosine, and the third by tryptophan. Furthermore, there are 2 different mutases that convert chorismate into prephenate. One of them is inhibited by phenylalanine, and the other by tyrosine.

3. Concerted feedback control. Here, the first common step (A ⟶ B) is inhibited if high quantities of both *Y* and *Z* are simultaneously present. In contrast with enzyme multiplicity mechanism, a high level of either *Y* or *Z* alone does not appreciably inhibit the A ⟶ B step. Like the two preceding control schemes, here also *Y* inhibits the C ⟶ D step and *Z* inhibits C ⟶ F step.

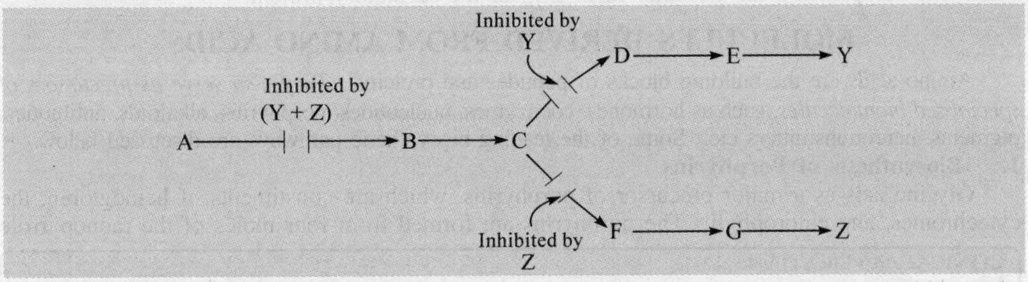

An example of the concerted feedback control is the inhibition of bacterial aspartyl kinase by threonine and lysine, the two final products.

4. Cumulative feedback control. In this mechanism, the first common step (A ⟶ B) is partially inhibited by each of the final products, which act independently of the other. Suppose that a high level of *Y* decreased the rate of the A ⟶ B step from 100 to 70 s^{-1} and that *Z* alone decreased the rate from 100 to 30 s^{-1}. Then, the rate of the A ⟶ B step in the presence of high levels of *Y* and *Z* would be ($0.7 \times 0.3 \times 100\ s^{-1}$) or 21 s^{-1}.

An important example of cumulative feedback control is in the bacterial synthesis of glutamine from glutamate, NH_4^+ and ATP which is regulated by *glutamine synthetase*. **Earl Stadtman** showed that this enzyme controls the flow of nitrogen. The amide group of glutamine is a source of nitrogen in the biosynthesis of a variety of compounds such as tryptophan, histidine, glucose-6-phosphate, CTP and ATP. Glutamine synthetase is cumulatively inhibited by each of these final products of glutamine metabolism, as well as by alanine and glycine. The enzymic activity of glutamine synthetase is switched off almost completely when all final products are bound to the enzyme.

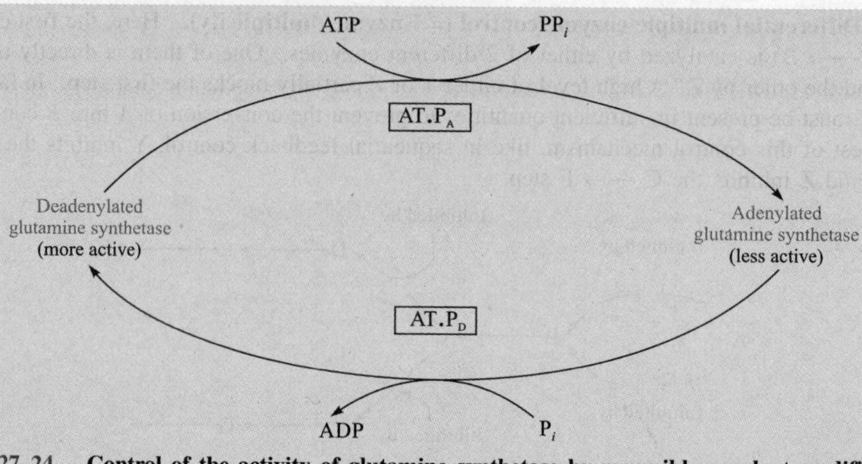

Fig. 27–24. Control of the activity of glutamine synthetase by reversible covalent modification

The activity of glutamine synthetase is also regulated by reversible covalent modification– the attachment of an AMP unit by a phosphodiester bond to the hydroxyl group of a specific tyrosine residue in each subunit. There are two forms of glutamine synthetase; the more active deadenylated form and the less active adenylated form. The *adenylated form is more susceptible to cumulative feedback inhibition than the deadenylated form.* The covalently-attached AMP unit can be removed from the adenylated enzyme by phosphorolysis. Adenylation and deadenylation both are catalyzed by the same enzyme, *adenylyl transferase* (AT). It has been observed that the specificity of adenylyl transferase is controlled by a regulatory protein (designated *P*), which can exist in 2 forms, P_A and P_D and (Fig. 27–24). Adenylation is catalyzed when the enzyme AT forms a complex with one form (P_A) of the regulatory protein. The same enzyme (AT) catalyzes deadenylation when it is complexed with the other form (P_D) of the regulatory protein.

MOLECULES DERIVED FROM AMINO ACIDS

Amino acids are the building blocks of peptides and proteins. *They also serve as precursors of specialized biomolecules*, such as hormones, coenzymes, nucleotides, porphyrins, alkaloids, antibiotics, pigments, neurotransmitters etc. Some of the leading biosynthetic pathways are discussed below.

1. Biosynthesis of Porphyrins

Glycine acts as a major precursor of porphyrins, which are constituents of hemoglobin, the cytochromes, and chlorophyll. The porphyrins are formed from four moles of the monopyrrole

CLINICAL IMPLICATIONS 27–1.

Porphyrias: Diseases That Result From Defective Heme Biosynthesis

Porphyrias are rare acquired disorders caused by a deficiency of an enzyme in the heme biosynthetic pathway. Porphyrias are classified as hepatic or erythropoietic, depending on the primary organ affected whether it is liver or erythrocyte precursors, respectively. Erythropoietic porphyrias account for about 85% of all porphyrias; hepatic porphyrias account for only 15%. *All types of porphyrias, except a few hepatic ones, are inherited.* Three porphyrias are known as acute porphyrias and can be a cause of emergency admissions with abdominal paint; they also cause neuropsychiatric symptoms. These are:

1. ***Acute intermittent porphyria*** is caused by the deficiency of hydroxymethylbilane synthase, an enzyme converting porphobilinogen (PBG) to a linear tetrapyrrole; in this disorder the concentrations of 5-aminolevulinate (5-ALA) and BPG increase in plasma and urine.
2. ***Hereditary coproporphyria*** is due to a defect in the conversion of coproporphyrinogen III to protoporphyrinogen III (coprooxidase).
3. ***Variegate porphyria*** is the third form of acute porphyria; its clinical manifestations are very similar to AIC.

Other porphyrias, such as **porphyria cutanea tarda,** present clinically as the sensitivity of skin to light (photosensitivity) which may cause disfiguration and scarring. Also, the pathway is inhibited by lead (Pb) at the stage of porphobilinogen synthase.

derivative, porphobilinogen (Fig. 27–25). In the first reaction, glycine reacts with succinyl-CoA to yield α-amino-β-ketoadipate, which is then decarboxylated to produce δ-aminolevulinate. This reaction is catalyzed by *δ-aminolevulinate synthase*, a PLP enzyme in mitochondria. Two moles of δ-aminolevulinate condense to form porphobilinogen, the next intermediate. This dehydration reaction is catalyzed by *α-aminolevulinate dehydrase*. Four moles of porphobilinogen come together to form protoporphyrin, through a series of complex enzymatic reactions. The iron atom is incorporated after the protoporphyrin has been assembled.

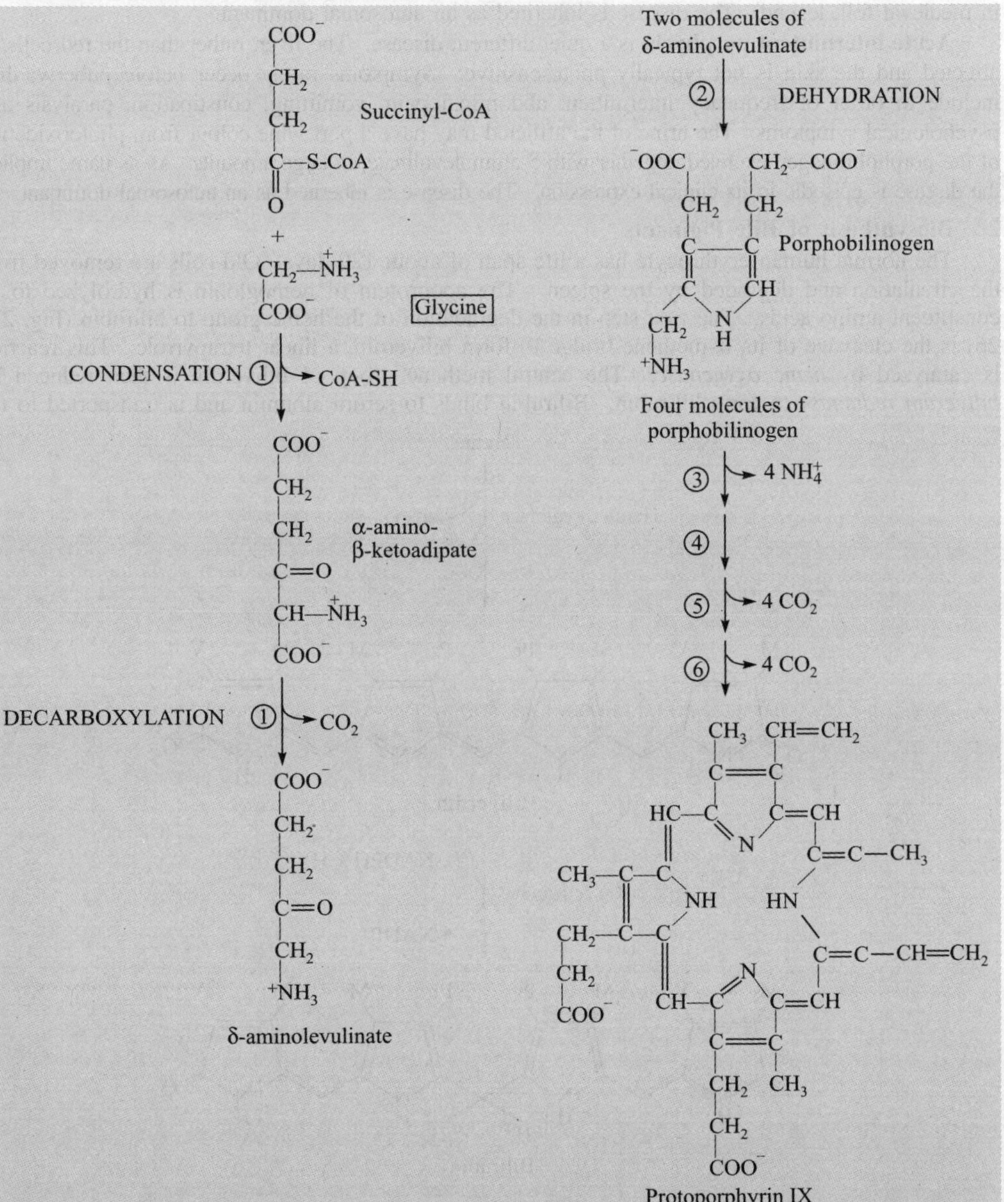

Fig. 27–25. Biosynthesis of protoporphyrin IX, the porphyrin of hemoglobin and myoglobin The atoms furnished by glycine are shown in *boldface*. The remaining carbon atoms are derived from the succinyl group of succinyl-CoA. The enzymes involved in the numbered reactions are: ① δ-aminolevulinate synthase, ② porphobilinogen synthase, ③ uroporphyrinogen synthase, ④ uroporphyrinogen III cosynthase ⑤ uroporphyrinogen decarboxylase, and ⑥ coproporphyrinogen oxidase.

In humans, genetic defects of certain enzymes of this pathway lead to the accumulation of specific porphyrin precursors in body fluids and in the liver. These genetic diseases are known as porphyrias. In **congenital erythropoietic porphyria**, which affects mainly erythrocytes, there is an accumulation of uroporphyrinogen I, the useless symmetric isomer of a precursor of protoporphyrin. It stains the urine red and causes the teeth to fluoresce strongly in UV light and the skin to be abnormally sensitive to light. Because insufficient heme is synthesized, patients with this disease are anemic, shy away from sunlight and have a propensity to drink blood. This condition may have given rise to the vampire myths in medieval folk legend. The disease is inherited as an autosomal dominant.

Acute intermittent porphyria is a quiet different disease. The liver, rather than the red cells, is affected and the skin is not typically photosensitive. Symptoms rarely occur before puberty; they include in order of frequency intermittent abdominal pain, vomitting, constipation, paralysis and psychological symptoms. The urine of the afflicted may have a port wine colour from photooxidation of the porphobilinogen excreted, together with 5-aminolevulinate, in large amounts. As its name implies, the disease is episodic in its clinical expression. The disease is inherited as an autosomal dominant.

2. Biosynthesis of Bile Pigments

The normal human erythrocyte has a life span of about 120 days. Old cells are removed from the circulation and degraded by the spleen. The apoprotein of hemoglobin is hydrolyzed to its constituent amino acids. The first step in the degradation of the heme group to bilirubin (Fig. 27–26) is the cleavage of its α-methene bridge to form biliverdin, a linear tetrapyrrole. This reaction is catalyzed by *heme oxygenase*. The central methene group of biliverdin is then reduced by *biliverdin reductase* to form bilirubin. Bilirubin binds to serum albumin and is transported to the

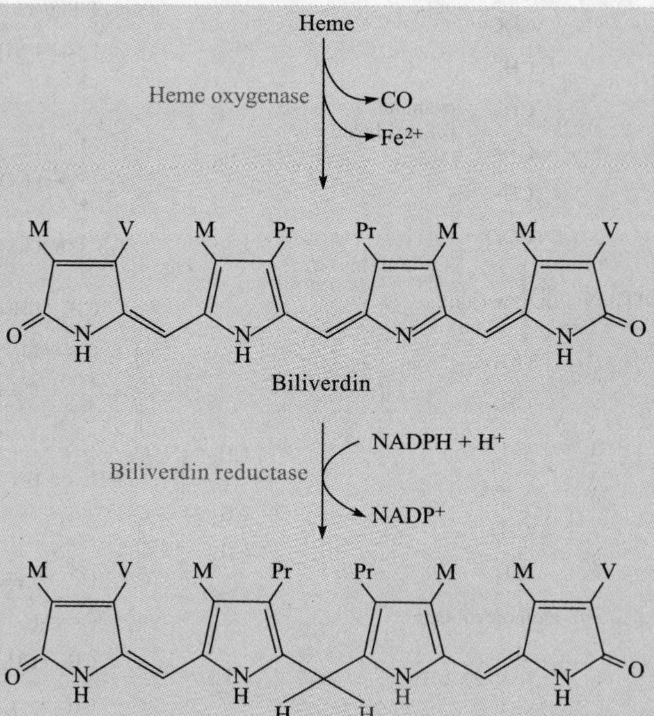

Fig. 27–26. Biosynthesis of billrubin from heme group of porphyrin (= Degradation of heme to billirubin)

For ease of comparison, these structures are shown in linear form, rather than in their correct stereochemical conformations.

[M = methyl; Pr = propionyl]

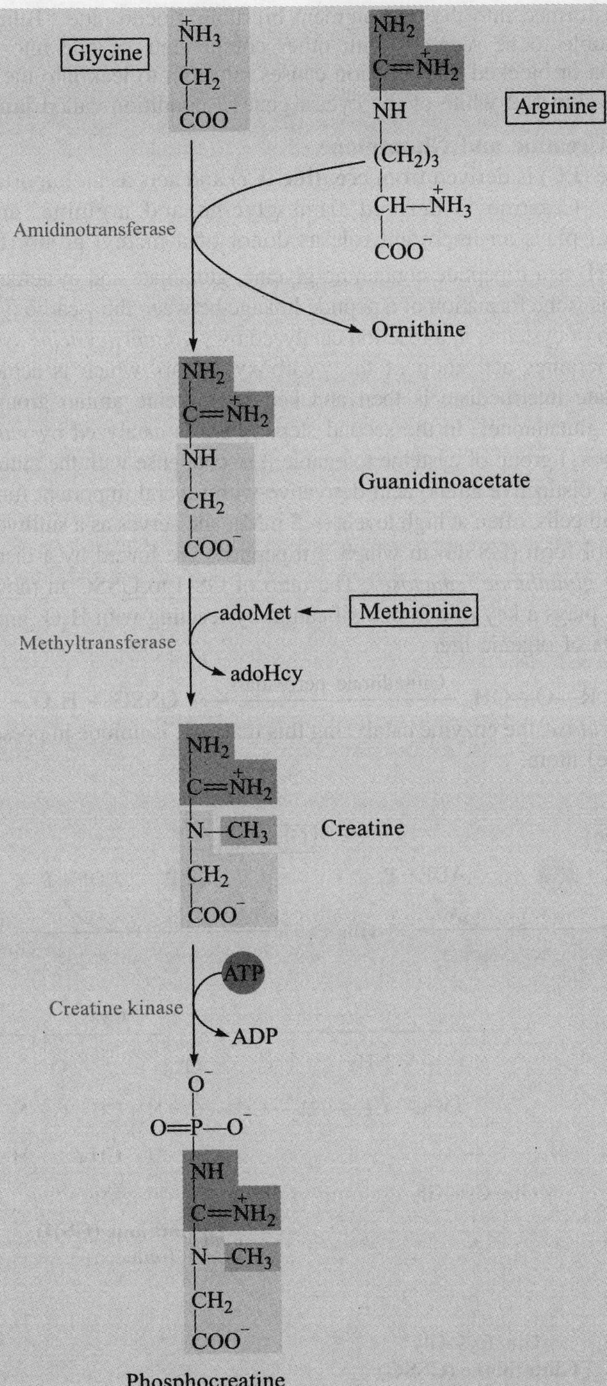

Fig. 27–27. Biosynthesis of creatine and phosphocreatine

Note that creatine is made from 3 amino acids: glycine, arginine and methionine. This pathway shows the versatility of amino acids as precursors in the biosynthesis of other nitrogenous biomolecules.

liver, where it is transformed into the bile pigment bilirubin glucuronide. Bilirubin glucuronide is sufficiently water-soluble to be secreted with other components of bile into the small intestine. Impaired liver function or blocked bile function causes bilirubin to leak into the blood. This results in a yellowing of the skin and white of the eye, a general condition called **jaundice**.

3. Biosynthesis of Creatine and Glutathione

Phosphocreatine (PC) is derived from **creatine** (Cr) and acts as an important energy reservoir in skeletal muscle. Creatine is derived from glycine and arginine, and methionine (as *S*-adenosylmethionine) plays an important role as donor of a methyl group (Fig. 27–27).

Glutathione (GSH) is a tripeptide containing glycine, glutamate and cysteine (Fig. 27–28). The first step in its synthesis is the formation of a peptide linkage between the γ-carboxyl group of glutamate and the α-amino group of cysteine, in a reaction catalyzed by *γ-glutamylcysteine synthetase*. Formation of this peptide bond requires activation of the γ-carboxyl group, which is achieved by ATP. The resulting acyl phosphate intermediate is then attacked by cysteine amino group. This reaction is feedback-inhibited by glutathione. In the second step, which is catalyzed by *glutathione synthetase*, ATP activates the carboxyl group of cysteine to enable it to condense with the amino group of glycine. Glutathione is a highly distinctive amino acid derivative with several important functions. Glutathione is present in virtually all cells, often at high levels (~ 5 mM), and serves as a sulfhydryl buffer. It cycles between a reduced thiol form (GSSG) in which 2 tripeptides are linked by a disulfide bond. GSSG is reduced to GSH by *glutathione reductase*. The ratio of GSH to GSSG in most cells is more than 500. Glutathione also plays a key role in detoxification by reacting with H_2O_2 and organic peroxides, the harmful byproducts of organic life.

$$2GSH + R—O—OH \xrightarrow{\text{Glutathione peroxidase}} GSSG + H_2O + R—OH$$

Glutathione peroxidase, the enzyme catalyzing this reaction, is unique in possessing a covalently-attached selenium (Se) atom.

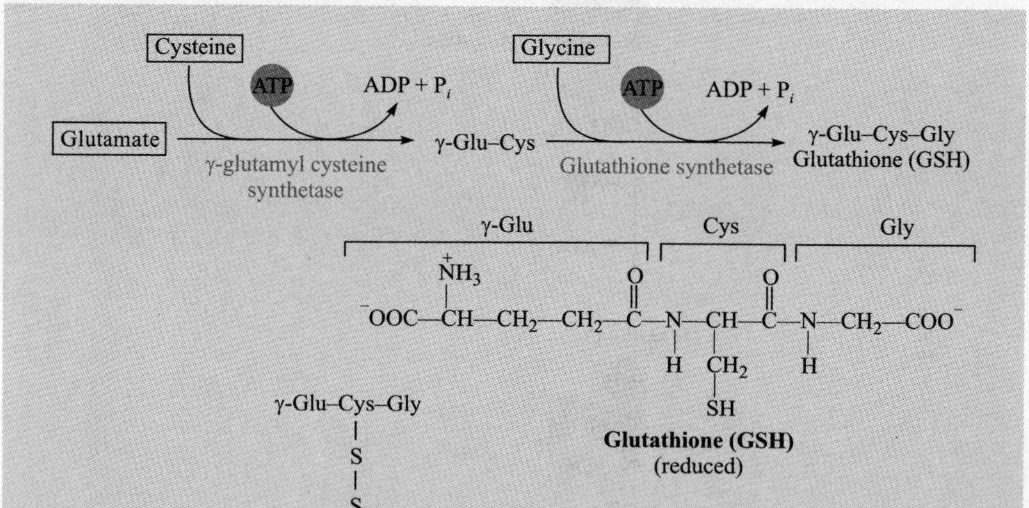

Fig. 27–28. **Biosynthesis and structure of glutathione**

Also shown is the oxidized form of glutathione (GSSG), which contains 2 moles of glutathione linked by a disulfide bond.

4. Biosynthesis of Neurotransmitters

The small diffusible molecules which communicate nerve impulses across most synapses are called **neurotransmitters**. Many important neurotransmitters are primary or secondary amines, derived from amino acids in simple pathways. Besides, certain polyamines, that are conjugated with DNA, are derived from the nonprotein amino acid, ornithine.

The synthesis of some neurotransmitters is illustrated in Fig. 27–29. **Tyrosine** gives rise to a family of catecholamines that comprises **dopamine, norepinephrine** and **epinephrine.** Levels of

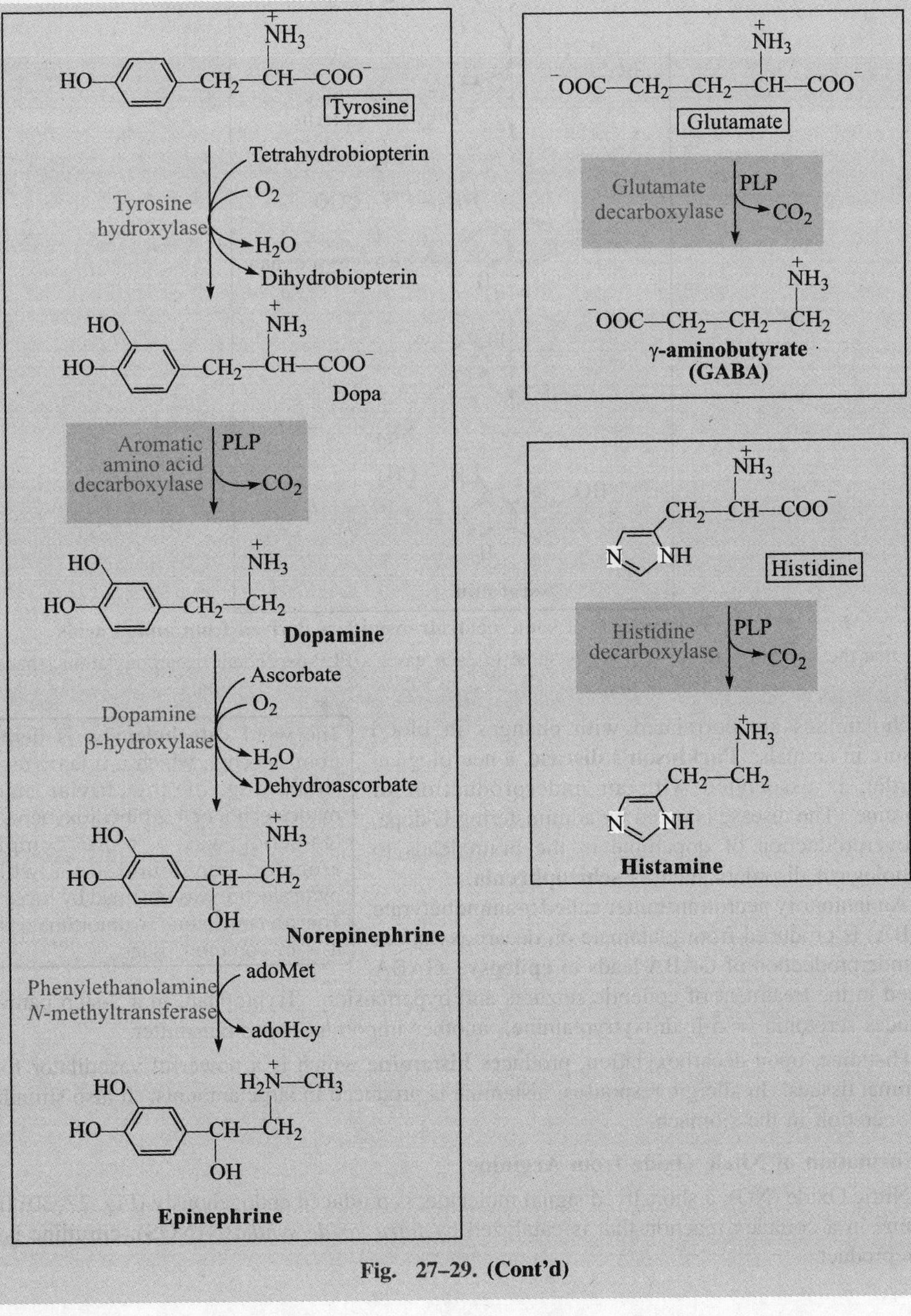

Fig. 27–29. (Cont'd)

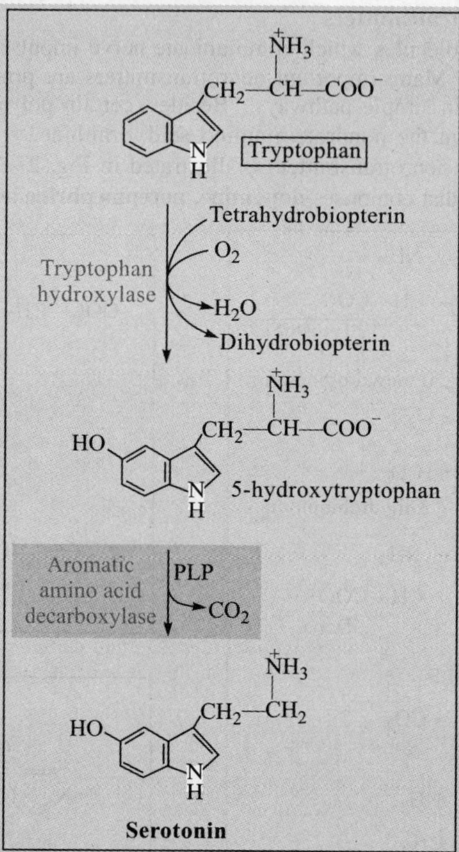

Fig. 27–29. Biosynthesis of some neutrotransmitters derived from amino acids

Note that the key biosynthetic step is the same in each case: a PLP-dependent decarboxylation (shaded in *pink*).

catecholamines are correlated with changes in blood pressure in animals. **Parkinson's disease**, a neurological disorder, is associated with an underproduction of dopamine. The disease is cured by administering L-dopa. An overproduction of dopamine in the brain leads to psychological disorders such as **schizophrenia**.

> The word **catecholamine** is derived from catechol, which is a fairly recent shortening of the trivial name pyrocatechol or 1,2-dihydroxybenzene (catechol was a more complex aromatic compound from which pyrocatechol was obtained by heating). The nomenclature is unfortunate, but it is in common usage.

An inhibitory neurotransmitter called γ-aminobutyrate (GABA) is produced from glutamate on decarboxylation. An underproduction of GABA leads to **epilepsy**. GABA is used in the treatment of epileptic seizures and hypertension. Tryptophan, in a 2-step pathway, produces serotonin (= 5-hydroxytryptamine), another important neurotransmitter.

Histidine, upon decarboxylation, produces **histamine** which is a powerful vasodilator found in animal tissues. In allergic responses, histamine is produced in large amounts. It also stimulates acid secretion in the stomach.

5. Formation of Nitric Oxide from Arginine

Nitric Oxide (NO), a short-lived signal molecule, is produced endogenously (Fig. 27–30) from arginine in a complex reaction that is catalyzed by *nitric oxide synthase* (NOS); citrulline is the other product.

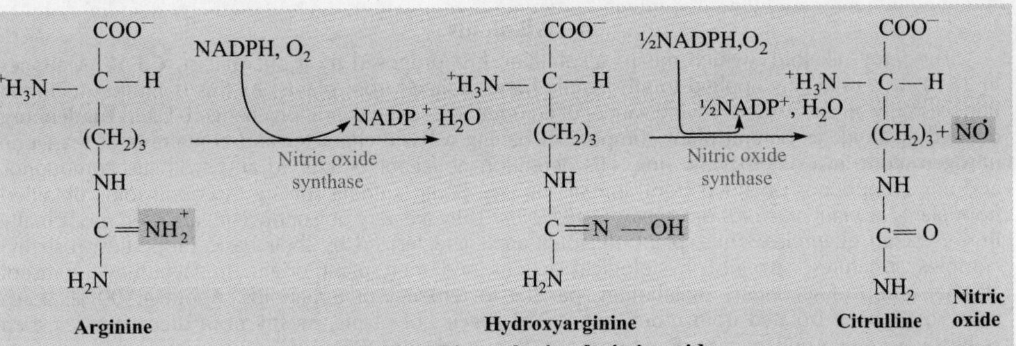

Fig. 27–30. **Biosynthesis of nitric oxide**

Note that both steps are catalyzed by *nitric oxide synthase*. The nitrogen of the NO is derived from the guanidino group of arginine.

Nitric oxide, a free-radical gas, is an important messenger in many vertebrate signal transduction processes. It diffuses freely across membranes but has a short life span, less than a few seconds, because it is highly reactive. Hence, *nitric oxide is well suited to serving as a transient signal molecule within cells and between adjacent cells.*

6. Biosynthesis of Lignin, Tannin and Auxin

The aromatic amino acids, phenylalanine, tyrosine and tryptophan are precursors of many plant substances. **Lignin**, a complex rigid polymer, is derived from phenylalanine and tyrosine. *It is second only to cellulose in abundance in plant tissues*. Phenylalanine and tyrosine also give rise to many natural products such as **tannins** that inhibit oxidation in wines.

Tryptophan gives rise to plant growth hormone, **indole-3-acetate** (IAA) or **auxin** (Fig. 27–31). This molecule regulates some important physiological processes.

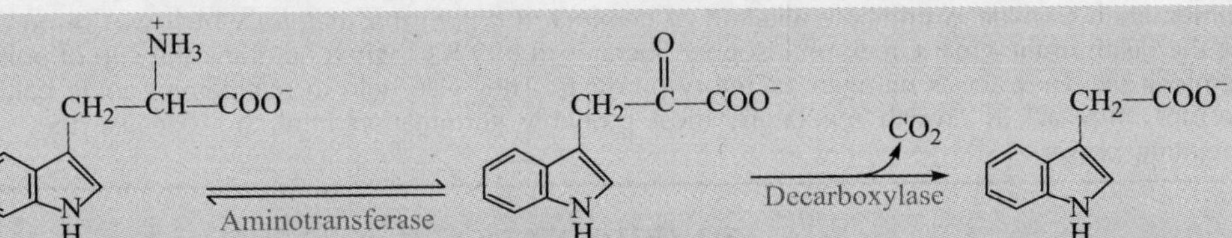

Fig. 27–31. **Biosyntheis of indole-3-acetate**

7. Biosynthesis of Papaverine from Tyrosine

The metabolism of most alkaloids is poorly known. Their chemical structure in many cases reveals their close relationship to amino acids such as phenylalanine, tyrosine, tryptophan, lysine or proline. As an example, the structure of **papaverine**, one of the alkaloids of opium, is given in Fig. 27–32. Note that this compound could be formed rather simply from two molecules of dihydroxyphenylalanine with the elimination of both carboxyls and one amino group. The actual pathway is not known.

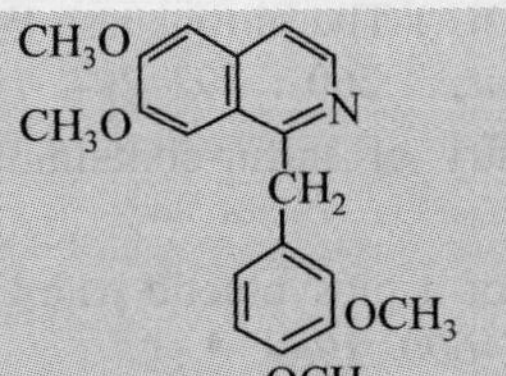

Fig. 27–32. **Chemical structure of papaverine**

Alkaloids

The term 'alkaloid' (which means alkali-like), first proposed by a pharmacist, **C.F.W. Meissner** in 1819, was originally applied to all organic bases isolated from plants. As this definition included unnecessarily an extraordinary wide variety of compounds, so the definition changed. Later, **Landenburg** defined alkaloids as natural plant compounds having a basic character and containing at least one nitrogen atom in a heterocyclic ring. His definition noticeably excluded any synthetic compounds and any compounds obtained from animal sources. Some authors specify those alkaloids obtained from plants as *plant alkaloids* or *vegetable alkaloids*. They are very poisonous, but are used medicinally in very small quantities. Thus, plant alkaloids are characterized by their basic properties, (usually) complex structures, specific physiological actions and their plant origin. In fact, they represent another group of secondary metabolites, parallel to terpenes or terpenoids. About 4,500 alkaloids have so far been isolated from more than 4,000 species of plants, mostly from dicot families such as Papaveraceae, Apocynaceae, Papilionaceae, Ranunculaceae, Rubiaceae, Rutaceae and Solanaceae : the first one being unusual in that almost all its members contain alkaloids. Among monocot families, they are found in Liliaceae and Gramineae (=Poaceae). Alkaloids are generally found in the roots, bark, stem, leaves and seeds of these plants and are stored in the vacuoles (hence, they do not appear in young cells until they are vacuolated) as salts of various plant organic acids, such as acetic, citric, oxalic, tartaric etc.

The alkaloids are usually colourless, crystalline, nonvolatile solids, which are insoluble in water but are soluble in ethanol, ether, chloroform etc. Some alkaloids are liquid which are soluble in water (e.g., coniine and nicotine) and a few are coloured (e.g., berberine is yellow). Most alkaloids have a bitter taste and are optically active (levorotatory). Alkaloids form insoluble precipitates with solutions of phosphotungstic acid, phosphomolybdic acid, picric acid etc; many of these have definite crystalline shapes, hence may be used to help in their identification. The alkaloids are stored in sites other than the site of their synthesis. For example, the alkaloid **nicotine** is synthesized in tobacco roots but is transported and stored in the leaves. Some alkaloids are even modified at the storage sites. Basically, the alkaloids are synthesized from various amino acids (ornithine, lysine, tyrosine, phenylalanine, asparagine, tryptophan, proline, glutamic acid) or their derivatives. Because of their physiological effects on human systems, alkaloids have been used in pharmacy since time immemorial. Coniine is principle alkaloid of hemlock (*Conium maculatum*) which was responsible for the death of the great Greek philosopher Socrates in 399 B.C. when he drank the cup of poison-hemlock oil. They act as nitrogen excretory products. They also help in maintaining ionic balance. Besides, they act as growth regulators, most probably germination inhibitors because they have chelating power.

REFERENCES

1. **Baldwin E :** An Introduction to Comparative Biochemistry. *4th ed. Cambridge University Press, New York. 1964.*
2. **Bender DA :** Amino Acid Metabolism. *2nd ed. Wiley-Interscience, New York, 1985.*
3. **Bondy PK, Rosenberg LE (editors) :** Metabolic Control and Disease. *8th ed., W.B. Saunders, Philadelphia. 1980.*
4. **Bredt DS, Snyder SH :** Nitric oxide, a novel neuronal messenger. Neuron. **8 :** *3–11, 1992.*
5. **Burris RH :** Nitrogenases. *J. Biol. Chem.* **266 :** *9339-9342, 1991.*
6. **Chan MK, Kim J, Rees DC :** The nitrogenase FeMo-Cofactor and P-cluster pair: 2.2 Å resolution studies. *Science.* **260 :** *792–794, 1993.*
7. **Cooper AJL :** Biochemistry of sulfur-containing amino acids. *Ann Rev. Biochem.* **52 :** *187–222, 1983.*
8. **Crawford IP :** Evolution of a biosynthetic pathway: The tryptophan paradigm. *Ann. Rev. Microbiol.* **43 :** *567–600, 1989.*
9. **Cunningham EB :** Biochemistry: Mechanisms of Metabolism. *McGraw-Hill Book Company, New York. 1978.*

10. **Granick S, Beale SI :** Hemes, chlorophylls, and related compounds: Biosynthesis and metabolic regulation. *Adv. Enzymol.* **46 :** *33–203, 1978.*
11. **Jordan PM (editor) :** Biosynthesis of Tetrapyrroles. *Elsevier. 1991.*
12. **Kim J, Rees DC :** Nitrogenase and biological nitrogen fixation. *Biochemistry.* **33 :** *389–397, 1994.*
13. **Leeper FJ :** The biosynthesis of porphyrins, chlorophylls, and vitamin B_{12}. *Nat. Prod. Rep.* **6 :** *171–199, 1989.*
14. **Meister A :** Biochemistry of the Amino Acids. *2nd ed., Vols. 1 and 2. Academic Press Inc., New York. 1965.*
15. **Meister A, Anderson ME :** Glutathione. *Ann. Rev. Biochem.* **52 :** *711–760, 1983.*
16. **Miflin BJ (editor) :** Amino acids and derivatives. In Biochemistry of Plants. *Vol. 5. Academic Press Inc., New York. 1981.*
17. **Miles EW :** Structural basis for catalysis by tryptophan synthase. *Adv. Enzymol. Mol. Biol.* **64 :** *93–172, 1991.*
18. **Mora J, Palacios R (editors) :** Glutamine: Metabolism, Enzymology and Regulation. *Academic Press Inc., New York. 1979.*
19. **Nathan C :** Nitric oxide as a secretary product of mammalian cells. *FASEB J.* **6 :** *3051–3064, 1992.*
20. **Orme-Johnson WH :** Molecular basis of biological nitrogen fixation. *Ann. Rev. Biophys. Chem.* **14 :** *419–459, 1985.*
21. **Postgate J :** Trends and perspectives in nitrogen fixation research. *Adv. Microbiol. Physiol.* **30 :** *1–22, 1989.*
22. **Rhee SG, Chock PB, Stadtman ER :** Regulation of *Escherichia coli* glutamine synthetase. *Adv. Enzymol. Mol. Biol.* **62 :** *37–92, 1989.*
23. **Scriver CR, Beaudet AL, Sly WS, Valle D (editors) :** The Metabolic Basis of Inherited Diseases. *6th ed. McGraw-Hill Information Services Company, Health Sciences Division, New York. 1989.*
24. **Shemin D :** An illustration of the use of isotopes: The biosynthesis of porphyrins. *BioEssays.* **10 :** *30–35, 1989.*
25. **Stocker R, McDonagh AF, Glazer AN, Ames BN :** Antioxidant activities of bile pigments: Biliverdin and bilirubin. *Meth-Enzymol.* **186 :** *301–309, 1990.*
26. **Stocker R, Yamamoto Y, McDonagh AF, Glazer AN, Ames BN :** Bilirubin is an antioxidant of possible physiologic importance. *Science.* **235 :** *1043–1046, 1987.*
27. **Umbarger HE :** Amino acid biosynthesis and its regulation. *Ann. Rev. Biochem.* **47 :** *533–606, 1978.*
28. **Walsh C :** Enzymatic Reaction Mechanisms. *W.H. Freeman and Company, New York. 1979.*

PROBLEMS

1. Write a balanced equation for the synthesis of alanine from glucose.
2. Isovaleric acidemia is an inherited disorder of leucine metabolism caused by a deficiency of isovaleryl CoA dehydrogenase. Many infants having this disease die in the first month of life. The administration of large amounts of glycine sometimes leads to marked clinical improvement. Propose a mechanism for the therapeutic action of glycine.
3. Most cytosolic proteins lack disulfide bonds, whereas extracellular proteins usually contain them. Why?

4. The synthesis of δ-aminolevulinate takes place in the mitochondrial matrix, whereas the formation of porphobilinogen takes place in the cytosol. Propose a reason for the mitochondrial location of the first step in heme synthesis.
5. Which of the 20 amino acids can be synthesized directly from a common metabolic intermediate by a transamination reaction?
6. For the following example of a branched pathway, propose a feedback inhibition scheme that would result in the production of equal amounts of Y and Z.

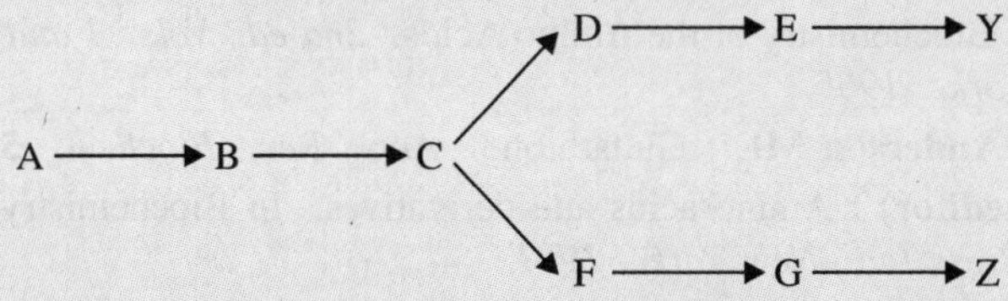

7. The first common step (A → B) is partly inhibited by both of the final products, each acting independently of the other. Suppose that a high level of Y alone decreased the rate of the A → B step from 100 to 60 s^{-1} and that a high level of Z alone decreased the rate from 100 to 40 s^{-1}. What would the rate be in the presence of high levels of both Y and Z?
8. How might increased synthesis of aspartate and glutamate affect energy production in a cell? How would the cell respond to such an effect?
9. The mitochondrial form of carbamoyl phosphate synthetase is allosterically activated by *N*-acetylglutamate. Briefly describe a rationale for this effect.
10. Most bacterial mutants that require isoleucine for growth also require valine. Why ? Which enzyme or reaction would be defective in a mutant requiring only isoleucine (not valine) for growth?
11. Describe a series of allosteric interactions that could adequately control the biosynthesis of valine, leucine, and isoleucine.
12. Identify carbon atoms, by number, that are incorporated from this structure into the compounds listed in (*a*) to (*d*).

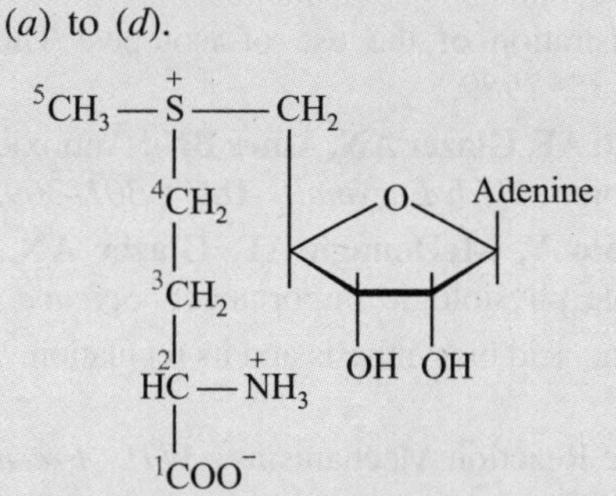

(*a*) Creatine phosphate
(*b*) Ethylene
(*c*) Glycine betaine
(*d*) Epinephrine

13. Why is phenylketonuria resulting from dihydropteridine reductase deficiency a more serious disorder than PKU resulting from phenylalanine hydroxylase deficiency?
14. Tyrosine is normally a nonessential amino acid, but individuals with a genetic defect in phenylalanine hydroxylase require tyrosine in their diet for normal growth. Explain.
15. Write the net equation for the synthesis of the nonessential amino acid aspartate from glucose, carbon dioxide, and ammonia.

ANSWERS TO PROBLEMS

1. Glucose + 2 ADP + 2 P_i + 2 NAD^+ + 2 glutamate ⟶
   2 alanine + 2 α-ketoglutarate + 2 ATP + 2 NADH + 2 H_2O + 2 H^+
2. The administration of glycine leads to the formation of isovalerylglycine. This water-soluble conjugate, in contrast with isovaleric acid, is excreted very rapidly by the kidneys.
3. The cytosol is a reducing environment, whereas the extracellular milieu is an oxidizing environment.
4. Succinyl CoA is formed in the mitochondrial matrix.
5. Alanine from pyruvate; aspartate from oxaloacetate; glutamate from α-ketoglutarate.
6. Y could inhibit the C → D step, Z could inhibit the C → F step, and C could inhibit A → B. This scheme is an example of sequential feedback inhibition. Alternatively, Y could inhibit the C → D step, Z could inhibit the C → F step, and the A → B step would be inhibited only in the presence of both Y and Z. This scheme is called concerted feedback inhibition.
7. The rate of the A → B step in the presence of high levels of Y and Z would be 24 s^{-1} ($0.6 \times 0.4 \times 100\ s^{-1}$).
8. Synthesis from oxaloacetate and α-ketoglutarate would deplete the citric acid cycle, which would decrease ATP production. Anaplerotic reactions would be required to replenish the citric acid cycle.
9. *N*-acetylglutamate is an intermediate in ornithine biosynthesis. Activity of the urea cycle requires both ornithine and carbamoyl phosphate. If insufficient carbamoyl phosphate is available, ornithine will accumulate, and this could cause accumulation of the precursor, *N*-acetylglutamate. This accumulation acts as a signal to stimulate carbamoyl phosphate synthesis to increase urea cycle flux.
10. Because the same enzymes are involved in comparable steps of both isoleucine and valine biosynthesis. Threonine dehydratase.
11. Threonine → α-ketobutyrate: inhibited by isoleucine

    α-ketoisovalerate + acetyl-CoA → β-isopropyl malate: inhibited by leucine

    Control of valine synthesis is more complicated, because three of the enzymes are involved in synthesis of all three amino acids. One could look for cumulative feedback inhibition– by valine, isoleucine, and leucine– of the first committed reaction:

    Pyruvate + hydroxyethyl-TPP ⟶ α-acetolactate
12. (*a*) 5 (*c*) 5
    (*b*) 3,4 (*d*) 5
13. Because a pteridine reductase deficiency would impair all tetrahydroiopterin-dependent reactions, which include the synthesis of catecholamines, serotonin, and nitric oxide, as well as tyrosine.
14. If phenylalanine hydroxylase is defective, the biosynthetic route to Tyr is blocked and Tyr must be obtained from the diet.
15. Glucose + 2 CO_2 + 2 NH_3 ⟶ 2 aspartate + 2 H^+ + 2 H_2O

CONTENTS

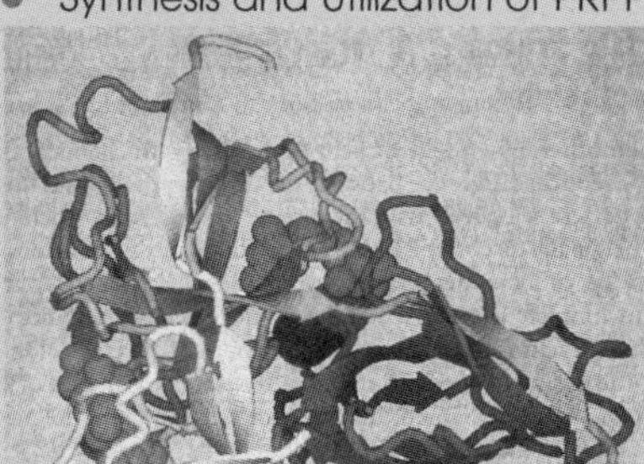

Crystal structure of *E. coli* dUTPase

One of the three subunits of the homotrimer is shown in *white*, one in *light blue*, and one in *dark blue*. Each subunit contains a C-terminal arm that wraps about an adjacent subunit. A competitive inhibitor, dUDP (shown in *orange*), is bound at each of the three catalytic sites and each site is formed at a junction between subunits.

[*Courtesy: Elle Cedergren-Zeppenzauer amd Gunilla Larsson*]

CHAPTER 28

Biosynthesis of Nucleotides

An ample supply of nucleotides is essential for many life processes. Purine and pyrimidine nucleotides participate in many critical cellular functions. Their metabolic roles vary widely from serving as monomeric precursors of RNA and DNA to serving as second messengers. Besides, some nucleotide derivatives such as UDP-glucose participate in biosynthetic processes such as the formation of glycogen. In addition to being an universal currency of energy, ATP also acts as the donor of phosphoryl groups transferred by protein kinases.

THE TWO BIOSYNTHETIC PATHWAYS

Two types of pathways lead to the synthesis of nucleotides: the *de novo* pathways and the salvage pathways (Fig. 28–1).

(a) ***de novo* (from scratch) pathways:** In these, the nucleotide bases are assembled from simple compounds. The framework for a *pyrimidine* base is assembled first and then attached to ribose. In contrast, the framework for a *purine* base is synthesized piece-by-piece directly onto a ribose-based structure. These pathways comprise a small number of elementary reactions that are repeated with variation to produce different nucleotides, as might be expected for pathways that appeared very early in evolution. *de novo* pathways are essentially identical throughout the biological world.

(b) **Salvage** (or **rescuing**) **pathways:** In these, preformed bases are recovered and reconnected to a ribose unit.

In fact, these pathways recycle the free bases and nucleosides released from nucleic acid breakdown. In essence, these pathways involve the utilization of preformed purine and pyrimidine compounds that would otherwise be lost to biodegradation.

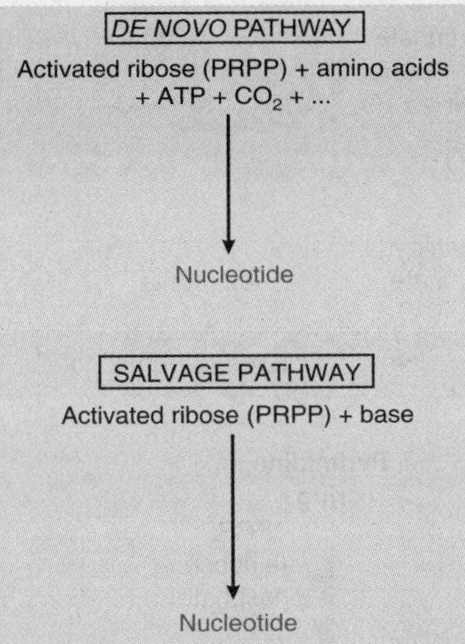

Fig. 28–1. *de novo* and salvage pathways

In *de novo* synthesis, the base itself is synthesized from simpler starting substances, including amino acids. ATP hydrolysis is required for *de novo* synthesis. In salvage pathway, a base is reattached to a ribose, activated in the form of 5-phosphoribosyl-1-pyrophosphate (PRPP).

Both *de novo* and salvage pathways lead to the synthesis of *ribonucleotides*. However, DNA is built from *deoxyribonucleotides*. Consistent with the notion that RNA preceded DNA in the course of evolution, all deoxyribonucleotides are synthesized from the corresponding ribonucleotides. The deoxyribose sugar is generated by the reduction of ribose within a fully-formed nucleotide. Furthermore, the methyl group that differentiates the thymine of DNA from the uracil of RNA is added at the last step in the pathway.

Two other features deserve mention. *First*, there is evidence, especially in the *de novo* purine pathway, that the enzymes are present as large, multienzyme complexes in the cell, a recurring theme in our discussion on metabolism. *Second*, the cellular pools of nucleotides, other than ATP, are quite small, perhaps 1% or less of the amounts required to synthesize the cell's DNA. Hence, the cells must continue to synthesize nucleotides during nucleic acid synthesis, and in some cases nucleotide synthesis may limit the rates of DNA replication and transcription. Because of the importance of these processes in the meristematic (dividing) cells, agents that inhibit nucleotide synthesis have become especially important to modern medicine.

de novo PYRIMIDINE NUCLEOTIDE SYNTHESIS

The common pyrimidine nucleotides are cytidine 5′-monophosphate (CMP; cytidylate) and uridine 5′-monophosphate (UMP; uridylate), which contain the pyrimidines cytosine and uracil, respectively. In *de novo* synthesis of pyrimidines, the 6-membered pyrimidine ring is synthesized first and then it is attached to ribose to form a pyrimidine nucleotide (Fig. 28–2). Pyrimidine rings are assembled from bicarbonate, aspartic acid, and ammonia. Although ammonia can be used directly, it is usually produced from the hydrolysis of the side chain of glutamine.

The first step in *de novo* pyrimidine biosynthesis is the synthesis of carbamoyl phosphate from bicarbonate and ammonia in a multistep process, requiring the cleavage of 2 molecules of ATP. This reaction is catalyzed by *carbamoyl phosphate synthetase* (CPS). An analysis of the structure of CPS reveals two homologous domains, each of which catalyzes an ATP-dependent step (Fig. 28–3).

In the first step of the carbamoyl phosphate synthesis, bicarbonate is phosphorylated by ATP to form carboxyphosphate and ADP. Ammonia then reacts with carboxyphosphate to form carbamic

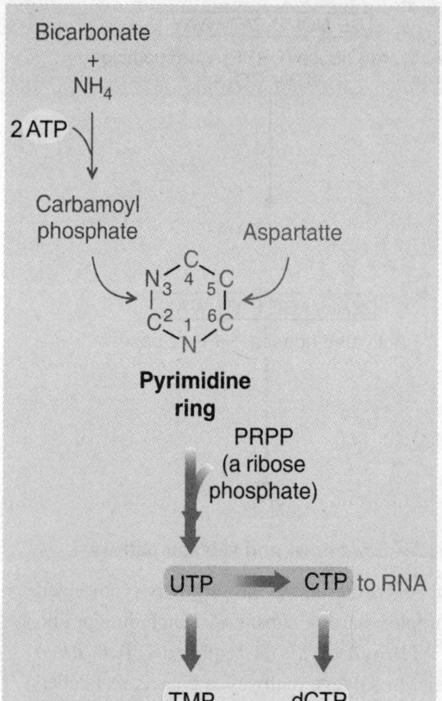

Fig. 28–2. *de novo* pathway for pyrimidine nucleotide synthesis

The C-2 and N-3 atoms in the pyrimidine ring come from carbamoyl phosphate, whereas the other atoms of the ring come from aspartate.

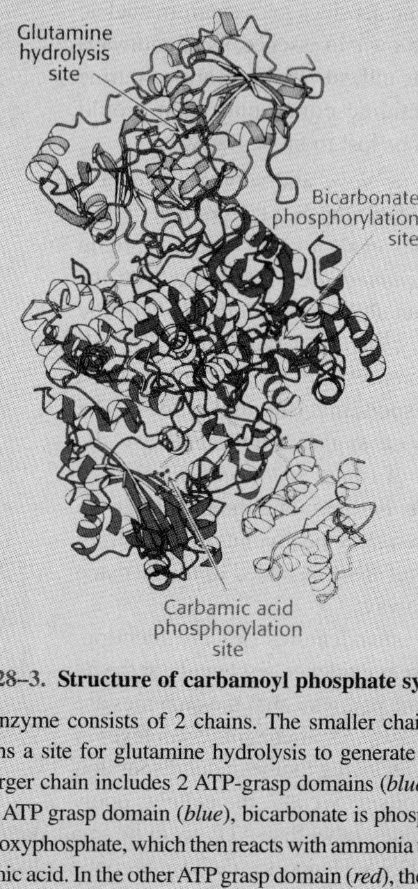

Fig. 28–3. Structure of carbamoyl phosphate synthetase

This enzyme consists of 2 chains. The smaller chain (*yellow*) contains a site for glutamine hydrolysis to generate ammonia. The larger chain includes 2 ATP-grasp domains (*blue* and *red*). In one ATP grasp domain (*blue*), bicarbonate is phosphorylated to carboxyphosphate, which then reacts with ammonia to generate carbamic acid. In the other ATP grasp domain (*red*), the carbamic acid is phosphorylated to produce carbamoyl phosphate.

acid and an inorganic phosphate. The active site for this reaction lies in a domain formed by the amino-terminal third of CPS. This domain forms a structure, called an *ATP-grasp fold*, that surrounds ATP and holds in an orientation suitable for nucleophilic attack at the γ phosphoryl group. Proteins

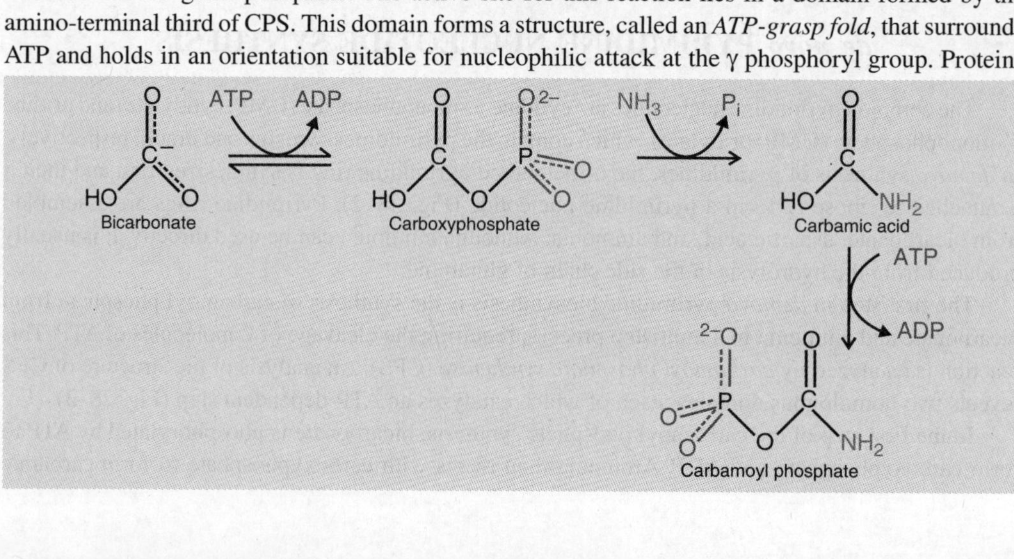

containing ATP-grasp folds catalyze the formation of carbon–nitrogen bonds through acylphosphate intermediates and are widely used in nucleotide biosynthesis. In the last step catalyzed by cabamoyl phosphate synthetase, carbamic acid is phosphorylated by a second molecule of ATP to form carbamoyl phosphate. This reaction takes place in a second ATP-grasp domain within the enzyme. The active sites leading to carbamic acid formation and carbamoyl phosphate formation are very similar, revealing that this enzyme evolved by gene duplication. Indeed, duplication of a gene encoding an ATP-grasp domain followed by specialization was central to the evolution of nucleotide biosynthetic processes.

Carbamoyl phosphate synthetase primarily uses glutamine as a source of ammonia. In this case, a second polypeptide component of the carbamoyl phosphate synthetase enzyme hydrolyzes glutamine to form ammonia and glutamate. The active site of the glutamine-hydrolyzing component of carbamoyl phosphate synthetase contains a catalytic dyad comprising a cysteine and histidine residue (Fig. 28–4). Such a catalytic dyad, reminiscent of the active site of cysteine proteases, is conserved in a family of amidotransferases, including CTP synthetase and GMP synthetase.

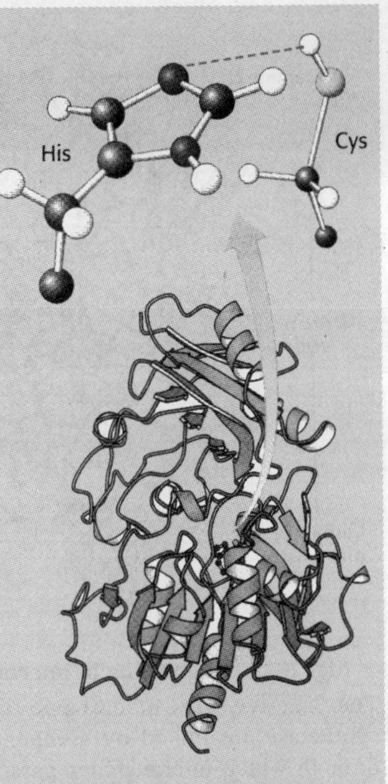

Fig. 28–4. The ammonia-generation site in the smaller domain of carbamoyl phosphate synthetase

This smaller domain contains an active site for the hydrolysis of the side chain carboxamide of glutamine to generate ammonia. The key residues in this active site include a cysteine residue and a histidine residue.

Channeling of Intermediates in Carbamoyl Phosphate Synthetase

Carbamoyl phosphate synthetase contains 3 different active sites, which are separated from one another by a total of 80 Å. (Fig. 28–5). Intermediates produced at one site move to the next site without leaving the enzyme, *i.e.*, they move by means of substrate channeling (similar to the process as occurs in tryptophan synthetase). The ammonia produced in the glutamine-hydrolysis active site travels 45 Å through a channel within the enzyme to reach the active site at which carboxypeptidase has been generated. The carbamic acid generated at this site diffuses an additional 35 Å through an extension of the channel to reach the site at which carbamoyl phosphate is generated. This channeling serves 2 functions:

(*a*) intermediates generated at one active site are captured with no loss caused by diffusion, and

(*b*) labile intermediates, such as carboxyphosphate and carbamic acid (which decompose in less than 1 second at pH 7), are protected from hydrolysis.

It may, however, be noted that in animals the carbamoyl phosphate required in urea synthesis is produced in *mitochondria* by **carbamoyl phosphate synthetase I**, whereas the carbamoyl phosphate required in pyrimidine biosynthesis is produced in the *cytosol* by a different form of the enzyme, **carbamoyl phosphate synthetase II.**

The *de novo* pyrimidine nucleotide biosynthesis (Fig. 28–6) proceeds in a somewhat different manner from purine nucleotide synthesis, the 6-membered pyrimidine ring is made first and then

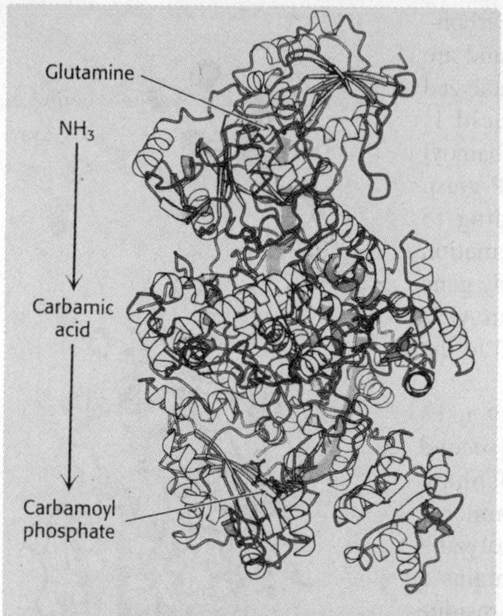

Fig. 28–5. Channeling of intermediates

The 3 active sites of carbamoyl phosphate synthetase are linked by a channel (*yellow*) through which intermediates pass. Glutamine enters one active site, and carbamoyl phosphate, which includes the nitrogen atom from the glutamine side chain, leaves another 80 Å away.

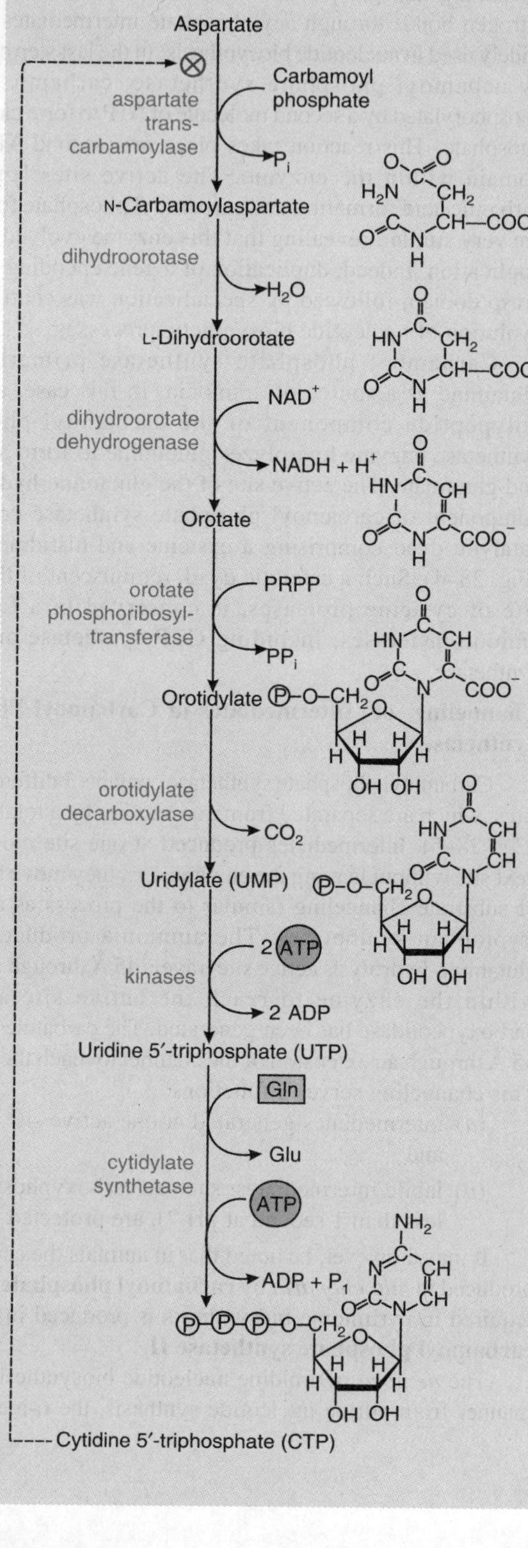

Fig. 28-6. *de novo* synthesis of pyrimidine nucleotides: biosynthesis of UTP and CTP *via* orotidylate

The pyrimidine is constructed from carbamoyl phosphate and aspartate. The ribose 5-phosphate is then added to the completed pyrimidine ring by orotate phosphoribosyltransferase. The first step in this pathway (not shown here; see page 6) is the synthesis of carbamoyl phosphate from CO_2 and NH_4^+, catalyzed in eukaryotes by carbamoyl phosphate synthetase II. The activities of carbamoyl phosphate synthetase II, aspartate carbamoyltransferase and dihydroorotase, are on a trifunctional protein (called CAD); the activities of orotate phosphoribosyltransferase and orotidylate decarboxylase are on a bifunctional protein (called UMP synthase).

attached to ribose 5-phosphate. Required in this process is carbamoyl phosphate (CAP), also an intermediate in the urea cycle.

Carbamoyl phosphate reacts with **aspartate** to yield *N*-carbamoylaspartate (CAA) in the first committed step of pyrimidine biosynthesis. This reaction is catalyzed by *aspartate transcarbamoylase* (ATCase). **Carbamoylaspartate** then removes a molecule of water thus bringing about closure of the pyrimidine ring (cyclization) to form L-dihydroorotate (DHOA); the reaction being catalyzed by *dihydroorotase*. **Dihydroorotate** is then oxidized to orotate (OA), a reaction in which NAD^+ is the ultimate electron acceptor. In eukaryotes, the first 3 enzymes in this pathway– *c*arbamoyl phosphate synthetase II, *a*spartate transcarbamoylase, and *d*ihydroorotase– are part of a single trifunctional protein. This protein, known by the acronym CAD, contains 3 identical polypeptide chains (each of M_r 2,30,000), each with active sites for all 3 reactions. This suggests that large, multienzyme complexes may be the rule in this pathway.

Once **orotate** is formed, the ribose 5-phosphate side chain, provided once again by 5-phosphoribosyl-1-pyrophosphate (PRPP), is attached to yield orotidylate (OMP), (PRPP is synthesized from ribose 5-phosphate, formed by the pentose phosphate pathway, by the addition of pyrophosphate from ATP). The enzyme that catalyzes this addition reaction, *orotate phosphoribosyltransferase,* is homologous to a number of other phosphoribosyltransferases that add diferent groups to PRPP to form the other nucleotides. **Orotidylate** is then decarboxylated to form uridylate (UMP), a major pyrimidine nucleotide that is a precursor to RNA. The reaction is catalyzed by *orotidylate decarboxylase. This enzyme is one of the most proficient enzymes known*. In its absence, decarboxylation is extremely slow and is estimated to take place once every 78 million years; with the enzyme present, it takes place approximately once in a second, a rate enhancement of 10^{17}-fold! The activities of orotate phosphoribosyltransferase and orotidylate (OMP) decarboxylase are found on a bifunctional protein, called *UMP synthase*. A defect in this bifunctional protein that affects either phosphoribosyltransferase activity or decarboxylase activity leads to a rare clinical condition known as ***hereditary orotic aciduria***.

Uridylate (UMP) is then phosphorylated to uridine triphosphate (UTP) by *kinases*. In fact, nucleoside monophosphates are converted into nucleoside triphosphates in stages. First, nucleoside monophosphates are converted into diphosphates by specific *nucleoside monophosphate kinases* that utilize ATP as the phosphoryl-group donor. For example, UMP is phosphorylated to UDP by *UMP kinase*:

$$\text{UMP + ATP} \rightleftharpoons \text{UDP + ADP}$$

Nucleoside diphosphates and triphosphates are interconverted by *nucleoside diphosphate kinase*, a ubiquitous enzyme with broad specificity, in contrast with the monophosphate kinases. X and Y can represent any of several ribonucleosides or even deoxyribonucleosides:

$$\text{XDP + YTP} \rightleftharpoons \text{XTP + YDP}$$

UTP can be transformed into cytidine triphosphate (CTP) by the replacement of a carbonyl group by an amino group. Like the synthesis of carbamoyl phosphate, this reaction requires ATP and uses glutamine as the source of the amino group. **CTP** formation is catalyzed by *cytidylate synthetase*, which displays homotropic sigmoidal kinetics. CTP, the product, is a negative effector of the reaction (Fig. 28–7). By regulating CTP synthetase in this way, cells maintain an appropriate ratio of UTP and CTP for cellular functions and RNA synthesis.

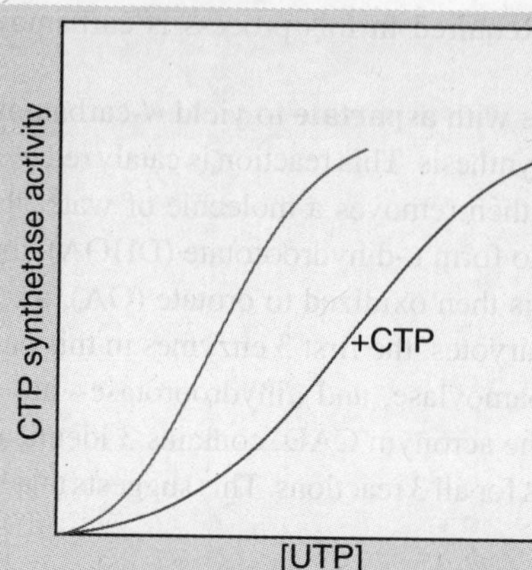

Fig. 28-7. Regulation of CTP synthetase

CLINICAL IMPLICATIONS 28–1.

Hereditary Orotic Aciduria

Hereditary orotic aciduria results from a defect in *de novo* synthesis of pyrimidine nucleotides. This genetic disease in characterized by severe anemia, growth retardation, and increased quantities of orotic acid excretion. The biochemical basis for orotic aciduria is a defect in one or both of the activities (orotate phosphoribosyltransferase or orotidylate decarboxylase) associated with *UMP synthase*, the bifunctional protein. Although a very rare disease, but the understanding of the metabolic basis for this disorder has led to successful treatment of this ailment. Patients are fed uridine, which leads not only to reversal of the hematologic problem but also to decreased formation of orotic acid. Uridine is taken up by cells and converted by *uridine phosphotransferase* to uridine monophosphate (UMP) that is sequentially converted to UDP and then to UTP. UTP formed from exogenous uridine, in turn, inhibits carbamoyl phosphate synthetase II (CPS II), the major regulated step in the *de novo* pathway. Consequently, orotic acid synthesis *via* the *de novo* pathway is markedly decreased to essentially normal levels. Since ATP is also the substrate for CTP synthesis, uridine treatment serves to replenish both the UTP and CTP cellular pools. In effect, then, exogenous uridine bypasses the defective UMP synthase and supplies cells with UTP and CTP required for nucleic acid synthesis and other celluar functions.

SALVAGE PATHWAYS FOR PYRIMIDINES

Pyrimidines are "salvaged" by conversion to nucleotides by *pyrimidine phosphoribosyltransferase*. The general reaction is:

$$\text{Pyrimidine + PRPP} \longrightarrow \text{Pyrimidine nucleoside 5'-monophosphate + PPi}$$

The enzyme from human erythrocytes utilizes orotate, uracil, and thymine as substrates but not cytosine. These salvage reactions divert the pyrimidine bases from the degradative pathway to the nucleotide level. As a pyrimidine base becomes available to cells, there are competing reactions that will either result in degradation and excretion or reutilization of the bases. For example, when normal liver is presented with uracil, it is rapidly degraded to β-alanine, whereas in proliferating tumour cells uracil is converted to UMP. This is the result of the availability of PRPP, enzyme levels, and metabolic state of the cells.

Mammalian cells do not appear to possess efficient means of salvaging *free* pyrimidine bases to their respective pyrimidine nucleotides. However, they do have active salvage pathways for converting the pyrimidine nucleosides– uridine, cytidine, and thymidine– to their respective nucleotides (Fig. 28–8). The enzyme required for *de novo* pyrimidine biosynthesis, *orotate phosphoribosyltransferase*, is capable of salvaving orotic acid to orotidylate (= orotidine monophosphate, OMP), but in a strict sense orotic acid is not considered a complete pyrimidine base. The orotate phosphoribosyltransferase cannot use normal pyrimidine bases as substrates, although it is capable of converting **allopurinol** (= 4-hydroxypyrazolopyrimidine) to a nucleotide in which the ribosyl phosphate is attached to the N_1 of the pyrimidine ring of that drug (Fig. 28–9). The anticancer drug, **5-fluorouracil (FUra)** is also phosphoribosylated by orotate phosphoribosyltransferase. 2′-deoxycytidine is phosphorylated by a distinct enzyme called *deoxycytidine kinase* that can also phosphorylate deoxyadenosine and deoxyguanosine.

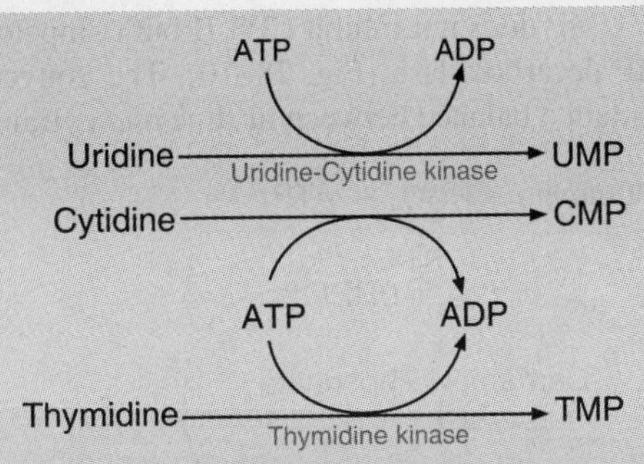

Fig. 28–8. The pyrimidine nucleoside kinase reactions responsible for the formation of the respective pyrimidine nucleoside monophosphates

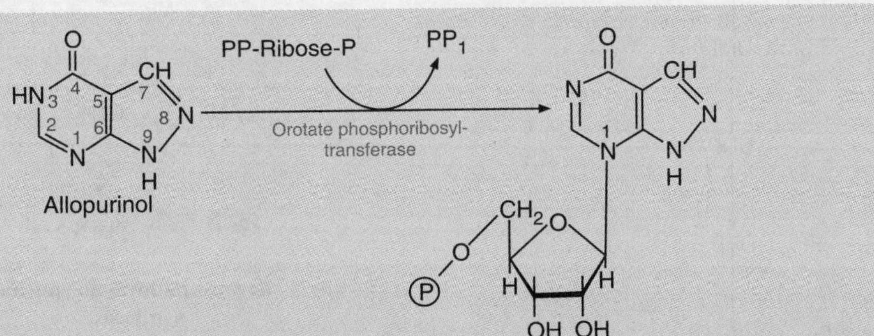

Fig. 28–9. Formation of allopurinol-1-ribonucleotide by the phosphorylation of allopurinol in a reaction catalyzed by orotate phosphoribosyltransferase

REGULATION OF PYRIMIDINE NUCLEOTIDE SYNTHESIS

Regulation of the rate of pyrimidine nucleotide synthesis in **bacteria** occurs in large part through the regulatory enzyme aspartate transcarbamoylase (ATCase), which catalyzes the first reaction in the sequence and is inhibited by cytidine 5′-triphosphate (CTP), the end product of the sequence. The bacterial ATCase molecule consists of 6 catalytic subunits and 6 regulatory subunits. The catalytic subunits bind the substrate molecules, and the allosteric subunits bind the allosteric inhibitor, CTP. The entire ATCase molecule, as well as its subunits, exists in 2 conformations, active and inactive. When CTP is not bound to the regulatory subunits, the enzyme is maximally active. As CTP accumulates and binds to the regulatory subunits, they undergo a change in conformation. This change is transmitted to the catalytic subunits, which then also shift to an inactive conformation.

Regulation to pyrimidine nucleotide synthesis in **mammalian cells** occurs at the carbamoyl phosphate synthetase II (CPS II) step. CPS II is a cytosolic enzyme and distinct from the mitochondrial carbamoyl phosphate synthetase I (CPS I), which utilizes ammonia as the amino donor and is activated by *N*-acetylglutamate. CPS II is inhibited by UTP, an end product of the pathway, and is activated by PRPP. *CPS II is the only source of carbamoyl phosphate in extrahepatic tissues*. However, in liver, under stressed conditions in which there is excess ammonia, CPS I generates carbamoyl phosphate in mitochondria, which ends up in the cytosol and serves as a substrate for pyrimidine nucleotide synthesis. This pathway serves to detoxify excess ammonia. Elevated levels of orotic acid are excreted as a result of ammonia toxicity in humans. This points to CPS II as being the major regulated activity in pyrimidine nucleotide metabolism.

UMP does not inhibit CPS II but competes with orotidine 5′-monophosphate (OMP) to inhibit OMP decarboxylase (Fig. 28–10). The conversion of UTP to CTP is also regulated so that cells maintain a balance between uridine and cytidine nucleotides.

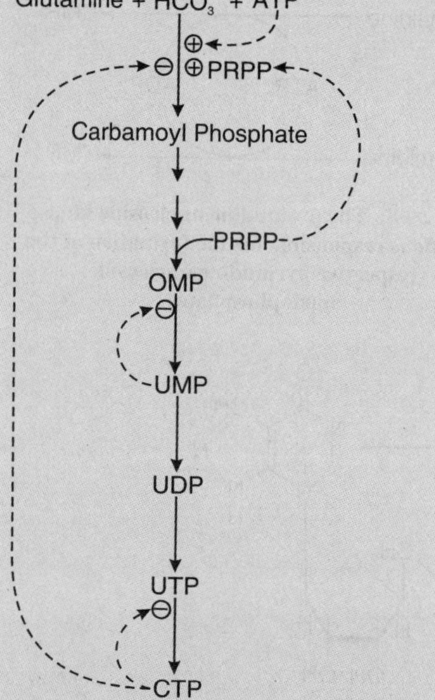

Fig. 28–10. Regulation of pyrimidine nucleotide synthesis

Solid arrows represent enzyme-catalyzed reactions and *dotted arrows* represent inhibition by the products of the reactions.

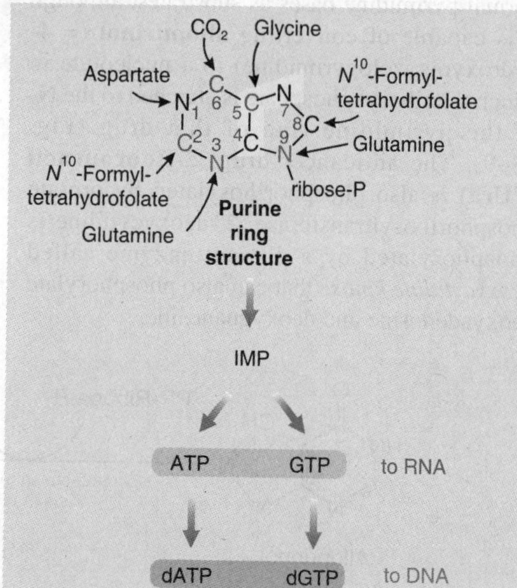

Fig. 28–11. *de novo* pathway for purine nucleotide synthesis

The information regarding the origin of the ring atoms of purines was obtained from tracer experiments with ^{14}C- or ^{15}N-labeled precursors. Formate is supplied in the form of N^{10}-formyltetrahydrofolate (10-formyl-THF).

de novo PURINE NUCLEOTIDE SYNTHESIS

The two parent purine nucleotides of nucleic acids are adenine 5′-monophosphate (AMP; adenylate) and guanosine 5′-monophosphate (GMP; guanylate), which contain the purines, adenine and guanine, respectively. The reactions of *de novo* purine nucleotide biosynthesis were identified in the 1950s, in the laboratories of **John Buchanan** and **G. Robert Greenberg**. Elucidation of the pathway began with the realization that birds excrete most of their excess nitrogen compounds in the form of *uric acid*, an oxidized purine. Thus, researchers were able to identify low-molecular-weight precursors to purines by administering isotopically-labeled compounds to pigeons, crystallizing uric acid from the droppings, and, by selective chemical degradation, determining which positions were labeled by which precursors. Fig. 28-11 shows the general plan of the *de novo* purine nucleotide synthesis along with the origin of the carbon and nitrogen atoms of the purine ring system, as determined by them. At that time, 10-formyltetrahydrofolate was not known, but compounds such as formate or serine labeled in the hydroxymethyl carbon readily labeled C-2 and C-8 of uric acid.

Purines are synthesized *de novo*, beginning with simple starting materials such as amino acids and bicarbonate (Fig. 28–11). *Unlike the case for pyrimidines, the purine bases are assembled already attached to the ribose ring.*

The *de novo* pathway for purine nucleotide synthesis, in a stepwise series of 11 reactions, leads to the synthesis of inosine 5′-monophosphate (IMP). IMP is the precursor for both adenosine 5′-monophosphate (AMP) and guanosine 5′-monophosphate (GMP). All enzymes involved in this synthesis are found in the cytosol. However, not all cells (*e.g.*, red cells) are capable of *de novo* purine nucleotide synthesis. Hydrolysis of ATP is required to drive several reactions in this pathway. Overall, the *de novo* pathway for purine nucleotide synthesis is expensive in terms of moles of ATP utilized per mole of IMP synthesized. This pathway is highly regulated by AMP and GMP; IMP is not normally found to any extent in cells. The pathway discussed here (Fig. 28–12), is identical in all organisms, with the exception of one step (Step ⑥) that differs in higher eukaryotes, as noted below.

Step ① : *de novo* purine biosynthesis like pyrimidine biosynthesis, requires 5-phosphoribosyl 1-pyrophosphate (PRPP), but for purines, PRPP provides the foundation on which the bases are constructed step-by-step. The initial committed step (Step①) is the displacement of pyrophosphate by ammonia, rather than by a preassembled base, to produce 5-phospho-β-D-ribosylamine (PRA), which carries the β configuration at C-1 of the sugar, as do all the common nucleotides. An inversion of configuration thus occurs in this reaction, as the amido nitrogen displaces the pyrophosphate moiety. 5-phosphoribosylamine is highly unstable, with a half-life of 30 seconds at pH 7.5. The purine ring is subsequently built up on this structure.

In the formation of 5-phosphoribosylamine, the N—C bond is formed that will ultimately be the *N*-glycosidic bond of the purine nucleotide. *Glutamine phosphoribosyl amidotransferase* catalyzes this reaction.

Glutamine phosphoribosyl amidotransferase comprises 2 domains: the first is homologous to the phosphoribosyltransferases in salvage pathways, whereas the second produces ammonia from glutamine by hydrolysis. However, this glutamine-hydrolysis domain is distinct from the domain that performs the same function in carbamoyl phosphate synthetase. In glutamine phosphoribosyl amidotransferase, a cysteine residue located at the amino terminus facilitates glutamine hydrolysis. To prevent wasteful hydrolysis of either substrate, the amidotransferase assumes the active configuration only on binding of both PRPP and glutamine. As is the case with carbamoyl phosphate synthetase, the ammonia generated at the glutamine-hydrolysis active site passes through a channel to reach PRPP without being released into solution. – additional steps are required to assemble the purine ring. Remarkably, the first steps are analogous reactions. Most of these steps are catalyzed by enzymes with ATP-grasp domains that are homologous to those in carbamoyl phosphate synthetase. *Each step consists of the activation of a carbon-bound oxygen atom (typically a carbonyl oxygen atom) by phosphorylation, followed by the displacement of a phosphoryl group by ammonia or an amine group acting as a nucleophile (Nu).*

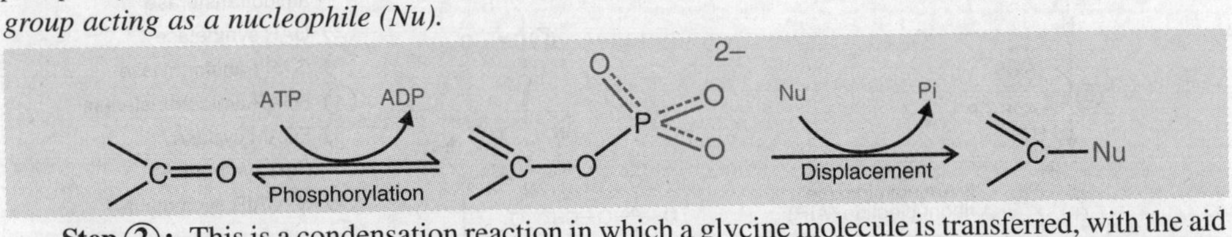

Step ②: This is a condensation reaction in which a glycine molecule is transferred, with the aid of ATP, to the nitrogen of phosphoribosylamine. In fact, a carboxylate group of a glycine residue is activated by phosphorylation and then coupled to the amino group of phosphoribosylamine, forming **glycinamide ribonucleotide (GAR)**. Thus, a new amide bond is formed while the amino group of glycine is free to act as a nucleophile in the next reaction (step③). This reaction is catalyzed by *GAR synthetase*.

Step ③. In this reaction a formyl group from N^{10}-formyltetrahydrofolate is activated and then attached to the newly-added amino group of glycine to form **formylglycinamide ribonucleotide (FGAR).** The reaction is catalyzed by *GAR transformylase*. In some organisms, two distinct enzymes can catalyze this step. One enzyme transfers the formyl group from N^{10}-formyl tetrahydrofolate. The other enzyme activates formate as formyl phosphate, which is added directly to the glycine amino group.

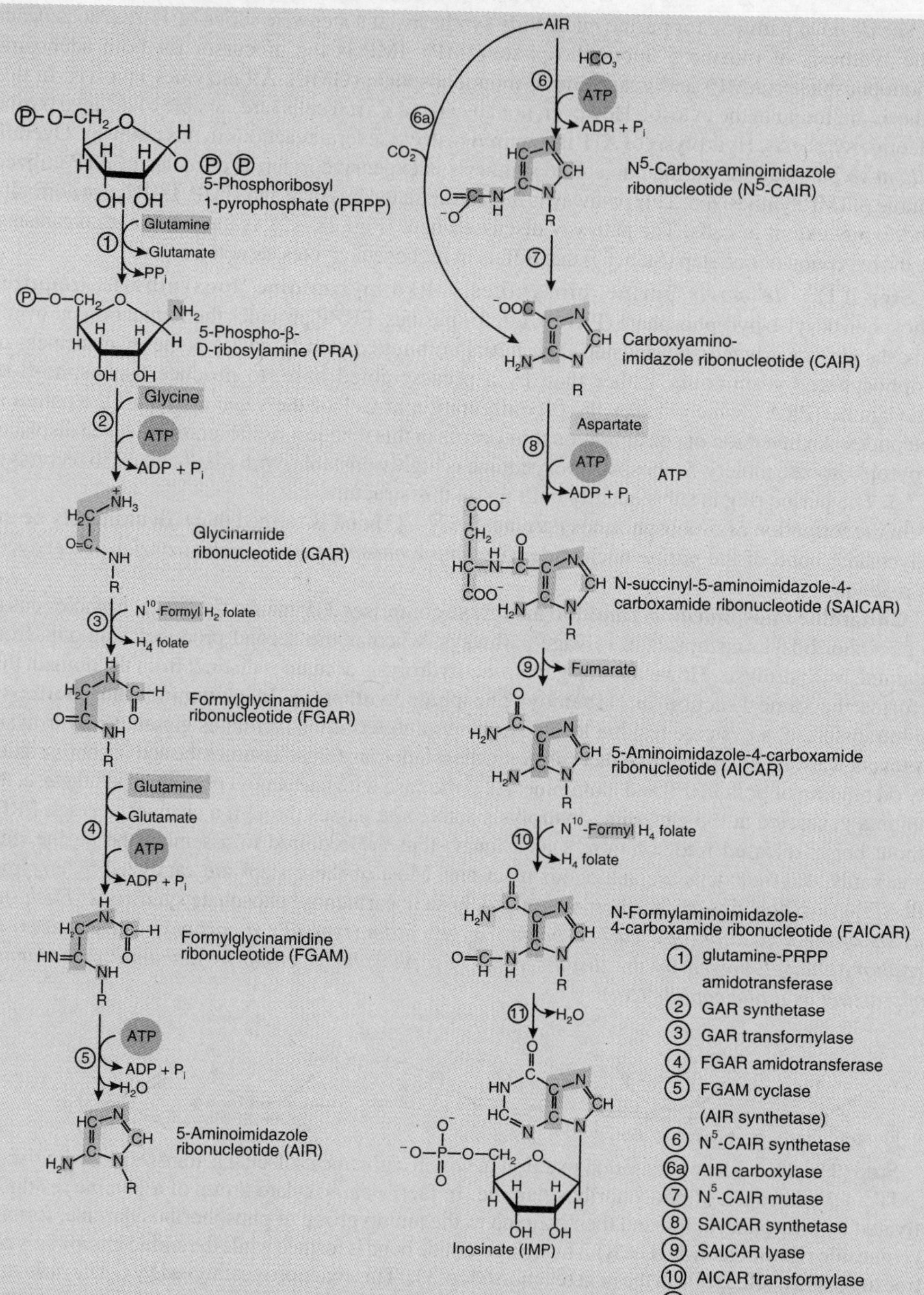

Fig. 28–12. *de novo* synthesis of purine nucleotides: formation of the purine ring of inosinate (IMP) Each addition to the purine ring is shaded differently to show the origin of atoms. After (step ②), R symbolizes the 5-phospho-D-ribosyl group on which the purine ring is built. Formation of 5-phosphoribosylamine (Step①) is the first committed step in purine synthesis. Abbreviations are given for most intermediates to simplify the naming of the pathway enzymes. (Step⑥a) is the alternative path from AIR to CAIR occurring in higher eukaryotes.

Step ④. In this reaction, an inner group is activated and then converted into an amidine, (**formylglycinamidine ribonucleotide, FGAM**), by the addition of ammonia derived from glutamine. The reaction is catalyzed by an ATP-dependent enzyme, *FGAR amidotransferase*.

Step ⑤. In this reaction, FGAM undergoes dehydration and a ring closure (cyclization) takes place to yield the 5-membered imidazole ring of the purine nucleus, as **5-aminoimidazole ribonucleotide (AIR)**. Although this cyclization is likely to be favourable, thermodynamically, a molecule of ATP is consumed to ensure irreversibility. The familiar pattern is repeated: a phosphoryl group from the ATP molecule activates the carbonyl group and is displaced by the nitrogen atom attached to the ribose molecule. Cyclization is, thus, an intramolecular reaction in which the nucleophile and phosphate-activated carbon atom are present within the same molecule. At this point, 3 of the 6 atoms needed for the second ring in the purine structure are in place. The enzyme *FGAM cyclase (= AIR synthetase)* catalyzes this reaction.

Step ⑥. In this reaction, bicarbonate is activated by phosphorylation and then attacked by the exocyclic amino group. This carboxylation reaction, which is catalyzed by N^5*-CAIR synthetase*, is unusual in that it does not require biotin, but instead uses the bicarbonate, as stated above, generally present in aqueous solutions. The product formed is **N^5-carboxyaminoimidazole ribonucleotide (N^5-CAIR)**.

Step ⑦. This is a rearrangement reaction in which the carboxylate from the exocyclic amino group is transferred to position 4 of the imidazole ring, forming **carboxyaminoimidazole ribonucleotide (CAIR);** the reaction being catalyzed by N^5*-CAIR mutase*.

It is noteworthy that steps ⑥ and ⑦ are found only in bacteria and fungi. Higher eukaryotes, including humans, do not require ATP for this step and in them the 5-aminoimidazole ribonucleotide (the product of step ⑤) is carboxylated directly to carboxyaminoimidazole ribonucleotide (which is the product of step ⑦) in one step instead of two (Step ⑥a); bicarbonate apparently attaches directly to the exocyclic amino group and is then transferred to the imidazole ring. The enzyme catalyzing this reaction (Step (6a)) is *AIR carboxylase*.

Step ⑧. In this reaction, the imidazole carboxylate is phosphorylated again and the phosphate group is displaced by the amino group of aspartate, forming, ***N*-succinyl-5-aminoimidazole-4-carboxamide ribonucleotide (SAICAR);** the reaction being catalyzed by *SAICAR synthetase*. The process is quite analogous to the synthesis of argininosuccinic acid in the urea cycle, where ATP is required as an energy source. In the present case, however, ATP is cleaved to ADP and Pi.

Thus, a 7-step process (from step ② to ⑧) links glycine, formate, ammonia, bicarbonate, and aspartate to form an intermediate that contains all but two of the atoms necessary for the formation of the purine ring.

Step ⑨. In this reaction, fumarate, an intermediate in the citric acid cycle, is eliminated, leaving the nitrogen atom from aspartate joined to the imidazole ring, yielding **5-aminoimidazole-4-carboxamide ribonucleotide (AICAR)**, the intermediate shown to accumulate in sulfonamide-treated bacteria. *SAICAR lyase* catalyzes this reaction. The reaction is quite analogous to the cleavage of arginosuccinic acid in the urea cycle.

Step ⑩. This is another transformylase reaction (besides step ③) where a single-carbon formyl group from N^{10}-formyltetrahydrofolate is added to the nitrogen atom of the imidazole ring to form the final intermediate, **N-formylaminoimidazole-4-carboxamide ribonucleotide (FAICAR)**; obviously, the enzyme catalyzing this reaction being *AICAR transformylase*. In effect, this step acquires one final carbon atom needed for the formation of the 6-membered ring of purine.

Step ⑪. Finally, in this second ring-closure reaction (the first one being step ⑤), an internal condensation takes place so that the first purine compound **inositate (IMP)** is formed from FAICAR with the loss of water. The enzyme catalyzing this reaction is *IMP synthase*. Inositate does occur free in biological materials, but of course is not a component of RNA or DNA.

The enzymes of IMP synthesis, like those in the tryptophan (Trp) and histidine (His) pathways, appear to be organized as large, multienzyme complexes in the cell. In eukaryotic cells, ranging from yeasts to fruitflies (*Drosophila*) to chickens, steps ①, ③ and ⑤ (refer Fig. 28–9) are catalyzed by a multifunctional protein. An additional multifunctional protein catalyzes steps ⑩ and ⑪. In humans, a multifunctional enzyme combines the activities of AIR carboxylase and SAICAR synthetase (Steps ⑥a and ⑧). In bacteria, these activities are found on separate proteins, but a large, noncovalent complex may exist in these cells. The channeling of reaction intermediates from one enzyme to the next permitted by these complexes is probably especially important for unstable intermediates such as 5-phosphoribosylamine (PRA).

Some *salient features* of IMP synthesis need be pointed out:

1. 5′-phosphoribosyl-1-pyrophosphate (PRPP) is synthesized from ribose 5-phosphate, generated by the pentose phosphate pathway (PPP).
2. The equivalent of 6 mole of ATP is utilized per mole of IMP synthesized.
3. The formation of 5-phosphoribosylamine (PRA) (Step ①) is the committed and regulated step.
4. In the formation of PRA, the N—C bond is formed that will ultimately be the *N*-glycosidic bond of the purine nucleotide.
5. There are no known regulated steps between PRA and IMP.
6. Tetrahydrofolate serves as a "C_1" carrier (N^{10}-formyltetrahydrofolate) in this pathway.
7. In steps ③ and ⑩, tetrahydrofolate is regenerated.
8. 5-aminoimidazole ribonucleotide (AIR) carboxylase, which in higher eukaryotes catalyzes the reaction in which CO_2 is used to introduce carbon-6 of the ring (Step ⑥), is not a biotin-dependent carboxylase.

If we express IMP synthesis pathway (Fig. 28–9), in terms of the overall reaction, we can write:

$$2NH_3 + 2 \text{ formic acid} + CO_2 + \text{glycine} + \text{aspartic acid} + \text{ribose 5-phosphate} \longrightarrow \text{Inosinic acid} + \text{fumaric acid} + 9H_2O$$

The energy required to accomplish this process is, of course, provided by 9 ATP molecules, including 2 required for glutamine formation and 2 for formate activation. All but one of these ATP molecules are cleaved to produce ADP and Pi. Thus, we have another example of the means by which

$$9\text{ ATP} + 9\text{ H}_2\text{O} \longrightarrow 8\text{ ADP} + 8\text{ Pi} + \text{AMP} + \text{PPi}$$

the energy-richness of ATP can be utilized in discrete reactions to accomplish a biosynthetic sequence requiring energy. The equations, as written, would apply only in those organisms which can activate free formate. In higher vertebrates, the one-carbon unit is derived from serine.

By way of comparison, the purine nucleotides possess a 6-membered ring (with 2 nitrogen atoms and 3 double bonds) fused to the 5-membered imidazole ring (with 2 nitrogen atoms and 1 double bond) whereas pyrimidine nucleotides possess only the 6-membered heterocyclic ring, akin to that present in the purine nucleus. These nuclotides have chemical and physiologic properties similar to those of pyrimidine nucleotides. Although the purine nucleus is more complex and its synthetic pathway longer than that of the pyrimidine structure, the two share several common precursors. PRPP, glutamine, CO_2, and aspartate are required for the synthesis of all purine and pyrimidine nucleotides. For all purine nucleotides and for the thymidine nucleotides, tetrahydrofolate derivatives are also necessary. There is, however, one striking difference between the synthesis of purine nucleotides and that of pyrimidine nucleotides, namely, that in the synthesis of pyrimidine nucleotides the pyrimidine base is formed and attachment of the ribose phosphate moiety is delayed until the later steps of the pathway, whereas the synthesis of the purine nucleotides commences with ribose phosphate as an integral part of the earliest precursor molecule.

BIOSYNTHESIS OF AMP AND GMP FROM IMP

(= Purine Nucleotide Interconversions)

Inosinic acid (IMP) represents a branch point in purine nucleotide synthesis. Fig. 28–13 depicts the conversion of this compound both to adenosine 5′-monophosphate (AMP) and guanosine 5′-monophosphate (GMP); both conversions being a two-step process.

A. Synthesis of AMP

IMP first participates with aspartic acid and GTP to form **adenylosuccinic acid**, a substituted succinic acid. The reaction involves the transfer of an amino group from Asp to IMP. Note that the nucleoside triphosphate derivative of one purine (guanine) is required for the formation of another purine nucleotide. This reaction is catalyzed by *adenylosuccinate synthetase* and is strictly analogous to Step (8) of IMP synthesis pathway but with a crucial difference that GTP rather than ATP is the source of high-energy phosphate in synthesizing adenylosuccinic acid. In the second step, nonhydric cleavage of adenylosuccinic acid yields **AMP** and **fumaric acid**. This reaction is catalyzed by *adenylosuccinate lyase* and is analogous to step ⑨ of IMP synthesis pathway.

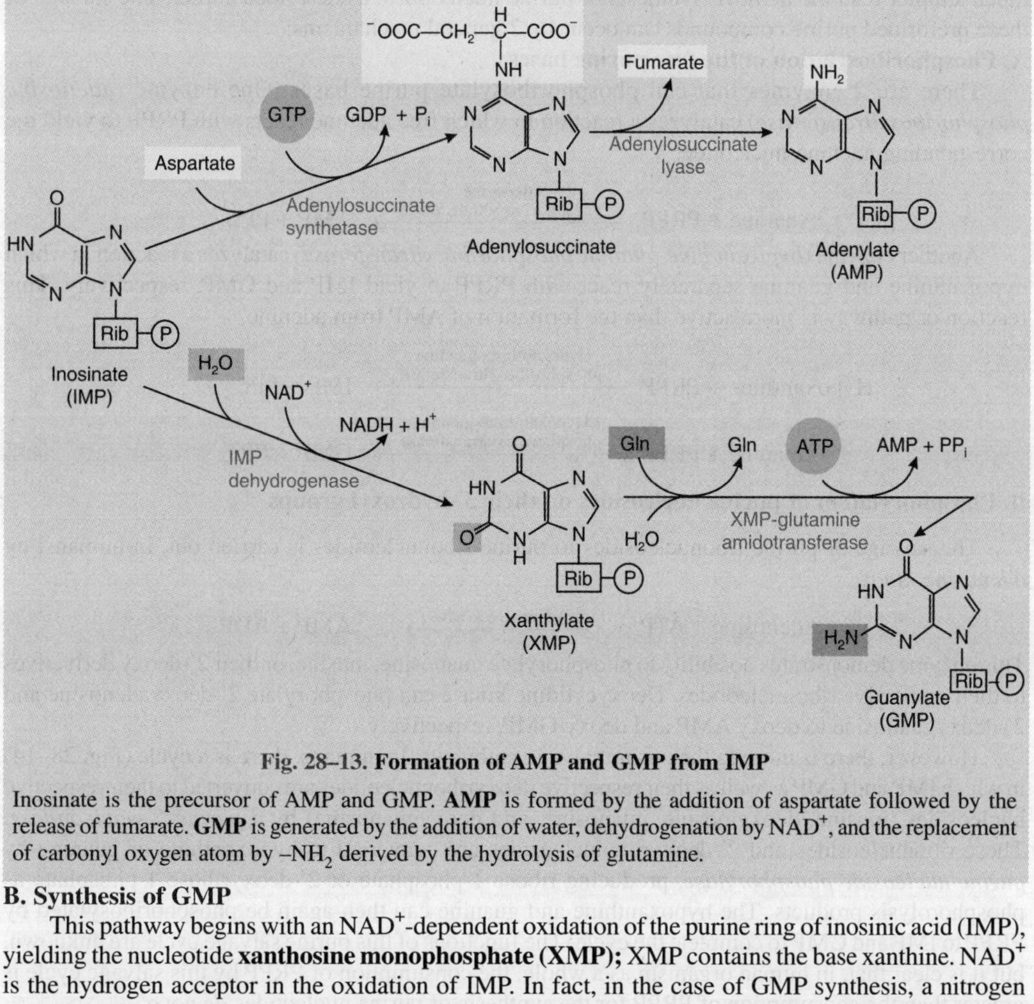

Fig. 28–13. Formation of AMP and GMP from IMP

Inosinate is the precursor of AMP and GMP. **AMP** is formed by the addition of aspartate followed by the release of fumarate. **GMP** is generated by the addition of water, dehydrogenation by NAD^+, and the replacement of carbonyl oxygen atom by $-NH_2$ derived by the hydrolysis of glutamine.

B. Synthesis of GMP

This pathway begins with an NAD^+-dependent oxidation of the purine ring of inosinic acid (IMP), yielding the nucleotide **xanthosine monophosphate (XMP);** XMP contains the base xanthine. NAD^+ is the hydrogen acceptor in the oxidation of IMP. In fact, in the case of GMP synthesis, a nitrogen atom must be introduced in position 2 of the purine nucleus. To do this, the carbon atom at that position in IMP must first be oxidized to the higher oxidation state in order to acquire the amino group which is already at that oxidation level. The enzyme *IMP dehydrogenase* catalyzes this reaction. In the next step, XMP is activated by the transfer of an AMP group (rather than a phosphoryl group) from ATP to the oxygen atom in the newly-formed carbonyl group. Ammonia generated by the hydrolysis of glutamine, then displaces the AMP group to form **GMP**, in a reaction catalyzed by *XMP-glutamine amidotransferase*. It may, thus, be seen that the synthesis of AMP requires GTP, whereas the synthesis of GMP requires ATP.

Note from Fig. 28–9 that the energy to drive the aspartate transfer reaction comes not from ATP but from GTP. This may represent a way to control the proportions of IMP that go to adenine and guanine nucleotide synthesis. GTP accumulation would tend to promote the pathway toward adenine nucleotides. Also, because the conversion of XMP to GMP is ATP-dependent, accumulation of ATP could promote guanine nucleotide synthesis.

SALVAGE PATHWAYS FOR PURINES

Free purine and pyrimidine bases are constantly released in cells during metabolic degradation of nucleotides. Free purines are, in large part, salvaged and reused to make nucleotides, in a pathway much simpler than the *de novo* synthesis of purine nucleotides, as described earlier. The salvage of these preformed purine compounds can occur by 2 general mechanisms:

A. Phosphoribosylation of the free purine bases

There are 2 enzymes that can phosphoribosylate purine bases. One enzyme (*adenosine phosphoribosyltransferase*) catalyzes a reaction in which free adenine reacts with PRPP to yield the corresponding adenine nucleotide:

$$\text{Adenine + PRPP} \xrightarrow[\text{phosphoribosyltransferase}]{\text{Adenosine}} \text{AMP + PPi}$$

Another enzyme (*hypoxanthine-guanine phosphoribosyltransferase*) catalyzes a reaction in which hypoxanthine and guanine separately react with PRPP to yield IMP and GMP, respectively. This reaction or pathway is more active than the formation of AMP from adenine.

$$\text{Hypoxanthine + PRPP} \xrightarrow[\text{phosphoribosyltransferase}]{\text{Hypoxanthine-guanine}} \text{IMP + PPi}$$

$$\text{Guanine + PRPP} \xrightarrow[\text{phosphoribosyltransferase}]{\text{Hypoxanthine-guanine}} \text{GMP + PPi}$$

B. Phosphorylation of purine nucleosides on their 5′-hydroxyl groups

The salvage of purine ribonucleosides to purine ribonucleotides is carried out, in humans, by *adenosine kinase*.

$$\text{Adenosine + ATP} \xrightarrow{\text{Adenosine kinase}} \text{AMP + ADP}$$

This enzyme demonstrates no ability to phosphorylate guanosine, inosine, or their 2′-deoxy derivatives to their respective ribonucleotides. Deoxycytidine kinase can phosphorylate 2′-deoxyadenosine and 2′-deoxyguanosine to deoxy AMP and deoxy GMP, respectively.

However, there is more to the purine salvage pathways. In humans, there is a cycle (Fig. 28–14) in which IMP and GMP as well as their respective deoxyribonucleotides are converted to their respective nucleosides (inosine, deoxyinosine, guanosine, and deoxyguanosine) by a *purine 5′-nucleotidase.* These ribonucleosides and 2′-deoxyribonucleosides are converted to hypoxanthine or guanine by *purine nucleoside phosphorylase*, producing ribose 1-phosphate or 2′-deoxyribose 1-phosphate as phosphorolysis products. The hypoxanthine and guanine can then again be phosphoribosylated by PRPP to IMP and GMP to complete the cycle. The functions of this purine salvage cycle are unknown, but it is clear that, in human organism as a whole, the consumption of PRPP by this salvage cycle is greater than the consumption of PRPP for the synthesis of purine nucleotides *de novo*.

There is a lateral pathway of this cycle which involves the conversion IMP to AMP (refer Fig. 28–13), with the subsequent conversion of AMP to adenosine. The latter is probably catalyzed by the same purine 5′-nucleotidase that hydrolyzes IMP to inosine. The adenosine, so produced, is then either salvaged directly back to AMP *via adenosine kinase*, or is converted to inosine by the enzyme *adenosine deaminase*. Quantitatively, the functoin of this inosine loop is less important than the previously-described cycle. Quantitatively, the action of adenosine deaminase is an important process, especially for the immune system.

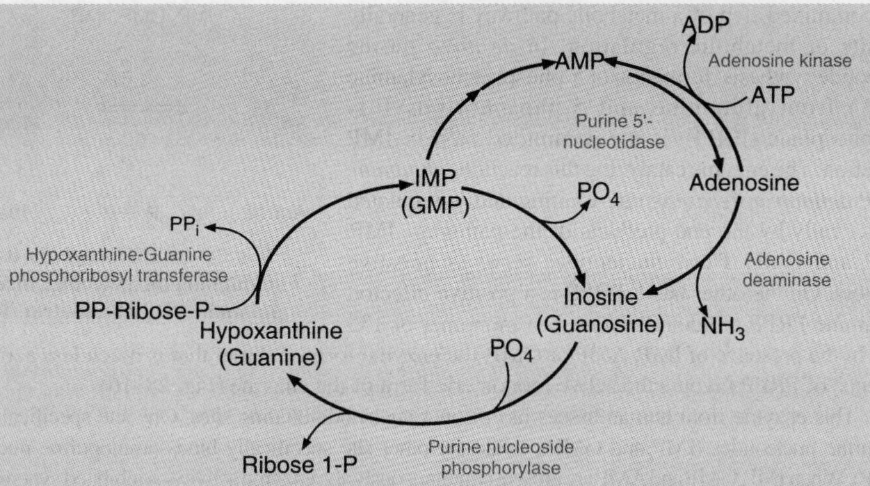

Fig. 28–14. The purine salvage cycles involving the interconversion of AMP, IMP and, to a lesser extent, GMP to their respective ribonucleosides and their eventual reconversion to purine ribonucleotides
Deoxyadenosine, deoxyinosine, and deoxyguanosine share the same pathways except that deoxyguanosine can be directly phosphorylated to deoxy GMP.

The salvage of the free purine adenine by *adenine phosphoryltransferase* seems to prevent the xanthine oxidase-mediated oxidation of adenine to 2,8-dihydroxyadenine. Dihydroxyadenine is a highly insoluble product which appears as kidney stones in patients devoid of adenine phosphoribosyltransferase activity. The free adenine for such a salvage process probably derives from the hydrolysis of *5′-methylthioadenosine*, a by-product of the synthesis of polyamines from S-adenosylmethionine.

REGULATION OF PURINE NUCLEOTIDE SYNTHESIS

The synthesis of purine nucleotides in living beings is controlled by feedback inhibition at several sites (Fig. 28–15).

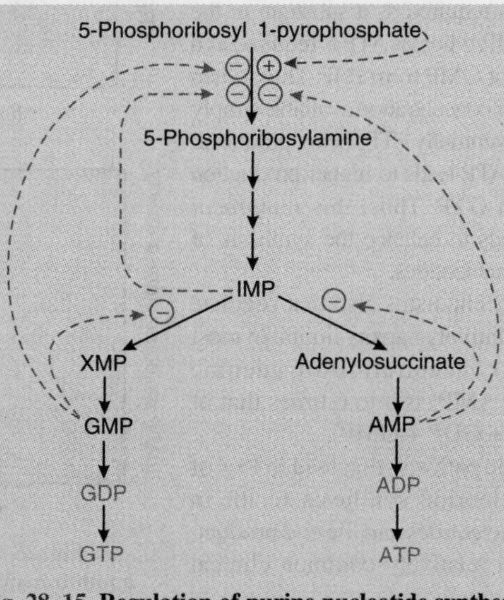

Fig. 28–15. Regulation of purine nucleotide synthesis

1. The committed step of a metabolic pathway is generally the site of metabolic regulation. In *de novo* purine nucleotide synthesis, formation of 5-phosphoribosylamine (PRA) from glutamine and 5-phosphoribosyl 1-pyrophosphate (PRPP) is the committed step in IMP formation. The enzyme catalyzing this reaction, *glutamine PRPP amidotransferase* is rate-limiting and is regulated allosterically by the end products of the pathway– IMP, AMP, and GMP. These nucleotides serve as negative effectors. On the other hand, PRPP is a positive effector. Glutamine PRPP amidotransferase is a monomer of 135 kDa. In the presence of IMP, AMP or GMP, the enzyme forms a dimer that is much less active. The presence of PRPP favours the active monomeric form of the enzyme (Fig. 28–16).

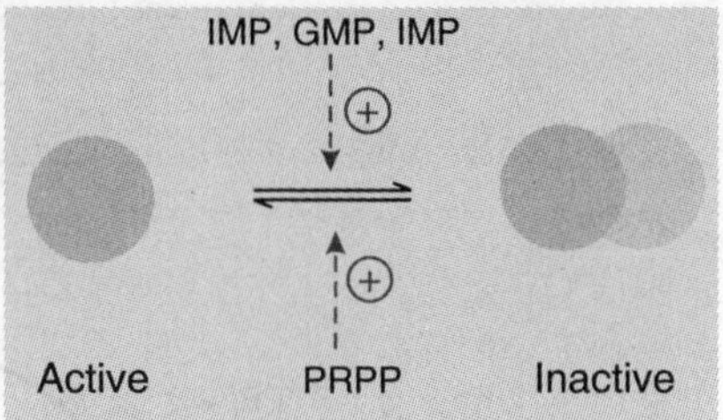

Fig. 28–16. Effects of allosteric modulators on molecular forms of glutamine PRPP amidotransferase

This enzyme from human tissues has distinct nucleotide-binding sites. One site specifically binds oxypurine nucleotides (IMP and GMP), while the other site specifically binds aminopurine nucleotides (AMP). When IMP, GMP and AMP are present simultaneously, the enzyme activity is inhibited synergistically. Glutamine-PRPP amidotransferase displays hyperbolic kinetics with respect to glutamine as the substrate and sigmoidal kinetics with respect to PRPP (Fig. 28–17). Since the intracellular concentration of glutamine varies relatively little and is close to the K_m of the enzyme, the glutamine concentration has little effect in regulating IMP synthesis. The intracellular concentration of PRPP, on the contrary, varies widely and can be 10 to 100 times less than the K_m for PRPP. Consequently, the concentration of PRPP plays an important role in regulating synthesis of purine nucleotides.

2. Inositate (IMP) is the branch point in the synthesis of AMP and GMP. *The reactions leading away from inositate are sites of feedback inhibition*. AMP inhibits the conversion of inositate into adenylo-succinate, its immediate precursor. Similarly, GMP inhibits the conversion of inositate into xanthylate (XMP), its immediate precursor.

3. As already noted, GTP is required as a substrate in the synthesis of AMP from IMP, whereas ATP is required as a substrate in the synthesis of GMP from IMP. Thus, when GTP is in excess, the higher concentration available simply leads to more AMP (and eventually ATP) being produced. Conversely, an excess of ATP leads to higher production of GMP and henceforth GTP. Thus, this *reciprocal substrate relationship* tends to balance the synthesis of adenine and guanine ribonucleotides.

There must be other mechanisms also that regulate the ATP/GTP ratio within relatively narrow limits. In most cells, the total cellular concentration of adenine nucleotides (ATP + ADP + AMP) is 4 to 6 times that of guanine nucleotides (GTP + GDP + GMP).

Defects in the metabolic pathway that lead to loss of regulation of purine nucleotide synthesis result in overproduction of purine nucleotides and the end product, uric acid. This results in a relatively common clinical condition called **gout**.

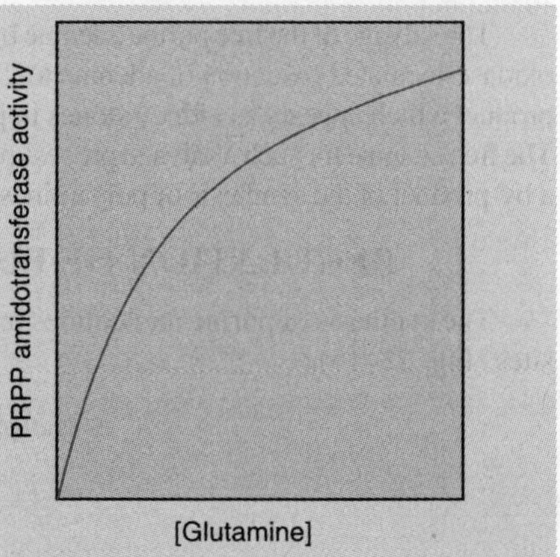

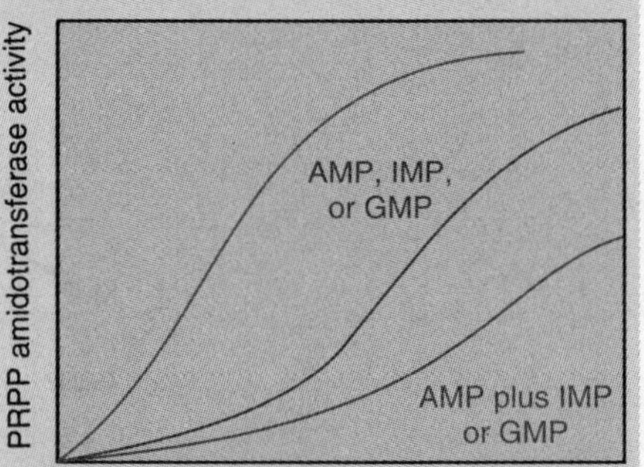

Fig. 28–17. Glutamine-PRPP amidotransferase activity as a function of glutamine or PRPP concentrations

CLINICAL IMPLICATIONS 28–2.

Gout

Gout is a disease of the joints caused by an elevated concentration of uric acid in the blood and tissues. The joints become inflammed, painful, and arthritic, owing to the abnormal deposition of sodium urate crystals. The kidneys are also affected, as excess uric acid is deposited in the kidney tubules. There is a large sex difference in the occurrence of gout: only about 5% of the cases occur in females. In many of the cases there appears to be a familial incidence. Hyperuricemia is frequently observed in the asymptomatic male relatives of gouty individuals. Eating and drinking purine-rich foods are apt to stimulate acute gouty attacks in susceptible individuals. Because such foods include "rich" items such as liver, sweetbreads, and wine, gout is historically linked with an excess of 'high living'. Hyperuricemia from overproduction of uric acid *via* the *de novo* pathway can be distinguished from hyperuricemia that results from kidney disease or excessive cell death (*e.g.*, increased degradation of nucleic acids from radiation therapy). Feeding of ^{15}N glycine to a patient who is an overproducer will result in uric acid excreted in urine that is enriched in ^{15}N at the N-7 of uric acid while in a patient who is not an overproducer, there will be no enrichment of ^{15}N in uric acid from these patients.

Gout results either from overproduction of purine nucleotides, leading to excessive uric acid synthesis, or from impaired uric acid excretion through the kidneys. Several specific enzymatic defects can lead to excessive purine synthesis, (Fig. 28–18). These biochemical defects are:

1. ***Increased PRPP synthetase activity:*** Elevated activity of PRPP synthetase results in increased intracellular concentrations of PRPP. PRPP acts as a positive effector of glutamine–PRPP amidotransferase, leading to increased flux through the *de novo* pathway since activity of the rate-limiting step is markedly increased.
2. ***Mutations in PRPP amidotransferase:*** Mutations render PRPP amidotransferase less sensitive to feedback inhibition by purine nucleotides. This loss of control also increases flux through the major regulated step.

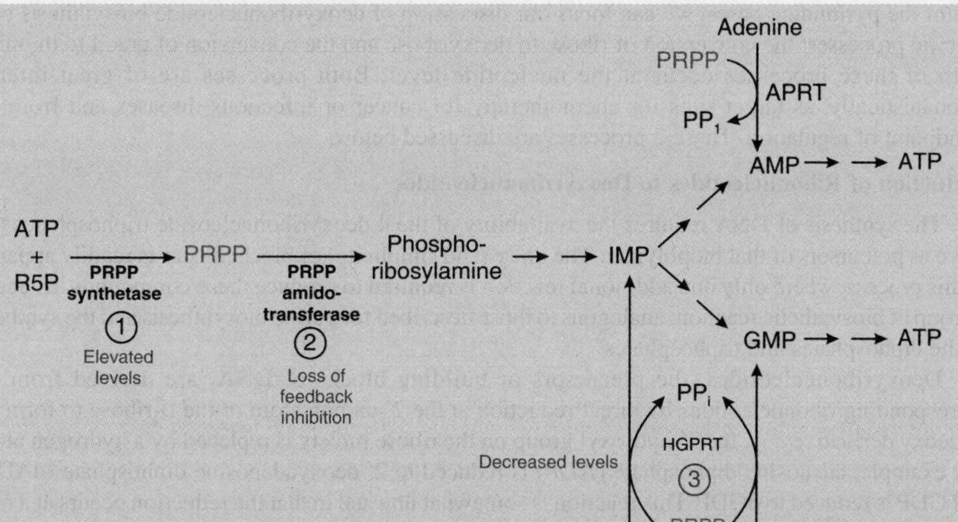

Fig. 28–18. Enzymatic abnormalities in 3 types of gout

Each of the 3 enzyme abnormalities shown here elevates the rate of *de novo* purine nucleotide biosynthesis, resulting in hyperuricemia and gout.
[HGPRT = Hypoxanthine-guanine phosphoribosyltransferase APRT = Adenine phosphoribosyltransferase]

3. ***Decreased HGPRTase activity:*** Partial decrease in HGPRTase activity has 2 follouts with respect to the *de novo* pathway for purine nucleotide synthesis. First, since there is decreased salvage of hypoxanthine and guanine, PRPP is not consumed by the HGPRTase reaction and PRPP can activate glutamine–PRPP amidotransferase activity. Second, with decreased salvage of hypoxanthine and guanine, IMP and GMP are not formed *via* this pathway so that regulation of the PRPP amidotransferase step by IMP and GMP as negative effectors is compromised.

Besides the above-mentioned enzymatic abnormalities, gout may also result from impaired uric acid excretion. Patients with some forms of glycogen storage disease are gouty. Evidently, prolonged hypoglycemia causes accumulation of organic acids (lactate and the like), and this accumulation interferes with tubular secretion of uric acid in the kidney. Gout is also a consequence of cancer therapy, presumably resulting from an overload of purines caused by nucleic acid degradation after death of tumour cells.

de novo DEOXYRIBONUCLEOTIDE SYNTHESIS

Most cells contain 5 to 10 times as much RNA as DNA. Moreover, ribonucleotides have multiple metabolic roles, while deoxyribonucleotides serve only as constituents of DNA. Henceforth, most of the carbon that flows through nucleotide synthetic pathways goes into ribonucleoside triphosphate (rNTP) pools. However, the relatively small fraction that is diverted to the synthesis of deoxyribonucleoside triphosphates (dNTPs) is of paramount importance in the life of the cell. dNTPs are used almost exclusively in the biosynthesis of DNA. Consequently, there are especially close regulatory relationships between DNA synthesis and dNTP metabolism– more so than seen between other macromolecular biosyntheses and the pathways that provide their precursors. The overall pathway of deoxyribonucleoside triphosphate (dNTP) biosynthesis is shown in Fig. 28–19.

Recalling that DNA differs chemically from RNA in the nature of the sugar and in the identity of one of the pyrimidine bases, we can focus our discussion of deoxyribonucleotide biosynthesis on 2 specific processes: the conversion of ribose to deoxyribose and the conversion of uracil to thymine. Both of these processes occur at the nucleotide level. Both processes are of great interest mechanistically, as target sites for chemotherapy for cancer or infectious diseases and from the standpoint of regulation. These 2 processes are discussed below.

Reduction of Ribonucleotides to Deoxyribonucleotides

The synthesis of DNA requires the availability of the 4 deoxyribonucleoside triphosphates that serve as precursors of that biopolymer. The unity (and simplicity) of biochemistry is readily apparent in this process, where only one additional reaction is required to produce these compounds instead of a group of biosynthetic reaction, analogous to those described for purine biosynthesis and the synthesis of the diphosphates and triphosphates.

Deoxyribonucleotides, the precursors or building blocks of DNA, are derived from the corresponding ribonucleotides by direct reduction at the 2′-carbon atom of the D-ribose to form the 2′-deoxy derivative, *i.e.,* the 2′-hydroxyl group on the ribose moiety is replaced by a hydrogen atom. For example, adenosine diphosphate (ADP) is reduced to 2′-deoxyadenosine diphosphate (dADP), and GDP is reduced to dGDP. This reaction is somewhat unusual in that the reduction occurs at a non-activated carbon; no closely analogous reactions are known. The general reaction, which is summarized in Fig. 28–20, is catalyzed by *nucleoside 5′-diphosphate* (also called *ribonucleotide reductase*), an enzyme discovered by **Peter Reichard** in 1998.

The reduction of the D-ribose portion of a ribonucleoside diphosphate to 2′-deoxy-D-ribose (Fig. 28–21) requires a pair of hydrogen atoms, which are ultimately donated by NADPH *via* an intermediate hydrogen-carrying protein called **thioredoxin** (This ubiquitous protein serves a similar redox function in photosynthesis and other processes). Thioredoxin is a small (108 amino acids) protein ($M_r \cong 12,000$) with 2 thiol (—SH) groups in the sequence Cys—Gly—Pro—Cys; these pairs of —SH groups carry

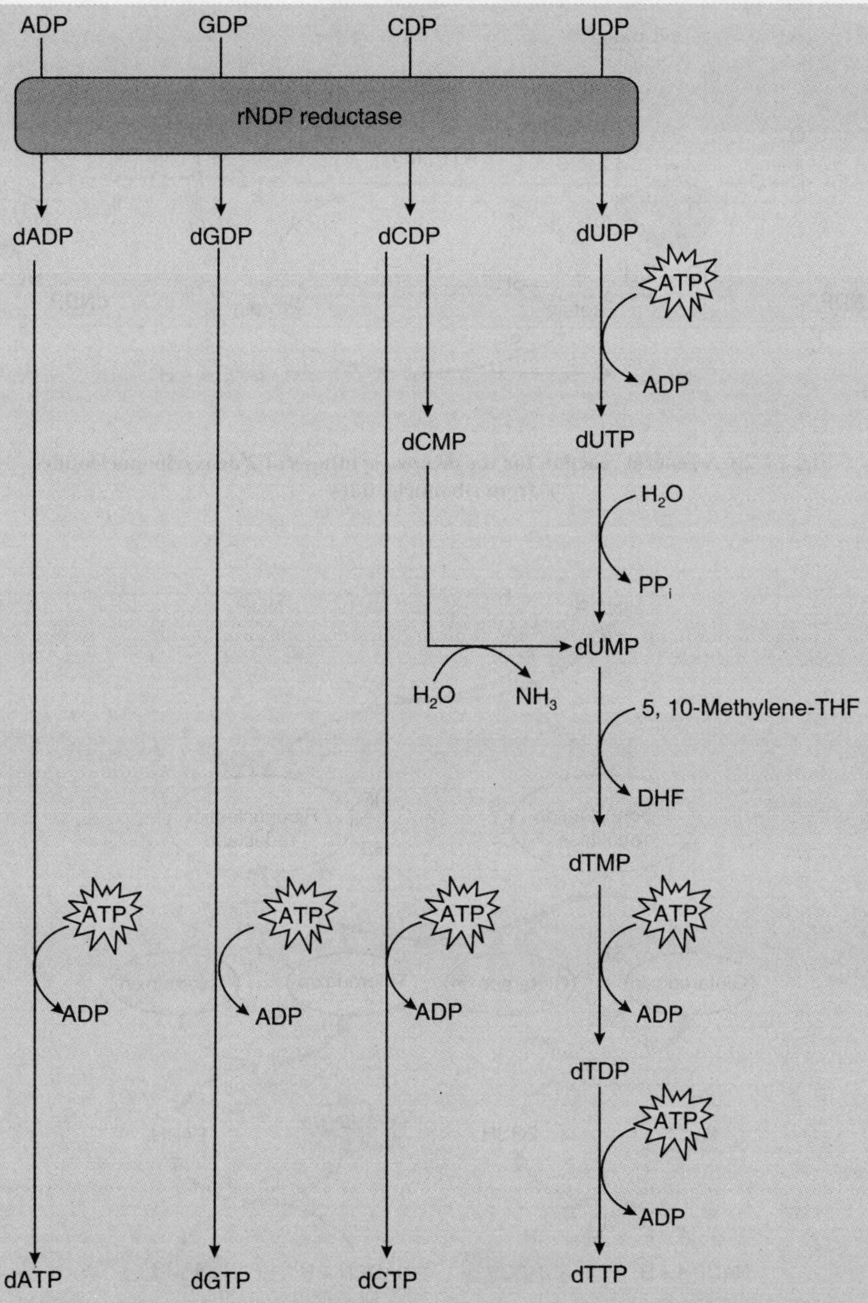

Fig. 28–19. An overview of the biosynthesis of deoxyribonucleoside triphosphate (dNTP)

hydrogen atoms from NADPH to the ribonucleoside diphosphate. Its oxidized (disulfide) form is reduced by NADPH in a reaction catalyzed by *thioredoxin reductase*, and reduced thioredoxin is then used by ribonucleotide reductase to reduce the nucleoside diphosphates (NDPs). A second source of reducing equivalents for ribonucleotide reductase is glutathione (GSH). GSH serves as the reductant for a protein closely related to thioredoxin, **glutaredoxin** (named so because of its ability to be reduced by glutathione), which then transfers the reducing power to ribonucleotide reductase.

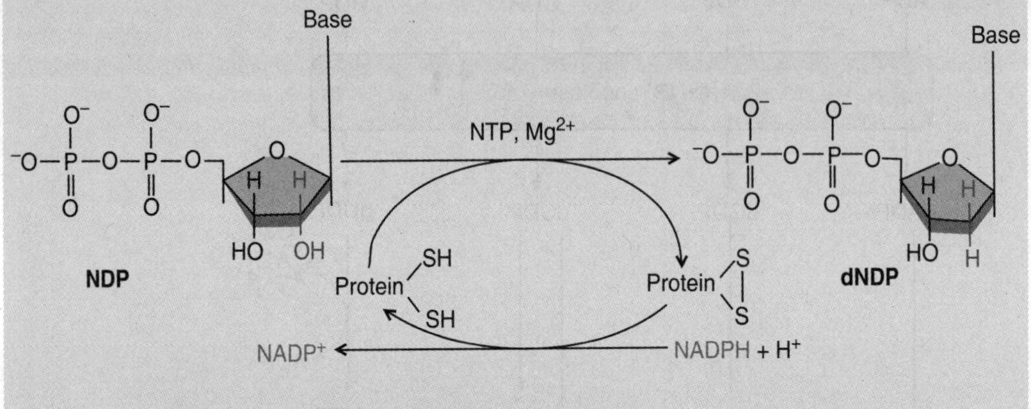

Fig. 28–20. A general reaction for the *de novo* synthesis of 2′deoxyribonucleotides from ribonucleotides

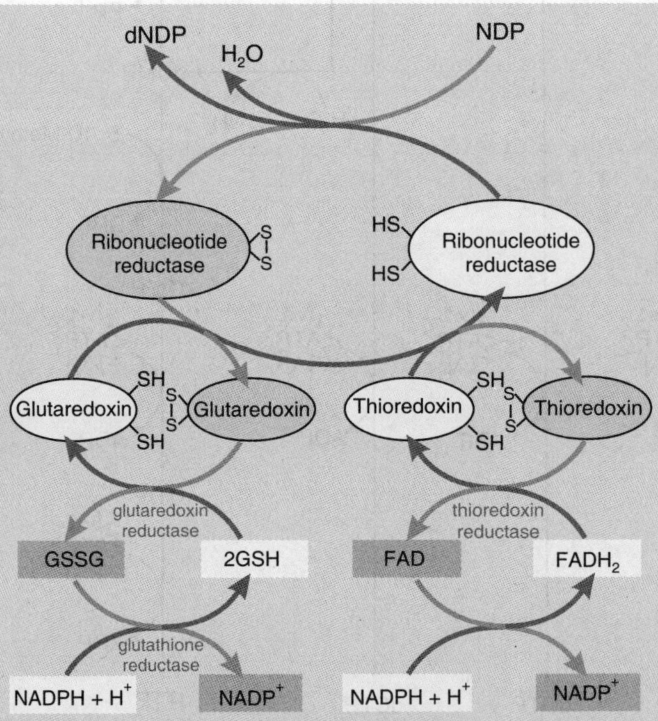

Fig. 28–21. Reduction of ribonucleotides to deoxyribonucleotides by ribonucleotide reductase

Electrons are transmitted (*blue* arrows) to the enzyme from NADPH by either glutaredoxin (***a***) or thioredoxin (***b***). The sulfide groups in glutaredoxin reductase are contributed by 2 molecules of bound glutathione (GSH; GSSG indicates oxidized glutathione). Note that thioredoxin reductase is a flavoenzyme, with FAD as prosthetic group.

Structure of Ribonucleotide Reductase

The enzyme **ribonucleotide reductase** is notable in that its reaction mechanism provides the *best-characterized example of the involvement of free radicals in biochemical transformations,* once thought to be rare in biological systems. This enzyme from *Escherichia coli* (Fig. 28–22) and most eukaryotes (including mammals) is an $\alpha_2\beta_2$ tetramer. The *E.coli* enzyme contains 2 nonidentical protein subunits (heterodimer), neither of which has enzyme activity. The 2 subunits are:

(*a*) **R1**, with two identical α polypeptide chains with *M*r of 87,000 each (*i.e.*, an 87-kd dimer), and

(*b*) **R2**, consisting of two identical β polypeptide chains with M_r of 43,000 each (*i.e.*, an 43-kd dimer).

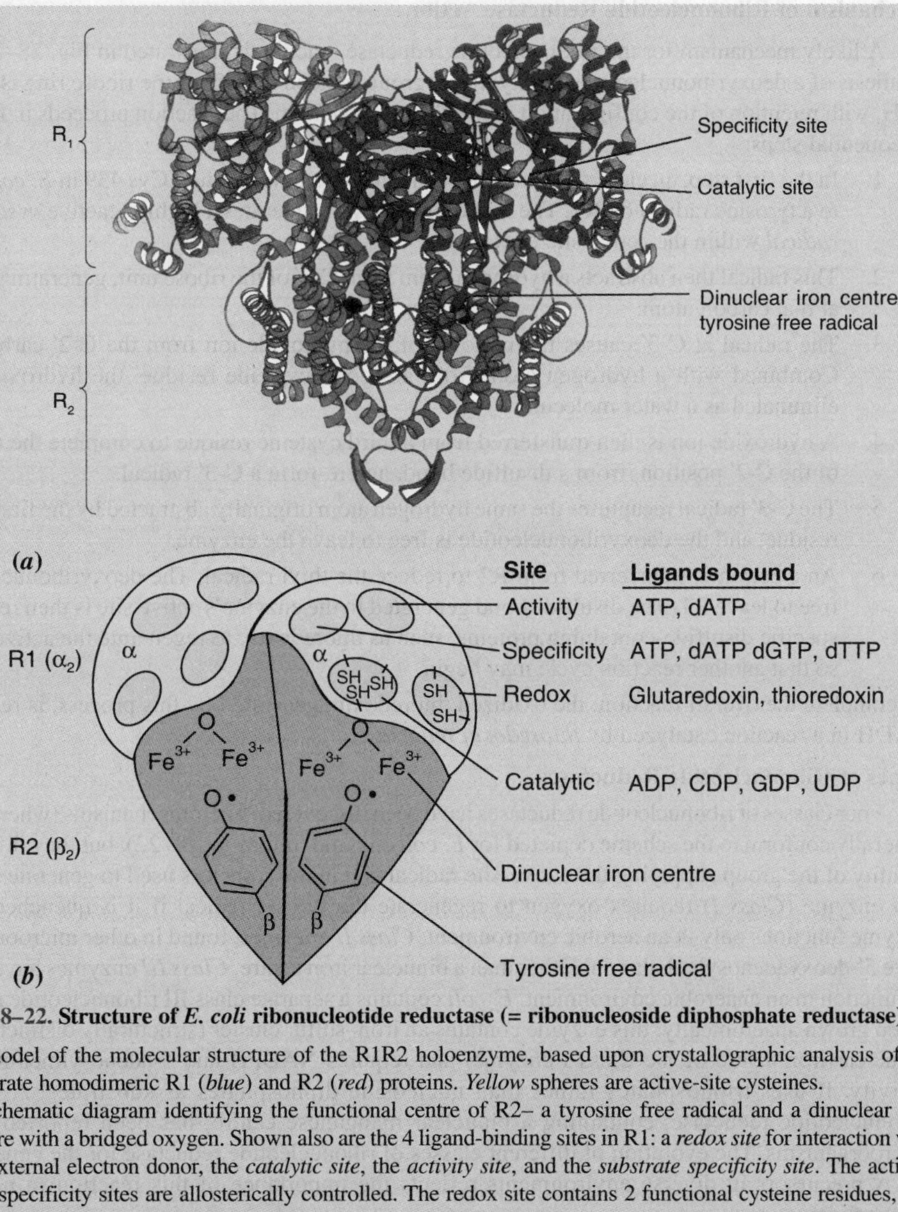

Fig. 28–22. Structure of *E. coli* ribonucleotide reductase (= ribonucleoside diphosphate reductase)

(*a*) A model of the molecular structure of the R1R2 holoenzyme, based upon crystallographic analysis of the separate homodimeric R1 (*blue*) and R2 (*red*) proteins. *Yellow* spheres are active-site cysteines.

(*b*) A schematic diagram identifying the functional centre of R2– a tyrosine free radical and a dinuclear iron centre with a bridged oxygen. Shown also are the 4 ligand-binding sites in R1: a *redox site* for interaction with an external electron donor, the *catalytic site*, the *activity site*, and the *substrate specificity site*. The activity and specificity sites are allosterically controlled. The redox site contains 2 functional cysteine residues, and the catalytic site contains 3 Cys residues.

(*Adapted from Uhlin U and Eklund H, 1994; © 1994 Macmillan Magazines Ltd.*)

The catalytic site lies in the large (R1) subuit. Within this site are 3 cysteine (Cys) residues which are conserved among different rNDP reductases. Two of the cysteine thiols are *redox-active*, so called because they undergo cyclic oxidation and reduction during the reaction. The third cysteine evidently functions as part of a free radical mechanism, as described below. The R2 dimer contains an unusual tyrosine free radical that is involved in the reaction and an oxygen atom bridging 2 ferric ions. This *dinuclear iron centre* stabilizes the free radical. The R1 protein contains 2 classes of regulatory sites. Finally, the R1 protein contains an additional pair of redox-active thiol groups, which interact with an external reductive cofactor. The mammalian enzyme is similar in structure.

Mechanism of Ribonucleotide Reductase Action

A likely mechanism for the ribonucleotide reductase reaction is illustrated in Fig. 28–23. In the synthesis of a deoxyribonucleotide, the hydroxyl group bonded to C-2′ of the ribose ring is replaced by H, with retention of the configuration at the C-2′ carbon atom. The reaction proceeds in following 6 sequential steps:

1. In the first step, an electron is transferred from a cysteine residue (Cys 439 in *E. coli*) on R1 to a tyrosine radical on R2. The loss of an electron generates a highly reactive *cysteine thiyl radical* within the active site of R1.
2. This radical then abstracts a hydrogen atom from C-3′ of the ribose unit, generating a radical at that carbon atom.
3. The radical at C-3′ causes the removal of the hydroxide ion from the C-2′ carbon atom. Combined with a hydrogen atom from a second cysteine residue, the hydroxide ion is eliminated as a water molecule.
4. A hydroxide ion is then transferred from a third cysteine residue to complete the reduction of the C-2′ position, from a disulfide bond, and re-form a C-3′ radical.
5. The C-3′ radical recaptures the same hydrogen atom originally abstracted by the first cysteine residue, and the deoxyribonucleotide is free to leave the enzyme.
6. An electron is transferred from R2 to reduce the thiyl radical. The deoxyribonucleotide is free to leave R1. The disulfide bond generated in the enzyme's active site is then reduced by specific disulfide-containing proteins, such as thioredoxin, to regenerate the active enzyme so that another reaction cycle may begin.

To complete the overall reaction, the oxidized thioredoxin, generated by this process, is reduced by NADH in a reaction catalyzed by *thioredoxin reductase*.

Types of Ribonucleotide Reductases

Four classes of ribonucleotide reductases have been discovered. Their mechanisms (where known) generally conform to the scheme depicted for *E. coli* enzyme (refer Fig. 28–22), but they differ in the identity of the group supplying the active-site radical and in the cofactors used to generate it. The *E. coli* enzyme (*Class I*) requires oxygen to regenerate the tyrosyl radical if it is quenched, so this enzyme functions only in an aerobic environment. *Class II* enzymes, found in other microorganisms, have 5′-deoxyadenosylcobalamin, rather than a binuclear iron centre. *Class III* enzymes have evolved to function in an anaerobic environment. *E. coli* contains a separate class III ribonucleotide reductase when grown anaerobically; this enzyme contains an iron–sulfur cluster (structurally distinct from the binuclear iron centre of the Class I enzyme) and requires NADPH and *S*-adenosylmethionine for activity. It uses triphosphates rather than nucleoside diphosphates as substrates. A *Class IV* ribonucleotide reductase, containing a binuclear manganese centre, has been reported in some microorganisms. The evolution of different classes of ribonucleotide reductase for the generation of DNA precursors in diverse environments reflects the importance of this reaction in nucleotide metabolism.

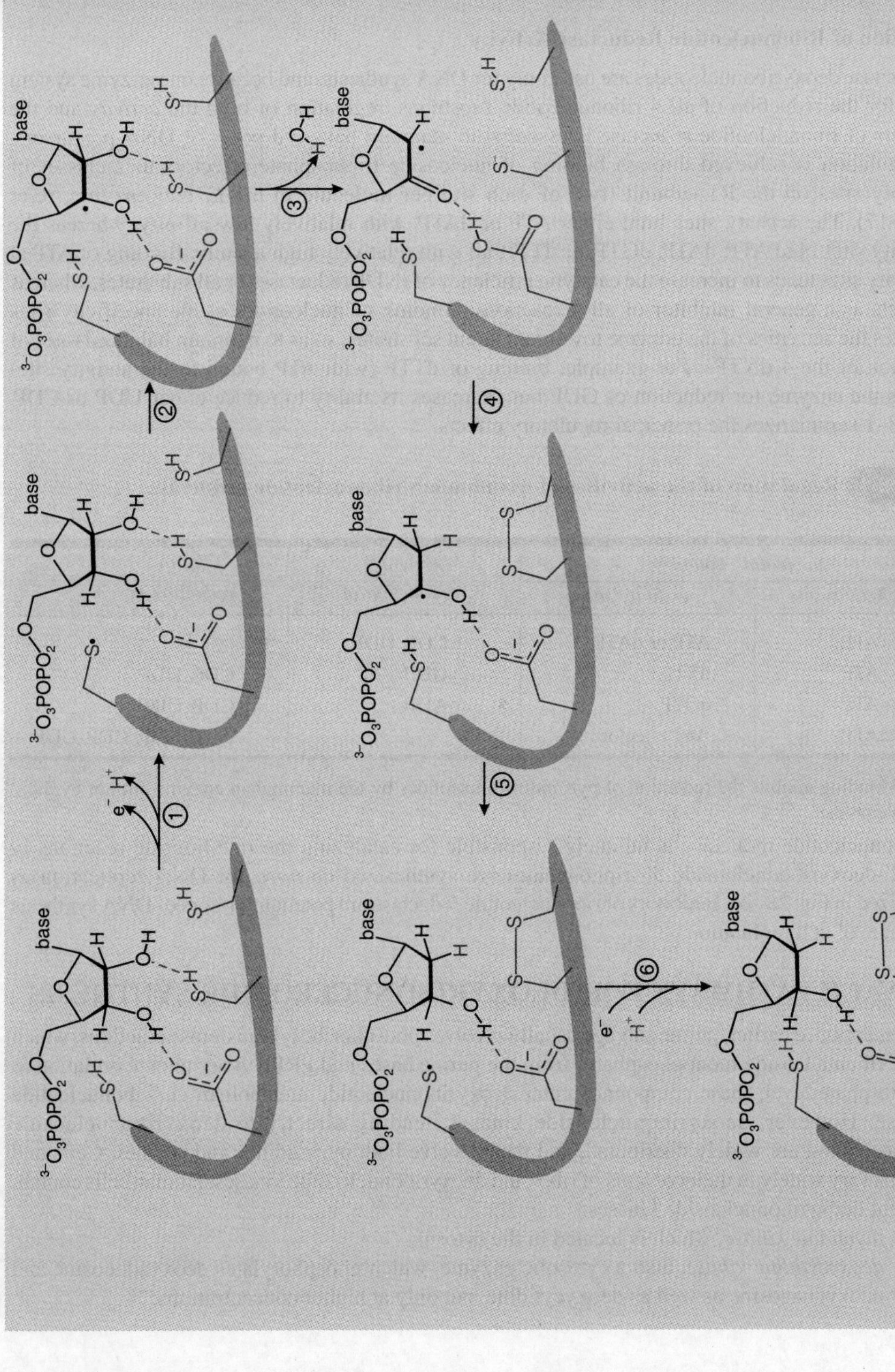

Fig. 28–23. Proposed mechanism for ribonucleotide reductase
See text for the 6 reaction steps.

Note that the reduction of ribonucleotides to deoxyribonucleotides is a difficult reaction chemically, likely to require a sophisticated catalyst. The existence of a common protein enzyme framework for this process strongly suggests that proteins joined the RNA world before the evolution of DNA as a stable storage form for genetic information.

Regulation of Ribonucleotide Reductase Activity

Because deoxyribonucleotides are used only for DNA synthesis, and because one enzyme system is used for the reduction of all 4 ribonucleotide substrates, regulation of both the *activity* and the *specificity* of ribonucleotide reductase is essential to maintain balanced pools of DNA precursors. This regulation is achieved through binding of nucleoside triphosphate effectors to 2 classes of regulatory sites on the R1 subunit (two of each site per molecule in the *E. coli* enzyme; refer Fig. 28–17). The activity sites bind either ATP or dATP, with relatively low affinity, whereas the specificity sites bind ATP, dATP, dGTP, or dTTP, all with relatively high affinity. Binding of ATP at the activity sites tends to increase the catalytic efficiency of rNDP reductase for all substrates, whereas dATP acts as a general inhibitor of all 4 reactions. Binding of nucleotides at the specificity sites modulates the activities of the enzyme toward different substrates, so as to maintain balanced rates of production of the 4 dNTPs. For example, binding of dTTP (with ATP bound in the activity site) activates the enzyme for reduction of GDP but decreases its ability to reduce either UDP or CDP. Table 28–1 summarizes the principal regulatory effects.

Table 28-1 Regulation of the activities of mammalian ribonucleotide reductase

Nucleotide bound in		*Activates reduction of*	*Inhibits reduction of*
Activity site	*Specificity site*		
ATP	ATP or dATP	CDP, UDP	
ATP	dTTP	GDP	CDP, UDP
ATP	dGTP	ADP	CDP, UDP*
dATP	Any effector		ADP, GDP, CDP, UDP

*dGTP binding inhibits the reduction of pyrimidine nucleotides by the mammalian enzyme but not by the *E. coli* enzyme.

Ribonucleotide reductase is uniquely responsible for catalyzing the rate-limiting reactions in which 2′-deoxyribonucleoside 5′-triphosphates are synthesized *de novo* for DNA replication as summarized in Fig. 28–24. Inhibitors of ribonucleotide reductase are potent inhibitors of DNA synthesis and, hence, of cell replication.

SALVAGE PATHWAYS FOR DEOXYRIBONUCLEOTIDE SYNTHESIS

As mentioned earlier, purine salvage usually involves phosphoribosyltransferase reactions, which generate ribonucleoside monophosphates from the purine bases and PRPP. After phosphorylation to the diphosphate level, these compounds enter deoxyribonucleotide metabolism *via* ribonucleotide reductase. However, deoxyribonucleoside kinases, leading directly to deoxyribonucleoside monophosphates, are widely distributed, and they involve both pyrimidines and purines. Cells and organisms vary widely in their contents of ribo- and deoxyribonucleoside kinases. Human cells contain 4 different deoxyribonucleoside kinases:

1. *thymidine kinase*, which is located in the cytosol;
2. *deoxycytidine kinase*, also a cytosolic enzyme, which phosphorylates deoxyadenosine and deoxyguanosine as well as deoxycytidine, but only at higher concentrations;

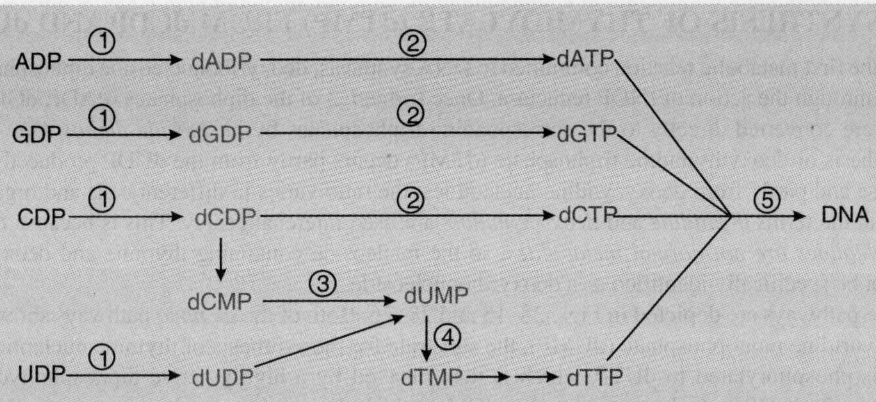

Fig. 28–24. Role of ribonucleotide reductase in DNA synthesis

The enzymes catalyzing the reactions are: (1) ribonucleotide reductase, (2) nucleoside 5′-diphosphate kinase, (3) deoxycytidylate deaminase, (4) thymidylate synthase, and (5) DNA polymerase.

3. *deoxyguanosine kinase*, which is mitochondrial in origin; and
4. *a mitochondrial isoform of thymidine kinase*, which has a broader substrate specificity than the cytosolic enzyme, acting also upon deoxycytidine and deoxyuridine.

Nowadays, several nucleoside analogues are being used/tested in the treatment of cancer and several viral diseases. Invariably, these drugs must be converted to deoxyribonucleotides in order to be effective, and this has focused attention upon the deoxyribonucleoside kinases. As an instance, a side effect of *3′-azido-2′,3′-dideoxythymidine,* AZT (Fig. 28–25) the first drug to receive approval for treating HIV infections, is cardiotoxicity (damage to heart muscle). The major route for azidothymidine utilization involves the mitochondrial isoform of thymidine kinase, also called TK2. Evidence suggests that deoxyribonucleotides of azidothymidine interfere with mitochondrial function, possibly by inhibiting mitochondrial DNA replication or transcription, and this is probably the basis for the cardiotoxicity.

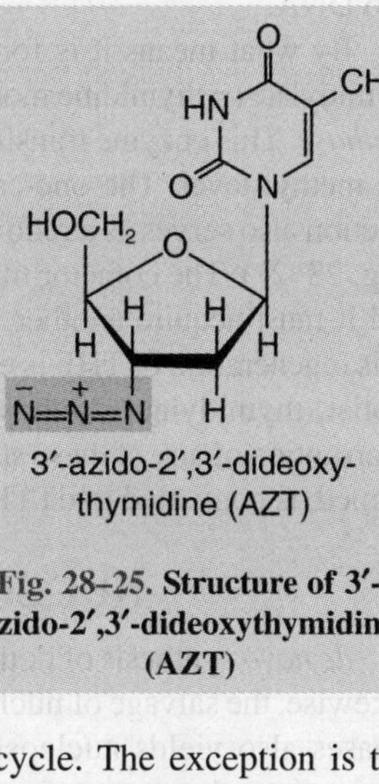

Fig. 28–25. Structure of 3′-azido-2′,3′-dideoxythymidine (AZT)

Of the 4 deoxyribonucleoside kinases, 3 are synthesized constitutively, being produced at constant rates throughout the cell cycle. The exception is the cytosolic thymidine kinase (TK1), the expression of which is highest at the time of DNA replication; in this respect, TK1 resembles enzymes of *de novo* deoxyribonucleotide synthesis, such as ribonucleotide reductase. For reasons not yet understood, TK1 salvages exogenous thymidine extremely efficiently. Experiments with radiolabeled precursors show that dTTP derived from salvage synthesis is usually incorporated into DNA in preference to thymidine nucleotides generated by *de novo* synthesis. This is the basis for the widespread technique of estimating rates of DNA replication by measuring incorporation of radiolabeled thymidine into DNA. However, an accurate measurement of the rate of DNA synthesis from thymidine incorporation data requires measuring the specific radioactivity of the labeled dTTP pool. Nevertheless, because of the great efficiency with which thymidine is utilized in most cells, simple measurements of thymidine incorporation often yield fairly accurate estimates of DNA replication rates.

BIOSYNTHESIS OF THYMIDYLATE (dTMP) FROM dCDP AND dUMP

In the first metabolic reaction committed to DNA synthesis, deoxyribonucleoside diphosphates are formed through the action of rNDP reductase. Once formed, 3 of the diphosphates (dADP, dGDP, and dCDP) are converted directly to the corresponding triphosphates by *nucleoside diphosphate kinase*. Biosynthesis of deoxythymidine triphosphate (dTMP) occurs partly from the dUDP produced *via* the reductase and partly from deoxycytidine nucleotides; the ratio varies in different cells and organisms. Note that the terms *thymidine* and *deoxythymidine* are used interchangeably. This is because *thymine ribonucleotides are not normal metabolites*, so the nucleoside containing thymine and deoxyribose need not be specifically identified as a deoxyribonucleoside.

The pathways are depicted in Figs. 28–15 and 28–26. Both of the *de novo* pathways shown lead to deoxyuridine monophosphate (dUMP), the substrate for the synthesis of thymine nucleotides: (*1*) dUDP is phosphorylated to dUTP, which is then cleaved by a highly active diphosphohydrolase, *dUTPase*, (*2*) dCDP is dephosphorylated to dCMP, which then undergoes deamination to dUMP by an aminohydrolase called *dCMP deaminase*. The latter reaction represents a branch point for pyrimidine dNTP synthesis; the enzyme requires dCTP as an allosteric activator and is inhibited by dTTP. *E. coli* and some other bacteria use a different route to dUMP. Deamination of dCTP is carried out at the triphosphate level by *dCTP deaminase*, and the resultant dUTP is cleaved by *dUTPase* to dUMP and PPi. The latter reaction must be efficient to keep dUTP pools low and prevent incorporation of uridylate into DNA.

By what means it is formed, deoxyuridylate (dUMP) serves as a substrate for the formation of thymidylate (= thymidine monophosphate, dTMP); this methylation reaction is catalyzed by *thymidylate synthase*. This enzyme transfers a one-carbon unit at the methylene level of oxidation, and reduces it to the methyl level. The one-carbon donor is N^5-N^{10}-methylenetetrahydrofolate, which in this unusual reaction also serves as a redox cofactor, to give 7,8-dihydrofolate (DHF) as the second reaction product (Fig. 28–27). The cofactor must then be reduced to tetrahydrofolate (THF) by *dihydrofolate reductase*, and it must acquire another methylene group, most commonly *via serine hydroxymethyltransferase*. This regeneration of THF is essential for the many processes that require THF. In plants and at least one protist, thymidylate synthase and dihydrofolate reductase reside on a single bifunctional protein. Interruption of any of these steps of the cycle interferes with thymine nucleotide formation. dTMP, once formed, is converted to dTTP by 2 successive phosphorylations.

NUCLEOSIDE AND NUCLEOTIDE KINASES

de novo synthesis of both pyrimidine and purine nucleotides yield nucleoside 5′-monophosphates. Likewise, the salvage of nucleobases *via* the phosphoribosyltransferases or nucleosides *via* nucleoside kinases also yields nucleoside 5′-monophosphate. This is particularly important in cells such as *erythrocytes* that cannot form nucleotides *de novo*.

Besides nucleoside kinases, there are nucleotide kinases that convert a nucleoside 5′-monophosphate to nucleoside 5′-dihosphate and nucleoside 5′-diphosphates to nucleoside 5′-triphosphates. These are important reactions since most reactions in which nucleotides function require nucleoside 5′-triphosphate (primarily) or nucleoside 5′-diphosphate.

Nucleoside kinases show a high degree of specificity with regard to the base and sugar moieties. There is also substrate specificity in nucleotide kinases. Mammalian cells also contain, in high concentration, nucleoside diphosphate kinase that is relatively nonspecific for either phosphate donor or phosphate acceptor in terms of pyrimidine or purine base or the sugar. This reaction proceeds as follows:

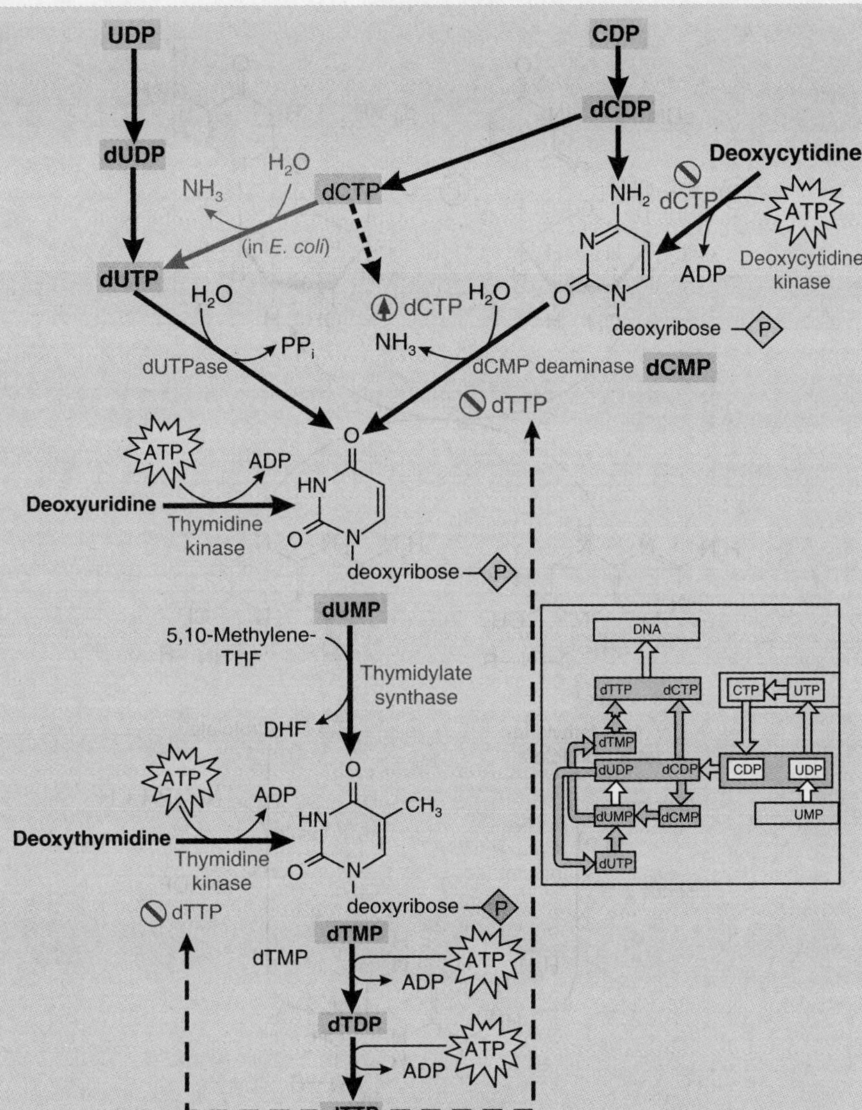

Fig. 28–26. *de novo* and salvage synthetic pathways to thymine nucleotides
The *de novo* pathways start with UDP or CDP, shown at the top.

Since ATP is present in the highest concentration and most readily regenerated on a net basis *via* glycolysis or oxidative phosphorylation, it is probably the major donor for these reactions.

NUCLEOTIDE COENZYME SYNTHESIS

Nucleotides are important constituents not only of RNA and DNA, but also of a number of key biomolecules which are frequently encountered in the study of many biochemical processes. Nicotinamide adenine dinucleotide (NAD), flavin adenine dinucleotide (FAD) and coenzyme A (CoA) serve as important coenzymes in intermediary metabolism. These are synthesized by a variety of mammalian cell types.

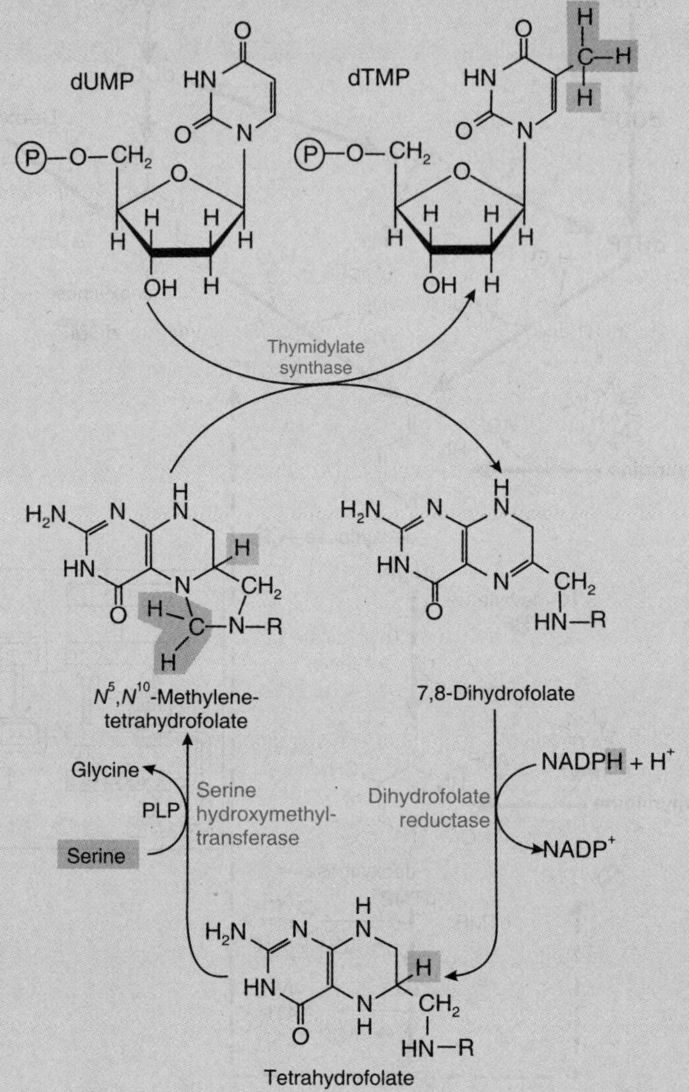

Fig. 28–27. Conversion of dUMP to dTMP by thymidylate synthase and dihydrofolate reductase
Serine hydroxymethyltransferase is required for regeneration of the N^5,N^{10}-methylene form of tetrahydrofolate. In the synthesis of dTMP, all 3 hydrogens of the added methyl group are derived from N^5,N^{10}-methylenetetrahydrofolate (*pink* and *grey*).

Pyridine Nucleotides

Nicotinamide adenine dinucleotide (NAD) is synthesized by 3 different pathways starting from tryptophan, nicotinate (= niacin) or nicotinamide (= niacinamide). When tryptophan is in excess of the amount needed for protein synthesis and serotonin synthesis, it is used for NAD synthesis. This situation is not likely in most normal diets. Consequently, niacin is required in the diet. The coenzyme NAD contains an AMP moiety, which is not directly involved in the functional part of the molecule since electron transfer in NAD occurs *via* the niacin ring. Biosynthesis of NAD$^+$ by any of the 3 pathways (refer Fig. 28–28 for the pathway starting from nicotinamide) requires utilization of 5'-phosphoribosyl 1-pyrophosphate (PRPP) as the ribose 5-phosphate donor. **Nicotinamide adenine dinucleotide phosphate ($NADP^+$)** is derived by phosphorylation of NAD^+. NAD^+ is used not only as

$$NAD^+ + ATP \xrightarrow{Mg^{2+}} NADP^+ + ADP$$

a cofactor in oxidation–reduction reactions but also as a substrate in ADP-ribosylation reactions (*e.g.*, DNA repair and pertussis toxin poisoning). These reactions lead to the turnover of NAD^+. The end product of NAD^+ degradation is 2-pyrolidone-5-carboxamide, which is excreted in urine.

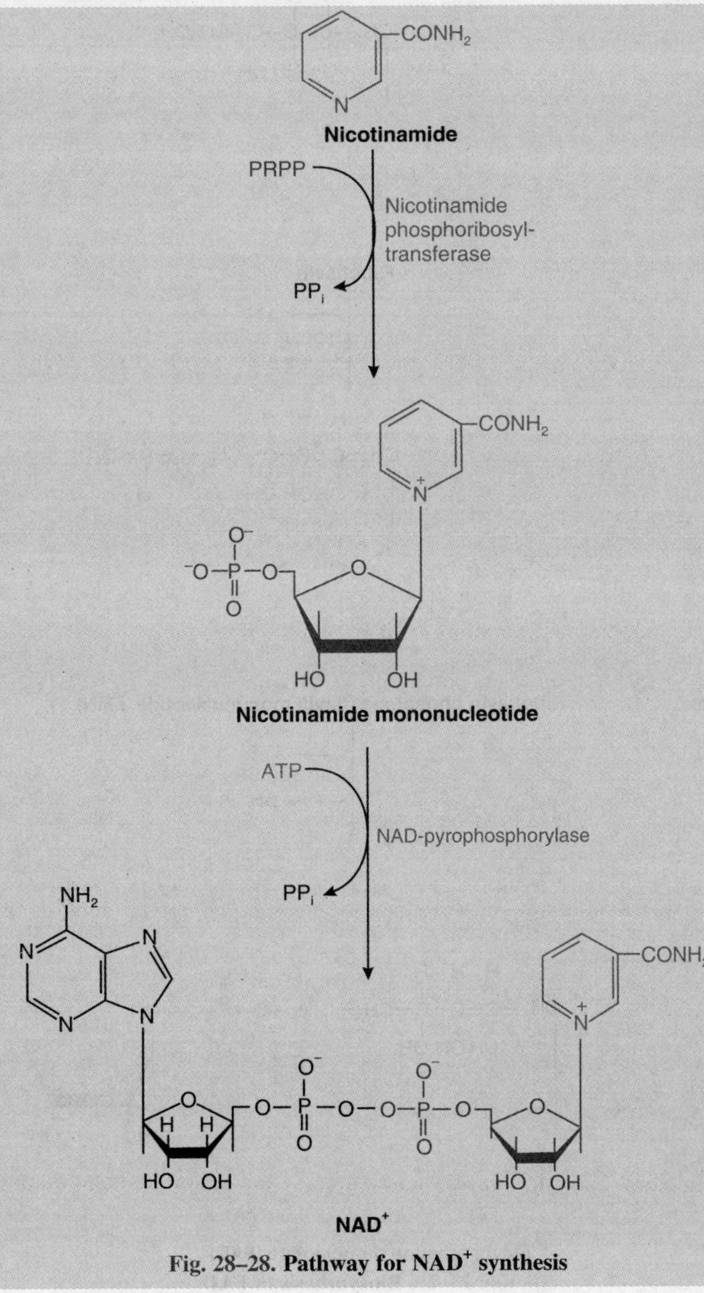

Fig. 28–28. Pathway for NAD^+ synthesis

Flavin Nucleotides

Riboflavin [6,7-dimethyl-9-(1′-D-ribityl) isoalloxazine] is an essential dietary constituent for mammals. It functions in a mono- or dinucleotide form as the prosthetic group of a number of enzymes, such as FMN and FAD. **Flavin adenine dinucleotide FAD;** Fig. 28–29 is formed from riboflavin and 2 molecules of ATP. Riboflavin is phosphorylated by a molecule of ATP to give riboflavin 5′-phosphate (also called **flavin mononucleotide, FMN**). FAD is then generated from FMN by the transfer of an AMP moiety from a second molecule of ATP. Like NAD^+, here also the AMP moiety is not directly involved in the functional part of the molecule since electron transfer in FAD occurs *via* the niacin ring.

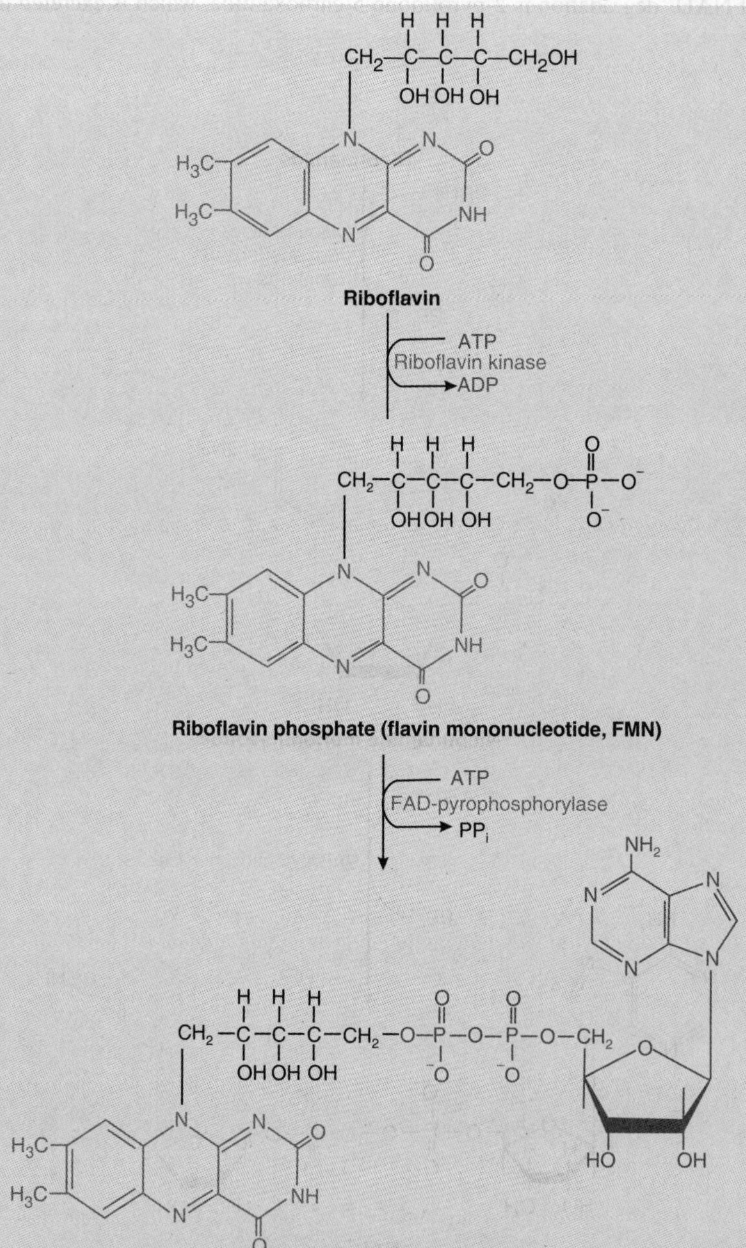

Fig. 28–29. Biosynthesis of FAD

Coenzyme A

The synthesis of coenzyme A or CoA (Fig. 28–30) occurs from pantothenic acid (= pantoyl-β-alanine) which is also a constituent of CoA. Pantothenic acid occurs in nature in combination with β-

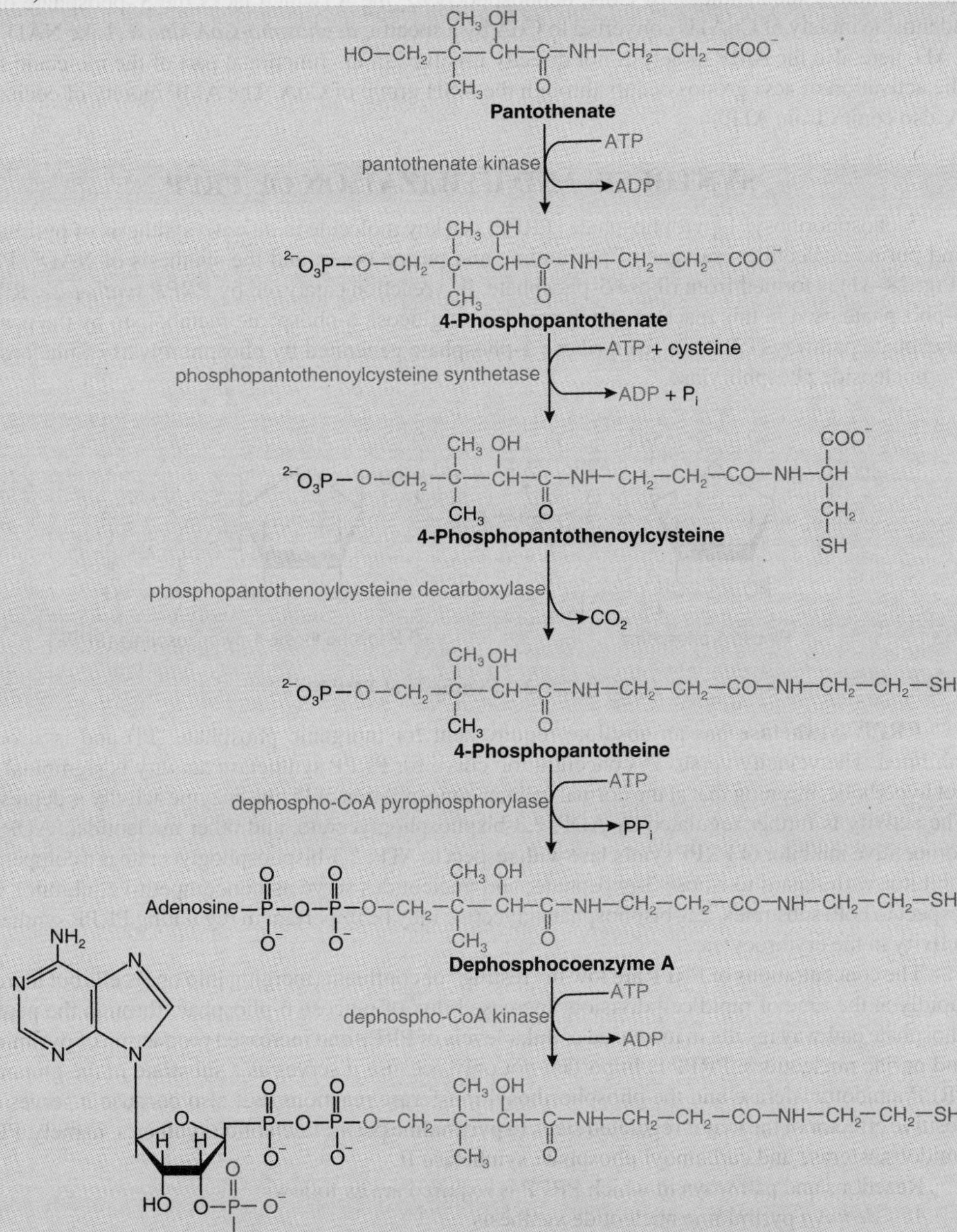

Fig. 28–30. Biosynthesis of coenzyme A (CoA)

A common feature of the biosyntheses of NAD^+, FAD and CoA is *the transfer of the AMP moiety of ATP to the phosphate group of a phosphorylated intermediate*. The pyrophosphate formed in these condensations is then hydrolyzed to orthophosphate. As in many other biosyntheses, *much of the thermodynamic driving force comes from the hydrolysis of the released pyrophosphate.*

mercaptoethylamine (cysteamine) as pantetheine or *Lactobacillus bulgaricus factor* (LBF), so called because it is an essential nutrient for this and certain other microorganisms. Pantetheine is an intermediate in the pathway of CoA formation in mammalian liver and some microorganisms. In the last reaction of the synthesis of CoA, dephosphocoenzyme A (which lacks the 3′-phosphate of the adenosine moiety of CoA) is converted to CoA by a specific *dephospho-CoA kinase*. Like NAD^+ and FAD, here also the AMP moiety is not directly involved in the functional part of the molecule since the activation of acyl groups occurs through the – SH group of CoA. The AMP moiety of coenzyme A also comes from ATP.

SYNTHESIS AND UTILIZATION OF PRPP

5-phosphoribosyl 1-pyrophosphate (PRPP) is a key molecule in *de novo* synthesis of pyrimidine and purine nucleotides, salvage of pyrimidine and purine bases, and the synthesis of NAD^+. PRPP (Fig. 28–31) is formed from ribose 5-phosphate, in a reaction catalyzed by *PRPP synthetase*. Ribose 5-phosphate used in this reaction is generated from glucose 6-phosphate metabolism by the pentose phosphate pathway (P.P.P.) or from ribose 1-phosphate generated by phosphorolysis of nucleosides *via* nucleoside phosphorylase.

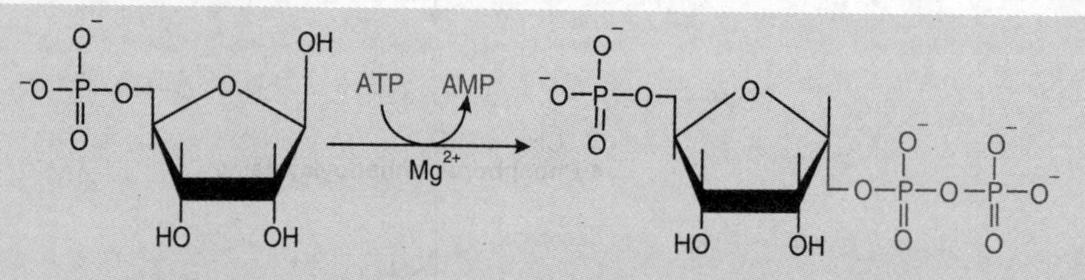

Fig. 28–31. Synthesis of PRPP

PRPP synthetase has an absolute requirement for inorganic phosphate (Pi) and is strongly inhibited. The velocity versus Pi concentration curve for PRPP synthetase activity is sigmoidal and not hyperbolic, meaning that at the normal cellular concentration of Pi, the enzyme activity is depressed. The activity is further regulated by ADP, 2,3-bisphosphoglycerate, and other nucleotides. ADP is a competitive inhibitor of PRPP synthetase with respect to ATP; 2,3-bisphosphoglycerate is a competitive inhibitor with regard to ribose 5-phosphate; and nucleotides serve as noncompetitive inhibitor with respect to both substrates. 2,3-bisphosphate glycerate may be important in regulating PRPP synthetase activity in the erythrocytes.

The concentrations of PRPP are low in "resting" or confluent (merging into one) cells but increase rapidly at the time of rapid cell division. Increased flux of glucose 6-phosphate through the pentose phosphate pathway results in increased cellular levels of PRPP and increased production of pyrimidine and purine nucleotides. PRPP is important not only because it serves as a substrate in the glutamine PRPP amidotransferase and the phosphoribosyltransferase reactions, but also because it serves as a positive effector of the major regulated steps in pyrimidine purine nucleotide synthesis, namely, PRPP amidotransferase and carbamoyl phosphate synthetase II.

Reactions and pathways in which PRPP is required are as follows:

1. *de novo* pyrimidine nucleotide synthesis:

$$\text{PRPP + orotate} \longrightarrow \text{OMP + PPi}$$

2. Salvage of pyrimidine bases:

$$\text{P RPP + uracil} \longrightarrow \text{UMP + PPi}$$

3. *de novo* purine nucleotide synthesis:

$$\text{PRPP + glutamine} \longrightarrow \text{5-phosphoribosylamine + glutamate + PPi}$$

4. Salvage of purine bases:

PRPP + hypoxanthine (guanine) $\longrightarrow$ IMP (GMP) + PPi

PRPP + adenine $\longrightarrow$ AMP + PPi

5. NAD^+ synthesis:

PRPP + nicotinate $\longrightarrow$ nicotinate mononucleotide + PPi

PRPP + nicotinamide $\longrightarrow$ nicotinamide mononucleotide + PPi

PRPP + quinolinate $\longrightarrow$ nicotinate mononucleotide + PPi

REFERENCES

1. **Arnér ESJ, Eriksson S :** Mammalian deoxyribonucleoside kinases. *Pharmacol Ther.* **67 :** *155–186, 1995.*
2. **Begley TP, Appleby TC, Ealick SE :** The structural basis for the remarkable proficiency of orotidine 5′-monophosphate decarboxylase. *Curr. Opin. Struct. Biol.* **10 :** *711–718, 2000.*
3. **Blakley RL :** Eukaryotic dihydrofolate reductase. *Adv. Enzymol. Relat. Areas Mol. Biol.* **70 :** *23–102, 1995.*
4. **Blakley RL, Benkovic SJ :** Folates and Pterins. *Vol. I. Academic Press, Inc., New York. 1984.*
5. **Brown KA, Kraut J :** Exploring the molecular mechanism of dihydrofolate reductase. *Faraday Discuss.* **1992 :** *217–224, 1992.*
6. **Buchanan JM, Hartman SC :** Enzymic reactions in the synthesis of Purines. *Advances in Enzymol.* **21 :** *199–261, 1959.*
7. **Carreras CW, Santi DV :** The catalytic mechanism and structure of thymidylate synthase. *Annu. Rev. Biochem.* ***64*** **:** *721–762, 1995.*
8. **Chargaff E, Davidson JN (editors) :** The Nucleic Acids, *Vol. III, Academic Press, Inc., New York. 1960.*
9. **Denenberg P :** Thymidylate synthase– A target enzyme in chemotherapy. *Biochim. Biophys. Acta.* **473 :** *73–92, 1977.*
10. **Davidson JN :** The Biochemistry of the Nucleic Acids. *5th ed., John Wiley & Sons, Inc., New York, 1965.*
11. **Elledge S, Zhou Z, Allen JB :** Ribonucleotide reductase : Regulation, regulation, regulation. *Trends Biochem. Sci.* **17 :** 119–123, 1992.
12. **Henderson JF :** Regulation of purine biosynthesis. *Monograph No. 170. American Chemical Society. 1972.*
13. **Henderson JF, Peterson ARP :** Nucleotide Metabolism : An Introduction. *Academic Press, Inc., New York, 1973.*
14. **Holmgren A :** Thioredoxin and glutaredoxin systems. *J. Biol. Chem.* **264 :** *13963–13966, 1989.*
15. **Hutchinson DW :** Nucleotides and Coenzymes. *John Wiley & Sons, Inc., New York, 1964.*
16. **Jordan A, Reichard P :** Ribonucleotide reductases. *Annu. Rev. Biochem.* **67 :** *71–98, 1998.*
17. **Kappock TJ, Ealick SE, Stubbe J :** Modular evolution of the purine biosynthetic pathway. *Curr. Opin. Chem. Biol.* **4 :** *567–572, 2000.*
18. **Kornberg A :** Pyrophosphorylases and phosphorylases in biosynthetic reactions. *Advances in Enzymol.* **18 :** *191–240, 1957.*
19. **Kornberg A, Baker TA :** DNA Replication. *2nd edition. Freeman, San Francisco. 1972.*
20. **Lee L, Kelly RE, Pastra-Landis SC, Evans DR :** Oligomeric structure of the multifunctional protein CAD that initiates pyrimidine biosynthesis in mammalian cells. *Proc. Natl. Acad. Sci. USA.* **82 :** *6802–6806, 1985.*

21. **Levdikov VM, Barynin VV, Grebenko AI, Melik-Adamyan WR, Lamzin VS, Wilson KS :** The structure of SAICAR synthase : An enzyme in the *de novo* pathway of purine nucleotide biosynthesis. *Structure* **6 :** *363–376, 1998.*

22. **Liang PH, Anderson KS :** Substrate channeling and domain-domain interactions in bifunctional thymidylate synthase-dihydrofolate reductase. *Biochemistry.* **37 :** *12195–12205, 1998.*

23. **Licht S, Gerfen GJ, Stubbe J :** Thiyl radicals in ribonucleotide reductases. *Science.* **27 :** *477–481, 1996.*

24. **Miller GP, Benkovic SJ :** Stretching exercises: Flexibility in dihydrofolate reductase catalysis. *Chem. Biol.* **5 :** *R105–R113, 1998.*

25. **Raushel FM, Thoden JB, Reinhart GD, Holden HM :** Carbamoyl phosphate synthetase: A crooked path from substrates to products. *Curr. Opin. Chem. Biol.* **2 :** *624–632, 1998.*

26. **Reichard P :** The enzymic synthesis of pyrimidines. *Advances in Enzymol* **21 :** *263–294, 1959.*

27. **Reichard P :** The evolution of ribonucleotide reduction. *Trends Biochem. Sci.* **22 :** *81–85, 1997.*

28. **Schmidt G :** Metabolism of nucleic acids. *Ann. Rev. Biochem.* **33 :** *667–728, 1964.*

29. **Schweitzer BI, Dicker AP, Bertino JR :** Dihydrofolate reductase as a therapeutic target. *FASEB J.* **4 :** *2441–2452, 1990.*

30. **Sigoillot FD, Berkowski JA, Sigoillot SM, Kotsis DH, Guy HI :** Cell cycle-dependent regulation of pyrimidine biosynthesis. *J. Biol. Chem.* **278 :** *3403–3409, 2009.*

31. **Smith JL, Zaluzec EJ, Wery J-P, Niu L, Switzer RL, Zalkin H, Satow Y :** Structure of the allosteric regulatory enzyme of purine biosynthesis. *Science.* **264 :** *1427–1433, 1994.*

32. **Smith LH Jr :** Pyrimidine metabolism in man. *N. Engl J Med.* **288 :** *764, 1973.*

33. **Stubbe J :** Ribonucleotide reductases in the twenty-first century. *Proc. Natl. Acad. Sci. USA.* **95 :** *2723–2724, 1998.*

34. **Stubbe J :** Ribonucleotide reductases : The link between an RNA and a DNA world? *Corr. Opin. Struc. Biol.* **10 :** *731–736, 2000.*

35. **Stubbe J, Riggs-Gelasco P :** Harnessing free radicals : Formation and function of the tyrosyl radical in ribonucleotide reductase. *Trends Biochem. Sci.* **23 :** *438–443, 1998.*

36. **Uhlin U, Eklund H :** Structure of ribonucleotide reductase protein R1. *Nature.* **370 :** *533–539, 1994.*

37. **Underwood M:** Diagnosis and management of gout. *BMJ.* **332 :** *1315–1319, 2006.*

38. **Villfranca JE, Howell EE, Voet DH, Stobel MS, Ogden RC, Abelson JN, Kraut J :** Directed mutagenesis of dihydrofolate reductase. *Science.* **222 :** *782–788, 1983.*

39. **Weber G, Nagai M, Natsumeda Y, Ichikawa S, Nakamura H, Eble JN, Jayaram HN, Zhen WN, Paulik E, Hoffman R :** Regulation of *de novo* and salvage pathways in chemotherapy. *Adv. Enzymol. Regul.* **31 :** *45–67, 1991.*

40. **Zalkin H, Dixon JE :** De novo purine nucleotide biosynthesis. *Prog. Nucleic Acid Res. Mol. Biol.* **42 :** *259–287, 1992.*

PROBLEMS

1. Write a balanced equation for the synthesis of PRPP from glucose through the oxidative branch of the pentose phosphate pathway.

2. Leukemia is a neoplastic (cancerous) proliferation of white blood cells. Clinicians are currently testing deoxycoformycin, an adenosine deaminase inhibitor, as a possible antileukemic agent. Why might one expect this therapy to be effective?

3. If thymine nucleotides are degraded by the same enzymes as those catabolizing uridine nucleotides, give the structure of the thymine metabolite that corresponds to β-ureidopropionate.

4. ATP is synthesized primarily by energy metabolism, whereas other nucleoside triphosphates are formed from the action of nucleoside diphosphate kinase. What additional pathway exists for GTP synthesis?

5. Write a balanced equation for the synthesis of orotate from glutamine, CO_2, and aspartate.

6. Amidotransferases are inhibited by the antibiotic azaserine which is an analogue of glutamine. Which intermediates in purine biosynthesis would accumulate in cells treated with azaserine?

7. Under what conditions might one expect a deficiency of hypoxanthine-guanine phosphoribosyltransferase to affect the rate of *pyrimidine* nucleotide biosynthesis? How might one estimate the rate of pyrimidine nucleotide biosynthesis in living animals or people?

8. If a cell capable of *de novo* synthesis of purine nucleotides has adequate AMP but is deficient in GMP, how would the cell regulate synthesis to increase the concentration of GMP, *i.e.*, [GMP]? If both AMP and GMP were present in appropriate concentrations, what would happen?

9. If a cell were unable to synthesize PRPP, what process would likely be directly impaired?

10. Which therapeutic agent works by impairing *de novo* purine synthesis?

11. Suppose that a person is found who is deficient in an enzyme required for IMP synthesis. How might this person be treated?

12. Bacterial growth is inhibited by sulfanilamide and related sulfa drugs, and there is a concomittant accumulation of 5-aminoimidazole-4-carboxamide ribonucleotide. This inhibition is reversed by the addition of *p*-aminobenzoate. Propose a mechanism for the inhibitory effect of sulfanilamide.

13. What major biosynthetic reactions utilize PRPP?

14. Some bacteria require *p*-aminobenzoate in the culture medium for normal growth, and their growth is severely inhibited by the addition of sulfanilamide. Moreover, in the presence of this drug, 5-aminoimidazole-4-carboxamide ribonucleotide (AICAR) accumulates in the culture medium. These effects are reversed by the addition of excess *p*-aminobenzoate.

    (*a*) what is the role of *p*-aminobenzoate in these bacteria?

    (*b*) Why does AICAR accumulate in the presence of sulfanilamide?

    (*c*) Why are the inhibition and accumulation reversed by addition of excess *p*-aminobenzoate?

15. Propose a mechanism for the conversion of 5-formamidoimidazole-4-carboxamide ribonucleotide into inositate.

16. Name some important functions of the nucleotides.

17. How are the two purine nucleotides found in RNA formed?

18. Succinate uniformly labeled with ^{14}C is added to cells actively engaged in pyrimidine biosynthesis. Propose a mechanism by which carbon atoms from succinate could be incorporated into a pyrimidine. At what positions is the pyrimidine labeled?

19. Under starvation conditions, organisms can use proteins and amino acids as sources of energy. Deamination of amino acids produces carbon skeletons that can enter the glycolytic pathway

and the citric acid cycle to produce energy in the form of ATP. Nucleotides, on the contrary, are not similarly degraded for use as energy-yielding fuels. What observations about cellular physiology support this statement? What aspect of the structure of nucleotides makes them a relatively poor source of energy?

20. Purine biosynthesis is allowed to take place in the presence of [^{15}N]aspartate, and the newly-synthesized GTP and ATP are isolated. What positions are labeled in the two nucleotides?

21. Who gets gout?

22. Why does gout strike the big toe?

23. What diseases are related to obesity?

ANSWERS TO PROBLEMS

1. Glucose + 2 ATP + 2 $NADP^+$ + $H_2O \longrightarrow$ PRPP + CO_2 + ADP + AMP + 2 NADPH + 3 H^+

2. It might lead to adenosine and deoxyadenosine accumulation, as in adenosine deaminase deficiency. This would cause dATP to accumulate and shut off all activities of ribonucleoside diphosphate reductase.

3. 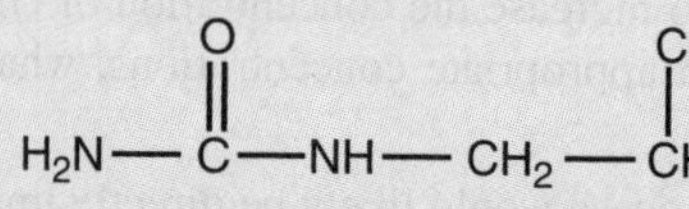

4. Succinyl-CoA synthetase.

5. Glutamine + aspartate + CO_2 + 2 ATP + NAD^+

$\longrightarrow$ Orotate + 2 ADP + 2 Pi + glutamate + NADH + H^+

6. PRPP and formylglycinamide ribonucleotide.

7. Decreased flux through HGPRT could cause the substrates to accumulate, including PRPP. This could increase pyrimidine synthesis at the orotate phosphoribosyltransferase reaction, if PRPP levels are normally subsaturating for that enzyme. Orotate incorporation into nucleotide pools or nucleic acids would give a reasonable estimate of the rate of *de novo* pyrimidine nucleotide synthesis.

8. AMP would partially inhibit *de novo* synthesis of IMP by its allosteric inhibition of PRPP amidotransferase. The IMP formed would be directed toward GMP because AMP is an inhibitor of its own synthesis from IMP. If both AMP and GMP are adequate, the synergistic effect of the two on PRPP amidotransferase would severely inhibit *de novo* synthesis.

9. Ribose 5-phosphate synthesis. PRPP is formed from ribose 5-phosphate in an irreversible reaction.

10. Methotrexate.

11. Inosine or hypoxanthine could be administered.

12. There is a deficiency of N^{10}-formyltetrahydrofolate. Sulfanilamide inhibits the synthesis of folate by acting as an analogue of *p*-aminobenzoate, one of the precursors of folate.

13. PRPP is the activated intermediate in the synthesis of phosphoribosylamine in the *de novo* pathway of purine formation; of purine nucleotides from free bases by the salvage pathway; of orotidylate in the formation of pyrimidines; of nicotinate ribonucleotide; of phosphoribosyl ATP in the pathway leading to histidine; and of phosphoribosylanthranilate in the pathway leading to tryptophan.

14. (*a*) Compound *p*-aminobenzoate is a component of N^5, N^{10}-methylenetetrahydrofolate, the cofactor involved in the transfer of one-carbon units.

(*b*) In the presence of sulfanilamide, a structural analogue of *p*-aminobenzoate, bacteria are unable to synthesize tetrahydrofolate, a cofactor necessary for converting AICAR to FAICAR; thus AICAR accumulates.

(c) The competitive inhibition by sulfanilamide of the enzyme involved in tetrahydrofolate biosynthesis is overcome by the addition of excess substrate (*p*-aminobenzoate).

15. The $-NH_2$ group attacks the carbonyl carbon atom to form a tetrahedral intermediate. Removal of a proton leads to the elimination of water to form inosinate.

16. The various functions carried out by nucleotides are: serve as monomeric units of nucleic acids; act as physiological mediators; serve as sources of chemical energy; and serve as structural components of coenzymes.

17. GMP and AMP are both formed from the first purine nucleotide, IMP, in a branched pathway.

18. Succinate $\longrightarrow$ malate $\longrightarrow$ oxaloacetate by the citric acid cycle. Oxaloacetate $\longrightarrow$ aspartate by transamination, followed by pyrimidine synthesis. Carbons 4, 5 and 6.

19. The organisms do not store nucleotides to be used as fuel, and they do not completely degrade them, but rather hydrolyze them to release the bases, which can be recovered in salvage pathways. The low C:N ratio of nucleotides makes them poor sources of energy.

20. N-1 in both cases, and the amine group linked to C-6 in ATP.

21. The average gout sufferer is an overweight, middle-aged man. The first attack can hit by age 30, although it is more common in the 40's, 50's and beyond. Premenopausal women rarely get gout. Ten men suffer from gout for every one woman.

22. More than half of gout attacks strike the big toe. Gout is a painful form of arthritis caused by uric acid crystals that settle in the joints. As they accumulate, a gout attack occurs. Scientists speculate that gravity is to blame for uric acid buildup in the big toe.

23. Obesity is linked to a number of diseases, *viz.*, heart disease, type 2 diabetes, high cholesterol, high blood pressure, stroke, gallbladder disease, liver disease, osteoarthritis, gout, pulmonary problems and certain cancers.

CONTENTS

CHAPTER

29

Degradation of Nucleotides

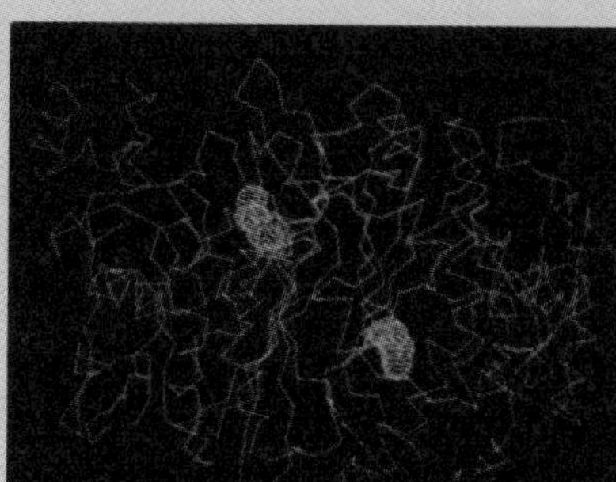

Structure of the dimeric enzyme thymidylate synthase from *Lactobacillus casei*

[*Courtesy : L.W. Hardy, J. Finer-Moore, D.V. Santi, and R.M. Stroud, 1987*]

DEGRADATION OF PYRIMIDINE NUCLEOTIDES

The turnover of nucleic acids results in the release of pyrimidine and purine nucleotides. Degradation of pyrimidine nucleotides follows the pattern shown in Fig. 29–1. Pyrimidine nucleotides are converted to nucleosides by *nonspecific* phosphatases. Cytidine and deoxycytidine are deaminated to uridine and deoxyuridine, respectively by the enzyme, *pyrimidine nucleoside deaminase*. Another enzyme, *uridine phosphorylase* catalyzes phosphorolysis of uridine and deoxyuridine to uracil, and that of deoxythymidine to thymine.

The catabolism of pyrimidines occurs mainly in the liver. The degradation of ^{14}C-labeled pyrimidines (uracil and thymine) was elucidated mainly by injection of large quantities of these substances and of possible intermediates, followed by examination of urinary excretion products. Some studies have also been performed by incubation of these substances with liver slices and with individual enzymes. The catabolism of pyrimidines results in the production of a series of highly soluble end products. This is in contrast with the production of sparingly soluble uric acid and sodium urate by purine catabolism. Because of the relatively high solubility of the intermediates, there are few known derangements of pyrimidine breakdown. Fig. 29–2 shows the pathways for the degradation fo uracil and thymine. It is blelieved that cystosine and methylcytosine are deaminated first to yield uracil and thymine, respectively.

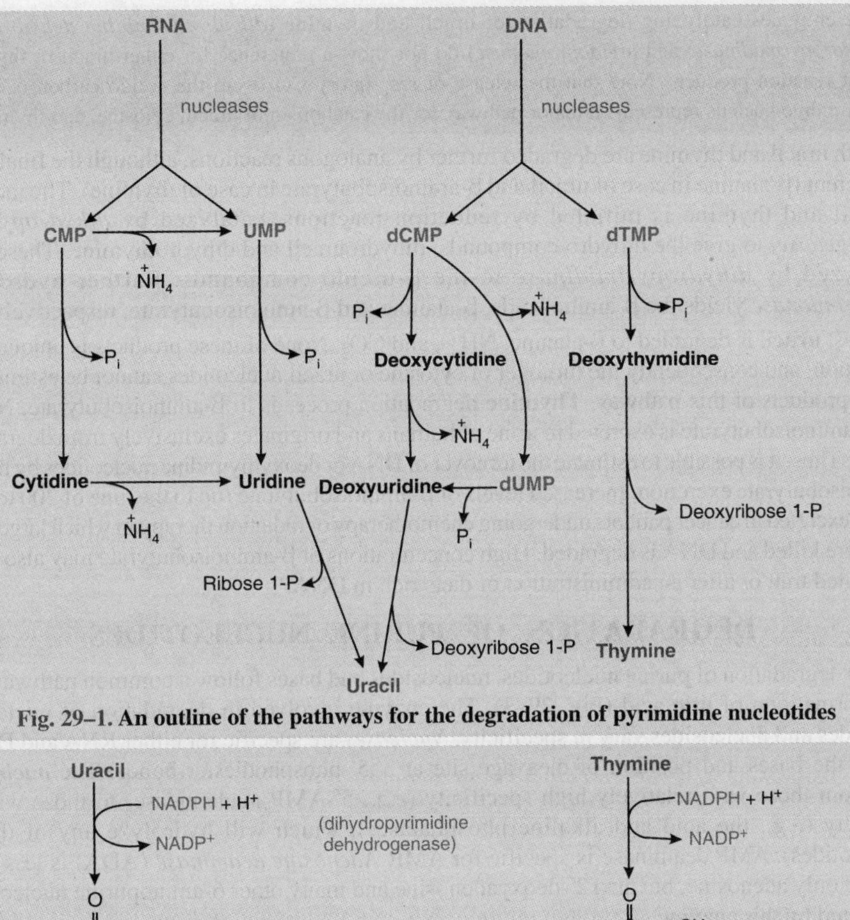

Fig. 29–1. An outline of the pathways for the degradation of pyrimidine nucleotides

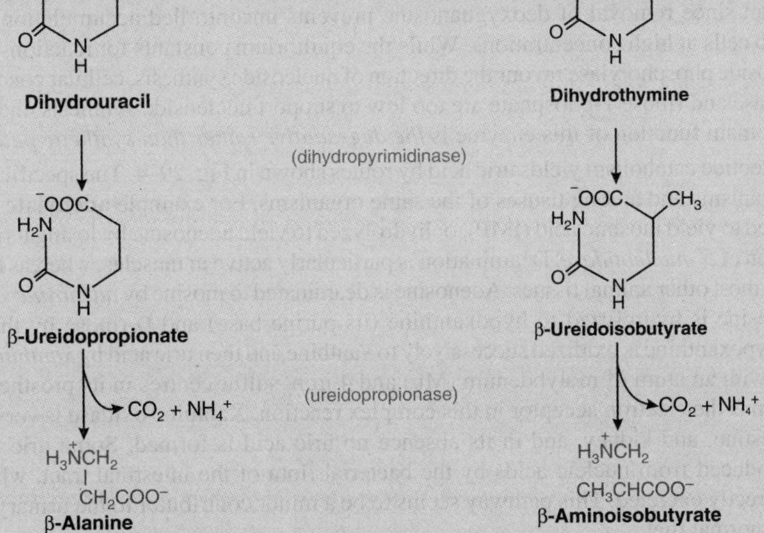

Fig. 29–2. Degradation of uracil and thymine to their end products, β-alanine and β-aminoisobutyrate

The enzymes catalyzing degradation of uracil and thymine (*dihydropyrimidine dehydrogenase, dihydropyrimidinase,* and *ureidopropionase*) do not show a preference for either uracil or thymine or their reaction products. Note that the release of respiratory CO_2 from the ureido carbon (C_2) of the pyrimidine nucleus represents a major pathway for the catabolism of uracil, cytosine, and thymine.

Both uracil and thymine are degraded further by analogous reactions, although the final products are different (β-alanine in case of uracil and β-aminoisobutyrate in case of thymine). The metabolism of uracil and thymine is initiated by reduction reactions, catalyzed by *dihydropyrimidine dehydrogenase*, to give the dihydro compounds, dihydrouracil and dihydrothymine. These are then hydrolyzed by *dihydropyrimidinase* to the β-ureido compounds. Further hydrolysis by *ureidopropionase* yields the β-amino acids, β-alanine and β-aminoisobutyrate, respectively.

Thus, **uracil** is degraded to β-alanine, NH_4^+, and CO_2. None of these products is unique to uracil degradation, and consequently the turnover of cytosine or uracil nucleotides cannot be estimated from the end products of this pathway. **Thymine** degradation proceeds to β-aminoisobutyrate, NH_4^+, and CO_2. β-aminoisobutyrate is excreted in urine of humans and originates exclusively from degradation of thymine. Thus, it is possible to estimate the turnover of DNA or deoxythymidine nucleotides by measuring β-aminoisobutyrate excretion. Increased levels of β-aminoisobutyrate (up to the tune of 200 to 300 mg/day) are excreted in cancer patients undergoing chemotherapy or radiation therapy in which large numbers of cells are killed and DNA is degraded. High concentrations of β-aminoisobutyrate may also be due to an inherited trait or after an administration of diets rich in DNA.

DEGRADATION OF PURINE NUCLEOTIDES

The degradation of purine nucleotides, nucleosides and bases follow a common pathway leading to the formatioon of uric acid (Fig. 29–3). The enzyme involved in degradation of nucleic acids, nucleotides and nucleosides vary in specificity. *Nucleases* are specific for either RNA and DNA, and also for the bases and position of cleavage site at 3′,5′-phosphodiester bonds. The *nucleotidases* range from those with relatively high specificity (*e.g.*, 5′-AMP nucleotidase) to those with broad specificity (*e.g.*, the acid and alkaline phosphatases), which will hydrolyze any of the 3′- or 5′-nucleotides). AMP deaminase is specific for AMP. *Adenosine deaminase* (ADA) is less specific, since not only adenosine, but also 2′-deoxyadenosine and many other 6-aminopurine nucleosides are deaminated by this enzyme.

Deoxyinosine and deoxyguanosine are also natural substrates for purine nucleoside phosphorylase. This is important since removal of deoxyguanosine prevents uncontrolled accumulation of dGTP, which is toxic to cells at high concentrations. While the equilibrium constants for reactions catalyzed by purine nucleoside phosphorylase favour the direction of nucleoside synthesis, cellular concentrations of free purine base and ribose 1-phosphate are too low to support nucleoside synthesis under normal conditions. The main function of this enzyme is *the degradative rather than synthetic pathway.*

Purine nucleotide catabolism yields uric acid by routes shown in Fig. 29–4. The specific pathways vary among organisms and among tisuses of the same organisms. For example **adenylate (AMP)** is either deaminated to yield inosinic acid (IMP), or hydrolyzed to yield adenosine by losing its phosphate through the action of *5′-nucleotidase*. Deamination is particularly active in muscles, whereas hydrolysis predominates in most other animal tissues. Adenosine is deaminated to inosine by *adenosine deaminase* (ADA), and inosine is hydrolyzed to hypoxanthine (its purine base) and D-ribose by the enzyme nucleosidase. Hypoxanthine is oxidized successively to xanthine and then uric acid by *xanthine oxidase*, a flavoenzyme with an atom of molybdenum (Mb) and 4 iron–sulfur centres in its prosthetic group. Molecular oxyen is the electron acceptor in this complex reaction. Xanthine oxidase is very active in liver, small intestine, and kidney, and in its absence no uric acid is formed. Some uric acid may, however, be produced from nucleic acids by the bacterial flora of the intestinal tract, whence it is absorbed and directly excreted. This pathway seems to be a minor contributor to the urinary uric acid of persons on a normal diet.

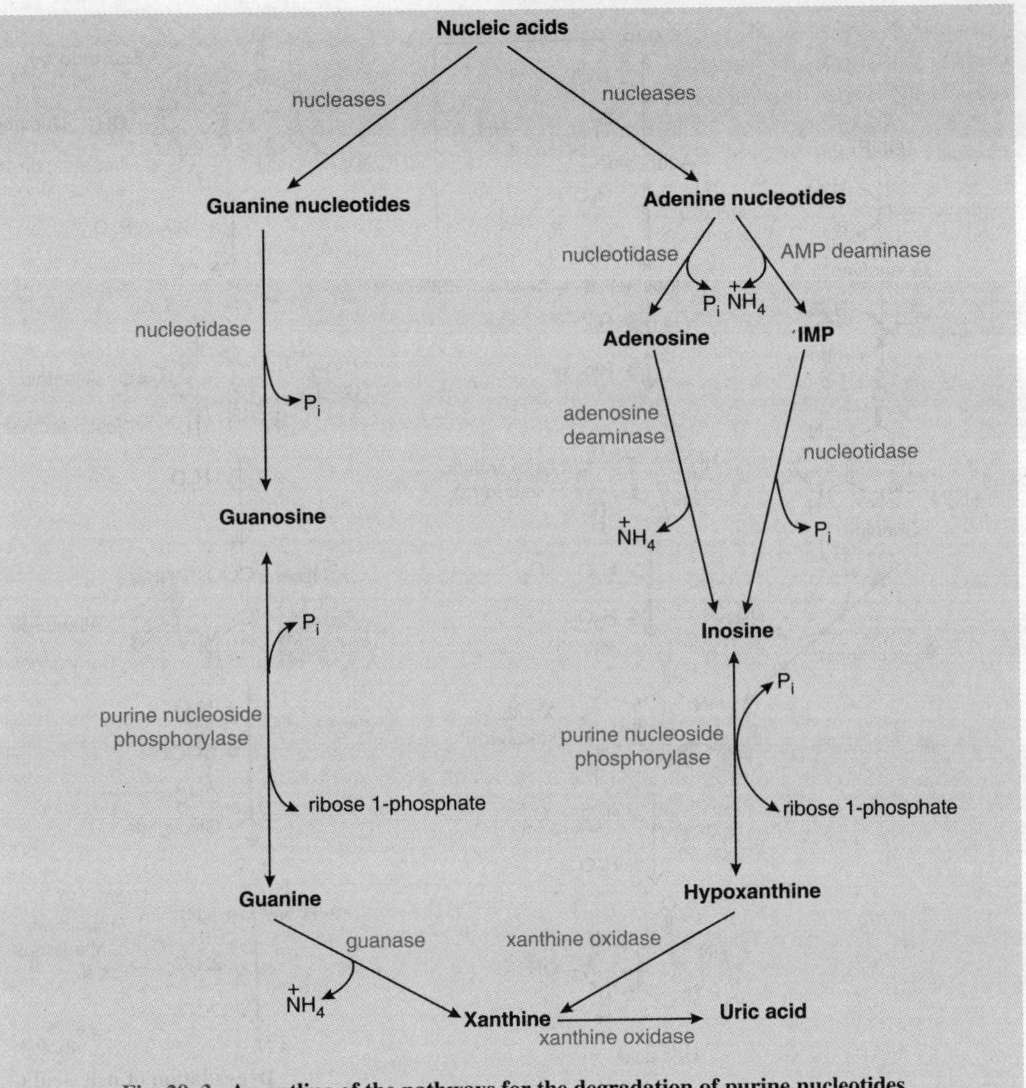

Fig. 29–3. An outline of the pathways for the degradation of purine nucleotides

Purine nucleoside phosphorylase catalyzes the following reversible reactions:

$$\text{Inosine + Pi} \rightleftharpoons \text{hypoxanthine + ribose 1-phosphate}$$
$$\text{or} \quad \text{Guanosine + Pi} \rightleftharpoons \text{guanine + ribose 1-phosphate}$$
$$\text{or} \quad \text{Xanthosine + Pi} \rightleftharpoons \text{xanthine + ribose 1-phosphate}$$

Guanylate (GMP) catabolism also yields uric acid as end product. GMP is first hydrolyzed to yield guanosine, which is then cleaved to free guanine and D-ribose by the enzyme *nucleosidase*. Guanine undergoes hydrolytic removal of its amino group to yield xanthine, by *guanine deaminase*, an enzyme found abundantly in mammalian brain and liver. Xanthine is converted to uric acid by *xanthine oxidase*.

Uric acid is the excreted end product of purine metabolism in primates (including man), birds, reptiles and insects. A healthy adult human excretes uric acid at a rate of about 0.6 g/24 h; the excreted product arises partly from ingested purines and partly from turnover of the purine nucleotides of

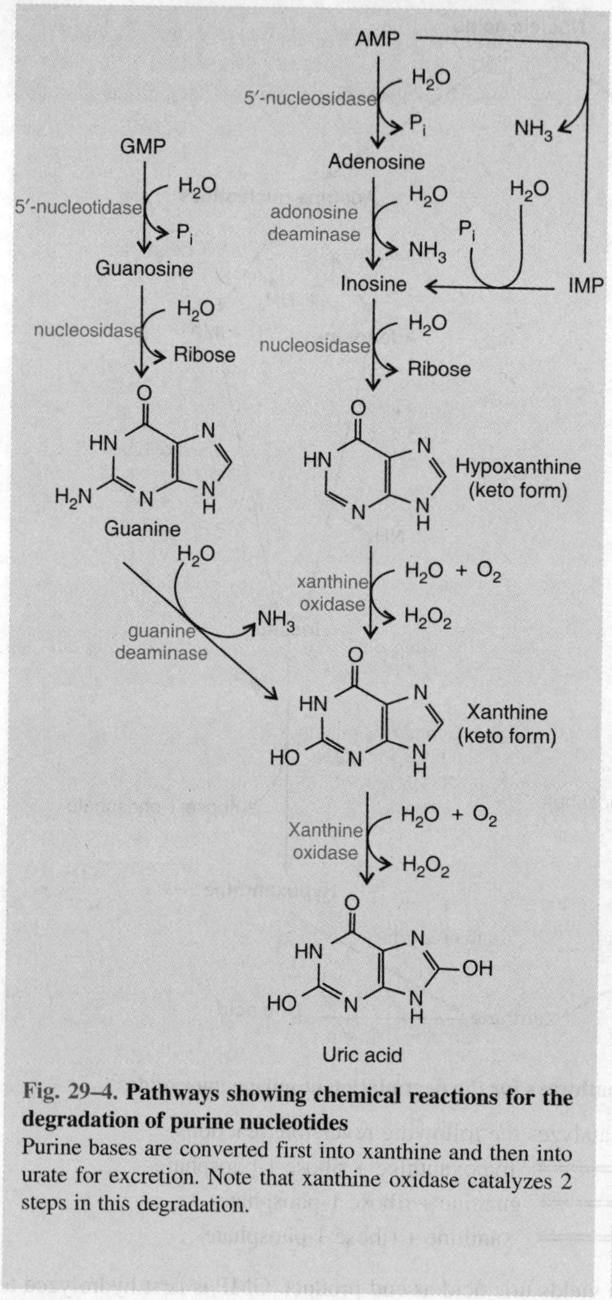

Fig. 29–4. Pathways showing chemical reactions for the degradation of purine nucleotides

Purine bases are converted first into xanthine and then into urate for excretion. Note that xanthine oxidase catalyzes 2 steps in this degradation.

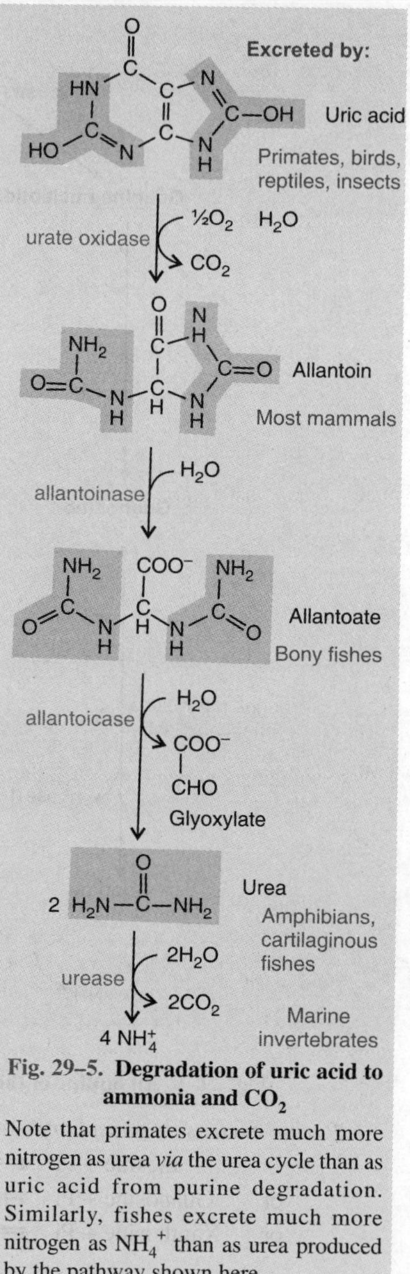

Fig. 29–5. Degradation of uric acid to ammonia and CO_2

Note that primates excrete much more nitrogen as urea *via* the urea cycle than as uric acid from purine degradation. Similarly, fishes excrete much more nitrogen as NH_4^+ than as urea produced by the pathway shown here.

nucleic acids. However, most mammals other than primates, some reptiles (turtles) and gastropod mollusks further oxidize uric acid to allantoin by the action of *urate oxidase* and then to allantoic acid by the enzyme *allantoinase*. Allantoic acid is either excreted (in bony fishes) or further catabolized to urea by *allantoicase* (in amphibians, most fishes, and fresh-water lamellibranch mollusks) or to ammonia by *urease* (in some marine invertebrates and crustaceans). Fig. 29–5 shows the pathway from uric acid to ammonia and CO_2.

Because of the degradation of purines is much less complete in higher animals, it is apparent that certain enzymes have been lost during animal evolution, *e.g.*, urate oxidase (= uricase), allantoinase, allantoicase, and urease.

CLINICAL DISORDERS OF PYRIMIDINE METABOLISM

The various disorders of pyrimidine metabolism are listed in Table 20–1. As described earlier, the end products of pyrimidine metabolism, unlike those of purine metabolism, are highly water-soluble compounds such as CO_2, ammonia, β-alanine and propionate. Hence, under circumstances of pyrimidine overproduction, clinically detectable abnormalities are rarely evident. In cases of hyperuricemia associated with severe PRPP overproduction, there is concomitant overproduction of pyrimidine nucleotides with increased excretion of compounds such as β-alanine. Because of the requirement for N^5, N^{10}-methylenetetrahydrofolate for thymidylate (dTMP) synthesis, disorders for folate and vitamin B_{12} metabolism result in deficiencies of TMP (in case of vitamin B_{12} deficiency, by an indirect mechanism).

Table 29–1. Inherited disorders of pyrimidine metabolism and their associated enzyme abnormalities

Clinical disorder	*Defective enzyme*	*Nature of the defect*	*Characteristics of clinical disorder*	*Inheritance pattern*
β-aminoisobutyric aciduria	Transaminase	Deficiency	No symptoms; frequent in Orientals	Autosomal recessive
Orotic aciduria, type I	Orotate phosphoribosyltransferase and orotidylate decarboxylase	Deficiency	Orotic acid crystalluria, failure to thrive, and megaloblastic anemia. Remission with oral uridine	Autosomal recessive
Orotic aciduria, type II	Orotidylate decarboxylase	Deficiency	Orotidinuria and orotic aciduria, megaloblastic anemia. Remission with oral uridine	Autosomal recessive
Orotic aciduria	Ornithine transcarbamoylase	Deficiency	Protein intolerance, hepatic encephalopathy, and mild orotic aciduria	X-linked recessive

A summary of the interconversions of purine derivatives (Fig. 29–6) shows that reactions may occur at the level of nucleotides, nucleosides, or with the free purines. Not all these reactions necessarily occur in every tissue or in every species, but there is evidence for each pathway shown. Of the pathways given, only 2 are known to be reversible by the same enzymic system. The reaction pathways include 4 types of reactions: *oxidative, hydrolytic, pyrophosphorolytic* and *phosphorolytic reactions*. The first three types of reaction may be considered to be essentially irreversible.

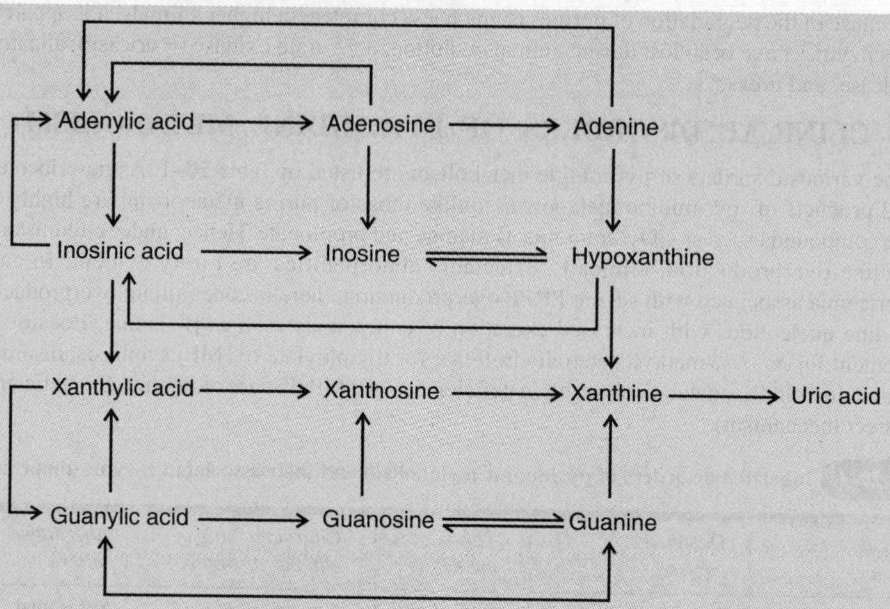

Fig. 29–6. Summary of interconversions of purine derivatives

In most instances, the biosynthetic pathways and the enzymes involved differ from the degradative reactions; this is indicated in the figure by separate pathways of the arrows.

Uric acid production and excretion proceed at a relatively constant rate in men when the diet is free of purines. This uric acid is derived from the endogenous purine metabolism and, as with many body constituents, reflects a steady state in which the rates of purine synthesis and purine degradation, measured by the excretion of uric acid, are approximately equal. Some protein foods (*e.g.,* milk, cheese, and eggs) are low in purines, whereas foods rich in nucleoproteins (*e.g.*, liver and pancreas) are high in purines. The concentration of uric acid in normal plasma is about 2–6 mg/100 ml, with an average for adults of about 3.5 mg/100 ml for females and of about 4.5 mg/100 ml for males. These levels are elevated in gout, a disease in which large amounts of the sparingly soluble monosodium urate are deposited as *tophi* in cartilage. Uric acid deposits may also be found as *calculi* in the kidney, with resultant renal damage. A large fraction of all renal calculi consists of the sparingly soluble uric acid and its salts.

This property of insolubility of uric acid and its urate salts is advantageous to egg-laying animals, because it provides a route for disposition of excess nitrogen in a closed environment: the waste material simply precipitates *in situ*. However, the insolubility of urates can present difficulties in mammalian metabolism. In humans about 3 individuals in 1,000 suffer from hyperuricemia– chronic elevation of blood uric acid levels well beyond normal limits. Although the biochemical reasons for hyperuricemia vary, the condition goes by the single clinical name of **gout.** Prolonged or acute elevation of blood urate leads to its precipitation, as crystals of sodium urate, in the synovial fluid of joints. These precipitates cause inflammation, resulting in painful arthritis which, if untreated, leads ultimately to severe degeneration of joints.

Gout is effectively treated by a combination of nutritional and drug therapy. Foods especially rich in nucleotides and nucleic acids, such as liver or glandular products, are withdrawn from the diet. Major alleviation of the symptoms is provided by the drug **allopurinol** (Fig. 29–7), a structural analogue of hypoxanthine that strongly inhibits *xanthine oxidase*, the enzyme that catalyzes the conversion of purines to uric acid. Allopurinol is a substrate of xanthine oxidase, which converts allopurinol to oxypurinol (= alloxanthine). Oxypurinol inactivates the reduced form of the enzyme by remaining tightly bound in its active site. When xanthine oxidase is inhibited, the excreted products of purine metabolism are xanthine and hypoxanthine, which are more water-soluble than uric acid and hence more readily excreted than is uric acid, and also less likely to form crystalline deposits. Allopurinol was developed by **Gertrude Elion** (LT, 1918–1999) and **George Hitchings** (LT, 1905–1998) who also developed *acyclovir* (used against AIDS) and other purine analogues used in cancer chemotherapy.

β-aminoisobutyric aciduria is an autosomal, recessively inherited disorder which is prevalent among the Oriental races. The excretion of β-aminoisobutyric acid, which is the precursor of both laboratory animals and in humans, is increased in leukemia as well as after the body has been subjected to x-irradiation. This is undoubtedly a reflection of increased destruction of cells and their DNA. A familial occurrence of an abnormally high excretion of β-aminoisobutyric acid has also been observed in otherwise normal individuals. High execretors result only when the trait is homozygous. It is interesting to note that about one-fourth of tested persons of Chinese or Japanese ancestry consistently excreted large quantities of β-aminoisobutyric acid. Although little is known about the mechanism whereby β-aminoisobutyric acid is degraded in humans, an enzyme which catalyzes the reversible transamination reaction has been identified in pig kidney. β-aminoisobutyric acid is converted to methylmalonic semialdehyde and thence to propionic acid, which in turn proceeds to succinate.

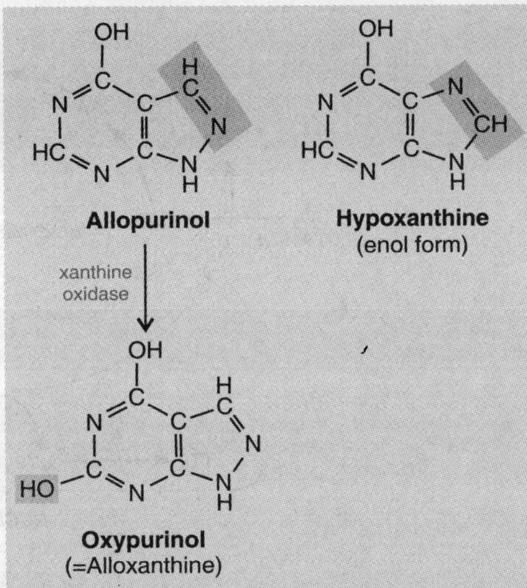

Fig. 29–7. Allopurinol, an inhibitor of xanthine oxidase

Note that hypoxanthine is the normal substrate of xanthine oxidase. Only a slight alteration in the structure of hypoxanthine (shaded *pink*) yields the medically effective enzyme inhibitor, allopurinol. At the active site, allopurinol is convertd to oxypurinol, a strong competitive inhibitor that remains tightly bound to the reduced form of the enzyme.

It should be noted that the initial steps in the degradation of pyyrimidine nucleotides, including the removal of the sugar phosphate moiety by hydrolysis of the N-glycosidic bond, are similar to reversing the latter steps of the synthetic pathway. For **pseudouridine**, which is formed *in situ* in tRNA by a rearrangement reaction, there is no mechanism to catalyze the hydrolysis or phosphorolysis of this unusual nucleoside to its respective pyrimidine base, uracil. As a result, pseudouridine is excreted unchanged in the urine of normal persons. When increased nucleic acid turnover occurs in patients with leukemia or lymphoma, there is a markedly increased urinary excretion of pseudouridine. This compound is highly soluble and by itself produces no disease.

Two types of **hereditary orotic aciduria** have been reported (Fig. 29–8): type I and type II. Their characteristics are:

Type I: This is the more common type, although still rare. In it, both orotate phosphoribosyltransferase and orotidylate decarboxylase are missing in all cell types tested. The patients are pyrimidine auxotrophs. They are readily treated with uridine. As infants, these patients fail to thrive and exhibit megaloblastic anemias and orange crystalluria (orotic acid). Unless treated with a source of pyrimidine nucleosides, they succumb to infections. In such type I patients, orotic acid is the major abnormal excretory product. In the erythrocytes of such patients, the specific catalytic activities of aspartate transcarbamoylase and dihydroorotase were found to be greatly enhanced but returned to normal upon treatment of the patient with oral uridine. These observations suggest that one or more end products of the pathway are normally responsible for the maintenance of these enzyme activities at a regulated level. In a deficient state when the cells are deprived of the end products of this pathway, there is a derepression, probably a coordinate one, of at least those 2 enzymes.

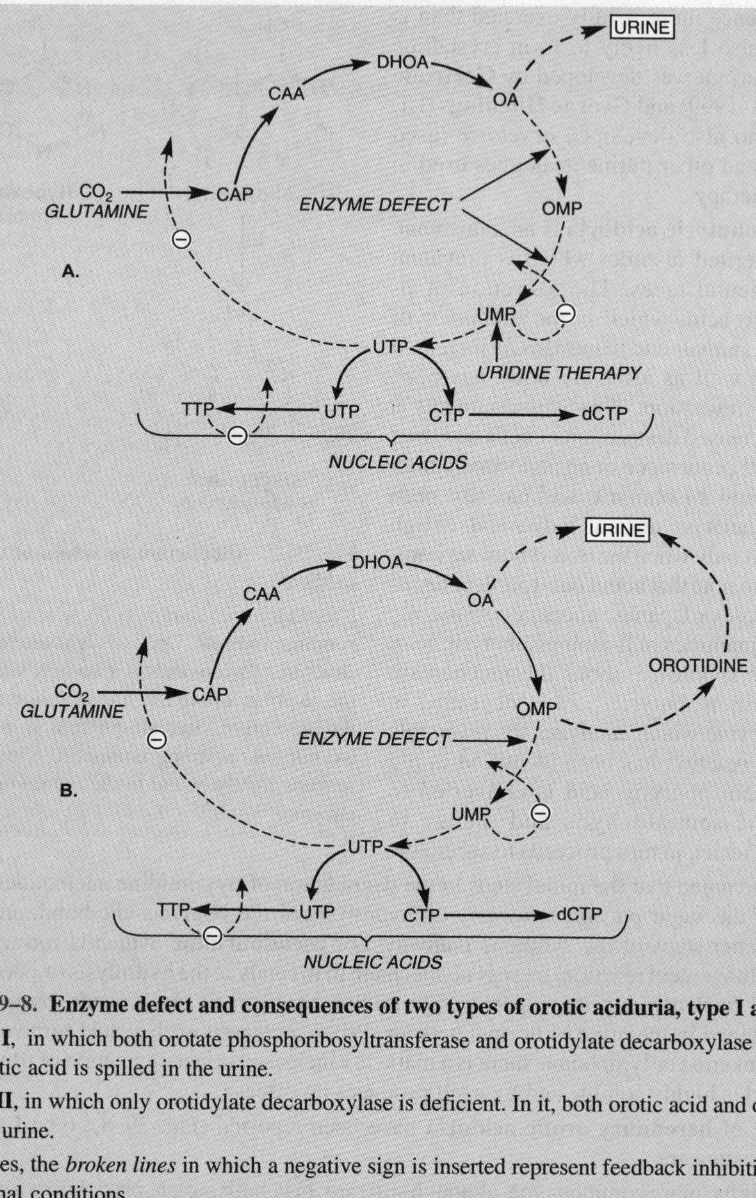

Fig. 29–8. Enzyme defect and consequences of two types of orotic aciduria, type I and type II

A. Type I, in which both orotate phosphoribosyltransferase and orotidylate decarboxylase are deficient. In it, orotic acid is spilled in the urine.

B. Type II, in which only orotidylate decarboxylase is deficient. In it, both orotic acid and orotidine appear in the urine.

In both types, the *broken lines* in which a negative sign is inserted represent feedback inhibition which exists under normal conditions.

[CAP = carbamoyl phosphate, CAA = carbomoyl aspartic acid, DHOA = dihydroorotic acid, OA = orotic acid]

(Adapted from: Smith LH Jr: Pyrimidine Metabolism in Man. N. Engl. J. Med. **288** *: 764, 1973.)*

Type II: This less common type is due to a deficiency only of orotidylate decarboxylase. In such a patient, orotidine is the major excretory product, although some orotic acid is also excreted.

The enzymology of this pathway suggests that there is a common protein molecule shared by carbamoyl phosphate synthetase, aspartate transcarbamoylase, and dihydroorotase, and another shared by orotate phosphoribosyltransferase and orotidylate decarboxylase.

Increased excretion of orotic acid, uracil, and uridine has been described in patients deficient in ornithine transcarbamoylase, a liver mitochondrial enzyme responsible for an early step in urea and arginine biosynthesis. In these patients, there is apparently mitochondrial carbamoyl phosphate

(CAP) accumulation in response to the enzyme blockage. The mitochondrial CAP diffuses into the cytosol to be utilized as a substrate for *de novo* synthesis of pyrimidine nucleotides. The excess production of orotic acid (OA) is then manifested as orotic aciduria, which usually occurs in a mild degree and appears without crystal formation but increases upon the ingestion of foodstuffs such as various types of meat that contain large amounts of nitrogen.

At least 2 drugs can result in orotic aciduria. One such drug, which is widely used clinically, is **allopurinol** (4-hydroxypyrazolopyrimidine), a purine analogue which directly inhibits xanthine oxidase. It can be phosphoribosylated by orotate phosphoribosyltransferase, thereby competitively inhibiting the phosphoribosylation of orotic acid. Furthermore, the unusual nucleotide formed inhibits orotidylate decarboxylase, producing orotic aciduria and orotidinuria. In humans, at least, the pyrimidine pathway appears to readjust itself to this inhibitioin so that the organism is only transiently starved for pyrimidine nucleotides during the early stages of treatment. Another drug, **6-azauridine**, after conversion to 6-azauridylate, is a competitive inhibitor of orotidylate decarboxylase, inducing high rates of excretion of orotic acid and orotidine as a result.

CLINICAL DISORDERS OF PURINE METABOLISM

The end products of purine metabolism, unlike those of pyrimidine metabolism, are sparingly water-soluble compounds. Thus, in circumstances where purine overproduction occurs, clinically detectable abnormalities are frequently evident (Table 29–2).

Table 29–2. Inherited disorders of purine metabolism and their associated enzyme abnormalities

Clinical disorder	*Defective enzyme*	*Nature of the defect*	*Characteristics of clinical disorder*	*Inheritance pattern*
Gout	PRPP synthetase	Superactive	Purine overproduction and overexcretion	X-linked recessive
Gout	PRPP synthetase	Resistane to feedback inhibition	Purine overproduction and overexcretion	X-linked recessive
Gout	PRPP synthetase	Low Km for ribose 5-phosphate	Purine overproduction and overexcretion	X-linked recessive?
Gout	HGPRTase	Partial deficiency	Purine overproduction and overexcretion	X-linked recessive
Lesch-Nyhan syndrome	HGPRTase	Complete deficiency	Purine overproduction and overexcretion; cerebral palsy and self-mutilation	X-linked recessive
Immune deficiency	Adenosine deaminase	Severe deficiency	Combined (T cell + B cell) immunodeficiency, deoxyadenosinuria	Autosomal recessive
Immune deficiency	Purine nucleoside phosphorylase	Severe deficiency	T cell deficiency, inosinuria, deoxyinosinuria, guanosinuria, deoxyguanosinuria, hyperuricemia	Autosomal recessive
Renal lithiasis	Adenine phosphoribosyltransferase	Complete deficiency	2,8-dihydroxyadenine renal lithiasis	Autosomal recessive
Xanthinuria	Xanthine oxidase	Complete deficiency	Xanthine renal lithiasis, hypouricemia	Autosomal recessive

The predominant form of uric acid is determined by the pH of its milieu (*e.g.*, blood, urine, cerebrospinal fluid). The pK of the N^9 proton is 5.75, and that of N^1 proton is 10.3. Thus, under physiologic conditions (*i.e.*, at the usual pH of physiologic fluids) only uric acid and its monosodium salt, sodium urate, are found. In a fluid where the pH is less than 5.75, the predominant molecular species will be uric acid. In a fluid at pH 5.75, the concentration of sodium urate will equal that of uric acid. At a pH greater than 5.75, sodium urate will predominate in the solution.

The miscible urate pool in the body is reflected by the sodium urate concentration in the serum. When this level exceeds the solubility of sodium urate in serum, a condition referred to as *hyperuricemia* is created. In it, the serum becomes supersaturated and crystals of sodium urate may precipitate. These crystals can collect and deposit in soft tissues, particularly in or about joints. These sodium urate deposits are referred to as *tophi*. Accumulation of sodium urate crystals in the tissues, can lead to an acute inflammatory reaction called **acute gouty arthritis**. The chronic inflammatory changes induced by the deposition of sodium urate tophi can generate **chronic gouty arthritis**, which results in joint destruction.

In water, uric acid (the protonated form of urate) is only 1/17th as soluble as sodium urate. An aqueous solution becomes saturated with uric acid when its concentration is greater than 80 mg/dl. Because the pH of urine of normal persons generally is below the pK of uric acid (*i.e.*, 5.75), the predominant form of urate in the urine is as uric acid, which is the highly insoluble form. Uric acid becomes the predominant form once the urine is acidified to a pH of less than 5.75, a process which occurs in the distal tubule and collecting ducts of the kidney. Therefore, more stones of the urinary collecting system are uric acid. The precipitation of uric acid stones can be prevented to a considerable degree by alkalinization of the urine to ensure that sodium urate, the more soluble form, will predominate.

As mentioned earlier, high serum levels of urate induce gout, a disease in which salts of urate crystallize and damage joints and kidneys. The average serum level of urate in humans is close to the solubility limit. Contrary to this, prosimians (*e.g.*, lemurs) have ten-fold lower levels. A striking increase in urate levels occurred in the evolution of primates. What is the selective advantage of a urate level so high that it wobbles on the brink of gout in many people? It turns out that urate has a markedly beneficial action. *Urate is highly effective scavenger of reactive oxygen species.* Indeed, urate is about as effecive as ascorbate (vitamin C) as an antioxidant. The increased levels of urate in humans, as compared to prosimians and other lower primates, may contribute significantly to the longer life span of humans and to lowering the incidence of human cancer.

Fig. 29–9. An electronmicrograph showing sodium urate crystals

Joints and kidneys are damaged by these crystals in gout.

(Courtesy: Dr. James McGuire)

The **sodium urate crystals** (Fig. 29–9) are acicular (needle-shaped) and intensely negatively birefringent (optically anisotropic). Thus, when viewed through a polarizing microscope, they can be distinguished from other types of crystals. If the synovial (or joint) fluid of a patient shows polymorphonuclear leucocytes containing crystals whose colour is yellow when viewed with their long axis parallel to the plane of polarized light and blue when perpendicular to the plane of light, then sodium urate crystals are present. This diagnoses gout. It should be noted, however, that calcium pyrophosphate crystals, which are found in synovial fluid, are positively birefringent and can be responsible for a syndrome known as **pseudogout**.

Purine metabolism also involves a disorder called **hyperuricemia** (elevated blood uric acid). Individuals showing hyperuricemia can be divided into 2 groups:

(a) those with normal urate excretion rates, where renal disorder is responsible for elevated serum urate. This renal disorder is somewhat analogous to an elevated threshold wherein excretion of the normal amount of urate formed daily requires that the level of serum urate be elevated to "flow over the dam", so to speak.

(b) those excreting excessive quantities of total urates either secondary to other diseases (*e.g.*, malignancy, psoriasis) or due to known defects is certain enzymes (*e.g.*, PRPP synthetase, HGPRTase, glucose 6-phosphatase).

Certain identifiable enzyme defects with regard to purine metabolism do occur in humans. Such defects include abnormalities of PRPP synthetase (feedback resistance and enhanced enzyme acivity), HGPRTase deficiencies (complete deficiency as in Lesch–Nyhan syndrome or incomplete deficiency), and glucose 6-phosphate deficiency (as in von Gierke's disease).

The **Lesch–Nyhan syndrome** (complete HGPRTase deficiency) is a compulsive X-linked recessive disorder characterized by cerebral palsy with choreoathetosis and spasticity, a bizarre syndrome of self-mutilation, and severe overproduction hyperuricemia but without any neurologic manifestations. There also exist male patients with partial deficiencies of HGPRTase which is attributed to a different mutation of the same gene. These males have severe overproduction hyperuricemia but usually are without significant neurologic signs and symptoms. Purine overproduction by patients deficient in hypoxanthine-guanine phosphoribosyltransferase (HGPRTase) is related to the increased intracellular concentrations of PRPP. Increased PRPP levels seem to result from the sparing of PRPP by the deficient salvage pathway.

The basis of purine overproduction and hyperuricemia in **von Gierke's disease** is purportedly secondary to the enhanced activity of the hexose monophosphate shunt and thus enhanced production of ribose 5-phosphate, from which PRPP is synthesized. However, patients with glucose 6-phosphatase deficiency also have chronic lactic acidosis and thus have decreased secretion of urate contributing to the accumulation of total body urates.

Hypouricemia (decreased blood uric acid) is due either to enhanced excretion or to decreased production of urate and uric acid. Dalmatian dogs, although possessing uricase acivity like rest other dogs, are not capable of reabsorbing completely the filtered uric acid in their kidneys. Hence, they excrete urate and uric acid in amounts that are excessive in respect to their serum urate levels. Similar defect in humans are exceeding rare.

A deficiency of the enzyme *purine nucleoside phosphorylase* (PNP) is associated with hypouricemia because persons deficient in this enzyme are not capable of producing hypoxanthine and guanine from inosine and guanosine, respectively. Consequently, excessive quantities of purine nucleosides are excreted in their urine. Two of the purine nucleosides excreted in significant quantities, guanosine and dexoyguanosine, have limited solubility and can, hence, result in **renal lithiasis**.

Two immunodeficiency diseases associated with deficiencies of purine metabolizing enzymes have been described in the recent past. **Adenosine deaminase (ADA) deficiency** is associated with a severe combined immunodeficiency disease in which both thymus-derived lymphocytes (T cells) and bone marrow-derived lymphocytes (B cells) are sparse and dysfunctional. Another disease, **purine nucleoside phosphorylase (PNP) deficiency** is associated with a severe T cell deficiency with apparent normal B cell function, a much milder form of immunodeficiency. Both of these diseases are inherited as an autosomal recessive disorders. The molecular bases for the immune dysfunctions seem to involve the intracellular accumulation of the triphosphates of the deoxyribonucleoside substrates, deoxyguanosine and deoxyadenosine, of purine nucleoside phosphorylase and adenosine deaminase, respectively. These deoxynucleoside triphosphates, deoxy GTP and deoxy ATP, are capable of allosterically inhibiting ribonucleotide reductase and thereby depleting cells, such as T cells, of the precursors of DNA synthesis, particularly deoxy CTP.

Purine deficiency states are rare in humans. These are restricted to circumstances attributable primarily to deficiencies of folic acid and perhaps of vitamin B_{12} when the latter results in a secondary deficiency of folate derivatives.

COMPOUNDS THAT INTERFERE WITH CELLULAR NUCLEOTIDE METABOLISM: CHEMOTHERAPEUTIC AGENTS

The *de novo* synthesis of pyrimidine and purine nucleotides is essential for normal cell replication, maintenance and function. Regulation of these pathways is important since diseases spring up from defects in the regulatory enzymes. Many compounds, both synthetic and natural products of plants, bacteria or fungi are structural analogues of the bases or nucleosides used in metabolic reactions. These compounds are specific inhibitors of enzymes involved in nucleotide synthesis or interconversions and have proved useful in the treatment of various clinical disorders. Such compounds, called *chemotherapeutic agents*, are generally classified as antimetabolites, antifolates, glutamine antagonists, antiviral agents and other agents.

1. Antimetabolites

Antimetabolites are usually structural analogues of pyrimidine and purine bases or nucleosides that interfere with the most specific metabolic reactions. They include *allopurinol* for the treatment of gout and hyperuricemia; *acyclovir* and *ganciclovir* for the treatment of herpes viral infections; *6-mercaptopurine* and *6-thioguanine* for the treatment of acute leukemia; and *azathioprine* for immunosuppression in patients with organ transplants. A study of the mechansim of action of these drugs has led to a better understanding of normal nucleotide metabolism in humans. *Nucleotides are poorly transported into cells because of the negative phosphate* charge, so most of the compounds, discussed below, are introduced into cells as nucleosides or nucleoside derivatives, where they are acted upon initially by nucleoside kinases. After uptake and conversion to nucleotides, these compounds interfere with metabolism in diverse ways. Some of these antimetabolites are described hereunder.

Allopurinol has already been described in this chapter under 'Degradation of Purine Nucleotides'.

Both **acyclovir** (= acycloguanosine) and **ganciclovir** (Fig. 29–10) are used in treating herpes viral infections. The herpesviruses are large DNA-containing viruses whose genomes encode several enzymes, including *deoxypyrimidine kinase*. Originally discovered as a thymidine kinase, the enzyme also phosphorylates dTMP. The extremely broad specificity of the nucleoside kinase activity of this enzyme is quite important. The thymidine analogue, **5-iododeoxyuridine**, used to treat herpes simplex eye infections, is readily phosphorylated by deoxypyrimidine kinase and incorporated into DNA in place of thymidylate. Recently, both acyclovir and its relative ganciclovir have been shown also to be phosphorylated by viral deoxypyrimidine kinases, even though they are analogues of a purine nucleoside, deoxyguanosine. Both acyclovir and ganciclovir are now being used for topical and systemic treatment of herpes viral infections. All 3 analogues are converted ultimately to the 5′-triphosphate, which interferes with DNA replication. Uninfected cells do not efficiently phosphorylate acyclovir and ganciclovir, and they phosphorylate iododeoxyuridine weakly, so DNA replication and, hence, virus growth are inhibited selectively in the infected cells.

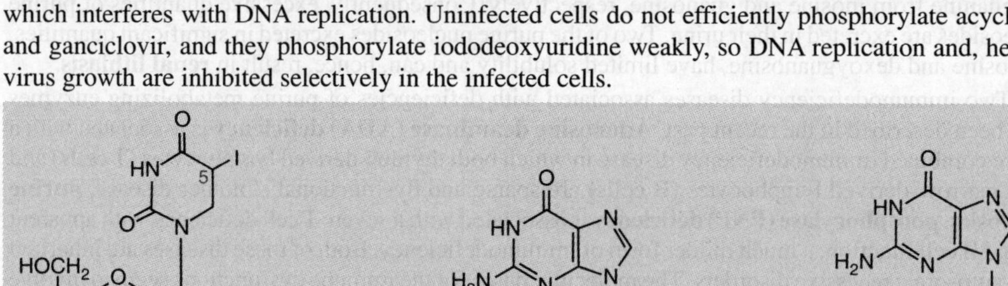

Fig. 29–10. **Three structural analogues used for the treatment of herpes viral infections**

CLINICAL IMPLICATIONS 29–1.

Lesch–Nyhan Syndrome

Lesch–Nyhan syndrome (named after its two discoverers, both 20th century US pediatricians, **Michael Lesch** and **William L. Nyhan**) is characterized clinically by hyperuricemia, excessive uric acid production, and neurological problems such as spasticity, mental retardation and self-mutilation. It springs up under circumstances of almost complete deficiency of HGPRTase activity. The gene for HGPRTase activity is situated on the Y chromosome, hence the deficiency is virtually to males. There are over 100 disease-related mutations defined in the HGPRTase gene from Lesch–Nyhan patients. These have led to the loss of HGPRTase protein, loss of HGPRTase activity etc.

The role of HGPRTase is to catalyze reactions in which hypoxanthine and guanine are converted to nucleotides. The hyperuricemia and excessive uric acid production that occur in patients with Lesch–Nyhan syndrome are easily explained by the lack of HGPRTase activity. Hypoxanthine and guanine are not salvaged, leading to increased intracellular pools of PRPP and decreased levels of IMP or GMP. Both of these factors promote *de novo* synthesis of purine nucleotides *without* regard for proper regulation of this pathway at PRPP amidotransferase level.

It is not understood why a severe defect in this salvage pathway leads to bizarre neurologic symptoms. Adenine phosphoribosyltransferase activity in these patients is normal or elevated. With this salvage enzyme, presumably the cellular needs for purine nucleotides could be met by conversion of AMP to IMP, followed by the conversion of IMP to GMP if the cell's *de novo* pathway were not functioning. The normal tissue distribution of HGPRTase activity perhaps could explain the neurological symptoms. The brain (frontal lobe, basal ganglia, and cerebellum) has 10 to 20 times the level of enzyme activity found in liver, spleen, or kidney and from 4 to 8 times that found in erythrocytes. Individuals who have primary gout with excessive uric acid formation and hyperuricemia do not exhibit neurological problems. It is argued that products of purine degradation (hypoxanthine, xanthine, and uric acid) cannot be toxic to the central nervous system (CNS). However, it is possible that these metabolites are toxic to the developing CNS or that lack of enzyme leads to an imbalance in the concentrations of the purine nucleotides at critical times during development. If the IMP dehydrogenase activity in the brain were extremely low, lack of HGPRTase could lead to decreased amounts of intracellular GTP due to decreased salvage of guanine. Since GTP is a precursor of tetrahydrobiopterin, a required cofactor in the biosynthesis of neurotransmitters, low concentrations of GTP during development could be the triggering factor in the observed neurological manifestations.

Treatment of Lesch–Nyhan patients with the drug **allopurinol** will decrease the amount of uric acid formed, relieving some of the problems caused by sodium urate deposits. However, since the Lesch–Nyhan patient has a marked reduction in HGPRTase activity, hypoxanthine and xanthine are not salvaged, PRPP is not consumed, and consequently *de novo* synnthesis of purine nucleotides is not shut down. There is no treatment for neurological problems. These patients usually die from kidney failure, resulting from high sodium urate deposits.

The Lesch–Nyhan syndrome demonstrates that the salvage pathway for the synthesis of IMP and GMP is not gratuitous. Moreover, this syndrome also reveals that *the abnormal behaviour such as self-mutilation and extreme hostility can be caused by the absence of merely a single enzyme*.

CLINICAL IMPLICATIONS 29–2.

Immunodeficiency Disease

Two distinct immune deficiency diseases are associated with defects in the *adenosine deaminase* (ADA) and *purine nucleoside phosphorylase* (PNP), respectively. These enzymes are involved in the degradative pathways leading to the formation of uric acid. Natural substrates for ADA are adenosine and deoxyadenosine while natural substrates for PNP are inosine, deoxyinosine, guanosine and deoxyguanosine. A deficiency in ADA is associated with a severe combined immunodeficiency involving both T cell and B cell functions. PNP deficiency is associated with an immunodeficiency involving T cell functions with the sparing of effects on B cell functions. In neither case is the mechanism(s) by which the lack of these 2 enzymes leads to immune dysfunction known.

However, in ADA-deficient patients, intracellular concentrations of dATP and *S*-adenosylhomocysteine are greatly increased. Several hypotheses have been proposed to explain the biochemical consequences of a lack of ADA:

1. High concentrations of dATP inhibit ribonucleotide reductase activity and consequently inhibit DNA synthesis.
2. Deoxyadenosine inactivates *S*-adenosylhomocysteine hydrolase leading to decreased *S*-adenosylmethionine required for methylation of bases in RNA and DNA.
3. Increased concentrations of adenosine result in increased cAMP levels.

It is possible that each of these mechanisms contributes to the overall effect of immune dysfunction. However, there is not a suitable explanation for the specificity of the effects on only T cells and B cells.

Treatments of children with ADA deficiency have included blood transfusions, bone marrow transplantation, enzyme replacement therapy with ADA-polyethylene glycol conjugate (ADA-PEG), and most recently gene therapy. Each of these treatments have drawbacks. Blood transfusions produce problems of "iron overload" and safety of the source. Bone marrow transplantation, while curative, requires a suitably-matched donor. Enzyme replacement therapy with ADA-PEG has been the most successful to date, but the treatment requires constant monitoring of ADA levels and frequent injections of very costly ADA-PEG. Gene therapy presents the hope for the future. There are strong indications in gene therapy trials that the ADA gene has been successsfully transfected into stem cells of ADA-deficient children.

Some other antimetabolites (Fig. 29–11) are discussed below to show (a) the importance of *de novo* synthetic pathways in normal cell metabolism, (*b*) that regulation of these pathways occurs *in vivo*, and (*c*) the concept of the requirement for metabolic activation of the drugs.

6-mercaptopurine (6-MP) is a useful antitumour drug in humans. Its cytotoxic activity is related to the formation of 6-mercaptopurine ribonucleotide by the tumour cell. Utilizing PRPP and HGPRTase, 6-mercaptopurine ribonucleoside 5′-monophosphate accumulates in cells and is a negative effector of PRPP amidotransferase, the committed step in the *de novo* pathway. This nucleotide also acts as an inhibitor of the conversion of IMP to AMP at the adenylosuccinate synthetase step and IMP to GMP at the IMP dehydrogenase step. Since 6-MP is a substrate for xanthine oxidase and is oxidized to 6-thiouric acid, allopurinol is generally administered to inhibit degradation of 6-MP and to potentiate the antitumour properties of 6-MP.

5-fluorouracil (Fura) is an analogue of uracil. *Fura is not the active species*. It must be converted by cellular enzymes to the active metabolites, 5-fluorouridine 5′-triphosphate (FUTP) and 5-fluoro-2′-deoxyuridine 5′-monophosphate (FdUMP). FUTP is efficiently incorporated into RNA and once incorporated inhibits maturation of 45S precursor rRNA into the 28S and 18S species and alters splicing of pre-mRNA into functional mRNA. FdUMP is a potent and specific

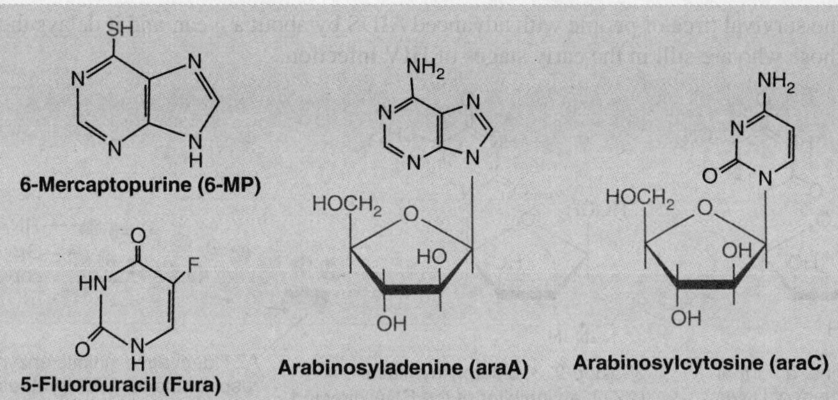

Fig. 29–11. Structure of 4 antimetabolites used against cancer and other diseases

inhibitor of thymidylate synthase. In the presence of tetrahydrofolate, FdUMP and thymidylate synthase, a ternary complex is formed that results in covalent binding of FdUMP to thymidylate synthase. This inhibits dTMP synthesis and leads a "thymineless death" for cells.

Arabinosyladenine (araA), also called adenine arabinoside, is one of the earliest antiviral drugs approved for use in the human beings. It is now being used to treat viral encephalitis, a neurological disease caused by another member of the herpesvirus famly. Unlike acyclovir, araA is phosphorylated to the triphosphate level by cellular kinases. The triphosphate, araATP, is a selective inhibitor of DNA polymerases encoded by herpesviruses. Thus, araA selectively interferes with viral DNA replication, even though all cells, infected and healthy, form the triphosphate. Because araA is susceptible to degradation by adenosine deaminase (ADA), its effectiveness can be increased when administered with an inhibitor of the latter enzyme.

Arabinosylcytosine (araC), also called cytosine arabinoside, is arabinose analogue of deoxycytidine. It is used in the treatment of several forms of human cancer. It must be metabolized by cellular enzymes to arabinosylcytosine 5′-triphosphate (araCTP) to exert its cytotoxic effects. araCTP competes with dCTP in the DNA polymerase reaction and araCMP is incorporated into DNA. This inhibits synthesis of the growing DNA strand. Clinically, the efficacy of araC as an antileukemic drug correlates with the concentration of araCTP that is achieved in the tumour cell, which in turn determines the amount of araCMP incorporated into DNA. Formation of araCMP *via* deoxycytidine kinase appears to be the rate-limiting step in activation to araCTP.

There are other analogues also which are being used to combat acquired immune deficiency syndrome (AIDS) caused by human immunodeficiency virus (HIV). One such drug is **3′-azido-2′, 3′-dideoxythymidine (AZT)** which is also called zidovudine. AZT (Fig. 29–12), a nucleoside analogue with an azido group on the sugar, was first synthesized in 1964 by **Jerome P. Horwitz**. It failed as an anticancer drug (the purpose for which it was made), but in 1985 it was found to be a useful treatment for AIDS.

It is, in fact, the first drug approved for treatment of HIV infection. AZT can be phosphorylated into the triphosphate form and competes with dTTP for incorporation into the reverse transcript. Once incorporated, it terminates the growing chain of the transcript because the azido group on the 3′-carbon of the sugar is not a substrate for nucleotide addition. AZT is much less efficient at competing with the more accurate cellular DNA polymerases, providing a therapeutic opportunity in which the effect is primarily upon viral replication. Nevertheless, the side effects include *toxicity* to bone marrow, which contains rapidly dividing cells, and *myopathy* that might be related to toxicity to mitochondria, which contain their own DNA polymerase, pol γ. However, AZT can increase the survival time of people with advanced AIDS by about a year, and it delays the onset of AIDS in those who are still in the early stages of HIV infection.

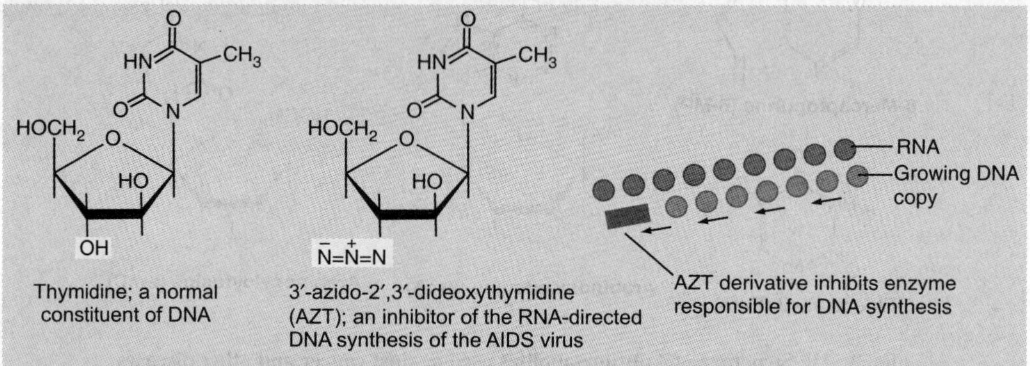

Fig. 29–12. Medical application of biochemistry, taking AZT as an example

It is, however, interesting to note that the 2 antiviral agents (acyclovir and AZT) demonstrate the diversity of responses required for selectivity. In one case, enzyme activity encoded by the viral genome is mandatory for activation of the drug (acyclovir); in the second case, although cellular enzymes activate AZT, the viral gene product (HIV-DNA polymerase) is the selective target.

Other nucleoside analogues– 2′, 3′-dideoxycytidine (ddC), 3′-thiacytidine (3TC), 2′,3′-didehydro-3′-deoxythymidine (d4T) and 2′,3′-dideoxyinosine (ddI)– act by conversion to the corresponding triphosphate, which is incorporated into DNA but then blocks further replicative chain elongation because of the absence of a 3′-hydroxyl terminus. All 4 analogues (Fig. 29–13) have been approved for use in treating human HIV infections and both AZT and 3TC are components, along with HIV protease inhibitors, of the three-drug "cocktails" recently credited with long-term remission of HIV infections.

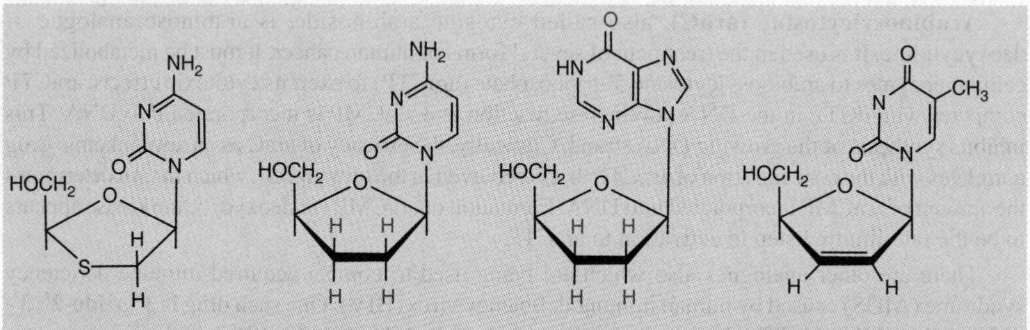

Fig. 29–13. Four other nucleoside analogues used against HIV infections

2. Antifolates

Antifolates are compounds which interfere with the formation of tetrahydrofolate (H_4 folate) from dihydrofolate (H_2 folate) or H_2 folate from folate by inhibition of dihydrofolate reductase. **Methotrexate (MTX)**, a close structural analogue of folic acid, is used as an antitumour agent in the treatment of human cancers. MTX and folate (Fig. 29–14) differ from each other at 2 places:

(i) the OH group present at C-4 in folate is repalced by an NH_2 group in methotrexate.

(ii) the H atom present at N-10 in folate is replaced by a CH_3 group in methotrexate.

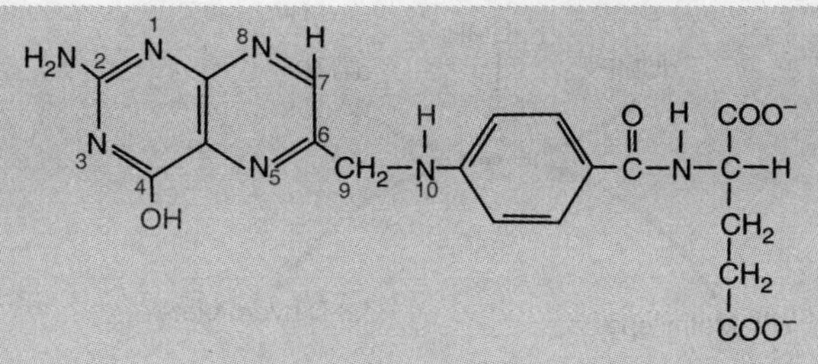

Folic acid, FA

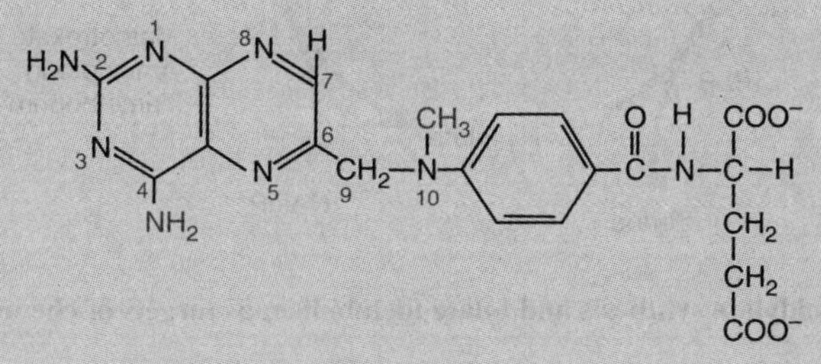

Methotrexate, MTX
(=amethopterin)

Fig. 29–14. A comparison of the structures of folic acid and methotrexate

MTX is an inhibitor of dihydrofolate reductase (Fig. 29–15). This folate analogue acts as a competitive inhibitor; the enzyme binds methotrexate with about 100 times higher affinity than dihydrofolate. MTX specifically inhibits dihydrofolate reductase with a *K*i in the range of 0.1 nM. The reactions inhibited are shown below:

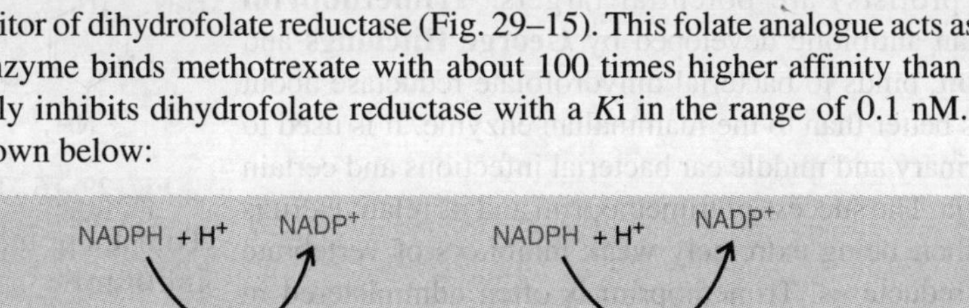

MTX at very low concentrations is cytotoxic to mammalian cells in culture. The inhibition of dihydrofolate reductase by MTX results in the lowering of both ribonucleoside 5′-triphosphate and 2′-deoxyribonucleoside 5′-triphosphate intracellular pools. The effects can be prevented by addition of deoxythymidine and hypoxanthine to the culture medium. Reversal of the MTX effects by thymidine and hypoxanthine indicates that MTX causes depletion of deoxythymidine and purine nucleotides in the cells.

During thymidylate synthesis, N^5,N^{10}-methylenetetrahydrofolate is converted to 7,8-dihydrofolate; the N^5, N^{10}-methylenetetrahydrofolate is regenerated in 2 steps. *This cycle is a major target of several chemotherapeutic agents*. Fluorouracil and methotrexate are important therapeutic agents. In cells, **fluorouracil** is converted to FdUMP, which inhibits thymidylate synthase. **Methotrexate**, a structural analogue of tetrahydrofolate, inhibits dihydrofolate reductase.

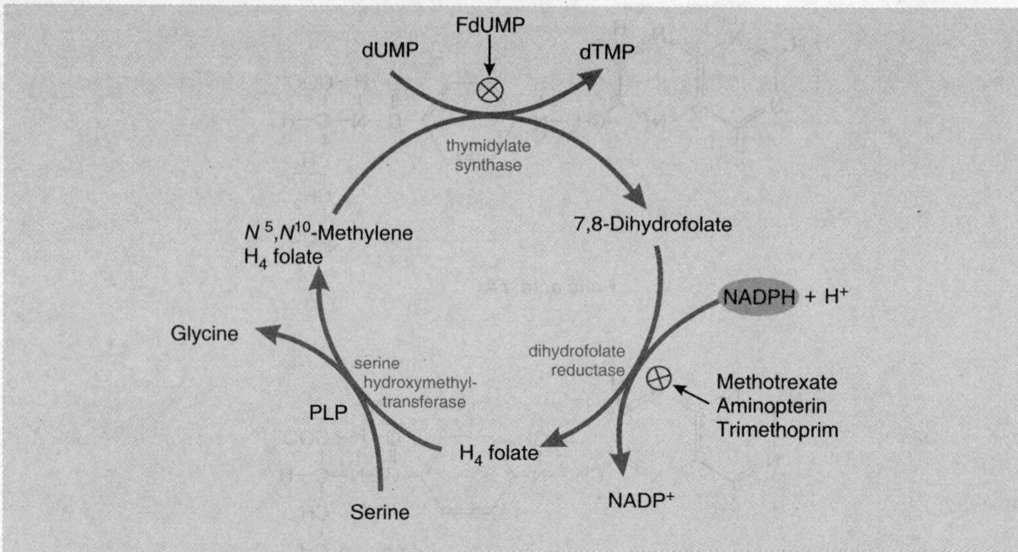

Fig. 29–15. Thymidylate synthesis and folate metabolism as targets of chemotherapy

Aminopterin is another important folate analogue. It is identical to methotrexate except that it lacks the methyl group at N-10. Its mechanism of action is similar to that of methotrexate.

The medical potential of inhibitors of nucleotide biosynthesis is not limited to cancer treatment. All fast-growing cells (including bacteria and protists) are potential targets. **Trimethoprim** (Fig. 29–16), an antibiotic developed by **George Hitchings** and **Gertrude Elion**, binds to bacterial dihydrofolate reductase about 1,00,000 times better than to the mammalian enzyme. It is used to treat certain urinary and middle ear bacterial infections and certain forms of malaria. The success of trimethoprim and its relative drugs derives from their being extremely weak inhibitors of vertebrate dihydrofolate reductases. Trimethoprim is often administered in conjunction with a sulfonamide drug to inhibit the *synthesis* of folate and hence to block sequential steps in the same pathway.

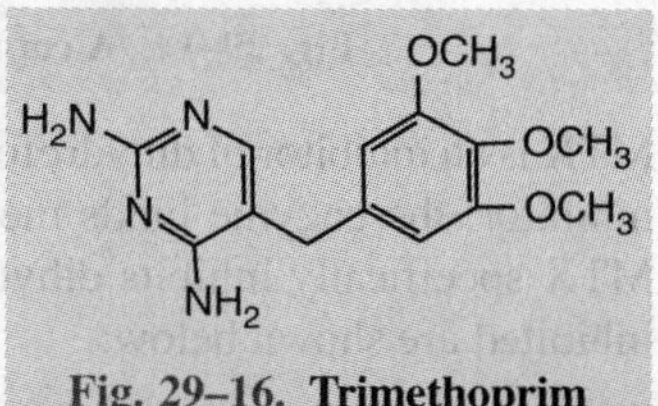

Fig. 29–16. Trimethoprim

Trimethoprim, a tight-binding inhibitor of bacterial dihydrofolate reductase, was developed as an antibiotic.

3. Glutamine Antagonists

Many reactions in mammalian cells utilize glutamine, which serves as the amino group donor. On the contrary, bacteria primarily utilize NH_3 as the amino donor in a similar reaction. These amidation reactions are critical in *de novo* synthesis of purine nucleotide (N-3 and N-9), synthesis of GMP from IMP, formation of cytosolic carbamoyl phosphate, synthesis of CTP from UTP, and synthesis of NAD^+.

Compounds that inhibit these reactions are referred to as *glutamine antagonists* (Fig., 29–17). **Azaserine** (*O*-diazoacetyl-L-serine) and 6-diazo-5-*oxo*-L-norleucine (DON), which were first isolated from cultures of *Streptomycin*, are very effective inhibitors of glutamine transferases, *i.e.*, the enzymes that utilize glutamine as the amino donor. Since azaserine and DON inactivate the enzymes involved, addition of glutamine alone will not reverse the effects of either of these 2 drugs. It would necessitate that many metabolites such as guanine, cytosine, hypoxanthine (or adenine), and nicotinamide be provided to bypass the many sites blcoked by these glutamine antagonists. As expected from the fact that so many key steps are inhibited by azaserine and DON, these agents are extremely toxic and not of clinical use. **Acivicin**, another glutamine analogue, shows promise as a cancer therapeutic agent.

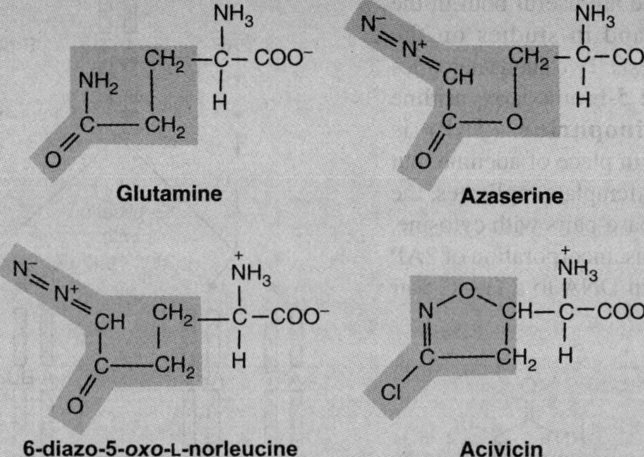

Fig. 29–17. Structure of 3 glutamine antagonists

These analogues of glutamine interfere in a number of amino acid and nucleotide biosynthetic pathways. Structure of *glutamine* is shown for comparison.

4. Cell Growth-Inhibiting Agents

There are certain chemical compounds that inhibit cell growth by interfering with nucleotide metabolism. Two such chemicals, hydroxyurea and tiazofurin (Fig. 29–18), are discussed here. Tumour cells treated with **hydroxyurea** show a specific inhibition of DNA synthesis with little or no inhibition of RNA or protein synthesis. Hydroxyurea inhibits ribonucleotide reductase to block reduction of CDP, UDP, GDP and ADP to the corresponding 2′-deoxyribonucleoside 5′-diphosphates. Toxicity results from depletion of 2′-deoxyribonucleoside 5′-triphosphates required for DNA replication. Clinical use of hydroxyurea as an antitumour agent is limited because of its rapid rate of clearance and the high drug concentration required for effective inhibition. However, hydroxyurea has recently been utilized in the treatment of sickle-cell anemia.

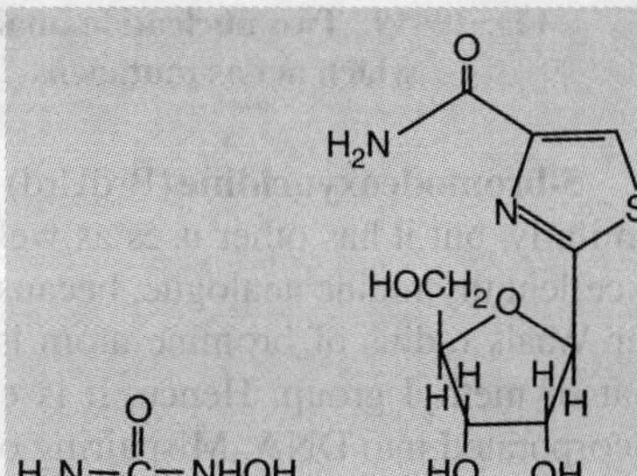

Fig. 29–18. Structure of 2 cell growth-inhibiting agents

Another compound, **tiazofurin** is converted by cellular enzymes to the NAD^+ analogue, *tiazofurin adenine dinucleotide* (TAD). TAD inhibits IMP dehydrogenase, the rate-limiting enzyme in GTP synthesis. As a result of IMP dehydrogenase inhibition, the concentration of GTP is markedly decreased. It is interesting that, although there are many dehydrogenases in the cells that utilize NAD^+ as a substrate, IMP dehydrogenase is the enzyme that is most effected. This could be because IMP dehydrogenase catalyzes a rate-limiting step in a pathway and is quantitatively limiting in concentration.

These clinically useful drugs serve as examples in which the knowledge of basic biochemical pathways and mechanisms lead to generation of effective drugs.

NUCLEOTIDE ANALOGUES AND MUTAGENESIS

Some nucleotide analogues are excellent mutagenic agents. These are useful both in the isolation of mutants and in studies on the mechanism of mutagenesis. Two such analogues are 2-aminopurine and 5-bromodeoxyuridine (Fig. 29–19). **2-aminopurine** (2AP) is incorporated into DNA in place of adenine, but when a 2AP-containing template replicates, the analogue occasionally base-pairs with cytosine, rather than thymine. Thus, incorporation of 2AP changes an A—T pair in DNA to a G—C pair (Fig. 29).

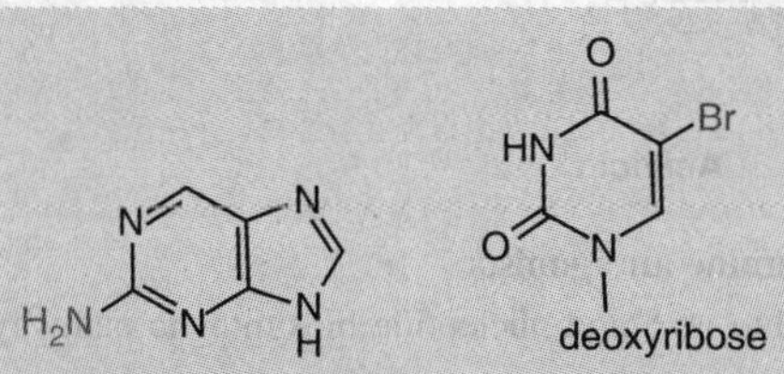

2-Aminopurine (2AP) **5-Bromodeoxyuridine (BrdUrd)**

Fig. 29–19. Two nucleotide analogues which act as mutagens

5-bromodeoxyuridine (BrdUrd) functions similarly, but it has other uses as well. It is an excellent thymidine analogue, because the van der Waals radius of bromine atom is close to that of methyl group. Hence, it is efficiently incorporated into DNA. Mispairing of BrdUrd with a deoxyguanosine residue on replication of a BrdUrd-containing template can lead to mutagenesis by changing an A—T base pair to a G—C base pair (Fig. 29–20). Alternatively, BrdUTP can compete with dCTP for incorporation opposite G in the template (not shown in the figure).

As the bromine is much heavier than a methyl group, it imparts an increased density to the substituted DNA, thereby providing a physical basis for separating replicating DNA from non-replicating one. Finally, radiobiologists use BrdUrd as a radiosensitizing agent. Bromo-dUMP residues in DNA debrominate readily when the substituted DNA is irradiated with UV or near-UV light. This process generates free radicals, which cause various kinds to damage to the DNA structure.

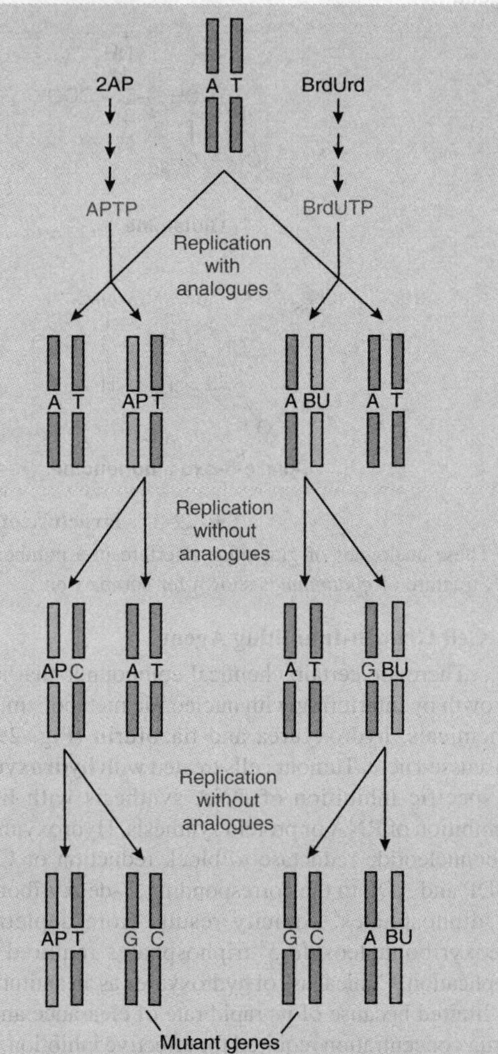

Fig. 29–20. Mechanisms of mutagenesis by nucleotide analogues

2-aminopurine (2AP) is converted by salvage pathways to dAPTP, the dNTP analogue of dATP. BrdUrd is converted to BrdUTP, the dNTP analogue of dTTP. The first round of replication occurs in the presence of the analogue, and the second and third rounds occur in its absence. Replication of only the analogue-containing duplex is shown in the second and third rounds. AP is a 2-aminopurine nucleotide residue in DNA, and BU is a bromodeoxyuridine nucleotide residue. In the pathways shown, both analogues change an A—T base pair to a G—C base pair (*red* letters). Other pathways can occur as well.

REFERENCES

1. **Chargaff E, Davidson JN (editors) :** The Nucleic Acids. *vol. III, Academic Press, Inc., New York. 1980.*
2. **Davidson JN :** The Biochemistry of the Nucleic Acids. *5th ed., John Wiley & Sons, Inc., New York. 1965.*
3. **Elion GB :** The purine path to chemotherapy. *Science.* **244 :** *41–47, 1989.*
4. **Gutman AB :** The biological significance of uric acid. *Harvey Lectures.* **60 :** *35–55, 1964–1965.*
5. **Hardy LW, Finer-Moore JS, Montfort WR, Jones MO, Santi DV, Stroud RM :** Atomic structure of thymidylate synthase: Target of rational drug design. *Science.* **235 :** *448–455, 1987.*
6. **Henderson JF, Paterson ARP :** Nucleotide Metabolism : An Introduction. *Academic Press, Inc., New York. 1973.*
7. **Hutchinson DW :** Nucleotides and Coenzymes. *John Wiley & Sons, Inc., New York. 1964.*
8. **Jackson RC (editor) :** Anti-folate drugs– Past and future perspectives (Series : Cancer Drug, Discovery and Development). *Humana Press, Inc., Totowa, N.J. 451 pages, 1999.*
9. **Kornberg A, Baker TA :** DNA Replication. *2nd edition. Freeman, San Francisco. 1972.*
10. **Mitsuya H (editor) :** Anti-HIV nucleosides: Past, Present, and Future. *R.G. Landes, Georgetown, Tex. 1997.*
11. **North TW :** Chemotherapy, antiviral agents. *In Encyclopedia of Human Biology. vol 2,* 395–402. *Academic Press, Inc., New York. 1992.*
12. **Nyhan WL :** The recognition of Lesch–Nyhan syndrome as an inborn error of purine metabolism. *J. Inherited Metab. Dis.* **20 :** *171–178, 1997.*
13. **Resta R, Thompson LF :** SCID : The role of adenosine deaminase deficiency. *Immunol. Today.* **18** : *371–374, 1997.*
14. **Scriver CR, Beaudet AL, Valle D, Sly WS, Childs B, Kinzler LW, Vogelstein B (editors) :** The Metabolic and Molecular Bases of Inherited Disease. *8th edition. McGraw-Hill Professional, New York. 2001.*
15. **Sculley DG, Dawson PA, Emerson BT, Gordon RB :** A review of the molecular basis of hypoxanthine-guanine phosphoribosyltransferase (HPRT) deficiency. *Hum. Genet.* **90 :** *195–207, 1992.*
16. **Seegmiller JE :** Contributions of Lesch–Nyhan syndrome to the understanding of purine metabolism. *J. Inherited Metab. Dis.* **12 :** *184–196, 1989.*
17. **Smith LH Jr :** Pyrimidine metabolism in man. *N. Engl. J. Med.* **288 :** *764, 1973.*
18. **Stanbury JB, Wyngaarden JB, Frederickson DS (editors) :** The Metabolic Basis of Inherited Disease. *4th edition. McGraw-Hill Book Company, New York. 1978.*
19. **Webster DR, Becroft DMO, Suttle DP :** Hereditary orotic aciduria and other disorders of pyrimidine metabolism. In Scriver CR, Beaudet AL, Sly WS, Valle D (Editors): The Metabolic and Molecular Bases of Inherited Disease. *7th edition, McGraw-Hill, New York. 1799–1837, 1995.*
20. **Wyngaarden JB, Kelley WN :** Gout and Hyperuricemia. *Grune & Stratton. 1976.*

PROBLEMS

1. Allopurinol is sometimes given to patients with acute leukemia who are being treated with anticancer drug. Why is allopurinol used?
2. Describe the metabolic fate of 5-bromodeoxyuridine. Would you expect BrdUrd to *inhibit* DNA replication. Briefly explain your answer.

3. Hereditary orotic aciduria is characterized by severe anemia, growth retardation, and higher levels of orotic acid excretion. The defect may be in either orotate phosphoribosyltransferase, orotidine decarboxylase, or both. The preferred treatment for this disease is dietary uridine, which reverses the anemia and decreases the formation of orotic acid. What are the elements involved in the effectiveness of the dietary treatment?
4. Mammalian cells can become resistant to the lethal action of methotrexate (MTX) by the selective survival of cells containing increases in dihydrofolate reductase gene copy number, so that intracellular levels of the enzyme become very high. What other biochemical or genetic changes in cells could cause them to become resistant to methotrexate?
5. Many patients with glucose 6-phosphatase deficiency have high serum levels of urate. Hyperuricemia can be induced in normal people by the ingestion of alcohol or by strenuous exercise. Propose a common mechanism that accounts for these findings.
6. Xanthine oxidase treated with allopurinol results in the formation of a new compound that is an extremely potent inhibitor of the enzyme. Propose a structure for this compound.
7. Gout is a disease characterized by hyperuricemia from an overproduction of purine nucleotides *via* the *de novo* pathway. In Lesch–Nyhan syndrome, the specific cause is a severe deficiency of HGPRTase. Allopurinol is used in the treatment of gout to reduce the production of uric acid. In Lesch–Nyhan syndrome, the decrease in uric acid is balanced by an increase in xanthine plus hypoxanthine in blood. In other forms of gout, the decrease in uric acid is greater than the increase in xanthine plus hypoxanthine. Explain this difference in the two forms of gout.
8. The diazo compound *O*-(2-diazoacetyl)-L-serine, known also as azaserine, is a powerful inhiitor of glutamine amidotransferases. If growing cells are treated with azaserine, what intermediates of nucleotide biosynthesis would accumulate? Explain.
9. Allopurinol, an inhibitor of xanthine oxidase, is used to treat chronic gout. Explain the biochemical basis for this treatment. Patients treated with allopurinol sometimes develop xanthine stones in the kidneys, although the incidence of kidney damage is much lower than in untreated gout. Explain this observation in the light of the following solubilities in the urine: uric acid, 0.15 g/L; xanthine, 0.05 g/L; and hypoxanthine, 1.4 g/L.
10. The uric acid is synthesized as an end product of
    (*a*) Protein metabolism (*c*) Carbohydrate metabolism
    (*b*) Lipid metabolism (*d*) Purine metabolism

ANSWERS TO PROBLEMS

1. These patients have a high level of urate because of the breakdown of nucleic acids. Allopurinol prevents the formation of kidney stones and blocks other deleterious consequences of hyperuricemia by preventing the formation of urate.
2. $$\text{BrdUrd} \xrightarrow[\text{ATP}]{\text{TK}} \text{Br-dUMP} \xrightarrow[\text{ATP}]{} \text{Br-dUDP} \xrightarrow[\text{ATP}]{\text{Nucleoside diphosphate kinase}} \text{Br-dUTP}$$
3. It is common for an exogenous agent to require conversion to an active form; in this case the uridine is "salvaged" to the monophosphate and ultimately to the triphosphate. The cell is deficient in UTP and CTP because the conversion of orotic acid is blocked so the exogenous uridine provides a bypass around the block. Orotic acid formation is decreased since UTP inhibits carbamoyl phosphate synthetase II, the control enzyme.
4. Mutation in DHFR gene that makes the enzyme resistant; mutation in transcriptional or translational control mechanism that causes overproduction of normal DHFR; transport defect that causes failure of methotrexate to be taken up by cell.

5. The cytosolic level of ATP in the liver falls and that of AMP rises above normal in all three conditions. The excess AMP is degraded to urate.

6. An oxygen atom is added to allopurinol to form alloxanthine.

7. Not only is uric acid production directly inhibited, *de novo* synthesis is as well, thus reducing production of xanthine and hypoxanthine. In Lesch–Nyhan syndrome, the only effect is the direct inhibition of xanthine oxidase. In non-Lesch–Nyhan gout, hypoxanthine and xanthine are salvaged to IMP and XMP and inhibit PRPP amidotransferase.

8. 5-phosphoribosyl-1-pyrophosphate; this is the first NH_3 acceptor in the purine biosynthetic pathway.

9. Treatment with allopurinol has two consequences:

(*a*) It inhibits conversion of hypoxanthine to uric acid, causing accumulation of hypoxanthine, which is more soluble and more readily excreted; this alleviates the clinical problems associated with AMP degradation.

(*b*) It inhibits conversion of guanine to uric acid, causing accumulation of xanthine, which is less soluble than uric acid; this is the source of xanthine stones. Because the amount of GMP degradation is low relative to AMP degradation, the kidney damage caused by xanthine stones is less than that caused by untreated gout.

9. (*d*) Purine metabolism

CONTENTS

Transfer RNA-binding sites

Three tRNA-binding sites are present on the 70S ribosome. They are called the A (for aminoacyl), P (for peptidyl), and E (for exit) sites. Each tRNA molecule contacts both the 30S and the 50S subunit.

CHAPTER

30

Biosynthesis of Proteins

GENERAL CONSIDERATIONS

Proteins are the end products of most metabolic pathways. A typical cell requires thousands of different protein molecules at any given moment. These must be synthesized in response to the cell's current requirements, transported (or targetted) to the appropriate cellular location and ultimately degraded when the cell no longer requires them. In fact, *the protein synthesis is well understood than protein targetting or degradation.*

Protein synthesis is the most complex of all the biosynthetic mechanisms. In eukaryotic cells, protein synthesis requires the participation of over 70 different ribosomal proteins ; 20 or more enzymes to activate the amino acid precursors; about 15 auxiliary enzymes and other specific protein factors for initiation, elongation and termination of polypeptides; perhaps 100 additional enzymes for the final processing of different kinds of proteins and 40 or more types of transfer and ribosomal RNAs (**Lehninger**, **Nelson** and **Cox**, 1993). Thus, about 300 different macromolecules are required to synthesize polypeptides or proteins. Many of these macromolecules are organized into the complex 3-`D' structure of the ribosome to carry out stepwise translocation of the mRNA as the polypeptide is assembled. Despite this great complexity, proteins are made at exceedingly high rates. For example, in *E. coli* cell at 37°C, a complete polypeptide chain of 100 amino acid residues is synthesized in about 5 seconds. The synthesis of different proteins in each cell is strictly regulated so that only the required number of molecules of each protein is made under certain metabolic conditions. The targetting and degradative processes keep pace with synthetic process so that an appropriate concentration of proteins is maintained in the cell.

Of the total chemical energy used for all the biosynthetic processes, a cell utilizes as much as 90% for protein synthesis only. In *E. coli*, the numbers of different types of proteins and RNA molecules involved in protein synthesis are similar to those in eukaryotic cells. When totalled, the 20,000 ribosomes, 1,00,000 related protein factors and enzymes and 2,00,000 tRNAs present in a typical bacterial cell (with a volume of 100 nm^3) can account for more than 35% of the cell's dry weight.

The transformation of hereditary information in biological systems falls into 3 categories (Fig. 30–1):

1. **General (or Information) transfers.** These are those that can occur in all cells and are of 3 types: DNA, RNA and protein syntheses.

2. **Special transfers.** These are those that occur in cells only under special circumstances and are also of 3 types: RNA replication, reverse transcription and DNA translation.

3. **Forbidden (or Unknown) transfers.** These are those that have never been detected experimentally or predicted theoretically. They too are of 3 types: protein to DNA, protein to RNA and protein to protein.

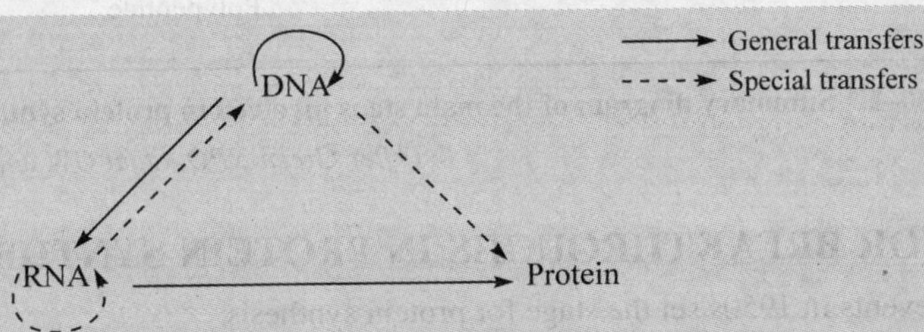

Fig. 30–1. Transfer of genetic information in biological systems

The central dogma of molecular biology has been depicted here. Forbidden transfers are not diagrammed.

In fact, information in the gene is contained in the base sequence of one of the strands of the double helix of DNA. Messages are transcribed onto a strand of RNA which serves as a template for the specific transfer RNA units carrying the amino acids for protein synthesis. The sequence of step may be shown schematically as:

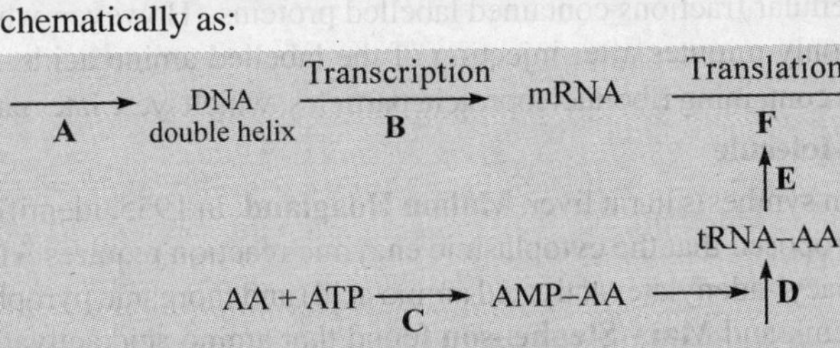

In the above scheme, some possible points of control have been indicated by the inclusion of letters A through F (depicted in *boldface*).

Let us take an overview of protein synthesis. A protein is synthesized in the amino-to-carboxyl direction by the sequential addition of amino acids to the carboxyl end of the growing peptide chain. The activated precursors are aminoacyl-tRNAs, in which the carboxyl group of an amino acid is joined to the 3′–OH of a transfer RNA (tRNA). The linking of an amino acid to its corresponding tRNA is catalyzed by an *aminoacyl-tRNA synthetase*. This activation reaction, which is analogous to the activation of fatty acids, is driven by ATP. For each amino acid, there is at least one kind of tRNA and activating enzyme.

Protein synthesis is a continuous anabolic process and takes place intracellularly. There is no *evidence to suggest that proteins can replicate. Self replication seems to be confined to nucleic acids exclusively.* The protein synthesis, in its simplest form, takes place as depicted in Fig. 30–2.

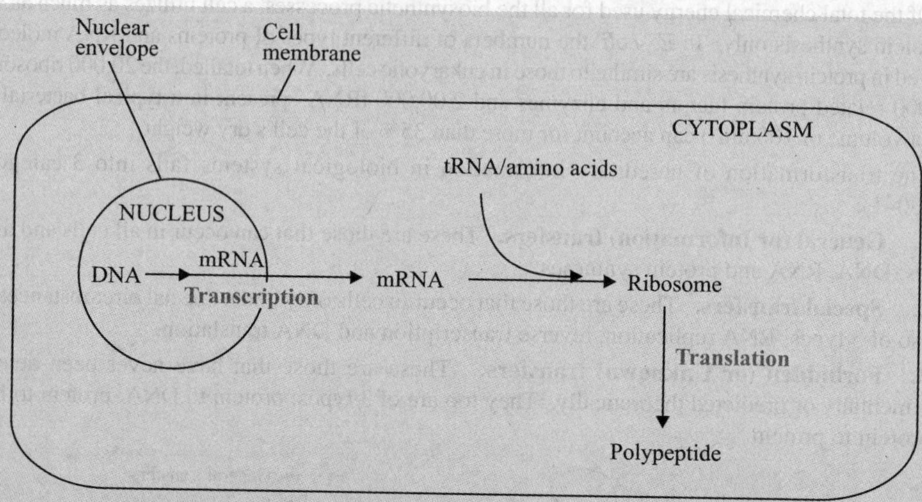

Fig. 30–2. Summary diagram of the main steps involved in protein synthesis

(After Green NPO, Stout GW and Taylor DJ, 1984)

MAJOR BREAKTHROUGHS IN PROTEIN SYNTHESIS

Three major events in 1950s set the stage for protein synthesis:

(a) Ribosomes as the Site of Protein Synthesis

In the early 1950s, **Paul Zamecnik** *et al* conducted an experiment to find out the site of protein synthesis. They injected radioactive amino acids into rats and then after different time intervals, the liver was removed, homogenized and fractionated by centrifugation. The subcellular fractions were then examined for the presence of radioactive protein. After few hours or days of the labelled amino acids injected, all the subcellular fractions contained labelled proteins. However, when the liver was removed and fractionated only minutes after injection of the labelled amino acids, labelled protein was found only in a fraction containing ribonucleoprotein particles, which were later named *ribosomes.*

(b) RNA as a Receptor Molecule

From studies on protein synthesis in rat liver, **Mahon Hoagland**, in 1955, identified the activated form of amino acids. He proposed that the cytoplasmic enzymic reaction requires ATP and an amino acid and produces an aminoacyl-adenylate (activated amino acid) and inorganic pyrophosphate. Later, in 1957, Hoagland, Zamecnik and **Mary Stephenson** found that amino acid activation also utilizes a small RNA (tRNA), found in the soluble fraction of the cell, to form aminoacyl-tRNAs; the reaction being catalyzed by aminoacyl-tRNA synthetases. In other words, the role for a nucleic acid as a receptor molecule in amino acid activation was, hence, identified.

$$\text{Amino acid} + \text{ATP} \rightleftharpoons \text{AMP}\sim\text{amino acid} + PP_i$$
$$\text{AMP}\sim\text{amino acid} + \text{tRNA} \rightleftharpoons \text{tRNA}\sim\text{amino acid} + \text{AMP}$$

(c) Adaptor Hypothesis

The third major advance in protein synthesis took place when **Francis H.C. Crick** (1953) found out how the genetic information that is coded in the 4-letter language of nucleic acids is translated

PAUL CHARLES ZAMECNIK

(LT, 1912–2009)

Zamecnik was an American biochemist who played a central role in the early history of molecular biology. He was born in Cleveland, Ohio. He attended Dartmouth College, majored in chemistry and zoology, and received his AB degree in 1933. Later, he attended Harvard Medical School and received his MD degree in 1936. Between 1936 and 1939, he worked at Collis P. Huntington Memorial Hospital in Boston, Harvard Medical School, and Lakeside Hospital in Cleveland. During his Lakeside Hospital internship, Zamecnik became interested in how cells regulate growth, and hence, in protein chemistry. Then he went to the Carlsberg Laboratory in Copenhagen to work with **Dr. Kai Linderström-Lang**. But because of World War II, he and his wife **Mary Connor** (whom he married in 1936) had to return to Boston, where he became an assistant physician at the Huntington Memorial Hospital, studying the toxic factors involved in traumatic shock for a wartime office of Scientific Research and Development Project led by Huntington director **Joseph Aub**. After a year in New York at the Rockefeller Institute for Medical Research studying protein synthesis with **Max Bergmann**, he returned to Harvard in 1942, to join the Faculty of Medicine at Harvard Medical School where he became an instructor and then Professor of Medicine, where he served until retirement as the Collis P. Huntington Professor of Oncologic Medicine, Emeritus in 1979.

He conducted, in the early 1950s, an experiment to find out the site of protein synthesis, along with his colleagues. He injected radioactive amino acids into rats and then after different intervals, the liver was removed, homogenized and fractionated by centrifugation. The subcellular fractions were then examined for the presence of radioactive protein. After few hours or days of the labelled amino acids injected, all the subcellular fractions contained labelled proteins. However, when the liver was removed and fractionated only minutes after injection of the labelled amino acids, labelled protein was found only in a fraction containing ribonucleoprotein particles, which were later named *ribosomes*. Later in 1957, Zamecnik along with **Mahlon Hoagland** and **M. Stephenson** found that amino acid activation also utilizes a small RNA (*i.e.*. transfer RNA, tRNA), found in the soluble fraction of the cell, to form aminoacyl-tRNAs; the reaction being catalyzed by aminoacyl-tRNA synthetases. In other words, the role for a nucleic acid as a receptor molecule in amino acid activation was, hence, identified.

$$\text{Amino acid + ATP} \rightleftharpoons \text{AMP} \sim \text{amino acid + PP}i$$

$$\text{AMP} \sim \text{amino acid + tRNA} \rightleftharpoons \text{tRNA} \sim \text{amino acid + AMP}$$

After retiring from Harvard Medical School, Zamecnik continued his research at the Worcester Foundation for Biomedical Research with his former colleague **Dr. Hoagland**. When the foundation merged with the University of Massachusetts Medical School in 1997, Zamecnik moved his laboratory to the Massachusetts General Hospital (MGH), where he continued to work until several weeks before his death. Through his later work at MGH, Zamecnik is credited as an **inventor of antisense therapeutics.** Throughout his career, Zamecnik earned over a dozen US patents for his therapeutic techniques. He also studied the application of synthetic oligonucleotides (antisense hybrids) for chemotherapeutic treatment of drug-resistant and XDR tuberculosis in his later years.

Zamecnik authored/coauthored 210 peer-reviewed scientific articles. He won many distinguished awards, including the **National Cancer Society National Award** in 1968, **National Medal of Science** in 1991, and the first-ever **Lasker Lifetime Achievement Award** in 1996. Zamecnik was also member of several societies/associations, some of which are as follows:

1. U.S. National Academy of Sciences
2. American Society of Biological Chemistry
3. American Association for Cancer Research (President, 1964–1965)
4. Association of American Physicians
5. Danish Academy of Sciences and Letter

Zamecnik died on October 27, 2009, at his home in Boston. He was 97 years old. Paul Zamecnik is generally regarded as the *founder of antisense therapy*.

into the 20-letter language of proteins. He opined that tRNA must serve the role of an adaptor, one part of the tRNA molecule binding a specific amino acid and some other part of the tRNA recognizing a short nucleotide sequence (a trinucleotide) in mRNA coding for that amino acid. This idea, termed as adaptor hypothesis, was soon verified. The tRNA adaptor 'translates' the nucleotide sequence of an mRNA into the amino acid sequence of a polypeptide. This overall mRNA-guided process of protein synthesis is often referred to as **translation**. As might be expected, translation is a more complex process than either the replication or the transcription of DNA, which takes place within the framework of a common base-pairing language. In fact, translation necessitates the coordinated interplay of more than a hundred kinds of macromolecules.

Adaptor is also spelt as **adapter.**

CENTRAL DOGMA OF MOLECULAR GENETICS

The 3 different disciplines– genetics, physics and biochemistry– have contributed a lot to the advancement of knowledge of genetical information pathways. The discovery of the double-helical structure of DNA by **Watson** and **Crick** (1953) epitomized the contributions of these 3 fields. *Genetics* contributed the concept of coding by genes. *Physics* made possible the determination of molecular structure of DNA by x-ray diffraction analysis. *Biochemistry* revealed the chemical composition of DNA.

The structure of DNA itself suggested how DNA might be copied so that the information contained therein could be transmitted from one generation to the next. Understanding of how the information contained in DNA was converted into functional proteins became possible through the discovery of messenger RNA, transfer RNA and the genetic code. All these advancements led to the understanding of the process of protein synthesis through the **central dogma of molecular genetics**, which postulates that genetic information flows from nucleic acids to protein. This concept, first forwarded by Francis H.C. Crick in 1958, consists of 3 steps or processes (Fig. 30–3). The *first step* is **replication**, *i.e.*, the copying of parental DNA to form daughter DNA molecules having identical nucleotide sequences. The *second step* is **transcription** (= copying), the process in which parts of coded genetic message in DNA are copied precisely in the form of RNA. The RNA so formed is called messenger RNA (mRNA) as it contains the information or 'message' as to the sequence of nucleotides present in the original DNA molecule. *Transcription does not involve a*

The term **'dogma'** (meaning literally 'a body of beliefs') is a misnomer here. It was introduced by **Francis H.C. Crick** (1958) at a time when little evidence supported these ideas. The 'dogma' is now a well-established principle.

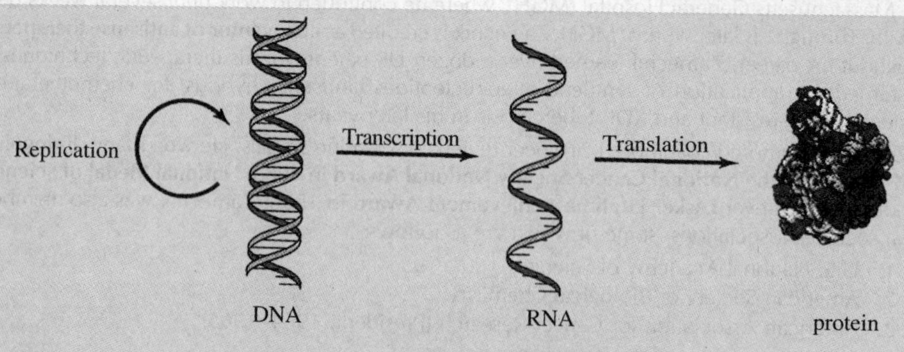

Fig. 30–3. Central dogma of molecular genetics

The principle depicts the general pathways of information flow *via* the processes of replication, transcription and translation.

change of code since DNA and the resulting mRNA are complementary. Transcription may also be defined as DNA-dependent RNA synthesis. The third step is translation, in which the genetic message coded in mRNA is translated, on the ribosomes, into a protein with a specific sequence of amino acids. In other words, reproduction of a primary polypeptide chain according to specification of mRNA is called translation. Apparently, *translation involves a change of code from nucleotide sequences to amino acid sequences.* Thus, according to this central dogma, the flow of information is one way or **unidirectional**, *i.e.*, from DNA the information is transferred to RNA (mRNA) and from RNA to protein. In other words,

'DNA makes RNA makes protein'.

However, in 1968, Barry Commoner suggested a **circular** flow of information, *i.e.*, DNA transcribes RNA, RNA translates into proteins, and proteins synthesize RNA and RNA synthesizes DNA (Fig. 30–4). But there is no evidence whatsoever for the synthesis of RNA from protein. If this is proved, one would have to believe in Lamarckism.

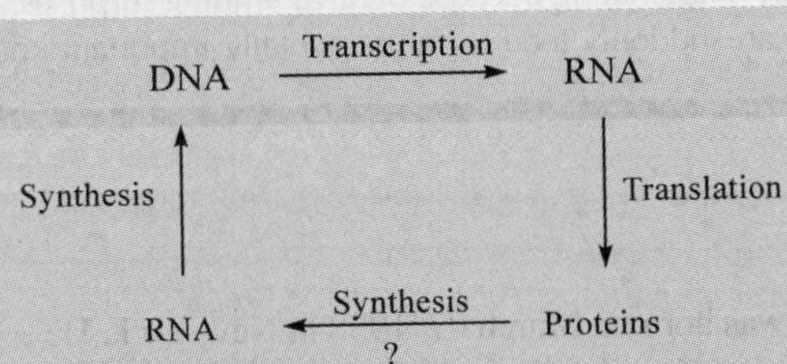

Fig. 30–4. Circular flow of information in DNA and protein syntheses

Note that this circular flow concept is untenable as there is no proof for the synthesis of RNA from protein.

Interestingly, the existence of enzymes **RNA-dependent DNA polymerases** (or **reverse transcriptases**) in RNA viruses was predicted by an American biochemist **Howard Temin** in 1962. And the enzymes were ultimately demonstrated to be present in such viruses by Temin and his compatriot **David Baltimore** both in 1970 but independently. Their discovery aroused much interest particularly because it constituted molecular proof that genetic information can sometimes flow 'backward' from RNA to DNA. The RNA viruses containing reverse transcriptases are also known as retroviruses (*retro*[L] = backward). This exciting finding gave rise to the concept of **central dogma reverse** which means that the sequence of information flow is not necessarily from DNA to RNA to protein but can also take place from RNA to DNA (Fig. 30–5).

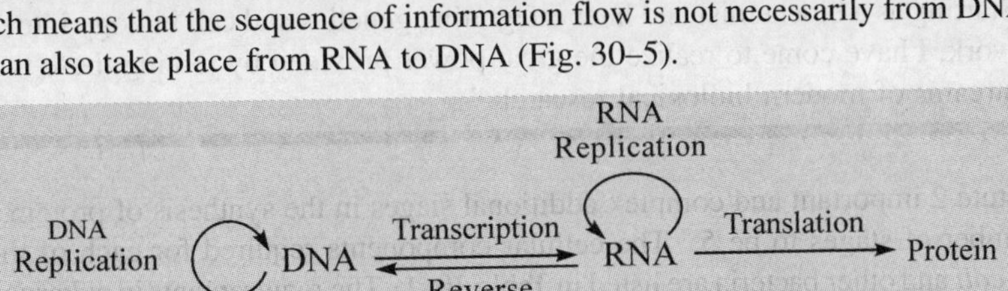

Fig. 30–5. The central dogma reverse of molecular genetics

Note that the diagram is the extension of the central dogma to include RNA-dependent synthesis of RNA and DNA.

PHASES OF PROTEIN SYNTHESIS

Like DNA and RNA, the synthesis of polymeric protein biomolecules can be distinguished into 3 stages: initiation, elongation and termination of polypeptide chain. Besides these 3 stages or phases, the activation of amino acids and the post-translational processing of the completed polypeptide

HOWARD MARTIN TEMIN
(LT, 1934–1994)

A US virologist who was born on December 10, 1934 at Philadelphia. He graduated from Swarthmore in Pennsylvania in 1955. A protege of Renato Dulbecco, one of the major influences in animal virology, Temin worked with him at the California Institute of Technology (CALTECH) in the late 1950s and discovered *reverse transcriptase* in 1970 (independently of Baltimore who also discovered the enzyme, the same year). He spent most of his career at the University of Wisconsin at Madison. Both Temin and **D. Baltimore** along with another fellow countryman **R. Dulbecco**, working in London, shared the prestigious **Nobel Prize in Medicine or Physiology for 1975** for their work on *cancer-forming viruses*. Temin died on February 09, 1994 at Madison. His thoughtful expression about conducting research and integrating related branches was:

> "Scientific research is fascinating because of never-ending surprises. Top quality work on any subject can suddenly lead to fundamentally important knowledge."

DAVID BALTIMORE
(born, 1938)

A US virologist, who was born on March 07, 1938 in New York. He graduated from Swarthmore in Pennsylvania, in 1960. A protege of **Renato Dulbecco**, one of the major influences in animal virology, Baltimore worked with him at the Salk Institute, San Diego, in the early 1960s and *discovered reverse transcriptase* in 1970 (independently of **Temin** who also discovered the enzyme, the same year). He spent most of his career at the Massachusetts Institute of Technology (MIT). Baltimore shared the **1975 Nobel Prize for Physiology or Medicine** with **R. Dulbecco** and **H.M. Temin** for their work linking viruses with the development of some cancerous tumours. His thoughtful expression about conducting research and integrating related branches is:

> "Starting as a biochemist and later integrating genetics and cell biology into my work, I have come to realize the great power provided by joining these three streams of modern biological research."

chain constitute 2 important and complex additional stages in the synthesis of protein, thus making the total number of stages to be 5. The cellular components required for each of the 5 stages in *Escherichia coli* and other bacteria are listed in Table 30–1. The requirements in eukaryotic organisms are quite similar.

An overview of these stages will help a better understanding of the discussion that follows.

Stage 1: Activation of amino acids

This stage takes place in the cytosol (and not on the ribosomes). During this stage, each of the 20 amino acids is covalently attached to a specific tRNA at the expense of ATP energy; the reactions being catalyzed by a group of Mg^{2+}-dependent activating enzymes called *aminoacyl-tRNA synthetases*, each specific for one amino acid and its corresponding tRNAs. Aminoacylated tRNAs are usually referred to as being 'charged'.

Stage 2 : Initiation

During this stage, the mRNA, bearing the code for the polypeptide to be synthesized, binds to the smaller ribosomal subunit. This is followed by the binding of the initiating aminoacyl-

Table 30–1. Cellular components required for the 5 major stages of protein synthesis in *Escherichia coli*

Stage	*Essential components*
1. Activation of amino acids	20 amino acids 20 aminoacyl-tRNA synthetases 32 or more tRNAs ATP Mg^{2+}
2. Initiation	mRNA N-formylmethionyl-$tRNA^{fMet}$ Initiation codon in mRNA (*i.e.,* AUG) 30S ribosomal subunit 50S ribosomal subunit Initiation factors (IF-1, IF-2, IF-3) GTP Mg^{2+}
3. Elongation	Functional 70S ribosome (initiation complex) Aminoacyl-tRNAs specified by codons Elongation factors (EF-Tu, EF-Ts, EF-G) GTP Mg^{2+}
4. Termination and release	Termination codon in mRNA Release factors (RF-1, RF-2, RF-3)
5. Folding and posttranslational processing	Specific enzymes, cofactors, and other components for removal of initiating residues and signal sequences, additional proteolytic processing, modification of terminal residues, and attachment of phosphate, methyl, carboxyl, carbohydrate, or prosthetic groups.

tRNA and the large ribosomal subunit to form an initiation complex. The initiating aminoacyl-tRNA base-pairs with the mRNA codon AUG that signifies the beginning of the polypeptide chain. This process, which requires GTP, is promoted by specific cytosolic proteins called *initiation factors*.

Stage 3 : Elongation

The polypeptide chain now increases in length by covalent attachment of successive amino acid units, each carried to the ribosome and correctly positioned by its tRNA, which base-pairs to its corresponding codon in the mRNA. Elongation is promoted by cytosolic proteins called *elongation factors*. The binding of each incoming aminoacyl-tRNA and the movement of the ribosome along the mRNA are facilitated by the hydrolysis of 2 moles of GTP for each residue added to the growing polypeptide chain.

Stage 4 : Termination and release

The completion of the polypeptide chain is signalled by a termination codon present in the mRNA. The polypeptide chain is then released from the ribosome with the help of proteins called *release factors*.

Stage 5 : Folding and processing

In order to attain its biologically active form, the polypeptide chain must fold into its proper 3-

dimensional conformation. Before or after folding, the new polypeptide may undergo enzymatic processing to carry out certain functions:

(*a*) to remove one or two amino acids from the amino terminus,

(*b*) to add acetyl, phosphate, methyl, carboxyl, or other groups to certain amino acid residues,

(*c*) to cleave the protein molecule proteolytically, or

(*d*) to attach oligosaccharides or prosthetic groups to the protein molecule.

Much of the post translational modification of proteins occurs in the Golgi apparatus.

THE TWO KEY COMPONENTS IN PROTEIN SYNTHESIS

The two key components of protein synthesis are: ribosomes and transfer RNA.

A. Ribosome

The ribosomes are ribonucleoprotein cellular particles and can be regarded as the organelles of protein synthesis, just as mitochondria are the organelles of oxidative phosphorylation. A ribosome is a highly specialized and complex structure. The best-characterized ribosomes are those of *E. coli.* Each *E. coli* cell contains 15,000 or more ribosomes, which make up about 25% of the dry weight of the cell. The *E. coli* ribosomes (MW = 2.7×10^6 or 2,500 kdal) have a diameter of about 21 nm (1 nm = 10 Å) and are made up of about 66% ribosomal RNA and 34% proteins. Since these particles are rich in

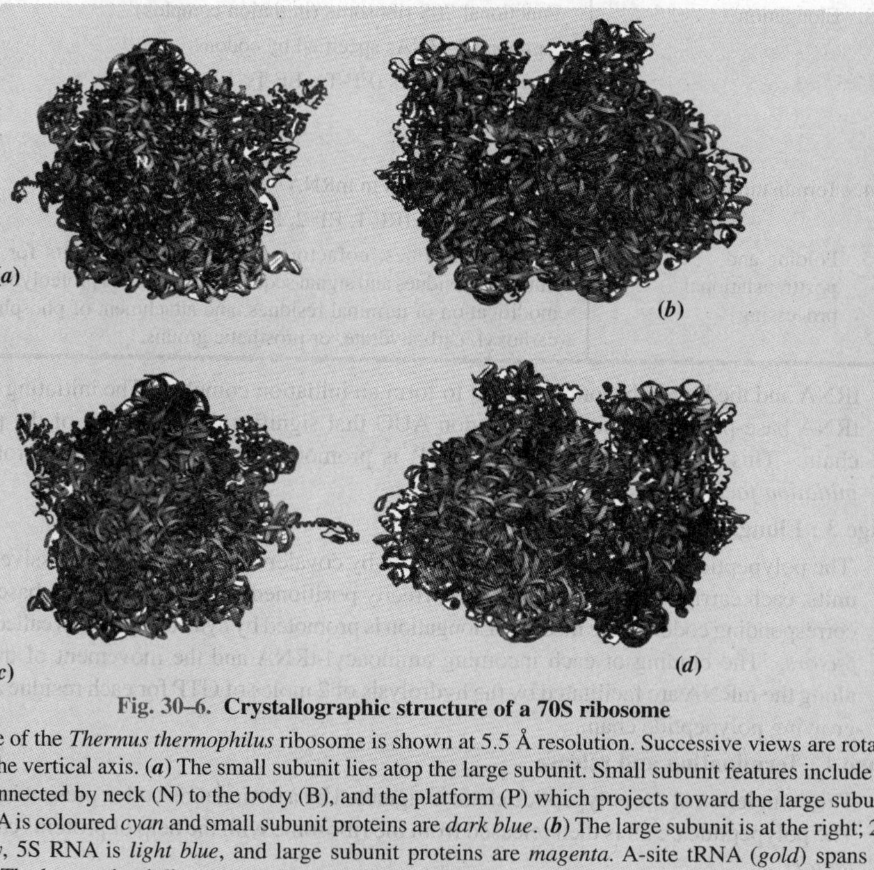

(*a*) (*b*) (*c*) (*d*)

Fig. 30–6. Crystallographic structure of a 70S ribosome

The structure of the *Thermus thermophilus* ribosome is shown at 5.5 Å resolution. Successive views are rotated 90° around the vertical axis. (***a***) The small subunit lies atop the large subunit. Small subunit features include the head (H), connected by neck (N) to the body (B), and the platform (P) which projects toward the large subunit. The 16S RNA is coloured *cyan* and small subunit proteins are *dark blue*. (***b***) The large subunit is at the right; 23S RNA is *grey*, 5S RNA is *light blue*, and large subunit proteins are *magenta*. A-site tRNA (*gold*) spans the subunits. (***c***) The large subunit lies on top with the stalk protruding to the left. (***d***) The large subunit is at the left, and elements of E- and A-side tRNAs are visible in the subunit interface.

(*Courtesy : Drs. A Baucom and H Noller, 2001*)

RNA content, these were called 'ribosomes'. These have a sedimentation coefficient of 70S when Mg^{2+} ion concentration in the solution is 0.01 M. If this concentration is lowered to 0.001 M, the 70S ribosomes dissociate into 50S and 30S particles and this phenomenon is reversible, if the Mg^{2+} ion concentration is again raised.

The detailed 3-dimensional structure of the bacterial ribosomes has been worked out using crystallography (Figs. 30–6 and 30-7) at different resolutions. The bacterial ribosomes consist of 2 subunits of unequal size (Fig. 30–8), the larger having a sedimentation coefficient of 50S and the smaller of 30S. The 50S subunit contains one mole of 5S rRNA (120 nucleotides), one mole of 23S rRNA (2,904 nucleotides) and 36 proteins (MW = 5,000 – 25,000). The 30S subunit contains one mole of 16S rRNA (152 nucleotides) and 21 proteins (MW = 8,000 – 26,000). The proteins are designated by numbers. Those in the large subunit are numbered L1 to L36 (L for large) and those in the smaller subunit S1 to S21 (S for small). Their molecular weight range between 6,000 to 75,000. The rRNAs appear to serve as a framework to which the ribosomal proteins are bound. Each of the 57 (36 + 21) proteins in the bacterial ribosome is believed to play a role in the synthesis of polypeptides, either as an enzyme or as a structural component in the overall process.

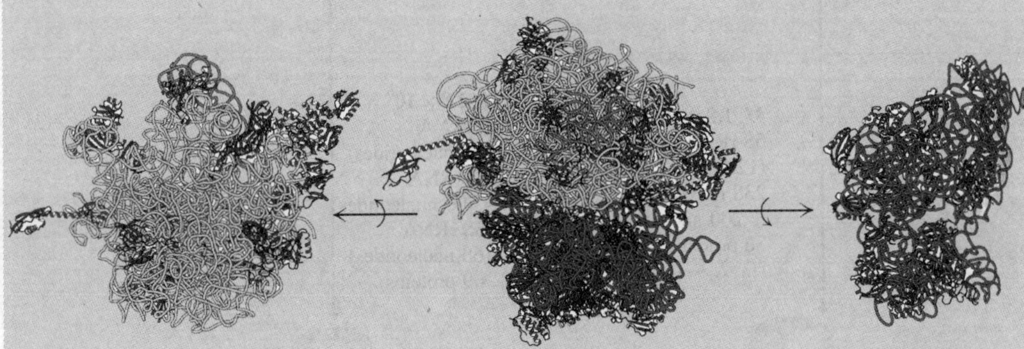

Fig. 30–7. The ribosome at high resolution

Detailed models of the ribosome based on the results of x-ray crystallographic studies of the 70S ribsome and the 30S and 50S subunits. 23S RNA is shown in *yellow*, 5S RNA in *orange*, 16S RNA in *green*, proteins of the 50S subunit in *red* and proteins of the 30S subunit in *blue*.

The two ribosomal subunits have irregular shapes. The 3-dimensional structure of the 30S and 50S subunits of *E. coli* ribosomes (Fig. 30–8) has been deduced from x-ray diffraction, electron microscopy, and other structural methods. The two oddly-shaped subunits fit together in such a way that a cleft is formed through which the mRNA passes as the ribosome moves along it during the translation process and the newly-formed polypeptide chain emerges from this cleft (Fig. 30–9).

The ribosomes in the cytoplasm of eukaryotic cells (other than mitochondrial and chloroplast ribosomes) are substantially larger and more complex than bacterial ribosomes (Fig. 30–8). They have a diameter of about 23 nm and a sedimentation coefficient of 80S. They also have 2 subunits, which vary in size between species but on an average are 60S and 40S. The small subunit (40S) contains a single 18S rRNA molecule and the large subunit (60S) contains a molecule each of 5S, 5.8S and 28S rRNAs. Altogether, eukaryotic ribosomes contain over 80 different proteins. Thus, an eukaryotic ribosome contains more proteins in each subunit and also has an additional RNA (5.8S) in the larger 60S subunit. The ribosomes of mitochondria and chloroplasts are different from those in the cytoplasm of eukaryotes. They are more like bacterial ribosomes. In fact, there are many similarities between protein synthesis in mitochondria, chloroplasts, and bacteria.

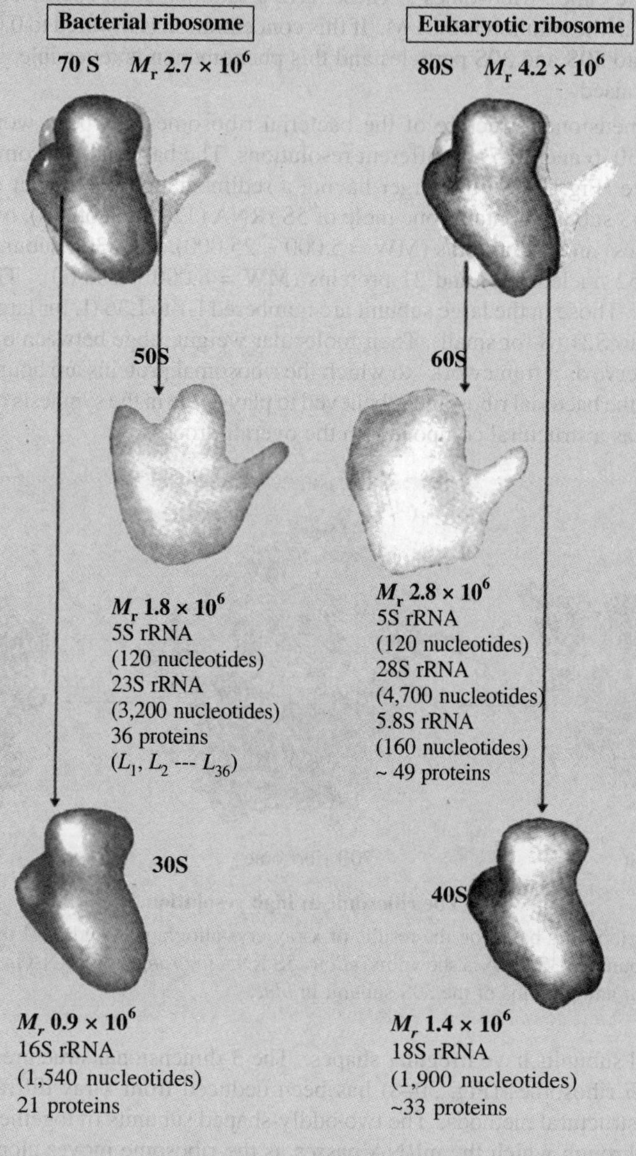

Fig. 30–8. Composition and mass of ribosomes in prokaryotes and eukaryotes

The ribosomal subunits are identified by their S (Svedberg unit) values, sedimentation coefficients that refer to their rate of sedimentation in a centrifuge. The S values (sedimentation coefficients) are not necessarily additive when the two subunits are combined because rates of sedimentation are affected by shape as well as mass.

Polypeptide synthesis takes place on the head and plateform regions of the 30S subunit and the upper half of the 50S subunit (translational domain). The mRNAs and tRNAs attach to the 30S subunit, and the peptidyl transferase site (where peptide bond formation occur) is associated with the central protuberance of the larger 50S subunit (Fig. 30–10).

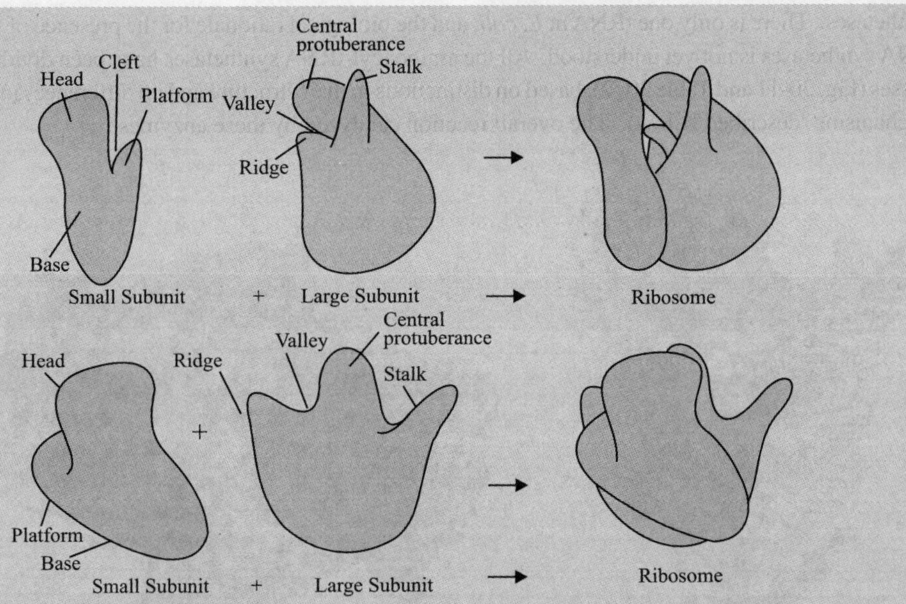

Fig. 30-9. Three-dimensional model of *E.coli* ribosome and its components shown in two different orientations, (A) and (B)

(*Redrawn from Scientific American, 245, 1981*)

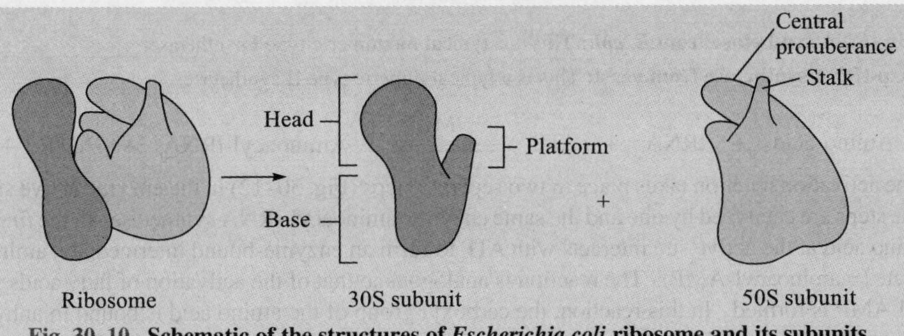

Fig. 30–10. Schematic of the structures of *Escherichia coli* ribosome and its subunits

The arrow indicates the cleft between the subunits.

(*Adapted from James A. Lake, 1985*)

B. Transfer RNA (tRNA)

The chemical structure and the role of tRNA has already been described on pages 384–389.

ACTIVATION OF AMINO ACIDS

A. Two Classes of Aminoacyl-tRNA Synthetases

As noted earlier, activation of amino acids takes place in the cytosol and in it the 20 different amino acids are esterified to their corresponding tRNAs by aminoacyl-tRNA synthetases. In most organisms, there is generally one aminoacyl-tRNA synthetase (also called aminoacyl-tRNA ligase or simply activation enzyme) for each amino acid. However, for amino acids that have 2 or more corresponding tRNAs, the same aminoacyl-tRNA synthetase usually aminoacylates all of them. However, in *E. coli*, the only exception to this rule is lysine, for which there are two aminoacyl-tRNA synthetases. There is only one tRNA in *E. coli*, and the biological rationale for the presence of two Lys-tRNA synthetases is not yet understood. All the aminoacyl-tRNA synthetases have been divided into 2 classes (Fig. 30-11 and Table 30–2), based on distinctions in their structure and on differences in reaction mechanisms (described below). The overall reaction catalyzed by these enzymes is:

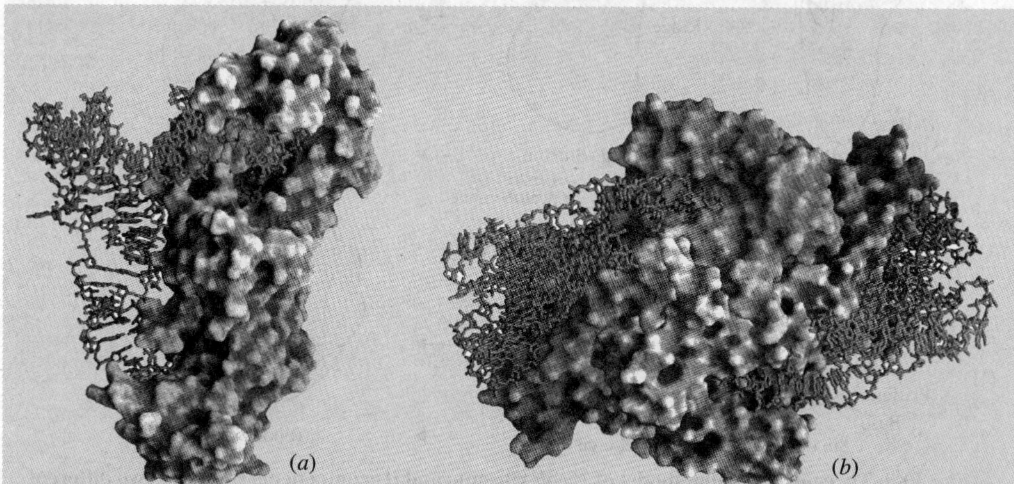

Fig. 30–11. Aminoacyl-tRNA synthetases
Both synthetases are complexed with their cognate tRNAs (*green stick* structures). Bound ATP (*red*) pinpoints the active site near the end of the aminoacyl arm.

(*a*) **Gln-tRNA synthetase from E. coli:** This is a typical monomeric type I synthetase.
(*b*) **Asp-tRNA synthetase from yeast:** This is a typical dimeric type II synthetase.

$$\text{Amino acid} + \text{tRNA} + \text{ATP} \xrightleftharpoons{Mg^{2+}} \text{Aminoacyl-tRNA} + \text{AMP} + PP_i$$

The activation reaction takes place in two separate steps (Fig. 30–12) in the enzyme active site but both the steps are catalyzed by one and the same enzyme, aminoacyl-tRNA synthetase. In the *first step*, the amino acid at the active site interacts with ATP to form an enzyme-bound intermediate, aminoacyl adenylate (= aminoacyl-AMP). The reaction is analogous to that of the activation of fatty acids where an acyl-AMP is formed. In this reaction, the carboxyl group of the amino acid is bound in anhydride linkage with the 5′-phosphate group of the AMP with displacement of pyrophosphate. In the *second step*, the aminoacyl group is transferred from enzyme-bound aminoacyl-AMP to its

Table 30–2. Two classes of aminoacyl-tRNA synthetases

Class I	*Class II*
Arginine	Alanine
Cysteine	Asparagine
Glutamic acid	Aspartic acid
Glutamine	Glycine
Isoleucine	Histidine
Leucine	Lysine
Methionine	Phenylalanine
Tryptophan	Proline
Tyrosine	Serine
Valine	Threonine

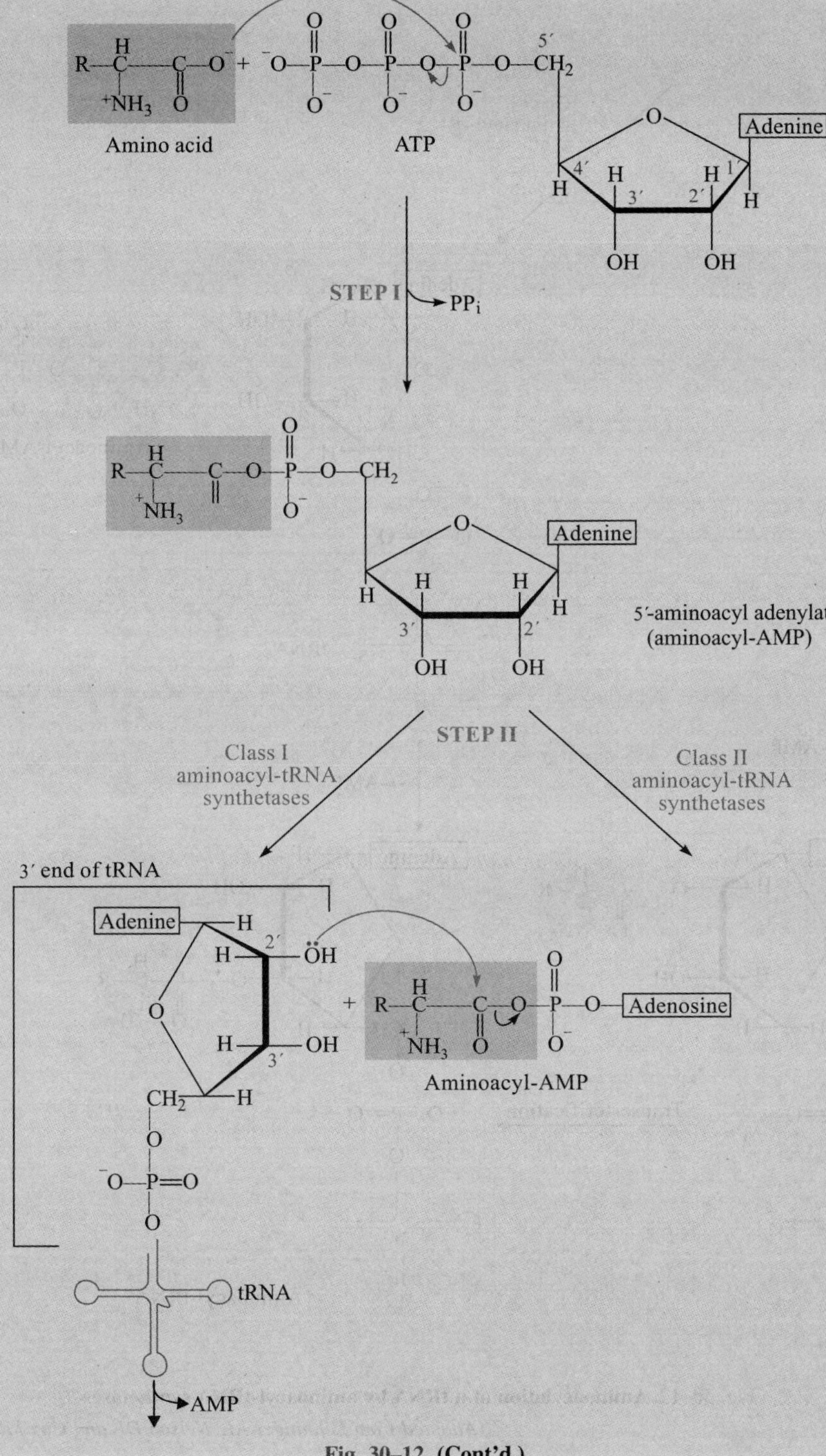

Fig. 30–12. (Cont'd.)

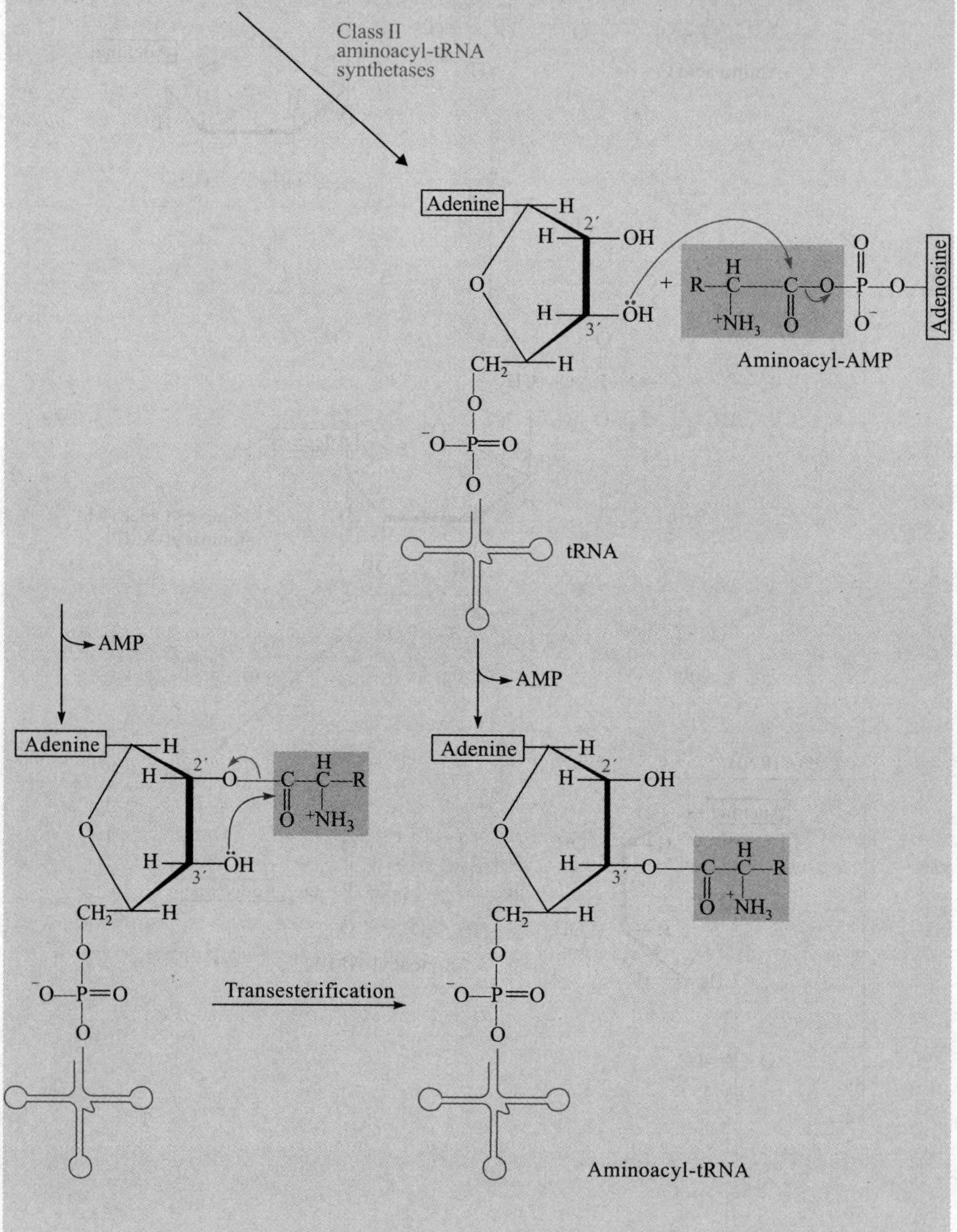

Fig. 30–12. Aminoacylation of a tRNA by aminoacyl-tRNA synthetases

(Adapted from Lehninger AL, Nelson DL and Cox MM, 2005)

corresponding specific tRNA. The mechanism of this step is somewhat different for the 2 classes of aminoacyl-tRNA synthetases (Fig. 30–12) For **class I enzymes**, the aminoacyl group is transferred initially to the 2′-hydroxyl group of the 3′-terminal adenylate residue, then moved to the 3′-hydroxyl group by a transesterification reaction. For **class II enzymes**, the aminoacyl group is transferred directly to the 3′-hydroxyl group of the terminal adenylate. The reason for this mechanistic difference between the two enzyme classes is not known. The resulting ester linkage between the amino acid and the tRNA (Fig. 30–13) has a high standard free energy of hydrolysis ($\Delta G^{0\prime} = -29$ kJ/mol). The pyrophosphate formed in the activation reaction undergoes hydrolysis to phosphate by inorganic phosphatase. Thus, *two* high-energy phosphate bonds are consumed in the activation of each amino acid molecule. One of the bonds is consumed in forming the ester linkage of aminoacyl-tRNA, whereas the other is consumed in driving the reaction forward. This renders the overall reaction for amino acid activation essentially irreversible:

$$\text{Amino acid + tRNA + ATP} \xrightarrow{Mg^{2+}} \text{Aminoacyl-tRNA + AMP} + 2\,P_i$$
$$\Delta G^{0\prime} \approx -29 \text{ kJ/mol}$$

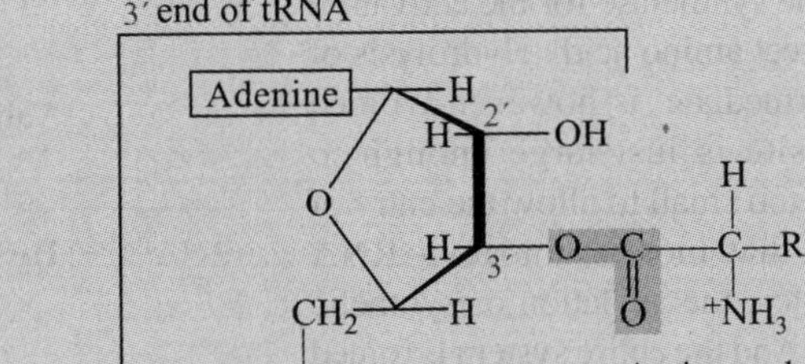

Fig. 30–13. General structure of aminoacyl-tRNAs

The aminoacyl group is esterified to the 3′-position of the terminal adenylate residue. The ester linkage, that both activates the amino acid and joins it to the tRNA, is shaded *pink.*

B. Proofreading by Some Aminoacyl-tRNA Synthetases

The aminoacylation of tRNA performs two functions: the activation of an amino acid for peptide bond formation and attachment of the amino acid to an adaptor tRNA which directs its placement within a growing polypeptide. In fact, the identity of the amino acid attached to a tRNA is not checked on the ribosome and attaching the correct amino acid to each tRNA is, henceforth, essential to the fidelity of protein synthesis as a whole.

The correct translation of genetic message depends on the high degree of specificity of aminoacyl-tRNA synthetases. These enzymes are highly sensitive in their recognition of the amino acid to be activated and of the prospective tRNA acceptor. A very high specificity is indeed necessary during the Stage I (*i.e.*, activation of amino acids), in order to avoid errors in the biosynthesis of proteins, because once the aminoacyl-tRNA is formed, there is no longer any control mechanism in the cell to verify the nature of the amino acid and it is therefore not possible to reject an amino acid which would have been incorrectly bound. Consequently, the amino acid would be erroneously incorporated in the protein molecule. Therefore, the aminoacyl-tRNA synthetases, *in vivo*, must either not commit any errors or be able to rectify them.

The tRNA molecules, that accept different amino acids, have differing base sequences, and hence they can be easily recognized by the synthetases. But the synthetases must, in particular, be able to distinguish between two amino acid of very similar structure; and some of them have this capacity. For example, isoleucine (Ile) differs from valine (Val) only in having an additional methylene (—CH_2—) group. The additional binding energy contributed by this extra —CH_2— group favours the activation of isoleucine (to form Ile-AMP) over valine by a factor of 200. The concentration of valine *in vivo* is about 5 times that of isoleucine, and so valine would be mistakenly incorporated in place of isoleucine 1 in every 40 times.

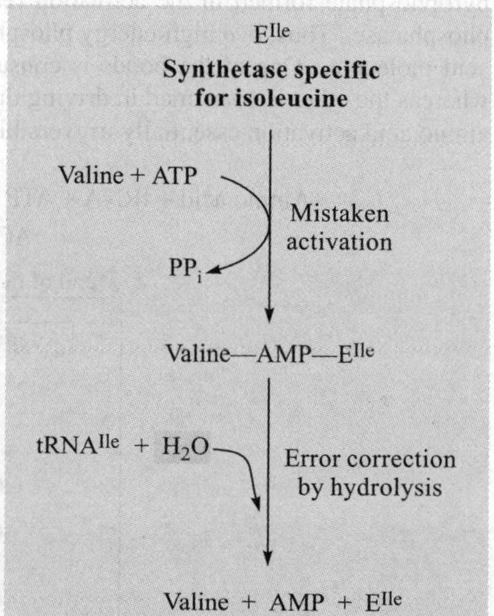

Fig. 30–14. Correction mechanism or proofreading on hydrolytic site, the "second filter"

The entry of tRNA specific for isoleucine induces hydrolysis of valyl-AMP.

However, the observed error frequency *in vivo* is only 1 in 3,000, indicating that there must be subsequent editing steps to enhance fidelity. In fact, the Ile-tRNA synthetase corrects its own errors, *i.e.*, in the presence of $tRNA^{Ile}$, the Val-AMP formed is hydrolyzed (but not Ile-AMP), thus preventing an erroneous aminoacylation (*i.e.*, a misacylation) of $tRNA^{Ile}$ (Fig. 30–14). Furthermore, this hydrolytic reaction frees the synthetase for the activation and transfer of Ile, the correct amino acid. Hydrolysis of Ile-AMP, the desired intermediate, is however avoided because the hydrolytic site is just large enough to accomodate Val-AMP, but too small to allow the entry of Ile-AMP. Thus, we see that most aminoacyl-tRNA synthetases contain two sites: the acylation or synthetic site and the hydrolytic site. And the entire system is forced through two successive "filters", rather than one, whereby increasing the potential fidelity by a power of 2. The *first filter* is the **synthetic site** on **synthetases** which brings about the initial amino acid binding and activation to aminoacyl-AMP. The *second filter* is the separate active site or **hydrolytic site** on synthetases which catalyzes deacylation of incorrect aminoacyl-AMPs. The synthetic site rejects amino acids that are larger than the correct one because there is insufficient room for them, whereas the hydrolytic site destroys activated intermediates that are smaller than the correct species. Hydrolytic proofreading is central to the fidelity of many aminoacyl-tRNA synthetases, as it is to DNA polymerases. In addition to proofreading after formation of the aminoacyl-AMP intermediate, most aminoacyl-tRNA synthetases are also capable of hydrolyzing the ester linkage between amino acids and tRNAs in aminoacyl-tRNAs. This hydrolysis is greatly accelerated for incorrectly-charged tRNAs, providing yet a *third filter* to further enhance the fidelity of the overall process. In contrast, in a few aminoacyl-tRNA synthetases that activate amino acids that have no close structural relatives, little or no proofreading occurs; in these cases, the active site can sufficiently discriminate between the proper amino acid and incorrect amino acids. *Proofreading is costly in energy and time and hence is selected in the course of evolution only when fidelity must be enhanced.*

The overall error rate of protein synthesis (~ 1 mistake per 10^4 amino acids incorporated) is not nearly as low as for DNA replication, perhaps because a mistake in a protein is erased by destroying the protein and is not passed onto future generations. This degree of fidelity is sufficient to ensure that most proteins contain no mistakes and that the large amount of energy required to synthesize a protein is rarely wasted.

C. The "Second Genetic Code"

An individual aminoacyl-tRNA synthetase must be specific not only for a single amino acid but for a certain tRNA as well. Discriminating among several dozen tRNAs is just as important for the overall fidelity of protein synthesis as is distinguishing among amino acids. The interaction between aminoacyl-tRNA synthetases and tRNAs has been referred to as the **"second genetic code"**, to reflect its crucial role in maintaining the accuracy of protein synthesis. The "coding" rules are apparently more complex than those in the "first" code.

D. Direction of the Growth of Polypeptide Chain

The direction of synthesis of polypeptide chain was determined by **Howard Dintzis** (1961) who followed, as a function of time, the course of incorporation of tritiated leucine (^{3}H-leucine) in the α and β chains of the hemoglobin, synthesized by reticulocytes (immature erythrocytes) in suspension (at low temperature to slow down protein synthesis). At regular intervals (between 4 and 60 minutes), he isolated hemoglobin, separated the α and β chains, subjected them to a hydrolysis by trypsin and fractionated the peptides obtained. *Leucine was chosen because it occurs frequently along both the α- and β-globin chains.* It was found that there was more radioactivity in the peptides near the carboxyl end than in those near the amino end. Thus, there was a gradient of increasing radioactivity from the amino to the carboxyl end of each chain (Fig. 30–15). The carboxyl end was the most heavily labelled because it was the last to be synthesized. Hence, the direction of chain growth is always from the amino to carboxyl end. In other words, *the proteins are synthesized in the amino-to-carboxyl direction.*

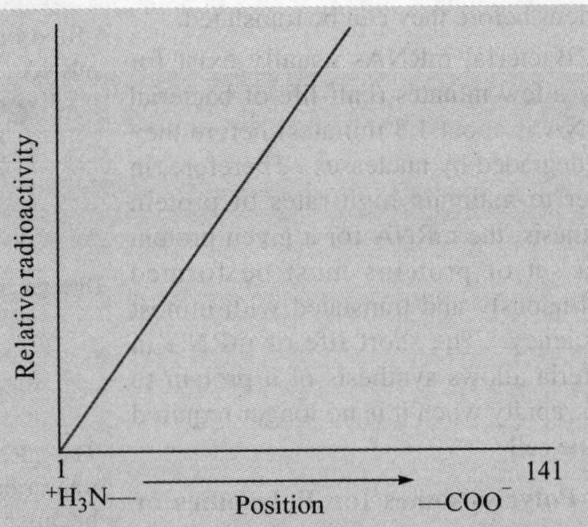

Fig. 30–15. Distribution of 3H-leucine in a chains of hemoglobin synthesized after exposure of reticulocytes to tritiated leucine

Note that each a chain of hemoglobin contains 141 amino acid residues. The higher radioactivity of the carboxyl ends, relative to the amino ends, indicates that the carboxyl end of each chain was synthesized last.

(*After Lubert Stryer, 1995*)

INITIATION OF PROTEIN SYNTHESIS

A. Translation of Messenger RNA in 5′ ⟶ 3′ Direction

The direction of reading of mRNA was determined by using **synthetic polynucleotide** as the

$$^{5'}A - A - A - (A - A - A)_n - A - A - C^{3'}$$

template in a cell-free protein-synthesizing system. The triplet AAA codes for lysine (Lys), whereas AAC codes for asparagine (Asn).

The **polypeptide product** was:

$$^{+}H_3N - Lys - (Lys)_n - Asn - C(=O)O^-$$

Because asparagine was the carboxyl terminal residue, the codon AAC was the last to be read. Hence, *the direction of translation of mRNA is 5′ ⟶ 3′.*

B. Coupling between Transcription and Translation in Bacteria

In bacteria, transcription of mRNA is closely coupled to its translation. Messenger RNAs are synthesized (or transcribed) in direction and are translated also in the same direction. As such, the ribosomes begin translating the end of the mRNA before its synthesis (or transcription) is complete

(Fig. 30–16). This is possible because the mRNA in bacteria does not have to be transported from a nucleus to the cytoplasm before encountering ribosomes. The situation is somewhat different in eukaryotes, where newly-transcribed mRNAs must be transferred out of the nucleus before they can be translated.

Bacterial mRNAs usually exist for only a few minutes (half-life of bacterial mRNA is about 1.5 minutes) before they are degraded by nucleases. Therefore, in order to maintain high rates of protein synthesis, the mRNA for a given protein or a set of proteins must be formed continuously and translated with utmost efficiency. The short life of mRNA in bacteria allows synthesis of a protein to stop rapidly when it is no longer required by the cell.

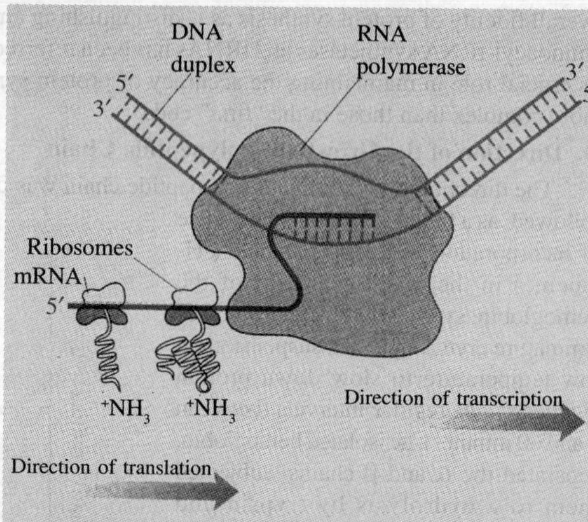

Fig. 30–16. The coupling of transcription and translation in bacteria

The mRNA is translated by ribosomes while it is still being transcribed from DNA by RNA polymerase. This is possible because the mRNA in bacteria does not have to be transported from a nucleus to the cytoplasm before encountering ribosomes. In this schematic diagram, the ribosomes are depicted as smaller than the RNA polymerase. In reality, the ribosomes ($M_r = 2.7 \times 10^6$) are an order of magnitude larger than the RNA polymerase ($M_r = 3.9 \times 10^5$).

C. Polyribosomes (or Polysomes or Ergosomes)

Many ribosomes can simultaneously translate an mRNA molecule. This parallel synthesis markedly increases the efficiency of utilization of the mRNA. The group of ribosomes bound to an mRNA molecule is called a **polyribosome** or simply **polysome** or rarely an **ergosome**. The ribosomes in this unit operate independently, each synthesizing a complete polypeptide chain. In optimal conditions for protein synthesis, there is an average one ribosome for every 80 nucleotides. In general, there are less than 10 ribosomes on the eukaryotic mRNA (which are usually *monocistronic*) while there can be several tens on prokaryotic mRNAs (which are often *polycistronic*). The polyribosomes synthesizing hemoglobin (which contains about 145 amino acids per chain, or about 500 nucleotides per mRNA) typically consist of 5 ribosomes. Ribosomes closest to the 5′ end of the messenger RNA have the smallest polypeptide chains, whereas those nearest the 3′ end of the mRNA have the longest chains that have almost finished synthesis (Fig. 30–17). Ribosomes dissociate into 30S and 50S subunits after the polypeptide chain is released. Polyribosomes, thus, allow rapid translocation of a single message.

D. N-formylmethionine as Initiator of Protein Synthesis in Bacteria

The simplest possibility as to how the protein synthesis starts may be that the first three nucleotides of each mRNA serve as the first codon; no special start signal would then be needed. However, the experimental evidences indicate that translation does not start immediately at the 5′ end of mRNA. In fact, the first translated codon is usually more than 25 nucleotides away from the 5′ end. Furthermore, many mRNA molecules in prokaryotes are *polycistronic*, *i.e.*, they code for 2 or more polypeptide chains. These facts and findings have led to the conclusion that *all known mRNA molecules contain signals that signify the beginning and end of each encoded polypeptide chain.*

An important finding was that nearly half of the amino-terminal residues of proteins in *E. coli* are methionine, yet this amino acid is uncommon at other positions of the polypeptide chain.

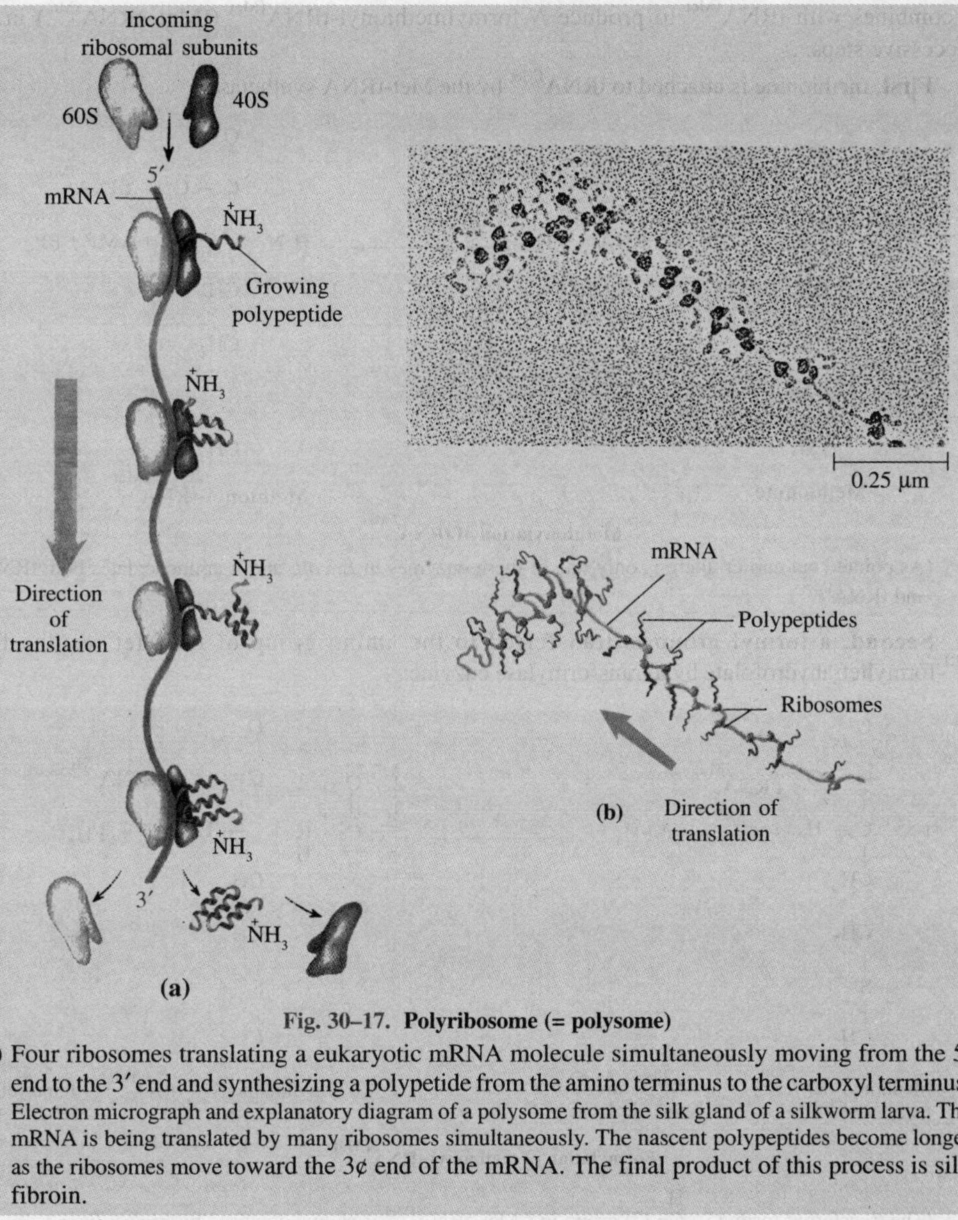

Fig. 30–17. Polyribosome (= polysome)

(a) Four ribosomes translating a eukaryotic mRNA molecule simultaneously moving from the 5′ end to the 3′ end and synthesizing a polypetide from the amino terminus to the carboxyl terminus.

(b) Electron micrograph and explanatory diagram of a polysome from the silk gland of a silkworm larva. The mRNA is being translated by many ribosomes simultaneously. The nascent polypeptides become longer as the ribosomes move toward the 3¢ end of the mRNA. The final product of this process is silk fibroin.

Furthermore, the amino end of nascent proteins is usually modified. These facts provided the clue that a derivative of methionine participates in initiation of polypeptide synthesis. In fact, *protein synthesis in bacteria starts with N-formyl methionine* (*fMet*).

Although there is only one codon (AUG) for methionine, there are 2 tRNAs for methionine in all organisms and both have a 3′ U–A–C 5′ anticodon that recognizes the unique 5′ A–U–G 3′ codon for methionine. One tRNA, designated as 'internal' tRNA or $tRNA^{Met}$, acts as an adaptor for methionine residues required within the polypeptide chain; the second tRNA, designated as initiator tRNA or $tRNA^{fMet}$ (the superscript f indicates that methionine attached to the initiator tRNA can be formylated) has a unique role in the initiation of the polypeptide chain synthesis. As already stated, the starting amino acid residue at amino-terminal is *N*-formylmethionine (fMet).

It combines with $tRNA^{fMet}$ to produce *N*-formylmethionyl-$tRNA^{fMet}$ (fMet–$tRNA^{fMet}$) in two successive steps:

First, methionine is attached to $tRNA^{fMet}$ by the Met-tRNA synthetase:

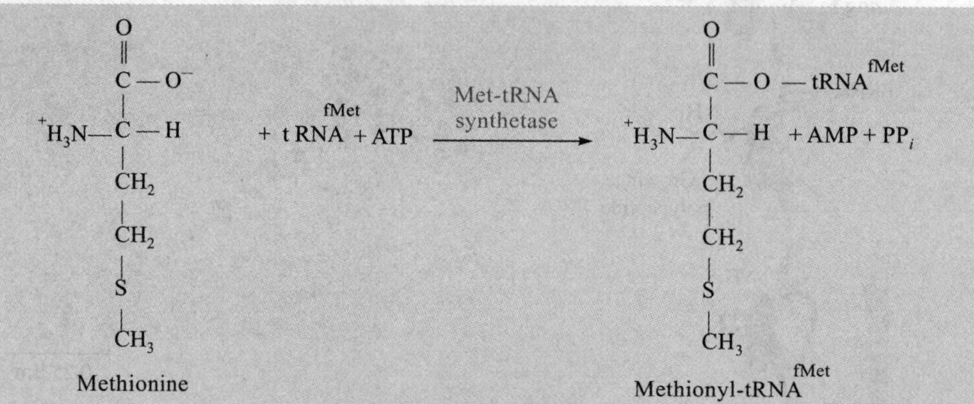

Methionylation of $tRNA^{fMet}$

(As pointed out earlier, there is only one of these enzymes in *E. coli*, and it aminoacylates both $tRNA^{Met}$ and $tRNA^{fMet}$.)

Second, a formyl group is transferred to the amino group of the Met residue from N^{10}-formyltetrahydrofolate by a transformylase enzyme:

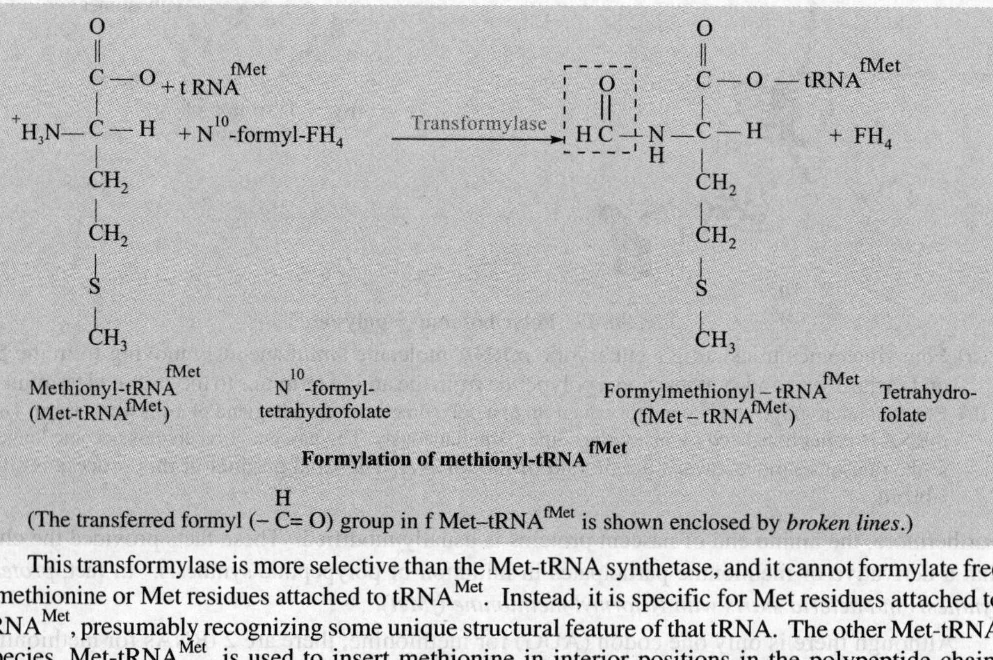

Formylation of methionyl-$tRNA^{fMet}$

(The transferred formyl (– HC= O) group in f Met–$tRNA^{fMet}$ is shown enclosed by *broken lines*.)

This transformylase is more selective than the Met-tRNA synthetase, and it cannot formylate free kmethionine or Met residues attached to $tRNA^{Met}$. Instead, it is specific for Met residues attached to $tRNA^{Met}$, presumably recognizing some unique structural feature of that tRNA. The other Met-tRNA species, Met-$tRNA^{Met}$, is used to insert methionine in interior positions in the polypeptide chain. Blocking of the amino group of methionine by the *N*-formyl group not only prevents it from entering interior positions but also allows fMet-$tRNA^{fMet}$ to be bound at a specific initiation site on the ribosome that does not accept Met-$tRNA^{Met}$ or any other aminoacyl-tRNA. In prokaryotes, the *N*-formylation of the Met residue on the tRNA seems to deceive the P site, leaving the A site free (refer the following Section for the two sites).

In eukaryotic cells, however, all peptides synthesized by cytosolic ribosomes begin with a Met residue (in contrast to fMet in bacteria), but again a specialized initiating tRNA is used that is distinct from the $tRNA^{Met}$ used at interior positions. Interestingly, one of the two types of methionyl-$tRNA^{Mets}$ of eukaryotes can be formylated by the *E. coli* formylase. It appears that, during the course of evolution, eukaryotes lost the ability to produce the enzyme, but one of the two methionyl-tRNAs retained the vestigial ability to be formylated. In contrast, polypeptides synthesized by the ribosomes in the mitochondria and chloroplasts of eukaryotic cells begin with *N*-formylmethionine. This and other similarities in the protein-synthesizing machinery of these organelles and bacteria strongly support the view that mitochondria and chloroplasts originated from bacterial ancestors symbiotically incorporated into the precursors of eukaryotic cells at an early stage of evolution.

E. The Three Steps of Initiation Process

The following description of the second stage of protein synthesis, *i.e.*, initiation (and also of elongation and termination stages to follow) pertains to the protein synthesis in bacteria; the process is not well-understood in eukaryotes. The initiation of polypeptide synthesis in bacteria requires 7 components (Table 30–1): the 30S ribosomal subunit (which contains 16S rRNA), the mRNA coding for the polypeptide to be made, the initiating fMet-$tRNA^{fMet}$, a set of 3 proteins called initiation factors (IF-1, IF-2, IF-3), GTP, the 50S ribosomal subunit, and Mg^{2+}. The formation of initiation complex takes place in 3 steps (Fig. 30–18):

(*i*) In the **first step**, the 30S ribosomal subunit binds initiation factor 3 (IF-3), which prevents the 30S and 50S subunits from combining prematurely to form a dead-end 70S complex, devoid of mRNA. The mRNA binds to the 30S subunit in such a manner that the initiation codon (AUG) binds to a precise location of the 30S subunit.

Each ribosome has 2 binding sites for tRNA: the **aminoacyl** or **A site** for the incoming aminoacyl-tRNA and the **peptidyl** or **P site** for the growing polypeptidyl-tRNA. The A site is also called acceptor site, while the P site is also called **donor** or **D site**. Both the 30S and 50S subunits contribute to the characteristics of each site. The initiating AUG codon resides in the P site, which is the only site to which fMet-$tRNA^{fMet}$ can bind. However, fMet-$tRNA^{fMet}$ is the exception: during the subsequent elongation stage, all other incoming aminoacyl-tRNAs, including the Met-$tRNA^{Met}$ that binds to interior AUGs, bind to the A site. The P site is the site from which the 'uncharged' tRNAs leave during elongation of the polypeptide chain.

(*ii*) In the **second step** of initiation process, the complex consisting of 30S subunit, IF-3, and mRNA now forms a still larger complex by binding IF-2 which already is bound to GTP and the initiating fMet-$tRNA^{fMet}$. The anticodon of this tRNA (UAC) pairs correctly with the initiation codon in this step. The protein factor IF-1, however, stimulates the activities of both IF-2, and IF-3.

(*iii*) In the **third step**, this large complex combines with the 50S ribosomal subunit. Simultaneously, the GTP, which is bound to IF-2, is hydrolyzed to GDP and P*i*. Both GDP and P*i* are released and IF-2 and IF-3 also dissociate from the ribosome.

These 3 steps in the initiation of bacterial protein synthesis result in a functional 70S ribosome called the **initiation complex**, containing the mRNA and the initiating fMet-$tRNA^{fMet}$. The correct binding of the fMet-$tRNA^{fMet}$ to the P site in the complete 70S initiation complex is assured by two points of recognition and attachment: the codon–anticodon interaction involving the initiating AUG fixed in the P site and binding interactions between the P site and the fMet-$tRNA^{fMet}$. The initiation complex is now ready for the elongation process.

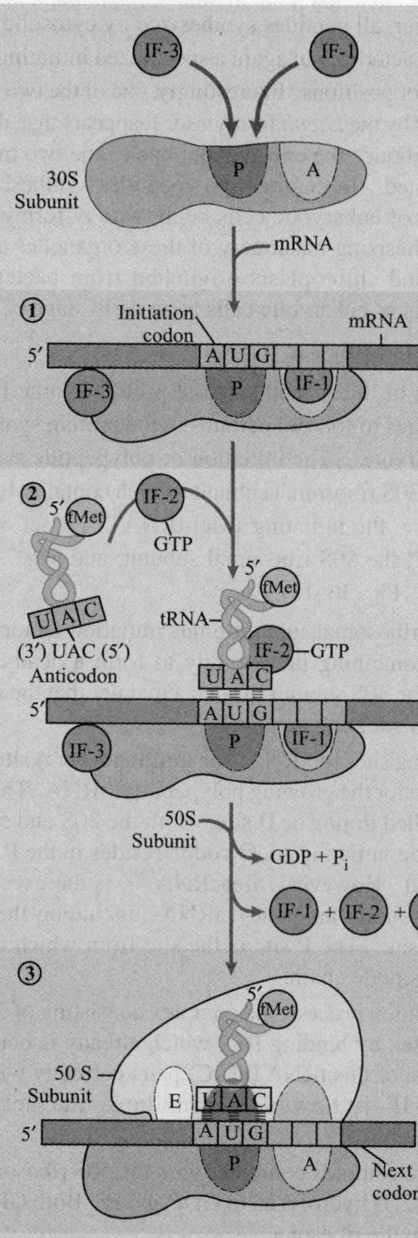

Fig. 30–18. Formation of the initiation complex in bacteria

The complex forms in three steps at the expense of the hydrolysis of GTP to GDP and P_i. IF-1, IF-2, and IF-3 are initiation factors. P designates the peptidyl site, A the aminoacyl site, and E the exit site. Here the anticodon of the tRNA is oriented 3′ to 5′, left to right.

One of the important differences between the protein synthesis in prokaryotes and eukaryotes is the presence of 9 eukaryotic initiation factors, most of them abbreviated as eIF (*e* for eukaryotic). One of these factors, called cap binding protein or CBPI, binds to the 5′ cap of mRNA. CPBI facilitates formation of a complex between the mRNA and the 40S ribosomal subunit. The mRNA is then scanned to locate the first AUG codon, which signals the start of the reading frame. Many additional initiation factors are required in this *mRNA scanning reaction* and also in assembly of the complete 80S initiation complex in which the initiating Met-tRNA^{Met} and mRNA are bound and are ready for elongation to proceed. The roles of various initiation factors in protein synthesis in both prokaryotes and eukaryotes are summarized in Table 30–3.

The terms **prokaryotes** and **eukaryotes** were first proposed by **Edward Chatton** in 1937.

Table 30–3. Protein factors required for initiation of translation in bacterial and eukaryotic cells

Factor	*Function*
Bacterial	
IF-1	Prevents premature binding of tRNAs to A site
IF-2	Facilitates binding of fMet-tRNA^{fMet} to 30S ribosomal subunit
IF-3	Binds to 30S subunit; prevents premature assosication of 50S subunit; enhances specificity of P site for fMet-tRNA^{fMet}
Eukaryotic	
eIF2	Facilitates binding of initiating Met-tRNA^{Met} to 40S ribosomal subunit
eIF2B, eIF3	First factors to bind 40S subunit; facilitate subsequent steps
eIF4A	RNA helicase activity removes secondary structure in the mRNA to permit binding to 40S subunit; part of the eIF4F complex
eIF4B	Binds to mRNA; facilitates scanning of mRNA to locate the first AUG
eIF4E	Binds to the 5′ cap of mRNA; part of the eIF4F complex
eIF4G	Binds to eIF4E and to poly(A) binding protein (PAB); part of the eIF4F complex
eIF5	Promotes dissociation of several other initiation factors from 40S subunit as a prelude to association of 60S subunit to form 80S initiation complex
eIF6	Facilitates dissociation of inactive 80S ribosome into 40S and 60S subunits

*The prefix "e" identifies these as eukaryotic factors.

F. Shine–Dalgarno Sequence

For the production of a 30S initiation complex in the first step of translation initiation, the 3′ terminal region of the 16S RNA component of the 30S ribosomal subunit is the binding site for the mRNA. This 3′ terminal region of 16S rRNA component is pyrimidine-rich and hydrogen-bonds, in an antiparallel fashion, with a complementary sequence of purine-rich region on the 5′ side (upstream) of the initiating AUG codon on mRNA (Fig. 30–19). This purine-rich region, consisting of 4 to 9 residues upstream from the AUG codon on mRNA, is called **Shine–Dalgarno sequence** and is needed for the interaction of the mRNA with its *ribosomal binding site*. This particular pattern of mRNA-16S rRNA binding aligns the initiating AUG codon of the message (or the protein to be formed) for the proper binding with the anticodon (UAC) in the 30S–fMet-tRNa^{fMet}complex. Thus, the specific AUG codon where fMet-tRNA^{fMet} is to be bound is thereby distinguished from interior methionine codons (AUGs) by its proximity to the Shine–Dalgarno sequence in the mRNA. This very well explains as to how the single codon, 5′-AUG serves to identify both the starting *N*-formylmethionine (or methionine in the case of eukaryotes) and those methionine residues that occur in interior

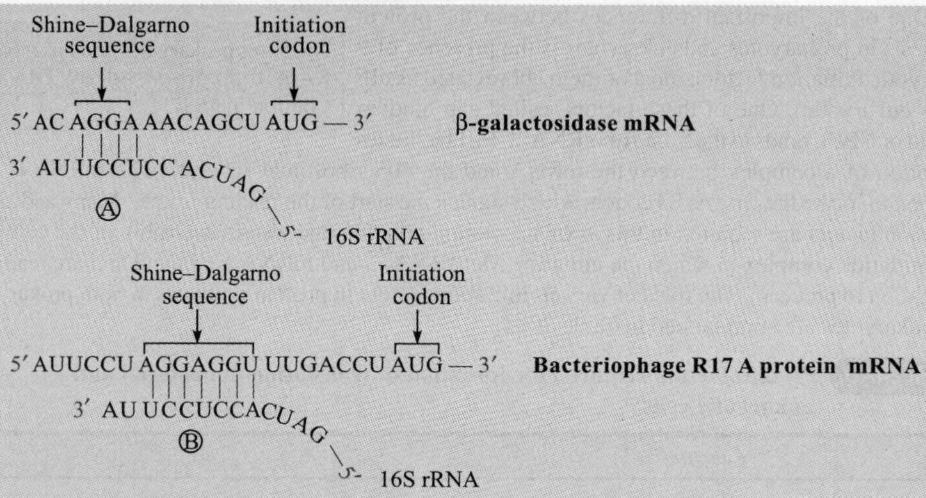

Fig. 30–19. Binding sequence of two mRNAs (Shine–Dalgarno sequence) which base-pairs with a sequence near the 3′ end of the 16S rRNA

The portions of the two mRNA transcripts shown here code for β-galactosidase A and the A protein of R17 phage B. The AUG codon defines the start of the polypeptide chain. Shine–Dalgarno sequence pairs with 16S rRNA, whereas the initiator codon pairs with fMet-tRNA^{fMet}, *i.e.*, initiator tRNA.

positions in the polypeptide chains. In a polycistronic mRNA, containing 2 or more messages, there is a ribosomal binding site to the 5′ side of each message; hence, such mRNAs can accommodate as many initiation complexes as there are messages to be translated. It may, thus, be inferred that two kinds of interactions determine where protein synthesis starts: the pairing of mRNA bases with the 3′ end of 16S rRNA, and the pairing of the initiator codon on mRNA with the anticodon of fMet initiator tRNA.

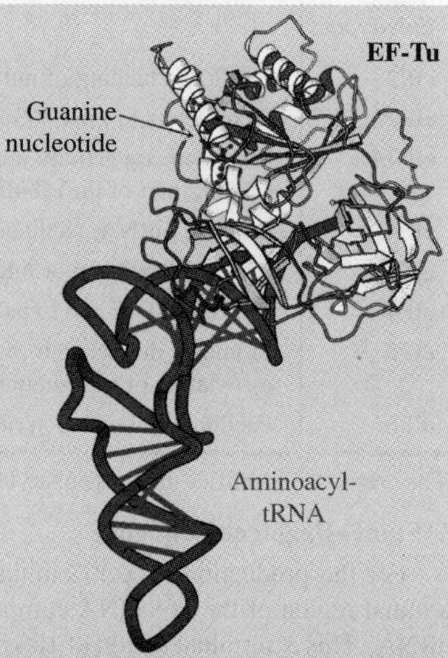

Fig. 30-20. The structure of a complex between elongation factor Tu (EF-Tu) and an aminoacyl-tRNA

The amino terminal domain of EF-Tu is a P-loop NTPase domain, similar to those in other G proteins.

ELONGATION OF THE POLYPEPTIDE CHAIN

The third stage of protein synthesis is elongation, *i.e.*, the stepwise addition of amino acids to the polypeptide chain. The process of elongation requires 4 components: the initiation complex (described above), the next aminoacyl-tRNA specified by the next codon in the mRNA, a set of 3 soluble cytosolic proteins called *elongation factors* (EF-Tu, EF-Ts, EF-G), and GTP. Three steps take place in the addition of each amino acid residue, and this cycle is repeated as many times as there are amino acid residues to be added. The three steps involved (Figs. 30–22, 30–23, 30–24) are:

Step 1: Codon recognition (= Binding of an aminoacyl-tRNA at site A of ribosome)

In this first step (Fig. 30–22), the next aminoacyl-tRNA (aa-tRNA) is first bound to a complex of EF-Tu (Tu for temperature-unstable) (Fig. 30–20) containing a molecule of bound GTP. The resulting aminoacyl-tRNA – EF-Tu · GTP complex is then bound to the A site of the 70S initiation complex. The GTP is hydrolyzed to GDP and P_i, and an EF-Tu · GDP complex leaves the 70S ribosome. The bound GDP is dissociated when the EF-Tu · GDP complex binds to the second elongation factor, EF-Ts (Ts for temperature-stable), and EF-Ts is subsequently released when another molecule of GTP becomes bound to EF-Tu. EF-Tu containing bound GTP is now ready to pick up another aminoacyl-tRNA and deliver it to the A site of the ribosome (Fig. 30–21).

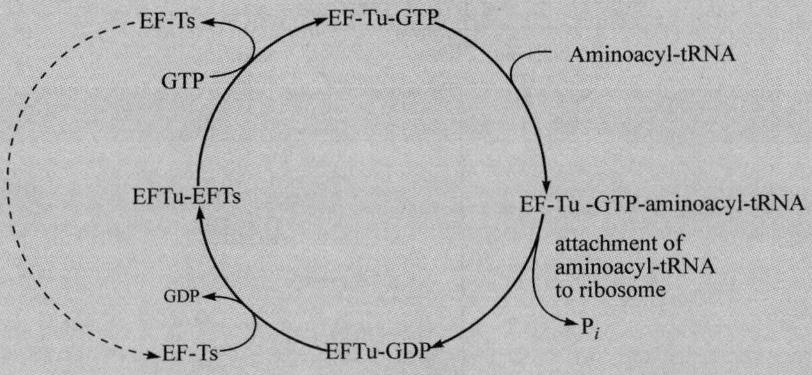

Fig. 30.21. Reaction cycle of elongation factor Tu (EF-Tu)
The diagram shows the regeneration of EF-Tu. GTP complex.

It is worthnoting that *EF-Tu does not interact with fMet–tRNA$_f$*. Hence, this initiator is not delivered to the A site. On the contrary, Met-tRNA$_m$, like all other aminoacyl-tRNAs, does bind to EF-Tu. These findings point out that *the internal AUG codons are not read by the initiator tRNA*. Conversely, initiation factor 2 recognizes fMet-tRNA$_f$ but no other tRNA. The GTP-GDP cycle of EF-Tu is reminiscent of that of transducin in vision, the stimulatory G protein in hormone action and the *ras* protein in growth control. Indeed, the amino-terminal domain (the G domain) of EF-Tu is structurally similar to the GTP-binding subunit of these signal-transducing proteins. The other two domains of the tripartite EF-Tu are distinctive– they mediate interactions with aminoacyl-tRNA and the ribosome.

Step 2 : Peptide bond formation

In the second step, a new peptide bond is created between the amino acids bound by their tRNAs to the A and P sites on the ribosome (Fig. 30–24). During the process, the initiating *N*-formylmethionyl group from its tRNA is transferred to the amino group of the second amino acid now in the A site. The α-amino group of the amino acid in the A site displaces the tRNA in the P site to form the peptide bond. This reaction produces a dipeptidyl-tRNA in the A site and the now 'uncharged' (or deacylated) tRNA^{fMet} remains bound to the P site. The peptide bond formation was till now being referred to have been catalyzed by **peptidyl transferase** (also called **peptide synthase**). But, in 1992, **Harry F. Noller** *et al* discovered that this activity was catalyzed not by a protein but *by a highly conserved domain of 23S rRNA that forms the peptidyl transferase active site*. This has added another critical biological function for ribozymes.

Peptidyl Transferase: A Ribozyme, not A Typical Enzyme

Peptidyl transferase is the activity responsible for peptide bond formation during protein synthesis. This enzyme activity catalyzes the reaction between the amino group of the aminoacyl-tRNA in the A site and the carboxyl carbon of the peptidyl-tRNA in the P site, forming a peptide bond from an ester bond. The activity is located in the ribosome, but none of the ribosomal proteins has the capacity to catalyze this reaction. Crystal structures of the bacterial 50S ribosome have shown that the peptidyl-transferase centre is composed entirely of rRNA, and although the requirement for proteins or specific amino acids for positioning the tRNAs or in stabilizing rRNA structure cannot be excluded, the catalytic activity resides in the ribosomal RNA.

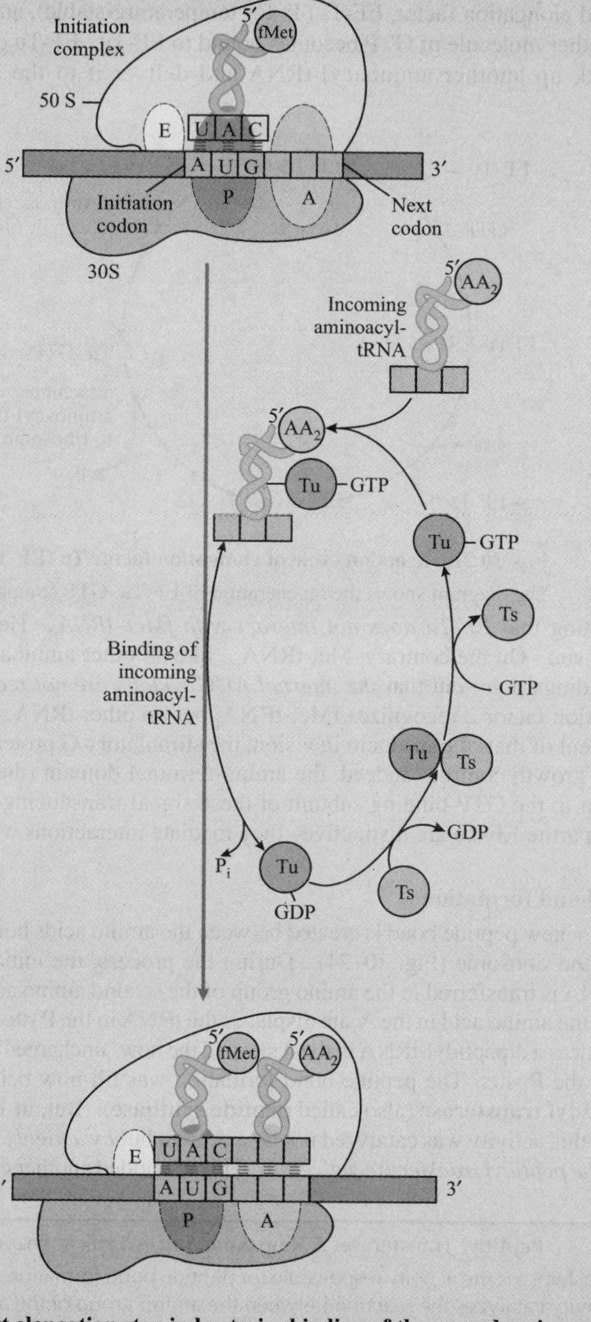

Fig. 30–22. First elongation step in bacteria: binding of the second aminoacyl-tRNA

The second aminoacyl-tRNA enters the A site of the ribosome bound to EF-Tu (shown here as Tu), which also contains GTP. Binding of the second aminocyl-tRNA to the A site is accompanied by hydrolysis of the GTP to GDP and P_i and release of the EF-Tu-GDP complex from the ribosome. The bound GDP is released when the EF-Tu-GDP complex binds to EF-Ts, and EF-Ts is subsequently released when another molecule of GTP binds to EF-Tu. This recycles EF-Tu and makes it available to repeat the cycle.

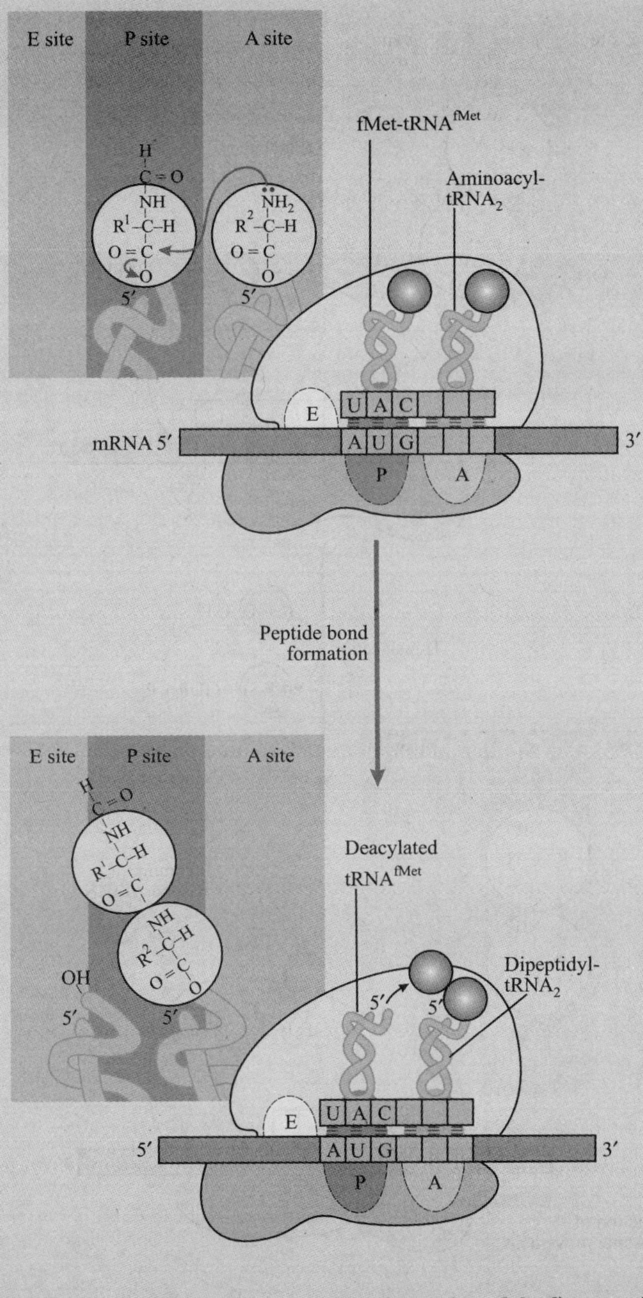

Fig. 30–23. Second elongation step in bacteria: formation of the first peptide bond

The peptidyl transferase catalyzing this reaction is the 23S rRNA ribozyme. The *N*-formylmethionyl group is transferred to the amino group of the second aminoacyl-tRNA in the A site, forming a dipeptidyl-tRNA. At this stage, both tRNAs bound to the ribosome shift position in the 50S subunit to take up a hybrid binding state. The uncharged tRNA shifts so that its 3′ and 5′ ends are in the E site. Similarly, the 3′ and 5′ ends of the peptidyl tRNA shift to the P site. The anticodons remain in the A and P sites.

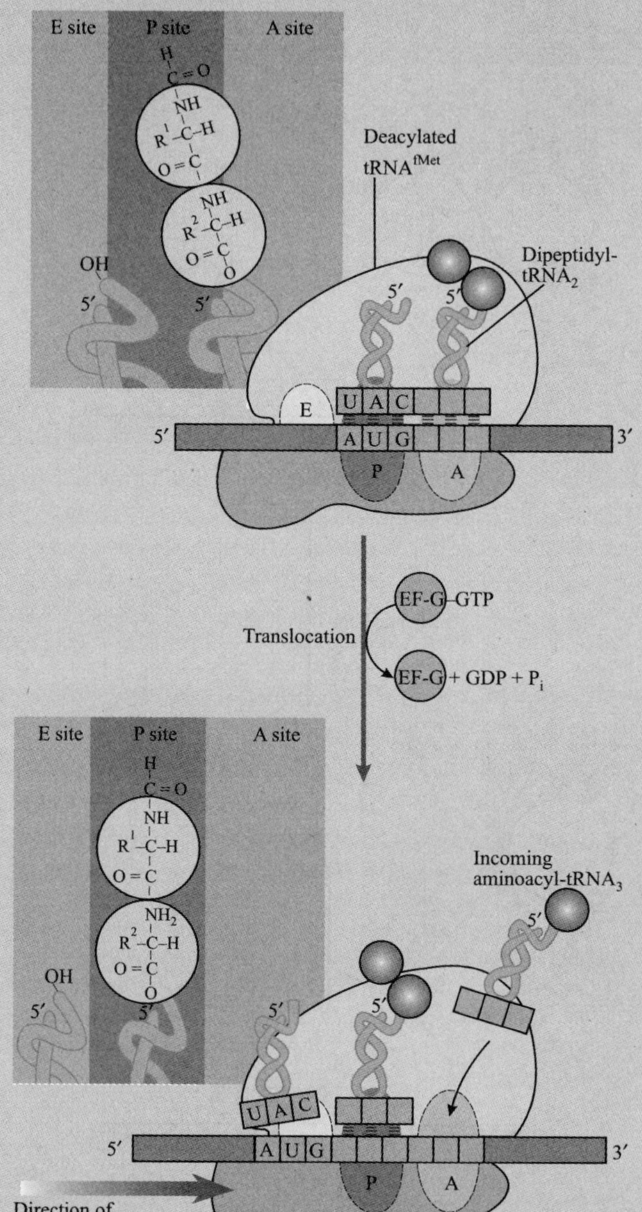

Fig. 30–24. Third elongation step in bacteria: translocation

The ribosome moves one codon toward the 3′ end of the mRNA, using energy provided by hydrolysis of GTP bound to EF-G (translocase). The dipeptidyl-tRNA is now entirely in the P site, leaving the A site open for the incoming (third) aminoacyl-tRNA. The uncharged tRNA dissociates from the E site, and the elongation cycle begins again.

The dipeptidyl-tRNA present at A site is now translocated to P site. Two *models* are now available to explain for the translocation of dipeptidyl-tRNA from A to P site. These are:

(*a*) **Two sites (A, P) model**. According to this model, deacylated tRNA is released from P site, and with the help of one GTP mole and an elongation factor, EF-G, the peptidyl-tRNA is translocated from A to P site. Thus, according to this model, tRNA is either entirely in the A site or entirely in the P site.

(*b*) **Three sites (A, P, E) model**. According to this newer model, initially the aminoacyl end of the tRNA bound to A site moves to the P site on 50S subunit at the time of peptide transfer (before translocation), but only later during translocation, the anticodon end of this tRNA moves from A to P site on 30S subunit. Only this later step requires action of elongation factor, EF-G. In this model, thus, there is an intermediate state, when anticodon of tRNA is still on A site (on 30S subunit), while the aminoacyl end occupies P site (on 50S subunit). A third tRNA binding site, E (exit) was also recognized, through which tRNA leaves the ribosome. E site is found mainly on 50S subunit, and deacylated tRNA interacts with it through its CCA sequence at the end. The movement of deacylated tRNA from P to E site also occurs in 2 steps: the aminoacyl end moves to E site during peptide bond formation, and the anticodon end leaves the P site (on 30S subunit) only during translocation. This three site model suggests that *tRNAs interact mainly at their ends*. It also assumes that a tRNA remains bound to ribosome at one or the other of its two ends (aminoacyl end, anticodon end) and does not leave ribosome at both its end simultaneously (either on A or P site). The two-step transfer of tRNAs from A to P site and from P to E site could result from reciprocating motions of the 2 subunits of a ribosome. It means that *50S and 30S subunits move alternately rather than simultaneously*.

Step 3 : Translocation of peptidyl-tRNA from A to P site

In the third step of elongation cycle (Fig. 30–24), 3 movements take place:

(*a*) the ribosome moves by the distance of one codon (*i.e.*, three nucleotides) towards the 3′ end of mRNA,

(*b*) because the dipeptidyl-tRNA is still attached to the second codon of the mRNA, the movement of the ribosome shifts the dipeptidyl-tRNA from the A site (on the 30S subunit) to the P site (on the 30S subunit), and

(*c*) the deacylated tRNA is released from the initial P site (on the 30 S subunit) back into the cytosol.

The result is that the third codon of the mRNA is now in the A site and the second codon in the P site. This shift of the ribosome along the mRNA requires **elongation factor G, EF-G** (called so since it was isolated as *G*TP-requiring factor; also earlier called **translocase**) and the energy provided by hydrolysis of another molecule of GTP. EF-G, like IF-2 and EF-Tu, cycles between a GTP and a GDP form. The GTP form of EF-G is the one that drives the translocation step. After translocation, the A site is empty, ready to bind an aminoacyl-tRNA to start another round of elongation. The filling of the A site induces the release of deacylated tRNA from the E site; *the A and E sites cannot be simultaneously occupied.*

Thus, the ribosome with its attached depeptidyl-tRNA and mRNA is now ready for another elongation cycle to attach the third amino acid residue. This occurs precisely in the same way as the addition of the second residue. Thus, for each amino acid residue added to the chain, two GTPs are hydrolyzed to GDP and P_i. The ribosome moves from codon to codon along the mRNA toward the end, adding one amino acid residue at a time to the growing chain.

The polypeptide chain always remains attached to the tRNA of the last amino acid to have been inserted. This continued attachment to a tRNA is the chemical glue that makes the entire process work. The ester linkage between the tRNA and the carboxyl terminus of the polypeptide activates the terminal carboxyl group for nucleophilic attack by the incoming amino acid to form a new peptide bond (second step of elongation). At the same time, this tRNA represents the only link between the growing polypeptide and the information in the mRNA. A 3-dimensional, x-ray structure of the entire gamut of ribosome, mRNA, tRNA and the nascent polypeptide chain or the **'protein assembly'**, as it is called, is presented in Fig. 30–25.

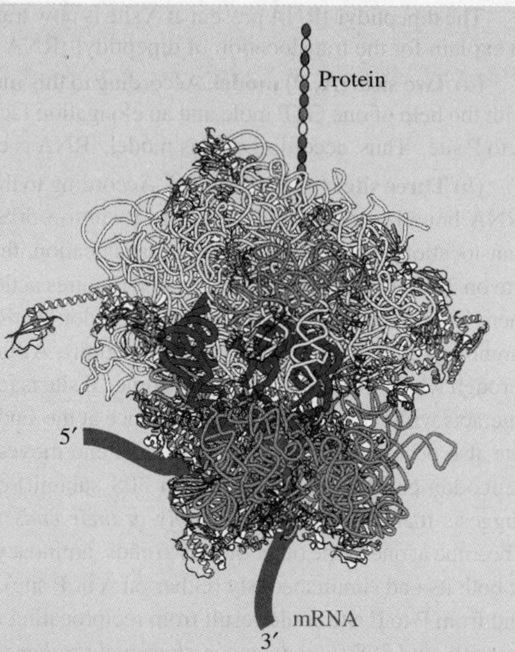

Fig. 30–25. Protein assembly

The ribosome is a factory for the manufacture of polypeptides. Amino acids are carried into the ribosome, one at a time, connected to transfer RNA molecules (*blue*). Each amino acid is joined to the growing polypeptide chain, which detaches from the ribosome only once it is completed. This assembly line approach allows even very long polypeptide chains to be assembled rapidly and with impressive acuracy.

(Courtesy: Doug Martin)

GTPase Rate of EF-Tu as the Pace Setter of Protein Synthesis

The GTPase activity of EF-Tu makes an important contribution to the rate and fidelity of the overall process of protein synthesis. The EF-GTP complex exists for a few milliseconds, and the EF-Tu·GDP complex also exists for a similar period before it dissociates (Fig. 30–26). Both these intervals provide an opportunity for the codon-anticodon interactions to be verified (*i.e.*, proofread). An incorrect aminoacyl-tRNA usually leaves the ribosome during one of these intervals, whereas the correct one stays bound. The correct aminoacyl-tRNA interacts strongly with mRNA in both states, but an incorrect one does not. In effect, the codon–anticodon interaction is scrutinized twice in different ways to achieve higher accuracy, just as proofreading of a manuscript by two readers for an hour each is much more efficient than two hours of proofreading by one. This proofreading mechanism verifies only that the proper codon-anticodon pairing has taken place. The identity of the amino acids attached to tRNAs is not checked at all on the ribosome. This was demonstrated experimentally in 1962 by two groups led by **Fritz Lipmann** and **Semour Benzer**. This finding also provided timely proof for **Crick's adaptor hypothesis** (Fig. 30–27).

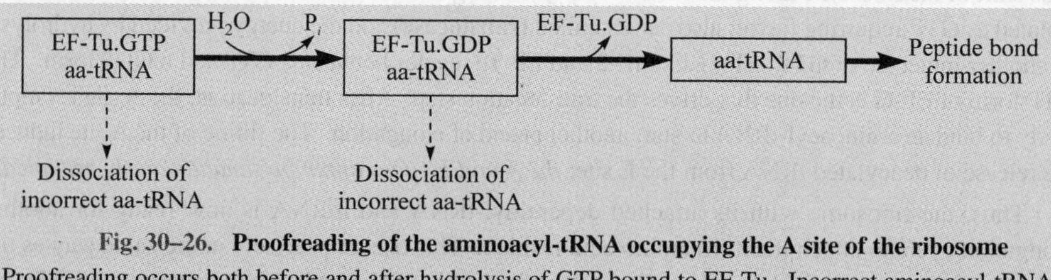

Fig. 30–26. Proofreading of the aminoacyl-tRNA occupying the A site of the ribosome

Proofreading occurs both before and after hydrolysis of GTP bound to EF-Tu. Incorrect aminoacyl-tRNAs dissociate from the A site (steps shown by *broken lines*), whereas the correct aminoacyl-tRNA stays bound and a peptide bond is formed (*thick arrows*).

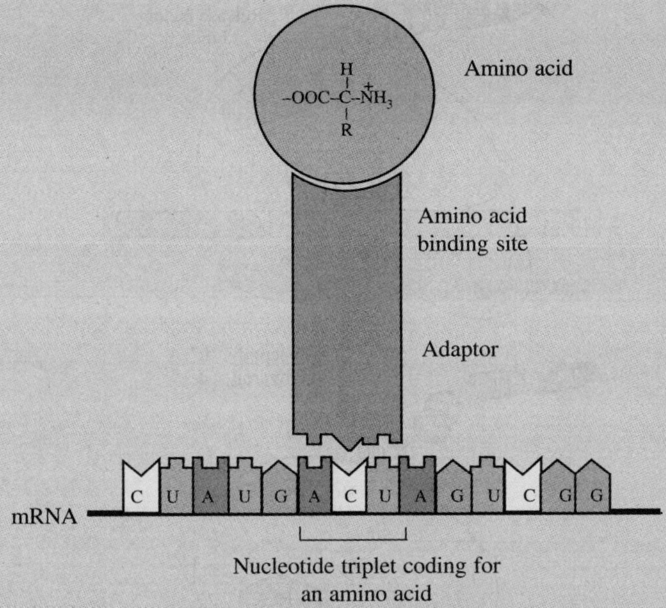

Fig. 30–27. Crick's adaptor hypothesis

The hypothesis asserts that the amino acid is covalently bound at the 3' end of a tRNA molecule and that a specific nucleotide triplet elsewhere in the tRNA interacts with a particular triplet codon in mRNA through hydrogen bonding of complementary bases.

The probability *p* of forming a protein without any error depends on *n*, the number of amino acid residues, and , the frequency of inserting a wrong amino acid:

$$p = (1 - \varepsilon)^n$$

It is apparent from Table 30–4 that an error frequency of 10^{-2} would be intolerable, even for small proteins. An ε of 10^{-3} would actually lead to the error-free synthesis of a 300-residue protein (~ 33 kd) but not of a 1000-residue protein (~ 110 kd). Thus, an error-frequency of the order of no more than 10^{-4} per residue is needed to effectively produce the larger proteins. In fact, the observed values of ε are close to 10^{-4}. *An error frequency of about 10^{-4} per residue has been selected in the course of evolution to produce the greatest number of functional proteins in the shortest time.*

Table 30–4. Accuracy of protein synthesis

Frequency of inserting an incorrect amino acid	*Probability of synthesizing an error-free protein with*		
	*100 AARs**	*300 AARs*	*1000 AARs*
10^{-2}	0.366	0.049	0.000
10^{-3}	0.905	0.741	0.368
10^{-4}	0.990	0.970	0.905
10^{-5}	0.999	0.997	0.990

*AARs = amino acid residues

TERMINATION OF POLYPEPTIDE SYNTHESIS AND RELEASE OF POLYPEPTIDE CHAIN

The process of elongation continues till the ribosome adds the last amino acid, completing the polypeptide coded by the mRNA. Termination, the 4th stage of polypeptide synthesis, is signalled by one of the 3

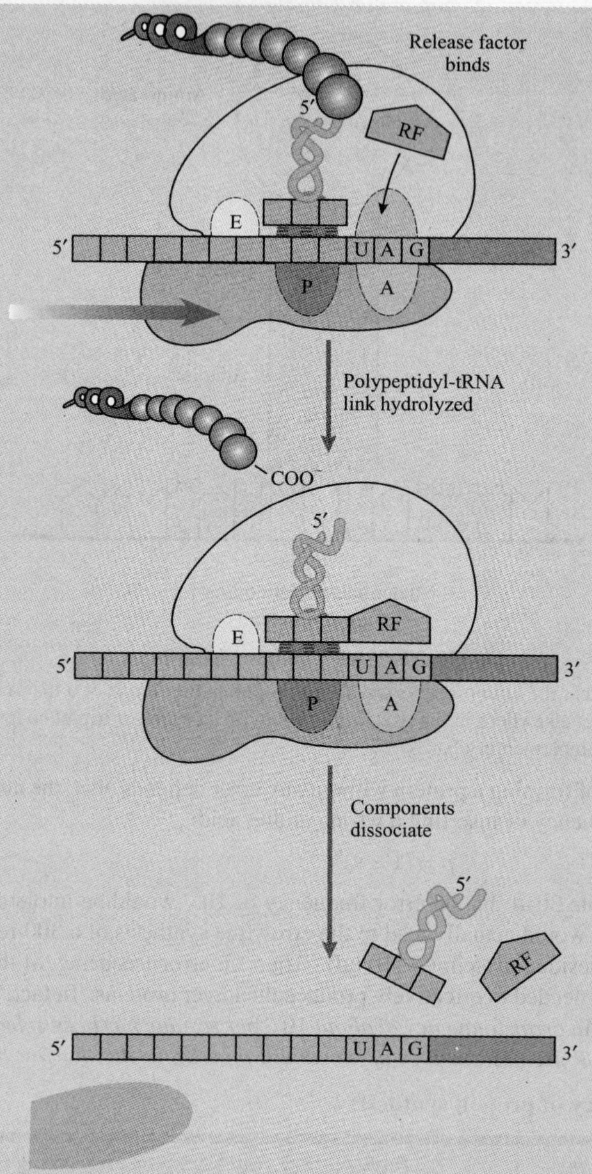

Fig. 30–28. Termination of protein synthesis in bacteria

Termination occurs in response to a termination codon in the A site. First, a release factor, RF (RF-1 or RF-2, depending on which termination codon is present), binds to the A site. This leads to hydrolysis of the ester linkage between the nascent polypeptide and the tRNA in the P site and release of the completed polypeptide. Finally, the mRNA, deacylated tRNA, and release factor leave the ribosome, and the ribosome dissociates into its 30S and 50S subunits.

termination codons or stop signals in the mRNA (UAA, UAG, UGA) immediately following the last amino acid codon (Fig. 30–30). Normal cells do not contain tRNAs with anticodons complementary to these stop signals. Instead, stop codons are recognized by release factors (RF_1, RF_2 and RF_3), which are proteins. RF_1 recognizes the termination codons UAA and UAG, and RF_2 recognizes UGA and UAA. First, either RF_1 or RF_2 (depending on which codon is present) binds at a termination codon to the A site of the mRNA. This leads, in the second step, to the activation of peptidyl transferase so that it hydrolyzes the bond between the polypeptide and the tRNA in the P site. *The specificity of peptidyl transferase is*

Fidelity of Translation

To guarantee the accuracy of protein synthesis, mechanisms have evolved to ensure selection of the correct amino acid for acylation and for proofreading of already-charged tRNAs. One such mechanism is found in the enzymes responsible for attaching an amino acid to the correct tRNA. The **aminoacyl-tRNA synthetases** have the ability not only to discriminate between amino acids before they are attached to the appropriate tRNA, but also to remove amino acids that are attached to the wrong tRNA. Thus, the aminoacyl-tRNA synthetases have the ability of proofreading. In addition, these enzymes must discriminate between the multitude of tRNAs and be able to pair the correct tRNA with the appropriate amino acid. These abilities exhibited by the synthetases are accomplished by a series of hydrogen-bonding interactions between the enzyme and the amino acid, and between the enzyme and the tRNA. These mechanisms combine to ensure the accurate transfer of information from mRNA to protein.

altered by the release factor so that water rather than an amino group is the acceptor of the activated peptidyl moiety. The completed polypeptide is, thus, detached and leaves the ribosome. In the third and final step, the mRNA, deacylated tRNA and release factor leave the ribosome and the ribosome itself dissociates into 30S and 50S components, which are now ready to start a new cycle of polypeptide synthesis. The specific function of RF_3 has not been decisively established. It is probably involved in the release of the completed polypeptide chain from its attachment site on the last tRNA molecule. In eukaryotes, however, a single release factor eRF recognizes all 3 termination codons.

The completed polypeptide chain has a formylmethionyl-amino terminal. Such a protein is called *nascent protein.* Before it becomes functional, the formylmethionyl-amino terminal must be removed which is accomplished by enzymes in 2 steps:

Step I :

$$\text{N-formylmethionyl peptide} \xrightarrow{\text{A specific deformylase}} \text{Methionyl peptide + Formic acid}$$

Step II :

$$\text{Methionyl peptide} \xrightarrow{\text{A specific aminopeptidase}} \text{Peptide + Methionine}$$

Possible Role of GTP in Translation

Whereas ATP is selectively used for the activation of amino acids, *GTP is the only nucleoside triphosphate that is utilized for ribosomal protein synthesis.* In fact, GTP is needed for each of the 3 phases of polypeptide synthesis. It is doubtful that the hydrolysis of GTP, that repeatedly occurs during polypeptide synthesis, provides energy to drive reactions, since there is no evidence for the nucleotide's utilization to form covalent bonds. Its main role appears to be in the noncovalent binding of the various factors to a ribosome, *e.g.*, EFT and RFs. One explanation is that GTP, in each of the translation phases, allows for the specific ribosomal binding of individual elongation factors and that its hydrolysis to GDP and P*i* is required for the release of the bound factors, freeing them for reparticipation in protein synthesis.

FOLDING AND PROCESSING OF POLYPEPTIDE CHAIN

In the fifth and final step of protein synthesis, the newly-formed peptide chain is folded and processed into its biologically-active form. At some point of time, during or after protein synthesis, the polypeptide chain spontaneously assumes its native conformation by forming sufficient number of hydrogen bonds and van der Waals, ionic, and hydrophobic interactions. In this way, the linear (or one dimensional) genetic message encoded in mRNA is converted into the 3-dimensional structure of the protein. However, there are some other nascent proteins which undergo one or more processing reactions called **posttranslational modifications,** for their conversion to the active forms.

Only 20 amino acids are encoded genetically and incorporated during translation. Posttranslational modification of proteins, however, leads to the formation of 100 or more different amino acid derivatives in proteins. Modification may be permanent or highly reversible. The amounts of modified amino acids may be small, but they often play a major functional role in proteins. Examples are listed in Table 30–5.

Table 30–5 Modified amino acids in proteins*

Amino acid	*Modifications found*
Amino terminus	Formylation, acetylation, aminoacylation, myristoylation, glycosylation
Carboxyl terminus	Methylation, glycosyl-phosphatidylinositol anchor formation, ADP-ribosylation
Arginine	*N*-Methylation, ADP-ribosylation
Asparagine	*N*-Glycosylation, *N*-methylation, deamidation
Aspartic acid	Methylation, phosphorylation, hydroxylation
Cysteine	Cystine formation, selenocysteine formation, palmitoylation, linkage to heme, *S*-glycosylation, prenylation, ADP-ribosylation
Glutamic acid	Methylation, γ-carboxylation, ADP-ribosylation
Glutamine	Deamidation, cross-linking, pyroglutamate formation
Histidine	Methylation, phosphorylation, diphthamide formation, ADP-ribosylation
Lysine	*N*-acetylation, *N*-methylation, oxidation, hydroxylation, cross-linking, ubiquitination, allysine formation, biotinylation
Methionine	Sulfoxide formation
Phenylalanine	β-Hydroxylation and glycosylation
Proline	Hydroxylation, glycosylation
Serine	Phosphorylation, glycosylation, acetylation
Threonine	Phosphorylation, glycosylation, methylation
Tryptophan	β-Hydroxylation, dione formation
Tyrosine	Phosphorylation, iodination, adenylation, sulfonylation, hydroxylation

* The listing is not comprehensive and some of the modifications are very rare. Note that no derivatives of alanine (Ala), glycine (Gly), isoleucine (Ile) and valine (Val) have been identified in proteins.

(*Adapted from: Krishna RG, Wold F: Post-translational modification of proteins. In Meister A (Editor): Advances in Enzymology. vol.* **67**: *265–298, Wiley-Interscience, New York. 1993.*)

Such modifications occur in both eukaryotes and prokaryotes and include the following:

1. N-terminal and C-terminal Modifications

All polypeptides begin with a residue of *N*-formylmethionine (in bacteria) or methionine (in eukaryotes). However, the formyl group, the terminal methionine residue, and often additional N-terminal or C-terminal residues must be removed enzymatically before they convert into the final functional proteins. The formyl group at the N-terminus of bacterial proteins is hydrolyzed by a *deformylase*. One or more N-terminal residues may be removed by *aminopeptidases*. In about half of the eukaryotic proteins, the amino group of the N-terminal residue is acetylated after translation. The C-terminal residues are also sometimes modified.

2. Loss of Signal Sequences

In certain proteins, some (15 to 30) amino acid residues at the N-terminus play a role in directing the protein to its ultimate destination in the cell. Such **signal sequences**, as they are called, are ultimately removed by specific *peptidases*.

3. Modification of Individual Amino Acids

Certain amino acid side chains may be specifically modified. For instance, the hydroxyl groups of certain serine, threonine, tyrosine residues of some proteins undergo enzymatic phosphorylation by ATP [Fig. 29(*a*)]; the phosphate groups add negative charge to these polypeptides. The functional significance of this modification varies from one protein to the other. For example, the milk protein *casein* has many phosphoserine groups, which function to bind Ca^{2+}. Given that Ca^{2+} and phosphate, as well as amino acids, are required by suckling young, casein provides three essential nutrients. The phosphorylation and dephosphorylation of the OH group of certain serine residues regulate the activity of some enzymes, such as *glycogen phosphorylase*.

Sometimes, additional carboxyl groups are added to Asp and Glu residues of some proteins. For instance, the blood clotting protein prothrombin contains many γ-carboxyglutamate residues [Fig. 30–29 (*b*)] in its N-terminal region. These groups bind Ca^{2+} which is required to initiate the clotting mechanism.

In some proteins, certain lysine residues are methylated enzymatically [Fig. 30–29(*c*)]. Monomethyl- and dimethyllysine residues are present in some muscle proteins and in *cytochrome c*. *Calmodulin* of most organisms contains one trimethyllysine residue at a specific position. In other proteins, the carboxyl groups of some Glu residues undergo methylation, which removes their negative charge.

Some proline and lysine residues in *collagen* are hydroxylated.

4. Formation of Disulfide Cross-links

Some proteins after acquiring native conformations are often covalently cross-linked by the formation of disulfide bridges between cysteine residues (refer Fig. 9–14). These cross-links help to protect the native conformation of the protein molecule from denaturation in an extracellular environment that is quite different from that inside the cell.

5. Attachment of Carbohydrate Side Chains

In glycoproteins, the carbohydrate side chains are attached covalently during or after the synthesis of polypeptide chain. In some glycoproteins, the carbohydrate side chain is attached enzymatically to Asn residues (*N*-linked oligosaccharides), in others to Ser or Thr residues (O-linked oligosaccharides). Many proteins that function extracellularly contain oligosaccharide side chains.

6. Addition of Prosthetic Groups

Many prokaryotic and eukaryotic proteins require for their activity covalently-bound prosthetic groups. These groups become attached to the polypeptide chain after it leaves the ribosome. The two significant examples are the covalently-bound biotin molecule in *acetyl-CoA carboxylase* and the heme group of *cytochrome c*.

7. Addition of Isoprenyl Groups

Many eukaryotic proteins are isoprenylated; a thioester bond is formed between the isoprenyl group and a cysteine residue of the protein. The isoprenyl groups are derived from pyrophosphate intermediates of the cholesterol biosynthetic pathway, such as farnesyl pyrophosphate (Fig. 30–30). Proteins so modified include the products of the *ras* oncogenes and proto-oncogenes, *G proteins*, and proteins called *lamins*, found in the nuclear matrix.

8. Proteolytic Trimming

Many proteins (*insulin, collagen*) and proteases (*trypsin, chymotrypsin*) are initially synthesized as larger, inactive precursor proteins. These precursors are proteolytically trimmed to produce their final, active forms. Some animal viruses, notably poliovirus, synthesize long polycistronic proteins from one long mRNA molecule. These protein molecules are subsequently cleaved at specific sites to provide the several specific proteins required for viral function.

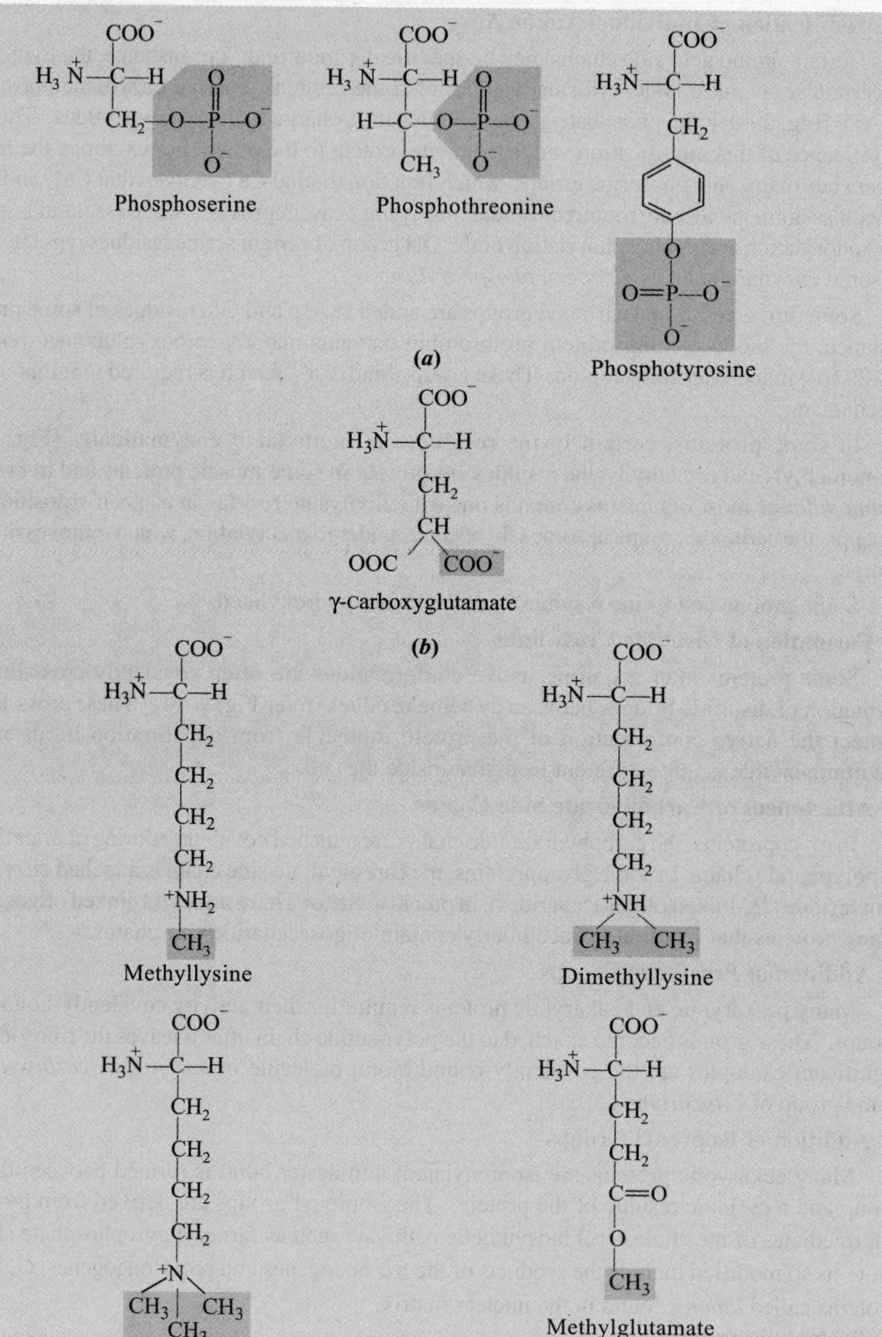

Fig. 30–29. Some modified amino acid residues

(*a*) Three phosphorylated amino acids (*b*) A carboxylated amino acid (*c*) Four methylated amino acids

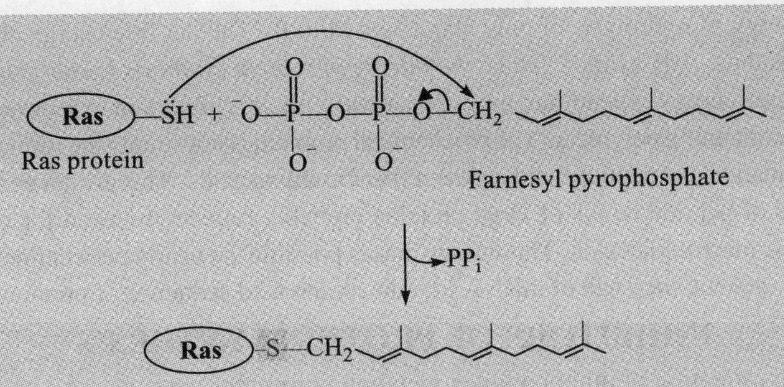

Fig. 30–30. Farnesylation of a cysteine residue on a protein

The *ras* protein is the product of the *ras* oncogene. The thioether linkage is shown *red*.

A reduced ability to carry out a posttranslational modification results in a disorder called *multiple sulfatase deficiency*.

CLINICAL IMPLICATION 30–1.

Absence of Posttranslational Modification: Multiple Sulfatase Deficiency

A variety of biological molecules are sulfated; for example, glycosaminoglycans (GAGs), steroids and glycolipids. Ineffective sulfation of the GAGs, chondroitin sulfate and keratan sulfate of cartilage, results in major skeletal deformities.

Degradation of sulfated molecules depends on the activity of several related sulfatases, most of which are located in lysosomes. Multiple sulfatase deficiency is a rare lysosomal storage disorder that combines features of metachromatic leukodystropohy and mucopolysaccharidosis. Affected individuals develop slowly and from age 2 lose the abilities to stand, sit or speak; physical deformities and neurological deficiencies develop and death before age 10 is usual. Biochemically, multiple sulfatase deficiency is characterized by severe lack of all the sulfatases. Contrarily, deficiencies in individual sulfatases are also known, and several distinct diseases are linked to single-enzyme defects.

Multiple sulfatase deficiency arises from a defect in posttranslational modification that is common to all sulfatase enzymes and is necessary for their enzymatic activity. In each case, a cysteine residue of the enzyme is normally converted to 2-amino-3-oxopropionic acid; the $-CH_2SH$ side chain of cysteine becomes a –CHO (aldehyde) group, which may react with $-NH_2$ or –OH groups of the enzyme, a cofactor, and so on. Fibroblasts from individuals with multiple sulfatase deficiency catalyze this modification with much lowered efficiency, and the unmodified sulfatases are catalytically inactive.

ENERGY REQUIREMENTS FOR PEPTIDE BOND FORMATION

The formation of a peptide bond in a small peptide requires the utilization of ATP. In the case of ribosome-dependent peptide bond synthesis, the equivalent of 4 ATPs are required per peptide bond formed. The break-up of these 4 ATP equivalents is as follows:

(*a*) *Two moles of ATP* are required for the activation of each amino acid to form an aminoacyl-tRNA.

(*b*) *One mole of GTP* is cleaved to GDP and P_i during the first step of elongation, *i.e.*, binding of an aminoacyl-tRNA to the A site.

(*c*) *Another mole of GTP* is hydrolyzed in the translocation of the peptidyl-tRNA from the P to A site.

Additional ATPs are used each time incorrectly-activated amino acids are hydrolyzed by the deacylation activity of some *aminoacyl-tRNA synthetases*. Henceforth, *a total of at least 4 high-energy bonds is ultimately required for the formation of each peptide bond of the completed polypeptide chain.*

This represents an exceedingly large thermodynamic 'push' in the direction of synthesis: at least $4 \times 30.5 = 122$ kJ/mol of phosphodiester bond energy is required to produce a peptide bond having a standard free energy of hydrolysis of only about –21 kJ/mol. The net free energy change in peptide bond synthesis is thus –101 kJ/mol. Thus, *the fidelity in protein synthesis is energetically expensive.* Although, this large energy expenditure may appear wasteful, it is important to remember that proteins are information-containing polymers. The biochemical problem is not simply the formation of a peptide bond, but the formation of a peptide bond between *specific* amino acids. This greater energy requirement for the synthesis of peptide bonds of large proteins probably reflects the need for greater *sequence specificity* in these macromolecules. This energy makes possible the nearly perfect fidelity in biological translation of the genetic message of mRNA into the amino acid sequence of proteins.

INHIBITORS OF PROTEIN SYNTHESIS

As in the investigation of other complex metabolic processes, specific inhibitors have played a major role in separating the steps in the biosynthesis of nucleic acids and proteins. Some of these inhibitors are synthetic compounds; others were first isolated from the culture filtrates of various organisms as antibiotics in efforts to control infectious diseases or to inhibit the growth of malignant tissue.

Protein synthesis is a pivotal function in cellular physiology and as such it is a primary target of a number of naturally-occurring antibiotics and toxins. Most of these antibiotics inhibit protein synthesis in bacteria (*i.e.*, prokaryotes) but are relatively harmless to eukaryotic cells. The inhibitory effects of antibiotics on bacterial growth led to their application in studies designed to determine the intermediary steps in protein synthesis. *Nearly every step in protein synthesis can be specifically inhibited by one antibiotic or the other* (Table 30–5).

> **Antibiotics** are substances produced by bacteria or fungi that inhibit the growth of other organisms. They are known to inhibit a variety of essential biological processes, such as DNA replication (e.g., novobiocin), transcription (e.g., rifamycin) and bacterial wall synthesis (e.g., penicillin). However, *the majority of known antibiotics, including a great variety of medically-useful substances, block translation.* This situation is presumably a consequence of the translation machinery's enormous complexity, which makes it vulnerable to disruption in many ways.

Some known inhibitors of protein synthesis are described below (Fig. 30–31):

1. Streptomycin. Streptomycin, which was discovered by **Selman Wakesman** in 1944, is a medically important member of a family of antibiotics known as *aminoglycosides* that inhibit prokaryotic ribosomes in a variety of ways. It is a highly basic trisaccharide and, at higher concentrations, interferes with the binding of fMet-tRNA to ribosomes and thereby prevents the correct initiation of protein synthesis. And at relatively low concentrations, streptomycin also leads to a misreading of the genetic code on the mRNA and inhibit initiation of the polypeptide chain. If poly U is the template, Ile (AUU) is incorporated in addition to Phe (UUU). An extensive series of experiments revealed that a single protein in the 30S subunit, namely protein S12, is the determinant of streptomycin sensitivity.

2. Tetracyclines. Tetracycline and its derivatives are broad-spectrum antibiotics that inhibit protein synthesis by blocking the A site on the ribosome so that the binding of aminoacyl-tRNAs is inhibited; the nascent polypeptide chain remains in the P site and can react normally with pyromycin, another antibiotic inhibitor.

3. Chloramphenicol, CAP (= Chloromycetin). Chloramphenicol, the first of the "broad-spectrum" anibiotics, inhibits peptidyl transferase activity on the large subunit of prokaryotic ribosomes. However, its clinical uses are limited to only severe infections because of its toxic side effects, which are caused, at least in part, by the chloramphenicol sensitivity of mtochondrial ribosomes. It is a classic inhibitor of protein synthesis in bacteria and acts, at relatively low concentrations on bacterial (also mitochondrial and chloroplast) ribosomes by blocking peptidyl transfer by interfering with the interactions of ribosomes with A site-bound aminoacyl-tRNAs, but does not affect cytosolic protein synthesis in eukaryotes. Of the various possible optical isomers, only the D (–) *threo* form shows significant inhibitory activity.

Table 30–5. Inhibitors of protein or RNA synthesis

Antibiotics	*Organisms*	*Site of action*	*Specific action**
Streptomycin	Prokaryotes	30S	Inhibits movement of initiation complex and causes misreading of mRNA
Tetracycline	Prokaryotes	30S	Blocks recognition site and inhibits binding of aminoacyl-tRNAs at A site
Chloramphenicol	Prokaryotes	50S	Inhibits the peptidyl transferase activity on 70S ribosomes
Cycloheximide	Eukaryotes	60S	Inhibits the peptidyl transferase activity on 80S ribosomes
Erythromycin	Prokaryotes	50S	Binds to the 50S subunit and inhibits translocation reaction on ribosomes
Puromycin	Prokaryotes, Eukaryotes	50S, 60S	Causes premature chain termination by acting as analogue of aminoacyl-tRNA
Fusidic acid	Prokaryotes, Eukaryotes	50S, 60S	Blocks the release of EF-G and GDP during elongation cycle
Rifamycin	Prokaryotes	Transcription	Blocks intiation of RNA chains by binding to RNA polymerase
Streptoligidin	Prokaryotes	Transcription	Inhibits elongation of the peptide chain
Actinomycin D	Prokaryotes, Eukaryotes	—	Binds to DNA and blocks the movement of RNA polymerase
Anisomycin	Eukaryotes	—	Blocks the peptidyl transferase reaction on ribosomes
α-amanitin	Eukaryotes	—	Blocks mRNA synthesis

* The ribosomes of eukaryotic mitochondria (and chloroplasts) often resemble those of prokaryotes in their sensitivity to inhibitors. Therefore, some of these antibiotics can have a deleterious effect on human mitochondria.

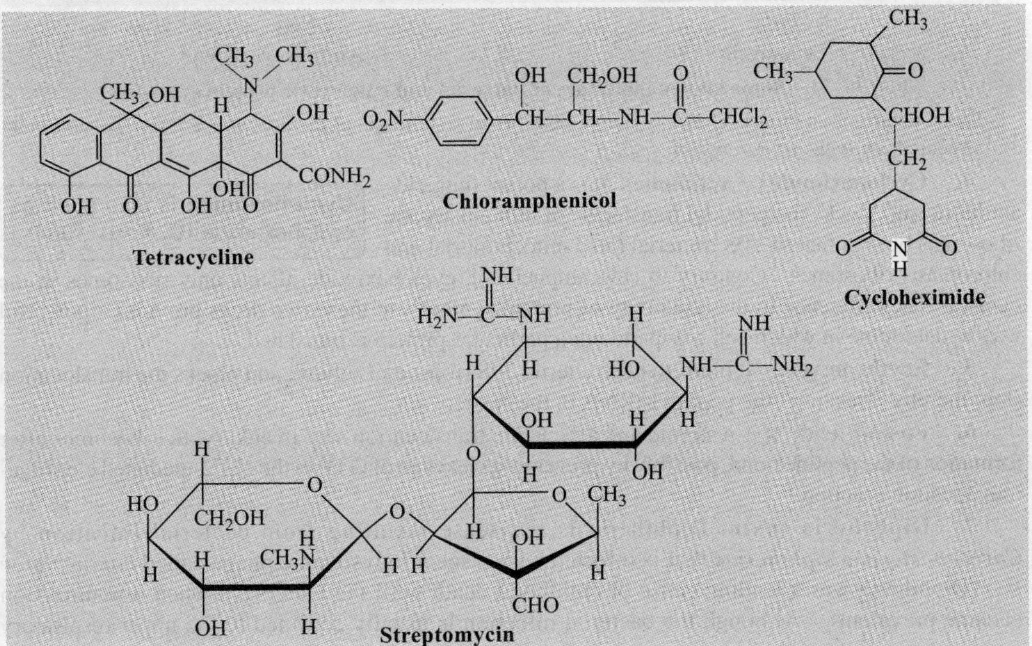

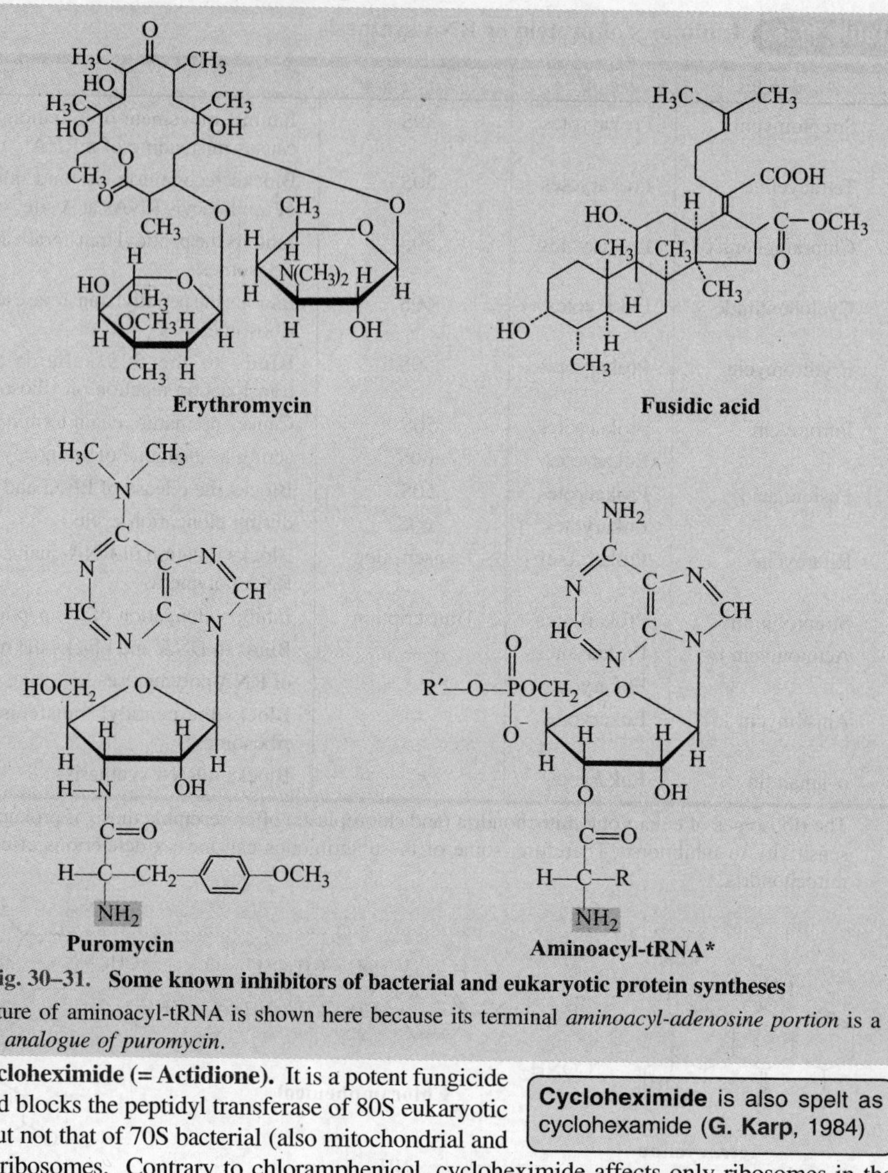

Fig. 30–31. Some known inhibitors of bacterial and eukaryotic protein syntheses

* The structure of aminoacyl-tRNA is shown here because its terminal *aminoacyl-adenosine portion* is a structural *analogue of puromycin*.

4. Cycloheximide (= Actidione). It is a potent fungicide antibiotic and blocks the peptidyl transferase of 80S eukaryotic ribosomes but not that of 70S bacterial (also mitochondrial and chloroplast) ribosomes. Contrary to chloramphenicol, cycloheximide affects only ribosomes in the cytosol. The difference in the sensitivity of protein synthesis to these two drugs provides a powerful way to determine in which cell compartment a particular protein is translated.

> **Cycloheximide** is also spelt as cyclohexamide (**G. Karp**, 1984)

5. Erythromycin. It binds to the bacterial 50S ribosomal subunit and blocks the translocation step, thereby "freezing" the peptidyl-tRNA in the A site.

6. Fusidic acid. It is a steroid and affects the translocation step in eukaryotic ribosomes after formation of the peptide bond, possibly by preventing cleavage of GTP in the eEF2-mediated cleavage-translocation reaction.

7. Diphtheria toxin. Diphtheria is a disease resulting from bacterial infection by *Corynebacterium diphtheriae* that is infected with a specific lysogenic phage called *corynephage β*. (Diphtheria was a leading cause of childhood death until the late 1920s when immunization became prevalent). Although the bacterial infection is usually confined to the upper respiratory tract, the bacteria secreate a phage-endoded protein, called *diphtheria toxin*, that is responsible for the disease's lethal effects. *Diphtheria toxin specifically inactivates the eukaryotic elongation factor eEF-2, thereby inhibiting eukaryotic protein synthesis.*

Diphtheria toxin acts in a particularly interesting way. It is a monomeric 535-residue protein that is cleaved past Arg residues 190, 192 and 193 by trypsin and trypsin-like enzymes. This hydrolysis occurs around the time diphtheria toxin encounters its target cell, yielding 2 fragments A and B, which nevertheless remain linked by a disulfide bond. The fragment B binds to a specific receptor on the plasma membrane of susceptible cells, whereupon it facilitates A fragment's cytosolic uptake *via,* receptor-mediated endocytosis (free fragment A is devoid of toxic activity). The intracellular reducing environment then cleaves the disulfide bond linking the A and B fragments.

Within the cytosol the fragment A catalyzes the ADP-ribosylation of eEF-2 by NAD^+, thereby inactivating this elongation factor. Since the fragment A acts catalytically, *one molecule is sufficient*

eEF-2 + NAD^+
(active)

↓ diphtheria toxin

ADP-ribosyl-eEF-2 + Nicotinamide + H^+
(inactive)

to ADP-ribosylate all of a cell's eEF-2s, which halts protein synthesis and kills the cells. Only a few micrograms of diphtheria toxin are, hence, sufficient to kill an unimmunized individual.

Diphtheria toxin specifically ADP-ribosylates a modified His residue on eEF-2 known as diphthamide:

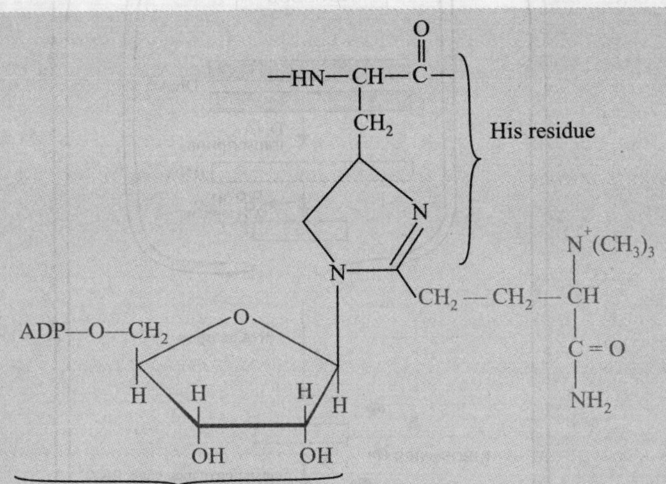

ADP-ribosylated diphthamide

Diphthamide occurs only in eEF-2 (not even in its bacterial counterpart, EF-G), which accounts for the specificity of diphtheria toxin in exclusively modifying eEF-2. Since diphthamide occurs in all eukaryotic eEF-2s, it probably is essential to eEF-2 activity.

8. Ricin or Abrin. It is an extremely toxic protein and inactivates the 60S subunit of eukaryotic ribosomes.

9. Puromycin. It is one of the best-understood inhibitory antibiotics and is obtained from a mould *Streptomyces alboniger*. Puromycin (Fig. 30–31) is a structural analogue of the terminal aminoacyl-adenosine portion of aminoacyl-tRNA. It binds to the A site on the ribosome and inhibits the entry of aminoacyl-tRNA. Furthermore, puromycin contains an α-amino group which, like the one on aminoacyl-tRNA, forms a peptide bond with the carboxyl group of the growing peptide chain in a reaction catalyzed by peptidyl transferase. The product is a peptide having a covalently-attached puromycin residue at its carboxyl end. Peptidyl puromycin then dissociates from the ribosome. Thus, normal chain growth is aborted, and incomplete peptide chains bearing carboxyl-terminal puromycin are released from the ribosomes. Puromycin has been used to ascertain the functional state of ribosomes. In fact, the concept of A and P sites resulted from the use of puromycin to ascertain the location of peptidyl-tRNA. When peptidyl-tRNA is in the A site (before translocation), it cannot react with puromycin.

EUKARYOTIC PROTEIN SYNTHESIS

The eukaryotic protein synthesis, which has been in essence diagrammatically represented in Fig. 30–32, follows essentially the same pattern as that described for *Escherichia coli* (the bacterium most ideal for *in vitro* experimentation) with certain distinct differences, already mentioned at places of their reference. These may, however, be summed up as follows:

1. Chromatin Material. In prokaryotes, there is no distinct nucleus bound by a nuclear membrane. As such, the two processes of transcription and translation take place *simultaneously* in

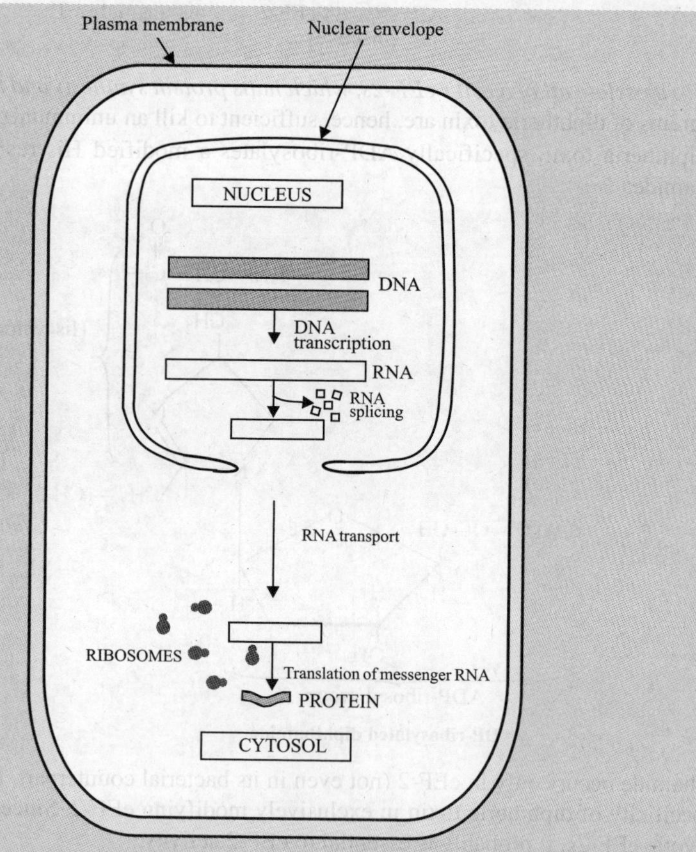

Fig. 30–32. Protein synthesis (DNA → RNA → Protein) in eukaryotes

Eurkaryotic cells have evolved many membrane-bound compartments that separate their various chemical reactions so as to make them more efficient, and the nucleus is one such compartment. The nuclear envelope keeps functional ribosomes out of the nucleus and thus prevents RNA transcripts from being translated into protein until they have been processed (spliced) and transported and transport steps are interposed between DNA transcription and RNA translation.

the cell; whereas in eukaryotes, transcription of mRNA takes place inside the nucleus and followed by its translocation in the cytoplasm, *i.e.*, the two processes take place *successively*. In other words, in eukaryotic cells, transcription and translation are separated in time and space; whereas in prokaryotes, they are not. However, in both transcription occurs at the site of DNA representing a gene. Prokaryotes lack a nuclear envelope. Since mRNAs are synthesized (or transcribed) in a $5' \rightarrow 3'$ direction, and translation also proceeds in the same direction, there can be, in prokaryotes, a coupling between transcription and translation, *i.e.*, translation of a mRNA can begin even before its synthesis is completed. Contrarily, in eukaryotes, such a coupling is not possible because the mRNAs are synthesized in the nucleus (where the DNA is localized) and must then cross the nuclear membrane and reach the cytoplasm before translation begins.

2. **Ribosomes.** The eukaryotic ribosomes are larger than those of the prokaryotes. They consist of a 60S larger subunit and a 40S smaller subunit, come together to form an 80S particle having a mass of 4,200 kd, in comparison to 2,700 kd for the prokaryotic 70S ribosome. The 40S subunit contains an 18S RNA that is homologous to the prokaryotic 16S RNA. The 60S subunit contains 3 RNAs: the 5S and 28S are homologous to the prokaryotic 5S and 23S; its 5.8S is unique to eukaryotes and has no counterpart in prokaryotes.

3. **Initiator tRNA.** Methionine initiates eukaryotic protein synthesis; whereas the starting amino acid residue at the amino-terminus in prokaryotes is N-formylmethionine. However, as in prokaryotes, a special tRNA participates in initiation. This aminoacyl-tRNA is called Met-$tRNA_f$ or Met-$tRNA_i$ (the subscript f indicates that it can be formylated in vitro, and i stands for initiation).

4. **Initiation Complexes.** The most enigmatic difference to date, however, is with regard to the number of initiation factors needed for the formation of the eukaryotic initiation complex. Whereas the *E. coli* system requires 3 factors (IF1, IF2, IF3), the eukaryotic initiation process requires at least 9 (eIF1, eIF2, eIF3, eIF4A, eIF4B, eIF4C, eIF4D, eIF4E, eIF5). This list does not include factors that enhance the activity of eIF2 (Co-eIF2 factors) or those that enhance the dissociation of 80S ribosomes into their subunits (ribosome-dissociation factors). Also, in prokaryotes, the smaller ribosomal subunit (30S) first combines with mRNA and the 30S-mRNA complex then unites with fMet–$tRNA^{fMet}$ - (Fig. 30–33). But in eukaryotic protein synthesis, the smaller ribosomal subunit (40S) associates with Met–$tRNA^{fMet}$ without the help of mRNA. The scheme for the assemblage of the eukaryotic initiation

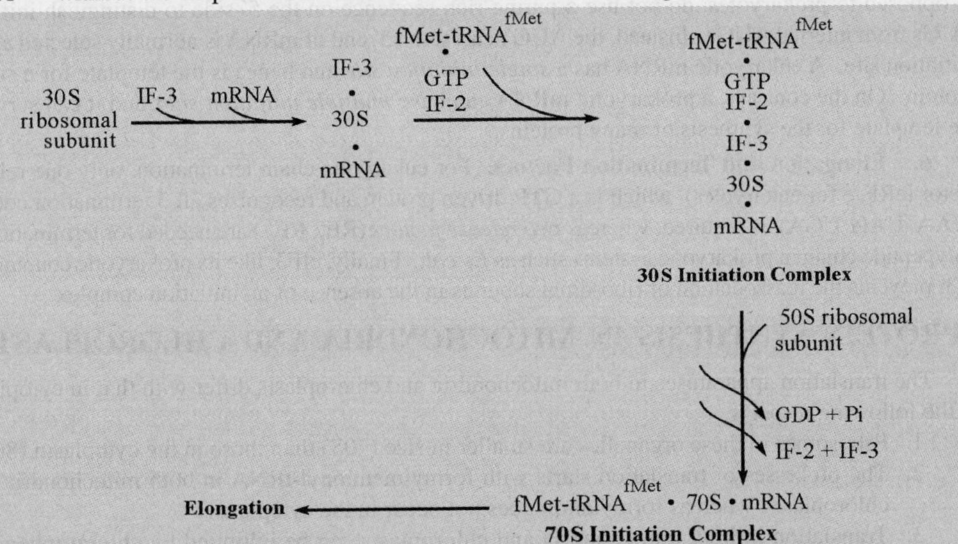

Fig. 30–33. Proposed assembly of the initiation complex for prokaryotic protein synthesis

The initiation factor 1 (IF-1) plays a role in stimulating activities of IF-2 and IF-3.

complex is presented in Fig. 30–34. The step-by-step process is similar to that described for the formation of the *E. coli* complex, *i.e.*, complexing of an initiating tRNA, the smaller ribosomal subunit, an mRNA, and, finally, the larger ribosomal subunit. With its requirement for 6 of the 9 initiation factors, the addition of mRNA to the complex is the most complex molecular step in the process. ATP hydrolysis is also required for mRNA addition. Thus, *for the eukaryotic system, both ATP and GTP are necessary, whereas the comparable E. coli process requires only GTP.*

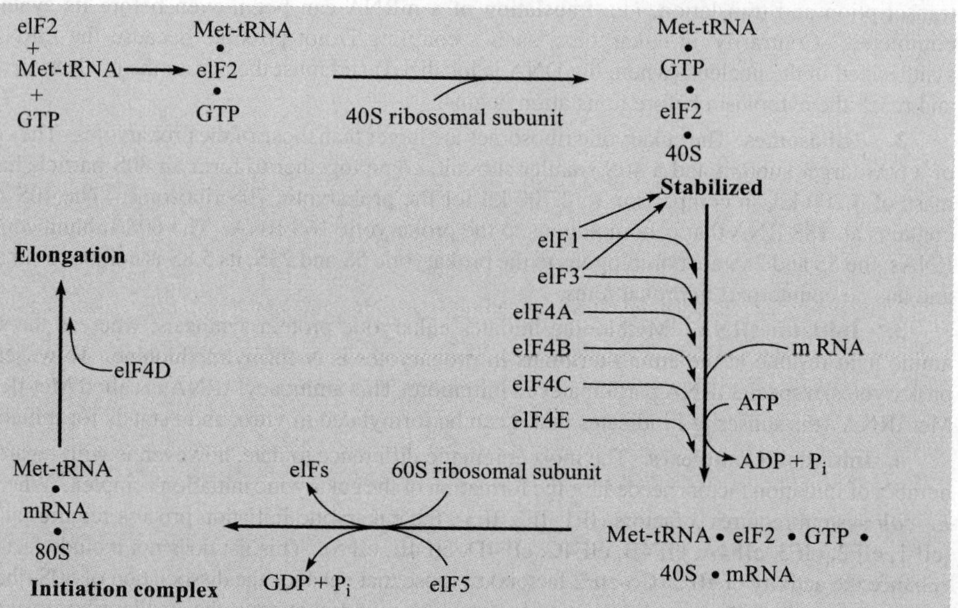

Fig. 30–34. Proposed assembly of the initiation complex for eukaryotic protein synthesis

5. Start Signals. The initiating codon is eukaryotic protein synthesis is always AUG. Eukaryotes, as opposed to prokaryotes, do not use a purine-rich sequence on the 5′ side to distinguish initiator AUGs from internal AUGs. Instead, the AUG nearest the 5′ end of mRNA is normally selected as the initiation site. A eukaryotic mRNA has a *single initiation site* and hence is the template for a single protein. On the contrary, a prokaryotic mRNA can have *multiple initiation sites* and it can serve as the template for the synthesis of many proteins.

6. Elongation and Termination Factors. For eukaryotic chain termination, only one release factor (eRF, e for eukaryotes), which is a GTP-driven protein and recognizes all 3 termination codons (UAA, UAG, UGA), is required; whereas *two release factors* (RF_1, RF_2) are needed for termination of polypeptide chain in prokaryotic systems such as *E. coli*. Finally, eIF3, like its prokaryotic counterpart IF3, prevents the reassociation of ribosomal subunits in the absence of an initiation complex.

PROTEIN SYNTHESIS IN MITOCHONDRIA AND CHLOROPLASTS

The translation apparatuses in both mitochondria and chloroplasts differ with that in cytoplasm in the following respects:

1. Ribosomes in these organelles are smaller in size (70S) than those in the cytoplasm (80S).
2. The processes of translation starts with formylmethionyl-tRNA in both mitochondria and chloroplasts, whereas formylation does not occur in the cytoplasm.
3. Translation in both mitochondria and chloroplasts can be inhibited by chloramphenicol, as in bacteria, since the 70S ribosomes are sensitive to chloramphenicol; on the contrary,

the translation in cytoplasm is inhibited by cycloheximide since 80S ribosomes are sensitive to cycloheximide.

The **mitochondria** contain circular DNA molecules which differ from nuclear DNA in size, base composition, synthesis and turnover rate. The DNA is located in many discrete regions in the mitochondrial matrix, which codes for fewer proteins. The mitochondria contain ribosomes of 70S type and the ribosomal species resemble those of prokaryotes, rather than their counterparts in the cytoplasm which are 80S type. Mitochondria also possess many organelle-specific tRNAs and aminoacyl-tRNA synthetases, which catalyze the attachment of amino acids to tRNA molecules. The mitochondrial tRNA hybridizes with mitochondrial DNA and hence may be made within the organelle on mitochondrial DNA template. The enzymes DNA and RNA polymerases, capable of acting on mitochondrial DNA, are also found in mitochondria.

The mitochondrial genome of various eukaryotic cells (esp, yeast cells and human cells) has been extensively studied in the recent past. The mitochondrial genome of humans, which is a circular double-stranded DNA, comprises 16,569 base pairs (as opposite to 78,000 base pairs in the case of mitochondrial genome of the yeast *Saccharomyces cerevisiae*) and codes for the following:

1. two rRNAs with 1,559 and 954 nucleotides respectively (3,200 and 1,660 nucleotides in yeast)
2. twenty two tRNAs (24 in yeast). This number of tRNAs is definitely smaller than the minimum number of tRNAs required to 'read' the 61 'sense' codons, taking into consideration the wobble hypothesis, *i.e.*, 31 tRNAs (plus the initiation tRNA).
3. about thirteen polypeptides which are part of oligomeric membrane proteins: cytochrome oxidase, apocytochrome b, ATPase, NADH dehydrogenase. The cytochrome oxidase contains seven different types of polypeptides, of which 3 are synthesized on mitochondrial ribosomes and the remaining 4 on the cytoplasmic ribosomes.

It has been observed that human mitochondrial genes are not separated by intercistronic sequences (punctuation is carried out by the tRNA genes) and that the mRNAs contain neither the leader sequence nor the tail sequence. Moreover, there seems to be only one promoter in each strand of human mitochondrial DNA, which suggests that the genome is transcribed completely and symmetrically. In the yeast mitochondrion, on the contrary, transcription seems to take place from several different promoters.

Although the coding sequences in both the genes of apocytochrome b in human and yeast mitochondria are of the same size, there is no intron in the human mitochondrial gene as opposed to 5 introns in yeast. Besides, while in man, there is neither leader sequence in 5′ nor tail sequence in , in yeast there is a leader sequence of about 1,000 nucleotides and a tail sequence of about 50 nucleotides.

Amazingly, it is also revealed that the genetic code used by the mitochondria is different from the normal genetic code which was thought to be universal. For example, the UGA codes for tryptophan (instead of acting as nonsense codon) and the AUA codes for methionine (instead of coding for isoleucine). Also, AUU and AUA can be used as initiation codons (instead of being codons for isoleucine), and AGA and AGG can be used either as termination codons in the case of vertebrates or as codons for serine in the case of *Drosophila* (instead of being codons for arginine). Surprisingly, the CUN (N = any nucleotide) codes for tryptophan in yeast (instead of coding for arginine).

The mitochondrial genome of plants is also circular but larger than that of either mammals (ca 16 kbp) or yeasts (ca 78 kbp) and that its size varies, according to the species, from 200 to 2,000 kbp. In fact, only about 15 genes of proteins have been identified and sequenced at present: these are some genes coding for subunits of cytochrome oxidase, NADH dehydrogenase, ATPase and cytochrome b, and some other genes coding for proteins of the small ribosomal particles S_2, S_3, S_4, etc. A characteristic feature of plant mitochondria is that a number of mitochondrial tRNAs are coded by the mitochondrial genome but the other mitochondrial tRNAs are coded by the nuclear genome and are imported into the mitochondrion. Moreover, a heterogeneity is observed in the size of mitochondrial DNA molecules of the same plant. This is due to the presence, in the mitochondrial genome, of *repeated sequences* which can generate sub-genomic DNA molecules by recombination.

Recently, it has been shown that in plant mitochondria, there exist a phenomenon of correction of genetic information at the level of mRNAs, called RNA editing. It consists in modifying some cytosines and uracils and thus changing some codons. This results in the synthesis of a protein whose amino acid sequence is different from the sequence which could be deduced from that of the gene, but has more similarity in the sequence of amino acids of the corresponding mitochondrial protein in other organisms such as mammals and yeasts.

The **chloroplasts**, like mitochondria, also possesses a level of autonomy. The chloroplast DNA is much larger than mitochondrial DNA and hence codes for a greater number of proteins. Each chloroplast has more than one DNA molecule and the number varies with the size of the organelle and the type of organism. As opposed to mitochondria, the size of the chloroplast genomes, which are also circular, does not vary considerably from one species to another and is about 120 to 190 kbp. In most of the species studied, the chloroplast DNA comprises a region of about 10 to 25 kbp present as two copies in opposite orientations (*i.e., inverted repeats*). These inverted repeats contain the genes of the rRNAs and are separated by a large single copy region and a small single copy region.

Recently, the complete sequence of the chloroplast genome of two plants namely tobacco, *Nicotiana tabaccum* (155 kbp) and a liverwort, *Marchantia* (120 kbp) was determined. The chloroplast DNA codes for the following:

1. the chloroplast 23S, 16S, 5S and 4.5S rRNAs
2. thirty chloroplast tRNAs
3. nineteen chloroplast ribosomal proteins (11 proteins of the 30S particle + 8 proteins of the 50S particle)
4. the translation initiation factor, IF-1
5. three subunits of the RNA polymerase
6. two proteins of photosystem I, five proteins of photosystem II, six polypeptides forming part of the ATP-synthase complex, and three proteins involved in electron transport.

Besides the genes for these 40 or so identified proteins, the chloroplast genome contains some 40 *open reading frames* (ORF) which could also code for polypeptides, but most of the chloroplast proteins are coded by the nuclear genome.

Ellis (1975) studied protein synthesis in the isolated chloroplasts of a leguminous plant, pea. The isolated chloplasts could synthesize following proteins:

(*a*) large subunit of Fraction I protein
(*b*) five unidentified proteins of the internal lamellar system
(*c*) two or three unidentified polypeptides of the envelope.

Ellis also demonstrated that while the large subunit of Fraction I protein is synthesized under the influence of chloroplast DNA, the small subunit is synthesized in the cytoplasm under the influence of nuclear DNA. Small subunit is then transported to the chloroplast where it associates with large subunit to give rise to Fraction I protein (Fig. 30–35).

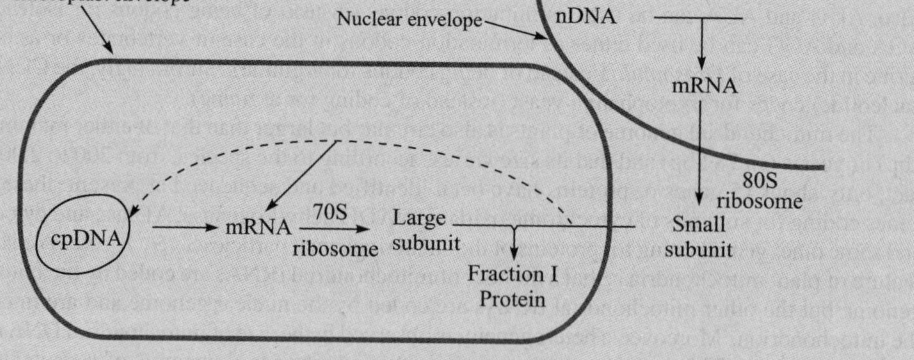

Fig. 30–35. A model of cooperativity between nuclear DNA (nDNA) and chloroplast DNA (cpDNA) in pea

(Redrawn from Ellis RJ, 1975)

SALIENT FEATURES OF RIBOSOMAL PROTEIN SYNTHESIS

The protein synthesis carried on ribosomes, which is summarily represented in Fig. 30–36, is a highly complex, anabolic process and occurs continuously within the cells, *i.e.*, intracellularly. Certain salient features of ribosomal protein synthesis are:

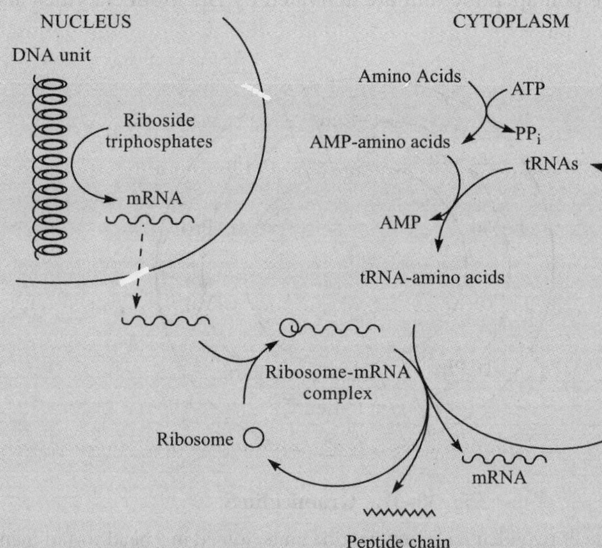

Fig. 30–36. A summary of the process of protein synthesis

(*Adapted from Fairley JL and Kilgour GL, 1974*)

1. *DNA molecule never leaves the chromosome* during the course of eukaryotic protein synthesis. It acts through the intermediary of mRNA.

2. The process of protein synthesis is *extremely accurate* and *highly fool-proof*. In fact, there is less than one error for every 10,000 amino acids polymerized into protein, *i.e.*, the error rate is about 0.01%.

3. The mechanism of protein synthesis is *cyclical*. Proteins are linear polymers of many monomer units. Therefore, this type of structure can be synthesized by a mechanism in which monomer units are added sequentially to the growing polymer, with essentially the same events of each step. It would be less efficient and certainly wasteful of genetic material if an entirely separate mechanism were used to add each amino acid onto the nascent polypeptide.

BIOSYNTHESIS OF SHORT PEPTIDES (= ANTIBIOTIC POLYPEPTIDE SYNTHESIS)

Certain short peptides are not synthesized by ribosomes and a different mechanism of peptide bond formation operates in them. The biosynthesis of short peptides that are not under genetic control, insofar as determination of the sequence of their amino acid residues is concerned, is known in at least two instances.

A. Synthesis of Gramicidin S

Gramicidin S, an antibiotic produced by certain strains of *Bacillus*

Gramicidin S is also spelt as gramacidin S.

brevis, is a cyclic decapeptide made up of two identical pentapeptides joined head-to-tail (Fig. 30–37). A characteristic feature of this molecule is that it contains a D-amino acid (phenylalanine) and a nonprotein amino acid ornithine, which is used in urea cycle but not in protein synthesis.

Gramicidin S biosynthesis requires a much simpler synthetic apparatus consisting only of two enzymes, E_I and E_{II}. D-phenylalanine is activated by E_{II}, whereas the remaining 4 amino acids (L-Pro, L-Val, L-Orn, L-Leu) of the pentapeptide unit are activated by E_1. Both enzymes also participate in peptide-bond formation.

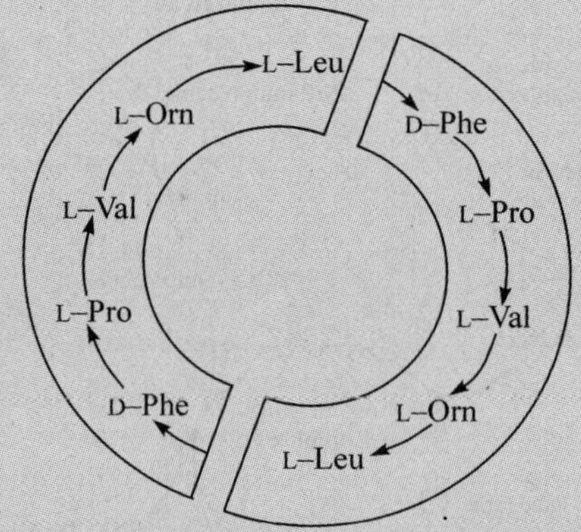

Fig. 30–37. Gramicidin S

This is a cyclic peptide made of two identical pentapeptide units, joined in a head-to-tail manner. The arrows denote the polarity of the polypeptide chain (amino to carboxyl direction).

In the biosynthetic process, amino acids are activated by the formation of enzyme-bound thioesters. The activated amino acids are attached to a sulfhydryl (– SH) group of E_I or E_{II}, instead of the terminal hydroxyl group of a tRNA:

$$\text{Amino acid} + \text{ATP} \rightleftharpoons \text{Aminoacyl-AMP} + \text{PP}i$$

$$\text{Aminoacyl-AMP} + \text{E–SH} \rightleftharpoons \text{E–S}\sim\overset{\overset{\displaystyle O}{\|}}{C}\text{–}\underset{\underset{\displaystyle R}{|}}{\overset{\overset{\displaystyle H}{|}}{C}}\text{–}NH_3^+ + \text{AMP}$$

Thioester

L-Proline, L-valine, L-ornithine and L-leucine form thioester linkages with specific sulfhydryls of E_I when they are incubated in the presence of ATP. Similarly, D-phenylalanine forms a thioester bond with E_{II} in the presence of ATP.

The synthesis of gramicidin S starts with the interaction of E_I and E_{II}. The D-phenylalanine residue on E_{II} is transferred to the imino (> NH) group of the L-proline residue on E_I to form a dipeptide:

$$E_{II}\text{–S}\sim\overset{\overset{\displaystyle O}{\|}}{C}\text{–Phe–}NH_2 + E_I\text{–S}\sim\overset{\overset{\displaystyle O}{\|}}{C}\text{–Pro–NH} \rightleftharpoons E_{II}\text{–SH} + E_I\text{–S}\sim\overset{\overset{\displaystyle O}{\|}}{C}\text{–Pro–Phe–}NH_2$$

Activated dipeptide

The subsequent reactions require only E_I. The activated carbonyl group of the proline residue in the dipeptide reacts with the amino group of the valine residue on the same enzyme to yield a tripeptide. This process is repeated with ornithine and then with leucine to produce an enzyme-bound pentapeptide.

The growing peptide is transferred to a different sulfhydryl each time a peptide bond is formed.

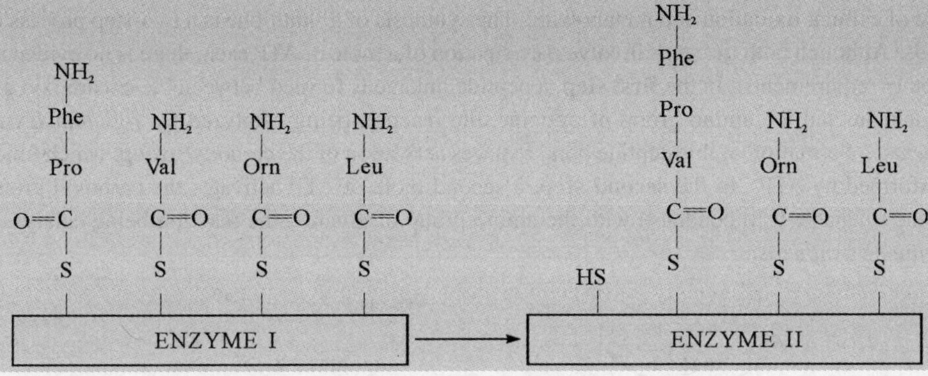

Finally, the activated pentapeptides attached to two different E_I molecules react with each other to form cyclic gramicidin S.

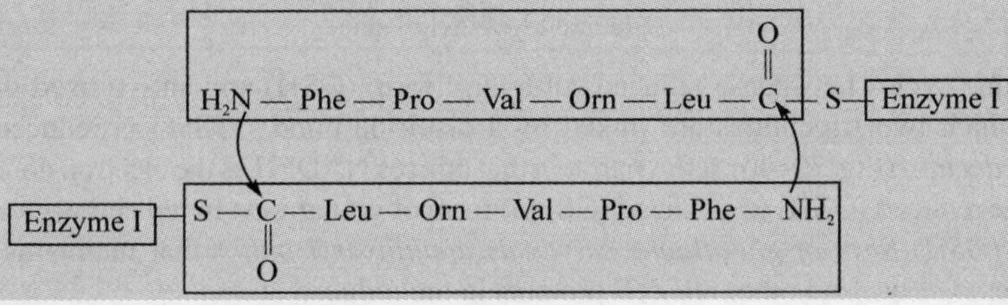

Two worthnoting features of this biosynthetic pathway are:

(*a*) The amino acid sequence of gramicidin S is determined by the spatial arrangement and specificity of the enzymes I and II. There is at least one protein subunit for each peptide bond formed. Therefore, *this mode of synthesis is uneconomical as compared to the ribosomal protein synthesis.* With the result, a peptide containing more than about 15 amino acid residues is not synthesized by this mechanism.

(*b*) *The synthesis of gramicidin S resembles that of fatty acid*, because the activated intermediates in both processes are thioesters. Furthermore, E contains a covalently-attached phosphopantetheine residue. This thiol probably carries the growing peptide chain from one site to the next in E_I. *This antibiotic polypeptide synthesis may be a surviving relic of a primitive mechanism of protein synthesis used early in evolution* (**Fritz Lipmann**, 1971). *Ribosomal protein synthesis may have evolved from fatty acid synthesis.*

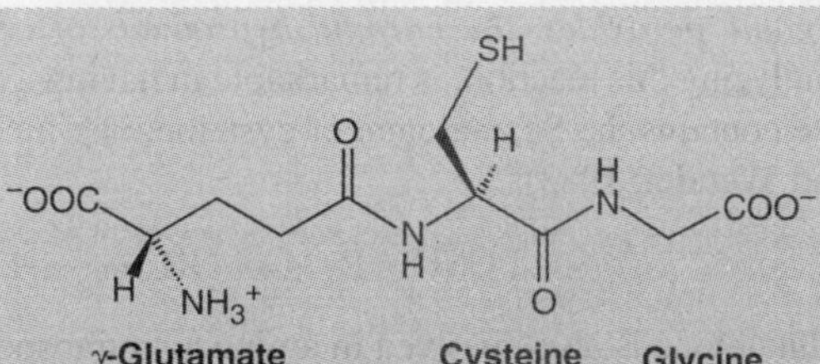

Fig. 30-38. Glutathione, GSH

This tripeptide consists of a cysteine residue flanked by a glycine residue and a glutamate residue that is linked to cysteine by an isopeptide bond between glutamate's side-chain carboxylate group and cysteine's amino group.

B. Synthesis of Glutathione

Glutathione (Fig. 30–38), a tripeptide containing a sulfhydryl group, is a highly distinctive amino acid derivative performing a number of important physiological functions. It was isolated in 1921 by Frederick Gowland Hopkins. The discovery of this sulfur-containing compound gave a great impetus to the study of the complicated nature of cellular oxidation and metabolism. The synthesis of glutathione is a two-step process (Fig. 30–39). Although both the steps involve participation of a mole of ATP each, there is no evidence for any other requirements. In the **first step**, a peptide linkage is formed between the α-carboxyl group of glutamate and the amino group of cysteine, the reaction being catalyzed by *γ-glutamylcysteine synthetase*. Formation of this peptide bond requires activation of the carboxyl group, which function is performed by ATP. In the **second step**, a second mole of ATP activates the carboxyl group of cysteine to enable it to condense with the amino group of glycine, the reaction being catalyzed by *glutathione synthetase*.

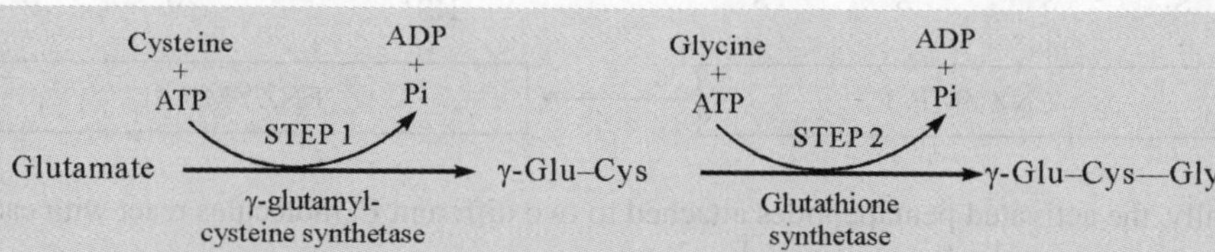

Fig. 30–39. Synthesis of glutathione, GSH
(γ-glutamylcysteinylglycine)

Glutathione cycles between a reduced sulfhydryl form (GSH) and an oxidized disulfide form (GSSG) in which two tripeptides are linked by a disulfide bond. GSSG is reduced to GSH by *glutathione reductase* (Fig. 30–40), a flavoprotein that utilizes NADPH as the electron donor. Normally, the ratio of the reduced (GSH) to oxidized (GSSG) form of glutathione in red cell is greater than 500. *The reduced (GSH) form of glutathione serves as a sulfhydryl buffer* that maintains the cysteine residues of hemoglobin and other red-cell proteins in the reduced state.

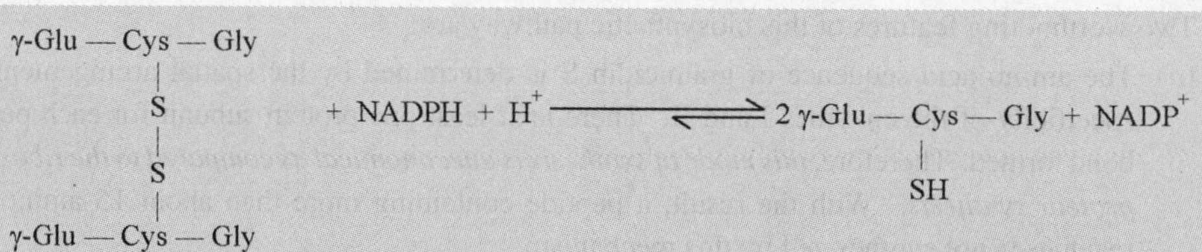

The reduced form of glutathione also plays a key role in detoxication by reacting with H_2O_2 and organic peroxides, the harmful byproducts of aerobic life. Glutathione peroxidase, the enzyme catalyzing this reaction, is remarkable in having a covalently-attached selenium (Se) atom. Its active site contains the Se analogue of cysteine, in which selenium has replaced sulfur (**Epp**, **Ladenstein** and **Wendel**, 1983).

$$2\,GSH + R—O—OH \longrightarrow GSSG + H_2O + ROH$$

Glutathione is also involved in amino acid transport across cell membranes.

EVOLUTION OF PROTEIN SYNTHESIS

The mechanism underlying protein synthesis seems extremely complex. Protein synthesis in the present-day organisms centres on the ribosome, which consists of proteins arranged around a core of ribosomal RNA (rRNA) molecules. How did rRNA molecules come to play such a dominant role in the ribosome?

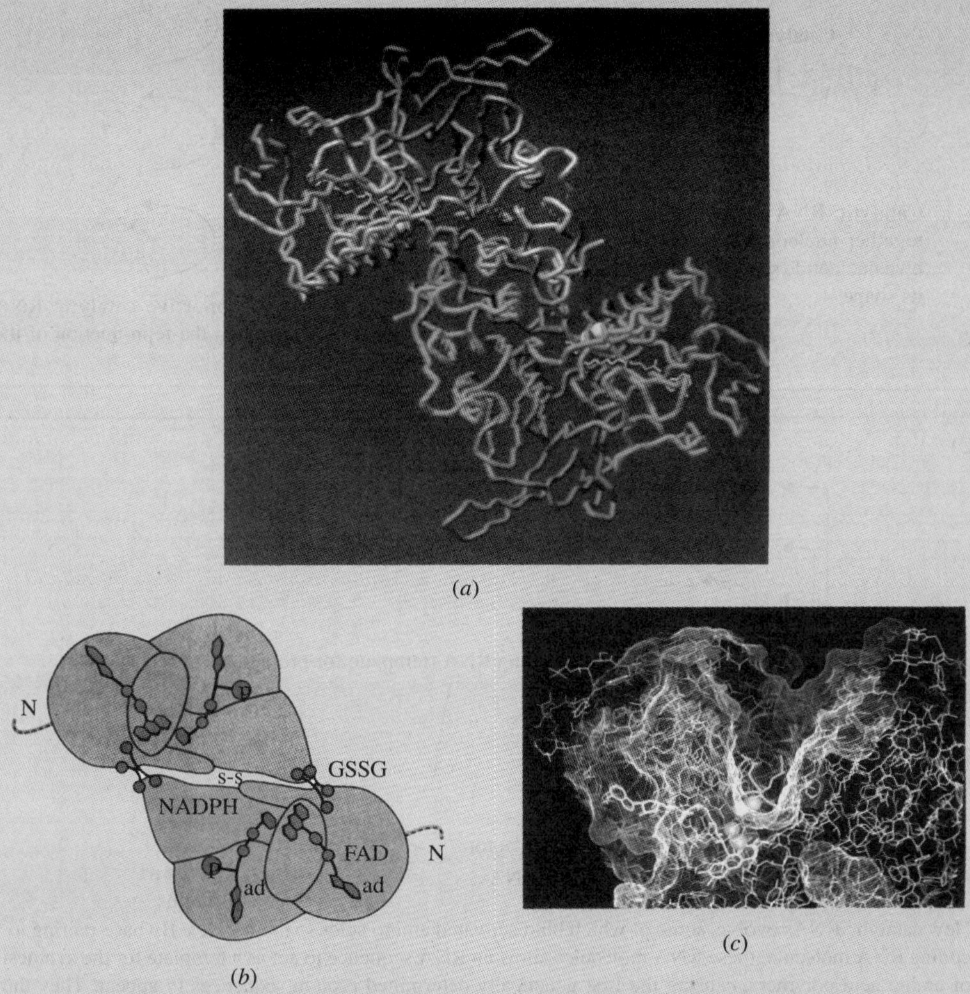

Fig. 30–40. The x-ray structure of the dimeric enzyme glutathione reductase, as viewed along the molecule's twofold axis of symmetry

(*a*) **The C_α backbone with the two identical subunits shown in different colours.** The S atoms of the redox-active disulfides are represented by *yellow* spheres and the FAD prosthetic groups are shown in *orange* (each flavin residue is near an active disulfide group).

(*b*) **An interpretive diagram of Part *a* showing how each subunit is organized into five domains.** The 18-residue N-terminal domain (*dashed lines*) is not visible in the x-ray structure presumably because it is flexibly linked to the rest of the protein. The binding sites of NADPH and GSSG [not shown in Part a], as well as those of FAD, are indicated. The two subunits are covalently linked by a disulfide bridge across the molecular twofold axis.

(*c*) **The active site region of glutathione reductase showing the FAD and GSSR positions.** Acidic residues (Asp and Glu) are *red*, basic residues (Arg, Lys, and His) are *blue* and all other residues are *white*. The dot surface, which is colored according to the nearest residue, represents the protein's solvent-accessible surface. The FAD has thick *yellow* bonds, the GSSR has thick *green* bonds, and the redox-active sulfur atoms of both the enzyme (*below*) and substrate (*above*) are represented by *green* spheres.

[*Courtesy: (a) Arthur Olson, (b) EF Pai and GE Schulz, and (c) John Kuriyan*]

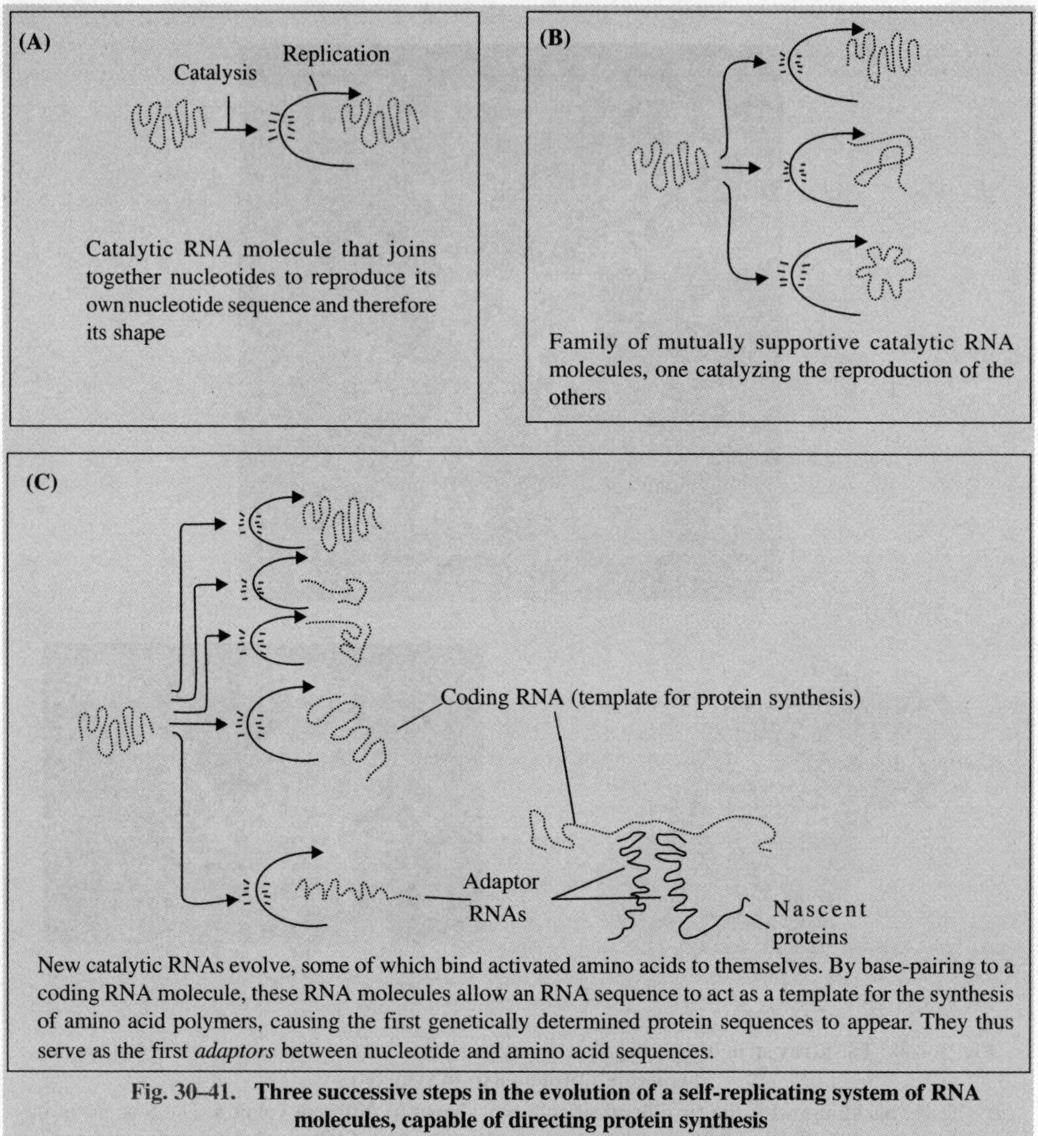

Fig. 30–41. Three successive steps in the evolution of a self-replicating system of RNA molecules, capable of directing protein synthesis

(*Adapted from Alberts B, Bray D, Lewis J, Raff M, Roberts K and Watson JD, 1994*)

Prior to the discovery of mRNA in the early 1960s, it was presumed that the large amount of RNA in ribosomes served a "messenger" function, carrying genetic information from DNA to proteins. However, it is now well known that all of the ribosomes in a cell contain an identical set of rRNA molecules that have no such information role. In bacterial ribosomes, rRNA molecules act as a catalyst in protein synthesis. As mentioned earlier, the major rRNA of the large ribosomal subunit appears to be the peptidyl transferase; besides, the rRNA of the small ribosomal subunit forms short base-paired helix with the initiation site sequence on bacterial mRNA molecules. Similarly, a variety of specific base-pair interactions form between tRNA molecules and bacterial rRNAs. This suggests that complex set of interactions take place that depend on the tertiary structure of the rRNA.

The process of protein synthesis relies heavily on a large number of proteins that are bound to the rRNAs in a ribosome. The complexity of a number of interacting components has made biochemists despair as to the understanding of the pathway by which protein synthesis evolved. The discovery that RNA molecules can act as enzymes, however, has given a new direction of viewing the pathway. Perhaps, early biological reactions used RNA molecules rather than protein molecules as catalysts. In the most primitive cells, tRNA molecules on their own may have formed catalytic surfaces that allowed them to bind and activate specific amino acids without requiring aminoacyl-tRNA synthetase enzymes. Similarly, rRNA molecules may have served by themselves as the entire "ribosome", folding up in complex ways to produce a complex set of surfaces that guided pairing of tRNA with mRNA codons as well as catalyzed the polymerization of the tRNA-linked amino acids (Fig. 30–41). During the course of evolution, individual proteins have been added to this complex machinery, each one making the process more accurate and efficient or adding regulatory controls. In this view, the large amount of RNA in present-day ribosomes is a remnant of a very early stage in evolution, before proteins dominated biological catalysis.

REFERENCES

1. **Altman S (editor) :** Transfer RNA. *MIT Press, Cambridge. 1978.*
2. **Altman S :** Transfer RNA processing enzymes. *Cell.* **23 :** *3, 1981.*
3. **Arnstein HRV (editor) :** Synthesis of amino acids and proteins vol. 7. *In* MTP International Review of Sciences. Biochemistry Section. *University Park Press, Baltimore, Md., 1975.*
4. **Aspirion D (editor) :** Processing of RNA. *CRC Press, Boca Raton. 1984.*
5. **Basavappa R, Sigler PB :** The 3 Å crystal structure of yeast initiator tRNA: Functional implications in initiator/elongator discrimination. *EMBO J.* **10 :** *3105–3111, 1991.*
6. **Berchtold H, Reshetnikova L, Reiser COA, Schirmer NK, Sprinzl M, Hilgenfeld R :** Crystal structure of active elongation factor Tu reveals major domain rearrangements. *Nature.* **365 :** *126–132, 1991.*
7. **Bermek E (editor) :** Mechanisms of Protein Synthesis. Structure-Function Relations, Control Mechanisms, and Evolutionary Aspects. *Springer-Verlag, New York. 1985.*
8. **Björk GR, Ericson JU, Gustafsson CED, Hagervall TG, Jönsson YH, Wikström PM :** Transfer RNA modification. *Ann. Rev. Biochem.* **56 :** *263–268, 1987.*
9. **Blobel G *et al* :** Translocation of proteins across membranes: the signal hypothesis and beyond. *Symp. Soc. Exp. Biol.* **33 :** *9–36, 1979.*
10. **Bradshaw RA :** Protein translocation and turnover in eukaryotic cells. *Microbiol. Rev.* **55** : *543–560, 1989.*
11. **Brimacombe R, Stöffler G, Wittman HG :** Ribosome structure. Ann. Rev. Biochem. **47 :** *217, 1978.*
12. **Burbaum JJ, Schimmel P :** Structural relationships and the classification of aminoacyl-tRNA synthetases. *J. Biol. Chem.* **266 :** *16965–16968, 1991.*
13. **Calender R, Gold L (editors) :** Sequence Specificity in Transcription and Translation. *Alan R. Liss, New York. 1985.*
14. **Carter CW Jr. :** Cognition, mechanism and evolutionary relationships in aminoacyl-tRNA synthetases. *Ann. Rev. Biochem.* **62 :** *715–748, 1993.*
15. **Caskey CT :** Peptide chain termination. *Trends Biochem. Sci.* **5 :** *234, 1980.*
16. **Cavarelli J, Moras D :** Recognition of tRNAs by aminoacyl-tRNA synthetases. *FASEB J.* **7 :** *79–86, 1993.*
17. **Cech TR :** RNA as an enzyme. *Sci. Amer.* **255(5) :** *64–75, 1986.*

18. **Chambliss G (editor) :** Ribosomes: Structure, Function and Genetics. *University Park Press, Baltimore. 1980.*

19. **Chambon P :** Eukaryotic nuclear RNA polymerases. *Ann. Rev. Biochem.* **44 :** *613, 1975.*

20. **Choe S, Bennett MJ, Fujii G, Curmi PM, Kantardjieff KA, Collier RJ, Eisenberg D :** The crystal structure of diphtheria toxin. *Nature.* **357 :** *216–222, 1992.*

21. **Chu C-Y, Rana TM :** Small RNAs: regulators and guardians of the genome. *J. Cell. Phys.* **213 :** *412–419, 2007.*

22. **Clark B :** The elongation of protein synthesis. *Trends Biochem. Sci.* **5 :** *207, 1980.*

23. **Collins FS, Morgan M, Patrinos A :** The Human Genome Project: lessons learned from large-scale biology. *Science.* **300 :** *286–290, 2003.*

24. **Craigen WJ, Lee CC, Caskey CT :** Recent advances in peptide chain termination. *Mol. Microbiol.* **4 :** *861–865, 1990.*

25. **Dahlberg AE :** The functional role of ribosomal RNA in protein synthesis. *Cell.* **57 :** *525–529, 1989.*

26. **Freifelder D :** Molecular Biology. *2nd edition. Jones and Bartlett. 1987.*

27. **Frydman J :** Folding of newly translated proteins *in vivo*: the chaperones. *Annu. Rev. Biochem.* **70 :** *603–647, 2001.*

28. **Gesteland RF, Atkins JF (editors) :** The RNA World. *Cold Spring Harbour Laboratory Press, New York. 1993.*

29. **Gluzman Y (editor) :** Eucaryotic Transcription. *Cold Spring Harbour Press, New York. 1985.*

30. **Grunberg-Manago M, Gros F :** Initiation mechanism of protein synthesis. Prog. Nucleic Acid Res. *Mol. Biol.* **20 :** *209, 1977.*

31. **Gualerzi CO, Pon CL :** Initiation of mRNA translation in prokaryotes, *Biochemistry.* **29 :** *5881–5889, 1990.*

32. **Hardesty B, Kramer G :** Structure, Function and Genetics of Ribosomes. *Springer-Verlag, New York. 1986.*

33. **Hershey JWB :** Overview: Phosphorylation and transport control. Enzyme. **44 (1-4) :** *17–27, 1990.*

34. **Hill WE, Dahlberg E, Garrett RA, Moore PB, Schlessinger D, Warner JR (editors) :** The Ribosome: Structure, Function, and Evolution. *The American Society for Microbiology, Washington, DC. 1990.*

35. **Holmes WW, Platt T, Rosenberg M :** Termination of transcription in *E.coli. Cell.* **32 :** *1029, 1983.*

36. **Hunt T :** The initiation of protein synthesis. *Trends Biochem. Sci.* **5 :** *178–181, 1980.*

37. **Hunt T, Prentis S, Tooze J (editors) :** DNA makes RNA makes Protein. *Elsevier, New York. 1983.*

38. **Hunter T, Karin M :** The regulation of transcription by phosphorylation. *Cell.* **70 :** *375–387, 1992.*

39. **Imal J, Yashiroda H, Maruya M, Yahara I, Tanaka K :** Proteasomes and molecular chaperones: cellular machinery responsible for folding and destructions of unfolded proteins. *Cell Cycle.* **2 :** *585–590, 2003.*

40. **International Human Genome Sequencing Consortium :** Finishing the euchromatic sequence of the human genome. *Nature.* **431 :** *931–945, 2004.*

41. **Jacques N, Dreyfus M :** Translation initiation in *Escherichia coli*: Old and new questions. *Mol. Microbiol.* **4 (7) :** *1063–1067, 1990.*

42. **Jiménez A :** Inhibitors of translation. *Trends Biochem. Sci.* **1** : *28–30, 1988.*
43. **Jukes T :** Evolution of the amino acid code: Interferences from mitochondrial codes. *J. Mol. Evol.* **19** : *219, 1983.*
44. **Kirkwood TBL, Rosenberger RF, Galas DJ (editors) :** Accuracy in Molecular Processes. Its Control and Relevance to Living Systems. *Chapman and Hall. 1986.*
45. **Kjeldgaard M, Nissen P, Thirup S, Nyborg J :** The crystal structure of elongation factor EF-Tu from Thermus aquaticus in the GTP conformation. *Structure* **1** : *35–50, 1993.*
46. **Kozak M :** Evolution of the 'Scanning Model' for initiation of protein synthesis in eucaryotes. *Cell.* **22** : *7, 1980.*
47. **Kozak M :** Comparison of initiation of P.S. in procaryotes, eucaryotes, and organelles. *Microbiol. Rev.* **47** : *1–45, 1983.*
48. **Kozak M :** Regulation of translation in eukaryotic systems. *Ann. Rev. Cell Biol.* **8** : *197–225, 1992.*
49. **Kurland CG :** Translational accuracy and the fitness of bacteria. *Ann. Rev. Genet.* **26** : *29–50, 1992.*
50. **Lake JA :** The ribosome. *Sci. Amer.* **245(2)** : *84–97, 1981.*
51. **Losick R, Chamberlin M (editors) :** RNA Polymerases. *Cold Spring Harbour, New York. 1976.*
52. **Maitra U, Stringer EA, Chaudhuri A :** Initiation factors in protein biosynthesis. *Ann. Rev. Biochem.* **51** : *869–900, 1982*
53. **Merrick WC :** Overview: Mechanism of translation initiation in eukaryotes. *Enzyme.* **44(1-4)** : *7–16, 1990.*
54. **Merrick WC :** Mechanism and regulation of eukaryotic protein synthesis. *Microbiol. Rev.* **56(2)** : *291–315, 1992.*
55. **Moldave K (editor) :** RNA and Protein Synthesis. *Academic Press Inc., New York. 1981.*
56. **Moldave K :** Eukaryotic protein synthesis. *Ann. Rev. Biochem.* **54** : *1109–1149, 1985.*
57. **Moldave K, Grossman L (editors) :** Nucleic Acids and Protein Synthesis. Parts G and H. Methods Enzymol. Vols. 59 and 60. *Academic Press Inc., New York. 1979.*
58. **Moller W :** Functional aspects of ribosomal proteins. *Biochimie.* **73** : *1093–1100, 1991.*
59. **Moore P :** Ribosomes. *Curr. Opin. Struct. Biol.* **1** : *258–263, 1991.*
60. **Moras D :** Structural and functional relationships between aminoacyl-tRNA synthetases. *Trends Biochem. Sci.* **17** : *159–164, 1992.*
61. **Nathans D :** Puromycin inhibition of protein synthesis: Incorporation of puromycin into peptide chains. *Proc. Natn. Acad. Sci., USA.* **51** : *485, 1964.*
62. **Nierhaus KH :** The allosteric three-site model for the ribosomal elongation cycle: Features and future. *Biochemistry.* **24** : *4997–5008, 1990.*
63. **Noller HF :** RNA structure: reading the ribosome. *Science.* **309** : *1508–1514, 2005.*
64. **Noller HF :** Structure of ribosomal RNA. *Ann. Rev. Biochem.* **53** : *119, 1984.*
65. **Noller HF :** Ribosomal RNA and translation. *Ann. Rev. Biochem.* **60** : *191–227, 1991.*
66. **Noller HF :** tRNA-rRNA interactions and peptidyl transferase. *FASEB J.* **7** : *87–89, 1993.*
67. **Noller HF :** Peptidyl transferase: protein, ribonucleoprotein or RNA? *J. Bacteriol.* **175(17):** *5297–5300, 1993.*
68. **Nomura M :** The control of ribosome synthesis. *Sci. Amer.* **250(1)** : *102–114, 1984.*
69. **Nomura M, Tissieres A, Lengyel P (editors) :** Ribosomes. *Cold Spring Harbour, New York. 1974.*
70. **Normanly J, Abelson J :** Transfer RNA identity. *Ann. Rev. Biochem.* **58** : *1029–1049, 1989.*

71. **Olsen GJ, Woese CR :** Ribosomal RNA: A key to phylogeny. *FASEB J. 7 : 113–123, 1993.*
72. **Perkins ND :** Post-translational modifications regulating the activity and function of the nuclear factor kappa B pathway. *Oncogene.* **25 :** *6717–6730, 2006.*
73. **Powers T, Noller HF :** Evidence for functional interaction between elongation factor Tu and 16S ribosomal RNA. *Proc. Nat. Acad. Sci.* **90 :** *1364–1368, 1993.*
74. **Proud CG :** Protein phosphorylation in translational control. *Curr. Top. Cell Regul.* **32 :** *243–369, 1992.*
75. **Rich A :** Polyribosomes. *Sci. Amer.* **209(6) :** *44–53, 1963.*
76. **Rich A :** Transfer RNA: Structure and biological function. *Trends Biochem. Sci.* **3 :** *34, 1978.*
77. **Rich A, Kim SH :** The three-dimensional structure of transfer RNA. *Sci. Amer.* **238(1) :** *52–62, 1978.*
78. **Riis B, Rattan SI, Clark BF, Merrick WC :** Eukaryotic protein elongation factors. *Trends Biochem. Sci.* **15(11) :** *420–424, 1990.*
79. **Roberts JM :** Turning DNA replication on and off. Curr. *Opin. Cell Biol.* **5 :** *201–206, 1993.*
80. **Rodina MV, Beringer M, Wintermeyer W :** How ribosomes make peptide bonds. *Trends Biochem. Sci.* **32 :** *20–26, 2006.*
81. **Sachs AB :** Messenger RNA degradation in eukaryotes. *Cell.* **74 :** *413–421, 1993.*
82. **Samuel CE :** The eIF-2 alpha protein kinases, regulators of translation in eukaryotes from yeasts to humans. *J. Biol. Chem.* **268 :** *7603–7606, 1993.*
83. **Schimmel P :** Parameters for the molecular recognition of transfer RNAs. *Biochemistry.* **28 :** *2747–2759, 1989.*
84. **Schimmel P, Söll D, Abelson J (editors) :** Transfer RNA: Structure, Properties and Recognition. *Cold Spring Harbour, New York. 1979.*
85. **Sentenac A :** Eukaryotic RNA polymerases. CRC Crit. *Rev. Biochem.* **18 :** *31–91, 1985.*
86. **Shatkin AJ :** mRNA cap binding proteins: Essential factors for initiating translation. *Cell.* **40 :** *223–224, 1985.*
87. **Söll D:** The accuracy of aminoacylation– ensuring the fidelity of the genetic code. *Experimentia.* **46(11-12) :** *1089–1096, 1990.*
88. **Söll D, Abelson J, Schimmel P (editors) :** Transfer RNA: Biological Aspects. *Cold Spring Harbour, New York. 1980.*
89. **Spirin AS :** Ribosomal translocation: Factors and Models. Prog. Nucleic Acid Res. *Mol. Biol.* **32 :** *75, 1985.*
90. **Sprin AS :** Ribosome Structure and Protein Biosynthesis. *Benjamin/Cummings, Menlo Park, California. 1986.*
91. **Sprinzl M :** Elongation factor Tu: A regulatory GTPase with an integrated effector. *Trends Biochem. Sci.* **19 :** *245–250, 1994.*
92. **Stewart PR, Letham DS (editors) :** The Ribonucleic Acids. *2nd ed. Springer-Verlag, New York. 1977.*
93. **Szekely M :** From DNA to Protein. *John Wiley and Sons, New York. 1980.*
94. **Tai P-C, Wallace BJ, Davis BD :** Streptomycin causes misreading of natural messenger by interacting with ribosomes after initiation. *Proc. Nat. Acad. Sci.* **75 :** *275–279, 1978.*
95. **Varshavsky A :** The N-end rule. *Cell.* **69 :** *725–735, 1992.*
96. **Von Hippel PH, Bear DG, Morgan WD, McSwiggen JA :** Protein–nucleic acid interactions in transcription: A molecular analysis. *Ann. Rev. Biochem.* **53 :** *389, 1984.*

97. **Walsh CT :** Posttranslarional Modification of Proteins: Expanding Nature's Inventory. *3rd ed. Roberts & Company, Colorado. 2007.*
98. **Watson JD :** The involvement of RNA in the synthesis of proteins. *Science.* **140 :** *17–26, 1963.*
99. **Yonath A :** Three dimensional crystals of ribosomal particles. *Trends Biochem. Sci.* **9 :** *227, 1984.*
100. **Young RA :** RNA polymerase II. *Ann. Rev. Biochem.* **60 :** *689–715, 1991.*
101. **Zamecnik P :** The machinery of protein synthesis. *Trends Biochem. Sci.* **9 :** *464–466, 1984.*

PROBLEMS

1. What is the smallest number of molecules of ATP and GTP consumed in the synthesis of a 200-residue protein, starting from amino acids? Assume that the hydrolysis of PP_i is equivalent to the hydrolysis of ATP for this calculation.
2. EF-Tu, a member of the G-protein family, plays a crucial role in the elongation process of translation. Suppose that a slowly hydrolyzable analogue of GTP were added to an elongating system. What would be the effect on rate of protein synthesis?
3. What is the nucleophile in the reaction catalyzed by peptidyl transferase? Write out a plausible mechanism for this reaction.
4. Which protein in G-protein cascades plays a role similar to that of elongation factor Ts?
5. Eukaryotic elongation factor 2 is inhibited by ADP ribosylation catalyzed by diphtheria toxin. What other G proteins are sensitive to this mode of inhibition?
6. The E site may not require codon recognition. Why?
7. Suppose that the probability of making a mistake in translation at each translational step is a small number, δ. Show that the probability, p, that a given protein molecule, containing n residues, will be completely error-free is $(1 - \delta)^n$.
8. Assume that the translational error frequency, δ, is 1×10^{-4}.
   (*a*) Calculate the probability of making a perfect protein of 100 residues.
   (*b*) Repeat for a 1,000-residue protein.
9. A given sequence of bases in an mRNA will code for one and only one sequence of amino acids in a polypeptide, if the reading frame is specified. From a given sequence of amino acid residues in a protein such as cytochrome *c*, can we predict the base sequence of the unique mRNA that coded for it? Give reasons for your answer.
10. The chemical mechanisms used to avoid errors in protein synthesis are different from those used during DNA replication. DNA polymerases utilize a 3′→5′ exonuclease proofreading activity to remove mispaired nucleotides incorrectly inserted into a growing DNA strand. There is no analogous proofreading function on ribosomes; and, in fact, the identity of amino acids attached to incoming tRNAs and added to the growing polypeptide is never checked. A proofreading step that hydrolyzed the last peptide bond formed when an incorrect amino acid was inserted into a growing polypeptide (analogous to the proofreading step of DNA polymerases) would actually be chemically impractical. Why ? (Hint : Consider how the link between the growing polypeptide and the mRNA is maintained during the elongation phase of protein synthesis; see Figs. 30–23 and 30–24).
11. Although the Shine–Dalgarno sequences vary considerably in different genes, they include examples like GAGGGG that could serve as code– in this case, for Glu–Gly. Does this imply that the sequence Glu–Gly cannot ever occur in a protein, lest it be read as a Shine–Dalgarno sequence? Speculate.

ANSWERS TO PROBLEMS

1. Two hundred molecules of ATP are converted into AMP + 400 P_i to activate the 200 amino acids, which is equivalent to 400 molecules of ATP. One molecule of GTP is required for initiation, and 398 molecules of GTP are needed to form 199 peptide bonds.
2. The rate would fall because the elongation step requires that the GTP be hydrolyzed before any further elongation can take place.
3. The nucleophile is the amino group of the aminoacyl-tRNA. This amino group attacks the carbonyl group of the ester of peptidyl-tRNA to form a tetrahedral intermediate, which eliminates the tRNA alcohol to form a new peptide bond.
4. EF-Ts catalyzes the exchange of GTP for GDP bound to EF-Tu. In G-protein cascades, an activated 7TM receptor catalyzes GTP–GDP exchange in a G protein.
5. The α subunits of G proteins are inhibited by a similar mechanism in cholera and whooping cough.
6. Because the tRNA is going to be released from the E-site, codon recognition and pairing with the anticodon are in no way advantageous and may, in fact, be detrimental to smooth release.
7. If δ is the probability that an error is committed at each step, then 1-δ is the probability that an error has *not* been made in any one step. The probability that no error has been made in *any* of the *n* steps is then $(1 - \delta)^n$. This is the probability and the protein is entirely error-free.
8. (*a*) 0.990.

   (*b*) 0.904. In this case, nearly 10% of the proteins would contain one or more errors.
9. No; because nearly all the amino acids have more than one codon (*e.g.*, Leu has six), any given polypeptide can be coded for by a number of different base sequences. However, because some amino acids are encoded by only one codon and those with multiple codons often share the same nucleotide at two of the three positions, *certain parts* of the mRNA sequence encoding a protein of known amino acid sequence can be predicted with high certainty.
10. The amino acid most recently added to a growing polypeptide chain is the only one covalently attached to a tRNA and hence is the only link between the polypeptide and the mRNA that is encoding it. A proofreading activity would sever this link, halting synthesis of the polypeptide and releasing it from the mRNA.
11. The Shine–Dalgarno sequence, like the AUG that is read uniquely as fMet, occurs near the 5′ end of the message. The implication is that, in forming the initiation complex, only a region near the 5′ end of mRNA can bind. That is, a Shine–Dalgarno sequence occurring by chance within a message could not bind at this site to form an initiation complex. Other explanations may be possible.

CHAPTER

31 Protein Targeting and Degradation

CONTENTS

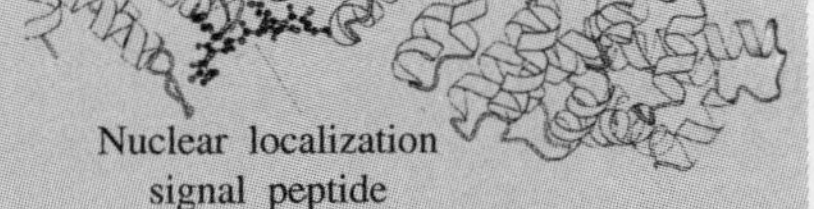

The structure of the nuclear localization signal-binding protein α-karyopherin (or α-importin) with a nuclear localization signal peptide bound to its major recognition site

GENERAL CONSIDERATIONS

Protein turnover– that is synthesis and degradation– occurs constantly in eukaryotic cells but it is a highly selective process with different rates of turnover for various proteins. Turnover of proteins can control the level of certain enzymes, furnish amino acids in times of need and degrade faulty or damaged proteins that are generated during synthesis or arise from deleterious activities in the cell. Nascent proteins contain signals that determine their ultimate destination. A newly synthesized protein in the prokaryotic *Escherichia coli* cell, for example, can stay in the cytosol or it can be sent to the plasma membrane, the outer membrane, the space between them, or the extracellular medium.

The eukaryotic cell is made up of many structures, compartments and organelles, each with specific functions requiring different types of proteins and enzymes. The synthesis of most of these proteins begins on free ribosomes in the cytosol. Therefore, eukaryotic cells must direct proteins to internal sites such as lysosomes, mitochondria, chloroplasts, nucleus etc. How then is sorting accomplished? In eukaryotes, a key choice is made soon after the synthesis of a protein begins. A ribosome remains free in the cytosol unless it is directed to the endoplasmic reticulum (ER) by a signal sequence in the protein being synthesized. Nascent polypeptide chains formed by ribosomes are translocated across the ER membrane. In the lumen of the ER, many of them are glycosylated and modified in other ways. These are then transported to Golgi complex where they are further modified. Finally, they are sorted for delivery to lysosomes, secretory vesicles, and the plasma membrane. Transported proteins are carried by vesicles that bulge out from donor compartments and fuse with target compartments. *The signals used to target eukaryotic proteins for transfer across the ER membrane are ancient*, for bacteria also use similar sequences or signals for sending proteins to their plasma membrane and to secrete them.

The transported proteins must reach their assigned cellular locations. This is of utmost importance because mistakes in transport can severely affect cellular metabolism and the cumulative effect may prove fatal to an organism. For instance, *I-cell disease*, a rare human disease, is characterized by export from the cell of at least 8 enzymes which should be transported to lysosomes.

FREE AND MEMBRANE-BOUND RIBOSOMES

In eukaryotic cells, one may distinguish free ribosomes in the cytosol and ribosomes bound to the membranes of the endoplasmic reticulum (ER). The **cytosolic ribosomes** are responsible for the a synthesis of the proteins which will be released in the cytosol, whereas the **membrane-bound ribosomes** (Fig 31–1) synthesize 3 major classes of proteins: secretory proteins (which are secreted outside the cell), lysosomal proteins and proteins spanning the plasma membrane. *ER amounts to about half of the total membrane of a cell*. The region of ER that binds ribosomes is called the **rough ER** because of its beaded appearance in comparison to the **smooth ER,** which is devoid of ribosomes. Virtually all integral membrane proteins of the cell, barring those located in the mitochondrial and chloroplast membranes, are formed by ribosomes bound to the ER. George Palade's (1975-Nobel Laureate) pioneering studies on the mechanism of secretion of zymogens by the pancreatic acinar cells opened a new field of enquiry– protein targeting– and delineated the pathway taken by the secretory proteins.

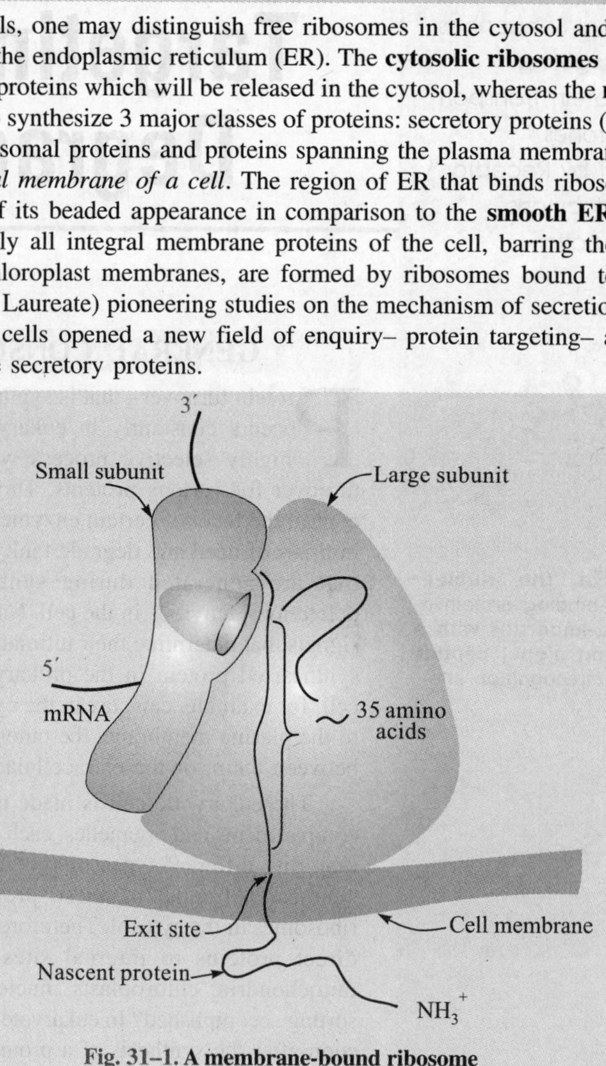

Fig. 31–1. A membrane-bound ribosome

(*After James A. Lake, 1981*)

		Cleavage site: ↓	
Human influenza virus A :	Met Lys Ala Lys Leu Leu Val Leu Leu Tyr Ala Phe Val Ala Gly	↓	Asp Gln --
Human preproinsulin* :	Met Ala Leu Trp Met Arg Leu Leu Pro Leu Leu Ala Leu Leu Ala Leu Trp Gly Pro Asp Pro Ala Ala Ala	↓	Phe Val --
Bovine growth hormone :	Met Met Ala Ala Gly Pro Arg Thr Ser Leu Leu Leu Ala Phe Ala Leu Leu Cys Leu Pro Trp Thr Gln Val Val Gly	↓	Ala Phe --
Bee promellitin :	Met Lys Phe Leu Val Asn Val Ala Leu Val Phe Met Val Val Tyr Ile Ser Tyr Ile Tyr Ala	↓	Ala Pro --
Drosophila glue protein :	Met Lys Leu Leu Val Val Ala Val Ile Ala Cys Met Leu Ile Gly Phe Ala Asp Pro Ala Ser Gly	↓	Cys Lys --
Zea maize protein 19:	Met Ala Ala Lys Ile Phe Cys Leu Ile Met Leu Leu Gly Leu Ser Ala Ser Ala Ala Thr Ala	↓	Ser Ile --

Fig. 31–2. Amino-terminal signal sequences of some eukaryotic proteins (secretory and plasma membrane), directing translocation into the endoplasmic reticulum

The hydrophic core (in *yellow*) is preceded by one or more basic residues (*blue*). Note the presence of polar and short-side-chain residues immediately preceding (to the left of, as shown here) the cleavage sites (indicated by *red arrows*).

* Insulin is synthesized as preproinsulin and the prefix *pre* refers to the 24-residue signal sequence, preceding the cleavage site. Note that leaving Arg, Gly and Asp, the remaining 21 residues have hydrophobic side chains (a higher percentage than is found in most signal sequences).

Studies conducted on the protein-synthesizing activities of ribosomes in cell-free systems confirmed that *the membrane-bound ribosomes and free cytosolic ribosomes are intrinsically identical.* In fact, free ribosomes from the cytosol were isolated and then added to rough ER membranes that had been stripped of their ribosomes. This reconstituted system actively synthesized secretory proteins when supplied with the proper mRNAs and other soluble factors. Similarly, ribosomes isolated from the rough ER were fully active in synthesizing proteins that are normally freed into the cystosol. Whether a particular ribosome is free or attached to the rough ER depends only on the kind of protein it is synthesizing.

SIGNAL HYPOTHESIS

The pathways by which proteins are sorted and transported to their proper cellular locations are referred to as *protein targeting pathways*. A characteristic feature of these targeting pathways (with the exception of cytosolic and nuclear proteins) is the presence of a short amino acid sequence at the amino terminus of a newly synthesized polypeptide called the **signal sequence** or **signal peptide**. In many cases, the targeting capacity of particular signal sequences has been confirmed by fusing the signal sequence from one protein, say protein A, to a different protein B, and showing that the signal directs protein B to the location where protein A is normally found. The signal sequence, whose function was first postulated by **David Sabatini** and **Günter Blobel** (1970), directs a protein to its proper location in the cell and is removed by a *signal peptidase* during transport or when the protein reaches its final destination. Obviously, the signal sequence is absent in the protein once secreted.

At present, the signal sequences of more than 100 secretory proteins from a wide variety of eukaryotic species have been determined; some of which have been presented in Fig. 31–2. A well-defined consensus sequence such as the TATA box guiding the initiation of transcription, is not evident. However, signal sequences do exhibit certain common characteristics:

1. They range in length from 13 to 36 amino acid residues.
2. The amino terminal part of the signal contains at least one or more positively charged amino acid residues, preceding the hydrophobic sequence.
3. A sequence of highly hydrophobic amino acids (10 to 15 residues long) forms the centre of the signal sequence. Ala, Val Leu, Ile and Phe residues are common in this region.
4. There is present a region of more polar short sequence (of about 5 residues) at the carboxyl terminus, upstream the cleavage site. The amino acid residues having short side chains (*esp*, Ala) predominate in this region at positions closest to the cleavage site.

However, in certain secretory and plasma membrane proteins, the signal sequence is not situated at the amino terminus. These proteins contain an **internal signal sequence** that serves the same role. For example, in the case of ovalbumin, the sequence is located between residues 22 and 41 and is critical for the transfer of nascent albumin across the ER membrane.

In 1975, **George Palade**, at the Rockefeller Institute in New York, demonstrated that proteins with these signal sequences are synthesized on ribosomes attached to the ER membrane. The overall pathway, summarized in Fig. 31–3, proceeds in following 8 steps:

1. First of all, the ribosomal subunits assemble in an initiation complex at the initiation codon and begin protein synthesis.

2. Later, a proper signal sequence appears early in the synthetic process because it is at the amino terminus of the nascent polypeptide.

3. Then, this signal sequence and the ribosome itself are rapidly bound by a large rod-shaped complex called **signal recognition particle (SRP).** This binding event halts elongation and the signal sequence has completely emerged from the ribosome. The SRP receptor is a heterodimer of α (M*r* 69,000) and β (M*r* 30,000) subunits and consists of a 305-nucleotide RNA (called 7 SL-RNA) and 6 different proteins, with a combined molecular weight of 3,25,000. One protein subunits binds directly to the signal sequence, inhibiting elongation by sterically blocking entry of aminoacyl-tRNAs and inhibiting peptidyl transferase.

4. The ribosome-SRP complex with the incomplete polypeptide is bound by two receptors (ribosome receptor and SRP receptor) present on the cytosolic face of the ER. For transport of a polypeptide into the ER lumen, the signal sequence attaches to the **SRP receptor**. The hydrophobicity of the signal sequence is postulated to be the molecular key for the polypeptide's interaction with the ER membrane, which is also a hydrophobic structure. The second recognition site, **ribosome receptor**, serves to anchor the organelle (ribosome) to the ER membrane. The interaction between the signal sequence and the ER membrane is believed to open a channel in the membrane through which the polypeptide is transported into the ER lumen. Thus, the molecular instructions for transport into the ER (in the form of a hydrophobic sequence) are furnished by the polypeptide.
5. The SRP dissociates and is recycled.
6. Protein synthesis then resumes, along with translocation of the polypeptide chain into the lumen of the ER. The nascent polypeptide is delivered to a **peptide translocation complex** in the ER. The translocation complex feeds the growing polypeptide chain into the lumen of the ER in a reaction driven by the energy of ATP.
7. The signal sequence is cleaved by a membrane enzyme, **signal peptidase** which is located on the lumenal side of the ER.

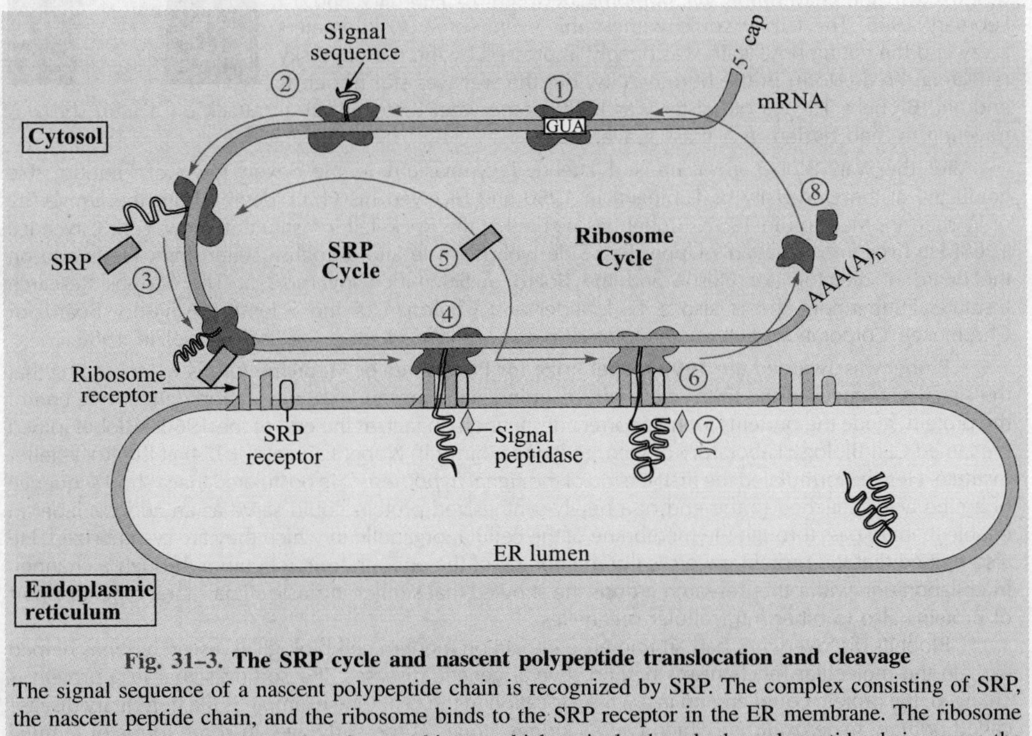

Fig. 31–3. The SRP cycle and nascent polypeptide translocation and cleavage

The signal sequence of a nascent polypeptide chain is recognized by SRP. The complex consisting of SRP, the nascent peptide chain, and the ribosome binds to the SRP receptor in the ER membrane. The ribosome is then transferred to the translocation machinery, which actively threads the polypeptide chain across the ER membrane. SRP released from its receptor is free to bind another emerging signal sequence. The steps conform to the description in the text.

(Adapted from Lehninger AL, Nelson DL and Cox MM, 1993)

8. Once the complete protein has been synthesized, the ribosome dissociates from the ER and is recycled.

The proteins to be secreted and the lysosomal proteins completely pass through the membrane of the ER. On the contrary, other proteins must form part of a membrane. Such proteins, in the lumen of the ER, are modified in several ways. Besides the removal of signal sequences, polypeptide chains fold and disulfide bonds form. Many proteins are also glycosylated.

As a result of about 20 years of strenuous work, **Günter Blobel** formulated, in 1980, general principles for the sorting and targeting of proteins to particular cell compartments. Each protein carries in its structure the information needed to specify its proper location in the cell. Specific amino acid sequences (**topogenic signals**) determine whether a protein will pass through a membrane into a particular organelle, become integrated into the membrane, or be exported out of the cell. In essence, the **signal hypothesis** may be summarized below:

GÜNTER BLOBEL

(born, 1936)

Blobel, a cell and molecular biologist at the Rockefeller Institute in New York, was born on May 21, 1936 in Waltersdorf in the Prussian Province of Lower Silesia. In January 1945, his family fled from native Silesia from the advancing Red Army. On their way to the West, they passed through the beautiful old city of Dresden, which left deep impressions in the young boy. Only days after their stay, Dresden was destroyed in the catastrophic air bombing between 13 February and 15 February 1945. The family could witness this event some 30 kilometers away and the young boy again was deeply impressed by the red night sky reflecting the firestorm in the burning city, but the war was still not at its end and Blobel's 19-year-old sister was killed some weeks later in an air attack on a train she was traveling in, and buried in a mass grave.

After the war, Blobel grew up and attended gymnasium in the Saxon town of Freiburg. He graduated at the University of Tübingen in 1960 and received his Ph.D. degree from the University of Wisconsin, Madison in 1967. Blobel has worked at the Rockefeller Institute in New York City since 1968. He lives in Manhattan's Upper East Side with his wife and 3 English setter dogs. He is also on the board of directors for Nestlé and the Board of Scientific Governors at The Scripps Research Institute. Furthermore, he is also a Co-founder and Chairman of the Scientific Advisory Board for Chromocell Corporation. He was appointed to the Howard Hughes Medical School in 1986.

Blobel was awarded the **1999 Nobel Prize for Physiology or Medicine** for his work proving that the signal sequences in the form of a chain of amino acids present either as a short 'tail' at one end of the protein, guide the proteins to their correct destination. In fact, at the end of the 1960s, Blobel joined the famed Cell Biology Laboratory of **George Palade** (himself, Nobel Laureate, 1974) at the Rockefeller Institute. Here he formulated the first version of the *signal hypothesis*. He postulated that a short sequence of amino acids attached to the end of a newly-synthesized protein could serve as an address label to enable them to pass through the membrane of the cellular organelle in which they are sysnthesized. He also opined that the protein traverses the membrane of the endoplasmic reticulum through a channel. In collaboration with other research groups, he showed that similar intrinsic signals target the transport of proteins also to other intracellular organelles.

Blobel's discovery has had an immense impact on modern cell biological research. It has helped explain the molecular mechanisms behind several genetic diseases. If a sorting signal in a protein is changed, the protein could end up in a wrong location in the cell. One example is the hereditary disease called *primary hyperoxaluria*, which causes kidney stones at the early age. In some forms of *familial hypercholesterolemia*, a very high level of cholesterol in the blood is due to deficient transport signals. Other hereditary diseases, *e.g.*, *cystic fibrosis* are caused by the fact that proteins do not reach their proper destination.

Blobel is also well known for his direct and active support for the rebuilding of **Dresden** in Germany, becoming, in 1994, the founder and of the nonprofit "Friends of Dresden, Inc." He *donated all of the Nobel award money* to the restoration of Dresden, in particular for the rebuilding of the *Frauenkirche* (completed in 2005) and the building of a new *synagogue*. In Leipzig, he pursued a rebuilding of the *Paulinerkirche* which had been blown up by the communist regime of East Germany in 1968, arguing:

"This is much more than a church– this is a shrine of German cultural history, connected to the most important names in German cultural history."

Besides the Nobel Prize, Blobel also received many awards and honours; a few of them are as follows:

1. Gairdner Foundation International Award (1982)
2. Louisa Gross Horowitz Prize from Columbia University (1987)
3. Albert Lasker Award for Basic Medical Research (1993)
4. Ciba Drew Award in Biomedical Research (1995)
5. Mayor's Award for Excellence in Science and Technology (1997)
6. Pontifical Academy of Sciences (2001)

Blobel's famous quote reads as follows:

"You are one, they are many. If you don't work hard night and day, they will crush you like a roach."

Proteins which are to be exported out of the cell are synthesized by ribosomes, associated with the ER. The genetic information from DNA is transferred *via* RNA. This information determines how the amino acids build up the proteins. First, a signal peptide is formed as a part of the protein. With the help of binding proteins, the signal peptide directs the ribosome to a channel in the ER. The growing protein chain penetrates the channel, the signal peptide is cleaved and the completed protein is released into the lumen of ER. The protein is subsequently transported out of the cell.

In fact, the signal hypothesis explains how new polypeptides scheduled for intracellular transport are routed into the ER lumen. The signal hypothesis was originally proposed for the transport of secretory proteins. But it is also applicable to storage proteins. An important feature of this hypothesis is that the membrane transport of the protein depends on the simultaneous protein synthesis by the membrane-bound ribosomes, thus causing the polypeptide to migrate through the tunnel in the endoplasmic reticular membrane. Thus, it may be emphasized that the signal hypothesis is both correct and universal, since the various processes associated with it operate in the same way in yeast, plant and animal cells.

GLYCOSYLATION OF PROTEINS AT THE LEVEL OF ER

While the individual polypeptides are in the ER lumen, the biochemical processes start for their cellular distribution. Whereas hydrophobicity provides the first molecular instructions for intracellular transport, it is **glycosylation** (*i.e.*, addition of carbohydrates) that establishes the molecular patterns acquired by polypeptides to continue their intracellular routing (Armstrong, 1989). The addition of oligosaccharide units, which convert polypeptides into glycoproteins, commences in the ER lumen and continues when they are transported from the ER to the Golgi apparatus. The particular oligosaccharide unit, attached to a glycoprotein, furnishes the molecular instructions for its cellular destination. Acquisition of oligosaccharide units by polypeptides may be compared to the assignment of Pin Codes to mailing addresses with each type of oligosaccharide, representing a distinct Pin Code.

Glycosylation begins soon after a nascent polypeptide enters the ER lumen. Carbohydrates bind to either the amide group of an asparagine (Asn) or the hydroxyl group of a serine (Ser) or threonine (Thr). Oligosaccharides attached to asparaginyl residues are referred to as **N-linked** and those to seryl or threonyl residues as **O-linked.** The following discussion will make it amply clear that, in the case of N-linked glycosylation, the molecular instructions dictating which oligosaccharide unit a protein will attain reside in the sequence and composition of the protein. For example, which asparaginyl residues will be glycosylated and which of the diverse oligosaccharides it will bear.

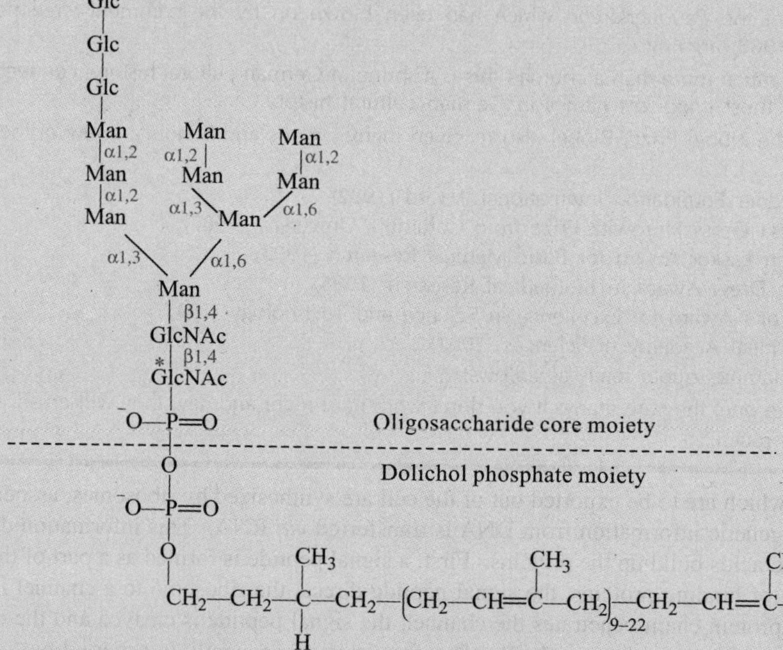

Fig. 31–4. Activated oligosaccharide core

Dolichol phosphate is a highly hydrophobic lipid carrier, whose terminal phosphate group is the site of attachment of the activated oligosaccharide. Note that the first carbohydrate, *N*-acetylglucosamine, GlcNAc (indicated by a *red asterisk*) is added to the dolichol phosphate moiety as a phosphorylated derivative.

A. Core Glycosylation

Carbohydrate processing in the ER is called **core glycosylation** to distinguish it from *terminal glycosylation* (described in the subsequent Section), which takes place in the Golgi complex. In the ER lumen, an N-linked oligosaccharide is not added to a polypeptide by a series of one-carbohydrate addition, but instead as an intact unit, called the *common oligosaccharide* core (Fig. 31–4), consisting of 14 residues (2 N-acetylglucosamine + 9 mannose + 3 glucose residues). However, this oligosaccharide core is constructed by the successive addition of single monosaccharide units to dolichol phosphate (Fig 31–4). Dolichol is an unusually long-chain lipid, containing from 9 to 22 isoprene units. Phosphorylation of dolichol at the nonolefinic end produces dolichol phosphate. Dolichol phosphate is used to carry activated sugars in the membrane-associated synthesis of glycoproteins and some polysaccharides.

When the oligosaccharide core is completely synthesized on dolichol phosphate moiety, the whole structure is now called the *activated* oligosaccharide core (Fig. 31–4). The synthesis of activated oligosaccharide core and its transfer to the protein in the ER is depicted in Fig. 31–5. Once this oligosaccharide core is completely synthesized, it is enzymatically transferred en bloc from dolichol phosphate to a specific asparagine residue of the growing polypeptide chain. The enzyme *transferase* is located on the lumenal face of the ER and thus does not catalyze glycosylation of cytosolic proteins. An asparagine residue can accept the oligosaccharide only if it is a part of an Asn-X-Ser or Asn-X-Thr sequence, and if it is sterically accessible to the transferase. Dolichol

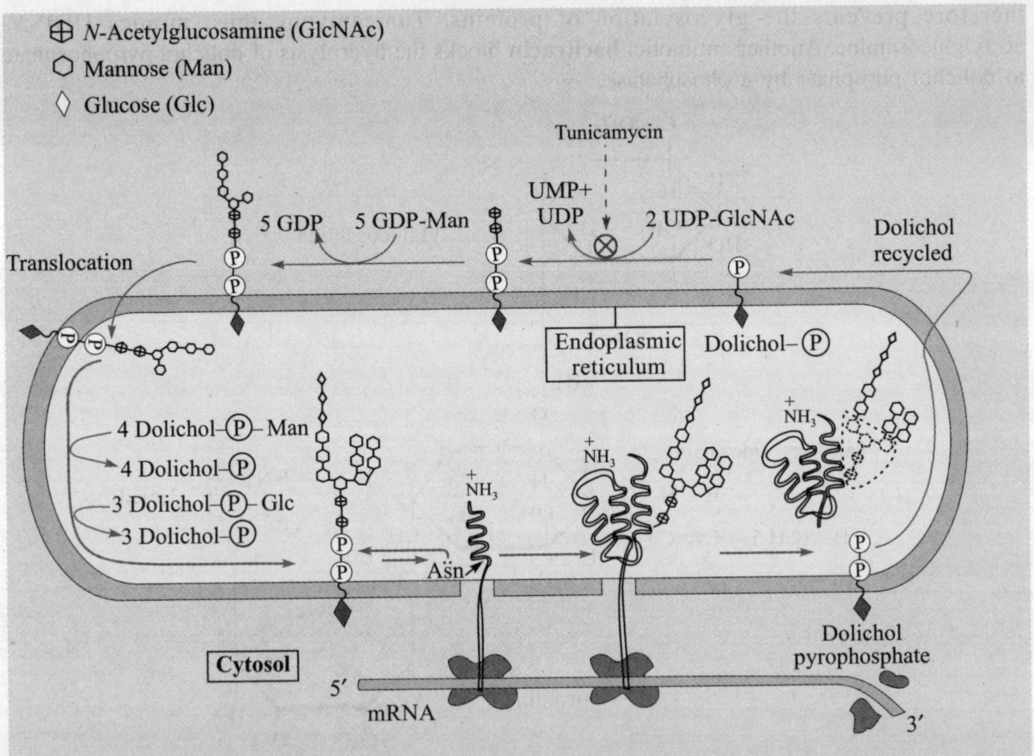

Fig. 31–5. Synthesis of the oligosaccharide core of glycoproteins and its transfer to the protein in the endoplasmic reticulum

The oligosaccharide core is built up in a series of steps as shown. The first few steps occur on the cytosolic face of the ER. Completion occurs within the lumen of the ER after a translocation step (upper left) in which the incomplete oligosaccharide is moved across the membrane. The synthetic precursors that provide additional mannose and glucose residues to the growing oligosaccharide in the ER lumen are themselves dolichol phosphate derivatives. Dolichol-phosphate-mannose and dolichol-phosphate-glucose are synthesized from dolichol phosphate and GDP-mannose or UDP-glucose, respectively. After it is transferred to the protein, the oligosaccharide core is further modified in the ER and the Golgi complex in pathways that differ for different proteins. The released dolichol pyrophosphate is recycled. The 5-sugar residues (shown in a dotted enclosure on lower right side) are retained in the final structure of all N-linked oligosaccharides.

(Adapted from Lehninger AL, Nelson DL and Cox MM, 1993)

pyrophosphate, released in the transfer of the oligosaccharide to the protein, is recycled to dolichol phosphate by the action of a *phosphatase*. After the transfer, the oligosaccharide core is trimmed (*i.e.*, carbohydrates removed) in the ER but all linked oligosaccharides retain a pentasaccharide core derived from the original 14-residue oligosaccharide. Trimming continues when the polypeptides are transferred to the Golgi apparatus. For some polypeptides, trimming produces the required oligosaccharide units; for others, trimming and subsequent addition of new carbohydrates are needed for these polypeptides to acquire their characteristic glycosylated patterns. It is in the Golgi apparatus that most of the final trimming and additions take place (called terminal glycosylation).

Several antibiotics interfere with one or more steps in the core glycosylation process. The best-characterized is **tunicamycin**, which blocks the first step (*i.e.*, addition of *N*-acetylglucosamine to dolichol phosphate). Tunicamycin (Fig. 31–6) is a hydrophobic analogue of UDP-*N*-acetylglucosamine which blocks the fixation of *N*-acetylglucosamine on dolichol phosphate, and therefore prevents the glycosylation of proteins. Tunicamycin, thus, mimics UDP-*N*-acetylglucosamine. Another antibiotic, **bacitracin** blocks the hydrolysis of dolichol pyrophosphate to dolichol phosphate by a phosphatase.

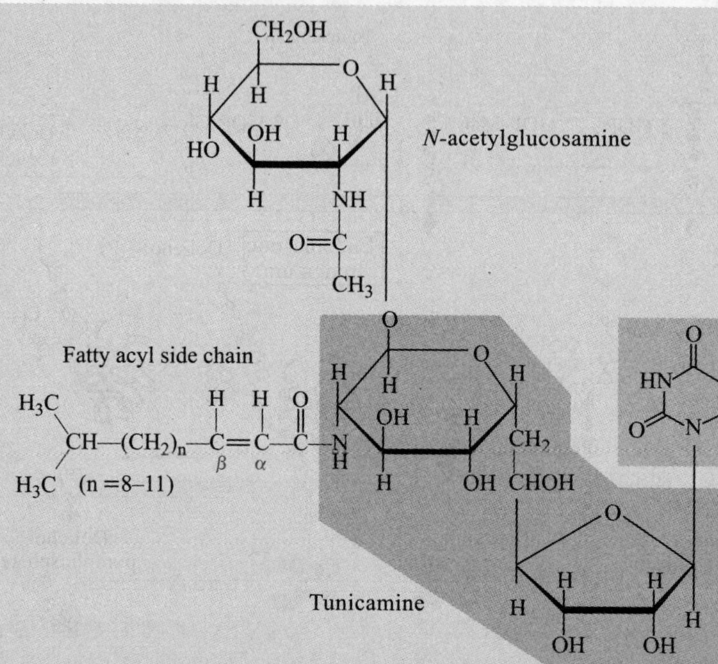

Fig. 31–6. Structure of tunicamycin

Tunicamycin is actually a family of antibiotics produced by (and isolated as a mixture from) *Streptomyces lysosuperficens*. They all contain uracil-*N*-acetylglucosamine, an 11-carbon aminodialdose called tunicamine and a fatty acyl side chain. The structure of the fatty acyl side chain varies in the different compounds within the family. In addition to the variation in length of the fatty acyl side chain, some homologues lack the isopropyl group at the end and/or α-, β-unsaturation.

B. Terminal Glycosylation

> The Golgi complex or Golgi apparatus is named for **Camillo Golgi** (1906-Nobel Laureate) who first detected the vesicles in 1898 by staining brain cells of a barn owl with silver salts.

Proteins are transported from the ER to the Golgi complex in transport vesicles (Fig. 31–7). Golgi complex (Fig. 31–7) is an asymmetric stack of flattened membranous sacs called **cisternae**. A typical mammalian cell has 3 or 4 cisternae, whereas many plant cells usually have about twenty. The Golgi (*pronounced* as GOAL-gee) is differentiated into (1) a *cis* compartment, the receiving end, which is closed to the ER; (2) *medial* compartments; and (3) a *trans* compartment, which exports proteins to various destinations. These compartments contain different enzymes and carry out distinctive functions. Different vesicles transfer proteins from one Golgi compartment to another and then to lysosomes, secretory vesicles, and the plasma membrane. The transport of proteins between the ER and Golgi, and between the Golgi and subsequent destinations is mediated by small (~ 50 to 100 nm in diameter) membrane-bound compartments called *transport vesicles* (or *transfer vesicles*). The Golgi complex performs two main roles. First, carbohydrate units of glycoproteins are altered and elaborated in the Golgi. O-linked sugar units are trimmed there, and N-linked ones are modified in many different ways. Second, the Golgi is the major sorting and packaging center of the cell. It sends proteins to lysosomes, secretory granules, or the plasma membrane according to signals encoded by their 3-dimensional structures (Fig. 31–7).

The carbohydrate moieties of glycoproteins are modified in each of the compartments of the Golgi complex (Fig. 31–8). In the *cis* compartment, 3 mannoses are removed from the oligosaccharide chains of proteins destined for secretion or for insertion in the plasma membrane. The carbohydrate moieties of glycoproteins targeted to the lysosomal lumen are modified differently (described later). In the medial compartments, 2 or more mannoses are removed, and 2 *N*-acetylglucosamines and a fucose are added. Finally, in the *trans* compartment, another N-acetylglucosamine is added, followed by galactose and sialic acid, to form a complex oligo-saccharide unit.

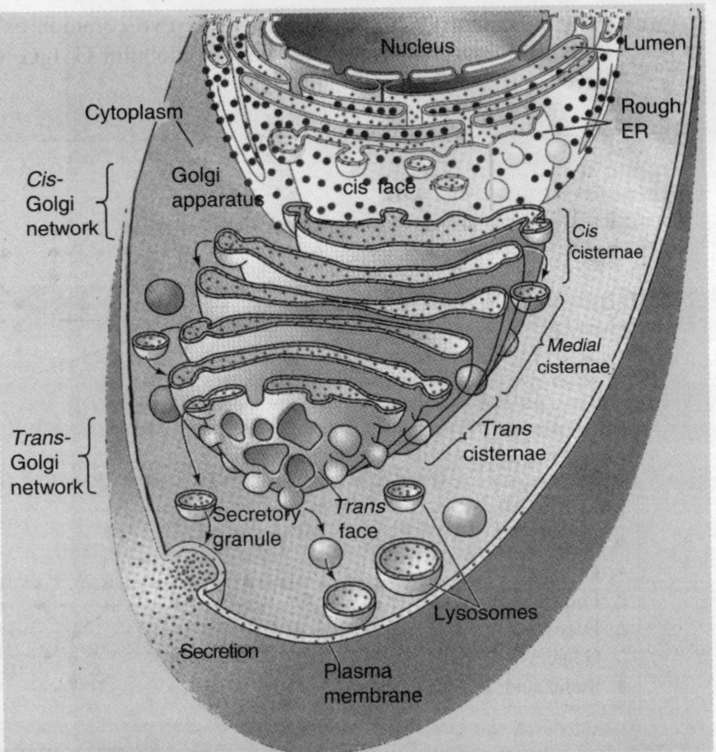

Fig. 31–7. Different pathways adopted by proteins destined for lysosomes, the plasma membrane or secretory granules

Most proteins destined for secretion or insertion into a membrane are synthesized by ribosomes (*blue dots*) attached to the rough endoplasmic reticulum (rough ER; *top*). As they are synthesized, the proteins (*red dots*) are either injected into the lumen of the endoplasmic reticulum or inserted into its membrane. After initial processing, the proteins are encapsulated in vesicles formed from endoplasmic reticulum membrane, which subsequently fuse with the *cis* Golgi network. The proteins are progressively processed according to their cellular destinations, in the *cis, medial,* and *trans* cisternae of the Golgi, between which they are transported by other membranous vesicles, Finally, in the *trans* Golgi network (*bottom*), the completed glycoproteins are sorted for delivery to the final destinations; for example, lysosomes, the plasma membrane, or secretory granules, to which they are transported by yet other vesicles.

Although the biochemical mechanisms involved in "sorting and packaging" is not fully understood, however, with respect to *N*-linked oligosaccharides, a unified concept about the types of units attached is developing. As a rule, *N*-linked oligosaccharides have the same *inner core* which is the branched pentasaccharide containing 3 mannose and 2 acetylglucosamine (Fig. 31–9). Apparently, trimming of the common oligosaccharide core can proceed to the level of the inner core.The *N*-linked oligosaccharides generally fall into one of the following two categories:

(*a*) **Simple mannose-rich units:** These possess the inner core either with short or long mannose oligosaccharides attached (chicken albumin) or with one or few carbohydrates attached (human immunoglobulin M, IgM).

(*b*) **Complex N-acetyllactosamine units:** These are oligosaccharides with *N*-acetylgalactosamine (disaccharide unit of galactose and *N*-acetylglucosamine) linked to the mannosyl residues of the inner core, since they generally have additional sialate (NAN) residues bonded to their galactosyl residues. The two common examples are the oligosaccharide units of human transferrin and immunoglobulin G, IgG, which, unlike the simple units, have been found only in animals.

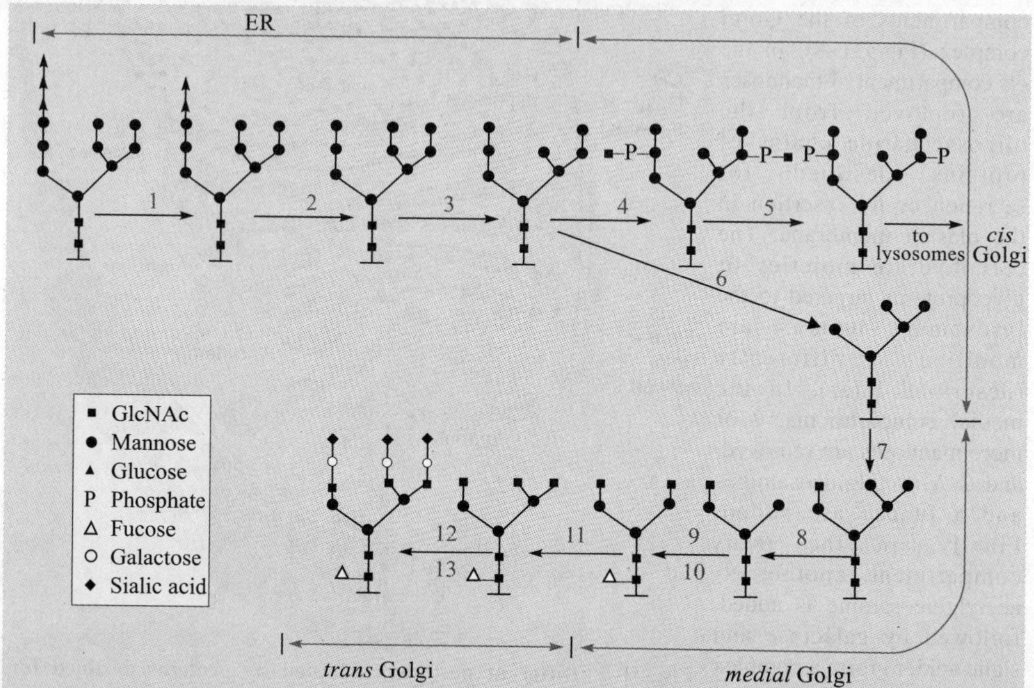

Fig. 31–8. Processing of asparagine-linked (or N-linked) oligosaccharides in the ER and the 3 compartments of the Golgi complex

Steps 4 and 5 apply only to proteins destined for delivery to lysosomes. The enzymes catalyzing all the 13 steps are:

1 glucosidase I
2 glucosidase II
3 ER-1,2-mannosidase
4 N-acetylglucosaminyl-phosphotransferase
5 phosphodiester glycosidase
6 Golgi mannosidase I
7 GlcNAc transferase I
8 mannosidase II
9 GlcNAc transferase II
10 fucosyl transferase
11 GlcNAc transferase IV
12 galactosyltransferase
13 sialyltransferase

GTP–GDP Cycle and the Signal Sequence

The signal sequence on the nascent polypeptide is protected rather sequestered by SRP until it is delivered to the translocation machinery on the ER membrane. The exact timing of the release of the polypeptide by SRP is achieved by a GTP–GDP cycle in the SRP receptor, which is an integral membrane protein consisting of two submits, α (68 kd) and β (30 kd). The binding of SRP-signal peptide to the receptor triggers the replacement of GDP (bound to the α subunit) by GTP (Fig. 31–10). The GTP form of the receptor firmly binds SRP, which loses its grip on the signal peptide. The released signal peptide quickly binds to the *translocon*, a multisubunit assembly of integral and peripheral membrane proteins, which act as translocation machinery. The α subunit of the receptor then hydrolyzes its bound GTP to GDP, which releases SRP. The delay in GTP

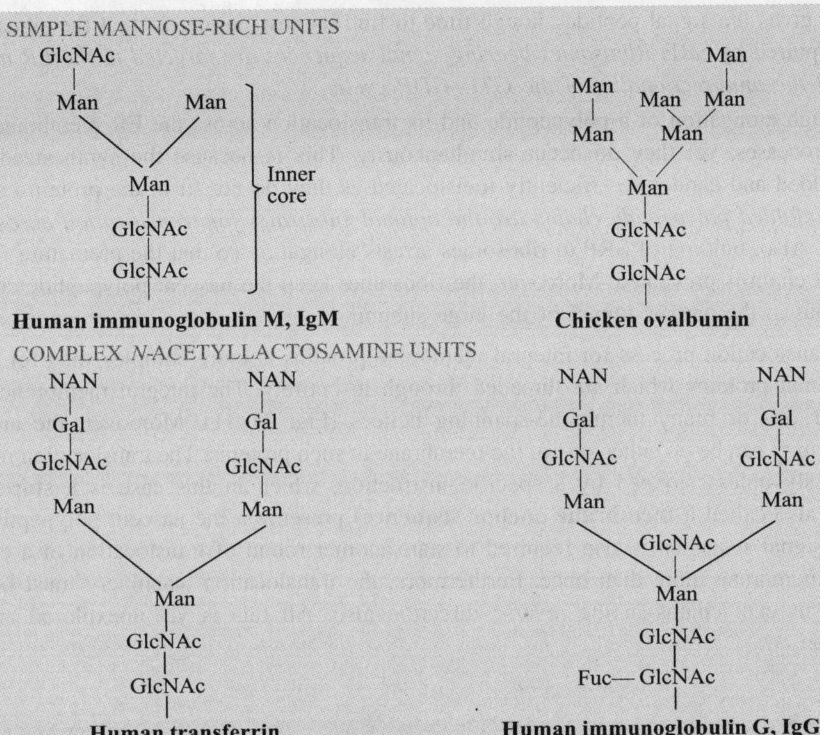

Fig. 31–9. Four oligosaccharide units of glycoproteins

Note that the two examples of complex units have 3 carbohydrates (galactose, sialate, fucose), not found in the common core.

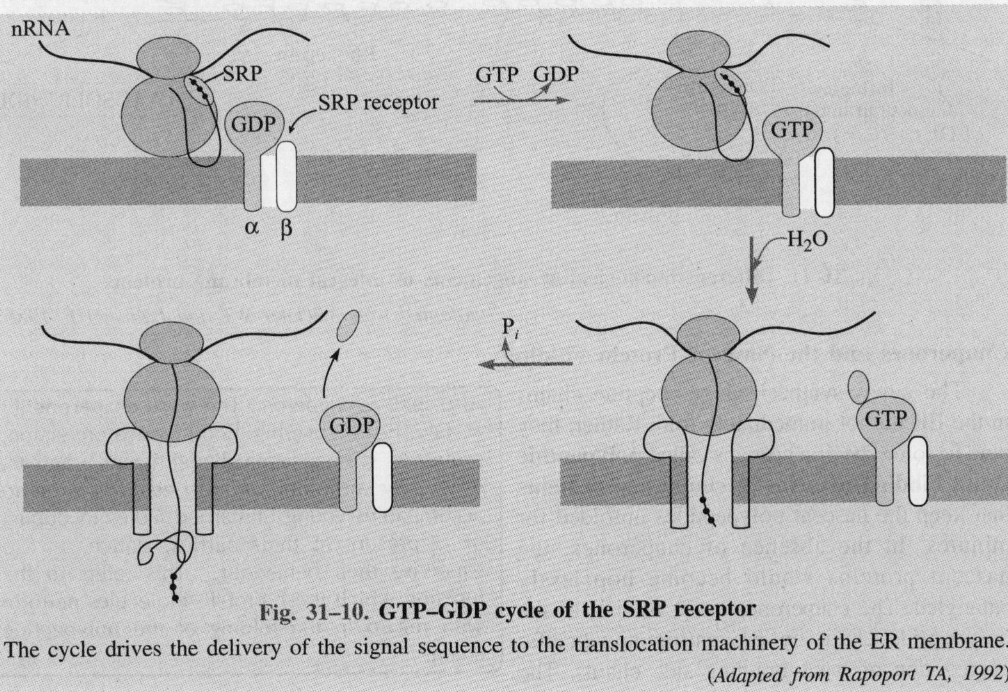

Fig. 31–10. GTP–GDP cycle of the SRP receptor

The cycle drives the delivery of the signal sequence to the translocation machinery of the ER membrane.

(Adapted from Rapoport TA, 1992)

hydrolysis gives the signal peptide enough time to find its new partner so that the signal peptide is not recaptured by SRP. *Ribosomes bearing signal sequences are targeted to the ER membrane because of the unidirectionality of the GTP–GDP cycle.*

Although elongation of a polypeptide and its translocation across the ER membrane are two separate processes, yet they do occur simultaneously. This is because the synthesized proteins become folded and cannot be efficiently translocated as they do not fit in the protein conducting channel. *Unfolded polypeptide chains are the optimal substrates for translocation across the ER membrane.* Also, binding of SRP to ribosomes arrests elongation so that the premature folding of the nascent chain is prevented. Moreover, the ribosomes keep the nascent polypeptide chain fully stretched out in the narrow tunnel of the large subunit.

The translocation process for integral membrane proteins is more complex than for secretory and lysosomal proteins which are threaded through in entirety. The integral membrane proteins have either one or many membrane-spanning helices (Fig. 31–11). Moreover, the amino and carboxy termini can be on either side of the membrane in such proteins. The translocation machinery acts restlessly unless stopped by a specific instruction, which in this case is a **stop transfer sequence** (also called **a membrane anchor sequence**) present on the nascent polypeptide chain. A second signal sequence is also required to start another round of translocation of a chain that spans the membrane more than once. Furthermore, the translocation machinery must be able to thread the nascent chains in the reverse direction also. All this is yet unexplored and needs investigation.

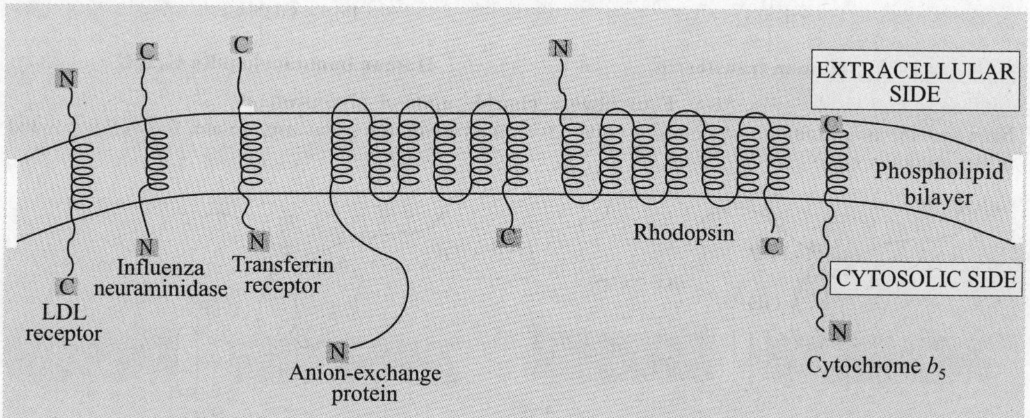

Fig. 31–11. Different topological arrangements of integral membrane proteins

(*Adapted from Wickner WT and Lodish HF, 1985*)

Chaperones and the Nascent Protein Folding

The newly-synthesized polypeptide chains in the ER do not immediately fold. Rather, they bind to some specific proteins called **polypeptide chain binding proteins** or **chaperone proteins** that keep the nascent polypeptides unfolded for minutes. In the absence of chaperones, the nascent proteins would become hopelessly entangled. The chaperones assist folding of the polypeptide chain by preventing nonspecific aggregation of weak-bonding side chains. The

> Also spelt as *chaperon*. The word **chaperone** [Fr < OFr, head, covering, hood (hence protection, protector) < chape: sea chape] means a person, especially an older or married woman who accompanies young unmarried ladies in public, or is present at their parties, dances etc to supervise their behaviour. This refers to the function which these protein molecules perform with regard to the folding of the polypeptide chain.

chaperone proteins were originally identified as members of "heat-shock" protein (hsp) family because they are induced in many cells when heat stress is applied and apparently help stabilize other proteins. When does a chaperone releases its bound nascent chain? Dissociation of chaperones from polypeptides is often coupled to ATP hydrolysis. In fact, *chaperone proteins are slow ATPases*. The ADP-chaperone complex (Fig. 31–12) has high affinity for unfolded peptides but not for the folded (or native) proteins. The binding of unfolded peptide to the chaperone induces the release of ADP and the entry of ATP into the chaperone's active site. The ATP-chaperone complex releases the peptide portion. The subsequent hydrolysis of bound ATP enables the chaperone to again bind an unfolded peptide. Thus, the interval between release and binding of the unfolded peptide chain is determined by the hydrolytic rate. In essence, chaperones buy time before proper folding proceeds.

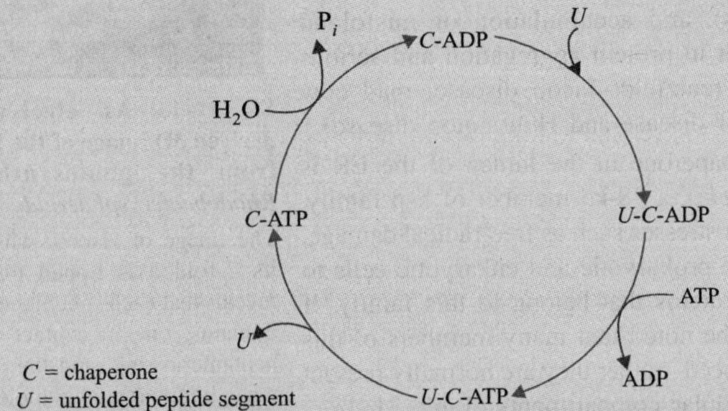

Fig. 31–12. Mechanism of chaperone action

Note that the chaperone action is driven by ATP.

Many proteins can independently and spontaneously generate their correct 3-'D' conformation. For example, fully denatured *pancreatic ribonuclease* can, under appropriate conditions, refold and generate correct disulfide bridges and full activity. Bacterial ribosomal subunits can reassemble from denatured RNA and protein components. In cells, folding of some proteins may be unaided

Table 31–1. Heat shock proteins

Family	Protein
Hsp60 (chaperon-60) family	
	GroEL (in bacterial cytosol)
	Hsp60 (in mitochondrial matrix)
	Rubisco binding protein (in chloroplasts)
Hsp70 (stress-70) proteins	
	Hsp70 (in mammalian cytosol)
	BiP (in ER of eukaryotes)
	Grp75 (in mitochondria)
	DnaK (in bacterial cytosol)
Hsp90 (stress-90) proteins	
	Hsp83 (in eukaryotic cytosol)
	Grp94 (in mammalian ER)
	Htgp (in bacterial cytosol)

(Adapted from Gething MJ and Sambrook J, 1992)

but many proteins attain their correct confirmation only with assistance of one or more of a group of protein chaperones. Chaperones reversibly bind hydrophobic regions of unfolded proteins and folding intermediates. They can stabilize intermediates, main proteins in an unfolded state to allow passage through membranes, help unfold misfolded segments, prevent formation of incorrect intermediates, and prevent inappropriate interactions with other proteins. All of these activities can assist a bound protein in attaining its functional, usually most compact conformation. Failure to fold correctly usually leads to rapid protein degradation (as in cystic fibrosis), and accumulation of misfolded proteins can result in protein aggregation and serious diseases (as in Creutzfeldt–Jacob disease, mad cow disease, Alzheimer disease and Huntington disease).

The major chaperone in the lumen of the ER is BiP (*binding protein*), a 78-kd member of hsp family. Besides heat, other stresses such as free-radical damage, also induce all the prokaryotic and eukaryotic cells to synthesize new proteins that belong to this family. If should, however, be noted that many members of this family are *not* induced– rather they are normally present in virtually all cellular compartments (Table 31–1).

The Hsp60 class of heat-shock protein from the photosynthetic bacterium *Rhodobacter sphaeroides* (Fig 31–13) consists of 14 identical ~60-kD subunits arranged to form two apposed rings of 7 subunits, each surrounding a central cavity.

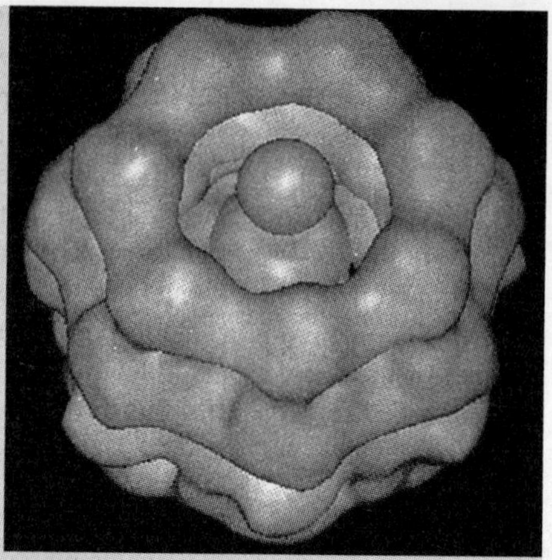

Fig. 31–13. An electron micrograph-derived 3D image of the Hsp60 chaperonin from the photosynthetic bacterium *Rhodobacter sphaeroides*

The image of Hsp60, which is viewed with its 7-fold axis tipped towards the viewer, reveals that each subunit consists of two major domains, one in contact with the opposing heptameric ring, and the other at the end of the cylindrical protein molecule. The spherical density occypying the protein's central cavity is thought to represent a bound polypeptide. The cavity presumably provides a protected microenvironment in which a polypeptide can progressively fold itself.

[*Courtesy*: *Helen Saibil and Steve Wood, Birbeck College, London*]

CLINICAL IMPLICATIONS 31–1.

Incorrect Posttranslational Modification and Premature Protein Degradation: Cystic Fibrosis

Cystic fibrosis (CF) is the most common autosomal recessive disease in Caucasians, occurring with a frequency of about 1 in 2,000 live births. It is a multiorgan disease with a principal manifestation being pulmonary obstruction; thick mucous secretions obstruct the small airways, allowing recurrent bacterial infections to lungs. Exocrine pancreatic dysfunction occurs early and leads to steatorrhea (fatty stool) in CF patients. Such patients have reduced Cl^- permeability, which impairs fluid and electrolyte secretion, leading to luminal dehydration. Diagnosis of CF is confirmed by a significant increase of Cl^- content of sweat of affected individuals in comparison to normal individuals.

The gene responsible for CF was identified in 1989 and over 400 mutations leading to CF have been found. The **CF gene** is 230 kb in length and includes 27 exons that encode a protein of 1480 amino acids. The protein known as the cystic fibrosis transmembrane conductance regulator (CFTR) is a member of a family of ATP-dependent transport proteins and it includes 2 membrane-spanning domains, 2 nucleotide-binding domains that interact with ATP, and 1 regulatory domain that includes several phosphorylation sites. CFTR functions as a cAMP-dependent choride channel and may itself regulate other channels; it is expressed in epithelial tissues. CF epithelia are characterized by defective electrolyte transport. The organs most strongly affected include the lungs, pancreas, and liver, and the most life-threatening effects, so to repeat, involve thick mucous secretions that lead to chronic obstructive lung disease and persistent infections of lungs.

In about 70% of affected individuals, there occurs a deletion of 3 nucleotides that results in deletion of a single amino acid, phenylalanine (Phe) at position 508, normally located in ATP-binding domain 1 on the cytoplasmic side of plasma membrane. As with many other CF mutations, the Phe 508-deletion protein does not fold correctly in the ER and is not properly glycosylated or transported to the cell surface. Instead, it is returned to the cytoplasm to be degraded within proteasomes.

The hsp70 class of heat-shock proteins, which includes BiP, is highly conserved in evolution– the amino acid sequences of the *E.coli* and human proteins are 50% identical. Hsp70 proteins (Fig. 31–14) consist of an ATPase domain and a peptide-binding domain. The ATPase moiety has 2 lobes with a cleft in between.

Fig. 31–14. Structure of Hsp70 proteins

ENVELOPE CARRIER HYPOTHESIS

The envelope carrier hypothesis is linked with the transport of proteins to mitochondria and chloroplasts. Initially, this hypothesis was put forth to explain the transport of smaller subunit of ribulose bisphosphate carboxylase-oxygenase (RUBISCO) enzyme from cytoplasm to chloroplasts. The hypothesis proposes that there are certain proteins in the chloroplastic envelope which can detect and bind with all chloroplast-bound proteins. The smaller subunit of RUBISCO can also bind with these protein in the envelope (Fig. 31–15). This subunit is synthesized as a slightly bigger precursor protein (MW = 20kd) than the actual size (MW = 16kd). The bigger precursor is transported through the chloroplastic envelope by the receptor protein, which also removes the extra segment from the protein through proteolytic action. The portion so removed is especially rich in acidic amino acids, Asp and Glu. Removal of this segment, which contains about 50 amino acid residues, induces a configurational change in the protein, leading to its transport across the chloroplastic envelope and into the stroma.

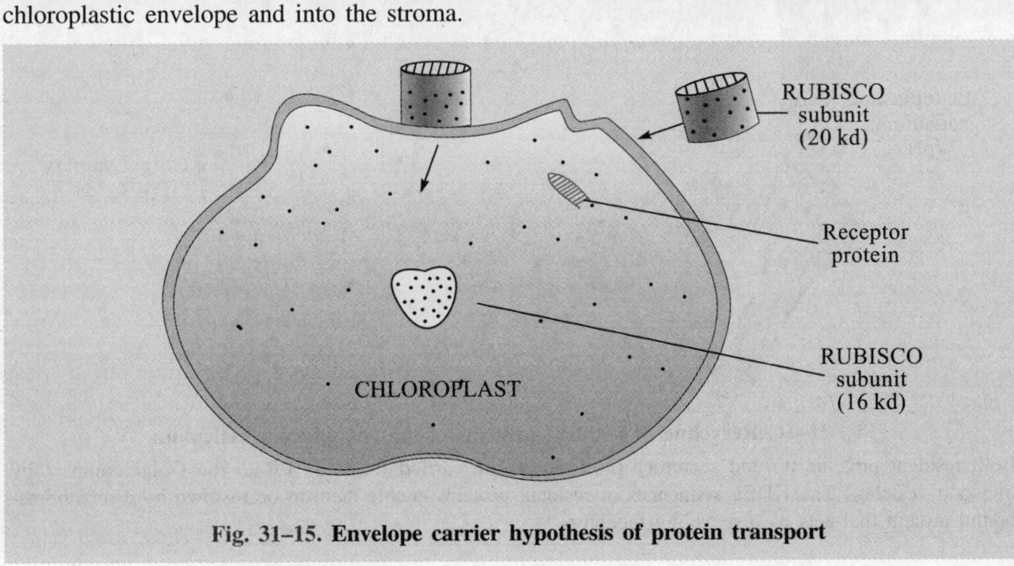

Fig. 31–15. Envelope carrier hypothesis of protein transport

PROTEINS WITH A CARBOXYL-TERMINAL KDEL SEQUENCE (= Recycling of Resident Proteins of the ER)

The ER is rich in chaperones and other proteins that help folding of the nascent peptides. What prevents the ER from losing these essential resident proteins ? In principle, either the resident or the secreted set of proteins could contain a distinguishing tag. However, the experimental evidences indicate that no special tag is needed for needed. A cytosolic protein given a signal sequence will emerge in the ER and then be efficiently secreted. *The resident proteins of the ER lumen carry a retention signal*. More than 50 resident ER proteins from vertebrates, plants, arthropods and nematodes have been shown to a carboxyl-terminal Lys-Asp-Glu-Leu (KDEL) sequence or a closely-related tetrapeptide sequence. In yeast, an HDEL (His-Asp-Glu-Leu) sequence does the same work. Annexation of KDEL sequence to the C-terminus of a secretory protein keeps it (the protein) in the ER of the higher eukaryotes. Conversely, removal of KDEL sequence from the C terminus of a resident ER protein changes its fate: the protein is secreted rather than retained in the ER. Thus, *KDEL sequences retain the resident proteins in the ER*.

In fact, a KDEL sequence does not block the departure of a proteins from the ER. Rather, it serves as a restoring or *retrieval tag*. On reaching the Golgi, KDEL proteins bind to membrane receptors that recognize their C terminal tail (Fig. 31–16). These protein-receptor complexes then incorporate into vesicles that return them to the ER. This recycling scheme operates only when the receptor has high affinity for KDEL in the Golgi but low affinity in the ER. The affinity of the KDEL for its receptor increases about 10 times when the pH is lowered from 7 to 5. As the lumen of the Golgi complex is more acidic than that of the ER, KDEL proteins are membrane-bound in the Golgi but free in solution is the ER. Tight binding in the Golgi ensures efficient retrieval, whereas weak binding in the ER gives resident proteins the freedom to promote the folding and modification of nascent proteins.

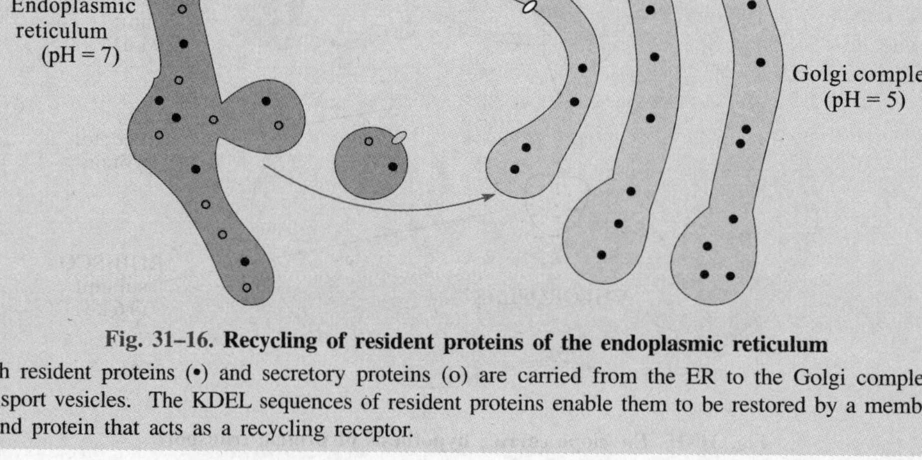

Fig. 31–16. Recycling of resident proteins of the endoplasmic reticulum

Both resident proteins (•) and secretory proteins (o) are carried from the ER to the Golgi complex by transport vesicles. The KDEL sequences of resident proteins enable them to be restored by a membrane-bound protein that acts as a recycling receptor.

A. Protein Transport to Lysosomes

Proteins from the Golgi to lysosomes are sent by a different kind of marker. In fact, a glycoprotein destined for delivery to lysosomes acquires a phosphoryl marker in the *cis* Golgi. This (*i.e.*, phosphorylation) is a two-step process (Fig. 31–17). In the first step, a *phosphorylase* adds a phospho-N-acetylglucosamine unit to the 6-OH group of a mannose; in the second step, a *phosphodiesterase* removes the added sugar to produce a mannose-6-phosphate residue in the core oligosaccharide. Phosphotransferase is a highly discriminating enzyme. It does not act on a mannose residue in an unattached oligosaccharide or on a short peptide containing such a unit. Rather, the phosphotransferase recognizes a 3-'D' structure (sometimes called a '**signal patch**') that is present only in glycoproteins destined to lysosomes.

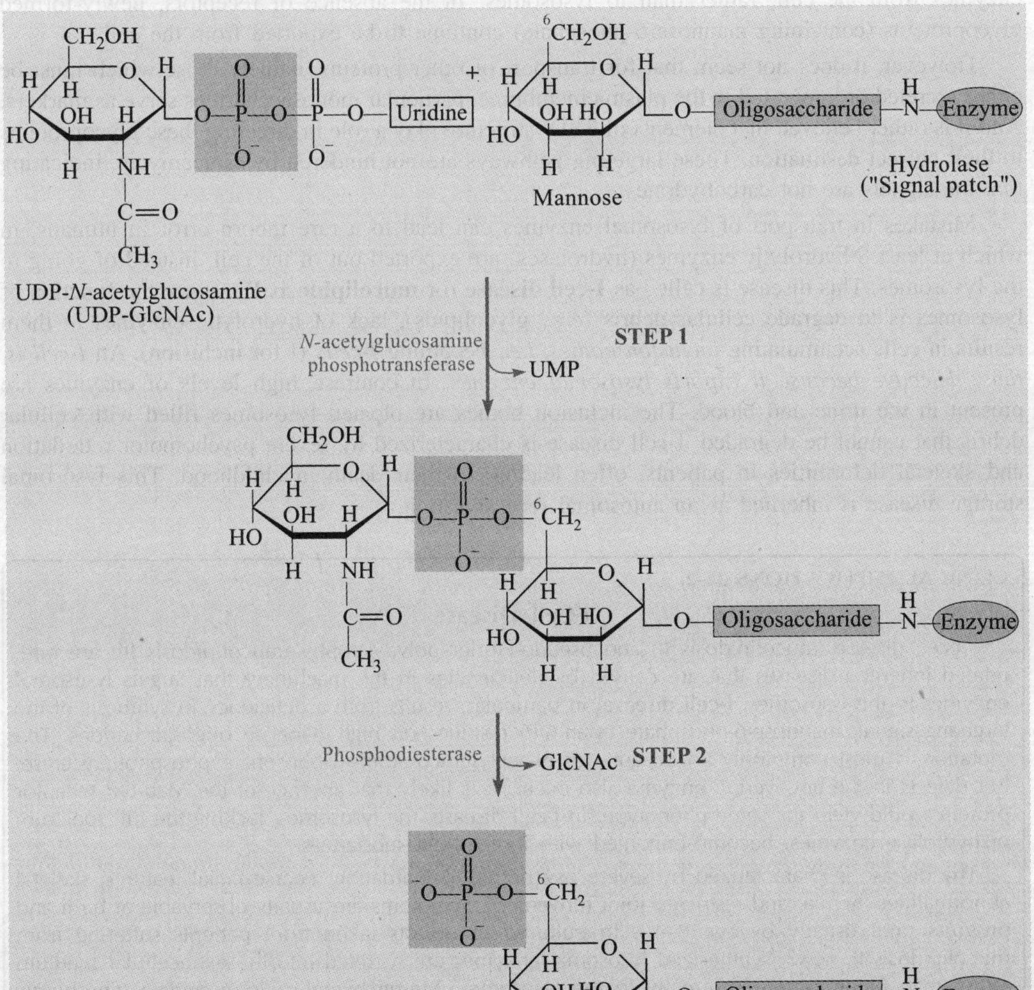

Fig. 31–17. Phosphorylation of mannose residues on lysosome-targeted enzymes such as hydrolases
Note that phosphorylation occurs in 2 steps. A phosphotransferase and a phosphodiesterase in the *cis* Golgi catalyze the addition of a phosphoryl tag. Mannose-6-phosphate, in fact, acts as a marker that normally directs many hydrolytic enzymes from the Golgi to lysosomes.

How is the lysosomal targeting accomplished by mannose-6-phosphate?. The Golgi membrane contains a receptor that specifically recognizes the mannose-6-phosphate residue and binds proteins marked by it. The vesicles containing this protein-receptor complex bud from the *trans* side of the Golgi. These vesicles then fuse with pre-lysosomal vesicles which are more acidic than the Golgi. A decrease in pH triggers dissociation of the marked glycoprotein from its receptor, which now returns to the Golgi. Pre-lysosomes mature into lysosomes by fusing with lysosomes and receiving their enzymes. The membrane-bound mannose-6-phosphate receptor returns to the Golgi by a different set of vesicles. This receptor is recycled so as to be used again and again. Lysosomal targeting is blocked by agents that make sorting vesicles less acidic. Chloroquine and ammonium chloride, for example, raise the pH of sorting vesicles and thus lead to the export of lysosomal enzymes from the cell, rather than to lysosomes. In the absence of receptors, newly-formed glycoproteins (containing mannose-6-phosphate) continue to be exported from the cell.

However, it does not seem that for transport of other proteins, namely those which must be either secreted or integrated in the plasma membrane, particular monosaccharides serve as markers. And it is rather believed that elements of 3-'D' structure play a role in directing these glycoproteins to their correct destination. These targeting pathways are not hindered by tunicamycin, indicating that the signals are not carbohydrates.

Mistakes in transport of lysosomal enzymes can lead to a rare inborn error in humans, in which at least 8 hydrolytic enzymes (hydrolases) are exported out of the cell, instead of going to the lysosomes. This disease is called as **I-cell disease** (or **mucolipidosis II**). Since the function of lysosomes is to degrade cellular debris (*e.g.*, glycolipids), lack of hydrolytic enzymes in them results in cells accumulating *inclusion bodies, i.e.*, becoming *I-cells* (I for inclusion). An *I-cell is, thus, defective because it exports lysosomal enzymes*. In contrast, high levels of enzymes are present in the urine and blood. The inclusion bodies are bloated lysosomes filled with cellular debris that cannot be degraded. I-cell disease is characterized by severe psychomotor retardation and skeletal deformities in patients, often leading to their death in childhood. This lysosomal storage disease is inherited as an autosomal recessive trait.

CLINICAL IMPLICATIONS 31–2.

I-Cell Disease

I-cell disease (mucolipidosis II) and pseudo-Hurler polydystrophy (mucolipidosis III) are rare, related inherited diseases that are caused by deficiencies in the machinery that targets lysosomal enzymes to the lysosomes. **I-cell disease,** in particular, results from a deficiency in synthesis of the targeting signal, mannose-6-phosphate (Man-6-P) residues on high-mannose oligosaccharides. The mutation is most commonly an absence of N-acetylglucosamine-1-phosphate phosphotransferase, but defects in the uncovering enzyme also occur. It is likely that absence of the Man-6-P receptor protein would yield the same phenotype. In I-cell disease, the lysosomes, lacking the full spectrum of hydrolase enzymes, become engorged with indigestible substances.

The disease is characterized by severe psychomotor retardation, coarse facial features, skeletal abnormalities, hernia, and restricted joint movement. Symptoms are usually observable at birth and progress until usually by age 8–10. In cultured fibroblasts taken from patients suffering from mucolipidosis II, newly-synthesized lysosomal enzymes are secreted into the extracellular medium rather than being targeted correctly to the lysosomes. Mesenchymal cells, especially fibroblasts, contain numerous membrane-bound vacuoles in the cytoplasm containing fibrillogranular material. These deposits are called inclusion bodies, and this is the origin of the name I-cell disease. This fatal congenital disorder provided the earliest connection between glycoprotein biosynthesis and human disease.

Pseudo-Hurler polydystrophy is a much milder form of the disease. Onset is usually delayed until the age of 2–4 years, the disease progresses more slowly, and patients survive into adulthood. Prenatal diagnosis of both diseases is possible, but there is as yet no definitive treatment.

B. Protein Transport to Mitochondria

The mitochondrial genome encodes all the RNA but only a very small number of mitochondrial proteins (about 12) with the result that the great majority of mitochondrial proteins is encoded by the nuclear genome, synthesized in the cytosol by free ribosomes and then imported into the mitochondria. Indeed about 10% of the proteins in a eukaryotic cell are imported into mitochondria. How do these majority proteins reach their mitochondrial destinations? The problem becomes more complex as mitochondrial proteins reside in 4 locations; the outer membrane, the inner membrane, the intermembrane space, and the matrix. **Gottfried Schatz** (1993) discovered that the import of a protein into the mitochondrion requires the presence of a particular sequence at its amino-terminal end, called the **presequence** or **matrix targeting sequence**. This sequence must be recognized by receptors situated on the external face of the outer membrane of the mitochondrion and leads to the import of the protein bearing it into the matrix. The presequences (Fig. 31–18) are typically 15 to 35 residues long. They are rich in positively-charged residues, and in serines and threonines. No consensus sequence has been found. In fact, matrix targeting sequences, like prokaryotic and eukaryotic signal sequences, are highly degenerate– about 20% of randomly-generated sequences allow proteins to enter mitochondria.

$^{+}H_3N$—Met—**Leu**—Arg—Thr—Ser—Ser—**Leu**—**Phe**—Thr—Arg—Arg—**Val**—Gln—Pro—Ser

—**Leu**—**Phe**—Arg—Asn—**Ile**—**Leu**—Arg—**Leu**—Gln—Ser—Thr

Fig. 31–18. A mitochondrial matrix targeting sequence

The 9 hydrophobic residues are shown in *boldface*. There are 5 basic residues (Arg) also.

The precursor proteins are imported into the mitochondrial matrix through mediation by proteins in the cystosol, outer and inner mitochondrial membranes, and the matrix. The fully folded proteins cannot enter mitochondria. Hence, chaperones play a key role in maintaining the precursor proteins in unfolded or partly folded state. A cytosolic *hsp70 protein* hands the precursor to an *import receptor* in the outer mitochondrial membrane (Fig. 31–19). The receptor-bound precursor then moves to a site where the outer and inner mitochondrial membranes adhere to one another. The precursor protein threads through a channel formed by several subunits from each membrane. Translocation is driven by the membrane potential across the inner mitochondrial membrane and by the ATP-powered **untate** of the nascent by mhsp70, a 70 kd heat-shock protein in the mitochondrial matrix. Finally, the amino-terminal signaling sequence is cleaved in the matrix by a metalloprotease to produce the mature protein.

The import of proteins into mitochondria (and plasts) is a post-translational process which may be performed *in vitro* by incubating organelles in presence of the protein precursor, and therefore differs from the transport of proteins through the membrane of the ER which takes place concomitantly with the translocation, *i.e.,* while the protein chain is being synthesized.

C. Protein Transport to Chloroplasts

Most proteins of chloroplasts, like those of mitochondria, are coded by the nuclear genome, synthesized by the cytosolic ribosomes and imported in the plast. But there are more possible localizations in the plast (6) than in the mitochondria: the outer membrane, the inner membrane, the intermembrane space, the stroma, the thylakoid membrane and the lumen of the thylakoid. The existence of the thylakoid membranes, separate from the inner membrane, gives rise to two more destinations than are present in mitochondria. As in the case of mitochondria, the import of proteins requires a **presequence** (also called **transit sequence** or **transit peptide**) at the amino-terminal end. Chloroplast presequences resemble mitochondrial presequences in being

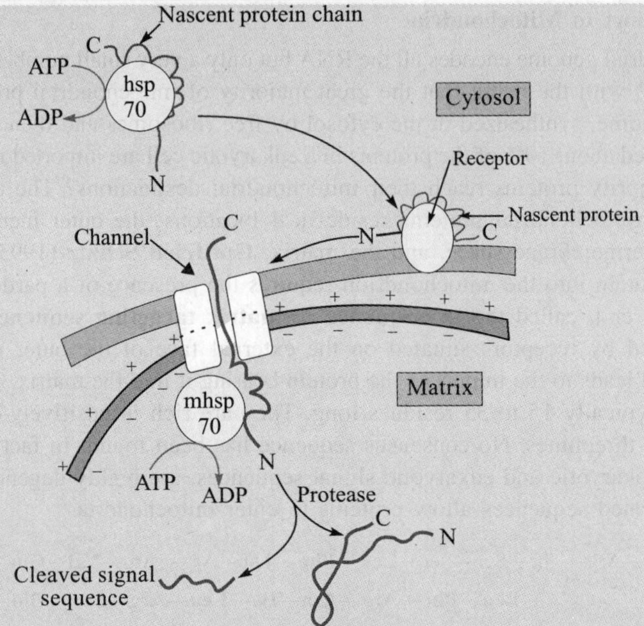

Fig. 31–19. Proposed mechanism for the delivery of a protein from the cytosol to the mitochondrial matrix

The nascent chains is delivered by a chaperone to a receptor in the outer mitochondrial membrane. The nascent protein threads through a channel that traverses both membranes at a contact site. A transmembrane potential across the inner mitochondrial membrane is essential for transport. The matrix-trageting sequence is removed by protease.

(*After Gottfried Schatz, 1993*)

positively-charged and rich in hydroxylated residues (*i.e.*, serine and threonine). This chloroplast presequence is cleaved during the transport process through stroma and thylakoids. Transport across the outer and inner chloroplast membranes is powered by ATP hydrolysis, whereas transport across the thylakoid is driven by the pH gradient. The presequences of proteins destined for the thylakoid lumen appear to contain two signals (Fig. 31–20). The amino-terminal signal leads to the import of the precursor protein into the chloroplast stroma. This part of the presequence (*first signal*) is cleaved in the stroma, or en route to it, exposing a *second signal* that directs the translocation of the modified precursor across the thylakoid membrane. A protein targeted to the stroma lacks this second signal, which contains a hydrophobic core reminiscent of bacterial and ER signal sequences.

D. Protein Transport to Peroxisomes

Peroxisomes are small membrane-bound compartments, present in most eukaryotic cells. These organelles contain *oxidases* that produce H_2O_2 and *catalase* which decomposes H_2O_2 to water and oxygen. Peroxisomes perform a variety of functions. They bring about detoxication. They also catalyze the first two steps in the synthesis of plasmalogens. In humans, β-oxidation of fatty acids longer than C_{18} occurs primarily in the peroxisomes, rather than in mitochondria. In plants, similar organelles (also called **glyoxysomes**) play a key role in the recycling of phosphoglycolate, which is generated by the oxygenase action of RUBISCO. Peroxisomes and glyoxysomes are together known as **microbodies**.

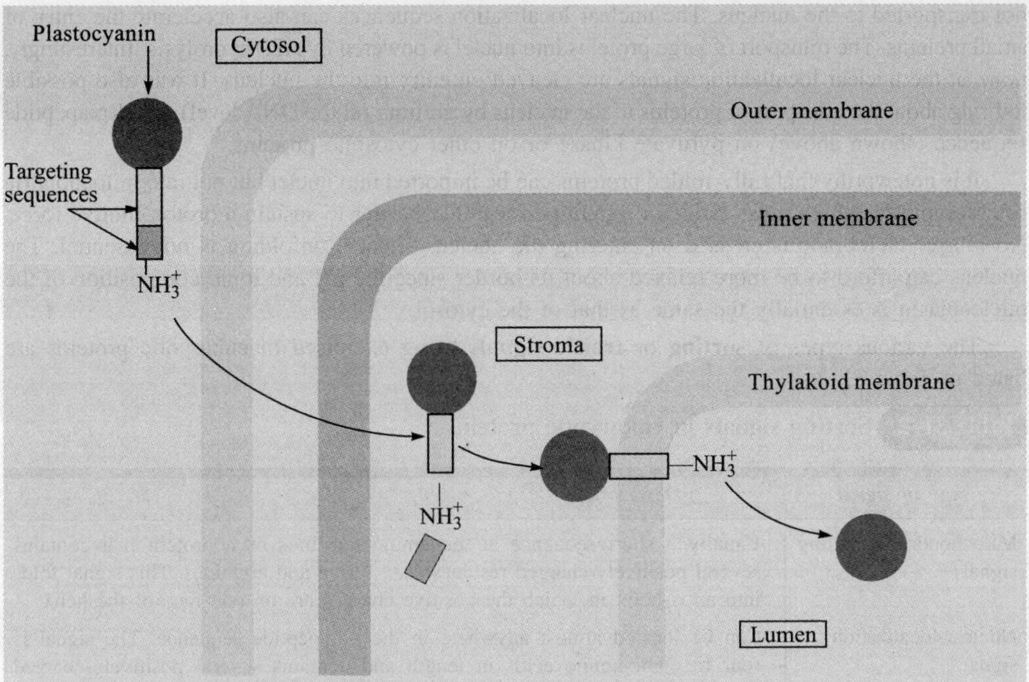

Fig. 31–20. Targeting of the plastocyanin to the thylakoid lumen of chloroplasts by the sequential action of two amino-terminal sequences

The first sequence enables plastocyanin to enter the stroma and the second (exposed by proteolysis) enables it to cross the thylakoid membrane.

(After Smeeckens, Bauerle, Hagemann, Keegstra and Weisbeek, 1986)

Both peroxisomes and glyoxysomes are devoid of DNA. The soluble proteins in their matrix are imported from the cytosol. The targeting signal for many peroxisomal matrix proteins is simply a carboxyl-terminal Ser-Lys-Phe or SKF (the one-letter symbol for the 3 amino acids) **tripeptide sequence**. Mutation of the SKF sequence of a cytosolic precursor protein blocks its import into peroxisomes. The SKF signal, in contrast with mitochondrial and chloroplast import signals, is *not* cleaved. Two other differences are its brevity and carboxyl-terminal location.

E. Protein Transport to Nucleus

All nuclear proteins (*esp*, histones, DNA polymerases, RNA polymerases and all proteins participating in the replication of DNA and transcription) are synthesized in the cystosol by free ribosomes and must pass through the nuclear envelope of eukaryotes comprising an outer membrane and an inner membrane. This transport, for the small proteins (*e.g*, histones), seems to take place through nuclear pores of 70 Å diameter; but for larger proteins (> 90 kd), a short peptide sequence (or signal sequence) appears to be necessary. For example, the T antigen of SV 40 virus is a protein of molecular weight 92,000 daltons (or 92 kd) that regulates the replication and transcription of viral DNA. And studies have shown that the transport of this large protein, depends on the presence of a **nuclear localization sequence**, containing five consecutive positively-charged residues (shown in *red*):

-Pro-Lys-Lys-Lys-Arg-Lys-Val-
128

A change of even a single amino acid residue can render this sequence inactive. As an instance, T antigen containing Thr or Asn in place of Lys at residue 128 stays in the cystol and not transported to the nucleus. The nuclear localization sequences can also accelerate the entry of small proteins. The transport of large proteins into nuclei is powered by ATP hydrolysis. Interestingly, none of the nuclear localization signals are cleaved on entry into the nucleus. It was also possible to bring about the transport of proteins to the nucleus by grafting (at the DNA level) this heptapeptide sequence (shown above) on pyruvate kinase or on other cytosolic proteins.

It is noteworthy that fully-folded proteins can be imported into nuclei but not into mitochondria or chloroplasts, which must maintain a tight permeability barrier to sustain a proton-motive force. No bilayer membrane is crossed on entering the nucleus– hence, unfolding is not essential. The nucleus can afford to be more relaxed about its border since the pH and ionic composition of the nucleoplasm is essentially the same as that of the cytosol.

The various types of **sorting** or **traffic signals** being exercised in eukaryotic proteins are listed in Table 31–2.

Table 31–2. Sorting signals in eukaryotic proteins

Type of signal	*Description*
Mitochondrial-targeting signal	Usually a short sequence at the amino terminus or a protein that contains several positively-charged residues (*i.e.*, lysine and arginine). This signal folds into an α-helix in which the positive charges are on one face of the helix.
Nuclear-localization signal	Can be located almost anywhere in the polypeptide sequence. The signal is four to eight amino acids in length and contains several positively-charged residues and usually one or more prolines.
Peroxisomal-targeting signal	A specific sequence of three amino acids, serine–lysine–leucine, which is usually located near the carboxyl terminus of the protein.
SRP signal (sorting to the ER)	A sequence of ~20 amino acids near the amino terminus that is composed of mostly nonpolar amino acids.
ER retention signal	A sequence of four amino acids, lysine–aspartic acid–glutamic acid–leucine, which is located at the carboxyl terminus of the protein.
Golgi retention signal	A sequence of 20 hydrophobic amino acids that forms a transmembrane domain flanked by positively-charged residues.
Lysosomal-targeting signal	A patch of amino acids within the polypeptide sequence. Positively-charged residues within this patch are thought to play an important role. This patch causes lysosomal proteins to be covalently modified to contain a mannose-6-phosphate residue that directs the protein to the lysosome.
Destination of a cellular protein	*Type of signal the protein contains within its amino acid sequence*
Cytosol	No signal required
Mitochondrion	Mitochondrial-targeting signal
Nucleus	Nuclear-localization signal
Peroxisome	Peroxisomal-localization signal
ER	SRP signal and an ER retention signal
Golgi	SRP signal and a Golgi retention signal
Lysosome	SRP signal and a lysosomal targeting signal
Plasma membrane	SRP signal and the protein contains a hydrophobic transmembrane domain that anchors it in the membrane
Secretion	SRP signal. No additional signal required

BACTERIAL SIGNAL SEQUENCES AND PROTEIN TARGETING

Protein targeting is not confined to eukaryotes. Bacteria also target proteins to destinations encoded in their sequences. A Gram-negative microorganism such as *E. coli* can translocate nascent proteins to the inner and outer membranes, the periplasmic space between the membranes or (rarely) the extracellular medium (secretion) as depicted in Fig. 31–21. *As in eukaryotes, translocation is not mechanistically coupled to chain elongation*. This targeting uses signal sequences (also called **leader sequences**) at the amino terminus of the proteins, much like those found on eukaryotic proteins targeted to the ER (Fig. 31–22). These signal sequences are usually 16 to 26 residues long. Though diverse, the prokaryotic signal sequences have a positively-charged amino-terminal region, and a helix-breaking segment. Like eukaryotes, the signal sequence is usually cleaved by a signal peptidase at the helix-breaking site. Another similarity is that polypeptide chain elongation and translocation usually take place at about the same time but are not mechanistically coupled.

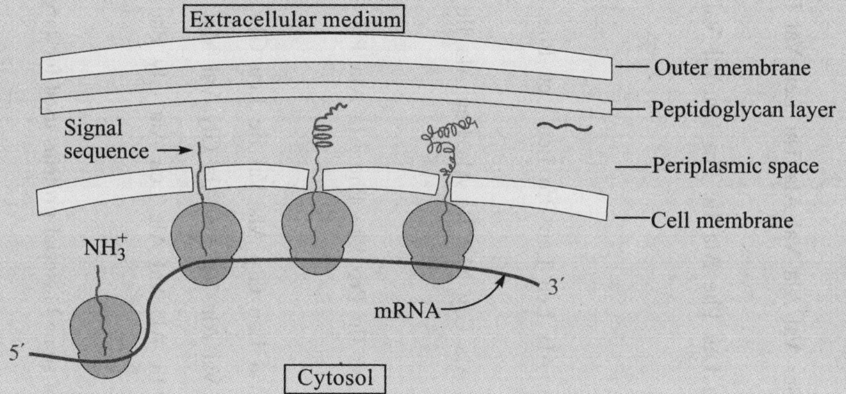

Fig. 31–21. Schematic of the synthesis of noncytosolic proteins by ribosomes bound to the plasma membrane in prokaryotes

A signal sequence (shown by *bold line*) on the nascent chain directs the ribosomes to the plasma membrane and enables the protein to be translocated. The translocation machinery is *not* depicted in this schematic diagram.

Translocation of a polypeptide chain through the cell membrane of *E. coli* is catalyzed by a soluble chaperone and a membrane-bound, multisubunit translocase (Fig. 31–23). The bacterium contains a major chaperone, *SecB protein* which keeps nascent chains in unfolded or partially folded state to enable them to traverse the membrane. SecB presents the nascent chain to Sec A, a peripheral membrane component of the translocase. SecA works in unison with SecY and SecE, the membrane-embedded portion of the translocase. Two forms of free energy (ATP and proton-motive force) drive protein translocation in *E. coli*. SecA is an ATPase– the ATP state has high affinity for the protein to be translocated, whereas the ADP state has low affinity. Portions of the nascent chain are successively handed from from SecA to SecY-SecE channel as a result of many cycles of ATP hydrolysis. The proton-motive force across the cell membrane then drives the threading of the nascent chain through the membrane.

Inner membrane proteins

Cleavage site

Phage fd, major coat protein: Met Lys Lys Ser Leu Val Leu Lys Ala Ser Val Ala Val Ala Thr Leu Val Pro Met Leu Ser Phe Ala ↓ Ala Glu --

Phage fd, minor coat protein: Met Lys Lys Leu Leu Phe Ala Ile Pro Leu Val Val Pro Phe Tyr Ser His Ser ↓ Ala Glu --

Periplasmic proteins

Alkaline phosphatase: Met Lys Gln Ser Thr Ile Ala Leu Ala Leu Leu Pro Leu Leu Phe Thr Pro Val Thr Lys Ala ↓ Arg Thr --

Leucine-specific binding protein: Met Lys Ala Asn Ala Lys Thr Ile Ile Ala Gly Met Ile Ala Leu Ala Ile Ser His Thr Ala Met Ala ↓ Asp Asp--

β-lactamase of pBR322: Met Ser Ile Gln His Phe Arg Val Ala Leu Ile Pro Phe Phe Ala Ala Phe Cys Leu Pro Val Phe Ala ↓ His Pro --

Outer membrane proteins

Lipoprotein: Met Lys Ala Thr Lys Leu Val Leu Gly Ala Val Ile Leu Gly Ser Thr Leu Leu Ala Gly ↓ Cys Ser --

LamB: Leu Arg Lys Leu Pro Leu Ala Val Ala Val Ala Ala Gly Val Met Ser Ala Gln Ala Met Ala ↓ Val Asp --

OmpA: Met Met Ile Thr Met Lys Lys Thr Ala Ile Ala Ile Ala Val Ala Leu Ala Gly Phe Ala Thr Val Ala Gln Ala ↓ Ala Pro --

Fig. 31–22. Amino-terminal signal sequences of some bacterial proteins used for targeting to different locations

Basic amino acids (*blue*) near the amino terminus and hydrophobic core amino acids (*yellow*) are highlighted. The cleavage sites denoting the ends of the signal sequences are marked by *red* arrows. Note that the inner bacterial cell membrane is where phage fd coat proteins and DNA are assembled into phage particles. *OmpA* is outer membrane protein A; *LamB* is a cell surface receptor protein for bacteriophage lambda.

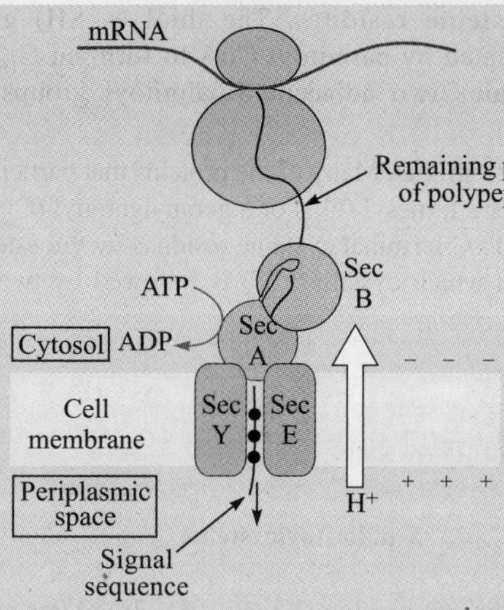

Fig. 31–23. Schematic diagram showing the interplay of Sec proteins in protein translocation across the cell membrane

Proton-motive force powers the unidirectional translocation of the polypeptide from the cytosolic to the periplasmic side of the membrane

Some proteins that are translocated through one or more membranes to reach their final destinations are maintained in a distinct "*translocation-competent*" *conformation* until this process is complete. The functional conformation is assumed after translocation, and proteins purified in this final form are now longer capable of translocation. Available evidences indicate that the translocation conformation is stabilized by a specialized set of proteins in all bacterial cells. These bind to the protein to be translocated while it is being synthesized, preventing it from folding into its final 3-'D' structure. In *E. coli*, a protein called trigger factor (M_r 63,000) appears to facilitate the translocation of at least one outer membrane protein through the inner membrane.

EUKARYOTIC PROTEIN TRANSPORT ACROSS MEMBRANES

The eukaryotic organisms employ different strategies to send proteins to the *cytosolic face* of either the plasma membrane or a compartment membrane. A fatty acyl or prenyl group is attached covalently to a soluble cytosolic protein. This membrane-anchoring may be accomplished in following 4 ways (**Lubert Stryer**, 1995):

1. Myristoylation at the N terminus. The amino terminus of many proteins is acylated with a myristoyl (C_{14}) or a similar fatty acyl group; the acyl group being donated by myristoyl-CoA. This reaction is catalyzed by *N-myristoyl transferase*. The N-terminal residue (residue 1) must be glycine, residue 5 is usually serine or threonine, and residues 6 and 7 are typically basic. Myristoylation enables a modified protein to interact with a membrane receptor or the lipid bilayer itself.

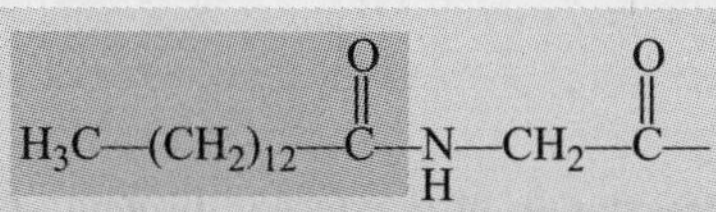

***N*- myristoylglycine**

2. Palmitoylation of cysteine residues. The thiol (– SH) group of some cysteine residues in proteins can be acylated by palmitoyl-CoA to form an C_{16} *S*-palmitoyl derivative. Rhodopsin, for example, contains two adjacent *S*-palmitoyl groups that act as membrane anchors.“

3. **Farnesylation at the C terminus.** Many of the proteins that participate in signal transduction and protein targeting contain either a farnesyl (C_{15}) or a geranylgeranyl (C_{20}) unit at their C terminus. These prenyl groups are attached to C-terminal cysteine residues by thioester linkages. Farnesylation occurs at **CaaX sequences** in which cysteine (C) is followed by two

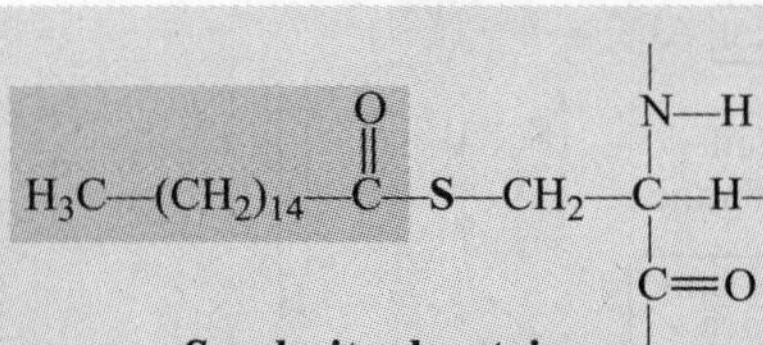

***S*-palmitoylcysteine**

aliphatic residues (*a*) and a C-terminal residue (X) (Fig. 31–24). After attachment of the C_{15} unit to this cysteine, the aaX residues are proteolytically removed and the new terminal carboxylate group is methylated. Thus, a highly hydrophobic C terminus is fashioned by a series of modifications. The ras protein does not insert in the plasma membrane unless it is farnesylated. In fact, unfarnesylated ras is unable to transduce growth signals.

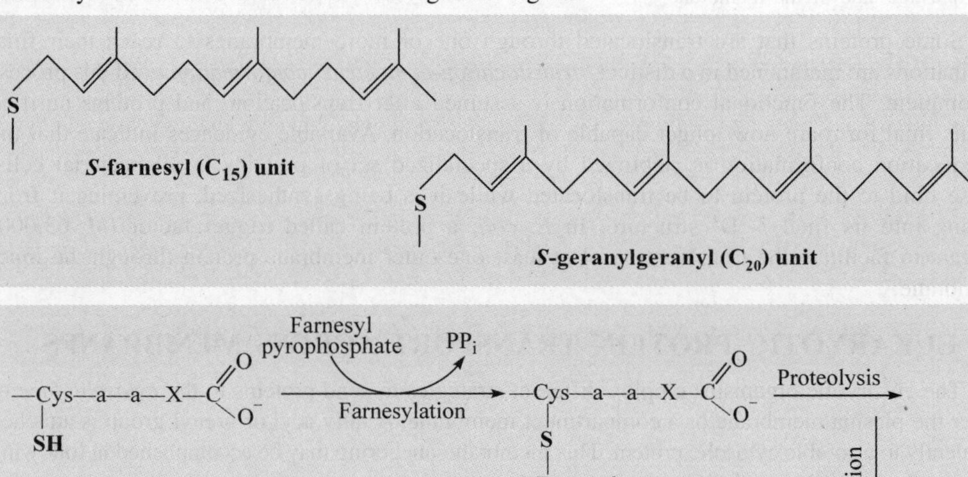

***S*-farnesyl (C_{15}) unit**

***S*-geranylgeranyl (C_{20}) unit**

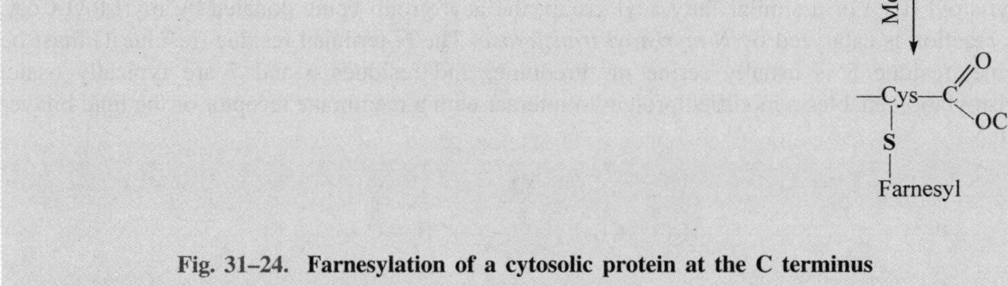

Fig. 31–24. Farnesylation of a cytosolic protein at the C terminus

Farnesylation is followed by trimming of 3 C-terminal residues and methylation of the terminal carboxylate.

4. Geranylgeranylation at the C terminus. When the C-terminal sequence is CC, CXC, or CCXX, a geranylgeranyl (C_{20}) unit, rather than a farnesyl unit, becomes attached to one or both cysteines. The rab family of small GTP-binding proteins, which participate in membrane targeting, are geranylgeranylated. The attachment of this highly hydrophobic prenyl unit is necessary for membrane binding.

PROTEIN IMPORT BY RECEPTOR-MEDIATED ENDOCYTOSIS

Specific proteins are imported into a cell by their binding to receptors in the plasma membrane and their inclusion into vesicles. Such a process is called **receptor-mediated endocytosis** and has a great number of biological applications:

1. It is a means of delivering essential metabolites to cells. For instance, the low-density lipoprotein (LDL) carrying cholesterol is taken up by the LDL receptor in the plasma membrane and internalized.
2. Endocytosis regulates responses to many protein hormones and growth factors. Epidermal growth factor and nerve growth factor are taken into the cell and degraded together with their receptors.
3. Proteins destined for degradation are taken up and delivered to lysosomes for digestion. Phagocytes, for example, have receptors that enable them to take up antigen-antibody complexes.
4. Receptor-mediated endocytosis is employed by many viruses and toxins to gain entry into cells, as exemplified by the ingenious mode of entry and departure of Semliki Forest Virus (SFV), a membrane-enveloped virus.
5. Disorders of receptor-mediated uptake can lead to diseases, such as some forms of familial hypercholesterolemia.

Cell-surface Receptors and Clathrin

The **cell-surface receptors**, which mediate endocytosis, are transmembrane glycoproteins. They have a large extracellular domain and a small cytosolic region and contain either one (*e.g.*, asialoglycoproteins) or two (*e.g.*, transferrin) transmembrane helices (Fig. 31–25). Many of the receptors are located in specialized regions of the plasma membrane called *coated pits*. The cytosolic side of these pits has a thick coat of *clathrin*, a protein designed to form lattices around membranous vesicles. Many receptors (such as those for LDL, transferrin, asialogycoproteins, insulin) congregate in coated pits; others (such as the receptor for epidermal growth factor) cluster there after binding their cognate protein.

Clathrin (Fig. 31–26) is a trimeric protein, consisting of 3 heavy chains (H; Mr = 180,000) and 3 light chains (L; Mr = 35,000). The $(HL)_3$ clathrin unit (8 S; Mr = 650,000) is organized as a three-legged structure, called a *triskelion*. The carboxy termini of the 3 heavy chains (each about 500 Å long) come together at a vertex. A bend in the heavy chain divides it into a proximal arm, closest to the vertex, and a distal arm. Each of 3 the light chains is aligned with the proximal arm of a heavy chain. Many clathrin assemble into closed shells having a polyhedral structure. The polyhedra are made of both pentagons and hexagons [Fig. 31–26(b)]. A single edge of a pentagon or hexagon is made of parts of four triskelions, 2 proximal arms and 2 distal arms. The flexibility of a triskelion is important in enabling it to fit into a pentagon or hexagon.

Receptor-mediated Endocytosis

Receptor-mediated endocytosis (Fig. 31–27) begins with the binding of certain proteins (such as LDL, transferrin, peptide hormones etc) to receptors on the outer face of the plasma membrane.

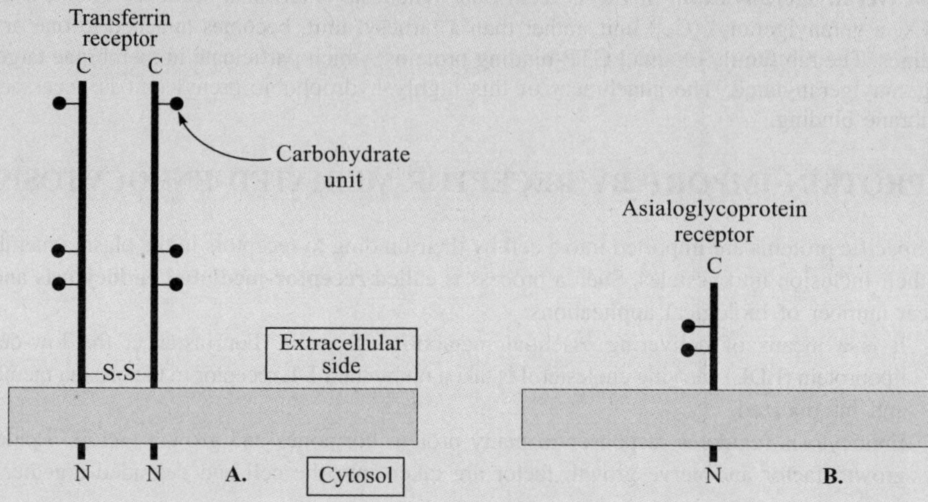

Fig. 31–25. Schematic of two cell-surface receptors that are internalized at coated pits

A. Transferrin receptor **B.** The asialoglycoprotein receptor

The short N-terminal tails of these receptors are critical for internalization.

The receptors are concentrated in invaginations of the membrane called coated pits, which are coated on their cytosolic face with a lattice made up of the protein clathrin. The clathrin lattice grows up until a complete membrane-bound endocytic vesicle, with a diameter of about 80 nm, buds off the plasma membrane and moves into the cytosol. The endocytic vesicle then rapidly loses its clathrin shell by uncoating enzymes and fuses with an endosome. The endosomes, in turn, fuse with one another to form bigger vesicles, ranging between 200 and 600 nm. The membrane ATPases present in the endosomes lower the pH, so that the receptors dissociate from their target proteins. Proteins and receptors then follow separate paths, their fates varying according to the system (Table 31–3). The protein transferrin transports iron from sites of absorption and storage to sites of utilization. Two Fe^{3+} ions are bound to the protein which contains two similar domains. The protein devoid of iron is called *apotransferrin*. Transferrin, but not apotransferrin, binds to a dimeric receptor (Fig. 31–25). The low pH within the endosome causes dissociation of Fe^{3+} from transferrin. The acidity lowers the affinity of transferrin for Fe^{3+} more than a millionfold. However, apotransferrin remains bound to the receptor.

Sorting then takes place: part of the vesicle bearing apotransferrin bound to the receptor pinches off and proceeds towards plasma membrane, whereas the remaining Fe^{3+} is stored in ferritin in the cytosol. When the pinched off vesicle fuses with the plasma membrane, apotransferrin is released from the receptor because of the sudden increase in pH. Apotransferrin has little affinity for the receptor at pH 7.4. Thus, *pH changes are used twice to drive the transferrin transport cycle*: first to release iron from transferrin in the endosome, and then to discharge apotransferrin into the extracellular fluid. *The cycle takes about 16 minutes*: 4 minutes for the binding of transferrin, 5 minutes for transport to endosomes, and 7 minutes for the return of the iron carrier and the receptor to the cell surface. Toxins (diphtheria toxin, cholera toxin) as well as viruses (influenza virus) enter cells by receptor-mediated endocytosis.

PROTEIN DEGRADATION

Proteins are constantly being degraded in all cells so as to prevent the buildup of the abnormal or unwanted proteins and to facilitate the recycling of amino acids. *Degradation is a*

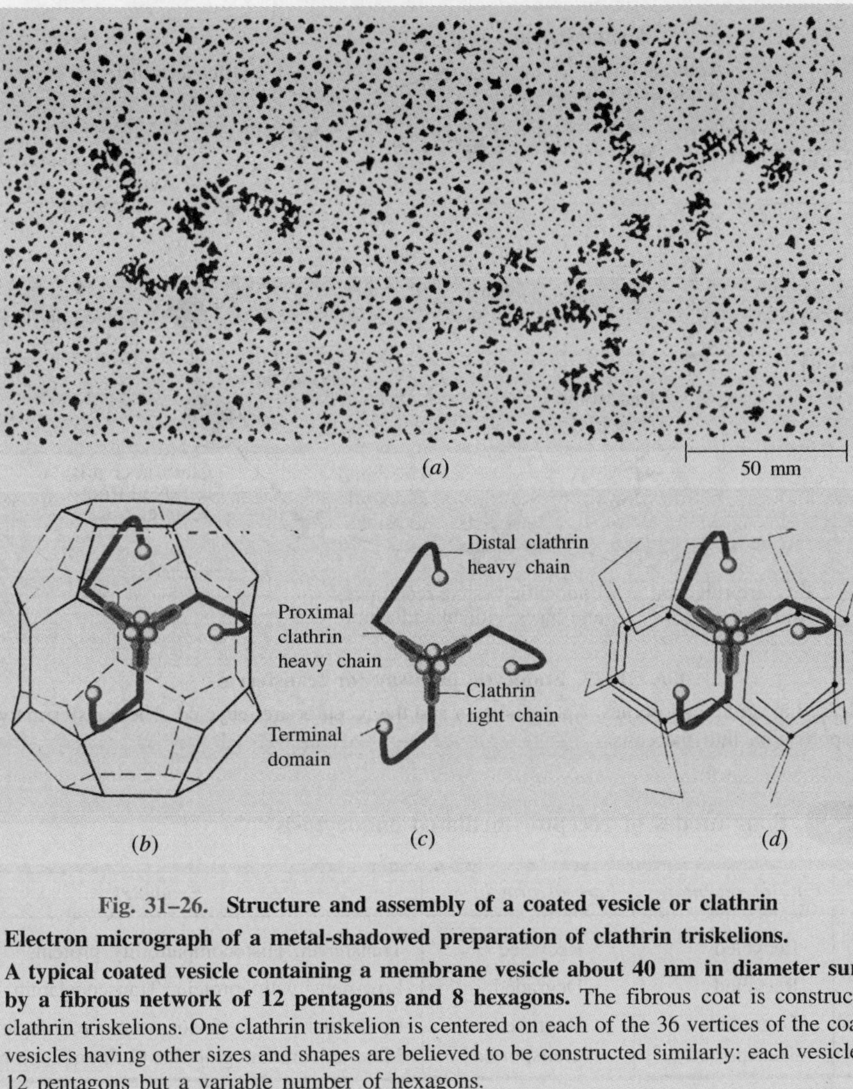

Fig. 31–26. Structure and assembly of a coated vesicle or clathrin

(*a*) **Electron micrograph of a metal-shadowed preparation of clathrin triskelions.**

(*b*) **A typical coated vesicle containing a membrane vesicle about 40 nm in diameter surrounded by a fibrous network of 12 pentagons and 8 hexagons.** The fibrous coat is constructed of 36 clathrin triskelions. One clathrin triskelion is centered on each of the 36 vertices of the coat. Coated vesicles having other sizes and shapes are believed to be constructed similarly: each vesicle contains 12 pentagons but a variable number of hexagons.

(*c*) **Details of a clathrin triskelion.** Each of three clathrin heavy chains is bent into a proximal arm and a distal arm. A clathrin light chain is attached to each heavy chain, most likely near the centre.

(*d*) **An intermediate in the assembly of a coated vesicle, containing 10 of the final 36 triskelions, illustrates the packing of the clathrin triskelions.** Each of the 54 edges of a coated vesicle is constructed of two proximal and two distal arms intertwined. The 36 triskelions contain 36 × 3 = 108 proximal and 108 distal arms, and the coated vesicle has precisely 54 edges.

[*Courtesy: (a) Ernst Ungewickell and Daniel Branton, 1981, and (b), (c) and (d) Nathke IS, Heuser J, Lupas A, Stock J, Turck CW and Brodsky EM, 1992 and Darnell J, Lodish H and Baltimore D, 1986*]

selective process. The lifetime of any particular protein is regulated by proteolytic systems meant for this purpose as opposed to proteolytic events that might occur during post-translational processing. The proteins differ markedly in their half-lives from half a minute to many hours or even days in eukaryotes. Most proteins are turned over rapidly in relation to the lifetime of a cell,

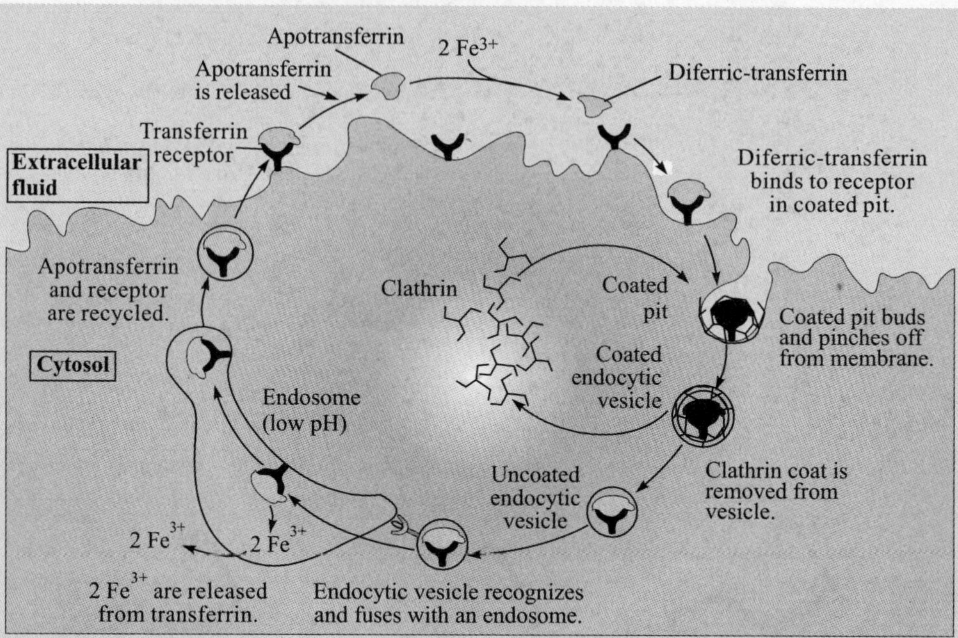

Fig. 31–27. Endocytic pathway for transferrin

Iron is released in acidic endosomes. Apotransferrin and the receptor are recycled. The transferrin cycle, thus, transports iron into the cells.

Table 31–3. Four modes of receptor-mediated endocytosis*

Mode	*Fate of receptor*	*Fate of protein*	*Examples*
1	Recycled	Recycled	Transferrin, Histocompatibility proteins
2	Recycled	Degraded	Low-density lipoprotein, Transcobalamin II
3	Degraded	Degraded	Epidermal growth factor, Immune complexes
4	Transported	Transported	Maternal immunoglobulin G, Immunoglobulin A

* As is evident from the table, the protein transferrin and its receptor both are eventually recycled, whereas LDL is degraded after the associated cholesterol has been delivered to its destination, but its receptor is recycled. Epidermal growth factor and its receptor are both degraded, whereas immunoglobulin A and its receptor are both transported. Hence, it may be concluded that *sorting decisions are made in endosomes*.

although a few stable proteins (hemoglobin, for example) can last for the life span of a cell, *i.e.*, about 110 days for an erythrocyte. The rapidly-degraded proteins include:

a. the defective proteins because of incorrect amino acid(s) insertion during synthesis, or because of a damage occurring during normal cell functioning, and

b. also many enzymes that act at key regulatory points in metabolic pathways.

Defective proteins and proteins with short half-lives are generally degraded in both bacteria and eukaryotes by *ATP-dependent cytosolic systems*. In vertebrates, however, a second system also operates in lysosomes which serves to recycle membrane proteins, extracellular proteins and proteins with long half-lives.

A. Protein Degradation in Prokaryotes

In *Escherichia coli*, many proteins are degraded by an ATP-*dependent protease* called La. The ATPase is activated only in the presence of defective proteins or those destined for rapid turnover. Two ATP moles are hydrolyzed for each peptide bond cleaved. The precise molecular function of ATP hydrolysis during peptide-bond cleavage is unclear.

B. Protein Degradation in Eukaryotes

i. Ubiquitin Proteolytic System

In eukaryotes, the ATP-dependent pathway is quite different. A key component in this system is a small protein (76 amino acids, 8.5 kd) called **ubiquitin**, so named because of its presence throughout the eukaryotic kingdoms. *Ubiquitin is one of the most highly conserved protein in evolution*, so much so that yeast and human ubiquitins differ at only 3 of 76 residues.

In vitro studies on reticulocytes (red blood cells) have established that the C-terminal glycine of ubiquitin becomes covalently linked to the -amino group of lysine residues of proteins destined for destruction. The energy for the formation of these **isopeptide bonds** (*iso* because ε- rather than α-amino groups are partners) come from ATP. This *ATP-dependent pathway* involves 3 steps, each catalyzed by a separate enzyme (Fig. 31–28). The **first step** involves activation of ubiquitin and then its linkage to the *ubiquitin activating enzyme* (E-SH).

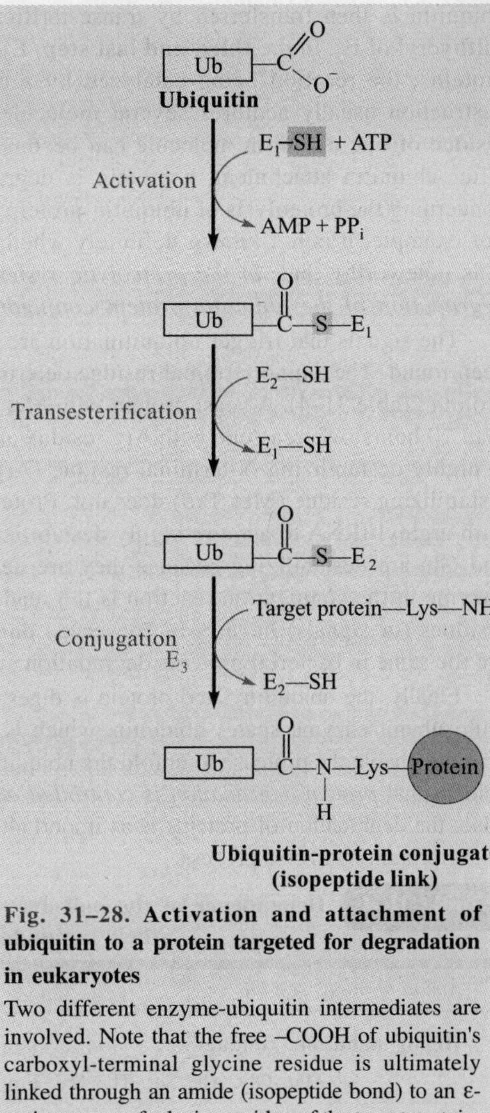

Fig. 31–28. Activation and attachment of ubiquitin to a protein targeted for degradation in eukaryotes

Two different enzyme-ubiquitin intermediates are involved. Note that the free –COOH of ubiquitin's carboxyl-terminal glycine residue is ultimately linked through an amide (isopeptide bond) to an ε-amino group of a lysine residue of the target protein.

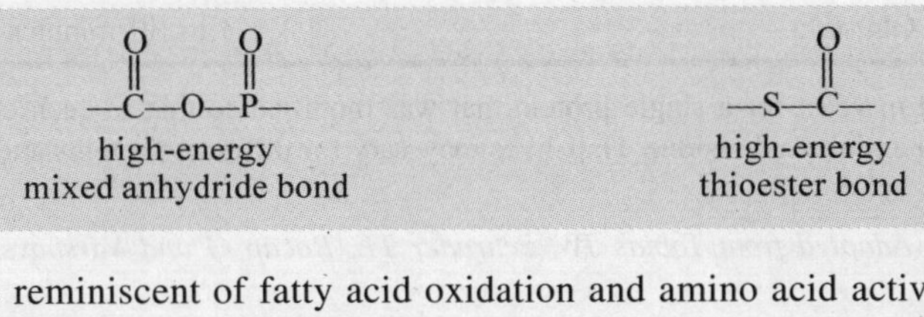

This step is reminiscent of fatty acid oxidation and amino acid activation. Activation requires ATP and formation of AMP ~ ubiquitin (an enzyme-bound intermediate) which, because of its mixed anhydride bond, is a high-energy molecule. Then, ubiquitin bonds to a specific sulfhydryl group of E_1–SH, establishing a high-energy thioester bond. In the **second step**, activated ubiquitin is then transferred by transesterification to a ubiquitin carrier (E_2–SH), *i.e.*, to a sulfhydryl of E_2. In the **third and last step**, E_2–SH donates the ubiquitin moiety to an acceptor protein ; the reaction being catalyzed by a conjugation enzyme (E_3). A protein tagged for destruction usually acquires several molecules of ubiquitin. The ε-amino group of a lysine residue of one ubiquitin molecule can become linked to the terminal carboxylate of another. After ubiquitin attachment, a protein is degraded, or, interestingly, deubiquitinated. Details concerning the proteolysis of ubiquitin-protein conjugates, which requires ATP, remain unclear. For example, it is not known definitely whether ubiquitin is degraded, released or reutilized. It is noteworthy that, *in the proteolytic system, ATP is required for both the formation and degradation of the ubiquitin-protein conjugate*.

The signals that trigger ubiquitination are also not all understood, but one simple one has been found. The amino terminal residue determines to a large extent the half-life of a cytosolic protein (Table 31–4). A yeast protein with Met residue at its N terminus has a half-life of more than 20 hours, whereas one with Arg residue at this position has a half-life of about 2 minutes. A highly destabilizing N-terminal residue (Arg, Leu) favours rapid ubiquitinylation, whereas a stabilizing residue (Met, Pro) does not. Proteins with an N-terminal Asp or Glu residue react with arginyl-tRNA to acquire highly destabilizing Arg as their new N terminus. Similarly, Asn and Gln are destabilizing because they are deamidated to Asp and Glu, respectively. The E_3 enzyme in the conjugation reaction is the reader of the N-terminal residues. These N-terminal residues (or signals) have been conserved during billions of years of evolution : the signals are the same in bacterial proteins degradation systems and in the human ubiquitination pathway.

Finally, the ubiquitinylated protein is digested by a 26 *S protease complex*. This ATP-driven multisubunit enzyme spares ubiquitin, which is then recycled. ATP hydrolysis is repeated many times to enable the protease to unfold the ubiquitinylated protein and gradually digest it. It is, thus, evident that *protein degradation is controlled and conducted by sophisticated molecular devices*. Also, the degradation of proteins is as important to a cell's survival in a changing environment as is the protein synthetic process.

Table 31–4. Dependence of the half-lives of cytosolic yeast proteins on the nature of their amino-terminal residue

Amino-terminal residue	*Half-life** ($t_{1/2}$)
Highly stabilizing residues	
Met, Gly, Ala, Ser, Thr, Val, Pro, Gly	>20 hours
Intrinsically destabilizing residues	
Arg, His, Leu, Ile, Lys, Phe, Trp, Tyr	2 to 30 minutes
Destabilizing following chemical modification	
Asn, Asp, Gln, Glu	3 to 30 minutes

* Half-lives were measured in yeast for a single protein that was modified so that in each experiment it had a different N-terminal amino acid residue. Half-lives may vary for different proteins and in different organisms.

(*Adapted from Tobias JW, Schrader TE, Rocap G and Varshavsky A, 1991*)

ii Polyubiquitin System

DNA sequencing of eukaryotic genes coding for ubiquitin unexpectedly revealed that the protein is synthesized initially as a polyubiquitin precursor molecule, which is then processed to

> ***Xenopus laevis*** is, in fact, a species of frog, not a toad, which is frequently used in studies of early vertebrate development.

produce individual ubiquitin molecules. For example, polyubiquitin of yeast (Fig. 31–29) contains 6 exact repeats of the 76-amino acid sequence of ubiquitin which are arranged, without spacing, in a *head-to-tail sequence*. The repeats are joined directly by the C-terminal glycine of one ubiquitin to the N-terminal methionine of the next. The only variation is the C-terminal residue of the precursor molecule which is asparagine (Asn). *Polyubiquitin organization is a common feature of eukaryotic coding sequences for ubiquitin*; only the number of repeats differ, e.g., human sequence has 9 repeats. The single residue difference at the C-terminus is also found in other eukaryotic precursor molecules, e.g., human and chicken polyubiquitin have valine and tyrosine, respectively. It has been suggested that the nonubiquitin residue at the C-terminus prevents polyubiquitin from participating in the protein-conjugating reactions before processing. However, the *Xenopus laevis* (a South African clawed toad) precursor has only ubiquitin sequences, *i.e.*, no variant C-terminus residue.

```
 1     76    1     76    1     76    1     76    1     76    1     76
Met——Gly  Met——Gly  Met——Gly  Met——Gly  Met——Gly  Met——Gly —Asn
        ↑         ↑         ↑         ↑         ↑
```

Fig 31–29. Polyubiquitin precursor protein of yeast

The 3-dimensional x-ray structures of ubiquitin and tetraubiquitin are presented in Figs. 31–30 and 31–31, respectively.

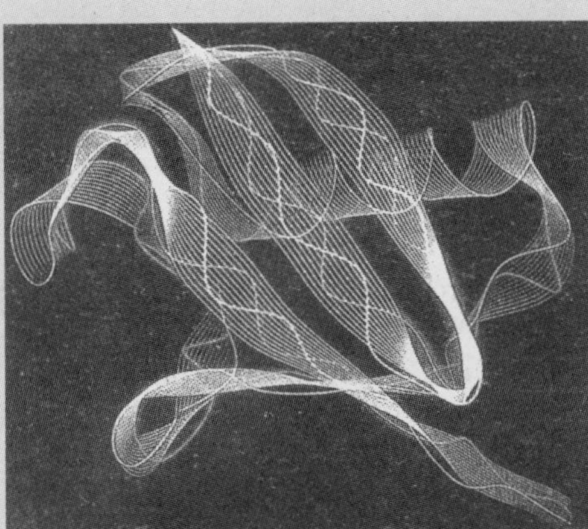

Fig. 30–30. The ribbon model of the 3–'D'structure of ubiquitin

The *white* ribbon represents the polypeptide backbone and the *red* and *blue* curves, respectively, indicate the directions of the carbonyl and amide groups.

[*Courtesy of Charles Bugg*]

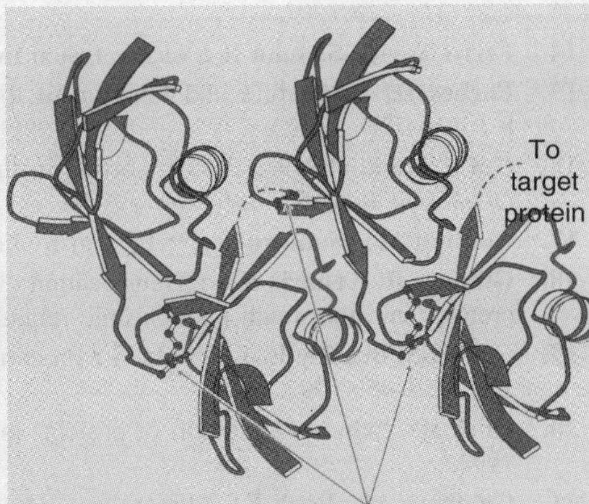

Fig. 30–31. Ribbon model of the 3–'D' structure of tetraubiquitin

Four ubiquitin molecules are linked by isopeptide bonds. This unit is the primary signal for degradation when linked to a target protein.

REFERENCES

1. **Alberts B, Bray D, Lewis J, Raff M, Roberts K, Watson JD :** Molecular Biology of the Cell. *Garland. 1994*
2. **Asano N :** Glycosidase inhibitors: update and perspectives on practical use. *Glycobiology.* **13 :** *93R–r104R, 2003.*

3. **Bachmair A, Finlay D, Varshavsky A :** In vivo half-life of a protein is a function of its amino-terminal residue. *Science.* **234** : *179–186, 1986.*

4. **Bennett MK, Scheller RH :** A molecular description of synaptic vesicle membrane trafficking. *Ann Rev. Biochem.* **63** : *63–100, 1994.*

5. **Bradshaw RA :** Protein translocation and turnover in eukaryotic cells. *Microbiol. Rev.* **55** : *543–560, 1989.*

6. **Blobel G, Walter P, Chang CN *et al* :** Translocation of proteins across membrane: the signal hypothesis and beyond. *Symp. Soc. Exp. Biol.* **33** : *9–36, 1979.*

7. **Blobel G, Dobberstein B :** Transfer of proteins across membrane. *J. Cell Biol.* **67** : *835–851, 1985.*

8. **Choe S, Bennett MJ, Fujii G, Curmi PM, Kantardjieff KA, Collier RJ, Eisenberg D** : The crystal structure of diphtheria toxin. *Nature.* **357** : *216–222, 1992.*

9. **Ciechanover A :** The ubiquitin-proteasome proteolytic pathway. *Cell.* **79** : *13–21, 1994.*

10. **Clarke S :** Protein isoprenylation and methylation at carboxyl-terminal cysteine residues. *Ann. Rev. Biochem.* **61** : *355–386, 1992.*

11. **Dahms NM, Lobel P, Kornfeld S :** Mannose 6-phosphate receptors and lysosomal enzyme targeting *J. Biol, Chem.* **264** : *12115–12118, 1989.*

12. **Dautry-Varsat A, Lodish HF :** How receptors bring proteins and particles into cells. *Sci. Amer.* **250 (5)** : *51–58, 1984.*

13. **Dingwall C :** Transport across the nuclear envelope: Enigmas and explanations. BioEssays. **12** : *213–218,1991.*

14. **Ferro-Novick S, Jahn R :** Vesicle fusion from yeast to man. *Nature.* **370** : *191–193, 1994.*

15. **Forbes DJ :** Structure and function of the nuclear pore complex *Ann. Rev. Cell Biol.* **8** : *495–527, 1992.*

16. **Gal S, Raikhel NV :** Protein sorting in the endomembrane system of plant cells. *Curr. Opin. Cell Biol.* **5** : *636–640, 1993.*

17. **Gething MJ, Sambrook J :** Protein folding in the cell. *Nature.* **355** : *33–45, 1992.*

18. **Gilmore R, Kellaris KV :** Translocation of proteins across and intergration of membrane proteins into the rough endoplasmic reticulum. *Ann. N.Y. Acad. Sci.* **674** : *27–37,1992.*

19. **Glick BS, Beasley EM, Schatz G :** Protein sorting in mitochondria. *Trends Biochem. Sci.* **17** : *453–459,1992.*

20. **Glick BS, Schatz G :** Import of proteins into mitochondria. *Ann. Rev. Genet.* **25** : *21–44, 1991.*

21. **Goldberg AL, Rock KL :** Proteolysis, proteasomes and antigen presentation. *Nature.* **357** : *375–379, 1992.*

22. **Goldstein JL, Brown MS, Anderson RGW, Russell DW, Schneider WJ :** Receptor-mediated endocytosis: Concepts emerging from the LDL receptor system. *Ann. Rev. Cell Biol.* **1** : *1–39, 1985.*

23. **Hall A :** Ras-related proteins. *Curr. Opin. Cell Biol.* **5** : *265–268, 1993.*

24. **Hammond C, Breakman I, Helenius A :** Role of N-linked oligosacchride recognition, glucose trimming, and calnexin in glycoprotein folding and quality control. *Proc. Nat. Acad. Sci.* **91** : *913–917, 1994*

25. **Hart GW :** Glycosylation. *Curr. Opin Cell Biol.* **4** : *1017–1023, 1992*

26. **Hartmann E, Rapoport TA, Lodish HF :** Predicting the orientation of eukaryotic membrane-spanning proteins. *Proc. Nat. Acad. Sci.* **86** : *5786–5790, 1989.*

27. **Hartmann E, Sommer T, Prehn S, Gorlich D, Jentsch S, Rapoport TA :** Evolutionary conservation of components of the protein translocation complex. *Nature.* **367** : *654–657,1994.*

28. **Hendrick JP, Hartl F-U :** Molecular chaperon functions of heat-shock proteins. *Ann. Rev. Biochem.* **62** : *349–384,1993.*

29. **Hershko A, Ciechanover A :** The ubiquitin system for protein degradation. *Ann. Rev. Biochem.* **61** : *761–807,1992.*

30. **Hill WE, Dahlberg E, Garrett RA, Moore PB, Schlessinger D, Warner JR (editors)** : The Ribosome: Structure, Function and Evolution *The American Society for Microbiology, Washington, DC. 1990.*

31. **Hinshaw JE, Carragher BO, Milligan RA :** Architecture and design of the nuclear pore complex. *Cell.* **69** : *1133–1141,1992.*

32. **Hirschberg CB, Snider MD :** Topography of glycosylation in the rough endoplasmic reticulum and Golgi apparatus. *Ann. Rev. Biochem.* **56** : *63, 1987*

33. **Hurt EC, van Loon APGM :** How proteins find mitochondria and intramitochondrial compartments. *Trends Biochem. Sci.* **11** : *204–207, 1986.*

34. **Jentoff N:** Why are proteins O-glycosylated ? *Trends Biochem. Sci.* **15** : *291–294, 1990*

35. **Knight JS, Madueno F, Gray JC :** Import and sorting of protein by chloroplasts. *Biochem. Soc. Trans.* **21** : *31–36, 1993*

36. **Machamer C :** Targeting and retention of Golgi membrane proteins. *Curr. Opin. Cell Biol.* **5** : *606–612, 1993.*

37. **Marquardt T, Denecke J :** Congenital disorders of glycosylation: review of their molecular basis, clinical presentations and specific therapies. *Eur. J. Pediatr.* **162** : *359–379, 2003.*

38. **McKay DB :** Structure and mechanism of 70-kDa heat-shock related proteins. *Adv. Prot. Chem.* **44** : *67–98, 1993.*

39. **Mellman I, Simons K :** The Golgi complex: In vitro veritas? *Cell.* **68** : *829–840, 1992.*

40. **Miller JD, Wilhelm H, Gierasch L, Gilmore R, Walter P :** GTP binding and hydrolysis by the signal recognition particle during initiation of protein translocation. *Nature.* **366** : *351–354, 1993.*

41. **Neupert W, Hill R (editors) :** New Comprehensive Biochemistry: Membrane Biogenesis and Protein Targeting *Vol. 22 Elsevier, Amsterdam. 1992.*

42. **Nunnhari J, Walter P :** Protein targeting to and translocation across the membrane of the endoplasmic reticulum. Curr. Opin. *Cell Biol* **4** : *573–580, 1992.*

43. **Nuoffer C. Balch WE :** GTPases: Multifunctional molecular switches regulating vesicular traffic. *Ann. Rev. Biochem.* **63** : *949–990, 1994.*

44. **Palade G :** Intracellular aspects of the process of protein system. *Science.* **189** : *347–358, 1975.*

45. **Palmer DJ, Helms JB, Beckers CJ, Orci L, Rothman JE :** Binding of coatomer to Golgi Membranes requires ADP-ribosylation factor. *J. Biol. Chem.* **268** : *12083–12089, 1993.*

46. **Pastan I, Willingham MC (editors) :** Endocytosis. *Plenum. 1985.*

47. **Pearse BM, Robinson MS :** Clathrin, adaptors, and sorting, *Ann. Rev. Cell Biol.* **6** : *151–171,1990.*

48. **Pelham HR :** The secretion of proteins by cells. *Proc. Roy. Soc. Lond. Sec. B* **250** : *1–10, 1992.*

49. **Peters J-M :** Proteosomes: Protein degradation machines of the cell. *Trends Biochem. Sci.* **19** : *377–382, 1994.*

50. **Pfanner N, Rasson J, vander Klei IJ, Neupert W :** A dynamic model of the mitochondrial protein import machinery. *Cell.* **68** : *999–1002, 1992.*

51. **Pfeffer SR :** GTP-binding proteins in intracellular transport. *Trends Biochem. Sci.* **2** : *41–46, 1992.*

52. **Pfeffer SR, Rothman JE :** Biosynthetic protein transport and sorting by the endoplasmic reticulum and Golgi. *Ann, Rev. Biochem.* **56** : *829–852, 1987.*

53. **Pryer NK, Wuestehube LJ, Schekman R :** Vesicle-mediated protein sorting. *Ann. Rev. Biochem.* **61** : *471–516, 1992.*

54. **Pugsley AP :** The complete general secretory pathway in gram-negative bacteria. *Microbiol. Rev.* **57** : *50–108, 1993.*

55. **Randall LL, Hardy SJS :** Export of protein in bacteria. *Microbiol. Rev.* **48** : *290–298, 1984.*

56. **Rapoport TA :** Transport of proteins across the endoplasmic reticulum membrane *Science.* **258** : *931–936, 1992.*

57. **Rothman JE :** The compartmental organization of the Golgi apparatus. *Sci. Amer.* **253 (3)** : *74–89, 1985.*

58. **Rothman JE (editor) :** Methods in Enzymology. vol. 219: Reconstitution of Intracellular Transport. *Academic Press, Inc., New York. 1992.*

59. **Rothman JE :** Mechanisms of intracellular protein transport. *Nature.* **372** : *59–67, 1994.*

60. **Schafer WR, Rine J :** Protein prenylation: Genes, enzymes, targets, and functions. *Ann. Rev. Genet.* **26** : *209–237, 1992.*

61. **Schatz G :** The protein import machinery of mitochondria. *Protein Sci.* **2** : *141–146, 1993.*

62. **Schmidt GW, Mishkind ML :** The transport of proteins into chloroplasts. *Ann. Rev. Biochem.* **55** : *879–912, 1986.*

63. **Silver PA :** How proteins enter the nucleus. *Cell.* **64** : *489–497, 1991.*

64. **Simon SM, Blobel G :** A protein-conducting channel in the endoplasmic reticulum. *Cell.* **65** : *371–380, 1991.*

65. **Simon SM, Blobel G :** Signal peptides open protein-conducting channels in *E coli. Cell.* **69** : *677–684, 1992.*

66. **Simons K, Garoff H, Helenius A :** How an animal virus gets into and out of its host cell. *Sci. Amer.* **246 (2)** : *58–66, 1982.*

67. **Smeeckens S, Weisbeek P, Robinson C :** Protein transport into and within chloroplasts. *Trends Biochem. Sci.* **15** : *73–76, 1990.*

68. **Spirin AS :** Ribosomal translocation: Factors and models. *Nucleic Acid Res. Mol. Biol.* **32** : *75, 1985.*

69. **van Heijne G :** Signal sequences: the limits of variation. *J. Mol. Biol.* **184** : *99–105, 1985.*

70. **van Heijne G :** Protein targeting signals. *Curr. Opin Cell Biol.* **2** : *604–608,1990.*

71. **van Meer G :** Transport and sorting of membrane lipids. *Curr. Opin. Cell Biol.* **5** : *661–673, 1993.*

72. **Varshavsky A :** The N-end rule. *Cell.* **69** : *725–735, 1992.*

73. **Ward WHJ :** Diphtheria toxin: A novel cytocidal enzyme. *Trends Biochem. Sci.* **12** : *28–31, 1987.*

74. **Welch W, Georgopoulos C :** Heat-shock proteins or chaperones. *Ann. Rev. Cell Biol.* **9** : *601–634, 1993.*

75. **Wickner W, Driessen AJ, Hartl F-U :** The enzymology of protein translocation across the Escherichia coli plasma membrane. *Ann. Rev. Biochem.* **60** : *101–124, 1991.*

76. **Wickner WT, Lodish HF :** Multiple mechanisms of protein insertion into and across membranes. *Science.* **230 :** *400–407, 1985.*

77. **Wolin SL :** From the elephant to *E.coli:* SRP-dependent protein targeting. *Cell.* **77:** *787–790, 1994.*

PROBLEMS

1. I-cell (inclusion bodies) disease results from a defect in the enzyme that transfers *N*-acetylglucosamine phosphate to proteins containing high mannose-type oligosaccharides. What enzymes would be deficient in I-cell patients? How does the disease name correspond to the problem in these patients?
2. What is the significance of the following structures? A positively-charged amphiphilic α-helix, a cluster of lysine and arginine residues, and a carboxy-terminal Ser-Lys-Leu (SKL) sequence?
3. All are membrane-bound organelles except
   (*a*) Sphaerosome
   (*b*) Ribosomes
   (*c*) Mitochondria
   (*d*) Lysosomes
4. Golgi body is found in
   (*a*) RBC of man
   (*b*) All cells except RBC and bacterial cell
   (*c*) Bacterial cell
   (*d*) All the cells
5. Which organelle is concerned with protein synthesis, mechanical support and enzyme transport?
6. Smallest cell organelle is
   (*a*) Mitochondria
   (*b*) Lysosome
   (*c*) Microfilament
   (*d*) Ribosome
7. Mitochondria are not found in
   (*a*) Young erythrocytes
   (*b*) Mature erythrocytes
   (*c*) Cells
   (*d*) Tissues

ANSWERS TO PROBLEMS

1. Multiple lysosomal enzymes are severely deficient in the lysosomes and are actually found in abnormally high levels in sera and other body fluids. Mannose 6-phosphate is the signal to target enzymes to lysosomes. The first step is addition of *N*-acetylglucosamine phosphate to high-mannose oligosaccharides. Subsequent cleavage, leaving the phosphate behind, produces the mannose 6-phosphate signal. Lack of lysosomal enzymes means that undegraded products accumulate in lysosomes– thus forming *inclusion bodies*.
2. These structure are all targeting signals to send proteins to particular subcellular particles. The three indicated are targets for mitochondria, nucleus, and peroxisomes, respectively.
3. (*b*) Ribosomes
4. (*b*) All cells except RBC and bacterial cell
5. Endoplasmic reticulum
6. (*d*) Ribosome
7. (*b*) Mature erythrocytes

CONTENTS

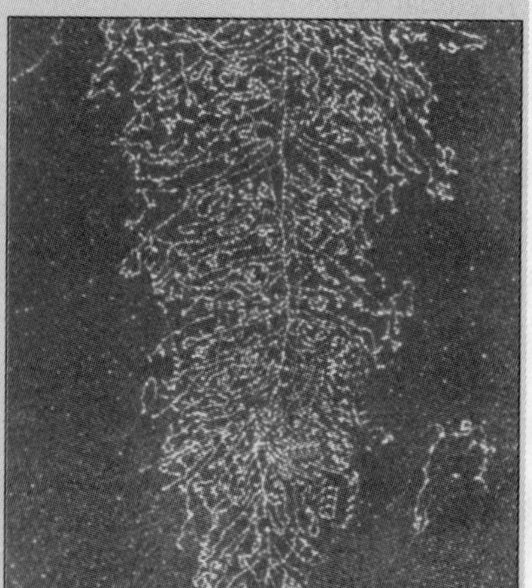

Clusters of messenger RNA molecules (magnified 6700 times) form a fernlike structure around a backbone of DNA molecules, which are undergoing transcription. And it is on these mRNA molecules that the entire genetic dictionary is written.

CHAPTER

32

Genetic Code

GENERAL CONSIDERATIONS

As DNA is a genetic material, it carries genetic informations from cell to cell and from generation to generation. At this stage, an attempt will be made to determine that in what manner the genetic informations are existed in DNA molecule ? Are they written in articulated or coded language on DNA molecule? If in the language of codes, what is the nature of genetic code ?

A DNA molecule is composed of three kinds of moieties: (*i*) phosphoric acid, (*ii*) deoxyribose sugar, and (*iii*) nitrogen bases. The genetic informations may be written in any one of the three moieties of DNA. But the poly-sugar-phosphate backbone is always the same, and it is therefore unlikely that these moiteies of DNA molecule carry the genetic informations. The nitrogen bases, however, vary from one segment of DNA to another, so the informations might well depend on their sequences. *The sequences of nitrogen bases of a given segment of DNA molecule, actually has been found to be identical to linear sequence of amino acids in a protein molecule.* The proof of such a **colinearity** between DNA nitrogen base sequence and amino acid sequence in protein molecules was first obtained from an analysis of mutants of head protein of bacteriophage T_4 (**Sarabhai** *et al*, 1964) and the A protein of tryptophan synthetase of *Escherichia coli* (**Yanofski** *et al*, 1964). *The colinearity of protein molecules and DNA polynucleotides has given the clue that the specific arrangement of four nitrogen bases (e.g., A, T, C and G) in DNA polynucleotide* *chains somehow determines the sequence of amino acids in protein molecules.* Therefore, these four DNA bases can be considered as four alphabets of DNA molecule. All the genetic information, therefore, should be written by these four alphabets of DNA. Now the question arises that whether the genetic informations are written in articulated language or coded language ? If genetic informations might have occurred in an articulated language, the DNA molecule might require various alphabets, a complex system of grammer and ample amount of space on it. All of which might be practically impossible and troublesome too for the DNA. Therefore, it was safe to conclude for molecular biologists that genetic informations were existed in DNA molecule in the form of certain special language of code words which might utilize the four nitrogen bases of DNA for its symbols. Any coded message is commonly called **cryptogram.**

NATURE OF THE GENETIC CODE

Earlier, **Gamow**, the well-known nuclear physicist, proposed that the genetic code consists of three nitogenous (N) bases and the adjacent triplets overlap. This meant that at any particular point the same N-base occurs three times in a vertical manner instead of one which is expected on the basis of colinear model. This hypothesis, however, was not accepted on the following grounds:

1. In the overlapping model only certain amino acids can follow certain others. After the first amino acid in a protein is coded, the next two and for that matter the remaining amino acids in the protein are partially predetermined. If the first code is CAG, then the next must begin with AG and the third one with G.
2. Mutation involving a change in one base, according to this hypothesis, must involve three amino acids.

In order to find the **arrangement of codons**, in later experiments, it was found that when a change occurs due to a mutation, it is confined only to one amino acid. For instance, when sickle cell anemia occurs, only one amino acid, namely glutamic acid is changed into valine, the two adjacent amino acids remaining unaffected. Further research showed that the codons are arranged in a linear order. This explains as to why the change in one involves only one amino acid and not three; if Gamow's hypothesis were correct, change of one nitrogenous base would have involved 3 amino acids.

GEORGE GAMOW
(LT, 1904–1968)

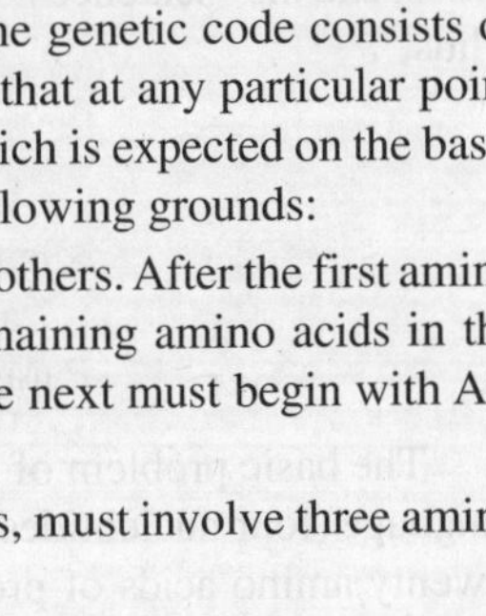

A Russian-born US nuclear physicist and cosmologist who was born on March 4, 1904, in Odessa in the erstwhile USSR. He is famous for his work related to the *evolution of stars*, and for the *advocacy of the Big Bang theory of the Universe*. In genetics, his work paved the way for the subsequent discovery of the role of DNA. In 1928,, Gamow derived a quantum-mechanical formula. This gave a non-zero probability of two charged particles to overcome their mutual electro-static repulsion and come very close to each other. This quantum-mechanical probability is universally known as the ""Gamow factor". The Gamow factor was then used to derive the rate at which nuclear reactions would proceed at high temperatures.

He is perhaps best known for his popular science writings, designed to introduce to the nonspecialist such diffuse subjects as relativity and cosmology. His popular writings include:

1. The Creation of the Universe
2. A Planet Called Earth
3. A Star Called the Sun
4. One, Two, Three ... Infinity

For his achievements as a popularizer of science, Gamow was awarded the **1956 Kalinga Prize** by UNESCO. He died on August 19, 1968 in Boulder, Colorado, USA.

The sequence of bases that encodes a functional protein molecule is called a **gene.** And the **genetic code** is the relation between the base sequence of a gene and the amino acid sequence of the polypeptide whose synthesis the gene directs. In other words, the specific correspondence between a set of 3 bases and 1 of the 20 amino acids is called the genetic code.

J.D. Burke (1970) defined genetic code in the following words,

> " The genetic code for protein synthesis is contained in the base sequence of DNA. ... The genetic code is a code for amino acids. Specifically, it is concerned with what codons specify what amino acids."

The genetic code is the key that relates, in Crick's words, "...the two great polymer languages, the nucleic acid language and the protein language."

The "letters" in the "language" were found to be the bases; the "words" (codons) are groups of bases; and the "sentences" and "paragraphs" equate with groups of codons (**Eldon J. Gardner**, *1968*). Thus,

Letters	≡	Bases
Words	≡	Groups of bases (*i.e.*, codons)
Sentences and Paragraphs	≡	Groups of codons

The basic problem of such a genetic code is to indicate how information written in a four-letter-language (four nucleotides or nitrogen bases of DNA) can be translated into a twenty-letter-language (twenty amino acids of proteins). *The group of nucleotides that specifies one amino acid is a code word or codon.* The simplest possible code is a **singlet code** (a code of single letter) in which one nucleotide codes for one amino acid. Such a code is inadequate for only four amino acids could be specified. A **doublet code** (a code of two letters) is also inadequate because it could specify only sixteen (4 × 4) amino acids, whereas a **triplet code** (a code of three letters) could specify sixty four (4 × 4 × 4) amino acids. Therefore, it is likely that there may be 64 triplet codes for 20 amino acids. *The possible singlet, doublet and triplet codes, which are customarily represented in terms of "mRNA language", (mRNA is a complementary molecule which copies the genetic informations during its transcription)* can be illustrated as in Fig. 32–1. Larger than three letter units would seem wasteful and evidence already accumulated suggests that such larger units are unlikely.

The first experimental evidence in support to the concept of triplet codes was provided by **Crick** and coworkers in 1961. During their experiment, when they added or deleted single, or double base pairs in a particular region of DNA of T_4 bacteriophages of *E. coli*, they found that such bacteriophages ceased to perform their normal functions. However, bacteriophages with addition or deletion of three base pairs in DNA molecule, had performed normal functions. From this experiment, they concluded that *a genetic code is in triplet form because the addition of one or two nucleotides has put the reading of the code out of order*, while the addition of third nucleotide resulted in a return to the proper reading of the message.

A strong evidence in favour of triplet coding units is derived from determinations of the coding ratio. **Coding ratio** is equal to the number of nucleotides in the mRNA (or nucleotide pairs in the double-stranded DNA) divided by the number of amino acid residues of the resultant polypeptide chain. This expresses the number of nucleotides per coding unit. In each of the several genes so far studied, the number of nucleotide pairs of DNA has been estimated by genetic techniques and compared with the number of amino acid residues of the protein synthesised by the gene. All estimates give coding ratios close to three, indicating that three nucleotides compose a coding unit (triplet code). A wide variety of genetic experiments is consistent with a triplet code. The conclusion is inescapable that sequences of

Singlet Code (4 Words)

A
G
C
U

Doublet Code (16 Words)

AA	AG	AC	AU
GA	GG	GC	GU
CA	CG	CC	CU
UA	UG	UC	UU

Triplet Code (64 Words)

AAA	AAG	AAC	AAU
AGA	AGG	AGC	AGU
ACA	ACG	ACC	ACU
AUA	AUG	AUC	AUU
GAA	GAG	GAC	GAU
GGA	GGG	GGC	GGU
GCA	GCG	GCC	GCU
GUA	GUG	GUC	GUU
CAA	CAG	CAC	CAU
CGA	CGG	CGC	CGU
CCA	CCG	CCC	CCU
CUA	CUG	CUC	CUU
UAA	UAG	UAC	UAU
UGA	UGG	UGC	UGU
UCA	UCG	UCC	UCU
UUA	UUG	UUC	UUU

Fig. 32–1. Singlet, doublet and triplet codes of mRNA

3 bases in the mRNA molecule (triplet) contain coded information for the various amino acids. Such a triplet is called a **codon.** Each triplet codon specifies only one particular amino acid and the position of the codon in the mRNA molecule specifies the position of the amino acid in a polypeptide chain. As stated above, with four bases, 64 triplet codons (4^3) are possible. The genetic dictionary, thus, consists of 64 words, each made of a specific sequence of 3 out of 4 letters of the genetic alphabet. Any letter may occur more than once in a codon. *An example from the English language may be used to explain this*. Consider a 4-letter word "SEAT". Out of the four letters of this word, you can make many 3-letter words, each of which conveys a definite meaning, *e.g.*, "SEA", "SEE", "SET", "SAT", "EAT", "ASS", "ATE" and "TEA". The genetic code has now been experimentally deciphered and perfected by the combined efforts of many biochemists, notably **Marshall Warren Nirenberg** and **Har Gobind Khorana**, who were awarded the 1968 Noble Prize for their work, along with **Robert Holley** who was the first scientist to determine the nucleotide sequence of several tRNAs.

THE GENETIC CODE

The genetic language consists of only four letters contained in the word "GACU". These four letters can be combined to form 64 genetic words, each consisting of 3 letters. Each triplet word (codon) has a specific meaning which the cell understands. It codes for a particular amino acid. The genetic code, as at present known, is shown in Fig. 32–2. It shows the base sequences of the various codons (triplets) and against each codon is given the amino acid that it codes. Just as different combinations of different words make different sentences, each having a specific meaning,

FIRST BASE OF CODON (5′ end)	SECOND (OR MIDDLE) BASE OF CODON: U	C	A	G	THIRD BASE OF CODON (3′ end)
U	UUU Phe	UCU Ser	UAU Tyr	UGU Cys	U
	UUC Phe	UCC Ser	UAC Tyr	UGC Cys	C
	UUA Leu	UCA Ser	UAA †stop ochre	UGA †stop opal	A
	UUG Leu	UCG Ser	UAG †stop amber	UGG Trp	G
C	CUU Leu	CCU Pro	CAU His	CGU Arg	U
	CUC Leu	CCC Pro	CAC His	CGC Arg	C
	CUA Leu	CCA Pro	CAA Gln	CGA Arg	A
	CUG Leu	CCG Pro	CAG Gln	CGG Arg	G
A	AUU Ile	ACU Thr	AAU Asn	AGU Ser	U
	AUC Ile	ACC Thr	AAC Asn	AGC Ser	C
	AUA Ile	ACA Thr	AAA Lys	AGA Arg	A
	AUG* Met/start	ACG Thr	AAG Lys	AGG Arg	G
G	GUU Val	GCU Ala	GAU Asp	GGU Gly	U
	GUC Val	GCC Ala	GAC Asp	GGC Gly	C
	GUA Val	GCA Ala	GAA Glu	GGA Gly	A
	GUG* Val/start	GCG Ala	GAG Glu	GGG Gly	G

Fig. 32–2. The genetic code dictionary

The genetic code comprises 64 codons, which are permutations of four bases taken in threes. Note the importance of sequence: three bases (A,C, and G, for example), each used once per triplet codon, gives six permutations. ACG, AGC, GAC, GCA, CAG and CGA which code for threonine, serine, aspartic acid, alanine, glutamine and arginine, respectively. The 64 triplet codons are listed in the 5′ ⟶ 3′ direction in which they are read. Against each condon is written the name of the amino acid encoded by it. The first genetic code ever to be established was the codon (UUU) for phenylalanine (Phe). AUG and GUG encode methionine, which initiates most polypeptide chains. All other amino acids, except tryptophan (which is encoded only by UGG), are represented by 2 to 6 triplets. The 3 triplets UAA, UAG and UGA are termination signals and do not encode any amino acids. UGA rarely codes for selenocysteine (SeCys).

To read code dictionary: A codon consists of three nucleotides read in the sequence indicated by the column heads. For example, ACU codes threonine. The first letter, A, is read in the first-letter column; the second letter, C, from the second-letter column; and the third letter, U, from the third-letter column. Each codon is recognized by a corresponding anticodon sequence on a tRNA molecule. Some tRNA molecules recognize more than one codon sequence but always for the same amino acid. Most amino acids are encoded by more than one codon. For example, threonine is encoded by four codons (ACU, ACC, ACA, and ACG), which differ from one another only in the third position.

* Chain initiation codons

† Chain termination codons

similarly different sequences of codons on mRNA specify different proteins, each with a specific sequence of amino acids. Although the genetic information resides in DNA, the terms genetic code and codon are used with reference to mRNA because mRNA is the nucleic acid which directly determines the sequence of amino acids in a protein. This expression of genetic information in the amino acid sequence of proteins by mRNA is called **translation.** *The DNA-RNA-Protein code may be expressed as under:*

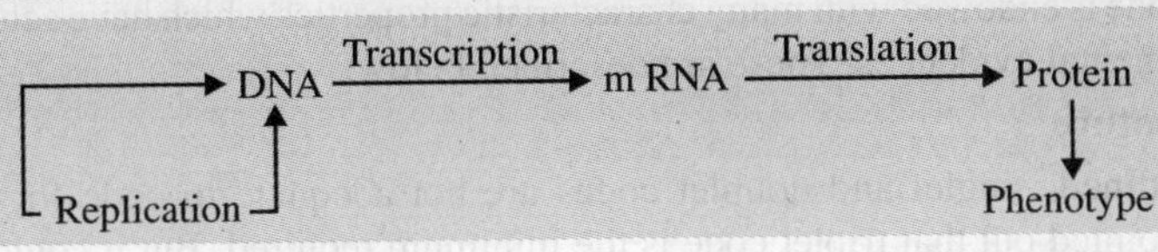

The synthesis of cellular proteins takes place in the joining together of several amino acids to form a linear polypeptide chain of variable length. There are 20 different amino acids which are commonly found in protenis (hence called protein amino acids) and which take part in their synthesis. The mRNA codons for these 20 protein amino acids, as can be deduced from Fig. 32–2, are given in Fig. 32–3.

Amino acid	*Abbreviation*	*mRNA codons*		*Total no. of codon(s)*
		Common bases	*Complete codon(s)*	
1. Alanine	Ala	GC–	GCU, GCC, GCA, GCG	4
2. Arginine	Arg	CG–	CGU, CGC, CGA, CGG	6
		AG–	AGA, AGG	
3. Asparagine	Asn	AA–	AAU, AAC	2
4. Aspartic acid	Asp	GA–	GAU, GAC	2
5. Cysteine	Cys	UG–	UGU, UGC	2
6. Glutamic acid	Glu	GA–	GAA, GAG	2
7. Glutamine	Gln	CA–	CAA, CAG	2
8. Glycine	Gly	GG–	GGU, GGC, GGA, GGG	4
9. Histidine	His	CA–	CAU, CAC	2
10. Isoleucine	Ile	AU–	AUU, AUC, AUA	3
11. Leucine	Leu	UU–	UUA, UUG	6
		CU–	CUU, CUC, CUA, CUG	
12. Lysine	Lys	AA–	AAA, AAG	2
13. Methionine	Met	AU–	AUG	1
14. Phenylalanine	Phe	UU–	UUU, UUC	2
15. Proline	Pro	CC–	CCU, CCC, CCA, CCG	4
16. Serine	Ser	UC–	UCU, UCC, UCA, UCG	6
		AG–	AGU, AGC	
17. Threonine	Thr	AC–	ACU, ACC, ACA, ACG	4
18. Tryptophan	Trp	UG–	UGG	1
19. Tyrosine	Tyr	UA–	UAU, UAC	2
20. Valine	Val	GU–	GUU, GUC, GUA, GUG	4
Terminator triplets	**Trm**	UA–	UAA, UAG	3
		UG–	UGA	
			Total	**64**

Fig. 32–3. Proteinogenic amino acids and their mRNA codons

CHARACTERISTICS OF THE GENETIC CODE

The genetic code is endowed with many characteristic properties which have actually been proved by definite experimental evidences. These are described below:

1. Triplet nature

As earlier outlined, singlet and doublet codes are not adequate to code for 20 amino acids; therefore, it was pointed out that triplet code is the minimum required. But it could be a quadruplet code or of a higher order. As pointed out above, in a triplet code of 64 codons, there is an excess of (64 – 20) = 44 codons and, therefore, more than one codons are present for the same amino acid. This excess will be still greater if more than three-letter words are used. In a *quadruplet code* there will be 4^4 $(4 \times 4 \times 4 \times 4) = 256$ possible codons. An account of the 20 amino acids along with their corresponding codons is presented below:

2	amino acids (Met, Trp)	... have 1 codon each	=	2
9	amino acids (Asn, Asp, Cys, Gln, Glu, His, Lys, Phe, Tyr)	... have 2 codons each	=	18
1	amino acid (Ile)	... has 3 codons	=	3
5	amino acids (Ala, Gly, Pro, Thr, Val)	... have 4 codons each	=	20
3	amino acids (Arg, Leu, Ser)	... have 6 codons each	=	18
		3 terminator codons	=	3
20	Amino acids			64

A closer scrutiny of the **Genetic Dictionary** (Fig. 32–2) has revealed the emergence of certain trends or **patterns of the genetic code.** These are:

1. Amino acids with similar structural properties tend to have related codons. Thus, aspartic acid codons (GAU, GAC) are similar to glutamic acid codons (GAA, GAG); the difference being exhibited only in the third base (toward 3′ end).
2. Similarly, the codons for the aromatic amino acids phenylalanine (UUU, UUC), tyrosine (UAU, UAC) and tryptophan (UGG) all begin with uracil (U).
3. The first two bases of all the 4 codons assigned to each of the 5 amino acids are similar : GC for alanine, GG for glycine, CC for proline, AC for threonine and GU for valine.
4. All codons with U in the second position specify hydrophobic amino acids (Ile, Leu, Met, Phe, Val).
5. All codons with A in the second position specify the charged amino acids, except Arg.
6. All the acidic (Asp, Glu) and basic (Arg, Lys) amino acids have A or G as the second base.

2. Degeneracy

The code is degenerate which means that the same amino acid is coded by more than one base triplet. Degeneracy, as used here, does not imply lack of specificity in protein synthesis. It merely means that a particular amino acid can be directed to its place in the peptide chain by more than one base triplets (**Eldon J. Gardner**, 1968). For example, the three amino acids arginine, alanine and leucine each have six synonymous codons. A non-degenerate code would be one where there is one to one relationship between amino acids and the codons, so that from the 64 codons, 44 will be useless or nonsense codons. It has been definitely shown that there are no nonsense codons. The codons which were initially called nonsense codons were later shown to mean stop signals.

The code degeneracy is basically of 2 types: partial and complete. In *partial degeneracy*, the first two nucleotides are identical but the third (*i.e.,* 3′ base) nucleotide of the degenerate codon differs; for example, CUU and CUC code for leucine. *Complete degeneracy* occurs when any of the 4 bases can take third position and still code for the same amino acid; for example, UCU, UCC, UCA and UCG all code for serine.

Degeneracy of genetic code has certain biological advantages. For example, it permits essentially the same complement of enzymes and other proteins to be specified by the microorganisms varying widely in their DNA base composition. Degeneracy also provides a mechanism of minimizing mutational lethality.

Degeneracies occur frequently in the third letter of the codon. Exceptions are, however, arginine (Arg), leucine (Leu) and serine (Ser) which have 2 groups of codons or triplets, which differ in either the first base only (Arg, Leu) or in both the first and second bases (Ser).

3. Nonoverlapping

The genetic code is nonoverlapping, *i.e.*,the adjacent codons do not overlap. A nonoverlapping code means that the same letter is not used for two different codons. In other words, no single base can take part in the formation of more than one codon. Fig. 32–4 shows that an overlapping code can mean coding for four amino acids from six bases. In actual practice, six bases code for not more than two amino acids. As an illustration, an end-to-end sequence of 5′ UUUCCC 3′ on mRNA will code only 2 amino acids, *i.e.*, phenylalanine (UUU) and proline (CCC).

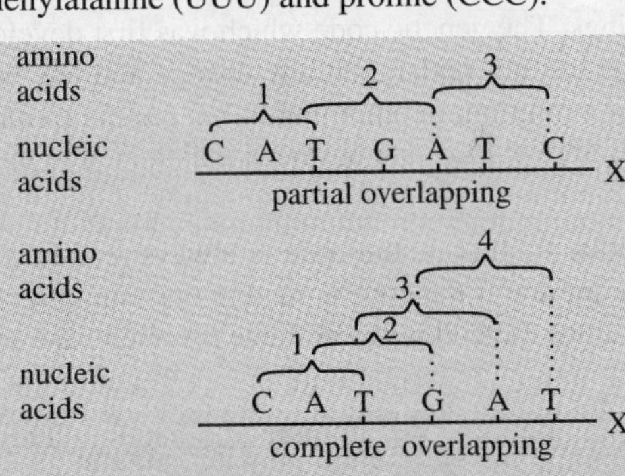

Fig. 32–4. Overlapping of codons due to one letter or two letters

4. Commaless

There is no signal to indicate the end of one codon and the beginning of the next. The genetic code is commaless (or comma-free). A commaless code means that no codon is reserved for punctuations or the code is without spacers or space words. There are no intermediary nucleotides (or commas) between the codons. In other words, we can say that after one amino acid is coded, the second amino acid will be automatically coded by the next three letters and that no letters are wasted for telling that one amino acid has been coded and that second should now be coded (Fig. 32–5).

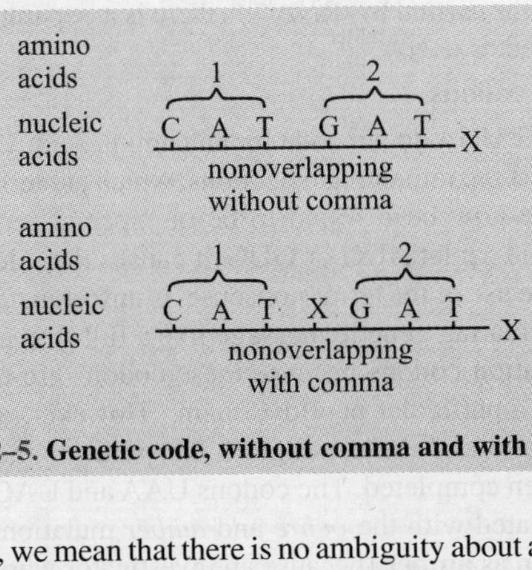

Fig. 32–5. Genetic code, without comma and with comma

5. Nonambiguity

By non-ambiguous code, we mean that there is no ambiguity about a particular codon. A particular codon will always code for the same amino acid. In an ambiguous code, the same codon could code for two or more than two different amino acids. Such is not the case. *While the same amino acid can be coded by more than one codon (the code is degenerate), the same codon shall not code for two or more different amino acids (non-ambiguous).* But sometimes the genetic code is ambiguous, that is, same codon may specify more than one amino acid. For example, UUU codon usually codes for phenylalanine but in the presence of streptomycin, may also code for isoleucine, leucine or serine.

6. Universality

The genetic code applies to all modern organisms with only very minor exceptions. Although the code is based on work conducted on the bacterium *Escherichia coli* but it is valid for other organisms. This important characteristic of the genetic code is called its **universality.** It means that the same sequences of 3 bases encode the same amino acids in all life forms from simple microorganisms to complex, multicelled organisms such as human beings. Consider any codon. It codes for the same amino acid from the smallest organism to the largest, plant or animal. Thus, UUU codes for phenylalanine and GUC for valine in all living things, from amoeba to ape, bacteria to the banyan tree, and from cabbage to kings. The genetic code which was first developed in the bacteria about 3 billion (320 crore) years ago has not undergone any change and has been preserved in its almost original form in the course of evolution. In other words, *the code is a conservative one, i.e.*, the code was fixed early in the course of evolution and has been maintained to the present day.

7. Polarity

The genetic code has polarity, that is, the code is always read in a fixed direction, *i.e.*, in the $5' \rightarrow 3'$ direction. It is apparent that if the code is read in opposite direction (*i.e.*, $3' \rightarrow 5'$), it would specify 2 different proteins, since the codon would have reversed base sequence:

Codon :	UUG	AUC	GUC	UCG	CCA	ACA	AGG
Polypeptide :	→ Leu	Ile	Val	Ser	Pro	Thr	Arg
	Val	Leu	Leu	Ala	Thr	Thr	Gly ←

8. Chain initiation codons

The triplets AUG and GUG play double roles in *E. coli.* When they occur in between the two ends of a cistron (intermediate position), they code for the amino acids methionine and valine, respectively in an intermediate position in the protein molecule. But when they occur immediately after a terminator codon, they act as "chain initiation" (C.I.) signals or "starter codons" for the synthesis of a polypeptide chain. It has also been shown that the initiating methionine molecule should be found in the formylated state. This makes a distinction between the initiating methionine and the methionine at internal position. The methionine when required at internal position should not be formylated. Also while formyl methionine is carried by $tRNA^{fMet}$, there is a separate species of tRNA for internal methionine and it is designated as $tRNA^{mMet}$.

9. Chain termination codons

The 3 triplets UAA, UAG, UGA do not code for any amino acid. They were originally described as **nonsense codons,** as against the remaining 61 codons, which are termed as **sense codons.** The so-called nonsense codons have now been found to be of "special sense". When any one of them occurs immediately before the triplet AUG or GUG, it causes the release of the polypeptide chain from the ribosome. Hence, the use of the term 'nonsense' is unfortunate. These *special-sense codons* perform the function of punctuating genetic message like a full stop at the end of a sentence. They are also called chain termination codons because these codons are used by the cell to signal the natural end of translation of a particular peptidyl chain. However, their inclusion in any mRNA results in the abrupt termination of the message at the point of their location even though the polypeptide chain has not been completed. The codons UAA and UAG were discovered in bacteria and were respectively associated with the *ochre* and *amber* mutations. Hence, UAA is also called **ochre** and UAG is also known as **amber** (because an investigator who studied the properties of this codon belonged to the Bernstein family, and Bernstein means amber in German). UGA is also called

opal. They resulted in the formation of incomplete polypeptide chains. UGA is the usual terminator codon in all cases.

Most of these principles are illustrated by the following analogy. Consider a sentence (gene) in which the words (codons) each consist of 3 letters (bases).

THE BIG RED FOX ATE THE EGG

(Here the spaces separating the words have no physical significance; they are only present to indicate the reading frame). The deletion of the 4th letter, which shifts the reading frame, changes the sentence to

THE IGR EDF OXA TET HEE GG

so that all words past the point of deletion are unintelligible (specify the wrong amino acids). An insertion of any letter, however, say an X in the 9th position,

THE IGR EDX FOX ATE THE EGG

restores the original reading frame. Consequently, only the words between the two changes (mutations) are altered. As in this example, such a sentence might still be intelligible (the gene could still specify a functional protein), particularly if the changes are close together. Two deletions or two insertions, no matter how close together, would not suppress each other but just shift the reading frame. However, three insertions, say X, Y, and Z in the 5th, 8th and 12th positions, respectively, would change the sentence to

THE BXI GYR EDZ FOX ATE THE EGG

which, after the third insertion, restores the original reading frame. The same would be true of three deletions. As before, if all three changes were close together, the sentence might still retain its meaning.

Crick and **Brenner** did not unambiguously demonstrate that the genetic code is a triplet code because they had no proof that their insertions and deletions involved only single nucleotides. Strictly speaking, they showed that a codon consists of $3r$ nucleotides where r is the number of nucleotides in an insertion or deletion. Although it was generally assumed at the time that $r = 1$, proof of this assertion had to await the elucidation of the genetic code.

DECIPHERING THE GENETIC CODE OR CODON ASSIGNMENT

The early studies that led to the breaking of the genetic code are serendipitous. Following the discovery of mRNA, the immediate question was: How is it read? It was clear that a triplet code was the minimum required to account for all 20 amino acids, but many kinds of triplet codes are possible. Some conceivable ones are shown in Fig. 32–6. An *overlapping code* like that shown in Fig. 32–6(*a*) would be space-saving. This possibility was eliminated, however, by the observation that most mutations result in the change of a single amino acid. If the codons overlapped, there should be a significant number of cases in which two adjacent residues are modified by mutation of the "overlap" residue. Furthermore, an overlapping code would lead to statistical regularities between neighbouring amino acid residues in proteins—that is, some amino acids would be neighbours more often than others—and this has never been observed. Another possibility was that the code was *punctuated*. That is, as shown in Fig. 32–6 (*b*), some base or bases might serve as "spacing" between code words.

> Serendipitous is adjective of the word serendipity which means occurrence of events by chance in a fortunate way. The word **'serendipity'** was coined ca 1754 by **Horace Walpole** after *The Three Princes of Serendip* (Serendip, a former name for Sri Lanka), a Pers fairy tale in which the princes make such discoveries.

We may here briefly describe the techniques which ultimately resulted in determining the code words for each of 20 proteinogenic amino acids. Most of our current knowledge of the general aa₁ aa₃ aa₅ aa₇ aa₉ aa₁₁ aa₁₃ aa₁₅

U C |C |G G |A|A A |C |U C |C| U U|U| C A|G| C C |C| U C |A| U C

aa₂ aa₄ aa₆ aa₈ aa₁₀ aa₁₂ aa₁₄ aa₁₆

(*a*) Overlapping code: There will be statistical regularities between adjacent amino acid residues. Point mutations will be able to change two amino acid residues.

aa₁ aa₂ aa₃ aa₄ aa₅ aa₆

U C C | G |G A A|A| C U C|C|U U U|C| A G C|C|C U C|A|U C

(*b*) Punctuated code: Deletions of four nucleotides (or multiples thereof) will restore the reading frame.

aa₁ aa₂ aa₃ aa₄ aa₅ aa₆ aa₇ aa₈

U C C |G G A| A A C|U C C|U U U|C A G |C C C|U C A|U C

(*c*) Unpunctuated code: Deletions of three nucleotides (or multiples thereof) will restore the reading frame. This is the actual form of the code.

Fig. 32–6. Three conceivable types of genetic codes

Early research on the nature of the code quickly showed that a nonoverlapping, unpunctuated code (*c*) fit all experimental observations.

nature of the genetic code and the nucleotide composition were obtained from the following *four main types of experimental approaches*:

A. Assignment of Codons with Unknown Sequences

1. Polyuridylic Acid Method : This method for establishing the genetic code consists of acids in the protein molecule synthesized under its influence. The sequence analysis for proteins is now a standard biochemical technique but unfortunately the determination of the base sequence of large nucleic acid molecules (like mRNA) is yet to be achieved. Biochemists, therefore, adopted indirect methods to crack the genetic code. A cell-free amino acid incorporating system and the polymerizing enzyme were known which could polymerize single ribonucleoside triphosphates or their mixtures to produce long nucleotide chains of random base sequence. *Initial breakthrough in this direction was achieved by* **Marshall Nirenberg** and **Heinrich Matthaei** *in 1961 who adopted the following technique*. Using the enzyme *polynucleotide phosphorylase*, they prepared a synthetic polyribonucleotide containing only one kind of base and added it to a cell-free amino acid-incorporating system from *E. coli*. In this way, they prepared, for example, a polyuracil (UUU...UUU) to which they added different amino acids labelled with C^{14}. They found that this poly-U system specifically stimulated the formation of a polypeptide which contained only the amino acid *phenylalanine*. The conclusion was drawn that the code word for the amino acid phenylalanine was a sequence of three uracil nucleotides (UUU) in mRNA. This was the first demonstration that synthetic polyribonucleotide could act as a messenger for polypeptide synthesis. Later on, **Nirenberg** and **Severo Ochoa** became deeply involved in this problem. Subsequently, they found that poly A gave polylysine and poly C gave polyproline. Therefore, AAA was assigned to *lysine* and CCC to *proline*. Similar experiments with poly G were not successful, because it attains secondary structure and thus cannot attach to ribosomes. The following three codons thus could be assigned, using homopolymers of RNA, without any difficulty: UUU = *phenylalanine;* AAA = *lysine;* CCC = *proline*.

2. Copolymers Method : In this method, Nirenberg used mixtures of two or more ribonucleoside disphosphates and with the help of the enzyme mentioned above, polyribonucleotides were prepared. Thus, using UDP and CDP in the ratio of 3:1, he obtained a polynucleotide which contained the following triplets in the descending order of frequency: UUU, UUC, UCU and CUU. The triplets containing 2Cs and IU were the least frequent. With such a poly-UC system, **Nirenberg**

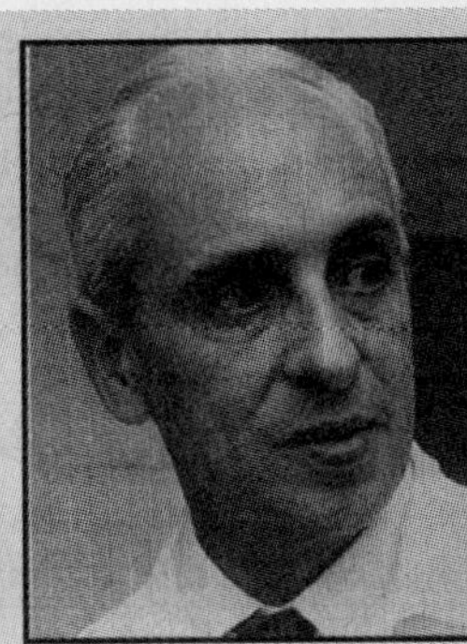

SEVERO OCHOA
(LT, 1905–1993)

A Spanish-born US biochemist who was educated in his home country, graduating from Malaga University in 1921 and then obtaining his MD from Madrid in 1929. He worked in Germany with **Otto Meyerhof** on muscle biochemistry and then moved *via* England to the USA, joining New York University in 1942. During the 1950s, his research centred on the enzymes that utilize the energy held in the energy-rich phosphate bonds of ATP. This led in 1955 to his discovery, with **Marianne Grunberg-Manago,** of *polynucleotide phosphorylase,* the enzyme that was subsequently used to prepare synthetic mRNA molecules for elucidation of the genetic code. Ochoa received the **1959 Nobel Prize for Physiology or Medicine** with **A. Kornberg** for his first synthesis of a nucleic acid (**synthetic RNA**). In 1985, he returned to Madrid University as Professor of Biology.

His advice to the science students regarding conduction of research reads as follows:

"My advice to students of science is that if they have an urge to do research they should do it by all means. Nothing should stand in the way of a strong wish to devote Life to Science."

"If you have the urge to do scientific research get the proper training and by all means do it; nothing else is likely to give you so much satisfaction and, above all such a sense of fulfilment."

obtained a polypeptide containing the amino acids phenyalanine and serine in the ratio 3:1. He concluded that the code word for serine contains 2Us and 1C. This method, however, did not give the exact sequence of the three bases. Ochoa also used this technique and determined the code words for most of the amino acids.The exact sequence of the three bases for any amino acid was not known because of these polynucleotides had random base sequence. The assigning of the exact triplet for every amino acid was not possible at this stage.

In an experiment, **Nirenberg** and his coworkers used RNA synthesized by using two or more bases. For instance, if only A and C are used, poly AC will consist of eight possible codons, namely AAA, AAC, ACA, CAA, CCA, CAC, ACC and CCC. The proportion of these eight codons, in the synthetic RNA can be calculated if the known quantities of A and C are used for the synthesis of poly AC. For instance, if A : C = 5 : 1, (5/6 is A and 1/6 is C), the calculated relative proportions of eight codons on random basis would be as given in Table 32–1.

Table 32–1. The relative proportions of different codons in mRNA formed due to bases A : C taken in 5 : 1 ratio.

Base composition	*Codon and probability*	*Ratio using*	
		minimum (as 1)	*maximum (as 100)*
(1) 3A	1. AAA 5/6 × 5/6 × 5/6 = 125/216	125	100
(2) 2A1C	2. AAC 5/6 × 5/6 × 1/6 = 25/216	25	20
	3. ACA 5/6 × 1/6 × 5/6 = 25/216	25	20
	4. CAA 1/6 × 5/6 × 5/6 = 25/216	25	20
(3) 1A2C	5. CCA 1/6 × 1/6 × 5/6 = 5/216	5	4
	6. CAC 1/6 × 5/6 × 1/6 = 5/216	5	4
	7. ACC 5/6 × 1/6 × 1/6 = 5/216	5	4
(4) 3C	8. CCC 1/6 × 1/6 × 1/6 = 1/216	1	0.80

The calculated relative proportions of codons were compared with the proportions in which different amino acids were present in the polypeptides synthesized using poly AC. For instance, if an amino acid is 1/5th of lysine (coded by AAA), we can say that it should be coded by one of the three possible 2A1C codons (AAC, ACA or CAA). Similar reasoning would allow assignments of 1A2C as well as 3C. However, using this technique, it was not possible to assign the three codons of the category 2A1C (*i.e.*, AAC, ACA, CAA) to three amino acids, since these will be present in equal quantities. Therefore, the codons were initially assigned only with respect to base composition, ignoring the sequences of the bases in codons, as done in the above example. The assignments are given in Table 32–2.

Table 32–2. Codon assignments derived due to use of A : C (5 : 1) in the synthesis of mRNA

Amino acids	*Codon composition*
(1) lysine	3A
(2) asparagine, glutamine and threonine	2A1C
(3) histidine, proline and threonine	1A2C
(4) proline	3C

B. Assignment of Codons with Known Sequences

1. Binding Technique : Marshall W. Nirenberg and **Philip Leder** in 1964 found that if a synthetic trinucleotide for a known sequence (with known bases at 5' end and 3' end) is used with

MARSHALL WARREN NIRENBERG
(LT, 1927–2010)

Nirenberg, a US biochemist and geneticist of Jewish origin, was born on April 10, 1927 in New York City. His parents were **Harry** and **Minerva Nirenberg**. He developed rheumatic fever as a boy, so the family moved to Orlando, Florida to take advantage of the subtropical climate there. He developed on early interest in biology. In 1948, he received his B.S. degree, and in 1952, a M.S. in Zoology from the University of Florida at Gainesville, where he was also a member of the *Pi Lambda Phi Fraternity*. His dissertation for the Master's thesis was an ecological and taxonomic study of caddis flies (Trichoptera). He received his Ph.D. in Biochemistry from the University of Michigan, Ann Arbor in 1957.

He began his post doctoral work at the National Institutes of Health (NIH) in 1957, as a fellow of the American Cancer Society in what was then called the National Institute of Arthritis and Metabolic Diseases. In 1959, he became a research assistant at the NIH and began to study steps that relate DNA, RNA, and protein. Nirenberg's groundbreaking experiments advanced him to become the Head of the Section of Biochemical Genetics in 1962 in the National Heart Institute (now the National Heart, Lung, and Blood Institute), where he remained a laboratory chief until his dearth. In 1961, he was married to **Perola Zaltman**, a chemist from the University of Brazil, Rio de Janeiro, who also worked at NIH and died in 2001. Nirenberg then married **Myrna M. Weismann,** Ph.D., Professor of Epidemiology and Psychiatry at Columbia University College of Physicians and Surgeons, in 2005. He had 4 stepchildren: **Susan Weissman** of Illinois, **Judith Weismann** of New York, **Sharon Weismann** of Connecticut, and **Jonathan Weismann** of California.

In 1961, Nirenberg and **Heinrich Matthaei** reported the first breakthrough with regard to finding out the 3-letter code words for each aming acid. They incubated synthetic polyuridylate, poly(U), with an *E. coli* extract, GTP, ATP, and a mixture of the 20 amino acids in 20 different tubes, each tube containing a different radioactively-labeled amino acid. Because poly(U) mRNA is made up of many successive UUU triplets, it should promote the synthesis of a polypeptide containing only the amino acid encoded by the triplet UUU. A radioactive polypeptide was indeed formed in only one of the 20 tubes, the one containing radioactive phenylalanine. They, therefore, concluded that the triplet codon UUU encodes phenylalanine. The same approach revealed that polycytidylate, poly(C), encodes a polypeptide containing only proline, and polyadenylate, poly(A), encodes polylysine.

Polyguanylate, poly(G), did not generate any polypeptide in this experiment because it spontaneously forms tetraplexes that cannot be bound by ribosomes. Using a variety of artificial mRNAs made by polynucleotide phosphorylase from different starting mixtures of ADP, GDP, UDP, and CDP, investigators soon identified the base compositions of the triplets coding for almost all the amino acids. Although these experiments revealed the base composition of the coding triplets, they could not reveal the *sequence* of the bases.

In 1964, Nirenberg and **Philip Leder** achieved another experimental breakthrough. Isolated *E. coli* ribosomes would bind a specific aminoacyl-tRNA in the presence of the corresponding synthetic polynucleotide messenger (By convention, the identity of a tRNA is indicated by a superscript, such as $tRNA^{Ala}$, and the aminoacylated tRNA by a hyphenated name: alanyl-$tRNA^{Ala}$ or Ala-$tRNA^{Ala}$). For example, ribosomes incubated with poly(U) and phenylalanyl-$tRNA^{Phe}$ (Phe-$tRNA^{Phe}$) bind both RNAs, but if the ribosomes are incubated with poly(U) and some other aminoacyl-tRNA, the aminoacyl-tRNA is not bound, because it does not recognize the UUU triplets in poly(U). Even trinucleotides could promote specific binding of appropriate tRNAs, so these experiments could be carried out with chemically-synthesized small oligonucleotides. With this technique, researchers determined which aminoacyl-tRNA bound to about 50 of the 64 possible triplet codons. For some codons, either no aminoacyl-tRNA or more than one would bind. Another method was needed to complete and confirm the entire genetic code; this work was admirably done by **H.G. Khorana**, at about the same time.

The period between 1961 and 1962 is often referred to as the **"coding race"** because of the competition between the labs of Nirenberg at NIH and Nobel Laureate **Severo Ochoa** at the New York University Medical School, who had a massive staff. Faced with the possibility of helping the first NIH scientist win a Nobel Prize, many NIH scientists put aside their own work to help Nirenberg in deciphering the mRNA codons for amino acids. **Dr. DeWitt Stetten Jr.**, Director of the National Institute of Arthritis and Metabolic Diseases, called this period of collaboration "NIH's finest hour". Nirenberg's later research focused on Neuroscience, neural development and the homeobox genes.

Nirenberg was awarded the **National Medal of Science in 1964.** He shared with **Robert W. Holley** and **H.G. Khorana** the **1968 Nobel Prize for Physiology or Medicine** for his major contributions toward *deciphering the genetic code.* In the same year, together with **H.G. Khorana**, he was awarded the **Louisa Gross Horowitz Prize** from the Columbia University. The **U.S. President Lyndon B. Johnson** awarded him the **National Medal of Honor**, also in 1968. He was elected to the American Philosophical Society in 2001. He died on January 15, 2010, from cancer after several months of illness.

ribosome and a particular aminoacyl-tRNA (tRNA having its own specific amino acid attached), these will form a complex, provided the used codon codes for the amino acid attached to the given aminoacyl tRNA.

$$\text{Codon}_1 + \text{Ribosome} + \text{AA}_1 + \text{tRNA} \rightarrow \text{Ribosome-Codon}_1\text{-AA}_1\text{-tRNA}_1$$

In a process such as above, if given AA_1 is used with a given codon, and the formation of the complex is detected, this would prove that the given codon codes for the given amino acid.

It was also observed that while the free AA-tRNA passed through nitrocellulose membrane, the **ribosome-codon-AA-tRNA complex** adsorbs on such a membrane. If in a particular mixture only one of the amino acids is made radioactive, then the presence or absence of the radioactivity on the nitrocellulose membrane will show whether there is a relationship between the codon and the amino acid which was made radioactive. For instance, 20 samples of a mixture of all 20 amino acids may be taken and in each sample one amino acid is made radioactive in such a manner that each and every amino acid is made radioactive in one sample or the other, and no two samples have same radioactive amino acid. A particular sample would be then known by its radioactive amino acid. Now tRNAs and ribosomes are mixed with each sample and same codon is used for complex formation in all 20 cases. When the mixture is poured on the nitrocellulose membrane, radioactivity on membrance will be observed only when the radioactive amino acid is taking part in the formation of the complex. Since in each sample the radioactive amino acid is known, it would be possible to detect the amino acid coded by a given codon by the presence of radioactivity on the membrane. Such a treatment was given by Nirenberg and his coworkers to all the 64 synthetic codons, and their respective amino acids were identified. The binding of AA-tRNA was not equally efficient in all cases. Therefore, the sequences of bases in only about 45 codons could be worked out by this method.

2. Repetitive Sequencing Technique : This method of confirming the genetic code is the most direct method and was devised by the Nobel prize winner **Prof. Har Gobind Khorana**. This method consists of *in vitro* chemical synthesis of short segments of DNA of known base sequence with the help of DNA polymerase. From this synthetic DNA, a polyribonucleotide (RNA) of strictly

HAR GOBIND KHORANA

(LT, 1922–2011)

Khorana, an Indian-born US biochemist, was born on January 9, 1922 to Hindu parents in British India, now part of Pakistan. His father was the village "patwari" (taxation official). He was home schooled by his father until High School. He earned his B.Sc. in 1943 and M.Sc. in 1945, both from Punjab University, Lahore. In 1945, he began studying at the University of Liverpool and received Ph.D. from there in 1948. He continued his postodoctoral studies in Zürich from 1948 to 1949. Subsequently, he spent two years at Cambridge University. Khorana married **Esther Elizabeth Sibler,** of Swiss origin, in 1952. Also in 1952, he went to the University of British Columbia, Vancouver and in 1960 moved to the University of Wisconsin, Madison. Khorana became a naturalized citizen of the United States in 1966, and subsequently received the **National Medal of Science**. Khorana shared with **Robert W. Holley** and **Marshall W. Nirenberg** (both from the United States) the **1968 Nobel Prize for Physiology or Medicine** for his major contribution toward *deciphering the genetic code.* In 1970, he and his research team synthesized the first artificial copy of a yeast gene. Khorana and Nirenberg were also awarded the **Louisa Gross Horowitz Prize** from the Columbia University, in the same year. In 1970, he became the Alfred P. Sloan Professor of Biology and Chemistry at the Massachusetts Institute of Technology (MIT) where he worked until retiring in 2007. He was also a member of the Board of Scientific Governors at The Scripps Research Institute.

Khorana developed chemical methods to synthesize polyribonucleotides with defined, repeating sequences of 2 to 4 bases. The polypeptides produced by these mRNAs had one or a few amino acids in repeating patterns. These patterns, when combined with information from the random polymers used by **Marshall Nirenberg** and colleagues, permitted unambiguous codon assignments. The copolymer $(AC)_n$,for example, has alternating ACA and CAC codons: ACACACACACACACA. The polypeptide synthesized on this messenger contained equal amounts of threonine (Thr) and histidine (His). Given that a histidine codon has one A and two Cs, CAC must code for histidine and ACA for threonine.

Consolidation of the results from many experiments permitted the assignment of 61 of the 64 possible codons. The other 3 (UAA, UAG, UGA) were identified as termination codons, in part because they disrupted amino acid coding patterns when they occurred in a synthetic RNA polymer. Meanings for all the triplet codons were established by 1966 and have been verified in many different ways. The cracking of the genetic code is regarded as one of the most important scientific discoveries of the 20th century.

The University of Wisconsin, Madison, the Government of India (DBT, Department of Biotechnology), and the Indo-US Science and Technology Forum jointly created the **Khorana Program**, in 2007. The mission of the Khorana Program is to build a seamless community of scientists, industrialists, and social entrepreneurs in the United States and India. The program is focused on 3 objectives: (*a*) providing graduate and undergraduate students with a transformative research experience, (*b*) engaging partners in rural development and food security, and (*c*) facilitating public-private partnerships between the U.S. and India. In 2009, Khorana was hosted by the Khorana Program and honoured at the 33rd Steenbock Symposium in Madison, Wisconsin. Khorana died of natural causes on November 9, 2011 in Concord, Massachusetts, at age 89. A widower, he was survived by his children, **Julia** and **Dave.**

defined base sequence is transcribed under the catalytic influence of RNA polymerase. *A polypeptide is then synthesised under the direction of RNA as represented in* Fig. 32–7. The amino acid sequence of the polypeptide so formed is then determined and correlated with the base sequence of DNA and RNA.

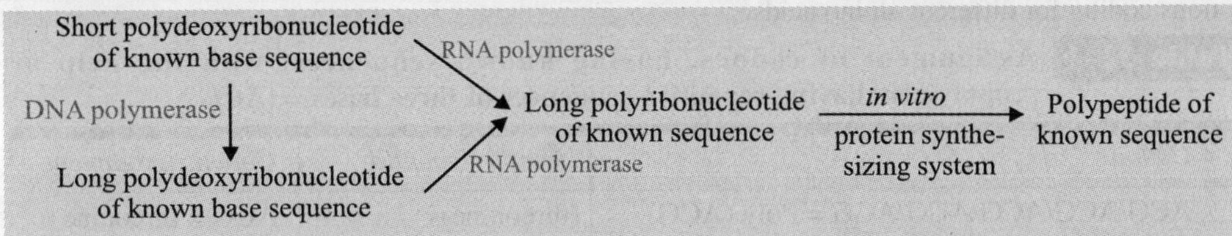

Fig. 32–7. Synthesis of polypeptide under the direction of RNA

Using homopolynucleotides (all bases the same) as templates could yield only a few code words. Many more words were deciphered after Khorana developed methods for synthesizing polyribonucleotides with different but repeating structures. In the example shown in Fig. 32–8, the repeating sequence $(AAG)_n$ was found to give 3 different homopolymers : polylysine, polyarginine and polyglutamic acid. This finding not only confirmed the importance of the reading frame but also showed that AAG, AGA and GAA must be codons for these amino acids. However, the experiment did not reveal which codon corresponded to which amino acid; further experiments were needed to discriminate among the possible matches.

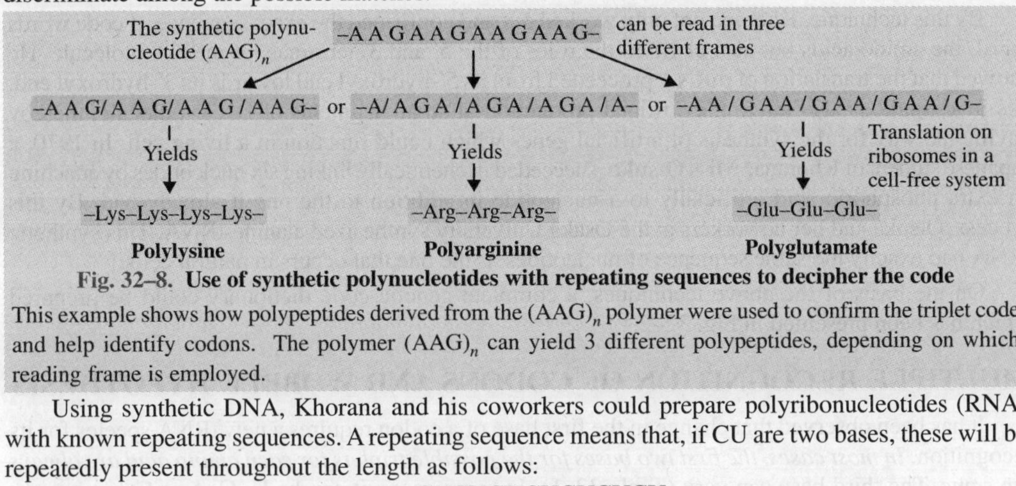

Fig. 32–8. Use of synthetic polynucleotides with repeating sequences to decipher the code

This example shows how polypeptides derived from the $(AAG)_n$ polymer were used to confirm the triplet code and help identify codons. The polymer $(AAG)_n$ can yield 3 different polypeptides, depending on which reading frame is employed.

Using synthetic DNA, Khorana and his coworkers could prepare polyribonucleotides (RNA) with known repeating sequences. A repeating sequence means that, if CU are two bases, these will be repeatedly present throughout the length as follows:

CUCUCUCUCUCUCU

In a similar manner, if ACU are three bases, they will be present repeatedly as follows:

ACUACUACUACU

Such copolymers will direct the incorporation of amino acids in a manner which can be theoretically predicted. For instance, in $(CU)_n$ =(CUC/UCU/CUC/UCU), only two codons are possible and these are CUC and UCU. Moreover, these codons are present in alternating sequence. The result would be that the polypeptide formed would have only two amino acids in alternating sequences. These two amino acids can be assigned to the two codons (Table 32–3).

Table 32–3 **Assignment of codons, having known sequences, with the help of copolymers having repetitive sequences of two bases**

Copolymers	*Codons*	*Amino acids*	*Codons*
$(CU)_n$	CUC/UCU/CUC	leucine/serine	CUC/UCU
$(UG)_n$	UGU/GUG/UGU	cysteine/valine	UGU/GUG
$(AC)_n$	ACA/CAC/ACA	threonine/histidine	ACA/CAC

We may similarly consider a repeating sequence of three bases *e.g.*, $(ACG)_n$. Depending upon where the reading is started, three kinds of homopolypeptides are expected (Table 32–4). Actual codon assignment *i.e,.* to find out which of the three codons codes for which amino acid would depend upon the previous information available regarding the composition of bases in different codons coding for different amino acids.

Table 32–4 Assignment of codons, having known sequences, with the help or copolymers having repetitive sequeuces of three bases = $(ACG)_n$

Codons	*Homopolypeptide*	*Codon assignment*
ACG/ACG/ACG/ACG/ACG = Poly (ACG)	(threonine)*n*	ACG = threonine
A/CGA/CGA/CGA/CGA/CGA = Poly (CGA)	(arginine)*n*	CGA = arginine
AC/GAC/GAC/GAC/GAC = poly (GAC)	(aspartic acid)*n*	GAC = aspartic acid

These studies of **Khorana** and his coworkers with chemically-defined messengers proved very conclusively that (*a*) the base sequence in DNA specifies the sequence of amino acids in proteins, (*b*) the information contained in DNA is conveyed through RNA, and (*c*) the genetic code is triplet and non overlapping in nature. The results obtained from these polypeptide synthesis in combination with the results obtained from the binding technique of **Nirenberg** and coworkers led to the deciphering and establishment of the genetic code.

By this technique, Khorana not only worked out and confirmed the exact sequence of code words for all the amino acids but also clarified the roles of the 5′ and 3′ terminals of mRNA molecule. He showed that the translation of mRNA proceeded from its 5′-hydroxyl end towards its 3′-hydroxyl end. His investigations also led to the artificial production of small segments of DNA molecule, thereby paving the way for the synthesis of artificial genes which could function in a living cell. In 1970, a Japanese student of Khorana, **Miss Otsuka**, succeeded in chemically linking six nucleotides by attaching an extra phosphoric acid artificially to a nucleotide in addition to the one it already had. By this process, Otsuka and her coworkers at the Osaka University synthesized alanine-tRNA. This synthetic tRNA had exactly the same sequence of nucleotides as the one that occurs in natural cells.

On the basis of the above techniques, a complete genetic code dictionary could be prepared which has been presented in Fig. 32–2.

MULTIPLE RECOGNITION OF CODONS AND WOBBLE HYPOTHESIS

It has been observed that change in the first base of a codon requires a new tRNA species for its recognition. *In most cases, the first two bases for the possible triplets for each amino acid are always the same.* The third base can vary (Table 32–1). In some cases, it can be U, C, A or G *but in most cases it can be either the pair U and C or the pair A and G.* These relationships probably enable in some way the anticodon triplet in tRNA to recognize the codon triplet in mRNA. It has been found that each tRNA anticodon recognizes the several mRNA codons for one amino acid. Generally, one species of tRNA can recognize three codons which have the last base U, C and A as a group. *Codons ending in the base G are mostly specific for species of tRNA.* To explain the capacity of the third anticodon base to recognize several complementary bases in the codon (redundancy), **Crick** (1965) postulated the **wobble hypothesis**. According to this hypothesis, the first two bases of the anticodon are strictly standard for the first two constant bases of the codon and pair strongly with them. The third base of the codon is not so specific in its base pairing and may *wobble* (*pair loosely*) in pairing with the corresponding base in the anticodon. As a result of this, each tRNA recognizes several codons for its amino acid. Hence, the number of tRNAs needed for the translation of genetic code is considerably less than the number of codons. He discovered that if U is present at first position of anticodon, it can pair with either A or G at the third position of codon. Similarly, when C or A occurs in the 5′ position of the anticodon, it can pair only with G or U, respectively, in the 3′ position of a codon. Transfer RNAs containing either G or U in the 5′ (or wobble) position of the anticodon can each pair with 2 different codons, whereas an inosine, (the deaminated form of adenine), in this position produces a tRNA that can pair with 3 codons differing in the 3′ base. The pairing relationships between first base of anticodon and third base of codon are given in Table 32–5.

Table 32–5 Wobble base-pairing hypothesis

*Anticodon base** *(first base)*	*Codon base* *(third base)*
U	A, G
C	G only
A	U only
G	U, C
I (Inosine, resembles G)	U, C, A

* The first base of anticodon pairs with the third base of codon.

Fig. 32–9 presents examples of a standard base pair and two wobble base pairs.

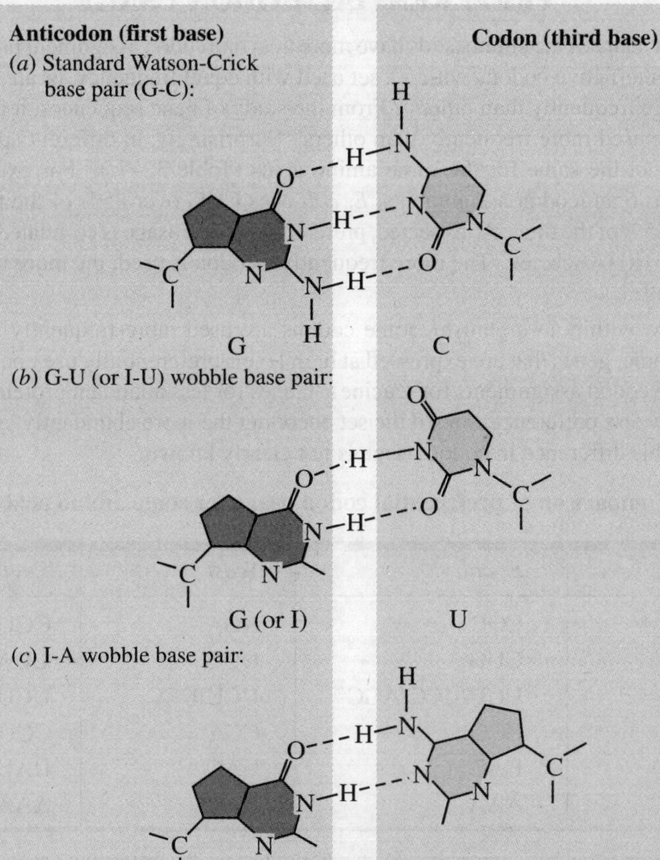

Fig. 32–9. Examples of standard and wobble base pairs

The standard (*a*) and wobble (*b* and *c*) base pairs are formed between the first base in the anticodon and the third base in the codon.

This kind of wobbling allows economy of the number of tRNA molecules, since several codons meant for same amino acid are recognized by the same tRNA. For instance, (i) anticodon IGC can recognize codons GCU, GCC and GCA (coding for alanine), (*ii*) IGA can recognize UCU, UCC and UCA (coding for serine), and (*iii*) IAC can recognize GUU, GUC and GUA (coding for valine). *Block explained multiple codon recognition on the basis of tautomeric shift model.* In some cases, the third base of the anticodon is an unusual base instead of one of the four common bases (Table 32–6).

Table 32–6. Codons and anticodons for some amino acids

Amino acid	*Alanine*	*Serine*	*Isoleucine*	*Phenylalanine*	*Tyrosine*	*Valine*
Codon (5′–3′)	U GCC A G	U UCC A G	U AUU A	U UUU	U UAC	U GUU A G
Anticodon (3′–5′)	CGI	AGI	UAI	AAMG	APSUG	CAI

PREFERENTIAL CODON USAGE

The fact that many of the amino acids have more than one codon assignment presents the question : Are each of the alternative codons within a set used with equal frequency, or are some of the codons in a set used more frequently than others ? From the study of gene sequences, it is now apparent that some codons are used more frequently than others. Surprisingly, in different species, the preferred codon usage is not the same for the same amino acids (Table 32–7). For example, even though leucine (Leu) has 6 anticodon assignments, *E. coli* uses CUG over 85% of the time, whereas yeast uses UUG over 85% of the time. As expected, preferential codon usage is correlated with the abundance of the respective tRNA species. The more frequently a codon is used, the more that tRNA species is present in the cell.

Additionally, within an organism, some codons are used more frequently for certain types of genes. For example, genes that are expressed at high levels preferentially use one codon in a set (*e.g.*, only one of the 6 codon assignments for leucine). Genes for less abundant proteins use a larger set of codons and show less preference toward the set encoding the more abundantly synthesized proteins. The reason for this difference in codon usage is not clearly known.

Table 32–7. Comparison of preferential codon usage for some amino acids

Codon for	*E. coli*	*Yeast*	*Euglena chloroplast*
Arginine (6)*	CGC	AGA	CGU/CGC/AGA
Leucine (6)	CUG	UUG	UUA/UUG/CUU
Serine (6)	UCU/UCC/AGC	UCU/UCC	UCU/UCA/AGU
Proline (4)	CCG	CCA	CCU/CCA
Tyrosine (4)	UAC	UAC	UAU/UAC
Lysine (2)	AAA	AAG	AAA

* The number in parenthesis represents the number of codons for that amino acid.

Preferred codons are those used more than 85% of the time. For example, UCU, UCC and AGC are used more than 85% of the time for serine in *E. coli.*

Another observation relating to codon usage is that codons with the forms NCG and NUA (n = any nucleotide) are avoided. It is speculated that it may be related to the frequent methylation of cytosine in CG dinucleotides and the tendency for methyl-CG to mutate by deamination to TG. However, this reasoning would apply only to eukaryotes because DNA methylation is minimal in prokaryotes.

Preferential codon usage may have some practical applications when it comes to transferring genes between different organisms and species. For example, the insulin gene from humans has been put into fast-growing *E. coli* to produce human insulin cheaply and in bulk.

MUTATIONS AND GENETIC CODE

Most of the work earlier discussed in this chapter deals with the study of genetic code in cell-free systems. Therefore, it could be questioned whether or not this information would apply to the living systems. With the study of certain mutants, it was possible to show that *the genetic code deciphered using cell-free systems applies to the living systems also.*

There are two kinds of mutations which have played a very significant role in the study of the genetic code. These are (*i*) frameshift mutations and (*ii*) base substitutions.

A. Frameshift Mutations

The genetic message, once initiated at a fixed point, is read in a definite frame in a series of three-letter words. The framework would be disturbed as soon as there is a deletion or addition of one or more bases. When such frame shift mutations were intercrossed, in certain combinations, they gave wild type. It was concluded that one of them was deletion and the other an addition, so that the disturbed order of the frame due to one mutation will be restored by the other (Fig. 32–10).

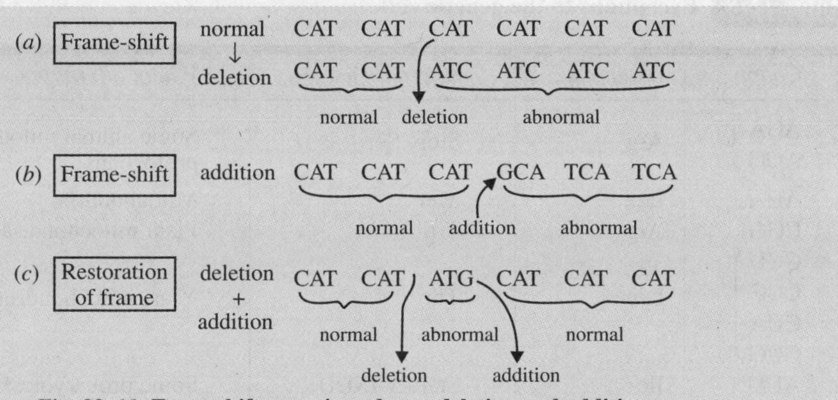

Fig. 32–10. Frameshift mutations due to deletion and addition

B. Base Substitutions or **Amino Acid Replacements**

If in a mRNA at a particular point, one base is replaced by another without any deletion or addition, the meaning of one codon containing this altered base will ordinarily change and in place of a particular amino acid at a particular position in a polypeptide, another amino acid will be incorporated. Such mutations have been studied in detail in a protein enzyme known as *tryptophan synthetase*. By working out the altered amino acid sequences in the polypeptides, some conclusions regarding the possible changes in the base-sequence in RNA can be easily made if the dictionary of genetic code is available. *This has been done in tryptophan synthetase and also in case of hemoglobin, using sickle cell anemia.* From the analysis of many different species, including bacteria, protozoa, fungi, plants and animals, researchers have found that the GC is nearly universal only a few rare exceptions to the GC have been noted. These are described in Table 32–8.

CLINICAL IMPLICATIONS 32–1.

Sickle Cell Anemia: Mutation of the Genetic Code

Sickle cell anemia (SCA) is an example of a disease in which a single nucleotide change within the coding region of the gene for the b chain of hemoglobin A, the major form of adult hemoglobin, yields an altered protein that has impaired function. The mutation that causes this disease is a single-nucleotide change in a codon that normally specifies glutamate (GAG) and which now produces a codon that specifies valine (GUG). Under conditions of low oxygen tension, this single amino acid change causes the protein to polymerize into rod-shaped structures, resulting in deformation of red blood cells (becoming sickle-shaped) and in altered flow properties of the cells in vessels and capillaries. When hemoglobin is not bound to oxygen, the conformation of the protein exposes a hydrophobic patch, allowing the mutant proteins with valine to interact more readily and form rods. This substitution of an amino acid with an acidic side chain for an amino acid with a nonpolar, hydrophobic side chain is termed a **nonconservative mutation.** Conservative mutations of one amino acid by another with similar physical and chemical properties may have less severe consequences, e.g., an Arg $\longrightarrow$ Lys or Asp $\longrightarrow$ Glu mutation.

Only individuals homozygous for hemoglobin S (HbS), a variant form of the normal adult hemoglobin (HbA_1), exhibit the disease. Individuals heterozygous for HbS have approximately 50% HbA_1 and 50% HbS in their RBCs and do not exhibit symptoms of sickle-cell anemia disease except under conditions of hypoxia (deficiency of oxygen).

Table 32–8. Exceptions to the genetic code

Codon	*Usual use*	*Alternate use*	*Where alternate use occurs*
AGA, AGG	Arg	Stop, Ser	Some animal mitochondria, some protozoans
AUA	Ile	Met	Mitochondria
CGG	Arg	Trp	Plant mitochondria
CUU, CUC, CUA, CUG	Leu	Thr	Yeast mitochondria
AUU	Ile	Start(*N*-fMet)	Some prokaryotes[a]
CUG	Val		
UUG	Leu		
UAA, UAG	Stop	Glu	Some protozoans
UGA	Stop	Trp	Mitochondria, mycoplasmas
		Selenocysteine	*E. coli**

* Depends on context of message, other factors.

NEW GENETIC CODES

A. Genetic Code in Mitochondria

Earlier we described that the genetic code is universal. It was also believed that the genetic code does not undergo any kind of evolution and, therefore, should be static. However, during the last few years, it was shown that variations in genetic code are found in mitochondria, particularly studied in yeast and mammals. In the case of yeast mitochondria (Fig. 32-11), UGA codes for tryptophan, although in the nuclear genes, UGA is a termination codon. Similarly, the codons beginning with CU represent Thr instead of Leu and the AUA codon represents Met instead of Ile. It has also been shown that although a new genetic code may exist in mitochondria, but mitochondria in all organisms may not have the same genetic code. For instance, UGA in mitochondria does not always code for tryptophan. In the gene coding for maize cytochrome oxidase subunit II, UGA codon is not present and tryptophan is coded by codon CGG (coding for arginine in the nuclear genes) codes for methionine in mitochondria of mammals, *Xenopus*, yeast and *Drosophila,* but not in *Neurospora* and *Aspergillus*.

First Position (5′ end)	Second Position: U		C		A		G		Third Position (3′ end)
U	UUU	Phe AAG	UCU	Ser AGU	UAU	Tyr AUG	UGU	Cys ACG	U
	UUC		UCC		UAC		UGC		C
	UUA	Leu AAU*	UCA		UAA	STOP	UGA	Trp ACU*	A
	UUG		UCG		UAG		UGG		G
C	CUU	Thr GAU	CCU	Pro GGU	CAU	His GUG	CGU	Arg GCA[b]	U
	CUC		CCC		CAC		CGC		C
	CUA		CCA		CAA	Gln GUU*	CGA		A
	CUG		CCG		CA G		CGG		G
A	AUU	Ile UAG	ACU	Thr UGU	AAU	Asn UUG	AGU	Ser UCG	U
	AUC		ACC		AAC		AGC		C
	AUA	Met UAU[a]	ACA		AAA	Lys UUU*	AGA	Arg UCU*	A
	AUG		ACG		AAG		AGG		G
G	GUU	Val CAU	GCU	Ala CGU	GAU	Asp CUG	GGU	Gly CCU	U
	GUC		GCC		GAC		GGC		C
	GUA		GCA		GAA	Glu CUU*	GGA		A
	GUG		GCG		GA G		GGG		G

Fig. 32-11. The genetic code of yeast mitochondria

The codons (5′ ⟶ 3′) are at the *left* and the anticodons (3′ ⟶ 5′) are at the *right* in each box.
* designates U in the 5′ position of the anticodon that carries the —$CH_2NH_2CH_2COOH$ grouping on the 5′ position of the pyrimidine.

[a] Two tRNAs for methionine have been found. One is used in initiation and one is used for internal methionines.

[b] Although an Arg tRNA has been found in yeast mitochondria, the extent to which the CGN codons are used is not clear.

(Adapted from S G Bonitz et al, 1980)

It has also been shown that in mitochondria, a number of single tRNA species with U in the wobble position (first base of anticodon pairing with the third base in codon) read all four codons in a family. It is also shown that there are only 22 tRNAs in mitochondria as against 55 tRNAs in universal code, and these are adequate for reading 60 codons, the remaining four being termination codons.

B. Genetic Code in Ciliate Protozoa

In 1986, a different genetic code was shown to be present in ciliate protozoa (*Mycoplasma capricolum*). In this genetic code, codons UAA and UAG specify glutamine instead of stop signals. In future, more such cases may be discovered, showing diversity in the genetic code.

OVERLAPPING GENES

Earlier, we stated that the genetic code is nonoverlapping– each ribonucleotide in an mRNA is part of only one triplet. However, this characteristic of the code does not rule out the possibility that a single mRNA may have multiple initiation points for translation. If so, these points could theoretically create several different reading frames within the same mRNA, thus specifying more than one polypeptide. This concept of overlapping genes is illustrated in Fig. 32–12(*a*).

That this might actually occur in some viruses was suspected when phage $\phi \times 174$ was carefully investigated . The circular DNA chromosome consists of 5386 nucleotides, which should encode a maximum of 1795 amino acids, sufficient for 5 or 6 proteins. However, this small virus in fact synthesizes 11 proteins consisting of more than 2320 amino acids. A comparison of the nucleotide sequence of the DNA and the amino acid sequences of the polypeptides synthesized has clarified the appearent paradox. At least four cases of multiple initiation have been discovered, creating overlapping genes (Fig. 32–12(*b*)).

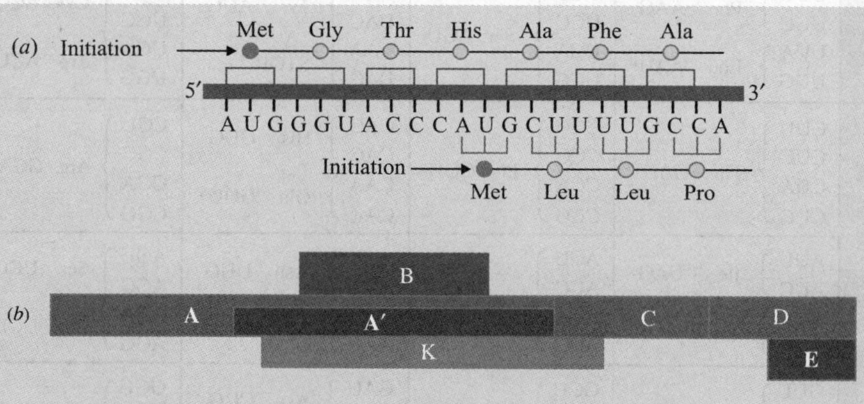

Fig. 32–12. An illustration of the concept of overlapping genes

(*a*) An mRNA sequence initiated at 2 different AUG positions out of frame with one another will give rise to 2 distinct amino acid sequences.

(*b*) The relative positions of the sequences encoding 7 polypeptides of the phage $\phi \times 174$.

The sequences specifying the K and B polypeptides are initiated with separate reading frames within the sequence specifying the A polypeptide. The K sequence overlaps into the adjacent sequence specifying the C polypeptide. The E sequence is out of frame with, but initiated in that of the D polypeptide. Finally, the A′ sequence, while in frame, begins in the middle of the A sequence. They both terminate at the identical point. In all, seven different polypeptides are created from a DNA sequence that might otherwise have specified only three (A, C and D).

A similar situation has been observed in other viruses, including phage G4 and the animal virus SV 40. Like $\phi \times 174$, phage G4 contains a circular single-stranded DNA molecule. The use of overlapping reading frames optimizes the use of a limited amount of DNA present in these small viruses. However, such an approach to storing information has a distinct disadvantage in that a single mutation may affect more than one protein and thus increase the chances that the change will be deleterious or lethal. In the case we just discussed, a single mutation at the junction of genes A and C could affect three proteins (the A, C, and K proteins). It may be for this reason that overlapping genes are not common in other organisms.

EVOLUTION OF THE GENETIC CODE

Two types of theories have been putforth to explain the evolution of the genetic code.

A. Mechanistic models

These models suggest that the original code was based upon some physico-chemical relationship between the amino acid and its codon. **Woese** (1967–68) had proposed that the original code might have been determined by a stereochemical fit between an amino acid and its codon, the nucleic acid acting as a physical template for assembly of the amino acids. Initially, the code might commence in an autocatalytic cycle in which a polynucleotide and a polypeptide assisted each other to replicate, so that their relationship was reciprocal, only subsequently becoming unidirectional. Nucleotides and amino acids involved under primitive conditions were not the same as those employed today and this creates difficulty in confirming the primitive code. **Lacey** and **Pruit** (1969) have demonstrated the formation of a complex between poly-L-lysine and mononucleotides. It is a controversial matter whether an amino acid could interact with its codon with sufficient stereochemical specificity to fit to account for the development of the code. Nevertheless, it is an attractive hypothesis.

B. Stochastic models

These models do not depend on physico-chemical relationship. It is presumed that the initial set of codon assignments was free to vary. Selective forces would modify it in the course of evolution and continued unchanged in all organisms descended from a single inbreeding population. Pure mechanistic and pure stochastic models differ in the nature of selective forces responsible for the evolution of the code.

Sonneborn (1965) proposed a mutational-buffer theory. According to him, mutations resulting from substitution of a single nucleotide by another tend to exert a deleterious effect upon the cell and if the cells can reduce the burden of harmful mutations, they are likely to be at a selective advantage. This theory is purely stochastic as it demands that the codons for similar amino acids should be related to each other. This model suffers from following two drawbacks:

1. It does not account for the universality of the code.
2. The model requires codons to change their meanings during the course of evolution.

Woese (1965, 1967, 1968) proposed an alternative model which suggests that the selective forces in evolution operate to minimise errors made in the translation of codons into amino acids. *This is not a purely stochastic hypothesis as it depends to some extent upon mechanistic features involved in translation.* Woese argues that translation is highly evolved today but primitive cells must have had different systems. No experimental evidence is, however, possible to confirm the primitive translation machinery. Woese further suggests that evolution would favour a codon dictionary in which more error-prone codons are represented by amino acids which are functionally less important. He observed that the probability of misreading occurring at each of the three nucleotides in a codon are 100 : 10 : 1 for positions III : I : II respectively. Thus, the third position, being most error-prone, is the most degenerate so that codons differing in the base at this position tend to be assigned to related amino acid. The second base is the least error-prone and is probably least involved in such restraints.

Crick (1968) has put forward a model in which the code might start in primitive form, a small number of triplets coding for only a few amino acids and assignment of codons to amino acids would become nonsense and probably lethal, on switching to the triplet code. Nobody knows which amino acids were present in the initial code. It seems probable that the present complex species might not have existed under prebiotic conditions. Through a series of changes, the nature of which is difficult to fathom, the final code would evolve as new amino acids replaced some of the primitive ones. This is possible if the new amino acids are related and the organism codes for a few rather crude and simple proteins. As the number and complexity of proteins increased, a situation might arise when substitution of new amino acids may prove lethal and at this point the code would be 'frozen'. This theory is termed a '*frozen accident model*'. But in order to explain universality, it is again necessary to postulate that this must have happened at an early stage of evolution where all cells descended from one population were involved.

REFERENCES

1. **Barrell BG, Banker AT, Drouin J :** A different genetic code in human mitochondria. *Nature* **282** : *189–194, 1979.*
2. **Barrell BG, Air G, Hutchinson C :** Overlapping genes in bacteriophage ϕ × 174. *Nature* **264** : *34–40, 1976.*
3. **Chambron P :** Split genes. *Scientific American. 244 (May) : 60–71, 1981.*
4. **Crick FHC Barnett L, Brenner S, Watts-Tobin RJ :** General nature of the genetic code for proteins. *Nature,* **192** : *1227–1232., 1961.*
5. **Crick FHC :** Codon–anticodon pairing : The wobble hypothesis. *J.Mol. Biol.* **19** : *548–55, 1966 b.*
6. **Crick FHC :** The genetic code. *Sci. America (Oct)* **207** : *66–77, 1962.*
7. **Crick FHC :** The genetic code III. *Sci. Amer. (Oct.)* **215** : *55–63, 1966a.*
8. **Gamow G :** Possible relation between DNA and protein structures. *Nature* **173** : *318, 1954.*
9. **Jukes TH :** The genetic code. *Am. Sci.* **51** : *227–245, 1963.*
10. **Khorana HG :** Polynucleotide synthesis and the genetic code. *Harvey Lectures.* **62** : *79–105, 1967.*
11. **Khorana HG *et al* :** Polynucleotide synthesis and the genetic code. *Cold Spring Harb. Sym.* **31** : *39–49, 1967.*
12. **Knight RD, Freeland SJ, Landweber LF :** Selection, history and chemistry : the three faces of the genetic code. *Trends Biochem. Sci.* **24(6)** : *241–247, 1999.*
13. **Lewin B :** Genes VI. *6th ed. Oxford University Press, New York. 1997.*
14. **Nirenberg MW :** The genetic code : II *Scientific American. (March issue)* **190** : *80–94, 1963.*
15. **Nirenberg MW, Leder P :** RNA code words and protein synthesis. *Science* **145** : *1399–1407, 1964.*
16. **Ochoa S :** Synthetic polynucleotides and the genetic code (Symposium on Genetic Mechanics). *Fed. Proc.* **22** : *62–74, 1963.*
17. **Singer M, Berg P :** Genes and Genomes. *University Science Books, Mill Valley, CA . 1991.*
18. **Smithe MA *et al* :** Direction of reading of the genetic message, II *Proc. Nat, Acad. Sci.* **55** : *141–147, 1966.*
19. **Szathmary, E :** What is the optimum size for the genetic alphabet. *Proc. Natl. Acad. Sci.* **89** : *2614–2618, 1992.*
20. **Woese CR :** The Genetic Code. *Harper & Row, New York. 1967.*
21. **Wong JT :** Evolution of the genetic code. *Microbiol. Sci.* **5(6)** : *174–181, 1988.*
22. **Ycas M :** The Biological Code. *North Holland, Amsterdam. 1969.*

PROBLEMS

1. (*a*) Write the sequence of the mRNA molecule synthesized from a DNA template strand having the sequence

5′-ATCGTACCGTTA-3′

(*b*) What amino acid sequence is encoded by the following base sequence of an mRNA molecule ? Assume that the reading frame starts at the 5′ end.

5′-UUGCCUAGUGAUUGGAUG-3′

(*c*) What is the sequence of the polypeptide formed on addition of poly (UUAC) to a cell-free protein-synthesizing system?

2. The code word GGG cannot be deciphered in the same way as can UUU, CCC, and AAA, because poly(G) does not act as a template. Poly(G) forms a triple-stranded helical structure. Why is it an ineffective template?

3. Synthetic RNA molecules of defined sequence were instrumental in deciphering the genetic code. Their synthesis first required the synthesis of DNA molecules to serve as a template. **Har Gobind Khorana** synthesized, by organic-chemical methods, two complementary deoxyribonucleotides, each with nine residues: $d(TAC)_3$ and $d(GTA)_3$. Partly overlapping duplexes that formed on mixing these oligonucleotides then served as templates for the synthesis by DNA polymerase of long, repeating double-helical DNA chains. The next step was to obtain long polyribonucleotide chains with a sequence complementary to only one of the two DNA strands. How did he obtain only poly(UAC)? Only poly(GUA)?

4. In a nonoverlapping triplet code, each group of three bases in a sequence ABCDEF... specifies only one amino acid– ABC specifies the first, DEF the second, and so forth– whereas, in a completely overlapping triplet code, ABC specifies the first amino acid, BCD the second, CDE the third, and so forth. Assume that you can mutate an individual nucleotide of a codon and detect the mutation in the amino acid sequence. Design an experiment that would establish whether the genetic code is overlapping or nonoverlapping.

5. Proteins generally have low contents of Met and Trp, intermediate ones of His and Cys, and high ones of Leu and Ser. What is the relation between the number of codons of an amino acid and its frequency of occurrence in proteins. What might be the selective advantage of this relation.

6. **Crick, Barnett, Brenner,** and **Watts-Tobin**, in their studies of frameshift mutations, found that either three pluses or three minuses restored the correct reading frame. If the code were a sextuplet (consisting of six nucleotides), would the reading frame be restored by either of the preceding combinations?

7. When repeating copolymers are used to form synthetic mRNAs, dinucleotides produce a single type of polypeptide that contains only two different amino acids. On the other hand, using a trinucleotide sequence produces three different polypeptides, each consisting of only a single amino acid. Why? What will be produced when a repeating tetranucleotide is used ?

8. In studies using repeating copolymers, ACA ... incorporates threonine and histidine, and CAACAA ... incorporates glutamine, asparagine, and threonine. What triplet code can definitely be assigned to threonine?

9. In the triplet-binding technique, radioactivity remains on the filter when the amino acid corresponding to the triplet is labeled. Explain the basis of this technique.

10. (*a*) Shown here is a theoretical viral mRNA sequence:

5′–AUGCAUACCUAUGUGACCCUUGGA–3′

Assuming that it could arise from overlapping genes, how many different polypeptide, sequences can be produced ? Using Figure 32–2, what are the sequences?

(*b*) A base substitution mutation that altered the sequence in (*a*) eliminated the synthesis of all but one polypeptide.

The altered sequence is shown here:

5′–AUGCAUACCUAUGUGACCCUUGGA–3′

Using Fig. 32–2, determine why?

11. Define the process of transcription. Where does this process fit into the central dogma of molecular genetics?

12. Describe the structure of RNA polymerase in bacteria. What is the core enzyme? What is the role of the sigma subunit.

13. In a mixed copolymer experiment, messengers were created with either 4/5C: 1/5A or 4/5A: 1/5C. These messages yielded proteins with the following amino acid compositions.

4/5C : 1/5A		4/5A : 1/5C	
Proline	63.0 percent	Proline	3.5 percent
Histidine	3.0 Percent	Histidine	3.0 percent
Threonine	16.0 percent	Threonine	16.6 percent
Glutamine	3.0 percent	Glutamine	13.0 percent
Asparagine	3.0 percent	Asparagine	13.0 percent
Lysine	0.5 percent	Lysine	50.0 percent
	98.5 percent		99.1 percent

Using these data, predict the most specific coding composition for each amino acid.

14. What would be the effect on reading frame and gene function under the following conditions?

(*a*) Three bases were inserted together in the middle of the gene.

(*b*) Three bases were deleted from the gene.

(*c*) One base was inserted and another one deleted five bases downstream of the insertion.

15. You know the amino acid sequence of a protein coded for by a gene in *E. coli* and the very end of the sequence is:

Pro–Trp–Ser–Glu

You find a mutant of this gene and the preceding sequence of the protein has changed to:

Pro–Gly–Val–Lys–Met–Arg–Val

Explain what has happened. What is your prediction as to the effect on protein function?

16. How does aminoacyl-tRNA synthetase^{Leu} recognize only the tRNA^{Leu} family (of which there are six) to specifically attach leucine to the proper tRNA?

17. Frameshift mutations frequently cause the translated protein to terminate downstream of the mutation. How do you explain this phenomenon?

18. The only amino acid that has 3 codons for its formation is:

(*a*) Leucine (*c*) Methionine

(*b*) Isoleucine (*d*) Tryptophan

ANSWERS TO PROBLEMS

1. (*a*) 5′–UAACGGUACGAU–3′.

(*b*) Leu-Pro-Ser-Asp-Trp-Met.

(*c*) Poly(Leu-Leu-Thr-Tyr).

2. Only single-stranded RNA can serve as a template for protein synthesis.
3. Incubation with RNA polymerase and only UTP, ATP, and CTP led to the synthesis of only poly(UAC). Only poly(GUA) was formed when GTP was used in place of CTP.
4. These alternatives were distinguished by the results of studies of the sequence of amino acids in mutants. Suppose that the base C is mutated to C′. In a nonoverlapping code, only amino acid 1 will be changed. In a completely overlapping code, amino acids 1, 2 and 3 will all be altered by a mutation of C to C′. The results of amino acid sequence studies of tobacco mosaic virus mutants and abnormal hemoglobins showed that alterations usually affected only a single amino acid. Hence, it was concluded that the *genetic code is nonoverlapping*.
5. Highly abundant amino acid residues have the most codons (*e.g.*, Leu and Ser each have six), whereas the least-abundant amino acids have the fewest (Met and Trp each have only one). Degeneracy (*a*) allows variation in base composition and (*b*) decreases the likelihood that a substitution for a base will change the encoded amino acid. If the degeneracy were equally distributed, each of the 20 amino acids would have three codons. Both benefits (*a* and *b*) are maximized by the assignment of more codons to prevalent amino acids than to less-frequently used ones.
6. No. The the term "reading frame" refers to the number of bases contained in each codon. The reason that (+ + +) or (– – –) restored the reading frame is because the code is triplet. By having the (+ + +) or (– – –), the translation system is "out of phase" until the third "+" or "–" is encountered. If the code contained six nucleotides (a sextuplet code), then the translation system is "out of phase" until the sixth "+" or "–" is encountered.
7. Assume that you have introduced a copolymer (ACACACAC ....) to a cell-free protein synthesizing system. There are two possibilities for establishing the reading frames: ACA, if one starts at the first base and CAC if one starts at the second base. These would code for two different amino acids (ACA = threonine; CAC = histidine) and would produce repeating polypeptides which would alternate *Thr-His-Thr-His ... or His-Thr-His-Thr...*

   Because of a triplet code, a trinucleotide sequence will, once initiated, remain in the same reading frame and produce the same code all along the sequence regardless of the initiation site. Given the sequence CUACUACUACUA, notice the different reading frames producing three different sequences, each containing the same amino acid.
8. From the repeating polymer ACACA ... one can say that threonine is either CAC or ACA. From the polymer CAACAA ... with ACACA ..., ACA is the only codon in common. Therefore, threonine would have the codon ACA.
9. The basis of the technique is that if a trinucleotide contains bases (a codon) which are complementary to the anticodon of a charged tRNA, a relatively large complex is formed which contains the ribosome, the tRNA, and the trinucleotide. This complex is trapped in the filter whereas the components by themselves are not trapped. If the amino acid on a charged, trapped tRNA is radioactive, then the filter becomes radioactive.
10. (*a*) Starting from the 5′ end and locating the AUG triplets, one finds two initiation sites leading to the following two sequences:

    Met-His-Tyr-Glu-Thr-Leu-Gly

    Met-Arg-Pro-Leu-Gly.

    (*b*) In the shorter of the two reading sequences (the one using the internal AUG triplet), a UGA triplet was introduced at the second codon. While not in the reading frames of the longer polypeptide (using the first AUG codon), the UGA triplet eliminates the product starting at the second initiation codon.

11. The central dogma of molecular genetics and to some extent, all of biology, states that DNA produces, through transcription, RNA, which is "decoded" (during translation) to produce proteins.
12. RNA polymerase from *E. coli* is a complex, large (almost 500,000 daltons) molecule composed of subunits (α, β, β′, σ) in the proportion α2, β, β′, σ for the holoenzyme. The β subunit provides catalytic function while the sigma (σ) subunit is involved in recognition of specific promoters. The core enzyme is the protein without the sigma.
13. Proline : C_3, and one of the C_2A triplets

    Histidine : one of the C_2A triplets

    Threonine : one C_2A triplet, and one A_2C triplet

    Glutamine : one of the A_2C triplets

    Asparagine : one of the A_2C triplets

    Lysine : A_3
14. A gene is responsible for the amino acid sequence of a protein, which may or may not be an enzyme or part of an enzyme. The gene also controls the amount of the protein synthesized at any one time.
15. After synthesis, the product of RNA polymerase II, messenger RNA, is capped with 7-methylguanosine, a poly(A) tail is added, and introns are removed.
16. Met–Pro–Gly–Asn–Thr–Ala–Gly–Asn–Ser

    Met–Pro–Gly–Lys–His–Ser–Arg–Glu
17. Overlapping codes affect two or more protein products during a mutational event, which may be more severe. Overlapping coding sequences allow compaction of information. More information can be stored in a shorter sequence.
18. (*b*) Isoleucine

CONTENTS

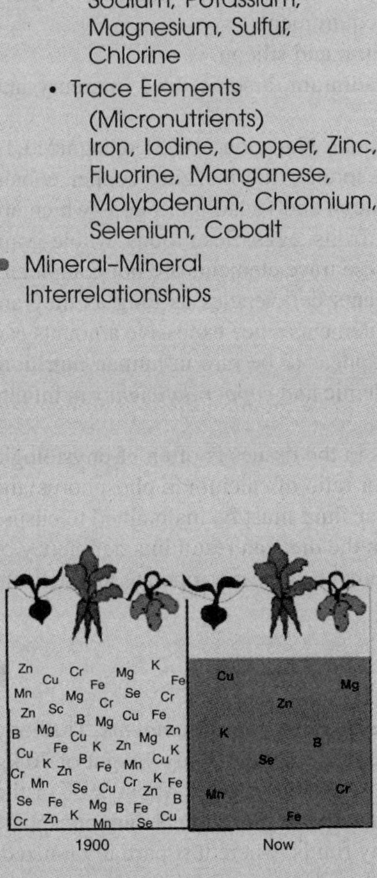

Plants may look the same but they now have far fewer minerals.

CHAPTER

33 Mineral Metabolism

INTRODUCTION

The minerals form only a small portion of the total body weight, nonetheless they are of great importance in the vital economy. They form only 7% of the composition of human body. The body contains about 24 minerals: *calcium, phosphorus, sodium, potassium, magnesium, iron, iodine, manganese, zinc, copper, cobalt, chromium, molybdenum, selenium, fluorine, chlorine, bromine, sulfur, aluminium, arsenic, nickel, tin, silicon* and *vanadium*. These minerals perform following important functions:

(*a*) as constituents of bones and teeth, *e.g.*, calcium, phosphorus and magnesium.

(*b*) as constituents of body cells of soft tissues such as muscles, liver etc, *e.g.*, phosphorus.

(*c*) as soluble salts which give to the body fluids and cell contents, their composition and stability which are both essential for life, *e.g.*, sodium, potassium, chlorine and phosphorus.

(*d*) maintenance of acid-base equilibrium, *e.g.*, sodium and potassium.

(*e*) some specific functions such as formation of hemoglobin (iron and copper); formation of thyroxine (iodine); constituent of an enzyme, carbonic anhydrase and a hormone, insulin (zinc); constituent of a vitamin, B_{12} (cobalt); and constituent of amino acids, cysteine and methionine (sulfur).

Many of these minerals are widely distributed in foods so that a well-balanced diet will supply them in sufficient quantities. The mineral elements present in the animal body may be classified into 2 groups:

1. **Principal mineral elements (= Macronutrients):** These elements occur in living tissues in comparatively large amounts. There are 7 essential elements: calcium, phosphorus, sodium, potassium, magnesium, sulfur and chlorine. They constitute 60 – 80% of all the inorganic material in the body.
2. **Trace elements (= Micronutrients):** These elements occur in living tissues in small amounts *i.e.,* in traces. They may be subdivided into 3 groups, according to their dietary requirements in higher animals:
   (*a*) ***Essential trace elements*:** iron, iodine, copper, zinc, fluorine manganese, cobalt, molybdenum, selenium, chromium and fluorine. All the essential trace elements are toxic when their intake is sufficiently in excess of dietary requirements.
   (*b*) ***Possibly essential trace elements*:** nickel, tin, vanadium and silicon.
   (*c*) ***Nonessential trace elements*:** aluminium, boron, cadmium, arsenic, lead, mercury and germanium.

The WHO Expert Committee on Trace Elements in Human Nutrition (1973) recognized 14 trace elements as being essential for human life. These are: iron, iodine, fluorine, zinc, copper, cobalt, chromium, manganese, molybdenum, selenium, nickel, tin, silicon and vanadium. Foods which are good sources of trace elements include green leafy vegetables, fruits, eggs, flesh foods, whole-grain cereals and sea foods. The functions in the body of many of these trace elements are not completely understood. Humans are not likely to suffer from trace-elements deficiencies as long as they are omnivorous. Trace elements should *not* be used as dietary supplements since excessive amounts can have adverse effects. Overall, deficiencies of trace elements appear to be rare in human nutrition. Exceptions are *iodine deficiency* in areas in which goitre is endemic and *copper deficiency* in infants on diets of unsupplemented milk.

The **balance** (*i.e.*, the ratio of one to another) of the ions in the tissues is often of physiologic importance. For example, normal ossification demands a proper ratio of calcium to phosphorus; the normal ratio between potassium and calcium in the extracellular fluid must be maintained to ensure the normal activity of muscle; and an excess of molybdenum in the diet can result in a deficiency of copper. Thus, both the principal mineral elements and trace elements appear to operate in a delicate balance among themselves.

CALCIUM

Distribution

Among all the minerals present in human body, calcium occurs in maximum amounts. The body of the infant at birth contains about 25–35 g of calcium while the adult with a body weight of 70 kg contains about 1,000–1,200 g of calcium (*i.e.*, about 1.5% of the body weight). About 99% of the calcium in the body is present in the skeleton where it is maintained as deposits of calcium phosphate in a soft fibrous matrix. The remaining 1% is present in the body fluids, where it is partially ionized. The calcium other than that in the bones and teeth is distributed as shown in Table 33–1.

Table 33–1. Distribution of calcium in body fluids or tissues

Fluid/Tissue	*mg/dl or 100 g*	*mEq/L*
Serum	9–11	5
Cerebrospinal fluid	4.5–5	2
Muscle	70	
Nerve	15	

Sources

Of the common foods, milk and milk products are undoubtedly the richest sources of calcium (Table 33–2). Skimmed milk powder contains 1.37% calcium. Though sesame seeds are rich in calcium (1.45%), the calcium is present as calcium oxalate mostly in the skin. The skin is usually removed before consumption. The dehusked sesame seed is only a fair source of calcium (0.15%). Green leafy vegetables are one of the cheapest natural sources of calcium containing about 0.34–1.13%. Small fish eaten along with bones is an excellent source of calcium (1.8%). 'Ragi' (*Eleusine coracana*) is the cheapest source of calcium with about 0.33%. Other less important sources of calcium are egg yolk, beans, lentils, nuts, figs, cabbage, turnip greens and cauliflower.

Table 33–2. Important dietary sources of calcium

Foodstuff	*Calcium content (mg/100 g)*	*Foodstuff*	*Calcium content (mg/100 g)*
Milk and milk products:		***Oilseeds and nuts:***	
Milk, cow's	120	Sesame seed	1450
Milk, buffalo	210	Sesame seed (without skin)	150
Milk, goat	170	***Cereals:***	
Curds (from cows or buffalo milk)	120–210	Ragi	330
Milk powder, whole	950	***Green leaf vegetables:***	
Milk powder, skimmed	1370	Sesbania leaves	1130
Cheese	790	Carrot leaves	340
Khoa (from whole buffalo milk)	650	Amaranth leaves	500
		Drumstick leaves	460
		Curry leaves	810
		Fish:	
		Small fish, dried	1800

Requirements

Surveys have shown that a large majority of population in the developing countries consume diets which provide 300–500 mg of calcium. Metabolic studies have shown that they maintain positive calcium balance on such diets. The FAO/WHO Expert Group on calcium requirements (1962) considered the question of calcium requirements and suggested 'practical allowances' for calcium which will be adequate to meet the needs of different age groups (refer Table 33–3). The ICMR Nutrition Expert Group (1981) adopted the above recommendations with slight modifications.

Table 33–3. Recommended intake of calcium (mg/day)

Age group	*FAO/WHO (1962)*	*ICMR (1981)*
Infants (0 – 12 months)	500–600	500–600*
Children (1 – 9 years)	400–500	400–500
Children (10 – 15 years)	600–700	600–700
Adolescents (16 – 19 years)	500–600	500–600
Adults (man and woman)	400–500	400–500
Pregnancy and Lactation	1,000–1200	1,000

* applies to artificially-fed infants. Breast-fed infants need no additional supplement.

HENRY CLAPP SHERMAN

(LT, 1875–1955)

Sherman, a U.S. biochemist, was born on October 16, 1875 in a farm near Ash Grove, Virginia. He graduated from Maryland Agriculture College and worked as an assistant till 1895. He was a research fellow in chemistry at Columbia University from 1895 to 1897 and received his Ph.D. degree. From 1897 to 1899, he worked with **Prof. Wilbur O. Atwater** (LT, 1844–1907) at Wesleyan University. He returned to Columbia University, in 1899, as lecturer and soon became associate professor and professor. Sherman's interest in nutrition embraced many aspects of science such as **energy metabolism**, protein and calcium requirements and practical problems of food selection and distribution. His classical work on **protein and calcium requirements of human beings** is quite well known. Another important contribution is his long term studies over several generations with rats on two diets containing varying amounts of milk. This study showed that the diet containing larger amounts of milk was more nutritious and promoted better growth and health in rats. He also developed a unit, called **Sherman unit**, which is a unit of minimum protective dose of vitamin C. It is the minimum amount of vitamin C that, fed daily, will protect a 300-g guinea pig from scurvy for 90 days; it is equivalent to 0.5–0.6 mg of ascorbic acid. His name is also associated with 2 more units: **Sherman–Bourquin unit** of vitamin B_2 and **Sherman–Munsell unit** of vitamin A. He wrote a classical book 'The Chemistry of Food and Nutrition', which has undergone several editions. He is author or coauthor of a few other books– 'Food Products', 'The Essentials of Nutrition' and 'The Science of Nutrition'. He died in 1955.

Absorption and Retention

Calcium is absorbed *incompletely* by an active transport process occurring mainly in the upper small intestine. The process is regulated by 1, 25-dihydroxycholecalciferol, a metabolite of vitamin D that is produced in the kidney in response to low plasma Ca^{2+} concentrations. Thus, Ca^{2+} absorption is adjusted to body needs. Absorption is facilitated by vitamin D, lactose and protein. On a high-protein diet, about 15% of the dietary calcium is absorbed, compared with 5% absorption on a low-protein diet. Calcium is well absorbed at the normal pH of the intestinal contents. If the contents become alkaline, Ca^{2+} absorption is lowered due to the formation of insoluble tricalcium phosphate. An increase in acidophilic flora (*e.g.*, *Lactobacillus*) is recommended to lower the pH, which favours Ca^{2+} absorption.

Absorption of calcium is inhibited by a numebr of dietary factors that cause formation of insoluble calcium salts in the intestine, including phytate (*e.g.*, cereal grain), oxalates (*e.g.*, spinach), and phosphates. When fat absorption is impaired, much free fatty acid is present which reacts with Ca^{2+} to form insoluble calcium salts of fatty acids which are excreted in the feces. Presence of excess of fibre in the diet interferes with the absorption of calcium.

Studies have been conducted by many nutritionists on the absorption and retention of calcium in human subjects. The absorption and retention of calcium depends on (*a*) calcium intake, (*b*) presence of interfering substances, and (*c*) presence of vitamin D. On normal intake of calcium, the retention varies widely from 10 to 20%, depending on the diet and the age of the subject, as shown in Table 33–4.

It is apparent from the above table that the infants fed on breast milk receive only 103 to 208 mg calcium daily and retain about 80 mg, while infants fed on cow's milk receive daily 600 to 900 mg calcium and may retain about 120 to 150 mg. Infants fed on human milk adapt to the cow intake and calcium absorption on human milk is about 2 or 3 times as high as on cow's milk. When calcium intake is low, the percentage absorption is high.

Table 33–4. **Retention of calcium in humans**

Subjects	*Source of calcium in diet*	*Calcium intake (mg)*	*Calcium retention (%)*
Intfants (1–12 months)	Human milk	103–208	40–70
	Cow's milk	600–900	15–30
Preschool children	Cow's milk	400–1200	18–25
	$CaHPO_4$	350–550	18
Early school age	Cow's milk	750–1300	21
School children	Rice diet	259–277	4–15
	Ragi diet	1151	20
	Pearl-millet diet	479	19
	Sorghum diet	441	17

Functions

The matrix of the bone in which calcium is deposited has a unique structure essential to normal calcification. The major inorganic constituent of bone consists of a crystalline form of calcium phosphate which closely resembles the mineral hydroxyapatite. In addition, however, the bone contains an appreciable amount of noncrystalline, amorphous calcium phosphate. It appears that this amorphous material is predominant in early life but is replaced in later life by the crystalline apatite. Bone is in a dynamic state, constantly being formed and resorbed– more rapidly during early development and at a slower and declining rate during adult life. It is estimated that in adult males about 700 mg of calcium enter and leave the bones each day.

About 15% of the noncollagen protein of bovine bone consists of the calcium-binding protein called **osteocalcin** (mol. wt. 6,800) which contains 3 mol of γ-carboxyglutamic acid per mole of enzyme (**Price** *et al,* 1976). Bone conains approximately one molecule of osteocalcin for each microcrystal of hydroxyapatite. Osteocalcin may be involved in the processes of bone mineralization and in the pathologic calcification of soft tissues, such as arteries.

As stated earlier, about 1% of calcium, which is present in the body fluids, is in part ionized. It is, indeed, this calcium which is of great significance in blood coagulation, in maintaining the normal excitability of the heart, muscles, and nerves, and in the differential aspects of membrane permeability.

Metabolism

The blood cells contain very little calcium. Most of the calcium is, hence, present in the plasma, where it exists in 3 *forms* or *fractions* (Fig. 33–1):

(*a*) ionized (or **diffusible**) **calcium:** This is present to the tune of 50% of the total plasma calcium of 8.8–10.4 mg/100 ml (= 2.2–2.6 m mol/L), *i.e.,* about 4.4–5.2 mg. The ionized (or free) calcium content is controlled by vitamin D, parathyroid hormone and calcitonin. A normal ionized calcium content is essential for normal calcification of bone.

(*b*) protein-bound (or **nondiffusible**) **calcium:** This accounts to 40% of the total plasma protein, *i.e.,* about 3.6–4.0 mg. It is mainly bound to negatively-charged albumin.

(*c*) complexed calcium: The remaining 10% amount of the total plasma protein, *i.e.,* about 0.8–1.2 mg is present complexed with citrate, phosphate and bicarbonate.

All of these forms of calcium in the serum are in equilibrium with one another. In the usual determination of calcium, all 3 fractions are measured together.

Endogenous Loss of Calcium

When the calcium intake is negligible, there will be depletion of body calcium. This is known as **endogenous loss** of body calcium. The endogenous calcium is excreted in the urine, feces and sweat. In practice, the total endogenous loss of calcium is assessed as the loss of body calcium when the intake of dietary calcium is zero and this is obtained by extrapolation from the regression of retention on intake estimated at 2 or more intervals. The losses so estimated vary from 150 – 300 mg calcium/day in adults. The isotopic studies, conducted by **Bauer** *et al* (1957), showed the endogenous losses in an infant to be 109 mg/day. It may be assumed that the endogenous calcium losses in children will be between the values for infants and adults.

Calcium Balance

Most of the calcium (70–90%) eliminated from the body is excreted in the feces. The fecal calcium concentration is positively correlated with calcium intake. In fact, most of the calcium in the feces is unabsorbed calcium and a very small amount is excreted into the intestine after absorption.

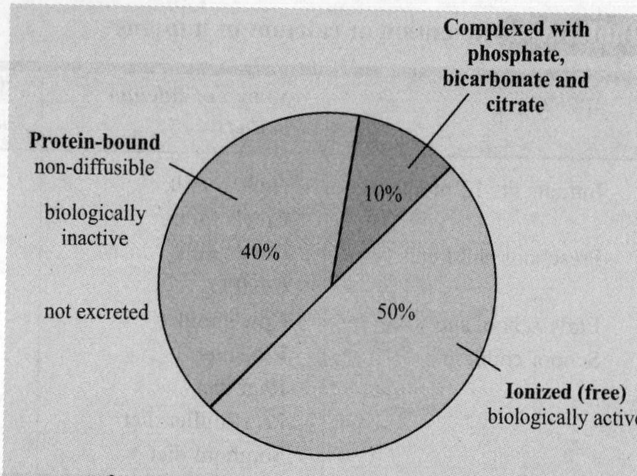

Total Ca^{2+} = Ionized Ca^{2+} + Protein-bound Ca^{2+} + Complexed Ca^{2+}
(2.2–2.6 mmol/L) (1.1–1.3 mmol/L) (0.9–1.0 mmol/L) (0.2–0.3 mmol/L)

Fig. 33–1. The 3 forms of circulating calcium

A relatively small quantity of absorbed calcium is excreted in urine. In man, about 10 g of calcium are filtered in a 24-hour period by the renal glomeruli, but only about 200 mg (0.2 g) appear in the urine. The **calcium balance** (the difference between the quantity of calcium ingested and that excreted in urine and the feces) is *positive* during growth, pregnancy, lactation and in normal adults. The excretion of calcium continues on a Ca-deficient diet when the body will be in *negative* calcium balance.

The average daily loss of calcium in sweat is about 15 mg. Strenuous physical exercise increases the loss by way of the sweat even during periods of low intake of calcium. In tropical climates also, the loss of calcium in sweat will be appreciable. Under severe sweating conditions, the daily calcium loss in adults may vary from 40 to 120 mg. This may account for the apparent retention of calcium in adults, as calcium loss in sweat is usually not determined in calcium balance studies.

Calcium x Phosphorus Product in Serum

The plasma phosphorus content (Table 33–5) in normal infants and children (5.2–6.4 mg) is higher than that found in normal adults (2.6–4.2 mg). In the serum, the product of Ca x P is over 50 in normal infants and children, and about 30–40 in normal adults. The Ca x P product is very much reduced in children suffering from rickets (20 to 30) and in adults suffering from osteomalacia (14 to 24). A high Ca x P product is very important for normal ossification of bone. If it is low, ossification does not take place.

Table 33–5. Calcium and phosphorus content of plasma in normal and vitamin D-deficient subjects

Subjects	*Plasma calcium (mg/100 ml)*	*Plasma phosphorus (mg/100 ml)*	*Ca x P product*
Normal			
Infants and children	9–11	5.2–6.4	50–60
Adults	9–11	2.6–4.2	30–40
Vitamin D-deficient			
Infants and children (Rickets)	7–8	3–4	20–30
Adults (Osteomalacia)	7–8	2–3	14–24

Calcium : Phosphorus Ratio in Diet

Diets having distorted Ca : P ratios have been used to produce experimental rickets in animals. Observations on several species have indicated that during rapid growth and calcification, the diet should have a Ca : P ratio 1 : 1. When the calcium is required only for maintenance as in adults, the calcium requirement is lower, both absolutely and relatively to the phosphorus requirement. Thus, the Ca : P ratio is highest at the earliest stage (1 : 1), decreases with the attainment of adulthood (1 : 2) and then, in the case of the female, increases again (1 : 1) in the latter part of pregnancy and during lactation. In the normal adult mammal, the Ca : P ratio of the whole body is a little under 1 : 1 and that of the bone is a little over 2 : 1.

Disease States

Calcium metabolism has profound effects on the human body as a whole. It affects following disease states:

(*a*) Hypoparathyroidism and hyperparathyroidism: Vitamin D increases the absorption of calcium and phosphorus in the intestines and also promotes calcification of bones. It also promotes mobilization of calcium from bones. The parathyroid hormone (parathormone, PTH) regulates calcium level in blood and calcium metabolism in bones. It acts directly on the bone releasing calcium from it and thus raising the plasma calcium level. PTH also increases the excretion of phosphates by the kidney. In **hypoparathyroidism**, such as occurs after operative removal of the parathyroid glands, the concentration of the serum calcium may drop below 7 mg/100 ml and tetany may result. There is a concomitant increase in serum phosphate and a decrease in urinary phosphates. The urinary calcium is extremely low as well. The symptoms of tetany can be relieved temporarily by injecting PTH.

In **hyperparathyroidism**, which is caused by hyperactive, hyperplastic, or adenomatous parathyroid glands, the followng signs develop: an increase in serum calcium level (hypercalcemia) from the normal level of 9–11 mg to 16–22 mg/100 ml, decrease in serum phosphate, decreased renal tubular reabsorption of phosphate, increased phosphatase activity, rise in urinary calcium and phosphorus from bone decalcification, and dehydration. Spontaneous fractures may also occur. These signs are due to increased renal losses of phosphorus, causing a decrease in serum phosphate which elicits an increase in calcium to maintain the Ca : P ratio. The extra calcium and phosphorus is lost from soft tissues and from bone by increased osteoclastic (bone-destroying) activity. The activities of calcitonin (CT) are opposed to those of parathormone: (*a*) It inhibits resorption and mobilization of Ca and P from bone and (*b*) It causes a lowering serum calcium content. The influence of vitamin D, PTH and calcitonin on calcium and phosphorus metabolism are shown in Fig. 33–2.

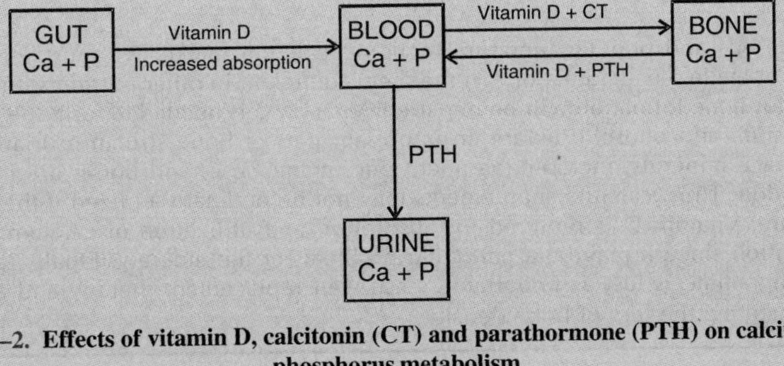

Fig. 33–2. Effects of vitamin D, calcitonin (CT) and parathormone (PTH) on calcium and phosphorus metabolism

(*b*) Osteoporosis: Osteoporosis is a skeletal disease that has often been regarded as possibly related to inadequate intake of calcium. It has become increasingly evident, however, that such a simple etiologic relationship is not tenable. No relationship has been established between bone loss and calcium intake; furthermore, intakes of calcium about 1,500 mg/day do not seem to be 'protective', nor are intakes below 300 mg/day unequivocally associated with bone tissue loss. There remains a possibility, however, that high protein intake associated with low calcium intake may play a role in the pathogenesis of osteoporosis since calcium losses can be significant when protein intake is high. Were this dietary situation to continue for prolonged periods, there would result a considerable loss in total body calcium.

CLINICAL IMPLICATION 33–1.

Calcium Metabolism and Osteoporosis

Based on the TV ads and wide variety of calcium-fortified foods on the market, it would be easy to assume that all an older woman needs to prevent osteoporosis is a diet rich in calcium. However, that may be like closing the barn door after the horse has left. There is a strong consensus that the years from age 10 to 35, when the bone density is reaching its maximum, are the most important for reducing the risk of osteoporosis. The maximum bone density obtained during these years is clearly dependent on both calcium intake and exercise, and dense bones seldom become seriously depleted of calcium following manopause. Unfortunately, most women consume far too little calcium during these years. The RDA for calcium is 1300 mg/day (4 or more glasses of milk a day) for women from age 11 to 18, 1000 mg/day (3 or more glasses of milk a day) for women from 19 to 50, and 1200 mg/day (4 glasses of milk a day) for women over 50. But the average calcium intake for women in this age range is only about 500 mg/day. Thus, it is evident that increased calcium intake should be encouraged in this group.

With regard to the postmenopausal women, the 1994 National Institutes of Health (NIH) consensus panel on osteoporosis recommended that such women consume up to 1500 mg calcium per day, but this recommendation has been vigorously criticised by other diet experts. In fact, calcium balance studies have shown that many postmenopausal women need 1200–1500 mg calcium per day to maintain a positive calcium balance (more calcium coming in than is lost in the urine), but that does not necessarily mean that the additional calcium will be stored in their bones. All of those studies have been complicated by the discovery that calcium intake may have varied effects on different types of bones. Calcium intakes in the range of 1000–1500 mg/day appear to slow the decrease in density of cortical bone, such as that found in the hip, hand, and some parts of the forearm. Similar doses, however, appear to have little or no effect on loss of density from the trabecular bone found in the spine, wrist, and other parts of the forearm. At least some of the confusion in the earlier studies appear to have resulted from differences in the site used for measurement of bone density. Thus, the effect of high calcium intakes alone on decreasing bone loss in postmenopausal women remains controversial at present. It is clear that elderly women should be getting the RDA for calcium in the diet. With the recent concern about the fat content of dairy products, calcium intakes in this group appear to be decreasing rather than increasing.

While the ads and popular literature focus on calcium intake, we must remember that bones are not made of calcium alone. If the diet is deficient in other nutrients, the utilization of calcium for bone formation will be impaired. Vitamin C is needed to form the bone matrix and magnesium and phosphorus are an important part of bone structure. Viamin K and a variety of trace minerals, including copper, zinc, manganese, and boron are important for bone formation. Thus, calcium supplements may not be optimally utilized if the overall diet is inadequate. Vitamin D is required for absorption and utilization of calcium. It deserves special mention since it may be a particular problem for the elderely. Finally, an adequate exercise programme is just as important as estrogen replacement therapy and an adequate diet for preventing the loss of bone density.

(*c*) Rickets: This disease is characterized by faulty calcification of bones due to a low vitamin D content of the body, a deficiency of Ca and P in the diet, or a combination of both. Usually the serum phosphate concentration is low or normal except in renal disease (where it is elevated) and the serum calcium remains normal or may be lowered. There is an increase in fecal phosphate and calcium because of poor absorption of these elements, with the simultaneous decrease in urinary phosphate and calcium. Another characteristic of rickets is an increase in alkaline phosphatase activity.

(*d*) **Renal rickets** (or, more accurately designated, **familial hypophosphatemic rickets**): This disease is inherited as an X-linked dominant trait. Affected males exhibit hypophosphatemia and severe rickets; heterozygous females, as opposed to males, tend to have higher serum phosphorus concentrations and less severe bone disease. Hypophosphatemic rickets is characterized biochemically not only by lowered serum phosphorus but also by hyperphosphaturia (increased loss of phosphorus in the urine) and a reduced intestinal absorption of Ca and P. The disease is supposed to be caused by defective transport of phosphate by the intestine and by the renal tubules. Vitamin D in ordinary dosages does not relieve the symptoms in renal rickets; thus, it has sometimes been referred to as **vitamin D-resistant rickets**.

(*e*) **Decrease of serum calcium:** In severe renal disease, the serum calcium may decrease– in part because of increased losses into the urine. However, except for circumstances contributing to malabsorption of calcium from the intestine, there is no clinical syndrome attributable exclusively to inadequate dietary intake of calcium.

PHOSPHORUS

Distribution

Phosphorus is found in every cell of the human body. An adult human contains about 400 – 700 g of phosphorus as phosphates. A major part of this (about 80% of the total) is present in combination with calcium in the bones and teeth. About 10% is combined with proteins, lipids and carbohydrates, and in other compounds in blood and muscle. The remaining 10% is widely distributed in various chemical compounds. The distribution of phosphorus in the body is shown in Table 33–6.

Table 33–6. Distribution of phosphorus in body fluids or tissues

Fluid/Tissue	*mg/dl or 100 g*	*mEq/L*
Blood	40	
Serum (inorganic)		
Children	4–7	1.3–2.3
Adults	3–4.5	0.9–1.5
Muscle	170–250	
Nerve	360	
Bones and Teeth	22,000	

Phosphorus is present in the body as inorganic salts of phosphoric acid or in combination with oranic compounds.

(*a*) **Inorganic phosphorus:** present as calcium phosphate in bones and teeth, and as phosphates of Na and K in soft tissues or body fluids.

(*b*) **Organic phosphorus:** present as (*a*) phospholipids (*e.g.*, lecithin, cephalin), (*b*) nucleoproteins and nucleic acids, (*c*) creatine phosphate, ATP and ADP and coenzymes I, II and cocarboxylase, and (*d*) hexose phosphates, triosephosphates and glycerophosphates.

Sources

Phosphorus is widely distributed in foodstuffs. Almost all foods contain phosphorus. The important sources are milk, eggs, meat and fish. A greater part of phosphorus present in cereals, pulses, nuts and oilseeds exists in the form of phytic acid (a compound of inositol and phosphoric acid) or phytin (Ca and Mg salts of phytic acid). Data regarding the phosphorus contents of some foods are given in Table 33–7. Since the distribution of calcium and of phosphorus in foods is very similar, an adequate intake of calcium generally ensures an adequate intake also of phosphorus.

Table 33–7. Important dietary sources of calcium*

Foodstuff	*Phosphorus content (100 g)*	*Foodstuff*	*Phosphorus content (100 g)*
Milk and milk products		Rice, raw, milled	0.11
Milk, cows	0.09	Rice, parboiled, homepounded	0.22
Milk, buffalo	0.13	Rice, parboiled, milled	0.17
Milk powder, skimmed	1.00	Wheat, whole	0.32
Milk powder, whole	0.73	Wheat flour (maida)	0.09
Cheese	0.52	***Pulses***	
Khoa (from whole buffalo milk)	0.42	Bengal gram	0.31
		Black gram	0.37
Flesh foods		Cow gram	0.49
Egg, hen's	0.22	Field bean, dry	0.45
Egg, duck's	0.26	Green gram	0.28
Fish	0.41	Horsegram	0.39
Mutton	0.24	Lentil	0.25
Liver, sheep	0.38	Peas, dried	0.30
Cereals		Red gram	0.26
Bajra (Cambu)	0.35	Soybean	0.69
Barley	0.23	***Nuts and oilseeds***	
Cholam (Jowar)	0.28	Almond	0.49
Italian millet	0.29	Cashewnut	0.45
Maize (corn)	0.33	Coconut, fresh	0.24
Oatmeal	0.38	Gingelly (sesame) seeds	0.57
Ragi	0.27	Groundnut	0.39
Rice, raw, husked	0.23		
Rice, raw, homepounded	0.21		

* The phosphorus present in milk and animal foods is available to greater extent than that present in cereals, pulses, nuts and oilseeds.

Requirements

The requirements of phosphorus depend on the availability of phosphorus present in diets. Phosphorus present in cereals and legumes are available to a lesser extent (as it is present in the form of phytic acid) than that present in milk, meat, eggs and fish. Consequently, phosphorus requirements of persons consuming mainly cereal-based diets, will be greater than those consuming large quantities of milk, meat, eggs and fish. The optimal Ca : P ratio for infants and children is 1 : 1 and for adults 1 : 2. The Ca : P ratios of diets consumed in different countries vary widely from 1 : 1.5 to 1:3. Metabolic studies conducted on these diets have revealed that the subjects maintained positive calcium and phosphorus balance. Phosphorus requirements have not been specifically considered by FAO/WHO Committees. However, NRC (USA) has recommended an allowance for phosphorus equal to that for calcium (*i.e.,* Ca : P ratio of 1 : 1) for all age groups except infants. With respect to young infants, it should be noted that the Ca : P ratio in cow's milk is about 1.2 : 1, whereas in human milk it is 2 : 1. Because an intake of phosphorus such as is provided in cow's milk may contribute to the occurrence of hypocalcemic tetany during the first week of life, it is now recommended that in early infancy the Ca : P ratio of the diet should be 1.5 : 1.

Functions

Phosphorus performs following important functions:

1. It is necessary for the formation of bones and teeth.
2. It is a constituent of phospholipids– lecithin and cephalin, which are integral parts of the cell structure and also act as intermediates in fat transport and metabolism.
3. It is essential for carbohydrate metabolism as phosphorylation of glycogen requires inorganic phosphorus and phosphoric esters such as adenylic acid, adenylpyrophosphate and creatine phosphate.
4. It is a constituent of certain coenzymes, *e.g.*, coenzyme I and cocarboxylase which take part in the enzyme systems concerned in the oxidation of carbohydrates, fats and proteins.
5. It is an essential constituent of nucleic acids and nucleoproteins which are integral parts of the cell nuclei.

Metabolism

Phosphorus plays an important role in almost all metabolisms. However, the metabolism of phosphorus is, in large part, related to that of calcium. The Ca : P ratio in the diet affects the absorption and excretion of these elements. If either element is given in excess, excretion of the other is increased. The optimal ratio is 1:1 when the intake of vitamin D is adequate. An increase in carbohydrate metabolism, such as during absorption of carbohydrate, is accompanied by a temporary decrease in serum phosphate. A similar decrease may occur during absorption of some fats. In diabetes mellitus, there is a lower concentration of organic phosphorus but a higher concentration of inorganic phosphorus in the serum.

Phosphorus is absorbed in the small intestine as inorganic phosphates. Phosphorus present in organic combination, such as phytic acid should be hydrolyzed to inorganic phosphates before absorption. Since the enzyme *phytase* is not present in human digestive juices, phytin phosphorus is absorbed only to a very slight extent in humans. Phosphorus present in animal foods such as milk, meat and eggs is absorbed to a greater extent than that present in cereals and legumes as the latter exists mostly in the form of phytic acid. The kidney is the major pathway of excretion of the phosphorus absorbed. The retention of phosphorus in children kept on different diets has been reported to vary widely from 10 to 40%. The retention of phosphorus depends on the following factors:

(*a*) the quantity of phosphorus ingested,

(*b*) the calcium content of the diet,

(*c*) the form in which phosphorus exiss in the diet, and

(*d*) the vitamin D intake.

Disease States

Phosphate retention is a prominent cause of the acidosis in severe renal disease and the resultant elevated serum phosphorus also contributes to the lowered serum calcium. Blood phosphorus levels are also high in hypoparathyroidism. A relationship of phosphorus metabolism to growth hormone (GH) is possibly indicated by the fact that growing children usually have high blood phosphorus levels and that in acromegaly an elevation of the blood phosphorus also occurs.

Blood phosphorus levels are low in hyperparathyroidism and in spruce and celiac disease. A low blood phosphorus together with an elevated alkaline phosphatase is also characteristic feature in patients with an inherited or acquired renal tubular defect in the reabsorption of phosphate. Such cases embrace familial hypophosphatemic rickets (so-called renal rickets), De Toni–Fanconi syndrome, and an inherited form of osteomalacia in adults called Milkman's syndrome. The greatly enhanced excretion of phosphate in the urine of these patients distinguishes them from those in whom a deficiency of vitamin D is the cause of the low serum phosphorus and the accompanying defects in the calcification of bone.

SODIUM

Distribution

An adult human body contains about 100 g of sodium ion. About 1/3rd of the total sodium content of the body is present in the inorganic portion of the skeleton. However, most of the sodium is found in the extracellular fluids of the body (Table 33–8).

Table 33–8. Distribution of sodium in body fluids or tissues

Fluid/Tissue	*mg/dl or 100 g*	*mEq/L*
Whole blood	160	70
Plasma	330	143
Cells	85	37
Muscle tissue	60–110	
Nerve tissue	312	

Sources

The main source of sodium is the sodium chloride (NaCl) used in cooking and seasoning; besides, ingested foods also contain additional sodium. It is estimated that about 10 g of NaCl (4 g of sodium) is, thus, ingested each day. In addition to salted foods, sodium-rich foods include bread, cheese, clams, oysters, crackers, wheat germ and whole grains. Foods such as carrots, cauliflower, celery, eggs, legumes, milk, nuts, spinach, turnips, oatmeal, and radishes are also relatively high in sodium. Data regarding sodium (and potassium) content of some foods are given in Table 33–9.

Table 33–9. Sodium and potassium contents of some foods (mg / 100 g)

Foodstuff	*Sodium*	*Potassium*
Cereals		
Barely	4	179
Maize (corn)	6	290
Rice, raw milled	3	110
Rice, parboiled milled	9	161
Wheat flour, refined	14	101
Wheat, whole	18	349
Cereal products		
Bread (without salt)	10	65
Bread (with added NaCl)	420	46
Biscuits (with added NaCl)	520	67
Corn flakes (with added NaCl)	870	120
Pulses		
Black gram	75	643
Chick pea	72	386
Cowpea	95	323
Field bean	89	402
Green gram	49	643
Lentil	29	402
Red gram	66	482
Peas, dry	20	640

Oilseeds and nuts		
Almonds	6	856
Cashewnut, roasted	41	425
Cashewnut, salted	428	422
Groundnut, roasted	24	686
Groundnut, salted	420	675
Milk and milk products		
Butter, unsalted	4	25
Butter, salted	840	25
Cheese, processed (with NaCl added)	740	100
Milk, cow's	50	140
Milk, dried	420	1260
Eggs, meat and fish		
Eggs, hen (whole)	122	129
Meat, lamb's	70	290
Fish	61	304
Pork, raw	65	390
Pork, cured (with NaCl added)	930	326
Vegetables		
Cabbage, raw	20	233
Carrots, raw	47	341
Cauliflower, raw	13	295
Potato, raw	4	503
Radish	18	322
Spinach	71	470
Fruits		
Apple, raw	1	110
Cherries	2	191
Cranberry juice	1	60
Grapes	3	158
Grape fruit	1	136
Lemon	1	104
Orange	1	200
Papaya	3	234
Pears	2	130

Requirements

The requirement of sodium in the form of NaCl in adults varies from as low as 5 to as high as 20 g/day, depending upon climate, occupation and physical activity. In temperate regions, the NaCl intake ranges between 5 and 8 g while in tropical countries the intake ranges between 10 and 20 g. People engaged in hard work will need more. The intake of sodium from the foodstuffs may vary from 100 to 200 mg as the common foods are poor sources of sodium.

A maximum NaCl intake of about 5 g/day may be recommended for adults without a history of hypertension. This is about half the daily amount which is ordinarily consumed. Furthermore, an intake 5 g of NaCl per day is 10 times the amount at which adequate NaCl balance can apparently be maintained. For persons with a family history of hypertension, **Dahl** (1958) has recommended a diet containing no more than 1 g of NaCl per day. Such a diet helps to lower blood pressure in subjects suffering from hypertension.

Since the daily NaCl consumption is higher than the requirements, the excess NaCl consumed is excreted in urine, feces and sweat. The obligatory (irreducible) daily losses of sodium in an adult, as determined by **Dahl** (1958) are: urine, 5 – 35 mg; feces, 10 – 125 mg; skin (not sweating visibly), 25 mg; the total losses being 40 – 185 mg/day. The minimum daily sodium need can be met by a daily intake of about 250 mg of sodium (equivalent to 525 mg of NaCl).

The most variable loss of sodium occurs by way of the sweat. Ordinarily, sweat contains 20 – 50 mEq sodium per litre. Excessive losses of sweat resulting from heavy exercise, environmental heat, or high fever can lead to sodium losses of as much as 350 mEq/day. However, the loss of NaCl in the sweat can be minimized during prolonged exposure to high temperatures if a period of a few days is allowed for adaptation. It is suggested that whenever a water intake of more than 4 L is required to replace sweat loss, extra NaCl should be provided. The need will vary from 2 g sodium chloride per litre of extra water loss up to an extra 7 g/day for persons doing heavy work under hot conditions.

Sodium is readily absorbed mainly in the ileum, so that the feces contain very little except in diarrhea, when much of the sodium excreted into the intestine in the course of digestion escapes reabsorption.

Functions

Sodium ion exists in the body in association with following anions: chloride, bicarbonate, phosphate, lactate and proteinate. It performs following functions:

1. Sodium is the major component of the cations of the extracellular fluid.
2. It regulates the acid-base equilibrium of the body.
3. It maintains the osmotic pressure of body fluids and thus protects the body against excessive fluid loss.
4. If preserves the normal irritability of muscle and the permeability of the cells.
5. It plays an important part in the absorption of monosaccharides and the amino acids from the small intestine.
6. Na^+ ion also plays an important role in initiating and maintaining the heart beat.

Metabolism

The metabolism of sodium is influenced by the adrenocortical steroids. In adrenocortical insufficiency, a decrease of serum sodium and an increase in sodium excretion occur. In chronic renal disease, particularly when acidosis coexists, sodium depletion may occur due to poor tubular reabsorption of sodium as well as to loss of sodium in the buffering of acids.

Unless the individual is well adapted to a high environmental temperature, extreme sweating may cause the loss of considerable sodium in the sweat; muscular cramps of the extremities and abdomen, headaches, nausea, and diarrhea may develop.

Disease States

The levels of serum sodium may not reflect accurately the total body sodium. Thus, a low concentration of serum sodium (**hyponatremia**) may develop if patients are fed with large quantities of salt-free fluids. This obviously is not an indication of actual depletion of body sodium but rather the effect of overhydration. A similar situation prevails in edematous states such as cirrhosis or congestive heart failure (CHF), wherein low serum sodium is frequently observed although the total body sodium may actually be excessive. Clinical conditions causing hyponatremia are (*a*) prolonged vomiting and diarrhea resulting in excessive loss of digestive juices rich in Na^+ ion, (*b*) chronic renal disease with acidosis due to poor reabsorption of sodium in the tubules, and (*c*) adrenocortical deficiency leading to Addison's disease. Severe hyponatremia will cause severe dehydration, decrease in blood volume, decrease in blood pressure, and circulatory failure.

Increased serum sodium (**hypernatremia**) is rare. It may occur as a result of rapid administration of sodium salts, or may be due to hyperactivity of the adrenal cortex as in Cushing's syndrome, or may be due to prolonged treatment with corticotropin (ACTH), cortisone, deoxycorticosterone or some of the sex hormones. The most common cause of hypernatremia, however, is rapid loss of water, such as in the dehydration associated with diabetes insipidus. Occasionally, hypernatremia may also follow excessive sweating. This is so because sweat is a hypotonic salt solution; consequently, loss of this fluid endengers a loss of water at a rate exceeding a loss of salt insofar as the ratio of water to salt in the body fluids is concerned. The main symptoms of hypernatremia are: increased retention of water in the body, increase in blood volume, and increase in blood pressure.

Addison's disease, which is characterized by an increased sodium loss because of adrenocortical insufficiency, is ameliorated during pregnancy presumably because of the production of steroid hormones, which cause sodium retention. It has also been shown that the placenta elaborates hormones, with sodium-retaining efects. And it is believed that these hormonal substances are responsible for the sodium and water retention, accompanied by rapid gains in weight, commonly observed in certain stages of pregnancy.

POTASSIUM

Distribution

Potassium is present largely within different cells. It is present in small amounts in the extracellular fluids. An adult human body contains about 250 g of potassium. Of this about 90% is present in the cells of various tissues and red blood cells, the remainder 10% is present in the extracellular fluids (Table 33–10).

Table 33–10. Distribution of potassium in body fluids or tissues

Fluid/Tissue	*mg/dl or 100 g*	*mEq/L*
Whole blood	200	50
Plasma	20	5
Cells	440	112
Muscle tissue	250–400	
Nerve tissue	530	

Sources

Potassium occurs widely in foodstuffs. A high potassium content is found in the following foods (300 to 600 mg/serving): chicken, beef liver, pork; dried apricots, dried peaches, bananas, orange and pineapple juices; yams, winter squash, broccoli, potatoes and Brussels sprouts. There are other foods rich in potassium but also rich in sodium. Since in many situations the need for high potassium intake parallels that for a low sodium intake, these foods are of less value as sources of potassium. Data regarding potassium content of some foods have already been given in Table 33–9.

Requirements

The normal daily intake of potassium from the diet and drinking water is between 2–4 g, depending on the composition of the diet. This element is so widely distributed in foodstuffs that a deficiency is unlikely except in the pathologic states.

Functions

Potassium performs following functions:

1. It is the main cation in the cells and plays an important role in the regulation of the acid-base balance in the cell.
2. It is an essential constituent of the extracellular fluid (ECF). Though the level is small (20 mg/100 ml) it influences the activity of muscle.
3. It is also essential for the growth and builup of tissues, as every 6.25 g tissue protein synthesized is accompanied by retention of 117 mg of potassium in the tissues.
4. Potassium is needed for the synthesis of glycogen. Every one gram of glycogen synthesized is accompanied by retention of 14 mg of potassium.
5. During muscular contraction, potassium is lost from the muscle into the ECF and during the recovery phase, K^+ ions are withdrawn by the muscle cells from the ECF.

Metabolism

Variations in extracellular potassium influence the activity of striated muscles so that paralysis of skeletal muscle and abnormalities in conduction and activity of cardiac muscle occur. The body maintains a state of potassium equilibrium by adjusting the excretion of potassium to its intake. Most of the ingested potassium (95%) is excreted in urine and a small part in feces. Although potassium is excreted into the intestine in the digestive fluids (Table 33–11), much of this is later reabsorbed. The kidney is the principal organ of excretion for potassium. Not only is potassium filtered by the kidney in the glomeruli but it is also secreted by the tubules. The excretion of potassium is markedly influenced by changes in acid-base balance as well as by the activity of the adrenal cortex. The capacity of the kidney to excrete potassium is so great that hyperkalemia (elevated serum potassium) will not occur, even after the ingestion or intravenous injection of relatively large quantities of potassium, if kidney function is unimpaired. This is not so when urine production is insufficient. Potassium should *not* be given intravenously until circulatory collapse, dehydration, and renal insufficiency have been corrected.

Table 33–11. Volume and composition of blood plasma, gastrointestinal secretions, and sweat

Fluid	*Average volume (ml/24 hours)*	*Electrolyte concentrations (mEq/L)*			
		Na^+	K^+	Cl^-	HCO_3^-
Blood plasma		135–150	3.6–5.5	100–105	24.6–28.8
Gastric juice	2500	31–90	4.3–12	52–124	0
Bile	700–1000	134–156	3.9–6.3	83–110	38
Pancreatic juice	> 1000	113–153	2.6–7.4	54–95	110
Small bowel (Miller-Abbott suction)	3000	72–120	3.5–6.8	69–127	30
Ileostomy					
Recent	100–4000	112–142	4.5–14	93–122	30
Adapted	100–500	50	3	20	15–30
Cecostomy	100–3000	48–116	11.1–28.3	35–70	15
Feces	100	< 10	< 10	< 15	< 15
Sweat	500–4000	30–70	0-5	30–70	0

(*Adapted from the Tables of: Lockwood, 1949 and Randall, 1952*)

Disease States

As potassium occurs widely in foods, there is a little likelihood of its deficiency. However, the two common disease states which develop are described hereunder:

(*a*) Hypokalemia (low serum potassium): Potassium deficiency is likely to develop under following *clinical conditions*:

1. prolonged diarrhea and vomiting resulting in continued loss of digestive juices
2. intravenous administration of potassium-free fluids/solutions to replace digestive juice lost by prolonged vomiting
3. chronic wasting diseases with malnutrition
4. prolonged negative nitrogen balance
5. overactivity of adrenal cortex (Cushing's syndrome) which causes increased excretion of body potassium in urine
6. prolonged use of diuretics such as acetazolamide and chlorothiazide
7. during treatment of heart failure with digitalis
8. during treatment of diabetic coma with insulin

Potassium deficits often become apparent only when water and sodium have been replenished in an attempt to correct dehydration and acidosis or alkalosis. **Darrow** states that changes in acid-base balance involve alterations in both intracellular and extracellular fluids and that the normal reaction of the blood cannot be maintained without a suitable relation between the body contents of sodium, potassium, chloride, and water. *Familial periodic paralysis* is a rare disease in which potassium is withdrawn from the extracellular fluid and retained in the cells, thus lowering extracellular concentration.

The **symptoms** of hypokalemia include muscle weakness, irritability, paralysis, tachycardia, and dilation of the heart with gallop rhythm and changes in the electrocardiogram (ECG) of the heart. It is, however, important to note that a potassium deficit may not be reflected in lowered (less than 3.5 mEq/L) extracellular fluid concentrations until late in the process. Thus, the serum potassium is not an accurate indicator of the true status of potassium balance.

The **treatment** of hypokalemia consists in giving the patient intravenously a solution containing 25 mEq of potassium (KCl, 1.8 g)/L after adequate urine flow has been established. A daily maintenance dose of at least 50 mEq of potassium (KCl, 3.6 g) intravenously is probably necessary for most patients, with additional amounts to cover excessive losses, as from gastrointestinal drainage, up to 150 mEq of potassium per day. The potassium salts may be added to saline solutions or to dextrose solutions. However, whenever possible the correction of a potassium deficit by the oral route is preferred. For adults, 4–12 g of KCl (as 1–2% solution) per day in divided doses is recommended.

(*b*) Hyperkalemia (elevated serum potassium)**:** Marked elevation of serum potassium occurs in patients under the following *clinical conditions:* renal failure, advanced dehydration, shock, adrenal insufficiency (Addison's disease) due to decreased excretion of potassium by kidney, and intravenous administration of excessive amount of potassium salts.

The **symptoms** of hyperkalemia are mainly cardiac and central nervous system (CNS) depression. The heart signs include brachycardia and weak heart sounds, followed by peripheral vascular collapse, leading to cardiac arrest. Other common symptoms include mental confusion, weakness, numbness, tingling of the extremities, weakness of respiratory mucles, and flaccid paralysis of the extremities.

The **treatment** of hyperkalemia consists in administering the patient desoxycorticosterone acetate (Doca).

The sodium-potassium pump

The high concentration of gradients of Na^+ and K^+ that exist across the cell membrane are maintained by the activity of an energy-requiring pump that transports Na^+ out of the cell in exchange for K^+. The energy for pumping is provided by ATP generated during metabolic reaction in the cell. In the erythrocyte with most reliable data, three Na^+ are pumped outward and two K^+ inward for each molecule of ATP hydrolyzed to ADP and inorganic phosphate.

The enzyme which is involved in this active transport of Na^+ and K^+ is **adenosine triphosphatase** (Na^+-K^+ ATPase) or **'transport ATPase'**. This enzyme was discovered by **Skou** (1965) and is activated by Mg^{2+} ions. The requirement of the enzyme for Na^+ is absolute, but other ions, including Li^+, can substitute for K^+ to some extent. The enzyme Na^+-K^+ ATPase (M.W. 2,50,000 – 3,00,000) consists of 2 large subunits of MW 1,00,000 – 1,30,000 each and 1 smaller subunit, possibly a glycoprotein, of MW 55,000. Following *evidences* support the active involvement of Na^+-K^+ ATPase in Na^+-K^+ transport:

(*a*) Tissues with high transport activity, including nervous and secretory tissues, possess high Na^+-K^+ ATPase activity.

(*b*) Both Na^+ and K^+ transport and ATPase activity are inhibited specifically by cardiac glycosides, *e.g.*, ouabain.

(*c*) The Na^+-K^+ ATPase in erythrocytes displays the same asymmetry as the Na^+ pump because Na^+ and ATP activate transport from the inside surface and K^+ and ouabain affect them from the outside.

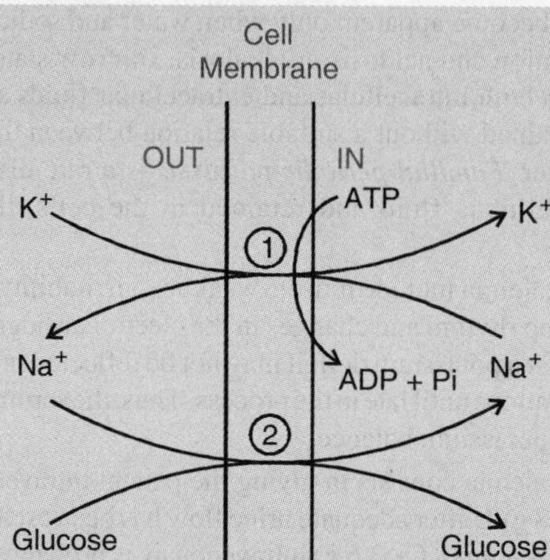

Fig. 33–3. The sodium gradient hypothesis of glucose transport linked to Na^+-K^+ transport

① represents the Na^+-K^+ transport system

② represents a passive carrier for the cotransport of Na^+ and glucose

(Courtesy: Crane RK, Fed. Proc. 24: 1000, 1965)

The activity of the Na^+ - K^+ ATPase appears to be linked to the uptake of a variety of solutes by tissues, using carrier molecules that facilitate the specific and compulsory cotransport of Na^+ and the solute molecule (**Crane**, 1965). When the external Na^+ concentration is maintained much higher than the internal Na^+ concentration by the Na^+-K^+ pump, Na^+ and solute bound to the carrier will move inward, down the Na^+ concentration gradient (Fig. 33–3). The **Crane hypothesis** was originally proposed to account for the Na^+-dependent transport of glucose during the process of glucose absorption from the intestine to the blood. Other solutes that appear to be transported in a similar manner in a variety of tissues include amino acids, iodide, thiamine, and uracil (**Schultz**, 1970). The energy stored in the Na^+ concentration gradient can, thus, be used to bring about the active transport of solutes without the requirement of separate and direct coupling of each of these systems to ATP.

MAGNESIUM

Distribution

Magnesium is present in all body cells. Most of the Mg of the body is in the skeleton. The adult human body contains about 21 g of magnesium. About two-third (*i.e.*, 67%) of this is combined with calcium and phosphorus in the complex salts of bone; the remaining amount is present in the soft tissues and body fluids. *Magnesium is one of the principal cations of soft tissue*. The whole blood contains 2–4 mg/dl (1.7–3.4 mEq/L). The serum contains less than half that in the blood cells (1.94 mEq/L). This is, in contrast to calcium, almost all of which is in the serum. The cerebrospinal fluid (CSF) contains about 3 mg/dl (2.4 mEq/L). The magnesium content of muscle is about 21 mg/ 100 g wet weight.

Sources

Human milk contains about 4 mg magnesium per dl; cow's milk contains about 12 mg/dl. Derivatives of cocoa, various nuts, soybeans and some seafoods are relatively rich in magnesium

(100 to as high as 400 mg/100 g). Whole grains and raw dried beans and peas may contain 100–200 mg magnesium per 100 g. Amounts of Mg^{2+} are significantly reduced during processing of food items.

Data regarding magnesium content of some foods are given in Table 33–12.

Table 33–12. Magnesium content of some foods (mg/100 g)

Foodstuff	*Magnesium*	*Foodstuff*	*Magnesium*
Cereals		***Milk and milk products***	
Barley	127	Butter, unsalted	10
Maize (corn)	144	Butter, salted	10
Rice, raw milled	48	Cheese, processed (with NaCl added)	48
Rice, parboiled milled	68	Milk, cow's	13
Wheat flour, refined	55	Milk, dried	102
Wheat, whole	139	***Eggs, meat and fish***	
Cereal products		Eggs, hen (whole)	11
Bread (without salt)	38	Meat, lamb's	21
Bread (with added NaCl)	27	Fish	24
Biscuits (with added NaCl)	42	Pork, raw	24
Corn flakes (with added NaCl)	80	Pork, cured (with NaCl added)	200
Pulses		***Vegetables***	
Black gram	185	Cabbage, raw	13
Chick pea	139	Carrots, raw	23
Cowpea	230	Cauliflower, raw	24
Field bean	196	Potato, raw	22
Green gram	189	Radish	15
Lentil	94	Spinach	88
Red gram	133	***Fruits***	
Peas, dry	124	Apple, raw	8
Oilseeds and nuts		Cherries	14
Almonds	257	Cranberry juice	9
Cashewnut, roasted	126	Grapes	13
Cashewnut, salted	129	Grape fruit	12
Groundnut, roasted	156	Lemon	11
Groundnut, salted	152	Orange	11
		Papaya	12
		Pears	7

Requirements

Magnesium is indispensable in the diet. In humans, the magnesium requirements for different age groups are as follows:

Infants and pre-school children	:	100–150 mg/day
Older children	:	150–200 mg/day
Adult women	:	300 mg/day
Adult men	:	350 mg/day

In the diet of Oriental peoples, there is a preponderance of foods high in magnesium, so that their intake of this mineral may approximate 6–10 mg/kg/d. The diet of Western peoples, however, provides less than 5 mg/kg/d for most adults (an average of 250–300 mg/d). It is claimed that several dietary constituents may interfere with retention or may increase the requirement of magnesium. Examples are calcium, protein and vitamin D. Alcohol is said to increase magnesium loss from the body. Since the Western diet may be high in calcium, protein and vitamin D and because alcohol is more commonly ingested by Western peoples, some nutritionists suggest that the optimal daily intake of magnesium under these conditions may be as high as 7–10 mg/kg/d.

Functions

Magnesium performs many function in the human body; some of which are as follows:

1. Mg is required for many enzyme activities, particularly those involving ATP, forming an ATP-Mg^{2+} complex.
2. Mg is found in certain enzymes, *e.g.*, cocarboxylase, which decarboxylase pyruvic acid.
3. It is required as a cofactor for oxidative phosphorylation.
4. It is required for neuromuscular transmission.

The vital role of magnesium in cellular metabolism is exercisd more by way of enzyme activation, *i.e.*, as metal–enzyme complexes rather than as metalloenzymes. The metal seems to activate all enzyme systems which catalyze the transfer of phosphate from ATP (GTP) to a substrate, or from a phosphorylated compound to ADP. Since ATP is involved in the activation of acetate, formate, sulfate, in oxidative phosphorylation, in muscular contraction, in group transfers and in active transport of nutrients across cellular membranes, the important role of magnesium is implicit in all these functions. In enzyme reactions involving ATP, the actual substrate is assumed to be ATP–Mg complex. In these systems, the enzyme activities are optimum when Mg : ATP ratio is 1 or more than 1. Magnesium may also interact directly with the enzyme protein and stimulate its activity. The list of Mg-activated enzymes is very large and includes: alkaline phosphatase, hexokinase, fructokinase, phosphofructokinase phosphopyruvic and acetyl transphosphorylases, and the enzymes which phosphorylate vitamins and their cofactors. The metal is also required for the activation of adenyl cyclase systems and for the activity of cyclic AMP-dependent kinases like those involved in phosphorylation of nucleotides and in glycogen metabolism.

Meabolism

The metabolism of magnesium is similar to that of calcium and phosphorus. Absorption and excretion of Mg from the intestine has been measured in human subjects with the aid of isotopic ^{28}Mg. On an *average* diet (20 mEq Mg/day), 44.3% of the ingested radioisotope was absorbed. On a *low* magnesium diet (1.9 mEq/day), 75.8% was absorbed; whereas on a *high* magnesium intake (47 mEq/day), only 23.7% was absorbed. The rate and duration of absorption of the ingested magnesium indicated that most was absorbed from the small intestine and little or none from the colon. Absorption of magnesium from the intestine did not appear to be related to the status of Mg stores in the body. In the first 48 hours after administration of radioactive Mg, only about 10% of the amount absorbed was excreted in the urine. Thus, renal conservation of body magnesium is excellent; the average urinary magnesium content is only about 6 - 20 mEq/L. A testicular hormone aldosterone increases the renal clearance of magnesium as it does also the excretion of potassium. Absorbed or administered Mg^{2+} is rapidly excreted in the urine. However, the major daily excretion occurs in the feces and represents unabsorbed magnesium. Ingestion of magnesium also increases excretion of calcium in both feces and urine.

An antagonistic effect between magnesium and calcium has been observed in certain experiments. The intravenous injection of Mg in a quantity sufficient to raise the Mg^{2+} ion concentration in the serum to about 20 mg/dl (normal value, 2.4 mg/dl) results in immediate and profound anesthesia together with paralysis of voluntary muscles. The intravenous injection of a corresponding amount of calcium results in an instantaneous reversal of this effect. It is, hence, suggestd that these 2 cations (Mg^{2+}, Ca^{2+}) are exerting differing effects on cell permeability. In the case of magnesium, there is about 10 times as much as of this element in the cells as in the extracellular fluid (ECF). As an instance, in plasma there is an average of 2.4 mg/dl; in muscle cells, 23 mg/100 g. This differential distribution between plasma and muscle cells is *not* observed with calcium, but it is especially prominent in the case of sodium and potassium as well as magnesium. Apparently, magnesium and potassium are normally concentrated within the cell and sodium without. An alteration in this relationship is followed by profound physiologic changes. In this connection it is of interest to note that both magnesium and potassium excretion by the kidney are increased to the same extent by aldosterone.

Deficiency States

In **children**, it is possible that magnesium deficiency exists in some cases of kwashiorkor as the serum Mg level has been reported to be low. Magnesium deficiency may contribute to some extent to the apathy and weakness found in kwashiorkor.

Spontaneous nutritional deficiency of Mg^{2+} in **adult humans** is extremely unlikely. However, excessive loss of this element may occur in diarrhea and become evident if such patients are rehydrated with fluids which do not contain Mg^{2+}. As the serum $[Mg^{2+}]$ declines to about 1.0 mEq/L, a syndrome resembling *delirium tremens* may be precipitated, *i.e.*, semicoma, tremor, carpopedal spasm and a general tetany-like neuromuscular irritability with marked susceptibility to auditory, mechanical and visual stimuli. Administration of Mg^{2+} effects a prompt improvement.

Magnesium deficiency may also occur in chronic alcoholics, cirrhosis of liver, toxemias of pregnancy, protein–energy malnutrition and malabsorption syndrome (**Kundu**, 1970). The principal clinical features attributed to Mg deficiency are: irritability, tetany, hyperreflexia and occasionally hyporeflexia.

Parenteral administration of large amounts of magnesium salts results in marked depression of both the central nervous system and peripheral neuromuscular activity. This depressive action can be antagonized by calcium; an animal anesthetized with magnesium returns to consciousness and behaves normally a few seconds after intravenous administration of sufficient calcium salt. There is no satisfactory explanation for either the magnesium narcosis or the calcium antagonism. Low serum (Mg^{2+}) leads to tetany resembling that of hypocalcemia. Thus, while a low concentration of either of these ions in the blood leads to similar functional changes, an excess of calcium antagonizes rather than synergizes the effects of excessive magnesium.

In renal failure, magnesium tends to rise in the serum. Indeed, this might contribute to the somnolence and weakness characteristic of the uremic state. Thus, it is of interest to note that some 14 patients with uremia and associated depression of the CNS had mean erythrocyte levels of 8.84 ± 1.71 mEq/L (upper limit of normal, 6.0 mEq/L), and plasma levels of 3.17 ± 1.30 mEq/L. It has been concluded that elevated serum or plasma levels are reliable evidence of a total body excess of magnesium. On the contrary, a deficit of magnesium may not be readily apparent from measurement of the serum level; in this instance, erythrocyte levels may be more informative.

SULFUR

Distribution

Sulfur is present in all cells of the human body, primarily in the cell protein in the form of 2 sulfur-containing amino acids, cysteine and methionine (The metabolism of sulfur and nitrogen, thus, tends to be associated). Sulfur is also present in small amounts in certain other organic compounds, viz., glutathione, thiamine, biotin, coenzyme A (CoA), lipoic acid, heparin, taurocholic acid, ergothioneine, sulfolipids, and chondroitin sulfate. *Sulfur in the form of inorganic sulfates is not present in the diet to any appreciable extent and is not utilized by the body.* Small amounts of inorganic sulfates, with Na and K, are present in blood and other tissues.

Sources

The main sources of sulfur for the body are the amino acids cysteine and methionine. Sulfate ions are only poorly absorbed from the intestine, and for this reason magnesium sulfate is useful as a saline cathartic. Elemental sulfur or inorganic sulfur is not known to be utilized for amino acid synthesis. Sulfate ions are incorporated into the 2 mucopolysaccharides– heparin and chondroitin sulfate– after 'activation' of sulfate.

Requirements

The intake of sulfur is mainly in the form of cysteine, cystine and methionine present in proteins. Only very small amounts of sulfur are contributed by other compounds present in the diet. The daily sulfur intake will vary with the protein intake.

Functions

The *cysteine* residues of proteins have important functions, both in protein structure and in enzymic activity. Specific sulfhydryl groups of cysteine residues in some enzyme molecules are essential for catalytic activity, which is inhibited when these residues are modified by combination with alkylating reagents, heavy metal compounds, or oxidizing agents.

Methionine is the principal donor of methyl groups in the body. The 'activated' form of methionine, *S*-adenosylmethionine, functions as a precursor in the synthesis of a large number of methylated compounds involved in intermediary and detoxification mechanisms.

Coenzyme A and lipoic acid are utilized for the synthesis of high-energy acyl-thioester intermediates such as acetyl-CoA and *S*-acetyl lipoate.

Metabolism

In the catabolism of sulfur-containing amino acids, the sulfur is oxidized mostly to sulfates and excreted in the urine. The output of this urine sulfur varies with the protein intake. The different forms of sulfur excreted (Table 33–13) are as follows:

Table 33–13. Sulfur intake and excretion on diets containing varying amounts of proteins in adult human*

Form of sulfur	*Low protein (26 g)*	*Moderate protein (80 g)*		*High protein (145 g)*
Sulfur intake from proteins	0.98	2.85	as per cent	3.88
Urine sulfur			of total sulfur	
Total sulfur	0.86	2.65	100	3.55
Inorganic sulfur	0.64	2.16	82	2.82
Ethereal sulfur	0.11	0.18	7	0.36
Neutral sulfur	0.11	0.31	11	0.37

* Sulfur expressed as SO_3

(*a*) **Total sulfur:** The total sulfur in the urine is derived mostly from sulfur present in proteins. Hence, it varies with the protein intake.

(*b*) **Inorganic** (or **sulfate**) **sulfur:** This is the completely oxidized sulfur precipitated from urine when barium chloride ($BaCl_2$) is added. It is roughly proportionate to the ingested protein with a ratio of 5 : 1 between urine nitrogen and inorganic sulfate (expressed as SO_3). Together with the total urinary nitrogen, this fraction of urine sulfur is an index of protein catabolism.

(*c*) **Ethereal sulfur** (or **conjugated sulfates**): This fraction represents roughly 10% of the total sulfur and includes the organic combinations of sulfur excreted in the urine. Examples are the phenol and cresol sulfuric acids, indoxyl and skatoxyl sulfuric acids, and other sulfur conjugates formed in detoxification. The ethereal sulfur is, in part, derived from protein metabolism; but in indican and some of the phenols, putrefaction in the intestine also takes place. After hydrolysis with hot HCl, the ethereal sulfates may be precipitated with $BaCl_2$.

(*d*) **Neutral sulfur:** This fraction represents the sulfur which is incompletely oxidized, such as that which is contained in cystine, taurine, thiocyanate, or sulfides. It does not vary with the diet to the same extent as the other fractions do. Neutral sulfur is determined as the difference between the total sulfur and the sum of the inorganic and ethereal sulfur.

CHLORINE

Distribution

The element chlorine is present as a component of sodium chloride. The distribution of the element chlorine (as chloride ion) is presented in Table 33–14. The chloride concentration in cerebrospinal fluid (CSF) is higher than that in other body fluids, including the gastrointestinal secretions.

Table 33–14. Distribution of chloride in body fluids or tissues

Fluid/Tissue	*mg/dl or 100 g*	*mEq/L*
Whole blood	250	70
Plasma or serum	365	103
Cells	190	53
Cerebrospinal fluid	440	124
Muscle tissue	40	
Nerve tissue	171	

Sources

All minerals except sodium chloride (NaCl) are usually present in sufficient amounts in a well-balanced diet. *Sodium chloride is the only mineral which is taken in more or less pure form in addition to the amount present in natural foods.*

Requirements

In the diet, the chloride occurs almost entirely as NaCl, and hence the intake of chloride is satisfactory as long as sodium intake is adequate. In general, both the intake and output of this element are, in fact, inseparable from those of sodium. On low-salt diets, both the chloride and sodium in the urine drop to low levels. Sodium chloride requirements depend on the climate, occupation, and the salt content of the diet. *Foods of animal origin contain more NaCl than those of vegetable origin.* The sodium chloride requirements for persons living in the tropical and temperate climates are given in Table 33–15. A perusal of the table points out that the NaCl requirements, in general, are more under tropical climate than under temperate climate.

Table 33–15. NaCl requirements under different climates

Age group and activity/condition	*Temperate*	*Tropical*
Children	3–5	5
Adolescents	6–8	10–25
Adults: light work	5	10
hard work	8	15–20

very hard work	15	25–30
Women : Pregnancy – first half	6	10
second half	3	5
Lactation	10	15

Absorption and Excretion

A healthy adult excretes about 10–15 g of NaCl in the urine daily. All this is derived from the salt taken in with food. Sodium chloride is also excreted in sweat. During excessive sweating in hot climates, while doing hard work, the loss of NaCl in sweat may vary from 10 to 20 g daily. During fasting or on salt-free diets, the excretion of chlorides in urine may be decreased or only traces of NaCl may be found. Absence of NaCl in the urine is an indication for adding more salt (NaCl) in the diet. Sodium chloride is stored in the subcutaneous tissue. During deprivation of NaCl, the store is consumed.

Functions

The element (as chloride ion) is essential in water balance and osmotic pressure regulation as well as in acid-base equilibrium. In maintaining the acid-base equilibrium, chloride plays a special role in the blood by the action of the chloride shift. In gastric juice, chloride is also of special importance in the production of hydrochloric acid (HCl).

Metabolism

Abnormalities of sodium metabolism are generally accompanied by abnormalities in chloride metabolism. When losses of sodium are excessive, as in diarrhea, profuse sweating, and certain endocrine disturbances, chloride deficit is likewise observed.

In loss of gastric juice by vomiting or in pyloric or duodenal obstruction, there is a loss of chloride more than that of sodium. This leads to a decrease in plasma chloride, with a compensatory increase in bicarbonate and a resultant hyperchloremic alkalosis. In Cushing's syndrome or after the administration of an excess of corticotropin (ACTH) or cortisone, hypokalemia with an accompanying hypochloremic alkalosis may also be observed. Some chloride is lost in diarrhea because the reabsorption of chloride in the intestinal secretions is impaired.

Deficiency States

Men doing hard work in hot humid climates, such as in mines, suffer from **heat cramps**, *i.e.* intense and painful contractions of skeletal muscle. This is due to NaCl deficiency caused by loss of NaCl from the body by excessive sweating. This loss may be about 2–5 g of NaCl/hour of very hard work or 10–20 g/day. Heat cramps may be prevented by adding NaCl (0.3 to 0.5%) to drinking water. Drinking pure water without replacing the salt loss aggravates the deficiency.

Excess States

Consumption of excessive amounts of NaCl causes **edema** in protein deficiency and increases blood pressure in patients with hypertension.

IRON

Distribution

Iron is of great importance in human nutrition. A 70-kg adult human contains about 4–5 g of iron (Table 33–16). A greater part (60–70%) of this amount is present in hemoglobin and myoglobin, and the remaining amount is stored in tissues such as bone marrow, liver, and reticuloendothelial system. Most of the body iron exists in complex forms bound to protein either as porphyrin or heme compounds or as ferritin and transferrin. Free inorganic iron occurs in the body only in very small amounts. The hemoprotein and flavoprotein enzymes also contain iron.

Table 33–16. Distribution of iron-containing compounds in the normal human adult

Compound	*Total in body (g)*	*Iron content (g)*	*Percent of total iron in the body*
Iron porphyrins (= Heme compounds)			
Hemoglobin	900	3.0	60–70
Myoglobin	40	0.13	3–5
Heme enzymes			
Cytochrome c	0.8	0.004	0.1
Catalase	5.0	0.004	0.1
Other cytochromes	—	—	—
Peroxidase	—	—	—
Nonporphyrin iron compounds			
Siderophilin (= Transferrin)	10	0.004	0.1
Ferritin	2–4	0.4–0.8	15.0
Hemosiderin	—	—	—
Total available iron stores		1.2–1.5	
Total iron in the body		4.0–5.0	

Sources

Iron is widely distributed in foodstuffs. Following are the important sources of iron:

(a) *Animal sources*: Liver, meat, fish and eggs are most important sources of readily-available iron, in all advanced countries. Not only this, they also increase the absorbability of iron in plant foods at the same time. *The iron content of milk is low in all mammalian species*. The average iron content of human milk is less than 1.0 µg/ml (**Picciano** ***et al***, 1976).

(*b*) *Plant soruces*: Cereals, pulses and vegetables, especially green leafy vegetables are the most important sources of iron in the diets of a large majority of the population of India and other developing countries. Other important sources are nuts, oilseeds, jaggery, and dried fruits such as raisin, apricots and dates. But the bioavailability of iron in plant foods is low owing to the presence of phytates and oxalates which interfere with iron absorption.

(*c*) *Other sources*: Significants amount of iron may be derived from cooking in iron vessels.

In a nutshell, it may be said that the best dietary sources of iron are "organ meats": liver, heart, kidney and spleen. Other good sources are egg yolk, whole wheat, fish, oysters, clams, nuts, dates, figs, beans, asparagus, spinach, molasses and oatmeal. Although the habitual Indian diet contains seemingly adequate amounts of iron, its bioavailability is limited. Data regarding the iron content of foods are given in Table 33–17.

Table 33–17. Iron content of some foods

Foodstuffs	*Iron (mg/100 g)*	*Foodstuffs*	*Iron (mg/100 g)*
Cereals	**2.0–8.8**	***Green leafy vegetables***	**3.9–21.4**
Bajra	8.8	Amaranthus, tender	21.4
Barley	3.7	Coriander	10.0
Cholam (*Sorghum*)	6.2	Drumstick	7.0
Maize, yellow	2.1	Mint	15.6
Oatmeal	3.8	Radish leaves	4.8

Ragi	5.4	Spinach	5.0
Rice, paraboiled milled	3.7	***Meat and fish, and eggs***	**2.1–6.3**
Rice, raw milted	2.8	Egg	2.1
		Fish	2.3
Legumes	**3.8–11.3**	Liver, goat	6.3
Bengal gram	8.9	Mutton, goat	2.5
Black gram	9.8	***Nuts and oil seeds***	**1.7–15.0**
Cow gram	3.8	Cashewnut	5.0
Field bean, dry	5.0	Groundnut	1.7
Green gram	8.4	Sesame seeds	15.0
Red gram	8.8	***Miscellaneous foods***	
Soybean	11.3	Jaggery	11.4

Requirements

The need for iron in the human diet varies greatly at different ages and under different circumstances. It is determined by the requirements for tissue growth and hemoglobin synthesis and the replacement needs due to iron losses in urine, feces and sweat, and, in the female, the additional losses in menstruation, gestation and lactation. The need for iron is greatest during the first 2 years of life, during the period of rapid growth and hemoglobin increase in adolescence, and throughout the childbearing period in women. Iron requirements are also influenced by the availability of iron present in foods. Iron present in cereals, legumes and green leafy vegetables is available to a lesser extent (due to the presence of phytates and oxalates) than that present in eggs, meat and fish. In view of this, iron requirements of persons consuming a predominantly cereal-based diet will be greater than those consuming large quantities of meat and eggs. Iron requirements suggested by different expert groups are given in Table 33–18.

Table 33–18. Recommended dietary allowances for iron (mg/day)

Age group	*ICMR Expert Group, India (1968)*	*NRC, USA* (1968)*	*Expert Group, UK (1969)*
Birth–1 year	1 mg/kg	6–15	6
1–6 years	15–20	15	7–8
7–12 Years	15–20	10	10–13
13–18 years: Boys	25	18	15
Girls	35	18	15
Men	24	10	10
Women: Normal	32	18	20
Pregnancy	40	18	15
Lactation	32	18	15

* It should be noted that these requirements for iron take into account the low amount of iron actually absorbed from orally-ingested iron. The RDA of 10 mg/d for adult males is readily obtained from the normal diet in the USA, which provides about 6 mg of iron/1000 kcal. However, the RDA for females (18 mg/d), based on 2000 kcal/d, is difficult to obtain from dietary sources without further iron fortification of foods.

Clinical studies have indicated that an additional 0.8–1 mg of iron per day is needed to maintain normal hemoglobin concentration in children and adolescents to provide for increased stores of iron in the growing body. The average loss of iron in the healthy adult male is estimated to be only about 1 mg/d, ranging from 0.4–2 mg depending upon the iron content of exfoliated cells. In adult women, the average loss of blood during a menstrual period is 35–70 ml, which represents a monthly loss of 16–32 mg of iron, or an additional average loss of 0.5–1 mg/d. This amount of iron is easily obtained from the diet. For women with excessive menstrual blood loss and a resultant chronic iron-deficiency anemia, a supplement of 100 mg of iron per day (as ferrous sulfate) is sufficient to produce a maximal response toward correction of anemia. Pregnancy increases the requirement to about 3.6 mg/d. In the latter part of pregnancy, iron requirement may reach 4 mg/d. It is clear that during growth, pregnancy and lactation, when the demand for hemoglobin formation is increased, additional iron is needed in the diet.

Assuming a 10–15% efficiency of iron absorption, the Food and Nutrition Board has set a recommended dietary allowance of 10 mg/d for normal adult males and 15 mg/d for menstruating females. For pregnant females this allowance is raised to 30 mg/d. While 10 mg/d of iron can easily be obtained from a normal diet, 15 mg is marginal at best and 30 mg can almost never be obtained.

Absorption

A strange and probably unique feature of the metabolism of iron is that *it occurs in what is virtually a closed system.* Under normal conditions, very little iron is absorbed, the amounts excreted in the urine are minimal, and a high proportion of the total body iron is continuously redistributed throughout the body in several metabolic circuits. As there is no way to excrete excess iron, its absorption from the intestine must be controlled if it is not to accumulate in the tissues to the toxicity level. In the ordinary/normal diet, 10–20 mg of iron are taken daily, but less than 10% of this amount is excreted. Infants and children absorb comparatively high percentage of iron from foods than adults. Iron-deficient children absorb twice as much from foods as normal children. Iron deficiency in infants can usually be attributed to dietary inadequacy.

Most of the iron in foods occurs in the ferric (Fe^{3+}) state either as ferric hydroxide or as ferric organic compounds. In acidic medium, these compounds are broken down into free ferric ions or loosely-bound organic iron. The HCl secreted from the stomach and the organic acids of the foods are both important for this purpose. The reducing substances in foods, sulfhydryl (–SH) groups in cysteine, and ascorbic acid convert ferric iron to the reduced (ferrous, Fe^{2+}) state. In this form, iron is more soluble and should, therefore, be more readily absorbed. Iron absorption is enhanced by *protein*, probably due to the formation of low-molecular-weight digestive products (peptides, amino acids) which can form soluble iron chelates. Inorganic iron also forms soluble complexes with normal gastric juice. Heme iron enters the mucosal cell without being released from the porphyrin ring. In humans, dogs and rats, heme is broken down in the mucosa and the iron appears in plasma transferrin. A diet high in *phosphate* causes a decrease in the absorption of iron since compounds of iron and phosphate are insoluble. Conversely, a diet very low in phosphates markedly increases iron absorption.

It has been demonstrated that the absorption of iron is enhanced by the simultaneous administration of *ascorbic acid* (vitamin C). It is reasonable to suppose that the effect is related to the reducing action of ascorbic acid. It has been demonstrated in normal and anemic human subjects that vitamin C increases the absorption of iron, but the effective amounts were very large, 500–1,000 mg. Infants on a normal diet did not absorb iron better if they were given an extra 100 mg of ascorbic acid per day. It does not seem likely that amounts of vitamin C ordinarily ingested would affect the absorption of iron.

Phytic acid, the hexaphosphoric acid of inositol, is a common constituent of plant parts that are used for food. It is conspicuous as a constituent of the bran of cereals. Many of the salts of phytic acid (phytates) have a low solubility, and phytate has been implicated as a deterrant to the absorption of metals, principally of calcium and iron. It has been shown that the response of serum iron to large amounts of dietary iron taken with bread and jam was less if sodium phytate had been added to the bread. In a similar experiment, it was demonstrated that sodium phytate given with test meals decreased the absorption of iron. The absorption of iron from ferric phytate is very low (2 – 5%). It has been demonstrated that anemic patients can utilize some of the iron from very large doses of iron phytate.

Oxalic acid, present in certain vegetables, forms insoluble iron oxalate and prevents the absorption of dietary iron.

The rate of iron absorption is influenced by a number of other factors such as iron status of the subjects, composition of the diet, and the presence of various disorders of the gut, such as celiac disease, tropical spru and tropical enteropathy which decrease iron absorption. Absorption also increases during growth and pregnancy. When the red blood corpuscles (RBCs) are broken down, the liberated iron is not excreted but is reutilized in the formation of new RBCs.

Absorption of iron occurs mainly in the duodenum. Impaired absorption of iron is observed in patients who have had subtotal or total removal of the stomach or in patients who have sustained surgical removal of major portion of the intestine. Iron absorption is also reduced in various malabsorption syndromes, such as steatorrhea.

Nearly all the iron released from the mucosal cell enters the portal blood, mostly in the ferrous state (Fig. 33–4). In the plasma, ferrous (Fe^{2+}) is oxidized rapidly to the ferric (Fe^{3+}) state and is then incorporated into a specific iron-binding protein, **transferrin**. This protein can bind 2 atoms of Fe^{3+} per molecule of prtein to form a red ferric–protein complex. Transferrin (also called **siderophilin**) is a glycoprotein of molecular weight 76,000 and contains 5.3% carbohydrate. The carbohydrate consists of 2 identical branched side chains, each branch ending in a molecule of N-acetyl-neuraminic acid (**Jamieson** *et al*, 1971).

The release of iron from the mucosal cell is facilitated by a low degree of transferrin saturation by iron. **Ceruloplasmin** (= **ferroxidase I**), a blue copper-binding plasma protein, exerts a catalytic activity (serum ferroxidase) in plasma to convert Fe^{2+} into Fe^{3+}, and it may, thus, promote the rate of iron incorporation into transferrin. Human serum also contains a yellow cuproprotein (**ferroxidase II**) which catalyzes the oxidation of ferrous ions (**Topham** and **Frieden**, 1970).

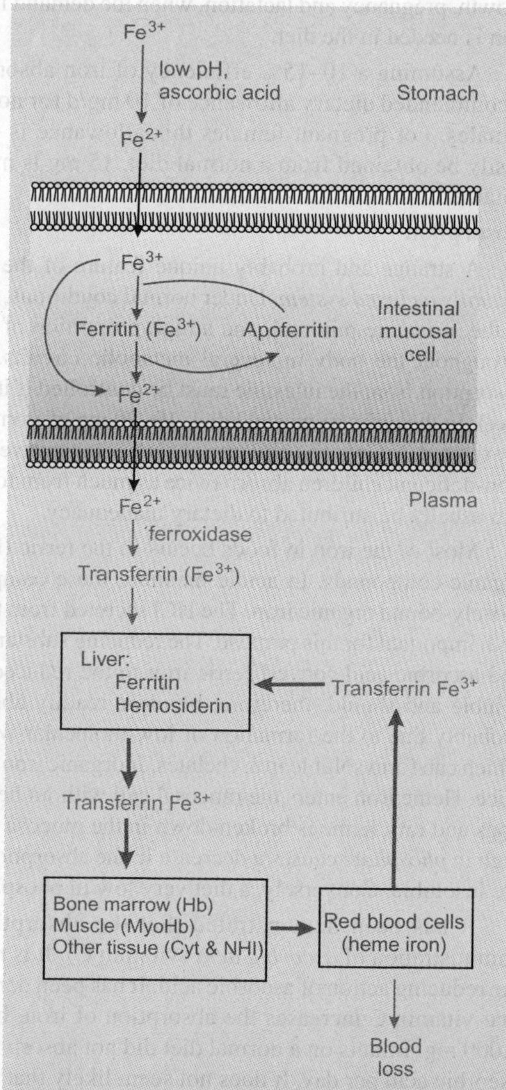

Fig. 33–4. An overview of iron absorption and metabolism

Red arrows indicate that most of the body's iron is efficiently reutilized by the pathway shown.

[Hb = hemoglobin, MyoHb = myoglobin, Cyt = cytochromes, NHI = nonheme iron]

Under normal circumstances, almost all of the iron bound to transferrin is taken up rapidly by the bone marrow. It appears that only the reticulocytes are capable of utilizing the Fe^{3+} bound to transferrin, although both the reticulocytes and the mature erythrocytes can take up unbound ferric iron. Thus, transferrin in some manner diverts the plasma iron into those cells which are actively making hemoglobin. The normal content of protein-bound iron (BI) in the plasma of males is 120–140 μg/dl; in females, this value is 90–120 μg/dl. However, the total iron-binding capacity (TIBC) is about the same in both sexes, *i.e.*, 300–360 μg/dl. This indicates that normally only about 30–40% of the iron-binding capacity of the serum is utilized for iron transport and that the iron-free siderophilin (*i.e.*, the unsaturated iron-binding capacity, UIBC) is therefore about 60–70% of the total.

Excretion

The body exercises amazing economy in the conservation and use of iron present in the body. The *red blood cells* destroyed each day yield about 25 mg of iron. This is conserved and stored in the body in the form of ferritin and reused for the formation of new hemoglobin. A major portion (over 90%) of the iron present in the *feces* consists of unabsorbed food iron. The quantity of metabolic iron present in the feces of an adult is about 0.3–0.5 mg/day as determined by radioactive iron technique. This iron is derived from the digestive juices and desquamated mucosal cells. The amount of iron excreted in *urine* may vary from 0.1 to 0.3 mg/day. There is a loss of iron in the *sweat*; this is called dermal loss. Cell-rich sweat may contain 1.15–1.61 mg Fe/litre while cell-free sweat may possess 0.34–0.44 mg Fe/litre in healthy men and women. The iron loss through *menstruation* ranges between 0.3 and 1.0 mg when calculated on a daily basis. Thus, the total daily loss of iron in men may range from 1 to 2 mg and in menstruating women from 2 to 3 mg. Hence, women need more iron than men. Recent studies have shown that contrary to earlier belief, iron losses are not increased in subjects working in hot and humid climates (WHO, 1975).

Storage

As the excretory capacity of the body for iron is very limited, the body is able to avoid iron intoxication by regulating the absorption of iron in the intestines. A greater part (95 – 98%) of iron administered orally as ferrous salt (100 mg iron) is not absorbed and is excreted in the feces. On the contrary, if iron is administered parenterally, all the iron is stored in the liver as **ferritin**.

The liver of an adult male contains about 700 mg of iron in the form of ferritin. Any excess iron which cannot be stored as ferritin is accumulated in the liver in the form of **hemosiderin**. (colloidal iron oxide as granules). The iron in the form of hemosiderin is not utilized by the body and excess deposition of hemosiderin affects adversely the function of the liver.

Functions

Iron is present in hemoglobin, myoglobin, cytochromes and many other oxidative enzymes (refer Table 33–19). As a constituent of these enzymes, iron performs a number of important functions, some of which are as follows:

1. As a component of hemoglobin and myoglobin, iron is required for O_2 and CO_2 transport.
2. As a component of cytochromes and nonheme iron (NHI) proteins, it is required for oxidative phosphorylation.
3. As a component of the essential lysosomal enzyme myeloperoxidase, it is required for proper phagocytosis and killing of bacteria by neutrophils.
4. As a component of catalse, it is required for the decomposition of H_2O_2 into H_2O and O_2.
5. As a component of various metalloproteins (such as succinic dehydrogenase, NADH dehydrogenase, and xanthine oxidase), it is essential for the oxidation of succinic acid, reduced nicotinamide, and xanthine, respectively.

Table 33–19. Functions of the oxidative enzymes containing iron

Enzymes	*Functions*
Cytochromes	
Cytochrome b	Carrier between dihydroubiquinone and cytochrome c_1
Cytochrome c_1	Carrier between cytochrome b and cytochrome
Cytochrome c	Carrier between cytochrome c_1 and cytochrome oxidase
Cytochrome (a + a_3) (cytochrome oxidase)	**Oxidases** hydrogen and electrons from reduced cytochrome c to H_2O
Hydroperoxidases	
Catalase	$H_2O_2 \longrightarrow H_2O + O_2$
Peroxidase	O_2 + Diphenols $\longrightarrow H_2O_2$ + Quinone
Tryptophan pyrrolase	Tryptophan + $O_2 \longrightarrow$ Formylkynutriene
Metalloflavoproteins	
Succinic dehydrogenase (Fe)	Succinate $\xrightarrow{-2H}$ Fumarate
NADH dehydrogenase (Fe)	NADH $\xrightarrow{-H}$ NAD
Aldehyde dehydrogenase (Fe, MO)	Aldehydes $\longrightarrow$ Carboxylic acids
Xanthine oxidase (Fe, MO)	Xanthine $\longrightarrow$ Hypoxanthine Hypoxanthine $\longrightarrow$ Uric acid

Metabolism

The 2 storage forms of iron, ferritin and hemosiderin, act as an internal iron reserve to protect against sudden losses of iron by bleeding. Ferritin is found not only in the intestine but also in the liver (about 700 mg), spleen and bone marrow. As mentioned earlier, if iron is administered parenterally in amounts which exceed the capacity of the body to store it as ferritin, it accumulates in the liver as hemosiderin, a form of colloidal iron oxide in association with protein. The iron content of hemosiderin is 35% by weight. Ferritin stores iron in a soluble form; it consists of a spectrum of molecules with varying affinities for iron. It, therefore, acts as both a short-term and a medium-term iron reserve. The mobilization of liver ferritin may require a ferric to ferrous transformation, involving the ferrireductase enzyme (**Osaki** and **Sirivech**, 1971). In the human adult, about 0.2 – 0.4% of serum iron is present as serum ferritin. Under most conditions, the concentration of serum ferritin is a satisfactory indicator of the iron stores of the body.

The level of iron in the plasma is the result of a dynamic equilibrium in which iron from the plasma is redistributed in several metabolic circuits, each circuit terminating in the return of iron to the plasma iron pool. In the quantitatively-important circuit, iron is used for hemoglobin synthesis, involving the transfer of plasma iron to bone marrow, the red cell, the ageing red cell, and back to the plasma (**hemoglobin cycle**). Studies of the turnover of iron, using isotopic ^{59}Fe, indicate that about 27 mg are utilized each day, 75% of this for the formation of hemoglobin. About 20 mg are obtained from the breakdown of RBCs, a very small amount from newly-absorbed iron, and the remainder from the iron stores. Normally, there is a rather slow exchange of iron between the plasma and the storage iron; in fact, following an acute hemorrhage in a normal individual, the level of iron in the plasma may remain low for weeks, a further indication that mobilization of iron from the storage depots is a slow process.

The outline of iron metabolism is depicted in Fig. 33–5.

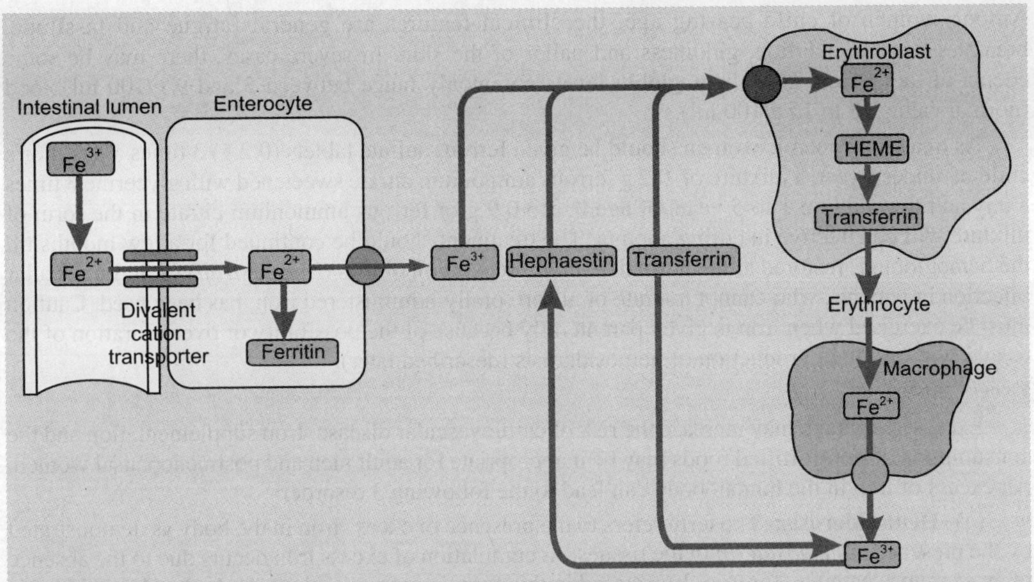

Fig. 33–5. Iron metabolism

Dietary iron is absorbed in the intestine and either stored in enterocytes as ferritin or transported out to plasma. In plasma, iron remains bound to transferrin. It is taken up by other cells, *e.g.*, erythrobalasts, through the mediation of the membrane transferrin receptor. In erythroblasts, iron is incorporated into heme, and then hemoglobin, Old erythrocytes are degraded by macrophages in the reticuloendothelial system. Liberated iron in released from cells and recycled bound to transferrin. Note that the dietary iron is in the ferric (Fe^{3+}) form. This is reduced to ferrous ion (Fe^{2+}) at the intestinal brush border. The transported form of iron is ferric again, but the form incorporated into heme is ferrous.

Deficiency States

Three stages of iron deficiency have been recognized by the World Health Organization (WHO) in its Technical Report, 1975. These are:

(*a*) *First stage*: It is characterized by the decreased storage of iron.

(*b*) *Second stage*: It begins after iron stores are exhausted.

(*c*) *Third stage*: This is the apparent/observable state of iron deficiency.

The single most effective tool for evaluating the iron status of different population groups is by measurement of serum ferritin.

Deficiency of iron causes anemias of hypochromic microcytic type. **Iron-deficiency anemia**, as it is called, is the most prevalent nutritional disorder among children, adolescent girls and expectant and nursing mothers worldwide, especially in all developing countries. Young children and pregnant females need enough iron for a continuing increase in blood volume. Menstruating females lose iron through blood loss and lactating females through production of lactoferrin, a glycoprotein closely homologous to transferrin. Thus, iron-deficiency anemia is primarily a problem for these population groups. Iron-deficiency anemia is also a problem with the elderly due to poor dietary intake and increased frequency of achlorhydria. A deficiency of iron may result from inadequate intake (*e.g.*, a high cereal diet, low in meat) or inadequate absorption (*e.g.*, gastrointestinal disturbances such as diarrhea, achlorhydria, steatorrhea, or intestinal disease, after surgical removal of the stomach, or after extensive intestinal resection) as well as from excessive loss of blood.

The weaned infants and young children with iron-deficient anemia are dull and inactive and show pallor of the skin. The appetite is poor and the growth and development of children are retarded due to low food intake. There is a tendency among children below 3 years to eat mud.

Among women of child-bearing age, the clinical features are general fatigue and lassitude, breathlessness on exertion, giddiness and pallor of the skin. In severe cases, there may be some edema of the ankles. Their hemoglobin levels commonly range between 5 and 9 g/100 ml blood (normal value: 13 to 15 g/100 ml).

As treatment, anemic women should be given ferrous sulfate tablets (0.2 g) 3 times a day. For a child of under 1 year, a mixture of 0.2 g ferrous ammonium citrate sweetened with glycerine 3 times a day and for children 1 to 5 years of age 0.4 to 0.9 g of ferrous ammonium citrate in the form of mixture will be effective in curing anemia. The treatment should be continued for a few months till the hemoglobin is restored to the normal level. A preparation of iron (*iron dextran*) for intramuscular injection in patients, who cannot tolerate or absorb orally-administered iron, has been used. Caution must be exercised when iron is given parenterally because of the possibility of oversaturation of the tissues with resultant production of hemosiderosis (described later).

Excess States

Excess iron intake may increase the risk of cardiovascular disease. Iron supplementation and the consumption of iron-fortified foods may be inappropriate for adult men and postmenopausal women. An excess of iron in the human body can lead to the following 3 disorders:

(a) Hemosiderosis: The term refers to the presence of excess iron in the body as demonstrated by the presence of *hemosiderin* in the tissues. Accumulation of excess iron occurs due to the absence of an excretory pathway for iron. Iron stored in this form is not available to the body. Hemosiderosis is observed in patients with aplastic or hemolytic anemia who have received many blood transfusions over a period of years. The existence in some individuals of an excessive capacity for the absorption of iron from the intestine has been detected by studies with radioactive iron, ^{59}Fe. Such individuals absorb 20 – 45% of an administered dose of the labeled iron; a normal subject absorbs ony 1.5–6.5%. The anomaly in iron absorption may be inherited. In such patients, a very large excess (as much as 40–50 g) of iron accumulates in the tissues after many years.

(b) Hemochromatosis: This is hemosiderosis accompanied by a bronzed pigmentation of the skin. Presumably because of the toxic effects of the unbound iron in the tissues, there may be liver damage with signs of cirrhosis, diabetes mellitus and a pancreatic fibrosis. The condition, which develops rarely, is sometimes referred to as '*bronze diabetes*'. As might be expected, the unsaturated iron-binding capacity of the serum of the patient with hemochromatosis is very low. Thus, whereas the iron-binding proteins of the serum in a normal individual are only about 30% saturated, in patients with hemochromatosis they are about 90% saturated, due to the excess absorption of iron from the intestine. Hemochromatosis is also occasionally seen in hemolytic anemias and liver disease.

CLINICAL IMPLICATIONS 33–2.

Hemochromatosis

Hemochromatosis is an autosomal recessive disorder resulting from the increased absorption of iron. It is the most common inherited disorder in persons of North European ancestry. Iron accumulates in heart, liver and pancreas and can cause liver cirrhosis, hepatocellular carcinoma, diabetes, arthritis and heart failure. In the classic form of hemochromatosis, the mutated gene encodes the protein known as *hereditary hemochromatosis protein* (HFE), which is structurally similar to class I major histocompatibility antigens. It is now known that mutations of other proteins can lead to a very similar clinical picture.

Iron metabolism is unique in that it operates largely as a closed sysem, with iron stores being efficiently reutilized by the body. Iron losses are minimal (less than 1 mg/day), but iron absorption is also minimal under the best of conditions. Iron usually occurs in foods in the ferric form bound to protein or organic acids. Before absorption can occur, the iron must be split from these carriers (a process that is facilitated by the acid secretions of the stomach) and reduced to the ferrous form (a process that is enhanced by ascorbic acid). Only 10% of the iron in an average mixed diet is usually absorbed, but the efficiency of absorption can be increased to 30% by severe iron deficiency. Iron absorption and metabolism are both summarized in Fig. 33–4.

The regulation of iron absorption is largely carried out by the intestinal epithelial cells, although the exact mechanism of this 'control' is not known. As one time, the so-called **'mucosal block' theory**, advanced by **Granic**, was considered to account for the intestinal control of iron absorption. According to this theory, the amount of an iron-binding protein, *apoferritin*, in the mucosal cells was the controlling factor. Ferrous ion, once within the mucosal cells, is oxidized to the ferric state and then combined with apoferritin to form the iron-containing protein *ferritin*. It was believed that the binding capacity of apoferritin for iron limited the further absorption of iron; when saturated with iron, no further uptake of iron could occur.

However, more recent observations render it unlikely that ferritin is involved as a major regulator of iron absorption. Iron taken into the mucosal cell is bound to one or more specific carriers which appear to regulate its passage across the cell to the blood. It is probable that iron entering the cell in excess of that which can be bound by the carrier system becomes incorporated into ferritin. Intestinal ferritin appears therefore to act *as a storage compound for iron* rather than as a controller of absorption. In normal rats, only, a small part of the iron taken up by the mucosal cells is transferred to the blood. Most of the iron, is lost to the lumen of the gut when the cells are shed from the intestinal villi. Experimental support for the existence of nonferritin control of iron absorption has been provided by studies in mice with X-linked inherited defect in iron absorption.

(c) Nutritional siderosis: This is an acquired siderosis of dietary origin and has been commonly observed among the Bantu peoples of South Africa, hence also called '*Bantu siderosis*'. It is believed to be caused by the fact that the natives consume a diet which is very high in corn and thus low in phosphorus and that their foods are cooked in iron pots; their daily iron intake may be as high as 100 mg or more. The combination of a low-phosphate diet and a high intake of iron enhances absorption of iron sufficiently to produce siderosis with accompanying organ damage as described above. Iron deficiency anemias, common among pregnant women in other areas of the world, are virtually unknown among the Bantus. Examination of the livers of Bantus who died of accidents revealed the presence of large amounts of iron (more than 100 mg/100 g fresh liver).

IODINE

Distribution

The adult human body contains about 50 mg of iodine.

Sources

In all human communities, about 90% of the iodine intake is obtained from the food consumed; the remainder is provided by the water (**Clements**, 1976). Marine foods– sea fish and shellfish– are the best sources of iodine and the oils of these fish are even richer in this element. The edible flesh of sea fish and shellfish may contain from 300 to 3,000 μg l/kg on the fresh basis, compared with 20 – 40 μg/kg for freshwater fish. Anadromous fish, which spend part of their life in the sea and part in the fresh water, contain intermediate levels of iodine. The iodine content of other foods (in μg l/kg on the fresh edible basis) are as follows: fresh fruits, 20; leafy and root vegetables (excluding spinach which is usually much higher), 30; cereal grains, 40; meats, 40; milk and milk products, 40; and hen's eggs, 90. There is evidence that certain vegetables such as cabbage, cauliflower and radish contain *glucosinolates* (= thioglucosides) which are potential goitrogens. Consumption of large quantities of these goitrogenic substances may lead to the development of goitre by making the iodine present in the food unavailable to the body.

The iodine content of the drinking water is a valuable index of the iodine content of the rocks and soils of a region (and hence of the locally-grown foods), and has been correlated with the incidence of goitre in many areas. Two instances may be cited to illustrate this relationship:

1. In **Latvia**, the iodine content of the water supplies of goitrous areas was found to range from 0.1 to 2.0 μg/litre and that of the nongoitrous areas from 2.0 to 15 μg/litre.
2. In **England**, the drinking water in Somerset villages, where the incidence of goitre in the human population was assessed at 56%, was shown to average 2.9 μg/litre compared with 8.2 μg/litre in Suffolk villages where the goitre incidence was only 3%.

Requirements

One of the earliest published estimates of human requirements of iodine was made by **von Fellenberg**, who compared the iodine intakes of individuals from goitrous and nongoitrous areas and suggested that less than 20 μg was required daily by an adult. This value is considerably lower than almost all subsequent estimates. Calculations based on average daily urine losses of iodine give an adult human requirement of 100–200 μg/day; the results of several iodine balance studies indicate that equilibrium or positive balance can be achieved at iodine intakes ranging between 44 and 162 μg/day. Estimates derived from the daily rate of thyroxine decay indicate that the amount of thyroxine supplied to the circulation to maintain normal metabolic activity in an adult man fluctuates between 200 and 400 μg/day (this is equivalent to 130–260 μg/day of iodine). However, these latter calculations make the erroneous assumption that there is no reutilization of iodine from the breakdown of thyroxine. From an assessment of a variety of estimations, **Elmer** places the optimum iodine requirement of humans at 100–200 μg/day. This estimate is an acordance with the requirement arrived at from the urinary excretion study and includes many of the estimates made from balance experiments. The use of iodized salt regularly will provide more than this. The daily requirement of iodine is also normally met by well-balanced diets and drinking water except in hilly regions where the food and water may be deficient in iodine. However, the need for iodine is increased in adolescence and in pregnancy and lactation.

The recommended dietary allowances (RDAs) of National Research Council, USA (1980) for iodine and other trace elements are given in Table 33–20.

Table 33–20. Recommended dietary allowances of trace elements*

Trace element	*Infants*		*Children*		*Adolescents*		*Adults*
	0–0.5 year	0.6–1 year	1–3 years	4–6 years	7–10 years	11–18 years	>18 years
Iodine	0.04	0.05	0.07	0.09	0.12	0.15	0.15
Copper	0.05–0.07	0.07–1.0	1.0–1.5	1.5–2.0	2.0–2.5	2.0–3.0	2.0–3.0
Zinc	3	5	10	10	10	15	15
Fluorine	0.1–0.5	0.2–1.0	0.5 – 1.5	1.0–2.5	1.5–2.5	1.5–2.5	1.5–4.0
Manganese	0.5–0.7	0.7–1.0	1.0 – 1.5	1.5–2.0	2.0–3.0	2.5–5.0	2.5–5.0
Molybdenum	0.03–0.06	0.04–0.08	0.05 – 0.1	0.06–0.15	0.1–0.3	0.15–0.5	0.15–0.5
Chromium	0.01–0.04	0.02–0.06	0.02 – 0.08	0.03–0.12	0.05–0.2	0.05–0.2	0.05–0.2

* All values are in mg/day.

(*Source: National Research Council, U.S.A., 1980*)

Absorption

Dietary iodine is efficiently absorbed and transported to the thyroid gland where it is stored and used for the synthesis of the thyroid hormones– triiodothyronine and thyroxine. These hormones function in regulating the basal metabolic rate (BMR) of adults and the growth and development of children.

Functions

Iodine was discovered in 1811 by **Courtois** in the ash of seaweeds. Its presence in the thyroid gland was first shown by **Beumann** in 1895. Triiodothyronine and thyroxine, the two iodine-containing hormones, accelerate cellular reactions in practically every organ and tissue of the body; this is reflected in an increase in basal metabolic rate. The ability of these hormones to accelerate growth and O_2 consumption of the whole organism is accompanied by increased activity of most enzymic systems that have been studied. For example, an injection of these hormones accelerates glucose oxidation *in vivo* and increases the activities of liver glucose 6-phosphatase, glucose 6-phosphate dehydrogenase and TPNH-cytochrome *c* reductase measured *in vitro* in broken cell preparations. Indeed, more than 100 enzymic systems have been reported to be altered in activity by thyroxine administration. However, few if any of these effects are likely to reflect a direct action of the hormone on the enzymes.

Metabolism

Thyroxine and triiodothyronine undergo deamination, deiodination, and conjugation. Deiodination occurs largely, but not solely, in extrahepatic tissues, and the other two reactions are prominent in liver, although not limited to this organ.

The liver participates in the metabolism of thyroxine and triiodothyronine in 2 ways: (*a*) it removes the amino group and deiodinates the hormones, and (*b*) it regulates blood levels of hormone by enterohepatic circulation. As much as 40% of an injected dose of thyroxine is found in the liver within 1 minute after administration. A small portion appears in the bile as iodide and as the administered hormone.

The liver also conjugates iodothyronines, with formation of glucosiduronides by linkage with the phenolic hydroxyl group of the thyronines. These conjugates are excreted in the bile. Because of the hydrolytic activity of intestinal enzymes and transmucosal transport of the liberated hormones, excretion of conjugated forms *via* the bile is not a significant route for their net disposal. The mechanism responsible for intestinal transport of the iodinated thyronines is similar to that for active transport of amino acids but may, in addition, involve re-formation of the glucosiduronides in the intestinal mucosa. Decarboxylation of the thyronines by bacteria may occur in the intestine.

Deiodination of the thyroidal hormones and their corresponding thiopyruvic acids in liver is catalyzed by *thyroxine dehalogenase*. Only trace amounts of the deiodinated thyronines are excreted in the urine. When thyronine was incubated with rat liver and kidney slices, the products formed were tyrosine, 3′-hydroxythyronine, 3,4-dihydroxyphenylalanine, thyroacetic acid, and *p*-hydroxyphenylpyruvic acid.

Thus, the liver can metabolize the iodinated thyroid hormones by deiodination, removal of amino groups, and disruption of the thyronine nucleus. Also, thyroidal compounds, as well as certain of the metabolites, can be conjugated with glucuronic acid and channeled into an enterohepatic circulation, or, in specific instances, conjugated with sulfate or O-methylated.

Deficiency States

Deficiency of sufficient amount of iodine in the diet leads to **goitre**, a disease characterized by the enlargement of the thyroid gland (the gland might weigh as much as 200–500 g or even more as opposed to a normal weight of about 25 g). Histological examination reveals diffuse overgrowth of the glandular tissue, known as general hypoplasia of the glands. The vesicles contain little or no colloid. A simple hyperplastic-colloid goitre, by definition, should not affect the thyroid state of the individual, *i.e.*, those with this goitre are euthyroid up to the age of 30–40 years, when other changes (described below) may occur. Until these changes occur, the main disadvantage of this condition is an aesthetic one, which may reduce the opportunities of goitrous individuals for employment, marriage etc.

When goitre occurs in a significant number of people in a defined geographic area, it is known as **endemic goitre**. Endemic goitre is considered to be of public health importance when 5% or more of the girls aged 12 to 14 years show Grade I enlargement of the thyroid gland, *i.e.*, gland usually visible with head thrown back and neck fully extended and gland enlarged on palpation (**Clements**, 1976). The public health importance of endemic goitre lies in the high incidence of myxedema, cretinism, deafmutism, thyrotoxicosis, mental retardation and other abnormalities in the endemic areas. Elimination of endemic goitre is, therefore, of great public health importance.

The progressive destruction of epithelial elements in the thyroid gland by the overgrowth of noncellular components and the formation of cysts and large colloid spaces gradually reduces the capacity of the gland to produce the thyroidal hormones and there is a slow onset of mild forms of **myxedema**. This becomes apparent in people over 40 years of age and occurs more often in females than the males.

According to the WHO Report on endemic goitre, there were about 200 million goitrous people in the world in 1960. India has the world's biggest 'goitre belt' in the sub-Himalayan region, extending from Ladakh, Jammu and Kashmir in the west, and passing through Himachal Pradesh, parts of Uttar Pradesh, Bihar, West Bengal, Sikkim, Bhutan, Assam and the North-eastern states of Arunachal Pradesh,

EDWIN BRET HART

(LT, 1874–1953)

Hart was born on December 25, 1874 near Sandusky, Ohio. He graduated from Michigan University with chemistry as a major subject. While at Michigan, he was assistant to **Dr. Campbell**, Professor of Analytical Chemistry. In 1897, he joined the New York State Experimental Station at Geneva. He proceeded to Germany in 1900 and worked with **Prof. A. Kossel** on proteins. He returned to Geneva in 1902 and stayed for 4 years at the Experimental Station. Here, he made valuable contributions in: (*a*) the study of chemical changes in cheese during ripening, and (*b*) elucidating the importance of phosphorus compounds in plants and animals.

In 1906, Hart was appointed as Head of the Department of Agricultural Chemistry in the Experimental Station, Wisconsin. He made noteworthy contributions on the **role of various minerals in nutrition**. He discovered that iodized salt will prevent goitre in human beings and animals. In 1928, he made the important discovery that copper is essential for the incorporation of iron in hemoglobin. He also showed that other trace elements such as manganese, zinc and cobalt are essential for nutrition of animals. He also showed the **toxic effects of fluorides in animals**. He guided 46 students for their Ph.D. degree. He retired from service in 1945 but continued to visit the laboratory till his death on March 11, 1953.

Meghalaya and Nagaland to Burma in the east (**Gopalan**, 1977). The average prevalence rate is around 40%. About 9 million people were estimated to be affected by goitre in India (**Duraiswamy**, 1969).

However, the prevalence rate of goitre can be slowly declined by increasing the amount of iodine, available to the thyroid gland, to the estimated daily requirements of 100 – 150 g/day for adults. This is achieved by the prevention of goitre in children and adolescents who did not have a goitre at the time when the preventive measures were introduced. Under ordinary circumstances, the quantities of additional iodine provided by the addition of iodide or iodate to a variety of foodstuffs do not bring about a significant reduction in size of an already well-established goitre. Four methods are practised to enhance the iodine intake among goitrous people. These are: use of iodized salt, iodized bread, potassium/sodium iodide tablets and iodized oil.

(*a*) **Iodized salt:** Common salt fortified with small quantities of sodium or potassium iodate has been widely used in reducing the prevalence of goitre. The degree of iodization varies from country to country (Table 33–21). The rate of iodization ranges from 1:1,000 to 1:1,00,000; the average being between 1:25,000 and 1:1,00,000. A global survey of the use of iodized salt by **Lowenstein** revealed that legislation had been enacted in 23 countries providing for the compulsory use of iodized salt. In 11 countries, the legislation covers the whole country; and in 12 countries, it applies to limited, defined areas in the country. Follow-up surveys in some of these countries suggest that iodized salt can reduce the goitre rate by 10 to 95% depending on the degree of iodine deficiency and the acceptance of the iodized salt. Iodization of salt is the most economical, convenient and effective means of mass prophylaxis in endemic areas. In European countries, Australia, and New Zealand, and North America free-flowing table salt is iodized with potassium iodide (KI) but in the countries of Asia, Africa, and Latin America the less refined salt is iodized with potassium iodate, which is more stable under the particular conditions of storage and the iodine in it is equally available for hormone synthesis.

(*b*) **Iodized bread:** The method of providing additional iodine in bread has been in use in parts of Netherlands and in Canberra, Australia for more than 15 years and in the whole of Tasmania, Australia for 3 years. In population where bread is universally consumed this appears to be an effective measure, judging by the results of the 3-year trial in Tasmania. In the Netherlands and Canberra, the iodine is provided through appropriately-enriched salt used in the bread formula. In Tasmania, iodine is provided as potassium iodate– a bread improver.

Table 33–21. Recommended level of iodine in salt in various countries

Country	*Stipulated level of iodine compound in salt*
United States	1 in 10,000 as KI
Canada	1 in 10,000 as KI or NaI
New Zealand	1 in 20,000 as KI or NaI
Argentina	1 in 30,000 as KI
England and Wales (table salt)	1 in 40,000 as KI
Mexico	1 in 66,000 as KI
England and Wales (all salt)	1 in 1,00,000 as KI
Switzerland	1 in 1,00,000 as KI or NaI
Netherlands	1 in 1,00,000 as KI
Italy	1 in 1,00,000 as KI
Yugoslavia	1 in 1,00,000 as KI
Poland	1 in 2,00,000 as KI
India	1 in 2,00,000 as KIO_3

To illustrate, the degree of iodization of the table and cooking salt in Switzerland is 1 part in 1,00,000 parts; in the United States it is 1 part in 10,000 parts, calculated as KI. On the assumption that the mean daily intake of salt is 10 g, this would supply 760 µg/day in the United States, and 76 µg/day in Switzerland.

(*c*) **Sodium (or potassium) iodide tablets:** In the original prophylactic trial in the USA at Akron, Ohio, Marine used tablets of sodium iodide. Tablets containing 10 mg of potassium iodide were distributed to infants, children, and young adolescents in Tasmania for 16 years before the introduction of iodized bread and proved highly effective when their regular distribution and consumption could be maintained.

(*d*) **Iodized oil:** In areas where salt iodization cannot be employed, intramuscular (IM) injection of iodized oil has been used for preventing endemic goitre. In fact, since 1957 a group of people (varying in number from 3,000 to 5,000) inhabiting a highly goitrous area of New Guinea have been given periodic IM injections of iodized oil (BP1953) as a prophylactic procedure. It is cheap, long-acting and relatively free from side-effects. Iodized poppyseed oil (ethiodized oil) containing 37% of iodine has also been successfully used. The dose recommended is 1–2 ml for adults and smaller doses for infants and children to be given once in every 3 years (WHO, 1974).

Thyrotoxicosis has been repeatedly reported after the introduction of iodide intervention programmes in populations in which endemic goitre has existed for long periods. It was described during trials with salt iodization and occasionally as a sequel to the administration of iodized oil. The most extensive and severe epidemic (called jodbasedow) on record occurred in the island state of Tasmania, Australia. **Jodbasedow**, which was originally defined by **Kocker** in 1910, is iodine-induced thyrotoxicosis affecting mostly elderly people who have lived for a major part of their life in iodine-deficient areas and who usually have had a goitre for many years. Jodbasedow occurred in a significant number of adults of all stages in Tasmania following the addition of iodate to bread and the widespread use of *ionophores* as sanitizers in the food industry during the 1970s. Mass medication of populations in which endemic goitre was a serious public health problem had been in force in many countries during the 1980s. Apart from the Tasmanian epidemic, relatively few outbreaks of jodbasedow have been reported. Investigators have suggested that the circumscribed area and the type of health sevices available in Tasmania might partly account for this observation. A certain prevalence rate of jodbasedow may be the inevitable outcome of increasing the iodine intake of a population which has had a low intake for health prophylactic programme to prevent goitre.

COPPER

Distribution

The adult human body contains 100 to 150 mg of copper; about 64 mg are found in the muscles, 23 mg in the bones, and 18 mg in the liver, which contains a higher concentration of copper than that of any of the other organs studied. It is of interest to note that the concentration of copper in the fetal liver is 5 – 10 times higher than that in the liver of an adult. Although copper is present in all tissues, but brain, liver, heart and kidney contain the highest amounts. Both the blood cells and serum contain copper; but the copper content of the red blood cell is constant, while that of the seurm is highly variable, averaging about 90 μg/dl.

The copper in plasma occurs in 2 main forms: firmly-bound and loosely-bound. The firmly-bound copper, comprising 80 – 95% of the total plasma copper, consists of *ceruloplasmin* while 60% of the RBC copper is present in a protein called *erythrocuprein*. The loosely-bound copper is known as 'direct-reacting' copper because it reacts freely with copper-complexing reagents (*e.g.*, dithizon and diethyldithiocarbamate) and is loosely bound to the protein, probably serum albumin. The albumin-bound copper may represent copper in transport. One of the important brain proteins called *cerebrocuprein* contains about 3% copper.

Sources

Copper is widely distributed in foods. Most diets provide the amount needed per day to maintain copper balance. The mixed diet of the later period of infancy meets or exceeds the value of 0.05 mg/kg/d suggested as needed by infants at this time. The richest sources of dietary copper are nuts, some shellfish, liver, kidney, raisins and dried legumes. Cow's milk is a poor source of copper, with concentrations of 0.015–0.18 mg/L. Human milk ranges from 1.05 mg/L at the beginning of lactation to 0.15 mg/L at the end. Copper contents of some foods are given in Table 33–22. The important sources of copper in the diet in India are cereals and pulses. In diets consumed in western countries, meat and eggs are important sources of copper in addition to cereals and vegetables.

Table 33–22. Copper content of some foods (mg/100 g)

Food	*Copper content*	*Food*	*Copper content*
Cereals		Coriander leaves	0.53
Bajra (pearl millet)	0.55	Lettuce	0.08
Kaffir corn (*Sorghum*)	0.55	Spinach	0.08
Maize (corn)	0.19	***Other vegetables***	
Rice, raw milled	0.23	Ash gourd	0.17
Wheat flour (refined)	0.32	Brinjal	0.17
Wheat, whole	0.49	Cucumber	0.10
Pulses		Cauliflower	0.05
		French beans	0.21
Bengal gram (Chickpea)	0.98	Peas, green	0.23
Black gram	0.72	Pumpkin	0.20
Cow gram	0.75	Tomato	0.19
Green gram	1.13	***Fruits***	
Peas, dried	0.85	Apple	0.13
Red gram	1.25	Banana	0.40
Nuts and oilseeds		Grapes	0.08
Almonds	0.14	Orange	0.08
Cashewnut	0.22	Peaches	0.06
Groundnut (peanut)	0.27	Pears	0.08

Coconut, dry	0.08	Pineapple	0.21
Roots and tubers		Papaya	0.20
Beet root	0.20	***Milk and eggs***	
Carrot	0.13	Milk, cow's	0.22
Potato	0.20	Egg, whole (hen's)	0.25
Tapioca	0.15	***Meat and fish***	
Green leafy vegetables		Fish	0.23
Cabbage	0.08	Mutton	0.27
Celery	0.30	Beef	0.22
		Liver beef	2.80

Requirements

The human requirement of copper (refer Table 33–20) has been studied by balance experiments. A daily allowance of 2.5 mg of copper has been suggested for adults; infants and children require about 0.05 mg/kg body weight. This is easily supplied in average diets, which contain 2.5 to 5 mg of copper. Even poor diets provide enough copper for human needs.

Absorption, Retention and Excretion

Studies conducted with stable and radioactive copper have shown that orally-administered copper is absorbed only to the extent of 2 – 5%. Only about 2 to 10% of the copepr present in the diets is absorbed and retained in the body. Copper absorption may depend on metallothionein, since excess intake of either copper or zinc interferes with the absorption of the other. The absorbed copper is transported by the blood to different tissues. Copper is stored mainly in the liver. Appreciable amounts are also found in the brain, kidney and heart.

Since copper in the plasma is largely bound to protein, it is not readily excreted in the urine. Most of it is excreted *via* the bile into the intestine and expelled with feces. Copper administered in small amounts by injection is stored in the body.

Functions

Several enzymes contain copper (Table 33–23). The activities of all these enzymes are reduced in copper deficiency. Some of these important enzymes and the reactions they catalyze are given below:

Table 33–23. Enzymes containing copper

S.N.	*Enzyme*	*Reaction it catalyzes*
1.	Ferroxidase I (= Ceruloplasmin)	$Fe^{2+} \longrightarrow Fe^{3+}$
2.	Ferroxidase II	$Fe^{2+} \longrightarrow Fe^{3+}$
3.	Superoxide dismutase (erythrocuprein, cerebrocuprein and hepatocuprein)	$2O_2^- + 2H^+ \longrightarrow H_2O_2 + O_2$
4.	Cytochrome c oxidase	$2H_2 + O_2 \longrightarrow 2H_2O$
5.	Monoamine oxidase	Aliphatic amine ⟶ Aldehyde + NH_3
6.	Diamine oxidase	Aliphatic diamine ⟶ Aminoaldehyde + NH_3
7.	Dopamine β-hydroxylase	Dopamine ⟶ Norepinephrine
8.	Lysyl oxidase	α amino group of lysine ⟶ Aldehyde
9.	Phenolase (= Tyrosinase)	Tyrosine ⟶ Dopa ⟶ Melanin pigment
10.	C_{18}, Δ^9-desaturase	Stearic acid ⟶ Oleic acid

1. **Ferroxidase I (= ceruloplasmin),** present in plasma, bone marrow and liver, is involved in the oxidation of ferrous to ferric ion. The oxidation of ferrous iron to ferric and reduction of ferric to ferrous is essential for the transport of iron from the intestine to blood and then to

bone marrow, liver and other organs and incorporation of hemoglobin. Ceruloplasmin (MW, 1,51,000), a copper-binding plasma protein, contains 0.34% copper, or about 8 atoms of copper per mole. Normal plasma contains about 30 mg of ceruloplasmin per dl.

2. **Ferroxidase II** also functions in a manner similar to that of ferroxidase I.
3. Another copper-containing enzyme is the cytosolic **superoxide dismutase** which catalytically scavenges the toxic free radical *superoxide ion* (O_2^-) generated during aerobic metabolism. Cytosolic superoxide dismutase (molecular weight, 32,000) consists of 2 identical subunits and contains one Cu^{2+} and one Zn^{2+} per subunit (**Fridovich**, 1975). This enzyme converts superoxide radical formed from oxygen in some oxidation or reduction reactions to H_2O_2 according to the following reaction:
$$2O_2^- + 2H^+ \longrightarrow H_2O_2 + O_2$$
The H_2O_2 formed is borken down by catalase to H_2O and O_2:
$$2H_2O_2 \longrightarrow 2H_2O + O_2$$
For example, autoxidation of Hb to metHb in the RBC results in the formation of superoxide radical, O_2^- by transfer of a single electron.

   Proteins identical to superoxide dismutase were earlier isolated from several different sources and, in the absence of known enzymic function, were given a variety of names, including *erythrocuprein* (from red blood cells), *hepatocuprein* (from liver), and *cerebrocuprein* (from brain). A superoxide dismutase protein of markedly different amino acid sequence and containing manganese instead of copper and zinc is present in the matrix space of mitochondria.
4. **Cytochrome oxidase** is essential in the electron-transport system leading to the reaction of hydrogen removed from substrates with molecular oxygen to form water.
5. The enzyme **monoamine oxidase** converts aliphatic amine to the corresponding aldehyde and NH_3.
6. **Diamine oxidase** catalyzes a reaction in which aliphatic diamine is converted to amino aldehyde and NH_3.
7. **Dopamine hydroxylase**, present in adrenal medulla, is a copper enzyme which converts dopamine to norepinephrine by adding oxygen to the β-carbon atom.
8. **Lysyl oxidase** is involved in the oxidation of ε-amino group of lysine to aldehyde group which reacts with lysine to form desmosine, the crosslinking group of collagen and elastin.
9. **Tyrosinase**, which also contains copper, is present in melanocytes and melanoma tissues of human beings and animals. This enzyme is responsible for melanin formation. The skin of human albinos has no detectable tyrosinase activity.
10. **C_{18}, Δ^9-desaturase** is responsible for converting stearic acid (a C_{18} saturated fatty acid) to oleic acid (a C_{18} monounsaturated fatty acid). This may be responsible for the fact that the dietary stearic acid does not have the cholesterol-raising property of the other saturated fatty acids.

Metabolism

Experiments using labeled copper (^{64}Cu) have established that copper is found to be largely associated with the albumin fraction of the plasma immediately after its ingestion. A decline in plasma radioactivity then occurs as copper attached to albumin is distributed to cuproproteins in the liver and elsewhere; a secondary rise in plasma radioactivity then takes place as copper incorporated into liver ceruloplasmin is released to the blood. In Wilson's disease (see below), this secondary rise does not occur.

Deficiency States

Either deficiency or excess of copper is very rare. However, copper deficiency has been observed in infants receiving only milk; the prime manifestation is a microcytic, nonchromic **anemia**. Also, a macrocytic anemia, alleviated by copper administration, has been seen in infants who had been ingesting milk fortified with iron and vitamin B_{12}. Anemia appears to be due to a defect in iron metabolism because the copper-containing enzyme *ferroxidase* is necessary for conversion of iron from the Fe^{2+} state (in which form it is absorbed) to the Fe^{3+} state (in which form it can bind to the plasma protein transferrin).

In copper deficiency, iron absorption by the mucosal cells appears to be unaffected, but release of iron into the plasma is impaired. Synthesis of ceruloplasmin– and therefore the movement of iron from ferritin stores to the plasma– is decreased, and **hypoferremia** results. Consequently, heme synthesis is depressed and anemia develops in all species when the deficiency is severe or prolonged.

Wilson's disease (= hepatolenticular degeneration), a rare autosomal recessive disorder, is associated with abnormalities in the metabolism of copper. In this disease, the liver and the lenticular nucleus of the brain contain abnormally large amounts of copper, and there is excessive urinary secretion of copper and low levels of copper and of ceruloplasmin in the plasma. Patients suffering from Wilson's disease absorb about 50% of the ingested copper as compared to 2–5% in normal persons. There is also defect in ceruloplasmin formation. Hence, the copper absorbed is deposited in various tissues such as brain, liver, kidney and in Descemet's membrane in the eye. Copper deposition in Descemet's membrane of the eye causes a golden-brown, yellow or green ring round the cornea. This lesion is called *Kayser–Fleischer ring* and occurs in all cases of Wilson's disease. These organs may be damaged by poisoning of certain enzymes by copper or by necrosis followed by fibrosis (or gliosis in the brain). A general aminoaciduria also occurs in this disease.

Almost all patients with Wilson's disease have less than 23 mg of ceruloplasmin/dl of serum, which may be taken to represent the lower limit of the normal range. In some patients, ceruloplasmin can hardly be detected. Hence, in Wilson's disease, much of the plasma copper remains loosely bound to serum albumin, and the copper can hence be more readily transferred to tissues such as the brain and liver or to the urine. The 'direct-reacting' fraction of the serum copper is not decreased in these patients and may actually be increased. Thus, the total serum copper may appear to be normal or only slightly decreased.

If the degree of copper in the liver becomes excessive, cirrhosis may develop. Accumulation of copper in the kidney may give rise to renal tubular damage, which leads to increased excretion of amino acids and peptides and, occasionally, glucose as well. It, thus, appears that many metabolic activities such as excessive copper absorption from the intestine, inadequate excretion of copper *via* the intestine, and synthesis of an abnormal liver protein with high affinity for copper may be factors in the genesis of Wilson's disease.

There are 3 *clinical types* of Wilson's disease: hepatic type, cerebral type, and hepatic and cerebral type. These are described below:

1. *Hepatic type*: There is a progressive hepatic cirrhosis of a coarse nodular type, leading gradually to portal hypertension and eventually to hepatic failure. Juvenile cirrhosis should always suggest the possibility of Wilson's disease. Jaundice may precede it. It is sometimes hemolytic in type.
2. *Cerebral type*: The key manifestations are neurolygic abnormalities arising from dysfunction of the lenticular region of the brain. Necrosis and sclerosis of the corpus structures cause basal ganglion syndromes in adolescents. Most common in choreo-athetosis but Parkinsonism may be caused according to the site mainly affected. There is usually cortical involvement leading to progressive dementia in which loss of emotional control is a feature.
3. *Hepatic and cerebral type*: In some patients, both hepatic and cerebral symptoms may be present.

Wilson's is a progressive disease and proves fatal within a few years after the appearance of clinical signs. Recently, some copper-binding drugs, such as *disodium calcium versenate, dimercaprol* and *penicillamine* have been found to be useful in improving the clinical condition and in preventing the pathological changes in the tissues; the latter (penicillamine) being the most effective of the three. These drugs bind (chelate) the ionic copper present in the tissues and copper is excreted as the *chelated* copper compounds in the urine.

Excess States

Hypercupremia may reflect excessive intake of copper which may result from eating food prepared in copper cooking vessels, or it may be associated with several acute and chronic infections, such as leukemia, Hodgkin's disease, severe anemia, hemochromatosis, myocardial infarction (MI) and hyperthyroidism (**Standstead *et al*,** 1970).

ZINC

Distribution

Zinc is widely distributed in the tissues of the body. The human adult, weighing 70 kg, contains 1.4 to 2.3 g of zinc. Zinc is present in large amounts in the skin, bones and teeth, and the prostate gland; in other tissues, it is present in small amounts (Table 33–24). Skin alone contains about 20% of the total body zinc.

Table 33-24. Zinc content of some tissues in humans

Tissue	*Zinc (mg/100 g)*	*Tissue*	*Zinc (mg/100 g)*
Prostate	86.01	Heart	3.3
Skin, epidermis	70.5	Spleen	3.1
Skin, dermis	12.6	Pancreas	2.9
Bones and teeth	15.0–25.0	Testes	1.7
Kidney	5.5	Lungs	1.5
Muscles	5.4	Brain	1.4
		Adrenal	1.2

Zinc is present in greater concentration in RBC than in the plasma. The zinc contents of human blood, plasma and RBC are 0.8 mg, 0.12 mg and 1.44 mg/100 ml, respectively. Most of zinc in RBC is present in the enzyme *carbonic anhydrase*.

The main zinc content of hair of normal Egyptians was 103.3 ppm (parts per million) and that of zinc-deficient dwarfs was 54.1 ppm. However, oral zinc sulfate therapy alleviated the signs of zinc deficiency and increased the zinc level in hair to 121.1 ppm.

Sources

Zinc content of some foods are given in Table 33–25. A perusal of the table points out that the seafoods (oysters and herrings, in particular) are very rich sources of zinc while flesh foods are good sources. Cereals, pulses, nuts and oilseeds, and vegetables and fruits are fair sources. Although zinc is widely distributed in foodstuffs– both animal and vegetable–, its bioavailability in vegetable foods is low due to the presence of phytates which interfere in its absorption.

Table 33–25. Zinc content of some foods (mg/100 g)

Foodstuff	*Zinc content*	*Foodstuff*	*Zinc content*
Cereals		***Flesh foods***	
Barley	2.7	Beef	3.5
Bread, whole wheat	2.4	Clams	2.0
Bread, white	1.2	Egg, whole	1.8
Corn, whole	2.5	Egg, yolk	3.4
Corn, flour (de-germed)	1.2	Herrings	110
Oat meal	8.2	Liver, beef	5.2
Rice, raw milled	1.5	Liver, pork	6.5
Rice, parboiled milled	1.9	Oysters	160
Wheat, whole	3.5		
Wheat, bran	14.0		

Pulses		***Milk***	
Pulses	3.5	Milk, cow	0.8
Peas	3.2		
Nuts and oilseeds		***Vegetables***	
Nuts and oilseeds	3.5	Beets	2.8
Peanut	2.2	Cabbage	0.8
Fruits		Carrots	1.2
Cherries, canned	0.5	Lettuce	0.4
Oranges, Pears, canned	0.2	Potato	0.2
		Spinach	0.5

Requirements

Zinc daily requirements of humans (refer Table 33–20) vary with age: infants, 3–5 mg; children, 10 mg; adolescents, 15 mg; and adults, 15 mg. An additional 15 mg during pregnancy and 10 mg during lactation is also required. Most human diets provide these amounts.

Absorption, Retention and Excretion

Only a small fraction (5–10%) of zinc present in the diet is absorbed by the human body. Zinc is absorbed maily in the small intestine, especially from the duodenum. **Evans** *et al* (1975) have shown that a low-molecular-weight zinc-binding ligand (or factor) is secreted by the pancreas and this factor forms a complex with zinc and helps in the absorption of zinc. High intakes of phosphorus and calcium and phytic acid interfere with the absorption of zinc. When radioactive zinc is given orally to animals, about 70% is lost in the feces, 0.3% is lost in the urine, and the remaining is retained in the body. However, upon intravenous administration of zinc, only 20% is excreted in feces, 0.25% is lost in the urine, and the remaining is retained in the body. On an average, of a zinc intake of about 10 mg/d by a healthy adult human, about 9 mg is lost in the feces, 0.5 mg is excreted in the urine, and the rest 0.5 mg is retained in the body. Endogenous zinc is secreted into the small intestine in the pancreatic juice, which contains carboxypeptidase.

The most rapid accumulation and turnover rate of zinc occurs in the soft tissues, especially the pancreas, liver, kidney and spleen. Liver cytosol contains zinc- and copper-binding proteins similar to or identical with metallothionein. In addition to its function in cellular heavy metal detoxication (see Cadmium), this protein may hence serve as a storage protein for zinc– analogous to the role of ferritin in iron storage. Relatively large quantities of zinc are deposited in bones, but these stores do not move into rapid equilibrium with the rest of the organism. The body pool of biologically-available zinc appears to be small and to have a rapid turnover.

Functions

Zinc is an essential component of a number of enzymes present in animal tissues, including alcohol dehydrogenase, alkaline phosphatase, carbonic anhydrase, procarboxypeptidase, and cytosolic superoxide dismutase (which also contains copper). Acetazolamide inhibits **carbonic anhydrase** activity by binding to the zinc atom present in the enzyme molecule. The retina contains a zinc metalloenzyme called **retinene reductase**, which is required for the reconstitution of retinene (= vitamin A aldehyde) during the rhodopsin cycle. Zinc is necessary to maintain normal concentrations of vitamin A (= retinol) in plasma. Zinc may be required for mobilization of vitamin A from the liver. When vitamin A concentrations in the plasma are below normal and are unresponsive to therapy with vitamin A, zinc supplementation may be effective.

Insulin forms complexes with zinc, which makes it possible for crystalline *zinc-insulin* to be prepared during insulin purification. Zinc adhering to the insulin molecule increases the duration of insulin action when injected into the body. Zinc-insulin complexes are also present in the β cells of the pancreas and available evidences suggest that zinc is used in these cells to store and release insulin as and when needed (**Davies**, 1972). Research conducted by **Babior** *et al* (1973) suggests that superoxide ions produced by leucocytes have a bactericidal function and the zinc-containing **superoxide dismutase** may therefore serve to regulate superoxide production in these cells.

Zinc is also essential for the normal growth, reproduction and life span of animals and has a beneficial effect on processes such tissue repair and wound healing. The influence of zinc on these physiologic processes may be related to the observation that depressed activity of **thymidine kinase**, essential in DNA synthesis and cell division, is an early metabolic defect in zinc deficiency. Thus, the element zinc is also involved in the biochemical processes which regulate protein and nucleic acid syntheses and turnover.

Metabolism

Porces and his associates, in 1966, found a significantly increased rate of wound healing when zinc sulfate, 50 mg three times a day, was administered to young men following surgery for pilonoidal sinuses. Several other workers have subsequently demonstrated improved wound healing in men following severe burns and major operations as a result of zinc sulfate therapy. The enhancement of wound healing by zinc may be due to increased collagen synthesis brought about by zinc.

Zinc is also present in *gustin,* a salivary polypeptide that appears to be necessary for normal development of taste buds. Zinc is also required for optional cytokine production by monocytes and T cells. Thus, zinc deficiencies are also associated with impaired immune function.

Deficiency States

In view of the wide distribution of zinc in foods, it seems unlikely that zinc deficiency will occur in human beings eating an otherwise adequate diet. However, certain deficiency states develop in case of malnutrition or undernourishment.

The occurrence of specific zinc deficiency in human beings in some parts of Egypt and Iran has been reported by **Prasad** in 1976. The subjects (adolescent boys, aged 16 – 20 years) suffered from a syndrome of growth failure (**dwarfism**) and retarded genital development (**hypogonadism**). Bone development was also retarded. The zinc content of blood plasma was low. Lab assessment of endocrine status of these patients revealed **hypopituitarism.** Therapeutic trials with zinc sulfate ($ZnSO_4$) along with the hospital diet brought about remarkable increase in height and genital development.

A physiological function for zinc in normal taste sensation was first reported by **Henkin** *et al* (1971) who found that **hypogeusia** (loss of taste acuity) and **dysgeusia** (perverted taste or pica) that occur in some adults respond to oral zinc therapy. A year later, **Hambidge** *et al* reported that poor growth, loss of appetite and loss of taste acuity in young malnourished children in Denver was associated with subnormal hair zinc levels. Supplementation of the diet with small amounts of $ZnSO_4$ normalised the taste acuity (the ability to think, see, or hear clearly), increased the hair zinc levels and improved the growth.

Severe zinc deficiency is seen primarily in alcoholics (especially if they have cirrhosis), patients with chronic renal disease or severe malabsorption diseases, and occasionally in people after long-term parenteral nutrition (TPN). The most characteristic early symptom of zinc deficient patients on TPN is dermatitis.

Evidence for a conditioned deficiency of zinc in patients with so-called postalcoholic cirrhosis has been presented by **Vallee** *et al* in 1959. Patients with Laennec's cirrhosis had a mean concentration of zinc of 66 ± 19 μg/dl, against a normal value of 120 ± 19 μg/dl. There was also increased urinary excretion of zinc in cirrhotic patients. When studied at autopsy, the livers of patients who had died of cirrhosis were found to have significantly less zinc than liver tissues from patients who had died of nonhepatic disease. It was suggested that the zinc deficiency in the cirrhotic patients was due to secondary factors related to the underlying disease which act to render a normal intake of zinc inadequate.

A clear-cut case of human zinc deficiency disorder is a rare congenital defect called **acrodermatitis enteropathica** (AE). Both AE and experimental zinc deficiency have clinical features in common. These are: anorexia, diarrhea, stunting, wasting, skin desquarmation with ulceration, a reduction in lymphoid tissue, and an increased susceptibility to infections. When the relationship of these clinical features to plasma zinc concentration in patients was examined shortly after admission to hospital (on an intake of 0.1 mg Zn/kg/d), the correlations with skin ulceration, stunting and edema were significant (**Golden** and **Golden**, 1979). The other cardinal feature of both conditions is increased susceptibility to infections.

FLUORINE

Distribution

Flourine is one of the most active elements of the halogen group. It is never found in the free state in nature. About 95% of the fluorine (in the form of fluoride) in the human body is found in bones and teeth.

Sources

There are 2 main sources of fluoride to humans: drinking water and food. *Drinking water* containing fluoride at a concentration of 1 ppm is enough for nutritional needs. When the drinking water in a given area contains less than that amount, it is regarded as a commendable public health practice to add fluoride to the water supply (**fluoridation**) in order to raise its concentration to an appropriate level. In the long-term studies of the results of fluoridation, it has been shown that the incidence of dental caries in younger age groups (14–16 years) was reduced by about 65%. When the water supply contains 1 ppm fluoride, 1–2 mg/day of fluoride will be ingested depending on the water intake in all forms, either as such or in beverages made with added water. This level of intake of fluoride is considered adequate.

The fluoride content of drinking water in India is about 0.5 mg/litre, but in fluorosis-endemic areas, the natural waters have been found to contain as much as 3–12 mg of fluorides per litre (**Gopalan**, 1974). Fluorides also occur in traces in many *foods*, but some foods such as sea fish, cheese and tea are rich in fluorine. The average man may ingest about 1 mg of fluoride daily from drinking water. In addition to this, the daily diet may provide 0.25–0.35 mg of fluorine (WHO, 1974).

Requirements

Since drinking water is the main source of fluorine to man, a concentration of 0.5 to 0.8 mg/litre in water is considered a safe limit in India. However, in temperate climates where the intake of water is low, the optimum level of fluorine in drinking water is accepted as 1–2 mg/litre. The requirements of fluorine are given in Table 33–20.

Absorption and Excretion

In humans, fluoride is normally accumulated in small quantities only in the bones and teeth. The amount of fluoride present in these tissues depends on several factors such as the duration of fluoride intake and the age at which intake occurs. The amounts of fluoride in the soft tissues are generally very low and do not increase with age. Soluble fluorides in drinking water are rapidly absorbed. Fluoride is initially distributed in the same manner as chloride and remains mostly in the extracellular water. Most of the fluoride that escapes retention by the bones and teeth is excreted rapidly into the urine.

Functions

Fluoride is essential for the normal mineralization of bones and formation of dental enamel. The protective effect of fluoride in tooth development is most important during infancy and early childhood because these are the periods that coincide with tooth development, although the caries-preventive activity of fluoride continues into adulthood. In addition, appropriate levels of dietary fluoride decrease the incidence of osteoporosis, a defect in the maintenance of bone structure that ocucrs in older adults– notably in women after menopause.

Metabolism

Based on *in vitro* studies in aqueous systems, it has been postulated that fluorine may be required in biological calcification. The precipitation of hydroxyapatite crystals from solutions containing calcium and phosphate ions is enhanced by the presence of fluorine, although a very low fluorine concentration (0.0002– 0.002 ppm) is sufficient for precipitation. Defective mineralization has not been demonstrated in animals deprived of fluorine, possibly because fluorine cannot be adequately removed from animal systems.

Deficiency States

Fluorine is often called a two-edged sword. Ingestion of large amounts is associated with dental and skeletal fluorosis, and inadequate amounts with dental caries. **Dental caries** (syn. **saprodontia**) in human beings, especially in children, develops in areas where the fluoride content of drinking water is low (less than 0.5 ppm). It is a localized, progressively destructive disease of the teeth that starts at the external surface (usually the enamel) with the apparent dissolution of the inorganic components by organic acids that are produced in immediate proximity to the tooth by the enzymatic action of masses of microorganisms (in the bacterial plaque) on carbohydrates. The initial demineralization is followed by an enzymatic destruction of the protein matrix with subsequent cavitation and direct bacterial invasion. In the dentin, demineralization of the walls of the tubules is followed by bacterial invasion and destruction of the organic matrix.

Dental caries may be prevented by fluoridation of community water supplies (to 1 mg/litre) where the fluoride intake from water and other sources for the given population is below optimum levels. Dental caries can also be significantly reduced by topical application of fluoride preparations to the teeth. Fluoride appears to exert an inhibitory effect on enzymes of cariogenic bacteria. Physicochemical changes in the tooth enamel may also be important. Once the teeth have been fully formed, fluoride is without effect on dental caries.

Excess States

In many parts of the world where drinking water contains excessive amounts of fluorine (10–45 ppm), **dental fluorosis** (syn. **mottled enamel**) develops in persons, especially young children. The enamel of their teeth loses its lustre and becomes rough. The chalkwhite patches develop on the affected teeth. This is an early sign of dental fluorosis. The white patches later become yellow and sometimes brown or black. Calcification is deficient, and cement substance may be lacking. In severe cases, loss of enamel is accompanied by 'pitting' which gives the tooth a corroded appearance. Mottling is best seen on the incisors of the upper jaw. It is almost entirely confined to the permanent teeth and develops only during their period of formation.

Chronic fluorine intoxication, through drinking water containing excessive amounts of fluorine (over 10 ppm) or among workers handling fluoride-containing minerals, results in a disease **skeletal fluorosis**. In such persons, especially the older people, the disease affects the bones, tendons and ligments. This is followed by pain and stiffness of the back, and later of joints of both limbs and limitation of neck movements. In addition, the ligaments of the spine become calcified, producing a 'poker back'. Early detection of the disease may escape till radiological help is sought. Radiological changes are quite characteristic– there is formation of new bone (exostosis) and calcification of tendons and ligaments as well as interosseous membranes. The affected persons become crippled and cannot perform simple daily tasks, such as bending, squatting etc, as the joints become stiff.

Toxic fluorosis can be prevented only by removing fluoride from the water supplies by treatment with activated carbon or by some other suitable absorbents.

Fluoride is very toxic in large doses. A lethal dose may be as small as 0.5 in a child, but doses many times as large are not always fatal. Fluoride is a powerful inhibitor of many magnesium-activated enzymes concerned in phosphate metabolism, including the glycolytic enzyme *enolase*. Fluoroacetate acts as a powerful inhibitor of the citric acid cycle (CAC) after metabolic conversion ("lethal synthesis") to fluorocitrate, which blocks aconitase activity.

MANGANESE

Distribution: The body of a 70-kg normal man contains about 10–20 mg manganese. It is distributed throughout the body tissues and fluids. The kidney and liver are the chief storage organs, followed by bones, pancreas and pituitary. Human blood contains very low amounts of manganese– 4 to 20 μg/dl.

Sources: Nuts and whole grains are rich sources of manganese; vegetables and fruits are good sources. Meats, poultry, seafoods and fish are poor sources.

Requirements: Daily requirements for manganese are given in Table 33–20. Human adults, on an average, require 2.5–5 mg daily.

Absorption and Excretion: Mitochondria are the principal intracellular sites of manganese uptake. Only 3–4% of the manganese present in the diet is absorbed while most of it is excreted into the intestine by way of the bile. Manganese usually occurs in traces in the urine.

Functions: Manganese is essential for normal bone structure, reproduction and the normal functioning of the central nervous system. Manganese ions may be involved in activating glycosyltransferases associated with the synthesis of glycoproteins, for example prothrombin. The enzyme pyruvate carboxylase, involved in glyconeogenesis, and the superoxide dismutase present in the matrix space of mitochondria both contain tightly-bound manganese. Manganese is also intimately bound to arginase of liver and activates many enzymes, *e.g.*, phosphoglucomutase, cholinesterase, the oxidative β-ketocarboxylases, certain peptides, and muscle adenosinetriphosphatase.

Metabolism: *In vitro*, manganese or magnesium ions activate isocitrate dehydrogenase and several phosphotransferase enzymes, and manganese ions strongly inhibit lipid peroxidation reactions. In human serum, manganese is bound to a specific β-globulin.

Deficiency States: An evident deficiency of manganese is not known to occur in humans, suggesting that the average daily dietary intake of 2.5–5 mg is adequate. However, manganese deficiency causes **skeletal defects** which can probably be explained by the observation that manganese ions activate glycosyltransferase enzymes involved with the synthesis of the mucopolysaccharides of cartilage.

Excess States: Ingestion of excessive quantities of manganese appears to interfere with absorption of iron, thus causing an **anemia** which is readily prevented by increasing the dietary iron.

MOLYBDENUM

Distribution: Molybdenum is present in all the tissues of human body. It's concentrations in animal tissues are normally very low in comparison to those of manganese. The highest concentrations of molybdenum are usually found in liver and kidney.

Sources: Besides liver and kidney, pulses, cereal grains and some green leafy vegetables are good sources of molybdenum; other vegetables and fruits are poor sources.

Requirements: The average diets provide adequate amounts of molybdenum. Recommended daily dietary allowances are given in Table 33–20. On an average, the infants and 0.05 mg, children 0.20 mg, adolescents and adults 0.35 mg.

Absorption and Excretion: About 50–70% of the total molybdenum intake is absorbed and half of the absorbed molybdenum is excreted in urine. An adequate protein intake seems to be essential for the excretion of molybdenum. At a high protein consumption (2.5 g/kg), the retention of metal was low, but was increased fivefold at a lower protein intake (0.7 g/kg). Thus, there is a direct relationship between protein intake and molybdenum excretion.

Functions: Molybdenum is an essential component of the molybdoflavoproteins, *xanthine oxidase* and *aldehyde oxidase* and the molybdohemoprotein, *sulfite oxidase* of animal tissues. It is also present in the enzymes *nitrate reductase* in plants and *nitrogenase*, which functions in nitrogen fixation by microorganisms. Traces of molybdenum are required for the maintenance of normal levels of xanthine oxidase in animal tissues. In humans, a genetic deficiency of xanthine oxidase (xanthinuria) and of sulfite oxidase has been recognized. Xanthinuric individuals do not appear to suffer any ill effects, but the single reported case of human sulfite oxidase deficiency proved fatal. Sulfite oxidase catalyzes the detoxification of sulfur dioxide and sulfite to sulfate.

Metabolism: Epidemiological studies in Hungary and New Zealand and experiments with mice suggest a beneficial effect of molybdenum on incidence and severity of dental caries. In this disease, molybdenum can augment the well-established effect of fluoride. A complex relationship exists between the intake of molybdenum, copper, and sulfate and the copper status of the animal (**Underwood**, 1977).

Deficiency States: It has proved difficult to induce molybdenum deficiency diseases experimentally, which suggests that many organisms have an extremely low requirement for molybdenum. However, deficiency of molybdenum has been linked with **mouth and esophageal cancer**.

Excess States: Excess absorption of molybdenum has been shown to produce **bony deformities**. Excess dietary molybdenum may also induce symptoms of **copper deficiency**.

CHROMIUM

Distribution: Chromium is widely distributed in human and animal tissues, although in extremely small amounts. The total chromium content of the adult human body is estimated to be less than 6 mg. Most adult tissues contain 0.02–0.04 ppm of chromium on dry basis. The blood contains about 0.009 to 0.055 ppm of chromium.

Sources: Food grains, cereals, vegetables and fruits are poor sources of chromium.

Requirements: Average human diets seem to meet the chromium daily requirements. Recommended dietary allowances (RDAs) of National Research Council, USA for chromium are given in Table 33–20.

Absorption, Retention and Excretion: Absorbed or injected chromium is excreted mainly in the urine, with small amounts being lost in bile and feces. Tissue uptake is quite rapid and plasma is cleared of a dose of chromium within 3 or 4 days. The bones, spleen, testis and epididymis retained more chromium after 4 days than the heart, lungs, pancreas or brain. Chromium is mobilized from the tissues in response to glucose administration.

Functions: There is suggestive evidence that chromium plays a role in relation to carbohydrate and insulin function (**Beaton** *et al*, 1976). Chromium is a component of the low-molecular-weight protein called *chromodulin*. Chromodulin potentiates the effects of insulin, presumably by facilitating insulin binding to cell receptor sites.

Metabolism: Studies have revealed that trivalent chromium may potentiate the action of insulin in promoting utilization of glucose, a conclusion drawn from the observation that impairment of glucose tolerance occurs in animals maintained on low-chromium diets and that the diminished glucose tolerance can be reversed by supplementation of the diet with chromium. There is also evidence to suggest that chromium is involved in glucose tolerance in humans (**Davies**, 1972).

Deficiency States: Chromium deficiency is characterized by impaired growth disturbances in glucose, protein, and lipid metabolism. Rats fed on diets low in protein and in chromium (less than 0.1 ppm chromium) developed **corneal opacity.** The opacity was ameliorated when the diet was supplemented with 5 ppm of chromium.

Excess States: Oral administration of excessive amounts (50 ppm) of chromium has been reported to produce **growth depression** and liver and kidney damages in some experimental animals.

SELENIUM

Distribution

Selenium is widely distributed in the animal body and is found in highest concentrations in the renal cortex, pancreas, pituitary, and liver. *Most of the body selenium is quite labile.* Transfer from a selenium-sufficient to a selenium-deficient diet induces a rapid loss of selenium from the tissues and a marked decrease in tissue glutathione peroxidase activities, especially in young animals.

Sources

Selenium is one of the few nutrients not removed by the milling of flour and is usually thought to be present in adequate amounts in the diet. The amount of selenium in foodstuffs is highly variable, due largely to differences in soil selenium content in the areas, where food is grown.

Requirements

The RDA for selenium in an adult is between 0.05 and 0.2 mg. The average diets contain adequate amounts of selenium.

Functions

Selenium is an essential component of the enzyme *glutathione peroxidase* isolated from sheep red blood cells, has a molecular weight of about 84,000. The enzyme contains one atom of selenium per protein subunit of about 21,000 MW (*i.e.*, 0.34% of selenium). The enzyme catalyzes the oxidation of reduced glutathione to oxidized glutathione, as shown below:

$$2GSH + H_2O_2 \longrightarrow GSSG + 2H_2O$$

Reduced glutathione protects membrane lipids and other cell constituents (*e.g.*, hemoglobin) against oxidative damage by destroying hydrogen peroxide and fatty acid hydroperoxides (of general structure ROOH) through reactions catalyzed by glutathione peroxidase, as follows:

$$2GSH + ROOH \longrightarrow GSSG + ROH + H_2O$$

Oxidized glutathione is regenerated by the activity of the enzyme glutathione reductase.

Selenium is involved in a variety of other important biologic processes including immune mechanisms, ubiquinone biosynthesis, and mitochondrial ATP biosynthesis (**Frost** *et al*, 1975). In addition to glutathione peroxidase, animal tissues also contain selenium that is incorporated into proteins, presumably as selenium analogues of the sulfur-containing amino acids.

Metabolism

Some of the disorders of animals induced by dietary means are responsive either to selenium or to vitamin E (tocopherol), indicating that a close relationship exists between the two nutrients. However, certain diseases are apparently caused by a deficiency that responds specifically to one nutrient but not to the other. The role of selenium in hydroperoxide destruction through glutathione peroxidase activity (described above) serves to clarify the interrelationships between vitamin E and cysteine (Cys) as a precursor of glutathione.

If vitamin E prevents fatty acid hydroperoxide formation, and the sulfur-containing amino acids (as precursors of glutathione) and selenium are involved in peroxide breakdown, all of these nutrients would obviously lead to a similar biochemical result, *i.e.*, lowering in the tissues of the concentrations of peroxides or products induced by them. Certain tissues or subcellular components that are inherently low in glutathione peroxidase would not be affected by selenium but would still be protected by vitamin E, which acts as an antioxidant by a mechanism not involving glutathione peroxidase.

Selenium has long been known to inhibit alcoholic fermentation by yeast and the activity of certain respiratory enzymes, possibly by forming complexes with essential enzyme sulfhydryl (—SH) groups.

Deficiency States

Studies indicate that human selenium deficiency may ocur in **protein–energy malnutrition** (**Standstead** *et al*, 1970).

Excess States

Selenium is usually thought of as an element with pronounced toxic properties. When selenium is present in the diet in concentrations of about 5–15 ppm, it is highly **toxic to animals**, as demonstrated in particular with ruminants maintained on pastures where the soil is rich in selenium. However, below about 3 ppm, selenium improves growth and is effective in combating several animal diseases. The clinical studies have suggested that supplementation with greater than RDA levels of selenium may reduce the risk of certain cancers.

COBALT

Distribution

The total human body content of cobalt is in the average of 1.1 mg. Cobalt is widely distributed throughout the body. The liver, kidneys and bones are the richest sources.

Sources

Most of the cobalt is present in cyanocobalamin (vitamin B_{12}).

Requirements

Cobalt requirements, if any, appear to be met by traces of cobalt found in foods. In fact, cobalt must be supplied in the diet of humans and other monogastric species in the physiologically-active form as vitamin B_{12}. The ruminants are capable of synthesizing their requirements of vitamin B_{12} whereas humans are not.

Absorption and Excretion

Cobalt, administered orally as a soluble salt, is poorly absorbed and is, hence, largely excreted in the feces. Injected isotopic cobalt is eliminated rapidly and almost completely by the kidneys into the urine.

Functions

Cobalt, as already mentioned, is a constituent of vitamin B_{12}, which is necessary for normal red blood cell formation. It is uncertain whether inorganic cobalt plays any separate essential biologic role. Enzymes requiring B_{12} for activity include methylmalonyl CoA mutase, methyltetrahydrofolate oxidoreductase, homocysteine methyltransferase, and ribonucleotide reductase.

Metabolism

Patients suffering from pernicious anemia due to vitamin B_{12} deficiency excrete large amounts of methylmalonic acid in the urine owing to a defective mutase reaction.

Deficiency States

Nutritional anemia in cattle and sheep, living in cobalt-poor soil areas, can be treated successfully with cobalt. Microorganisms in the rumens of these animals utilize dietary cobalt to synthesize vitamin B_{12}. The dietary cobalt, incorporated in vitamin B_{12} synthesized by rumen organisms, is available to the ruminants while that synthesized in the cecum and large intestines cannot be absorbed. However, there have been no reports of the production of cobalt deficiency in nonruminating animals including humans, all of which require dietary vitamin B_{12}.

Excess States

The pharmacologic doses of inorganic cobalt salts, administered to many species including humans, induce a vast increase in the number of erythrocytes in the circulation (**polycythemia**). But the amounts required are so large that the effect probably results from tissue hypoxia. The polycythemic effect of cobalt led to the use of cobalt as a nonspecific erythropoietic stimulant in man. Cobalt salts have been used in the treatment of the anemias of nephritis and infection, and several reports have appeared of hemopoietic response to cobalt, in addition to iron, in children and in pregnant women. The amounts required to elicit these responses are so large (20–30 mg cobalt/day) that serious toxic manifestations, including thyroid hyperplasia, myxedema, and congestive heart failure (CHF) in infants can occur. Hence, *cobalt occupies a very restricted place in the management of human anemias*.

Cobalt has been accused as the precipitating factor in several outbreaks of severe cardiac failure in heavy beer drinkers. Cobalt was suspected because of the high incidence of polycythemia, thyroid epithelial hyperplasia, and colloid depletion noted in the fatalities, in addition to the congestive heart failure (CHF). Cobalt salts had been added to the beer, *a practice abandoned long ago*, to improve its foaming qualities at concentrations of 1.2 to 1.5 ppm cobalt. At such concentrations, the consumption of 2 pints a day would supply approximately 8 mg of cobalt sulfate, an amount well below that which can be taken with impunity by normal persons. In fact, up to 300 mg a day of cobalt salts have been used therapeutically without cardiotoxic effects. It appears that high cobalt and high alcohol intakes are both necessary to induce the distinctive cardiomyopathy plus a third factor which may be low dietary protein, or thiamine deficiency.

MINERAL–MINERAL INTERRELATIONSHIPS

Studies carried out by several workers have shown that certain minerals affect the metabolism of other minerals. The various interactions are as follows:

1. **Copper and Iron:** Copper is essential for the incorporation of iron in hemoglobin.
2. **Calcium and Phosphorus:** The presence of excess of calcium in the diet affects adversely the intestinal absorption of phosphorus and vice versa.
3. **Phosphorus, Magnesium and Iron:** Presence of excessive amounts of phosphates and phytates in the diet interferes with the absorption of magnesium and iron.
4. **Copper and Molybdenum:** The presence of excess of molybdenum prevents the absorption of dietary copper and causes copper deficiency in experimental animals.
5. **Nickel, Iron and Copper:** Signs of nickel deprivation were more severe when dietary iron was low or the signs of moderate iron deficiency were more severe when dietary nickel was deficient. Signs of copper deficiency were more severe in nickel-supplemented than in nickel-deprived rats (**Nielson** *et al*, 1979).

REFERENCES

1. **Carnes WH :** Role of copper in connective tissue metabolism. *Fed. Proc.* **30 :** 995, 1971.
2. **Comar CL, Bronner F :** Mineral Metabolism. Vol. 2, *Academic Press, Inc., New York, 1964.*
3. **Cotzias GC :** Manganese in health and disease. *Physiol. Revs.* **38 :** *503–532, 1958.*
4. **Crane RK :** Na^+-dependent transport in the intestine and other animal tissues. *Fed. Proc.* **24:** *1000, 1965.*
5. **Dahl JL, Hokin LE :** Sodium–potassium adenosine triphosphatase. *Ann. Rev. Biochem.* **43 :** 327, 1974.
6. **Dahl LK :** Salt intake and salt need (2 parts). *N. Engl. J. Med.* **258 :** *1152, 1205, 1958.*
7. **Davies IJT :** The Clinical Significance of Essential Biological Metals. *Heinemann. 1972.*
8. **Deane N, Smith HW :** The distribution of sodium and potassium in man. *J. Clin. Invest.* **31 :** *197, 1952.*
9. **Dollman PR :** Biochemical basis for the manifestations of iron deficiency. *Annu. Rev. Nutr.* **6 :** *13, 1986.*
10. **EI-Youseff M :** Wilson disease. *Mayo. Clin. Proc.* **78 :** *1126–1136*, *2003.*
11. **Evans GW, Dubois RS, Hambidge KM :** Wilson's disease**:** Identification of an abnormal copper-binding protein. *Science*. **181 :** *1175, 1973.*
12. **Frost DV, Lish PM :** Selenium in biology. *Annu. Rev. Pharmacol.* **15 :** *259, 1975.*
13. **Ganther HE :** Selenium metabolism, selenoproteins and mechanisms of cancer prevention: complexities with thioredoxin reductase. *Carcinogenesis*. **20 :** *1657, 1999.*

14. **Irving JT :** Calcium Metabolism. *Metheun & Co. Ltd., London. 1957.*
15. **Jacobs A, Worwood M (editors) :** Iron in Biochemistry and Medicine. *Academic Press, Inc., New York. 1974.*
16. **Mertz W :** Chromium in human nutrition: a review. *J. Nutr.* **123 :** *626, 1993.*
17. **Mertz W, Comatzer WE (editors) :** Newer Trace Elements in Nutrition. *Marcel Dekker, Inc., New York. 1971.*
18. **Monier-Williams GW :** Trace Elements in Food. *John Wiley & Sons, Inc., New York. 1949.*
19. **Munro HN, Linder MC :** Ferritin: Structure, biosynthesis and role in iron metabolism. *Physiol. Rev.* **58 :** *317, 1978.*
20. **Nordian BEC :** Calcium, Phosphorus and Magnesium Metabolism. *Churchill, Livingstone. 1976.*
21. **Prasad AS (editor) :** Trace Elements in Human Health and Disease, Vols. I and II. *Academic Press, Inc., New York, 1976.*
22. **Roholm K :** Fluorine Intoxication. *H.K. Lewis, London. 1973.*
23. **Schneider BD, Leibold EA :** Regulation of mammalian iron homeostasis. *Curr. Opin. Clin. Nutr. Metab. Care.* **3 :** *267–273, 2000.*
24. **Schwarz K :** Recent dietary trace elements research, exemplified by tin, fluorine, and silicon. *Fed. Proc.* **33 :** *1748, 1974.*
25. **Stadtman T :** Selenium biochemistry. *Science,* **183 :** *915, 1974.*
26. **Underwood EJ :** Trace Elements in Human and Animal Nutrition. 4th ed., *Academic Press, Inc., New York. 1977.*
27. **Vallee BL** ***et al*** **:** Zinc metabolism in hepatic dysfunction. *Ann. Intern. Med.* **50 :** *1077, 1959.*
28. **Vincent JB :** The biochemistry of chromium. *J. Nutr.* **130 :** *715, 2000.*

PROBLEMS

1. Absorption of foods is inhibited by the excess of which mineral(s)?
2. Why the excess of dietary protein causes rapid excretion of calcium?
3. Why is the risk of nutritional deficiency high in young children?
4. Unsupplemented diets of populations living in inland areas may normally be deficient in which mineral?
5. What is the essential component of the enzyme, glutathione peroxidase?
6. Serum calcium levels are usually normal in spite of suboptimal calcium absorption and vitamin D deficiency. Serum calcium is being maintained:
   (*a*) by low parathyroid hormone (parathormone, PTH) levels inhibing calcium excretion
   (*b*) by an increase in calcitonin
   (*c*) by increased bone resorption stimulated by elevated PTH
   (*d*) because the kidney reabsorbs from the urine
7. What do minerals do?

ANSWERS TO PROBLEMS

1. Dietary zinc and dietary copper. These two probably share the same protein, *metallothionein*.
2. This is a concern about a diet that is very much higher than required in protein.
3. Rapid growth in children causes high demands for iron.
4. Iodine. The problem would be rare in areas whose population commonly uses iodized salt.
5. Selenium, which is bound to a cysteine residue.
6. (*c*) Parathormone promotes bone resorption and inhibits calcium excretion.
7. Minerals play an important role in blood formation and clotting, regulation of heart rate, nerve and muscle activity, bone formation, and digestion. Macrominerals, like calcium, are needed in large amounts. Trace minerals, like boron, are needed in small amounts.

PART

IV

Hormone Biochemistry

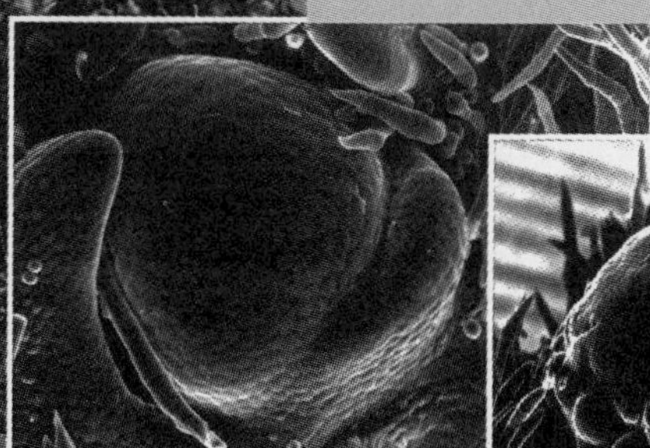

In a flowering apex, sepals, petals, stamens and carpels are produced instead of leaves. The arrangement of these is specific to each species and is determined by hormones and developmental genes.

(*Left*) The weed cocklebur (*Xanthium strumarium*)

(*Middle*) A highly magnified shoot tip showing bumplike "initials"

(*Right*) Shoot tip showing flower production

"Life– human life included– is the outcome of an elaborate organization based on trivial ingredients and ordinary forces"

– *G.E. Palade : Nat. Acad. of Sciences, Proceedings, 1964*

CONTENTS

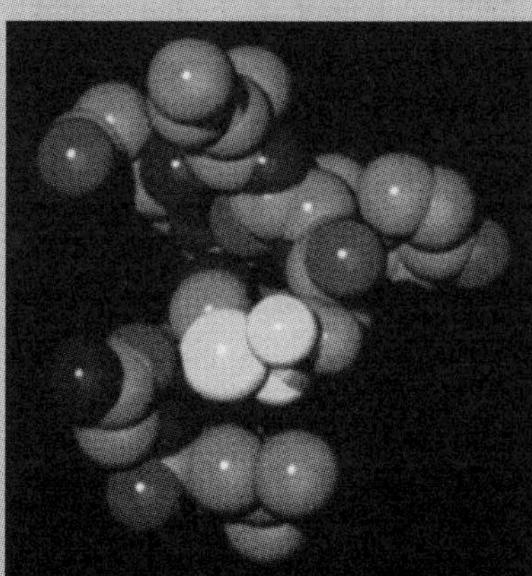

Space-filling model of ocytocin
This hormone stimulates the contraction of smooth muscles, especially those of uterus, thus facilitating childbirth.

CHAPTER

34

Animal Hormones

DEFINITION

The term hormone (*hormao*[G] = to excite) was first used by **William M. Bayliss** and his brother-in-law **Ernest H. Starling**, both of London University College, in 1904, who showed that a chemical substance (*secretin*) from the intestine could stimulate the action of a pancreatic secretion. These substances were then called as 'chemical messengers'. **Went** and **Thimann** (1937) defined a hormone as *"a substance which, produced in any one part of an organism, is transferred to another part and there influences a specific physiological process."* The tissues or organs where they are produced are called as **effectors** and those where they exert their influence as **targets.** These have low molecular weight and diffuse readily. As they are readily oxidized, their effects do not remain permanent unless these are supplied continuously.

Based on their site on action, the hormones are of two types: local and general. The *local hormones,* obviously, have specific local effects, whence their nomenclature. These may be exemplified by acetylcholine, secretin, cholecystokinin etc. The *general hormones*, on the other hand, are secreted by specific endocrine glands and are transported in the blood to cause physiologic actions at points remote from their place of origin. A few of the general hormones affect almost all cells of the body, *e.g.*, growth hormones (GH) and thyroid hormones ; whereas other general hormones, however, affect specific tissues far more

SIR WILLIAM MADDOCK BAYLISS

(LT, 1860–1924)

Bayliss, an English physiologist of London University, was born on May 2, 1860 in Wolverhampton, Staffordshire. He gained a B.Sc. from London University. He did MA and DSc in physiology from Wadham College, Oxford. In 1893, Bayliss married **Gertrude Ellen Starling**, the sister of his collaborator **Ernest H. Starling**.

Bayliss, in collaboration with his brother-in-law **E.H. Starling,** introduced the term *hormone* to describe substances that, when secreted from one part of the body, have a specific effect on another. In 1902, they discovered *secretin*, one of the first hormones to be identified. Later (in 1904), they showed that secretin could stimulate the action of a pancreatic secretion. These substances were then called as 'chemical messengers'. Bayliss also introduced the *saline injection* for cases of surgical shock during World War I (1914–1918). Both Bayliss and Starling discovered **peristalsis** of the intestine. The **Bayliss effect** is named after him. He was also involved in the **Brown Dog affair**, successfully suing **Stephen Coleridge** for libel over accusations he made about Bayliss's vivisection work.

Bayliss was elected a **Fellow of the Royal Society** (F.R.S.) in 1903. He jointly delivered their Croonian lecture in 1904 and was awarded their Royal Medal in 1911 and their Copley Medal in 1919. He was knighted, for his contributions to medicine, in 1922. Bayliss died in London on August 27, 1924.

The **Bayliss and Starling Society** was founded in 1979 as a forum for scientists with research interests in central and autonomic peptide function. Both Bayliss and Starling are called 'founders of endocrinology'.

ERNEST HENRY STARLING

(LT, 1866–1927)

Starling, an English physiologist of London University, was born on April 17, 1866. He worked mainly at University College, London, although he also worked for many years in Germany and France. In 1891, Starling married **Florence Amelia Woolridge**, the widow of **Leonard Charles Woolridge**, who had been his physiology teacher at Guy's and died at the age of 32. She was a great support to Starling as a sounding board, secretary, and manager of his affairs as well as mother of their 4 children. His main collaborator in London was his brother-in-law, **Sir William M. Bayliss**.

Starling, in collaboration with **Sir W.M. Bayliss**, introduced the term *hormone* to describe substances that, when secreted from one part of the body, have a specific effect on another. In 1902, they discovered *secretin*, one of the first hormones to be identified. Later (in 1904), they showed that secretin could stimulate the action of a pancreatic secretion. These substances were then called as 'chemical messengers'. Starling also did research into the heart, demonstrating the *Starling equilibrium* in 1896– the balance between hydrostatic pressure, causing fluids to flow out of the capillary membrane, and osmotic pressure, causing the fluids to be absorbed from the tissues into the capillary. Both Starling and Bayliss discovered **peristalsis** of the intestine. He is also known for his involvement, along with Bayliss, in the **Brown Dog affair**, a controversy relating to vivisection. Starling is most famous for developing the **"Frank–Starling law of the heart",** presented in 1915 and modified in 1919.

Starling was elected **Fellow of the Royal Society** (F.R.S.) in 1899. He died an May 2, 1927 during a sea voyage to the Carribean, and was buried in Kingston, Jamaica two days later.

The **Bayliss and Starling Society** was founded in 1979 as a forum for scientists with research interests in central and autonomic peptide function. Both Bayliss and Starling are called 'founders of endocrinology'.

than other tissues, *e.g.*, adrenocorticotropin (a hormone secreted from adenohyprophysis and stimulating the adrenal cortex) and ovarian hormones (affecting the uterine endometrium).

Non-specificity and Cross Effects: Hormones are not sepcific for the organisms in which they are produced and may influence bodily processes in other individuals also. Adrenalin, for example, also influences protozoans and crustaceans besides man and other vertebrates. Their cross effects have also been found between plants and animals. For example, auxin, a growth-promoting plant hormone, can also stimulate a protozoan, *Euglena viridis*. Likewise, some animal hormones also stimulate growth in root tips of many plants after decapping.

Differences with Nutrient Materials and Enzymes: A distinction between hormones and **nutrient materials** may be made, as the former are utilized in extremely minute quantities in comparison to the latter. For example, a 5×10^{-10} molar solution of ocytocin (about 1 mg in 2,000 litres of solution) will cause the contraction of uterine muscles.

They also differ from **enzymes**, which are essential to initiate and continue reactions, in that the hormones cannot initiate reactions but can influence the rate at which they proceed. Furthermore, the hormones are usually consumed in the process of growth, whereas the enzymes are not.

GENERAL FUNCTIONS

The hormones conduct a wide variety of functions ranging from growth, vegetative and sexual development, cellular oxidation to thermal production and the metabolism of carbohydrates, proteins and fats. The various functions performed by hormones may, in general, be discussed under following heads :

1. Regulatory or homeostatic function. The hormones have regulatory effects on the composition of the body fluids, the rate of gaseous exchange and the activity of the vascular system and the central nervous system (CNS). There always exists a high degree of precision and constancy in the composition of the body fluids in a normal individual for the conduction of various activities. Such an environment within the cell has been termed the *millieu intérieur* (or internal enivirnoment) in 1857 by **Claude Bernard**, a French physician. Throughout his research, he had been impressed by the way in which organisms were able to regulate physiological parameters, such as body temperature and water content, and maintain them within fairly narrow ranges. This concept of self-regulation leading to physiological stability was summed up by Bernard in the now classic statement, '*La fixité du milieu intérieur est la condition de la vie libre.*' (The constancy of the internal environment is the condition of the free life). In other words, it means that for an organism to function optimally, its component cells must be surrounded by a medium of closely-regulated composition. Bernard went on to distinguish between the *external environment* in which organisms live and the *internal environment* in which the individual cells live (in mammals, this is tissue, or interstitial, fluid). He realized the importance of conditions in the latter being continuously stable. He concluded that an organism is the sum of its constituent cells and the optimum functioning of the whole depends upon the optimum functioning of its parts.

For performing various metabolic functions, an organism should maintain a normal, constant internal state or **homeostasis** (*homoios*^{G} = same or similar ; *stasis*^{G} = state or standing), a term coined by an American physiologist, **Walter B. Cannon** in 1932. Homeostasis can be defined as the tendency to maintain uniformity or stability in the internal environment of the organism and to maintain the normal composition of the body fluids. In other words, homoeostasis is the maintenance of a constant internal environment in the face of changes in the external environment. Hormones play an important and decisive role in homeostatic regulation of internal mileu.

2. Permissive function. Not only does each endocrine gland affect a number of processes, but these glands also affect the functioning of one another. Thus certain hormones require the presence (or 'permission') of another hormone for the expression of their activity. This helps in maintaining a perfect hormonal balance. Derangements of this balance, either clinical or experimental, lead to a variety of metabolic aberrations.

CLAUDE BERNARD
(LT, 1813–1878)

Claude Bernard, a French physiologist, is often regardeed as the **father of experimental medicine**. He had parents who were vineyard workers. From his early experiences, Bernard retained an enthusiasm for the Beaujolais region of his native country and for the countryside in general, throughout a working life that kept him in a Paris laboratory. He trained as an apothecary, and later as a doctor, but was no more than an average student. He never worked as a medical practitioner, but did make his mark in experimental medicine.

Bernard's work centred on the **physiology of mammals**. Working with dogs and rabbits he established that:

(*a*) an enzyme is present in the gastric juice.
(*b*) the digestion of dietary carbohydrates to sugar occurs prior to their absorption.
(*c*) the digestion of fat involves bile and pancreatic juices.
(*d*) glycogen and sugar are interconverted in the liver.
(*e*) respiration produces heat in all body tissues.

In 1851, Bernard reported the existence of the vasomotor nerves. Following the work of a US army physician **William Beaumont** (LT, 1785–1853), he opened artificial fistules in animals to study their digestive systems. Bernard was a 'modern' researcher, combining experimental skills with an appreciation of the theory behind his work. He believed in the **constancy of the *milieu intérieur*** ("internal environment"), which is the extracellular fluid bathing the cells. He pointed out that it is through the *milieu intérieur* that foods and wastes and gases are exchanged and through which chemical messengers are distributed. He wrote :

"The living organism does not really exist in the external environment (the outside air or water) but in the liquid *milieu intérieur*... that bathes the tissue elements."

He is best remembered for his idea that animal life is dependent upon a constant internal environment– that cells function best in a narrow range of conditions of solute potential and temperature, and when bathed in a constant concentration of chemical constituents. Truly , Bernard was one of the most influential of nineteenth century physiologists.

His famous quote on Science reads as follows:

"Observation is a passive science, experimentation an active science."

JAMES EFRAIM LOVELOCK
(Born, 1919)

Lovelock, an independent English chemist, inventor, environmentalist and futurologist, was born on July 26, 1919 in Letchworth Garden City in Hertfordshire, England to working-class parents who were strong believers in education. **Nell**, his mother, started work at 13 in a pickle factory. His father, **Tom**, had served 6 months hard labour for poaching in his teens and was illiterate until attending technical college. The family moved to London where his dislike of authority made him, by his own account, an unhappy pupil at Strand School. Lovelock could not afford to go to university after school, something which he believes helped prevent him becoming over-specialized and aided the development of Gaia theory. He worked at a photography firm, attending Birbeck College during the evenings, before being accepted for chemistry at the University of Manchester although he could only pay for 2 years of the 3-year course. Lovelock worked at a Quaker farm before a recommendation from his professor led to him taking up a Medical Research Council post working on ways of shielding soldiers from burns. Lovelock refused to use the shaved and anesthetized rabbits that were used as burn victims, and *exposed his own skin to heat radiation instead*, an experience he describes as "exquisitely painful". His student status enabled temporary deferment of military service during WW II, but he registered as a conscientious objector. He later abandoned this position in the light of Nazi atrocities and tried to enlist for war service, but was told that his medical research was too valuable for this to be considered. In 1948,

Lovelock received a Ph.D. degree in medicine at the London School of Hygiene and Tropical Medicine. Within the United States, he has conducted research at Yale, Baylor College of Medicine, and Harvard University. Lovelock works out of a barn-turned -laboratory on the Devon/Cornwall border.

A lifelong inventor, Lovelock had invented an ingenious device that detected tiny amounts of chemicals. His device was widely used to measure trace amounts of pollutants, such as the insecticide DDT. He has also created and developed many scientific instruments, some of which were designed for NASA in its programme of planetary exploration. It was while working as a consultant for NASA that Lovelock developed the Gaia hypothesis, for which he is widely known.

The **Gaia hypothesis** proposes that the living and nonliving parts of the Earth form a complex interacting system that can be thought of as a single organism. The hypothesis postulates that the biosphere has a regulatory effect on the Earth's environment that acts to sustain life.

While the Gaia hypothesis was readily accepted by many in the environmentalist community, it has not been widely accepted within the scientific community. Among its more famous critics are the evolutionary biologists **Richard Dawkins**, **Ford Doolittle**, and **Stephen Jay Gould**–notable, given the diversity of this trio's views on other scientific matters. These (and other) critics have questioned how natural selection operating on individual organisms can lead to the evolution of planetary-scale homeostasis. Lovelock has responded to these criticisms with models such as Daisyworld, that illustrate how individual-level effects can translate to planetary homeostasis, under the right circumstances.

In the early 1960s, NASA (National Aeronautics and Space Administration) invited Lovelock to the Jet Propulsion Laboratory, California, to help design instruments for detecting life on Mars (In fact, the space agency was especially keen to examine Martian soil for signs of life or conditions that could support life. But analyzing the Martian soil would require new instruments specially designed for the project). Lovelock's fascination with the search for extraterrestrial life and his insatiable curiosity soon led him beyond this technical chore to reconsider the much greater question of life on Mars. What if Martian or other life did not look or act anything like terrestrial life? How could life be described so that scientists could recognize it even in an alien form.

Lovelock approached these questions by considering the concept of entropy. *Entropy is a measure of disorder in systems.* A *system* is any collection of interacting parts or objects. If any system– whether a solar system or a dormitory– is left alone, its disorder (or entropy) will increase. Organisms, by contrast, create order and thus *decrease* local entropy. Plants, for example, convert CO_2 and water into highly structured leaves, stems and flowers. And animals convert amino acids and simple fats into muscle and brain tissue. This decrease in local entropy exacts a cost, however. It requires energy from the sun. And the decrease in entropy is only local. The universe as a whole is running down; its entropy is increasing.

A sure sign of life on Mars, then, Lovelock realized, would be a reduction in local entropy. But what in the world, he wondered, does a "reduction in entropy" look like? He also wondered how an alien being would recognize life on Earth. And it dawned on him that an alien would notice the peculiar chemical composition of Earth's atmosphere. Earth's atmosphere is loaded with unlikely combinations of chemicals. For example, the atmosphere contains both methane and oxygen. Methane is a simple C-1 hydrocarbon made by bacteria that gives swamp gas its smell. Methane and oxygen readily react with one another to form carbon dioxide and water.

$$CH_4 + 2\,O_2 \longrightarrow CO_2 + 2\,H_2O$$

This reaction constantly destroys methane. If no living organisms were present, methane would soon disappear from Earth's atmosphere. But bacteria in marshes, rice paddies, and cow's stomachs constantly make more methane, so the gas persists.

Lovelock realized that the Earth's chemically unlikely atmosphere could be seen as an extension of our planet's biosphere– the system of living things that covers the Earth. Just as a tortoise's shell is part of its body, the Earth's atmosphere is part of the biosphere. Lovelock envisioned the atmosphere, the oceans and soils, as constituting a single, immense living individual. He further hypothesized that this immense organism acts to preserve itself. Lovelock, at the suggestion of novelist **William Golding**, named this entity *Gaia,* after the Greek Earth goddess. He reasoned that the biosphere regulates the temperature and chemistry of the atmosphere, just as many animals regulate their blood temperature and chemistry. Such interrelation is called *homeostasis.*

While the biologists found the idea amusing, most also found it untenable. Despite Lovelock's impressive credentials as an inventor, scientists viewed him as a crank. Evolutionary biologists dismissed Gaia, arguing that the biosphere cannot be considered as an organism because it neither reproduces nor evolves. Lovelock now denies that he insisted the Gaia to be understood as an organism. He says Gaia is more like a *superorganism*, a group of individuals that cooperate almost like various parts of the body.

However, in 1974, **Lynn Margulis**, a distinguished professor of biology at the Amherst campus of the University of Massachusetts, began enthusiastically promoting a version of the Gaia concept– to the horror of her scientific colleagues. According to her, Gaia is not a superorganism but a tightly integrated giant ecosystem, the highest level of organization of life. Margulis also proposed a related idea on the cellular level that is now widely accepted. She proposed that the mitochondria and chloroplasts of eukaryotic cells were once free-living bacteria that came to live permanently inside other prokaryotic bacteria. In other words, we eukaryotes are a combination prokaryotes that began cooperating millions of years ago. Such a cooperative arrangement is an example of *symbiosis,* any association between two organisms. Margulis asserts that the different organisms of our biosphere cooperate in similar ways to create a functioning ecosystem. The view that organisms cooperate as well as compete is fundamental to the view that our biosphere is a tightly-integrated whole.

Lovelock invented the **electron capture detector**, which ultimately assisted in discoveries about the persistence of CFCs (chlorofluorocarbons) and their role in stratospheric ozone depletion. After studying the operation of the Earth's sulfur cycle, Lovelock and his colleagues developed the CLAW hypothesis as a possible example of biological control of the Earth's climate.

After the development of his electron capture detector, in the late 1960s, Lovelock was the first to detect the widespread **presence of CFCs in the atmosphere**. He found a concentration of 60 parts per trillion (ppt) of CFC-11 over Ireland and, in a partially self-funded expedition in 1972, went on to measure the concentration of CFC-11 from the Northern Hemisphere to the Antarctic aboard the research vessel RRS Shackleton. He found the gas in each of the 50 air samples that he collected but, not realising that the breakdown of CFCs in the stratosphere would release chlorine that posed a threat to the ozone layer, concluded that the level of CFCs constituted "no conceivable hazard". He has since stated that he meant "no conceivable toxic hazard".

However, the experiment did provide the first useful data on the ubiquitous presence of CFCs in the atmosphere. The damage caused to the ozone layer by the photolysis of CFCs was later discovered by **Sherwood Rowland** and **Mario Molina**. After hearing a lecture on the subject of Lovelock's results, they embarked on research that resulted in the first published paper that suggested a link between stratospheric CFCs and ozone depletion in 1974, and later shared the 1995 Nobel Prize in Chemistry (with **Paul Crutzen**) for their work.

Lovelock has become concerned about the threat of **global warming** from the greenhouse effect. In 2004, he caused a media sensation when he broke with many fellow environmentalists by pronouncing that "only nuclear power can now halt global warming". In his view, nuclear energy is the only realistic alternative to fossil fuels that has the capacity to both fulfill the large-scale energy needs of humankind while also reducing greenhouse emissions. He is an open member of Environmentalists for Nuclear Energy. In 2005, against the backdrop of renewed UK government interest in nuclear power, Lovelock again publicly announced his support for nuclear energy, stating, "I am a Green, and I entreat my friends in the movement to drop their wrong-headed objection to nuclear energy". Although these interventions in the public debate on nuclear power are recent, his views on it are longstanding. In this 1988-book *The Ages of Gaia*, he states:

> "I have never regarded nuclear radiation or nuclear power as anything other than a normal and inevitable part of the environment. Our prokaryotic forbears evolved on a planet-sized lump of fallout from a star-sized nuclear explosion, a supernova that synthesized the elements that go to make our planet and ourselves".

In his 2006-book *The Revenge of Gaia*, Lovelock argues that the lack of respect humans have had for Gaia, through the damage done to rainforests and the reduction in planetary biodiversity, is testing Gaia's capacity to minimize the effects of the addition of greenhouse gases in the atmosphere. This eliminates the planet's negative feedbacks and increases the likelihood of homeostatic positive feedback potential associated with runaway global warming. Similarly, the warming of the oceans is extending the oceanic thermocline layer of tropical oceans into the Arctic and Antarctic waters, preventing the rise of oceanic nutrients into the surface waters and eliminating the algal blooms of phytoplankton on which oceanic food chains depend. As phytoplankton and forests are the main ways in which Gaia draws down greenhouse gases, particularly CO_2, taking it out of the atmosphere, the elimination of this environmental buffering will see, according to Lovelock, most of the Earth becoming uninhabitable for humans and other life-forms by the middle of 21st century, with a massive extension of tropical deserts.

In his most recent book *The Vanishing Face of Gaia,* he rejects scientific modeling that disagrees with the scientific findings that sea levels are rising faster, and Arctic ice is melting faster, than the models predict and he suggests that we may already be beyond the tipping point of terrestrial climate into a permanently hot state. Given these conditions, Lovelock expects human civilization will be hard-pressed to survive. He expects the change to be similar to the *Paleocene–Eocene Thermal Maximum* when atmospheric concentration of CO_2 was 450 ppm. At that point, the Arctic Ocean was 23°C and had crocodiles in it, with the rest of the world mostly scrub and desert.

Writing in the British newspaper *The Independent* in January 2006, Lovelock argues that, as a result of global warming, "billions of us will die and the few breeding pairs of people that survive will be in Arctic where the climate remains tolerable" by the end of the 21st century. He has been quoted in *The Guardian* that 80% of human will perish by 2100 AD, and this climate change will last 1,00,000 years. According to Lovelock, by 2040, the world population of more than 6 billion will have been culled by floods, drought and famine. Indeed "[t]he people of Southern Europe, as well as South-east Asia, will be fighting their way into countries such as Canada, Australia and Britain". ... "By 2040, parts of the Sahara desert will have moved into the middle Europe. We are talking about Paris–as far north as Berlin. In Britain we will escape because of our oceanic position".

He asserts that if you take the Intergovernmental Panel on Climate Change (IPCC) predictions, then by 2040 every summer in Europe will be as hot as it was in 2003– between 110°F and 113°F. It is not the death of people that is the main problem, it is the fact that the plants cannot grow– there will be almost no food grown in Europe. He further predicts that the average temperature in temperate regions will increase by as much as 8°C and by up to 5°C in the tropics, leaving much of the world's land uninhabitable and unsuitable for farming, with northernly migrations and new cities created in the Arctic. He foresees much of Europe will become uninhabitable having turned to desert and Britain will become Europe's "life-raft" due to its stable temperature caused by being surrounded by the ocean.

Lovelock partly retreated from this position in a September 2007 address to the World Nuclear Association's Annual Symposium, suggesting that climate change would stabilize and prove survivable, and that the Earth itself is in "no danger" because it would stabilize in a new state. Life, however, might be forced to migrate *en masse* to remain in habitable climes.

In a March 2010 interview with *The Guardian* newspaper, he said that even the best democracies agree that when a major war approaches, democracy must be put on hold for the time being. And the climate change, in his opinion, may be an issue as severe as a war.

3. **Integrative function.** The integrative function of the hormones is reflected in the fact that they support the role of nervous system. However, the integrative properties of the endocrine system are slow and steady whereas those of the nervous system are rapid. This close tie between the two systems has led to the emergence of a new discipline of science called *neuroendocrinology.*

WALTER BRADFORD CANNON
(LT, 1871–1945)

Cannon, an American scientist, was George Higginson Professor of Physiology, Emeritus at Harvard University Medical School from 1906 till 1942. He introduced the first radio-opaque agent, barium (later, he used barium salts) into the gut of a living animal to create a 'shadow' of the alimentary tract so that the mechanics of digestion could be observed. He devised this technique first in the year 1896, the year after **Roentgen** discovered x-rays. The technique has been used in diagnostic radiography ever since. He developed the concept originated by French physiologist **Claude Bernard.** Developed out of his studies of the nervous system and reactions to stress, he described the ceaseless balancing and rebalancing of physiological processes that maintain stability and restore the normal state when it has been disturbed. And he gave it the name, **homeostasis.** He spent his life in the study of homeostatic mechanisms. The term soon flooded the medical literature of 1930s. Physicians spoke of getting their patients back into homeostasis. Even politicians and sociologists saw what they considered to be deep nonphysiological implications. Cannon enjoyed this broadened application of the concept and later suggested that democracy was the form of government that took the homeostatic middle course. Despite the enduring importance of the homeostasis concept, Cannon never received the Nobel Prize – *one of several acknowledged oversights of the Nobel Committee.* In his terminal year (*i.e.*, 1945), Cannon expressed his ideas about scientific research in his autobiography, *The way of an Investigator.* This absorbing book describes the resourceful career of a homespun man whose life embodied the traits that favour successful research.

(Photo courtesy: Fulton JF and Wilson LG, 1966)

4. Morphogenetic function. The hormones govern the ontogenetic development of an individual from the embryonic to the adult state.

An account of **animal hormones** is given below.

INVERTEBRATE HORMONES

Among invertebrates, evidence of hormone control is less satisfactory in the acoelomate and pseudocoelomate animals. However, studies of endocrine function of the coelomate vertebrates have been mainly confined to the arthropods, annelids, molluscs and echinoderms; the arthropods (*esp.*, the insects and crustaceans) are the best-studied group.

1. Hormones from Coelenterata

According to **Burnett** and **Diehl** (1964), the hyopostome of an adult *Hydra* possesses neurosecretory cells in abundance. This is the region of new growth. Buds fail to develop if detached from the parent body prior to the secretion of neurosecretory cells in the hypostome. With the onset of sexual maturity, the neurosecretory cells disappear. Thus, growth and reproduction are antagonistic processes. The *neurosecretory growth hormone*, as it may be called, is credited with the following two functions :

(*a*) It activates cell proliferation.

(*b*) It causes interstitial cells to develop into somatic structures such as nematocysts.

In the absence of this hormone, growth ceases and interstitial cells form gametes.

2. Hormones from Annelida

Investigations conducted during the past four decades make it clear that neurosecretions play an important role in the regulation of various processes in annelids including seasonal swarming of distinct sexual forms, hermaphroditism and viviparity. In the polychaete worm *Nereis*, many neurosecretory cells have been localized in the brain and other ganglia. These cells govern diverse processes such as growth, metamorphosis, sexual development and reproductive behaviour.

The studies regarding regeneration of *Nereis diversicolor* have been conducted by **Clark** *et at* in 1962. The worm grows by adding new segments until it has produced about 50 segments; thereafter growth is mostly due to increase in segment size up to 90 segments. Segment proliferation then ensues which is regulated by hormones synthesized in the supra-oesophageal ganglia. Removal of these ganglia in the young ones ceases segment proliferation, which can be resumed after implanting ganglia from other young worms. Besides, old worms, which have stopped growing, can be induced to grow by implanting ganglia from young worms and not from other old worms. Amputation of a part of the worm (with less than 60 segments) activates neurosecretory cells to elaborate hormones which provoke segment proliferation. It is, however, not known whether the growth-promoting and the regeneration-promoting hormones are identical (**Jenkin**, 1970)

3. Hormones from Arthropoda

Among invertebrates, insects and crustaceans furnish excellent examples of hormone secretion. These hormones govern many metabolic processes including growth, development, reproduction, colour adaptation etc.

A. Hormones from Insecta. In insects, the growth after emergence from the egg is characterized by a process of metamorphosis during which larva passes to the adult phase via pupal stage. During these changes, the insect undergoes a process of moulting or ecdysis (*ekdysis*[G] = a getting out). The processes of moulting and metamorphosis are controlled by hormones secreted by the following three organs of endocrine nature (refer Fig. 34–1).

(*a*) neurosecretory cells of corpus cardiacum

(*b*) prothoracic gland

(*c*) corpus allatum.

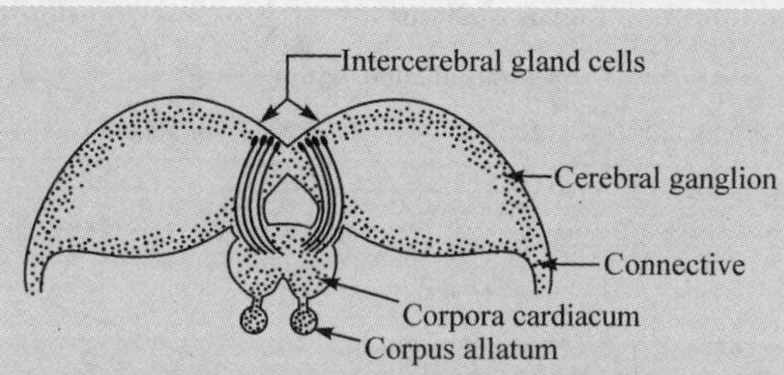

Fig. 34–1. Some of the endocrine glands of an insect

(*After Bodenstein*)

The neurosensory cells of the corpus cardiacum produce a secretion which stimulates ecdysial or prothoracic gland to secrete a hormone named **ecdysone** or **prothoracic gland hormone, PGH** (Fig. 34–2). This induces moulting. Based on its biological function, this hormone has been variously called as moulting hormone, pupation hormone or metamorphosis hormone. Chemically, it is a steroid.

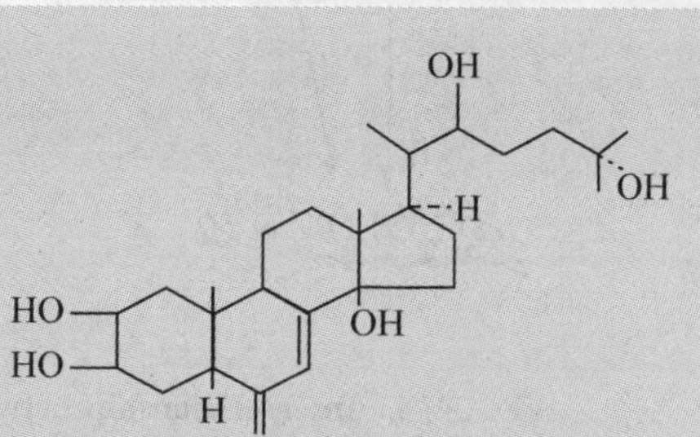

Fig. 34–2. α-ecdysone or moulting hormone, MH

In the larva, corpus allatum secretes another hormone called **status quo hormone (SQH)** or **neotenin** (Fig. 34–3). This delays metamorphosis of immature insects by maintaining the juvenile (= larval) character of the growing insect for a longer period. Henceforth, this hormone is variously called as **juvenile hormone, larval hormone** or **inhibitory hormone**. It was isolated in pure form from the giant silkworm moth, *Hyalophora cecropia* by **Röller** *et al* in 1965. It is a derivative of farnesoic acid, a compound that is related to intermediates in steroid biosynthesis. Juvenile hormone may be regarded as an acyclic sesquiterpenoid-type of compound. SQH remains active only in the presence of ecdysone. In later stages, the activity of SQH decreases so that the larva leads to adult stage by ecdysone activity. Besides SQH, the corpus allatum also produces other secretions which control colour changes and reproduction.

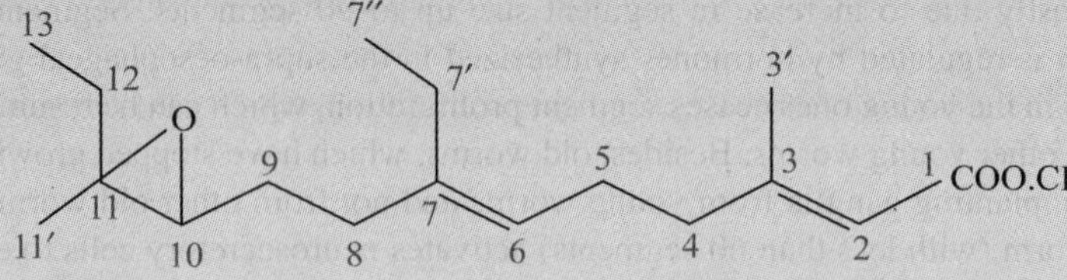

Fig. 34–3. Status quo hormone, SQH or juvenile hormone, JH

(methyl-10-epoxy-7-ethyl,-3, 11-dimethyltrideca-2, 6-dienoate)

The endocrine as well as neural control of growth and moulting in a moth has been diagramatically represented in Fig. 34–4.

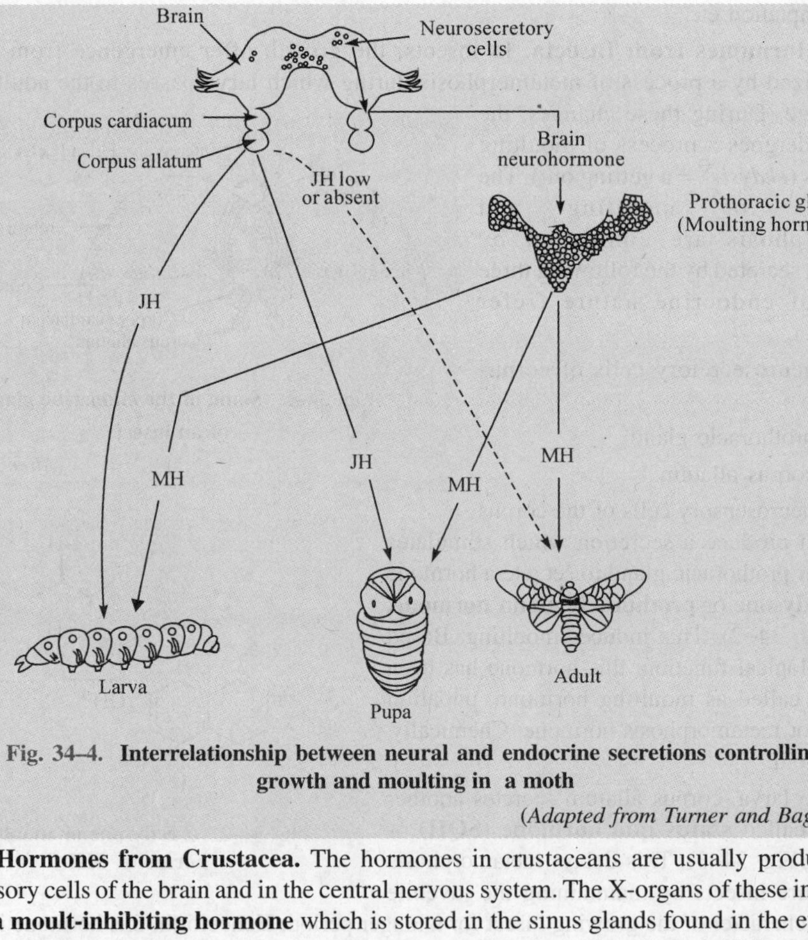

Fig. 34–4. Interrelationship between neural and endocrine secretions controlling growth and moulting in a moth

(Adapted from Turner and Bagnara, 1971)

B. Hormones from Crustacea. The hormones in crustaceans are usually produced in the neurosensory cells of the brain and in the central nervous system. The X-organs of these invertebrates produce a **moult-inhibiting hormone** which is stored in the sinus glands found in the eye stalks or in the head in species without eye stalks. This hormone has been thought to inhibit moulting as in the absence of the X-organs (or the sinus glands), the crustaceans moult more frequently than the normal individuals. The X-organs also secrete another hormone, **moult-promoting hormone** which stimulates the moulting gland or Y-organ.

The chromatophores in crustaceans contain pigments concerned with the body colour. The relative dispersion and concentration of these pigments is controlled by **colour-change hormones.** Two types of colour-change hormones may be recognized:

(a) *lightening hormones*– induce concentration of the dark pigments (red, black), thereby bleaching the body colour of the animal.

(b) *darkening hormones*– cause dispersal of these pigments within the chromotophores.

The crustaceans can, thus, change their body colour in harmony with their environment.

4. Hormones from Mollusca

In the cephalopod *Octopus*, which is one of the greatest evolutionary achievements, the brain (subpeduncular lobes) controls a pair of optic glands through the action of inhibitory nerves. The optic glands are located on the optic stalks, one on either side of the central part of the supraesophageal brain (see Fig. 34–5). The brain, in turn, receives environmental hints through the eyes. The optic

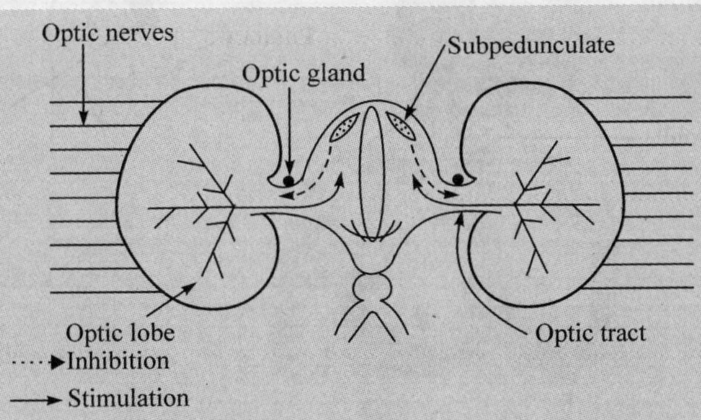

Fig. 34–5. Position of the optic glands in *Octopus*

glands dominate the reproductive endocrinology in a manner somewhat comparable to the vertebrate pituitary. The secretions from the optic gland promote vitellogenesis in females and spermatogenesis in males. They also govern the differentiation of the female reproductive tract. As a contrast, the male sex differentiation is under direct control of testicular hormones (Wells and Wells, 1969). As a corollary, *the endocrine controls in Octopus parallel the vertebrate pattern* but are much simpler in that the neurosecretions and ovarian hormones seem to be absent.

Among gastropods, *Lymnaea stagnalis* has been most carefully studied. Here the well-established endocrine organs are the dorsal bodies. There are small discrete bodies on the cerebral ganglia and develop from glial cells and the perineurium of the cerebral ganglia. Their secretions stimulate vitellogenesis and the differentiation of the female genital tract. The functions carried out by the dorsal bodies are more diverse than those conducted by the cephalopod optic glands (**Bonga**, 1972).

5. Hormones from Echinodermata

According to **Chaet** (1967), in the starfishes, a simple system of hormones regulates sexual maturation and spawning. Two such hormones recognized are:

(a) *gonad-stimulating substance* (*GSS*)– It is a low molecular weight (2,000) protein which is synthesized in the radial nerves.

(b) *maturation-inducing substance* (*MIS*)– It is 1-methyl adenine and is produced by the ovarian follicles.

It is well known that spawning reactions of starfishes are synchronized and that the presence of eggs or sperms in the water stimulates other individuals to spawn. The endocrinological explanation of this mass spawning is simple. A neurosecretion released by the radial nerves acts on the follicle cells to induce the synthesis of MIS, which triggers ovulation, release of gametes and reproductive behaviour.

VERTEBRATE HORMONES

Most glands of the body release their secretions by means of ducts. These are called as the **exocrine** (*exos*[G] = outside ; *krinein*[G] = to separate) **glands** or **duct glands**, *e.g.*, salivary, sebaceous, sweat glands etc. Other glands in the body, on the contrary, produce chemical substances which are

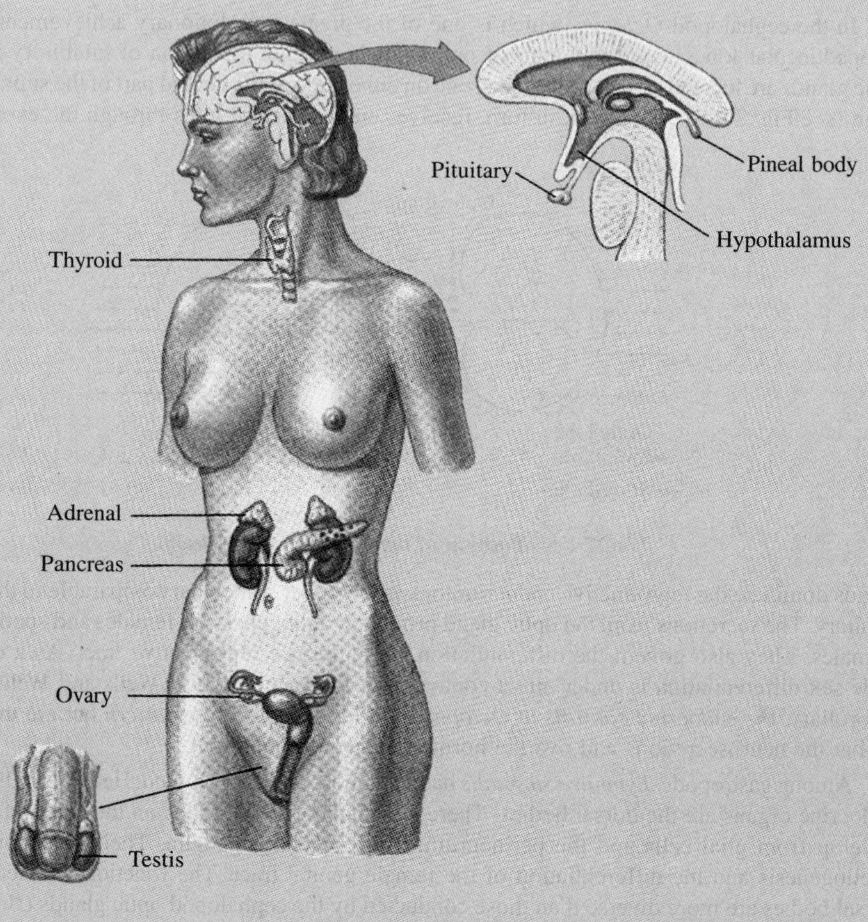

Fig. 34–6. Location of the major endocrine glands in human body

The hypothalamus regulates the anterior pituitary, which regulates the hormonal secretions of the thyroid, adrenals, and gonads (ovary in the female and testis in the male).

(*After Lehninger AL, 1984*)

released directly into the nearby blood and lymph vessels which carry them to various organs (target or effector organs) where they exert their characteristic effect. These glands have been termed as the **endocrine** (*endon*[G] = within) **glands** or **ductless glands** and their secretions are called as the **hormones**. The science of the study of structure and function of these endocrine glands and their secretions (*i.e.*, hormones) is designated as **endocrinology**, a term introduced by **Pende**.

A noteworthy point about the endocrines is their similarity, except in minor details, in all the vertebrates. They are highly vascular as the blood not only acts as a vehicle for transport of their

VERTEBRATE HORMONES

STEROID HORMONES

C_{18} STEROIDS

1. **Ovarian Hormones**
   β-estradiol
   Estriol
   Estrone

C_{19} STEROIDS

2. **Testicular Hormones**
   From testes
   Testosterone
   Androsterone
   Dehydroepiandrosterone

   From adrenal gland
   Androst-4-ene-3,17,dione
   Androst-4-ene-3,11,17-trione

C_{21} STEROIDS

3. **Adrenal Cortical Hormones**
   Mineralocorticoids
   Aldosterone
   Deoxycorticosterone

   Glucocorticoids
   Cortisone
   Cortisol
   Corticosterone

4. **Corpus Luteal Hormone**
   Progesterone

PEPTIDE HORMONES

1. **Hormones of the Pancreas**
   Insulin
   Glucagon

2. **Hormones of the Hypophysis**
   Pars distalis
   Thyrotropin, TSH
   Corticotropin, ACTH
   Gonadotropins, GTH
   FSH
   LH
   LTH
   Somatotropin, SH

   Pars intermedia
   Intermedins, MSH
   α-MSH
   β-MSH

   Pars nervosa
   Ocytocin or pitocin
   Vasopressin or pitressin

3. **Hormone of the Parathyroid**
   Parathormone, PTH

4. **Hormones of the Gastrointestinal Tract**
   Gastrin
   Secretin
   Cholecystokinin
   Pancreozymin
   Enterogastrone
   Enterokrinin
   Hepatocrinin
   Duicrinin
   Villikinin
   Parotin

5. **Hormone of the Corpus Luteum**
   Relaxin

AMINO ACID HORMONES

1. **Thyroidal Hormones**
   Triiodothyronine, T_3
   Tetraiodothyronine T_4

2. **Adrenal Medullary Hormones**
   Adrenalin
   Noradrenalin

secretions but also serves as a source of the chemical raw materials from which these hormones are synthesized. Ontogenetically, all these glands are derived from the epithelial tissues. The amount of hormones produced by the endocrines in 24 hours measures fractions of a milligram. The principal endocrine glands (refer Fig. 34–6) include:

(*a*) two in the head region— the *pineal* and the *pituitary,*

(*b*) three in the neck region— the *thymus,* the *thyroid* and the *parathyroid,* and

(*c*) four in the abdominal region— the *pancreas,* the *gastrointestinal mucosa,* the *adrenals* and the *gonads.*

Barring the parathyroid, all the remaining 8 glands are found in all the vertebrates. There is, however, controversy regarding the endocrinal nature of the two glands, the pineal and the thymus. But it is customary to include them in the endocrine system.

A gland may be **temporary**, lasting only for a limited period and then dying away, *e.g.*, thymus. Some glands are, however, **seasonal** being periodic and recurrent, *e.g.*, placenta and corpus luteum.

Chemically, a hormone may be any kind of organic molecule. Most known hormones are either **steroids** or **peptides** with usually high molecular weights. A third group of hormones, which is less common, consists of **amino acid derivatives** (or phenolic derivatives) with relatively low molecular weights. Thus, three categories of hormones may be recognized: steroids, peptides and amino acid derivatives. A classification of vertebrate hormones, based on their chemical composition, is outlined on page 843.

The majority of animal homones are water-soluble, and are derived from amino acids, small peptides, or larger proteins. Because these molecules are hydrophilic, and often large, they cannot easily pass through the hydrophobic lipid layer of the cell membrane. Instead, they interact with receptor molecules on the membrane surface; the hormones themselves do not enter the cell. Other hormones, such as the steroid and thyroid hormones, are hydrophobic in nature, and as such can readily diffuse through the cell membrane and bind to specific receptor inside the cell.

STEROID HORMONES

These include the sex hormones and the hormones from adrenal cortex. These are synthesized in mammals by the ovary (or testis), adrenal cortex, corpus luteum and the placenta. The activity of sex hormones appears to be controlled by the hormones secreted by the anterior lobe of the hypophysis (= adenohypophysis). Because of this, the sex hormones are, sometimes, referred to as *secondary sex hormones* and the hormones of the adenohypophysis, which are of proteinaceous nature, are called as *primary sex hormones*. Three types of sex hormones are recognized:

(*a*) the estrogens (female or ovarian or follicular hormones)

(*b*) the androgens (male or testicular hormones)

(*c*) the gestogens (corpus luteal hormones).

The sex hormones are concerned with the sexual processes and the development of secondary characteristics which differentiate males from females. The adrenal cortical hormones perform a variety of important functions related to cell metabolism.

Based on the number of carbon atoms present in the molecule, the steroid hormones may be named as C_{18}, C_{19} or C_{21} steroids.

C_{18} STEROIDS

1. Ovarian Hormones

Structure. Mammalian ovary contains ovarian follicles and corpus lutea. Hormones produced mainly in the follicles are known as estrogens (*oistros*[G] = a gadfly, hence sting or frenzy). "Estrogen is a generic term for a substance that induces estrus, which is a cyclic phenomenon of the female reproductive system. The stages and timings differ in various species but, in general, first a *proestrus* period occurs, during which the follicle repens and the organs of reproduction develop. This is followed by *estrus*, the period of heat, in which the female will receive the male. Ovulation takes place toward the end of estrus, either spontaneously or, as in rabbit, after mating. Then follows a period of retrogression of the accessory reproductive organs and a period of sexual inactivity" (**Orten** and **Neuhaus**, 1979).

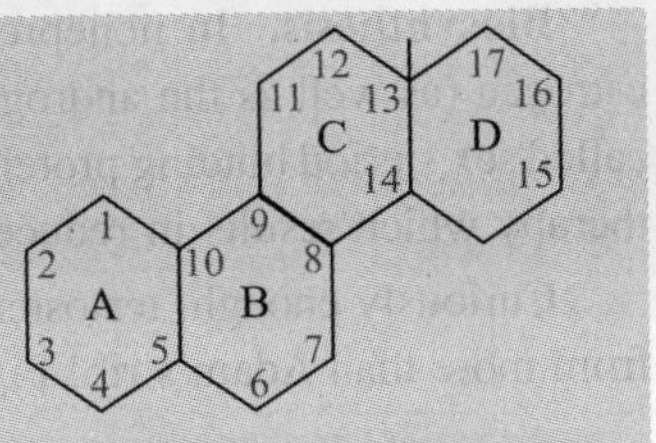

Fig. 34–7. Estrane (parent hydrocarbon of estrogens)

Chemically, the estrogens are derivatives of a C_{18} hydrocarbon, estrane (Fig. 34–7).

The three compounds of this group (Fig. 34–8) with hormonal activity are :

1. β-estradiol (= dihydrotheelin), $C_{18}H_{24}O_2$
2. Estriol (= theelol), $C_{18}H_{24}O_3$
3. Estrone (= theelin), $C_{18}H_{22}O_2$

Adolf Friedrich Johann Butenandt
(LT, 1903–1995)

A German chemist who shared the **1939 Nobel Prize for Chemistry** with **L.S. Ruzicka** for his work on *isolating and determining the chemical structures of the human sex hormones.* His main contributions include:

(*a*) isolation of estrone (1929)
(*b*) isolation of androsterone (1931), about 15 mg from 15,000 litres of urine
(*c*) isolation and synthesis of progesterone (1934)
(*d*) commercial production of testosterone
(*e*) nature and position of the side chain of cholesterol

Estrone is the first known member of the sex hormones and was isolated by **Adolf Butenandt** and **Edward A. Doisy** independently in 1929 from the urine of pregnant women. A year later, the estriol was isolated from human pregnancy urine by **Marrian**. Later, the estradiol was also isolated.

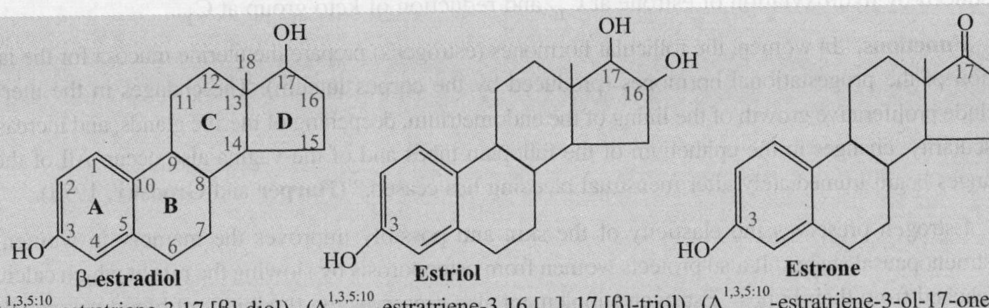

β-estradiol ($\Delta^{1,3,5:10}$-estratriene-3,17 [β]-diol) **Estriol** ($\Delta^{1,3,5:10}$-estratriene-3,16 [ ], 17 [β]-triol) **Estrone** ($\Delta^{1,3,5:10}$-estratriene-3-ol-17-one)

Fig. 34–8. Principal ovarian hormones

All these are characterized by the absence of a CH_3 group at carbon 10 and by the aromatic nature of ring A, making the OH group phenolic in character. Of all these, β-estradiol is most potent physiologically, estrone less potent and estriol is least active. Their relative potencies are 50 : 5 : 1, respectively. Although ovary is the chief source of estrogens, they are in smaller amounts also produced by the testis and the adrenal cortex.

Estrogen production is highest when a woman is young and slows down with age, giving rise to menopausal symptoms. The symptoms vary with each women. By 40's, women enter perimenopause, when menstruation becomes less regular, skin becomes dryer, hair turn brittle and sparser; women may feel a loss of libido and may suffer fluctuations in mood. Menopause follows, on an average between 45 and 50 years of age. Heavy women have an advantage over the slim ones. Their fat cells manufacture a form of estrogen called estrone, even after estrogen from ovaries shuts off.

Biosynthesis. In nonpregnant females, estrogen is mainly synthesized in the ovary. The estrogen (as well as the androgen) are, in part, transported by binding to a specific plasma protein called sex steriod binding protein SBT. The amount of this protein increases in pregnancy or estrogen therapy which results in reduced androgenic action.

Curiously enough, testosterone, a male hormone, is the precursor of estrogens. Eve was made from more than Adam's rib! Fig. 34–9 depicts the probable pathway of estrogen synthesis.

Cholesterol
↓
Testosterone
↓
19-hydroxytestosterone
↙ ↘
19-Oxotestosterone → Estradiol-17
19-Oxotestosterone ↓
10 -carboxy-17 -hydroxyestr-4-ene-3-one ⟶ 19-nortestosterone

Fig. 34–9. Biosynthesis of estrogen

Metabolism. Most of the metabolic reactions of the estrogens (*i.e*, the interconversion reactions of all the three forms) take place in the liver as follows:

Estradiol-17β	Estrone ⟶	16 α-hydroxyestrone
↓	↓	↓
2-hydroxyestradiol-17β ⟶	2-hydroxyestrone	Estriol
↓		
2-methoxyestradiol-17β ⟶	2-methoxyestrone	

Estriol is the principal estrogen found in the placenta and urine of pregnant females. It is produced by hydroxylation of estrone at C_{16} and reduction of keto group at C_{17}.

Functions. "In women, the follicular hormones (estrogens) prepare the uterine mucosa for the later action of the progestational hormones (produced by the corpus luteum). The changes in the uterine include proliferative growth of the lining of the endometrium, deepening of uterine glands, and increased vascularity; changes in the epithelium of the fallopian tubes and of the vagina also occur. All of these changes begin immediately after menstrual bleeding has ceased." (**Harper** and **Grodsky**, 1973).

Estrogen preserves the elasticity of the skin and possibly improves the memory in women at postmenopausal stages. It also protects women from osteoporosis by slowing the rate at which calcium is leached from their bones. Estrogen supplement also preserves the flexibility of blood vessels, thus helping to prevent cardiac diseases. Nowadays, estrogen in combination with the hormone progestin is considered an important tool for helping women remain healthy. The combination is known as **hormone replacement therapy (HRT)**, which has indeed become the closest in medicine to a woman's elixir of youth. HRT, when used by menopausal women, relieves hot splashes, dry sweats and vaginal dryness. However, the long-term use of HRT or estrogen therapy hightens the risk of ovarian cancer.

Victoria Luine (1997) of the Rockefeller University, New York has cleary shown the effect of estrogen on two brain areas, commonly associated with memory and learning, the hippocampus and cerebral cortex. She found that ovariectomized female rats given estrogen had more of an enzyme called *choline acetyltransferase* (ChAT) in the hippocampus and cerebral cortex than control animals did. ChAT enchances the working of the cells in basal forebrain because ChAT makes

Hormone Replacement Therapy

Women normally reach menopause at 45–55 years of age. The absence of ovarian follicles leads to an absence of estrogen and inhibin secretion, resulting in a marked rise in pituitary FSH (follicle-stimulating hormone). There are vasomotor symptoms of estrogen deficiency (*flushing*) associated with menopause. Long-term post-menopausal estrogen deficiency is known to increase the rate of bone loss, leading to osteoporosis, and also alters lipoprotein metabolism in a way that increases the risk of cardiovascular disease. Human replacement therapy (HRT) with an estrogen preparation was standard practice in most developed countries both to relieve the acute symptoms and to provide prophylaxis against the long-term risks. However, recent studies have shown that HRT increases a woman's risk for breast cancer, cardiovascular disease, stroke and pulmonary embolism. Although HRT did have a modest effect on preventing osteoporosis, there are newer drugs, the *bisphosphonates*, that are just as effective. The current recommendations are that HRT should only be used as short-term therapy for vasomotor symptoms that do not respond to other treatment.

acetylcholine (ACh) which helps nervous communication with other nerve cells. In a nutshell, *estrogen enchances brain power.* Estrogens could also protect brain cells from toxins (**Simpkins, 1994**). Estrogens can also act as an antioxidant soaking up highly reactive molecules called free radicals which can kill a cell by fracturing its membrane lipids, proteins and DNA. Estrogen is also being widely perceived to have a significant effect (of improving verbal memory) on patients with Alzheimer's disease.

The estrogens are also effective in the development of *secondary sex characters in females.* These are listed in Table 34–1.

Table 34–1. Body changes at puberty in girls (=Secondary sex characters in females)

Characters	*Changes*
1. External genitalia	Enlargement of uterus and vagina; Widening of pelvis.
2. Internal genitalia	Periodic vaginal bleeding that occurs with the shedding of the uterine mucosa (i.e., menstruation).
3. Voice	Larynx retains its prepubertal proportions, i.e., small in size; Voice stays high-pitched.
4. Hair growth	Less body hair and more scalp hair; Hair line on scalp resembles that of a child and does not recede anterolaterally (Fig. 30–10); Hair appear in axillae (axillary hair) and around vagina; Pubic hair have a characteristic female pattern, i.e., flat-topped; Hair on face absent i.e., no beard.
5. Mental	Less aggressive; Passive attitude; Interest in opposite sex less pronounced.
6. Body conformation	Narrow shoulders and broad hips, which are popularly called as 'hip pads'; Thighs that converge and arms that diverge, i.e., a wide carrying angle; Distribution of fat in the breasts and buttocks takes place, leading to their enlargement, the breasts also have high pigmentation in the areola which becomes even more intense during the first pregnancy; Muscles not pronounced.
7. Skin	Sebaceous gland secretions become more fluid and thus counter the effect of testosterone and inhibit formation of comedones ('black-heads') and acne (a hard, red inflamed pimple)
8. Weight gain	Tend to gain weight from the waist down.

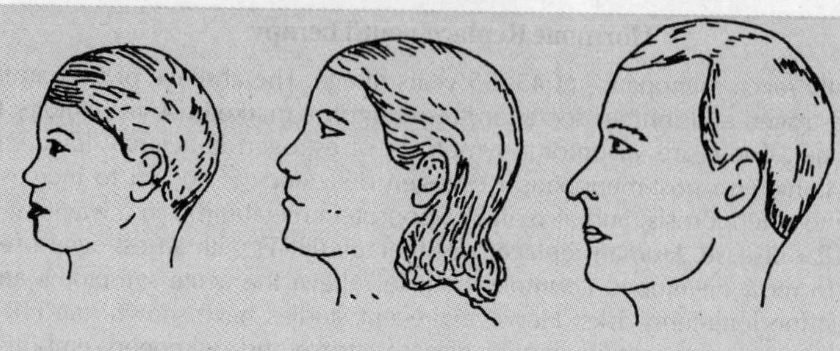

Fig. 34–10. Hair line pattern of child, woman and man

The hair line of the women is like that of the child, whereas that of the man is indented in the lateral frontal region.
(*Adapted from Grenlich et al, 1942*)

Estrogens also influence to a great deal (*a*) the inorganic metabolism of Ca and P and (*b*) the organic metabolism of proteins and lipids.

Castration (*castratus*[L] = to prune). Removal of the ovary in females is known as ovariectomy. In females, castrated prior to puberty (*pubertas*[L] = of ripe age, adult), both the menstrual as well as the reproductive cycles never appear. The typical pelvic enlargement fails to occur and the pubic and axillary hair become scanty. Post-pubertal castration results in suspension of menstrual cycle and atrophy of uterus and vagina. Also, mammary glands become involuted and osteoporosis gradually appears.

> **Castration** is the act of pruning in any of its senses.

> **Puberty** is the state of physical development at which persons are first capable of begetting or bearing children. In law, the age of puberty is usually fixed at 14 in the male and 12 in the female.

Stilbesterol. Stilbesterol is a synthetic product with marked estrogenic properties. It is an amino acid derivative and obviously does not resemble estrogens in chemical structure. However, it produces practically all the physiologic effects that estradiol does. Its diethyl derivative, diethylstilbesterol (Fig. 34–11), is more potent physiologically. Stilbesterol is administered orally and in some cases certain unpleasant side effects spring up. However, if the dosage is controlled carefully, these effects can be alleviated.

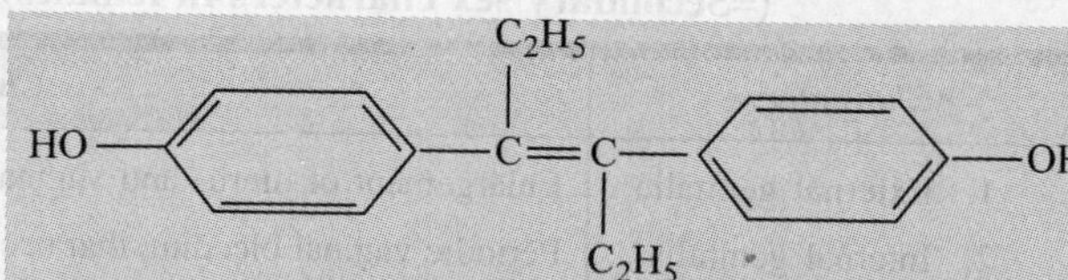

Fig. 34–11. Diethylstilbesterol

C_{19} STEROIDS

2. Testicular Hormones

Structure. These hormones are secreted mainly by the testes, the male reproductive organs and are called as androgens (*andros*[G] = male). Chemically, these are derivatives of a C_{19} hydrocarbon, *androstane* (Fig. 34–12).

There are many hormones secreted from testes with androgenic activity. The three important ones (Fig. 34–13) are:

1. Testosterone, $C_{19}H_{28}O_2$
2. Androsterone, $C_{19}H_{30}O_2$
3. Dehydroepiandrosterone, $C_{19}H_{25}O_2$

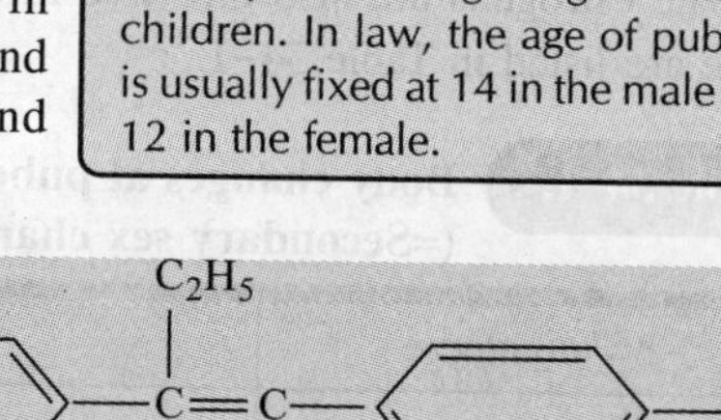

Fig. 34–12. Androstane
(parent hydrocarbon of androgens)

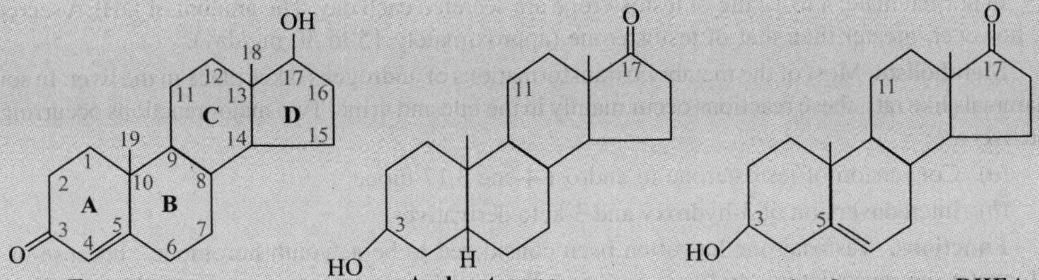

Testosterone (3-keto-17-hydroxy-Δ^4-androstene)　　Androsterone (3 -hydroxy-5 -androstan-17-one)　　Dehydroepiandrosterone, DHEA

Fig. 34–13. Principal testicular hormones

Androsterone was first isolated by **Adolf Butenandt** *et al* in 1931 from male urine (about 15 mg from 15,000 litres of urine).

Testosterone is most potent of all these and dehydroepiandrosterone is least active. The relative potency ratio of these three forms is 20 : 7 : 1. Testosterone has a tendency to rise during late summer and early fall to peak in September. DHEA production peaks between ages 25 and 30 and wanes with age. Restoring DHEA levels to peak is said to boost the immune system.

A few testicular hormones are also produced by the adrenal gland. The structure of two such hormones is given in Fig. 34–14.

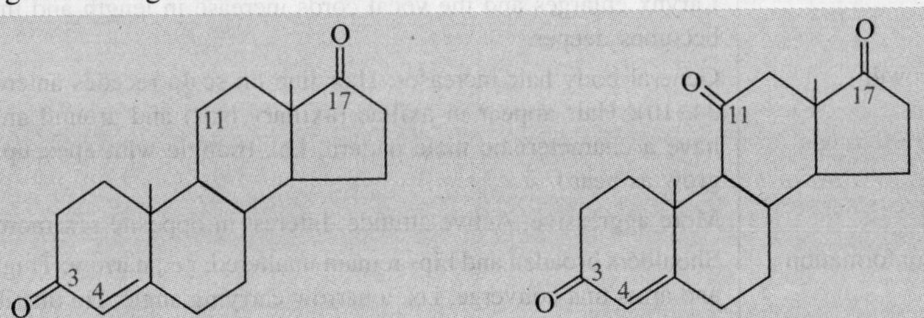

Androst-4-ene-3,17-dione　　Androst-4-ene-3,11,17-trione

Fig. 34–14. Two testicular hormones produced from adrenal cortex

Biosynthesis. In Fig. 34–15, biosynthesis of testosterone from cholesterol has been depicted. The principal male hormone, testosterone, is synthesized by the Leydig cells of the testes from cholesterol through pregnenolone, progesterone and hydroxyprogesterone. The latter is then converted to a C_{19} ketosteroid called androstenedione which is the immediate precursor of testosterone. It is presumed that the same sequence of events also takes place in the adrenal gland, ovary and placenta.

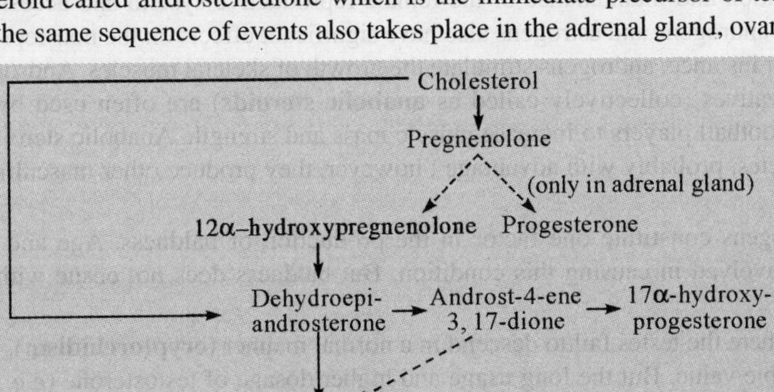

Fig. 34–15. Biosynthesis of testosterone

In addition to testosterone, adrostenedione and DHEA are also synthesized in the testes, although in amounts far less than that of testosterone.

In normal male, 4 to 12 mg of testosterone are secreted each day. The amount of DHEA secreted is, however, greater than that of testosterone (approximately 15 to 50 mg/day).

Metabolism. Most of the metabolic transformations of androgens takes place in the liver. In some mammals like rats, these reactions occur mainly in the bile and urine. Two major reactions occurring in the liver are:

(*a*) Conversion of testosterone to androst-4-ene-3,17-dione.

(*b*) Interconversion of 3-hydroxy and 3-keto derivatives.

Functions. Testosterone has often been considered to be a 'youth horomone', because of its effects on the musculature, and it is occasionally used for treatment of persons who have poorly developed muscles. Because of the ability of testosterone to increase the size and strength of bones, it is often used in old age to treat osteoporosis. Like estrogens, the androgens are also responsible for the development of secondary sex characters in males. These are listed in Table 34–2.

Table 34–2. Body changes at puberty in boys (= Secondary sex characters in males)

Characters	*Changes*
1. External genitalia	Penis increases in length and width; Scrotum becomes pigmented and rugose.
2. Internal genitalia	Seminal vesicles enlarge and secrete– they also begin to form fructose ; Prostate and bulbourethral glands also enlarge and secrete.
3. Voice	Larynx enlarges and the vocal cords increase in length and thickness; Voice becomes deeper.
4. Hair growth	General body hair increases; Hair line on scalp recedes anterolaterally (Fig. 34–10); Hair appear in axillae (axillary hair) and around anus; Pubic hair have a characteristic male pattern, i.e., triangle with apex up; Hair on face grow as beard.
5. Mental	More aggressive; Active attitude; Interest in opposite sex more pronounced.
6. Body conformation	Shoulders broaden and hips remain unaltered, i.e., narrow; Thighs that diverge and arms that converge, i.e., a narrow carrying angle; No distribution of body fat in the chests and buttocks; Muscles enlarge, leading to a muscular body contour.
7. Skin	Sebaceous gland secretion thickens and increases (predisposing to acne).
8. Weight gain	Tend to gain weight in the abdominal region.

Androgens regulate the activities of the male reproductive system; estrogens stimulate the growth, maturation and maintenance of the female reproductive system and accessory sex tissues. However, both androgens and estrogens also have significant effects on the non-reproductive tissues of the body. For instance, androgens stimulate the growth of skeletal muscles. Androgens and certain androgen derivatives (collectively called as **anabolic steroids**) are often used by weight lifters, wrestlers and football players to increase muscle mass and strength. Anabolic steroids are also used by female athletes, probably with advantage ; however, they produce other masculinizing effects as well.

The androgens constitute one factor in the production of baldness. Age and inheritence are other factors involved in causing this condition. But baldness does not ensue without androgenic stimulation.

In cases where the testes fail to descend in a normal manner (**cryptorchidism**), the testosterone is of considerable value. But the long usage and higher dosage of testosterone (*e.g.*, 25 mg per day for 4–6 weeks) in individuals often leads to atrophy of the sperm. The effect can, however be reversed by discontinuing the treatment for a similar period. Testosterone also controls the libido, and also the development of muscle mass and bone density.

Castration. Prepubertal castration in males results in :

1. Ossification of long-bone epiphyses
2. Increased adiposity
3. Lipid deposition leading to slight extrusion of the chest *i.e.,*enlargement of the male breast (gynecomastia)
4. Depressed growth of hair on the face and chest
5. Suppressed male organ (penis)
6. Diminished muscular growth.

Castration after puberty, leads to almost similar changes described above but less intensified.

C_{21} STEROIDS

3. Adrenal Cortical Hormones

Secretory gland. The adrenals (*ad* = at; *renal* = pertaining to kidneys)or suprarenal glands in all tetrapods are a pair of glands, so named because of their position very close to or at the top of the kidneys. Each of the two adrenals among mammals is actually a '*double gland*' and is composed of 2 distinct parts : namely an outer barlike covering called the *cortex*, surrounding an inner corelike dark-coloured mass called the *medulla*. The cortex is derived from the mesodermal glandular tissue and the medulla originates from the cells of neural crest. Both these parts secrete hormones which differ from each other chemically as well as physiologically. Hence, these 2 components are discussed separately. In man, the adrenals are two small structures sitting like 'cocked hats' over the apical end of the kidneys (refer Fig. 34–16) and each gland weighs about 3 grams.

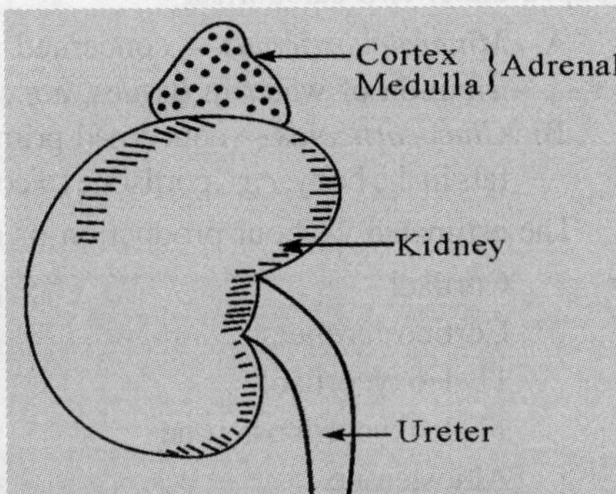

Fig. 34–16. The adrenal gland in man
(*After D. Marsland, 1964*)

Histologically, the adrenal cortex is made up of three layers (Fig. 34–17):

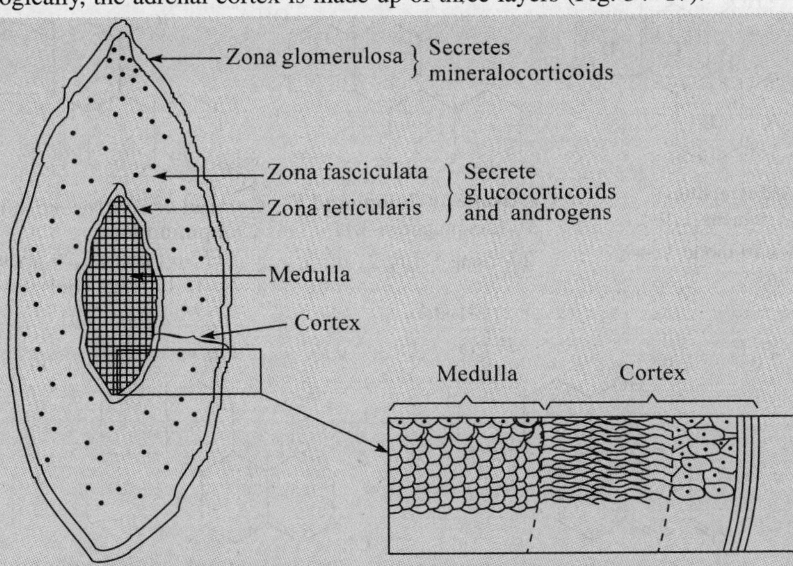

Fig. 34–17. Microscopic structure of the adrenal gland

(*a*) an outer narrow *zona glomerulosa*, believed to be the site of biosynthesis of the mineralocorticoid hormones.

(*b*) a middle, comparatively broader *zona fasciculata*, responsible for the production of glucocorticoid hormones and the adrenal androgens.

(*c*) an inner narrow *zona reticularis*, secreting glucocorticoids along with the middle zone.

When there is prolonged stimulation of the adrenal cortex by adrenocorticotropic hormone (ACTH), the middle and inner zones both hypertrophy; but a total lack of ACTH causes these two zones to atrophy almost entirely, leaving the outer zona glomerulosa partially intact. On the other hand, enhanced aldosterone production causes hypertrophy of the zona glomerulosa, while the other two zones remain almost unaffected.

Structure. Adrenal cortex secretes some 40-50 closely related C_{21} steroids, collectively called as corticosteroids (refer Fig. 34–18). From physiological viewpoint, the corticosteroids may be grouped under two categories:

A. *Mineralocorticoids*— concerned primarily with the transport of electrolytes and the distribution of water in tissues, *e.g.*, aldosterone and deoxycorticosterone.

B. *Glucocorticoids*— concerned primarily with the metabolism of carbohydrates, proteins and fats in the body, *e.g.*, cortisone (= compound E), cortisol (= hydrocortisone) and corticosterone.

The estimated 24-hour production of major compounds of human adrenal gland is:

Cortisol	8-24 mg
Corticosterone	1.5–4 mg
11-deoxycortisol	0.5 mg
11-deoxycorticosterone	0.2 mg
Aldosterone	0.04–0.2 mg
18-hydroxycorticosterone	0.15–0.45 mg
18-hydroxy-11-deoxycorticosterone	*ca* 0.1 mg

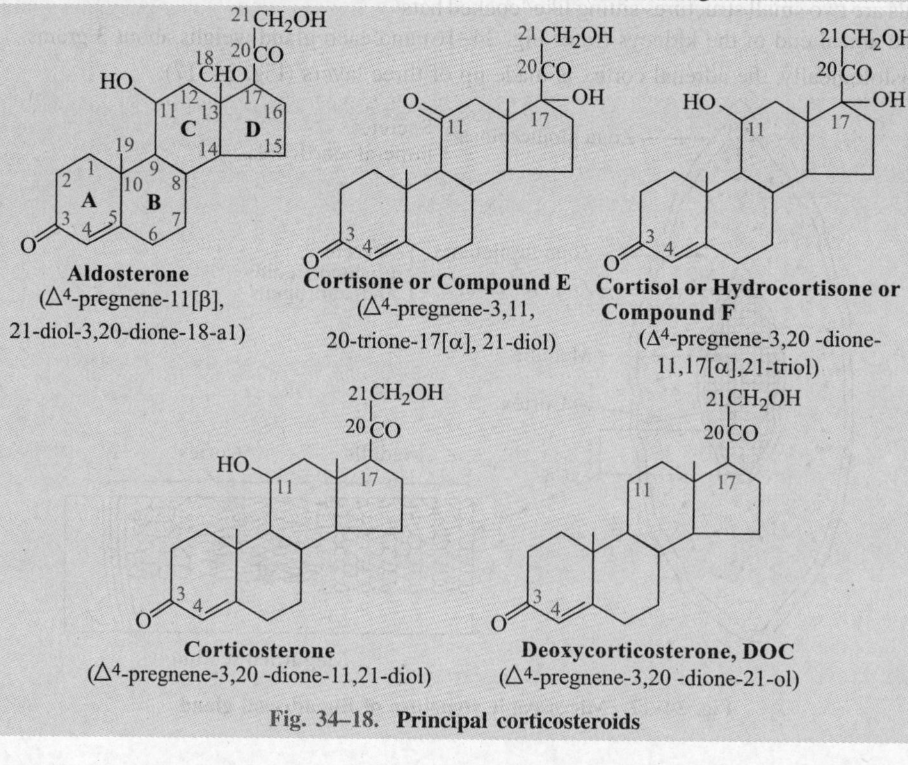

Aldosterone (Δ^4-pregnene-11[β], 21-diol-3,20-dione-18-al)

Cortisone or Compound E (Δ^4-pregnene-3,11, 20-trione-17[α], 21-diol)

Cortisol or Hydrocortisone or Compound F (Δ^4-pregnene-3,20 -dione-11,17[α],21-triol)

Corticosterone (Δ^4-pregnene-3,20 -dione-11,21-diol)

Deoxycorticosterone, DOC (Δ^4-pregnene-3,20 -dione-21-ol)

Fig. 34–18. Principal corticosteroids

However, during periods of stress, the amount of cortisol released during 24 hours may increase even up to 300 mg. The release of cortisol by the adrenal cortex has a diurnal rhythm. More is released during the day than during the night. Their structure is given in Fig. 34–18.

Aldosterone is 30 times more active than deoxycorticosterone. Deoxycorticosterone, in its turn, is 4 times more potent than cortisone and cortisol in maintenance of life. Corticosterone is least active in this regard.

Functions. A *mineralocorticoid*, aldosterone is chiefly concerned with water-salt balance of the body. It stimulates the reabsorption of Na^+ ion from the kidney tubules and as such regulates NaCl contents of the blood. This also causes excretion of K in the urine. Aldosterone is also more potent in maintaining the life of adrenalectomized animals.

Glucocorticoids, on the contrary, govern many other processes. They perform the following physiological functions:

1. Influence the carbohydrate metabolism firstly by increasing release of glucose from the liver and secondly by promoting the transformation of amino acids to carbohydrates.
2. Inhibit protein synthesis in muscle tissues.
3. Control eosinophil cells of the blood.
4. Regulate lipogenesis.
5. Reduce the osteoid matrix of bone, thus favouring osteoporosis (weak bones) and heavy loss of calcium from the body.
6. Decrease immune responses associated with infection and anaphylaxis (immunosuppressive effects).
7. Cause increased secretion of hydrochloric acid and pespinogen by the stomach and that of trypsinogen by the pancreas (exocrine secretory effects).
8. Cause retention of sodium(and water) and loss of potassium to some extent. In this respect, it resembles aldosterone in action.

Hypoadrenocorticism. A decrease in the amount of corticosteroids in the body (hypoadrenocorticism) leads to the decreased metabolic rate, excessive pigmentation, loss of appetite (anorexia), muscular weakness, deficiency of blood (anemia), eosinophilia and decreased blood sugar (hypoglycemia) with fasting.

Hyperadrenocorticism. The excessive supply of adrenal cortical steroids (hyperadrenocorticism) results from cortical cell tumours which may arise in or outside the adrenal gland. Oversecretion of cortisol in man leads to a rare disease, **Cushing's syndrome** (Fig. 34–19), after its discoverer, **Harvey Williams Cushing**. The most common cause of the symptoms of Cushing's syndrome is the prolonged administration of glucocorticoids for medical treatment and henceforth the disease is also known by its synonym *glucocorticoid excess*. The syndrome is characterized by profound disturbance of carbohydrate, protein, fat and calcium metabolism. There occurs mobilization of fat from the lower part of the body, with the concomitant extra deposition of fat in the thoracic region. The obesity becomes visible on the neck (*buffalo hump*) and on the face (*moon face*). Weakness and muscle wastings with marked osteosis become evident. Hypertension, pigmentation of the hair and excessive growth of hair are other symptoms.

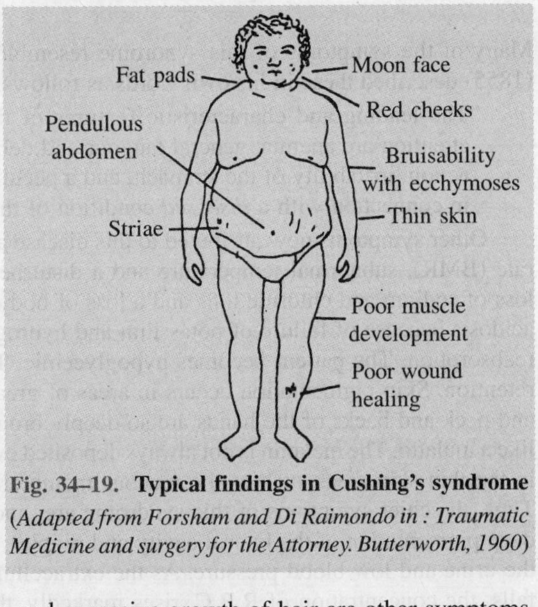

Fig. 34–19. Typical findings in Cushing's syndrome

(*Adapted from Forsham and Di Raimondo in : Traumatic Medicine and surgery for the Attorney. Butterworth, 1960*)

In men, there is impotence, and in women, amenorrhea and masculinization. Thus, Cushing's syndrome resembles somewhat adrenogenital syndrome.

Hypersecretion of aldosterone leads to a marked Na^+ and water retention, resulting in edema and hypertension causing heart failure.

The adrenal cortex also produces androgenic steroids known as *adrenosterones*. Their hypersecretion has effects varying according to the age and sex of the patient. In adult female, it leads to **adrenal virilism.** In it menstruation stops, breasts atrophy, hair on breast and face develop and the voice deepens. In all, the adult woman becomes masculine. In adult males, there occurs excessive hair growth, enlargement of the sex organ and increased sexual desire. However, in children excessive supply of adrenosterones results in precocious development of sex organs and the secondary sexual characters.

Adrenal decortication. Removal of adrenal cortex (*adrenalectomy*) leads to a fatal human disease known as **Addison's disease** (Fig. 34–20), named after its discoverer Thomas Addison.

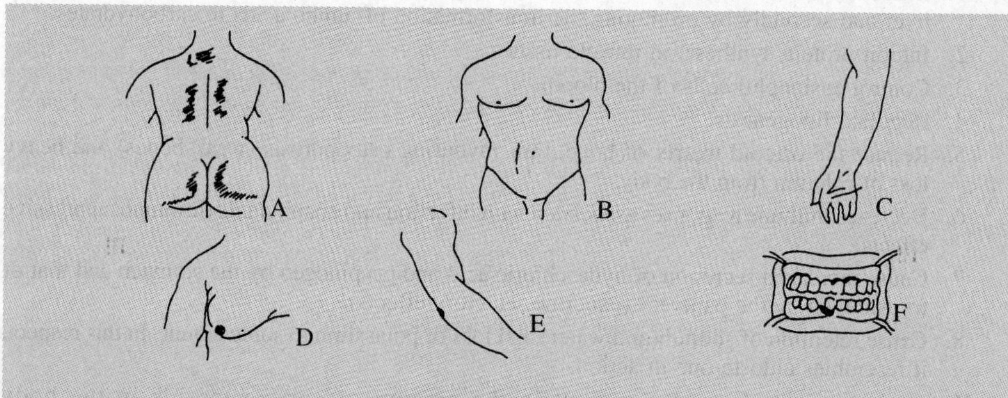

Fig. 34–20. Pigmentation in Addison's disease

(A) Tan and vitiligo (B) Pigmentation of scars (C) Pigmentation of skin creates (D) Darkening of areolas (E) Pigmentation of pressure points (F) Pigmentation of the gums.

(Adapted from Forsham and Di Raimondo, 1960)

Many of the symptoms of this syndrome resemble those of adrenalectomized animals. **Addison** (1855) described these in his own words as follows:

> "The leading and characteristic features of the morbid state to which I would direct attention are anemia, general langour and debility, remarkable feebleness of the heart's action, irritability of the stomach, and a peculiar change of colour in the skin, occurring in connection with a *diseased* condition of the 'supra-renal' capsules"

Other symptoms now attributed to this disease are low blood pressure, lowered basal metabolic rate (BMR), subnormal temperature and a disturbed water and electrolyte balance. This includes loss of sodium and chloride ions and a loss of body water. The person develops hyperkalemia and acidosis because of failure of potassium and hydrogen ions to be secreted in exchange for sodium reabsorption. The patient becomes hypoglycemic. The kidneys are also affected, resulting in urea retention. Skin pigmentation occurs in areas of greatest normal pigmentation. Frequently, the face and neck and backs of the hands are so deeply bronzed as to cause the afflicted individual to look like a mulatto. The melanin is not always deposited evenly but occasionally in blotches and especially in the thin skin areas such as the mucous membranes of the lips and the thin skin of the nipples. Thus, the chief symptoms of this syndrome are: anorexia, emesis (= vomiting), diarrhea, anemia, deep pigmentation of the buccal cavity and nipples, rapid loss of weight, excessive loss of NaCl in the urine and low blood pressure. As the extracellular fluid becomes depleted, the plasma volume falls, the concentration of R.B.C. rises markedly, the cardiac output lowers down and the patient dies in shock. The grip of Addisonian patients on their life is tenuous and any stress, infection, cold or even noise can precipitate a crisis leading to death.

Addisonian patients may be cured by giving extracts of the adrenal cortex. This has been very successful since its beginning in about 1929. Because this is substitution therapy, like most endocrine therapy, the constant administration of potent extracts is essential. **Loeb** (1939) has, however, demonstrated that NaCl is of immense value to Addisonian patients as it corrects the electrolyte and water imbalance in them. The administration of deoxycorticosterone (usually as acetate) or aldosterone also cures the disease.

4. Corpus Luteal Hormones

Structure. The hormones secreted by the ovarian bed, corpus luteum are collectively called as **gestogens** or **progestins**. The principal gestogen is progesterone (the two pregnenolones, 20α-OH and 20β-OH are other hormones secreted by the corpus luteum). Progesterone (Fig. 34–21) is a C_{21} steroid and is secreted by the corpus luteum during the second half of the menstrual cycle. This was first isolated in pure form by Adolf Butenandt *et al* (1934) from corpus lutea of pregnant sows. It has also been isolated from adrenal cortical extracts. But its presence in the adrenal tissue is a consequence of its role as an intermediate in the biosynthesis of the typical adrenal cortical hormones. Chemically, progesterone is one of the pregnane derivatives and lacks the ketol group. Its molecular formula is $C_{21}H_{30}O_2$. It closely resembles deoxycorticosterone in structure. It is, therefore, not surprising to find progesterone with certain adrenocortical properties, *viz.*, those influencing salt and water. Indeed, it serves as a procursor of the steroidal adrenocorticoids. It is soluble in most organic solvents and in vegetable oils but is insoluble in water.

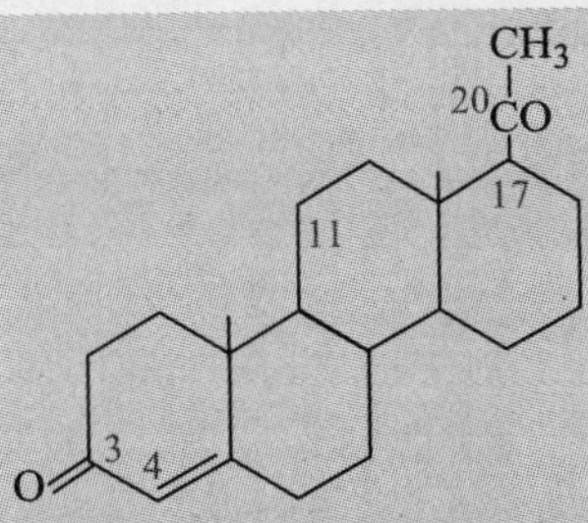

Fig. 34–21. Progesterone or "pregnancy hormone"
(Δ^4-pregnene-3,20-dione or 3,20-diketo-$\Delta^{4,5}$-pregnene, *cis*)

The 3-dimensional structures of some steroid hormones determined by x-ray crystallography is presented in Fig. 34–22.

Biosynthesis. Progesterone is synthesized in the corpus luteum, placenta and the adrenal cortex from its immediate precursor, pregnenolone by a combined dehydrogenase and isomerase reaction. Unlike testosterone and estradiol, the progesterone is bound in plasma to the corticosteroid-binding globulin (CBG).

Metabolism. The metabolic fate of progesterone has been studied by injecting C^{14} - labelled hormone. About 75% of the injected progesterone is translocated to the intestine *via* bile and passed out in feces. A 20-hydroxy compound, pregnanediol, is the main urinary product of progesterone metabolism in human beings and rabbits. Such a conversion of progesterone into pregnanediol is not carried out in rat liver.

Functions. Progesterone has manifold functions:

1. The primary function of progesterone is to promote the proliferation of uterine mucosa so that the latter may receive the fertilized ovum. It, thus, serves for implantation of the fertilized ovum. If pregnancy ensues, continued secretion of progesterone is essential for completion of term.
2. It brings mammary glands to full maturity during gestation (pregnancy) for their onward use in breast-feeding by the newborn (Fig. 34–23)
3. It also maintains the uterus quiescent during pregnancy (*i.e.*, inhibits contraction of the uterus) and also prevents further release of eggs from the ovary.
4. If given between 5th and 25th day of the normal menstrual cycle, progesterone exerts an antiovulatory effect. This is the basis for the use of certain progestins as oral contraceptive agents.

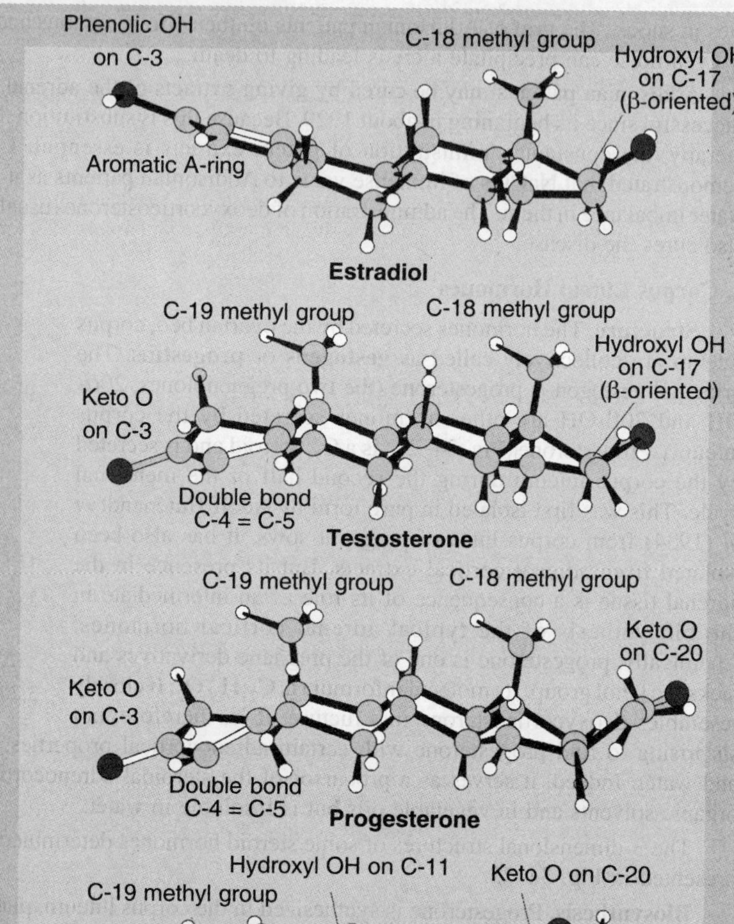

Fig. 34–22. Ball-and-stick representation of some steroid hormones

Details of each structure are labelled. In aldosterone, the acetal grouping is

$$R{-}CH\begin{matrix} OR_1 \\ OR_2 \end{matrix}$$

and the hemiketal grouping is

$$\begin{matrix} R_1 & & OR_3 \\ & C & \\ R_2 & & OH \end{matrix}$$

,where R_1, R_2 and R_3 refer to different substituents.

(Courtesy: Glusker JP, 1979)

5. It serves as precursor of cortisol and corticosterone in the adrenal glands.
6. It is used as an oral contraceptive, either on its own or combined with estrogen, to suppress ovulation or implantation.

Fig. 34–23. A mother breastfeeding her baby

Progesterone brings mammary glands to full maturity and ocytocin helps in the ejection of milk which is suckled by the baby during infancy.

Fig. 34–24. The effect of hormones on sexual reproduction in humans

Hormones control gamete production, puberty, pregnancy, birth and lactation in humans.

Oral Contraception

Oral contraceptives usually contain an estrogen and a progestin. Taken orally, the levels of these steroids increase in blood to a level where secretion of FSH and LH is repressed. Consequently, the gonadotropic hormone levels in blood fall and there is insufficient FSH to drive the development of the ovarian follicle. As a result, the follicle does not mature and ovulation cannot occur. In addition, any corpora lutea cannot survive because of low LH levels. In essence, the ovarian cycle ceases. The uterine endometrium thickens and remains in this state, however, because of elevated levels of estrogen and progestin. Pills without the steroids (*placebos*) are inserted in the regimen at about the 28th day and, consequently, blood levels of the steroids fall dramatically and menstruation occurs. When oral contraceptive steroids are resumed, the blood levels of estrogen and progestin increase again and the uterine endometrium thickens. This sequence creates a "false" cycling because of the occurrence of menstruation at the expected time in the cycle. The ovarian cycle and ovulation are suppressed by the oral contraceptive based on the negative feedback effects of estrogen and progestin on the secretion of the anterior pituitary gonadotropes.

It is also possible to provide contraception by implanting in the skin *silicone tubes* containing progestins. The steroid is slowly released, providing contraception for up to 3 to 5 years.

[**Relaxin.** Corpus luteum also secretes another hormone (or group of hormones) called relaxin or uterine-releasing factor (URF). It was discovered by Hisaw in 1926. Relaxin is a polypeptide of molecular weight about 9,000. It has not yet been obtained in pure form and, consequently, its structure has remained undetermined as yet. It is produced during pregnancy. The hormone is active only when injected into an animal which is in normal estrus (sexual heat). A wide variety of mammals including pregnant women, cows, rabbits and dogs secrete this hormone. Relaxin causes relaxation (hence the name) of the ligaments of the symphysis pubis in the estrus rat and guinea pig. This effect is, however, least pronounced in pregnant woman. It performs two other functions:

I. softening of the cervix of the pregnant woman at the time of delivery.
II. inhibition of the uterine motility.]

We, thus, see that many aspects of sexual reproduction in humans are controlled by hormones, including gamete production, puberty, pregnancy, birth and lactation (Fig. 34–24).

Prohormones or Hormogens

Some polypeptide hormones, like certain enzymes, are synthesized in an inactive form, *i.e.*, as prohormones or hormogens. Prohormones are examples of proteins with extracellular functions, such as the enzymes chymotrypsin and trypsin, whose biological activities are dormant until activated by *peptidases*. Table 34–3 lists some of the known prohormones.

Table 34-3. Some prohormones

Prohormone	*Source*
Proinsulin	Pancreas– β cells
Proparathormone	Parathyroid
Angiotensinogen	Liver
Progastrin	Stomach

PEPTIDE HORMONES

1. Hormones of the Pancreas

Secretory gland. The pancreas (from pankreas : *pan*[G] = all, *kreas*[G] = flesh ; referring to the fleshy nature of the tissue) is both an exocrine and endocrine gland. It is situated transversely below and behind the stomach between the curve of duodenum and spleen. It is a compact and lobulated organ. It weighs about half a pound and resembles an elongated cone lying on its side. The broad end or '*head*' of the pancreas is located next to a curve of the duodenum, the part of the small intestine just beyond the stomach. The gland tapers off to the left in the direction of the spleen and left kidney and ends in a portion called 'tail' (Fig. 34–25).

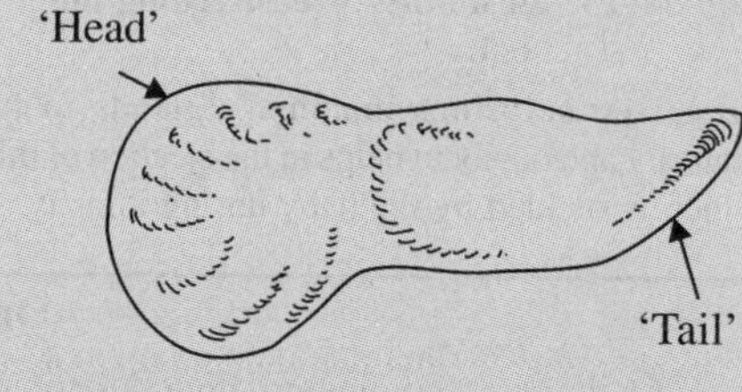

Fig. 34–25. Morphologic structure of pancreas

Barring cyclostomes, the pancreas in all the vertebrates is composed of two types of cells (Fig. 34–26).

(*a*) the glandular cells or *acinar* or *acini* (exocrine), which make up the bulk of the pancreatic tissues and secrete digestive juices into the duodenum by the pancreatic duct.

(*b*) the polygonal cells or *islets of Langerhans* or *islet tissue* (endocrine), which do not have any means for emptying their secretions externally but instead pour their secretions (*i.e.*, insulin and glucagon) directly into the blood. These were discovered by **Langerhans** in 1867, hence so named.

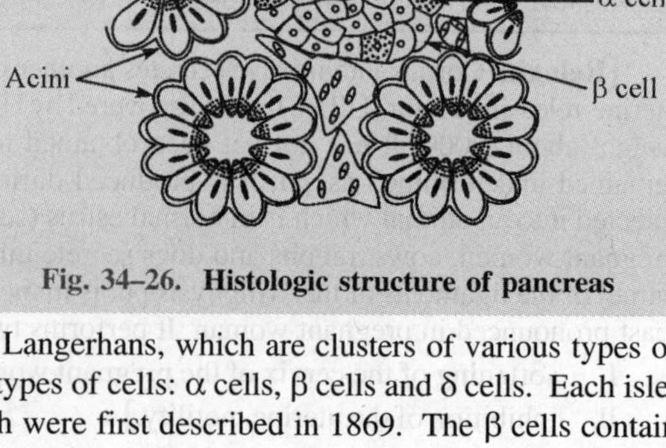

Fig. 34–26. Histologic structure of pancreas

Each pancreas has about 1,00,000 islets of Langerhans, which are clusters of various types of cells. These islets in mammals contain 3 major types of cells: α cells, β cells and δ cells. Each islet contains between 1,000 and 2,000 **β cells,** which were first described in 1869. The β cells contain granules which are insoluble in alcohol and manufacture a hormone insulin, store it and eventually release it directly into the bloodstream at the appropriate times. These amazing cells (*i.e.*, β cells) are also capable of measuring the blood glucose level within seconds to within a range of 2 mg %. Using this information, they can determine how much insulin is needed, and, within a minute or so, secrete the precise amount of required insulin. The **α cells** contain granules which are soluble in alcohol and produce another hormone, glucagon. The α-cells tend to be arranged about the periphery of the islets. The existence of αcells is dubious in cyclostomes and urodeles. Of the two cell types, the α cells predominate in the reptiles and birds, whereas in amphibians and mammals, the β cells are more abundant. Teleost fishes, however, often exhibit seasonal variations between the two cell types. However, a 3rd cell type **δ cells** secrete another hormone, *somatostatin*.

The islets of Langerhans are also known to contain one more type of cells designated as **D** or **$\alpha_1$cells.** These cells probably secrete *gastrin* which is a recognized gastrointestinal hormone and is derived from pyloric mucosa.

FREDERICK SANGER
(born, 1918)

A British biochemist who was born on August 13, 1918 at Rendcombe, England. He received his doctorate in 1943 on the metabolism of lysine from Cambridge University. During his service tenure, he preferred working in a laboratory with his own hands to teaching and administrative jobs. Sanger has been responsible for two of the most important technical advances of the past 55 years. First, in the early 1950s, he developed a method for determining the amino acid sequence of a polypeptide, and put it to use by working out the sequence of amino acids in insulin. This work, completed in 1955, provided the first real proof that the amino acids in a protein are present as a constant, genetically-determined sequence. This gave molecular biologists the confidence to tackle other problems such as the genetic code. Then, in 1977, Sanger perfected a rapid method for sequencing DNA, opening the door to precise examination of gene structure and organization. Both advances have been invaluable and both were recognized by Nobel Committee. Sanger is *one of the two scientists to have ever won Nobel Prizes twice in the same subject: first in 1958* for determining the amino acid sequence of (bovine) insulin on his own, and *later in 1980* for determining the base sequences in nucleic acids, along with **Paul Berg** and **Walter Gilbert**, both from USA, thus both times in chemistry; Sanger's contribution was to refine the techniques for determining the structure of more complex proteins [the other being an American physicist **John Bardeen**– in 1956 and 1972, both time jointly]. Sanger has spent his entire career at the University Department of Biochemistry and the Medical Research Council Laboratory for Molecular Biology at Cambridge, U.K. The principle behind his success, in his own words, is:

"If the planned experiment doesn't work, don't worry, start planning the next experiment".

Insulin

Structure. Insulin (*insula*^{L} = island) was first isolated in 1922 from the pancreas of dogs by **Banting** and **Best**, both of the University of Toronto, Canada. They also demonstrated the curative effect of pancreatic extract in dogs ailing with diabetes mellitus. Abel and his associates (1926) obtained insulin in crystalline form (Fig. 34–27) and also demonstrated its protein nature. *Insulin is, in fact, the first hormone to be recognized as a protein.* The chemical structure of insulin has been determined by **Sanger** and his coworkers at Cambridge, England in a painstaking 10-year (1944-54) study. It has a molecualr formula, $C_{254}\ H_{377}\ O_{75}\ N_{65}\ S_6$ (consult Chapter 9 for structure). Just to recall, bovine insulin consists of 51 amino acid residues dispersed in two chains. The two polypeptide chains are held together by cross linkages of two disulfide bonds. The acidic chain A contains 21 residues and the peptide chain B having 30 residues. It has a molecular weight of 5,733 and is isoelectric at pH 5.4. Human insulin, however, has a molecular weight of 5,808. Insulin is destroyed by alkali but is relatively stable in acid solutions. Reduction of the disulfide bond results in a loss of biologic activity. Zinc is always found with this hormone but is not a part of the insulin molecule.

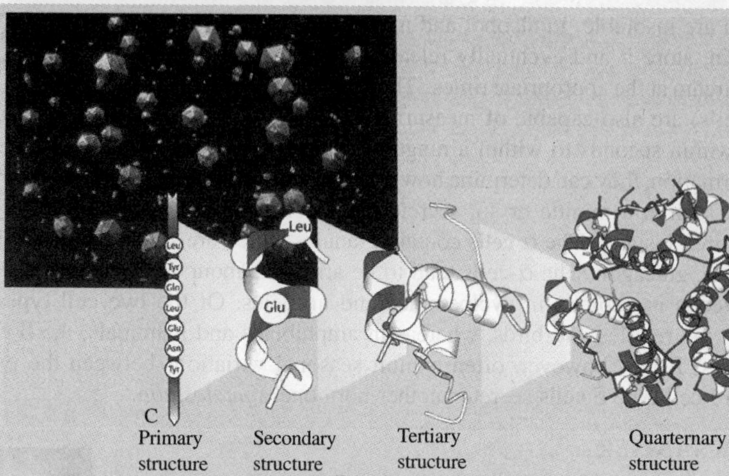

Fig. 34–27. Crystals of human insulin

Insulin is a protein homone, crucial for maintaining blood sugar at appropriate levels. (Below) Chains of amino acids in a specific sequence (the *primary structure*) define a protein like insulin. These chains fold into well-defined structures (the *tertiary structure*)– in this case a single insulin molecule. Such structures assemble with other chains to form arrays such as the complex of six insulin molecules shown at the *far right* (the *quartenary structure*). These arrays can often be induced to form well-defined crystals (photo at *left*), which allows determination of these structures in detail.

[Courtesy: (Left) Alfred Pasieka/Peter Arnold]

It has been established that the insulins from various species differ only slightly, although their biologic activities are identical. **Sanger** and **Smith** (1957) have determined the amino acid sequence of insulins from 5 different species. They found the sequence to be identical except the identity and sequence of only three amino acids at positions 8, 9 and 10 in the chain A (refer Table 34–4).

SIR FREDERICK GRANT BANTING

(LT, 1891–1941)

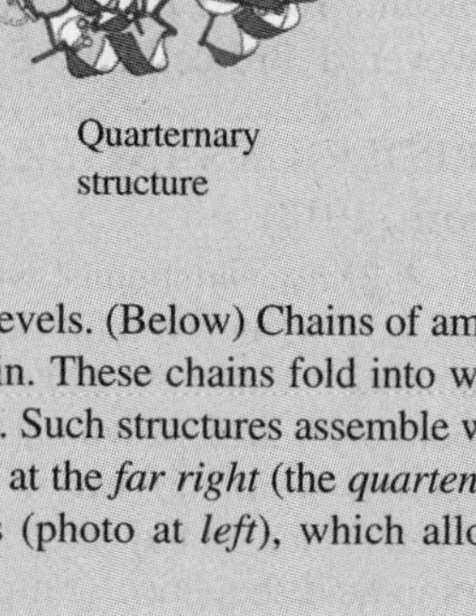

Banting, a Canadian physician, was born on November 14, 1891, in a farmhouse near Alliston, Ontario. He was the youngest of 5 children of **William Thompson Banting** and **Margaret Grant**. He attended public and high schools in Alliston. He attempted to enter the army but was refused due to poor eyesight. He then attended the University of Toronto in the Faculty of Divinity but soon transferred to Medicine. He received his M.B. degree in 1916 and enlisted in the Canadian Army Medical Corps, which had a need for medicos in World War I. He was wounded at the Battle of Cambrai in 1918. Despite his injuries, he helped other wounded men for 16 hours, until another doctor told him to stop. He was awarded the **Military Cross**, in 1919, for heroisn.

Banting returned to Canada after the war and briefly took up general practice in London, Ontario. Returning to Toronto, he studied Orthopaedic Medicine and, during 1919–1920, was Resident Surgeon at the Hospital for Sick Children. During 1920–1921, he continued his general practice, while teaching orthopaedics and anthropology part-time at the University of Western Ontario in London. From 1921 to 1922, he lectured at the University of Toronto, receiving his M.D. degree in 1922, along with a Gold Medal.

An article he read about the pancreas piqued Banting's interest in **diabetes**. Research by **Naunyn, Minkowski, Schafer** and others suggested that diabetes resulted from a lack of a protein hormone secreted by the islets of Langerhans in the pancreas. Schafer had named this hormone *insulin*. Insulin was thought to control the metabolism of sugar; its lack led to an increase of sugar in the blood which was then extracted in urine. Attempts to extract insulin from ground-up pancreas cells were unsuccessful, likely because of the destruction of the insulin by the proteolytic enzyme of the pancreas. The challenge was to find a way to extract insulin from the pancreas prior to it being destroyed.

Moses Baron published an article on experimental closure of the pancreatic duct by ligature which further influenced Banting's thinking. The procedure caused deterioration of the cells of the pancreas which secrete trypsin but left the islets of Langerhans intact. Banting realized that this procedure would destroy the trypsin-secreting cells but not the insulin. Once the trypsin-secreting cells died, insulin could be extracted from the islets of Langerhans. Banting discussed this approach with his physiology professor, **J.J.R. Macleod**, at the University of Toronto. Macleod provided experimental facilities and the assistance of one of his students, **Dr. Charles Best**. Banting and Best began the production of insulin, already discovered in 1916 by Romanian physiologist **Nicolae Paulescu**, by this means.

Banting was appointed Senior Demonstrator in Medicine at the University of Toronto, in 1922. The following years (*i.e.*, in 1923), he was elected to the new Banting and Best Chair of Medical Research, endowed by the *Legislature of the Province of Ontario*. At the Banting and Best Institute, he researched **silicosis, cancer**, and the **mechanisms of drowning**.

Banting, along with C.H. Best, is credited with curing dogs ailing with *diabetes mellitus* by giving them pancreatic extract containing insulin. Later, insulin purified from pig or cow pancreas was successfully tried on humans but it produced immunogenic reactions in them. However, in 1982, using genetic engineering (recombinant DNA technology), insulin was manufactured and marketed for human use. Thus, *insulin became the first hormone product of genetic engineering*. The new recombinant insulin has the exact structure of human insulin and hence will not produce immunogenic reactions. Sir Banting (and not Best) was awarded the **1923 Nobel Prize in Medicine or Physiology** for demonstrating the curative effect of insulin, along with his physiology professor **J.J.R. Macleod,** also of the University of Toronto. As of September 2012, Banting, who received the Nobel Prize at age 32, remains the youngest Nobel Laureate in the area of Physiology or Medicine.

Prior to the award of the Nobel Prize in 1923, Banting received the *Reeve Prize* of the University of Toronto, in 1922. In 1923, the Canadian Parliament granted him a Life Annuity of $ 7,500. In 1928, Banting gave the Cameron Lecture in Edinburgh. He was a member of numerous medical academies and societies in Canada and abroad, including the British and American Physiological Societies, and the American Pharmacological Society. He was knighted as a **Knight Commander of the Order of British Empire** (KBE), in 1934. In May 1935, he was elected a **Fellow of the Royal Society** (F.R.S.). A *flame of hope* was lit in 1989 as a tribute to Dr. Banting and all the people that have lost their lives to diabetes. The flame will remain lit until there is a cure for diabetes. The flame is located at Sir Frederick Banting Square in London, Ontario, Canada.

Banting developed an interest in **painting** beginning around 1921, while he was in London, Ontario. He became friends with Group of Seven artists **A.Y. Jackson** and **Lawrence Harris**, sharing their love of the rugged Canadian landscape. In 1927, he made a sketching trip with Jackson to the St. Lawrence in Quebec. They also made painting expeditions to Great Slave Lake, Walsh Lake, Georgian Bay, French River and Sudbury District.

Banting married twice. His first marriage was to **Marion Robertson**, in 1924; they had one child, **William** (born, 1928). They divorced in 1932 and Banting married **Henrietta Ball**, in 1937.

In 1938, Banting's interest in **aviation medicine** resulted in his participation with the Royal Canadian Air Force (RCAF) in research concerning the physiological problems encountered by pilots operating high-altitude combat aircraft. During the World War II, he investigated the problems of aviators, such as "blackout" (syncope). On February 21, 1941, Banting was killed in an airplane crash in Newfoundland. He was *en route* to England to conduct operational test on the Franks flying suit developed by his colleague, **Wilbur Franks**.

In 1994, Banting was inducted into the **Canadian Medical Hall of Fame**. Banting's name is immortalised in the yearly Banting Lectures, given by an expert in diabetes, and by the creation of the **Banting and Best Department of Medical research** of the University of Toronto; Banting Memorial High School in Alliston; Sir Frederick Banting Secondary School in London, ON; and Ecole Banting Middle School in Coquitlam, BC. **Banting House**, his house in London, Ontario, was declared a National Historic Site of Canada, in 1997. The crater Banting on the Moon is also named after his brother. In 2004, he was nominated as one of the 10 "Greatest Canadians" by viewers of the Canadian Broadcasting Corporation. When the final votes were counted, Banting finished 4th behind **Tommy Douglas, Terry Fox** and **Pierre Trudeau.**

An oil painting of Sir Frederick Banting in 1925 by **Tibor Polya**, is now in the possession of the National Portrait Gallery of Canada.

CHARLES HERBERT BEST
(LT, 1899–1978)

Best, an American-Canadian physiologist, was born in West Pembroke, Washington County, Maine, on February 27, 1899. He was the son of **Luella Fisher** and **Herbert Huestis Best**, Canadians from Nova Scotia. He moved to Toronto, Ontario in 1915 where he started studying for B.A. degree at the University College, University of Toronto. In 1918, he enlisted in the Canadian Army serving with the 2nd Canadian Tank Battalion. After, the World War I, he completed his degree in physiology and biochemistry. As a 22-year-old medical student at the University of Toronto, he worked as an assistant to **Dr. Frederick Banting** and played a major role in the discovery of the pancreatic hormone **insulin**– one of the more significant medical advances, enabling an effective treatment for diabetes.

In the spring of 1921, Banting traveled to Toronto to visit **John J.R. Macleod**, professor of physiology at the University of Toronto, and asked Macleod if he could use his laboratory. Macleod was initially sceptical, but eventually agreed before leaving on holiday for the summer. Before leaving for Scotland, he supplied Banting with 10 days for experiment and 2 medical students, **Charles Best** and **Clarke Noble**, as lab assistants. Since Banting only required one assistant, Best and Noble flipped a coin to see which would assist Banting first. Banting won and took the first shift. Loss of the coin toss proved unfortunate for Noble, given that Banting decided to keep Best for the entire summer and eventually shared half of his Nobel Prize money and a large part of the credit for the discovery of insulin. Had Noble won the toss, his career might have taken a different path.

In 1923, the Nobel Prize Committee honoured Banting and his physiology professor J.J.R. Macleod with the **Nobel Prize in Physiology or Medicine** for the *discovery of insulin*, ignoring **Charles H. Best**. This incensed Banting who then chose to share half of the prize money with Best.

In 1924, Best married **Margaret Hooper Mahon** in Toronto. They had 2 sons. He succeeded Macleod as professor of physiology at the University of Toronto, in 1929. During World War II (1944–1948), he was influential in establishing a Canadian programme for securing and using dried human blood serum. In his later years, he was an advisor to the Medical Research Committee of the United Nations World Health Organization.

Best had a fair share of awards and honours to his credit. In 1948, he was elected a Foreign Honorary Member of the American Academy of Arts and Sciences. In 1967, he was made a **Companion of the Order of Canada** in recognition for "his contribution to medicine, particularly as co-discoverer of insulin". He was a commander of the Civil Division of the Order of the British Empire and was made a member of the Order of the Companions of Honour, in 1971, "for services to Medical Research". He was a **Fellow of the Royal Society of London** (F.R.S.), the Royal Society of Canada and was the first Canadian to be elected into the **Pontifical Academy of Sciences.** In 1994, he was inducted into the **Canadian Medical Hall of Fame**. In 2004, he was inducted into the **National Inventors Hall of Fame**. Dr. Charles Best Secondary School in Coquitlam, British Columbia, C.H. Best West Elementary School in Burlington, Ontario, and C.H. Best East Middle School in Toronto, Ontario, are named in his honour. Best died on March 31, 1978.

Table 34–4. Differences in amino acid sequences in insulin of various species

Species	*Amino acid present in chain A at position*		
	8	9	10
Beef or Cattle	Ala	Ser	Val
Pig or Hog	Thr	Ser	Ile
Sheep	Ala	Gly	Val
Horse	Thr	Gly	Ile
Whale	Thr	Ser	Ile

Dorothy Crowfoot Hodgkin and coworkers of Oxford have shown that porcine and presumably also bovine insulin (the types used for injection in man) are hexameric molecules composed of three subunits (Fig. 34–28). Each subunit, in turn, is a dimer composed of two polypeptide chains, A and B. This multicomponent molecule of insulin, thus, is a triangular ring like structure consisting of three tilted, football like dimers around two central zinc atoms in the core. The hexamer is held together by a variety of forces. The ends of the overlapping dimers are apparently joined by interlocking phenylalanine groups. There is also a hydrogen bonding between dual glutamic acid groups midway in the dimer. The two chains of the monomers are linked by hydrogen and hydrophobic interactions and by two disulfide bridges between the chains. The role of two zinc atoms in the quaternary structure is yet to be ascertained.

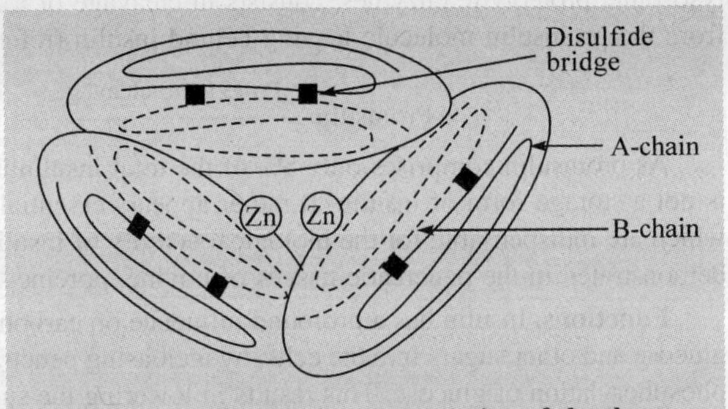

Fig. 31–28. Schematic representation of the three-dimensional structure of insulin

[The zinc atoms are apparently involved in the formation of the quaternary structure]

(Adapted from Orten and Neuhaus, 1970)

Biosynthesis. β-cells of the pancreas synthesize insulin by the ribosomes of the endoplasmic reticulum. Previously, it was suggested that the two chains of insulin are synthesized independently and later these combine by disulfide bonds. But now **Donald F. Steiner** *et al* (1967) have shown that it is formed from its precursor, *proinsulin*. Proinsulin has been isolated and purified from pancreatic extracts. It is a linear protein with 84 amino acid residues and has a molecular weight of about 9,100.

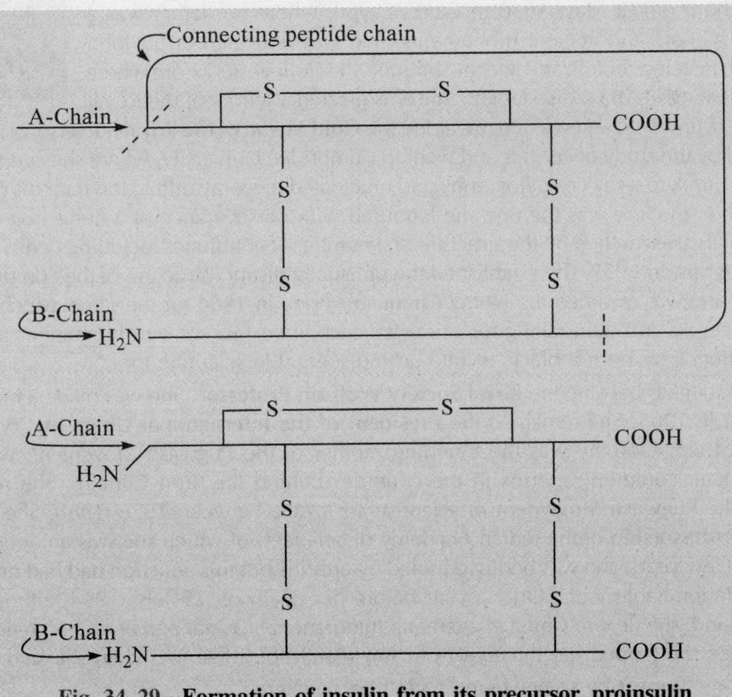

Fig. 34–29. Formation of insulin from its precursor, proinsulin

The transformation of proinsulin to insulin takes place in the granules and not in the endoplasmic reticulum where synthesis of proinsulin takes place. The conversion, which is brought about by lysosomal proteolytic enzymes, consists in cleavage of a 33 amino acid-connecting peptide chain from the proinsulin molecule leaving behind insulin (refer Fig. 34–29).

$$\text{Proinsulin} \xrightarrow{\text{Proteolytic cleavage}} \text{Insulin} + \text{Peptide}$$

As proinsulin comprises only 5% of the total insulinlike protein of the islets of Langerhans, it is not a storage form of insulin. It rather appears essential in the formation of the disulfide bonds which are indispensable for the biological activity of insulin. The formation of proinsulin has been demonstrated in the pancreatic tissues of bovine, porcine, rat and man as well.

Functions. Insulin has a profound influence on carbohydrate metabolism. It facilitates entry of glucose and other sugars into the cells, by increasing penetration of cell membranes and augmenting phosphorylation of glucose. This results in lowering the sugar content of the blood – a fact leading to its common name, **hypoglycemic factor**, which may be abbreviated as **hG-factor.**

Insulin administration promotes protein synthesis (proteogenesis) by assisting incorporation of amino acids into proteins. This effect is not dependent on glucose utilization. At the same time, it also acts as an *antiproteolytic agent*, *i.e.*, discourages excessive breakdown of tissue protein. This action is similar to that of GH and testosterone.

DOROTHY CROWFOOT HODGKIN

(LT, 1910–1994)

An outstanding British scientist of the 20th century who was endowed with a brilliant mind and an iron will to succeed and in many ways proved herself a trendsetter. One of her students is **Margaret Thatcher**, the 'Iron Lady' and ex Prime Minister of Great Britain. **Dorothy Mary Crowfoot** (her name before she married) was born on 12 May, 1910 in Cairo (Egypt), where her father was a supervisor of schools and ancient monuments. She was educated at Sir John Leman School, Beccles, Suffolk and later at Sommerville College at Oxford where she, after graduating in 1931, also taught. She was elected a fellow of the Royal Society at age 37 (in 1947) and was also awarded the **Gold Medal of the Royal Society** in 1956. Dorothy was fascinated by the study of crystals and went to Cambridge University, where she eventually joined the faculty. She took to x-ray crystallography and elucidated the x-ray diffraction patterns of pepsin and insulin. Insulin's structure was the one she laboured with perseverance of a hedgehog (*Erinaceus*, a mammal). She also researched on the structure of a number of antibiotics including penicillin (in 1945) and cephalosporins. In 1957, she established the unusually distinct structure of the complex molecule of vitamin B_{12} and was awarded the **Nobel Chemistry Prize in 1964** for the same. Much of her work required pioneering and painstaking experimental technique that, as one biographer put it, "transformed crystallography from a black art into an indispensable scientific tool."

In 1960, Hodgkin became the **Royal Society Wolfson Professor.** She was made a member of the Order of Merit in 1965. She remained the **President of the International Union of Crystallography** from 1969 to 1975. Dorothy was the founding mother of the *Pugwash Movement*, which helped scientists maintain communication with the countries behind the 'Iron Curtain'. She remained the **President of the Pugwash Movement** of scientists for peace between 1976–1988. She held the **Sir C.V. Raman Professorship** of the Indian Academy of Sciences, of which she was an honorary fellow. During the last few years, she was bodily crippled by arthritis but this affliction had had no debilitating effect on her indomitable will. Only a year before her death on 29 July, 1994, although she was wheelchair-bound, she flew to China to attend an International Crystallography Conference. Such was Hodgkin! Once she helped her three-year-old son search for a lost toy when she said:

"It must be somewhere; it can't be nowhere."

Synthesis of lipids (*lipogenesis*) is also stimulated by administration of insulin. Insulin, thus, promotes uptake by body cells (esp., muscle, liver, adipose cells) of free glucose and of amino acids by muscles; it is essentially anabolic in its action.

> One of the present authors (JL Jain) proposes the abbreviation **'hG-factor'** for the hypoglycemic factor insulin, where the small letter 'h' justifiably denotes 'small' or hypo function. It should, however, not be abbreviated as 'HG- factor', which is aptly kept reserved for the hyperglycemic factor glucagon, where the capital letter 'H' signifies 'big' or hyper function.

Insulin also influences the inorganic metabolism *esp*., that of phosphate and potassium. Insulin administration lowers the blood phosphate level and facilitates absorption of inorganic phosphate by the cells. This phosphate appears within the cells as ATP. A similar mechanism operates in potassium uptake.

It may be said, in general, that insulin promotes anabolic processes (synthesis of glycogen, fatty acids and proteins) and inhibits catabolic ones (breakdown of glycogen and fat).

Insulin deficiency. The deficiency of insulin caused either by inadequate insulin production or by accelerated insulin destruction, leads to *diabetes mellitus* in man. This disease is characterized by:

1. an increase in blood sugar or glucose (*hyperglycemia*) from a normal value of 80 mg/100 ml of plasma to abnormal value ranging between 150—200 mg/100 ml of plasma.
2. the appearance of sugar in the urine (*glycosuria*); with the result, the victim's urine tastes sweet.
3. an increase in concentration of ketone bodies in the blood (*ketonemia*) and in the urine (*ketonuria*).
4. the excretion of large quantities of urine (*polyuria*), frequently at night (*nocturia*), leading to dehydration.
5. the excessive drinking of water (*polydipsia*) on account of an unrelenting thirst.
6. the excessive eating (*polyphagia*) due to feeling constant hunger. This is because the tissues cannot utilize glucose normally, even though they need fuel. Consequently, the body wastes away, despite a large intake.
7. the lack of energy (*asthenia*) which is apparently caused mainly by loss of body protein.

CLINICAL IMPLICATIONS 34–1.

Diabetes Mellitus

The Greek physician Arteus (ca A.D. 200) apparently was the first to call insulin deficiency disease as **diabetes** (to follow through; urine), referring to a large volume of urine passed in it. Regarding the disease, he wrote: "The epithet diabetes has been assigned to the disorder, being something like passing of water by a siphon." He perceptively characterized diabetes as "being a melting-down of the flesh and limbs into urine". Later, the Latin term **mellitus** (sweetened or honey-like) was added to it and dates from the time when urine was tested by tasting, and the urine in this condition is sweet to the taste. Used alone, diabetes is taken to mean diabetes mellitus, by far the most common (80–90%) of the two types of diabetes (*diabetes mellitus* and *diabetes insipidus*). Normally, some sugar (glucose) is in the blood at all times. These terms, glucose and sugar, are used interchangeably, although with regard to diabetes we almost always mean glucose. The level of glucose is usually between 60 and 120 milligrams per cent (mg %), which means that there are 60 to 120 milligrams of glucose in each 100 cubic centimeters (cc) of blood. Milligrams per cent is sometimes expressed as milligrams per deciliter (mg/dL). *Diabetes mellitus is characterized by a level of glucose in the blood that is above normal, most of the time.*

Glucose is the major fuel of the body. We ingest a great deal of glucose each day, as carbohydrates (starches) are made of glucose. When we eat glucose, it gets into the bloodstream and ultimately into each of the individual cells of the body to provide them with energy. Glucose cannot just flow into the cells, however. Cells are enclosed in membranes that separate what is inside the cell from what is outside. Somehow, the cells must be told that the glucose waiting outside the cell should be allowed in. That is what insulin does. It is needed to signal to the cells that they should allow the glucose, that normally is in the blood, to penetrate their outer layer. If the glucose cannot enter the cells, it "backsup" in the blood. The level of glucose in the blood increases and a state of diabetes is produced. Such a condition occurs if there is a lack of insulin for whatever reason. Perhaps the pancreas cannot produce enough insulin.

Diabetes mellitus is of 2 types : type I (insulin-dependent) and type II (noninsulin-dependent). The use of Roman numerals probably dignifies the importance of this classification.

Type I (Insulin-dependent diabetes mellitus, IDDM) : IDDM occurs because the insulin producing β-cells are destroyed and there is not enough insulin produced. This form of diabetes usually occurs at an early age and requires the injection of insulin to make up for the failure of the pancreas to produce enough insulin. In the past, there was treatment that had some measure of success. This was *relative starvation.* People with IDDM had very short careers. The former synonyms for this disease are *brittle diabetes* or *juvenile-onset diabetes*.

Type II Noninsulin-dependent diabetes mellitus, NIDDM) : NIDDM is caused by a relative insulin insufficiency due to insulin resistance– the inability of the insulin to tell the cells to use glucose– plus insufficient insulin to overcome this resistance. This is more common of the 2 types of diabetes, in which the body fails to utilize insulin properly. The result is swings in blood levels that affect daily as well as long-term health. Regarded as a disease of middle or old age until recently, type II diabetes is now being seen increasingly in obese teenagers and even in some obese young children. Patients with NIDDM often lived for many years, if they heroically reduced their weight to live with the small amount of insulin that might be available. In a sense, this was *organised starvation*. NIDDM is also known by the former names, *stable diabetes* or *adult-onset diabetes*. In case of type II diabetes, it may be said that heredity may load the cannon, but stress or obesity pulls the trigger.

Table given below draws comparison between type I and type II diabetes.

Comparison of two types of diabetes mellitus

Item	*Type I (IDDM)*	*Type II (NIDDM)*
1. Percentage of all people with diabetes	20%	80%
2. Dependence on insulin	Dependent	Not dependent
3. Age of discovery	In children or in adults, usually under 40	In adults, usually over 40
4. Cause of diabetes	Reduced or none insulin production	Insulin resistance and relative or absolute deficiency of insulin
5. Pancreatic insulin content	0	> 5% of normal
6. Weight	Often underweight or normal weight	Usually overweight but 20% are normal weight
7. Primary insulin resistance	Minimal	Marked
8. Anti-islet antibodies	85%	> 5%
9. Concordance rate of identical twins for diabetes mellitus	25–50%	~ 100%
10. Condition when discovered	Usually moderately to severely ill	Often not ill at all or having mild symptoms
11. Acute complications	Ketoacidosis	Nonketotic, hyperosmolar, hyperglycemic comma, not usually prone to ketoacidosis.
12. Usual treatment	Insulin, eating plan, exercise	Diet, exercise, insulin, if needed oral agents

The control of diabetes, however, depends on the triad of food, activity and insulin or oral agents. All of these, however, must be balanced properly by the person with diabetes.

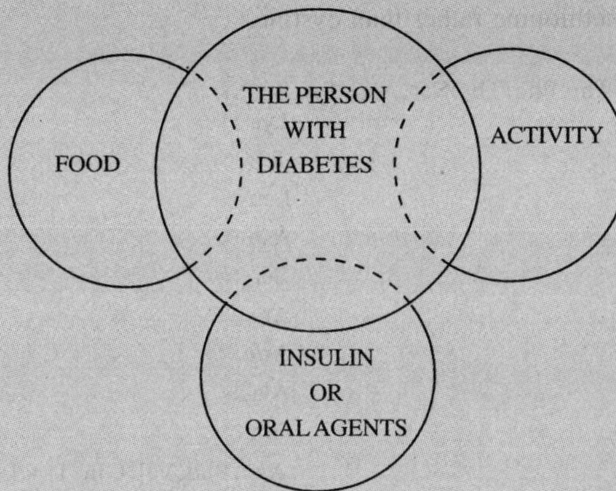

Diabetes and its controlling triad

The diabetic person gradually becomes weaker and loses weight due to the failure of glucose utilization by the body despite a voracious appetite. Diabetic comma ensues, the plasma volume decreases and the kidney fails. This eventually leads to death. In England, the disease for centuries was approximately called the "***pissing evil***".

Since insulin is a protein, it is not amazingly digested and thus inactivated by the proteolytic enzymes, pepsin and trypsin. Hence, insulin has no effect when taken orally and must be, therefore, administered parenterally in diabetic patients.

Parenteral is any route outside the gastrointestinal tract, including subcutaneous, intramuscular, intraperitoneal and intravenous injections or infusions. In practice, parenteral feeding is done intravenously.

Hypoglycemic agents. There are many hypoglycemic drugs, effective when taken orally, that control diabetes. Certain of these drugs, which belong to the group sulfonamide, are tolbutamide (orinase), chlorpropamide (diabinese) and tolazamide (tolinase).

In 1978, **Riggs** and his collaborators of the City of Hope National Medical Centre, California, have been able to produce human insulin by using artificial genes which make bacteria produce it. The scientists first synthesized a gene that dictates insulin production. This gene was then inserted into molecules of a harmless laboratory strain of human bacteria called *Escherichia coli*. These bacterial molecules are called **plasmids** or **DNA rings.** In the plasmids, the genes were joined to clusters of regulatory genes called **lac operons** which react to the presence of milk sugar (lactose) by switching on the bacteria's protein-making apparatus. When an experimental colony of E. coli was salted with the new genes and plasmids, it obeyed the command to make protein. The protein in this case was obviously insulin since that was the DNA blueprint in the bacteria's doctored genetic material. This finding, however, might benefit the world's 100 million diabetics by providing them better and probably cheaper insulin. Besides, about 5% of the diabetics are allergic to the animal insulin now in use and the man-made carbon copy of human insulin will be a boon to them. The greatest achievement of this new research, however, is that it has opened up large vistas in a new era of biology since any protein can be *tailor-made* by this process. The cell proteins include hormones, antibodies, enzymes etc., and these can now be produced at will.

Glucagon

Structure. Glucagon was first isolated in crystalline form by **Behrens** and others. This peptide hormone has a molecular weight of 3,482 and is isolectric at pH 8. It has 29 amino acid residues (of

15 different types) arranged in a linear row. Histidine is the N-terminal amino acid and threonine, the C-terminal amino acid. Unlike insulin, it contains no cystine, proline or isoleucine, but possesses methionine and tryptophan in appreciable amounts (Fig. 34–30). The small amount of sulfur present is, thus, in the form of methionine rather than cystine.

His.Ser.Gln.Gly.Thr.Phe.Thr.Ser.Asp.Tyr.Ser(1–11).Lys(12).Tyr(13).Leu(14).Asp(15).Ser(16).Arg(17).Arg(18).Ala(19).Gln(20).Asp.Phe.Val.Gln.Trp.Leu.Met.Asn.Thr(21–29)

1	His
2	Ser
3	Gln
4	Gly
5	Thr
6	Phe
7	Thr
8	Ser
9	Asp
10	Tyr
11	Ser
12	Lys
13	Tyr
14	Leu
15	Asp
16	Ser
17	Arg
18	Arg
19	Ala
20	Gln
21	Asp
22	Phe
23	Val
24	Gln
25	Trp
26	Leu
27	Met
28	Asn
29	Thr

Fig. 34–30. Structure of glucagon

Functions. Like insulin, glucagon also influences carbohydrate metabolism but in an opposing way (refer Fig. 34–31). Glucagon (as also the other hormone, epinephrine) activates the enzyme adenyl cyclase which converts ATP to cyclic AMP. The latter compound activates phosphorylase *b* kinase which, in its turn, activates phosphorylase *b* to yield phosphorylase *a*. This releases glucose-l-phosphate from glycogen of liver. Glucose-l-phosphate then yields free glucose in blood, whereby increasing blood sugar contents. It is because of this reason that the hormone is also termed as **hyperglycemic factor** or **HG-factor**.

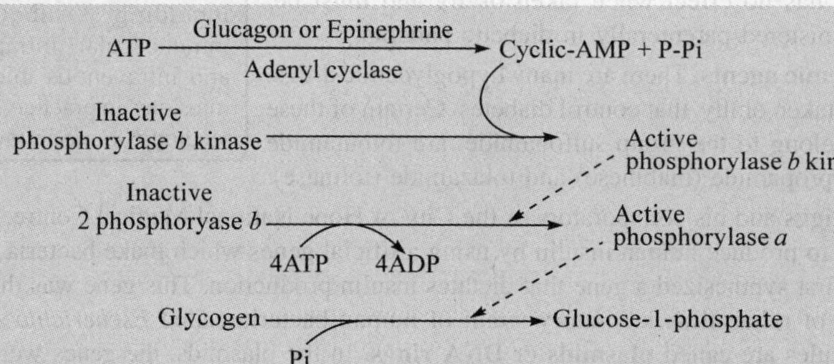

Fig. 34–31. Role of glucagon and epinephrine in glycogen synthesis

(*After Fairley JL and Kilgour GL 1966*)

In contrast to epinephrine, glucagon does not cause an increase in blood pressure. Therefore, glucagon and not epinephrine has found clinical applications and is administered in patients with acute hypoglycemia.

Acting in the liver, it stimulates glycogenolysis (glycogen breakdown) and gluconeogenesis (production of glucose from noncarbohydrate sources such as proteins and fats). The former function is similar to that of ACTH and epinephrine.

Glucagon also affects lipid metabolism by accelerating ketogenesis and inhibiting synthesis of fatty acids.

Glucagon has a catabolic action on proteins. Its administration in the body results in excretion of enough nitrogen and phosphorus, in decrease of liver tissues and in loss of body weight.

Hyperglycemic agent. Overdosages of insulin, given to diabetic patients, often result in acute hypoglycemia. This may be cured by giving crystalline glucagon (in the form of glucagon hydrochoride) either intramuscularly or intravenously. Response may be observed within 10–15 minutes after administration of 0.5 to 1.0 mg of glucagon.

2. Hormones of the Hypophysis or Pituitary Gland

Secretory gland. Hypophysis (meaning, undergrowth) is so named because of its location below the brain as an undergrowth. Its synonym pituitary gland is, however, misleading as the gland is not concerned with the secretion of mucus or phlegm (*pituita*[L] = phlegm), as was thought previously. This is a composite unpaired small, but essential, ovoid gland and is no larger than the end of the little finger (*i.e.,* pea-sized). It is located at the base of the brain and lies below the diencephalon in a depression of basisphenoid bone of the skull called sella turcica. It is a complex structure formed of ectodermal outgrowth of the mouth cavity and downgrowth of the infundibulum (= *pituitary stalk*).

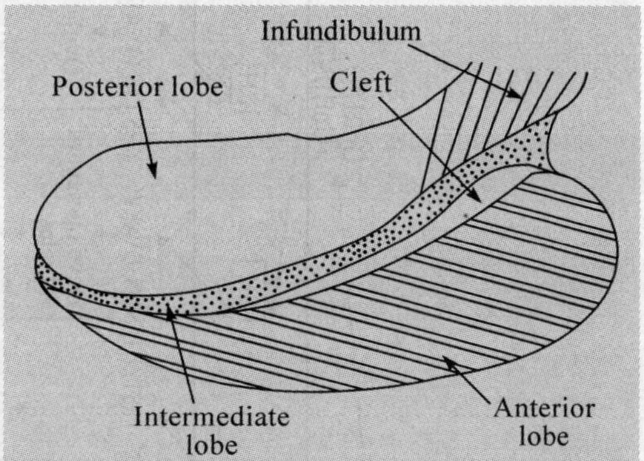

Fig. 34–32. The hypophysis in man
(*After D. Marsland, 1964*)

The human hypophysis is a reddish- grey oval structure and measures about 10 mm in diameter. Its average weight varies in the two sexes: 0.5—0.6 g in males and 0.6—0.7 g in females. It consists of 3 lobes (Fig. 34–32):

(*a*) an anterior richly vascular largest lobe, pars distalis or adenohypophysis.

(*b*) an intermediate relatively avascular smallest lobe, pars intermedia.

(*c*) a posterior neural lobe, pars nervosa or neurohypophysis.

While all the three parts are important, it is the anterior lobe that seems to be essential to life.

This gland plays perhaps the most dominant role as it secretes hormones which govern the secretion of other endocrine glands (like thyroid, adrenal and gonads) and also its secretions have a direct effect on the metabolism of nonendocrine tissues (refer Fig. 34–33). It has, therefore, aptly been described as the '*master gland*' of the system or the 'master (= *conductor*) *of endocrine orchestra*'. However, the view held by some biochemists is that since the hypophysis is subservient to the nervous system and some of the other endocrine glands, it is erroneous to call this gland as the master gland of the body.

Pars distalis or *Adenohypophysis*

The anterior lobe is the largest and shares about 70% of the total weight of hypophysis. The adenohypophysis originates from an embryonic invagination of the pharyngeal epithelium called Rathke's pouch. This explains the epithelioid nature of its cells. It consists of glandular epithelial cells of varying shapes and sizes, arranged in columns separated by sinusoids containing blood. In general, there is one type of cell for each type of hormone that is produced in this gland. These various cell types can be differentiated from one another on the basis of special staining techniques (the only likely exception to this is that the same cell type may secrete both luteinizing hormone and follicle-stimulating hormone). Using acid-base histological stains, only 3 types of cells may be differentiated in this region:

(*a*) *acidophils* or *α cells*— these stain strongly with acidic dyes (acid fuchsin) and secrete growth hormone and prolactin.

(*b*) *basophils* or *β cells*—these stain strongly with basic dyes (aniline blue) and secrete luteinizing hormone, follicle-stimulating hormone and thyroid-stimulating hormone.

(*c*) *neutrophils* or *chromophobe cells*— these do not stain with either and are believed to secrete adrenocorticotropic hormone.

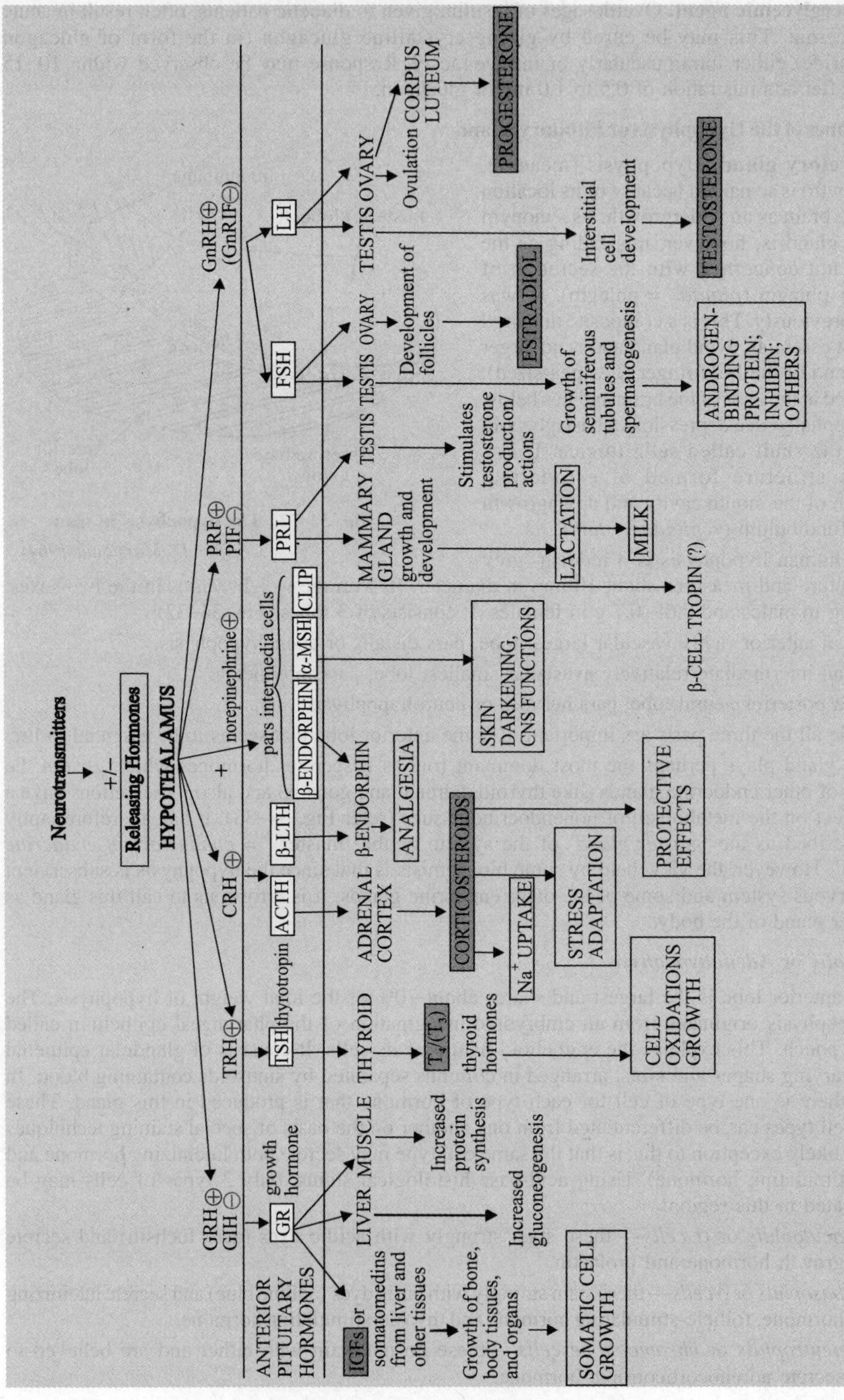

Fig. 34–33. Overview of anterior pituitary hormones with hypothalamic releasing hormones and their actions

The anterior lobe produces hormones which govern the production of hormones secreted by other glands. These hormones are called as **tropins** (*tropos*[G] = turning) or **trophic** (*trophikos*[G] = nursing) **hormones**. Evidences available indicate that the rate of secretion of a trophic hormone is inversely proportional to the concentration, in the blood, of the hormone with which it is related. For example, a high blood level of thyroid hormone tends to inhibit the secretion of TSH from the adenohypophysis and a low level causes an increased production of it.

Neurohormones— The secretion of thyrotropin (as of almost all other pituitary hormones) is controlled by the hormones (or factors) released from hypothalamus, a region of the brain immediately proximal to the pituitary. These hormones are called as **hypothalamo-releasing hormones** or **hypothalamic factors.** These have been classified as *neurohormones*, *i.e.*, those produced by the nerve cells. These are unlike *neurohumors* (*e.g.*, acetylcholine, serotonin, norepinephrine etc) which are released at nerve endings and activate the adjacent nerve bodies. The neurohormones, on the contrary, are released into the blood and activate cells a little far from their point of release. These are introduced into the capillary of the hypothalamo-hypophyseal portal system at the floor of the hypothalamus called median eminence. In addition, the release of anterior pituitary hormone may be inhibited by *a release inhibiting* factor which passes down the same hypothalamo-hypophyseal portal veins.

The various hypothalamic factors controlling the release of pituitary hormones have been listed in Table 34–5.

Table 34–5. Pituitary hormones and the hypothalamic factors controlling their secretion

S.N.	*Pituitary Hormones*	*Hypothalamic Releasing Factors**
1.	Thyrotropin, TSH	Thyrotropin-releasing factor, TRF
2.	Corticotropin, ACTH	Corticotropin-releasing factor, CRF
3.	Follicle-stimulating hormone, FSH	Follicle-stimulating hormone-releasing factor, FSH-RF
4.	Luteinizing hormone, LH	Luteinizing hormone-releasing factor, LH-RF
5.	Prolactin, PL	Prolactin-releasing factor, PRF
6.	Growth hormone, GH	Growth hormone-releasing factor, GH-RF
7.	Melanocyte-stimulating hormone, MSH	Melanocyte-stimulating hormone-releasing factor, MRF

* Also, sometimes called hormones, so in some books, the abbreviations may have an H instead of an F at the end.

(*After Schally et al, 1973*)

Some 30 tropins are secreted from the anterior lobe. The four important tropins are described below.

1. Thyrotropin or thyroid-stimulating hormone, TSH. It is a glycoprotein with molecular weight about 30,000. Each molecule has 8-9 cystine residues and the disulfide groups are present as intrachain linkages rather than interchain linkages.

In general, it stimulates the activity of thyroid gland and enhances the rate of certain reactions such as:

(*a*) removal of iodide from blood by thyroid

(*b*) conversion of iodide to thyroid hormones

(*c*) release of hormonal iodine from thyroid.

The release of thyrotropin is controlled by another hormone from hypothalamus called thyrotropin-releasing factor, TRF.

2. Corticotropin or adrenocorticotrophic hormone, ACTH. Corticotropin (Fig. 34–34) is a straight chain polypeptide with a molecular weight of about 4,500 and consists of 39 amino acid residues in mammals like man, ox, sheep and pig. The most potent segment of activity is from residue 15 to 18. ATCH molecule in these species differs from each other only in the constituents present from residue 25 to 33. Thus far, no differences in residues 1-24 and 34-39 have been reported.

ATCH, in general, has a stimulatory effect on the hormone-producing capacity of the adrenal cortex. ATCH administration leads to accelerated gluconeogenesis (= neoglucogenesis) with accompanied retardation of protein synthesis in all tissues except liver. It also possesses an intrinsic melanocyte-stimulating activity, causing darkening of the skin in a manner similar to that of another hormone, MSH.

Oversecretion of ACTH results in Cushing's disease, already described earlier.

Certain peptides found in hypothalamus and also in neurohypophysis have ACTH-releasing activity. These have been termed as corticotropin-releasing factor, CRF.

```
 1   2   3   4   5   6   7   8   9   10
Ser.Tyr.Ser.Met.Glu.His.Phe.Arg.Trp.Gly.Lys 11
                                         Pro 12
                                         Val 13
                                         Gly 14
                                       { Lys 15
             Most potent segment      {  Lys 16
             of activity              {  Arg 17
                                       { Arg 18
                                         Pro 19
                                         Val 20
                                         Lys 21
                                         Val 22
                                         Tyr 23
                                         Pro 24
                                         Asp 25
                                         Ala 26
                                         Gly 27
                                         Glu 28
                                         Asp 29
                                         Gln 30
                                         Ser.Ala.Glu.Ala.Phe.Pro.Leu.Glu.Phe
                                          31  32  33  34  35  36  37  38  39
```

Fig. 34–34. Structure of human adrenocorticotropin

3. Gonadotropins or gonadotrophic hormones, GTH. These hormones control the development and functioning of the gonads which remain dormant until the age of 12-14 years in the human beings. Damage to certain areas of the hypothalamus greatly decreases the secretion of gonadotrophic hormones by the anterior pituitary. If this occurs prior to puberty, it causes typical eunuchism. The damage often causes simultaneous overeating because of its effect on the feeding centre of the hypothalamus. Consequently, the person develops severe obesity along with the eunuchism. This condition is called **adiposogenital syndrome** or **Frohlich's syndrome** or **hypothalamic eunuchism**. Three gonadotropins are known:

A. Follitropin or follicle-stimulating hormone, FSH. It is a glycoprotein that contains galactose, mannose, galactosamine, glucosamine, sialic acid, fucose and uronic acid. It has a molecular weight of about 30,000 in man. In human females, it induces the growth of graafian follicles resulting in an increased weight of the ovary. In males, however, FSH promotes spermatogenesis by stimulating the development of seminiferous tubules, thus leading to the formation of a large number of spermatocytes.

The release of this hormone is controlled by another hormone from hypothalamus called follicle-stimulating hormone-releasing factor, FSH-RF.

B. Luteinizing hormone, LH or interstitial cell-stimulating hormone, ICSH. It is a peptide hormone with molecular weight of about 26,000 (in man) or 100,000 (in swine). It lacks tryptophan but has a high content of cystine and proline. Each molecule contains 10 glucosamine and 3 galactosamine residues. In females, LH is concerned with the ripening and rupturing of ovarian follicles, which later transform into corpus lutea. It also induces the development of interstitial cells of both the ovaries and the testes– a fact responsible for its nomenclature.

The secretion of LH is controlled by luteinizing hormone-releasing factor, LH-RF, a secretion from hypothalamus. The long-acting analogue of LH-RF has been found useful in the treatment of precocious puberty in females by **Florence Comite** *et al* (1983), the American researchers at the National Institute of Health. Puberty is initiated by pulsed nocturnal secretions of gonadotropins that result from the release of the hormone LH-RF by the hypothalamus gland in the brain. The administration of LH-RF analogue "initially stimulated but subsequently inhibited" the release of LH and FSH, the two sex hormones that initiate puberty.

C. Luteotropin or luteotrophic hormone, LTH. Because of its broad spectrum of effects on vertebrates in general, *luteotropin is the most versatile of all the adenohypophyseal hormones.* It is also the first anterior pituitary hormone to be obtained in pure form. This is also a peptide hormone with 198 amino acid residues and a molecular weight of about 23,500. It has 3 disulfide bonds between cysteine residues at 4-11, 58-173 and 190-198 (**Li** *et al*, 1971). *It differs from FSH and LH in that it contains no carbohydrate*. It is thermolabile and is destroyed by tryptic digestion. Its secretion is increased by suckling. In pure form, it has no growth-promoting, thyrotropic, diabetogenic, adrenotropic or gonadotropic activities. However, in association with estrogen, luteotropin promotes the growth of the mammary glands (mammogenesis) and also induces secretion of milk (lactation) at the time of child birth (parturition). Henceforth, this hormone is variously called as *prolactin*, PL or *lactogenic hormone* or *mammotrophic hormone*, MH. Its prolonged use in preindustrial societes can lead to birth spacings of 3–4 years through disruption of the menstrual cycle (*amenorrhea*). It also stimulates glucose uptake and lipogenesis. Along with androgens, it causes the development of secondary male sex characters. In rat, at least, prolactin also has gonadotrophic activities in that it maintains functional corpora lutea in hypophysectomized animals. It also acts as an anabolic agent

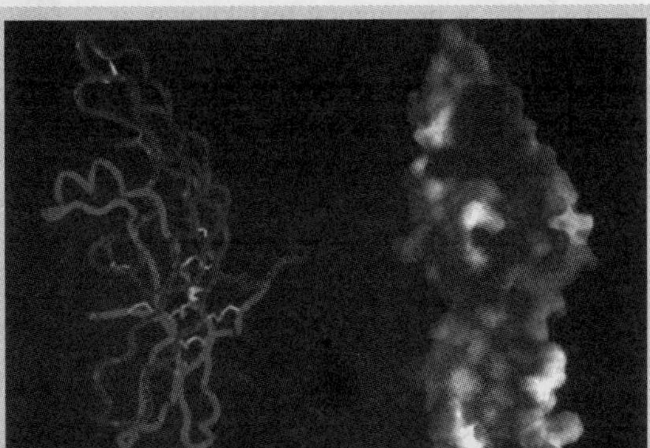

Fig. 34–35. The x-ray structure of human chorionic gonadotropin (hCG)

(*a*) A C_α diagram showing the α subunit in *red* and the β subunit in *blue* with disulfide linkages in *yellow*. The pseudo-twofold axis relating the two subunits is horizontal.

(*b*) The surface electrostatic potential of the hCG as viewed towards its receptor-binding surface. The surface is coloured according to its electrostatic potential, with the most positive regions *dark blue* and the most negative regions *dark red*. The view is related to that in Part (*a*) by a 60º rotation about the vertical axis.

[*Courtesy: Hao Wu and Wayne Hendrickson, Columbia University.*]

mimicking the effects of growth hormone, although it is less active in this respect. In pigeons, however, it stimulates enlargement of crop glands and formation of 'crop milk', and is also necessary for maintenance of incubation (which itself stimulates its secretion) by both sexes. In fact, prolactin is credited with performing some more than 80 functions and it is for the same reason that it has been jocularly termed as a "*jack-of-all-trades*." According to Nicoll and Bern (1971), these various functions of prolactin fall under 5 major categories : reproduction, osmoregulation, growth, integument and synergistic effect with the steroid hormones.

[**N.B.**—A hormone chorionic gonadotropin, although of placental origin, resembles hypophyseal hormones in its biological effects. It was previously recognized as anterior pituitary like factor (APL factor). *Human chorionic gonadotropin, hCG* (Fig 34–35) is a glycoprotein with molecular weight of about 30,000. The carbohydrate moiety contains some 6 components, *viz*., D-galactose, D-mannose, *N*-acetylglucosamine, *N*-acetylgalactosamine, L-fucose and *N*-acetylneuraminic acid. Their sequence is not fully known. *N*-acetylneuraminic acid is essential to the biologic activity of this hormone. Apparently, all the common amino acids are present in the protein moiety, proline being especially high (about 20%).

> The name **anterior pituitary like factor** is, however, inaccurate because the hormone is not effective in hypophysectomized rats whereas the hypophyseal hormones are.

As to its biological role, hCG factor supplements hypophysis in maintaining growth of the corpus luteum during pregnancy. This hormone, when administered, stimulates Leydig tissue and hence the male accessory organ.

The hormone appears sharply after pregnancy in the urine. This fact has become the basis for a pregnancy test known as *Aschheim-Zondek test*. This involves injecting urine or an alcoholic precipitate of urine from a woman into immature female rats. Urine from a pregnant woman containing this hormone will cause rupturing (hemorrhage) of ovarian follicles within 48 hours.

In another pregnancy test known as *Friedman test*, a virgin rabbit is used and the test urine is injected into an ear vein. After 24 hours, the ovaries are examined for ruptured or hemorrhagic follicles.]

4. Somatotropin or somatotrophic hormone, STH or growth hormone, GH. Somatotropin obtained from human hypophysis (Fig. 34–36) is a protein with molecular weight 27,000 (41,600 in pig) and an isoelectric point 4.9. It has 190 amino acids and consists of 2 disulfide bridges between adjacent cysteine residues at 53-164 and 181–188 (**Niall**, 1971; **Li**, 1972). The N-terminal and C-terminal residues are both phenylalanine.

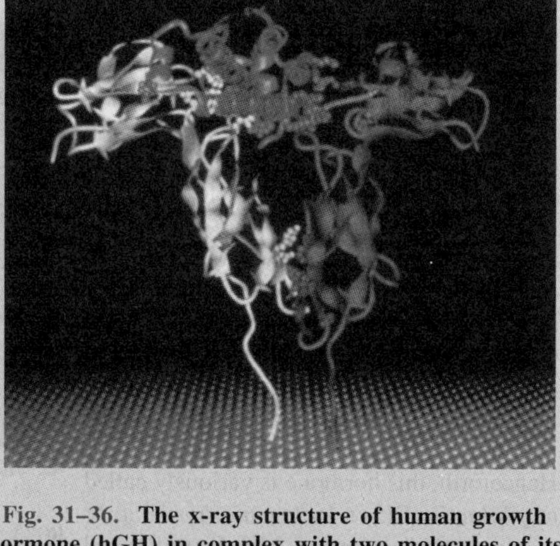

Fig. 31–36. The x-ray structure of human growth hormone (hGH) in complex with two molecules of its receptor's extracellular domain (hGHbp)

The proteins are shown in ribbon form, with the two hGHbp molecules, which together bind one molecule of hGH, *green* and *blue* and with the hGH *red*. The side chains involved in intersubunit interactions are shown in *space-filling form*. The *yellow pebbled surface* represents the cell membrane, through which the C-terminal ends of the hGHbp molecules are shown penetrating as they do in the intact hGH receptor.

(*Courtesy: Abraham de Vos and Anthony Kossiakoff, Genentech Inc., South San Francisco, California.*)

Unlike other adenohypophyseal hormones, the various effects of somatotropin are not due to its influence on other endocrine glands. It acts rather directly upon various tissues to produce diverse effects. *It is, therefore, not a true tropic hormone*. The various metabolic activities particularly attributed to this hormone are listed below.

(*a*) It affects the rate of skeletal growth and gain in body weight. In adult animals with closed epiphyses, STH stimulates chondrogenesis followed by ossification.

(*b*) It causes abnormal increase in blood sugar by producing degenerative changes in islets of Langerhans (*diabetogenic effect*).

(*c*) It stimulates the growth of the islets of Langerhans (*pancreatotropic effect*).

(*d*) It controls the production of fat in the body and its deposition in the liver (*ketogenic effect*).

(*e*) It prevents the fall of muscle glycogen in fasting and hypophysectomized animals.

(*f*) It also stimulates milk secretion in cows and also the growth of mammary glands in hypophysectomized rats (*galactopoietic effect*).

(*g*) STH is also known to cause adrenal enlargement (*corticotropic effect*). The adrenal enlargement can be greatly augmented by simultaneous treatment with low doses of thyroxine, thus indicating the participation of thyroid gland along with the hypophysis in this action.

It is now recognized that as people live to their 70s and beyond, the effect of declining hormones may contribute significantly to chronic, debilitating and costly illnesses. Among the well-known **attributes of ageing** that hormone loss may bring about are loss of muscle mass and strength, an increase of body fat, particularly fat around the abdomen, a weakening of the bones, a decline in immune responses and a general loss of energy. In 1990s, attention has now been diverted to the study of growth hormone and other trophic hormones, substances that promote growth or maintenance of tissues. They may, at least, have a promise for halting or reversing degenerative changes in bones, muscles, nerves and cartilage.

Pituitary diabetes— A general increase in the secretion of all the adenohypophyseal hormones causes elevated blood glucose concentration. This condition is clinically designated as pituitary diabetes. It, however, differs from diabetes mellitus, which results from insulin deficiency, in the following respects :

I. In pituitary diabetes the rate of glucose utilization by the cells is only moderately depressed whereas in diabetes mellitus almost no utilization of glucose takes place.

II. Many of the side effects that result from reduced carbohydrate metabolism in diabetes mellitus are, however, lacking in pituitary diabetes.

Pars Intermedia

In human beings, the intermediate lobe is poorly developed but is much larger and much more functional in some lower animals. It is, however, absent in certain mammals such as whale, Indian elephant and armadillo.

The intermediate lobe secretes **melanocyte** (or melanophore)-**stimulating hormones, MSH** or **melanotrophins.** These are also known as **intermedins**. In chicken and whales, where intermediate lobe is absent, MSH function is taken over by adenohypophyseal hormones. In mammals, MSH is a vestigial hormone. The structure of 2 types of intermedins, α-MSH and β-MSH is given in Fig. 34–37.

Acetyl-Ser. Tyr. Ser. Met. Glu. His. Phe. Arg. Trp. Gly. Lys. Pro. Val amide
1 2 3 4 5 6 7 8 9 10 11 12 13

α–**MSH** (from monkey)

Asp. Glu. Gly. Pro. Tyr. Arg. Met. Glu His. Phe. Arg. Trp. Gly. Ser. Pro. Pro. Lys. Asp
1 2 3 4 5 6 7 8 9 10 11 12 13 14 15 16 17 18

β–**MSH** (from monkey)

Fig. 34–37. Two types of intermedins

[Note that the strucutre of α-MSH is identical with the first 13 residues of ACTH except that in α-MSH terminal serine residue is acetylated.]

α-MSH contains 13 amino acid residues with N-terminal residue acylated and the C-terminal valine is in the amide form. β-MSH has been found to possess 18 amino acid residues with aspartic acid residue at both N- and C-terminals. Thus, in α-type the terminal groups are blocked whereas the β-type has both the end groups free. *The α-MSH has greater biologic activity than β-MSH.*

MSH causes dispersal of black pigment (melanin) in melanophore cells of certain lower vertebrates. This leads to darkening of the skin. This hormone is particularly used in colour adaptation in lower vertebrates. The animal becomes pale on a light background and dark on a dark background. Its function in higher vertebrates including man is not known but it is, however, thought to participate in darkening the areas already pigmented. In humans, it also appears to raise the general excitability of neurones in the central nervous system, probably affecting learning.

Pars Nervosa or Neurohypophysis

The neurohypophysis develops from an outgrowth of the hypothalamus. This explains the presence of glial type cells in the gland. The posterior pituitary secretes 2 hormones : ocytocin and vasopressin. These were separated by **Kamm** and others.

> **Ocytocin** is often called oxytocin. However, the name ocytocin is preferred since its Greek origin (meaning *quick birth*) tallies with it and also it avoids confusion with the prefix *oxy* = oxygen-containing, as the hormone refers in no way to oxygen.

1. Ocytocin (*ocy*[G] = quick ; tokos[G] = birth) or **pitocin.** It is a nonapeptide amide. A disulfide bond (—S—S—) is present to link the two cysteine residues present in the molecule (refer Fig. 9–15). Ocytocin preparations from man, cow and pig are identical. Besides mammals, it is also found in chondrichthyes, lung fishes, amphibians and aves.

Ocytocin stimulates the contraction of smooth muscles, *esp*., those of uterus, thus facilitating coitus and childbirth. Commercial form of ocytocin is frequently used to induce 'labour'. In general, it also causes contraction of other smooth muscles like those of intestine, urinary bladder and the ducts of mammary glands resulting in milk ejection. It is, therefore, also called as *milk-let-down-ejection factor*. Ocytocin levels are increased by suckling which is necessary for the continued formation of milk by the breasts. Optimum milk secretion lasts for about 8 to 10 months after which it gradually falls and ultimately ceases.

2. Vasopressin or **pitressin.** It is also a cyclic nonapeptide amide and resembles ocytocin except that isoleucine is replaced by phenylalanine and leucine by arginine (Fig. 34–38). The hormone was synthesized, in 1953, by **Vincent du Vigneaud** and colleagues from U.S.A. and du Vigneaud received the Nobel Prize in 1955 for the first synthesis of a polypeptide hormone.

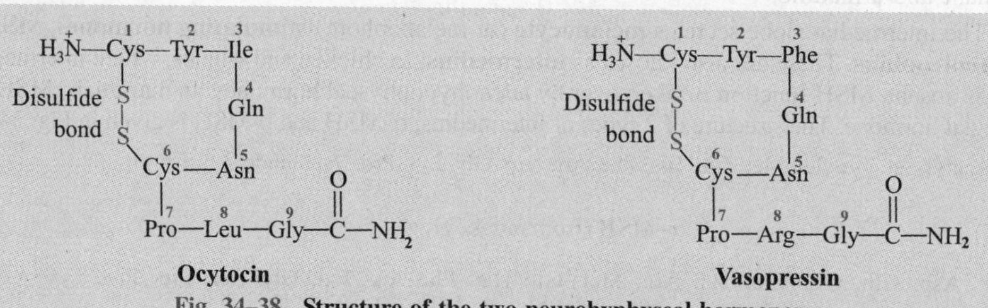

Fig. 34–38. Structure of the two neurohyphyseal hormones

Vasopressin (VP), obtained from pig, differs from that of human beings (shown above) in arginine being replaced by lysine at position 8. It is, therefore, better called as *lysine vasopressin* and that from human beings as *vasopressin.*

VINCENT du VIGNAUD

(LT, 1901–1978)

du Vignaud was a US biochemist who first worked out the structure of vitamin B_7 (*biotin*) in 1942. He graduated from Schurz High School in 1918. He then began studying chemistry at the University of Illinois at Urbana-Champaign. He was influenced by the lectures of **Carl Shipp Marvel**. After receiving his M.S. in 1924, he joined DuPont. He married **Zella Zon Ford** on June 12, 1924. Restarting his academic career in 1925, he joined the group of **John R. Murlin** at the University of Rochester for his Ph.D. thesis. He was graduated in 1927 with his work, *The Sulfur in Insulin*. Later, he became professor at the University of Illinois. He next went to George Washington University Medical School in Washington D.C. in 1932 and to Cornell Medical College, New York City in 1938, where he stayed until his *emeritation* in 1967. Following that retirement, he held a position at Cornell University in Ithaca, New York.

His career was characterized by an interest in sulfur, proteins, and especially peptides. Even before his famous work on elucidating and synthesizing ocytocin (also called oxytocin) and vasopressin, he had established a reputation for work on insulin, biotin, transmethylation and penicillin. He also began a series of *structure–activity relationships* for ocytocin and vasopressin, perhaps the first for peptides. That work culminated in the publication of a book entitled, *A Trail of Research in Sulfur Chemistry and Metabolism and Related Field.* He was awarded the **1955 Nobel Prize for Chemistry** in recognition of his synthesis of the two nonapeptide pituitary hormones, *ocytocin* and *vasopressin* (1953–1954). In 1974, he suffered from a stroke which ended his academic career. One year after his wife's death in 1977, he died.

Vasopressin regulates many functions:

I. Circulatory or pressor action. Vasopressin causes a rise in blood pressure by contraction of peripheral blood vessels. It is, thus, a vasopressor substance and for this reason it is termed 'vasopressin'. It has been used in surgical shock as an adjuvant in elevating blood pressure. It may also be used at the time of delivery to overcome uterine inertia. The effect is similar to that caused by adrenalin. However, the rise in blood pressure caused by vasopressin lasts much longer than that caused by adrenalin which wears off in a few minutes.

II. Antidiuretic action. It brings about a reduction in the urine volume by causing renal tubules to withhold more water. Consequently, the urine passed is rich in sodium chloride, phosphate and total nitrogen. It was earlier thought that this antidiuretic effect was due to a different hormone called *antidiuretic hormone*, ADH. But it is now established that the two hormones (vasopressin and ADH) are one and the same. Vasopressin, therefore, finds use against persons suffering from **diabetes insipidus**, a disease characterized by excretion of large quantities of urine (*polyuria*) and a marked thirst (*polydipsia*). The urine specific gravity remains almost constant between 1.002 and 1.006. The urine output becomes 4 to 6 litres a day but can be sometimes as high as 12 to 15 litres a day, depending mainly on the amount of water taken by the patient. Furthermore, the rapid loss of fluid in the urine creates a constant thirst which keeps the water flushing throughout the body. The patient, thus, has a tendency to become dehydrated. But this tendency is quite well offset by the increased thirst. The disease may be controlled by administration of posterior pituitary extracts subcutaneously or even by nasal instillation.

The term **'insipidus'** (= tasteless) dates from the time when the only method of testing urine was to taste it after diluting, and in this condition the urine is tasteless. Diabetes mellitus is caused by impaired renal absorption of water, whereas diabetes mellitus develops due to the deficiency of insulin caused either by inadequate insulin production or by accelerated insulin destruction.

Table 34–6. Summarizes the biological effects of the two neurohypophyseal hormones.

Table 34–6. Comparison of the biological effects of ocytocin and vasopressin

S. N.	*Biological Effect*	*Ocytocin*	*Vasopressin*
1.	Water diuresis	no effect	inhibits
2.	Blood pressure	slightly lowers	raises
3.	Coronary arteries	slightly dilates	constricts
4.	Intestinal contractions	?	stimulates
5.	Uterine contractions	stimulates	stimulates
6.	Ejection of milk	stimulates	slightly stimulates

Hypopituitarism. Insufficient secretion of pituitary hormones (or hypopituitarism) may occur as a result of pituitary tumours or atrophy of the gonad. Hypoactivity of this gland may lead to the following disorders:

1. **Dwarfism.** It refers to the arrested growth of the individuals (Fig 34–39). It is of 2 types:
   (*a*) *Lorain type*— short-statured individuals with a head large in comparison to the rest of the body ; usually intelligent but unattractive.
   (*b*) *Fröhlich type*— obesity and arrested sexual development; mentally below normal and usually lethargic.
2. **Panhypopituitarism.** It is caused because of the destruction of the gland, thus leading to cessation of all hypophyseal functions.
3. **Pituitary myxedema.** It is caused due to the lack of TSH and produces symptoms similar to those described for primary hypothyroidism (see page 1143).

Fig. 34–39. Two physiologic conditions arising from abnormal pituitary secretion

Here we see dwarfism and gigantism both. Both conditions arise from abnormal pituitary secretions.

Hyperpituitarism. It refers to the overproduction of hypophyseal hormones. Hypersecretion of this gland leads to **gigantism** during childhood or adolescence, *i.e.*, before closure of the epiphyses. The disease is characterized by overgrowth of the bones, especially at joints, leading to a tall individual with 2 to 2.5 M height. The limbs usually become disproportionately large.

In human adults with closed epiphyses, excessive secretion of pituitary hormones causes **acromegaly** (*acron*[G] = extremity). The chief symptoms of this disease are (Fig. 34–40):

1. consistent overgrowth of the bones of face, hands and feet (*acral* parts ; hence the term acromegaly) so that the patient often complains of having have to change globes and shoes frequently as they no longer fit.
2. protrusion of the lower jaw (*prognathism*). Overgrowth of the malar, frontal, and basal bones combines with prognathism to produce the coarse facial features called *acromegalic facies.*

CLINICAL IMPLICATIONS 34–2.

Obesity

Obesity is a chronic disease of excessive fat deposition in the body. It is the condition of a person having excessive weight (or being overweight) for his height, build, and age because of a buildup of fat. Being overweight is not a disease, obesity is. And the line differentiating the two is quite thin. Obesity predisposes to or is associated with numerous diseases including diabetes (type II), heart diseases, stroke, arthritis, atherosclerosis, high blood pressure and certain cancers. It is "that state in which the accumulation of reserve fat becomes so extreme that the functions of the organism are interfered with". It can be quantitated by measuring skinfold thickness with calipers. Obesity is most readily the result of long-term overconsumption of energy, with resultant excessive increase in body mass. However, there may be contributory malfunction in the individual's energy homeostasis. It may arise from either overeating or diminished utilization or a combination of both. Obesity results from a low basal metabolic rate (BMR) in obese and therefore, even with a normal intake of food, an excess of calorie is available. The BMR diminishes slightly as the individuals grow older but often the food consumption is not decreased proportionally. As a consequence, obesity may result. Obesity is usually due to an excess intake of food. Appetite may be influenced by a variety of factors such as psychologic disturbances; hypothalamic, pituitary, or other brain lesions; and hyperinsulinism. Obesity may also develop from an increase in number or in size of fat cells, adipocytes. Adipocytes increase in number when caloric intake is increased, especially in the gestational months and during the first year of life. The obese may become resistant to insulin, resulting in an increase in the levels of circulating insulin. Insulin decreases lipolysis and increases fat synthesis and uptake.

Obesity may become evident at any stage, but it appears most frequently in the first year of life, at 5 to 6 years of age, and during adolescence. The adiposity in the mammary regions of **boys** is often indicative of breast development and hence an embarrassing feature. The abdomen tends to be pendulous, and white or purple striae are often present. The external genitalia of boys appear disproportionately small but actually are most often of average size; the penis is often embedded in the pubic fat. Puberty may occur early, with the result that the ultimate height of the obese may be less than that of their slower-maturing peers. The development of external genitalia is normal in the majority of **girls,** and menarche is usually not delayed. The obesity of the extremities is usually greater in the upper arm and thigh and is sometimes limited to them. The hands may be relatively small and the fingers tapering. Obesity remains an enigma today as it has for many centuries– the earliest recorded case of human obesity dating back to the **Willendorf Stone Age Venus** of about 22,000 B.C.

3. bowing of the spine (*kyphosis*)
4. overgrowth of the body hair
5. thickening of the soft tissues of nose, lips and forehead
6. enlargement of the visceral organs such as lungs, heart, liver and spleen
7. increased sexual activity in the beginning which is ultimately followed by atrophy of the gonads. This leads to impotence in man and amenorrhea in woman.

A classical example of acromegaly is that of **Akhenaton** (Fig. 34–41), the Pharaoh who ruled Egypt in the years 1379–1362 BC. His predominant facial features strongly suggest that he suffered from acromegaly. Acromegalic persons usually have enlarged facial & chiral features (Fig. 30–42).

Hypophysectomy. The effects of hypophysectomy (removal of hypophysis) appear to be almost entirely due to the loss of adenohypophysis. Removal of only the neurohypophysis exerts no striking dysfunctions. Hypophysectomy, in general, leads to:

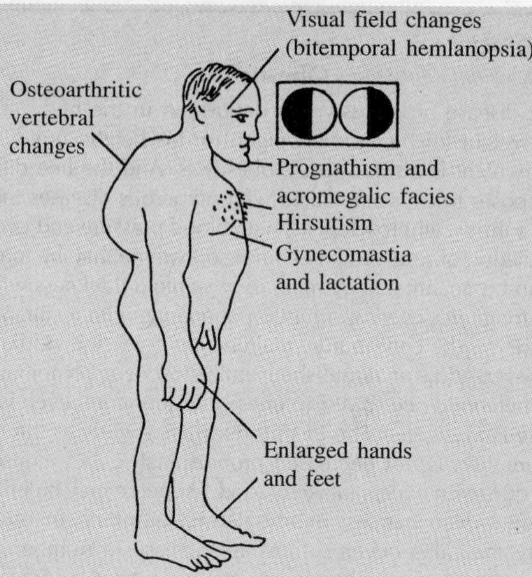

Fig. 31–40. Typical finding in acromegaly

1. gonadal atrophy in either sex
2. atrophy of the thyroid
3. atrophy of the adrenal cortex
4. loss of body tissue with some reversion to younger characters, *i.e.*, appearance of juvenile hair.

Fig. 34–41. The ancient carving of the Egyptian Pharaoh Akhenaton, who ruled from 1379-1362 B.C.

His characteristic enlarged features strongly suggest that he suffered from acromegaly. It may be the oldest known case. **Akhenaton** is famous for his following quote: "True wisdom is less presuming than folly. The wise man doubteth often, and changeth his mind; the fool is obstinate, and doubteth not; he knoweth all things but his own ignorance ... To be satisfied with a little, is the greatest wisdom; and he that increaseth his riches, increaseth his cares; but a contended mind is a hidden treasure, and trouble findeth it not."

(*Courtesy: Agytisches Museum. Staadtliche Museen Preussicher Kulturbesitz, Berlin, Germany*)

Fig. 34–42. An acromegalic person

Sudden increases in the production of growth hormone after maturity produce growth in only certain body parts, such as those of the face and hands. The condition is called acromegaly.

3. Hormone of the Parathyroid

Secretory gland. The parathyroids were first discovered by **Sandström** in 1880. In the human beings, there are usually four small parathyroid glands so closely associated with the thyroid gland that they remained undiscovered for some time. Each parathyroid gland is a reddish or brownish, oval body, measuring about 5–7 mm in length, 3–5 mm in width and 1–2 mm in thickness. The four glands together weigh only 0.1 to 0.2 gm. The glands have a macroscopic appearance of dark brown fat, therefore these are difficult to locate and hence often removed during thyroidectomy (removal of the thyroid). Of the four parathyroids, two lie embedded in the thyroid and are called as *internal parathyroids* ; the other two lie close and behind the thyroid and are known as *external parathyroids.* There may be fewer than four or as many as eight parathyroids. The extra ones are scattered along the trachea and are called as accessory parathyroids. Occasionally, the parathyroids are located in the anterior mediastinum or, rarely, in the posterior mediastinum. The parathyroids are, however, lacking in the fishes.

Histologically, the parathyroid of the adult human being consists mainly of chief cells (= principal cells) and oxyphil cells (refer Fig. 34–43). The oxyphil cells are usually absent in young human beings. The chief cells are concerned with the secretion of the parathyroid hormone. The oxyphil cells are rich in mitochondria but lack glycogen. Their function is uncertain. They are regarded as probably aged chief cells that still secrete some hormones.

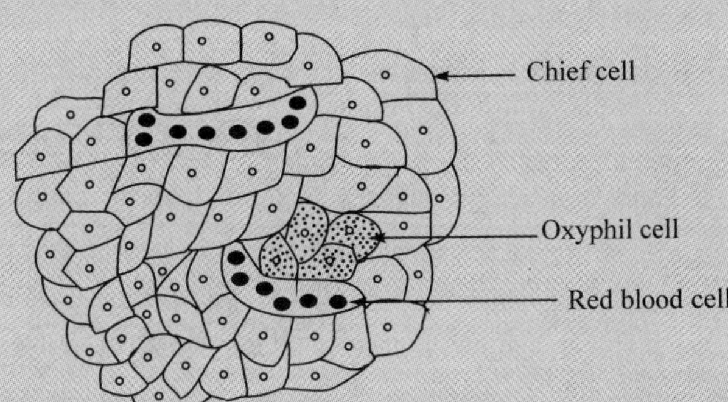

Fig. 34–43. Macroscopic structure of a parathyroid gland

Structure. Parathyroids secrete a hormone called parathyroid hormone or **parathormone, PTH**; it is also known as **Collip's hormone.** PTH has been isolated in pure form. It is a linear polypeptide consisting of 84 amino acid residues (Fig. 34–44) and has a molecular weight of about 7,000 to 8,500. **Potts** and others (1971) have, however, indicated that the physiologic activity of this hormone on both skeletal and renal tissues is contained within the first 34 amino acids from the N-terminal of the chain. As is also true for many other hormones (such as α-MSH and corticotropin), the PTH can be cleaved to form smaller but still active units. Oxidation inactivates the parathormone rapidly.

Functions. The principal sites of parathyroid action are bones, kidney and gastrointestinal tract. Following physiological functions are attributed to this hormone:

1. *Bone resorption–* It exerts a direct influence on the metabolism of bone, leading to an increased release of bone Ca^{2+} into the blood. The exact mechanism behind this phenomenon is not truly known. It has, however, been suggested that the hormone stimulates the production of citric acid in the bone tissues and an increased concentration of citrate ions leads to the removal of phosphate from calcium phosphate, the bone material. The bone is, thus, made soluble.

2. *Renal reabsorption of calcium–* In the kidney, parathormone affects renal tubular reabsorption of calcium and reabsorption or secretion of phosphate. It increases the elimination of calcium and phosphorus in the urine. It is interesting to note that the secretion of this hormone is controlled by Ca^{2+} ion concentration of the blood itself. As the Ca^{2+} ion concentration increases, PTH secretion decreases tending to preserve the original condition. This affords an excellent example of feedback mechanism of metabolic control.

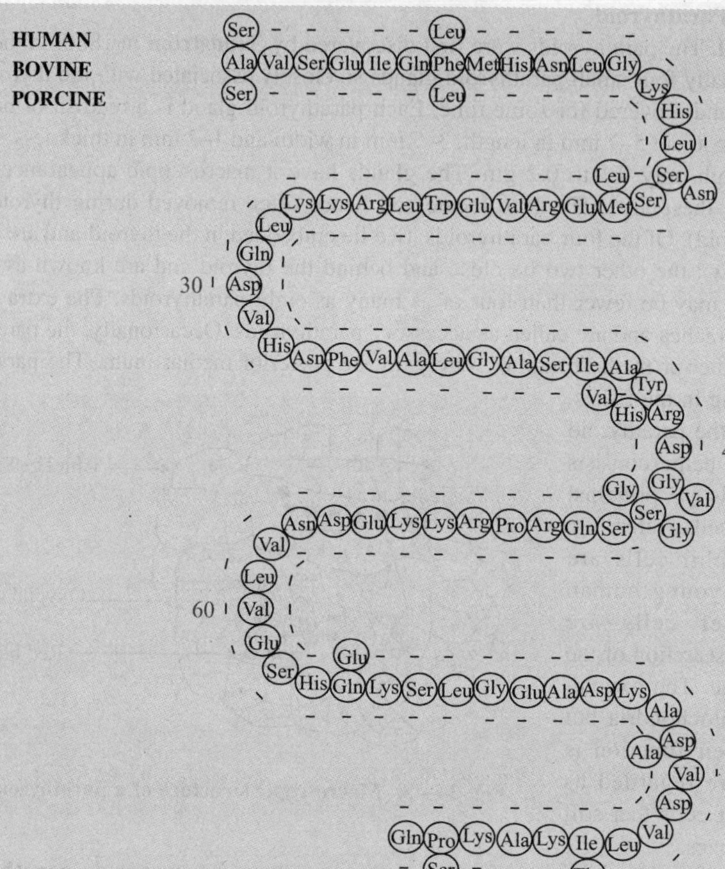

Fig. 34–44. The structure of bovine parathyroid hormone

Identical amino acid residues in porcine and human parathyroid hormone are indicated by *dashes*, whereas different amino acid residues are identified above or below the corresponding residue in bovine hormone.

(*Courtesy: Keutman et al, 1977*)

3. *Increase in osteoelastic activity*– Parathormone increases the number of osteoclasts, which break down the bone matrix and release calcium into the blood. It also augments growth of the connective tissue.

4. *Calcium homeostasis*– An optimum Ca^{2+} concentration is necessary for various functions of the body, *viz.*, normal transmission of impulses, the contraction of the muscles, the formation of the bones, the coagulation of the blood etc.

5. *Vitamin D activation*– Parathormone converts vitamin D to its active form, which increases calcium absorption in the intestine.

Henceforth, a precise endocrine mechanism has been evolved so as to ensure a stable concentration of this ion. The three hormones of importance in this process of calcium homeostasis, as it is called, are vitamin D, PTH and calcitonin. The vitamin D can be classified as a hormone since it is manufactured in the skin (which is an organ) which it is exposed to light. An interesting finding is the observation that in the presence of vitamin D, the parathormone sitmulates the release of calcium accumulated by mitochondria.

Hypoparathyroidism. Undersecretion of PTH causes a decrease in Ca contents of the blood from the normal 10 mg per 100 ml to 7 mg per 100 ml (hypocalcemia) which leads to excessive contraction of the muscles (convulsions). In fact, convulsions occur when the calcium is further decreased to 4 mg per 100 ml of plasma. As the calcium decreases in the blood, there is a decrease in the urine. However, during this period, the phosphorus in plasma increases from a normal 5 mg per 100 ml to 9 mg per 100 ml and even higher (hyperphosphotemia). These changes develop into a fatal disease called **muscular twitchings** or **tetany**. It is characterized by locking up of the jaw, rapid breathing, increased heart beat, rise in temperature and ultimately death due to asphyxia. The signs of tetany in man include **Chvostek's sign**, a quick contraction of the ipsilateral facial muscles elicited by tapping over the facial nerve at the angle of the jaw; and **Trousseau's sign**, a spasm of the muscles of the upper extremity that causes flexion of the wrist and thumb with extension of the fingers (Fig. 34–45). Tetany can be relieved either by the administration of a soluble calcium salt or of PTH.

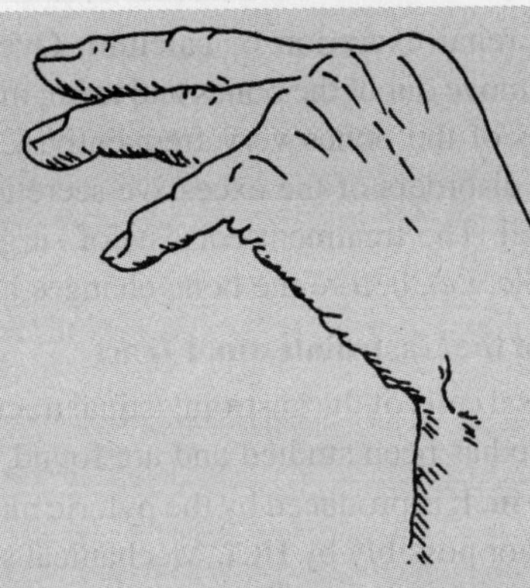

Fig. 31–45. Position of the hand in hypocalcemic (Trousseau's sign)

(*After Aub. Drawn from a photograph in Adams: Physical Diagnosis, 14th eddition, Williams and Wilkins, 1958*)

Hyperparathyroidism. An increase in PTH production is usually due to a tumour of the gland (parathyroid adenoma). Oversecretion of PTH in man results in a cystic bone disease variously called as **osteitis fibrosa cystica** or **von Recklinghausen's disease** or **neurofibromatosis.** It is an autosomal dominant condition. The disease is characterized by increased calcium contents of the blood (hypercalcemia) usually up to 20 mg calcium per 100 ml plasma, decreased phosphate concentration

CLINICAL IMPLICATIONS 34–3.

Neurofibromatosis

Neurofibromatosis is one of the most common autosomal dominant genetic disorders in man, produced due to oversecretion of parathormone. It occurs roughly in 1 of 4,000 live births. It is seen equally in every racial and ethnic group throughout the world. There is virtually a complete dominant penetrance of the gene localized at the centromeric region of chromosome 17. The gene responsible for the disease, isolated in 1990, is huge and actually includes 3 smaller genes. This is only the second time that *nested genes* have been found in humans. This is a tumour-suppressor gene active in controlling cell division. When it mutates, a benign tumour develops. At birth or later, the affected individual may have 6 or more large tan spots on the skin. Such spots may grow in size and number and become darker. Small benign tumours (or lumps), called *neurofibromas,* may occur under the skin or elsewhere. Neurofibromas are made up of nerve cells and other cell types. Occasionally, the upper jaw is also affected. This genetic disorder shows *variable expressivity*. In most cases, the symptoms are mild and patients live a normal life. In some cases, however, the effects are severe. Skeletal deformities, including a large head, become evident; eye and ear tumours can lead to blindness and hearing loss. Many children with neurofibromatosis have learning disabilities and are hyperactive. In the most extreme form of neurofibromatosis, the connective tissue of the bones and the lining of the nerves produce disfiguring tumorous growths. A devastating expression of the gene occurred in **Joseph Merrick**, a severely-deformed late 19th century Londoner, whose tragic life became the subject of a contemporary play and movie, *The Elephant Man*. However, researchers today believe that Merrick actually suffered from a much rare disorder called **Proteus syndrome**.

and increased renal excretion of calcium. Overproduction of parathormone causes calcium and phosphorus to move out of the bones and teeth, making them soft and fragile. Such patients, therefore, suffer fractures of the bones very frequently. Cysts in the bones are another characteristic of this disease. Other disorders of the excessive secretion include the hemorrhages in the stomach and the intestine as well. The treatment consists of surgically removing the gland. This should be done as early as possible, *i.e.*, before the bone changes have become irreversible.

4. Hormones of the Gastrointestinal Tract

The mucosal cells of the gastrointestinal tract, GIT secrete a number of hormones. A few of them, whose structure has been studied and are found to be proteinaceous in nature, are described below.

1. **Gastrin.** It is produced by the pyloric mucosa, which is apparently stimulated by the proteins present in food or possibly by HCl. Mechanical stimulation caused by distention of the stomach also results in the secretion of gastrin. This hormone is absorbed into the blood stream and is carried to the fundic cells, causing them to secrete HCl actively. The hormone stimulates the gastric glands to pour out more gastric juice. Gastrin is a heptadecapeptide (Fig. 34–46) with a molecular weight 2,100. It dialyzes through a membrane and has an isolectric point of 5.5.

1 (Glu) … 10 (Glu) … 13 (Gly) … 17 (Phe)

Glu.Gly.Pro.Trp.Leu.Glu.Glu.Glu.Glu.Glu.Ala.Tyr.Gly.Trp.Met.Asp.Phe–NH_2

(with HSO_3 attached to Tyr)

Fig. 34–46. Gastrin

2. **Secretin.** *It is the first compound to be designated as hormone*. It is formed by the upper intestinal mucosa and is liberated by HCl present in the acid chyme (semifluid material from the stomach). It is carried by the bloodstream to the pancreas which it stimulates. Secretin, thus, causes the flow of pancreatic juice rich in bicarbonate. This occurs even if the nerves supplying the pancreas are cut. Hence, it is a true hormonal action. Secretin also stimulates the flow of bile and intestinal juices, although to a lesser extent. It has been obtained in crystalline form. Secretin (Fig. 34–47) is a polypeptide of 27 amino acid residues (14 of which are identical to those found in glucagon) with molecular weight of 3,056. The molecule has no structural homology with gastrin or cholecystokinin.

1 (His) … 10 (Leu) … 20 (Gln) … 27 (Val)

His.Ser.Asp.Gly.Thr.Phe.Thr.Ser.Glu.Leu.Ser.Arg.Leu.Arg.Glu.Gly.Ala.Arg.Leu.Gln.

Arg.Leu.Leu.Gln.Gly.Leu.Val—NH_2

Fig. 34–47. Secretin

3. **Cholecystokinin, CCK.** It is also secreted by the upper part of the small intestine. It sitmulates contraction of the gall bladder so as to release its contents into the duodenum. Cholecystokinin (Fig. 30–48) is a polypeptide with 33 amino acid residues and a molecular weight of 3,883. It is, however, noteworthy that the last 5 amino acids towards the C-terminal in the gastrin and cholecystokinin are exactly the same. It is in this terminal portion of these hormones that the principal activity resides.

1 (Lys) … 27 (Tyr) … 30 (Trp) … 33 (Phe)

Lys.(Ala, Gly, Pro, Ser).Arg.Val.(Ile, Met, Ser). Lys.Asn.(Asn, Gln, His, Leu_2, Pro, Ser_2).

Arg.Ile.(Asp, Ser).Arg.Asp.Tyr.Met.Gly.Trp.Met.Asp.Phe—NH_2

(with HSO_3 attached to Tyr)

Fig. 34–48. Cholecystokinin

4.. Pancreozymin, PZ. It is another polypeptide hormone elaborated by the intestinal mucosa. Along with secretin, it functions to stimulate the release of pancreatic juice, which is rich in enzymes or their zymogens as well as in bicarbonate. The enzymes or zymogens include trypsinogen, lipase, carboxypeptidase etc. The release of pancreozymin is believed to be brought about in the presence of any one of a variety of compounds such as peptone, casein, dextrin, maltose, lactose, saline water and also even distilled water. Pancreozymin has been obtained in pure form. It is thermostable and is not destroyed by acid but it is destroyed by alkali. Like cholecystokinin, it also contains 33 amino acid residues. The 5 C-terminal amino acid residues are the same as those of gastrin and cholecystokinin. **Jorpes** (1968) believes that the two gastrointestinal hormones, cholecystokinin and pancreozymin are identical. They are now thought to be a single factor and represented as CCK-PZ.

5.. Enterogastrone. It is present in duodenal mucosa and is secreted from it only when fat (derived from food) reaches the duodenum. The hormone reaches the stomach where it inhibits gastric secretion and motility of the stomach. There is secretion of lesser volume of juice with a lower concentration of HCl and a smaller amount of pepsin. The effect is to permit digestion to be more completely accomplished.

6. Enterokrinin. It is also isolated from intestinal mucosa and controls the intestinal secretion. It does not influence pancreatic secretion but may increase the volume as well as the enzyme concentrations of succus entericus. It is distinct from secretin in that it stimulates the secretion of both fluid and enzymes by the intestinal mucosa. It is also proteinaceous in nature.

7. Hepatocrinin. It is related to enterokrinin and is believed to stimulate the secretion of dilute, low-salt type of bile juice.

8. Duicrinin. It controls the secretion of Brunner's glands which are located in the submucosa of the upper duodenum.

9. Villikinin. It is secreted by the intestinal mucosa. It stimulates movements of the intestinal villi and hence accelerates intestinal absorption.

10. Parotin. Ito (1953) has claimed that the salivary glands elaborate parotin, a protein hormone. It has multifarious effects : stimulates calcification of the teeth, decreases the calcium content of the blood and increases the phosphorus level of the serum.

Although evidences exist to indicate the presence of all the humoral agents mentioned above, yet gastrin, secretin and cholecystokinin are the only three gastrointestinal hormones which have been fully established with regard to their biochemical and physiological behaviour.

AMINO ACID HORMONES

1. Thyroidal Hormones

Secretory gland. *The thyroid is the largest endocrine gland in the body.* It was first described by **Whartonin** in 1659 who gave it the descriptive |name, thyroid because of its resemblance to a shield (*thyreoides*[G] = shield-shaped). In man, the gland consists of two lobes on either side of and anterior to the trachea just below the larynx. The two lobes are connected across the ventral surface of the trachea with a narrow bridge called isthmus, making the entire gland more or less H-shaped in appearance (refer Fig. 34–49). The isthmus crosses infront of the 2nd, 3rd and 4th tracheal ring. In the adult, the gland weighs about 25 to 30 gm. The thyroid gland receives blood flow about 5 times its own weight per minute. It has a blood supply as rich as that of any other area of the body barring probably the adrenal cortex. The thyroid is presumed to be homologous with the endostyle of the early vertebrates.

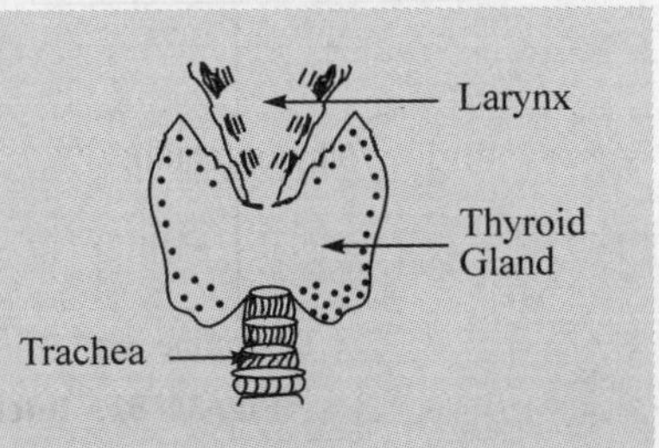

Fig. 34–49. The thyroid in man

(After D. Marsland, 1964)

Histologically, the thyroid gland (refer Fig. 34–50) is composed of a large number of tiny closed vesicles called *follicles*, 150 to 300 microns in diameter. The follicles are held together by areolar tissue and are surrounded by a rich network of capillaries. Each follicle is lined with a single peripheral layer of columnar or cuboidal *epithelioid cells* that secrete into the interior of the cells. Its lumen is filled with a secretory substance called colloid. The major constituent of colloid is a large protein called thyroglobulin, which contains the thyroid hormones. Once the secretion has entered the follicles, it must be absorbed back through the follicular epithelium into the blood before it can perform its function in the body. The thyroid gland is, thus, unique amongst the endocrine glands in that *it stores its hormone as a colloid in small vesicles in the gland.* The other endocrine glands, however, store their hormones in the cells themselves.

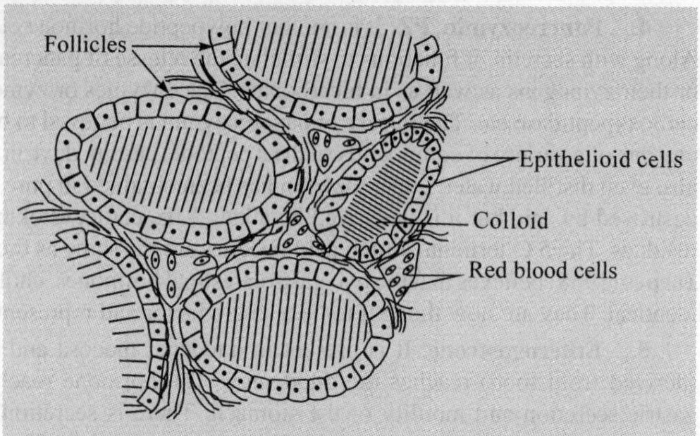

Fig. 34–50. Histologic structure of thyroid gland

As age advances, the thyroid activity tapers off. That's why elderly persons feel colder, since their body does not produce sufficient heat.

Structure. Thyroid contains large amounts of elemental iodine which is bound to a protein named iodothyroglobulin or simply thyroglobulin. It is a glycoprotein with a molecular weight of about 650,000 and iodine content from 0.5 to 1.0%. This protein represents the storage form of the hormone in the gland.

Evidences available at present indicate that thyroglobulin is hydrolyzed, in the presence of thyrotropin, to release thyroxine (= 3, 5, 3′,5′-tetraiodothyronine) in the blood (Fig. 34–51). The release of thyrotropin is, in turn, controlled by the level of thyroxine in the blood.

Thyroxine is one of the earliest recognized hormones. It was so named and isolated first by **Kendell** in 1915. **Harington** and **Barger** (1925) established its chemical formula. It is an iodine-containing aromatic amino acid and closely resembles tyrosine in structure. Diiodotyrosine is believed to be the precursor of thyroxine.

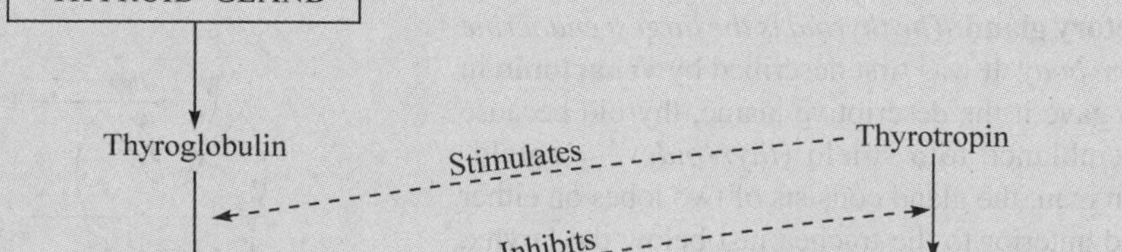

Fig. 34–51. Interdependence of thyroxine and thyrotropin

Besides thyroxine, 3,5,3′-triodothyronine is also produced from enzymatic hydrolysis of thyroglobulin. It is 5 to 10 times more potent in biologic activity than thyroxine. This may, possibly, be due to the fact that triiodothyronine is bound loosely by serum proteins and hence diffuses much more rapidly into the tissues. It is present in the blood in much smaller quantities and persists for a much shorter time than does thyroxine. The structure of triiodothyronine and tetraiodothyronine is given in Fig. 34–52.

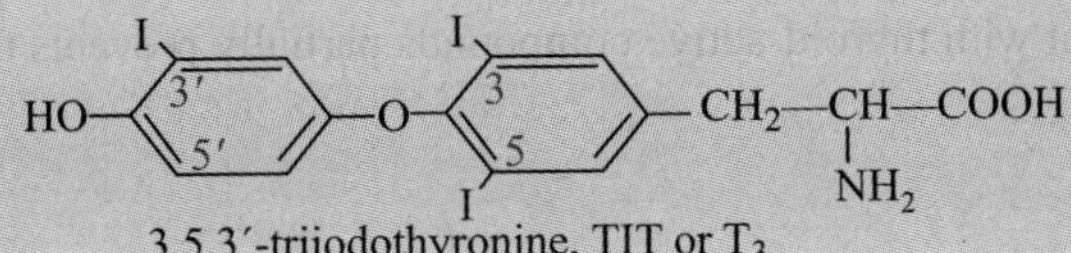

3,5,3′-triiodothyronine, TIT or T_3

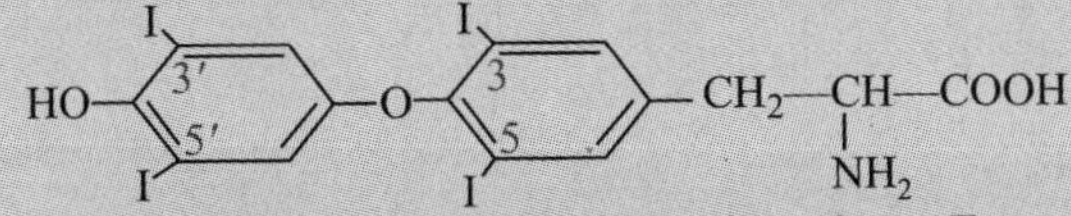

3,5,3′,5′-tetraiodothyronine or thyroxine or T_4

Fig. 34–52. Thyroidal hormones

Functions. Thyroid hormones have widespread effects on the ossification of cartilage, the growth of teeth, the contours of the face and the proportions of the body. They also carry out following functions :

1. They bring about deamination reactions in the liver.
2. They also carry on deiodination in the extrahepatic tissues.
3. These influence oxidative phosphorylation by altering the permeability of the mitochondrial membrane.
4. Their presence accelerates metamorphosis in amphibians. This is so a sensitive test that tadpole has been widely used for the assay of the potency of these hormones.
5. These may increase the level of cytochrome *c* in the tissues.

It is, thus, apparent that these hormones affect the general metabolism, regardless of the nature of its specific activity. It is for this reason that the thyroid gland has rightly been called as the '*pace setter*' of the endocrine system.

Thyroid inhibitors or goitrogens. Certain substances act as antithyroid agents by inhibiting the production of thyroxine. This is explained by the fact that these compounds prevent the oxidation of iodide, which means that the iodine cannot be used for the synthesis of thyroxine. Some important thyroid inhibitors (Fig. 34–53) are 6-propyl-2-thiouracil, 2-thiouracil, thiourea and the sulfonamides. Thiouracil itself is relatively toxic ; less so are propylthiouracil and thiourea. Of these, propylthiouracil has been used most extensively. It is 3 to 5 times more active than thiouracil. The antithyroid drugs or the goitrogens, as they are also called, are also found in common foods such as cabbage, turnips, spinach, peas etc.

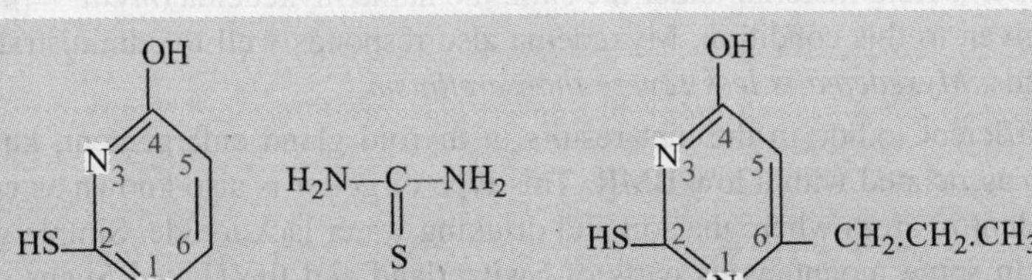

2-thiouracil (lactim form)

Thiourea

6-propyl-2-thiouracil (lactim form)

Fig. 34–53. Some thyroid inhibitors

Hypothyroidism. Underactivity of the thyroid may result from two causes : a degeneration of thyroid cells or a lack of sufficient iodide in the diet. The disease that results from thyroid cell degeneration is cretinism in children and myxedema in adults. **Cretinism** (=feeble-mindedness) is characterized by dwarfism and mental suppression. The cretinic children have infantile features

(Fig. 34–54). They possess a large head and an apathetic face ; their teeth erupt late and the speech is retarded. Early treatment with thyroid-active compounds partially prevents this disease.

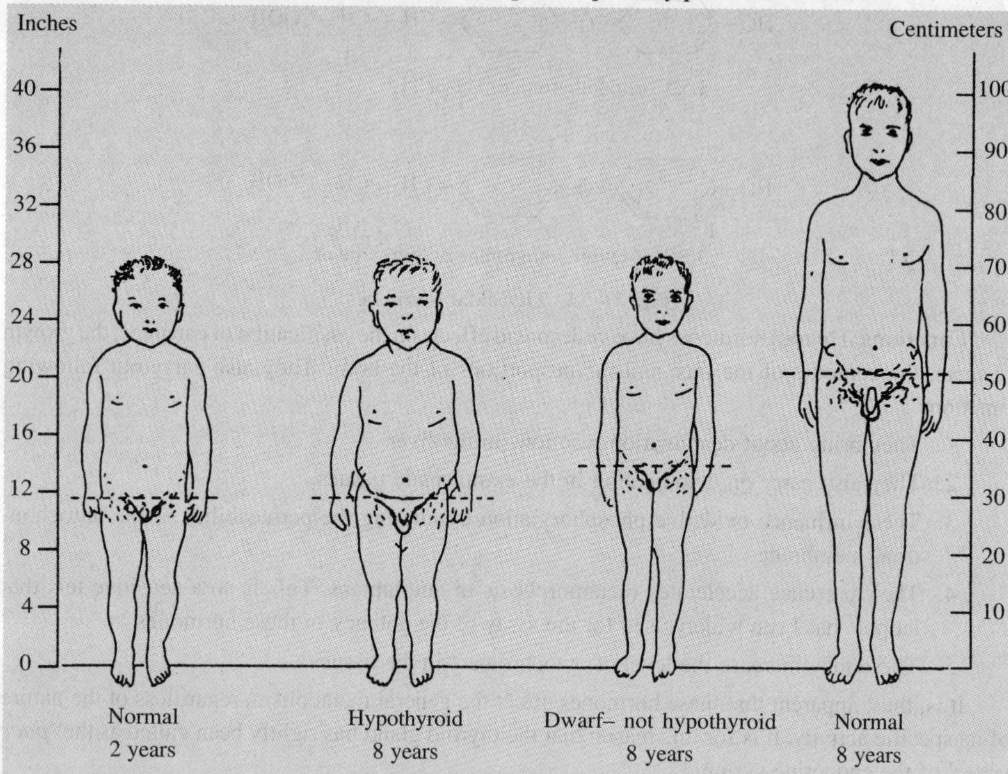

Fig. 34–54. Normal and abnormal growth

Hypothyroid dwarfs retain their infantile proportions, whereas dwarfs of the constitutional type and of the hypopituitary type have proportions characteristic of their chronologic age.

(*Adapted from Wilkins: The Diagnosis and Treatment of Endocrine Disorders in Childhood and Adolescence, 34 ed., Thomas. 1966*)

Myxedema is characterized by an abnormally low basal metabolic rate (BMR). In it, the adults become mentally lethargic and possess thick puffy skin (edema) and dry hair. The patient shows bagginess under the eyes and swelling of the face. The hair thin on the eyebrows and scalp. As there is deposition of semi-fluid material under the skin, the name myxedema (*myxa*[G] = mucus; *oidema*[G] = swelling) is given to this condition. Myxedema also responds well to administration of thyroid-active compounds. *Myxedema is less severe than cretinism.*

Lack of sufficient iodide in the diet results in thyroid gland enlargement, known as **simple goiter.** It is also associated with a low BMR. This type of goiter is also known as **endemic goiter**, since it is prevalent in areas where the soil and drinking water lack iodide. Simple goiter was once fairly common in some mountainous parts of Switzerland and the United States, where soil and water are deficient in iodine compounds.

Hyperthyroidism or **Thyrotoxicosis.** Abnormally high activity of this gland may occur due to either oversecretion of the gland or an increase in size of the gland. Swelling of the gland results in an **exophthalmic goiter** (Fig. 34–55), characterized by protrusion of the eye balls. In it, the BMR increases considerably above the normal figures; 80% above normal is not unusual. Consequently, appetite is increased in hyperthyroid individuals. In spite of this, they lose weight and often feel hot because of the increased heat production. Their pulse rate is also increased and excessive sweating occurs. Other symptoms of this disease are dilated pupils, mental excitement, irritability and cardiac dilatation. Hyperthyroid individuals are, in general, characterized by an above-normal rate of many physiological activities. The clinical syndrome is generally termed **Graves' disease**, after its discoverer **Robert James Graves**. **Basedow's disease** and **thyrotoxic exophthalmos** are other names of this disease. Not surprisingly, people with hyperactive thyroids (*i.e.*, with Graves' disease), thus, show many symptoms opposite those in hypothyroidism.

Hyperthyroidism can be cured by surgical removal of the thyroid (thyroidectomy), treatment with x-rays, injection of radioactive iodide (^{131}I) or by treating with antithyroid drugs or with agents like thiocyanate or perchlorate which compete with iodide for the uptake mechanism. Propylthiouracil is being particularly used against the Graves' disease.

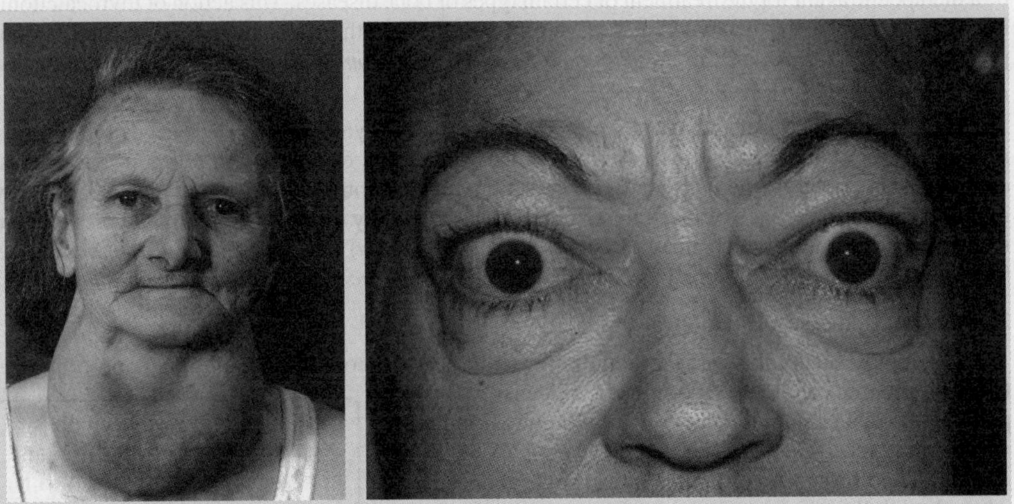

(*a*) (*b*)

Fig. 34–55. Graves' disease

Note the goitre (*a*) and exophthalmos (*b*).

Thyrocalcitonin or Calcitonin

A quite different type of hormone which has effects of decreasing calcium ion concentration in the blood has been discovered by **Copp** *et al* in 1961 as an impurity in the commercial parathyroid extracts. They termed this hypocalcemic factor as calcitonin (CT). *Later researches revealed that this factor originated in the thyroid, rather than in the parathyroids.* This substance was, henceforth, also named as thyrocalcitonin (TCT). It is secreted by the parafollicular cells or C-cells in the interstitial tissue between the follicles of the human thyroid gland. Later, it was discovered that CT is secreted by the ultimobranchial glands of fishes, amphibians, reptiles and birds. In mammals, however, these glands do not exist as such but have become incorporated into either the parathyroids or the thyroid. The parafollicular cells of human thyroid glands are remnants of the ultimobranchial glands of lower animals.

Thyrocalcitonin (Fig. 34–56) is a large, unbranched polypeptide containing 32 amino acids and has a molecular weight of about 3,600. It is unique in having no isoleucine and lysine residues. Its amino acid sequence has been determined by **Foster** in 1968.

Thyrocalcitonin acts by causing a transfer of calcium from blood into bone either by increasing calcification of the bones or by diminishing decalcification or by both processes. In other words, it

S—————————————S
Cys—Gly—Asn—Leu—Ser—Thr—Cys—Met—Leu—Gly—Thr—Tyr—Thr—Gln—Asp—
1 7 10
Phe—Asn—Lys—Phe—His—Thr—Phe—Pro—Glu—Thr—Ala—Ile—Gly—Val—Gly—
20 30
Ala—Pro—NH_2
32

Fig. 34–56. Human calcitonin

Note the presence of an intrachain disulfide bond between two cysteinyl residues placed at positions No.1 and No. 7.

rapidly inhibits calcium withdrawal from bones. This characteristic property, however, promises it to be a therapeutic agent for the treatment of certain types of bone diseases. This action of thyrocalcitonin is counterbalanced by the hypercalcemic hormone, the parathormone which is secreted by the parathyroids. *TCT, thus, aids in the maintenance of calcium homeostasis.*]

Effect of thyroid hormones on the gonads:

For the normal sexual activity to occur, the secretion of thyroidal hormones needs to be almost normal– neither too little nor too more. In the male, deficiency of the thyroid hormones may cause complete loss of libido whereas their hypersecretion, on the contrary, leads to impotence. Similarly, in the female also, the lack of thyroid hormones leads to greatly diminished libido and often results in menstrual bleeding which may be excessive (*menorrhagia*) and frequent (*polymenorrhea*). A hyperthyroid female exhibits greatly reduced bleeding (*oligomenorrhea*) and sometimes, *amenorrhea.*

2. Adrenal Medullary Hormones

As already stated, the adrenal medulla forms the central core of adrenal gland and originates from the neural canal. It is composed of densely packed polyhedral cells containing chromaffin granules. It is highly vascular and receives 6–7 ml of blood per gram of tissue per minute. The chromaffin granules store large quantities of adrenal medullary hormones.

Adrenalin and **noradrenalin** are also spelt as adrenaline and noradrenaline, respectively (**J.H. Green**, 1978).

Structure. Adrenal medulla, whose secretion is under nervous control, produces two hormones (Fig. 34–57): (*a*) epinephrine or adrenalin ($C_9H_{13}O_3N$)in 1897 and (*b*) norepinephrine or noradrenalin ($C_8H_{11}O_3$ N). *Epinephrine was the first hormone to be isolated in the crystalline form.* The isolation was, however, done by **Abel** in 1897. It has been produced synthetically by **Stoltz**. Chemically, these two hormones are catecholamines (= dihydroxy-phenylamines) and are closely related to tyrosine and phenylalanine. Norepinephrine, however, differs from epinephrine structurally in having a hydrogen atom in place of the methyl group. The commonly available epinephrine, therefore, is a mixture of these two hormones (usually 10—20% norepinephrine) and its effects are a resultant of the combined actions of these two hormones. Since these hormones

The word **'nor'** in norepinephrine (and noradrenalin) was originally coined to indicate nitrogen (**N**) without (**O**-Ohne) a radical (**R**), in this case a methyl group.

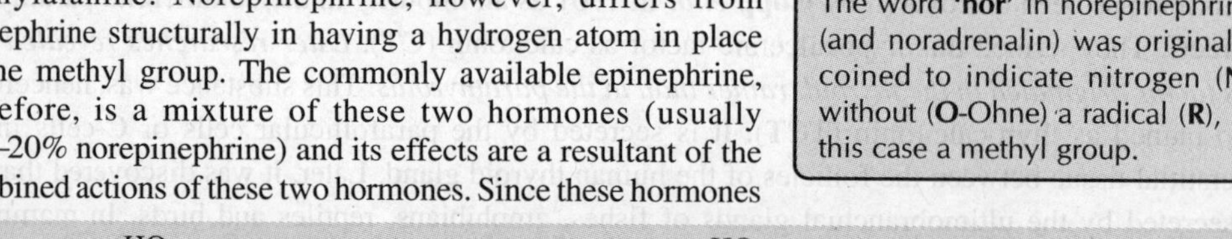

Epinephrine (Adrenalin)

Norepinephrine (Noradrenalin or arterenol)

Fig. 34–57. Adrenal medullary hormones

possess an asymmetric carbon atom, two stereoisomers are possible for each one of them. The naturally-occurring epinephrine is the L-isomer and is levorotatory. It is 15 times more active than the D-form. Similarly, the natural D (-) form of norepinephrine is about 20 times more potent than the unnatural isomer. Their relative proportion in the adrenal medulla differs from species to species (refer Table 34–7).

Table 34–7. Percentage composition of the two adrenal medullary hormones in various organisms

S.N.	*Organism*	*Epinephrine (Adrenalin)*	*Norepinephrine (Noradrenalin)*
1.	Man	75-90	25-10
2.	Cat	50	50
3.	Whale	10	90

Functions. In general, the adrenal medullary hormones reinforce the functions performed by the sympathetic nervous system. Although both these hormones exert similar effects in regulating carbohydrate metabolism and blood pressure, yet epinephrine is more closely related to carbohydrate metabolism and norepinephrine to blood pressure.

Epinephrine conducts a wide variety of functions, which are as follows:

1. It promotes glycogenolysis in muscles and liver, resulting in an increase of blood glucose level and an increased lactic acid formation in muscles. These changes are then followed by an increase in oxygen consumption.
2. It causes an increase in blood pressure because of arteriolar vasoconstriction of the skin and splanchnic vessels.
3. It brings about an increase in the heart rate and in the cardiac output.
4. It causes dilation of vessels (= *vasodilation*) of skeletal muscles, corona and the viscera. This results in an increase of blood flow in these areas.
5. It relaxes the muscles of gastrointestinal tract and bronchials of the lungs but causes contraction of the pyloric and ileocecal sphincter muscles.
6. It causes constriction of the muscles at the base hair. This makes the hair stand on end which, in many animals, may serve to make them look bigger and fiercer than they actually are.
7. It may also enhance memory during periods of excitement.
8. It also serves in cases of emergency. Under emotional stress, fear or anger, it is secreted in the bloodstream and the blood is shifted from the viscera to the brain and the muscles so that the individual becomes ready for fight. It is for this reason that the adrenals are frequently referred to as the '*emergency glands*' or the '*glands of flight*, fright and fight' and the two adrenal medullary hormones as **'emergency hormones'**.

Norepinephrine, on the other hand, does not relax bronchiolar muscles and has little effect on cardiac output. It augments both systolic and diastolic blood pressure.

The physiological activities performed by these two hormones have been compared in Table 34–8.

Adrenal demedullation. Despite the varied and definite physiologic effects of its characteristic hormones, the adrenal medulla does not appear to be essential to life. Hence, removal of only the medullary portion of the adrenal gland leads to no specific physiologic disorder. This is because the autonomous nervous system may take over in its absence. Consequently, the exact importance of the adrenal medulla is really undetermined. However, certain tumours of the medullary cells result in **pheochromocytoma**, characterized by hypertension and ultimately leading to death due to coronary insufficiency and pulmonary edema. Treatment is to remove the tumour surgically.

Table 34–8. Comparison of the effects of intravenous infusion of epinephrine and norepinephrine

S.N.	*Physiological Activities*	*Epinephrine* (Adrenalin)*	*Norepinephrine (Noradrenalin)*
1.	Heart rate	+	–
2.	Cardiac output	+ + +	–, 0
3.	Systolic blood pressure	+ + +	+ + +
4.	Diastolic blood pressure	+, –, 0	+ +
5.	Total peripheral resistance	–	+ +
6.	Blood glucose	+ + +	+, 0
7.	Blood lactate	+ + +	+, 0
8.	Central nervous system action	–	0

* + = increase; – = decrease; 0 = no change

(Modified from Goldenberg, 1951)

CLINICAL IMPLICATIONS 34–4.

Pheochromocytoma

Pheochromocytoma is a tumour of the adrenal medulla. Both norepinephrine and epinephrine may be secreted: norepinephrine causes hypertension by activating α1-adrenoceptors on vascular smooth muscle, and epinephrine increases heart rate by activating β1-adrenoceptors on the heart muscle. Hypertension may be paroxysmal and severe, leading to stroke or heart failure. Although pheochromocytoma is a rare cause of hypertension, comprising only about 1% of cases, it is very important to remember it, as the condition is dangerous and often amenable to surgical cure.

Diagnosis of pheochromocytoma is made by measuring catecholamines in plasma or urine, or their metabolites, such as metanephrines and vanillylmandelic acid, in urine. The tumour is usually localized by radiologic techniques such as nuclear magnetic resonance (NMR) or computed tomography (CT) scanning.

PARAHORMONES OR TISSUE HORMONES

In addition to the hormones discussed in the preceding pages, a number of others have been found to possess hormonal properties. These compounds are effective at low concentrations; unlike global hormones, they are not transported between tissues in the blood, but act on the tissue in which they are produced. They are, hence, also called as **tissue hormones**. They are at best local hormones because they are short-lived and are *paracrine* rather than endocrine in nature, but they fit the definition of hormone. They alter the activities of the cells in which they are synthesized and of adjoining cells. The nature of these effects may vary from one type of cell to another, in contrast with the more uniform actions of global hormones such as insulin or glucagon. Four such hormones (or groups of hormones), whose hormonal function has been established, are: *melatonin, renal hormones, eicosanoids and opiate peptides.*

1. Pineal Hormone or Melatonin

Secretory gland : The pineal gland or epiphysis (*epi*^{G} = upon ; *phyein*^{G} = to grow) is a tiny conical or pine-like organ (hence its nomenclature, pineal) on the dorsal side of brain in vertebrates, lying deep in the groove between the cerebellum and cerebral hemispheres. The gland was long associated with the folklore of a mysterious "third eye". As the gland works round the clock, **Bruce Fellman** (1987) calls this as a '*clockwork gland*'. Pineal has a back-dated history. **Herophilus** (4th century BC) called this as the **'sphincter of thought'** the mind's valve. **René Descartes**, a 17th century French physiologist called it as **'the seat of the rational soul'.** Scientists of the early 20th century, wearied by a vain search for the gland's function, cast it as a neurological vestige, an appendix of the brain. The pineal, however, was not easily relegated to the cerebral scrap heap. In the early 1960s **Richard Wurtman** of the Massachusetts Institute of Technology (MIT), USA, aptly coined a phrase to describe the gland's function, **"a neurological transducer"** : a system that converts a nerve-type signal– one set off by dark and light– into an endocrine signal– that of a hormone whose levels rise and fall in the bloodstream.

Nearly all modern fish, amphibians reptiles, birds and mammals have a pineal. Even the few exceptions– notably alligators, crocodiles and armadillos– have cells that act like the gland. And the organ itself has occupied most vertebrates' skulls since the first backboned creatures joined the earth's fossil record 500 million years ago. A structure as this ubiquitous and ancient is surely not just going along for the evolutionary ride. In man, this gland is very well developed in children and shrinks after the 7th year and is reduced to a little knot of tissue after puberty. It weighs less than a button. Many functions have been attributed to this gland but they, however, remain to be verified.

Structure. Pineal secretes a hormone called **melatonin** which inhibits the secretion of an adenohypophyseal hormone called luteinizing hormone. Melatonin is synthesized (Fig. 34–58) from its precursor, *serotonin* under the influence of two enzymes, N-acetyltransferase (NAT) and hydroxyindole-o-methyltransferase (HIOMT) which are present in rich quantity in the pineal gland. Serotonin was first isolated and crystallized by Page and others in 1948. In fact, like the clockwork, the pineal manufactures melatonin only in the night. However, during daytime, it does not sit idle and converts the amino acid tryptophan (Trp) into serotonin which accumulates and is stored in an inactive form until night. Neurotransmitters (such as norepinephrine) are released by the nerve cells connected to the pineal and cause increased concentration in the gland of the enzymes NAT and HIOMT. These enzymes convert serotonin into melatonin which then flows into the bloodstream. Thus, *exposure to light slows while exposure to darkness stimulates the production of melatonin in the pineal gland.* Serotonin and melatonin, thus, demonstrate the inverse circadian rhythms in their production.

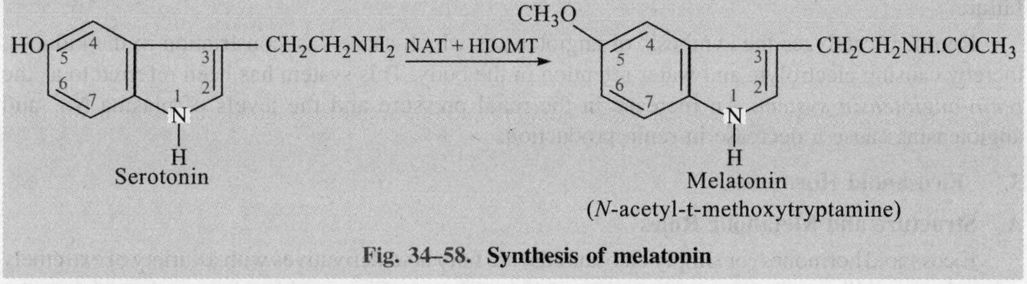

Fig. 34–58. Synthesis of melatonin

David Klein argues that the pineal is a slave and it is under the control of a master clock. The master clock is a concentration of specialized nerve cells found in the hypothalamus, close to where the optic nerve crosses. Called the suprachiasmatic nuclei (SCN), this structure was also identified late, like the pineal gland but its function was a mystery. **Klein** has found that the SCN turns melatonin production on and off in the pineal by directing the nightly increase in NAT.

Functions. In humans, the gland's rhythm may affect the sleep cycle. **Lieberman** and **Vollrath** domonstrated that melatonin makes people drowsy. Hence, nowadays melatonin is being consumed as a sleep aide by insomniacs and people suffering from jet-lag, and also by those having trouble in adaptation to work at odd hours. The hormone is also being used for prevention of cancer and for strengthening the immune system. This *'miracle hormone'* keeps a person using it, away from growing old. As melatonin production diminishes naturally with age, it has been linked to a possible role in the onset of puberty. Studies have also indicated that melatonin may regulate menstrual cycles in women and sperm production in men. Melatonin also controls pigmentation changes; it triggers aggregation of the pigment melanin into melanophores in the skin, causing the skin to turn pale.

The removal of pineal gland (*pinealectomy*) in animals causes the gonads to regress. Many animals tie their breeding activities to the seasons, but after a pinealectomy they suffer from reproductive asynchrony, losing touch with nature's calender. Birds also abandon daily activity rhythms, some eventually losing the urge to migrate.

2. Renal Hormones

The kidney secretes two hormones, erythropoietin and renin.

> **Erythropoietin** is also spelt as erythropoetin (**W.S. Hoar,** 1975).

Erythropoietin or **erythrocyte stimulating factor, ESF.** It is secreted by the kidney. Its secretion is stimulated by tissue anoxia and also by androgenic hormones and cobalt. In fact, erythropoietin is secreted as an inactive protein called renal erythropoietin factor (REF) which is enzyme like in behaviour. The REF converts a plasma globulin to the active erythropoietin.

$$\text{Plasma globulin} \xrightarrow{\text{REF (Enzyme) ?}} \text{Active erythropoietin}$$

Erythropoietin has been prepared from the plasma of anemic sheep and also from the urine of anemic human beings. It is now prepared using genetically-engineered cell cultures. It is a glycoprotein with 8–12% total hexose and has a molecular weight of about 60,000. Its molecule contains all the common amino acids except methionine. Its activity is much retarded by proteolytic enzymes and also by the antibiotic actinomycin-D.

Erythropoietin stimulates the differentiation of the stem cell (*hemohistioblast*) of the bone marrow into the erythroid series, increasing the numbers of proerythroblasts in the bone marrow. This is followed by increases in other nucleated erythrocytes and finally, by increases in reticulocytes (erythrocyte precursor cells) and mature erythrocytes, in the peripheral circulation. Recent studies indicate that the earliest effect of erythropoietin is stimulation of the synthesis of a very large RNA (150s) by bone marrow cells. Erythropoietin is administered therapeutically in cases of kidney failure.

Renin. It catalyzes the synthesis of angiotensins which cause vasoconstriction in the kidneys, thereby causing electrolyte and water retention in the body. This system has been referred to as the *renin-angiotensin system.* An increase in the renal pressure and the levels of plasma Na^+ and angiotensins cause a decrease in renin production.

3. Eicosanoid Hormones

A. Structure and Metabolic Roles

Eicosanoid hormones (or simply eicosanoids) are fatty acid derivatives with a variety of extremely potent hormonelike actions on various tissues of vertebrates. Eicosanoids, in general, are known to be involved in reproductive function; in the inflammation, fever and pain; in the formation of blood clots and the regulation of blood pressure; in gastric acid secretion; and in a variety of other human processes.

Eicosanoids are all derived from 20-carbon polyunsaturated fatty acid, **arachidonic acid** (20 : 4; 5, 8, 11, 14), from which they take their general name (*eikosi*[G] = twenty). There are 3 classes of eicosanoids (or the *signal molecules*, as they are also called): prostaglandins, thromboxanes, and prostacyclins and leukotrienes (Fig. 34–59).

Prostaglandins

Kurzrok and **Lieb** (1930), for the first time, observed that the human semen is able to bring about strong muscular contraction or relaxation when placed in the uterus. This is due to the presence in the semen of a number of structurally-related compounds collectively termed prostaglandins. The name prostaglandin (abbreviated as PG) was first given by a Swedish chemist **U.S. von Euler** in

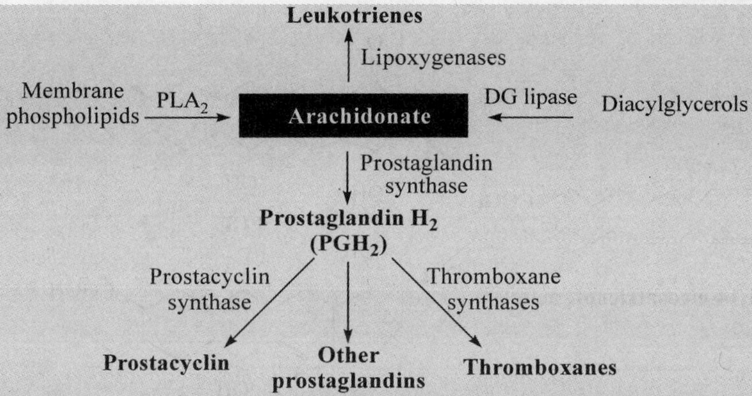

Fig. 34–59. Arachidonic acid and some of its eicosanoid derivatives

In response to certain hormonal signals, *phospholipase* A_2 releases arachidonic acid (arachidonate at pH 7) from membrane phospholipids. Arachidonic acid then serves as a precursor of various eicosanoids. *Prostaglandin synthase* catalyzes the first step in a pathway leading to prostaglandins, prostacyclins and thromboxanes. *Lipoxygenase catalyzes* the initial step in a pathway leading to leukotrienes.

1935 to this lipid-soluble acidic substance, initially found by him in human seminal fluid and in extracts from ovine vesicular glands. Although these were originally found in the seminal fluid of man and other species (hence the nomenclature) but now prostaglandins have been found to occur in a wide variety of mammalian tissues including brain, spinal cord, thymus, lungs, pancreas, kidneys, menstrual fluid and placenta. However, *semen remains one of the richest sources of prostaglandins as yet.* The purification and structural characterization of various prostaglandins was subsequently carried out largely by **S. Bergstöm**, **J. Sjovall** and **B. Samuelsson**, also in Sweden.

Chemically, the prostaglandins resemble prostanoic acid (Fig. 34–60). They are hydroxy derivatives of the polyunsaturated C-20 cyclic fatty acids. In its molecule, the carbon atoms 8 to 12 are involved in the formation of a 5-carbon ring called cyclopentane ring. Variations in the double bonds and in the hydroxyl and ketone groups give rise to prostaglandins that can be divided into 9 groups designated as A through I (and accordingly, the prostaglandins are designated as PGA through PGI). The PG, obviously, stands for prostaglandin and the third capital letter, in most cases, indicates the type of the substituents found on the hydrocarbon chain. In fact, two groups of prostaglandins were originally recognized: **PGE** which is ether-soluble (hence the nomenclature,

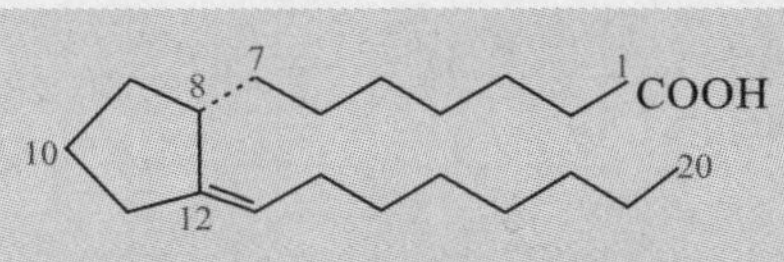

Fig. 34–60. Prostanoic acid

ULF SVANTE VON EULER

(LT, 1905–1983)

von Euler, a Swedish physiologist, was born in Stockholm. He received an MD from the Karolinska Institute, where he remained as a member of the faculty after receiving his degree. He first *identified prostaglandins*– from semen– in the early 1930s, and named them (in 1935) for their source, the prostate gland. By the time it was realized that all cells except red blood cells synthesize prostaglandins, their name had become entrenched.

He remained **President of the Nobel Foundation** from 1965. He discovered noradrenalin and identified its function as a chemical intermediate in nerve transmission. He shared with **Julius Axelrod** and **Sir Bernard Katz** the **1970 Nobel Prize in Medicine or Physiology** for their independent work on the *chemistry of the transmission of nerve impulses.* He died on March 9, 1983.

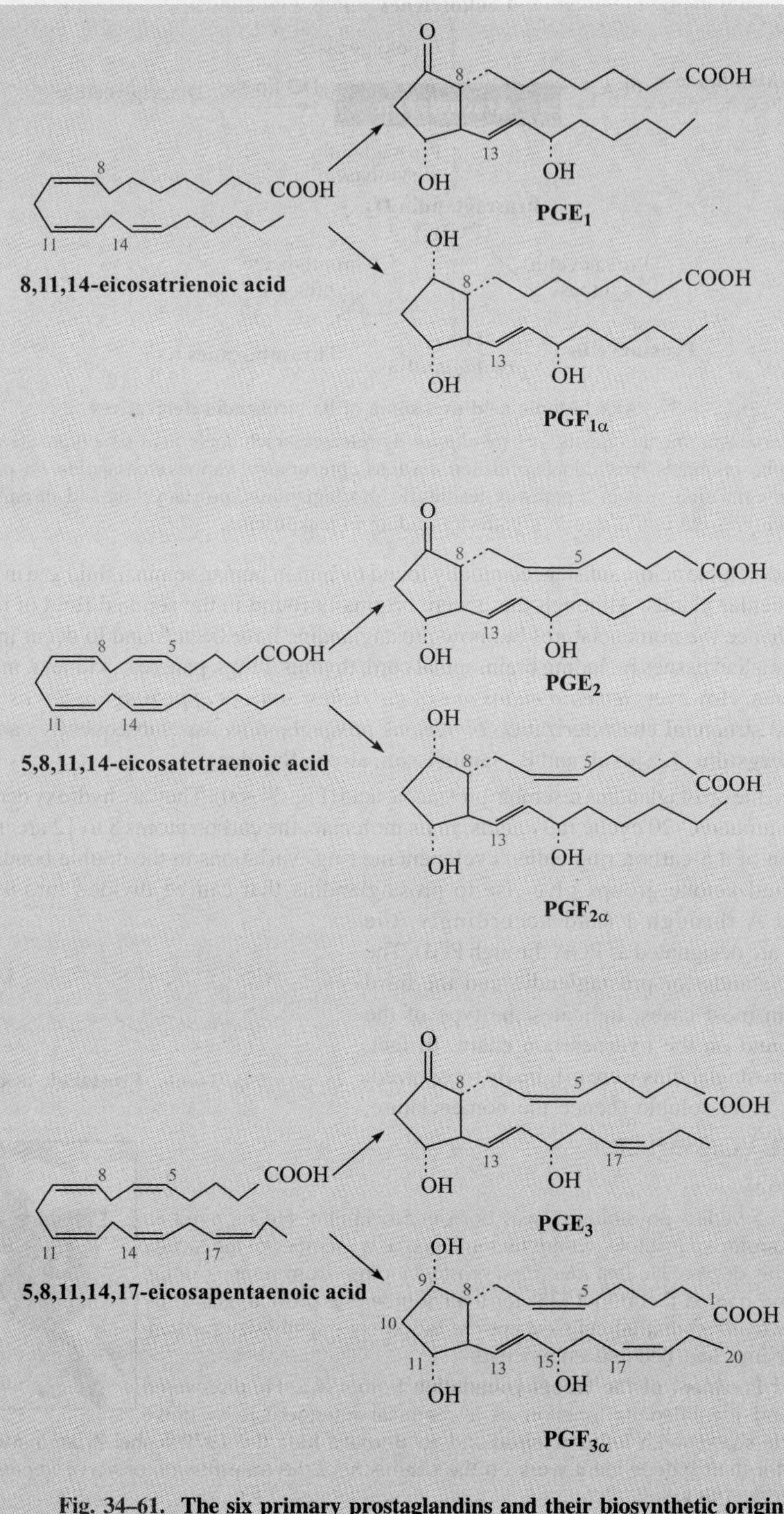

Fig. 34–61. The six primary prostaglandins and their biosynthetic origins

E from *e*ther) and has a keto group at C_9, and **PGF** which is phosphate-buffer-soluble (hence the nomenclature, F from *fosfat* in Swedish) and has a hydroxyl group at C_9. There are 3 compounds from each of these two groups, arising from eicosanoic (*i.e.*, C_{20}) fatty acids with 3, 4 or 5 double bonds. These are the 6 primary prostaglandins and are abbreviated as PGE_1, PGE_2, PGE_3, $PGF_{1\alpha}$, $PGF_{2\alpha}$ and $PGF_{3\alpha}$. In the abbreviations of the prostaglandins, the number 1, 2, 3 is added as subscript to indicate the number of carbon-carbon double bonds outside the ring. When there is more than one member, the group is subdivided into α, β, γ etc. These occur in most cells. Their structure appears in Fig. 30–61 along with their biosynthetic origin. In addition to these, there are several secondary prostaglandins that are derived from PGE types through enzymic conversions. The x-ray structure of PGH_2 synthase from sheep seminal vesicles is presented in Fig. 34–62.

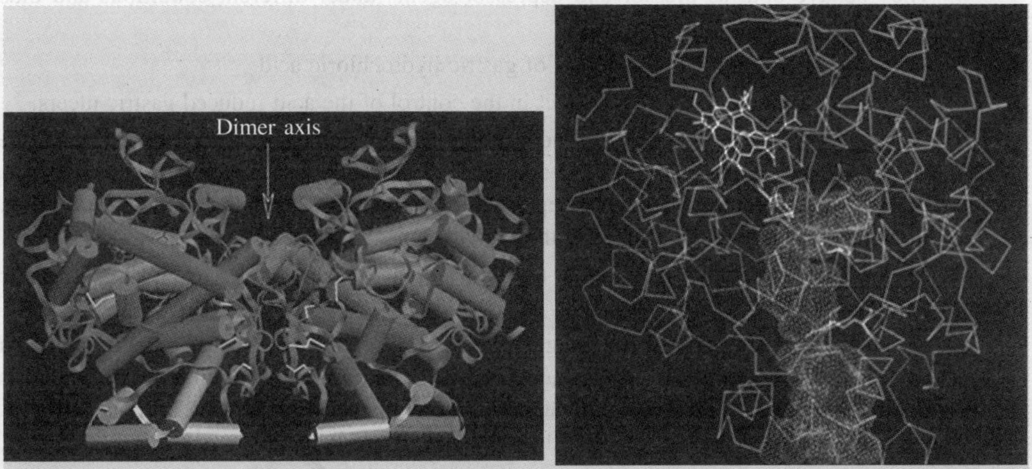

(*a*) (*b*)

Fig. 34–62. The x-ray structure of PGH_2 synthase from sheep seminal vesicles

(*a*) **A diagram of the dimer as viewed along the dimer interface with its twofold axis of symmetry vertical.**

The EGF module, the membrane-binding motif, and the catalytic domain are colored *green, tan,* and *blue,* respectively, whereas the heme is *red* and the five disulfide bonds in each subunit are *yellow*.

(*b*) **A C_α diagram of a PGH_2 subunit (*green*), the left subunit in Part *(a)* as viewed from 30° to the left.** The peroxidase active site is located above the heme (*pink*). The hydrophobic channel, which penetrates the subunit from the membrane-binding motif at the bottom of the figure to the cyclooxygenase active site below the heme, is represented by its van der Waals surface (*blue dots*). The three residues in the channel that are shown in *yellow* are, from top to bottom: Tyr 385, which forms a transient radical during the cyclooxygenase reaction; Ser 530, which is acetylated by aspirin; and Arg 120, which forms an ion pair with the NSAID flurbiprofen when it binds in the channel.

(*Courtesy: Michael Garavito, University of Chicago*)

Prostaglandins have a half-life of 5 minutes or less, hence they are destroyed very rapidly in the body. They have, therefore, *a high turnover rate.* The short half-life of these tissue hormones is thought to ensure their transient and limited response at the intermediate site of production. From a medical viewpoint, they are potentially the most revolutionary therapeutic substances yet discovered. Prostaglandins affect smooth muscles and blood pressure and often the activities of individual prostaglandins oppose one another. For example, prostaglandin E_2 (PGE_2) dilates blood vessels and bronchi, and prostaglandin $F_{2\alpha}$ ($PGF_{2\alpha}$) constricts these smooth muscle tissues.

The prostaglandins perform a wide variety of biologic activities:

1. As mentioned earlier, they bring about contraction or relaxation of the smooth muscles of the uterus, *esp*., at the time of ovulation. This may be due to a chelation of calcium ions. As little as 1 ng/ml can cause contraction of the smooth muscles. They, thus, resemble ocytocin in this regard. They are modulators of hormone action.
2. They lower down blood pressure.
3. They inhibit lipolysis in adipose tissue, possibly by inhibiting the conversion of ATP to cyclic AMP and inhibition of platelet aggregation. The prostaglandins, thus, have the opposite effect of epinephrine, norepinephrine, glucagon and corticotropin on the release of fatty acids from adipose tissue.
4. They behave both as pressor and depressor agents under different conditions and thus affect the cardiovascular system.
5. They appear to control the secretion of gastric hydrochloric acid.
6. They also have some beneficial effect in the control of the acid-induced gastric ulcers.
7. Prostaglandins are best known for their effects on reproductive system. There seems to be a strong link between male fertility and seminal prostaglandin content. Human semen is rich in prostaglandins, which when deposited in vagina through coitus, facilitate conception. Thus, low prostaglandin content in the human semen is related to infertility.
8. They are effective inducers of labour and abortion in pregnant women and also in domestic animals.
9. Recent work indicates that the prostaglandins are also involved in the inflammatory reaction and pain. Anti-inflammatory drugs such as aspirin, in part, act by inhibiting the synthesis of prostaglandins. However, paracetamol (an analgesic drug like aspirin) is not anti-inflammatory as it does not inhibit the synthesis of prostaglandins.

Like hormones, they may be released into the blood but, unlike them, are only effective over short distances. The widespread distribution of prostaglandins and their capacity to carry out varied metabolic effects have, however, led some to question the propriety of their being called hormones.

Thromboxanes and Prostacyclins

Thromboxanes (TXAs) and prostacyclins (PGIs) are structurally-related compounds that arise from a nascent prostaglandin. In compounds of both these categories, carbons 8 and 12 are joined and an oxygen atom is added to form the six-membered ring (*cf* prostaglandins, where a five-membered ring is formed). **Thromboxane A_2** (TXA_2) was first isolated, in 1975, by **Samuelsson** *et al* from blood platelets (also known as thrombocytes, hence the nomenclature). The following year, **John Vane** and colleagues at the Royal College of Surgeons, London identified yet another type of eicosanoid, **prostacyclin I_2** (PGI_2), which is produced primarily in vascular tissues, *i.e.*, blood vessels.

The striking similarity and diversity in the physiological roles of thromboxanes and prostacyclins displays a critical balance required for the normal functioning in the body. TXA_2 and PGI_2 are medically important examples of how such a balance operates *in vivo*. TXA_2 is a highly effective vasoconstrictor (blood vessel constrictor) and platelet aggregator ; conversely, PGI_2 is a potent vasodilator and inhibitor of platelet aggregation. Platelets are the blood cells that first appear and aggregate at the site of injury to produce a temporary plug that serves as a base on which the strong fibrin clot ultimately forms. However, for maintenance of normal blood flow, TXA_2-induced aggregation of platelets would quickly prove fatal. Thus, a vital opposing role of PGI_2 is to prevent platelets from aggregating on blood vessel walls, a site of PGI_2 production.

KARL SUNE DETLOF BERGSTRÖM

(LT, 1916–2004)

Bergström, a Swedish biochemist of Karolinska Institute, was born on January 10, 1916 in Stockholm. He shared the **1982 Nobel Prize for Physiology or Medicine** with **Bengt I. Samuelsson** and **John R. Vane** for their research work on eicosanoids (prostaglandins). Bergström's contribution was to identify the main members of the prostaglandin family. Bergström had a fair share of awards and honours to his credit. He was elected a member of the Royal Swedish Academy of Sciences in 1965, and its President in 1983. In 1965, he was also elected a member of the Royal Swedish Academy of Engineering Sciences. In 1966, he was elected a Foreign Honorary Member of the American Academy of Arts and Sciences. In 1985, he was appointed member of the Pontifical Academy of Sciences. He was an honorary member of the International Academy of Science. He died on August 15, 2004. Sune Bergstörm was the father of the evolutionary geneticist **Svante Pääbo** and of the businessman **Rurik Bergström** (both born 1955).

BENGT INGEMAR SAMUELSSON

(born, 1934)

Samuelsson, a Swedish biochemist of Karolinska Institute, was born on May 21, 1934 in Halmstad in southwest Sweden. He studied at the Stockholm University where he became a professor in 1967. In 1975, he was awarded the Louisa Gross Horowitz Prize from Columbia University together with **Sune K. Bergström**. He shared the **1982 Nobel Prize for Physiology or Medicine** with **Sune K. Bergström** and **John R. Vane**, for his work on the control of prostaglandins. Discussing the role of **prostaglandins** in the body, Samuelsson explained:

"It is a control system for the cells that participates in many biological functions. There are endless possibilities of manipulating this system in drug development."

His research interests were originally in cholesterol metabolism with importance to reaction mechanisms. Following the structural work on prostaglandins along with **Sune Bergström**, he was interested mainly in the transformation products of arachidonic acid. This has led to the identification of endoperoxides, thromboxanes and the leukotrienes, and his group has chiefly been involved in studying the chemistry, biochemistry, and biology of these compounds and their function in biological control systems. This research has implications in numerous clinical areas, especially in thrombosis, inflammation, and allergy. Between 1981 and 1995, about 3,000 papers per year were published that specifically used the expression "prostaglandins", or related terms such as "prostacyclins", "leukotrienes", and "thromboxanes", in their labels and titles.

SIR JOHN ROBERT VANE

(LT, 1927–2004)

Vane, a British pharmacologist, was born on March, 1927 in Tardebigge, Worcestershire to **Maurice Vane** and **Frances Vane**. He grew up in suburban Birmingham. He attended a local state school from age 5, before moving on to King Edward's School in Edgbaston, Birmingham. An early interest in chemistry was to prove the inspiration for studying the subject at the University of Birmingham in 1944. During his undergraduate studies, Vane became disenchanted with chemistry but still enjoyed experimentation. When **Maurice Stacey**, the Professor of Chemistry at Birmingham, was asked by **Harold Burn** to recommend a student to go to Oxford and study pharmacology, Vane jumped at, rather grabbed, the chance and moved to Burn's department in 1946. Under Burn's guidance, Vane found motivation and enthusiasm for pharmacology writing:

"[the] laboratory gradually became the most active and important centre for pharmacological research in the U.K. and the main school for training of young pharmacologists."

John Vane married in 1948 to **Elizabeth Daphne Page** and had 2 daughters. He completed a B.Sc. in pharmacology and briefly went to work at the University of Sheffield, before coming back to Oxford to complete his doctorate in 1953 under **Geoffrey Dawes**. After completing his Ph.D., Vane worked as an assistant professor in the Department of Pharmacology at Yale University before moving back to the United Kingdom to take up a post as a senior lecturer in the Institute of Basic Medical Sciences at the University of London, in 1955.

Vane held a post at the University of London for 18 years, progressing from senior lecturer to Professor of Experimental Pharmacology in 1966. During that time, he developed certain bioassay techniques and focused his research on both angiotensin-converting enzyme (ACE) and the actions of aspirin, eventually leading to the understanding of the relationship between aspirin and the prostaglandins. In 1973, Vane left academia and took up the position as Director of Research of the Wellcome Foundation, taking a number of his colleagues with him who went on to form the Prostaglandin Research Department. Under the leadership of **Salvador Moncada,** this group continued important research that eventually led to the discovery of prostacyclin. In 1985, Vane returned to academic life and set up the *William Harvey Research Institute* at the Medical College of St. Bartholomew's Hospital (now Barts and The London School of Medicine and Dentistry). At the Institute, Vane's work focused on selective inhibitors of COX-2, and the interplay between nitric oxide and endothelin in the regulation of vascular function.

Vane shared the **1982 Nobel Prize for Physiology or Medicine** with Sune **K. Bergström** and **Bengt I. Samuelsson** for their isolation and analysis of prostaglandins. Vane's research provided the means of measuring prostaglandin activity in tissue. Vane was also awarded the **Lasker Award** for the discovery of prostacyclin and was knighted in 1984 for his contributions to science. He died on November 19, 2004 in Princess Royal University Hospital, Kent, from long-term complications arising from leg and hip fractures he sustained in May of that year.

Unlike other eicosanoids, PGI_2 is not metabolized during passage through the lungs. Thus, TXA_2 and PGI_2 are continuously engaged in a 'tug of war' with respect to platelet aggregation (Fig. 34–63).

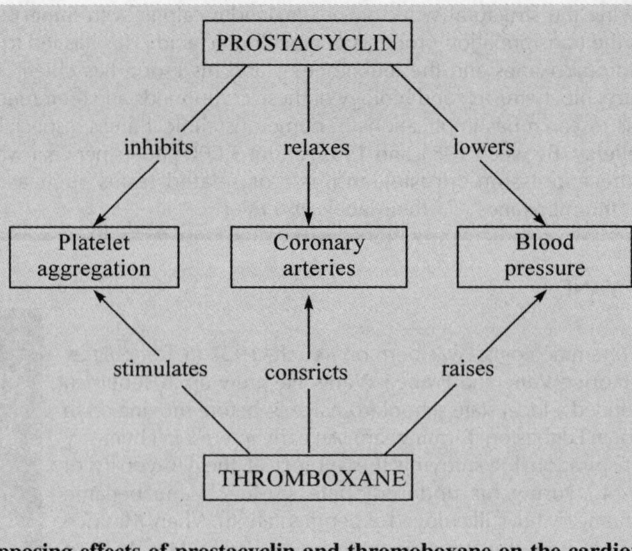

Fig. 34–63. Opposing effects of prostacyclin and thromoboxane on the cardiovascular system

Leukotrienes

Leukotrienes (LTs), first found in leukocytes, are cysteinyl-containing derivatives of arachidonic acid with a series of three conjugated double bonds in its molecule (Fig. 34–64), hence their nomenclature.

Leukotriene B$_4$ (LTB$_4$) has two hydroxyl groups at C-5 and C-12 and three conjugated double bonds at C-6, C-8 and C-10. An additional double bond is present between C-14 and C-15. **Leukotriene C$_4$** (LTC$_4$) contains the tripeptide glutathione (γ-Glu-Gly-Cys) covalently bonded to a derivative of arachidonic acid. **Leukotriene D$_4$** (LTD$_4$) possesses the dipeptide, Gly-Cys (Glu residue is eliminated) and **leukotriene E$_4$** (LTE$_4$), the amino acid Cys (Gly residue eliminated).

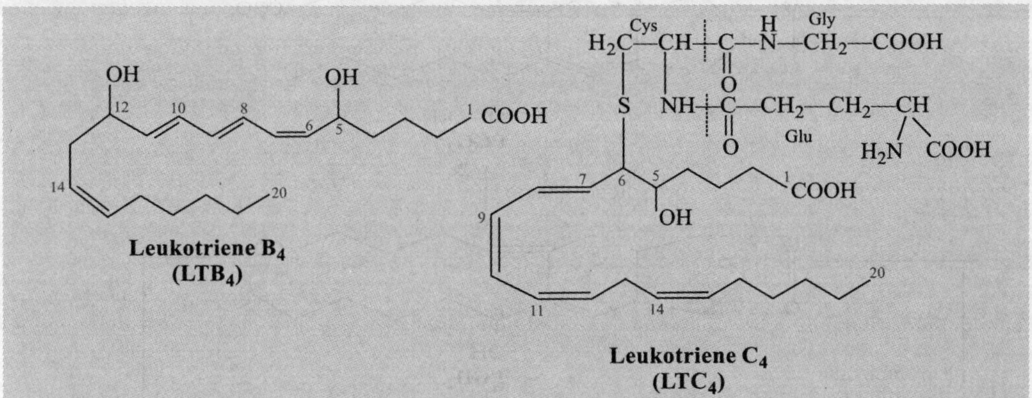

Fig. 34–64. Two common leukotrienes

Neutrophils make one class of leukotrienes to alter mobility and act as chemotactic agents. *Mast cells* make another class, formerly known as **slow-reacting substances**, which is responsible for bronchial constriction and other anaphylactic allergic reactions.

Leukocytes are powerful biological signals; for example, they induce contraction of the muscle lining the airways to the lung. They stimulate chemotaxis of white blood cells and regulate inflammation by increasing permeabilities of small blood vessels. They also cause a slow and persistent contraction in the smooth muscle of blood vessels. Overproduction of leukotrienes causes asthmatic attacks and also stimulates mucus secretion.

B. Biosynthesis

Various eicosanoids are produced in different cell types by different synthetic pathways, and have different target cells and biological activities. While prostaglandins are made everywhere in the body, synthesis of thromboxanes and leukotrienes occurs in restricted locations. Both compounds are synthesized in platelets, neutrophils and the lung. Some amount of thromboxane production also takes place in the brain.

Fig. 34–65 depicts various reactions involved in the biosynthesis of 3 types of eicosanoids from arachidonic acid. Arachidonic acid is generated from phospholipids by the action of *phospholipase A$_2$* (PLA$_2$), or from diacylglycerol by the action of a *lipase*. The biosynthesis of most eicosanoids starts at arachidonate, which has 4 double bonds at C$_5$, C$_8$, C$_{11}$ and C$_{14}$. In the first key reaction, arachidonate gets converted to prostaglandin H$_2$ (PGH$_2$) by the enzyme *prostaglandin synthase*, which is made up of two components, cyclooxygenase (COX) and hydroperoxidase. This is a two-step reaction. In *first step*, cyclooxygenase component of prostaglandin synthase catalyzes the addition of one mole of oxygen to C-9 of arachidonate and of a second mole to C-15. The bond formation between C-8 and C-12 accompanying this oxygenation produces the 5-membered endoperoxide

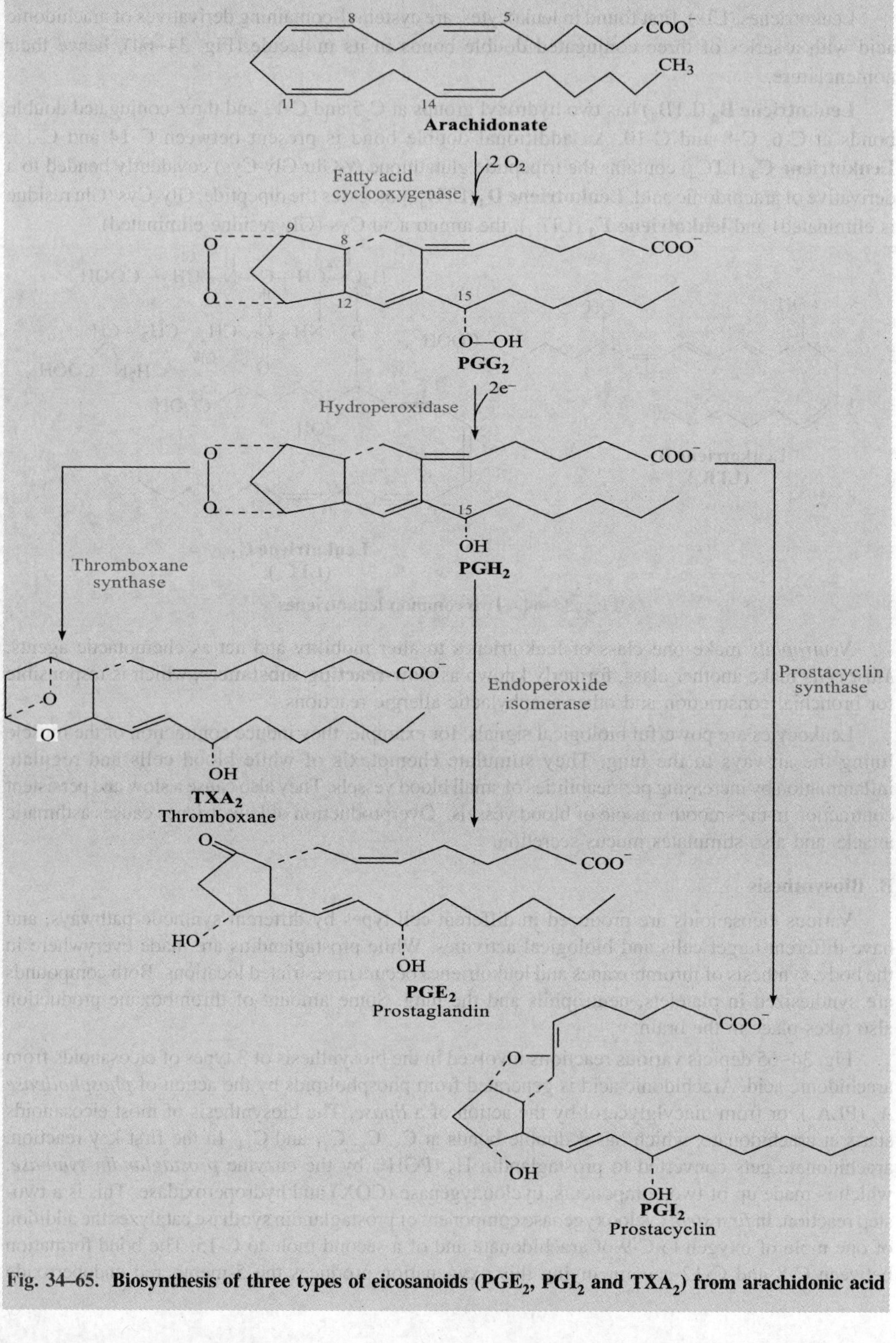

Fig. 34–65. Biosynthesis of three types of eicosanoids (PGE_2, PGI_2 and TXA_2) from arachidonic acid

ring structure, characteristic of eicosanoids. The compound so formed is called as **prostaglandin G_2** (PGG_2). The 4 oxygen atoms introduced into PGG_2 come from 2 moles of oxygen. In the second step, the hydroperoxidase component of prostaglandin synthase, then, catalyzes a two-electron reduction of the 15-hydroperoxy group of PGG_2 to a 15-hydroxyl group, producing **prostaglandin H_2** (PGH_2). The highly unstable PGH_2 is rapidly transformed into other prostaglandins, prostacyclins and thromboxanes. In fact, the biochemical fate of the PGH_2 synthesized is determined by tissue-specific enzymes. For example, in a tissue producing prostaglandin E_2, the enzyme endoperoxide isomerase is present and converts PGH_2 into **prostaglandin E_2** (PGE_2). The synthesis of PGE2 from arachidonate was the first-ever pathway elucidated for eicosanoid production and was discovered by **Bengt Samuelsson** and his associates in 1964.

Inhibitors of Prostaglandin Synthesis

The *nonsteroidal antiinflammatory drugs* (NSAIDs), such as aspirin, indomethacin, and phenylbutazone, decrease inflammation, pain and fever through inhibition of the synthesis of prostaglandins, by blocking the first stage whereby arachidonic acid is converted to the common prostaglandin precursor, PGG_2, by a cyclooxygenase enzyme. There are 2 forms of the enzyme. Cyclooxygenase 1 (COX-1) is mainly constitutive and found in platelets, stomach, and kidney. Cyclooxygenase 2 (COX-2) is mainly inducible and responsible for the synthesis of inflammatory prostaglandins. Aspirin (=acetylsalicyclic acid) covalently modifies and irreversibly inactivates both forms of the enzyme by acetylation. Aspirin is more potent against COX-1 than COX-2. Other NSAIDs inactivate cyclooxygenase but do so by binding noncovalently to the enzyme instead of acetylating it; they are called "non-aspirin NSAIDs". Most NSAIDs inhibit COX-1 more than COX-2. These drugs have undesirable side-effects; aplastic anemia can result from phenylbutazone therapy. COX is an important target for pharmacological agents.

Steroidal antiinflammatory drugs like hydrocortisone, prednisone, and betamethasone block prostaglandin release by inhibiting phospholipase A_2 activity so as to interfere with mobilization of arachidonic acid. The rate-limiting step in the synthesis of prostaglandins is the release of arachidonic acid from membrane phospholipid stores in response to phospholipase A_2 activation.

Phospholipid

Phospholipase A_2 ↓ ← inhibited by antiinflammatory steroids

Arachidonic acid

Cyclooxygenase ↓ ← inhibited by aspirin, indomethacin, phenylbutazone

Prostaglandins

Sites of action of inhibitors of prostaglandin synthesis

Control of the biosynthesis of prostaglandins is poorly understood, but, in general, prostaglandin release seems to be triggered by hormonal or neural excitation, or muscular activity. For example, histamine increases prostaglandin concentration in gastric perfusates. Also, prostaglandins are released during labour and after cellular injury (e.g., platelets exposed to thrombin, lungs irritated by dust).

The wondrous drug aspirin has been used for centuries to decrease inflammation, pain and fever. Its mode of action was an enigma until **John Vane**, in 1975, discovered that *aspirin inhibits the synthesis of prostaglandins by inactivating prostaglandin synthase*. Specifically, aspirin (acetylsalicylate) irreversibly inhibits the cyclooxygenase activity of this enzyme by acetylating a specific serine hyydroxyl group (Fig. 34–66). Aspirin is a potent antiinflammatory agent because it blocks the first step in the synthesis of prostaglandins. This drug is also widely used to prevent excessive blood clotting, which can lead to heart attacks and strokes. Aspirin is antithrombotic also because it blocks the formation of thromboxane A_2 (TXA_2), a potent aggregator of blood platelets.

Aspirin

The chemical name of aspirin is *acetylsalicylic acid*, $CH_3CO_2C_6H_4COOH$. The name 'aspirin' is derived from a plant genus *Spirea* (from a Latin word which means spiral, which refers to the fact that most members of this genus are used for making wreaths and garlands, which are coiled or somewhat spiral in shape), whose members contain salicylic acid, and thus are its natural sources. However, salicylic acid also occurs in many other plants such as jasmine, patridge berry, legumes (pea, bean, clover), grasses (wheat, rye, sugarcane) and many trees (beech, birch, olive, poplar, willow). The chemical term salicylic acid itself comes from *Salix*, the Latin name for willow tree. The active ingredient in willow bark is a bitter glycoside called **salicin.** Salicin was first identified by the Italian chemists, **Fontana** and **Brugnatelli** in 1826, and isolated in pure form by the French pharmacist, **Henri Leroux** in 1829. On hydrolysis, salicin yields glucose and salicylic alcohol. The latter gets converted in the body to **salicylic acid**, which actually brings the fever down. Salicylic was first successfully extracted, in 1838, by an Italian chemist **Rafelle Piria** (the second name of this chemist, as one can probably note, has an earie similarity to the plant genus *Spirea*). Later in 1874, the same German scientist **Adolf Kolbe**, who had earlier synthesized acetic acid, came out with an economical procedure for the laboratory synthesis of salicylic acid as well. In pure form, however, salicylic acid is too corrosive to the tissues in that it peels off the upper horny layers of the skin and that is why it is an excellent agent for the removal of corns, callosities and warts. But taken orally, it produces severe vomiting and sometimes even comma. For internal administration, however, it had to be taken in the form of a salt. The most significant and potent salt, **acetylsalicylic acid** was, however, produced in 1899 by **Felix Hoffmann** (LT, 1868–1946) who also looked to its usefulness for a compelling reasons to try the same on his father suffering from rheumatoid arthritis and both remedies of pain (sodium, phenyl salicylate), available at that time, having had failed to give him relief. The drug fortunately worked on him. **Heinrich Dreser**, Director of the Friedrich Bayer & company, gave it the tradename **'aspirin'**. Aspirin is widely used to suppress platelet function in humans at risk of coronary heart disease, or deep vein thrombosis, and to treate patients that have suffered a heart attack or stroke.

Inhibition of the cyclooxygenase blocks the formation of prostaglandin H_2, PGH_2 (Fig. 30–66). PGH_2, which is produced in the platelets, is also the precursor of thromboxane A_2 (TXA_2); the reaction being catalyzed by the enzyme *thromboxane synthase*. **Prostacyclin I_2** (PGI_2) is synthesized from PGH_2 and the reaction is mediated by *prostacyclin synthase*. Thus, we see that the tissues are differently endowed with enzymes that transform endoperoxides into sepcific types of eicosanoids.

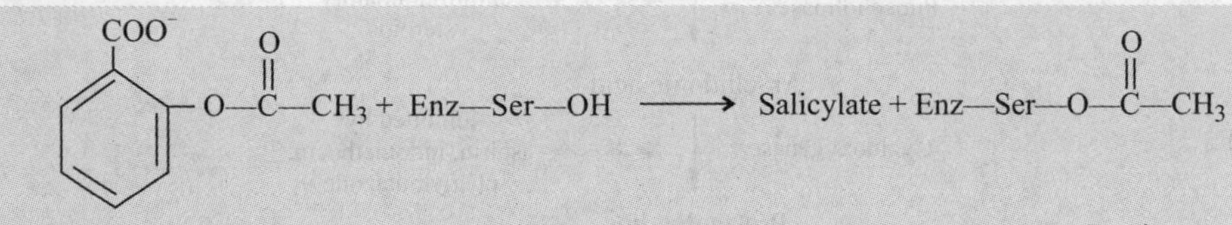

Aspirin (Acetylsalicylate) — Active enzyme — Inactive enzyme

Fig. 34–66. Inactivation of prostaglandin synthase by aspirin

Leukotrienes are made from arachidonate by another pathway, beginning with the addition of oxygen to C-5 of arachidonate; the reaction being catalyzed by *lipoxygenase*. This reaction is not affected by antiinflammatory drugs.

4. Opiate Peptides

Opiate peptides are compounds that bind to specific receptors and exhibit hormone-like actions. The various opiate peptides fall under 3 families, enkephalins, dynorphins and endorphins, which are generated in the body by the action of protease.

(*a*) **Enkephalins.** These consist of 5 to 7 amino acids and are derived from a precursor, *proenkephalin*. These have been isolated from extracts of brain and pituitary and exhibit morphine-like properties.

(*b*) **Dynorphins.** These consist of 10 to 17 amino acids and are derived from a precursor, *prodynorphin*. These are endogenous compounds made in several locations, but the major locations are brain, pituitary, adrenal medulla and peptidergic neurons.

(*c*) **Endorphins.** These consist of 16 to 27 amino acids and are derived from a precursor, propiomelanocortin (POMC). These have also been isolated from extracts of brain and pituitary and exhibit morphine-like properties.

It has been suggested that there are at least 3 types of receptors, δ, μ and *k*, through which these drugs mediate their physiological effects.

Though these opoids produce major effects on the central nervous system, they also function as neurotransmitters and neurohormones to modulate neurotransmission. Some of the effects that opiate peptides produce are analgesia, drowsiness, nausea, vomiting, respiratory depression, decreased gastrointestinal motility and modulations of endocrine and autonomic nervous systems.

VASOACTIVE PEPTIDES

There are three groups of well defined peptides that possess vasoactive properties. These are: *neurohypophyseal hormones* (ocytocin and vasopressin), *angiotensins* and *kinins*. Ocytocin and vasopressin have already been described on pages 1132 through 1134. The remaining two groups of vasoactive peptides are described below.

1. Angiotensins

Goldblatt (1947) demonstrated that the partial occlusion of the renal artery in a dog results in permanent hypertension. Obviously, slowing the circulation causes the formation of a substance that produces vasoconstriction. The presence of such a substances has been demonstrated in the blood from such ischemic kidneys. The mechanism behind this operates as follows (Fig. 34–67). An enzyme renin is formed in the kidney and is released into the blood. It is a proteinase and acts on *angiotensinogen*, a plasma protein synthesized in the liver, to split off a decapeptide called *angiotensin I*. This decapeptide, in turn, is acted upon by a peptidase called converting enzyme, present in serum, to form an octapeptide called *angiotensin II*. Another peptidase, termed angiotensinase, produced by kidney, hydrolyzes angiotensin II to produce a heptapeptide called *angiotensin III*. Angiotensinase, thus, serves as a balancing antipressor agent. The sequence of biochemical steps leading to its release is as follows:

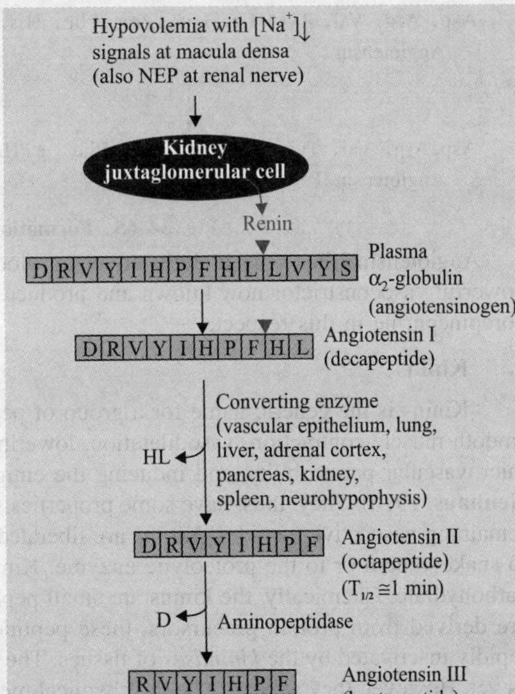

Fig. 34–67. Renin–angiotensin system
One-letter amino acid symbols are given in Appendix VIII.
[NEP = norepinephrine]

Angiotensinogen (an α_2-globulin of plasma) —Renin→ Angiotensin I (a decapeptide) —Angiotensin-converting enzyme→ Angiotensin II (an octapeptide)

Angiotensin II —Angiotensinase→ Angiotensin III (a heptapeptide)

Various synonyms exist in the literature for the above substances. Angiotensinogen has been variously called as hypertensinogen or renin activator. Likewise, angiotensin II is also known as hypertensin or angiotonin and angiotensinase as hypertensinase.

The two angiotensins, I and II are peptides and their amino acid sequence (refer Fig. 34–68) was determined by **Page** and coworkers in 1960.

Asp. Arg. Val. Tyr Ile. His. Pro. Phe. His. Leu. Leu. Val. Tyr. Ser–Glycoprotein
Angiotensinogen (?)

↓ Renin

Asp. Arg. Val. Tyr. Ile. His. Pro. Phe. His. Leu + Leu. Val. Tyr. Ser–Glycoprotein
Angiotensin I

↓ Peptidase

Asp.Arg. Val. Tyr. Ile. His. Pro. Phe + His. Leu
Angiotensin II

Fig. 34–68. Formation of angiotensin II

Angiotensin I has only a slight effect on blood pressure, whereas angiotensin II is the most powerful vasoconstrictor now known and produces hypertension. It is 200 times more active than norepinephrine in this respect.

2. Kinins

"Kinins is the generic name for a group of peptides with potent biologic activities in causing smooth muscle contraction, vasodilatation, lowering of blood pressure, increasing blood flow and microvascular permeability, and inducing the emigration of granulocytic leukocytes" (**Orten** and **Neuhaus**, 1970). They, thus, have some properties similar to and other properties different from the remaining vasoactive peptides. Kinins are liberated from plasma protein called *kininogen* exposed to snake venom or to the proteolytic enzyme. Kininogen is an α_2-globulin containing about 18% carbohydrate. Chemically, the kinins are small peptides of 9 to 11 amino acid residues. Since they are derived from protein precursors, these peptides contain only protein amino acids. They are rapidly inactivated by the *kininases* of tissues. The biological significance of the kinins is not clear as yet. However, they appear to be the chemical mediators of inflammation. They are also powerful hypotensive agents. *Bradykinin* and *kallidin* are typical examples of kinins.

Bradykinin. It is formed from its precursor *bradykininogen*, an α_2-globulin in serum, in the presence of an enzyme kallikrein. Bradykininogen has been found in a variety of tissues. Heart contains the most, followed by liver, kidney and brain. The amount of bradykininogen in the blood is 6 to 8 μg per ml. Bradykinin is a nonapeptide with the amino acid sequence shown in Fig. 34–69.

Arg. Pro. Pro. Gly. Phe. Ser. Pro. Phe. Arg

Fig. 34–69. Bradykinin

Bradykinin is released by active sweat and salivary glands. It is a powerful vasodilator that enormously increases the blood flow locally and thus promotes secretion of sweat and saliva. But its role as a vasodilator is mainly conjectural.

Kallidin. Kallidin (Fig. 34–70) is a decapeptide and contains lysine, in addition, on the N-terminal of the peptide chain of bradykinin. Hence, it is also called *lysylbradykinin.*

Lys. Arg. Pro. Pro. Gly. Phe. Ser. Pro. Phe. Arg

Fig. 34–70. Kallidin

HORMONE FROM THYMUS

The thymus is present in all jawed vertebrates. It is a flat, pinkish, bilobed structure, located in the chest behind the sternum. The gland arises as a proliferation of the gill pouch epithelium which becomes infiltrated with lymphocytes. It grows rapidly to acquire a large size in the young animals and atrophies in the adults. In man, the gland reaches its greatest development at the age of 14 to 16, after which it atrophies because of the activity of the sex cells. Thymus is, however, absent in the hagfishes and in the lamprey; it is represented by a group of cells beneath the gill epithelium. This is called as a *prothymus*.

Structure. Purification of thymus extracts have yielded an active glycopeptide.

Functions. Certain biochemical functions are attributed to this gland. These are as follows:

(*a*) It is the primary source of lymphocytes in the mammals. The lymphocytes are responsible for the immunological functions of the body by producing antibodies.

> **Antibodies** are substances that help the animals to combat with the invading bacteria and viruses.

(*b*) It helps the body to distinguish between its own tissue proteins and the foreign proteins such as those present in tissues transplanted from another animal. This has been inferred from the fact that when the thymus is removed at birth, skin grafts in animals persist for a long period. However, when the thymus is left intact, the graft or the transplanted tissue is rejected.

(*c*) It is concerned with accelerating growth and as such metamorphosis is postponed.

Thymectomy. Removal of the thymus gland has resulted in the following dysfunctions:

1. reduction in lymphocyte population (*lymphopenia*)
2. atrophy of the lymphoid tissue
3. loss of immunological competence.

The renal hormones, the prostaglandins, the angiotensins, the kinins and the thymus hormone are all sometimes collectively referred to as **parahormones** or **tissue hormones** as many of them act very locally. They are all produced endogenously by specialized groups of cells. These are different from hormones as they produce their effects without entering the bloodstream. Moreover, their secretion is controlled by hormones, metabolic products, enzymes etc.

PHEROMONES OR 'SOCIAL' HORMONES

Definition. The pheromones or smell signals, as they are also called, are species-specific chemical substances or 'odours' which are released from animals (insects to mammals) into the environment and evoke behavioural, developmental or reproductive responses. The term '*pheromones*' was first used by **Karlson** and **Luscher** to designate active substances that mediate humoral correlations among individuals of a given species. *The pheromones work on other members of the same species (i.e., they are species-specific) whereas the hormones proper confine their activities to the animal which produces them.* Unlike the hormones, which are produced in glands to act on target tissues *internally*, the pheromones are produced in glands and discharged *externally* to influence other members of the same species. The pheromones, for the same reason, have also been christened as `social hormones'. These include substances responsible for olfactory attraction between the sexes, alarm substances which warn other members of the species of danger and 'markers' which establish territories and trials or mark rich sources of food. They are well known sex-attractants also. These peculiar 'odours' or 'scents' are elaborated by different epithelial glands in the epidermis, the oro-anal and the urinogenital regions.

Structure. Chemically, they belong to a diverse category of compounds ranging from amino acids to lipids, alcohols and organic acids. They are highly volatile and can be effective at minute concentrations. Female silkworm moth (*Bombax mori*) secretes a sex-attractant which excites male moths (Fig 34–71). The sex-attractant of the silk moth (Fig. 34–72) is a long-chain alcohol ($C_{16}H_{30}O$) with two double bonds at C_{10} and C_{12} (**Adolf Butenandt, Hecker** and **Stamm**).

Fig. 34–71. The male silkworm moth (*Bombax mori*)

The female moth releases sex attractants such as the pheromone bombykol into the air in minute quantities. The male is extremely sensitive to these pheromones, because his antennae are large and crammed with many specialist receptors. This male has about 17,000 receptors in each antenna that respond only to bombykol.

In fact, **Adolf Butenandt** and his colleagues obtained only 12 mg of the compound from half a million glands of the silkworm moth. Only as low as 10^{-18}g per millilitre of this substance placed in the vicinity of the male silk moth is enough to induce flutter, dancing movements and other symptoms of sexual excitation. The signals could be detected even 1.5 kilometres away by the male silk moth making it one of the most active bioligical substances known. The *trans*-12 isomer requires 10^{12} times the quantity for equal effectiveness as an attractant.

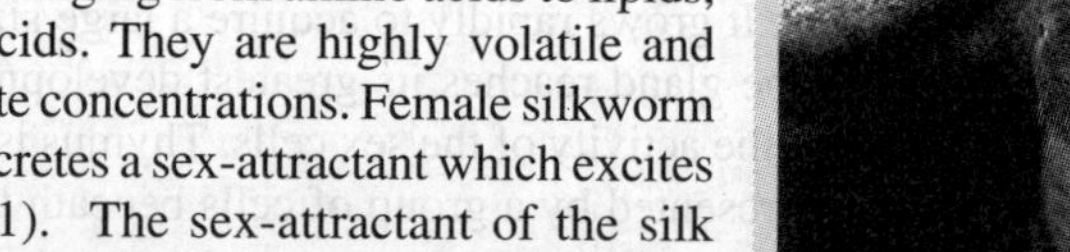

Fig. 34–72. Sex-attractant of the silk moth

(*trans*-10-*cis*-12-hexadecadien-1-ol)

The fall army worm utilizes *cis*-9-tetradecen-1-ol (Fig. 34–73) as the sex-attractant. This alcohol could be derived through biological reduction of myristoleic acid.

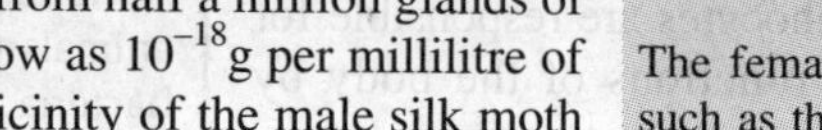

Fig. 34–73. Sex-attractant of the fall army worm, bombykol

Insects also produce alarm pheromones. Many of these are hydrocarbon, oxidized hydrocarbon, also hydrocarbon, or terpenoid in nature. The chemical structure of 4 such pheromones is given in Fig. 34–74, along with the name of the insect which produces them.

Types. Based on their functional aspect, the pheromones have been grouped under two categories.

1. *Releaser pheromones*—They initiate specific patterns of behaviour. They serve as powerful sex-attractants, mark territories or trails, initiate alarm reactions or bring about aggregation of individuals.
2. *Primer pheromones*—They trigger physiological changes in endocrine activity, *esp*., related to sexual maturation, growth or metamorphosis.

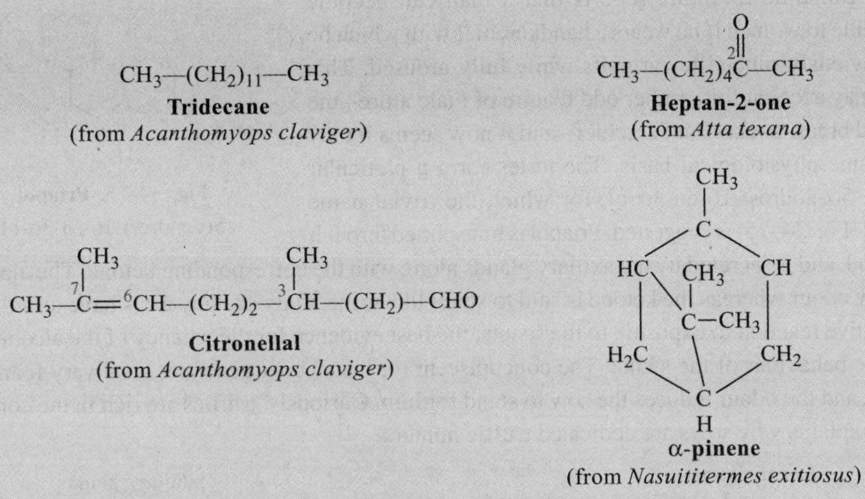

Fig. 34–74. **Four alarm pheromones from insects**

Examples. Many of the insects live in a world dominated by odours which greatly affect their social life. Two such interesting examples are given below.

In the honeybee colony, the queen bee secretes from its mandibular glands a pheromone called '*queen substance*' which inhibits the growth of the ovaries in the worker bees. The latter obtain this substance by occasionally licking queen's body. If, however, an accident occurs to the queen or else she is removed from the hive, the ovaries of the worker bees start developing. The 'queen substance' has been isolated and synthesized by **Englemann** in 1970. Chemically, it is 9-oxodecenoic acid.

The female gypsy moths, *Porthetria dispar* cannot fly and have to depend, for luring the winged males for copulation, on sex-attractants which they secrete. Only very little amount of this substance (*i.e.*, 1×10^{-12} µg) is sufficient to attract males from some distance.

Many social insects produce alarm substances which excite other members of the species to attack intruders. The honeybee, for instance, injects a pheromone called 2-heptanone in the body of the victim it stings. This marks the invader so that its chances of getting more stings increase (**Free** and **Simpson**, 1968).

Pigs have been known to detect truffles burried as deep as 3 feet below the ground by scent alone. It has been discovered that truffles contain a pig sex pheromone called *5α-androst-16-en-3α-ol*. This might also explain why we like the fungus which is said to taste like a cross between musk, nuts and ozone. The above steroid is synthesized by human males in the testes and secreted by axillary sweat glands!

A human baby even when 2-days old has already learnt to recognize the odour of the breast of the mother, given off probably by the areola of the breast. Thus, if two wads of cloth– one infiltrated with the odour of the mother's breast and the other with the odour of the breast of another woman– are placed on either side of an infant's head, the child instinctively turns towards the wad with the odour of its mother's breast.

A survey, held in 1982, at a `Taste and Smell' clinic, Georgetown (U.S.A.) has shown that there is a strong link between sex and smell. According to this survey, sexual desire is conveyed by a hormonal substance secreted by the body. Synthetically-produced *rostenol* can give the right fragrance to make love.

In an attempt to find whether a link between sex and smell in humans exists, studies have been conducted on two possible pheromones, androstenone and copulins. **Androstenone**, a hormone found in perspiration from the armpit and genitals is found more prevalent in men than in women. **Copulins** are estrogen-dependent fatty acids found in vaginal secretions of some women. Earlier

French romantic literature reports that a man can become irresistible to women if he wears a handkerchief with which he has previously rubbed his armpits while fully aroused. This belief may account for a rather odd feature of male attire– the exposed breast pocket handkerchief– and it now seems it may have some physiological basis. The testes form a particular steroid, 5α-androst-16-en-3α-ol, for which the trivial name **priapol** (Fig. 34–75) is suggested. Priapol is transported through the blood, and is secreted by the axillary glands along with the corresponding ketone. The alcohol has a musky odour whereas the ketone is said to smell like urine. Only sketchy trials have tested humans for positive reaction to exposure to the scents; the best evidence for the potency of the alcohol comes from the behaviour of the swine. The concupiscent (or lusty) boar generates a salivary foam rich in priapol, and the odour induces the sow to stand for him. Curiously, truffles are rich in the compound, which explains why sows are dedicated truffle hunters.

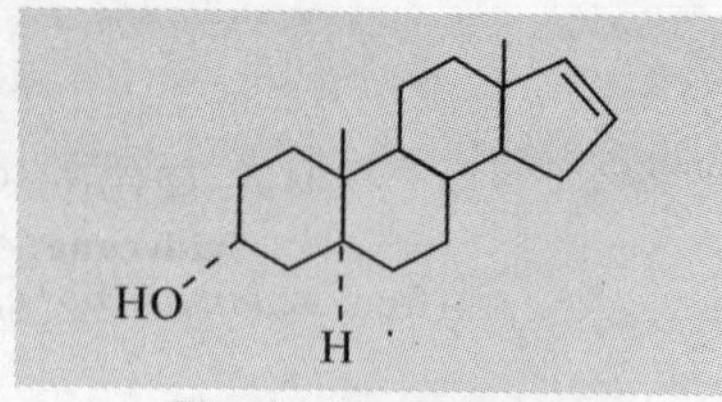

Fig. 34–75. Priapol
(5α-androst-16-en-3α-ol)

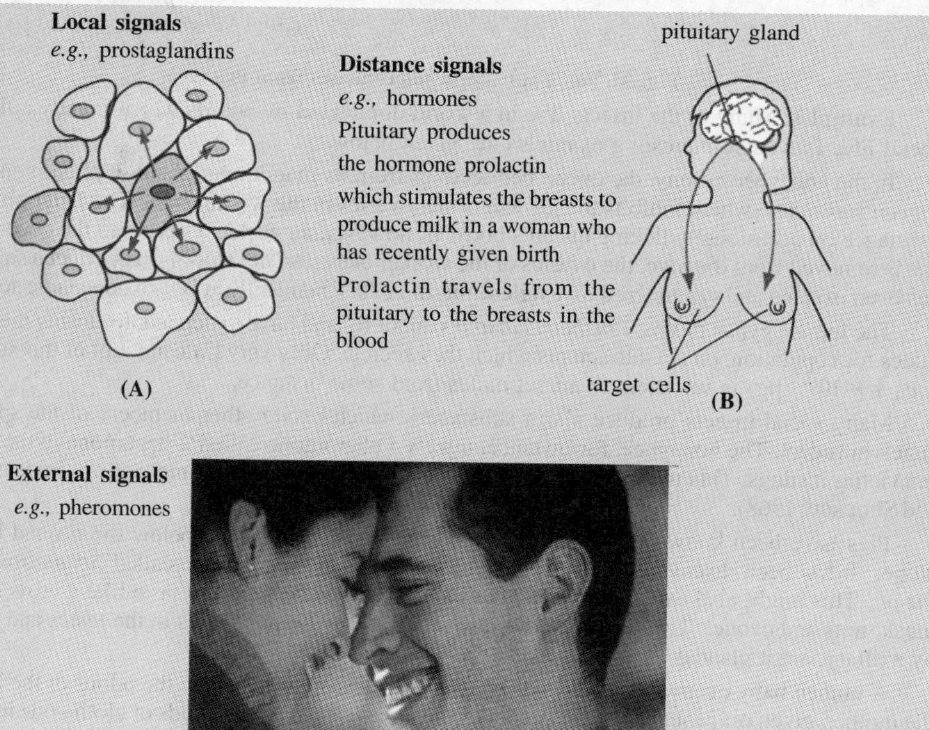

Fig. 34–76. A summary of the different types of chemical signalling that exist within and between organisms

(A) Local signals include neurotransmitters such as acetylcholine and local messengers such as prostaglandins, endorphins and histamines.

(B) Distant signals such as hormones allow communication between different organs, *via* the blood.

(C) Externally-produced messengers allow communication between organisms.

Dr. George Dodd, the father of the psychology of perfumery, has the world's only laboratory for sensitivity to smell. He has chemically identified and synthesized all the human pheromones. Dr. Dodd (1996) has also developed a synthetic human pheromone booster, the *Pheromone Factor*. He has discovered that the 7 families of human pheromones correspond to the aroma of the foods traditionally considered aphrodisiac : truffles, caviar, shellfish, champagne, beer, ripe cheese and vintage wine. Currently, an attempt is being made to synthesize pheromones to alleviate problems related to stress, sleep, diet and smoking.

A general emerging trend is now to classify the various chemical signals, that exist within and between organisms, into 3 discrete categories : local signals, distant signals and external signals (Fig 34–76).

MECHANISMS OF HORMONE ACTION

The function of different hormones is to *control* the activity of levels of target tissues. To achieve this, the hormones may alter either the permeability of the cells or they may activate some other specific cellular mechanism. Although the exact site of action of any hormone is not established, five *general sites* have been proposed.

A. Hormonal Action at Cyclic Nucleotides Level. Many hormones exert their effect on cells by first causing the formation of a substance, cyclic 3′, 5′-adenosine monophosphate (Fig. 34–77) in the cell. Once formed, the cyclic AMP causes the hormonal effects inside the cell. Thus, *cyclic AMP acts as an intracellular hormonal mediator*. It is also frequently referred to as the *second messenger* for hormone mediation; the *first messenger* being the original hormone itself.

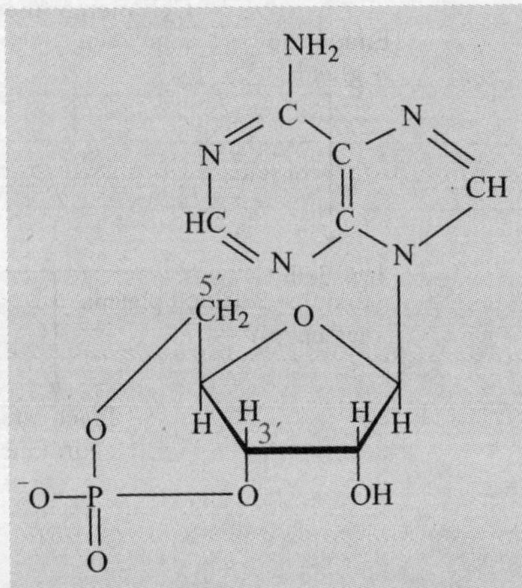

Fig. 34–77. Cyclic 3′, 5′-adenosine monophosphate, cAMP

The effects of cyclic AMP on the action of a hormone was first described by **Earl W. Sutherland** and **T.W. Rall** in 1960. They found that the effect of epinephrine on hepatic glycogenolysis (breakdown of glycogen) is a result of the conversion of inactive phosphorylase *b* into an active form phosphorylase *b* by cyclic AMP (Note that *glycogen phosphorylase* is often referred to simply as phosphorylase– so honoured because it was the first phosphorylase to be discovered; the shortened name has persisted in common usage and in the literature). Epinephrine was found to activate the enzyme, adenyl cyclase which, in turn, converts ATP to cAMP. Besides epinephrine, other hormones like glucagon, parathormone, ACTH, TSH, ICSH, LH, α-MSH and vasopressin are now known to have a stimulatory effect on cAMP levels. Several hormones, on the contrary, decrease cAMP levels and thus produce an opposite effect. These include insulin, melatonin

EARL WILBUR SUTHERLAND JR.
(LT, 1915–1974)

Sutherland, an American pharmacologist and physiologist, was awarded the **1971 Nobel Prize in Physiology or Medicine**, for his discoveries in *cellular signal transduction*. He showed that hormones such as adrenalin goad phosphorylase into action, indirectly, by increasing the production of another biochemical called cyclic AMP (cAMP). The latter is known as the `second messenger,' the first messenger being the hormone itself. He, thus, demonstrated the role of cAMP in the way that hormones affect the bodily organs.

and the prostaglandins. From the many names of hormones given above, it appears that hormone action not mediated by cAMP may be an exception rather than the rule.

Fig. 34–78 depicts, in a schematic way, the effect of cAMP on hormone action. The cell contains receptor for hormones in the plasma membrane. The stimulating hormone acts at the plasma membrance of the target cell and combines with a specific receptor for that particular type of hormone. The specificity of the receptor determines which hormone will affect the target cell. The combination of the hormone with its receptor leads to the activation of the enzyme, adenyl cyclase, which is also bound to the plasma membrane. The portion of the adenyl cyclase that is exposed to the cytoplasm causes immediate conversion of cytoplasmic ATP into cAMP. The reaction representing cAMP synthesis may, thus, be written as:

$$\text{ATP} \xrightarrow{Mg^{2+}} \text{Cyclic AMP} + PP_i + H^+$$

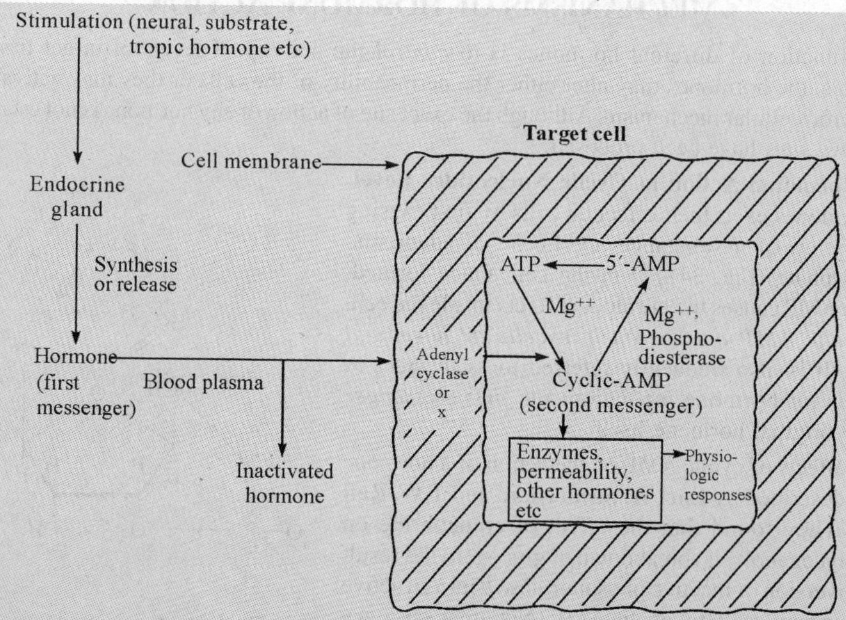

Fig. 34–78. Role of cyclic AMP in hormone actions

(*Modified from R.W. Butcher, 1968*)

The reaction is slightly endergonic and has a $\Delta G°'$ value of about 1.6 kcal/mol. The cAMP then acts inside the cell to initiate a number of cellular functions before it itself is destroyed. The various functions initiated include:

(*a*) activating the enzymes
(*b*) altering the cell permeability
(*c*) synthesizing the intracellular proteins
(*d*) contracting or relaxing the muscles
(*e*) releasing other hormones (*third messengers*).

It should, however, be emphasized that *what cAMP does in a particular effector cell is determined by the cell itself, rather than by cAMP.*

Cyclic AMP is, however, destroyed (or inactivated) by a specific enzyme called *phosphodiesterase*, which hydrolyzes it to AMP. Like adenyl cyclase, the phosphodiesterase is present in practically all tissues.

$$\text{Cyclic AMP} + H_2O \xrightarrow{Mg^{2+}} \text{AMP} + H^+$$

This reaction is highly exergonic, having a $\Delta G°'$ value of about –12 kcal/mol. Cyclic AMP is a very stable compound unless hydrolyzed by a specific phosphodiesterase.

An important feature of the second messenger model is that *the hormone need not enter the cell and its impact is made at the cell membrane*. The biological effects of the hormone are mediated inside the cell by cAMP rather than by the hormone itself.

cAMP and the Protein Kinases. Cyclic AMP elicits many of its effects by activating protein kinases. Protein kinases are ubiquitous in nature and are activated by cAMP at extremely low concentrations of 10^{-8} M. These kinases molecule the activities of different proteins in different cells by phosphorylating them. The enzyme *protein kinase* (Fig. 34–79) consists of two subunits : a catalytic subunit and a regulatory subunit which can bind cAMP. In the absence of cAMP, the catalytic and regulatory subunits form a complex that is enzymatically inactive. In the presence of cAMP, however, the complex disintegrates, freeing the catalytic subunit which now becomes catalytically active. The regulatory subunit binds cAMP to form a complex. Thus, the binding of cAMP to the regulatory subunit relieves its inhibition of the catalytic subunit. The cAMP acts as an allosteric effector.

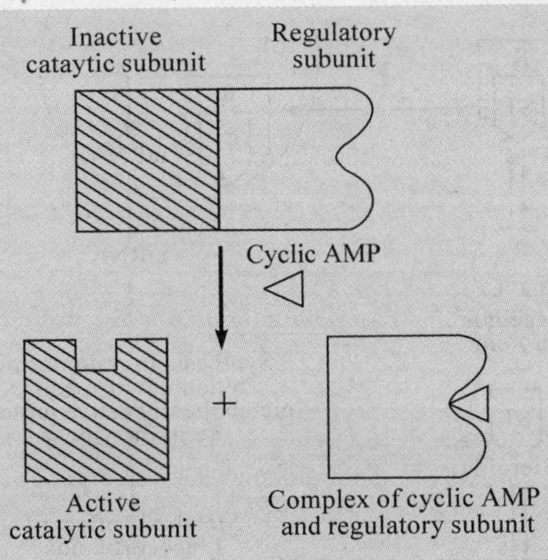

Fig. 34–79. Activation of protein kinase by cAMP

[Cyclic AMP activates protein kinases by dissociating the complex of the catalytic and regulatory subunits.]

(Adapted from Lubert Stryer, 1975)

cAMP as an Ancient Hunger Signal. It is interesting to note that cAMP has a long evolutionary history as a regulatory molecule. In bacteria, it stimulates the transcription of certain genes. Thus, in these microorganisms, it acts as 'hunger signal'. The word signifies an absence of glucose and leads to the synthesis of enzymes that can exploit other energy sources. In mammalian liver and muscle cells, cAMP has retained its ancient role as a hunger signal. But here cAMP acts by stimulating a protein kinase, rather than by enhancing the transcription of certain genes as is the case in bacteria. Moreover, cAMP has become a second messenger in higher organisms.

The querry emerging apparently from this discussion is that why was cyclic AMP chosen during evolution to be a second messenger? This is because of the following three important reasons :

1. cAMP is derived from ATP which is an omnipresent molecule.
2. cAMP is very stable except for the presence of phosphodiesterase.
3. cAMP has a number of active groups that can bind it tightly to the receptor proteins such as the regulatory subunit of the protein kinase is muscles.

Other Intracellular Hormonal Mediators. It has been postulated that, besides cAMP, other types of intracellular hormonal mediators also exist.

1. One almost-certain mediator is **cyclic guanosine monophosphate** (= cyclic GMP). Cyclic GMP is a nucleoside similar to cAMP and is found in most tissues. It can probably catalyze some intracellular functions in a manner similar to that of cAMP.
2. Another type of intracellular hormonal mediator is a group of compounds referred to as

prostaglandins (refer page 1150). These substances frequently cause intracellular inhibition, in contrast to the activation usually caused by cAMP.

B. Induction of Enzyme Synthesis at the Nuclear Level. A second major mechanism by which the hormones, *esp.*, the steroidal and thyroidal ones, act is to cause synthesis of proteins in the target cell. These proteins are presumably the enzymes which, in turn, activate other functions of the cells. The mechanism behind the **steroidal hormones** is depicted in Fig. 34–80. The sequence of events is as follows:

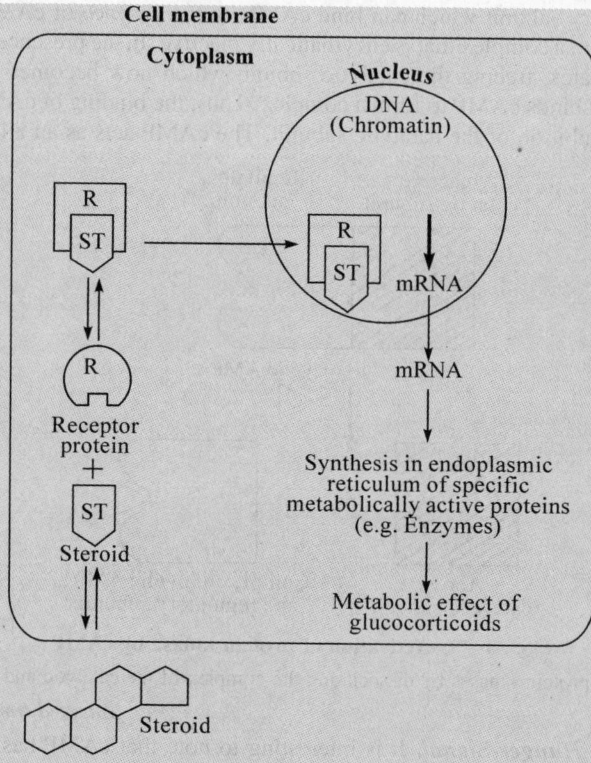

Fig. 34–80. Mechanism of action of steroidal hormones

ST = Steroid ; R = specific receptor protein

[The dissimilar shapes of R are intended to represent different conformations acquired by this protein.]

(*Adapted from Baxter and Forsham, 1972*)

1. The steroidal hormone enters the cytoplasm of the target cell where it binds with a specific, high- affinity receptor protein.
2. The receptor protein- hormone complex, so formed, then diffuses into (or is transported into) the nucleus, where it reacts with the nuclear chromatin.
3. Somewhere along this route, the receptor protein is structurally altered to form a smaller protein with low molecular weight. Or else the steroid hormone is transferred to a second smaller protein.
4. The combination of the small protein and hormone is now the active factor that stimulates the specific genes to form messenger RNA (mRNA) in the nucleus.
5. The mRNA diffuses into the cytoplasm where it accelerates the translation process at the ribosomes to synthesize new proteins.

It is, however, noteworthy that a direct chemical reaction of the hormone with DNA or RNA polynucleotide is not likely. Instead, the hormone must first combine with a specific receptor protein and it is this combination that acts on DNA chromatin. It is possible that the chromatin proteins may influence hormonal activity by modifying the ability of the receptor complex to bind with DNA.

To cite an example, the aldosterone, one of the mineralocorticoids secreted by adrenal cortex, enters the cytoplasm of the renal tubular cells. These tubular cells contain its specific receptor protein and hence above sequence of events follows. After about 45 minutes, the proteins begin to appear in the renal tubular cells that promote sodium reabsorption from the tubules and potassium secretion into the tubules. This characteristic delay, of about 45 minutes, in the final action of this steriod hormone is in marked contrast to the almost instantaneous action of some of the peptide hormones.

The **thyroidal hormones** act similarly to enchance RNA and enzyme synthesis but may do so by directly binding with the sepcific receptor proteins present in the nuclear chromatin. The receptors present in the cytoplasm are less effective in this regard.

C. Stimulation of Enzyme Synthesis at Ribosomal Level. In the case of some hormones, the activity is at the level of translation of information carried by the mRNA on the ribosomes to the production of enzyme protein. For example, the ribosomes taken from animals, which have been given growth hormone, have a capacity for protein synthesis in the presence of normal mRNA.

D. Direct Activation at the Enzyme Level. It has been experimentally observed that treatment of the intact animal (or of isolated tissue) with some hormones results in a change in enzyme behaviour which is not related to *de novo* synthesis. The cell membrane is usually required for such activity. Henceforth, it is possible that activation of a membrane receptor might be an initial step in hormone action.

E. Hormone Action at the Membrane Level. Many hormones appear to transport a variety of substances, including carbohydrates, amino acids and nucleotides, across cell membranes. These hormones, in fact, bind to cell membranes and cause rapid metabolic changes in the tissues. Catecholamines (epinephrine and norepinephrine) and many protein hormones stimulate different membrane enzyme systems systems by direct binding to specific receptors on cell membrane rather than in the cytoplasm.

A schematic representation of the two principal mechanisms of action involving water soluble hormones and steroid hormones is presented in Fig. 34–81.

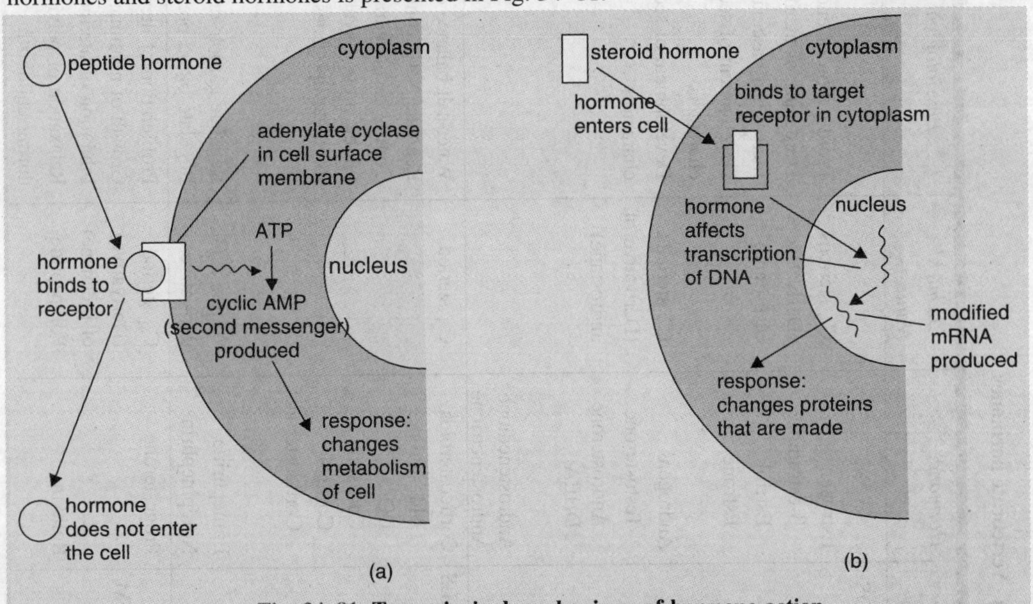

Fig. 34–81. Two principal mechanisms of hormone action

(a) Water-soluble hormones work *via* a second messenger which is formed inside the cell when they bind to receptors on the cell surface.

(b) When steroid hormones bind to a receptor inside the cytoplasm, they form a complex similar to an enzyme–substrate complex. This enters the nucleus of the cell and binds directly to the DNA, interfering with the cell's ability to read some of its genes. A steroid hormone can either switch on or switch off protein synthesis of particular genes. The overall effect of hormone action is different for each steroid hormone.

Table 34–9. Vertebrate hormones

S.No.	Sectretory Organ	Hormone	Chemical Nature	Physiological Role	Hypofunction	Hyperfunction	Effect of Removal of Secretory Organ
1.	GONADS						
	Ovary	Estrogens β-estradiol Estriol Estrone	C_{18} steriods (Derivative of estrane)	Proliferation of endometrium Deepening of uterine glands; Increase in vasculature; Development of secondary sex characters	—	—	Suspension of menstrual and reproductive cycles; Atrophy of uterus & vagina; Involuted mammary glands
	Testes	Androgens Testosterone Androsterone DHEA Androstenedione* Androstenetrione	C_{19} steriods (Derivative of androstane)	Development of secondary sex characters	—	—	Ossification of long-bone epiphyses; Increased adiposity; Extrusion of chest; Suppressed penis
2.	ADRENALS Cortex	Corticosteroids Aldosterone DOC Cortisone Cortisol Corticosterone	C_{21} steriods	Water-salt balance of the body; Regulate NaCI contents of blood; Control carbohydrate metabolism; Inhibit protein synthesis; Regulate lipogenesis	Decreased metabolism; Excessive pigmentation; Anorexia; Anemia; Hypoglycemia	Cushing's syndrome	Addison's disease
	Medulla	Epinephrine Norepinephrine	Amino acid derivatives	Regulate carbohydrate metabolism; Regulate blood pressure	—	—	Pheochromocytoma
3.	CORPUS LEUTEUM	Progesterone	C_{21} steriod (Derivative) of pregnane)	Proliferation of uterine mucosa; Growth of mammary glands; Inhibition of uterine contraction	—	—	—
		Relaxin	Polypeptide	Relaxation of pelvic ligaments during childbirth	—	—	—

Table 34–9. (Cont'd)

S.N.	Secretory Organ	Hormone	Chemical Nature	Physiological Role	Hypofunction	Hyperfunction	Effect of Removal of Secretory Organ
4.	ISLETS OF LANGERHANS						
	β-cells	Insulin	Polypeptide – 51 AARs†	Lowers sugar level in blood; Promotes protein synthesis; Promotes lipid synthesis	Diabetes mellitus	—	—
	α-cells	Glucagon	Polypeptide – 29 AARs	Increases sugar level in blood Catabolic action on proteins; Accelerates ketogenesis	Hypoglycemia	—	—
5.	HYPOPHYSIS						
	Pars distalis	Thyrotropin	Polypeptide	Stimulates thyroid activity	Dwarfism	Gigantism (in children)	Atrophy of gonads, thyroid and adrenal cortex; Loss of body tissue
		Corticotropin	Polypeptide	Stimulates adrenal cortex	Panhypo-pituitarism		
		Gonadotropins					
		FSH	Polypeptide	Induces growth of graafian follicles	Pituitary myxedema	Acromegaly (in adults)	
		LH	Polypeptide	Induces ripening of ovarian follicles			
		LTH	Polypeptide	Lactation during parturition			
		Somatotropin	Polypeptide	Controls skeletal growth			
	Pars intermedia	Intermedins	Polypeptide	Darkening of the skin			
	Pars nervosa	Ocytocin	Polypeptide –9 AARs	Stimulates contraction of smooth muscles *esp.* of uterus			
		Vasopressin	Polypeptide –9 AARs	Causes a rise in blood pressure; Antidiuretic action			

Table 34–9. (Cont'd)

S.N.	*Secretory Organ*	*Hormone*	*Chemical Nature*	*Physiological Role*	*Hypofunction*	*Hyperfunction*	*Effect of Removal of Secretory Organ*
6.	GASTROINTESTINAL TRACT	Secretin Pancreozymin Gastrin	Polypeptide Polypeptide Polypeptide	Stimulates flow of pancreatic juice Stimulates flow of pancreatic juice Stimulates flow of gastric	—	—	—
7.	PARATHYROID	Parathormone	Polypeptide –84 AARs	Bone resorption; Renal reabsorption of calcium	Tetany	Osteitis fibrosa	—
8.	THYROID	TIT or T_3	Amino acid derivative	Deamination reactions in the liver;	Cretinism (in children)	Exophthalmia;	—
		Thyroxine, T_4	Amino acid derivative	Deiodination in extrahepatic tissues	Myxedema (in adults)	Goiter	

* Secreted from adrenal gland

† AARs = Amino acid residues

REFERENCES

1. **Aaronson SA :** Growth factors and cancer. *Science.* **254 :** *1146–1153, 1991.*
2. **Arnaud CD Jr., Tenenhouse AM, Rasmussen H :** Parathyroid hormone. *Ann. Rev. Physiol.,* **29 :** *349–372, 1967.*
3. **Austin CR, Short RV :** Mechanisms of Hormone Actions. Vol. 7 in the Reproduction in Mammals Series. *Cambridge University Press, New York. 1980.*
4. **Austin LA, Health, H :** Calcitonin : Physiology and pathophysiology. *N. Engl. Jour. Med.,* **304 :** *269, 1981.*
5. **Axelrod J, Weinshilbaum R :** Catecholamines. *N. Engl. Jour. Med.,* **287 :** *237, 1972.*
6. **Bahl OP :** Human chorionic gonadotropin, its receptor and mechanism of action. *Fed. Proc.,* **36 :** *2119, 1977.*
7. **Bailey JM (editor) :** Prostaglandins, Leukotrienes and Lipoxins. *Plenum Press, New York. 1985.*
8. **Barrington EJW :** An Introduction to General and Comparative Endocrinology. *Oxford Univ., Press, Fair Lawn. 1963.*
9. **Barrington EJW, Dockray GT :** Gastrointestinal hormones. Jour. Endocrinol., **69 :** *299. 1976.*
10. **Baxter JB, Funder JW :** Hormone receptors. *N. Engl. Jour. Med.,* **301 :** *1149, 1979.*
11. **Beers RF Jr, Bassett EG (editors) :** Polypeptide Hormones (12th Miles International Symposium). *Raven Press, New York. 1980.*
12. **Belfore F :** Frontiers in Diabetes. Vol. 2. New Events and Facts in Diabetes. *S. Karger, London. 1982.*
13. **Bergström S :** The Prostaglandins. Recent Progress. *Recent Progr. Horm. Res.,* **22 :** *153–175, 1966.*
14. **Bernal J, Refetoff S :** The action of thyroid hormone *Clin. Endocrinol.,* **6 :** *227, 1977.*
15. **Berridge MJ, Irvine RF :** Inositol phosphates and cell signalling. *Nature.* **341 :** *197–205, 1989.*
16. **Blackwell RE, Guillemin R :** Hypothalamic control of adenohypophyseal secretions. *Ann. Rev. Physiol.,* **35 :** *357, 1973.*
17. **Bliss M :** The Discovery of Insulin. *Paul Haris Edinburgh Publishing. 304 pp, 1983.*
18. **Bradshaw RA, Frazier WA :** Hormone receptors as regulators of hormone action, *Cur. Topics Cell Regul.* **12 :** *1–35, 1977.*
19. **Bransome ED Jr :** Adrenal cortex. *Ann. Rev. Physiol.,* **30 :** *171, 1968.*
20. **Butt WR :** Hormone Chemistry. *D. Van Nostrand Co., Ltd., London. 1967.*
21. **Caldeyro-Barcia R, Heller H (editors) :** Oxytocin. *Pergamon Press. New York. 1961.*
22. **Chapra IJ :** New insights into metabolism of thyroid hormones : Physiological and clinical implications. *Progr. Clin. Biol. Res.* **74 :** *67, 1981.*
23. **Chord IT :** The posterior pituitary gland. *Clin. Endocrinol.,* **4 :** *89, 1975.*
24. **Cooper DS :** Hyperthyroidism.: *Lancet.* **363 :** *459–468, 2003.*
25. **Crapo L :** Hormones : The Messengers of Life. *W.H. Freeman and Company, New York. 1985.*
26. **Crowley VEF:** Overview of human obesity and central mechanisms regulating energy homeostasis. *Ann. Clin. Biochem.* **45 :** *245–255, 2008.*
27. **Cuatrecasas P, Hollenberg MD :** Membrane receptors and hormone action. *Adv. Protein Chem.,* **30 :** *251, 1977.*
28. **Daneman D :** Type I diabetes. *Lancet.* **367 :** *847–858, 2006.*

29. **Dattani M, Preece M :** Growth hormone deficiency and related disorders: insights into causation, diagnosis and treatment. *Lancet.* **363 :** *1977–1987, 2004.*
30. **Dorfman RI :** Steroid Hormones. Elsevier Publishing Co., New York. 1974.
31. **Eckel RH :** Nonsurgical management of obesity. *N. Engl. J. Med.* **358 :** *1941–1950, 2008.*
32. **Edelman IS :** Mechanism of action of steroid hormones. *Jour. Steroid Biochem.,* **6 :** 147, 1975.
33. **Fellman B :** A clockwork gland. *Frontline. Issue 11–24th July, 1987.*
34. **Foa P :** Glucagon. *Charles C. Thomas Pub., Springfield, Ill., 1962.*
35. **Frantz AG :** Prolactin. *N. Engl. Jour. Med.,* **298 :** *201, 1978.*
36. **Fritz IB (editor) :** Insulin Action. *Academic Press, Inc., New York. 1972.*
37. **Gaillard PJ, Talmadge RV, Budy AM (editors) :** The Parathyroid Glands : Ultrastructure, Secretion and Function, *University of Chicago Press, Chicago. 1965.*
38. **Green JH :** Basic Clinical Physiology, *3 rd ed. ELBS and Oxford University Press, London. 1978.*
39. **Grodsky GM, Forsham PH :** Insulin and the pancreas. *Ann. Rev. Physiol.,* **28 :** *347–380, 1966.*
40. **Hall K :** Relaxin. *Jour. Reprod. Fertility.* **1 :** *368–384, 1960.*
41. **Harris GW, Donovan BT :** The Pituitary Gland. *3 vols. University of California Press, Berkley. 1966.*
42. **Henderson J :** Ernest Starling and 'hormones': an historical commentary. *J. Endocrinol.* **18:** *5–10, 2005.*
43. **Hinman JW :** Prostaglandins. *Ann. Rev. Biochem.,* **41 :** *161–178, 1972.*
44. **Hirsch PF, Munson PL :** Thyrocalcitonin. *Physiol. Rev.,* **49 :** *548, 1969.*
45. **Hoar WS :** General and Comparative Physiology. *Prentice Hall of India. 1975.*
46. **Hofmann K :** Chemistry and function of polypeptide hormones. *Ann. Rev. Biochem.,* **31 :** *213-246, 1962.*
47. **Hou W, Schubert MI :** Gastric secretion. *Curr. Opin. Gastroenterol.* **12 :** *593–598, 2006.*
48. **Imura H, Nakai Y :** Endorphins in pituitary and other tissues. *Ann. Rev. Physiol.* **43 :** *265, 1981.*
49. **Jackson IMD :** Thyrotropin-releasing hormone. *N. Engl. Jour. Med.,* **306 :** *145, 1982.*
50. **Jensen EV :** On the mechanism of estrogen action. *Perspectives Biol. Med.,* **6 :** *47–60, 1962.*
51. **Karlson P :** Chemistry and biochemistry of insect hormones. *Angew. Chem. Intern. Ed. Engl.,* **2 :** *175–182, 1963.*
52. **Karlson P (editor) :** Mechanisms of Hormone Action. *Academic Press, Inc., New York. 1965.*
53. **Kopelman PG :** Obesity as a medical problem. *Nature.* **404 :** *635–643, 2000.*
54. **Krall LP, Beaser RS :** Joslin Diabetes Manual. *12th ed. Lea & Febiger, Philadelphia, London. 1989.*
55. **Krebs EG :** Role of the cyclic AMP-dependent protein kinase in signal transduction. *JAMA,* **262 :** *1815–1818, 1989.*
56. **Kryston LJ, Shaw RA, Schwager P (editors) :** Endocrinology and Diabetes. *Grune and Stratton. Inc., New York. 1975.*
57. **Kunos G (editor) :** Adrenoreceptors and Catecholamine Action. *Wiley. 1981.*
58. **Lands WE :** Biosynthesis of Prostaglandins. *Ann. Rev. Nutrit.* **11 :** *41–60, 1991.*

59. **Lebovitz HE :** Etiology and pathogenesis of diabetes mellitus. *Pediatr. Clin. North Am.* **31 :** *525, 1984.*

60. **Lee JB (editor) :** Prostaglandins. *Elsevier. 1982.*

61. **Lerner AB, Lee TH (editors) :** Melanocyte-stimulating hormones. *Vitamins and Hormones.* **20 :** *337–346, 1962.*

62. **Li CH (editor) :** Hormonal Proteins and Peptides. *13 vols. Academic Press, Inc., New York. 1973–1987.*

63. **Linder ME, Gilman AG :** G Proteins. Sci. Amer. **267 (July) :** *56–65, 1992.*

64. **Litwick G (editor) :** Biological Action of Hormones. *vol. 2, Academic Press, Inc., New York. 1972.*

65. **Mainwaring WI :** The mechanism of action of androgens. *Monogr. Endocrinol.* **10 :** *1, 1977.*

66. **Makin HLJ (editor) :** Biochemistry of Steroid Hormones. *2nd ed. Blackwell Scientific, Boston. 1984.*

67. **Malins J (editor) :** Clinical Diabetes Mellitus. *Eyre & Spottiswood (Publishers) Ltd., London. 1968.*

68. **Mann J, McAuley K :** Carbohydrates: is the advice to eat less justified for diabetes and cardiovascular health. *Curr. Opin. Lipidol.* **18 :** *9–12, 2007.*

69. **Martini L, Ganong WF (editors) :** Neuroendocrinology. *2 vols., Academic Press, Inc., New York. 1966, 1967.*

70. **Melmed S :** Acromegaly. *N. Engl. J. Med.* **355 :** *2558–2573, 2006.*

71. **Moudgal NR (editor) :** Gonadotropins and Gonadal Function. *Academic Press, Inc. New York, 1974.*

72. **Neville AM, Ó Hare MJ :** The Human Adrenal Cortex. *Springer-Verlag. 1982.*

73. **Newell P J, Bertagna X, Grossman A, Nieman L :** Cushing's syndrome. *Lancet.* **367 :** *1605–1617, 2006.*

74. **Pastan I :** Cyclic AMP. *Sci. Amer.* **227 (2) :** *97–105, 1972.*

75. **Pierce JG, Parsons TF :** Glycoprotein hormones. *Ann. Rev. Biochem.* **50 :** *465, 1981.*

76. **Pike JE :** Prostaglandins. *Sci. Amer.* **225 (5) :** *84–92, 1971.*

77. **Pilkis SJ, Park CR :** Mechanism of action of insulin. *Ann. Rev. Pharmacol.,* **14 :** *365, 1974.*

78. **Pincus G, Thimann KV, Astwood EB (editors) :** The Hormones : Physiology, Chemistry and Applications. *5 vols., Academic Press, Inc., New York. vol. I, 1948 ; vol. II, 1950 ; vol. III, 1955 ; vol. IV, 1964 ; vol. V, 1964.*

79. **Pitt-Rivers R, Trotter WR (editors) :** The Thyroid Gland. 2 *vols., Butterworth, Inc., Washington. 1964.*

80. **Rasmussen H :** The cycling of calcium as an intracellular messenger. *Sci. Amer.* **261 (October) :** *66–73, 1989.*

81. **Rasmussen H, Pechet MM :** Calcitonin. *Sci. Amer.,* **223 :** *49, 1970.*

82. **Roberts CGP, Ladenson PW :** Hypothyroidism. *Lancet.* **362 :** *793–803, 2004.*

83. **Robinson GA, Butcher RW, Sutherland EW :** Cyclic AMP. *Acadmic Press, Inc., New York. 1971.*

84. **Roth RA, Cassells DJ :** Insulin Receptor : Evidence That It Is a Protein Kinase. *Science.* **219 :** *299, 1983.*

85. **Samuelsson B *et al* :** Prostaglandins and thromboxanes. *Ann. Rev. Biochem.,* **47 :** *997, 1978.*

86. **Saxena BB, Beling CG, Gandy HM :** Gonadotropins. *John Wiley and Sons, Inc., New York. 1972.*

87. **Schally AV, Arimura A, Kastin AJ :** Hypothalamic regulatory hormones. *Science.* **179 :** *341, 1973.*

88. **Schauer R :** The mode of action of hormones. *Angew. Chem. Intern. Ed. Engl.,* **11 :** *7–16, 1972.*

89. **Schneider HJ, Aimaretti G *et al* :** Hypopituitarism. *Lancet.* **369:** *1461–1470, 2007.*

90. **Smith WL :** Nutritionally essential fatty acids and biologically indispensable cyclooxygenases. *Trends Biochem. Sci.* **33 :** *27–37, 2008.*

91. **Snyder SH :** The molecular basis of communication between cells. *Sci. Amer.* **253 (October) :** *132–141, 1985.*

92. **Stadel JM, DeLean A, Lefkowitz RJ :** Molecular Mechanisms of Coupling in Hormone Receptor–Adenylate Cyclase Systems. *Adv. Enzymol.* **53 :** *1, 1982.*

93. **Steiner DF :** Insulin Today. *Diabetes.* **26 :** *322–340, 1977.*

94. **Sterling K, Lazarus JH :** The thyroid and its control. *Ann. Rev. Physiol.,* **39 :** *349, 1977.*

95. **Stern MP :** Diabetes and cardiovascular disease. The 'common soil' hypothesis. *Diabetes.* **44 :** *369–374, 1995.*

96. **Stryer L, Bourne HR :** G Proteins : A family of signal transducers. *Ann. Rev. Cell Biol.,* **2 :** *391, 1986.*

97. **Stumvoll M, Goldstein BJ, von Haeften TW :** Type 2 diabetes: principles of pathogenesis and therapy. *Lancet.* **365 :** *1333–1346, 2005.*

98. **Sutherland EW :** Studies on the mechanism of hormone action. *Science.* **177 :** *401–408, 1972.*

99. **Tager HS, Steiner DF :** Peptide hormones. *Ann. Rev. Biochem.,* **43 :** *509–538, 1974.*

100. **Tang W-J, Gilman AG :** Adenylyl cyclases. Cell. **70 :** *869–872, 1992.*

101. **Terry LC, Martin JB :** Hypothalamic hormones : Subcellular distribution and mechanism of release. *Ann. Rev. Pharm. Tox.,* **18 :** *111, 1978.*

102. **Turner CD :** General Endocrinology. *3rd ed., W.B. Saunders Co., Philadelphia. 1960.*

103. **von Euler US, Heller H (editors) :** Comparative Endocrinology. 2 *vols., Academic Press, Inc., New York. 1963.*

104. **Werner SC, Nauman JA :** The thyroid. *Ann. Rev. Physiol.,* **30 :** *213, 1968.*

105. **Williams CM :** The Juvenile Hormone. *Scientific American, 1958.*

106. **Williams RH (editor) :** Textbook of Endocrinology. *7th ed., W.B. Saunders Co., Philadelphia. 1985.*

107. **Wilson JD, Foster DW (editors) :** Williams Textbook of Endocrinology. *8th ed., W.B. Saunders Company, Philadelphia. 1992.*

108. **Wurtman RJ, Axelrod J :** The Pineal Gland. *Sci. Amer.,* **213 :** *50–60, 1965.*

109. **Young WC (editor) :** Sex and Internal Secretions, *3rd ed., 2 vols.* The *Williams and Wilkins Co., Baltimore. 1961.*

110. **Yudkin JS :** Inflammation, obesity, and the metabolic syndrome. *Horm. Metab. Res.* **39 :** *707–709, 2007.*

111. **Zimmermann MB, Jooste PL, Pandav CS :** Iodine-deficiency disorders. *Lancet.* **372 :** *1152–1168, 2008.*

112. **Zimmet P, Alberti KGMM, Shaw J :** Global and societal implications of the diabetes epidemic. *Nature.* **414 :** *782–787, 2001.*

PROBLEMS

1. The hormone progesterone contains two ketone groups. At pH 7, which side chains of the receptor might form hydrogen bonds with progesterone?
2. Ingesting large amounts of glucose before a marathon might seem to be a good way of increasing the fuel stores. However, experienced runners do not ingest glucose before a race. What is the biochemical reason for their avoidance of this potential fuel ? (**Hint:** Consider the effect of glucose ingestion on the level of insulin.)
3. Insulin-dependent diabetes is often accompanied by hypertriglyceridemia, which is an excess blood level of triacylglycerides in the form of very low density lipoproteins. Suggest a biochemical explanation.
4. The hormone glucagon signifies the starved state, yet it inhibits glycolysis in the liver. How does this inhibition of an energy-production pathway benefit the organism?
5. Sildenafil (Viagra) is a drug widely used to treat male impotence. Sildenafil exerts its effect by inhibiting a cGMP phosphodiesterase isozyme (PDE5) that is especially prevalent in smooth muscle. Interestingly, certain airlines restrict pilots from flying for 24 hours after using sildenafil. Suggest a reason for this restriction.
6. How would you determine whether an inability to produce cortisol in response to stress was caused by a problem in the hypothalamus, the anterior pituitary, or the adrenal cortex?
7. Steroidal anti-inflammatory drugs inhibit prostaglandin synthesis in at least two ways– inhibition of phospholipase A_2 and inhibition of cyclooxygenase. Why are the new generation of antiinflammatory drugs called COX2 inhibitors (Celebrex, Vioxx) better tolerated than the older drugs?
8. What is the relationship between 7-dehydrocholesterol and 1α, 25-dihydroxycholecalciferol?
9. Marathon runners preparing for a race engage in "carbo loading" to maximize their carbohydrate reserves. This involves eating large quantities of starchy foods. Why is starch preferable to candy of sugar-rich foods?
10. Ketone bodies are exported from liver for use by other tissues. Because many tissues can synthesize ketone bodies, what enzymatic property of liver might contribute to its special ability to export these compounds?
11. During a "fight or flight" situation, the release of epinephrine promotes glycogen breakdown in the liver, heart, and skeletal muscle. The end product of glycogen breakdown in the liver is glucose. In contrast, the end product in skeletal muscle is pyruvate.
    (*a*) Why are different products of glycogen breakdown observed in the two tissues?
    (*b*) What is the advantage to the organism during a "fight or flight" condition of having these specific glycogen breakdown routes?
12. Certain malignant tumours of the pancreas cause excessive production of insulin by the β cells. Affected individuals exhibit shaking and trembling, weakness and fatigue, sweating, and hunger. If this condition is prolonged, brain damage occurs.
    (*a*) What is the effect of hyperinsulinism on the metabolism of carbohydrate, amino acids, and lipids by the liver?
    (*b*) What are the causes of the observed symptoms? Suggest why this condition, if prolonged, leads to brain damage.
13. Thyroid hormones are intimately involved in regulating the basal metabolic rate. Liver tissue of animals given excess thyroxine shows an increased rate of O_2 consumption and increased heat output (thermogenesis), but the ATP concentration in the tissue is normal. Different explanations have been offered for the thermogenic effect of thyroxine. One is that excess thyroid hormone causes uncoupling of oxidative phosphorylation in

mitochondria. How could such an effect account for the observations? Another explanation suggests that the thermogenesis is due to an increased rate of ATP utilization by the thyroid-stimulated tissue. Is this a reasonable explanation? Why?

14. What are the possible advantages in the synthesis of hormones as prohormones or preprohormones?
15. Which of these hormones inhibits FSH production in a female mammal?
    (*a*) estrogen
    (*b*) adrenalin
    (*c*) thyroxine
    (*d*) testosterone
    (*e*) luteinizing hormone
16. Does it make any difference when you take a thyroid hormone?
17. During pregnancy, the chief source of progesterone is the:
    (*a*) ovary
    (*b*) placenta
    (*c*) pituitary gland
    (*d*) mammary gland
    (*e*) uterus
18. Name two animal hormones that are peptides (or proteins) and steroids.
19. Name two animal hormones that are antagonistic in their effects.
20. Why are diabetics more sensitive to skin diseases?
21. On the basis of their physical properties, hormones fall into one of two categories: those that are very soluble in water but relatively insoluble in lipids (*e.g.*, epinephrine) and those that are relatively insoluble in water but highly soluble in lipids (*e.g.*, steroid hormones). In their role as regulators of cellular activity, most water-soluble hormones do not penetrate into the interior of their target cells. The lipid-soluble hormones, by contrast, do penetrate into their target cells and ultimately act in the nucleus. What is the correlation between solubility, the location of receptors, and the mode of action of the two classes of hormones?
22. Which is the most widely-used drug in the world?
    (*a*) Aspirin (*b*) Morphine (*c*) Paracetamol (*d*) Progesterone
23. Acetylcholine is a:
    (*a*) Neurotransmitter (*b*) B-vitamin (*c*) Enzyme (*d*) Protein
24. Name the gland which acts both as an 'endocrine' as well as an 'exocrine' gland.
    (*a*) Pancreas (*b*) Pituitary (*c*) Thyroid (*d*) Adrenal
25. Can obesity affect life expectancy in humans?
26. Why is edible salt iodized?
27. The gland that degenerates during adolescene is:
    (*a*) Thymus (*b*) Liver (*c*) Pituitary (*d*) Pineal
28. Excess production of thyroxine leads to:
    (*a*) Diabetes (*b*) Cretinism (*c*) Gigantism (*d*) Exophthalmic goitre
29. Which is the hormone that lowers the concentration of calcium in the blood?
    (*a*) Somatostatin (*b*) Ocytocin (*c*) Calcitonin (*d*) Androgens
30. Which of the following contains iodine?
    (*a*) Insulin (*b*) Thyroxine (*c*) Ocytocin (*d*) Adrenalin

31. Which of the following is called 'emergency hormone'?
(*a*) Antidiuretic hormone
(*b*) Insulin
(*c*) Adrenalin
(*d*) Thymosin

32. Deficiency of which of the following hormones is the cause of diabetes insipidus?
(*a*) Vasopressin
(*b*) Thyroxine
(*c*) Follicle-stimulating hormone
(*d*) Calcitonin

33. The endocrine gland which controls some other gland is:
(*a*) Thyroid
(*b*) Pancreas
(*c*) Pituitary
(*d*) Adrenal

34. Juvenile hormone is also known as:
(*a*) Thymosin
(*b*) Renin
(*c*) Testosterone
(*d*) Estrogen

35. How much does diabetes affect one's lifespan?

36. What are some natural diabetic remedies?

37. What are the 4 enemies of blood sugar?

38. What are the risks of taking aspirin?

ANSWERS TO PROBLEMS

1. Potential hydrogen-bond donors at pH 7 are the side chains of the following residues: arginine, asparagine, glutamine, histidine, lysine, serine, threonine, tryptophan, and tyrosine.

2. A high blood-glucose level would trigger the secretion of insulin, which would stimulate the synthesis of glycogen and triacylglycerols. A high insulin level would impede the mobilization of fuel reserves during the marathon.

3. Lipid mobilization can occur so rapidly that it exceeds the ability of the liver to oxidize the lipids or convert them into ketone bodies. The excess is reesterified and released into the blood as VLDL.

4. A role of the liver is to provide glucose for other tissues. In the liver, glycolysis is used not for energy production but for biosynthetic purposes. Consequently, in the presence of glucagon, liver glycolysis stops so that the glucose can be released into the blood.

5. The phosphodiesterase in photoreceptor cells is homologous to PDE5 and is inhibited to some extent by sildenafil, leading to visual side effects.

6. The end organ is adrenal cortex. If cortisol is increased in response to an injection of ACTH, the adrenal cortex is functioning properly. An intravenous infusion of CRH (corticotropic-releasing hormone) should be able to reach the anterior pituitary. If ACTH increases in response to this, the pituitary is functioning so the problem is probably with the hypothalamus. If ACTH does not increase in response to CRH, the anterior pituitary is not functioning.

7. There are two cyclooxygenases. COX1 is a constitutively expressed enzyme that catalyzes the synthesis of required prostaglandins, for example, protection of the stomach lining and blood coagulation processes. COX2, however, is induced as part of the inflammatory response. NSAIDs (nonsteroidal antiinflammatory drugs like *aspirin*) inhibit both COX1 and COX2, so stomach irritation is common. COX2 drugs inhibit production of only the inflammatory prostaglandins.

8. 1α,25-dihydroxycholecalciferol [1,25 $(OH)_2D_3$] is the active form of vitamin D. Ultraviolet light acting on the skin converts 7-dehydrocholesterol to cholecalciferol. This compound must be hydroxylated in liver to 25-hydroxycholecalciferol and subsequently in kidney to yield the active $1,25(OH)_2D_3$.

9. It is presumed that starch increases blood glucose levels less than simple sugars do. Thus, there is less stimulation of insulin secretion. Insulin would tend to retard energy mobilization from intracellular stores– something not desirable during a marathon.

10. Liver contains low levels of the enzyme that synthesizes acetoacetyl-CoA from acetoacetate, ATP, and CoA-SH. Therefore, when liver synthesizes ketone bodies, they cannot readily be activated for catabolism within the hepatocyte. Instead, they are released and ultimately utilized by other tissues.

11. **(*a*)** Heart and skeletal muscle lack the enzyme glucose-6- phosphate phosphatase. Any glucose-6-phosphate that is produced enters the glycolytic pathway, and under O_2-deficient conditions is converted into lactate *via* pyruvate.

(*b*) Phosphorylated intermediates cannot escape from the cell, because the membrane is not permeable to charged species. In a "fight or flight" situation, the concentration of glycolytic precursors needs to be high in preparation for muscular activity. The liver, on the other hand, must release the glucose necessary to maintain the blood glucose level. Glucose is formed from glucose-6-phosphate and passes from the liver cells to the bloodstream.

12. **(*a*)** Excessive utilization of blood glucose by the liver, leading to hypoglycemia; shutdown of amino acid and fatty acid catabolism

(*b*) Little circulating fuel is available for ATP requirements. Brain damage results because glucose is the main source of fuel for the brain.

13. Thyroxine acts as an uncoupler of oxidative phosphorylation. Uncouplers lower the P/O ratio, and the tissue must increase respiration to meet the normal ATP demands. Thermogenesis could also be due to the increased rate of ATP utilization by the thyroid-stimulated tissue, because the increased ATP demands are met by increased oxidative phosphorylation and thus respiration.

14. Because prohormones and preprohormones are inactive, they can be stored in quantity in secretory granules. Rapid activation is achieved by enzymatic cleavage in response to an appropriate signal.

15. **(*a*)** estrogen.

16. **(*a*)** Yes. Take it on an empty stomach and do not eat immediately afterward since food alters its absorption.

17. **(*b*)** placenta.

18. *Peptide (or protein) hormones*: Insulin and Growth hormone

Steroid hormones: Testosterone and Estrogen

19. Insulin and glucagon.

20. In diabetes mellitus, which is a disorder of the carbohydrate metabolism, glucose cannot enter the cells but remains concentrated in the blood instead. Thus, the cells starve in the midst of plenty. The increased amount of sugar in the blood also brings about adverse changes in the membrane transport system of the cells. The blood circulation is also impaired. A net result of all these changes is that opportunistic bacteria etc flourish and skin infections especially at the extremities (toes etc) are common. Increased incidence of skin infection is only one of a group of complications that may set in due to diabetes. Skin is easily affected as it is the most easily accessible part of the body.

21. Because of their low solubility in lipid, water-soluble hormones cannot penetrate the cell membrane; they bind to receptors on the outer surface of the cell. In the case of epinephrine, this receptor is an enzyme that catalyzes the formation of a second messenger (cAMP) *inside* the cell. In contrast, lipid-soluble hormones can readily penetrate the hydrophobic core of the cell membrane. Once inside the cell, they can act on their target molecules or receptors directly.

22. (*a*) Aspirin

23. (*a*) Neurotransmitter

24. (*a*) Pancreas

25. A new study suggests that being obese during adulthood decreases lifespan by 3 years. Obesity is defined as having a body mass index (BMI) of 30 or more. Obesity risks are similar to smoking risks

26. Iodine is an essential nutrient which is required in minute amounts for normal growth of animals and humans. Iodine deficiency results in impairment of the functioning of the thyroid gland, which produces 'thyroxine' hormone. *Since thyroxine controls the rate of body metabolism and growth, iodine deficiency leads to a variety of disorders like goitre in children or adults and impaired mental functioning, poor intellectual performance and muscular disorders in children.* Women with iodine deficiency give birth to stillborn babies. Iodization of salt offers the most convenient way of combating iodine deficiency. Firstly, because salt is an essential ingredient of our diet and secondly, one can never take too much of it. So by iodization of common salt, a minimum intake of iodine can be ensured for the entire population, thereby minimizing the incidence of iodine deficiency.

27. (*a*) Thymus

28. (*d*) Exophthalmic goitre

29. (*c*) Calcitonin

30. (*b*) Thyroxine

31. (*c*) Adrenalin

32. (*a*) Vasopressin

33. (*c*) Pituitary

34. (*a*) Thymosin

35. If one is older than 40, having type 1 or 2 diabetes, *ages his heart by 15 years*. Controlling blood sugar by regular exercise and a low-glycemic diet is essential to the health of diabetics. These same habits can help prevent the onset of diabetes.

36. A large-scale review of clinical trials in Iran revealed that 3 food substances exhibited powerful blood-sugar control for type 2 diabetes: sweet potatoes, fenugreek and milk thistle.

37. Blood glucose levels rise when diabetics consume the 'four whites': white potatoes, white rice, white flour and white sugar.

38. A new study has revealed that those who take aspirin regularly are at an increased risk of developing the more severe "wet" form of age-related macular degeneration. Those who took the most aspirin were at the greatest risk of developing macular degeneration.

CONTENTS

The tomatoes (*above*) are the same variety and the same age as those shown *below*.

The deterioration and spoilage shown by the normal variety (*above*) is normal. The tomatoes, shown *below*, have been genetically modified and stay fresher for longer. They have been given a gene which blocks the normal production of **ethene**, a plant hormone which promotes ripening.

CHAPTER 35

Plant Hormones

DEFINITION

Thimann (1948) designated the plant hormones by the term *'phytohormones'* in order to distinguish them from animal hormones. He defined a phytohormone as *"an organic compound produced naturally in higher plants, controlling growth or other physiological functions at a site remote from its place of production and active in minute amounts."*

This definition includes a variety of compounds, besides those responsible for growth curvatures in organs like *Avena* coleoptile. For example, it embraces the hormones which induce flowering, wound-healing and also those vitamins which act as growth factors.

A definition of plant hormones with still wider scope has been given by **Johannes van Overbeek** (1950). According to him, the plant hormones are defined as *"organic compounds which regulate plant physiological process— regardless of whether these compounds are naturally occurring and/or synthetic; stimulating and/or inhibitory; local activators or substances which act at a distance from the place where they are formed."*

The migratory nature of hormones has been specifically emphasized by **Meirion Thomas** (1956) who stated that "all hormones are migratory correlating substances or correlators which play an essential part in the integration of plant behaviour."

Three types of plant hormones are usually recognized. These are *auxins, gibberellins* and *cytokinins*. These were discovered in the early decades of the twentieth century, in 1930's and in 1960's respectively. Naturally, the knowledge accumulated on auxins and gibberellins is far greater than that gathered for cytokinins.

AUXINS

Definition

Kögl and **Haagen-Smit** (1931) introduced the term 'auxin' (*auxein*[G] = to grow or to increase) for designating those plant hormones which are specially concerned with cell enlargement or the growth of the the shoots. An auxin may, thus, be defined as "an organic substance which promotes growth (*i.e.*, irreversible increase in growth) along the longitudinal axis when applied in low concentrations to shoots of the plants freed as far as practicable from their own inherit growth-promoting substances. Auxins may, and generally do, have other properties but this one is critical" (**Thimann**, 1948). This definition is nowadays widely accepted among plant physiologists.

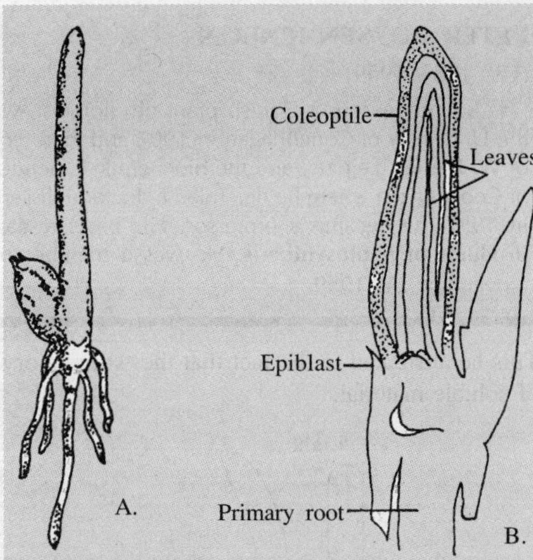

Fig. 35–1. Structure of oat coleoptile

Oat Coleoptile and the Auxins

Since long the action of auxin has been clearly demonstrated in the leaf sheath or coleoptile of oat plant (*Avena sativa*). The coleoptile (Fig. 35–1) is a tubular structure with a conical top and encloses the first-formed leaf in it. A transection of the coleoptile reveals that it consists of an epidermal layer and a few parenchyma cells with two vascular bundles running longitudinally.

The experimental utility of oat coleoptile lies in the fact that the process of cell elongation can be easily studied over here. After four days, the primary leaf breaks through the coleoptile at its tip and the organ stops growing. Thus, cell divisions cease relatively early in the development of oat coleoptile and further growth in length takes place as a result of cell elongation only. Elongation is confined to a region about 1 cm below the tip of the coleoptile. The rate of elongation is most rapid (about 1 mm per hour) when the seedling is about 3 days old. At this time, the coleoptile may measure about 3 cm in length. It is mostly at this stage that it is experimentally used for conducting growth experiments.

Hormone Concept

The presence of growth-regulating hormones in plants was first suggested by **Julius von Sachs** in 1980. He proposed that there were certain *'organ-forming substances'* in plants which were produced in the leaves and translocated downward in plant body.

Also in 1880, **Charles Darwin**, an evolutionist, studied the effect of unilateral light on plant movements. While conducting his experiments on canary grass (*Phanaris canariensis*), he found that if the coleoptile tip is provided light from one side only (*i.e.*, unilateral illumination), the tip would bend towards light. In the absence of illumination, however, no curvature could be induced (refer Fig. 35–2 for this and other subsequent experiments).

The material nature of the hormones was first conclusively demonstrated by a Dane, **Peter Boysen-Jensen** (1910). He first cut off (or decapitated) the coleoptile tip a few millimeters from the apex, then put a block of gelatin on the decapitated stump and ultimately replaced the cut tip on the gelatin block. Upon unilateral illumination, the coleoptile showed curvature towards light.

PETER BOYSEN-JENSON

(LT, 1883–1959)

Boysen-Jensen, a Danish plant physiologist, was born on January 18, 1883. He graduated from the University of Copenhagen in 1907 and then worked there. In 1909, he worked in the laboratories of **W. Pfeffer** (Leipzig) and the biochemist **E. Schulze** (Zürich). In 1910, he returned to the University of Copenhagen where he defended a doctoral dissertation on sugar conversion during plant respiration. In 1927, he became a professor. His basic researches are on **plant hormones, tropisms**, and the **products of photosynthesis**. He was a member of the Danish Academy of Sciences. He died on November 21, 1959.

This he attributed to the fact that the 'stimulatory effect' can diffuse through gelatin and is a type of soluble material.

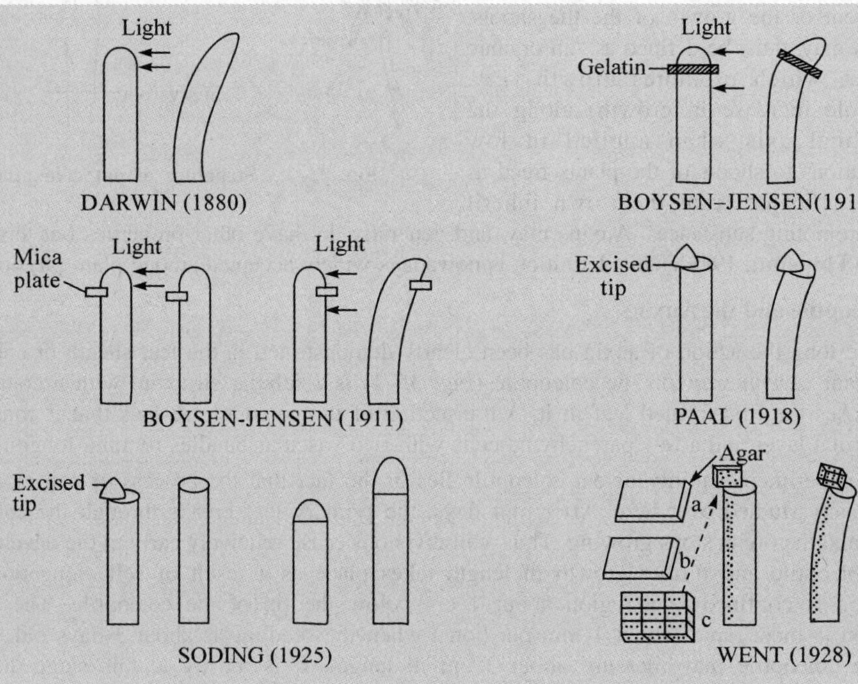

Fig. 35–2. Major experiments in the discovery of auxins

A year later (*i.e.*, in 1911) in another experiment, he made a transverse slit halfway through the coleoptile below the tip on one side and inserted a piece of mica in the slit. Upon unilateral illumination from the side opposite the slit, no phototropic curvature was observed. If, however, the operated coleoptile is illuminated on the side of the slit, the phototropic bending towards the source of light is seen. From this was inferred that the stimulus for bending passes down the dark side of the coleoptile.

In 1918, the Hungarian plant physiologist, **Arpad Paál** of the University of Budapest in an experiment decapitated coleoptile and replaced the tip eccentrically. He found that the coleoptile showed a negative curvature, *i.e.*, bent away from the side with the tip even in the dark. Paal, therefore, concluded that the material substance (called as '*correlation carrier*' by him) diffuses from the tip downwards and stimulates growth of the cells below the tip.

Soding (1925) found that decapitation of a coleoptile markedly retards the rate of cell division. But reheading of the tip on the cut stump resulted in resumption of vertical growth due to cell elongation.

The task of isolation, extraction and bioassay of these growth-promoting substances was admirably done by a Dutch botanist, **Frits W. Went**, in 1928. He worked in his father's laboratory at the University of Utrecht in the Netherlands. He placed numerous freshly-cut coleoptile tips on an agar block for some time. Later, this block was divided into small rectangular blocks (*a*, *b* and *c*). These agar blocks were, then, placed eccentrically on decapitated coleoptiles for 2 hours in dark. In all cases, the coleoptiles showed curvature towards the side opposite to the agar blocks (*i.e.*, negative curvature). Went also devised a method for the bioassay (to be described later) of these substances.

FRITS WARMOLT WENT

(LT, 1903–1990)

Went, a Dutch biologist, was born on May 18, 1903 at Utrecht. His father **F.A.F.C. Went** was himself a famous Dutch botanist. Warmolt graduated from the University of Utrecht, Holland in 1927 with a dissertation on the effects of plant hormone called **auxin**. He then worked as a Plant Pathologist in the research labs of the Royal Botanical Garden, in Buitenzorg, Dutch East Indies, now Bogor, Indonesia from 1927 to 1933. Later, he took a position at CalTech in Pasadena, California, first researching plant hormones. His interest gradually shifted to environmental influences on plant growth. He played an important role in the **development of synthetic plant hormones**, which then became the basis of much of the agricultural chemical industry.

Funded by generous donors, Went constructed a series of greenhouses at CalTech in which he could vary light conditions, humidity, temperature, air quality and other variables. In 1949, this led to him to construct a large new complex of climate-controlled rooms called the *Earhart Plant Research Laboratory*, also known as the **"phytotron"**. Here he produced fundamental research of the effects of air pollution on plant growth.

In 1958, Went was appointed Director of the Missouri Botanical Garden and Professor of Botany at Washington University in St. Louis, at a point where he had become a world-recognized authority on plant growth. He moved from Pasadena to St. Louis with his wife **Catharina** and their two children, **Hans** and **Anneka**. After the opening of the *Climatron,* the world's first geodesic dome greenhouse, Went's vision of a renewed Missouri Botanical Garden eventually came into conflict with that of its Board of Trustees, and he resigned as Director in 1963. After 2 years as simply Professor of Botany at Washington University, in 1965 he then became Director of the Desert Research Institute at the University of Nevada-Reno, where he continued his research on desert plants for the remainder of his career. He died on May 1, 1990 in Little Valley, Nevada.

Cholodny in Russia also worked on similar lines independently and arrived at the same conclusions.

The availability of a technique of extracting the hormone opened up new vistas in this field and within a short span of about 8 years (from 1928 to 1936) extensive work was done. With the result, three different hormones (auxin *a*, auxin *b* and heteroauxin) were identified and their characteristics noted mainly on account of the efforts of **Kögl** and his colleagues in Holland.

Extraction of Auxins

The two forms of auxins (free and bound) appear to be in a dynamic state as there are many examples where the bound auxin is released in free state during extraction. With the result, the strictly separate measurement of free and bound auxins is often difficult. There are, however, two methods commonly employed for auxin extraction.

A. Diffusion method. It was devised by **Went** (1928) at Utrecht. In this method the growing tip (or other organ to be tested), under conditions of low transpiration, is severed (or cut) and is then placed on an agar block (usually of 1.5 concentration) for about an hour or so. During this period, the auxin diffuses from the cut tip into the agar block.

This method, though simple, has some major *drawbacks* and is, henceforth, not widely used.

(*a*) Excessive transpiration may prevent the accumulation of the auxin in the agar block.

(*b*) Severing the tip results in lowering the amount of auxin from the cut surface.

(*c*) The method cannot be widely adopted on account of the presence of growth inhibitors in many green plants.

B. Solvent extraction method. Here, the tissues are grinded in some organic solvents like chloroform, ether, ethyl alcohol or even water and the liquid is then filtered. The auxin is separated from the filtrate by chromatographic technique.

This method is widely employed for the extraction of auxins, *esp.*, the bound auxins. But this one also suffers from certain *drawbacks*.

(*a*) Use of chloroform as a solvent causes slow accumulation of chlorine which is a toxic substance and probably an auxin inactivator too.

(*b*) Diethyl ether, if used as a solvent, brings about oxidation of the auxin in the presence of a spontaneously formed peroxide. This can, however, be avoided if the solvent is distilled with ferrous sulfate and calcium oxide before use.

(c) During auxin extraction, a new auxin may be produced which may thus contaminate the auxin to be extracted. The difficulty can be overcome by employing Gustafson's technique which involves the boiling of plant material for about a minute prior to extraction.

Bioassay of Auxins

The term *bioassay* refers to determining the amount of active substance present in the plant tissues. In the various methods employed for bioassay, the activity of the auxin is in general determined by making it available in certain concentration to a seedling (or an ovary or a root) and the degree of either acceleration or inhibition of growth is recorded.

One of the most commonly employed methods of bioassay of auxins, as devised by **Went** (1928), is known as **Avena curvature test**. The test involves the following steps (Fig. 35–3).

Went found that the degree of angular curvature of coleoptile tip is proportional, within limits, to the concentration of auxin present in the agar block. This fact he made the basis of his *Avena test*.

ARIE JAN HAAGEN-SMIT

(LT, 1900–1977)

Haagen-Smit, a Dutch chemist, was born on December 22, 1900 at Utrecht. He attended the Rijks Hoogere Burger School in Utrecht. He graduated from the University of Utrecht in 1922 with a major in Organic Chemistry. He passed M.A. in 1926 and did Ph.D. in 1929. His work was on terpenes, a hydrocarbon found in plants. His dissertation is titled 'Investigations in the Field of Sesquiterpenes'.

He stayed at the University of Utrecht from 1929 to 1935 as chief assistant. He became an expert in plant-derived chemicals, particularly **auxins**, a hormone. He was invited to lecture at Harvard University in 1936 by **Kenneth Thimann**, himself an authority on plant hormones. He was appointed as associate professor by California Institute of Technology in 1937 by **Thomas Hunt Morgan**, and professor in 1940, becoming one of the 'Dutch Mafia' at CalTech (The other member of the mafia was **Frits Warmolt Went**). Two other scientists and he jointly published a paper on **traumatic acid**, a wound-healing hormone, in the journal *Science* in 1939. He remained the Director of the Plant Environmental Laboratory at CalTech from 1965 to 1971.

Haagen-Smit is best known for linking the **smog** in Southern California to automobiles and is hence known by many as the **"father" of air pollution control**. After serving as an original board member of the Motor Vehicle Pollution Control Board, formed in 1960 to combat the smog, Dr. Haagen-Smit became the California Air Resources Board's first chairman in 1968. Shortly before his death of ling cancer (on March 17, 1977 in Pasadena, California), the Air Resources Board's El Monte Laboratory was named after him.

Kögl and **Haagen-Smit** (1931) used *Avena test* as the unit of measurement. It is termed *Avena Einheit* and is abbreviated as A.E. One **A.E.** is defined as the amount of auxin present in an agar block (2 × 2 × 1 cm) which produces a curvature of 10° to a decapitated *Avena* coleoptile when placed eccentrically on it for 90 minutes.

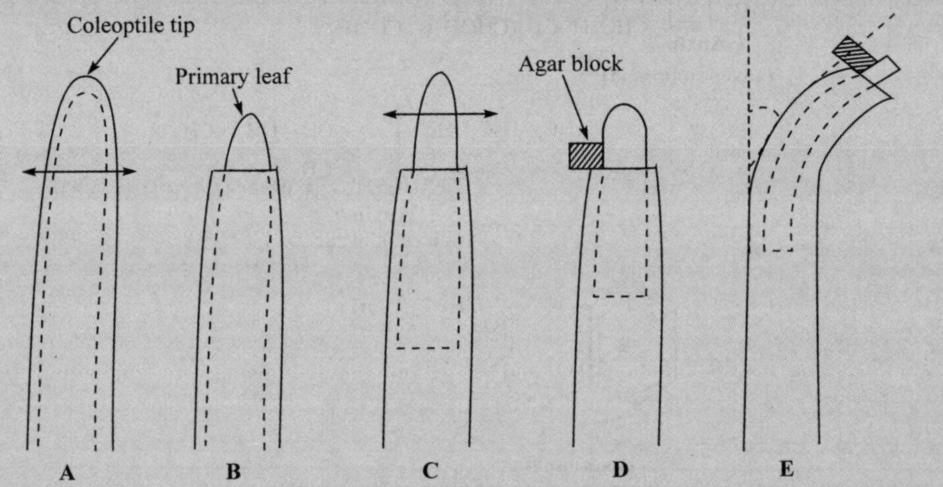

Fig. 35–3. Steps involved in Avena curature test

A. Decapitation of the coleoptile tip containing the primary leaf.
B. Pulling out of the exposed primary leaf so that its elongation may not dislocate the agar block.
C. Cutting the tip of primary leaf.
D. Affixing the agar block containing the auxin unilaterally to the cut tip.
E. Reading the angle resulting from movement of auxin into the side of coleoptile on which agar block is placed.

Biochemistry of Auxins

Application of *Avena* test to a wide variety of substances led to the discovery of auxins in human urine. **Kögl and Haagen-Smit** (1931) isolated 40 mg of auxin (which they named as *auxin a*) from as much as 33 gallons of human urine. Three years later, Kögl and his associates isolated two more compounds with auxin activity. These were named as *auxin b* and *heteroauxin* and were obtained from corn germ oil and human urine respectively. Their structure is shown in Fig. 35–4.

> **1 gallon** = 3.78533 litres (U.S. units)
> = 4.54596 litres (British units)

1. **Auxin a, auxentriolic acid.** It occurs at the meristematic apices (buds and growing leaves) both in the free state and bound to plasma proteins. It is a weak acid and is soluble in water, alcohol, ether and chloroform. It is stable in acid solutions but decomposes in alkaline soultions (*i.e.*, acid-stable and alkali-labile).

2. **Auxin b, auxenolonic acid.** It is present in corn germ oil, other vegetable oils, malt and a fungus called *Rhizopus*. It is also a weak acid and is soluble in water, alcohol, ether and chloroform. It is both acid-labile and alkali-labile. This as well as auxin *a* both are derivatives of cyclopentene.

3. **Heteroauxin, indole-3-acetic acid.** It is of universal occurrence in plants and is also synthesized by microorganisms including certain bacteria, yeasts and fungi like *Rhizopus*. It is resistant to alkalies whereas destroyed by acids and undergoes rapid decomposition on heating.

Unlike the first two, it can be easily synthesized in the laboratory. Chemically, it is a monobasic acid of a relatively simple structure.

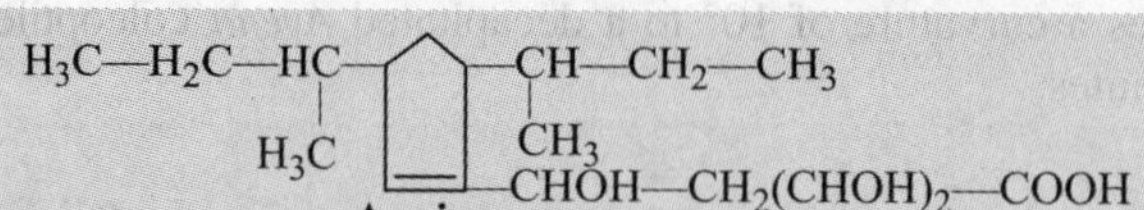

Auxin *a*

(auxentriolic acid)

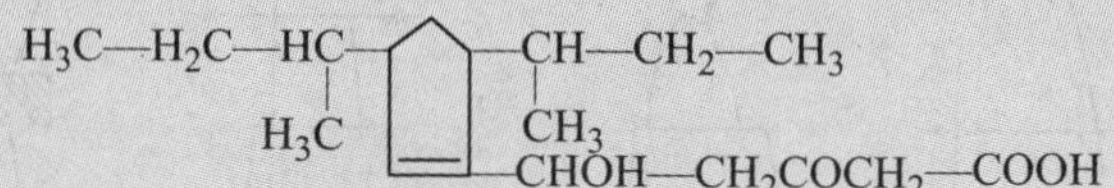

Auxin *b*

(auxenolonic acid)

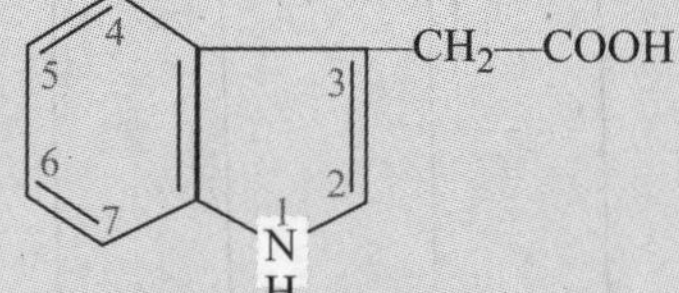

Heteroauxin

(indole-3-acetic acid, IAA)

Fig. 35–4. Natural hormones

Natural Auxins

Besides the above-mentioned auxins, certain other compounds have recently been shown to occur in plants. These show similar behaviour in terms of their effect on growth, although usually less intense. The principal naturally-occurring auxins in plants that have been definitely identified, isolated, purified and their chemical structure determined are all indole derivatives, (refer Fig. 35–5), as mentioned below:.

1. Indole-3-acetic acid, IAA
2. Indole-3-acetonitrile, IAN
3. Indole-3-acetaldehyde, IAc
4. Ethylindoleacetate
5. Indole-3- pyruvic acid, IPyA
6. Indole-3-ethanol, IEtOH

Controversy exists regarding whether ethylindoleacetate occurs naturally or is produced during extraction as an artifact. The presence of indole-3-ethanol is also not conclusively established.

Non-indole compounds, *e.g.*, some fatty acids may also possess auxin-like properties. But they have not yet been properly characterized.

The position of the side chain in the ring structure in indole-3-acetic acid (IAA) appears to be highly specific for activity, since 1-, 2- and 4-indole acetic acids are only very slightly active in bioassay. Substitution with halogens, like fluorine and chlorine, can result, however, in very active derivatives, *e.g.*, the 4-chloro and 6-chloro compounds. The replacement of an aromatic CH of IAA by N can give rise either to 7-aza (indole-3) acetic acid, a synthetic auxin (see Fig. 35–6 for the structure of this and other synthetic auxins) which is less active in the *Avena* test or to another synthetic auxin, indazole-3-acetic acid which has activity equal to IAA. Indole-3-acetonitrile (IAN) may be converted into either IAA or IAc, depending on whether –C ≡ N is replaced by –COOH or –CHO, respectively. Replacement of the imino group (–NH) in IAA by a S atom produces another synthetic auxin called thianaphthen-3-indoleacetic acid which is quite

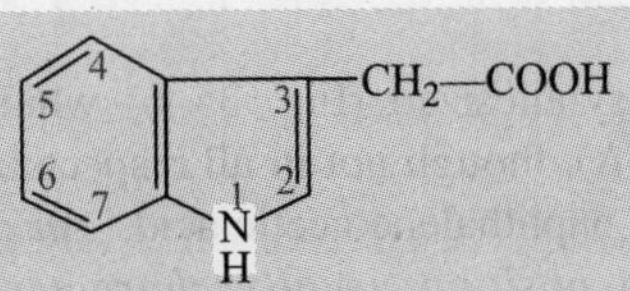

Indole-3-acetic acid, IAA
(175)

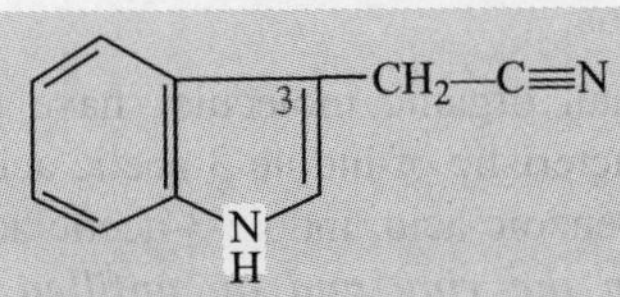

Indole-3-acetonitrile, IAN
(156)

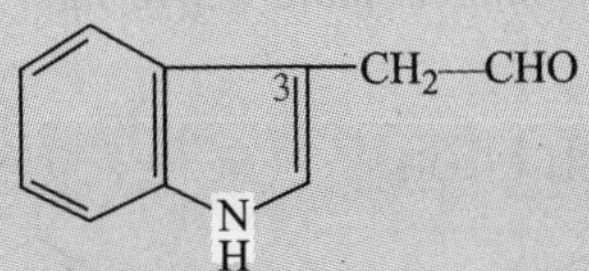

Indole-3-acetaldehyde, IAc
(159)

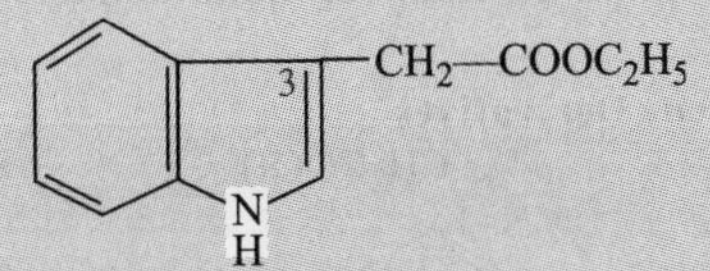

Ethylindoleacetate
(203)

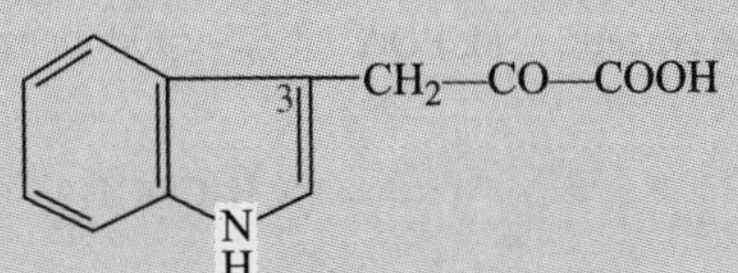

Indole-3-pyruvic acid, IPyA
(203)

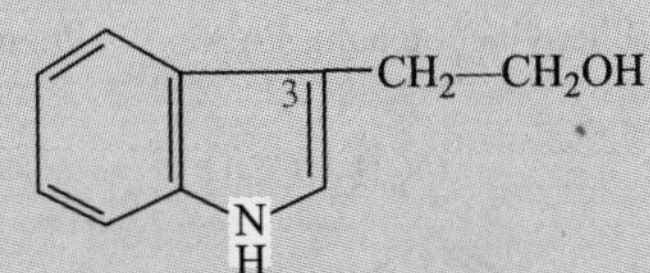

Indole-3-ethanol, IEtOH
(161)

Fig. 35–5. Some natural auxins (IAA and its derivatives)
Bracketed numbers underneath each compound represent its molecular weight.

active in various biossay tests (**Allsopp**, 1965). Indole-3-propionic acid (IPA) is less active than IAA, but indole-3-butyric acid (IBA) is less active than IAA, but indole-3-butyric acid (IBA) is more active than IPA; however, in some tests as in the case of rooting of cuttings, IBA tests as in the case of rooting of cuttings, IBA is even as active as or more active than IAA. It is noteworthy that the side chains having even number of C atoms are more active than those having an odd number. It is hypothesized that the side chain is oxidized to IAA before the compound shows a biological activity.

Indoleacetonitrile (IAN), a nitrile derivative of IAA, is a neutral substance and has been obtained in crystalline form from many plant materials. It is believed that the nitrile derivative (*i.e.*, IAN) as such is inactive and has to be converted to IAA in order to be active. IAN, upon alkali hydrolysis, yields IAA. IAN promotes growth of only those plant organs (like *Avena* coleoptile, bean seeds, tomato ovary etc) which possess the enzyme *indoleacetonitrilase*, while tissues (like pea roots)which cannot convert IAN to IAA because of the absence of indoleacetonitrilase, are unreactive to IAN. However, in higher concentrations, IAN is inhibitory like IAA. In addition to acting as an auxin precursor, IAN is also known to be an activator or booster of IAA responses.

Ethylindoleacetate is the ethyl ester derivative of IAA and has been isolated from many plant tissues. Both IAN and ethylindoleacetate seem to be more active than IAA, in those tissues which are capable of converting them into IAA. IAN, ethylindoleacetate and **indoleacetaldehyde**, **IAc** (the aldehyde derivative of IAA), all three exist in nature as auxin precursors and all of which can be converted into IAA.

Synthetic Auxins

Several small organic molecules have been synthesized (Fig. 35–6) which show biologic properties characteristic of indole-3-acetic acid (IAA), though not in all respects. They are usually derivatives of benzoic acid, indole-3-acetic acid or naphthalene acetic acid. *An acidic side chain, unsaturation in the ring and an unfilled ortho position usually characterize many active molecules*; although no general rule can be framed as some thiocarbamates, which do not satisfy most of the requirements, are quite active. *The D-isomer is more active than the L-isomer.*

Benzene Derivatives

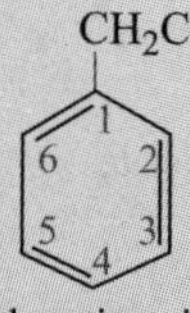

Phenylacetic acid, PAA

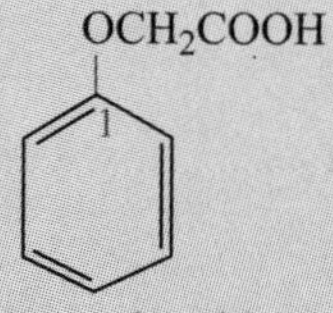

Phenoxyacetic acid, POA

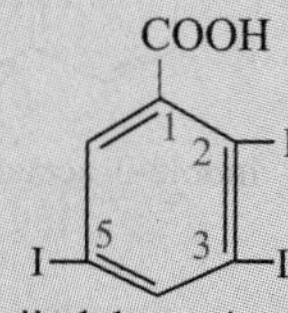

2,3,5-triiodobenzoic acid, TIBA

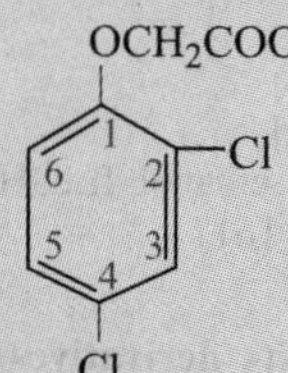

2,4-dichlorophenoxyacetic acid, 2,4-D

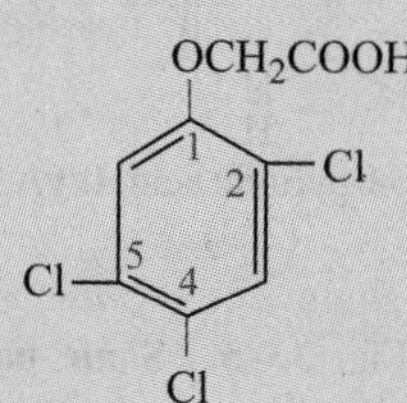

2,4,5-trichlorophenoxyacetic acid, 2,4,5-T

Indole Derivatives

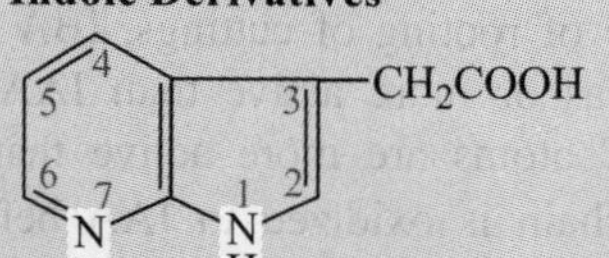

7-aza (indole-3) acetic acid

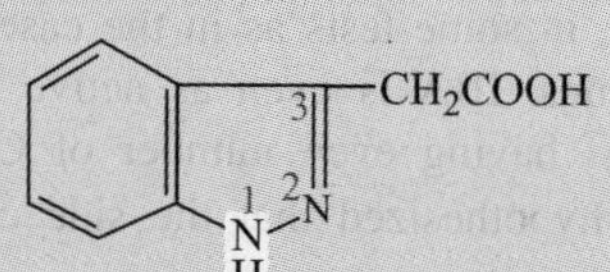

Indazole-3-acetic acid

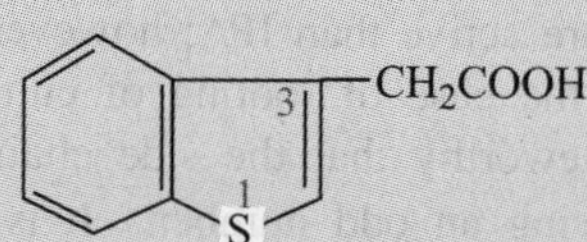

Thianaphthen-3-indoleacetic acid

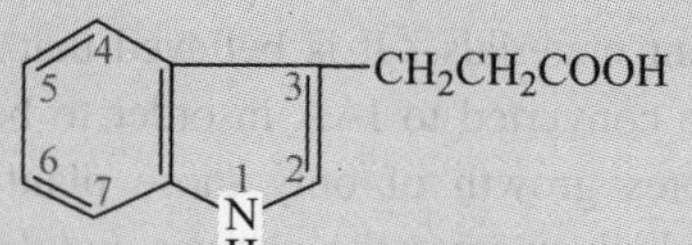

Indole-3-propionic acid, IPA

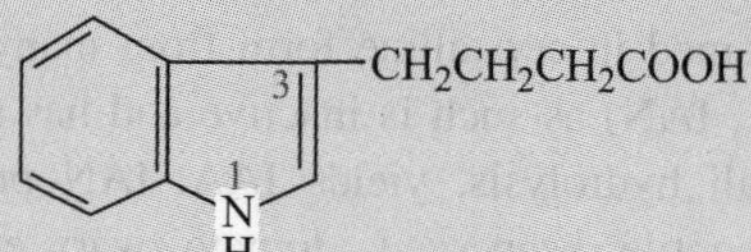

Indole-3-butyric acid, IBA

Naphthalene Derivatives

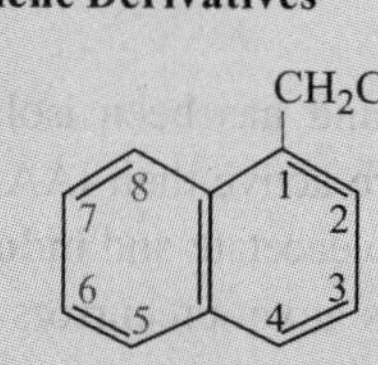

1-naphthaleneacetic acid, NAA

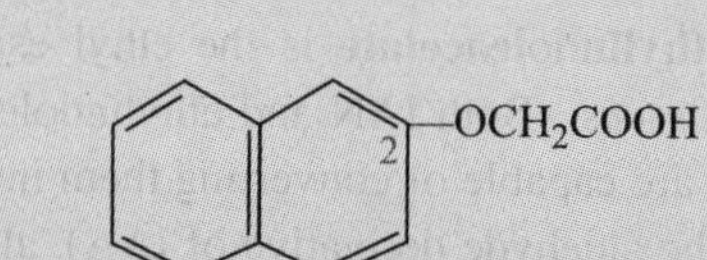

2-naphthoxyacetic acid, NOA

Fig. 35–6. Some synthetic auxins

Cis-cinnamic acid is an auxin whereas *trans*- cinnamic acid behaves as an antiauxin. The 3-'D' structure of the molecule and the spatial relationship between the side chain and the aromatic ring, if present, determine biologic activity. Some of the potent synthetic auxins are TIBA, 2-4-D, 2,3,5-T, NAA and NOA. Phenylacetic acid (PAA) is, however, a weak auxin.

System of Carbon Numbering in Naphthalene

A molecule of naphthalene consists of two benzene rings joined together. The various carbon atoms in naphthalene (as also in other organic compounds) are numbered using either the Arabic numerals (*Arabic system*) or Greek alphabets (*Greek system*), as shown below:

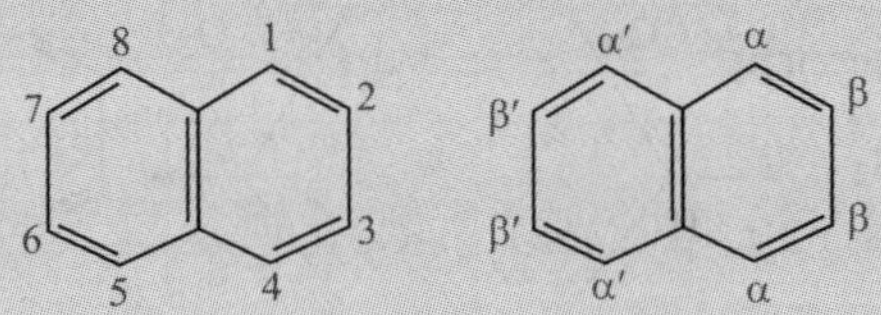

Two systems of carbon numbering in naphthalene

Positions of carbon atoms numbered 1, 4, 5, and 8 in Arabic system are similar; hence, they all are called as α in Greek system. However, to differentiate between these two pairs of C atoms (1–4 pair of one ring and 5–8 pair of another ring) in the Greek system, carbon atoms 5 and 8 of the second ring are numbered by putting a prime (′) sign on the right of α as superscript, thus numbering them as α′. Similarly, position of carbon atoms numbered 2, 3, 6 and 7 in Arabic system are similar; hence they all are called as β in Greek system. By the same reasoning, carbon atoms numbered 6 and 7 are designated by β′.

The various positions of carbon atoms, denoted by Arabic numerals, have been given specific names in Greek system. These are as follows:

Position 1, 2 is known as *ortho*.
Position 1, 3 is known as *meta*.
Position 1, 4 is known as *para*.
Position 1, 5 is known as *ana*.
Position 1, 6 is known as *epi*.
Position 1, 7 is known as *kata*.
Position 1, 8 or 4, 5 is known as *peri*.
Position 2, 6 or 3, 7 is known as *amphi*.
Position 2, 7 or 3, 6 is known as *pros*.

Biogenesis (= Synthesis) of Auxins

Folke Skoog (1937) of the University of Wisconsin, for the first time, experimentally proved that tryptophan is an auxin precursor in higher plants. Since then a wide variety of plant tissues (like leaf, stem, buds, coleoptile, ovary, pollen, embryo, endosperm and callus tissue) have been shown to convert tryptophan to indoleacetic acid (**Larsen**, 1951). In fact, it is quite probable that all living plant tissues may have the capability of bringing about this conversion. Although tryptophan is the primary precursor of IAA, about half a dozen indole compounds have been found to serve as potential precursors of IAA.

For the enzymic production of IAA, a plausible hypothesis has been suggested by **Wildman**, **Ferri** and **Bonner** in 1946. According to this hypothesis (Fig. 35–7), tryptophan, liberated by hydrolysis of proteins, undergoes either oxidative amination first and then decarboxylation or vice versa to yield indoleacetaldehyde (IAc). IAc is then oxidized to yield the free auxin. Indoleacetaldehyde, thus, acts as an intermediate metabolite as well as an immediate precursor of IAA. The capability of living plant tissues to form auxin from tryptophan has been demonstrated for spinach (**Wildman** *et al*, 1947), pineapple (**Gordon** and **Nieva**, 1949) and many others. The three essential conditions for IAA synthesis are the presence of light, zinc and an enzyme system.

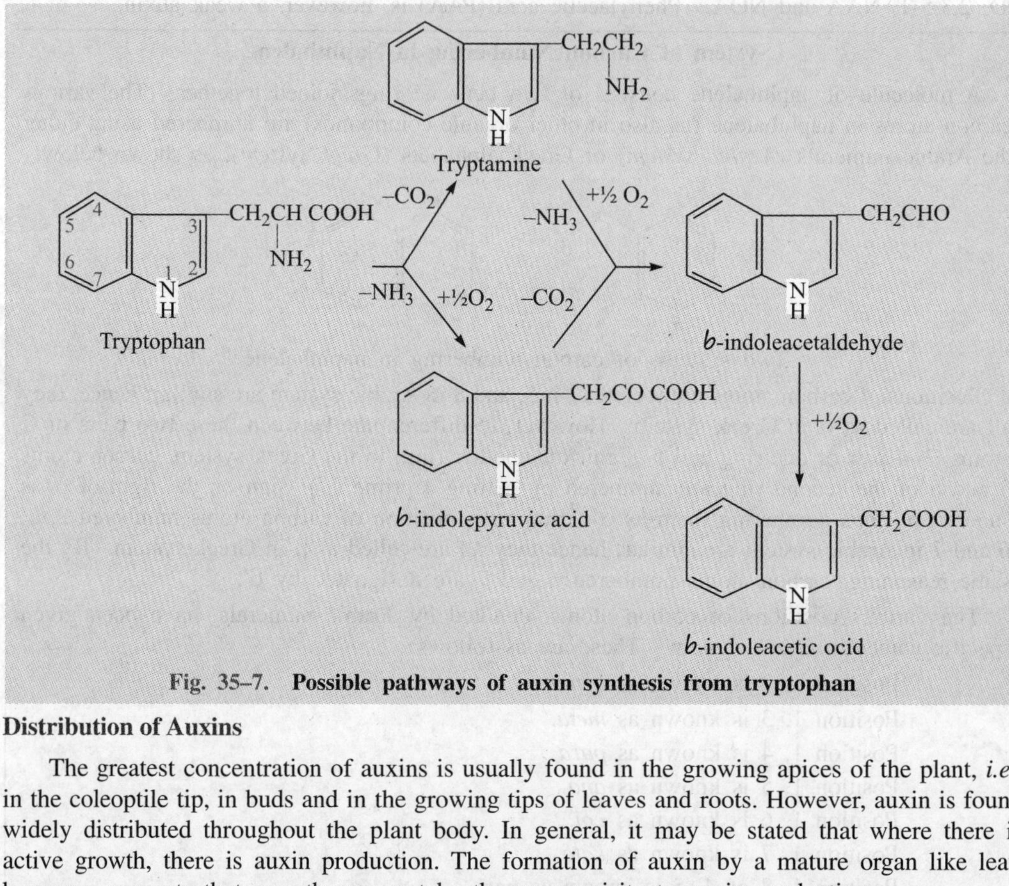

Fig. 35–7. Possible pathways of auxin synthesis from tryptophan

Distribution of Auxins

The greatest concentration of auxins is usually found in the growing apices of the plant, *i.e.*, in the coleoptile tip, in buds and in the growing tips of leaves and roots. However, auxin is found widely distributed throughout the plant body. In general, it may be stated that where there is active growth, there is auxin production. The formation of auxin by a mature organ like leaf, however, suggests that growth may not be the pre-requisite to auxin production.

Thimann (1934) studied the distribution of auxin, in detail, in etiolated *Avena* coleoptile (Fig. 35–8). He found that the concentration of auxin drops as one progresses from the coleoptile tip to its base; the highest concentration being at the tip and the lowest at the base. If one progresses further from the base of the coleoptile along the root, there is a steady increase in auxin content till a maximum is reached in the root tip. Of the two maximal values, that for the stem tip is much higher than that for the root tip.

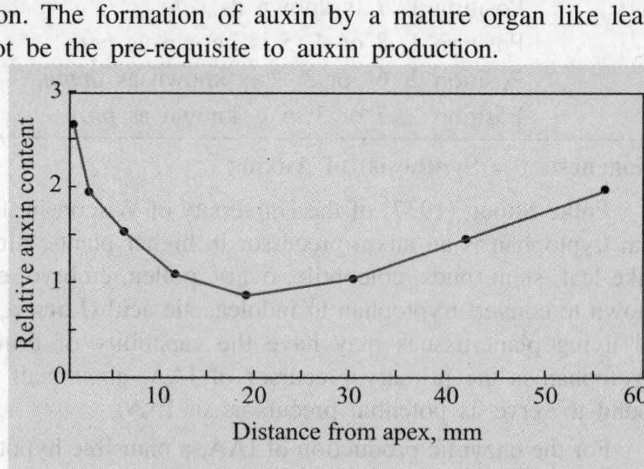

Fig. 35–8. Distribution of auxin in an etiolated *Avena* seedling

(*After Thimann KV, 1934*)

Thimann and **Skoog** (1934), while working on *Vicia faba* seedlings grown in light, found the concentration of auxins in various organs in the following descending order:

Apical buds > Young leaves > Mature leaves

The amount of diffusible auxin per hour for these organs was found to be approximately in the ratio of 12 : 2 : 1.

Van Overbeek (1947) studied the distribution of both free and bound auxins in pineapple. He found that large quantities of free auxin occurred in apical buds and lowest amount in mature leaves. For bound auxin, however, the condition was found to be reverse.

A few other examples of auxin concentration studies are given in Table 35–1.

Table 35–1. Amount of auxin present in different organs of some plants

S.N.	*Plant*	*Organ*	*Maximum Concentration Found (in μg IAA equivalent)*
1.	Corn	Endosperm	105, 000
2.	Lily	Stem tip	83,900
3.	Oat	Grain	1,000
4.	Rice	Endosperm	250
5.	Turnip	Seed	250

Concentration of Auxins

The concentration of auxins, which has a profound influence on growth changes, varies from organ to organ (Fig. 35–9). The effect of auxin concentration on the shoots is quite different from

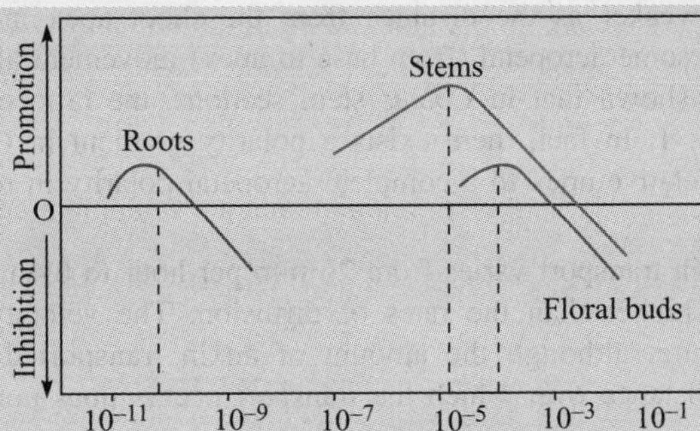

Fig. 35–9. Graph showing the interrelationship between growth promotion (or growth inhibition) and various plant organs such as roots, stems and floral buds

The optimum concentrations for growth promotion were found to be between 10–11 and 10–9 for roots, 10–8 and 10–3 for stems and 10–5 and 10–3 for floral buds.

(*After Leopold AC and Thimann KV, 1949*)

that on the roots. Higher concentration of auxin, which has growth-stimulatory effect on the shoots, is growth-inhibitory for the roots ; the latter growing better at much lower concentration. Thus, *in general, the optimum range of concentration for elongation in stems is much higher than for the roots*. The stems grow best at an auxin concentration of 1.0 mg/litre, whereas the optimum concentration for the roots is of 0.001 mg/litre.

Translocation (= Movement) of Auxins

The auxins are transported in plants from one organ to the other. The usual direction of auxin transport is downward but when added to the soil these are absorbed by the roots and carried upward along with transpiration stream to various plant organs. The prevailing downward movement takes place through the living phloem cells whereas the upward movement occurs through the dead xylem elements.

The most striking characteristic of auxin movement is its almost strict *basipetal* (from apex to base) *polarity*. This has been demonstrated in various plant organs such as oat coleoptiles (**Went** and **White**, 1939), petioles of leaves and herbaceous and woody strems (**Oserkovsky**, 1942). The actual proof of polar transport was furnished by **Went** (Fig. 35–10). He showed that if an agar block containing auxin be affixed to the morphologically upper end of a coleoptile segment and a block of pure agar to the lower end, auxin would move and collect in the agar block at the lower end. But if the coleoptile segment is inverted, no translocation of auxin would occur. The translocation of auxins in oat takes place through the parenchyma tissue.

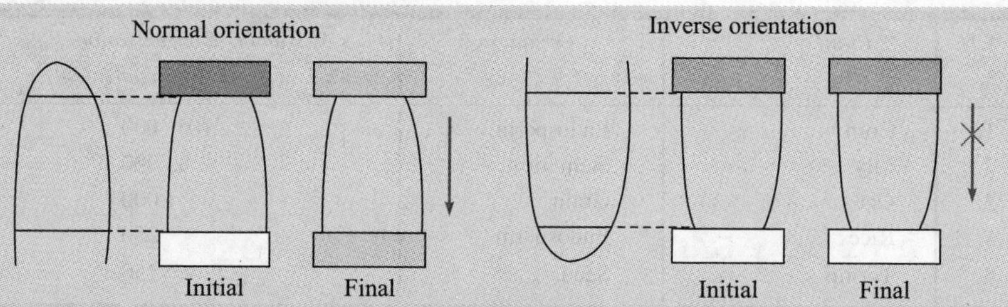

Fig. 35–10. Experiment demonstrating basipetal movement of auxin in oat coleoptile

(*After Went FW, 1935*)

In *Coleus*, however, **Leopold** and **Guernsey** (1953) have shown that the basipetal polarity becomes progressively weaker as the distance from the shoot apex increases. Further, in the flowering stems there is some acropetal (from base to apex) movement also even at the stem tip. **Jacobs** (1961) has also shown that in *Coleus* stem sections, the ratio of basipetal to acropetal transport of auxin is 3 : 1. In fact, there exists a polarity gradient in *Coleus* from a complete basipetal polarity in vegetative apex to a complete acropetal polarity in root apex with a gradual transition in between.

The velocity of auxin transport varies from 26 mm per hour to 6.4 mm per hour (**Rajgopal**, 1967). These rates are higher than the rates of diffusion. The velocity of auxin transport is unaffected by temperature, although the amount of auxin transported is proportional to the temperature. Also, the distance over which the transport occurs does not influence the velocity.

Mechanism of Auxin Action

The action of auxins, like growth itself, seems to be a complex of many functions. Although the mechanism of auxin action may, in part, be attributed to each of these functions, none of these can account for *in toto* the multifarious effects of these growth substances. The various views put forward to explain auxin behaviour may be, for convenience, grouped under following five headings:

1. Molecular reaction theories. Skoog (1942) expressed the view that the auxin may act like a coenzyme and serves as a point of attachment for some substrate onto an enzyme regulating growth. The molecular configuration of the auxin affects the activity by altering the fit and functioning of this molecular union. The higher auxin concentrations would inhibit growth owing to separate molecules combining with the enzyme and the substrate.

Muir *et al* (1949) advanced the hypothesis that the auxins (*esp.*, phenoxy acids) may combine with some material (*e.g.*, protein) in the cell at two points, the ortho position of the ring and the acid group of the side chain

Foster and his associates (1952) put forward a theory of auxin action by *2-point attachment*.

It is presumed that the enzyme is attached to a substrate to form an enzyme-substrate complex and that the complex may, then, dissociate to produce the end product of the reaction (*i.e.*, growth) and regenerate the enzyme. Considering the enzyme as the material with which auxin reacts, the theory may thus be expressed as:

$$E + S \rightleftharpoons ES \longrightarrow \text{Growth} + E$$

[where E = auxin receptor; S = auxin; ES = complex]

The inhibitory effect of high auxin concentration on growth may well be explained by this hypothesis. This would result from two auxin molecules becoming attached to the receptor substrate, one at each of the points of attachment and each preventing the functioning of the other (Fig. 35–11).

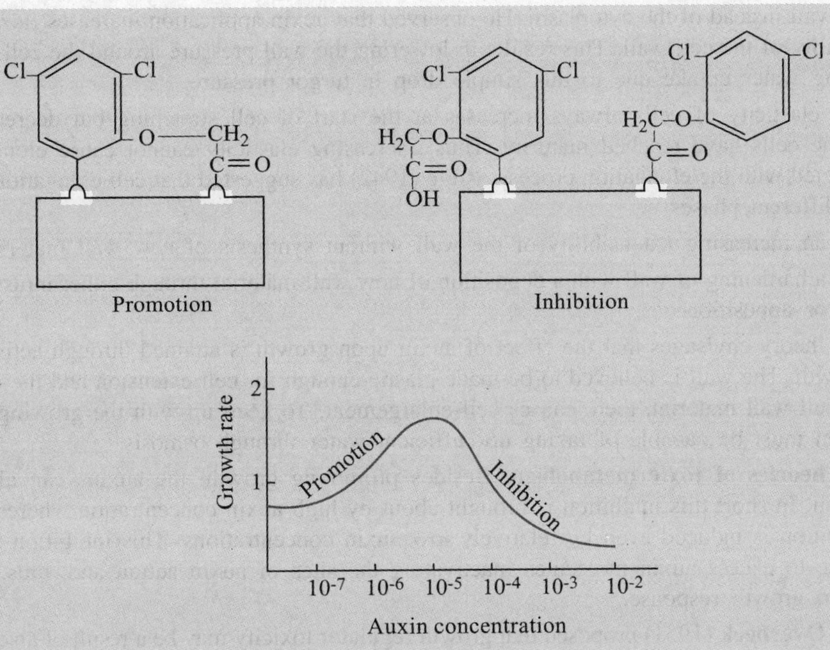

Fig. 35–11. Effect of auxin concentration on growth

2. Theories of enzymatic effects. The fact that growing tissues, upon treatment with auxin, show an increased activity of a number of enzymes has proved to be the basis of these theories. **Northen** (1942) pointed out that the auxin causes decrease in cytoplasmic viscosity and also brings about dissociation of the cytoplasmic proteins. The latter effect would result in an increase in water permeability and also in the osmotic value of the cytoplasm. These effects ultimately lead to enhanced enzymic activity.

Burger and **Avery** (1942) demonstrated that some *dehydrogenases*, under certain conditions, could be stimulated by auxin. **Thimann** (1951), however, found that the auxins act as agents protecting certain growth enzymes from destruction rather than as substances activating enzymes.

3. Theories of osmotic effects. During the process of growth, the cell increases in volume due to water uptake. The uptake of water occurs due to changes in the cytoplasm itself (*esp.*, changes in the osmotic value) or due to changes in permeability of the cell wall and the cell membranes.

Czaja (1935), for the first time, stated that the auxin may increase the volume of the cell which could result directly in water uptake and growth. **Van Overbeek** (1944), however, pointed out that growth is not necessarily associated with an increase in osmotic value.

Commonor and his associates (1942–43) suggested that since water uptake in growth may be linked with respiration, the process of growth may be explained as due to the osmotic uptake of water which is, in its turn, activated by respiratory uptake of salts.

Two important *objections* put against this theory are as follows:

(*a*) In some instances (*e.g.*, potato slices), growth cannot be a function of salt accumulation causing an increase in osmotic value (**Van Overbeek**, 1944).

(*b*) **Reinders** (1942) points out that the auxin-induced growth can occur in pure water in the absence of salt uptake. She describes water uptake as dependent upon oxidative metabolism.

Thus, water uptake is not the cause but the consequence of growth (Burström, 1953).

4. Theories of cell wall effects. Hyen (1940) attributed growth to the dynamic function of the cell wall instead of the cytoplasm. He observed that auxin application increases flexibility and extensibility of the cell wall. This results in lowering the wall pressure around the cell wall, thus permitting water uptake due to this simple drop in turgor pressure.

The elasticity of cells always increases at the start of cell stretching but decreases again before the cells have reached maturity. Thus, increasing elasticity cannot cause elongation but is connected with the elongation process. **Ruge** (1942) has suggested that cell elongation proceeds in two different phases:

(*a*) an increasing extensibility of the wall without synthesis of new wall material.

(*b*) a hardening of wall with a deposition of new wall material through either intussusception or apposition.

The theory envisages that the effect of auxin upon growth is attained through activating cell wall growth. The wall is believed to be made plastic enough for cell extension and the deposition of new cell wall material, then, causes cell enlargement. To keep up with the growing wall, the cytoplasm must be capable of taking up sufficient water through osmosis.

5. Theories of toxic metabolism. Besides promoting growth, the auxins can also inhibit elongation. In short this inhibition is brought about by high auxin concentration whereas in roots the inhibition is induced even by relatively low auxin concentrations. This inhibition is thought to be due to excess auxin molecules inactivating the sites of auxin action and, thus, checking maximum growth response.

Van Overbeek (1951) proposed that growth regulator toxicity may be a result of an alternation of metabolism in such a manner that unsaturated lactones are accumulated in plant tissues. These are toxic to plants when applied in higher concentrations. Such toxic compounds may accumulate in plant tissues following application of such hormones as 2,4-D.

Thus, it may be concluded that the mechanism of auxin action remains unsolved, although the various theories put forward provide some clues regarding various functions which the auxins perform.

Physiological Roles of Auxins

It was previously thought that the sole function of auxins was to promote cell enlargement. But the work done in later years has proved them to be deeply associated with a variety of functions. In some cases they act as a stimulating agent, in others as an inhibitory agent and in still others as a necessary participant in the growth activity of other phytohormones such as gibberellins and cytokinins. The various growth processes in which the auxins (both natural and synthetic) play their role are discussed below:

1. Cell elongation. It is usually considered that cell elongation occurs only in the presence of auxins and also that the rate of elongation is directly proportional to the amount of auxin applied provided no other factors are limiting. But relatively high concentrations usually exert inhibitory effect on this phase of growth.

Auxin as Herbicide

Some commercial weedkillers are chemically similar to auxin and have similar effects. They cause affected plant cells to elongate and the plant grows. The exact mechanism in not yet understood, but the auxin seems to interfere with DNA transcription and RNA translation. The amount of auxin applied in a weedkiller is far greater than the amount produced within the plant, and so the rate of growth produced is much greater than normal. The plant cannot sustain this rate of growth: it becomes weakened, unable to reproduce, and then it dies. The auxin like herbicides have a much greater effect on dicots than on monocots, such as grasses. This is partly because dicot leaves have a larger surface area than monocot leaves and so absorb more herbicide (refer adjoining figure).

Auxin-like herbicides as efficient means of getting rid of weeds like dock

As already discussed, the various plant organs like roots, buds and stems all react in a comparable way to auxins : their growth being promoted by relatively low and inhibited by relatively high auxin concentrations. Elongation of roots is promoted only at very low concentrations ; at higher concentrations their growth is retarded. Stems and coleoptiles respond similarly except that optimum range of concentrations for elongation is much higher than for roots. Flowers require still higher concentration for growth.

Auxins also play a significant role in the elongation of petiole, mid rib and major lateral veins of the leaves. Thus, adenine favours enlargement in detached leaves of radish and pea. Similarly, coumarin has been shown to promote expansion of leaves in some plants.

An osmotic equilibrium exists in a cell where the turgor pressure developed is counterbalanced by the wall pressure acting in opposite direction. Regarding the mechanism of cell elongation, it is thought that auxins stimulate cell elongation by modifying certain conditions responsible for this equilibrium (**Devlin**, 1969). These modifications include:

(*a*) an increase in osmotic contents of the cell

(*b*) an increase in permeability of the cell to water

(*c*) a decrease in wall pressure

(*d*) an increase in wall synthesis and

(*e*) an inducement of specific RNA and protein synthesis.

2. Cambial activity. In the spring season, the trees exhibit growth by developing buds which later on open. This is then followed by elongation of the young stems. This resumption of growth by cambial cells is activated by the auxins which move basipetally in the stems from developing buds. **Snow** (1935) has shown that a steady supply of auxin a at 1/10,00,000 mg per hour (or of IAA at 1/5,00,000 mg per hour) from a gelatin block, upon affixing it to the cut end of a decapitated shoot of sunflower (*Helianthus annuus*) seedling, stimulated meristematic activity of the cambium.

The suggestion by **Jost** (1940) seems to imply that the major function of hormones that migrate from the developing apex of the epicotyl is to activate the differentiation of procambial strands.

3. Callus formation and galls. Besides acting as stimulants of cell elongation, the auxins may also activate cell division. This may be illustrated by applying 1% IAA in lanolin paste to a debladed petiole of a bean plant. This causes prolific division of parenchyma cells resulting in the formation of a swelling or callus tissue at a point where the auxin in applied. The amount of callus tissue formed is directly proportional to the concentration of IAA applied (**Ropp**, 1950).

Bezerinck, in as early as 1885, investigated the production of cecidomid galls in *Poa nemoralis.* He compared gall production to callus formation and thought of a substance coming from the larval body as a causative agent. He regarded this action of animal substance as analogous to that of the inner causes which lead to the formation of roots in normal plants.

4. Rooting of stem cuttings (= Formation of adventitious roots). It is a common observation that the presence of buds on a cutting favours development of roots when the lower end is dipped in a suitable rooting medium. Developing buds are effective in accelerating root formation. Young leaves also favour the initiation of roots on the cuttings. These observations led to the suggestion that the root formation is favoured by the auxins which are synthesized in the buds and young leaves and are later translocated to the basal part of the cutting.

Two techniques are usually employed to introduce auxins into the cuttings. In one, *"dry form" method*, the auxin is first mixed with an inert powder such as talc in the proportion of 500 to 2,000 parts of the auxin to 10,00,000 parts of the talc. Later, the basal end of the cutting is first dipped in water and then in powder before immersion into the rooting medium.

In the other, *"quick dip" method*, the basal end of the cutting is dipped for about 5 seconds into a relatively concentrated solution of auxin (5,000 to 10,000 ppm in water or 5% ethyl alcohol) before immersing in the rooting medium.

The auxins most commonly employed for this purpose are IAA, NAA, 2,4-D, naphthalene acetamide (NAd) etc.

Auxin-induced rooting is not only of academic interest but also of enormous horticultural value as it helps propagation of certain plants by cuttings. Although some success has been achieved in bean and alike plants, yet certain other plants like apples do not respond to it (**Thimann** and **Behnke**, 1947).

5. Apical dominance. It has been generally observed that so long as the apical bud is intact on the plant, the growth of the lateral buds remains suppressed. Upon removal of the apical bud, the lateral bud nearest the apical bud establishes its dominance over the remaining buds, causing them to become inactive again. *This inhibitory effect of a terminal bud upon the development of the lateral buds is called apical dominance and produces a cone-shaped plant.* This is why a gardener keeps on trimming the hedge occasionally in order to obtain a denser growth. Plants that are tall and unbranched exhibit strong influence of apical dominance than those which are short and branched.

> The characteristic conical shape of most **Christmas trees** results from apical dominance. *Almost all Christmas-tree-shaped trees are conifers*. Ever since **U.S. President Franklin Pierce** (a politician, otherwise known for putting stickum on the backs of postage stamps) first had a Christmas tree in the White House in 1856; people have been choosy about these trees, preferring ones that are 2 metres high and 1.3 metres wide at the base. Scotch pine (*Pinus sylvestris*) is the most popular Christmas tree in India, the United States and elsewhere. It is a fast-grower and takes only 8 years to grow to a height of 2 metres. Although the tree has long needles, but they stay on the tree much longer than those of balsam fir (*Abies balsamia*) or spruce (*Picea* sp.), the previous best-sellers.

The relation of apical dominance with the auxin supply was first reported by **Skoog** and **Thimann** (1934). They demonstrated that when agar block containing auxin *b* or IAA was kept on the decapitated shoot of broad bean (*Vicia faba*), the lateral buds, as might be expected, resulted in the usual suppression of growth as if the terminal buds were present. But when the same decapitated shoot was reheaded with an agar block containing no auxin, these lateral buds resumed growth. Similar results were also obtained with field-grown tobacco plants using NAA as the auxin.

Although the exact mechanism behind this growth correlation is not yet known, the best explanation for this has been furnished by **Snow** (1939, 40). According to him, under the influence of auxin, a growth inhibitor is formed that is responsible for the inhibition of growth of the lateral buds. The growth inhibitor is synthesized in some unknown manner when the auxin moves in its usual downward direction. The theory lends support from the fact that certain growth inhibitors have chemical structure not much different from that of the auxins. It may, thus, be visualized that such a synthesis of growth inhibitor from the auxin may take place in the tissues.

The knowledge of apical dominance has been utilized practically in solving the problem of storage of the potatoes. Potatoes, stored for some time, sprout and become sweet in taste, thus causing financial loss to the grower as the sweet taste is disliked by its consumers. But spraying the potatoes with auxins like indole butyric acid (IBA) and NAA would prevent sprouting (or in other words, would prolong dormancy) by inhibiting the development of buds or 'eyes' ; the effect persisting for as long a period as 3 years. Although such a treatment is in the interest of the breeder but certainly not in that of the consumer.

6. Delay (or inhibition) of abscission of leaves. The abscission of leaves can be delayed or inhibited by the application of auxins on the surface of the lamina or on the cut surface of a debladed petiole. The controlling behaviour of the auxins on the abscission was first noted by **Laibach** (1933) who showed that the extract of orchid pollinia is capable of preventing the leaf fall. Since then, enough work has been carried out in this direction. **Addicott** and **Lynch** (1955) have proved conclusively the delaying effect of IAA on the abscission of various plant organs.

As to the mechanism of abscission, it has been suggested that the leaf fall is retarded by the basipetal migration of a hormone from the blade to the base of the petiole. Removal of the leaf blade eliminates the supply of hormone to the abscission zone and thus induces leaf fall.

The correlation between the amount of diffusible auxin and the age of the organ concerned (Fig. 35–12) was demonstrated by **Shoji** *et al* in 1951. They found high auxin contents in the young leaf blades of bean plant as compared to their petioles. But on ageing, the auxin contents in the leaf blade and the petiole fall almost to the same level.

7. Flowering. A flowering hormone, *florigen*, is produced in the leaves under correct light and dark period. It moves first down the petiole and then up the stem to the growing apex where it causes the development of floral buds in place of vegetative buds (**Cajlachjan**, 1936). Auxins are useful in modifying flowering in one of the following ways:

A. ***Altering earliness***— **Leopold** and **Guernsey** (1953) conducted experiments on oat, corn, barley, peas etc., and found that when their seeds were treated with auxins followed by low temperature treatment at 4°C for 5–15 days, they produced quantitative increase in earliness.

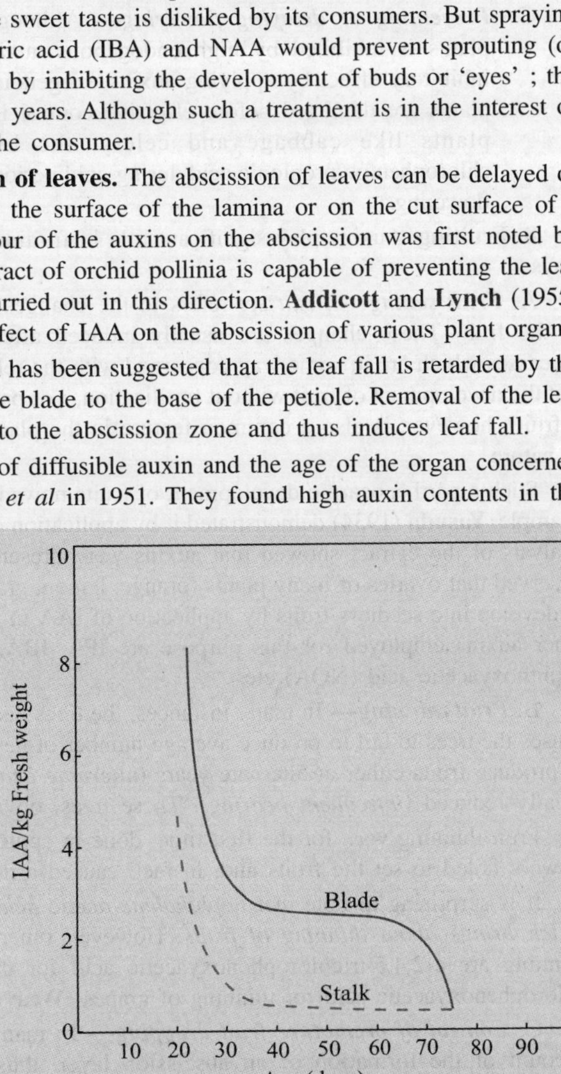

Fig. 35–12. Efect of ageing of the petiole and the leaf blades of bean plant on the auxin contents of these organs

(After Shoji et al, 1951)

Chakravarti (1955) also produced earliness and greater yield in mustard when auxin was applied before chilling. No gains were, however, obtained when the auxin was applied after chilling.

B. ***Inducing flowering***— The flowering in pineapple has been successfully promoted in Hawaii islands by treating the fields with sodium naphthalene acetate at some time during growing period. In Caribbean areas, however, 2,4-D is more commonly used.

Leopold and **Thimann** (1949) have, likewise, found low concentrations of auxins such as NAA or IAA as effective in inducing flowering in barley and Teosinte. Furthermore, 2,4-D has been successfully used in control of flowering in sweet potato.

C. ***Preventing or delaying flowering***— Although natural flowering of many species of plants is inhibited by high endogenous auxin concentrations, attempts to attain this inhibitory effect by applying auxins exogenously (*i.e.*, from outside) usually fail as the plants are damaged before effective control is obtained. But experimentally in certain plants like cabbage and celery, the bolting is prevented by applying *p*-chlorophenoxypropionic acid during cold periods which would otherwise normally induce flowering.

8. Fruiting. Auxins play significant role in fruiting by modifying it in one of the following ways:

A. ***Fruit setting***— *Fruit set* refers to the changes in the ovary leading to the development of the fruit. These changes are usually induced after pollination and fertilization— the two processes which are in some way concerned with the release of some stimulus of hormone nature. But the development of fruit without fertilization, *i.e.*, parthenocarpy (*parthenos*^{G} = virgin; *carpos*^{G} = fruit), however, is also a common feature in the plant world and henceforth occurs frequently in nature.

Such a parthenocarpic development of fruits nowadays has also been induced artificially. For example, **Yasuda** (1934) demonstrated it by application of pollen extracts to cucumber flowers. An analysis of the extract showed that auxins were present in it. Later, **Gustafson** (1936, 39) also observed that ovaries of many plants (orange, lemon, grape, banana, tomato etc.) could be induced to develop into seedless fruits by application of IAA in lanolin paste to their stigmas. The various other auxins employed for this purpose are IPA, IBA, α-NAA, phenoxyacetic acid (POA), α-naphthoxyacetic acid (NOA) etc.

B. ***Fruit thinning***— In many instances, the trees bear extensively large number of fruits. This causes the trees to fail to produce average number of new flower buds. Such trees, therefore, have to produce fruits either at alternate years (*alternate bearing*) or if yearly, the number of fruits is greatly reduced (*infrequent bearing*). These trees, obviously, require thinning.

Fruit thinning was, for the first time, done in apple when naphthalene acetic acid applied to flowers failed to set the fruits and, in fact, caused a decrease in fruit set.

It is surprising to note that *naphthalene acetic acid appears to be the only successful auxin which brings about thinning of fruits*. However, other examples of auxins employed for fruit thinning are a-2,4,5-trichlorophenoxyacetic acid for thinning of pears (**Griggs**, 1951) and *p*-chlorophenoxyacetic acid for thinning of grapes (**Weaver** and **Winkle**, 1952).

C. ***Control of premature fruit dropping***— In many fruit trees, the unripe fruits fall off on account of the formation of an abscission layer, thus causing serious losses in yield to the gardeners. This problem has now been successfully overcome in many cases like apples by the application of auxins which prevent the formation of abscission layer and thus check preharvest drop of the fruits. Besides apples, such as control has also been induced in citrus fruits (like oranges and lemons) using 2,4-D and 2,4,5-trichlorophenoxyacetic acid as auxins.

D. ***Improving fruit quality***— The various processes like colouration, softening, sweetening and ripening are all involved in improving the quality of the fruit.

Auxin effects on fruit colouration are most pronounced in apples where use of 2,4, 5-trichlorophenoxyacetic acid has greatly increased red pigments. 2,4-D when applied to bananas hastened the process of ripening as the auxin facilitates the conversion of starch to sugars. By injecting 2,4-D, IBA or maleic hydrazide, the accumulation of sugars was reported in sugarcane.

9. Increase in respiration. James Bonner (1953), for the first time, recognized that auxins stimulate the process of respiration. And as such a direct relation between growth due to auxin treatment and the rate of respiration has been found. *Greater the growth, higher is the rate of respiration.*

Such a knowledge about the behaviour of auxins has been applied in controlling the development of weeds which grow obnoxiously in the crop fields. This has been successfully achieved by spraying 2,4-D which acts as a *weed killer*. In fact, this hormone, which operates only on broad-leaved herbaceous plants, increases the rate of respiration so much so that the plants die of over-oxidation and exhaustion. And fortunately most of the weeds are broad-leaved dicots and the crop plants are usually narrow-leaved monocots which, henceforth, escape destruction on spraying 2,4-D.

Another hormone 2,4-dichloropropionic acid (or dalapon), however, destroys graminaceous weeds.

> The defoliation in Vietnam forests was protested during the war by many botanists, but the problems have proved to be more serious than the destruction of the forests. Synthesis of 2,4,5-T, a component of Agent Orange, also produces **dioxin** (2,3,7,8-tetrachlorodibenzo-*para*-dioxin). *Dioxin is one of the most toxic synthetic chemicals known*. It is also known to be carcinogenic as well as endocrine disruptor in nature and is highly persistent in the environment, besides being extremely resistant to chemical or physical breakdown. Thousands of U.S. pilots and Vietnamese citizens exposed to Agent Orange (which is often contaminated with dioxin) have had higher frequencies of miscarriages, birth defects, leukemia and other types of cancer. Although the **Environmental Protection Agency** (EPA), in 1979, banned the use of 2,4,5-T in the United States, its effects continue to plague many Vietnamese citizens, especially the veterans.

10. Increased resistance to frost damage. In parsnip, the tops resist damage by frost on treatment with 2,4,5-T. Similarly, the application of 2,4,5-T in apricot fruits before the onset of frost resulted in less damage than the untreated fruits.

11. Great weapon of war. Auxins when applied in greater concentrations on enemy crop fields by air may cause devastation of land and thus form the basis of what is called *biological warfare*.

The synthetic auxins (such as 2,4-D, 2,4,5-T and NAA) have been widely used since 1940s. Their use has decreased production costs by reducing the amount of labour and mechanical weeding, needed to grow and effectively harvest a crop. Unfortunately, the effects of synthetic auxins have not all been positive. Most of the negative effects can be traced to a defoliant, once used in Vietnam War, by the code name Agent Orange. **Agent Orange** was a 1 : 1 mixture of 2,4-D and 2,4,5-T that was sprayed throughout the jungles of Vietnam and followed by napalm bombs. This resulted in the destruction of hundreds of square kilometres of Vietnam forests.

GIBBERELLINS

The gibberellins are another class of compounds whose minute quantities profoundly stimulate the growth of many plants.

Discovery

The gibberellins were discovered in an interesting and incidental way. In early part of the twentieth century, Japanese farmers noted that some plants in rice fields were taller, thinner and paler than the normal plants; had longer and narrower leaves markedly overgrowing their unaffected neighbours; and were sometimes devoid of fruits too. They named this disease as "*bakanae*", meaning foolish asseedlings. **Sawada** (1912) suggested that the disease is due to a 'substance' secreted by a parasitic ascomycetous fungus, *Gibberella fujikuroi* (the perfect form, occurring only occasionally; the imperfect form is *Fusarium moniliforme*), in infecting the diseased plants. This suggestion was experimentally supported by Ewiti Kurosawa (1926) who demonstrated that sterile filtrates of the fungus could initiate symptoms of bakanae disease in healthy rice seedlings. Later in 1939, **Yabuta** and **Hayashi** isolated this growth promoting substance in crystalline form and named it as gibberellin A, which has now been shown as a mixture of many growth promoters, collectively known as **gibberellins.**

Since that time, gibberellins and allied substances have been found in higher plants also by **Mitchell** *et al* (1951), **West** and **Phinney** (1957) and **Sumiki** and **Kawarada** (1961).

Definition

A *gibberellin* (abbreviated as GA, for gibberellic acid) may be defined as a compound which is active in gibberellin bioassays and possesses a gibbane ring skeleton (refer page 1206). There are, however, other compounds (like kaurene) which are active in some of the assays but do not possess a gibbane ring. Such compounds have been called *gibberellin-like* rather than gibberellins.

Isolation, Distribution and Biosynthesis

About 29 gibberellins were previously isolated and their chemical structures known. These

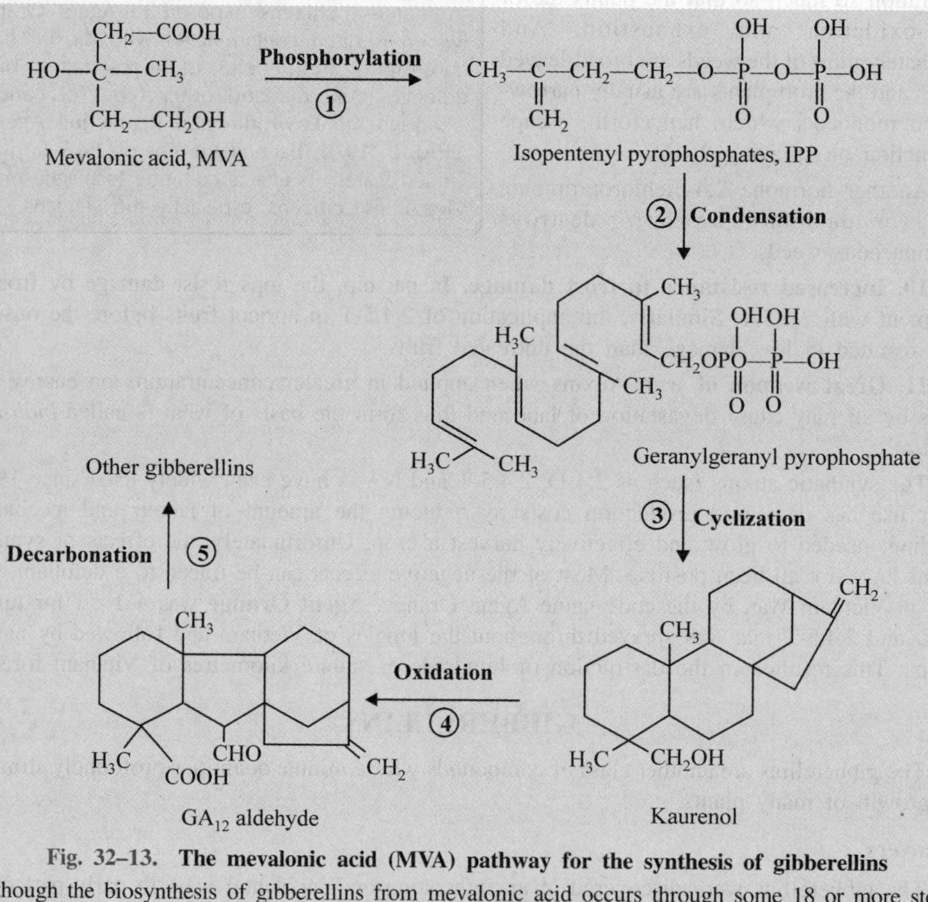

Fig. 32–13. The mevalonic acid (MVA) pathway for the synthesis of gibberellins

Although the biosynthesis of gibberellins from mevalonic acid occurs through some 18 or more steps, 5 key steps culminating in GA production have been outlined here. Mevalonate, in its turn, is produced from acetate.

(*Adapted from Moore R, Clark WD and Stern KR, 1995*)

have been named as gibberellin A_1 (GA_1), gibberellin A_2 (GA_2) and so on up to gibberellin A_{29} (GA_{29}). Of these, Cross *et al* (1961) have isolated 6 gibberellins from the fungus, *Fusarium moniliforme* and designated them as GA_1, GA_2, GA_3, GA_4, GA_7 and GA_9. The same year, **MacMillan** *et al* isolated 3 gibberellins from bean seeds and named them as GA_5, GA_6 and GA_8. The GA_{10} and GA_{13} have been discovered by **Mulholland** (1963). All these compounds are sometimes referred to as constituting the **gibberellin A series**. Till date, more than 80 gibberellins have been isolated from various plant sources.

Although the gibberellins were originally isolated from a fungus, but now they have been shown to be present in almost all the groups of plant kingdom including angiosperms, gymnosperms, ferns, mosses and algae but are unknown in bacteria. For example, GA_1, and GA_5 have been isolated from immature seeds of *Phaseolus vulgaris* by **West** and **Phinney** (1959). Although all the organs of the flowering plants contain gibberellins, but the highest level has been detected in seeds. Young leaves and roots are also rich in them. It may, thus, be generalized that *rapidly growing and developing regions of the plant possess higher concentrations of gibberellins.*

The gibberellins can exist in more than one form within the plant. **Hashimoto** and **Rappaport** (1966) suggested that the esterified forms of gibberellins (*i.e.*, neutral gibberellins) act as reservoir of active gibberellins. The active acidic form may be drawn from the neutral form as and when needed. In addition, bound forms of gibberellins also exist (**Mc Comb**, 1961).

Many angiospermous plants have now been used as bioassay for gibberellins and gibberellin-like substances. A few of them are *Avena sativa* (leaf section), *Pisum sativum* (intact seedling), *Triticum vulgare* (excised coleoptile) and *Rudbeckia bicola* (resetted plants).

Gibberellins are synthesized *via* the mevalonic acid (MVA) pathway. In fact, the biosynthesis of GA_3 from MVA proceeds *via* 18 or more steps or intermediates and about 15 related compounds. The biosynthetic studies were conducted using cell-free systems of *Gibberella fujikuroi* and systems from immature seeds of *Echinocystis macrocorpa* and *Cucurbita maxima*. In essence, the MVA pathway (Fig. 35–13) comprises of 4 major steps for the formation of C_{20} gibberellins, from which C_{19} gibberellins (such as GA_3) are produced by decarbonation, the 5th major step.

Many commercial compounds inhibit the synthesis of gibberellins. These inhibitors (Fig. 35–14), which are called **growth retardants**, include B-Nine, Cycocel (CCC), Phosphon D and Amo-1618. Growth retardants inhibit stem elongation whereby producing stunted plants. Growth retardants are frequently sprayed on growing chrysanthemums to produce flowers with thicker, sturdier stalks.

$HOOC—CH_2—CH_2—CO—NH—N(CH_3)_2$

N-dimethylaminosuccinamic acid
(B-Nine)

$Cl—CH_2—CH_2—N^+(CH_3)_2—CH_3Cl^-$

2-chloroethyltrimethylammonium chloride
or chlorocholine chloride or cycocel
(CCC)

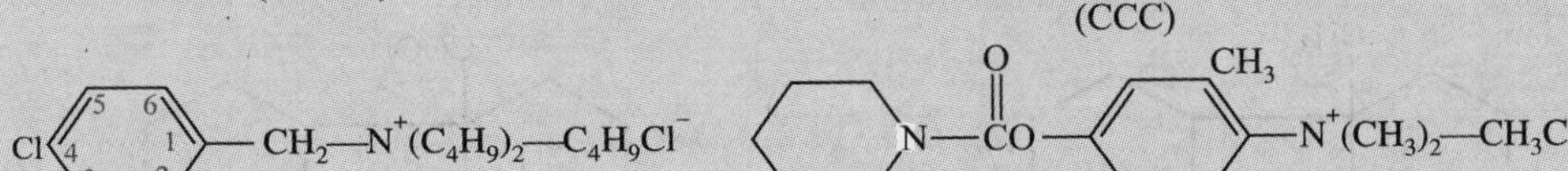

2, 4-dichlorobenzyltributyl-
phosphorium chloride
(Phosphon D)

2-isopropyl-4-dimethylamino-5-methylphenyl-
1-piperidine-carboxylate methyl chloride
(Amo 1618)

Fig. 35–14. Some growth retardants for the synthesis of gibberellins

Unlike the transport of IAA and other auxins, the transport of gibberellins is not polar: it moves in all directions (*i.e.*, *nonpolarly*) in the xylem and phloem.

Chemistry

The chemical structure of the gibberellins was established by **Cross** *et al* in 1961. They showed that the gibberellins (Fig. 35–15) are a group of closely related compounds with one or

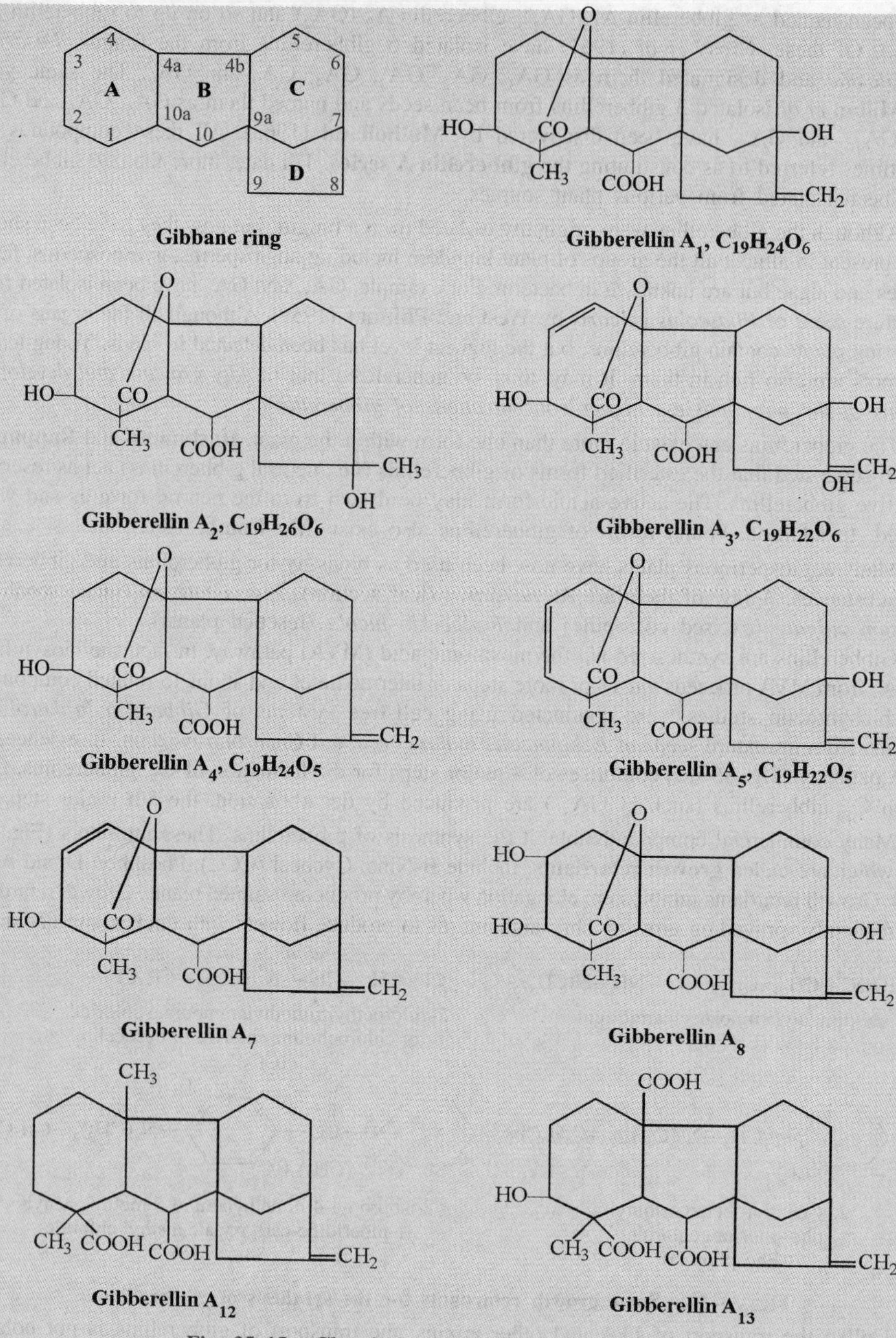

Fig. 35–15. Chemical structure of some gibberellins

Note the minor but significant differences between any two gibberellins. Differences in biologic potency are sometimes associated with presence or absence of double bonds, particular position of the OH groups(s), presence of the lactone ring and the number of carboxyl groups. Also note that while most of the gibberellins have only one carboxyl group, GA$_{12}$ and GA$_{13}$ have two and three carboxyl groups, respectively.

more carboxyl groups that impart acidic properties to the molecule and possess a common feature, the *gibbane ring skeleton*. The gibbane ring consists of a carbon skeleton with 4 interlocking rings, designated as A, B, C and D. They are, hence, described as tetracarbocyclic compounds. Of the 29 gibberellins, previously isolated, 19 are C_{19} compounds and the other 10 have 20 carbon atoms each. Eighteen gibberellins are monocarboxylic, 7 are dicarboxylic and 4 are tricarboxylic acids. The various gibberellins differ from each other in the number and position of the functional groups present in the molecule. In *gibberellin* A_1, the functional groups are a carboxyl, one ethylenic double bond, two alcoholic hydroxyl groups (one secondary and the other tertiary), a saturated lactone and one methyl group. *Gibberellin* A_3 differs from GA_1 in the presence of one more ethylenic double bond in the ring A. It is, thus, more unsaturated than GA_1.

The *gibberellin* A_2 *and gibberellin* A_4 both have structures similar to that of GA_1 except for the difference in position of the tertiary hydroxyl group and the absence of a double bond in GA_2 and in the absence of tertiary hydroxyl group in GA_4. The gibberellin A_5 is, in fact, a dehydrogibberellin A_1 where the secondary hydroxyl group is eliminated from ring A, making the compound more unsaturated.

Gibberellin A_3 has been usually shown to be biologically most active followed by GA_1, GA_4 and GA_2 in descending order of their activity:

$$GA_3 > GA_1 > GA_4 > GA_2$$

Certain gibberellins have been found in plant tissues to occur in bound or conjugated form with other compounds; *free gibberellins* can be released from '*bound ones*' by enzymatic treatment with *emulsin*. In pea seedlings during the first few days of growth, the necessary gibberellin needed for this phase of growth, may partly be released from bound ones. Among the bound forms of gibberellins, those deserving mention are:acetyl-GA_3, D-glycopyranosyl-GA_3, D-glucopyranosyl-GA_8, D-glucopyranosyl-GA_{27} and D-glucopyranosyl-GA_{28}.

Physiological Roles

Gibberellins may be regarded as natural phytohormones on account of their wide range of distribution in plants and specificity of response of individual flowering plants to the exogenously applied gibberellins. The gibberellins, however, play important roles in the following processes:

1. Genetic dwarfism. In certain plants, dwarfism is caused by the mutation of a single gene. Such individuals are called '*single gene dwarfs*'. In these plants, dwarfism is due to shortening of internodes rather than a decrease in the number of internodes. Application of gibberellins on such dwarfs causes them to elongate so much as to become indistinguishable from the tall normal plants. Elongation of the stem, in fact, takes place due to an elongation in the internodes rather than an increase in the number of internodes. Thus, genetic dwarfism has been successfully overcome by gibberellin A_3 treatment in many single gene dwarf mutants like *Pisum sativum, Vicia faba and Phaseolus multiflorus* (**Brian** and **Hemming**, 1955). The gibberellins, thus, make most plants grow taller by causing the internodes to elongate considerably (Fig. 35–16).

Fig. 35-16. Wheat stems, cut lengthwise, to show the nodes
The 4 plants on the *left*, treated with gibberellic acid, have internode lengths much greater than the internodes of the two untreated on the *right*.

Two views have been put forward regarding the mechanism of control of dwarfism by gibberellins:

(*a*) It is due to the lack of endogenous gibberellins in dwarf plants or if at all present, they are in traces as to have no effect.

(*b*) A natural inhibitor is present in those plants which retard growth. And the gibberellin, when applied, nullifies the effect of this inhibitor.

2. Bolting and flowering. '*Rosette plants*' are characterized by their profuse leaf development and retarded internodal growth. But prior to the reproductive phase, there occurs striking elongation in the internode so that the plant attains 5 to 6 times the original height. Treatment of these `rosette' plants with gibberellins, under conditions that would normally maintain the rosette form, induces them to bolting (or shoot elongation) and flowering (**Lang**, 1957). By regulating the amount of gibberellin applied, it is also possible to separate shoot elongation from flowering; *with low dosages of gibberellins, the plant will bolt but not flower* (**Phinney** and **West**, 1961).

It is, therefore, not amazing to find a direct correlation between the amount of gibberellin present and the habit of the plant, whether rosetted or bolted. Native gibberellin-like substances are found in higher concentrations in the bolted forms than in the nonbolted ones. This has been experimentally demonstrated in quite a few plants including the biennial *Hyoscyamus niger* by **Lang** (1957) and the cold-requiring plant *Chrysanthemum morifolium* and a long-day plant *Rudbeckia speciosa* by **Harada** and **Nitsch** (1959).

As far as the use of gibberellins in agriculture is concerned, it may be possible to grow cold-requiring plants in warm countries and long-day plants in short-day conditions at lower altitudes.

Gibberellic acid (GA) hastened flowering and improved the flower yield in *Coriandrum sativum* (coriander). This was accomplished by a decline in starch content and an increase in reducing sugars, as well as enhanced amylase activity. It was inferred that GA_3 hastened flowering probably through its influence on carbohydrate metabolism (**Amrutavalli**, 1979).

3. Light-induced inhibition of stem growth. Light-grown plants reveal suppressed stem growth than the dark-grown (or etiolated) plants, indicating that light has an inhibitory effect on stem elongation. But this inhibitory effect of light on stem elongation can be reversed at least in some plants (like *Pisum sativum*) by the application of gibberellins on these plants. This clearly suggests that endogenous gibberellin is the limiting factor in stem elongation.

Lockhart (1961) has given a possible explanation for it. According to him, exposure to light lowers the level of available gibberellins present in the plant. The lowered available gibberellin contents then, in turn, decrease the plasticity of cell walls, thus inhibiting stem growth. The theory has, however, not won the universal support on account of the following *drawbacks*:

(*a*) Stem elongation is also induced in mustard seedlings, grown in dark, upon application of gibberellin.

(*b*) In some plants, gibberellin-stimulated stem growth has been found to be partially due to enhanced cell division and has nothing to do with cell wall plasticity.

(*c*) Germination of the seeds of *Lactuca sativa* is not only promoted by gibberellins but by red light too.

4. Parthenocarpy. Like auxins, the gibberellins are also capable of inducing parthenocarpic fruit-set. *Gibberellins are, in fact, more efficient than the auxins in inducing parthenocarpy.* For example, **Wittwer** and **Bukovac** (1957) have found gibberellin to be about 500 times more effective than IAA in inducing parthenocarpy in tomatoes. Moreover, there are cases where auxins have failed to induce parthenocarpy while gibberellins are effective, as shown experimentally for apples (**Davison**, 1960) and stone fruits (**Crane** *et al*, 1960).

Gibberellin-induced parthenocarpy has been reported in many plants such as *Cucumis sativus* (cucumber), *Solanum melongena* (brinjal) and *Zephyranthes* sp. Whether the production of parthenocarpic fruits is a direct action of gibberellins or an interaction with the natural auxins of the plant has not been conclusively proved.

5. Breaking dormancy of seeds. The light-sensitive seeds (lettuce, tobacco) show poor germination in dark and on exposure to light their germination starts vigorously. But when these seeds are treated with GA_3, the light requirement is alleviated and they germinate in dark.

6. Breaking dormancy of buds. In temperate areas, the buds produced in winter remain dormant until the next spring due to very low temperature. The dormancy in such cases is overcome by gibberellin treatment. Thus, GA_3 treatment to birch buds has replaced the light requirement for breaking dormancy (**Eagles** and **Wareing**, 1964). Gibberellins are also capable of breaking dormancy in potato tubers.

7. Role in abscission. GA_3 treatments have shown accelerated rate of abscission in explants of bean (**Chatterjee** and **Leopold**, 1964) and of *Coleus* (**Gupta** and **Kaushik**, 1969).

8. Stimulation of enzyme activity in cereal endosperm. Yomo (1960) and Paleg (1960) working independently showed that the gibberellins applied exogenously could stimulate *amylase* activity in isolated barley endosperm. It was then shown that it is the aleurone layer of the endosperm which is sensitive to the gibberellin. Subsequent researches by **Paleg** (1964) and **Varner** (1964) revealed that GA treatment of isolated aleurone can cause release of the enzymes, *amylase* and *proteinase*. Finally, **Jacobson** and **Varner** (1967) showed that the two enzymes (amylase and proteinase) induced by GA treatment arise through *de novo* synthesis. These enzymes participate in the breakdown of the stored starch to simple sugars. These sugars are then translocated to the growing embryo where they provide energy for growth.

> ***de novo*** is a Latin phrase, meaning anew.

9. Sex expression. Gibberellins are also capable of altering the sex of the flowers. **Galun** (1959) could induce maleness by foliar application of GA_3 to the female flowers of *Cucumis*. Also, the antheridia have been induced to develop in many fern gametophytes by GA_3 treatment.

10. Juvenility. Many plants in their life cycle exhibit two different stages of growth: a juvenile stage and an adult stage. For example, in a species of eucalyptus (*Eucalyptus globulus*), the juvenile leaves are oppositely-placed and are shorter, softer and almost obcordate with emarginate apex; while the leaves of the adult stages are spirally-arranged and are larger, harder and falcate with acute apex (Fig. 35–17). Gibberellins may help determine whether a particular part of a plant

Fig. 35–17. Photographs of two growth stages of *Eucalyptus globulus* leaves

A. Juvenile stage B. Adult stage

Juvenile and mature leaves are shaped differently. Juvenile leaves are softer and opposite each other. Adult leaves are hard and spirally arranged. The difference in the leaves of 2 stages is also exhibited in their anatomy. The juvenile leaves have palisade cells only on the upper surfaces, whereas the adult leaves contain palisade parenchyma on both sides.

is juvenile or adult. For instance, the buds of adult branches usually develop only into adult branches, but treating them with gibberellin causes them to grow into juvenile branches.

Relationship between Auxins and Gibberellins

Auxins and gibberellins are similar to each other in that both promote cell elongation, flowering and parthenocarpy. These, however, differ form each other in many of the physiological activities. These differences are listed in Table 35–2.

Table 35–2. Differences between auxins and gibberellins

S. N	*Physiological activity*	*Auxins*	*Gibberellins*
1.	Transport polar	Yes	No
2.	Promote root initiation	"	"
3.	Inhibit root elongation	"	"
4.	Delay leaf abscission	"	"
5.	Inhibit lateral buds	"	"
6.	Induce callus formation	"	"
7.	Promote epinastic responses	"	"
8.	Control dwarfism	No	Yes
9.	Promote seed germination and the breaking of dormancy	"	"
10.	Promote bolting and flowering in non-vernalized biennials and in long-day plants	"	"

(*Adapted from Galston and Purves, 1960*)

Accumulated evidences indicate that auxins and gibberellins act both independently and together, depending upon the type of plant and the conditions under which the plant grows. The fact whether the auxins and the gibberellins interact or not is not conclusively proved.

CYTOKININS (= KININS)

Discovery and Nomenclature

Auxins and gibberellins, besides inducing cell elongation, also do promote cell division under certain conditions. But this behaviour of them is an exception rather than a rule. However, there exist in plants many substances inducing cell division. For example, **Van Overbeek** *et al* (1941) found coconut milk as an active stimulant of cell division. Later, in 1955 **Carlos Miller** *et al* isolated a *"cell-division-stimulating factor"* from yeast DNA. It was named as kinetin because of its amazing power to stimulate cell division (cytokinesis) in the presence of an auxin. In subsequent years, many other compounds promoting cell division have been synthesized. **Miller** and his associates (1956) have grouped all such compounds including kinetin under a generic name kinin. **D.S. Leetham** (1963) of New Zealand proposed the term cytokinins for such substances. This term is the most acceptable one.

Fairley and **Kilgour** (1966), however, prefer to use the term **'phytokinins'** for such substances in order to distinguish them from the peptide hormones of animal gastrointestinal tract.

Definition

Skoog, Strong and **Miller** (1965) have defined cytokinins as chemicals which, regardless of their activities, promote cytokinesis (cell division) in cells of various plant organs.

FOLKE KARL SKOOG
(LT, 1908–2001)

Skoog, a leading plant physiologist, was born on July 15, 1908 in Holland, Sweden. He decided to take up residence in the United States during a trip to California in 1925, and became a U.S. Citizen 10 years later. He received Bachelor of Science degree in chemistry from the California Institute of Technology (CalTech), in 1932. He competed, and finished sixth in heat 2, in the 1500– meter race during the 1932 Summer Olympics. In 1936, he received his PhD in Biology from CalTech for his work done with auxin, a plant hormone. While at CalTech, he worked closely with prominent plant physiologists **Kenneth V. Thimann** and **Frits W. Went** (known for their work and discovery of auxins). Between 1937 and 1941, Skoog was on staff at Harvard University, and from 1941 to 1944 at John Hopkins University. He served as a chemist and technical representative of the U.S. Army in Europe from 1944 to 1946. In 1947, he joined the University of Wisconsin, Madison, U.S.A. as Professor of Botany and remained there for 32 years.

While at Wisconsin, he discovered a major new class of plant hormones, the **cytokinins**, which stimulate the division of plant cells, and regulate plant growth and development. During the cultivation of tobacco pith cells, Skoog and his associates observed that the incorporation of adenine (a purine base) in the culture medium enhances callus growth. Likewise, DNA extracts and Herring sperm whale also favoured callus initiation and callus growth of tobacco pith cells. This led to the isolation of a compound from autoclaved Herring sperm whale DNA, called **kinetin** (6-furfurylaminopurine). Though kinetin enhances cell division in pith cells, it does not present in this form in plant cells. Hence, they concluded that this product was formed during autoclaving of whale DNA. It is produced due to dehydration and migration of a deoxyribose moiety from the 9-position of adenine to the N^6-position (**Miller** *et al*, 1956). Skoog also showed that a number of cytokinins occur naturally, and at least one of these compounds occurs in every organism tested, from bacteria to humans. With colleague **Nelson J. Leonard** at the University of Illinois, Skoog synthesized and tested hundreds of compounds for cytokinin activity, and established the principles that govern relationships between plants structure and activity.

Skoog was also a pioneer in investigating how to control the formation of roots, stems and leaves from undifferentiated cells in plant tissue cultures. Tissues could be made to develop as undifferentiated mass of cells, or to become roots, stems or a combination of roots and stems, resulting in complete plants. **Skoog** and **Miller** (1957) demonstrated, for the first time, that a ratio of auxin and cytokinin can control the root, shoot and callus formation. Skoog's theory that plant development is controlled by hormone levels and other factors, led to the realization that whole plants can be generated from cultured cells. This laid the groundwork for the production of transgenic plants and other advances in biotechnology.

In 1962, Skoog and a graduate student **Toshio Murashige** published what is probably the best-known paper in Plant Tissue Culture; in a futile attempt to discover a yet-unknown plant growth regulator in tobacco juice for his doctoral thesis, Murashige and Skoog instead developed a greatly improved salt base for the sterile culture of tobacco. Now referred to as **Murashige and Skoog medium**, the final paper (*Murashige T and Skoog FK:* A revised medium for rapid growth and bioassays with tobacco tissue cultures. *Physiol. Plant,* ***18*** *: 100–127, 1962.)* is considered one of the most often-cited papers in biology. Now 50 years after the work, *M & S salt base* remains an essential component in plant tissue culture.

Skoog played an active role in a number of biological societies. In 1970, he was elected a foreign member of the Royal Swedish Academy of Sciences. He served as President of the American Society of Plant Physiologists, the Society of Developmental Biology, the American Society of General Physiologists, and the International Plant Growth Substances Association. He was a member of the American Academy of Arts and Sciences, and the National Academy of Sciences. He was awarded the **National Medal of Science** in 1991 during a ceremony at the White House. He died on February 15, 2001 after a protracted illness, at age 92.

Fox (1969) has defined cytokinins as chemicals composed of one hydrophilic adenine group of high specificity and one lipophilic group without specificity.

Isolation, Distribution and Biosynthesis

Although kinetin does not occur in nature but other kinins are found occurring widely, if not universally, in plants. The naturally-occurring kinins do not occur free in nature but are normally bound to a pentose sugar, ribose and sometimes to an inorganic phosphate, the ribonucleotide.

Fruits and endosperm are the richest sources of kinins. Coconut milk and corn endosperm possess the active substance. Substances with cytokinin activity have also been reported in tomato juice, in floral extracts of apples and pears and also in cambial tissues of certain plants. A *kinetin-like substance* is also present in peach embryo (**Powell** and **Pratt**, 1964) and sunflower root exudate (**Kende**, 1964).

Diphenylurea and many of its derivatives have cytokinin activity. Several synthetic cytokinins are available. These include benzimidazole and 6-benzyladenine. Adenine also has some cytokinin activity.

In angiosperms, the cytokinins are synthesized mostly in roots and probably originate at the root tips. Whether the shoots also synthesize cytokinins or else receive their cytokinin requirement from the roots is not certain. Contrary to what was first believed, the cytokinins are not breakdown

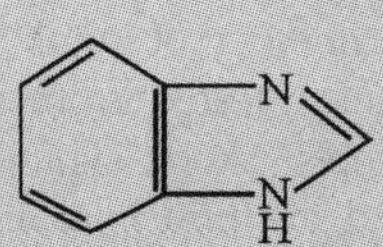

Benzimidazole

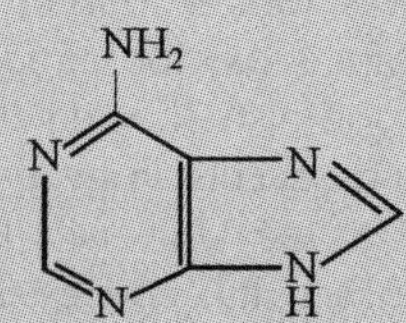

Adenine

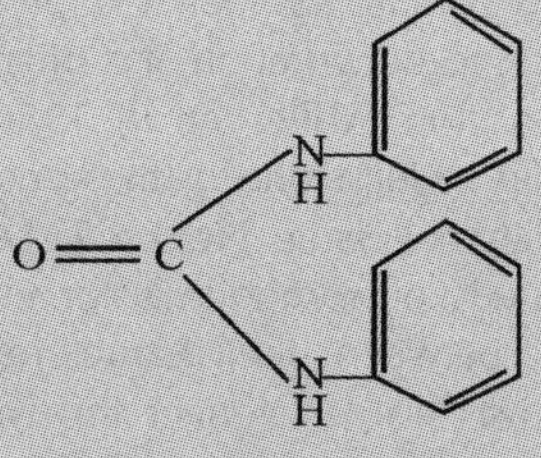

Diphenylurea

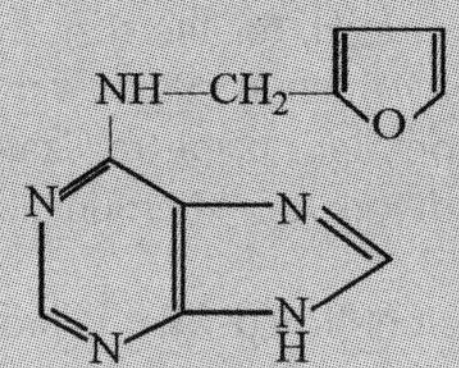

6-furfurylaminopurine
(Kinetin or K)

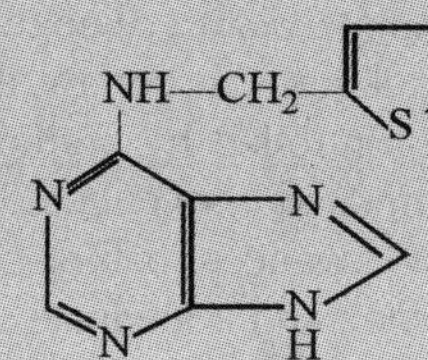

6-(2-thenylamino) purine

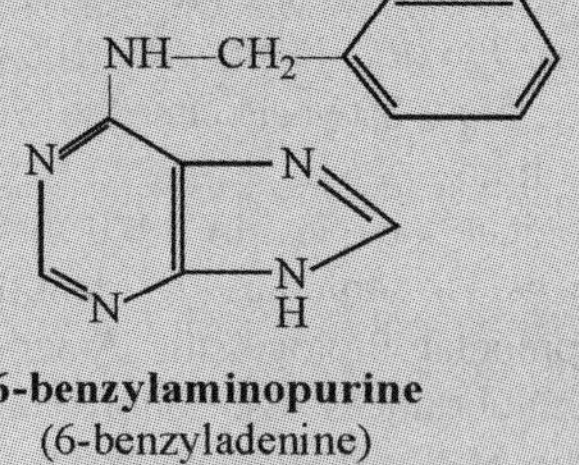

6-benzylaminopurine
(6-benzyladenine)

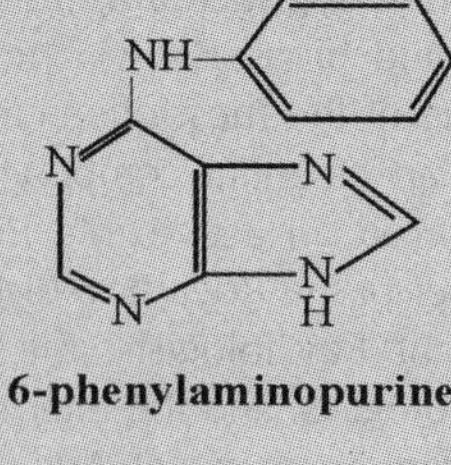

6-phenylaminopurine

Fig. 35–18. Structural formulae of some kinins

Note that all cytokinins are structurally related to the purine derivative, adenine, *i.e.*, they have a side chain rich in carbon and hydrogen and attached to nitrogen protruding from the top of the adenine ring.

products of DNA. Rather, they are made *via* the mevalonate pathway, the same pathway used to make gibberellins. Like gibberellins, the cytokinins move nonpolarly in xylem, phloem, and parenchyma cells.

Chemistry

Chemically, *kinetin* ($C_{10}H_9ON_5$) is 6-furfurylaminopurine. It is formed from deoxyadenosine which is a degradation product of DNA (**Hall** and **de Ropp**, 1955). The structural formulae of kinetin and its 3 structural analogues are given in Fig. 35–18. All these substances promote cell division.

Apart from the above-mentioned kinins, Letham (1963) successfully isolated a cytokinin in pure crystalline form from immature maize grains. It was named as zeatin (Fig. 35–19) and identified as 6-(4-hydroxy 3-methylbut-*trans*-2-enyl) aminopurine. *Zeatin is more powerful than any other known cytokinin probably because of the presence of a highly reactive allylic OH group in its side chain.* Zeatin riboside (**Letham**, 1966) and zeatin ribotide (**Letham**, 1966; **Miller**, 1967) also occur naturally in plants. A *cis*-ribosyl zeatin and a *ms*-ribosyl zeatin have also been extracted from plant tissue.

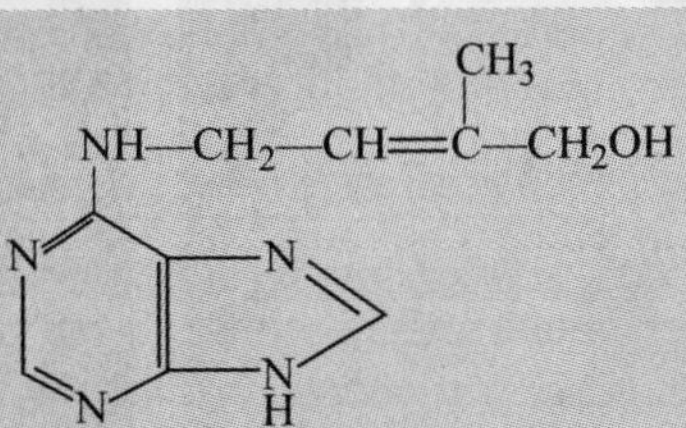

Fig. 35–19. Zeatin
[6-(4-hydroxy 3-methylbut-*trans*-2-enyl) aminopurine]

Fleissner and **Borek** (1962) have described compounds such as N^6-methylaminopurine and N^6, N^6-dimethylaminopurine. These are widespread in plants and have cell division stimulating property. Later, a cytokinin called N^6-purine was isolated from serine-transfer RNA of yeast cells by **Hall** and others in 1966.

Physiological Roles

Certain physiological processes which are influenced by the cytokinins *esp*., kinetin are given below:

1. Cell division. *Kinins are notable for their stimulatory effect on cell division.* Using tobacco pith cultures (Fig. 35–20), **Skoog** and **Miller** (1957) found that, in addition to IAA, kinetin is also needed for growth. The growth response is much more pronounced when both IAA and kinetin are used together in right ratio of concentrations. When either of them is used alone, a little response is produced which is due to the presence of small amounts of endogenous kinetin-like substances and IAA, already present in the tissues.

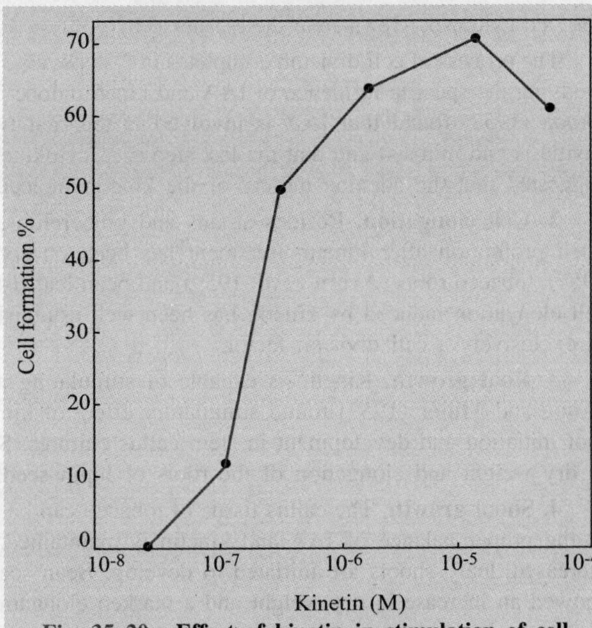

Fig. 35–20. Effect of kinetin in stimulation of cell division in tobacco pith cultures
[2mg/litre of IAA was also present in the medium]
(*After Skoog and Miller, 1957*

If a mixture of cytokinin and auxin is added to unspecialized cells (Fig. 35–21), they will begin to differentiate. A high cytokinin to auxin ratio will lead to the formation of shoots, buds and leaves while a low cytokinin to auxin ratio will lead to root formation. The one treatment followed by the other provides a means of forming small plantlets. Such *in vitro* culture methods have become widely adopted for the rapid propagation of new plant varieties. The technique allows growers to produce very large numbers of plants, quite rapidly and in a small space. The alternative conventional propagation methods can take several years and cover a large area of land.

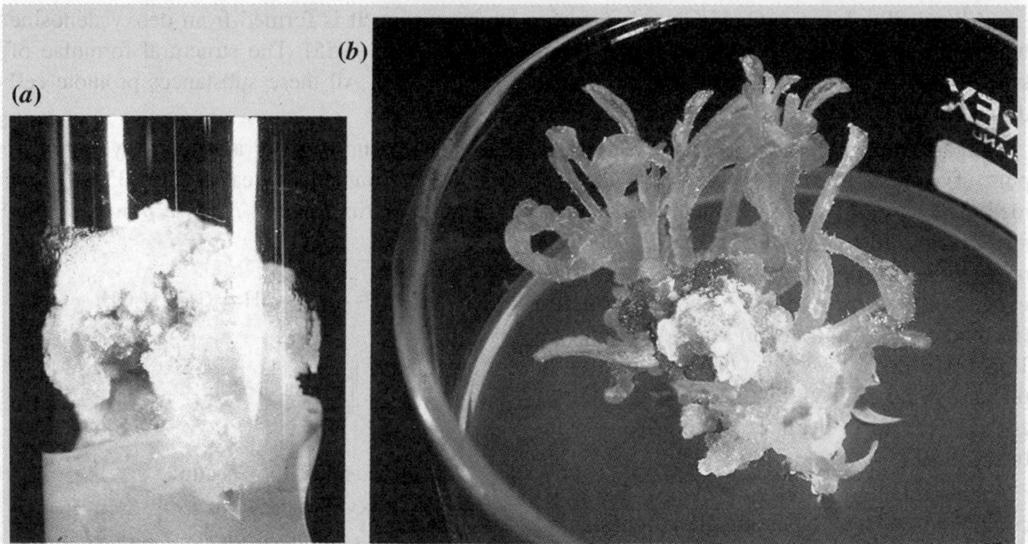

Fig. 35–21. Callus formation and production of plants from callus cells

(*a*) If a small piece of pith from a shoot is placed on agar in aseptic conditions, it will grow into a mass of unspecialized cells called callus.

(*b*) Plants can be grown from these callus cells.

The process of cell division completes in 3 steps, *viz.*, DNA synthesis, mitosis and cytokinesis. Studying the specific influence of IAA and kinetin alone on any of these 3 steps, **Patau**, **Das** and **Skoog** (1957) found that IAA is involved in the first two steps of cell division (i.e., in DNA snythesis and mitosis) and that the last step (*i.e.*, cytokinesis) is controlled by kinetin. It has been suggested that the adenine moiety of the kinetin molecule is essential for cell division.

2. Cell elongation. Besides auxins and gibberellins, kinetin also promotes cell elongation. Such promotion after kinetin treatment has been observed in tobacco pith cultures (**Glasziou**, 1957), tobacco roots (**Arora** *et al*, 1959) and bean leaf tissues (**Powell** and **Griffith**, 1960). Since cell elongation induced by kinetin has been well established, the kinetin should not be regarded as exclusively a cell division factor.

3. Root growth. Kinetin is capable of stimulating as well as inhibiting root development. **Skoog** and **Miller** (1957) found stimulatory effect of kinetin, when applied along with IAA, on root initiation and development in stem callus cultures. Similarly, kinetins also induced increase in dry weight and elongation of the roots of lupin seedlings (**Fries**, 1960).

4. Shoot growth. The callus tissue of tobacco can be kept in an undifferentiated state so long as the proper balance of IAA and kinetin is maintained. If, however, the amount of kinetin is increased, leafy shoots are initiated to develop. Bean seedlings, soaked in kinetin solution, also showed an increase in dry weight and a marked elongation of stem and petioles (**Miller**, 1956).

5. Organogenesis. Cytokinins can cause organogenesis (*i.e.*, the formation of organs) in a variety of tissue cultures. For instance, **Skoog** and **Miller** (1957) observed that tobacco pith callus can be made to develop either buds or roots by changing the relative concentrations of kinetins and auxins. High kinetin and low auxin contents result in the production of buds. In reverse condition (high auxin and low kinetin), however, the roots appear on the pith.

The kinins also stimulate the production of buds in leaf segments of various plants such as *Saintpaulia ionantha, Bryophyllum* sp *and Begonia* sp.

In addition to the root and shoot differentiation, the cytokinins also bring about other morphogenetic responses. These are:

(a) maturation of proplastids into plastids

(*b*) differentiation of tracheids

(*c*) induction of parthenocarpy

(*d*) induction of flowering

6. Counteraction of apical dominance. As discussed earlier (refer page 1200), the auxins emanating from the apical bud inhibit the growth of lateral buds (apical dominance). **Wickson** and **Thimann** (1958) studied the antagonistic effect of auxin and kinetin in apical dominance using pea stem sections in culture solutions. They found, as might be normally expected, that the growth of lateral buds is inhibited when the culture medium contained IAA and is uninhibited when the culture medium does not contain IAA. They further noted that addition of kinetin, along with IAA, stimulates the growth of lateral buds.

The above workers also conducted experiments with entire shoots, *i.e.*, with the apical bud intact. As long as the apical bud is present, the lateral buds do not develop but removal of the apical bud leads to the stimulation of growth of the lateral buds. If, however, the intact shoot is soaked in kinetin solution, the inhibition of lateral buds is checked to a great extent or, in other words, the lateral buds tend to develop, although less vigorously, as if the apex of the shoot has been cut off. The above findings point out towards the possibility of controlling apical dominance by maintaining a proper balance of concentrations between IAA and the endogenous kinetin-like substances.

Studies conducted in subsequent years by **Sachs** and **Thimann** (1964, 67) and **Panigrahi** and **Audus** (1966) also indicated that the cytokinins are strong promoters of lateral bud growth.

7. Breaking dormancy of seeds. Cytokinins are also effective in breaking seed dormancy in lettuce, tobacco, white clover and carpet grass. **Thimann** (1963) suggested that the site of cytokinin action in such cases is the cotyledon. Furthermore, the inhibitory effect of infrared light on germination of lettuce seeds is also alleviated by kinetin treatment.

The seeds of parasites such as Striga asiatica require the presence of host plant for germination. But when treated with kinetin, the seeds germinate even in the absence of their host.

8. Delay of senescence (= Richmond–Lang effect). The term *senescence* refers to the ageing of the leaves which is associated with the loss of chlorophyll and the breakdown of proteins. **Richmond** and **Lang** (1957) showed that the senescence in the detached leaves of Xanthium could be postponed for many days by kinetin treatment. This effect of kinetin in retarding senescence (or ageing) is known as *Richmond-Lang effect.* According to **Mothes** and **Engelbracht** (1961), the cytokinins have the ability to attract certain substances including auxins and to prevent the movement of leaf components out of the treated area. However, the mobilizing effect of cytokinin may actually induce senescence in others parts of the plant. **Osborne** (1962) suggested that the high protein content in kinetin-treated areas is probably due to enhanced protein synthesis than their breakdown. The protein synthesis, in its turn, is dependent on RNA synthesis, a process governed by kinetins. It may, however, be emphasized that the cytokinin-induced delay in leaf senescence occurs only in detached leaves; cytokinins have little or no effect on senescence in attached organs. Leaf senescence is also delayed by the formation of adventitious roots. As the roots are rich in cytokinins, the transport of these cytokinins from roots to leaves could account for the delayed senescence.

A correlation between the age of the leaf and the kinetins has been established. Mature leaves of tobacco respond more vigorously to kinetin treatment in delaying senescence than the young leaves.

Cytokinins are sometimes used commercially to maintain the greenness of excised plant parts, such as cut flowers. However, their use on edible crops such as broccoli is banned in some countries. This is possible because any compound like cytokinin that resembles a nucleic acid component is automatically a suspected carcinogen.

9. Role in abscission. Cytokinins can accelerate as well as retard the process of abscission in leaf petioles depending on the site of their application (**Osborne** and **Moss**, 1963). In explant petioles of *Coleus blumei*, **Gupta** and **Kaushik** (1969) reported accelerated abscission on kinetin application."

10. Effects on cotyledons. Cytokinins promote cellular division and expansion in cotyledons. Cellular expansion results from cytokinin-induced increases in wall plasticity that do not involve wall acidification. Cytokinins also increase the amount of sugars (especially glucose and fructose) in cells, which may account for the osmotic influx of water and the resulting expansion of cytokinin-treated cells in cotyledons.

OTHER NATURAL GROWTH HORMONES IN PLANTS

In addition to the 3 well-established categories of plant hormones described above, some other compounds with hormonal actions have been identified in plants which fall under 4 categories: ethylene, traumatic acid, calines and vitamins. Their account follows.

1. ETHYLENE (= ETHENE)

Discovery

During the 1800s, the city streets of Germany were illuminated by lamps that burned "illuminating gas". Soon after these lamps were installed, city residents made a curious observation: plants growing near the lamps had short thick stems and leaves falling from most of them. The mystery was solved in 1901 by a Soviet plant physiologist, **Dimitry Neljubow** who identified ethylene as the combustion product of "illuminating gas" that was responsible for defoliation and stunted growth of plants growing near the lamps. He also showed that only micro quantities are needed to bring about these effects, *i.e.*, only 0.06 ppm of ethylene.

> **ppm** is part(s) per million;
> 1 ppm = 1 millilitre in 1,000

Later in 1910, an annual report submitted to the Japanese Department of Agriculture recommended that *oranges be not stored with bananas*, because oranges released something that caused premature ripening of the bananas; this "something" was in 1934 identified by **R. Gane** as ethylene which is made by plants. Subsequent researches showed that ethylene has all the characteristics which warrant its inclusion under plant hormones, *i.e.*, it is made in one part of a plant and transported to another, where it induces a physiological response. Thus was discovered the *gaseous plant hormome: ethylene*. Later, the presence of ethylene was shown in certain fungi (*Penicillium digitatum, Alternaria citri*) and in the leaves, flowers and fruits of many higher plants; its recognition as a natural plant hormone was confirmed by **Pratt** and **Goeschl** only in 1969.

Distribution and Biosynthesis

All parts of angiospermous plants produced ethylene but especially large amounts are released into the air by roots, the shoot apical meristem, nodes, senescing flowers and ripening fruits (for example, the dark flecks on a ripening banana peel are concentrated pockets of ethylene). Because most ethylene-induced effects result from ethylene in the air, *the effects of ethylene can be contagious:* ethylene made by one "bad" (*i.e.*, overripe) apple can "spoil" (*i.e.*, induce rapid ripening of) an entire bushel of apples. Ethylene also occurs in minute quantities in city gas and in tail gases of blast furnaces. It is a volatile gas of peculiar odour and is sparingly soluble in water but a little more in ethanol and ether. It is inflammable and hence the ignition of a mixture of ethylene with air leads to explosion.

$$H_2C{=}CH_2$$

Ethylene, ETH

Ethylene is made from methionine, a S-containing amino acid (Fig. 35–22). Its synthesis, which requires O_2, is inhibited by CO_2 When plants are placed in pure CO_2 (or O_2-free air), ethylene synthesis decreases dramatically. Biosynthesis begins with methionine which reacts with ATP to form a compound known as *S*-adenosylmethionine (SAM). SAM is then split into two different molecules, one of which contains a ring consisting of 3 carbon atoms (1-aminocyclopropane-1-carboxylic acid, ACC). This compound is then converted into ethylene, carbon dioxide and ammonia by enzymes on the tonoplast. The reaction forming ACC is the step of the pathway that is affected by treatments, such as high auxin concentrations, air pollution, wounding and damage. These stimulate ethylene production by plant tissues.

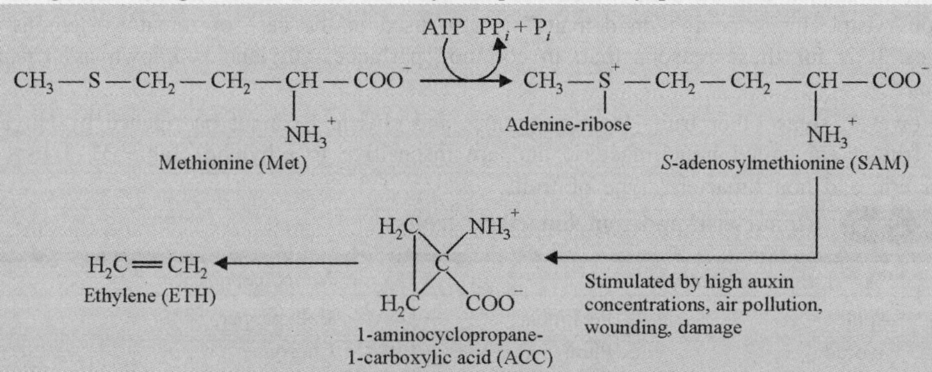

Fig. 35–22. Biosynthesis of ethylene from methionine

Note that a cyclic compound, 1-aminocyclopropane-1-carboxylic acid (ACC) is the immediate precursor of ethylene.

Physiological Roles

1. Stimulates fruit ripening. The ancient Chinese knew that fruits would ripen faster if placed in a room containing burning incense. The factor responsible for this hastened ripening was not heat, but ethylene released as the incense burned. The stimulation of fruit ripening by ethylene is a consequence of many ongoing processes such as:

(*a*) the breakdown of chlorophyll and synthesis of other pigments; for example, apples changing from green to red during ripening,

(*b*) fruit softening due to breakdown of cell walls by cellulase and pectinase, and

(*c*) conversion of starches and acids to sugars.

And ethylene stimulates each of these processes, leading ultimately to fruit ripening.

Fig. 32–23. The oranges infected with the fungus *Penicillium digitatum*

The fungus itself produces large amounts of ethylene, which accelerates the post-harvest maturation of oranges, making them more vulnerable to infection.

We often say that 'one rotten apple spoils the rest of the barrel' because pieces of fruit near to a rotten one start to go bad quickly. This is because a damaged or fungus-infected fruit starts to produce ethene. And in a closed barrel, the concentration of ethene is high enough to quickly trigger the ripening process in neighbouring fruits making them more vulnerable to infection (Fig. 35–23).

Some fruits (such as tomatoes and apples) show a conspicuous increase in respiration just before fruit ripening. This increase in respiration is called a **climacteric**, and fruits that display it are referred to as **climacteric fruits.** The climacteric begins just after a huge increase (up to a 100-fold) in ethylene production. Thus, *the climacteric and fruit ripening are both triggered by ethylene.*

Fruit growers often take advantage of ethylene's capability of stimulating fruit ripening for making fruits available for sale out-of-season. For instance, many apples are plucked in September and October when they are green and immature. These are then stored in rooms containing air that has small amounts (1–3%) of O_2 large amounts (5–10%) of CO_2 and no ethylene. As these conditions inhibit protein synthesis, the fruits can be stored without the fear of their being ripened. When these unripe fruits are needed for sale, producers expose them to normal air containing 1 ppm of ethylene, which is enough to induce climacteric and ripening of fruits. Thus the "fresh" apples one buys in March are the ones harvested in September/October of the previous year. This "ripening on demand" is also used in the case of tomatoes, lemons and oranges. It is for these reasons that, in common parlance, ethylene is known as *'ripening hormone'*.

However, some other fruits (such as grapes and cherries) cannot be ripened by ethylene. Such fruits are called nonclimacteric and are insensitive to ethylene. Table 31–3 lists the climacteric and nonclimacteric type of fruits.

Table 35–3. Climacteric and nonclimacteric fruits

*Climacteric**		*Nonclimacteric**
Apple	Persimon	Bell pepper
Avocado	Plum	Cherry
Banana	Tomato	Citrus
Contaloupe		Grape
Cherimoya		Pineapple
Fig		Snap beam
Mango		Strawberry
Olive		Watermelon
Peach		
Pear		

* Note that the term **'climacteric'** can be used either as a noun, as in "most fruits exhibit a climacteric during ripening" or as an adjective, as in "a climacteric rise in temperature". The term **'nonclimacteric'** however, is used only as an adjective.

2. Promotes flowering. Although ethylene inhibits flowering in most species, but induces it in a few plants including mangoes, pineapples and some ornamentals. The Filipino mango growers and the Peurto Rican pineapple growers, who knew this effect long ago, set bonfires near their crops. The fires produced ethylene which initiated flowering of their plants. Nowadays, the pineapple growers in Hawaii produce pineapple fruits round the year by spraying plants with ethepon. Ethepon splits under neutral and alkaline conditions to release ethylene.

3. Hastens leaf abscission. Abscission zone in leaves causes the increased production of ethylene which triggers the breakdown of middle lamella, thus leading to the initiation of abscission (Fig. 35–24). This effect is also utilized by horticulturists to minimize the harvesting period of such fruits as cherries, grapes and blueberries. These fruits are sprayed with ethepon to coordinate abscission, thereby allowing growers to harvest their crops in shorter periods of time.

4. Induces leaf epinasty. When a plant's roots are kept submerged in water for long periods, water fills the intercellular spaces. Since these spaces are the primary routes of gas exchange

Fig. 35–24. Photograph of a plant showing leaf fall

Leaf abscission is strongly influenced by hormones, *esp*., ethylene.

with the atmosphere, the submerged roots become waterlogged and anaerobic. The symptoms of waterlogging (*e.g*., leaf chlorosis, shorter and thicker shoots and wilting) can also be induced by placing roots in O_2-free air. Because O_2 is required to produce ethylene, its synthesis is greatly checked in roots of waterlogged plants. The small amount of ethylene that is made in these roots is trapped, where it accumulates and eventually stimulates the activity of enzymes like cellulase and pectinase. These enzymes break down the cell walls. This leads to the formation of many intercellular spaces, characteristic of hydrophytes. Concomittantly, the ethylene precursors in the shoot are also converted to ethylene, which causes parenchyma cells on the upper side of the petiole to expand and point the leaf down, a physiologic response called epinasty.

5. Controls stem elongation. Mechanical disturbances such as shaking decrease stem elongation. This effect, which is called **thigmomorphogenesis**, is mediated by ethylene. Mechanical disturbances enhance ethylene production several times. Ethylene so produced causes cells to arrange their cellulose microfibrils *longitudinally.* This lengthwise reinforcement inhibits cellular elongation, causing cells to elongate radially; this leads to the formation of short and thick stems. This effect is opposite to that of auxin, which causes cells to orient their microfibrils *transversely*, thereby accounting for cellular elongation.

6. Determines sex expression. Both ethylene and gibberellins determine the sex of flowers on monoecious plants, *i.e*., plants having male and female flowers on the same individual. As an instance, cucumber (*Cucumis sativus*) buds, on treatment with ethylene, become carpellate (⊡) flowers, whereas those treated with gibberellins become staminate (⊠) flowers. Correspondingly, buds that ultimately become flowers produce more ethylene than do buds that become male flowers.

Ethylene *versus* Auxin

IAA stimulates ethylene production, thereby linking the responses of these two hormones. But ethylene does not account for all the effects induced by applying IAA. For example, IAA's stimulation of cellular elongation and the formation of lateral roots occur independently of ethylene. Similarly, leaf epinasty, decreased elongation of roots and shoots, and determination of sex are responses to ethylene application rather than IAA.

2. TRAUMATIC ACID

Most plants form callus when they are injured. This observation of **Gottlieb Haberlandt** (1913) led to postulate that the injured cells secrete a '**wound hormone**' which induces the neighbouring uninjured cells to become meristematic and divide till the injured part is healed up. Later, the active substance was isolated and identified as traumatic acid by **Bonner** and **English** in 1938. Traumatic acid is an open-chain dicarboxylic acid with a single double bond.

$$COOH—CH{=}CH—(CH_2)_8—COOH$$

Traumatic acid or **Traumatin** or **Nekrohormone**

Traumatic acid is specifically effective in inducing cell division in bean pods. It, however, seems to be of confined activity as it has failed to develop meristematic activity in a number of plant tissues including tobacco pith. **Davies** (1949), henceforth, regards this as a specific wound hormone for bean pods.

3. CALINES

Some indirect evidences accumulated during the past few decades have established beyond doubt the presence of some hormones which are needed for initiating the action of auxins on roots, stems and leaves. For example, the presence of leaves and buds is necessary for a better rooting caused by auxin application. These hormones have been collectively called as **calines** or **formative hormones**. The various postulated calines are:

A. *Rhizocaline*— It is a root forming hormone. It is produced by the leaves and transported down the stem.

B. *Caulocaline*—It is a stem forming hormone. It is synthesized mostly in the roots and translocated up in the stem.

C. *Phyllocaline*— It stimulates the development of mesophyll in the leaves. It is synthesized probably in the cotyledons from where it moves to its site of action.

None of these calines has been isolated from plants as yet.

4. VITAMINS

Their account follows in the Chapters 38 and 39.

GROWTH INHIBITORS

Introduction. While all the hormones discussed so far possess the power of growth promotion, the plants also do possess a few substances which inhibit growth. Although auxins when present in higher concentrations also inhibit growth, the growth inhibitors do so *irrespective of their concentration.* They retard such processes as seed germination, root and stem elongation and bud opening. As a matter of fact, the growth inhibitors act as chemical check upon plants preventing the seeds from germinating and the buds from opening under unfavourable conditions. For example, the sprouting of lateral buds during cold weather is inhibited by the presence of growth inhibitors produced within the plant body during the previous growing season.

Characteristics. The growth inhibitors are, in general, characterized by some common features (**Addicott** and **Lyon**, 1969; **Milborrow**, 1969). These are:

1. The amount of growth inhibitors decreases during active growth period of plants and increases during the period of growth suppression.

2. They counteract the activities of growth promoters.
3. They inhibit the growth of various isolated organs and tissues.
4. They do not evoke the strong stimulatory effects specific for auxins, gibberellins and kinins.
5. They are actively synthesized in green tissues and are found associated with auxins, gibberellins and kinins. However, they alone are accumulated in the absence of these growth substances in senescent and resting organs.
6. They are accumulated in woody plants during the period of dormancy.

Types. There are two types of growth inhibitors:

A. *Phenolic inhibitors*— These are better known and are widespread in plant organs. These are derivatives of either benzoic acid (*e.g.*, *p*-hydroxybenzoic, salicylic and gallic acids) or cinnamic acid (e.g., p-coumaric acid) and as well as coumarin.

B. *Abscisic acid and other substances*— Besides abscisic acid, these also include pyridine derivatives (*e.g.*, fusarinic and picolinic acids), quinones and flavonoids (*e.g.*, naringenin, phloridzin).

The structure of some common growth inhibitors is given in Fig. 35–25.

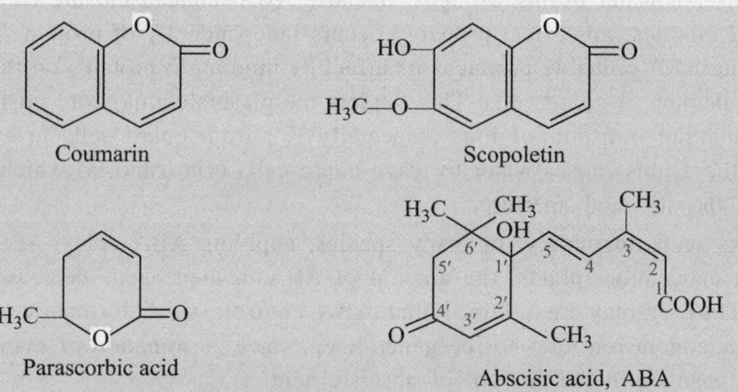

Fig. 35–25. Some common growth inhibitors

ABSCISIC ACID or ABSCISIN II or DORMIN

Discovery

Most of the effects first discovered for plant hormones were stimulatory. For example, IAA stimulates cellular elongation and cytokinins stimulate cell division. But near the end of the decade 1940s, **Torsten Hemberg** of Sweden reported that dormant buds of ash and potato contained inhibitors (rather than stimulators) that blocked the effects of IAA. When the buds germinated, the amount of these inhibitors decreased. Later, **Eagles** and **Wareing** (1963) isolated an inhibitor from the birch (*Betula pubescens*) leaves held under short day conditions. When this substance was reapplied to the leaves of birch seedlings, apical growth was completely arrested. As this substance induced dormancy, they named it as **dormin**. Later in 1965, **Ohkuma** *et al* isolated an inhibitor from cotton fruits and named it **abscisin II**. The same year, Cornforth and his associates isolated a growth inhibitor from sycamore and pointed out that both dormin and abscisin II are identical. Abscisin II is peculiar in that it is effective in much lower concentration than phenolic inhibitors and is accumulated under short day conditions. This compound was later named as abscisic acid (abbreviated as ABA)– *an unfortunate name*, because subsequent research has shown that *ethylene rather than* *abscisic acid controls abscission*. Abscisic acid (from *abscissus*[L] = cut off) is sometimes called *stress hormone.*

Distribution and Biosynthesis

Abscisic acid occurs in angiosperms and gymnosperms but apparently not in liverworts. ABA in plants is made from carotenoids. Once synthesized, ABA moves throughout a plant in xylem, phloem and parenchyma. *Like gibberellins and cytokinins, abscisic acid moves nonpolarly. There are no synthetic abscisic acids.*

Chemistry

Abscisic acid (Fig. 35–25) is a sesquiterpene with molecular formula $C_{15}H_{20}O_4$. Its molecule contains a corboxyl groups, 4 double bonds and an identical number of methyl groups.

Physiological Roles

1. Closure of stomata. It is a known fact that during drought, leaves synthesize large amounts of ABA which causes stomata to close. Thus, ABA acts as a messenger and enables plants to conserve water during drought. Because ABA-induced closure of stomata occurs within 1 to 2 minutes, this effect probably occurs independently of protein synthesis. As to its mechanism, ABA probably produces its effect by binding to proteins on the outer surface of the plasmalemma of guard cells. This renders the plasmalemma more positively-charged. thereby stimulating transport of ions (especially K^+) from guard cells to epidermal cells. The loss of these ions causes water to leave guard cells (via osmosis) which then collapse, thus closing the stomatal aperture.

2. Delays seed dormancy. In many species, applying ABA delays seed germination. Similarly, in many other plants, the amount of ABA in their seeds decreases, when seeds germinate. Thus, it may be inferred that ABA controls seed dormancy in some cases. However, this conclusion may not be generalized, since germination of many seeds occurs without any changes in the amount of abscisic acid.

3. Controls bud dormancy. Bud dormancy was previously thought to be controlled solely by ABA. But when leaves are treated with radioactive ABA, no radioactivity has been detected in buds. This suggests that, besides ABA, bud dormancy is probably also influenced by cytokinins and IAA-induced synthesis of ethylene.

4. Counteracts the effects of other hormones. ABA counteracts the stimulatory / inhibitory effects of other hormones. For example,

(*a*) ABA inhibits cell growth promoted by IAA.

(*b*) ABA inhibits amylase produced by seeds treated with gibberellin.

(*c*) ABA promotes chlorosis that is inhibited by cytokinins.

This may be due to the fact that *ABA is a* Ca^{2+} *antagonist* and its inhibition of the stimulatory effects of IAA and cytokinin may be due to its interference with Ca^{2+} metabolism. Although ABA usually inhibits growth, it is not toxic to plants as are inhibitors of RNA/ protein synthesis. ABA often decreases gene acctivity, but there are instances of ABA stimulating genes. For example, ABA stimulates the synthesis of mRNAs for storage proteins in developing wheat grains.

MORPHACTINS

Recently, a group of growth substances called morphactins (meaning morphologically active substances) has diverted the attention of physiologists on account of its regulatory effect on the growth and development of plants (**Merck**, 1970). *The role of morphactins is usually of inhibitory nature.*

Chemically, the morphactins are derivatives of fluorene-9-carboxylic acid (Fig. 35–26) and are absorbed through the seeds, roots, and leaves. Their action is systematic but slow and can be transported both acropetally and basipetally. Considering from phylogenetic viewpoint, the morphogenic action of morphactins starts with the complex brown algae, the mosses and ferns and is exhibited well in higher plant groups. *Unicellular plants and simple filaments are, however, not affected by morphactins.*

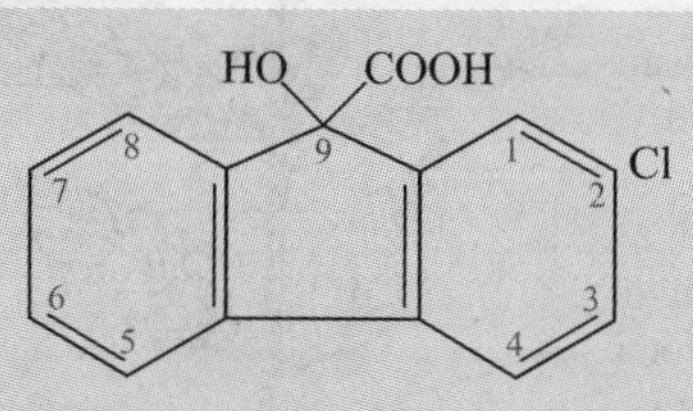

Fig. 35–26. Chlorofluorenol or morphactin

Although morphactins are much alike to abscisic acid in action, yet some of their effects are highly specific to them. Some such effects are:

1. development of dwarf, bushy habit due to shortening of internodes.
2. inhibition of sprouting of rosette plants.
3. complete abolition of polarity in plants.
4. inhibition of mitosis in apical meristems.
5. stimulation of cell division in cambium of intact plants and cuttings.
6. reinforcement of the apical dominance in tap roots so that the formation of lateral roots is strongly inhibited.
7. abolition of geotropic and phototropic responses.
8. reduction of apical dominance in the main shoot so that the formation of branches is promoted, resulting in a broomlike growth of the plant.
9. change in orientation of the spindle axes of dividing cells.

OLIGOSACCHARINS AND OTHER PLANT HORMONES

Plant hormones are not only restricted to those mentioned above. For example, fragments of the cell wall called oligosaccharins control plant growth, differentiation, reproduction and defense against disease and hence function as plant hormones. However, oligosaccharins typically differ from other plant hormones because they elicit specific effects:

1. Different oligosaccharins can induce cultured cells to form undifferentiated callus roots, shoots, or flowers in a variety of plants.
2. Oligosaccharins that inhibit flowering and promote vegetative growth in one species have the same effect on other species.
3. Oligosaccharins induce effects by impressively small amounts: about 100 to 1,000 times less oligosaccharin than IAA or cytokinin to induce a response.

Recent researches have revealed the presence of a variety of other compounds that act as hormones in different groups of plants. For example:

(*a*) Yams contain **batasins** which induce dormancy of bulbils (vegetative reproductive structures) that form from lateral bulbs.

(*b*) **Brassosteroids** are plant hormones present in tea, bean and rice plants that stimulate the growth of stems. Their chemical structure resembles that of ecdysone, an insect moulting hormone.

HORMONAL INTERACTIONS

It is very rarely, if ever, that the plant hormones work alone; rather *plant growth and development usually result from interactions of plant hormones* (Fig. 35–27).

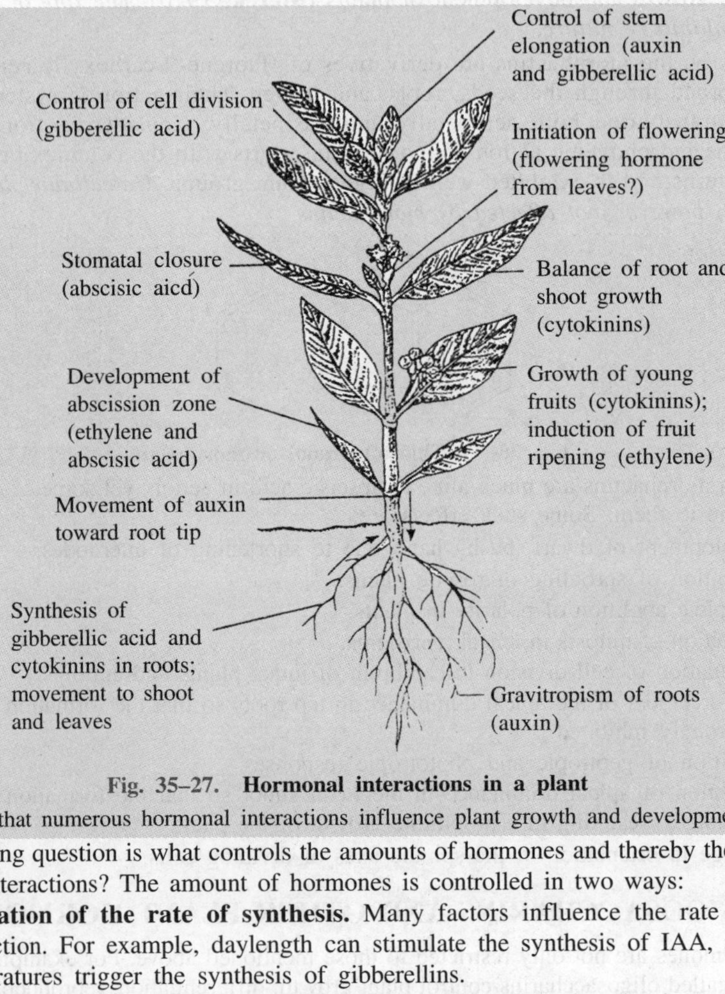

Fig. 35–27. Hormonal interactions in a plant

Note that numerous hormonal interactions influence plant growth and development.

An intriguing question is what controls the amounts of hormones and thereby their ratios and the resulting interactions? The amount of hormones is controlled in two ways:

(*a*) **Regulation of the rate of synthesis.** Many factors influence the rate of hormone production. For example, daylength can stimulate the synthesis of IAA, and the cold temperatures trigger the synthesis of gibberellins.

(*b*) **Regulation of the rate of breakdown or inactivation.** Inactivation of a hormone usually takes place either by its oxidation or by its conjugation with other compound.

PLANT HORMONES *VERSUS* ANIMAL HORMONES

Since plant hormoes were discovered later than animal hormones, early research was based on the assumption that plant and animal hormones had basically similar physiological effects. However, there are important differences in plant and animal hormones (Table 35–4).

1. There is no evidence to indicate that the fundamental actions of plant and animal hormones are the same.
2. Plant hormones are not made in tissues specialized for hormone production, whereas animal hormones are produced in specialized organs, mostly in endocrine glands.
3. Plant hormones do not have definite target areas, whereas animal hormones do act at target .areas.
4. Animal hormones are usually very specific in their actions, while plant hormones seldom have such specific effects.

Table 35–4. Comparison of animal and plant hormones

Characteristic	*Animal hormones*	*Plant hormones*
Site of production	Specific endocrine glands specialized for homone production.	Produced by actively metabolizing tissues that have other functions.
Target tissues	Each hormone acts on a specific target tissue or organ.	Each hormone acts on a variety of tissues.
Number of hormones	Many, each with a specific function.	Relatively few, each with a variety of functions.
Primary function	Affect homeostasis and are regulatory in action; effects are reversible.	Affect growth and development; effects are permanent.

LESLIE JOHN AUDUS

(LT, 1911–2011)

Audus, a British botanist, was born on December 9, 1911 in Isleham in the fens of Cambridgeshire. He was educated at Soham Grammar School, and in 1929 was awarded a scholarship to Downing College, Cambridge. He was Royal Air Force Volunteer Reserve survivor of the Japanese internment camps of World War II. He cultured yeast to feed and save the lives of his fellow prisoners of war (POWs). In the mid-1950s, he did research on **plant growth regulators**.

Audus was Professor Emeritus at Bedford College, London. He wrote the books *Plant Growth Substances* (1959 and 1972) and *Spice Island Slaves* (1996). He was editor of the 1964-textbook, *The Biochemistry and Physiology of Herbicides*. Between 1965 and 1974, he was an editor of the *Journal of Experimental Botany*. He died on May 5, 2011, missing completing his century by a few months.

Despite these differences, botanists have traditionally suggested that plant hormones function like animal hormones; that is, the response elicited by a plant hormone is determined by the concentration of the hormone. Certain responses *can* be obtained by applying hormones to various parts of the plants, but this does not necessarily mean that such responses are naturally controlled by the hormone. As a result, some botanists now question the propriety of plant hormones being called as hormones, in a sense as applied to animal hormones. These botanists, therefore, refer the plant hormones as **plant growth regulators** and suggest that plant hormones are intergrating agents that are necessory for, but do not *control*, the response. For example, cytokinins regulate the conversion of buds from a dormant to nondormant state, but do not control the subsequent growth of the bud. Thus, cytokinins are necessary for bud growth , but do not control it. According to this perspective, "the response elicited by a hormone is determined not by the amount of hormone, but rather by the sensitivity of the tissue to the hormone" (**Moore, Clark** and **Stern**, 1995).

REFERENCES

1. **Abeles FB :** Ethylene in Plant Biology. *Academic Press, Inc., New York. 1973.*
2. **Addicott FT (editor):** Abscisic Acid. *Praeger, New York. 1983.*
3. **Addicot FT, Smith OE, Lyon JL :** Some physiological properties of abscisin II. *Plant Physiol.* **40** : *Supple. XXVI, 1965.*
4. **Albersheim P, Darvill AG :** Oligosaccharins. *Scientific American* **253** : *58–64, 1985.*
5. **Audus LJ :** Plant Growth Substances. *2nd ed. Interscience Publishers, New York. 1959.*
6. **Audus LJ :** Plant Growth Substances. vol. 1. Chemistry and Physiology. *Leonard Hill (Books) Ltd., London. 1973.*
7. **Brian PW :** The Gibberellins as hormones. *Int. Rev. Cytol.* **19** : *229–266, 1966.*
8. **Carr DJ :** Plant Growth Substances. *Springer-Verlag. 1972.*
9. **Chrispeels MJ, Varner JE :** Hormonal control of enzyme synthesis– On the mode of action of gibberellic acid and abscisin in aleurone layers of barley. Pl*ant Physiol.* **42** : *1008,* 1967.

10. **Cline M :** Apical dominance. Bot. *Rev.* **57** : *318–350, 1991.*
11. **Eagles CF, Wareing PF :** The role of growth substances in the regulation of bud dormancy. *Plant Physiol.* **17** : *697, 1964.*
12. **Evans ML :** The action of auxin on plant cell elongation. CRC Crit. Rev. Plant Sci. **2** : *317–365, 1984.*
13. **Galston AW :** Life Processes of Plants. *W.H. Freeman and Co., New York. 1994.*
14. **Galston AW, Purves WK :** The mechanism of action of auxin. *Ann. Rev. Plant Physiol.* **11** : *239–276, 1960.*
15. **Helgeson :** Cytokinins. *Science.* **161** : *974, 1968.*
16. **Huddart H, Smith RJ, Langton PD, Hetherington AM, Mansfield TA :** Is abscisic acid a universally active calcium antagonist? *New Phytologist.* **104** : *161–173, 1986.*
17. **Kefeli VI :** Natural Plant Growth Inhibitors and Phytohormones. *Dr. W. Junk Publishers, Hague. 1978.*
18. **Kefford :** Natural plant regulators. *Science.* **142** : *1495, 1963.*
19. **Kende H :** Ethylene biosynthesis. *Ann. Rev. Plant Physiol. Plant Mole. Biol.* **44** : *283–308, 1993.*
20. **Kuraishi S, Muir RM :** Effect of the Gibberellin. *Science.* **137** : *760, 1962.*
21. **Kuraishi S, Muir RM :** The relationship of gibberellin and auxin in plant growth. *Plant Cell Physiol.* **5** : *61, 1964.*
22. **Lang A :** Gibberellins : Structure and metabolism. *Ann. Rev. Plant Physiol.* **21** : *537–570, 1970.*
23. **Leopold AC :** Auxins and Plant Growth. *Univ. of California Press, Los Angeles. 1955.*
24. **Leopold AC :** Auxin uses in the control of flowering and fruiting. *Ann. Rev. Plant Physiol.* **9** : *281–310, 1958.*
25. **Leopold AC :** Plant Growth and Development. *McGraw Hill Book Co., New York. 1964.*
26. **Letham DS :** Zeatin, a factor inducing cell division isolated from Zea mays. *Life Sciences.* **2** : *569, 1963.*
27. **Letham DS :** Chemistry and physiology of kinetin-like compounds. *Ann. Rev. Plant Physiol.* **18** : *349, 1967.*
28. **Mc Fadden CH, Keeton WT :** Biology– An Exploration of Life. *1st ed. W.W. Norton & Company Inc., New York. 1995.*
29. **Miller C :** Kinetin and related compounds in plant growth. *Ann. Rev. Plant Physiol.* **12** : *395–408, 1961.*
30. **Moore R, Clark WD, Stern KR :** Botany. *1st ed. Wm. C. Brown Publishers, Dubuque IA. 1995.*
31. **Nickell LG :** Plant Growth Regulators– Agricultural Uses. *Springer-Verlag, New York. 1982*
32. **Ohkuma K, Addicott FT, Smith OE, Thiessen WE :** The structure of abscisin II. *Tetrahedron Letters.* **29** : *2529, 1965.*
33. **Paleg LG :** Physiological effects of gibberellins. *Ann. Rev. Plant Physiol.* **16** : *291–322, 1965.*
34. **Phillips IDJ :** Introduction to the Biochemistry and Physiology of Plant Growth Hormones. *McGraw Hill Book Co., New York. 1971.*
35. **Ryan CA, Farmer EE :** Oligosaccharide signals in plants : A current assessment. *Ann. Rev. Plant Physiol. Plant Mole. Biol.* **42** : *651–674, 1991.*
36. **Salisbury FB, Ross CW :** Plant Physiology. *4th edition. Wadsworth, Belmont, CA. 1992.*
37. **Shaw G, Wilson DV :** A synthesis of zeatin. *Proc. Chem. Soc. (London). 231, 1964.*
38. **Sircar SM :** Plant Hormone— Research in India. *Indian Council of Agricultural Research, New Delhi. 1971.*
39. **Sisler EC, Y and SF :** Ethylene, the gaseous hormone. *BioScience.* **33** : *233–338, 1984.*
40. **Skoog F, Strong FM, Miller CO :** Cytokinins. *Science.* **148** : *532, 1965.*
41. **Steward FC, Krikorian AD :** Plants, Chemicals and Growth. *Academic Press, Inc., New York. 1971.*

42. **Taiz L, Zeiger E :** Plant Physiology. *2nd ed. Sinauer Associates, Inc., Publishers, Sunderland, Massachusetts. 1998.*
43. **Thimann K V :** Plant Growth Substances : Past, Present and Future. *Ann. Rev. Plant Physiol.* **14** : *1–18, 1963.*
44. **Thomas H, Stoddart, JL :** Leaf senescence. *Ann. Rev. Plant Physiol.* **31** : *83–111, 1980.*
45. **Trewavas A :** How do plant growth substances work ? Plant, *Cell and Environment.* **14** : *1–12, 1991.*
46. **Wareing PF, Good JEG, Manuel J :** Some possible physiological roles of abscisic acid. Proc. 6th Inter. *Conf. Plant Growth Substances, Ottawa. 1968.*
47. **Wareing PF, Phillips IDJ :** Growth and Differentiation in Plants. *3rd edition. Pergamon Press. Elmsford, New York. 1981.*
48. **Went FW, Thimann KV :** Phytohormones. *The Macmillan Co., New York. 1937.*
49. **Wilkins MB (editor) :** Physiology of Plant Growth and Development. *McGraw Hill Book Co., New York. 1969.*

PROBLEMS

1. Describe two general chemical features that distinguish plant homones from animal hormones.
2. Name two plant hormones that are antagonistic in their effects.
3. The primary effect of gibberelllins is on:
   (*a*) mitosis
   (*b*) meiosis
   (*c*) cell enlargement
   (*d*) flowering
   (*e*) root growth
4. Which part of a plant produces auxins?
   (*a*) Root hair
   (*b*) Meristematic regions of stem and root
   (*c*) Fruits
   (*d*) Leaves
5. Which of the following accelerates cell division in plants?
   (*a*) Auxins
   (*b*) Cytokinins
   (*c*) Gibberellins
   (*d*) Florigens
6. The plant part which produces cytokinin is:
   (*a*) Flowers
   (*b*) Leaves
   (*c*) Stem
   (*d*) Roots
7. How do fruits ripen?
8. How does calcium carbide ripen mangoes?

ANSWERS TO PROBLEMS

1. (*a*) Most plant hormones are terpenoid compounds.
   (*b*) There are no known peptide hormones in higher plants. Steroid-like compounds have been described in plants, but it is not clear that they act like steroid hormones.
2. Gibberellin and abscissic acid
3. (*c*) cell enlargement
4. (*b*) Meristematic regions of stem and root
5. (*b*) Cytokinins

6. (*d*) Roots
7. Unripe fruits are typically hard, green and have a sour or astringent taste due to the presence of organic acids such as malic, citric and tartaric. They have high amounts of polysaccharides and low amounts of proteins. *Ethylene* gas that triggers the ripening process emanates from fruits and causes rapid and dramatic changes. The green colour of the fruits disappears and red- or yellow-coloured pigments such as anthocyanins and carotenoids appear giving the fruits distinct colours. The polysaccharides are broken down to smaller sugars which give the fruits sweet taste. The fruits also begin to soften. Volatile substances are produced during ripening which give many fruits their distinct aroma.
8. For reasons of safe transportation of fruits, mangoes, bananas etc are picked before they ripen fully. Slightly green harvested mangoes are subjected to small containers of calcium carbide (CaC_2) with a plastic covering. CaC_2 reacts with the moisture in the air to release acetylene (= ethyne) gas, which like ethylene– a chemical substance produced by fruits to accelerate the normal process of fruit maturation– is a ripening hormone. Industrial-grade calcium carbide contains arsenic and phosphorus, and its use for the purpose of ripening is illegal in many countries.

PART

V

Nutrition Biochemistry

Fruits and vegetables are nutritious. Besides containing carbohydrates proteins, lipids, minerals and some fibres, they also possess appreciable amounts of vitamins which are *sine quo non* for life.

"What drives life... is a little electric current, kept up by sunshine."

— *Albert Szent-Györgyi : Chemistry of Muscular Contraction, 1960.*

CONTENTS

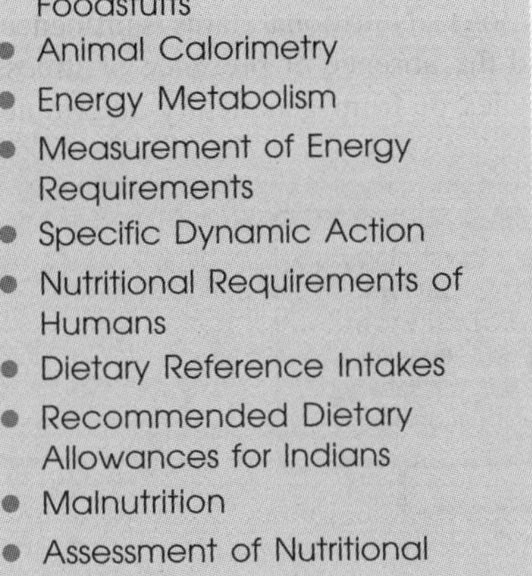

A simplified chart of the targeted nutrition

CHAPTER

36 Principles of Nutrition

"One cannot think well, love well, sleep well, if one has not dined well."

–Virginia Woolf

NUTRITION

The terms food and nutrition are sometimes used synonymously, but that is not strictly correct. **Food** is a composite mixture of various substances, the quantity of which may vary from a fraction of a gram in certain cases, to hundreds of grams in others. The term foodstuff is defined as "anything which can be used as food" (**Brock**, 1961). **Nutrition**, on the contrary, is defined as the composition and quantity of food intake and the utilization of the food intake by living organisms. Since the process of food utilization is biochemical, the major thrust of this chapter and the latter two chapters shall be on a discussion of the basic nutritional concepts. Nutrition attracts more than its share of controversy in our society, and a thorough understanding of nutrition almost demands an understanding of the issues behind these controversies. Study of human nutrition can be divided into 3 areas:

1. Undernutrition: This is the condition which results when insufficient food is eaten over a prolonged period of time. In extreme cases, it is called *starvation.* Undernutrition is particularly a serious problem in the underdeveloped countries because of the intake of food, deficient in many nutrients, by their inhabitants. In developed countries, nutritional deficiency diseases are now quite rare.

2. Overnutrition: This is the pathological state resulting from the consumption of excessive quantity of food over a prolonged period of time. Overnutrition is the main concern of the developed countries. For example, current estimates suggest that between 15 % and 30% of the US population is obese, and obesity is known to have a number of serious health problems, such as obesity, atheroma, diabetes, hypertension and CVDs.

3. Optimal (or **ideal) nutrition:** This is a concept that has meaning only in an affluent society. Only when food supply becomes abundant enough so that deficiency diseases are a rarity does it become possible to consider long-range effects of nutrition on health.

From biochemical viewpoint, nutrition is an essential interaction of an organism with the environment. Nutrients, apart from providing energy for survival, signal the organism in diverse ways, and this in turn affects their usage and storage. Nutrition underpins health and affects susceptibility to disease; both malnutrition and obesity put an organism at risk. The factors that determine nutritional status of an individual are: the environment, the phase of the life cycle, the level of physical activity and the presence of disease (Fig. 36–1). Nutritional status is influenced by the availability of food, its palatability and variety, and the absence or presence of illness. Nutritional deficiencies may result from dietary inadequacies or from genetically-determined metabolic errors.

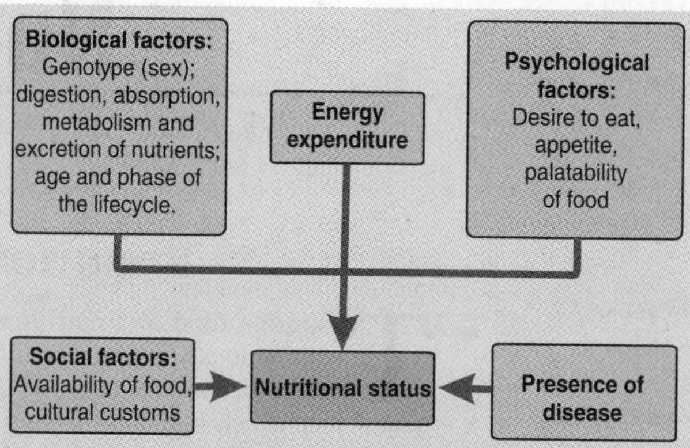

Fig. 36–1. Factors that determine the state of nutrition

Food intake is controlled by *hunger* (a desire to eat) and *appetite* (a desire for a particular food). The main centres regulating appetite are located in the hypothalamus in the central nervous system (CNS). The appetite is regulated by the hypothalamic arcuate and paraventricular nuclei. The brain regulates energy homeostasis and is also the primary regulator of body weight (Fig. 36–2). According to the lipostatic model of energy homeostatis, signals controlling energy intake originate from the adipose tissue and are sent to the CNS. These signals are mediated by the adipokine leptin and by insulin. In response, the brain sends efferent signals through the complex network of neuropeptides. These signals regulate hunger and appetite. The neurons in the arcuate nucleus express 2 neuropeptides: the catabolic proopiomelanocortin (POMC), and an anabolic neuropeptide Y/agouti-related protein (NPY/AgRP). POMC is cleaved yielding melanocortins, such as α-MSH, that decrease food intake. NPY/AgRP links to neurons expressing melanin-concentrating hormone (MCH) and orexins A and B. They, in turn, are involved in the control of food intake by acting on brainstem neurons. These neurons connect with the brain cortex (the satiety centre) to promote hunger and to stimulate yet another set of hormones such as thyreoliberin (TRH), corticoliberin (CRH) and ocytocin. TRH increases thermogenesis and food intake, whereas CRH decreases food intake and, through the sympathetic activity, increases energy expenditure. Further signals that control food intake are conveyed by gastrointestinal peptides such as glucagon, cholecystokinin, glucagon-like peptide, amylin and bombesin-like peptide. Ghrelin, secreted by the stomach, stimulates NPY/AgRP-expressing neurons. *Ghrelin is the only known appetite-stimulating peptide.* Gastric stretch also affects food intake. Finally, hypoglycemia decreases the activity of the satiety centre.

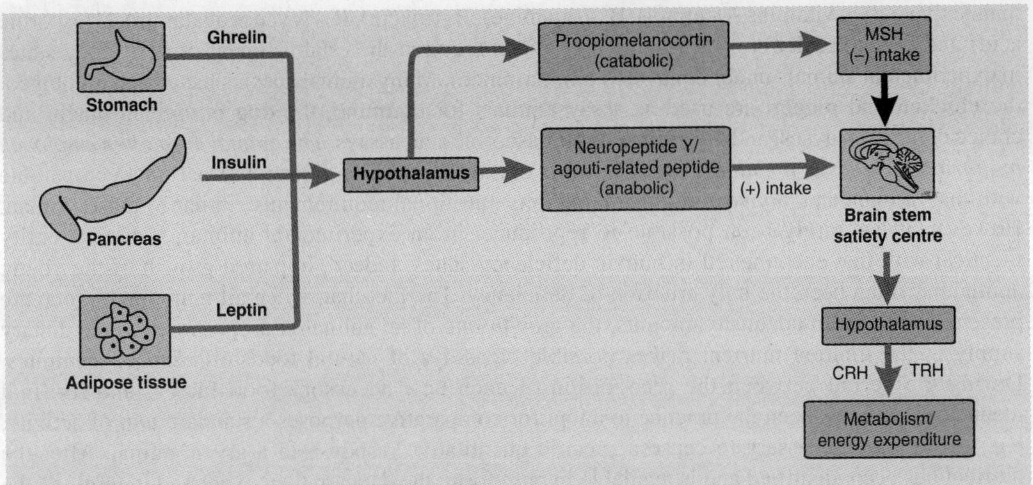

Fig. 36–2. Regulation of food intake

Regulation of food intake is accomplished by signals generated in the adipose tissue, pancreas, stomach, and brain. The hypothalamus translates signals related to the energy balance into eating behaviour through secretion of a range of neuropeptides. The (+) sign means action leading to increase of appetite and food intake, and the (–) sign to a decrease.

[MSH = Melanocyte-stimulating hormone; CRH = Corticoliberin; TRH = Thyreoliberin]

The main nutrients are carbohydrates (including fibres), proteins, fats, vitamins, and minerals. Carbohydrates, proteins, fats, fibres, and some minerals are *macronutrients.* Vitamins and trace elements are *micronutrients.* The functions of various nutrients are summarized in Fig. 36–3. The caloric values of the main nutrients are given in Table 36–1 (see page 1235).

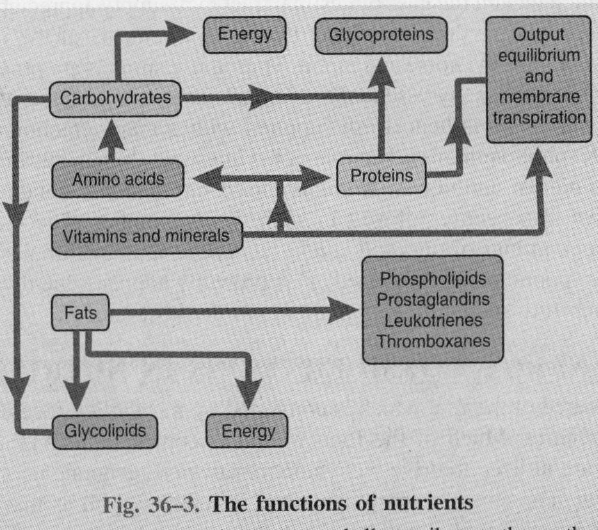

Fig. 36–3. The functions of nutrients

All main classes of nutrients can be used to make energy, and all contribute to the synthesis of more complex compounds. The main role of vitamins and minerals is their participation in enzymatic reactions.

ASSAY ANIMALS FOR NUTRITION STUDIES

Recognition of the role of nutrition in the etiology of certain human diseases led to a search for the specific nutrients which would prevent these diseases. Among the nutrients recognized in this manner have been vitamins A (retinol), B_1 (thiamine), B_5 (niacin), B_{12} (cyanocobalamin), C (ascorbic acid) and D (cholecalciferol). These investigations required the establishment of deficiency states in experimental animals under controlled circumstances. Many animal species have been employed: the **chicken** and **pigeon** are used as assay animals for thiamine; the **dog** is used in niacin and cholecalciferol analyses; and the **guinea pig** for ascorbic acid assays. *The animal most often employed in nutrition studies is the* ***albino rat***. The rat is omnivorous, has a gastrointestinal tract comparable with that of man, and, above and overall, exhibits nutritional requirements similar to those of man. However, it has rarely been possible to reproduce, in an experimental animal, a disease entity identical with that encountered in human deficiency states. Indeed, impaired growth of the young animal has often been the only criterion of deficiency. The fact that, when all nutrients but one are present in the diet in adequate amounts, the growth rate of an animal is proportional to the dietary supply of the limiting nutrient makes possible bioassays of natural foodstuffs and concentrates. During the period between the recognition of each new accessory food factor and its final identification, it has been the practice to adopt for comparative purposes a standard *unit* of activity, *e.g.*, the amount necessary to cause a specific quantitative response in a given animal. After the nutrient has been identified and is available in pure form, the data are then expressed in terms of the weight of the nutrient.

Microorganisms have been extremely useful in the nutritional investigations. Vitamins B_3 (pantothenic acid), B_7 (biotin) and inositol were first recognized as required for the initial growth of certain strains of yeast. Vitamin B_9 (folic acid) was first established as an essential factor for the growth of several strains of lactobacilli several years before its importance in mammalian nutrition was discovered. Vitamin B_{12} (cyanocobalamin), the antipernicious anemia factor, was identified in one laboratory because it is essential for the growth of *Lactobacillus lactis*. Many species of **bacteria, moulds, yeasts** and **fungi** have been found for which one or more of the nutrients known to be essential to humans, are also growth factors, thus providing a basis for rapid quantitative microbiological assays for these substances.

The bacteria which inhabit the gastrointestinal tract may supply appreciable amounts of vitamins to the host. This accounts for the simpler nutritional requirements of the ruminants and of those species with large ceca, *e.g.*, the horse and rabbit. Thus, these animals are provided with fermentation mechanisms which produce many essential nutritional factors. To a lesser degree, this occurs also in the intestine of man, who is henceforth supplied with a major fraction of his requirement for biotin and vitamin K (phylloquinone). The role of the intestinal flora in nutrition has been established by inclusion in the diet of antibiotics. Animals raised under sterile conditions with bacteria-free intestinal tracts have also been employed in such type of studies. The inclusion of tetracycline antibiotics in the stock rations of pigs and chicks has been found to stimulate their growth, with an increased weight per pound of ingested feed. This probably happens due to the establishment of an intestinal flora which furnishes utilizable nutrients to the host animal.

ENERGY CONSIDERATIONS IN NUTRITION

Food is the source of the fuel which is converted by metabolic processes of the body into the energy for vital activities. Much of the food we eat is converted to ATP and other high-energy compounds, which are utilized to drive biosynthetic pathways, generate nerve impulses, and power muscular contraction. The unit of energy contained in foods as well as that involved in metabolic activities is most often expressed as a unit of heat, the calorie. A **calorie** is defined as the amount of heat required to raise the temperature of 1 gram of water by 1°C (from 15°C to 16°C). This amount of heat (a small calorie, written with a small c and abbreviated *cal*) is exceedingly small, so that in nutrition it is common to use instead the large calorie (written with a capital C and abbreviated *Cal*) or **kilocalorie**, which is equal to 1,000 calorie. The accepted international unit of energy is now the **joule** (10^7 ergs). One joule is equal to the energy expended when 1 kilogram is moved to a distance of 1 meter by applying a pressure of 1 newton. The interrelationship between kilocalorie and joule is as follows:

1 kilocalorie = 4184 joules = 4.184 kilojoules (kJ)

1 kJ = 0.239 kcal

The combustion of a foodstuff in the presence of oxygen results in the production of heat. The amount of heat thus produced can be measured in a bomb calorimeter. By this technique, the **caloric values** of a foodstuff can be determined. The **caloric value** (or **fuel value**) of the 3 principal foodstuffs (carbohydrate, protein and fat), determined originally by burning in a bomb calorimeter, is given in Table 36–1.

Table 36–1. Fuel values of foods*

Type of foodstuff	*Kilocalories per gram*	
	In bomb calorimeter	*In the body†*
Carbohydrate	4.1	4 (= 17 kJ)#
Protein	5.6	4 (= 17 kJ)
Fat	9.4	9 (= 38 kJ)

* The fuel values given in this table are average values, since variations occur within each class; for example, monosaccharides do not have exactly the same caloric content as polysaccharides.

† It is customary to round off the energy value of foods as utilized in the body to the figures given in this table. These figures also correct for the efficiency of the digestion of foods.

\# The energy value of foods may be described in kilojoules, but for total daily intake or expenditure, *megajoules* are more suitable. [1 kcal = 4.184 kJ; 1 kJ = 239 cal]

When utilized in the body, carbohydrate and fat are completely oxidized to carbon dioxide and water, as they are in the bomb calorimeter also. Proteins, however, are not burned completely since the major end product of protein metabolism, urea, still contains some energy which is not available to the body. For this reason, the energy value of protein in the body (4.1 kcal/g) is less than that obtained in the bomb calorimeter.

The total energy requirement of an individual is made up of 2 main components: basal energy and the energy required for the actual physical activties. A small allowance is also made for wastes of all kinds.

(*a*) ***Basal metabolism*:** It is the energy expenditure of a person who is relaxed and comfortable in the morning soon after awakening and 14 hours after the last meal. This is the energy required for such vital functions as respiration, blood circulation and maintenance of body temperature. Basal energy requirements are closely related to body surface area. The reported values of basal metabolic rate (BMR) in India are : 38.1 kcal for normal males and 32.9 kcal for normal females per square metre of body surface per hour (**Narsinga Rao**, 1969).

For estimating the energy needs, the concept of a reference man and reference women is universally adopted. An *Indian reference man* is defined as one who is between 20 – 39 years of age and weighs 55 kg. He is free from diseases and physically fit for active work (ICMR, 1981). He has a body surface area of 1.40 sq. metres. On this basis, an Indian reference man expends during a period of 8 hours' sleep a net of (8 × 1.62 × 38.1) or 493 kcal (2 MJ) of energy; and an *Indian reference woman*, a net of (8 × 1.40 × 32.9) or 368 kcal (1.5 MJ) of energy. This is the first step in the calculation of the total energy requirements of a person.

(*b*) ***Energy expenditure "off-work"*:** In addition to basal metabolism, a person expends energy for daily activities which are nonoccupational, such as sitting, standing, dressing, undressing, walking etc. For all these activities, energy is required. It has been estimated that an Indian reference man may expend 1220 kcal (4.9 MJ), and an Indian reference woman about 826 kcal (3.3 MJ) for 8 hours' off -work.

(*c*) ***Energy expenditure for work*:** Additional calories are required for the performance of daily work. The ICMR (1960) suggested following values for each type of work:

Kind of work	*Kcal/kg/hour*
Light	1.7
Moderate	2.5
Heavy	5.0

(*d*) ***Fecal loss of energy*:** In any computation of energy requirement, necessary allowance for fecal loss of energy should be made. An allowance of 10% is recommended in India (**Narsinga Rao**, 1969).

The energy requirements for Indians recommended by the Nutrition Expert Committee of the ICMR (1981) are given in the Table 36–2. These can be used as guidelines for planning food requirements.

Table 36–2. Recommended daily energy requirements*

	Activity	*Kcal/day*	*MJ*
Man (55 kg)	Light work	2,400	10.00
	Moderate work	2,800	11.70
	Heavy work	3,900	16.30
Woman (45 kg)	Light work	1,900	7.95
	Moderate work	2,200	9.20
	Heavy work	3,000	13.10
Pregnancy	2nd and 3rd trimester	+ 300	1.25
Lactation	First 6 months	+ 550	2.30
	6 to 12 months	+ 400	1.68
Infants	Under 3 months	120/kg	
	3 to 5 months	115/kg	
	6 to 8 months	110/kg	
	9 to 11 months	105/kg	
	Average 1st year	112/kg	
Children	1 to 3 years	1,220	5.1
	4 to 6 years	1,720	7.2
	7 to 9 years	2,050	8.6
	10 to 12 years (males)	2,420	10.1
	10 to 12 years (females)	2,260	9.5
	13 to 15 years (males)	2,660	11.1
	13 to 15 years (females)	2,300	9.5
	16 to 18 years (males)	2,820	11.7
	16 to 18 years (females)	2,200	9.2

* A perusal of the table shows that children above the age 13 years need as much energy as adults. This is because they show a good deal of physical activity, almost equal to hard work by adults. This is also the age when puberty sets in and there is a spurt in growth and an increase in metabolic rate. This fact should be borne in mind in planning dietaries for children and adolescents.

It may be added that energy requirements decrease with age because of a fall in BMR and a decrease in physical activity in most persons. The FAO/WHO Committee (1973) suggested that after the age of 40 years, requirements should be reduced by 5% per decade until the age of 60, by 10% between 60 and 70, and by a further 10% after the age of 70 years.

When the energy requirement of a group of people consisting of adult men and women, and children is to be calculated, it is usual to assess their needs in terms of those of the average man. The needs of an average man are reckoned as 1.0 coefficient or *Man Value* or *consumption unit*. The energy requirement of the rest of the members of the group are adjusted according to the following *scale of coefficients* (**Gopalan** *et al,* 1974):

Person		*Coefficient*	
Adult male	Sedentary worker	1.0	
	Moderate worker	1.2	
	Heavy worker	1.6	
Adult female	Sedentary worker	0.8	
	Moderate worker	0.9	
	Heavy worker	1.2	
12–21 years			1.0
9–12 years			0.8
7–9 years			0.7
5–7 years			0.6
3–5 years			0.5
1–3 years			0.4

As an illustration, for a family consisting of father and mother doing sedentary work, and children aged 14, 8 and 6– the total 'adult man values' or consumption units would be 1.0 + 0.8 + 1.0 + 0.7 + 0.6 = 4.1.The minimum daily energy (calorie) requirement of this family would be 2400 × 4.1 = 9840 kcal.

The heat generated by the body in the course of metabolism of foodstuffs maintains the body temperature. Warm-blooded animals, such as birds and mammals, have heat-regulating mechanisms that either increase heat production or radiate or otherwise dissipate excess heat, depending on the temperature of their external environment. When the external temperature rises above the normal body temperature (37ºC), evaporation from the surface of the body becomes the only mechanism available for cooling the body.

RESPIRATORY QUOTIENTS OF FOODSTUFFS

The **respiratory quotient**, RQ (also called **respiratory exchange rate,** RER) is the ratio of the volume of CO_2 eliminated to the volume of O_2 utilized in oxidation. The RQ values are different for the 3 principal foodstuffs:

A. **Carbohydrates:** The complete oxidation of glucose, for example, may be represented as follows:

$$C_6H_{12}O_6 + 6\,O_2 \longrightarrow 6\,CO_2 + 6\,H_2O$$

The RQ for carbohydrate is therefore:

$$\frac{CO_2}{O_2} = \frac{6}{6} = 1$$

B. **Fats:** Fats have a lower RQ because the oxygen content of their molecule in relation to the carbon content is quite low. As a result, they require more oxygen from the outside. The oxidation of tristearin, for example, may be represented as follows:

$$2\,C_{57}H_{110}O_6 + 163\,O_2 \longrightarrow 114\,CO_2 + 110\,H_2O$$

$$\frac{CO_2}{O_2} = \frac{114}{163} = 0.7$$

C. **Proteins:** The oxidation of proteins cannot be so readily expressed because their chemical structure is quite variable. However, by indirect methods, the RQ of proteins has been calculated to be approximately 0.8.

D. **Mixed diets under varying conditions:** In mixed diets containing varying proportions of carbohydrate, protein, and fat, the RQ is about 0.85. As the proportion of carbohydrate metabolized is increased, the RQ approaches closer to 1. When carbohydrate metabolism is impaired, as in diabetes, the RQ is lowered. Treatment with insulin is followed by an elevation in the RQ. A high carbohydrate intake, as used in fattening animals, will result in an RQ exceeding 1. This rise is caused by the conversion of much of the carbohydrate, an oxygen-rich substance, to fat, an oxygen-poor substance; a relatively small amount of oxygen is required from the outside, and the ratio of CO_2 eliminated to the oxygen taken in (*i.e.*, the RQ) will be considerably elevated. A reversal of the above process, *i.e.*, the conversion of fat to carbohydrate, would lower the RQ below 0.7. This has been reported but has not been generally confirmed.

ANIMAL CALORIMETRY

Calorimetry deals with the measurement of the energy requirements of the body under various physiologic conditions and of the fuel values of foods which supply this energy. The heat generated by the body in the course of the metabolism of foodstuffs maintains the body temperature. **Homoiotherms** ('warm-blooded' animals), such as birds and mammals, have heat-regulating mechanisms that either increase heat production or radiate or otherwise dissipate excess heat, depending on the temperature of their external environment. When the external temperature rises above the normal body temperature (*i.e.*, 37ºC), evaporation from the surface of the body becomes the only mechanism available for cooling the body.

Since all of the energy produced in the body is ultimately dissipated as heat, measurement of the heat production of an animal is a way to estimate its energy expenditure. Following are the 2 methods of accomplishing this:

(*a*) **Direct calorimetry:** In it, the subject is placed in an insulated chamber; his heat production is measured directly by recording the total amount of heat transferred to a weighed quantity of water circulating through the calorimeter. The O_2 intake, the CO_2 output, and the nitrogen excretion in the urine and feces are also measured during the entire period of observation. However, direct calorimetry is attended by considerable technical difficulties.

(*b*) **Indirect calorimetry:** Here, by measuring gas exchange and determining the respiratory quotient, energy metabolism studies are considerably simplified and thus rendered applicable to field studies and to clinical analysis.

The use of the indirect calorimetry for calculating the total energy output and the proportions of various foodstuffs being burned may be illustrated by the following example:

The subject utilized at a rate of 412.6 L/d and eliminated 348.2 L of CO_2 in the same period. The urinary nitrogen for the day was 12.6 g. Because of the incomplete metabolism of protein, the gas exchange is corrected for the amount of protein metabolized; a nonprotein RQ (nonprotein portion of the total RQ) is thus obtained. One g of urinary nitrogen represents the combustion of an amount of protein which would require 5.92 L of oxygen and would eliminate 4.75 L of CO_2. With these data, the values for various parameters can be derived at.

A. Calculation of the nonprotein RQ

Multiply the amount of urinary nitrogen by the number of litres of oxygen required to oxidize that amount of protein represented by 1 g of urinary nitrogen.

$$12.6\,\text{g} \times 5.92 = 74.59\,\text{L}$$

Multiply the amount of urinary nitrogen by the number of litres of carbon dioxide which result from this oxidation.

$$12.6\,\text{g} \times 4.75 = 59.85\,\text{L of } CO_2$$

Thus, 74.59 L of the total oxygen intake was used to oxidize protein, and 59.85 L of the CO_2 eliminated was the product of this oxidation. The remainder was used for the oxidation of carbohydrates and fats. Therefore, to determine the nonprotein RQ, subtract these values for protein from the totals for the day.

$$\text{Oxygen}: \quad 412.6 - 74.59 = 338.01\,\text{L}$$

$$\text{Carbon dioxide}: \quad 348.2 - 59.85 = 288.35\,\text{L}$$

$$\frac{CO_2}{O_2} = \frac{288.35}{338.01} = 0.85\ \text{(nonprotein RQ)}$$

B. Conversion of the nonprotein RQ to grams of carbohydrate and fat metabolized

Reliable tables have been worked out which give the proportions of carbohydrate and of fat metabolized at various RQs. According to the Table 36–3, when the nonprotein RQ is 0.85, 0.580 g of carbohydrate and 0.267 g of fat are metabolized per litre of oxygen used. Therefore, to determine the total quantities of carbohydyrate and fat used, multiply these figures by the number of litres of oxygen (derived from the nonprotein RQ) consumed during the combustion of carbohydrate and fat.

$$338.01 \times 0.580\ \text{g} = 196.04\ \text{g carbohydrate}$$

$$338.01 \times 0.267\ \text{g} = 90.24\ \text{g fat}$$

C. Determination of the amount of protein metabolized

Each gram of urinary nitrogen represents the oxidation fo 6.25 g of protein. Therefore, to determine the amount of protein metabolized, multiply the total urinary nitrogen by 6.25.

$$12.6\ \text{g} \times 6.25 = 78.75\ \text{g of protein}$$

Table 36–3. The significance of the nonprotein respiratory quotient as regards the heat value of 1 L of oxygen, and the relative quantity in calories of carbohydrate and fat consumed*

Nonprotein respiratory quotient	One litre of oxygen is equivalent to: *Grams* — *Carbohydrate*	*Fat*	*Calories*
0.707	0.000	0.502	4.686
0.71	0.016	0.497	4.690
0.72	0.055	0.482	4.702
0.73	0.094	0.465	0.714
0.74	0.134	0.450	4.727
0.75	0.173	0.433	4.739
0.76	0.213	0.417	4.751
0.77	0.254	0.400	4.764
0.78	0.294	0.384	4.776
0.79	0.334	0.368	4.788

0.80	0.375	0.350	4.801
0.81	0.415	0.334	4.813
0.82	0.456	0.317	4.825
0.83	0.498	0.301	4.838
0.84	0.539	0.284	4.850
0.85	0.580	0.267	4.862
0.86	0.622	0.249	4.875
0.87	0.666	0.232	4.887
0.88	0.708	0.215	4.899
0.89	0.741	0.197	4.911
0.90	0.793	0.180	4.924
0.91	0.836	0.162	4.936
0.92	0.878	0.145	0.948
0.93	0.922	0.127	4.961
0.94	0.966	0.109	4.973
0.95	1.010	0.091	4.985
0.96	1.053	0.073	4.998
0.97	1.098	0.055	5.010
0.98	1.142	0.036	5.022
0.99	1.185	0.018	5.035
1.00	1.232	0.000	5.047

* Table of **Zuntz** and **Schumberg**, as modified by **Lusk**, and later by **McClendon**

(Reproduced from: Bodansky M: Introduction to Physiological Chemistry. 4th edition. Wiley. 1938.)

D. Calculation of the total heat production

Multiply the quantity in grams of each foodstuff oxidized by the caloric value of that food to obtain the heat productioin due to its combustion. The sum of these caloric values equals the total heat production of the diet.

$$\begin{aligned} \text{Carbohydrate} &: 196.04 \text{ g} \times 4 \text{ kcal} = 784.16 \text{ kcal} \\ \text{Protein} &: 78.75 \text{ g} \times 4 \text{ kcal} = 315.00 \text{ kcal} \\ \text{Fat} &: 90.24 \text{ g} \times 9 \text{ kcal} = 812.16 \text{ kcal} \\ \text{Total heat production} &= 1911.32 \text{ kcal} \end{aligned}$$

ENERGY METABOLISM

The total heat production or energy expenditure of the body is the sum of that required merely to mainain life (*i.e.*, **basal metabolism**), together with such additional energy as may be expended for any additional activities. The lowest level of energy production consonant with life is the **basal metabolic rate (BMR).** For measuring the BMR, following conditions should be adhered to:

1. The subject/patient must be in a post-absorptive state, *i.e.*, should have had nothing orally for the past 12 hours.
2. The subject must be mentally and physically relaxed immediately prior to the test. For this, usually half-an-hour of bed rest is used, although ideally the patient should not arise from bed after a prolonged rest.
3. The subject must be in a recumbent position during the test.
4. Patient must be awake while undergoing test.
5. The environmental temperature of the vicinity should be between 20 and 25°C.

A. Factors Affecting Basal Metabolism

Following principal factors influence energy expenditure:

1. ***Surface area***: The surface area of an individual is directly related to the rate of heat loss by the body– the greater the surface area, the greater the rate of heat loss. While it may seem surprising, a lean individual actually has a greater surface area, and thus a greater energy requirement, than an obese individual of the same weight. The basal metabolic rates (BMRs) of different individuals, when expressed in terms of surface area (m^2), are remarkably constant. In general, however, smaller individuals have a higher rate of metabolism per unit of surface area than larger individuals.
2. ***Age:*** Age may reflect 2 factors: growth and lean muscle mass. In infants and children, more energy expenditure is required for rapid growth, which is reflected in a higher BMR (rate of energy utilization in the resting state). In adults (even lean adults), muscle tissue is gradually replaced by fat and water during the ageing process, resulting in a 2% decrease in BMR per decade of adult life. In the newborn, the BMR is low; it rises to maximum at age 5 or 6 (50 – 53 kcal/m^2), after which the rate begins to decline (at age 21, it is between 36 and 41 kcal/m^2), continuing into old age. There is, however, a relative rise just before puberty.
3. ***Sex***: Women normally tend to have a lower basal metabolic rate than men due to a smaller percentage of lean muscle mass and the effects of female hormones on metabolism. Interestingly, the BMR of females declines more rapidly between the age group 5–17 than that of males.
4. ***Racial groups***: Different racial groups exhibit variations in the BMR. For example, the BMR of Oriental women living in the U.S. average 10% below the standard BMR for the American women of the same age; the BMRs of adult Chinese are equal to or below the lower limit of normal for Occidentals; high values (33% above normal) have been reported for Eskimos living in the region of Baffin Bay.
5. ***Climate***: The basal metabolic rate is lower in warm climates.
6. ***Nutrition state***: In starvation or semistarvation, the BMR can decrease up to 50% because energy intake itself has an inverse relatonship to expenditure. This is of great survival value in cases of genuine starvation, but not much help to the person who wishes to lose weight on a calorie-restricted diet.
7. ***Activity level***: The effect of activity levels on energy requirements is obvious. However, most people overemphasize the immediate, as opposed to the long-term, effects of exercise. For example, one would need to jog for over an hour to burn up the calories present in one piece of apple pie. The effect of a regular exercise programme on energy expenditure can be quite beneficial. Regular exercise increases BMR, allowing calories to burn up more rapidly 24 hours a day. A regular exercise programme should be designed to increase lean muscle mass and should be repeated 3 to 5 days a week. For an elderly or infirm individual, even daily walking may help to increase BMR slightly.
8. ***Hormone levels***: The hormones also affect metabolism. *Thyroxine* is the most important of the hormones in this respect, and the principal use of calorimetry in clinical practice is in the diagnosis of thyroid disease. The basal metabolic rate is lowered in hypothyroidism and increased in hyperthyroidism. Changes in the BMR are also noted in pituitary disease. Other than thyroxine, the only hormone which has a direct effect on the rate of heat production is *epinephrine*, although the effect of epinephrine is rapid in onset and brief in duration. The effects of epinephrine and cortisol probably explain, in part, why severe stress and major trauma significantly increase energy requirements. Tumours of the adrenal (pheochromocytoma) cause an elevation in the BMR. In adrenal insufficiency (Addison's disease), the basal metabolism is subnormal, whereas adrenal tumours and Cushing's disease may produce a slight increase in the metabolic rate.

B. Measurement of Basal Metabolism

In clinical practice, the BMR can be estimated with sufficient accuracy merely by measuring the oxygen consumption of the patient for 2 six-minute periods under basal conditions. This is corrected to standard conditions of temperature and barometric pressure. The averge oxygen consumption for the 2 periods is multiplied by 10 to convert it to an hourly basis (6 × 10 = 60 minutes) and then multiplied by 4.825 kcal, the heat production represented by each litre of oxygen consumed. This provides the heat production of the patient in kilocalories per hour. This is corrected to kilocalories per square metre of body surface per hour by dividing kilocalories per hour by the patient's surface area. The **simple formula** for calculating the surface area is:

$$\underset{\text{(in cm)} \times 2}{\text{Circumference of midthigh}} \times \underset{\text{(in cm)}}{\text{Height}} = \underset{(\text{in cm}^2)}{\text{Surface area}}$$

The classical formula, **DuBois' surface area formula**, is as follows:

$$A = H^{0.725} \times W^{0.425} \times 71.84$$

where, A = surface area in cm^2

H = height in cm

W = weight in kg

(Surface area in cm^2/10,000 = Surface area in m^2)

In practice, a nomogram which relates height and weight to surface area is used. It is based on the Du Bois formula.

EUGENE FLOYD DuBois

(LT, 1882–1959)

DuBois, a metabolist, was born on June 4, 1882 in West New Brighton States, Island, New York. He studied at Milton Academy, near Boston. He studied humanities at Harvard. He joined Columbia College of Physicians and Surgeons in 1902 and graduated in 1906. He spent 6 months in Berlin working in a Path Lab, then returned to New York and completed his term as House Physician. Later, he went to Germany and worked in Krau's Clinic on the energy requirements in diabetes mellitus. In 1909, DuBois returned to New York. In 1911, he was appointed Medical Director of the Russel Sage Institute of Pathology. He held this post till his retirement.

The first study he carried out was to determine the value of high calorie diets in typhoid fever. Later, he carried out energy metabolism studies in various other diseases such as those of thyroids, which have become classical. He also carried out studies in various other diseases such as those of thyroids, which have become classical. He also carried out studies on the basal metabolism of normal subjects. He evolved a formula (called **DuBois formula**) for calculating a person's body surface area from his weight and height:

$$A = 71.84\, W^{0.425}\, H^{0.725}$$

where, A = surface area (*in* cm^2)

W = weight (*in kg*)

H = height (*in cm*)

Jointly with a U.S. physician **Joseph C. Aub** (LT, 1890–1973), DuBois devised a table (called **Aub-DuBois table**) of basal metabolic rates in calories per square metre of body surface per hour or day for different ages. He published a classical book 'Basal Metabolism in Health and Disease'. He died on February 12, 1959.

C. Calculation of Basal Metabolic Rate

The normal BMR for an individual of the patient's age and sex is obtained from standard tables. The patient's actual BMR is expressed as a + or – percentage of the normal. An example will clarify this concept.

A man, aged 35 years, 170 cm tall and weighing 70 kg, consumed an average of 1.2 L of oxygen (corrected to normal temperature and pressure: 0ºC, 760 mm Hg) in a six-minute period.

$$1.2 \times 10 = 12 \text{ L of oxygen/h}$$

$$12 \times 4.825 = 58 \text{ kcal/h}$$

$$\text{Surface area} = 1.8 \text{ m}^2 \text{ (from Du Bois' formula)}$$

$$\text{Basal metabolic rate} = 58 \text{ kcal}/1.8$$

$$= 32 \text{ kcal/m}^2\text{/h}$$

The normal BMR for this patient, by reference to the Du Bois standards, is 39.5 kcal/m^2/h. His BMR, which is below normal, is then reported as:

$$\frac{39.5 - 32}{39.5} \times 100 = 18.98\% \quad \text{or} \quad -18.98$$

A basal metabolic rate between –15 and +20% is considered normal. In hyperthyroidism, the basal metabolic rate may exceed +50 to +75%. The basal metabolic rate in an hypothyroidic individual may be –30 to –60%.

MEASUREMENT OF ENERGY REQUIREMENTS

The metabolic rate increases with activity. Maximal increases occur durign exercise– as much as 600 to 800% over BMR. The energy requirement over a 24-hour period, therefore, will vary considerably in accordance with the physical activity in which the subject is engaging himself. Examples of energy expenditure of mature men and women engaged in light occupations are given in Table 36–4.

Table 36–4 Daily energy expenditures of mature men and women in light occupations*

Activity category	*Time* (h)	*Man, 70 kg* *Rate* (kcal/min) †	*Man, 70 kg* *Total* (kcal [kJ]) †	*Woman, 58 kg* *Rate* (kcal/min)	*Woman, 58 kg* *Total* (kcal [kJ])
Sleeping, reclining	8	1.0 – 1.2	540 (2270)	0.9–1.1	440 (1850)
Very light Seated and standing activities, auto and truck driving, laboratory work, typing, playing musical instruments, sewing, ironing	12	Up to 2.5	1300 (5460)	Up to 2.0	900 (3780)
Light Walking on level 2.5–3 mph, tailoring, pressing, garage work, electrical trades, carpentry, restaurant trades, cannery workers, washing clothes, shopping with light load, golf, sailing, table tennis, volleyball	3	2.5–4.9	600 (2520)	2.0–3.9	450 (1890)
Moderate Walking 3.5–4 mph, plastering, weeding and hoeing, loading and stacking bales, scrubbing floors, shopping with heavy load, cycling, skiing, tennis, dancing	1	5.0 – 7.4	300 (1260)	4.0–5.9	240 (1010)

Heavy Walking with load uphill, tree felling, work with pick and shovel, basketball, swimming, climbing, football	0	7.5–12.0		6.0–10.0
Total	**24**		**2740 (11,510)**	**2030 (8530)**

* The data are from **Durnin JVGA** and **Passmore R**: *Energy, Work and Leisure*. Heineman. 1967.

† kcal = kilocalories (Cal); kJ = kilojoules = 4.2 kcal

(*Reproduced from: Recommended Dietary Allowances. 8th edition. National Academy of Sciences. 1974.*)

It is suggested that the daily energy requirement of a moderately active person might be increased by about 300 kcal over the totals given in Table 36–4. For very active persons such as those engaged in heavy athletic pursuits, military personnel in training, or workers in heavy construction, the allowance might be increased by 600–900 kcal/d.

In attempts to regulate body weight, prime emphasis must be placed on restriction of excess caloric intake. Weight loss cannot be achieved under any conditions as long as caloric intake exceeds expenditure, including provision for growth, if applicable. However, exercise as a means to assist in weight loss or to control weight is useful in addition to caloric restriction.

SPECIFIC DYNAMIC ACTION

The **specific dynamic action** (SDA) of a foodstuff is the extra heat production, over and above the caloric value of a given amount of food, which is produced when this food is used by the body. For example, when an amount of **protein** which contains 100 kcal (25 g) is metabolized, the heat production in the body is not 100 kcal but 130 kcal. This extra 30 kcal is the product of the specific dynamic action of the protein. In the body, a 100 kcal portion of **fat** produces 113 kcal, and a100 kcal portion of **carbohydrate** produces 105 kcal. The origin of this extra heat is not clear, but it is attributable to the activity of the tissues metabolizing these foodstuffs.

The SDA of each food, as given above, is obtained when each foodstuff is fed separately; but when these foods are taken in a mixed diet, the dynamic effect of the whole diet cannot be predicted by merely adding the individual effects of each foodstuff in accordance with its contribution to the

GRAHAM LUSK
(LT, 1866–1932)

Lusk was born on February 15, 1866 in Bridgeport, Connecticut. He graduated from the Columbia University, in 1887. He spent the next 4 years in Germany. He worked with **Carl von Voit** (LT, 1831–1908) at Munich and received his Ph.D. degree in 1891 from the University of Munich. On his return to U.S.A., Lusk was appointed instructor of physiology at Yale University, in 1891. He became assistant professor in the following year and Professor of Physiology, in 1895. He later moved over to the University Medical College, New York city as Professor of Physiology in 1898, which post he held for 11 years. In 1911, he was appointed Professor of Physiology at Cornell University, which post he continued for 21 years.

Lusk did pioneering work on the production of glucose from amino acids using phorhizinised dogs. He also built a **calorimeter** on the model of Atwater-Benedict-Rosa calorimeter for studying energy metabolism in dogs. Lusk also discovered the phenomenon of **specific dynamic action** (SDA) **of foods,** which is the extra heat produced by the organism, over and above the basal heat production, as a result of food ingestion. He made important contributions to explain the basis of SDA. He also wrote a remarkable book entitled 'The Science of Nutrition' in 1906. Its revised editions appeared successively, the last one being in 1928. Lusk died in 1932. Among his numerous collaborators are: **JR Murlin, AI Ringer, EF DuBois, RJ Anderson, HJ Deuel** and **WH Chambers.**

diet, *e.g.*, 30% of the caloric value of protein, 13% of the caloric value of fat, and 5% of the caloric value of cabrohydrate. According to **Forbes**, the dynamic effects of beef muscle protein, glucose, and lard (fat), when each was fed separately, were 32, 20 and 16%, respectively, of their caloric content. But when the glucose and protein were combined, the dynamic effect was 12.5% less than predicted from the sum of their individual effects; combining lard, glucose, and protein produced 22% less dynamic effect than predicted; a glucose-lard combination was 35% less, and a protein-lard mixture 54% less than calculated from the individual percentages of each foodstuff.

These observations are of importance because they indicate that the high specific dynamic action of protein can be reduced depending on the quantities of other foodstuffs in the diet. Forbes' data show that fat (lard) has a greater influence on SDA than does any other nutrient, *i.e.*, fat decreases the SDA more than any other nutrient.

In calculating the total energy requirement for the day, it is customary to add 10% to the total caloric requirement to provide energy for the specific dynamic action, *i.e.*, expense of utilization of the foods consumed. This figure may be too large, particularly if the fat content of the diet is high.

NUTRITIONAL REQUIREMENTS OF HUMANS

Several difficulties are encountered in studies of human nutrition. The most serious of these are the impracticality of feeding human subjects rations consisting exclusively of chemically-purified materials for comparatively longer periods and the uncertainty concerning the qualitative and quantitative activities of the intestinal flora. It has been particularly difficult to establish the quantitative requirements for many nutrients known, qualitatively, to be essential.

Nitrogen balance studies have yielded values for the amounts of the essential **amino acids** required for short periods of time for maintenance of nitrogen balance in human adults. However, the amino acid requirements for growth have not been established for the human beings. Balance studies have also been employed to establish quantitative requirements for elements such as calcium, potassium, and iron. Perhaps the most striking result of these studies is the extreme variability found within the population. The amount of **calcium** required to maintain one individual in balance may be 2 or 3 times that which is sufficient for another. A similar situation exists with respect to **iron**. The quantitative requirements for **vitamins** are even less securely established. The recommended dietary allowances (RDAs), established for various foodstuffs, represent dietary levels which, in clinical experience, are known *not* to lead to deficiency; they are not intended to represent minimal requirements but are objectives which at present are thought desirable to seek in planning diets for normal individuals.

DIETARY REFERENCE INTAKES

Dietary reference intakes (DRIs) are quantitative estimates of nutrient intakes to be used for planning and assessing diets for healthy people. The Food and Nutrition Board of the National Academy of Sciences, National Research Council, has collected the best available data on the quantities of various nutrients required by normal persons of varying ages and body sizes, and in different physiologic states. The first recommendations derived from such data were published in 1943. At intervals of approximately 5 years since that time, recommendations for vitamins B_6 (pyridoxine), B_9 (folacin), B_{12} (cyanocobalamin), and vitamin E (tocopherol) as well as for phosphorus, iodine, magnesium, and zinc have been added to the recommendations previously made for calories, protein, calcium, iron, and vitamins A (retinol), B_1(thamine), B_2 (riboflavin), B_5 (niacin), C (ascorbic acid), and D (cholecalciferol). Note that the Dietary Reference Intakes (DRIs) have replaced the Dietary Reference Values (DRVs) used previously in the UK, and the Recommended Dietary Allowances (RDAs) used in the USA. Note that DRIs are intended for healthy people. Also, a DRI established for any one nutrient presupposes that requirements for other ones are being met.

In assessing quantitative standards for nutrient intake, the Food and Nutrition Board considers the amount of nutrients and food components required for preventing deficiency diseases and, where the data are definitive, for promoting optimal health. The first step in this process is determining the **estimated average requirement** (EAR), which is the amount of nutrient estimated to meet the nutrient requirement of half of the healthy individuals in a life-style and gender group. The **recommended dietary allowance** (RDA), or in the UK the **reference nutrient intake** (RNI), is normally set at 2 standard deviations above the EAR and is assumed to be the dietary intake amount that is sufficient to meet the nutrient requirement of nearly all (97–98%) of healthy individuals in a group. If a nutrient is considered essential but the experimental data are considered inadequate for determining an EAR, an **adequate intake** (AI) is set rather than an RDA. The AI is based on observed or experimentally-determined approximations of nutrient intake by a group of individuals. For example, the AI for young infants is based on the daily mean nutrient intake supplied by human milk for healthy, full-term infants who are exclusively breast-fed. Lastly, for most nutrients, the Food and Nutrition Board sets a **Tolerable Upper Intake Level** (UL). The UL is defined as the highest level of daily nutrient intake that is likely to pose no risks of adverse health effects to almost all individuals in the general population. The RDAs, AIs and ULs are designed to be of use in planning and evaluating diets for individuals. The EAR is designed to be used in setting goals for nutrient intake and assessing the prevalence of inadequate intake in a population group.

Dietary intake is not easy to assess. Available date are based on (sometimes incomplete) population surveys. Sets of values derived from these describe suggested minimal average, and adequate intakes of particular nutrients. Different values are used in different countries, and there has been a degree of confusion and overlap between various definitions, as seen above. Currently, the estimates of nutrient intake are based on the dietary reference intakes (DRIs), the sets of values that describe the intake of a given nutrient in a population (Fig. 36–4). The DRIs include: EAR, RDA, AI, and UL. The EAR and RDA values are given for a nutrient or, if these cannot be determined, AI is used.

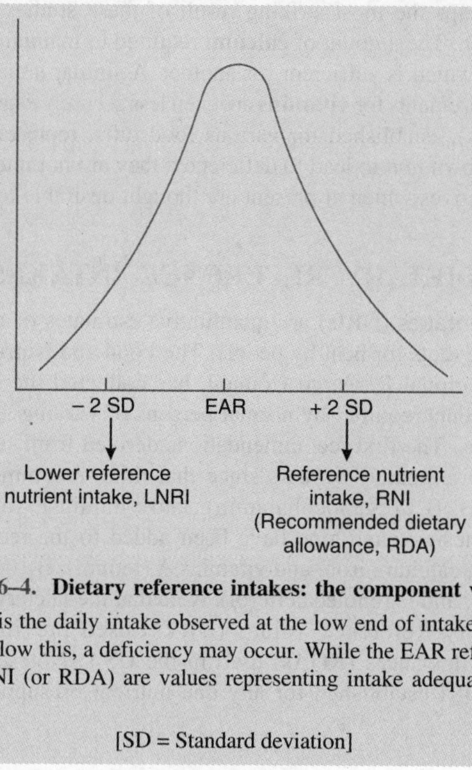

Fig. 36–4. Dietary reference intakes: the component values

The LNRI, used in the UK, is the daily intake observed at the low end of intake distribution in a population (about 2%). If intake falls below this, a deficiency may occur. While the EAR reflects the intake adequate for half of a population, the RNI (or RDA) are values representing intake adequate for the great majority of individuals.

[SD = Standard deviation]

The DRI determinations are relatively easy to make for those nutrients associated with dramatic deficiency diseases; for example, vitamin C and survey. In other instances, more indirect measures must be used, such as tissue saturation or extrapolation from animal studies. There are no criteria that can be used for all micronutrients, and there are always some uncertainty and debate as to the correct criteria. The criteria are changed by new research. The Food and Nutrition Board normally meets every 6–10 years to consider currently available information and update its recommendations.

Unless the purposes of the RDAs are kept in mind, it might be concluded that the allowances are too high. The recommendations are designed to set levels suitable for maintenance of good nutrition in the majority of healthy individuals. Thus, they are proposed as a means of fixing goals in planning food supplies and interpreting food intakes of groups of people. It follows that the allowances may be higher than required for some individuals. Indeed, the allowances should not be interpreted as specific requirements for individuals unless it is understood that the stated amounts of each nutrient may, in fact, be too high for a given person. Therefore, the diet of an individual may not necessarily be regarded as deficient if it does not equal in every respect the recommendations specified.

Recommended dietary allowances (RDAs) serve a useful general guide in evaluating adequacies of individual diets. However, the RDAs have several limitations that should be kept in mind. Important *limitations* are the following:

1. RDAs are designed to meet the needs of healthy people and do not take into account special needs arising from infections, metabolic disorders, or chronic diseases.
2. Since present knowledge of nutritional needs is incomplete, there may be unrecognized nutritional needs. To provide for these needs, the RDAs should be met from as varied a selection of foods as possible. No single food can be considered, even if it meets the RDA for all known nutrients. This is important, especially in light of the current practice of fortifying foods of otherwise low nutritional value.
3. RDAs may not define the "optimal" level of any nutrient, since optimal amounts are difficult to define. Because of information suggesting that optical intake of certain micronutrients (*e.g.*, vitamins A, C and D) may reduce heart disease and cancer risk, the RDAs for these nutrients have recently been increased slightly; however, some experts feel that the current RDAs may not be sufficient to promote optimal health.

In order to simplify the concept of an adequate diet for normal individuals, foodstuffs have been arranged into 4 groups, each of which makes a major contribution to the diet. It is recommended that some food from each group be taken daily, as specified below:

1. **Milk group:** For children, 3 or more 8 oz glasses; for teenagers, 4 or more glasses; for adults, 2 or more glasses. Cheese, ice cream, and other milk products can supply part of the milk recommended.
2. **Meat group:** Two or more servings of meat, fish, poultry, eggs, or cheese. Dry beans, peas, and nuts are alternates, although they cannot substitute entirely for meat, fish, poultry, or eggs.
3. **Vegetables and fruits:** Four or more servings, including some dark green or yellow vegetables, citrus fruits or tomatoes. Some uncooked vegetables should be included.
4. **Breads and cereals:** Four or more servings of enriched or whole-grain breads and cereals.

RECOMMENDED DIETARY ALLOWANCES FOR INDIANS

The Nutrition Advisory Committee of Indian Research Fund Association, now **Indian Council of Medical Research**, recommended for the first time dietary allowances for Indians to energy, proteins, iron, calcium, vitamin A, thiamine, ascorbic acid and vitamin D. These recommendations were revised in 1958 in respect of energy and protein. The Nutrition Expert Group of the Indian Council of Medical Research revised, in 1968, the earlier recommendations of nutrient allowances for Indians in respect of all nutrients, except energy. This Committee also considered the desirable levels of intake of additional nutrients like folic acid, vitamin B_{12}, vitamin D and fat. In making these recommendations, the Committee was guided by the safe allowances of various nutrients suggested by the expert groups of the Food and Agricultural Organisations and the World Health Organization and data on nutrient requirement of Indians. These recommendations were again revised, in 1978, by an Expert Group of the I.C.M.R. and allowances were updated and revised in respect of several nutrients. Nutrient allowances for Indians were again examined in detail by an Expert Committee of the I.C.M.R. in April 1988 and detailed recommendations with respect to safe levels of intake of all nutrients were made. The basis of these recommendations is discussed in detail in the Report of the Committee. Here, the recommended intakes are given in Table 36–5 and the main points on which they are based and guidelines to use them are described below briefly.

1. **Body weight:**
   (*a*) ***Reference man and woman:*** On the basis of height of Indian adult, body weight of 60 kg and 50 kg for reference man and reference woman, respectively, should be used in arriving at RDA, instead of 55 kg and 40 kg used earlier.
   (*b*) ***Growing infants and children:*** For them, the desirable body weights are to be used. In the absence of Indian data, NCHS standards will be used for the time being.
2. **Energy allowances:**
   (*a*) Energy requirement is expressed in terms of basal metabolic rate (BMR) values which are derived from the body weight of Indians using the equation given by FAO, 1980 and the values reduced by 5% to allow for the lower BMR of Indians.
   (*b*) Energy allowances for children and adolescents are for normally growing healthy Indian children. Allowance is made for the age and no correction for actual body weight to children need to be made.
3. **Protein allowances:** Protein allowances are given in terms of mixed vegetable proteins contained in Indian diets, the NPU of which is assumed to be 65. Adult protein requirement is 1 g/kg.
4. **Fat requirement:** Fat requirement was examined in great detail and more realistic estimates of minimum intakes have been given. In arriving at the requirements, the total invisible fat content of cereal-based diets eaten in India and the minimum EFA requirement of various groups were considered.
5. **Minerals:**

   Iron: Iron requirements have been arrived at using the factorial approach taking into account the basal loss in case of men, basal loss + menstrual loss in case of women and basal loss + growth requirement in case of children. Dietary iron requirements are given at levels of absorption of 5%, 3% and 2% applicable respectively to the diet based on rice, based on mixed cereals (rice + wheat) and based wholly on wheat or millets.

   Calcium and phosphorus: Because of close relationship between Ca and P, allowances for both these nutrients are given. The desirable ratio of Ca : P is 1 : 5 in case of infants and 1 : 1 in case of other groups.

   Trace elements: Desirable daily intakes of some trace elements have been suggested for an adult: chromium 65 μg, copper 2.2 mg, manganese 5.5 mg, zinc 15.5 mg and molybdenum 500 μg.

Table 36–5. Recommended dietary allowances for Indians

Group	Particulars	Body Wt/kg	Net energy KCal	Protein g/d	Fat g/d	Calcium mg/d	Iron mg/d	Vit. A µg/d Retinol	Vit. A µg/d β-carotene	Thiamine mg/d	Riboflavin mg/d	Nicotinic acid mg/d	Pyridoxine mg/d	Ascorbic acid mg/d	Folic acid µg/d	Cyanocobalamin µg/d
Man	Sedentary work		2350							1.2	1.4	16				
	Moderate work	60	2700	60	15	400	28	600	2400	1.4	1.6	18	2.0	40	100	1
	Heavy work		3200							1.6	1.9	21				
Woman	Sedentary work		1800							0.9	1.1	12				
	Moderate work		2100	50	15	400	30	600	2400	1.1	1.3	14	2.0	40	100	1
	Heavy work		2450							1.2	1.5	16				
	Pregnant woman	50	+ 300	+15	25	1000	38	600	2400	+0.2	+0.2	+2	2.5	40	400	1
	Lactation															
	0–6 months		+550	+25						+0.3	+0.3	+4				
					35	1000	30	950	3800				2.5	80	150	1.5
	6–12 months		+ 400	+ 18						+0.2	+0.2	+3				
Infants	0–6 months	5.4	118/kg	2.05/kg						59 µg/kg	71 µg/kg	780 µg/kg	0.3			
						500								25	25	0.2
	6–12 months	8.6	108/kg	1.65/kg				300	1200	54µg/kg	65 µg/kg	710 µg/kg	0.4			
Children	1–3 years	12.1	1125	23	20		12	400	1600	0.6	0.7	7.0	0.9			
	4–6 years	18.2	1600	31		400	18	400	1600	0.8	1.0	11				
	7–9 years	25.2	1925	41	15		25	600	2400	1.0	1.2	13				
Boys	10–12 years	33.5	2150	53	15		28	600	2400	1.1	1.3	14	1.6			
Girls	10–12 years	35.0	1950	55	15	600	20			1.0	1.2	13		40	100	0.2–1.0
Boys	13–15 years	46.8	2400	71	15		43	600	2400	1.2	1.4	16				
Girls	13–15 years	47.8	2050	67	15	600	28			1.0	1.2	14				
Boys	16–18 years	56.1	2600	79	15		50	600	2400	1.3	1.6	17	2.0			
Girls	16–18 years	49.7	2050	65	15	500	30			1.0	1.2	14				

(Courtesy: Indian Council of Medical Research, 1989)

6. **Vitamins:**

Vitamin A: The recommended level of vitamin A (retinol) for adult is 600 μg. In terms of beta-carotene, it will be 2,400 μg, the conversion factors to be used in case of retinol and beta-carotene is as follows:

1 μg beta-carotene = 0.25 μ retinol

If diet contains both retinol and beta-carotene, its vitamin A content can be expressed as follows:

(*a*) Retinol content (μg) = μg beta-carotene × 0.25, it retinol and beta-carotene are expressed in μg.

(*b*) Retinol content (μg) = vitamin A IU × 0.3 + beta-carotene IU × 0.15, if vitamin A and beta-carotene are given in International Units (IU).

Thiamine, riboflavin and nicotinic acid: The daily allowances for these vitamins are related to energy intake. The basic allowances per 1,000 KCal are: thiamine 0.5 mg, riboflavin 0.6 mg and niotinic acid 6.6 mg niacin equivalent. Niacin allowance takes into account the contribution of tryptophan also, assuming that 60 mg dietary tryptophan yielded 1 mg of niacin in the body.

$$\text{Niacin equivalents} = \text{Niacin content} + \frac{\text{Tryptophan content}}{60} \ (\text{mg})$$

Folic acid: Allowance for folic acid is in terms of free folate (L. casein activity) present in foods. RDA of folate in pregnancy will be 300 μg in addition to normal requirement of 100 μg. Since such high levels of folate cannot be obtained, therapeutic supplementation becomes necessary in pregnancy. Folic acid requirement can be expressed on the basis of body weight also, *i.e.*, 3 μg/kg.

Vitamin B_{12}: Recommended intake of vitamin B_{12} is 1 μg/day entirely derived from animal foods. This provides justificatioin for including some animal foods in our daily diet.

Vitamin C: Earlier recommendation of 40 mg was retained. In case of infants, the RDA was changed from 20 to 25 mg/day.

Vitamin D: No dietary recommendations for vitamin D was made since adequate vitamin D can be obtained through exposure to sunlight. In case of conditions of inadequacy of vitamin D, medicinal supplements can be given.

Vitamin E: An intake of 0.8 mg of vitamin E per g of polyunsaturated fatty acids was suggested.

Vitamin B_6: Earlier (1978) recommendations were retained.

MALNUTRITION

Malnutrition is defined as "a pathological state resulting from a relative or absolute deficiency or excess of one or more essential nutrients." Malnutrition is a gradual decline in nutritional status, which in its more advanced stages leads to a decrease in functional capacity and other complications. *It is a man-made disease*. It is a disease of the human societies. It begins quite commonly in the womb and ends in the grave. Not much is known about the dimensions of the problem. However, according to Food and Agriculture Organization's (FAO's) report, about 1/5th of world's population is malnourished. What makes the situation most serious is that malnutrition's main victims are children under the age of 15. But children under-5 are hit the hardest. Malnutrition also affects a large group of older individuals. In the U.K., up to 40% of patients admitted to hospitals are undernourished. In adition to malnutrition, specific deficiencies such as those of vitamin D, iron , and vitamin C may occur. On a global scale, the 5 principal nutritional deficiency diseases that are being accorded the highest priority action are: kwashiorkor, marasmus, xerophthalmia, nutritional anemias and endemic goitre (WHO, 1977). These diseases represent the tip of the "iceberg" of malnutrition; a much larger population are affected by 'hidden' malnutrition which is not easy to diagnose. The effects of malnutrition are summarized in Table 36–6. Globally, malnutrition contributes to 54% of the 11.6 million deaths annually among children below 5 years of age.

Table 36–6 Consequences of protein–calorie malnutrition

Consequences of protein–calorie malnutrition
Decreased protein synthesis
Decreased activity of Na^+/K^+-ATPase
Decreased glucose transport
Fatty liver, liver necrosis, liver fibrosis
Depression, apathy, mood changes
Hypothermia
Compromised ventilation
Compromised immune system: impaired wound healing
Risk of wound breakdown
Decreased cardiac output
Decreased renal function
Loss of muscle strength
Anorexia
Decreased mobility

The effects malnutrition on the community are both direct and indirect. The *direct* effects are the occurrence of frank and subclinical nutrition deficiency diseases such as kwashiorkor, marasmus, vitamin and mineral deficiency diseases. The *indirect* effects are a high morbidity and mortality among young children (nearly 50% of total deaths in the developing countries occur among children under 5 years of age, as compared to about 5% in the developed countries), retarded physical and mental growth and development (which may be permanent), lowered vitality of the people leading to lowered productivity and reduced life expectancy. Malnutrition predisposes to infection and infection to malnutrition; and the morbidity arising from as a result of complications from such infectious diseases as tuberculosis and gastroenteritis is considerable. The high rate of maternal mortality, stillbirth, and low birth weight are all associated with malnutrition.

In the more developed countries of the world, nutritional problems are somewhat different. Overnutrition is encountered much more frequently than undernutrition. The health hazards from overnutrition are a high incidence of obesity, diabetes, hypertension, cardiovascular and renal diseases, and disorders of liver and gall bladder. Thus, it is apparent that the consequences of malnutrition are ominous.

The point at which malnutrition starts is usually defined as a reduction in body weight below 80% of the Harvard 50th percentile; this corresponds approximately to the Harvard 3rd percentile. The Nutrition Subcomittee of the Indian Academy of Pediatrics (IAP) has classified malnutrition into 3 degrees:

(a) *First-degree malnutrition*: weight between 80 and 71% of the Harvard standard

(b) *Second-degree malnutrition*: weight between 70 and 61% of the Harvard standard

(c) *Third-degree malnutrition*: weight below 60% of the Harvard standard

The Harvard Reference Standard or median or 50th percentiles are given in a number of pediatrics textbooks. These standards allow international comparisons.

Despite significant advancements in understanding of nutrition and the fact that no one need be malnourished, malnutrition is increasingly prevalent among the world's population. Nor is it entirely true that the roots of malnutrition are solely economic; in many cases ignorance is an equally important factor. Beriberi, kwashiorkor and xerophthalmia may occur on a large scale; severe rickets, pellagra, and scurvy are less frequent.

Surveys of nutritional habits conducted on peoples in underdeveloped areas frequently indicate an appalingly low intake of nutrients. This does not, however, always result in the appearance of deficiency diseases. The explanation for this, in fact, lies in one of the basic concepts of nutrition, the *balanced diet*. In both men and experimental animals, it is known that a balanced dietary mixture, even if restricted in *total quantity*, is preferable to a regime that is adequate in most respects but badly deficient in only *one* nutrient. Inadequate quantity may lead to delayed growth, anemia, or hypoproteinemia, but the imbalanced diet results in pellagra, beriberi, scurvy etc. For example, the rat fed a diet completely lacking in B vitamins survives for a surprising length of time, while a deficiency of thiamine alone may be fatal in a few weeks. Genetic differences among individuals may drastically change the quantitative, or even qualitative, composition of the balanced dietary mixture desirable for optimal health. Thus, galactosemic children must eschew lactose, phenylketonuric individuals must ingest a minimum of phenylalanine, and the diabetics must carefully balance dietary carbohydrate and lipid.

Improved methods of laboratory diagnosis allow detection of early deficiencies of many of the essential nutrients. Application of these techniques, particularly in large-scale nutritional surveys of populations, has helped to improve the nutritional status of the Western World while revealing the enormous magnitude of the problem of undernourishment of peoples living in the equatorial belt.

Relatively few investigations of conditions for *optimal nutrition* have been reported; even the criteria for optimal nutrition have not been established. The criterion generally employed in experimental nutrition has been the weight response of young rats. However, it is not established that ingestion of a diet which permits maximum weight gains through childhood and adolescence contributes to long and healthy life. Indeed, available experimental evidence suggests that diets which provide the minimal intake of essential nutrients compatible with health also promote longevity. Should this prove generally to be the case, many of the recommendations that have been given earlier may require revision.

Malnutrition and Infection

It is generaly recognized that frequent occurrence of infectious diseases such as diarrhea, dysentery, or malaria affects adversely the health of the individuals. In the developing countries, infection and infestation are important factors involved in the causation of severe malnutrition among preschool children consuming inadequate and ill-balanced diets. Some of the ill effects of infection and infestation are:

(*a*) decreased food intake due to poor appetite,

(*b*) decreased absorption of nutrients due to diarrhea, vomiting or long-continued micro-hemorrhage, and

(*c*) increased needs of nutrients for human host due to increase in basal metabolism and the breakdown of tissue proteins.

The nutritionally-relevant infections and infestations are of 3 types: bacterial, viral and parasitic.

1. **Bacterial infections:** These increase the requirements for proteins and other nutrients. Gastroenteritis causes high mortality among malnourished infants and preschool children in the tropics. Those who survive develop severe forms of malnutrition such as kwashiorkor or marasmus. Another condition known as 'weaning diarrhea' is caused by bacteria which is not ordinarily pathogenic in a well-nourished child. Weaning diarrhea occurs extensively among weaned infants in the developing countries and is responsible for precipitating the occurrence of kwashiorkor or marasmus in malnourished children.

2. **Viral infections:** Measles cause high mortality among malnourished children. Those who survive develop severe protein–calorie malnutrition. Malnourished who suffer from frequent attacks of influenza are also prone to develop protein–calorie malnutrition.
3. **Parasitic infections:** Malarial parasites cause hemolytic anemia and depletion of protein reserves. Hookworm infection causes constant loss of blood and hence anemia. Ascariasis interferes with the absorption of nutrients.

Malnutrition and Possible Defence Mechanism

Malnutrition affects adversely various defence mechanisms in the human body. Some of these are as follows:

1. **Antibody formation:** In experimental animals, deficiencies of various vitamins of the B-complex, vitamin A, vitamin C and proteins have all been shown to interfere with antibody formation. In man, antibody formation is low in deficiencies of protein, vitamin B_3 (panthothenic acid), vitamin B_6 (pyridoxine) and multiple factors associated with grossly inadequate diet.
2. **Phagocytic activity:** Severe undernutrition or protein depletion in experimental animals leads to atrophy of liver, spleen, bone marrow, and lymphoid tissue from which phagocytes originate. Accumulated evidences all indicate that deficiencies of vitamins A, B_1, B_2, B_6, B_9, and C reduce the activity of phagocytes.
3. **Tissue integrity:** Dietary inadequacies decrease resistance of the tissues to infection by causing pathological changes in epithelial and other cells. The gross epithelial lesions resulting from nutrient deficiencies include (*a*) megaplastic hyperkeratosis, (*b*) dermatitis, cheilosis, and angular stomatitis resulting from ariboflavinosis, (*c*) the characteristic dermatosis and mucosal atrophy of pellagra, (*d*) spongy gums and subcutaneous hemorrhages of scurvy, and (*e*) atrophy of skin and gastrointestinal mucosa of severe protein deficiency.
4. **Intestinal flora:** There is sufficient proof that changes in the normal intestinal flora induced by diet can influence susceptibility to some intestinal pathogens. In kwashiorkor, there is a large increase in the total count of intestinal bacteria and a tendency for the bacteria of lower intestinal tract to appear in the upper intestinal tract, as is common with patients suffering from sprue. Further, intestinal organisms, which are nonpathogenic in normal human subjects, may become pathogenic in acute diarrheal diseases of malnourished children.
5. **Resistance of infection:** Lysozyme, which can destroy pathogenic microorganisms, is excreted in tears and sweat. Their activity has been reported to be reduced in human subjects ailing from vitamin A (retinol) deficiency. Vitamin B_3 (pantothenic acid) deficiency has been reported to reduce the content of properdin (euglobin) which appears to function by combining selectively with bacterial polysaccharides of high molecular weight.

Malnutrition and Mental Development

Only within the latter half of the 20th century, studies had been carried out by some nutritionists on the effects of malnutrition and undernutrition on the development and functioning of the central nervous system (CNS) in experimental animals. These studies have clearly shown that malnutrition *per se* is capable of producing physical, biochemical and physiological changes in the CNS. The effects are more marked and less transient when the episode of malnutrition occurs at the time of most rapid growth of the brain when the increment is due mainly to cellular hyperplasia. Changes in nucleic acid and protein content of human brains of marasmic infants as reported by **Winick** (1968, 1969) also established malnutrition as the cause of alteration. Since the correlation among structural, biochemical and behavioural changes cannot be done at present, the ultimate meaning of these findings is difficult to assess.

It has been experimentally found out that the periods of maximum brain growth and cell division, as well as the period of myelination is not retarded or postponed by undernutrition. These are processes which must occur at certain chronological times. It nutrition is sufficiently restricted to retard growth at these times, then the physiological processes of brain development themselves are quantitatively restricted and on subsequent restoration of an adequate diet, the brain has no capability for making good the deficit. In other words, the developing brain has a once and for all opportunity to accomplish certain important developmental processes. These must be completed at the most appropriate under the best conditions. Otherwise, there will be persisting deficits into adult life. This situation has obvious implications for the human species.

Malnutrition and Intelligence and Behaviour in Children

Recent studies by several group of workers have shown that malnutrition can affect significantly the intelligence and learning capacity of children. **Cabak** and **Najadanvic** (1965) tested a group of 36 Serbian children who had suffered from malnutrition in their preschool years and found about half of them scored below the accepted limit of normal intelligence. Similarly, **Cravioto** *et al* (1966) reported that children who have recovered from severe malnutrition proved very much inferior to normal well-nourished children of the same age group with regard to their mental development and learning capacity.

Liang and coworkers (1967) also found very low intelligence quotients in malnourished Indonesian children showing signs of vitamin A deficiency as compared with control healthy children from the same locality. **Cravioto** *et al* (1968) also carried out a systematic study of neuro-integrative development and intelligence in school children who recovered from severe protein–calorie malnutrition in early childhood. Following results were inferred upon:

(*a*) the experimental children attained significantly lower scores in the verbal and performance scales of intelligence tests,

(*b*) their visual-kinesthetic intersensory integration was significantly inferior, and

(*c*) they showed lag in development of audio-visual competence as compared with a control group of well-nourished children of the same age group.

Assessment of Malnutrition

There are 3 increasingly stringent criteria for measuring malnutrition:

1. Dietary intake studies: These studies are usually based on a 24-hour recall and are less stringent. The 24-hour recalls tend to overestimate the number of people with deficient diets. Also, poor dietary intake alone is not a problem in developed countries unless the situation is compounded by enormously increased need.

2. Biochemical assays: These are a more useful indicator of the nutritional status of an individual. At their best, they indicate subclinical nutritional deficiencies, which can be treated before deficiency diseases develop. However, all biochemical assays are not equally valid, an unfortunate fact that is not well recognized. Changes in biochemical parameters due to stress need to be interpreted with caution. The distribution of many nutrients in the body changes dramatically in a stress situation such as illnes, injury, and pregnancy. A drop in level of a nutrient in one tissue compartment (usually blood) need not signal a deficiency or an increased requirement. It could simply reflect a normal metabolic adjustment to stress.

3. Clinical symptoms: The appearance of clinical symptoms is the most stringent criterion. However, it is desirable to intervene long before symptoms became apparent.

Dietary surveys are seldom a valid indication of general malnutrition unless the average intake for a population group falls significantly below the estimated average requirement (EAR) for one or more nutrients. However, by looking at the percentage of people within a population group who have suboptimal intake, it is possible to identify high-risk population groups that should be monitored more closely. Biochemical assays can definitely identify subclinical cases of malnutrition where nutritional intervention is desirable, provided:

(*a*) the assay has been shown to be reliable,

(*b*) the deficiency can be verified by a second assay, and

(*c*) there is no unusual stress situation that may alter micronutrient distribution.

In assessing nutritional status, it is important for the clinician to be aware of those population groups at risk, the most reliable biochemical assays for monitoring nutritional status, and the symptoms of deficiencies if they should occur.

ASSESSMENT OF NUTRITIONAL STATUS

The **nutritional status** of a community is the sum of nutritional status of the individuals who form that community. The principal aim of the nutritional assessment of a community is "to map out the magnitude and geographical distribution of malnutrition as a public health problem, to discover and analyze the ecological factors that are responsible and to plan and put into effect measures not only for the control and eradication of malnutrition but also subsequent maintenance of good health" (**Jelliffe**, 1966).

The problem of assessing the nutritional status of population groups in different countries has been engaging the attention of Public Health Nutrition workers for a long time. The first attempt to deal with this problem on an international scale was made by the Health Organization of the League of Nations in 1932 by convening a conference of Nutrition Experts to consider the best methods available for detecting different degree of malnutrition (L.O.N. Hlt. Org., 1933). In 1939, the Technical Commission of the Nutrition of the Health Organization of the League of Nations published a monograph containing procedures for conducting nutrition surveys. In 1951, the FAO/WHO Joint Expert Committee prepared a report on the assessment of the nutritional status of the population. This report described in broad terms the application of anthropometric, clinical, and dietary methods for assessing nutritional status. The Interdepartmental Committee on Nutrition for National Defence (ICNND) of USA issued a manual outlining procedures for conducting nutrition surveys of Armed Forces in the Near and Far East (ICNND, 1963). This publication describes the various methods in detail and has been used for studing the nutritional status of population groups in various countries.

Types of Nutritional Surveys

There are 2 types of nutritional surveys: longitudinal and cross-sectional.

(*a*) **Longitudinal surveys:** In these, selected members or families in a community are kept under continually systematic surveillance for at least one year. Longitudinal studies provide valuable information not only about the true annual *incidence* of malnutrition but also show seasonal variations caused by climate, the availability of food, community activities and other factors.

(*b*) **Cross-sectional surveys:** In these studies, a satisfactory sample of the population is examined for *prevalence* of malnutrition. Cross-sectional studies do not provide information about most of the acute conditions, and also do not reflect seasonal variations.

In nutritional surveys, it is not necessary to examine all the persons in a given community. Examination of a random and representative sample of the population comprising of all ages and both sexes in the different socioeconomic groups is sufficient.

Methods of Nutritional Assessment

The assessment of nutritional status is accomplished by carrying out the following investigations:

(*a*) **Clinical examination:** It is the most essential part of all nutritional surveys. It is also the simplest, the most practical, and undoubtedly the soundest means of ascertaining the nutritional status of a group of individuals. There are a number of physical signs, some specific and others nonspecific, known to be associated with states of malnutrition. The person is examined from head to foot in good illumination for presence or absence of these signs.

(*b*) **Anthropometric (Body) measurements:** Body measurements such as height, weight etc are important tools in the evaluation of nutritional status of individuals or groups. Although genetically determined, anthropometric measurements are strongly influenced by nutrition. Correctly recorded and interpreted, they reflect the patterns of growth and physical state of individuals. Following basic anthropometric measurements are made in nutritional surveys:

1. *Height, weight and age: Height* should be taken in a standard position without footwear and is recorded to the nearest half-centimetre. *Weight* should be taken with minimum clothing and recorded to the nearest quarter-kilogram. *Age* assessment is equally important to record the age of the individuals correctly for proper interpretation of heights and weights. The international convention of expressing age at the last birthday should be followed.

   Weight for age has been used as an index of malnutrition. *Height for age* is very important, especially in the case of children. However, in adults since maximum height is already attained, height loses its value as an indicator of nutritional status. Height for age gives a picture of the past nutritional history. But this index is also age-dependent. The index *weight for height* is independent of age. It is regarded as an index of current nutritional status.

   Body weight in relation to height is the most commonly used measurement in nutritional assessment. Relationship between the two is expressed as the body mass index (BMI) calculated according to the following formula:

$$\text{BMI} = \text{Weight (kg)} / \text{Height (cm)}^2$$

   The BMI is used to categorize the nutritional status as shown in Table 36–7. The BMI alone, however , should not be used as a sole/definitive indication of the nutritional status.

Table 36–7. Body mass index (BMI) and nutritional status

Body mass index	*Interpretation*
<18.5	Malnourished
18.5–20	Underweight
21–25	**Desirable**
26–30	Overweight
>30	Obese

The ABC of Emergency Treatment

Eating and drinking, like breathing, connect living organisms with the environment. To survive we need oxygen, water and nutrients. One can survive without oxygen for minutes only, without water, the survival time is days. With these two supplied, a human can survive without food for between 60 and 90 days.

These considerations determine the urgency of treatment when things go wrong. Re-establishment of oxygen supply and circulating volume is the first priority (**ABC of resuscitation**: **A**irway, **B**reathing, **C**irculation). Repletion of lost fluid and electroloytes is also necessary within hours to days, depending on the state of the patient. Provision of other nutrients becomes important as soon as the life-saving measures have been taken. The *rule of thumb* is that patients unable to eat would need nutritional support if they have been (or will be) unable to take food for 5–7 days. This period is shorter in hypercatabolic individuals.

2. *Arm circumference:* The left upper arm is measured while hanging freely at its midpoint. The midpoint of the upper arm is assessed by measuring the distance between the acromial process of the scapula and the olecranon process of the ulna and taking the midpoint of that distance. The tape is placed gently but firmly round the midpoint, avoiding compression of the soft tissues. The arm circumference is considered a useful and practical means of assessing the protein–energy deficiency in childhood. Unfortunately, there is lack of suitable standards for comparison.
3. *Skinfold thickness:* Skinfold measurements are sometimes used for assessing the nutritional status (**Goel** *et al*, 1971). The sites commonly used for skinfold measurements are over the triceps and subscapular region. The skin used is a *skin callipers* which is designed to measure the thickness of the skin and subcutaneous fat. It provides a measurement of considerable in assessing the amount of fat and henceforth the reserve of calories in the body.

The other measurement used in nutritional assessment is the *waist-to-hip ratio.* The simple measurement of waist circumference correlates with the amount of visceral fat and is used in the diagnosis of metabolic syndrome.

(*c*) **Laboratory examinations:** The laboratory tests are usually performed with regard to hemoglobin, stools and urine. *Hemoglobin estimation* is the most important laboratory test that is carried out in nutritional surveys. Hemoglobin level is a useful index of the overall state of nutrition, irrespective of its significance in anemia. An RBC count and a hematocrit determination are also conducted. *Stool* should be examined for the ova of hookworm (*Ancylostoma duodenale*), roundworm (*Ascaris lumbricoides*) and other parasites. *Urine* should be examined for sugar, albumin, and casts.

(*d*) **Biochemical examinations:** There are many biochemical tests which may be applied to measure individual nutrient concentration and to establish malnutrition, especially in its preclinical stages. Biochemical tests are time-consuming. Also, they cannot be applied on a large scale, as for example in the nutritional assessment of the whole community. However, they are useful aids in the diagnosis of specific diseases. Some of the biochemical tests applied in nutritional surveys are given in Table 36–8.

Table 36–8. Biochemical tests applicable to nutritional surveys

Nutritional deficiency	*Biochemical test*
Protein	Total serum protein
	Serum albumin
	Urinary urea
Vitamin A (Retinol)	Serum vitamin A
	Serum carotene
Vitamin B_1 (Thiamine)	Urinary thiamine
	Blood pyruvate
Vitamin B_2 (Riboflavin)	Urinary riboflavin
Vitamin B_5 (Niacin)	Urinary *N*-methyl nicotinamide
Vitamin C (Ascorbic acid)	Serum ascorbic acid
	Urinary ascorbic acid
Vitamin D (Cholecalciferol)	Serum alkaline phosphatase (in young children)
Iron	Hemoglobin
	Serum iron
Iodine	Urinary iodine
	Tests for thyroid function

(*Adapted from: WHO: Techn. Rep. Set. No. 258, 1963.*)

More detailed analysis includes assessment of total body water, analysis of body bioelectrical impedance and the mesurements of lean body mass using dual-energy x-ray absorptiometry (DEXA). Some of these measurements allow calculation of variables such as body fat content and composition. Functional measurements such as grip strength or peak expiratory flow are also relevant to nutritional assessment. Full assessment involves measurements of vitamins and trace metals. This is particularly important in patients who remain on long-term parenteral nutrition.

(*e*) **Dietary examination:** The value of nutritional assessment is greatly enhanced when it is supplemented by a diet survey. A diet survey provides information about the amount and types of food consumed by the people and brings out dietary inadequacies as judged by the available standards. A diet survey may be carried out by one of the following methods:

1. *Weighment of raw foods:* This is the method widely employed in India as it is practicable and, if properly carried out, is fairly accurate. The survey team visits the households and weighs all food that is going to be cooked and eaten as well as that which is wasted or discarded. The duration of the survey may vary from 1 to 21 days, but commonly 7 days, which is called 'one dietary cycle'.
2. *Weighment of cooked foods:* This is the best method but not easily acceptable among the people.
3. *Oral questionnaire method:* This method is useful in carrying a diet survey of a large number of people in a short time. Inquiries are made about the nature and quantity of foods eaten during the previous 1 to 2 days. If properly carried out, this method can give reliable results.
4. *Checking of stock inventory:* This method is widely used in the western countries. The housewife is required to make entries of all the purchases during the period of enquiry. In countries where the literacy rate of women is quite low, this method is not applicable.

The data that is collected by the diet surveys is analyzed: (*a*) for the mean intake of foods in terms of cereals, pulses, vegetables, fruits, milk, meat, fish and eggs and (*b*) for the mean intake of calories, proteins, fats, carbohydrates, vitamins and minerals per adult man value or consumption unit.

(*f*) **Study of vital statistics:** The study of vital statistics provides an indirect means of nutritional assessment of human groups. The parameters used for this purpose are the infant mortality rate, neonatal mortality rate, stillbirth rate, perinatal mortality rate, mortality in the age group 1–4 years, and life expectancy. These rates are influenced by nutritional status. It is well known that these rates are low in well-nourished population. Lastly, hospital statistics particularly those relating to kwashiorkor, marasmus, keratomalacia, parasitic infestations etc can be of some value in nutritional surveys.

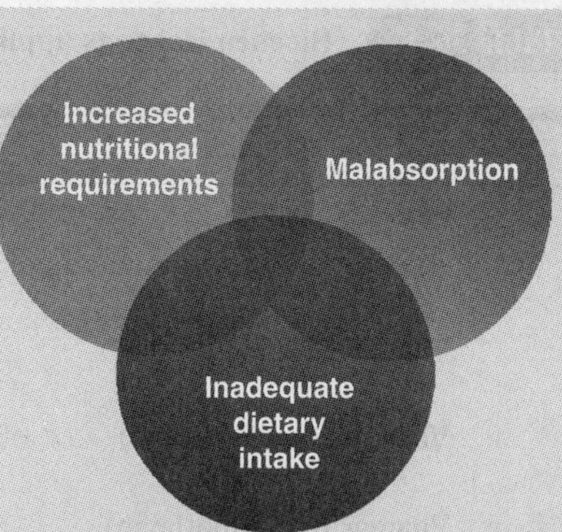

Fig. 36–5. Schematic representation of the 3 important risk factors in determining nutritional status

A person in the periphery would have very low risk of any nutritional deficiency, whereas those in the *green, orange, purple,* or *centre areas* would be much more likely to experience some symptoms of nutritional deficiencies.

Nutritional Status in Medical Practice

The process of evaluating the nutritional status of a patient is an overwhelming task. It is perhaps best to recognize that there are 3 factors that can lead to nutritional deficiencies: poor diet, malabsorption, and increased nutritional need. Only when 2 or 3 components overlap in the same person (Fig. 36–5) do the risks of symptomatic deficiencies become significant. For example, *infants* and *young children* have increased needs for iron, calcium, and protein. Dietary surveys show that many of them consume diets inadequate in iron and some consume diets that are low in calcium. Protein is seldom a problem unless the children are being raised as strictly vegetarians. Thus, the chief nutritional concerns for most children are iron and calcium. *Teenagers* tend to consume diets low in calcium, magnesium, and vitamins A, B_6 and C. Of all these nutrients, their needs are particularly high for calcium and magnesium during the teenage (13–19) years, so these are the nutrients of greatest concern. *Young women* are likely to consume diets low in iron, calcium, magnesium, zinc, and vitamins B_6 (pyridoxine) and B_9 (folic acid)– and all of these nutrients are needed in greater amounts during pregnancy and lactation. *Adult women* often consume diets low in calcium, yet they may have a particularly high need for calcium to prevent rapid bone loss. Finally, the *elderly* have unique nutritional needs and tend to have poor nutrient intake due to restricted income, loss of appetite, miserness, and loss of the ability to prepare a wide variety of foods. They are also more prone to suffer from malabsorption problems and to use multiple prescription drugs that increase nutrient needs (see Table 36–9).

Table 36–9. Drug–nutrient interactions

Drug	*Potential nutrient deficiencies*
Alcohol	Vitamin B_1 (Thiamine)
	Vitamin B_6 (Pyridoxine)
	Vitamin B_9 (Folic acid)
Anticonvulsants	Vitamin B_9 (Folic acid)
	Vitamin D (Cholecalciferol)
	Vitamin K (Phylloquinone, Flavinoquinone)
Cholestyramine	Fat-soluble vitamins (A, D, E, K)
	Iron
Corticosteroids	Vitamin D (Cholecalciferol)
	Calcium
	Zinc
	Potassium
Diuretics	Potassium
	Zinc
Isoniazid	Vitamin B_6 (Pyridoxine)
Oral contraceptives and estrogens	Vitamin B_6 (Pyridoxine)
	Vitamin B_9 (Folic acid)
	Vitamin B_{12} (Cyanocobalamin)

Illness and metabolic stress often cause increased demand or decreased utilization of certain nutrients. For example, diseases leading to fat malabsorption cause a particular problem with absorption of calcium and fat-soluble vitamins. Other malabsorption diseases can result in deficiencies of many nutrients depending on the particular malabsorption disease. Liver and kidney disease can prevent activation of vitamin D and storage or utilization of many other nutrients including vitamins A, B_9, and B_{12}. Severe illnesses or trauma increase the need for calories, protein, and possibly some micronutrients such as vitamin C and certain B vitamins. Long-term use of many drugs in the treatment of chronic disease states can affect the need for certain micronutrients. Some of these drugs are summarized in Table 36–9.

REFERENCES

1. **Albritton EC (editor) :** Standard Values in Nutrition and Metabolism. *Saunders. 1954.*
2. **Brock JF :** Recent Advances in Human Nutrition. *Little, Brown and Company, Boston. 1961.*
3. **Cooper LF, Barber EM, Mitchell HS :** Nutrition in Health and Disease. *J.B. Lippincott Company, Philadelphia. 1960.*
4. **Crampton EW, Lloyd LE :** Fundamentals of Nutrition. *W.H. Freeman and Company, San Francisco. 1960.*
5. **Cravioto J, DeLcardie ER, Birch :** Nutrition, growth and neurointegrative development: An experimental and ecologic study. *Paediatrics.* **38** *: 319–372, 1966.*
6. **Davidson S, Passmore R :** Human Nutrition and Dietetics, *4th ed., Williams & Wilkins. 1969.*
7. **Dobbing J :** Vulnerable periods in developing brain. In Davison AN, Dobbing J (editors)' Applied Neurochemistry. *Blackwell, Oxford. 1968.*
8. **Food and Nutrition Board of the National Academy of Sciences :** Towards Healthful Diets. *U.S. Government Printing Office, Washington DC. 1980.*
9. **Food and Nutrition Board of the National Academy of Sciences :** Recommended Daily Allowances. *10th edition, Washington DC. 1989.*
10. **Food and Nutrition Board of the National Academy of Sciences :** Dietary reference intakes for thiamin, riboflavin, niacin, vitamin B_6, folate, vitamin B_{12}, pantothenic acid, biotin, and choline. *1999.*
11. **Food and Nutrition Board of the National Academy of Sciences :** Dietary reference intakes for vitamin C, vitamin E, selenium, and carotenoids. *2000.*
12. **Indian Council of Medical Research :** Recommended Dietary Intakes for Indians. *ICMR, New Delhi. 1981.*
13. **Jelliffe DB :** The Assessment of the Nutritional Status of the Community. *WHO, Geneva. 1966.*
14. **Kritchevsky D :** Dietary Guidelines. The rationale for intervention. *Cancer.* **72** *: 1011, 1993.*
15. **National Research Council :** Diet, Nutrition and Cancer. *National Academy Press, Washington. 1982.*
16. **Rau P :** Nutrition. *National Institute of Nutrition, Hyderabad. July 1977.*
17. **Scrimshaw NS, Gordon JE (editors) :** Malnutrition, Learning and Behaviour. *M.I.T. Press, Boston, USA. 1967.*
18. **Scrimshaw NS, Taylor CE, Gordon JE :** Interactions of Nutrition and Infection. *World Health Organization, Geneva. 1968.*
19. **Sebrell WH (editor) :** Control of Malnutrition in Man. *American Public Health Association, New York. 1960.*
20. **Sebrell WH Jr :** The new recommended dietary allowances. *Nutr. Rev.* **26** *: 355, 1968.*
21. **Truswell AS :** Evolution of dietary recommendations, goals, and guidelines. *Am. J. Clin. Nutr.,* **45** *: 1060, 1987.*
22. **Williams SR :** Nutrition & Diet Therapy. *2nd ed., Mosby. 1973.*
23. **Wohl MG, Goodhart RS (editors) :** Modern Nutrition in Health and Disease. *5th ed., Lea & Febiger, Philadelphia. 1973.*
24. **World Health Organization :** The Health Aspects of Food and Nutrition. *WHO, Regional Office, Manila. 1972.*
25. **Yates AA, Schlicker SA, Suitor CW :** Dietary reference intakes: the basis for the recommendation for calcium and related nutrients, B vitamins, and choline. *J. Am. Diet. Assoc.,* **98** *: 699, 1998.*

PROBLEMS

1. Of the two people with approximately the same weight, the one with the higher basal energy requirement would most likely be:
   (*a*) taller
   (*b*) female, if the other were male
   (*c*) older
   (*d*) under less stress
2. Basal metabolic rate (BMR):
   (*a*) is not influenced by energy intake.
   (*b*) increases in response to starvation.
   (*c*) may decrease up to 50% during starvation periods.
   (*d*) increases in direct proportion to daily energy expenditure.
3. An excessive intake of calories:
   (*a*) usually does not have adverse metabolic consequences.
   (*b*) leads to metabolic changes that are usually irreversible.
   (*c*) frequently leads to elevated serum levels of free fatty acids, cholesterol, and triacylglycerols.
   (*d*) is the only component of obesity.
4. Recommended dietary allowances (RDAs):
   (*a*) are standards for all individuals.
   (*b*) include all nutritional needs.
   (*c*) define optimal levels of nutrients.
   (*d*) are useful only as general guides in evaluating the adequacy of diets.
5. The estimated average requirement (EAR) of a nutrient is:
   (*a*) the same as the RDA of that nutrient.
   (*b*) an amount that should meet the requirement of half of the healthy individuals of a particular group.
   (*c*) based on the observed nutrient intake of a particular group.
   (*d*) the highest level of nutrient deemed to pose no risk or adverse health effects on the particular population.
6. Who is a vegivore?
7. What is organic food? Is it really healthier?
8. What are the best cancer-fighting foods?
9. How common are genetically-modified (GM) foods? What is wrong with them?
10. What are superfoods?

ANSWERS TO PROBLEMS

1. (*a*) A taller person with the same weight would have a greater surface area.
   (*b*) Males have higher energy requirements than females.
   (*c*) Energy requirements decrease with age.
   (*d*) Stress, probably because of the effects of epinephrine and cortisol, increase energy requirements.
2. (*c*) This is part of the survival mechanism in starvation. (*a*), (*b*) BMR decreases when energy intake decreases. (*d*) BMR as defined is independent of energy expenditure. Only when

the exercise is repeated on a daily basis so that lean muscle mass is increased does BMR also increase.

3. (*c*) Probably because an increased number of and/or size of adipose cells will contain fewer or insensitive insulin receptors. (*a*) Excess caloric intake will lead to obesity if continued long enough. (*b*) Fortunately, most of the changes accompanying obesity can be reversed if weight is lost. (*d*) Inadequate exercise and a genetic component also play roles in obesity.
4. (*d*) RDAs should meet the needs of about 97% of the population but do not take into account special requirements such as those arising from chronic diseases which often change dietary requirements. (*a*) RDAs are designed for most individuals; however, exceptions do occur. (*b*) Some nutritional needs may be unknown; the requirements for all known nutrients are not even clear. (*c*) Optimal levels of nutrients are hard to define; it depends on the criterion for optimal.
5. (*b*) This is the first step in defining nutritional guidelines. (*a*) The RDA is set at 2 standard deviations above the EAR. (*c*) This is called the adequate intake (AI). (*d*) This is the total upper intake level (UL).
6. A vegivore is a person who craves or has a special fondness for vegetables. This need not be because the person has an aversion to eating meat but solely due to an intense craving for vegetables. To a vegivore, meat production may only seem like an accompaniment to add flavour to the dish. He/she will find the vegetable dish more satiating than the meat dish.
7. Organic food is the food produced without synthetic pesticides or fertilizers and antibiotics or growth hormones. A new Stanford study has taken the nutrition world by surprise, revealing that they found little difference in health benefits between organic and conventional foods. Only the mineral phosphorus was higher in organic produce. Organic foods were not found to be 100% chemical-free.
8. The top caner-fighting foods include: garlic, tomatoes, green tea, cruciferous vegetables like broccoli, fish (especially cold-water fatty fish) and fruit.
9. About 75% of processed foods contain GM ingredients. There are now 8 major GM food crops on the market: corn, soy, cottonseed (for oil), canola (for oil), sugar from sugarbeets, Hawaiian papaya, some varieties of zucchini, and crookneck squash. There is plenty of wrong with GM foods. They have built-in pesticides. Hamsters-fed GM soy had a 5-fold infant mortality rate. The 3rd generation of hamsters on a GM diet were almost all sterile.
10. Superfoods are extremely nutrient-dense foods. For maximum nutritional benefits, include them in your diet. The top-20 list of superfoods include: apples, blueberries, broccoli, quinoa, dark chocolate, sea vegetables, garlic, avocados, parsley, wild salmon, beans, kale, green tea, pumpkin seeds, microplants, oats, sweet potatoes, walnuts, yogurt, and fermented foods.

CONTENTS

Fruits and vegetables are the key sources of various components of foods.

CHAPTER

37 Elements of Nutrition

"Eating good food is the best of all types of material happiness."

–Chanakya

COMPONENTS OF AN ADEQUATE DIET

There are 6 major components of the diet. **Carbohydrate, fat** and **protein** yield energy, provide for growth, and maintain tissue subjected to wear and tear. **Vitamins, minerals** and **water**, although they do not yield energy, are essential parts of the chemical mechanisms for the utilization of energy and for the synthesis of various necessary metabolites such as hormones and enzymes. The minerals are also incorporated into the structure of the tissue and, in solution, play an important role in acid-base equilibrium.

Protein, fats, and carbohydrates are often termed *proximate principles*. Together with water, they form the main bulk of food. The human body is built up from these 6 constituents and has a widely differing composition (refer Table 37–1).

Table 37–1. Approximate composition of human body

Dietary constituent	*Per cent composition*
Water	63
Proteins	17
Fats	12
Minerals	7
Carbohydrates	1
Vitamins	Traces

Energy for physiologic processes is provided by the combustion of carbohydrate, fat and protein. The daily energy requirement or the daily caloric need is the sum of the basal energy demands plus that required for the additional work of the day. During periods of growth, pregnancy, or convalescence, extra calories are also needed. While all 3 major nutrients yield energy to the body, carbohydrate and, to a lesser extent, fat are physiologically the most economical sources. Protein serves primarily to provide for tissue growth and repair; but if the caloric intake from other foods is inadequate, it is burned for energy. The caloric requirement of persons, with respect to energy, protein and minerals, of varying ages and under various physiologic conditions, are tabulated in Table 37–2.

Table 37–2. Recommended daily dietary allowances*

	Age (years)	*Weight* (kg)	(lb)	*Height* (cm)	(in.)	*Energy* (kcal)†	*Protein* (g)
Infants	0.0–0.5	6	14	60	24	kg × 117	kg × 2.2
	0.5–1.0	9	20	71	28	kg × 108	kg × 2.0
Children	1–3	13	28	86	34	1300	23
	4–6	20	44	110	44	1800	30
	7–10	30	66	135	54	2400	36
Males	11–14	44	97	158	63	2800	44
	15–18	61	134	172	69	3000	54
	19–22	67	147	172	69	3000	54
	23–50	70	154	172	69	2700	56
	51+	70	154	172	69	2400	56
Females	11–14	44	97	155	62	2400	44
	15–18	54	119	162	65	2100	48
	19–22	58	128	162	65	2100	46
	23–50	58	128	162	65	2000	46
	51+	58	128	162	65	1800	46
Pregnant						+ 300	+30
Lactating						+500	+20

* The allowances are intended to provide for individual variations among most normal persons as they live in the USA under usual environmental stresses. Diets should be based on a variety of common foods in order to provide other nutrients for which human requirements have been less well defined. See text for more detailed discussion of allowances and of nutrients not tabulated.

† Kilojoules (kJ) = 4.2 × kcal

(*Reference: Recommended Dietary Allowances. 8th rev. ed. Food and Nutrition Board, National Research Council– National Academy of Sciences, 1974.*)

Obesity is now a major nutritional problem in the world, especially in many western countries. It occurs in about 15% of the adolescent population. Of the total cases of obesity, about 90% are due to overeating; the remaining 10% are due to endocrine disturbances. Obesity in infants and children may be a predisposing factor for adult obesity. The main features of obesity are overweight and fatness. Persons are said to be overweight when they are 10% or more above the accepted standards. The hazards of obesity include increased incidence of diabetes mellitus, arterial hypertension, atheroma and atheros arteriosclerosis, cerebrovascular disease, diseases of gall bladder, and increased morbidity and mortality. Accumulated evidences indicate that some of the hazards of obesity are partly related to serum triglyceride and cholesterol levels. These tend to be elevated on diet rich in saturated fats.

As **obesity** is almost always the result of excess consumption of calories, its treatment is, henceforth, directed at reducing the caloric intake from carbohydrate and fat but maintaining the protein, vitamin, mineral, and water intake at normal levels. An adequate diet containing not more than 800–1,000 kcal should be maintained and normal or increased amounts of energy expended until the proper weight is reached.

Carbohydrate and fat 'spare' protein and thus make it available for various anabolic processes. Carbohydrate is somewhat more efficient at protein sparing than fat– presumably because carbohydrate can be used as an energy source by almost all tissues whereas fat cannot. This "*protein-sparing action*" of carbohydrates and fats is particularly important in the nutrition of patients, especially those being fed parenterally, when it is difficult or even impossible for them to take in enough calories. If the caloric intake is inadequate, giving proteins orally or amino acids intravenously is a relatively inefficient way of supplying energy because the primary function of proteins is tissue synthesis and repair, and not energy production.

The amounts of each of the 3 energy-yielding components in the diet of persons world over, especially in the developed countries, have changed markedly during the 20th century. Most noteworthy is the decline in *carbohydrate* intake and a corresponding increase in the intake of *fats*. The total *protein* content of the diet has remained consistent at an amount that supplies approximately 12% of the total caloric intake per day. Carbohydrate and fat now supply about the same amount of energy in the diet in developed countries, *i.e.*, 46 and 42%, respectively. It is of interest to note that the intake of starches (complex sugar) has declined from 43% during the first decade of the 20th century to 28% of the total energy contained in the diet of persons during the last quarter of the 20th century. Furthermore, the intake of simple sugars has increased to a point where they now supply 17 to 18% of the total energy supplied in the daily food consumed. The increase in fat intake is attributable mainly to the use of more separated fats and oils and, most significantly, to increased consumption of meat.

CARBOHYDRATES

Carbohydrates are composed of carbon, hydrogen and oxygen. They are the first and most efficient sources of energy for vital processes. They are also essential for the oxidation of fats and for the synthesis of certain nonessential amino acids. The 3 principal sources of carbohydrates in foods are: sugars, starches and cellulose. Cereal grains, potatoes and rice, the staple foods of most countries, are the main sources of dietary carbohydrates. Indeed, such carbohydrates furnish 60–80% of the total caloric intake when there is a scarcity of proteins and fats. The chief metabolic role of dietary carbohydrates is for energy production. Any carbohydrate in excess of that needed for energy is converted to glycogen and triacylglycerol for long-term storage. The body can adapt to a wide range of dietary carbohydrate levels. *Carbohydrate-rich diets* result in higher steady-state levels of glucokinase and some of the enzymes involved in the pentose phosphate pathway and triacylglycerol synthesis. *Carbohydrate-poor diets* result in higher steady-state levels of some of the enzymes involved in gluconeogenesis, fatty acid oxidation, and amino acid catabolism. *Glycogen stores* are also affected by the carbohydrate content of the diet.

The optimum quantity of carbohydrate in balanced diets is placed between 50 to 70% of the total energy intake. Most Indian diets contain excessive amounts of carbohydrate, providing as much as 90% of the total energy intake. The carbohydrate reserve of a human adult is only about 500 g. When a man is fasting, this reserve is rapidly exhausted.

While dietary carbohydrates provide a source of energy and the substrates for many synthetic pathways, animals and man can be maintained on diets devoid of carbohydrate. However, since dietary calories may be provided most cheaply in this form, carbohydrate supplies about half of the calories of most human dietaries. A carbohydrate-poor ration must provide calories in the form of lipid since protein intake is limited by the feeling of satiety provided by relatively small quantities of protein. Subjects transferred from ordinary mixed rations to a lipid-rich diet rapidly develop **ketosis**. After a few weeks, the ketosis abates and may disappear entirely. In fact, some preformed carbohydrate is necessary as a preventive measure against ketosis. It is considered that a minimum of 5 g of carbohydrate per 100 kcal of the total diet is necessary to prevent the onset of ketosis, but prevention

Carbohydrate Loading and Endurance Athletes

The practice of carbohydrate loading dates back to observations made in the early 1960s that endurance during vigorous exercise was limited primarily by muscle glycogen stores. Of course, glycogen stores are not the sole energy source for muscle. Free fatty acids are present in the blood during vigorous exercise and are utilized by muscle along with the muscle glycogen stores. Once the glycogen stores are exhausted, however, muscle cannot rely entirely on free fatty acids without tiredness. This is probably related to the fact that muscle becomes increasingly hypoxic during vigorous exercise. While glycogen stores are utilized equally well with or without oxygen, fatty acids can only be utilized in presence of oxygen, *i.e.*, aerobically. Under anaerobic conditions, fatty acids cannot provide ATP rapidly enough to serve as the sole energy source.

Thus, the practice of carbohydrate loading to increase glycogen stores was devised for track and other endurance athletes. Originally, it was thought that it would be necessary to trick the body into increasing glycogen stores. The original carbohydrate loading regimen consisted of a 3–4 day period of heavy exercise while on a low-carbohydrate diet, followed by 1–2 days of light exercise while on a high-carbohydrate diet. The initial low-carbohydrate-high-energy demand period caused a depletion of muscle glycogen stores. Apparently, the subsequent change to a high-carbohydrate diet resulted in a slight rebound effect, with the production of higher-than-normal levels of insulin and growth hormone (somatotropin). Under these conditions, glycogen storage was favoured and glycogen stores reached almost twice the normal amounts. This practice did increase endurance significantly.

While the technique clearly worked, the athletes often felt lethargic and irritable during the low-carbohydrate phase of the regimen, and the high-fat diet ran counter to current health recommendations. Fortunately, recent studies indicate that regular consumption of a high-complex-carbohydrate low-fat diet during training increases glycogen stores without the need for trickling the body with certain dietary changes. Current recommendations are for endurance athletes to consume a high-carbohydrate diet (with emphasis on complex carbohydrates) during training. Then carbohydrate intake is increased further (to 70% of calories) and exercise tapered off during the 2–3 days just prior to an athletic event. This increases muscle glycogen stores to levels comparable to the previously described carbohydrate loading regimen.

of the untoward effects of fasting or of high-fat diets will require 50–100 g of digestible carbohydrate per day. Indeed, the Eskimo normally lives on a high-lipid, low-carbohydrate ration. This problem is significant in the diabetic individual who depends on lipids for his calories, since the extent of carbohydrate utilization will depend on the amount of insulin administered. The relative tendencies of various foodstuffs to elicit or prevent ketosis are expressed as their ketogenic or antiketogenic activity. As an aid in the management of diabetes mellitus, diets are planned to contain a balance of ketogenic and antiketogenic foods.

Very rarely, individuals cannot utilize **sucrose**. Those with a genetic lack of *fructokinase* exhibit a benign fructosuria. However, those *who phosphofructose aldolase* cannot utilize fructose 1-phosphate formed by fructokinase. Accumulation of fructose 1-phosphate results in a bizarre syndrome of headache and acute abdominal distress.

The most common nutritional problems involving carbohydrates are those of **carbohydrate intolerance**. The common form of carbohydrate intolerance is *diabetes mellitus*, caused either by subnormal insulin production or lack of insulin receptors. This causes an intolerance to glucose and sugars that are readily converted to glucose. For years, the American Diabetes Association has recommended diets that were low in fat and high in complex carbohydrates and fibre for diabetics. The logic behind this recommendation was that diabetics are prone to hyperlipidemia with risk of heart disease, and low-fat diets seemed likely to reduce risk of hyperlipidemia and heart disease. In addition, numerous chemical studies had suggested that the high-fibre content of these diets improved control of blood sugar. However, this recommendation has proved to be unsuitable since making dietary recommendations for population groups, rather than individuals, is difficult. Also, it is very difficult to make any major changes in dietary composition without changing other components of the diet.

Lactase insufficiency is also a common disorder of carbohydrate metabolism. It is most prevalent among blacks, Asians, and Hispanics. Without the enzyme *lactase*, dietary lactose is not significantly hydrolyzed or absorbed. It remains in the intestine where it draws water into the gut through osmosis and serves as a substrate for conversion to lactic acid and CO_2, by intestinal bacteria. The end result is bloating, flatulence, and diarrhea – all of which can be avoided simply by eliminating milk and milk products from the diet. The rare infants whose intestines are devoid of lactase grow poorly and may have severe diarrhea. Transfer to a synthetic milk containing sucrose results in prompt improvement.

CLINICAL IMPLICATIONS 37–1.

Lactase Deficiency

Intestinal disaccharidase deficiencies are encountered relatively frequently in humans. Deficiency can be present in one enzyme or several enzymes for a variety of reasons, such as genetic defect, physiological decline with age, or the result of 'injuries' to the mucosa. Of the disaccharidases, *lactase* is the most common enzyme with an absolute or relative deficiency, which is experienced as **milk intolerance**. The consequences of an inability to hydrolyze lactose in the upper small intestine are inability to absorb lactose and bacterial fermentation of ingested lactose in the lower small intestine. Bacterial fermentation results in the production of gas (distention of gut and flatulence) and osmotically active solutes that draw water into the intestinal lumen (diarrhea). The lactose in yogurt has already been partially hydrolyzed during the fermentation process of making yogurt. Thus, individuals with lactase deficiency can often tolerate yogurt better than unfermented dairy products. Lactase is commercially available to pretreat milk so that the lactose is hydrolyzed.

LIPIDS

Because of its high fuel value, triacylglycerols (= fats) are an important component of the diet. Furthermore, the palatability of foods is generally enhanced by their fat contents. As a form of energy storage in the body, fat has more than twice the value of protein or carbohydrate. Fatty acids can be used directly as sources of energy by all of the cells of the body; indeed, skeletal muscle and the myocardium appear to utilize fatty acids preferentially as a source of fuel. Exceptions to the otherwise extensive use of fatty acids by the cells of most tissues are the erythrocytes and the cells of the central nervous system, although in total starvation, after a brief period of adaptation, the brain can use ketones formed from fatty acids or certain amino acids. Fat in the body provides support to viscera such as heart, kidney and intestine; fat beneath the skin provides insulation against cold.

The human requirement for fat is not precisely known. Fats contribute in human nutrition as components of the essential fatty acids (EFAs), linoleic and arachidonic acids. Dietary fats also serve as the major carriers of the fat-soluble vitamins. With the exception of these specific contributions of fats to the diet, it appears that there is no absolute need for lipids in the diet. The specific contributions of fats can be supplied by the daily consumption of 15–25 g of appropriate lipids. However, at ordinary levels of intake, 20–25% of the calories of the diet might be derived from fats; for a 3,000 kcal intake, this would represent 66–83 g of fat. When higher caloric levels are consumed, more fat is likely to be included. Suggested level of fat intake are given in Table 37–3.

Table 37–3. Suggested intake of dietary fat

Person	*Fat* (energy per cent)	*Essential fatty acid* (energy per cent)
Infants	—	6
Young children	25	5–6
Adults: Man and Woman	20*	3
Pregnant woman	—	4.5
Lactating mother	—	6

* About half of this (*i.e.,* 10%) will come from invisible fat present in the foods.

Fats such as 'ghee', butter, 'vanaspati' and vegetable oils are known as **visible fats**, because it is easy to estimate their intake in the daily diet. But fats present in foods such as cereals, pulses, nuts, vegetables, milk and milk products, fish, meat and eggs are unseen, and difficult to estimate. These are called *invisible fats.* The invisible fats furnish about 9% of energy in Indian diets.

Fats are directly utilized by many tissues of the body as an energy source and, as phospholipids, are an important constituent of membranes. Excess fat in the diet can only be stored as triacylglycerol. *As with carbohydrates, the body adapts to a wide range of fat intakes*. However, problems arise at the extremes (either high or low) of fat consumption. *At the low end*, **essential fatty acid (EFA) deficiency** may become a problem. The fatty acids linoleic and linolenic cannot be made by the body and, thus, are essential components of the diet. These EFAs are needed for maintaining the function and integrity of membrane structure, for fat metabolism and transport, and for the synthesis of prostaglandins and related compounds. The most characteristic symptom of EFA deficiency is a scaly dermatitis or **phrenoderma**. *At the high end* of the scale, the excess dietary fat causes elevation of serum lipids and, thus, an increased risk of heart disease. Recent studies point out that high-fat intakes are linked with increased risk of colon, breast, and prostate cancer. However, it is not yet certain whether the cancer risk is associated with fat intake *per se*, or with the excess calories associated with a high-fat diet. Animal studies suggest that polyunsaturated fatty acids (PUFAs) of the ω-6 series may be more tumorigenic than other unsaturated fatty acids. The conjectural explanation is that prostaglandins derived from the ω-6 fatty acids may stimulate tumour progression.

Evidence suggests that **hyperlipemia** and, hence, **atherosclerosis** may occur as a result of a *relative* deficiency in polyunsaturated fatty acids. Deposition of lipid plaques on the intima of arteries (**atheroma**) appears to occur most readily when the serum lipids are elevated in concentration. The material deposited has a composition similar to that of the serum lipids at the time of deposition. The concentration of plasma lipids is determined, in part, by the composition of the diet. In general, variation in the amount of dietary cholesterol, as in normal diets, is without significant influence. When greater quantities are included in experimental diets, the consequences vary with the amount of cholesterol and the composition of the diet. Large quantities (2% of the diet) evoke a massive lipemia; triglyceride is mobilized from adipose tissue and deposited in the liver. Lesser quantities result in marked lipemia, including high concentrations of cholesterol itself with cocomitant fall in the serum concentration of polyunsaturated fatty acids in the various lipid fractions. In contrast, a diet extremely low in lipid results in prompt decline in all plasma lipid fractions, although a lipid-free, high-carbohydrate regime may elicit an increase in serum triglycerides. Thus, it is now well established that the level of serum cholesterol is decreased by the ingestion of polyunsaturated fatty acids and increased by saturated fatty acids as well as by dietary cholesterol itself. While there continues to be some disagreement about the relative importance of dietary cholesterol as compared to other factors known to influence the blood cholesterol level, there is a growing body of evidence supporting the idea that over a certain range of dietary intake, the amount of cholesterol ingested does exert a substantial influence on the levels of cholesterol in the serum.

Exercise significantly decreases serum cholesterol and triglyceride levels while total unsaturated fatty acids of serum appear to increase. It is perhaps significant also that diets which increase serum lipid concentrations have been reported to result in increased blood clot strength and delayed fibrinolysis. Thus, in addition to the complex nutritional effects, genetic tendencies, emotional variability, exercise, and even anatomical relationships must be significant in the control of serum lipid concentrations. Moreover, it is not established that reduction in serum lipids will delay atherosclerosis or prevent coronary artery disease in man.

PROTEINS

Proteins are of greatest importance in human nutrition. They are complex organic nitrogenous compounds. Proteins are composed of carbon, hydrogen, oxygen, nitrogen and sulfur in varying amounts. Some proteins also contain phosphorus and iron and occasionally other elements. Proteins differ from carbohydrates and fats in that they contain nitrogen, this amounting to about 16%.

Protein carries a certain mystique as a "body building" food. While protein is an essential structural component of all cells, it is equally important for maintaining essential secretions such as digestive enzymes and peptide or protein hormones. Protein is also needed for the synthesis of plasma proteins, which are essential for maintaining osmotic balance. Excess proteins, consumed in the diet, are treated as a source of energy, with the glycogenic amino acids being converted to glucose and the ketogenic amino acids being converted to fatty acids and keto acids. Both types of amino acids are eventually converted to triacylglycerol in adipose tissue if fat and carbohydrate supplies are already adequate to meet energy requirements. Thus, for most of the humans, the only body building food obtained from high-protein diets is in adipose tissue. It is also interesting to note that proteins, besides performing other important functions, also constitute the most important sources of nitrogen, sulfur, and phosphorus for the body.

It has been popular to say that the body has no storage depot for protein, and thus adequate dietary protein must be supplied with every meal. However, actually speaking, this is not quite accurate. While there is no separate class of "storage" protein, there is a certain percentage of body protein that undergoes a constant process of breakdown and resynthesis. In the *fasting state*, the breakdown of this store of body protein is enhanced, and the resulting amino acids are utilized for glucose production, for the synthesis of nonprotein nitrogenous compounds, and for the synthesis of the secretory and plasma proteins. Even in the *fed state*, some of these amino acids are utilized for energy production and as biosynthetic precursors. Thus, the turnover of body protein is a normal process– and an essential feature of what is called nitrogen balance.

Nitrogen Balance

A healthy adult eating a varied and plentiful diet is generally in 'nitrogen balance'. **Nitrogen balance** is a state where the amount of nitrogen ingested (chiefly in the form of protein) each day is balanced by the amount excreted (chiefly in the form of undigested protein in the feces, and urea and ammonia in urine), resulting in no net change in the amount of body nitrogen. In the well-fed condition, excreted nitrogen comes mostly from digestion of excess protein or from normal protein turnover (= synthesis and degradation of protein). Under some conditions, the body is either in negative or positive nitrogen balance. In **negative nitrogen balance** (Fig. 37–1d), more nitrogen is excreted than ingested. This occurs in starvation and certain diseases. During starvation, carbon chains of amino acids from proteins are need for gluconeogenesis; ammonia released from amino acids is excreted mostly as urea and is not reincorporated into protein. A diet deficient in an essential amino acid also leads to a negative nitrogen balance, since body proteins are degraded to provide the deficient essential amino acid, and the other amino acids liberated are metabolized. It also occurs in injury when there is net destruction of tissue and in major trauma or illness when the body's adaptive response causes increased catabolism of body protein stores. Negative nitrogen balance may also exist in senescence. **Positive nitrogen balance** (Fig. 37–1a, b, c) occurs in growing children, who are increasing their body weight and incorporating more amino acids into proteins than they break down. Positive nitrogen balance also occurs in pregnant women, during refeeding after starvation, and in convalescing adults.

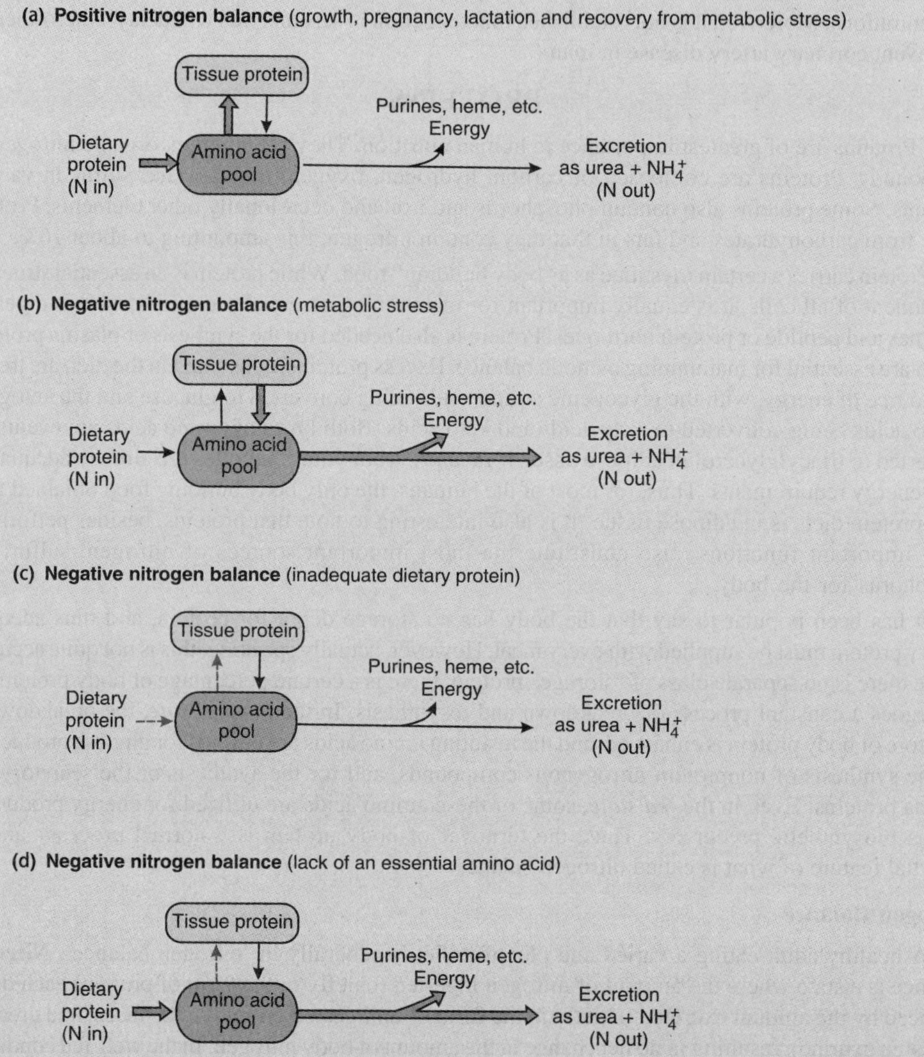

Fig. 37–1. Schematic representations of the metabolic interrelationships involved in determining nitrogen balance

Each figure represents the nitrogen balance resulting from a particular set of metabolic conditions. The dominant pathways in each situation are indicated by *heavy red arrows*.

Protein Requirement

Protein requirements vary from individual to individual. Many factors such as age, gender, physiologic state, infections, worm infestations, emotional disturbances and stress situations can affect a person's protein requirements. The protein requirement also varies with the net protein utilization (NPU) of the dietary protein. If the NPU is low, the protein requirement is high and vice versa. The NPU of the protein of Indian diets varies between 50 and 80. All estimates of protein requirement are valid only when energy requirements are fully met. If the total energy intake is insufficient, some dietary protein will be diverted to provide energy.

Protein requirements are, customarily, expressed in terms of body weight. The Indian Council of Medical Research (ICMR), in 1981, has recommended 1.0 g protein/kg body weight for an Indian adult, assuming a NPU of 65 for the dietary proteins. The ICMR recommendations for daily protein requirements for various age groups are given in Table 37–4.

Table 37–4. Recommended daily protein allowances

Age (years)	*Daily allowance in terms of dietary protein*	
	g/kg body weight	*Total requirement (g)*
Men (55 kg)	1.0	55
Women (45 kg)	1.0	45
Pregnancy (2nd and 3rd trimester)		+14
Lactation (0 to 6 months)		+25
Infants		
0 to 3 months	2.3*	
3 to 6 months	1.8*	
6 to 9 months	1.8†	
9 to 12 months	1.5 †	
Children		
1 to 3 years	1.83	22
4 to 6 years	1.56	29
7 to 9 years	1.35	36
Males		
10 to 12 years	1.24	43
13 to 15 years	1.10	52
16 to 18 years	0.94	53
Females		
10 to 12 years	1.17	43
13 to 15 years	0.95	43
16 to 18 years	0.88	44

* in terms of milk protein

† partly vegetable protein

A minimal amount of protein is indispensable in the diet to provide for the replacement of tissue protein, which constantly undergoes destruction and resynthesis. This is often spoken of as the **wear and tear quota**. The protein requirement is considerably increased, however, by the demands of growth, increased metabolic activity (as in infection with fever), in burns, blood loss, surgery, proteinuria, diabetes and after trauma. The recommended intake of protein for individuals of various age groups (refer Table 37–2) presume that the caloric demand is adequately supplied by other foods so that the ingested protein is available for tissue growth and repair. The requirement for protein in the diet is, however not only quantitative; there is also an important qualitative aspect since the metabolism of protein is inextricably connected with that of its constituent amino acids. Certain amino acids are called **indispensable** or **essential amino acids** in the sense that they must be obtained preformed and cannot be synthesized by the animal body (Table 37–5). The remainder, the so-called **dispensable** or **nonessential amino acids**, are also required by the organism since they are found in the protein of the tissues, but they can apparently be synthesized, presumably from α-keto acids by amination.

The list of essential amino acids varies with the animal species tested; for example, the chicks require *glycine*. It may also vary with the physiologic state of the animal. For some amino acids (*e.g.*, *arginine*), the rate of synthesis may be too slow to supply fully the needs of the animal. *Histidine* in the diet is also necessary to maintain growth during childhood. Such amino acids are said to be **relatively indispensable**. The nutritive value of a protein is now known to be dependent on its content of essential amino acids. Examples of incomplete proteins are *gelatin* (which lacks tryptophan) and *zein* of corn (which is poor in both tryptophan and lysine). Such incomplete proteins are unable to support growth if given as the sole source of protein in the diet.

As pointed out earlier, 10 of the 20 amino acids contained in natural proteins are indispensable nutrients for the human species. *Cystine* can spare as much as 80– 0% of the *methionine* requirement by supplying some needs for organically-bound sulfur, and *tyrosine* can spare 70–75% of the need for *phenylalanine*. This is so because normal metabolism of methionine and phenylalanine involves their conversion in whole or in part to cystine or tyrosine, respectively.

Table 37–5. Essential and nonessential amino acids

Essential	*Nonessential*
Arginine[1]	Alanine
Histidine	Aspartate
Isoleucine	Asparagine
Leucine	Cysteine
Lysine	Glutamate
Methionine[2]	Glutamine
Phenylalanine[3]	Glycine
Threonine	Proline
Tryptophan	Serine
Valine	Tyrosine

[1] *Arginine* is synthesized by mammalian tissues, but the rate is insufficient to meet the need during growth.

[2] *Methionine* is required in large amounts to produce cysteine if the latter is not supplied adequately by the diet.

[3] *Phenylalanine* is needed in large amounts to form tyrosine if the latter is not supplied adequately by the diet.

Amino acid requirements for infants (**Holt**, 1965), for children (**Nakagawa**, 1964), for men (**Rose**, 1957), and for women (**Leverton**, 1959) have been estimated (refer Table 37–6). The values shown for adults are higher than those previously quoted because they have been adjusted upward by 30% to allow for individual variability. A perusal of Table 36–6 indicates that the amino acid requirements for the human infants are, in general, higher than those for the adult when expressed on the basis of body weight. It is also important to note that *histidine* is to be considered an essential amino acid for requirements of growth in the young infant.

Table 37–6. Estimated amino acid requirements of man

Amino acid	*Requirement (per kg of body weight), mg/d*			*Amino acid pattern for high-quality proteins (mg/g of protein)*
	*Infant (3–6 months)**	*Child (10–12 years)*	*Adult*	
Histidine	33	?	?	17
Isoleucine	80	28	12	42
Leucine	128	42	16	70
Lysine	97	44	12	51
Total S-containing amino acids	45	22	10	26
Total aromatic amino acids	132	22	16	73
Threonine	63	28	8	35
Tryptophan	19	4	3	11
Valine	89	25	14	48

* Two grams per kilogram of body weight per day of protein of the quality listed in the last column would meet the amino acid needs of the infant.

(Reproduced from: Recommended Dietary Allowances. 8th edition. National Academy of Sciences, 1974.)

WILLIAM CUMMING ROSE

(LT, 1887–1985)

Rose, an American nutritionist, was born in Greenville, South Carolina, USA on April 4, 1887. He received his B.S. degree from Davidson College in 1907 and worked as an assistant in Physiological Chemistry under **Prof. Lafayette Benedict Mendel** (LT, 1872 – 1935) at Yale University from 1908–1911. During this period, he worked on protein and amino acid nutrition and was awarded the Ph.D. degree in 1911. He was appointed Professor of Biological Chemistry at the College of Medicine, University of Texas from 1914–1922. In 1922, he went to the University of Illinois as professor of Physiological Chemistry, a title which was changed to professor of Biochemistry in 1936. From 1922 to 1955, he transformed his department into a centre of excellence for the training of biochemists. He retired from the University of Illinois in 1955. During this period, 56 students received Ph.D. degree and 34 students M.S. degrees under his guidance.

Rose began his classical researches on **'Amino Acid Nutrition'** in 1930. He and his colleagues attended to rear rats on a synthetic diet containing a mixture amino acids known to be present in casein at that time. They found that the animals did not grow. This led them to conclude that there must be some other amino acid in casein essential for the growth of rats. Hence, they tried for the isolation and identification of the unknown amino acid and succeeded in isolating a new essential amino acid, **threonine** (Thr) in 1936. They also made the important discovery that incorporation of 0.6% of the new amino acid (threonine), along with amino acid mixture, promoted excellent growth in rats. Next, they proceeded to investigate the essential nature of different amino acids for growth, by omitting one amino acid at a time from the above mixture. These studies led to the discovery of essentiality of the following 10 amino acids for the growth of rats: arginine, histidine, isoleucine, leucine, lysine, methionine, threonine, tryptophan and valine. They also made the remarkable discovery that incorporation of the above 10 amino acids in place of casein in a synthetic diet would promote good growth in rats indicating that the rest 10 amino acids need not be provided in the diet and these can be synthesized in the body. These were termed 'nonessential amino acids'. Rose was the author/co-author of over 120 scientific publications. Rose died in Urbana, Illinois on September 25, 1985 at age 98.

Rose received many awards and honours in his lifetime. From 1939 to 1941, he became President of the American Society of Biological Chemists. From 1945 to 1946, he became President of the American Institute of Nutrition. In 1952, he was awarded **Willard Gibbs Medal of the American Chemical Society**. In 1966, he received **National Medal of Science**.

Amino acids serve as constituents of a diverse variety of proteins. Besides, many amino acids have specific functions in metabolism. They include the role of methionine as a methyl donor, cystine as a source of thiol (–SH) groups, and tryptophan as a precursor of niacin, arginine and the urea cycle.

It should also be apparent that, in assessing protein requirements, consideration must be given to the intake of the essential amino acids rather than only to that of whole proteins of varying essential amino acid content. Although it is not commonly appreciated, there is virtually no evidence that intakes of protein exceeding requirements in healthy adults have any nutritional benefit. On the contrary, low-protein diets tend to be limited in their content of foods of animal origin and thus low in important trace nutrients. They are also generally less palatable.

In fact, the normal adult protein requirements depend on the diet. Assuming adequate calorie intake and a 75% efficiency of utilization (which is typical of mixed protein in the average US diet), the recommended protein intake is 0.8 g/kg body wt day^{-1}. This amounts to about 58 g protein day^{-1} for a 72-kg (160-lb) man and about 44 g protein day^{-1} for a 55-kg (120 - lb) woman. These recommendations would need to be increased on a vegetarian diet if overall efficiency of utilization were less than 75%.

Evaluation of Protein Quality

A knowledge of the amino acid content of protein is not sufficient for the evaluation of protein quality. Information is also required about the digestibility and suitability to meet the protein needs of the body. A large number of methods have been proposed by many nutritionists for the evaluation of protein quality. Briefly, these are as follows:

1. **Digestibility coefficient (DC):** It is the percentage of ingested protein absorbed into the blood stream after the process of digestion is complete.
2. **Biological value (BV):** The term was first introduced by **Thomas**, in 1909. It is the percentage of nitrogen retained from out of the nitrogen absorbed from the diet.
3. **Protein efficiency ratio (PER):** It is the gain in weight of young animals per unit weight of protein consumed.
4. **Net protein utilization (NPU):** Miller and Bender first used this term, in 1955. It is a product of digestibility coefficient and biological value divided by 100.

THOMAS BURR OSBORNE

(LT, 1859–1929)

Osborne was born in New Haven, Connecticut on August 5, 1859. He graduated from Yale University in 1881. He obtained his Ph.D. degree in analytical chemistry from the same university in 1885. He joined as Analytical Chemist in the Scheffield Scientific School of Yale University, in 1886. In 1887, Osborne undertook studies on proteins of plant seeds. His work on the isolation and determination of amino acid composition of plant proteins has now become classical and represents highly original contributions in the field of proteins. Osborne, in collaboration with **Prof. Lafayette Benedict Mendel** (LT, 1872–1935) started his classical studies on the nutritive value of isolated proteins, in 1909. They both developed an original growth method for assaying the nutritive value of proteins. They expressed the **nutritive value of the protein** as a ratio (called *protein efficiency ratio,* PER) the growth (in g) to protein (in g) consumed. This method has now become classical and, with slight modifications, is widely used at present for assaying the nutritive value of proteins. Both Osborne and Mendel showed that:

(*a*) *Zein,* a protein isolated from maize, will not promote growth in albino rats and when the amino acids lacking in zein (*i.e.,* tryptophan and lysine) were added, good growth occurred in rats.

(*b*) Rats grow at a slow rate on a diet containing *gliadin* (a protein, isolated from wheat) as the sole source of protein but when lysine (which is present in small amounts in gliadin) was added, good growth occurred.

Thus, they both discovered that tryptophan and lysine are essential for the growth of rats. Osborne also published a monograph on 'Vegetable Proteins'. He died in 1929.

Data regarding the DC, BV and NPU are presented in Table 37–7

Table 37–7. Protein content, DC, BV and NPU of some common foods

Name of Food	*Protein content (%)*	*DC (%)*	*BV (%)*	*NPU (%)*
Animal foods				
Egg, hen's	13.3	98	98	96
Milk, cow's	3.5	95	85	81
Meat	19.8	96	82	79
Fish	21.5	96	80	77
Cereals				
Rice, milled	7.0	93	70	65
Wheat, whole				
Maize	11.8	85	60	51
Pulses				
Bengal gram	22.5	84	62	52
Green gram	24.0	85	58	49
Red gram	22.3	83	56	46
Black gram	24.0	82	54	45
Nuts and oil seeds				
Coconut, fresh	4.5	82	67	56
Soybean	40.0	86	64	55
Sesame	18.3	80	60	48
Groundnut	26.7	92	54	50

Biological Value of Proteins

The term **biological value of protein** refers to the relative nutritional value of individual proteins as compared to a standard and includes a factor relating to the digestibility of a protein source as well as to its amino acid composition. Various procedures are employed to ascertain this value. By these procedures, proteins which lack any of the essential amino acids will have no biological vlaue as compared with a standard protein preparation such as *lactalbumin*, which is known to be qualitatively complete and readily digestible. It is apparent that if the amino acid compositions of 2 proteins are generally similar except that protein *A* contains only half as much leucine as does protein *B*, twice as much of *A* is needed to meet the leucine requirement and the biological value of *B* is greater than that of *A* when the leucine supply of the diet is a limiting factor. *In general, proteins of animal origin have greater biological value than the proteins of plant origin*. Casein, lactalbumin, mixed muscle proteins, and mixed egg proteins are approximately equal in value.

Most **animal proteins** generally contain all essential amino acids in about the quantities needed by the human body. **Vegetable proteins,** on the contrary, often lack one or more essential amino acids and may, in some cases, be more difficult to digest. Vegetable proteins are not only generally nutritionally inferior to those of animal origin, but proteins are also present in relatively smaller concentrations in plant materials. In general, plant proteins are lower in lysine, methionine and tryptophan content and are less readily digestible than the animal proteins. Even so, vegetarian diets can provide adequate protein provided enough extra protein is consumed to provide sufficient quantities of the essential amino acids and/or 2 or more different proteins are consumed together, which complement each other in amino acid content. For example, if corn (which is deficient in lysine) is combined with legumes (which are deficient in methionine but rich in lysine), the efficiency of utilization for the combination of the two vegetable proteins approaches that of animal protein.

Vegetarian Diets and Protein–Energy Requirements

One of the important problems of a purely *vegetarian diet*, in contrast to a *lacto-ovo vegetarian diet*, is the difficulty in getting sufficient calories and protein. The potential caloric deficit results since the caloric densities of fruits and vegetables are much less than the meats they replace (30–50 cal/100 g) (150–300 cal/100 g). This is because of 3 reasons:

1. most plant products contain much less protein (1–2 g of protein/100 g versus (15–20 g/100 g);
2. most plant protein is of low biological value; and
3. some plant proteins are not completely digested.

Actually, well-defined vegetarian diets usually provide enough calories and protein for the average adult. Indeed, the reduced caloric intake may be beneficial because strict vegetarians do tend to be lighter than their nonvegetarian counterparts.

However, whereas an adult male may require 0.8 g of protein and 40 cal kg^{-1} of body weight, a young child may require two- to threefold of that amount. Likewise, a pregnant woman needs an additional 10 g of protein and 300 cal day^{-1} and a lactating woman an extra 15 g of protein and 500 cal day^{-1}. Thus, both young children and pregnant and lactating women run a risk of protein–energy malnutrition. Children of vegetarian mothers generally have a lower birth weight than children of mothers consuming a mixed diet. Similarly, vegetarian children generally have a slower rate of growth through the first 5 years, but generally catch up by age of 10.

However, it is possible to provide sufficient calories and protein even for these high-risk groups provided the diet is adequately planned. Three principles should be followed to design a *calorie-protein-sufficient vegetarian diet* for young children:

1. whenever possible, include eggs and milk in the diet; they are both excellent souces of calories and high-quality protein;
2. include liberal amounts of those vegetable foods with high-caloric density in the diet, including nuts, grains, dried beans and dried fruits; and
3. include liberal amounts of high-protein vegetable foods that have complementary amino acid composition.

It used to be thought that these complementary proteins must be present in the same meal. Recent animal studies, however, suggest that a meal low in (but not devoid of) an essential amino acid may be supplemented by adding the limiting amino acid at a subsequent meal.

It may be noted that proteins differ in their 'biological value' depending on their content of essential amino acids. The proteins of eggs, dairy products, kidney and liver have high biological values because they contain all of the essential amino acids and may be termed *excellent proteins*. *Good quality proteins*, which are somewhat less efficient in supplying amino acids, include shellfish, soybeans, peanuts, potatoes, and the muscle tissue of meats, poultry, and fish. *Fair proteins* are those of cereals and most root vegetables. The proteins of most nuts and legumes are of poor biological value and hence termed *poor quality proteins*. It is important to emphasize, however, that 2 or more proteins– in themselves only 'poor' or 'fair' in quality– may have a 'good' biologic value when taken together because they may complement each other in supplying the necessary amino acids.

Amino Acid Imbalance

Certain amino acids when present in excessive amounts have been found to affect the health and growth of experimental animals. This is called **amino acid imbalance.** In the latter half of the 20th century, pellagra was reported in India where great millet (*Sorghum vulgare*) is the staple cereal. This is attributed to an excess of leucine in which appears to interfere in the conversion of tryptophan to niacin.

Protein–Energy Malnutrition

The most common form of malnutrition in the world today is **protein–energy malnutrition (PEM).** It occurs frequently among infants and young children between 1 to 3 years of age. It is not only an important cause of childhood morbidity and mortality, but leads to permanent impairment of physical and possibly of mental growth of those who survive. In developing countries, inadequate intake of protein and energy is very common, especially in infants and children. These nutritional deficiency states (or PEM cases) develop into either of the 2 clinical syndromes: kwashiorkor and marasmus; they represent the tip of the 'iceberg'. The great majority of cases of PEM are the 'intermediate' ones, *i.e.*, the mild and moderate cases, which go frequently unrecognized.

Kwashiorkor (a Bantu word, meaning *displaced child*, since it occurs in infants displaced from the breast by younger siblings) is undoubtedly the world's major health problem, and, as population growth outstrips the food supply of the developing nations, this problem will be exacerbated. The disease, which is prevalent in economically undeveloped areas such as tropical America, Central and South Africa, and India, is characterized by growth retardation, anemia, hypoproteinemia frequently with edema, fatty infiltration of the liver with ensuing fibrosis, and, in young Negroes, red to light-brown hair. Often there is atrophy of the acinar tisue of the pancreas that results in diarrhea and steatorrhea. The lack of pancreatic digestive secretions makes available even the small amount of dietary protein. Moreover, a renal lesion develops which leads to strikingly increased urinary excretion of free amino acids. If the diet contains much iron, the intestinal mucosa fails to restrict its absorption, and iron may accumulate in the liver. In general, kwashiorkor appears in children who are fed almost exclusively on a gruel of a cereal such as plantain, taro, millet, cassava (manioc), or, most frequently, corn. In contrast to milk, which contains 5.4 g of protein per 100 cal, these diets provide less than 2g of protein per 100 cal and are deficient also in other essential dietary factors. The mortality rate in untreated children has been reported to be 30–90%; these children are unusually susceptible to intercurrent infection and frequently succumb to acute diarrhea.

CICELY DELPHINE WILLIAMS

(LT, 1893–1992)

Williams was born at Kew Park, Darliston, Westmoreland on December 2, 1893. She was educated in Jamaica, partly at Woolmer's Girls' School. During World War I in 1914, she started to take First Aid and nursing classes and thought seriously about studying medicine. In 1916, after her father's death, she decided to go to Oxford, her father's alma mater. She was one of the women admitted because there were so few male students during the war. She took her degree in Medicine at Oxford in 1923. After serving as House Physician at South London Hospital for women and children, she worked as a full-time Doctor in Queen's Hospital for children and later in King's College Hospital. She obtained the diploma in Tropical Medicine and Hygiene from the London University, in 1929. The same year, she was appointed to the British Colonial Medical Service in the Gold Coast (now Ghana), West Africa where she established clinics and hospitals and improved record keeping. Here, she later made her classic observations on the common and often fatal condition **kwashiorkor**. She learned that kwashiorkor meant the sickness the older child gets when the next baby is born. This seemed to indicate that, when they were no longer breastfed, children were not receiving enough to eat. She quickly published her diagnosis of kwashiorkor as a protein-deficiency disease, which attracted the attention of medical fraternity. In 1931, she wrote a paper entitled 'Deficiency Disease of Infants' based on her work for the Gold Coast Colony Annual Medical Report. In 1933, she published her classic paper 'Nutritional Disease of Children Associated with a Maize Diet'. She showed that the disease can be cured by a well-balanced diet rich in proteins and other nutrients. She concluded 'As maize was the only source of supplementary food, some amino acid or protein deficiency cannot be excluded as a cause'. She also found that milk could cure the disease. In 1935, she published another classical paper 'Kwashiorkor– A nutritional disease of children associated with a maize diet'.

After 7 years of service in West Africa, Williams was transferred to Malaya as paediatrician in the College of Medicine, Singapore. During 1936 – 1941, she carried out extensive studies on different diseases occurring among Malayan children. In December 1941, her work was interrupted due to invasion of Malaya by Japanese forces. She was kept in a P.O.W. (Prisioner of War) camp, for about three-and-a-half years. She was the first woman made incharge of the maternity and child welfare services, to look after the health of women and children in the camp.

After her release from the P.O.W. camp, which brought her near to death due to terrible conditions and bad treatment, Dr. Williams returned to work in Malaya as Advisor in Child Health. Later, she was appointed as Head of the section on 'Maternal and Child Health' in the World Health Organization (WHO), Geneva. She campaigned vigorously against the substitute for breastfeeding in Third-World countries. From 1955–1957, she was senior lecturer in Nutrition in London University where she investigated 'Toxemias of Pregnancy'. In 1959, she joined the Faculty of Nutrition of American University of Beirut. As advisor in Training Programmes for the Family Planning Association of the U.K., she toured extensively in 70 countries lecturing and teaching. She had to her credit more than 50 research papers. She was the recipient of many honours and awards. She was elected Honorary Member of the American Academy of Paediatricians, in 1973. A special number highlighting her contributions was published in the November 1973 Issue of the journal *Nutrition Reviews*. **Dr. William Darby** in his tribute has observed:

> 'The lessons from her writings provide not only instructions in the art, the science, and the sociology of medicine, but an inspirational guide to life.'

She died in England in 1992, at age 99. She was also an active member of the Medical Association for the prevention of war. Indeed, she is one of the many outstanding Jamaican women who have received recognition for their contributions to the world's peoples.

The problem can be solved by a combination of population control, improved agricultural practices, and general economic development. Nutritionally, the principal problem is the low lysine content of the above-mentioned cereals and of wheat flour. Supplementation with synthetic lysine could significantly ameliorate this situation, as could introduction of recently-developed corn strains containing the genes *opaque-2* or *floury-2*. Whereas ordinary corn protein contains 2.8 g lysine per 100 g, protein from strains positive for *opaque-2* contains 4.7 g per 100 g. However, supplementation of cereal diets with fish protein concentrate could also prevent kwashiorkor.

In **marasmus**, there is a generalized deficiency in food intake, most notably in respect to both protein and energy (in kwashiorkor, however, there is inadequate intake of protein with adequate energy intake). Although marasmus is commonly associated with inadequate food intake for a variety of reasons, it may also represent a form of starvation secondary to such diseases as cystic fibrosis, celiac disease, or overwhelming infections. Other causes of marasmus are prematurity, mental deficiency, and gastrointestinal diseases involving severe diarrhea and malabsorption. The marasmic infant will have a thin, wasted appearance and will be small for his/her age.

The *diagnostic features* of kwashiorkor and marasmus are given in Table 37–8.

Table 37–8. Principal features of kwashiorkor and marasmus

Features	*Kwashiorkor*	*Marasmus*
Essential features		
Age of maximum incidence	12–48 months	6–18 months
Edema	Lower legs, sometimes face or generalized	None
Wasting	Less obvious; sometimes fat, blubbery	Marked; 'all skin and bone'
Muscle wasting	Sometimes	Severe
Cholesterol	Lowered	Normal
Growth retardation (in terms of body weight)	Less than in marasmus	Severe
Blood urea	Lowered	Normal
Mental changes	Usually present	Usually good
Response to adrenalin	Lowered	Exaggerated
Variable features		
Appetite	Usually poor	Usually good
Diarrhea	Often (past or present)	Often (past or present)
Skin changes	Often, diffuse pigmentation	Usually none
Hair changes	Often sparse– straight and silky; dyspigmentation– greyish or reddish	Texture may be modified
Moon face	Often	None
Hepatic enlargement	Frequent	None

In general, the *principles of treatment* of PEM are the following:

(a) *Adequate diet:* The aim is to supply what has been lacking in the diet, *viz.,* energy and proteins. It is generally agreed that 125–150 kcal and 3–4 g of good quality protein per kg of body weight per day will cover the child's needs. Mild to moderate cases have been treated successfully on and outpatient basis.

(b) *Treatment of infections*

(c) *Health education:* The parents and others taking care of the patient should be trained as to how to prevent a relapse.

Children can be effectively rehabilitated at home by educating the mother and convincing her that cure can be effected by simple modifications in the child's home diet without external food supplements and within the economic constraints of the family. This concept is known as **nutritional rehabilitation** which was originally formulated by **Bengoa** and has since been developed and modified in various countries. The National Institute of Nutrition (NIN), Hyderabad has formulated an *energy-protein-rich mixture* to treat PEM at home level. The mixture consists of the following:

Whole wheat, roasted		40 g
Bengal gram		16 g
Groundnut		10 g
Jaggery		20 g
	Total	86 g
Energy		330 kcal
Protein		11.3 g

Signs of protein deficiency in the adult are not so clear cut as in the child. This is presumably because the demands of growth for protein are no longer present and the results of deficiency are not so obviously manifest. There is evidence that in areas where protein deficiency syndromes of children are endemic, the adult population is also protein-deficient. The effects of protein deficiency in the adult may show itself in the form of loss of weight, reduced subcutaneous fat, anemia, greater susceptibility to infection, frequent loose stools, general lethargy, incapacity to sustained hard work, delay in healing of wounds, and edema.

Excess Protein–Energy Intake

The sum of all the caloric requirements in man and other mammals, including the basal requirement plus increments for muscle work, other forms of work, and specific dynamic action (SDA), is met by the caloric yield of the diet. If the diet provides excess calories, the differece will appear in the form of stored chemical potential energy, *i.e.*, as **depot lipid**. Other things being equal, a surplus of 9 cal may be expected to reuslt in the deposition of 1 g of additional fat. *Thus, it is apparent that the commonest cause of obesity is overeating*. Specifically in developed countries, people consume larger amounts of protein than needed to maintain nitrogen balance. Although a healthy adult can consume large amounts of protein (about 100 g per day) with no apparent harm, the problem is with the effects of high-protein intake on calcium requirements. Studies conducted suggest that high-protein intake increases urinary loss of calcium and may accelerate bone demineralization associated with ageing.

Overeating plays an important role in many individuals, as does inadequate exercise, but there is also a strong genetic component as well. An **obesity gene** in mice and humans that appears to regulate obesity through effects on both appetite and deposition of fat has been identified. This gene produces a protein called *leptin*. Initially, it was thought that leptin might be a useful therapy for obesity. However, it is now apparent that most obese persons already overproduce leptin and that it has deleterious as well as beneficial effects.

One striking clinical feature of overweight individuals is a marked elevation of serum free fatty acids, cholesterol, and triacylglycerols irrespective of the dietary inake of fat. Why is this? Obesity is obviously associated with an increased number and/or size of **adipose tissue cells**. These cells overproduce factors such as leptin and **tumour necrosis factor α** (TNF-α), which cause cellular resistance to insulin by interfering with autophosphorylation of the insulin receptor (IR) and the subsequent phosphorylation of the insulin receptor substrate-1 (IRS-1). Initially, the pancreas maintains glycemic control by overproducing insulin. Thus, many obese individuals with apparently normal blood glucose control have a syndrome characterized by insulin resistance of the peripheral tissue and high concentrations of insulin in the circulation. This hyperinsulinemia appears to stimulate the sympathetic nervous system, leading to sodium and water retention and vasoconstriction, which increases blood pressure.

The insulin resistance in adipose tissue results in enhanced activity of the hormone-sensitive lipase, which along with increased mass of adipose tissue is probably sufficient to explain the increase in circulating free fatty acids. These excess free fatty acids are carried to the liver and converted to triacylglycerol and cholesterol. Excess triacylglycerol and cholesterol are released as very low density lipoprotein (VLDL) particles, leading to higher circulating levels of both triacylglycerol and cholesterol. This also results in a decrease in high density lipoprotein (HDL) particles, for reaosns not very clear. Eventually, the capacity of the pancreas to overproduce insulin declines, which leads to higher fasting blood sugar levels and decreased glucose tolerance. Because of these metabolic changes, obesity is a primary risk factor in coronary heart disease, hypertension, and diabetes mellitus. This is nutritionally significant because all of these metabolic changes are reversible. Undoubtedly obesity materially decreases life expectancy in the older age groups.

Quite often reduction to ideal body weight is the single most important aim of nutritional therapy. The therapeutic restriction of caloric intake below the total caloric requirement necessarily leads to loss of body lipid, although, owing to transient water retention, this may not be immediately reflected in changes in body weight. It is, however, worthnoting that weight reduction regimes result in a metabolic consequence of obesity. Obesity can lead to increased retention of both sodium and water. As the fat stores are metabolized, they produce water (which is denser than the fat), and the water may largely be detained. In fact, some individuals may actually observe short-term weight gain on certain diets, even though the diet is working perfectly well in terms of breaking down their adipose tissue. This metabolic fact of life can be psychologically devastating to dieters, who expect quick results for all their sacrifice.

DIETARY FIBRES

Dietary fibres are defined as the plant polysaccharides and lignin which are resistant ot hydrolysis (*i.e.,* cannot be broken down) by the digestive enzymes in human beings. It is incorrect, however, to assume that fibre is indigestible since some fibres are, in fact, at least partially broken down by the intestinal bacteria. The major types of fibre and their properties are summarized in Table 37–9.

Table 37–9 Major types of fibres and their properties

Types of fibre	*Major source in diet*	*Chemical properties*	*Physiological effects*
Cellulose	Unrefined cereals Bran Whole wheat	Nondigestible Water-insoluble Absorbs water	Increases stool bulk Decreases intestinal transit time Decreases intercolonic pressure
Hemicellulose	Unrefined cereals Some fruits and vegetables Whole wheat	Partially digestible Usually water-insoluble Absorbs water	Increases stool bulk Decreases intestinal transit time Decreases intercolonic pressure
Lignin	Woody parts of vegetables	Nondigestible Water-insoluble Absorbs organic substances	Increases stool bulk Bind cholesterol Bind carcinogens
Pectin	Fruits	Digestible Water-soluble Mucilaginous	Decreases rate of gastric emptying Decreases rate of sugar uptake Decreases serum cholesterol
Gums	Dried beans Oats	Digestible Water-soluble Mucilaginous	Decreases rate of gastric emptying Decreases rate of sugar uptake Decreases serum cholesterol

(Adapted from: Devlin, Thomas M, 2002)

Data regarding the *dietary fibre content of foods* have been reported by **Southgate** (1978) and **van Soest** (1978). These are presented in Table 37–10.

Table 37–10. Fibre contents of some foods

Foodstuffs	*Total fibre (g/100 g)*	*Components of total fibre (as % of total fibre)*		
		Noncellulosic polysaccharide	*Cellulose*	*Lignin*
Cereals				
Whole meal flour (100%)	13.5	74	20	6
Brown flour (90–95%)	8.7	76	18	6
White flour (70–72%)	3.5	80	19	1

Wheat bran	30.1		75	16	9
Oatmeal	7.7		82	12	6
Rice, milled	2.7		78	22	—
Rice bran	28.5		74	17	9
Rye, whole meal	12.7		71	14	15
Legumes and nuts					
Peanut (without skin)	9.3		70	18	12
Peas, dried	32.5		62	36	2
Vegetables					
	(A)	(B)			
Brussel sprouts	4.2	35.5	72	25	3
Cabbage	3.4	29.4	62	25	13
Peas, frozen	7.8	37.1	69	27	2
Runner beans	2.9	26.4	52	42	7
Onions	1.3	18.1	74	26	Traces
Carrots	2.9	28.4	60	40	Traces
Parsnips	4.9	38.4	58	40	2
Turnips	2.2	25.6	62	36	2
Fruits					
Apple, flesh	1.4	9.2	76	23	1
Banana	1.8	6.8	74	21	5
Guava	3.6	14.4	60	35	5
Pears	2.4	14.7	54	28	19
Peaches	1.5	9.6	65	15	19
Tomato	1.4	12.2	47	32	21

(A) on fresh basis; (B) on moisture-free (= dry) basis

These days, the growing realization is that the dietary fibres play a very important role in the incidence of diseases of metabolism. The credit for attributing low fibre content of diets to the large incidence of certain diseases of metabolism, colon and cancer goes to **Himsworth** (1935, 1949), **Cleaves** (1956), **Burkitt** (1975) and **Trowell** (1975). **Cellulose** and most **hemicelluloses** increase stool bulk and decrease transit time. They decrease intercolonic pressure and appear to play a beneficial role with regard to diverticular diseases. They dilute out the potential carcinogens and speed up their transit through the colon; thus, they may also play a role in reducing the risk of colon cancer. **Lignins** not only enhance the bulk of the food but they also adsorb organic substances such as cholesterol so as to lower plasma cholesterol concentration. Mucilaginous fibres, such as **pectin** and **gums**, tend to form viscous gels in the stomach and intestine and slow the rate of gastric emptying, thus slowing the rate of absorption of many nutrients. The most important clinical role of these fibres is to slow the rate at which carbohydrates are digested and absorbed. Thus, both the rise in blood sugar and the subsequent rise in insulin levels are significantly decreased if these fibres are ingested along with carbohydrate-containing foods.

Water-soluble fibres (pectins, gums, some hemicelluloses, and storage polysaccharides) also help to lower serum cholesterol levels in most people. Whether this is due to their effect on insulin levels (insulin stimulates cholesterol synthesis and export) or to other metabolic effects (perhaps caused by end products of partial bacterial digestion) is not conclusively known. Vegetables, wheat and most grain fibres are the best sources of the water-insoluble cellulose, hemicellulose and lignin. Fruits, oats, and legumes are the best sources of the water-soluble fibres. Obviously, a balanced diet should include food sources of both soluble and insoluble fibres.

It has long been recognized that ingestion of some foods, *e.g.*, pulses, onions etc, produces greater degree of flatus in human beings than other foods, *e.g.*, rice, wheat etc. Various studies conducted to identify the chemical nature of flatus-producing components in foods have shown that the flatus-producing factors consist mostly of oligosaccharides, and other nonavailable carbohydrates. Among the oligosaccharides, the trisaccharide *raffinose* and the tetrasaccharides *stachyose* and *scorodose* do commonly occur in foods such as garlic, onion etc. These are not digested by the enzymes present in the human gastrointestinal tract (GIT) and pass on along with undigested food to the colon. They are hydrolyzed and fermented by the bacterial enzymes in the colon, producing flatus. Besides, consumption of large amounts of legumes is also known to cause flatus.

Stagnation of the feces may give rise to carcinogenic compounds by the action of the intestinal flora on steroids present in the stools. Presence of fibre, however, may help to prevent cancer by rapid transit and elimination of the feces or may help to bind cancer-producing (carcinogenic) chemicals and make them ineffective (**Walker**, 1978; **Rubio** *et al,* 1979).

Studies conducted by various nutritionists have shown that fibre increases excretion of bile acids in stools. **Portman** *et al* (1958) reported that cellulose added to semipurified diet caused a small increase in the excretion of bile acids. This observation was later confirmed by **Eastwood** and **Boyd**, in 1967. Incorporation of bagasse and Metamucil (a hydrophilic colloid) caused a marked increase in the excretion of bile acids in stools. **Kritchevsky** and **Story** (1974) found that the dietary fibre present in human foods absorbed, to varying extent, bile acids.

Studies carried out by **McCance** and **Widdowson** (1954) showed that the weight of feces in children fed on whole meal bread was very much higher than that fed on white bread. This is evidently due to the high fibre content of the whole meat bread. **Subramanyam** *et al* (1955) reported that the dry weight of feces in adults fed on whole ragi flour diet was twice as much as that obtained from those fed on rice diet. This is due to the fact that 'ragi' (*Eleusine coracana*) contains large amounts (13–16%) of dietary fibre. An intake of 50 g of wheat bran daily has been reported to double the fecal weight in adult human (**Flock** and **Fuchs**, 1978).

COMPOSITION OF MACRONUTRIENTS IN THE DIET

The 3 major components of the diet (carbohydrate, fat and protein) are required and consumed in larger amounts; hence, they are designated as **macronutrients**. There are, however, relatively few instances of their deficiencies in the diet. Thus, much of the interest in recent years has focused on whether there is an ideal diet composition consistent with good health. Heart disease, stroke, and cancer kill lacs of persons each year across the globe. It is hoped that many of these deaths could be prevented with a balanced or prudent diet.

Diet Composition and Serum Cholesterol

There is a close association between heart disease and the concentration of serum cholesterol and triacyglycerol in the diet. The controversies centred around the dietary control of cholesterol levels illustrate perfectly the trap one falls into by trying to look too closely at each individual component of the diet instead of the diet as a whole. For example, there at least 4 dietary components that can be identified as having an effect on serum cholesterol: cholesterol itself, polyunsaturated fatty acids (PUFAs), saturated fatty acids (SFAs), and fibre. It would seem that the more cholesterol one eats, the higher the serum cholesterol should be. However, cholesterol synthesis is tightly regulated and decreases in dietary cholesterol have relatively little effect on serum cholesterol levels. One can obtain a more significant reduction in cholesterol and triacylglycerol levels by increasing the ratio of PUFA/SFA in the diet. Finally, some plant fibres, especially the water-soluble fibres appear to decrease cholesterol levels significantly.

Saturated fatty acids inhibit receptor-mediated uptake of low density lipoproteins (LDLs). Palmitic acid (saturated, C16) raises serum cholesterol levels while stearic acid (saturated, C18) is neutral. Polyunsaturated fatty acids lower both LDL and HDL cholesterol levels, while oleic acid (monounsaturated, C18) appears to lower LDL without affecting HDL levels. Furthermore, the ω-3 and ω-6 polyunsaturated fatty acids have slightly different effects on lipid profiles. However, these mechanistic complications do not significantly affect dietary recommendations. Most foods rich in standard fats contain both palmitic and stearic acids and are atherogenic. Since oleic acid lowers LDL levels, olive oil (and possibly peanut oil) may be considered as beneficial as polyunsaturated oils.

PUFAs AND RISK FACTORS FOR HEART DISEASE

Recent researches confirm that reduction of elevated serum cholesterol levels can lessen the risk of heart disease. They have rekindled interest in the effects of diet on serum cholesterol levels and other risk factors for heart disease. One of the most important dietary factors regulating serum cholesterol levels is the ratio of polyunsaturated fats (PUFAs) to saturated fats (SFAs) in the diet. A recent discovery is that different types of polyunsaturated fatty acids have different effects on lipid metabolism and on other risk factors for heart disease. There are 2 families of polyunsaturated essential fatty acids:

1. **the n-6 (= ω-6) or linoleic family:** These have a double bond after the 6th carbon atom counting from the terminal methyl (ω) carbon atom; for example linoleic acid.

$$^{1}CH_3—(CH_2)_3—(^{5}CH_2—^{6}CH=CH)_n—(CH_2)_m—COOH$$

Basic formula of the linoleic acid series

2. **the n-3 (= ω-3) or linolenic family:** These have a double bond after the 3rd carbon atom counting from the terminal methyl (ω) carbon atom; for example linolenic acid.

$$^{1}CH_3—(^{2}CH_2—^{3}CH=CH)_n—(CH_2)_m—COOH$$

Basic formula of the linolenic acid series

Recent clinical studies have shown that the ω-6 PUFAs (chief dietary source is linolenic acid from plants and vegetable oils) primarily decrease serum cholesterol levels, with only modest effects on serum triacylglycerol levels. The ω-3 PUFAs (chief dietary source is eicosapentaenoic acid from certain ocean fish and fish oils) cause modest increases in serum cholesterol levels and significantly lower serum triacylglycerol levels. The mechanism behind these effects on serum lipid levels are not known.

The ω-3 PUFAs have yet another unique physiological effect that may decrease the risk of heart disease: they decrease platelet aggregation. The mechanism of this effect is slightly clear. Archidonic acid (ω-6 family) is known to be a precursor of thromboxane A_2 (TXA_2), which is a potent proaggregating agent, and prostaglandin I_2 (PGI_2), which is a weak antiaggregating agent. The ω-3 PUFAs are thought to act by either of the 2 mechanisms:

(*a*) Eicosapentaenoic acid (ω-3 family) may be converted to thromboxane A_3 (TXA_3), which is only weakly proaggregating and prostaglandin A_3 (PGA_3), which is strongly antiaggregating. Thus, the balance between proaggregation and antiaggregation would be shifted toward a more antiaggregating condition as the ω-3 PUFAs displace ω-6 PUFAs as a source of precursors to the thromboxanes and prostaglandins.

(*b*) The ω-3 PUFAs may also act by simply inhibiting the conversion of arachidonic acid to TXA_2.

The unique potential of eicosapentaenoic acid and other ω-3 PUFAs in reducing the risk of heart disease is being tested in numerous clinical trials.

Effects of Refined Carbohydrate in the Diet

Much of the nutritional dispute in the area of carbohydrate intake centres around the amount of refined carbohydrate in the diet. Until recently, simple sugars (primarily sucrose) were blamed for almost for every ill from tooth decay to heart disease and diabetes. In the case of tooth decay, these assertion were truly correct. In the case of heart disease, however, the association is more obscure.

The situation with respect to diabetes is probably even less direct. Whereas restriction of simple sugars is often desirable in patients who already are diabetic, recent studies show less than expected correlation between the type of carbohydrate ingested and the subsequent rise in serum glucose levels (Tabe 37–11). Ice cream, for example, causes a much smaller increase in serum glucose levels than either potatoes or whole-wheat bread. This is because other components of food– such as protein, fat, and the soluble fibres– are much more important than the type of dietary carbohydrate in determining how rapidly glucose will enter the bloodstream.

Table 37–11. Glycemic index* of some selected foods

Grain and cereal products		**Root vegetables**	
Bread (white)	69 ± 5	Beets	64 ± 16
Bread (whole wheat)	72 ± 6	Carrots	92 ± 20
Rice (white)	72 ± 9	Potato (white)	70 ± 8
Sponge cake	46 ± 6	Potato (sweet)	48 ± 6
Breakfast cereals		**Dried legumes**	
All bran	51 ± 5	Beans (kidney)	29 ± 8
Cornflakes	80 ± 6	Beans (soy)	15 ± 5
Oatmeal	49 ± 8	Peas (blackeye)	33 ± 4
Shredded wheat	69 ± 10	**Fruits**	
Vegetables		Apple (golden delicious)	39 ± 3
Sweat corn	59 ± 11	Banana	62 ± 9
Frozen peas	51 ± 6	Oranges	40 ± 3
Dairy products		**Sugars**	
Ice cream	36 ± 8	Fructose	20 ± 5
Milk (whole)	34 ± 6	Glucose	100
Yogurt	36 ± 4	Honey	87 ± 8
		Sucrose	59 ± 10

* **Glycemic index** is defined as the area under the blood glucose response curve for each food expressed as a percentage of the area after taking the same amount of carbohydrate as glucose (mean: 5–10 individuals).

(*Data adapted from: Jenkins DA et al: Glycemic index of foods: a physiological basis for carbohydrate exchange. Am. J. Clin. Nutr.* ***34*** *: 362, 1981.*)

An Increase of Fibre Content in the Diet

The dietary fibres have a profound beneficial effect on the human metabolism. Hence, most nutritionists recommend an enhanced amount of dietary fibres in the balanced or prudent diet. The current fibre content of the diet globally ranges between 10 to 15 g/day. Most nutritionists feel that an increase to about 25 to 30 g/day would be safe and beneficial. Since different types of fibres have different physiolgoical roles, the increase in fibre intake should come from a wide variety of fibre sources– including fresh fruits, vegetables and legumes as well as the more popular sources of fibre (which are primarily cellulose and hemicellulose).

A Prudent Diet

The Senate Select Committee on Human Nutrition, USA spearheaded a movement wherein several private and government groups were asked to recommend composition of an ideal diet, specifically for the Americans. The Committee, after enough deliberations and thorough study, first published its *Dietary Goals for the United States,* in 1977. This Committee recommended that the

American Public reduce consumption of total calories, total fat, saturated fat, cholesterol, simple sugars and salt to "ideal" goals more compatible with good health (Fig. 37–2).

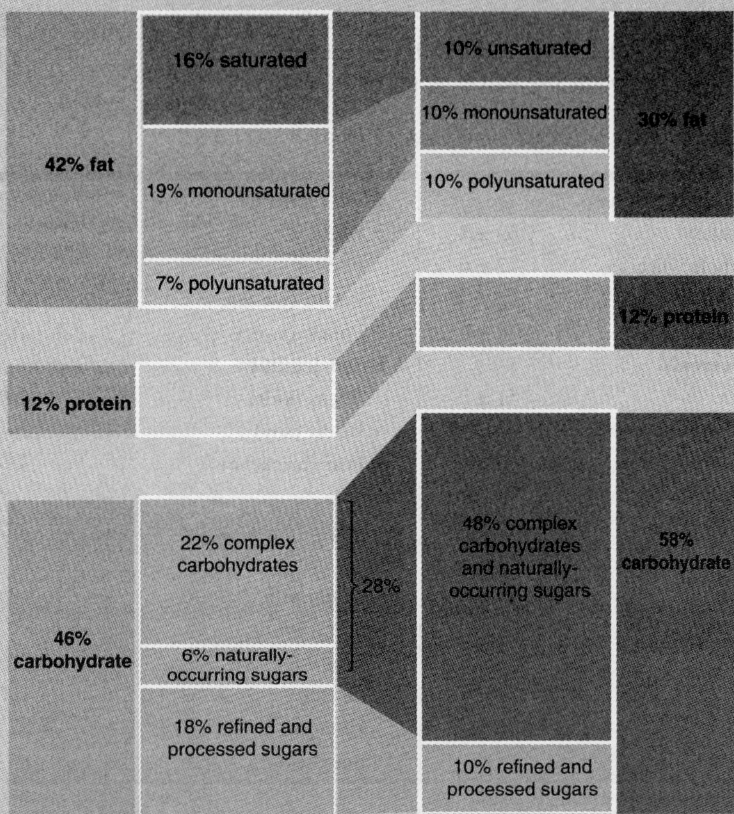

Fig. 37–2. Graphical comparison of the composition of the current U.S. diet and the dietary goals for the U.S. population, as suggested by the Senate Select Committee on Human Nutrition

(Adapted from: Dietary Goals for the United States 2nd ed., Washington, DC: U.S. Government Printing Office, 1977.)

In 1972, the United States Department of Agriculture (USDA), after consulting recommendations made by various scientific bodies, has designed revised recommendations for a **balanced** or "**prudent**" **diet** (Fig. 37–3). How valid is the scientific basis of the recommendations for a prudent diet? An important argument against such recommendations is that we do not have enough information to set concrete goals. We might be creating some problems while solving others. For example, the goals of reducing total and saturated fat in the diet are best met by replacing animal protein with vegetable protein. This might reduce the amount of available iron and vitamin B_{12} (cyanocobalamin) in the diet. It is also quite clear that the same set of guidelines do not apply for every individual.As an example, exercise is known to raise serum HDL cholesterol and obesity is known to elevate serum cholesterol and triacylglycerols and reduce glucose tolerance. Thus, a very active individual who maintains ideal body weight can likely tolerate higher fat and sugar intakes than an obese individual.

On the "pro" side, however, it clearly can be argued that all of the dietary recommendations are aimed at reducing nutritional risk factors in the general population. Besides, similar diets have been consumed by the people the world over and their ancestors with no apparent harm.

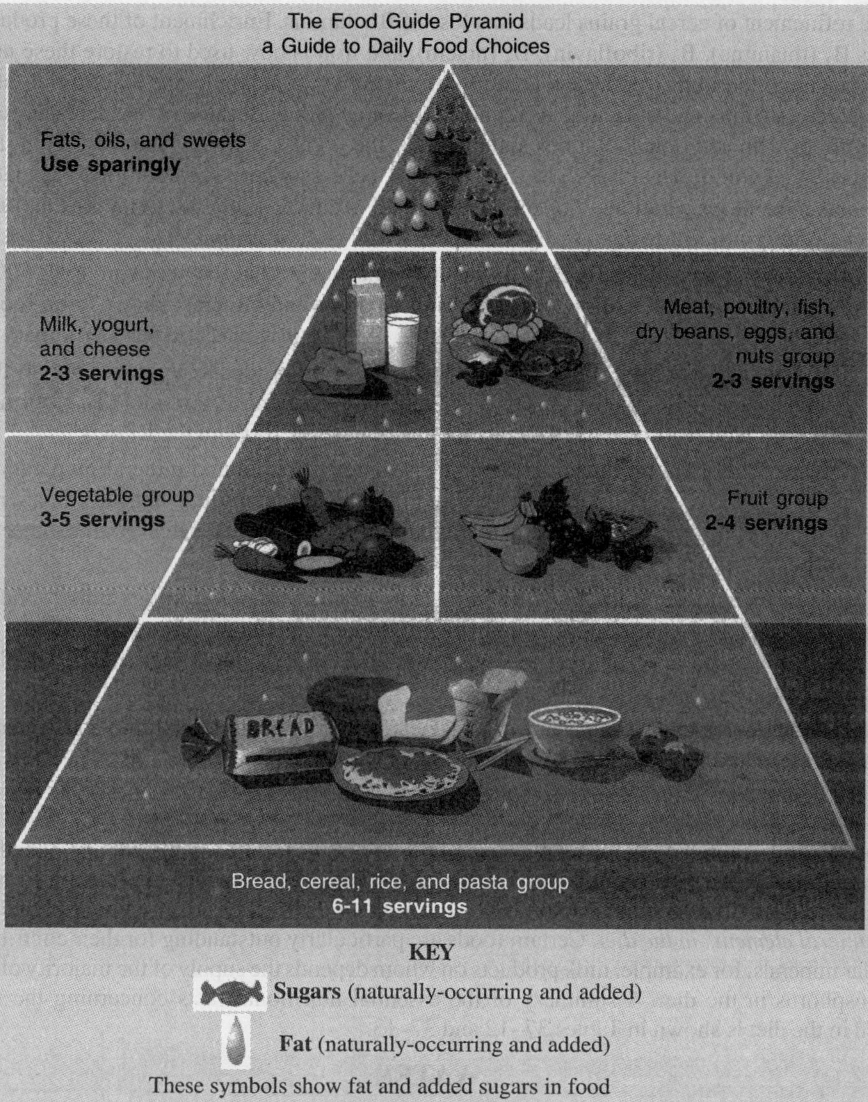

Fig. 37–3. Graphical representation of USDA recommendations for a balanced diet

This illustrates current recommendation on healthy eating. The pyramid is an outline of what to eat each day. In is not a rigid prescription, but a general guide that lets you choose a healthfull diet that is right for you. The pyramid calls for eating a variety of foods to get the nutrients you need and at the same time the right amount of calories to maintain or improve you weight. The pyramid also focuses on fat because most American diets are too high in fat, especially saturated fat.

(Adapted from: HG Bulletin #252. Washington, DC: U.S. Government Printing Office,1992.)

VITAMINS

The history, nomenclature, chemistry, physiological functions and deficiency diseases of the vitamins have been discussed in Chapters 38 and 39. Normal individual on an adequate diet can obtain all of the required vitamins from natural foods; no supplementation with vitamin concentrates is necessary. In disease states in which digestion and assimilation are impaired or the normal requirement for the vitamins is increased, they must, of course, be supplied in appropriate quantities from other sources.

The refinement of cereal grains leads to a loss of B vitamins. Enrichment of these products with vitamins B_1 (thiamine), B_2 (riboflavin), B_5 (niacin), and iron is now used to restore these nutrients. Other foods are often improved by the addition of vitamins; for example, the addition of vitamin D (cholecalciferol) to milk and of vitamin A (retinol) to oleomargarine. Because of the increasing tendency to fortify foods with additional vitamins and minerals, the Food and Drugs Administration (FDA) of the United States Public Health Service has found it necessary to issue additional regulations on vitamin and mineral preparations. For this purpose, all products containing vitamins and minerals have been divided into following 3 categories:

1. **Ordinary foods:** These are products supplemented so that they contain up to 50% of the Recommended Daily Allowance (RDA) of vitamins and minerals. Any natural food which already contains more than 50% of the RDA is, of course, classified as an ordinary food.
2. **Dietary supplements:** With 3 exceptions, these are products supplemented so that they contain 50–150% of the RDA. The 3 exceptions are: vitamins A (retinol), B_9 (folic acid) and D (cholecalciferol). The upper limits for these are set at 100% of the RDA.
3. **Drugs:** These are products supplemented wherein vitamin and mineral preparations with potencies above 150% of the RDA exceed food requirements and, henceforth, are regarded as appropriate only for the treatment of vitamin–mineral deficiencies or some other medical purpose.

Because of their toxicity, the FDA has directed particular attention to vitamins A and D. It is recommended that any preparation of vitamin A with more than 10,000 International Units (IU) per dosage unit or any preparation of vitamin D with more than 400 IU per dosage unit be supplied only on prescription.

The summaries of the essential nutritional facts concerning the fat-soluble and water-soluble vitamins are presented in Tables 37–12 and 37–13.

MINERALS

The minerals, though they form only a small portion of the total body weight, are nonetheless of great importance in the vital economy of living beings. Their requirements and functions for each as known till date are discussed in Chapter 33. *Fruits, vegetables and cereals are the principal sources of the mineral elements in the diet*. Certain foods are particularly outstanding for their contribution of particular minerals; for example, milk products on whom depends the supply of the majority of calcium and phosphorus in the diet. A summary of the essential nutritional facts concerning the minerals required in the diet is shown in Tables 37–12 and 37–13.

WATER

Water is not a food, but since it is ordinarily consumed in the diet, it is included as one of its components. The water requirements, its structure, properties and the functions it performs in the body are discussed in Chapter 2.

BALANCED DIET

A **balanced diet** may be defined as one which contains the various groups of foodstuffs such as energy-yielding foods, body-building foods and protective foods in the correct proportions so that an individual is assured of obtaining the minimum requirements of all the nutrients. The components of a balanced diet differ according to age, sex, physical activity, economic status and the physiologic state (such as pregnancy, lactation) of the individual. If the dietary lacks one or some of these constituents or elements, malnutrition results and the human body becomes limited in physical and mental development, reveals a propensity to illness and has a low capacity for work.

Table 37–12. The fat-soluble vitamins in nutrition

Nomenclature	Important sources	Stability	Functions	Metabolism
Vitamin A Vitamin A_1: Retinol Vitamin A_2: 3-dehydroretinol *Antixerophthalmic factor; 'Bright eyes' vitamin*	Fish-liver oils, liver, butter, cream, whole milk, whole-milk cheese, egg yolk. Dark green leafy vegetables, yellow vegetables and fruits, fortified margarine. Principal source in foods is as the provitamin beta-carotene. The vitamin itself occurs rarely in foods and is confined to lipids of animal tissues.	Insoluble in water: fat-soluble. Associated with lipid in foods. Stable to heat by usual cooking methods. Destroyed by oxidation, drying, and very high temperatures.	Maintenance of the integrity of epithelial tissue. Constituent of visual purple (rhodopsin) of retinal cells. Essential to growth, particularly of skeleton and other connective tissues.	*Absorption:* From gastrointestinal tract, vitamin A follows pathway of fats; consequently, any impairment in fat absorption impairs absorption of vitamin A. *Storage:* In liver (95%). *Deficiency:* Dry, keratinized epithelium; night blindness; xerophthalmia; arrested growth.
Vitamin D Vitamin D_2: Ergocalciferol Vitamin D_3: Cholecalciferol *Antirachitic factor; 'Sunshine' vitamin*	Fish-liver oils, fortified milk, activated sterols, exposure to sunlight. Very small amounts in butter, liver, egg yolks.	Fat-soluble. Relatively stable to heat and oxidation.	Increases absorption of calcium and phosphorus from the intestine. Essential to ossification. Influences handling of phosphate by kidneys. Active form in tissues: 1.25-dihydroxycholecalciferol.	*Absorption:* From intestine with fats, bile salts essential; can be synthesized in skin by activity of ultraviolet light on provitamin (D_3). *Storage:* Chiefly in liver, *Deficiency:* Rickets in children; tetanic convulsions in infants with severe deficiency.
Vitamin E Mainly α-tocopherol; also β-, γ-, δ-tocopherol. *Antisterility factor*	Plant tissues; oils of wheat germ, rice germ, cottonseed; green leafy vegetables, nuts, legumes, milk, eggs; muscle meats, fish.	Fat-soluble. Not affected by heat or acid. Oxidized in rancid fats and in the presence of lead and iron salts, alkali, ultraviolet light.	Exerts antioxidant effect to protect other vitamins in foods. May have auxiliary function in tissue respiration. In experimental animals, in association with other factors, prevents certain types of liver necrosis. No demonstrated function in human nutrition except in some infants immediately postpartum, particularly prematures. Need may be related to unsaturated fatty acid intake.	*Absorption:* Similar to other fat-soluble vitamins. Transfer *via* placenta is limited; mammary gland transfer better; hence, breast milk is effective source for infants. *Deficiency:* May occur in malabsorptive states associated with lipid malabsorption.
Vitamin K Vitamin K_1: Phylloquinone Vitamin K_2: Flavinoquinone *Antihemorrhagic vitamin; Coagulation vitamin*	Green leaves, such as alfalfa, spinach, cabbage; liver. Synthesis in intestine by activity of microorganisms is probably most important source. Dietary deficiency unlikely.	Fat-soluble. Unstable to alkali and light. Fairly stable to heat.	Molecular action is to catalyze reactions of carboxylation of γ-carbon atoms of glutamic acid residues in certain proteins. The resulting carboxylated residues permit the affected proteins to carry more Ca^{2+} cations. Clinically, K deficiency is most evident by diminished production of prothrombin by liver. The vitamin is also required for activity of several blood clotting (thromboplastic) factors. Carboxylation catalytic function is noted also in bone and other mineralized tissues, indicating a more extensive metabolic role of vitamin K than only its connection with blood clotting factors.	*Absorption:* Similar to other fat-soluble vitamins. *Utilization:* Presence of bile needed; impairment in fat absorption seriously affects vitamin K absorption. Produced by intestinal microorganisms; therefore, antibiotic suppressive therapy may induce vitamin K deficiency. *Storage:* Limited amount in liver. *Deficiency:* Hypoprothrombinemia with resultant prolonged blood clotting; uncontrollable hemorrhage in newborn.

Table 37–13. The water-soluble vitamins in nutrition

Nomenclature	Important sources	Stability	Functions	Metabolism
Vitamin B_1 Thiamine or Aneurin *Antiberiberi vitamin; Antineurtic factor*	Lean pork; liver, heart, kidney; brewer's yeast, wheat germ, whole-grain or enriched cereals and breads; soybeans, legumes, peanuts, milk.	Soluble in water. Stable in slightly acid solution. Quickly destroyed by heat in neutral or alkaline solution. Sulfite quickly destroys thiamine.	Constitutent of enzyme systems of tissues, particularly in connection with decarboxylation (*e.g.*, pyruvic and ketoglutaric acids). Deficiency affects mainly peripheral nervous system, gastrointestinal tract, and cardiovascular system.	*Absorption:* Readily absorbed in aqueous solutions from both small and large intestines. *Storage:* Limited; hence, daily supply needed. *Excretion:* Excess excreted to some extent in perspiration; also in urine. *Deficiency:* Anorexia, gastrointestinal atony and constipation, beriberi (including polyneuritis, cardiac failure, and edema). Requirement increased by high carbohydrate intake; also in fever, hyperthyroidism, pregnancy, lactation.
Vitamin B_2 Riboflavin or Lactoflavin *'Yellow enzyme'*	Milk, powdered whey; liver, kidney, heart, meats; eggs; green leafy vegetables; dried yeast, enriched foods (flour, bread). Cereals low; germination of oats, wheat, barley, and corn increases content.	Sparingly soluble in water. Quickly decomposed by ultraviolet or visible light; very sentitive to alkali. Relatively resistant to heat in acid media.	Constituent of tissue respiratory enzyme systems, as well as some enzymes (flavoproteins) involved in amino acid and lipid metabolism.	*Absorption:* May require phosphorylation in intestinal mucosa. *Storage:* Limited in the body. *Excretion:* Excess excreted in urine. *Deficiency:* Cheilosis, seborrheic dermatitis of face, magenta tongue, certain functional and organic eye disorders.
Vitamin B_3 Pantothenic acid *Filtrate factor; Yeast factor*	Organ meats (liver, kidney), lean beef; egg yolk; peanuts; broccoli, cauliflower, cabbage; whole-grains, cereal bran; skim milk; fruits, sweet potatoes.	Easily destroyed by heat and alkali. Stable in neutral solution.	Constituent of coenzyme A which participates in synthesis and breakdown of fatty acids, synthesis of cholesterol and steroid hormones, utilization of pyruvate and acetate, reactions of acetylation, metabolism of some amino acids, synthesis of heme for hemoglobin and cytochromes.	*Storage:* Limited in the body. *Excretion:*Urine. *Deficiency:* Gastrointestinal symptoms, skin symptoms, anemia, and impairment in functions of adrenal cortex in experimental animals.
Vitamin B_5 Niacin and Niacinamide *Anti-pellagra vitamin; Pellagra-preventive factor*	Liver, kidney, lean meat, fish (salmon), poultry; whole-grain or enriched cereals, and breads; some leafy green vegetables, tomatoes; peanuts, brewer's yeast, tryptophan in proteins. Most fruits and vegetables are poor sources of niacin.	Soluble in water. Relatively stable to heat, oxidation, and light. Relatively stable to acid and alkali.	Constituent of 2 coenzymes (NAD, NADH) which operate as hydrogen and electron transfer agents in respiration. Tryptophan normally contributes to niacin supply (60 mg tryptophan equivalent to 1 mg niacin).	*Storage:* Limited in the body. *Excretion:* Urine, mainly as methylated derivatives. *Deficiency:* Pellagra with gastrointestinal, skin, and neurologic changes.
Vitamin B_6 Pyridoxine, Pyridoxal, and Pyridoxamine *Antidermatitis factor*	Wheat germ; meat, liver, kidney; whole-grain cereals, soybeans, peanuts, corn, yams; brewer's yeast. Synthesis by activity of microorganisms.	Fairly stable to heat, but sensitive to ultraviolet light and oxidation.	Pyridoxal phosphate is a prosthetic group of enzymes that decarboxylate tyrosine, arginine, glutamic acid, and certain other amino acids. Essential to transulfuration and in conversion of tryptophan to niacin; also as a coenzyme in transamination. Participates in metabolism of essential fatty acids. Essential in synthesis of porphyrins (*e.g.*, heme for hemoglobin and cytochromes).	*Absorption:* Intestinal bacteria synthesize some pyridoxine, and it is absorbed from the intestine. *Storage:* Limited in the body. *Excretion:* Urine. *Deficiency:* Hypochromic macrocytic anemia, lesions of central nervous system evidenced by epileptiform seizures and encephalographic changes, particularly in infants.

Table 37–13. (Cont'd.)

Nomenclature	Important sources	Stability	Functions	Metabolism
Vitamin B_7 Biotin *Antiegg white injury factor; Coenzyme R*	Required by various animal species but of questionable need for humans. If they are in fact needed, the amounts required are very small and are probably synthesizable in tissues or provided by intestinal flora.			
Vitamin B_9 Folic acid or Folacin *Liver Lactobacillus casei factor*	Liver, kidney; yeast; fresh green leafy vegetables, cauliflower. Synthesis by activity of intestinal microorganisms.	Easily oxidized in acid medium and sunlight. Labile to heat. (similar to thiamine)	Involved in transfer and utilization of the single-carbon moiety; participates in synthesis of purines, thymine, and methyl groups; has specific role in metabolism of histidine and well-demonstrated role in hematopoiesis.	*Excretion:* Excess in both urine and feces. *Deficiency:* May produces macrocytic anemia with concurrent glossitis, gastrointestinal lesions, diarrhea, and intestinal malabsorption (sprue). Deficiency in pregnancy not uncommon.
Vitamin B_{12} Cyanocobalamin *Antipernicious anemia factor*	Foods of animal origin; liver, kidney, muscle meat, eggs, milk, cheese. No significant amounts in higher plants. Synthesis within intestine by activity of microorganisms	Labile to heat, acids, alkali, and light.	Involved in purine and pyrimidine metabolism, synthesis of nucleic acid (DNA), maturation of red blood cells, methionine metabolism, and transmethylation. Contains cobalt, which is the only known function for this element.	*Absorption:* From ileum, but requiring intrinsic factor and hydrochloric acid contributed by stomach. *Storage:* Principally in liver, for long periods. *Excretion:* In feces (represents unabsorbed vitamin). *Deficiency:* Macrocytic anemia or pernicious anemia with degenerative changes in gastric mucosa, characteristic lesions in nervous system (combined system disease).
Vitamin C Ascorbic acid *Antiscorbutic vitamin; Cevitamin*	Citrus fruits, tomatoes, strawberries, cantaloupe, cabbage, broccoli, kale, potatoes, green peppers, salad greens.	Soluble in water. Most easily destroyed of all vitamins– by heat, air, alkali, enzymes. Acid inhibits destruction. Copper accelerates destruction. Cooking generally reduces vitamin C content of food. Consumption of some uncooked foods essential to ensure adequate intake.	Maintains normal interecellular material of cartilage, dentine, and bone. Probably has specific role in synthesis by activity on proline hydroxylation. Association with oxidation–reduction systems in tissues. Metabolism of some amino acids, *e.g.,* tyrosine, proline.	*Storage:* Large quantity in adrenal cortex. With exception of muscle, tissues with high metabolic activity have increased concentrations. *Excretion:* Urine. *Deficiency:* Mild– petechial hemorrhages. Severe– loosening of teeth, lesions of gums, poor wound healing, easily fractured bones, scurvy.

(Modified from: Harper HA: Review of Physiological Chemistry. 17th edition. Lange Medical Publications, California. 1979.)

The characteristics of the 3 types of foods, which constitute a balanced diet, are as follows:

A. **Energy-yielding foods:** These are foods rich in carbohydrates and also fats. They may be broadly divided into 2 groups: (*a*) cereals, roots and tubers, and (*b*) pure carbohydrates and fats. *Cereals* provide, in addition to energy, the greater part of the proteins, certain minerals and vitamins in the diets of the low-income groups in the developing nations. *Roots and tubers* also provide some amounts of proteins, minerals and vitamins while pure *carbohydrates and fats* provide only energy.

B. **Body-building foods:** These are foods rich in proteins. These may be broadly divided into 2 groups: (*a*) milk, egg, meat and fish which are rich in proteins of high biological value, and (*b*) pulses, oilseeds and nuts and low-fat oilseed flours which are rich in proteins of medium nutritive value.

C. **Protective foods:** These are foods rich in proteins, vitamins and minerals. These may also be broadly classified into 2 groups: (*a*) foods rich in vitamins, minerals and proteins of high biological value; *e.g.*, milk, eggs, fish and liver, and (*b*) foods rich in certain vitamins and minerals only; *e.g.*, green leafy vegetables and some fruits. Diets in India are generally poor in respect of protective foods.

The main functions of food are: (*a*) provision of energy, (*b*) body building and repair, and (*c*) maintenance and regulation of tissue functions.

Balanced diets may be planned for all sections of the society including persons of high-, medium- and low-income group. Accordingly, 3 different types of balanced diet are recognized:

1. **Balanced diets at low cost:** These diets include small amounts of milk, eggs, meat, fish and fats and liberal amounts of cereals, pulses, nuts and given leafy vegetables.
2. **Balanced diets at moderate cost:** Such diets include moderate amounts of milk, eggs, meat, fish, fruits and fats and liberal amounts of cereals, pulses, nuts and green leafy vegetables.
3. **Balanced diets at high cost:** These diets include liberal amounts of costly foods such as milk, eggs, fish and fruits and moderate quantities of cereals, pulses, nuts and fats.

Balance diets at low, moderate and high costs for adult males and for normal, pregnant and lactating women, as suggested by the Indian Council for Medical Research (ICMR) Nutrition Expert Group 1968, are given in Tables 37–14 to 37–19.

Table 37–14. Balanced diets (g/caput/day) for adult male (Low cost)

Foodstuffs	*Sedentary work*		*Moderate work*		*Heavy work*	
	V	*NV*	*V*	*NV*	*V*	*NV*
Cereals	400	400	475	475	650	650
Pulses	70	55	80	65	80	65
Green leafy vegetables	100	100	125	125	125	125
Other vegetables	75	75	75	75	100	100
Fruits	30	30	30	30	30	30
Milk	200	100	200	100	200	100
Fats and oils	35	40	40	40	50	50
Sugar and jaggery	30	30	40	40	55	55
Meat and fish	–	30	–	30	–	30
Eggs	–	30	–	30	–	30
Groundnut, roasted	–	–	–	–	50*	50*

*An additional 25 g of fats and oils can be included in the diet in place of groundnut.

(*Source: Nutrition Expert Group, I.C.M.R., 1968*)

Table 37–15. Balance diets (g/caput/day) for adult male (Moderate cost)

Foodstuffs	Sedentary work		Moderate work		Heavy work	
	V	NV	V	NV	V	NV
Cereals	350	350	425	425	600	600
Pulses	70	55	80	65	80	65
Green leafy vegetables	100	100	125	125	125	125
Other vegetables	75	75	100	100	100	100
Roots and tubers	75	75	100	100	100	100
Fruits	60	60	60	60	60	60
Milk	600	400	600	400	600	400
Fats and oils	35	40	40	40	50	50
Sugar and jaggery	30	30	40	40	55	55
Meat and fish	–	60	–	60	–	60
Eggs	–	30	–	30	–	30
Groundnut, roasted	50	50	50	50	50	50

(*Source: Nutrition Expert Group, I.C.M.R., 1968*)

Table 37–16. Balanced diets (g/caput/day) for adult male (High cost)

Foodstuffs	Sedentary work		Moderate work		Heavy work	
	V	NV	V	NV	V	NV
Cereals	300	300	375	375	550	550
Pulses	70	50	80	60	80	60
Green leafy vegetables	100	100	100	100	100	100
Other vegetables	75	75	75	75	100	100
Roots and tubers	75	75	100	100	100	100
Fruits	200	200	200	200	200	200
Milk	1000	600	1000	600	1000	600
Cheese	40	–	40	–	40	–
Meat and fish	–	100	–	100	–	100
Eggs	–	60	–	60	–	60
Groundnut, roasted	–	–	–	–	50	50

(*Source: Nutrition Expert Group, I.C.M.R., 1968*)

Table 37–17. Balanced diets for normal, pregnant and lactating women (Low cost)

Foodstuffs	*Normal women*						*Additional allowances during*	
	Sedentary work		*Moderate work*		*Heavy work*		*Pregnancy*	*Lactation*
	V (g)	*NV (g)*	*V (g)*	*NV (g)*	*V (g)*	*NV (g)*	*(g)*	*(g)*
Cereals	300	300	350	350	475	475	50	100
Pulses	60	45	70	55	70	55	–	10
Green leafy vegetables	125	125	125	125	125	125	25	25
Other vegetables	75	75	75	75	100	100	–	–
Roots and tubers	50	50	75	75	100	100	–	–
Fruits	30	30	30	30	30	30	–	–
Milk	200	100	200	100	200	100	125	125
Fats and oils	30	25	35	40	40	45	–	15
Sugar and jaggery	30	30	30	30	40	40	10	20
Meat and fish	–	30	–	30	–	30	–	–
Eggs	–	30	–	30	–	30	–	–
Groundnut	–	–	–	–	40*	40*	–	–

* An additional 25 g of fats and oils can be included in the diet in place of groundnut.

(*Source: Nutrition Expert Group, I.C.M.R., 1968*)

Table 37–18. Balanced diets for normal, pregnant and lactating women (Moderate cost)

Foodstuffs	*Normal women*						*Additional allowances during*			
	Sedentary work		*Moderate work*		*Heavy work*		*Pregnancy*		*Lactation*	
	V (g)	*NV (g)*	*V (g)*	*NV (g)*	*V (g)*	*NV (g)*	*V (g)*	*NV (g)*	*V (g)*	*NV (g)*
Cereals	260	250	310	300	440	425	–	–	110	100
Pulses	60	50	60	50	60	50	20	–	40	–
Green leafy vegetables	100	100	100	100	100	100	–	–	–	–
Other vegetables	75	75	75	75	100	100	–	–	–	–
Roots and tubers	50	50	75	75	100	100	–	–	–	–
Fruits	60	60	60	60	60	60	50	50	50	50
Milk	400	250	400	250	400	250	400	200	600	300
Fats and oils	30	35	35	40	40	45	–	–	20	20
Sugar and jaggery	30	30	30	30	40	40	–	–	20	20
Meat and fish	–	60	–	60	–	60	–	25	–	40
Eggs	–	30	–	30	–	30	–	–	–	–
Groundnut	50	30	50	30	50	30	–	–	–	–

(*Source: Nutrition Expert Group, I.C.M.R., 1968*)

Table 37–19. Balanced diets for normal, pregnant and lactating women (High cost)

Foodstuffs	*Normal women*						*Additional allowances during*			
	Sedentary work		*Moderate work*		*Heavy work*		*Pregnancy*		*Lactation*	
	V	*NV*	*V*	*NV*	*V*	*NV*	*V*	*NV*	*V*	*NV*
	(g)	*(g)*	*(g)*	*(g)*	*(g)*	*(g)*	*(g)*	*(g)*	*(g)*	*(g)*
Cereals	230	220	280	270	410	400	–	–	70	70
Pulses	60	50	60	50	60	50	–	–	–	–
Green leafy vegetables	100	100	100	100	100	100	–	–	–	–
Other vegetables	75	75	75	75	100	100	–	–	–	–
Roots and tubers	50	50	75	75	100	100	–	–	–	–
Fruits	100	100	100	100	100	100	50	50	50	50
Milk	600	400	600	400	600	400	500	200	700	300
Cheese	40	–	40	–	40	–	30	–	40	–
Fats and oils	30	35	35	40	40	40	–	–	20	20
Sugar and jaggery	30	30	30	30	40	40	–	–	20	20
Meat and fish	–	100	–	100	–	100	–	25	–	40
Eggs	–	60	–	60	–	60	–	–	–	–
Groundnut	50	30	50	30	50	30	–	–	–	–

(Source: Nutrition Expert Group, I.C.M.R., 1968)

THERAPEUTIC NUTRITION

Therapeutic nutrition is concerned with the nutritional requirements of patients suffering from different diseases and prescribing the right type of diets for them. Many illnesses caused by infection, injuries and metabolic disorders affect the food intake of the patients leading to malnutrition and undernutrition. For this purpose, it is necessary to gather information on the following:

(*a*) the underlying causes of the disease
(*b*) the possible duration of the disease
(*c*) the food habits of the patient
(*d*) the nature of dietary changes required

Before formulating the diet, the information must be obtained with regard to the effect of illness on food acceptance and the psychological and emotional factors confronting the patient. The common complaints of most patients are lack of appetite and dislike for some types of foods. Hence, the diet prescribed should be to the liking of the patient. Care should be taken to avoid foods to which the patient has strong aversion. Also, the diet prescribed should be within the means of the patient. Overall, the *objectives of the diet theory* are as follows:

1. Correction of the existing dietary deficiencies and to maintain the patient in good nutritional status
2. Formulation of the diet to meet the needs of the patient taking into consideration his food habits
3. Education of the patient regarding the need for adherence to the prescribed diet

Acceptability of Modified Therapeutic Diets

Many diseases have a profound effect on appetite and food acceptance. Illnesses, in general, cause anorexia, nausea and dislike of even foods commonly eaten by the patient. The quantity of food consumed will be small. This will involve frequent feeding with small amounts of the diet. The therapeutic diet should be acceptable to the patient and should conform to the diets habitually consumed by the patient. As far as possible, the diets should include the food items commonly consumed by the patient. A real problem will arise if wheat has to be completely eliminated from the diet of the patient (whose staple food is wheat) suffering from malabsorption as *gluten* present in wheat aggravates the clinical condition of the patient. In such cases, the patient should be asked to choose rice and corn as substitutes for wheat. Similarly, diabetics will have to forego a part of the cereal consumed and a greater part of sugar consumed and eat other foods in its place. Such changes will bring about resistance from the patients. They should be advised that, in the interest of their health, they have to accept the modified diets. The patient should be asked to suggest recipes from the food items allowed so that the prepared food items will be acceptable to them.

Modification in Diet Consistency

Diets are usually modified with regard to consistency to suit the needs of the patient. The 3 common modifications are: soft diet, clear-fluid diet and full-fluid diet. Further, for patients who cannot consume diets or fluids orally, tube feeding will have to be restored to.

(*a*) **Soft diet:** This diet is given in acute infection, gastrointestinal disorders and in post operative cases. The diet is soft in consistency, easily digested, and contains very little fibre, spices and condiments. The composition of the soft diet is presented in Table 37–20.

Table 37–20. Composition of soft diet (g/caput/day) (Calories 2000–2200 Kcal; Proteins 60–70 g)

Foodstuffs	*Vegetarian*	*Nonvegetarian*
Milled cereals and cereal products (milled rice, white flour, bread, biscuits and breakfast cereals)	300	300
Daal	50	30
Milk	1000	600
Meat and fish sausages	—	100
Egg	—	30
Cheese (soft and cheddar)	80	30
Tender vegetables	50	50
Potato	100	100
Tender leafy vegetables	100	100
Fruits (Banana, Apples, Oranges)	100	100
Fats and oils (Butter and vegetable oil)	30	30
Sugar	80	80

(*b*) **Full-fluid diet:** This diet is prescribed when the patient is unable to chew or swallow solid foods. It consists of milk, eggs, cereal porridge, soups, fruit juices and gelatin dessert. This diet provides the minimum requirements of proteins, calories, vitamins and minerals. The composition of the full-fluid diet is given in Table 37–21.

Table 37–21. Composition of full-fluid diet (g/caput/day)

Foodstuffs	*Vegetarian*	*Nonvegetarian*
Semolina and corn flour	50	50
Dhal flour	30	—
Milk	1000 ml	800 ml
Meat	—	50
Egg	—	One
Fruit Juice	800	800
Butter	2 tablespoons	2 tablespoons
Sugar	100	100

(*c*) **Clear-fluid diet:** This diet is prescribed for patients with marked intolerance to foods as manifested by nausea, vomiting, loss of appetite etc. It consists of tea with lemon and sugar, fat-free meat broth, carbonated beverages, fruit juice, glucose water, cereal water and plain gelatin. The clear-fluid diets are usually given to the patient for 1 or 2 days till he is able to retain and digest a more liberal liquid diet. At the beginning, the amount of fluid given to the patient is restricted to about 40–8 ml/hr, the quantity gradually being increased to 100–120 ml/hr as the patient recuperates. Clear-fluid diets are meant to provide some calories and make up the water loss in the body. Table 37–22 gives the composition of a clear-fluid diet.

Table 37 – 22. Composition of clear-fluid diet

Foodstuffs	*Amount*
Orange juice	1,000 ml
Glucose water	1,000 ml
Barley water (with glucose)	1,000 ml
Tender coconut water	1,000 ml

Modification of Diets in Different Nutrients

Diets will have to be modified in one or more nutrients depending on the nature of the disease the patient is suffering from. This is quite necessary as the metabolism of the individual changes in different diseases with regard to one or more nutrients. Some such examples are:

(*a*) inefficient utilization of carbohydrates in *diabetes mellitus*,

(*b*) inability of the kidney to excrete sodium chloride in *nephritis*,

(*c*) increased production and inefficient elimination of uric acid in *gout*, and

(*d*) increase in energy metabolism and in the catabolism of tissue proteins in *fever*.

The simplest modification of diets is in the treatment of *allergy* where the foods responsible for the allergic reactions are eliminated. Modification in diets in other diseases may involve change in different constituents such as:

1. bland diets, omitting condiments and spices
2. low residue (fibre) or high residue (fibre) diets
3. high protein or low protein diets
4. high calorie or low calorie diets

5. high carbohydrate or low carbohydrate diets
6. low sodium diets
7. low purine diets

In case the patient cannot consume food orally, then administration of nutrients through parenteral route or through nasogastric tube is essential to avoid starvation and loss of proteins and other nutrients from the tissues. The *types of changes required in the diets* in various diseases are briefly discussed below:

(*a*) **Modifications in carbohydrate content:** High-carbohydrate diet may be indicated in Addison disease, liver diseases and in preoperative conditions. Hypoglycemia may be present in Addison disease, while in the other two conditions, adequate glycogen storage is of considerable value in the therapy of these conditions. Restricted carbohydrate diet is essential in the treatment of diabetes mellitus.

(*b*) **Modifications in fibre content:** Diets rich in fibre (or residue) are prescribed for the treatment of constipation while low-fibre diets are essential in the treatment of several gastrointestinal disorders, such as peptic ulcer, ulcerative colitis, celiac disease, diarrhea and dysentery.

(*c*) **Modifications in protein content:** High-protein diets (i.e., those about twice the actual requirements) with restriction in other nutrients are prescribed in a wide variety of ailments such as protein–calorie malnutrition, cirrhosis of liver, peptic ulcer, nephrosis, and celiac disease. Low protein or complete withdrawal of proteins may be necessary in hepatic coma, acute uraemia etc.

(*d*) **Modifications in fat content:** Moderately-high-fat diet is used in the treatment of severe under-nutrition. Restricted or low-fat diet may be necessary in the treatment of steatorrhea, malabsorption syndrome and diseases of the liver.

(*e*) **Modifications in calorie content:** Diets with increased calorie value are used for the treatment of the patients who are markedly underweight and also for patients with increased calorie requirements as in fever, infections and hyperthyroidism. Low-calorie diets are used for the treatment of obesity, cardiovascular diseases (CVDs), acute uraemia and hepatic coma.

(*f*) **Modifications in mineral content:** High-calcium diet is essential in the treatment of rickets and osteomalacia, while a diet restricted in calcium and phosphorus is desirable in renal calculi. Sodium-restricted diets are essential in the treatment of cardiac failure and hypertension. Restriction in sodium chloride (NaCl) intake is essential in kidney diseases.

(*g*) **Modification in vitamin content:** An increase in the content of vitamins can be easily achieved by the addition of synthetic vitamins. This is essential as most of the therapeutic may be partially lacking in one or more vitamins.

(*h*) **Modifications in Purine content:** Diets low in purine contents are prescribed in the treatment of gout, while diets low in purines and oxalic acid are prescribed in renal calculi.

NUTRITION FOR HEALTH IN INDIA

The **Indian Council of Medical Research** (ICMR), New Delhi, the apex body in India for the planning, formulation, coordination, implementation and promotion, of biomedical research, is one of the oldest medical research bodies in the world. With a modest beginning in 1911 under the name the **Indian Research Fund Association** (IRFA), which after independence was redesignated as ICMR, this prestigious body today enjoys a pan-India presence with 31 dedicated research institutes / centres across the major disciplines. In fact, ICMR has played a key role in the development of public health activities in the country, in the field of nutrition. Research efforts of the council almost eliminated the first vitamin deficiency disease called **beriberi** from the country in the mid-20th century with a simple intervention: 'use a varied diet, do not restrict the diet too rigidly to machine-milled rice.'

Similarly, ICMR studies busted the **protein–energy malnutrition** (PEM) myth convincingly. Studies of preschool children were made to gauge their energy and protein intakes. It indeed turned out to be a surprising revelation that the protein concentration of their diets was nearly adequate contributing to about 10% of their energy. However, the average energy intake was grossly deficient. This analysis paved the way for a new hypothesis that *the malnutrition among underprivileged children was due to energy deficiency and not solely due to protein deficiency, thus busting the protein myth*. Large-scale community studies were initiated to test this hypothesis and the view was soon endorsed by the United Nations Organizations like WHO and UNICEF.

Iron deficiency leading to nutritional anemia is most frequently encountered among the malnourished population of India. After a series of studies on the stability and bioavailability of iron in common salt fortified with iron employing different iron compounds and wide range of stabilizing agents, a successful process for fortification of salt with iron was developed. At present, technologies of *double fortified salt* (DFS) and *fortification of wheat flour* with iron and other essential nutrients have been transferred to industries.

In yet another important development, studies on administration of massive dose of vitamin A to avoid **nutritional blindness** among children, paved the way for a nationwide vitamin A prophylaxis programme with its inclusion in the Fourth Five-Year Plan. The ICMR had also evaluated the National Anemia Prophylaxis Programme and the shortcomings of the programme were brought out.

Similarly, studies on **goitre** control programmes in coastal Andhra Pradesh established that there was irregular supply of iodized salt, lack of iodine content in salt, and absence of quality control measures to ensure iodine content in salt.

REFERENCES

1. **Albanese A (editor) :** Protein and Amino Acid Nutrition. *Academic Press, Inc., New York. 1959.*
2. **Anderson JW :** Fiber and health: an overview. *Am. J. Gastroenterol.* **81 :** *892, 1986.*
3. **Berry EM :** Dietary fatty acids in the management of diabetes mellitus. *Am. J. Clin. Nutr.* **66 :** *99IS, 1997.*
4. **Block RJ, Weiss KW :** Amino Acid Handbook. *Thomas. 1956.*
5. **Boden G :** Free fatty acids, insulin resistance, and type 2 diabetes mellitus. *Proc. Assoc. Am. Physicians. 111, 241, 1999.*
6. **Bondy PK, Rosenberg LE :** Duncan's Diseases of Metabolism. *7th edition. Saunders, Philadelphia. 1974.*
7. **Bray GA (editor) :** Obesity in Perspective. *DHEW Publications (NIH), Washington DC. 1974.*
8. **Bray GA :** Obesity– a disease of nutrient or energy balance? *Nutr. Rev.* **45 :** *33, 1987.*
9. **Burton BT (editor) :** Heinz Handbook of Nutrition. *2nd ed., McGraw-Hill Book Company, New York. 1965.*
10. **Champakam S, Srikantia SG, Gopalan C :** Kwashiorkor and mental development. *AM. J. Clin. Nutr.* **21 :** *844–852, 1968.*
11. **Dulloo AG, Miller DS :** Obesity: a disorder of the sympathetic nervous system. *World Rev. Nutr. Diet.* **50 :** *1, 1987.*
12. **Eastwood M :** Dietary fibre and the risk of cancer. *Nutr. Rev.* **45 :** *193, 1987.*
13. **Francis DEM :** Diets for Sick Children. *Blackwell. 1975.*
14. **Goldsmith G :** Nutritional Diagnosis. *Charles C. Thomas, Publisher, Springfield III, 1959.*
15. **Gorlin R :** The biological actions and potential clinical significance of dietary ω-3 fatty acids. *Arch. Intern. Med.* **148 :** *2043, 1988.*
16. **Griswold RM :** The Experimental Study of Foods. *Houghton Mifflin Company, Boston. 1962.*
17. **Grundy SM :** Monounsaturated fatty acids, plasma cholesterol, and coronary heart disease. *Am. J. Clin. Nutr.* **45 :** *1168, 1987.*

18. **Holt LE Jr, Snyderman SE :** Protein and amino acid requirements of infants and children. *Nutr. Abstr. & Revs.* **35 :** *1–12, 1965.*
19. **James AT, Lovelock JE :** Essential fatty acids and human disease. *Brit. Med. Bull.* **14 :** *262–266, 1958.*
20. **Kisselbah A, Schetman G :** Polyunsaturated and saturated fat, cholesterol, and fatty acid supplementation. *Diabetes Care.* **11 :** *129, 1988.*
21. **Kurtz RC :** Nutrition in Gatro-intestinal Disease. *Churchill Livingstone, London. 1981.*
22. **Lasanga L (editor) :** Obesity: Causes, Consequences and Treatment. *Medican Press, New York. 1974.*
23. **Mattson FH, Erickson BA, Gligman AM :** Effect of dietary cholesterol on serum cholesterol in man. *Am. J. Clin. Nutr.* **25 :** *589, 1972.*
24. **Maxwell MH, Waks AU :** Obesity and hypertension. *Bibl. Cardiol.* **41 :** *29, 1987.*
25. **Miller JCB :** Importance of glycemic index in diabetes. *Am. J. Clin. Nutr.* **59 (suppl.) :** *7475, 1994.*
26. **Pi-Sunyer FX :** Health implications of obesity. *Am. J. Clin. Nutr.* **53 :** *15955, 1991.*
27. **Proudfit FT, Robinson CH :** Normal and Therapeutic Nutrition. *16th edition. The MacMillan Co., New York. 1978.*
28. **Rose WC :** The amino acid requirements of adult man. *Nutr. Abstr. Rev.* **27 :** *631, 1957.*
29. **Shankar S, Lanza E :** Dietary fibre and cancer prevention. *Hematol. Oncol. Clin. North Am.* **5 :** *25, 1991.*
30. **Simopoulos AP :** Obesity and carcinogenesis: historical perspective. *Am. J. Clin. Nutr.* **45 :** *271, 1987.*
31. **Simopoulos AP :** Omega-3 fatty acids in health and disease and in growth and development. *Am. J. Clin. Nutr.* **54 :** *438, 1991.*
32. **U.S. Department of Agriculture :** The Food Guide Pyramid. Stock No. HSG-252. In Hyatsville MD : *Human Nutrition Information Service, 1992.*
33. **Welsh CW :** Enhancement of mammary tumorigenesis by dietary fat: review of potential mechanisms. *Am. J. Clin. Nutr.* **45 :** *191, 1987.*
34. **Wohl MG, Goodhart RS (editors) :** Modern Nutrition in Health and Disease. *Lea and Febiger, Philadelphia, USA. 1976.*
35. **Wolever TMS, Jenkins DJA, Jenkins AL, Josse RG :** The glycemic index: methodology and clinical implications. *Am. J. Clin. Nutr.* **54 :** *846, 1991.*

PROBLEMS

1. The primary effect of the consumption of excess protein beyond the body's immediate needs will be:
   (*a*) excretion of the excess as protein in the urine
   (*b*) an increase in the 'storage pool' of protein
   (*c*) an increased synthesis of muscle protein
   (*d*) an increase in the amount of adipose tissue
2. Which of the following individuals would most likely be in nitrogen equilibrium?
   (*a*) A normal adult male
   (*b*) A normal, pregnant female
   (*c*) A growing child
   (*d*) An adult male recovering from surgery

3. Kwashiorkor is:
   (*a*) characterized by a thin, wasted appearance
   (*b*) an inadequate intake of food of any type
   (*c*) an inadequate intake of total calories but a specific deficiency of protein
   (*d*) an adequate intake of total protein but a deficiency of the essential amino acids
4. A complete replacement of animal protein in the diet by vegetable protein:
   (*a*) would be satisfactory regardless of the nature of the vegetable protein used.
   (*b*) would reduce the total amount of food consumed for the same number of calories.
   (*c*) might reduce the total amount of iron and vitamin B_{12} available.
   (*d*) could not satisfy protein requirements.
5. Which of the following statements about dietary fibre is/are correct?
   (*a*) Water-soluble fibre helps to lower serum cholesterol is most people.
   (*b*) Mucilaginous fibre slows the rate of digestion and absorption of carbohydrates.
   (*c*) Insoluble fibre increases stool bulk and decreases transit time.
   (*d*) All of the above are correct.
6. Which one of the dietary regimens would be *most* effective in lowering serum cholesterol?
   (*a*) restrict dietary cholesterol
   (*b*) increase the ratio of polyunsaturated to saturated fatty acids
   (*c*) increase fibre content
   (*d*) restrict cholesterol, increase PUFA/SFA ratio, increase fibre content
7. For diabetics:
   (*a*) the only carbohydrate that must be eliminated in the diet is sucrose
   (*b*) fibre increases the rate at which carbohydrate is digested and absorbed
   (*c*) not all carbohydrate foods raise blood glucose levels at the same rate because the glycemic index of all foods is not the same.
   (*d*) a vegetarian diet is the only appropriate choice
8. A 120-lb woman is consuming a diet with adequate total calories and 44 g of protein per day. The protein is exclusively from vegetable sources, primarily corn-based. What would be her state of nitrogen balance?
9. Are high-protein diets a good way to lose weight?
10. Why is it so important to eat plenty of fruits and vegetables?
11. What is the best grain for getting protein?
12. What is the safest fruit to eat?
13. What is gluten-free diet?
14. What is auto-eating.

ANSWERS TO PROBLEMS

1. (*d*) Excess protein is treated like any other excess energy source and stored (minus the nitrogen) eventually as adipose tissue lipid. (*a*) Protein is not found in normal urine except in very small amounts. The excess nitrogen is excreted as $NH_4^+$and urea, whereas the excess carbon skeletons of the amino acids are used as energy sources. (*b*), (*c*) There is no discrete storage form of protein, and although some muscle and structural protein is expendable, there is no evidence that increased intake leads to generalized increased protein synthesis.

2. (*a*) A normal adult male would be in nitrogen equilibrium. (*b*), (*c*), (*d*) Although normal, pregnancy is also a period of growth, requiring positive balance as does a period of convalescence.
3. (*c*) This is the definition of kwashiorkor. (*a*), (*b*) These are the characteristics of marasmus. (*d*) This would lead to negative nitrogen balance but does not have a specific name.
4. (*c*) This would reduce the amount of fat, especially saturated fat, but could also reduce the amount of necessary nutrients that come primarily from animal sources. (*a*), (*d*) It is possible to satisfy requirements for all of the essential amino acids completely if vegetables with complementary amino acid patterns, in proper amounts, are consumed. (*b*) The protein content of vegetables is quite low, so much larger amounts of vegetables would have to be consumed.
5. (*d*) These each illustrate the different properties and roles of the common types of fibres.
6. (*d*) Any of the measures alone would decrease serum cholesterol slightly, but to achieve a reduction of more than 15% requires all three, *i.e.*, restrict cholesterol, increase PUFA/SFA ratio, and increase fibre content.
7. (*c*) Foods vary in their glycemic index. Bread and rice raise blood glucose levels more rapidly than does sucrose. (*a*) See (*c*). (*b*) One of the benefits of fibre is that it decreases the rate of carbohydrate absorption. (*d*) An appropriately designed vegetarian diet is perfectly acceptable but is certainly not the only choice.
8. The 44 g protein / 120 lb is 0.8 g protein kg^{-1} day^{-1}. However, this is inadequate because pure vegetable protein is less than 75% efficient in utilization. Also, heavy dependence on one protein (corn is deficient in lysine) would likely lead to a deficiency of one or more essential amino acids. Therefore, this woman would be in negative nitrogen balance.
9. High-protein diets put a strain on kidneys, and make one dehydrated. A high-protein diet is one that consists of 30% or more protein. If one eats a lot of protein, one must drink a lot of water.
10. They help protect a person against heart disease, high blood pressure, stroke, cancer, diabetes, constipation, diverticulosis, cataract and macular degeneration.
11. That honour goes to **quinoa**. It is a 'complete protein', meaning that it contains all 9 amino acids. Usually thought of as a grain, quinoa is actually a seed from the goosefoot plant. Quinoa originated in the Andean region of Peru, Colombia, Equador and Bolivia. It was domesticated 3,000 to 4,000 years ago. It is closely related to beets, spinach and tumbleweed. Considered a superfood, quinoa is higher in protein and fibre than brown rice.
12. Pineapple (*Annona squamosa*) is the safest fruit to eat because it has such a thick skin, which protects it from pesticides, bugs and surface bacteria. Bananas and mangoes are also safe to eat because of the thickness of their skins.
13. Celiac disease symptoms are triggered by gluten, a protein found in wheat, barley and rye. Oats can also present a problem. Avoiding these grains is the first step in managing symptoms of celiac disease.
14. Eating without actually feeling hungry is auto-eating. Most often, it is women who do it. They many not even be aware they are consuming so many calories. A study found that boredom can lead women to overeat, even slip into an auto-eating mode. Men also auto-eat but women are more at risk, especially if they are upset. *As much as a quarter of the food eaten each day can be contributed to auto-eating*. The term was coined by nutritionist **Mary Strugar.**

CONTENTS

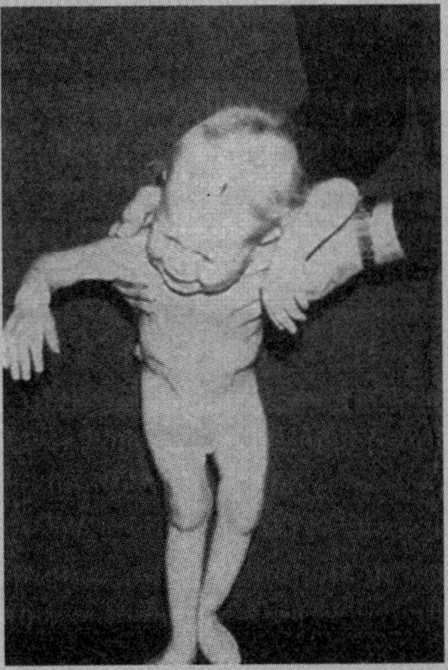

A rachitic child

Pictured above is a 21/2-year-old boy with severe rickets, which is due to the deficiency of vitamin D in the body. The disease was once common in cold climates where heavy clothing blocks the ultraviolet (UV) component of sunlight necessary for the production of vitamin D_3 in skin. Cholecalciferol, (vitamin D_3) is produced in the skin by UV irradiation of 7-dehydrocholesterol, the provitamin form. In the liver, a hydroxyl group is added at C-25; in the kidney, a second hydroxylation at C-1 produces the active hormone, 1,25-dihydroxycholecalciferol. This hormone regulates the metabolism of Ca^{2+} in kidney, intestine, and bone. Dietary vitamin D prevents rickets.

CHAPTER

38

Fat-soluble Vitamins

HISTORICAL RESUME

Vitamins have a back-dated history. The dreaded disease **scurvy** was one of the prevalent diseases in Europe during 15th and 16th centuries. The disease is said to have afflicted the crusadors. Scurvy was reported by **Vasco de Gama** during his sea voyages and **Jacques Cartier** in 1535 had reported death of about 25% of his sailing crew due to scurvy. On the recommendation of **Sir James Lancaster**, an English privateer, the ships of East India Company in 1601 carried oranges and lemons to prevent scurvy. In 1757, **James Lind**, a British naval surgeon, stated that fresh fruits and vegetables alone are effective to protect the body from various maladies and urged the inclusion of lemon juice in the diet of sailors to prevent scurvy. And some 40 years later, the Admiralty took his advice. After limes were substituted for lemons in 1865, British sailors began to be known as "limeys." Similarly, **rickets** was also attributed to faulty diets and **Guérin** (1838) produced it experimentally in puppies to prove the dietary connection.

In 1887, **Admiral Takaki**, Director-General of the Medical Services in Japan, demonstrated that another scourge **beriberi** could be prevented by enriching the diet with meat, vegetables and milk and at the same time decreasing the amount of milled rice in the case of Japanese sailors. Later, **Eijkman** (1897), a Dutch physician, found that experimental beriberi could be

CHRISTIAAN EIJKMAN
(LT, 1858–1930)

Eijkman, a Dutch physician, was born on August 11, 1858 at Nijerk in Poland. He studied medicine at the University of Amsterdam. In 1883, he took up a post of army medical officer in Dutch East Indies. In 1886, a commission was sent by the Dutch government to study the cause of beriberi in Dutch East Indies. Eijkman collaborated with the work of the commission. At that time, all diseases were thought to be caused by mircoorganisms. When the microorganism that caused beriberi could not be found, the team left the East Indies. Eijkman stayed behind to become the director of a new bacteriological laboratory. He went back to Holland, in 1896, due to illness. The same year, he accidentally discovered the cause of beriberi when he noticed that chickens being used in the laboratory had developed symptoms characteristic of the disease *polyneuritis*. He found that the symptoms had developed when a cook had started feeding the chickens rice meant for hospital patients (*i.e*, the polished rice). The symptoms disappeared when a new cook resumed feeding them chicken-feed containing unpolished rice. Later, it was recognized that thiamine (**Vitamin B_1**) is present in rice hulls but not in polished rice. For this work (*i.e.*, the discovery of **antiberiberi factor**), Eijkman shared the **1929 Nobel Prize in Medicine or Physiology**, with **Frederick G. Hopkins**. He died in 1930.

induced in hens when fed with polished rice without bran. Such hens could be cured by giving them the rice polishings. **Eijkmann**, for a time, believed that the rice polishings contained something that neutralized the beriberi toxin in the polished rice. In 1906, however, **Frederick Gowland Hopkins** ascribed the diseases such as scurvy and rickets to the lack of some '*dietary factors*'. Six years later (*i.e.*, in 1912), **Hopkins** in collaboration with **Casimer Funk** of Poland, who was working at the Lister Institute of London, suggested the **vitamin theory** which postulates that '*specific diseases such as beriberi, scurvy and rickets are each caused by the absence from the diet of a particular nutritional factor.*'

FREDERICK GOWLAND HOPKINS
(LT, 1861–1947)

Hopkins, a British biochemist, was born in Eastbourne, East Sussex, England on June 20, 1861. He qualified for the fellowship of the Institute of Chemistry and was employed as a chemist in a Medical Jurist Laboratory at Guy's Hospital in London for some years. During this period, he studied medicine and took his M.B. degree from the London University, in 1894. He was then appointed as a demonstrator in Physiology. In 1898, he was invited by **Sir Michael Foster** to develop teaching and research in Chemical Physiology at Cambridge. He accepted the invitation and spent the rest of his scientific career at Cambridge. In 1901, Hopkins along with **S.W. Cole** isolated **tryptophan** (Trp). Later, Hopkins and **Willcock** showed that Trp is an essential amino acid and zein (a protein) lacking in Trp could be improved nutritionally by supplementing with this amino acid.

Between 1906 and 1912, Hopkins and his team at Cambridge University studied accessory food factors. In his experiment, two groups of rats were used. *Group 1* rats were fed on an artificial diet of purified casein (milk protein), starch, cane sugar, lard and salts. *Group 2* rats were fed the same diet but with a milk supplement. Those on the artificial diet (Group 1) stopped growing and lost body mass; those with the milk supplement (Group 2) grew normally, even though the milk contributed only 4% of the total food eaten.

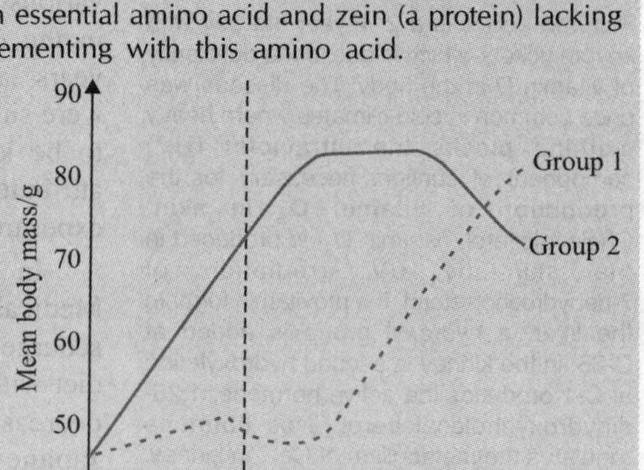

Hopkin's greatest work was in the **identification and isolation of vitamins**. He worked out the importance of vitamins in the diet in the early 1900s. His classical experiment (described in the graph, given above) helped people understand the need for vitamins as additional food factors in a healthy diet. Basically, he showed that the milk-deficient rats did not grow properly. He suggested that the milk contained an **'accessory food factor'**, or 'vital amine'– hence the name vitamin, which the rats need to stay healthy. Later work by American researchers showed this compound to be fat-soluble. We now know this compound to be vitamin A. In 1912, Hopkins published his classical paper on the importance of 'accessory factors' in the growth of animals. Hopkins also discovered **glutathione**, in 1912, and studied its role in tissue oxidation.

In collaboration with a U.S. chemist **Stanley R. Benedict** (LT, 1884–1936), and a 20th century French pathologist **Laurent Cole, Hopkins** developed a reagent (called **Benedict-Hopkins-Cole reagent**) which consists of magnesium glyoxylate made from a mixture of oxalic acid and magnesium and used for testing proteins for the presence of tryptophan.

Hopkins shared the **1929 Nobel Prize in Medicine or Physiology** for his discovery of essential nutrient factors now known as vitamins– needed in animal diets to maintain health, along with **Christiaan Eijkman** of The Netherlands for his discovery of vitamin B_1 (thiamine). He established a world-famous School of Biochemistry at Cambridge. Many leading scientists the world over have worked in his laboratory. His students included such great names as **Adair, Baldwin, Chain, Dixon, Elvehjem, Gale, Green, Haldane, Harris, King, Krebs, Lehmann, Martin, Mellanby, Moore, Needham, Neuberg, Peters, Synge and Szent-Györgi.** It is really a tribute to Hopkins that some of his students have been conferred the coveted Nobel Prize for their contributions. He died on May 16, 1947 at Cambridge. Hopkins was truly one of the great founders of Biochemistry.

CASIMIR FUNK
(LT, 1884–1967)

Funk was born on February 23, 1884 in Warsaw, Poland. His father was a physician, who had several research papers in clinical medicine to his credit. He studied biology in Geneva, Switzerland and later specialized in organic chemistry at Berne. He obtained the doctorate degree in organic chemistry in 1904. He joined the department of biochemistry under **Prof. Gabriel Bertrand** at the Pasteur Institute in Paris and carried out work on amino acids and bases. Two years later, he moved to Emil H. Fischer's laboratory in Berlin. Here, he worked with **Emil Abderhalden**, who was Fischer's assistant, on some aspects of amino acids and proteins. In 1907, he was appointed biochemist at Municipal Hospital in Wiesbaden and later moved to the University Hospital in Berlin. Not seeing much hope for a bright future, Funk moved to the Lister Institute in London, in 1910. In 1911, he published a paper on the synthesis of dihydroxyphenylalanine. He was assigned by the Director of the Institute to work on the substance responsible for causing beriberi. **Sir Charles J. Martin**, Director of the Lister Institute was of the opinion that an essential amino acid was removed when rice was polished while **W.L. Braddon**, an English physician working in Malaya, was of the opinion that beriberi was caused by a toxin.

Funk discarded both the amino-acid deficiency and toxin theories and started to look for an unknown material present in rice polishing but absent in polished rice. In 1912, he succeeded in preparing concentrates rich in antiberiberi factor first from rice polishings and later from yeast. Funk stated that 'the curative substance is a pyrimidine base analogous to uracil and thymine...' In the following year, he isolated and identified **nicotinic acid** (vitamin B_5) from rice bran extract but he did not realize that this was a vitamin [Only 25 years later, in 1937, **Elvehjem** isolated nicotinamide (an amide of vitamin B_5) and identified it with the pellagra-preventive vitamin]. Funk coined the term **'vitamine'**. In 1912, Funk, in collaboration with **F.G. Hopkins**, suggested the **vitamin theory**. The same year, he published a paper entitled 'The aetiology of deficiency diseases'. In this paper, Funk stated 'the deficient substances which are of the nature of organic bases we will call Vitamines and we will speak of a beriberi and scurvy vitamine'. In 1914, he published a book 'The Vitamine'. Funk became an US citizen in 1920. In 1923, he returned to Poland to direct the State Institute of Health but returned to the United States permanently when World War II broke out. He died in 1967. According to his coworker, friend and biographer **Benjamin Harrow**, Funk's imagination was sometimes more that of a poet than of a scientist and often his ideas were ahead of what was warranted by the results of scientific studies.

Funk, for the first time, also isolated the dietary factor from rice polishings which acted as antiberiberi substance. Since this factor was an amine and necessary to life, **Funk** at the suggestion of **Dr. Max Nierenstein** introduced the term 'vitamine' (*vita*[L] = life) to denote it. Since subsequent studies showed that not all these substances are amines, the terminal letter 'e' was dropped from its spelling at the suggestion of **Sir. J.C. Drummond** (1919), who also proposed their **alphabetical nomenclature**. In fact, the various vitamins have no structural resemblance to each other, but because of a similar general function in metabolism, they are studied together. Although these molecules serve nearly the same roles in all forms of life, but higher animals have lost the capacity to synthesize them.

DEFINITION

The vitamin concept has undergone extensive revisions during the history of biochemistry. However, **Franz Holfmeister's** (LT, 1850–1922) definition– 'vitamins are substances which are indispensable for the growth and maintenance of the animal organism, which occur both in animals and plants and are present in only small amounts in food'– still holds good, although it has been interpreted in various ways.

> ***Sine qua non*** is a Latin phrase, which means an indispensable condition.

The term '**vitamin**', in its modern sense, usually refers to *the substances distinct from major components of food, required in minute quantities (i.e., oligodynamic in nature) and whose absence causes specific deficiency diseases.* As the living organisms cannot synthesize most of these compounds, a steady supply of them is *sine qua non* for life. Their ultimate source is the plant or bacterial world.

Differences with hormones. Indeed, for some time, it was believed that the distinction between vitamins and hormones was no longer tenable. But there exists a fundamental difference between these two classes of active substances : *the hormones are regulatory substances whereas the vitamins are accessory nutrients.* The vitamins also differ from the hormones in that they are supplied to the body from some external source (*i.e.,* chiefly from the food ingested) whereas the hormones are synthesized within some part of the body of an organism. Moreover, most vitamins and some hormones are involved either directly or indirectly in enzyme systems in order to carry out biochemical functions. Some vitamins are also known to be coenzymes.

A comparison of enzymes, hormones and vitamins is presented in Table 38–1.

Table 38–1. Differences among enzymes, hormones and vitamins

Enzymes	*Hormones*	*Vitamins*
1. These are specific catalytic substances.	These are regulatory substances.	These are accessory nutrients.
2. All enzymes are proteinaceous in nature.	Hormones are steroids (estradiol, testosterone) or peptides (insulin, ocytocin) or amino acid derivatives (thyroxine, adrenalin).	Vitamins belong to diverse groups chemically.
3. These are synthesized by the animal cells.	These are synthesized within some part of the body.	These are not synthesized by most animal cells but supplied to the body from outside source (mainly as food). Plants can, however, synthesize them.
4. These can initiate and continue reactions.	These cannot initiate reactions but can influence the rate at which they proceed.	—
5. These are not consumed during growth.	They are consumed during growth.	—
6. These are involved in coenzyme syntems.	These are not involved in coenzyme systems.	These are involved in coenzyme systems.

7. These operate in biological systems. *Examples*: Pepsin, Trypsin, Urease, Diastase, Lipase, Catalase, Maltase, Alcohol dehydrogenase, Cytochrome oxidase.	Only some hormones are involved in enzyme systems. *Examples*: Estradiol, Testosterone, Relaxin, Progesterone, Insulin, Thyrotropin, Ocytocin, Secretin, Intermedin, Thyroxine, Adrenalin.	Most vitamins are involved in enzyme systems. *Examples*: Fat-soluble vitamins A, D, E, K; Water-soluble vitamins B, C.

GENERAL CHARACTERISTICS

The vitamins are characterized for some general facts, which are listed below:

1. Vitamins are of widespread occurrence in nature, both in plant and animal worlds.
2. All common foodstuffs contain more than one vitamin.
3. The plants can synthesize all the vitamins whereas only a few vitamins are synthesized in the animals.
4. Human body can synthesize some vitamins, *e.g.*, vitamin A is synthesized from its precursor carotene and vitamin D from ultraviolet irradiation of ergosterol and 7-dehydrocholesterol. Some members of the vitamin B complex are synthesized by microorganisms present in the intestinal tract. Vitamin C is also synthesized in some animals such as rat.
5. Most of vitamins have been artificially synthesized.
6. All the cells of the body store vitamins to some extent.
7. Vitamins are partly destroyed and are partly excreted.
8. Vitamins are nonantigenic.
9. Vitamins carry out functions in very low concentrations. Hence, the total daily requirement is very small.
10. Vitamins are effective when taken orally.
11. Synthetically-made vitamins are just as nutritionally good as natural vitamins.
12. Old people need about the same amounts of vitamins as young people.

CLASSIFICATION

In 1913, **McCollum** and **Davis** described a lipid-soluble essential food factor in butter fat and egg yolk. In 1915, a water-soluble factor in wheat germ necessary for the growth of young rats was also described. Since then, two categories of vitamins are usually recognized: fat-soluble and water-soluble. These two groups discharge rather different functions.

ELMER VERNER McCOLLUM

(LT, 1879–1967)

McCollum was born in March 1879 in a farmhome near Ft. Scott, Kansas. He received his B.A. degree in 1903 and M.S. degree in chemistry in 1904 from the University of Kansas. He did Ph.D. from the Yale University, in 1906, under **Prof. T.B. Johnson**. Immediately after this, he joined as a research fellow under **Prof. L.B. Mendel** (LT, 1872–1935). In 1907, on the advice of Prof. Mendel, he accepted a position of assistantship under **Prof. E.B. Hart** (LT, 1874 –1953) at the Wisconsin Agricultural Experimental Station and remained on this post till 1917. During this period, McCollum made the discovery of the presence of **'fat-soluble A'** in 1913 and **'water-soluble B'** in 1914. He also developed the biological assay for foods using experimental animals. In 1917, he was appointed as Professor and Head of the Department of Chemical Hygiene and Public Health at The John Hopkins University. He continued in this position till his retirement in 1946. During this period, more than 150 research papers were published. The investigations included studies on **trace elements** and **vitamins**. The deficiency symptoms of several trace elements were also studied by McCollum and his associates.

McCollum became a world leader in nutrition. He published, in 1918, his famous book 'The Newer Knowledge of Nutrition', whose fifth and the last edition appeared in 1939. He could the term **'protective foods'** for foods rich in proteins, vitamins and minerals. With **J.E. Becker** (later Mrs. McCollum), he wrote a small popular book 'Food, Nutrition and Health'. During his period of retirement, he wrote a monumental book 'A History of Nutrition' (1957). He died on November 15, 1967 at the ripe age of 88 years. McCollum was truly one of the great pioneers in nutrition.

A. Fat-soluble vitamins. These are oily substances, not readily soluble in water and their biochemical functions are not well understood. They contain only carbon, hydrogen and oxygen. Their examples are vitamins A, D, E and K. They, however, play more specialized roles in certain group of animals and in particular type of activities. For instance, they function in the formation of a visual pigment (vitamin A), in the absorption of calcium and phosphorus from the vertebrate intestine (vitamin D), in protecting mitochondrial system from inactivation (vitamin E) or in the formation of a blood clotting factor in vertebrates (vitamin K). The individual fat-soluble vitamins bear a closer resemblance to each other chemically. In fact, the 4 fat-soluble vitamins can be regarded as lipids. Vitamins A, E and K are terpenoids, and vitamin D is a steroid. All four are isoprenoid compounds, since they are synthesized biologically from units of isoprene (for structure, refer page 319), a building block of many naturally-occurring oily, greasy or rubbery substances of plant origin. Unlike the water- soluble vitamins (B and C), fat-soluble vitamins can be stored in the body, *e.g.*, an adult's liver can store enough vitamin A to last several months or longer. However, because *fat-soluble vitamins are storable*, their excessive intakes can result in toxic conditions (hypervitaminoses).

B. Water-soluble vitamins. Most of these are universally vitamins since they perform the same general functions wherever they occur. Besides C, H and O, they also contain nitrogen. They are catalytic factors and as such form vital links in the chains of biochemical reactions characteristic of all living objects. For instance, thiamine is required whenever sugars are oxidized aerobically to release energy. The individual water-soluble vitamins bear no closer resemblance to each other chemically. The biochemical or coenzyme function of nearly all of these is known. The common water-soluble vitamins are vitamins of B complex such as B_1 through B_{12} (vitamins B_4, B_8, B_{10} and B_{11}, however, do not exist) and the vitamin C. Choline, inositol, *p*-aminobenzoic acid, bioflavonoids and α-lipoic acid are frequently included in this category. Many nutritionists, however, do not consider them as true vitamins, although their dietary deficiencies in animals lead to the development of characteristic symptoms. Moreover, none of them except α-lipoic acid is a part of the coenzyme system. The B-series of vitamins, being water-soluble and excretable, are *required daily in meagre amounts* (in milligrams or even less) for the normal growth and good health of humans and many other organisms. It is virtually impossible to 'overdose' on them.

STORAGE OF VITAMINS IN THE BODY

The vitamins can be stored in the body to a slight extent. The liver cells are, however, rich in certain **fat-soluble vitamins**. For instance, the amount of *vitamin A* contained in the liver is sufficient enough to meet its requirement without any additional intake for about 6 months. Similarly, the quantity of *vitamin D* stored ordinarily in the liver is sufficient to maintain a person without any additional intake of vitamin D for about 2 months. The storage of *vitamin K* is, however, relatively slight.

The **water-soluble vitamins** are stored even in lesser amounts in the cells. Evidently, in cases of deficiency of *vitamin B compounds*, clinical symptoms appear rather early, that is within a few days. Similarly, absence of *vitamin C* can induce deficiency symptoms within a few weeks. Vitamin C is stored in the adrenal cortex.

DAILY HUMAN REQUIREMENTS OF VITAMINS

Vitamins are no substitute for food. They are supplements to be taken along with a good diet for dealing with deficiencey. The requirement of vitamins varies considerably depending on the nature of the individual consuming them. Some general facts regarding vitamin requirement may be listed below:

1. Greater the size of the individual, higher are his vitamin needs.
2. Younger ones require higher quantities of vitamins than do the elders.
3. The vitamin requirements increase when a person performs exercise.
4. During ailments, the vitamin requirements are ordinarily enhanced.
5. Under certain specific conditions of metabolic disorders when the vitamins cannot be properly utilized, the requirement for one or more specific vitamins is at extreme.

Table 38–2. Recommended daily dietary allowances*

	Age[†]	Weight	Height	Energy	Protein	Fat-soluble vitamins			Water-soluble vitamins						
	years	kg	cm	kcal	g	A μg[‡]	D_3 μg	E mg	B_1 mg	B_2 mg	B_5 mg	B_6 mg	B_9 μg	B_{12} μg	C mg
INFANTS	0.0–0.5	6	60	650	kg × 2.2	420	10	3	0.3	0.4	6	0.3	30	0.5	35
	0.5–1.0	9	71	970	kg × 2.0	400	10	4	0.5	0.6	8	0.6	45	1.5	35
	1–3	13	90	1300	23	400	10	5	0.7	0.8	9	0.9	100	2.0	45
CHILDREN	4–6	20	112	1700	30	500	10	6	0.9	1.0	11	1.3	200	2.5	45
	7–10	28	132	2400	34	700	10	7	1.2	1.4	16	1.6	300	3.0	45
	11–14	45	157	2700	45	1000	10	8	1.4	1.6	18	1.8	400	3.0	45
	15–18	66	176	2800	56	1000	10	10	1.4	1.7	18	2.0	400	3.0	60
MEN	19–22	70	177	2900	56	1000	7.5	10	1.5	1.7	19	2.2	400	3.0	60
	23–50	70	178	2700	56	1000	5	10	1.4	1.6	18	2.2	400	3.0	60
	51+	70	178	2400	56	1000	5	10	1.2	1.4	16	2.2	400	3.0	60
	11–14	46	157	2200	46	800	10	8	1.1	1.3	15	1.8	400	3.0	50
	15–18	55	153	2100	46	800	10	8	1.1	1.3	14	2.0	400	3.0	60
WOMEN	19-22	55	163	2100	44	800	7.5	8	1.1	1.3	14	2.0	400	3.0	60
	23–50	55	163	2000	44	800	5	8	1.0	1.2	13	2.0	400	3.0	60
	51+	55	163	1800	44	800	5	8	1.0	1.2	13	2.0	400	3.0	60
Pregnant				+300	+30	+200	+5	+2	+0.4	+0.3	+2	+0.6	+400	+1.0	+20
Lactating				+500	+20	+400	+5	+3	+0.5	+0.5	+5	+0.5	+100	+1.0	+40

* The recommended allowances can be attained with a variety of common foods, including those nutrients also for which human requirements have been less well defined.

† All the entries for the age range represent allowances for the midpoint of the specified age range. However, in the case of age range 23–50 years, the reference is to the man and woman at age 25 years.

‡ Retinol equivalents: 1 retinol equivalent = 1 μg retinol or 6 μg β-carotene.

Note that there is no recommended dietary allowances (RDAs) for vitamin K, vitamin B_3 (pantothenic acid), and vitamin B_7 (biotin) as intestinal bacteria synthesize the body's requirement.

(Adapted from Food and Nutrition Board, National Academy of Sciences— National Research Council, Revised 1980)

6. Growing children require comparatively high quantities of vitamin D.
7. During pregnancy and lactation, the vitamin D requirement by the mother is greatly enhanced.
8. The requirements of vitamin B complex (*esp.*, that of vitamin B_1) are increased under conditions of greater utilization of carbohydrates.

The daily requirement of any vitamin (refer Table 38–2) for any individual is not a fixed quantity and varies according to the rate of metabolism. *In general, in all cases of high metabolic activity (such as heavy muscular work, during pregnancy and lactation and in growing children), the vitamin requirement is proportionately high.* Normally, a man doing ordinary work can obtain enough vitamin from his balanced diet.

The intestinal organisms may synthesize vitamins in significant amounts and play a vital role in regulating the quantity of vitamin available to the organism. Most of the vitamins of B complex group (such as thiamine, riboflavin, nicotinic acid, pyridoxine, biotin, folic acid) and vitamain K are some of the vitamins synthesized by the intestinal organisms. These may be absorbed to varying extents and utilized. This fact renders rather 'inaccurate' the figures for daily requirement of the various vitamins. Certain organisms, however, destroy vitamins. Supplementing the diet with certain antibiotics and sulfa drugs enhances the growth of these organisms.

In measuring human requirements of vitamins, certain units have been used. In the beginning, these were arbitrarily fixed and were mainly based on the amount necessary to check avitaminosis in animals under standard conditions. With the passage of time, the various vitamins were synthesized and so it became possible to base the unit on the weight of purified preparations.

AVITAMINOSES (Deficiencies of Vitamins)

A lack or deficiency of one or more vitamins leads to characteristic deficiency symptoms in man. Multiple deficiencies caused by the lack of more than one vitamin are, however, more common in human beings. Vitamin deficiencies occur rather frequently in certain parts of the world for socioeconomic reasons. Avitaminosis may be of following 2 types:

A. ***primary*** or ***direct dificiency***— This arises due to inadequate intake of vitamins resulting from chronic alcoholism, dietary fads etc.

B. ***secondary*** or ***'conditioned' deficiency***— This arises due to other factors such as malabsorption, increased excretion, allergies, anorexia, gastrointestinal disorders etc.

Vitamin deficiency, whether primary or secondary, leads to:

(*a*) a gradual decrease in tissue levels of the vitamin(s) deficient,
(*b*) a biochemical lesion,
(*c*) an anatomic lesion, and
(*d*) finally cellular pathology and disease.

This sequence is schematically represented in Fig. 38–1.

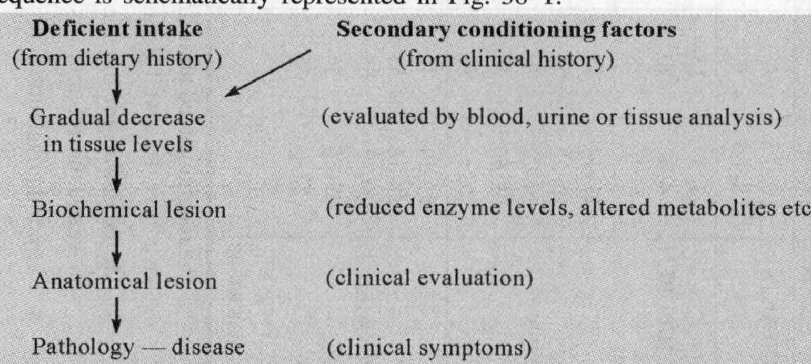

Fig. 38–1. Sequence of events occurring in a tyical avitaminosis

An account of **fat-soluble vitamins** is given below.

VITAMIN A

A. History. It was first recognized as an essential nutritional factor by **Elmer McCollum** in 1915 and then isolated from fish-liver oil by **Holmes** in 1917. On account of its established role in the visual process, it is often called as **antixerophthalmic factor** or the **"bright eyes" vitamin**. It was first synthesized in 1946 by **Milas.**

B. Occurrence. *Liver oils of various fishes are the richest natural sources of vitamin A.* Shark and halibut contain maximum amount whereas the cod-liver has lowest amount. Depending upon the species of fish and the time of year of catch, the fish livers contain 2,000 to 100,000 I.U. per gram of vitamin A. The amount present in human liver is much less (500 to 1,000 I.U.per gram). However, polar bear liver is an extremely concentrated source of vitamin A. Other noteworthy sources are butter, milk and eggs and, to a lesser extent, kidney. In its provitamin form (*i.e.*, as carotenes) it is supplied by all pigmented (particularly yellow) vegetables and fruits such as carrots, pumpkins, cantaloupes, turnips, peppers, peas, sweet potatoes, papayas, tomatoes, apricots, peaches, plums, cherries, mangoes etc. Yellow corn is the only cereal containing significant amounts of carotene. New-born infants have low quantities of vitamin A content that is rapidly augmented because colostrum and breast milk furnish large amounts of the vitamin. The milk of well-fed mothers, however, contains sufficient amounts of vitamin A (in ester form) for the infant's need. However, vitamin A is absent from vegetable fats and oils (olive oil, linseed oil, groundnut oil). It is added to margarine during its manufacture from these oils.

The β-carotene content of some items are presented in Table 38–3.

Table 38–3. β-carotene content of some fruits and vegetables

Fruit / Vegetable	*β–carotene (μg/100 g)*
Mango (ripe)	300
Orange	210
Tomato	190
Drumstick leaves	7,000 – 8000
Coriander leaves	7,000
Carrot	1,500 – 3,000
Radish leaves	3,600
Mint leaves	1,800
Cabbage	1,300

Table 38–4. presents vitamin A content of some animals including those of polar regions and fishes.

Table 38–4. Vitamin A content of some animal foods

Source	*Biological name*	*Vitamin A (I.U./g of food)*
Weddel seal	*Leptonychotes weddelli*	444
Ox liver*		550
Cod liver*		600
Southern elephant seal	*Mirounga leonina*	1,160
Antarctic husky	*Canis familiaris*	10,570
Halibut liver oil*		30,000
Arctic bearded seal	*Erignathus barbatus*	12,000–14,000
Polar bear	*Thalarctos maritimus*	24,000–35,000

* Commonly recommended as source of vitamin A

Vitamin A originates in marine algae, and then passes up the food chain to reach the large carnivorous animals. Toxic levels of vitamin A may accumulate in the livers of a wide range of creatures such as Polar bears, seals, porpoises, dolphins, sharks, whales, Arctic foxes and huskies. Even a small meal of southern Australian seal liver, say 80g, may produce illness in man. Most of the foods recommended as source of vitamin A contain well below the toxic levels of vitamin A, but one– Halibut liver oil– contains dangerously high amounts of vitamin A.

As regards the vitamin A contents of **polar animals**, one can see that, in reality, very little quantities of livers of these animals are required to kill a human being: 30–90 g of polar bear liver or halibut liver, 80–240 g of bearded seal liver and 100–300 g of Antarctic husky liver is enough to kill a human being.

C. Structure. Vitamin A is found in two forms A_1 and A_2. The carotenoids that give rise to vitamin A in animal body is named as provitamin A. These inculde α-, β- and γ-carotenes and cryptoxanthin. β-*carotene is most potent of all these forms*. A molecule of β-carotene is made of eight 5-carbon isoprenoid units, linked to form a long chain of 40 carbon atoms with an ionone ring at each end. It is an orange-red hydrocarbon and upon hydrolysis yields 2 moles of vitamin A_1 (Fig. 38–2). During hydrolysis, a cleavage occurs at the mid point of the carotene in the polyene chain connecting the 2 β-ionone rings. This conversion usually takes place in livers of fishes and mammals. Vitamin A_1 is a complex primary alcohol called retinol, having the empirical formula, $C_{20}H_{29}OH$. The terminal hydroxyl group is ordinarily esterified. It contains β-ionone ring.

> **α- and γ-carotenes** as well as cryptoxanthin, however, yield only 1 molecule of vitamin A since in them one of the two rings present in their molecule differs in structure from that of vitamin A.

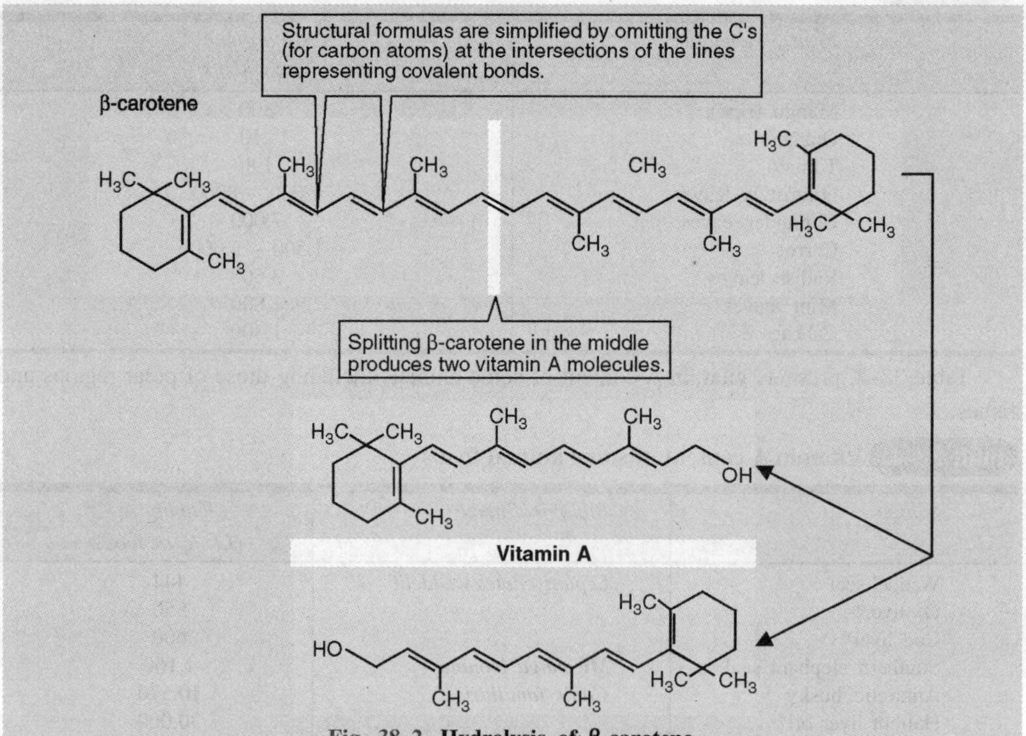

Fig. 38–2. Hydrolysis of β-carotene

The carotenoid β-carotene is symmetrical around the central double bond; when split, β-carotene becomes two vitamin A molecules. The simplified structural formula is standard chemical shorthand for large structures with many carbon atoms.

Another form of vitamin A, present in fresh-water fishes, is known as vitamin A_2 (Fig. 38–3). It differs from vitamin A_1, which is found in salt-water fishes, in possessing an additional conjugate double bond between carbon atoms 3 and 4 of the β-ionone ring. Its potency is 40% that of vitamin A_1.

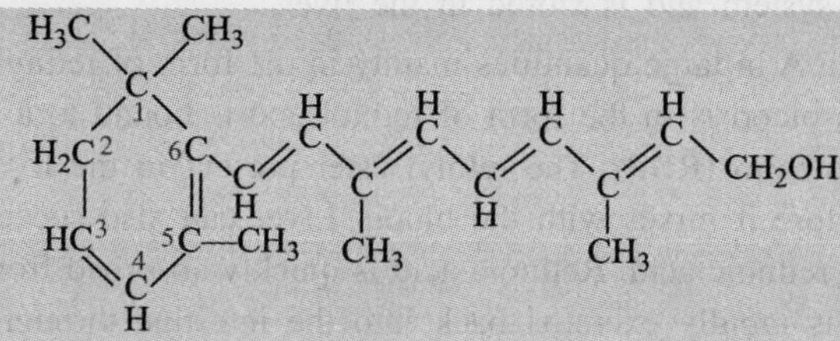

Fig. 38–3. Vitamin A_2 or 3-dehydroretinol
(all *trans*)

D. Properties. Ordinarily retinol is a viscid, colourless oil but by careful fractionation it has also been isolated as pale yellowish needles. It gives a characteristic absorption band in ultraviolet (UV) spectrum at 328 *m*μ. It is soluble in fat and fat solvents but insoluble in water. Loss of vitamin A in cooking, canning and freezing of foodstuffs is small; oxidizing agents, however, destroy it. It is destroyed on exposure to UV light. Vitamin A is relatively unstable in air unless protected by antioxidants including vitamin E.

E. Metabolism. In the tissues, the metabolic transformation of retinol is carried out by enzymes. The dietary β-carotene is split into retinal (Fig. 38–4) by an enzyme of the intestine. Retinal is

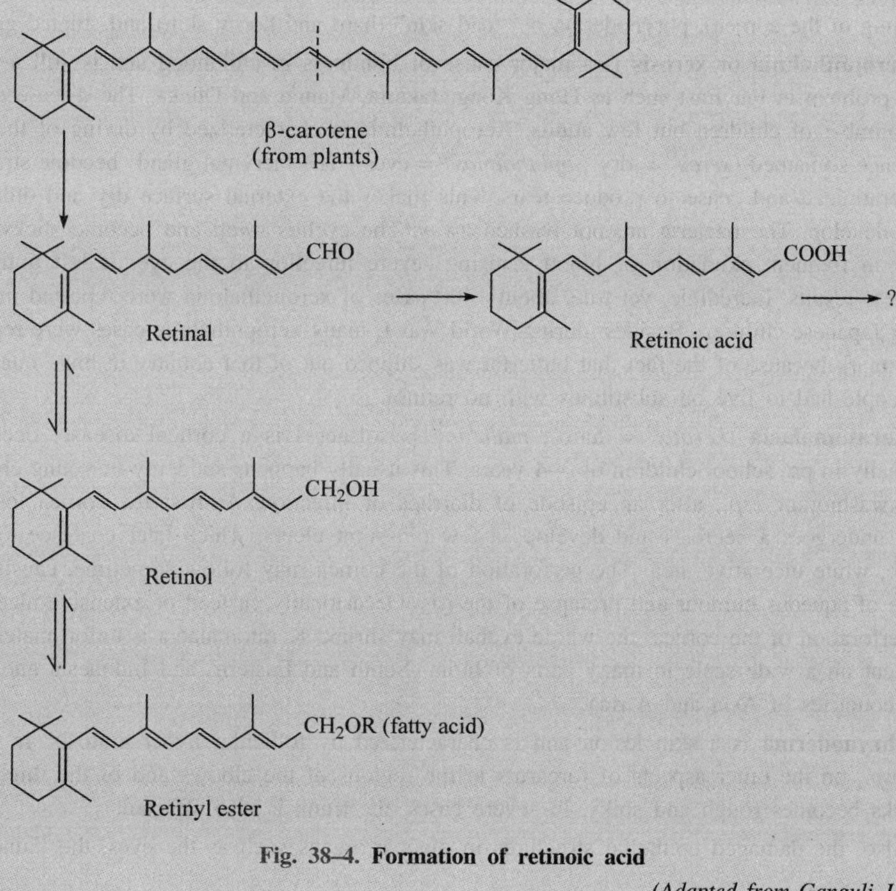

Fig. 38–4. Formation of retinoic acid

(Adapted from Ganguli J, 1973)

then reduced by another enzyme to retinol which, in turn, is converted to retinyl ester by reacting with a fatty acid like palmitic. The retinyl ester, on the contrary, is enzymatically hydrolyzed to produce retinol which is re-esterified with palmitic acid to produce retinyl ester. This is absorbed through the lymphatic system and is stored in the liver.

Liver stores vitamin A in large quantities mainly in the form of retinyl ester. But the vitamin A that circulates in the blood is in the form of retinol and is bound to a specific carrier protein called retinol binding protein (RBP). The retinyl ester present in the liver, therefore, has to be converted to retinol before it mixes with the blood. Liver can also successively convert retinol to retinal and retinal to retinoic acid. Retinoic acid is quickly absorbed from the intestine through the portal system and is rapidly excreted back into the intestine through the bile.

Vitamin A helps maintain the epithelial cells of the skin and the linings of the digestive, respiratory and genito-urinary systems. These linings play a protective role against cancer-causing agents (or carcinogens), viruses and bacteria and are rendered vulnerable by a deficiency of vitamin A. Vitamin A guards against cancer by protecting cell walls from undesirable oxidation, and scavenging the products of oxidation— free radicals, which are linked to the development of cancer.

F. Deficiency. *Vitamin A is perhaps the most important as it affects the various metabolic processes in the body. It has profound effect on epithelial structures, in general.* Vitamin A deficiency leads to the onset of many diseases like nyctalopia or night blindness (inability to see in night), xerophthalmia (scaly condition of the delicate membrane covering the eyes), keratomalacia (softening of the cornea), phrynoderma or "toad skin" (hard and horny skin) and stunted growth.

Xerophthalmia or **xerosis** is a major cause of blindness in childhood and is still a major health problem in Far East such as Hong Kong, Jakarta, Manila and Dhaka. The disease affects large number of children but few adults. Xerophthalmia is characterized by drying of the eyes and hence so named (*xeros*^{G} = dry ; *ophthalmos*^{G} = eyes). The lacrymal glands become stratified and keratinized and cease to produce tears. This makes the external surface dry and dull. The ulcers develop. The bacteria are not washed away. The eyelids swell and become sticky. This results in frequent exudation of blood, causing severe infection to the eye. If left untreated, blindness results. Incredible, yet true, about 1400 cases of xerophthalmia were reported in 1904 among Japanese children. Besides, during World War I, many xerophthalmic cases were reported in Denmark because of the fact that butterfat was shipped out of that country in huge quantities and people had to live on substitutes with no retinol.

Keratomalacia (*kerato*^{G} = hair ; *malakia*^{G} = softness) is a corneal disease, occurring maximally in pre-school children of 3–4 years. This usually happens suddenly in young children with kwashiorkor *esp.*, after an episode of diarrhea or infection. At first the cornea loses its lustre, undergoes a necrosis and develops a few pin-point ulcers, which later coalesce to form a large, white ulcerative area. The perforation of the cornea may follow sometime, causing the release of aqueous humour and prolapse of the iris. Occasionally, instead of extensive ulceration and perforation of the cornea, the whole eyeball may shrink. Keratomalacia is unfortunately still prevalent on a wide scale in many parts of India (South and Eastern) and Indonesia and some other countries of Asia and Africa.

Phrynoderma is a skin lesion and is characterized by follicular hyperkeratosis. In it, the skin *esp.*, on the outer aspects of forearms in the regions of the elbows and of the thighs and buttocks becomes rough and spiky. In severe cases, the trunk is also affected.

Also, the damaged epithelial structures in diverse organs such as the eyes, the kidneys or the respiratory tract often become infected. It is for this reason that vitamin A has been called an '*anti-infection*' *vitamin*. The vaginal epithelium may become cornified, and epithelial metaplasia of the urinary tract may contribute to pyuria and hematuria. Increased intracranial pressure with wide separation of cranial bones at the sutures may occur.

In the alimentary tract, the deficiency of vitamin A causes damage to the intestinal mucosa, resulting in diarrhea.

Vitamin A is an important factor in tooth formation. In its deficiency, there is a defective formation of enamel so that the dentin is exposed. Evidently, sound tooth formation does not occur. Vitamin A deficiency may result in retardation of mental and physical growth and in apathy. Anemia with or without hepatosplenomegally is usually present.

Other deficiency diseases that have been attributed to vitamin A are atrophy of the testes and disturbances of the female genital organs.

According to a report, presented at a meeting of the National Seminar on Corneal Disasters held in New Delhi (1981), India accounts for about one-third of the total blind population of the world, that is about 9 million blind people and that every year about 25,000 children go blind in India due to malnutrition.

Researches conducted by the National Institute of Nutrition (NIN), Hyderabad (1978), show that about 30% of the blind in India lost their sight before they were 21. NIN studies have also revealed that about 10% of the school children belonging to the poorer socioeconomic groups show signs of vitamin A deficiency. It is however, worth-mentioning that the vitamin A deficiency is mainly a result of ignorance rather than the non-availability of food or of the resources to acquire the food as is the case with protein calorie malnutrition. According to NIN researches, if 40 g of green leafy vegetables are included in the existing diet without any other changes, the child will get the necessary vitamin A.

G. Hypervitaminosis A. Vitamin A is less toxic to man and other animals when it is taken in large doses over a long period of time. The children receiving overdosages of 500,000 units of vitamin A per day exhibit tender swellings over the bones, limited motion and definite hyperosteoses.

Human adults consuming 500,000 units or more show nosebleed, weakness, headache, anorexia and nausea. Continuous excessive intake of the vitamin is dangerous because it results in fragile bones and abnormal fetal development. In the case of plant carotenoids, a dietary oversupply (for example, eating too many carrots on a daily basis) results in **carotenosis**, a condition readily diagnosed by yellowing of the skin. These toxic effects develop presumably due to the fact that viatmin A, like vitamin D, is not readily excreted and consequently tissue levels may build up dangerous concentrations.

H. Human requirements. The recommended dietary allowance (RDA) of vitamin A is about 5,000 International Unit (I.U.). Growing children, adults and pregnant and lactating mothers require high doses of up to 8,000 I.U. It is also possible that some individuals require more than the minimal requirement due to either faulty absorption or some other reason. For vitamin A, the World Health Organization (WHO) has chosen as one International Unit the biologic activity of 0.000344 mg (0.344 μg) of synthetic vitamin A-acetate, which is equivalent to 0.30 μg of retinol. In fact, one μg of β-carotene, the provitamin form, gets converted to about 0.167 μg of retinol, the true vitamin form. Another way of putting it is that the retinol equivalent of 1 μg of β-carotene is 0.167. A diet consisting of 1/2 pint of milk, 1 ounce of butter and an adequate amount of

1 pint	=	0.568 litre (British Units)
	=	0.473 litre (U.S. Units)
1 ounce	=	27.09 gm

green vegetables or carrots daily is sufficient enough to meet minimal requirement. The optimal requirement of this vitamin is met with by taking 1 pint of milk and cod liver oil daily.

Fatality Due to Excessive Intake of Vitamin A

A three-man team of explorers from the Australasian Antarctic Expedition started their expedition to explore Antarctica in January, 1912. The team was led by Douglas Mawson and the other two members were Lt. BES Ninnis and Xavier Mertz, a Swiss scientist. Disaster struck on December 14, 1912, when Ninnis fell into a very deep pit and died. With him also went precious food supplies. With most of their food gone, Mawson and Mertz decided to return to their base at Commonwealth Bay, which is at the shores of Antarctica. From here, they could take the ship back to their country. But Commonwealth Bay was about 315 miles from where they were stationed. Covering that distance in the inhospitable surroundings of Antarctica would have taken them weeks, and they had only 10 days' food left with them. They had 6 huskies with them (Huskies are Eskimo dogs, used as ponies in Antarctic region). Both knew that sooner or later they would have to eat those dogs to remain alive.

So they did kill the huskies and ate their flesh, but the flesh was stringy (*i.e.*, fibrous) and they could not eat it. But they found the liver softer and easier to eat, so they took generous quantities of liver. Mertz was a near vegetarian; he could not eat the stringy flesh, so he took more liver than Mawson. Little did he realize that he was taking fatal amounts of vitamin A in this form. On New Year's Eve, Mertz began to feel ill. Next day, he complained of stomachache. Few days later, both men began displaying typical symptoms of vitamin A poisoning, although Mawson was affected less. Their skin was falling off their bodies in strips and their hair was dropping out in handfuls. A week later, Mertz fell into a delirious sleep and never woke. His was the first known case of death due to overdose of vitamin A. Mawson, however, survived, and ultimately did return to Commonwealth Bay.

I. Treatment. For xerophthalmia, the treatment consists in giving 1,500 μ*g*/kg/24 hr of vitamin A orally for 5 days and then continued with intramuscular injection of 7,500 μ*g* of vitamin A in oil daily until recovery occurs.

J. Vitamin A and the vision. George Wald (Physiology or Medicine Nobel Laureate, 1967) of Harvard University has made major contributions to our understanding of the role of vitamin A in visual process. The retina of the human eye contains 2 types of receptor cells, *rods* and *cones*. Animals which have vision only in bright light ("*day vision*") like pigeons have only cones while animals which can see in night or dim light ("*night vision*") like owls have only rods. Thus, the rods are concerned with seeing at low illumination and the cones are responsible for colour vision.

(*a*) Rod vision. The rods contain a photosensitive visual pigment known as **rhodopsin**. It is a conjugated protein and upon illumination splits into a protein called opsin and a carotenoid called retinene$_1$ (Fig. 38–5). It is actually this reaction which may initiate an enzymic reaction responsible for visual mechanism. The retinene$_1$, released by bleaching of rhodopsin, is reduced to vitamin A$_1$ by NADH in the presence of retinene1 reductase, an enzyme present in the retina. Vitamin A$_1$ of mammals is in all-*trans* form like the retinene liberated by bleaching rhodopsin. The isomerization of all-trans retinene1 and vitamin A$_1$ to Δ"-*cis* forms is catalyzed by the enzyme *retinene isomerase*. The major site of this reaction is the liver. Opsin in dark reacts with Δ"-cis retinene1 to regenerate rhodopsin. The series of reactions leading from vitamin A1 to rhodopsin constitutes the major events in '*dark adaptation*'. And the reactions leading to bleaching of rhodopsin constitute what is known as '*light adaptation*'.

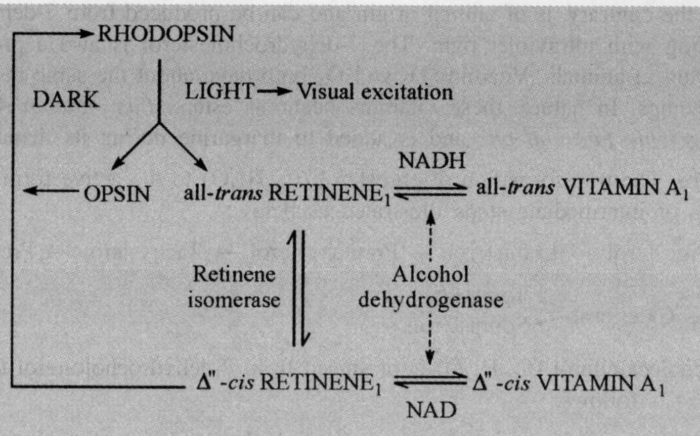

Fig. 38–5. Visual cycle in vertebrates

Whereas the retina of mammals, birds, frogs and marine fishes contain rhodopsin, all fresh-water fishes have another visual pigment **porphyropsin** in their retina. **Wald** has shown that porphyropsin undergoes the same cyclic changes on bleaching and regeneration as in the case of rhodopsin except that here retinene A_2 and vitamin A_2 come in picture.

***(b)* Cone vision.** According to **Young-Helmholtz** trichromatic theory, there are present at least 3 closely related pigments in cones. These are **iodopsin, chlorolabe** and **erythrolabe**. The last two pigments absorb actively in the green and the red portions of the spectrum respectively. When light strikes the retina, it bleaches one or more of these pigments based upon the quality of light. The pigments are converted to all-trans retinene and the protein, opsin. This reaction produces the nerve impulse which is read as blue, green or red based on the pigment affected.

VITAMIN D

A. History. The first demonstration of the existence of vitamin D was shown by **Elmer McCollum** in 1922 who found that cod liver oil was effective in preventing rickets, a disease induced in rats by providing low calcium diet. On account of its preventive action on rickets, vitamin D is often called as **antirachitic factor**. It is also known as '**sunshine vitamin**' as its provitamin form present in human skin is easily converted to the active form by irradiation with ultraviolet light. At least 10 different compounds are known to have antirachitic properties and are designated as D_2, D_3 etc, but the two, namely, vitamin D_2 (ergocalciferol) and vitamin D_3 (cholecalciferol) are more important. Vitamin D_3 was, however, first isolated by **Brockmann** and others.

> The **designation D_1** is not used, however, because D_1 was the designation for the first crystalline material, which later turned out to be a mixture.

B. Occurrence. The best natural sources of vitamin D are the liver oils of many fishes such as cod and halibut. The flesh of oily fishes (*e.g.*, sardine, salmon, herring) is also excellent source. Egg yolks are fairly good but milk, butter and mushrooms are poor. The diets of infants may contain only small amounts of vitamin D; cow's milk contains only 0.1 to 1 μg/quart (1 μg = 40 IU). Cereals, vegetables and fruits contain only negligible amounts. Eggs and yolks contain 3 to 10 μg/g. Most marketed cow's milk is fortified with 10 μg of vitamin D per quart and most commercially-prepared milks for infant formulae are also fortified. Vitamin D_2 is of plant origin and is produced commercially by irradiation with ultraviolet light of a provitamin known as ergosterol which is found in plants, especially in ergot (hence, so named) and yeast.

Vitamin D_3, on the contrary, is of animal origin and can be produced from 7-dehydrocholesterol also by irradiating with ultraviolet light. The 7-dehydrocholesterol is also a provitamin found naturally occurring in animals. Vitamins D_2 and D_3 both have about the same degree of activity in the human beings. In nature, these vitamins occur as esters. *Like vitamin A, vitamin D is absent from vegetable fats and oils* and is added to margarine during its manufacture.

C. Structure. The transformation of ergosterol ($C_{28}H_{44}O$) to the active form D_2 takes place through a series of intermediate steps illustrated as below:

Ergosterol → Lumisterol → Protachysterol → Tachysterol → Precalciferol → Calciferol → Toxisterol / Suprasterols

Similarly, cholecalciferol ($C_{27}H_{44}O$) is produced from 7-dehydrocholesterol through a series of intermediaries as follows:

7-dehydrocholesterol→ Lumisterol$_3$ → Tachysterol$_3$ → Precalciferol$_3$ → Cholecalciferol

During the activation of the provitamins, the ring B is cleaved between carbon atom 9 and 10 to produce vitamins D_2 and D_3 (Fig. 38–6).

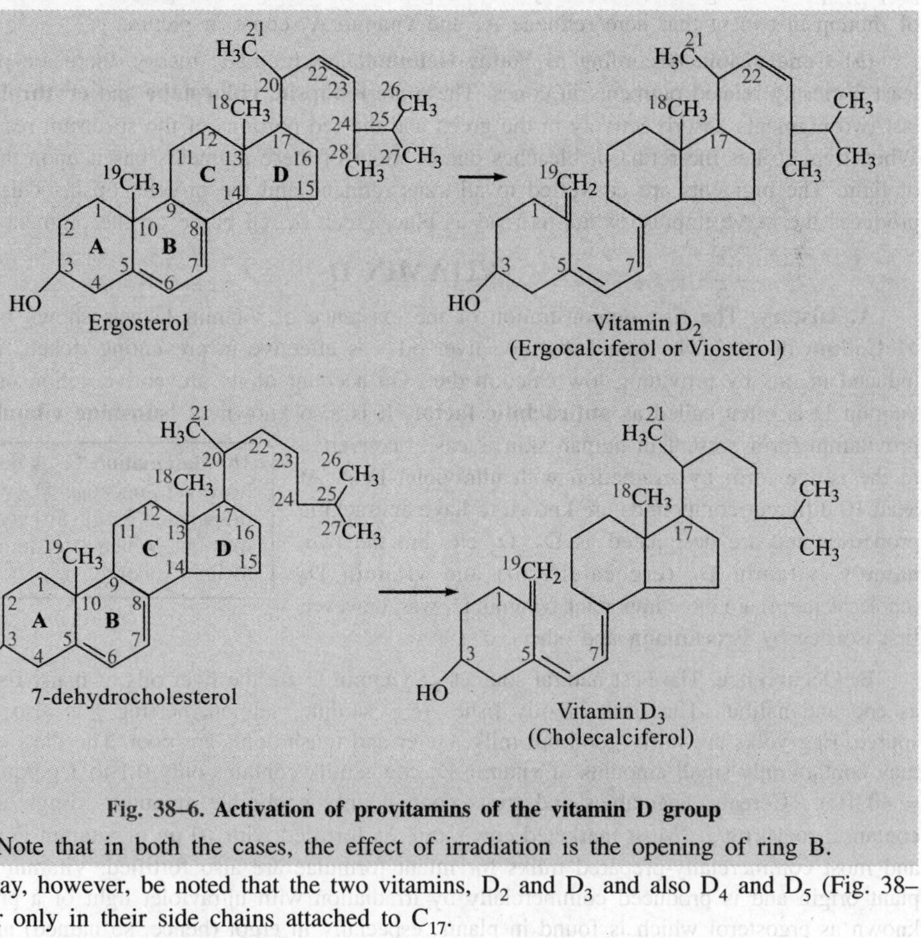

Fig. 38–6. Activation of provitamins of the vitamin D group

Note that in both the cases, the effect of irradiation is the opening of ring B.

It may, however, be noted that the two vitamins, D_2 and D_3 and also D_4 and D_5 (Fig. 38–7), differ only in their side chains attached to C_{17}.

Fig. 38–7. Side chains of vitamins D_4 and D_5

Other forms of vitamin D may be obtained by irradiation of other sterols. Vitamin D_4, for example, is produced by irradiation of 22-dihydroergosterol. Its potency is 50–70% that of vitamin D_2.

However, researches conducted by **DeLuca** *et al* (1968) indicate that the biologically active form of vitamin D_3, present in animals such as rat, has a slightly different structure. It is identified as 25-hydrocholesterol (Fig. 38–8). It is a more polar compound and has an additional OH group at C_{25}. It is one and a half times more potent than the vitamin in curing ricket. Also, it stimulates bone metabolism and intestinal calcium transport more rapidly than the vitamin. It is synthesized

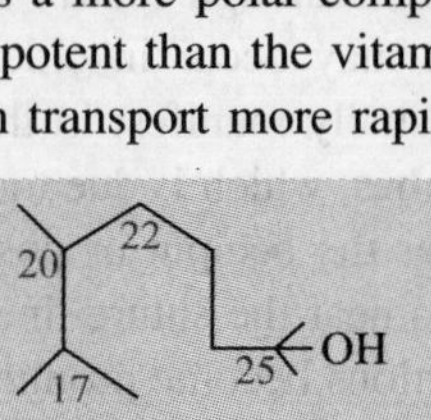

Fig. 38–8. Side chain of 25-hydrocholesterol or 25–hydroxycholecalciferol

in the liver and is then converted into *1,25-dihydroxycholecalciferol* [$1,25(OH)_2D_3$] in the kidneys. Cholecalciferol and its derivatives are **seco-steroids**, *i.e.,* ring A is not rigidly fused to ring B. Because of its conformational mobility, ring A of a seco-steroid exists in two equilibrated conformers. Now as the *in vitro* mode of action of vitamin D is understood, it has been proposed that *1,25(OH)2D$_3$ is actually a hormone and not a vitamin* as it fits the traditional description of a hormone. $1,25(OH)_2D_3$ is produced in the kidneys (organ) and transported by the blood to intestinal mucosa and bone (target tissues), where it functions in the processes for the absorption, reabsorption and mobilization of calcium and phosphate ions. In conjunction with parathormone and calcitonin, it has a major role in homeostasis of Ca and P in the body's fluids and tissues. Now *because a hormone-receptor protein has been identified for $1,25(OH)_2D_3$, the status of vitamin D as a hormone is well established.*

D. Properties. Vitamin D is a white and almost odourless crystalline substance, soluble in fat and fat solvents. It is fairly heat resistant and also relatively resistant to oxidation. It is not affected by acids and alkalies.

E. Metabolism. The provitamin D_3 can be synthesized within the human body so that it may, in fact, not be required in the diet. This may, henceforth, not be treated as a vitamin. In the past when man lived mainly outdoors and with minimum clothing, there was no hindrance for the penetration of ultraviolet light from the sun to convert it into the active form. In the far northern areas, however, the amount of light is not adequate for conversion and as such fish liver oils serve as excellent source of vitamin D in these areas. The increased need of this vitamin is usually felt in growth and in pregnancy to provide for the needs of the fetus.

Vitamin D plays an important role in calcification of bones and teeth. It encourages the absorption, into the blood, of calcium salts and phosphates. Calcium passage across duodenum occurs mainly by diffusion and active transport of Ca^{2+} occurs across the ileal mucosa. Both these processes are related in deficiency of vitamin D. The subsequent release of bound calcium is also markedly stimulated by vitamin D but only in the presence of parathyroid hormone. *On the whole, the function of vitamin D is to cause increased absorption, longer retention and better utilization of calcium and phosphorus in the body.*

There is considerable difference in the potency of these 2 common forms of vitamin D. For example, vitamin D_2 is a powerful antirachitic agent for man and for the rat but not for the chicken. Vitamin D_3, on the contrary, is much more potent for the chicken than either for man or the rat.

F. Deficiency. The most characteristic symptom of vitamin D deficiency is the childhood disease known as rickets. Deficiency of it in human adults leads to osteomalacia, a condition that might also be termed "adult rickets".

Rickets (origin doubtful; false derivation from *rachitis*^{G}, from *rhachis*^{G} = spine + *-itis*^{G} = inflammation; also said to have been derived from an old English word, *wrickken* = to twist) is primarily a disease of growing bones. In it, the deposition of inorganic materials on the matrix of bones (*i.e.*, calcification) fails to occur, although matrix formation continues. Rickets is unusual below the age of 3 months (mo). It may occur in older children with malabsorption. Clinical manifestations of rickets in children usually manifest in the first year or in the second year. One of the early signs of rickets is **craniotabes**, which is due to thickening of the outer table of the skull and is detected by pressing firmly over the occiput or posterior parietal bones. A ping-pong ball like sensation will be felt. Craniotabes near the suture line may, sometimes, be present in normal premature infants. Costochondral junctions become prominent to give appearance of a beaded ribs, the **rachitic rosary**. Thickening of the wrists and ankles are other early evidences of osseous changes. Increased sweating, especially around the head, may also be present.

CLINICAL IMPLICATIONS 38–1

Rickets

As described by **Daniel Webster** in 1645, in rickets, " .... the whole bony structure is as flexible as softened wax, so that the flaccid and enervated legs can hardly support the superposed weight of the body; hence the tibia, giving way beneath the overpowering weight of the frame, bend inwards; and for the same reason the legs are drawn together at their tops; and the back, by reason of the bending of the spine, sticks out in a hump in the lumbar region .... the patients in their weakness cannot (in the most severe stages of the disease) bear to sit upright, much less stand .....".

Signs of advanced rickets are easily identified. These are listed below:

1. Head: Craniotabes may obliterate before the end of the 1st year, although the rachitic process continues. The softness of the skull may result in flattening and, at times, permanent asymmetry of the head. The anterior fontenel is larger than normal; its closure may be delayed until after the 2nd year of life. The central parts of the parietal and frontal bones are usually thickened, forming prominences or bosses, which give the head a box-like appearance (**catput quadratum**).

2. Thorax: The sides of the thorax become flattened, and the longitudinal grooves develop posterior to the rosary. The sternum with its adjacent cartilage projects forward leading to protruding chest (*pigeonbreast*). Along the lower border of the chest develops a horizontal depression (*Harrison groove*), which corresponds with the costal insertions of the diaphragm.

3. Spinal column: Moderate degree of lateral curvature (*scoliosis*) is common and a *kyphosis* (= increased convexity in the region of thoracic spine) may appear in the dorsolumbar region when sitting. *Lordosis* (= forward curvature of the lumbar spine) may be seen in the erect position.

4. Pelvis: The pelvic entrance is narrowed by a forward projection of the promontory; the exit, by a forward displacment of the caudal part of the sacrum and coccyx. In the female, these changes, if they become permanent, add to the hazards of childbirth and necessitate caesarean section.

5. Extremities: The epiphyseal enlargement at the wrists and ankles becomes more noticeable. Bending of the softened shafts of the femur, tibia and fibula results in bowlegs (*knock-knees*) ; the femur and the tibia may also acquire an anterior convexity. *Coxa vara* is sometimes the result of rickets.

Deformities of the spine, pelvis and legs results in reduced stature, **rachitic dwarfism**.

6. Ligaments: Relaxation of ligaments helps to produce *deformities* and partly accounts for knock-knees, weak ankles, kyphosis and scoliosis.

7. Muscles: The muscles are poorly developed and lack tone. As a result, the rachitic children are late in standing and walking. Abdomen becomes protuberant (*potbelly*) because of marked hypotonia of abdominal wall muscles, visceroptosis and lumbar lordosis.

8. Sense organ: Avitaminosis D in early infancy results in *bilateral lamellar cataracts*.

9. Dentition: Eruption of temporary teeth may be delayed in rachitic children. The first tooth in such babies appears between 6th and 9th month, at which time it has appeared in half of the normal babies. In deficiency of vitamin D, the formation of teeth becomes defective and leads to the development of *dental caries*.

Chemical analysis of the bones of rachitic children reveals the presence of low inorganic and high organic and water contents in them. The ratio of calcium to phosphorus (C/P), however, remains constant. In the blood serum, there is usually a normal content of calcium but the phosphate content is reduced (1.5–3.5 mg/dL), against a normal value of 4.5–6.5 mg/dL in healthy infants. Vitamin D deficiency is also accompanied by generalized aminoaciduria, a decrease of citrate in bone and its increased urinary excretion, decreased ability of the kidneys to make an acid urine, phosphaturia, and, occasionally, mellituria. The parathyroid glands hypertophy in rickets.

Rickets in itself is not a fatal disease but complications and intercurrent infections such as pneumonia, tuberculosis, and enteritis are more likely to cause death in rachitic children than in normal children.

Rickets is most prevalent where climate or custom prevents individuals from exposure to sun, whereby checking vitamin D production by irradiation of the skin. In the seventeenth century. this disease was so common in England that it used to be known as "*English disease*". A study conducted by the Indian Social Institute, New Delhi (1981) shows that about 168 children per 1,000 of live births die of rickets in India.

In **osteomalacia** (*osteon*^{G} = bone ; *malakia*^{G} = softness), the action of bones is essentially like that in rickets. However, the bones become softer than the rachitic bones and the C/P ratio does not remain constant. The loss of calcium is greater than that of phosphorus and there is a relative gain in magnesium content. The disease is prevalent in India, China and Arab, particularly among women because of the custom that keeps them indoor and also prevents them from exposure to sun. This is particularly true of Bedouin Arab women who are clothed so that only their eyes are exposed to sunlight. The serum calcium is reduced, sometimes to such an extent that tetany develops.

Vitamin D_3 deficiency also leads to a disease called **idiopathic steatorrhea** or **celiac disease**. Like osteomalacia, the disease is characterized by demineralization of the bones which may result in deformities or dwarfism. In fact, celiac disease is *indirectly* a vitamin D deficiency because the primary abnormality appears to be, in part, a fatty diarrhea. The fat is not absorbed in the intestine and is passed out in stool along with calcium salts and vitamin D.

G. Hypervitaminosis D. Overdosing of calciferol to the children and adults as well produces demineralization of bone. Serum concentrations of both calcium and phosphate are greatly raised, resulting in metabolic calcification of many soft tissues and the formation of renal calculi. The latter disorder may block the renal tubules causing hydronephrosis.

The use of very high doses of vitamin D is not danger-free, however. The toxic effects caused by excess dosage include anorexia, nausea, polyuria, weakness, headache etc. The toxicity is due to the diminished excretion of this vitamin, rather than its storage in the liver. *The water-soluble vitamins, on the contrary, if given in excess pass out immediately in the urine and are henceforth nontoxic.* Much of the whole milk available in urban areas and evaporated milk are fortified with vitamin D concentrate so that 1 quart of fresh, whole milk or a cane of evaporated milk contains the required amount (*i.e.*, 10 μg).

H. Human requirements. Vitamin D requirement is greatly influenced by the amount of ultraviolet light to which the individual is exposed. Half an hour of direct sunlight on the cheeks of a baby each day is sufficient to generate the minimal daily requirement of vitamin D. For adults also, exposure to sunlight for 30 minutes a day is believed to satisfy the daily requirement (about 10 μ*g* or 400 IU) of vitamin D. As effective UV rays do not penetrate glass windows, exposure to sun through window glass is of little importance. Smoke also hinders the progress of these rays and as such city sunshine is not much beneficial. It is for these and some other reasons that vitamin D should be included in the diet. This is particularly true for older people. The recommended daily allowance of vitamin D is 400 I.U. for infants, pregnant women and lactating mothers. For adults, 400 units are adequate. One International Unit is defined as the biologic activity of 0.025 μ*g* of pure crystalline vitamin D_3.

VITAMIN E

A. History. The presence of this active principle was first demonstrated in vegetable oils by **Evans** and **Mattill** independently in 1920. This was designated as vitamin E or **antisterility factor** on account of the development of sterility in animals in its absence. In 1936, two compounds with vitamin E activity were isolated from wheat germ oil by **Evans** and his associates and given the name, α- and β-tocopherol (*tokos*^{G} = childbirth; *pheros*^{G} = to bear; *ol* = an alcohol). Subsequently, five other tocopherols were obtained from various cereal grains like wheat germ, corn oil, rice etc.

B. Occurrence. The tocopherols are of widespread occurrence in many plant oils such as wheat germ, rice, corn, cottonseed, soybean, coconut and peanut *but not olive oil.* Whole grain cereals (fortified breakfast cereals) and margarine butter are a good source of vitamin E. They are also present in small amounts in meat, milk, eggs, leafy vegetables (such as cabbage and lettuce) and some fruits. *Fish liver oils, so abundant in vitamin A and D, are devoid of vitamin E.* Of all the tocopherols discovered so far, the α-form has the widest distribution and greatest biologic activity. The relative *biologic potencies* of various tocopherols are:

α-tocopherol—100

β-tocopherol—25

γ-tocopherol—19

The vitamin E content of some oils are presented in Table 38–5.

Table 38–5. Vitamin E content of some oils

Oil	*Vitamin E (mg/100 g)*
Groundnut	261
Wheatgerm	225
Soybean	166
Linseed	110
Palm	56
Mustard	32

C. Structure. Vitamin E is the collective name for a group of closely related lipids called tocopherols. The tocopherols are derivatives of 6-hydroxychroman (also known as tocol) bearing an isoprenoid side chain at carbon 2. The structure of α-tocopherol ($C_{29}H_{50}O_2$) is given in Fig. 38–9.

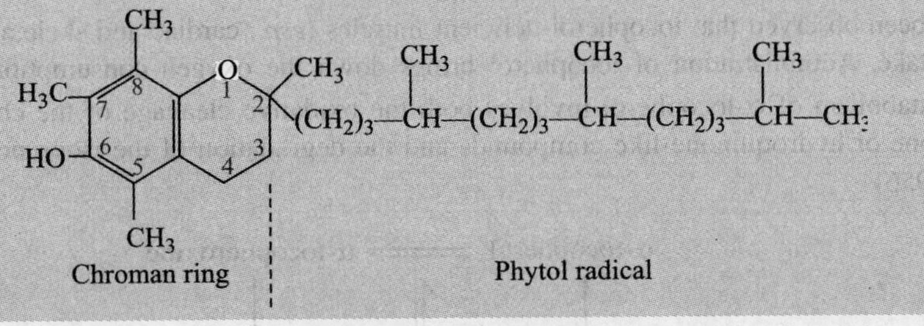

Fig. 38–9. α-tocopherol
(5,7,8-trimethyl tocol)

The various tocopherols differ from each other in substituents on carbons 5, 7 and 8. These substituents are methyl groups and hydrogen atoms. a-tocopherol contains 3 methyl groups whereas other tocopherols are short one or two methyl groups on the aromatic ring (refer Table 38–6).

Table 38–6. Composition of the variable segment of different tocopherols

Types of Tocopherol		*Substituents at Carbon Atoms*		
		5	7	8
Alfa	α	CH_3	CH_3	CH_3
Beta	β	CH_3	H	CH_3
Gamma	γ	H	CH_3	CH_3
Delta	δ	H	H	CH_3
Epsilon	∈	CH_3	H	H
Zeta	ζ	CH_3	CH_3	H
Eta	η	H	CH_3	H

It is noteworthy that the presence of all 3 methyl groups attached to the benzene ring is necessary for full activity. d-tocopherol has but one methyl group and is almost without activity. A slight change in the structure of the tocopherols, for example shortening the side chain, may greatly diminish their physiologic activity.

D. Properties. Vitamin E is a light yellow oil. It is resistant to heat (up to 200°C) and acids but acted upon by alkalies. It is easily but slowly oxidized and is destroyed by UV rays. The tocopherols are excellent antioxidants. They prevent other vitamins presents in food (*e.g.,* vitamin A) from oxidative destruction. It is found in the nonsaponifiable fraction of the vegetable oils.

E. Metabolism. *Tocopherols act as antioxidants, i.e.,* they can prevent the oxidation of various other easily oxidized substances such as fats and vitamin A. It is for this reason that they are commercially added to foods to retard their spoilage. It may be recalled that vitamin A is essential for reproduction. *Whereas the beneficial action of vitamin A is mainly on the ectoderm and endoderm, that of vitamin E is on the mesodermal tissue*. But, very likely, vitamin E influences all the 3 germinal layers of the embryo by preventing the too rapid destruction of vitamin A. Certain substances such as phenols and vitamin C (ascorbic acid) stimulate the antioxidant property of vitamin E.

In fact, the biochemical activity of tocopherol lies in its capacity to protect mitochondrial system from inactivation by fat peroxides. Thus, in mitochondria obtained from vitamin E-deficient animals, a marked deterioration in activity is found due to peroxidation of unsaturated fatty acids which are usually present in these particles. Addition of vitamin E prevents this deterioration by acting as antioxidant for peroxidation.

It has been observed that tocopherol-deficient muscles (*esp.*, cardiac and skeletal) show a high oxygen uptake. Administration of tocopherol brings down the oxygen consumption to normal.

The catabolism of α-tocopherol involves both the oxidative cleavage of the chroman ring to yield quinone or hydroquinone-like compounds and the degradation of the isoprenoids side chain (**Simon**, 1956)

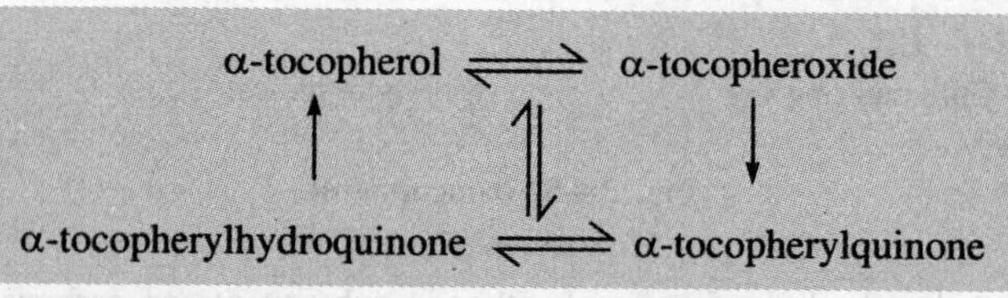

F. Deficiency. The characteristic symptoms of experimentally-induced vitamin E deficiency vary from animal to animal. In mature female rats, **sterility** develops because of reabsorption of fetus after conception while in males, the germinal epithelium of the testes degenerates and the spermatozoa become nonmotile.

Avitaminosis E in herbivorous animals like rabbits and guinea pigs leads to acute **muscular dystrophy** (atrophy of muscle fibres), which ultimately results in creatinuria ; young chicks exhibit capillary damage and **encephalomalacia** ; hen eggs show **low hatchability** and monkeys reveal **hemolytic anemia**. There is, however, little evidence that man is ever short of vitamin E.

Finally, *as is true for almost all the vitamins, avitaminosis E prevents normal growth.* It also sometimes causes degenatation of the renal tubular cells.

G. Human requirements. Vitamin E is not a problem in human nutrition because it is ubiquitous in foods. However, the minimum daily requirement of vitamin E for adults is 30 I.U. for men and 25 I.U. for women. The pregnant and lactating mothers, however, require 30 I.U. daily. For infants and children,. the vitamin E requirement is at the rate of 1 to 1.25 I.U. per kilogram of body weight. One International Unit of dl-α-tocopherol is equivalent to the biologic activity of 1.1 mg of pure compound or 0.67 mg of d-α-tocopherol.

VITAMIN K

A. History. Henrik Dam (1929), a Danish investigator, found that newly-hatched chicks, fed on artificial diets, develop hemorrhage, a fatal disease characterized by prolonged blood-clotting period. The term vitamin K (K for Danish *koagulations*) was then proposed by **Dam** (1934) himself

CARL PETER HENRIK DAM

(LT, 1895–1976)

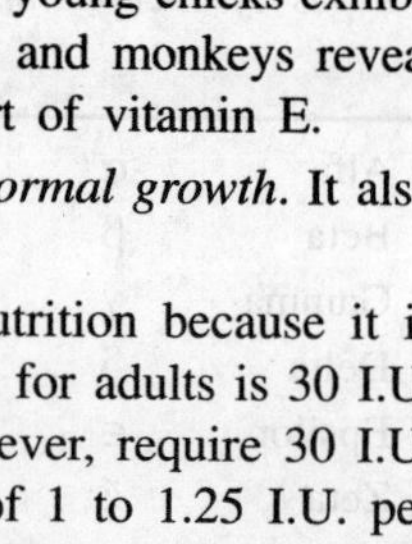

Dam was a Danish biochemist who *proposed the term vitamin K* (*K* for Danish koagulations), in 1934, to designate the active factor which cured or prevented hemorrhage, a disease found in newly-hatched chicks fed on artificial diets. He also isolated vitamin K_1 from alfalfa (1939) and investigated the physiological properties of vitamin K. Dam was on a lecture tour of the United States and Canada when the Nazis occupied his homeland, Denmark, in 1940. He spent the war years at the University of Rochester, and it was during this period that he received his **Nobel Prize (1943) for Physiology or Medicine** along with E.A. Doisy.

to designate the active factor which cured or prevented this disease. On account of its blood-clotting power, it is also called as **antihemorrhagic factor** or **coagulation vitamin.**

Of the 2 naturally-occurring forms of this vitamin, vitamin K_1 was first isolated by Dam *et al* from alfalfa in 1939 and the other form, vitamin K_2 from fish meal by **Doisy** *et al*, also in 1939.

B. Occurrence. Vitamin K_1 occurs in green vegetables like spinach, alfalfa, cabbage etc. Fruits and cereals are poor sources. Vitamin K_2 is found in some intestinal bacteria. A rich source of K_2 is putrefied fish meal. Their relative *biologic potencies* are:

vitamin K_1— 100
vitamin K_2— 80

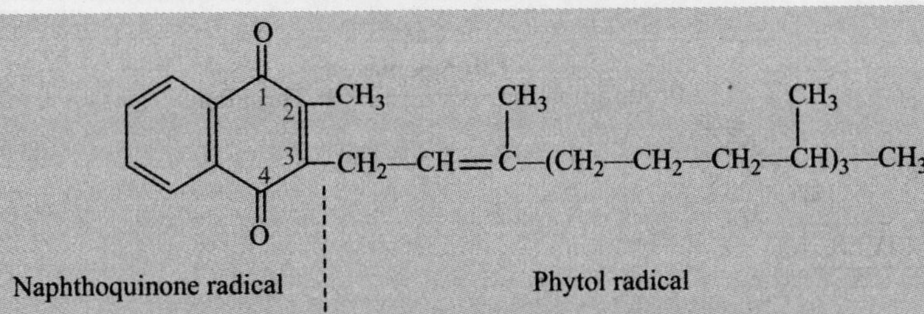

Vitamin K_1 or **phylloquinone** or **phytonadione**
(2-methyl-3-phytyl-1, 4-naphthoquinone)

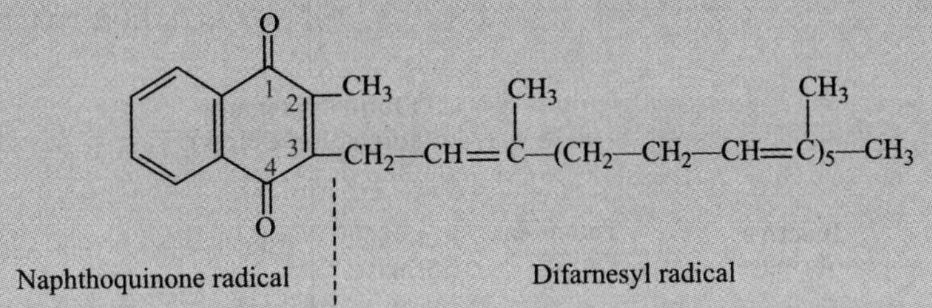

Vitamin K_2 or **flavinoquinone** or **farnoquinone** or **menaquinone**
(2-methyl-3-difarnesyl-1,4-naphthoquinone)

Fig. 38–10. Two vitamins of K group

C. Structure. Chemically, the two forms of vitamin K (Fig. 38–10) are derivatives of quinones and differ from each other in the composition of their side chain present at carbon 3 of the naphthoquinone ring. It is a phytol radical in vitamin K_1 ($C_{31}H_{46}O_2$) and a difarnesyl radical in vitamin K_2 ($C_{41}H_{56}O_2$). Vitamin K_1, found in plants, has 4 isoprene units in its side chain whereas vitamin K_2, found in animals, contains in its side chain 6 isoprene units, each with a double bond.

Analogues— The various analogues of naphthoquinone, however, have also been shown to possess vitamin K activity for animals. This is due to their structural resemblance. The common examples are menadione and phthiocol (Fig. 38–11).

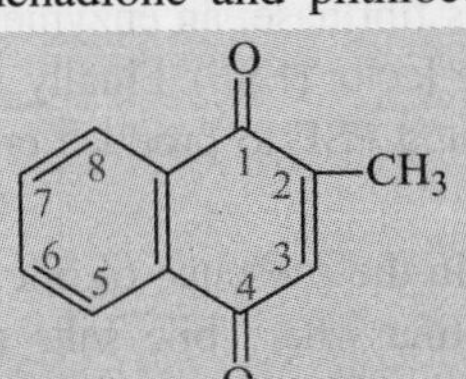

Menadione or **Vitamin K_3**
(2-methyl-1,4-naphthoquinone)

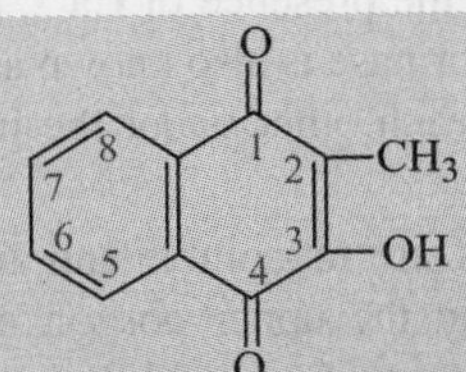

Phthiocol
(2-methyl-3-hydroxy-1,4-naphthoquinone)

Fig. 38–11. Two analogues of vitamin K

Menadione, which is sometimes referred to as vitamin K_3, is twice as potent as vitamin K_1. It is soluble in oil, sparingly so in water and is not oxidized in air when protected from light. Its diphosphate ester is water-soluble and is widely used clinically.

D. Properties. Vitamin K_1 is a yellow viscid oil but vitamin K_2 is a yellowish crystalline solid. It is sensitive to light and is, therefore, kept in dark bottles. It is destroyed by irradiation, strong acids, alkalies and oxidizing agents.

E Metabolism. Vitamin K plays an essential role in the biosynthesis of prothrombin— a blood plasma protein needed in the process of blood clotting and produced in liver. The process of blood coagulation may be summarized as below.

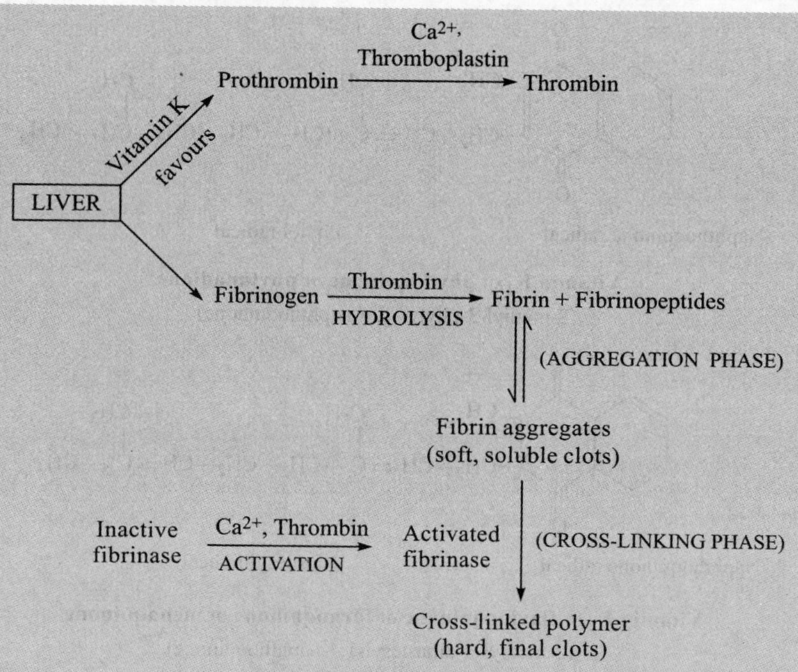

Fig. 38–12. Probable mechanism of blood coagulation

[Note the role of vitamin K in the synthesis of prothrombin which is a precursor of thrombin, the latter has dual function of (*a*) hydrolyzing fibrinogen to fibrin, and (*b*) activating fibrinase of FSF, which brings about clotting.]

The formation of the blood clot (refer Fig. 38–12) is caused by the enzymic hydrolysis of the soluble plasma protein, fibrinogen to the insoluble protein, fibrin and fibrinopeptides. This transformation is catalyzed by an enzyme, *thrombin*. Thrombin itself is not present in the blood but is produced from its precursor, prothrombin in the presence of Ca^{2+} ions and another protein called *thromboplastin* (= thrombokinase). In the next step, fibrin forms soft, fibrous networks (or soft clots). Then in the presence of Ca^{2+}, thrombin activates fibrinase, an enzyme precursor found in blood plasma. Fibrinase is also known as fibrin-stabilizing factor (FSF). Finally, fibrous networks of fibrin link with each other under the influence of activated FSF to produce cross-linked fibrin (or hard clots).

The vitamins K are fat-soluble and are absorbed only in the presence of bile. As a result, the absorption occurs in the upper portion of the small intestine where bile salts are present. The avitaminosis K may occur where bile is prevented from entering the intestinal tract. This is true for most of the fat-soluble vitamins but is, in particular, important in the case of vitamin K because of its blood clotting action.

Like vitamin E and coenzyme Q, the vitamin K has also been ascribed a role in electron transport system (ETS) and oxidative phosphorylation in mitochondria. The vitamin K_1 and K_2 both activate electron transport in the succinate oxidase of cardiac muscle preparations that have been made inactive by treating with isooctane. The bacterial extracts or liver mitochondria, when antagonists irradiated, require vitamin K for oxidative phosphorylation. This suggests a possible role of vitamin K in oxidative phosphorylation. The specific site of action is believed to occur between NADH and cytochrome *b*. It has been suggested that a phosphate ester of vitamin K, upon oxidation, transfers phosphate to ADP to form ATP.

Antagonists— Two antagonists of vitamin K are dicoumarol and warfarin (Fig. 38–13) ; both antagonists prevent blood clotting. Dicoumarol was first isolated from mouldy clover hay. It is often given to patients, who have suffered heart attacks caused by blood clots, as as preventive measure against further clotting in the blood vessels. Warfarin (name derived from the initials of the *W*isconsin *A*lumni *R*esearch *F*oundation, which sponsored the research on the compound) is a synthetic analogue of vitamin K. It is extremely poisonous to rats, causing death by internal bleeding. Ironically, this potent rodenticide is also a valuable anticoagulant drug for the treatment of human patients in whom excessive bleeding is dangerous– surgical patients and victims of coronary thrombosis.

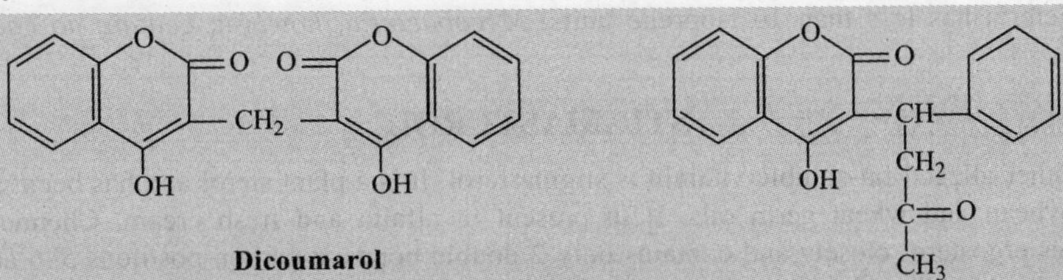

Dicoumarol

[3,3′–methylenebis (4-hydroxy-1,2-benzopyrone)]

Warfarin

Fig. 38–13. Two vitamin K antagonists

Note that both dicoumarol and warfarin lack an isoprenoid side chain.

In 1988, by introducing the antioxidant vitamin E, **David Gershon** and his colleagues at the Technion–Israel Institute of Technology, have succeeded in reducing cell damage and increasing the life span of nematode worms. The damage body cells sustain is due to oxidation, an underlying mechanism of ageing, which also damages the disposal system. Paradoxically, the oxygen we depend on for life is a source of our age-associated decline in function. Similar researches conducted on humans might shed light on how to intervene and retard ageing in them.

F. Deficiency. Deficiency of vitamin K causes loss of blood-clotting power. The infants may also show signs of vitamin K deficiency by developing **hemorrhage**. This disease persists by the time the bacteria develop in the intestine. Administration of this vitamin to pregnant mothers before parturition decreases the onset of this disease. In man, however, avitaminosis K results in **steatorrhea** with diminished intestinal absorption of lipids.

In general, vitamin K deficiency is rarely found in higher animals as this is provided by food and also synthesized by intestinal bacteria.

G. Human requirements. *There is seldom a lack of sufficient vitamin K in human beings.* As such, no standard requirement has been set.

COENZYME Q

Crane *et al* (1959) have demonstrated that coenzyme Q and certain other ubiquinones are components of mitochondrial lipids. These substances serve as electron transport agents and are involved in the formation of ATP at a cytochrome *a*. **Roles** (1967) has classified the coenzyme Q group as vitamins because of their ability to cure (or protect against) vitamin E deficiency in several animal species. Some of the coenzymes also participate in electron transport and/or oxidative phosphorylation.

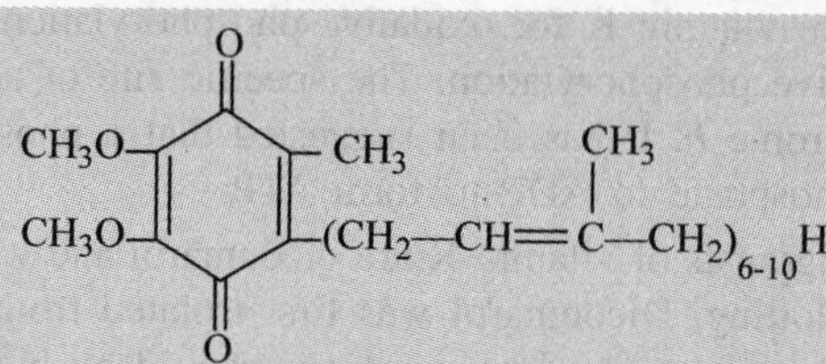

Fig. 38–14. Coenzyme Q

Various homologues of coenzyme Q, containing 6 to 10 isoprene units, have been isolated from various microbes, chloroplasts of green plants and mitochondria of animals. These have the same quinonoid nucleus (Fig. 38–14) but differ in the number of isoprenoid units in the side chain. For example, coenzyme Q from animal source has 10 isoprene units and is, henceforth, called coenzyme Q_{10} or *ubiquinone*$_{50}$ (50 carbon atoms, *i.e.*, 10 isoprene units, in the side chain), whereas the one from bacteria has less than 10 isoprene units. *Mycobacteria, however, contain no coenzyme Q.*

STIGMASTEROL

Another alleged fat-soluble vitamin is stigmasterol. It is a plant sterol and has been isolated from soybean and wheat germ oils. It in present in alfalfa and fresh cream. Chemically, it resembles ergosterol closely and contains only 2 double bonds at carbon postitons 5–6 and 22–23 (Fig. 38–15). The absence of stigmasterol causes stiffness of the wrists and elbows of the guinea pigs. For this reason, it is commonly called as **'antistiffness factor'**. The muscles atrophy and become streaked.

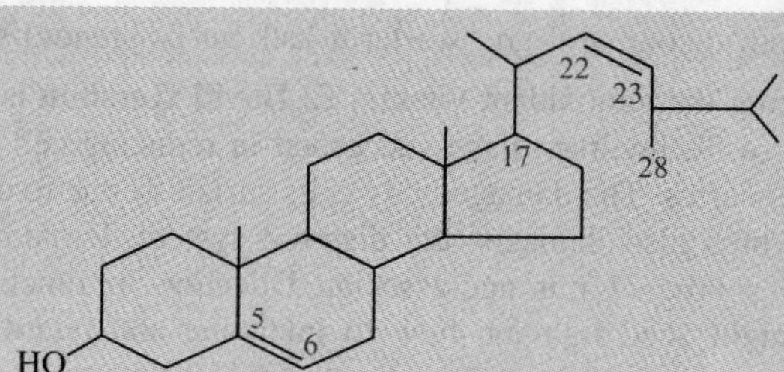

Fig. 38–15. Stigmasterol

Table 38–7 summarizes some characteristic features of various fat-soluble vitamins.

Table 38–7. Fat-soluble vitamins

Vitamin	*Common name(s)*	*Chemical name*	*Sources*	*Metabolic functions*	*Deficiency diseases**	*Daily requirement of man*
A	Antixerophthalmic factor or 'Bright eyes' vitamin or Anti-infection vitamin	A_1— Retinol, $C_{20}H_{30}O$; A_2— 3-dehydroretinol, $C_{20}H_{28}O$	Fish liver oils, butter, milk, eggs and kidneys; Pigmented (esp., yellow) vegetables and fruits	Vasual cycle; Membrane integrity	Demyelinization; Nyctalopia; Xerophthalmia; Keratomalacia; Phrynoderma	5,000 I.U.
D	Antirachitic factor or 'Sunshine' vitamin	D_2— Ergocalciferol, $C_{28}H_{44}O$; D_3— Cholecalciferol, $C_{27}H_{44}O$	Fish liver oils, egg yolks, milk, butter and mushrooms	Calcification of bones and teeth	Rickets (in children); Osteomalacia (in adults); Celiac disease	400 I.U.
E	Antisterility factor	α-tocopherol, $C_{29}H_{50}O_2$; β-tocopherol, $C_{28}H_{48}O_2$; γ-tocopherol, $C_{28}H_{48}O_2$	Plant oils like wheat germ, rice cottonseed, corn, soyabean etc; Also meat, milk and eggs	Act as antioxidants; Control O_2 consumption; Participate in nucleic acid metabolism	Sterility in rats; Muscular dystrophy in rabbits and guinea pigs; Encephalomalacia in young chicks	30 I.U.
K	Antihemorrhagic factor or Coagulation vitamin	K_1— Phylloquinone, $C_{31}H_{46}O_2$; K_2— Farnoquinone, $C_{41}H_{56}O_2$	Green vegetables like spinach, alfalfa, cabbage etc.; Also fruits and cereals	Biosynthesis of prothrombin; Oxidative phosphorylation; Electon transport system	Hemorrhage (in infants); Steatorrhea (in adults)	0.001 mg

* Deficiency diseases have been mentioned in relation to man, if not stated otherwise.

REFERENCES

1. **Baker H :** Clinical Vitaminology. *Interscience Publishers, Inc., New York. 1968.*
2. **Briggs MH (editor) :** Vitamins in Human Biology and Medicine. *CRC Press, Boca Raton, Fla. 1981.*
3. **Briggs MH (editor) :** Recent Vitamin Research. *CRC Press, Boca Raton, Fla, 1984.*
4. **British Medical Journal :** The Nutrition of Man. *Vol. 37, 1981.*
5. **Davidson S, Passmore R, Brock JF, Truswell AS (editors) :** Human Nutrition and Dietetics. *6th ed. Churchill Livingstone Ltd., Edinburgh. 1975.*
6. **DeLuca HF :** The Vitamin D Story. *FASEB J.* **2** : *224–236, 1988.*
7. **Dyke SF :** The Chemistry of the Vitamins. *Interscience Publishers, Inc., New York. 1965.*
8. **Fairfield KM, Fletcher RH :** Vitamins for chronic disease prevention in adults: scientific review. *JAMA.* **287** : *3116–3126, 2002.*
9. **Fletcher RH, Fairfield KM :** Vitamins for chronic disease prevention in adults: clinical applications. *JAMA.* **287** : *3127–3129, 2002.*
10. **Ganguly J :** The "British Eyes" Vitamin. *Science Today. 30–33, 1973.*
11. **Goodwin TW :** The Biosynthesis of Vitamins and Related Compounds. *Academic Press, Inc., New York. 1963.*
12. **Goodman DS :** Vitamin A metabolism. *Fed. Proc.* **39** : *2716, 1980.*
13. **György P, Pearson WN (editors) :** The Vitamins. *Vols. 6 and 7. Academic Press, Inc., New York. 1967.*
14. **Harris LJ :** Vitamins in Theory and Practice. *Cambridge University. Press, Cambridge. 1965.*
15. **Harris LJ :** The Discovery of Vitamins : in J. Needham's (ed) : The Chemistry of Life. *150–170. Cambridge University. Press, New York. 1970.*
16. **Hauschka PV, Lian BJ, Gallop PM :** Vitamin K and mineralization. *Trends Biochem. Sci.* **3** : *75, 1978.*
17. **Holman WIM :** Distribution of vitamins within the tissues of common foodstuffs. *Nutr. Abstr. Revs.,* **26** : *227–304, 1956.*
18. **Isler O, Wiss O :** Chemistry and biochemistry of the K vitamins. *Vitamins and Hormones.* **17** : *54–91, 1959.*
19. **Jones G :** Eating fruit and vegetables. *BMJ.* **326** : *888, 2003.*
20. **Keys A :** The caloric requirement of adult man. *Nutr. Abstr. & Revs.,* **19** : *1–21, 1949.*
21. **Kleiber M :** The Fire of Life. *Wiley, New York. 1961.*
22. **Leung AKC :** Carotenemia. *Adv. Paediatr.* **34** : *223, 1987.*
23. **Loomis WF :** Rickets. *Sci. Amer.* **223** : *76–91, 1970.*
24. **Machlin LJ (editor) :** Vitamin E. A Comprehensive Treatise. *M.Dekker. 1980.*
25. **Marks J :** The Vitamins in Health and Disease. *Little Brown & Co., Boston. 1969.*
26. **McCormick DB, Wright LD :** Vitamins and coenzymes. In the series Methods in Enzymology., Vols. 62, 66 and 67. *Academic Press, Inc., New York. 1979, 1980.*

27. **National Research Council :** Recommended Dietary Allowances. *National Academy of Sciences (U.S.), 1980.*

28. **Nicolaysen R, Eeg-Larsen N :** Biochemistry and physiology of vitamin D. *Vitamins and Hormones* **11** : *29–61, 1953.*

29. **Pett LB :** Vitamin requirements of human beings. Vitamins and Hormones. **13** : *214–238, 1955.*

30. **Readings From Scientific American :** Human Nutrition. *Freeman, San Francisco. 1978.*

31. **Reichel H, Koeffler HP, Norman AW :** The role of vitamin D endocrine system in health and disease. *N. Engl. J. Med.* **320** : *980, 1989.*

32. **Roels OA :** Present knowledge of vitamin A. *Nutr. Rev.* **24** : *129, 1966.*

33. **Sebrell WH (editor) :** Control of Malnutrition in Man. *Amer. Public Health Asso., New York. 1960.*

34. **Sebrell WH Jr, Harris RS (editors) :** The Vitamins : Chemistry, Physiology, Pathology and Methods. *2nd ed. Vols. 1-5. Academic Press, Inc., New York. 1967–1972.*

35. **Vasington FD, Reichard SM, Nason A :** Biochemistry of vitamin E. *Vitamins and Hormones.* **20** : *43–88, 1960.*

36. **Wagner AF, Folkers K :** Vitamins and Coenzymes. *Interscience Publishers. Inc., New York. 1964.*

37. **Wasserman RH, Corradino RA :** Metabolic roles of vitamin A and D. *Ann. Rev. Biochem.,* **40** : *501, 1971.*

38. **Wasserman RH, Taylor AN :** Metabolic roles of vitamin D, E, and K. *Ann. Rev. Biochem.,* **41** : *179, 1972.*

39. **Yetgin S, Ozsoylu S, Raucan S *et al* :** Vitamin D-deficiency rickets and myelofibrosis. *Science.* **170** : *404, 1970.*

PROBLEMS

1. In contrast to water-soluble vitamins, which must be a part of our daily diet, fat-soluble vitamins can be stored in the body in amounts sufficient for many months. Suggest an explanation for this difference based on solubilities.
2. In the presence of warfarin, an analogue of vitamin K, several proteins of the blood coagulation pathway are ineffective because they cannot bind Ca^{2+} efficiently. Why?
3. What is the chemical reaction in which vitamin K participates? How is this reaction involved in blood coagulation and bone formation?
4. If one takes vitamin E to protect his heart, how much should he take?
5. Will taking vitamins give one extra energy?
6. When taking vitamin E, is it okay to take other medicine?
7. Should one take multivitamin pills containing vitamin D every day as one gets older?
8. How much exposure to sunlight a person needs to supply his body with enough vitamin D?
9. Vegetables oils are often fortified with:

   (*a*) vitamin A (*c*) vitamin E

   (*b*) vitamin D (*d*) vitamin K

10. Why should one take coenzyme Q_{10}?
11. One claim put forth by purveyors of health foods is that vitamins obtained from natural sources are more healthful than those obtained by chemical synthesis. For example, it is claimed that pure L-ascorbic acid (vitamin C) obtained from rose hips is better for you than pure L-ascorbic acid manufactured in a chemical plant. Are the vitamins from the two sources different? Can the body distinguish a vitamin's source?
12. How much vitamin D a person needs daily to help prevent osteoporosis?
13. Is it safe to take very high doses of vitamin D?
14. Vitamin A is the essential dietary factor for the formation of:
    (*a*) rhodopsin (*c*) hemoglobin
    (*b*) biliverdin (*d*) biotin
15. Which of the following is both a vitamin and a hormone?
    (*a*) ascorbic acid (*c*) thiamine
    (*b*) calciferol (*d*) riboflavin
16. Does your skin ever lose its ability to make vitamin D from sunlight?
17. When is a good time to take your multivitamin/mineral tablet?
18. If a person takes a daily multivitamin and supplements, can he get away with less fruits, vegetables and grains?
19. Can one take vitamins with thyroid pills?
20. The most nutritive substance on Earth is:
    (*a*) Spinach (*c*) *Laminaria*
    (*b*) *Spirulina* (*d*) Soybean
21. What are the best food sources of vitamin D?
22. How powerful is turmeric?
23. What is the "forgotten vitamin"?
24. *Seanol* is the name of the compound derived from *Ecklonia cava*, a seaweed that grows off the coast of Japan and Korea. Why is seanol the most powerful antioxidant known?

ANSWERS TO PROBLEMS

1. Water-soluble vitamins are more rapidly excreted in the urine and are not stored effectively. Fat-soluble vitamins have very low solubility in water and are excreted only very slowly by the kidney.
2. The proteins affected are those that have γ-carboxyglutamyl (Gla) residues, which are excellent calcium chelators. The formation of Gla is a posttranslational modification catalyzed by a carboxylase whose essential cofactor is vitamin K. Warfarin prevents the reduction of vitamin K epoxide formed during the carboxylation back to the dihydroquinone form, which is necessary for the reaction.
3. Vitamin K is necessary for carboxylation of specific glutamic acid residues in certain proteins to form γ-carboxyglutamic acid residues. In blood coagulation, this step is required for the conversion of preprothrombin to prothrombin. In bone formation, this is required to form the calcium-binding residues of the protein osteocalcin.
4. The usual recommended dose is 400 IU once a day.
5. Despite what many people believe, vitamins pills or vitamin shots do not give you extra energy.

6. No. If you take vitamin E while you are taking the anticlotting drug coumadin (warfarin), you could develop dangerous internal bleeding.
7. Yes, older people need extra vitamin D because they absorb less of it from their food and their skin makes less of it.
8. Ten to fifteen minutes of sunlight on the face and arms. Three times a week is usually enough.
9. (*b*) vitamin D
10. Coenzyme Q_{10} (CoQ_{10}) is a longevity supplement. Recent animal studies have revealed that animals given CoQ_{10} experienced a 24% increase in lifespan. That translates to 9 years in human lifespan. Be sure to take the ubiquinol form of CoQ_{10}. It is more absorbable.
11. The vitamins from the two sources are identical; the body cannot distinguish the source.
12. Ordinarily 400 I.U., but if one is over 50, 600 I.U. and if over 70, 800–1,000 I.U. a day
13. No. 5,000 I.U. a day can cause permanent kidney damage, but taking up to 2,000 I.U. a day is usually safe.
14. (*a*) rhodopsin
15. (*b*) calciferol (vitamin D)
16. Yes, as one grows older, his skin makes less and less vitamin D and one needs to take more orally.
17. After eating. This shows its movement through the intestive allowing more time for the pill to dissolve completely and be absorbed.
18. No. These foods contain thousands of substances that have not been identified yet and that they fight cancer and other health problems.
19. Not, if the vitamin pills contain iron. The iron can keep the thyroid pills from being adequately absorbed.
20. (*b*) *Spirulina*
21. Plant sources provide you with vitamin D_2, an inferior form of the vitamin. Only animal sources, such as salmon, mackerel, sardines, egg yolks and raw milk can provide you with vitamin D_3, the superior form of the vitamin.
22. Turmeric is one of the nature's most powerful antioxidants. It is 5 to 8 times stronger than vitamin E, and also stronger than vitamin C. It is 3 times more powerful than grape seed or pine-bark extract.
23. Vitamin K is set to explode on the nutritional scene like vitamin D did 15 to 20 years ago. Up to 99% of the population could be deficient in this vital nutrient. Lettuce, broccoli and spinach are good K sources. The best food source is *natto beans*, if you can handle the taste. It is also available in capsule form.
24. It all comes down to the number of molecular rings. The more rings, the more antioxidant activity. Resveratol has 2 rings, green tea has 4 rings. Seanol has 8 rings, allowing it to trap far more free radicals than any other antioxidant. Also, seanol is both water- and fat-soluble.

CONTENTS

Crystals of ascorbic acid (vitamin C), viewed under polarized light

CHAPTER

39

Water-soluble Vitamins

VITAMIN B COMPLEX

Originally, vitamin B referred to a vitamin whose deficiency causes beriberi in man and polyneuritis in birds. Later, Goldberger's researches on pellagra led to the view that vitamin B consisted of at least 2 factors: a heat-labile *antiberiberi factor* and a comparatively heat-stable *antipellagra factor*. Some called the former factor as vitamin B_1 and the latter as vitamin B_2. But the later researches conducted by **Richard Kuhn**, Conrad Elvehjem and others have established the fact that vitamin B complex, as represented by yeast, rice bran and liver extracts, contains still other factors. At present, the vitamin B complex is known to consist of a group of at least 13 components usually named as B1, B_2, B_3 etc. But to prevent confusion, their chemical names are now frequently used. The various members of the vitamin B complex are not related either chemically or physiologically, yet they have many features in common:

(*a*) All of them except lipoic acid are water-soluble.

(*b*) Most of them, if not all, are components of coenzymes that play vital roles in metabolism (refer Table 39–1).

(*c*) Most of these can be obtained from the same source, *i.e.,* liver and yeast.

(*d*) Most of them can be synthesized by the intestinal bacteria.

Table 39–1. Coenzyme derivatives of water-soluble vitamins

Vitamin		*Coenzyme form*
Vitamin B_1	(Thiamine)	Thiamine pyrophosphate (TPP)
Vitamin B_2	(Riboflavin)	Flavin mononucleotide (FMN) Flavin adenine dinucleotide (FAD)
Vitamin B_3	(Pantothenic acid)	Coenzyme A (CoA)
Vitamin B_5	(Niacin)	Nicotinamide adenine dinucleotide (NAD) Nicotinamide adenine dinucleotide phosphate (NADP)
Vitamin B_6	(Pyridoxine)	Pyridoxal phosphate (PALP), Pyridoxamine phosphate (PAMP)
Vitamin B_7	(Biotin)	Biocytin
Vitamin B_9	(Folic acid)	Tetrahydrofolic acid (THFA)
Vitamin B_{12}	(Cyanocobalamin)	Deoxyadenosyl cobalamin
Vitamin C	(Ascorbic acid)	Not known

VITAMIN B_1

A. History. Thiamine was the first member of the vitamin B group to be identified and hence given the name vitamin B_1. Thiamine was first isolated by **Jansen** (1949) in Holland and **Adolf Windaus** in Germany. On account of its curing action against beriberi, it is commonly known as **antiberiberi factor.** It is also known as **antineuritic factor** or heat-labile factor. In Europe, it is also designated **aneurin.**

Thiamine is also spelt as thiamin (**Frank B. Armstrong**, 1989; **Albert L. Lehninger, David L. Nelson** and **Michael M. Cox**, 1993).

B. Occurrence. *Thiamine is found practically in all plant and animal foods.* Cereals, heart, liver and kidney are excellent sources of it. In cereals, the outer layers of seeds are especially

BAREND COENRAAD PETRUS JANSEN
(LT, 1884–1962)

Jansen was born on April 1, 1884 in Zwolle, Holland. He received his Ph.D. degree in 1912. He was appointed lecturer in Physiological Chemistry at the University of Amsterdam. His researches centered around studies on intestinal digestion and metabolism of amino acids. His major contribution was the **biosynthesis of urea** from amino acids by the liver *in vitro* which he published in 1915. He accepted the post of Chief Chemist in the Pharmaceutical Chemistry Division in an Institute in Dutch East Indies as he was keen to work on the **isolation of antiberiberi vitamin** (vitamin B_1). He developed a biological assay method for antiberiberi vitamin using a small bird called bondols (*Mina maja*). Jansen evolved a method for the purification of vitamin B_1 from rice polishings extract and succeeded in isolating vitamin B_1 in crystalline form. The work, which was published in 1926, aroused worldwide interest as vitamin B_1 was the first vitamin to be obtained in a crystalline form. He also developed a fluorimetric method for the assay of vitamin B_1 by oxidizing it to thiochrome, which gives fluorescence.

Jansen went back to Holland in 1928 and continued his researches in the field of nutrition as professor of Physiological Chemistry in the University of Amsterdam. The Netherlands Institute of Nutrition was also shifted to his department. He continued his researches on various problems such as nutrition survey of pregnant women, relationship of nutrition to dental caries, and development of methods for the assay of vitamins. He was a member of several National and International Committees on Nutrition. He edited a 'Text Book of Nutrition' which was published in 1959. He died on October 18, 1962.

rich in thiamine (Fig. 39–1). In yeasts and animal tissues, however, it is present mainly as its coenzyme, thiamine pyrophosphate (TPP). Milk also contains thiamine, although in relatively low amounts. The milling of wheat flour lowers the thiamine content considerably, sometimes to the extent of even 80%. Consequently, wheat flour is usually enriched with thiamine at many places. Furthermore, improper cooking (*esp.*, when the water in which foods are cooked is discarded) loses thiamine content. This is because of the solubility of thiamine in water. Therefore, it is desirable to use the "cook water" for soups and sauces. Thiamine is easily destroyed by heat in neutral or alkaline media. Because the covering of the grains of cereals contains most of the vitamin, polishing reduces its availability. Canning processes are, however, not particularly destructive.

Fig. 39–1. Crystals of thiamine (vitamin B_1) Thiamine pyrophosphate (TPP) is a cofactor in a number of enzymes.

C. Structure. The chemical structure of thiamine (Fig. 39–2) was determined in 1935 by **Robert R. Williams** and his associates in the United States and its chemical synthesis was achieved soon thereafter. Thiamine ($C_{12}H_{17}N_4OS$) is 2,5-dimethyl-6-aminopyrimidine bonded through a methylene linkage to 4-methyl-5-hydroxyethyl-thiazole. Thus, pyrimidine and thiazole are the two moieties present in its molecule. The pyrimidine is unique in that it is the only natural pyrimidine containing an alkyl group at C_2. Also, *with the possible exception of penicillin, thiamine is the only natural compound which contains a thiazole group*. It is interesting to note that plants can use a mixture of pyrimidine and thiazole compounds in place of thiamine itself. On the other hand, all the animals except pigeon require the complete vitamin.

ROBERT RUNNELS WILLIAMS

(LT, 1886–1965)

Robert Williams, an Indian-born U.S. biochemist, was born on February 16, 1886 in Nellore (India) to Baptist missionaries. He returned to the United States when he was 10-years old. He got a postgraduate degree at the University of Chicago, in 1908. He spent some time teaching in the Philippines. After returning to the United States, he worked for Bell Telephone Laboratories from 1915, until his retirement in 1945.

Robert first isolated vitamin B_1 (**thiamine**) from rice polishings in 1933 and worked out its molecular structure in 1935. Its chemical synthesis was achieved soon thereafter, in 1936. The monies that Robert received from his patents on the chemical synthesis of thiamine were used to establish the **Williams–Waterman Fund for the Combat of Dietary Disease**, which supported nutritional research and field programmes to abolish malnutrition.

Among his awards were the **Elliott Cresson Medal** in 1940 and the **Perkin Medal** in 1947. A resident of Summit, New Jersey, Robert died there at the age of 79 on October 2, 1965. Robert R. Williams was the elder brother of **Roger John Williams**, who first isolated vitamin B_3 (pantothenic acid). Although both shared similar interests in the B-vitamins and nutrition, their research careers were independent of each other and never involved collaboration.

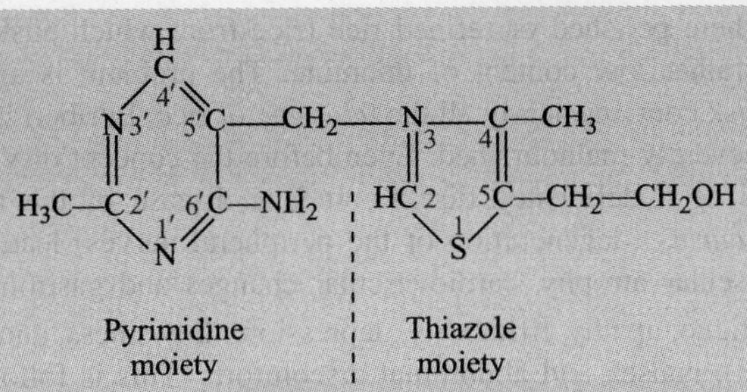

Fig. 39–2. Vitamin B_1 or thiamine or aneurin

D. Properties. Thiamine is a white crystalline substance, readily soluble in water, slightly so in ethyl alcohol but insoluble in ether and chloroform. Its odour resembles that of a yeast. The aqueous solution is optically inactive. Thiamine is destroyed at elevated temperature, unless the *p*H is low. It can stand short boiling up to 100°C. Hence, it is only partly lost in cooking or canning processes. Long boiling or boiling with alkali destroys it. But it is stable in acid medium. On oxidation, it produces *thiochrome*, which gives fluorescence.

E. Metabolism. The requirement of this vitamin is increased under high metabolic conditions such as fever, increased muscular activity, pregnancy and lactation and also under surgery and stress. A correlation also exists between the type of food taken and the vitamin B_1 requirement. Fats and proteins reduce while carbohydrates increase the amount of this vitamin required in the daily diet. Thiamine absorption decreases with gastrointestinal or liver disease.

Raw seafoods (*e.g.*, fishes and molluscs) contain an enzyme, *thiaminase* which destroys thiamine in the body. People consuming such foods may, therefore, reveal symptoms of thiamine deficiency. Thiaminase cleaves the thiamine molecule between the pyrimidine and thiazole rings.

If thiamine is administered in human body, a part of it is excreted or recovered in the urine and a part is converted to pyramin by the enzyme, thiaminase. Besides thiaminase, certain flavonoids of nonenzymic nature also work against thiamine. These have been shown to be present in ferns and certain higher plants.

Thiamine is phosphorylated with ATP to form thiamine pyrophosphate (TPP), which is also called diphosphothiamine (DPT).

$$\text{Thiamine} \xrightarrow{\text{ATP} \curvearrowright \text{AMP}} \text{Thiamine pyrophosphate}$$

TPP, in association with lipoic acid, forms the prosthetic group, *cocarboxylase* for the enzyme *carboxylase*. TPP participates in many reactions, such as decarboxylation of α-keto acids, notably pyruvic and α-ketoglutaric and transketolation.

F. Deficiency. Vitamin B_1 deficiency leads to polyneuritis in animals and beriberi in human beings.

Polyneuritis in birds renders them unable to fly, walk or even stand. Rats develop, among other symptoms, a brachycardia (slowing of the heart rate).

Beriberi (*beri*[Singhalese] = weakness, which here means *I cannot*, symbolizing the incapacitated condition created by thiamine deficiency) has been and continues to be a serious human health

CLINICAL IMPLICATIONS 39–1

Beriberi

Jacobus Bonitus, a Dutch physician working in Java, had first described beriberi, in 1630, as follows: "A certain very troublesome affliction, which attacks men, is called by the inhabitants Beri-beri (which means sheep). I believe those, whom this same disease attacks, with their knees shaking and the legs raised up, walk like sheep. It is a kind of paralysis, or rather Tremor: for it penetrates the motion and sensation of the hands and feet indeed sometimes of the whole body ..."

problem in Far East where polished or refined rice (rice from which husk has been removed) is eaten. The rice has a rather low content of thiamine. The problem is aggravated if the rice is polished because the husk contains nearly all the thiamine of rice. Beriberi is also occasionally seen in alcoholics who are severely malnourished. Even before the concept of vitamins was developed, beriberi was described as a deficiency disease. It is a disease of the nervous system and is characterized by *polyneuritis* (degeneration of the peripheral nerves) leading to partial *paralysis* of the extremities, muscular atrophy, cardiovascular changes and gastrointestinal disorders.

At first, there is fatigue, apathy, irritability, depression, drowsiness, anorexia (loss of appetite), insomnia (sleeplessness), nausea and abdominal discomfort. This is followed by symptoms like peripheral neuritis with tingling, burning paresthesias of the toes and feet; decreased tendon reflexes; loss of vibration sense; tenderness and cramping of leg muscles; congestive heart failure and psychic disturbances. There may be ptosis of the eyelids and atrophy of the optic nerve. Hoarseness due to paralysis of the laryngeal nerve is a typical sign. Muscular atrophy and tenderness of the nerve trunks are followed by ataxia, loss of coordination, and loss of deep sensation. Paralytic symptoms are more common in adults than in children. Finally, the major symptoms may follow one of the following 3 courses (and accordingly beriberi is of 3 types):

(*a*) Symptoms involving nervous system, causing ***dry beriberi*** **:** In it, the child may appear plump but is pale, flabby, listless and dyspneic; the heart beat is rapid and the liver enlarged.

(*b*) Symptoms associated with edema and effusions, leading to ***wet beriberi*** **:** In it, the child is undernourished, pale and edematous and has dyspnea, vomiting and tachycardia. The skin appears waxy. The urine may contain albumin and casts.

(*c*) Symptoms involving heart, resulting in ***acute pernicious beriberi*** **:** In it, the lesions may be found principally in the heart, peripheral nerves, subcutaneous tissue and serous cavities. The heart is enlarged, especially to the right and there is fatty degeneration of the myocardium. Generally edema or edema of the legs, serous effusions, and venous engorgement may be seen. Lesions in the brain include vascular dilatation and hemorrhage. Finally, death ensues due to heart failure.

Often the symptoms characteristic of more than one of these 3 types of beriberi appear simultaneously in individuals causing ***mixed beriberi***. Although beriberi is caused due to thiamine avitaminosis, it is usually associated with deficiencies of other vitamins. This is true of all vitamin B complex-deficiency conditions in man.

G. Human requirements. The daily recommended dietary allowances are 1.2–1.4 mg for men and 1.0 mg for women. Pregnant and lactating mothers, however, require up to 1.5 mg daily. The thiamine requirement for infants is between 0.2 and 0.5 mg daily.

H. Treatment. If beriberi occurs in breast-fed infant, both the mother and child should be treated with thiamine. In such cases, the daily dose for adults is 50 mg and for children 10 mg or more. Oral administration is effective until gastrointestinal disturbances prevent absorption. Thiamine should be instilled intramuscularly or intravenously to children with cardiac failure.

VITAMIN B_2

A. History. Riboflavin or vitamin B_2 was first isolated in 1879 from milk whey which is an essential dietary factor for rats. Since it was first isolated from milk, vitamin B_2 is also known as **lactoflavin.** Originally, it was also known as **ovoflavin** (*from eggs*) and **hepatoflavin** (*from liver*). Its synthesis was done by **Richard Kuhn** and **Paul Karrer.** It is popularly called as the **"yellow enzyme"** because of its colour.

B. Occurrence. In nature, it occurs almost exclusively as a constituent of one of the

RICHARD KUHN
(LT, 1900–1967)

A Austria-born German chemist who received his doctorate in 1922 on enzymes. He was awarded the **1938 Nobel Prize for Chemistry** for his work on carotenoids and vitamins but its acceptance was forbidden by the German government. After the war, he received the gold medal and diploma of the prize. His main contributions include:

1. the separation of carotenoids (α and β isomers) from carrots and thus isolated the third isomer (γ *carotene*)
2. the structure of *lycopene*
3. the synthesis of *riboflavin*

two flavin coenzymes, namely, flavin mononucleotide (FMN) and flavin adenine dinucleotide (FAD). Milk, cheese, eggs, liver, kidney, heart and brewer's yeast are excellent sources of this vitamin. Cow's milk contains about 5 times as much riboflavin as human milk. Leafy vegetables are good sources. They are usually richer in riboflavin than they are in thiamine. Fruits and most root vegetables contain moderate quantities. Whole grains, cereals and milled flour contain low riboflavin content. The riboflavin contents in cereals, however, increase strikingly during germination. The ordinary cooking processes do not affect the riboflavin content of the food. Roasted or boiled meat retains about 75% of the vitamin. *It is only very rarely that vitamin* B_2 *is present in free or uncombined state as in retina and spleen.* Fermentation residues from alcohol manufacture probably offer the richest large supplies.

C. Structure. Riboflavin ($C_{17}H_{20}N_4O_6$) belongs to a class of water-soluble pigments called *lyochromes.* A molecule of thiamine (Fig. 39–3) consists of a sugar alcohol, D-ribitol, attached to a chromogenic dimethyl isoalloxazine ring at position number 9.

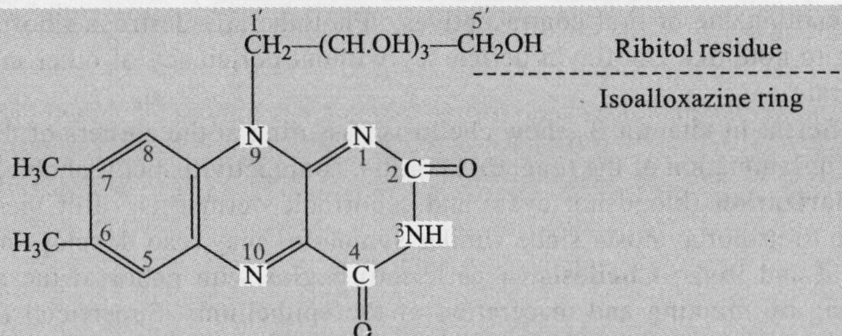

Fig. 39–3. Vitamin B_2 or riboflavin or lactoflavin
[6,7-dimethyl-9-(1′-D-ribityl) isoalloxazine]

D. Properties. Riboflavin is a bright orange-yellow crystalline powder. It is soluble in water and ethanol but insoluble in ether and chloroform. It is stable to heat and acids but is easily decomposed by alkalies and exposure to light. The aqueous solution exhibits yellow-green fluorescence. It stands ordinary cooking and canning. On exposure to light, the ribityl residue splits off, forming a compound *lumiflavin* in alkaline solution and *lumichrome* in acidic or neutral solution.

E. Metabolism. Riboflavin is synthesized by all green plants, most bacteria, yeasts and moulds. *Ashbya gossypii*, an yeast, produces it in such large amounts that riboflavin crystals are formed in the culture medium. *Animals have, so far, not been shown to synthesize riboflavin.* In man, the ingested riboflavin is largely passed out as such or as its coenzyme, the FMN.

Experiments with plant tissues have suggested that riboflavin and flavoproteins may play a significant role in phototropic curvature of various plant organs (**Galston**, 1950).

Riboflavin is essential for growth and tissue respiration; it may have a role in light adaptation and is required for conversion of pyridoxine to pyridoxal phosphate.

When riboflavin is phosphorylated in the presence of an enzyme, *flavokinase*, it gets converted to FMN which is essential in the biosynthesis of fats.

$$\text{Riboflavin} + \text{ATP} \xrightleftharpoons{\text{Flavokinase}} \text{FMN} + \text{ADP}$$

FMN may undergo a further reaction with ATP, in the presence of an enzyme found in yeast and animal tissues, to produce FAD. It is a chief constituent of electron transport system (ETS). A decrease in the amount of FAD, therefore, would severely hamper the efficiency of ETS.

$$\text{FMN} + \text{AMP} \rightleftharpoons \text{FAD} + \text{PP}$$

The coenzymes undergo reversible oxidation-reduction in the presence of their enzymes and a suitable substrate.

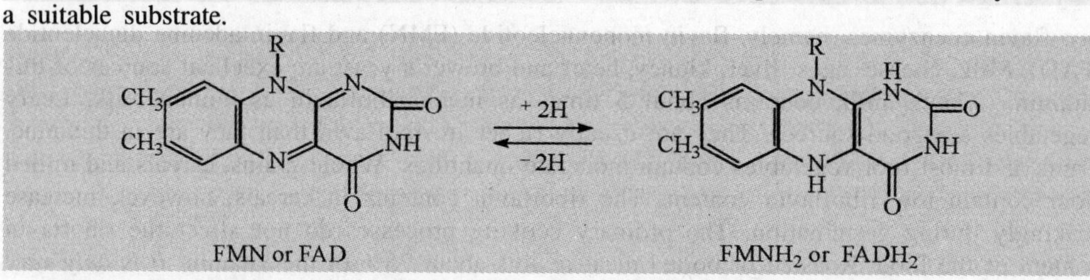

The flavoenzymes play a key role in cell metabolism. They function in accepting hydrogen atoms from reduced pyridine nucleotides. They have been shown to participate in the enzymic oxidation of glucose, fatty acids, amino acids and purines.

F. Deficiency. Riboflavin deficiency is usually caused by inadequate intake. Faulty absorption may contribute in patients with biliary atresia or hepatitis or in those receiving probenecid, phenothiazine or oral contraceptives. Phototherapy destroys riboflavin content. It is interesting to note that riboflavin deficiency without deficiency of other member of the B complex is rare.

Persons deficient in vitamin B_2 show cheilosis (fissuring at the corners of the mouth and lips), glossitis (inflammation of the tongue), keratitis, conjunctivitis photophobia, lacrimation, **corneal vascularization** (bloodshot eyes) and seborrheic dermatitis. But these symptoms are not specific to ariboflavinosis since similar symptoms may also develop in the absence of nicotinic acid and iron. **Cheilosis** (= perle′che) begins with pallor at the angles of the mouth, following by thinning and maceration of the epithelium. Superficial fissures often covered by yellow crusts develop in the angles of the mouth and extend radially into the skin for distances upto 2 cm. In **glossitis**, the tongue is smooth, and loss of papillary structure occurs. A normocytic and normochronic anemia with bone marrow hyperplasia is common.

However, patients suffering from pellagra and beriberi are usually also deficient in riboflavin content.

G. Human requirements. The minimum daily requirement of riboflavin varies from 0.6 to 1.7 mg for children and adults. During pregnancy and lactation, the women require up to 2.0 mg daily.

H. Treatment. Ariboflavinosis may be prevented by a diet that contains adequate amounts of milk, eggs, leafy vegetables, and lean meats. Treatment consists in oral administration of 3-10 mg of riboflavin daily. If no response occurs within a few days, intramuscular injections of 2 mg of riboflavin in saline solution may be administered 3 times in a day.

VITAMIN B_3

A. History. This was first isolated by **Roger J. Williams** in 1938 from yeast and liver concentrates. On account of its wide distribution, he named it as pantothenic acid (*pantos*[G] =

ROGER JOHN WILLIAMS

(LT, 1893–1988)

Roger Williams, an Indian-born American biochemist, was born in Ootacamund (India) of American missionary parents on August 14, 1893. At age 2, his family returned to the U.S., where he grew up in Kansas and California. Roger married **Hazel Elizabeth Wood** in 1916; they raised three children. After Hazel's death in 1952, he married **Mabel Phyllis Hobson** the next year. His formal education culminated in a Ph.D. degree (*magna cum laude*) from the University of Chicago in 1919. He taught at the University of Texas at Austin, beginning in 1940. While studying the nutrition of yeast cells, he discovered that *aneurin* (now named vitamin B_1 or thiamine), first described as a vitamin for humans, also promoted the growth of yeasts. This important finding opened the way for the use of microorganisms (yeast and bacteria) in nutritional research, which greatly accelerated their discovery and the analysis of their metabolic effects. Later again using yeast cells, Roger discovered and isolated vitamin B_3 and named it **pantothenic acid** (because of its wide distribution: *pentos*[G] = everywhere) and found a method for its chemical synthesis. Later, he concentrated vitamin B_9 and named it **folic acid** because it was first obtained from spinach leaves (*folium*[L] = leaf).

At the University of Texas, he founded and directed the Clayton Foundation Biochemical Institute (now called the **Biochemical Institute**) from 1940 to 1963, when he retired from that position. More vitamins and their variants were discovered in this laboratory than in any other laboratory in the world. Following his retirement from the Directorship of the Institute, he concentrated on human nutrition as his central field of interest, while writing several widely-read books. In 1971, he became Emeritus Professor of Chemistry at the University of Texas and retired from that position in 1986, at age 92. After his 80th birthday (in 1973), he continued to be actively involved, writing and editing several important books dealing with aspects of human nutrition and education. In his brief 1954-autobiography, Roger wrote that he considered his most important to be the study of individual human differences. His ideas on this subject are embodied in the books *The Human Frontier* and *Free and Unequal*, as well as his later book *Alcoholism: The Nutritional Approach*.

Common differences in nutritional needs formed the basis of a **vitamin formula** Roger developed to diminish the craving for alcohol in people biochemically susceptible to alcoholism. In a course he taught, 'The Biochemical and Physiological Bases of Individual Human Differences', Roger recounted his experience after publishing a book on this research. Because he had written that alcoholic people who got their individual nutritional needs fully met could drink socially without bingeing, he was assailed by people associated with *Alcoholics Anonymous*. Humble by nature, Roger allowed himself to be convinced that the social side of alcoholism was also of major importance. He withdrew his book and destroyed the printing. Roger never attempted to profit from this research; he gave this vitamin formula away to a number of pharmaceutical companies. A version modified by one of his former colleagues, **Dr. Donald R. Davis**, is sold under the name *Vitamin & Mineral Insurance Formula,* by Bronson Laboratories.

In 1941, for his discovery of pantothenic acid, he received **Mead Johnson Award of the American Institution of Nutrition** and the Chandler Medal of Columbia University. Prof. Williams was a member of the National Academy of Sciences, and was **President of the American Chemical Society** in 1957. He received honorary D.Sc. degrees from Columbia University, Oregon State University, and the University of Redlands, his alma mater. In 1972, he served as a member of President Nixon's Advisory Panel on Heart Disease.

Williams was an avid fisherman, golfer, walker and fan of University of Texas at Austin athletics. He died of pnemonia in an Austin nursing home on February 20, 1988, at age 94, and is buried in Austin Memorial Park. His papers are in the University of Texas archives. Roger J. Williams was the younger brother of **Robert Runnels Williams**, who first established the structure of vitamin B_1 (thiamine). Although both shared similar interests in the B-vitamins and nutrition, their research careers were independent of each other and never involved collaboration.

everywhere). The coenzyme form of this vitamin (coenzyme A or CoA-SH) was isolated and its structure determined by **Fritz A. Lipmann.** The chemical synthesis of this coenzyme was, however, described by **Khorana** in 1959. This vitamin is sometimes called as **filtrate factor** or the **yeast factor**.

B. Occurrence. Although widespread in nature, yeast, liver and eggs are the richest sources of it. The vegetables (potatoes, sweet potatoes, cabbage, cauliflower, broccoli) and fruits (tomatoes, peanuts) and also the skimmed milk, wheat bran, whole milk and canned salmon are some of the less important sources. In most animal tissues and microorganisms, it occurs as its coenzyme.

C. Structure. Pantothenic acid ($C_9H_{17}NO_5$) is an amide of pantoic acid (α, γ-dihydroxy-β, β-dimethyl butyric acid) and β-alanine (Fig. 39–4).

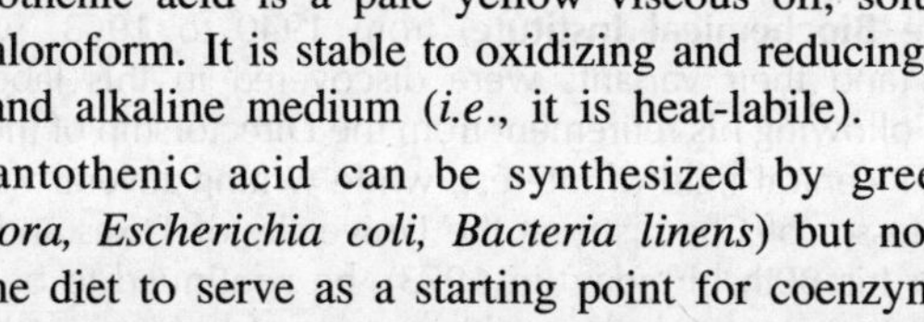

Fig. 39–4. Vitamin B_3 or pantothenic acid
(Pantoyl-β-alanine)

D. Properties. Pantothenic acid is a pale yellow viscous oil, soluble in water and ethyl acetate but insoluble in chloroform. It is stable to oxidizing and reducing agents but is destroyed by heating in an acidic and alkaline medium (*i.e.*, it is heat-labile).

E. Metabolism. Pantothenic acid can be synthesized by green plants and various microorganisms (*Neurospora, Escherichia coli, Bacteria linens*) but not by mammals. Hence, this must be present in the diet to serve as a starting point for coenzyme A (CoA). Coenzyme A is richly found in the liver and in poor quantities in the adrenals. There may be as much as 400 mg of CoA per kilo of liver. It functions in acetylation reactions. In order to be effective, CoA must be present in the form of acetyl-CoA. It may arise in many ways but the most common way of its production is that CoA, in the presence of ATP, acetate and a suitable enzyme, is converted into acetyl CoA. The overall reaction may be shown in 3 steps:

1. ATP + Enzyme $\rightleftharpoons$ Adenylic acid-Enzyme + Pyrophosphate
2. Adenylic acid-Enzyme + CoA $\rightleftharpoons$ CoA-Enzyme + Adenylic acid
3. CoA-Enzyme + Acetate $\rightleftharpoons$ Acetyl-CoA + Enzyme

The acetyl groups may be transferred to an acetyl acceptor in the presence of a suitable acceptor. This may occur in two ways: either the acetyl group is attached to the accepting group at the carbonyl end (head reaction) or at the methyl end (tail reaction).

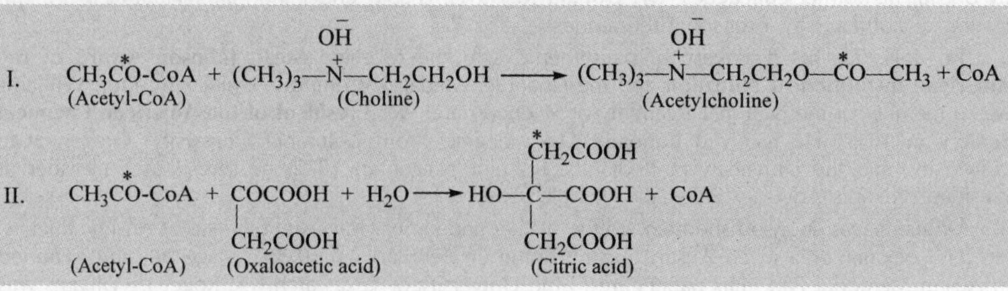

The only known metabolic fate of vitamin B_3 is its participation in the formation of the biologically important coenzyme A. It functions as a thioester of carboxylic acids.

F. Deficiency. A deficiency of pantothenic acid leads to depigmentation of the hair in rats, pigs and dogs and to depigmentation of feathers in chicks. Atrophy of the adrenal cortex with necrosis and hemorrhage may also occur in animals including rat. Corneal changes consisting of vascularization, thickening and opacity may be seen.

In human beings, no definite deficiency syndrome has been ascribed to pantothenic acid, probably because of the ubiquitous nature of this vitamin and because of the fact that a little amount of this vitamin can perhaps be synthesized in the body. Its correlation with **achromotrichia** (premature greying of the hair) has been described in the case of man, sometimes. But it seems too much to hope that grey hair can be averted by attention to diet ; rather it appears we must expect to go grey in spite of this vitamin.

G. Human requirements. The dietary allowance dose has not been officially worked out. Yet, 5—10 mg per day of vitamin B_3 has been suggested.

VITAMIN B_5

A. History. Vitamin B_5 refers to nicotinic acid and was named as **pellagra-preventive (PP) factor** by an Austrian-American physician of the U. S. Public Health Service, **Joseph Goldberger** (ca 1920) because of its curing action on pellagra (After Goldberger's death, vitamin B_5 was sometimes called **vitamin G** in his honour). The vitamin role on nicotinic acid was first recognized by **Conrad Elvehjem** and **Dilworth Wayne Woolley** of Wisconsin University in 1937. As this vitamin has a curing action against blacktongue disease in dogs, it is also called as **antiblacktongue factor.** It was first

Although **Dilworth Wayne Woolley** (LT, 1914–1966) was blind since young adulthood, undaunted, he pursued a science career that established him as one of the most prominent biochemists in the United States.

JOSEPH GOLDBERGER
(LT, 1874–1929)

Goldberger was born on July 16, 1874 in Giralt, Hungary. His parents emigrated to USA when he was 7-year old. He graduated in medicine in 1895 and stood first in the Intern Examination. In 1899, he joined the U.S. Public Health Service. For about 15 years till 1914, Goldberger worked on several infectious diseases such as yellow fever, typhoid fever, dengue fever and typhus fever. In 1914, he was asked to undertake an investigation on **pellagra** which was widely prevalent in the southern states of USA. He proceeded in his careful and methodical way to study the diets of persons suffering from pellagra and those free from the disease. He found the diets consumed by persons suffering from pellagra consisted mostly of biscuits, hominy grits, corn meal syrup and fat pork with little or no milk, meat, fish and eggs. Next, Goldberger studied the effect of addition of milk and meat for 2 years in orphanages on the incidence of pellagra. He made the remarkable discovery that all subjects suffering from pellagra were cured and no new cases occurred during the 2-year period among the subjects receiving milk and meat. He, thus, proved beyond doubt that pellagra is caused by faulty diet and could be cured by addition of milk and meat to the diet. In 1920, he reached the conclusion that the pellagra-preventing power of milk and meat was due to the correction of a deficiency of amino acids or some unknown vitamin or a combination of the two.

It is really remarkable that as early as 1921, Goldberger's associate, **Tanner** found that tryptophan (Trp) was effective in curing pellagra. In 1925, Goldberger and Tanner found that dried brewyer's yeast could cure and prevent pellagra and they called the curative principal as **pellagra-preventive factor** (P-P factor), which is identified as vitamin B_5. They found, using dogs as experimental animals, that the disease *blacktongue* in dogs produced by feeding on a maize diet was cured by brewyer's yeast. In 1928, Goldberger published a paper on the pellagra-preventive value of some common foods, determined using dog as the experimental animal. By the end of 1928, he fell ill and died on January 17, 1929 at the early age of 55.

ARNOLD CONRAD ELVEHJEM

(LT, 1901–1962)

Elvehjem was born on May 27, 1901 in McFarland, Wisconsin. He majored in agricultural chemistry. He became a teaching assistant immediately after obtaining bachelor's degree in 1923 and remained in that position until he received his Ph.D. degree in 1927. His graduate work was done under **Prof. Edwin B. Hart** (LT, 1874–1953). In 1929, he went to Cambridge and worked on cellular enzymes for a year in the laboratory of **Prof. F.G. Hopkins** (LT, 1861–1947). In 1930, Elvehjem was appointed as assistant professor at the University of Wisconsin. He was promoted as associate professor in 1932 and professor in 1936. He continued his work on **trace elements**– iron, copper, zinc, manganese and molybdenum, and on **B-complex vitamins**. In 1937, he made the great discovery of **isolating nicotinamide** from liver and identifying it as the antipellagra vitamin. Another discovery he made was that tryptophan served as a precursor of nicotinic acid, and tryptophan and nicotinic acid were interchangeable in the nutrition of animals. This work led to the discovery of **amino acid imbalance.** Elvehjem and **Krehl** reported, in 1955, that addition of gelatin to 9% casein diet lacking in nicotinic acid (= niacin) resulted in a severe depression in growth of rats that was prevented by tryptophan or nicotinic acid.

Elvehjem was the author or coauthor of over 800 research publications and 89 graduate students obtained their Ph.D. degrees partly or wholly under his direction. He was a member of many National and International Expert Committees on Nutrition. When Prof. Hart retired, Elvehjem was appointed as Chairman of the Department. Two years later, he was made the Dean of the Graduate School. In 1958, Dean Elvehjem became the President of the University of Wisconsin. He proved a great success in this position. He died suddenly in Madison on July 27, 1962.

isolated by **Funk** in 1911. As the name `nicotinic acid' might mislead some people into thinking that tobacco is nutritious, nicotinic acid has been given the alternative official name *niacin* for public use.

B. Occurrence. Nicotinic acid is widely distributed in nature in plant and animal tissues mainly as its amide called nicotinamide (commercially called **niacinamide**, to avoid any misassociation with the alkaloid nicotine of tobacco). It is an essential component of mammalian diet. As dietary tryptophan can be converted, in restricted quantities, to niacin in the body, it can partially substitute for niacin, although other sources of vitamin B_5 are necessary. *Niacin is most abundantly found in yeast.* Liver, lean pork, salmon, poultry and red meat, and also fortified white bread, groundnut and sunflower meals are also good sources, but most cereals contain only small amounts of it. Most vegetables and fruits are poor sources of it. Milk, cheese and eggs, which contain very little or practically no niacin, are good pellagra-preventive foods because of their high content of tryptophan. Since a number of stable vegetable articles of diet are not particularly rich in nicotinamide, the vegetarian's diet may be lacking in this vitamin. Nicotinamide, is present as a constituent in two pyridine nucleotide coenzymes namely NAD and NADP (previously called as DPN and TPN respectively). Since niacin is stable to heating and oxidation, there are only small losses in cooking. Like thiamine, most of vitamin B_5 is lost in the milling process. It can be prepared by oxidizing nicotine with a variety of agents, or, more cheaply, from pyridine or quinoline.

C. Structure. *Niacin* ($C_6H_5O_2N$) *is simplest of all the known vitamins.* It is a pyridine derivative (Fig. 39–5).

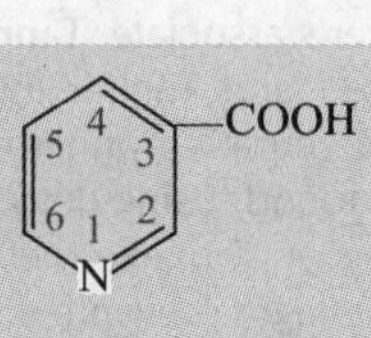

Nicotinic acid or **niacin**
(pyridine-3-carboxylic acid)

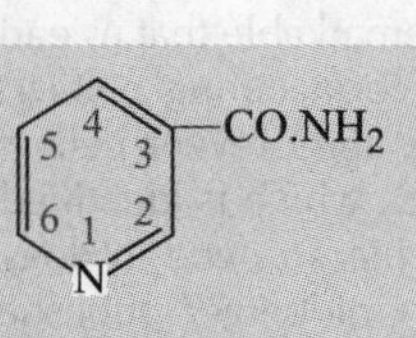

Nicotinamide or **niacinamide**
(pyridine-3-carboxylic acid amide)

Fig. 39–5. Vitamin B_5 and its amide

D. Properties. Niacin is a white crystalline substance. It is soluble in ethyl alcohol but is less soluble in ether and benzene than nicotinamide. It is heat-stable. Nicotinamide, when pure, occurs as white needle-like crystals. It is soluble in water and is stable in air and heat.

E Metabolism. The conversion of niacin to niacinamide takes place in the kidney and brain slices and also in the liver slices, if glutathione is present. Nicotinamide is synthesized by amidation of nicotinic acid adenine dinucleotide and subsequent degradation of NAD thus formed.

The niacin in man and other animals is derived from the amino acid tryptophan, which also cures pellagra. The conversion of tryptophan to nicotinic acid in the body takes place through a series of intermediate steps, which are represented below:

Tryptophan ⟶ Kynurenine ⟶ 3-hydroxykynurenine

3-hydroxyanthranilic acid ⟶ ... ⟶ Quinolinic acid ⟶ Nicotinic acid

Nicotinamide undergoes methylation in mammalian liver to produce N′-methyl nicotinamide which is oxidized to give corresponding 6-pyridone. In many plant seeds, nicotinic acid is, however, converted to trigonelline.

The nicotinic acid and its amide both are necessary for the growth of various microorganisms. Pyridine-3-sulfonic acid and its amide (Fig. 39–6) both prevent such growth which can be resumed by the addition of these vitamins. The relationship of these growth inhibitors to the two vitamins is not much different with the relationship of *p*-aminobenzoic acid to sulfanilamide (refer page 460).

Pyridine-3-sulfonic acid **Pyridine-3-sulfonamide**

Fig. 39–6. Antagonists of vitamin B_5 and its amide

The two coenzyme forms of this vitamin, NAD and NADP, carry out 2 important functions in the tissues:

(*a*) Oxidation of alcohols, aldehydes, amino acids and hydroxy-carboxylic acids

(*b*) Reduction of the flavin coenzymes

F. Deficiency. A deficiency of niacin causes pellagra in man and blacktongue in dogs. **Pellagra** (of Italian origin, *pellis* = skin; *agra* = rough) is characterized by 3 "Ds", namely **d**ermatitis of the exposed parts, **d**iarrhea and **d**ementia. The early symptoms of pellagra are vague. Anorexia, lassitude, fatigue, burning sensations, numbness and dizziness may be prodromal symptoms. Their manifestation in children who have parasites or chronic disorders may be particularly severe. The most characteristic manifestations are the cutaneous ones, which may develop abruptly or insidiously and may be elicited by irritants, esp., by intense light. They first appear as symmetric erythema of the exposed surfaces that may resemble sunburn. The lesions are usually sharply demarcated from the healthy skin around them, and their distribution may change very often. The lesions on the hands sometimes have the appearance of a glove (*pellagrous glove*), and similar demarcations are sometimes seen on the foot and leg (*pellagrous boot*) or around the neck (*Casal necklace*). The healed parts of the skin may remain pigmented. The cutaneous lesions are sometimes preceded by stomatitis, glossitis, vomiting or diarrhea. Swelling and redness of the tip of the tongue and its lateral margins may be followed by intense redness of the entire tongue and of the papillae and even ulceration. Nervous symptoms include depression, disorientation, insomnia and delirium. The histologic changes in the nervous system occur relatively late in the disease and consist of patchy areas of demyelinization and degeneration of ganglion cells.

The classic symptoms of pellagra are usually not pronounced in infants and children. Anorexia, irritability, anxiety and apathy are common in "*pellagra families*". They may also have sore tongues and lips and the skin is usually dry and scaly. Diarrhea and constipation may alternate and a moderate secondary anemia may occur. Pellagral children often have symptoms characteristic of other nutritional deficiency diseases. As coffee (*Coffea arabica*) is particularly rich in niacin, the heavy coffee drinkers usually do not develop pellagra. Other factors like thiamine-deficiency also seem to be responsible for this disease. Unbelievably, more than 1,20,000 cases of pellagra were reported in the United States in 1927, mainly among poor people with unvaried diets. Incredible as it may seem, over 600 deaths were attributed to pellagra, in 1948. Pellagra is greatly aggravated in persons kept on a corn diet (as natives of Africa) because corn is very much deficient of tryptophan.

The canine **blacktongue disease** leads to complete loss of appetite. The inner surfaces of the lips and cheeks develop pustules ; the pustules may also develop on the thorax and abdomen. Intensive salivation and bloody diarrhea are other symptoms.

G. Human requirements. The recommended daily allowance of nicotinic acid is between 8 and 15 mg for children, between 15 and 20 mg for men and between 13 and 15 mg for women. Pregnant and lactating mothers may require up to 20 mg daily.

H. Treatment. Children respond quickly to antipellagral therapy. A well-balanced diet should be augmented with 50-300 mg/day of niacin; 100 mg may be given intravenously in acute cases or in cases of poor intestinal absorption. The diet should be supplemented with other vitamins, especially with other members of B complex group. Sun exposure should be avoided during the active phase; the skin lesions may be covered by applying soothers. The diet of the cured pellagrin should be supervised continuously to prevent recurrence.

VITAMIN B_6

A. History. The name vitamin B_6 was suggested by **Albert Szent-Györgyi** (1934) to designate substances, other than thiamine and riboflavin, which cured a dermatitis (acrodynia) in rats. It was, henceforth, also named as **adermin** or **antidermatitis factor**. Vitamin B_6 group includes 3 compounds: pyridoxine, pyridoxal and pyridoxamine. Pyridoxine was first isolated, in 1938, from yeast and liver. Later, **Snell** (1942) discovered the other two compounds.

B. Occurrence. The B_6 vitamins are widely distributed in nature in plant and animal tissues. They are especially rich in cereals (wheat, rice), peas, turnip greens, brussels sprouts, carrots, potatoes, sweet potatoes, bananas, avocados, watermelons and yeasts. B_6 vitamins are also found in egg yolk, salmon, chicken, fish, beaf, pork and liver. Pyridoxine is adequately available in human and cow's milk. Pyridoxal (PAL) and pyridoxamine (PAM) also occur in nature as their coenzymes, namely, pyridoxal phosphate (PALP) and pyridoxamine phosphate (PAMP), respectively.

C. Structure. All the 3 forms of vitamin B_6 (Fig. 39–7) are derivatives of pyridine, C_5H_5N and differ from each other in the nature of substituent at position 4 of the ring. All the 3 forms are readily interconvertible biologically.

D. Properties. Pyridoxine ($C_8H_{11}O_3N$) is a white crystalline substance and is soluble in water and alcohol and slightly so in fat solvents. It is sensitive to light and ultraviolet irradition. It is resistant to heat (*i.e.*, heat-stable) in both acidic and alkaline solutions but its two allies pyridoxal and pyridoxamine are destroyed at high temperatures (*i.e.*, heat-labile).

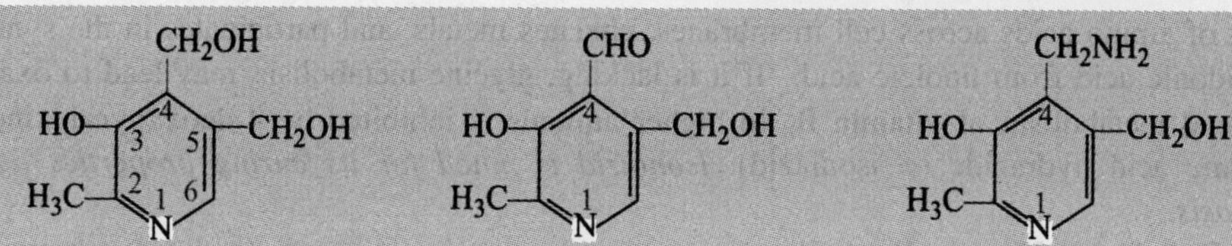

Pyridoxine [2-methyl-3-hydroxy-4, 5-di (hydroxymethyl)-pyridine] **Pyridoxal** **Pyridoxamine**

Fig. 39–7. Vitamins of B_6 group

E. **Metabolism.** The various forms of vitamin B_6 serve as growth factors to a number of bacteria. In addition, the 3 forms (pyridoxine, pyridoxal, pyridoxamine) are converted to pyridoxal-5-phosphate (Fig. 39-8), which acts as a coenzyme in various enzymic reactions involved in amino acid metabolism such as transamination, decarboxylation and racemization and in the metabolism of glycogen and fatty acids. It is also essential in the metabolism of hydroxy amino acids, sulfur-containing amino acids and also tryptophan.

Pyridoxal or its phosphate derivative also possibly acts as a carrier in the active transport of amino acids across cell membranes.

Pyridoxine can be converted to either pyridoxal or pyridoxamine (Fig. 39–8) but neither of them can be changed to pyridoxine. All these three can be detected in the urine after ingestion although 4-pyridoxic acid is the most important excretion product quantitatively. It is for this reason that when administered in the human body, about 90% of pyridoxine is oxidized to pyridoxic acid and excreted in human urine in this form.

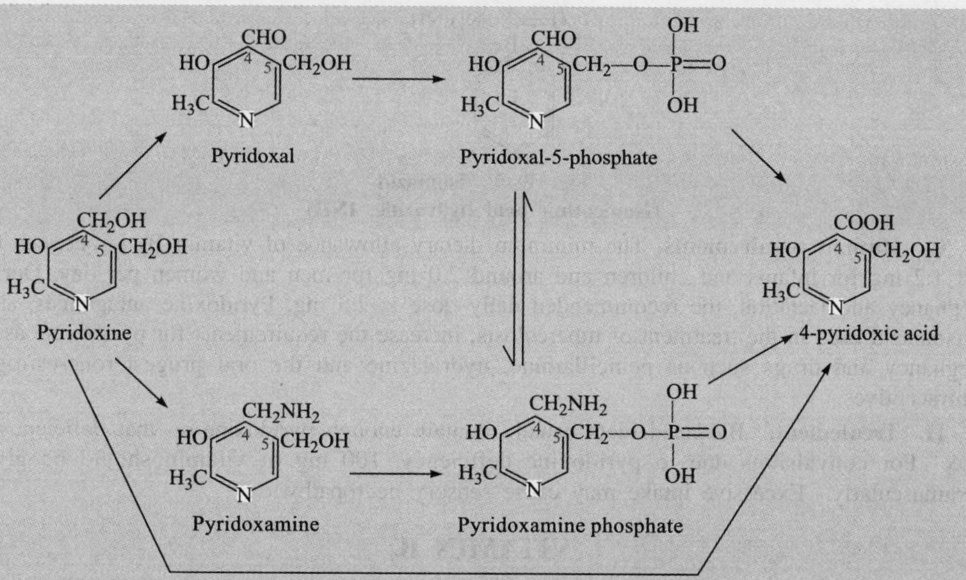

Fig. 39–8. Interrelationship between pyridoxine and its derivatives

B_6 vitamins are also essential for the breakdown of kynurenine. When this does not happen, xanthurenic acid appears in the urine. In addition, adequate functioning of the nervous system depends on pyridoxine, deficiency of which leads to seizures and to peripheral neuropathy.

Pyridoxal phosphate (PALP) is the coenzyme for both glutamic decarboxylase and γ-amino-butyric acid transaminase; each is essential for normal brain metabolism. It participates in active transport of amino acids across cell membranes, chelates metals, and participates in the synthesis of arachidonic acid from linoleic acid. If it is lacking, glycine metabolism may lead to oxaluria.

Normal metabolism of vitamin B_6 in higher animals is inhibited by 4-deoxypyridoxine and isonicotinic acid hydrazide (= isoniazid). *Isoniazid is noted for its curing properties against tuberculosis*.

F. Deficiency. Vitamin B_6 deficiency or apyridoxosis in rats leads to the development of acrodynia, a disease of dermatitis on ears, mouth and tail and accompanied by edema and scaliness of these structures. Dogs and chick develop anemia and nervous lesions in apyridoxosis.

In human infants, vitamin B_6 deficiency results in **convulsions, anemia, dermatitis** and **gastrointestinal disorders** such as nausea and vomiting. However, this deficiency is rare. Moreover, tryptophan metabolism is also disturbed. In adults, the vitamin B_6 deficiency is normally not found because the intestinal bacteria are capable of synthesizing vitamin B_6.

In B_6-deficient anemia, the RBCs are microcytic and hyperchromic. There are increased serum iron concentrations, saturation of iron-binding protein, hemosiderin deposits in bone marrow and liver, and failure of iron utilization for hemoglobin synthesis.

Diseases with malabsorption, such as celiac syndrome, may contribute to vitamin B_6 deficiency.

A syndrome resembling vitamin B_6 deficiency, as observed in animals, has also been reported in man during the treatment of tuberculosis with high doses of the drug *isoniazid* (Fig. 39–9). Only 2-3% of patients receiving conventional doses (2–3 mg/kg) of isoniazid developed neuritis; 40% of patients receiving high doses (20 mg/kg) developed neuropathy. The symptoms were alleviated by the administration of pyridoxine. Thus, 50 mg of pyridoxine per day completely prevented the development of neuritis. It is believed that isoniazid forms a hydrazone complex with pyridoxal, resulting in partial activation of the vitamin. *Isoniazid, thus, is a potent antagonist of vitamin B_6.*

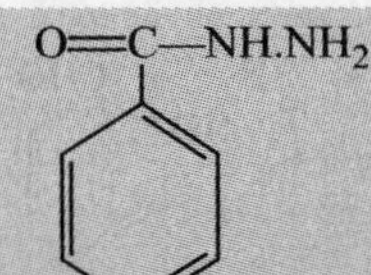

Fig. 39–9. Isoniazid (Isonicotinic acid hydrazide, INH)

G. Human requirements. The minimum dietary allowance of vitamin B_6 is between 0.2 and 1.2 mg for infants and children and around 2.0 mg for men and women per day. During pregnancy and lactation, the recommended daily dose is 2.5 mg. Pyridoxine antagonists, such as isoniazid used in the treatment of tuberculosis, increase the requirements for pyridoxine as do pregnancy and drugs such as penicillamine, hydralazine and the oral progesterone-estrogen contraceptives.

H. Treatement. Balanced diets usually contain enough pyridoxine so that deficiency is rare. For convulsions due to pyridoxine deficiency, 100 mg of vitamin should be given intramuscularly. Excessive intake may cause sensory neuropathy.

VITAMIN B_7

A. History. In 1935, **Fritz Kögl**, a Dutch biochemist, isolated in crystalline form from 250 kg of dried egg yolks about 1 mg of a *'bios'* factor (growth promoting factor) necessary for yeast and named it as "biotin". Four years later, **Szent-Györgyi** *et al* conclusively proved that biotin is synonymous to the **"antiegg white injury factor"** which is responsible for the cure of egg white injury, induced in rats and other animals by feeding them with raw egg white. The raw egg white contains a biotin-antagonist protein, *avidin*, which combines with biotin in a firm linkage to form a compound that cannot be absorbed by the intestine and is, therefore, excreted. It is also called as **coenzyme R** because it is a growth factor for the nitrogen-fixing bacterium, *Rhizobium.*

B. Occurrence. Biotin has a wide range of distribution both in the animal and the vegetable kingdoms. Yeast, liver, kidney, milk and molasses are among the richest sources; peanuts and eggs have lesser amounts. Biotin occurs in nature usually in combined state as biocytin (Fig. 39–10). It is a bound form of biotin, linked as a peptide with the amino acid lysine.

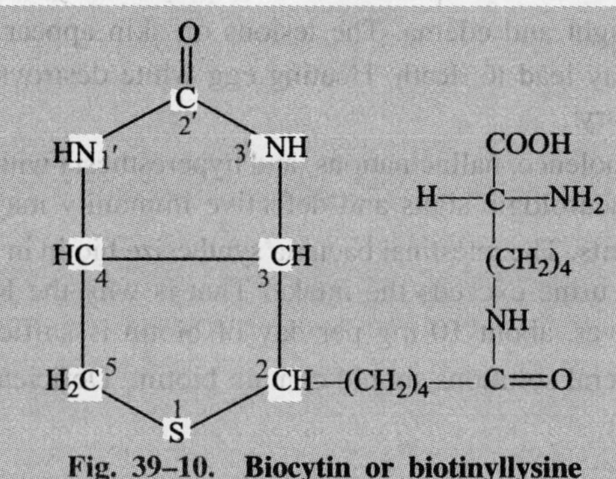

Fig. 39–10. Biocytin or biotinyllysine
(ε-N-biotinyl-L-lysine)

C. Structure. The structure of biotin ($C_{10}H_{16}O_3N_2S$) was worked out by **Vincent du Vigneaud** in 1942. Biotin (Fig. 39–11) has an unusual structure and consists of a fused imidazole and thiophene ring with a fatty acid side chain. Two forms of biotin can exist, *allobiotin and epibiotin.* Biotin is optically active. Only the (+) biotin is active; the DL-biotin is half as active as the naturally occurring biotin. The oxybiotin, in which S atom of biotin is replaced with an O atom, has some activity. *Biotin and thiamine are the only sulfur-containing vitamins isolated to date.*

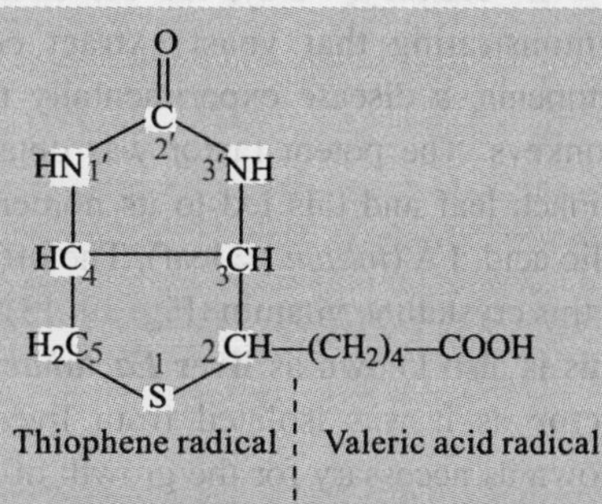

Fig. 39–11. Vitamin B_7 or biotin or vitamin H
(2′-keto-3,4-imidazolido-2-tetrahydrothiophene-*n*-valeric acid)

D. Properties. Biotin crystallizes as long needles. It is soluble in water and ethyl alcohol but is insoluble in chloroform and ether. It is heat-stable and is resistant to both acids and alkalies. It has a melting point of 230°C.

E. Metabolism. This vitamin serves as a prosthetic group for many enzymes. These biotin-containing enzymes catalyze the fixation of CO_2 into organic molecules, thus bringing about carboxylation. The carbon dioxide is carried as a carboxyl group attached to one of the ureidonitrogen atoms of biotin, forming *N-carboxybiotin complex* (Fig. 39–12). They also bring about synthesis of fatty acids such as oleic acid.

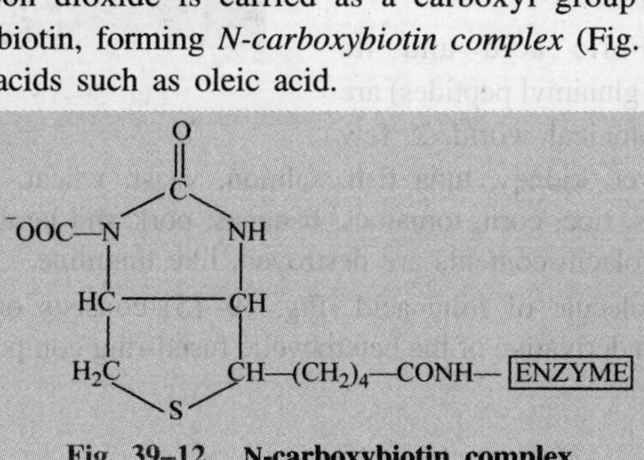

Fig. 39–12. N-carboxybiotin complex

F. Deficiency. In most animals including man, intestinal bacteria synthesize appreciable amounts of biotin. It is because of this reason that biotin-deficiency in human beings, fed on biotin-free diets, cannot be produced. However, biotin-deficiency may be induced by sterilization of intestine and by feeding with raw egg white. Avidin, the egg white protein, inactivates biotin by eliminating it from an otherwise complete diet. Such a deficiency in man leads to **dermatitis**, loss of hair, decrease in weight and edema. The lesions on skin appear with changes in posture and gait. These disorders may lead to death. Heating egg white destroys the avidin and prevents the so-called egg white injury.

Brawny dermatitis, somnolence, hallucinations, and hyperesthesia with accumulation of organic acids are common. Other neurologic signs and defective immunity may occur.

G Human requirements. The intestinal bacteria synthesize biotin in such appreciable amounts that the amount excreted in urine exceeds the intake. That is why the RDA for this vitamin has not been established. However, about 10 mg per day of biotin is sufficient for an adult.

H. Treatment. Parenteral solutions should contain biotin. Deficient patients respond to oral administration of 10 mg.

VITAMIN B_9

A. History. Day, for the first time, showed the existence of this nutritional factor by demonstrating that yeast extract could cure cytopenia, a disease experimentally induced in monkeys. The potent factor was obtained from spinach leaf and this led to its nomenclature as folic acid, FA (*folium*^{L} = leaf). The official name of this crystalline vitamin (Fig. 39–13) is **folacin.** This is also known as **liver *Lactobacillus casei* factor** as it was isolated from liver and was shown as necessary for the growth of lactic acid bacteria. **Hogan** called this as vitamin B_c.

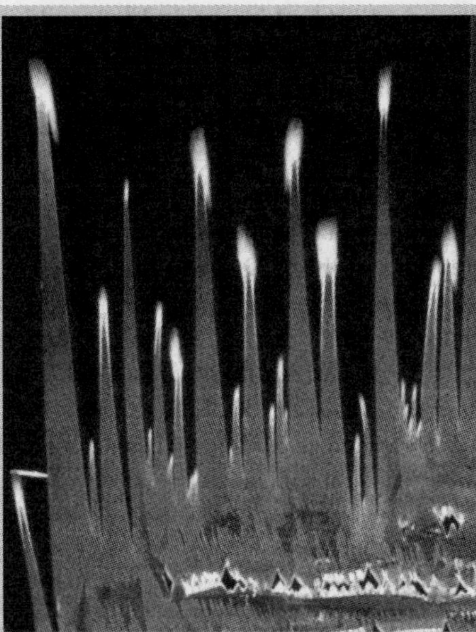

Fig. 39–13. Folic acid crystal

However, a number of other compounds or factors (Fig. 39–14), having similar or different biochemical functions but closely related to folic acid, were isolated from different sources. These are *fermentation Lactobacillus casei factor, Streptococcus lactis R (SLR) factor, B_c conjugate and citrivorum factor, CF.*

B. Occurrence. Folic acid and its derivatives (tri- and hepta-glutamyl peptides) are widely distributed in biological world. A few important sources are liver, kidney, tuna fish, salmon, yeast, wheat, dates and spinach. Root vegetables, sweet potatoes, rice, corn, tomatoes, bananas, pork and lamb contain little folid acid. With improper cooking, folacin contents are destroyed, like thiamine.

C. Structure. A molecule of folic acid (Fig. 39–15) consists of 3 units: glutamic acid, *p*-aminobenzoic acid and a derivative of the heterocyclic fused-ring compound pterin. Its molecular formula is $C_{19}H_{19}O_6N_7$.

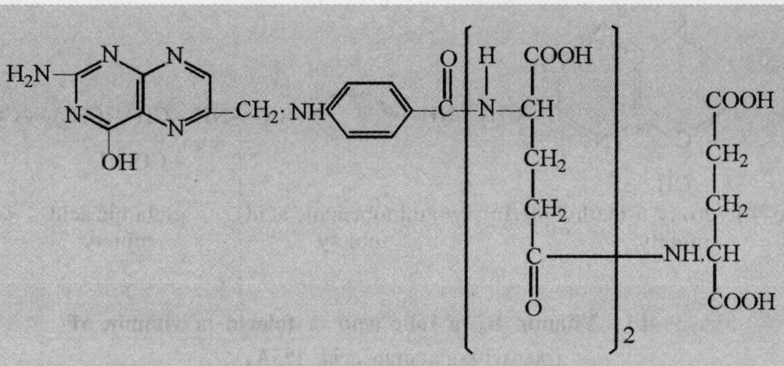

Fermentation *Lactobacillus casei* factor
(pteroyl-γ-glutamyl-γ-glutamylglutamic acid)

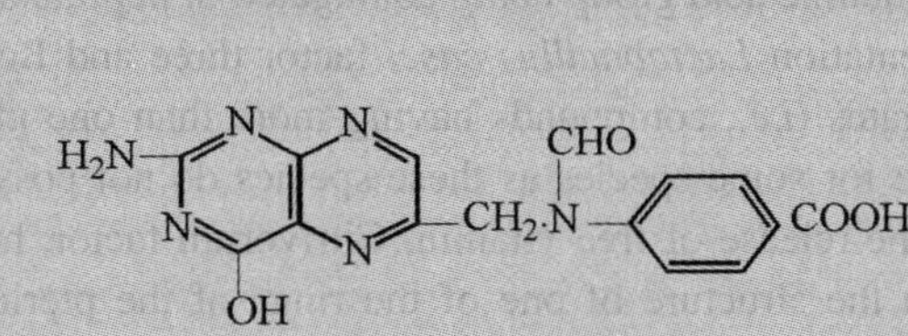

***Streptococcus lactis* R (SLR) factor or rhizopterin**
(N^{10}-formylpteroic acid)

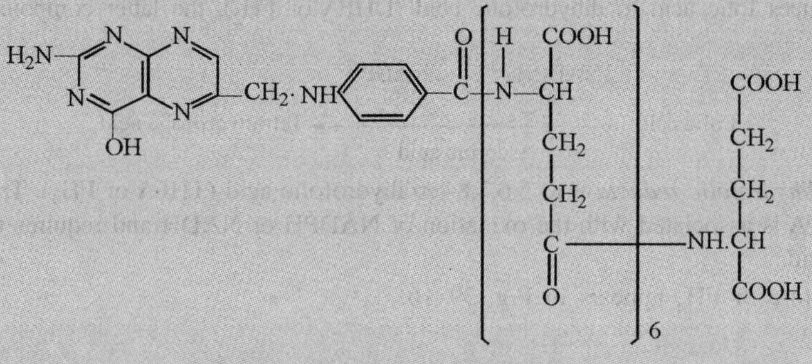

B_c conjugate
(pteroylhexaglutamylglutamic acid)

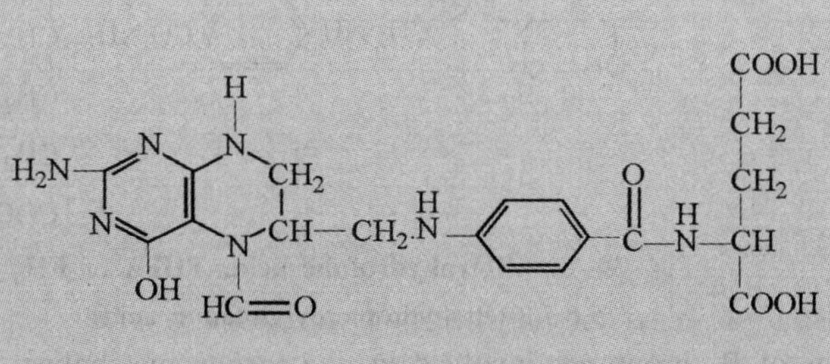

Citrivorum factor, CF or folinic acid
(5-formyl-5,6,7,8-tetrahydropteroylglutamic acid or N^5-formyl-FH_4)

Fig. 39–14. The four structural homologues of vitamin B_9

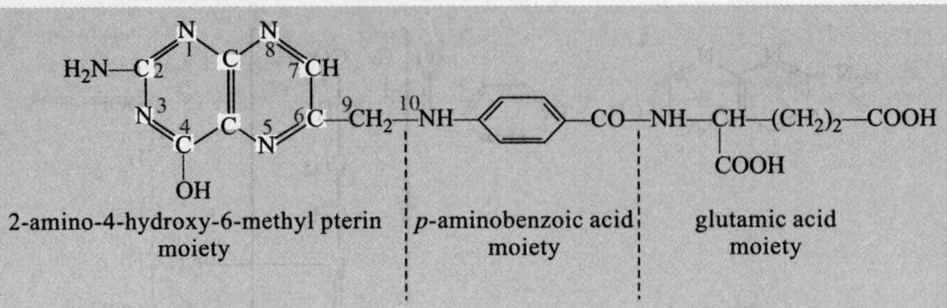

Fig. 39–15. Vitamin B_9 or folic acid or folacin or vitamin M
(pteroyl-L-glutamic acid, PGA)

The various vitamins of B_9 group differ from each other in the number of glutamic acid groups present; the additional glutamic acid group being conjugated in peptide linkages. For example, folic acid contains one, fermentation *Lactobacillus casei* factor three and Bc conjugate seven glutamic acid groups. The conjugates (*i.e.*, compounds having more than one glutamic acid groups in the molecule) are ineffective for some species as these species do not possess the enzyme *conjugase* which is necessary for the release of free vitamin. Citrivorum factor, however, differs from other vitamins of B_9 group in the structure of one of the rings of the pterin moiety.

D. Properties. Folic acid is a yellow crystalline substance, slightly soluble in water but insoluble in fat solvents. It is stable to heat in alkaline or neutral solutions only. It is inactivated by sunlight.

E. Metabolism. The reduction products of folic acid act as coenzymes. An enzyme, *folic reductase*, reduces folic acid to dihydrofolic acid (DHFA or FH_2), the latter compound is further

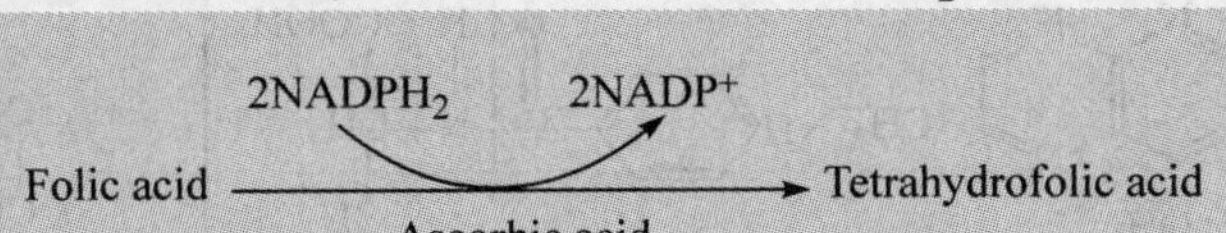

reduced by *dihydrofolic reductase* to 5,6,7,8-tetrahydrofolic acid (THFA or FH_4). The formation of FH_4 from FA is associated with the oxidation of NADPH or NADH and requires the presence of ascorbic acid.

The structure of FH_4 appears in Fig. 39–16.

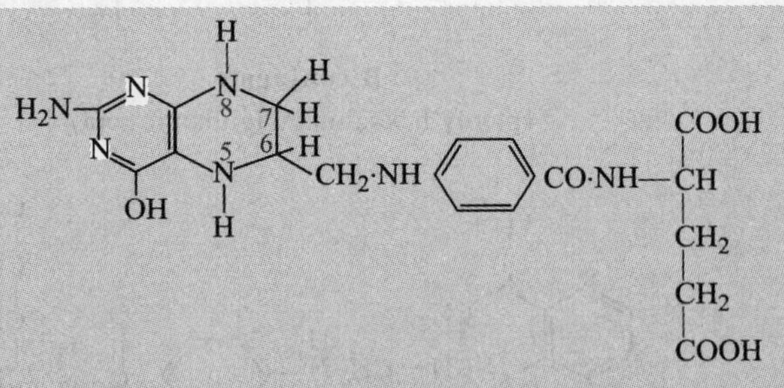

Fig. 39–16. Tetrahydrofolic acid, THFA or FH_4
(5,6,7,8-tetrahydropteroylglutamic acid)

The vitamins of B_9 group are involved in *one-carbon* metabolism in a way similar to the *two-carbon* metabolism in which CoA is involved. THFA acts as an acceptor of a one-carbon unit either from formate (in which case 5,6,7,8-tetrahydrofolic acid is formed) or from formaldehyde (in which case 5-hydroxymethyl-5,6,7,8-tetrahydrofolic acid is formed). THFA is also involved in the transfer of the methyl group and in the utilization of single carbons (formate) in the synthesis of serine, methionine, thymine, purines, choline and inosinic acid.

Folic acid, in conjunction with ascorbic acid, also appears to be related to tyrosine metabolism.

Citrivorum factor (CF) is 5 formyl derivative of tetrahydrofolic acid and is so named because it supports the growth of *Leuconostoc citrivorum*. CF is about one thousand times more potent biologically than folic acid. Its chemical name is folinic acid. During the conversion of FA to CF, vitamin B_{12} and ascorbic acid are also required. Citrivorum factor (and also folic acid to a lesser degree) are concerned in the production of an agent that stimulates the formation of normal RBCs.

It is interesting to note that the bacteria, which require PABA for growth, also utilize FA with almost equal ease.

Recent studies have shown that folic acid provides protection against Alzheimer's disease.

Folic acid is, however, essential for lactation in rats and hatchability of eggs in chicks, turkeys and guinea pigs. Rats, dogs and probably man do not need folic acid because the intestinal bacteria synthesize sufficient quantity of this vitamin.

F. Deficiency. In chicks, a lack of this factor leads to **anemia.** Rats develop **achromotrichia** (failure in normal pigmentation of the hair). The monkeys show **macrocytic anemia** (anemia characterized by the presence of giant RBCs), leukopenia, diarrhea and **edema** (retention of water by skin tissues).

On a worldwide basis, deficiency of folic acid is believed to be the most common form of vitamin undernutrition. In man, the folic acid deficiency leads to **megaloblastic anemia, glossitis** and gastrointestinal disorders. Pregnant women and infants are also particularly vulnerable. Folic acid deficiency is a major feature of **tropical sprue**, in which there is a general deficiency in absorption of many nutrients from the small intestine.

Folic acid has been successfully used in the treatment of certain macrocytic anemias such as those developed in sprue and anemias of pellagra, pregnancy and infancy. However, the long-held hope that it would cure pernicious anemia (caused by avitaminosis B_{12}) has not been held true since it fails to cure the neurological lesions of the disease.

Folic Acid (Vitamin B_9) Deficiency

The occurrence of folic acid in practically all natural foods makes deficiency difficult, and apparently a normal person accumulates more than adequate reserves of this vitamin. For pregnant women the situation is very different. Needs of the fetus for normal growth and development include constant, uninterrupted supplies of coenzymes (in addition to amino acids and other cell constituents). Deficiency of folic acid, however, is not uncommon. It may result from limited diets, especially when the food is cooked at high temperatures for long periods, which destroys the vitamin. Intestinal diseases, notably **celiac disease**, are often characterized by folic-acid deficiency caused by malabsorption. Inability to absorb folate is rare. Folate deficiency is usually seen only in newborns and produces symptoms of **megaloblastic anemia**. All B_7-deficient patients respond to continuous therapy, although permanent damage appears to be caused by delayed or inadequate treatment.

A classic experiment was carried out by a physician, *apparently serving as his own experimental subject,* to study the human requirements for folic acid. His own diet consisted of foods (boiled repeatedly to extract the water-soluble vitamins) to which vitamins (and minerals) were added, barring folic acid. Symptoms attributable to folate deficiency did not appear for 7 weeks, altered appearance of blood cells and formiminoglutamate excretion were seen only at 13 weeks, and serious symptoms (irritability, forgetfulness, and macrocytic anemia) appeared only after four months (*i.e.*, about 17 weeks). Neurological symptoms were alleviated within 2 days after folic acid was added to the diet; the blood picture became normal more slowly. Recently, folate deficiency has been implicated in *spina bifida*.

G Human requirements. The daily dietary allowance of folic acid is 0.1 mg for infants, 0.2 mg for children and 0.4 mg for adult men and women. Pregnant mothers may, however, require up to 0.8 mg per day.

VITAMIN B_{12}

A. History. In 1926, two American physicians, **George Minot** and **William Parry Murphy** discovered that patients suffering from pernicious anemia could be cured by feeding them with about half a pound of liver a day. This landmark in medicine brought them Nobel Prize in 1934. In 1929, **Castle** suggested that gastric juice contained a factor (intrinsic factor) that, together with a factor present in the food (extrinsic factor), is responsible for the cure of pernicious anemia. This **antipernicious anemia factor (APA factor)** was, later, isolated in crystalline form in 1948 independently by **E. Lester Smith** in England and by **Edward Rickes** and **Karl Folkers** in the United States. It was then named as vitamin B_{12} or cyanocobalamin. It is the last B-vitamin to be isolated and is also known as *Factor X* or *L.L.D. factor*. The coenzyme form of this vitamin (deoxyadenosyl cobalamin or cobamide coenzyme) was first isolated by **Barker** of California. Coenzyme B_{12} has been called a *"biologic Grignard reagent"*.

> **Cyanocobalamin** is also spelt as cyanocobalamine (**Albert L. Lehninger**, 1984)

B. Occurrence. *Vitamin B_{12} has been found only in animals*; the chief source is liver, although it is also present in milk, meat, eggs, fish, oysters and clams. Animal tissues contain it in varying amounts as shown in Table 39–2.

Table 39–2. Amount of vitamin B_{12} in animal tissues

S. N.	*Source of vitamin B_{12}*	*Amount present (in millionth gm per 100 gm of fresh weight)*
1.	Liver (Ox)	134
2.	Liver (Herring)	34
3.	Meat (Fish)	25
4.	Meat (Beef)	8
5.	Cheese	3.6
6.	Egg yolk	1.3

Under certain dietary conditions, vitamin B_{12} may be synthesized by the intestinal microorganisms. *In general, cyanocobalamin is not present in plant foods* except in *Spirulina,* a blue-green alga. However, it occurs in foods bound to proteins and is apparently split off by proteolytic enzymes.

Animals and plants are unable to synthesize this vitamin. *Cyanocobalamin is unique in that it appears to be synthesized only by microorganisms especially anaerobic bacteria.* However, a process of producing vitamin B_{12} from waste products has been developed, in 1977, by the department of Chemical Engineering of the Indian Institute of Technology, Chennai.

C. Structure. The structure of vitamin B_{12} (Fig. 39–17), one of the most complex known, has been established, in 1957, by **Dorothy Crowfoot Hodgkin** (Nobel Laureate, 1964). Cyanocobalamin ($C_{63}H_{88}O_{14}N_{14}$ P Co) is a pigment alike to the tetrapyrrole ring structure of the porphyrins, *e.g.*, chlorophyll and haem. A unique feature of this vitamin (and other related compounds) is the presence, in its molecule, of an atom of a heavy metal cobalt in the trivalent state. *No other cobalt-containing organic compound has been found in nature*. The cobalt atom is centrally-situated and is surrounded by a macrocyclic structure of 4 reduced pyrrole rings (A, B, C and D) collectively called as corrin. It may be noted from the structural formula that the

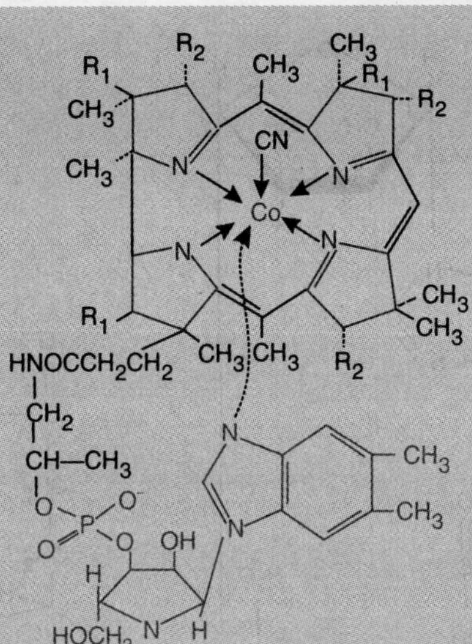

Fig. 39–17. Structure of vitamin B_{12} or cyanocobalamin

6 coordinate valences of the cobalt atom (Co^{2+})are satisfied by the 4 nitrogens of the reduced tetrapyrrole, a nitrogen atom of 5, 6-dimethylbenzimidazole and a cyanide ion. Two of the pyrrole rings, namely A and D, are directly linked to each other and the corrin has lower degree of unsaturation with only 6 double bonds. The other two pyrrole rings, namely B and C, are joined through a single methene carbon. Another distinct feature of the vitamin B_{12} molecule is the presence of a loop of the isopropanol, phosphate, ribose and 5,6-dimethyl-benzimidazole in that order, the end of the loop being attached to the central cobalt atom.

Many compounds with vitamin B_{12} activity have been isolated from natural sources. Cyanocobalamin is the most common form and is sometimes also written as vitamin B_{12a}. In other forms, cyanide ion is replaced by other ions, *e.g.*, by hydroxyl ion in hydroxocobalamin (also designated as vitamin B_{12b}), by nitrite ion in nitrocobalamin (also designated as vitamin B_{12c}) etc. The latter two, B_{12b} and B_{12c} can be converted to vitamin B_{12a} by treatment with cyanide.

The structure of *vitamin B_{12} coenzyme* (= 5′-deoxyadenosyl cobalamin) is similar to that of cobalamin except that here the CN group is replaced by adenosine and the linking with cobalt atom taking place at 5′ carbon atom of the ribose of adenosine (Fig. 39–18). *Vitamin B_{12} coenzyme is the only known example of a carbon-metal bond in a biomolecule.*

D. Properties. Vitamin B_{12} (molecular weight, ca 1,500) is a deep red crystalline substance. It is soluble in water, alcohol and acetone but not in chloroform. It is levorotatory. It is stable to heat in neutral solutions but is destroyed by heat in acidic or alkaline solutions.

E. Metabolism. Vitamin B_{12} is converted to coenzyme B_{12} by extracts from microorganisms supplemented with ATP.

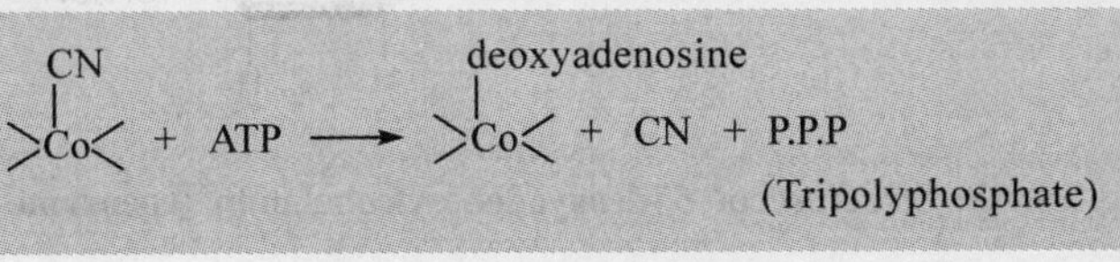

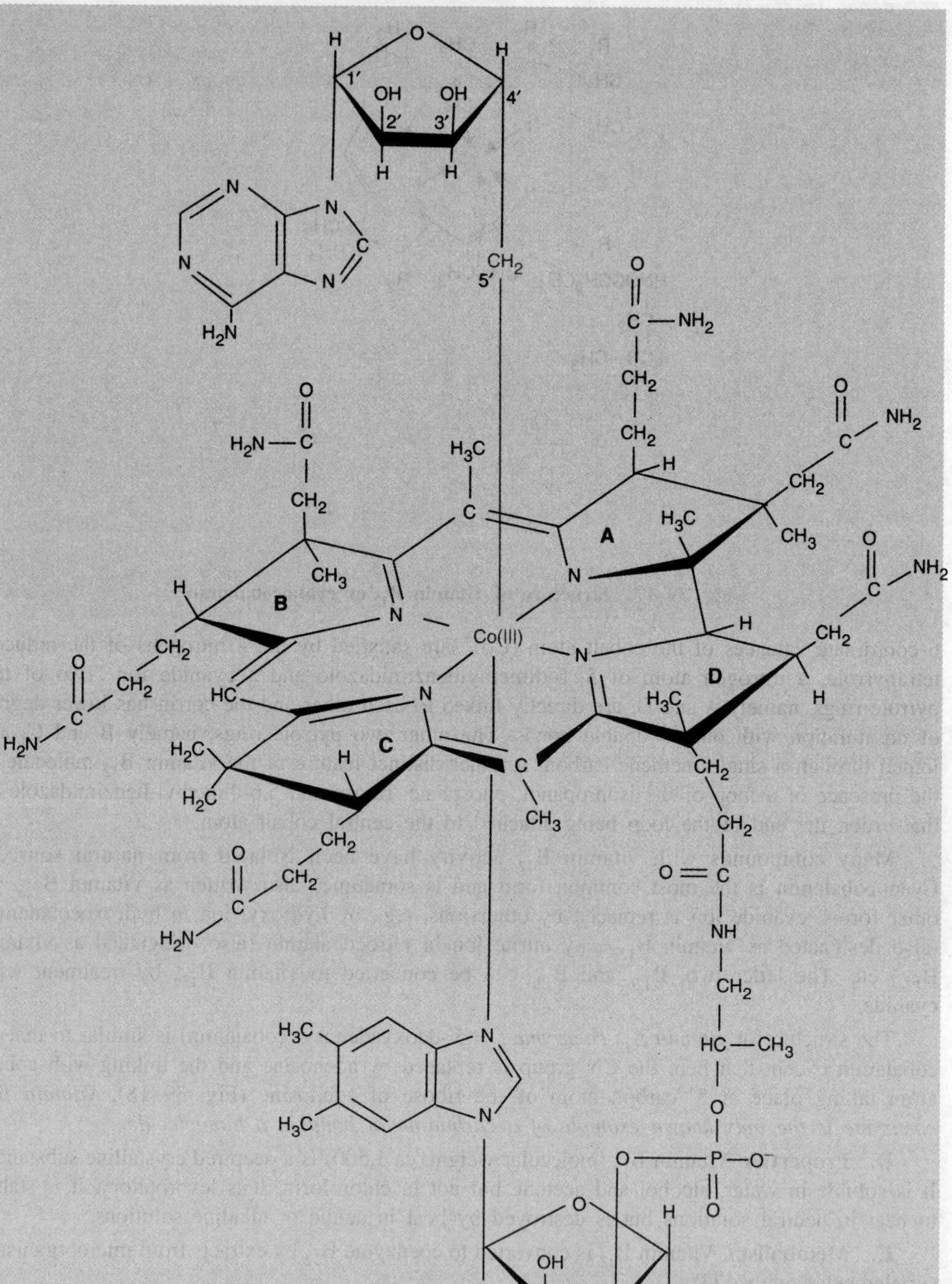

Fig. 39–18. Structure of 5′-deoxyadenosylcobalamin (coenzyme B_{12})

Coenzyme B_{12} is associated with many biochemical reactions:

(*a*) *1,2 shift of a hydrogen atom*: Coenzyme B_{12} catalyzes 1,2 shift of a hydrogen atom from one carbon atom of the substrate to the next with a concomitant 2,1 (reverse) shift of some other group, *e.g.*, hydroxyl, alkyl etc.

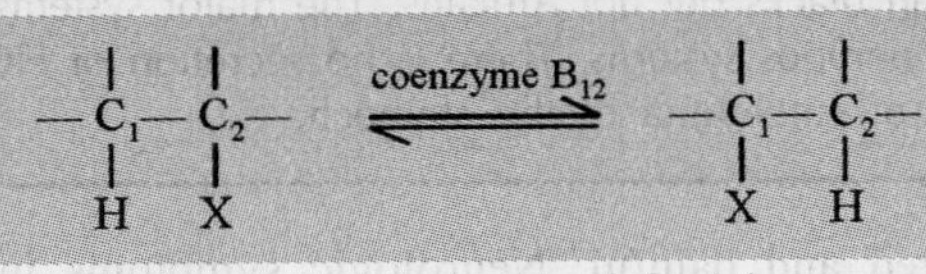

Conversion of methylmalonyl-CoA to succinyl-CoA is an example of 1,2-shift.

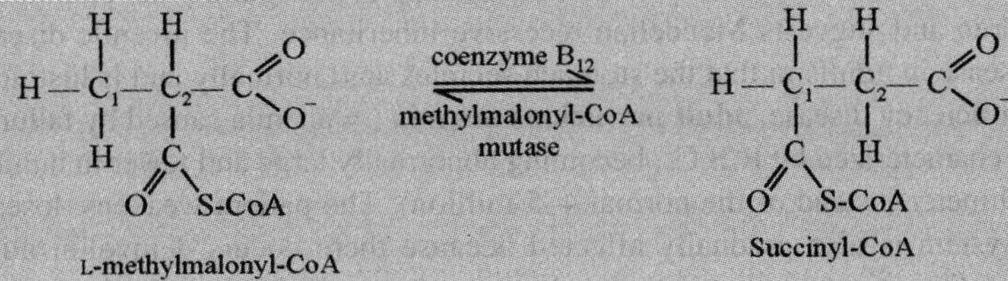

(*b*) *Carrier of a methyl group*: Coenzyme B_{12} also serves as a carrier of a methyl group, obtained from N^5-methyltetrahydrofolate, to the appropriate acceptor molecule. In the reaction, a methyl group occupies the 5-deoxyadenosyl coordination position of coenzyme B_{12}. Methylation of homocysteine to produce methionine is an example of such reaction. The reaction is catalyzed by *homocysteine methyltransferase*. An absence of vitamin B_{12} inhibits the reaction and leads to the buildup of N^5-methyltetrahydrofolate (N^5-methyl-FH_4). This is known as the **tetrahydrofolate trap.**

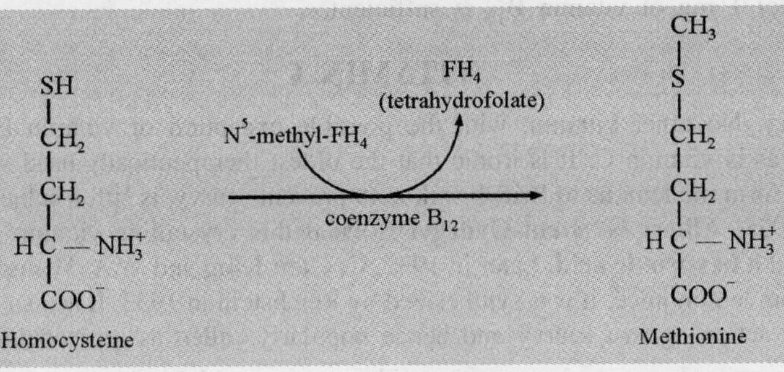

(*c*) *Isomerization of dicarboxylic acids*: Coenzyme B_{12} is associated with isomerization of dicarboxylic acids, *e.g.*, glutamic acid into β-methyl-aspartic acid.

(*d*) *Dismutation of vicinal diols*: Coenzyme 12 also catalyzes dismutation of vicinal diols to the corresponding aldehydes, *e.g.*, propane-1,2-diol into propionaldehyde.

Vitamin B_{12} is also needed for the biosynthesis of methyl groups from 1-carbon precursors and for the synthesis of thymidine and other deoxyribosides. It also functions in protein synthesis and in the activation of SH enzymes. Cyanocobalamin also affects myelin formation.

F. Deficiency. A nutritional deficiency of this vitamin is usually not observed on account of its ubiquitous (= widespread) nature in foodstuffs. Thus, most cases of deficiency stem from failure to absorb the vitamin. However, deficiency may be observed in individuals who abstain from all animal products including milk and eggs, *i.e.*, those who are strict vegetarians.

The rare disease **juvenile** (or **congenital) pernicious anemia** springs up due to an inability to secrete gastric intrinsic factors. The symptoms of this disease become prominent at 9 months to 10 years of age. As the anemia becomes severe, irritability, anorexia and listlessness occur.

CLINICAL IMPLICATIONS 39-2.

Pernicious Anemia

Pernicious anemia was described in 1855 by **Dr. Thomas Addison** and called pernicious because it is a fatal disease. The disease pernicious anemia affects those of Northern European ancestry including American blacks but not Africans. The major systems affected by this disease are hematopoietic and central nervous systems. There is no secretion of HCl in the stomach, and there is an increased incidence of carcinoma of the stomach.

The tongue is smooth, red and painful. Neurologic symptoms include ataxia, paresthesias, hyporeflexia, clonus, Babinski responses and coma. Consanguinity is common in parents of affected children and suggests Mendelian recessive inheritance. The juvenile disease differs from the typical disease in adults in that the stomach secretes acid normally and is histologically normal. This typical deficiency disease, **adult pernicious anemia** (= anemia caused by failure of erythrocyte formation), is characterized by R.B.Cs. becoming abnormally large and fewer in number (1–3 million per cubic millimeter instead of the normal 4–5 million). The patient weakens, loses its weight and the nervous system is also gradually affected because there occurs demyelinization of the large nerve fibres of the spinal cord. All these changes ultimately lead to death.

G Human requirements. The recommended daily allowance of vitamin B_{12} is 2 to 4 μg for children and 5 μg for men and women. Pregnant and lactating mothers require 8 μg and 6μg daily.

H. Treatment. The excessive secretion of methylmalonic acid in the urine is a reliable and sensitive index of vitamin B_{12} deficiency. The physiologic need for vitamin B_{12} is 1-5 μg/24 hr, and hematologic responses have been observed with these small doses. If there is evidence of neurologic involvement, 1 mg should be injected intramuscularly daily for a minimum of 2 weeks. Maintenance therapy is necessary throughout patient's life; monthly intramuscular administration of 1 mg of vitamin B_{12} is sufficient.

VITAMIN C

A. History. No other vitamin, with the possible exception of vitamin E, is as generally misunderstood as is vitamin C. It is ironic that the oldest therapeutically-used vitamin, furnished in 1750s in the form the lemons to British sailors to prevent scurvy, is still a subject of controversy. However, in 1928, **Albert G. Szent-Györgyi** isolated this crystalline vitamin from the paprika plant and named it **hexuronic acid.** Later in 1932, **C. Glen King** and **W.A. Waugh** in United States isolated this from lemon juice. It was synthesized by Reichstein in 1933. It is also called **cevitamin.** It has a curing action against scurvy and hence popularly called as **antiscorbutic factor.**

B. Occurrence. *In general, ascorbic acid is not as widely distributed as other vitamins.* Among plants, it is present in all fresh fruits and vegetables. *The richest source of vitamin C, known uptodate, is the acerola fruit (Malpighia punctifolia).* The fruit yeilds 1,000–4,000 mg of ascorbic acid per 100 g of edible matter. Citrus fruits (such as orange, lemon, lime), gooseberry, pineapple, guavas, tomatoes, melons, raw cabbage and green pepper are also rich sources of it. New potatoes contain relatively large amounts. Dried legumes and cereals contain very little vitamin C. Dry seeds, in general, are devoid of it but during their sprouting, the vitamin appears. Woody tissues also lack it. Vitamin C is synthesized by most mammals, but not by primates (such as apes, man) and guinea pigs which acquire it from their diets. In animals, the vitamin occurs in tissues and various glands or organs such as liver, adrenals, thymus, corpus luteum etc. Meat contains relatively low concentration. Human milk is 3 to 4 times richer in vitamin C contents than cow's milk. *Vitamin C is, however, absent from fish, fats and oils.* It is also not present in

CHARLES GLEN KING

(LT, 1896–1988)

King, a U.S. biochemist, was born on October 22, 1896 in Entiat, Washington to **Charles Clement King** and **Mary Jane Bookwalter**. He entered Washington State University early, as his local one-room school did not have a 12th grade. He was a member of Lambda Chi Alpha. World War I interrupted his college studies, where he served in the 12th infantry, a machine gun company. He did not receive his B.S. in chemistry until 1918. He immediately departed for the University of Pittsburgh, earning his M.S. in 1920 and Ph.D. in 1923. He remained in Pittsburgh as professor until 1942. In 1943, he became the first Scientific Director of the *Nutrition Foundation, Inc.*, which worked to promote scientific and public health research, both in the U.S. and internationally. Later, he became Professor in the Institute of Nutritional Sciences, Columbia University and President of the International Union of Nutritional Sciences.

King's contribution to the science of nutrition revolves around his **isolation of vitamin C** in 1931–1932 by studying the antiscorbutic activities of guinea pigs with preparations of lemon juice. He, along with **W.A. Waugh**, succeeded in isolating and identifying vitamin C from lemon juice in 1932. **Albert Szent-Györgyi** was conducting similar research at the University of Szeged in Hungary, focusing on hexuronic acid. The chemical identity of King's active substance was almost identical to Szent-Györgyi's hexuronic acid, but the research of **S.S. Silva** had declared the hexuronic acid was not vitamin C. However, within 2 weeks of each other in the spring of 1932, King first, and then Szent-Györgyi, published articles declaring that vitamin C and hexuronic acid were indeed the same compound. Szent-Györgyi would later (in 1937) win a Nobel Prize for his part in the discovery, and controversy remains over whether both men deserve equal credit. Surely a neglect on the part of the Nobel Committee! King later established the important functional role of vitamin B.

Throughout his 40-year research career, King made many significant contributions in the areas of fats, enzymes, and vitamins. He authored over 200 articles on good nutritional practices and the positive effects of vitamins. Apart from his work with the Nutrition Foundation, King's public service activities involved creation of the USDA's *Plant, Soil and Nutrition Laboratory* in Ithaca, NY. He helped establish the *Food and Nutrition Board,* dealing with food and nutrition problems in military and civilian populations, beginning in World War II and continuing through 1970. He also helped create the *Food Protection Committee*, the *Recommended Dietary Allowances* (RDAs), the *Protein Advisory Group*, and the *International Union of Nutritional Sciences*. King officially retired from the Nutrition Foundation in 1963, only to begin a second career as Associate Director of the Institute of Nutrition Sciences and a consultant to the Rockefeller Foundation. His notable awards include **John Scott Award** and **Nicholas Appert Award**. King died on January 23, 1988, at age 91.

or required by microorganisms. Since this vitamin is a good reducing agent, it is lost under oxidizing conditions like aeration and heating. Thus, many cooked and canned foods contain little ascorbic acid. It is also found in the combined form as *ascorbigen*. The adrenal and lenses have particularly high contents of vitamin C.

The infant is usually born with adequate stores of ascorbic acid if the mother's intake has been adequate; the ascorbic acid content of cord blood plasma is 2-4 times greater than that of maternal plasma. Under these conditions, the breast milk contains ca 4–7 mg/dL of ascorbic acid and is an adequate source of ascorbic acid.

The vitamin C content of some food items are presented in Table 39–3.

Table 39–3 Vitamin C content of some fruits and vegetables

Fruit / Vegetable	*Vitamin C (mg/100 g)*
Amla	600–700
Guava	200–300
Lime	65
Papaya	55–60
Lemon	40–50
Orange	30
Tomato	25–30
Drumstick leaves	200–250
Bathua leaves	200–250
Radish leaves	80
Fenugreek leaves	50
Palak leaves	30

C. Structure. The structure of ascorbic acid ($C_6H_8O_6$) was established mainly by **Haworth**. It is a derivative of a hexose called L-gulose. Chemically, it is 1-threo-2,4,5, 6-pentoxyhexen-2-carboxylic acid lactone (Fig. 39–19). Although ascorbic acid is a small molecule when compared with DNA, RNA or proteins, its metabolic impact is no less considerable. The dienolic group consisting of hydroxyls on C2 and C3 with a double bond between them invests the ascorbic acid molecule with redox property.

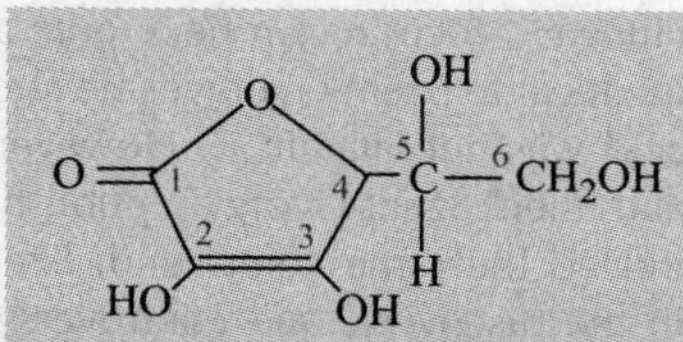

Fig. 39–19. Vitamin C or ascorbic acid

D. Properties. Ascorbic acid is a colourless and odourless crystalline substance, slightly sour in taste and optically active. *Only the L-isomer has antiscorbutic properties*. It is soluble in water and alcohol but practically insoluble in chloroform, solvent ether and light petroleum. It is readily oxidized, particularly in the presence of copper and iron but not of aluminium. It is for this reason that the foods cooked in copper utensils lose ascorbic acid quickly. This vitamin is also rapidly destroyed by alkalies but is fairly stable in weak acid solutions. Therefore, baking soda has a deleterious effect but steam cooking destroys very little amount of ascorbic acid. Drying of vegetables and also their storage results in a loss of their ascorbic acid. However, freezing has no detrimental effect on this vitamin. Citrus fruit juices and tomato juice may be canned with but little loss of ascorbic acid. On account of its easily oxidizable nature, the ascorbic acid is a powerful reducing agent.

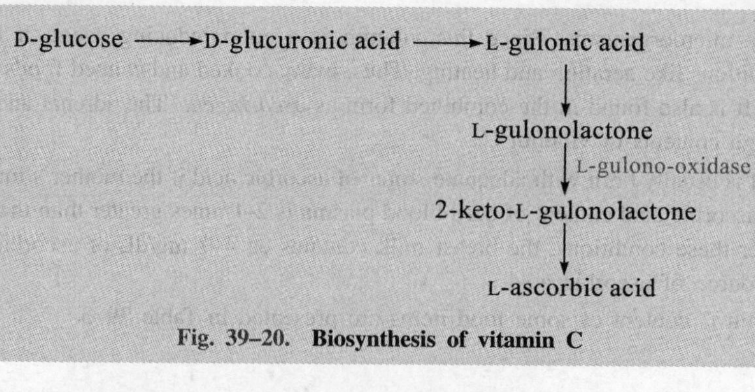

Fig. 39–20. Biosynthesis of vitamin C

E Metabolism. Higher plants (like pea) and all known mammals except man, the primates and guinea pig can synthesize ascorbic acid from L-gulonolactone. The rat, for example, is resistant to scurvy. In animals, the liver and adrenals (cortical portion) are the main sites of synthesis. The biosynthesis of ascorbic acid in animals takes place according to the scheme as depicted in Fig. 39–20. The scheme also probably applies to the plants. *Man (and higher primates, guinea pigs and fruit-eating bats) lack the enzyme L-gulono-oxidase and as such are incapable of synthesizing ascorbic acid.*

Ascorbic acid can be readily oxidized (Fig. 39–21) to dehydroascorbic acid in the presence of metal ions. Dehydroascorbic acid is a much more powerful electron donor than even ascorbic acid by virtue of its unpaired electron. It is, in fact, the free radical form of ascorbic acid. Dehydroascorbic acid can be reduced, in the presence of H_2S or cysteine, back to ascorbic acid. The reduced form (*i.e.*, L-ascorbic acid) predominates in the plasma and also apparently in tissues at a ratio of about 15 : 1 of the oxidized form (*i.e.*, dehydro-L-ascorbic acid). Both of these are biologically active and are equally potent in carrying out their metabolic functions. When dehydro-L-ascorbic acid is hydrated, 2,3-diketo-L-gulonic acid is formed which is biologically inactive and cannot be converted back to either of the active forms in the body. Since the hydration reaction takes place automatically in the neutral medium, the oxidation of ascorbic acid, in other words, means its biologic inactivation. These reactions have been shown to occur *in vivo* in man and guinea pigs.

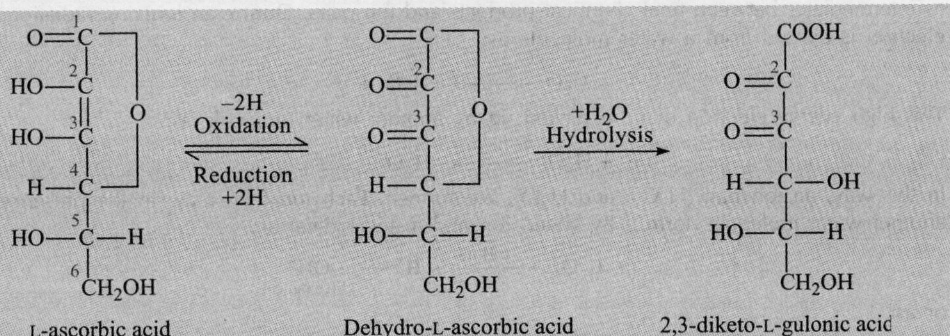

Fig. 39–21. Metabolism of vitamin C

Ascorbic acid functions in a number of enzymatic activities. A major function of ascorbic acid is the formation of tissue collagen or '*intracellular cement substance*'. In fact, ascorbic acid appears to be essential to the activity of the enzyme *collagen proline hydroxylase*, which catalyzes the conversion of proline to hydroxyproline. Hydroxyproline (Hyp) is found exclusively in collagen and is vital in maintaining the tertiary structure of this major vertebrate protein, *i.e.*, collagen.

Recent researches have established the role of ascorbic acid in the conversion of folic acid to a physiologically active form, tetrahydrofolic acid.

Ascorbic acid also plays a key role in tyrosine metabolism. One of the steps in tyrosine metabolism is the oxidation of *p*-hydrophenylpyruvic acid to homogentisic acid. The vitamin C protects the enzyme *p-hydrophenylpyruvic acid oxidase* from inhibition by excess substrate.

Ascorbic acid is also involved in electron transport in the microsomal fraction. However, *in none of the biological oxidation systems, ascorbic acid has been shown to act as a specific coenzyme.*

Vitamin C is found concentrated in certain parts of human body such as brain and the white blood cells. This body "pool" amounts to roughly 1,500 mg for a man and slightly more for a woman, which is normally enough for about a month's need. However, illness or stress can substantially decrease the body's vitamin C reserves, as also can smoking, drinking, and a variety of drugs such as aspirin. Vitamin C plays an important role in our body's wondrous immune system. It may enhance the body's production of interferon, prostaglandins, T-lymphocytes and immunoglobulins-weapons of the body's self-defence arsenal. However, huge doses of ascorbic acid can leach calcium and other needed minerals out of the body. It can act as a diuretic and laxative.

The fact that in old age the deficiency of vitamin C is frequently observed point to a possible role of vitamin C as an anti-ageing agent. Free radicals are the major cause of ageing. These are oxygen molecules which lose an electron in the course of circulating through the blood-stream. These highly reactive molecules try to regain chemical stability by 'snatching' electrons from other molecules, a process which causes much damage. While normal metabolic processes produce some free radicals, their number increases by tissue injuries from infection, toxins, reduced blood flow, excessive exercise and environmental hazards like radiation, heat, and cold. And *the antioxidants like vitamins A, C and E neutralize these free radicals by 'donating' electrons.*

Free Radicals

Free radicals are natural by-products of metabolism and are involved in the body's defence against microorganisms. But, if in excess, they damage body cells and tissues and, thus, play a role in degenerative disorders such as heart ailments and cancer. Free radicals are highly reactive, unstable molecules that normally attack cellular proteins, lipids and even DNA. It may be an atom or groups of atoms containing an unpaired electron. Usually free radicals are formed in radiation as intermediates between final chemical products and ion pairs. During an ionizing radiation, an electron is ejected from a water molecule as:

$$H_2O \longrightarrow H_2O^+ + e^-$$

This high energy electron may be picked up by another water molecule as:

$$e^- + H_2O \longrightarrow H_2O^-$$

In this way, an ion pair, H_2O^+ and H_2O^-, are formed. Each ion then may, in the presence of another water molecule, form a hydrogen ion and a free radical as:

$$H_2O^+ \xrightarrow{\bar{c}\, H_2O} H^+ + \underset{\text{(Free radical)}}{OH^\circ}$$

or as:

$$H_2O^- \xrightarrow{\bar{c}\, H_2O} \underset{\text{(Free radical)}}{H^\circ} + OH^-$$

The H^+ and OH^- will combine to form water. The H° and OH° free radicals are very reactive. In fact, many of them react to form H_2O_2. In cells containing catalase and peroxidases, hydrogen peroxide formation may not be significant. In the absence of such enzymes, hydrogen peroxide formation in cells may be important biologically. Free radicals may also react with oxygen to enhance the effect of radiation. Free radicals may be formed from nearly any cellular component which ionizes to contribute to the indirect effect of radiation. They are also formed in the body as a result of exposure to smoking, pollution and sunlight.

Vitamin C (and also vitamins A and E) have long been known to be beneficial for the skin. While vitamins A and E work by exfoliating the skin's surface cells, vitamin C works from the inside by boosting collagen production and repair. It also inhibits the excess production of melanin ($C_{17}H_{98}O_{33}N_{14}S$) which leads to a tan and hyperpigmentation. Some dermatologists encourage the use of vitamin C as an anti-inflammatory agent as its topical application speeds up the skin's healing process, reducing redness and irritation caused by sun exposure.

Ascorbic acid plays an important role in germination, growth, metabolism and flowering of plants (**Chenoy JJ**, 1962, 1967–1973). During germination, embryo axis has higher ascorbic acid content as well as higher rate of ascorbic acid utilization compared with those of the endosperm or the cotyledons. Ascorbic acid stimulates amylase, protease and RNAase activity and RNA content in various crops including gram (*Cicer arietinum*). Free radical content of the endosperm / cotyledon has been shown to be higher during the initial stages of germination as compared to that in the embryo axis, suggesting an active participation of free radicals in the process of energy flow for transport of metabolites from storage organ to the embryo axis. Increase in free radical content of the embryo axis, at later stages of germination is highly suggestive of the important role of free radical in the biosynthesis of macromolecules and other cell constituents for seedling growth. Further, it was established that the ascorbic acid turnover is appreciably higher during the reproductive phase of differentiation in many thermophobes (wheat, barley, oat), as well as in many thermophytes (maize, sesame). During the period of reproductive differentiation, the free radical content is enhanced.

F. Deficiency. Avitaminosis C leads to **scurvy,** which may occur at any stage but is rare in the newborn infant. The majority of cases appear in infants 6-24 months (mo) of age. Breast-fed infants are protected as the breast milk contains adequate amounts of vitamin C. Clinical manifestations require time to develop. However, after a variable period of vitamin C depletion, contain vague symptoms of irritability, tachypnea (very rapid respiration), digestive disturbances and loss of appetite appear. The main symptoms which later develop are listed below:

1. *Tender bones*: There is evidence of general tenderness, esp., noticeable in the legs when the infant is picked up or when the diaper is changed. The legs assume the typical "*frog-like position*", in which the hips and knees are semiflexed with the feet rotated outward. This may be mistaken for paralysis and is hence aptly called **pseudo-paralysis.**
2. *Edematous swellings*: These develop along the shafts of the legs and in some cases a subperiosteal hemorrhage can be palpated at the end of the femur.
3. *Petechial hemorrhages*: The capillaries become brittle and burst, thus giving rise to red and purple spots (or *petechiae*) over the body. Petechiae may be seen in the skin and mucous membranes. Hematuria, melena, and orbital or subdural hemorrhages may be found.
4. *Bleeding gums*: Changes in the gums, most noticeable when the teeth are erupted, are characterized by bluish-purple, spongy swellings of the mucous membrane, usually over the upper incisors (Fig. 39–22).

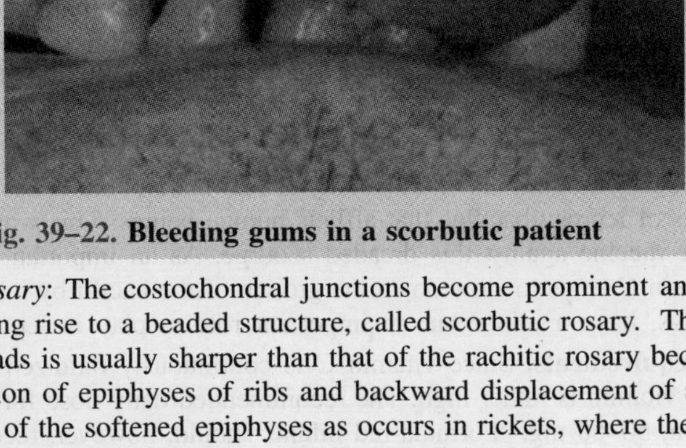

Fig. 39–22. Bleeding gums in a scorbutic patient

5. *Scorbutic rosary*: The costochondral junctions become prominent and appear sharp and angular, giving rise to a beaded structure, called scorbutic rosary. The angulation of the scorbutic beads is usually sharper than that of the rachitic rosary because it is produced by a separation of epiphyses of ribs and backward displacement of sternum rather than by widening of the softened epiphyses as occurs in rickets, where the prominence of the costochondral junction is dome-shaped and semicircular.

CLINICAL IMPLICATIONS 39–3.

SCURVY

A vivid description of scurvy, a dietary deficiency disease, was given by **Jacques Cartier** in 1536, when it afflicted his men when they were exploring the Saint Lawrence River:

"Some did lose all their strength, and could not stand on their feet ... Others also had their skins spotted with spots of blood of a purple colour: then did it ascend up to their ankles, knees, thighs, shoulders, arms and necks. Their mouths became stinking, their gums so rotten, that all the flesh did fall off, even to the roots of the teeth, which did also almost all fall out".

The need for fruits like oranges, lemons and limes was first discovered in the seventeenth century when sailors on long sea voyages developed scurvy after many months of living on dried and stale meat.

But it was not until the 1940s that anyone showed that the substance in fruit which prevented these symptoms was vitamin C. In 1940, a young American surgeon, **John Crandon**, *experimented on himself* to investigate the effects of a diet deficient in ascorbic acid. After 3 months of deficiency, a skin cut healed normally, but after 6 months another cut would not heat at all. We now know that vitamin C deficiency leads eventually (but not immediately) to a severe reduction in the body's ability to produce collagen and intercellular 'cement'. Any damage to the skin or other tissues cannot therefore be repaired.

Today, some scientists think that some vitamins, including vitamin C may help protect the body against cancer. There is no hard evidence for this, but many people take vitamin supplements, as just in this case.

6. *Delayed wound healing*: The wound healing is delayed or, in many cases, even does not occur because of the failure of the cells to deposit collagen fibrils. The healed wounds may even breakdown.
7. *Cessation of bone growth*: The bones cease to grow. The cells of growing epiphyses continue to proliferate but no new matrix is laid down between the cells. Consequently, bones fracture easily at the point of growth because they fail to ossify. Moreover, when an already ossified bone fractures in a scorbutic individual, the osteoblasts cannot secrete a new matrix for the deposition of new bone. With the result, the fractured bone does not heal.
8. *"Sicca" syndrome*: Swollen joints and follicular hyperkeratosis may develop, as well as the "sicca" syndrome of Sjögren, which is usually associated with collagen disorders and includes xerostomia, keratoconjunctivitis sicca, and enlargement of the salivary glands.
9. *Anemia*: Anemia may reflect inability to utilize iron or impaired folic acid metabolism.
10. *Pyrexia*: Low-grade fever is usually present in scorbutic children.

Infants 6–12 months of age, who are fed on processed milk only, are very susceptible to this disease (*infantile scurvy*). Adult cases appear less frequently. Elderly bachelors and widowers who have to cook their own foods are especially prone to the development of vitamin C deficiency– a syndrome termed *'bachelor scurvy'*. Food faddists may also develop scurvy if their diet lacks fruits and vegetables.

In 1975, **Prof. Olaf Skinsnes** of the University of Hawaii, Honolulu, has succeeded in obtaining almost pure cultures of leprosy bacillus that afflicts human beings. This work has, however, raised the possibility of a vaccine against this dreaded scourge. As an important sidelight of his work, it appears that vitamin C tends to slow down the growth of the bacilli by inhibiting enzyme action. Vitamin C may, thus, have a minor but an important role to play in leprosy treatment.

G. Human requirements. Since vitamin C is continuously oxidized in the body, the daily requirement of this vitamin is rather high. The recommended daily dose for children is 40 mg and for men and women, 50–60 mg. Formula-fed infants should, however, receive even lower doses, *i.e.,* 30 mg of ascorbic acid daily. Lactating mothers should take higher doses, *i.e.*, 100 mg daily. According to a report published by the British Nutrition Foundation, the RDA for vitamin C on which most nutritionists base diets, is far too low. The 30 mg daily that is usually recommended in most countries is way below the United States at 60 mg, Germany at 75 mg and Russia at 100 mg. Ante diluvian though it sounds, the figure of 30 mg is based on the amount of vitamin C needed to prevent that scourge of seafarers, scurvy. The Nutrition Foundation believes that a more contemporary approach to vitamin C would be to consider at what level it actually promotes good health. Besides, vitamin C requirement in humans may vary with the time of day and time of year. Early autumn, for example, is when most poeple's vitamin C levels are at their lowest.

H. Treatment. Scurvy is prevented by a diet rich in ascorbic acid; citrus fruits and juices are excellent sources. The administration of orange juice or tomato juice daily will quickly produce healing but ascorbic acid is preferable. The daily therapeutic dose is 100-200 mg or more, orally or parenterally.

CHOLINE

A. History. Choline is an essential component of the diet of animals and is, therefore, usually included among the vitamins. **Best**, for the first time, pointed out the role of choline in nutrition. He also showed that choline prevented the development of fatty livers in depancreatized dogs.

B. Occurrence. Choline is widely distributed. The richest source is egg yolk. Liver, kidney, meats, cereals and many vegetables such as beans and peanuts are other good sources. It is an important constituent of lecithins.

> A **quaternary ammonium compound** is that in which N bears four organic substituents.

C. Structure. Choline ($C_5H_{15}O_2N$) is a quaternary ammonium compound, where out of the 4 H atoms, one is replaced by hydroxyethyl group and the other three by 3 methyl groups (Fig. 39–23).

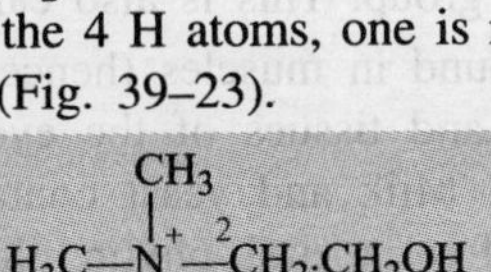

Fig. 39–23. Choline
(trimethyl-2-hydroxyethylammonium hydroxide)

D. Properties. Choline is water-soluble and has very strong basic properties.

E. Metabolism. Choline can be synthesized in the body by methylation of ethanolamine and, therefore, strictly speaking, this is not a vitamin.

Choline may function in many ways:

(*a*) It is an important constituent of phospholipids like lecithin.

(*b*) It undergoes esterification with acetyl-CoA to form acetylcholine. This is an endergonic reaction, the energy being derived from ATP. Acetylcholine is responsible for the transmission of nerve impulses in the central nervous system (CNS).

$$\text{Choline} + \text{Acetyl-CoA} \xrightarrow{\text{ATP} \quad \text{ADP}} \text{Acetylcholine} + \text{CoA}$$

(*c*) It acts as an important methyl group donor in intermediary metabolism.

(*d*) It is an important lipotropic agent and participates in the mobilization of fat from the liver. Its absence, henceforth, causes accumulation of fat in the hepatic tissues.

Using mutants of *Neurospora*, Horowitz has shown that the inability of the fungus to synthesize choline is due to a deficiency in the formation of an intermediate compound, N-monomethyl aminoethanol. The synthesis involves the following steps:

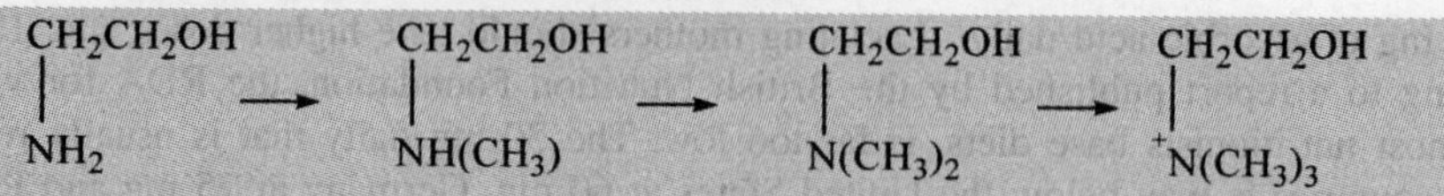

Aminoethanol — *N*-monomethyl aminoethanol — Dimethyl aminoethanol — Choline

F. Deficiency. In the deficiency of choline, puppies develop **anorexia**, hens do not lays eggs and mice do not lactate normally. A low-choline diet also develops hemorrhages of the kidneys and eyes, in addition to **fatty livers**, in young rats.

No definite symptoms of choline deficiency have been established in man. However, alcoholic cirrhosis of the liver in man is largely a result of dietary deficiency of many lipotropic agents, of which choline is an important example.

G. Human requirements. In the case of choline, the dietary requirement for human beings has not been established.

INOSITOL

A. History. Woolley (1940) discovered that the mice, when fed on a synthetic diet containing all the known vitamins even, failed to grow and their hair growth was arrested. The addition of pantothenic acid, the absence of which may also cause hair changes, however, proved futile. Neither biotin nor *p*-aminobenzoic acid could also cure them. The curative effects were, however, obtained by the addition of phytin (obtained from cereal grain) or inositol (isolated from liver). Thus, it was established as a vitamin of B group. This is also called as mouse **antialopecia factor.**

> **Phytin** is a mixed calcium and magnesium salt of inositol hexaphosphate (= phytic acid), which is found in plants.

B. Occurrence. Inositol is found in muscles (hence, its nomenclature as **muscle sugar**), liver, kidneys, brain, erythrocytes and tissues of the eye. Among plants, it occurs in furits, vegetables, whole grains and nuts. Milk and yeast contain appreciable quantities. Inositol is found in nature in at least 4 forms: free inositol, phytin, phosphatidylinositol and a nondialyzable complex. Inositol containing phosphatide or phosphoinositide (= lipositol of Woolley) has been isolated in pure form from soyabeans and is also known to be present in brain and spinal cord.

C. Structure. Inositol, $C_6H_{12}O_6$ or better $C_6H_6(OH)_6$, is a carbocyclic hexahydric alcohol. It has 9 possible stereoisomers, of which only one myoinositol (Fig. 39–24), found in muscles, is biologically active and happens to be a symmetric, optically inactive meso-form.

> Myoinositol was formerly called **mesoinositol**. This nomenclature is, however, vague since 7 of the 9 isomers are meso forms.

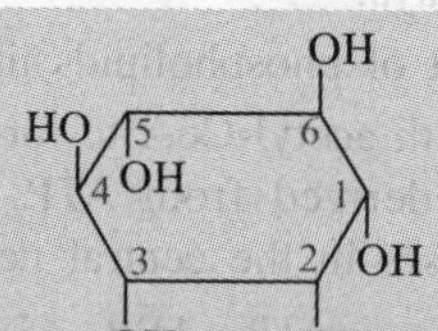

Fig. 39–24. **Myoinositol** or **mesoinositol** (hexahydroxycyclohexane)

D. Properties. *Although not a sugar, inositol is sweet in taste*. This is, in fact, a common property to many polyatomic alcohols including glycerol. Inositol is soluble in water.

E. Metabolism. Inositol as phosphoinositide helps in transport processes in cells.

Inositol stimulates the growth of many microorganisms such as *Saccharomyces cerevisiae* and *Neurospora.*

It also acts as a lipotropic agent and prevents the formation of fatty livers.

Possibly, it is an intermediate between carbohydrates and aromatic substances.

F. Deficiency. Inositol deficiency results in **retarded growth** and a peculiar hairlessness in mice. Lack of inositol also causes insufficient lactation in experimental animals. Deficiency of inositol, however, does not occur in man.

G Human requirements. The amount of inositol needed by man is not known.

PARA-AMINOBENZOIC ACID

A. History. Ansbacher demonstrated that depigmentation of the hair (achromotrichia) in mouse could be cured by feeding rice polishings or by adding *p*-aminobenzoic acid (PABA) to the diet. PABA also appeared essential for the growth of rat and chick and also for bacterial multiplication.

B. Occurrence. PABA is widely distributed in nature. The good sources are liver, yeast, rice bran and whole wheat. PABA occurs in conjugated form as a part of folic acid and its derivatives (see pages 1351 and 1352).

C. Structure. The structural formula of *p*-aminobenzoic acid appears on page 360.

D. Properties. PABA is a white crystalline substance. It is sparingly soluble in cold water but freely soluble in hot water and alcohol.

E. Metabolism. Woods, for the first time, observed that PABA blocks the bacteriostatic properties of sulfanilamide (refer page 360). PABA is synthesized from shikimic acid *via* chorismic acid.

F. Deficiency. The deficiency of PABA affects adversely the growth and the maintenance of a normal fur coat in rats.

G. Human requirements. There is, at present, no evidence that PABA is an essential dietary factor for man.

ALPHA-LIPOIC ACID

A. History. Lipoic acid is a relatively newly-discovered factor. This factor supports the growth of a number of bacteria and protozoa. It has been variously called as **pyruvate oxidation factor (POF)** or **acetate replacement factor** or **protogen. Lester J. Reed** *et al* (1951) isolated this factor in crystalline form from the insoluble residue of liver. When first isolated, lipoic acid was believed to be a B-vitamin because of its coenzyme function. However, the current opinion is in favour of treating lipoic acid as a *pseudovitamin* since it is synthesized by most animals.

Such small quantities of **lipoic acid** are present in tissues that 10 tons of water-insoluble liver residue were used by **Reed** *et al* to obtain about 30 mg of the biomolecule.

B. Occurrence. It is found in many biologic materials including yeast and liver.

C. Structure. α-lipoic acid ($C_8H_{14}O_2S_2$) is a cyclic disulfide, derived from 6, 8-dimercapto-*n*-caprylic acid (Fig. 39–25).

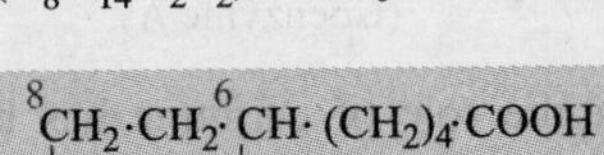

Fig. 39–25. **α-lipoic acid** or **thioctic acid**
(6,8-dithiooctanoic acid)

D. Properties. Lipoic acid is an exception of the vitamins of B series in that *it is fat-soluble rather than water-soluble*.

E Metabolism. Lipoic acid acts as a catalytic agent for the oxidative decarboxylation of pyruvic acid and α-ketoglutaric acid by certain microorganisms. It is probably a coenzyme or part of a coenzyme, called as lipothiamide pyrophosphate (LTPP), for this reaction.

F. Deficiency. A deficiency of lipoic acid usually does not occur since it is synthesized by most animals.

G. Human requirements. α-lipoic acid is not an essential factor of the diet.

CARNITINE

A. History. Carnitine has long been known as a constituent of meat extractives. Its vitamin nature was first recognized when it was shown to be an essential food factor of certain insects such as the yellow mealworm, *Tenebrio molitor*. It was, however, first isolated from muscles.

B. Occurrence. Carnitine is widely distributed in most tissues including plants, animals and microorganisms. It occurs in the free state and also bound to lipid.

C. Structure. Carnitine (Fig. 39–26) is a betaine (pronounced as 'bay-tah-een').

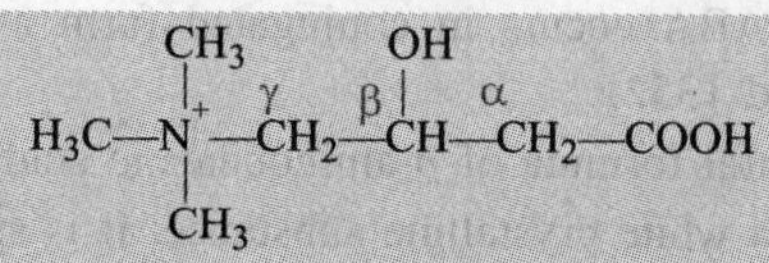

Fig. 39–26. Carnitine

(β-hydroxy-γ-butyrobetaine)

D. Metabolism. Fatty acids are activated on the outer mitochondrial membrane whereas they are oxidized in the mitochondrial matrix. Since the inner mitochondrial membrane is impermeable to long-chain acyl CoA molecules, a special transport mechanism is needed for them. It has been suggested that *carnitine acts as a carrier of the activated long-chain fatty acids across the inner mitochondrial membrane*. The acyl group is transferred from the S atom of acyl CoA to the OH group of carnitine to produce acyl carnitine, which traverses the inner mitochondrial membrane. In the mitochondrial matrix, the acyl group from acyl carnitine is transferred back to CoA so as to regenerate acyl CoA and free the carnitine. As the value of K is near 1, the O-acyl bond of carnitine is a high-energy bond. These transacylation reactions are reversible and are catalyzed by *fatty acyl CoA : carnitine fatty acid transferase*.

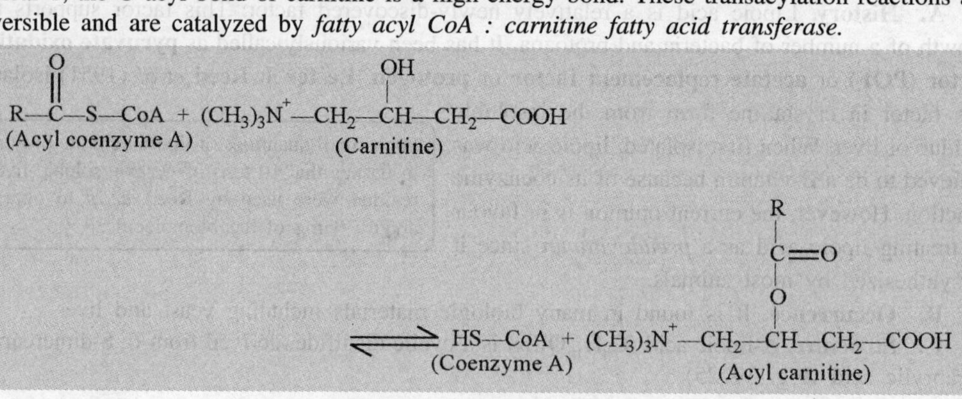

The postulated mechanism for the role of carnitine has been outlined in Fig. 39–27.

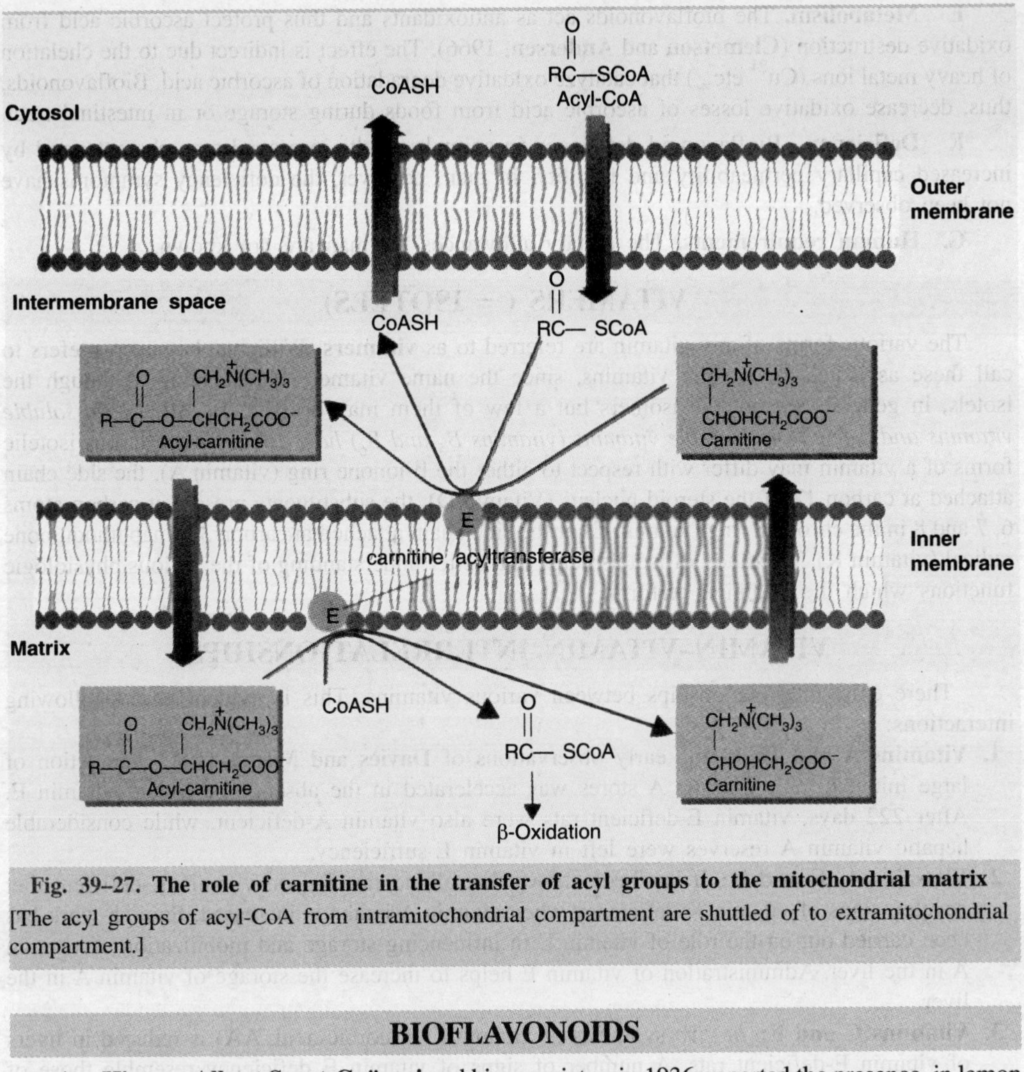

Fig. 39–27. The role of carnitine in the transfer of acyl groups to the mitochondrial matrix

[The acyl groups of acyl-CoA from intramitochondrial compartment are shuttled of to extramitochondrial compartment.]

BIOFLAVONOIDS

A. History. Albert Szent-Györgyi and his associates, in 1936, reported the presence, in lemon peel, of a material which they named **citrin**. It consisted of a mixture of flavonoids and was shown to be associated with the maintenance of normal capillary permeability and fragility. The active principle in citrin was found to be **hesperidin**. It shows physiologic roles similar to those exhibited by other structurally-related compounds such as flavanones, flavones and flavonols. The term **vitamin P** (for *p*ermeability) was at first assigned to this group of compounds. They are now more commonly referred to as **bioflavonoids.**

B. Occurrence. The bioflavonoids are widely distributed in nature. *They are always of plant origin.* They are present in the juice, peel and pulp of citrus fruits, in tobacco leaves, in buckwheat (*Fagopyrum esculentum*), in grapes and in many other fruits and vegetables.

C. Structure. Hesperidin is 5,3′-dihydroxy-4′-methoxy-7-rhamnoglucosidoflavanone. Its aglycone hesperitin, rutin (5,7,3′,4′- tetrahydroxy-3-glucorhamnosidoflavone) and its aglycone, quercitin all have comparable physiologic roles.

D. Properties. The bioflavonoids are water-soluble.

E. **Metabolism.** The bioflavonoids act as antioxidants and thus protect ascorbic acid from oxidative destruction (**Clemetson** and **Andersen**, 1966). The effect is indirect due to the chelation of heavy metal ions (Cu^{2+} etc.,) that catalyze oxidative degradation of ascorbic acid. Bioflavonoids, thus, decrease oxidative losses of ascorbic acid from foods during storage or in intestinal tract.

F. **Deficiency.** Bioflavonoid deficiency in animals results in a syndrome characterized by increased capillary permeability and fragility. In man, however, the deficiency symptoms have not been observed.

G. **Human requirements.** The dietary allowances for man are not known.

VITAMERS (= ISOTELS)

The various forms of any vitamin are referred to as **vitamers**. Williams, however, prefers to call these as isotels or isotelic vitamins, since the name vitamer is misleading. Although the isotels, in general, are not the isomers but a few of them may be isomers. *All the fat-soluble vitamins and a few water-soluble vitamins (vitamins B_5 and B_6) have isotels.* The various isotelic forms of a vitamin may differ with respect to either the β-ionone ring (vitamin A), the side chain attached at carbon 17 of the steroid nucleus (Vitamin D), the substituents present at carbon atoms 6, 7 and 8 in the chroman ring (vitamin E) or the side chain attached at carbon 3 of naphthoquinone radical (vitamin K). The study of isotels helps in a better understanding of the various physiologic functions which the vitamins perform.

VITAMIN–VITAMIN INTERRELATIONSHIPS

There exist interrelationships between various vitamins. This is evident in the following interactions:

1. **Vitamins A and E:** In the early observations of **Davies** and **Moore** (1941), depletion of large initial hepatic vitamin A stores was accelerated in the absence of dietary vitamin E. After 222 days, vitamin E-deficient rats were also vitamin A-deficient, while considerable hepatic vitamin A reserves were left in vitamin E-sufficiency.
2. **Vitamins A, E and K:** It has been shown that albino rats fed on vitamin K-deficient diet develop signs of vitamin K deficiency when vitamin A intake is increased. Enough work has been carried out on the role of vitamin E in influencing storage and mobilization of vitamin A in the liver. Administration of vitamin E helps to increase the storage of vitamin A in the liver.
3. **Vitamins C and E:** *In vivo* synthesis of vitamin C (ascorbic acid, AA) is reduced in livers of vitamin E-deficient rats. A number of signs of vitamin E deficiency resemble those of scurvy. For example, in the pig, hemorrhages are often observed. In the rat, skin ulcerations, suggesting defective collagen synthesis, are seen after a prolonged deficiency of vitamin E. These scorbutic-like symptoms might be explained by a defect in ascorbic acid synthesis in vitamin E-deficient animals *in vivo*. **Chen** and **Barnes** (1976) have reported that plasma levels of both total and reduced AA are lower in vitamin E-deficient rats or guinea pigs than in controls.
4. **Vitamins C, B_1 and B_2:** The albino rat can synthesize ascorbic acid. This synthetic capacity, however, is diminished considerably in thiamine (vitamin B_1) and riboflavin (vitamin B_2) deficiencies. This indicates that both these vitamins play an important role in the biosynthesis of AA. A depletion of AA content of tissues of rats occurred when the animals were fed on diets deficient in vitamin B_1 or B_2.
5. **Vitamins C and B_6: Selivanova** *et al* (1963) reported that the urinary excretion of 4-pyridoxic acid (the major metabolite of pyridoxine, the vitamin B_6) was elevated with an increased intake of AA. **Baker** *et al* (1964) followed whole blood ascorbic acid levels in

normal adult men subjected to depletion and repletion of pyridoxine. Whole blood AA level progressively fell during the vitamin B_6 depletion phase and returned to normal level upon repletion with pyridoxine. In a subsequent study on vitamin C deficiency in adult men, the urinary excretion of free pyridoxine was measured. During the vitamin C-depletion period, a 3½-fold increase occurred in the free urinary pyridoxine, even though the subjects were receiving a constant daily intake of 2.5 mg of pyridoxine. Upon repletion with AA, urinary free pyridoxine levels progressively fell until they returned to the levels observed during the control period.

6. **Vitamins C and B_9:** Anemia is observed in vitamin C-deficient patients. A normochromic, normocytic, or macrocytic type of anemia is generally observed, although megaloblastic anemia has been reported in some patients. The presence of megaloblastic anemia in some scorbutic patients has given rise to the consideration that either a dietary deficiency of folic acid (vitamin B_9) existed or that folate metabolism was impaired in vitamin C deficiency (**Cox** *et al*, 1966). In some instances, the megaloblastic anemia was effectively treated with AA alone, while in other reports, supplements of folic acid were required. In 1975, **Stokes** *et al* reported megaloblastic anemia in a patient with scurvy.
7. **Vitamins B_1 and B_2:** Thiamine deficiency is often accompanied by disturbance of riboflavin metabolism, causing considerable excretion of riboflavin in urine.
8. **Vitamins B_1, B_2 and B_3:** Administration of thiamine, riboflavin and pantothenic acid (vitamin B_3) offered partial protection from spontaneous seizures in vitamin B_6-deficient experimental animals.
9. **Vitamins B_1, B_2 and B_5:** The livers of rats fed on thiamine-deficient or riboflavin-deficient diets contained smaller amounts of niacin (vitamin B_5) than normal controls. Treatment of pellagra with niacin precipitates symptoms of thiamine and riboflavin deficiencies indicating that niacin deficiency is accompanied by secondary deficiencies of thiamine and riboflavin.
10. **Vitamins B_2 and B_3:** A deficiency of pantothenic acid (vitamin B_3) affects adversely the metabolism of riboflavin from the liver. Administration of pantothenic acid to deficient animals increases the mobilization of riboflavin from liver and raises the levels of riboflavin in blood.
11. **Vitamins B_2 and B_6:** Riboflavin deficiency in the rat showed the uptake of labeled pyridoxine (vitamin B_6) into the liver and decreased the conversion of pyridoxine to its metabolites. Pyridoxine phosphate oxidase in the livers of riboflavin-deficient rats was only 15% that of control animals. Riboflavin deficiency also lowered pyridoxal kinase activity but did not influence pyridoxal phosphate phosphatase activity. The results suggest that the slower rate of pyridoxal phosphate (PALP) synthesis in riboflavin deficiency may be due to an impairment in the pyridoxine phosphate oxidase reaction as well as to a reduction in the pyridoxal kinase activity (**Lakshmi** and **Bamji**, 1974). Lakshmi and Bamji, in 1976, further studied vitamin B_6 metabolism in human subjects who showed both biochemical and clinical signs of riboflavin deficiency. Normal and deficient subjects were administered pyridoxine, and their blood PALP levels followed. Subjects with riboflavin deficiency signs had significantly lower blood levels of PALP when compared to the control subjects. Clinical and biochemical deficiency signs as well as the ability to convert pyridoxine to PALP were all improved by riboflavin administration.
12. **Vitamins B_3 and B_7:** Biotin (vitamin B_7) deficiency may be aggravated by a simultaneous pantothenic acid deficiency. The addition of biotin to the diet not only protects the animals from a biotin deficiency but also reduces the severity of the symptoms of pantothenic acid deficiency (**Banister**, 1976).
13. **Vitamins B_6 and B_{12}:** Vitamin B_6 deficiency has been reported to cause an impairment in vitamin B_{12} (cyanocobalamin) absorption in the rat, resulting in a lowering of vitamin B_{12} levels in the serum and in reduced vitamin B_{12} stores in the liver (**Ranke** *et al*, 1960).

14. **Vitamins B_9 and B_{12}:** For the conversion of folic acid (vitamin B_9) to folinic acid (citrivorum factor, CF), vitamin B_{12} is essential. Folic acid therapy of patients suffering from sprue, causes an increase in plasma vitamin B_{12} content. Vitamin B_{12} deficiency causes a rise in unconjugated folates and a marked depletion of intracellular conjugated folates.

VITAMIN–MINERAL INTERRELATIONSHIPS

Certain vitamins influence the metabolism of some minerals. Following vitamin–mineral interrelationships exist:

1. **Vitamin B_2 and iron:** Ferrireductase (known also as NADH-FMN oxidoreductase) has been proposed for the mobilization of iron from ferritin. Riboflavin deficiency was reported to decrease hepatic iron mobilization and to impair the absorption of dietary iron. It did not appear to interfere with iron-induced biosynthesis of ferritin (**Zaman** and **Verwilghen**, 1977).
2. **Vitamin C and iron:** Food iron is absorbed by the intestinal mucosa from 2 separate pools of heme and nonheme iron (NHI). Heme iron, derived from hemoglobin and myoglobin, is well absorbed and relatively little affected by other foods eaten in the same meal. On the contrary, the absorption of NHI, the major dietary pool, is greatly influenced by meal composition. Ascorbic acid (AA) is a powerful enhancer of NHI absorption and can reverse the inhibiting effect of such substances as tea and calcium/phosphorus. Its influence may be less pronounced in meals of high iron availability– those containing meat, fish, or poultry. The enhancement of iron absorption from vegetable meals is directly proportional to the quantity of AA present. The absorption of soluble inorganic iron added to a meal increases in parallel with the absorption of NHI, but AA has a much smaller effect on insoluble iron compounds, such as ferric oxide or ferric hydroxide, which are the common food contaminants. Ascorbic acid facilitates iron absorption by forming a chelate with ferric iron at acid pH that remains soluble at the alkaline pH of the duodenum (**Sayers** *et al*, 1974).
3. **Vitamin C, iron and cadmium:** Toxic levels of dietary cadmium (5 – 200 ppm) interferred with iron absorption and produced iron deficiency. Supplements of iron (particularly, the divalent form) and ascorbic acid protected against the cadmium. With very low levels of dietary cadmium (0.06–0.08 ppm) typical of those in the diets of humans, supplements of iron and AA markedly decreased cadmium concentrations in the liver, kidney and small intestine. Iron deficiency changed the distribution of cadmium within the body (**Jacobs** *et al*, 1969).
4. **Vitamin C, lead and mercury:** Iron will alleviate lead toxicity but ascorbic acid is ineffective. Ascorbic acid will alleviate mercury toxicity, but iron exhilarates this condition (**Hill**, 1979).
5. **Vitamin D, Calcium and Phosphorus:** In vitamin D deficiency, the absorption of calcium and phosphorus from the intestines is very low. Administration of vitamin D (cholecalciferol) corrects the defect.
6. **Vitamin E and selenium:** An organic compound of selenium (known as *factor 3*) appears to function in the same way as vitamin E (tocopherol) in preventing necrosis of the liver in animals fed on low-protein diets. Factor 3 also prevents lipid peroxidation occurring in tissues of vitamin E-deficient animals.

Table 39–4 summarizes some characteristic features of various water-soluble vitamins.

Table 39–4. Water-soluble vitamins

Vitamin	*Common name(s)*	*Chemical name*	*Sources*	*Metabolic functions*	*Deficiency diseases(s)**	*Daily requirement of man*
B_1	Antiberiberi factor or Antineuritic factor	Thiamine or Aneurin, $C_{12}H_{17}ON_4S$	All plant and animal foods; Cereals, heart, liver, kidney and milk	Thiamine pyrophosphate (TPP) as coenzyme in many decarboxylation reactions	Beriberi (in man); Polyneuritis (in birds)	1.4–1.7 mg
B_2	'Yellow enzyme'	Riboflavin or Lactoflavin, $C_{17}H_{20}O_6N_4$	Milk, liver, kidney, heart and green vegetables; Occur almost exclusively as a constituent of either FMN or FAD	Phototropic curvature of plant organs; FAD as a cofactor in respiration; Also in bioluminescence	Glossitis; Cheilosis; Corneal vascularization; Dermatitis	1.4–1.7 mg
B_3	Filtrate factor or Yeast factor	Pantothenic acid, $C_9H_{17}O_5N$	Yeast, liver and eggs; Also fruits, vegetables and skimmed milk; In most animal tissues, it occurs as its coenzyme	Participates in the formation of coenzyme A	Adrenal cortical insufficiency in man; Depigmentation of hair in rats, pigs and dogs; Depigmentation of feathers in chicks	5.0–10.0 mg
B_5 (or G)	Pellagra-preventive factor or Antiblacktongue factor	Niacin, $C_6H_5O_2N$ Niacinamide, $C_6H_6ON_2$	Yeast, liver, meat, poultry, vegetables and fruits; Milk and eggs devoid of niacin	As a constituent in two pyrimidine nucleotide coenzymes, NAD and NADP	Pellagra (in man); Blacktongue (in dogs)	15–20 mg
B_6	Antidermatitis factor	Pyridoxine, $C_8H_{11}O_3N$ Pyridoxal, $C_8H_9O_3N$ Pyridoxamine, $C_8H_{12}O_2N_2$	Cereal grams, yeast, egg yolk and meat ; Pyridoxal and pyridoxamine also occur as their phosphates	Serve as growth factor to a number of bacteria; Also act as a carrier in active transport of amino acids across cell membranes	Convulsions and anemia in human infants; Acrodynia in rats; Anemia and nervous lesions in dogs and chicks	2.0 mg

Table 39–4. (Cont'd)

Vitamin	*Common name(s)*	*Chemical name*	*Sources*	*Metabolic functions*	*Deficiency diseases(s)**	*Daily requirement of man*
B_7 (or H)	Antiegg white injury factor or Coenzyme R	Biotin, $C_{10}H_{16}O_3N_2S$	Yeast, liver, peanuts and eggs	As a prosthetic group for many enzymes which bring about carboxylation and synthesis of fatty acids	Very rare; Sterilization of intestine, however, may lead to dermatitis, loss of hair and edema	25 mg
B_9 (or M)	Liver Lactobacillus casei factor	Folic acid or folacin, $C_{19}H_{19}O_6N_7$	Liver, kidney, yeast and wheat	Enzymatic synthesis of purines, pyrimidines and amino acids	Megaloblastic anemia and Tropical sprue (in man); Macrocvtic anemia (in monkey)	1.0–2.0 mg
B_{12}	Anti-pernicious anemia factor (APA factor)	Cyanocobalamin, $C_{63}H_{88}O_{14}N_{14}PCo$	Animals and microorganisms only; Plants devoid of it	Nucleic acid metabolism	Pernicious anemia	1.0 mg
C	Antiscorbutic factor or Cevitamin	Ascorbic acid or Cevitamic acid, $C_6H_8O_6$	Freash fruits and vegetables- Acerola fruit, citrus fruit, tomatoes, green pepper and new potatos	Reducing agent; Biosynthesis of adrenal steriod hormines; Synthesis of colagen	Scurvy	75–100 mg
P	Bioflavonoids	Hesperidin	Of plant origin– juice, peel and pulp of citrus fruits, in grapes etc.	As antioxidant; permeability and fragility	Hemorrhage Maintenance of capillary	Not known (in animals)
Choline	–	Trimethyl-2-hydroxyethyl-ammonium hydroxide	Widely distributed; Richest source is egg yolk; Liver, kidney, meat and cereals are other good sources.	Part of phospholipids; A lipotropic agent; Acts as a methyl group donor.	Anorexia (in puppies); Fatty liver (in rats)	Not known

*Deficiency diseases have been mentioned in relation to man, if not stated otherwise.

REFERENCES

1. **Bhutiani S :** The anti-scurvy vitamin C. *Science Reporter. 257–260, 1973.*
2. **Blakley RL :** The Biochemistry of Folic Acid and Related Pteridines. *Elsevier, North Holland, Amsterdam. 1969.*
3. **Blakley RL, Whitehead VM :** Folates and Pterins. *Vol. 1-3, John Wiley and Sons, New York. 1984–86.*
4. **Briggs MH (editor) :** Vitamins in Human Biology and Medicine. *CRC Press, Boca Raton, Fla. 1981.*
5. **Briggs MH (editor) :** Recent Vitamin Research. *CRC Press, Boca Roton, Fla. 1984.*
6. **British Medical Journal :** The Nutrition of Man. *Vol. 37, 1981.*
7. **Brown GM, Reynolds JJ :** Biogenesis of the water-soluble vitamins. *Ann. Rev. Biochem.* **32** : *419–462, 1963.*
8. **Burns JJ (editor) :** Vitamin C. *Ann. N. Y. Acad. Sci.*, **92** : *1, 1961.*
9. **Chenoy JJ *et al* :** Role of free radicals in germination. *Sci. & Cult.* **36** : *690 – 693, 1969.*
10. **Chick H :** The causation of pellagra. *Nutr. Abstr. and Revs.*, **20** : *523–547, 1951.*
11. **Ciba Foundation Study Group :** Mechanism of Action of Water-soluble Vitamins. *Little Brown and Co., Boston, 1962.*
12. **Cinnamon AD, Beaton JR :** Biochemical assessment of vitamin B_6 status in man. *Am. J. Clin. Nutr.* **26** : *96, 1970.*
13. **Daniel LJ :** Inhibitors of the vitamins of the B-complex. *Nutr. Abstr. and Revs.*, **31** : *1–31, 1961.*
14. **Darby WJ, McNutt KW, Todhunter EN :** Niacin. *Nutr. Rev.* **33** : *289, 1975.*
15. **Dolphin D :** Vitamin B_{12}. *Vols. 1 and 2. Wiley, New York. 1981.*
16. **Dolphin D, Poulson R, Avramovic O (editors) :** Vitamin B_6 Pyridoxal Phosphate, Parts A and B. Chemical, Biochemical and Medical Aspects. *John Wiley and Sons, New York. 1986.*
17. **Gubler CJ, Fujiwara W, Dreyfus PM :** Thiamine. *Wiley. 1976.*
18. **Jaenicke L :** Vitamin and coenzyme function : Vitamin B_{12} and folic acid. *Ann. Rev. Biochem.*, **33** : *287–312, 1964.*
19. **Jansen BCP :** The physiology of thiamine ; *Vitamins and Hormones.* **7** : *84–110, 1949.*
20. **Kuksis A, Mookerjea S :** Choline. *Nutr. Rev.*, **36** : *201, 1978.*
21. **Leitch I, Hepburn A :** Pyridoxine, metabolism and requirement. *Nutr. Abstr. and Revs.*, **31** : *389–401, 1961.*
22. **Levine M :** New concepts in the biology and biochemistry of ascorbic acid. *N. Engl. J. Med.* **314** : *892, 1986.*
23. **Lucock M :** Is folic acid the ultimate functional food component for disease prevention? *BMJ.* **328** : *211–214, 2004.*
24. **Luhby AL, Cooperman JM :** Folic acid deficiency in man and its interrelationships with vitamin B_{12} metabolism. *Advances in Metabolic Disorders.* **1** : *264–333, 1964.*
25. **Martell AE :** Reaction pathways and mechanism of pyridoxal catalysis. *Adv. Enzymol.* **53** : *163, 1982.*
26. **McCormick DB, Wright LD :** Vitamins and Coenzymes. In the series Methods in Enzymology. Vols. 62, 66 and 67. *Academic Press, Inc., New York. 1979, 1980.*
27. **Meiklejohn AP :** Physiology and biochemistry of ascorbic acid. V*itamins and Hormones.* **11** : *62–96, 1953.*
28. **Mistry SP, Dakshinamurti K :** The biochemistry of biotin. *Vitamins and Hormones.* **22** : *1–56, 1965.*
29. **National Research Council :** Recommended Dietary Allowances. *National Academy of Sciences (U.S.), 1980.*
30. **Obermayer M, Lynen F :** Structure of biotin enzymes. *Trends Biochem. Sci.* **1** : *169, 1976.*

31. **Padhi SC :** Vitamin C and old age. *Science Reporter. 589–590, 1973.*
32. **Plaut GWE, Smith CM, Alworth WL :** Biosynthesis of water-soluble vitamins. *Ann. Rev. Biochem.,* **43** : *899, 1970.*
33. **Readings from Scientific American :** Human Nutrition. *Freeman, San Francisco. 1978.*
34. **Reynolds RD, Leklem JE (editors) :** Vitamin B_6 : Its Role in Health and Disease. *Alan R. Liss, New York. 1985.*
35. **Rivlin RS :** Hormones, drugs and riboflavin. *Nutr. Rev.* **37** : *241, 1979.*
36. **Scriver CR :** Vitamin B_6 deficiency and dependency in man. *Am. J. Dis. Child.* **113** : *109, 1967.*
37. **Smith EL :** Vitamain B_{12}. 3rd ed. *John Wiley and Sons, Inc., New York. 1965.*
38. **Staudinger HA :** Ascorbic acid. *Trends Biochem. Sci.,* **3** : *211–212, 1978.*
39. **Stokstad ELR, Koch J :** Folic acid metabolism. *Physiol. Revs.,* **17** : *83–116, 1967.*
40. **Terroine T :** Physiology and biochemistry of biotin. V*itamins and Hormones.* **20** : *1–42, 1960.*
41. **Thakur RS :** The versatile vitamin B_{12}. *Science Reporter. 579–583, 1973.*
42. **Vitale JJ :** Present knowledge of folacin. Nutr. *Rev.,* **24** : *289, 1966.*
43. **Wagner AF, Folkers K :** Vitamins and Coenzymes. *Interscience Publishers, Inc., New York. 1964.*
44. **Wagner F :** Vitamin B_{12} and related compounds. *Ann. Rev. Biochem.,* **35** : *405–434, 1966.*
45. **Weiner M, Van Eys J :** Nicotinic Acid. *Marcel Dekker, New York. 1983.*
46. **Wolstenholme GEW, O'Connor M (editors) :** Thiamine Deficiency : Biochemical Lesions and Their Chemical Significance. *J & A, Churchill Ltd., London. 1967.*
47. **Wood HG, Barden RE :** Biotin enzymes. *Ann. Rev. Biochem.* **46** : *385, 1977.*

PROBLEMS

1. Individuals with thiamine deficiency display a number of characteristic neurological signs: loss of reflexes, anxiety, and mental confusion. Suggest a reason why thiamine deficiency is manifested by changes in brain function.
2. The neurological disorders seen in vitamin B_{12} deficiency are caused by progressive demyelination of nervous tissue. How does lack of B_{12} interfere with formation of the myelin sheath?
3. What should a person eat to get all the vitamin C his body can use?
4. Scurvy is to vitamin C as fatty liver is to:
   (*a*) vitamin K
   (*b*) pantothenic acid
   (*c*) riboflavin
   (*d*) choline
5. If maize is the staple diet, the deficiency to develop is likely to be that of:
   (*a*) nicotinic acid
   (*b*) cholecalciferol
   (*c*) pyridoxine
   (*d*) thiamine
6. Macrocystic anemia can result from the deficiency of:
   (*a*) folic acid
   (*b*) vitamin B_2
   (*c*) inositol
   (*d*) vitamin B_{12}
7. If you are a complete vegetarian, do you miss any vitamins?

8. Flour is often enriched with:
(*a*) vitamin A (*c*) vitamin B_{12}
(*b*) vitamin B_1 (*d*) vitamin E

9. Can B vitamins cut the risk of developing Alzheimer's disease?

10. Rain drops contain which of the following vitamins:
(*a*) vitamin A (*c*) vitamin B_{12}
(*b*) vitamin B_2 (*d*) vitamin D

11. Can vitamins help in the treatment of depression?

12. Why do doctors prescribe vitamin tablets along with antibiotics?

13. Can vitamin B_6 cure carpal tunnel syndrome in humans?

14. Which microbe is highly sensitive to the presence of biotin (vitamin H)?
(*a*) *Lactobacillus vugraicus*
(*b*) *Loctobacillus casei*
(*c*) *Bacillus anthracis*
(*d*) *Micrococcus orthobacter*

15. Egg has a pride of place among the foods of animal origin. It is considered as a balanced meal that comes ready in a natural pack. Yet eating raw egg is considered not good for health. Why?

16. Deficiency of which vitamin causes bleeding tendencies?

17. Name the scientist who advocated massive doses of vitamin C as a preventive measure for common cold and also cancer.

18. What is wrong with synthetic vitamins?

19. What are the best sources of vitamin C?

20. What is good about oranges?

21. What needs to be stored in dark bottles?
(*a*) biotin (*c*) riboflavin
(*b*) nicotinic acid (*d*) retinol

ANSWERS TO PROBLEMS

1. Glucose is the primary fuel of the brain, and the brain is particularly sensitive to any change in the availability of glucose for energy production. A key reaction in glucose catabolism is the thiamine pyrophosphate-dependent oxidative decarboxylation of pyruvate to acetyl-CoA, and thus a thiamine deficiency reduces the rate of glucose catabolism.

2. One of the reactions in which vitamin B_{12} participates is conversion of methylmalonyl CoA to succinyl CoA (a step in the catabolism of valine and isoleucine). Methylmalonyl CoA is a competitive inhibitor of malonyl CoA in fatty acid biosynthesis, necessary for the maintenance of the myelin sheath. Secondly, methylmalonyl CoA can be used in fatty acid synthesis leading to formation of branched-chain fatty acids, which might disrupt normal membrane structure.

3. Five servings of fruits and vegetables a day will give one plenty of vitamin C (200 mg) – you would not need many supplements.

4. (*d*) choline

5. (*a*) nicotinic acid

6. (*a*) folic acid

7. Yes, B_{12} is found only in animal products. To compensate, get B_{12} from a supplement or fortified products such as soy milk.

8. (*b*) vitamin B_1
9. A new study showed that people with high homocysteine levels were twice as likely to develop Alzheimer's as those with low levels. To cut the risk of Alzheimer's, take a multivitamin that has B_6, B_{12} and folic acid. Wheat germ is a good source of B vitamins and folic acid.
10. (*c*) vitamin B_{12}
11. Possibly. Vitamin B_{12} and folate deficiencies may cause symptoms of depression and may also cause a poor response to antidepressants.
12. Doctors prescribe vitamin tablets along with any antibiotic prescription to prevent our body from suffering vitamin deficiency. The antibiotics are powerful chemicals that can kill microorganisms, chiefly bacteria. When ingested these drugs kill not only the disease-causing bacteria but also the harmless ones normally residing in our intestines. Called the 'intestinal microflora', these friendly bacteria convert some of the food constituents into vitamin precursors which are absorbed into the body through intestine. But for this help some of the essential vitamins like those of vitamin-B complex will not be available for us, for our body lacks the machinery to convert food constituents into vitamin precursors. Use of powerful antibiotics kills these helpful bacteria and curtail the supply of vitamin precursors to the body. Therefore, in order to keep the body supplied with usual dose of vitamins, doctors prescribe readymade easily absorbable vitamin-B tablets along with antibiotics. Vitamins are also believed to help in early recovery from the ailments.
13. Some experts opine that more than 90% of carpal tunnel cases can be cured by vitamin B_6. Take 100 mg daily for 90 days to feel relief from symptoms. Do not take more than 100 mg of B_6 daily for extended periods of time, as nerve damage may result.
14. (*b*) *Lactobacillus casei*
15. Raw egg white contains a protein named *avidin*. Avidin binds with biotin. Thus, the vitamin is not available for uptake by the body and biotin deficiency occurs.
16. Vitamin C (= ascorbic acid)
17. **Linus Carl Pauling**
18. Nature intended food to be consumed in its *whole form*. Synthetic vitamins are fractions or isolated pieces of the whole. The human body does not properly absorb vitamins in their synthetic form. Although typically cheap and easily available in supermarkets, the synthetic vitamins often give you massive quantities of some vitamins and inadequate amounts of others.
19. According to one estimate, camu camu berry, which grows in the South American rain forest, is the best source of vitamin C. It contains 30 to 60 times as much vitamin C as an orange. Other top vitamin C-source fruits are: kiwi (98 mg), papaya (85 mg), and oranges and lemons (53 mg).
20. One orange gives you 100% of your RDA of vitamin C. Oranges and all citrus fruits contain flavonoids that are able to cross the blood-brain barrier and help protect the brain from oxidative stress.
21. (*c*) riboflavin

PART VI

Analytical Biochemistry

A selection of preparative ultracentrifuge

The sample tubes of the swinging bucket rotors (*rear*) are hinged so that they swing from the vertical to the horizontal position as the rotor starts spinning, whereas the sample tubes of the other rotors have a fixed angle relative to the rotation axis.

(*Courtesy : Beckman Instruments, Inc.*)

"There is no higher or lower knowledge but only one flowing out of experimentation"

–Leonardo da Vinci (LT, 1452–1519)

CONTENTS

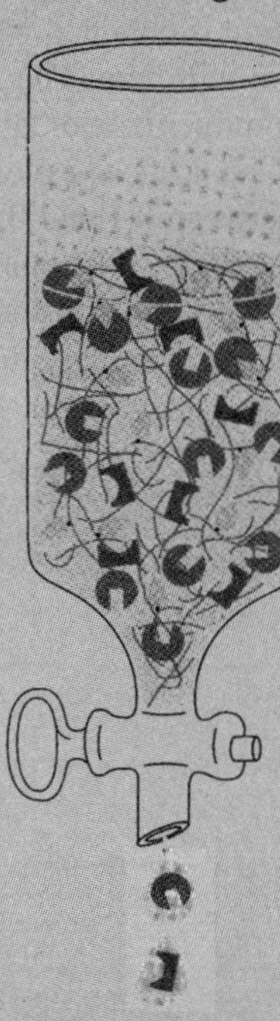

The separation of macromolecules by affinity chromatography

The cutout squares, semicircles, and triangles are schematic representations of ligand-binding sites on macromolecules. Only those ligand-binding sites represented by the orange circles with triangle cutouts specifically bind to the chromatographic matrix-anchored ligands (*yellow*).

CHAPTER

40

Biochemical Techniques

"Theory is good thing but a good experiment lasts forever."

–*Kapitza*

OBSERVATIONS ON TISSUES

It is often desirable to conduct biochemical studies on an individual organ or individual cells rather than on the body as a whole. But since the growing of single cells in multicelled organisms is not easy and also that after many generations cells from specialized tissues tend to revert to a more primitive type, the biochemist is forced to study either intact organs or preparations from these organs. He can also follow another technique of breaking up the tissues and studying the behaviour of a broken cell preparation.

PERFUSION

This is the simplest way to study intact cells organized in tissues. Here the whole organs is perfused *in situ* or *in vitro* with blood or an isotonic saline solution. The solution is kept recirculating and is also reoxygenated. Arrangements are also made to remove CO_2 from the system. The entire preparation is kept at 37°C by means of a water jacket. In such experiments the tissue itself can be analyzed at the end of perfusion and the changes noted and compared with reference to the composition of the organ before perfusion. The method, however, suffers from certain *drawbacks*:

(*a*) During perfusion experiments, one is sure to lose certain of the regulatory mechanism such as hormonal and nervous control which operate upon the organ in its normal locus.

(*b*) Moreover, inferences drawn from such experiments are always those obtained under artificial conditions in which the tissue is operating.

TISSUES SLICES

This techinque has been much used in the past for organs such as liver, kidney and brain and also for certain plant tissues like roots and tubers. Here thin slices (approx. 50 μ thick) of tissues are cut. Many of the cells are, of course, damaged but some between 50–70% remain intact. Slicing also offers sufficient surface to the bathing fluid to permit adequate exchange of materials and waste products so that the tissues remain viable for several hours. These slices are then suspenced in buffered isotonic saline solutions in airtight vessel to prevent evaporation. Excess O_2 is given by filling the large airspace, present above the liquid, with pure O_2. The vessel is maintained at 37°C and is frequently shaken to promote diffusion. The changes in cellular composition are then studied.

The two serious *drawbacks* with this techniques are as follows :

(*a*) The cells are feared to 'die' early during such operations. Henceforth, it is best to conduct the experiments for the shortest possible time but consistent with accuracy in analysis of the various changes.

(*b*) Although a fraction of cell membranes is cut in slicing, most cellular constituents still remain contained within the cells.

HOMOGENIZATION

In this the cells are broken up completely so that a homogeneous mixture is obtained which can be adequately oxygenated. The suspension medium can also be changed at will here. The mincing of the material nowadays is done using either a Potter-Elvehjme homogenizer or a blender (Fig. 40–1). **Potter-Elvehjem homogenizer** consists of a glass tube into which fits a pestle of glass or of polytetrafluoroethylene. The clearance between tube and pestle is between 0.1 and 0.15 mm. The pestle is driven by a motor at 2,000 rpm. A **blender**, on the other hand, consists of a glass vessel fitted with rotating knives. In both the instruments, the tissue is roughly chopped and suspended in a relatively large volume of suitable medium. In order to prevent local heating, the glass tube must be immersed in ice.

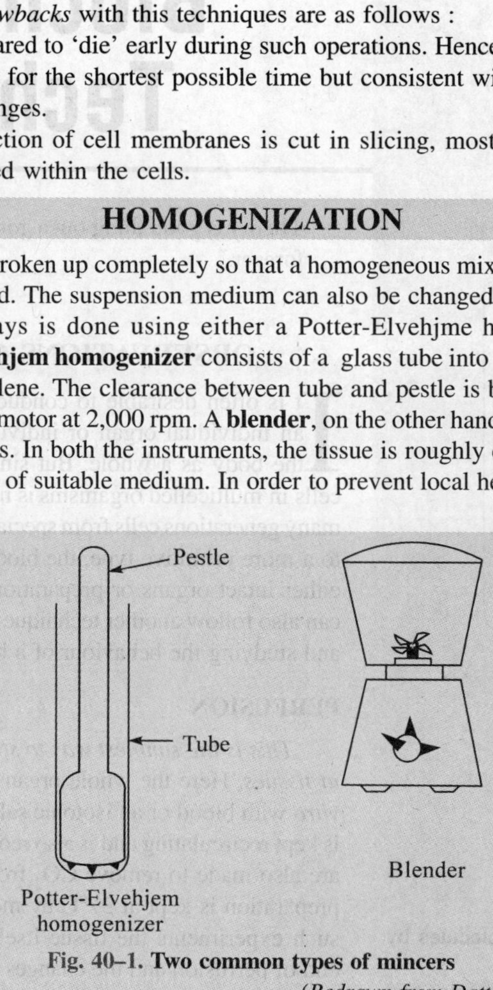

Fig. 40–1. Two common types of mincers

(*Redrawn from Datta SP and Ottaway JH, 1965*)

After homogenization, the various cell components–nuclei,mitochondria, microsomes etc., can be separated by differential centrifugation (*see* subsequent section). The various fractions so obtained can be identified by their biochemical composition and, to a limited extent, by histological staining. The intracellular organelles can, however, be observed under electron microscope.

This procedure has proved especially useful in studies concerned with determining the location of chemical processes within the cell.

DIFFERENTIAL CENTRIFUGATION

The centrifugal force exerted on a particle in the solution is expressed in multiples of the force exerted by gravity. *The centrifugal force is proportional to the radius of the centrifugal head and to the square of angular velocity*. Hence, it is more convenient to use relatively small heads rotating at high speeds. Thus, a head of approximately 10 cm diameter rotating at about 40,000 rpm will produce nearly 100,000 g. At such high speeds, the head is run in a vacuum to prevent the heat produced by air friction. The tube containing the homogenate is usually held at an angle to the axis of rotating to keep the path of particles through the solution as short as possible (Fig. 40–2).

The broken cell preparation is loosely referred to as cell-free preparation or '**homogenate**'. Properly prepared, a homogenate is the sum of all the cell components obtained by rupturing the cell membrane.

The homogenate is first diluted and then centrifuged at a low speed to remove nuclei etc. The supernatant liquid is then poured off and centrifuged at a higher speed to remove next fraction. The

Supernatant is that part of the cell material not sedimented by a centrifugal force known to sediment any subcellular structure.

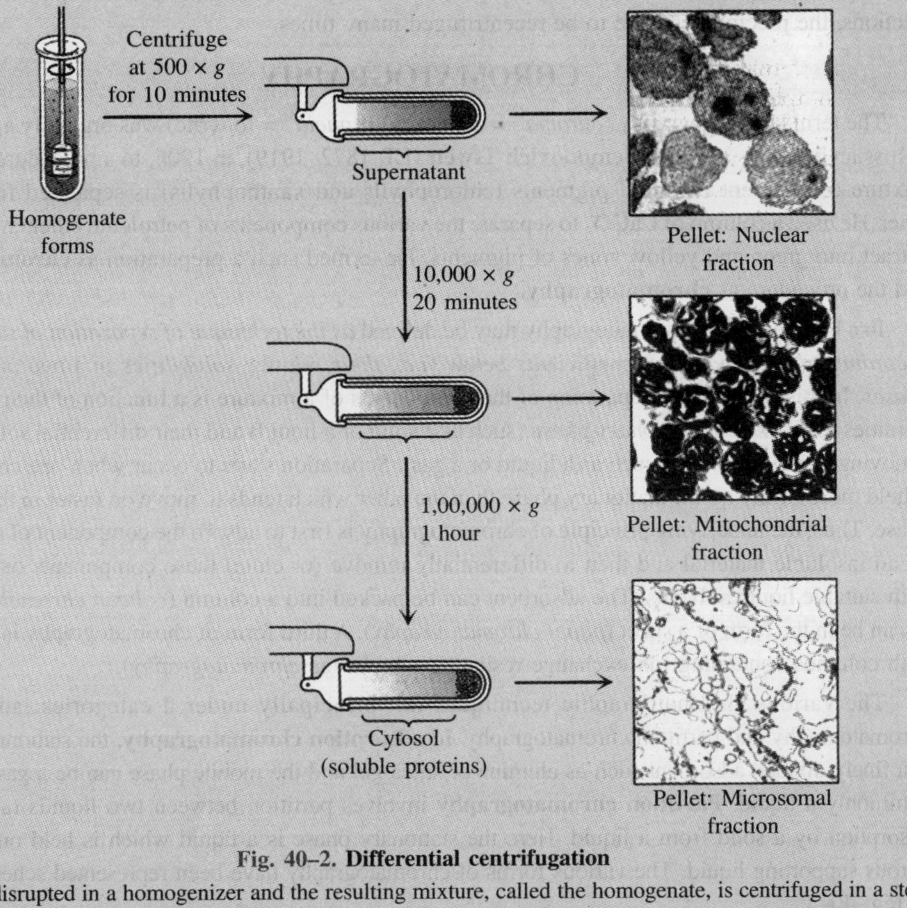

Fig. 40–2. Differential centrifugation

Cell are disrupted in a homogenizer and the resulting mixture, called the homogenate, is centrifuged in a step-by-step fashion of increasing centrifugal force. The denser material will form a pellet at lower centrifugal force than will the less-dense material. The isolated fractions can be used for further purification.

(Photos courtesy : S Fleischer and B Fleischer)

supernatant is then again contrifuged at still a higher speed and fraction separated. The removal of particles of a particular size depends on the magnitude of the field and also on the time for which the field is applied (refer Table 40–1).

Table 40–1. Separation of cell organelles by differential centrifugation

*Field**	*Time*	*Structure(s) separated*
700 g	10 min	Nuclei, cell membranes
5,000 g	10 min	Mitochondria, lysosomes
57,000 g	60 min	Microsomes, lysosomes
150,000 g	30 min	Ribosomes
Unsedimented		Cell sap, some ribosomes

* These centrifugal fields refer to homogenates in 0.25 M sucrose.

(*After Datta SP and Ottaway JH, 1969*)

In practice, a mixture of fractions is obtained on single centrifugation. In order to obtain pure fractions, the precipitates have to be recentrifuged many times.

CHROMATOGRAPHY

The term chromatography (*chroma*[G] = a colour; *graphein*[G] = to write) was orginally applied by a Russian botanist, **Michael Semonovich Tswett** (LT, 1872–1919), in 1906, to a procedure where a mixture of different coloured pigments (chlorophylls and xanthophylls) is separated from each other. He used a column of $CaCO_3$ to separate the various components of petroleum ether chlorophyll extract into green and yellow zones of pigments. He termed such a preparation as **chromatogram** and the procedure as **chromatography**.

In a broader sense, chromatography may be defined *as the technique of separation of substances according to their partition coefficients below (i.e., their relative solubilities in) two immiscible phases*. In this method, the separation of the components of a mixture is a function of their different affinities for a fixed or *stationary phase* (such as a solid or a liquid) and their differential solubility in a moving or *mobile phase* (such as a liquid or a gas). Separation starts to occur when one component is held more firmly by the stationary phase than the other which tends to move on faster in the mobile phase. Thus, the underlying principle of chromatography is first to adsorb the component of a mixture on an insoluble material and then to differentially remove (or elute) these components one by one with suitable liquid solvents. The adsorbent can be packed into a column (*column chromatography*) or can be in the form of a sheet (*paper chromatography*). A third form of chromatography is obtained with columns containing ion exchange resins (*ion exchange chromatography*).

The various chromatographic techniques fall principally under 2 categories: adsorption chromatography and partition chromatography. In **adsorption chromatography**, the stationary phase is a finely divided adsorbent such as alumina or silica gel and the mobile phase can be a gas or more commonly a liquid. **Partition chromatography** involves partition between two liquids rather than adsorption by a solid from a liquid. Here the stationary phase is a liquid which is held on an inert porous supporting liquid. The various forms of chromatography have been represented schematically in Fig. 40–3.

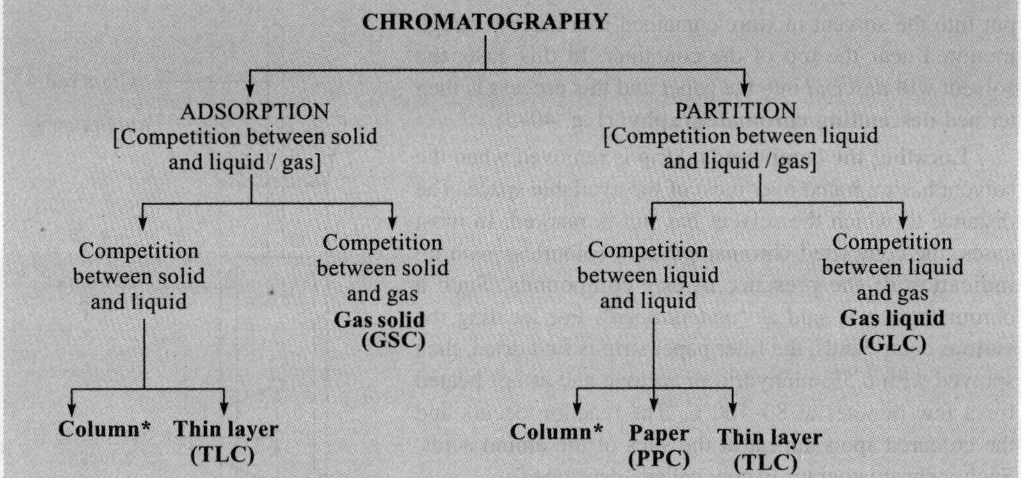

Fig. 40–3. Various forms of chromatography

* Other than gas chromatography

A few chromatographic techniques are described below.

PAPER PARTITION CHROMATOGRAPHY, PPC

Two Russian workers, **Izmailov** and **Schraiber** (1938) discovered this important technique. *This method is especially useful for the detection and separation of amino acids.* Here the filter paper strips are used to support a stationary water phase while a mobile organic phase moves down the suspended paper strip in a cylinder. Separation is based on a liquid-liquid partition of the compounds. Thus, this is essentially a form of partition chromatography between two liquid phases, although adsorption to the paper may also take place.

In this method, a drop of solution containing a mixture of amino acids (or other compounds) to be separated is applied at a marked point, about 3 cm from one end of a strip of filter paper. Whatman No. 1 paper is most frequently used for this purpose. The chemical composition of Whatman filter paper No. 1 is:

Component	Percentage
α-cellulose	98–99%
β-cellulose	0.3–1%
Pentosans	0.4–0.8%
Ash	0.07–0.1%
Ether soluble matter	0.015–0.1%

Whatman No. 3 MM is a thick paper and is best employed for separating large quantities of material; the resolution is, however, inferior to whatman No.1. For a rapid separation, Whatman Nos. 4 and 5 are convenient, although the spots are less well-defined. In all cases, the flow rate is faster in the 'machine direction', which is normally noted on the box containing the papers. The paper may be impregnated with a buffer solution before use or chemically modified by acetylation. Ion exchange papers are also available.

The filter paper is then dried and 'equilibrated' by putting it into an air-tight cylidrical jar which contains an aqueoues solution of a solvent. The most widely applicable solvent mixture is *n*-butanol : acetic acid : water (4 : 1 : 5), which is abbreviated as BAW. The end of the filter paper nearest the applied drop is inserted into the solvent mixture at the bottom of the jar, taking care that the marked point of application remains well above the level of the solvent in the jar. The paper is suspended in such a manner so that it hangs freely without touching the sides of the container. Thus, the solvent will *ascend* into the paper and this process is, therefore, termed **ascending chromatography** (Fig. 40–4).

Alternatively, the same end of the filter paper may be put into the solvent mixture contained in a narrow trough mounted near the top of the container. In this case, the solvent will *descend* into the paper and this process is then termed **descending chromatography**. (Fig. 40–5).

Locating the compounds. Strip is removed when the solvent has migrated over most of the available space. The distance to which the solvent has run is marked. In most cases, the completed chromatogram is colourless with no indication of the presence of any compounds. Such a chromatogram is said as '*undeveloped*'. For locating the various compounds, the filter paper strip is first dried, then sprayed with 0.5% ninhydrin in acetone and at last heated for a few minutes at 80–100°C. The reaction occurs and the coloured spots appear at the sites of the amino acids. Such a chromatogram is now called '*developed*'.

However, if the solution chromatographed is one derived from a tracer experiment, the compounds will be radioactive and can be located either by using a Geiger counter or by placing the paper strip against a sheet of x-ray film. The β-rays from the radioactive compounds will expose the film. The negative, on developing, will show dark spots at the site of radioactive compounds. This method has been used most successfully in tracing the carbon pathway in photosynthesis.

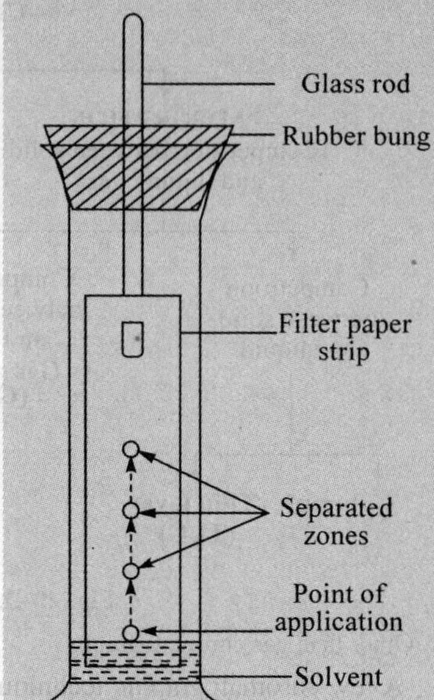

Fig. 40–4. One-dimensional paper chromatography (Ascending type)

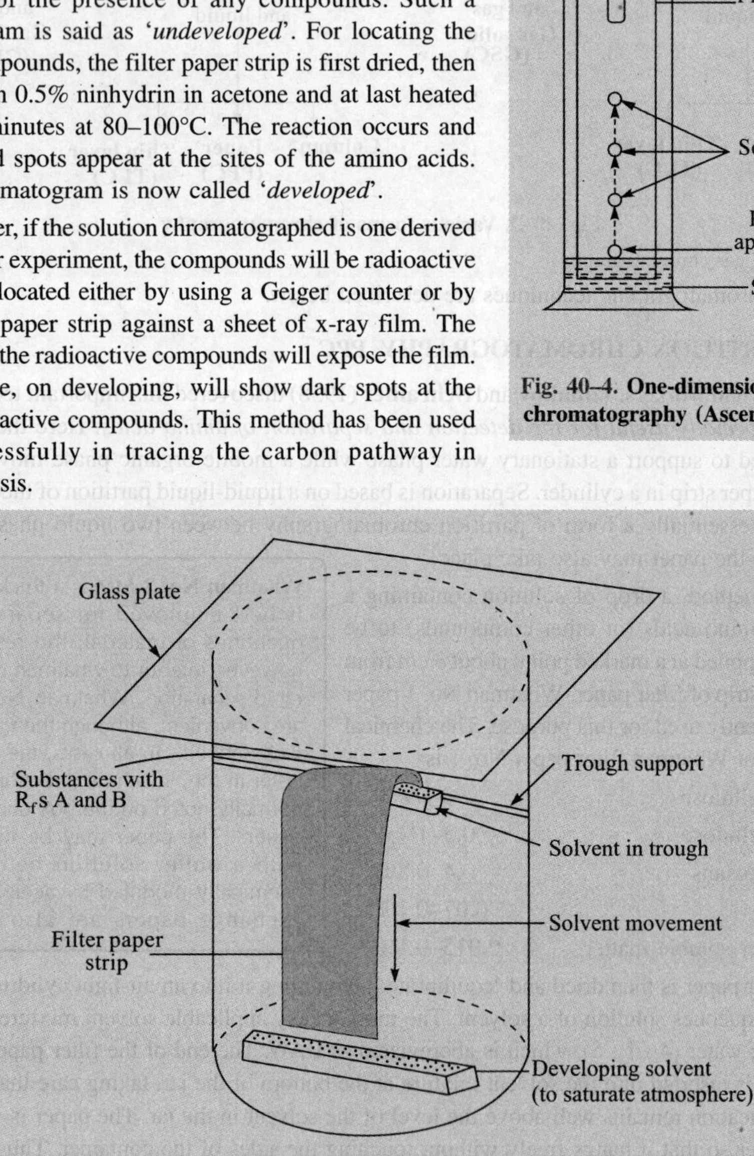

Fig. 40–5. One-dimensional paper chromatography (Descending type)

In paper chromatography, the stationary cellulose phase is more polar than the mobile organic phase. Amino acids with large nonpolar side chains (leucine, isoleucine, phenylalanine, tryptophan, valine, methionine, tyrosine) magrate farther in *n*-butanol ; acetic acid : water (4 : 1 : 5) than those with shorter nonpolar side chains (proline, alanine, glycine) or with polar side chains (threonine, glutamic acid, serine, arginine, aspartic acid, histidine, lysine, cysteine). This reflects the greater relative solubility of polar molecules in the hydrophilic stationary phase and on nonpolar molecules in organic solvents.

Identifying the compounds. In chromatography, the distance travelled by a particular component of a mixture (or solute) is used to identify it. The ratio of the distance travelled by a component (*i.e.*, amino acid) to that travelled by the solvent front, both measured from the marked point of application of the mixture, is called the **resolution front** (R_f) value for that component. Thus,

$$R_f = \frac{\text{Distance from origin run by the compound}}{\text{Distance from origin run by the solvent}}$$

The R_f value for a given compound varies with the type of solvent used (refer Table 40–2, for amino acids). *Similar R_f values, however, do not necessarily mean the identical compounds as different compounds may have identical R_f values in a given solvent system.* Henceforth, different solvent systems are used to identify the compounds by chromatography. It is preferable to chromatograph known amino acid standards along with the unknown mixture so that the R_f values be easily compared and the amino acids be safely identified.

Table 40–2. R_f values of amino acids in various solvent systems*

S. N.	*Amino acid*	*Phenol : Water*	*Collidine : Water*	*Butanol : Acetic Acid : Water, BAW*
1.	Glycine	0.36	0.26	0.26
2.	Alanine	0.55	0.32	0.38
3.	Valine	0.72	0.43	0.60
4.	Leucine	0.80	0.55	0.73
5.	Isoleucine	0.83	0.53	0.72
6.	Serine	0.30	0.30	0.27
7.	Threonine	0.43	0.32	0.35
8.	Cystine	0.24	0.11	0.08
9.	Methionine	0.74	0.53	0.55
10.	Proline	0.88	0.34	0.43
11.	Aspartic acid	0.22	0.23	0.24
12.	Glutamic acid	0.23	0.27	0.30
13.	Phenylalanine	0.83	0.54	0.68
14.	Tyrosine	0.55	0.59	0.45
15.	Tryptophan	0.71	0.59	0.50
16.	Histidine	0.62	0.30	0.20
17.	Arginine	0.54	0.17	0.20
18.	Lysine	0.41	0.11	0.14

* Each amino acid has a characteristic R_f value for a given solvent.

(Adapted from Fruton JS and Simmonds S, 1958)

Quantitation of the compounds. Quantitative analysis of the amino acids may be accomplished by cutting out each spot, eluting (or removing) the compound with a suitable reagent and performing a calorimetric (ninhydrin) or chemical (nitrogen) analysis. Alternatively, the filter paper strip may be sprayed with ninhydrin and heated so that the coloured spots indicating the location of amino acids may develop. The colour densities of these spots may be measured with a recording transmittance or reflectance photometer device.

The chromatographic technique described above allows *vertical* separation of amino acid and is termed as **one-dimensional paper chromatography**.

For obtaining better separation of components and to improve their quantitation, a **two-dimensional paper chromatography** is applied (Fig. 40–6). In this technique, a square sheet of filter paper, rather than a strip is taken. The test sample is applied to the upper left corner and chromatographed for some hours with one solvent mixture (*e.g.*, *n*-butanol : acetic acid : water). After drying to remove this solvent, the paper is tuned through 90° and chromatographed in a second solvent mixture (*e.g.*, collidine : water). This technique thus allows both *vertical* and *horizontal* separation of the amino acids.

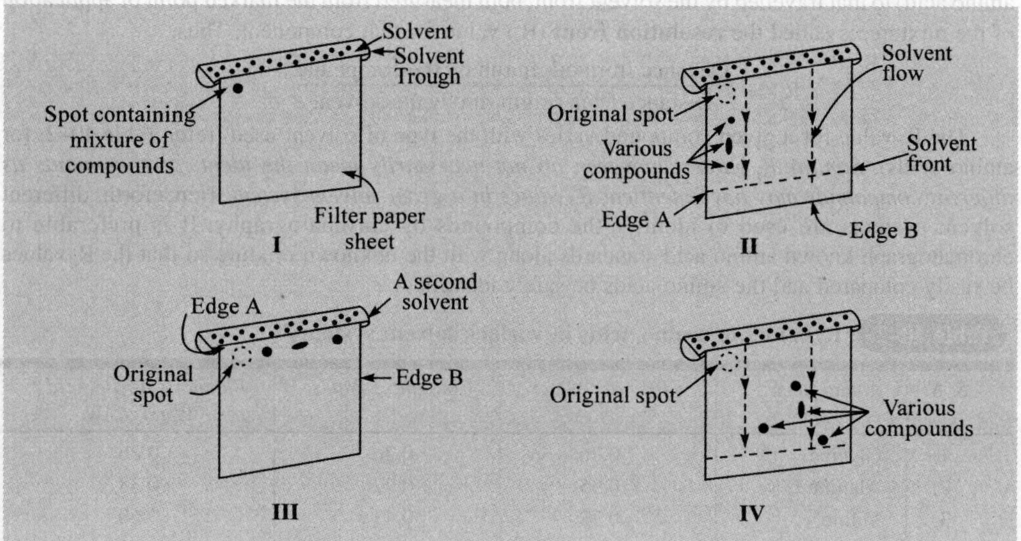

Fig. 40–6. Two dimensional (or '2-D') paper chromatography (Descending type)

Forces in operation. The movement of the solute molecules on the chromatogram depends on the net result of a number of forces operating in the system. These forces are of 2 types: propelling and retarding.

(*a*) ***Propelling forces–*** These include the capillary force and the solubility force of the solvent. The Whatman paper is made up of numerous fibrils which are placed very close to each other, thus forming a network of capillaries. The solute rises through these capillaries as a result of **capillary force**. *The smaller the bore of the capillaries, the greater is the height to which the solute rises.*

The **solubility forces of the solvent** refers to the capacity of the solvent to dissolve the solute. The rise of solute also depends on its solubility in the solvent being used. *The greater the solubility of the solute in the solvent, the greater is the height to which it rises in the chromatogram.*

(*b*) ***Retarding forces–*** Concurrent with the forces of propulsion, certain retarding forces also operate in the system which try to drag the solute molecules from moving in either direction. These retarding forces include the gravitational force and the partition force. Under the ascending chromatography, the solute molecules have to move against the **gravitational force** which acts from below and tends to retard the movement of solute molecules in the upward direction. Under the descending chromatography, obviously, the gravitational force does not figure as a retarding force, rather it assists in propulsion and, henceforth, acts as a propelling force.

The **partition force** refers to the force between liquid and liquid molecules. The interstices between the fibrils on the Whatman paper are occupied by the solvent molecules and the solute molecules, for their movement, have to displace them. Should the spaces not been filled with the solvent molecules, the movement of the solute molecules through them would have been much more facilitated.

Thus, the movement of the solute molecules on the Whatman paper in either direction is the net result of the interaction between various forces of propulsion and retardation. Movement of the solute molecules is exhibited only when the propelling forces exceed the retarding forces in magnitude.

THIN LAYER CHROMATOGRAPHY, TLC

Thin layer chromatography is adsorption chromatography performed on open layers of adsorbent materials supported on glassplates. *This technique combines many of the advantages of paper chromatography with those of column chromatography*. Here a thin uniform film of adsorbent (like silica gel or alumina powder) containing a binding medium (like calcium sulfate) is spread onto a glass plate. The thin layer is allowed to dry at room temperature and is then activated by heating in an oven between 100° to 250°C. The activated plate is then placed flat and samples spotted with micropipettes carefully on the surface of the thin layer. After the solvent has evaporated, the plates are placed vertically in a glass tank containing a suitable solvent. Within a short time (5 to 30) minutes), the various components get separated by the solvent rising through the thin layer. The glass plate is then taken out from the tank, allowed to dry and then the spots are detected by spraying the plate with a variety of reagents.

Superiority of TLC– The superiority of TLC over paper chromatography lies in the following facts:

(*a*) Because of the inorganic nature of the adsorbent (supporting medium), concentrated sulfuric acid spray followed by heating may be used to develop (or locate substances on) the chromatogram by charring and rendering visible any spots of organic nature.

(*b*) Moreover, amino acid mixtures, which require 18 hours for separation on paper, require as little as 3 hours using cellulose TLC.

(*c*) The advantage of this technique also lies in the choice of the adsorbents which allow separation not possible on paper.

Thus, *the speed, efficiency and sensitivity of TLC has made this technique most valuable to the biochemists*. Lipids including sterols may be neatly separated by TLC on alumina. Also after localization of these substances on the glass plate with reagents, selected spots of the plate's surface may be scrapped off and the compounds isolated by extraction of the alumina powder with suitable solvents.

ION EXCHANGE CHROMATOGRAPHY, IEC

Ion exchange chromatography is in wide use for the separation of amino acids. It employs synthetic resins such as a strongly acidic cation exchanger, **Dowex-50** and a strongly basic anion exchanger, **Dowex-1**. The former is a polystyrene sulfonic acid and the latter a polystyrene quaternary ammonium salt.

Two common cellulose derivatives, carboxymethyl cellulose (CMC) and diethylaminoethyl cellulose (DEAE) are successfully used in protein purification. CMC is a cationic derivative and DEAE, an anionic derivative. An even more versatile substance for the separation of protein is a cross-linked dextran called **sephadex**. When packed in a column and equilibrated with buffer, the polysaccharide acts as a molecular sieve for protein molecules. Five grades of sephadex are available. Sephadex G-25 beads retain molecules with molecular weights up to 5,000. In other words, the 'exclusion limit' of G-25, is 5,000. The exclusion limits of the remaining four grades, G-50, G75, G-100 and G-200– are 10,000, 50,000, 100,000 and 200,000 respectively.

Thus, electrostatic attraction of oppositely charged ions on a polyelectrolyte surface forms the basis of ion exchange chromatography. In principle, the ionized groups of the ion exchange materials form salts with ions of opposite charge and they exchange these ions for others when the relative salt concentrations in the solution are varied. Ion exchange chromatography is especially useful in separation and purification of nucleotides owing to their low molecular weights and the presence of ionizable groups.

In practice, the protein mixture is passed through a column of the modified cellulose contained in a glass tube and allowed to become attached to the column material. Increasing salt concentration and/or buffers of varying pH values are then passed through the column. Proteins will be washed off the column at different times depending on their molecular structure, net charge and side group. The essential process is depicted in Fig. 40–7.

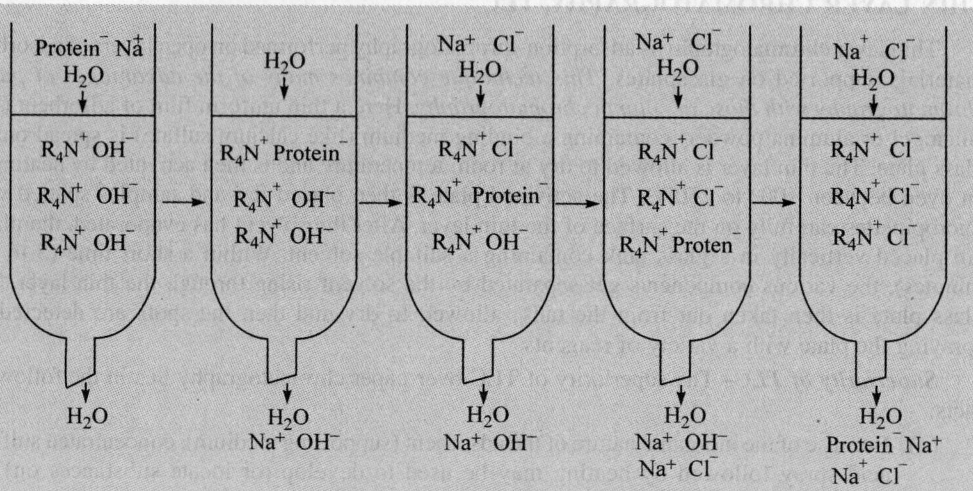

Fig. 40–7. Ion exchange chromatography

The various processes occurirng as protein is absorbed by and eluted from an ion exchange column are illustrated here. The ion exchange material is assumed to be an insoluble polymer having free substituted amine groups.

(Adapted from Fariley JL and Kilgour GL, 1966)

ISOTOPIC TRACER TECHNIQUE

In the studies of metabolic transformation of a particular substance, the primary objective of the biochemists remains to observe the fate of that substance *in vivo* under experimental conditions which cause minimum physiological disturbance to the test organism. Isotopic tracer technique has been most successfully used for such type of studies. Here one or more of the atoms in metabolite under study is "labelled" by means of one of the rare or artificially produced isotopes and its path followed while frequently testing the intermediary compounds at intervals.

The nucleus of each elementary species is characterized by an **atomic number**, which is equal to the number of protons and is also equal to the number of electrons around the nucleus when the atom exists in an electrically neutral state. All atoms of a particular element have the same atomic number. Each nucleus is also characterized by what is called as the **mass number** and which is equal to the total number of protons and neutrons contained therein. The difference between the two numbers is obviously the number of neutrons contained in the nucleus. *The various isotopes of an element have the same atomic number but different atomic mass.* Conventionally, the atomic number is written as subscript that precedes the elementry symbol and the mass number as superscript that follows the symbol. For example, nitrogen (atomic number 7) is found in nature both with a mass of 14 (designated as ${}_7N^{14}$) and with a mass of 15 (designated as ${}_7N^{15}$). Similarly, carbon with atomic number 6 has different mass number of 11, 12, 13 and 14 and the various isotopes are accordingly written as ${}_6C^{11}$, ${}_6C^{12}$, ${}_6C^{13}$ and ${}_6C^{14}$. Being implicit in the symbol, the atomic number is often not written. Thus, the two common isotopes of carbon are simply written as C^{12} and C^{14}.

Types of Isotopes. Isotopes are of 2 types– stable and radioactive. The *stable isotopes* are those whose nuclei do not undergo spontaneous decomposition. N^{14}, N^{15}, O^{18}, C^{13}, S^{33} and S^{34} belong to this category. Atoms whose nuclei decompose spontaneously with the emission of radiations are termed *radioactive isotopes.* C^{14} , S^{35}, P^{32} and H^3 belong to this group.

Stable Isotopes. The stable isotopes of biological importance are available in enriched concentrations. The important ones are H^2, N^{15}, C^{13} and O^{18}. These atoms occur in nature in the following relative abundance:

H^1/H^2 – 99.98/0.02

N^{14}/N^{15} – 99.63/0.37

C^{12}/C^{13} – 98.09/1.10

O^{16}/O^{18} – 99.80/0.20

The concentration of these stable isotopes, 0.37% of N^{15} for example, is called the *normal abundance.* The concentration of a heavy isotope is usually measured as *atom per cent excess, i.e.,* the amount in per cent by which the isotope exceeds its normal abundance. As an illustration, if a sample of nitrogen gas contains 3% N^{15} (and 97% N^{14}), the concentration of N^{15} in this sample is said to be (3.0–0.37) or 2.63 atom per cent excess.

The stable isotopes are measured quantitatively with a **mass spectrometer**. Devised by **Francis William Aston** in 1919, this instrument has now been greatly improved. In fact, this apparatus converts the uncharged atoms or molecules into positively charged ions (cations). These are then accelerated into the field of a powerful magnet. In the magnetic field, the ions will be deflected to an extent corresponding to their mass. *The path of the heaviest particles is bent least and that of the lightest particles is bent most*. The relative amounts of the ions of different mass is then determined by collecting them on a plate and measuring the current produced.

This technique for measuring stable isotopes is less sensitive and more tedious than that for measuring radioactive isotopes. However, since no useful radioactive isotopes of oxygen and nitrogen exist, the mass spectrometer remains an essential tool for studies involving these elements.

Table 40–3. Properties of useful radioactive isotopes

S.N.	*Isotope*	*Half-life**	*Type of radiation*	*Energy of radiation (in million electronvolts, mev)*
1.	Cl^{36}	3.1×10^5 years	β	0.714
2.	C^{14}	5760 years	β	0.155
3.	H^3	12.5 years	β	0.018
4.	Co^{60}	5.271 years	β	–
			γ	–
5.	Fe^{55}	2.94 years	x-rays	–
6.	Ca^{45}	152 days	β	0.255
7.	S^{35}	87.1 days	β	0.167
8.	I^{125}	60 days	γ	0.035
9.	Fe^{59}	45 days	β	0.46
			γ	0.27
10.	P^{32}	14.3 days	β	1.712
11.	I^{131}	8.1 days	β	0.610
			γ	0.280
12.	Na^{24}	15 hours	β	1.39
			γ	–
13.	K^{42}	12.5 hours	β	3.550

* **Half-life** is the time required for the loss of 50% of radioactivity. In other words, the time taken for half the atoms in a quantity of a radioactive substance to disintegrate is known as the half-life. The half-life is given in terms of whatever time unit is most appropriate. It can be anything from years to seconds.

Radioactive Isotopes. *The radioactive isotopes are usually more useful as tracers than the stable isotopes since the analytical methods for their measurement are extremely sensitive.* The biochemically-important radioactive isotopes are H^3, C^{14}, P^{32}, S^{35} and Ca^{45}. All these emit β-rays and their nuclei, upon disintegration, produce electrons. The β-rays interact with the molecules through which they traverse causing ionization of the molecules. It is this ionization property which is used to measure quantitatively the amount of radioactive isotope present. Table 40–3 lists some properties of useful radioactive isotopes.

The most widely-used apparatus for the determination of radioactive isotopes is **Geiger–Müller counter or G–M tube** (Fig. 40–8). The functioning of this apparatus is based on the ability of the emitted radiation to ionize atoms. It consists of a large, round tube forming cathode with a fine wire stretched in the centre as anode. The fine wire is maintained at a high potential (1,000–2,500 volts) with respect of the outer cathode. The tube is filled with an easily ionized gas such as helium or argon and an organic quenching substance such as ethanol. The voltage as well as the gas filling is so adjusted that normally no current flows. The open end of the tube is covered with an extremely thin

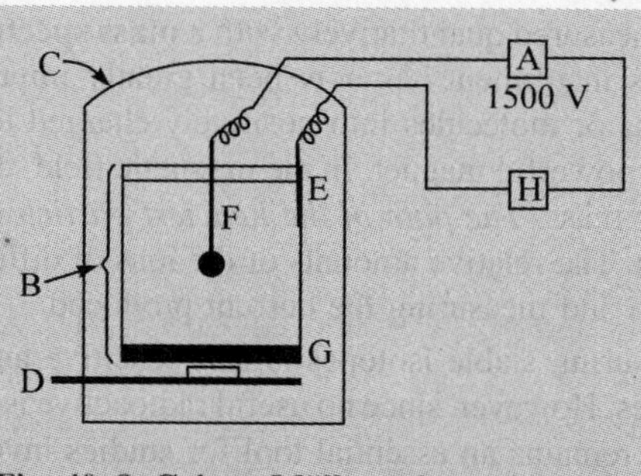

Fig. 40–8. Geiger–Müller tube and a scaler

[A = source of potential, B = Geiger–Müller tube, C = leadshield, D = sample pan holder, E = cathode shield, F = anode wire, G = mica window, H = scaler]

(After Cowgill and Pardee, 1957)

window of mica (1.5 to 2.0 mg/cm^2) or synthetic plastic. The radioactive material which is usually a solid is placed beneath this window. The radioactive particle enters the tube and ionizes the gas molecules with a subsequent release of a shower of electrons. These free electrons are then accelerated to the positive wire. As they progress through the gas, additional molecules are ionized. Thus, the tube becomes momentarily conductive. The resultant electrical pulse is collected in an electronic computing machine (called scaler) which records the number of such pulses in a predetermined time.

> The **radioactive sample** to be counted is, however, in some instances a gas and is actually included in the filling of the counter tube.

The radioactivity of a sample is given in terms of its *specific radioactivity, i.e.*, the number of counts per minute (*cpm*) per unit weight (milligram or micromole etc). The relationship of this quantity to the number of disintegrations per unit time depends on the counting efficiency of the system. The absolute unit of disintegration rate is the *curie*. A curie is defined as the amount of emitter which exhibits 3.7×10^{10} disintegrations per second (*dps*). More common units are a millicurie, mc (10^{-3} curie) and a microcurie, μc (10^{-6} curie).

The Geiger–Muller technique is being rapidly replaced nowadays by an alternative method called **scintillation counting**. It involves transformation of β or γ radiation into ultraviolet or visible light by the use of a *phosphor*. The phosphors are often highly aromatic substances such as *p*-terphenyl and have the property of emitting a light flash (or a scintillation) when they absorb radiation from a radioactive compound. The scintillations are detected by a sensitive photomultiplier and the resultant electrical pulses are collected in a scaler. *The scintillation technique has proved particularly useful for isotopes which emit low energy β particles such as H^3, C^{14}* etc.

Although extremely sensitive, this method also suffers from certain *drawbacks*. They are:

(*a*) quenching of light flashes by coloured samples, and

(*b*) limited solubility of highly polar compounds in nonpolar solvents.

Therapeutic Applications of Radioactive Emissions. Radioactive emissions, particularly of γ-rays, are frequently used to kill unwanted tissues. Some common examples are the radioactive sterilization of surgical accessories, partial destruction of over-active thyroids by I^{131} and the irradiation of tumours by Co^{60} and by Au^{198} or Ra^{226}.

Neutron Activation. The radioactive isotopes are made usually by bombarding a suitable target with neutrons in an atomic pile. But sometimes it becomes also possible to bombard a biological sample directly so that a little element is converted into a radioactive isotope. Thus, O^{18} has been converted into F^{18} by neutron activation. Of special interest was the dramatic example of bombardment of a single hair of the **Emperor Napoleon Bonaparte**. The hair was, in fact, taken at the end of his life and is preserved since then. The hair, after this treatment, was found to possess a radioactive isotope of arsenic which could only have come from arsenic already present in the hair. This finding was evidenced to the fact that Napoleon died of *arsenic poisoning*.

Arsenic Poisoning

Most forms of arsenic are toxic, but the trivalent form (arsenite as AsO_2^-) is much more toxic than the pentavalent form (arsenate or $HAsO_4^{2-}$). Less ATP is produced whenever arsenate substitutes for inorganic phosphorus (P_i) in biological reactions. Arsenate competes for P_i-binding sites on enzymes, resulting in the formation of arsenate esters that are unstable. Arsenite works by a completely different mechanism, involving the formation of a stable complex with enzyme-bound lipoic acid.

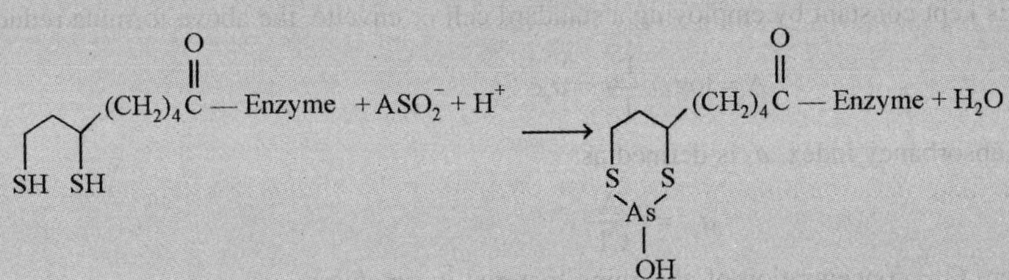

For the most part, arsenic poisoning is explained by inhibition of those enzymes that require lipoic acid as a coenzyme. These include **pyruvate dehydrogenase, α-ketoglutarate dehydrogenase,** and **branched-chain α-keto acid dehydrogenase.** Chronic arsenic poisoning from well-water contaminated with arsenical pesticides or through the efforts of a murderer is best diagnosed by determining the concentration of arsenic in the hair or fingernails of the victim. About 0.5 mg of arsenic would be found in a kilogram of hair from a normal individual. The hair of a person chronically exposed to arsenic could have 100 times as much.

SPECTROPHOTOMETRY

Principle. When a beam of incident light of intensity, I_o passes through a solution (Fig. 40–9), a part of it is reflected (I_r), a part absorbed (I_a) and rest transmitted (I_t), *i.e.*,

$$I_o = I_r + I_a + I_t$$

In colorimetric methods, I_r is eliminated because the measurement of I_o and I_t will be sufficient to determine I_a. For this purpose, the amount of light reflected (or I_r) is kept constant by using cells that have identical properties. I_o and I_t are then measured.

The mathematical relationship between the amount of light absorbed and the concentration of a substance can be shown by the following two fundamental laws, on which the spectrophotometry is based.

A. Lambert's law (= Bouguer's law). This law states that the amount of light absorbed is directly proportional to the length or *thickness* of the solution under analysis. Thus,

$$A = \log_{10} \frac{I_o}{I_t} = a_s b$$

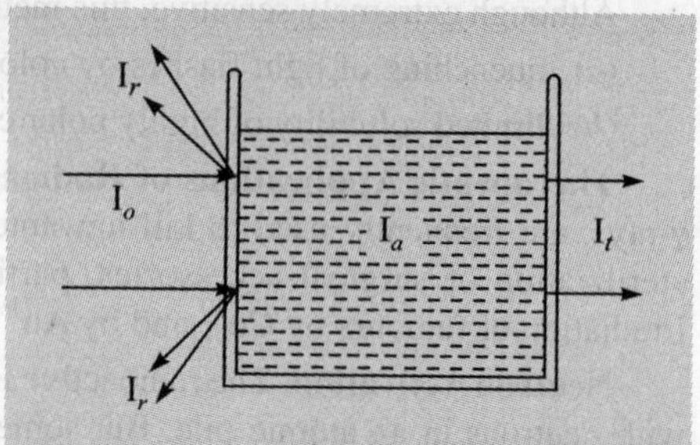

Fig. 40–9. Distribution of incident light passing through a solution contained in a cell

[where A = absorbancy

a_s = absorbancy index characteristic for the solution

b = length or thickness of the medium.]

B. Beer's law. This law states that the amount of light absorbed is directly proportional to the *concentration* of the solute in solution. Thus,

$$\log_{10} \frac{I_o}{I_t} = a_s c$$

[where c = concentration of solute in solution.]

The combined **Beer–Lambert Law** then becomes:

$$\log_{10} \frac{I_o}{I_t} = a_s bc$$

If b is kept constant by employing a standard cell or cuvette, the above formula reduces to:

$$A = \log_{10} \frac{I_o}{I_t} = a_s c$$

The absorbancy index, a_s is defined as:

$$a_s = \frac{A}{CL}$$

[where C = concentration of absorbing material in gms/litre

L =distance in cms travelled by the light in solution.]

If one wishes to express the light absorption in terms of the molar concentration of the absorbing material, the molar absorbancy index, a_m will be equal to:

$$a_m = a_s M$$

[where M = molecular weight of the absorbing material.]

Spectrophotometer. A spectrophotometer has two fundamental parts : a source of radiant light and a monochromator. Fig. 40–10 outlines Beckman spectrophotometer. It consists of a prism. This disperses the radiant energy into a spectrum. A slit is also fixed which selects a narrow portion of the spectrum. The standard cell or cuvette is placed in a light-tight unit. The incident light strikes the standard cell and emergent light passes into a photocell. The photocell changes the emerging light energy into measureable electrical energy.

Applications. The technique finds many applications:

(*a*) It can be used in determining the concentration of a compound by measuring the optical density, provided the absorbancy index, a_s is known.

(*b*) The course of a reaction can be determined by measuring the rate of formation or disappearance of a light-absorbing compound.

(*c*) A compound can be identified by determining its absorption spectrum in the visible and ultraviolet region of the spectrum.

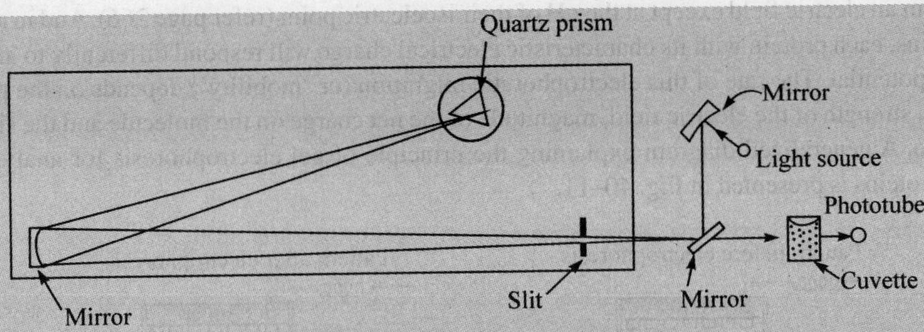

Fig. 40–10. Beckman spectrophotometer

(Adapted from Conn EE and Stumpf PK, 1966)

ELECTROPHORESIS

Principle. Migration of ions in an electric field at a definite pH is called **electrophoresis**. This method was developed by **Arne W.K. Tiselius** in 1937 and is based on the principle that the proteins

ARNE WILHELM KAURIN TISELIUS

(LT, 1902–1971)

Tiselius, a Swedish physical biochemist, was born on August 10, 1902, in Stockholm. After the early loss of his father, the family moved to Gothenburg where he went to school, and after graduation at the local "Realgymnasium" in 1921, he studied at the University of Uppsala, specializing in chemistry. He became research assistant in The Svedberg's laboratory in 1925 and obtained his doctor's degree in 1930 on a thesis "The moving-boundary method of studying the electrophoresis of proteins" and was appointed Docent (Assistant Professor) in Chemistry from 1930 on. He was married (1930) with **Ingrid Margareta (Greta) Delén**, daughter of city judge **Per Dalén** of Gothenburg. They had two children: **Eva** (born, 1932) and **Per** (born, 1934). During the years 1931–1935, Tiselius published a number of papers on diffusion and absorption phenomena in naturally occurring base-exchanging zeolites, and these studies were continued during a year's visit to H.S. Taylor's laboratory in Princeton with support of a Rockefeller Foundation Fellowship. On his return to Uppsala, Tiselius resumed his interest in proteins and the application of physical methods to biochemical problems, in general. This led to a much improved method of electrophoretic analysis which he refined in subsequent years. In 1938, a special research professorship was established for Tiselius through a donation to the University of Uppsala by **Major Herbert Jacobsson** and his wife. Some space was put at the disposal of the new professor in the Institute of Physical Chemistry (Prof. The Svedberg). In 1946, biochemistry was established as an independent department and in 1950–1952 obtained a new building, the present Institute of Biochemistry.

Under the leadership of Tiselius, this institute has contributed to the development and improvement of a number of useful methods in biochemistry, such as electrophoresis, chromatography, phase partition, gel filtration etc. These methods and others have been applied to studies of large-molecular weight-substances, chiefly proteins and enzymes, but also polysaccharides (dextran) and nucleic acids. Tiselius won the **1948 Nobel Prize in Chemistry** for the discovery of proteins in blood serum and for the development of electrophoresis as a technique for studying proteins, notably those of the blood.

Tiselius took an active part in the reorganization of scientific research in Sweden in the years following World War II. Thus, he was **Chairman of the Swedish Natural Science Research Council** (1946–1950), and Chairman of the Research Committee of the Swedish Cancer Society (1951–1955). He was **President of the International Union of Pure and Applied Chemistry** (1951–1955), became **Vice President of the Nobel Foundation** in 1947 and **President** from 1960 to 1964. He also served as member of the Nobel Committee for Chemistry since 1946. The lunar crater Tiselius was named in his honour. He died of a heart attack on October 29, 1971 in Uppsala.

migrate in an electric field except at the pH of their isoelectric point (refer page 266). And in a mixture of proteins, each protein with its characteristic electrical charge will respond differently to an applied electric potential. The rate of this electrophoretic migration (or 'mobility') depends on the pH of the medium, strength of the electric field, magnitude of the net charge on the molecule and the size of the molecule. A generalized diagram explaining the principle of gel electrophoresis for analyzing and sizing proteins is presented in Fig. 40–11.

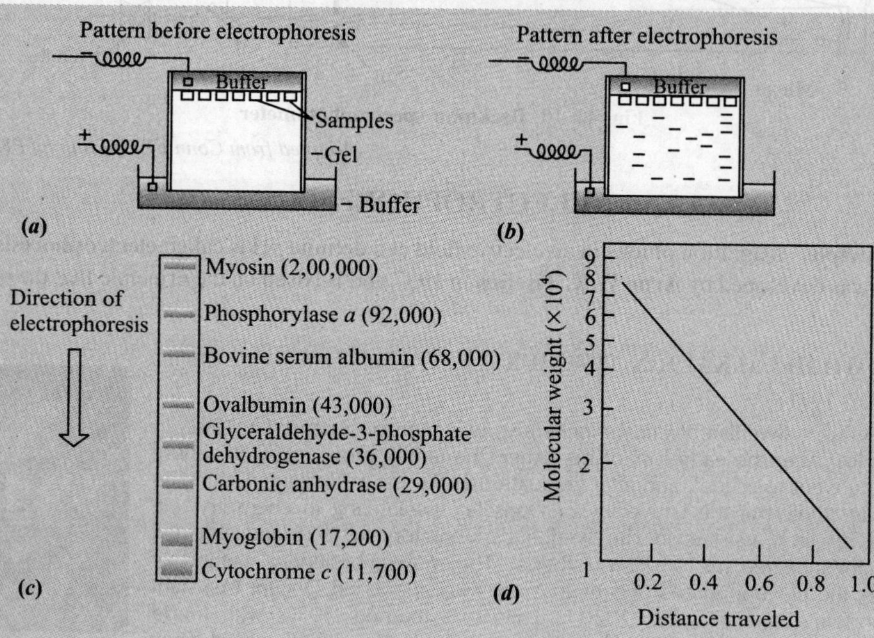

Fig. 40–11. Gel electrophoresis for analyzing and sizing proteins

(*a*) Apparatus for slab-gel electrophoresis. Samples are layered in the little slots cut in the top of the gel slab. Buffer is carefully layered over the samples, and a voltage is applied to the gel for a period of usually 1–4 h.

(*b*) After this time, the proteins have moved into the gel at a distance proportional to their electrophoretic mobility. The pattern shown indicates that different samples were layered in each slot.

(*c*) Results obtained when a mixture of proteins was layered at the top of the gel in phosphate buffer, pH 7.2, containing 0.2% SDS. After electrophoresis, the gel was removed from the apparatus and stained with Coomassie Blue. The protein and its molecular weight are indicated next to each of the stained bands.

(*d*) The logarithm of the molecular weight against the mobility (distance traveled) shows an approximately linear relationship.

(Source: Data of K Weber and M Osborn)

Electrophoresis apparatus. Tiselius electrophoresis apparatus consists of compartmented cells forming a U tube, connected to an anode and a cathode compartment. The protein solution is placed at the bottom of the U tube. The U tube is kept immersed in a water bath at 4°C to minimize convection currents and the movement of the proteins is visualized by a Schlieren optical system.

Consider that the protein solution is composed of 3 components, A, B and C and that these components carry charges of the same sign (either + or –) but of different magnitude. Before the electric field is applied, the solution will be homogeneous from ascending boundary to descending boundary. On the flow of current, the 3 components will separate. If the magnitude of the electric charge is in the order A > B > C, the components will separate as shown in Fig. 40–12. The faster

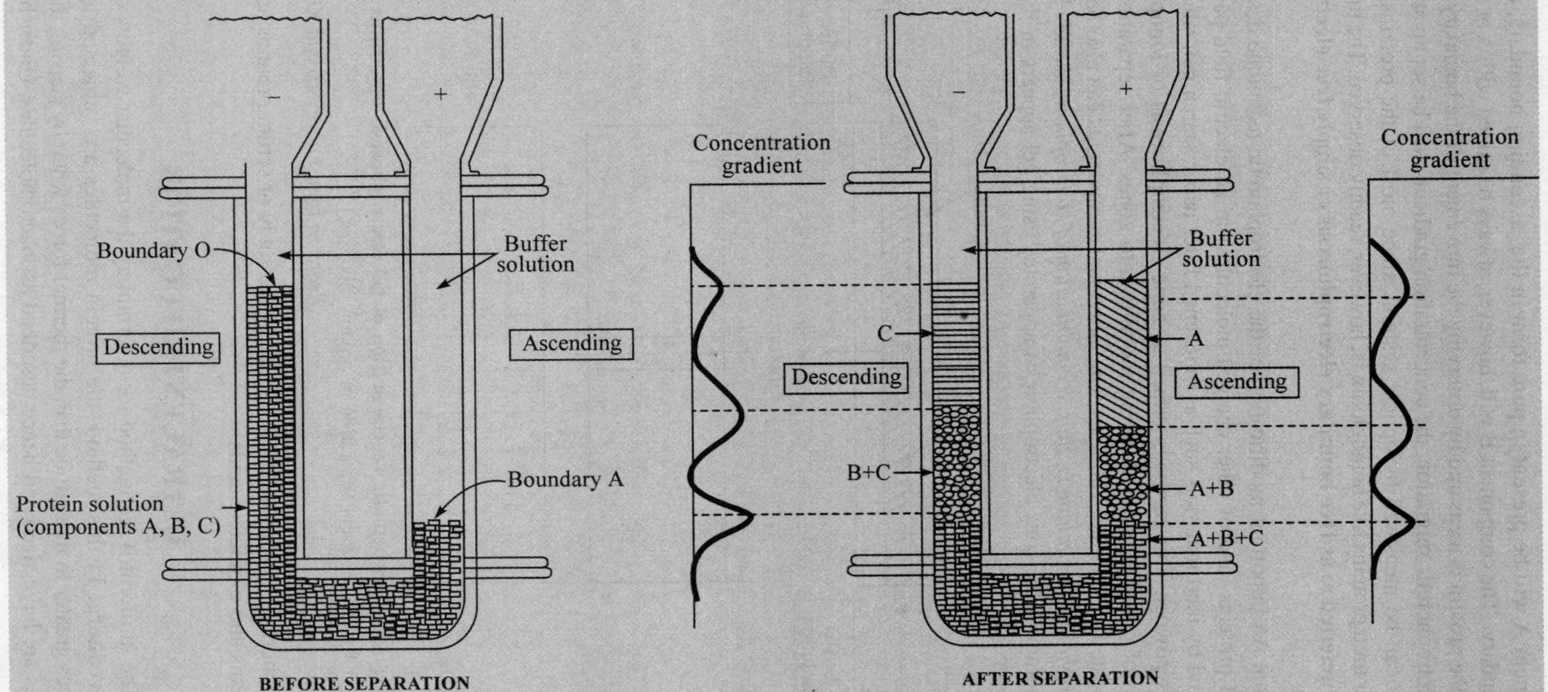

Fig. 40–12. U tube of an electrophoresis cell before and after electrophoretic separation

moving component A will be present in pure form in the ascending boundary, as will C in the descending boundary. The component B will, however, always be mixed with A or with C. The rate of migration of the protein is measured by observing the movement of the boundary as a function of time. During electrophoretic migration, the concentration gradients will be set up at the boundaries. These gradients can be measured by optical systems since most of the proteins are colourless. Migration of coloured proteins like hemoglobin is, however, readily observed. The method described above is usually referred to as **free boundary electrophoresis** or simply **free electrophoresis** (*i.e.*, in free solution).

Modification. An important modification of the electrophoretic technique described above is the migration of proteins (and of other charged molecules) in an electric field passing through a solution supported by inert materials such as moistened filter paper, starch gel, silica gel, cellulose sponges or glass powder. This method, termed as **zone electrophoresis** or **ionophoresis**, allows ready separation of components of different mobility into zones. After separation, the different zones may be located by staining with dyes. The individual components can also be extracted from these zones. *Ionophoresis technique is, thus, a combination of electrophoresis and chromatography.* A schematic diagram of the zone electrophoresis apparatus using gel appears in Fig. 40–13.

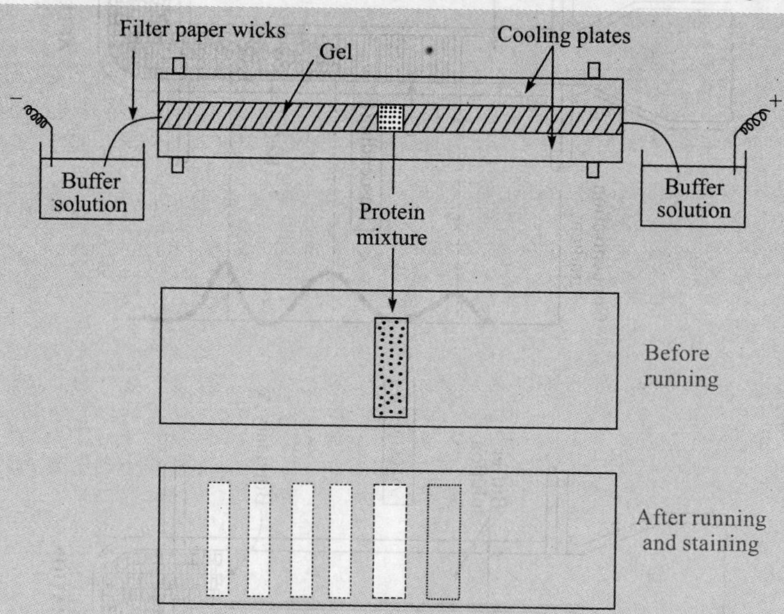

Fig. 40–13. Schematic representation of gel electrophoresis apparatus

[The two lower figures represent separation of protein mixtures.]

(*Redrawn from Fairley JL and Kilgour GL, 1966*)

Zone electrophoresis has proved much useful for the study of serum proteins and the cleavage products of proteins and of nucleic acids.

ULTRACENTRIFUGATION

Principle. The ultracentrifuge method for determining the molecular weights of proteins was developed by **Svedberg**. In this method, the protein molecules are subjected to gravitational (centrifugal) forces greater in magnitude than the thermal forces whereby causing them to diffuses. Protein molecules are large-sized and hence described as macromolecules. Under high centrifugal

THEODOR SVEDBERG

(LT, 1884–1971)

Svedberg, a pioneer Swedish physical chemist, was born at Fleräng, Valbo, in the county of Gävleborg, Sweden on August 30, 1884, as the son of **Elias Svedberg**, a works manager, and his wife, **Augusta, née Alstermark**. He attended the Köping School, the Örebro High School and Gothenburg Modern School, where he passed his student examination in December 1903, before matriculating and beginning his lifelong association with Uppsala University in January 1904. He gained his Bachelor of Arts degree in 1905, his Master's degree in 1907 and his Doctor of Philosophy in 1908.

Svedberg accepted a post as assistant in the Chemical Institute at Uppsala in 1905 and in 1907 he was given the additional position of lecturer in chemistry in the university. He obtained a special appointment in 1909 as lecturer and demonstrator of physical chemistry and in 1912 he was elected Professor of Physical Chemsitry, University of Uppsala. He was made emeritus in 1949, since when he has been Director of the Gustaf Werner Institute for Nuclear Chemistry in the University.

Svedberg married four times: 1909, **Andrea Andreen**, a medical graduate; in 1916, **Jane Frodi**; **Ingrid Blomquist** in 1938; and in 1948, **Margit Hallén**. He had twelve children, six sons and six daughters.

Svedberg's work has mainly been concerned with colloids and macromolecular compounds. In his doctoral thesis *Studien zur Lehre von den kolloiden Lösungen* (1908), now considered as a classic, he described a new method of producing colloid particles and gave convincing evidence of the validity of the theory founded by **Albert Einstein** and **Marian von Smoluchowski** on the Brownian motion, thereby providing definite proof of the existence of molecules. With numerous collaborators, he has studied the physical properties of colloids, such as their diffusion, light absorption, and sedimentation, from which could be concluded that the gas laws could be applied to disperse systems. For the study of sedimentation, he constructed his well-known ultracentrifuge, where large molecules in solution, such as proteins, carbohydrates, and high polymers have been investigated. In the ultracentrifuge the large molecules are put under the influence of centrifugal forces up to about 10^6 gravity. The findings were related to molecular size and shape. He showed that the molecules of certain pure proteins are all of one size and demonstrated the use of the ultracentrifuge to show the presence of contaminants. Later, he turned his attention to the problems of nuclear chemistry and radiation biology. He has also studied photographic processing, and assisted in photographing the famous *Codex argenteus* in colour, which made this the source of the Gothic language available for research. During the Second World War, he worked out a **method for making synthetic rubber**.

He was awarded the **John Ericsson Medal** (1942), the **Berzelius Medal** (1944), and the **Medal of the Franklin Institute** (1949). For this work on the disperse systems, he received the **1926 Nobel Prize in Chemistry**. He was a member of the *Swedish Academy of Sciences*, the Halle Academy, The Chemical Society (London), the Indian Academy of Science, the American Philosophical Society (Philadelphia), The New York Academy of Sciences, The Royal Society (London) and the National Academy of Science (Washington). The **unit Svedberg** (symbol S), a unit of time amounting to 10–13 s or 100 fs, is named after him, as well as the **The Svedberg Laboratory** in Uppsala. He died on February 25, 1971.

forces, these molecules can be made to sediment toward one end of a centrifuge tube. By means of photographic and optical systems similar to those used in electrophoresis, it is possible to follow the rate of sedimentation. The rate of sedimentation for a protein under a certain centrifugal force depends on the density, shape and size of the molecule.

The fundamental equation, devised by **Svedberg**, for the molecular weight (M) of a protein is:

$$M = \frac{RTs}{D(1 - Vp)}$$

where,
R = gas constant
T = absolute temperature
s = sedimentation constant
D = diffusion constant in cm_2/sec
V = partial specific volume of the protein
p = density of the solvent.

The rate of sedimentation is usually expressed in terms of sedimentation constant, s which has the dimensions of time per unit of gravitational field and which usually lies between 1×10^{-13} and 200×10^{-13} sec. For convenience, the sedimentation constant of 1×10^{-13} is referred to as 1 Svedberg unit (S) and all sedimentation constants are then expressed in Svedberg units. S value (or Svedberg unit) refers to the sedimentation coefficient of the RNA; the larger the number, the more rapidly the molecule moves through a field of force during centrifugation, and (for a group of chemically similar molecules) the larger the size of the molecule. The 28S, 18S, 5.8S and 5S RNAs consist of nucleotide lengths of about 5,000, 2,000, 160 and 120 nucleotides, respectively. A *Svedberg unit* is, thus, defined as the velocity of sedimenting molecule per unit of gravitational field or 1×10^{-13} cm/sec/dyne/g. The S value of an organelle or macromolecule is related to its molecular weight and shape. Typical S values are 1.83 for cytochrome *c*, 4.4 for bovine serum albumin and 185 for TMV. For the sake of uniformity, the sedimentation coefficient is customarily corrected to the value that would be obtained at 20 ºC in a solvent with the density and viscosity of pure water. This is symbolized $S_{20,\omega}$. Table 40–4 and Fig. 40–14 indicate the values of $S_{20,\omega}$ in Svedbergs for a variety of biological materials.

Table 40–4. Physical constants of some proteins

Protein	*Molecular mass (kD)*	*Partial specific volume, $V_{20,\omega}$ ($cm^3 . g^{-1}$)*	*Sedimentation coefficient, $S_{20,\omega}$ (S)*
Lipase (milk)	6.7	0.714	1.14
Ribonuclease A (bovine pancreas)	12.6	0.707	2.00
Cytochrome *c* (bovine heart)	13.4	0.728	1.71
Myoglobin (horse heart)	16.9	0.741	2.04
α-chymotrypsin (bovine pancreas)	21.6	0.736	2.40
Diphtheria toxin	70.4	0.736	4.60
Cytochrome oxidase (*P. aeruginosa*)	89.8	0.730	5.80
Lactate dehydrogenase H (chicken)	150	0.740	7.31
Catalase (horse liver)	222	0.715	11.20
Fibrinogen (human)	340	0.725	7.63
Glutamate dehydrogenase (bovine liver)	1015	0.750	26.60
Turnip yellow mosaic virus protein	3013	0.740	48.80

(Adapted from Smith MH, 1970)

The diffusion coefficient or constant D may be defined as the quantity of material that diffuses per second across a surface 1 cm^2 in area. The partial specific volume V is equal to the reciprocal of the density of the molecule. *For most proteins, however, the value of V is in the range of 0.70 to 0.75.*

Ultracentrifuge. Using the above principle, **Svedberg** in 1925 devised an instrument called ultracentrifuge (Figs. 40–15 and 40–16). By this instrument, it is possible to get information on the purity of a protein, its molecular weight and approximate dimensions. Two types of ultracentrifuges are now known:

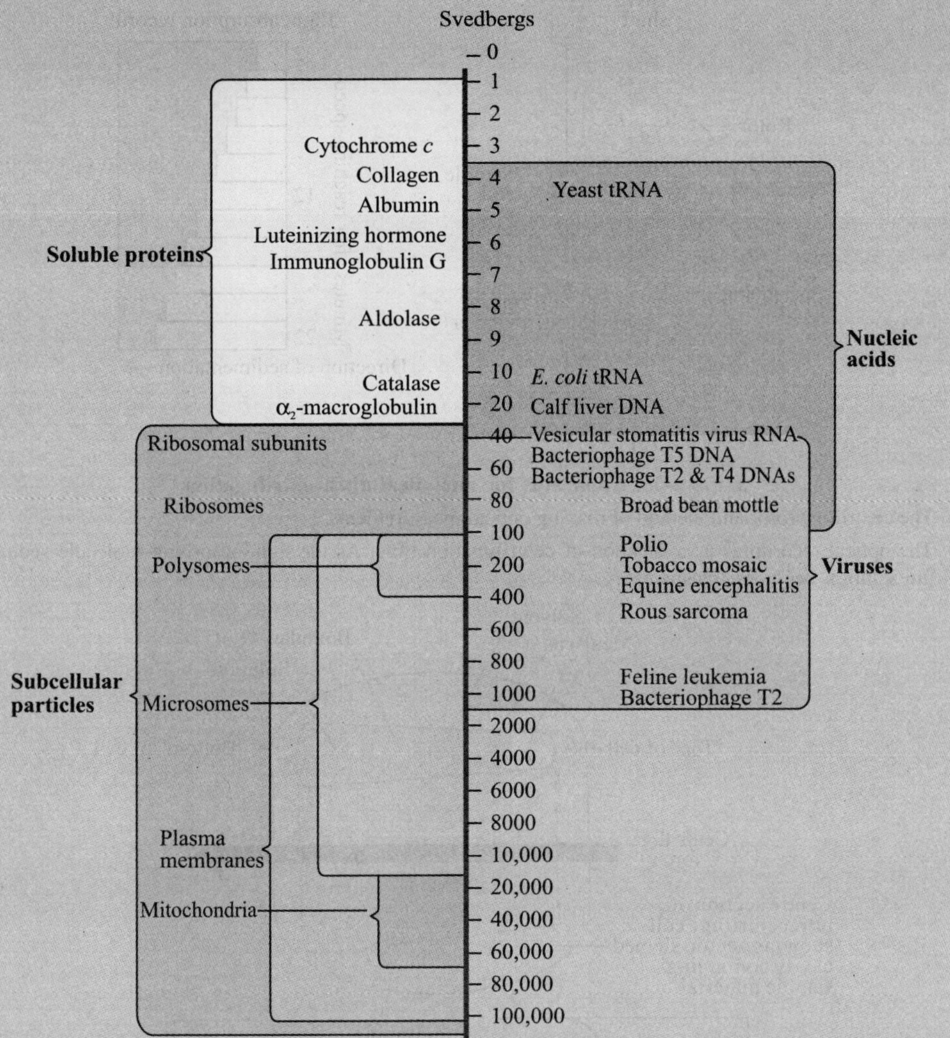

Fig. 40–14. The sedimentation coefficients in Svedbergs for some biological materials

(*Courtesy: Beckman Instruments, Inc.*)

(*a*) *Low speed type*, with centrifugal forces of the order of 100 to 8,000 times gravity and having the maximum speed of about 40,000 *rpm*.

(*b*) *Oil-turbine type*, with centrifugal forces of the order of 8,000 to 400,000 times gravity and having the maximum speed of about 80,000 *rpm*.

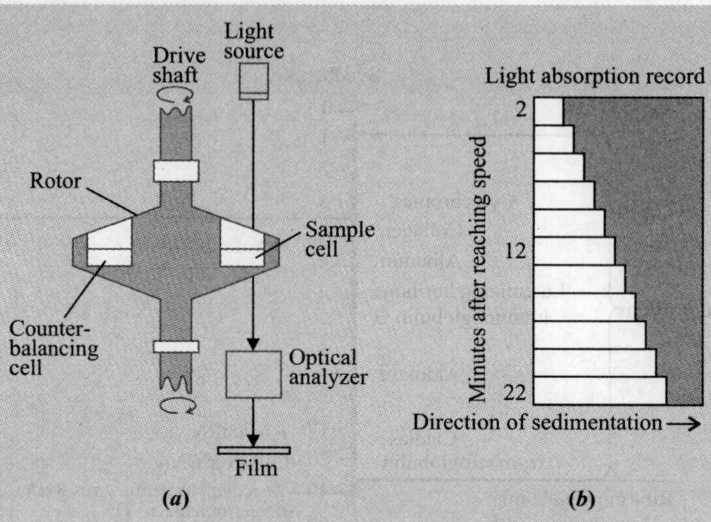

Figl 40–15. Apparatus for analytical ultracentrifugation

(*a*) The centrifuge rotor and method of making optical measurements.

(*b*) The optical recordings as a function of centrifugation time. As the light-absorbing molecule sediments, the solution becomes transparent.

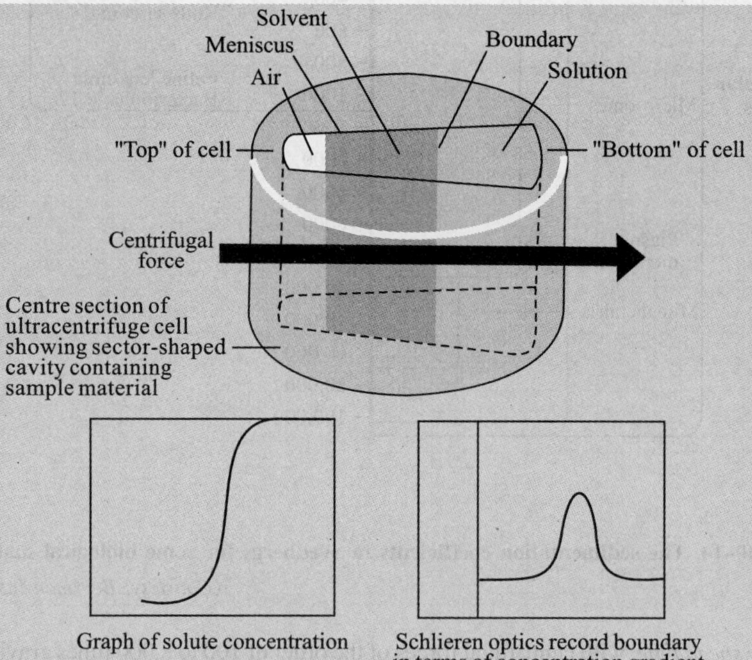

Fig. 40–16. Centre section of ultracentrifuge cell

The solute molecules which are initially evenly distributed are forced towards the bottom of the cell by the centrifugal force. This migration creates at the top of the cell a region that is without solute molecules and contains only solvent molecules. A boundary is set up in the cell between solvent and solution in which concentration varies with the distance from the axis of rotation. The measurement of boundary movement, which represents its movement of protein molecules, is the basis of analytical ultracentrifugation.

(*Courtesy: Beckman Instruments, Inc.*)

POLYMERASE CHAIN REACTION

Polymerase chain reaction (PCR) is a technique developed by an American **Kary B. Mullis** in 1985 while at the Cetus Corporation, for which he received the 1993-Nobel Prize in Chemistry. It is a repetitive, extremely simple yet immensely powerful technique for isolating and amplifying a specifically desired DNA sequence through a series of reactions in a test tube (*in vivo*), *provided that short sequences either side of it are known*. This information is used to synthesize two oligonucleotide primers that are added to a denatured (single-stranded) DNA sample. The technique is shown in the figure given below. A PCR reaction contains *(1)* the target double-stranded DNA, *(2)* two short single-stranded DNA primers (known to anneal upwards of the 3′-end of the DNA sequence to be amplified) that hybridize to flanking sequences on opposite strands of the target DNA, *(3)* all the 4 deoxyribonucleotides, and *(4)* a DNA polymerase.

PCR (Fig. 40–17) consists of 3 basic steps:

(*a*) ***Denaturation***: The DNA to be amplified is denatured into single strands. This DNA does not have to be purified and can come from any number of sources, including genomic DNA, forensic samples such as dried blood or semen, dried samples stored as part of medical records, single hair, saliva-licked postage stamps, mummified remains and fossils. The reaction mixture (containing the DNA to be amplified) is heated to 95°C for a short time period (ca 15–30 sec) to denature the target DNA into single strands that can act as templates for DNA synthesis.

(*b*) ***Primer annealing***: The mixture is rapidly cooled to a defined temperature (somewhere between 50 and 60ºC) which allows the two primers (synthetic oligonucleotides, 15–30 nucleotides long) to bind to the sequences on each of the two strands flanking the target DNA. Thus, the primers serve as starting points for synthesizing new DNA strands complementary to the target DNA. This *cooling* or *annealing temperature* depends upon the GC content of the sequence and is calculated carefully to ensure that the primers bind only to the desired DNA sequences. One primer binds to each strand. The two parental strands do not reanneal with each other because the primers are in large excess than the parental DNA.

(*c*) ***Elongation*** (***DNA synthesis***): A key to the process is that DNA polymerase requires a priming sequence upwards of the 3′-end of the DNA sequence to extend the whole complementary strand in a 5′-wards direction. Because the reaction muxture periodically gets heated to high temperature, PCR depends upon using a *heat-stable* DNA polymerase which is usually taken from the extreme thermophilic bacterium T*hermus* aq*uaticus* (the polymerase from this bacterium being called Taq polymerase), thus allowing automation of the process. The thermostable nature of *Taq* polymerase gives it an advantage over the *E. coli* DNA polymerase which is thermolabile. However, one disadvantage of *Taq* polymerase is that it does not possess proofreading properties, thus mistakes in synthesis occur at a rate of about 1 in 1,000 bases incorporated. These mistakes are amplified in subsequent cycles, thus producing copies that are not true to the original. Other heat-stable polymerases have subsequently been discovered that possess proofreading properties, but they are usually more expensive, and investigators use them only when they need an accurate PCR-generated copy of a gene.

In this third step, the temperature of the reaction mixture is raised to about 72°C and kept at this temperature for a pre-set period of time to allow Taq polymerase, which is added to the mixture, to elongate each primer by copying the single-stranded templates. Thus, at the end of this incubation, both single-stranded (ss) template strands have been made partially double-stranded (ds). The new strand of each double-stranded DNA extends for a variable distance downstream.

KARY BANKS MULLIS

(born, 1944)

Mullis, a multifaceted American biochemist and author, was born in Lenoir, North Carolina, near the Blue Ridge Mountains, on December 28, 1944. His family had a background in farming in this rural area. As a child, Mullis recalls, he was interested in observing organisms in the countryside. He grew up in Columbia, South Carolina, where he attended Dreher High School. Mullis earned a Bachelor of Science degree in chemistry from the Georgia Institute of Technology in Atlanta in 1966, during which time he got married and started a business. He then received a Ph.D. in biochemistry from the University of California, Berkeley in 1972; his research focused on synthesis and structure of proteins. Following his graduation, Mullis became a postdoctoral fellow in pediatric cardiology at the University of Kansas Medical School, going on to complete two years of postdoctoral work in pharmaceutical chemistry at the University of California, San Francisco.

After receiving his Ph.D., Mullis left science to write fiction, but quite and became a biochemist at a medical school in Kansas City. He then managed a bakery for two years. Mullis returned to science at the encouragement of friend **Thomas White**, who later got Mullis a job with the biotechnology company *Cetus Corporation* of Emeryville, California. Mullis worked as a DNA chemist at Cetus for seven years; it was there, in 1983, that Mullis invented his prize-winning improvements to the polymerase chain reaction. In fact, in the spring of 1983, according to Mullis, he was driving his vechicle late one night with his girlfriend, who was also a chemist at Cetus, when he had the idea to use a pair of primers to bracket the desired DNA sequence and to copy it using **DNA polymerase**, a technique which would allow a small strand of DNA to be copied almost an infinite number of times. Cetus took Mullis off his usual projects to concentrate on PCR full-time. Mullis succeeded in demonstrating PCR on December 16, 1983. In his Nobel Prize lecture, he remarked that the success didn't make up for his girlfriend breaking up with him shortly before: "I was sagging as I walked out to my little Silver Honda Civic. Neither [assistant] **Fred**, empty Beck's bottles, nor the sweet smell of the dawn of the age of PCR could replace Jenny. I was lonesome." He received a $10,000 bonus from Cetus for the invention.

Other Cetus scientists, including **Randall Saiki** and **Henry Erlich**, were placed on PCR projects to work on developing HIV and other tests utilizing PCR. Saiki generated the needed data and authored the first paper to include utilization of the technique, while Mullis was still working on a paper that would describe PCR itself.

A complication at that point was that the DNA polymerase used was destroyed by the high heat used at the start of each replication cycle and had to be replaced. In 1986, Mullis started to use *Thermophilus aquaticus* (Taq) DNA polymerase to amplify segments of DNA. The Taq polymerase was heat-resistant and would only need to be added once, thus making the technique dramatically more affordable and subject to automation. This has created revolutions in biochemistry, molecular biology, genetics, medicine and forensics.

Mullis has also **invented a UV-sensitive plastic** that changes color in response to light, and most recently has been working on an approach for mobilizing the immune system to neutralize invading pathogens and toxins, leading to the formation of his current venture, Altermune LLC.

After leaving Cetus in 1986, Mullis served as director of molecular biology for Xytronyx, Inc. in San Diego for two years. Mullis has consulted on nucleic acid chemistry for multiple corporations. In 1992, Mullis founded a business with the intent to sell pieces of jewelry containing the amplified DNA of decreased famous people like **Elvis Presley** and **Marilyn Monroe**. Mullis is also a member of the USA Science and Engineering Festival's Advisory Board. He is a researcher at the Children's Hospital Oakland Research Institute in Oakland, California.

In recognition of his improvement of the polymerase chain reaction (PCR) technique, he shared the **1993 Nobel Prize in Chemistry** with **Michael Smith** and earned the Japan Prize in the same year. The process was first described by **Kjell Kleppe** and 1968-Nobel Laureate **Har Gobind Khorana**, and allows the amplification of specific DNA sequences. The improvements made by Mullis allowed PCR to become a central technique in biochemistry and molecular biology, described by The New York Times as "highly original and significant, virtually dividing biology into the two epochs of *before P.C.R.* and *after P.C.R.*" Since winning the Nobel Prize, Mullis has been criticized in The New York Times for promoting ideas in areas in which he has no expertise. He has promoted AIDS denialism, climate change denial and his belief in astrology.

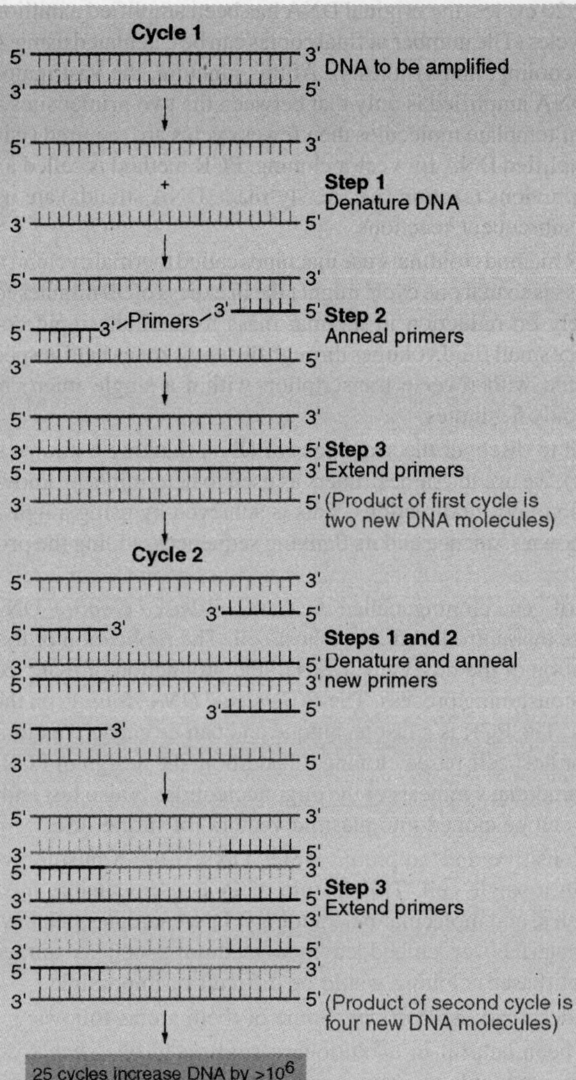

Fig. 40–17. The technique of polymerase chain reaction (PCR)

The PCR can be used to amplify a specific DNA sequence if the flanking sequences are known. The flanking DNA sequence gives information for the chemical synthesis of the primers. In the polymerase chain reaction, the target DNA is denatured into single strands; each strand is then annealed to a short, complementary primer. The primers are synthetic oligonucleotides that are complementary to sequences flanking the region to be amplified. DNA polymerase and nucleotides extend the primers in the 3' direction, using the single-stranded DNA as a template. The result is a double-stranded DNA molecule with the primers incorporated into the newly-synthesized strand. In a second PCR cycle, the products of the first cycle are denatured into single strands, primers are annealed, and DNA polymerase then synthesizes new strands. Each cycle is repeated 20–30 times, doubling the amount of DNA each time.

Thus, in the second cycle, the 4 strands denature, bind primers which are still present and are extended. No other reactants need to be added. The 3 steps are repeated once more for a third cycle and so on for a set number of additional cycles. By the third cycle, some of the PCR products represent DNA sequence only between the two primer sites and the sequence does not extend beyond these sites. As more and more reaction cycles are carried out, this type of dsDNA molecule becomes the majority species present. After 20 cycles, the original DNA has been amplified a million-fold and this rises to a billion-fold after 30 cycles (The number of final copies can be calculated using 2^n, where *n* is the number of cycles of heating, cooling, and synthesis). At this juncture, the vast majority of the products are identical in that the DNA amplified is only that between the two primer sites. If the process is started with a large number of template molecules then fewer cycles are required (with less mutation risk) to produce sufficient amplified DNA for vector cloning. PCR method is called a chain reaction, because the products of each previous reaction (*i.e.,* newly-made DNA strands) are used as reactants (*i.e.,* as template strands) in subsequent reactions.

Traditional PCR methods routinely use machines called thermal cyclers (= *thermocyclers*), which have high thermal masses so that one cycle might take in excess of 90 minutes to complete. New micro-fabricated systems rely on reduction in thermal mass to facilitate rapid cooling and heating, and continuous flow of very small fluid volumes through different zones held at specific temperatures. PCR can, thus, be integrated with reverse transcription within a single microchip, 20 cycles of DNA amplification lasting only 5 minutes.

Should we wish to discover the sequences of DNA flanking a known sequence (*for which no primers are available*), we use the same primers as used to determine the known sequence, but in such a way that they face in opposite directions. This is achieved by using a ligase to circularize dsDNA which includes the known sequence and its flanking sequences, adding the primers, and then carrying out normal PCR.

In one method of gene cloning, called the *host cell-based cloning*, DNA of interest is inserted into a vector, which is then introduced into a host cell. The replication of the vector within the host cell, and the proliferation of the host cells, leads to the production of many copies of the DNA. This is a tedious and time-consuming process. The *PCR-based DNA cloning*, on the contrary, does not use host cells and vectors. The PCR is a fast technique and can be carried out in a few hours, rather than the weeks required for host cell-based cloning. In addition, the design of PCR primers uses computer software, and the commercial synthesis of the oligonucleotides is also fast and economical. If desired, the products of PCR can be cloned into plasmid vectors for further use.

PCR is very sensitive and amplifies target DNA from vanishingly small DNA samples, including the DNA in a single cell. This feature of PCR is invaluable in several areas, including genetic testing, forensics, and molecular paleontology. DNA samples that have been partly degraded, mixed with other materials, or embedded in a medium (such as amber) can be used where conventional (host cell-based) cloning would be difficult or impossible.

PCR has widespread *applications*. Some of them are as follows:

1. PCR has been helpful in evolutionary studies. In one such study, using the PCR, the researchers were able to obtain enough DNA to sequence a portion of the DNA from a woolly mammoth, freed from its icy grave of 40,000 years. The DNA had been highly degraded, but sufficient template DNA remained to support the PCR. Researchers, using PCR, were able to obtain enough mitochondrial DNA from the extinct species to eventually determine the entire sequence. This DNA sequence was then compared to present-day elephant's mitochondrial DNA to ascertain homology. The findings showed that the woolly mammoth and elephant are very closely related.
2. The PCR can also be used to detect the presence of extremely small quantities of DNA, such as DNA found in disease-causing virus. For example, an individual may suspect that he/she has the HIV-1 virus (human immunodeficiency or AIDS virus) but may test negative using conventional methods (a false-negative). With a modification of the PCR method (HIV is an RNA virus, so it must be converted to DNA first), one copy of the virus can be replicated for easy detection.

3. By using suitable primers, it is possible to use PCR to produce point mutations, deletions and insertions of target DNA which greatly facilitates the analysis of gene expression.
4. Because of its extreme sensitivity, PCR has proved invaluable in forensic medicine, enabling amplification of minute traces of genetic material for DNA fingerprinting. It is even possible to use PCR to amplify the DNA from a single human hair or a microscopic drop of blood left at the scene of a crime to allow detailed characterization.
5. PCR is used for characterizing medically-important DNA samples. For example, in screening for human genetic diseases, it is rapidly replacing the use of RFLPs. The PCR screen is based on the analysis of microsatellites. The method is very fast, reliable and uses very small amounts of clinical material.
6. The 'pooled DNA samples' collected by sampling whole environments, notably soil, when analyzed by PCR, can reveal the existence of unnoticed taxa. In the case of bacteria, so few bacterial species have been cultured that such culture-independent PCR (ciPCR) have more than doubled the estimated number of prokaryotic phyla and have revealed the existence of bacteria-sized eukaryotes (nano- and picoeukaryotes).

Although PCR is a valuable technique, it does have *limitations*. Some information about the nucleotide sequence of the target DNA must be known, and even minor contamination of the sample with DNA from other sources can create problems. For example, cells shed from the skin of a laboratory worker while performing PCR can contaminate samples gathered from a crime scene or taken from fossils. Such contamination makes it difficult to obtain accurate results. PCR technique must always be run with carefully designed and appropriate controls. The PCR is usually used to amplify sequences of less than 1,500 bases long, but with some modifications of the procedure, segments as long as 40,000 bases can be amplified.

RADIOIMMUNOASSAY

Radioimmunoassay (RIA) is a technique developed by **Rosalyn S. Yalow** and **Solomon A. Berson**. It is an immunologic method for detecting trace amounts of a biomolecule, based on its capacity to displace a radioactive form of the molecule from combination with its specific antibody. This procedure uses competitive inhibition between a radioisotope-labeled antigen or other substance and unlabeled antigen for antiserums, binding as a means to quantify the agent of interest; any method for detecting or quantitating antigens or antibodies using radiolabeled reactants. The method is extraordinarily sensitive as minute quantities of enzymes, hormones, or other substances can be assayed.

Hormone-specific antibodies (Fig. 40–18) are the key to the radioimmunoassay. Purified hormone, injected into rabbits, elicits antibodies that bind to that hormone with very high affinity and specificity. When a constant amount of isolated antibody is incubated with a mixed amount of the radioactively-labeled hormone, a certain fraction of the radioactive hormone binds to the antibody. If, in addition to the radiolabeled hormone, unlabeled hormone is also present, the unlabeled hormone competes with and displaces some of the labeled hormone from its binding site on the antibody. This binding competition can be quantified by reference to a standard curve obtained with known amounts of unlabeled hormone. The degree to which labeled hormone is displaced from antibody is a measure of the amount of unlabeled hormone in a sample of blood or tissue extract. By using very highly radioactive hormone, researchers can make the assay sensitive to even picograms of hormone.

Radioimmunoassay (RIA) can be used to measure a multitude of substances found in tiny quantities in fluids within and outside of organisms (such as viruses, drugs and hormones). The list of current possible uses is endless, but specifically, RIA allows blood donations to be screened

ROSALYN SUSSMAN YALOW
(LT, 1922–2011)

Yalow was born in 1922 in Manhattan to **Simon** and **Clara (née Zipper) Sussman**. She attended *Walton High School*. Knowing how to type, she won a part-time position as secretary to **Dr. Rudolf Schoenheimer**, a leading biochemist at *Columbia University's* College of physicians and Surgeons. Not believing that any good graduate school would admit and provide financial support to a woman, she took a job as a secretary to **Michael Heidelberger**, another biochemist at Columbia, who hired her on the condition that she studied stenography. She graduated from *Hunter College* in Juanary 1941.

In mid-February of 1941, she received an offer of a teaching assistantship in physics at the *University of Illinois at Urbana-Champaign* with the primary reason being that World War II commenced and many men went off to war and the university decided to offer scholarships for women rather than shut down. That summer, she took two tuition-free physics courses under government auspices at *New York University*. At the *University of Illinois*, she was the only woman among the department's 400 members, and the first since 1917. She married fellow student **Aaron Yalow**, the son of a rabbi, in June 1943. They had two children and kept a kosher home. Yalow earned her Ph.D. in 1945.

After graduating, Yalow joined the Bronx Veterans Administration Hospital to help set up its radioisotope service. There she collaborated with **Solomon Berson** to develop **radioimmunoassay** (RIA). RIA is a radioisotope tracing technique that allows the measurement of tiny quantities of various biological substances in human blood as well as a multitude of other aqueous fluids. RIA testing relies on the creation of two reagents. One reagent is a *molecule* that is the product of *covalently bonding* a radioactive isotope atom with a molecule of the target. The second reagent is an antibody which specifically chemically reacts with the target substance. The measurement of target signal is done using both reagents. They are mixed with the fluid containing an unknown concentration of target to be measured. The radioactive atom supplies a signal that can be monitored. The target supplied from the unknown concentration solution displaces the radiolabelled target bond to the antibody. Originally used to study *insulin levels in diabetes mellitus*, the technique has since been applied to hundreds of other substances– including *hormones*, *vitamins* and *enzymes*– all too small to detect previously. *Despite its huge commercial potential, Yalow and Berson refused to patent the method.* In 1968, Yalow was appointed Research Professor in the Department of Medicine at *Mount Sinai Hospital*, where she later became the Solomon Berson Distinguished Professor at Large.

In 1975, Yalow and Berson (who had died in 1972) were awarded the **AMA Scientific Achievement Award**. The following year, she became the first female recipient of the **Albert Lasker Award for Basic Medical Research**. She was elected a Fellow of the American Academy of Arts and Science, in 1978. Yalow received the **National Medal of Science**, in 1988.

In 1977, she received the **Nobel Prize in Chemistry**, together with **Roger Guillemin** and **Andrew V. Schally**, for her role in devising the *radioimmunossay* (RIA) technique that by measuring substances in the human body, that made possible the screening the blood of donors for such diseases as hepatitis among other uses.

Until the time of her death, she continued to reside in the same house in *Riverdale* that she and her husband purchased after she began working at the Bronx Veterans Administration Medical Center in the 1940s. Her husband, Dr. Aaron Yalow, died in 1992. Rosalyn Yalow died on May 30, 2011, aged 89, in *The Bronx* from undisclosed causes.

for various types of hepatitis. The technique can also be used to identify hormone-related health problems. Further, RIA can be used to detect in the blood many foreign substances including some cancers. Finally, the technique can be used to measure the effectiveness of dose levels of antibiotics and drugs. A newer variation of this technique is the *enzyme-linked immunosorbent assay* (ELISA).

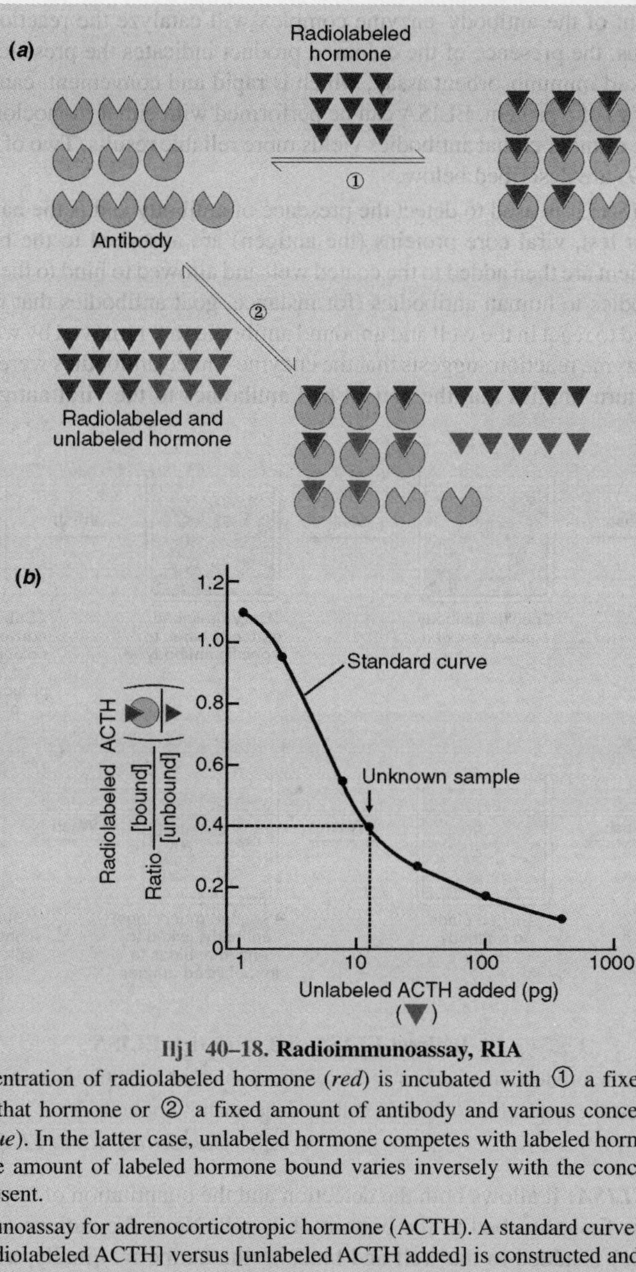

Ilj1 40–18. Radioimmunoassay, RIA

(*a*) A low concentration of radiolabeled hormone (*red*) is incubated with ① a fixed amount of antibody specific for that hormone or ② a fixed amount of antibody and various concentrations of unlabeled hormone (*blue*). In the latter case, unlabeled hormone competes with labeled hormone for binding to the antibody; the amount of labeled hormone bound varies inversely with the concentration of unlabeled hormone present.

(*b*) A radioimmunoassay for adrenocorticotropic hormone (ACTH). A standard curve of the ratio [bound] to [unbound radiolabeled ACTH] versus [unlabeled ACTH added] is constructed and used to determine the amount of (unlabeled) ACTH in an unknown sample. If an aliquot containing an unknown quantity of unlabeled hormone gives, say, a value of 0.4 for the ratio [bound]/[unbound] (see *arrow*), the aliquot must contain about 20 pg of ACTH.

ELISA (*E*NZYME-*L*INKED *I*MMUNO*S*ORBENT *A*SSAY)

A highly sensitive *in vitro* binding assay in which an enzyme and its substrate (rather than a radioactive substance) serve as the indicator system. In this method, an enzyme, which reacts with a colourless substrate to produce a coloured product, is covalently linked to a specific antibody that recognizes a target antigen. If the antigen is present, the antibody–enzyme complex will bind to it, and the enzyme component of the antibody–enzyme complex will catalyze the reaction generating the coloured product. Thus, the presence of the coloured product indicates the presence of the antigen. Such an enzyme– linked immunosorbent assay, which is rapid and convenient, can detect even less than a nanogram (10^{-9} g) of a protein. ELISA can be performed with either monoclonal or polyclonal antibodies, but the use of monoclonal antibodies yields more reliable results. Two of the several types of ELISA (Fig. 40–19) are described below.

A. *Indirect ELISA*: It is used to detect the presence of antibody and is the basis of the test for HIV infection. In *that* test, viral core proteins (the antigen) are absorbed to the bottom of a well. Antibodies from a patient are then added to the coated well and allowed to bind to the antigen. Finally, enzyme-linked antibodies to human antibodies (for instance, goat antibodies that recognize human antibodies) are allowed to react in the well and unbound antibodies are removed by washing. Substrate is then applied. An enzyme reaction suggests that the enzyme-linked antibodies were bound to human antibodies, which in turn implies that the patient had antibodies to the viral antigen.

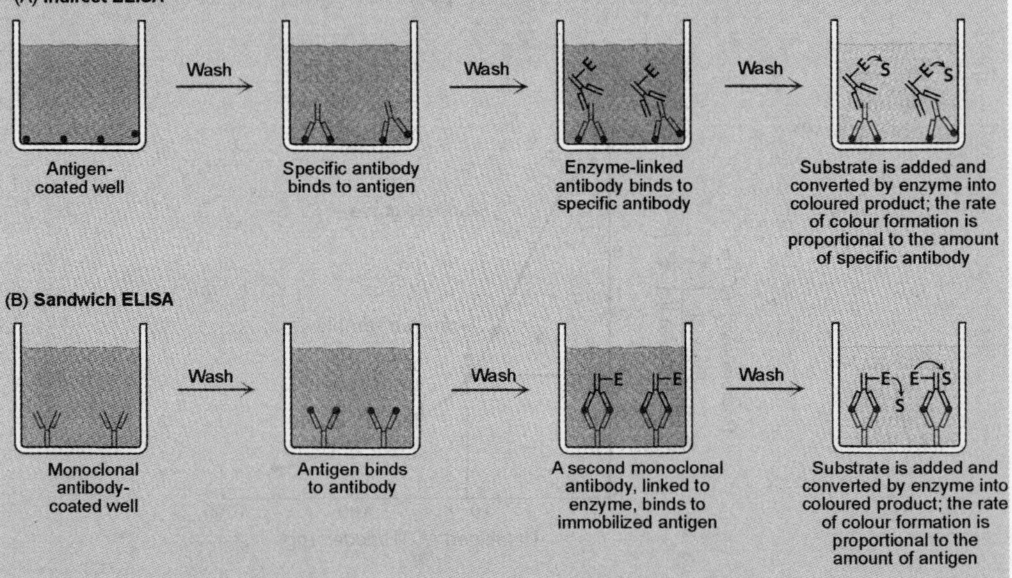

Fig. 40–19. Indirect ELISA and sandwich ELISA

(A) In indirect ELISA, the production of colour indicates the amount of an antibody to a specific antigen.

(B) In sandwich ELISA, the production of colour indicates the quantity of antigen.

(*After Goldsky RA, Kindt TJ, Osborne BA, 2000*)

B. *Sandwich ELISA*: It allows both the detection and the quantitation of antigen. Antibody to a particular antigen is first absorbed to the bottom of a well. Next, the antigen (or blood or urine containing the antigen) is added to the well and binds to the antibody. Finally, a second, different antibody to the antigen is added. This antibody is enzyme-linked and is processed as described for indirect ELISA. In this case, the extent of reaction is directly proportional to the amount of antigen present. Consequently, it permits the measurement of small quantities of antigen.

X-RAY CRYSTALLOGRAPHY

X-ray crystallography is a technique pioneered between 1912 and 1915 by **William Lawrence Bragg**, who developed the theory, and his father **William Henry Bragg,** who designed the instruments; both receiving the **1915 Physics Nobel Prize**, the only father-son duo to have ever

SIR WILLIAM LAWRENCE BRAGG

(LT, 1890–1971)

Bragg, an Australian-born British physicist and x-ray crystallographer, was born on March 31, 1890 in North Adelaide. His father, **William Henry Bragg**, was Elder Professor of Mathematics and Physics at the *University of Adelaide*. Shortly after starting school aged 5, William Lawrence Bragg fell from his tricycle and broke his arm. His father, who had read about Röntgen's experiments in Europe and was performing his own experiments, used the newly-discovered x-rays and his experimental equipment to examine the broken arm. This is the first recorded surgical use of x-rays in Australia.

Bragg was a brilliant student. After beginning his studies at *St Peter's College, Adelaide* in 1904, he went to the University of Adelaide at age 14 to study mathematics, chemistry and physics, graduating in 1908. In the same year, his father accepted the Cavendish Chair of Physics at the *University of Leeds*, and brought the family back to England. Bragg entered *Trinity College, Cambridge* in the autumn of 1909 and received a major scholarship in mathematics, *despite taking the exam while in bed with pneumonia*. After initially excelling in mathematics, he transferred to the physics course in the later years of his studies, and graduated with first class honours in 1911. In 1914, Bragg was elected to a Fellowship at Trinity College– a Fellowship at a Cambridge college involves the submission and defence of a thesis.

Bragg is most famous for his law on the diffraction of X-rays by crystals. **Bragg's x-ray diffraction law** makes it possible to calculate the positions of the atoms within a crystal from the way in which an x-ray beam is diffracted by the crystal lattice. He made this discovery in 1912, during his first year as a research student in Cambridge. He discussed his ideas with his father, who developed the x-ray spectrometer in Leeds. This tool allowed many different types of crystals to be analyzed.

He received jointly with his father Sir William Henry Bragg, the **Nobel Prize for Physics in 1915**, *"For their services in the analysis of crystal structure by means of X-ray"* an important step in the development of x-ray *crystallography*. He was **knighted** in 1941. *To date, Lawrence Bragg is the youngest Nobel Laureate at age 25.* He was the **Director of the Cavendish Laboratory**, Cambridge, when the epochal discovery of the structure of DNA was made by **James D. Watson** and **Francis Crick**, in February 1953.

Bragg's research work was interrupted by both World War I and World War II. During both wars, he worked on sound ranging methods for locating enemy guns. In this work, he was aided by **William Sansome Tucker**, **Harold Roper Robinson** and **Henry Harold Hemming**. For his work during WW I, he was warded the **Military Cross** (MC) and appointed **Officer of the Order of the British Empire** (OBE).

Between the wars, from 1919 to 1937, he worked at the *Victoria University of Manchester* as Langworthy Professor of Physics. After World War II, he returned to Cambridge, splitting the *Cavendish Laboratory* into research groups. He believed that "the ideal research unit is one of six to twelve scientists and a few assistants".

In 1948, he became interested in the structure of proteins and was partly responsible for creating a group that used physics to solve biological problems. He played a part in the 1953 discovery of the *structure of DNA*, in that he provided support to **Francis Crick** and **James D. Watson** who worked under his aegis at the Cavendish. Bragg's original announcement of the discovery of the structure of DNA was made at a Solvay conference on proteins in Belgium on 8 April 1953, but went unreported by the press. He then gave a talk at Guys Hospital Medical School in London on Thursday, 14 May 1953, which resulted in an article by **Ritchie Calder** in The News Chronicle of London on Friday, 15 May 1953, entitiled "Why You Are You. Nearer Secret of Life."

Bragg was gratified to see that the x-ray method that he developed forty years before was at the heart of this profound insight to the nature of life itself. At the same time at the Cavendish, **Max Perutz** was also doing his Nobel-Prize-winning work on the structure of hemoglobin. Bragg subsequently successfully lobbied for, and nominated, **Crick**, **Watson** and **Maurice Wilkins** for the 1962 Nobel Prize in Physiology or Medicine; Wilkins' share recognized the contribution made by researchers (using x-ray crystallography) at *King's College London* to the determination of the structure of DNA. Among those researchers was **Rosalind Franklin**, whose "photograph 51" showed that DNA was a *double helix*, not a triple helix as **Linus Pauling** had proposed. Franking died before the prize (which only goes to living people) was awarded. Lawrence Bragg died on July 1, 1971.

SIR WILLIAM HENRY BRAGG

(LT, 1862–1942)

Henry Bragg, a British physicist, chemist, mathematician and active sportsman, uniquely shared a Nobel Prize with his son **William Lawrence Bragg**– the **1915 Nobel Prize in Physics**: *"For their services in the analysis of crystal structure by means of X-ray"*. The mineral **Braggite** is named after him and his son. He was **knighted** in 1920.

Bragg was born at *Westward* near *Wigton, Cumberland,* the son of **Robert John Bragg**, a merchant marine officer and farmer, and his wife **Mary née Wood**, a clergyman's daughter. When Bragg was seven years old, his mother died, and he was raised by his uncle, also named **William Bragg**, at *Market Harborough*, Leicestershire. He was educated at the Old Grammar School there, at *King William's College* on the *Isle of Man*, and having won an exhibition [scholarship], at *Trinity College, Cambridge*. He graduated in 1884 as third wrangler, and in 1885 was awardwd a "first" in the mathematical tripos.

In 1885, (at the age of 23), Bragg was appointed (Sir Thomas) **Elder Professor of Mathematics** and Experiment Physics at the *University of Adelaide*, and started work there early in 1886. Being a skilled mathematician, at that time he had limited knowledge of physics, most of which was in the form of applied mathematics he had learnt at Trinity. Also at that time, there were only about a hundred students doing full courses at Adelaide, of whom less than a handful belonged to the science school, whose deficient teaching facilities Bragg improved by apprenticing himself to a firm of instrument makers. Bragg was an able and popular lecturer; he encouraged the formation of the student union, and the attendance, free of charge, of science teachers at his lectures.

Bragg's interest in physics developed, particularly in the field of electromagnetism. In 1985, he was visited by **Ernest Rutherford**, en route from New Zealand to Cambridge; this was the commencement of a lifelong friendship. "The turning-point in Bragg's career came in 1904 when he gave the presidential address to section A of the *Australasian Association for the Advancement of Science* at Dunedin, New Zealand", on "Some Recent Advances in the Theory of the Ionization of Gases". This idea was followed up "in a brilliant series of researches" which, within three years, earned him a fellowship of the *Royal Society of London*. This paper was also the origin of his first book *Studies in Radioactivity* (1912). Soon after the delivery of his 1904 address, some *radium bromide* was made available to Bragg for experimentation. In December 1904, his paper "On the Absorption of α Rays and on the Classification of the α Rays from Radium" appeared in the *Philosophical Magazine*, and in the same Issue a paper "On the Ionization Curves of Radium", written in collaboration with his student **Richard Kleeman,** also appeared.

At the end of 1908, Bragg returned to England. During his 23 years in Australia, "he had seen the number of students at the University of Adelaide almost quadruple, and had a full share in the development of its excellent science school."

Whilst in Adelaide, Bragg played tennis and golf, and as a founding member of the *North Adelaide* and *Adelaide University Lacrosse Clubs*, contributed to the introduction of *lacrosse* to South Australia and was also the secretary of the Adelaide University Chess Association. He also met **Gwendoline née Todd**, a skilled water-colour painter, whom he married in 1889. Their first son, **William Lawrence**, was born in North Adelaide in 1890.

Bragg occupied the Cavendish Chair of Physics at the *University of Leeds* from 1909. He continued his work on x-rays with much success. He invented the **x-ray spectrometer** and with his son, *William Lawrence Bragg*, then a research student at Cambridge, founded the new science of x-ray *crystallography*, the analysis of crystal structure using x-ray diffraction.

From 1914, both father and son contributed to the war effort; W.H. Bragg was connected with submarine detection, at Aberdour on Forth and at Harwich, and returned to London in 1918 as a consultant to the admiralty.

In 1915, father and son were jointly award the Nobel Prize in Physics for their studies, using the x-ray spectrometer, of x-ray spectra, x-ray diffraction, and of crystal structure. Ten years later, their volume *X-Rays and Crystal Structure* (1995) had reached a fifth edition.

Bragg was appointed Quain Professor of physics at University College London in 1915, but did not take up his duties there until after World War I. While Quain Professor at London, he continued his work on crystal analysis.

From 1923, he was Fullerian Professor of chemistry at the Royal Institution and director of the Davy Faraday Research Laboratory. This institution was practically rebuilt in 1929–30 and, under Bragg's directorship, many valuable papers were issued from the laboratory. In 1919, 1923 and 1925, he was invited to deliver the Royal Institution Christmas Lecture on *The World of Sound; Concerning the Nature of Things;* and *Old Trades and New Knowledge,* respectively.

received a Nobel Prize and **Lawrence**, who was 25 years old at the time, the youngest-ever Nobel Laureate. It is a technique for determining the structure of crystals or molecules (such as nucleic acid or protein) through the way in which they scatter x-rays. The technique (Fig 40–20) involves directing, rather 'firing', a narrow beam of x-rays through a crystalline sample and recording the diffracted x-rays on a photographic plate. The diffraction pattern consists of a pattern of spots on the plate, and the crystal structure can be worked from the positions and intensities of the diffraction spots. Since x-rays are scattered by electrons, it is possible to draw up *electron density maps* from the diffraction patterns showing the positions of atoms. Small atoms such as hydrogen with few electrons do not show up well, so their position can be hard to determine. Loose quasicrystalline structures such as cell walls and cell membranes are sometimes analyzed this way. Among the early successes of x-ray crystallography were the determination of the structures of:

(*a*) the protein α-helix by **Linus Carl Pauling**, in 1951

(*b*) DNA by **Rosalind Franklin, Maurice Hugh Frederick Wilkins, Francis Harry Compton Crick** and **James Dewey Watson,** in 1953

(*c*) vitamin B_{12} by **Dorothy Crowfoot Hodgkin,** in 1957

(*d*) myoglobin by **John Cowdery Kendrew** at a resolution of 6Å in 1957, 2Å in 1959 and 1.4Å in 1962

(*e*) hemoglobin by **Max Ferdinand Perutz** and **John Cowdery Kendrew,** in 1959

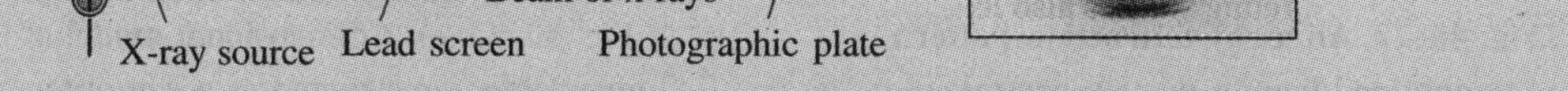

Figl 40–20. X-ray crystallography

X-ray crystallography revealed the basic helical nature of DNA structure. The positions of atoms in DNA can be inferred by the pattern of diffraction of x-rays passed through it, although the task requires tremendous skill

X-RAY DIFFRACTION

It is the diffraction of x-rays by a crystal. The wavelengths of x-rays are comparable in size to the distances between atoms in most crystals, and the repeated pattern of the crystal lattice acts like a diffraction grating for x-rays. Because of interference between the waves scattered by different atoms, the scattering occurs in directions characteristic of the spatial arrangement of the unit cells of the crystal, while the relative intensity of the different beams reflects the structure of the unit cell itself. Thus, a crystal of suitable type can be used to disperse x-rays in a spectrometer (Fig. 40–21). *X-ray diffraction is also the basis of x-ray crystallography.*

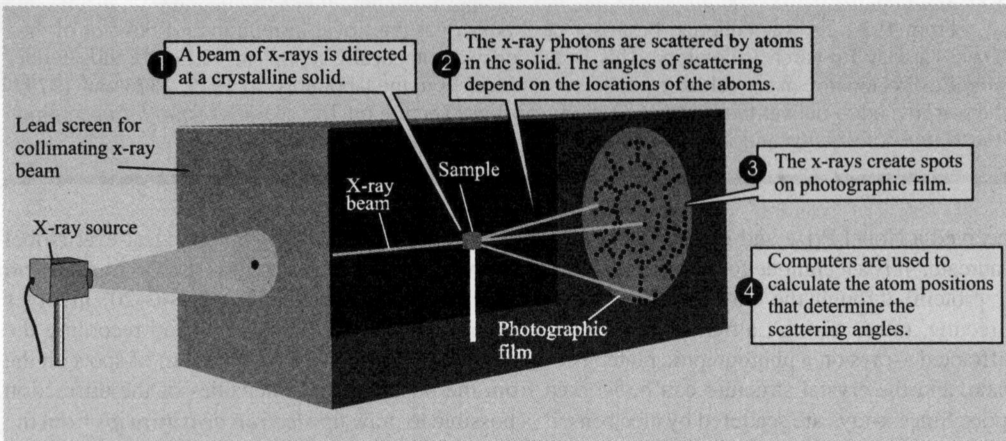

Fig. 40–21. An x-ray diffraction experiment

In fact, when a beam of x-rays is fired at a crystal of purified protein, the atoms in that protein diffract (bend) the x-rays, producing a specific pattern on a photographic plate, as shown in the Figs. 40–22 and 40–23. As the Fig. 40–22 shows, the shape of a protein is not immediately obvious from the information produced, but a trained scientist with a computer can produce an accurate 3-dimensional model of the molecule, as shown in the Fig. 40–23.

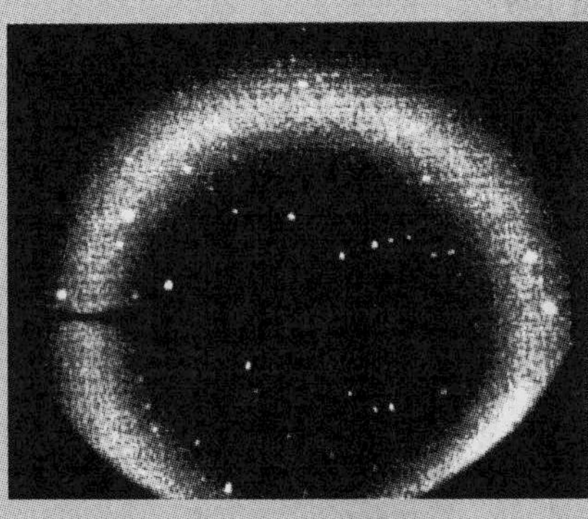

Fig. 40–22. An x-ray diffraction pattern obtained from a protein crystal

Fig. 40–23. A 3-dimensional model of myoglobin, constructed as a computer graphic image with information obtained by x-ray crystallography

NUCLEAR MAGNETIC RESONANCE

It is a powerful analytical technique for finding the structures of carbon compounds. It consists of the absorption of electromagnetic radiation (radio waves) at a suitable precise frequency by certain atomic nuclei placed in an external strong and stable magnetic field. This results in a change of orientation of the nuclei which respond to the magnetic field like miniature bar magnets. The name of the technique 'nuclear magnetic resonance' (NMR)summarizes its key features:

nuclear— The technique detects nuclei of atoms such as hydrogen-1 (protons).

magnetic— The nuclei detected by the technique are the ones which are like tiny magnets which can line up either in the same direction or in the opposite direction to an external magnetic field.

resonance — The absorption of energy is from radiowaves with the frequency corresponding to the size of the energy jump as the nuclei flip from one alignment in a magnetic field to the other.

The samples for analysis are dissolved in a solvent with no hydrogen-1 atoms. Also in the solution is some tetramethyl silane (TMS) which is a standard reference compound which produces an absorption peak well away from the sample peaks. The tube with the sample is supported in a strong magnetic field. The operator turns on an oscillator which produces radiation at radio frequencies (rf). The rf detector records the intensity of the signal as the oscillator scans across a range of wavelengths (Fig. 40–24).

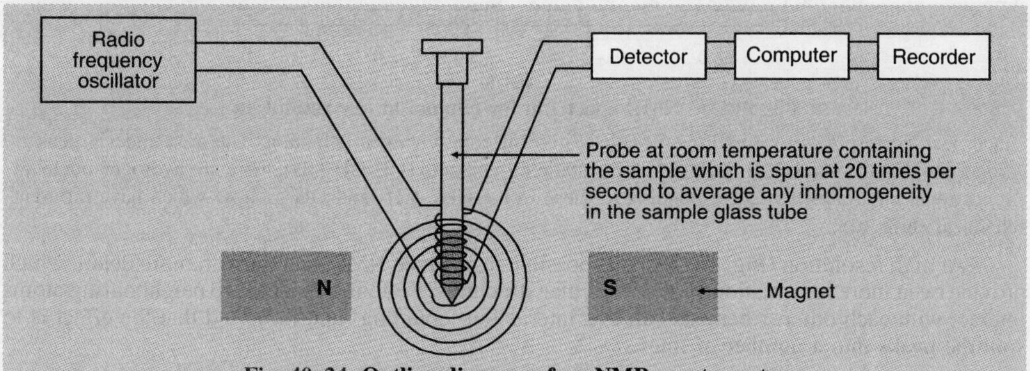

Fig. 40–24. Outline diagram of an NMR spectrometer

The recorder (Fig. 40–25) prints out a spectrum with peaks wherever the sample absorbs radiation strongly. The zero on the scale is fixed by the absorption of hydrogen atoms in the reference chemical TMS. The distances of the sample peaks from zero are called their '*chemical shifts*' (δ).

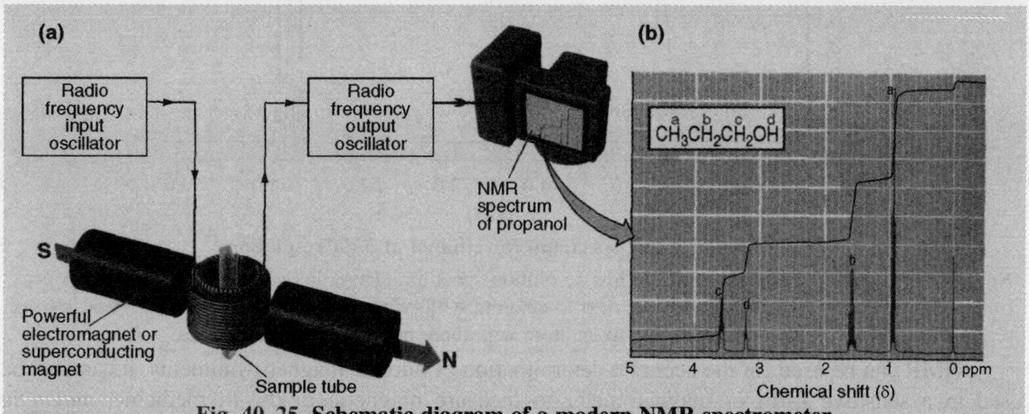

Fig. 40–25. Schematic diagram of a modern NMR spectrometer

(a) A modern NMR spectrometer is a highly automated, computer-controlled instrument.

(b) The spectrum pictured here is of propanol, showing the 4 different types of protons present in the molecule.

At low resolution (Fig. 40–26), the 3 peaks obtained are in the ratio of 3:2:1. The peaks also have a fine structure caused by interaction between spins in the molecule.

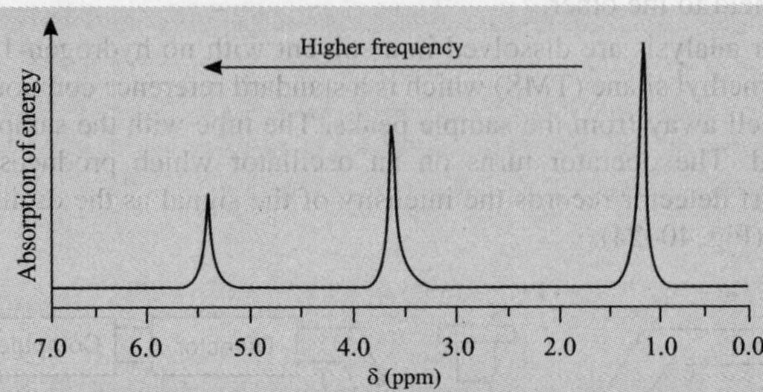

Fig. 40–26. NMR spectrum for ethanol at *low* resolution

Each peak corresponds to a hydrogen nucleus in a different chemical situation. The area under a peak is proportional to the number of nuclei in each situation. In ethanol (CH_3CH_2OH), there are hydrogen nuclei in 3 electronic environments (corresponding to those in the CH_3, CH_2 and OH groups) which have different chemical shifts (d).

At high resolution (Fig. 40–27), it is possible to produce NMR spectra with more detail, which provide even more information about molecular structures. Protons connected to neighbouring atoms interact with each other. Chemists call this interaction 'coupling' and they find that the effect is to split the peaks into a number of lines.

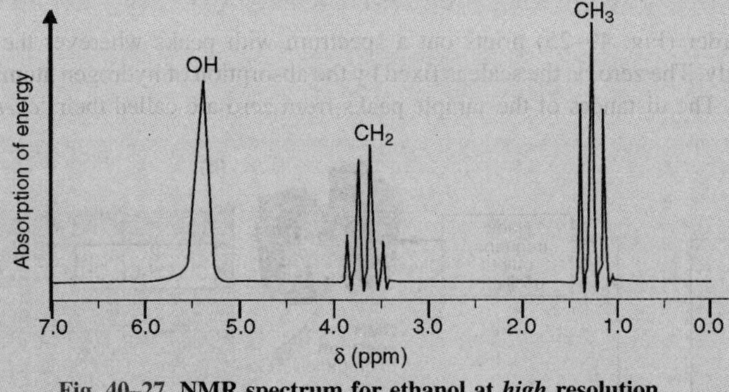

Fig. 40–27. NMR spectrum for ethanol at *high* resolution

Note the extra peaks compared with the low-resolution spectrum. This illustrates the 'n + 1' rule. A peak from protons bonded to an atom which is next to an atom with two protons splits into 3 lines. A peak from protons bonded to an atom which is next to an atom with three protons splits into 4 lines.

NMR can be used for the accurate determination of nuclear magnetic moments. It can also be used in a sensitive form of magnetometer to measure magnetic fields. In medicine, *magnetic resonance imaging* (MRI) has been developed, in which images of soft tissue are produced by magnetic resonance technique. This technique is useful for locating tumours and tissue abnormalities.

The main application of NMR is as a technique for chemical and biochemical analysis and structure determination, known as *NMR spectroscopy*. It depends on the fact that the electrons in a molecule shield the nucleus to some extent from the field, causing different atoms to absorb at slightly different frequencies or at slightly different fields for a fixed frequency (as already mentioned, such effects are known as chemical shifts). There are two methods of NMR spectroscopy:

(*a*) ***Continuous wave (CW) NMR.*** In it, the sample is subjected to a strong field, which can be varied in a controlled way over a small region. It is irradiated with radiation at a fixed frequency, and a detector monitors the field at the sample. As the field changes, absorption corresponding to transitions occur at certain values, and this causes oscillations in the field, which induce a signal in the detector.

(*b*) ***Fourier transform (FT) NMR.*** This uses a fixed magnetic field and the sample is subjected to a high-intensity pulse of radiation covering a range of frequencies. The signal produced is analyzed mathematically to give the NMR spectrum. The most common nucleus studied is ^{1}H. Other nuclei can also be used for NMR spectroscopy (*e.g.*, ^{13}C, ^{14}N, ^{19}F) although these generally have lower magnetic moment and natural abundance than hydrogen and produce weaker signals. The spectrum produced is characteristic of the molecule absorbing the radiation.

NMR can be used analytically, *e.g.*, for estimation of water and oils in foodstuffs. The technique is also used to detect artery-clogging plaques, changes in brain chemistry etc. NMR is not recommended for pregnant women and cannot be used on people with artificial pacemakers or metal joints.

NORTHERN BLOTTING

The name '*Northern*' is given to this technique by way of analogy to another similar, earlier-discovered technique called 'Southern blotting' and is a geographical pun! Even though Southern blotting was named after **Edwin M. Southern**, Northern blotting was not named after anyone called Northern! It was originally termed *reverse-Southern blotting* and later *Northern blotting*. This technique is similar to southern blotting but is used to analyze RNA rather than DNA. The DNA in every human cell contains copies of all the genes present in the human genome. However, only about 15% are active in any particular cell. The remaining 85% are inactive. These are not transcribed and do not lead to the synthesis of protein. Furthermore, different genes are active in different cell types. For example, the genes that are active in a muscle cell are very different from those active in red blood cells. The active genes reflect the very different protein compositions of these cells. Muscle cells contain proteins such as actin and myosin which are required for contraction and the genes for these proteins are very active. In contrast, in red blood cells, the major protein present is hemoglobin. Consequently, the globin genes which encode the polypeptides that constitute hemoglobin are very active.

Northern blotting is used to identify which genes are active in different cell types. The methods used for Northern blotting is similar to Southern blotting. The main difference is with respect to the starting material. For Northern blotting (Fig. 40–28), RNA (rather than chromosomal DNA) is isolated from cells. This contains messenger RNA (mRNA) derived from all the active genes in that cell type. The mRNAs are of different sizes, depending on the size of the protein they encode. The mRNAs that encode proteins present in high concentrations in a cell are also more abundant than those encoding proteins present in low concentrations. The RNA is separated by agarose gel electrophoresis. Each mRNA migrates to a position on the gel determined by its size, with the larger transcripts near the top and the smaller transcripts near the bottom. Like for Southern blotting, the gel is blotted and the membrane is hybridized with a probe specific for the mRNA being investigated. The membrane is then washed and exposed to x-ray film. Development of the film reveals usually a single band corresponding to the mRNA recognized by the probe. The position of the band relative to the top of the membrane can be used to estimate the length of the mRNA and the intensity or darkness of the band is a measure of how much of that mRNA was present in the original cells. Thus, mRNAs of genes that are expressed at low levels appear as light or faint bands and very abundant mRNAs appear as dark or intense bands.

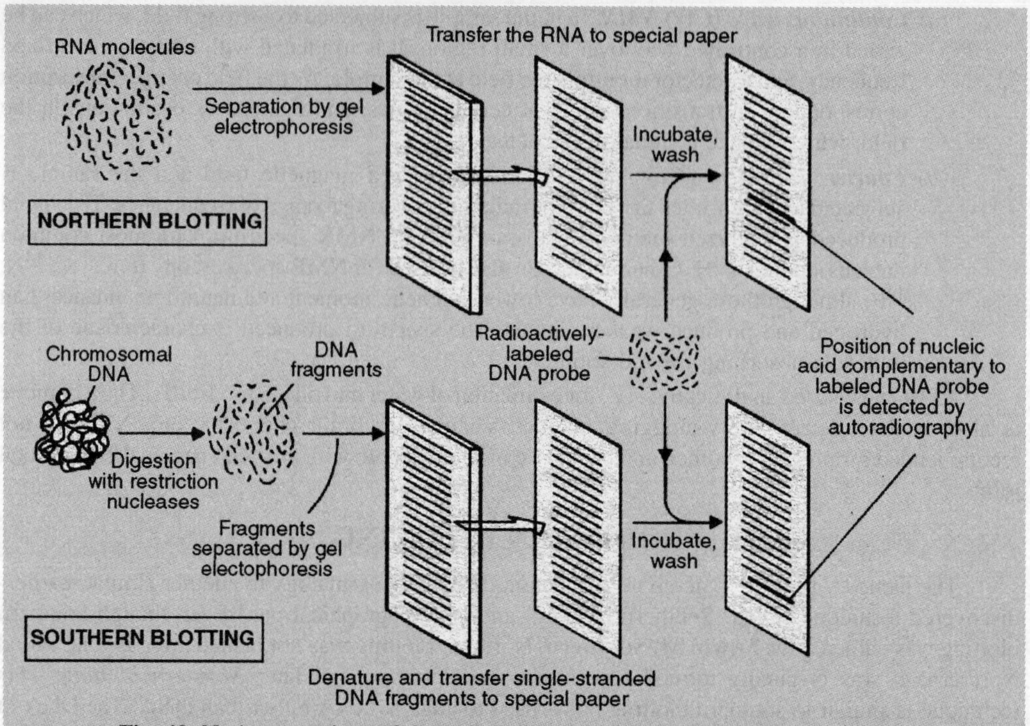

Fig. 40–28. A comparison of the techniques of Northern and Southern blotting

Northern blotting is useful in following ways:

(*a*) Northern blots provide information about the expression of specific genes and are used to study patterns of gene expression in embryonic and adult tissues.

(*b*) Northern blots also detect alternatively-spliced mRNAs and multiple types of transcripts from a single gene and are used to derive other information about transcribed mRNAs.

(*c*) If marker RNAs of known size are run in an adjacent lane, the size of gene's mRNA can be measured.

(*d*) In addition, the amount of transcribed RNA present in the cell or tissue being studied is related to the density of the RNA band on the film. Measuring the density of the band gives the relative transcriptional activity of genes in different cells, tissues, and organisms.

Fig. 40–28 also provides a comparison of the Northern and Southern blotting techniques.

SOUTHERN BLOTTING

This technique was devised by a US biologist **E.M. Southern**, born 1938, of the University of Oxford; hence, so named. It is an analytical technique used to detect DNA molecules and is typically used to analyze the structure of genes. The basis for a Southern blotting experiment is that two DNA fragments will bind to each other only if they have complementary sequences. In the experiment, the labelled strands from the cloned gene will pair specifically with complementary DNA strands, even if the complementary strands are found within a mixture of many other DNA pieces.

In this procedure (Fig. 40–29), the DNA for analysis is purified from cells and is obtained as large fragments of chromosomal DNA, typically 20,000 base pairs or more in length. The first step in Southern blotting involves *digestion of the DNA* with a restriction enzyme. These enzymes, which are derived from bacteria, cut DNA molecules specifically at each point where a DNA sequence of

SIR EDWIN MELLOR SOUTHERN

(born, 1938)

Southern, a molecular biologist, was born on June 7, 1938 in Burnley, Lancashire and educated at *Burnley Grammar School*. He had a brother named **John Southern**. He went on to ready Chemistry at the *University of Manchester* (BSc Hons., 1958). He continued as a graduate student (then Demonstrator, 1963) in the Department of Chemistry, *University of Glasgow*, where he was awarded his Ph.D., in 1962. Southern is Professor of Biochemistry (Emeritus) at the *University of Oxford* and a fellow of *Trinity College*.

He invented the Southern blot technique which is used for *DNA* analysis and was routinely used for genetic fingerprinting and *paternity testing* prior to the development of microsatellite markers for this purpose. The procedure is also frequently used to determine the number of copies of a gene in the genome. The concepts of the Southern blot were used in the development and creation of the modern microarray slide, which is a promising experimental tool. The northern blot is a similar procedure for RNA, playing off the Southern name. The *western blot* is a further pun on the Southern blot, but is an important research tool in protein detection.

Southern is also the Founder, Chairman and Chief Scientific Officer of **Oxford Gene Technology** (OGT), a company that developed DNA microarray technology. OGT won a 1999 *patent infringement* lawsuit against *Affymetrix* based on his *patent* holdings in microarray technology. He is also the Founder and Chairman of a Scottish charity, **The Kirkhouse Trust**, which aims to promote education and research in the Natural Sciences, particularly the biological and medical sciences. In addition, the **Edina Trust** was founded to promote science in schools. *These charities are financed using royalty income from licensing microarray technology*.

In 1990, Southern was one of the winners of the Gairdner Foundation International Award. In 1998, he was awarded the **Royal Medal** of the Royal Society of London. He was made a **Knight Bachelor** in the June 2003 Birthday Honours for services to the development of DNA microarray technologies. In 2005, he was awarded the prestigious **Albert Lasker Award for Clinical Medical Research**, jointly with **Alec Jeffreys** of the University of Leicester, for his invention of the Southern blot, now a common laboratory procedure, when he was working at the University of Edinburgh.

4–8 bases recognized by the enzyme occurs. Digestion of the chromosomal DNA produces thousands of DNA fragments ranging in size from just a few bases to several thousand bases. These DNA fragments are then subjected to electrophoresis in an agarose gel (gel electrophoresis). This separates them according to size with the large fragments near the top of the gel and the small fragments near the bottom. After electrophoresis, the gel is soaked in an alkaline solution that denatures the DNA to single strands making them suitable for hybridization. The pH is then neutralized. The next stage involves transferring the denatured fragments out of the gel and onto a membrane made from nylon (or sometimes nitrocellulose) where they become accessible for analysis using a probe. Transfer is achieved by *blotting*. The gel is placed on a platform above a dish containing a buffer. A wick made from filter paper runs from the platform into the buffer reservoir. A sheet of membrane is placed on top of the membrane. The gel is blotted for several hours during which time the buffer in the reservoir is drawn by capillary action from the wick through the gel and the membrane into the stack of paper towels. The flow of buffer causes the DNA fragments to pass out of the gel and onto the membrane above, where they become attached. After several hours, a replica of the pattern of fragments in the gel forms on the membrane. At this point, the blot is dismantled and the membrane containing the fragments is removed. The membrane is treated to firmly attach the DNA fragments either by baking it at 80°C or by exposing it to ultraviolet radiation.

The next stage involves incubating the membrane with a labelled probe. This is called *hybridization* and is carried out at a temperature and in a buffer that favour the formation of hybrids between the probe and fragments of DNA bound to the membrane whose sequence is complementary to the probe. If hybridization is done at very high temperatures (namely, 65°C) or at high salt (NaCl)

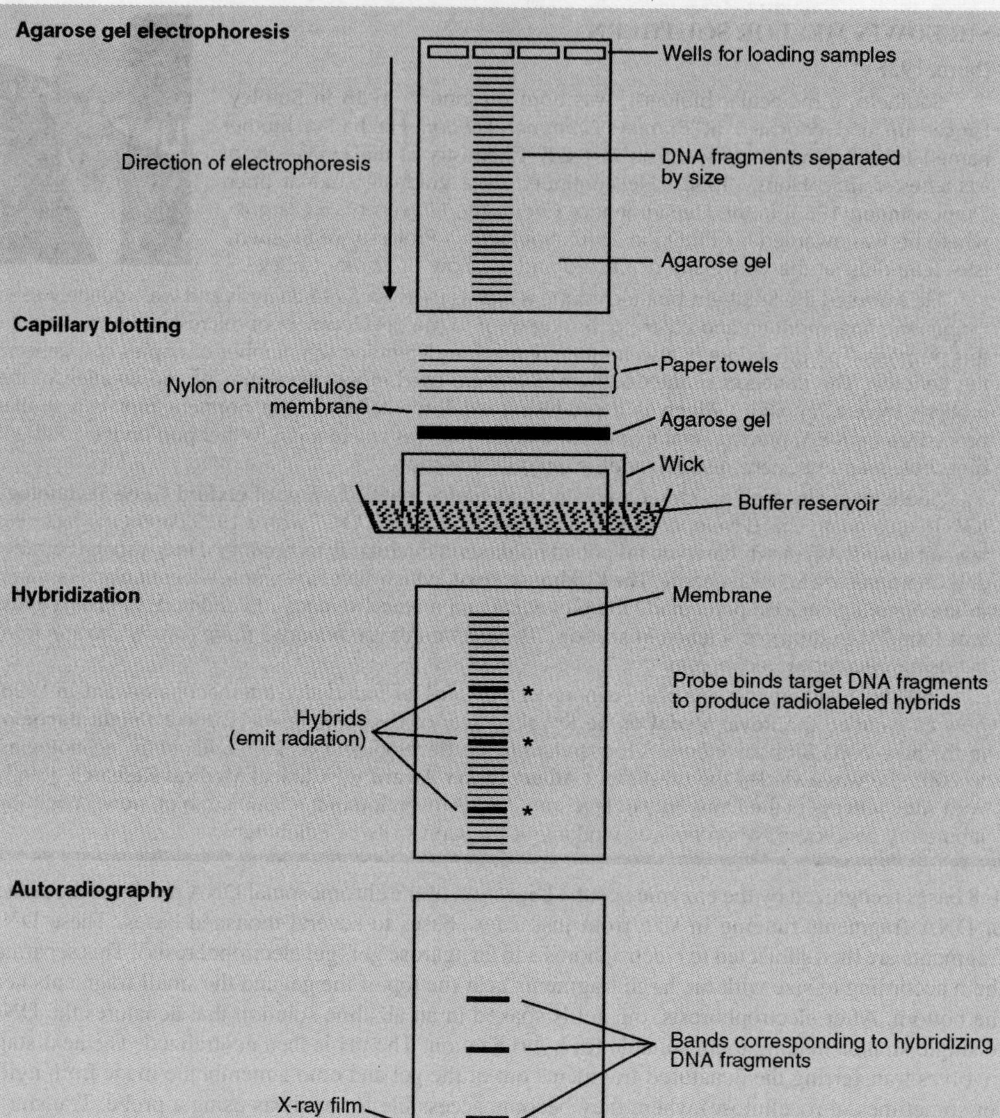

Fig. 40–29. Southern blotting technique

The procedure shown is the original method using capillary action to blot the DNA bands from the gel to the nitrocellulose membrane. *Electrolyte transfer is now often used instead.*

concentrations, then the probe DNA and chromosomal fragment must be very complementary– nearly a perfect match– to hybridize. This condition is called *high stringency*. Conditions of high stringency are used to detect a match between the cloned gene and a chromosomal DNA fragment. However, if the temperature and/or ionic strength are lower, genes that are similar but not necessarily identical may hybridize to the probe. This condition is called *low stringency*. Conditions of low stringency are used to detect homologous genes with DNA sequences that are similar but not identical to the cloned gene being used as a probe. The membrane is then washed with a buffer to remove probe that is bound nonspecifically so that only labelled probe bound to target sequences remains. If the probe is labelled with radioactivity, the membrane is placed in the dark against a sheet of x-ray film. After several hours, development of the film reveals one or more dark bands which correspond to the position of fragments bound by the probe. The length of the fragments can be calculated from their position relative to marker DNA molecules of known length. In this way, the structure of individual genes can be characterized in terms of the number of hybridizing fragments and their sizes. When a probe for a new gene is isolated, one of the first experiments carried out is to use Southern blotting to analyze the structure of the gene in the chromosomal DNA.

Southern blotting has several *uses*:

(*a*) It can determine the copy number of a gene within the genome of an organism. For example, Southern blotting has revealed that rRNA genes are found in multiple copies within a genome whereas many other structural genes are unique.

(*b*) It can also detect small gene deletions that cannot be distinguished under the light microscope.

(*c*) Southern blots also detect rearrangements and duplications in genes associated with human genetic disorders and cancers.

(*d*) The technique can also be used to reveal abnormalities in the structure of genes. For example, about 20% of individuals with the inherited bleeding disorder, hemophilia, show alterations in the structure of the gene for the blood-clotting protein, Factor VIII. These alterations can be used to identify relatives who are carriers of hemophilia and may need genetic counseling.

(*e*) A common use of Southern blotting is to identify gene families. A *gene family* is a group of 2 or more genes derived from the same ancestral gene. The members of a gene family have similar, but not identical, DNA sequences; they are called homologous genes. Southern blotting can distinguish the homologous members of a gene family. Similarly, Southern blotting can identify homologous genes among different species.

REFRENCES

1. **Alexander P, Lundgren HP :** A Laboratory Manual of Analytical Methods of Protein Chemistry. *Vols. 1–5, Pergamon Press, Oxford. 1966–69.*
2. **Andrews AT :** Electrophoresis: Theory, Techniques and Biochemical and Clinical Applications. *2nd ed. Clarendon Press, Oxford (U.K.). 1986.*
3. **Bailey JL :** Techniques in Protein Chemistry. *2nd ed. Elsevier, Amsterdam. 1967.*
4. **Bergmeyer HU :** Principles of Enzymatic Analysis. *Verlag Chemie. New York. 1978.*
5. **Bier M (editor) :** Electrophoresis: Theory, Methods and Applications. *Vols. I and II. Academic Press, Inc., New York. 1958, 1967.*
6. **Blattner FR *et al* :** The complete genome sequence of *Escherichia coli* K12. *Science.* **277 :** *1453–1462, 1997.*
7. **Bowen TJ :** An Introduction to ultracentrifugation. *Wiley-Interscience. New York. 1970.*
8. **Bragg L :** X-ray Crystallography. *Sci. Amer.,* **219 :** *58, 1968.*
9. **Bruening G, Criddle R, Preiss J, Rudert F :** *Biochemical Experiment. Wiley-Interscience, New York. 1970.*
10. **Campbell ID, Dwek RA :** Biological Spectroscopy. *Benjamin/cummings, Menlo Park. 1984.*
11. **Chaykin S :** Biochemistry Laboratory Techniques. *Wiley Eastern Private Ltd., New Delhi. 1970.*
12. **Cooper TG :** The Tools of Biochemistry. *Wiley, New York. 1977.*
13. **Determan H :** Gel Chromatography. *Springer-Verlag, Berlin. 1969.*

14. **Dayl Z :** Separation Methods. *Elsevier, Amsterdam. 1984.*
15. **Edman P, Begg G :** A protein sequenator. *Eur. J. Biochem.* **1 :** *80–91, 1967.*
16. **Erlich HA, Arnheim N :** Genetic analysis using the polymerase chain reaction. *Annu. Rev. Genet.* **26 :** *469–506, 1992.*
17. **Frais F :** Practical Biochemistry. *Butter Worths, London. 1972.*
18. **Francis GE, Mulligan W, Wormall A :** Isotopic Tracers. *The Athlone Press, London. 1954.*
19. **Freifelder D :** Physical Biochemistry. *2nd ed. W.H. Freeman, San Francisco. 1982.*
20. **Giddings JC, Keller RA (editors) :** Advances in Chromatography. *8 vols., Marcel Dekker, Inc., New York. 1966–1969.*
21. **Grossman L, Moldave K :** Methods in Enzymology. *Vol. 12*–Nucleic Acids. *Academic Press, Inc., New York. 1968.*
22. **Hames BD, Rickwood D (editors) :** Gel Electrophoresis of Proteins. *2nd ed., IRL Press, Oxford, Washington, D.C. 1980.*
23. **Harborne JB :** Phytochemical Methods. *Ist ed., Chapman and Hall, London. 1973.*
24. **Heftmann E (editor) :** Chromatography. *2nd ed., Reinhold Publishing Corp., New York. 1967.*
25. **Hudson L, Hay PC :** Practical Immunology. *Blackwell Scientific Publications, Oxford. 1980.*
26. **James A, Prichard M :** Practical Physical Chemistry. *3rd ed., Longmans, London. 1975.*
27. **Karlin S, Campbell AM, Mrazek J :** Comparative DNA analysis across genomes. *Annu. Rev. Genet.* **32 :** *185–225, 1998.*
28. **Kates M :** Techniques in Lipidology. *North Holland, Amsterdam. 1972.*
29. **Maniatis T, Fritsch EF, Sambrook J :** Molecular Cloning: A Laboratory Manual. *2nd ed. Cold Spring Harbor, New York. 1989.*
30. **McCleary BV, Matheson NK :** Enzymatic analysis of polysaccharide structures. *Adv. Carbohydr. Chem. Biochem.* **44 :** *147–276, 1986.*
31. **Morell V :** Huntington's gene finally found. *Science.* **260 :** *28–29, 1993.*
32. **Morris C, Morris P :** Separation Methods in Biochemistry, *Pitman, London. 1963.*
33. **Mullis K, Ferre F, Gibbs R (editor) :** The polymerase chain reaction. *Boston Birkhäuser. 1994.*
34. **Parish JH :** Principles and Practice of Experiments with Nucleic Acids. *Longmans, London. 1972.*
35. **Phizicky EM, Fields S :** Protein–Protein interactions: Methods for detection and analysis. *Microbiol. Rev.* **59 :** *94–123, 1995.*
36. **Plummer DT :** An Introduction to Practical Biochemistry. *3rd ed., McGraw-Hill, London. 1987.*
37. **Rickwood D, Hames BD (editors) :** Gel Electrophoresis of Nucleic Acids. *IRL Press, Oxford, Washington, D.C. 1982.*
38. **Rommens JM *et al* :** Identification of the cystic fibrosis gene: Chromosome walking and jumping. *Science.* **245 :** *1059–1079, 1989.*
39. **Schachman HK :** Ultracentrifugation in Biochemistry. *Academic Press, Inc., New York. 1959.*
40. **Segel IH :** Biochemical Calculations. *John Wiley and Sons, Inc., New York. 1968.*
41. **Senko MW, McLafferty FW :** Mass spectrometry of macromolecules. *Annu. Rev. Biophys. Biomol. Struct.* **23 :** *763–785, 1994.*
42. **Shaw DJ :** Electrophoresis *Academic Press, Inc., New York. 1969.*
43. **Smith I :** Chromatographic and Electrophoretic Techniques. *4th ed., Vol. I–Chromatography. Vol II– Electrophoresis. William Heinemann Ltd., London. 1976.*

44. **Stahl E :** Thin-layer Chromatography. A Laboratory Handbook. *2nd ed., Springer-Verlag, Berlin. 1969.*
45. **Stein WH, Moore S :** Chromatography. *Sci. Amer.* **184 (3) :** *35, 1951.*
46. **Svedberg T, Pedersen KO :** The Ultracentrifuge. *Oxford Univ. Press., New York. 1940.*
47. **Touchstone JC :** Practice of Thin Layer Chromatography. *Wiley-Interscience. 1978.*
48. **Umbreit WW, Burris RH, Stauffer JF :** Manometric Techniques and Related Methods for the Study of Tissue Metabolism. *4th ed., Burgess Publishing Co., Minneapolis. 1964.*
49. **van Holde, KE :** Physical Biochemistry. *2nd ed., Englewood Cliffs, N. J., 1971.*
50. **van Holde KE, Johnson WC, Ho P-S :** Principles of Physical Biochemistry. *Prentice Hall, N.J. 1998.*
51. **van Vunakis H, Langone JJ (editors) :** Immunological methods. *Methods in Enzymology. vol. 70. Academic Press, New York.*
52. **Wait R :** Strutural analysis of carbohydrates by mass spectrometry. *Adv. Carbohydr. Anal.* **1 :** *335–440, 1991.*
53. **Wang CH, Wills DL :** Radiotracer Methodology in Biological Science. *Prentice-Hall, New Jersey. 1965.*
54. **Watson JD, Gilman M, Witkowski J, Zooler M :** Recombinant DNA. *2nd ed. Scientific American Books, New York. 1992.*
55. **Watt IM :** The Principles and Practice of Electron Microscopy. *Cambridge University Press, Cambridge, England. 1985.*
56. **Wilson K, Goulding KH (editors) :** A Biologist's Guide to Principles and Techniques of Practical Biochemistry. *3rd ed., Edward Arnold (Publisher's) Ltd., London. 1986.*
57. **Williams VR, Williams HB :** Basic Physical Chemistry for the Life Sciences. *W.H. Freeman and Co., San Francisco. 1967.*
58. **Wu R, Grossman L, Moldave K (editors) :** Recombinant DNA Methodology. *Academic Press, San Diego. 1989.*
59. **Yalow RS :** Radioimmunoassay: A probe for the fine structure of biological systems. *Science.* **200 :** *1236–1240, 1978.*

PROBLEMS

1. The absorbance A of a solution is defined as

$$A = \log_{10} (I_0/I)$$

in which I_0 is the incident light intensity and I is the transmitted light intensity. The absorbance is related to the molar absorption coefficient (extinction coefficient) ϵ (in M^{-1} cm^{-1}), concentration c (in M), and path length l (in cm) by

$$A = \epsilon lc$$

The absorption coefficient of myoglobin at 580 nm is 15,000 M^{-1} cm^{-1}. What is the absorbance of a 1 mg ml^{-1} solution across a 1-cm path ? What percentage of the incident light is transmitted by this solution?

2. Tropomyosin, a 93-kd muscle protein, sediments more slowly than does hemoglobin (65 kd). Their sedimentation coefficients are 2.6S and 4.31S, respectively. Which structural feature of tropomyosin accounts for its slow sedimentation?
3. The relative electrophoretic mobilities of a 30-kd protein and a 92-kd protein used as standards on an SDS-polyacrylamide gel are 0.80 and 0.41, respectively. What is the apparent mass of a protein having a mobility of 0.62 on this gel?
4. Suppose that an enzyme is dissociated into 4 identical subunits and that you want to test for the enzymatic activity of the individual subunits. However, you must be sure that there are no tetramers remaining in the sample.

(*a*) What chromatographic procedure would you choose to free the monomers from the tetramers?

(*b*) How would you know where the tetramer would be if it were present?

5. The density of DNA in CsCl is approximately 1.7 g/cm^3 and that of most proteins is approximately 1.3 g/cm^3. What would you expect to be the density of a typical bacteriophage which is 50% protein and 50% DNA?
6. A solution of a substance at a concentration of 32 μg/ml, having a MW of 423, has an absorbance of 0.27 at 540 nm measured with a 1-cm light path. What is the molar extinction coefficient at 540 nm, assuming that Beer's law is obeyed at an absorbance of 0.27?
7. Details of the citric acid cycle were worked out by the use of:
   (*a*) x-ray crystallography
   (*b*) radioactive carbon compounds
   (*c*) ultracentrifugation
   (*d*) electron microscopy
8. In two-dimensional chromatography and electrophoresis, such as is employed in finger-printing technique, does it matter which is done first? Explain.
9. A mixture of proteins is subjected to electrophoresis in 3 polyacrylamide gels, each having a different pH value. In each gel five bands are seen.
   (*a*) Can one reasonably conclude that there are only 5 proteins in the mixture. Explain.
   (*b*) Would the conclusion be different if a mixture of DNA fragments was being studied?
10. Which would have a higher sedimentaion coefficient, a rigid rod or flexible rod, both having the same radius and same mass?

ANSWERS TO PROBLEMS

1. A 1 mg/ml solution of myoglobin (17.8 kd) corresponds to 5.62×10^{-5} M. The absorbance of a 1-cm path length is 0.84, which corresponds to an I_0/I ratio of 6.96. Hence, 14.4% of the incident light is transmitted.
2. Tropomyosin is rod shaped, whereas hemoglobin is approximately spherical.
3. 50 kd.
4. (*a*) Gel chromatography or gel electrophoresis.
   (*b*) Add proteins of known molecular weight and obtain a standard curve in which either volume of the effluent (for chromatography) or the distance migrated (for electrophoresis) is plotted against molecular weight.
5. (1.7 +1.3)/2 = 1.5 g/cm^3.
6. The molarity of the solution is (1/423) = 0.032. The molar extinction coefficient is 0.27/(0.032/423) = 3,569.
7. (*b*) radioactive carbon compounds
8. No. To move an object *x* units north and the *y* units east brings the object to the same place reached by moving it first *y* units east and then *x* units north.
9. (*a*) No, since differences in shape can affect mobility, *i.e.*, two proteins differing in both molecular weight and shape might have the same electrophoretic mobility.
   (*b*) Yes.
10. A flexible rod encounters less friction when moving through a fluid.

APPENDICES

Only because water is so cohesive can trees raise it hundreds of feet off the ground. Individual California coast redwoods (*Sequoia sempervirens*) may grow to heights of 360 feet, taller than a 25-story building.

Appendix is "an addition to a document or book (generally at the end) which supplements or illustrates the text"

[*Source : (one of the meanings) : The New International **Webster's Dictionary** of the English Language, Trident Press International*]

APPENDIX I

SELECTED BIBLIOGRAPHY

1. **Abeles RH, Frey PA, Jenckes WP :** Biochemistry. *Jones and Bartlett, Boston. 1992.*
2. **Abraham C, Schepartz B :** Biochemistry. *4th edition. W. B. Saunders Co., Philadelphia and London. 1967.*
3. **Alberts B, Bray D, Lewis J, Raff M, Roberts K, Watson JD :** Molecular Biology of the Cell. *3rd edition Garland Publishing, Inc., New York./London. 1994.*
4. **Applin D :** Key Science (Biology). *2nd edition. Stanley Thornes (Publishers) Ltd., Cheltenham. 1997.*
5. **Armstrong FB :** Biochemistry. *3rd* edition. *Oxford University Press, New York. 1989.*
6. **Baldwin E :** Dynamic Aspects of Biochemistry. *4th edition. Cambridge Univ. Press, Cambridge, U.S.A. 1963.*
7. **Baldwin E :** The Nature of Biochemistry. *2nd edition. Cambridge Univ. Press, Londan. 1967.*
8. **Behrman RE :** Nelson Textbook of Pediatrics. *14th edition. 1992.*
9. **Bennet TP, Frienden E :** Modern Topics in Biochemistry. *The McMillan Co., New York. 1969.*
10. **Berg JM, Tymoczko JL, Stryer L :** Biochemistry. *5th edition. W. H. Freeman and Company, New York. 2002.*
11. **Bhagavan NV :** Biochemistry. *Lippincott. 1975.*
12. **Bonner JE :** Plant Biochemistry. *Academic Press, Inc., New York. 1966.*
13. **Boyle M, Indge B, Senior K :** Human Biology. *1st edition. Collins Educational, London. 1999.*
14. **Brown TA :** Genetics: A Molecular Approach. *3rd edition. Nelson Thorns Ltd., Cheltenham, U.K. 1998.*
15. **Burke JD :** Cell Biology. *The Williams & Wilkins Company, Baltimore (U.S.A.). 1970.*
16. **Campbell MK :** Biochemistry. *2nd edition, Saunders College Publishing, Philadelphia. 1995.*
17. **Chamberlain EN, Ogilvie CO :** Symptoms and Signs in Clinical Medicine. *9th edition. John Wright & Sons Ltd. Bristol.*
18. **Conn EE, Stumpf PK, Bruening G, Dio RH :** Outlines of Biochemistry. *5th edition. John Wiley and Sons, Singapore. 1987.*
19. **Conoley C, Hills P :** Chemistry. *1st edition. Collins Educational, London. 1998.*
20. **Cunningham EB :** Biochemistry: Mechanism of Metabolism. *McGraw-Hill, New York. 1978.*
21. **Darnell JE, Lodish HF, Baltimore D :** Molecular Cell Biology. *2nd edition. Scientific American Books, New York. 1990.*
22. **Datta SP, Ottaway JH :** Biochemistry. *2nd edition. Bailliere Tindall and Cassell, London. 1969.*

23. **Devlin RM :** Plant Physiology. *3rd edition. Van Nostrand Reinhold Co., New York. 1975.*
24. **Devlin TM (Editor) :** Text Book of Biochemistry with Clinical Correlations. *5th edition. Wiley-Liss, New York .2002.*
25. **Doby G :** Plant Biochemistry. *Interscience Publishers, London and New York. 1965.*
26. **Dow J, Lindsay G, Morrison J :** Biochemistry, Molecules, cells and the Body. *Addison-Wesley. 1996.*
27. **Edwards NA, Hassall KA :** Biochemistry and Physiology of the cell. *McGraw-Hill Book Company (UK) Ltd., London. 1980.*
28. **Fairley JL, Kilgour GL :** Essentials of Biological Chemistry. *2nd edition. Van Nostrand Reinhold Co., New York. 1966.*
29. **Fasman GD (editor) :** Handbook of Biochemistry and Molecular Biology. *3rd edition. CRC, Press, Boca Raton, Fla. 1976.*
30. **Finar IL :** Organic Chemistry. *Vol. I (6th edition), Vol II (5th edition). Longman, London. 1973, 1975.*
31. **Flickinger CJ, Brown JC, Kutchai HC, Ogilvie JW :** Medical Cell Biology. *W.B. Saunders Company, Philadelphia. 1979.*
32. **Florkin M, Stotz EH (editors) :** Comprehensive Biochemistry. *Amer. Elsevier Publishing Co., Inc., New York. 1962-67.*
33. **Freeland PW :** Problems in Theoretical Advanced Level Biology. *Ist edition. Hodder & Stoughton, London. 1985.*
34. **Freifelder DM :** Molecular Biology and Biochemestry: Problems and Applications. *W.H. Freeman and Company, San Francisco. 1978.*
35. **Fruton JS, Simmonds S :** General Biochemistry. *2nd edition, John Wiley and Sons, Inc., New York. 1958.*
36. **Ganong WF :** Review of Medical Physiology. *Lange. 1977.*
37. **Green JH :** Basic Clinical Physiology. *3rd edition. ELBS and Oxford University Press, London. 1978.*
38. **Harper HA, Rodwell VW, Mayes PA :** Review of Physiological Chemistry. *17th edition. Lange Medical Pub., California. 1979.*
39. **Hartman PE, Suskind SR :** Gene Action. *2nd edition. Prentice-Hall, Inc., Englewood Cliffs, N.J., U.S.A. 1969.*
40. **Jain JL, Jain S, Jain N :** Dictionary of Biochemistry. *1st (Multicolour) edition. S. Chand & Company Ltd., New Delhi. 2012.*
41. **Jain JL, Jain S, Jain N :** Elementary Biochemistry. *3rd (Multicolour) edition. S. Chand & Company Ltd., New Delhi. 2014.*
42. **Jain JL, Jain S, Jain N :** Fundamentals of Biochemistry. *7th (Multicolour) edition. S. Chand & Company Ltd., New Delhi. 2013.*
43. **Karlson, P :** Introduction to Modern Biochemistry. *4th edition. Academic Press, Inc., New York. 1975.*
44. **Karp G :** Cell and Molecular Biology: Concepts and Experiments. *3rd edition. John Wiley & Sons, Inc., New York. 2002.*
45. **Kingsman SM, Kingsman AJ :** Genetic Engineering. *Ist edition. Blackwell Scientific Publications, Oxford/London. 1988.*

46. **Kornberg A *et al* :** Reflections on Biochemistry. *Pergamon Press. 1976.*

47. **Nelson DL, Cox MM :** Lehninger Principles of Biochemistry. *4th edition. W.H. Freeman and Company, New York. 2005.*

48. **Mahler HR, Cordes EH :** Basic Biological Chemistry. *2nd edition. Harper and Row, Publishers Inc., New York. 1968.*

49. **Mallette MF, Clagett CO, Phillips AT, McCarl RL :** Introductory Biochemistry. *1st edition. Robert E. Krieger Publishing Co., New York. 1979.*

50. **Marsland D :** Principles of Modern Biology. *4th edition. Hold, Rinehart and Winston, New York. 1964.*

51. **Mathews CK, van Holde KE, Ahern KG :** Biochemistry. *3rd Edition. Pearson Education, Inc., California.* 2000.

52. **Mazur A, Harrow B :** Textbook of Biochemistry. *10th editon. W.B. Saunders Co., Philadelphia/London. 1971.*

53. **McElroy WD :** Cellular Physiology and Biochemistry. *Englewood Cliffs, N.J., U.S.A. 1961.*

54. **McGilvery RW, Goldstein :** Biochemistry: A Functional Approach. *3rd edition. W. B. Saunders Co., London/Toronto. 1983.*

55. **Mertz ET :** Elementary Biochemistry. *3rd edition. Burgess Pub. Co., Minneapolis. 1967.*

56. **Metzler DE :** Biochemistry. *Academic Press, Inc., New York. 1977.*

57. **Orten JM, Neuhaus OW :** Biochemistry. *8th edition. The C.V. Mosby Co., Saint Louis. 1970.*

58. **Rawn JD :** Biochemistry. *Neil Patterson Publishers, Burlington, N.C. 1990.*

59. **Rethel FJ :** Concepts in Biochemistry. *McGraw-Hill Book Co., New York. 1967.*

60. **Rosenberg E, Cohen IR :** Microbial Biology. *Ist edition. Holt-Saunders, Japan. 1983.*

61. **Routh JI, Eyman DP, Burton DJ :** General, Organic and Biochemistry. *2nd edition. W.B. Saunders Co., Philadelphia. 1976.*

62. **Schleif R:** Genetics and Molecular Biology. *Addison–Wesley, Reading, A. 1986.*

63. **Smith EL, Hill RL, Lehman IR, Lefkowitz RJ, Handler P, White A :** Principles of Biochemistry: General Aspects. *7th edition. McGraw-Hill International Book Co., London/ Tokyo 1983.*

64. **Smith EL, Hill RL, Lehman IR, Lefkowitz RJ, Handler P, White A :** Principles of Biochemistry: Mammalian Biochemistry. *7th edition. McGraw-Hill International Book Co., London/Tokyo. 1983.*

65. **Stephenson WK :** Concept in Biochemistry. A programmed Text. Reprint. *Wiley Eastern Pvt. Ltd., New Delhi. 1973.*

66. **Street HE :** Plant Metabolism. *Pergamon Press Ltd., Headington Hill Hall, Oxford, 1966.*

67. **The Royal Society :** Quantities, Units and Symbols. *London. 1975.*

68. **Tobin AJ, Dusheck J :** Asking About Life. *2nd edition. Harcourt College Publishers, Philadelphia, 2001.*

69. **Umland JB, Bellama JM :** General Chemistry. *2nd edition. Brooks/Cole Publishing Company, Albany. 1996.*

70. **Vander AJ, Sherman JH, Luciano DS :** Human Physiology. *6th edition. WCB McGraw-Hill, Boston. 1994.*

71. **Varmus H, Weinberg RA :** Genes And the Biology of Cancer. *1st edition. W.H. Freeman & Company, New York. 1993.*

72. **Voet D, Voet JG :** Biochemistry. *2nd edition. Wiley, New York. 1995.*

73. **Vogel AI :** Textbook of Quantitative Inorganic Analysis. *3rd edition, Longmans, Green & Co., London. 1961.*

74. **Watson JD, Hopkins NH, Roberts JW, Steitz JA, Wiener AM :** Molecular Biology of the Gene. *4th edition. The Benjamin/Cummings Publishing Company Inc., California. 1987.*

75. **Whitten KW, Davis RE, Peck ML :** General Chemistry. *5th edition. Saunders College Publishing, Forth Worth. 1996.*

76. **Wiel JH :** General Biochemistry. *6th edition. Wiley Eastern Limited, New Delhi. 1990.*

77. **West ES, Todd WR, Mason HS, van Bruggen JT :** Text-book of Biochemistry. *4th edition. Macmillan Co., 1966.*

78. **Wood WB, Wilson JH, Benbow RM, Hood LE :** Biochemistry. A Problems approach. *2nd edition. Benjamin/Cummings, Menlo Park, California. 1981.*

79. **Woods RA :** Biochemical Genetics. *2nd edition. Chapman and Hall, London. 1980*

80. **Yudkin M, Offord R :** Comprehensible Biochemistry. *Longman. 1973.*

81. **Zubay GL, Parson WW, Vance DE :** Principles of Biochemisty, *1st edition, Wm. C. Brown Publishers, Dubuque. 1995.*

APPENDIX II

GREEK ALPHABET*

Greek letter	*Greek name*	*English equivalent*
Α, α	Alpha	a
Β, β	Beta	b
Γ, γ	Gamma	g
Δ, δ	Delta	d
Ε, ε	Epsilon	ĕ
Ζ, ζ	Zeta	z
Η, η	Eta	ē
Θ, θ	Theta	th
Ι, ι	Iota	i
Κ, κ	Kappa	k
Λ, λ	Lambda	l
Μ, μ	Mu	m
Ν, ϖ	Nu	n
Ξ, ξ	Xi	x
Ο, ο	Omicron	ŏ
Π, π	Pi	p
Ρ, ρ	Rho	r
Σ, σ	Sigma	s
Τ, τ	Tau	t
Υ, υ	Upsilon	u
Φ, ϕ	Phi	ph
Χ, χ	Chi	ch
Ψ, ψ	Psi	ps
Ω, ω	Omega	ō

* Greek letters are preferably used to indicate the positions of substituents in compounds described by trivial names, *e.g.*, α-hydroxypropionic acid (lactic acid). They are used to indicate the *class* of compound, *e.g.*, γ- and δ-lactones. The use of Greek letters to indicate the positions of substituents in a chain should, however, be avoided. *Instead, the numerals should be used.*

APPENDIX III

EXPONENTIAL NOTATION

Exponential notation is used to deal with very large and very small numbers. An **exponent** is a symbol written to the right and above a number or mathematical expression; it indicates number of times that figure is to be multiplied by itself. An exponent is also called the **power** of a number.

(*a*) *For numbers greater than one*, the exponents is positive and is equal to the number of zeroes following the one.

$$10 = 10^1 = \text{ten}$$
$$100 = 10^2 = \text{one hundred}$$
$$1000 = 10^3 = \text{one thousand}$$
$$10000 = 10^4 = \text{ten thousand}$$
$$1000000 = 10^6 = \text{one million}$$
$$1000000000 = 10^9 = \text{one billion}$$
$$1000000000000 = 10^{12} = \text{one trillion}$$
$$1000000000000000 = 10^{15} = \text{one quadrillion}$$

(*b*) *For numbers smaller than one*, the exponent is negative and is equal to the number of places the one is to the right of the decimal point.

$$.1 = 10^{-1} = 1 \text{ tenth}$$
$$.01 = 10^{-2} = 1 \text{ hundredth}$$
$$.001 = 10^{-3} = 1 \text{ thousandth}$$
$$.0001 = 10^{-4} = 1 \text{ ten-thousandth}$$
$$.000001 = 10^{-6} = 1 \text{ millionth}$$
$$.000000001 = 10^{-9} = 1 \text{ billionth}$$
$$.000000000001 = 10^{-12} = \text{one trillionth}$$
$$.000000000000001 = 10^{-15} = \text{one quadrillionth}$$

Any large or small number may be expressed as the product or a more convenient-sized number and a power of 10. Thus, 357 may be expressed as $3.57 \times 100 = 3.57 \times 10^2$. Similarly, $357{,}000{,}000 = 357 \times 10^6$, or 3.57×10^8.

In decimals smaller than one, the decimal point is moved to the right, and the resulting number is multiplied by a negative power of 10 equal to the number of places the decimal point was moved : $.0357 = 3.57 \times 10^{-2}$ or 35.7×10^{-3}; $.0000357 = 3.57 \times 10–5$ or 35.7×10^{-6}.

APPENDIX IV

THE INTERNATIONAL SYSTEM OF UNITS (Systéme International d'Unités)

The International system of units is a metric system of units derived from the *metre, kilogram and second* (m.k.s. system) in place of the *centimetre, gram and secont* (the old c.g.s system). These SI units (Systéme International d'Unités) were approved in 1960 by The 11th General Conference of Weights and Measures and sub-sequently amended and are being adopted by the scientific laboratories throughout the world. They are the legal standard in many countries. The S.I. units are a *coherent system of units* so that if two unit quantities are multiplied or divided, then the answer is the unit of the resultant quantity. In this way, the number of multiples and submultiples of units now in use, will be reduced.

A. Basic Units

This system is now based on 8 base units in relation to 8 basic kinds of quantity. All other units have been derived from basic units---The primary unit of **length** is the *metre*. A metre is the distance between two lines, under standard conditions, engraved on a prototype preserved in the Bureau International des Poids et Measures, Sévres, France.

The primary unit of **mass** is the *kilogram*, which is the mass of the prototype preserved in the same place. The primary unit of **time** is the *second*, which is defined as the duration of 9 192 631 770 periods of the radiation corresponding the ground state of caesium-133 atom.

The 8 Base Units

S.N.	*Physical quantity*	*Unit*	*Symbol*
1.	Length	Metre	m
2.	Mass	Kilogram*	kg
3.	Time	Second	s
4.	Amount of substance	Mole	mol
5.	Thermodynamic temperature	Kelvin	K
6.	Electric current	Ampere	A
7.	Luminous intensity	Candela	cd
8.	Katalytic amount	Katal	kat

* For more than a century, all measurements of weight have been defined in relation to a lump of metal sitting in Paris. The "international prototype" kilogram has been at the heart of trade and scientific experiment since 1889, but now experts want to get rid of it.

The kilogram is still defined as the mass of a piece of platinum which is kept in a laboratory safe of the International Bureau of Weights and Measures in Sévres, France. In fact, the kilogram is the only base unit in the International System of Units (SI) still defined by a physical object, *i.e.,* a cylinder of platinum-irridium. It is about 39 mm high, 39 mm in diameter, cast by **Johnson Matthey** in Hatton Garden in 1889, delivered to the International Committee on Weights and Measures in Sévres shortly afterwards, polished and adjusted to be made equal in mass to the mass of the old French kilogram of the archives which dates from the time of the French Revolution. Then in 1889, it was adopted by the first general conference for weights and measures as the international prototype of the kilogram.

The international prototype of kilogram

One problem with using a lump of metal to define such a basic quantity as the kilogram is that it is liable to change over time. Measurements over the past century have shown that the international prototype has lost around 50 micrograms, around the weight of a grain of sand. "Why should it [the current standard] be stable? It is a piece of platinum cast in London 130 years ago, full of holes, full of hydrogen", said **Terry Quinn**, ex-director of the International Bureau of Weights and Measures. "What is on the surface, it's impossible to know. There are all sorts of surface layers of hydrocarbons". he further adds.

However, the stability of the kilogram is crucial as there are other scientific measures derived from the kilogram. There are:

Newton: Unit of force required to accelerate one kilogram at one metre per second squared.

Pascal: Unit of pressure equal to one newton per unit metre squared.

Hence, the scientists met on January 25, 2011 at the Royal Society in London, Britain's national academy of science, to discuss how to bring the kilogram into the 21st century, by defining this basic unit of measurement in terms of the fundamental constants of nature, rather than a physical object.

B. Derived SI Units

Besides the above basic units, there are also a number of derived SI units, obtained by appropriate combination of these basic units. For convenience, these derived units are given special names. Each unit has an agreed symbol (a capital letter or an initial capital letter if it is named after a scientist, otherwise the symbol consists of one or two lowercase letters). Decimal multiples of the units are indicated by a set of prefixes; when possible a prefix representing 10 raised to a power that is usually a multiple of three is often used.

Some Derived SI Units*

S. N.	*Physical quantity*	*Unit*	*Symbol*	*Definition*
1.	Frequency	Hertz	Hz	s^{-1}
2.	Force	Newton	N	$kg\ m\ s^{-2}$
3.	Energy	Joule	J	$kg\ m^2\ s^{-2}$
4.	Pressure	pascal	Pa	$kg\ m^{-1}\ s^{-2}$
5.	Power	Watt	W	$kg\ m^2 s^{-3}$
6.	Electric charge	Coulomb	C	A s
7.	Electric potential	Volt	V	$kg\ m^2\ s^{-3}\ A^{-1}$
8.	Electric resistance	Ohm	Ω	$kg\ m^2\ s^{-3}\ A^{-2}$
9.	Electric capacitance	Farad	F	$A^2\ s^4\ kg^{-1}\ m^{-2}$
10.	Customary temperature	Degree Celsius	°C	°C = K – 273.15

* None of these units take the plural form, so that 2 volts is written as 2 V, not 2 Vs and 3 metres is written as 3 m, not 3 ms.

C. Prefixes for the SI Units

Sometimes units may be too large or too small and in such cases in order to avoid writing too many zeros, a prefix is placed before the symbol of the unit. The recommended multiples or fractions of a unit change mostly by 1,000 each time (refer Table, given on the next page). Thus, 0.000025 mol is written 25 µmol and 12,500 m is written 12.5 km.

Prefixes Denoting Decimal Factors for the SI Units*

Magnitude	*Prefix*	*Symbol*
Multiples		
10^{33} (decillion)	vendeka	V
10^{27} (octillion)	xenna	X
10^{24} (septillion)	yotta	Y
10^{21} (sextillion)	zetta	Z
10^{18} (quintillion)	exa	E
10^{15} (quadrillion)	peta	P
10^{12} (trillion)	tera	T
10^{9} (billion/gillion)	giga	G
10^{6} (million)	mega	M
10^{3} (thousand)	kilo	k
10^{2} (hundred)	hecto	h
10^{1} (ten)	deca	da
Fractions		
10^{-1} (tenth)	deci	d
10^{-2} (hundredth)	centi	c
10^{-3} (thousandth)	milli	m
10^{-6} (millionth)	micro†	μ
10^{-9} (billionth/gillionth)	nano	n
10^{-12} (trillionth)	pico	p
10^{-15} (quadrillionth)	femto	f
10^{-18} (quintillionth)	atto	a
10^{-21} (sextillionth)	zepto	z
10^{-24} (septillionth)	yocto	y
10^{-27} (octillionth)	xenno	x
10^{-33} (decillionth)	vendeko	v

* These prefixes are applied to metric and other units. For example, a micrometer (μm) is 10^{-6}; a picoliter (pl) is 10^{-12}; and a kilogram (kg) is 10^{3} grams. Also applied to seconds, units, mols, hertz, volts, farads, ohms, curies, equivalents, osmols etc. Combinations of prefixes are no longer allowed, so that *n* is used instead of mμ and *p* instead of μμ.

† Formerly called *micron*, with the same abbreviation, *i.e.*, μ. A micron is a unit of measurement of length or distance.

$$1\mu = 0.001 \text{ mm} = 0.000039 \text{ in.}$$

D. Non-SI Units

Although the SI system of units is coming into use the world over, there are a few recommended non-SI units which are still frequently used. However, these will be abandoned in time but probably not until after the useful life of this text.

Non-SI Units Still in Use

S.N.	*Unit*	*Symbol*	*Definition*
1.	Angström*	Å	10^{-10} m; 10–8 cm; 10^{-7} mm; 10^{-4} μ
2.	Atmosphere	atm	760 mm Hg
3.	Calorie *a.* international	cal_{IT}	4.1868 J
	b. thermochemical	cal	4.184 J
4.	Dyne	dyn	10^{-5} N
5.	Erg	erg	10^{-7} J
6.	Gauss	G	10^{-4} T (testa) = Vsm^{-2} (volt)
7.	Litre	l	10^{-3} m^3 = dm^3
8.	Micron	μ	10^{-6} m = μm
9.	Millimicron	mμ	10^{-9} m = nm

* No longer used, nanometer used instead.

E. Units Used in Conjunction with SI

The principal changes relevant to physiology and medicine are as follows :

1. *Litre* (*l*). The unit of volume is the cube metre (m^3), but since this is too large a unit for practical use, the cubic decimetre (dm^3) is frequently used. An alternative name for the cubic decimetre is the litre. A litre is the volume occupied by one kilogram of pure water at 4°C under a presence of 760 mm of mercury.

$$1 \text{ litre} = 1 \text{ dm}^3 = 10^{-3} \text{ m}^3$$

$$1 \text{ millilitre} = 1 \text{ cm}^3 = 10^{-6} \text{ m}^3$$

$$1 \text{ microlitre} = 1 \text{ mm}^3 = 10^{-9} \text{ m}^3$$

The terms millilitre and microlitre will, however, be abandoned in time.

2. *Gram* (*g*). The basic units of mass is the kilogram, but the gram is used as an elementary unit and also in association with prefixes (μg, mg) until a name is adopted for the basic unit of mass.

3. *Minute, hour and year*. The basic SI unit of time is the second but the common units of time such as minute, hour and year are frequently used for convenience.

4. *Kilopascal* (*kPa*). Pressure is the force exerted, divided by the area over which it acts. In SI units, the basic unit of pressure is the newton per square metre (N/m^2 or Nm^{-2}). It is this unit that is also known as the pascal (Pa). The pascal is too small a pressure for use in clinical medicine and the unit of 1,000 pascals or kilopascal (kPa) is, therefore, used. The relationship between the kilopascal and mm Hg is that:

$$1 \text{ kPa} = 7.5 \text{ mm Hg}$$

$$0.133 \text{ kPa} = 1 \text{ mm Hg}$$

Since blood pressure is usually measured using a mercury manometer, it seems likely that blood pressure will continue to be expressed in mm Hg for many years to come.

Central venous pressures are frequently measured in centimetres of water. The conversion of kilopascals is:

10 cm of water = 1 kPa

5. *Kilojoule* (*kJ*). The unit of heat and energy that has been used in nutrition is the Calorie (kilocalorie). This is the amount of heat required to raise 1 kilogram of water 1°C. But this does not form part of the SI units system.

The work done when the force of one newton acts through a distance of one metre is termed the joule (J). This is also the work done when one ampere of electricity flows through a resistance of one ohm for one second. One joule per second is, thus, equal to one watt. The joule has been adopted as the unit of both heat and energy in the SI units system. Like the pascal, it is too small a unit for clinical use and the kilojoule (kJ), corresponding to 1,000 joules and the megajoule (MJ), corresponding the 1,000,000 joules are therefore used.

The conversion from Calories to kilojoules is:

1 Calorie = 4.2 kJ

6. *Millimoles per litre.* The concentration of a substance in body fluids is expressed in *millimoles per litre* (instead of g/100 ml) provided that its molecular weight is known. When the molecular weight cannot be acurately determined in the case of the plasma proteins), concentrations are expressed in *grams per litre*.

F. Definition of Standards and Equivalents

Standard	*Abbreviation*	*Equivalent*
Meter	m	1,650,763.73 wavelengths in vacuo of the unpertured transition $2p_{10}$ -$5d_5$ in Kr^{86}
Kilogram	kg	mass of the international kilogram at Sévres, France
Second	sec	9, 192, 631, 770 vibrations of the unper turbed hyperfine transition 4, 0–3, 0 of the fundamental state $^2S_{1/2}$ in Cs^{133}*
Thermochemical calorie	cal	4.1840 joules
Litre	1	0.001, 000, 028 $meter^3$
Inchin.	0.0254 meter	
Pound (avdp.)	1b	0.453, 592, 37 kg

*There is no measurable difference between this and the previous standard of time, 1/31, 556, 925. 9747 of the tropical year at 12^h ET, 0 January 1990. For this reason and because even more accurate maser standards may soon be available, the Cs standard was adopted provisionally rather than "permanently".

HINTS FOR USING THE METRIC SYSTEM

1. The metric system uses symbols rather than abbreviations. Hence, do not place a period (or full stop) after metric symbols (*e.g.*, 1 g, not 1 g.). Use a period after a symbol only at the end of a sentence, but always use a period with the symbol for inches (in.)
2. Metric symbols are always singular; they do not have a plural form (*e.g.*, 20 km, not 20 kms).
3. Capital initial letters are never used for units, except when named after famous scientists, such as N (Newton), J (Joule) and W (Watt).
4. Do not mix units or symbols (*e.g.*, 5.3 m, not 5 m 300 mm).
5. Symbols combined in a quotient can be written as, for example, metre per second or ms^{-1}. The use of the solidus (strope, /) is restricted to indicating the unit of a variable, such as temperature /ºC.
6. Except for degree Celsius, always leave a space between a number and a metric symbol (*e.g.*, 70 mm, not 70mm; 10°C, not 10° C).
7. The comma is no longer used to separate groups of 3 digits, but a space is left instead (*e.g.*, 236 892 714 not as 236,892,714.
8. The raised decimal point is not correct. The internationally-accepted decimal sign is placed level with the feet of the numeral (*e.g.*, 3.182, not 3·182).
9. Express measurements in units that require only a few decimal places. For example, 0.4 m is more easily manipulated and understood than 400,000,000 nm.
10. Familiarize yourself with manipulations within the metric system. Work within one system, and do not convert back and forth between the metric and English (or the old c.g.s.) systems.
11. Use a zero before a decimal point when the number is less than one (*e.g.*, 0.25 m, not .25 m)
12. When measuring water, the metric system offers an easy and common conversion from volume measured in litres to volume measured in cubic metres to mass measured in grams : 1 ml = 1 cm^3 = 1 g.

APPENDIX V

COMPARISON OF METRIC AND OTHER UNITS

Units of Length

1 centimetre	= 0.3937 inch	1 inch	= 2.5400 centimetres
1 mitre	= 3.2808 feet	1 foot	= 0.30480 metre
1 metre	= 1.09361 yards	1 yard	= 0.91440 metre
1 kilometre	= 0.62137 mile	1 mile	= 1.60934 kilometres

Units of Area

1 square cm	= 0.1550 sq. in.	1 square in.	= 6.4516 sq. cm.
1 square metre	= 10.7638 sq. ft.	1 square ft.	= 0.9290 sq. m.
1 square metre	= 1.1960 sq. yds.	1 square yd.	= 0.83613 sq. m.
1 square km.	= 0.38610 sq. mile	1 square mile	= 2.5900 sq. km.
1 hectare	= 2.47105 acres	1 acre	= 0.40469 hectare

Units of Volume

1 cubic cm. (c.c.)	= 0.061024 cubic in.	1 cubic in.	= 16.3871 c.c.
1 cubic m.	= 35.3144 cubic ft.	1 cubic ft.	= 0.028317 cubic m.
1 cubic m.	= 1.3079 cubic yds.	1 cubic yd.	= 0.7645 cubic m.

Measures of Liquid Capacity

British Units (or Imperial Units)

pennyweight = 3.858 carats
1 gallon = 4 quarts = 8 pints = 32 gills
1 gallon = 4.54596 litres
1 litre = 0.2200 gallon = 1.7598 pints

U.S. Units

1 gallon = 4 quarts = 8 pints = 32 gills
1 gallon = 3.78533 litres
1 litre = 0.26418 gallon = 2.1134 pints

Apothecaries' Units (British)

1 gallon = 8 pints = 160 fluid ounces
1 fluid ounce = 8 fluid drachms = 24 scruples = 480 minims
1 gallon = 4.54596 litres
1 litre = 0.2200 gallon = 35.196 fluid ounces

Apothecaries' Units (U.S.)

1 gallon = 8 pints = 128 fluid ounces
1 fluid ounce = 8 fluid drachms = 480 minims
1 gallon = 3.78533 litres
1 litre = 0.26418 gallon = 33.81504 fluid ounces

Units of Mass

Avoirdupois Weight. This is a British and American system of weights, based on a pound of 16 ounces.

1 hundredweight = 4 quarters = 8 stones = 112 pounds
1 pound = 16 ounces = 256 drachms = 7,000 grains
1 grain = 0.064799 g
1 gram = 15.4323 grains
1 hundredweight = 50.80238 kg
1 kilogram = 0.0198461 hundredweight
1 millier = 1 metric ton = 10 quintals
1 quintal = 10 myriagram = 100 kilogram = 0.09842059 ton
1 kilogram = 2.204621 pounds
1 pounds = 0.4535926 kilogram

Troy Weight. This is a system of weights, used for gems and precious metals (gold, silver) and also for drugs.

1 pound = 12 ounces = 240 pennyweights
1 pennyweight (dwt.) = 6 carats = 24 grains
1 gram = 0.3215 ounce = 0.643 pennyweight
= 3.858 carats
1 carat = 0.25920 gram

Apothecaries' Weight. This is a system of weights, used in pharmacy.

1 pound = 12 ounces = 96 drachms
1 drachm = 3 scruples = 60 grains
1 gram = 0.03215 ounce = 0.2572 drachm = 0.7716 scruple
1 scruple = 1.29598 gram

APPENDIX VI

MATHEMATICAL SIGNS AND SYMBOLS

$+$	plus; positive
$-$	minus; negative
$\pm$	plus or minus; positive or negative
$\mp$	minus or plus; negative or positive
$\times$, $\cdot$	multiplied by
$\div$, :	divided by
=, ::	equal to; equals
$\backsimeq$, $\cong$	approximately equal to; congruent
$\neq$	not equal to
$\equiv$	identical to; identically equal to
$\not\equiv$	not identical to
$>$	greater than
$\not>$	not greater than
$\gg$, »	much greater than
$<$	less than
$\not<$	not less than
$\ll$, «	much less than
$\geqq$, $\geq$	greater than or equal to (or, is no less than)
$\leqq$, $\leq$	less than or equal to (or, is no greater than)
$\sim$	of the order of; similar to; equivalent to
$\approx$	similar to
Σ	sum of
Π	product of
$\bar{x}$	the average value of x
$\surd$	square root; root; radical
π	pi (= 3.14159295)
$^\circ$	degree(s)
∞	infinity
$\propto$	proportional to
Δ	difference
$\therefore$	therefore
$\because$	since; because
:	is to (in ratios)
$\angle$	angle
L, $\perp$	right angle
$\parallel$	parallel; parallel to
$\perp$	perpendicular to
$\int$	indefinite integral
!	factorial
$'$	minutes of arc; feet; prime
$''$	seconds of arc; inches; double prime

APPENDIX VII

THE RELATIVE SIZES OF CELLS, MOLECULES, AND MULTICELLULAR ORGANISMS

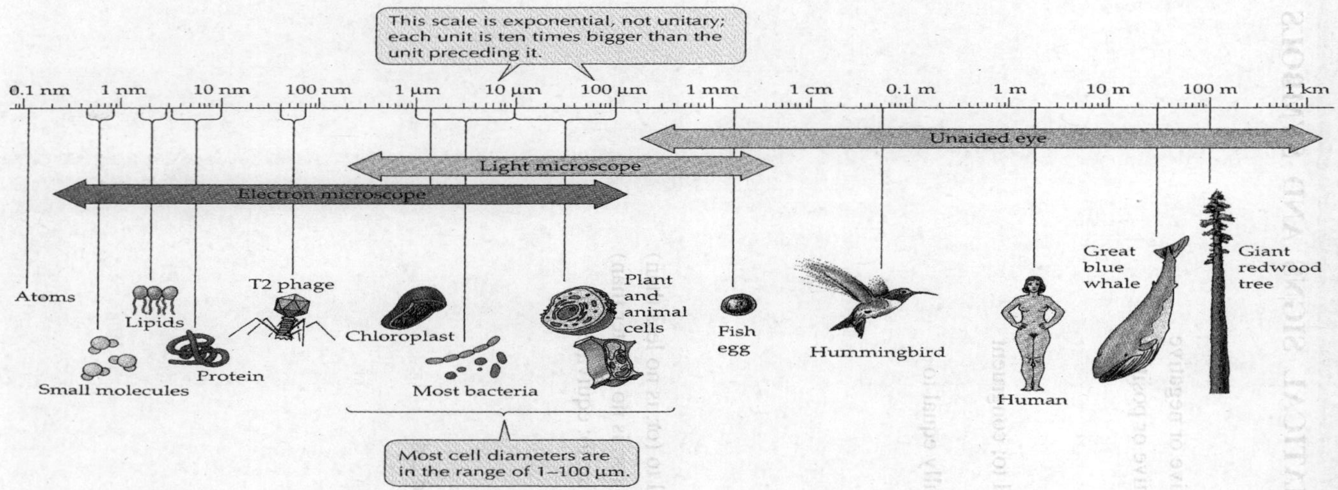

APPENDIX VIII

LIST OF ABBREVIATIONS AND SYMBOLS

The abbreviations are used to conserve space. They have become a part of the nomenclature of modern biochemistry. These are used in the current biochemical literature to such an extent that some publications are rendered unintelligible by the numerous abbreviations. They should, henceforth, be used cautiously. The abbreviations are justifiably used in the case of high molecular weight compounds (such as **proteins**) where the individual components are indicated by such symbols. The monomers are abbreviated by lowercase letters. The abbreviations of **metabolites** are mostly superfluous. *The names of the enzymes should never be abbreviated.* Although the use of abbreviations, in general, should be discouraged, they have their own value and are, therefore, listed here as a useful reference.

1. CARBOHYDRATES

Abe	Abequose
Ara	Arabinose
Fru	Fructose
Fru-1-P, F1P	Fructose-1-phosphate
Fru-6-P, F6P	Fructose-6-phosphate
Fru-1-P, 6-P, FDP	Fructose-1, 6-diphosphate
Fuc	Fucose
Gal	Galactose
GalN	D-galactosamine
GalNAc	*N*-acetyl-D-galactosamine
Glc	Glucose
GlcA	Gluconic acid
GlcN	D-glucosamine
GlcNAc, NAG	*N*-acetyl-D-glucosamine
Glc-1-P, G1P	Glucose-1-phospahate
Glc-6-P, G6P	Glucose-6-phosphate
GluUA	Glucuronic acid
IdoUA	Iduronic acid
Lyx	Lyxose
Mal	Maltose
Man	Mannose
Mur	Muramic acid
Mur*N*Ac, NAM	*N*-acetyl-muramic acid
Neu	Neuraminate
Neu*N*Ac, NANA (or Sia)	*N*-acetyl-D-neuraminic acid (or Sialic acid)
Rha	Rhamnose
Rib	Ribose
dRib	Deoxyribose
Rul	Ribulose
Xul	Xylulose
Xyl	Xylose

Glucose may not be abbreviated as Glu since this symbol is already reserved for an amino acid, *glutamic acid*. Also, the symbol G cannot be used for glucose since this is often used to denote a nitrogenous base, *guanine* in the nucleic acids.

2. AMINO ACIDS

Three-letter abbreviations are used in text books to designate amino acid residues in peptide sequences but *not to designate free amino acids*. **One-letter symbols** were devised for computer storage and retrieval of sequence date but are also being used for tabulation of long sequences.

Three-letter abbreviation†	*One-letter symbol* ‡	*Amino acid*
Ala	A	Alanine
Arg	R	Arginine
Asn, Asp-NH_2*] Asx	N] B	Asparagine
Asp	D]	Aspartic acid
Cys, Cys-SH*	C	Cysteine
Gln, Glu-NH_2*] Glx	Q] Z	Glutamine
Glu	E]	Glutamic acid
Gly	G	Glycine
His	H	Histidine
Ile	I	Isoleucine
Leu	L	Leucine
Lys	K	Lysine
Met	M	Methionine
Phe	F	Phenylalanine
Pro	P	Proline
Ser	S	Serine
Thr	T	Threonine
Trp	W	Tryptophan
Tyr	Y	Tyrosine
Val	V	Valine

† Note that the abbreviations for amino acids are the first letters of their names except tryptophan (Trp), isoleucine (Ile), asparagine (Asn) and glutamine (Gln).

‡ Also note that barring 3 letters (J, O, U), all the remaining 23 letters of English alphabet have been utilized in giving one-letter symbol to the standard amino acids. The symbols for the small amino acids are the first letters of their names (*e.g.*, A for alanine, G for glycine and L for leucine); the other symbols have been agreed upon by convention. However, the letter X is used for indicating an undetermined or nonstandard amino acid.

* Not preferred

The 3-letter abbreviations for the **derivatives of amino acids** are:

AdoMet, SAM	*S*-adenosylmethionine
Cys-SO_3H	Cysteic acid
Cys-S-S-Cys	Cystine
fMet	N-formylmethionine
Hyl	Hydroxylysine
Hyp	Hydroxyproline
Pgl, pGlu, <Glu	Pyroglutamic acid

3. PROTEINS

ACP	**Acyl carrier protein**
CBG	**Corticosteroid-binding globulin**
GSH	**Glutathione (reduced form)**
GSSG	**Glutathione (oxidized form)**
Hb	**Hemoglobin (deoxygenated)**
HbA	**Adult hemoglobin**
HbCO	**CO hemoglobin**
HbF	**Fetal hemoglobin**
HbO_2	**Oxyhemoglobin**
HDL	**High density lipoprotein**
HGPRT	**Hypoxanthine-guanine phosphoribosyl-transferase**
IDL	**Intermediate density lipoprotein**
IF, IFN	**Interferon**
IgG	**Immunoglobulin G**
IP_3	**Inositol triphosphate**
ITP	**Inosine triphosphate**
LDL	**Low density lipoprotein**
Mb	**Myoglobin (deoxygenated)**
MbCO	**CO myoglobin**
MbO_2	**Oxymyoglobin**
Met Mb	**Methemyoglobin**
NHI	**Nonheme iron protein**
RBP	**Retinol-binding protein**
TBG	**Thyroxine-binding globulin**
UCP	**Uncoupling proteins**
VHDL	**Very high density lipoprotein**
VLDL	**Very low density lipoprotein**

4. LIPIDS

Cer	Ceramide
DAG	Diacylglycerol
EFA	Essential fatty acids (refer to linoleic, linolenic and arachidonic acids)
FA	Fatty acid
FFA	Unesterified free fatty acid (same as NEFA)
HCC	25-hydroxycholecalciferol, an active metabolite of vitamin D_3
LTB_4	Leukotriene B_4
LTC_4	Leukotriene C_4
LTD_4	Leukotriene D_4
LTE_4	Leukotriene E_4
NEFA	Unesterified (= nonesterified) free fatty acid (same as FFA)
PG	Prostaglandin
PGE	Prostaglandin (ether-soluble)
PGE_2	Prostaglandin E_2
PGF	Prostaglandin (phosphate-buffer-soluble)
PGG_2	Prostaglandin G_2
PGH_2	Prostaglandin H_2
PGI_2	Prostaglandin I_2
TXA_2	Thromboxane A_2

5. NUCLEIC ACIDS AND RELATED COMPOUNDS

A. Nitrogenous bases

A, Ade	Adenine
B	5-bromouracil
C, Cyt	Cytosine
G, Gua	Guanine
MC	Methylcytosine
T, Thy	Thymine
U, Ura	Uracil

B. Nitrogenous base derivatives

DiHu	5, 6-dihydrouracil
DiMeA	6-dimethyladenine
2-DiMeG	2-dimethylguanine
HMC	5-hydroxymethylcytosine
6-IPA	6-N-isopentenyladenine
6-MeA	6-methyladenine
5-MeC	5-methylcytosine
1-MeG	1-methylguanine
4-ThioU	4-thiouracil
Ψ, ψU	Pseudouridine

C. Nucleosides

*Abbreviations**	*Systemic name*	*Trivial name*
Ribonucleosides		
AR, Ado	Adenine ribonucleoside	Adenosine
CR, Cyd	Cytosine ribonucleoside	Cytidine
GR, Guo	Guanine ribonucleoside	Guanosine
TR, Thd	Thymine ribonucleoside	Thymidine
UR, Urd	Uracil ribonucleoside	Uridine
Deoxyribonucleosides		
AdR	Adenine deoxyribonucleoside	Deoxyadenosine
CdR	Cytosine deoxyribonucleoside	Deoxycytidine
GdR	Guanine deoxyribonucleoside	Deoxyguanosine
TdR	Thymine deoxyribonucleoside	Deoxythymidine
UdR	Uracil deoxyribonucleoside	Deoxyuridine

* In the case of ribonucleosides, the second set of abbreviations is preferred.

D. Nucleoside phosphates (= nucleotides)

Nucleosides		*Abbreviations for nucleotides**		
Name	*Abbreviation*†	*Monophosphate*	*Diphosphate*	*Triphosphate*
Adenosine	A, Ado	AMP	ADP	ATP
Cytidine	C, Cyd	CMP	CDP	CTP
Guanosine	G, Guo	GMP	GDP	GTP
Inosine	I, Ino	IMP	IDP	ITP
Thymidine	T, Thd	TMP	TDP	TTP
Uridine	U, Urd	UMP	UDP	UTP

* These abbreviations are used for 5′-mono-5′-di-and 5′-triphosphates. The 3′-phosphates are, however, denoted by a number; for example, adenosine -3'-monophosphate is abbreviated as 3′-AMP or Ado-3′-P, adenosine 3′-triphosphate as 3′-ATP or Ado-3′ P_3.

† The first set of abbreviations is used to describe the long sequence in oligonucleotides. These letters are, otherwise, reserved for the nitrogenous bases.

E. Related compounds

AICAR	5-aminoimidazole-4-carboxamide ribonucleotide
AZT	3′-azido-3′-deoxythymidine
CAIR	Carboxyaminoimidazole ribonucleotide
cDNA	Complementary DNA
DHU, UH_2	Dihydrouridine
DNA	Deoxyribonucleic acid
mDNA	Inosine
MtDNA	Mitochondrial DNA
RNA	Ribonucleic acid
hnRNA	Heterogeneous nuclear RNA
mRNA	Messenger RNA or template RNA
nRNA	Ribosomal RNA
ns	Nucleoside(s)
nt	Nucleotide(s)
Pu	Purine base (in a nucleotide sequence)
Py	Pyrimidine base (in a nucleotide sequence)
R	Purine nucleoside
rRNA	Ribosomal RNA
snRNA	Small nuclear RNA
snRNP	Small nuclear ribonucleoprotein
sRNA	Soluble RNA (now replaced by tRNA)
SSB	Single-strand binding protein
tRNA	Transfer RNA (formerly called as sRNA)
X	Any nucleoside
Y	Pyrimidine nucleoside

6. ENZYMES

ACE	Angiotensin-converting enzyme
ADA	Adenosine deaminase
ADH	Alcohol dehydrogenase
ACTase (or ATCase)	Aspartate carbamoyltransferase (or Aspartate transcarbamoylase)
ALP	Alkaline phosphatase
ALT	Alanine aminotransferase
APRT	Adenosine phosphoribosyl transferase
AST	Aspartate aminotransferase
ATPase	Adenosine triphosphatase
AVP	Arginine vasopressin (same as antidiuretic hormone)
Bio	Biotin
CA	Carbonic anhydrase
CAT	Catalase
CDK	Cyclin-dependent kinase
ChAT	Choline acetyltransferase
ChE	Cholinesterase
COX	Cyclooxygenase-1
CPK, CK	Creatine phosphokinase
DNase	Deoxyribonuclease
E, Enz	Enzyme
EC (followed by numbers)	Enzyme commission number (IUB system)
EF	Elongation factor
eu	enzyme unit
FSH	Fibrin-stabilizing factor (or fibrinase)
GDH	Glutamic dehydrogenase
GK	Glucokinase
GOT	Glutamate-oxaloacetate transaminase
HGPRT	Hypoxanthine-guanine phosphoribosyl transferase
HIOMT	Hydroxyindole-o-methyl transferase
IF	Initiation factor
LDH	Lactate dehydrogenase
LPL	Lipoprotein lipase
LPS	Lipopolysaccharide
MDH	Malate dehydrogenase
PABA	p-aminobenzoic acid
PFK	Phosphofructokinase
RF	Releasing factor
RNase	Ribonuclease
RNR	Ribonucleotide reductase
SOD	Superoxide dismutase
SRP	Signal recognition particle
XO	Xanthine oxidase

7. COENZYMES

Acetyl-CoA	Acetyl-coenzyme A
Acyl-CoA	General symbol for an organic compound. coenzyme A ester
CoA	Coenzyme A ("A" stands for acyl activation), used in names of compounds
CoA-SH	Coenzyme A, used in reactions
CoF	Coenzyme F ('F" stands for formyl and formaldehyde activation)
CoQ	Coenzyme Q (now replaced by Q)
DHFA, FH_2	Dihydrofolic acid
DPN^+	Diphosphopyridine nucleotide (oxidized form), now replaced by NAD^+
DPNH	Diphosphopyridine nucleotide (reduced form), now replaced by NADH
DPT	Diphosphothiamine (same as TPP)
FAD	Flavin adenine dinucleotide (oxidized form)
$FADH_2$	Flavin adenine dinucleotide (reduced form)
FMN	Flavin mononucleotide (oxidized form)
$FMNH_2$	Flavin mononucleotide (reduced form)
HS-CoA	Reduced coenzyme A
LIP (S_2)	Lipoic acid
LTPP	Lipothiamide pyrophosphate
Malonyl-CoA	Malonyl-coenzyme A
NAD^+	Nicotinamide adenine dinucleotide (oxidized form), formerly called as DPN^+ or coenzyme I or coezymase
NADH	Nicotinamide adenine dinucleotide (reduced form), formerly called as DPNH
$NADP^+$	Nicotinamide adenine dinucleotide phosphate (oxidized form), formerly called as TPN^+ or coenzyme II
NADPH	Nicotinamide adenine dinucleotide phosphate (reduced form), formerly called as TPNH
PAL, PL	Pyridoxal
PALP, PLP	Pyridoxal phosphate
PAMP	Pyridoxamine phosphate
PEP	Phosphoenolpyruvate
Q, UQ	Ubiquinone (formerly called as CoQ)
THFA, FH_4	Tetrahydrofolic acid
TPN^+	Triphosphopyridine nucleotide (oxidized form), now replaced by $NADP^+$
TPNH	Triphosphopyridine neucleotide (reduced form), now replaced by NADPH
TPP, ThPP	Thiamine pyrophosphate (same as DPT) ; formerly called as cocarboxylase
UDP	Uridine diphosphate

8. ANIMAL HORMONES

ACTH	Adrenocorticotrophic hormone (= corticotropin)
ADH	Antidiuretic hormone (= vasopressin)
cAMP	Adenosine 3', 5'-cyclic monophosphate (= 3', 5'-adenylic acid)
CCK	Cholecystokinin
cGMP	Guanosine 3', 5'-cyclic monophosphate
Compound A	11-dehydrocorticosterone
Compound B	Corticosterone
Compound E	Cortisone
Compound F	Cortisol
Compound S	11-deoxycortisol
CRF	Corticotropin-releasing factor
CT	Calcitonin (same as TCT)
DEA, DHEA	Dehydroepiandrosterone
DHT	Dihydrotestosterone
DOC	Deoxycorticosterone
DOCA	Deoxycorticosterone acetate
ESF	Erythrocyte stimulating factor (= erythropoietin)
FSH	Follicle-stimulating hormone (= follitropin)
FSH-RF	Follicle-stimulating hormone-releasing factor
GH	Growth hormone (same as STH)
GH-RF	Growth hormone-releasing factor
GSS	Gonad-stimumating substance (in echinoderms)
GTH	Gonadotrophic hormones (= gonadotropins)
HCG	Human chorionic gonadotropin
HCS	Human chorionic somatotropin (same as HPL)
hG-factor	Hypoglycemic factor (*i.e.*, insulin)
HG-factor	Hyperglycemic factor (*i.e.*, glucagon)
HPL	Human placental lactogen (same as HCS)
5-HT	5-hydroxytryptamine (*i.e.*, serotonin)
ICSH	Interestitial cell-stimulating hormone (same as LH)
In	insulin
JH	Juvenile hormone (same as SQH)
LH	Luteinizing hormone (same as ICSH)
LH-RF	Luteinizing hormone-releasing factor
LTH	Luteotrophic hormone (same as MH or PL)
MH	Mammotrophic hormone (same as LTH or PL) ; also moulting hormone (same as PGH)
MIS	Maturation-inducing substance (in echinoderms)
MRF	Melanocyte-stimulating hormone-releasing factor
MSH	Melanocyte-stimulating hormone

PG	Prostaglandin
PGH	Prothoracic gland hormone (= ecdysone)
PL	Prolactin (same as LTH or MH)
PRF	Prolactin-releasing factor
PTH	Parathormone (= Collip's hormone)
PZ	Pancreozymin
REF	Renal erythropoietic factor
SQH	Status quo hormone (same as JH)
STH	Sometotrophic hormone or somatotropin (same as GH)
TCT	Thyrocalacitonin (same as CT)
Thx, T_4	Thyroxine (= 3, 5, 3', 5'-tetraiodothyronine)
TIT, T_3	3, 5, 3'-triiodothyronine
TRF	Thyrotropin-releasing factor
TSH	Thyroid stimulating hormone (= thyrotropin)
URF	Uterine-relaxing factor (= relaxin)
VP	Vasopressin

9. PLANT HORMONES

ABA	**Abscisic acid**
AE	**Avena Einheit**
CCC	**Chlorocholine chloride (also called cycocel)**
2,4-D	**2, 4-dichlorophenoxyacetic acid**
ETH	**Ethylene**
GA	**Gibberellic acid**
IAA	**Indole-3-acetic acid**
IAc	**Indole-3-acetaldehyde**
IBA	**Indole-3-butyric acid**
IPA	**Indole-3-propionic acid**
IPyA	**Indole-3-pyruvic acid**
K	**Kinetin (= 6-furfurylaminopurine)**
NAA	**1-naphthaleneacetic acid**
NAd	**1-naphthalene acetamide**
NOA	**2-naphthoxyacetic acid**
PAA	**Phenyl acetic acid**
POA	**Phenoxyacetic acid**
2,4,5-T	**2,4,5-trichlorophenoxyacetic acid**

10. VITAMINS

APA factor	Antipernicious anemia factor
CR	Citrivorum factor
DHF, FH2	Dihydrofolate (or H_2 folate)
FA	Folic acid (same as PGA)
LA	Lipoic acid
PGA	Pteroyl-L-glutamic acid (same as FA)
PABA	Para-aminobenzoic acid
POF	Pyruvate oxidation factor, refers to α-lipoic acid
SLR factor	Streptococcus lactis R factor
THF, FH_4	Tetrahydrofolate (or H_4 folate)
Vitamin A_1	Retinol
" A_2	3-dehydroretinol
" B_1	Thiamine
" B_2	Riboflavin
" B_3	Pantothenic acid
" B_5	Niacin
" B_6	Pyridoxine
" B_7	Biotin
" B_9	Folacin
" B_{12a}	Cyanocobalamin
" B_{12b}	Hydroxocobalamin
" B_{12c}	Nitrocobalamin
" C	Ascorbic acid
" D_2	Ergocalciferol
" D_3	Cholecalciferol
" E	Tocopherol
" K_1	Phylloquinone
" K_2	Farnoquinone
1,25 $(OH)_2D_3$	1, 25-dihydroxycholecalciferol

11. MISCELLANEOUS

A	Ampere
A	Absorbance
AA	Amino acid
Å	Angström units(s), 10^{-10} m, 0.1 nm
A^-	General symbol for anion
a	atto, a prefix denoting one million millionth, *i.e.*, 10^{-18}
Ab	Absorbance; Antibody
Ag	Antigen
atm	Atmosphere : 1 atm = 760 torr = mean atomspheric pressure at sea level
ACH, ACh	Acetylcholine
ADC	AIDS-dementia complex
AHF	Antihemophilic factor
AI	Angiotensin I
AIDS	Acquired immune deficiency syndrome
AT-10	Dehydrotachysterol (= antitetany compound 10)
BMR	Basal metabolic rate
BOH	General symbol for an alkali
bp	Base pair
bw	Body weight
C	centigrade
c	centi, a prefix denoting one hundredth, *i.e.*, 10^{-2}
Cal	1,000 calories ; kilocalorie
cal	calorie (gram calorie)
CaM	Calmodulin
cap	Capillary
cc	cubic centimetres
cer	Ceramide
Ci	curie
Chl	Chlorophyll
CM-cellulose	o-(carboxymethyl) cellulose
CNS	Central nervous system
conc.	Concentration
cpm	counts per minute
cps	cycles per second (same as Hz)
Cr	Creatinine
CrP	Creatinine phosphate
CSF	Cerebrospinal fluid
cyt	cytochrome
D	Geometric isomer of L form of chemical compound
D	Diffusion Coefficient
d	diem, day(s)
d	2′-deoxyribo-

d	density
Da	Dalton
DEAE-cellulose	o-(diethylaminoethyl) cellulose
DFP	Diisoprpyl fluorophosphate
DHAP	Dihydroxyacetone phosphate
DHF	Dihydrofolate
DIPF	Diisopropylphosphofluoride
DIT	Diiodotyrosine (same as I_2tyr)
DNFB	Dinitrofluorobenzene (same as FDNB)
DNP	2,4-dinotrophenol
Dol	Dolichol
Dol-P	Dolichol phosphate
DOPA, dopa	3,4-dihydroxyphenylalanine
dps	disintegrations per second
DPT	Diphosphothiamine
E	Electrical potential
e, e^-	Electron
Ea	Energy of activation
ECF	Extracellular fluid
ECM	Extracellular matrix
EDTA	Ethylenediminetetraacetic, acid, a reagent used to chelate divalent metals
EF	Elongation factor
ELISA	Enzyme-linked immunosorbent assay
EM	Electron microscopy
ER	Endoplasmic reticulum
ETS	Electron transport system
Eq	Equivalent(s) ; also equation
FA	Female
F	Fahrenheit
f	Frictional coefficient
FA	Fatty acid
Fd	Ferredoxin
FDA	Food and Drug Administration
FDNB	Fluorodinitrobenzane (same as DNFB)
FP	Flavoprotein
FSF	Fibrin-stabilizing factor
g, gm	grams (s)
g	unit of force, 1 *g* is equal to the force of gravity on the earth's surface
G	giga, a prefix denoting thousand million, 10^9 (American) or million million, 10^{12} (British)
GABA	γ-aminobutyric acid
GAG	Glycosaminoglycan
GIT	Gastrointestinal tract

GLC	Gas liquid chromatography
G3P, GAP	Glyceraldehyde-3-phospahte
GSC	Gas solid chromatography
^3H	Tritium
h, hr	Hour(s)
HA	General symbol for an acid
Hb	Hemoglobin
HbO_2	Oxyhemoglobin
Hct	Hematocirt
HIV	Human immuno deficiency virus
HPLC	High-performance liquid chromatography
HRT	Hormone replacement therapy
5-HT	5-hydroxytryptamine
Hz	Hertz, unit of frequency : 1 cycle per second = 1 hertz
ICF	Intracellular fluid
IDDM	Insulin-dependent diabetes mellitus
IdUA	L-iduronic acid
IEP, pI	Isolectric point
Ig	Immunoglobulin
IgG	Immunoglobulin G
IL	Interleukin (IL-1 to IL-29)
IMP	Inosine monophosphate
INH	Isonicotinic acid hydrazide (= isoniazid)
IR	Infrared light
Ityr	Monoiodotyrosine (same at MIT)
I_2tyr	Diiodotyrosine (same as DIT)
IU	International Unit of hormone activity
IUB	International Union of Biochemistry
K	Dissociaton constant
K_a	Acid dissociation constant
k	Kilo, a prefix denoting thousand, *i.e.*, 10^3
k	rate constant
kb	kilobase
kbp	kilobase pair(s)
kcal (Cal)	Kilocalorie (1,000 calories)
kD	kilo Dalton
K_{eq}	Equilibrium constant
αKG	α-ketoglutarate
K_i	Inhibitor constant
K_m	Michaelis constant (substrate concentration producing half-maximal velocity)
L	Litre
L	Geometric isomer of D form of chemical compound
LATS	Long-acting thyroid stimulator, now replaced by TSI
LCAT	Lecithin: cholesterol acyltransferase

log	Logarithm to the base 10
LSD	Lysergic acid diethylamide
M	Male
M	Molarity (mols/litre)
M	mega, a prefix denoting million, *i.e.*, thousand thousand or 10^6
m	milli, a prefix denoting one thousandth or 10^{-3}; also metre(s)
MAB	Monoclonal antibody
Mb	Myoglobin
MbO_2	Oxymyoglobin
MDR	Multidrug resistance
mev	million electron volt
MHC	Major histocompatibility complex
MI	Myocardial infarction
min	minute(s)
MIT	Monoiodotyrosine (same as Ityr)
mm Hg	Millimeters of mercury
mo	Month(s)
mol	Mole
MS	Mass spectrometry
MSUD	Maple syrup urine disease
MW, mol wt	Molecular weight
N	Normality (of a solution)
n	nano, a prefix denoting one billionth, *i.e.*, one thousand millionth or 10^{-9}
NDP	Any nucleoside diphosphate
NF	Neurofibromatosis; Nuclear factor
NIDDM	Noninsulin-dependent diabetes mellitus
nm	nanometer (10^{-9} metre)
NMR	Nuclear magnetic resonance
NPN	Nonprotein nitrogen
NSAID	Nonsteroidal anti-inflammatory drug
NTP	Any nucleoside triphosphate
OAA	Oxaloacetate
OD	Optical density
P	Phosphate (radical)
P	Pressure
p	pico, a prefix denoting one trillionth, *i.e.*, one thousand billionth or 10^{-12}
~P	High-energy phosphate
Pa	Pascal
PA	Phosphatidic acid

PAGE	Polyacrylamide gel electrophoresis
PC	Phosphatidyl choline; Pyruvate carboxylase
PC, PPC	Paper partition chromatography
PEP	Phosphoenolpyruvate
PCr	Phosphocreatine
PCR	Polymerase chain reaction
PDE	Phosphodiesterase
PDH	Pyruvate dehydrogenase
PE	Phosphatidyl ethanolamine
PFK	Phosphofructokinase
2-PG	2-phosphoglycerate
3-PG	3-phosphoglycerate
PGK	Phosphoglycerate kinase
PGM	Phosphoglucomutase
pH, *p*H	Negative logarithm of the hydrogen ion concentration of a solution
P_i	Inorganic phosphate
PK	Pyruvate kinase
*p*K	Negative logarithm of the equilibrium constant for a chemical reaction
PL	Phospholipase
PNS	Peripheral nervous system
pO_2	Partial pressure of oxygen
PP, PP_i	Inorganic pyrophosphate
PP...P_i	Inorganic polyphosphates
PQ	Plastoquinone (same as Q)
PRL	Prolactin
PrP	Prion protein
PRPP	5-phosphoribosyl 1-pyrophosphate
PS	Phosphatidyl serine
Q	Ubiquinone (= plastoquinone)
QH_2	Ubiquinol (= plastoquinol)
R	General symbol for remainder of a chemical formula, *e.g.*, an alcohol is R—OH ; also gas constant
R_f	Resolution front
Rb	Ribosome
RBC, rbc	Red blood cell(s) or erythrocyte(s)
RDA	Recommended dietary allowance
RIA	Radioimmunoassay
RNS	Reactive nitrogen species
ROS	Reactive oxygen species
rpm	revolutions per minute
rps	revolutions per second
RQ	Respiratory quotient
RT	Room temperature

s	**Second(s)**
S	**Substrate**
S	**Svedberg unit(s)**
SAM	***S*-adenosylmethionine**
SCID	**Severe combined immunodeficiency disease**
SD	**Standard deviation**
SDS	**Sodium dodecyl sulfate**
sec	**second(s)**
SER	**Smooth endoplasmic reticulum**
S_f	**Svedberg unit of floatation**
SGOT	**Serum glutamic-oxaloacetic transaminase (= aspartate aminotransferase)**
SGPT	**Serum glutamic-pyruvic transaminase (= alanine aminotransferase)**
SH	**Sulfhydryl**
sq cm	**square centimetre(s)**
SRP	**Signal recognition particle**
std.	**Standard**
STP	**Standard temperature and pressure (273° absolute, 760 mm Hg)**
Str	**Streptomycin**
T	**Absolute temperature**
T	**tera, a prefix denoting trillion, *i.e.*, million million or 10^{12}**
T_3	**Triiodothyronine**
T_4	**Thyroxine**
TB	**Tuberculosis**
TBW	Total body water
TCA	Tricarboxylic acid cycle
TEAE-cellulose	o-(triethylaminoethyl) cellulose
Tet	Tetracycline
TF	Transcription factor (with qualifier)
TG	Triacylglycerol
THF	Tetrahydrofolate
TLC	Thin layer chromatography
T_m	Melting temperature (= denaturation temperature)
TMV	Tobacco mosaic virus
TNF	Tumour necrosis factor
Tris	Tris (hydroxymethyl) aminomethane, a buffer (tromethamine
TSI	Thyroid-stimulating immunoglobin, formerly called as LATS
U	Unit(s)
U	International Unit of enzyme activity
UV	Ultraviolet light
V	Volume

V	**Volt**
V_o	**Initial velocity**
V_{max}	**Maximal velocity**
WBC, wbc	**White blood cell(s) or leukocyte(s)**
WHO	**World Health Organization**
wk	**Week(s)**
XMP	**Xanthosine monophosphate**
Yr	**Year(s)**
μ	**micro, a prefix denoting one millionth, *i.e.*, one thousand thousandth or 10^{-6}**
Δ	**Change in (Example : ΔV = change in volume) ; In steroid nomenclature, Δ followed by a number (*e.g.*, Δ^4) indicate the position of a double bond**
ΔG	**Free-energy change**
$\Delta G^{o\prime}$	**Standard free energy change**
ΔH	**Enthalpy change**
ΔS	**Entropy change**
[ ]	**Concentration of**
$[\alpha]_D{}^t$	**Specific rotation**
λ	**Wavelength**
Z	**Net charge**

APPENDIX IX

THE NOBEL PRIZES

Alfred Bernhard Nobel
The originator of the Nobel Prizes, the world's most venerated annual awards

The Nobel Prizes are among the world's most venerated awards. Originally, they were a gesture of peace by a man whose discovery had unintentionally added to the destructive forces of warfare.

Alfred Bernhard Nobel (LT, 1833–1896), one of the most imaginative of the 19th century inventors, is famous as originator of the Nobel Prizes, the world's most venerated annual awards. He was an engineer, a chemist, a philanthropist, a litterateur, an eccentric genius, and a linguist. He was born on October 21, 1833 in Stockholm, Sweden as fourth son to **Immanuel** and **Caroline Nobel**. His family was descended from **Olof Rudbeck**, the best-known technical genius of Sweden's 17th century era as a great power in northern Europe. His father **Immanuel Nobel** was an engineer and inventor who built bridges and buildings in Stockholm. His mother **Andriette Nobel**, came from a wealthy family. In connection with his construction work, Immanuel Nobel also experimented with different techniques for blasting rocks. The naval mines designed by Immanuel Nobel were simple devices consisting of submerged wooden casks filled with gunpowder. Anchored below the surface of the Gulf of Finland, they effectively deterred the British Royal Navy from moving into firing range of St. Petersburg during the Crimean War (1853–1856). Immanuel Nobel was also a pioneer in arms manufacture and in designing steam engines. Nobel, who remained unmarried and had no children, was closely associated with **Countess Bertha Kinsky**, who was a prominent Austrian pacifist.

By the age of 17, Alfred Nobel was fluent in Swedish, Russian, French, English and German. His primary interests were in English literature and poetry as well as in chemistry and physics. Alfred's father, who wanted his sons to join his enterprise as engineers, disliked Alfred's interest in poetry and found his son rather introverted. In order to widen Alfred's horizons, his father sent him abroad for further training in chemical engineering. During a two-year period, Alfred Nobel visited Sweden, Germany, France and the United States. In Paris, the city he came to like best, he worked in the private laboratory of **Professor T. J. Pelouze**, a famous chemist. There he met the young Italian chemist **Accanio Sobrero** who, three year earlier, had invented **nitroglycerine**, the oily substance that was 25 times more explosive than gunpowder. Nitroglycerine was produced by mixing glycerine with sulfuric and nitric acids. It was considered too dangerous to be of any practical use. Although its explosive power greatly exceeded that of gunpowder, the liquid would explode in a very unpredictable manner if subjected to heat and pressure. Alfred Nobel became very interested in nitroglycerine and how it could be put to practical use in construction work. He also realized that the safety problems had to be solved and a method had to be developed for the controlled detonation of nitroglycerine.

After his return to Sweden in 1863, **Alfred Nobel** concentrated on developing nitroglycerine ($C_3H_5O_9N_3$) as an explosive. Several explosions, including one (1864) in which his younger brother **Emil** and several other persons were killed, convinced the authorities that nitroglycerine production was exceedingly dangerous. They forbade further experimentation with nitroglycerine within the Stockholm city limits and Alfred Nobel had to move his experimentation to a barge anchored on Lake Malaren. Alfred was not discouraged and right in 1864, he was able to start mass production of nitroglycerine. To make the handling of nitroglycerine safer, Alfred Nobel experimented with different additives. He soon found that mixing nitroglycerine with diatomaceous earth (*i.e.,* silica) would turn the liquid into a paste which could be shaped into rods of a size and form suitable for insertion into drilling holes. He called this mixture **dynamite** (*dynamis*^{G} = power). The explosive power of dynamite was demonstrated by Nobel on July 14, 1867 at a quarry in Redhills, Surrey, UK. Dynamite had a clear advantage over other explosives because it could be transported easily and handled without fear. It became an overnight success and was adapted to applications in mining, tunnel construction, and bridge and road building. Before long, it was being used in armaments on the battlefield. Nobel patented dynamite in 1867 (UK) and 1868 (US). Later, he invented **gelignite** and patented it in 1876. He then turned his attention to military explosives, where the main requirement was for a smokeless powder, and achieved success with his combination of nitroglycerine with gun cotton known as **ballistite** (patented in 1888).

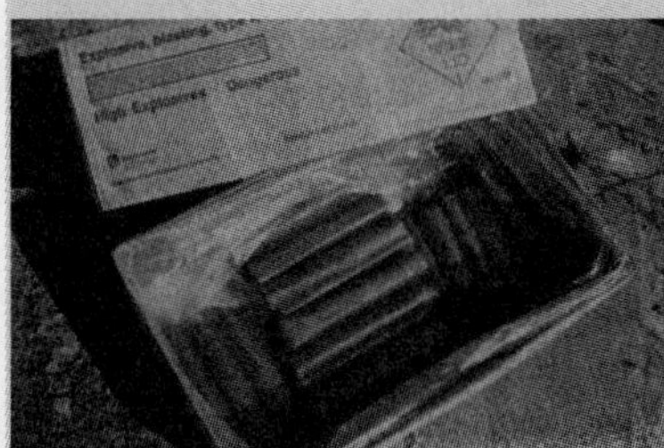

Dynamite contains nitroglycerine.

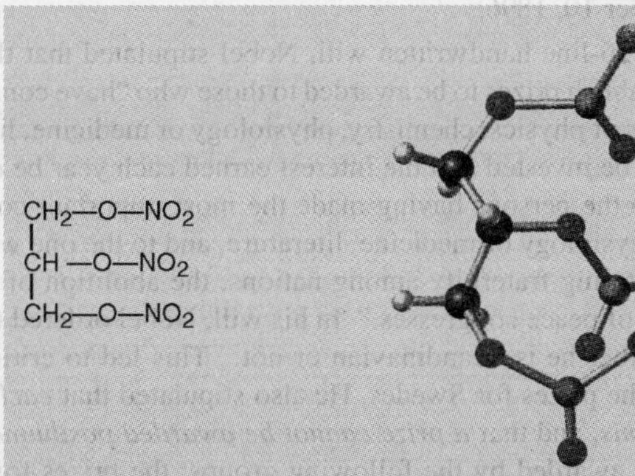

Structural formula Ball-and-stick model

Nitroglycerine

To be able to detonate the dynamite rods, Nobel also invented a detonator (blasting cap) which could be ignited by lighting a fuse. He also invented blasting gelatin and smokeless powder. These inventions were made at the same time as the diamond drilling crown and the pneumatic drill came into general use. Together, these inventions drastically reduced the cost of blasting rock, drilling tunnels, building canals and many other forms of construction work. The market for dynamite and detonating caps grew very rapidly and Alfred Nobel also proved himself to be a very skillful entrepreneur and businessman. By 1865, his factory in Krümmel near Hamburg, Germany was exporting nitroglycerine explosives to other countries in Europe, America and Australia, Over the years, he founded factories and laboratories in some 90 different places in more than 20 countries.

Nobel soon amassed a fortune through control of several European companies that produced dynamite. A lifelong pacifist, he wished his explosives to be used solely for peaceful purposes, and was much embittered by their military use. Money never brought Nobel happiness, nor release him from loneliness and distrust. It is certain that the awards he instituted reflect his lifelong interest in the fields of Physics, Chemistry, Physiology or Medicine, and Literature. However, there is also abundant evidence that his friendship with his one-time secretary Countess Bertha inspired him to establish the prize for peace. When Nobel first proposed a prize for world peace in 1883, it is interesting that he suggested it be given for the first 30 years only. He was optimistic that the world would realize the ravages of war and prize the peace dividend. Coincidentally, Hitler's putsch in Munich began just then.

Although Nobel lived in Paris, much of his life he was constantly traveling. **Victor Hugo** at one time described him as "Europe's richest vagabond". He focused on the development of explosives technology as well as other chemical inventions, including such materials as synthetic rubber and leather, artificial silk, etc. By the time of his death in 1896, he had 355 patents which made Nobel a wealthy man. Alfred Nobel's greatness lay in his ability to combine the penetrating mind of the scientist and inventor with the forward-looking dynamism of the industrialist. Nobel was very interested in social and peace-related issues and held what were considered radical views in his era, He had a great interest in literature and wrote his own poetry and dramatic works. The Nobel Prizes became an extension and a fulfillment of his lifetime interests. Many of the companies founded by Nobel have developed into industrial enterprises that still play a prominent role in the world economy; for example, Imperial Chemical Industries (ICI), Great Britain; Société Centrale de Dynamite, France; and Dvno Industries, Norway. Toward the end of his life, he acquired the company AB Bofors in Karlskoga, where Björkborn Manor became his Swedish home. By 1895, Alfred Nobel had developed angina pectoris, and he died of cerebral hemorrhage at his village in San Remo, Italy, on December 10, 1896.

In his 26-line handwritten will, Nobel stipulated that the bulk of his estate ($ 9,200,000) be used to establish prizes to be awarded to those who "have conferred the greatest benefits on mankind in the fields of physics, chemistry, physiology or medicine, literature and peace." He instructed that the money be invested and the interest earned each year be divided into five equal portions "to be awarded to the persons having made the most important contributions in the fields of chemistry, physics, physiology or medicine, literature, and to the one who had done the most or the best work toward fostering fraternity among nations, the abolition of standing armies, and the holding and promotion of peace congresses." In his will, Nobel ordered that "the most worthy shall receive the prize, whether he is Scandinavian or not." This led to criticism of him being unpatriotic for not reserving the prizes for Swedes. He also stipulated that *each prize can be shared by no more than three persons*, and that *a prize cannot be awarded posthumously*. Nobel instructed that the prizes were to be awarded by the following groups: the prizes for chemistry and physics by the Royal Swedish Academy of Sciences, for physiology or medicine by its medical equivalent, the Karolinska Institute in Stockholm, for literature by the Swedish Academy, and for peace by a five-person committee elected by the Norwegian Parliament. The deliberations are secret, and appeals cannot be made about the decisions.

The executors of his will were two young engineers, **Ragnar Sohlman** and **Rudolf Lilljequist**. They set about forming the Nobel Foundation as an organization to take care of the financial assets left by Nobel for this purpose and to coordinate the work of the Prize-Awarding Institutions. This was not without its difficulties since the will was contested by his relatives, in a futile bid, for 3 years and also questioned by authorities in various countries.

After his death, the Governments of Sweden and Norway established *Nobel Memorial Prizes* in 5 categories: Chemistry, Physics, Physiology or Medicine, Literature and Peace (The first Nobel Prizes were awarded in 1901, and since then these are being uninterruptedly awarded each year on December 10– the anniversary of Nobel's death). A sixth category, Economics, was established by the Central Bank of Sweden in 1968 and first awarded in 1969 (Curiously, enough, there is no provision for a Nobel Prize in Mathematics). All the prizes are awarded in Stockholm except for the Peace prize, which is awarded in Oslo. Each laureate receives a medallion, a scroll, and all or part of a cash award, currently valued at 10 million Kronor (*i.e.,* Rs. 6.35 crore) per category. Until 1980, the Nobel medal was made of 23-carat gold and weighed about 175 grams. Since then, the medallion is being made of 18-carat green-gold plated with 24-carat gold.

Only speculations can be made about the reasons for Nobel's establishment of the prizes that bear his name. He was reticent about himself, and he confided in no one about his decision in the months preceding his death. The most plausible assumption is that a bizarre incident in 1888 may have clicked the idea that culminated in his bequest for the Nobel Prizes. In fact, that his brother **Ludwig Nobel**, who himself made his millions in oil, had died while staying in Cannes, France. The French newspapers reported Ludwig's death but confused him with Alfred; one paper sported the headline "*Le marchand de la mort est mort*" ("The merchant of death is dead"). While other papers made critical comments on his explosive contribution for wartime destruction of man by man. Perhaps, Nobel established the prizes to avoid precisely the sort of posthumous reputation aired by the newspapers. *Thus, the millionaire Alfred also provides the rare instance of a celebrity who read his own obituaries.*

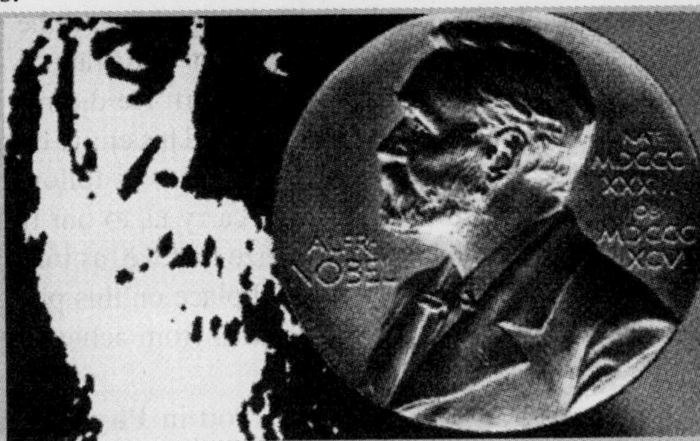

The Nobel Medallion

In the background (on the *left*) is the sketch of Rabindranath Tagore, the 1913 Nobel Laureate for Literature from India.

Not surprisingly, then, a few countries have dominated in the Nobel stakes. Germany was first, and won the lion's share of science prizes until the 1930s, but the onset of Nazism and the flight of refugees to the U.S. brought about a clear American lead that has continued ever since. But even after World War II, many North American laureates have been immigrants. Up to 1986, United States citizens had won 47 physics prizes, 57 medicine and 29 chemistry prizes. Only in chemistry did Germany come anywhere near, with 24, while in physics and medicine/physiology Britian came a poor second with 20 in each, and Germany third with 15 and 11, respectively.

The Nobel Prizes have been awarded 44 times to women, between 1901 and 2011. Only one woman, **Madam Curie**, has been honoured twice, with the 1903 Prize in Physics and the 1911 Nobel in Chemistry.

Nobel often used to say:

"If I have a thousand ideas and only one turns out to be good, I am satisfied."

The first Nobel Prizes were announced and awarded in 1901. Among the recipients were:

1. **Wilhelm Konrad Roentgen** (LT, 1845–1923), a German physicist, for his discovery of x-rays which heralded the age of modern physics and revolutionized diagnostic medicine.
2. **Jacobus Henricus van't Hoff** (LT, 1852–1911), a Dutch physical chemist, for formulating the laws of chemical dynamics and of osmotic pressure.
3. **Emil Adolf von Behring** (LT, 1854–1917), a German biologist and the founder of the science of immunology, for his work on serum therapy, especially in developing a vaccine against tetanus and diptheria.
4. **Jean Henri Dunant** (LT, 1828–1910), a Swiss humanitarian, for peace. He was founder of the Red Cross, in 1864 (now Red Cross and Red Crescent).

Till 2012, seven Indians have been honoured with the coveted Nobel Prize, embracing all the 6 categories, in which the award is presented. They are:

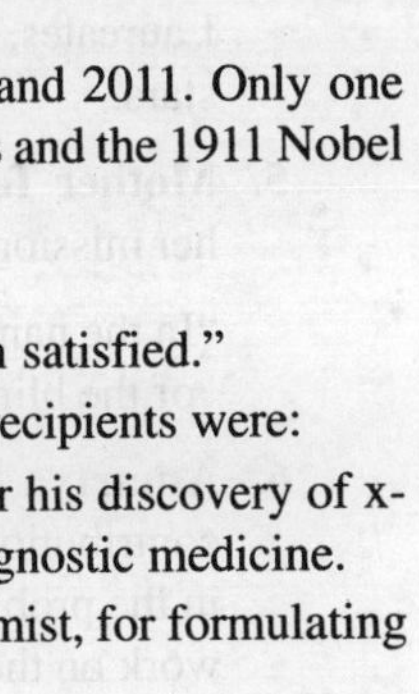

1. **Rabindranath Tagore** (LT, 1861–1941), in 1913 for **Literature** for his classical work *Gitanjali* (1910). He was a Bengali poet, short-story writer, song composer, playwright, essayist, and painter (He published several books of poetry in the 1880s and completed *Manasi* (1890), a collection that marks the maturing of his genius), He was awarded a Knighthood in 1915, but he repudiated it in 1919 as a protest against the Amritsar Massacre.

2. **Chandrashekhar Venkat Raman** (LT, 1888–1970), in 1930 in **Physics**, for his laws on light scattering, better known as 'Raman effect' which is considered as an optical analogue of the Compton effect. Raman had a obssession for crystals, and added to his collection whenever he went abroad. Once, when bringing home a huge collection of these rare crystals, he faced the problem of excess baggage at the airport. He did not have enough money to pay the surcharge, so he surrendered his entire baggage but took the crystals home. Raman's famous quote reads as follows:

   " We need a spirit of victory, a spirit that would carry us to our rightful place under the sun, a spirit which will recognize that we, as inheritors of a proud civilization, are entitled to a rightful place on this planet. If that indomitable spirit arises, nothing can hold us from achieving our rightful destiny".

3. **Har Gobind Khorana** (LT, 1922–2011), in 1960 in **Physiology or Medicine**, for deciphering genetic code . In 1970, he and his research team synthesized the first artificial copy of a yeast gene.

4. **Subrahmanyan Chandrasekhar** (LT, 1910–1995), in 1983 in **Physics** (along with William Alfred Fowler) for his theory on black holes. He is best known for what was later called as 'Chandrasekhar limit'. It is interesting to note that two of Subrahmanyan's former students Tsung Dao Lee and Chen Ning Yang got 1957 Nobel Prize for discovering violations of the principle of parity. And the teacher, in a strange quirk of fate, had to wait for 26 years to join his students in the pantheon of Nobel Laureates, for their investigations into the ageing and ultimate collapse of stars.

5. **Mother Teresa** (LT, 1910–1997), in 1979 for **Peace**, in recognition of her missionary activities. She received this coveted prize by proclaiming:

   "In the name of the hungry, of the naked, of the crippled, of the homless, of the blind, in their name, I accept the award.

6. **Amartya Sen** (born, 1933), in 1998 in **Economic Science**, for his contributions to welfare economics and social choice and for his interest in the problems of society's poorest members. Sen is best known for his work on the causes of famine, which led to the development of practical solutions for preventing/limiting the effects of real/perceived shortages of food. Regarding public resources, he laments:

   "The amount of public resources going into providing healthcare for all is extraordinarily small."

7. **Venkatraman Ramakrishnan** (born, 1952), in 2009 in **Chemistry** (along with the American Thomas A. Steitz and the Isreali Ada E. Yonath), for their study of the structure and function of the ribosome. In the August 26, 2000 issue of **Nature**, Ramakrishnan and his colleagues published the structure of the small ribosomal subunit (305) of **Thermus thermophilus**, a heat stable bacterium related to the one found in the Yellowstone hot springs. Presently, he is working as a Senior Scientist and Group Leader in the Structural Studies Division of the MRC Laboratory of Molecular Biology in Cambridge.

APPENDIX X

A CHRONOLOGY OF BIOCHEMISTRY

Certain key discoveries greatly influence the development of a particular discipline of study. This appendix presents some of the major landmarks in the progress of biology, biochemistry in particular and the individuals whose names are commonly associated with them. It is very difficult to appraise the historical development of any field of study. One investigator often gets the credit for an important discovery when many others should share in the honour. Many 'unlucky' others go in history unnoticed, while some of them get recognition posthumously. No one individual has a monopoly on ideas; advances in science are built on the work of many minds. Especially in recent years, important discoveries have resulted from the efforts of teams of investigators. Furthermore, because knowledge of a topic accrues over a period of time, it is often difficult to decide in which year the "discovery" *actually* occurred. In this brief outline, the student may be able to see some of the relationships that exist between one discovery and another. The discoveries are not completely isolated, as they may sometimes appear. One may also notice that fundamental discoveries in a particular branch of biology tend to be grouped fairly close together chronologically because that particular interest may have dominated the thought of the biological investigators at that time.

Biochemistry had its earliest origin in speculations of the role of air in the utilization of food and on the nature of fermentation. **Leonardo da Vinci** (1452–1519) was one of the first to compare animal nutrition to the burning of a candle. This sort of line of reasoning was further developed by **van Helmont** in 1648. However, the real history of biochemistry began during the late 18th century when the science of chemistry began to take form.

18th Century

1770–1774 **Joseph Priestley**– discovered oxygen and showed that it was consumed by animals and produced by plants.

1770–1786 **Karl Wilhelm Scheele**– isolated glycerol from olive oil, as well as citric, malic, lactic and urin acids from natural sources.

1773 **Rouelle**– isolated urea from urine.

1779–1796 **Ingen-Housz**– showed that light is required for oxygen production by green plants; also proved that plants use carbon dioxide.

1780–1789 **Antoine Laurent Lavoisier**– demonstrated that animals require oxygen; recognized that respiration is oxidation; and first measured oxygen consumption by a human subject.

1783 **Lazaro Spallanzani**– deduced that protein digestion in the stomach is a chemical rather than mechnical process.

19th Century

1804 **John Dalton**– enunciated the atomic theory.

1804 **De Saussure**– carried out the first balance sheet for the stoichiometry of gas exchanges in photosynthesis.

1806 **Vauguelin** and **Robiquet**– first isolated an amino acid, asparagine (Asn).

1815 **Biot**– discovered optical activity.

1828 **Friedrich Wöhler**– synthesized the first organic compound from inorganic components: urea from lead cyanate and ammonia.

1830–1840 **Justus von Liebig**– developed techniques of quantitative analysis and applied them to biological systems.

1833	**Payen and Persoz–** purified diastase (amylase) of wheat, showed it to be heat-labile, and postulated the central importance of enzymes in boilogy.
1837	**Jöns Jacob Berzelius–** postulated the catalytic nature of fermentation.
1838	**Mattias J. Scheleiden and Theodore Schwann–** erunciated the cell theory.
1838	**Geradus Johannes Mulder–** carried out the first systematic studies of proteins; also proposed the name "proteine", later "protein," for the basic constituents of protoplasmic materials.
1842	**Herman von Helmholtz and J. R. von Mayer–** enunciated the First Law of thermodynamics and its applicability to living organisms.
1847	**Dubrunfaut–** degraded starch to maltose by diastase.
1850–1855	**Claude Bernard–** isolated glycogen from the liver and showed that it is converted into blood glucose.
1854–1864	**Louis Pasteur–** proved that fermentation is caused by microorganisms and demolished the spontaneous generation hypothesis. famous for his aphorism : *omne vivum e vivo* (every living thing from the living).
1857	**Kölliker–** discovered mitochondria ("sarcosomes") in muscle.
1859	**Charles Darwin–** published 'Origin of Species' and developed the concept of natural selection as a factor in evolution.
1860	**Claude Bernard–** developed the concept of constancy of the internal environment (homeostasis).
1862	**Sachs–** proved that starch is a product of photosynthesis.
1864	**F. Hoppe-Seyler–** first crystallized a protein, hemoglobin which he also renamed.
1866	**Gregor Johann Mendel–** published his experiments leading to the principles of independent segregation and assortment of genes, thus formulated the first two laws of heredity.
1869	**Friedrich Miescher–** isolated DNA for the first time.
1877	**Friedrich Wilhelm Kuhne–** coined the term 'enzyme' and distinguished enzymes from bacteria.
1879	**Albrecht Kossel–** isolated nucleoproteins in the heads of fish sperm (Nobel Laureate, 1910)
1886	**MacMunn–** discovered histohematins, later renamed cytochromes.
1887	**Emil Fischer–** described structural patterns of proteins (Nobel Laureate, 1902).
1890	**Altmann–** described procedures for staining mitochondria, studied their distribution, and postulated them as having metabolic and genetic autonomy.
1893	**Ostwald–** proved enzymes are catalysts.
1894	**Emil Fischer–** demonstrated the specificity of enzymes and the lock-and-key relationship between enzyme and substrate.
1895	**Wilhelm Conrad Roentgen–** observed that a new form of penetrating radiation, which he named x-rays, was produced when cathode rays (electrons) hit a metal target (Nobel Laureate, 1901).
1897	**Bertrand–** christened the term 'coenzyme'.
1897	**Eduard Buchner and Hans Buchner–** showed that cell-free extracts of yeast can ferment sugars to form CO_2 and $C_2H_5.OH$, laying the foundations of enzymology.
1897	**Christiaan Eijkman–** proved that beriberi is a dietary deficiency disease, and that a water-soluble component of rice polishings can cure it (Nobel Laureate, 1929).

20th Century

1901–1904 **J. Takmine and Aldrich**, and also **J.J.Abel**– first isolated a hormone, epinephrine. **Stoltz**– achieved its synthesis.

1902 **Emil fischer and Hofmeister**– demonstrated that proteins are polypeptides.

1903 **Carl Neuberg**– first used the term 'boichemistry'.

1904 **William M. Bayllis and Ernest H. Starling**– demonstrated the action of secretin, a hormone from the mucosa of the stomach; this marked the real birth of the science of endocrinology.

1905 **Knoop**– deduced the β-oxidation of falty acids.

1905 **Harden and Young**– showed the reqquirement of phosphate in alcoholic fermentation and isolated the first coenzyme, coxymase, later shown to be NAD.

1906 **Michael Tswett**– invented column chromatography, passing petroleum extracts of plant leaves through columns of powered chalk.

1906 **Frederick Gowland Hopkins**– explained dietary bya biochemical investigation of the lack of essential amino acids in the diet (Nobel Laureate, 1929).

1907 **Fletcher Gowland Hopkins**– showed that lactic acid is formed during anaerobic muscle contraction.

1909 **Soren Sörensen**– showed the effect of pH on enzyme action.

1911 **Casimer Funk**– isolated crystals with vitamin B activity and coined the term 'vitamine'.

1912 **Batelli and Stern**– discovered dehydrogenases.

1912 **Carl Neuberg**– proposed a chemical pathway for fermentation.

1912 **Warburg**– postulated a respiratory enzyme for the activation of oxygen; also showed the requirement of iron in respiration.

1912–1922 **Wieland**– showed the activation of hydrogen in dehydrogenation reactions.

1913 **Leonor Michaelis and Maud Menten**– proposed a kinetic theory of enzyme action.

1913 **Willstätter and Stahl**– isolated and studied chlorophyll.

1914 **Edward C. Kendall**– isolated the hormone, thyroxine, which was later (in 1927) artificially synthesized by Harington.

1917 **Elmer Mc Collum**– showed that zerophatalmia in rats is due to lack of vitamin A.

1919 **Francis W. Aston**– discovered isotopes (Nobel Laureate, 1922).

1920 **firz Went**– demonstrated how auxim affects plant growth.

1921 **Frederick Gowland Hopkins**– isolated glutathione

1921 **Loewi, Otto and H.H. Dale**– isolated acetylcholine (Nobel laureates, 1936).

1922 **Ruzicka**– recognized isoprene as the building block of many natural products, *i.e.*, propounded isoprene rule.

1922 **Elmer McCollum**– showed that lack of vitamin D causes rickets.

1922 **Otto Warburg**– devised monometric methods for studying metabolism of living cells (Nobel Laureate, 1931).

1925 **Briggs and John B.S. Haldane**– made important refinements in the theory of enzyme kinetics.

1925 **Walter Norman Howorth**– formulated sugars as pyranoses.

1925 **David Keilin**– coined the term 'cytochrome'.

1925–1926 **George R. Minot, George William Murphy** and **G.H. Whipple**– discovered liver treatment of pernicious anemia (Nobel Laureates, 1934)

1925–1930 **Levene**– elucidated the structure of mononucleotides and showed that they are building blocks of nucleic acids.

1925–1930 **Theodore Svedberg**– developed the first analytical ultracentrifuge and used it to estimate the MW of hemoglobin as 68,000.

1926 **Fritz Went**– demonstrated how auxin affects plant growth.

1926 **James B. Sumner**– first obtained crystals of an enzyme, uresase from extracts of jack beans (*Canavalia ensiformis*) proved it to be a protein (Nobel Laureate, 1946).

1926 **Hisaw**– discovered a hormone, relaxint.

1926 **Jansen** and **Donath**– isolated vitamin B_1, thiamine from rice polishings.

1927 **Adolf Windaus**– showed ergosterol as a precursor of vitamin D.

1927 **P. Eggleton, G.P. Eggleton, C.H. Fiske** and **Y. Subbarow**– demonstrated the role of phosphagen (phosphocreatine) in muscular contraction.

1928 **Euler**– isolated carotene and showed it to have vitamin A activity.

1928 **Albert Szent-Györgyi**– isolated vitamin C (ascorbic acid) from paprika plant (Nobel laureate, 1937).

1928 **H. Wieland** and **Adolf Windaus**– established the structure of cholesterol molecule (Noble Laureates, 1927, 1928)

1928–1933 **Warburg**– deduced the iron– porphyrin nature of the respiratory enzyme.

1929 **K. Lohamann, C. Fiske** and **Y. Subbarow**– isolated a labile phosphate (*i.e.*, ATP) and phosphocreatine from muscle extracts.

1929 **A. Butenandt** and **E.A. Doisy**– the first-ever isolation of a sex hormone (estrone) by these two investigators, independently (Doisy, Nobel laureate, 1943).

1930–1933 **John H. Northrop**– isolated crystalline pepsin and trypsin and proved their protein nature.

1930–1935 **Edsall** and **von Muralt**– isolated myosin from muscle.

1931 **Linus Carl Pauling**– published his first essays on "The Nature of the Chemical Bond", detailing the rules of covalent bonding.

1931 **Englehardt**– discovered that phosphorylation is coupled to respiration.

1932 **C. Glen King** and **W.A. Waugh**– isolated ascorbic acid from lemon juice.

1932 **Warburg** and **Christian**– discovered the 'yellow enzyme' (vitamin B_2 or riboflavin), which is a flavoprotein.

1932 **A. Bethe**– developed the concept of ectihormone (pheromone)

1933 **Hans Adolf Krebs** and **Kurt Hanseleit**– discovered the urea cycle.

1933 **Keilin**– isolated cytochrome *c* and reconstituted electron transport in particulate heart preparations.

1933 **Gustave Embden** and also **Otto Meyerhof**– demonstrated crucial intermediates in the chemical pathway of glycolysis and fermentation.

1933 **Arne W.K. Tiselius**– introduced electrophoresis for separating proteins in solution.

1933 **M. Goldblatt and U.S. von Euler**– discovered prostaglandins.

1933 **George Wald**– discovered the presence of vitamin A in the retins of the eye (Nobel lautrate, 1967).

1934 **V. B. Wogglesworth**– discovered the role of corpus allatum gland in insect metamorphosis.

1934 **Carl P. Henrik Dam** and **Edward A. Doisy**– isolated and synthesized vitamin K (Nobel Laureates, 1943).

1935	**Rose**– discovered theromine, the last essential amino acid to have been recognized.
1935	**Robert R. Williams** ***et al***– deduced the structure of vitamin B_1 (thiamine).
1935	**Wendell M. Stanley**– first crystallized a virus, tobacco mosaic virus (Nobel laureate, 1946).
1935	**Schoneheimer** and **Rittenberg**– first used isotopes as tracers in the study of intermediary metabolism of carbohydrates and lipids.
1935	**B. Hansteöm**– discovered the presence of the X-organs in crusaceans.
1935–1946	**Edward C. Kendall** and **P.S. Hench**– dicovered a hormone, cortisone (Nobel Laureates, 1950).
1935–1936	**Warburg** and **Van Euler**– isolated and determined the structure and action of pyridine lucleotides.
1937	**Hans Adolf Krebs**– postulated the citric acid cycle or Krebs cycle (Nobel laureate, 1953).
1937	**Lohman** and **Schuster**– showed that thiamine is a component of the prosthetic group of pyruvate carbomylase.
1937	**Carl F. Cori** and **Gerti T. Cori**– began their incisive studies of glycogen phosphorylase.
1937	**G. W. M. Findley** and **F.O. MacCollum**– discovered interferon.
1937	**P. Köenig** and **Arne W. K. Tiselius**– developed electrophoresis.
1938	**Theodore Svedberg**– developed ultracentrifuge (Nobel laureate, 1926).
1938	**Hill**– found that cell-free subpensions of chloroplasts yield O_2 when illuminated in the presence of an electron acceptor.
1938	**Braunstein** and **Kritzmann**– discovered transmination reactions.
1938	**Beherens**– employed differential centrifugation to separate nuclei and cytoplasm from liver cells, a technique further developed for the fractionation of cell organelles by Claude, Brachet, Hogeboom and others in the 1940s and early 1950s.
1939–1941	**Fritz A. Lipmann**– postulated the central role of ATP in the energy transfer cycle.
1939–1942	**Engelhardt** and **Ljubimova**– discovered ATPase activity of myosin.
1939–1946	**Albert Szent-Györgyi**– discovered actin and myosin and elucidated the role of ATP in muscular contraction.
1940	**George Beadle** and **Edward Tatum**– proposed the one gene-one enzyme hypothesis.
1940–1943	**Albert Claude**– isolated and studied a mitochondrial fraction from liver.
1941	**W.T. Astbury**– obtained the first x-ray diffraction pattern of DNA.
1941	**Carl F. Cori** and **Gerti T. Cori**– elucidated lactic acid metabolic cycle (Nobel Laureates, 1947).
1942	**D. McClean** and **L.M. Rowlands**– discovered the enzyme hyaluronidase in mammalian sperm.
1942	**Konrad S. Bloch** and **Rittenberg**– discovered that acetate is the precursor of cholesterol.
1942	**A.J.P. Martin** and **R.L.M. Synge**– devolped partition chromatography, leading to paper chromatography two years later.
1943	**Chance**– first applied sensitive spectrophotometric methods to enzyme– substrate interactions.
1943	**A. Claude**– isolated various cell constituents (such as mitochondria, microsomes and nuclei) by differential; centrifugation (Nobel Laureate, 1974).

1943–1947 **Leloir** and **Munoz**– demonstrated fatty acid oxidation in cell-free systems.

Albert L. Lehninger– showed the requirement of ATP and the stoichiometry of fatty acid oxidation.

1944 **Ostwald T. Avery, Colin M. Macleod** and **Maclyn McCarty**– demonstrated that bacterial transformation is caused by DNA, *i.e.*, DNA is genetic material.

1944 **Selman Waksman**– discovered streptomycin.

1945 **Keith R. Porter**– analyzed the fine structure of endoplasimic reticulum.

1947 **P. Holtz**– discovered norepinephrine (=noradrenalin).

1947–1950 **Fritz A. Lipmann** and **Kaplan**– isolated and characterized coenzyme A.

1948 **Leloir**– discovered the role of uridine nucleotides in carbohydrate biosynthesis.

1948 **Hogeboom, Schneider** and **Palade**– refined the differential centrifugation method for cell fractionation.

1948 **Alexander Ogston**– proposed a 3-point attachment hypothesis for citrate behaviour.

1948 **Melvin Calvin** and **Benson**– discovered that phosphoglyceric acid is an early intermediate in photosynthetic CO_2 fixation.

1948–1950 **Eugene P. Kennedy** and **Albert L. Lehninger**– discovered that the citric acid cycle, falty acid oxidation and oxidative phosphorylation take place in mitochondria.

1948–1950 **Loomis** and **Fritz A. Lipmann**– deduced the action of uncoupling agents.

1949 **Albert Szent-Györgyi**– showed that isolated myofibrils from skeletal muscle cells contract upon addition of ATP.

1949 **U.S. von Euler**– isolated and identified norepinephrine as the neurotransmitter in the synpathetic nervous system (Nobel laureate, 1970).

1950–1953 **Erwin Chargaff**– discovered that the base pairing ratios in DNA were always the same, whatever the organism.

1951 **Rosalind Franklin**– obtained high-quality x-ray diffraction studies of DNA, showing that it has a helical structure.

1951 **Fritz A Lipmann** and **Feodor Lynen**– elucidated 'active' acetate *i.e.*, acetyl-coenzyme A.

1951 **Brakke**– used density-gradient centrifugation in sucrose solutions to purify a plant virus.

1951 **Albert L. Lehninger**– demonstrated that oxidative phosphorylation is coupled to electron transport in respiratory chain.

1951 **Linus Carl Pauling** and **Robert Corey**– proposed the structure of helical conformation of a chain of L-amino acids- the α-helix, and the structure of the β-sheet, both of which were later found in many proteins (Pauling, Twice Nobel Laureate, 1954 and 1962).

1952 **George E. Palade**– analyzed the fine structure of the mitochondrion (Nobel Laureate, 1974).

1952–1953 **Palade, Porter** and **Sjöstrand**– perfected thin sectioning and fixation methods for electron microscopy of intracellular structures.

1952–1954 **Paul Zamecnik** *et al*– discovered that ribonucleoprotein particles, later named ribosomes are the site of protein synthesis, also developed the first cell-free system to carry out protein synthesis.

1953 **James Dewey Watson** and **Francis Harry Compton Crick**– proposed the double-helix model of DNA, based on x-ray diffraction patterns obtain by Rosalind Franklin and Maurice Wilkins. (Watson, Crick and Wilkins, Nobel Laureates, 1962)

1953 **V. du Vigneaud**– carried out the first laboratory synthesis of the peptide hormones ocytocin and vasopressin.

1953 **Horecker, Dickens** and **Racker**– elucidated the 6-phosphogluconate pateway of glucose catabolism.

1954 **Christian de Duve**– isolated lysosomes, and later peroxisomes by centrifugation (Nobel laureate, 1974).

1954 **Arnon et al**– discovered photosynthetic phosphorylation.

1954 **Max F. Perutz** *et al*– developed heavy-atom methods to solve the phase problem in protein crystallography.

1954 **Hugh Huxley**– proposed the sliding filament model for muscular contraction.

1955 **Frederick Sanger**– completed the analysis of amino acid sequence of bovine insulin, the first protein to the sequenced (Twice Nobel Laureate, 1958 and 1980).

1955 **Kennedy and Weiss**– described the role of CTP in the biosynthesis of phosphatidylcholine.

1955–1957 **Arthur Kornberg** and **Severo Ochoa**– synthesized DNA (Kornberg) and RNA (Ochoa) artificially outside of cells, *i.e.*, in vitro (Nobel Laureates, 1959)

1956 **Umbarger**– reported that the end product isoleucine inhibits the first enzyme in its biosynthesis from threonine.

1956 **Dorothy Crowfoot Hodgkin**– determined the structure of coenzyme B_{12}.

1956 **Ingram**– showed that normal and sickle-cell hemoglobin differ in a single amino acid residue.

1956 **Christian B. Anfinsen** and **White**– concluded that the three-dimensional conformation of proteins is specified by their amino acid sequence.

1956 **Luis F. Leloir**– determined the pathway to uridine diphosphate glucose (UDPG).

1957 **Earl Wilbur Sutherland** and **T.W. Rall**– discovered cyclic adenosine-3′5′ monophosphate (=cyclic AMP), an intracellular mediating agent, found in all living animal tissues (Sutherland, Nobel Laureate, 1971).

1957 **Hans Adolf Krebs** and **H.L. Kornberg**– discovered the glyoxylate cycle.

1957 **Mathew S. Meselson, Frank W. Stahl** and **Vinograd**– developed equilibrium density-gradient centrifugation in $CsCl_2$ solutions for separating nucleic acids.

1957 **Arthur Kornberg**– purified DNA polymerase from *E. coli* and used it to make DNA from nucleotides in a test tube.

1957 **Melvin Calvin**– elucidated chemical pathways in photosynthesis by using radioactive carbon 14 (Nobel Laureate, 1961)

1957 **Hoagland, Zamecnik** and **Stephenson**– isolated tRNA and determined its function.

1958 **A.B. Learner**– discovered the hormone melatonin in the pineal gland.

1958 **A.F.J. Butenandt** *et al*– chemically analyzed the first pheromone-the sex attractant substance of silk moths (*Bombyx mori*) and named it as bombykol

1958 **Weiss, Hurwitz** and **Stevens**– discovered DNA-directed RNA polymerase.

1958 **Meselson** and **Stahl**– using radioisotopes of nitrogen, demonstrated that DNA is replicated by a semiconservative mechanism.

1959 **Wakil** and **Ganguly**– reported that malonyl-CoA is a key intermediate in fatty acid biosynthesis.

1959–1960 **Rosalyn Yalow** and **S.A. Berson**– developed the technique of radiommunoassay (Yalow, Nobel Laureate, 1977).

1960s–1970s **S. Bergström, B. Samuelsson** and **J. Vane**– characterized the prostalandins, the chemical transmitters of intracellular and intercellular signals, Bergström is credited with isolating prostaglandins and determining their structures; Samuelsson with determining their biosynthesis and metabolism; and Vane for discovering prostacyclin (Nobel Laureates, 1982).

1960 **John C. Kendrew**– described the first detailed 3-'D' structure of a protein (sperm whale myoglobin) to a resolution of 0.2 nm.

Max F. Perutz– proposed a lower-resolution structure of larger protein hemoglobin.

1960 **William Moore** and **Stanford Stein**– determined the amino acid sequence of ribonuclease.

1961 **J.F.A. Miller**– established the function of the thymus gland, which was long known as a transitory organ.

1961 **Marshall W. Nirenberg** and **J.H. Matthaei**– deciphered the genetic code (Nirenberg, Nobel Laureate, 1968).

1961 **Marshall W. Nirenberg** and **Severo Ochoa**– decoded the base code of nucleic acids.

1961 **Francois Jacob** and **Jacques Monod**– presented a model for the regulation of gene activity called operon model (Nobel Laureates, 1965).

1961 **Peter Mitchell**– proposed chemiosmotic coupling hypothesis of ATP formation.

1961 **Harold Copp** *et al*– discovered a hormone, calcitonin.

1961 **Cross et al**– isolated gibberllins and elucidated their chemical structure.

1961 **François Jacob, Jacques Monod** and **Changeux**– proposed a theory of the function and mechanism of allosteric enzymes.

1961 **Marmur** and **Doty**– discovered DNA renaturation.

1962 **Efraim Racker et al**– isolated F_1 component of ATP synthetase from mitochondria and reconstituted oxidative phosphorylation in submitochondrial vesicles.

1963 **D.S. Letham**– coined the term 'cytokinins; also isolated zeatin.

1963 **C.F. Eagles** and **P.R. Wareing**– isolated a plant growth inhibitor, dormin.

1963 **John Cairns**– demonstrated the existence of circular DNA in a bacterium.

1965 **Robert W. Holley** *et al*– determined for the first time the base sequence of a nucleic acid (Nobel Laureate, 1968).

1966 **Marshall W. Nirenberg, Severo Ochoa** and **Arthur Kornberg**– elucidated the genetic code.

1966 **David C. Phillips** *et al*– described the 3-'D' structure of lysozyme, the first enzyme to be analyzed in detail.

1966 **Marshall W. Nirenberg** and **Har Gobind Khorana**– cracked the genetic code, *i.e.*, presented a complete dictionary of nucleotide triplets (Nobel Laureates, 1968).

1966 **Maizel**– introduced the use of sodium dodecylsulfate (SDS) for high-resolution electrophoresis of protein mixtures.

1966 **Francis Harry Compton Crick**– proposed the wobble hypothesis.

1966 **Gilbert** and **Muller-Hill**– isolated the lac repressor.

1967 **Waclaw Szybalski**– determined that only one strand of DNA is transcribed.

1967 **Gellert**– discovered DNA ligase, the enzyme used to join DNA fragments together.

1968 **Glomset**– proposed the theory of reverse cholesterol transport in which HDL is involved in the return of cholesterol to the liver.

1968 **Meselson and Yuan**– discovered the first DNA restriction enzyme. Shortly after, **Smith and Wilcox** discovered the first restriction enzyme that cuts DNA at a specific sequence.

1969 **Max Delbrück, Alfred D. Hershey** and **Salvadore E. Luria**– worked for the genetics of viruses suggesting the reproductive patterns.

1969 **Zubay** and **Lederman**– developed the first cell-free system for studying the regulation of gene expression.

1970 **David Baltimore, Howard M. Temin** and **Renatto Dulbecco**– studied the interaction between tomour viruses and the genetic material of the cell; discovered the reverse transcription (Baltimore and Temin, Nobel Laureates, 1975).

1970 **Gerald M. Edelman** and **Rodney R. Porter**– worked out the structure of gamma globulin (Nobel Laureates, 1972).

1970 **Smith** and **Wilcox**– isolated the first restriction enzyme, *Hin*dll, which can cut DNA.

1971 **W.Y. Cheung**– discovered calmodulin.

1971 **John Robert Vane**– discovered that aspirin blocks the biosynthesis of prostaglandins.

1971–1973 **A.M. Cormack** and **G.N. Hounsfield**– invented the computer-assisted tomography (CAT) scanner (Nobel Laureates, 1979).

1972 **P.B. Woodard** and **A. Eschenmoser**– synthesized vitamin B_{12}.

1972 **Paul Berg** and **Berg Boyer**– produced the first recombinant DNA molecules.

1972 **Michael S. Brown** and **Joseph L. Goldstein**– discovered the LDL receptor on cell surfaces; also explained the intracellular regulation of cholesterol metabilism (Nobel Laureates, 1985).

1973 **Stanley Cohen, Annie Chang, Berg Boyer** and **Helling**– showed that DNA could be inserted and cloned inside a bacterium.

1975 **Dobberstein** and **Günter Blobel**– discovered the mechanism for protein translocation across the ER membranes in a cell-free system– the signal hypothesis (Blobel, Nobel Laureate, 1999).

1976 **Walter Gilbert, A. Maxam,** and **Frederick Sanger**– devised methods for determining base sequence in DNA (Gilbert and Sanger, Nobel Laureates, 1980).

1976 **J. Michael Bishop** and **Harold E. Varmus**– discovered oncogenes (Nobel Laureates, 1989)

1976 **Kim, Rich and Klug**– described the detailed 3-'D' structure of tRNA determined by x-ray diffraction.

1977 **Frederick Sanger**– developed a method for sequencing DNA.

1977 **Genentech**– the first genetic engineering company founded, using recombinant DNA to make pharmaceuticals.

1977 **Roger C.L. Guillemin** and **Andrew V. Schally**– synthesized peptide hormones.

1977 **Roslin Institute, Edinburgh**– first mammal cloned. Dolly, the sheep, fuels controversy about genetic research.

1977 **McGarry, Mannaerts** and **Foster**– discovered that malonyl-CoA is a potent inhibitor of β oxidation.

1977 **Nishizuka** ***et al***– reported the existence of protein kinase C.

1977–1978 **Holmes** and **Klug**– determined the structure of TMV.

Harrison and **Rossman**– determined the structure of 2 small spherical viruses.

1978 **Susumu Tonegawa**– demonstrated DNA splicing for an immunoglobulin gene.

1981 **Steitz**– determined the structure of CAP protein.

1982 **Stanley B. Prusiner**– discovered a new biological principle of infection, under the name 'prion theory' (Nobel Laureate, 1997).

1982–1984 **Sidney Altman** and **Thomas R. Cech**– dicovered RNA catalysis (Nobel laureates, 1989).

1983 **Luc Montagnier**– discovered HIV, the cause of acquired immodeficiency syndrome (AIDS).

1984 **Alec Jeffreys**– developed DNA fingerprinting.

1984 **Rothman** ***et al***– reconstituted Golgi vesicle trafficking *in vitro* using a cell-free system.

1984 **Schwartz** and **Cantor**– developed pulsed field gel electrophoresis for the separation of very large DNA molecules.

1984 **Hartmut Michel, Johann Deisenhofer** and **Robert Huber**– determined the structure of the bacterial photosynthetic reaction centre.

1985 **Kary Banks Mullis**– invented the polymerase chain reaction (PCR), a means of amplifying tiny amounts of DNA (Nobel Laureate, 1993).

1988 **Elion** and **Hitchings**– designed and synthesized therapeutic purines and pyrimidines.

1989 **Francis Collins**– identified the gene CFTR, responsible for cystic fibrosis, on chromosome 7.

1989 **Snyder** ***et al***– purified and reconstituted the inositol-1,3,4-P_3 receptor.

1991 **Roger Beachy**– demonstrated that plants can acquire resistance to viral pathogens.

1995 **J. Roca**– described the role of DNA topoisomerases in DNA replication.

1996 **S.C. West**– described the role of DNA helicases in DNA replication.

1997 **Roslin Institute, Edinburgh**– first mammal cloned. Dolly, the sheep, fuels controversy about genetic research.

2000 **D.A. Clayton**– studied transcription and replication of mitochondrial DNA.

21st Century

2002 **C. Kelleher** ***et al***– studied biochemistry of telomerase enzyme.

GLOSSARY

A grove of *Acacia* trees, from a plain

> *"Exactness cannot be established in the arguments unless it is first introduced into the definitions."*
>
> *–Jules Henri Poincare (LT, 1854–1912)*

GLOSSARY

In studying any science, one must become familiar with its vocabulary. This involves acquiring a clear understanding of the technical terms of science. In general, these terms have been devised to enable scientists the world over to understand each other when discussing a structure, a process, a theory, or any other scientific method. Furthermore, the use of a scientific term often makes it possible to avoid lengthy descriptions and explanations. It is important, therefore, for a student/learner to make every effort to learn a new term the first time it is presented. If he does this, he will have made a major stride in learning the subject.

The technical names both in boldface and italics used in the text are defined in the glossary. However, some of the words, both boldfaced and italicized, in the text are not included here because they are sufficiently defined or otherwise understood as used in the text. The following glossary defines terms according to the way they are used in this book. In some cases, the terms have additional meanings which have also been provided.

For an exhaustive account and better understanding of these (and other related) terms, the readers are advised to consult our another muticolour and illustrative book entitled 'DICTIONARY OF BIOCHEMISTRY', published, in 2012, by S. Chand & Company Ltd., New Delhi.

A

Abaxial (*ab*[L] = away) **:** Away from the axis

Abiotic : Something that is nonliving and never has been alive.

ABO blood group : A classification of blood which is based on natural variation in human blood types. There are 4 groups; A, B, AB and O, each classified by a particular combination of antigen(s) on the red blood cells (RBCs) and naturally-occurring antibodies in the blood plasma. The relative frequency of the 4 ABO groups (A, B, AB and O) differs widely between races. It is 41, 9, 4 and 46 in the same order for UK; 31, 28, 7 and 34 for China; and 57, 0, 0 and 43 for Australian aborigines. Antigens and antibodies of the same type cause agglutination when mixed, resulting in difficulties in blood transfusion. Although possessing no A or B antigens, group O individuals have an H-antigen, which is a precursor to the A and B types. H, A and B antigens are found also in human body secretions such as saliva and semen, often a useful fact in forensic tests. Inheritance of grouping is controlled by a single autosomal gene on chromosome 9 with 3 major alleles A, B and O (sometimes written as I^A, I^B and I^O). The 4 blood groups were mainly indentified and named by **Karl Landsteiner** in 1901.

Abortion (*abortum*[L] = miscarried) **:** The termination of pregnancy before the fetus reaches the stage of viability, which is approximately 20 to 28 weeks of gestation. Alternatively, the spontaneous or induced expulsion of a fetus before it becomes viable outside the uterus or womb.

Abscess (*abscessus*[L], from *ab* = away + *cedere* = to go) **:** A circumscribed collection of pus.

Abscisic acid (ABA) : A plant hormone (growth regulator) associated with water stress and the inhibition of growth; in ageing leaves, it is partly responsible for abscission (hence also called *abscisin*) and in buds and seeds, it causes induction of dormancy (hence also called *dormin*).

Abscission (*ab*[L] = away + *scisso*[L] = dividing) **:** In vascular plants, the detachment (or dropping) of leaves, flowers, fruits or stems at the end of a growing season, as the result of formation of a corky layer of young thin-walled cells, the abscission zone, at the base which rupture under stress, *e.g.*, wind.

Absorbance (A_λ) **:** A dimensionless number that indicates how well a solution of a substance absorbs light of a given wavelength. It is defined as the negative logarithm of the fraction of light of wavelength λ that passes through a sample of the solution; its value depends on the length of the light path, the concentration of the solution, and the extinction coefficient of the substance at that wavelength.

Absorption (*absorbere*[L] = to swallow down) : The movement of water and of substances dissolved in water into a cell, tissue or organism.

Absorption spectrum : The range of photons that a given atom or molecule is capable of absorbing, depending on the electron energy levels available in the atom or molecule.

Accessory pigment : A pigment such as a carotenoid (β-carotene) or chlorophyll *b* that captures light energy and transfers it to chlorophyll *a*; an accessory pigment, in fact, increases the percentage of the photons of sunlight that are harvested.

Acetylcholine (ACh) : The most important of the numerous chemical neurotransmitters responsible for the passing of nerve impulses across synaptic junctions. It is found both in the brain and in the peripheral nervous system.

Acetylcholinesterase : An enzyme found in cholinergic synapses and that removes the leftover acetylcholine from the synaptic cleft at the neuromuscular junction after the last impulse. It is one of the fastest acting enzymes in the vertebrate body.

Acetyl-coenzyme A (Acetyl-CoA) : A two-carbon water-soluble organic acid whose hydroxyl group has been replaced with coenzyme A; common to the metabolism of all major types of food, forming a step leading to the Krebs cycle.

Acid : Any substance that dissociates to form H^+ ions (protons) when dissolved in water.

Acidosis : A body condition in which there is excessive acidity in body fluids, normally regulated by the kidney; opposite of *alkalosis*.

Acquired immune deficiency syndrome (AIDS) : An infectious and usually fatal human disease caused by a retrovirus, HIV, which attacks T-cells. The virus multiplies within and kills individual T-cells, until no T-cells remain, leaving the affected individual helpless in the face of microbial infections because his/her immune system is now incapable of putting a defense against them.

Acridines : A group of planar heterocyclic compounds that can intercalate between stacked bases in DNA and cause frameshift mutations.

Acromegaly (*acron*[G] = tip, extremity + *megas*[G] = big) : A chronic disease characterized by enlargement of extremities, *i.e.,* head, hands and feet; caused by oversecretion of the growth hormone by pituitary.

Acropetal (*acron*[G] = apex) : (1) In the case of plant structures (*e.g.,* leaves and flowers), their production one after another from the base of the stem to the apex. (2) In the case of substanes (*e.g.,* water), their movement from the base of a plant to the apex.

Actin (*actis*[G] = ray) : One of the two major globular proteins that make up myofilaments in all eukaryotic cells (the other is myosin). It provides the cell with mechanical support and plays major roles in determining cell shape and cell movement. The monomeric form is sometimes called globular or G–actin; the polymeric form is filamentous or F–actin.

Actinomycin D : A substance that inhibits the transcription of RNA from DNA. When isolated from soil bacteria and used pharmaceutically, it acts as an antibiotic.

Action spectrum : The spectrum of light that elicits a particular response.

Activation : The process by which the regulatory protein binds to DNA and turns on the transcription of specific genes.

Activation energy ($\Delta G^{\neq}$) : Extra energy that must be possessed by atoms or molecules in addition to their ground-state energy in order to undergo a particular chemical reactivon. Activation energy can be applied externally as heat but this is inappropriate for living organisms. Instead, they rely on biological catalysts (enzymes) which decrease the activation energy needed for the reaction to take place.

Active centre : (1) The part of an enzyme molecule that interacts with and binds the substrate, forming an enzyme-substrate complex. **(2)** The part of an antibody molecule that interacts with and binds the antigen, forming an antibody/antigen complex.

Active site : An area of enzyme surface which has a shape complementary to a particular substrate, enabling the enzyme and substrate to become temporarily bonded to form an enzyme-substrate complex. It is often a cleft or pocket in the surface of the enzyme.

Active transport : The transport of a substance across a biological membranae by protein carrier molecules by a mechanism that can work against a concentration or electrochemical gradient (*i.e.*, to a region of higher concentration). It always requires the expenditure of cellular energy. Compare *facilitated transport, passive transport*.

Adaptation (*adaptare*^{L} = to fit) **:** Any peculiarity of structure, physiology, or behaviour that promotes the likelihood of an organism's survival and reproduction in a particular environment. Alternatively, the process by which organisms are modified to function in a given environment.

Adaxial (*ad*^{L} = toward) : Toward the axis or (in the case of a leaf) facing the stem.

Addison's disease (named after **Thomas Addison**, an English physician) : A disease caused by a deficiency of adrenocorticosteroid hormones, produced by the cortical cells of the adrenal gland; the major symptoms of the disease are lowered blood pressure, lowered blood-sugar levels, reduced kidney function, loss of weight, extreme muscular weakness and a brownish pigmentation of the skin and mucous membranes.

Adenine : An organic molecule composed of two carbon-nitrogen rings. It is a purine component of nucleic acids and necleotides. Adenine always forms complementary base-paring with thymine.

Adenosine : A nitrogen-containing compond consisting of an aderine base attached to a ribose sugar

Adenosine phosphates : A group of organic phosphates including adenosine monophosphate (AMP), adenosine diphosphate (ADP) and adenosine triphosphate (ATP). They function in phosphate transfer in the cell, particularly in the transfer of the high-energy phosphate bonds of ADP and ATP. ATP is the most directly utilizable source of energy of the cell.

Adenylation : In cells; the transfer of an adenyl moiety from ATP to another molecule. Some enzymes are regulated by reversible adenylylation.

Adenyl cyclase : An enzyme which catalyzes the formation of cyclic AMP from ATP by the removal of pyrophosphate.

Adhesion (*adhaerere*^{L} = to stick to) **:** The molecular attraction exerted between the surfaces of unlike bodies in contact, as in water molecules to the walls of the narrow tubes that occur in plants.

Adipocyte : A fat cell; cell that is specialized for storing triacylglycerols and for releasing them to the blood in the form of fatty acids and glycerol as required.

Adipose tissue : A fatty connective tissue, the matrix of which contains large, closely-packed, fat-filled cells; occurs either round the liver and kidneys where it stores energy, or in the dermis of the skin where it stimulates the body from heat loss.

Adrenal gland : An endocrine organ consisting of a central medulla which secretes adrenaline and noradrenaline, and an outer cortex which secretes adrenal cortical hormones called corticosteroids. The two parts are closely associated in mammals, but are sometimes separated into distinct organs in other vertebrates, *e.g.*, fish . In mammals there is a pair of adrenal glands situated anteriorly to the kidneys, other vertebrates have more than two adrenals.

Adrenalin (*ad*^{L} = above; *renal* = kidney) **:** Hormone released by chromaffin cells (in the medulla of adrenal gland) and by some neurons in response to stress. It prepares the body for emergency action leading to "fight or flight" responses, which include increased heart rate, blood pressure and blood sugar levels, widening of the pupils, and erection of the hair.

Adrenocorticotrophic hormone (ACTH) : A small protein hormone secreted by anterior lobe of the hypophysis; controls the secretion of other hormones of the adrenal cortex.

Adsorption : The adhesion (attachment), in a very thin layer, of the molecules of gases, dissolved substances, or liquids to the surfaces of solid bodies with which they are in contact. In *physical adsorption*, molecules are held by van der Waal's forces of attraction; in *chemical adsorption*, there is exchange or sharing of electrons.

Aerobe : Any organism (typically a microorganism) requiring free oxygen for respiration and life.

Aerobic (*aer*^{G} = air + *bios*^{G} = life) : Oxygen requiring.

Aerobic pathway : A metabolic pathway, at least one step of which is an oxidation/ reduction reaction that depends on oxygen gas as an electron acceptor. It includes the citric acid cycle and pyruvate oxidation.

Aerobic respiration : A type of cellular respiration that requires oxygen. Glucose is broken down to release energy in a series of steps.

Agar : A gelatinous polysaccharide which, on forming a gel with water and allowed to solidify, is used extensively as a culture medium for the growth of bacteria and fungi; derived from certain red algae (*Gelidium, Gracilaria, Gigartina, Ahnfeltia*).

Agglutination : A clumping together of cells, usually as a result of reaction between specific antigens and antibodies in blood and lymph, forming a natural defence against foreign materials, including bacterial cells.

Aglycon : The nonsugar part of a glycoside.

Albinism : The inability to produce melanin. Albinism is a recessive mutation, but mutations in several different gene can result in this phenotype.

Albumin (*albus*^{L} = white) : A protein constituent of blood, sometimes found in the urine.

Aldose : A simple monosaccharide in which the carbonyl group comes at the end of the chain and thus represents an aldehyde group, *e.g.*, glucose; compare *ketose*.

Aleurone : A proteinaceous material in the form of small grains found in the outer layer of the endospern of many grains. Aleurone releases large quantities of hydrolytic enzymes (amylases, proteases, nucleases) for digestion of the endosperm by the growing embryo.

Alkaloid : A small but complex organic substance in which at least one nitrogen is part of a ring, having strong basic properties and produced in some families of plants (*e.g.*, Papaveraceae) as a defense against herbivores. Examples include caffeine, nicotine, cocaine, morphine, colchicine, quinine and strychnine.

Alkalosis : A body condition in which there is excessive alkalinity in body fluids; opposite of *acidosis*.

Alkane : A compound of carbon and hydrogen that has only single covalent bonds, *e.g.*, ethane (CH_3– CH_3).

Alkene : A hydrocarbon with one or more carbon–carbon double bonds, *e.g.*, ethylene (CH_2= CH_2).

Alkyl group : General term for a group of covalently-linked carbon and hydrogen atoms, such as methyl (–CH_3) or ethyl (–CH_2CH_3) groups. These groups can be formed by removing a hydrogen atom from an alkane.

Allantoin : A heterocyclic end product of purine catabolism in some reptiles, and mammals other than primates.

Allele (originally *allelomorph, allelon*^{G} = of one another) : Either of a pair of contrasting Mendelian characters such as round peas and wrinkled peas; also applied to one of the two or more differing forms (or mutants) of the same gene. In a diploid cell, each gene will have normally two alleles, each occupying the same position (locus) on homologous chromosomes. Alleles are detected only when the differences in phenotype can be detected.

Allele frequency : The relative proportion of a particular allele among individuals of a population. In other words, the total set of alleles of a given gene. Not equivalent to *gene frequency*, although the two terms are sometimes confused.

Allergen : Any antigenic substance that initiates a strong immune response in a particular individual.

Allergy : The overreaction of the immune response of the body to minute traces of foreign substances (antigens). The reaction is usually visible in the form of rashes, itching, breathing difficulties etc.

Allosteric interaction (*allos*[G] = other + *stereos*[G] = shape) **:** The change in shape that occurs when an activator or inhibitor binds to an enzyme. These changes result when specific, small molecules bind to the enzyme, molecules that are not substrates of that enzyme.

Allozyme : Distinct allelic forms of an enzyme that can be physically separated by electrophoresis.

Alpha-cell (α-cell) : A cell in the islets of Langerhans that produces the hormone glucagon, which causes liver cells to release stored glucose and fat cells to break down triglycerides.

Alpha-helix (α-helix) : Common structural motif of proteins in which a linear sequence of amino acids folds into a right-handed helix, which are stabilized by internal hydrogen bonds between backbone atoms.

Alzheimer's disease : A type of dementia seen in the elderly, probably resulting from a deficiency of acetylcholine, which leads to degeneration and death of neurons.

Amide : Molecule containing a carbonyl group linked to an amine. Adjacent amino acids in a protein molecule are linked by amide groups.

Amino acids (*Ammon*[G], referring to the Egyptian sun god, near whose temple ammonium salts were first prepared from camel dung) **:** Organic molecules containing both an amino group(s) and a carboxyl group(s). Those that serve as building blocks of proteins are alpha-amino acids, since both the amino and carboxyl groups are linked to the same carbon atom. They have the general formula, R – $CHNH_2$ – COOH. They are amphoteric, *i.e.*, act as acids or bases, if the pH is shifted.

Amniocentesis : A technique for genetic testing of a fetus by isolating fetal cells from amniotic fluid. Amniocentesis is often used for determining the fetal karyotype.

Amino group : A weakly basic functional group derived from ammonia (NH_3) in which one or more hydrogen atoms are replaced by another atom. In aqueous solution, it can accept a proton and carry a positive charge.

Aminoacyl-transfer RNA (aa-tRNA) : The molecule produced when an amino acid is activated into its aminoacyl form and attached to its specific transfer-RNA molecule, the whole process being catalyzed by a specific aminoacyl-tRNA synthetase enzyme.

Aminopeptidase : An enzyme that catalyzes the sequential hydrolysis of amino acids in a polypeptide chain from the N-terminal.

Amino terminus (N-terminus) : The end of a polypeptide chain that carries a free or unreacted amino group. A ribosome synthesizes a polypeptide in the direction from N-terminus to C-terminus.

Aminotransferases : Enzymes that catalyze the transfer of amino groups from α-amino to α-keto acids; also called *transaminases.*

Amphipathic : For a molecule, the property of having both hydrophilic and hydrophobic proteins. Usually one end or side of the molecule is hydrophilic and the other end or side is hydrophobic.

Ampholyte : A substance whose molecules have both acidic and basic groups.

Amphoteric : A chemical that can act both as an acid and as a base, *e.g.*, amino acids.

Amylase (*amylon*^{G} = starch + *-ase* = enzyme suffix) : A digestive enzyme that breaks up starches and other carbohydrates into sugars; found in the saliva of most mammals; the amount of amylase can also vary, being high when meat is eaten; previously called *ptyalin*.

Amylopectin : A highly branched polymer of more than 50,000 molecules of alpha-glucose; forms a major (80 – 85%) component of natural starches.

Amylose : An unbranched (or linear) polymer of up to 50,000 molecules of alpha-glucose; forms a minor (15 – 20%) component of natural starches.

Anabolism (*ana*^{G} = up + *bolein*^{G} = to throw) : A constructive process in which complex molecules are synthesized from simpler ones; consumes rather than produce cellular energy; includes processes such as photosynthesis and assimilation; opposite of *catabolism*.

Anaemia : (*an*^{G} = not + *haima*^{G} = blood) A deficiency in the number of red blood cells, their volume, or the hemoglobin content.

Anaerobe : An organism capable of living in the absence of free oxygen.

Anaerobic (*an*^{G} = without + *aer*^{G} = air + *bios*^{G} = life) : Any process that can occur in the absence of air or, more precisely, without molecular oxygen, *e.g.*, glycolysis and fermentation.

Anaesthesia (*an*^{G} = not + *aisthesis*^{G} = sensation) : Insensibility

Analgesic : A substance reducing pain without causing unconsciousness.

Analogy : In comparative anatomy, the relationship of different structures serving similar functions, for example, gills and lungs, both means of respiration.

Angiosperm (*angein*^{G} = vessel + *sperma*^{G} = seed) : Any plant whose seeds are born in a fruit; an informal name for flowering plant (Division – Anthophyta or Magnoliphyta)

Angström (named in honour of **Anders J. Angström**, a spectroscopist) : A unit used for measuring lengths shorter than 100 Å such as those of atoms and molecules (nanometer, nm is used for longer dimensions); 1 Å = 10^{-10} meter or 10^{-8} centimeter or 0.1 nanometer.

Anion (*aiion*^{G} = to go up) : A negatively charged ion that is attracted to the anode during electrolysis.

Annual (*annus*^{L} = year) : A plant that germinates from seed, grows to maturity and produces new seed all within one year or growing season. Since the life cycle duration is so short, annuals are usually herbaceous rather than woody, *e.g.*, *Capsella* and groundsel.

Anorexia (*an*^{G} = not + *orexis*^{G} = appetite) : Loss of appetite

Antagonism : (1) The inhibiting or nullifying action of one substance or organism on another, *e.g.*, the antibiotic effect of penicillin. **(2)** The normal opposition between certain muscles.

Anterior (*ante*^{L} = before) : Situated before or toward the front. In animals, the head end of an organism.

Anthocyanin : Any red or blue water-soluble pigment that is a flavonoid; they are the primary pigmets of red and blue plant parts (*e.g.*, flowers, fruits and leaves).

Antibiotic : The naturally occurring or synthetic product of a living organism (usually a microbe) that inhibits the growth of another organism (usually another microbe), *e.g.*, penicillin or streptomycin. *Unfortunately, most antibiotics are not lethal to viruses.*

Antibody (*anti*^{G} = against) : A highly specific protein produced in the blood by B lymphocytes in response to a foreign molecule (antigen) and released into the bloodstream. Often binds to the foreign molecule or cell extremely tightly, thereby inactivating it or marking it for destruction by phagocytosis or complement-induced lysis. Each antibody protein is composed of 4 subunits, two heavy chains and two light chains.

Anticoagulant : A substance that hinders agglutination or clotting of blood cells.

Anticodon : A three-nucleotide sequence at the end of a transfer RNA (tRNA) molecule, that is complementary to the three-nucleotide codon on a messenger RNA (mRNA) molecule; the anticodon is matched to a specific amino acid that is covalently attached to the tRNA molecule.

Antidiuretic hormone (ADH) : *See* **vasopressin.**

Antigen (*anti*[G] = against + *genos*[G] = origin) : A large, foreign molecule (such as a protein or polysaccharide) that stimulates its host lymphocytes to proliferate and secrete specific antibodies that bind to the foreign molecule, labelling it as foreign and destined for destruction.

Antimetabolite : A substance that is a structural analogue of a normal metabolite or otherwise resembles it and that interferes with the utilization of the metabolite by the cell.

Antiparallel : Refers to double-stranded DNA, in which the direction of each strand is opposite to its complementary strand.

Antiport : Membrane carrier protein that transports two different ions or small molecules across a membrane in opposite directions, either simultaneously or sequentially. Compare *symport.*

Antisense strand : In DNA, the antisense strand of a gene is the one that does not contain a coding sequence for a molecule of RNA; the antisense strand is not transcribed.

Antiseptic : A substance that prevents infection in a wound. *Antisepsis* is carried out by disinfection or sterilization using nontoxic, noninjurious substances, and has the effect of killing or inactivating microbes which cause infection.

Antiserum : Serum that contains a high concentration of antibodies against a particular antigen.

Antithetic theory : A theory of the alternation of generations that visualizes a gradual decrease in gametophyte and increase in sporophyte as parts of the life cycle of most plants. In fact, it is the theory of an increase in importance of one generation at the expense of another.

Antitoxin : Antibody molecule that unites specifically with toxin molecule to neutralize it.

Anuria (*an*[G] = not + *ouron*[G] = urine) : Failure to secrete urine

Amyloplast (*amylon*[G] = starch + *plastos*[G] = formed) : A type of plant organelle found particularly in storage organs such as the potato tuber; stores starch in a unit membrane.

Apical : Describes the tip of a cell, a structure, or an organ. The apical surface of an epithelial cell is the exposed free surface, opposite to the basal surface. The basal surface rests on the basal lamina that separates the epithelium from other tissue.

Apical dominance : The influence exerted by a terminal bud in suppressing the growth of lateral buds.

Apical meristem (*apex*[L] = top + *meristos*[G] = divided) : A region of active cell division that ocurs at or near the tips of the plant axis (*i.e.*, root and shoot apices in a vascular plant).

Apoenzyme : The basic or protein part of a holoenzyme.

Aqueous : Pertaining to water, as in an aqueous solution.

Aromatic : Refers to a molecule that contains carbon atoms in a ring, linked through alternating single and double bonds. Often a molecule related to benzene.

Arthritis (*arthron*[G] = joint + *-itis* = inflammation) : Inflammation of a joint.

Artifact : An image seen in a microscope that results from the coagulation or precipitation of materials that had no structure in the living cell.

Ascorbic acid ($C_6H_8O_6$) **:** A water-soluble vitamin, designated as vitamin C; not as widely distributed as other vitamins: among plants, it is present in all fresh fruits and vegetables, and acerola fruit (*Malpighia punctifolia*) being the richest source and among animals, it occurs in liver, adrenals, thymus etc and absent from fish, fats and oils; a colourless, odourless crystalline substance, slightly sour in taste and is easily oxidizable and hence a powerful reducing agent; functions in the biosynthesis of the adrenal steroid hormones and of collagen, also plays a key role in tyrosine metabolism. Avitamimosis C leads to scurvy, a disease characterized by petechial hemorrhages in the skin, mucous membrane and degenerative changes in the cartilage and bone matrices; ascorbic acid acts as a vitamin in man but rats and most other mammals can synthesize their own supply from D-glucose; also called *antiscorbutic factor* or *cevitamin.*

Asphyxia (a^G = not + *sphyxis*^{G} = pulse) : Suffocation, lack of oxygen.

Aspirin : An analgesic that relieves pain without loss of consciousness; chemically known as acetylsalicylic acid.

Assimilation : The transformation of food into the living substance, protoplasm.

Association constant (Ka) : Measure of the association of a complex. For the binding equilibrium $A + B \rightleftharpoons AB$, the association constant is given by [AB]/[A][B], and it is larger the tighter the binding between A and B.

Asthenia (a^G = not + *sthenos*^{G} = strength) : Loss of strength

Atom (*atomos*^{G} = indivisible) **:** A core (nucleus) of protons and neutrons surrounded by an orbiting cloud of electrons. The chemical behaviour of an atom is largely determined by the distribution of its electrons, particularly the number of electrons in the outermost shell. Alternatively, an atom may also be defined as a unit of an element that cannot be further subdivided without losing the quality of an element.

Atomic mass : The atomic mass of an electron consists of the combined weight of all its protons and neutrons.

Atomic number : It is the number of protons in the nucleus of an atom. In an atom that does not bear an electric charge (that is, one that is not an ion), the atomic number is also equal to the number of electrons.

Atomic weight : It is the sum of the relative weights of the protons, neutrons, and electrons of an atom.

ATP phosphohydrolase (ATPase) : One of a large class of enzymes that catalyze a process involving the hydrolysis of ATP. The energy so released is used to actively transport ions or other solutes against their concentration gradient.

ATP synthase : Enzyme complex in the inner membrane of the mitochondrion and the thylakoid membrane of a chloroplast that catalyzes the formation of ATP from ADP and inorganic phosphate during oxidative phosphorylation and photosynthesis, respectively. Also present in plasma membrane of bacteria.

Atrophy (a^G = not + *trophe*^{G} = nourishment) : Wasting or the reduction in size of an organism or tissue mass, often after disuse.

Autocatalysis : A reaction that is catalyzed by one of its products, creating a positive feedback (self-amplifying) effect on the reaction rate.

Autoclave : (1) A strong metallic vessel, gastight when closed, using steam under pressure at temperatures in excess of 100°C for sterilization. **(2)** A sealed vessel in which chemical reactions can occur at high pressure.

Autolysis : The breakdown of tissues, usually after death, by their own enzymes.

Autoradiography : A technique in which a radioactive object produces an image of itself on a photographic film. The image is called an autoradiograph or autoradiogram.

Autosome (*autos*^{G} = self + *soma*^{G} = body) **:** A chromosome not directly involved in determination of sex.

Autotroph (*autos*^{G} = self + *trophikos*^{G} = food) **:** An organism that produces its own food, usually by photosynthesis; *virtually all plants are autotrophs.* Compare *heterotroph.*

Auxin (*auxein*^{G} = to grow) **:** A plant hormone (growth regulator) that controls cell elongation, among other effects. Of the various auxins, indole-3-acetic acid (IAA) is the most commonly-used auxin.

Axil (*axilla*^{L} = armpit) **:** The upper angle between a branch or leaf and the stem from which it arises.

B

B-DNA : The naturally-occurring form of DNA duplexes *in vivo*, that is the same as the model proposed by **Watson** and **Crick** in 1953.

B lymphocyte : A type of lymphocyte that recognizes invading pathogens much as T-cells do but instead of attacking the pathogens directly, it marks them for destruction by the nonspecific body defenses; also called as *B cell*.

Bacillus, *plural* **bacilli** (*baculus*^{L} = rod) **:** A straight or rod-shaped bacterium; for example, *Bacillus subtilis*.

Bactericide : A substance causing death to bacteria; for example, an antibiotic.

Bacteriophage (*bakterion*^{G} = little rod + *phagein*^{G} = to eat) **:** A type of virus that infects and destroys bacterial cells and often replicate within them; extensively used in virus and DNA studies. Also called a *phage*.

Bacterium, *plural* **bacteria** (*bakterion*^{G}, diminutive of *baktron*^{G} = a staff) **:** The simplest cellular organism. Its cells are smaller and prokaryotic in nature, and they lack internal organization.

Bar : A unit of pressure; one bar is the atmospheric pressure of air at sea level and room temperature.

Barbiturate : Any ureide such as amytal, seconal, phenobarbital etc; have depressant effect on the central nervous system (CNS), usually producing sleep.

Bark : All the tissues of a stem or root, exterior to the vascular cambium.

Basal : Situated near the base. The basal surface of a cell is opposite the apical surface.

Basal metabolic rate (BMR) : The minimal rate of metabolism in a resting organism in an environment with a temperature the same as its own body heat, whilst not digesting or absorbing food. The rate is commonly expressed in terms of energy per unit surface area per unit time, usually as $kJm^{-2}h^{-1}$; also called *basal metabolic level (BML)*.

Base : Any substance that combines with H^+ ions (protons) present in a solution and thereby increasing the number of hydroxyl ions.

Base pair (bp) : Two nucleotides in a DNA or an RNA molecule that are paired by hydrogen bonds, for example, G pairs with C, and A pairs with T or U.

Basipetal : (1) In the case of plant structures (*e.g.*, stem), their production one after the other from the apex down to the base, so that the oldest are at the apex. **(2)** In the case of substances (*e.g.*, auxins), their movement away from the apex to lower parts of the plant.

Basophil : A white blood cell containing granules that rupture and release histamine that enhances the inflammatory response; closely related to a mast cell and important in causing allergic responses.

Batesian mimicry (after *Henry W. Bates*, an English naturalist) **:** A situation in which a palatable or nontoxic organism resembles another kind of organism that is distasteful or toxic. Both species exhibit warning colouration.

Benign : Nonmalignant, as in a growth.

Benzene : An organic cyclic compound composed of a six-membered carbon ring having 3 double bonds; occurs as part of many biological molecules.

Beriberi (*beri*^{Singhalese} = weakness) **:** A human disease caused by vitamin B_1 (thiamine) deficiency; characterized by wasting of muscles, paralysis of the extremities, mental confusion and sometimes heart failure.

Beta-carotene : An orange pigment that is composed of 8 isoprene units; it occurs in most plants as an accessory pigment to photosynthesis; a precursor of vitamin A.

Beta-cell (β-cell) : A cell in the islets of Langerhans that secretes the hormone insulin when a person eats, storing glucose to be used later.

Beta-oxidation (β-oxidation) : A sequence of biochemical reactions that oxidize fatty acids into a series of two-carbon compounds that are converted to acetyl-CoA.

Biennial (*biennium*[L] = a two-year period) : A plant that normally requires two years (*i.e.*, two growing seasons) to complete its life cycle and then dies usually it produces a rosette of leaves the first year, and flowers and undergoes fruiting the second year. Such plants are herbaceous rather than woody, *e.g.*, *Campanula* and carrot.

Bile : A thick brown-green fluid secreted by the liver which is alkaline in its reactions, containing bile salts, bile pigments, cholesterol and inorganic salts; contains no enzymes; the secretion of bile from the liver is stimulated by the hormone *secretin*.

Bile pigments : The pigments secreted in bile, bilirubin and biliverdin, which result from the breakdown of hemoglobin in red blood corpuscles, giving the bile its colouration which in turn affects the colour of the feces.

Bile salts : The sodium salts secreted in bile, sodium taurocholate and sodium glycocholate, which greatly lower surface tension and are important in emulsifying fats.

Binomial system (*bi*[L] = twice + *nomos*[G] = usuage, Law) : A system of nomenclature that uses two words: the first word designates the genus (*generic* name), and the second word signifies the species (*specific* name). Both words are italicized or underlined, if handwritten.

Bioassay : A technique in which the presence of a chemical is quantified by using a part of or an entire living organism, rather by carrying out chemical analysis.

Biochemical : Organic and inorganic chemicals that occur in living organisms and are involved in the processes of life.

Biodegradation : The breaking down of inorganic and organic susbstances by biological action, a process usually involving bacteria and fungi, which are known as saprobionts when the substrate is biological.

Biodeterioration : The unwanted breakdown of materials such as foodstuffs, surface coatings, rubber, lubricants, by microorganisms, resulting in significant financial losses in many industries.

Bioenergetics : The energy relationships of living organisms.

Biogenesis (*bios*[G] = life + *gen*[G] = origin) : The production of organisms from other, parental organisms.

Biological warfare : The use of living organisms, particularly microbes, or their products, to induce illness or death in a population, usually during wartimes.

Bioluminescence : The emission of light from living organisms as a result of internal oxidative changes.

Biomass (*bios*[G] = life + *maza*[G] = lump or mass) : The collective dry weight of all the organisms in a population, area, or sample.

Biopoiesis : The production of living from nonliving material.

Biosphere : The world of living organisms; the narrow zone, where land, water and atmosphere meet, that supports life.

Biota : The living organisms (both plants and animals) present in a specific region or area, ranging in size from a small.

Biotic : Pertaining to living organisms.

Biotin ($C_{10}H_{16}O_3N_2S$) : A water-soluble vitamin of the B-complex group and designated as vitamin B_7; has a wide range of distribution : yeast, liver, kidney, milk and molasses are among the richest sources; also occurs in nature in its combined form as *biocytin*; heat-stable and is resistant to both acids and alkalies; serves as a prosthetic group for many enzymes which bring about carboxylation and synthesis of fatty acids; a deficiency (rare in humans) of biotin causes dermatitis, and decrease in weight and edema; also called as *antiegg white injury factor* or *coenzyme R*.

Blanching : Treatmeant with heat for a short period of time to destroy cellular enzymes.

Bohr effect : The effect of pH on oxygen binding by hemoglobin, by which a decrease in pH causes a decrease in oxygen affinity. The effect promotes both the release of O_2 from hemoglobin in the tissues and the release of CO_2 from the blood to the air in the lungs.

Bolting : An unusual lengthening of plant stems, due to elongation of cells, which can be induced by a group of plant hormones called gibberellins, producing a stem with long internodes.

Bond : A force holding atoms together in a chemical compound; the principal types of bond are covalent, ionic, and hydrogen.

Bond energy : Strength of the chemical linkage between two atoms, measured by the energy in kilocalories or kilojoules needed to break it.

Bradycardia (*bradys*[G] = slow + *kardia*[G] = heart) : Slow action of the heart.

Bromellin : A protease obtained from ripe pineapples.

Brown fat : A special fat layer found between the neck and shoulders of some mammals, *e.g.*, bats and squirrels; brown fat is heavily vascularized and has many mitochondria, the latter giving it its brown colour due to the presence of mitochondrial cytochrome oxidase; enables the production of large amounts of heat, particularly after hibernation.

Buffer : A chemical substance which has the capacity to bond to H^+ ions, removing them from solution when their concentration begins to rise and releasing H^+ ions when their concentration begins to fall. In this way, buffers stabilize the pH of biological solutions and are thus important in maintaining homeostasis. As an example, hemoglobin maintains a stable pH in the erythrocyte.

C

Calciferol : Vitamin D_2, a fat-soluble antirachitic vitamin.

Calcitonin (*calcem*[L] = lime) : A polypeptide hormone, secreted by both thyroid and parathyroid glands, that lowers the calcium content of the blood.

Callose : A complex carbohydrate in sieve tubes of sieve tube members; callose is especially abundant in injured sieve tubes.

Callus : A parenchymatous tissue formed as an overgrowth of a wound or in tissue culture.

Calmodulin : An ubiquitous protein that is activated when it binds to calcium ions (Ca^{2+}); calmodulin activates enzymes in membranes; as much as 2% of the plasma membrane may be calmodulin.

Calorie, Cal (*calorie*[L] = heat) : A unit of heat which is equal to 1,000 calories. One Calorie (capital "C") is the amount of energy in the form of heat required to raise the temperature of 1 kilogram of water by 1°C. A slice of apple pie contains about 365 Cal (Capital "C").

calorie, cal (*calorie*[L] = heat) : Also a unit of heat which is equal to 1/1,000 = 0.001 Calories. One calorie (small "c") is the amount of heat energy required to raise the temperature of 1 gram (1 cm^3) of water by 1°C. Used as a unit of energy content or output, but now largely superseded by the S.I. unit joule (1 cal = 4.12 J).

Calvin cycle (after **Melvin Calvin**, an American chemist) : A major enzymatically-catalyzed metabolic pathway in which CO_2 is reduced to 3-phosphoglyceraldehyde (a C-3 compound) and the CO_2 acceptor (ribulose-1,5-bisphosphate) is regenerated.

Cancer : An unrestrained invasive cell growth. A tumour or cell mass resulting from uncontrollable cell division.

Cap : The altered 5' end of eukaryotic mRNA that includes a 7-methylguanosine.

Capsid : Protein coat of a virus, formed by the self-assembly of one of more protein subunits into a geometrically regular structure.

Carbohydrase : Any enzyme that catalyzes the hydrolysis of carbohydrates.

Carbohydrate : General term for sugars and related compounds containing carbon, hydrogen and oxygen, usually with the empirical formula $(CH_2O)_n$.

Carbon fixation : A process by which green plants incorporate carbon atoms from atmospheric carbon dioxide into sugars. It is the second stage of photosynthesis.

Carbonic anhydrase : An enzyme that accelerates the reaction between CO_2 and water to form carbonic acid in the erythrocytes.

Carbonyl group : A pair of atoms consisting of a carbon atom linked to an oxygen atom by a double bond (C = O).

Carboxyl group : A carbon atom linked both to an oxygen atom by a double bond and to a hydroxyl group $(-\mathrm{C}{\substack{/\!\!/\mathrm{O} \\ \backslash \mathrm{OH}}})$. Molecules containing a carboxyl group are weak (carboxylic) acids.

Carboxyl terminus (C-terminus) : The end of a polypeptide chain that carries a free or unreacted α-carbonyl group.

Carboxylase : An enzyme capable of splitting off CO_2 from the carboxyl group (COOH) of certain organic acids.

Carcinogen (*karkinos*[G] = a crab, symbolizing cancerous growth + *–gen* = producing) : An agent, such as a chemical or a form of radiation, that causes cancer.

Carcinoma (*karkinos*[G] = a crab) : A malignant growth from epithelial tissues that cover the body surface and line the intestine and other internal organs; the most common form of human cancer; carcinomas account for about 90% of malignant tumours.

Carnitine A low-molecular-weight lysine derivative or betaine that shuttles fatty acids through the inner mitochondrial membrane to the matrix. The fatty acyl moiety is transferred from CoA to carnitine for transit through the membrane and is then transferred back to CoA; the carnitine released on the matrix side of the membrane is shuffled back for reuse.

Carotenoids : A group of yellow/orange fat-soluble accessory pigments that are derived from 8 isoprene units linked together; subdivided into carotenes (orange) and xanthophylls (yellow); the most widespread carotenoid in plants is beta-carotene.

Carotenols : Any of a class of yellow plastid carotenoid pigments, differing from the carotenoids by having oxygen in the molecules; also called as *xanthophylls*.

Carrageenan : A slimy polysaccharide, consisting mostly of a specific mixture of α-galactose sulfates and found in certain red algae (*Chondrus*, for example); of economic importance as a gel, used as a food, and as a stabilizer in ice cream and other products; also spelt as *carrageenin*.

Carrier protein : Membrane transport protein that binds to a solute and transports it across the membrane by undergoing a series of conformational changes.

Cartilage : A form of connective tissue composed of cells (called chondrocytes) embedded in a matrix rich in type II collagen and chondroitin sulfate.

Castration (*castratus*[L] = to prune) : The act of pruning in any of its senses.

Catabolism (*katabole*[G] = throwing down) : A degradative process in which complex molecules are broken down into simpler ones; includes processes such as respiration and digestion; opposite of anabolism; also spelt as *katabolism*.

Catalase : An iron-containing enzyme found in tissues such as liver and potato tubers; catalyzes the breakdown of toxic hydrogen peroxide into water and oxygen and works by lowering (or reducing) the activation energy required from 80 kJ to 10 kJ; has a very high turnover number of about 6 million.

Catalysis (*katalysis*[G] = dissolution + *lyein* = to loosen) : The enzyme-mediated process in which the subunits of polymers are held together and their bonds are stressed.

Catalyst (*kata*[G] = down + *lysis* = a loosening) : A substance that changes the rate of a chemical reaction (usually accelerating it), without itself undergoing a change. Enzymes are biological catalysts and are proteinaceous in nature.

Catecholamine : Any catechol-derived compound (such as adrenalin or dopamine) which exerts an action similar to that of sympathetic nervous system; closely related to tyrosine and phenylalanine; chemically known as *dihydroxyphenylamines*.

Cell (*cella*[L] = a chamber or small compartment) : The basic organizational unit of all organisms; the smallest unit of life; composed of a nuclear region containing the hereditary apparatus within a larger volume called the cytoplasm, bounded by a lipid membrane.

Cell cycle : Reproductive cycle of the cell; the orderly sequence of events by which the cell duplicates its contents and divides into two.

Cell membrane : *See* **plasma membrane**.

Cell plate : A flattened membrane-bound structure that forms from the fusion of vesicles at the equator of the spindle apparatus during early telophase of mitosis in plants and some algae; when mature , the cell plate becomes the middle lamella.

Cell sap : The contents of a cell found within the vacuoles. It may include a variety of substances such as water, pigments, sugars, and inorganic substances.

Cell theory : The theory, now universally accepted, that organisms are cells or are composed principally of cells.

Cell wall : A mechanically strong extracellular matrix deposited by a cell outside its plasma membrane and is composed mainly of cellulose secreted by the protoplasm of the cell. It is prominent in most plants, bacteria, algae, and fungi. *Not present in most animal cells.*

Cellobiase : An enzyme facilitating the hydrolysis of cellobiose into glucose; one of the enzymes active in the digestion of cellulose.

Cellobiose ($C_{12}H_{22}O_{11}$) : A disaccharide consisting of two molecules of glucose; a hydrolytic product of cellulose.

Cellulase : An enzyme facilitating the hydrolysis of cellulose to cellobiose and ultimately to glucose; used particularly in the softening or digestion of plant cell walls; produced in large quantities in the abscission layer formed in leaf stalks of higher plants, causing a weakening of cell walls prior to leaf fall.

Cellulose : A complex carbohydrate (polysaccharide) found in the cell walls of plants; consists of long chains of covalently-linked glucose units; undoubtedly the most abundant of all biomolecules in the biosphere; although a chemically-inert material but provides tensile strength in plant cell walls.

Central dogma of molecular genetics : The hypothesis (based on Weismannism) that genetical information flows only in one direction, from DNA to RNA to protein, and not in the opposite direction; refers to how genes work to make proteins; each protein-coding gene is transcribed into a molecule of mRNA, which is translated into a sequence of amino acids that comprise a polypeptide (*i.e.*, a protein).

Centripetal : Developing from the outside toward the centre; said of an inflorescence in which the lower or outer flowers bloom first.

Chaperone : A protein that helps other proteins avoid misfolding pathways that produce inactive or aggregated states.

Chargaff's equivalence rules (after **Erwin Chargaff**, an American biochemist) **:** The observation that, in all natural DNA molecules, the amount of adenine is always equal to the amount of thymine, and the amount of guanine is always equal to the amount of cytosine.

Chelate : Combine reversibly, usually with high affinity, with a metal ion such as iron, calcium, or magnesium.

Chelation : The binding of a metal ion to an organic molecule from which it can later be released. The process enables plants to take up metal ions such as iron that are not readily available in free state.

Chemical bond : The force that holds atoms together in molecules.

Chemiosmosis : A biochemical process in which energy from electrons powers the movement of protons across the bacterial membrane, a process that leads to ATP formation. This cellular process is responsible for almost all of the ATP harvested from eaten food and for all the ATP produced by photosynthesis.

Chemosynthesis : A process by which a chemical source of energy, instead of light, is used by certain bacteria such as the hydrogen bacteria and the nitrifying bacteria, in making carbohydrates out of CO_2 and water.

Chemotherapy : The use of chemical substances to combat disease caused by microorganisms; the term is often extended to include cancer treatment by chemicals.

Chiral (*cheir*^{G} = hand) : With respect to a molecule or other object, the property of being nonsuperimposable on its mirror image. An atom that makes a molecule chiral (such as carbon with 4 different substituents) is called asymmetric or chiral atom or centre of chirality.

Chitin : A tough, horny, resistant, nitrogenous polymer of high molecular weight, commonly found in the exoskeleton of arthropods but occurring also in the cell walls of many fungi, and in a few other animals and protists.

Chloramphenicol (CAP) : A bacteriostatic antibiotic produced by the ascomycetous fungus, *Streptomycin*; inhibits protein synthesis in a variety of organisms.

Chlorophyll : A light-absorbing green pigment found in chloroplasts; a methyl-phytol ester of the chlorophyllins; responsible for trapping light energy in the primary events of photosynthesis.

Chloroplast (*chloros*^{G} = green + *plastos*^{G} = moulded) : A complex, energy-producing, cytoplasmic organelle containing bacteria-like elements with vesicles containing chlorophyll; chloroplasts occur in the cells of aboveground parts of plants and are the site of photosynthesis.

Chlorosis : A yellowing of plant leaves caused by lack of chlorophyll pigment due to mineral deficiency (*e.g.*, magnesium, iron) or disease (*e.g.*, virus yellows) which results in a decrease in photosynthetic rate.

Cholecalciferol : One of at least 10 different fat-soluble group of vitamins, collectively designated vitamin D group; of these, 2 vitamins, vitamin D_2 (ergocalciferol) and vitamin D_3 (cholecalciferol) are physiologically more important; vitamin D_2 is of plant origin and produced commercially by irradiation with UV light of a provitamin called ergosterol, whle D_3 is of animal origin and produced from 7-dehydrocholesterol also by irradiating with UV light; both D_2 and D_3 have the same degree of activity in the human beings; the best natural sources of vitamin D are the liver oils of many fishes such as cod and halibut; the flesh of oily fishes (sardine, salmon, herring) is also an excellent source; a white, almost odourless, crystalline substance, fairly heat-resistant and not affected by acids and teeth and encourages the absorption, into the blood, of calcium salts and phosphorus; avitaminosis D results in rickets in children and osteomalacia in human adults, particularly women after many pregnancies; also called *antirachitic factor* or *'sunshine' vitamin*.

Cholecystitis (*chole*^{G} = bile + *kystis*^{G} = bladder + *–itis*^{G} = inflammation) : Inflammation of gall bladder.

Cholecystokinin (CCK) : A polypeptide hormone secreted by the upper part of the small intestine; contains 33 amino acid residues with a molecular weight 3,883; it stimulates contraction of the gall bladder so as to release its contents into the duodenum; both cholecystokinin (CCK) and pancreozymin (PZ) are now considered to be a single factor under the name *cholecystokinin–pancreozymin (CCK–PZ)*.

Cholesterol : A lipid molecule with a characteristic four-ringed steroid structure (called a sterane nucleus); occurs in the plasma membranes of animal cells, but not in plants; precursor of the animal steroid hormones and bile acids.

Choline : An organic base which is a constituent of acetylcholine.

Chondriosome : *See* **mitochondrion**.

Chromaffin cell : A cell that stores adrenalin in secretory vesicles and secretes it in times of stress when stimulated by the nervous system.

Chromatid (*chroma*^{G} = colour + *-id*^{L} = daughters of) : One of the two daughter strands of a duplicated chromosome, formed by DNA replication, that is still joined by a single centromere to the other daughter strand.

Chromatin (*chroma*[G] = colour) **:** The complex of DNA, histones and nonhistone proteins found in the nucleus of a eukaryotic cell; the material of which eukaryotic chromosomes are made; that part of the cell nucleus which becomes deeply stained with basic dyes.

Chromatography (*chroma*[G] = colour + *graphein*[G] = to write) **:** A biochemical technique in which a mixture of substances is separated by charge, size, or some other property by allowing it to partition between a mobile phase and a stationary phase.

Chromosome (*chroma*[G] = colour + *soma*[G] = body) **:** A generally threadlike on rodlike structure within a cell, composed essentially of protein and nucleic acid, containing the hereditary units or genes; a cell usually contains a definite number of chromosomes. Chromosomes contain, or consist of, a linear sequence of genes.

Chronic disease : A disease that lingers for a long time, rarely reaches a climax, and disappears slowly.

Chylomicron : A type of lipoprotein which is largest in size (180-500 nm diameter) and has very low density ($d < 0.95 g/cm^3$) because of a high triacylglycerol contents (about 85%) in it; synthesized in the smooth ER of epithelial cells lining the small intestine and serves to transport dietary lipids in the circulation.

Chymase : *See* **rennin.**

Cirrhosis (*kirrhos*[G] = tawny-yellow) **:** Fibrosis, generally of the liver.

Cis and trans isomers : *See* **geometric isomers**.

Cisterna, *plural* **cisternae :** A flattened membrane-bound tube or saclike region, as found in the endoplasmic reticulum or Golgi apparatus.

Cistron : The smallest unit of DNA that must be intact to code for the amino acid sequence of a polypeptide; thus, the coding part of a gene, minus 5′ and 3′ untranslated sequences and regulatory elements.

Citric acid : A six-carbon organic acid that is converted to isocitric acid in the second step of the citric acid cycle.

Citric acid cycle, CAC : A central metabolic pathway found in all aerobic organisms and by which acetyl-CoA, derived from food molecules, is oxidized in mitochondria to CO_2; each of the citric acid cycle also forms one ATP by substrate-level phosphorylation, reduces one NAD^+ to NADH, and reduces one ubiquinone to ubiquinol; also known as *Krebs cycle* or *tricarboxylic acid (TCA) cycle*.

Clathrin : A protein that assembles into a polyhedral cage on the cytoplasmic side of a membrane so as to form a clathrin-coated pit, which buds off to form a clathrin-coated vesicle.

Clitoris : An erectile structure occurring ventral to the uterus in female mammals, which is homologous with the penis in the male.

Clone (*klon*[G] = twig) **:** A population of cells or organisms formed by repeated (asexual) division from a common cell or organism. Also used as a verb: "to clone a gene" means to produce many copies of a gene by repeated cycles of replication.

Coagulation : The separation or precipitation of suspended particles from a dispersed state.

Coated pit : Invagination of the plasma membrane associated with a bristlelike layer of protein on its cytoplasmic surface. Pinches off to form a coated vesicle in the process of endocytosis.

Coated vesicle : Small membrane-bound organelle formed by the pinching off a coated region of membrane. Some coats are made of clathrin, whereas others are made from other proteins.

Cobalamin : *See* **cyanocobalamin**.

Cocarboxylase : The coenzyme of carboxylase.

Codon (*code*[L] = book + *-on*) **:** A sequence of three nucleotides in a DNA or messenger RNA molecule that represents the instruction for incorporation of a specific amino acid or of a stop signal into a growing polypeptide chain; of the 64 possible codons, 61 are codes for amino acids and 3 are stop codons.

Coenzyme : A dialyzable, nonprotein prosthetic group of an enzyme; like enzymes, coenzymes are not altered or used up in the reaction and can be used many times; most of them are derived metabolically from vitamins; NAD^+ and coenzyme A are examples of coenzyme.

Coenzyme A : A derivative of the B-complex vitamin, pantothenic acid (vitamin B_3); a small molecule, used in the transfer of the acyl (CH_3CO–) group.

Coenzyme Q : *See* **ubiquinone.**

Cofactor : An inorganic, dialyzable nonprotein prosthetic group of enzyme-catalyzed reactions.

Cognitive (*cognoscere*[L] = to know) **:** Thinking. Using the mind.

Coiled coil : Especially stable rodlike protein structure formed from two α helices coiled around each other.

Colchicine : A poisonous alkaloid, derived from the corms of a monocot plant, *Colchicum autumnale*; interferes with spindle formation during nuclear and cell division, thus inducing polyploidy due to nondisjunction.

Coleoptile : In monocotyledons of the grass type, a nonchlorophyllous covering over the young foliar leaves in the growing stem tip of the seedling.

Coleorrhiza : A structure of the grass embryo that is similar to the coleoptile but located around the radicle of yound seedlings.

Collagen : A fibrous protein abundant in glycine and proline that is a major component of the extracellular matrix and connective tissues; exists in many forms : *type I*, the most common, is found in skin, tendon, and bone; *type II* is found in cartilage; *type IV* is present in basal laminae.

Colloid : A mixture of two substances which are immiscible, but where the particles of one are too small to settle out, and so remain suspended indefinitely; colloids are common in cells; colloid particles measure between 0.000001 to 0.0001 mm in diameter; does not diffuse through cell membranes; a liquid colloid is called a *sol*, and a solid colloid, a *gel*.

Coma (*koma*[G] = deep sleep) **:** Deep unconsciousness.

Combinatorial : Describes any process that is governed by a specific combination of factors (rather than by any single factor), with different combinations having different effects.

Competitive inhibition : A form of enzyme control in which an inhibitor molecule, very similar in structure to the normal substrate of an enzyme, becomes reversibly bound to the active site, thus reducing the quantity of enzyme available. However, if excess substrate is present, the inhibitor can be forced out by the substrate molecule which takes its place and the reaction proceeds. Compare *noncompetitive inhibition*.

Complementary DNA, cDNA : DNA that is made by reverse-transcribing mRNA into its DNA complement and therefore lacking the introns that are present in genomic DNA; used to determine the amino acid sequence of a protein by DNA sequencing or to make the protein in large quantities by cloning followed by expression; the collection of vector-cloned cDNA fragments of an organism are its *cDNA library*.

Complex : Assembly of molecules that are held together by noncovalent bonds; protein complexes perform most cell functions.

Compound : A substance composed of two or more elements.

Concentration gradient : The concentration difference of a substance as a function of distance. In a cell, a greater concentration of its molecules in one region than in another.

Configuration : The spatial arrangement of an organic molecule that is conferred by the presence of either (a) double bonds, around which there is no freedom of rotation, or (b) chiral centres, around which substituent groups are arranged in a specific sequence. Configurational isomers cannot be interconverted without breaking one or more covalent bonds.

Conformation : The spatial arrangement of substituent groups that are free to assume different positions in space, without breaking any bonds, because of the freedom of bond rotation.

Conjugate acid-base pair : A proton donor and its corresponding deprotonated species; for example, acetic acid (donor) and acetate (acceptor).

Conjugate redox pair : An electron donor and its corresponding electron acceptor form; for example, Cu^+ (donor) and Cu^{2+} (acceptor) or NADH (donor) and NAD^+ (acceptor).

Conjugated protein : A protein containing one or more prosthetic groups.

Consensus sequence : A most typical form of a sequence that occurs with minor variations in a group of related DNA, RNA, or protein sequences. The consensus sequence shows the nucleotide or amino acid most often found at each position. The preservation of a consensus sequence implies that the sequence is functionally important.

Conservative replication : A form of DNA replication, where both strands of the parent DNA are transferred to one daughter molecule, whereas the other molecule has 2 newly-synthesized strands. Compare *semi-conservative replication*, where one strand of the parent molecule ends up in each of the progeny molecule.

Constitutive enzyme : An enzyme that is made all the time at a constant rate, unaffected by inducers.

Cooley's disease (named after the American paediatrician, **Thomas B. Cooley**, LT 1871–1945) **:** A type of human anemia in which there is a deficiency of either α or β hemoglobin chains. Various causes have been found for the condition, including a recessive mutant allele for β chain deficiency that is present in high frequencies in areas, particularly found at sea borders, *e.g.*, mediterranean, often associated with high incidence of mosquito activity. Also called *thalassemia*.

Cori cycle : The metabolic cycle by which lactate produced by tissues engaging in an aerobic glycolysis (such as exercising muscles) is regenerated to glucose in the liver and returned to the tissues *via* the bloodstream.

Correlation : The tendency of 2 variables to vary together.

Correlation coefficient : A statistic that measures the strength of association of 2 variables.

Corticotropin : *See* **adrenocorticotrophic hormone.**

Cortisol : An adrenocortical steroid with effects similar to cortisone; also called *hydrocortisone*.

Cortisone : A glucocorticoid hormone secreted by the adrenal cortex whose function is to combat stress; causes shrinkage of lymph nodes and lowers the WBC count, reduces inflammation, promotes healing, and stimulates gluconeogenesis.

Cotyledon (*kotyledon*^{G} = a cup-shaped hollow) **:** (1) An embryonic leaf-like storage organ in angiosperm seeds. Monocot embryos have one cotyledon, and dicots have two. (2) A part of the mammalian placenta on which a tuft of villi occurs, particularly in ruminants.

Coupled reactions : Two chemical reactions that have a common intermediate and thus a means of energy transfer from one to the other.

Covalent bond (*co*-^{L} = together + *valare*^{L} = to be strong) **:** A stable chemical bond formed by the sharing of one or more pairs of electrons among the atoms in a molecule.

Creatinine : The nitrogenous waste material of muscle creatine.

Cresols : Phenol derivatives containing methyl groups.

Cretinism (*cretin*^{Fr} = from Swiss patois + *crestin*, from *Christianum*^{L} = a Christian creature) **:** Disease resulting from thyroid cell degeneration in children; characterized by dwarfism and mental suppression.

Crista, *plural* **cristae** (Latin, for *crest*) **:** (1) In mitochondria, the infoldings of the inner mitochondrial membrane, which forms a series of "shelves" containing the electron transport chains involved in ATP formation. (2) A sensory structure in the inner ear.

Crossing over : An essential feature of meiosis occurring during prophase when nonsister chromatids exchange portions of DNA strand. Alternatively, it is an exchange of corresponding segments between chromatids of homologous chromosomes.

Curie (named in honour of Madame **Curie**) : The basic unit of radioactive decay; an amount of radioactivity equivalent to that produced by 1g of radium, namely 2.22×10^{12} distintegrations per minute.

Cuticle (*cutis*^{L} = skin) **:** A waxy layer found on the outer walls of epidermal cells, especially of leaves, floral parts and fruits in many plants.

Cutin (*cutis*^{L} = skin) **:** A waxy, waterproof substance that is the chief ingredient of the cuticle of a plant; consists of hydroxylated fatty acids that are linked together in a complex array.

Cyanocobalamin ($C_{63}H_{88}O_{14}N_{14}PCo$) : A water-soluble vitamin of the B complex group and designated as vitamin B_{12}; occurs only in animals: the chief source being liver, although also present in milk, meat, eggs, fish, oysters etc; its chemical structure, one of the most complex known, was established by Dorothy C. Hodgkin who was awarded Nobel prize for the same in 1964; a cobalt–containing, deep–red, crystalline substance, soluble in alcohol and acetone but not in chloroform; functions in nucleic acid metabolism, in the formation of red blood cell and acts as a carrier of methyl group; a nutritional deficiency of cyanocobalamin leads to pernicious anemia especially among old people, which is characterized by RBCs becoming abnormally large and fewer; also called simply as *cobalamin* or *antipernicious anemia (APA) factor*.

Cyclic AMP, cAMP : A second messenger within cells that is generated from ATP in response to hormonal stimulation of cell-surface receptors; cAMP acts as a signalling molecule by activating A-kinase; it is hydrolyzed to AMP by phosphodiesterase.

Cytochrome oxidase : An enzyme acting as the last hydrogen electron carrier in the electron transport system, receiving an electron from cytochrome and passing it on to oxygen, with the formation of water.

Cytochromes : Coloured heme-containing metalloproteins that serve as electron carriers during cellular respiration, photosynthesis, and other oxidation-reduction reactions; usually designated as cytochromes a, b, and c.

Cytokinins : A group of hormones (growth regulators) that promote growth by stimulating cell division; also called *kinins*.

Cytoplasm (*kytos*^{G} = hollow vessel + *plasma*^{G} = anything moulded) **:** The semifluid living contents of a cell that are contained within its plasma membrane but, in the case of eukaryotic cells, outside the nucleus; it contains sugars, amino acids, proteins, and the organelles (in eukaryotes) such as mitochondria, endoplasmic reticulum, Golgi bodies etc.

Cytosine : An organic molecule composed of one carbon-nitrogen ring. It is a pyrimidine component of nucleic acids and nucleotides. Cytosine always forms complementary base-pairing with guanine.

Cytoskeleton (*kytos*^{G} = hollow vessel + *skeleton*^{G} = a dried body) **:** An organized network of protein filaments in the cytoplasm of all eukaryotic cells that maintain the shape of the cell and anchor organelles (such as nucleus) to fixed locations; consists mostly of actin filaments, microtubules and intermediate filaments.

Cytosol : The fluid medium that is located inside a cell but outside the nucleus and organelles (for eukaryotes) or the nucleoid (for prokaryotes); a semisolid concentrated solution or suspension. Alternatively, it may be defined as the continuous aqueous phase of the cytoplasm with its dissolved solutes; excludes the organelles such as mitochondria.

Cytosome : The cytoplasm of a cell.

D

Dalton (after *John Dalton*, a British chemist) : A unit of molecular mass; approximately equal to the weight of a single hydrogen atom (1.66×10^{-24}g).

Datum, *plural* **data** : An observation used as a basis for inference or induction.

de novo **pathway** : Pathway for synthesis of a biomolecule, such as a nucleotide, from simple precursors, as distinct from a salvage pathway.

Deamination : The enzymatic removal of amino group from a biomolecule, such as amino acids or nucleotides. Deamination of adenine or cytosine causes base-pairing changes.

Decarboxylation : The removal of carbon dioxide from a biomolecule, as in the conversion of oxalosuccinic acid (C_6) to α-ketoglutaric acid (C_5) in the third step of the Krebs cycle.

Decomposition : **(1)** The break-up of a chemical substance into two or more simpler substances. **(2)** The breakdown of organic material by microbes.

Deficiency disease : Any condition exhibiting abnormalities produced by lack of a particular component in the diet of animals and plants; examples include beriberi (due to deficiency of vitamin B_1), scurvy (vitamin C), rickets (vitamin D) and kwashiorkor (protein) among humans; among plants, deficiency of magnesium causes chlorosis of the leaves.

Degenerate code : A code in which one amino acid is specified by more than one codon.

Dehydration reaction : Water-losing. The process in which a hydroxyl (OH) group is removed from one subunit of a polymer and a hydrogen (H) group is removed from the other subunit.

Dehydrogenases : The oxidizing enzymes catalyzing the removal of pairs of hydrogen atoms from their substrates.

Delirium (*delirium*[L], from *de* = away from + *lira* = furrow) : Mental excitement with confusion and sometimes hallucinations.

Denaturation : Partial or complete change in conformation of a polypeptide, protein, or nucleic acid caused by heating or by exposure to chemicals and usually resulting in the loss of biologic property; usually a irreversible process.

Denitrification : The breaking down of nitrates to nitrites and ultimately to free gaseous nitrogen by soil organisms, especially denitrifying bacteria.

Deoxyribonucleic acid, DNA : A polynucleotide having a specific sequence of deoxyri-bonucleotide units covalently joined through 3′,5′-phosphodiester bonds; two of the polymers wind around each other, like the outside and inside rails of a winding staircase; serves as a carrier of genetic information.

Deoxyribonucleotide : A nucleotide containing 2-deoxy-D-ribose as its pentose component.

Deoxyribose : A 5-carbon sugar with one oxygen atom less than the related sugar ribose; a component of deoxyribonucleic acid.

Depurination : Cleavage of the glycosidic bond between C–1′ of deoxyribose and a purine base in DNA; used in Maxam-Gilbert sequence analysis.

Desaturases : Enzymes that catayze the introduction of double bonds into the hydrocarbon portion of fatty acids.

Desiccation : The process by which a substance is dried out and the moisture removed; desiccation is often carried out in a desiccator, which contains a substance which will take up water, *e.g.*, calcium chloride.

Detergent : A type of small amphipathic molecule that tends to coalesce in water, with its hydrophobic tails buried and its hydrophilic heads exposed; when dissolved in water, a detergent acts as a cleansing agent for the removal of grease by altering the interfacial tension of water with other liquids or solids; widely used to solubilize membrane proteins; powerful detergents are used to break up oil spillages at sea.

Detoxification : The process by which poisonous substances are rendered less harmful; for example, the liver converts ammonia into the less toxic compound urea *via* ornithine cycle, and hydrogen peroxide is split into water and oxygen by the enzyme catalase.

Dextrin : A polysaccharide carbohydrate that may form an intermediate step in the hydrolysis of insoluble starch to soluble glucose which is ready for cell respiration, translocation or further synthesis. Dextrins possess adhesive properties and are used as adhesives on paper products.

Dextrorotatory isomer : A stereoisomer that rotates the plane of plane-polarized light to the right or clockwise; represented by the symbol *d* or (+).

Dextrose ($C_6H_{12}O_6$) : A monosaccharide sugar of wide occurrence in plants; also known as *glucose*.

Diabetes insipidus (*diabetes*[G], a siphon, from *dia* = through + *be nai* = to go; *insipidus*[L] = tasteless) : A metabolic disorder resulting due to the failure of the pituitary to secrete antidiuretic hormone (ADH); characterized by an increase in the amount of urine excreted (polyuria) and an increased thirst (polydipsia)); less common than diabetes mellitus.

Diabetes mellitus (*mellitus*[L] = honeyed) : A metabolic disease resulting from insulin deficiency; characterized by a failure in glucose transport from the blood into cells at normal glucose concentrations, with the result excess sugar appears in the blood and urine, associated with thirst and loss of body weight; more common than diabetes insipidus.

Diacylglycerol : Lipid produced by the cleavage of inositol phospholipids in response to extracellular signals; composed of two fatty acid chains linked to glycerol; serves as a signalling molecule to help active protein kinase C.

Dialysis : A process by which small molecules can be separated from larger ones using a semipermeable membrane (*e.g.*, collodion) to contain the larger molecules but which allows the smaller molecules to pass through into the excess water on the other side. The kidney functions by means of this principle, which is also the basis for kidney machines used in cases of kidney disease or failure.

Diarrhoea (*diarrhoia*[G], from *dia* = through + *rhoia* = flowing) : Increased frequency and fluidity of the stools.

Diastase : An enzyme mixture, common in seeds such as barley; involved in starch hydrolysis; the mixture contains *amylases* for conversion of starch to maltose and *maltase* for conversion of maltose to glucose.

Dichlorodiphenyltrichloroethane (DDT) : A chlorinated hydrocarbon which acts as a powerful insecticide with long-lasting effects; DDT was the first major insecticide in use; although cheap to manufacture, DDT has produced adverse ecological consequences; DDT has produced adverse ecological consequences; DDT's lack of biodegradability and the fact that it tends to accumulate in fatty tissues has resulted in its transfer from one consumer to another up the food chain, becoming concentrated at each step. One effect of this has been to endanger the top carnivorous birds whose eggshells have become paper-thin because DDT has prevented the mobilization of calcium in the oviduct, so reducing the reproductive potential of many rare species.

Dicotyledon : A class of flowering plants characterized by having two cotyledons in their embryo, leaves with reticulate venation, and flower parts in twos, fours or fives; also called *dicot*.

Dictyosome : *See* **Golgi body**.

Dielectric constant : A dimensionless constant that expresses the screening effect of an intervening medium on the interaction between two charged particles. Every medium (such as a water solution or an intervening portion of an organic molecule) has a characteristic dielectric constant.

Differential centrifugation : Separation of cell organelles or other particles of different size by their rates of sedimentation in a centrifugal field.

Diffusion (*diffundere*[L] = to pour out) : The net movement of molecules from a region of higher concentration to a region of lower concentration as a result of random, spontaneous molecular motions; diffusion tends to distribute molecules uniformly throughout a medium.

Digestion (*digestio*[L] = separating out, dividing) : The enzymatic process by which food is changed chemically into materials, which are soluble and diffusible, that the cells can assimilate, store, oxidize, or use as nourishment.

Dihydroxyphenylalanine (DOPA) : A precursor in the biochemical pathway leading to melanin formation in animals; DOPA is not metabolized in individuals with albinism.

Denitrification : The breaking down of nitrates to nitrites and ultimately to free gaseous nitrogen by soil organisms, especially denitrifying bacteria.

Dioecious (*di*[G] = two + *eikos*[G] = house, dwelling) : Having male and female flowers on separate plants of the same species; mulberry and willow are common examples of dioecious species.

Dioxin : A chemical byproduct of the manufacture of certain herbicides such as 2,4,5-trichloro-phenoxyacetic acid; chemically known as 2,3,7, 8-tetrachlorodibenzo-para-dioxin (TCDD); one of the most toxic synthetic chemicals known.

Dipeptide : An organic compound consisting of two amino acids, the $-NH_2$ group of one amino acid being united with the –COOH group of the other.

Diphosphopyridine nucleotide (DPN) : *See* **nicotinamide adenine dinucleotide (NAD^+).**

Diphtheria (*diphthera*[G] = skin, leather) : An infection with *Corynebacterium diphtheriae*, generally affecting the pharyngeal area.

Dipole : A molecule having both positive and negative charges.

Diprotic acid : An acid having two dissociable protons.

Disaccharide, $C_{12}H_{22}O_{11}$ (*di*[G] = two + *sakcharon*[G] = sugar) : A carbohydrate consisting of two monosaccharide units that are linked by a covalent bond; for example, sucrose (table sugar) is a disaccharide formed by linking a molecule of glucose to a molecule of fructose.

Disease : An abnormality of an animal or plant caused by a pathogenic organism that affects performance of the vital functions and usually gives diagnostic symptoms.

Disinfectant : A chemical substance used to control microorganisms on a lifeless object.

Dismutation : A reaction in which 2 identical substrate molecules have different fates; particularly, a reaction in which one of the substrate molecules is oxidized and the other reduced.

Dissociation constant (Ka) : Measure of the tendency of a complex to dissociate. For the binding equilibrium $A + B \rightleftharpoons AB$, the dissociation constant is given by [A][B]/[AB], and it is smaller the tighter the binding between A and B.

Disulfide bond : A type of covalent bond formed between two sulfur atoms of two cysteine molecules in the same protein; disulfide bonds strengthen the tertiary structure of proteins.

Diterpene : A compound that consists of four isoprene units linked together; gibberellins are examples of diterpenes.

Diuresis : The increased output of watery urine by the kidneys.

DNA fingerprint : The use of highly variable regions of the DNA sequence to identify an individual.

DNA library : Collection of cloned DNA molecules, representing either an entire genome (*genomic library*) or DNA copies of the mRNA produced by a cell (*cDNA library*).

DNA ligase : An enzyme that joins adjacent nucleotides together by catalyzing the formation of sugar-phosphate bonds in a strand of DNA.

DNA polymerase : An enzyme that catalyzes DNA replication only at the OH ends of DNA strands.

DNA sequencing : Determination of the order of nucleotides in a DNA molecule.

DNA topoisomerases : *See* **topoisomerases.**

Domain : A distinct structural unit of a polypeptide, which may be encoded separately by a specific exon; domains may have separate functions and may fold as independent, compact units. Large globular proteins often consist of several domains, which are connected to each other by stretches of relatively extended polypeptide.

Dopamine ($C_8H_{11}O_2N$) : The decarboxylation product of dihydroxyphenylalanine (DOPA).

Dormancy : A state in which seeds and other structures (such as underground stems) reduce their metabolic activities to a minimum level during unfavourable conditions (low temperature, drought etc.) so as to survive until conditions improve.

Dormin : *See* **abscisic acid.**

Dorsal (*dorsum*[L] = the back) : **(1)** In animals, the part that normally occurs uppermost. The back of an animal is called the *dorsal surface*. The dorsal side is normally directed upwards (dorsal fin) but backwards in primates in the upright position. **(2)** In plants, the part situated on the side of an organ that is directed away from the axis, for example, the upper surface of a leaf; opposite of *ventral*.

Dose : The known amount of chemical or other treatment received by an organism.

Double bond : A covalent bond sharing two pairs of electrons.

Double helix : The natural coiled conformation of two complementary, antiparallel DNA chains.

Double reciprocal plot : A plot of $1/V_0$ versus $1/[S]$, which allow a more accurate determination of V_{max} and K_m than a plot of V_0 versus [S]; also called the *Lineweaver-Burk plot.*

Drosophila melanogaster **:** A species of small dipterous fly, commonly called a fruit fly; extensively used in genetical studies of development.

Drug : (1) Any substance used as an ingredient in medical preparations. (2) Any substance that affects the normal body functions.

Dwarfism : A form of body malfunction in which the adult individual does not reach the normal height and may sometimes have other abnormalities. Such conditions can be due to a deficiency of *growth hormone (GH)* secreted by the anterior pituitary, or to cartilage abnormalities due to genetical defects (*achondroplasia*) as seen in typical circus dwarfs.

Dysentery : A severe disorder of the ileum and colon caused by the bacterium *Shigella dysenteriae* (and many other species), resulting in abdominal cramps, diarrhea, and fever. The disease is spread by 'food, feces, fingers and flies', and can be controlled by sanitary precautions.

Dyspepsia (*dys*[G] = bad + *peptos*[G] = cooked) **:** Deranged digestion.

Dyspnea : Laboured breathing, with breathlessness.

2,4-dichlorophenoxyacetic acid (2,4-D) : A growth substance widely used as a herbicide.

E

e value : The solar energy present on other planets expressed as a percentage of the earth's solar energy.

E_0 values : A numerical series indicative of the redox potential of molecules; protons are accepted by a molecule from any other molecule with a more positive E_0 value.

Ecdysis (*ekdysis*[G] = stripping off) **:** The shedding of the outer covering or skin of certain animals, especially the shedding of the exoskeleton by arthropods, usually in the preadult stage.

Ecdysone : A hormone secreted by the thoracic gland of insects which brings about the moulting of the cuticle and subsequent growth (ecdysis). It raises the metabolic rate and increases the build-up of proteins from amino acids in growing tissue. Also called as *moulting hormone.*

Edema : A swelling of the body tissues caused

by the capillary blood vessels passing out water into the surrounding tissues, and so increasing the intercellular fluid content; in Old English, also spelt as *oedema*.

Effector : A structure or organ, such as a muscle, that brings about an action as a result of a stimulus received through a receptor which can come from the central nervous system or from a hormone.

Egestion : The evacuation of feces or unused food substances from the body.

Ejaculation : The process by which semen is expelled from the penis by strong muscular contractions of the urethral wall.

Elastin : The protein found in elastin fibres.

Electrochemical gradient : Driving force that causes an ion to move across a membrane due to the combined influence of a difference in its concentration on the two sides of the membrane and the electrical charge difference across the membrane.

Electrolyte : A compound with ionic bonds, forming ions and capable of transmitting electric current in solution.

Electromagnetic spectrum : The entire range of wavelengths of electromagnetic radiation, most of which are not detectable by the human eye except in the *visible spectrum* from about 400–700 nm wavelength. Wavelengths shorter than the visible spectum contain large quantities of energy which can be harmful to the living beings.

Electron : A minute negatively-charged particle in an atom. The negative charge of one electron exactly balances the positive charge of one proton.

Electron acceptor : An atom or molecule that receives electrons readily (and thereby becoming reduced) in an oxidation-reduction reaction.

Electron carrier : A protein, such as a flavoprotein or a cytochrome, that can reversibly gain and lose electrons; functions in the transfer of an electron from a donor molecule to an acceptor molecule.

Electron donor : A molecule that readily gives up electrons (and thereby becoming oxidized) in an oxidation-reduction reaction.

Electron microscope (EM) : An instrument that uses an electron beam to magnify images of specimens.

Electron transport system (ETS) : A collective term describing a sequence of membrane–associated electron carriers, generated by the citric acid cycle, that use the energy from electron flow to transport protons against a concentration gradient across the inner mitochondrial membrane.

Electrophile : An electron-deficient group with a strong tendency to accept electrons from an electron-rich group (nucleophile).

Electrophoresis : Movement of charged solutes in response to an electric field; a technique often used to separate mixtures of ions, proteins, or nucleic acids.

Element : A chemical substance that cannot be separated into different substances by ordinary chemical methods; comprises atoms of a single kind.

Elongation factors : Specific nonribosomal proteins required for the addition of amino acids to growing polypeptide chains on ribosomes.

Emendation : The correction of a previously published misspelt scientific name.

Emphysema: The pulmonary disorder involving overdistention and destruction of the air spaces in the lungs.

Enantiomers : Stereoisomers that are nonsuperimposable mirror images of each other. The enantiomers of a compound rotate polarized light in opposite direction, hence also called as *optical isomers*.

Encephalitis (*enkephalos*^{G} = brain, from *en* = in + *kephale* = head + *–itis* = inflammation) : Inflammation of the brain.

Endemic : Refers to organisms or disease whose occurrence is limited to a particular geographical area, such as an island.

Endergonic reaction (*endon*^{G} = within + *ergon*^{G} = work) : A chemical reaction that requires an input of energy before it can proceed (that is, for which ΔG is positive); endergonic reactions never occur spontaneously; compare *exergonic reaction*.

Endocrine gland (*endon*[G] = within + *krinein*[G] = to separate or distinguish) : A ductless gland producing hormonal secretions that pass directly into the bloodstream or lymph.

Endocytosis (*endon*[G] = within + *kytos*[G] = cell) : The process by which the plasma membrane invaginates and forms vesicles (endosomes) whose contents from outside of the cell can be brought into the cell.

Endonuclease : An enzyme that hydrolyzes the interior phosphodiester bonds of a nucleic acid; that is, it acts at points other than the terminal bonds.

Endopeptidase : *See* **protease.**

Endoplasmic reticulum, ER (*endon*[G] = within + *plasma*[G], from cytoplasm; *reticulum*[L] = network) : An extensive network of double membranes distributed throughout the cytosol of eukaryotic cells; portions that are studded with ribosomes are called *rough ER*, and other portions with fewer ribosomes are called *smooth ER*; concerned with the synthesis of lipids and membrane-bound proteins.

Endorphins : A class of endogenous brain peptides that exert analgesic effects in the CNS by binding to opiate receptors. These consist of 16-27 amino acids and are derived from a precursor proopiomelanocortin (POMC). They exhibit actions similar to morphine.

Endoskeleton (*endon*[G] = within + *skeletos*[G] = hard) : In vertebrates, an internal scaffold of bone to which muscles are attached.

Endosperm (*endon*[G] = within + *sperma*[G] = seed) : A nutritive, storage tissue characteristic of the seeds of angiosperms that develops from the fusion of a male nucleus with the polar nuclei in the embryo sac; the endosperm is either digested by the growing embryo or retained in the mature seed to nourish the germinating seedling.

Endothelial (*endon*[G] = within + *thele*[G] = nipple) : Describes the innermost layer of tissue that lines the heart, blood vessels and lymph vessels of vertebrates.

Endothermic reaction : A chemical reaction that takes up heat (that is for which ΔH is positive), as in photosynthesis.

End-product inhibition : *See* **feedback inhibition.**

Energy : The capacity of a body or system to do work; *potential energy* is at rest, as in a chemical compound; *kinetic energy* is in any of many forms of motion, such as light radiation or muscular contraction. In plants and animals, the energy is stored in ATP (short-term storage), and starch and fat (long-term storage).

Energy charge (a term coined by Daniel Atkinson in 1970) : It is a quantity that indicates the state of a cell's energy reserves. It is equal to the cell's reserves of the free energy sources, ATP and ADP (taking into account that ADP stores less free energy than ATP) divided by the total supply of ATP and its breakdown products, ADP and AMP. Thus,

$$\text{Energy charge} = \frac{[\text{ATP}] + \frac{1}{2}[\text{ADP}]}{[\text{ATP}] + [\text{ADP}] + [\text{AMP}]}$$

Energy charge : The fractional degree to which the ATP/ADP/AMP system is filled with high-energy phosphate groups.

Enterocrinin : A gastrointestinal hormone that controls the secretion of intestinal juice.

Enterogastrone : A hormone secreted by the mucosa of the duodenum that decreases gastric secretions and movement in response to the ingestion of fat.

Enterokinase : An enzyme secreted by the wall of the small intestine, whose function is to catalyze the conversion of inactive trypsinogen in the pancreatic juice to active trypsin.

Enterotoxin : A toxin affecting the gastrointestinal tract in humans.

Enthalpy (*H*) : The heat content of a system. A thermodynamic property of a substance given by $H = U + pV$, where U is the internal *energy*, p the *pressure*, and V the *volume*.

Entropy, *S* (*en*[G] = in + *tropos*[G] = change in manner) : A thermodynamic quantity that expresses (or measures) the amount or degree of disorder of a system; the higher the entropy, the more the disorder; a measure of energy that has become so randomized and uniform in a system that the energy is no longer available to do work.

Environment : The surroundings of any organism, including the medium, substrate, climatic conditions, other organisms, light and pH.

Enzyme (*enzymos*[G] = leavened, from *en* = in + *zyme* = leaven) : A biomolecule, either protein or RNA, that catalyzes a specific chemical reaction; it does not affect the equilibrium of the catalyzed reaction; remains unaltered in the process.

Epidemic : The occurrence of many cases of a disease within an area.

Epilepsy (*epilepsis*[G], from *epi* = upon + *lambanein* = to seize) : A nervous condition due to abnormalities in the brain cortex that results in seizures ranging from a sense of numbness in certain body areas (*petit mal*) to extreme muscular convulsions and fits (*grand mal*). Epileptics exhibit large, abnormal brain waves, which can be detected on an electro-encephalogram (EEG).

Epimerases : Enzymes that catalyze the reversible interconversion of the two epimers.

Epimers : Two stereoisomers differing in configuration at one asymmetric centre, in a compound having two or more asymmetric centres.

Epinephrine : *See* **adrenalin**.

Epiphysis : The ossified part of the end of a mammalian limb bne or vertebra which, during growth, is separated by a plate of cartilage from the rest of the ossified bone. When growth is complete, the epiphysis fuses with the rest of the bone.

Epithelium (*epi*[G] = on + *thele*[G] = nipple) : A thin layer of cells forming a tissue that covers the internal and external surfaces of the body.

Equilibrium : The state of a system in which no further net change is occurring; the free energy is at a minimum.

Equilibrium constant (Keq) : A ratio of forward and reverse rate constants for a reaction and equal to the association constant; a constant characteristic for each chemical reaction.

Eradicant : Any biocide (*e.g.*, a fungicide) used to cure an established infection.

Ergosterol ($C_{28}H_{43}OH$) : A white crystalline sterol that occurs in small amounts in the fats of animals; converted into vitamin D_2 (calciferol) by the action of ultraviolet radiation; m.p. 163°C.

Erythema (*erythema*[G] = a flush) : Redness of the skin.

Erythrocyte (*erythros*[G] = red + *kytos*[G] = hollow vessel) : A red blood cell of vertebrates, containing large amounts of hemoglobin; transports oxygen and carbon dioxide to and from tissues; during the process of maturation, an erythrocyte loses its nucleus and mitochondria, and its endoplasmic reticulum is reabsorbed; also called *red blood corpuscle*.

***Escherichia coli* :** A common rodlike bacterium normally found in the small intestine (colon) of humans and other mammals; the most well-studied organism; widely used in biomedical research; also called *E. coli*.

Essential amino acids : Amino acids that cannot be synthesized by humans (and other vertebrates) and must be obtained from the diet.

Essential fatty acids : The group of polyunsaturated fatty acids produced by plants, but not by humans; required in the human diet.

Ester : A molecule formed by the condensation reaction of an alcohol group with an acidic group; most phosphate groups are esters.

Esterases : A class of enzymes involved in the hydrolysis (digestion) of fats and other esters.

Estradiol : An estrogen produced by the follicle cells of the vertebrate ovary; promotes estrus, the development of the endometrium, and stimulates ICSH secretion; in Old English, also spelt as *oestradiol*.

Estrogen (*oestros*[G] = frenzy + *genos*[G] = origin) : A hormone produced by the ovary of the female vertebrate; also produced by the female placenta and in small quantities by the adrenal cortex and the male testis. Estrogen maintains the female secondary sex characters and is involved in the repair of the uterine wall after menstruation.

Estrous cycle : In mammals, the periodic cycle in which periods of estrus correspond to ovulation events; in Old English, also spelt as *oestrous cycle*.

Estrus (*oestrus*[L] = frenzy) : The period of maximum female sexual receptivity. Associated with ovulation of the egg. Being "in heat".

Ethylene, ETH ($CH_2 = CH_2$) : A gaseous plant hormone (growth regulator) that inhibits elongation in most growing tissues and promotes leaf abscission and fruit ripening and other physiological responses in some plants. Plants cells produce ethylene from the amino acid methionine. Also known as *ethene*.

Etiolation : The abnormal elongation of stems caused by insufficient light; etiolated stems are pale yellow due to lack of chlorophyll, have long internodes and rudimentary leaves.

Etiology : The study of causes (usually of a disease); in Old English, also spelt as *aetiology*.

Euchromatin (eu^G = good + $chroma^G$ = colour) **:** Lightly staining portion of chromatin, not easily visible by light microscopy; "normal" chromatin, as opposed to the more condensed heterochromatin.

Eukaryote (eu^G = good + $karyon^G$ = kernel) **:** Living organism composed of one or more cells with a distinct membrane-bound nucleus, multiple chromosomes and internal organelles; includes all forms of life except viruses and bacteria (prokaryotes).

Exergonic reaction (ex^L = out + $ergon^G$ = work) **:** A chemical reaction that proceeds with the release of free energy (that is, for which ΔG is negative); exergonic reactions occur spontaneously; compare *endergonic reaction*.

Euthanasia : The act of painless killing to relieve human suffering from an incurable disease.

Exocytosis : An active process in which vesicles containing excretory or secretary materials are actively carried to the periphery of the cell, and released to the outside when the vesicle membrane fuses with the cell membrane; opposite of endocytosis.

Exon : A region in the coding sequence of a gene that is translated into protein (as opposed to an intron, which is not). It is, thus, a sequence of DNA within a gene that codes for an amino acid sequence. The name comes from the fact that exons are the only parts of an RNA transcript that are present outside the nucleus. Compare *intron*.

Exonuclease : An enzyme that removes a terminal nucleotide (3' or 5') in a polynucleotide chain; exonucleases remove the nucleotides in a successive way, one by one, and are highly specific in their action.

Exopeptidase : A type of protein-splitting enzyme that hydrolyzes the terminal peptide bonds rather than those bonds within the chain; such enzymes complete the digestion of protein prior to absorption into the bloodstream; opposite of endopeptidase.

Exophthalmos (ex^G = out + $ophthalmos^G$ = eye) **:** Prominence of the eyes.

Exoskeleton : A skeleton present on the outside of an organism as in arthropods or molluscs. Some vertebrates possess an exoskeleton in addition to an endoskeleton, for example, armadillos and turtles. The exoskeleton may lie outside the epidermis (as in arthropods) or inside (as in vertebrates, such as scaly fish, tortoises etc).

Exothermic reaction : A chemical reaction that releases heat (that is, for which ΔH is negative), as in respiration.

Experiment : The test of a hypothesis. A successful experiment is one in which one or more alternative hypotheses are demonstrated to be inconsistent with experimental observation and are thus rejected.

Explant : An actively dividing plant tissue that can be induced to produce callus tissue in tissue culture.

Exponent : The number or quantity placed as a superscript to the right of another number or quantity, indicating how many times the number is to be multiplied by itself. For example, 10^4 means that 10 is to be multiplied by 10 four times and is hence equal to $10 \times 10 \times 10 \times 10 = 10{,}000$.

3' end : The end of a nucleic acid that lacks a nucleotide bound at the 3′ position of the terminal residue.

5' end : The end of a nucleic acid that lacks a nucleotide bound at the 5′ position of the terminal residue.

F

Facilitated transport : The transport of a substance across a biological membrane in response to a concentration or electrochemical gradient (*i.e.*, in the direction of lowest concentration) where the movement is facilitated by membrane pores or by specific transport proteins; also called *facilitated diffusion*. Compare *active transport, passive transport*.

Faeces (*faeces*^{L}, plural of *faex* = dregs) **:** The bodily waste material that is formed in the large intestine and eliminated *via* the anus. Faeces contain a mixture of excretory material from the liver (*e.g.*, bilirubin which gives the faeces their characteristic colour), food material which has passed straight through the gut, dead bacteria, dead cells and mucus.

Fastidious : Having special requirement.

Fat : An organic compound used as a food and consisting of carbon, hydrogen and oxygen, with relatively more carbon and hydrogen and less oxygen than a carbohydrate; glycerol esters of fatty acids, insoluble in water but soluble in ether and chloroform; found in almost all organisms; abundant in plant seeds, and are also found in roots, stems and leaves, forming about 5% of the total dry weight; in animals, fats are stored in specialized cells making up adipose tissue.

Fatigue : Exhaustion in muscles resulting from excretion or overstimulation following a period of activity.

Fat-soluble vitamin : Any of several vitamins, including vitamins A, D, E and K that are soluble in organic solvents but insoluble in water.

Fatty acids, $C_nH_{2n+1}COOH$: A long-chain aliphatic carboxylic acid found in natural fats and oils; also a component of membrane phospholipids and glycolipids; used as a major source of energy during metabolism.

Fauna : The total of all animals of a specified region or time.

Fecundity : The number of young ones produced by an organism during the entire course of its life.

Feedback inhibition : A regulatory mechanism in which a biochemical pathway is regulated by the amount of the product that the pathway produces; also known as *end-product inhibition*.

Fermentation (*fermentum*^{L} = ferment) **:** A process by which energy is obtained from organic compounds without the use of oxygen as an electron acceptor; yields lactate, ethanol, or some other simple product.

Ferredoxin : An important iron-containing protein acting as an electron carrier in the electron transport system that operates in the light reactions of photosynthesis, especially in noncyclic photophosphorylation.

Ferritin : A conjugated, electron-dense protein concerned in the absorption of iron through the intestinal mucosa; serves as a storage protein for iron in the liver and spleen.

Fertilization (*ferre*^{L} = to bear) **:** The union of male and female gametes to form a zygote which then subsequently develops into a new organism.

Fetus (Latin for *pregnant*) **:** An animal embryo during the later stages of its development in the womb. In humans, a developing individual is referred to as a fetus from the end of the second month of gestation until birth; also spelt as *foetus*.

Fever (*febris*^{L} = fever, through O.E. féfor) **:** Rise of body temperature above normal, or diseases characterized by this phenomenon.

Fibrin : *See* **fibrinogen.**

Fibrinogen : A large soluble protein found in blood plasma that is formed in the liver and is converted to insoluble fibrin during the process of blood clotting; also known as *fibrin*.

Fibrosis (*fibra*^{L} = a fibre) **:** Formation of scar tissues.

Fibrous proteins : Insoluble, elongated proteins that serve in protective or structural role. Compare *globular proteins*.

First law of thermodynamics : The law stating that in all processes, the total energy of the universe remains constant. It is neither be created nor destroyed. It is, therefore, possible to account for any change in the internal energy of a system ΔE by an exchange of heat (q) and/or work (w) with the surroundings.

$$\Delta E = q - w$$

First-order reaction : A reaction in which the rate of reaction is directly proportional to the concentration of one of the reactants, either product or substrate. Compare *second-order reaction.*.

Fixative : A chemical reagent such as formaldehyde or osmium tetroxide used to preserve cells for microscopy. Samples treated with these reagents are said to be "fixed", and the process is called *fixation*.

Flaccid (*flaccid*[L], from *flaccus* = flabby) : Relaxed; flabby.

Flagellum, *plural* **flagella** (*flagellum*[L] = whip) : A fine, long, threadlike organelle protruding the surface of a cell. Bacterial flagella have much simpler structure than eukaryotic flagella, which are similar to cilia (a cilium is a small flagellum); common in protists and motile gametes; used in locomotion and feeding.

Flatulence (*flatus*[L] = a blowing) : Excess of gas in the digestive tract.

Flavin nucleotides : Nucleotide coenzymes (FMN and FAD) containing riboflavin.

Flavin-linked dehydrogenases : Dehydrogenases requiring one of the riboflavin coenzymes, FMN or FAD.

Flavone : A water-soluble yellowish pigment related to the anthocyanins.

Flavonoid : Any phenylpropanoid-derived compound that is linked to 3 acetate units and condensed into a multiple-ringed structure; the most common flavonoid is *rutin*; flavonoids also include *naringin*, which is a bitter substance in grapefruits.

Flavoprotein : An enzyme containing a flavin nucleotide as a tightly bound prosthetic group; acts as an intermediate carrier in respiratory chains between dehydrogenases and cytochromes.

Flocculation : The formation of jellylike masses of coagulated material.

Flora : The total of all plants of a specified region or time.

Florigen : A hypothetical flowering hormone, that may be produced in leaves and moves to the bud to stimulate flowering; florigen has never been identified or isolated.

Fluorescence : The property of giving out light when molecules are excited by incident light; emitted light is always of a shorter wavelength than the incident light.

Flux : With reference to a chemical pathway, the rate (in moles per unit time) at which reactant 'flows through' the pathway to emerge as a product. The term can be used for the rate at which particles undergo any process in which they either flow or can be thought of metaphorically as flowing.

Folic acid ($C_{19}H_{19}O_6N_7$) : A water-soluble vitamin of the B complex group and designated as vitamin B_9; widely distributed in biological world; the important sources are liver, kidney, yeast and wheat; a yellow crystalline substance, insoluble in fat solvents and inactivated by sunlight; carries out enzymatic synthesis of pyrimidines, purines and amino acids; on a worldwide basis, deficiency of folic acid is believed to be the most common form of vitamin undernutrition: in man, folic acid deficiency leads to megaloblastic anemia, glossitis and gastrointestinal disorders; also called *liver Lactobacillus casei factor* or *vitamin M*.

Follicle (*folliculus*[L] = small ball) : **(1)** In a mammalian ovary, one of the spherical chambers containing an oocyte. **(2)** A dry fruit formed by a single carpel splitting along a line, usually ventral, to liberate its seeds.

Follicle-stimulating hormone (FSH) : A glycoprotein hormone secreted by the anterior lobe of the pituitary gland that stimulates the growth of the ovarian follicles and oocytes in the ovary, and spermatogenesis in the seminiferous tubules of the testis.

Food : An organic substance (carbohydrate, protein, or fat) furnishing energy or building material for protoplasm.

Food poisoning : An acute disorder of the gut caused by food contaminated with bacteria or their toxins (*e.g.*, botulism) or by some chemical.

Formalin : A 37% solution of formaldehyde in water.

Free energy (G) : The component of the total energy of a system that can do work at constant temperature and pressure and defined by the equation, $G = H - TS$, where H is the heat content (*enthalpy*), T the *thermodynamic temperature*, and S the *entropy*; takes into account changes in both energy and entropy.

Free radical : *See* **radical**.

Free radicals : Natural byproducts of metabolism that are involved in the body's defence against microorganism. But, if in excess, they damage body cells and tissues and, thus, play a role in degenerative disorder such as heart ailments and cancer. Free radicles are highly reactive, unstable molecules that normally attach cellular proteins, likes and even DNA. It may be an atom or groups of atoms containing an unpaired electron, usually free radicals are formed in radiation as intermediates free radicals are formed in radiation as intermediates between final chemical products an ion pairs. During an ionizing radiation, an electron is ejected from a water molecule as:

$$H_2O \longrightarrow H_2O^+ + e^-$$

This high energy electron may be picked up by another water molecule as:

$$e^- + H_2O \longrightarrow H_2O^-$$

In this way, an ion pair, H_2O^+ and H_2O^-, are formed. Each ion then may, in the presence of another water molecule, form a hydrogen ion and a free radial as:

$$H_2O^+ \xrightarrow[\text{(free radical)}]{\bar{c}\,H_2O} H^+ + OH^o$$

or as: (free radicals)

$$H_2O^- \xrightarrow{\bar{c}\,H_2O} H^0 + OH^-$$

(free radicals)

The H^+ and OH^- will combine to form water. The H^o and OH^o free radicals are very reactive. In fact, many of them react to form H_2O_2. In cells containing catalase and peroxidases, hydrogen peroxide formation may not be significant. In the absence of such enzymes, hydrogen peroxide formation in cells may be important biologically. Free radicals may also reach with oxygen to enhance the effect of radication. Free radicals may be formed from nearly any cellular component which ionizes to contribute to the indirect effect of radiation. They are also formed in the body as a result of exposure to smoking pollution and sunlight.

Free-energy change (ΔG) : The amount of free energy released (negative ΔG) or absorbed (positive ΔG) in a reaction at constant temperature and pressure.

Fructose ($C_6H_{12}O_6$) : A monosaccharide found in many plants; also called *fruit sugar*.

Fugacious : (of plant organs) Withering or falling quickly.

Fumaric acid : A C-4 organic acid that takes on a molecule of water and becomes malic acid in the 7th step of the Krebs cycle, the reaction being catalyzed by fumarate hydratase.

Functional group : The specific atom or group of atoms that confers a particular chemical property on a biomolecule.

Fungicide : Any substance such as *benomyl* or *captan* that kills fungi.

Furanose : A simple sugar containing the 5-membered furan ring.

G

G proteins : A large family of heterotrimeric proteins that are important intermediaries in cell-signaling pathways. Usually activated by the binding of a hormone or other signaling ligand to seven-pass transmembrane receptor protein. They are called G proteins because binding of GTP and GDP is essential for their functioning.

Galactosaemia : A rare inborn error of metabolism in which the breast-fed human infant is literally poisoned by mother's milk. The affected infants are unable to metabolize the milk sugar galactose, which normally is converted to glucose ready for oxidation and the release of energy. Instead, affected infants store the galactose in various tissues including the brain, resulting in severe malnutrition along with mental retardation. Galactosaemia is due to blockage of the step galactose-1-phosphate to glucose-1-phosphate because the enzyme uridyltransferase is absent or inactive.

Galactose : A monosaccharide carbohydrate that does not occur freely in nature, but is combined with glucose to form lactose, a disaccharide sugar found in milk.

Gamete (Greek, for *wife*) **:** A haploid reproductive cell that fuses with another gamete to form a diploid zygote; the female is an *egg* and the male gamete is a *sperm*; in certain kinds of algae and fungi, however, the gametes are neither male nor female.

Gamma-aminobutyric acid (GABA) : One of many neurotransmitters that vertebrate nervous systems use, each with specific receptors on postsynaptic membranes. GABA opens the channel, which leads to the exit of positively-charged potassium ions and a more negative interior.

Ganglion, *plural* **ganglia** (Greek, for *swelling*) **:** A group of nerve cells forming a nerve centre in the peripheral nervous system.

Gangliosides : Sphingolipids, containing complex oligosaccharides as head groups; found in the plasma membrane of eukaryotic cells and especially abundant in nervous tissue.

Gastric juice : The fluid secreted by glands of the stomach, containing pepsin, renin, and hydrochloric acid.

Gastrin : A polypeptide hormone produced by gastrin cells of the pyloric gland that regulates the synthesis of HCl by the parietal cells of the gastric pits.

Geiger-Müller counter, G-M tube (named after **Hans Geiger**, LT, 1882–1947) **:** A widely-used apparatus for the determination of radioactive isotopes.

Gel : The gellylike solid state of a colloid, as milk, as distinct from the liquid *sol*.

Gel electrophoresis : A technique by which nucleic acids or proteins are separated in a gel that is placed in an electric field.

Gel filtration : A chromatographic procedure for the separation of a mixture of molecules on the basis of size; based on the capacity of porous polymers to exclude solutes above a certain size.

Gene (*genos*[G] = birth, race) **:** The basic unit of heredity. A sequence of DNA nucleotides on a chromosome that codes for a polypeptide or RNA molecule and thus determines of an individual's inherited traits. A gene can be defined either genetically on physically. The genetic test for a gene is the *cis-trans* test or complementation test. In physical terms, the gene is defined as the coding region of DNA that determines a protein product.

Gene amplification : A process in which many copies are made of some genes at one time, while other genes are not replicated. The replicated genes enable enhanced manufacture of product in a short time.

Gene bank : (1) A collection of clones containing all the genes of a particular organism, such as *Escherichia coli*. **(2)** A collection of many lines of a particular crop plant, used as a genetic resource by plant breeders.

Gene cloning : The technique of genetic engineering in which specific genes are excised from host DNA, inserted into a vector plasmid and introduced into a host cell, which then divides to produce many copies (clones) of the transferred gene.

Gene expression : Transcription and, in the case of proteins, translation to yield the product of a gene; a gene is expressed when its biological product is present and active.

Gene frequency : The frequency with which individuals in a population possess a particular gene; often confused with *allele frequency*.

Gene pool : The sum total of all the genes of all breeding individuals in a population at a particular time, represented by their gametes.

Gene splicing : The enzymatic attachment of one gene, or part of a gene, to another; also called *splicing*.

Genetic code : The "language" of the genes. The set of triplet code words in DNA (or mRNA) that code for the amino acids of proteins. Of the 64 possible codons, 61 are the codes for amino acids, and the remaining being termination codons that are not translated. With a few minor exceptions, all living beings use the same code, *i.e.*, the genetic code is *universal*.

Genetic counselling : The process of identifying parents at risk, for producing children with genetic defects and of assessing the genetic state of early embryos.

Genetic engineering : A collective term for the techniques of transferring genes from one organism to another and multiplying them; synonymous with *recombinant DNA technology.*

Genetic map : A diagram showing the relative sequence and position of specific genes along a chromosome.

Genome (*genos*[G] = offspring + *oma*[L] = abstract group) : Total genetic information encoded in a cell or an organism, or a virus.

Genomic library : The set of fragments of an organism's genome that are cloned in a virus or bacterial plasmid.

Genotype (*genos*[G] = offspring + *typos*[G] = form) : The genetic constitution of an organism, as distinct from its physical (or visible) characteristics actually developed during its life history, or phenotype; also used to refer to the set of alleles at a single gene locus. Compare *phenotype.*

Genus, *plural* **genera** (Latin, for *race*) : A taxonomic category between family and species; includes one or more closely-related species; it forms the first part of the binomial of the scientific name of organism and is written or printed with an initial capital letter.

Geometric isomers : Isomers related by rotation about a double bond; also called *cis and trans isomers.*

Geraniol ($C_{10}H_{17}OH$) : A liquid terpene alcohol; present either free or as an ester, in many essential oils; b.p. 107°C.

Germicide : A substance that kills microorganisms.

Germination (*germinare*[L] = to sprout) : The resumption of growth and development by a spore, seed or other structure.

Gestation : The development of the embryo in the uterus of mammals. The gestation period is the time elapsing from fertilization and the implantation of the embryo to birth. This takes 60 days in the domestic cat, 9 months in humans, and 18 months in the Indian elephant.

Gibberellin (*Gibberella*, a genus of fungi) : A type of plant hormone, produced in the apical regions of shoots and roots; controls stem elongation and seed germination, in particular but, unlike auxins, does not inhibit root growth.

Gigantism (*gigas*[G] = giant) : A rare human condition in which excess production of growth hormone by the anterior pituitary gland during childhood and adolescence causes over-elongation of bones, producing a pituitary giant. Compare *dwarfism.*

Gland (*glandis*[L] = acorn) : Any of several organs in the body, such as exocrine or endocrine, that secrete substances for use in the body. Glands are composed of epithelial tissue.

Gliadin : A simple protein, belonging to the group 'prolamines'; a storage protein in the grains of wheat.

Globular protein : Any protein with an approximately rounded shape (the value of axial ratio being between 1 and 4) such as insulin and ribonuclease; contrasts with highly elongated, *fibrous proteins,* such as collagen and fibrinogen.

Globulins : A group of proteins that are soluble in salt solution and coagulated by heat; occur in blood plasma and antibodies, and are the main proteins of plant seeds.

Glucagon : A polypeptide hormone (of 29 amino acids) produced by the α-cells in the islets of Langerhans that raises blood sugar level by breaking down glycogen to glucose; has effect opposite to *insulin.*

Glucocorticoid : A steroid endocrine secretion produced by the adrenal cortex, influencing the metabolism of carbohydrates and proteins, *e.g.,* cortisol and corticosterone.

Glucogenic amino acids : Amino acids with carbon chains that can be metabolically converted into glucose or glycogen *via* gluconeogenesis.

Gluconeogenesis : The biosynthesis of a carbohydrate from simpler, noncarbohydrate precursors such as oxaloacetate or pyruvate.

Glucose ($C_6H_{12}O_6$) : The commonest simple sugar which is colourless, crystalline and soluble in water; m.p. 146°C; occurs in honey and sweet fruits; the instant source of energy in organisms; plays a major role in the metabolism of living cells; stored in polymeric form as glycogen in animal cells and as starch in plant cells; used in brewing, jam-making, confectionery etc; also known as *dextrose* or *grape sugar*.

Glucoside : An organic compound yielding glucose on hydrolysis.

Glutathione (GSH) : An autooxidizable, widely-distributed tripeptide containing glutamic acid, cysteine and glycine; involved in certain oxidations in the cell.

Glutelins : A composite term for a group of simple coagulable proteins isolated from plant seeds only; includes *glutelin* from wheat and *oryzenin* from rice.

Glycan : Another term for polysaccharide; a polymer of monosaccharide units joined by glycosidic bonds.

Glycerol : A sweet, syrupy trihydroxyalcohol, colourless and odourless; formed by the digestion of fats and fixed oils in the plants; parent compound of many small molecules in the cell, including phospholipids; also called *glycerine*.

Glycerophospholipid : A amphipathic lipid with a glycerol backbone; fatty acids are ester-linked to C-1 and C-2 of glycerol, and a polar alcohol is attached through a phosphodiester linkage to C-3.

Glycogen (*glykys*^{G} = sweet + *gen*^{G} = of a kind) **:** A white, amorphous, tasteless polysaccharide composed exclusively of glucose units and used to store energy in animal cells. Large granules of glycogen are especially abundant in liver and muscle cells. Also called "animal starch".

Glycolipid : A membrane lipid molecule with a short carbohydrate chain attached to a hydrophobic tail.

Glycolysis (*glykys*^{G} = sweet + *lyein*^{G} = to loosen) **:** The ubiquitous catabolic pathway, taking place in the cytosol, by which a molecule of glucose is oxidatively broken down into 2 molecules of pyruvic acid; the substrate-level phosphorylation of 2 molecules of ADP to ATP and the reduction of 2 molecules of NAD^+ to NADH occur for the breakdown of each molecule of glucose; constitutes the early stages of respiration.

Glycoprotein : Any protein with one or more covalently-linked oligosaccharide chains; includes most secreted proteins and most proteins exposed on the outer surface of the plasma membrane such as *extensin*.

Glycosaminoglycan (GAG) : A heteropoly-saccharide of two alternating units: one is either *N*-acetylglucosamine or *N*-acetylgalactosamine, the other is a uronic acid (usually glucuronic acid); hyaluronic acid and heparin are common examples; formerly called *mucopoly-saccharide*.

Glycoside : An organic compound yielding a sugar on hydrolysis. Glycosides containing glucose are called *glucosides*, those with galactose are called *galactosides*.

Glycosidic bond : Bonds between a sugar and another molecule (typically an alcohol, purine, pyrimidine, or sugar) through an intervening oxygen or nitrogen atom; the bonds are classified as *O*-glycosidic or *N*-glycosidic, respectively.

Glycosuria : The condition where glucose is excreted in the urine because the blood sugar level exceeds the normal (*hyperglycemia*). Glycosuria is one of the symptoms of diabetes.

Glycosylation : The addition of a carbohydrate to an organic molecule such as a protein.

Glyoxylate cycle : A variant of the citric acid cycle that converts acetyl-CoA into succinate and, eventually, new carbohydrate; occurs in bacteria and some plant cells.

Glyoxysome : A specialized peroxisome containing the enzymes of the glyoxylate cycle; found in cells of germinating oilseeds and seedlings that arise from them.

Goitre (*goitre*^{Fr}, from *goitreux*; *gutturosus*^{L}, from *guttur* = the throat) **:** An enlargement of the thyroid gland resulting from a deficiency of iodine in the diet.

Golgi body (after *Camillo Golgi*, an Italian physician) : A stack of flattened, membranous vesicles, often branched and present in the cytoplasm of eukaryotic cells; Golgi bodies are the sites where precursors of cell wall materials and other cellular components are assembled and prepared for secretion from the cell or are incorporated into the plasma membrane or organellar membranes; also known as *dictyosome*.

Golgi complex (after **Camillo Golgi**, an Italian physician) : A collective term for Golgi bodies.

Gonad (*gone*[G] = generation) : An organ in which gametes, male or female, are produced.

Grain : A simple, indehiscent, dry, single-seeded fruit in which the seed coat is fused to the pericarp over its entire surface, as in cereal grasses such as wheat, rice, maize etc; *grain is a caryopsis fruit*.

Gram molecular weight : The weight in grams of a compound that is numerically equal to its molecular weight; the weight of one mole.

Gram reaction : A method of differential staining of bacteria by treating them with a special iodine solution after they have been stained with Gentian violet. Certain species of bacteria (gram-positive) retain the purple dye and others (gram-negative) are decolourized, thus affording a basis for classification.

Granum, *plural* **grana** (Latin, for *grain* or *seed*) : In chloroplasts, stacks of membrane-bound disks called thylakoids; the thylakoids contain the chlorophylls and carotenoids and are the sites of the photochemical (*i.e.*, "light") reactions of photosynthesis.

Ground state : The normal, stable form of an atom or molecule; as distinct from the *excited state*.

Growth hormone (GH) : *See* **somatotrophic hormone (STH).**

Guanine ($C_5H_5ON_5$) (Sp. from *Quechua*, huanu, dung) : A purine base found in DNA and RNA; occurs in high concentration as a white crystalline base in guano (=bird manure), hence its nomenclature.

Gum : Any of a class of colloidal, water- soluble substance (chemically, a hemicellulose) that is exuded by or extracted from plants; gluey when wet but hardening on drying; consists of many kinds of monosaccharides. An example is *gum arabic* (extracted from *Acacia senegal*), which is a complex branched-chain hemicellulose consisting of the monosac- charides arabinose, galactose, glucose, and rhamnose.

H

Half-life : **(1)** For a chemical reaction, the time at which half of the substance has been consumed and turned into product. **(2)** In biochemistry, the time required for the disappearance or decay of one-half of a given component in a system. Alternatively, it may be defined as the time required for one-half of a population of molecules to be converted to another form; for example, conversion of ^{32}P to ^{31}P. Half-lives vary from isotope to isotope, some being less than a millionth of a second and some more than a million years; symbol, $T_{1/2}$; also called *half-time*.

Half-time : *See* **half-life**.

Homeothermy (*homeo*[G] = similar + *therme*[G] heat) : The maintenance of a stable body temperature independent of the environmental temperature by the organisms such as birds and mammals.

Heat-shock protein : Protein synthesized in response to an elevated temperature or other stressful treatment; usually helps the cell to survive the stress; also called *stress-response protein*.

Heat-labile : Destroyable by heat.

Heavy isotope : A stable atom in which there are more neutrons than in the normal isotope of the element, giving it a greater mass; for example, ^{15}N is the heavy isotope, ^{14}N is the common form.

HeLa cell : Line of human epithelial cells that grows vigorously in tissue culture that is grown as a standard in research laboratories all over the world; derived from a human cervical carcinoma obtained from *He*nrietta *La*cks, in 1951.

Helicase : An enzyme that breaks hydrogen bonds between complementary base pairs of DNA, thereby causing separation of 2 strands in a DNA molecule before replication.

Helix : Anything of a spiral shape; in biology, it refers to the shape of DNA molecules, which occur as double helices.

Heme : A complex organic ring structure, called a protoporphyrin, to which an iron atom is bound; occurs in cytochromes of all organisms and in the hemoglobin of animals; also spelt as *haem*.

Heme protein : A protein containing a heme as a prosthetic group.

Hemicellulose : Primarily a cell wall polysaccharide of variable composition and structure; less complex than cellulose and easily hydrolyzable to simple sugars; hemicellulose that is secreted by plants is also called a *gum*.

Hemoglobin (*haima*^{G} = blood + *globus*^{L} = a ball) **:** A globular heme protein in vertebrate red blood cells and in the plasma of many invertebrates that carries oxygen and carbon dioxide; heme group binds oxygen and carbon dioxide and as well as imparts red colour to the blood; also spelt as *haemoglobin*.

Hemolytic (*haima*^{G} = blood + *lytikos*^{G} = able to loosen) **:** An agent causing dissolution of the red blood corpuscles.

Hemophilia : An individual that lacks a clotting factor and consequently bleeds easily. Some of the genes for clotting factors are sex-linked.

Hemopoiesis : Generation of blood cells, mainly in the bone marrow; also called *hematopoiesis*.

Hemorrhage (*haima*^{G} = blood + *rhegnunai*^{G} = to break) **:** An escape of blood from the blood vessels, due to a wound or disease.

Henderson–Hasselbalch equation : An equation relating the pH, pK_a, and the ratio of the concentrations of the proton-acceptor (A^-) and proton-donor (HA) species in a solution. Alternatively, it is a formula for calculating the pH of a buffer solution.

Heparin : A mucopolysaccharide molecule produced in the liver that acts as an anticoagulant, inhibiting the transformation of prothrombin to thrombin, a vital step in blood clotting.

Hepatitis : A serious disorder of the liver that leads to severe jaundice, liver degeneration and even death. The condition is caused by two viruses : hepatitis A virus, which produces *infective hepatitis* transmitted by the intestinal-oral route, and hepatitis B virus, which produces *serum hepatitis* transmitted *via* infected blood or its products. Of the two viral types, type A usually has a shorter incubation period than type B.

Hepatocyte : The major cell type of liver tissue.

Herbicide : Any chemical that kills plants. Herbicides can be highly selective; for example, 2,4-D only kills dicotyledons (broad-leaved plants), leaving monocotyledons unharmed.

Heredity : Transmission of traits from one generation to the next.

Heterochromatin (*heteros*^{G} = different + *chroma*^{G} = colour) **:** Region of a eukaryotic chromosome that remains permanently condensed and therefore is not transcribed into RNA; stains darkly with Giemsa stain; most centromere regions are heterochromatic; easily visible by light microscopy. Heterochromatin is either constitutive or facultative. Constitutive heterochromatin is composed of repeated sequences of DNA. Facultative heterochromatin is a transient form of inactive DNA (an inactive X chromosome, for example) that is usually the result of methylation of cytosine.

Heterogeneous nuclear RNA (hn RNA) : The pool of primary RNA transcripts in the nucleus which are of various, usually large sizes. It is, in fact, the immediate product of transcription in an eukaryote, containing both introns and exons; also known as *pre-mRNA*.

Heteropolymer : A polymer that is made of more than one kind of monomer; for example, polypeptides and nucleic acids. Compare *homopolymer*.

Heteropolysaccharide : A polysaccharide containing more than one type of monosaccharide units.

Heterotroph (*heteros*^{G} = different; *trophikos*^{G} = food) : An organism that cannot synthesize its own food (*i.e.*, organic compounds) entirely from inorganic precursors but must consume at least some organic compounds made by other organisms. In particular, these organisms can not use CO_2 as a carbon source. Compare *autotroph*.

Hexose : A simple sugar with a backbone having six carbon atoms.

Hierarchy (*hieros*^{G} = sacred + *archos*^{G} = leader) : Refers to a graded system of classification in which each step of category includes all those below it; especially the Linnean hierarchy used in classification of organisms.

High-energy bond : A covalent bond that has a low activation energy and is broken easily and which on hydrolysis releases an unusually large amount of free energy under the conditions existing in a cell. A group linked to a molecule by such a bond is readily transferred from one molecule to another; common examples are the phosphodiester bonds in ATP and the thioester linkage in acetyl-CoA.

High-energy compound : A compound that on hydrolysis undergoes a large decrease in free energy under standard conditions.

Histamine ($C_5H_9N_3$) : A small molecule derived from the amino acid histidine, released from mast cells and basophils in allergic reactions; causes irritation, dilation of blood vessels, and contraction of smooth muscle.

Histone (*histos*^{G} = tissue) : The family of five very basic positively-charged, low-molecular-weight polypeptides, rich in arginine and lysine, that are tightly associated with DNA in the chromosomes of all eukaryotic cells. Histones form the core of nucleosomes, around which DNA is wrapped. The 5 major histones are represented as H1, H2A, H2B, H3 and H4.

Holoenzyme : A complete or intact, catalytically-active, enzyme consisting of a protein part, apoenzyme and a nonprotein part, prosthetic group, which may be an organic molecule (*coenzyme*) or a metal ion (*cofactor*). It is, thus, the entire molecular structure required to carry on an enzymatic function (for example, polymerase III, with all of its 7 subcomponents).

Homeosis : The transformation of one tissue or organ into another.

Homeostasis (*homeos*^{G} = similar + *stasis*^{G} = standing) : The maintenance of a relatively stable internal physiological environment in an organism, or steady-state equilibrium in a population or ecosystem, by regulatory mechanisms that compensate for changes in external circumstances.

Hominid (*homo*^{L} = man) : Human beings and their direct ancestors, *i.e.*, a member of the family, Hominidae; *Homo sapiens* is the only living member.

Homologous chromosome (*homologia*^{G} = agreement) : One of the two nearly identical versions of each chromosome. Chromosomes that associate in pairs in the first state of meiosis. In diploid cells, one chromosome of a pair that carry equivalent genes.

Homologous proteins : Proteins having sequences and functions similar in different species, for example, the hemoglobins.

Homology (*homologia*^{G} = agreement) : Similarity in structure of an organ or a molecule, reflecting a common evolutionary origin, specifically such a similarity in protein or nucleic acid sequence; structures related by homology are homologous and are called as homologues; contrasted with *analogy*– a similarity that does not reflect a common evolutionary origin.

Homoplasy : Resemblance among organisms not due to inheritance from a common ancestry (not homologous).

Homopolymer : A polymer that is made of only one kind of monomer; for example, starch which is made only of glucosyl units. Compare *heteropolymer*.

Hormone (*hormaein*^{G} = to excite) **:** A chemical substance (often a steroid or peptide), synthesized in minute quantities by an endocrine gland and carried in the blood to another tissue, where it acts as a messenger to regulate the function of the target tissue or organ.

Hormone receptor : A protein in, or on the surface of, target cells that binds a specific hormone and initiates the cellular response.

Human Genome Project (HPG) : An international effort to determine the sequence of nucleotides in the human genome.

Human immunodeficiency virus (HIV) : The virus responsible for acquired immunodeficiency syndrome (AIDS), a deadly disease that destroys the human system. HIV is a retrovirus (its genetic material is RNA) that is believed to have been introduced to humans from African green monkeys.

Hyaluronidase : An enzyme present in snake venom and bacteria that catalyzes the hydrolysis of hyaluronic acid, thus making it ineffective in stopping the spread of invading microbes and other toxic substances.

Hybrid : (*hybrida*^{L} = the offspring of a tame sow and a wild boar) – An individual that results from the crossing of dissimilar parents.

Hydration : Union of a chemical substance with water without chemical decomposition.

Hydrazine ($H_2N.NH_2$) : A fuming, strongly basic liquid; b.p. 113°C; a powerful reducing agent and is highly reactive; used in organic synthesis and as a rocket propellant, either alone or mixed with the dimethyl derivative.

Hydrocarbon : A compound that has only carbon and hydrogen atoms.

Hydrogen bond : A weak electrostatic chemical bond that forms between one electronegative atom (such as oxygen or nitrogen) and a hydrogen atom covalently linked to a second electronegative atom. Hydrogen bonds hold the 2 complementary strands of DNA together. Life would be impossible without this type of linkage.

Hydrolases : Enzymes that catalyze hydrolysis reactions; common examples are lipases, nucleases, phosphatases and proteases.

Hydrolysis (*hydro*^{G} = water + *lyse*^{G} = break) **:** Cleavage of a covalent bond, such as an anhydride or peptide bond, by the addition of elements of water, yielding two or more products. Common examples are conversion of maltose to glucose and dipeptide to 2 amino acids; essentially the reverse of *dehydration*.

Hydronium : The hydrated hydrogen ion (H_3O^+).

Hydropathy index : A scale that expresses the hydrophilic and hydrophobic tendencies of a chemical group.

Hydrophilic (*hydro*^{G} = water + *philic*^{G} = loving) **:** Polar or charged; describing molecules or groups that form enough hydrogen bonds with water and hence dissolve freely in water; sugars are hydrophilic; compare *hydrophobic*.

Hydrophobic (*hydro*^{G} = water + *phobos*^{G} = hating) **:** Nonpolar or uncharged; describing molecules or groups, which do not form hydrogen bonds with water and hence are not soluble in water (but are soluble in nonpolar solvents such as ether, acetone); lipids and hydrocarbons are generally hydrophobic; also called as *lipophilic*; compare *hydrophilic*.

Hydrophobic interactions : The association of nonpolar groups, or compounds, with each other in aqueous systems, driven by the tendency of the surrounding water molecules to seek their most stable (disordered) state.

Hydroxyl (–OH) : A chemical group consisting of a hydrogen atom linked to an oxygen atom, as in an alcohol.

Hyperchromic effect : The large increase in light absorption at 260 nm occurring as a double-helical DNA is melted (unwound).

Hyperemia : An increased blood flow to an organ or tissue.

Hyperplasia : An increase in tissue mass caused by an increase in cell number.

Hypertonic (*hyper*[G] = above + *tonos*[G] = tension) : Refers to the solution surrounding a cell that has a sufficiently high concentration of solutes to cause water to move out of the cell due to osmosis.

Hypertrophy (*hyper*[G] = too much + *trophe*[G] = nurture) : An excessive growth or development of an organ or tissue.

Hypophysectomy : The removal of the pituitary gland.

Hypothalamus (*hypo*[G] = under *thalamos*[G] = inner room) : The region of the brain under the thalamus that controls temperature, hunger, and thirst, and that produces hormones that influence the pituitary gland.

Hypothesis, *plural* **hypotheses** (*hypo*[G] = under + *tithenai*[G] = to put) : A proposal that might be true. No hypothesis ever proven correct. All hypotheses are provisional– proposals that are retained for the time being as useful but that may be rejected in the future if found to be inconsistent with new information. Altenatively, a hypothesis may be defined as a supposition put forward in explanation of observed facts. A hypothesis that stands the test of time– often tested and never rejected– is called a *theory*.

Hypotonic (*hypo*[G] = under + *tonos*[G] = tension) : Refers to the solution surrounding a cell that has a sufficiently low concentration of solutes to cause water to move into the cell due to osmosis.

I

Iatrochemistry : Medieval medical chemistry; early attempts at the application of drugs to medicine.

Imbibition : The adsorption of water onto the internal surfaces of materials; one of the methods by which root hair and other plant parts obtain water.

Immune response : Response made by the immune system of a vertebrate when a foreign substance or microorganism enters its body; the response is exhibited by generating antibodies and T-cells against a foreign specific antigen.

Immunoglobulin, I_g : An antibody protein generated against, and capable of binding specifically to, an antigen. Higher vertebrates have five classes of immunoglobulin– I_gA, I_gD, I_gE, I_gG and I_gM– each with a different role in the immune response.

In vitro (Latin, for "*in glass*") : A term used by biochemists to describe a process taking place in an isolated cell-free extract. Also used by cell biologists to refer to cells growing in culture usually taken in glass equipments, as opposed to in an organism (*in vivo*).

In vivo (Latin, for "*in life*") : Refers to a process taking place in an intact (*i.e.*, living) cell or an organism, as opposed to in a culture (*in vitro*).

Inbreeding : The mating of related individuals.

Indeterminate growth : Growth that is not inherently limited, as with a vegetative apical meristem that produces an unrestricted number of organs indefinitely.

Indole-3-acetic acid, IAA : A plant hormone (growth regulator) that influences cellular elongation, among other things.

Indole-3-acetonitrile (IAN) : A plant growth-regulating hormone.

Induced fit : A change in the conformation of an enzyme in response to substrate binding that renders the enzyme catalytically active.

Inducer : A signal molecule that, when bound to a regulatory protein, produces an increase in the expression of a given gene. It is an environmental agent that triggers transcription from an operon.

Inducible enzyme : An enzyme that is produced in response to the presence of its substrate (the inducer).

Infection : The invasion of tissues by micro-organisms with or without disease production.

Informational macromolecules : Biomolecules containing information in the form of specific sequences of different monomers; for example, many proteins, lipids, polysaccharides and nucleic acids.

Infusion : The liquid extract of any substance which has been soaked in water.

Inhibitor : A chemical whose binding alters the shape of a protein and shuts off enzyme activity.

Initiation complex : A complex of a ribosome with an mRNA and the initiating Met-$tRNA^{Met}$ or fMet-$tRNA^{fMet}$, the formtion of which begins polypeptide (or protein) synthesis.

Initiation codon : AUG (sometimes GUG in prokaryotes); codes for the first amino acid in a polypeptide sequence: *N*-formylmethionine in prokaryotes and methionine in eukaryotes; also called *start codon*.

Initiation factor : A protein that promotes the proper association of ribosomes with mRNA and is required for the initiation of protein synthesis.

Innate (*innatus*^{L} = born) **:** Describing a characteristic based partly or wholly on inherited gene differences.

Inorganic compound : A type of molecule that either lacks carbon or *contains carbon but not hydrogen*; carbon dioxide (CO_2) and water (H_2O) are examples of inorganic compounds.

Inorganic pyrophosphatase : An enzyme that hydrolyzes a molecule of inorganic pyrophosphate to yield two molecules of (ortho) phosphate; also known simply as *pyrophosphatase*.

Inositol phospholipids : *See* **phosphoinositides**.

Inositol : A cyclic molecule with 6 hydroxyl (OH) groups that forms the hydrophilic head group of inositol phospholipids.

Inositol phosphate : A mediator molecule produced during insulin receptor-induced enzyme activity.

Insane (*insanus*^{L}, from *in* = not + *sanus* = sane) **:** Unsound in mind or lunatic.

Insecticide : Any substance that is used to kill insects; for example, DDT or malathion.

Insulin (*insula*^{L} = an island, from the 'islets of Langerhans') **:** A peptide hormone secreted by the β-cells of islets of Langerhans in the pancreas that acts as a storage hormone. It enables the body to use sugar and other carbohydrates by regulating the body's sugar metabolism in animals. Insulin has three targets: the liver, the muscles and adipose tissue.

Interferon : In vertebrates, a class of glycoproteins produced naturally in virus-infected cells that inhibits viral multiplication.

Intermediary metabolism : In cells, the enzyme-catalyzed reactions that extract chemical energy from nutrient molecules and utilize it to synthesize and assemble cell components. It does not include nucleic acid and protein syntheses.

Internal membrane : An eukaryotic cell membrane other than the plasma membrane; for example, the membranes of the endoplasmic reticulum and Golgi apparatus.

Internode : The part of the plant stem between two successive nodes, where branches and leaves attach.

Intracellular digestion : Digestion occurring within a cell.

Intron (*intra*^{L} = within) **:** A noncoding sequence of nucleotides within a eukaryotic gene that is transcribed into an mRNA molecule but is then excised by RNA splicing before the gene is translated. These untranslated regions of DNA make up the bulk of most eukaryotic genes; also called as *intervening sequence*. Compare *exon*.

Inulin : A white, tasteless soluble polysaccharide, yielding fructose on hydrolysis and found especially in the roots, tubers, and rhizomes of the members of Compositae (*Dahlia, Helianthus, Taraxacum*).

Invertase : An enzyme which catalyzes the breakdown of sucrose by hydrolysis into glucose and fructose; also called as *saccharase*, *sucrase*, or β, *(+)-fructosidase*.

Iodine number : A measure of the degree of unsaturation (content of double bonds) of a product such as an oil or fat; it is expressed in grams of idodine absorbed by

100 g of the given oil or fat; also called as *Koettstorfer number* or *Hübl number*.

Ion : An atom in which the number of electrons does not equal the number of protons; an ion does carry an electrical charge. Alternatively, an ion is an electrically-charged atom or groups of atoms.

Ion channel : An integral membrane protein that provides for the regulated transport of a specific ion or ions, across a membrane.

Ion exchange resin : A polymeric that contains fixed charged groups; used in chromatographic columns to separate ionic compounds.

Ion product of water (K_w) : The product of the concentrations of H^+ and OH^- in pure water; $K_w = [H^+]\,[OH^-] = 1 \times 10^{-14}$ at 25°C.

Ionic bond : A noncovalent bond formed between two ions, one with a positive charge and the other with a negative charge, as a result of the attraction of opposite electrical charges.

Ionization : The process of spontaneous ion formation.

Ionizing radiation : A type of radiation such as *x* rays and gamma rays, that causes loss of electrons from some organic molecules, thus making them more reactive.

Ionophore : A small hydrophobic molecule that binds one or more metal ions and is capable of diffusing across a membrane, carrying the bound ion.

Iron–sulfur centre (Fe–S centre) : A prosthetic group of certain redox proteins involved in electron transfers; Fe^{2+} and Fe^{3+} is bound to inorganic sulfur and to cysteine groups in the protein.

Islets of Langerhans (after **Paul Langerhans**, a German anatomist) **:** The small, scattered endocrine glands in the pancreas that secrete the hormones, insulin and glucagon.

Isocitric acid : A C-6 organic acid that loses a molecule of CO_2 in the third step of the Krebs cycle, thereby being converted to α-ketoglutaric acid; also during this conversion, one molecule of NAD^+ is reduced to NADH.

Isoelectric point (pI) : The pH at which a solute has no net electric charge and thus does not move in an electric field.

Isoenzyme : Multiple forms of an enzyme that catalyze the same reaction but differ from each other in their amino acid sequence, substrate affinity, V_{max}, and/or regulatory properties; also called *isozyme*.

Isoforms : Multiple forms of the same protein that differ somewhat in their amino acid sequence.

Isomerases : Enzymes that catalyze the transformation of compounds into their positional isomers, *i.e.*, from D to L, or L to D forms.

Isomers (*isos*[G] = same; *meros*[G] = part) **:** Molecules that are formed from the same atoms in the same chemical linkages but have different three-dimensional conformations; alternatively, molecules with the same molecular formulae but with different structural formulae.

Isoprene : A small, unsaturated hydrocarbon containing 5 carbon atoms; chemically called 2-methyl-1,3-butadiene; a parent compound of isoprenoids such as retinoic acid and cholesterol.

Isoprene rule : Refers to the statement that nearly all the terpenoids are made of varying number of repetitive units (C_5H_8), called isoprene units.

Isothermal : Occurring at constant temperature.

Isotonic (*isos*[G] = equal + *tonos*[G] = tension) **:** Refers to the solution surrounding a cell that has the same concentration of solutes as in the cell.

Isotope (*isos*[G] = equal + *topos*[G] = place) **:** Different forms of an element that differ in atomic weight but are otherwise chemically identical to the naturally abundant from of the element; the various forms have the same number of protons but different numbers of neutrons; may be either stable or radioactive; used as tracers.

Isozyme : *See* **isoenzyme.**

J

Jaundice (M.E. jaundice from *jaunice*[Fr] = yellowness) : Condition characterized by yellowness of the skin and mucous membranes and the secretion of bile pigments in the urine, turning it dark yellowish.

Joint : The part of a vertebrate where one bone meets and moves on another.

Joule (J) : The amount of energy needed to move one kilogram through one meter with an acceleration of one metre per second per second; one joule is the energy delivered in one second by a one-watt power source; 10^7erg; approximately equal to 0.24 calories; a slice of apple pie contains about 1.5×10^6J.

Juvenile hormone (JH) : A hormone present in juvenile insects and secreted by the corpus allatum of the brain. So long as it is produced, the cuticle maintains the characteristics of the nymphal or larval form at each moult. Only when it ceases to be present or the level falls below a threshold value does the insect moult to the adult form. Also called *status quo hormone (SQH)*.

K

Karyotype (*karyon*[G] = kernel + *typos*[G] = stamp or print) : Full set of chromosomes of a cell arranged with respect to size, shape and number.

Kelvin scale : An absolute scale of temperature where a range of 1° kelvin (K) is equal to a range of 1°C.

0°K = –273.15°C or –459.67°F.

Keratins : Hard, fibrous, insoluble, sulfur-containing, structural or protective proteins consisting of parallel polypeptide chains in α-helical or β-conformations, found in the epidermis of vertebrates, mainly in the outermost layers of skin; can have several forms : in scales, feathers, hooves, horns, claws and nails, it is hard, while wool and hair are made up of a soft and flexible form.

Keratitis (*keras*[G] = horn + *-itis*[G] = inflammation) : Inflammation of the cornea.

Keto-enol tautomerism : A type of tautomerism that occurs in ketones as the result of the migration of a hydrogen atom from an alkyl group to the carbonyl group.

Ketogenic amino acids : Amino acids with carbon skeletons that can serve as precursors of the ketone bodies.

Ketone : An organic molecule containing a carbonyl group linked to two alkyl groups.

Ketone bodies : Water-soluble fuels normally exported by the liver but overproduced during fasting or in untreated diabetes mellitus; acetoacetate, D-β-hydroxybutyrate and acetone common examples.

Ketonuria (*ketone*[Eng] = variant of a acetone, from *acetum*[L] = vinegar + *ouron*[G] = urine) : The presence of ketone bodies in the urine.

Ketose : A simple monosaccharide in which the carbonyl group occurs within the chain and, hence, represents a ketone group, *e.g.*, fructose; compare *aldose*.

Ketosis : A condition in which the concentration of ketone bodies in the blood, tissues, and urine is abnormally high.

Kilocalorie : Unit of heat energy, equal to 1,000 calories.

Kilojoule : Standard unit of energy, equal to 1,000 joules, or 0.24 kilocalories.

Kinases : Enzymes that catalyze the phosphorylation of certain molecules by ATP.

Kinetic energy : The energy of motion; a solute that moves down its concentration gradient has kinetic energy.

Kinetics : The study of reaction rates.

Kinetin, K ($C_{10}H_9ON_5$) : A growth substance prepared from deoxyribonucleic acid; chemically known as 6-furfurylaminopurine; acts as a cytokinin, *i.e.*, promotes cell division in plants.

Kinetochore (*kinetikos*[G] = putting in motion + *choros*[G] = chorus) : A discoid complex of proteins that is bound on one side to a centromere and on the other side to a spindle fibre.

Kingdom : The highest grouping (taxon) in biological classification, such as animal kingdom (*Animalia*) and plant kingdom (*Plantalia*).

Kinins : A group of substances that promotes cell division; kinetin is one of the common kinins.

Krebs cycle : *See* **citric acid cycle.**

Kwashiorkor : The most widespread and serious human nutritional disease brought on by acute protein starvation. The disease is characterized by apathy, impaired growth, skin ulcers, swollen hands and feet and an enlarged liver. If left untreated, it is fatal. Kwashiorkor typically affects children in the early stages after weaning.

Kyphosis (a Greek word, meaning *humpback*) **:** Abnormal curvature of the vertebral column.

L

Lac operon : A cluster of genes encoding 3 proteins that bacteria use to obtain energy from the sugar lactose.

Lachrymal gland : The tear gland that lies below the upper eyelid of mammals and serves to moisten and cleanse the surface of the eye by secreting sterile and antiseptic liquid. Excess fluid is drained away through the lachrymal duct in the corner of the eye which leads eventually into the nasal cavity.

Lactase : An enzyme that splits the disaccharide lactose into galactose and glucose, secreted as part of the intestinal juice by glands in the wall of small intestine.

Lactation : The production of milk by the adult female mammal from the mammary gland, and with which it suckles its young. Lactation is induced by luteotrophic hormone (LTH), from the anterior pituitary gland.

Lactose : A disaccharide carbohydrate found in the milk of mammals; produced by condensation reaction between galactose and glucose; broken into its component monosaccharides by lactase; souring of milk is due to the coversion of lactose to lactic acid by microbes present in the air.

Lagging strand : One of the two newly-made strands of DNA found at a replication fork. The lagging strand is synthesized in the direction opposite to that in which the replication fork moves and is made in discontinuous lengths that are later joined covalently. Compare *leading strand.*

Lamella, *plural* **lamellae** (Latin, for *a little plate*) **:** A thin plate-like structure. In chloroplasts, a layer of chlorophyll-containing membranes. In bivalve mollusks, one of the two plates forming a gill. In vertebrates, one of the thin layers of bone laid concentrically around the Haversian canals.

Lateral meristem : Meristem that produces secondary tissue; the vascular cambium and cork cambium are examples of lateral meristem.

Latex : A milky, usually white, fluid found in the laticiferous ducts of certain plants and used commercially in the production of rubber, chicle and guttapercha.

Law of mass action : The law stating that the rate of any given chemical reaction is proportional to the product of the activities (or concentrations) of the reactants.

Leader sequence : (1) For an mRNA, the nontranslated sequence at the 5' end of the molecule that precedes the initiation codon. **(2)** For a protein, a short, noncoding, hydrophobic sequence of amino acids from amino terminal that signals the cellular fate or destination of a newly-synthesized protein, also called *signal sequence.*

Leading strand : One of the two newly-made strands of DNA found at a replication fork. The leading strand is synthesized in the same direction in which the replication fork moves and is made by continuous synthesis in the $5' \longrightarrow 3'$ direction. Compare *lagging strand.*

Lecithin : Any of a group of phospholipids, composed of choline, phosphoric acid, fatty acids and glycerol; found in animal and plant tissues.

Leprosy : A chronic loathsome disease characterized by mutilating and disfiguring lesions, with loss of sensation in fingers and toes; caused by infection with the bacterium *Mycobacterium leprae*; transmitted by contact between an affected area in the donor and skin abrasions in the recipient, although it is not highly contagious; treatment by sulfone drugs over long periods can produce gradual improvement.

Lesion : A localized area of diseased tissue.

Lethal dose (LD) : The amount of a treatment (*e.g.*, viral inoculation, insecticide) that induces death in a laboratory animal in a standard time. Because the response to treatment is often nonlinear, it is more conventional to measure the amount of a treatment that causes 50% mortality in a standard time, the so-called LD_{50}. In some tests, the treatment dosage is fixed and the duration of the treatment period is varied, producing an estimate of lethal time and an LT_{50}.

Leucocyte : A nucleated blood cell lacking hemoglobin; includes lymphocytes, neutrophils, eosinophils, basophils and monocytes; also called *white blood corpuscle*; also spelt *leukocyte*; compare erythrocyte.

Leukaemia (*leukos*^{G} = white + *haima*^{G} = blood) **:** Disease characterized by gross excess of white blood cells in body organs and often in the blood itself. Together with lymphomas, they account for about 8% of all malignant tumours.

Leukotrienes : A family of molecules derived from arachidonate by the lipoxygenase pathway and function as local hormones, primarily to promote inflammatory and allergic reactions (such as the bronchial constriction of asthma).

Levorotatory isomer : A stereoisomer that rotates the plane of plane-polarized light to the left or anticlockwise; represented by the symbol *l* or (–).

Leydig cells : The cells of the testes that secrete testosterone when stimulated by luteinizing hormone (LH); also called *interstitial cells*.

Ligament (*ligare*^{L} = to bind) **:** A band on sheet of elastic connective tissue that links bone to bone.

Ligand (*ligare*^{L} = to bind) **:** A small molecule that binds specifically to a larger one; for example, a hormone is the ligand for its specific protein receptor. The term can also be used to mean a chemical species that forms a coordination complex with a central atom which is usually a metal atom.

Ligase : An enzyme that joins together (ligates) two molecules in an energy-dependent process. DNA ligase, for example, links two DNA molecules together through a phosphodiester bond.

Ligation : The joining of 2 DNA molecules with a covalent bond.

Light chains : The two identical short strands of the 4 polypeptide chains of an antibody molecule.

Light compensation point : Light intensity at which photosynthesis equals respiration.

Light microscope : Any optical instrument that uses light to magnify images of specimens.

Light reactions : The reactions of photosynthesis that require light and cannot occur in the dark; also known as the *light-dependent reactions*.

Light year (ly) : The distance light travels in one year or 5.9×10^{12} miles.

Lignin : A complex phenylpropanoid polymer that makes cell walls stronger, more water-proof, and more resistant to pests, herbivores, and disease organisms; in case of the xylem of plants, lignin combines with cellulose to form lignocellulose; since lignin forms an impermeable barrier, the cells are dead.

Lineage analysis : Tracing the ancestry of individual cells in a developing embryo.

Lineweaver–Burk equation : An algebraic transform of the Michaelis-Menten equation, allowing determination of V_{max} and K_m by extrapolation of [S] to infinity.

Linkage : The condition of having genes on the same chromosome (linked); alleles of genes that are linked tend to be inherited together; the greater the linkage, the lower the frequency of recombination between the two loci.

Linoleic acid ($C_{17}H_{31}COOH$) : An unsaturated fatty acid which occurs in various vegetable oils, particularly linseed and cottonseed oils; contains 2 double bonds at C_9 and C_{12}.

Linolenic acid ($C_{17}H_{29}COOH$) : A yellow oily liquid unsaturated fatty acid, b.p. 229°C, which occurs in various vegetable oils, particularly linseed oil; contains 3 double bonds at C_9, C_{12} and C_{15}; once known as *vitamin F,* its function in this capacity is discredited.

Lipase (*lipos*^{G} = fat + *-ase* = suffix, used for enzymes) **:** An enzyme that hydrolyzes fats (= triacylglycerols) to fatty acids and glycerol. Under certain conditions, it can reverse this action and synthesize fats out of glycerol and fatty acids. The main source of lipase is the pancreatic juice.

Lipid (*lipos*^{G} = fat) **:** A loosely-defined group of small biomolecules that are insoluble in water but dissolve readily in nonpolar organic solvents, and contain fatty acids, sterols, or isoprenoid compounds. Oils, such as olive and coconut, as well as waxes, such as beeswax and earwax, are all lipids. One class, the phospholipids, forms the structural basis of biological membranes.

Lipogenesis : The formation of fats from nonfatty sources.

Lipoic acid ($C_8H_{14}O_2S_2$) : A cyclic disulfide vitamin for some microorganisms; exceptional in being fat-soluble, rather than water-soluble; an intermediate carrier of hydrogen atoms and acyl groups in α-keto acid dehydrogenases.

Lipophilic : *See* **hydrophobic**.

Lipoprotein : A lipid–protein conjugate that serves to carry water-insoluble lipids in the blood. The protein component alone is an apolipoprotein.

Liposome : An artificial phospholipid bilayer vesicle formed from an aqueous suspension of phospholipid molecules.

Locus *plural* **loci** (Latin, for *place*) **:** In genetics, the position of a gene on a chromosome. Different alleles of the same gene all occupy the same locus.

Longevity : Length of life.

Low-energy phosphate compound : A phosphory-lated compound with a relatively small standard free energy of hydrolysis.

Lumen : A cavity enclosed by an epithelial sheet (in a tissue) or by a membrane (in a cell).

Luteinizing hormone (LH) : A glycoprotein hormone produced by the anterior pituitary gland; in females, it brings about ovulation on the stimulus of increasing estrogen from the ovarian tissues and causes a change in the graafian follicle to form a corpus luteum; in males, it causes androgens to be secreted by the testis; also called *interstitial cell-stimulating hormone (ICSH).*

Luteol : A xanthophyll pigment, commonly found in leaves.

Luteotrophic hormone (LTH) : A protein hormone secreted by the anterior pituitary gland of vertebrates; most versatile of all the adenohypophyseal hormones, hence jocularly called as "jack-of-all-trades"; influences the onset of lactation, stimulates progesterone secretion and also stimulates maternal behaviour in all vertebrates; also called *luteotropin* or *mammotrophic hormone (MH)* or *prolactin (PL).*

Lycopene : A carotenoid pigment found in the fruit of tomato and of other members of the family Solanaceae.

Lymph (*lympha*^{L} = clear water) **:** In animals, a colourless, plasma-like transparent fluid (95% water), widespread in the body outside the blood circulatory system; derived from blood by filtration through capillary walls in the tissues; carries lymphocytes in a special system of ducts and vessels– the lymphatic vessels.

Lymphocytes (*lympha*^{G} = water + *kytos*^{G} = hollow vessel) **:** A subclass of leucocytes involved in the immune response when activated by a foreign molecule (an antigen); *B lymphocytes* develop in the bone marrow in mammals and are responsible for the production of circulating antibodies; *T lymphocytes* develop in the thymus and are responsible for cell-mediated immunity or kill foreign and virus-infected cells.

Lymphomas : Cancers of the lymph system. Together with leukemia, they account for about 8% of all malignant tumours.

Lysergic acid diethylamide (LSD) : An hallucinogenic drug prepared from lysergic acid.

Lysis : Destruction of a cell's plasma membrane or of a bacterial cell wall, releasing the cellular contents and killing the cell.

Lysogeny : State of a bacterium in which it carries the DNA of an inactive virus integrated into its genome. The virus can subsequently be activated to replicate and lyse the cell.

Lysosome (*lysis*[G] = a loosening + *soma*[G] = a body) : A membrane-bounded organelle in the cytoplasm of eukaryotic cells, containing many digestive enzymes which are most active at the acid pH found in the lumen of lysosomes; serves as a degrading and recycling centre for unneeded components.

Lysozyme : An enzyme that breaks down bacterial cell walls and provides protection against bacterial invasion in the skin, mucus membranes and many body fluids.

M

Macromolecule (*macros*[G] = large + *moliculus*[L] = a little mass) : A large molecule having a molecular weight in the range of a few thousand to many millions; refers specifically to polysaccharides, proteins, enzymes, lipids, and nucleic acids.

Malaise : A general feeling of illness.

Malaria : (*mal' aria*[It], for *mala aria* = bad air) : A febrile disease due to infection with specific parasites.

Malic acid : A C-4 acid that is oxidized by the reduction of NAD^+ to NADH in the eighth step of the Krebs cycle.

Malignant (*malignus*[L] = of an evil nature) : Describes tumours and tumour cells that are invasive and/or able to undergo metasis; a malignant tumour is a cancer.

Maltase : An enzyme which hydrolyzes maltose to glucose. In mamnals, it is produced in the crypt of Lieberkuhn in the small intestine and is present in the succus entericus. Maltase is also present in many seeds.

Maltose ($C_{12}H_{22}O_{11}$) : A hard, crystalline, soluble, reducing sugar, made up of two molecules of glucose; less sweet than canesugar ; formed in malt by the action of the enzyme diastase on starch; also known as *maltobiose* or *malt sugar*.

Mammal : Any animal of the class Mammalia, a group often regarded as the most highly evolved animals; characterized by the presence of hair, a diaphragm used in aerial respiration and mammary glands in females; have 3 living subclasses:

1. **Monotremata–** primitive egg-laying mammals such as the duck-billed platypus and *Echidna*, the spiny anteater.
2. **Marsupialia–** which transfer their young to pouches for the latter part of their early development, such as kangaroos.
3. **Eutheria–** which have a placenta, such as cat, cow, monkey, man etc.

Mammary gland : A gland present in female mammals that produces milk used to suckle their young; usually concentrated on the underbelly beneath the pelvic girdle; in most mammals, the size of the gland is determined by the state of the estrus cycle; probably evolved from a modified sweat gland and at least two are normally present (as in humans), though in many mammals more than two are developed (as in bitch).

Marrow : The soft tissue that fills the cavities of most bones and is the source of red blood cells.

Mass : In chemistry, the total number of protons and neutrons in the nucleus of an atom, approximately equal to the atomic weight.

Matrix : The aqueous contents of a cell or organelle (the mitochondria, for example) with dissolved solutes.

Medium : A substance on or in which microorganisms and other small organisms can be cultured. A medium can be liquid or solid. If solid, it frequently contains agar, a stiffening agent extracted from seaweed. Culture media can contain all necessary nutrients and trace elements for normal growth (a minimal medium) but can also be supplemented. For example, antibiotics can be added to test for antibiotic resistance in bacteria.

Medulla (Latin, for *marrow*) : The inner portion of an organ, in contrast to the cortex or outer portion, as in the kidney or adrenal gland. Also refers to the part of the brain that controls breathing and other involuntary functions, located at the top end of the spinal cord; also called *medulla oblongata*.

Megapascal (MPa) : A unit of pressure; one million (10^6) pascals; 1 MPa = 10 atmospheres of pressure; a car tyre is typically inflated to about 0.2 MPa, whereas the water pressure in home plumbing is 0.2 – 0.3 MPa.

Melanin : A dark-brown or black pigment found in skin or hair.

Melanism : A condition in which excess dark pigment produces dark colour or blackness in scales, skin or plumage.

Melanocytes : Cells that produce the pigment melanin which is responsible for the pigmentation of skin and hair; has a protective or camouflage function; also called *melanophores*.

Melatonin (from *mela* [nin] + [sero] *tonin*) : A hormone secreted by the pineal gland, whose removal causes the ovaries to undergo hypertrophy.

Membrane : Double layer of lipid molecules and associated proteins that encloses all cells and, in eukaryotic cells, many organelles.

Membrane channel : Transmembrane protein complex that allows inorganic ions or other small molecules to diffuse passively across the lipid bilayer.

Membrane protein : Protein that is normally closely associated with a cell membrane.

Membrane transport : Movement of molecules across a membrane *via* a specific membrane protein (a transporter).

Menarche : The onset of menstruation which occurs at puberty.

Menopause : The time at which women stop ovulating, with the result that the normal menstrual cycle no longer occurs. This is normally at about 45 to 50 years of age. Also called *climacteric*.

Menstruation (*mens*^{L} = *month*) : Periodic shedding of the blood-enriched lining of the uterus with accompanying bleeding when pregnancy does not occur in female mammals.

Menstrual cycle (*mens*^{L} = month) : Monthly cycle. The term used to describe the reproductive cycle, usually occurring every twenty-eight days, in women.The menstrual cycle in the primates is the cycle of hormone-regulated changes in the condition of the uterine lining, which is marked by the periodic discharge of blood and disintegrated uterine lining through the vagina, a process called *menstruation*.

Meristem (*merizein*^{G} = to divide) : A zone of unspecialized (and organized), dividing cells whose derivatives give rise to other tissues and organs of a flowering plant. Key examples are the root apical meristem and shoot apical meristem.

Mesophyll (*mesos*^{G} = middle + *phyllon*^{G} = leaf) : The photosynthetic parenchyma of a leaf, located within the epidermis. The vascular strands (veins) run through the mesophyll.

Messenger RNA (mRNA) : A class of RNA molecules, each of which is complementary to one strand of DNA and which passes from the nucleus to the cytoplasm; carries the genetic message of genes from the chromosomes to the ribosomes in the cytoplasm, where the message is translated into the amino acid sequence of a polypeptide.

Metabolic pathway : The sequence of enzymatic reactions followed in the formation of one substrate from another.

Metabolic rate : The rate at which an organism carries out metabolism, and which is closely linked to temperature. The relationship between metabolic rate and temperture can be expressed in terms of a value called Q_{10}.

Metabolic waste : Any waste substance that is produced during the metabolism of an organism, such as nitrogen in the form of urea.

Metabolic water : Water formed by a type of metabolism called catabolism in which complex molecules are broken down to release their stored energy, with water as a by-product. In certain insects and desert mammals, which feed primarily on dry seeds, the water conservation mechanisms are so efficient that metabolic water alone is sufficient to replace the normal water loss; 'free' water is not required in the diet.

Metabolism (*metabole*[G] = change) **:** The entire set of enzyme-catalyzed transformations of organic molecules in living cells; the sum total of all the processes involved in the building up and tearing down of protoplasm; the synthetic or building up processes constitute the anabolic phase of metabolism and the degradative or tearing down processes, the catabolic phase. A special energy-carrying molecule called ATP is involved in these processes.

Metabolite : A chemical intermediate in the enzyme-catalyzed reactions of metabolism.

Metalloprotein : A protein having a metal ion as its prosthetic group.

Metaplasia : The transformation of a tissue to another form.

Metaplasm : The lifeless constituents of protoplasm.

Metastasis, *plural* **metastases** (Greek, for *to place in another way*) **:** Spread of cancer cells from their site of origin to other sites in the body, forming new tumours there.

Methyl ($-CH_3$) : A hydrophobic chemical group derived from methane (CH_4).

Micelle : An aggregate of amphipathic molecules in water, with the nonpolar portions in the interior and the polar portions at the exterior surface, exposed to water.

Michaelis–Menten constant (K_m) : The substrate concentration (in moles/litre) at which the enzyme-catalyzed reactions proceed at one-half its maximum velocity.

Michaelis–Menten equation : The equation describing the hyperbolic dependence of the initial reaction velocity, V_0, on substrate concentration, $[S]$, in many enzyme-catalyzed reactions:

$$V_0 = \frac{V_{max}[S]}{K_m + [S]}$$

Michaelis–Menten kinetics : A kinetic pattern in which the initial rate of an enzyme-catalyzed reaction exhibits a hyperbolic dependence on substrate concentration.

Microbodies : Cytoplasmic, single-membrane-bounded vesicles, containing peroxide-forming and peroxide-destroying enzymes; generally derived from endoplasmic reticulum; includes lysosomes, peroxisomes and glyoxysomes.

Microsomes : Membranous vesicles formed by fragmentation of the endoplasmic reticulum of eukaryotic cells; recovered by differential centrifugation.

Mil : One thousandth (1/1,000) of an inch.

Milk : (1) A whitish, highly nutritive fluid secreted by the mammary glands in mammals which serves to nourish the young. **(2)** Any white fluid, such as coconut milk

Mimicry (*mimos*[G] = mime) **:** The resemblance in form, colour, or behaviour of certain organisms (mimics) to other more powerful (and hence less subject to predation) or more protected ones (models), which results in the mimics being protected in some way.

Miscarriage : The expulsion of a foetus before it is viable outside the womb.

Missense mutation : A mutation that alters a DNA codon so as to cause one amino acid in a protein to be replaced by a different one.

Mitochondrion, *plural* **mitochondria** (*mitos*[G] = thread + *chondrion*[G] = small grain) **:** A membrane-bounded, tubular or sausage-shaped organelle in the cytoplasm of eukaryotes; closely resemble the aerobic bacteria from which they were originally derived; contains the enzymes required for the citric acid cycle, fatty acid oxidation, electron transfer, and oxidative phosphorylation; acts as chemical furnaces of the cell and produces most of the ATP in eukaryotic cells; also called *chondriosome* or in common parlance as 'powerhouse' of the cell.

Mixed-function oxidase : An oxygenase enzyme that catalyzes a reaction in which two different substrates are oxidized, one by the addition of an oxygen atom from O_2 and the other by supplying two hydrogen atoms to reduce the remaining oxygen atom to H_2O.

Modulator : A metabolite that, when bound to the allosteric site of an enzyme, alters its kinetic characteristics.

Module : With reference to a protein or nucleic acid, a unit of structure or function that is used in a variety of different contexts.

Molar solution : One mole of solute dissolved in water to give a total volume of 1,000 mL.

Mole : A mole is the amount of a substance, in grams, that equals the combined atomic mass of all the atoms in a molecule of that substance.

Mole, mol (*moles*[L] = mass) **:** The atomic weight of a substance, expressed in grams. One mole of any substance is defined as the mass of 6.0222×10^{23} atoms.

Molecular genetics : Field of genetics devoted to the study of the biochemical mechanisms by which heredity information is stored in nucleic acid and transmitted to proteins.

Molecular weight (Mol. wt. or MW) : Numerically, the same as the relative molecular mass of a molecule,expressed in daltons ; alternatively, the sum of the atomic weight of the atoms in a molecule.

Molecule (*moliculus*[L] = a small mass) **:** A group of atoms joined together by covalent bonds; the smallest unit of a compound that displays the properties of that compound and cannot be further subdivided without losing the quality of the compound.

Monoamine : An amine containing one amino group especially one that functions as a neurotransmitter. Acetylcholine and norepinephrine are monoamines.

Monochromatic light : Light consisting of vibrations of the same or nearly the same frequency; light of one colour.

Monocotyledon (monocot) : A class of flowering plants characterized by having one cotyledon in their embryo, leaves with parallel venation and flower parts often in threes. Compare *dicotyledon*.

Monoecious (*monos*[G] = one + *eikos*[G] = house, dwelling) **:** Having male and female flowers on the same individual.

Monomer : Small molecular building block that can be linked to others of the same type to form a larger molecule (a polymer).

Monoprotic acid : An acid having only one dissociable proton.

Monosaccharide (*monos*[L] = one + *sakcharon*[G] = sugar) **:** A simple sugar that cannot be broken down by hydrolysis and having the general formula $(CH_2O)_n$, where n = 3 to 7; glucose and fructose are examples of monosaccharide. Monosaccharides are usually white, crystalline solids with a sweet taste and are generally soluble in water.

Monosodium glutamate (MSG), $HOOC.(CH_2)_2\ CH(NH_2).COONA$: A white soluble crystalline substance; used to intensify the flavour of foods; also called *sodium hydrogen glutamate* or *'taste powder'*.

Monoterpene : A compound that consists of two isoprene units linked together; menthol is an example of monoterpene.

Morphine ($C_{17}H_{19}O_3N$) : A white, crystalline, narcotic alkaloid drug obtained from opium (*Papaver somniferum*).

Morphogenesis (*morphe*[G] = form + *genesis*[G] = origin) **:** The development of form in the growth of an individual; the growth and differentiation of cells and tissues during development.

Mosaic : In genetics, an organism made of a mixture of cells with different genotypes.

Motif : An element of structure or pattern that recurs in many contexts ; specifically, a small structural domain that can be recognized in a variety of proteins.

Mucopolysaccharide : An older name for a *glycosaminoglycan*.

Mucoprotein : A complex of protein with polysaccharide.

Mucosa : Any epithelium that secretes mucus, such as the mucous membrane lining the alimentary canal.

Mucous membrane : The lining of the gut system and the urinogenital system of animals, consisting largely of moist epithelium overlying connective tissue. It is so called because of the presence in the epithelium of goblet cells that secrete mucus.

Mucus (*mucus*[L] = snivel) **:** Jellylike viscous, slimy substance secreted by mucous membranes.

Multienzyme system : A group of related enzymes participating in a given metabolic pathway.

Murein : The cross-linked mucopeptides that form the rigid framework of bacterial cell walls.

Muscle (*musculus*[L] = mouse) **:** The tissue in the body of humans and animals, made up of highly contractile cells, that can be contracted and relaxed to make the body move.

Muscular dystrophy (MD) : A disease characterized by the progressive wasting of muscles and eventual death. One type, called *Duchenne MD*, is controlled by a recessive sex-linked gene and consequently affects more boys than girls. The disease first shows itself between 1 and 6 years, progressing until the patient is confined to a wheelchair by the early teens, with death resulting by the late teens in most affected individuals. Other forms of MD are controlled by autosomal genes (both dominant and recessive) and hence they are equally frequent in boys and girls.

Mustard gas, $(CH_2CH_2Cl)_2S$: An oily liquid that has been used as a 'war gas'; destroyed by oxidizing agents, *e.g.*, bleaching powder; synonymous with *yperite* or *dichlorodiethyl sulfide*.

Mutagen (*mutare*[L] = to change) **:** A chemical capable of damaging DNA, thus causing mutation.

Mutant (*mutare*[L]= to change) **:** A mutated gene. An organism carrying a gene that has undergone a mutation.

Mutarotation : The change in specific rotation of a pyranose or furanose sugar or glycoside accompanying the equilibration of its α- and β- anomeric forms.

Mutases : Enzymes that catalyze the transposition of functional groups.

Mutation (*mutare*[L] = to change) **:** Heritable changes in the nucleotide sequence of genomic DNA (or genomic RNA, in the case of an RNA virus).

Myeloma : A cancerous tumour of the bone marrow.

Myoglobin : A relatively small (MW = 16,700), oxygen-binding, globular protein consisting of 153 amino acid residues in a polypeptide chain and one heme group; has molecular dimensions 45 × 35 × 25 Å; found in the muscles of vertebrates and some invertebrates (giving the muscles a deep red-brown colour due to the presence of iron-porphyrin or heme group); has a high affinity for oxygen.

Myosin (*myos*[G] = muscle + *in*[G] = belonging to) **:** A contractile protein; one of the two protein components of myofilaments (the other is actin). *Myosin II* is a very large protein that forms the thick filament of skeletal muscle that slide over actin filaments during contraction. *Myosin* I is smaller, more widely distributed, and not assembled into filaments.

Myxedema (*myxa*[G] = slime + *oidema*[G] = swelling **:** A hypothyroidal disease, characterized by an abnormally low basal metabolic rate (BMR), deposition of a semifluid material under the skin, an increase in subcutaneous fat and mental and physical sluggishness; also spelt as *myxoedema*.

N

Nail : A horny, keratinized layer protecting the distal end of each finger and toe in humans and most other primates; in other terrestrial vertebrates, the nail is shaped into a claw.

Native conformation : The biologically-active conformation of a macromolecule.

Nausea (*nausia*[G] = sea-sickness, from *naus* = a ship) **:** Feeling of sickness.

Nectar (Greek, *nektar*) **:** A sweet liquid rich in sugar and amino acids and found in many flowers; attracts insects and birds that bring about pollination.

Negative feedback : Regulation of a biochemical pathway achieved when a reaction product inhibits an earlier step in the pathway.

Neonatal : (of newborn offspring, particularly human) the first month of independent life.

Neontology : The study of life of recent living organisms. Compare *palaeontology*.

Neurohormones : Substances that are secreted from neurons and modulate the behaviour of target cells, which are often other neurons; unlike neurotransmitters, they do not act strictly across a synapse; most neurohormones are peptides.

Neuron (Greek, for *nerve*) **:** A nerve cell with long processes specialized to receive, conduct, and transmit signals in the nervous system; also spelt as *neurone*.

Neurotransmitter (*neuron*^{G} = nerve + *trans*^{G} = acroos + *mitere*^{G} = to send) **:** A small signaling compound (usually containing nitrogen) secreted from the terminal of a neuron and bound by a specific receptor protein in the next neuron; serves to transmit a nerve impulse; examples include acetylcholine, glutamate, GABA, glycine, and many neuropeptides.

Neutron (*neuter*^{L} = neither) **:** A subatomic particle located within the nucleus of an atom; similar to a proton in mass, but as its name implies a neutron is neuter and possesses no charge.

Nick : A single-stranded cut or break in a DNA molecule; nicking of DNA may form part of a DNA repair mechanism, as occurs after damage caused by ultraviolet light.

Nicotinamide adenine dinucleotide (NAD^+) : A coenzyme derived from the vitamin B_5 (nicotinic acid); participates in an oxidation reaction by accepting a hydride (H^-) ion from a donor molecule and the NADH, thus formed, is an important carrier of electrons for oxidative phosphorylation; formerly called as *diphosphopyridine nucleotide (DPN)*.

Nicotinamide adenine dinucleotide phosphate (NADP) : A coenzyme closely related to NAD^+; functions as an electron donor in many of the reduction reactions of biosynthesis, thus used extensively in biosynthetic, rather than catabolic, pathways; $NADP^+$ is the oxidized form of NADP, and $NADPH_2$ is the reduced form; formerly called as *triphosphopyridine nucleotide (TPN)*.

Nicotine : An alkaloid derived from tobacco.

Nipple : An conical projection at the centre of a mammary gland on which the milk ducts have outlets.

Nicotinic acid (C_6H5O_2N) : A water-soluble vitamin of the B complex group and designated vitamin B_5; widely distributed in nature in plant and animal tissues, mainly as its amide called *nicotinamide*; most abundantly found in yeast, although liver, meat and poultry are also good sources; milk and eggs are usually devoid of nicotinic acid; simplest in structure of all the known vitamins; a white crystalline substance, soluble in ethyl alcohol and stable in air and heat; acts as a constituent in two pyrimidine nucleotide coenzymes, NAD and NADP; avitaminosis B_5 leads to pellagra in man, which is characterized by 3 "Ds", namely dermatitis of the exposed parts, diarrhea and dementia; also called as *pellagra preventive (PP) factor* or *antiblacktongue factor*.

Ninhydrin reaction : A colour reaction given by amino acids and peptides on heating with ninhydrin; widely used for their detection and estimation.

Nitrification : The oxidation of ammonia and ammonium compounds into nitrites and nitrates through the action of nitrifying bacteria in the soil.

Nitrogen cycle : The cycling of various forms of biologically-available nitrogen through the plant, animal, and microbial worlds, and through the atmosphere and geosphere. Alternatively, the natural circulation of nitrogen between organic molecules in living organisms and inorganic molecules in the soil.

Nitrogen fixation : A biochemical process that reduces atmospheric nitrogen (N_2) to ammonia (NH_3) and hence into various nitrogen-containing metabolites by certain free-living and symbiotic bacteria; the process fixes the free, nonavailable, atmospheric nitrogen into nitrogen compounds and makes it available to the plant in its combined form.

Nitrogenase complex : An enzyme complex that converts (reduces) atmospheric nitrogen gas into ammonia in the presence of ATP.

Nocturnal (*nocturnus*^{L} = night) **:** Active primarily at night.

Node (*nodus*^{L} = knot) **:** The part of a stem where leaves and axillary buds arise.

Noncompetitive inhibition : A form of enzyme control in which the enzyme has 2 kinds of active site, one for an inhibitor, the other for the enzyme substrate. The inhibitor prevents catalytic activity of the enzyme by licking or even changing the shape of the substrate active site. The enzyme inhibition is not reversed by increasing the substrate concentration. Compare *competitive inhibition.*

Noncovalent bond : A chemical bond in which, in contrast to a covalent bond, no electrons are shared. Noncovalent bonds are relatively weak but they can sum together to produce strong, highly specific interactions between molecules.

Nonessential amino acids : Amino acids that can be made by humans and other vetebrates from simple precursors, and are thus not required in the diet.

Nonheme iron proteins : Proteins containing iron but no porphyrin groups; usually acting in oxidation–reduction reactions.

Nonhistone proteins : Proteins associated with chromosomes that are not histones. For example, they may be enzymes involved in DNA replication or structural components of the chromatin.

Nonpolar molecules : Molecules or groups that are poorly soluble in water (*i.e.*, hydrophobic); lack any asymmetric accumulation of positive or negative charge; also called as *apolar molecules.*

Nonreducing sugar : A sugar in which the carbonyl (anomeric) carbon is involved in the formation of a glycosidic bond and hence cannot undergo oxidation, such as sucrose, glycogen and inulin.

Nonsense codon : A codon that does not specify an amino acid, but signals the termination of a peptide chain; the three nonsense codons are: UAA, UAG and UGA; also called *stop codon* or *termination codon.*

Nonsense mutation : A mutation that creates an abnormal stop codon and thus causes translation to terminate prematurely; the resulting truncated protein is usually nonfunctional.

Nor : (1) A combining form of *normal.* **(2)** The prefix is also used to indicate the loss of a *methyl group, e.g.*, noradrenaline or the loss of a *methylene group* from a chain.

Noradrenalin : A transmitter substance produce at the nerve endings of adrenergic nerves giving effects very similar to those of adrenalin. After transmission, it is inactivated by monoaminooxidase in order to prevent a build-up. The action of noradrenalin can be inhibited by various drugs, such as mescaline, which produce hallucinatory effects.

Nuclear envelope : The double membrane surrounding the nucleus of eukaryotes; consists of outer and inner membranes perforated by nuclear pores.

Nuclear pore : Shallow depressions, like the craters of the moon, that are scattered over the surface of the nuclear envelope. Such pores contain many embedded proteins that act as molecular channels, permitting selected molecules to pass into and out of the nucleus.

Nucleases : Enzymes that hydrolyze the inter-nucleotide (phosphodiester) linkages of nucleic acids.

Nucleic acids : Biologically-occurring polynucleotides in which the nucleotide residues are linked in a specific sequence by phosphodiester bonds. The two types of nucleic acids are deoxyribonucleic acid (DNA) which is double-stranded, and ribonucleic acid (RNA) which is typically single-stranded.

Nucleoid : The large circular DNA molecule of a prokaryotic cell, along with its associated proteins; is supercoiled and forms a dense mass within the cell, and the term nucleoid often refers to the cell region occupied by this mass, but has no surrounding membrane; also sometimes called *bacterial chromosome*.

Nucleolus, *plural* **nucleoli** (Latin, for *small nucleus*) **:** A densely staining rounded structure in the nucleus of eukaryotic cells; a nucleus may contain one or more nucleoli; involved in ribosomal RNA (rRNA) synthesis and ribosome formation.

Nucleophile : An electron-rich group with a strong tendency to donate electrons to an electron-deficient nucleus (electrophile); the entering reactant in a bimolecular substitution reaction.

Nucleoplasm : The portion of a cell's contents enclosed by the nuclear membrane; also called as *nuclear matrix*.

Nucleoprotein : A compound of one or more proteins with a nucleic acid.

Nucleoside : A compound composed of a purine or pyrimidine base covalently linked to a pentose sugar, either a ribose (*ribonucleoside*) or a deoxyribose (*deoxyribonucleoside*).

Nucleosome (*nucleus*^{L} = kernel + *soma*^{L} = body) **:** The structural, beedlike packaging unit of a eukaryotic chromosome, composed of a short DNA strand wrapped around a core of histone proteins; the fundamental subunit of chromatin.

Nucleotide : A monomeric unit of nucleic acids, composed of a phosphate, a pentose sugar (ribose or deoxyribose) and a nitrogenous base (purine or pyrimidine); DNA and RNA are polymers of nucleotides. Alternatively, a nucleotide is a nucleoside with one or more phosphate groups joined in ester linkages to the sugar moiety.

Nucleus, *plural* **nuclei** (Latin, for *a kernel*, dim. fr. *nux* = nut) **:** In a eukaryotic cell, a prominent, membrane-bounded organelle that contains chromosomes; the nucleus is the repository of the genetic information that directs all activities of a living cell. In atoms, the central core, containing positively-charged protons and (in all but hydrogen) electrically-neutral neutrons.

Numerator : The number above the line in a vulgar fraction, *e.g.*, 5 in 5/16.

Nutrient : A substance derived from outside a cell and providing it with energy and structural material.

O

Occam's razor : *See* **Ockham's razor.**

Ockham's razor : (named after **William of Ockham**, died circa 1349) **:** A principle of logic that holds that the best explanation of an event is the simplest, using the fewest assumptions of hypothesis; also spelt as *Occam's razor*.

Ocytocin (*ocy*^{G} = quick + *tokos*^{G} = birth) **:** A hormone of the posterior pituitary gland that accelerates uterine contractions during menstruation and childbirth and stimulates lactation. Also spelt *oxytocin*.

Oestradiol : *See* **estradiol.**

Oestrogen : *See* **estrogen.**

Oestrous cycle : *See* **estrous cycle.**

Okazaki filaments (after **R. Okazaki**, the discoverer) **:** Short lengths of DNA produced on the lagging strand during DNA replication; usually about 1,000 –1,200 bp in length; they are rapidly joined by DNA ligase to form a continuous DNA strand.

Oleic acid, $C_{17}H_{33}COOH$ (*oleum*^{G} = oil) **:** A monoethenoid, oily liquid found in almost all natural fats; used in cosmetics, soaps and lubricating oils.

Oligomer (*oligos*^{G} = few, little) **:** A short polymer, usually consisting (in a cell) of amino acids (*oligopeptide*), sugars (*oligosaccharide*), or nucleotides (*oligonucleotide*); the definition of "short" is somewhat arbitrary, but usually less than 50 subunits.

Oligomeric proteins : A multisubunit protein having two or more identical polypeptide chains.

Oligonucleotide : A short polymer of nucleotides (usually less than 50) joined by phosphodiester bonds.

Oligopeptide : A few amino acids joined by peptide bonds.

Oligosaccharide : A carbohydrate composed of a small number of monosaccharide residues.

Oncogene (*oncos*[G] = tumour) **:** A cancer-causing gene; any of several mutant genes that cause cells to exhibit rapid and uncontrolled proliferation.

Oncogene theory (*oncos*[G] = tumour) **:** The hypothesis that cancer results from the action of a specific tumour-inducing *onc* gene.

One-gene/one-enzyme hypothesis : The hypothesis that genes produce their effects by specifying the structure of enzymes and that each gene encodes the structure of a single enzyme; put forward by George Beadle and Edward Tatum in 1941.

Ontogeny : The whole of the development of an organism from fertilization to the completion of life history.

Open reading frame (ORF) : A group of contiguous nonoverlapping nucleotide codons in a DNA or RNA molecule between an initiation codon and a termination codon. It represents the coding sequence for a polypeptide.

Open system : A system that exchanges matter and energy with its surroundings.

Operator : A short region of DNA in a bacterial chromosome that interacts with a repressor protein to control the expression (transcription) of adjaent gene(s).

Operon (*operis*[L] = work) **:** In bacterial chromosome, a group of contiguous functionally-related genes that are transcribed into a single mRNA molecule; although a common mode of gene regulation in prokaryotes, it is rare in eukaryotes except fungi.

Opiate : A narcotic substance derived from opium (*Papaver somniferum*).

Opsin : A protein that occurs in rodes and cones of the retina of the eye, which combines with retinal$_1$ or retinal$_2$ to form visual pigments.

Optical activity : The capacity of a substance to rotate the plane of plane-polarized light.

Optimum pH : The characteristic pH at which an enzyme has maximal catalytic activity.

Organ (*organon*[L] = tool) **:** A complex body structure, composed of one or more kinds of tissues and acting as a structural and functional unit, such as the liver, eye or leaf.

Organelle (*organella*[G] = little tool) **:** Membrane-bounded structures (except ribosomes) found in eukaryotic cells; nuclei, chloroplasts, mitochondria and ribosomes are all organelles; contain enzymes and other components required for specialized cell functions.

Organic compound : A type of molecule that contains carbon except carbon dioxide and the carbonates that have a mineral origin.

Organism : Any individual living creature; either unicellular or multicellular plant or animal.

Organogenesis : The period during embryonic development of an organism when the main body organs are formed.

Organophosphate : A phosphate with insecticidal properties; probably the most commonly used group of insecticides.

Ornithine cycle : *See* **urea cycle**.

Osmic acid : *See* **osmium tetroxide**.

Osmium tetroxide (OsO_4) : An inorganic compound, used as a fixative for cytological preparations that is characterized by the small amount of distortion it causes; also called *osmic acid.*

Osmoregulation : The maintenance of a constant internal solute concentration by an organism, regardless of the environment in which it lives.

Osmosis (*osmos*[G] = act of pushing, thrust) **:** The diffusion of water (or other solvent) through a differentially permeable membrane-a membrane that permits the free passage of water but not that of one or more solutes-into another aqueous compartment containing solute at a higher concentration.

Osmotic pressure : Pressure generated by the osmotic flow of water through a differentially permeable membrane into an aqueous compartment containing solute at a higher concentration; osmotic pressure is an indicator of how concentrated a solution is on the other side of a membrane from pure water; a molar solution of any nonelectrolyte has an osmotic pressure of 22.4 atm.

Osteoblast (*osteon*^{G} = bone + *blastos*^{G} = bud) : A bone-forming cell.

Osteocyte (*osteon*^{G} = bone + *kytos*^{G} = hollow vessel) : A mature osteoblast.

Ovary (*ovum*^{L} = egg) : (1) In animals, the organ that produces eggs. (2) In flowering plants, the enlarged, basal portion of a carpel (or of a cluster of fused carpels) which contains the ovule(s); the ovary matures to become the fruit.

Ovulation : The successful development and release of an egg by the ovary which then normally passes into the oviduct.

Ovule (*ovulum*^{L} = a little egg) : A structure in a seed plant that becomes a seed when mature.

Ovum, *plural* **ova** (Latin, for *egg*) : A mature egg cell; a female gamete. The human egg is about 0.14 mm in diameter, which is some 50,000 times larger than the human sperm.

Oxaloacetic acid (OAA) : A C-4 organic acid that is converted to citric acid by the addition of an acetyl group in the first step of the Krebs cycle; OAA is also the product of the CO_2 fixation of phosphoenolpyruvic acid in C_4 and CAM photosynthesis.

Oxidase : Any of a group of enzymes that promote the oxidation of a substrate with oxygen as the electron acceptor. Most oxidases are proteins with metallic groups attached.

Oxidation (*oxider*^{Fr} = to oxidize) : The loss of an electron during a chemical reaction from one atom to another; occurs simultaneously with reduction; the second stage of the ten reactions of glycolysis; opposite of *reduction*.

Oxidation–reduction reaction : A reaction in which electrons are transferred from a donor to an acceptor molecule; also called a *redox reaction*.

Oxidative phosphorylation : Phosphorylation of ADP to ATP that uses energy from a proton pump fuelled by the electron transport system. The process is the major means by which aerobic organisms obtain their energy from foodstuffs.

Oxidizing agent : The acceptor of electrons in an oxidation–reduction reaction; also called *oxidant*.

Oxidoreductase : One of a group of enzymes that catalyzes oxidation–reduction reactions.

Oxygenases : Enzymes that catalyze the incorporation of oxygen into a substrate.

Ozone (O_3) : A form of oxygen in the stratosphere when compared with ordinary oxygen (O_2), more effectively shields living organisms from intense ultraviolet radiation.

P

Pain : A conscious sensation produced in the brain and stimulated by pain receptors in, for example, the skin; pain has a protective function and often produces a reflex action in response.

Palaeontology : The study of fossil animals and plants. Compare *neontology*.

Palindrome : A segment of duplex DNA with a nucleotide sequence that is identical to its complementary strand when each is read in the same chemical direction; for example, TGAC in the following DNA segment:

5' XXX TGAC XXX 3'
3' XXX CAGT XXX 5'

Palisade parenchyma (*palus*^{L} = stake; *para*^{G} = beside + *en*^{G} = in + *chein*^{G} = to pour) : Parenchyma cells that are chlorophyllous and columnar, closely packed together, and located between the upper epidermis and the spongy parenchyma of a leaf.

Pancreas (*pan*^{G} = all + *kreas*^{G} = flesh) : In vertebrates, the large, principal digestive gland situated between the stomach and the small intestine that secretes a host of digestive enzymes and the hormones, insulin and glucagon.

Pancreozymin (PZ) : A polypeptide hormone elaborated by the intestinal mucosa; contains 33 amino acid residues; PZ is thermostable but alkali-labile; along with secretin, it functions to stimulate the release of pancreatic juice, which is rich in enzymes or their zymogens as well as in bicarbonate. Jorpes (1968) believes that the two gastrointestinal hormones, cholecystokinin (CCK) and pancreozymin (PZ), are identical. They are now believed to be a single factor and represented as CCK-PZ.

Pantothenic acid ($C_9H_{17}O_5N$) : A water-soluble vitamin of the B complex group and designated as vitamin B_3; although widespread in nature (hence so named, *pantos*[G] = everywhere), yeast, liver and eggs are richest sources; in most animals and microbes, it occurs as its coenzyme; participates in the formation of coenzyme A, which when bonded to acetic acid, forms acetyl-coenzyme A (acetyl-CoA); a deficiency of pantothenic acid causes adrenal cortical insufficiency and achromotrichia in man; also called *filtrate factor* or *yeast factor*.

Papain : A proteolytic enzyme obtained from the juice of the green fruit of the papaya or papaw tree (*Carica papaw*).

Paradigm : In biochemustry, an experimental model or example.

Paralysis (*paralyein*[G] = to loose from the side, to disable) : Loss of muscular power from a lesion of the nervous system.

Parasite (*para*[G] = beside + *sitos*[G] = food) : A symbiotic relationship in which one organism benefits and the other is harmed.

Parathyroid hormone, PTH (*para*[G] = beside + thyroid + hormone) : A hormone produced by the parathyroid glands that regulates the way the body uses calcium.

Parthenocarpy (*parthenos*[G] = virgin + *carpos*[G] = fruit) : The development of fruit without seeds due to the lack of pollination, fertilization or embryo development.

Parthenogenesis (*parthenos*[G] = virgin + *genesis*[Eng] = beginning) : The development of an adult from an unfertilized egg; a common form of reproduction in insects and some fruits.

Partition coefficient : A constant that expresses the ratio in which a given solute will be partitioned or distributed between two given immiscible liquids at equilibrium.

Parturition : The act of giving birth to an offspring.

Pascal (Pa) : The pressure unit (*i.e.*, energy per unit volume) used to measure water potential; one pascal equals the force of one newton per square meter; one atmosphere of pressure equals 1.0×10^5 Pa.

Passive transport : The movement of a substance across a biological membrane by molecular diffusion through the lipid bilayer; also called *passive diffusion*. Compare *active transport, facilitated transport*.

Pathogen (*pathos*[G] = suffering + *genesis*[Eng] = beginning) : An organism or other agent that causes disease, such as viruses or bacteria.

Pectin (*pektos*[G] = curdled, congealed, *pegnynai*[Fr] = to make fast or stiff) : A gluey polysaccharide that holds cellulose fibrils together; pectins are mostly polymers of galacturonic acid monomers with α-1,4 linkages. When heated, pectin forms a gel which can 'set', a feature used in the making of jams.

Penicillin : An antibiotic obtained from the fungus *Penicillium* that is toxic to a number of bacteria, both pathogenic and nonpathogenic; discovered by **Sir Alexander Fleming** in 1928.

Penis (Latin, for *tail*) : The male organ of copulation which conveys the sperm to the genital tract of the female; in mammals, it is also the male urinary organ; also called *phallus*.

Pentose : A simple sugar with a backbone containing five carbon atoms.

Pentose phosphate pathway (PPP) : A pathway that serves to interconvert hexoses and pentoses and is a source of reducing equivalents and pentoses for biosynthetic processes; present in most organisms; during this pathway, NADP is reduced fo NADPH, but no ATP is produced; also called the *phosphogluconate pathway* or *hexose monophosphate shunt*.

Pepsin : A proteolytic enzyme secreted in the inactive form (pepsinogen) by chief or peptic cells in the gastric pits of the stomach of vertebrates, and which breaks down proteins in acid solution into short polypeptide chains which are subsequently broken down further by peptidases.

Peptidase : An enzyme present in pancreatic juice that releases free amino acids from polypeptide chains. *Exopeptidases* split off terminal amino acids, whereas *endopeptidases* split links with the chain.

Peptide (*peptein*^{G} = to soften, digest) **:** Two or more amino acids covalently joined by peptide bonds.

Peptide bond : A substituted amide linkage between the α-amino group of one amino acid and the α-carboxyl group of another, with the elimination of a molecule of water.

Peptide hormone : A hormone that interacts with a receptor on the cell surface and initiates a chain of events within the cell by increasing the levels of secondary messengers.

Peptidoglycan : A major component of bacterial cell walls; generally consists of parallel heteropolysaccharides cross-linked by short peptides.

Peptidyl site : A tRNA binding site on the ribosome at which a peptide bond occurs.

Peptidyl transferase : An enzyme in the large ribosomal subunit that catalyzes the formation of a peptide bond between the amino acid at the end of a growing polypeptide and the next amino acid to be added to the chain.

Peptone : Any of a class of initial, water-soluble decomposition products of protein digestion; noncoagulable by heat and not capable of being precipitated with saturated ammonium sulfate.

Perennial (*per*^{L} = through + *annus*^{L} = a year) **:** A plant continuing to live from year to year; one that lives more than two years.

Perfusion : The passage of a liquid through an organ or tissue.

Peripheral nervous system (*peripherein*^{G} = to carry around) **:** All of the neurons and nerve fibres outside the central nervous system, including motor neurons, sensory neurons, and the autonomic nervous system.

Peristalsis (*peri*^{G} = around + *stellein*^{G} = to wrap) **:** The alternate contraction and relaxation of circular and longitudinal muscle which produces waves that pass along the intestine (and other tubular systems) of animals, moving the tube contents in one direction.

Permeases : *See* **transporters**.

Pernicious anemia : A servere condition in which there is a progressive decrease in the number of red blood cells together with an increase in their size, producing poor colour, weakness and gut disorders. The disease can be fatal but may be treated by dosing with vitamin B_{12}.

Peroxidase : An enzyme involved in the transfer of oxygen from peroxides to substances to be oxidized, such as catalase.

Peroxisome : A small, membrane-bounded spherical organelle in the cytoplasm of eukaryotic cells, and derived from smooth endoplasmic reticulum; contains peroxide-forming and peroxide-destroying enzymes.

Pesticide : Any agent that causes the death of a pest. The general definition is usually restricted to chemicals with pesticidal properties such as herbicides, insecticides, acaricides and fungicides. Pesticide application can produce many problems such as (*a*) destruction of organisms useful to man and (*b*) directly harmful effects to man, if used incorrectly.

pH (p stands for power of 10 and *H* stands for hydrogen ion concentration) **:** The number of grams of hydrogen ions per litre of solution. It is useful as a measure of the *acidity* of a solution and in this context is usually expressed in terms of pH = $\log_{10}$

$1/[H^+]$, where $[H^+]$ is the hydrogen ion concentration. Alternatively, pH is also defined as the negative logarithm of the hydrogen ion concentration of an aqueous solution. As pure water at ordinary temperature (of 25°C) dissociates slightly into hydrogen ions and hydroxyl ions ($H_2O = H^+ + OH^-$), the concentration of each type of ion being 10^{-7} mole per litre, the pH of pure water will therefore be $\log_{10} 1/10^{-7} = 7$; this figure is accordingly taken to represent neutrality on the pH scale. If acid is added to water, its hydrogen ion concentration will increase and its pH will therefore decrease. Thus, a pH below 7 indicates acidity and similarly a pH above 7 indicates alkalinity. In fact, the pH values extend from 0 to 14 : the lower the pH value, the higher the acidity or the more hydrogens the solution contains; 0 is maximum acidity, 14 is maximum alkalinity and 7 is neutrality. Since the pH is logarithmic, each of 1 pH unit means a 10-fold change in the number of hydrogen ions. The pH can be measured by using indicators which change colour with changing pH, or by electrical means using a pH meter. The term pH was introduced by a Danish biochemist, **Soren Sörensen** in 1909.

pH buffers : Chemical substances that maintain a constant pH in a solution.

Phage : *See* **bacteriophage.**

Phagocyte (*phagein*^G = to eat + *kytos*^G = hollow vessel) **:** A general term for a cell that kills invading cells by engulfing them; includes macrophages and neutrophils; capable of discriminating between different particles, for example phagocytic white blood cells will engulf only certain bacteria.

Phagocytosis (*phagein*^G = to eat + *kytos*^G = hollow vessel) **:** A form of endocytosis in which cells engulf organisms or fragments of organisms; prominent in carnivorous cells such as *Amoeba proteus*, and in vertebrate macrophages and neutrophils.

Phallus : *See* **penis.**

Phasic : Transient.

Phenocopy : Situation that occurs when an environment agent induces a phenotype that resembles a particular mutant phenotype.

Phenolic : Any compound that contains a fully unsaturated six-carbon ring that is linked to an oxygen-containing side group.

Phenotype (*phainein*^G = to show + *typos*^G = stamp or print) **:** The observable characteristics of a cell or an organism, either individually or collectively; the phenotype results from the interaction of the genotype of an organism with its environment. The interaction is that between nature and nurture. Variations due to nature are the inherited aspects of the organism, the genotype, while nurture denotes the (usually not inherited) effects of the environment upon the organism. Compare *genotype.*

Phenylketonuria (PKU) : An inborn error of metabolism in humans in which there is the inability to convert phenylalanine to tyrosine, due to the absence of a functional *phenylalanine hydroxylase* enzyme. The condition is controlled by the recessive allele of an autosomal gene present on chromosome 1. The effects of PKU are many and extremely serious. Most of the phenylketonurics (about 90%) have an IQ of less than 40 and thus are severely *mentally retarded* because of an abnormal accumulation of phenylalanine in their blood and other tissues which adversely affects their nervous system. There is *microcephally* (reduced brain size) in about 65% of the affected individuals and about 75% show '*tailor posture*', in which muscular hypertonicity causes contraction of the leg and arm muscles so that individuals sit cross-legged with their arms drawn into the body. The disease can be treated if discovered in very early childhood. Treatment consists of a diet low in phenylalanine (as some of this amino acid must be present in the diet since it is an essential amino acid).

Phenylpropanoid : A complex phenolic that has a three-carbon side chain; generally derived from the amino acids phenylalanine and tyrosine; myristicin, the main flavour ingredient of nutmeg, is a phenylpropanoid.

Pheromone (*phereinG* = to carry + [hor]*mone*) : A chemical substance used in communication between organisms of the same species, especially with regard to their reproductive readiness; found mainly in animals but they occur in some lower plant groups where a chemical is secreted into water by female gametes to attract male gametes; in animals, for example, pheromones are transmitted in the air, as in female emperor and eggar moths, which secrete a chemical that is attractive to males over long distances, or by a dog marking out his territory with urine; insect pheromones have been used to trap females of serious pests.

Phloem (*phloosG* = bark) : In vascular plants, a specialized complex tissue, basically composed of sieve elements, companion cells, parenchyma cells, fibres and sclereids; transports nutrients from the leaves downward in the plant.

Phosgene ($COCl_2$) : A colourless poisonous gas with a penetrating smell, resembling musty hay or green corn; heavier (3.43 times) than air, hence was used extensively in gas warfare in World War I; used as an intermediate in organic synthesis.

Phosphagen : A type of chemical found in the muscles of all animals, whose function is to pass on high-energy phosphate to ADP to form ATP. Phosphagens thus act as energy-storage molecules and are especially useful when cellular respiration is not providing sufficient ATP molecules, for example when sudden muscular activity takes place. Phosphagens are of 2 types; *creatine phosphate* found in vertebrates and echinoderms, and *arginine phosphate* found in many other invertebrates.

Phosphatase : An enzyme that removes a phosphate group from a protein by hydrolysis; for example, in the mammalian liver phosphorylated glucose can be broken down to glucose with a phosphatase system; also called *phosphoprotein phosphatase*.

Phosphate group (PO_4) : A chemical group commonly involved in high-energy bonds.

Phosphatidylinositol : An inositol phospholipid.

Phosphocreatine : A compound used in the production of ATP from ADP in muscles, being converted to creatine with the release of inorganic phosphorus.

Phosphodiester bond (–O–P–O–) : A covalent chemical bond formed when two hydroxyl groups are linked in ester linkage to the same phosphate group, such as adjacent nucleotides in RNA or DNA.

Phosphoenolpyruvic acid (PEP) : A high-energy compound which when dephosphorylated to pyruvic acid, gives rise to the synthesis of ATP from ADP in glycolysis.

Phosphogluconate pathway : *See* **pentose phosphate pathway**.

Phosphoinositides : One of a family of lipids containing phosphorylated inositol derivatives; although minor components of the plasma membrane, they are important in signal transduction in eukaryotic cells; also called *inositol phospholipids*.

Phospholipid (*phosphorosG* = light-bearer + *liposG* = fat) : A lipid composed of two fatty acids linked through glycerol phosphate to one of a variety of polar groups. One end of a phospholipid molecule is therefore strongly nonpolar (water-insoluble), whereas the other end is extremely polar (water-soluble). The two nonpolar fatty acids extend in one direction, roughly parallel to each other, and the polar alcohol group points in the other direction; the principal lipid molecule that is used to construct biological membranes; phospholipids in the blood are responsible for the transport and utilization of fats in the body.

Phosphoprotein phosphatase : *See* **phosphatase**.

Phosphoric acid : An important component of nucleic acids, connecting the pentose sugars to form a polynucleotide chain.

Phosphorolysis : Cleavage of a compound with phosphate as the attacking group; analogous to *hydrolysis*.

Phosphorylation : A reaction in which a phosphate group becomes covalently linked to another molecule. Such reactions occur regularly in biological systems. For example, in aerobic respiration glucose is phosphorylated in glycolysis; ADP is phosphorylated to ATP at the substrate-level and *via* the electron transport system ; in plants ATP is produced by photophosphorylation.

Photochemical reactions : The "light" reactions of photosynthesis; these reactions occur on the grana of chloroplasts and produce ATP and reduced NADP.

Photon (*photos*[G] = light) **:** The elementary particle of light and other electromagnetic radiation.

Photooxidation : A chemical reaction occurring as a result of absorption of light in the presence of oxygen.

Photophosphorylation (*photos*[G] = light + *phosphoros*[G] = bringing light) **:** The enzymatic formation of ATP from ADP coupled to the light-dependent transfer of electrons in photosynthetic cells.

Photoreduction : The light-induced reduction of an electron acceptor in photosynthetic cells.

Photorespiration : The light-dependent formation of glycolic acid in chloroplasts and its subsequent oxidation in peroxisomes; a process in which carbon dioxide is released without the production of ATP or NADPH and mitochondria are not involved; because it produces neither ATP nor NADPH, photorespiration acts to undo the work of photosynthesis.

Photosynthesis (*photos*[G] = light + *-syn*[G] = together + *tithenai*[G] = to place) **:** The process by which plants, algae, and some bacteria use the energy of sunlight to create, from carbon dioxide and water, the more complicated molecules (carbohydrates), that make up living organisms; occurs in chloroplasts and releases oxygen.

Phylloquinone : One of the 2 fat-soluble group of vitamins, collectively designated vitamin K group; these 2 vitamins are vitamin K_1 (phylloquinone) and vitamin K_2 (flavinoquinone); vitamin K_1 occurs in green vegetables like spinach, alfalfa, cabbage etc, while K_2 is found in intestinal bacteria, besides putrefied fish meal; vitamin K_1 is a yellow viscid oil, while vitamin K_2 is a yellowish crystalline solid; both are sensitive to light and hence kept in dark bottles; destroyed by irradiation, strong acids, alkalies and oxidizing agents; plays a key role in the biosynthesis of prothrombin, oxidative phosphorylation and electron transport system; avitaminosis K results in hemorrhage in infants and steatorrhea in human adults; also known as *antihemorrhagic vitamin* or *coagulation vitamin*.

Phylogeny (*phylon*[G] = race, tribe) **:** The evolutionary relationships among any group of organisms; often presented in chart form as a phylogenetic tree.

Phylum (*phylon*[G] = race, tribe) **:** A major taxonomic category between kingdom and class in animals; it is equivalent to *division* in plants.

Physiology (*physis*[G] = nature + *logos*[G] = a discourse) **:** The study of the function of cells, tissues, and organs.

Phytochrome : A group of blue-green, photoreceptive, proteinaceous pigments produced in plants and involved in phenomena such as photoperiodism, the germination of seeds, and leaf formation; occurs in 2 forms: P_{660}, which is biologically inert and absorbs red light and P_{725}, which is biologically active in that it stimulates enzymic reactions and absorbs far-red light. The conversion of one form into the other occurs simultaneously.

Phytohormones : Hormones produced by plants.

Phytol ($C_{20}H_{39}OH$) **:** A long-chain alcohol derived from the hydrolysis of chlorophyll; an acyclic diterpene with 2 chiral centres, b.p. 145°C.

Pigment (*pigmentum*[L] = paint) **:** A molecule that reflects and absorbs light at particular wavelengths.

Piloerection : The standing up of hair on the skin to increase thermal insulation.

Pineal body : *See* **pineal gland**.

Pineal gland (*pinus*^{L} = pine tree) **:** An outgrowth of the roof of the forebrain. The posterior part (epiphysis) has an endocrine function and secretes the hormone, melatonin. It appears to function in humans as a light-sensing organ and in a variety of other roles concerning sexual development. Exposure to light inhibits the production of melatonin. The anterior part forms an eye-like structure (third eye) in some lizards; it is vestigeal in the lamprey and absent in other vertebrates. Also called as *pineal body*.

Pinocytosis (*pinein*^{G} = to drink + *kytos*^{G} = cell) **:** A type of endocytosis in which soluble materials are taken up from the environment and incorporated into vesicles for digestion.

Pituitary (*pituita*^{L} = phlegm) **:** The major hormone-producing gland of the brain and is under the control of the hypothalamus; secretes hormones that promote growth, stimulate glands, and regulate many other bodily functions.

pK : The negative logarithm of an equilibrium constant.

Placebo : (1) Any inactive substance given to satisfy a patient's psychological need for medication. **(2)** A control in an experiment to test the effect of a drug.

Placenta, *plural* **placentae** (Latin, for *a flat cake*) **: (1)** In animals, a specialized organ, held within the womb in the mother across which she supplies the offspring with food, water and oxygen and through which removes wastes. **(2)** In plants, a part of the ovary wall on which ovules are borne.

Plasma (Greek, for *form*) **: (1)** The fluid of vertebrate blood; contains dissolved salts, metabolic wastes, hormones, and a variety of proteins, including antibodies and albumen; blood minus the blood cells. **(2)** The cellular protoplasm inside a plasma membrane.

Plasma membrane : A lipid bilayer with embedded proteins that control the cell's permeability to water and dissolved substances. Alternatively, the differentially permeable membrane that surrounds the cytoplasm of a cell and is next to the cell wall. Also called the *cell membrane* or the *plasmalemma* or *ectoplast*.

Plasma proteins : The proteins present in blood plasma.

Plasmalogen : A phospholipid with an alkenyl ether substituent on the C-1 of glycerol.

Plasmid (*plasma*^{G} = a form or something moulded) **:** A small fragment of circular double-stranded DNA that replicates independently of the bacterial chromosome; because of their ability to take up foreign DNA, bacterial plasmids are used as vectors for genetic engineering and research; many different types of plasmids exist, some conferring antibiotic resistance, and others, the ability to metabolize unusual organic compounds.

Plasmodesma, *plural* **plasmodesmata** (*plasma*^{G} = something moulded + *desma*^{G} = band) **:** A tiny, membrane-lined channel between adjacent cells in plants through which cytoplasmic connections extend; in some textbooks, *singular* plasmodesmus, *plural* plasmodesmi are used.

Plasmolysis : The shrinkage of plant cell contents due to loss of water, resulting in the cell membrane pulling away from the cell wall, leaving a fluid-filled space. Plasmolysis occurs when plant cells are placed in a hypertonic medium so that they lose water by osmosis.

Plastid : A cytoplasmic, self-replicating, large (3–6, μm in diameter) organelle in plants, bounded by a double membrane; colourless or pigmented, with definite physiological functions; the types are leucoplasts, chloroplasts and chromoplasts.

Platelets (dim. of *plattus*^{G} = flat) **:** Small enucleated cells that arise from cells called megakaryocytes in the bone marrow and are found in large numbers in the bloodstream; they help initiate blood clotting when blood vessels are injured; also known as *thrombocytes*.

Pleated sheet : The side-by-side hydrogen-bonded arrangement of polypeptide chains in the extended β-conformation.

Pneumonia (*pneumon*^G = lung) **:** Inflammation of the lungs, commonly due to pneumococcal infection.

Poikilothermy : Changeable temperature in an animal, dependent on environmental temperature, exercise, and other irregular factors. All animals apart from Aves and Mammals are poikilotherms.

Polar : Hydrophilic or "water-loving; desribing molecules or groups that are soluble in water.

Polar bond : Covalent bond in which the electrons are attracted more strongly to one of the two atoms, creating a polarized distribution of electric charge.

Polar molecule : A molecule with positively and negatively charged ends. One portion of a polar molecule attracts electrons more strongly than another portion, with the result that the molecule has electron-rich (–) and electron-poor (+) regions, giving it magnetlike positive and negative poles. Polar molecules are usually soluble in water. Water is one of the most polar molecules known.

Polarity : (1) In *biology*, establishment of poles of specialization at opposite ends of a cell, tissue, organ, or organism; for example, polarity leads to the differentiation of roots and shoots. **(2)** In *chemistry*, the nonuniform distribution of electrons in a molecule. **(3)** In *molecular biolgy*, the distinction between the 5' and 3' ends of nucleic acids.

Poliomyelitis : A paralytic disease in which cells of the central nervous system become destroyed by the polio virus, leading to crippling, although most infections are not serious and the patient usually recovers full health. The disease can be kept in check by the administration of vaccines, the most popular being a live, attenuated type.

Poly-A tail : A chain of adenylic acid molecules that is added to a molecule of RNA immediately after it has been transcribed and cleaved from its DNA template.

Polydipsia (*polys*^G many + *dipsa*^G = thirst) **:** Increased thirst.

Polymer (*polus*^G = many + *meris*^G = part) **:** A large molecule formed by the linking together of similar smaller molecules (monomers). The number of monomers in a polymer may range up to millions.

Polymerase : An enzyme that catalyzes the joining of DNA or RNA nucleotides.

Polymerase chain reaction (PCR) : A repetitive procedure that results in geometric amplification of a specific region of DNA by multiple cycles of DNA polymerization, each followed by a brief treatment to separate complementary strands; the PCR is cycled 30 or more times to produce million-fold amplification of the target DNA sequence; the free nucleotides are assembled in a nucleic acid chain in a test tube by enabling the activity of a bacterial DNA polymerase to bind them together.

Polymerization (*polus*^G = many + *meris*^G = part + *izein*^G = to combine with) **:** A process in which many small identical subunits (monomers) combine to one another chemically to form a long chain of a polymer molecule.

Polymorphism (*polys*^G = many + *morphe*^G form) **:** The presence of distinctly different, genetically-determined phenotypic characteristics within a population of a single species.

Polynomial (*polys*^G = many + *nomos*^G = naming) **:** Before Linnaeus, the propounder of binomial system of nomenclature, naming a genus by use of a cumbersome string of Latin words and phrases; the system is now abandoned.

Polynucleotide : A covalently-linked sequence of nucleotides in which the 3' hydroxyl of the pentose of one necleotide residue is joined by a phosphodiester bond to the 5' hydroxyl of the pentose of the next residue.

Polypeptide (*polys*^G = many + *peptein*^G to digest) **:** A general term for a long chain of amino acids linked end-to-end by peptide bonds. A protein is a long, complex polypeptide; the two terms are often used interchangeably.

Polyribosome : A complex of an mRNA molecule and two or more ribosomes; such an arrangement ensures that the mRNA is 'read' at the maximum speed; also known as *polysome* (poly + [ribo]some).

Polysaccharide (*polys*[G] = many + *sakcharon*[G] = sugar) **:** A linear or branched polymer of monosaccharide units linked by glycosidic bonds. Polysaccharides are insoluble, unsweet and are important as storage molecules (starch, glycogen, inulin) and as reinforcing materials (cellulose, chitin).

Polysome : *See* **polyribosome.**

Polyuria (*polys*[G] = many + *ouron*[G] = urine) **:** Increased urine output.

Porphyrin : A complex nitrogenous compound containing 4 substituted pyrroles covalently joined into a ring; often complexed with a centrally-located atom of a heavy metal; forms part of several important biological molecules; examples include the heme group of myoglobin and hemoglobin, chlorophyll (with magnesium) and cytochromes (with iron).

Posttranslational modification : Enzyme-catalyzed changes (or modifications) in a polypeptide chain after it is synthesized (or translated) from its mRNA; the various modifications include cleavage, glycosylation, phosphorylation, methylation, and prenylation.

Posterior (*post*[L] = after) **:** Situated behind or tail end of the body.

Posttranscriptional processing : The enzymatic processing of the primary RNA transcript, producing functional mRNA, tRNA and/or rRNA molecules.

Postulate : A basic or essential assumption; a set of postulates that address the same phenomenon can be taken together as a *theory*.

Potential difference : A difference in electrical charge on two sides of a membrane, caused by an unequal distribution of ions.

Potential energy : Any stored energy that can be released (or has the potential) to do work.

Pre-mRNA : *See* **heterogeneous nuclear RNA.**

Prenylation : Covalent attachment of an isoprenoid lipid group to a protein.

Primary structure : Sequence of units in a linear polymer, such as amino acid sequence of a protein or nitrogenous base sequence of a nucleic acid.

Primary transcript : The immediate RNA product of transcription before any posttranscriptional processing reactions.

Primase : An enzyme that catalyzes the formation of RNA oligonucleotides used as primers by DNA polymerase.

Primer : A short oligomer (of sugars or nucleotides, for example) to which an enzyme adds additional monomeric subunits.

Prion : An infectious agent that contains protein but no nucleic acid.

Probe : In genetic research, a fragment of labelled (radioactively or chemically labelled) DNA or RNA, containing a nucleotide sequence complementary to a gene or genomic sequence that one wishes to detect in a hybridization experiment.

Prochiral molecule : A symmetric molecule that can react asymmetrically with an enzyme having an asymmetric active site, generating a chiral product.

Progesterone ($C_{21}H_{30}O_2$) : A C_{21} steroid female sex hormone, secreted by the corpus luteum during the second half of the menstrual cycle; closely resembles deoxycorticosterone in structure; makes the lining of the uterus more receptive to a fertilized ovum, brings the mammary glands to full maturity during pregnancy, inhibits contraction of the uterus during pregnancy, and serves as a precursor of cortisol and corticosterone in the adrenal glands.

Projection formulae : A method for representing molecules to show the configuration of groups around chiral centres; also known as *Fischer projection formulae*.

Prokaryote (*pro*[G] = before + *karyon*[G] = *kernel*) **:** A unicellular organism with a single chromosome, no nuclear envelope, no membrane-bounded organelles and no mitosis or meiosis; prokaryotes include bacteria and cyanobacteria; also spelt as *procaryote*; compare *eukaryote*.

Prolactin (PL) : *See* **luteotrophic hormone**.

Promoter : A DNA sequence at which RNA polymerase may bind, leading to intiation of transcription.

Proofreading : The ability of DNA polymerases to remove mismatched nucleotides with 3′ to 5′ exonuclease activity during DNA synthesis.

Prostaglandins (from *prosta*[te] *gland* + – *in*) : A class of lipid-soluble, hormonelike, regulatory molecules derived from arachidonate and other polyunsaturated fatty acids by virtually all cells; stimulate contraction or expansion of smooth muscles and contraction of blood vessels, have also been used in the induction of labour and abortion.

Prostate gland : (*prostates*[G] = one standing in front) : A large gland surrounding the male urethra just below the bladder. Its secretions, which transport sperm cells, make up a large part of the semen. Androgens affect the size and secretion of the prostate gland, whose exact function is unknown.

Prosthetic group : A metal ion or an organic compound (other than an amino acid) that is covalently bound to a protein and is essential to its activity. For example, many enzymes contain metallic ions, as in carboxypeptidase which contains zinc; hemoglobin contains heme with an iron atom at the centre.

Protease (*proteios*[G] = primary + *-ase* = suffix for an enzyme) : An enzyme that breaks up proteins into amino acids by hydrolyzing some of their peptide bonds located *inside* the chain, but not at the ends; pepsin, erepsin, rennin and trypsin are proteases; also called *proteinase* or *proteolytic enzyme*.

Protein (*proteios*[G] = pre-eminent, primary) : The major macromolecular constituent of cells, consisting of carbon, hydrogen, oxygen, nitrogen, and usually sulfur and phosphorus; a linear polymer of amino acids linked together by peptide bonds in a specific sequence; various proteins exhibit 4 levels of structural organization: primary, secondary, tertiary and quaternary.

Protein kinase : Enzyme that transfers the terminal phosphate group of ATP to a specific amino acid of a target protein.

Protein targeting : The process by which newly-synthesized proteins are sorted and transported to their proper locations in the cell.

Proteinase : *See* **protease**.

Proteoglycan : A hybrid molecule consisting of one or more heteropolysaccharide (glyco-saminoglycan, GAG) chains joined to a core protein; the polysaccharide is the major component; an important component of the intercellular matrix.

Proteolytic enzyme : *See* **protease**.

Proteolysis : Degradation of a protein, usually by hydrolysis at one or more of its peptide bonds.

Proton : A subatomic particle in the nucleus of an atom that carries a positive charge. The number of protons determines the chemical character of the atom because it dictates the number of electrons orbiting the nucleus and available for chemical activity.

Proton acceptor : An anionic compound capable of accepting a proton from a proton donor, that is, a *base*.

Proton donor : The donor of a proton in an acid-base reaction, that is, an *acid*.

Proton-motive force : The electrochemical potential inherent in a transmembrane gradient of H^+ concentration; used in oxidative phosphorylation and photophosphorylation to drive ATP synthesis.

Protoplasm : A general term referring to the entire contents of a living cell.

Pruritis : Itching.

Psoriasis (from a Greek word, meaning *to have the itch*) : A noncontagious disease of the skin marked by scaly red patches, due probably to a disorder of the immune system.

Ptyalin : *See* **amylase.**

Puberty (*pubertas*[L] = of ripe age, adult) : The state of physical development at which persons are first capable of begetting or bearing children. In law, the age of puberty is usually fixed at 14 in the male and 12 in the female.

Pupation : The formation of puparium by an insect larva, the first step in metamorphosis.

Purine ($C_5H_4N_4$) : A nitrogenous heterocyclic base found in nucleotides and nucleic acids; contains a 6-membered pyrimidine ring fused to the 5-membered imidazole ring; the most common purine derivatives are adenine and guanine; purine derivatives always pair with pyrimidine derivatives in the two strands of DNA, ensuring a parallel-sided molecule.

Puromycin : An antibiotics that inhibits polypeptide synthesis by being incorporated into a growing polypeptide chain, causing its premature termination.

Pus : A yellowish fluid consisting of serum, white blood cells, bacteria and tissue debris formed during the liquiefaction of inflammed tissue.

Putrefaction : The anaerobic decomposition of organic substances, particularly proteins, brought about by bacteria and other microbes and which gives rise to foul-smelling products.

Pyorrhea (*pyon*[G] = pus + *rhoia*[G] = flowing) : Refers to inflammation of gums with purulent discharge.

Pyranose : A simple sugar containing the 6-membered pyran ring.

Pyrethrum : An insecticide prepared from the dried flowers of the chrysanthemum plant, the active ingredient being called *pyrethrin.*

Pyrexia (*pyrexis*[G] = fever) **:** Fever.

Pyridine nucleotide – A nucleotide coenzyme containing the pyridine derivative nicotinamide; NAD or NADP.

Pyridoxal phosphate : A coenzyme containing the vitamin pyridoxine (vitamin B_6); functions in reactions involving amino group transfer.

Pyridoxine ($C_8H_{11}O_3N$) : A water-soluble vitamin of the B complex group and designated as vitamin B_6; in fact, vitamin B_6 group includes 3 closely-related vitamins: pyridoxine, pyridoxal ($C_8H_9O_3N$) and pyridoxamine ($C_8H_{12}O_2N_2$); B_6 vitamins are widely distributed in nature in plant and animal tissues, *e.g.*, cereal grains (wheat, rice), yeast, eggyolk and meat; pyridoxal and pyridoxamine also occur in nature as their coenzymes, namely, pyridoxal phosphate (PALP) and pyridoxamine phosphate (PAMP), respectively; pyridoxine is a white crystalline substance, soluble in alcohol and sensitive to light and ultraviolet radiation; various forms of vitamin B_6 serve as growth factors to a number of bacteria; also act as a carrier in the active transport of amino acids across cell membranes; avitaminosis B_6 leads to convulsions and anemia in human infants; also known as *antidermatitis factor*.

Pyrimidine ($C_4H_4N_2$) : A nitrogenous heterocyclic base found in nucleotides and nucleic acids; contains a 6-membered ring only; the most common pyrimidine derivatives are cytosine, thymine and uracil; pyrimidine derivatives always pair with purine derivatives in DNA.

Pyrimidine dimer : A covalently-joined dimer of two adjacent pyrimidine residues in DNA, induced by absorption of ultraviolet light; most commonly derived from two adjacent thymines (a *thymine dimer*).

Pyrogallol, $C_6H_4(OH)_3$: A soluble phenol which, in alkaline solution, will absorb oxygen and is used to estimate the volume of oxygen in a gaseous sample; also called *trihydroxybenzene*.

Pyrophosphatase : *See* **inorganic pyrophosphatase.**

Pyrrole : A porphyrin building block that has a 5-membered heterocyclic structure and contains nitrogen.

Pyruvate dehydrogenase : The complex of enzymes that removes carbon dioxide from pyruvate; one of the largest known enzymes, containing 48 polypeptide chains.

Pyruvic acid : A C-3 compound that is the starting material of the citric acid cycle; it is also the end product of glycolysis where it is formed from glucose and glycerol.

Q

Q_{10} : A measure of the rate of increase in metabolic rate over a 10-degree range in temperature. Thus, if an organism has a metabolic rate at 10°C of T units, a rate of twice T units at 20°C, the $Q_{10} = 2$. A Q_{10} of 2 is the typical exponential increase in rate exhibited by enzymes up to a certain maximum rate, after which denaturation occurs faster than increase, causing an overall reduction of the rate.

Quantum : The ultimate unit of energy.

Quarantine : A forced stoppage of the transportation of diseased plants from one region to another.

Quart : Unit of capacity equal to one quarter of a gallon.

Quaternary structure : The three-dimensional structure of a multisubunit protein; particularly the manner in which the subunits fit together.

Quinone : Any of the various compounds derived from benzene.

R

R group : **(1)** Formally, an abbreviation denoting any alkyl group **(2)** Occasionally, used in a more general sense to denote virtually any organic substituent, such as a methyl group, a hydroxyl group, or a monosaccharide.

Racemic mixture : An equimolar mixture of the D and L stereoisomers of an optically-active compound.

Radial symmetry (*radius*[L] = a spoke of a wheel + *summetros*[G] = symmetry) : The regular arrangement of parts around a central axis so that any plane passing through the central axis divides the organism or an organ into two halves that are approximate mirror images; for example, human beings, starfish and tulip flowers.

Radiation : The electromagnetic energy that travels through empty space with the speed of light (2×10^8 ms^{-1}). All objects emit radiation, at room temperature mostly in the infrared range, whereas at high temperatures visible radiation is produced.

Radical : **(1)** A functional group of atoms recurring as part of various different molecules. **(2)** An atom or group of atoms possessing an unpaired electron; also called a *free radical*. **(3)** A leaf *seemingly* arising from the apical portion of a root.

Radicle (*radicula*[L] = root) : The part of the plant embryo that develops into the root.

Radioactive isotope : An isotopic form of an element with an unstable nucleus that stabilizes itself by emitting ionizing radiation; in fact, the nucleus tends to break up into elements with lower atomic numbers in a process called radioactive decay, such as carbon-14 which breaks up into a stable carbon-12.

Radioautography : *See* **autoradiography**.

Radioimmunoassay (RIA) : A sensitive and quantitative method for detecting trace amounts of a biomolecule, based on its capacity to displace a radioactive form of the molecule from combination with its specific antibody.

Raffinose : A trisaccharide found in barley, cottonseeds, beet roots and other parts of plants.

Ramachandran plot (named in honour of **Gopalasamudram Narayana Ramachandran,** LT, 1922–2001, an Indian biochemist) : A plot that constitutes a map of all possible backbone configurations for an amino acid in a polypeptide chain. The axes of the plot consist of the rotation angles of the 2 backbone bonds that are free to rotate (ϕ and ψ, respectively); each point ϕ, ψ on the plot, thus, represents a conceivable amino acid backbone configuration.

Random coil : A linear polymer that has no secondary or tertiary structure but instead is wholly flexible with a randomly-varying geometry. This is the state of a denatured protein or nucleic acid.

***Ras* protein** : One of a large family of GTP-binding proteins that help relay signals from cell-surface receptors to the nucleus; named for the *ras* gene, first identified in viruses that cause *rat* sarcomas.

Rate-limiting step : **(1)** Generally, the step in an enzymatic reaction with the greatest activation energy or the transition state of highest free energy. **(2)** The slowest step in a metabolic pathway.

Reaction : In chemistry, any process in which the arrangement of atoms into molecules is changed.

Reaction intermediate : Any chemical species in a reaction in a pathway that has a *finite* (definite) chemical lifetime.

Reading frame : A contiguous and nonoverlapping set of three nucleotide codons in DNA or RNA; an mRNA molecule can be read in any one of three reading frames.

Recapitulation theory : The rejected hypothesis, developed by **Thomas Huxley**, that *ontogeny repeats phylogeny.*

Receptor : Protein that binds a specific extracellular signalling molecule (ligand) and initiates a response in the cell. Cell surface receptors (acetylcholine receptor, insulin receptor) are located in the plasma membrane with their ligand-binding site exposed to the external medium. Intracellular receptors (steroid hormone receptors) bind ligands that diffuse into the cell across the plasma membrane.

Recombinant DNA : Any DNA molecule created in the laboratory by joining segments from two or more precursor DNA molecules; widely used in the cloning of genes, in the genetic modification of organisms and in molecular biology generally.

Recombinant DNA technology : *See* **genetic engineering**.

Recombination : A process in which chromosomes or DNA molecules are broken and the fragments are rejoined in new combinations. In *bacteria*, it is accomplished by the transfer of genes into cells, often in association with viruses. In *eukaryotes*, it is accomplished by reassortment of chromosomes during meiosis and by crossing-over.

Recon : The smallest genetic unit between two of which recombination can take place, effectively a DNA base.

Red blood corpuscle (RBC) : *See* **erythrocyte.**

Redox pair : An electron donor and its corresponding oxidized form; for example, NADH and NAD^+.

Redox potential : A quantitative measurement of the willingness of an electron carrier to act as a *red*ucing or *ox*idizing agent. Redox potential is measured in volts; the more negative the value, the better the carrier will act as a reducing agent. Thus, in an electron transport system the carriers are arranged in order of increasing redox potentials (negative to positive).

Redox reaction : *See* **oxidation–reduction reaction**.

Reducing agent : The electron donor in an oxidation–reduction reaction; also called *reductant.*

Reducing end : The end of a polysaccharide having a terminal monosaccharide moiety with a free anomeric carbon; the terminal residue can act as a reducing sugar.

Reducing equivalent : An amount of a reducing compound that donates the equivalent of 1 mole of electrons in an oxidation-reduction reaction. The electrons may be expressed in the form of hydrogen atoms.

Reducing power : The use of light energy to extract atoms from water.

Reducing sugar : A sugar in which the carbonyl (anomeric) carbon is not involved in the formation of a glycosidic bond and hence can undergo oxidation with metal hydroxides which are themselves reduced to lower oxides or free metals; common examples are glucose, fructose, lactose and maltose.

Reduction (*reduction*^{L} = a bringing back; originally, "bringing back" a metal from its oxide) **: (1)** In *biochemistry*, the gain of an electron during a chemical reaction from one atom to another, as occurs during the addition of hydrogen to a molecule or the removal of oxygen from it; occurs simultaneously with oxidation; reduction involves the addition of energy to one substance, which is coupled with the simultaneous removal of energy from another substance by oxidation. **(2)** In *genetics*, division of chromosomes in meiosis.

Reductionism : The approach of studying simpler components in order to understand the functions of complex systems.

Regulatory enzyme : An enzyme having a regulatory function through its capacity to undergo a change in catalytic activity by allosteric mechanisms or by covalent modification.

Regulatory gene : A gene that gives rise to a product involved in the regulation of the expression of another gene; for example, a gene coding for a repressor protein.

Regulatory sequence : A DNA sequence involved in regulating the expression of a gene; for example, a promoter or operator.

Regulon : A group of unlinked (non-adjacent) genes or operons that are all regulated by a common mechanism.

Relative molecular weight : Mass of a molecule expressed as a multiple of the mass of a hydrogen atom.

Release factors : *See* **termination factors**.

Releasing factors : Peptide hormones released by the hypothalamus that stimulate the release (secretion) of other hormones by the anterior pituitary (adenohypophysis).

Renaturation : Refolding of an unfolded (denatured) globular protein so as to restore native structure and protein function.

Rennin : An enzyme, present in the gastric juice, secreted by the gastric glands of the stomach wall that coagulates caseinogen in milk to form casein, which forms an insoluble curd (a calcium-casein compound) which is then attacked by pepsin. It is important particularly in young mammals because it increases retention time in the stomach, allowing for a more efficient digestion of the primary food source.

Replication : Synthesis of a daughter duplex DNA molecule identical to the parental duplex DNA.

Replication fork : Y-shaped region of a replicating DNA molecule at which the two daughter strands are formed and separate.

Replication origin : The point of initiation of DNA synthesis along the double helix; two replication forks form at the replication origin and move in opposite directions from one another during DNA synthesis.

Replicon : A block of DNA capable of replication (for example, a plasmid or a chromosome).

Replisome : The multiprotein complex that promotes DNA synthesis at the replication fork.

Repressible enzyme : In bacteria, an enzyme whose synthesis is inhibited when its reaction product is readily available to the cell.

Repression (*reprimere*[L] = to press back, keep back) **:** The state in which a gene is prevented from being transcribed, in response to a change in the activity of a regulatory protein, so that no protein is produced.

Repressor (*reprimere*[L] = to press back, keep back) **:** A protein that prevents transcription of mRNA from DNA by binding to the operator (a specific region of DNA) and so preventing RNA polymerase from attaching to the promoter.

Repetitive DNA : Sequences of DNA that occur in many copies in a genome; some sequences of repetitive DNA can occur in a million copies per nucleus.

Reserpine : An alkaloid extracted from *Rauwolfia serpentina*, that is used as a sedative and as an antihypersensitive agent to reduce hypertension.

Reserves : Any stored food supplies which may be drawn on in times of food shortage.

Residue : A general term for the unit of a polymer; for example, an amino acid within a polypeptide chain and a mononucleotide within a nucleic acid; the term reflects the fact that sugars, amino acids, and nucleotides lose a few atoms (generally the elements of water) when incorporated in their respective polymers.

Resin : A thick, translucent, combustible, organic fluid usually secreted into resin ducts in pines and many other seed plants.

Resolving power : The ability of a microscope to distinguish two lines as separate.

Respiration (*respirare*[L] = to breathe) **:** The catabolic process in which electrons are removed from nutrient molecules and passed through a chain of carriers to oxygen. Alternatively, a general term for any process in a cell in which the uptake of O_2 molecules is coupled to the production of CO_2.

Respiratory chain : The electron transport chain; a sequence of electron-carrying proteins that transfer electrons from substrates to molecular oxygen in aerobic cells.

Restriction endonuclease : A special kind of enzyme that can recognize and cleave both strands of DNA molecules into fragments at points within or near the specific site recognized by the enzyme; one of the basic tools of genetic engineering.

Restriction fragment : A segment of double-stranded DNA produced by the action of a restriction endonuclease on a larger DNA.

Restriction map : Diagrammatic representation of a DNA molecule indicating the sites of cleavage by various restriction endonucleases.

Retinene : The main carotenoid pigment found in the retina of the eye; turns yellow in light.

Retinoblastoma : A childhood cancer of the retina.

Retinol ($C_{20}H_{30}O$) : A fat-soluble vitamin and designated as vitamin A; liver oils of various fishes (shark, halibut, cod) are the richest natural sources of vitamin A; butter, milk, eggs, kidneys and pigmented (especially yellow) vegetables and fruits are other good sources; vitamin A is absent from vegetable fats and oils; found in two forms: Vitamin A_1 (retinol) and vitamin A_2 (3-dehydroretinol); retinol is a viscid colourless oil and is destroyed on exposure to UV light; participates in visual cycle and also helps maintain the epithelial cells of the skin and the linings of the digestive, respiratory and genito-urinary systems; avitaminosis A leads to nyctalopia (inability to see in night), xerophthalmia (drying of the eyes), keratomalacia (softening of the cornea), phrynoderma (hard and horny skin) and stunted growth; also called as *antixerophthalmic factor* or '*bright eyes*' *vitamin* or *anti-infection vitamin*.

Retrovirus (*retro*[L] = turning back) **:** An RNA virus containing a reverse transcriptase and that which replicates in a cell by first making a double-stranded DNA intermediate, which it can then insert into the cellular DNA as if it were a cellular gene. Human immunodeficiency virus (HIV) is a common example.

Reverse transcriptase : An RNA-dependent, DNA polymerase enzyme, present in retroviruses such as AIDS, that catalyzes the synthesis of a double-stranded DNA copy from a single-stranded RNA template molecule; in genetics, reverse transcriptase is used for making complementary DNA (cDNA) of eukaryotic genes.

Rhesus blood group : A form of human blood variation in which most of the world population (about 85% of the UK population, for example) possess an *Rh factor* (D-antigen) on the surface of red blood cells. Such people are described as being *Rh-positive*, those without the factor are *Rh-negative*. Unlike the ABO blood group, there is normally no rhesus antibody, unless an Rh-negative person is sensitized by a rhesus antigen, for example in a blood transfusion or when an Rh-negative mother receives blood cells from an Rh-positive fetus.The blood group factor is controlled by a single autosomal gene on chromosome 1, with 2 principal alleles; the allele for the Rh-factor is dominant. It was named for the Rhesus monkey in which it was first described.

Rhodopsin : A purple pigment found in the rods of the retina of the vertebrate eye. Lack of rhodopsin causes night blindness (*nyctalopia*).When bleached by light, rhodopsin liberates a protein called opsin and a yellow pigment called retinene. As a result of this reaction, energy is released which triggers off an action potential. Also called *scotopsin*.

Riboflavin : A water-soluble vitamin of the B complex group and designated as vitamin B_2; found in a wide variety of foods including heart, liver, kidney, milk and green vegetables; occurs almost exclusively as a constituent of either FMN or FAD; required in the metabolism of all animals, acting as a carrier in the electron transport system; persons deficient in vitamin B_2 may show glossitis, cheilosis, corneal vascularization and a typical dermatitis; also called '*yellow enzyme*' because of its colour.

Ribonuclease : A nuclease that catalyzes the hydrolysis of certain internucleotide linkages of RNA.

Ribonucleic acid, RNA : A polynucleotide having a specific sequence of ribonucleotide units covalently joined through 3',5'-phosphodiester bonds; molecules of RNA, which are made as complements of DNA segments called genes, function in protein synthesis; differentiated into 3 types: mRNA, rRNA and tRNA.

Ribonucleotide : A nucleotide containing D-ribose as its pentose component.

Ribose : A 5-carbon sugar with one oxygen atom more than the related sugar deoxyribose; a component of ribonucleic acid.

Ribosomal RNA (rRNA) : A class of RNA molecules serving as components of ribosomes and often distinguished by their sedimentation coefficient, such as 28 s rRNA or 5 s rRNA; participate in the synthesis of proteins.

Ribosome : A cell organelle composed of rRNAs and proteins (approximately 18 to 22 nm in diameter) that are arranged in two subunits, one large and one small; prokaryotes have ribosomes with 70 s size and mass and eukaryotes have larger ribosomes with 80 s size and mass; ribosome associates with mRNA and catalyzes the synthesis of proteins.

Ribozymes : Ribonucleic acid molecules with catalytic activities; RNA enzymes.

Ribulose ($C_5H_{10}O_5$) : A ketopentose sugar, found in syrup; plays important role in carbohydrate metabolism.

Ribulose bisphosphate (RBP) : A C-5 ketose that acts as a receptor of CO_2 in Calvin cycle.

Rickets : A vitamin D-deficiency disease in children, which is primarily a disease of growing bones; characterized by hunched back, beaded ribs, protruding chest, enlarged skulls, swollen joints and bow legs; more prevalent where climate or custom prevents individuals from exposure to sun, whereby checking vitamin D production by irradiation of the skin.

RNA polymerase : An enzyme that catalyzes the synthesis of an RNA molecule from ribonucleoside 5'-triphosphate precursors, using a strand of DNA or RNA as a template.

RNA processing : *See* **RNA splicing**.

RNA splicing : The trimming of larger primary RNA transcripts, in the nucleus, into smaller, coding sequences that are exported into the cytosol. Alternatively, a process in which intron sequences are excised from RNA molecules in the nucleus during formation of mRNA. Synonymous with *RNA processing*.

Rochelle salt, COOK $(CHOH)_2$ COONa . $4H_2O$: A white crystalline soluble salt, m.p. 70 – 80°C; used in the preparation of baking powder, Seidlitz powder etc.

Royal Society : An English scientific society of physicists, chemists and biologists founded in 1660s and in existence today.

Rubisco (short for ribulose bisphosphate carboxylase-oxygenase) : The enzyme that carries out carbon fixation in photosynthesis by adding CO_2 to ribulose-1,5-bisphosphate. It can also add O_2 in place of CO_2, initiating photorespiration.

Rut : The period of maximum testicular activity in male mammals; the term particularly applied to the period of sexual activity in deer; compare *estrus cycle*.

S

Saccharin, $C_6H_4SO_2CONH$: A white, crystalline, sparingly soluble solid with a m.p. 227°C; about 400 times sweeter than sugar; manufactured from toluene, C_6H_5. CH_3; used as an artificial sweetening agent which is noncalorific, *i.e.*, provides no energy, hence of no food value; may have harmful effects if used in excess; also used in the form of a sodium salt called saccharin sodium, $C_6H_4COSO_2NNa.2H_2O$.

Saccharase : *See* **invertase**.

Saccharomyces : An ascomycetous genus of unicellular fungi that reproduce asexually by budding or sexually by conjugation; economically important in brewing and baking, also widely used in genetic engineering and as simple model organisms in the study of eukaryotic cell biology.

Saccharose : *See* **sucrose**.

S-adenosylmethionine (adoMet) : An enzymatic cofactor involved in methyl group transfers.

Saliva : A viscous, transparent liquid containing water, salts, mucin and sometimes salivary amylase (previously called *ptyalin*); secreted by cells of the salivary glands; the quantity of saliva produced depends on the type of food being consumed: dry foods and acidic foods stimulate large quantities of nonviscous saliva while liquid foods such as milk stimulate small quantities of thick saliva.

Salivary gland : Any gland that secretes saliva; in humans, the salivary glands occur in 3 pairs, one in the cheek and two between the bones of the lower jaw.

Salmonella : A rodlike, motile, aerobic genus of bacteria; includes species that cause food poisoning.

Salt : A chemical compound formed when the hydrogen of an acid has been replaced by a metal; a salt is produced, together with water, when an acid reacts with a base; salts are named according to the acid and the metal from which the salts are derived; for example, *sodium chloride* is a salt derived from sodium and hydrochloric acid.

Salvage pathway : Synthesis of a biomolecule (such as a nucleotide) from intermediates in the degradative pathway for the biomolecule; a recycling pathway, as distinct from a *de novo* pathway.

Saponification : The hydrolysis of an ester; the term is often confined to the hydrolysis of an ester by an alkali, thus forming a salt (a soap, in the case of some of the higher fatty acids) and the free alcohol.

Sarcoma : Solid tumour or the cancer of connective tissue (muscle and bone). Sarcomas are the least frequent type of tumour, constituting about 2% of all malignant tumours.

Satellite DNA : DNA consisting of multiple tandem repeats of very short, simple nucleotide sequences; makes up 10 to 20% of genome of higher eukaryotes; usually identifiable by its unusual nucleotide composition; most often associated with the centromeric region; satellite DNA is not transcribed and has no known function.

Saturated fatty acid : A fatty acid containing a fully saturated (*i.e.*, having only single bonds) alkyl chain; palmitic and stearic acids are common examples of saturated fatty acids.

Schiff's reagent (named after **Hugo Schiff**, LT, 1834–1915) **:** A reagent consisting of the dye magenta, which has been decolourized with sulfur dioxide or sulfurous acid; used to test for aldehydes–the aldehydes oxidize the reduced form of the dye back to its original colour.

Scleroprotein : A class of complex, insoluble fibrous proteins (*e.g.*, keratin, collagen, elastin) that occur in the surface coatings of animals and form the framework binding cells together in animal tissues.

Scotopsin : *See* **rhodopsin**.

Scurvy (substantive use of adj. scurvy from scurf, influenced by *scorbut*^{Fr} = scurvy) **:** A disease characterized by petechial hemorrhages (bleeding) in the skin and mucous membrane, swelling of limbs, bleeding of gums and dropping of teeth; results from insufficient intake of vitamin C (ascorbic acid) in the diet; normal health being restored by administration of fresh fruits, particularly citrus fruits.

Scutellum : The single cotyledon of the grass embryo; the scutellum is specialized for absorbing nutrients from the endosperm as the seed germinates.

Sebaceous gland : One of many glands occurring in the skin; secretes oil or sebum into the hair follicles in mammals as a result of cell destruction; maintains an oily coating to the hair and contributes to its waterproofing.

Second : (1) The SI unit of time defined as the duration of 9 192 631 770 periods of the radiation corresponding to the transition between two hyperfine levels of the ground state of the caesium-133 atom. Symbol *s*. **(2)** A measure of angle: 1/60 of a minute.

Second law of thermodynamics : The law stating that in any physical or chemical process, the entropy (degree of disorder) of the universe tends to increase.

Second messenger : A small effector molecule that is formed in or released into the cytosol in response to an extracellular signal (first messenger) such as a hormone; helps to relay the signal to the interior of the cell; examples include cAMP, IP_3, and Ca^{2+}.

Second order reaction : A reaction in which 2 reactant molecules must come together for the reaction to occur. The reaction is called second-order because the reaction rate depends on the square of reactant concentration (for 2 molecules of the same reactant) or on the product of 2 reactant concentrations (for 2 different reactants). Compare *first-order reaction.*

Secondary metabolism : Pathways that lead to the production of specialized products which are not found in every living cell; usually have no known metabolic role in cells.

Secondary sex characters : Features of a male or female animal that develop when sexual maturity occurs; in humans, the *female secondary sex characters* are induced by estrogens at puberty and include mature genitalia, prominent breasts, high-pitched voice, passive attitude, less body hair and more scalp hair, and feminine body shape, particularly a broadening of the pelvis and fat deposition around the hips; the *male human secondary sex characters* are induced by androgens at puberty and include mature genitalia, normal chests, deeper voice, active attitude, more body hair and less scalp hair, and masculine body shape, particularly broadened shoulders and narrow hips; all these characters are excellent examples of sex limitation.

Secondary structure : Regular local folding pattern of a polymeric molecule, *i.e.*, residue-by-residue conformation of the backbone of a polymer; for example, the B- and Z-forms of DNA helix and the α helices and β pleated sheets in proteins.

Secondary metabolite : A product of microbial cells in culture when growth is slowing down. While having no obvious role in the cellular physiology of the producer, secondary metabolites are sometimes most useful to humans, for example as antibiotics.

Secretin : A peptide hormone of 27 amino acid residues and with molecular weight of 3,056; formed by the upper intestinal mucosa and liberated by HCl present in the acid chyme; causes the flow of pancreatic juice; also stimulates the flow of bile and intestinal juices; the first compound to be designated as hormone.

Secretory vesicle : Membrane-bounded organelle in which molecules destined for secretion are stored prior to release; also called *secretory granule* because darkly staining contents make the organelle visible as a small solid object.

Sedative : A drug that reduces nervousness and excitement.

Sedimentation : The process of separating an insoluble solid from a liquid in which it is suspended by allowing it to fall to the bottom of the containing vessel, with or without agitation or centrifuging.

Sedimentation coefficient : A physical constant specifying the rate of sedimentation of a particle in a centrifugal field under specified conditions.

Seed : A structure that develops from the mature ovule of a seed plant; contains an embryo and stored food enclosed by protective seed coat(s); a reproductive and dispersal unit of plant.

Seed coat : The outer layer of a seed; develops from the integuments of the ovule.

Selectively permeable membrane (*seligere*[L] = to gather apart + *permeare*[L] = to go through) : Refers to the ability of a cell membrane to allow passage across the membrane of some solutes, but not others; previously called as *semipermeable membrane*.

Selfish DNA : Refers to DNA that can perpetuate itself by semi-autonomous replication; transposons are considered to be selfish DNA because they can move copies of themselves to several sites in a genome.

Self-replication : Refers to the ability of DNA to make exact copies of itself.

Semen (Latin, for *seed*) **:** The ejaculate from the male reproductive organs which in mammals contains the sperms together with secretions from the seminal vesicles and prostate gland which are essential for maintaining the viability of the sperms.

Semiconservative replication : A form of DNA replication wherein half of each new double strand consists of one newly-synthesized strand and the other strand derived from the parent double helix. This is the way DNA replication actually occurs. Compare *conservative replication,* wherein both parent strands would end up in one progeny molecule.

Seminal fluid : The fluid in which the sperms are bathed.

Semipermeable membrane : *See* **selectively permeable membrane**.

Senescence : The process of growing old which occurs in all species and is characterized by a gradual slowing down of metabolism and breakdown of tissues, often accompanied by endocrinal changes.

Sense organ : Any receptor of external or internal stimuli.

Sense strand : For a gene, the DNA strand that is homologous to an RNA transcript of the gene, that is it carries the same sequence as the transcript, except with T in place of U. It is, thus, complementary to the strand that served as a template for the RNA.

Septicaemia : An infection of the bloodstream by a variety of pathogenic microorganisms, such as *Salmonella* and *Pseudomonas*, usually from a nonintestinal source, leading to fever, lesions in many body organs and even death; commonly known by the name *blood poisoning*.

Secretory granule : *See* **secretory vesicle**.

Serotonin : A pharmacologically-active compound, derived from tryptophan; acts as a vasodilator, increases capillary permeability, and causes contraction of smooth muscle.

Serum : Plasma with clotting factors removed.

Sex chromosome : A chromosome pair in eukaryotes that usually differ in size and shape and control sex determination. The presence of identical pairs is usually present in the female and nonidentical pairs in males, although the reverse also occurs; also called *allosome*; compare *autosome*.

Sex hormones : Hormones capable of stimulating the development of the reproductive organs and secondary sex characters in both sexes of mammals; three types of sex hormones are recognized: the estrogens (female hormones), the androgens (male hormones) and the gestogens (corpus luteal hormones); synthesized in mammals by the ovary (or testis), adrenal cortex, corpus luteum, and the placenta.

Shine-Dalgarno sequence : A sequence in an mRNA required for binding prokaryotic ribosomes.

SI units : Those units of measurement forming the *Systeme International d'Unites*, consisting of the metre, kilogram, second, mole, kelvin, ampere, candela and katal.

Sickle-cell anemia : A genetic human disease resulting from a hemoglobin mutation; caused by a homozygous allele coding for the β chain of hemoglobin; produces fragile erythrocytes, leading to anemia.

Signal recognition particles (SRPs) : Cytoplasmic particles that dock ribosomes on the surface of the ER if the nascent polypeptide is destined to be processed by the ER. The SRP recognizes and binds to a specific N-terminal signal sequence on the nascent polypeptide.

Signal sequence : *See* **leader sequence.**

Silk : A thread-like substance produced by the silkworm; composed mainly of the proteins, sericin and fibroin.

Simple protein : A protein yielding only amino acids on hydrolysis.

Single bond : A covalent bond that shares only one electron pair.

Small nuclear RNAs (snRNAs) : RNAs 100-300 nucleotides in length that are involved in the splicing reaction in eukaryotes and present in spliceosomes.

Soap : A compound of fatty acid and potassium or sodium hydroxide; soaps are important cleansing agents because of their emulsifying action.

Sol : The fluid state of a colloid.

Soluble : Refers to polar molecules that dissolve in water and are surrounded by a hydration shell.

Solute : The molecules dissolved in a solution.

Solution : A mixture of molecules, such as sugars, amino acids, and ions, dissolved in water.

Solvent : The most common of the molecules dissolved in a solution; usually a liquid, commonly water; the solvent is the larger part of the solution.

Somatic cell (*soma*[G] = body) **:** Any cell of a plant or animal other than a germ cell or germ-cell precursor.

Somatotrophic hormone (STH) : A anterior pituitary protein hormone with a molecular weight 27,000 and consisting of 190 amino acid residues; acts directly upon various tissues to produce diverse effects; affects the rate of skeletal growth and gain in body weight; stimulates chondrogenesis followed by ossification, in adult animals with closed epiphyses; also called *somatotropin* or *growth hormone (GH).*

Sorbitol, $CH_2OH\ (CHOH)_4CH_2OH$: A white, crystalline, sweet, soluble polyhydric alcohol with a m.p. 110°C (for the dextrorotatory compound); obtained from dextrose; used as a sugar substitute and in the manufacture of synthetic resins.

Species, *plural* also **species** (*speci*[L] = a kind or sort) **:** A basic category in the classification system between a genus and a variety; a group of plants or animals having in common one or more distinctive characters and being capable of interbreeding and reproducing their characters in their offspring, thus remaining relatively stable in nature; the scientific names of species are binomials consisting of a genus (generic) name and a specific epithet; unitary in evolution and ecological role.

Species-specific : Limited in reaction or effect to one species.

Specific activity : The number of micromoles (μ mol) of a substrate transformed by an enzyme preparation per minute per milligram of protein at 25°C; a measure of enzyme purity.

Specific gravity : The former term for the ratio of the density of a substance to that of water. As the word *specific* now has a different usage, the term *relative density* is now used for this concept.

Specific heat : The amount of energy (in joules or calories) needed to raise the temperature of 1 g of a pure substance by 1°C.

Specific rotation : The rotation, in degrees, of the plane of plane-polarized light (D-line of sodium) by an optically-active compound at 25°C, with a specified concentration and light path.

Specificity : The ability of an enzyme or receptor to discriminate among competing substrates or ligands.

Sperm (*sperma*[G] = sperm, seed) **:** A sperm cell or the male gamete.

Sphingolipids : An amphipathic lipid with a sphingosine backbone to which are attached a long-chain fatty acid and a polar alcohol.

Spliceosome : A large complex of nucleic acids and proteins in the nucleus that catalyzes the splicing reaction in eukaryotes.

Splicing : *See* **gene splicing**.

Spongy parenchyma : Parenchyma cells that are chlorophyllous and rounded, loosely arranged together, and located between the lower epidermis and the palisade parenchyma of a leaf.

Standard free-energy change (ΔG°) : The free-energy change for a reaction occurring under a set of standard conditions: temperature 298 K, pressure 1 atm or 101.3 kPa, and all solutes at 1M concentration; $\Delta G^{\circ\prime}$ denotes the standard free-energy change at pH 7.0.

Standard reduction potential (E_0') : The electromotive force exhibited at an electrode by 1M concentrations of a reducing agent and its oxidized form at 25°C and pH 7.0; a measure of the relative tendency of the reducing agent to lose electrons.

Standard state : A reference state, with respect to which thermodynamic quantities (such as chemical potentials) are defined. For substances in solution, standard state indicates 1M concentration at 1 atm pressure and 25° C.

Starch : The most important homopolysaccharide, acting as reserve food material of the higher plants; occurs in cereals, legumes, potatoes and other vegetables; consists of 2 components: an unbranched straight-chain component, *amylose* (15 – 20%) and a branched chain component, *amylose* (80 – 85%); a white, soft, unsweet, amorphous powder, insoluble in water and alcohol; breaks down into large fragments called *dextrins* on heating, especially in the presence of moisture; readily hydrolyzed by mineral acid with the final production of glucose.

Start codon : *See* **initiation codon.**

Steady state : A nonequilibrium state of a system through which matter is flowing and in which all components remain at a constant concentration.

Stearic acid (*stear*[G] = hard fat) **:** A C-18 saturated fatty acid; commonly found in animal and plant fats.

Steatorrhea (from two Greek words meaning *fat + a flow*, respectively) **:** A disease characterized by stools with increased fluidity due to excess of fat content.

Stereoisomers : Compounds that have the same molecular formula and the same structure, but differ only in spatial configuration of the atoms in the molecule.

Steroid (*stereos*[G] = solid + *ol*[L] = from oleum, oil) **:** A hydrophobic molecule related to cholesterol; derivatives of a fused and fully saturated ring system called cyclopentanoperhydr-ophenanthrene or sterane, which consists of 3 cyclohexane rings fused in nonlinear phenanthrene manner and a terminal cyclopentane ring; many important hormones such as estrogen and testosterone are steroids.

Steroid hormones : Hormones derived from cholesterol; C_{18} and C_{19} steroids promote the development of the female and male secondary sex characters, respectively; C_{21} steroids are concerned with the transport of electrolytes, with the metabolism of carbohydrates, proteins and fats and also with the implantation of the fertilized ovum.

Sterols : A class of lipids containing the steroid nucleus with hydroxyl groups, *i.e.*, they are hydroxy steroids; resist saponification; common example is cholesterol.

Stoichiometry : The part of chemistry dealing with the composition of substances; more particularly with the determination of combining proportions or chemical equivalents; also spelt as *stoicheiometry*.

Stop codons : RNA codons that signal a ribosome to stop translating an mRNA and to release the polypeptide. In the normal genetic code, there are 3 stop codons, : UAA, UAG and UGA.

Strain : A population of individuals that all share the same trait. Strains are maintained by interbreeding individuals within the same strain.

Streptomycin, $C_{21}H_{39}O_{12}N_7$: An antibiotic substance produced by the fungus, *Streptomyces*; effective against several types of disease bacteria, including some against which *Penicillin* is inactive; used in the treatment of tuberculosis.

Stroma (Greek, for *anything spread out*) **: (1)** In chloroplasts, the fluid matrix between the grana inside the chloroplast; does not include the contents within the thylakoid membranes; the site of the biochemical (*i.e.*, "dark") reactions of photosynthesis. **(2)** The connective tissue in which a glandular or other epithelium is embedded.

Structural gene : A gene or a region of DNA that codes for an RNA molecule and consequently the protein; as distinct from a regulatory gene that regulates gene expression.

Structural polysaccharide : A polysaccharide that holds cells and organisms together; *cellulose* is the most abundant structural polysaccharide in plants.

Suberin : A waxy waterproof substance that occurs in cork cells and in the cells of underground plant parts; consists of hydroxylated fatty acids that are linked together in a complex array.

Sublimation : The conversion of a solid direct into vapour, and subsequent condensation, without melting.

Substrate : **(1)** The specific compound acted upon by an enzyme molecule. **(2)** The medium on which an organism (especially a microorganism) can grow.

Substrate-level phosphorylation : Phosphorylation of ADP (or some other nucleoside 5'-diphosphate) coupled to the dehydrogenation of an organic substrate; results in ATP production; independent of the electron transport system used in oxidative phosphorylation.

Subunit : A polypeptide that combines with other polypeptides to comprise a multisubunit protein.

Succinic acid : A C-4 organic acid that is oxidized by the reduction of ubiquinone to ubiquinol in the sixth step of the Krebs cycle; the product of this oxidation is fumaric acid.

Succinyl-CoA : An acetylated C-4 organic acid that is converted to succinic acid by losing its acetyl-CoA group, thereby driving the substratelevel phosphorylation of one molecule of ADP to ATP in the fifth step of the Krebs cycle.

Sucrase : *See* **invertase.**

Sucrose ($C_{12}H_{22}O_{11}$) : A common, nonreducing, disaccharide made up of a molecule of glucose and one of fructose; obtained from sugarcane and sugar beet; used as a sweetening agent; also called *table sugar* or *saccharose*.

Sugar : A general term usually applied for any monosaccharide or disaccharide.

Sulfhydryl (–SH) : A chemical group containing sulfur and hydrogen which is present in the amino acid cysteine and other molecules; two sulfhydryl groups can join to produce a disulfide bond which is present in cystine; also called *thiol*.

Sulfonamide drugs : A group of organic compounds, containing the sulfonamide group, SO_2 NH_2 or its derivative; the group includes sulfanilamide, sulfadiazine, sulfadiazine, sulfathiazole and many others; some of the sulfonamides are sulfa drugs and act as powerful inhibitors of bacterial activity.

Supercoiled DNA : A region of DNA in which the double helix is further twisted on itself.

Supernatant : The clear liquid above a precipitate which has settled out.

Surface tension : A tautness of the surface of a liquid, caused by the cohesion of the liquid molecules. Alternatively, it is the attraction between a surface and a particle lying on it. Water has an extremely high surface tension.

Svedberg (S) : A unit of measure of the rate at which a particle sediments in a centrifugal field. It is equal to 10^{-13} second.

Sweat gland : A structure present in the skin of some mammals that secretes sweat containing NaCl as part of the cooling system. A typical human has about 2.5 million sweat glands. Dogs, cats and rabbits are amongst the mammals that do not sweat but use evaporation from the upper respiratory tract to lose excess body heat.

Symport : A form of cotransport in which a membrane carrier protein transports two solute species across the membrane in the same direction. Compare *antiport*.

Syndrome (*syn*[G] = together + *dramein*[G] = to run) **:** Refers to a disease characterized by a group of symptoms; often named after their discoverer, for example Down's syndrome, Klinefelter's syndrome and Turner's syndrome.

Synergism : A chemical phenomenon in which the combined activity of two or more compounds is greater than the sum of the individual activities. For example, auxin and cytokinin act synergistically in promoting DNA replication.

Synovial fluid : A viscous fluid contained within a membrane enclosing moveable joints such as the elbow and knee. The fluid lubricates the cartilages which make the surface contact between the bones at the joints.

Synthases : Enzymes that catalyze condensation reactions in which no nucleoside triphosphate is required as an energy source.

Synthetases : Enzymes that catalyze condensation reactions using ATP or another nucleoside triphosphate as an energy source.

System : An isolated collection of matter; all other matter in the universe apart from the system is called the *surroundings*.

T

T cell : A type of lymphocyte involved in cell-mediated immune responses and interactions with B cells; includes both cytotoxic T cells and helper T cells; also called *T lymphocyte*.

T lymphocyte : *See* **T cell.**

Tachycardia (*tachys*^{G} = fast + *kardia*^{G} = heart) : Rapid beating of the heart.

Tallow : The rendered hard fat of animals, particularly cattle and sheep; composed of long, saturated fatty acids.

Tannin : A bitter astringent organic substance; found widely in plant sap, particularly in bark, leaves and unripe fruits; used in the production of leather and ink.

Target cell : Any cell that responds to specific hormones.

Tartrazine : An azo dye that produces a yellow colour; widely used as a food additive (E102).

TATA box : Consensus sequence in the promoter region of many eukaryotic genes that binds a general transcription factor and hence specifies the position where transcription is initiated.

Tautomers (*tauto*^{G} = same, *meros*^{G} = part) : A term coined by Laar in 1225; structural isomers that differ in the location of their hydrogens and double bonds; aldehydes, ketones and other carbonyl compounds such as esters exhibit tautomerism.

Taxol : A drug obtained from Pacific yew, and also from a fungus that grows on the yew; has potential for treating certain forms of cancer.

Taxon, *plural* **taxa :** Any taxonomic category within the classification of organisms, such as class, order, family, genus or species.

Tay–Sachs disease : A common hereditary disorder, in which children who are apparently normal at birth show signs within 6 months of marked deterioration of brain and spinal cord. By the age of one year, the child can only lie helplessly, becoming mentally retarded, increasingly blind, and paralyzed. Death occurs between 3 and 4 years with no known survivors and no cure. The disease is caused due to a deficiency of the enzyme hexosaminidase A (= *N acetylgalactosaminidase*), which cleaves a specific bond (β-8 → 4) between an N-acetyl-D-galactosamine and a D-galactose residue in the polar head of the ganglioside, G_{M2}. In effect, G_{M2} is not degraded to G_{M1}. With the result, G_{M2} is accumulated in large amounts in the lysosomes, particularly in the brain cells, causing degeneration of the nervous system. The disease is controlled by the recessive allele of a gene located on chromosome 15, double recessives producing a deficient amount of the enzyme hexosaminidase A which leads to the accumulation of complex fatty substances in the central nervous system. Although the disease is rare in the population at large (1 in 3,00,000 births), it has a very high incidence (1 in 3,600 births) in Ashkenazic Jews.

Temperature : The degree of hotness or coldness, usually related to a zero at the melting point of ice (Celsius scale) or absolute zero (Kelvin scale).

Temperature quotient : *See* Q_{10}.

Template : A macromolecular mould or pattern for the synthesis of an informational macromolecule.

Tendon (*tenon*^{G} = stretch) : A bunch of parallel collagen fibres making up a band of connective tissue which serves to attach a muscle to a bone.

Tensile strength : The tensile (pulling) stress that has to be applied to a material to break it; measured as a force per unit area, *e.g.*, *newtons* per square metre, *dynes* per square centimetre, *pounds* or *tons* per square inch.

Teratogen : A chemical that causes developmental abnormalities in an organism.

Terminal transferase : An enzyme that catalyzes the addition of nucleotide residues of a single kind to the 3' end of DNA chains.

Termination codons : UAA, UAG and UGA; in protein synthesis, these three codons signal the termination of a polypeptide chain; also known as *stop codons* or *nonsense codons*.

Termination factors : Protein factors of the cytosol required in releasing a completed polypeptide chain from a ribosome; also known as *release factors*.

Termination sequence : A DNA sequence that appears at the end of a transcriptional unit and signals the end of transcription.

Terpenes : Organic hydrocarbons or hydrocarbon derivatives derived from recurring 5-C precursor units called *isoprene units*; terpenes produce some of the scents and tastes of plant products, for example, the scents of geranium leaves and pine needles; examples include menthol (2 isoprene units), β-carotene (8 isoprene units) and rubber (500 to 5,000 isoprene units).

Tertiary structure : The three-dimensional conformation of a polymer in its native folded state, especially that of a protein; in the case of protein, conformation (shape) of the molecule is maintained by disulfide bonds, ionic interactions or hydrophobic attraction between amino acids.

Testa : A protective coat around the seed, formed from the *integuments* of the ovule. At one area of the testa is the *hilum*, at one end of which is often found the *micropyle*. Sometimes, the testa is responsible for seed dormancy, a state which is broken when the coat ruptures.

Testis, *plural* **testes** (Latin, for *witness*) : In animals, the organ producing male gametes; in mammals a testicle. It also produces androgens, the male sex hormones.

Testosterone (*testis*^{G} = testicle + *steiras*^{G} = barren) : A steroid hormone secreted by the testes that is responsible for the development of secondary sex characters of males.

Tetany (*tètanie*^{Fr}, from *tetanos*^{G} = stretching) : An abnormal increase in nerve and muscle excitability resulting in spasms of the arms and legs, caused by a deficiency of parathyroid secretion, called parathormone (PTH) or Collip's hormone. Tetany can be relieved either by the administration of a soluble calcium salt or of PTH. Also called *muscular twitchings*.

Tetanus (*tetanos*^{G} = stretching) : A soil-borne disease produced by toxins from the bacterium *Closteridium tetani* which usually enters the body through a wound, producing spasm of the voluntary muscle, especially of the jaw (lockjaw). Bacterial tetanus can be treated by administering antitetanus serum containing ready-made antibodies, or by antitetanus vaccine which induces the formation of antibodies by the recipient.

Tetracycline : An antibiotic of the *broad spectrum type*, so called because of its effects on a wide variety of bacterial types. Since they can adversely affect host cells as well as bacteria, tetracyclines are prescribed medicinally with extreme caution.

Tetrahydrobiopterin : The reduced coenzyme form of biopterin.

Tetrahydrofolate : The reduced, active conezyme form of the vitamin B_9 (folic acid).

Thalassemia (from two Greek words, meaning *sea* + *blood*, respectively) : *See* **Cooley's disease**.

Theory (*theorein*[G] = to look at) **:** A well-tested hypothesis supported by a great deal of sound, convincing factual evidence, obtained as a result of properly-controlled experiments.

Thermoduric : Heat-enduring.

Thermodynamics (*therme*[G] = heat + *dynamis*[G] = power) **:** The study of transformations of energy, using heat as the most convenient form of energy measurements. There are two well-defined laws (first and second) of thermodynamics.

Thermoregulation : The control of body heat in homoiotherms.

Thiamine : A water-soluble vitamin of the B complex group and designated as vitamin B_1; found practically in all plant and animal foods, and cereals, heart, liver and kidney are excellent sources; a white crystalline substance and destroyed at high temperatures unless the pH is low; acts as a coenzyme in sugar breakdown by forming part of the NAD molecule; deficiency of thiamine results in a human disease called beriberi; also called *antiberiberi factor* or *antineuritic factor*.

Thiamine pyrophosphate : The active coenzyme form of vitamin B_1; involved in aldehyde transfer reactions.

Thioester bond : A high-energy bond formed by a condensation reaction between an acid (acyl) group and a thiol (–SH) group; present in acetyl-CoA and many enzyme-substrate complexes.

Thiols : A class of organic compounds of the general formula, RSH, with sulfur attached directly to carbon. They are the sulfur analogues of alcohols, containing SH instead of OH groups. Formely called *mercaptans*.

Threshold : The lowest value of any stimulus, signal, or agency that results in a specified effect or response, *e.g.*, threshold frequency. Below the threshold value, there is no response despite the application of a stimulus.

Thrombin : An enzyme formed in the blood of vertebrates that acts upon fibrinogen to form fibrin; it is hence, essential to the process of blood clotting; formed from a blood protein called prothrombin.

Thrombocytes : *See* **platelets**.

Thrombosis (Greek, for *a curdling*) **:** Stoppage of blood circulation by the formation of a clot (thrombus) of blood in a blood vessel; clotting.

Thromboxanes : A class of molecules derived from arachidonate and involved in platelet aggregation during blood clotting.

Thrombus (Greek, for *clot*) **:** A coagulation (clotting) that forms in a blood vessel and obstructs circulation.

Thylakoid (*thylakos*[G] = sac + *-oides*[G] = like) **:** A flattened, saclike, chlorophyll-bearing membrane in the chloroplast of a eukaryote; thylakoids are stacked on top of one another in arrangements called grana; carries out the light-gathering reactions of photosynthesis.

Thymine : A pyrimidine occurring in DNA but not in RNA; always base-pairs with a DNA purine base called adenine.

Thymine dimer : *See* **pyrimidine dimer**.

Thymus : An endocrine gland present in all jawed vertebrates, situated in the neck region of most vertebrates but close to the heart in mammals. In humans, it is a flat, pinkish, bilobed structure located in the chest behind the sternum. The gland reaches its greatest development at the age of 14 to 16, after which it atrophies because of the activity of sex cells. It produces lymphocytes which then move to lymph nodes. The thymus produces a hormone called *thymosin* which causes the lymphocytes to form antibody-producing plasma cells immediately after birth, but regress in adult animals.

Thyroglobulin : A protein containing and storing tetraiodothyronine (= thyroxine) and triiodothyronine in the thyroid gland.

Thyroid gland (*thyreoides*[G] = shield-shaped) **:** The largest endocrine gland in human body. The gland consists of two lobes near the trachea, connected across with a narrow bridge called *isthmus*, making the entire gland more or less H-shaped in appearance. Thyroid receives blood flow about 5 times its own weight per minute. Thyroid produces 2 hormones : tetraiodothyronine (= thyroxine) and triiodothyronine, of which the latter is 5 to 10 times more potent in biologic activity than the former. Thyroxine controls the basal metabolic rate (BMR) in humans. Undersecretion of thyroid leads to *cretinism* in children and *myxedema* in adults. Its oversecretion causes *exophthalmic goitre*, where the thyroid, and thus the neck, swells and the eyewalls protrude. Hyperthyroidism can be cured by surgical removal of the thyroid or by treating with antithyroid drugs.

Thyroid-stimulating hormone (TSH) : A hormone secreted by the anterior lobe of the pituitary gland which triggers the shedding of thyroxine from the colloidal follicles of the thyroid gland into the bloodstream. Excess thyroxine inhibits the production of TSH – an example of a negative feedback mechanism. Also called *thyrotropin*.

Thyrotropin : *See* **thyroid-stimulating hormone.**

Thyroxine (*thyros*[G] = shield) : An iodine-containing hormone secreted by the thyroid that increases metabolic rate and promotes growth.

Tincture of iodine : A 2% iodine solution in ethyl alcohol.

Tints : Colours that have the same hue but different saturation.

Tissue (*texere*[L] = to weave) : A group of cells organized into a structural and functional unit such as muscle and xylem; simple tissues are made up of similar cells, and complex tissues of different kinds of cells.

Tissue culture : A technique for growing and manipulating pieces of tissue in a liquid medium after their removal from the multicellular organism.

Titration curve : A plot of the pH versus the equivalents of base added during titration of an acid.

Titre : The concentration of a substance in solution, for example, the amount of specific antibody in a serum. The concentration is usually expressed as the reciprocal dilution. For example, when 1 : 250 gives a positive test and greater dilutions are negative, the titre is 250.

Tocopherols : A fat-soluble vitamin and designated as vitamin E; of widespread occurrence in many plant oils such as wheat germ, rice, corn, cottonseed, soybean but not olive oil; also present in small amounts in meat, milk, eggs, leafy plant and some fruits; fish liver oils are devoid of vitamin E; there are at least 7 types of tocopherols, designated as alpha (α) through eta (η), of which α-tocopherol is biologically most potent; α-tocopherol is a light yellow oil, resistant to heat (up to 200°C) and acids but acted upon by alkalies, easily but slowly oxidized and is destroyed by UV rays; act as antioxidants, control O_2 consumption and participate in nucleic acid metabolism; avitaminosis E leads to sterility in rats and muscular dystrophy in rabbits and guinea pigs; there is, however, little evidence that man is ever short of vitamin E; also called *antisterility factor*.

Tonoplast : The cytoplasmic membrane bordering a vacuole and controlling movements of ions into and out of the vacuole.

Topoisomerases : Enzymes that make reversible cuts in a double helical DNA molecule for the purpose of removing knots or unwinding excessive twists. They change the supercoiling of DNA helices by either allowing the superhelical torsion to relax (thus reducing the supercoiling) or adding more twists (thus increasing the supercoiling). Topoisomerase I makes one incision in double-stranded DNA and allows rotation of the other strand and reseals the end. Topoisomerase II makes double-stranded cuts, allows rotation and then reseals; also called *DNA isomerases*.

Torsion : 'Twisting' about an axis, produced by the action of two opposing couples acting in parallel planes.

Totipotent cell : A cell that has the capability to give rise to all structure or cell types.

Toxins : Proteins produced (or secreted) by some organisms and toxic to certain other species.

Trace element : A chemical element required by an organism in only trace amounts of less than 10^{-5} M. Absence can cause disease and death.

Tracer : A molecule or atom that has been labelled either chemically or radioactively so that it can be followed in a biochemical process or readily located in a cell or tissue.

Trachea, *plural* **tracheae** (Latin, for *windpipe*) : **(1)** In air-breathing vertebrates, the windpipe. **(2)** In insects, a tube distributing air to internal cells.

Transition State : A structure that forms transiently, in the course of a chemical reaction and has the highest free-energy of any reaction intermediate; a rate-limiting step in the reaction.

Tranquilizer : A drug used to reduce tension and anxiety, without impairing alertness or causing drowsiness.

Transaminases : *See* **aminotransferases**

Transamination : In the cell, the enzymatic transfer of an amino group from an α-amino acid to an α-keto acid. The keto acid becomes an amino acid and vice versa.

Transcript : A RNA product of DNA transcription.

Transduction : (1) Generally, the conversion of energy or information from one form to another **(2)** The transfer of DNA from one bacterium to another, using a virus as a vector.

Transcription (*trans*^{L} = across + *scribere*^{L} = to write) : Copying of one strand (sense strand) of DNA into a complementary sequence of bases in an mRNA molecule by the enzyme RNA polymerase; in eurkaryotes, transcription takes place in the nucleus. Compare *translation.*

Transfection : Introduction of a foreign DNA molecule into a eukaryotic cell; usually followed by expression of one or more genes in the newly-introduced DNA.

Transfer RNA, tRNA (*trans*^{L} = across + *ferre*^{L} = to bear or carry) : A class of small molecules (M.W. 25,000–30,000) that float free in the cytoplasm and each of which combines covalently with a specific amino acid and to a codon on messenger RNA and later transfers the amino acid to mRNA in ribosome for assembly into proteins.

Transferase : Any enzyme that catalyzes the transfer of a chemical group (such as amino, methyl or alkyl) from one substrate to another substrate.

Transformation (*trans*^{L} = across + *formare*^{L} = to shape) : **(1)** The incorporation of a piece of foreign DNA into the genome of a bacterial cell, causing the recipient to acquire a new phenotype. The process is important historically since, following transformation experiments by **Frederick Griffith** on *Pneumococcus* bacterium, DNA was shown to be the genetic material of cells by **Avery, MacLeod** and **McCarthy**. **(2)** In the case of cultured animal cells, the term usually refers to the acquisition of cancerlike properties following treatment with a virus or a carcinogen.

Transgenic : Refers to an organism that has genes from another organism incorporated within its genome as a result of recombinant DNA technology and can pass them on to successive generations.

Translation (*trans*^{L} = across + *latus*^{L} = that which is carried) : The process by which the sequence of nucleotides in a messenger RNA molecule directs the incorporation of amino acids into protein; occurs on a ribosome. Compare *transcription.*

Transition state : Structure that forms transiently in the course of a chemical reaction that has the highest free energy of any reaction intermediate; a rate-limiting step in the reaction.

Translational repressor : A repressor that binds to an mRNA, blocking translation.

Translocase : (1) An enzyme that catalyzes membrane transport **(2)** An enzyme that causes a movement, such as the movement of a ribosome along an mRNA.

Translucent : Permitting the passage of light in such a way that an object cannot be seen clearly through the substance; for example, an oily paper or a frosted glass.

Transplantation : The transfer of an organ or tissue from a donor to a recipient in need of a healthy organ or tissue. Kidney, lung, heart and liver transplants are in vogue these days. For successful transplantation to occur, similar tissues types must be involved and genetical similarities is one of the best ways of ensuring this.

Transporters : Proteins that span a membrane and transport specific nutrients, metabolites, ions, or proteins across the membrane; sometimes called *permeases*.

Transposition : The movement of a gene or set of genes from one site in the genome to another.

Transposon : A segment of DNA that can multiply and move spontaneously from one position in the genome to another; also called *transposable element*.

Traumatic acid : *See* **wound hormone**.

Triacylglycerol : An ester of glycerol with three molecules of fatty acids which may be saturated (in animal fats) or unsaturated (in vegetable oils); also called *triglyceride* or *neutral fat*.

Tricarboxylic acid cycle (TCA cycle) : *See* **citric acid cycle.**

Trichromatic theory : The theory that all colours can be produced by the mixing of blue, green and red.

Triglyceride : *See* **triacylglycerol**

Trihydroxybenzene : *See* **pyrogallol**.

Triiodothyronine (TIT or T_3) : A minor hormone of the thyroid gland that has the same function as that of tetraiodothyronine (T_4) but is 5 to 10 times more potent in biologic activity than T_4.

Triose : A simple sugar with a backbone containing three carbon atoms.

Triphosphopyridine nucleotide (TPN) : *See* **nicotinamide adenine dinucleotide phosphate.**

Triplet : Refers to a sequence of three nucleotides that together comprise a codon.

Trisaccharide : An oligosaccharide whose molecule has 3 linked monosaccharide moieties. Rhamninose, gentianose and raffinose are common examples.

Triterpene : A compound that consists of six isoprene units linked together; lanosterol and squalene are common examples of triterpenes.

Tritium, T : ^3_1H. A radioactive isotope of hydrogen with mass number 3 and atomic mass 3.016. The abundance of tritium in natural hydrogen is only one atom in 10^7, and its half-life is 12.5 years. It can, however, be made artificially in nuclear reactors, and tritiated compounds are used in radioactive tracing.

Tropic hormone : A peptide hormone that stimulates a specific target gland to secrete its hormone; for example, thyrotropin produced by the pituitary stimulates secretion of thyroxine by the thyroid gland; also called *tropin*.

Tropin : *See* **tropic hormone.**

Trypsin : A proteolytic enzyme produced by the pancreas; in the process of digestion, it breaks up proteins into amino acids; secreted as an inactive precursor trypsinogen, which is converted to trypsin by enterokinase secreted in the small intestine.

Tuberculosis (*tuberculum*[L] = a small swelling) **:** A disease caused by *Mycobacterium tuberculosis*.

Tumour (Latin, for *swollen*) **:** A mass of cells, growing in an uncontrolled manner.

Turgor pressure : The pressure developed as a result of diffusion of water into a living cell; pressure exerted against the cell wall.

Turnover number : The number of times an enzyme molecule transforms a substrate molecule per unit time, under conditions giving maximal activity at substrate concentrations that are saturating. For most enzymes, the turnover number falls between 1 to 10^4 per second. It is equivalent to the catalytic rate constant, k_{cat}.

Turpentine : The volatile, combustible component of resin.

Twist (*T*) **:** With respect to a DNA double helix, the total number of times the 2 strands of the helix cross over each other, excluding writhing. It is a measure of how tightly the helix is wound. See also *writhe.*

Tyndall effect (named after *John Tyndall*, LT, 1820–1893) **:** The scattering of light by particles of matter in the path of the light, thus making a visible 'beam' such as is caused by a ray of light illuminating particles of dust floating in the air of a room.

Tyrosinase : An oxidizing enzyme of the copper oxidase type, attacking the amino acid tyrosine.

Tyrosine kinase : An enzyme that transfers the terminal phosphate of ATP to a specific tyrosine residue on its target protein.

2,4,5-trichlorophenoxyacetic acid (2,4,5-T) : A plant hormone widely used as a herbicide to kill woody species.

U

Ubiquinol : The reduced form of ubiquinone; ubiquinol donates electrons to cytochrome b in the electron transport chain.

Ubiquinone : A lipid-soluble quinone; ubiquinone accepts electrons from electron donors like NADH and from the oxidation of fatty acids; also called *coenzyme Q*.

Ubiquitin : A small, highly conserved protein present in all eukaryotic cells that becomes covalently attached to lysines of other proteins. Attachment of a chain of ubiquitones tags a protein for intracellular proteolytic destruction in a proteasome.

Ultracentrifuge : A high-speed centrifuge that is capable of spinning 1,00,000 revolutions per minute (rpm), producing up to 10,00,000 g forces. The high speeds enable the separation of tiny particles, which are identified by the rate at which they move down the centrifuge tube. The units of rate are called Svedberg (S), after the inventor of the centrifuge.

Ultrafilteration : The technique of filtering a solution under pressure through a semipermeable membrane, which allows water and small solutes to pass through but retains macromolecules.

Ultrasound : A noninvasive procedure that uses sound waves to produce an image of the fetus but that harms neither the mother nor the fetus.

Ultraviolet (UV) radiations : Electromagnetic radiation in the region of 200 to 400 nm.

Umbilical cord : The cord that joins the embryo of placental mammals to the placenta consisting of blood vessels supported by connective tissue. The cord is usually severed at birth.

Uncoupling agent : A substance that uncouples phosphorylation of ADP from electron transfer; for example, 2,4-dinitrophenol (DNP).

Unicellular : Composed of a single cell.

uniport : A transport system that carries only one solute, as distinct from cotransport.

Universal acceptor : An individual with an AB blood type who may receive any type of blood.

Universal code : The genetic code which is exactly the same in all species.

Universal donor : A type O blood donor, whose blood is compatible with any individual's immune system.

Unsaturated fatty acid : A fatty acid containing one or more double bonds; oleic, linoleic, and linolenic are common examples of unsaturated fatty acids, containing one, two and three double bonds in their molecules, respectively.

Uracil : A pyrimidine found in RNA (during transcription) but not in DNA; always base-pairs with a purine base called adenine in DNA (during transcription) or RNA (during translation).

Urea cycle : A metabolic pathway in vertebrates, for the synthesis of urea from amino groups and carbon dioxide; occurs in the liver. The reaction is essentially:

$$2\ NH_3 + CO_2 \longrightarrow NH_2\ CONH_2 + H_2O$$

Also called *ornithine cycle*.

Urea, H_2NCONH_2 (*ouron*^{G} = urine) **:** An organic molecule formed in the vestebrate liver; the principal form of disposal of nitrogenous wastes by mammals.

Urease : An enzyme which hydrolyzes urea to ammonia and carbon dioxide.

Ureotelic : Excreting excess nitrogen in the form of urea.

Urethra (*ourein*^{G} = to urinate) **:** A tube which passes both semen (during ejaculation) and urine (during liquid elimination) in males but only urine in females.

Uric acid, $C_5H_4O_3N_4$: An organic compound belonging to the purine group; a colourless crystalline solid that is slightly soluble in water; occurs in very small amounts in the urine of some animals (reptiles and aves) as a breakdown product of amino acids and nucleic acids; being quite insoluble in water, it is thus nontoxic when released during embryonic development within the egg; also permits the removal of nitrogen with a minimum of water loss and is eliminated as a thick paste or even dry pellets; sodium and potassium salts of the acid are deposited in the joints in cases of gout.

Uricotelic : Excreting excess nitrogen in the form of uric acid (urate).

Urine (*ouron*^{G} = urine) **:** The liquid waste product of metabolism filtered from the blood by the kidneys. In mammals, elasmobranch fishes, amphibia, tortoises and turtles, nitrogen is excreted in the form of urea which, in humans, forms 2% of the urine on average.

Uterus (Latin, for *womb*) **:** A mammals, a chamber in which the developing embryo is contained and nurtured during pregnancy.

V

Vaccination : The injection of a harmless microbe into a person or animal to confer resistance to a dangerous microbe.

Vaccine (*vacca*^{L} = cow) **:** A suspension of dead or weakened bacteria or other pathogens which may be injected into an organism to immunize against the same species or kind of pathogen or its toxins. Vaccines are, thus, not quick-acting, but rely on the recipient to build up a supply of antibodies gradually. For example, the Salk polio vaccine contains attenuated viruses.

Vacuole (*vacuus*^{L} = empty) **:** A cavity or vesicle in the cytoplasm of a cell that is bound by a single membrane called tonoplast and contains a watery solution of pigments and waste products of cell metabolism called *cell sap*; typically found in plant cells.

Vagina (Latin, for *sheath*) **:** The membranous passage in female animals that leads to the mouth of the uterus; the female copulatory organ into which the penis is introduced during copulation.

van der Waals interactions : The weak bonds formed between electrically-neutral molecules or parts of molecules when they lie close together. Such interactions are common in the secondary and tertiary structure of protein.

Varicosities : Swellings that occur on blood vessels or nerve fibre.

Vascular bundle : In vascular plants, a strand of tissue containing primary xylem and primary phloem (and possibly procambium) that runs up through the roots, into the stems and out into the leaves; often enclosed by a bundle sheath; conducts water with dissolved minerals and carbohydrates throughout the plant body.

Vasoconstriction : A narrowing of the blood vessels, often in response to cold, which through a contraction of involuntary muscles in the walls of the vessels brought about by a stimulus from the sympathetic nervous system.

Vasodilation : The expansion of blood vessels by relocation of muscles, mainly controlled by the sympathetic nervous system.

Vasopressin : A posterior pituitary hormone that regulates the kidney's retention of water; also referred to as *antidiuretic hormone (ADH).*

Vector : (1) In cell biology, an agent (virus or plasmid) that can take up a foreign gene and transmit it into the genome of a target organism. For example, the *Anopheles* mosquito transmits the malarial parasite. **(2)** In plant reproduction, an animal that carries pollen (a pollinator) from a pollen sac to a stigma or to an ovule.

Velamen : Multiple epidermis covering the aerial roots of some orchids and aroids.

Venereal disease : Any contagious disease transmitted usually during sexual intercourse, such as gonorrhoea or syphilis.

Ventral (*venter*[L] = belly) **:** Situated toward the belly surface of an animal or that side which is normally directed downwards in the usual stance or resting position. In bipedal primates such as humans, the ventral side is the front, but would obviously be the underside if a four-legged gait were assumed; opposite of *dorsal*.

Vertebrate : An animal having a backbone made of bony segments called vertebrae, such as fish, amphibia, reptiles, birds and mammals. In addition, they are characterized by having a skull which surrounds a well-developed brain and a bony or cartilaginous skeleton.

Vertigo (*vertigo*[L] = a whirling, from *vertere* = to turn) **:** Dizziness.

Vesicle (*vesicula*[L] = a little bladder) **:** A small, membrane-bounded spherical organelle in the cytoplasm of a eukaryotic cell; created by weaving sheets of endoplasmic reticulum through the cell's interior.

Viral vector : A virus DNA altered so that it can act as a vector for recombinant DNA.

Virion : A complete virus particle outside its host cell.

Viroid : A plant-infecting viruslike particle with a single circular strand of RNA that is not associated with any protein.

Virulence : The collective properties of an organism that render it pathogenic to another one, the host.

Virus : An ultramicroscopic pathogenic particle, capable of passing through bacteriological filters; consists of nucleic acid (DNA or RNA) enclosed in a protein coat; capable of replicating within a living host cell only and spreading from cell to cell; infect cells of bacteria, plants and animals, and whilst viruses carry out no metabolism themselves, they are able to control the metabolism of the infected cell.

Virusoid : A particle similar to a viroid but located inside the protein coat of a true virus.

Viscosity : (1) The property of stickiness by which substances resist change or shape. (2) A measure of the ease with which layers of fluid pass each other.

Visible light : The range of colours from violet (380 nm) to red (750 nm) that a human eye can see.

Vitamin (*vita*[L] = vital for life + *amine* = of chemical origin) **:** An organic compound required by an organism, which cannot synthesize it in minute quantities for growth and activity; either fat-soluble (vitamins A, D, E and K) or water-soluble (vitamins of the B complex group and vitamin C); vitamins of the B complex group generally function as a component of coenzymes.

Vitamin A : *See* **retinol**.

Vitamin B complex : A group of at least 13 vitamins, designated by using Arabic numerals as subscript of B, such as B_1, B_2, B_3 etc; although not related either chemically or physiologically, B complex vitamins have many features in common: (a) all of them, except lipoid acid, are water-soluble, (b) most of them are components of enzymes that play vital roles in metabolism, (c) most of these can be obtained from the same source, *i.e.*, liver and yeast, and (d) most of them can be synthesized by the intestinal bacteria.

Vitamin B_1 : *See* **thiamine.**

Vitamin B_2 : *See* **riboflavin.**

Vitamin B_3 : *See* **pantothenic acid.**

Vitamin B_5 : *See* **niacin.**

Vitamin B_6 : *See* **pyridoxine.**

Vitamin B_7 : *See* **biotin.**

Vitamin B_9 : *See* **folic acid.**

Vitamin B_{12}: *See* **cyanocobalamin.**

Vitamin C : *See* **ascorbic acid.**

Vitamin D : *See* **ergocalciferol** and **cholecalciferol.**

Vitamin E : *See* **tocopherols.**

Vitamin F : See **linoleic acid.**

Vitamin G : *See* **riboflavin.**

Vitamin H : *See* **biotin.**

Vitamin K : *See* **phylloquinone** and **farnoquinone.**

Vitamin M : *See* **folic acid.**

Vitiligo (plural of *vitiligines*[L]) **:** Smooth light-coloured patches of the skin.

V_{max} : The maximum velocity of an enzymatic reaction when the binding site is saturated with substrate.

Vomitus (*vomitus*[L], from *vomere* = to vomit) **:** Substance vomited.

Vulgar fraction : Common fraction. A fraction expressed in terms of a numerator and a denominator, *e.g.*, 3/4.

W

Warfarin (C_{19} H_{16} O_4) : A colourless crystalline substance, m.p. 161°C; used as a rat poison.

Warm-blooded animal : *See* **homoiotherm.**

Water (H_2O) : A colourless, odourless and tasteless liquid which is most abundant of all other substances on earth; mother liquor of all forms of life as there is no life without water; life almost certainly originated in water, provides the medium for biological reactions to take place; accounts for about 70% or more of the weight of most organisms; unlike rest other substances, occurs in 3 states– solid, liquid and gaseous – at the same time; has a melting point 0°C, boiling point 100°C, surface tension 72.8 and dielectric constant 80.

Water potential : The potential energy of water to move down its concentration gradient; water potential is expressed in units of pressure instead of units of energy, because pressure is simpler to measure.

Water-soluble vitamins : All vitamins soluble in water; perform the same general functions wherever they occur; most of them act as components of coenzymes and as such form vital links in the chains of biochemical reactions characteristic of all living objects; being water-soluble and excretable, required daily in meagre amounts (in milligrams or even less) for the normal growth and good health of individuals; include vitamins of the B complex group (from B_1 through B_{12}), vitamin C, choline, inositol, PABA, bioflavonoids, lipoic acid etc.

Wavelength : The distance between two successive points at which the wave has the same phase. For example, visible light has a wavelength from 400 nm (violet) to 750 nm (red).

Wax (O.E. *weax*, meaning 'the material of the honey comb', reminding of the beeswax, the honeycomb is made of) **:** Esters of long-chain saturated and unsaturated fatty acids with long-chain monohydroxy alcohols; far less spread than fats; insoluble in water; usually inert due to their saturated nature of the hydrocarbon chain; acts as major food and storage lipids in marine organisms, such as whale, herring and salmon. The term is often loosely applied to solid, nongreasy, insoluble substances that soften or melt at fairly low temperatures, *e.g.*, paraffin wax.

Weed : Any plant growing in cultivated or otherwise utilized ground that competes for resources with a crop or desired vegetation or disfigures the area it grows; an economically, useless, unsightly, or undesired plant, of no particular kind; usually have a high viability and can use up disproportionate amounts of water, sunlight and nutrients.

Western blotting : A technique by which proteins are separated and immobilized on a paper sheet and then analyzed, usually by means of a labelled antibody.

Whey : The clear liquid portion of milk remaining after the protein curd has been removed.

White blood corpuscle (WBC) : *See* **leucocyte**.

Wild type : The normal unmutated form of an organism; the form found in nature (in the wild); wild type alleles are usually given a 'plus' symbol. Thus, the wild type allele of the vestigial wing mutation (vg) in *Drosophila* is vg^+.

Wobble hypothesis : The ability of a base in the third position of an anticodon to hydrogen-bond to different bases, so as to read more than one codon.

Wood sugar : *See* **xylose**.

Work : Energy where mechanical effort is involved; measured in joules. A joule is defined, in work terms, as a force of 1 newton moving its point of application through 1 meter.

Wound hormone, COOH – CH = CH – $(CH_2)_8$ – COOH : An open-chain dicarboxylic acid with a single double bond; a plant hormone that stimulates the development of a protective layer of tissue over a cut or wound, specifically in bean pods; also known as *traumatic acid or traumatin.*

Writhe (*W*) : With respect to a supercoiled DNA helix, the number of times the helix as a whole crosses over itself , *i.e.*, the number of superhelical turns that are present. See also *twist.*

X

Xanthophyll : A yellow carotenoid pigment, present in plant cells and sap.

***Xenopus laevis* :** A species of frog, not a toad; frequently used in studies of early vertebrate development; also used to test for pregnancy in women: injections of urine from a pregnant woman results in egg-laying in the frog.

X-ray crystallography : A technique for determining the three-dimensional arrangement of atoms in a molecule, based on the diffraction pattern of x-rays passing through a crystal of the molecule.

Xylem (*xylon*[G] = wood) **:** In vascular plants, a specialized complex tissue, composed primarily of vessels, tracheids, parenchyma and fibres; the location of xylem is different in roots and stem, and the area of xylem is greatly increased by secondary thickening; transports water and dissolved minerals from the roots upward in the plant.

Xylol, C_6H_4 $(CH_3)_2$: A liquid resembling toluene that occurs in coal tar; exists in 3 isomeric forms, a mixture of which boils at 137–140°C; used in the manufacture of dyes; also called *xylene* or *dimethylbenzene.*

Xylose, $C_5H_{10}O_5$: A colourless, crystalline pentose sugar with a m.p. 144°C; present in plant cell walls, especially in xylem; also called *wood sugar*.

Y

Year (yr) : A measure of time, commonly understood to be the time taken by the Earth to complete its orbit round the sun. The *civil year* has an average value of 365.2425 mean solar days; three consecutive years consisting of 365 days, the fourth or leap year of 366 days. *Century years do not count as leap years unless divisible by 400.*

Yeast : A common term for many families of unicellular fungi; includes species used for brewing beer and making bread, as well as pathogenic speciles; contains enzymes to convert grape sugar to wine and other products of fermentation.

Yolk (O.E. *geolu* = yellow) **:** The stored substance in egg cells, made up mainly of fat and protein granules; provides the embryo's primary food supply.

Z

Zeatin : A natural cytokinin isolated from the endosperm of immature corn (*Zea mays*) grains.

Zeaxanthin : *See* **zeaxanthol**.

Zeaxanthol : The chief yellow carotenoid pigment of corn; also called *zeaxanthin.*

Zein : A simple protein belonging to the group prolamines; stored in the kernels of corn (*Zea mays*).

Zero : Nought. The starting-point of any scale of measurement.

Zero-order reaction : A chemical reaction in which the rate is independent of the concentration of the reactants.

Zwitterion (German, for '*ions of both kind*') : A dipolar ion, with spatially-separated positive and negative charges.

Zygote (*zygotos*[G] = paired together) : A diploid (2N) cell that is formed by the fusion of a male and female gamete; a fertilized egg. A zygote may either develop into a diploid individual by mitotic divisions or undergo meiosis to form haploid (N) individuals that divide mitotically to form a population of cells.

Zymase : An enzyme complex present in yeast; acts on sugar with the formation of alcohol and carbon dioxide, *i.e.*, brings about alcoholic fermentation of sugar.

Zymogen : An inactive precursor of an enzyme, particularly by those concerned with protein digestion; for example, *pepsinogen*, the precursor of pepsin and *trypsinogen*, the precursor of trypsin; zymogens require activation energy to become functional; generally, prefix *pro-* or suffix *-ogen* is added to enzyme's name to denote its zymogen, such as prothrombin, proelastase, trypsinogen, fibrinogen etc.

INDICES

Competing for light, many tropical tree species evolved very tall trunks, sometimes 100 meters or more high. Woody vines called lianas grow up along their trunks to the crown of the tree, where there is enough light for the lianas to flower. More light and space are made for lianas when a tree falls.

Index is "an alphabetical list of names, subjects, titles etc., giving page numbers where reference is made and generally placed at the back of a book."

[*Source : (one of the meanings) The New International* ***Webster's Dictionary*** *of the English Language, Trident Press International*]

INDEX I

AUTHOR / SCIENTIST INDEX

Numbers in boldface type indicate references from the work of the author cited. Numbers followed by a D or T indicate the diagram or table, respectively, contributed by the author cited. Numbers followed by a B indicate the life history and contributions, presented in a box, of the author cited.

D

E

H

I

J

M

T

X

Y

Z

INDEX II

SUBJECT INDEX

Numbers in boldface type refer to pages on which the subject is illustrated or especially described. Numbers followed by 'CI' indicate the Clinical Implication pertaining to the subject cited.

A

B

C

E

F

G

H

I

M

P

Q

R

S

U

V

W

X

Y

Z